The Gun Digest® Book of

Modern

GUN VALUES

◆ For Modern Arms Made from 1900 to Present ◆

11th Edition

Edited by Ken Ramage

MODERN GUN VALUES STAFF

EDITOR

Ken Ramage, Editor
Firearms & DBI Books

CONTRIBUTING EDITORS

Handguns – Joseph Schroeder   Rifles – John Malloy
Shotguns – Larry Sterett   Commemoratives – Kevin Cherry

About Our Covers:

Front Cover: These guns are indicative of the period this book covers, 1900 to present. Clockwise from the top, a compact Glock in 40 S&W (*current production*); a pre-WWII vintage Iver Johnson safety Automatic Hammer (3rd Model) 5-shot revolver chambered 38 S&W; a Colt Government Model 45 ACP carried in WWII and a S&W K-22 revolver, with target hammer and trigger, which left the factory in the latter 70s.

Back Cover: *Top*, a Ruger classic – the 'old model' Super Blackhawk 44 Magnum. According to the paperwork accompanying the gun, this revolver was purchased in 1964 through a military base exchange on Okinawa. *Center*, the compact Glock. *Bottom*, another look at the 5-shot Iver Johnson.

Published by

700 E. State Street • Iola, WI 54990-0001
Telephone: 715/445-2214
Web: www.krause.com

Please call or write for our free catalog of publications. Our toll-free number to place an order or obtain a free catalog is 800-258-0929 or please use our regular business telephone, 715-445-2214.

Library of Congress Catalog Number: 2001086509

ISBN: 0-87349-249-8

Introduction

MODERN GUN VALUES continues its established role as an on-going research project within the field of modern arms collecting and valuing. It is as much a firearm identification guide – providing over 3000 photographs of 20th-century handguns, rifles, shotguns and commemoratives – as it is a guide to current used firearm values.

As a basic reference work, it provides specifications for most of the world's small arms manufactured in or imported into the United States since 1900, including many military small arms. Over 7000 firearms are documented and valued in these pages.

For some of the more widely collected firearms — such as Colt Single Action Armies, High Standards, Lugers and Mausers – our coverage includes most if not all of the most commonly encountered variations as well as some rarely found in even advanced collections. We also include other collectibles such as Nambus and Tokarevs.

Because of its historical base, MODERN GUN VALUES does include some prototypes, experimental pieces and one-of-a-kind developmental or evolutionary models which are very rare and seldom, if ever, seen on the used firearm market. These rarities are important and interesting to document but, as non-production firearms, next to impossible to price since they have little history on the used gun market.

New This Edition

We have added content features, including a heavily-illustrated color section, to the book to help the reader successfully inspect used guns, understand the NRA grades of condition and apply them to his own dealings.

Since many used guns invite maintenance and possibly do-it-yourself repair, we have included assembly/disassembly instructions for two widely available used arms – the SKS carbine and the Colt Police Positive revolver. These photo-illustrated instructions are excerpted from the popular *Gun Digest Book of Assembly/Disassembly* six-book series, by J. B. Wood.

The Reference section is new, too. Collectors, hobbyists and students of the gun may wish to contact collectors' organizations and research books in print that address their areas of interest. This information, like the gun values, is freshly updated for this edition.

How to Use This Edition

MODERN GUN VALUES is divided into four major sections - Handguns, Rifles, Shotguns and Commemoratives. Within each of these sections the individual firearms entries are listed in strict numerical model, then alphabetical, order. In addition, the firearms of larger manufacturers such as Smith & Wesson, Colt, Winchester, Marlin, Savage, etc. have been further delineated into action type - e.g. Semi-Automatic Pistols, Revolvers and Single Shots for handguns; and Semi-Automatic, Single Shot, Bolt-, Lever- and Pump-Action for rifles and shotguns.

Our page layout should also facilitate finding any given firearm. The shaded box found on each page's margin acts as a quick easy-to-use directory to the firearms contained on that page. Cross-references to firearms known by both model numbers as well as names – e.g. Beretta New Puma (See Beretta Model 70) - can also be found in this shaded directory. Other cross-references help the reader find manufacturers who have undergone reorganization and as a result acquired a new company name – e.g. Detonics (See New Detonics).

We have established a three-tiered pricing approach for each firearm based on the six *NRA Modern Condition Pricing Standards: New, Perfect, Excellent, Very Good, Good* and *Fair*. Each firearm has been evaluated to determine which three NRA standards most accurately reflect the condition in which the particular firearm is most likely to be found on the used market; the firearm is priced accordingly. For example, firearms currently on the retail market, or manufactured in the last four years, will most always be found in *New* or *Perfect* condition on the used market, while most older firearms will rarely be found in any condition exceeding *Very Good*. The three pricing levels we provide reflect current observations of prices seen at gun shows, in the gun shops or in the various periodicals, dealer's catalogs or at auction sales. Our prices reflect high gun show/low gun shop levels. For firearms currently manufactured and on the retail market, the Last Manufacturer's Suggested Retail price (LMSR) in bold is also provided. The *LMSR* reflects the price of the basic or standard grade of that particular firearm.

Keys to Determining Value

There are three primary factors that are key to establishing a firearm's value. These are *demand, availability* and *condition*.

Demand and Availability

One of the factors driving demand - and thus the value of a particular firearm - is availability. When demand exceeds availability, the price of the firearm increases. This pertains not only to collectibles but to shooting guns as well. For example, it is not unusual for a current-production firearm in limited supply to sell in the marketplace for prices considerably above the manufacturer's suggested list price. Even some Colt and Smith & Wesson models have enjoyed such run-ups. However, realize that in this case when supply finally catches up with demand, prices will plunge.

For collectibles, although the supply is fixed, demand plays a similar role in establishing value. Some older firearms may be in scarce supply, yet are not in demand by collectors. Without demand, regardless of availability, the dollar value of the firearm will be low.

Condition

Of the three factors used to establish value, condition is perhaps the most important and certainly the factor most frequently used by shooters as well as collectors for determining the price of a firearm. A "like new" example of a relatively common older Colt, Smith & Wesson, Luger or other arm popular with collectors can bring up to twice

---

NRA Modern Condition Standards

To understand the pricing structure of the 11th Edition, it is essential to establish a set of condition standards by which a firearm can be judged. We have adopted the well-respected and popular National Rifle Association's Modern Condition Standards as a guideline to the various grades of condition but have made slight modifications *(italics)* to further help readers determine condition degrees.

New: In same condition as current factory production, *with original box and accessories.*

Perfect: In new condition in every respect, *but may be lacking box and/or accessories.*

Excellent: *Near* new condition, used but little, no noticeable marring of wood or metal, bluing perfect (except at muzzle or sharp edges).

Very Good: In perfect working condition, no appreciable wear on working surfaces, *visible finish wear* but no corrosion or pitting, only minor surface dents or scratches.

Good: In safe working condition, minor wear on working surfaces, no corrosion or pitting that will interfere with proper functioning.

Fair: In safe working condition, but well worn, perhaps requiring replacement of minor parts or adjustments, no rust, but may have corrosion pits which do not render article unsafe or inoperable.

the price of another that easily ranks as "Excellent." On the other end of the scale, a scarce, popular collectible in "Poor" or even "Fair" condition usually will go begging at a fraction of its value in more acceptable shape.

Of course, neither condition nor pricing is ever absolute. There is no such thing as a fixed price. It's an axiom among gun enthusiasts that there are only two conditions any two collectors can agree upon and those are "New" and "Junk"; everything in between is highly subjective and subject to debate. For example, a shooter looking at a potential acquisition might rate a gun with little finish but mechanically tight and with a fine bore, "Excellent"; to the collector that same firearm would rate only a "good" or possibly "Very Good." An old Luger with most of its original blue but with the bore badly pitted from corrosive ammunition would rate an "excellent" from most collectors but only a "Fair" from a shooter. Collectors and shooters alike find the line between "Good" and "Very Good" or "Very Good" and "Excellent" a fuzzy one.

Other Keys Affecting Values
Other factors can play an important role in establishing the value of a particular firearm. Engraving, inlays, fancy stocks or grips, accessories such as after-market sights, grips or triggers, rare or odd calibers, nickel or stainless finishes, import markings, etc., all affect a firearm's value. Even magazine capacity can affect value. With the 1994 ban on all magazines with capacities of more than 10 rounds, those larger-capacity magazines on existing firearms grandfathered and still legal under the ban bring a premium over the same gun made today with a 10-shot magazine.

Pricing
Pricing, like condition, can also be highly subjective. It has been said that any price guide is obsolete even before it gets off press. To some degree this is true, though most firearms prices are not that volatile. However, any value guide is only a guide. In reality the only true established value for any given firearm is the amount of money mutually agreed upon between buyer and seller, and the actual monetary amount that changed hands in the transaction. A seller can ask any price he wishes for the firearm he is selling, but if no one will pay that price, then the value he has affixed to it is not meaningful. Even the manufacturer's suggested retail price is only a guideline, as many dealers are willing to negotiate price on at least their slower moving stock.

Remember that the values shown here should be taken as guidelines, not absolutes. When a rarely seen gun that you've been wanting for a long time shows up at a price somewhat higher than we indicate, it still may be worth buying because the pleasure of ownership is often well worth the higher price. Many collectors freely admit they rarely regret the firearms they buy, but all too often regret the ones they didn't buy. On the other hand, don't be too quick to buy a common gun that you'd like to have when you see it at market price, since one will likely show up sooner or later at a lower price.

Then there is also the perception of condition when pricing used firearms. Though most dealers try to be scrupulously fair in their catalogs or advertisements, a few are notorious for overstating condition. And even when you have the opportunity to examine a gun at a shop or a gun show, it is easy to overlook a mismatched serial number or a well-done repair. Gun shop prices tend to be higher than gun show prices, but there is also an advantage to dealing with an established business if a problem arises.

Restoration
For collectors in particular, there's another important aspect of condition that needs to be addressed and that is restoration. As the demand for many collectible arms has exceeded the supply of those in "acceptable" condition, many rare, and even not so rare, firearms have been restored. Pitted areas built up by welding, missing or damaged parts remade, obliterated markings re-rolled, grips and stocks with re-cut checkering, or the metal polished and refinished using the techniques of the original maker are some examples of the restoration work possible. A first-class restoration is an expensive proposition, but if properly done is often difficult to tell from original factory work.

But how does the value of a restored firearm compare with that of an original in like condition? A truly first-class job can bring close to the price of an original-condition example; however, very few restorations are that good. Generally, a very good restoration is usually worth at best half as much as a nice original. Two warnings: First, beware of restorations passed off as original. Second, be extra aware of a common model that's "restored" into a rare variation by modifying markings, barrel length or the like. When in doubt, ask an expert and refer to the NRA's Code of Ethics. Misrepresentation can be fraud.

NRA Code of Ethics
A listing of practices considered unethical and injurious to the best interests of the collecting fraternity.

1. The manufacture or sale of a spurious copy of a valuable firearm. This shall include the production of full-scale replicas of historic models and accessories, regardless of easily effaced modern markings, and it also shall include the rebuilding of any authentic weapon into a rarer and more valuable model. It shall not include the manufacture or sale of firearms or accessories which cannot be easily confused with the rare models of famous makers. Such items are: plastic or pottery products, miniatures, firearms of original design, or other examples of individual skill, plainly stamped with the maker's name and date, made up as examples of utility and craftsmanship, and not representative of the designs or models of any old-time arms maker.

2. The alteration of any marking or serial number, or the assembling and artificially aging of unrelated parts for the purpose of creating a more valuable or unique firearm with or without immediate intent to defraud. This shall not include the legitimate restoration or completion of missing parts with those of original type, provided that such completions or restorations are indicated to a prospective buyer.

3. The refinishing (bluing, browning, or plating) or engraving of any collector's weapons, unless the weapons may be clearly marked under the stocks or elsewhere to indicate the date and nature of the work, and provided the seller unequivocally shall describe such non-original treatment to a buyer.*

4. The direct or indirect efforts of a seller to attach a spurious historical association to a firearm in an effort to inflate its fair value; efforts to "plant" a firearm under circumstances which are designed to inflate the fair value.

5. The employment of unfair or shady practices in buying, selling, or trading at the expense of young and inexperienced collectors or anyone else; the devious use of false appraisals, collusion and other sharp practices for personal gain.

6. The use of inaccurate, misleading, or falsified representations in direct sales or in selling by sales list, catalog, periodical advertisement and other media; the failure to make prompt refunds, adjustments or other proper restitution on all just claims which may arise from arms sales, direct or by mail.

*When the NRA formulated this Code of Ethics many years ago restoration was rarer than it is today and some restoration was indeed marked. However, such marking is rarely if ever done today and restoration is not only considered ethical but desirable when appropriate. Furthermore, for many, a restorer's mark in even the most inconspicuous internal location would detract from originality.

Fortunately, many knowledgeable collectors and dealers are able to distinguish even the best restoration work from "factory original." In addition, in the current marketplace it is not at all unusual for an owner or seller to not only admit restoration, but to state with pride that a certain arm was restored by a specific well-known restorer.

Ethically, of course, the fact that restoration work has been done and to what extent should always be disclosed to a prospective purchaser. Not to do so has, in the case of some very valuable collectibles, resulted in expensive and embarrassing legal actions.

Modern Gun Values, Eleventh Edition

Table of Contents

Inspection Guide To Used Guns

by Patrick Sweeney

WHILE THE SATISFACTION of buying a new firearm, from the standpoint of warranty and features, appeals to many shooters and collectors, sometimes *"used"* is the only route. After all, how many new-in-the-box Winchester pre-64s still exist? Sometimes the only way to acquire the firearm model, or the firearm with the features you desire, is by buying it used. We all have budgets, and purchasing a used gun is much easier on them! Read on and learn how to buy "used" – safely.

If at all possible buy from an established dealer, with a track record and reputation. Even better, a dealer who has an in-house gunsmith who inspects all their used firearms and makes sure no lemons slip through. In the event one does, a reputable dealer will take it back or make it right.

What should the dealer warranty? The normal and expected performance and durability of that model firearm, and that he presented it correctly as to its features and performance. If you buy a plain old used 30-30 and find it shoots three- to four-inch groups at 100 yards, don't expect to be able to return it. If, however, it shoots those groups four feet to the left or right, you have every right to return it. An *as-new* benchrest rifle better do well under an inch with its provided ammo, or you may have cause to return it (assuming **you** can shoot that well). If the dealer doesn't have a written warranty, ask what the return policy is.

If you are not buying from a dealer, you have the standard business-school Latin to guide you – *Caveat Emptor:* "Let the buyer beware." One approach that some of my customers took – and more should have – was to have their purchase inspected by a pro. If you have any doubts about an attractive purchase, take it to a gunsmith and explain things. Don't just drop it off for a "strip and clean" and count on him (or her) uncovering hidden problems.

Come right out and explain: You just bought it, and you want it inspected for safety, durability, function and headspace. If there is a limited return time, the gunsmith needs to know in order to inspect it within the allotted time. Many gunsmiths are booked solid for months and may not get it back to you in time if you leave your purchase for what the 'smith understands is just a "regular cleaning." By explaining your inspection period time constraint you can get your new purchase back in time to meet the refund terms of the sale, should you need to return the gun.

General Inspection of a Used Firearm

To start, give the firearm in question a quick visual inspection. I call it the *"tire tracks and hammer marks"* look, and it is the same regardless of the type you are thinking of buying. The inspection of a firearm to determine the percentage of finish remaining is covered elsewhere. The purpose of this initial inspection is to uncover damage, repairs or abuse. Is the stock straight and clean? Is the barrel straight? Are the sights centered? Are there dents, scratches, cracks or repairs to be seen? Does the bluing have the right color? Are the barrel markings clean and crisp, or are they blurry or smeared? Is the barrel, the correct length? Is the muzzle uniform? Does the chambering marked on, the barrel match what the seller tells you it shoots? Try to get a "feel" for the history and typical condition of the gun you are looking at. Does it match the description of the one the seller is trying to sell you? A firearm that doesn't match what the seller describes is probably best left on the table.

Just because you are looking at a worn, *used-to-gray* rifle the seller describes as "the best he's seen" doesn't mean he's fibbing. If you are holding a Remington 700 in 308 Winchester – yes, he is. On the other hand, if you are holding a pre-'64 M-70 in 300 Savage – no, he isn't.

RIFLES

Open the action. With a light or reflector – and with the action open and bolt removed if appropriate – look down the bore. Clean, shiny and clear of obstructions, right? If not, let the bargaining begin!

While many rifles will shoot accurately with a slightly pitted bore, some won't – and all will require more frequent cleaning. Work the action and see if there are any binding spots or if the action is rough. Ask if you can dry fire it to check the safety.

Some people do not like to have any gun in their possession dry-fired; others don't care. If you cannot, you may have to pass on the deal. Or, you can assure the owner that you will restrain the cocking piece to keep the striker from falling.

Close the action and dry-fire it. How much is the trigger pull? Close the action, push the safety to *ON*, and pull the trigger. It should stay cocked. Let go of the trigger and push the safety *OFF*. It should stay cocked. Now, dry-fire it. Is the trigger pull different than it was before? If the pull is now lighter, the safety is not fully engaging the cocking piece, and you'll have to have someone work on it to make it safe. If the rifle fires at any time while manipulating the safety (even without your having touched the trigger) it is unsafe until a gunsmith repairs it.

While you were checking the safety, just what was the trigger pull? A very light trigger pull is not always bad, but may need adjustment. As an example, if you are handling a Remington 700 or Winchester 70, and the trigger pull is one pound, someone may have adjusted the trigger mechanism. If you are handling a Winchester '94 and the trigger pull is a pound, someone has been stoning the hammer or sear. On the first two, you or your gunsmith can adjust the weight back to normal ranges. On the '94 you may have to buy a new hammer or sear – or both – to get the pull back into the normal range.

Inspect the action and barrel channel. Is the gap between the barrel and the channel uniform? Or does the forearm bend right or left? Changes in humidity can warp a forearm and, if the wood touches the barrel, alter accuracy. The owner may be selling it because the accuracy has *"gone south,"* and not know that some simple bedding work can cure it.

Look at the action where it meets the stock. Is the wood/metal edge clean and uniform? Or do you see traces of epoxy bedding compound? Epoxy could mean a bedding job, and it could mean a repair of a cracked stock. Closely inspect the wrist of the stock, right behind the tang. Look for cracks and repairs.

Turn the rifle over and look at the action screws. Are the slots clean, or are they chewed up? Mangled slots indicates a rifle that has been taken apart many times – and at least a few of those times with a poorly-fitting screwdriver.

Remove the bolt if you can. If not, use a reflector or light to illuminate the bore. Is the bore clean and bright? Look at the bore near the muzzle. Do you see jacket fouling or lead deposits? Many an *"inaccurate"* rifle can be made accurate again simply by cleaning the

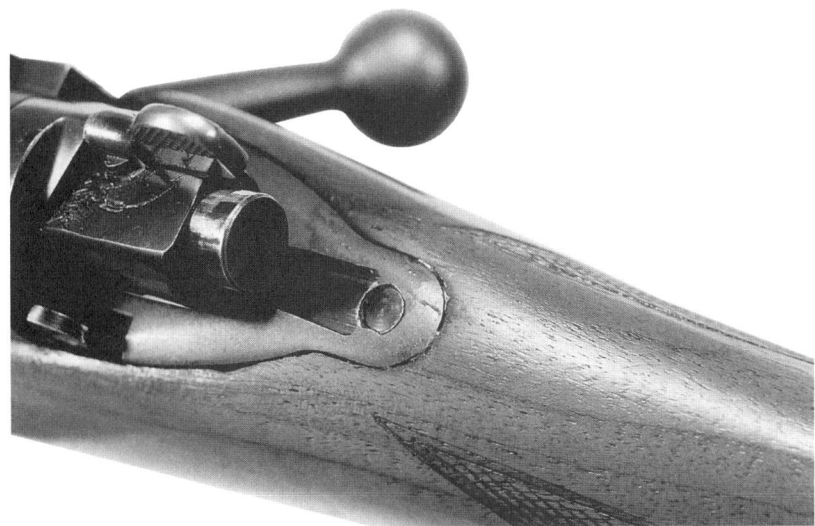

A rifle fired with a poorly-fitted stock, or one dried out from decades of storage, will often crack light behind the tang. Look closely for cracks or repairs.

Are the screw slots clean, or are they chewed up? This screw is just tolerable. Any worse and it would indicate abuse instead of 59 years of indifferent attention.

Check the edges of the stock where the action rests for signs of glass bedding. Bedding is not bad; in fact, it can be good. But don't pay collectors' prices for a working tool that has been modified.

jacket fouling out of the bore. While looking down the bore, hold the barrel so a vertical or horizontal bar in a window reflects down the bore. If the reflection of the bar has a 'break' in it, the barrel is bent. Sight down the outside of the barrel and see if you can spot it. A slightly bent barrel can still be

When inspecting the barrel, look for dents or creases. Also, inspect the rib (if there is one) to see if it has been dented. While a dented barrel or rib can be repaired so the damage is almost unnoticeable, you can still see evidence of the repair.

Remove the barrel to inspect the bore. Is the bore clean of

plastic? Is the choke clean? If not, swab them clean. If there are screw-in chokes, do they unscrew easily and smoothly? While you have the forearm off an autoloader, look at the gas system. Is it clean? Or is it crusty from powder residue? Powder residue can be wiped off, but rust requires more a vigorous remedy, and may leave the shot-gun as a non-cycling autoloader.

Double-Barrel Shotguns

Doubles require a different inspection. While looking over the barrels, look to see that the side or bottom ribs are smoothly attached along their length. A lifted rib that has been repaired will have a different appearance at the repair.

Open and close the action. Does the lever move smoothly into place, or do you have to push it the last fraction to fully close it? How far does the lever move? Levers are initially posi-tioned to not go fully to the cen-terline. As the action wears, the lever moves further and further, taking up the wear. A shotgun with a lever too far past the cen-terline may have been shot a great deal, or been taken apart and put back together with the lever mis-timed.

Next, check to see the barrels are tight. Often, the forearm will put enough pressure on the action to make it seem tight.

Be sure of the chambering, and be sure it is clearly marked. This Marlin was not a 38-55 whenit left the factory; but is now and is clearly so marked.

Consider rarity when assessing condition. A 30-06 with signs of honest use and hunting wear is a tool. This same rifle, were the barrel marked "300 Savage" or "35 Remington," would be a collector's dream.

accurate, but will walk its shots when it heats up. A severely bent barrel must be replaced.

SHOTGUNS

Pump-Action & Autoloading Shotguns

Pumps and autoloaders require the same safety check as rifles do, with a few additions. Safety on, pull the trigger, let go, safety off. Dry fire and see if the weight of the trigger pull changes.

Are screw-in chokes easy to remove? Or do you have to wrestle with them? A bulged choke may mean a ruined barrel. Always unscrew the chokes to make sure theywork as intended.

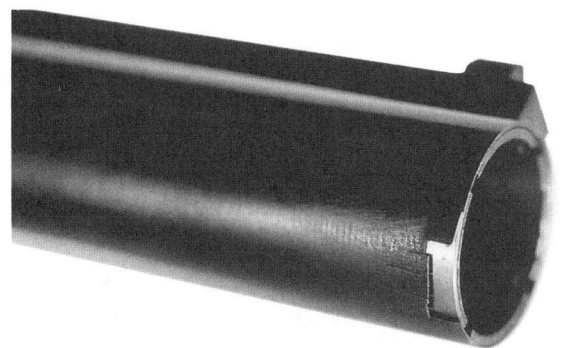

The screw-in chokes in this barrel, while functional, doom the barrel for resale. The wall was cut too thin on one side and chipped out. The owner will never be able to sell the barrel, for who would buy it? You might, if you neglected to check.

Remove the forearm and then check barrel tightness. Does the barrel assembly move or wobble when closed? Can you see the joint at the action changing size when you try to move the barrel? A loose barrel is an expensive repair, so be sure to check. Pull or twist the barrel in all three axes; attempt to move the rear side to side, lift as if you were opening the action (but not pushing the lever), and pull the barrels forward. The action should be as tight as a bank vault.

Next, the triggers. You'll need snap caps and the owner's permission. Insert the snap caps and close the action. Put the safety on. If the double is a twin trigger, check to make sure the safety blocks both triggers. If it is a single trigger, make sure the safety blocks the trigger when the barrel selector is set to each barrel in turn.

Push the safety to *OFF* and snap one of the barrels. Open the action *(keep your hand over the action to stop the snap cap from being launched across the room)* then close it and select the other barrel. Snap that barrel and open the action again, stopping the snap cap from being ejected. Both barrels work? Good. Close the action, snap one of the barrels, and then slap the butt of the shotgun with your hand. If the shotgun has a non-inertial trigger – like the Ruger Red Label – you can forego the slapping. Does the second barrel now fire when you pull the trigger? If so, the inertial trigger is working. If not, you may have to slap it harder, or the inertia weight needs adjusting.

HANDGUNS

Handguns come in two types: revolvers and autoloading pistols, and each has sub-types with their own peculiarities. The four types we'll cover are the **single-action revolver** and **double-action revolver, single-action autoloading pistol** and **double-action autoloading pistol**.

Single-Action Revolvers

The single-action revolver is known by many as the *cowboy* revolver. Your quick visual inspection of the exterior should start with the sights, to make sure they are straight, and the grips to make sure they are without cracks or dents. Also look at the exterior edge of the muzzle, and the corners of the frame, for signs of dropping. Bent sights and cracked grips indicate a dropped handgun. Dropping can bend the barrel, warp the frame or throw off the timing.

Hold the revolver up to the light, sideways, and look at the cylinder gap. There should be daylight, but not too much of it. The SA comes in the Colt pattern, old Ruger, and the new Ruger.

In Colts and old Rugers, open the loading gate, cock the hammer back to the *(half-cock)* notch that frees the cylinder, and rotate the cylinder. Look to see that it isn't loaded. On new Rugers, opening the loading gate frees the cylinder to rotate. To close up both action "systems," close the loading gate, cock the hammer and, with your thumb on the hammer spur, pull the trigger and ease the hammer forward while holding the trigger back. Check the cylinder for play.

For those who may not know, spinning the cylinder at high speed, or fanning the hammer

The wood on this shotgun can be repaired, but what caused it? If a previous owner used magnum shells in a non-magnum-capable gun, the action may be loose. Check the tightness of the barrels to the receiver.

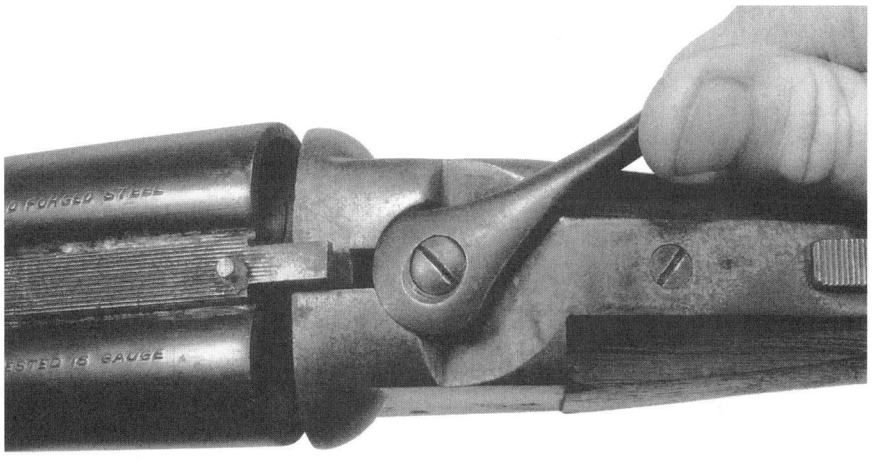

On doubles, check to see if the top lever is easy to move and the action easy to open.

are both considered abusive handling, and will likely end the sale before it starts. Don't do either!

Does the cylinder move back and forth? Called *endshake*, it can be easily fixed, but if there is too much it indicates a revolver that has seen a lot of use. On a Colt-pattern revolver it could mean that the cylinder has been replaced and not properly fitted, or the bushing is worn - or peened - from heavy loads.

Does it wobble from side to side? The cylinder stop may be worn, or the slots may be worn or too large. Look at the slots. If they have been abused, the edges will be chewed up. If they appear sharp and clean, the cylinder stop may be worn or its spring weak. A worn or abused cylinder is expensive, while a new cylinder stop or spring is relatively cheap.

Slowly cock the revolver, watching the cylinder. Does it come fully into position? Or do you have to push the cylinder around the last fraction of an inch to get it to lock? A cylinder failing to carry up will require a new hand – or require that the old one be "stretched."

Check each chamber. It isn't unusual for a revolver to have one chamber that has a slightly different timing on the *carry up* than the others do. Once you've checked *carry-up*, test the trigger pull. If the owner is leery of letting you dry fire, catch the hammer with your other hand each time you cock it and pull the trigger. Is the pull within normal limits? A heavy pull may indicate someone has fussed with the trigger – as would a very light pull.

While a trigger is relatively cheap, they can be salvaged only sometimes. The hammer is expensive, but you can often have the notch re-stoned (properly, of course) or in extreme cases, welded and re-cut. If the trigger pull has been "messed

Does the top lever go past center? As the locking surfaces wear the lever moves farther and farther. When it reaches the far side of the top strap, it needs to be refitted.

with," what was done? You can't tell without getting out a screwdriver set and disassembling the revolver there and then. You will have to either take the risk, or insist on a return/refund option if your gunsmith finds something too expensive to fix.

To continue inspecting the Single Action, open the loading gate, release and pull the center pin, roll the freed-up cylinder out of the frame *(to the right)* and inspect the front and rear of the cylinder. On the rear, is the bluing of the ratchet that the hand pushes against evenly worn white? *(A difficult inspection on a stainless or nickel gun, but you can see the wear if you look closely.)* Each chamber should be clean, their edges unmarred. On the front face of the cylinder, check to see if there are marks from the cylinder face rubbing against the rear of the barrel. A cylinder with *endshake* may rub. The rubbing may even be only partial. Don't worry unless the rubbing has been hard or extensive enough to have marred the face of the cylinder.

Look at the rear of the barrel. Is the end even and square

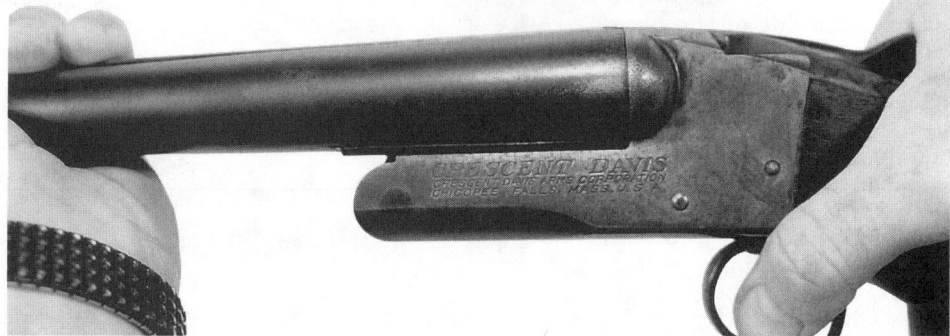

When doubles loosen their ribs and plates, the damage often starts at the muzzle. Check there first, and then work your way down the barrels.

You must remove the forearm before checking a double for tightness. If you don't, the forearm's support may mask any looseness present in the action.

to the bore? Or has someone been stoning or filing the rear face for some reason? Is the forcing cone clean and smooth?

A revolver that has been fired with lead alloy bullets will often have a forcing cone crusted with lead, even when the rest of the barrel is clean. A revolver that has seen a lot of jacketed magnum-level loads will show the wear in the forcing cone, the edges of which will be slightly rounded from the heat and abrasion.

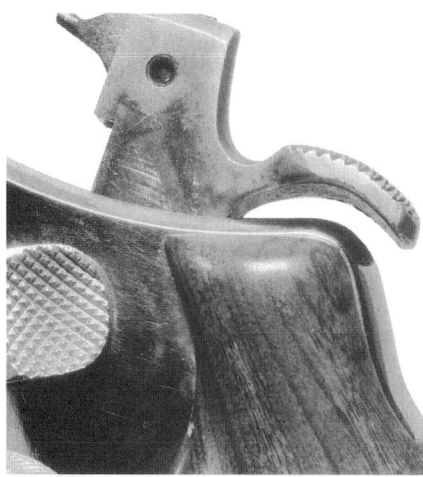

Hammer clearance on a revolver in single action mode is important. You can see here the hammer may bind on the frame if the spur is bent.

Look down the bore. Is it clean and are the lands and grooves smooth and shiny? A pitted bore means the barrel must be replaced. If you see a dark ring (or *donut*) that indicates a "ringed barrel," the barrel may still be accurate but will probably lead quickly. A bullet stopping partway down the bore, and then being jolted out by the next round fired, causes a "ringed" barrel. The bulge may not show on the outside.

Double-Action Revolvers

The quick exterior inspection should include the hammer spur. Dropping a DA revolver can bend the spur, keeping the hammer from being cocked. Your inspection will reveal this, so be prepared when you get there.

Push the cylinder latch and open the cylinder. Does the latch move smoothly? Does the cylinder move without binding or catching? Check by opening the cylinder at each of its six (or five, seven or eight) positions. A dropped DA revolver can have a bent center pin, and the bend will interfere with opening at only one chamber. On a DA revolver, opening and closing it Hollywood-style, by flicking the wrist, is flagrant handling abuse which will get it snatched out of your hands by many owners.

Check *carry-up*, both in single action and double action modes. You may have to ride your offhand thumb on the hammer as you slowly do the double-action check, to keep the hammer (and the trigger) from jerking to the end of the DA stroke and thus hiding improper carry-up. If the revolver has been dropped and the hammer spur bent, this is when you'll find out. A bent spur can still work fine in double action, but the hammer goes back farther in cocking for single action. A bent spur may bind against the frame and not allow the revolver to be cocked. With the hammer cocked, put your thumb behind it and give it a gentle push... no more than ten pounds worth. The hammer should stay cocked.

Years ago I had a run-in with a desk sergeant at a local police department about proper testing for push-off (*my home state of Michigan requires a safety inspection for the sale of a handgun*). He was pushing for all he was worth, with both thumbs, and rejecting every revolver my customers came in with. I finally had to bring in the S&W Armorers Manual, and

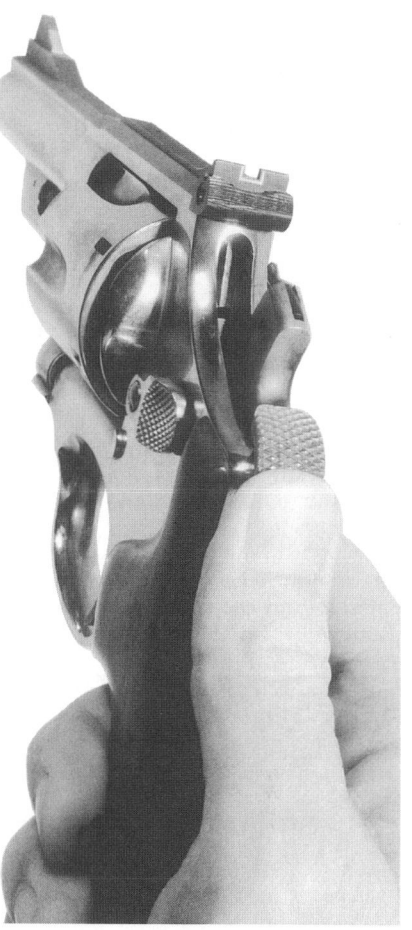

To check single action engagement, cock the hammer and push the hammer forward with one thumb. Again, ten pounds is all you need.

show him and his supervisor what the factory-accepted test was. If the revolver you are testing pushes off at ten pounds or less, the single-action notch is worn - or has been worked on. Depending on the remedy required, it may be

Does the cylinder unlatch smoothly and easily? Binding or requiring force to move is a bad sign, usually indicating a bent crane or bent center pin.

Close the cylinder, dry fire and hold the trigger back. Check the cylinder for wobble: front-to-back and side-to-side. Then release the trigger and try again. The cylinder shouldn't move at all when the trigger is held back, and only a little when released.

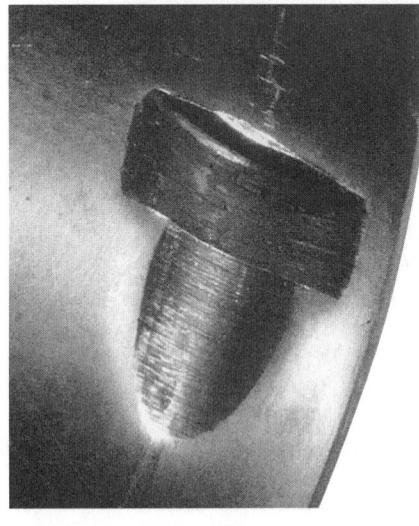

A peened cylinder locking slot indicates heavy use, either many rounds or magnum loads. Peened slots can't be fixed, and require a new cylinder, a major cost.

If you see beveling, look closely at the ejector star. Has the star been beveled, too? A proper job bevels the cylinder but not the ejector. A beveled ejector that improperly ejects (the empties will not be fully ejected) is a moderately expensive repair.

Inspect the forcing cone. Is it clean, with sharp edges? A revolver that has seen a lot of magnum loads, especially jacketed ones, will have an eroded forcing cone. A worn forcing cone can cause *spitting* and a loss of accuracy. A worn forcing cone can be fixed, but only by setting the barrel back and cutting a new cone in fresh steel. The gunsmith will also have to shorten the ejector rod and center pin, and will have to remove *endshake* to do the job properly and the cost will be moderate to moderately high.

Look down the bore. Clean, shiny and straight? Good. If it is pitted, or ringed from a bullet having been lodged in the bore, you'll need a new barrel.

The last check concerns the crane. The swing-out crane makes loading and unloading easier, but it is relatively fragile and can be bent by being dropped, or being flipped open Hollywood-style.

Gently close the cylinder, and see how much thumb pressure it takes to lock up. Does the cylinder swing into place and click shut without force? Great. Try it on all chambers, as a bent crane can be offset by other tolerances, and may be

removing *endshake* increases the gap beyond tolerances, you'll have to have the barrel set back, a moderately expensive fix.

Open the action and look at the front and rear of the cylinder. The front of the cylinder should not show rub marks from the rear of barrel. If it does, it is a sign of excessive *endshake*, which must be fixed. The rear of the chambers should have clean ninety-degree edges. Some shooters bevel the rear opening of the chambers to make speedloading faster and easier. Properly done, beveling does speed reloads but, improperly done, it can cause improper ejection.

expensive to fix; sometimes requiring a new hammer.

Check cylinder tightness with the trigger held back, as with the SA revolver, checking for play side-to-side and front-to-back. As on the SA revolver, side-to-side play can be caused by peened locking slots in the cylinder, which is expensive to repair. Or, it can be caused by a worn cylinder stop (less expensive), or a tired cylinder stop spring (cheap to fix). *Endshake* is a sign of use with heavy or magnum loads. *Endshake* is easy and inexpensive to fix by stretching the crane or installing shims, but both increase cylinder gap at the rear of the barrel. If

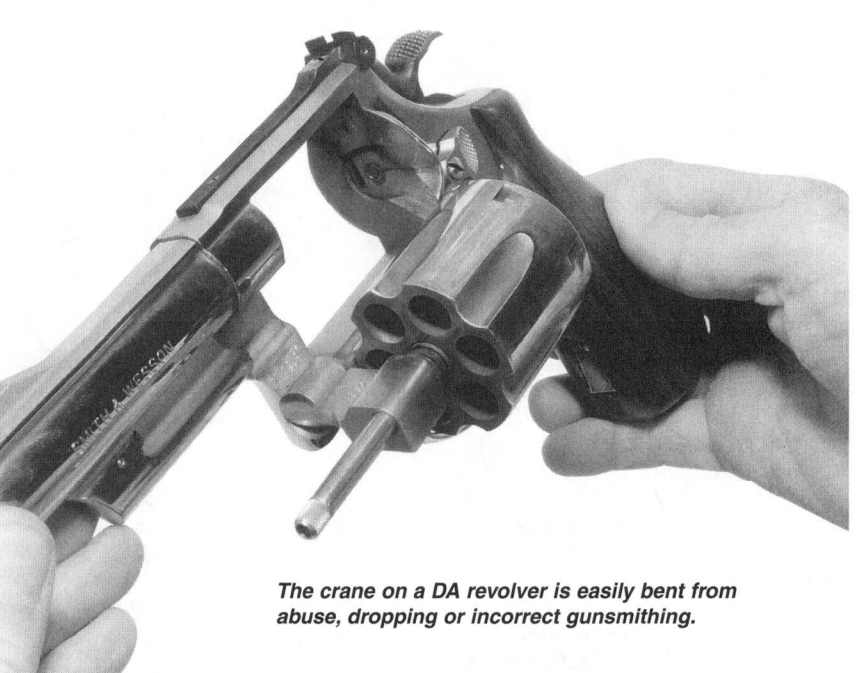

The crane on a DA revolver is easily bent from abuse, dropping or incorrect gunsmithing.

hidden on one or more chambers. If you find you need moderate thumb pressure to get the cylinder to lock in place, the crane may be bent.

A bent crane and its repair are brand-dependent. Rugers are so stoutly built that you need a ball-peen hammer to bend the crane. You also need one to straighten it. A S&W crane is more fragile and more sensitive to misalignment, but a simple job to straighten. The Colt system is less sensitive than the S&W, not as stout as the Ruger, and a more involved job to fix.

Autoloading Pistols, Single-Action

The icon of single-action autoloading pistols is the 1911 pistol. Of all handguns, this one is the most likely to be assembled from parts, played with, experimented upon - and had parts swapped in and out. Any used pistol requires a close inspection to ensure you don't end up with a pig in a poke.

On your exterior visual inspection, don't be put off by parts of different colors. The government never cared about matching the color of Parkerized parts on military-issue 45s, and many shooters through the years have come to favor deliberately two-toned pistols. It is not at all unusual

A dropped revolver can bend the center pin where it protudes into the frame. It cannot be straightened and must be replaced.

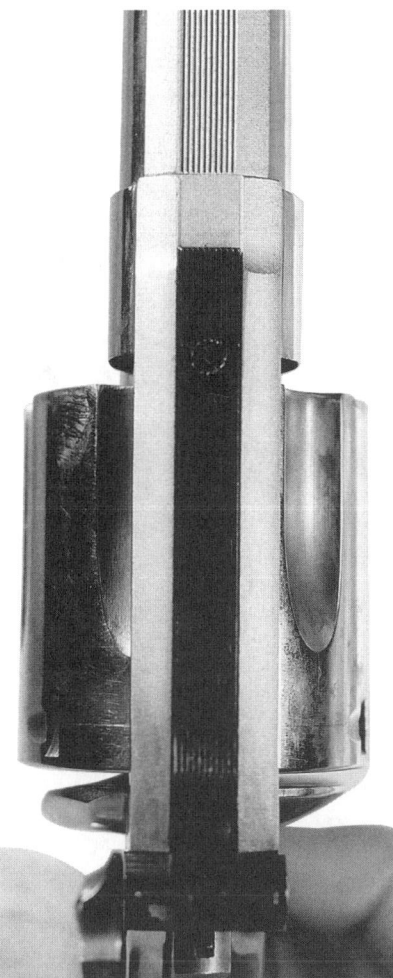

A revolver with a cylinder that won't fully "carry up" into position behind the barrel is dangerous. If it fires unlocked, the bullet won't be centered in the bore, and will split fragments out through the gap.

to find a 1911 with a blued slide and nickeled or stainless frame, or blued or Parkerized parts on a hard-chromed gun.

Check the muzzle end of the slide for dings and gouges indicating it has been dropped. Look at the magazine well. A dropped pistol can crack at the magazine well if the well has been beveled for fast magazine insertion. You may see a crack on the frame forward of the slide stop lever. Pay it no mind. A cracked dustcover on high-mileage auto-pistols is not rare. If you see the crack and the owner says it has never been shot, be suspicious. *Any crack in a slide is grounds for immediate rejec-*

tion. Cracked slides cannot be repaired, cannot be trusted, and must be replaced.

Give the pistol a brief visual check for signs of dropping, or tool marks from previous experimenting. Work the slide. Does it move smoothly? It should move its full travel without catching, binding or hesitating. A binding slide could be a bent slide, dented frame rails, or a mis-fit replacement barrel. All will be moderately expensive to fix. Or, it could simply be a replacement slide that was not fully lapped to fit – which is cheap to fix.

Flip the thumb safety up and down. It should move smoothly and snap from one setting to another. Check the grip safety. It should move in and out without binding, and its spring should snap it back out when released. A grip safety that doesn't move should set off alarm bells in your head. It was popular in competition circles a decade or more ago to pin down grip safeties so they would not move. A pistol with a pinned grip safety is probably a high-mileage competition gun that has seen tens of thousands of rounds. Even if it has seen only light use, you will have to have the grip safety unpinned and properly tuned.

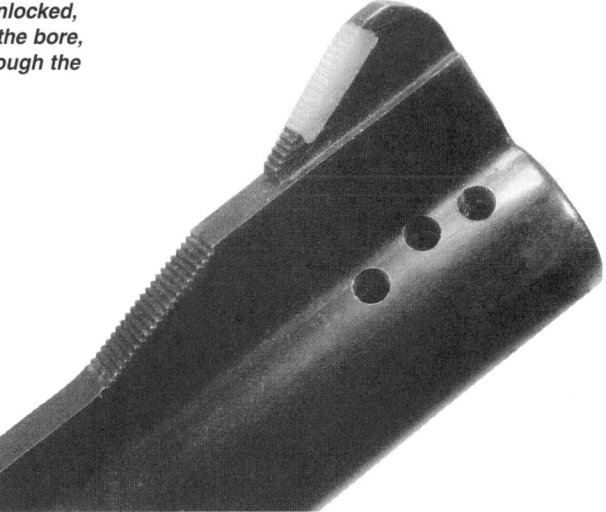

Some home gunsmithing is beyond the pale. These home-drilled "ports" on this revolver have ruined the barrel. If the seller won't subtract the cost of a new barrel and installation from the cost, pass it up.

Now check the function of the safeties. Happily, owners of the 1911 are much less prone to the *"don't dry fire"* attitude. Check to make sure the pistol isn't loaded, then cycle the slide and

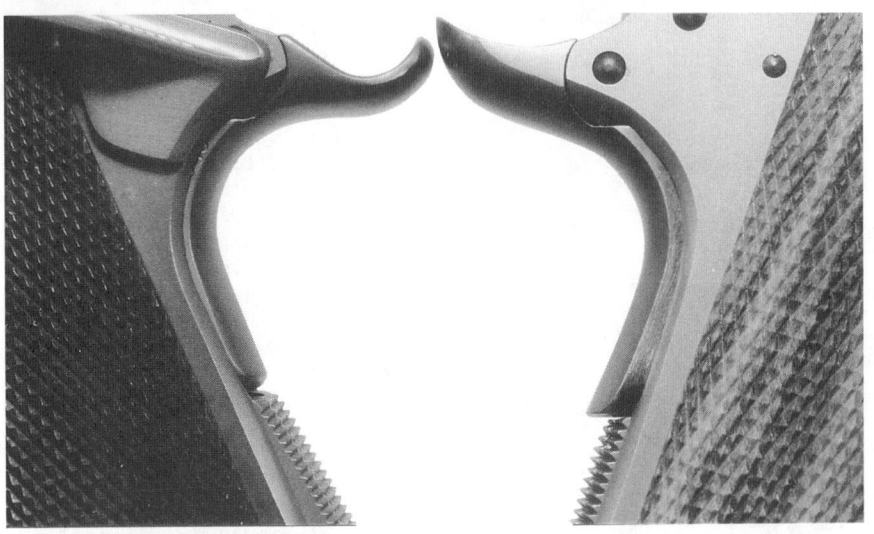

dry fire it. Hold the trigger back and work the slide. It should move smoothly. A pistol that is hard to cycle with the trigger held back could have disconnector problems – or an improperly adjusted trigger binding the disconnector. With the slide cycled back and forth, does the hammer stay cocked *(It better, or you will be facing expensive repairs)?* Next, push the thumb safety *ON*. Pull the trigger (using no more than ten pounds pressure), release the trigger and push the safety *OFF*. If the hammer falls, the safety isn't blocking the sear's movement. I've seen pistols that would fire when the safety was ON and the trigger was pulled. Not very safe and, potentially, an expensive repair.

The grip safety on the left has been pinned down, and doesn't work. Don't buy a 1911 with a pinned grip safety unless it can be unpinned and tested for function.

Does the thumb safety move smoothly, or do you need to force it? Forcing is bad, and indicates a poorly fitted thumb safety.

Once the safety is on, pull the trigger with about 10 pounds of force. Then push the safety off and listen to the sear.

If the hammer stays back, you now listen. Lift the pistol to your ear, and gently thumb back the hammer. If you hear nothing *(assuming you have properly worn hearing protection during all those years of shooting)* then the thumb safety is fine. If you hear a little metallic *"tink"* then the safety needs adjustment. If the safety blocks the sear - but not entirely - the sear can move minutely when you pull the trigger. The *"tink"* is the sear tip snapping back into the bottom of the hammer hooks when the spring pressure is released. If the thumb safety passes the "listen" test, you're on to the grip safety.

Cock the hammer and hold the pistol so you don't grip the grip safety. Pull the trigger. The test, and "listen," are the same as the thumb safety test, looking for the same problems. Now start looking for signs of abuse or experimentation.

Hold the slide partway back and look at the feed ramp. It should be clean and shiny. There should be a gap between the ramp on the frame and the ramp on the barrel. If someone has polished them to be an uninterrupted surface, they have decreased feeding reliabil-

To check the grip safety's function you have to hold the pistol so you don't depress the safety. Then pull the trigger.

To check the disconnector: dry fire, hold the trigger down and slowly cycle the slide.

ity. An improperly polished or ground ramp is expensive to fix.

Should you check barrel fit? Checking won't tell you much. The customary check is to press down on the chamber area to see if it moves, and having moved, if it springs back. The problem is, it doesn't tell you much. I've seen apparently loose pistols that shot quite accurately, and tight pistols that wouldn't shoot worth a darn.

There are some indications that something is amiss. If you are looking at a custom competition gun with a name-brand barrel fitted, and the fit is loose, be suspicious. The barrel may have been simply dropped in (with no attempt at properly fitting it), or it may have been shot tens of thousands of rounds until it wore loose.

If you have a pistol with a plain barrel, tightly fitted, and the front sight is very short, something is up. The barrel may be tight simply because the owner has fitted a long link to the barrel. In which case the link is propping the barrel up to be tight, and the front sight had to be shortened to get the sights to line up with the groups.

Lock the slide open and look down the bore. More so than many other pistols, the 1911 can be a high-mileage survivor.

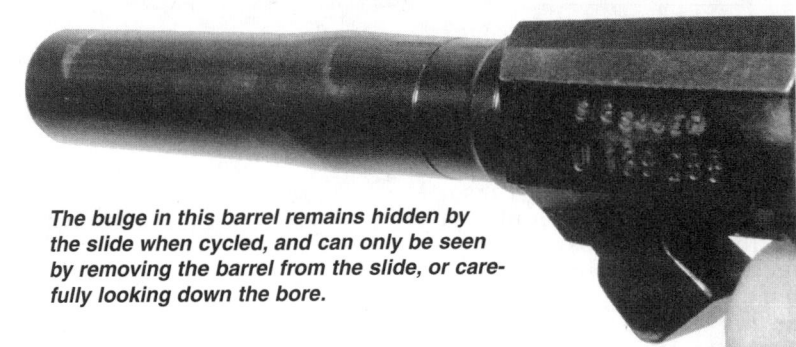

The bulge in this barrel remains hidden by the slide when cycled, and can only be seen by removing the barrel from the slide, or carefully looking down the bore.

On the subject of the cost of repairs to a 1911, the same symptoms can be cheap - or expensive - depending whether the parts involved merely need adjustment, or must be replaced. Accept a dysfunctional 1911 into your home only after careful consideration and acceptance of potentially high repair costs.

Autoloading Pistols, Double-Action

Your visual inspection for the DAs will be the same as with the 1911, except that more of the DAs will have alloy frames. You must take a closer look, especially at a police trade-in, to check for signs of dropping. If you have a pistol with worn bluing, but new grips, look closely. New grips go on only when the old ones are too far

ing grade" here, as you can't do more than cosmetic damage, even by throwing one into a cement mixer. Do the dry fire and slide cycle test just as you would with the 1911. Hammerless guns, or DA-only guns, where the hammer follows the slide down, obviously won't show you a cocked hammer to manually manipulate. Dry fire them, cycle the slide, and dry fire again.

The safety check is less involved than with the 1911, and is dependent on design. On Glocks, cycle the action and attempt to press the trigger back without depressing the centrally-mounted trigger safety. On DA guns, drop a pencil down the muzzle, eraser end first, and point up.

Push the safety lever to *SAFE,* or use the de-cocking lever. The pencil shouldn't move. Don't pay attention to vibrations. If the

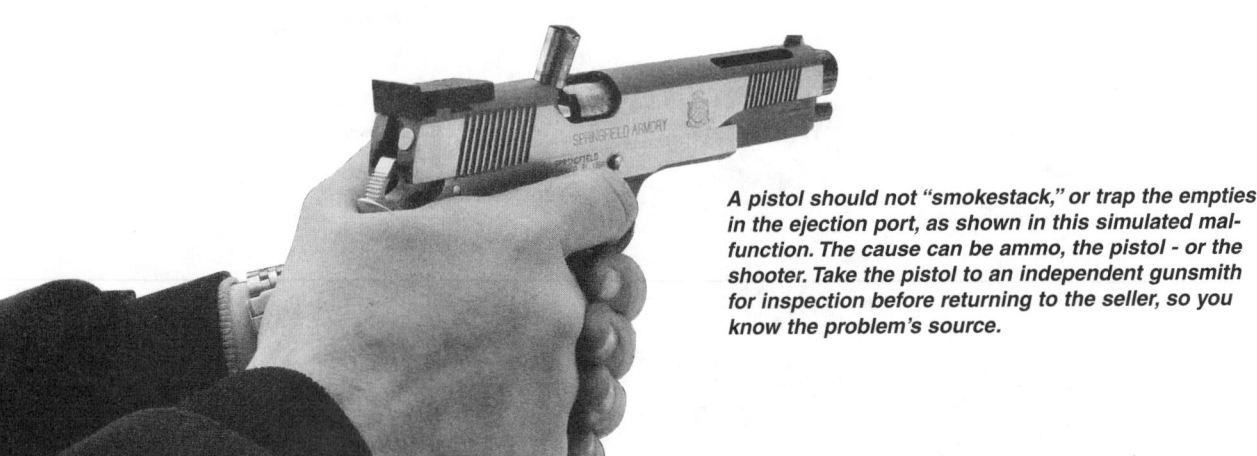

A pistol should not "smokestack," or trap the empties in the ejection port, as shown in this simulated malfunction. The cause can be ammo, the pistol - or the shooter. Take the pistol to an independent gunsmith for inspection before returning to the seller, so you know the problem's source.

Is the bore clean, or fouled with lead or copper? Is the muzzle worn from cleaning? Is there heavy brass "marking" behind the ejection port? Signs of high mileage are not a reason to pass, but if the pistol is offered as "new" or "like new" and you see signs of bore wear, hold on to your money.

gone to be presentable. Police guns get dropped, whacked into car doors and frames, door jambs, light poles, vending machines and seat belt buckles – and that is just when holstered!

Check the frame closely for cracks and signs of dropping, and pass on cracked frames. Glocks get an automatic "pass-

safety isn't blocking the firing pin, the pencil will get launched out and upwards.

Buying a used firearm can be rewarding, fun and educational. By taking a few precautions, and using the inspection procedures outlined, you can avoid buying a walnut and blue steel lemon. Have fun and stay safe! ◆

Evaluation of Arms Condition

by Tom Turpin

An accurate assessment of a used firearm's condition is necessary to establish the value range for the model. The author reviews examples that range from *NRA Perfect* to *NRA Fair* to guide you in evaluating the condition of modern used firearms.

One of the most popular semi-auto shotguns ever made, the Remington Model 1100. This one is a 20 gauge. The gun has been used considerably but has not been abused. It retains about 95% of its finish and is in *NRA Very Good* condition.

Perhaps best known for the wartime versions produced by Walther *(AC)*, Mauser *(BYF)* and Speewerke *(CYQ)*, the P-38 was also produced commercially after WWII by Walther. This one is unfired, with all the original paperwork and in *NRA Perfect* condition.

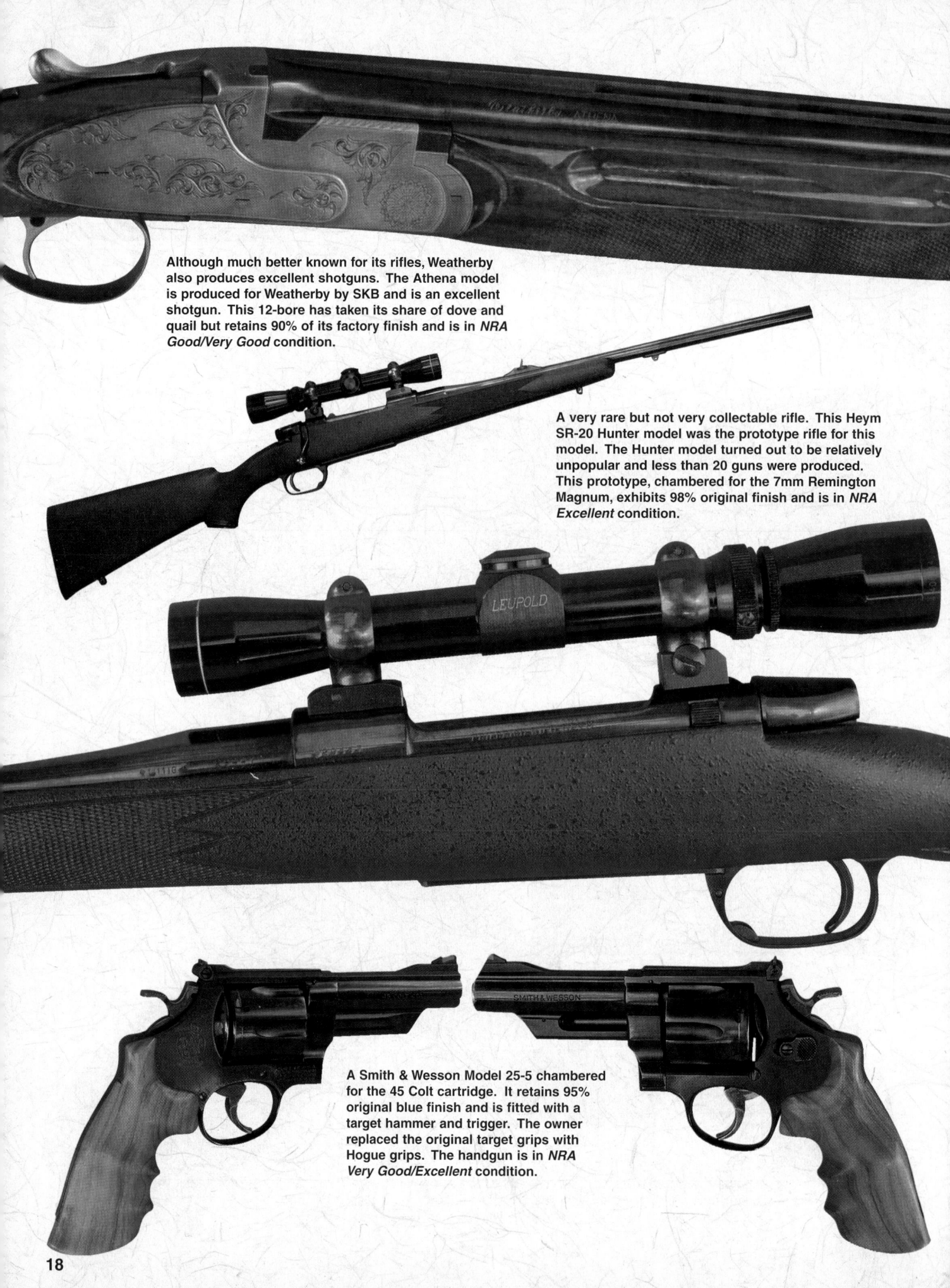

Although much better known for its rifles, Weatherby also produces excellent shotguns. The Athena model is produced for Weatherby by SKB and is an excellent shotgun. This 12-bore has taken its share of dove and quail but retains 90% of its factory finish and is in *NRA Good/Very Good* condition.

A very rare but not very collectable rifle. This Heym SR-20 Hunter model was the prototype rifle for this model. The Hunter model turned out to be relatively unpopular and less than 20 guns were produced. This prototype, chambered for the 7mm Remington Magnum, exhibits 98% original finish and is in *NRA Excellent* condition.

A Smith & Wesson Model 25-5 chambered for the 45 Colt cartridge. It retains 95% original blue finish and is fitted with a target hammer and trigger. The owner replaced the original target grips with Hogue grips. The handgun is in *NRA Very Good/Excellent* condition.

This 1911 Colt 45 auto has an interesting history. The frame is a commercial frame produced in about 1914 that is mated to a wartime Ithaca slide. The frame retains about 50% original blue finish while the slide retains 80% of the military Parkerized finish. The owner of the gun, when it was a complete commercial gun, was a young 2nd Lieutenant on active duty in Germany whose superiors frowned upon the blue finish. He had his company armorer swap the commercial slide for one of the wartime Parkerized slides. Whatever happened to the original slide is unknown. Due to its mix-and-match parts, the pistol would be graded as in *NRA Fair* condition.

A little known and quite rare Ruger Blackhawk. This one is a special version produced for the Arizona Rangers. I believe that approximately 210 pistols were produced. It is chambered for the 45 Colt cartridge and is marked Arizona Rangers on the barrel. It also features an outline of the state of Arizona on the top strap and a special serial number on the bottom of the grip strap. This one is 100% with all original paperwork and is rated in *NRA Perfect* condition.

A Smith & Wesson Model 18 Combat Masterpiece in 22 LR. This Model 18 is factory-fitted with a target hammer and trigger. The owner of the gun replaced the factory grips with a set made by Hogue. Retaining 95% of its original finish, it is in *NRA Very Good/Excellent* condition.

Remington Model 700 BDL Varmint rifle chambered for the 22-250 cartridge. One of the most popular bolt-action sporting rifles of all time, the Model 700 is commonly encountered. This one has 95% original finish and has been fitted with a set of Redfield scope mounts. It is rated *NRA Very Good/Excellent*.

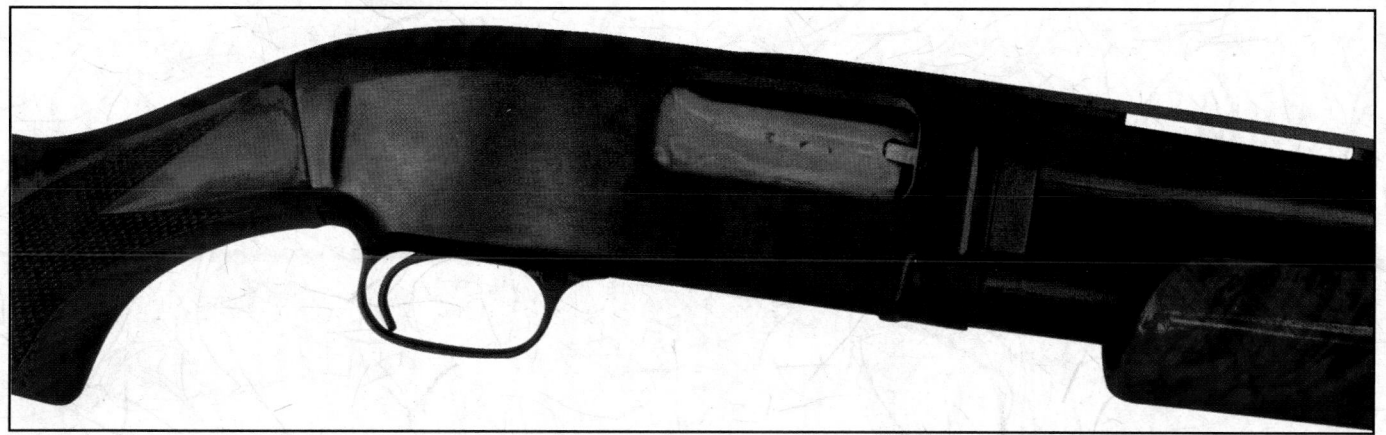

A 1950s version of the Winchester Model 12 pump-action shotgun. This gun is chambered for the 3-inch 12 gauge shell and is choked about as full as one can get. Although the gun is presently in 98% condition, I believe it has been refinished. The rib is not factory original and was added by Simmons, which most likely did the refinish job as well. If the gun was all original, it would merit an NRA Excellent condition rating. Refinished, and with a non-factory original rib, it would rate as *NRA Good*, at best.

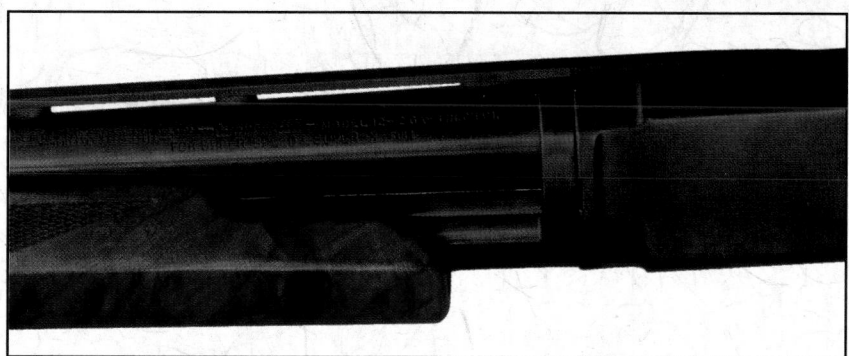

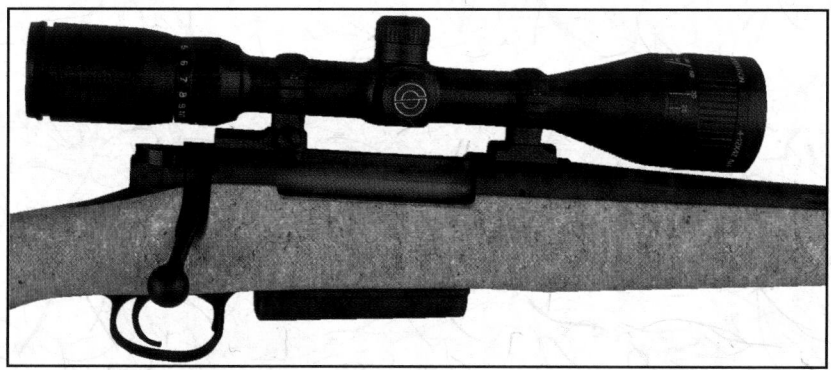

This rifle is an H-S Precision Pro-Hunter Model. It features a special aluminum bedding block system molded into the Kevlar-reinforced graphite stock and H-S cut-rifled barrel. Originally manufactured in Prescott, Arizona, they are now made in Rapid City, South Dakota. This rifle retains 99% finish and is in *NRA Excellent* condition.

This Winchester Model 94 was produced in 1905 and is chambered for the 38-55 cartridge. It features a 26-inch octagon barrel and has seen extensive use. Only about 45% of the original blue finish remains and the stock has been dinged and dented considerably. Still, it is an interesting rifle in *NRA Fair* condition.

The Sako Safari Model was not a smashing success and is not commonly encountered. This one is chambered for the 338 Winchester Magnum and although used, retains 90% of its factory finish. It is in *NRA Good/Very Good* condition.

This Ruger M-77 is a varmint model chambered for the 220 Swift cartridge. The Ruger folks added many desirable features to the M-77, including the tang safety and scope ring bases integral to the bridge of the action. This rifle has 95% original finish and is in *NRA Very Good/Excellent* condition.

The Walther Model PP was first produced in 1929 and, as far as I know is still in production. This German-made Walther is chambered for the 9mm *Kurz* (380 ACP). It retains 95% finish and has been used very little. It is in *NRA Excellent* condition.

A CZ-83, chambered for the 380 ACP, in *NRA Perfect* condition. The chambering must have given the CZ folks some problems as the box features an added sticker marked "Cal. 9mm Short," which was obviously cheaper than printing a new box. On the other hand, the slide of the gun is marked "9mm Browning," another European name for the 380 ACP.

An Illustrated Guide to Condition

by Joseph Schroeder

*To further clarify the standard classifications of "Excellent" through "Fair," we present the following photographs. Since "New" means just that—new, in the original box with all papers and accessories and "Perfect" means the gun is new but the original box and / or accessories may be missing, we do not show those classifications.
All other classifications are subjective, and subject to argument!*

EXCELLENT

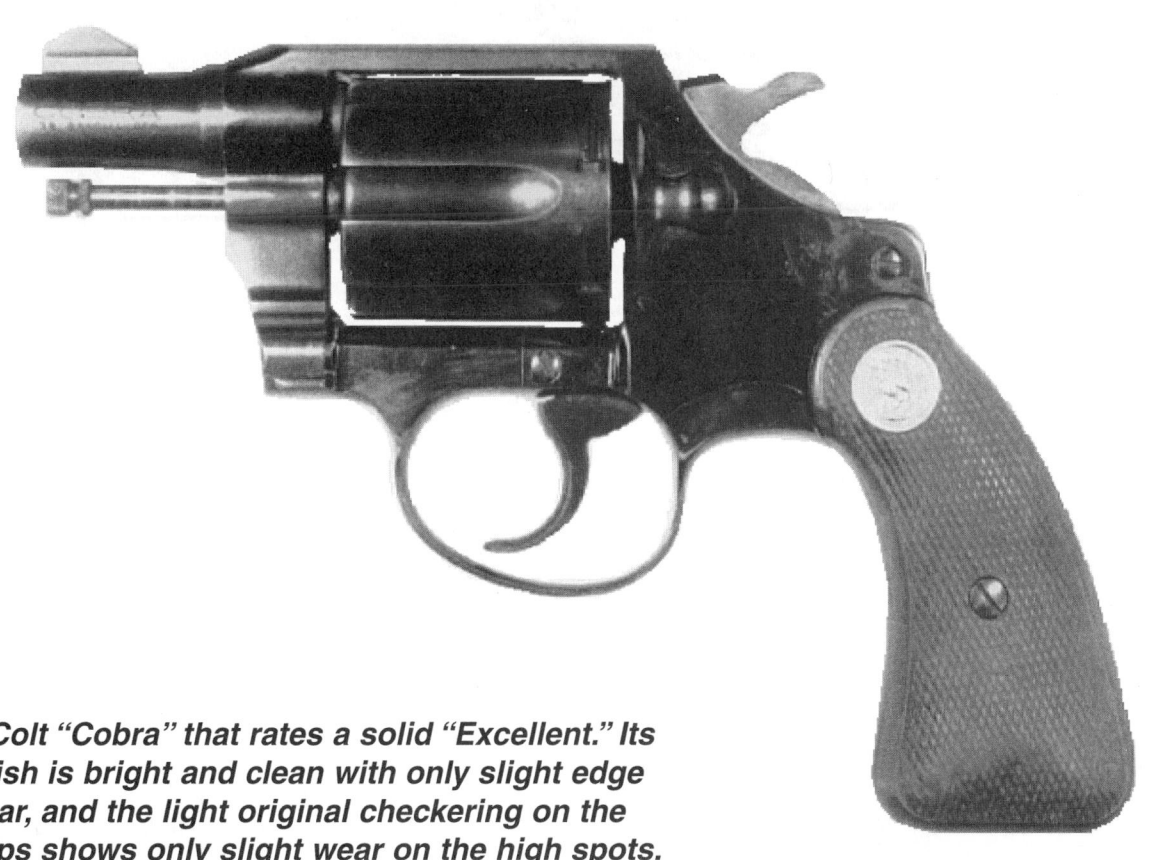

A Colt "Cobra" that rates a solid "Excellent." Its finish is bright and clean with only slight edge wear, and the light original checkering on the grips shows only slight wear on the high spots.

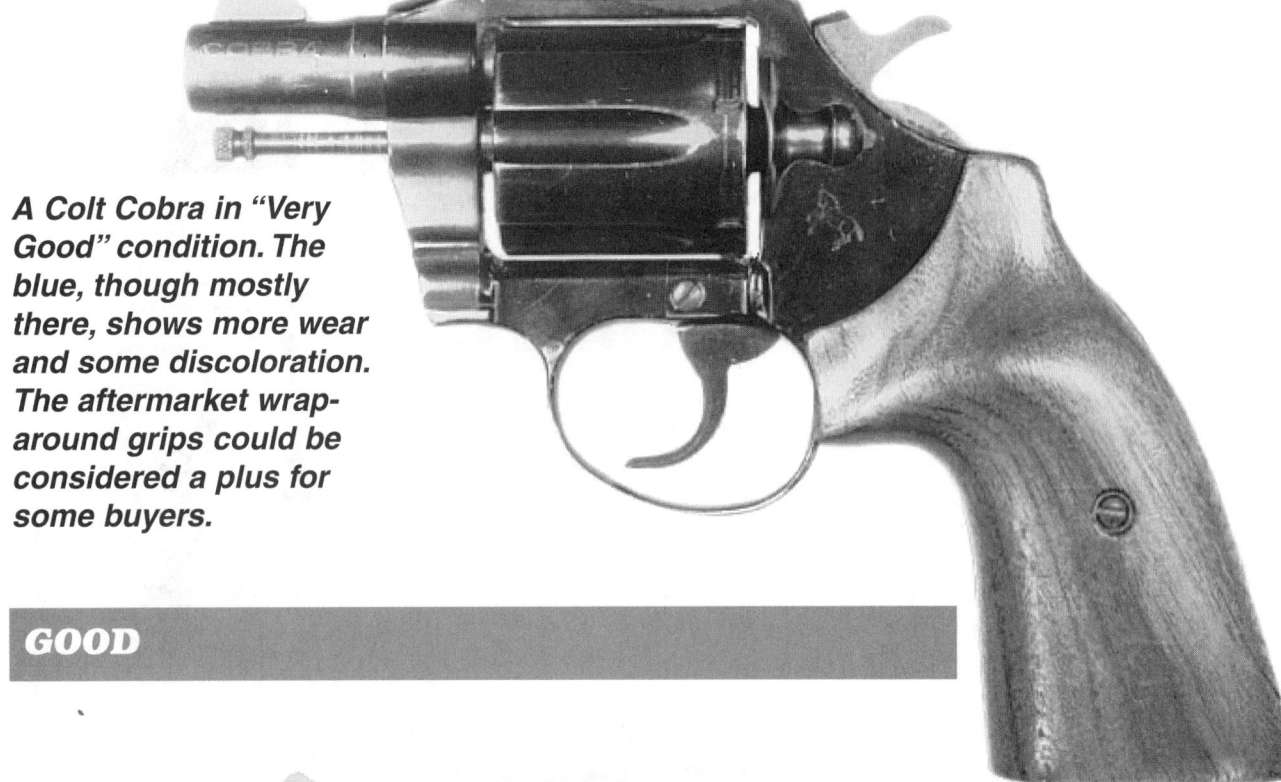

A Colt Cobra in "Very Good" condition. The blue, though mostly there, shows more wear and some discoloration. The aftermarket wrap-around grips could be considered a plus for some buyers.

GOOD

A Colt Cobra that would rate a high "Good." There is a good deal of blue wear with some light rust freckling, and the frame has been drilled just above the hammer for a hammer shroud (missing). The grips are so worn that the original checkering is entirely gone.

The cylinder of the "Good" condition Colt Cobra. Note the fingerprint etched into the cylinder by perspiration and the pits on the adjacent chamber.

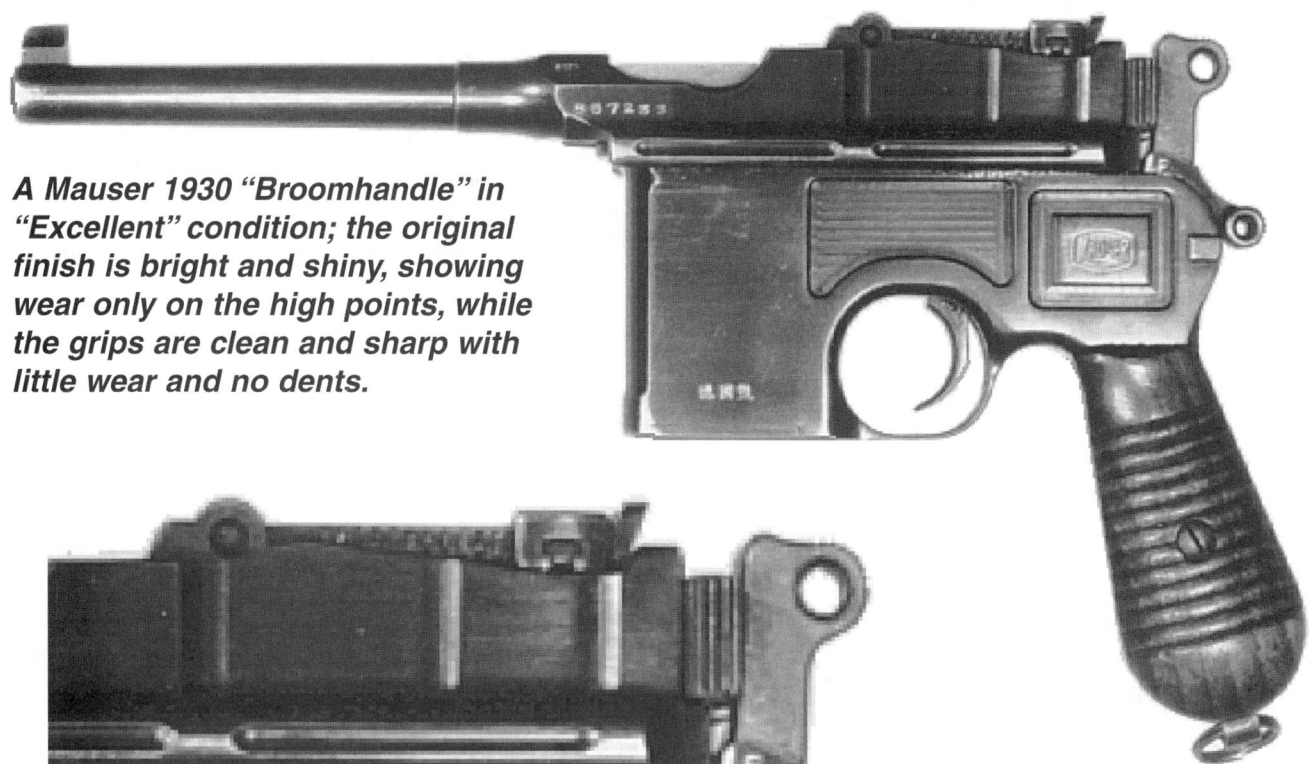

A Mauser 1930 "Broomhandle" in "Excellent" condition; the original finish is bright and shiny, showing wear only on the high points, while the grips are clean and sharp with little wear and no dents.

The left side of the "Excellent" Mauser 1930; note the tight fit of the magazine bottom. The Chinese characters translate to "Made In Germany" and adds 5 to 10 percent to the value as compared to the same model without markings.

VERY GOOD

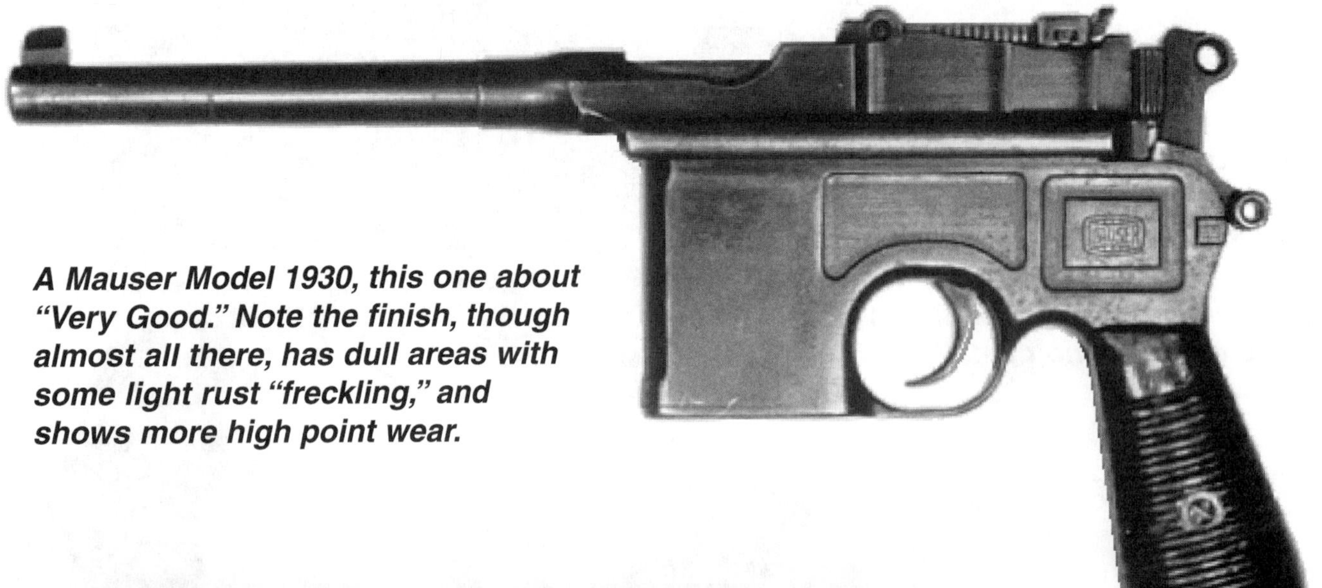

A Mauser Model 1930, this one about "Very Good." Note the finish, though almost all there, has dull areas with some light rust "freckling," and shows more high point wear.

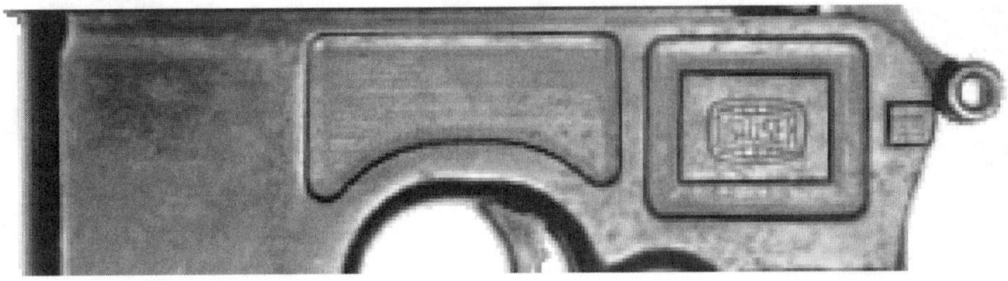

Though the grip grooves are still deep and sharp on this "Very Good" Model 1930, the wood does show the nicks and scratches that come from use.

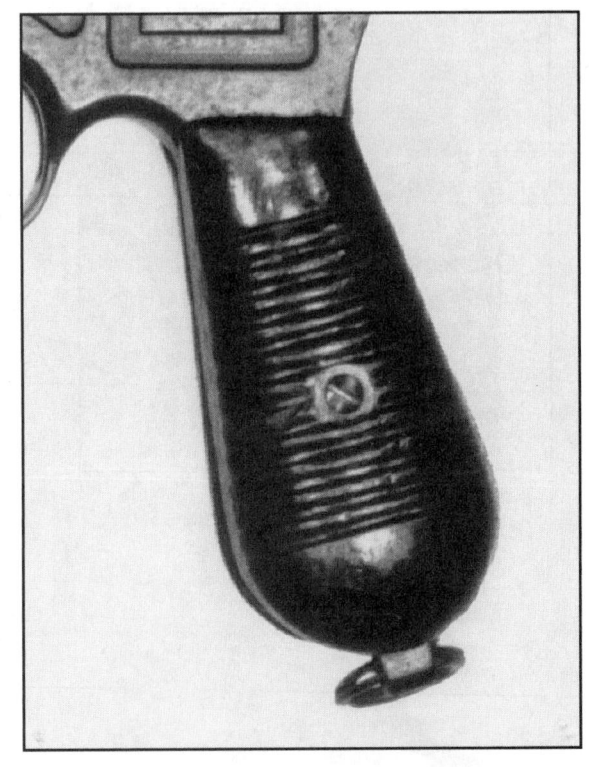

A Model 1930, this one about "Good." The finish is mostly gone, and there is pitting on the muzzle and around the grips. The grips themselves have some bad scars, with much wear.

The muzzle of the "Good" Mauser; in addition to the deep pitting, the muzzle crown is also worn almost flat.

IMPORT MARKINGS

Two examples of the import markings now required on all firearms imported into the U.S. The OBI OBNY under the Mauser chamber (right) indicated the importer was Oyster Bay Industries in Oyster Bay, New York. Oyster Bay also marked some Mausers by rolling their marking around the barrel (above). Guns without import markings, such as those brought back by GIs, bring premium from collectors.

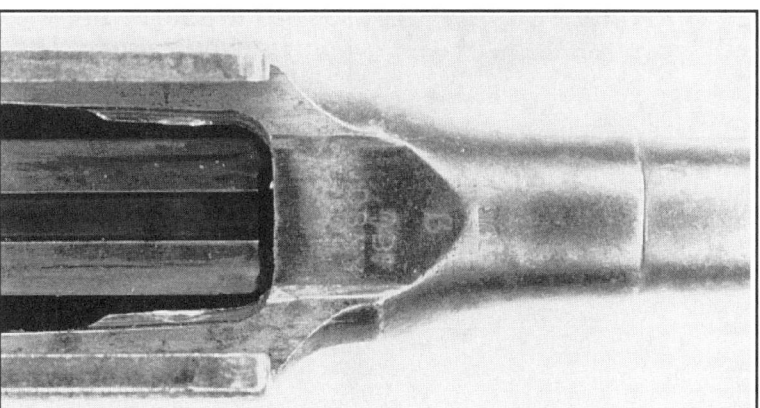

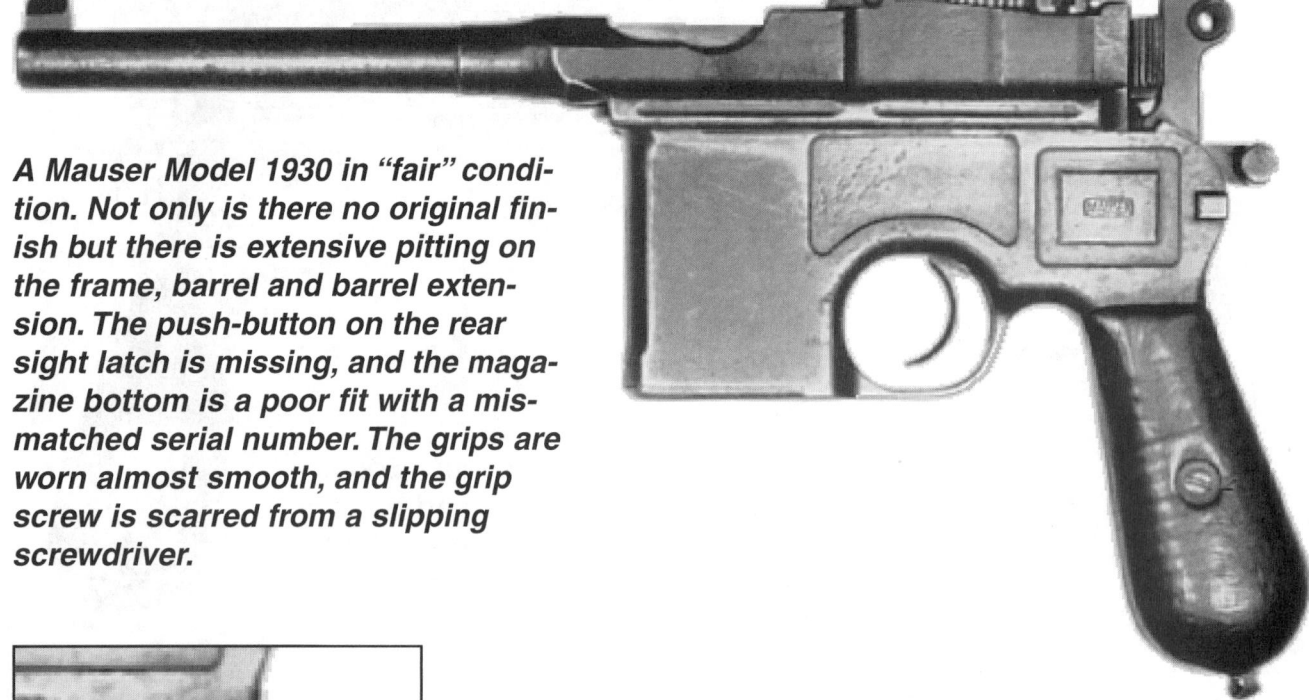

A Mauser Model 1930 in "fair" condition. Not only is there no original finish but there is extensive pitting on the frame, barrel and barrel extension. The push-button on the rear sight latch is missing, and the magazine bottom is a poor fit with a mismatched serial number. The grips are worn almost smooth, and the grip screw is scarred from a slipping screwdriver.

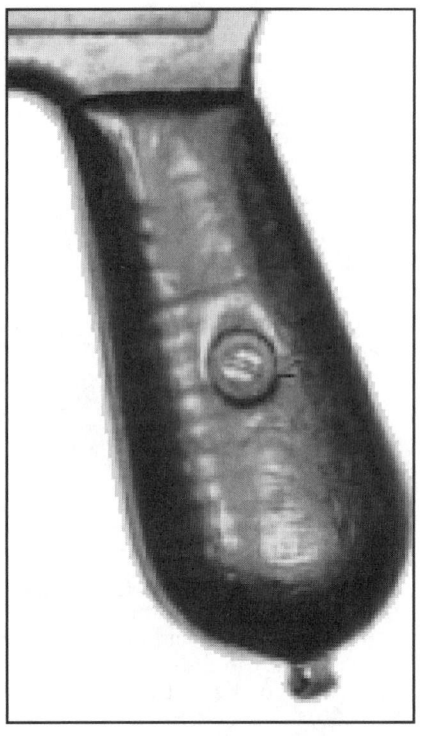

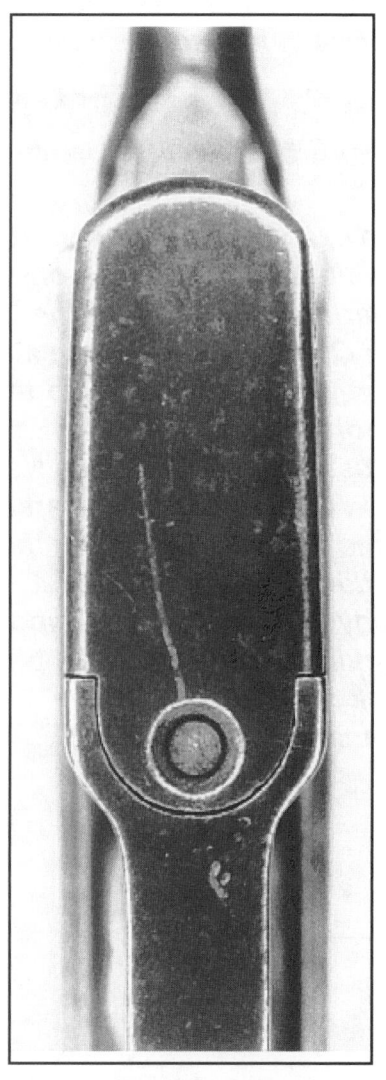

The underside of the "Fair" Mauser; the magazine bottom is a poor fit in the gun, showing an irregular gap between parts. The scratch is from careless use of a screwdriver during removal.

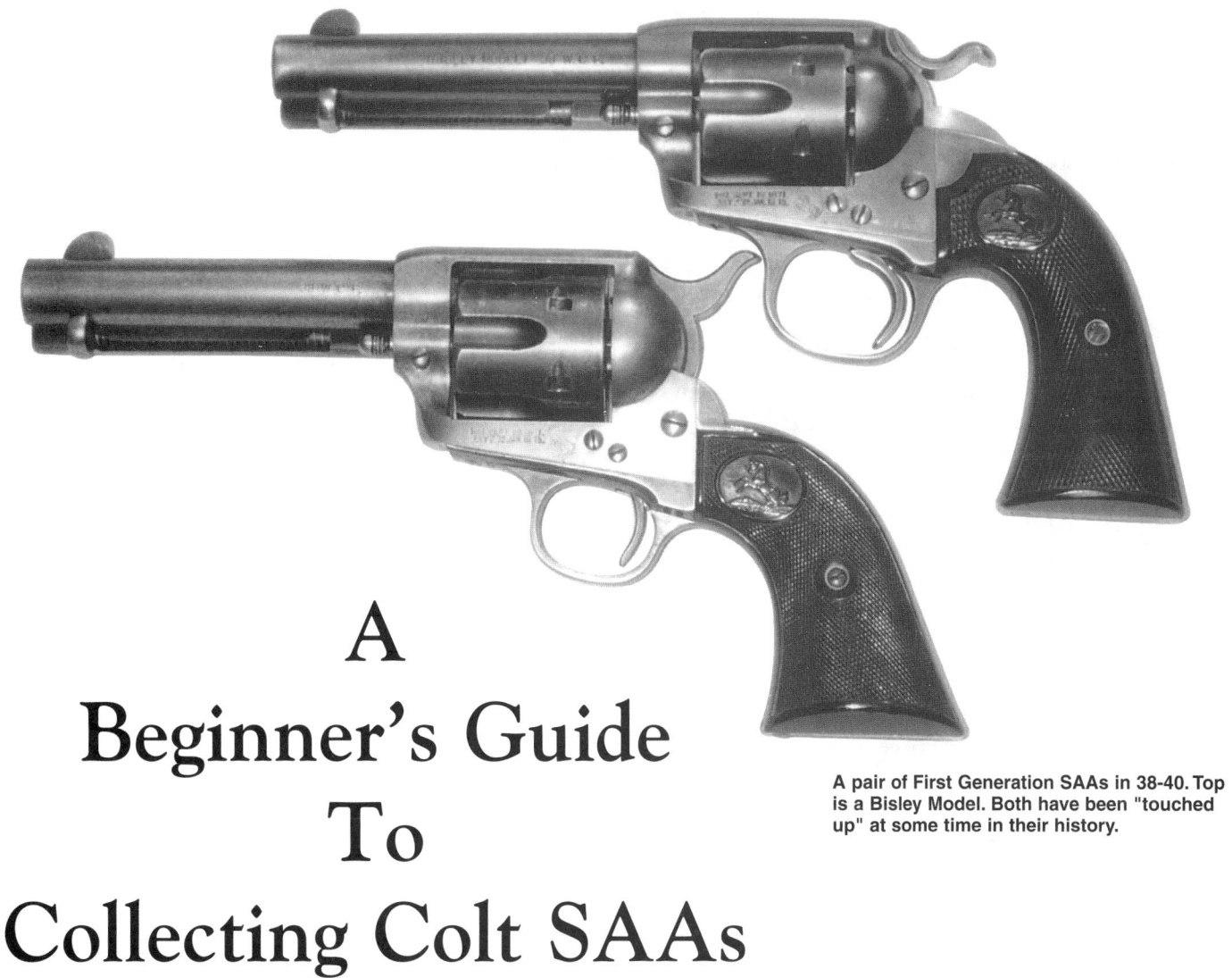

A pair of First Generation SAAs in 38-40. Top is a Bisley Model. Both have been "touched up" at some time in their history.

A
Beginner's Guide
To
Collecting Colt SAAs

By John A. Luchsinger, Jr.

THE COLT SINGLE Action Army (SAA) revolver is one of the most easily recognizable handguns of the last one hundred and twenty-five years. It was at Custer's Last Stand and was used by lawmen and desperadoes alike during the westward expansion of the United States. Not only was it popular in the United States, but Colt's heavy-framed SAA could be found in the holsters of adventurers worldwide. The history associated with the SAA helps to make it the most popular collectors' handgun of all time. Collecting Colt SAAs is an enjoyable and fascinating pastime, but it is not a task to be undertaken without advance preparation and research.

One might ask, "Why should I collect Colt Single Action Armies?" Well, for one thing, I can guarantee that you will not be able to collect all of the different variations. There are simply too many, and some types were produced in such low numbers that there are not

enough to go around. The second reason is economic. In the past twenty years, the value of SAAs has gone up at a rate to rival some Internet stocks. The 1979 GUN DIGEST lists a suggested retail price for a SAA of $336.95. Today, GUN DIGEST 2001 lists a suggested retail price of $1900. This price is just for a run-of-the-mill SAA!

The traditional way to look at the manufacture of Colt SAAs is by breaking them down into three generations, representing the three basic phases of manufacture by Colt. The First Generation SAAs were manufactured from 1873 to 1940. The Second Generation SAAs were manufactured from 1956 to 1975. The Third Generation SAAs began production in 1976 and continue to present.

When contemplating the purchase of any SAA, there are three things to consider: First, *the condition and originality* of the gun and box, if present; second, the *rarity* of the gun (how

many of that variation were manufactured originally and how many still exist); finally, and not to be underestimated in its importance, the *demand* for the SAA in question. All three considerations help to create the true market value of the individual gun.

Condition is critical in evaluating the value of a Colt SAA. What percentage of original finish does the gun have? Are the grips original to the gun? The key word here is **ORIGINAL**. I would take 30% original condition over 100 percent-refinished condition any day. Do the screws show approximately the same amount of wear as the rest of the gun? Shiny, blued screws when the rest of the gun is "gray" can be a dead giveaway that they have been replaced. Does the label of the box match the SAA that is in it? Is it even the correct type of box? A for-sale listing for a SAA may say new in box (NIB), but has the cylinder been turned? Is there a drag mark around the cylinder? All of these

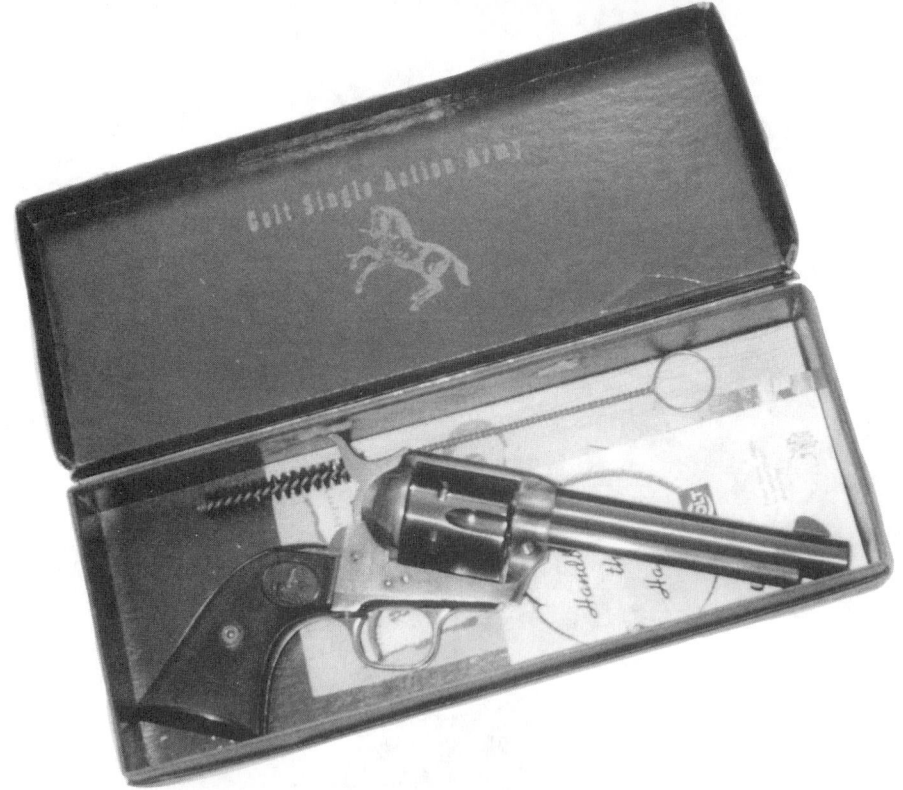

First year production (1956) Second Generation SAA in 38 Special with 5 1/2-inch barrel in original black box with paperwork and bore brush. They don't come much cleaner than this!

aspects – and many more – should be analyzed before a purchase is made.

A Second Generation SAA that might be worth two thousand dollars in 100% original condition in its original black box, would be worth less than half that amount if it has been refinished. The situation becomes even more complex if one is dealing with First Generation guns which have been around so long and have changed hands so many times

that many have, at some time or another, been improved/modified by their owners. It becomes the task of the prospective buyer to evaluate the gun and figure out whether it is, in fact, an "honest gun." But do not be afraid, because there are people and resources that can help you.

Colt provides a service whereby for a fee of $100 per gun, they will research a given serial number and provide a factory letter describing the original configuration of the

gun. All you have to do is send the serial number and a check for $100 to: Colt Historian, P.O. Box 1868, Hartford, CT 06144. The research service is offered for all generations of SAAs. This allows one to pass on a hundred year-old Colt SAA that is now nickel-plated with a 4 3/4-inch barrel in 45 Colt that was originally blued with a 7 1/2-inch barrel in 32-20 when it left the factory. The difference in value between an original "honest gun" and a parts gun can be astronomical.

Directly linked to **condition** is **rarity**, but the two are not mutually exclusive. Over its production period, Colt's SAA has been offered in the three standard barrel lengths of 4 3/4, 5 1/2, and 7 1/2 inches, but other barrel lengths have been available on special order. One common barrel length variant SAA is the Sheriff's Model, a type of SAA without an ejector rod under the barrel, which is usually, but not always shorter than standard. A 1st Generation SAA Sheriff's Model is a very rare gun. Even if it lacks all of its original finish, it is still a very valuable gun. A more common variant in the same poor condition would be worth much less. Once again the best way to play it safe is to obtain a letter authenticating the gun from Colt prior to purchase.

The First Generation SAA was manufactured in over thirty different chamberings with 45 Colt, 44-40, 38-40, 41 Colt, and 32-20 being the most common. Now, if I were contemplating the purchase of a First Generation SAA in – let's say – 357 Magnum, I would be very wary since records show that only 525 SAAs were manufactured in that chambering prior to the war. The number of replacement 357 barrels and cylinders manufactured between then and now is anyone's guess. With the money at stake being so high, I would not buy such a rare gun without a factory letter stating that the gun left the factory as a 357 Magnum. Watch out for the ones that are too good to be true; they probably are.

The final aspect that helps to determine the ultimate value of a SAA is the **current market demand** for the individual variant. The Colt New Frontier, which is basically a SAA with adjustable sights and Colt's Royal Blue (a few were nickel) finish, is a perfect example. Colt New Frontiers were manufactured both as Second and Third Generation guns, and although they were but a relatively small percentage of the overall SAA production, there has not been the same demand for them in the collec-

Another Second Generation SAA in 38 Special with 5 1/2-inch barrel. Note the pitting on the cylinder and hammer.

A pair of Third Generation New Frontiers in Colt Royal Blue finish and wood stocks. Top is a 45 Colt with 7 1/2-inch barrel. Bottom is a 44 Special with 5 1/2-inch barrel.

tors' market. They are, therefore, usually priced less than their more traditional SAA counterparts. The New Frontier is a fine gun, and some barrel lengths and calibers were manufactured in very low numbers. They just do not appear to have the same collector appeal at this time. If I were starting a collection of SAAs now, I think that I would concentrate on New Frontiers, since I could get the most bang for my collecting buck. But then again, I must not be representing the general SAA-collecting public.

The most important thing to do before you start collecting SAAs is to start collecting knowledge. You can begin by reading reference works on the SAA by acknowledged SAA experts such as R.L. Wilson, Donald Wilkerson, John Kopec and Keith Cochran, among others. Obtain a current copy of one – or several – of the well-known arms price guides: *Blue Book of Gun Values, Flayderman's Guide to Antique American Firearms...and their values, Modern Gun Values* and the *Standard Catalog of Firearms* and have it with

Third Generation Colt Sheriff's Model in 44-40 supplied with extra 44 Special cylinder in factory French-fitted case. Set is one of 1,150 so equipped (44 Special cylinder shown in gun).

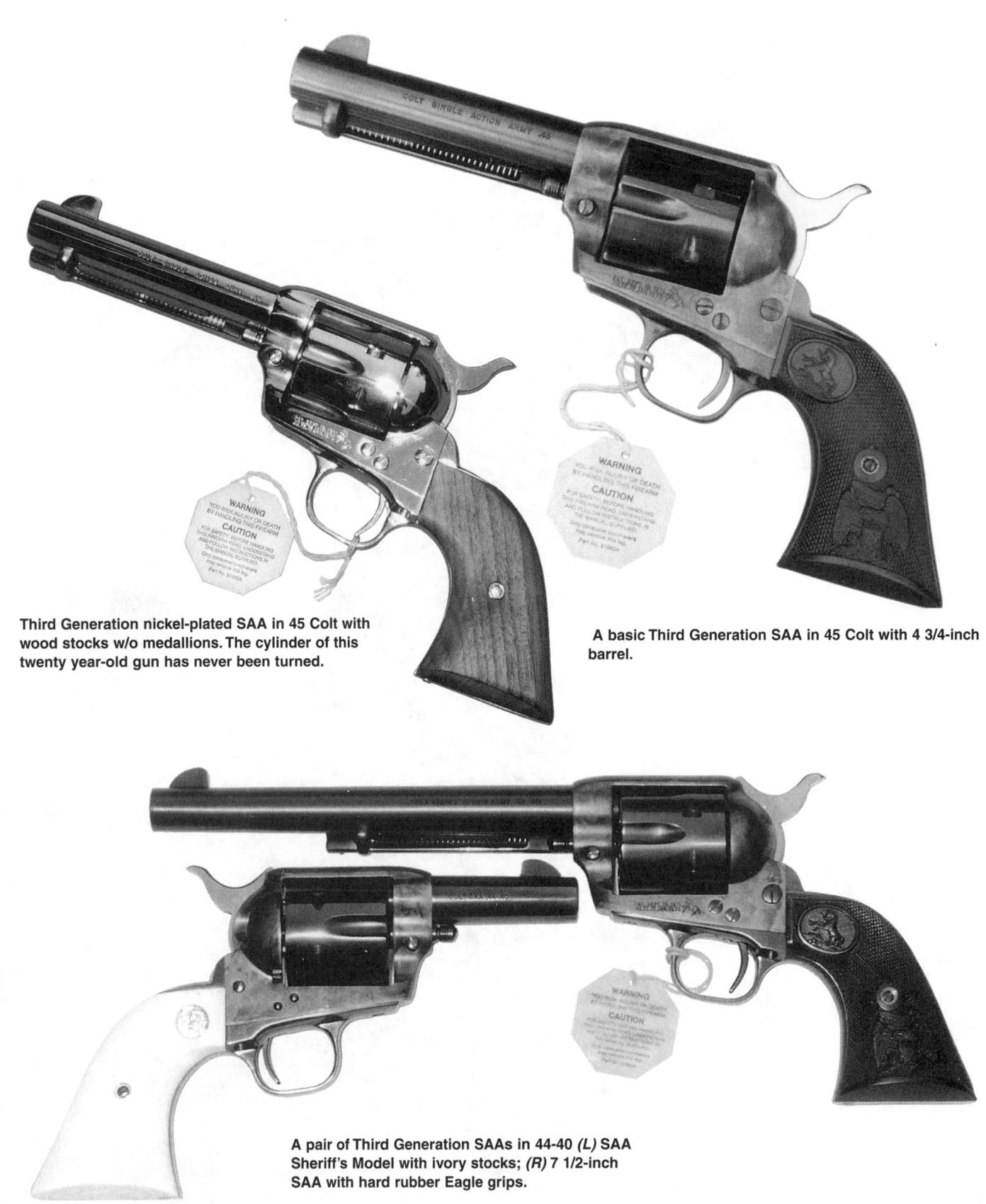

Third Generation nickel-plated SAA in 45 Colt with wood stocks w/o medallions. The cylinder of this twenty year-old gun has never been turned.

A basic Third Generation SAA in 45 Colt with 4 3/4-inch barrel.

A pair of Third Generation SAAs in 44-40 (L) SAA Sheriff's Model with ivory stocks; (R) 7 1/2-inch SAA with hard rubber Eagle grips.

you when you contemplate a purchase. The reference works will let you know what to look for and the price guides will give you general guidance on what an individual variant is worth. Remember that gun collecting is a very detail-intensive hobby, and those little details make the difference between a sow's ear and a silk purse. The more you know about Colt SAAs – and the more you know about evaluating guns in general – the better prepared you will be when that once-in-a-lifetime chance comes to buy that rare SAA of your dreams. ◆

BUYING USED CUSTOM RIFLES

by John A. Luchsinger Jr.

A PREVIOUSLY OWNED custom rifle may be the best of deals, or the worst of deals, depending on what one is looking for in a rifle. Often, one can purchase a used custom rifle for much less than having a new one made by the same maker. Custom rifles, in general, are built for individuals who want something special or unique in a rifle not obtainable in a production model. They come in all action types with bolt actions being the most common. These rifles can be both practical and collectible. For many years, custom rifles have been built by the most prestigious makers in the world – and in countless neighborhood gun shops and basements. Quality ranges from the apex of the rifle makers' art to simply abysmal. This article will concentrate on the high end of the spectrum.

The purchase of a used custom rifle should not be entertained without first considering your personal requirements for that rifle. Do you want to actually hunt with the rifle or do you simply want to possess a work of art? If you want to use the rifle for hunting, the rifle should fit you without significant modification. Length of pull, the distance from the trigger to the buttplate can vary greatly from one custom rifle to another. These rifles were, after all, designed to fit one individual. If the length of pull is

Steel trap-door buttplate on a Hoffman Arms Company Magnum Mauser in 375 H&H. Would you replace it with a recoil pad?

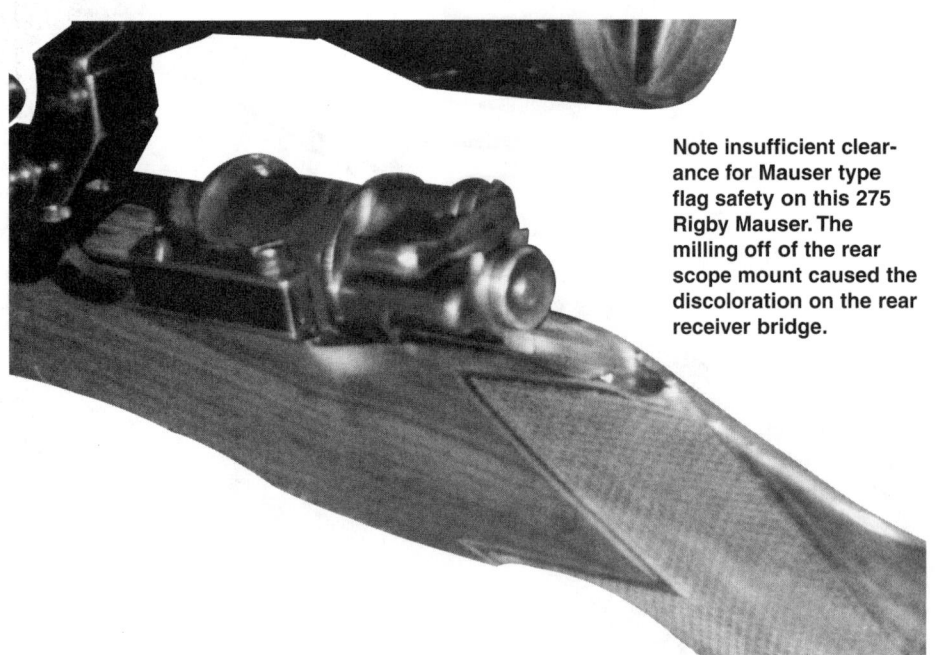

Note insufficient clearance for Mauser type flag safety on this 275 Rigby Mauser. The milling off of the rear scope mount caused the discoloration on the rear receiver bridge.

Chamberings such as the 30-06, 7mm Mauser, 375 H & H, and 416 Rigby, among others, are alive and well. They remain the choice of many knowledgeable hunters. Now, if you are the kind of person who has to hunt with the latest, high velocity magnum, these older custom guns are probably not for you. Regardless of chambering, if a custom rifle has a severely worn or neglected barrel, it is probably not a good choice for a hunting rifle purchase.

Type of sighting equipment also must be considered. What was acceptable a century ago might not be what one wants to use on that African safari of a lifetime. My Hoffman Mauser has an ancient, cloudy, two-power Noske scope – not exactly the latest in German optics – and a Howe-Whelen peep sight. Obviously, scopes can be changed, but in this case, replacement would be difficult due to the current, low mount configuration. Many custom rifles of the first part of the last century have only iron sights. In the case of a heavy rifle, designed for use against dangerous game at short range, this may not be a problem. On the other hand, most hunters

too short, you can have spacers added, but this isn't really an option if the rifle in question has a fitted steel or horn buttplate. Do you really want to add a recoil pad to a seventy-five year old collector's item? An example of this is my Hoffman Arms Company Magnum Mauser in 375 H&H. This rifle is fitted with a steel trap-door buttplate. I wouldn't even think about adding a recoil pad. If the length of pull is too long, a thinner recoil pad may be added (if one is present in the first place) or the stock may be shortened. Only you can know whether the finished rifle will be worth the effort and cost of modifying it.

Barrel length is another consideration in the purchase of a custom rifle for hunting. In the first part of the 20th century, it was quite common to fit extremely long barrels to custom rifles, a carryover from the black powder era. It is not uncommon to see both light and heavy rifles fitted with twenty-eight inch barrels. I have seen several English Mannlicher-actioned rifles, as well as numerous double rifles, which have these long barrels. In a double rifle, this may not be as great of a concern, since the action length of a double is relatively short. On the other hand, the 425 Magnum Westley Richards magazine rifle commonly was supplied with a 28-inch barrel. Many users of this rifle, such as the famous African hunter and author, John "Pondoro" Taylor, found these long barrels ungainly. On the other extreme, I have recently seen listed a custom 460 Weatherby Magnum with a 21-inch barrel for sale! Don't worry; there are plenty of custom rifles out there in the more frequently sought 22- to 26-inch barrel range.

Caliber selection is another point to consider when purchasing a used custom rifle. The grand old calibers of the past will kill game just as efficiently as they did a century ago.

Note the more contemporary side mount on this 275 Rigby Mauser with Schmidt & Bender scope. The rifle was originally made for Major P.A. Birkin in 1924.

The classic lines of the 275 Rigby Mauser are unmistakable.

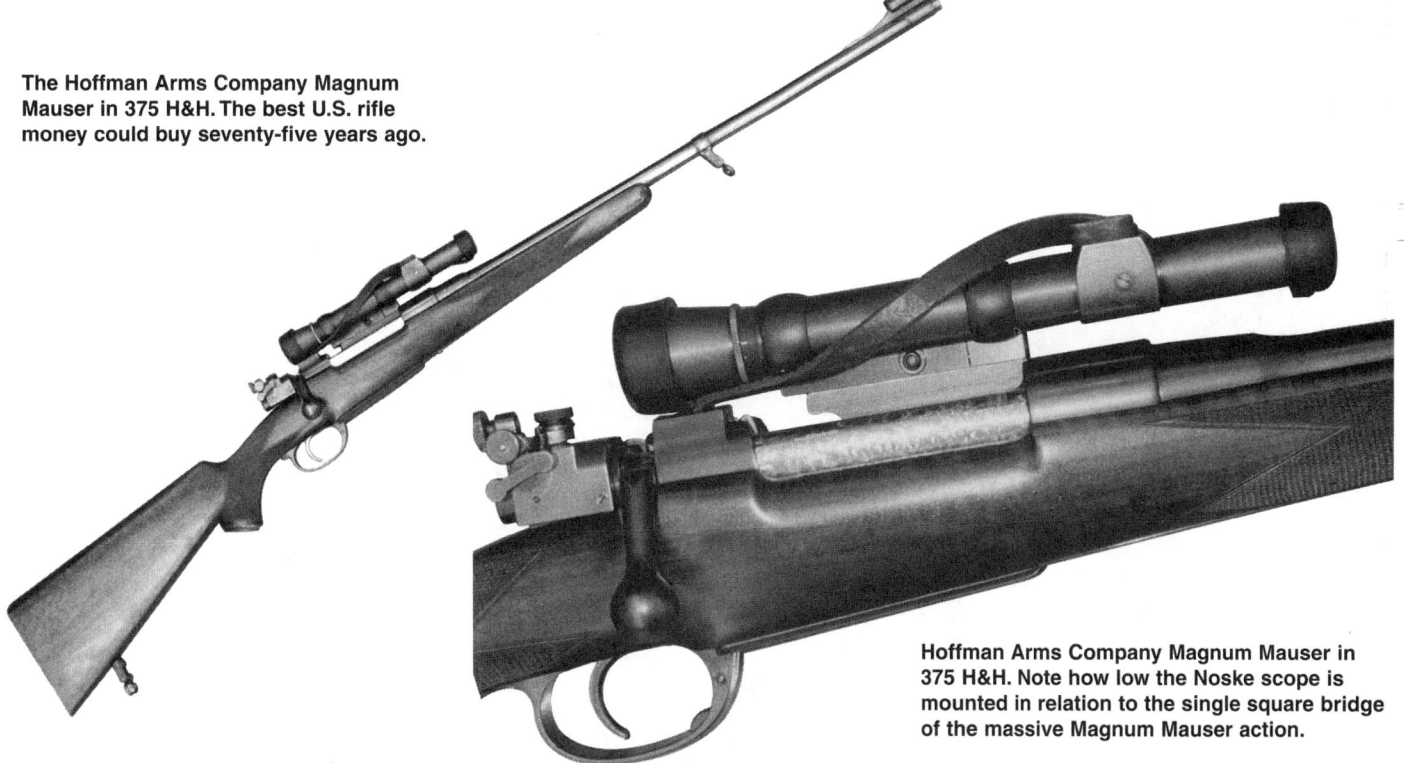

The Hoffman Arms Company Magnum Mauser in 375 H&H. The best U.S. rifle money could buy seventy-five years ago.

Hoffman Arms Company Magnum Mauser in 375 H&H. Note how low the Noske scope is mounted in relation to the single square bridge of the massive Magnum Mauser action.

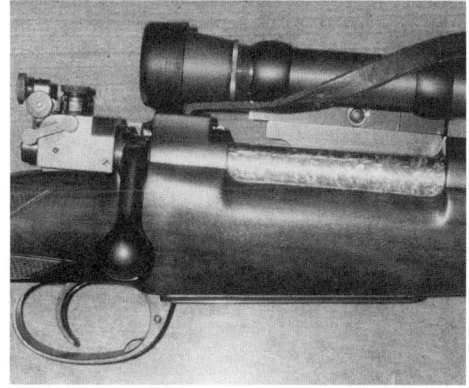

Note the extremely intricate Howe-Whelan peep sight located on the end of the bolt of the Hoffman Magnum Mauser. The safety is the lever located on the immediate right of the sight.

today prefer to have scope sights on their light-caliber rifles. I have a Holland & Holland Mauser in 240 Apex (the equivalent of the 243 Winchester), which has only iron sights. Would I drill and tap it or send it back to Holland & Holland for a scope to be mounted? The answer is no, since drilling and tapping would be a crime, and having a scope mounted by Holland & Holland would probably cost half again what the rifle is worth. In other words, I have a fine rifle with which I can hunt, but I would not take it on a hunt where shots at long range are anticipated.

Something as apparently simple as type and placement of the safety can become an issue with a custom rifle. There are many different types of safeties commonly found on custom rifles, including the traditional Mauser-type flag safety, the Winchester Model 70-type safety, and the shotgun-type tang safety. The Mauser-type flag safety on my 275 Rigby (7mm Mauser) sporting rifle cannot be engaged when the scope is mounted due to the positioning and type of scope. It is a candidate for the fitting of a relatively inexpensive (approximately $250) Model 70-type safety. My Hoffman Mauser has a unique side safety; basically a lever mounted on the side of the Howe-Whelan sight. For non-dangerous game, this unique type of safety arrangement is not a problem. But for dangerous game, I would have second thoughts about using a rifle with a safety so different from any other I have ever used. The choice of safety

type is up to you, but do not buy a hunting rifle if you are not comfortable with its safety operation.

If you simply want to collect custom rifles, with no real intention of using them for serious hunting, then the question becomes both simpler – and more complex. Custom rifles by the best makers can be very collectible and costly. Companies such as Griffin & Howe, Hoffman Arms Company, Rigby, Holland & Holland, Westley Richards, and countless others have produced some of the world's most well-crafted and beautiful rifles. These rifles are expensive. Not only is one paying for the workmanship that went into these rifles, but also for the maker's name on the barrel; with the name comes collectability – and therefore desirability.

As a boy growing up, I remember going to my neighborhood gun shop and seeing an entire collection of beautiful custom rifles for sale. A local gunsmith had crafted these rifles for one customer, who was then selling. These rifles, built on fine Mauser military actions, were stocked in the finest woods. Although extremely well crafted, these rifles were priced little higher than the factory-made rifles that stood in the racks next to them. The reason? The gunsmith who lovingly built these rifles, was and is, an unknown. These rifles will probably never be worth what it took to build them, simply because they were not built by a "name maker." A custom rifle may be well made and functional, but that does not mean it is necessarily collectible.

Whether a custom rifle has been significantly modified from its original specifications is another point to consider. My 275 Rigby Mauser falls into this category. Thanks to Mr. Paul Roberts, owner of the English shotgun- and rifle-making house of J. Roberts & Son, and former owner of John Rigby and Company, I know that my Rigby was scoped when Rigby made it. Unfortunately, at a later date, the rear scope base was rather crudely milled off, and a side mount was placed on the rifle with a more contemporary Schmidt & Bender scope. This type of modification is fine for a hunting rifle, but not nearly so desirable for a collector's piece.

One should also be on the lookout for poorly executed repairs, particularly in the small of the stock, where

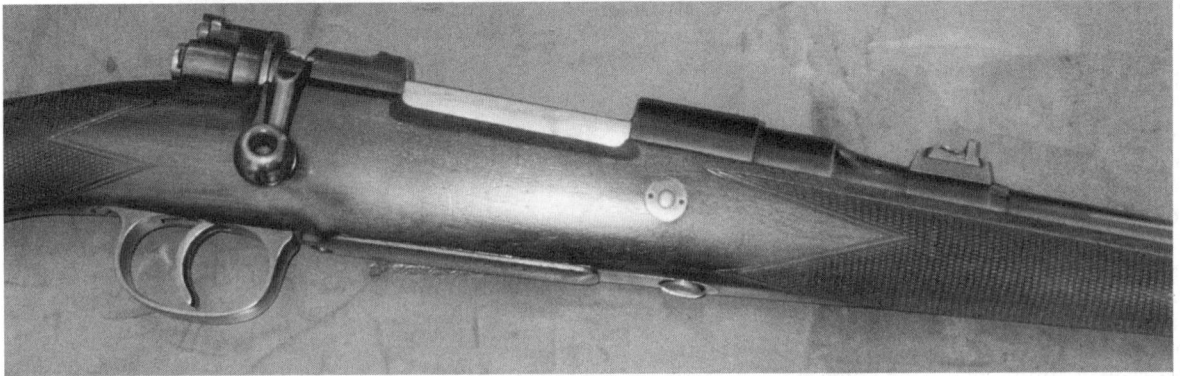

Holland and Holland rifle in 240 Apex. The thumb-screw located forward of the magazine box indicates this rifle is a takedown model.

A scope sight would spoil the clean lines of this rifle.

most cracks occur. As a boy, I remember seeing an original Griffin & Howe Springfield sporter in 30-06 at my neighborhood gun shop. This rifle had all of the bells and whistles: trap-door buttplate, receiver sight, and Griffin & Howe side mount with Zeiss scope. I believe it was priced at $650. It also had a beautiful walnut stock that was, unfortunately, cracked. Even more unfortunate, the crack had been repaired with a wood screw! Now, this rifle would probably give outstanding service in the field, but its poorly repaired stock greatly reduces its desirability to a collector. Properly restocking such a rifle would cost about as much as the rifle is currently worth. Before spending the big bucks for a used custom rifle, carefully consider whether you will be able to recover your money if you decide to sell.

There are many fine custom rifles out there just waiting to be purchased, but before you shell out those hard-earned greenbacks, consider what you want out of the rifle. I will probably never place an order for a fine custom rifle fitted to me, but I will continue to watch for bargains in the used custom gun marketplace. The pride of owner-ship, which comes with owning one of the works of the great gun makers, is priceless. ◆

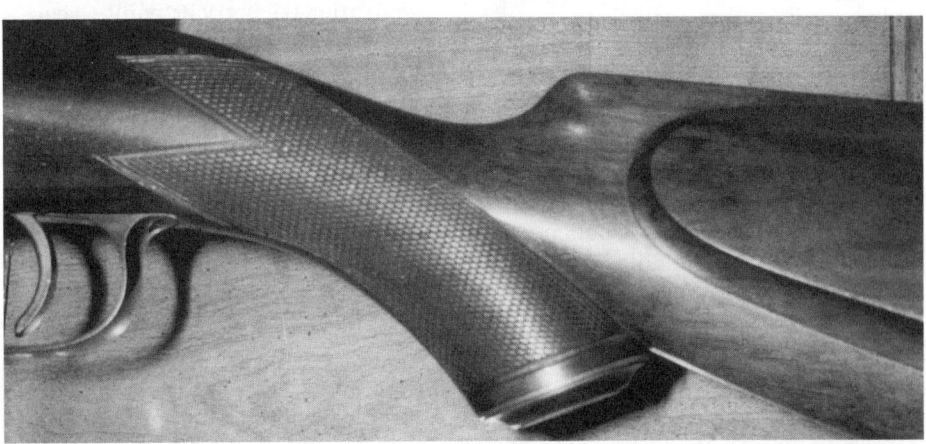

Fine stock and metal work are the hallmark of a Holland and Holland rifle. Note the metal pistol grip cap.

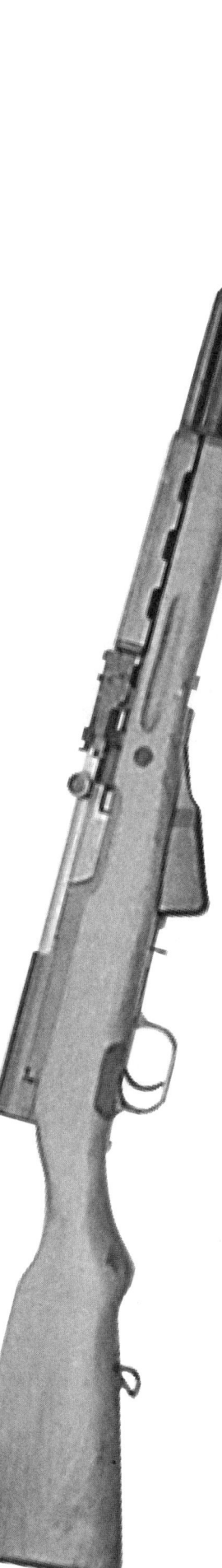

Russian SKS (Simonov)

Similiar/Identical Pattern Guns

The same basic assembly/disassembly steps for the Russian SKS (Simonov) also apply to the following guns:

Chinese Type 56   **Yugoslavian Model 59/66**

Date: Russian SKS (Simonov)
Origin: Russia
Manufacturer: Russian arsenals, and factories in China and other satellite nations
Cartridges: 7.62x39mm Russian
Magazine capacity: 10 rounds
Overall length: 40.2 inches
Barrel length: 20.47 inches
Weight: 8.5 pounds

Introduced in 1945, the Samozaryadnyi Karabin Simonova (SKS) was the first rifle chambered for the then-new 7.62mm "short" cartridge. The gun was made in large quantity, and it has been used at some time by every communist country in the world. Versions of it have been made in China, Yugoslavia, and elsewhere. While some of these variations may be different in small details, the mechanism is the same, and the instructions will apply.

Disassembly:

1. Cycle the action to cock the internal hammer. Turn the takedown latch up to vertical position, and pull it out toward the right until it stops. Take off the receiver cover toward the rear.

2. Move the bolt and recoil spring assembly back until it stops, and lift it off the receiver.

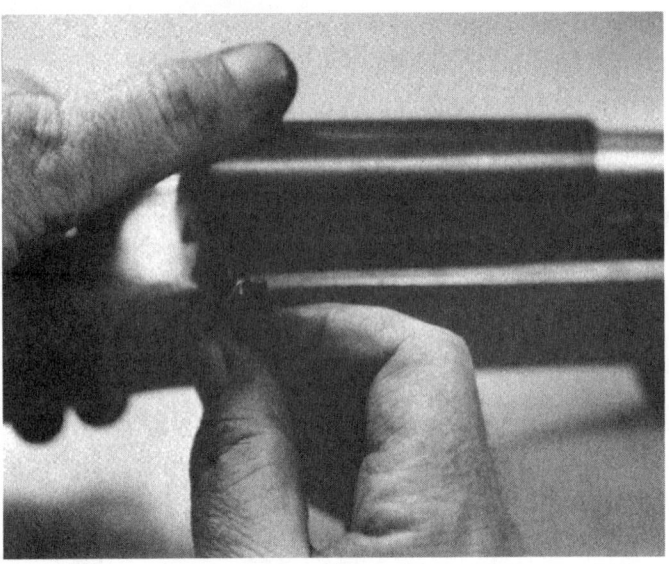

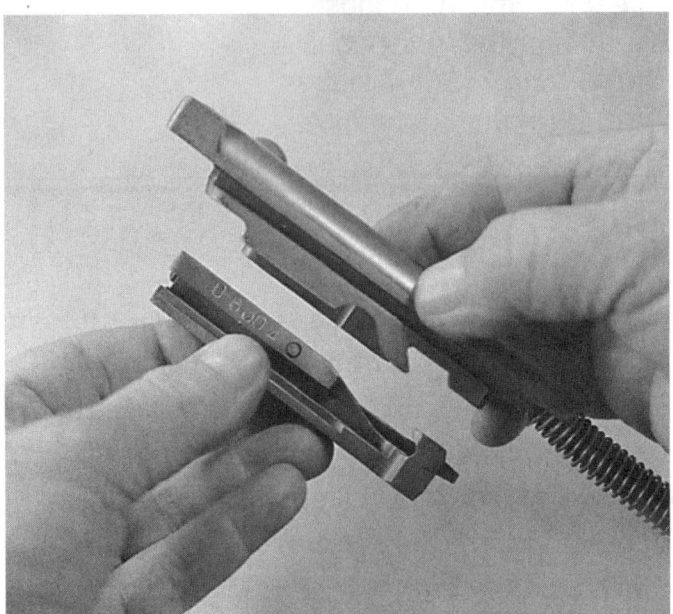

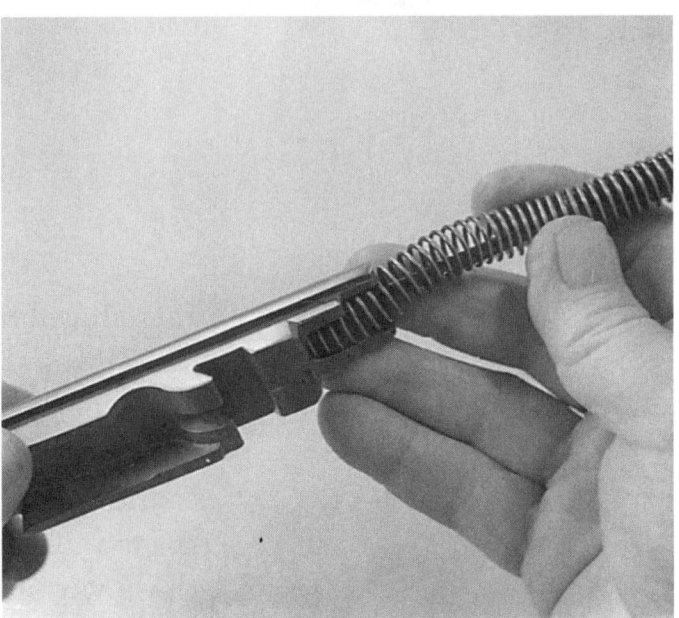

3. The bolt will probably be left in the receiver when the carrier and recoil spring unit are removed. If not, the bolt is simply lifted out of the carrier.

4. The captive recoil spring assembly is removed from the bolt carrier toward the rear.

5. If it is necessary to dismantle the spring assembly, rest the rear tip on a firm surface, pull back the spring at the front, and move the collar downward until it clears the button and take it off to the side. **Caution:** *Control the spring.*

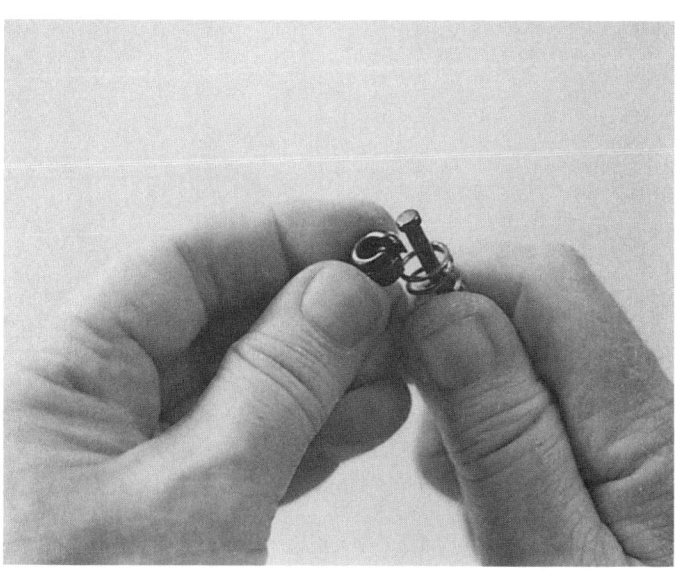

6. Push out the firing pin retainer toward the right.

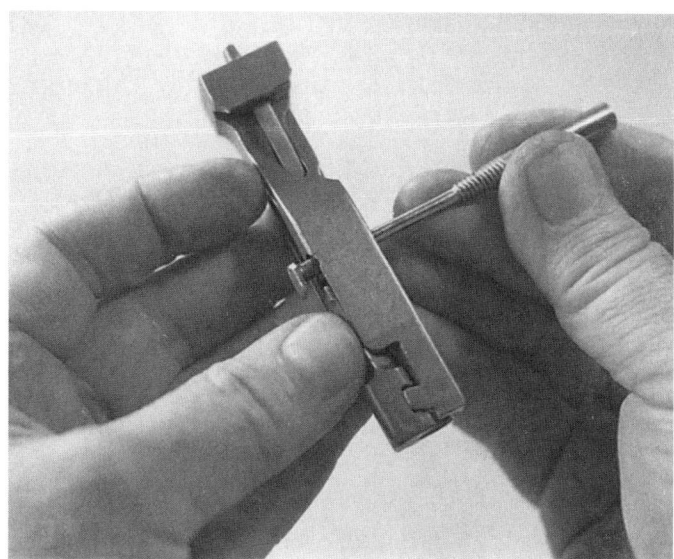

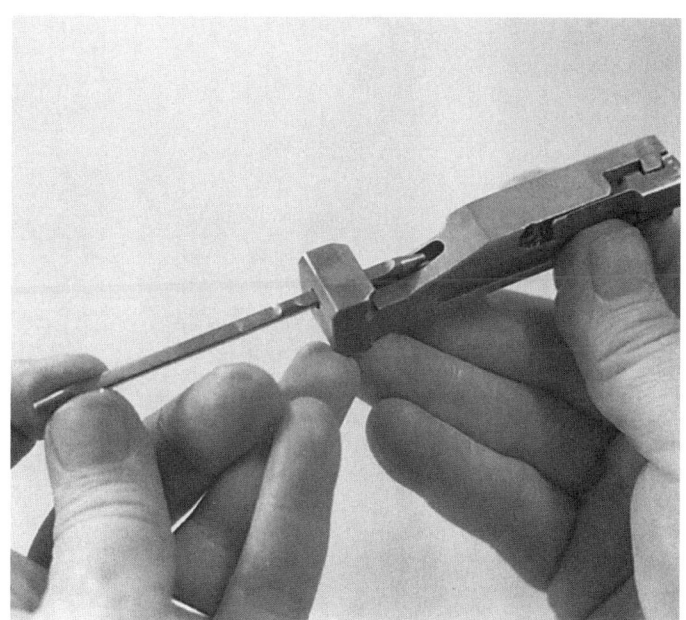

7. Remove the firing pin toward the rear.

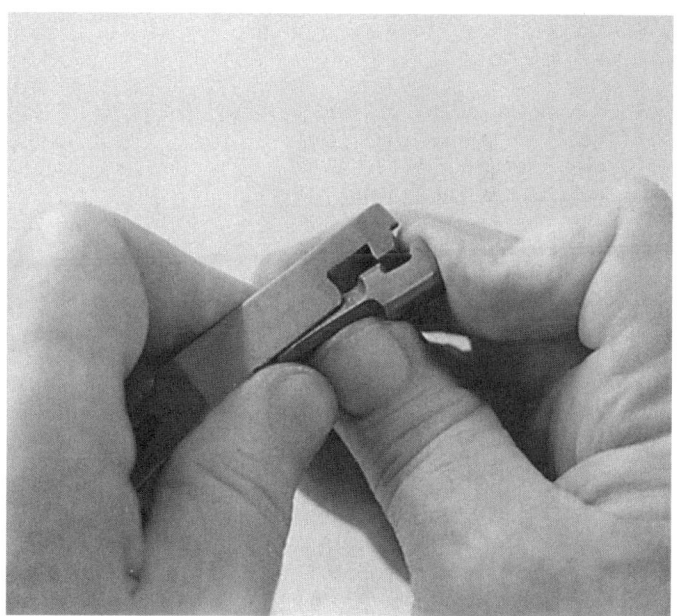

8. Push the extractor toward the rear, and tip it out toward the right for removal. The spring is mounted inside the rear of the extractor, and it will come off with it.

9. The takedown latch is retained by an internal cross pin. In normal takedown, it is best left in place. To get the latch out of the way for the remainder of takedown, push it back into its locked position.

10. Insert a drift through the hole in the head of the cleaning rod, lift it out of its locking recess, and remove it toward the front.

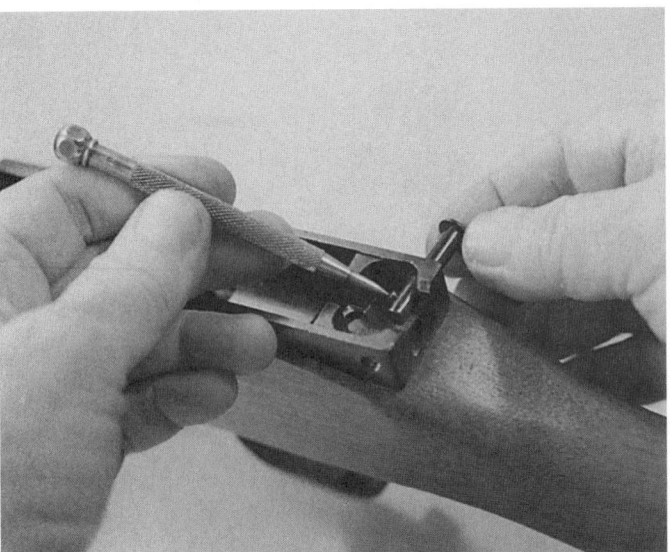

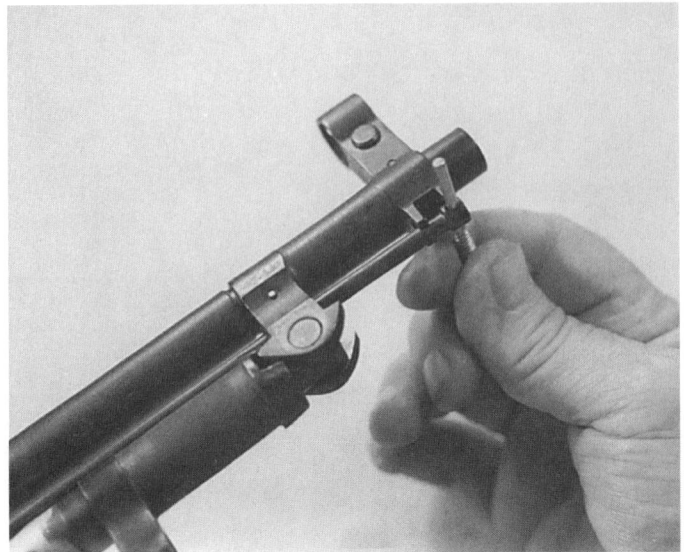

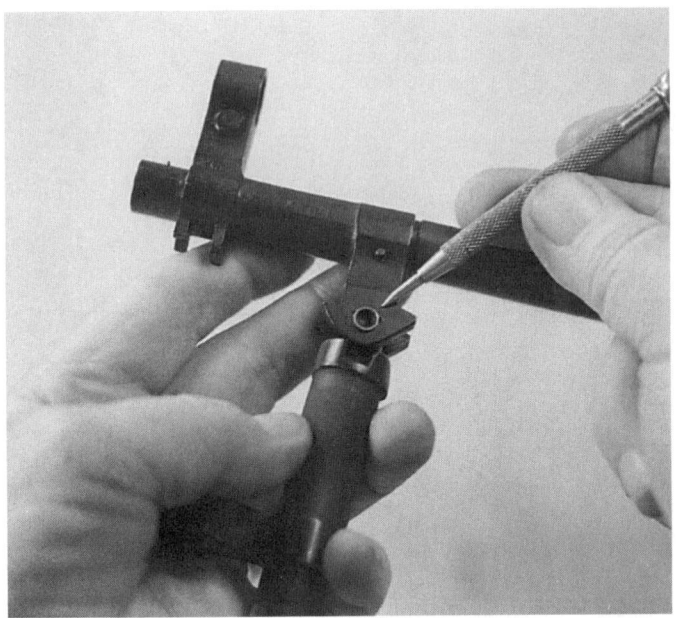

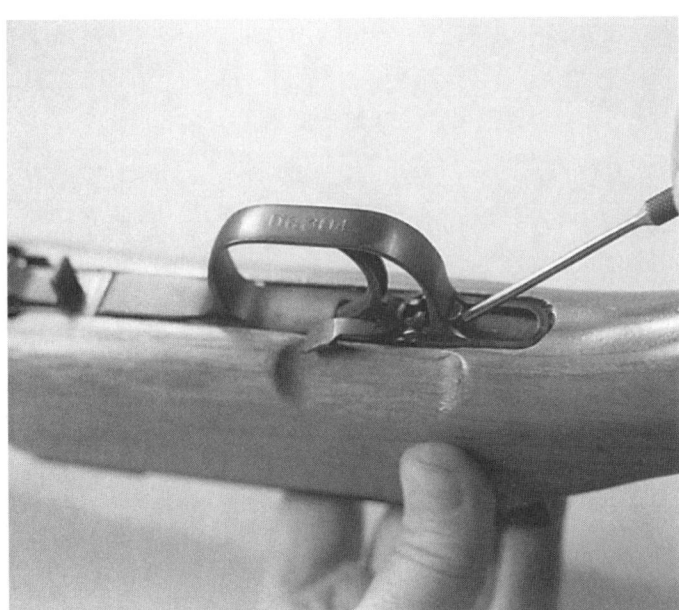

11. The bayonet hinge is often riveted in place. If removal is not necessary for repair, it is best left in place.

12. Be sure the internal hammer is in cocked position, and set the manual safety in on-safe position. Use a bullet tip or a suitable tool to push the guard latch foward.

13. When the latch releases, the guard will jump out slightly. Tip the guard away from the stock, move it toward the rear, and remove the guard unit.

14. Release the safety. Depress the disconnector, at the front of the hammer, about half way. Control the hammer, and pull the trigger. Ease the hammer down to fired position. **Caution:** *The hammer spring is powerful.*

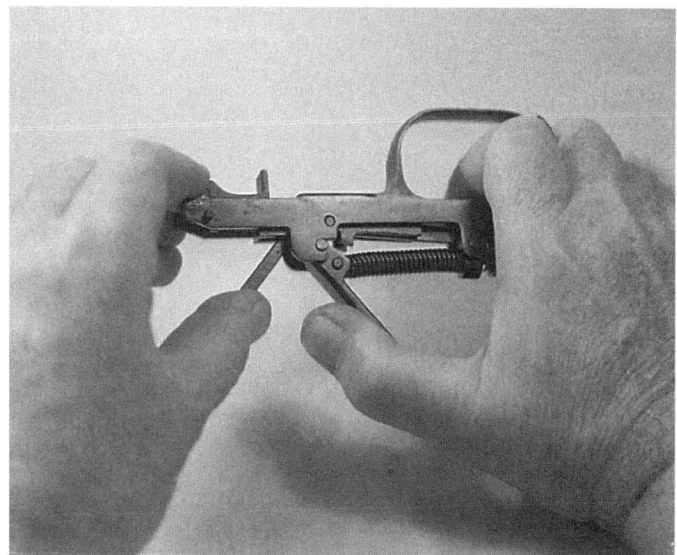

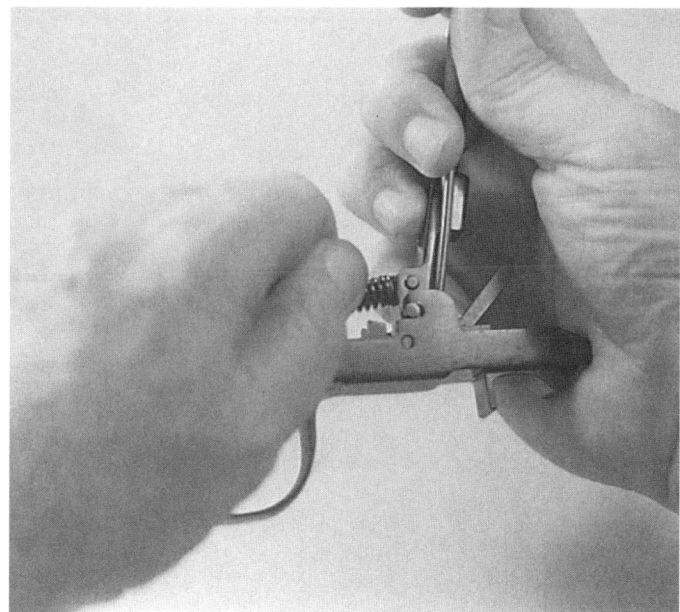

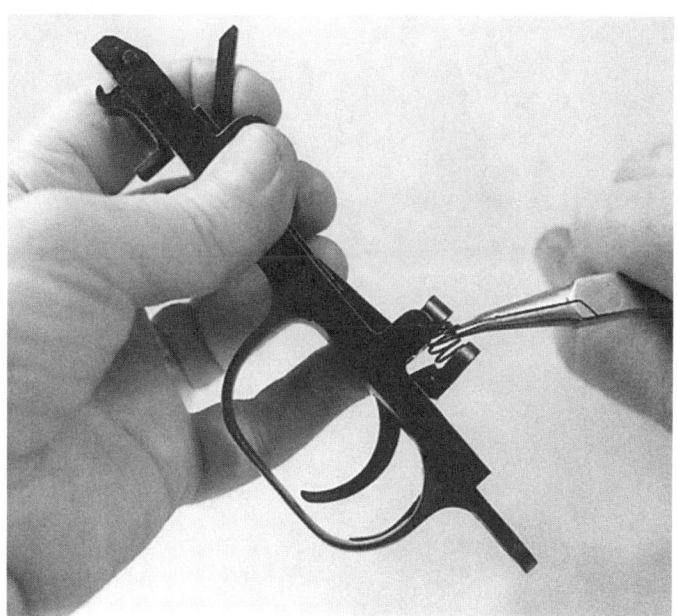

15. Insert a sturdy drift in front of the hammer, and lever it toward the rear until its pivot studs are clear of the hooks on the unit. **Caution:** *Keep a good grip on the hammer.* When it is clear, take off the hammer and its spring and guide. Another method is to grip the unit in a padded vise and use a bar of metal to apply pressure to the front of the hammer.

16. The trigger spring can be removed by gripping a forward coil with sharp-nosed pliers and compressing it slightly rearward, then tipping it out.

17. Pushing out the cross pin will free the disconnector for removal upward.

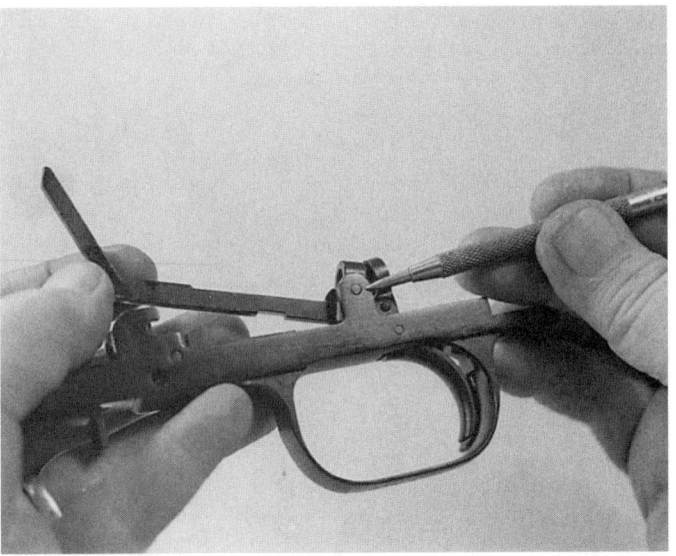

18. Removal of this cross pin will allow the rebound disconnector to be taken out.

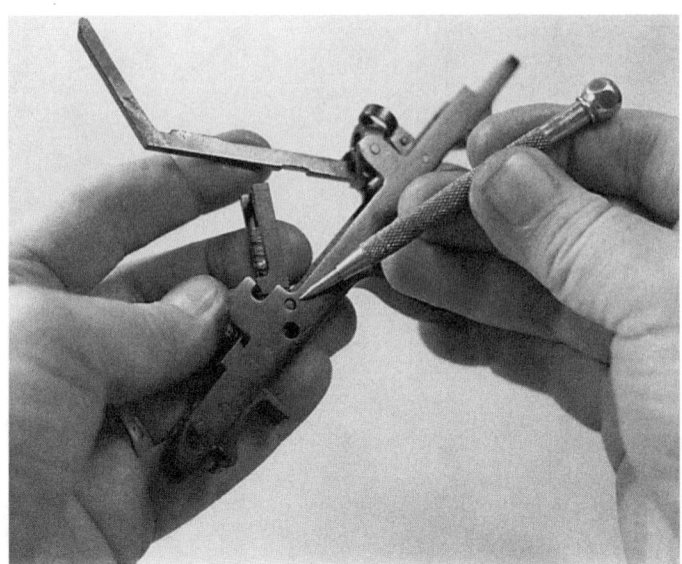

19. Drift out the trigger cross pin, and take out the trigger upward. The safety spring will also be freed for removal.

20. Drift out the safety-lever pin, and remove the safety.

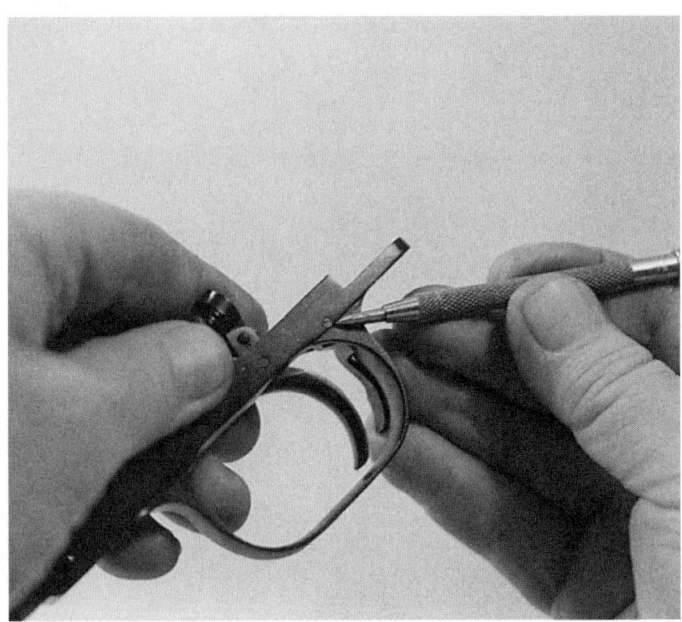

21. This cross pin at the front of the trigger guard assembly retains the magazine catch, the combination spring, and the sear.

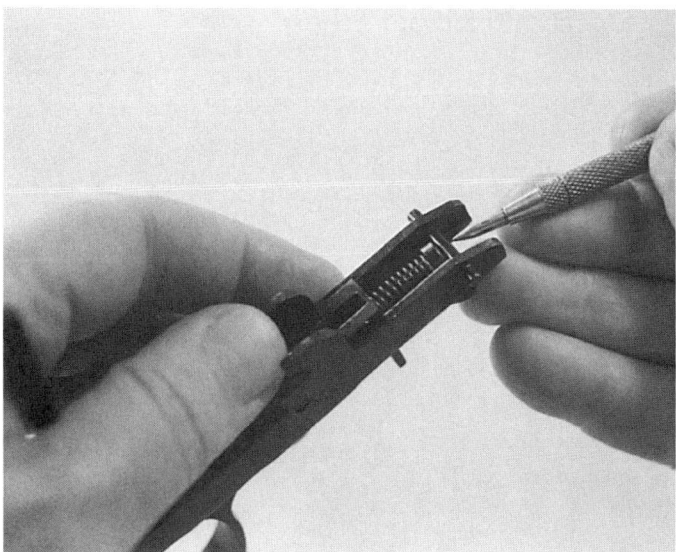

22. The trigger guard spring can be lifted out of its well in the stock.

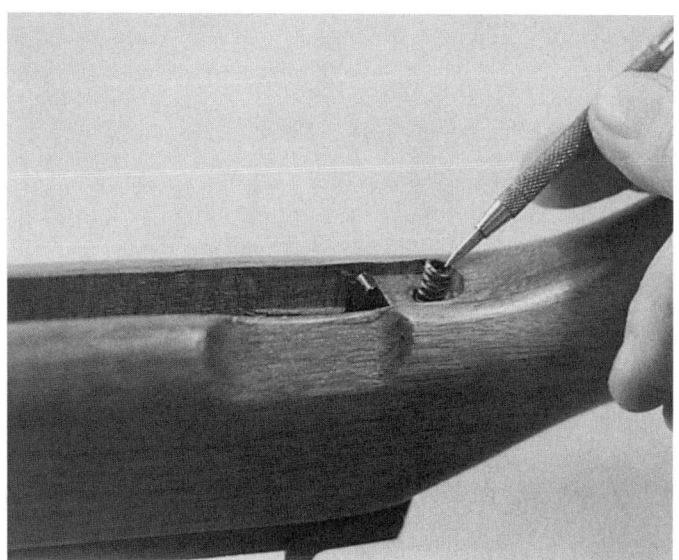

23. With a bullet tip or a non-marring tool, turn the handguard latch upward until it is stopped by its lower stud in the track. Lift the handguard and gas cylinder assembly at the rear, and remove it.

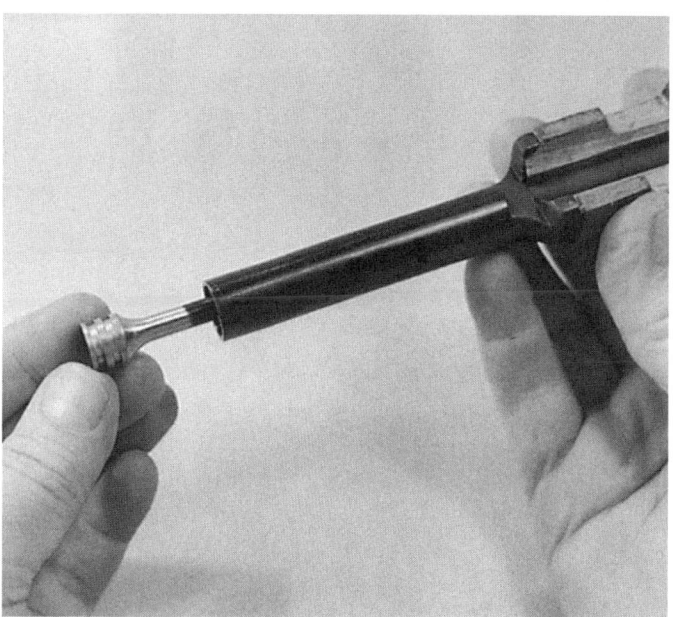

24. Remove the gas piston from the cylinder.

25. The gas port unit is retained on the barrel by a cross pin. In normal takedown, it is left in place.

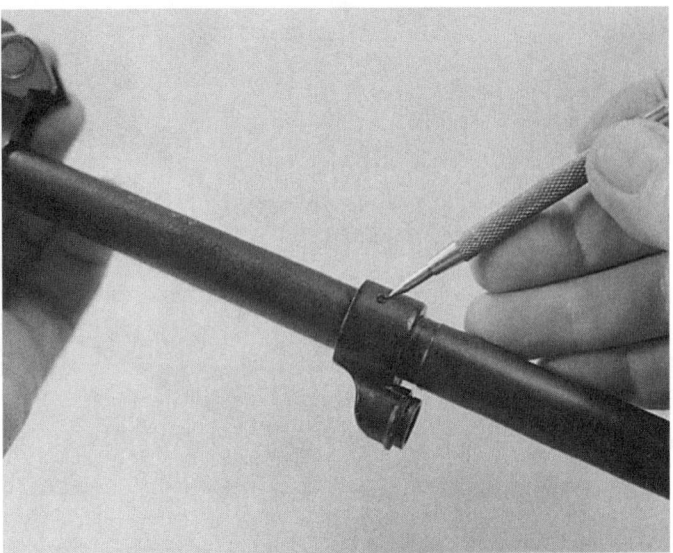

26. Drift out the stock end cap cross pin. Use a non-marring tool to nudge the end cap slightly forward.

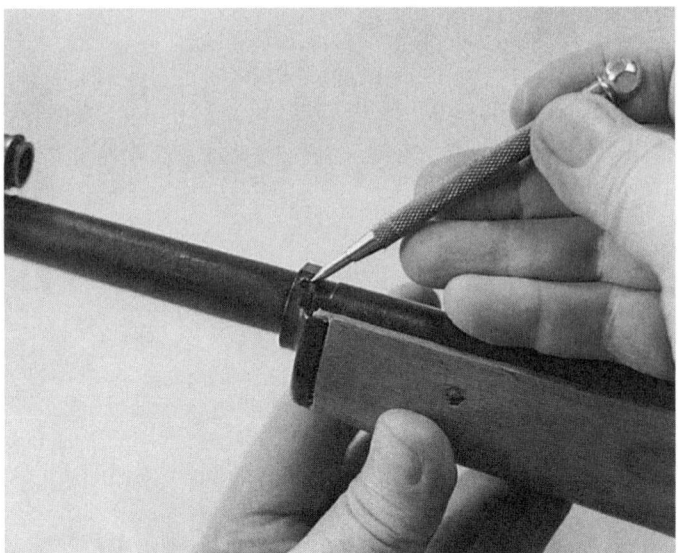

27. Grip the magazine assembly, and pull it rearward and downward for removal.

28. Remove the action from the stock.

29. The rear sight assembly is retained on the barrel by a cross pin, and is driven off toward the front. In normal takedown, the unit is left in place.

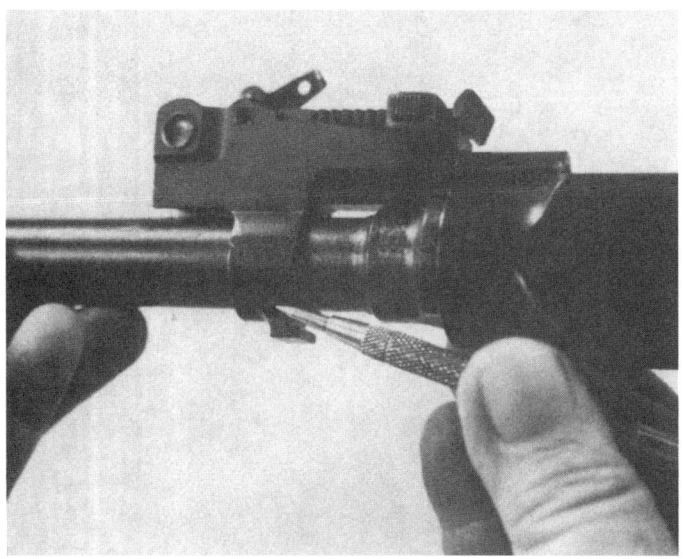

30. Drifting out this cross pin will allow removal of the bolt hold-open latch and its coil spring downward.

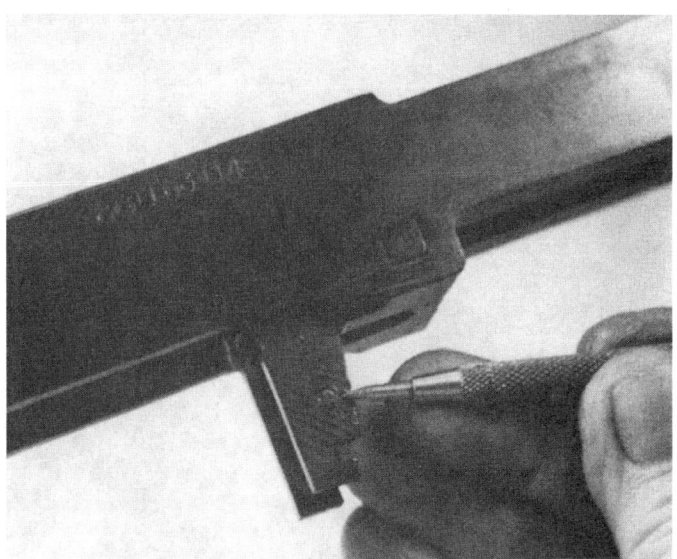

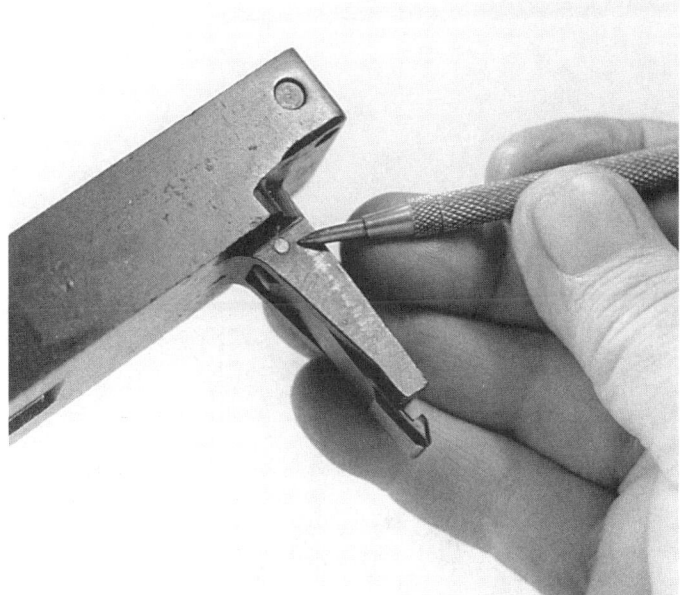

31. The trigger guard latch, which is its own spring, is retained by a cross pin. After removal of the pin, the latch is driven out downward.

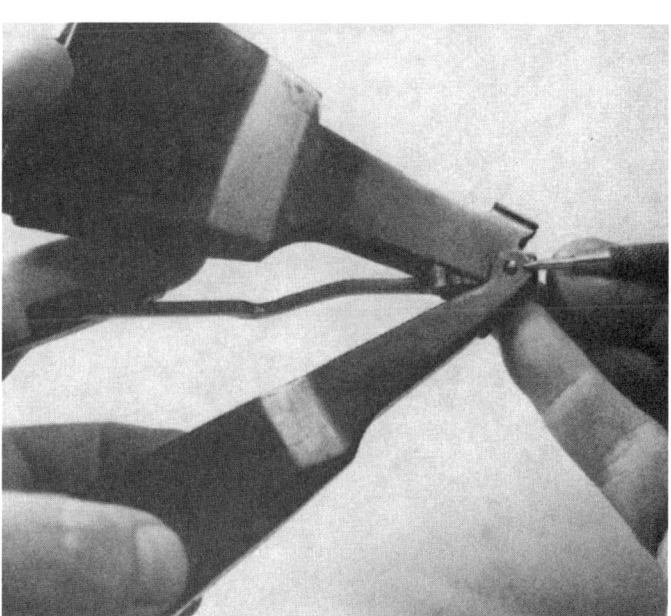

32. The magazine, the follower and its spring, and the magazine cover are joined by a cross pin at the front. The pin is riveted on both sides, and removal should be only for repair.

Reassembly Tips:

1. When the stock end piece has been nudged back into position, insert a drift to align the hole with the barrel groove.

2. For those who have disassembled the trigger group, here is a view of that unit with all of the parts properly installed.

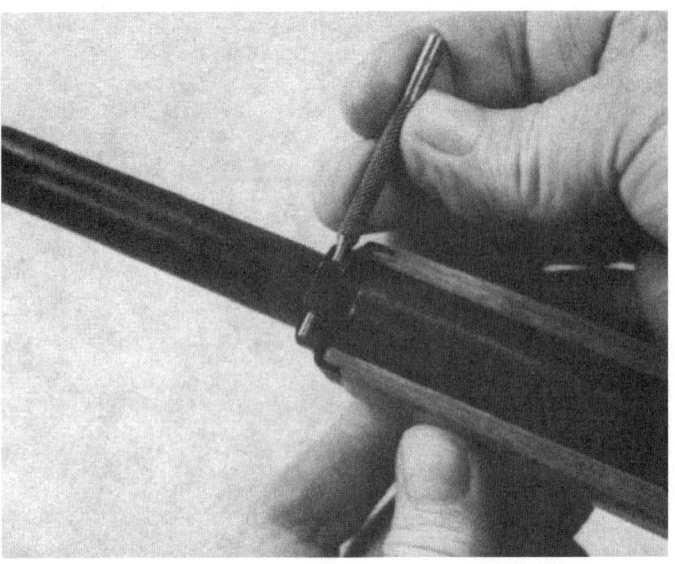

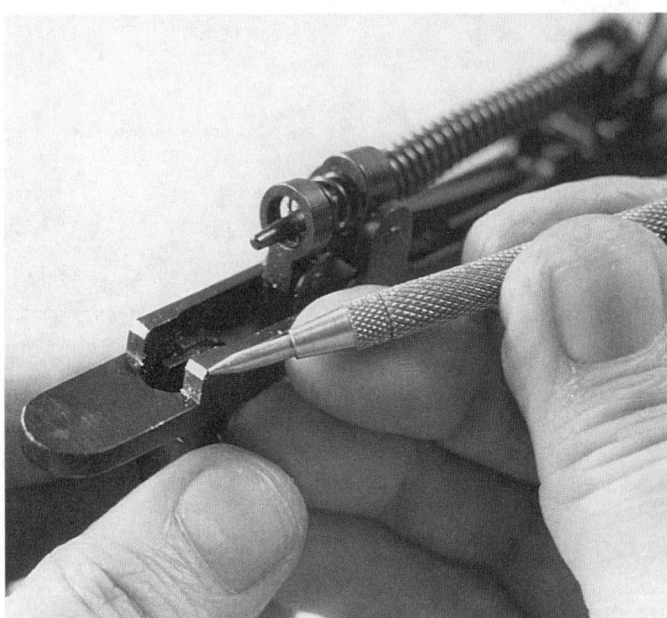

3. When replacing the trigger group in the gun, extremely sharp shoulders at the rear of the sub-frame may cause a seating problem. These can be easily beveled, as shown, with a file. Note: For installation of the trigger group, the manual safety must be in on-safe postion. Rest the top of the receiver on a firm surface as the guard unit is pressed into place.

4. When replacing the firing pin in the bolt, be sure the retaining shoulder is on top. Also, be sure the retainer is oriented to fit into its recess on the right side.

Colt Police Positive

Similar/Identical Pattern Guns

The same basic assembly/disassembly steps for the Colt Police Positive also apply to the following guns.

Colt Army Special	**Colt Banker's Special**	**Colt Detective Special**
Colt Official Police	**Colt Officer's Model Match**	**Colt Officer's Model Target**
Colt Police Positive Special	**Colt Police Positive Target**	**Colt Shooting Master**

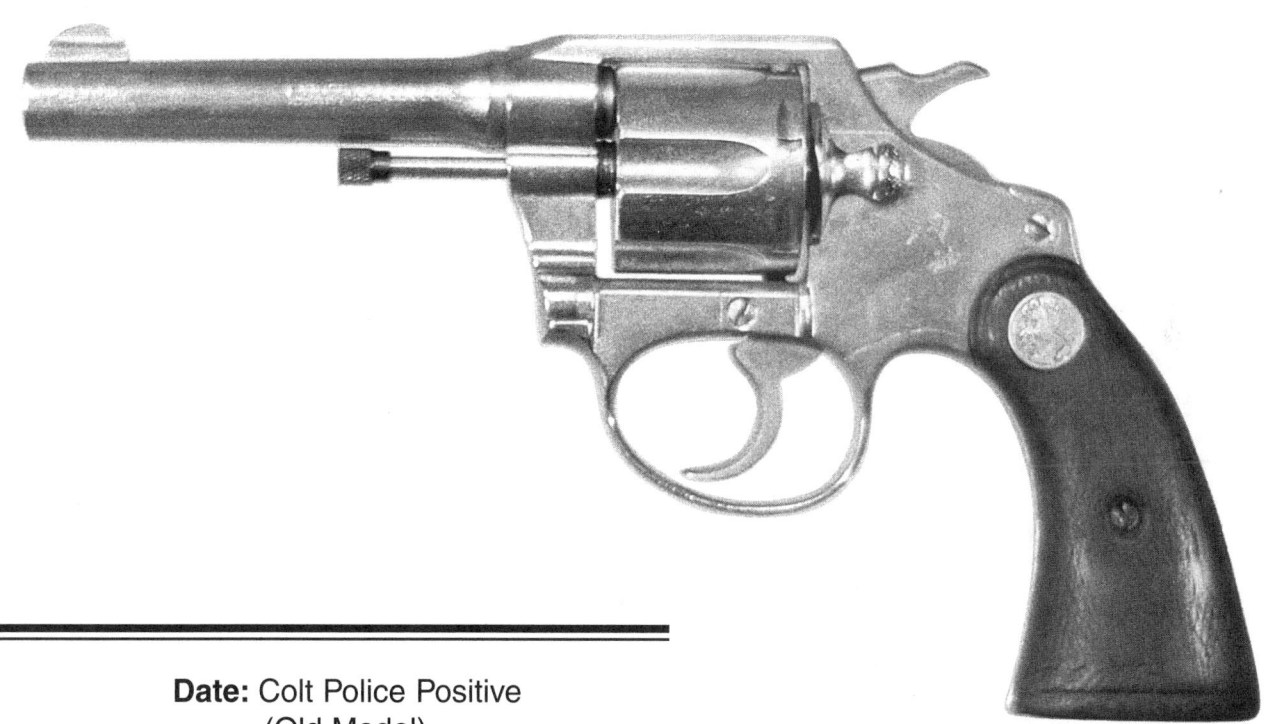

Date: Colt Police Positive
(Old Model)
Origin: United States
Manufacturer: Colt's, Hartford,
Connecticut
Cartridges: 38 Colt New Police
(38 S&W) and others
Magazine capacity: 6 rounds
Overall length: 8 1/2 inches
(with 4-inch barrel)
Barrel length: 2 1/2, 4, 5, and 6 inches
Weight: 20 ounces
(with 4-inch barrel)

The original Colt Police Positive was introduced in 1905 and its production life ran to 1947. Along the way there were several modifications, the most notable external change being a widening of the upper neck of the grip frame. The Police Positive Special, which began in 1907, was available in 38 Special, and was made until 1973. These guns are mechanically the same, and the takedown and reassembly instructions will apply to either one. This also applies to the Target models and the other variations listed in the cross-reference list.

Disassembly:

1. The crane retainer and its screw are located on the right side of the frame, just forward of the trigger. Remove the crane and its screw toward the right.

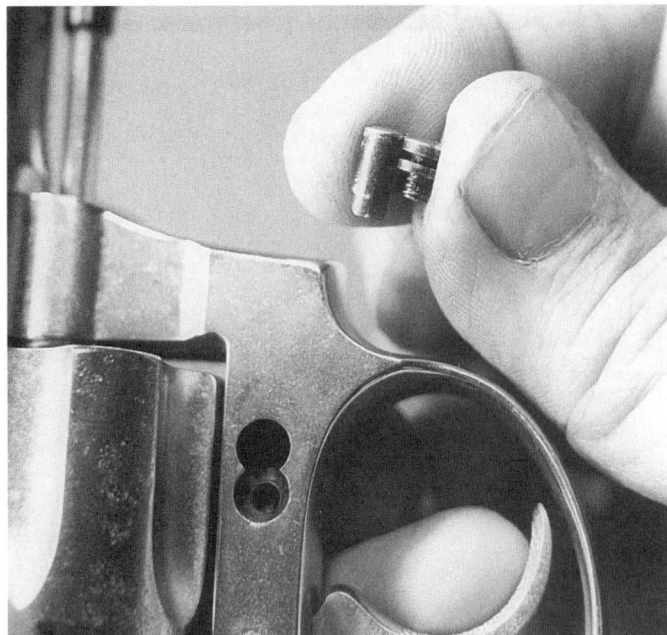

2. Move the crane forward out of the frame, and remove the crane and cylinder assembly toward the left.

3. Grip the ejector rod head with leather-padded smooth-jawed pliers, and unscrew it counter-clockwise (front view).

4. Brownells has a wrench that will fit the ejector/ratchet of the 38-caliber guns, but the one shown is a 32, so another tool must be used. Round-nosed wire-bending pliers will work, without burring the edges of the ejector. Unscrew the ejector/ratchet head counterclockwise (rear view).

5. Remove the ejector/ratchet head from the ejector rod, toward the rear.

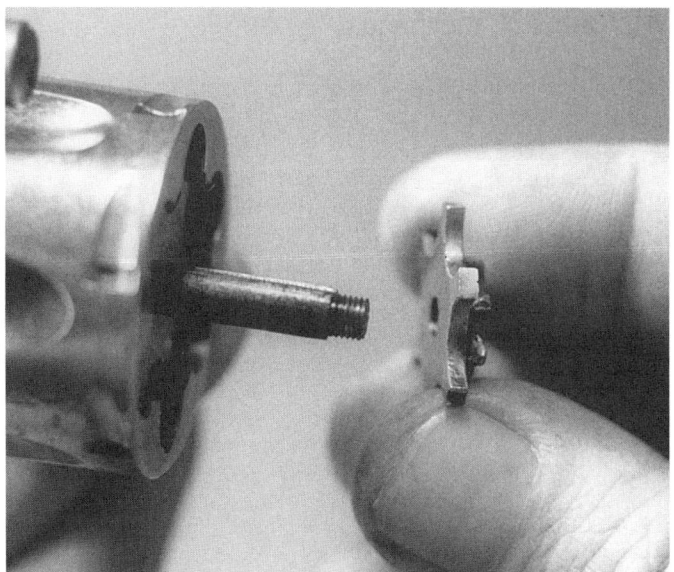

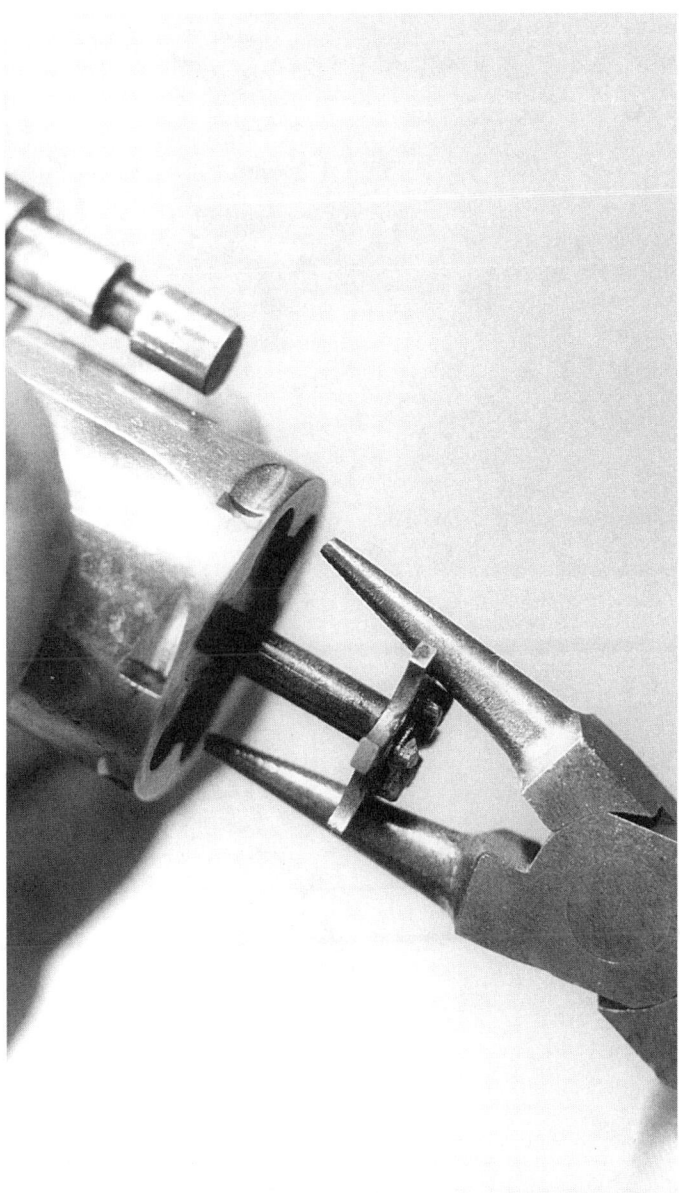

6. Remove the crane/ejector rod assembly from the front of the cylinder.

7. Use a standard Colt wrench to unscrew the ejector retainer nut from the rear of the cylinder arbor on the crane.

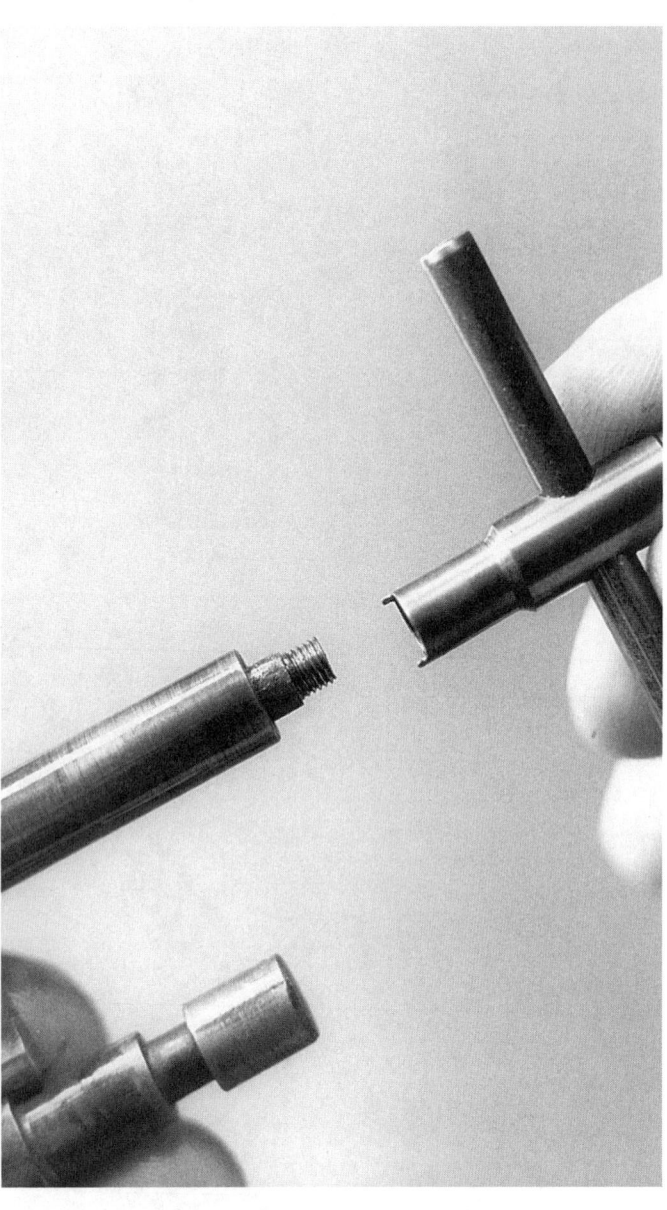

8. Remove the retaining nut toward the rear.

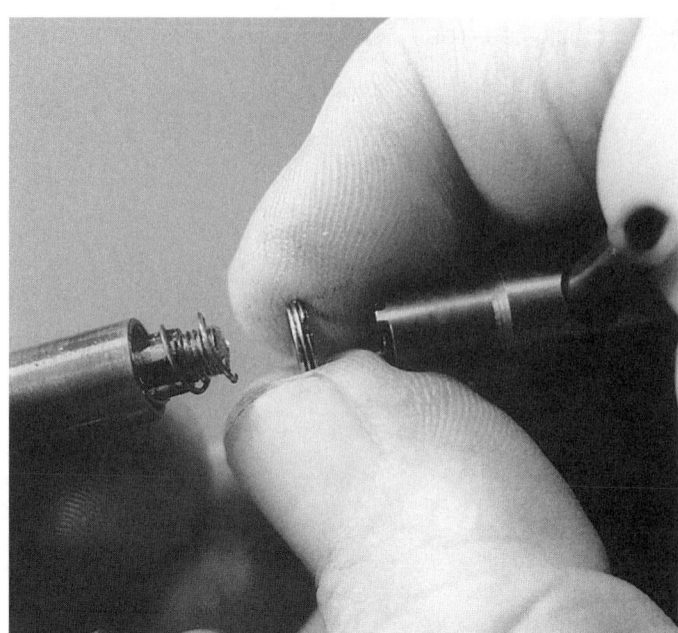

9. Remove the ejector rod and its spring toward the rear.

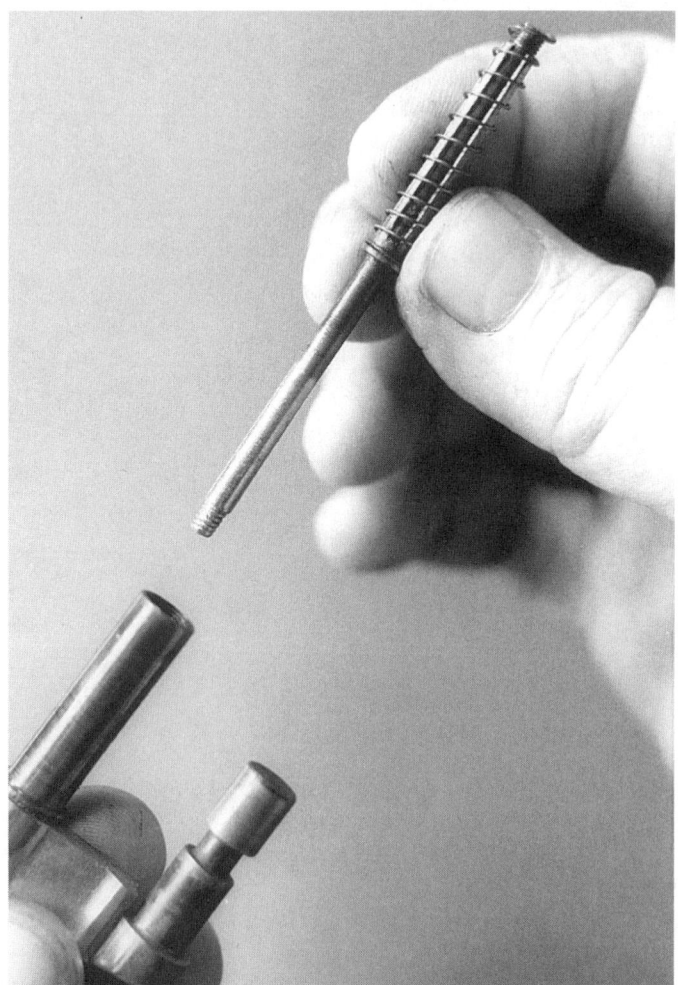

10. Remove the two sideplate screws on the left side of the frame. Hold the gun as shown, and tap the grip frame with a nylon mallet until the sideplate drops into the hand.

11. Remove the cylinder latch thumb piece and its spring and plunger from the sideplate.

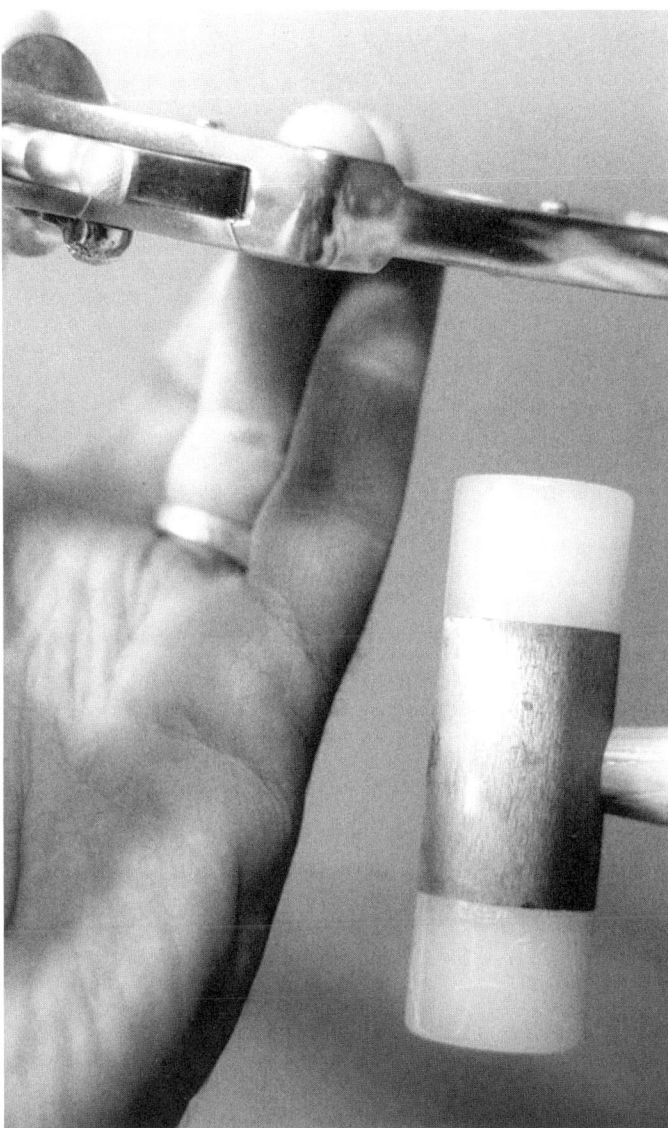

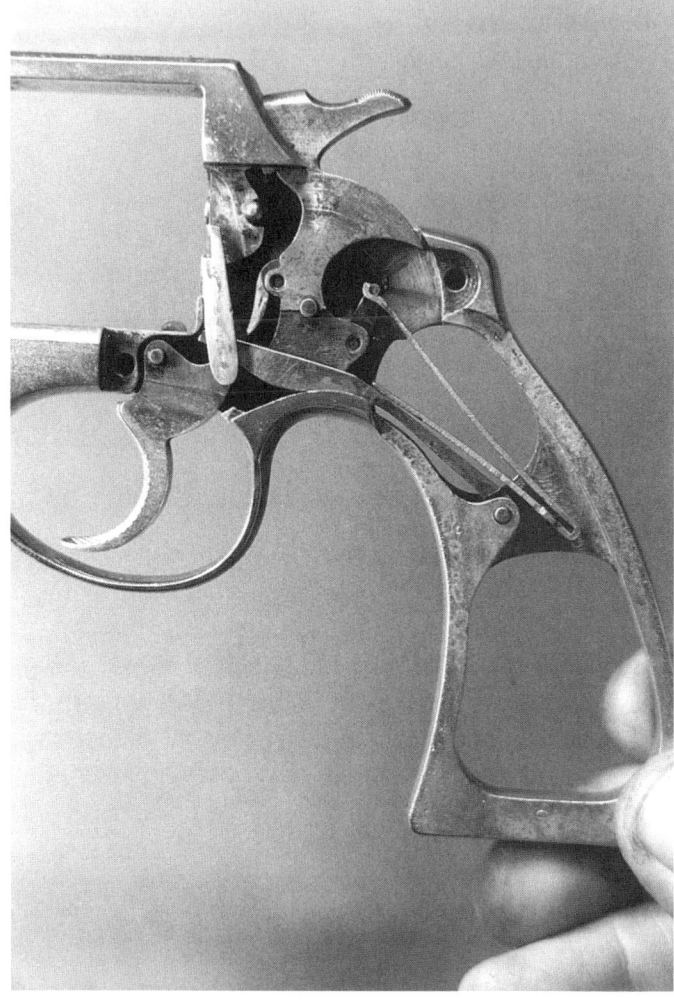

12. The internal parts are shown in proper order, prior to disassembly.

13. Grip the mainspring with smooth-jawed sharp-nosed pliers, and compress it slightly. Lift its rear end out toward the left, disengage the spring hooks from the hammer stirrup, and remove the spring toward the left.

14. Remove the cylinder hand toward the left.

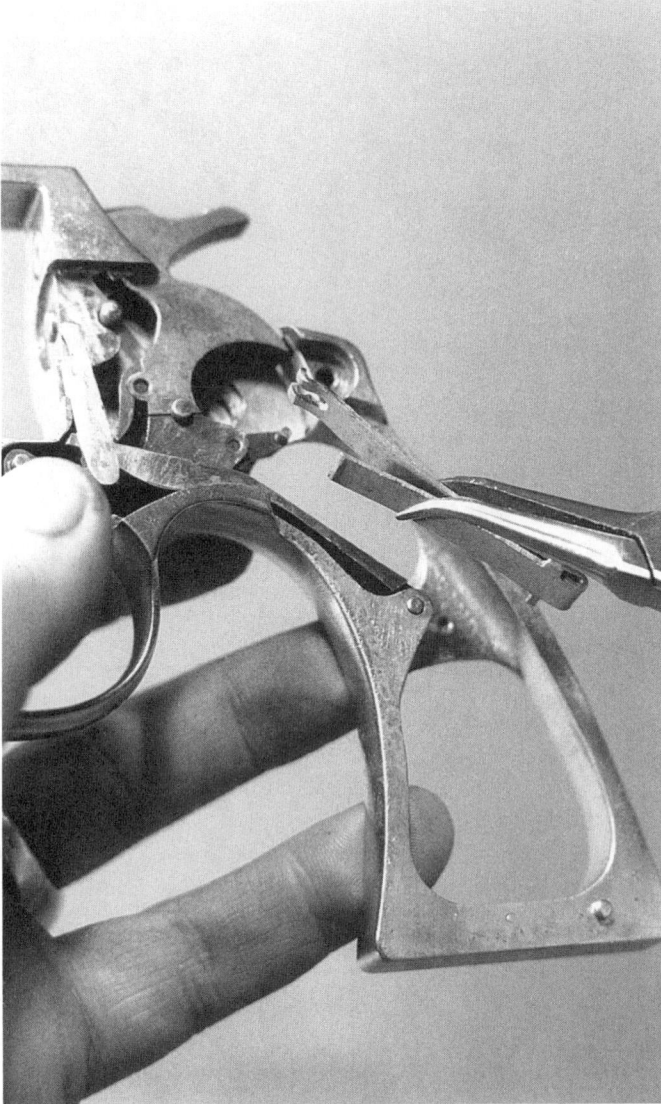

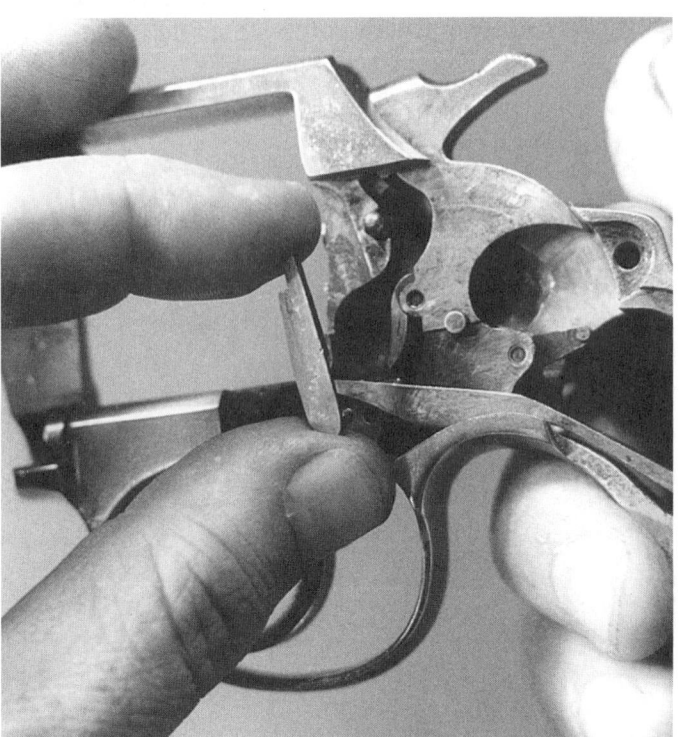

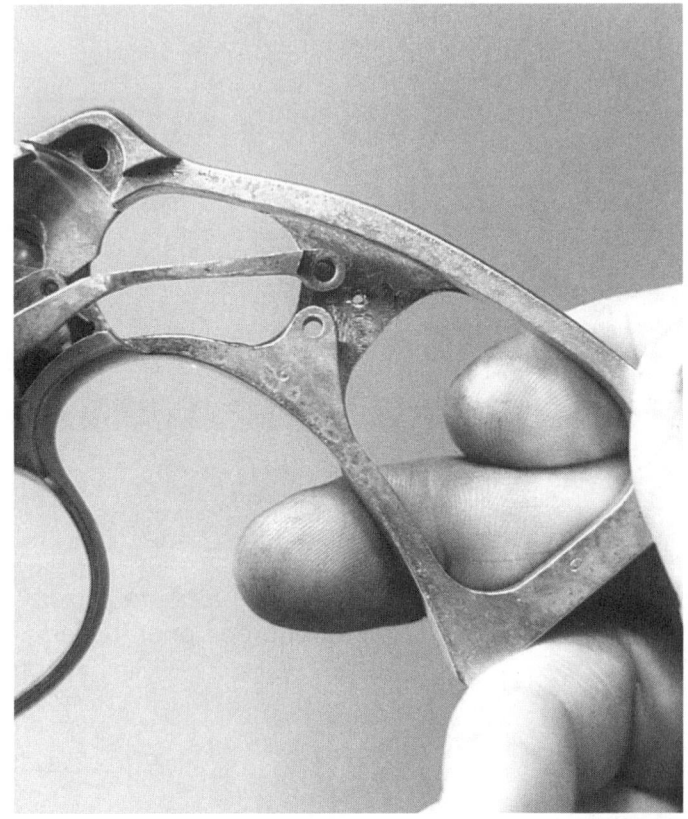

15. Drift out the rebound lever pin, located at the center of the grip frame. Move the rear loop of the rebound lever out of its recess in the grip frame, and remove the lever toward the left.

16. Pull the trigger to the rear, tipping the hammer back, and remove the hammer toward the left. Drifting out the cross-pin at the front of the hammer will allow removal of the double-action lever and its spring. The cross-pin at the lower rear of the hammer retains the hammer stirrup.

17. Remove the cylinder latch bolt toward the rear.

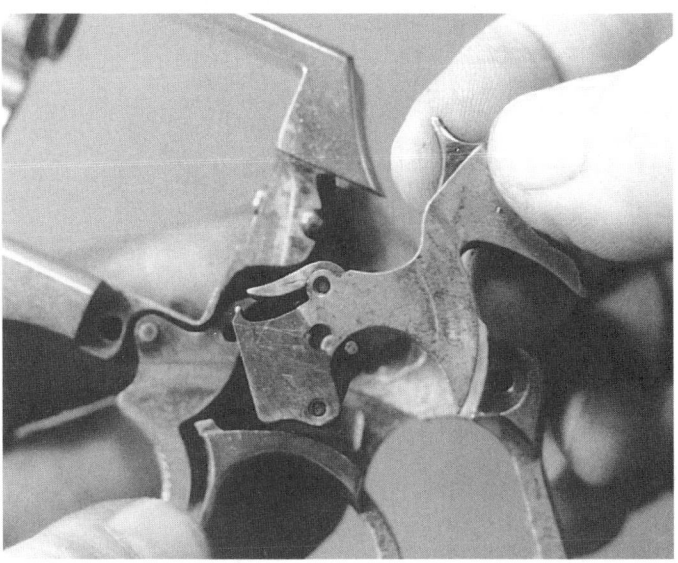

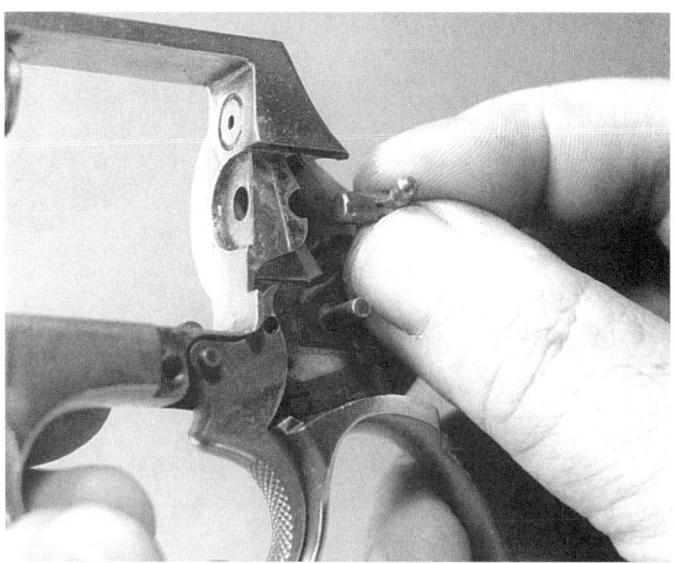

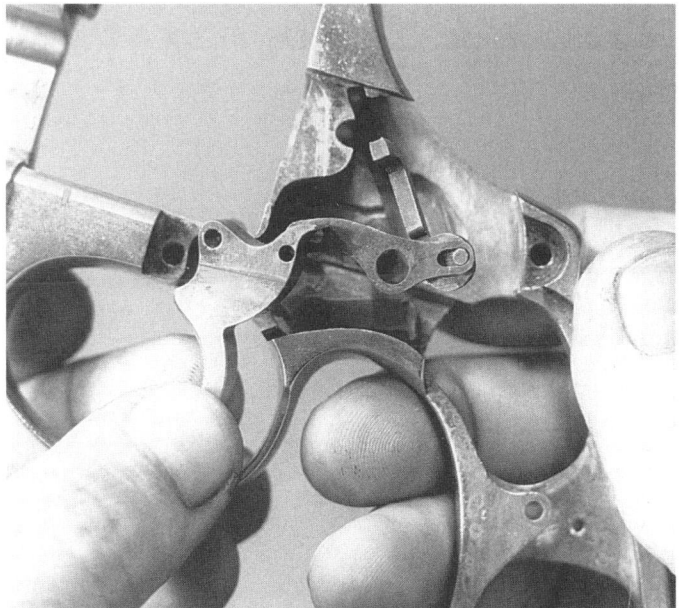

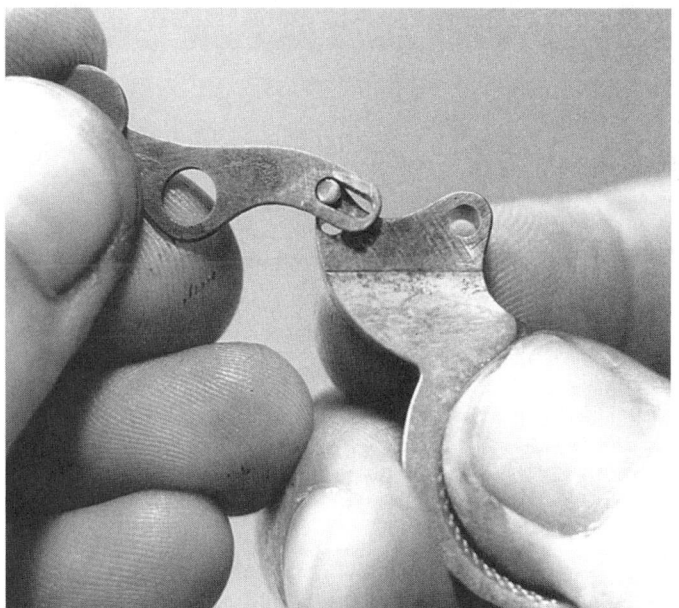

18. Remove the trigger, and the attached hammer block system, toward the left.

19. The hammer block lever is attached to the trigger by a stud on the right side, its head keyed into a slot in the lever. Move the stud back to the enlarged opening in the slot, and separate the lever from the trigger.

20. Separate the lever from the hammer block in the same way.

21. Remove the internal screw (arrow) that retains the cylinder stop. Move the front of the cylinder stop down out of its slot in the frame, and remove the stop and its spring toward the rear. Take care not to lose the small coil spring.

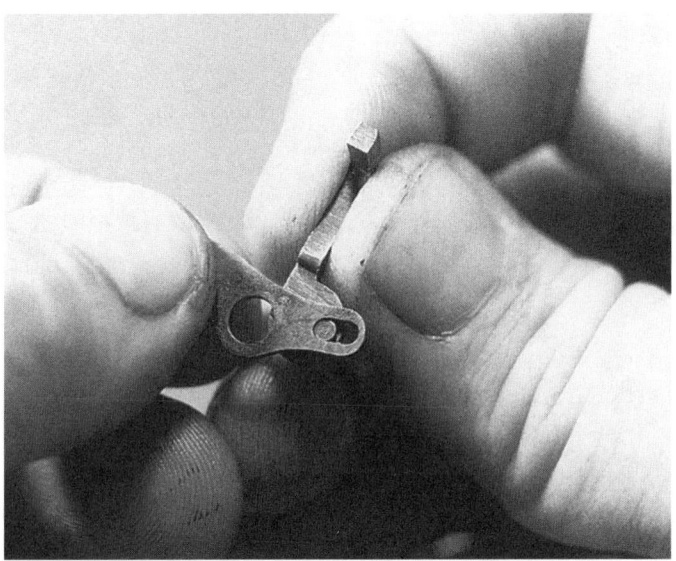

Reassembly Tips:

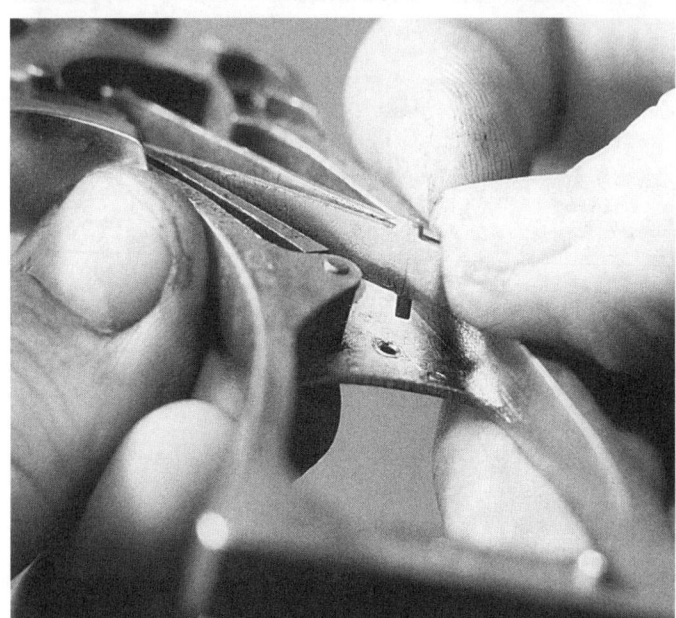

1. When replacing the hammer spring, engage the hooks with the stirrup first, then compress the spring and move it into place. Be sure the stud on the right side of the spring at the rear enters its hole in the grip frame.

2. When replacing the ejector/ratchet, note that one arm of the ejector has an index line at its tip, to be aligned with a corresponding line on the rear face of the cylinder.

GUNDEX®

A listing of all MODERN GUN VALUES, 11 TH EDITION firearms by manufacturer in numeric/alpha order.

Bergmann

Bergmann-Bayard

Bernardelli

HANDGUNS

Directory of Manufacturers

A.A. Arms
AP9
AP9 Target
P95

Accu-Tek
Model AT-9SS
Model AT-25SS
Model AT-25AL
Model AT-25SSB
Model AT-32SS
Model AT-32B
Model AT-40SS
Model AT-45SS
Model AT-380SS
Model AT-380 Lady
Model HC-380SS

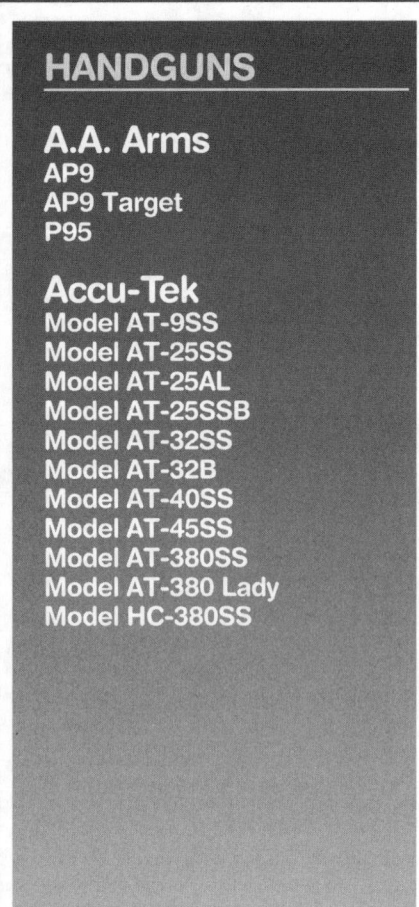

Accu-Tek AT-9SS

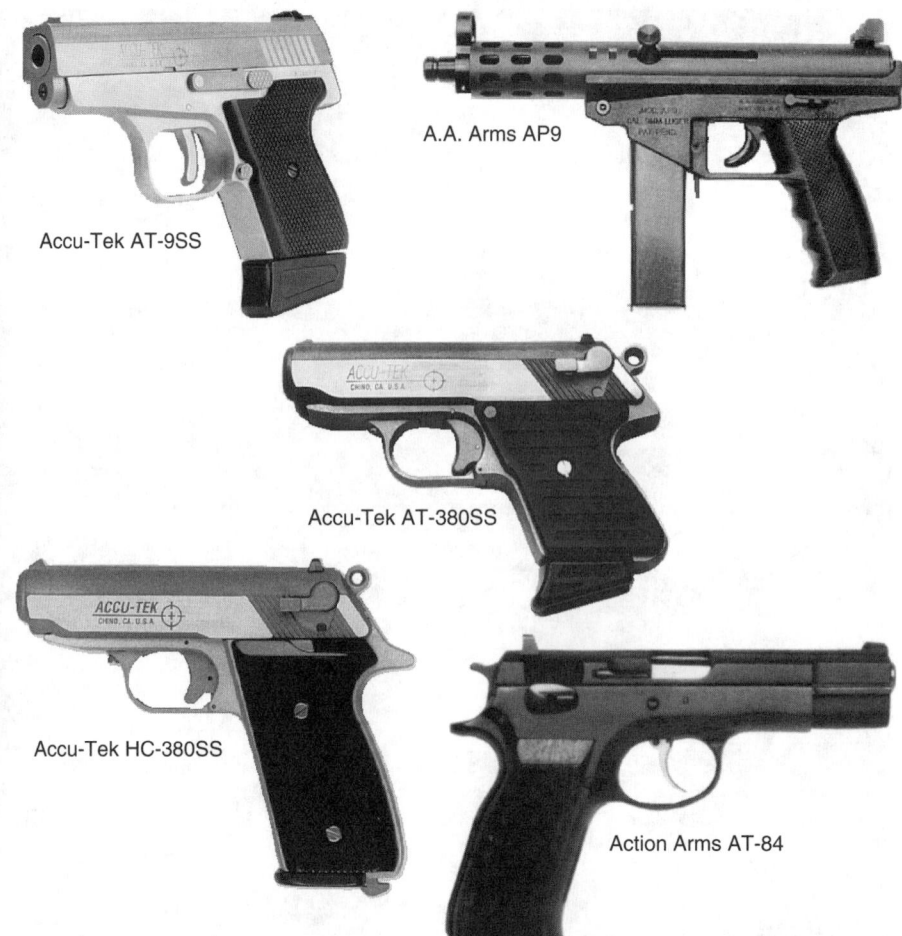

A.A. Arms AP9

Accu-Tek AT-380SS

Accu-Tek HC-380SS

Action Arms AT-84

A.A. ARMS AP9
Semi-automatic; 9mm Para.; 20-shot magazine; 5″ barrel; 11 ¹³/₁₆″ overall length; weighs 3 ½ lbs.; adjustable post front sight in ring, open fixed rear; vented barrel shroud; checkered plastic grips; lever safety; fires from closed bolt; 3″ barrel available; matte blue-black or nickel finish. Introduced 1988; no longer in production.
   Perf.: $200   **Exc.:** $175   **VGood:** $150
Nickel finish
   Perf.: $225   **Exc.:** $200   **VGood:** $175

A.A. Arms AP9 Target
Same specs as the AP9 except 12″ barrel; grooved forend; blue finish. No longer in production. Was marketed by Kimel Industries.
   Perf.: $375   **Exc.:** $325   **VGood:** $300

A.A. ARMS P95
Semi-automatic; 9mm Para.; 5-, 20-, 30-shot magazines; 5″ barrel; 11 ¹³/₁₆″ overall length; weighs 3 ½ lbs.; adjustable post front sight in ring, open fixed rear; checkered plastic grips; lever safety; fires from closed bolt; 3″ barrel available; matte blue-black or nickel finish. Introduced 1989; dropped 1991.
   Perf.: $200   **Exc.:** $175   **VGood:** $150
Nickel Finish
   Perf.: $225   **Exc.:** $200   **VGood:** $175

ACCU-TEK MODEL AT-9SS
Semi-automatic; double-action-only; 9mm Para.; 8-shot magazine; 3 ¼″ barrel; 6 ¼″ overall length; weighs 28 oz.; black checkered nylon grips; blade front sight, windage-adjustable rear; three-dot system; stainless steel construction; firing pin block with no external safeties. Made in U.S. by Accu-Tek. Introduced 1992.
   New: $250   **Perf.:** $200   **Exc.:** $150

ACCU-TEK MODEL AT-25SS
Semi-automatic; single action; 25 ACP; 7-shot magazine; 2 ¾″ barrel; 6″ overall length; weighs 28 oz.; blade front sight, windage-adjustable rear; checkered black composition grips; external hammer; manual thumb safety; firing pin block; trigger disconnect; satin stainless steel finish. Made in U.S. by Accu-Tek. Introduced 1991; no longer in production.
   Perf.: $150   **Exc.:** $125   **VGood:** $100

Accu-Tek Model AT-25AL
Same specs as AT-25SS except aluminum frame and slide with 11-oz. weight. Made in U.S. by Accu-Tek. Introduced 1991; no longer in production.
   Perf.: $135   **Exc.:** $110   **VGood:** $85

Accu-Tek Model AT-25SSB
Same specs as Model AT-25SS except matte black finish over stainless steel. Introduced 1991; no longer in production.
   Perf.: $150   **Exc.:** $125   **VGood:** $90

ACCU-TEK MODEL AT-32SS
Semi-automatic; single action; 32 ACP; 5-shot magazine; 2 ¾″ barrel; 6″ overall length; weighs 28 oz.; blade front sight, windage-adjustable rear; checkered black composition grips; external hammer; manual thumb safety; firing pin block; trigger disconnect. Satin stainless steel finish. Made in U.S. by Accu-Tek. Introduced 1991; still in production.
   Perf.: $175   **Exc.:** $150   **VGood:** $125

Accu-Tek Model AT-32B
Same specs as Model AT-32SS except matte black finish over stainless steel. Introduced 1991; still in production. **LMSR: $190**
   Perf.: $185   **Exc.:** $160   **VGood:** $135

ACCU-TEK MODEL AT-40SS
Semi-automatic; double-action-only; 40 S&W; 7-shot magazine; 3 ¼″ barrel; 6 ¼″ overall length; weighs 28 oz.; black checkered nylon grips; blade front sight, windage-adjustable rear; three-dot system; stainless steel construction; firing pin block with no external safeties. Made in U.S. by Accu-Tek. Introduced 1992.
   Perf.: $250   **Exc.:** $200   **VGood:** $175

ACCU-TEK MODEL AT-45SS
Semi-automatic; double-action-only; 45 ACP; 6-shot magazine; 3 ¼″ barrel; 6 ¼″ overall length; weighs 28 oz.; blade front sight, windage-adjustable rear; three-dot system; stainless steel construction; firing pin block with no external safeties. Made in U.S. by Accu-Tek. Introduced 1995.
   New: $250   **Perf.:** $200   **Exc.:** $150

ACCU-TEK MODEL AT-380SS
Semi-automatic; single action; 32 ACP, 380 ACP; 5-shot magazine; 2 ¾″ barrel; 5 ⅝″ overall length; weighs 16 oz.; blade front sight, windage-adjustable rear; black combat or wood stocks; external hammer; manual safety with firing pin block, trigger disconnect; black, chrome or chrome with black slide. Made in U.S. by Accu-Tek. Introduced 1990; still in production.
   Perf.: $150   **Exc.:** $125   **VGood:** $110

Accu-Tek Model AT-380 Lady
Same specs as Accu-Tek 380 except 380 ACP only; chrome finish; gray bleached oak stocks. Introduced 1990; dropped 1991.
   Perf.: $150   **Exc.:** $125   **VGood:** $100

ACCU-TEK MODEL HC-380SS
Semi-automatic; single action; 380 ACP, 10-shot magazine; 2 ¾″ barrel; 6″ overall length; weighs 28 oz.; checkered black composition grips; blade front sight, windage-adjustable rear; external hammer; manual thumb safety with firing pin and trigger disconnect; bottom magazine release; stainless finish. Made in U.S. by Accu-Tek. Introduced 1993; still in production.
   New: $200   **Perf.:** $175   **Exc.:** $150

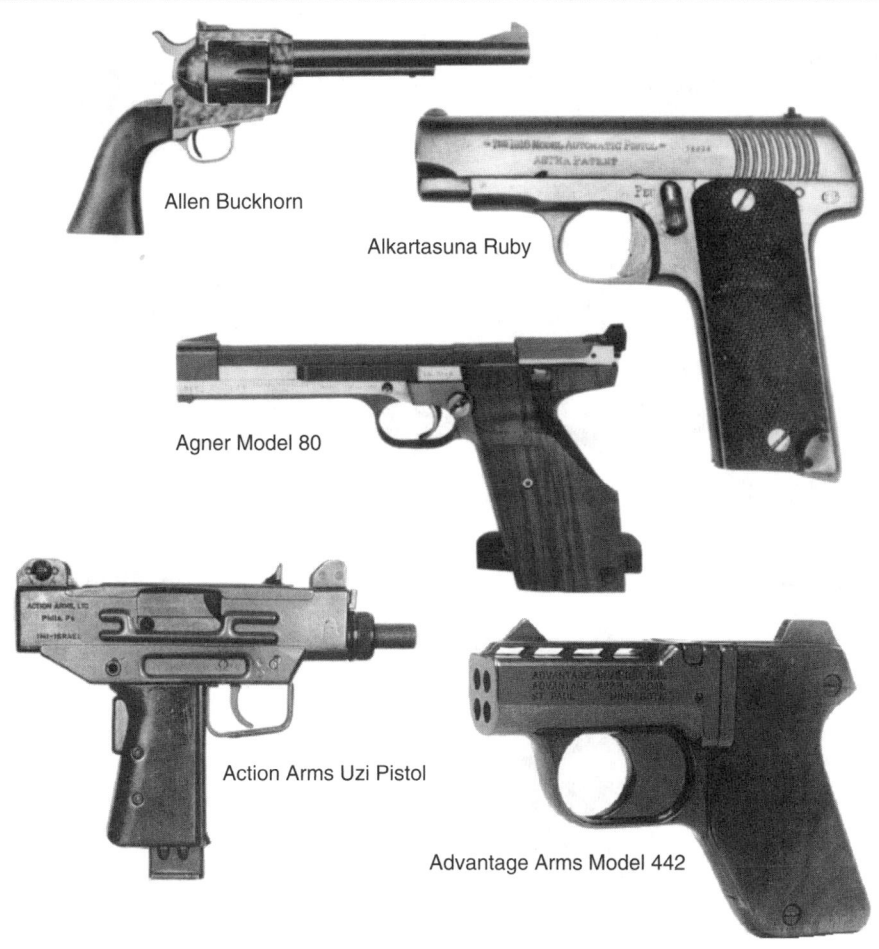

Allen Buckhorn

Alkartasuna Ruby

Agner Model 80

Action Arms Uzi Pistol

Advantage Arms Model 442

HANDGUNS

Action Arms
AT-84
AT-84P
AT-88S
AT-88H
AT-88P
Uzi Pistol

Advantage Arms
Model 442

Agner
Model 80

Air Match
500

A.J. Ordnance
Thomas 45

Alkartasuna
Ruby

Allen
1875 Army Outlaw
Buckhorn
Buckhorn Buntline Model

ACTION ARMS AT-84

Semi-automatic; double action; 9mm Para., 41 Action Express; 15-shot magazine (9mm), 10-shot (41 AE); 4 3/4" barrel; 8 1/16" overall length; weighs 35 1/2 oz.; drift-adjustable rear sight, blade front; polished blued finish. Made in Switzerland. Introduced 1987; no longer imported by Action Arms.

   Perf.: $500   **Exc.:** $450   **VGood:** $350

Action Arms AT-84P

Same specs as Model AT-84 except 13-shot (9mm Para.), 8-shot (41 AE); 3 11/16" barrel; weighs 32 oz. Prototype only.

ACTION ARMS AT-88S

Semi-automatic; double action; 9mm Para., 41 Action Express; 10-shot magazine; 4 3/4" barrel; 8 1/8" overall length; weighs 35 1/2 oz.; blade front sight, drift-adjustable rear; checkered walnut grips; polished blue finish; originally marketed with both barrels. Made in England. Introduced 1987; importation dropped 1990.

   Perf.: $550   **Exc.:** $500   **VGood:** $475

Action Arms AT-88H

Same specs as AT-88S except 10-shot (9mm Para.), 7-shot (41 Action Express) magazine; 3 1/2" barrel; 6 7/8" overall length; weighs 30 1/2 oz. Introduced 1989; importation dropped 1990.

   Perf.: $550   **Exc.:** $500   **VGood:** $475

Action Arms AT-88P

Same specs as AT-88S except 13-shot (9mm Para.), 8-shot (41 Action Express) magazine; 3 11/16" barrel; 7 5/16" overall length; weighs 32 oz. Introduced 1989; importation dropped 1990.

   Perf.: $550   **Exc.:** $500   **VGood:** $475

ACTION ARMS UZI PISTOL

Semi-automatic; single action; 9mm Para., 45 ACP; 20-shot magazine; 4 1/2" barrel; 9 1/2" overall length; weighs 3 1/2 lbs.; black plastic grip; post front sight with white dot, open fully click-adjustable rear, two white dots; semi-auto blowback action; fires from closed bolt; floating firing pin; comes in moulded plastic case. Imported from Israel by Action Arms. Introduced 1984; dropped 1993.

   Perf.: $925   **Exc.:** $750   **VGood:** $650

ADVANTAGE ARMS MODEL 442

Derringer; break-top action; 22 LR, 22 WMR; 4-shot; 2 1/2" barrel; 4 1/2" overall length; weighs 15 oz.; smooth walnut stocks; fixed sights; double-action trigger; rotating firing pin; spring-loaded extractors; nickel, blued or black finish. Introduced 1983; dropped 1985. Reintroduced 1989 by New Advantage Arms Corp.

   Perf.: $150   **Exc.:** $125   **VGood:** $100

AGNER MODEL 80

Semi-automatic; 22 LR; 5-shot magazine; 5 15/16" barrel; 9 1/2" overall length; weighs 36 oz.; adjustable French walnut stocks; fixed blade front sight, adjustable rear; safety key locks trigger, slide, magazine; dry fire button; right- or left-hand models. Made in Denmark. Introduced 1984; dropped 1987; was imported by Beeman.

   Perf.: $1000   **Exc.:** $900   **VGood:** $750

AIR MATCH 500

Single-shot; 22 LR; 10 7/16" barrel; weighs 28 oz.; match grip of stippled hardwood; left- or right-hand; match post front sight, adjustable match rear; adjustable sight radius; marketed in case with tools, spare sights. Made in Italy. Introduced 1984; dropped 1987; was imported by Kendall International Arms.

   Perf.: $500   **Exc.:** $450   **VGood:** $350

A.J. ORDNANCE THOMAS 45

Semi-automatic; double action; 45 ACP; 6-shot magazine; 3 1/2" barrel; 6 1/2" overall length; windage-adjustable rear sight, blade front; checkered plastic grips; blued finish; matte sighting surface. Introduced 1977; discontinued 1978.

   Perf.: $550   **Exc.:** $475   **VGood:** $375
Chrome finish
   Perf.: $800   **Exc.:** $600   **VGood:** $450

ALKARTASUNA RUBY

Semi-automatic; 32 ACP; 9-shot magazine; 3 3/8" barrel; 6 3/8" overall length; checkered hard rubber or wooden stocks; blued finish; fixed sights. Manufactured in Europe. Manufactured in Spain, 1917 to 1922; distributed primarily in Europe; some used by French army in World Wars I, II.

   Exc.: $150   **VGood:** $100   **Good:** $85

ALLEN 1875 ARMY OUTLAW

Revolver; single action; 357 Mag., 44-40, 45 Colt; 6-shot cylinder; 7 1/2" barrel; 13 3/4" overall length; weighs 44 oz.; notch rear sight, blade front; uncheckered walnut stocks; brass triggerguard; color case-hardened frame, rest blued. Copy of Model 1875 Remington. Made in Italy by Uberti. Introduced 1986; importation dropped 1987.

   Perf.: $300   **Exc.:** $250   **VGood:** $225

ALLEN BUCKHORN

Revolver; single-action; 44 Mag., 44 Spl., 44-40; 6-shot cylinder; 4 3/4", 6", 7 1/2" barrel; grooved rear sight, blade front; one-piece uncheckered walnut stocks; steel or brass backstrap, triggerguard; color case-hardened frame; blued cylinder, barrel. Made in Italy by Uberti. Formerly imported by Iver Johnson. Reintroduced 1986; importation dropped by Allen Fire Arms 1987.

   Perf.: $350   **Exc.:** $300   **VGood:** $275

Allen Buckhorn Buntline Model

Same specs as Buckhorn except 18" barrel. No longer imported by Allen.

   Perf.: $400   **Exc.:** $350   **VGood:** $325

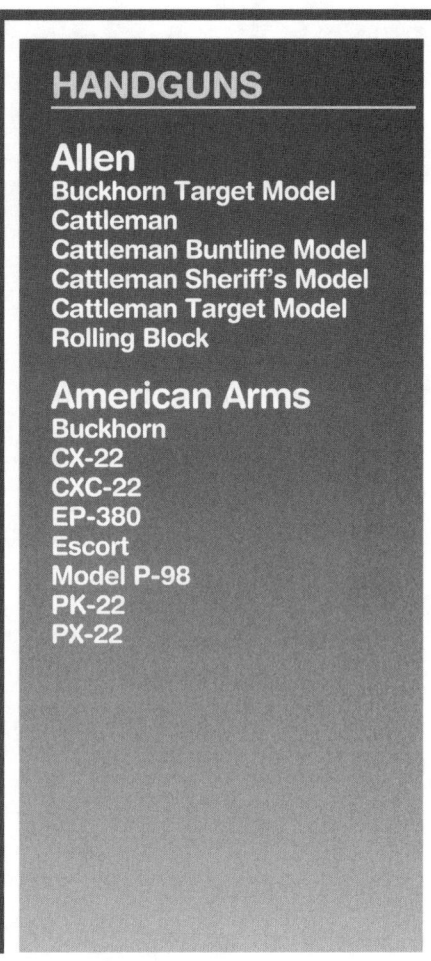

Allen
Buckhorn Target Model
Cattleman
Cattleman Buntline Model
Cattleman Sheriff's Model
Cattleman Target Model
Rolling Block

American Arms
Buckhorn
CX-22
CXC-22
EP-380
Escort
Model P-98
PK-22
PX-22

American Arms ZC-380

American Arms
TT-9mm Tokarev

American Arms CX-22

American Arms PK-22

American Arms Escort

Allen Buckhorn Target Model
Same specs as Buckhorn except flat-top frame; adjustable rear sight ramp front. No longer imported by Allen.
Perf.: $425   **Exc.:** $375   **VGood:** $350
Convertible cylinder model
Perf.: $475   **Exc.:** $425   **VGood:** $400

ALLEN CATTLEMAN
Revolver; single action; 22 LR, 22 WMR, 38 Spl., 357 Mag., 44-40, 45 Colt; 6-shot cylinder; 4 3/4", 5 1/2", 7 1/2" barrel; fixed groove rear sight, blade front; one-piece uncheckered walnut stocks; brass backstrap, triggerguard; blued barrel, cylinder; case-hardened frame; polished hammer flats. Made in Italy by Uberti. Formerly imported by Iver Johnson. Reintroduced 1986; no longer imported by Allen Fire Arms.
Perf.: $350   **Exc.:** $300   **VGood:** $275

Allen Cattleman Buntline Model
Same specs as Cattleman except 18" barrel; 357, 44-40, 45 Colt only. No longer imported by Allen.
Perf.: $425   **Exc.:** $375   **VGood:** $325

Allen Cattleman Sheriff's Model
Same specs as Cattleman except 3" barrel; 44-40, 45 Colt only. No longer imported by Allen.
Perf.: $350   **Exc.:** $300   **VGood:** $275

Allen Cattleman Target Model
Same specs as Cattleman except flat-top frame; fully-adjustable rear sight. No longer imported by Allen.
Perf.: $300   **Exc.:** $275   **VGood:** $250

ALLEN ROLLING BLOCK
Revolver; single shot; 22 LR, 22 WMR, 357 Mag., 45 Colt; 9 7/8" half-round, half-octagon barrel; 14" overall length; weighs 44 oz.; fully-adjustable rear sight, blade front; walnut stocks, forend; brass triggerguard; color case-hardened frame, blued barrel. Replica of 1871 target rolling block. Made in Italy by Uberti. Introduced 1986; no longer imported by Allen Fire Arms; currently imported by Uberti U.S.A.
Perf.: $350   **Exc.:** $300   **VGood:** $275

AMERICAN ARMS BUCKHORN
Revolver; single action; 44 Mag.; 4 3/4", 6", 7 1/2" barrel; 11" overall length; weighs 44 oz. (6" barrel); smooth walnut grips; blade front sight, groove rear; blued barrel and cylinder; brass triggerguard and backstrap. Imported from Italy by American Arms, Inc. Introduced 1993; still imported.
New: $350   **Perf.:** $300   **Exc.:** $275

AMERICAN ARMS CX-22
Semi-automatic; double action; 22 LR; 8-shot magazine; 3 5/16" barrel; 6 5/16" overall length; weighs 22 oz.; blade front sight, windage-adjustable rear; checkered black polymer grips; manual hammer-block safety; firing pin safety; alloy frame; blue-black finish. Resembles Walther PPK externally. Introduced 1990.
Perf.: $150   **Exc.:** $135   **VGood:** $125

American Arms CXC-22
Same specs as CX-22 except chromed slide. Introduced 1990; no longer produced.
Perf.: $150   **Exc.:** $135   **VGood:** $125

AMERICAN ARMS EP-380
Semi-automatic; double action; 380 ACP; 7-shot magazine; 3 1/2" barrel; 5 1/2" overall length; weighs 25 oz.; checkered wood stocks; fixed sights; slide-mounted safety; made of stainless steel. Manufactured in West Germany. Introduced 1988; no longer imported.
Perf.: $350   **Exc.:** $325   **VGood:** $300

AMERICAN ARMS ESCORT
Semi-automatic; double-action-only; 380 ACP; 7-shot magazine; 3 3/8" barrel; 6 1/8" overall length; weighs 19 oz.; soft polymer grips; blade front sight, windage-adjustable rear; stainless steel construction; chamber-loaded indicator. Marketed by American Arms, Inc. Introduced 1995.
New: $275   **Perf.:** $225   **Exc.:** $220

AMERICAN ARMS MODEL P-98
Semi-automatic; double action; 22 LR; 8-shot magazine; 5" barrel; 8 1/8" overall length; weighs 25 oz.; blade front sight, windage-adjustable rear; grooved black polymer grips; magazine disconnect safety; hammer-block safety; alloy frame; blue-black finish. Has external appearance of Walther P.38. Introduced 1969.
Perf.: $175   **Exc.:** $150   **VGood:** $135

AMERICAN ARMS PK-22
Semi-automatic; double action; 22 LR; 8-shot magazine; 3 5/16" barrel; 6 5/16" overall length; weighs 22 oz.; fixed rear sight; checkered plastic grips; slide-mounted safety; polished blue finish. From American Arms, Inc. Introduced 1989.
Perf.: $175   **Exc.:** $150   **VGood:** $135

AMERICAN ARMS PX-22
Semi-automatic; double action; 22 LR; 7-shot magazine; 2 15/16" barrel; 5 3/8" overall length; weighs 15 oz.; checkered black plastic stocks; fixed sights; polished blued finish. From American Arms, Inc. Introduced 1989.
Perf.: $175   **Exc.:** $150   **VGood:** $135

American Arms Regulator

American Arms P-98

American Derringer COP

American Arms PX-22

American Derringer Lady Derringer

HANDGUNS

American Arms
PX-25
Regulator
Sabre
Spectre
TT-9mm Tokarev
ZC-380

American Derringer
COP 357
COP Mini
DA 38
High Standard Derringer
Lady Derringer

AMERICAN ARMS PX-25

Semi-automatic; double action; 25 ACP; 7-shot magazine; 2 3/4" barrel; 5 3/8" overall length; weighs 15 oz.; checkered black plastic stocks; fixed sights; blued finish. Introduced 1990; no longer in production.

    Perf.: $175   **Exc.:** $150   **VGood:** $135

AMERICAN ARMS REGULATOR

Revolver; single action; 357 Mag., 44-40, 45 Colt; 4 3/4", 5 1/2", 7 1/2" barrel; 8 1/8" overall length (4 3/4" barrel); weighs 32 oz. (4 3/4" barrel); smooth walnut grips; blade front sight, groove rear; blued barrel and cylinder, brass triggerguard and backstrap. Imported from Italy by American Arms, Inc. Introduced 1992.

    New: $325   **Perf.:** $300   **Exc.:** $250
Dual cylinder (44-40/44 Spl. or 45 Colt/45 ACP), or DLX all-steel model
    Perf.: $375   **Exc.:** $325   **VGood:** $275

AMERICAN ARMS SABRE

Semi-automatic; double-action-only; 9mm Para., 40 S&W; 9-shot magazine (9mm Para.), 8-shot (40 S&W) magazine; 3 3/4" barrel; 6 13/16" overall length; weighs 26 oz.; black polymer composite grips; blade front sight, square notch windage-adjustable rear; short-recoil action; left-side safety and magazine catch; blue or stainless steel. Imported from Italy by American Arms, Inc. Introduced 1991; never in series production.

AMERICAN ARMS SPECTRE

Semi-automatic; double action; 9mm Para., 45 ACP; 30-shot magazine; 6" barrel; 13 3/4" overall length; weighs 4 lbs., 8 oz.; black nylon grip; fully-adjustable post front sight, fixed U-notch rear; blowback action fires from closed bolt; ambidex-trous safety and decocking levers; matte black finish; comes with magazine loading tool. For standard velocity ammunition only. Was marketed by American Arms, Inc.
9mm caliber
    Perf.: $450   **Exc.:** $350   **VGood:** $300
45 ACP caliber
    Perf.: $475   **Exc.:** $375   **VGood:** $325

AMERICAN ARMS TT-9MM TOKAREV

Semi-automatic; single action; 9mm Para.; 9-shot magazine; 4 1/2" barrel; 8" overall length; weighs 32 oz.; grooved plastic stocks; fixed sights; blued finish. Copy of Russian Tokarev, made in Yugoslavia. Introduced 1988; dropped 1990.

    Perf.: $125   **Exc.:** $100   **VGood:** $85

AMERICAN ARMS ZC-380

Semi-automatic; single action; 380 ACP; 7-shot magazine; 3 3/4" barrel; 6 1/2" overall length; weighs 26 oz.; checkered plastic stocks; fixed sights; polished, blued finish. Made in Yugoslavia. Introduced 1988; dropped 1990.

    Perf.: $250   **Exc.:** $225   **VGood:** $200

AMERICAN DERRINGER COP 357

Derringer; double-action-only; 4-shot; 38 Spl., 357 Mag.; 3 1/8" barrel; 5 1/2" overall length; weighs 16 oz.; rosewood grips; fixed sights; stainless steel construction. Made in U.S. by American Derringer Corp. Introduced 1990; no longer in production.

    Perf.: $275   **Exc.:** $250   **VGood:** $225

American Derringer Cop Mini

Same specs as COP 357 except 22 WMR; 2 7/8" barrel; 4 15/16" overall length; double action; auto-matic hammer-block safety; rosewood, walnut or other hardwood grips. Made in U.S. by American Derringer Corp. Introduced 1990; no longer in production.

    Perf.: $250   **Exc.:** $225   **VGood:** $200

AMERICAN DERRINGER DA 38

Derringer; double action; 38 Spl., 357 Mag., 9mm Para., 40 S&W; 3" over/under barrels; 4 13/16" overall length; weighs 14 1/2 oz.; rosewood, walnut or other hardwood grips; fixed sights; manual safety; made of satin-finished stainless steel, aluminum. Introduced 1989; still in production.

    Perf.: $275   **Exc.:** $225   **VGood:** $200

AMERICAN DERRINGER HIGH STANDARD DERRINGER

Derringer; double action; 22 LR, 22 WMR; 2-shot; 3 1/2" barrel; 5 1/8" overall length; weighs 11 oz.; hammer-block safety; blued finish. Direct copy of original High Standard derringer. Reintroduced 1990; no longer in production.

    Perf.: $150   **Exc.:** $135   **VGood:** $125

AMERICAN DERRINGER LADY DERRINGER

Derringer; over/under; 32 H&R Mag., 38 Spl., 22 LR, 22 WMR, 380 ACP, 357 Mag., 9mm Para., 45 ACP, 45 Colt/.410 shotshell; 2-shot; tuned action; scrimshawed synthetic ivory grips; deluxe grade polished; also deluxe engraved version with 1880s-style engraving. Marketed in fitted box. Introduced 1991; still in production.

    Perf.: $300   **Exc.:** $250   **VGood:** $185
Deluxe grade
    Perf.: $350   **Exc.:** $275   **VGood:** $200
Deluxe engraved
    Perf.: $650   **Exc.:** $500   **VGood:** $350

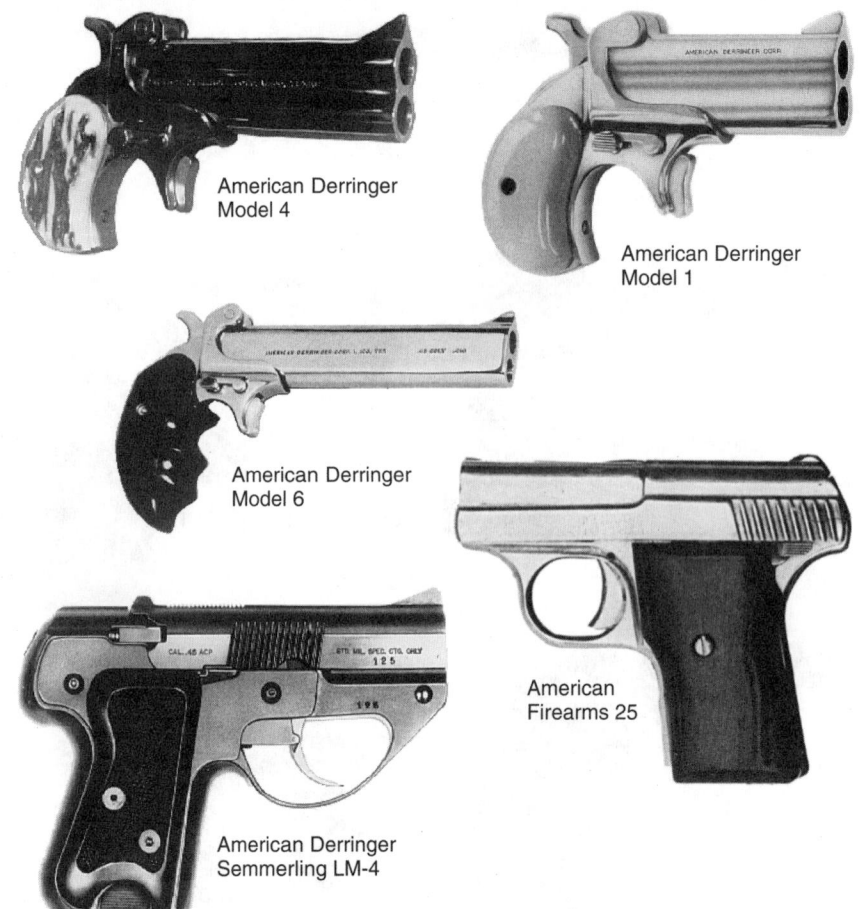

HANDGUNS

American Derringer
Model 1
Model 1 Texas Commemorative
Model 3
Model 4
Model 4 Alaskan Survival Model
Model 6
Model 7 Ultra Lightweight
Model 10 Lightweight
Model 11
Semmerling LM-4

American Firearms
25

American Derringer Model 4

American Derringer Model 1

American Derringer Model 6

American Firearms 25

American Derringer Semmerling LM-4

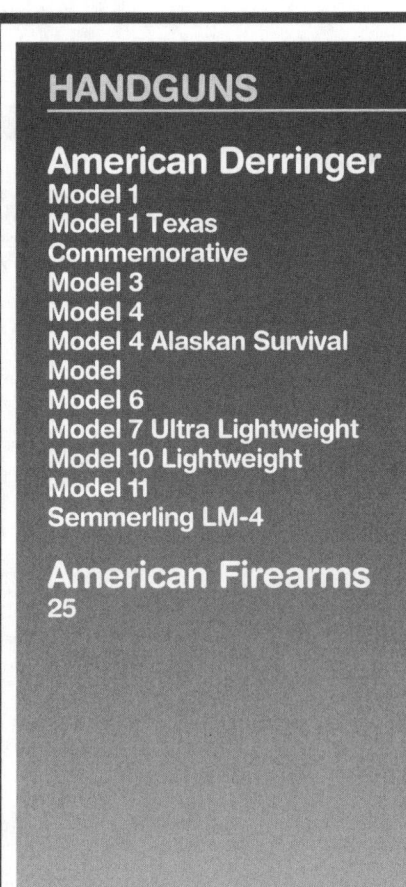

AMERICAN DERRINGER MODEL 1
Derringer; over/under; 22 LR, 22 WMR, 22 Hornet, 223 Rem., 30 Luger, 30-30, 32 ACP, 38 Super, 380 ACP, 38 Spl., 9x18, 9mm Para., 357 Mag., 41 Mag., 44-40, 44 Spl., 44 Mag., 45 Colt/.410; 2-shot; break action; 3″ barrel; 4$^{13}/_{16}$″ overall length; weighs 15 $^1/_2$ oz. (38 Spl.); rosewood, zebra wood stocks; blade front sight; stainless steel with high-polish or satin finish; manual hammer-block safety. Made in U.S. by American Derringer Corp. Introduced 1980; still in production.
Regular caliber
  Perf.: $275   **Exc.:** $225   **VGood:** $185
Large caliber
  Perf.: $325   **Exc.:** $275   **VGood:** $200
Engraved model
  Perf.: $1000   **Exc.:** $850   **VGood:** $600

American Derringer Model 1 Texas Commemorative
Same specs as Model 1 except solid brass frame; stainless steel barrel; rosewood grips; 38 Spl., 44-40 Win., 45 Colt. Made in U.S. by American Derringer Corp. Introduced 1987; still in production.
LMSR: $225
38 Special caliber
  Perf.: $250   **Exc.:** $200   **VGood:** $175
44-40 or 45 Colt caliber
  Perf.: $375   **Exc.:** $300   **VGood:** $225

AMERICAN DERRINGER MODEL 3
Derringer; single shot; 38 Spl.; 2 $^1/_2$″ barrel; 4 $^{15}/_{16}$″ overall length; weighs 8 $^1/_2$ oz.; rosewood grips; blade front sight; stainless steel construction; manual hammer-block safety. Made in U.S. by American Derringer Corp. Introduced 1985; no longer in production.
  Perf.: $100   **Exc.:** $85   **VGood:** $75

AMERICAN DERRINGER MODEL 4
Derringer; over/under; 357 Mag., 357 Max., 45-70/.410/3″, 45 Colt/.410/3″, 45 ACP, 44 Mag.; 2-shot; break-action; 4 $^1/_8$″ barrel; 6″ overall length; weighs 16$^1/_2$ oz.; stainless steel construction. Made in U.S. by American Derringer Corp. Introduced 1985; still in production.
  Perf.: $300   **Exc.:** $250   **VGood:** $200

American Derringer Model 4 Alaskan Survival Model
Same specs as Model 4 except upper barrel chambered for 45-70 or 44 Mag. Made in U.S. by American Derringer Corp. Introduced 1980; still in production.
  Perf.: $375   **Exc.:** $325   **VGood:** $250

AMERICAN DERRINGER MODEL 6
Derringer; over/under; .410/3″, 22 WMR, 357 Mag., 45 ACP, 45 Colt; 2-shot; 6″ barrels; 8 $^1/_4$″ overall length; weighs 21 oz.; rosewood stocks; manual hammer-block safety. Made in U.S. by American Derringer Corp. Introduced 1986; still in production.
  Perf.: $275   **Exc.:** $225   **VGood:** $200

AMERICAN DERRINGER MODEL 7 ULTRA LIGHTWEIGHT
Derringer; over/under; 22 LR, 22 WMR, 32 H&R Mag., 380 ACP, 38 Spl., 44 Spl.; 2-shot; break-action; 3″ barrel; 4 $^7/_8$″ overall length; weighs 7 $^1/_2$ oz.; high strength aircraft aluminum construction; rosewood, zebra wood stocks; blade front sight; stainless steel; manual hammer-block safety. Made in U.S. by American Derringer Corp. Introduced 1986; still in production.
  Perf.: $225   **Exc.:** $175   **VGood:** $135
44 Spl. caliber
  Perf.: $400   **Exc.:** $350   **VGood:** $325

AMERICAN DERRINGER MODEL 10 LIGHTWEIGHT
Derringer; over/under; 45 ACP, 45 Colt, 2-shot; break-action; 3″ barrels; 4 $^{13}/_{16}$″ overall length; weighs 10 oz.; aluminum frame; matte gray finish. Introduced 1990; still in production.
45 Colt
  Perf.: $300   **Exc.:** $250   **VGood:** $200
45 ACP
  Perf.: $250   **Exc.:** $225   **VGood:** $175

AMERICAN DERRINGER MODEL 11
Derringer; over/under; 38 Spl.; 2-shot; break-action; 3″ barrel; 4 $^{13}/_{16}$″ overall length; weighs 11 oz.; aluminum barrels; matte gray frame. Introduced 1990; no longer in production.
  Perf.: $200   **Exc.:** $150   **VGood:** $135

AMERICAN DERRINGER SEMMERLING LM-4
Derringer; manually-operated repeater; 9mm Para., 45 ACP; 7-shot (9mm), 5-shot (45 ACP) magazine; 3″ barrel; 5 $^1/_4$″ overall length; weighs 24 oz.; height 3 $^3/_4$″; width 1″; checkered plastic grips (blued finish), rosewood (stainless finish); open fixed sights. Was sold with manual, leather carrying case, spare stock screws, wrench. Made in U.S. by American Derringer Corp.
  Perf.: $2000   **Exc.:** $1800   **VGood:** $1300
Stainless steel
  Perf.: $2200   **Exc.:** $2000   **VGood:** $1600

AMERICAN FIREARMS 25
Semi-automatic; 25 ACP; 8-shot magazine; 2 $^1/_8$″ barrel; 4 $^7/_{16}$″ overall length; fixed sights; walnut grips; blued ordnance steel or stainless steel. Manufactured by American Firearms Co., Inc. Introduced 1966; dropped 1974.
  Perf.: $350   **Exc.:** $275   **VGood:** $225
Stainless steel
  Perf.: $400   **Exc.:** $325   **VGood:** $250

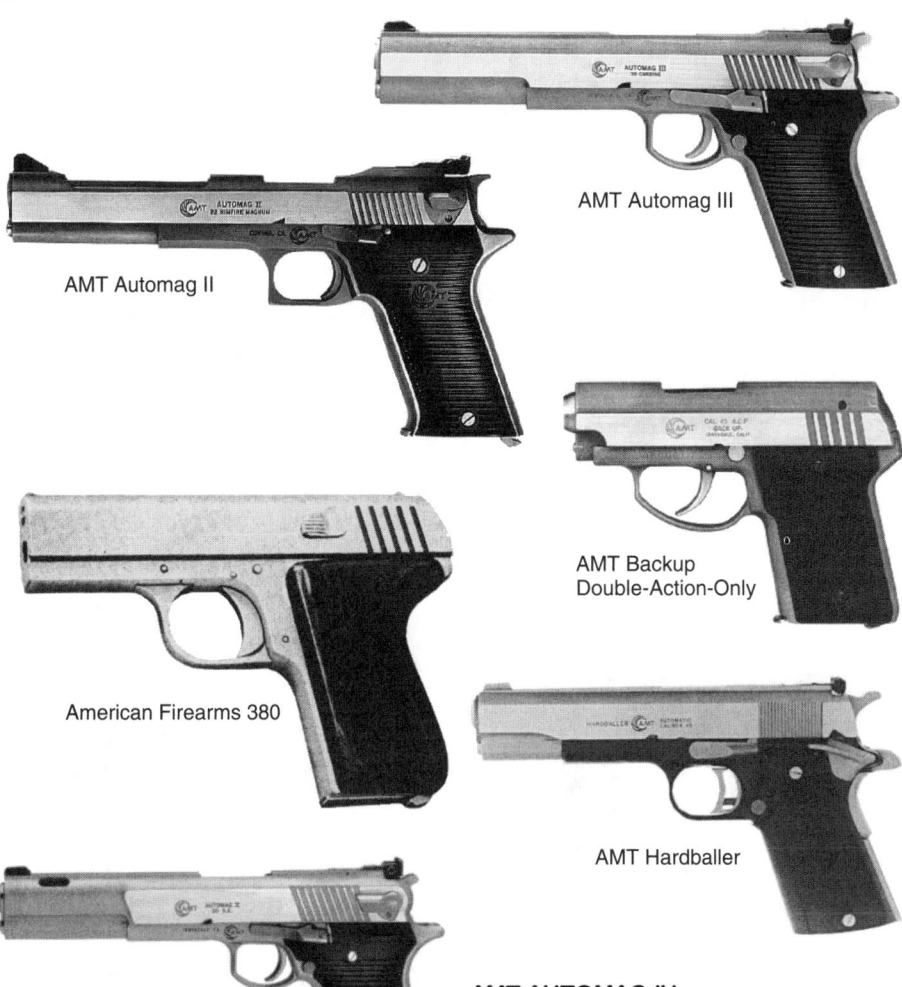

AMT Automag III

AMT Automag II

AMT Backup
Double-Action-Only

American Firearms 380

AMT Hardballer

AMT Automag V

HANDGUNS

American Firearms
380
Derringer

AMT
Automag II
Automag III
Automag IV
Automag V
Backup
Backup II
Backup Double-Action-Only
Bull's-Eye Target
Bull's-Eye Target Regulation
Combat Government
Hardballer

AMERICAN FIREARMS 380

Semi-automatic; 380 ACP; 8-shot magazine; 3 1/2" barrel; 5 1/2" overall; stainless steel; smooth walnut stocks. Limited manufacture 1972.
   Perf.: $400   **Exc.:** $350   **VGood:** $300

AMERICAN FIREARMS DERRINGER

Derringer; over/under; 38 Spl., 22 LR, 22 WMR; 2-shot; 3" barrel; fixed open sights; checkered plastic grips; stainless steel construction; spur trigger, half-cock safety. Introduced 1972; dropped 1974.
   Perf.: $225   **Exc.:** $175   **VGood:** $135

AMT AUTOMAG II

Semi-automatic; 22 WMR; 9-shot magazine, 7-shot (3 3/8" barrel); 3 3/8", 4 1/2", 6" barrel; 9 3/4" overall length; weighs 23 oz.; Millett adjustable rear sight, blade front; grooved black composition stocks; squared triggerguard; stainless steel construction; gas-assisted action; brushed finish on slide flats, rest sandblasted. Introduced 1987; still in production.
   Perf.: $300   **Exc.:** $250   **VGood:** $200

AMT AUTOMAG III

Semi-automatic; single action; 30 Carbine, 9mm Win. Mag.; 8-shot magazine; 6 3/8" barrel; 10 1/2" overall length; weighs 43 oz.; carbon fiber grips; blade front sight, adjustable rear; stainless steel construction; hammer-drop safety; brushed finish on slide flats, rest sandblasted. Made in U.S. by AMT. Introduced 1989; still in production.
   Perf.: $450   **Exc.:** $400   **VGood:** $300

AMT AUTOMAG IV

Semi-automatic; single action; 45 Winchester Mag.; 7-shot magazine; 6 1/2" barrel; 10 1/2" overall length; weighs 46 oz.; carbon fiber grips; blade front sight, adjustable rear; stainless steel construction with brushed finish. Made in U.S. by AMT. Introduced 1990; still in production.
   Perf.: $500   **Exc.:** $450   **VGood:** $375

AMT AUTOMAG V

Semi-automatic; single action; 50 AE; 5-shot magazine; 6 1/2" barrel; 10 1/2" overall length; weighs 46 oz.; carbon fiber grips; blade front sight, adjustable rear; stainless steel construction with brushed finish. Made in U.S. by AMT. Introduced 1990.
   Perf.: $750   **Exc.:** $650   **VGood:** $600

AMT BACKUP

Semi-automatic; double action; 22 LR, 380 ACP; 8-shot (22 LR), 5-shot (380 ACP) magazine; 2 1/2" barrel; 5" overall length; weighs 17 oz.; smooth wood stocks; fixed open sights; concealed hammer; manual grip safeties; blowback operated; stainless steel. 380 ACP introduced 1974; 22 introduced 1982; no longer in production.
   Perf.: $225   **Exc.:** $200   **VGood:** $150

AMT Backup II

Same specs as standard Backup except single action; 380 ACP only with 5-shot magazine; carbon fiber grips; weighs 18 oz. Introduced 1993.
   Perf.: $250   **Exc.:** $200   **VGood:** $175

AMT Backup Double-Action-Only

Same specs as standard Backup except 9mm Para., 38 Super, 40 S&W, 45 ACP; double-action-only; enlarged triggerguard; slide rounded at rear; 5-shot magazine. Made in U.S. by AMT. Introduced 1992.
   Perf.: $350   **Exc.:** $300   **VGood:** $275

AMT BULL'S-EYE TARGET

Semi-automatic; 40 S&W; 8-shot magazine; 5" barrel; 8 1/2" overall length; weighs 38 oz.; Millett adjustable rear sight; neoprene wrap-around grips; stainless steel construction; long grip safety; wide adjustable trigger; beveled magazine well. Introduced 1991; dropped 1992.
   Perf.: $350   **Exc.:** $300   **VGood:** $275

AMT Bull's-Eye Target Regulation

Same specs as Bull's-Eye Target except 22 LR; 5", 6 1/2", 8 1/2", 10 1/2", 12 1/2" bull or tapered barrel; adjustable sights; vent rib; wooden target grips. Manufactured 1986 only.
   Perf.: $375   **Exc.:** $300   **VGood:** $250

AMT COMBAT GOVERNMENT

Semi-automatic; single action; 45 ACP; 5" barrel; 8 1/2" overall length; stainless steel construction; extended combat safety; adjustable target-type trigger; flat mainspring housing; custom-fitted barrel bushing; fixed combat sights; checkered walnut grips. Introduced 1978; dropped 1980.
   Perf.: $325   **Exc.:** $275   **VGood:** $250

AMT HARDBALLER

Semi-automatic; 45 ACP; 5" barrel; 8 1/2" overall length; weighs 39 oz.; stainless steel construction; extended combat safety; serrated matted slide rib; long grip safety; beveled magazine well; grooved front, back straps; adjustable trigger; custom barrel bushing; adjustable combat sights; checkered walnut or wrap-around rubber grips. Introduced 1978; still in production.
   Perf.: $400   **Exc.:** $350   **VGood:** $300

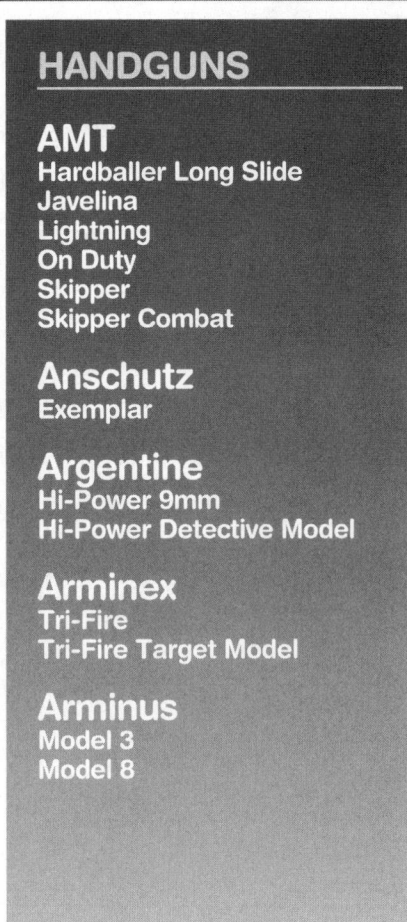

AMT
Hardballer Long Slide
Javelina
Lightning
On Duty
Skipper
Skipper Combat

Anschutz
Exemplar

Argentine
Hi-Power 9mm
Hi-Power Detective Model

Arminex
Tri-Fire
Tri-Fire Target Model

Arminus
Model 3
Model 8

AMT On Duty

AMT Lightning

Anschutz Exemplar

AMT Hardballer Long Slide

Argentine Hi-Power 9mm

AMT Skipper

AMT Hardballer Long Slide
Same specs as Hardballer except 7″ barrel; 10 1/2″ overall length; fully-adjustable micro rear sight. Introduced 1977; still in production. **LMSR: $595**
 Perf.: $425 Exc.: $375 VGood: $325

AMT JAVELINA
Semi-automatic; single action; 10mm Auto; 8-shot magazine; 7″ barrel; 10 1/2″ overall length; weighs 40 oz.; wrap-around rubber grips; blade front sight, Millett adjustable rear; all stainless construction, brushed finish. Made in U.S. by AMT. Introduced 1989; no longer in production.
 Perf.: $500 Exc.: $450 VGood: $400

AMT LIGHTNING
Semi-automatic; 22 LR; 10-shot magazine; 6 1/2″, 8 1/2″, 10 1/2″, 12 1/2″ tapered barrels; 5″, 6 1/2″, 8 1/2″, 10 1/2″, 12 1/2″ bull barrels; 10 3/4″ overall length (6 1/2″ barrel); weighs 45 oz.; checkered wrap-around rubber stocks; blade front sight, fixed rear; adjustable rear at extra cost; stainless steel; Clark trigger with adjustable stops; receiver grooved for scope; interchangeable barrels. Introduced 1984; dropped 1989.
 Perf.: $325 Exc.: $250 VGood: $175
Adjustable Sights
 Perf.: $350 Exc.: $275 VGood: $225

AMT ON DUTY
Semi-automatic; double action; 9mm Para., 40 S&W, 45 ACP; 15-shot (9mm), 11-shot (40 S&W), 9-shot (45 ACP) magazine; 4 1/2″ barrel; 7 3/4″ overall length; weighs 32 oz.; smooth carbon fiber grips; blade front sight, windage-adjustable rear; three-dot system; choice of DA with decocker or double-action-only; inertia firing pin, trigger disconnector safety; aluminum frame with steel recoil

shoulder, stainless steel slide and barrel. Made in U.S. by AMT. Introduced 1991; no longer in production.
 Perf.: $375 Exc.: $300 VGood: $275
45 ACP caliber
 Perf.: $425 Exc.: $350 VGood: $300

AMT SKIPPER
Semi-automatic; 45 ACP, 40 S&W; 7-shot magazine; 4 1/4″ barrel; 7″ overall length; weighs 33 oz.; Millet adjustable rear sight; checkered walnut grips; matte stainless steel construction. Introduced 1978; dropped 1984.
 Perf.: $325 Exc.: $275 VGood: $250

AMT Skipper Combat
Same specs as standard Skipper except fixed sights. Discontinued 1984.
 Perf.: $350 Exc.: $300 VGood: $275

ANSCHUTZ EXEMPLAR
Bolt action; 22 LR, 22 Hornet; 5-shot magazine; 10″, 14″ barrel; 17″ overall length; weighs 56 oz.; open-notch adjustable rear sight, hooded front on ramp; European walnut stock, stippled grip, forend; built on Anschutz Match 64 rifle action with left-hand bolt; Anschutz #5091 two-stage trigger; receiver grooved for scope mount. Made in Germany. Introduced 1987; still imported. **LMSR: $558**
 Perf.: $400 Exc.: $350 VGood: $300
22 Hornet
 Perf.: $800 Exc.: $750 VGood: $700

ARGENTINE HI-POWER 9MM
Semi-automatic; single action; 9mm Para.; 10-shot magazine; 4 21/32″ barrel; 7 3/4″ overall length; weighs 32 oz.; blade front sight, adjustable rear; checkered walnut grips; licensed copy of Browning Hi-Power. Imported from Argentina by Century International Arms, Inc. Introduced 1990; still imported.
 Perf.: $250 Exc.: $225 VGood: $200

Argentine Hi-Power Detective Model
Same specs as standard model except 3 13/16″ barrel; 6 15/16″ overall length; weighs 33 oz.; finger-groove, checkered soft rubber grips; matte black finish. Imported by Century International Arms, Inc. Introduced 1994; still imported.
 New: $275 Perf.: $250 Exc.: $200

ARMINEX TRI-FIRE
Semi-automatic; single action; 9mm Para., 38 Super, 45 ACP; 9-shot (9mm), 7-shot (45 ACP) magazine; 5″, 6″, 7″ barrel; 8″ overall length; weighs 38 oz.; smooth contoured walnut stocks; interchangeable post front sight, adjustable rear; slide-mounted firing pin safety; contoured backstrap; blued or electroless nickel finish; convertible by changing barrel, magazine, recoil spring. Introduced 1982; dropped 1987.
 Perf.: $700 Exc.: $550 VGood: $400
Presentation cased
 Perf.: $750 Exc.: $600 VGood: $500
With conversion unit
 Perf.: $850 Exc.: $700 VGood: $600

Arminex Tri-Fire Target Model
Same specs as Tri-Fire except 1″ longer slide; 6″ or 7″ barrel. Introduced 1982; dropped 1987.
 Perf.: $800 Exc.: $650 VGood: $550

ARMINIUS MODEL 3
Revolver; double action; 25 ACP; 5-shot; 2″ barrel; hammerless; safety catch; folding trigger. Produced by Friedrick Pickert, Arminius Waffenwerk.
 Perf.: $150 Exc.: $125 VGood: $100

ARMINIUS MODEL 8
Revolver; double action; 320, 32 ACP; 5-shot; 2″, 5 1/2″ barrel; hammerless; safety catch; folding trigger. Produced by Friedrick Pickert, Arminius Waffenwerk.
 Perf.: $150 Exc.: $125 VGood: $110

Astra A-75 Decocker

Arminex Tri-Fire

HANDGUNS

Arminus
Model 9
Model 9A
Model 10
Model TP1
Model TP2

Armscor
Model 38
Model 200P
Model 200SE
Model 200TC

Arm Tech
Derringer

Astra
A-60
A-70
A-75 Decocker
A-80

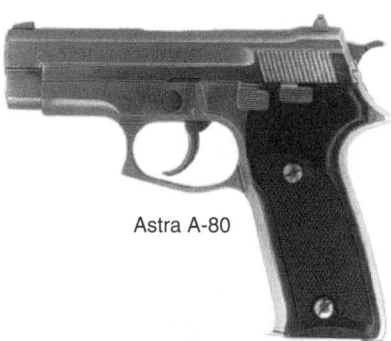
Astra A-80

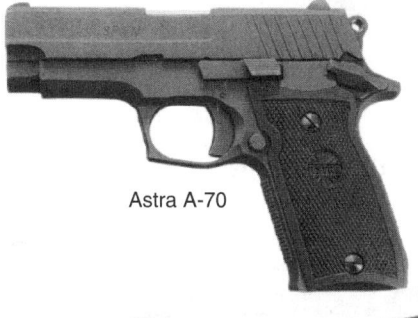
Astra A-70

Armscor Model 38

Armscor Model 200SE

ARMINIUS MODEL 9
Revolver; double action; 320, 32 ACP; 5-shot; 2 5/16" barrel; hammer model. Produced by Friedrick Pickert, Arminius Waffenwerk.
Perf.: $185   **Exc.:** $135   **VGood:** $110

Arminius Model 9A
Same specs as Model 9 except 3 1/8" barrel.
Perf.: $185   **Exc.:** $135   **VGood:** $110

ARMINIUS MODEL 10
Revolver; double action; 320, 32 ACP; 5-shot; 2 5/16" barrel; hammerless; safety catch; folding trigger. Produced by Friedrick Pickert, Arminius Waffenwerk.
Perf.: $165   **Exc.:** $145   **VGood:** $125

ARMINIUS MODEL TP1
Revolver; single shot; 22 LR; adjustable sights; drop-down action; hammer fired. Target pistol produced by Friedrick Pickert, Arminius Waffenwerk.
Perf.: $250   **Exc.:** $225   **VGood:** $175

Arminius Model TP2
Similiar to Model TP1 except double-set trigger; concealed hammer. Target pistol produced by Friedrick Pickert, Arminius Waffenwerk.
Perf.: $325   **Exc.:** $275   **VGood:** $225

ARMSCOR MODEL 38
Revolver; double action; 38 Spl.; 6-shot cylinder; 4" barrel; weighs 32 oz.; windage-adjustable rear sight, ramp front; checkered Philippine mahogany stocks; vent rib; polished blue finish. Made in the Philippines. Introduced 1986; no longer imported.
Perf.: $175   **Exc.:** $135   **VGood:** $110

ARMSCOR MODEL 200P
Revolver; double action; 38 Spl.; 6-shot cylinder; 2 1/2", 4" barrel; 8 7/8" overall length (4" barrel); weighs 26 oz.; ramp front sight, fixed rear; checkered mahogany or rubber grips; blued finish. Made in the Philippines. Introduced 1990; no longer imported.
Perf.: $175   **Exc.:** $135   **VGood:** $110

Armscor Model 200SE
Same specs as Model 200P except 22 LR, 22 WMR, 38 Spl.; 2", 3", 6"; overall length 9 1/4"(4" barrel); weighs 36 oz.; Introduced 1989; no longer imported.
Perf.: $200   **Exc.:** $175   **VGood:** $150

Armscor Model 200TC
Same specs as Model 200P except 22 LR, 22 WMR, 38 Spl.; full shroud barrel; adjustable rear sight; checkered wood grips. Introduced 1990; no longer imported.
Perf.: $225   **Exc.:** $200   **VGood:** $175

ARM TECH DERRINGER
Derringer; double-action-only; 22 LR, 22 WMR; 2 5/8" barrel; 4 5/8" overall length; weighs 19 oz.; fixed sights; hard rubber or walnut stocks; stainless steel; blued model available. Introduced 1983; no longer in production.
Perf.: $175   **Exc.:** $150   **VGood:** $125

ASTRA A-60
Semi-automatic; double action; 380 ACP; 13-shot magazine; 3 1/2" barrel; adjustable rear sight, fixed front; moulded plastic grips; slide mounted ambidextrous safety; blued steel only. Made in Spain. Introduced 1986; no longer imported.
Perf.: $300   **Exc.:** $250   **VGood:** $225

ASTRA A-70
Semi-automatic; double action; 9mm Para., 40 S&W; 8-shot (9 mm), 7-shot (40 S&W) magazine; 3 1/2" barrel; weighs 29 oz.; 6 1/2" overall length; checkered black plastic grips; blade front sight, windage-adjustable rear; all steel frame and slide; checkered grip straps and trig-

ger guard; nickel or blue finish. Imported from Spain by European American Armory. Introduced 1992; still imported. **LMSR: $360**
New: $300	**Perf.:** $250	**Exc.:** $225
Nickel finish
New: $325	**Perf.:** $300	**Exc.:** $250
Stainless steel
New: $325	**Perf.:** $300	**Exc.:** $275

ASTRA A-75 DECOCKER
Semi-automatic; double action; 9mm Para., 40 S&W, 45 ACP; 8-shot (9mm), 7-shot (40 S&W) magazine; 3 1/2" barrel; 6 1/2" overall length; weighs 29 oz.(Featherweight, 23 1/2 oz.); contoured pebble-grain grips; blade front sight, windage-adjustable rear; all steel frame and slide; checkered grip straps, triggerguard; nickel or blue finish; ambidextrous decocker system. Imported from Spain by European American Armory. Introduced 1993.
Blue finish
New: $300	**Perf.:** $250	**Exc.:** $200
Nickel finish
New: $325	**Perf.:** $275	**Exc.:** $225
Stainless finish
New: $400	**Perf.:** $350	**Exc.:** $325
Featherweight, 9mm Para.
New: $325	**Perf.:** $300	**Exc.:** $250

ASTRA A-80
Semi-automatic; double action; 9mm Para., 38 Super, 45 ACP; 15-shot (9mm, 38 Super), 9-shot (45 ACP); 3 3/4" barrel; 7" overall length; weighs 40 oz.; checkered black plastic stocks; square blade front sight, drift-adjustable square notch rear; loaded-chamber indicator; combat-style triggerguard; optional right-side slide release; automatic internal safety; decocking lever; blued or chrome finish. Imported from Spain by Interarms. Introduced 1982; dropped 1989.
Blue finish
Perf.: $350	**Exc.:** $300	**VGood:** $275
Chrome finish
Perf.: $375	**Exc.:** $325	**VGood:** $300

HANDGUNS

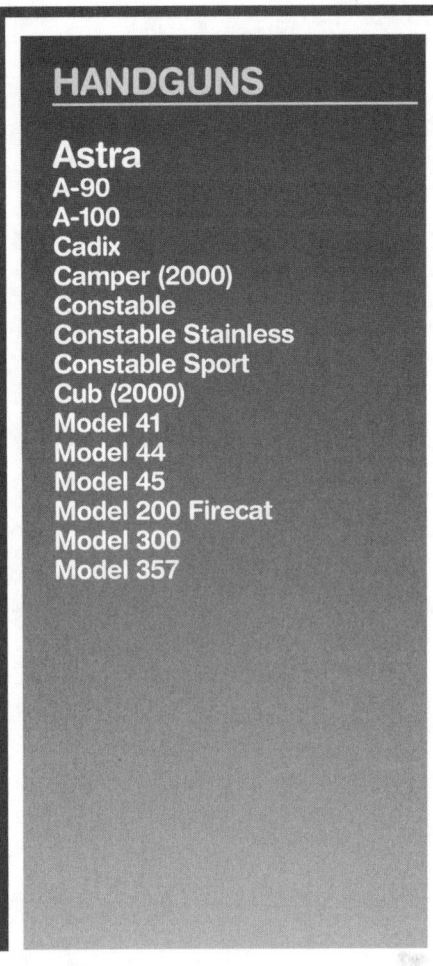

Astra
A-90
A-100
Cadix
Camper (2000)
Constable
Constable Stainless
Constable Sport
Cub (2000)
Model 41
Model 44
Model 45
Model 200 Firecat
Model 300
Model 357

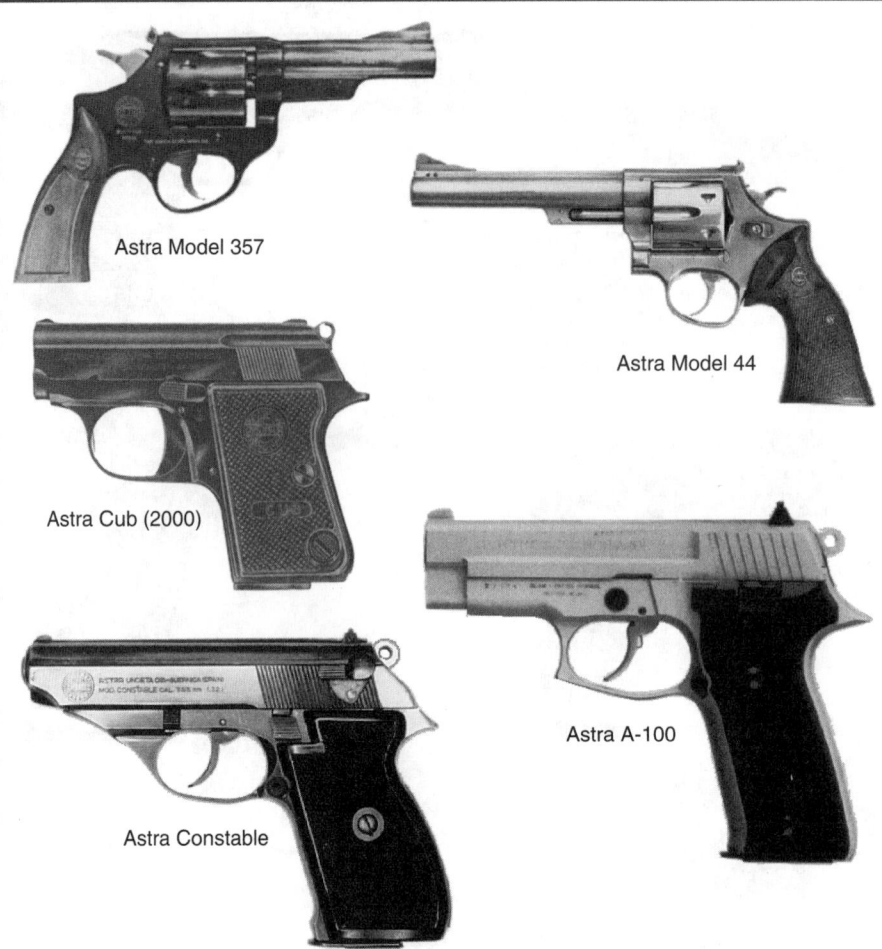

Astra Model 357

Astra Model 44

Astra Cub (2000)

Astra Constable

Astra A-100

ASTRA A-90

Semi-automatic; double action; 9mm Para., 45 ACP; 14-shot (9mm), 8-shot (45 ACP); 3 ³/₄″ barrel; 7″ overall length; weighs 48 oz.; checkered black plastic stocks; square blade front sight, square notch windage-adjustable rear; double or single action; loaded chamber indicator; optional right- or left-side slide release; combat-type triggerguard; auto internal safety; decocking lever. Made in Spain. Introduced 1985; dropped 1990.

   Perf.: $400   **Exc.:** $350   **VGood:** $300

ASTRA A-100

Semi-automatic; double action; 9mm Para., 40 S&W; 45 ACP; 10-shot (9mm, 40 S&W), 9-shot (45 ACP) magazine; 3 ¹⁵/₁₆″ barrel; 7 ¹/₁₆″ overall length; weighs 29 oz.; checkered black plastic grips; blade front sight, interchangeable rear blades for elevation, screw-adjustable for windage; decocking lever permits lowering hammer onto locked firing pin; automatic firing pin block; side button magazine release. Imported from Spain by European American Armory. Introduced 1993; still imported. **LMSR: $450**
Blue finish
   New: $375   **Perf.:** $325   **Exc.:** $300
Nickel finish
   New: $400   **Perf.:** $350   **Exc.:** $325

ASTRA CADIX

Revolver; double action; 22, 38 Spl.; 9-shot (22), 5-shot (38 Spl.); 4″, 6″ barrel; blued finish; adjustable rear sight, ramp front; checkered plastic stocks. Manufactured 1960 to 1968.
   Perf.: $150   **Exc.:** $135   **VGood:** $125

ASTRA CAMPER (2000)

Semi-automatic; 22 Short; 4″ barrel; 6 ¹/₄″ overall length; fixed sights; blued or chrome finish; plastic grips. Manufactured 1953 to 1960.
   Perf.: $300   **Exc.:** $250   **VGood:** $200

ASTRA CONSTABLE

Semi-automatic; double action; 22 LR, 32 ACP, 380 ACP; 10-shot (22 LR), 8-shot (32 ACP), 7-shot (380 ACP) magazine; 3 ¹/₂″ barrel; adjustable rear sight, fixed front; moulded plastic grips; exposed hammer; non-glare rib on slide; quick, no-tool takedown feature; blued or chrome finish except 32. Was imported by Interarms. Introduced 1969; dropped 1990.
   Perf.: $275   **Exc.:** $200   **VGood:** $175
Chrome finish or wood grips
   Perf.: $275   **Exc.:** $225   **VGood:** $200
Blue, engraved
   Perf.: $350   **Exc.:** $300   **VGood:** $275
Chrome, engraved
   Perf.: $375   **Exc.:** $325   **VGood:** $300

Astra Constable Stainless

Same specs as Constable except 380 ACP only; stainless steel finish. Manufactured 1986.
   Perf.: $300   **Exc.:** $275   **VGood:** $250

Astra Constable Sport

Same specs as Constable except 6″ barrel; blue finish; weighs 35 oz. Manufactured 1986-1987.
   Perf.: $300   **Exc.:** $250   **VGood:** $225

ASTRA CUB (2000)

Semi-automatic; 22 Short, 25 ACP; 6-shot magazine; 2 ¹/₂″ barrel; 4 ⁷/₁₆″ overall length; fixed sights; blued or chrome finish; plastic grips. Introduced 1957; still in production, but U.S. importation dropped 1968.
   Perf.: $175   **Exc.:** $150   **VGood:** $135
Chrome finish
   Perf.: $200   **Exc.:** $175   **VGood:** $150
Engraved
   Perf.: $300   **Exc.:** $250   **VGood:** $225

ASTRA MODEL 41

Revolver; double action; 41 Mag.; 6-shot; 6″ barrel only; 11 ³/₈″ overall length; weighs 2 ³/₄ lbs. Introduced 1980; no longer imported.
   Perf.: $275   **Exc.:** $250   **VGood:** $200

ASTRA MODEL 44

Revolver; double action; 44 Mag.; 8 ¹/₂″ barrel. Introduced 1980; importation dropped 1990.
   Perf.: $300   **Exc.:** $275   **VGood:** $250
Stainless, 6″ barrel
   Perf.: $350   **Exc.:** $300   **VGood:** $275

ASTRA MODEL 45

Revolver; double action; 45 Colt; 6-shot; 6″ barrel. Manufactured in Spain. Introduced 1980; no longer imported.
   Perf.: $250   **Exc.:** $225   **VGood:** $200

ASTRA MODEL 200 FIRECAT

Semi-automatic; 25 ACP, 6-shot magazine; 2 ¹/₄″ barrel; 4 ³/₈″ overall length; fixed sights; blued finish; plastic grips. Introduced in 1920; still in production; U.S. importation dropped 1968.
   Perf.: $200   **Exc.:** $165   **VGood:** $150

ASTRA MODEL 300

Semi-automatic; 32 ACP, 380 ACP; 4″ barrel; 5 ³/₈″ overall length; hard rubber grips. Introduced 1922; dropped 1947.
   Perf.: $450   **Exc.:** $400   **VGood:** $375
With Nazi proofmarks
   Perf.: $550   **Exc.:** $500   **VGood:** $475

ASTRA MODEL 357

Revolver; 357 Mag.; 6-shot swing-out cylinder; 3″, 4″, 6″, 8 ¹/₂″ barrel; integral rib, ejector rod; click-adjustable rear sight, ramp front; target hammer; checkered walnut grips; blued. Imported by Interarms. Introduced 1972; importation dropped 1990.
   Perf.: $225   **Exc.:** $185   **VGood:** $150
Stainless, 4″ barrel
   Perf.: $250   **Exc.:** $200   **VGood:** $175

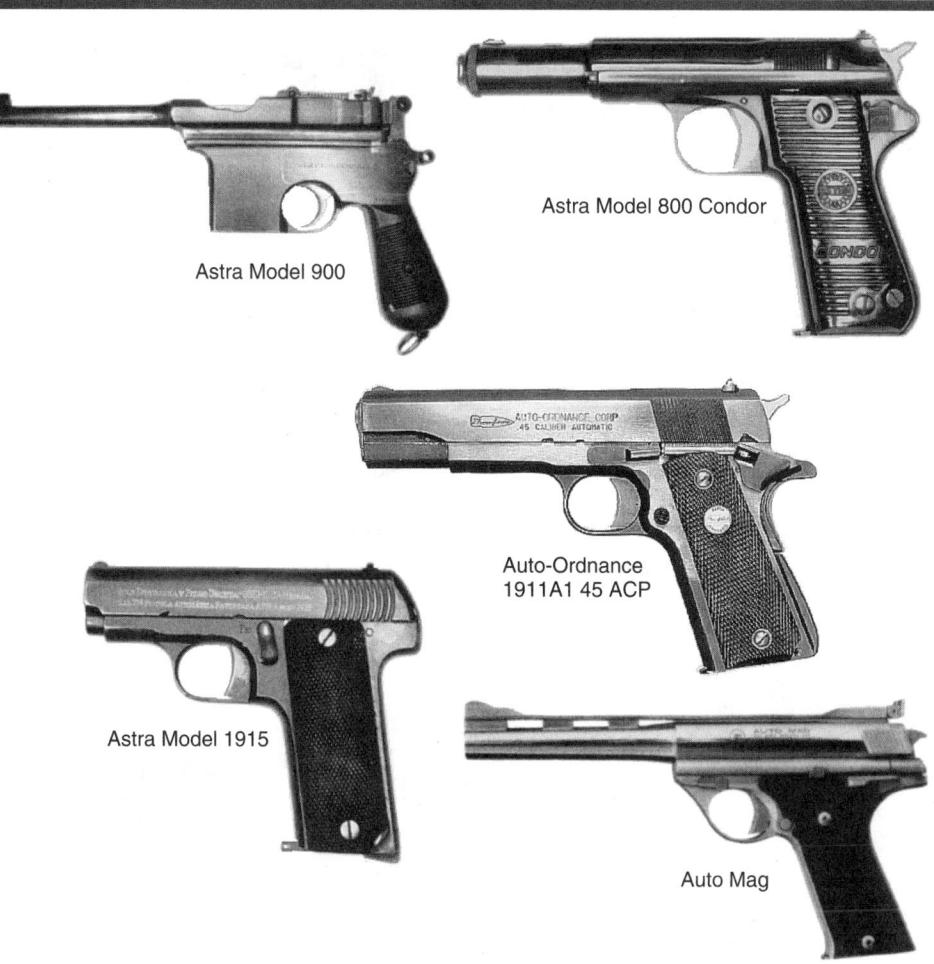

Astra Model 900

Astra Model 800 Condor

Astra Model 1915

Auto-Ordnance
1911A1 45 ACP

Auto Mag

Astra
Model 400
Model 600
Model 800 Condor
Model 900
Model 1911 25
Model 1911 32
Model 1915
Model 3000
Model 4000 Falcon
Victoria (See Astra Model 1915)

Auto Mag

Auto-Ordnance
1911A1
1911A1 40 S&W
1911A1 Competition Model

ASTRA MODEL 400
Semi-automatic; 9mm Bergmann-Bayard; 9-shot magazine; 6″ barrel; 10″ overall length; fixed sights; blowback action; some will also chamber and fire 9mm Para. and 38 ACP but it isn't recommended; blued finish; hard rubber or walnut grips. Introduced 1921; dropped 1946.
   Exc.: $225   **VGood:** $150   **Good:** $125
Navy Model
   Exc.: $250   **VGood:** $150   **Good:** $125
With Nazi proof marks
   Exc.: $500   **VGood:** $400   **Good:** $300

ASTRA MODEL 600
Semi-automatic; 9mm Luger; 8-shot; 5 1/4″ barrel, 8″ overall length; fixed sights; blued finish; hard rubber or walnut grips. Military, police issue. Introduced 1942; dropped 1946.
   Exc.: $250   **VGood:** $200   **Good:** $150
With Nazi proofmarks
   Exc.: $400   **VGood:** $350   **Good:** $300

ASTRA MODEL 800 CONDOR
Semi-automatic; 9mm Para.; 8-shot magazine; 5 5/16″ barrel; 8 1/4″ overall length; fixed sights; blued finish; grooved plastic grips; tubular-type design. Based on Model 400 design. Introduced 1958; dropped 1968. Few imported; produced primarily for European police, military use.
   Exc.: $1500   **VGood:** $1250   **Good:** $800

ASTRA MODEL 900
Semi-automatic; 7.63mm Mauser; 10-shot fixed magazine; 5 1/2″ barrel; 11 1/2″ overall length; adjustable rear sight, fixed front; small ring hammer on early models, larger hammer on later; grooved walnut grips; lanyard ring. Based on design of "broomhandle" Mauser, but has barrel, barrel extension as two parts rather than one as in German Mauser, different lockwork, etc. Introduced 1928; dropped 1940. Originally priced at $37; has collector value.
   Exc.: $1500   **VGood:** $700   **Good:** $350

ASTRA Model 1911 25
Semi-automatic; 25 ACP; 6-shot; 2″ barrel; 4 1/8″ overall length; fixed sights; horn grips; external hammer, loaded indicator on top; marked "Victoria Patent." Rare.
   Exc.: $400   **VGood:** $250   **Good:** $175

ASTRA MODEL 1911 32
Semi-automatic; 32 ACP; 7-shot; 3 3/16″ barrel; 5 1/2″ overall length; weighs 21 oz.; black horn grips; blued finish. Similar to the Browning Model 1903 with external hammer; marked "Victoria Patent." Manufactured until 1916.
   Exc.: $300   **VGood:** $200   **Good:** $150

ASTRA MODEL 1915
Semi-automatic; 32 ACP; 9-shot magazine; 3 1/4″ barrel; 5 3/4″ overall length; fixed sights; checkered wood or horn grips. Actually introduced in 1911 as the "Victoria," first marked "Astra" in 1915; dropped after WWI.
   Exc.: $175   **VGood:** $125   **Good:** $85

ASTRA MODEL 3000
Semi-automatic; 32 ACP; 380 ACP; 6-shot, 7-shot magazine; 4″ barrel; fixed sights; plastic grips; blued finish. Introduced 1948; dropped 1956.
   Perf.: $300   **Exc.:** $250   **VGood:** $200
Engraved
   Perf.: $350   **Exc.:** $300   **VGood:** $275

ASTRA MODEL 4000 FALCON
Semi-automatic; 22 LR, 32 ACP, 380 ACP; 10-shot (22 LR), 8-shot (32 ACP), 7-shot (380 ACP) magazine; 3 1/2″ barrel; 6 1/2″ overall length; thumb safety; exposed hammer; fixed sights; checkered black plastic grips; blued. Introduced 1956; U.S. importation dropped 1968.
   Perf.: $400   **Exc.:** $350   **VGood:** $300
22 LR
   Perf.: $500   **Exc.:** $450   **VGood:** $400
With conversion kit
   Perf.: $1000   **Exc.:** $800   **VGood:** $700

AUTO MAG
Semi-automatic; 357 Auto Mag., 44 Auto Mag.; 7-shot magazine; 6 1/2″ barrel; 11 1/2″ overall length; short recoil; rotary bolt system; stainless steel construction; checkered plastic grips; fully-adjustable rear sight, ramp front. Manufactured by Auto Mag. Corp. and TDE Corp. 1970 to 1975.
   Perf.: $2000   **Exc.:** $1850   **VGood:** $1600

AUTO-ORDNANCE 1911A1
Semi-automatic; 9mm Para., 38 Super, 10mm, 41 AE, 45 ACP; 7- shot (45 ACP), 8-shot (41 AE), 9-shot (others); 5″ barrel; 8 1/2″ overall length; weighs 39 oz.; blade front sight, windage-adjustable rear; parts interchangeable with original Colt Gov't. Model; blued, nonglare finish. Made in U.S. by Auto-Ordnance Corp. Introduced 1983; still in production except for 41 AE.
LMSR: $389
   Perf.: $300   **Exc.:** $275   **VGood:** $225

Auto-Ordnance 1911A1 40 S&W
Same specs as standard Auto-Ordnance 1911A1, except 4 1/2″ barrel; 7 3/4″ overall length; weighs 37 oz.; 40 S&W; 8-shot magazine; black rubber wrap-around grips; 3-dot sight system. Introduced 1991; no longer in production.
   Perf.: $350   **Exc.:** $325   **VGood:** $275

Auto-Ordnance 1911A1 Competition Model
Same specs as 1911A1 except 45 ACP; black textured, rubber wrap-around grips; fully-adjustable rear sight with three-dot system; machined compensator; Commander combat hammer; flat mainspring housing; low-profile magazine funnel; magazine bumper; high-ride beavertail grip safety; full-length recoil spring guide system; extended slide stop, safety and magazine catch; Videcki adjustable speed trigger; extended combat ejector. Made in U.S. by Auto-Ordnance Corp. Introduced 1994; still in production. **LMSR: $615**
   New: $500   **Perf.:** $450   **Exc.:** $425

HANDGUNS

Auto-Ordnance
1911A1 General Model
ZG-51 Pit Bull

Azul

Baer
1911 Bullseye Wadcutter
1911 Concept Series I-X
Pistols
1911 Custom Carry
1911 National Match Hardball
1911 Premier II

Baer 1911 Concept
Series I Pistol

Auto-Ordnance
ZG-51 Pit Bull

Azul

Baer 1911 Bullseye
Wadcutter

Baer Custom Carry

Auto-Ordnance 1911A1 General Model
Same specs as 1911A1 except 45 ACP; 4 1/2" barrel; 7 3/4" overall length; weighs 37 oz.; three-dot sight system; black textured, rubber wrap-around grips with medallion; full-length recoil guide; blue finish. Made in U.S. by Auto-Ordnance Corp. **LMSR: $428**

| New: $350 | Perf.: $325 | Exc.: $300 |

AUTO-ORDNANCE ZG-51 PIT BULL
Semi-automatic; single action; 45 ACP; 7-shot magazine; 3 1/2" barrel; 7 1/4" overall length; weighs 36 oz. Introduced 1989; still in production. **LMSR: $421**

| Perf.: $350 | Exc.: $300 | VGood: $275 |

AZUL
Semi-automatic; 7.63mm Mauser; 10-shot magazine; 5 1/2" barrel; 12" overall length; grooved wood grips. Close Spanish copy of Mauser Model 1896. Introduced 1930, dropped about 1935. Warning: Some Azuls have selector switches for full-auto fire and are illegal unless registered with the U.S. Treasury.

| Exc.: $1,000 | VGood: $750 | Good: $600 |

BAER 1911 BULLSEYE WADCUTTER
Semi-automatic; single action; 45 ACP; 7-shot magazine; 5" barrel; 8 1/2" overall length; weighs 37 oz.; Baer dovetail front sight with undercut post, low-mount Bo-Mar rear with hidden leaf rear; checkered walnut grips; polished feed ramp and barrel throat; Bo-Mar rib on slide; full-length recoil rod; Baer speed trigger; Baer deluxe hammer and sear; Baer beavertail grip safety with pad; flat mainspring housing checkered 20 lpi; blue finish. Made in U.S. by Les Baer Custom, Inc. Still in production. **LMSR: $1347**

| New: $1300 | Perf.: $1200 | Exc.: $1100 |

BAER 1911 CONCEPT SERIES I-X PISTOLS
Semi-automatic; single action; 45 ACP; 7-shot magazine; 5" barrel; 8 1/2" overall length; weighs 37 oz.; checkered rosewood grips; Baer dovetail front sight, Bo-Mar deluxe low-mount rear with hidden leaf; Baer forged steel frame, slide and barrel with Baer stainless bushing; slide fitted to frame; double serrated slide; Baer beavertail grip safety; checkered slide stop; tuned extractor; extended ejector; deluxe hammer and sear; match disconnector; lowered and flared ejection port; fitted recoil link; polished feed ramp, throated barrel; Baer fitted speed trigger; flat serrated mainspring housing; blue finish. Aluminum frame, stainless slide, BoMar and Novak sights and smaller Commanche frame sizes comprise the Series I-X. Made in U.S. by Les Baer Custom, Inc. Still in production. **LMSR: $1279**

| New: $1200 | Perf.: $1100 | Exc.: $1000 |

BAER 1911 CUSTOM CARRY
Semi-automatic; single action; 45 ACP; 7- or 10-shot magazine; 5" barrel; 8 1/2" overall length; weighs 37 oz.; checkered walnut grips; Baer improved ramp-style dovetailed front sight, Novak low-mount rear; Baer forged NM frame, slide and barrel with stainless bushing; fitted slide to frame; double serrated slide (full-size only); Baer speed trigger with 4-lb. pull; Baer deluxe hammer and sear; tactical-style extended ambidextrous safety; beveled magazine well; polished feed ramp and throated barrel; tuned extractor; Baer extended ejector; checkered slide stop; lowered and flared ejection port; full-length recoil guide rod; recoil buff. Made in U.S. by Les Baer Custom, Inc. Still in production. **LMSR: $1265**
Standard size, blued

| New: $1200 | Perf.: $1100 | Exc.: $1000 |
Standard size, stainless
| New: $1300 | Perf.: $1200 | Exc.: $1100 |

Commanche size, blued
| New: $1200 | Perf.: $1100 | Exc.: $1000 |
Commanche size, stainless
| New: $1300 | Perf.: $1200 | Exc.: $1100 |
Commanche size, aluminum frame, blued slide
| New: $1300 | Perf.: $1200 | Exc.: $1100 |
Commanche size, aluminum frame, stainless slide
| New: $1400 | Perf.: $1300 | Exc.: $1200 |

BAER 1911 NATIONAL MATCH HARDBALL
Semi-automatic; single action; 45 ACP; 7-shot magazine; 5" barrel; 8 1/2" overall length; weighs 37 oz.; Baer dovetail front sight with undercut post, low-mount Bo-Mar rear with hidden leaf rear; checkered walnut grips; Baer NM forged steel frame, double serrated slide and barrel with stainless steel bushing; slide fitted to frame; Baer match trigger with 4-lb. pull; polished feed ramp; throated barrel; checkered frontstrap; arched mainspring housing; Baer beveled magazine well; lowered, flared ejection port; tuned extractor; Baer extended ejector; checkered slide stop; recoil buff. Made in U.S. by Les Baer Custom, Inc. Still in production. **LMSR: $1130**

| New: $950 | Perf.: $875 | Exc.: $725 |

BAER 1911 PREMIER II
Semi-automatic; single action; 45 ACP; 7-shot magazine; 5" barrel; 8 1/2" overall length; weighs 37 oz.; Baer dovetail front sight with undercut post, low-mount Bo-Mar rear with hidden leaf rear; checkered rosewood grips with double diamond pattern; Baer NM forged steel frame and barrel with stainless steel bushing; slide fitted to frame; double serrated slide; aluminum speed trigger with 4-lb. pull; deluxe Commander hammer and sear; polished feed ramp; throated barrel; checkered frontstrap; flat mainspring housing; Baer beveled magazine well; lowered, flared ejection port; tuned, polished extractor; Baer

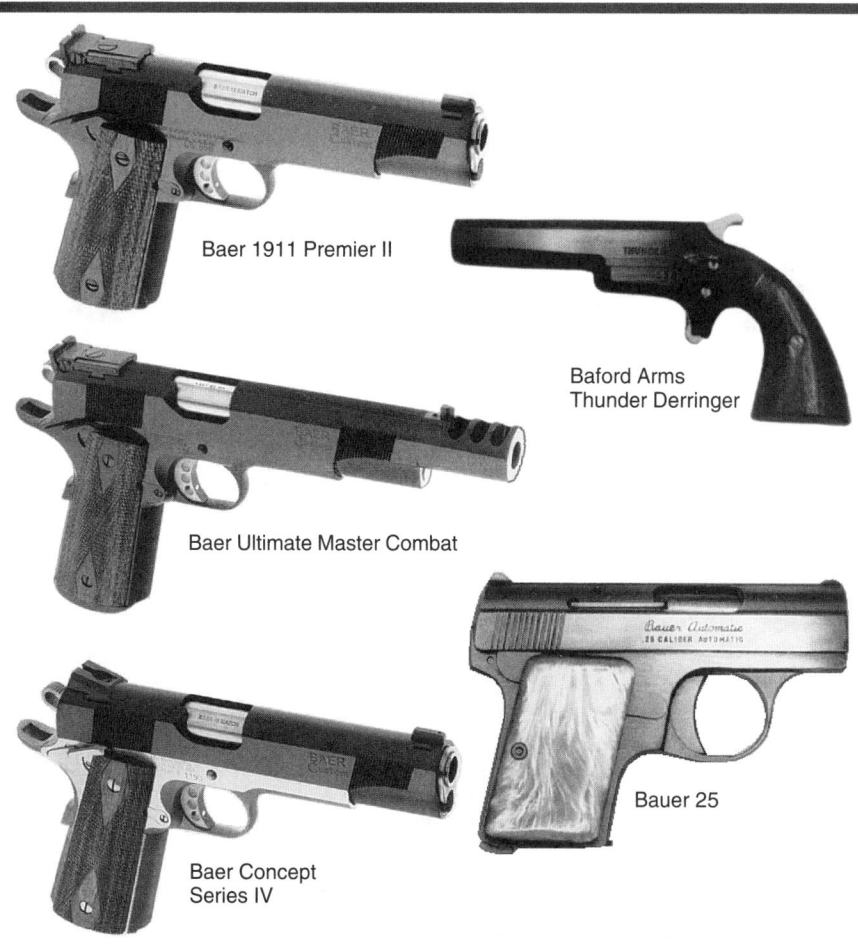

Baer 1911 Premier II

Baford Arms
Thunder Derringer

Baer Ultimate Master Combat

Bauer 25

Baer Concept
Series IV

HANDGUNS

Baer
1911 Prowler III
1911 Target Master
1911 Ultimate Master Combat
1911 Ultimate Master "Steel Special"

Baford Arms
Thunder Derringer

Bauer
25

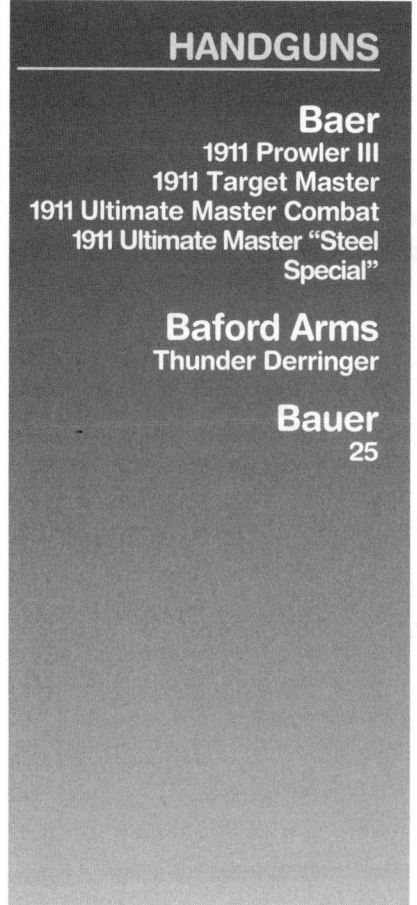

extended ejector; checkered slide stop. Made in U.S. by Les Baer Custom, Inc. Still in production. **LMSR: $1428**
Blued
   New: $1350   **Perf.:** $1200   **Exc.:** $975
Stainless steel
   New: $1400   **Perf.:** $1295   **Exc.:** $1100

BAER 1911 PROWLER III
Semi-automatic; single action; 45 ACP, 7- or 10-shot magazine; 5″ barrel; 8 1/2″ overall length; weighs 37 oz.; checkered rosewood grips with double diamond pattern; Baer dovetailed front sight, low-mount Bo-Mar rear with hidden leaf; Baer NM forged steel frame and barrel with stainless bushing; slide fitted to frame; double serrated slide; lowered, flared ejection port; tuned, polished extractor; Baer extended ejector; checkered slide stop; aluminum speed trigger with 4-lb. pull; deluxe Commander hammer and sear; beavertail grip safety with pad; beveled magazine well; extended ambidextrous safety; flat mainspring housing; polished feed ramp and throated barrel; 30 lpi checkered frontstrap; tapered cone stub weight and reverse recoil plug. Made in U.S. by Les Baer Custom, Inc. Still in production. **LMSR: $1795**
Standard or Commanche size, blued
   New: $1700   **Perf.:** $1625   **Exc.:** $1500
Standard or Commanche size, stainless
   New: $1800   **Perf.:** $1725   **Exc.:** $1600

BAER 1911 TARGET MASTER
Semi-automatic; single action; 45 ACP; 7-shot magazine; 5″ barrel; 8 1/2″ overall length; weighs 37 oz.; checkered walnut grips; Baer post-style dovetail front sight, low-mount Bo-Mar rear with hidden leaf; Baer NM forged steel frame; double serrated slide and barrel with stainless bushing; slide fitted to frame; standard trigger; polished feed ramp; throat-

ed barrel; checkered frontstrap; flat serrated mainspring housing; Baer beveled magazine well; lowered, flared ejection port; tuned extractor; Baer extended ejector; checkered slide stop; recoil buff. Made in U.S. by Les Baer Custom, Inc. Still in production. **LMSR: $1233**
   New: $1175   **Perf.:** $1095   **Exc.:** $950

BAER 1911 ULTIMATE MASTER COMBAT
Semi-automatic; single action; 45 ACP (others available); 10-shot magazine; 5″ barrel; 8 1/2″ overall length; weighs 37 oz.; checkered rosewood grips; Baer dovetail front sight, low-mount Bo-Mar rear with hidden leaf; Baer forged NM blued steel frame and double serrated slide; Baer triple port, tapered cone compensator; fitted slide to frame; lowered, flared ejection port; Baer reverse recoil plug; full-length guide rod; recoil buff; beveled magazine well; Baer Commander hammer and sear; Baer extended ambidextrous safety, extended ejector; checkered slide stop; beavertail grip safety with pad; extended magazine release button; Baer speed trigger. Made in U.S. by Les Baer Custom, Inc. Still in production. **LMSR: $1996**
Compensated, open sights
   New: $1900   **Perf.:** $1795   **Exc.:** $1650
Uncompensated "Limited" Model
   New: $1800   **Perf.:** $1695   **Exc.:** $1550
Compensated, with Baer optics mount
   New: $2300   **Perf.:** $2250   **Exc.:** $2100

BAER 1911 ULTIMATE MASTER "STEEL SPECIAL"
Same specs as Ultimate Master except scope and Baer scope mount; bushing-type compensator; fitted slide to frame; lowered, flared ejection port; Baer reverse recoil plug; two-piece guide rod. Designed for maximum 150 Power Factor; hard chrome finish. Still in production. **LMSR: $2570**
   New: $2500   **Perf.:** $2425   **Exc.:** $2350

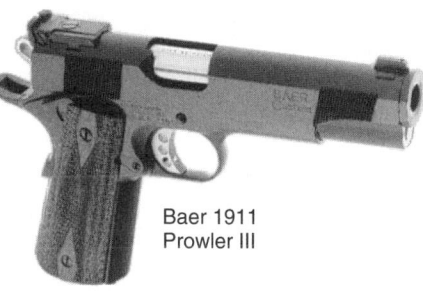

Baer 1911
Prowler III

BAFORD ARMS THUNDER DERRINGER
Derringer; 410 2 1/2″ shotshell/44 Spl.; insert sleeves available for 22 Short, 22 LR, 25 ACP, 32 ACP, 32 H&R Mag., 30 Luger, 380 ACP, 38 Super, 38 Spl., 38 S&W, 9mm Para.; barrel lengths vary with caliber inserts; weighs 8 1/2 oz., sans inserts; no sights; uncheckered walnut stocks; blued steel barrel, frame; polished steel hammer, trigger; side-swinging barrel; positive lock, half-cock safety; roll block safety. Introduced 1986; dropped 1988.
   Perf.: $125   **Exc.:** $110   **VGood:** $100
Extra barrel inserts
   Perf.: $75   **Exc.:** $65   **VGood:** $50

BAUER 25
Semi-automatic; 25 ACP; 6-shot magazine; 2 1/8″ barrel; 4″ overall length; stainless steel construction; fixed sights; plastic pearl or checkered walnut grips; manual, magazine safeties. Introduced 1973; dropped 1984.
   Perf.: $175   **Exc.:** $150   **VGood:** $125

HANDGUNS

Baikal
IJ-70
IJ-70HC
IJ-70HC, 380 ACP

Bayard
Model 1908 Pocket Automatic
Model 1923 Pocket Automatic
(Large Model)
Model 1923 Pocket Automatic
(Small Model)
Model 1930 Pocket Automatic

Beeman
P-08
P-08 Mini
SP Deluxe

Beholla
Pocket Auto

Benelli
B76
MP90S Match
MP95E Match

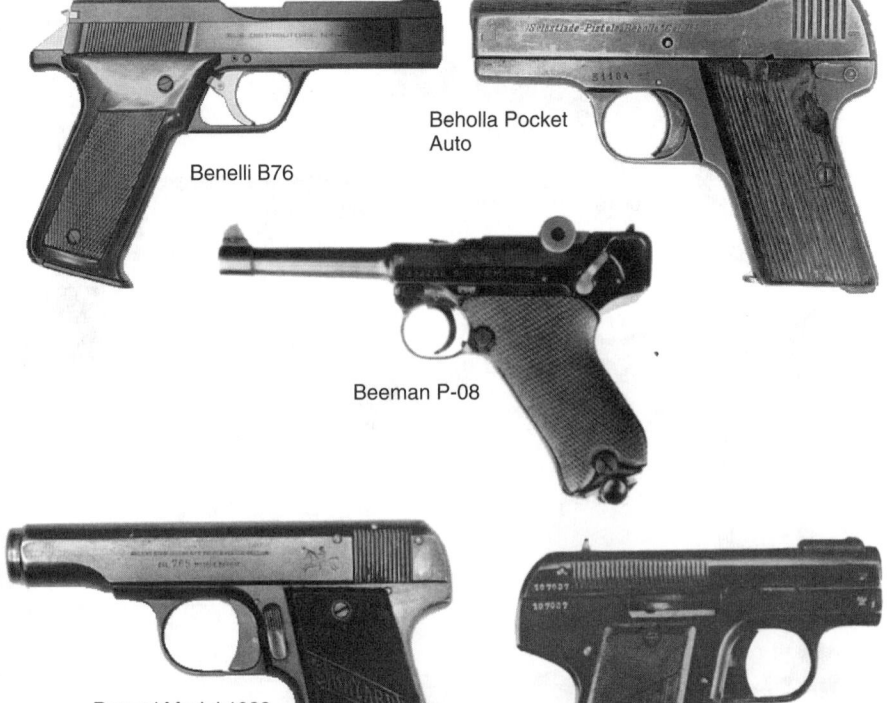

Benelli B76

Beholla Pocket Auto

Beeman P-08

Bayard Model 1923 (Large Model)

Bayard Model 1908 Pocket Automatic

BAIKAL IJ-70
Semi-automatic; double action; 9x18mm Makarov; 8-shot magazine; 4″ barrel; 6 1/4″ overall length; weighs 25 oz.; checkered composition grips; blade front sight, fully-adjustable rear; all-steel construction; frame-mounted safety with decocker. Comes with two magazines, cleaning rod, universal tool. Imported from Russia by Century International Arms, K.B.I., Inc. Introduced 1994; still imported. **LMSR: $199**

New: $125	**Perf.: $110**	**Exc.: $100**

Baikal IJ-70HC
Same specs as IJ-70 except 10-shot magazine. Still imported by K.B.I., Inc. **LMSR: $239**

New: $125	**Perf.: $110**	**Exc.: $100**

Baikal IJ-70HC, 380 ACP
Same specs as IJ-70HC except 380 ACP. **LMSR: $249**

New: $135	**Perf.: $125**	**Exc.: $110**

BAYARD MODEL 1908 POCKET AUTOMATIC
Semi-automatic; 25 ACP, 32 ACP, 380 ACP; 6-shot magazine; 2 1/4″ barrel; 4 7/8″ overall length; fixed sights, blued finish; hard rubber grips. Introduced 1908; dropped 1923.
25-caliber.

Exc.: $200	**VGood: $150**	**Good: $125**

32-caliber.

Exc.: $150	**VGood: $135**	**Good: $110**

380-caliber.

Exc.: $200	**VGood: $175**	**Good: $135**

BAYARD MODEL 1923 POCKET AUTOMATIC (LARGE MODEL)
Semi-automatic; 32 ACP, 380 ACP; 6-shot magazine; 3 5/16″ barrel; 4 5/16″ overall length; fixed sights; blued finish; checkered grips of hard rubber. Introduced 1923; dropped 1940.

Exc.: $175	**VGood: $150**	**Good: $125**

BAYARD MODEL 1923 POCKET AUTO-MATIC (SMALL MODEL)
Semi-automatic; 25 ACP only; 2 1/2″ barrel; 4 5/16″ overall length; fixed sights, blued finish; checkered grips of hard rubber. Scaled-down model of large Model 1923. Introduced 1923; dropped 1930.

Exc.: $175	**VGood: $150**	**Good: $135**

BAYARD MODEL 1930 POCKET AUTOMATIC
Semi-automatic; 25 ACP; 6-shot; 2 1/8″ barrel; 5 15/16″ overall length; fixed sights; blued finish; checkered hard rubber grips. Similar to small Model 1923 auto pistol with improvements in finish, internal mechanism. Introduced 1930; dropped 1940.

Exc.: $175	**VGood: $150**	**Good: $135**

BEEMAN P-08
Semi-automatic; 22 LR; 8-shot magazine; 4″ barrel; 7 3/4″ overal length; weighs 25 oz.; checkered hardwood stocks; fixed sights; based on original Luger design; Luger-type toggle action. Made in Germany. Introduced 1989; importation dropped 1990.

Perf.: $185	**Exc.: $165**	**VGood: $150**

Beeman P-08 Mini
Same specs as P-08 except 380 ACP; 5-shot magazine; 3 1/2″ barrel; weighs 22 1/2 oz.; magazine and sear disconnect safeties. Made in Germany. Introduced 1989; importation dropped 1990.

Perf.: $200	**Exc.: $185**	**VGood: $150**

BEEMAN SP DELUXE
Single shot; 22 LR; 8″, 10″, 11 3/16″, 15″ barrel; European walnut stocks; adjustable palm rest; blade front sight, adjustable notch rear; detachable forend, barrel weight; standard version has no forend or weight. Imported by Beeman. Introduced 1984; dropped 1987.

Perf.: $600	**Exc.: $550**	**VGood: $500**

BEHOLLA POCKET AUTO
Semi-automatic; 32 ACP; 7-shot magazine; 3″ barrel; 5 1/2″ overall length; fixed sights; grooved hard rubber or wooden stocks; blued finish. German-made. Manufactured 1915 to 1925.

Exc.: $175	**VGood: $135**	**Good: $100**

BENELLI B76
Semi-automatic; 9mm Para.; 8-shot magazine; 4 1/4″ chrome-lined barrel; 8 1/16″ overall length; weighs 34 oz. (empty); walnut stocks with cut checkering and high gloss finish; blade front sight with white face, windage-adjustable; fixed barrel, locked breech; stainless steel inertia firing pin and loaded chamber indicator; blued finish with internal parts hard chrome plated; all steel construction. Introduced 1979.

Perf.: $350	**Exc.: $250**	**VGood: $225**

BENELLI MP90S MATCH
Semi-automatic; single action; 22 Short, 22 LR, 32 S&W wadcutter; 5-shot magazine; 4 3/8″ barrel; 11 7/8″ overall length; weighs 39 oz.; stippled walnut match-type grips with fully-adjustable palm shelf, anatomically shaped; match type, blade front sight, fully click-adjustable rear; removable trigger fully-adjustable for pull and position; special internal weight box on sub-frame below barrel; comes with loading tool, cleaning rod. Imported from Italy by European American Armory. Introduced 1993; still imported. **LMSR: $1295**

Perf.: $1000	**Exc.: $900**	**VGood: $800**

BENELLI MP95E MATCH
Semi-automatic; single action; 22 LR, 32 S&W WC; 9-shot magazine; 4 3/8″ barrel; 11 7/8″ overall length; weighs 39 oz.; checkered walnut match-type grips, anatomically shaped; blade front sight, fully click-adjustable match-type rear; removable, adjustable trigger; special internal weight box on sub-frame below barrel; cut for scope rails. Imported from Italy by European American Armory. Introduced 1993; still imported. **LMSR: $550**
Blue finish

New: $500	**Perf.: $450**	**Exc.: $400**

Chrome finish

New: $550	**Perf.: $500**	**Exc.: $400**

Benelli MP95E Match

Beretta Model 70

Beretta Model 20

Beretta Model 76

Beretta Model 70S

Beretta Model 21A Bobcat

HANDGUNS

Beretta
**Brigadier Model
(See Beretta Model 951
Brigadier)
Cougar
(See Beretta Model 70S)
Featherweight
(See Beretta Model 948)
Jaguar
(See Beretta Model 71
Jaguar)
Lady Beretta
Model 20
Model 21A Bobcat
Model 70
Model 70S
Model 70T
Model 71
Model 71 Jaguar
Model 72 Jaguar
Model 76
Model 81
Model 82**

BERETTA LADY BERETTA
Semi-automatic; double action; 22 LR; 7-shot; 2 1/2″ barrel; 4 15/16″ overall length; weighs 11 1/2 oz.; fixed sights; thumb safety; half-cock safety; hinged tip-up barrel; similar to Model 21A except gold etching on top of frame and side of slide.
   Perf.: $200   **Exc.:** $175   **VGood:** $150

BERETTA MODEL 20
Semi-automatic; double action; 25 ACP; 8-shot magazine; 2 1/2″ barrel; 4 1/2″ overall length; weighs 11 oz.; fixed sights; plastic or walnut stocks. Made in Italy. Introduced 1984; dropped 1986.
   Perf.: $135   **Exc.:** $125   **VGood:** $110

BERETTA MODEL 21A BOBCAT
Semi-automatic; double action; 22 LR, 25 ACP; 7-shot (22 LR), 8-shot (25 ACP); 2 1/2″ barrel; 4 15/16″ overall length; weighs 11 1/2 oz.; fixed sights; thumb safety; half-cock safety; hinged tip-up barrel; blued, nickel or matte finish; checkered walnut or plastic grips. Introduced 1985; still in production. Made in the U.S.
LMSR: $244
   Perf.: $200   **Exc.:** $175   **VGood:** $135
Nickel finish
   Perf.: $200   **Exc.:** $185   **VGood:** $150
Matte finish
   Perf.: $175   **Exc.:** $150   **VGood:** $135
Engraved
   Perf.: $250   **Exc.:** $225   **VGood:** $200

BERETTA MODEL 70
Semi-automatic; double action; 32 ACP; 8-shot magazine; 3 1/2″ barrel; 6 5/16″ overall length; 23 1/2 oz.; fixed sights; plastic grips; crossbolt safety; blue or chrome finish. Replaced Model 948; has hold-open device and push-button magazine release. Alloy frame 32 ACP marketed in U.S. as the New Puma. Introduced 1958; dropped 1968.
   Perf.: $175   **Exc.:** $165   **VGood:** $150
Chrome finish
   Perf.: $185   **Exc.:** $175   **VGood:** $165

Beretta Model 70S
Same specs as Beretta Model 70 except 380 ACP, 22 LR; weighs 18 oz. (22 LR), magazine safety; steel frame; fixed sights (380 ACP); adjustable rear (22 LR); checkered plastic wrap-around grips with thumbrest; blued. Steel frame 380 ACP marketed in U.S. as the Cougar. Introduced 1977; dropped 1985.
   Perf.: $200   **Exc.:** $175   **VGood:** $165

Beretta Model 70T
Same specs as Beretta Model 70 except 9-shot magazine; 6″ barrel; 8 1/2″ overall length; adjustable rear sight, fixed front; slide stays open after last shot; checkered plastic wrap-around grips. Introduced 1969; dropped 1975.
   Perf.: $235   **Exc.:** $200   **VGood:** $175

BERETTA MODEL 71
Semi-automatic; single action; 22 LR; 8-shot magazine; 6″ barrel only; 8 3/4″ overall length; adjustable rear sight; wrap-around checkered plastic grips; lever thumb safety. Imported 1987 only.
   Perf.: $200   **Exc.:** $185   **VGood:** $165

BERETTA MODEL 71 JAGUAR
Semi-automatic; 22 LR; 8-shot magazine; 3 1/2″ barrel; windage-adjustable rear sight, fixed front; checkered plastic grips; blued. Marketed in U.S. as Jaguar. Introduced 1956; dropped 1968.
   Perf.: $200   **Exc.:** $185   **VGood:** $165

BERETTA MODEL 72 JAGUAR
Semi-automatic; 22 LR; 8-shot magazine; 3 1/2″, 6″ interchangeable barrels; windage-adjustable rear sight, fixed front; checkered plastic grips; blued finish. Introduced 1956; dropped 1968.
   Perf.: $225   **Exc.:** $200   **VGood:** $185

BERETTA MODEL 76
Semi-automatic; single action; 22 LR only; 10-shot magazine; 6″ barrel; 9 1/2″ overall length; adjustable rear sight, interchangeable blade front; non-glare, ribbed slide; heavy barrel; external hammer; checkered wrap-around plastic (76P) or wood (76W) grips; blued. Also known as New Sable. Introduced 1971; dropped 1985.
   Perf.: $350   **Exc.:** $300   **VGood:** $275
With wood grips
   Perf.: $400   **Exc.:** $350   **VGood:** $325

BERETTA MODEL 81
Semi-automatic; double action; 32 ACP; 12-shot magazine; 3 13/16″ barrel; 6 13/16″ overall length; blued finish; fixed sights; plastic or wood stocks. Introduced 1976; dropped 1981.
   Perf.: $375   **Exc.:** $250   **VGood:** $200
Nickel finish
   Perf.: $350   **Exc.:** $300   **VGood:** $250
With wood grips
   Perf.: $300   **Exc.:** $275   **VGood:** $225
Beretta Model 81BB
   Perf.: $300   **Exc.:** $275   **VGood:** $225

BERETTA MODEL 82
Semi-automatic; double action; 32 ACP; 9-shot straightline magazine; 3 13/16″ barrel; 6 13/16″ overall length; weighs 17 oz.; non-reversible magazine release; blued finish; fixed sights; wood grips. Introduced 1977; dropped 1984.
   Perf.: $275   **Exc.:** $250   **VGood:** $200
Nickel finish
   Perf.: $350   **Exc.:** $300   **VGood:** $250

Beretta
Model 84 Cheetah
Model 84F
Model 85 Cheetah
Model 85F
Model 86 Cheetah
Model 87 Cheetah
Model 87 Target
Model 89
Model 90
Model 92
Model 92S (Second Issue)
Model 92SB-P (Third Issue)
Model 92SB-P Compact

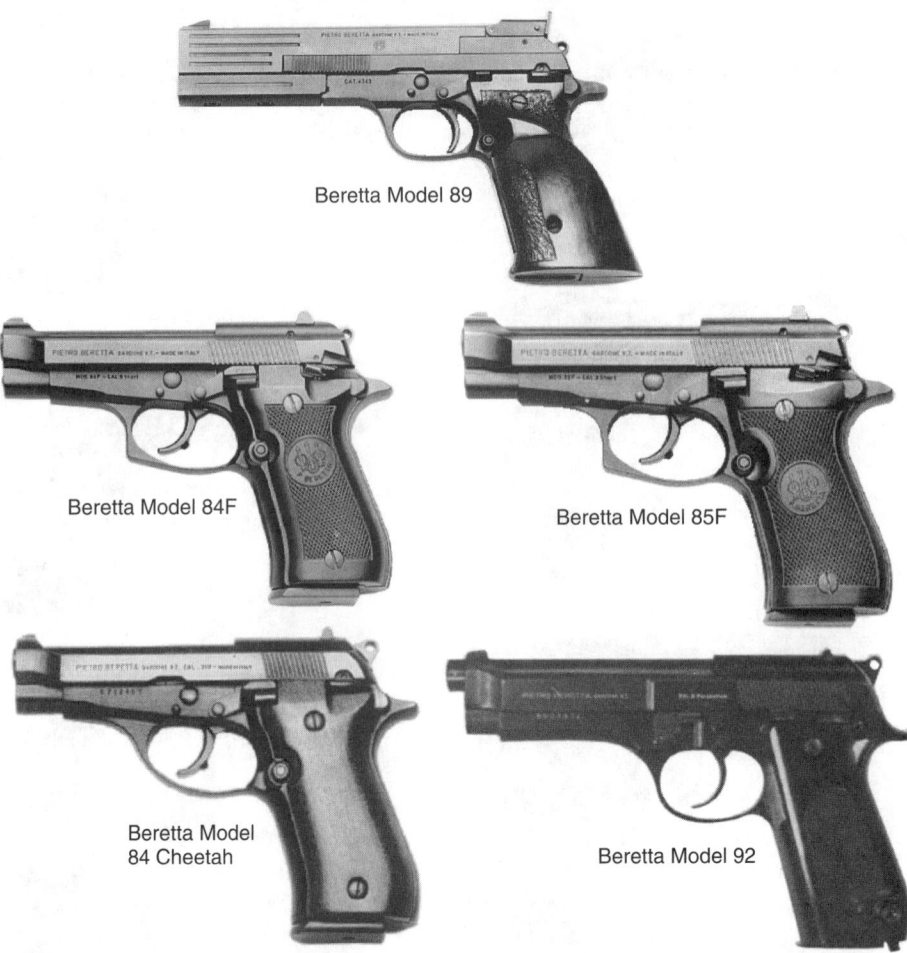

Beretta Model 89

Beretta Model 84F

Beretta Model 85F

Beretta Model
84 Cheetah

Beretta Model 92

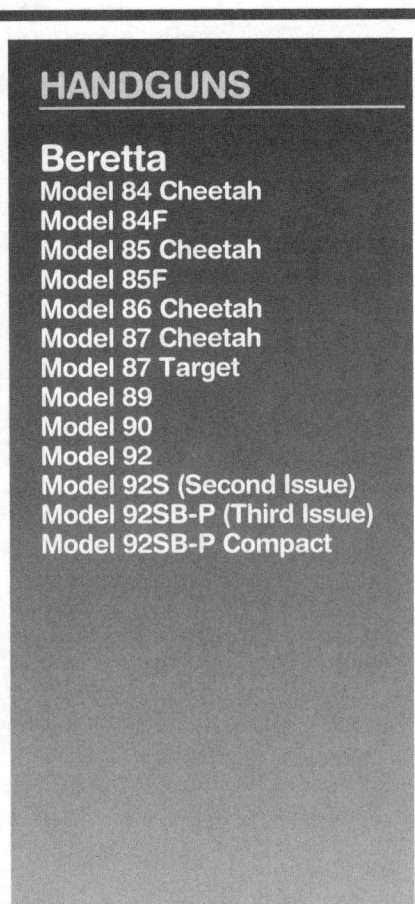

BERETTA MODEL 84 CHEETAH
Semi-automatic; double action; 380 ACP; 10-shot magazine; 3 7/8" barrel; 6 13/16" overall length; weighs 23 oz.; black plastic (84P) or wood (84W) grips; fixed front sight, drift-adjustable rear; blue finish; squared trigger guard; exposed hammer. Imported from Italy by Beretta U.S.A. Introduced 1977; no longer imported.

    Perf.: $400   **Exc.:** $325   **VGood:** $275
Blue, wood grips
    Perf.: $375   **Exc.:** $350   **VGood:** $300
Nickel, wood grips
    Perf.: $475   **Exc.:** $400   **VGood:** $325

Beretta Model 84F
Same specs as Model 84 except combat-style frame; grooved triggerguard; plastic grips; manual safety; decocking device; matte finish. Made in Italy. Introduced 1990; still in production.

    Perf.: $350   **Exc.:** $325   **VGood:** $275

BERETTA MODEL 85 CHEETAH
Semi-automatic; double action; 380 ACP; 8-shot straightline magazine; 3 7/8" barrel; 6 13/16" overall length; weighs 23 oz.; black plastic (85P) or wood (85W) grips; fixed front sight, drift-adjustable rear; non-reversible magazine release; blue finish; squared triggerguard; exposed hammer. Imported from Italy by Beretta U.S.A. Introduced 1977; still imported. **LMSR:** $486

    Perf.: $350   **Exc.:** $300   **VGood:** $250
Blue, wood grips
    Perf.: $375   **Exc.:** $325   **VGood:** $275
Nickel, wood grips
    Perf.: $425   **Exc.:** $375   **VGood:** $325

Beretta Model 85F
Same specs as Model 85 except combat-type frame; grooved triggerguard; plastic grips; matte black finish; manual safety; decocking device. Made in Italy. Introduced 1990; no longer in production.

    Perf.: $375   **Exc.:** $325   **VGood:** $275

BERETTA MODEL 86 CHEETAH
Semi-automatic; double action; 22 LR; 8-shot magazine; 4 7/16" barrel; 7 5/16" overall length; weighs 23 1/2 oz.; fixed sights; tip-up barrel; checkered walnut grips; matte finish; gold trigger. Importation began 1991.

    Perf.: $450   **Exc.:** $375   **VGood:** $300

BERETTA MODEL 87 CHEETAH
Semi-automatic; double action; 22 LR; 7-shot magazine; 3 7/8" barrel; 6 13/16" overall length; weighs 21 oz.; black plastic grips; fixed front sight, drift-adjustable rear; blue finish; squared trigger guard; exposed hammer. Imported from Italy by Beretta U.S.A. Introduced 1977; still imported.

    Perf.: $450   **Exc.:** $375   **VGood:** $300

Beretta Model 87 Target
Same specs as Model 87 except single action; 6" barrel with counterweight; wood grips. Made in Italy. Introduced 1977; importation began in 1986; no longer imported.

    Perf.: $500   **Exc.:** $425   **VGood:** $350

BERETTA MODEL 89
Semi-automatic; single-action only; 22 LR; 8-shot magazine; 6" barrel; 9 1/2" overall length; weighs 41 oz.; interchangeable front sight, fully-adjustable rear; semi-anatomical walnut grips; fixed monoblock barrel machined from forged steel block; aluminum alloy frame; manual safety on either side. Made in Italy. Introduced 1977. Importation began 1986; still imported.

    Perf.: $600   **Exc.:** $500   **VGood:** $400

BERETTA MODEL 90
Semi-automatic; double action; 32 ACP; 8-shot magazine; 3 5/8" barrel; 6 3/4" overall length; fixed sights, matted rib on slide; chamber-loaded indicator; external hammer; stainless steel barrel; moulded plastic wrap-around grips; blued. Introduced 1969; dropped 1982.

    Perf.: $250   **Exc.:** $200   **VGood:** $150

BERETTA MODEL 92
Semi-automatic, double action; 9mm Para.; 15-shot magazine; 5" barrel; 8 1/2" overall length; blued finish; fixed sights; plastic grips. Updated version of the Model 951. Introduced 1976; dropped 1981.

    Perf.: $600   **Exc.:** $500   **VGood:** $425

BERETTA MODEL 92S
(SECOND ISSUE)
Semi-automatic; double action; 9mm Para.; 15-shot magazine; 5" barrel; 8 1/2" overall length; fixed sights; plastic grips; blued finish. Variant of Model 92 with safety catch on side. Discontinued.

    Perf.: $550   **Exc.:** $500   **VGood:** $400

BERETTA MODEL 92SB-P
(THIRD ISSUE)
Automatic; double action; 9mm Para.; 16-shot magazine; 5" barrel; 8 1/2" overall length; weighs 34 1/2 oz.; blade front sight, windage-adjustable rear; black plastic or wood grips. Developed for U.S. Army pistol trials of 1979-1980. Introduced 1979; dropped 1985.

    Perf.: $450   **Exc.:** $400   **VGood:** $350
Wood grips
    Perf.: $500   **Exc.:** $450   **VGood:** $400

Beretta Model 92SB-P Compact
Same specs as Model 92SB-P except 4 5/16" barrel, 14-shot magazine; plastic grips. Introduced 1977; dropped 1985.

    Perf.: $500   **Exc.:** $450   **VGood:** $400
Wood grips
    Perf.: $525   **Exc.:** $475   **VGood:** $425

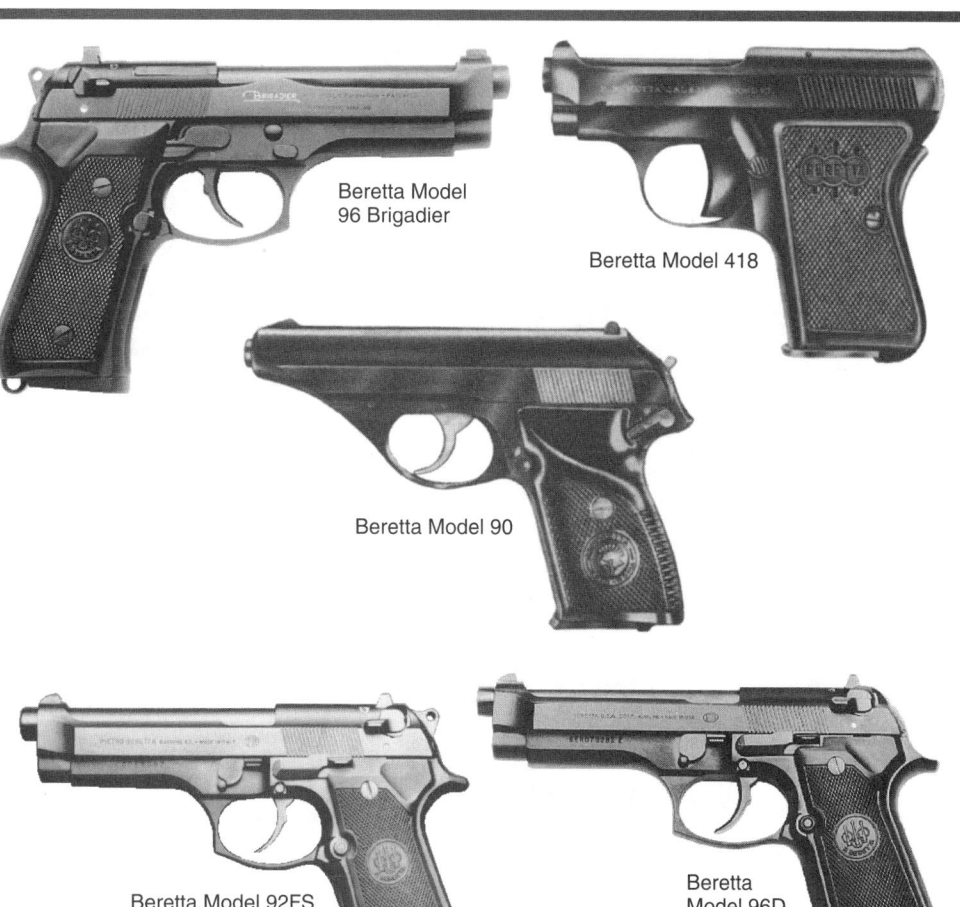

Beretta Model
96 Brigadier

Beretta Model 418

Beretta Model 90

Beretta Model 92FS

Beretta
Model 96D

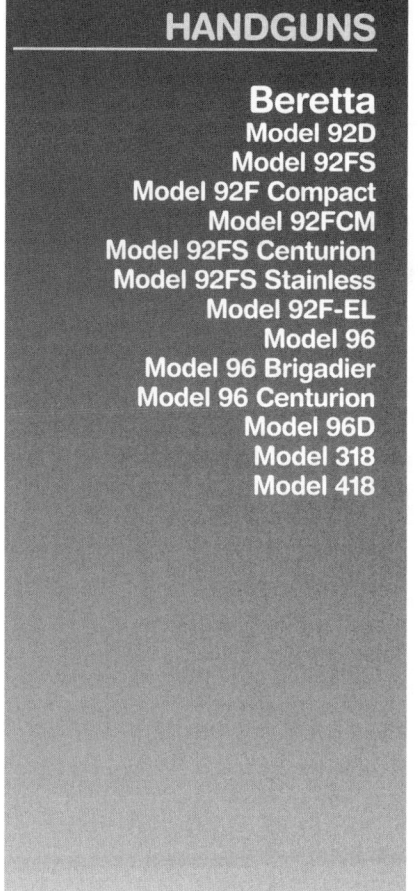

HANDGUNS

Beretta
Model 92D
Model 92FS
Model 92F Compact
Model 92FCM
Model 92FS Centurion
Model 92FS Stainless
Model 92F-EL
Model 96
Model 96 Brigadier
Model 96 Centurion
Model 96D
Model 318
Model 418

BERETTA MODEL 92D

Semi-automatic; double-action only; 9mm Para.; 10-shot magazine; 5" barrel; 8 1/2" overall length; weighs 34 oz.; bobbed hammer; no external safety; plastic grips; three-dot sights. Also offered with Trijicon sights. From Beretta U.S.A. Introduced 1992.
   New: $450   **Perf.:** $400   **Exc.:** $350

BERETTA MODEL 92FS

Semi-automatic; double action; 9mm Para.; 15-shot magazine; 5" barrel; 8 1/2" overall length; matte black finish; squared triggerguard; grooved front backstraps; inertia firing pin; extractor acts as chamber loaded indicator; wood or plastic grips. Adopted 1985 as official U.S. military sidearm. Introduced 1984; still imported by Beretta USA.
   Perf.: $550   **Exc.:** $500   **VGood:** $450
Stainless steel barrel and slide (92F)
   Perf.: $600   **Exc.:** $550   **VGood:** $400

Beretta Model 92F Compact

Same specs as Beretta Model 92FS except cut down frame; 4 3/8" barrel; 7 7/8" overall length; 13-shot magazine; weighs 3 1/2 oz.; Trijicon sights and wood grips optionally available. From Beretta U.S.A. Introduced 1989; no longer in production.
   Perf.: $500   **Exc.:** $450   **VGood:** $400
Wood grips
   Perf.: $525   **Exc.:** $475   **VGood:** $425

Beretta Model 92FCM

Same specs as Model 92F Compact except thinner grip; straightline 8-shot magazine; weighs 31 oz.; 1 1/4" overall width; plastic grips. From Beretta U.S.A. Introduced 1989; no longer in production.
   Perf.: $500   **Exc.:** $450   **VGood:** $400

Beretta Model 92FS Centurion

Same specs as Model 92 FS except 4 3/8" barrel shorter slide; Trijicon or three-dot sights; plastic or wood grips; 9mm Para. only. From Beretta U.S.A. Introduced 1992.
With plastic grips, three-dot sights
   New: $550   **Perf.:** $500   **Exc.:** $450
With wood grips
   New: $575   **Perf.:** $525   **Exc.:** $475

Beretta Model 92FS Stainless

Same specs as Model 92FS except for stainless steel satin finish; groove in slide rail; larger hammer pin head; modified to meet special-use military specs. Introduced 1990; still produced.
   Perf.: $550   **Exc.:** $500   **VGood:** $450

Beretta Model 92F-EL

Same specs as Model 92FS Stainless except gold trim on safety levers, magazine release, trigger and grip screws; gold inlaid Beretta logo top of barrel; P. Beretta signature inlaid in gold on slide; contoured walnut grips with Beretta logo engraved; high-polish blued finish on barrel, frame, slide. Introduced 1991.
   Perf.: $650   **Exc.:** $600   **VGood:** $500

BERETTA MODEL 96

Semi-automatic; double action; 40 S&W; 10-shot magazine; 5" barrel; 8 1/2" overall length; weighs 34 oz.; ambidextrous triple safety with passive firing pin catch; slide safety/decocking lever; trigger bar disconnect; plastic grips; blued finish. Introduced 1992.
   New: $550   **Perf.:** $500   **Exc.:** $400

Beretta Model 96 Brigadier

Same specs as Model 96 except removable front sight; reconfigured high slide wall profile; 10-shot magazine; three-dot sights system; 4 15/16" barrel; weighs 35 oz.; matte black Bruniton finish. From Beretta U.S.A. Introduced 1995; still in production.
   New: $600   **Perf.:** $550   **Exc.:** $450

Beretta Model 96 Centurion

Same specs as Model 96 except 4 3/8" barrel; shorter slide; Trijicon or three-dot sights; plastic or wood grips; 40 S&W only. From Beretta U.S.A. Introduced 1992.
   New: $550   **Perf.:** $500   **Exc.:** $400
With three-dot sights, plastic grips
   New: $650   **Perf.:** $600   **Exc.:** $500

Beretta Model 96D

Same specs as Model 96 except double-action-only; three-dot sights. From Beretta U.S.A. Introduced 1992.
   New: $450   **Perf.:** $400   **Exc.:** $325

BERETTA MODEL 318

Semi-automatic; 25 ACP; 8-shot magazine; 2 1/2" barrel; 4 1/2" overall length; weighs 14 oz.; blued finish, fixed sights; plastic grips. Sold in the U.S. as the Panther. Manufactured 1934 to 1946.
   Exc.: $275   **VGood:** $225   **Good:** $175

BERETTA MODEL 418

Semi-automatic; 25 ACP; 8-shot magazine; 3 1/2" barrel; 5 3/4" overall length; fixed sights; plastic grips; blued finish; grip safety redesigned to match contour of backstrap. Introduced 1946; dropped 1958.
   Exc.: $250   **VGood:** $175   **Good:** $150

Beretta
Model 420
Model 421
Model 935
(See Beretta Model 1935)
Model 948
Model 949 Tipo Olimpionico
Model 950B Jetfire
Model 950B Minx M2
Model 950CC Minx M4
Model 950BS
Model 951 Brigadier
Model 1915
Model 1919
Model 1923
Model 1931
Model 1934

Beretta Model 949
Tipo Olimpionico

Beretta Model 950BS
(2 1/2" barrel)

Beretta Model 951
Brigadier

Beretta Model 1931

Beretta
Model 950BS
(4" barrel)

Beretta Model 1923

BERETTA MODEL 420
Semi-automatic; 25 ACP; 8-shot magazine; 3 1/2" barrel; 5" overall length; fixed sights; plastic grips; chrome-plated, engraved version of the Model 418.
Exc.: $250   **VGood:** $200   **Good:** $150

BERETTA MODEL 421
Semi-automatic; 25 ACP; 8-shot magazine; 3 1/2" barrel; 5" overall length; fixed sights; plastic grips; gold-plated, elaborately engraved version of the Model 418.
Exc.: $500   **VGood:** $350   **Good:** $250

BERETTA MODEL 948
Semi-automatic; double action; 22 LR; 3 5/16", 6" barrel; fixed sights; hammer; grip displays "BERETTA" across top instead of earlier "PB" monogram. Sold in U.S. as Featherweight or Plinker.
Exc.: $200   **VGood:** $175   **Good:** $150

BERETTA MODEL 949 TIPO OLIMPIONICO
Semi-automatic; 22 Short; 6-shot magazine; 8 3/4" barrel; 12 1/2" overall length; target sights; adjustable barrel weight; muzzle brake; hand-checkered walnut thumbrest grips; blued. Introduced 1949; dropped 1968.
Exc.: $575   **VGood:** $400   **Good:** $300

BERETTA MODEL 950B JETFIRE
Semi-automatic; single action; 25 ACP; 7-shot magazine; 2 3/8" barrel; 4 1/2" overall length; weighs 9 1/2 oz.; tip-up barrel; rear sight milled in slide, fixed front; black plastic or wood grips; blue or nickel finish. Introduced 1950; not imported since 1968. Now being made in U.S.
Exc.: $125   **VGood:** $115   **Good:** $100

BERETTA MODEL 950B MINX M2
Semi-automatic; 22 Short; 6-shot magazine; 2 3/8" barrel; 4 1/2" overall length; rear sight milled in slide, fixed front; black plastic grips; blued. Introduced 1950.
Exc.: $125   **VGood:** $115   **Good:** $100

Beretta Model 950CC Minx M4
Same specs as Model M2 except 3 3/4" barrel. Introduced 1956. Not imported since 1968.
Exc.: $125   **VGood:** $115   **Good:** $100

BERETTA MODEL 950BS
Semi-automatic; single action; 22 Short, 25 ACP; 8-shot magazine; 2 1/2", 4" (22) barrel; 4 1/2" overall length; weighs 10 oz. (25 ACP); checkered plastic or walnut grips; fixed sights; thumb safety, half-cock safety; hinged barrel for single loading, cleaning; blued, nickel or matte finish. Made by Beretta USA. Introduced 1982; 25-caliber still in production.
Perf.: $150   **Exc.:** $135   **VGood:** $125

BERETTA MODEL 951 BRIGADIER
Semi-automatic; 9mm Para.; 8-shot magazine; 4 1/2" barrel; 8" overall length; external hammer; crossbolt safety; slide stays open after last shot; fixed sights; moulded grooved plastic grips; blued. Advertised originally as Brigadier model. Introduced 1951; no longer in production.
Exc.: $350   **VGood:** $250   **Good:** $175

BERETTA MODEL 1915
Semi-automatic; 32 ACP, 9mm Glisenti; 8-shot magazine; 6" overall length; 3 5/16" barrel; blued finish; fixed sights; grooved wooden grips. Smaller caliber manufactured 1915 to 1930; 9mm Glisenti, 1915 to 1918.
32 ACP
Exc.: $325   **VGood:** $250   **Good:** $175
9mm Glisenti
Exc.: $500   **VGood:** $350   **Good:** $250

BERETTA MODEL 1919
Semi-automatic; 25 ACP; 8-shot magazine; 3 1/2" barrel; 5 3/4" overall length; hammerless; grip safety on frame backstrap; fixed sights; wood (early models), pressed sheet steel or plastic grips; blued finish. Introduced in 1919, went through several modifications to be designated as Model 318; dropped 1939.
Exc.: $250   **VGood:** $175   **Good:** $150

BERETTA MODEL 1923
Semi-automatic; 9mm Glisenti; 8-shot magazine; 4" barrel; 6 1/2" overall length; 28 oz.; ring-type external hammer; blued finish; fixed sights; pressed steel or wood grips. Manufactured 1923 to 1936.
Exc.: $550   **VGood:** $375   **Good:** $275

BERETTA MODEL 1931
Semi-automatic; double-action; 32 ACP; 8-shot magazine; 3 1/2" barrel; 6" overall length; weighs 22 1/2 oz.; fixed sights; walnut grips; blued finish. Italian Navy issue bears medallion with "R/Anchor/M" emblem. Commercial issue has standard "PB" embossed black plastic grips.
Exc.: $750   **VGood:** $600   **Good:** $400

BERETTA MODEL 1934
Semi-automatic; 380 ACP; 7-shot magazine; 3 3/8" barrel; 5 7/8" overall length; weighs 28 1/8 oz.; fixed sights; plastic grips; thumb safety; blued or chrome finish. Official Italian service sidearm in WWII; military versions marked "RE", "RA" or "RM" for the army, air force and navy respectively. Police versions marked "PS" on rear of frame. Wartime version lacks quality of commercial model. Introduced 1934; no longer legally importable; still in production.
Exc.: $300   **VGood:** $275   **Good:** $200

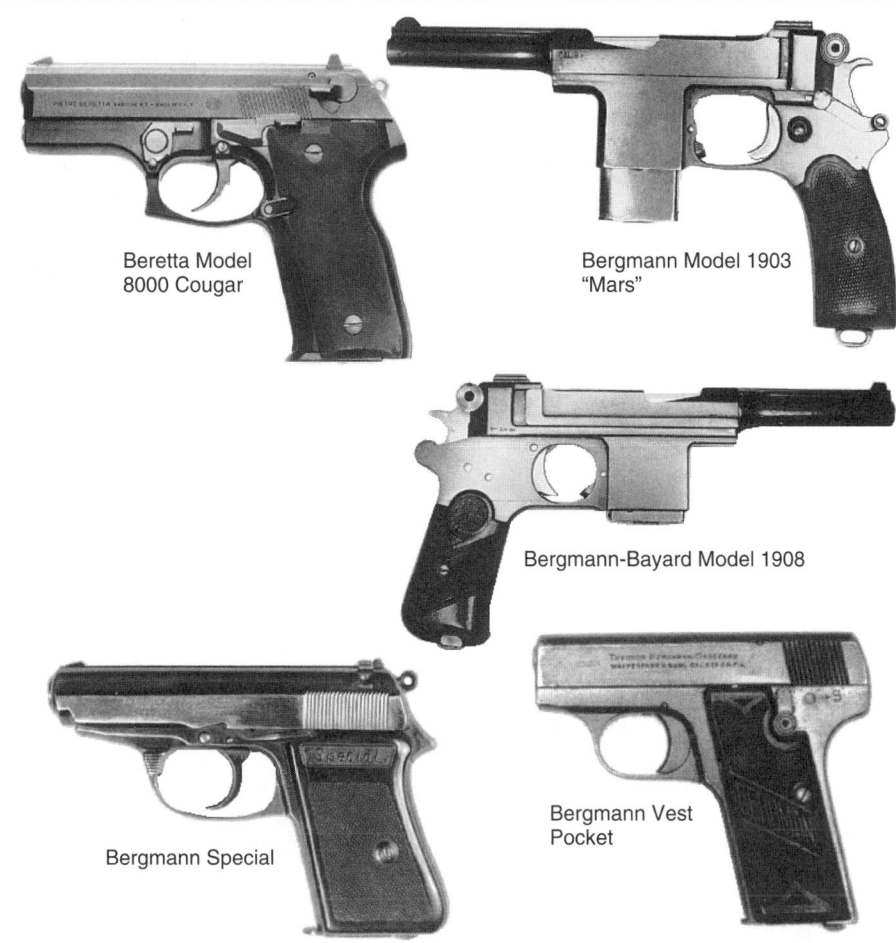

Beretta Model
8000 Cougar

Bergmann Model 1903
"Mars"

Bergmann-Bayard Model 1908

Bergmann Vest
Pocket

Bergmann Special

HANDGUNS

Beretta
Model 1935
Model 8000 Cougar
Model 8000D Cougar
Model 8040 Cougar
Model 8040D Cougar
New Puma (See Beretta
Model 70)
New Sable (See Beretta
Model 76)
Panther (See Beretta
Model 318)
Plinker (See Beretta
Model 948)

Bergmann
Model 1903 "Mars"
Special
Vest Pocket Models

Bergmann-Bayard
Model 1908
Model 1910/21

Bernardelli
Model 60
Model 68
Model 69 Target

BERETTA MODEL 1935

Semi-automatic; double action; 32 ACP; 7-shot magazine; 3 3/8" barrel; 5 7/8" overall length; fixed sights; plastic grips; thumb safety; blued or chrome finish. Italian service pistol during WWII; frame markings "RA" or "RM" on frame for the air force and navy respectively. Commercial version after 1945 known as Model 935.
Exc.: $300   VGood: $275   Good: $200

BERETTA MODEL 8000 COUGAR

Semi-automatic; double action; 9mm Para.; 10-shot magazine; 3 1/2" barrel; 7" overall length; weighs 33 1/2 oz.; textured composition grips; blade front sight, drift-adjustable rear; slide-mounted safety; exposed hammer; matte black Bruniton finish. Imported from Italy by Beretta U.S.A. Introduced 1994; still imported.
New: $525   Perf.: $450   Exc.: $375

Beretta Model 8000D Cougar

Same specs as Model 8000 except double-action-only trigger mechanism. Imported from Italy by Beretta U.S.A. Introduced 1994; still imported.
New: $550   Perf.: $500   Exc.: $400

BERETTA MODEL 8040 COUGAR

Semi-automatic; double action; 40 S&W; 10-shot magazine; 3 1/2" barrel; 7" overall length; weighs 33 1/2 oz.; textured composition grips; blade front sight, drift-adjustable rear; slide-mounted safety; exposed hammer; matte black Bruniton finish. Imported from Italy by Beretta U.S.A. Introduced 1994; still imported.
New: $575   Perf.: $550   Exc.: $525

Beretta Model 8040D Cougar

Same specs as Model 8040 except double-action-only trigger mechanism. Imported from Italy by Beretta U.S.A. Introduced 1994; still imported.
New: $550   Perf.: $525   Exc.: $500

BERGMANN MODEL 1903 "MARS"

Semi-automatic; 9mm Bergmann-Bayard; 6-, 10-shot magazine; 4" barrel; 10" overall length; checkered wood grips. Marked "Bergmann Mars Pat. Brev. S.G.D.G." on locking block top. About 1,000 made in Germany before rights were sold to Pieper in Belguim where, with a few minor changes, it was made as the Bergmann-Bayard.
Exc.: $4500   VGood: $3000   Good: $2000

BERGMANN SPECIAL

Semi-automatic; double action; 32 ACP; 8-shot magazine; 3 7/16" barrel; 6 3/16" overall length; checkered plastic wrap-around grips. Double-action competitor to Walther's Model PP, actually made by Menz but sold under Bergmann's name.
Exc.: $1200   VGood: $850   Good: $500

BERGMANN VEST POCKET MODELS

Semi-automatic; 25 ACP; 6-, 9-shot magazine; 2 1/8" barrel; 4 3/4" overall length; checkered hard rubber or wood (rare) grips. Both conventional and one-hand-cocked ("Einhand") models later marketed under the Lignose name (see Lignose).
Conventional
Exc.: $300   VGood: $200   Good: $125
"Einhand"
Exc.: $400   VGood: $250   Good: $150
32 or 380 "Einhand"
Catalogued but apparently never made.

BERGMANN-BAYARD MODEL 1908

Semi-automatic; 9mm Bergmann-Bayard; 6-shot magazine; 4" barrel; 10" overall length; checkered wood or hard rubber (rare) grips. Marked "ANCIENS ESTABLISSE-MENTS PIEPER" on side. Made under contract for the Spanish and Danish armies as well as commercial sale.
Exc.: $1250   VGood: $950   Good: $700
Spanish contract (tiny circle with three-line proofmark)
Exc.: $1500   VGood: $1250   Good: $750
Shoulder stock slot in backstrap
Exc.: $2000   VGood: $1500   Good: $950

BERGMANN-BAYARD MODEL 1910/21

Semi-automatic; 9mm Bergmann-Bayard; 6-shot magazine; 4" barrel; 10" overall length. Made under contract for Danish army but reworked post-WWI models with oversize wood or plastic grips; screw-retained instead of latched sideplate; stamped "1910/21" on side. After WWI when Pieper could no longer supply pistols to Denmark, the Danes produced about 2200 Model 1910/21 pistols in their own arsenals.
Exc.: $1100   VGood: $850   Good: $650

BERNARDELLI MODEL 60

Semi-automatic; 22 LR, 32 ACP, 380 ACP; 10-shot (22 LR), 9-shot (32 ACP), 7-shot (380 ACP); 3 1/2" barrel; 6 1/2" overall length; weighs 26 1/2 oz.; ramp front sight, adjustable white-outlined rear; checkered plastic grips, thumbrest; hammer-block slide safety; loaded chamber indicator; dual recoil buffer springs; serrated trigger, inertia-type firing pin. Made in Italy. Introduced 1978; no longer imported.
Perf.: $200   Exc.: $175   VGood: $150

BERNARDELLI MODEL 68

Semi-automatic; 22 Short, 22 LR; 6-shot; 2" barrel; 4 1/8" overall length; weighs 8 1/2 oz.; plastic grips; blued finish. A replacement for the Vest Pocket Model with rounded slide, loaded chamber indicator, optional extended magazine. Discontinued 1970.
Perf.: $175   Exc.: $150   VGood: $125

BERNARDELLI MODEL 69 TARGET

Semi-automatic; 22 LR; 10-shot magazine; 5 15/16" barrel; 9" overall length; weighs 38 oz.; interchangeable target sights; wrap-around hand-checkered walnut stocks; thumbrest; meets UIT requirements; manual thumb safety, magazine safety; grooved trigger. Made in Italy. Introduced 1987.
Perf.: $500   Exc.: $450   VGood: $400

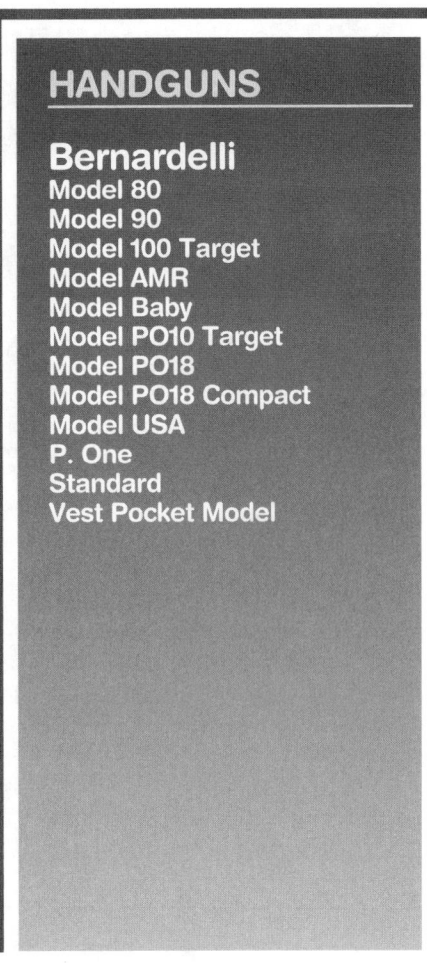

HANDGUNS

Bernardelli
Model 80
Model 90
Model 100 Target
Model AMR
Model Baby
Model PO10 Target
Model PO18
Model PO18 Compact
Model USA
P. One
Standard
Vest Pocket Model

Bernardelli
Model PO18

Bernardelli
Model USA

Bernardelli Vest
Pocket Model

Bernardelli
Model 80

Bernardelli
Model Baby

Bernardelli
Model 68

BERNARDELLI MODEL 80
Semi-automatic; 22 LR, 32 ACP, 380 ACP; 10-shot (22 LR), 8-shot (32 ACP, 380 ACP); 3 1/2" barrel; 6 1/2" overall length; adjustable rear sight, white dot front; plastic thumbrest stocks; blued finish;. Introduced 1968 as modification of Model 60 to meet U.S. import requirements.
Perf.: $175   **Exc.:** $150   **VGood:** $135

BERNARDELLI MODEL 90
Semi-automatic; 22 LR, 32 ACP; 10-shot (22 LR), 8-shot (32 ACP); 6" barrel; 9" overall length; adjustable rear sight, white dot front; plastic thumbrest stocks; blued finish. Introduced 1968; no longer imported.
Perf.: $200   **Exc.:** $175   **VGood:** $150

BERNARDELLI MODEL 100 TARGET
Semi-automatic; 22 LR; 10-shot magazine; 6" barrel; 9" overall length; adjustable rear sight, interchangeable front; checkered walnut thumbrest stocks; blued finish. Introduced in 1969 as Model 68; dropped 1983.
Perf.: $350   **Exc.:** $300   **VGood:** $275

BERNARDELLI MODEL AMR
Semi-automatic; single action; 22 LR, 32, 380 ACP; 6" barrel; target sights. Similar to the Model USA. Introduced 1989; dropped 1991. Was imported by Magnum Research.
Perf.: $350   **Exc.:** $300   **VGood:** $250

BERNARDELLI MODEL BABY
Semi-automatic; 22 Short; 22 LR; 2 1/8" barrel; fixed sights; plastic grips. This was the first of Bernardelli's 22 pistols and was largely the Vest Pocket Model altered to handle rimfire cartridges. Introduced 1949; discontinued 1968.
Perf.: $200   **Exc.:** $175   **VGood:** $150

BERNARDELLI MODEL PO10 TARGET
Semi-automatic; single action; 22 LR; 5- or 10-shot magazine; 6" barrel; weighs 40 1/2 oz.; walnut thumbrest stocks; fully-adjustable target-type sights; external hammer with safety notch; pivoted adjustable trigger; matte black finish. Made in Italy. Introduced 1989; importation dropped 1990.
Perf.: $650   **Exc.:** $600   **VGood:** $550

BERNARDELLI MODEL PO18
Semi-automatic; double action; 9mm Para; 16-shot magazine; 4 7/8" barrel; 6 3/16" overall length; weighs 36 1/3 oz.; low-profile combat sights; checkered, contoured plastic or optional walnut stocks; manual thumb safety, half-cock, magazine safeties; ambidextrous magazine release; auto-locking firing pin block safety. Made in Italy. Introduced 1987; still imported by Armsport. Was imported by Magnum Research, Inc.
Perf.: $450   **Exc.:** $400   **VGood:** $350
Walnut grips
Perf.: $500   **Exc.:** $450   **VGood:** $350

Bernardelli Model PO18 Compact
Same specs as standard PO18 except 4 1/8" barrel; 14-shot magazine. Introduced 1987; still imported by Armsport. Was imported by Magnum Research, Inc.
Perf.: $500   **Exc.:** $450   **VGood:** $400
Walnut stocks
Perf.: $550   **Exc.:** $500   **VGood:** $450

BERNARDELLI MODEL USA
Semi-automatic; 22 LR, 380 ACP; 10-shot (22 LR), 7-shot (380 ACP) magazine; 3 1/2" barrel, 6 1/2" overall length; weighs 26 1/2 oz.; ramp front sight; fully-adjustable white-outline rear; checkered plastic stocks, thumbrest; serrated trigger; inertia-type firing pin; hammer-block safety; loaded chamber indicator. Made in Italy. Introduced 1989; still imported by Armsport. Was imported by Magnum Research, Inc.
Perf.: $350   **Exc.:** $300   **VGood:** $275

BERNARDELLI P. ONE
Semi-automatic; double action; 9mm Para., 40 S&W; 16-shot (9mm), 10-shot (40 S&W) magazine; 4 7/8" barrel; 8 3/8" overall length; weighs 34 oz.; checkered black plastic grips; blade front sight, fully adjustable rear; three-dot system; forged steel frame and slide; full-length slide rails; reversible magazine release; thumb safety/decocker; squared triggerguard; blue/black finish. Imported from Italy by Armsport. Introduced 1994.
Perf.: $550   **Exc.:** $450   **VGood:** $400
Chrome finish
Perf.: $575   **Exc.:** $500   **VGood:** $450

Bernardelli P. One Practical VB Pistol
Same specs as P. One except 9x21mm; two- or four-port compensator; straight trigger; micro-adjustable rear sight. Imported from Italy by Armsport. Introduced 1994.
New: $1300   **Perf.:** $1200   **Exc.:** $1100

BERNARDELLI STANDARD
Semi-automatic; 22 LR; 10-shot magazine; 6", 8", 10" barrel; 13" overall length (10" barrel); target sights; adjustable sight ramp; walnut target grips; blued. Simply the Pocket Model adapted for 22 LR cartridge. Introduced 1949; no longer imported.
Perf.: $225   **Exc.:** $200   **VGood:** $175

BERNARDELLI VEST POCKET MODEL
Semi-automatic; 25 ACP; 5-shot; 8-shot extension magazine also available; 2 1/8" barrel; 4 1/8" overall length; no sights, but sighting groove milled in slide; plastic grips; blued finish. Introduced 1948; U.S. importation dropped 1968.
Exc.: $200   **VGood:** $175   **Good:** $135

Bersa Model 95
Series

Bersa Model 383

Bersa Model 223

Bersa Model 23

Bersa Model 83

Bersa Model 225

HANDGUNS

Bersa
Model 23
Model 83
Model 85
Model 86
Model 95 Series
Model 223
Model 224
Model 225
Model 226
Model 383

BERSA MODEL 23

Semi-automatic; double action; 22 LR; 10-shot magazine; 3 1/2″ barrel; 6 1/2″ overall length; weighs 24 1/2 oz.; walnut grips with stippled panels; blade front sight, notch windage-adjustable rear; three-dot system; firing pin, magazine safeties; blue or nickel finish. Made in Argentina; distributed by Eagle Imports, Inc. Introduced 1989; dropped 1994.
Blue finish

Perf.: $200	**Exc.:** $175	**VGood:** $150

Nickel finish

Perf.: $225	**Exc.:** $200	**VGood:** $175

BERSA MODEL 83

Semi-automatic; double action; 380 ACP; 7-shot magazine; 3 1/2″ barrel; 6 1/2″ overall length; weighs 25 3/4 oz.; stippled walnut grips; blade front sight, notch windage-adjustable rear; three-dot system; firing pin, magazine safety systems; blue or nickel finish. Imported from Argentina; distributed by Eagle Imports, Inc. Introduced 1989; dropped 1994.
Blue finish

Perf.: $200	**Exc.:** $175	**VGood:** $150

Nickel finish

Perf.: $225	**Exc.:** $200	**VGood:** $175

BERSA MODEL 85

Semi-automatic; double action; 380 ACP; 13-shot magazine; 3 1/2″ barrel; 6 1/2″ overall length; weighs 25 3/4 oz.; stippled walnut grips; blade front sight, notch windage-adjustable rear; three-dot system; firing pin, magazine safety systems; blue or nickel finish. Imported from Argentina; distributed by Eagle Imports, Inc. Introduced 1989; dropped 1994.
Blue finish

Perf.: $250	**Exc.:** $225	**VGood:** $200

Nickel finish

Perf.: $300	**Exc.:** $275	**VGood:** $250

BERSA MODEL 86

Semi-automatic; double action; 380 ACP; 13-shot magazine; 3 1/2″ barrel; 6 1/2″ overall length; weighs 22 oz.; wrap-around textured rubber grips; blade front sight, windage-adjustable rear; three-dot system; firing pin, magazine safeties; combat-style triggerguard; matte blue or satin nickel finish. Imported from Argentina; distributed by Eagle Imports, Inc. Introduced 1992; dropped 1994.
Blue finish

Perf.: $275	**Exc.:** $225	**VGood:** $175

Nickel finish

Perf.: $300	**Exc.:** $250	**VGood:** $200

BERSA MODEL 95 SERIES

Semi-automatic; double action; 380 ACP; 7-shot magazine; 3 1/2″ barrel; 6 1/2″ overall length; weighs 22 oz.; blade front sight, windage-adjustable rear; three-dot system; firing pin and magazine safeties; combat-style triggerguard; wrap-around textured rubber grips; matte blue or satin nickel. Imported from Argentina; distributed by Eagle Imports, Inc. Introduced 1992.
Matte blue finish

New: $200	**Perf.:** $175	**Exc.:** $150

Satin nickel finish

New: $225	**Perf.:** $200	**Exc.:** $175

BERSA MODEL 223

Semi-automatic; double action; 22 LR; 11-shot magazine; 3 1/2″ barrel; weighs 24 1/2 oz; blade front sight, square-notch windage-adjustable rear; checkered target-type nylon stocks; thumbrest; blowback action; squared-off triggerguard; magazine safety; blued finish. Made in Argentina. Introduced 1984; no longer imported.

Perf.: $200	**Exc.:** $175	**VGood:** $150

BERSA MODEL 224

Semi-automatic; single action; 22 LR; 11-shot magazine; 4″ barrel; weighs 26 oz; blade front sight, square-notch windage-adjustable rear; checkered target-type nylon stocks; thumbrest; blowback action; combat-type triggerguard; magazine safety; blued finish. Made in Argentina. Introduced 1984; no longer imported.

Perf.: $200	**Exc.:** $175	**VGood:** $150

BERSA MODEL 225

Semi-automatic; single action; 22 LR; 10-shot magazine; 5″ barrel; weighs 26 oz.; blade front sight, square-notch windage-adjustable rear; checkered target-type nylon stocks; thumbrest; blowback action; combat-type triggerguard; magazine safety; blued finish. Made in Argentina. Introduced 1984; no longer imported.

Perf.: $175	**Exc.:** $150	**VGood:** $135

BERSA MODEL 226

Semi-automatic; single action; 22 LR; 10-shot magazine; 6″ barrel; weighs 26 oz; blade front sight, square-notch windage-adjustable rear; checkered target-type nylon stocks; thumbrest; blowback action; combat-type triggerguard; magazine safety; blued finish. Made in Argentina. Introduced 1984; no longer imported.

Perf.: $175	**Exc.:** $150	**VGood:** $135

BERSA MODEL 383

Semi-automatic; single action; 380 ACP; 9-shot magazine; 3 1/2″ barrel; weighs 25 oz.; square-notch windage-adjustable rear sight, blade front; target-type black nylon stocks; blowback action; combat-type trigger guard; magazine safety; blued finish. Made in Argentina. Introduced 1984; no longer imported.

Perf.: $135	**Exc.:** $125	**VGood:** $110

HANDGUNS

Bersa
Model 383 DA
Thunder 9
Thunder 22
Thunder 380
Thunder 380 Plus

BF
Single Shot
Single Shot Ultimate
Silhouette

Bren Ten
Bren Ten
Pocket Model
Special Forces Model

BRNO
CZ 75
CZ 83
CZ 85

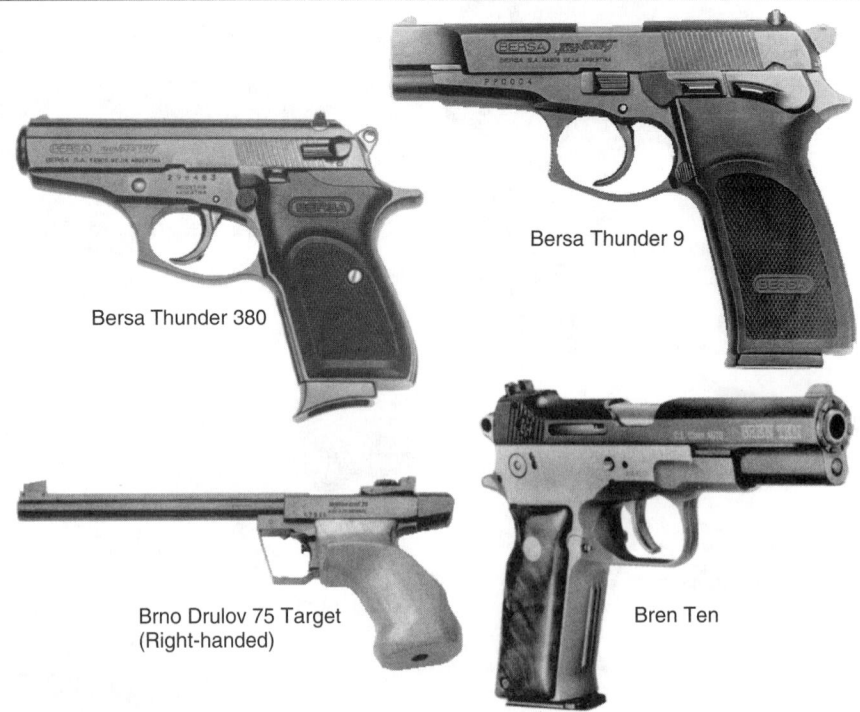

Bersa Thunder 9

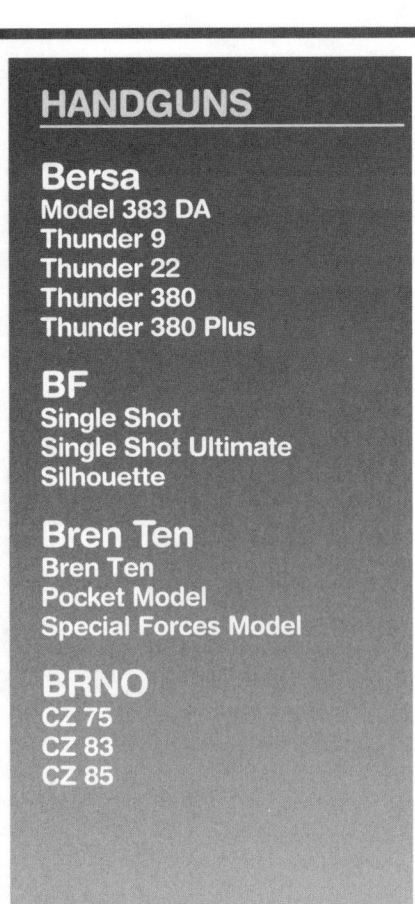

Bersa Thunder 380

Brno Drulov 75 Target
(Right-handed)

Bren Ten

BF Single Shot

Bersa Model 383DA
Same specs as Model 383 except double action.
Perf.: $150   **Exc.:** $135   **VGood:** $125

BERSA THUNDER 9
Semi-automatic; double action; 9mm Para.; 10-shot magazine; 4″ barrel; 7 3/8″ overall length; weighs 30 oz.; checkered black polymer grips; blade front sight, fully-adjustable rear; three-dot system; ambidextrous safety, decocking levers and slide release; internal automatic firing pin safety; reversible extended magazine release; adjustable trigger stop; alloy frame; link-free locked breech design; matte blue, satin nickel or duo-tone finish. Imported from Argentina by Eagle Imports, Inc. Introduced 1993; still imported. **LMSR: $474**
New: $400   **Perf.:** $350   **Exc.:** $300
Satin nickel finish
New: $450   **Perf.:** $400   **Exc.:** $350
Duo-tone finish
New: $425   **Perf.:** $375   **Exc.:** $325

BERSA THUNDER 22
Semi-automatic; double action; 22 LR; 10-shot magazine; 3 1/2″ barrel; 6 1/2″ overall length; weighs 24 oz.; black polymer grips; blade front sight, notch windage-adjustable rear; three-dot system; firing pin and magazine safeties; blue or nickel finish. Imported from Argentina; distributed by Eagle Imports, Inc. Introduced 1995; still imported. **LMSR: $309**
New: $225   **Perf.:** $200   **Exc.:** $175
Nickel finish
New: $250   **Perf.:** $225   **Exc.:** $200

BERSA THUNDER 380
Semi-automatic; double action; 380 ACP; 7-shot magazine; 3 1/2″ barrel; 6 1/2″ overall length; weighs 25 3/4 oz.; black rubber grips; blade front sight, notch windage-adjustable rear; three-dot system; firing pin and magazine safeties; blue, nickel or duo-tone finish. Imported from Argentina; distributed by Eagle Imports, Inc. Introduced 1995; still imported. **LMSR: $309**

New: $225   **Perf.:** $200   **Exc.:** $175
Nickel finish
New: $250   **Perf.:** $225   **Exc.:** $200
Duo-tone finish
New: $235   **Perf.:** $215   **Exc.:** $185

Bersa Thunder 380 Plus
Same specs as Thunder 380 except 10-shot magazine. Imported from Argentina; distributed by Eagle Imports, Inc. Introduced 1995; still imported. **LMSR: $368**
New: $250   **Perf.:** $200   **Exc.:** $175
Nickel finish
New: $300   **Perf.:** $250   **Exc.:** $225
Duo-tone finish
New: $275   **Perf.:** $225   **Exc.:** $200

BF SINGLE SHOT
Single shot; 22 LR, 357 Mag., 44 Mag., 7-30 Waters, 30-30 Win., 375 Win., 45-70; custom chamberings from 17 Rem. through 45-cal.; 10″, 10 3/4″, 12″, 15+″ barrel lengths; weighs about 52 oz.; custom Herrett finger-groove grip, forend; undercut Patridge front sight, 1/2-MOA match-quality fully-adjustable RPM Iron Sight rear; barrel or receiver mounting; drilled and tapped for scope mounting; rigid barrel/receiver; falling block action with short lock time; automatic ejection; air-gauged match barrels by Wilson or Douglas; matte black oxide finish standard, electroless nickel optional. Guns with RPM sights worth slightly more. Made in U.S. by E.A. Brown Mfg. Introduced 1988; still in production. **LMSR: $500**
Perf.: $450   **Exc.:** $400   **VGood:** $350

BF Single Shot Ultimate Silhouette
Same specs as standard model except 10 3/4″ heavy barrel; special forend; RPM rear sight; hooded front; gold-plated trigger. Made in U.S. by E.A. Brown Mfg. Introduced 1988; still in production. **LMSR: $529**
Perf.: $650   **Exc.:** $600   **VGood:** $550

BREN TEN
Semi-automatic; 10mm Auto; 11-shot magazine; 5″ barrel; 8 3/4″ overall length; textured black nylon stocks; adjustable, replaceable combat sights; double- or single-action; stainless steel frame; blued slide; reversible thumb safety, firing pin block. Made by Dornaus & Dixon. Introduced 1983; no longer in production.
Perf.: $1500   **Exc.:** $1200   **VGood:** $1000
Military/Police black matte finish model
Perf.: $1500   **Exc.:** $1200   **VGood:** $1000
Dual presentation 10mm and 45 ACP
Perf.: $3000   **Exc.:** $2500   **VGood:** $2200

Bren Ten Pocket Model
Same specs as Bren Ten except 4″ barrel; 7 3/8″ overall length; weighs 28 oz.; 9-shot magazine; chrome slide.
Perf.: $1400   **Exc.:** $1100   **VGood:** $900

Bren Ten Special Forces Model
Same specs as Pocket Model except standard grip frame; 11-shot magazine; weighs 33 oz.; black or natural light finish. Introduced 1984; no longer in production.
Perf.: $1500   **Exc.:** $1200   **VGood:** $1000
Light finish
Perf.: $1500   **Exc.:** $1200   **VGood:** $1000

BRNO CZ 75
Semi-automatic; double action; 9mm Para., 40 S&W; 15-shot magazine; 4 11/16″ barrel; 8″ overall length; weighs 35 oz.; windage-adjustable rear sight, blade front; checkered wood stocks; blued finish. Made in Czechoslovakia. Introduced 1986; still imported by Action Arms Ltd. **LMSR: $539**
Perf.: $400   **Exc.:** $350   **VGood:** $300

BRNO CZ 83
Semi-automatic; double action; 32 ACP, 380 ACP; 15-shot magazine (32), 13-shot (380); 3 11/16″ barrel; 6 11/16″ overall length; weighs 26 1/2 oz.; windage-adjustable rear sight, blade front; checkered black plastic stocks; ambidextrous magazine release, safety; matte blue or polished. Made in Czechoslovakia. Introduced 1987; still imported by Action Arms Ltd. **LMSR: $409**
Perf.: $300   **Exc.:** $275   **VGood:** $225

BRNO CZ 85
Semi-automatic; double action; 9mm Para., 40 S&W; 15-shot; 4 11/16″ barrel; 8″ overall length; weighs 35 oz.; ambidextrous slide release; safety lever; contoured composition stocks; matte finish on top of slide. Made in Czechoslovakia. Introduced 1986; still imported by Action Arms Ltd. **LMSR: $549**
Perf.: $400   **Exc.:** $375   **VGood:** $325

Bronco Model 1918

Browning Buck Mark Micro

Browning
BDA 380

Browning 380 Standard

Browning BDM

BRNO
Drulov 75 Target

Bronco
Model 1918

Browning
25 "Baby"
25 Lightweight
25 Renaissance Model
380 Standard
380 Renaissance
BDA
BDA 380
BDM
Buck Mark
Buck Mark Field 5.5
Buck Mark Micro
Buck Mark Micro Plus
Buck Mark Plus
Buck Mark Silhouette
Buck Mark Target 5.5

BRNO DRULOV 75 TARGET

Bolt-action; single shot; 22 LR; 10" barrel; 14 3/4" overall length; weighs 44 oz.; interchangeable blade front sight, micro-click fully adjustable rear; European walnut grips; all-steel construction; adjustable set trigger; blued finish. Made in Czechoslovakia. Introduced 1991; importation dropped 1992.
    Perf.: $275   **Exc.: $250**   **VGood: $200**

BRONCO MODEL 1918

Semi-automatic; 32 ACP; 6-shot magazine; 2 1/2" barrel; 5" overall length; blued finish; fixed sights; hard rubber stocks. Manufactured in Spain 1918 to 1925.
    Exc.: $175   **VGood.: $125**   **Good: $100**

BROWNING 25 "BABY"

Semi-automatic; 25 ACP; 6-shot magazine; 2 1/8" barrel; 4" overall length; fixed sights, hard rubber grips; blued finish. Introduced ca. 1936; dropped, not importable, 1968.
    Perf.: $350   **Exc.: $300**   **VGood: $250**

Browning 25 Lightweight

Same specs as Browning 25 except chrome-plated; polyester pearl grips; alloy frame. Introduced 1954; dropped 1968.
    Perf.: $325   **Exc.: $275**   **VGood: $250**

Browning 25 Renaissance Model

Same specs as standard Browning 25 except chrome-plated finish; polyester pearl grips; full engraving. Introduced 1954; dropped 1968.
    Perf.: $800   **Exc.: $700**   **VGood: $500**

BROWNING 380 STANDARD

Semi-automatic; 380 ACP, 32 ACP; 9-shot magazine; 3 7/16" barrel; 6" overall length; pre-'68 models have fixed sights, later adjustable; hard rubber grips; blued finish. Redesigned 1968; dropped 1974. Last models had 4 1/2" barrel; 7" overall length.
    Perf.: $300   **Exc.: $250**   **VGood: $200**

Browning 380 Renaissance

Same specs as standard Browning 380 Standard except chrome-plated finish; full engraving; polyester pearl grips. Introduced 1954; dropped 1981.
    Perf.: $800   **Exc.: $700**   **VGood: $600**

BROWNING BDA

Semi-automatic; 9mm Para., 38 Super, 45 ACP; 9-shot magazine (9mm, 38 Super), 7-shot (45 ACP); 4 7/16" barrel; 7 13/16" overall length; fixed sights; plastic stocks; blued finish. Manufactured by Sauer; same as SIG-Sauer P220. Imported 1977 to 1981.
9mm, 45 ACP
    Perf.: $500   **Exc.: $450**   **VGood: $350**
38 Super
    Perf.: $600   **Exc.: $550**   **VGood: $450**

BROWNING BDA 380

Semi-automatic; 380 Auto; 12-shot magazine; 3 13/16" barrel; 6 3/4" overall length; blade front sight, windage-adjustable rear; combination safety; de-cocking lever; inertia firing pin; uncheckered walnut grips. Manufactured in Italy. Introduced 1978; still in production. **LMSR: $615**
    Perf.: $400   **Exc.: $350**   **VGood: $300**

BROWNING BDM

Semi-automatic; double action; 9mm Para.; 15-shot magazine; 4 3/4" barrel; 7 7/8" overall length; weighs 31 oz; low-profile removable blade front sight, windage-adjustable rear; checkered black moulded composition grips, thumbrest on both sides; all-steel frame; matte black finish; two safety systems; mode selector for switching from DA auto to "revolver" mode. Introduced 1991; still produced. **LMSR: $595**
    Perf.: $450   **Exc.: $400**   **VGood: $350**

BROWNING BUCK MARK

Semi-automatic; 22 LR; 10-shot magazine; 5 1/2" barrel; 9" overall length; weighs 32 oz.; adjustable rear sight, ramp front; skip-line checkered moulded black composition stocks; all-steel construction; gold-colored trigger; matte blue finish. Introduced 1985; still in production. **LMSR: $250**
    Perf.: $200   **Exc.: $175**   **VGood: $150**

Browning Buck Mark Field 5.5

Same specs as Buck Mark except 9 7/8" barrel; matte blue finish only; hoodless ramp-type front sight, low profile rear; contoured or finger-groove walnut stocks. Introduced 1991; still in production. **LMSR: $400**
    Perf.: $325   **Exc.: $275**   **VGood: $200**

Browning Buck Mark Micro

Same specs as Buck Mark except 4" barrel; 16-click Pro Target rear sight; blue or nickel finish. From Browning. Introduced 1992; still in production. **LMSR: $250**
    Perf.: $225   **Exc.: $200**   **VGood: $175**
Nickel finish
    Perf.: $250   **Exc.: $225**   **VGood: $200**

Browning Buck Mark Micro Plus

Same specs as Buck Mark Micro except laminated wood grips. From Browning. Introduced 1992; still in production. **LMSR: $305**
    New: $250   **Perf.: $225**   **Exc.: $175**

Browning Buck Mark Plus

Same specs as Buck Mark except laminated wood stocks. Introduced 1985; still in production. **LMSR: $305**
    Perf.: $250   **Exc.: $200**   **VGood: $175**

Browning Buck Mark Silhouette

Same specs as Buck Mark except 9 7/8" heavy barrel; hooded sights; interchangeable posts; black, laminated wood grip and forend. Introduced 1987; still in production. **LMSR: $422**
    Perf.: $350   **Exc.: $300**   **VGood: $275**

Browning Buck Mark Target 5.5

Same specs as Buck Mark except 5 1/2" barrel with .900" diameter hooded sights; adjustable post front; contoured walnut target grips; matte blue or gold anodized finish; 10-shot magazine. Introduced 1990; still in production. **LMSR: $400**
    Perf.: $325   **Exc.: $300**   **VGood: $250**
Gold anodized
    Perf.: $350   **Exc.: $325**   **VGood: $300**

Browning
Buck Mark Unlimited Match
Buck Mark Varmint
Challenger
Challenger II
Challenger III
Challenger III Sporter
Challenger Gold Model
Challenger Renaissance
Model
Hi-Power
Hi-Power 40 S&W Mark III
Hi-Power Capitan
Hi-Power HP-Practical
Hi-Power Louis XVI
Hi-Power Renaissance
Medalist
Medalist Gold Medal
Medalist International

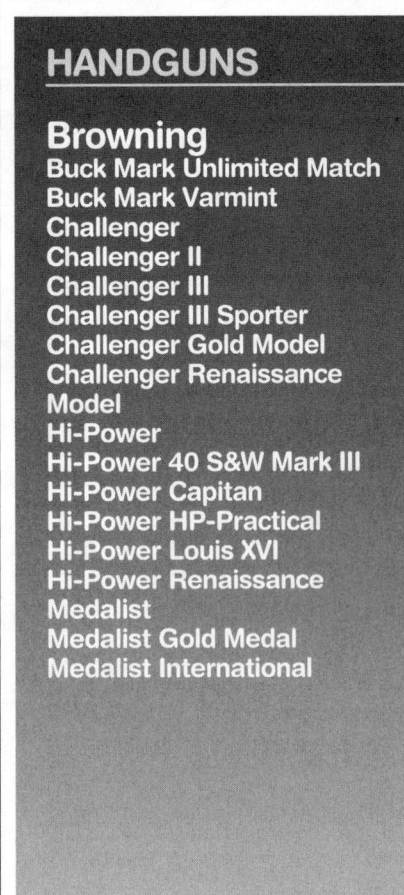
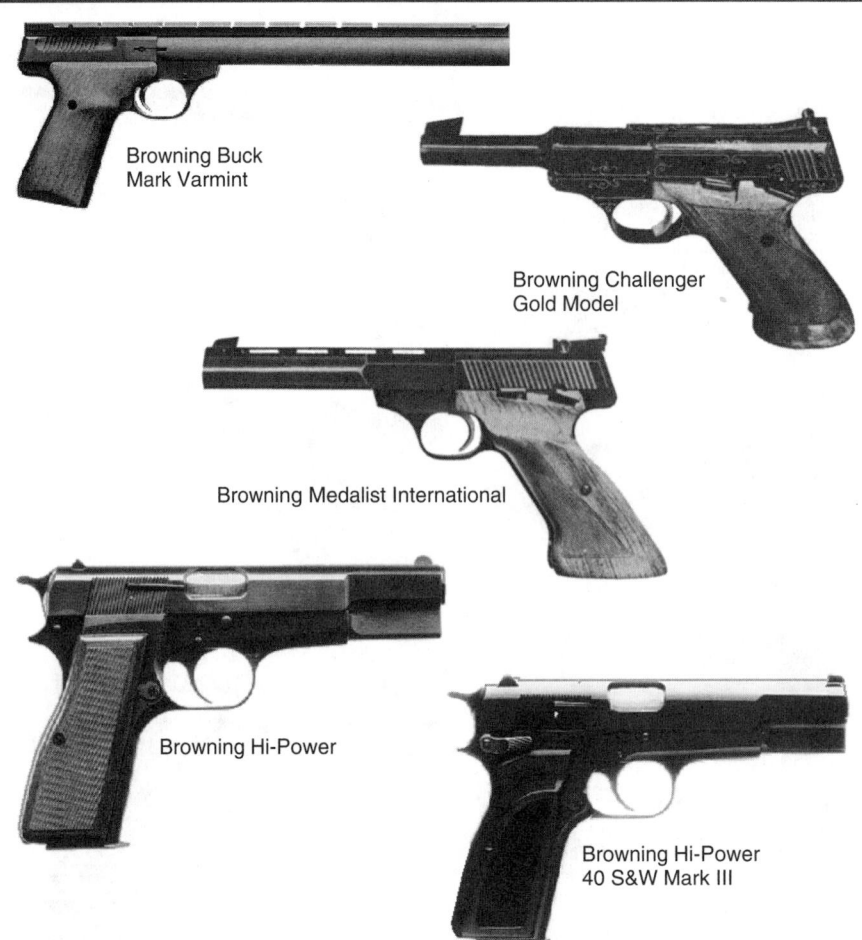

Browning Buck
Mark Varmint

Browning Challenger
Gold Model

Browning Medalist International

Browning Hi-Power

Browning Hi-Power
40 S&W Mark III

Browning Buck Mark Unlimited Match
Same specs as Buck Mark except 14" heavy barrel; 15" sight radius. Introduced 1991; still produced. **LMSR: $520**
   Perf.: $400   **Exc.:** $350   **VGood:** $300

Browning Buck Mark Varmint
Same specs as Buck Mark except 9 7/8" heavy barrel; full-length scope base; no sights; black laminated wood stocks, optional forend; weighs 48 oz. Introduced 1987; still in production. **LMSR: $380**
   Perf.: $300   **Exc.:** $250   **VGood:** $200

BROWNING CHALLENGER
Semi-automatic; 22 LR; 10-shot magazine; 4 1/2", 6 3/4" barrels; overall length 11 7/16" (6 3/4" barrel); screw-adjustable rear sight, removable blade front; hand-checkered walnut grips; gold-plated trigger; blued finish. Introduced 1962; dropped 1974.
   Perf.: $350   **Exc.:** $300   **VGood:** $275

Browning Challenger II
Same specs as Challenger except changed grip angle; impregnated hardwood stocks. Introduced 1975; replaced by Challenger III, 1982.
   Perf.: $225   **Exc.:** $200   **VGood:** $175

Browning Challenger III
Same specs as Challenger except 5 1/2" heavy bull barrel; 9 1/2" overall length; weighs 35 oz.; lightweight alloy frame; improved sights; smooth impregnated hardwood grips; gold-plated trigger. Introduced 1982; dropped 1985.
   Perf.: $200   **Exc.:** $175   **VGood:** $165

Browning Challenger III Sporter
Same specs as Challenger III except 6 3/4" barrel; 10 7/8" overall length; weighs 29 oz.; all-steel construction; blade ramped front sight, screw-adjustable rear, drift-adjustable front. Introduced 1982; dropped 1985.
   Perf.: $200   **Exc.:** $175   **VGood:** $165

Browning Challenger Gold Model
Same specs as Challenger except gold wire inlays in metal; figured hand-carved, checkered walnut grips. Introduced 1971; dropped 1974.
   Perf.: $1200   **Exc.:** $1000   **VGood:** $750

Browning Challenger Renaissance Model
Same specs as Challenger except satin nickel finish; full engraving; top-grade hand-carved, figured walnut grips. Introduced 1971; dropped 1974.
   Perf.: $1200   **Exc.:** $1000   **VGood:** $750

BROWNING HI-POWER
Semi-automatic; 9mm Para.; 13-shot magazine; 4 5/8" barrel; fixed sights; checkered walnut (pre-1986) or moulded grips; ambidextrous safety added in '80s; blued finish. Introduced 1954; still in production. **LMSR: $557**
   Perf.: $500   **Exc.:** $450   **VGood:** $350

Browning Hi-Power 40 S&W Mark III
Same specs as standard Hi-Power except 40 S&W; 10-shot magazine; weighs 35 oz.; 4 3/4" barrel; matte blue finish; low profile front sight blade, drift-adjustable rear; ambidextrous safety; moulded polyamide grips with thumbrest. Imported from Belgium by Browning. Introduced 1993; still imported. **LMSR: $525**
   New: $500   **Perf.:** $450   **Exc.:** $350

Browning Hi-Power Capitan
Same specs as standard Hi-Power except adjustable tangent rear sight authentic to early-production model; Commander-style hammer; checkered walnut grips; polished blue finish. Imported from Belgium by Browning. Reintroduced 1993; still imported. **LMSR: $660**
   New: $550   **Perf.:** $500   **Exc.:** $400

Browning Hi-Power HP-Practical
Same specs as standard Hi-Power except silver-chromed frame, blued slide; wrap-around Pachmayr rubber grips; round-style serrated hammer; removable front sight, windage-adjustable rear. Introduced 1981; still in production. **LMSR: $600**
   Perf.: $600   **Exc.:** $550   **VGood:** $450

Browning Hi-Power Louis XVI
Same specs as standard Hi-Power 9mm except fully engraved silver-gray frame, slide; gold-plated trigger; hand checkered walnut grips; issued in deluxe walnut case. Introduced 1981; no longer in production.
   Perf.: $950   **Exc.:** $750   **VGood:** $550

Browning Hi-Power Renaissance
Same specs as standard Hi-Power except chrome-plated finish; full engraving; polyester pearl grips. Introduced 1954; dropped 1982.
   Perf.: $1200   **Exc.:** $1000   **VGood:** $750

BROWNING MEDALIST
Semi-automatic; 22 LR only; 10-shot magazine; 6 3/4" barrel; 11 1/8" overall length; vent rib; full wrap-around grips of select walnut, checkered with thumbrest; matching walnut forend; left-hand model available; screw-adjustable rear sight, removable blade front; dry-firing mechanism; blued finish. Introduced 1962; dropped 1974.
   Perf.: $750   **Exc.:** $650   **VGood:** $500

Browning Medalist Gold Model
Same general specs as Browning Medalist except gold wire inlays; better wood in grip. Introduced 1963; dropped 1974.
   Perf.: $1600   **Exc.:** $1300   **VGood:** $1000

Browning Medalist International
Same specs as Medalist except sans forend; 5 15/16" barrel; 10 15/16" overall length; meets qualifications for International Shooting Union regulations. Introduced 1964; dropped 1974.
   Perf.: $600   **Exc.:** $550   **VGood:** $500

Browning Nomad

Browning Hi-Power
HP-Practical

Browning
Hi-Power Capitan

Browning Hi-Power
Louis XVI

BROWNING HANDGUN PRODUCTION DATES

Handgun	Introduced	Discontinued	Manufactured
Baby Browning 25 Caliber	1954	1968	Belgium
380 Auto Pistol	1954	1968	Belgium
380 Auto Pistol Adjustable Sights	1971	1974	Belgium
32 Caliber Auto Pistol	1968	1968	Belgium
Browning BDA (by Sig Sauer)	1977	1980	Germany
Nomad 22	1962	1974	Belgium
Challenger 22	1962	1974	Belgium
Medalist 22	1962	1974	Belgium
International Medalist	1977	1980	Belgium
Challenger II	1976	1982	US
Challenger III	1982	1984	US
Challenger III Sporter	1983	1985	
Buckmark	1985	To date	US
Buckmark Varmint	1987	To date	US
Buckmark Plus	1987	To date	US
Buckmark Silhouette	1987	To date	US
Buckmark Target 5.5	1990	To date	US
Buckmark Nickel	1991	To date	US
Buckmark 5.5 Field	1991	To date	US
Buckmark 5.5 Gold Target	1991	To date	US
Buckmark Unlimited Silhouette	1991	To date	US
Micro Buckmark	1992	To date	US
Micro Buckmark Nickel	1992	To date	US
Micro Buckmark Plus	1992	To date	US
Buckmark with Fingergroove	1992	To date	US
Buckmark 5.5 Nickel Target	1994	To date	US
Buckmark 5.5 Nickel Target Fingergroove	1994	To date	US
9mm Hi-Power	1954	To date	Belgium
9mm Renaissance Hi-Power	1954	1979	Belgium
9mm Silver Chrome Hi-Power	1981	1984	Belgium
9mm Silver Chrome Hi-Power	1991	To date	Belgium
9mm Nickel Hi-Power	1980	1984	Belgium
9mm Louis XVI Hi-Power	1980	1984	Belgium
9mm Hi-Power Centennial	1978	Ltd. Issue of 3500	Belgium
9mm Hi-Power Matte Finish	1985	To date	Belgium
9mm Classic	1985	Ltd. Issue of 5000	Belgium
9mm Gold Classic	1985	Ltd. Issue of 500	Belgium
9mm Hi-Power Mark III	1991	To date	Belgium
9mm Capitan	1993	To date	Belgium
9mm HP-Practical with Adjustable Sights	1993	To date	Belgium
40 S&W Hi-Power	1994	To date	Belgium
BDM Double Action 9mm	1991	To date	US
BDA 380	1977	To date	Britain/Italy

Browning Renaissance Set

Browning Medalist Renaissance Model
Same specs as Medalist except finely figured, hand-carved grips; chrome plating; full engraving. Introduced 1964; dropped 1974.
    Perf.: $2000   **Exc.:** $1700   **VGood:** $1200

BROWNING NOMAD
Semi-automatic; 22 LR; 10-shot magazine; 4 1/2″, 6 3/4″ barrel; 8 15/16″ overall length (4 1/2″ barrel); screw-adjustable rear sight, removable blade front; brown plastic grips; blued finish. Introduced 1962; dropped 1973.
    Perf.: $275   **Exc.:** $250   **VGood:** $225

BROWNING RENAISSANCE CASED SET
Includes Renaissance versions of Hi-Power 9mm, 380 Automatic, 25 ACP in a specially fitted walnut case. Oddly, value depends to a degree upon condition of the case. Introduced 1954; dropped 1968.
    Perf.: $3500   **Exc.:** $3000   **VGood:** $2500

HANDGUNS

Browning (FN)
Hi-Power 9mm
Model 1900
Model 1903
Model 1906
Model 1910
Model 1922

Bryco
Model 25
Model 38
Model 48
Model 58
Model 59

Budischowsky
TP-70

Butler
Derringer

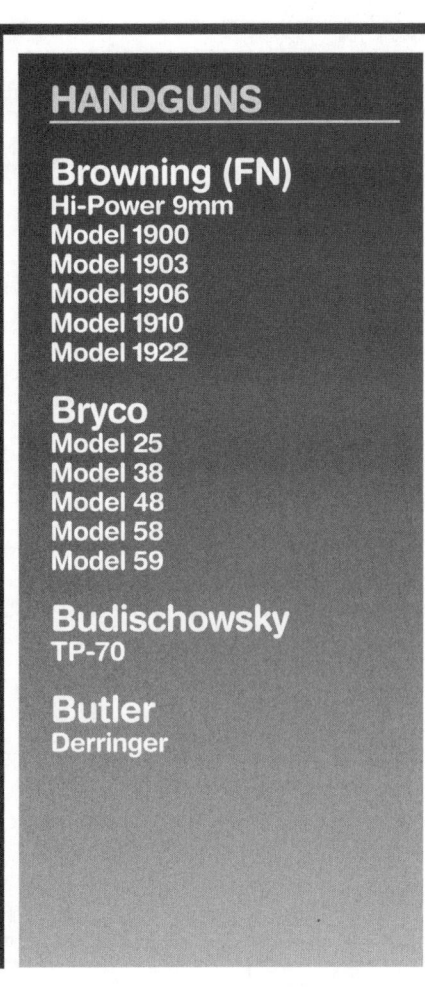

Browning (FN)
Hi-Power 9mm

Browning (FN)
Hi-Power 9mm

Browning (FN)
Model 1903

Browning (FN)
Model 1900

Browning (FN)
Model 1910

Bryco Model 48

BROWNING (FN) HI-POWER 9MM

Semi-automatic; 9mm Para.; 13-shot magazine; 4 5/8″ barrel; 7 3/4″ overall length; based on Browning-Colt 45 ACP; adjustable or fixed rear sight; checkered walnut grips; thumb, magazine safeties; external hammer with half-cock safety feature; blued finish. Pre-WWII-made and Nazi occupation marked models have collector value. Introduced 1935; still in production.
Pre-War Commercial Adjustable Sight
    Exc.: $950   **VGood:** $750   **Good:** $450
Pre-War Military Contact
    Exc.: $1000   **VGood:** $850   **Good:** $700
WWII Production (Fixed sights, Nazi proofed)
    Exc.: $500   **VGood:** $400   **Good:** $300
Post-War Production (Adjustable sights)
    Exc.: $750   **VGood:** $600   **Good:** $500
Post-War Production (Fixed sights)
    Exc.: $500   **VGood:** $400   **Good:** $300

BROWNING (FN) MODEL 1900

Semi-automatic; 32 ACP; 7-shot magazine; 4″ barrel; 6 3/4″ overall length; checkered hard rubber stocks; fixed sights; blued finish. Manufactured in Belgium, 1899 to 1910.
    Exc.: $300   **VGood:** $200   **Good:** $150

BROWNING (FN) MODEL 1903

Semi-automatic; 9mm Browning Long; 7-shot magazine; 5″ barrel; 8″ overall length; hard rubber checkered stocks; fixed sights; blued finish. Designed primarily for military use. Manufactured 1903 to 1939.
    Exc.: $450   **VGood:** $350   **Good:** $250

BROWNING (FN) MODEL 1906

Semi-automatic; 25 ACP; 6-shot magazine; 2″ barrel; 4 1/2″ overall length; weighs 12 1/2 oz.; fixed sights; blued or nickel finish; hard rubber grips. At about production number 100,000, a safety lever was fitted to left rear of frame. Almost identical to Colt Vest Pocket 25 semi-auto. Introduced 1906; dropped 1940.
    Exc.: $275   **VGood:** $225   **Good:** $175

BROWNING (FN) MODEL 1910

Semi-automatic; 32 ACP, 380 ACP; 7-shot magazine (32), 6-shot (380); 3 1/2″ barrel; 6″ overall length; weighs 20 1/2 oz.; hard rubber stocks; fixed sights; blued finish. Made in Belgium. Introduced 1910; still produced, but not imported.
    Exc.: $300   **VGood:** $225   **Good:** $175

BROWNING (FN) MODEL 1922

Semi-automatic; 32 ACP, 380 ACP; 9-shot magazine (32); 8-shot (380); 4 1/2″ barrel; 7″ overall; checkered hard rubber stocks; fixed sights; blued finish. Modified 1910 with longer barrel and slide and deeper grip to fulfill military contracts. Introduced 1922; dropped 1945.
    Exc.: $250   **VGood:** $200   **Good:** $150

BRYCO MODEL 25

Semi-automatic; 25 ACP; 6-shot magazine; 2 1/2″ barrel; 5″ overall length; weighs 11 oz.; resin-impregnated wood stocks; fixed sights; safety locks sear and slide; satin nickel, chrome or black Teflon finish. Introduced 1988; no longer in production.
    Perf.: $50   **Exc.:** $35   **VGood:** $25

BRYCO MODEL 38

Semi-automatic; 22 LR, 32 ACP, 380 ACP; 6-shot magazine; 2 13/16″ barrel; 5 5/16″ overall length; weighs 15 oz.; resin-impregnated wood stocks; fixed sights; safety locks sear and slide; satin nickel, chrome or black Teflon finish. Introduced 1988; still in production. **LMSR: $110**
    Perf.: $85   **Exc.:** $75   **VGood:** $65

BRYCO MODEL 48

Semi-automatic; 22 LR, 32 ACP, 380 ACP; 6-shot magazine; 4″ barrel; 6 11/16″ overall length; weighs 19 oz.; resin-impregnated wood stocks; fixed sights; safety locks sear and slide; satin nickel, chrome or black Teflon finish. Introduced 1988; still in production. **LMSR: $139**
    Perf.: $100   **Exc.:** $85   **VGood:** $65

BRYCO MODEL 58

Semi-automatic; single action; 9mm Para.; 10-shot magazine; 4″ barrel; 5 1/2″ overall length; weighs 30 oz.; black composition grips; blade front sight, fixed rear; striker-fired action; manual thumb safety; polished blue finish. Made in U.S. by Jennings Firearms. Introduced 1994; no longer in production.
    Perf.: $100   **Exc.:** $85   **VGood:** $75

BRYCO MODEL 59

Semi-automatic; single action; 9mm Para.; 10-shot magazine; 4″ barrel; 6 1/2″ overall length; weighs 33 oz.; black composition grips; blade front sight, fixed rear; striker-fired action; manual thumb safety; polished blue finish. Made in U.S. by Jennings Firearms. Introduced 1994; still in production. **LMSR: $169**
    New: $100   **Perf.:** $85   **Exc.:** $65

BUDISCHOWSKY TP-70

Semi-automatic; double action; 22 LR, 25 ACP; 6-shot magazine; 2 7/16″ barrel; 4 11/16″ overall length; fixed sights; all stainless steel construction; manual, magazine safeties. Introduced 1973; no longer in production.
    Exc.: $325   **VGood:** $275   **Good:** $200

BUTLER DERRINGER

Derringer; single shot; 22 Short; 2 1/2″ barrel; blade front sight, no rear; smooth walnut or pearlite grips; spur trigger; blued, gold, chrome finish. Introduced 1978; dropped 1980.
    Exc.: $65   **VGood:** $50   **Good:** $40

Budischowsky
TP-70

Century Model 100

Calico M-110

Charter Arms
Bulldog

Cabanas Single-Shot

Campo-Giro
Model 1913-16

Cabanas
Single-Shot

Calico
Model 100-P
(See Calico Model M-110)
Model M-110
Model M-950

Campo-Giro
Model 1913
Model 1913-16

Century
FEG P9R
FEG P9RK
F.N. Hi-Power
Model 100

Charter Arms
Bonnie & Clyde Set
Bulldog
Bulldog Pug
Bulldog Tracker
Explorer II

CABANAS SINGLE-SHOT

Single shot; 177 pellet or round ball; 9″ barrel; 19″ overall length; weighs 51 oz.; smooth wood stocks, thumbrest; ramped blade front sight, open adjustable rear; automatic safety; muzzle brake. Fires pellet or ball with 22 blank cartridge. Made in Mexico. Introduced 1987; still imported by Mandall Shooting Supplies. **LMSR: $140**
    Exc.: $75   **VGood:** $60   **Good:** $50

CALICO MODEL M-110

Semi-automatic; single action; 22 LR; 100-shot magazine; 6″ barrel; 17 15/16″ overall length; weighs 3 3/4 lbs. (loaded); moulded composition grip; adjustable post front sight, notch rear; aluminum alloy frame; flash suppressor; pistol grip compartment; ambidextrous safety. Uses same helical-feed magazine as M-100 Carbine. Made in U.S. by Calico. Introduced 1986 as Model 100-P; still in production. **LMSR: $359**
    Perf.: $400   **Exc.:** $350   **VGood:** $325

CALICO MODEL M-950

Semi-automatic; single action; 9mm Para.; 50- or 100-shot helical-feed magazine; 7 3/4″ barrel; 14″ overall length with 50-shot magazine; weighs 2 1/4 lbs. (empty); glass-filled polymer grip; fully-adjustable post front sight, fixed notch rear; ambidextrous safety; static cocking handle; retarded blowback action. Made in U.S. by Calico. Introduced 1989; dropped 1994.
    Perf.: $450   **Exc.:** $400   **VGood:** $375

CAMPO-GIRO MODEL 1913

Semi-automatic; 9mm Largo (Bergmann-Bayard); 8-shot magazine; 6 1/2″ barrel; 9 1/4″ overall length; weighs 32 oz.; fixed sights; checkered black horn grips; blued finish. Distinguished by magazine release lever behind triggerguard. Adopted by Spanish Army in 1913, superceded by 1913-16 Model after only 1300 made.
    Exc.: $1650   **VGood:** $1300   **Good:** $900

CAMPO-GIRO MODEL 1913-16

Semi-automatic; 9mm Largo; 8-shot magazine; 6 1/2″ barrel; 9 1/4″ overall length; weighs 35 oz.; fixed sights; checkered black horn or wood grips; blued finish. Magazine release at bottom of left grip. Total production 13,625.
    Exc.: $950   **VGood:** $700   **Good:** $500

CENTURY FEG P9R

Semi-automatic; double action; 9mm Para.; 10-shot magazine; 4 5/8″ barrel; 8″ overall length; weighs 35 oz.; checkered walnut grips; blade front sight, windage drift-adjustable rear; hammer-drop safety; polished blue finish; comes with spare magazine; blue or chrome finish. Imported from Hungary by Century International Arms. Still imported. **LMSR: $263**
    New: $250   **Perf.:** $200   **Exc.:** $150
Chrome finish
    New: $250   **Perf.:** $200   **Exc.:** $150

Century FEG P9RK

Same specs as the P9R except 4 1/8″ barrel; 7 1/2″ overall length; weighs 33 1/2 oz.; checkered walnut grips; fixed sights; 10-shot magazine. Imported from Hungary by Century International Arms, Inc. Introduced 1994; still imported. **LMSR: $290**
    New: $225   **Perf.:** $200   **Exc.:** $150

CENTURY F.N. HI-POWER

Semi-automatic; single action; 9mm Para.; 13-shot magazine; 4 11/16″ barrel; 7 3/4″ overall length; weighs 32 oz.; blade front, adjustable rear; checkered walnut grips; direct copy of Browning Hi-Power. Made in Argentina under license from FN Browning. Introduced 1990; no longer imported by Century International Arms, Inc.
    Perf.: $250   **Exc.:** $200   **VGood:** $175

CENTURY MODEL 100

Revolver; single action; 375 Winchester, 444 Marlin, 45-70; 6 1/2″, 8″, 12″ barrel; 15″ overall length (8″ barrel); weighs 6 lbs.; Millett adjustable square notch rear, ramp front sights; uncheckered walnut stocks; manganese bronze frame; blued cylinder, barrel; coil spring trigger mechanism. Imported by Century Gun Distributors. Introduced 1975; still in production. **LMSR: $2000+**
    Perf.: $2000   **Exc.:** $1650   **VGood:** $1250

CHARTER ARMS BONNIE & CLYDE SET

Revolvers; double action; 32 H&R Mag. (Bonnie) and 38 Spl. (Clyde); 6-shot magazine; 2 1/2″ barrels, fully shrouded; fixed rear sight, serrated ramp front; laminated grips; various colors; marked as "Bonnie" or "Clyde" on barrels; blued finish. Introduced 1989; no longer in production.
    Perf.: $400   **Exc.:** $350   **VGood:** $300

CHARTER ARMS BULLDOG

Revolver; double action; 44 Spl.; 5-shot cylinder; 2 1/2″, 3″ barrel; 7 1/2″ overall length; weighs 19 oz.; square notch fixed rear sight, Patridge-type front; checkered walnut or neoprene grips; chrome-moly steel frame; wide trigger and hammer, or spurless pocket hammer; blued or stainless finish. Introduced 1973.
    Perf.: $200   **Exc.:** $175   **VGood:** $150
Stainless Bulldog
    Perf.: $175   **Exc.:** $165   **VGood:** $135

Charter Arms Bulldog Pug

Same specs as Bulldog except 2 1/2″ shrouded barrel; 7 1/4″ overall length; weighs 19 oz.; notch rear sight, ramp front; wide trigger and hammer spur; shrouded ejector rod; blued or stainless steel finish. Introduced 1986; reintroduced 1993; still in production. **LMSR: $267**
    Perf.: $225   **Exc.:** $200   **VGood:** $165
Stainless Bulldog Pug
    Perf.: $225   **Exc.:** $200   **VGood:** $175

Charter Arms Bulldog Tracker

Same specs as Bulldog except 357 Mag.; 2 1/2″, 4″, or 6″ bull barrel; adjustable rear sight, ramp front; checkered walnut or Bulldog-style neoprene grips; square butt on 4″ and 6″ models; blued finish. Introduced 1986.
    Perf.: $175   **Exc.:** $150   **VGood:** $135

CHARTER ARMS EXPLORER II

Semi-automatic; 22 LR; 8-shot magazine; 6″, 8″ or 10″ barrel; 15 1/2″ overall length; blade front sight, open elevation-adjustable rear; serrated simulated walnut stocks; action adapted from Explorer AR-7 carbine; black, camo, gold or silver finish. Introduced 1980; discontinued 1986.
    Perf.: $100   **Exc.:** $85   **VGood:** $75

Charter Arms
Lady On Duty
Magnum Pug
Model 40
Model 79K
Off-Duty
Pathfinder
Pit Bull
Police Bulldog
Police Bulldog Stainless
Police Undercover
Target Bulldog
Target Bulldog Stainless
Undercover
Undercoverette

Charter Arms
Magnum Pug

Charter Arms
Target Bulldog

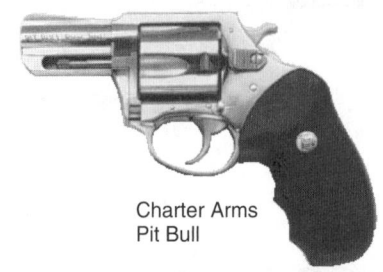

Charter Arms
Pit Bull

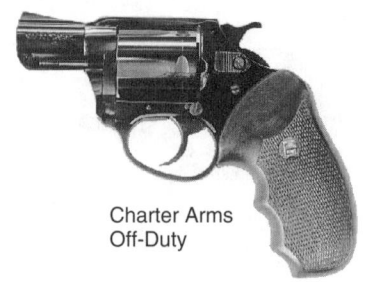

Charter Arms
Off-Duty

Charter Arms
Model 40

Charter Arms
Police Bulldog

CHARTER ARMS LADY ON DUTY
Revolver; double action; 22 LR, 22 WMR, 38 Spl.; 6-shot (22LR, 22 WMR), 5-shot (38 Spl.) magazine; 2″ barrel; 6 1/4″ overall length; weighs 17 oz.; rosewood-color checkered plastic grips; ramp-style front sight, fixed rear; comes with lockable plastic case, trigger lock; choice of spur, pocket or double-action hammer; blue or nickel finish. Made in U.S. by Charco, Inc. Introduced 1995; still in production. **LMSR: $219**
   New: $175   **Perf.:** $165   **Exc.:** $150

CHARTER ARMS MAGNUM PUG
Revolver; double action; 357 Mag.; 5-shot magazine; 2 1/4″ barrel; weighs 18 oz.; Bulldog or neoprene combat grips; ramp-style front sight, fixed rear; fully shrouded barrel; pocket or spur hammer; blue finish. Made in U.S. by Charco, Inc. Introduced 1995; still in production. **LMSR: $273**
   New: $200   **Perf.:** $175   **Exc.:** $165

CHARTER ARMS MODEL 40
Semi-automatic; double action; 22 LR; 8-shot magazine; 3 5/16″ barrel; weighs 2 11/2 oz.; fixed sights; stainless steel finish. Introduced 1984; discontinued 1986.
   Perf.: $250   **Exc.:** $200   **VGood:** $185

CHARTER ARMS MODEL 79K
Semi-automatic; double action; 32 ACP, 380 ACP; 7-shot magazine; 3 5/8″ barrel; 6 1/2″ overall length; weighs 24 1/2 oz.; fixed sights; hammer block; firing pin, magazine safeties; stainless steel finish. Imported from West Germany. Introduced 1984; no longer imported.
   Perf.: $300   **Exc.:** $275   **VGood:** $225

CHARTER ARMS OFF-DUTY
Revolver; double action; 22 LR, 38 Spl.; 5-shot magazine (38 Spl.); 2″ barrel; 6 3/4″ overall length; fixed sights; choice of smooth or checkered walnut, neoprene grips; all steel; matte black finish. Introduced 1984; reintroduced 1993; still in production. **LMSR: $200**
   New: $165   **Perf.:** $150   **Exc.:** $135
Stainless Off-Duty
   New: $225   **Perf.:** $200   **Exc.:** $175

CHARTER ARMS PATHFINDER
Revolver; double action; 22 LR, 22 WMR; 6-shot magazine; 2″, 3″, 6″ barrels; 7 3/8 ″ overall length; weighs 18 1/2 oz.; adjustable rear sight, ramp front; round butt; wide hammer spur and trigger; blued finish. Introduced 1971; no longer in production.
   Perf.: $175   **Exc.:** $165   **VGood:** $150
Square butt Pathfinder
   Perf.: $175   **Exc.:** $165   **VGood:** $150
Stainless Pathfinder with 3 1/2″ shrouded barrel
   Perf.: $175   **Exc.:** $165   **VGood:** $150

CHARTER ARMS PIT BULL
Revolver; double action; 9mm Federal, 38 Spl., 357 Mag.; 5-shot magazine; 2 1/2″, 3 1/2″, 4″ barrels; 7 1/4″ overall length; weighs 24 1/2 oz. (2 1/2″ barrel); blade front sight, fixed rear on 2 1/2″, adjustable on others; checkered neoprene grips; blued finish. Introduced 1989; discontinued 1991.
   Perf.: $225   **Exc.:** $200   **VGood:** $165
Stainless Pit Bull
   Perf.: $225   **Exc.:** $200   **VGood:** $175

CHARTER ARMS POLICE BULLDOG
Revolver; double action; 32 H&R, 38 Spl., 44 Spl.; 5-shot magazine (44) or 6-shot; 3 1/2″, 4″ barrel; fixed sights; neoprene grips; square butt (44 Spl.); blued finish. Introduced 1976.
   Perf.: $200   **Exc.:** $175   **VGood:** $150

Charter Arms Police Bulldog Stainless
Same specs as Police Bulldog except also offered in 357 Mag. (5-shot); shrouded barrel; square butt; stainless finish.
   Perf.: $200   **Exc.:** $175   **VGood:** $150

CHARTER ARMS POLICE UNDERCOVER
Revolver; double action; 32 H&R Mag., 38 Spl.; 6-shot magazine; 2″ barrel, shrouded; Patridge front sight, fixed square-notch rear; checkered walnut grips; pocket or spur hammer; blued finish. Introduced 1984; dropped 1989. Reintroduced 1993. **LMSR: $238**
   Perf.: $200   **Exc.:** $175   **VGood:** $165
Police Undercover Stainless
   Perf.: $225   **Exc.:** $200   **VGood:** $185

CHARTER ARMS TARGET BULLDOG
Revolver; double action; 357 Mag., 44 Spl.; 5-shot magazine; 4″ barrel; 9″ overall length; weighs 21 oz.; adjustable rear sight, blade front; square-butt walnut stocks; all-steel frame; shrouded barrel, ejector rod. Introduced 1986; discontinued 1988.
   Perf.: $225   **Exc.:** $200   **VGood:** $175

Charter Arms Target Bulldog Stainless
Same specs as Target Bulldog except also in 9mm Federal; 5 1/2″ barrel, fully shrouded; adjustable sights; square butt target grips; stainless finish.
   Perf.: $250   **Exc.:** $225   **VGood:** $200

CHARTER ARMS UNDERCOVER
Revolver; double action; 22 LR, 22 WMR, 32 S&W Long, 38 Spl.; 6-shot magazine (32), 5-shot (38 Spl.); 2″ (32), 3″ barrels; 6 3/4″ overall length; weighs 16 oz.; fixed rear sight, serrated ramp front; walnut standard or Bulldog grips; wide spur hammer and trigger, or pocket hammer (38 Spl.); blued. Introduced 1965.
   Perf.: $175   **Exc.:** $160   **VGood:** $135
Stainless with 2″ barrel only
   Perf.: $250   **Exc.:** $225   **VGood:** $175

Charter Arms Undercoverette
Same specs as Undercover except 32 S&W Long; 6-shot magazine; 2″ barrel; blued finish. Introduced 1969; discontinued 1981.
   Perf.: $150   **Exc.:** $135   **VGood:** $125

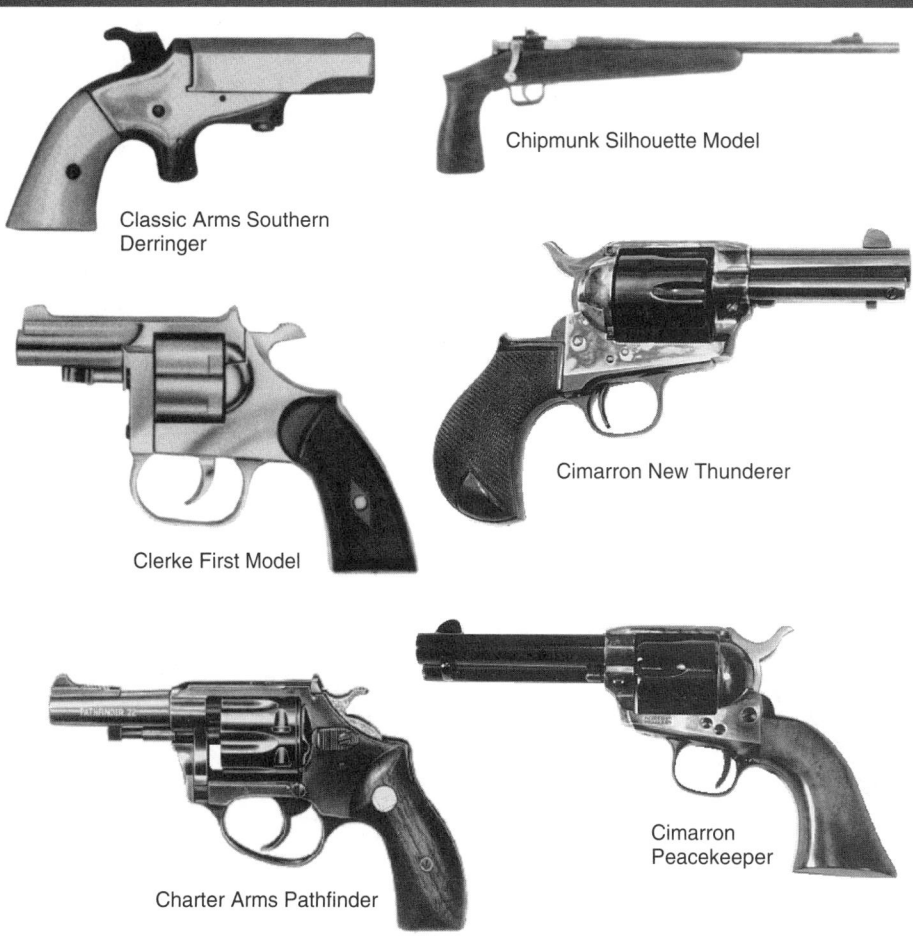

Classic Arms Southern Derringer

Chipmunk Silhouette Model

Cimarron New Thunderer

Clerke First Model

Charter Arms Pathfinder

Cimarron Peacekeeper

Chipmunk
Silhouette Model

Cimarron
1873 Peacemaker
1873 Peacemaker Sheriff's Model
1875 Remington
1890 Remington
New Thunderer
Peacekeeper
U.S. Artillery Model
U.S. Cavalry Model

Claridge Hi-Tec
Model L
Model S, T, ZL, ZT

Classic Arms
Southern Derringer
Twister

Clerke
First Model

CHIPMUNK SILHOUETTE MODEL

Bolt action; single shot; 22 LR; 14 7/8″ barrel; 20″ overall length; weighs 32 oz.; American walnut stock; post ramp front sight, peep rear; meets IHMSA 22 Unlimited competition rules. Introduced 1985; dropped 1990.
Perf.: $125   **Exc.:** $110   **VGood:** $100

CIMARRON 1873 PEACEMAKER

Revolver; single action; 22 LR, 22 WMR, 38 Spl., 357 Mag., 44-40, 45 Colt; 3″, 4″, 4 3/4″, 5 1/2″, 7 1/2″ barrel; 10″ overall length (4″ barrel); blade front, fixed or adjustable rear sight; walnut stocks; reproduction of original Colt design; various grades of engraving; old model blackpowder frame with "bullseye" ejector or New Model frame. Made in Europe. Introduced 1989; still imported by Cimarron Arms. **LMSR: $430**
Perf.: $350   **Exc.:** $300   **VGood:** $250
Engraved
Perf.: $800   **Exc.:** $700   **VGood:** $500

Cimarron 1873 Peacemaker Sheriff's Model

Same specs as 1873 Peacemaker except 44-40, 45 Colt; 3″, 4″ barrels. Introduced 1989; importation dropped 1990.
Perf.: $375   **Exc.:** $325   **VGood:** $250

CIMARRON 1875 REMINGTON

Revolver; single action; 357 Mag., 44-40, 45 Colt; 6-shot cylinder; 7 1/2″ barrel; 13 3/4″ overall length; weighs 44 oz.; blade front, notch rear sight; smooth walnut stocks; replica of 1875 Remington Single Action Army model; brass triggerguard; color case-hardened frame; rest blued or nickel finished. Made in Europe. Introduced 1989; no longer imported.
Perf.: $300   **Exc.:** $250   **VGood:** $225

CIMARRON 1890 REMINGTON

Revolver; single action; 357 Mag., 44-40, 45 Colt; 6-shot cylinder; 5 1/2″ barrel; 12 1/2″ overall length; weighs 37 oz; blade front, groove-type rear sight; American walnut stocks; replica of 1890 Remington Single

Action; brass trigger guard; rest blued or nickel finished; lanyard ring. Made in Europe. Introduced 1989; no longer imported.
Perf.: $300   **Exc.:** $250   **VGood:** $225

CIMARRON NEW THUNDERER

Revolver; single action, 357 Mag., 44 WCF, 44 Spl., 45 Colt; 6-shot; 3 1/2″, 4 3/4″ barrel, with ejector; weighs 38 oz. (3 1/2″ barrel); hand-checkered walnut grips; blade front, notch rear sight; Thunderer grip; color case-hardened frame with balance blued, or nickel finish. Imported by Cimarron Arms. Introduced 1993; still in production. **LMSR: $440**
New: $400   **Perf.:** $350   **Exc.:** $300

CIMARRON PEACEKEEPER

Revolver; single action; 357 Mag., 44 WCF, 44 Spl., 45 Colt; 6-shot; 3 1/2″, 4 3/4″ barrel, with ejector; weighs 38 oz. (3 1/2″ barrel); hand-checkered walnut grips; blade front, notch rear sight; Thunderer grip; color case-hardened frame with balance blued, or nickel finish. Imported by Cimarron Arms. Introduced 1993; no longer imported.
Perf.: $400   **Exc.:** $350   **VGood:** $325

CIMARRON U.S. ARTILLERY MODEL

Revolver; single action; 45 Colt; 5 1/2″ barrel; 11 1/2″ overall length; weighs 39 oz.; fixed sights; walnut stocks; Renaldo A. Carr Commemorative Edition. Made in Europe. Introduced 1989; still imported. **LMSR: $459**
Perf.: $400   **Exc.:** $350   **VGood:** $300

CIMARRON U.S. CAVALRY MODEL

Revolver; single action; 45 Colt; 7 1/2″ barrel; 13 1/2″ overall length; weighs 42 oz.; fixed sights; walnut stocks; original-type inspector and martial markings; color case-hardened frame and hammer; charcoal blued; exact copy of original. Made in Europe. Introduced 1989; still in production. **LMSR: $459**
Perf.: $400   **Exc.:** $350   **VGood:** $300

CLARIDGE HI-TEC MODEL L

Semi-automatic; single action; 9mm Para., 40 S&W, 45 ACP; 18-shot magazine; 7 1/2″ barrel; 15″ overall length; weighs about 3 lbs.; aluminum or stainless frame; telescoping bolt; floating firing pin locked by the safety; moulded composition grip; post elevation-adjustable front sight, open windage-adjustable rear. Made in U.S. by Claridge Hi-Tec, Inc. No longer produced.
Perf.: $500   **Exc.:** $450   **VGood:** $400

CLARIDGE HI-TEC MODEL S, T, ZL, ZT

Semi-automatic; single action; 9mm Para., 40 S&W, 45 ACP; 18-shot magazine; 5″, 9 1/2″, 7 1/2″ barrel; 12 3/4″ overall length (5″ barrel); weighs about 3 lbs.; aluminum or stainless frame; telescoping bolt; floating firing pin locked by the safety; moulded composition grip; post elevation-adjustable front sight, open windage-adjustable rear. Made in U.S. by Claridge Hi-Tec, Inc. No longer produced.
Perf.: $450   **Exc.:** $400   **VGood:** $350

CLASSIC ARMS SOUTHERN DERRINGER

Single shot; 22 LR, 41 rimfire; 2 1/2″ barrel; 5″ overall length; weighs 12 oz.; blade front sight; white plastic stocks; steel barrel; brass frame. Introduced 1982; dropped 1984.
Perf.: $75   **Exc.:** $60   **VGood:** $50

CLASSIC ARMS TWISTER

Derringer; over/under; 22 LR, 9mm rimfire; 3 1/4″ barrel; no sights; pearlite grips; rotating barrels; spur trigger. Originally marketed by Navy Arms. Introduced 1980; dropped 1984.
Perf.: $85   **Exc.:** $75   **VGood:** $60

CLERKE FIRST MODEL

Revolver; double action; 32 S&W, 22 LR, 22 Long, 22 Short; 5-shot swing-out cylinder (32), 6-shot (22); 2 1/4″ barrel; 6 1/4″ overall length; fixed sights; checkered plastic grips; blued, nickel finish. Introduced 1973; dropped 1975.
Perf.: $50   **Exc.:** $35   **VGood:** $25

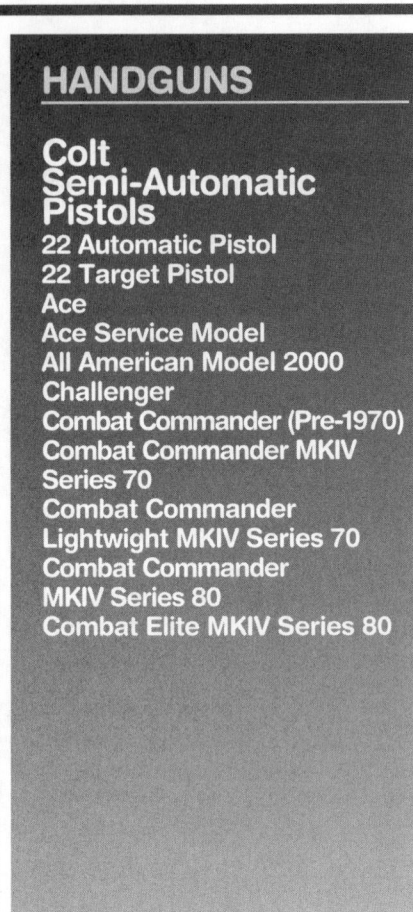

HANDGUNS

Colt Semi-Automatic Pistols

22 Automatic Pistol
22 Target Pistol
Ace
Ace Service Model
All American Model 2000
Challenger
Combat Commander (Pre-1970)
Combat Commander MKIV Series 70
Combat Commander Lightwight MKIV Series 70
Combat Commander MKIV Series 80
Combat Elite MKIV Series 80

Colt 22 Target Pistol

Colt Combat Commander (Pre-1970)

Colt Combat Commander MKIV Series 70

Colt Ace

Colt All American Model 2000

COLT SEMI-AUTOMATIC PISTOLS

COLT 22 AUTOMATIC PISTOL

Semi-automatic; single action; 22 LR; 10-shot magazine; 4 1/2" barrel; weighs 33 oz.; 8 5/8" overall length; textured black polymer grip; blade front sight, rear drift-adjustable for windage; stainless steel construction; ventilated barrel rib; single-action mechanism; cocked striker indicator; push-button safety. Made in U.S. by Colt's Mfg. Co. Introduced 1994; still produced.
  New: $220   **Perf.:** $200   **Exc.:** $180

Colt 22 Target Pistol

Same specs as the Colt 22 except 6" bull barrel; full-length sighting rib with lightening cuts and mounting rail for optical sights; fully-adjustable rear sight; removable sights for optics mounting; two-point factory adjusted trigger travel; stainless steel frame. Made in U.S. by Colt's. Introduced 1995; still produced.
  New: $275   **Perf.:** $225   **Exc.:** $200

COLT ACE

Semi-automatic; 22 LR only, standard or high velocity; 4 3/4" barrel; 8 1/4" overall length; no floating chamber; adjustable rear, fixed front sight; target barrel; hand-honed action. Built on same frame as Government Model 45 automatic. Introduced 1931; dropped 1947.
  Exc.: $2500   **VGood:** $1800   **Good:** $1100

COLT ACE SERVICE MODEL

Semi-automatic; 22 LR, standard or high velocity; 10-shot magazine; identical to Colt National Match. Specially designed chamber increases recoil four-fold to approximate that of 45 auto. Introduced 1937; dropped 1945; reintroduced 1978; dropped 1982. Collector interest affects value.
Old model
  Exc.: $3000   **VGood:** $2250   **Good:** $1500
New model
  Perf.: $600   **Exc.:** $500   **VGood:** $450

COLT ALL AMERICAN MODEL 2000

Semi-automatic; double-action-only; 9mm Para.; 15-shot magazine; 4 1/2" barrel; 7 1/2" overall length; weighs 29 oz. (polymer frame), 33 oz. (aluminum frame); checkered polymer grips; ramped blade front sight, drift-adjustable rear; three-dot system; moulded polymer or aluminum frame; blued steel slide; internal striker block safety. Made in U.S. by Colt's Mfg. Co., Inc. Introduced 1991; no longer produced.
Polymer frame
  Perf.: $650   **Exc.:** $400   **VGood:** $300
Aluminum frame
  Perf.: $700   **Exc.:** $500   **VGood:** $350

COLT CHALLENGER

Semi-automatic; 22 LR; 4 1/2", 6" barrel; 9" overall length (4 1/2" barrel); fixed sights; checkered plastic grips; blued finish. Same basic specs as third-issue Target Woodsman, with fewer features; slide does not stay open after last shot. Introduced 1950; dropped 1955.
  Perf.: $450   **Exc.:** $350   **VGood:** $250

COLT COMBAT COMMANDER (Pre-1970)

Semi-automatic; single action; 45 ACP, 38 Super, 9mm Para.; 7-shot (45 ACP), 9-shot (38 Super, 9mm); 4 1/4" barrel; 7 7/8" overall length; weighs 35 oz.; fixed sights; thumb grip safties; steel or alloy frame; grooved trigger; walnut grips; blue or satin nickel finish. Introduced 1950; dropped 1976.
45 ACP, 38 Super
  Exc.: $800   **VGood:** $600   **Good:** $400
9mm
  Exc.: $600   **VGood:** $450   **Good:** $300

COLT COMBAT COMMANDER MKIV SERIES 70

Semi-automatic; single action; 45 ACP, 38 Super, 9mm Para.; 7-shot (45 ACP), 9-shot (38 Super, 9mm); 4 1/4" barrel; 7 7/8"overall length; weighs 36 oz.; fixed blade front sight; square notch rear; all steel frame; grooved trigger; lanyard-style hammer; checkered walnut grips; blue or nickel finish. Introduced, 1970; dropped 1983.
  Perf.: $750   **Exc.:** $600   **VGood:** $400
Nickel finish
  Perf.: $800   **Exc.:** $650   **VGood:** $450

Colt Combat Commander Lightweight MKIV Series 70

Same specs as Combat Commander MKIV Series 70 except 45 ACP; weighs 27 oz.; aluminum alloy frame; Introduced 1970; dropped 1983.
  Perf.: $700   **Exc.:** $550   **VGood:** $450

COLT COMBAT COMMANDER MKIV SERIES 80

Semi-automatic; 45 ACP, 38 Super Auto, 9mm Para.; 7-shot (45 ACP), 9-shot (38 Super, 9mm); 4 1/4" barrel; 8" overall length; weighs 36 oz.; fixed blade front sight; grooved trigger, hammer spur; arched housing, grip and thumb safeties; steel frame; rubber combat grips; blued finish, except for satin nickel (no longer offered) and stainless steel 45 version. Introduced, 1979; called Commander until 1981; still in production.
  Perf.: $750   **Exc.:** $600   **VGood:** $450
Stainless steel finish
  Perf.: $800   **Exc.:** $650   **VGood:** $500
Satin nickel finish
  Perf.: $850   **Exc.:** $700   **VGood:** $550

COLT COMBAT ELITE MKIV SERIES 80

Semi-automatic; single action; 38 Super, 45 ACP; 7-, 8-shot magazine; 5" barrel; 8 1/2" overall length; weighs 39 oz.; high profile sights with 3-dot system; checkered rubber combat grips; stainless steel frame; ordnance steel slide, internal parts; beveled magazine well; extended grip safety. Introduced, 1986; still in production.
  Perf.: $700   **Exc.:** $600   **VGood:** $500

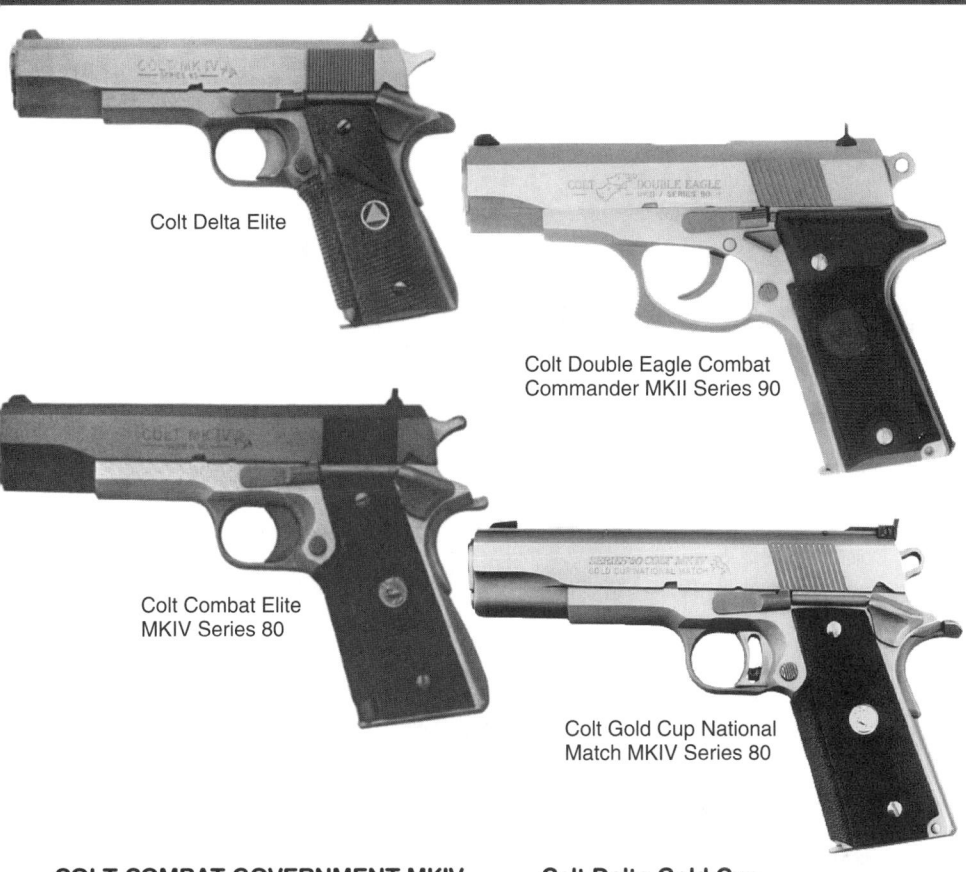

Colt Delta Elite

Colt Double Eagle Combat
Commander MKII Series 90

Colt Combat Elite
MKIV Series 80

Colt Gold Cup National
Match MKIV Series 80

Colt Semi-Automatic Pistols

Combat Government MKIV Series 70
Combat Government MKIV Series 80
Commander Lightweight MKIV Series 80
Delta Elite
Delta Gold Cup
Double Eagle Combat Commander MKII Series 90
Double Eagle MKII Series 90
Double Eagle Officer's Model MKIV Series 90
Gold Cup National Match MKIII
Gold Cup National Match MKIV Series 70
Gold Cup National Match MKIV Series 80

COLT COMBAT GOVERNMENT MKIV SERIES 70

Semi-automatic; single action; 45 ACP; 8-shot magazine; 5″ barrel; 8 ³/₈″ overall length; weighs 40 oz. Similar to Government Model except higher undercut front sight, white outlined rear; flat mainspring housing; longer trigger; beveled magazine well; angled ejection port; Colt/Pachmayr wrap-around grips; internal firing pin safety. Introduced 1973; no longer in production.

Perf.: $650   **Exc.:** $550   **VGood:** $450

COLT COMBAT GOVERNMENT MKIV SERIES 80

Semi-automatic; single action; 45 ACP; 5-shot magazine; 5″ barrel; 8 ¹/₂″ overall length; weighs 39 oz. Similar to Combat Government MKIV Series 70 with higher undercut front sight, white outlined rear; flat mainspring housing; longer trigger beveled magazine well; angled ejection port; Colt/Pachmayr wrap-around grips; internal firing pin safety; 45 with blue, satin or stainless finish; 9mm, 38 blue only. Introduced 1983; no longer in production.

Perf.: $600   **Exc.:** $500   **VGood:** $400

COLT COMMANDER LIGHTWEIGHT MKIV SERIES 80

Semi-automatic; 45 ACP, 38 Super, 9mm Para.; 4 ¹/₄″ barrel, 8″ overall length; 7-shot magazine (45 ACP), 9-shot (38 Super, 9mm); weighs 26 ¹/₂ oz.; basic design of Government Model auto, but of lightweight alloy, reducing weight; checkered walnut grips; fixed sights; rounded hammer spur; blued, nickel finish. Introduced 1983; still in production.

Perf.: $600   **Exc.:** $500   **VGood:** $400

COLT DELTA ELITE

Semi-automatic; 10mm Auto; 8-shot; 5″ barrel; 8 ¹/₂″ overall length; weighs 38 oz.; same general design as Government Model; 3-dot high profile combat sights; rubber combat stocks; internal firing pin safety; new recoil spring/buffer system; blued or matte stainless finish. Introduced 1987; still in production.

Perf.: $700   **Exc.:** $550   **VGood:** $450
Matte stainless
Perf.: $750   **Exc.:** $600   **VGood:** $500

Colt Delta Gold Cup

Same space as Delta Elite except Accro adjustable rear sight; adjustable trigger; wrap-around grips; stainless steel finish. Introduced 1989; still in production.

Perf.: $850   **Exc.:** $700   **VGood:** $600

COLT DOUBLE EAGLE COMBAT COMMANDER MKII SERIES 90

Semi-automatic; double action; 40 S&W, 45 ACP; 8-shot magazine; 4 ¹/₄″ barrel; 7 ³/₄″ overall length; weighs 36 oz.; blade front sight, windage-adjustable rear; 3-dot system; Colt Accro adjustable sight optional; stainless steel; checkered, curved extended triggerguard; wide steel trigger; decocking lever; traditional magazine release; beveled magazine well; grooved frontstrap; extended grip guard; combat-type hammer. Introduced 1991; still in production in 45 ACP.

Perf.: $600   **Exc.:** $500   **VGood:** $400
Accro sight
Perf.: $650   **Exc.:** $550   **VGood:** $450

COLT DOUBLE EAGLE MKII SERIES 90

Semi-automatic; double action; 45 ACP; 8-shot magazine; 4 ¹/₂″, 5″ barrel; 8 ¹/₂″ overall length; weighs 39 oz.; blade front sight, windage-adjustable rear; black checkered thermoplastic grips; stainless steel construction; matte finish; extended triggerguard; decocking lever; grooved frontstrap; beveled magazine well; rounded hammer; extended grip guard. Introduced 1989; still in production.

Perf.: $600   **Exc.:** $500   **VGood:** $400

COLT DOUBLE EAGLE OFFICER'S MODEL MKIV SERIES 90

Semi-automatic; double action; 45 ACP; 8-shot magazine; 3 ¹/₂″ barrel; 7 ¹/₄″ overall length; weighs 35 oz.; stainless steel construction; blade front, windage-adjustable rear sight with 3-dot system; black checkered thermoplastic grips; matte finish; extended trigger guard; decocking lever; grooved frontstrap; beveled magazine well; rounded hammer; extended grip guard. Introduced 1989; still in production.

Perf.: $600   **Exc.:** $500   **VGood:** $400
Lightweight model (25 oz.)
Perf.: $600   **Exc.:** $500   **VGood:** $400

COLT GOLD CUP NATIONAL MATCH MKIII

Semi-automatic; 45 ACP, 38 Spl.; 5″ match barrel; 8 ¹/₂″ overall length; weighs 37 oz.; Patridge front sight, Colt Elliason adjustable rear; arched or flat housing; wide, grooved trigger with adjustable stop; ribbed top slide; hand-fitted with improved ejection port. Introduced 1959; dropped 1970.

45 ACP
Exc.: $1000   **VGood:** $800   **Good:** $650
38 Spl.
Exc.: $1100   **VGood:** $850   **Good:** $700

COLT GOLD CUP NATIONAL MATCH MKIV SERIES 70

Semi-automatic; 45 ACP; 7-shot magazine; 5″ barrel; 8 ³/₈″ overall length; weighs 38 ¹/₂ oz; undercut front sight, Colt Elliason adjustable rear; match-grade barrel, bushing; long, wide grooved trigger; flat grip housing; hand-fitted slide; checkered walnut grips; blued finish. Introduced 1970; dropped 1983.

Exc.: $850   **VGood:** $750   **Good:** $600

COLT GOLD CUP NATIONAL MATCH MKIV SERIES 80

Semi-automatic; single action; 45 ACP; 8-shot magazine; 5″ barrel; 8 ¹/₂″ overall length; weighs 39 oz.; Patridge front sight, Colt-Elliason adjustable rear; match-grade barrel, bushing; long, wide trigger adjustable trigger stop; flat mainspring housing; hand-fitted slide; wider ejection port; checkered walnut grips; blue, stainless, bright stainless finish. Introduced 1983; still in production.

Perf.: $700   **Exc.:** $600   **VGood:** $500
Stainless finish
Perf.: $750   **Exc.:** $650   **VGood:** $550
Bright stainless
Perf.: $800   **Exc.:** $700   **VGood:** $600

HANDGUNS

Colt Semi-Automatic Pistols

Government Model MKIV
Series 70
Government Model MKIV
Series 80
Government Model 380 MKIV
Series 80
Huntsman, Targetsman
Junior
Model 1900
Model 1900 Army
Model 1900 USN
Model 1902 Military
Model 1902 Sporting
Model 1903 Pocket Hammer
Model 1903 Pocket
Hammerless (First Issue)

Colt Model 1905

Colt Huntsman

Colt Model 1902 Sporting

Colt Model 1900

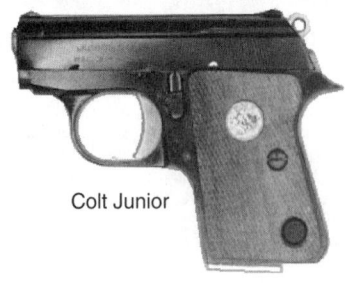

Colt Junior

COLT GOVERNMENT MODEL MKIV SERIES 70

Semi-automatic; single action; 45 ACP, 38 Super, 9mm Para.; 7-shot magazine; 5″ barrel; 8 3/8″ overall length; weighs 40 oz.; ramp front sight, fixed square notch rear; grip, thumb safeties; grooved trigger; accurizor barrel, bushing; blue or nickel (45 only) finish. Redesigned, redesignated Colt Model 1911A1. Introduced 1970; dropped 1983.

Perf.: $500   **Exc.:** $450   **VGood:** $400
Nickel finish
Perf.: $550   **Exc.:** $500   **VGood:** $400

COLT GOVERNMENT MODEL MKIV SERIES 80

Semi-automatic; single action; 45 ACP, 38 Super, 9mm Para.; 7-, 8-shot magazine; 5″ barrel; 8 1/2″ overall length; weighs 38 oz.; ramp front sight, fixed square notch rear; grip and thumb safeties; internal firing pin safety; grooved trigger; accurizor barrel bushing; checkered walnut grips; blue, nickel, satin nickel, bright stainless, stainless steel finish. Introduced 1983; still in production.

Blue, nickel, satin nickel finish
Perf.: $600   **Exc.:** $500   **VGood:** $425
Stainless steel, bright stainless finish
Perf.: $650   **Exc.:** $550   **VGood:** $475

COLT GOVERNMENT MODEL 380 MKIV SERIES 80

Semi-automatic; 380 ACP; 7-shot magazine; 3″ barrel; 6″ overall length; weighs 21 3/4 oz.; checkered composition stocks; ramp front sight, fixed square-notch rear; thumb and internal firing pin safeties; blue, nickel, satin nickel or stainless steel finish. Introduced 1983; still in production.

Perf.: $375   **Exc.:** $325   **VGood:** $275
Nickel, satin nickel finish
Perf.: $400   **Exc.:** $325   **VGood:** $275
Stainless steel finish
Perf.: $425   **Exc.:** $375   **VGood:** $325

COLT HUNTSMAN, TARGETSMAN

Semi-automatic; 22 LR; 4 1/2″, 6″ barrel; 9″ overall length; checkered plastic grips; no hold-open device; fixed sights (Huntsman), adjustable sights and 6″ barrel only (Targetsman). Introduced 1955; dropped 1977.
Huntsman
Exc.: $350   **VGood:** $275   **Good:** $225
Targetsman
Exc.: $450   **VGood:** $350   **Good:** $275

COLT JUNIOR

Semi-automatic; 22 Short, 25 ACP; 6-shot magazine; 2 1/4″ barrel; 4 3/8″ overall length; exposed hammer with round spur; checkered walnut stocks; fixed sights; blued. Initially produced in Spain by Astra, with early versions having Spanish markings as well as Colt identity; parts were assembled in U.S., sans Spanish identification after GCA '68 import ban. Introduced 1968; dropped 1973.
Spanish model
Exc.: $300   **VGood:** $250   **Good:** $200
U.S.-made model
Exc.: $325   **VGood:** $275   **Good:** $200

COLT MODEL 1900

Semi-automatic; 38 ACP; 7-shot magazine; 6″ barrel; 9″ overall length; spur hammer; plain walnut stocks; blued finish. Dangerous to fire modern high-velocity ammo. Was made in several variations. Introduced 1900; dropped 1903. Collector value.
Standard model
Exc.: $5000   **VGood:** $3500   **Good:** $2500
Altered sight safety
Exc.: $2200   **VGood:** $1400   **Good:** $1000

Colt Model 1900 Army

Same specs as Model 1900. Marked *U.S.* on left side of triggerguard bow, with inspector markings.
Exc.: $7500   **VGood:** $5500   **Good:** $4500

Colt Model 1900 USN

Same specs as Model 1900. Marked *USN* and Navy serial number on left side of frame, Colt serial number on right.
Exc.: $8500   **VGood:** $6500   **Good:** $4750

COLT MODEL 1902 MILITARY

Semi-automatic; 38 ACP; 8-shot magazine; 9″ overall length; 6″ barrel; fixed blade front, notched V rear sight; checkered hard rubber stocks; round-back hammer (changed to spur type in 1908); blued finish. Dangerous to fire modern high-velocity loads. Early production with serrations on front area of slide. U.S. Army marked pistols between serial numbers 15,001 and 15,200 have inspector stampings, slide serrations at front. Introduced 1902; dropped 1929.
Early production
Exc.: $2500   **VGood:** $1500   **Good:** $1200
Standard production
Exc.: $2000   **VGood:** $1250   **Good:** $1000
U.S. Army-marked serial numbers
Exc.: $5500   **VGood:** $4500   **Good:** $3500

COLT MODEL 1902 SPORTING

Semi-automatic; 38 ACP; 8-shot magazine; 6″ barrel; 9″ overall length; sans safety; checkered hard rubber grips; blade front sight, fixed notch V rear; round hammer. Not safe with modern loads. Manufactured 1902 to 1908.
Exc.: $1750   **VGood:** $1400   **Good:** $950

COLT MODEL 1903 POCKET HAMMER

Semi-automatic; 38 ACP; 4 1/2″ barrel; 7 1/2″ overall length; round back hammer changed to spur type in 1908; checkered hard rubber grips; blued finish. Not safe with modern ammo. Manufactured 1903 to 1929.
Exc.: $1450   **VGood:** $1000   **Good:** $650

COLT MODEL 1903 POCKET HAMMERLESS (First Issue)

Semi-automatic; 32 ACP; 8-shot magazine; 4″ barrel; 7″ overall length; fixed sights; internal hammer; grip safety; hard rubber stocks; blued or nickel finish; barrel takedown bushing. Manufactured from 1903 until 1908.
Exc.: $500   **VGood:** $400   **Good:** $300

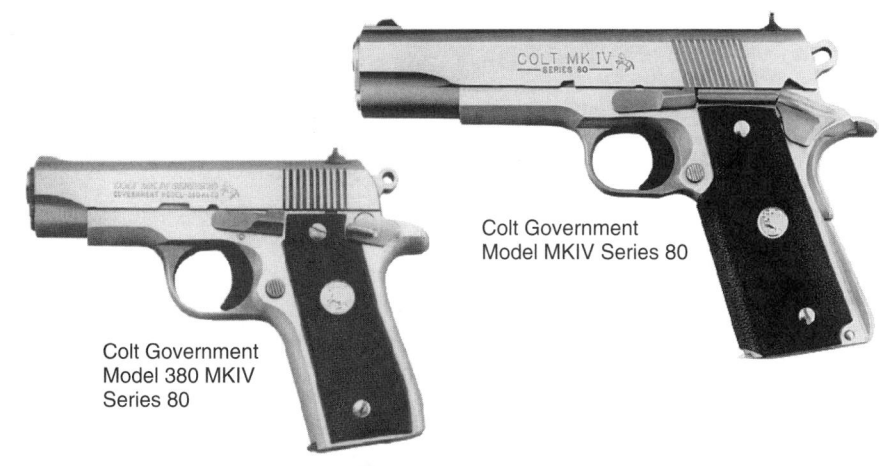

Colt Government
Model MKIV Series 80

Colt Government
Model 380 MKIV
Series 80

Colt Model 1902 Military

HANDGUNS

Colt Semi-Automatic Pistols

Model 1903 Pocket
Hammerless (Second Issue)
Model 1903 Pocket
Hammerless (Third Issue)
Model 1903 Pocket
Hammerless (Fourth Issue)
Model 1905
Model 1908 Pocket
Hammerless (First Issue)
Model 1908 Pocket
Hammerless (Second Issue)
Model 1908 Pocket
Hammerless (Third Issue)
Model 1911 Commercial
Model 1911 British 455
Model 1911 Military
Model 1911 Russian

Colt Model 1903 Pocket Hammerless (Second Issue)

Same specs as First Issue except 3 3/4" barrel. Made from 1908 until 1910.

Exc.: $475   **VGood:** $350   **Good:** $275

Colt Model 1903 Pocket Hammerless (Third Issue)

Same specs as First Issue except no barrel bushing. Introduced 1910; dropped 1926.

Exc.: $400   **VGood:** $325   **Good:** $200

Colt Model 1903 Pocket Hammerless (Fourth Issue)

Same specs as First Model except magazine safety. On all guns above serial number 468,097, safety disconnector prevents firing cartridge in chamber if magazine is removed. Manufactured from 1926 to 1945.

Exc.: $375   **VGood:** $300   **Good:** $200

U.S. Property marked

Exc.: $1000   **VGood:** $850   **Good:** $750

COLT MODEL 1905

Semi-automatic; single action; 45 ACP; 7-shot magazine; 5" barrel; snub-type and spur hammers; checkered, varnished walnut stocks; case-hardened hammer; blued finish; forerunner of the Model 1911 45 Auto; produced in standard, military test (with loaded chamber indicator, separate serial range 1-200) and shoulder-stocked models. Introduced 1905; dropped 1911.

Early production (s/n below 700)

Exc.: $3750   **VGood:** $2500   **Good:** $1800

Standard production

Exc.: $3500   **VGood:** $2250   **Good:** $1300

Military test model

Exc.: $8500   **VGood:** $6500   **Good:** $4500

With shoulder stock

Exc.: $12,500   **VGood:** $10,000   **Good:** $8500

COLT MODEL 1908 POCKET HAMMERLESS (First Issue)

Semi-automatic; 380 ACP; 7-shot magazine; 3 3/4" barrel; 7" overall length; fixed sights; hammerless; slidelock; grip safety; hard rubber stocks; blued finish. Same general design, specs as Model 1903 Pocket Hammerless. Manufactured from 1908 to 1911.

Exc.: $550   **VGood:** $450   **Good:** $350

Colt Model 1908 Pocket Hammerless (Second Issue)

Same specs as First Issue sans barrel bushing. Manufactured from 1911 to 1926.

Exc.: $450   **VGood:** $350   **Good:** $300

Colt Model 1908 Pocket Hammerless (Third Issue)

Same specs as Second Issue except safety disconnector installed on guns with serial numbers above 92,894. Introduced 1926; dropped 1945.

Exc.: $450   **VGood:** $350   **Good:** $300

U.S. Property marked

Exc.: $1500   **VGood:** $1250   **Good:** $1000

COLT MODEL 1911 COMMERCIAL

Semi-automatic; also known as Government Model; 45 ACP; 7-shot magazine; 5" barrel; weighs 39 oz.; checkered walnut grips; fixed sights; blued finish; commercial variations with letter "C" preceding serial number. Introduced 1912; dropped 1923 to be replaced by Model 1911A1. Collector value.

Early production (through s/n 4500)

Exc.: $3500   **VGood:** $2750   **Good:** $1500

Standard model

Exc.: $1200   **VGood:** $750   **Good:** $600

Colt Model 1911 British 455

Same specs as standard model but carries "Calibre 455" on right side of slide and broad arrow British ordnance punch; made 1915-1916; 11,000 manufactured. Collector value.

Exc.: $2000   **VGood:** $1600   **Good:** $1250

Colt Model 1911 Military

Same specs as the civilian version except serial numbers not preceded by the letter C; produced from 1912 to 1924; bright blue finish, early production; duller blue during war. Parkerized finish indicates post-war reworking and commands lesser value than original blue finish. Navy issue marked "Model of 1911 U.S. Navy" on slide.

Standard

Exc.: $2000   **VGood:** $1400   **Good:** $850

U.S. Navy marked

Exc.: $3750   **VGood:** $2500   **Good:** $1800

Remington-UMC (21,676)

Exc.: $2200   **VGood:** $1700   **Good:** $950

Springfield Armory (25,767)

Exc.: $2500   **VGood:** $2000   **Good:** $950

Colt Model 1911 Russian

Same specs as the standard version except Russian characters ANGLOZAKAZIVAT stamped on the slide. All were in 45 ACP, carrying serial numbers C-50000 through C-85000. This is an exceedingly rare collector item.

Exc.: $3000   **VGood:** $2250   **Good:** $1650

Colt Model 1911 Military
(Springfield Armory)

Colt Model 1903
Pocket Hammerless
(Early Model)

Colt Model 1903 Pocket
Hammerless (Fourth Issue)
U.S. Property marked

Colt Semi-Automatic Pistols
Model 1911A1
Model 1991A1
Model 1991A1 Commander
Model 1991A1 Compact
Mustang
Mustang Plus II
Mustang Pocket Lite
National Match
Officer's ACP MKIV Series 80

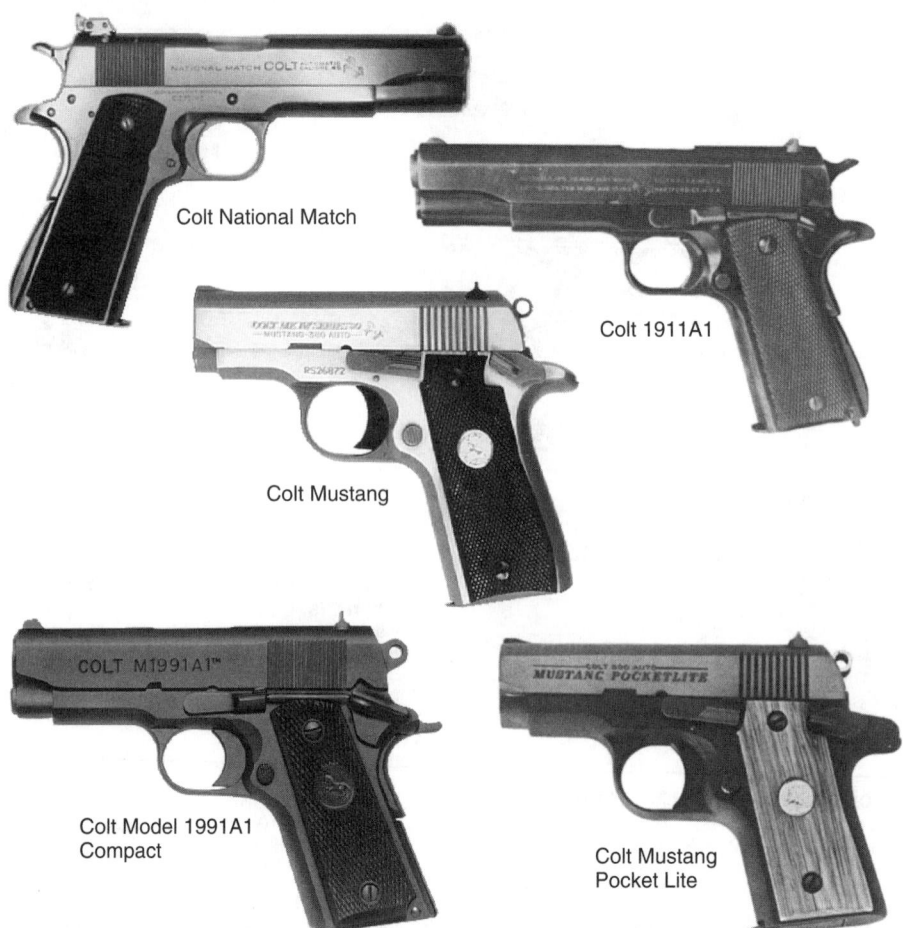

Colt National Match

Colt 1911A1

Colt Mustang

Colt Model 1991A1 Compact

Colt Mustang Pocket Lite

COLT MODEL 1911A1

Semi-automatic; single action; 45 ACP, 7-shot magazine; 5″ barrel; 8 1/2″ overall length; weighs 38 oz.; checkered walnut (early production) or brown composition (military version) grips; ramped blade front sight, fixed high-profile square notch rear; Parkerized finish. Same specs as Model 1911 except longer grip safety spur; arched mainspring housing; finger relief cuts in frame behind trigger; plastic grips. During WWII other firms produced the 1911A1 under Colt license, including Remington-Rand, Ithaca Gun Co., and Union Switch & Signal Co. These government models bear imprint of licensee on slide. In 1970, this model was redesigned and redesignated as Government Model MKIV Series 70; approximately 850 1911A1 guns were equipped with split-collet barrel bushing and BB prefix serial number which adds to the value. Modern version (marked "Colt M1911A1™") continues serial number range used on original G.I. 1911A1 guns and comes with one magazine and moulded carrying case. Introduced 1923; still produced.

Commercial model ("C" s/n)
Exc.: $850	**VGood:** $700	**Good:** $500

BB s/n prefix
Exc.: $950	**VGood:** $800	**Good:** $600

Military model
Exc.: $950	**VGood:** $650	**Good:** $500

Singer Mfg. Co. (500)
Exc.: $15,000	**VGood:** $12,000	**Good:** $8500

Union Switch & Signal (55,000)
Exc.: $1250	**VGood:** $850	**Good:** $700

Remington-Rand (1,000,000)
Exc.: $700	**VGood:** $600	**Good:** $500

Ithaca Gun Co. (370,000)
Exc.: $700	**VGood:** $600	**Good:** $500

45/22 conversion unit
Exc.: $650	**VGood:** $550	**Good:** $400

22/45 conversion unit
Exc.: $1500	**VGood:** $1000	**Good:** $800

COLT MODEL 1991A1

Semi-automatic; 45 ACP; 7-shot magazine; 5″ barrel; 8 1/2″ overall length; weighs 38 oz.; ramped blade front sight, fixed square notch high-profile rear; checkered black composition grips; Parkerized finish; continuation of serial number range used on original G.I. 1911A1 guns. Introduced 1991; still in production. **LMSR: $538**
New: $500	**Perf.:** $450	**Exc.:** $400

Colt Model 1991A1 Commander
Same specs as the Model 1991A1 except 4 1/4″ barrel; Parkerized finish. Comes in moulded case. Made in U.S. by Colt's Mfg. Co. Introduced 1993; still in production. **LMSR: $538**
New: $500	**Perf.:** $450	**Exc.:** $400

Colt Model 1991A1 Compact
Same specs as the Model 1991A1 except 3 1/2″ barrel; 7″ overall length; 3/8″ shorter height. Comes with one 6-shot magazine, moulded case. Made in U.S. by Colt's Mfg. Co. Introduced 1993; still produced. **LMSR: $538**
New: $500	**Perf.:** $450	**EXC.:** $400

COLT MUSTANG

Semi-automatic; 380 ACP; 5-, 6-shot magazine; 2 3/4″ barrel; 5 1/2″ overall length; weighs 18 1/2 oz.; steel frame; stainless, blued or nickel finish. Similar to Colt Government Model 380. Introduced 1987; still in production. **LMSR: $462**
Perf.: $350	**Exc.:** $300	**VGood:** $250

Nickel finish
Perf.: $375	**Exc.:** $325	**VGood:** $275

Stainless finish
Perf.: $375	**Exc.:** $325	**VGood:** $275

Colt Mustang Plus II
Same specs as Mustang except 7-shot magazine; weighs 20 oz.; composition grips; blued or stainless finish. Introduced 1988; still in production.
Perf.: $350	**Exc.:** $300	**VGood:** $250

Colt Mustang Pocket Lite
Same specs as Mustang except aluminum alloy frame; 2 3/4″ barrel; weighs 12 1/2 oz; blue or stainless finish. Introduced 1987; still in production.
Perf.: $350	**Exc.:** $300	**VGood:** $250

Stainless finish
Perf.: $375	**Exc.:** $325	**VGood:** $275

COLT NATIONAL MATCH

Semi-automatic; 45 ACP; 7-shot; 5″ barrel; 8 1/2″ overall length; weighs 37 oz.; adjustable rear, ramp front target sight; match-grade barrel; hand-honed action. Also available with fixed sights. Introduced 1933; dropped 1941.

Fixed sights
Exc.: $2750	**VGood:** $2000	**Good:** $1600

Target sights
Exc.: $3250	**VGood:** $2500	**Good:** $1850

COLT OFFICER'S ACP MKIV SERIES 80

Semi-automatic; single action; 45 ACP; 6-shot magazine; 3 1/2″ barrel; 7 1/4″ overall length; weighs 34 oz. (steel frame), 24 oz. (alloy frame); checkered walnut or rubber composite grips; blade front sight, square notch rear; 3-dot sight system; grooved trigger; flat mainspring housing; blued, matte, stainless or bright stainless finishes. Introduced 1985; still in production.
Perf.: $550	**Exc.:** $475	**VGood:** $350

Matte finish
Perf.: $500	**Exc.:** $425	**VGood:** $350

Stainless steel finish
Perf.: $600	**Exc.:** $525	**VGood:** $450

Bright stainless finish
Perf.: $650	**Exc.:** $575	**VGood:** $500

Alloy frame
Perf.: $550	**Exc.:** $450	**VGood:** $400

Colt Vest Pocket
Model 1908
Hammerless

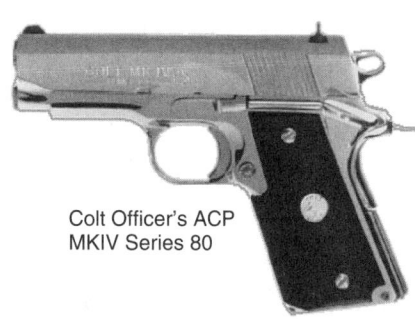

Colt Officer's ACP
MKIV Series 80

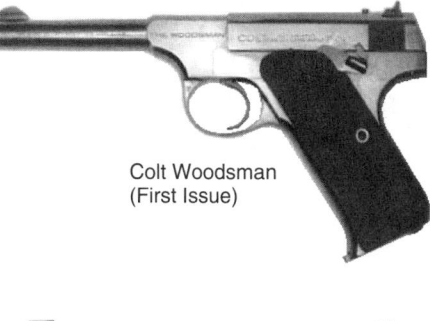

Colt Woodsman
(First Issue)

Colt Woodsman
(Second Issue)

Colt Woodsman
(Third Issue)
Match Target 6"

Colt Semi-Automatic Pistols

Super 38
Super 38 Match
Targetsman
(See Colt Huntsman, Targetsman)
Vest Pocket Model 1908 Hammerless
Woodsman (First Issue)
Woodsman (First Issue) Sport Model
Woodsman (First Issue) Match Target
Woodsman (Second Issue)
Woodsman (Second Issue) Sports Model
Woodsman (Second Issue) Match Target 4¹/₂"
Woodsman (Second Issue) Match Target 6"
Woodsman (Third Issue)
Woodsman (Third Issue) Sport Model
Woodsman (Third Issue) Match Target 4¹/₂"
Woodsman (Third Issue) Match Target 6"

COLT SUPER 38

Semi-automatic; double action; 38 Colt Super; 9-shot magazine; fixed sights standard, adjustable rear sights available. Same frame as 1911 commercial until 1970. Introduced in 1929. In 1937, the design of the firing pin, safety, hammer, etc. were changed and the model renamed the New Style Super 38. Still in production as Government Model Mark IV/Series 80.

   Exc.: $750   **VGood:** $600   **Good:** $500
Pre-WWI
   Exc.: $2500   **VGood:** $2000   **Good:** $1500

Colt Super 38 Match

Same specs as Colt Super 38 except hand-honed action; adjustable target sights; match-grade barrel. Manufactured from 1935 to 1941.

   Exc.: $4500   **VGood:** $3500   **Good:** $2500

COLT VEST POCKET MODEL 1908 HAMMERLESS

Semi-automatic; 25 ACP; 2" barrel; 4 ¹/₂" overall length; 6-shot magazine; hammerless; hard rubber or checkered walnut grips; fixed sights milled in top of slide; incorporates straight-line striker, rather than pivoting hammer, firing pin; slide-locking safety; grip safety magazine disconnector added in 1917 at serial number 141,000; blued, with case-hardened safety lever, grip safety, trigger, or nickel finished. Introduced 1908; dropped 1941.

   Exc.: $450   **VGood:** $375   **Good:** $300
Nickel finish
   Exc.: $500   **VGood:** $400   **Good:** $325
U.S. Property marked
   Exc.: $950   **VGood:** $750   **Good:** $600

COLT WOODSMAN (FIRST ISSUE)

Semi-automatic; 22 LR standard velocity; 10-shot magazine; 6 ¹/₂" barrel; 10 ¹/₂" overall length. Designation, "The Woodsman," added after serial number 34,000; adjustable sights; checkered walnut grips. Introduced 1915; replaced 1943. Collector value.

   Exc.: $350   **VGood:** $275   **Good:** $200

Colt Woodsman (First Issue) Sport Model

Same specs as First Issue Woodsman except adjustable rear sight; adjustable or fixed front; 4 ¹/₂" barrel; 8 ¹/₂" overall length; fires standard- or high-velocity 22 LR ammo. Introduced 1933; dropped 1943.

   Exc.: $300   **VGood:** $250   **Good:** $200

Colt Woodsman (First Issue) Match Target

Same specs as Woodsman First Issue except 6 ¹/₂" flat-sided barrel; 11" overall length; adjustable rear sight, blade front; checkered walnut one-piece extension grips; blued. Introduced 1938; dropped 1943.

   Exc.: $1800   **VGood:** $1250   **Good:** $800
U.S. Property marked
   Exc.: $2000   **VGood:** $1500   **Good:** $1200

COLT WOODSMAN (SECOND ISSUE)

Semi-automatic; 22 LR; 10-shot magazine; 6 ¹/₂" barrel; 10 ¹/₂" overall length; slide stop; hold-open device; heat-treated mainspring housing for use with high-velocity cartridges; heavier barrel; push-button magazine release on top side of frame. Introduced in 1932; dropped 1948.

   Exc.: $500   **VGood:** $350   **Good:** $275

Colt Woodsman (Second Issue) Sport Model

Same specs as the Second Issue Woodsman except 4 ¹/₂" barrel; 9" overall length; plastic grips. Introduced 1947; dropped 1955.

   Exc.: $550   **VGood:** $375   **Good:** $275

Colt Woodsman (Second Issue) Match Target 4 ¹/₂"

Same specs as Second Issue Woodsman except 4 ¹/₂" heavy barrel. Introduced 1947; dropped 1955.

   Exc.: $850   **VGood:** $650   **Good:** $550

Colt Woodsman (Second Issue) Match Target 6"

Same specs as Second Issue Woodsman except flat-sided 6" heavy barrel; 10 ¹/₂" overall length; 22 LR, standard- or high-velocity; checkered walnut or plastic grips; click-adjustable rear sight, ramp front; blued. Introduced 1947; dropped 1955.

   Exc.: $750   **VGood:** $600   **Good:** $500

COLT WOODSMAN (THIRD ISSUE)

Semi-automatic; 22 LR; 10-shot; 6 ¹/₂" barrel; longer grip; larger thumb safety; slide stop magazine disconnector; thumbrest; plastic or walnut grips; magazine catch on bottom of grip; click-adjustable rear sight, ramp front. Introduced 1955; dropped 1977.

   Exc.: $375   **VGood:** $325   **Good:** $250

Colt Woodsman (Third Issue) Sport Model

Same specs as Third Issue Woodsman except 4 ¹/₂" barrel. Introduced 1955; dropped 1977.

   Exc.: $400   **VGood:** $350   **Good:** $250

Colt Woodsman (Third Issue) Match Target 4 ¹/₂"

Same specs as Third Issue Woodsman except 4 ¹/₂" heavy barrel. Introduced 1955; dropped 1977.

   Perf.: $600   **Exc.:** $550   **VGood:** $500

Colt Woodsman (Third Issue) Match Target 6"

Same specs as Third Issue Woodsman except 6" heavy barrel. Introduced 1955; dropped 1977.

   Perf.: $500   **Exc.:** $450   **VGood:** $400

Colt 3rd Model Derringer

Colt Agent (Second Issue)

Colt 38 SF-VI

Colt Agent (First Issue)

Colt 357 Magnum

Colt Anaconda

COLT REVOLVERS & SINGLE SHOTS

COLT 3RD MODEL DERRINGER

Derringer; single shot; 41 rimfire; 2 1/2" pivoting barrel; varnished walnut grips; bronze frame, nickel- or silver-plated; blued or plated barrels. Introduced 1875; dropped 1912. (Note: Since 1959, Colt has produced this model intermittently in 22 caliber; it should not be confused with the 41 version.)

  Exc.: $1200   **VGood.:** $850   **Good.:** $725
S/N under 2000
  Exc: $2500   **VGood.:** $2000   **Good.:** $1200

COLT 38 SF-VI

Revolver; double action; 38 Spl.; 6-shot cylinder; 2" barrel; 7" overall length; weighs 21 oz.; checkered black composition grips; ramp front sight, fixed rear; new (for Colt) lockwork; made of stainless steel. Made in U.S. by Colt's Mfg. Introduced 1995; still in production. **LMSR: $408**

  New: $350   **Perf.:** $325   **Exc.:** $300

COLT 357 MAGNUM

Revolver; double action; 357 Mag.; 6-shot swing-out cylinder; 4", 6" barrel; 9 1/4" overall length (4" barrel); available as service revolver or in target version; latter with wide hammer spur, target grips; checkered walnut grips; Accro rear sight, ramp front; blued finish. Introduced 1953; dropped 1961.

  Perf.: $475   **Exc.:** $425   **VGood.:** $375
Target Model
  Perf.: $500   **Exc.:** $450   **VGood.:** $375

COLT AGENT (FIRST ISSUE)

Revolver; double action; 38 Spl.; 6-shot swing-out cylinder; 2" barrel; 6 3/4" overall length; weighs 14 oz.; minor variation of Colt Cobra with shorter stub grip for maximum concealment; Colt alloy frame, sideplate; steel cylinder, barrel; no housing around ejector rod; square butt; blued finish. Introduced 1955; dropped 1972.

  Perf.: $300   **Exc.:** $250   **VGood.:** $200

Colt Agent (Second Issue)

Same specs as Agent First Issue except 6 5/8" overall length; weighs 16 oz; ramp front sight, square notch rear; checkered walnut grips; grip design extends just below bottom of frame. Introduced 1973; dropped 1981.

  Perf.: $250   **Exc.:** $200   **VGood:** $175

Colt Agent (Third Issue)

Same specs as Agent First Issue except shrouded ejector rod; alloy frame; Parkerized-type finish. Introduced 1982; no longer in production.

  Perf.: $225   **Exc.:** $175   **VGood:** $150

COLT ANACONDA

Revolver; double action; 44 Rem. Mag., 45 LC; 6-shot cylinder; 4", 6", 8" barrel; 11 5/8" overall length; weighs 53 oz.; red insert front sight, white-outline adjustable rear; finger-grooved black neoprene combat grips; stainless steel construction; ventilated barrel rib; offset bolt notches in cylinder; full-length ejector rod housing; wide spur hammer. Introduced 1990; still produced. **LMSR: $612**

  New: $500   **Perf.:** $425   **Exc.:** $375

COLT ARMY SPECIAL

Revolver; double action; 32-20, 38 Spl., 41 Colt; 6-shot cylinder; 4", 4 1/2", 5", 6" barrels; 9 1/4" overall length (4" barrel); 41-caliber frame; hard rubber grips; fixed sights; blued or nickel finish. Not safe for modern 38 Special high-velocity loads in guns chambered for that caliber. Introduced 1908; dropped 1927.

  Exc.: $525   **VGood.:** $375   **Good.:** $250

COLT BANKER'S SPECIAL

Revolver; double action; 22 LR (with countersunk chambers after 1932), 38 New Police, 38 S&W; 6-shot swing-out cylinder; 2" barrel; 6 1/2" overall length; essentially the same as pre-1972 Detective Special, but with shorter cylinder of Police Positive rather than that of Police Positive Special; a few produced with Fitzgerald cutaway triggerguard; checkered hammer spur, trigger; blued or nickel finish. Low production run on 22 gives it collector value. Introduced 1926; dropped 1940.

22 blue finish
  Exc.: $1750   **VGood.:** $1250   **Good.:** $800
22 nickel finish
  Exc.: $2250   **VGood.:** $1600   **Good.:** $1000
38 blue finish
  Exc.: $850   **VGood.:** $650   **Good.:** $400
38 nickel finish
  Exc.: $1200   **VGood.:** $800   **Good.:** $500

COLT BISLEY

Revolver; single action; 455 Eley, 45, 44-40, 41, 38-40, 32-20; with trigger, hammer and grips redesigned for target shooting. Target model features target sights, flat-top frame. Bisley version of Colt SAA revolver first offered in flat-top target version in 1896 for use in target matches held at Bisley Common in England. Revolver's good performance encouraged Colt to offer it both in target and standard sighted versions until 1912. Bisley version featured a longer, more angled grip frame which placed the hand at almost a 90-degree angle to the bore. Parts, except for triggerguard, backstrap, grips, mainspring, hammer, and trigger, are totally interchangeable with standard Colt SAA. Standard calibers, barrel lengths, and finishes same as 1st Generation Colt SAA. Manufactured from 1894 to 1913; sought after by collectors.

Standard model
  Exc.: $7000   **VGood.:** $4500   **Good.:** $2500
Target model
  Exc.: $12,000   **VGood.:** $7500   **Good.:** $4000

HANDGUNS

Colt Revolvers & Single Shots

3rd Model Derringer
38 SF-VI
357 Magnum
Agent (First Issue)
Agent (Second Issue)
Agent (Third Issue)
Anaconda
Army Special
Banker's Special
Bisley

Colt Bisley

Colt Buntline Special
(2nd Generation)

Colt Army Special

Colt Cobra (First Issue)

Colt Camp Perry Model
(Second Issue)

Colt Cobra
(Second Issue)

Colt Revolvers & Single Shots

Buntline Special
Camp Perry Model
(First Issue)
Camp Perry Model
(Second Issue)
Cobra (First Issue)
Cobra (Second Issue)
Commando
Commando Special
Detective Special (First Issue)
Detective Special
(Second Issue)

COLT BUNTLINE SPECIAL

Revolver; single action; 45 Colt; 12″ barrel; case-hardened frame; hard rubber or walnut grips; designed after guns made as presentation pieces by author Ned Buntline. Introduced 1957; dropped 1992.
2nd Generation (1957-1975)
    Perf.: $1500   **Exc.:** $1000   **VGood:** $850
3rd Generation (1976-1992)
    Perf.: $1250   **Exc.:** $850   **VGood:** $700

COLT CAMP PERRY MODEL (FIRST ISSUE)

Revolver; single shot; 22 LR, with countersunk chamber for high-velocity ammo after 1930; 10″ barrel; 13 3/4″ overall length; built on frame of Colt Officers' Model; checkered walnut grips; hand-finished action; adjustable target sights; trigger, backstrap, hammer spur checkered; blued finish, with top, back of frame stippled to reduce glare. Chamber, barrel are single unit, pivoting to side for loading and extraction. Introduced 1926; dropped 1934.
    Exc.: $1500   **VGood:** $1200   **Good:** $750

Colt Camp Perry Model (Second Issue)

Same specs as First Issue except 8″ barrel; 12″ overall length; shorter hammer fall. Only 440 produced. Collector value. Introduced 1934; dropped 1941.
    Perf.: $3000   **Exc.:** $2000   **VGood:** $1500

COLT COBRA (FIRST ISSUE)

Revolver; double action; 22 LR, 22 New Police, 38 New Police, 38 Spl.; 6-shot swing-out cylinder; 2″, 3″, 4″, 5″ barrels; 3″, 4″, 5″ barrel styles special order; based on pre-1972 Detective Special, except frame, sideplate of high-tensile aluminum alloy; Coltwood plastic grips on later guns, checkered wood on early issues; square butt, early issue, replaced by round butt; optional hammer shroud; blue finish, matted on top, rear of frame; shrouded ejector rod (later issue). Introduced 1951; dropped 1972.
    Perf.: $350   **Exc.:** $300   **VGood:** $225
Nickel finish
    Perf.: $390   **Exc.:** $350   **VGood:** $275
22 LR
    Perf.: $425   **Exc.:** $350   **VGood:** $300

Colt Cobra (Second Issue)

Same specs as Cobra First Issue except integral protective shroud enclosing ejector rod. Introduced 1973; dropped 1981.
    Perf.: $400   **Exc.:** $300   **VGood:** $250

COLT COMMANDO

Revolver; double action; 38 Spl.; 6-shot cylinder; 2″, 4″, 6″ barrels; plastic grips; sand-blasted blue finish. Government contract issue made to military specs; a downgraded version of the Colt Official Police. Introduced 1942; dropped 1945.
    Exc.: $450   **VGood:** $325   **Good:** $250

COLT COMMANDO SPECIAL

Revolver; 38 Spl.; 6-shot cylinder; 2″ barrel; 6 5/8″ overall length; weighs 22 oz.; steel frame; combat grade finish; rubber grips. Identical to Detective Special. Introduced 1984; dropped 1987.
    Perf.: $475   **Exc.:** $375   **VGood:** $325

COLT DETECTIVE SPECIAL (FIRST ISSUE)

Revolver; double action; 38 Spl.; 6-shot swing-out cylinder; 2″, 3″ barrels; 6 3/4″ overall length; rounded butt introduced in 1934; blue or nickel finish. Introduced 1927; dropped 1936.
    Exc.: $575   **VGood:** $475   **Good:** $350
Nickel finish
    Exc.: $600   **VGood:** $500   **Good:** $375

Colt Detective Special (Second Issue)

Same specs as First Issue except 32 New Police, 38 New Police; heavier barrel; integral protective shroud enclosing ejector rod; frame, sideplate, cylinder, barrel, internal parts of high-tensile alloy steel; plastic or walnut Bulldog-type grips; fixed rear sight; notch milled in topstrap, serrated ramp front; checkered hammer spur; blue or nickel finish. Introduced 1947; dropped 1972.
    Perf.: $450   **Exc.:** $350   **VGood:** $275
Nickel finish
    Perf.: $475   **Exc.:** $375   **VGood:** $300

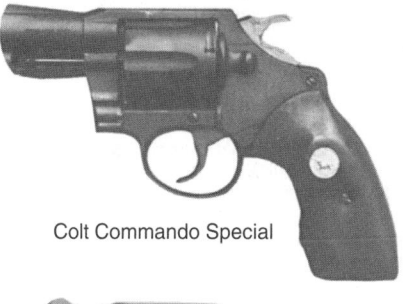

Colt Commando Special

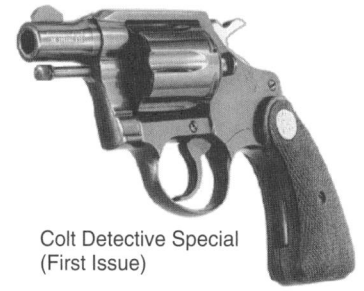

Colt Detective Special
(First Issue)

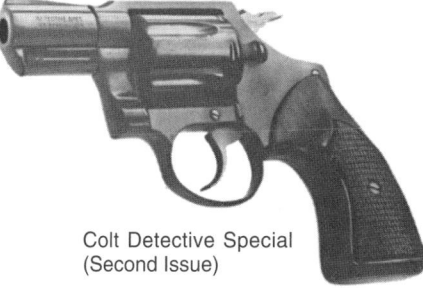

Colt Detective Special
(Second Issue)

Colt Revolvers & Single Shots

Detective Special (Third Issue)
Detective Special (Fourth Issue)
Diamondback
Frontier Scout
Frontier Scout Buntline
King Cobra
Lawman MKIII
Lawman MKV
Metropolitan MKIII
Model 1877 Lightning
Model 1877 Thunderer
Model 1878 Frontier

Colt Detective Special (Third Issue)

Colt Frontier Scout

Colt Detective Special (Fourth Issue)

Colt Lawman MKIII

Colt Diamondback

Colt 1877 Lightning

Colt Detective Special (Third Issue)

Same specs as Second Issue except 38 Spl.; shrouded ejector rod; wrap-around walnut grips. Introduced 1973; dropped 1986.

Perf.: $400   **Exc.:** $325   **VGood:** $275

Colt Detective Special (Fourth Issue)

Same specs as Third Issue except 2″ barrel only; weighs 22 oz.; 6 5/8″ overall length; black composition grips; ramp front sight, fixed square notch rear; glare-proof sights; grooved trigger; Colt blue finish. Made in U.S. by Colt's Mfg. Reintroduced 1993; still produced.

New: $350   **Perf.:** $290   **Exc.:** $250

COLT DIAMONDBACK

Revolver; double action; 38 Spl., 22 LR; 6-shot swing-out cylinder; 2 1/2″, 4″ (22 LR) barrel; vent-rib barrel; scaled-down version of Python; target-type adjustable rear sight, ramp front; full checkered walnut grips; integral rounded rib beneath barrel shrouds, ejector rod; broad checkered hammer spur. Introduced 1966; dropped 1986.

Perf.: $450   **Exc.:** $350   **VGood:** $275

Nickel finish

Perf.: $500   **Exc.:** $400   **VGood:** $325

22 LR

Perf.: $525   **Exc.:** $425   **VGood:** $350

COLT FRONTIER SCOUT

Revolver; single action; 22 LR, 22 Long, 22 Short; interchangeable cylinder for 22 WMR; 4 1/2″ barrel; 9 15/16″ overall length; originally introduced with alloy frame; steel frame, blue finish introduced in 1959; fixed sights; plastic or wooden grips. Introduced 1958; dropped 1971.

Alloy frame

Perf.: $425   **Exc.:** $375   **VGood:** $225

Blue finish, plastic grips

Perf.: $600   **Exc.:** $400   **VGood:** $300

Nickel finish, wood grips

Perf.: $450   **Exc.:** $375   **VGood:** $250

With 22 WMR cylinder

Perf.: $450   **Exc.:** $400   **VGood:** $275

Colt Frontier Scout Buntline

Same specs as Frontier Scout except 9 1/2″ barrel. Introduced 1959; dropped 1971.

Perf.: $500   **Exc.:** $425   **VGood:** $325

COLT KING COBRA

Revolver; double action; 357 Mag.; 6-shot cylinder; 2 1/2″, 4″, 6″ barrels; 9″ overall length (4″ barrel); weighs 42 oz.; adjustable white outline rear sight, red insert ramp front; full-length contoured ejector rod housing, barrel rib; matte finish; stainless steel construction. Introduced 1986; still in production.

Perf.: $350   **Exc.:** $250   **VGood:** $225

COLT LAWMAN MKIII

Revolver; double action; 357 Mag.; also chambers 38 Spl.; 2″, 4″ barrels; 9 3/8″ overall length (4″ barrel); choice of square, round butt walnut grips; fixed sights; blued, nickel finish. Introduced 1969; dropped 1983.

Perf.: $300   **Exc.:** $225   **VGood:** $175

Nickel finish

Perf.: $325   **Exc.:** $250   **VGood:** $200

Colt Lawman MKV

Same specs as Lawman MKIII except redesigned lockwork to reduce double-action trigger pull; faster lock time; redesigned grips; shrouded ejector rod (2″ barrel); fixed sights; solid rib. Introduced 1984; dropped 1985.

Perf.: $325   **Exc.:** $250   **VGood:** $200

COLT METROPOLITAN MKIII

Revolver; double action; 38 Spl.; 6-shot swing-out cylinder; 4″ barrel; designed for urban law enforcement; fixed sights; choice of service, target grips of checkered walnut; blued finish, standard; nickel, optional. Introduced 1969; dropped 1972.

Perf.: $350   **Exc.:** $300   **VGood:** $200

COLT MODEL 1877 LIGHTNING

Revolver; double action; 32 Colt; 6-shot cylinder; 4 1/2″ - 10″ barrels with ejector; 1 1/2″, 2 1/2″, 3 1/2″, 4 1/2″, 6″ sans ejector; fixed sights; blued or nickel finish; hard rubber grips. Manufactured 1877 to 1909.

Exc.: $1500   **VGood:** $850   **Good:** $600

COLT MODEL 1877 THUNDERER

Revolver; double action; 41 Colt; 6-shot cylinder; 4 1/2″-10″ barrels with ejector; 1 1/2″, 2 1/2″, 3 1/2″, 4 1/2″, 6″ sans ejector; fixed sights; blued or nickel finish; hard rubber grips. Manufactured 1877 to 1909.

Exc.: $1500   **VGood:** $750   **Good:** $550

COLT MODEL 1878 FRONTIER

Revolver; double action; 32-20, 38-40, 44-40, 45 Colt, 450, 455, 476 Eley; 6-shot cylinder; 4 3/4″, 5 1/2″, 7 1/2″ barrels with ejector; 3 1/2″, 4″ sans ejector; similar to Lightning model but with heavier frame; lanyard ring in butt; hard rubber grips; fixed sights; blued or nickel finish. Manufactured 1878 to 1905.

Exc.: $3500   **VGood:** $2500   **Good:** $1750

HANDGUNS
Colt Revolvers & Single Shots
Model 1905 Marine Corps
Model 1917 Army
Navy Model 1889
(First Issue)
Navy Model 1892-1903
(Second Issue)
New Army
(See Colt Navy Model
1892-1903 (Second Issue))
New Frontier
New Frontier Buntline Special
New Frontier 22
New Pocket Model
New Police

Colt Model 1905 Marine Corps

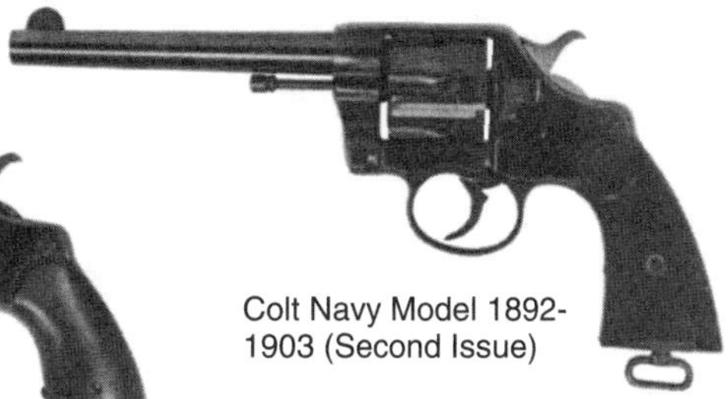

Colt Navy Model 1892-1903 (Second Issue)

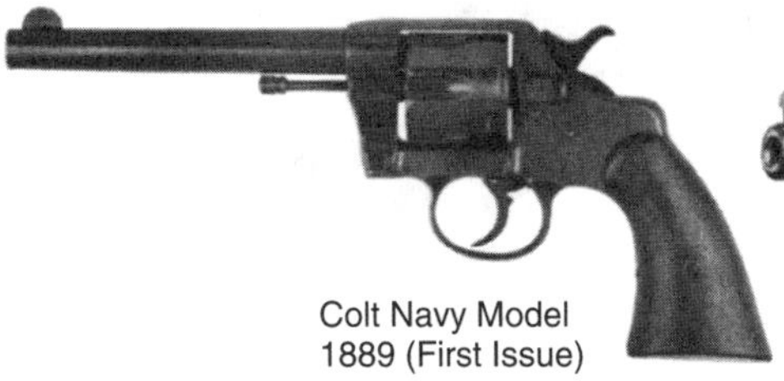

Colt Navy Model 1889 (First Issue)

Colt New Frontier 22

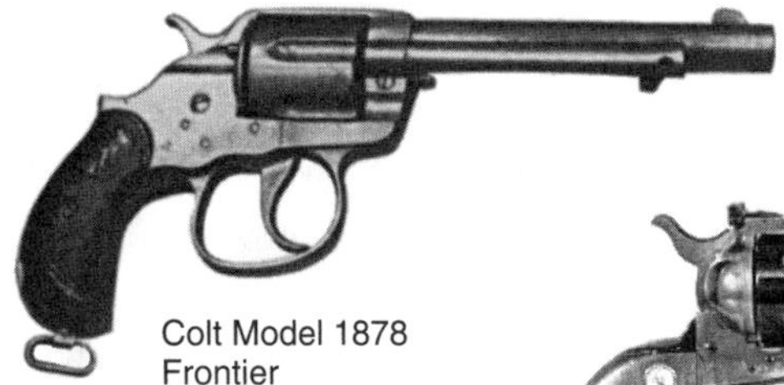

Colt Model 1878 Frontier

Colt New Frontier

Colt Buntline Special

COLT MODEL 1905 MARINE CORPS
Revolver; double action; 38 Spl., 38 Short, 38 Long; 6″ barrel; round butt; double cylinder notches; shorter flutes; double locking butt. Manufactured 1905 to 1909.
Exc.: $2500 **VGood:** $1500 **Good:** $1000

COLT MODEL 1917 ARMY
Revolver; double action; 6-shot swing-out cylinder; 45 Auto Rim cartridge can be fired in conventional manner; based upon New Service revolver to fire 45 ACP cartridge with steel half-moon clips; 10 13/16″ overall length; 5 1/2″ barrel; smooth walnut grips; fixed sights; dull finish. Should be checked for damage by corrosive primers before purchase. Introduced 1917; dropped 1925.
Exc.: $850 **VGood:** $550 **Good:** $400

COLT NAVY MODEL 1889 (FIRST ISSUE)
Revolver; double action; 38 Short Colt, 38 Long Colt, 41 Short Colt, 41 Long Colt; 6-shot left-revolving cylinder; 3″, 4 1/2″, 6″ barrels; fixed blade front sight, V-notched rear; Colt's first solid-frame, swing-out cylinder model; hard rubber or walnut grips; blued or nickel finish. Manufactured 1889 to 1894. In demand by collectors.
Exc.: $2000 **VGood:** $1000 **Good:** $800

Colt Navy Model 1892-1903 (Second Issue)
Same specs as First Issue except double cylinder notches, double locking bolt; lanyard loop on butt; added calibers; 38 Spl., 32-20; does not fire modern high-velocity loads safely. Also known as New Army, adopted by both Army and Navy. Manufactured 1892 to 1905.
Exc.: $1250 **VGood:** $1000 **Good:** $600

COLT NEW FRONTIER
Revolver; flat-top target version of Colt Single Action Army first introduced in 1962 during 2nd Generation production and continued until 1974. Serial numbers began at 3000NF and ended in 1974 with 7288NF. Calibers offered during 2nd Generation were 45 Colt, 44 Spl., 357 Mag., and 38 Spl. Standard barrel lengths 4 3/4″, 5 1/2″, 7 1/2″; a total of seventy-two 45 Colt Buntline New Frontiers manufactured with 12″ barrels. Finish was Colt's royal blue with color case-hardened frame. Two-piece walnut grips were standard. Manufacture of New Frontiers resumed in 1978 during 3rd Generation production and continued until 1985. Serial numbers began again at 01001NF and ran to approximately 17000NF by 1985. During 3rd Generation production, calibers offered were 45 Colt, 44 Spl., and 357 Mag., with 44-40 being added in 1981. Finishes were Colt's royal blue with color case-hardened frame, with fully nickel-plated, and fully-blued rarely found options. Barrel lengths were the standard 4 3/4″, 5 1/2″, and 7 1/2″.
2nd Generation
Perf.: $1000 **Exc.:** $750 **VGood:** $600
3rd Generation
Perf.: $850 **Exc.:** $650 **VGood:** $500

COLT New Frontier Buntline Special
Same specs as New Frontier except 45 Colt; 12″ barrel; flat-top frame; adjustable rear sight.
2nd Generation
Perf.: $2500 **Exc.:** $1850 **VGood:** $1250
3rd Generation
Perf.: $2000 **Exc.:** $1500 **VGood:** $1000

COLT NEW FRONTIER 22
Revolver; single action; 6-shot cylinder; furnished with dual cylinders for 22 LR, 22 WMR; 4 3/8″, 6″, 7 1/2″ barrels; 11 1/2″ overall length (6″ barrel); scaled-down version of New Frontier 45; target-type fully-adjustable rear sight, ramp front; checkered black plastic grips; flat topstrap; color case-hardened frame, rest blued. Introduced 1973; dropped 1975; reintroduced 1981; dropped 1982.
Perf.: $300 **Exc.:** $250 **VGood:** $200
7 1/2″ barrel
Perf.: $275 **Exc.:** $235 **VGood:** $180

COLT NEW POCKET MODEL
Revolver; double action; 32 Colt, 32 S&W; 6-shot cylinder; 2 1/2″, 3 1/2″, 5″, 6″ barrels; checkered hard rubber stocks; blued or nickel finish. Introduced 1893; dropped 1905. Collector value.
Exc.: $550 **VGood:** $400 **Good:** $350

COLT NEW POLICE
Revolver; double action; 32 Colt, 32 Colt New Police; 2 1/2″, 4″, 6″ barrels; built on New Pocket Model frame, with larger hard rubber grips; same sights as New Pocket Colt; blued or nickel finish. Manufactured 1896 to 1907.
Exc.: $550 **VGood:** $425 **Good:** $350

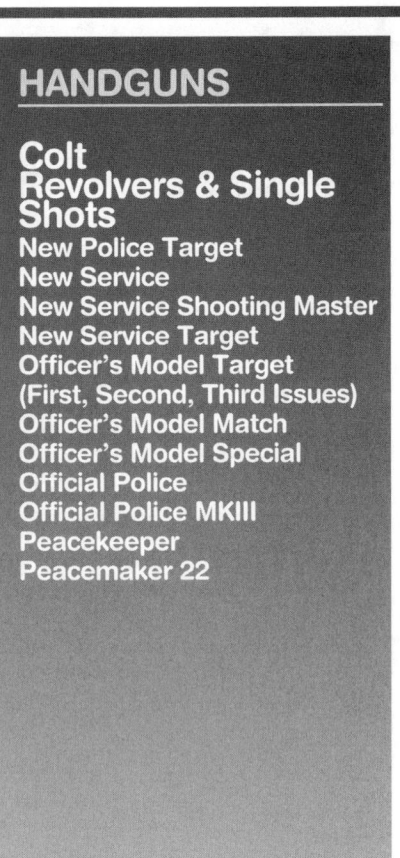

HANDGUNS

Colt Revolvers & Single Shots

New Police Target
New Service
New Service Shooting Master
New Service Target
Officer's Model Target
(First, Second, Third Issues)
Officer's Model Match
Officer's Model Special
Official Police
Official Police MKIII
Peacekeeper
Peacemaker 22

Colt New Police

Colt New Service
Shooting Master

Colt New Service

Colt Official
Police MKIII

Colt Officer's
Model Match

Colt Official Police

Colt New Police Target
Same specs as New Police model except target sights; 6" barrel only. Manufactured 1896 to 1905.
  Exc.: $100   **VGood:** $850   **Good:** $650

COLT NEW SERVICE
Revolver; double action; 38 Spl., 357 Mag., 38-40, 38-44, 44 Russian, 44 Spl., 44-40, 45 Colt, 45 ACP, 450 Eley, 455 Eley, 476 Eley; 2", 3 1/2" 4", 4 1/2", 5", 5 1/2", 6", 7 1/2" barrels; 6-shot swing-out cylinder; large frame. Special run in 45 ACP. During WWI was designated as Model 1917 Revolver under government contract. Fixed open notch rear sight milled in topstrap, fixed front; checkered hard rubber grips on commercial New Service; lanyard loop on most variations; blued, nickel finish. Introduced 1897; dropped 1944.
Commercial model
  Exc.: $1350   **VGood:** $1000   **Good:** $800
Military model
  Exc.: $1250   **VGood:** $1000   **Good:** $800
357 Mag.
  Exc.: $1200   **VGood:** $1000   **Good:** $875

Colt New Service Shooting Master
Same specs as New Service except 38 Spl., 357 Mag., 44 Spl., 45 ACP/Auto Rim, 45 Colt; 6" barrel; 11 1/4" overall length; deluxe target version; checkered walnut grips, rounded butt; windage-adjustable rear sight, elevation-adjustable front; blued. Introduced 1932; dropped 1941.
  Exc.: $1000   **VGood:** $900   **Good:** $700
45 ACP
  Exc.: $1900   **VGood:** $1700   **Good:** $1400
44 Spl.
  Exc.: $1900   **VGood:** $1700   **Good:** $1400

Colt New Service Target
Same specs as standard New Service model except windage-adjustable rear sight, elevation-adjustable front; 44 Spl., 44 Russian, 45 Colt, 45 ACP, 450 Eley, 455 Eley, 476 Eley; 5", 6", 7 1/2" barrels; flat topstrap; finished action; blued finish; hand-checkered walnut grips. Introduced 1900; dropped 1940.
  Exc.: $2000   **VGood:** $1500   **Good:** $1000

COLT OFFICER'S MODEL TARGET (FIRST, SECOND, THIRD ISSUES)
Revolver; double action; 38 Spl. (First Issue) 32 Colt, 38 Spl. (Second Issue) 22 LR (Third Issue); 6-shot cylinder; 6" barrel (38 Spl.); 4", 4 1/2", 5", 6", 7 1/2" barrels (32 Colt, 38 Spl., 22 LR); hand-finished action; adjustable rear target, blade front sight; blued finish. First Issue manufactured 1904-1908; Second Issue, 1908-1926; Third Issue, 1927-1949.
First Issue
  Exc.: $1000   **VGood:** $800   **Good:** $650
Second Issue
  Exc.: $800   **VGood:** $650   **Good:** $425
Third Issue
  Exc.: $750   **VGood:** $600   **Good:** $400

Colt Officer's Model Match
Same specs as Officer's Model Special, which it replaced, except checkered walnut target grips; tapered heavy barrel; wide hammer spur. Introduced 1953; dropped 1970.
  Perf.: $425   **Exc.:** $375   **VGood:** $300

Colt Officer's Model Special
Same specs as Second Issue Officer's Model Target except heavier barrel; ramp front, Coltmaster adjustable rear sight; redesigned hammer; checkered plastic stocks; 6" barrel; 22 LR, 38 Spl.; blued finish. Introduced 1949; dropped 1955.
  Exc.: $550   **VGood:** $425   **Good:** $375

COLT OFFICIAL POLICE
Revolver; double action; 22 LR, 32-20, 38 Spl., 41 Long Colt; 6-shot cylinder; 2", 6" heavy barrels (38 Spl. only); 4", 6" barrels (22 LR only); 4", 5", 6" (other calibers); checkered walnut grips, plastic grips on post-WWII models. Version made to military specs in WWII was called Commando model, had sand-blasted blue finish. Introduced 1927 as a replacement for Army Special; dropped 1970. Collector value.
  Perf.: $450   **Exc.:** $375   **VGood:** $300
Military model
  Perf.: $475   **Exc.:** $400   **VGood:** $350

COLT OFFICIAL POLICE MKIII
Revolver; double action; 38 Spl.; 4", 5", 6" barrels; 9 3/8" overall length; an old name, but a renewed design, incorporating coil mainspring in place of leaf spring; square butt, checkered walnut grips; fixed rear sight notch milled in topstrap, fixed ramp front; grooved front surface on trigger; checkered hammer spur. Introduced 1969; dropped 1975.
  Perf.: $325   **Exc.:** $275   **VGood:** $225

COLT PEACEKEEPER
Revolver; double action; 357 Mag.; 6-shot cylinder; 4", 6" barrel; weighs 42 oz. (6" barrel); red insert ramp front sight, white outline adjustable rear; rubber round-bottom combat grips; matte blue finish. Introduced 1985; dropped 1987.
  Perf.: $375   **Exc.:** $350   **VGood:** $300

COLT PEACEMAKER 22
Revolver; single action; dual cylinders for 22 LR, 22 WMR; 6-shot cylinder; 4 3/8", 6", 7 1/2" barrels; 11 1/4" overall length (6" barrel); scaled-down version of century-old Model 1873; rear sight notch milled into rounded topstrap; fixed blade front; color case-hardened frame, rest blued; black plastic grips. Introduced 1973; dropped 1975.
  Perf.: $325   **Exc.:** $275   **VGood:** $200
7 1/2" barrel
  Perf.: $350   **Exc.:** $300   **VGood:** $225

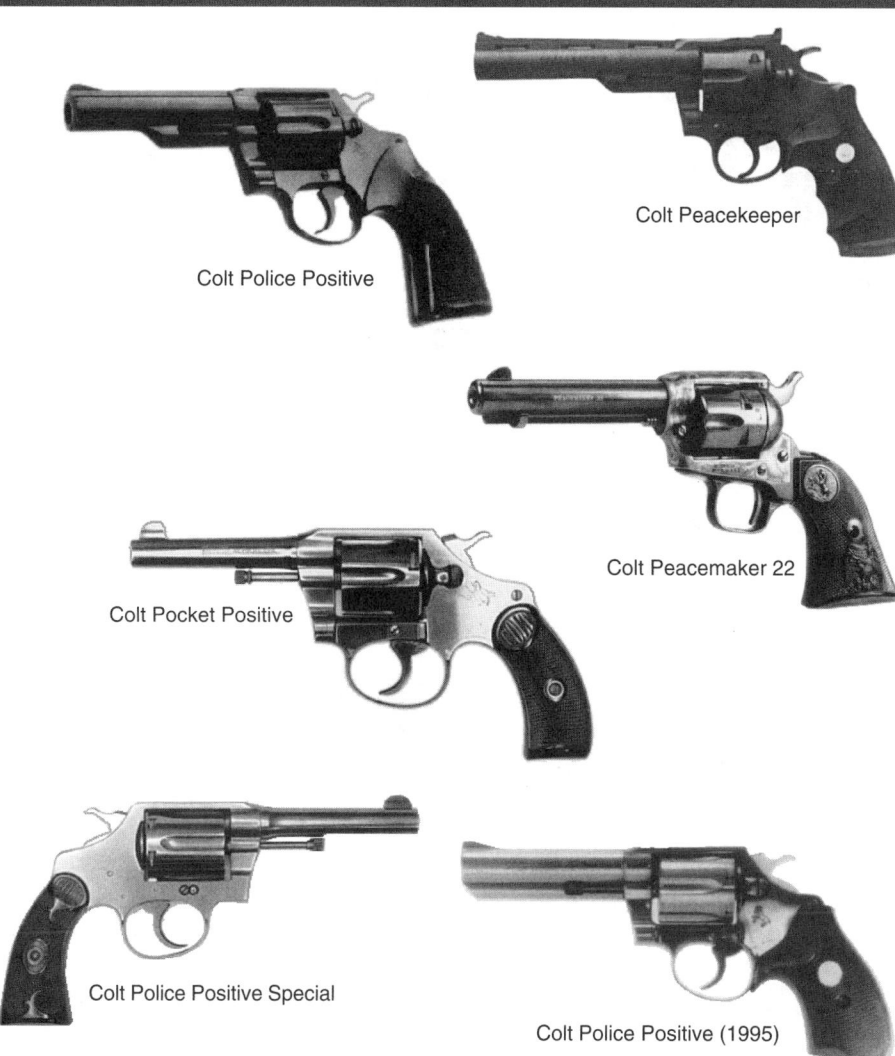

Colt Police Positive

Colt Peacekeeper

Colt Peacemaker 22

Colt Pocket Positive

Colt Police Positive Special

Colt Police Positive (1995)

HANDGUNS

Colt Revolvers & Single Shots
Pocket Positive
Police Positive
Police Positive Special
(First, Second, Third Issues)
Police Positive Target
Python
Python Stainless

Colt Python

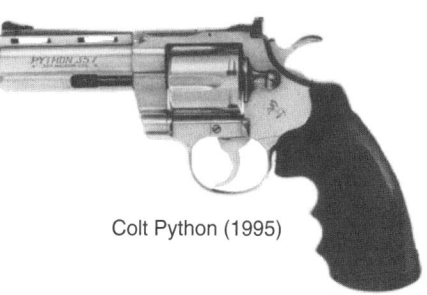

Colt Python (1995)

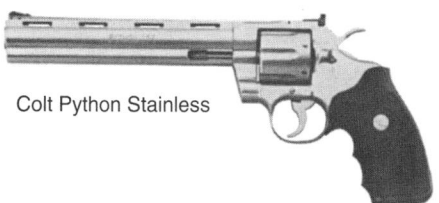

Colt Python Stainless

COLT POCKET POSITIVE

Revolver; double action; 32 Short Colt, 32 Long Colt, 32 Colt New Police (interchangeable with 32 S&W Long, S&W Short cartridges); 6-shot cylinder; 2″, 2 1/2″, 3 1/2″, 4″, 5″, 6″ barrels; 7 1/2″ overall length (3 1/2″ barrel); rear sight groove milled in top-strap, rounded front sight; positive lock feature; hard rubber grips; blued or nickel finish. Based upon New Pocket Model dropped in 1905. Introduced 1905; dropped 1940.

  Exc.: $335   **VGood:** $275   **Good:** $225
Nickel finish
  Exc.: $390   **VGood:** $310   **Good:** $260

COLT POLICE POSITIVE

Revolver; double action; 32 Short Colt, 32 Long Colt, 32 Colt New Police, 38 S&W; 6-shot swing-out cylinder; 2 1/2″, 4″, 5″, 6″ barrels; 8 1/2″ overall length (4″ barrel); fixed sights; checkered walnut, plastic or hard rubber grips; top of frame matted to reduce glare; blue or nickel finish; replaced New Police model. Introduced 1905; dropped 1943. Reintroduced 1995. **LMSR: $400**

  Exc.: $425   **VGood:** $375   **Good:** $325
Nickel finish
  Exc.: $500   **VGood:** $425   **Good:** $350

Colt Police Positive Special
(First, Second, Third Issues)

Same specs as standard Police Positive except lengthened frame to accommodate longer cylinder for 38 Spl., 32-20; also made in 32 and 38 Colt New Police; 4″, 5″, 6″ barrels; 8 3/4″ overall length (4″ barrel); blued or nickel. Introduced 1908; dropped 1970. Reintroduced 1995; still in production. **LMSR: $400**

  Perf.: $450   **Exc.:** $400   **VGood:** $350

Colt Police Positive Target

Same specs as standard Police Positive except 32 Short Colt, 32 Long Colt, 32 New Police, 32 S&W Short, 32 S&W Long, 22 LR. In 1932, cylinder was modified by countersinking chambers for safety with high-velocity 22 ammo; those with non-countersunk chambers should be fired only with standard-velocity 22 LR ammo. Has 6″ barrel; 10 1/2″ overall length, windage-adjustable rear sight, elevation-adjustable front; checkered walnut grips; blued; backstrap, hammer spur, trigger checkered. Introduced 1905; dropped 1941.

  Exc.: $600   **VGood:** $500   **Good:** $400

COLT PYTHON

Revolver; double action; 357 Mag., but will handle 38 Spl.; 6-shot swing-out cylinder; made first appearance in 6″ barrel later with 2 1/2″, 3″, 4″, 8″; checkered walnut grips contoured for support for middle finger of shooting hand; vent-rib barrel; ramp front sight, fully-adjustable rear; full-length ejector rod shroud; wide-spur hammer; grooved trigger; blued, nickel finish; hand-finished action. Introduced 1955; still in production. **LMSR: $815**

  Perf.: $550   **Exc.:** $475   **VGood:** $400
Nickel finish
  Perf.: $550   **Exc.:** $475   **VGood:** $400

Colt Python Stainless

Same specs as standard Python except stainless steel construction; 4″, 6″ barrel lengths only. Introduced 1983; still produced. **LMSR: $904**

  Perf.: $675   **Exc.:** $550   **VGood:** $475

HANDGUNS

Colt Revolvers & Single Shots

Single Action Army
Trooper
Trooper MKIII
Trooper MKV
Viper

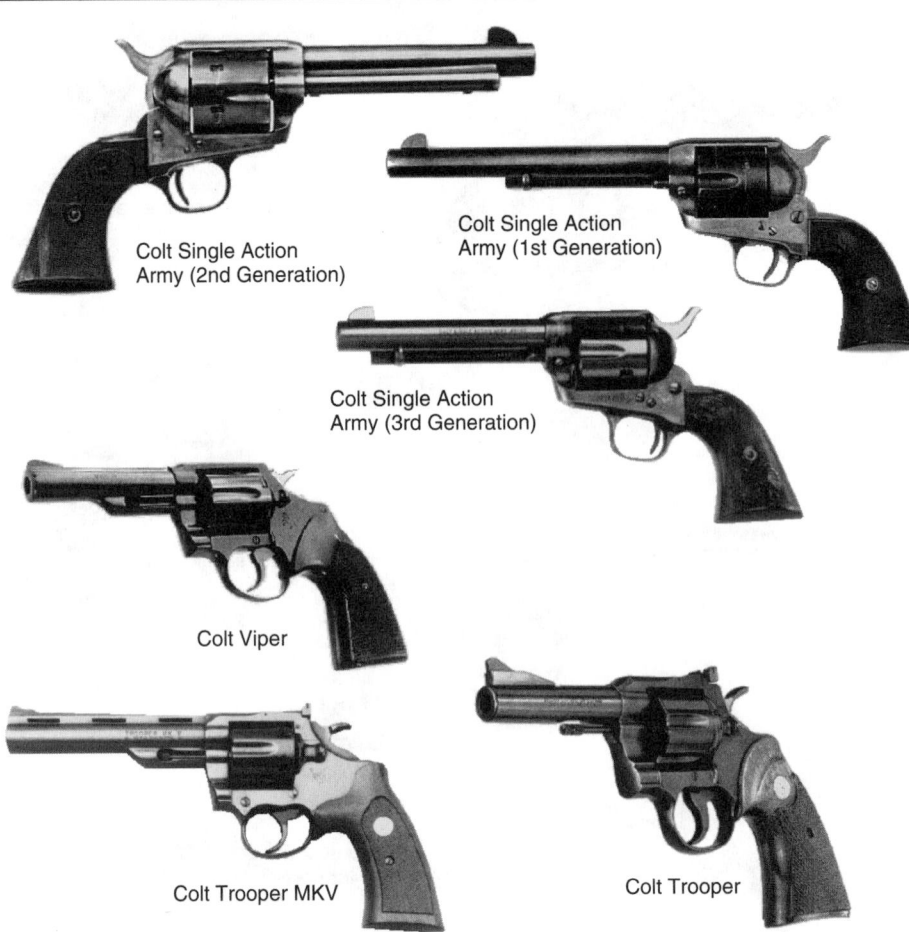

Colt Single Action Army (1st Generation)

Colt Single Action Army (2nd Generation)

Colt Single Action Army (3rd Generation)

Colt Viper

Colt Trooper MKV

Colt Trooper

COLT SINGLE ACTION ARMY

Colt Single Action Army also known as Peacemaker, or Frontier Six Shooter (44-40 caliber only). Originally introduced in 1873 and produced continuously until 1941. This production is run known as "1st Generation." The production run resumed in 1956 and stopped again in 1974 is known as "2nd Generation." Production known as "3rd Generation" began in 1976 and is still being manufactured.

1st Generation: Serial numbers began in 1873 with #1 and ended in 1941 with serial number 357859. The first Generation production included two frame styles: From 1873 until 1892, frames had screw angling in from front of frame to secure cylinder base pin; starting in 1892, and becoming standard by 1896, cylinder pin is secured by spring-loaded transverse catch. These frames are mistakenly called "blackpowder" and "smokeless powder" frames respectively, but it should be noted that Colt did not warranty any SAA revolvers for smokeless powder until 1900 at about serial number 192000. Rarer calibers and barrel lengths bring considerable premium over prices shown.

When **2nd Generation** production began in 1956, serial numbers started at #0001SA, and continued until approximately #74000SA. **3rd Generation** production started in 1976 and serial numbers resumed at #80000SA. When #99999SA was reached in 1978, the SA was made a prefix and numbers started once more at #SA00001. When #SA99999 was reached about 1993, the SA was split and numbers started once more at #S0001A.

Standard barrel lengths for all three generations have been 4 3/4" (introduced 1879), 5 1/2" (introduced 1875), and 7 1/2" (introductory length offered in 1873). During 1st Generation production, any barrel length could be special ordered. Examples are known ranging from 2 1/2" to 16".

During 2nd and 3rd Generation production, special runs were made as Sheriff's Model with 3" barrels and no ejector rod housings. During 2nd and 3rd Generation, Buntline Special versions with 12" barrels were

cataloged. A 3rd Generation "Storekeeper's Model" was also cataloged with 4" barrels and no ejector rod housings.

Standard finishes for all three generations have been color case-hardened frame with remainder blued, or fully nickel-plated. Some fully-blued SAAs have been offered by Colt from time to time. For 1st Generation, standard grips were one-piece walnut in early production and two-piece hard rubber in later production. Standard grips in 2nd Generation were hard rubber, and both hard rubber and two-piece walnut in 3rd Generation. Standard sights were groove down revolver's topstrap and blade front. However, a rare flattop target model was also offered during 1st Generation run.

Caliber options for 1st Generation revolvers numbered approximately thirty, ranging from 22 rimfire to 476 Eley. However, the top five in terms of numbers produced were 45 Colt, 44-40, 38-40, 32-20 and 41 Colt. 2nd Generation calibers in order of numbers produced were 45 Colt, 357 Magnum, 38 Special, 44 Spl.

3rd Generation calibers started in 1976 with 45 Colt and 357 Mag. In 1978, 44 Spl. was added, and in 1980 44-40 was included once more. In 1993, 38-40 was made an option again.

2nd Generation Buntline and Sheriff's Models made only in 45 Colt from 1957 until 1974. 3rd Generation Buntline and Sheriff's Models made in 45 Colt, 44-40, and 44 Spl. calibers from 1980 until 1985. So-called "blackpowder frame" model with screw angling in from front of frame to secure cylinder base pin re-introduced in 1984 and offered as custom feature until early 1990s.

During period 1984 to 1994, Colt SAAs could be ordered from Colt Custom Shop with almost any combination of features listed above. **LMSR: $1213**
1st Generation (s/n up to 357859, common models)
   Exc.: $6500 **VGood:** $4500 **Good:** $3000
2nd Generation (s/n #0001SA-#74000SA)
   Exc.: $1500 **VGood:** $1200 **Good:** $800
3rd Generation (s/n #80000SA-) **LMSR: $1213**
   Perf.: $900 **Exc.:** $800 **VGood:** $700

COLT TROOPER

Revolver; double action; 22 LR, 38 Spl., 357 Mag.; 6-shot swing-out cylinder; ramp front sight; choice of standard hammer, service grips, wide hammer spur, target grips. Has the same specs as Officer's Match model except 4" barrel. Introduced 1953; dropped 1969.
   Exc.: $450 **VGood:** $300 **Good:** $275
Target model
   Exc.: $475 **VGood:** $325 **Good:** $300

COLT TROOPER MKIII

Revolver; double action; 22 LR, 22 WMR, 38 Spl., 357 Mag.; 4", 6", 8" barrels; 9 1/2" overall length (4" barrel); rear sight fully-adjustable, ramp front; shrouded ejector rod; checkered walnut target grips; target hammer; wide target trigger; blued, nickel finish. Introduced 1969; dropped 1983.
   Perf.: $325 **Exc.:** $250 **VGood:** $200
Nickel finish
   Perf.: $350 **Exc.:** $275 **VGood:** $225

Colt Trooper MKV

Same specs as Trooper MKIII with redesigned lock work to reduce double-action trigger pull; faster lock time; redesigned grips; adjustable rear sight, red insert front; 4", 6", 8" vent-rib barrel; blued or nickel finish. Introduced 1984; dropped 1988.
   Perf.: $350 **Exc.:** $250 **VGood:** $200
Nickel finish
   Perf.: $375 **Exc.:** $300 **VGood:** $225

COLT VIPER

Revolver; double action; 38 Spl.; 6-shot cylinder; 4" barrel; 9" overall length; weighs 20 oz.; ramp front, fixed square-notch rear sights; checkered walnut wrap-around grips; lightweight aluminum alloy frame; blue or nickel finish. Uses Colt Cobra frame. Introduced 1977.
   Perf.: $375 **Exc.:** $325 **VGood:** $225
Nickel finish
   Perf.: $475 **Exc.:** $425 **VGood:** $300

Competitor
Single Shot

Coonan 357
Magnum

CZ Model 38

CZ Model 27

Competition Arms Competitor

HANDGUNS

Competition Arms
Competitor

Competitor
Single Shot

Coonan
357 Magnum
357 Magnum Classic
357 Magnum Compact Cadet
357 Magnum Compensated
Model

**Cumberland
Mountain Arms**
The Judge Single Shot

CZ
Model 22
Model 24
Model 27
Model 38
Model 50
Model 52

COMPETITION ARMS COMPETITOR

Single shot; 22 LR, 223, 7mm TCU, 7mm International, 30 Herrett, 357 Mag., 41 Mag., 454 Casull, 375 Super Mag.; other calibers on special order; 10″, 14″ barrels; adjustable open rear sight, ramp front; interchangeable barrels of blued ordnance or stainless steel; vent barrel shroud; integral scope mount. Introduced 1987; no longer in production.
10 1/2″ barrel
  Perf.: $350   **Exc.:** $300   **VGood:** $275
14″ barrel
  Perf.: $350   **Exc.:** $325   **VGood:** $300
Extra barrels
  Perf.: $100   **Exc.:** $85   **VGood:** $75

COMPETITOR SINGLE SHOT

Single shot; 22 LR through 50 Action Express, including belted magnums; 14″ barrel (standard), 10 1/2″ silhouette, 16″ (optional); weighs about 59 oz. (14″ barrel); 15 1/8″ overall length; rotary cannon-type action; ambidextrous synthetic grip (standard) or laminated or natural wood; ramp front sight, adjustable rear; actions cocks on opening; cammed ejector; interchangeable barrels, ejectors; adjustable single-stage trigger; sliding thumb safety and trigger safety; matte blue finish. Made in U.S. by Competitor Corp. Introduced 1988; still produced.
14″ barrel, standard calibers, synthetic grips
  Perf.: $350   **Exc.:** $300   **VGood:** $250
Extra barrels
  Perf.: $75   **Exc.:** $65   **VGood:** $50

COONAN 357 MAGNUM

Semi-automatic; 357 Mag.; 7-shot magazine; 5″ barrel; 8 5/16″ overall length; weighs 42 oz.; smooth walnut stocks; open adjustable sights; barrel hood; many parts interchangeable with Colt autos; grip, hammer, half-cock safeties. Introduced 1983; limited production.
  New: $650   **Perf.:** $500   **Exc.:** $400

Coonan 357 Magnum Classic

Same specs as 357 Magnum except 8-shot magazine; Teflon black two-tone or Kal-Gard finish; fully-adjustable rear sight; integral compensated barrel. Made in U.S. by Coonan Arms, Ins. Introduced 1983; still in production.
  New: $1250   **Perf.:** $1100   **Exc.:** $950

COONAN 357 MAGNUM COMPACT CADET

Semi-automatic; single action; 357 Mag.; 6-shot magazine; 3 15/16″ barrel; weighs 39 oz.; 7 7/8″ overall length; smooth walnut grips; interchangeable ramp front sight, windage-adjustable rear; linkless bull barrel; full-length recoil spring guide rod; extended slide latch. Made in U.S. by Coonan Arms, Inc. Introduced 1993; still in production.
  Perf.: $700   **Exc.:** $650   **VGood:** $500

Coonan 357 Magnum Compensated Model

Same specs as 357 Magnum except 6″ barrel with compensator. Introduced 1990; still in production.
  LMSR: $999
  New: $850   **Perf.:** $800   **Exc.:** $650

CUMBERLAND MOUNTAIN ARMS THE JUDGE SINGLE SHOT

Single shot; 22 Hornet, 22 K-Hornet, 218 Bee, 7-30 Waters, 30-30; 10″, 16″ barrel; walnut grip; bead on ramp front sight, open adjustable rear; break-open design; made of 17-4 stainless steel. Also available as a kit. Made in U.S. by Cumberland Mountain Arms. Introduced 1995; still produced.

CZ MODEL 22

Semi-automatic; 380 ACP; 8-shot magazine; 3 5/16″ barrel; 6″ overall length; weighs 21 7/8 oz.; made under license from Mauser; external hammer; rotating barrel lock system. Manufactured 1921 to 1923.
  Exc.: $400   **VGood:** $325   **Good:** $250

CZ MODEL 24

Semi-automatic; 380 ACP; 8-shot magazine; 3 1/2″ barrel; 6″ overall length; weighs 24 oz.; same general design as Model 22 except addition of magazine safety; about 190,000 made. Manufactured from 1924 to 1939.
  Exc.: $325   **VGood:** $275   **Good:** $200

CZ MODEL 27

Semi-automatic; 32 ACP; 8-shot magazine; 3 13/16″ barrel; 6 1/2″ overall length; weighs 25 oz.; vertical retracting grooves on slide; fixed sights; plastic grips; blued finish. Commercial and non-commercial models made until 1939 marked on top rib and left side of frame; post-1941 omit top rib marking; post-war models made until 1951 marked on slide. Manufactured from 1927 to 1951.
Pre-war
  Exc.: $300   **VGood:** $250   **Good:** $200
Post-1941
  Exc.: $325   **VGood:** $275   **Good:** $225
Post-war
  Exc.: $300   **VGood:** $250   **Good:** $200

CZ MODEL 38

Semi-automatic; double-action-only; 380 ACP; 8-shot magazine; 4 5/8″ barrel; 8 1/8″ overall length; weighs 32 oz.; plastic grips; fixed sights; blued finish. Also listed as CZ Pistole 39(t) during German occupation. Manufactured 1939 to 1945.
  Exc.: $375   **VGood:** $275   **Good:** $200

CZ MODEL 50

Semi-automatic; double action; 32 ACP; 8-shot; 3 3/4″ barrel; 6 1/2″ overall length; weighs 24 3/4 oz.; frame mounted safety catch; loaded chamber indicator. Issued to Czech police as Model VZ50. Manufactured from 1951 to approximately 1967.
  Exc.: $135   **VGood:** $110   **Good:** $100

CZ MODEL 52

Semi-automatic; 7.62 Tokarev; 8-shot; 4 3/4″ barrel; 8 1/4″ overall length; weighs 40 oz.; complex roller locking breech system. Manufactured from 1952 to approximately 1956.
  Exc.: $150   **VGood:** $125   **Good:** $100

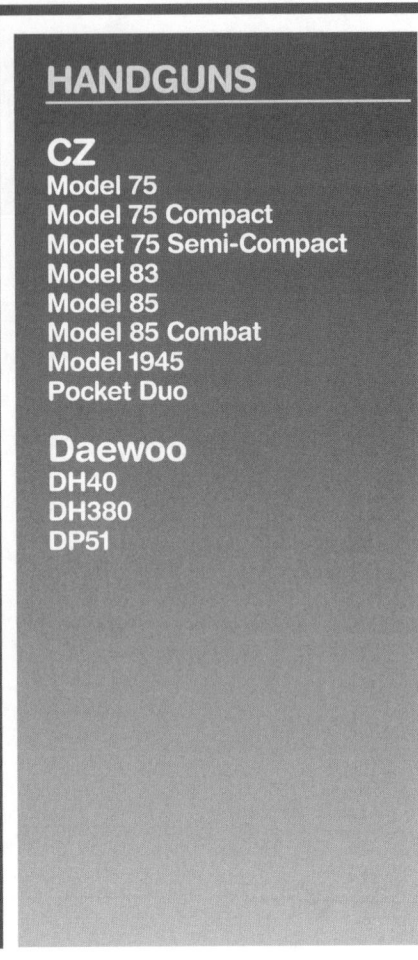

HANDGUNS

CZ
Model 75
Model 75 Compact
Modet 75 Semi-Compact
Model 83
Model 85
Model 85 Combat
Model 1945
Pocket Duo

Daewoo
DH40
DH380
DP51

Daewoo DP51C

CZ Pocket Duo

CZ Model 75

CZ Model 1945

CZ Model 83

CZ MODEL 75

Semi-automatic; double action or double-action-only; 9mm Para., 40 S&W; 15-shot magazine; 4 3/4" barrel; 8" overall length; weighs 35 oz.; all-steel frame; adjustable rear sight, square post front; checkered plastic; black polymer, matte or high-polish blued finish. Imported by Magnum Research.
Perf.: $400   **Exc.:** $350   **VGood:** $300

CZ Model 75 Compact

Same specs as Model 75 except 10-shot magazine; 3 15/16" barrel; weighs 32 oz.; removable front sight; non-glare ribbed slide top; squared and serrated triggerguard; combat hammer. Introduced 1993; still imported by Magnum Research.
Perf.: $400   **Exc.:** $350   **VGood:** $300

CZ Model 75 Semi-Compact

Same specs as Model 75 Compact with shorter slide and barrel on full-size CZ 75 frame; 10-shot magazine; 9mm Para. only. Introduced 1994; still imported by Magnum Research.
Perf.: $400   **Exc.:** $350   **VGood:** $300

CZ MODEL 83

Semi-automatic; double action; 32 ACP, 380 ACP; 3 13/16" barrel; 6 15/16" overall length; weighs 26 1/4 oz; adjustable rear, removable square post front sight; ambidextrous magazine release and safety; non-glare ribbed slide top; high-impact checkered plastic grips; blue finish.
Perf.: $300   **Exc.:** $250   **VGood:** $200

CZ MODEL 85

Semi-automatic; double action or double-action-only; 9mm Para., 40 S&W; 15-shot magazine; 4 3/4" barrel; 8" overall length; weighs 35 oz.; all-steel frame; ambidextrous slide release and safety levers; non-glare ribbed slide top; squared, serrated triggerguard; trigger stop to prevent overtravel; adjustable rear sight, square post front; checkered plastic; black polymer, matte or high-polish blued finish. Introduced 1986; still imported by Magnum Research.
Perf.: $450   **Exc.:** $400   **VGood:** $350

CZ Model 85 Combat

Same specs as the Model 85 except walnut grips; round combat hammer; fully-adjustable rear sight; extended magazine release; trigger parts coated with friction-free beryllium copper. Introduced 1992; still imported by Magnum Research.
Perf.: $500   **Exc.:** $450   **VGood:** $400

CZ MODEL 1945

Semi-automatic; double action; 25 ACP; 8-shot magazine; 2" barrel; 5" overall length; plastic grips; fixed sights; blued finish. Manufactured from 1945 to approximately 1960.
Exc.: $225   **VGood:** $175   **Good:** $145

CZ POCKET DUO

Semi-automatic; 25 ACP; 6-shot magazine; 2 1/8" barrel; 4 1/2" overall length; fixed sights; plastic grips; manufactured in Czechocslovakia; blue or nickel finish. Manufactured 1926 to 1960.
Exc.: $175   **VGood:** $150   **Good:** $125

DAEWOO DH40

Semi-automatic; double action; 40 S&W; 10-shot magazine; 4 1/8" barrel; 7 1/2" overall length; weighs 28 1/4 oz.; checkered composition grips; 1/8" blade front sight, adjustable rear; three-dot system; fast-fire mechanism; ambidextrous manual safety with internal firing pin lock; no magazine safety; alloy frame; squared triggerguard; matte black finish. Imported from Korea by Nationwide Sports Distributors. Introduced 1991; still imported.
New: $350   **Perf.:** $275   **Exc.:** $225

DAEWOO DH380

Semi-automatic; double action; 380 ACP; 8-shot magazine; 3 7/8" barrel; 6 3/4" overall length; weighs 23 oz.; checkered black composition thumbrest grips; 1/8" blade front sight, drift-adjustable rear; three-dot system; all-steel construction; blue finish; dual safety system with hammer block. Imported from Korea by Kimber of America; distributed by Nationwide Sports Dist. Introduced 1994; still imported.
Perf.: $300   **Exc.:** $250   **VGood:** $200

DAEWOO DP51

Semi-automatic; 9mm Para., 40 S&W; 12-shot (40 S&W), 13-shot (9mm) magazine; 4 1/8" barrel; 7 1/2" overall length; weighs 28 1/4 oz.; blade front, drift-adjustable square-notch rear sight; checkered composition grips; tri-action mechanism; ambidextrous manual safety, magazine catch; half-cock and firing pin block; alloy frame; matte black finish; square triggerguard. Made in Korea. Introduced 1991; still imported by Firstshot.
Perf.: $300   **Exc.:** $275   **VGood:** $225

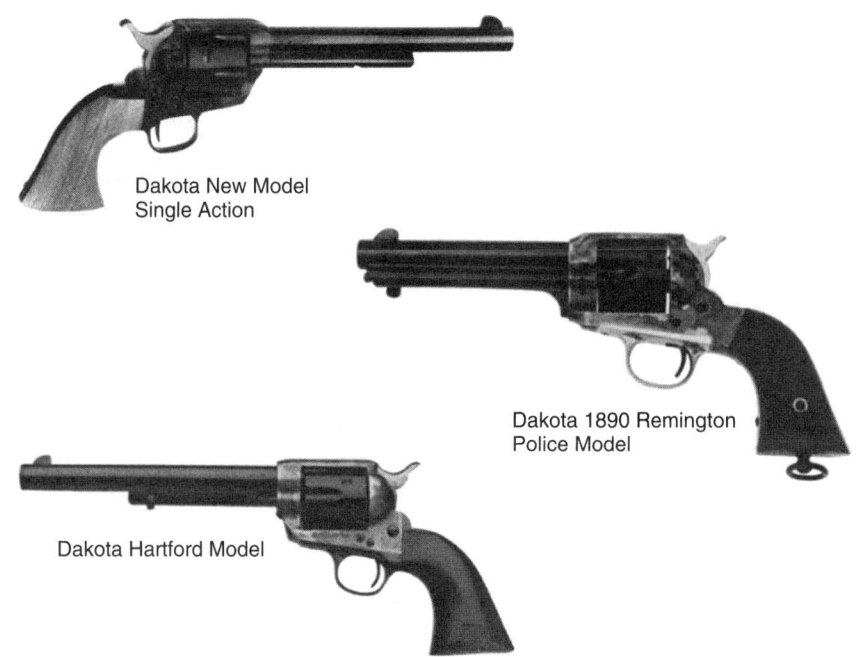

Dakota New Model Single Action

Dakota 1890 Remington Police Model

Dakota Hartford Model

HANDGUNS

Daewoo
DP51C
DP51S
DP52

Dakota
1873 Premier
1875 Remington
Outlaw Model
1890 Remington Police Model
Bisley Model
Hartford Model
Hartford Model Pinkerton
New Model Single Action
Old Model Single Action
Sheriff's Model

Daewoo DP51C
Same specs as the DP51 except 3 5/8" barrel; 1/4" shorter grip frame; flat mainspring housing; weighs 26 oz. Imported from Korea by by Kimber of America, Inc., distributed by Nationwide Sports Dist. Introduced 1995; still imported.
Perf.: $325   **Exc.:** $275   **VGood:** $225

Daewoo DP51S
Same specs as the DP51 except 3 5/8" barrel; weighs 27 oz. Imported from Korea by Kimber of America, Inc.; distributed by Nationwide Sports Dist. Introduced 1995.
Perf.: $300   **Exc.:** $275   **VGood:** $225

DAEWOO DP52
Semi-automatic; double action; 22 LR; 10-shot magazine; 3 7/8" barrel; 6 3/4" overall length; weighs 23 oz.; checkered black composition thumbrest grips; 1/8" blade front sight, drift-adjustable rear; three-dot system; all-steel construction; blue finish; dual safety system with hammer block. Imported from Korea by Kimber of America; distributed by Nationwide Sports Dist. Introduced 1994; still imported.
Perf.: $300   **Exc.:** $250   **VGood:** $200

DAKOTA 1873 PREMIER
Revolver; single action; 45 Colt; 6-shot cylinder; 4 5/8", 5 1/2" barrel; 10" overall length (4 5/8" barrel)); weighs 39 oz.; blade front, fixed rear sight; smooth walnut grips; reproduction of the 1873 Colt revolver with set-screw for the cylinder pin release; most parts said to interchange with the early Colts; blue finish, color case-hardened frame. Was imported by E.M.F.; no longer imported.
Perf.: $400   **Exc.:** $350   **VGood:** $250

DAKOTA 1875 REMINGTON OUTLAW MODEL
Revolver; single action; 357 Mag., 44-40, 45 Colt; 7 1/2" barrel; 13 1/2" overall length; weighs 46 oz.; blade front, fixed groove rear sight; uncheckered walnut stocks; authentic copy of 1875 Remington; color case-hardened frame; blued cylinder, barrel; steel backstrap; brass triggerguard; also made with nickel finish; factory engraving. Made in Italy. Introduced 1986; still imported by EMF.
Perf.: $400   **Exc.:** $350   **VGood:** $300
Nickel finish
Perf.: $450   **Exc.:** $400   **VGood:** $350
Engraved
Perf.: $500   **Exc.:** $450   **VGood:** $400

DAKOTA 1890 REMINGTON POLICE MODEL
Revolver; single action; 357 Mag., 44-40, 45 Colt; 5 1/2" barrel; 12" overall length; weighs 40 oz.; blade front, fixed groove rear sight; uncheckered walnut stocks; color case-hardened frame; blued cylinder, barrel; steel backstrap; brass trigger guard; lanyard ring in butt; also available in nickel finish, factory engraving. Made in Italy. Introduced 1986; currently imported by EMF.
Perf.: $400   **Exc.:** $350   **VGood:** $300
Nickel finish
Perf.: $450   **Exc.:** $400   **VGood:** $350
Engraved
Perf.: $500   **Exc.:** $450   **VGood:** $400

DAKOTA BISLEY MODEL
Revolver; single action; 22 LR, 22 WMR, 32-20, 32 H&R Mag., 357 Mag., 38-40, 44 Spl., 44-40, 45 Colt, 45 ACP; 4 3/4", 5 1/2", 7 1/2" barrels; 10 1/2" overall length (5 1/2" barrel); weighs 37 oz.; blade front, fixed groove rear sight; uncheckered walnut stocks; color case-hardened frame; blued barrel, cylinder; steel trigger guard, backstrap; Colt-type firing pin on hammer. Also available with nickel finish, factory engraving. Made in Italy. Introduced 1985; dropped 1991; reintroduced, 1993.
Perf.: $400   **Exc.:** $350   **VGood:** $300
Nickel finish
Perf.: $525   **Exc.:** $450   **VGood:** $400
Engraved
Perf.: $600   **Exc.:** $550   **VGood:** $500

DAKOTA HARTFORD MODEL
Revolver; single action; 22 LR, 32-20, 357 Mag., 38-40, 44 Spl., 45 Colt; 4 3/4", 5 1/2", 7 1/2" barrel; blade front, fixed rear sight; one-piece walnut stock; identical to original Colts; all major parts serial numbered with original Colt-type lettering, numbers; color case-hardened frame, hammer; various options. Made in Italy. Introduced 1990; still imported by EMF.
Perf.: $400   **Exc.:** $350   **VGood:** $300
Nickel plated
Perf.: $500   **Exc.:** $450   **VGood:** $400
Calvary or Artillery models
Perf.: $400   **Exc.:** $350   **VGood:** $300
Cattlebrand engraved
Perf.: $500   **Exc.:** $450   **VGood:** $400
Engraved nickel
Perf.: $600   **Exc.:** $550   **VGood:** $500

Dakota Hartford Model Pinkerton
Same specs as Hartford Model except 4" barrel; 45 Colt; bird's-head walnut grips; ejector tube. Introduced 1994; still in production.
Perf.: $400   **Exc.:** $350   **VGood:** $300

DAKOTA NEW MODEL SINGLE ACTION
Revolver; single action; 357 Mag., 44-40, 45 Colt; 6-shot; 4 3/4", 5 1/2", 7 1/2" barrel; blade front, fixed rear sight; one-piece walnut stock; Colt-type hammer, firing pin; nickel or color case-hardened forged frame; black nickel backstrap, triggerguard; also available in nickel finish. Made in Italy. Introduced 1991; currently imported by EMF.
Perf.: $300   **Exc.:** $250   **VGood:** $200
Nickel finish
Perf.: $350   **Exc.:** $300   **VGood:** $250
Engraved, nickel-plated
Perf.: $400   **Exc.:** $350   **VGood:** $300

DAKOTA OLD MODEL SINGLE ACTION
Revolver; single action; 22 LR, 32-20, 357 Mag., 38-40, 44 Spl., 44-40, 45 Colt; 6-shot cylinder; 4 5/8", 5 1/2", 7 1/2" barrel; blade front, fixed rear sight; one-piece walnut stocks; color case-hardened frame; brass grip frame, triggerguard; blued barrel cylinder; also available in nickel finish. Made in Italy. Introduced 1967; sporadic production. Was imported by EMF.
Perf.: $275   **Exc.:** $225   **VGood:** $200
Nickel finish
Perf.: $325   **Exc.:** $275   **VGood:** $225

DAKOTA SHERIFF'S MODEL
Revolver; single action; 32-20, 357 Mag., 38-40, 44 Spl., 44-40, 45 Colt; 6-shot; 3 1/2" barrel; blade front, fixed rear sight; one-piece walnut stock; blued finish. Made in Italy. Introduced 1994; was imported by EMF.
Perf.: $350   **Exc.:** $300   **VGood:** $250

Davis
Big-Bore Derringers
D-Series Derringers
Long-Bore Derringers
P-32
P-380

Desert Eagle
(See Magnum Research Desert Eagle)

Desert Industries
Double Deuce
Two-Bit Special
War Eagle

Detonics
Combat Master Mark I
Combat Master Mark V
Combat Master Mark VI
Combat Master Mark VII
Compmaster
Janus Scoremaster
(See Detonics Scoremaster)
Mark I
Mark II
Mark III

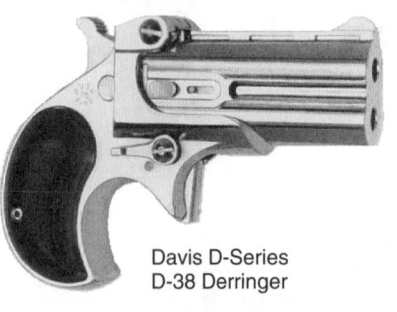

Davis D-Series
D-38 Derringer

Davis P-32

Desert Industries
War Eagle

Detonics Combat
Master Mark V

Desert Industries
Double Deuce

DAVIS BIG-BORE DERRINGERS
Derringer; over/under; 22 WMR, 32 H&R Mag., 38 Spl., 9mm Para.; 2-shot; 2 3/4" barrels; 4 9/16" overall length; weighs about 14 oz.; textured black synthetic grips; fixed sights; chrome or black teflon finish. Larger than the Davis D-Series models. Made in U.S. by Davis Industries. Introduced 1995; still produced.
 Perf.: $60 **Exc.:** $45 **VGood:** $35

DAVIS D-SERIES DERRINGERS
Derringer over/under; 22 WMR, 32 H&R, 38 Spl; 2-shot; 2 3/8" barrels; 4" overall length; weighs about 11 1/2 oz.; textured black synthetic grips; blade front, fixed notch rear sight; alloy frame; steel-lined barrels; steel breechblock; plunger-type safety with integral hammer block; chrome or black Teflon finish. Made in U.S. by Davis Industries. Introduced 1992; still produced.
 Perf.: $60 **Exc.:** $45 **VGood:** $35

DAVIS LONG-BORE DERRINGERS
Derringer over/under; 22 WMR, 32 H&R Mag., 38 Spl., 9mm Para.; 2-shot; 3 1/2" barrels; 5 3/8" overall length; weighs 16 oz.; textured black synthetic grips; fixed sights; chrome or black teflon finish. Larger than the Davis D-Series models. Made in U.S. by Davis Industries. Introduced 1995; still produced.
 Perf.: $75 **Exc.:** $60 **VGood:** $50

DAVIS P-32
Semi-automatic; double action; 32 ACP; 6-shot magazine; 2 13/16" barrel; 5 3/8" overall length; weighs 22 oz.; laminated wood stocks; fixed sights; chrome or black Teflon finish. Introduced 1986; still in production.
 Perf.: $70 **Exc.:** $60 **VGood:** $50

DAVIS P-380
Semi-automatic; double action; 380 ACP; 5-shot magazine; 2 13/16" barrel; 5 7/16" overall length; weighs 22 oz.; fixed sights; black composition grips; chrome or black Teflon finish. Introduced 1991; still in production.
 Perf.: $80 **Exc.:** $60 **VGood:** $50

DESERT INDUSTRIES DOUBLE DEUCE
Semi-automatic; double action; 22 LR; 6-shot; 2 1/2" barrel; 5 1/2" overall length; weighs 15 oz.; fixed groove sights; smooth rosewood grips; ambidextrous slide-mounted safety; stainless steel construction; matte finish. Introduced 1991; still in production.
 Perf.: $325 **Exc.:** $275 **VGood:** $225

DESERT INDUSTRIES TWO-BIT SPECIAL
Semi-automatic; double action; 25 ACP; 5-shot; 2 1/2" barrel; 5 1/2" overall length; weighs 15 oz.; fixed groove sights; smooth rosewood grips; ambidextrous slide-mounted safety; stainless steel construction; matte finish. Introduced 1991; still in production.
 Perf.: $325 **Exc.:** $275 **VGood:** $225

DESERT INDUSTRIES WAR EAGLE
Semi-automatic; double action; 380 ACP, 9mm Para., 40 S&W, 10mm Auto, 45 ACP; 10-shot, 14-shot (9mm), 8-shot (380 ACP); 4" barrel; fixed sights; rosewood grips; stainless steel construction; matte finish. Introduced 1991; still in production.
 Perf.: $700 **Exc.:** $650 **VGood:** $550

DETONICS COMBAT MASTER MARK I
Semi-automatic; 9mm Para., 38 Super, 45 ACP; 6-shot magazine (45 ACP); 3 1/2" barrel; 6 3/4" overall length; weighs 28 oz.; fixed or adjustable combat-type sights; checkered walnut stocks; throated barrel; polished feed ramp; self-adjusting cone barrel centering system; beveled magazine inlet; full-clip indicator; two-tone matte blue and stainless finish. Introduced 1977; discontinued 1992.
 Perf.: $700 **Exc.:** $600 **VGood:** $500

Detonics Combat Master Mark V
Same specs as Combat Master I except all matte stainless finish. Introduced 1977; discontinued 1985.
 Perf.: $600 **Exc.:** $550 **VGood:** $500

Detonics Combat Master Mark VI
Same specs as Combat Master Mark I except 451 Detonics Mag.; adjustable sights; polished stainless steel slide. Introduced 1977; discontinued 1989.
 Perf.: $950 **Exc.:** $750 **VGood:** $600

Detonics Combat Master Mark VII
Same specs as Combat Master Mark VI except no sights as special order option. Introduced 1977; discontinued 1982.
 Perf.: $850 **Exc.:** $750 **VGood:** $650

DETONICS COMPMASTER
Semi-automatic; 45 ACP; 7-, 8-shot magazine; 5", 6" barrel; Millet adjustable sights; checkered Pachmayr stocks; matching mainspring housing; Detonics recoil system; extended grip safety; ambidextrous safety; extended magazine release; stainless steel. Introduced 1988; discontinued 1992.
 Perf.: $1800 **Exc.:** $1400 **VGood:** $1200

DETONICS MARK I
Semi-automatic; 45 ACP; 6-shot magazine; 3 1/2" barrel; 6 3/4" overall length; weighs 29 oz.; fixed sights; adjustable rear sights; checkered walnut stocks; compact based on Model 1911; blued finish. Introduced 1977; discontinued 1981.
 Perf.: $450 **Exc.:** $400 **VGood:** $350

Detonics Mark II
Same specs as Mark I except satin nickel plated. Introduced 1977; discontinued 1979.
 Perf.: $500 **Exc.:** $450 **VGood:** $400

Detonics Mark III
Same specs as Mark I except hard chrome plated. Introduced 1977; discontinued 1979.
 Perf.: $500 **Exc.:** $400 **VGood:** $325

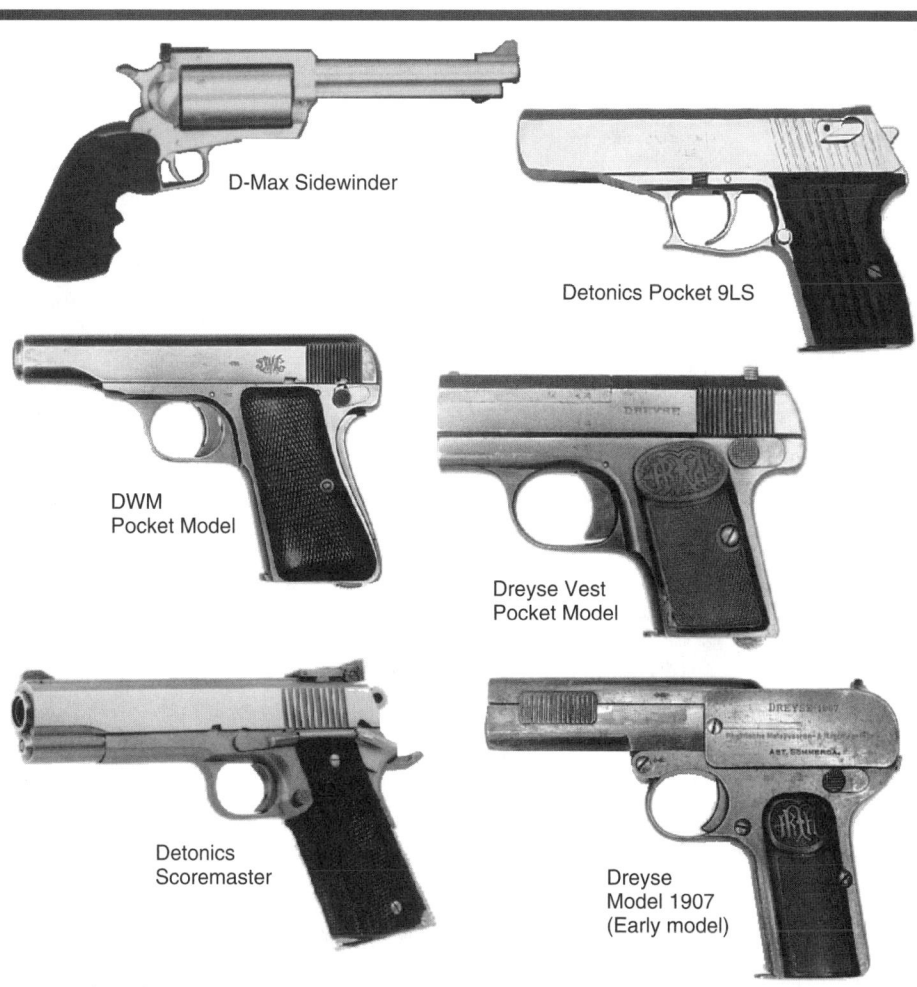

D-Max Sidewinder

Detonics Pocket 9LS

DWM
Pocket Model

Dreyse Vest
Pocket Model

Detonics
Scoremaster

Dreyse
Model 1907
(Early model)

Detonics
Mark IV
Military Combat MC2
Pocket 9
Pocket 9LS
Pocket 380
Power 9
Scoremaster
Servicemaster
Servicemaster II

D-Max
Sidewinder

Dreyse
Model 1907
Vest Pocket Model

DWM
Pocket Model

E.A.A.
Big Bore Bounty Hunter
Bounty Hunter

Detonics Mark IV

Same specs as Mark I except polished blue finish. Introduced 1977; discontinued 1981.
Perf.: $525 **Exc.:** $425 **VGood** $350

DETONICS MILITARY COMBAT MC2

Semi-automatic; 38 Super, 9mm Para., 45 ACP; 6-shot magazine; 3 1/2" barrel; 6 3/4" overall length; weighs 29 oz.; fixed sights; Pachmayr grips; camo pistol rug; matte finish. Introduced 1977; discontinued 1984.
Perf.: $600 **Exc.:** $550 **VGood** $475

DETONICS POCKET 9

Semi-automatic; double action; 9mm Para.; 6-shot; 3" barrel; 5 11/16" overall length; weighs 26 oz.; fixed sights; black Micarta stocks; triggerguard hook; snag-free hammer; captive recoil spring; matte stainless steel finish. Introduced 1985; dropped 1986.
Perf.: $475 **Exc.:** $400 **VGood** $350

Detonics Pocket 9LS

Same specs as Pocket 9 except 4" barrel; weighs 28 oz. Introduced 1986; discontinued 1986.
Perf.: $550 **Exc.:** $475 **VGood** $400

Detonics Pocket 380

Same specs as Pocket 9 except in 380 ACP; weighs 23 oz. Introduced 1986; discontinued 1986.
Perf.: $550 **Exc.:** $450 **VGood** $400

DETONICS POWER 9

Semi-automatic; double-action; 9mm Para.; 6-shot; 4" barrel; weighs 28 oz.; polished slide flats; fixed sights; black Micarta stocks; triggerguard hook; snag-free hammer; captive recoil spring; matte stainless steel finish. Introduced 1985; dropped 1986.
Perf.: $525 **Exc.:** $450 **VGood** $400

DETONICS SCOREMASTER

Semi-automatic; 45 ACP, 451 Detonics Mag.; 7-, 8-shot; 5", 6" barrel; 8 3/4" overall length (5" barrel); weighs 41 oz.; Millet adjustable sights; checkered Pachmayr stocks, matching mainspring housing; Detonics recoil system; extended grip safety; ambidextrous safety; extended magazine release; stainless steel. Introduced, 1983; discontinued, 1992.
Perf.: $950 **Exc.:** $800 **VGood** $700

DETONICS SERVICEMASTER

Semi-automatic; 45 ACP; 8-shot magazine; 4 1/4" barrel; 7 7/8" overall length; weighs 39 oz.; adjustable combat sights; Pachmayr rubber stocks; extended grip safety; thumb and grip safeties; stainless steel construction; matte finish. Introduced 1977; discontinued 1986.
Perf.: $750 **Exc.:** $600 **VGood** $500

Detonics Servicemaster II

Same specs as Servicemaster except polished slide flats. Introduced 1986; discontinued 1992.
Perf.: $800 **Exc.:** $650 **VGood** $550

D-MAX SIDEWINDER

Revolver; single action; 45 Colt/410 shotshell; 6-shot cylinder; 6 1/2", 7 1/2" barrel; 14 1/8" overall length (6 1/2" barrel); weighs 57 oz. (6 1/2" barrel); Hogue black rubber grips with finger grooves; blade on ramp front sight, fully-adjustable rear; stainless steel construction; removable choke for firing shotshells; grooved, wide-spur hammer; transfer bar ignition. Made in U.S. by D-Max, Inc. Introduced 1992; still produced.
Perf.: $700 **Exc.:** $650 **VGood** $600

DREYSE MODEL 1907

Semi-automatic; 32 ACP; 8-shot magazine; 3 1/2" barrel; 6 1/4" overall length; fixed sights; hard rubber checkered stocks; blued finish. Manufactured in Germany, 1907 to 1920.
Exc.: $200 **VGood** $150 **Good:** $100

DREYSE VEST POCKET MODEL

Semi-automatic; 25 ACP; 6-shot magazine; 2" barrel; 4 1/2" overall length; fixed sights; checkered hard rubber grips; blued finish. Manufactured 1909 to 1914.
Exc.: $200 **VGood** $175 **Good:** $135

DWM POCKET MODEL

Semi-automatic; 32 ACP; 3 1/2" barrel; 6" overall lengh; hard rubber checkered stocks; blued finish. Design resembles Browning FN Model 1910. Manufactured in Germany, 1920 to 1931.
Exc.: $600 **VGood** $500 **Good:** $425

E.A.A. BIG BORE BOUNTY HUNTER

Revolver; single action; 357 Mag., 44 Mag., 45 Colt; 6-shot cylinder; 4 1/2", 7 1/2" barrel; 11" overall length (4 5/8" barrel); weighs 2 1/2 lbs.; smooth walnut grips; blade front sight, grooved topstrap rear; transfer bar safety; three-position hammer; hammer-forged barrel; blue, chrome or case-hardened frame. Imported by European American Armory. Introduced 1992; still imported.
Blue or case-hardened frame
Perf.: $250 **Exc.:** $200 **VGood** $175
Chrome finish
Perf.: $300 **Exc.:** $250 **VGood** $200

E.A.A. BOUNTY HUNTER

Revolver; single action; 22 LR, 22 WMR; 6-shot cylinder; 4 3/4", 6", 9" barrel; 10" overall length (4 3/4" barrel); weighs 32 oz.; European hardwood grips; blade front sight, adjustable rear; blue finish only. From European American Armory Corp. Introduced 1991; no longer available.
Perf.: $75 **Exc.:** $65 **VGood** $50
4 3/4", barrel 22 LR/22 WMR combo
Perf.: $85 **Exc.:** $75 **VGood** $65
9" barrel, 22 LR/22 WMR combo
Perf.: $95 **Exc.:** $85 **VGood** $75

E.A.A.

EA22T Target
European 320
European 380/DA
European 380 Ladies
Windicator Standard Grade
Windicator Tactical Grade
Windicator Target Grade
Witness
Witness Compact
Witness Gold Team
Witness Silver Team

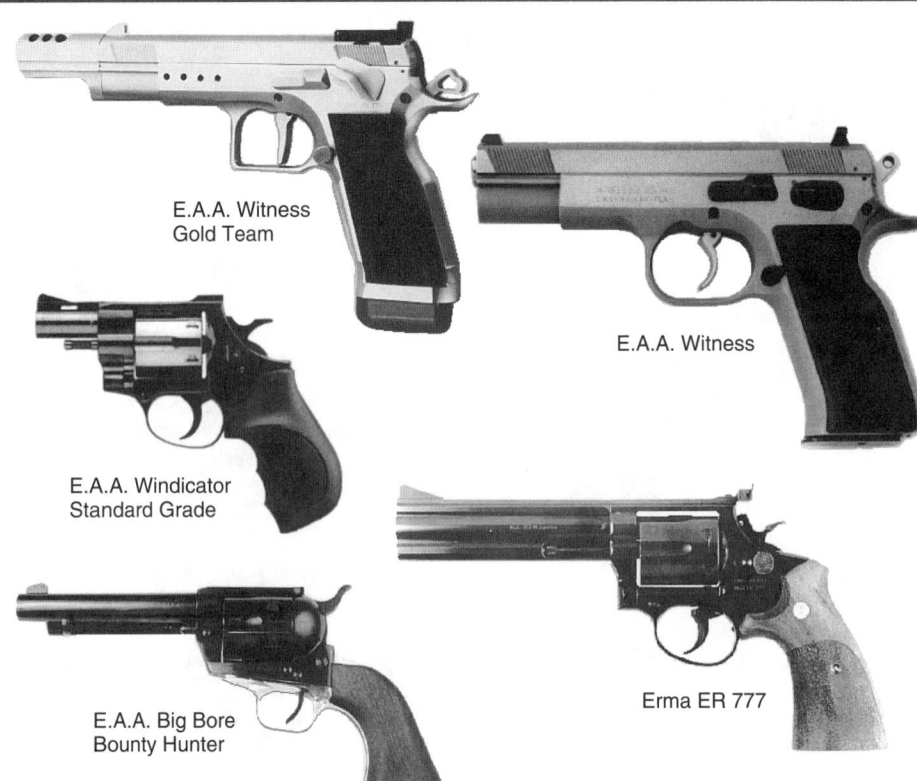

E.A.A. Witness
Gold Team

E.A.A. Witness

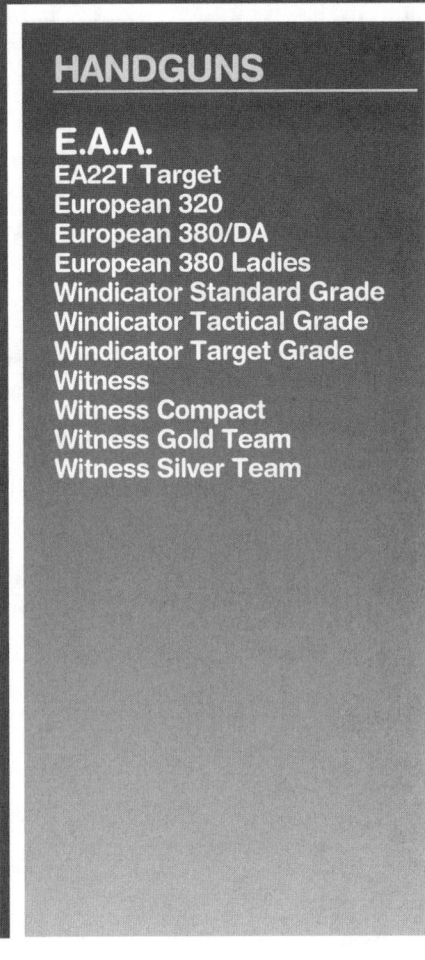

E.A.A. Windicator
Standard Grade

E.A.A. Big Bore
Bounty Hunter

Erma ER 777

E.A.A. EA22T TARGET

Semi-automatic; 22 LR; 12-shot magazine; 6" barrel; 9 1/8" overall length; weighs 40 oz.; ramped blade front sight, fully-adjustable rear; checkered walnut grips; thumbrest; blued finish. Made in Italy. Introduced 1991; no longer imported.

Perf.: $350 **Exc.:** $275 **VGood:** $200

E.A.A. EUROPEAN 320

Semi-automatic; 22 LR, 32 ACP; 380 ACP, 10-shot (22), 7-shot (32 ACP, 380 ACP); 3 7/8" overall length; weighs 26 oz.; fixed blade front sight, windage-adjustable rear; European hardwood grips; magazine, thumb and firing pin safeties; external hammer; safety-lever; chrome or blued finish. Made in Italy. Introduced 1991; still imported by European American Armory Corp. **LMSR:** $160

Perf.: $125 **Exc.:** $100 **VGood:** $85

E.A.A. EUROPEAN 380/DA

Revolver; double action; 380 ACP; 7-shot; 3 7/8" barrel; 7 3/8" overall length; 26 oz.; double-action trigger mechanism; steel construction; blue, chrome or blue/chrome finish. From European American Armory Corp. Introduced 1992; still imported. Blue finish

Perf.: $125 **Exc.:** $100 **VGood:** $85

Chrome finish or blue/chrome finish

Perf.: $135 **Exc.:** $115 **VGood:** $100

E.A.A. European 380 Ladies

Same specs as European 380/DA except blue or gold finish; ivory polymer grips. From European American Armory Corp.

Perf.: $200 **Exc.:** $175 **VGood:** $150

E.A.A. WINDICATOR STANDARD GRADE

Revolver; double action; 22 LR, 22 LR/22 WMR combo, 38 Spl.; 8-shot cylinder (22 LR, 22 LR/22 WMR); 6-shot cylinder (38 Spl.); 4", 6" barrel (22); 2", 4" barrel (38 Spl.); 8 13/16" overall length (4" barrel); weighs 38 oz.

(22, 4" barrel); rubber grips with finger grooves; blade front sight, fixed or adjustable (22); fixed sights (32, 38 Spl. only); swing-out cylinder; hammer-block safety; blue finish. Imported from Germany by European American Armory Corp. Introduced 1991.

38 Special, 2" barrel

Perf.: $150 **Exc.:** $125 **VGood:** $110

22 LR, 6" barrel, 38 Special, 4" barrel

Perf.: $140 **Exc.:** $120 **VGood:** $110

22 LR/22 WMR combo, 4" or 6" barrel

Perf.: $200 **Exc.:** $150 **VGood:** $135

E.A.A. WINDICATOR TACTICAL GRADE

Revolver; double action; 38 Spl.; 2", 4" barrel; fixed sights; compensator on 4"; bobbed hammer (DA only) on 2" model. Imported from Germany by European American Armory Corp. Introduced 1991; no longer imported.

2" barrel, bobbed hammer

Perf.: $200 **Exc.:** $175 **VGood:** $150

4" barrel, compensator

Perf.: $250 **Exc.:** $225 **VGood:** $200

E.A.A. WINDICATOR TARGET GRADE

Revolver; double action; 22 LR, 38 Spl., 357 Mag.; 8-shot cylinder (22 LR); 6-shot (38 Spl., 357 Mag.); 6" barrel; 11 13/16" overall length; weighs 50 oz.; walnut competition-style grips; blade front sight with three interchangeable blades, fully-adjustable rear; adjustable trigger with trigger stop and trigger shoe; frame drilled and tapped for scope mount; target hammer; comes with barrel weights, plastic carrying box. Imported from Germany by European American Armory Corp. Introduced 1991; still imported.

Perf.: $350 **Exc.:** $300 **VGood:** $250

E.A.A. WITNESS

Semi-automatic; 9mm Para., 40 S&W, 41 AE, 45 ACP; 16-shot (9mm), 12-shot (40 S&W), 11-shot (41 AE), 10-shot (45 ACP); 4 11/16" barrel; 8 1/8" overall length; weighs 35 3/8 oz.; undercut blade front, open adjustable rear sight; squared-off triggerguard; frame mounted safety; checkered rubber grips; blue, satin chrome finish.

9mm, blue finish

Perf.: $400 **Exc.:** $350 **VGood:** $300

9mm, satin finish

Perf.: $425 **Exc.:** $375 **VGood:** $325

40 S&W, blue finish

Perf.: $450 **Exc.:** $400 **VGood:** $350

40 S&W, satin finish

Perf.: $475 **Exc.:** $425 **VGood:** $375

45 ACP, blue finish

Perf.: $475 **Exc.:** $425 **VGood:** $375

45 ACP, satin finish

Perf.: $500 **Exc.:** $450 **VGood:** $400

41 AE

Perf.: $525 **Exc.:** $475 **VGood:** $425

E.A.A. Witness Compact

Same specs as Witness except 9mm, 40 S&W, 45 ACP; 10-shot (9mm), 9-shot (40 S&W), 8-shot (45 ACP); 3 5/8" barrel; 7 1/4" overall length; weighs 30 oz.; blue or hard chrome finish. Introduced 1995; still in production.

New: $375 **Perf.:** $300 **Exc.:** $250

45 ACP

New: $450 **Perf.:** $400 **Exc.:** $325

E.A.A. WITNESS GOLD TEAM

Semi-automatic; single action; 9mm Para., 9x21, 38 Super, 10mm, 40 S&W, 45 ACP; 10-shot magazine; 5 1/8" barrel; 9 5/8" overall length; weighs 41 1/2 oz.; checkered walnut competition-style grips; square post front sight, fully-adjustable rear; triple-chamber cone compensator; competition SA trigger; extended safety and magazine release; competition hammer; beveled magazine well; beavertail grip; hand-fitted major components; match-grade barrel; hard chrome finish. From E.A.A. Custom Shop. Marketed by European American Armory Corp. Introduced 1992; still offered.

Perf.: $1750 **Exc.:** $1500 **VGood:** $1250

E.A.A. WITNESS SILVER TEAM

Semi-automatic; single action; 9mm Para., 9x21, 38 Super, 10mm, 40 S&W, 45 ACP; 10-shot magazine; 5 1/8" barrel; 9 5/8" overall length; weighs 41 1/2 oz.; black rubber grips; square post front sight, fully-adjustable rear; double-chamber compensator; competition SA trigger; extended safety; oval magazine release; competition hammer; beveled magazine well; beavertail grip; hand-fitted major components; match-grade barrel; double-dip blue finish. Comes with Super Sight and drilled and tapped for scope mount. Built for the intermediate competition shooter. From E.A.A. Custom Shop. Marketed by European American Armory Corp. Introduced 1992; still offered.

Perf.: $850 **Exc.:** $750 **VGood:** $650

Erma ESP-85A
Junior Match

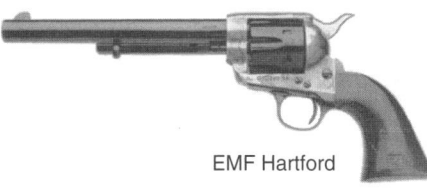

EMF Hartford

Erma ER 773

EMF 1894 TARGET BISLEY

Revolver; single action; 45 Colt; 5 1/2", 7 1/2" barrel; 11" overall length (5 1/2" barrel); weighs about 45 oz.; blade front, fixed rear sight; smooth walnut grips, specially shaped grips frame and triggerguard; wide spur hammer; blue or nickel finish. Imported by E.M.F. Introduced 1995; still imported.

Blue finish
  New: $600   **Perf.:** $525   **Exc.:** $475
Nickel finish
  New: $675   **Perf.:** $600   **Exc.:** $550

EMF DAKOTA 1875 OUTLAW

Revolver; single action; 357, 44-40, 45 Colt; 7 1/2" barrel; 13 1/2" overall length; weighs 46 oz.; smooth walnut grips; blade front, fixed groove rear sight; authentic copy of 1875 Remington with firing pin in hammer; color or case-hardened frame, blue cylinder, barrel; steel backstrap and brass triggerguard. Also offered in nickel finish, factory engraved. Imported by E.M.F.; still imported.

  New: $400   **Perf.:** $350   **Exc.:** $300
Nickel finish
  New: $450   **Perf.:** $400   **Exc.:** $375
Engraved
  New: $500   **Perf.:** $450   **Exc.:** $400
Engraved nickel
  New: $700   **Perf.:** $650   **Exc.:** $600

EMF DAKOTA 1890 POLICE

Revolver; single action; 357, 44-40, 45 Colt; 5 1/2" barrel; 12 1/2" overall length; weighs 40 oz.; smooth walnut grips; blade front, fixed groove rear sight; authentic copy of 1875 Remington with firing pin in hammer; color case-hardened frame, blue cylinder, barrel; steel backstrap and brass triggerguard; lanyard ring in butt. Also offered in nickel finish, factory engraved. Imported by E.M.F.; still imported.

  Perf.: $400   **Exc.:** $350   **VGood:** $300
Nickel finish
  Perf.: $550   **Exc.:** $500   **VGood:** $375
Engraved
  Perf.: $600   **Exc.:** $550   **VGood:** $500
Engraved nickel
  Perf.: $800   **Exc.:** $750   **VGood:** $600

EMF DAKOTA NEW MODEL

Revolver; single action; 357 Mag., 44-40, 45 Colt; 6-shot cylinder; 3 1/2", 4 3/4", 5 1/2", 7 1/2" barrel; 13" overall length (7 1/2" barrel); weighs 45 oz.; blade front, fixed rear sight; smooth walnut grips; color case-hardened frame, black nickel backstrap and triggerguard; also offered nickel plated. Imported by E.M.F.; still imported.

  Perf.: $350   **Exc.:** $300   **VGood:** $250
Nickel-plated
  Perf.: $450   **Exc.:** $375   **VGood:** $300

EMF HARTFORD

Revolver; single action; 22 LR, 357 Mag., 32-20, 38-40, 44-40, 44 Spec., 45 Colt; 6-shot cylinder; 4 3/4", 5 1/2", 7 1/2" barrel; 13" overall length (7 1/2" barrel); weighs 45 oz.; smooth walnut grips; blade front, fixed rear sight; identical to the original Colts with inspector cartouche on left grip, original patent dates and U.S. markings; all major parts serial numbered using original Colt-style lettering, numbering; bullseye ejector head and color case-hardening on frame and hammer. Imported by E.M.F. Introduced 1990; still imported.

  Perf.: $450   **Exc.:** $375   **VGood:** $300
Cavalry or Artillery model
  Perf.: $450   **Exc.:** $375   **VGood:** $300
Nickel-plated
  Perf.: $550   **Exc.:** $475   **VGood:** $400
Engraved, nickel-plated
  Perf.: $700   **Exc.:** $650   **VGood:** $550

EMF HARTFORD PINKERTON MODEL

Revolver; single action; 45 Colt; 6-shot cylinder; 4" barrel; 10" overall length; weighs 40 oz.; smooth walnut grips; blade front, fixed rear sight; bird's-head grip; identical to the original Colts with original patent dates and markings; all major parts serial numbered using original Colt-style lettering, numbering; bullseye ejector head and color case-hardening on frame and hammer. Imported by E.M.F. Introduced 1990; still imported.

  Perf.: $400   **Exc.:** $350   **VGood:** $300

ERMA ER 772

Revolver; double action; 22 LR; 6-shot; 6" shrouded barrel; 11 3/16" overall length; blade front, adjustable micro rear sight; sporting or adjustable match-type walnut stocks; solid rib; blued finish. Made in Germany. Introduced 1989.

  Perf.: $950   **Exc.:** $800   **VGood:** $700

ERMA ER 773

Revolver; double action; 32 S&W Long; 6-shot; 6" shrouded barrel; 11 3/16" overall length; weighs 46 oz.; blade front, adjustable micro rear sight; sporting or adjustable match-type walnut stocks; solid rib; blued finish. Made in Germany. Introduced 1989; still in production.

  Perf.: $800   **Exc.:** $650   **VGood:** $550

ERMA ER 777

Revolver; double action; 357 Mag.; 6-shot; 4", 5 1/2" shrouded barrel; 9 1/2" overall length (4" barrel); weighs 44 oz.; blade front, adjustable micro rear sight; checkered sport grips; solid rear; blued finish. Made in Germany. Introduced 1989; still imported.

  Perf.: $750   **Exc.:** $600   **VGood:** $500

ERMA ESP-85A SPORTING/MATCH TARGET

Semi-automatic; single action; 22 LR, 32 S&W Long; 8-shot (22 LR), 5-shot (32 S&W Long); 6" barrel; 10" overall length; weighs 40 oz.; checkered walnut thumbrest grips; adjustable target stocks optional; interchangeable blade front sight, micrometer fully-adjustable rear; interchangeable caliber conversion kit available; adjustable trigger, trigger stop. Imported from Germany by Precision Sales Int'l. Introduced 1988; match model still imported.

HANDGUNS

EMF
1894 Target Bisley
Dakota 1875 Outlaw
Dakota 1890 Police
Dakota New Model
Hartford
Hartford Pinkerton Model

ERMA
ER 772
ER 773
ER 777
ESP-85A Sporting/Match Target
ESP-85A Golden Target
ESP-85A Junior Match

Erma ESP-85A Target

22 LR
  Perf.: $1200   **Exc.:** $1000   **VGood:** $850
32 S&W Long
  Perf.: $1500   **Exc.:** $1300   **VGood:** $950
22 LR, chrome finish
  Perf.: $1300   **Exc.:** $1100   **VGood:** $900
22 LR conversion unit
  Perf.: $750   **Exc.:** $650   **VGood:** $600
32 S&W conversion unit
  Perf.: $1000   **Exc.:** $900   **VGood:** $800

Erma ESP-85A Golden Target

Same specs as the ESP-85A except high-polish gold finish on the slide; adjustable match stocks with finger grooves; comes with fully interchangeable 6" barrels for 22 LR and 32 S&W. Imported from Germany by Precision Sales International. Introduced 1994; no longer imported.

  Perf.: $2000   **Exc.:** $1850   **VGood:** $1500

Erma ESP-85A Junior Match

Same specs as the ESP-85A except 22 LR; blue finish only; stippled non-adjustable walnut match grips (adjustable grips optional). Imported from Germany by Precision Sales International. Introduced 1995; still imported.

  Perf.: $850   **Exc.:** $750   **VGood:** $700

Erma
KGP22 (See Erma KGP69)
KGP32 (See Erma KGP68A)
KGP38 (See Erma KGP68A)
KGP68A
KGP69
LA 22
RX 22

ESFAC
Four Aces Model 1
Four Aces Model 2
Four Aces Model 3
Four Aces Model 4
Little Ace Derringer
Pocket Pony

Falcon
Portsider

FAS
Model 601 Match
Model 602 Match
Model 603 Match
Model 607 Match

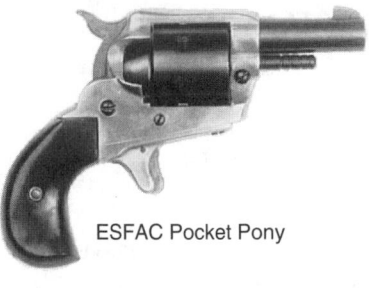

ESFAC Pocket Pony

FAS Model 607 Match

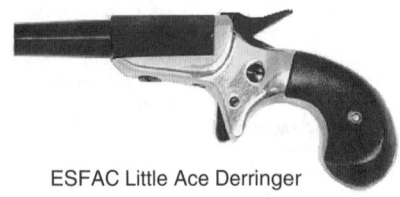

ESFAC Little Ace Derringer

Falcon Portsider

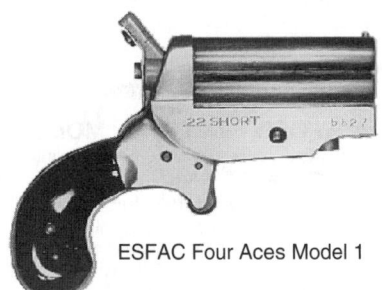

ESFAC Four Aces Model 1

FEG FP9

ERMA KGP68A
Semi-automatic; 32 ACP, 380 ACP; 6-shot; 3 1/2", 4" barrel; 6 3/4" overall length; weighs 20 oz.; adjustable blade front, fixed rear sight; checkered plastic or walnut grips; Luger look-alike; blued finish. Later models designated KGP32, KGP38. Introduced 1968. **LMSR: $795**
   Perf.: $450   **Exc.:** $350   **VGood:** $300

ERMA KGP69
Semi-automatic; 22 LR; 8-shot magazine; 3 3/4" barrel; 7 3/4" overall length; weighs 22 oz.; fixed sights; checkered plastic or walnut grips; Luger look-alike; blued finish. Later models designated KGP22. Introduced 1969.
   Perf.: $300   **Exc.:** $250   **VGood:** $200

ERMA LA 22
Semi-automatic; 22 LR; 10-shot magazine; 8 3/16", 11 3/4" barrel; adjustable target sights; checkered plastic grips; Luger look-alike; blued finish. Introduced 1964; discontinued 1967.
   Exc.: $350   **VGood:** $275   **Good:** $175

ERMA RX 22
Semi-automatic; double action; 22 LR; 8-shot magazine; 3 1/4" barrel, 5 9/16" overall length; fixed sights; plastic wrap-around grips; thumb safety; patented ignition safety system; polished blued finish. Assembled in USA. Introduced 1980; discontinued 1986.
   Perf.: $125   **Exc.:** $110   **VGood:** $100

ESFAC FOUR ACES MODEL 1
Derringer; 22 Short; 4-shot; 1 11/16" barrels (4); black plastic grips; round butt; post front sight; manganese bronze frame; spur trigger; barrel selector on hammer; blued steel barrels. Introduced 1973; maker ceased business 1974.
   Exc.: $175   **VGood:** $135   **Good:** $100

ESFAC Four Aces Model 2
Same specs as Model 1 except square-butt frame.
   Exc.: $175   **VGood:** $135   **Good:** $100

ESFAC Four Aces Model 3
Same specs as Model 2 except 2" barrels; 22 LR.
   Exc.: $150   **VGood:** $125   **Good:** $100

ESFAC Four Aces Model 4
Same specs as Model 3 except slightly longer overall length.
   Exc.: $160   **VGood:** $125   **Good:** $100

ESFAC LITTLE ACE DERRINGER
Derringer; single shot; 22 Short; 2" barrel; black plastic grips; no sights; manganese bronze frame; blued barrel; color case-hardened spur trigger, hammer. Introduced 1973; maker ceased business 1974.
   Exc.: $100   **VGood:** $75   **Good:** $50

ESFAC POCKET PONY
Revolver; single action; 22 LR; 6-shot; 1 3/8" barrel; fixed sights; black plastic grips; manganese bronze frame; blued barrel; color case-hardened trigger, hammer; non-fluted cylinder. Introduced 1973; maker ceased business 1974.
   Exc.: $150   **VGood:** $125   **Good:** $100

FALCON PORTSIDER
Semi-automatic; 45 ACP; 7-shot magazine; 5" barrel; 8 1/2" overall length; weighs 38 oz.; fixed combat sights; checkered walnut stocks; stainless steel construction; extended safety; wide grip safety; enlarged left-hand ejection port; extended ejector. Introduced 1986; dropped about 1989.
   Perf.: $500   **Exc.:** $450   **VGood:** $400

FAS MODEL 601 MATCH
Semi-automatic; 22 Short; 5-shot magazine; 5 1/2" barrel; 11" overall length; weighs 41 oz.; gas ported barrel; wrap-around adjustable walnut stocks; match blade front sight, fully-adjustable open notch rear; magazine inserts from top; adjustable, removable trigger group; single-lever take-down. Made in Italy. Introduced 1984; still imported by Nygord Precision Products. **LMSR: $1250**
   Perf.: $1000   **Exc.:** $825   **VGood:** $650

FAS MODEL 602 MATCH
Semi-automatic; single action; 22 LR; 5-shot magazine; 5 5/8" barrel; 11" overall length; weighs 37 oz.; walnut wrap-around grips in small, medium or large, or adjustable; match blade front, open notch fully-adjustable rear sight; magazine inserted from top; adjustable and removable trigger mechanism; single-lever takedown. Imported from Italy by Nygord Precision Products; no longer imported.
   Perf.: $850   **Exc.:** $750   **VGood:** $600

FAS MODEL 603 MATCH
Semi-automatic; single action; 32 S&W; 5-shot magazine; 5 5/8" barrel; 11" overall length; weighs 37 oz.; walnut wrap-around grips in sizes small, medium or large, or adjustable; blade front, open notch fully-adjustable rear match sight; magazine inserted from top; adjustable and removable trigger mechanism; single-lever takedown. Imported from Italy by Nygord Precision Products; still imported. **LMSR: $1175**
   Perf.: $900   **Exc.:** $800   **VGood:** $700

FAS MODEL 607 MATCH
Semi-automatic; single action; 22 LR; 5-shot magazine; 5 5/8" barrel; 11" overall length; weighs 37 oz.; walnut wrap-around grips in small, medium or large, or adjustable; blade front, open notch fully-adjustable rear match sight; magazine inserted from top; adjustable and removable trigger mechanism; single-lever takedown. Imported from Italy by Nygord Precision Products; still imported. **LMSR: $1175**
   Perf.: $900   **Exc.:** $800   **VGood:** $700

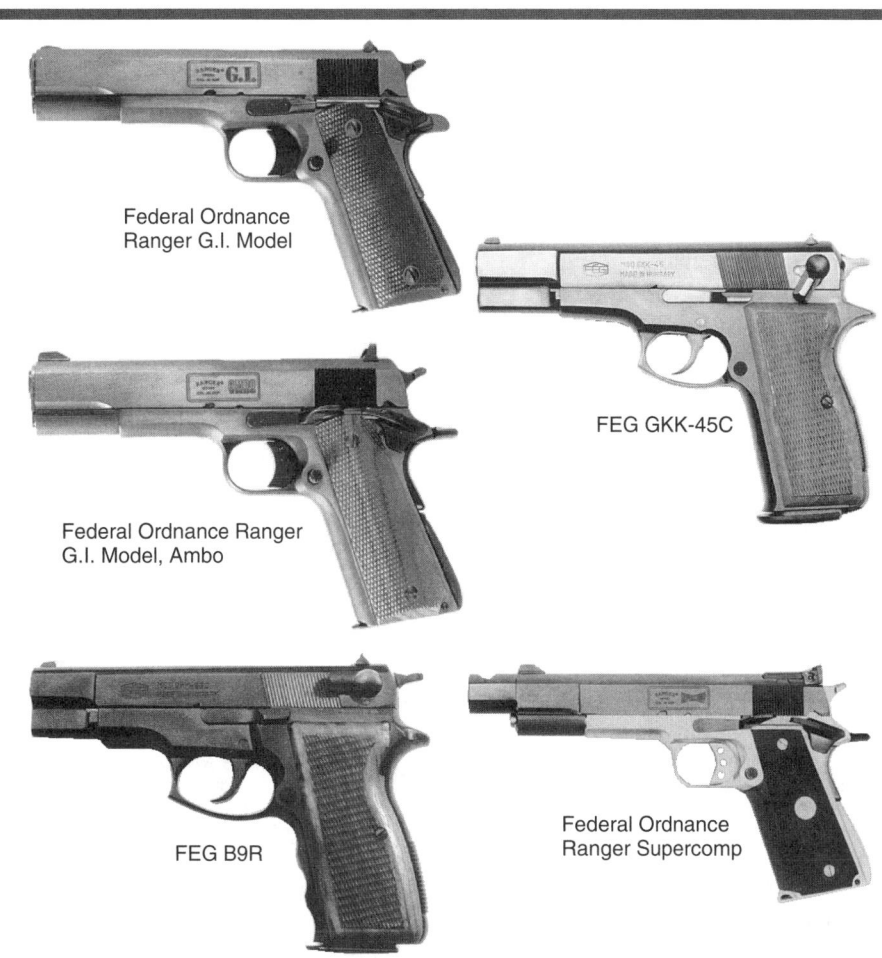

Federal Ordnance
Ranger G.I. Model

FEG GKK-45C

Federal Ordnance Ranger
G.I. Model, Ambo

FEG B9R

Federal Ordnance
Ranger Supercomp

Feather
Guardian Angel

Federal Ordnance
Ranger Alpha
Ranger G.I. Model
Ranger G.I. Model, Ambo
Ranger G.I. Model, Extended
Ranger G.I. Model, Lite
Ranger G.I. Model, Ten Auto
Ranger Supercomp

FEG
B9R
FP9
GKK-40C
GKK-45C

FEATHER GUARDIAN ANGEL
Derringer; 9mm Para., 38 Spl.; 3″ barrel; 5 1/2″ overall length; weighs 17 oz.; black composition stocks; fixed sights; stainless steel; matte finish; two-shot drop-in magazine. Introduced 1988; dropped 1989. New 22 LR/22 WMR model introduced 1990 with 2″ barrels; still in production. **LMSR: $120**
   Perf.: $100   **Exc.:** $80   **VGood:** $60

FEDERAL ORDNANCE RANGER ALPHA
Semi-automatic; single action; 38 Super, 10mm Auto, 45 ACP; 9-shot (38 Super), 8-shot (10mm Auto), 7-shot (45 ACP); 5″, 6″ barrel, ported or unported; 8 1/2″ overall length; weighs 42 oz.; wrap-around rubber grips; interchangeable front sight, fully-adjustable Peters Stahl rear; Peters Stahl linkless barrel system; polygonal rifling; extended grip safety, thumb safety, slide release, magazine release; high polish blue finish. Imported by Federal Ordnance. Introduced 1990; no longer imported.
5″ unported barrel, 38 Super, 45 ACP
   Perf.: $850   **Exc.:** $650   **VGood:** $600
5″ ported barrel, 38 Super 45 ACP or unported 10mm
   Perf.: $900   **Exc.:** $700   **VGood:** $650
5″, 6″ ported barrel, 10mm, 38 Super, 45 ACP
   Perf.: $950   **Exc.:** $750   **VGood:** $700
6″ ported barrel, 10mm
   Perf.: $950   **Exc.:** $750   **VGood:** $700

FEDERAL ORDNANCE RANGER G.I. MODEL
Semi-automatic; single action; 45 ACP; 7-shot magazine; 5″ barrel; 8 1/2″ overall length; weighs 38 oz.; checkered plastic grips; blade front sight, drift-adjustable rear; made in U.S. from 4140 steel and other high-strength alloys; barrel machined from a forged billet. From Federal Ordnance, Inc. Introduced 1988; no longer produced.
   Perf.: $350   **Exc.:** $300   **VGood:** $275

Federal Ordnance Ranger G.I. Model, Ambo
Same specs as Ranger G.I. Model except ambidextrous slide release, safety; extended grip safety, thumb safety; slide and magazine releases; weighs 40 oz. Introduced 1990; discontinued 1992.
   Perf.: $400   **Exc.:** $350   **VGood:** $300

Federal Ordnance Ranger G.I. Model, Extended
Same specs as Ranger G.I. Model except extended grip safety, thumb safety; slide and magazine releases; weighs 40 oz. Introduced 1990; discontinued 1992.
   Perf.: $425   **Exc.:** $375   **VGood:** $350

Federal Ordnance Ranger G.I. Model, Lite
Same specs as Ranger G.I. Model except aluminum alloy frame; Millet fixed sights; rubber wrap-around grips; lightened speed trigger; black anodized frame; blued steel slide. Introduced 1990; discontinued 1992.
   Perf.: $375   **Exc.:** $325   **VGood:** $300

Federal Ordnance Ranger G.I. Model, Ten Auto
Same specs as Ranger G.I. Model except 10mm Auto chambering. Introduced 1990; discontinued 1991.
   Perf.: $600   **Exc.:** $525   **VGood:** $450

FEDERAL ORDNANCE RANGER SUPERCOMP
Semi-automatic; single action; 10mm Auto, 45 ACP; 8-shot (10mm Auto), 7-shot (45 ACP); 6″ barrel; 9 1/2″ overall length; weighs 42 oz.; wrap-around rubber grips; ramped blade front sight, fully-adjustable, low-profile Ranger rear; Peters Stahl linkless barrel system with polygonal rifling and integral competition compensator; lightened speed trigger; beveled magazine well; ramped and throated barrel; blued slide, electroless nickel frame. From Federal Ordnance. Introduced 1990; no longer produced.
10mm
   Perf.: $1000   **Exc.:** $800   **VGood:** $700
45 ACP
   Perf.: $1000   **Exc.:** $800   **VGood:** $700

FEG B9R
Semi-automatic; 380 ACP; 10-shot magazine; 4″ barrel; 7″ overall length; weighs 25 oz.; blade front, drift-adjustable rear sight; hand-checkered walnut grips; hammer drop safety; grooved backstrap; squared trigger guard. Introduced 1993; still imported. **LMSR: $312**
   New: $290   **Perf.:** $250   **Exc.:** $200

FEG FP9
Semi-automatic; single action; 9mm Para.; 10-shot magazine; 5″ barrel; 7 7/8″ overall length; weighs 35 oz.; checkered walnut grips; blade front sight, windage-adjustable rear; full-length ventilated rib; polished blue finish; comes with extra magazine. Imported from Hungary by Century International Arms. Introduced 1993; still imported. **LMSR: $269**
   New: $225   **Perf.:** $200   **Exc.:** $175

FEG GKK-40C
Semi-automatic; double action; 40 S&W; 9-shot magazine; 4 1/8″ barrel; 7 3/4″ overall length; weighs 36 oz.; hand-checkered walnut grips; blade front sight, windage-adjustable rear; three-dot system; combat-type trigger guard; polished blue finish; comes with two magazines, cleaning rod. Imported from Hungary by K.B.I., Inc. Introduced 1995; still imported. **LMSR: $399**
   New: $325   **Perf.:** $300   **Exc.:** $250

FEG GKK-45C
Semi-automatic; double action; 45 ACP; 8-shot magazine; 4 1/8″ barrel; 7 3/4″ overall length; weighs 36 oz.; hand-checkered walnut grips; blade front sight, windage-adjustable rear; three-dot system; combat-type trigger guard; polished blue finish; comes with two magazines, cleaning rod. Imported from Hungary by K.B.I., Inc. Introduced 1995; still imported. **LMSR: $399**
   New: $325   **Perf.:** $300   **Exc.:** $250

FEG
GKK-92C
MBK-9HP
MBK-9HPC
P9R
PJK-9HP
PMK-380
SMC-22
SMC-380
SMC-918

F.I.E.
Arminus Model 222
Arminus Model 222B
(See F.I.E. Arminus Model 222)
Arminus Model 232
Arminus Model 232B
(See F.I.E. Arminus Model 232)
Arminus Model 357
Arminus Model 357T
Arminus Model 382TB
Arminus Model 384TB

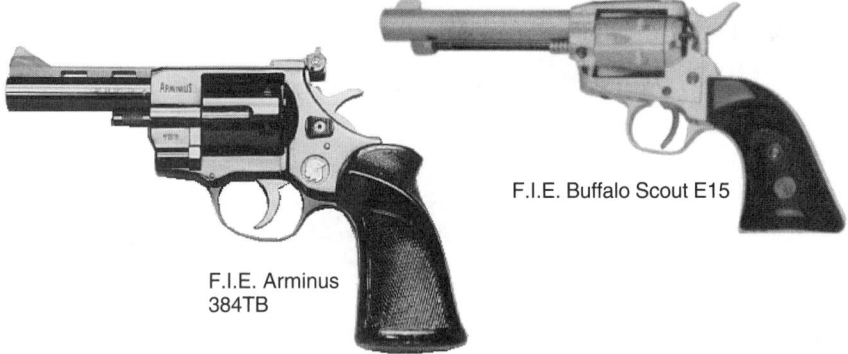

F.I.E. Buffalo Scout E15

F.I.E. Arminus 384TB

FEG PJK-9HP

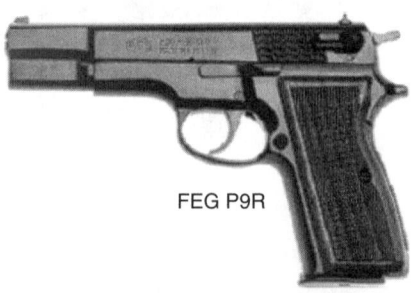

FEG P9R

FEG GKK-92C

Semi-automatic; double action; 9mm Para.; 14-shot magazine; 4″ barrel; 7 ³/₈″ overall length; weighs 34 oz.; hand-checkered walnut grips; blade front sight, windage-adjustable rear; slide-mounted safety; hooked trigger guard; finger-grooved frontstrap; polished blue finish. Imported from Hungary by K.B.I., Inc. Introduced 1992; no longer imported.

Perf.: $300   **Exc.:** $250   **VGood:** $200

FEG MBK-9HP

Semi-automatic; double action; 9mm Para.; 14-shot magazine; 4 ⁵/₈″ barrel; 7 ³/₈″ overall length; weighs 34 oz.; hand-checkered walnut grips; blade front sight, windage-adjustable rear; slide-mounted safety; hooked trigger guard; smooth frontstrap; polished blue finish. Imported from Hungary by K.B.I., Inc. Introduced 1992; no longer imported.

Perf.: $225   **Exc.:** $200   **VGood:** $185

FEG MBK-9HPC

Semi-automatic; double action; 9mm Para.; 14-shot magazine; 4″ barrel; 7 ¹/₄″ overall length; weighs 34 oz.; hand-checkered walnut grips; blade front sight, windage-adjustable rear; slide-mounted safety; hooked trigger guard; smooth frontstrap; polished blue finish. Imported from Hungary by K.B.I., Inc. Introduced 1992; no longer imported.

Perf.: $225   **Exc.:** $200   **VGood:** $185

FEG P9R

Semi-automatic; double action; 9mm Para.; 10-shot magazine; 4 ⁵/₈″ barrel; 7 ¹⁵/₁₆″ overall length; weighs 35 oz.; checkered walnut grips; blade front sight, windage-adjustable rear; slide-mounted safety; all-steel construction with polished blue finish; comes with extra magazine. Imported from Hungary by Century International Arms. Introduced 1993; still imported. **LMSR: $262**

New: $225   **Perf.:** $200   **Exc.:** $185

FEG PJK-9HP

Semi-automatic; single action; 9mm Para.; 10-shot magazine; 4 ³/₄″ barrel; 8″ overall length; weighs 32 oz.; hand-checkered walnut grips; blade front sight, windage-adjustable rear; three-dot system; polished blue or hard chrome finish; rounded combat-style serrated hammer; comes with two magazines and cleaning rod. Imported from Hungary by K.B.I., Inc. Introduced 1992; still imported. **LMSR: $349**

New: $275   **Perf.:** $225   **Exc.:** $200
Hard chrome finish
New: $300   **Perf.:** $250   **Exc.:** $225

FEG PMK-380

Semi-automatic; double action; 380 ACP; 7-shot magazine; 4″ barrel; 7″ overall length; weighs 21 oz.; checkered black nylon thumbrest grips; blade front sight, windage-adjustable rear; anodized aluminum frame; polished blue slide; comes with two magazines, cleaning rod. Imported from Hungary by K.B.I., Inc. Introduced 1992; no longer imported.

Perf.: $250   **Exc.:** $200   **VGood:** $175

FEG SMC-22

Semi-automatic; double action; 22 LR; 8-shot magazine; 3 ¹/₂″ barrel; 6 ¹/₈″ overall length; weighs 18 ¹/₂ oz.; checkered composition thumbrest grips; blade front sight, windage-adjustable rear; alloy frame, steel slide; blue finish; comes with two magazines, cleaning rod. Patterned after the PPK pistol. Imported from Hungary by K.B.I., Inc. Introduced 1994; still imported. **LMSR: $279**

New: $235   **Perf.:** $200   **Exc.:** $185

FEG SMC-380

Semi-automatic; double action; 380 ACP; 6-shot magazine; 3 ¹/₂″ barrel; 6 ¹/₈″ overall length; weighs 18 ¹/₂ oz.; checkered composition thumbrest grips; blade front sight, windage-adjustable rear; alloy frame, steel slide; blue finish; comes with two magazines, cleaning rod. Patterned after the PPK pistol. Imported from Hungary by K.B.I. Introduced 1994; still imported. **LMSR: $279**

New: $250   **Perf.:** $225   **Exc.:** $200

FEG SMC-918

Semi-automatic; double action; 9x18 Makarov; 6-shot magazine; 3 ¹/₂″ barrel; 6 ¹/₈″ overall length; weighs 18 ¹/₂ oz.; checkered composition thumbrest grips; blade front sight, rear adjustable for windage; alloy frame, steel slide; blue finish; comes with two magazines, cleaning rod. Patterned after the PPK pistol. Imported from Hungary by K.B.I. Introduced 1995; still imported. **LMSR: $279**

New: $250   **Perf.:** $225   **Exc.:** $200

F.I.E. ARMINIUS MODEL 222

Revolver; double action; 22 LR, 22 WMR; 8-shot; 2″ barrel; plastic or walnut grips; swing-out cylinder; blued or chrome finish. Discontinued 1985; reintroduced as F.I.E. Arminius Model 222B 1987; discontinued 1989.

Perf.: $125   **Exc.:** $110   **VGood:** $95

F.I.E. ARMINIUS MODEL 232

Revolver; double action; 32 S&W; 7-shot; 2″ barrel; fixed or adjustable sights; plastic or walnut grips; swing-out cylinder; blued finish. Discontinued 1985; reintroduced as F.I.E. Arminius Model 232B 1987; discontinued 1989.

Perf.: $100   **Exc.:** $85   **VGood:** $65

F.I.E. ARMINIUS MODEL 357

Revolver; double action; 357 Mag.; 6-shot; 3″, 4″, 6″ barrel; plastic or walnut grips; blued or chrome finish. Discontinued 1990.

Perf.: $175   **Exc.:** $150   **VGood:** $135
Chrome finish
Perf.: $185   **Exc.:** $175   **VGood:** $150

F.I.E. ARMINIUS MODEL 357T

Revolver; double action; 357 Mag.; 6-shot; 2″ barrel; plastic or walnut grips; blued or chrome finish. Discontinued 1984.

Perf.: $200   **Exc.:** $175   **VGood:** $135
Chrome finish
Perf.: $225   **Exc.:** $200   **VGood:** $150

F.I.E. ARMINIUS MODEL 382TB

Revolver; double action; 38 Spl.; 6-shot; 2″ barrel; plastic or walnut grips; blued or chrome finish. Discontinued 1985.

Perf.: $110   **Exc.:** $85   **VGood:** $75
Chrome finish
Perf.: $125   **Exc.:** $100   **VGood:** $85

F.I.E. ARMINIUS MODEL 384TB

Revolver; double action; 38 Spl.; 6-shot; 4″ barrel; plastic or walnut grips; blued or chrome finish. Discontinued 1990.

Perf.: $125   **Exc.:** $110   **VGood:** $85
Chrome finish
Perf.: $140   **Exc.:** $125   **VGood:** $100

F.I.E. Hombre

F.I.E. Model 722 TP
Silhouette Pistol

F.I.E. Little Ranger

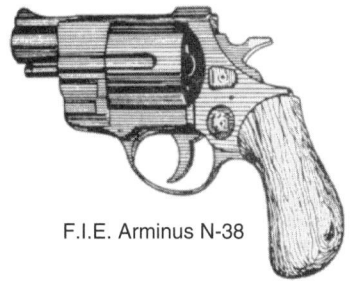

F.I.E. Arminus N-38

HANDGUNS

F.I.E.
Arminus Model 386TB
Arminus Model 522TB
Arminus Model 532TB
Arminus Model 722
Arminus Model 732B
Arminus Model N-38
Arminus Standard Model
Arminus Zephyr
Arminus Zephyr Lady
Buffalo Scout E15
Cowboy
Gold Rush
Hombre
KG-99 Pistol
Legend
Little Ranger
Model 722 TP Silhouette
Pistol

F.I.E. ARMINIUS MODEL 386TB
Revolver; double action; 38 Spl.; 6-shot; 6" barrel; plastic or walnut grips; blued or chrome finish. Discontinued 1990.
Perf.: $125   **Exc.:** $110   **VGood:** $85
Chrome finish
Perf.: $140   **Exc.:** $125   **VGood:** $100

F.I.E. ARMINIUS MODEL 522TB
Revolver; double action; 22 LR; 8-shot cylinder; 4" barrel; fixed sights; plastic or walnut grips; swing-out cylinder; blued finish. Discontinued 1990.
Perf.: $110   **Exc.:** $100   **VGood:** $85

F.I.E. ARMINIUS MODEL 532TB
Revolver; double action; 32 S&W; 7-shot cylinder; 4" barrel; adjustable sights; plastic or walnut grips; blued or chrome finish. Discontinued 1985.
Perf.: $125   **Exc.:** $100   **VGood:** $85
Chrome finish
Perf.: $135   **Exc.:** $110   **VGood:** $100

F.I.E. ARMINIUS MODEL 722
Revolver; double action; 22 LR; 8-shot cylinder; 6" barrel; fixed sights; plastic or walnut grips; swing-out cylinder; blued or chrome finish. Discontinued 1985.
Perf.: $110   **Exc.:** $100   **VGood:** $85
Chrome finish
Perf.: $125   **Exc.:** $110   **VGood:** $125

F.I.E. ARMINIUS MODEL 732B
Revolver; double action; 32 S&W; 7-shot cylinder; 6" barrel; fixed sights; blued finish. Discontinued 1988.
Perf.: $100   **Exc.:** $85   **VGood:** $75

F.I.E. ARMINIUS MODEL N-38
Revolver; double action; 38 Spl.; 6-shot cylinder; 2", 4" barrel; ramp front sight, fixed rear; Bulldog-style checkered plastic or walnut grips; swing-out cylinder; one-stroke ejection; blued or chrome finish. Marketed as Titan Tiger. Introduced 1978; discontinued 1990.
Perf.: $110   **Exc.:** $100   **VGood:** $85
Chrome finish
Perf.: $125   **Exc.:** $110   **VGood:** $100

F.I.E. ARMINIUS STANDARD MODEL
Revolver; double action; 22 LR, 22 WMR, 32 H&R Mag., 38 Spl.; 8-shot cylinder (rimfire), 6-shot swing-out cylinder (centerfire); 2", 4" barrel; 6 1/4" overall length (2" barrel); weighs 27 oz.; fixed sights; Bulldog-style checkered plastic or walnut grips; swing-out cylinder; one-stroke ejection; blued, chrome finish or gold-plated. Introduced 1989; discontinued 1990.
Perf.: $75   **Exc.:** $65   **VGood:** $50
Chrome finish
Perf.: $85   **Exc.:** $75   **VGood:** $60
Gold plated
Perf.: $85   **Exc.:** $75   **VGood:** $60

F.I.E. ARMINIUS ZEPHYR
Revolver; double action; 38 Spl.; 5-shot cylinder; 2" barrel; weighs 14 oz.; checkered grips; aluminum frame; blued finish. Introduced 1990; discontinued 1990.
Perf.: $125   **Exc.:** $110   **VGood:** $100

F.I.E. Arminius Zephyr Lady
Same specs as Zephyr except polymer ivory grips with scrimshaw; gold trim; gold case. Introduced 1990; discontinued 1990.
Perf.: $175   **Exc.:** $150   **VGood:** $135

F.I.E. BUFFALO SCOUT E15
Revolver; single action; 22 LR, 22 WMR; 6-shot cylinder; 4 3/4" barrel; 10" overall length; adjustable sights; black checkered plastic or uncheckered red walnut grips; sliding spring ejector; blued, chromed or blue/brass finish. Introduced 1978; discontinued 1990.
Perf.: $75   **Exc.:** $60   **VGood:** $50
Chrome finish
Perf.: $85   **Exc.:** $70   **VGood:** $60
Blue/brass finish
Perf.: $90   **Exc.:** $75   **VGood:** $65

F.I.E. COWBOY
Revolver; single action; 22 LR, 22 WMR; 6-shot cylinder; 3 1/4", 6 1/2" barrel; weighs 28 oz. (3 1/4" barrel); blade front sight, fixed rear; smooth nylon square butt grips; hammer-block safety; floating firing pin; blued finish. Introduced 1989; discontinued 1990.
Perf.: $75   **Exc.:** $65   **VGood:** $50
With combo cylinders
Perf.: $85   **Exc.:** $75   **VGood:** $60

F.I.E. GOLD RUSH
Revolver; single action; 22 LR, 22 WMR, 22 LR/22 WMR combo; 6-shot cylinder; 3 1/4", 4 3/4", 6 1/2" barrel; blade front, fixed rear sight; ivory-styled grips; gold band on barrel and cylinder; round or square butt; blue finish. Introduced 1989; discontinued 1990.
Perf.: $150   **Exc.:** $135   **VGood:** $125
With combo cylinders
Perf.: $200   **Exc.:** $175   **VGood:** $160

F.I.E. HOMBRE
Revolver; single action; 357 Mag., 44 Mag., 45 Colt; 5 1/2", 6", 7 1/2" barrels; 45 oz.; blade front, grooved backstrap rear sight; smooth walnut grips; medallion; color case-hardened frame; blue finish. Introduced 1970; discontinued 1990.
Perf.: $200   **Exc.:** $175   **VGood:** $150
Gold plated
Perf.: $225   **Exc.:** $200   **VGood:** $175

F.I.E. KG-99 PISTOL
Semi-automatic; 9mm Para.; 20-, 36-shot magazine; 3", 5" barrel; 12 1/2" overall length; weighs 46 oz.; fixed sights; military-type configuration. Introduced 1982; discontinued 1984.
Perf.: $500   **Exc.:** $425   **VGood:** $350

F.I.E. LEGEND
Revolver; single action; 22 LR, 22 LR/22 WMR combo; 6-shot cylinder; 4 3/4", 5 1/2" barrel; fixed sights; black checkered plastic or walnut grips; brass backstrap, trigger guard; case-hardened steel frame; blue finish. Introduced 1978; discontinued 1984.
Perf.: $110   **Exc.:** $100   **VGood:** $85
With combo cylinders
Perf.: $125   **Exc.:** $110   **VGood:** $100

F.I.E. LITTLE RANGER
Revolver; single action; 22 LR, 22LR/22WMR combo; 6-shot cylinder; 3 1/4" barrel; blade front, notch rear sight; American hardwood bird's-head grips; blue/black finish. Introduced 1986; discontinued 1990.
Perf.: $75   **Exc.:** $60   **VGood:** $50
With combo cylinders
Perf.: $100   **Exc.:** $75   **VGood:** $65

F.I.E. MODEL 722 TP SILHOUETTE PISTOL
Repeater; bolt action; 22 LR; 6-, 10-shot magazine; 10" barrel; 19" overall length; weighs 54 oz.; hooded front sight, micro-adjustable target rear; walnut-finished hardwood stock with stippled grip, forend; fully-adjustable match trigger; receiver grooved for scope mounting; marketed with two-piece high-base mount, both magazines. Made in Brazil. Introduced 1990; importation dropped 1990.
Perf.: $200   **Exc.:** $175   **VGood:** $150

F.I.E.
Model D38
Model D86
Spectre Pistol
SSP Model
SSP Model Lady
Super Titan II E32
Titan A27
Titan E27
Titan E38
Titan II E22
Titan II E22 Lady
Titan II E32
Titan Tiger
(See F.I.E. Arminus
Model N-38)
TZ-75
TZ-75 Series 88
TZ-75 Series 88
Compensated

F.I.E. Yellow Rose

F.I.E. Titan E27

F.I.E. Model D38

F.I.E. Texas Ranger

F.I.E. Super Titan II E32

F.I.E. Model D86

F.I.E. MODEL D38
Derringer; over/under; 38 Spl.; 2-shot cylinder; 2″ barrel; fixed sights; checkered plastic or walnut grips; spur trigger; chrome finish. Introduced 1979; discontinued 1985.
    Perf.: $60   **Exc.:** $50   **VGood:** $40

F.I.E. MODEL D86
Derringer; single shot; 38 Spl.; 3″ barrel; weighs 14 oz.; fixed sights; checkered black nylon or walnut grips; spur trigger; tip-up barrel; extractors; chrome or blued finish. Introduced 1986; dropped 1990.
    Perf.: $75   **Exc.:** $60   **VGood:** $50
Chrome finish
    Perf.: $85   **Exc.:** $70   **VGood:** $60

F.I.E. SPECTRE PISTOL
Semi-automatic; double action; 9mm Para., 45 ACP; 30-, 50-shot magazine; 6″ barrel; weighs 76 oz.; adjustable sights; military-type configuration. Introduced 1989; discontinued 1990.
    Perf.: $600   **Exc.:** $500   **VGood:** $400

F.I.E. SSP MODEL
Semi-automatic; single action; 32 ACP, 380 ACP; 3 1/8″ barrel; 6 1/4″ overall length; weighs 25 oz.; blade front, windage-adjustable rear sight; smooth European walnut grips; external hammer, magazine safety; blued, satin chrome finish. Introduced 1990; dropped 1990.
    Perf.: $100   **Exc.:** $85   **VGood:** $75
Satin chrome finish
    Perf.: $115   **Exc.:** $100   **VGood:** $85

F.I.E. SSP Model Lady
Same specs as SSP except scrimshawed polymer ivory grips; gold trimmed; gold case. Introduced 1990; discontinued 1990.
    Perf.: $200   **Exc.:** $175   **VGood:** $135

F.I.E. SUPER TITAN II E32
Semi-automatic; single action; 32 ACP, 380 ACP; 12-shot magazine (32 ACP), 11-shot (380 ACP); 3 7/8″ barrel; 6 3/4″ overall length; adjustable sights; walnut grips; blue finish. Introduced 1981; discontinued 1990.
    Perf.: $175   **Exc.:** $150   **VGood:** $125

F.I.E. TITAN A27
Semi-automatic; single action; 25 ACP; 6-shot magazine; 2 1/2″ barrel; 4 3/8″ overall length; fixed sights; checkered walnut stocks; all-steel construction; thumb, magazine safeties; exposed hammer; blued finish. Introduced 1978; discontinued 1988.
    Perf.: $110   **Exc.:** $85   **VGood:** $75

F.I.E. TITAN E27
Semi-automatic; single action; 25 ACP; 6-shot magazine; 4″, 6″ barrel; fixed sights; checkered walnut stocks; wide trigger; blue or chrome finish. Introduced 1977; dropped 1990.
    Perf.: $50   **Exc.:** $40   **VGood:** $35
Chrome finish
    Perf.: $55   **Exc.:** $45   **VGood:** $40

F.I.E. TITAN E38
Semi-automatic; single action; 25 ACP; 6-shot magazine; 4″, 6″ barrel; fixed sights; checkered walnut stocks; wide trigger; blue or chrome finish. Introduced 1990; dropped 1990.
    Perf.: $50   **Exc.:** $40   **VGood:** $35

F.I.E. TITAN II E22
Semi-automatic; single action; 22 LR; 10-shot; 3 7/8″ barrel; 6 3/4″ overall length; adjustable sights; walnut grips; magazine disconnector; firing pin block; standard slide safety; blued finish. Introduced 1978; discontinued 1990.
    Perf.: $110   **Exc.:** $85   **VGood:** $75

F.I.E. Titan II E22 Lady
Same specs as Titan II E22 except scrimshaw polymer ivory grips; blue/gold finish. Introduced 1990; discontinued 1990.
    Perf.: $145   **Exc.:** $125   **VGood:** $100

F.I.E. TITAN II E32
Semi-automatic; single action; 32 ACP, 380 ACP; 6-shot magazine; 3 7/8″ barrel; 6 3/4″ overall length; adjustable sights; checkered nylon thumbrest or walnut grips; magazine disconnector; firing pin block; standard slide safety; blue or chrome finish. Introduced 1978; discontinued 1990.
    Perf.: $175   **Exc.:** $150   **VGood:** $125
Chrome finish
    Perf.: $200   **Exc.:** $175   **VGood:** $135

F.I.E. TZ-75
Semi-automatic; double action; 9mm Para.; 15-shot magazine; 4 3/4″ barrel; 8 1/4″ overall length; weighs 35 oz.; undercut blade front, windage-adjustable open rear sight; walnut stocks or rubber grips; squared-off trigger guard; rotating slide-mounted safety; steel frame, slide; blue or chrome finish. Introduced 1983; dropped 1990.
    Perf.: $400   **Exc.:** $350   **VGood:** $275
Chrome finish
    Perf.: $425   **Exc.:** $375   **VGood:** $300

F.I.E. TZ-75 SERIES 88
Semi-automatic; double action; 9mm Para., 41 AE; 17-shot magazine (9mm Para.), 11-shot (41 AE); 4 3/4″ barrel; 8 1/4″ overall length; weighs 35 oz.; undercut blade front, removable rear sight; walnut or rubber grips; re-engineered version of the TZ-75 with frame mounted safety; matte blue, chrome finish or with blue slide/chrome frame. Introduced 1988; discontinued 1990.
    Perf.: $400   **Exc.:** $350   **VGood:** $325
Chrome finish
    Perf.: $425   **Exc.:** $375   **VGood:** $350
With blue slide/chrome frame
    Perf.: $400   **Exc.:** $375   **VGood:** $350

F.I.E. TZ-75 Series 88 Compensated
Same specs as TZ-75 Series 88 except 5 3/4″ compensated barrel; weighs 42 oz. Introduced 1990; discontinued 1990.
    Perf.: $650   **Exc.:** $600   **VGood:** $500
Chrome finish
    Perf.: $650   **Exc.:** $600   **VGood:** $500
With blue slide/chrome frame
    Perf.: $650   **Exc.:** $600   **VGood:** $500

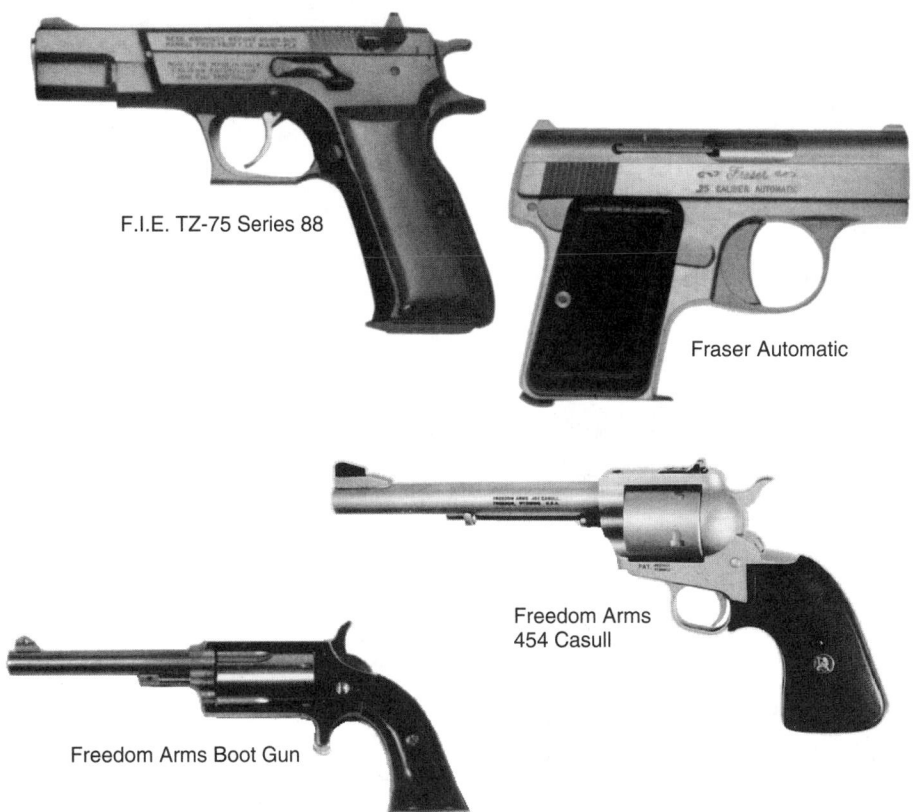

F.I.E. TZ-75 Series 88

Fraser Automatic

Freedom Arms
454 Casull

Freedom Arms Boot Gun

HANDGUNS

F.I.E.
TZ-75 Series 88 Government Model
TZ-75 Series 88 Ported
Texas Ranger
Yellow Rose

Fraser
Automatic

Freedom Arms
44 Magnum Casull
44 Magnum Silhouette
454 Casull
454 Casull Silhouette Model
Boot Gun
FA-S-22LR Mini Revolver
FA-S-22M Mini Revolver
Mini Revolver
Model 252 Silhouette

F.I.E. TZ-75 Series 88 Government Model

Same specs as TZ-75 Series 88 except 9mm Para.; 12-shot magazine; 3 1/2" barrel; 6 7/8" overall length; weighs 33 1/2 oz.; walnut grips; compact version of TZ-75 Series 88; blue, chrome or with blue slide/chrome frame. Introduced 1990; discontinued 1990.

Perf.: $400   **Exc.:** $325   **VGood:** $250
Chrome finish
Perf.: $425   **Exc.:** $350   **VGood:** $275
With blue slide/chrome frame
Perf.: $400   **Exc.:** $325   **VGood:** $250

F.I.E. TZ-75 Series 88 Ported

Same specs as TZ-75 Series 88 except 5" barrel and slide. Introduced 1990; discontinued 1990.
Perf.: $600   **Exc.:** $550   **VGood:** $400
Chrome finish
Perf.: $625   **Exc.:** $575   **VGood:** $450
Blue slide/chrome frame finish
Perf.: $600   **Exc.:** $550   **VGood:** $400

F.I.E. TEXAS RANGER

Revolver; single action; 22 LR, 22 LR/22 WMR combo; 6-shot cylinder; 3 1/4", 4 3/4", 6 1/2", 7", 9" barrels; blade front, notch rear sight; American hardwood stocks; blue/black finish. Introduced 1983; discontinued 1990.
Perf.: $100   **Exc.:** $75   **VGood:** $60
With combo cylinders
Perf.: $125   **Exc.:** $100   **VGood:** $85

F.I.E. YELLOW ROSE

Revolver; single action; 22 LR, 22 WMR; 6-shot cylinder; 4 3/4" barrel; 10" overall length; adjustable sights; walnut grips; sliding spring ejector; 24-karat gold plated. Introduced 1986; discontinued 1990.
Perf.: $150   **Exc.:** $135   **VGood:** $110
With scrimshawed ivory grips and walnut case
Perf.: $275   **Exc.:** $225   **VGood:** $200

FRASER AUTOMATIC

Semi-automatic; 25 ACP; 6-shot magazine; 2 1/4" barrel; 4" overall length; weighs 10 oz.; recessed fixed sights; checkered walnut or plastic pearl stocks; stainless steel construction; positive manual and magazine safeties; satin stainless, gold-plated or black QPQ finish. Introduced 1983; dropped 1986.
Exc.: $125   **VGood:** $115   **Good:** $100

FREEDOM ARMS 44 MAGNUM CASULL

Revolver; single action; 44 Mag.; 5-shot cylinder; 4 3/4", 6", 7 1/2", 10" barrel; 14" overall length (7 1/2" barrel); weighs about 59 oz.; blade front sight on ramp, fully-adjustable rear; Pachmayr rubber grips; Field Grade with matte stainless finish; Premier Grade with brushed stainless finish; impregnated hardwood grips, Premier Grade sights. Made in U.S. by Freedom Arms. Introduced 1992; still in production. **LMSR: $1261**
Field Grade
New: $1100   **Perf.:** $875   **Exc.:** $825
Premier Grade
New: $1500   **Perf.:** $1200   **Exc.:** $900

Freedom Arms 44 Magnum Silhouette

Same specs as the 44 Magnum except 10" barrel only; silhouette competition sights; Field Grade finish only; trigger over-travel screw; front sight hood. Made in U.S. by Freedom Arms. Introduced 1992; still produced. **LMSR: $1304**
New: $1400   **Perf.:** $1100   **Exc.:** $950

FREEDOM ARMS 454 CASULL

Revolver; single action; 454 Casull; 5-shot cylinder; 4 3/4", 6", 7 1/2", 10", barrels; weighs 50 oz. (7 1/2" barrel); impregnated hardwood grips; blade front sight, notch or adjustable rear; stainless steel; sliding bar safety. Introduced 1983; still in production.
Fixed sights
Perf.: $1250   **Exc.:** $950   **VGood:** $800
Adjustable sights
Perf.: $1300   **Exc.:** $1000   **VGood:** $850

Freedom Arms 454 Casull Silhouette Model

Same specs as 454 Casull except competition sights and trigger overtravel screw.
Perf.: $1350   **Exc.:** $1100   **VGood:** $950

FREEDOM ARMS BOOT GUN

Revolver; single action; 22 LR, 22 WMR; 3" barrel; 5 7/8" overall length; weighs 5 oz.; oversize grips; floating firing pin; stainless steel construction. Introduced 1982; dropped 1990.
Perf.: $250   **Exc.:** $200   **VGood:** $175
22 WMR
Perf.: $250   **Exc.:** $200   **VGood:** $175

FREEDOM ARMS FA-S-22LR MINI REVOLVER

Revolver; 22 LR; 5-shot; 1", 1 3/4", 3" barrels; blade front, notched rear sights; black ebonite grips; stainless steel construction; partial high polish finish; simulated ivory or ebony grips.
Perf.: $225   **Exc.:** $200   **VGood:** $165

FREEDOM ARMS FA-S-22M MINI REVOLVER

Revolver; 22 WMR; 4-shot; 1", 1 3/4", 3" barrels; stainless steel construction; blade front, notched rear sights; simulated ivory or ebony grips.
Perf.: $75   **Exc.:** $225   **VGood:** $200

FREEDOM ARMS MINI REVOLVER

Revolver; single action; 22 LR, 22 WMR; 5-shot; 4" (22 WMR), 1", 1 3/4" barrel (22 LR); 4" overall length; blade front sight, notched rear; black ebonite grips; stainless steel construction; marketed in presentation case. Introduced 1978; dropped 1990.
Perf.: $200   **Exc.:** $175   **VGood:** $165

FREEDOM ARMS MODEL 252 SILHOUETTE

Revolver; single action; 22 LR; 5-shot cylinder, 10" barrel; weighs 63 oz.; Patridge-type front sight, click-adjustable rear; Western-style black Micarta grips; constructed of stainless steel on 454 Casull frame; lightened hammer; two-point firing pin; extra 22 WMR cylinder available. Introduced 1991; still in production. **LMSR: $1432**
Perf.: $1000   **Exc.:** $850   **VGood:** $750
With extra WMR cylinder
Perf.: $1200   **Exc.:** $1000   **VGood:** $900

Freedom Arms
Model 252 Varmint
Model 353
Model 353 Silhouette
Model 555

Frommer
Baby Pocket Model
Liliput Pocket Model
Stop Pocket Model

FTL
Auto Nine
Pocket Partner

Galesi
Model 6
Model 9

Gamba
SAB G90 Standard
SAB G90 Competition
SAB G91 Compact
SAB G91 Competition
Trident Fast Action
Trident Match 900
Trident Super

Garcia
Regent

Galesi Model 9

Frommer Liliput
Pocket Model

Garcia/FI Model D

FTL Auto Nine

Freedom Arms Model 252 Varmint
Same specs as Model 252 Silhouette except 7 1/2″ barrel; weighs 59 oz.; black/green laminated hardwood grips; brass bead front sight, fully-adjustable rear; extra 22 WMR cylinder available. Introduced 1991; still produced. **LMSR: $1384**
Perf.: $1000   Exc.: $800   VGood: $700

FREEDOM ARMS MODEL 353
Revolver; single action; 5-shot; 4 3/4″, 6″, 7 1/2″, 9″ barrel; 14″ overall length (7 1/2″ barrel); weighs about 59 oz.; blade front sight on ramp, fully-adjustable rear; Pachmayr rubber grips (Field Grade), impregnated hardwood grips (Premier Grade); Field Grade with matte stainless finish; Premier Grade with brush stainless finish. Introduced 1992; still in production. **LMSR: $1216**
Field Grade
New: $1000   Perf.: $900   Exc.: $700
Premier Grade
New: $1300   Perf.: $1000   Exc.: $800

Freedom Arms Model 353 Silhouette
Same specs as Model 353 except 9″ barrel; silhouette competition sights; Field Grade finish and grips; trigger over-travel screw; front sight hood. Introduced 1992; still in production. **LMSR: $1304**
New: $1100   Perf.: $1000   Exc.: $850

FREEDOM ARMS MODEL 555
Revolver; single action; 50 AE; 5-shot cylinder; 4 3/4″, 6″, 7 1/2″, 10″ barrel; 14″ overall length (7 1/2″ barrel); weighs 50 oz.; removable blade front sight on ramp, adjustable rear; Pachmayr rubber grips on Field Grade; impregnated hardwood on Premier Grade; Field Grade with matte stainless finish; Premier Grade with brushed stainless finish. Introduced 1994; still produced. **LMSR: $1263**
Field Grade
New: $1100   Perf.: $1000   Exc.: $850
Premier Grade
New: $1300   Perf.: $1250   Exc.: $1000

FROMMER BABY POCKET MODEL
Semi-automatic; 32 ACP; 6-shot magazine; smaller version of Stop Pocket Model with 2″ barrel; 4 3/4″ overall length. Manufactured from 1919 to approximately 1922.
Exc.: $350   VGood: $275   Good: $200

FROMMER LILIPUT POCKET MODEL
Semi-automatic; blowback action; 25 ACP; 6-shot magazine; 2 1/8″ barrel; 4 5/16″ overall length; hard rubber stocks; fixed sights; blued finish. Manufactured 1921 to 1924.
Exc.: $250   VGood: $200   Good: $150

FROMMER STOP POCKET MODEL
Semi-automatic; 32 ACP, 380 ACP; 6-shot (380 ACP), 7-shot (32 ACP) magazine; 3 7/8″ barrel; 6 1/2″ overall length; external hammer; grooved wood or checkered hard rubber stocks; fixed sights; blued finish. Manufactured in Hungary, 1912 to 1920.
Exc.: $325   VGood: $250   Good: $175

FTL AUTO NINE
Semi-automatic; single action; 22 LR; 8-shot magazine; 2 1/8″ barrel; 4 5/16″ overall length; weighs 9 1/4 oz.; blade front sight, fixed notch rear; checkered black plastic grips; manual push-button safety; alloy frame; hard chromed slide and magazine; barrel support bushing. Made by Auto Nine Corp. Marketed by FTL Marketing Corp. Introduced 1985; no longer in production.
Perf.: $175   Exc.: $150   VGood: $125

FTL POCKET PARTNER
Semi-automatic; 22 LR; 8-shot magazine; 2 1/4″ barrel; 4 3/4″ overall length; weighs 10 oz.; internal hammer; fixed sights; checkered plastic stocks; all-steel construction; brushed blue finish. Introduced 1985; no longer in production.
Perf.: $125   Exc.: $100   VGood: $85

GALESI MODEL 6
Semi-automatic; 25 ACP; 6-shot magazine; 2 1/2″ barrel; 4 3/8″ overall length; plastic stocks; slide-top groove sights; blued finish. Manufactured in Italy. Introduced 1930; no longer in production.
Exc.: $250   VGood: $150   Good: $125

GALESI MODEL 9
Semi-automatic; 22 LR, 32 ACP, 380 ACP; 7-shot magazine; 3 1/4″ barrel; plastic stocks; fixed sights; blued finish. Introduced 1930; no longer in production.
Exc.: $300   VGood: $250   Good: $175

GAMBA SAB G90 STANDARD
Semi-automatic; double-action; 32 ACP, 9x18 Ultra, 9mm Para.; 15-shot magazine; 4 3/4″ barrel; weighs 33 oz.; windage-adjustable rear sight, undercut blade front; uncheckered European walnut stocks; squared trigger guard; blue or chrome finish. Made in Italy. Introduced 1986; discontinued 1990.
Perf.: $550   Exc.: $400   VGood: $350
Chrome finish
Perf.: $600   Exc.: $450   VGood: $300

Gamba SAB G90 Competition
Same specs as SAB G90 Standard except 9mm Para. only; cocked-and-locked operation; checkered walnut grips. Introduced 1990; discontinued 1990.
Perf.: $700   Exc.: $550   VGood: $400

GAMBA SAB G91 COMPACT
Semi-automatic; double action; 32 ACP, 9x18 Ultra, 9mm Para.; 12-shot magazine; 3 9/16″ barrel; 30 oz.; windage-adjustable rear sight, undercut blade front; squared trigger guard; blued or chrome finish. Introduced 1986; discontinued 1990.
Perf.: $350   Exc.: $300   VGood: $275
Chrome finish
Perf.: $375   Exc.: $325   VGood: $300

Gamba SAB G91 Competition
Same specs as SAB G91 Compact except 9mm Para.; cocked-and-locked operation; checkered walnut grips. Introduced 1990; discontinued 1990.
Perf.: $500   Exc.: $450   VGood: $300

GAMBA TRIDENT FAST ACTION
Revolver; double action; 32 S&W, 38 Spl.; 6-shot; 2 1/2″, 3″ barrel; 23 oz.; fixed sights; checkered walnut grips; blued finish. Discontinued 1986.
Perf.: $500   Exc.: $400   VGood: $325

Gamba Trident Match 900
Same specs as Trident Fast Action except 6″ heavy barrel; 35 oz.; adjustable sights; checkered walnut target grips; blued finish. Discontinued 1986.
Perf.: $700   Exc.: $600   VGood: $450

Gamba Trident Super
Same specs as Trident Fast Action except 4″ vent-rib barrel; 25 oz. Discontinued 1986.
Perf.: $550   Exc.: $450   VGood: $400

GARCIA REGENT
Revolver; 22 LR; 3″, 4″, 6″ barrel; 8-shot swing-out cylinder; fixed rear sight, ramp front; checkered plastic grips; blued. Introduced 1972; dropped, 1977.
Exc.: $100   VGood: $85   Good: $65

Glock Model
19 Compact

Gaucher GN1 Silhouette

Glock Model 24
Competition

Glock Model 21

Great Western Frontier

Garcia/FI
Model D

Gaucher
GN1 Silhouette
GP Silhouette
Randall Singleshot

Glock
Model 17
Model 17L Competition
Model 19 Compact
Model 20
Model 21
Model 22
Model 23 Compact
Model 24 Competition

Great Western
Derringer
Frontier

GARCIA/FI MODEL D
Semi-automatic; 380 ACP; 6-shot magazine; 3 1/8" barrel; 6 1/8" overall length; checkered American walnut stocks; windage-adjustable rear sight, blade front; blued finish; lanyard ring. Imported by Firearms International 1977 to 1979.
Perf.: $200   **Exc.:** $150   **VGood:** $125

GAUCHER GN1 SILHOUETTE
Single shot; bolt action; 22 LR; 10" barrel; 15 1/2" overall length; weighs 2 1/2 lbs.; blade front, open adjustable rear sight; Euopean hardwood stock; adjustable trigger. Made in France. Introduced 1990; still imported by Mandall Shooting Supplies. **LMSR: $525**
Perf.: $350   **Exc.:** $300   **VGood:** $275

GAUCHER GP SILHOUETTE
Single shot; bolt action; 22 LR; 10" barrel; 15 1/2" overall length; weighs 2 1/2 lbs.; hooded ramp front, open adjustable rear sight; stained harwood grips; matte chrome barrel; blued bolt, sights. Made in France. Introduced 1991; still imported by Mandall Shooting Supplies. **LMSR: $425**
Perf.: $275   **Exc.:** $250   **VGood:** $200

GAUCHER RANDALL SINGLESHOT
Bolt action; 9mm Para.; 16 1/2" overall length; weighs 36 oz.; bead front sight; European hardwood stock; blued finish. Made in France. Introduced 1990; discontinued 1992. Reportedly imported by Mandall Shooting Supplies.

GLOCK MODEL 17
Semi-automatic; double action; 9mm Para.; 17-, 19-shot magazine; 4 1/2" barrel; 7 7/16" overall length; weighs 24 oz.; white outline adjustable rear sight, dot on front blade; polymer frame; steel slide; trigger safety; recoil-operated action; firing pin safety; drop safety. Made in Austria. Introduced, 1986; still imported. **LMSR: $609**
Perf.: $400   **Exc.:** $350   **VGood:** $300

Glock Model 17L Competition
Same specs as Model 17 except compensated 6" barrel; lighter trigger pull; 26 oz. Made in Austria. Introduced 1988; still imported. **LMSR: $806**
Perf.: $600   **Exc.:** $550   **VGood:** $450

GLOCK MODEL 19 COMPACT
Semi-automatic; double action; 9mm Para.; 15-, 17-shot magazine; 4" barrel; 6 7/8" overall length; 22 oz.; white outline adjustable rear sight, dot on front blade; polymer frame; steel slide; trigger safety; recoil-operated action; firing pin safety; drop safety. Made in Austria. Introduced 1988; still imported. **LMSR: $609**
Perf.: $450   **Exc.:** $400   **VGood:** $350

GLOCK MODEL 20
Semi-automatic; double action; 10mm Auto; 15-shot magazine; 4 5/8" barrel; 28 oz.; white outline adjustable rear sight, dot on front blade; polymer frame; steel slide; trigger safety; recoil-operated action; firing pin safety; drop safety. Made in Austria. Introduced 1990; still imported. **LMSR: $670**
Perf.: $600   **Exc.:** $525   **VGood:** $450

GLOCK MODEL 21
Semi-automatic; double action; 45 ACP; 13-shot magazine; 4 5/8" barrel; 27 oz.; white outline adjustable rear sight, dot on front blade; polymer frame; steel slide; trigger safety; recoil-operated action; firing pin safety; drop safety. Made in Austria. Introduced 1991; still imported. **LMSR: $670**
Perf.: $600   **Exc.:** $500   **VGood:** $425

GLOCK MODEL 22
Semi-automatic; double action; 40 S&W; 15-shot magazine; 4 1/2" barrel; 7 7/16" overall length; 24 oz.; white outline adjustable rear sight, dot on front blade; polymer frame; steel slide; trigger safety; recoil-operated action; firing pin safety; drop safety. Made in Austria. Introduced 1990; still imported. **LMSR: $670**
Perf.: $475   **Exc.:** $425   **VGood:** $350

GLOCK MODEL 23 COMPACT
Semi-automatic; double action; 40 S&W; 13-shot magazine; 4" barrel; 6 7/8" overall length; 22 oz.; white outline adjustable rear sight, dot on front blade; polymer frame; steel slide; trigger safety; recoil-operated action; firing pin safety; drop safety. Made in Austria. Introduced 1991; still imported. **LMSR: $609**
Perf.: $500   **Exc.:** $450   **VGood:** $400

GLOCK MODEL 24 COMPETITION
Semi-automatic; double action; 40 S&W; 10-shot magazine; 6" barrel; 8 1/2" overall length; weighs 29 1/2 oz.; black polymer grips; blade front sight with dot, white outline windage-adjustable rear; long-slide competition model available as compensated or non-compensated; factory-installed competition trigger; drop-free magazine. Imported from Austria by Glock, Inc. Introduced 1994; still imported. **LMSR: $807**
New: $650   **Perf.:** $525   **Exc.:** $450
Compensated barrel
New: $700   **Perf.:** $575   **Exc.:** $500

GREAT WESTERN DERRINGER
Derringer; over/under; 38 S&W; 2-shot; 3" barrels; 5" overall length; checkered black plastic grips; fixed sights; blued finish. Replica of Remington Double Derringer. Manufactured 1953 to 1962.
Exc.: $275   **VGood:** $200   **Good:** $150

GREAT WESTERN FRONTIER
Revolver; single action; 22 LR, 22 Hornet, 38 Spl., 357 Mag., 44 Spl., 44 Mag., 45 Colt; 6-shot cylinder; 4 3/4", 5 1/2", 7 1/2" barrel lengths; grooved rear sight, fixed blade front; imitation stag grips; blued finish. Was sold primarily by mail order. Manufactured 1951 to 1962.
Exc.: $450   **VGood:** $400   **Good:** $300

HANDGUNS

Grendel
P-10
P-12
P-30
P-30L
P-30M
P-31

Guardian
Guardian-SS

Gunworks
Model 9 Derringer

Hammerli
Model 33MP
Model 100 Free Pistol
Model 101
Model 102
Model 103 Free Pistol
Model 104 Match

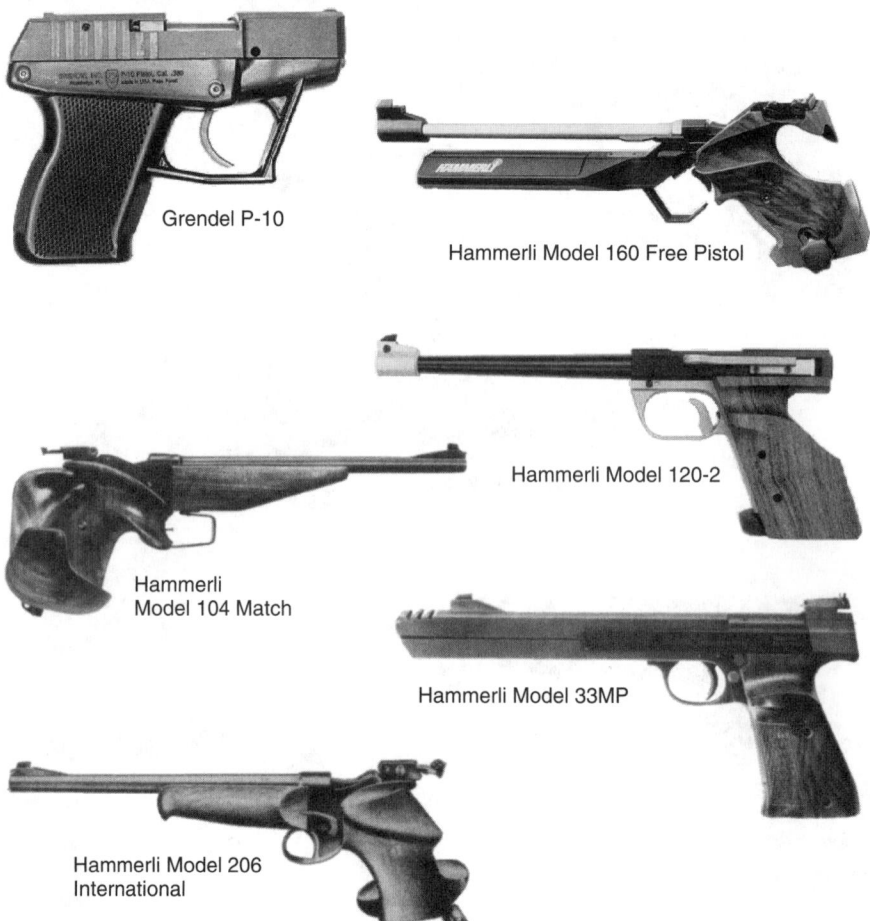

Grendel P-10

Hammerli Model 160 Free Pistol

Hammerli Model 120-2

Hammerli
Model 104 Match

Hammerli Model 33MP

Hammerli Model 206
International

GRENDEL P-10

Semi-automatic; double action; 380 ACP; 10-shot magazine; 3″ barrel; 5 5/16″ overall length; weighs 15 oz.; checkered polycarbonate metal composite grips; fixed sights; inertia safety hammer system; magazine loads from top; matte black, electroless nickel or green finish. Made in U.S. by Grendel, Inc. Introduced 1987; discontinued 1991.

    Perf.: $110   **Exc.:** $90   **VGood:** $75

GRENDEL P-12

Semi-automatic; double action; 380 ACP; 10-shot magazine; 3″ barrel; 5 5/16″ overall length; weighs 13 oz.; checkered DuPont ST-800 polymer grips; fixed sights; inertia safety hammer system; all-steel frame; grip forms magazine well and trigger guard; blue finish. Made in U.S. by Grendel, Inc. Introduced 1992; still produced. **LMSR: $175**

    New: $150   **Perf.:** $125   **Exc.:** $110
Electroless nickel finish
    New: $170   **Perf.:** $150   **Exc.:** $135

GRENDEL P-30

Semi-automatic; double action; 22 WMR; 30-shot magazine; 5″, 8″ barrel; 8 1/2″ overall length (5″ barrel); weighs 21 oz. (5″ barrel); checkered Zytel grips; blade front sight, fixed rear; blowback action with fluted chamber; ambidextrous safety; reversible magazine catch. Made in U.S. by Grendel, Inc. Introduced 1990; dropped 1994.

    Perf.: $175   **Exc.:** $150   **VGood:** $135

Grendel P-30L

Same specs as the P-30 except 8″ barrel only. Made in U.S. by Grendel, Inc. Introduced 1990; dropped 1994.

    Perf.: $200   **Exc.:** $175   **VGood:** $165

Grendel P-30M

Same specs as the P-30 except removable muzzlebrake. Made in U.S. by Grendel, Inc. Introduced 1990; dropped 1994.

    Perf.: $200   **Exc.:** $175   **VGood:** $150

GRENDEL P-31

Semi-automatic; 22 WMR; 30-shot magazine; 11″ barrel; 17 1/2″ overall length; 48 oz.; adjustable blade front, fixed rear; blowback action with fluted chamber; ambidextrous safety; muzzlebrake; scope mount optional; checkered black Zytel grip and forend; matte black finish. Made in U.S. by Grendel, Inc. Introduced 1991; no longer in production.

    Perf.: $300   **Exc.:** $250   **VGood:** $225

GUARDIAN-SS

Semi-automatic; double action; 380 ACP; 6-shot magazine; 3 1/4″ barrel; 6″ overall length; weighs 20 oz.; checkered walnut stocks; ramp front sight, windage-adjustable combat rear; narrow polished trigger; Pachmayr grips; blue slide; hand-fitted barrel; polished feed ramp; funneled magazine well; stainless steel. Introduced 1982; dropped 1985. Marketed by Michigan Armament, Inc.

    Exc.: $300   **VGood:** $250   **Good:** $200

GUNWORKS MODEL 9 DERRINGER

Derringer; over/under; 38/357 Mag., 9mm/9mm Mag.; bottom-hinged action; 3″ barrel; weighs 15 oz.; smooth wood stocks; Millett orange bar front sight, fixed rear; all steel; half-cock, through-frame safety; dual extraction; electroless nickel finish; marketed with in-pants holster. Introduced 1984, no longer in production.

    Exc.: $125   **VGood:** $110   **Good:** $100

HAMMERLI MODEL 33MP

Single shot; 22 LR; 11 1/2″ octagonal barrel; 16 1/2″ overall length; competition free pistol; micrometer rear sight, interchangeable front; European walnut stocks, forearm; Martini-type action; set trigger; blued finish. Manufactured in Switzerland 1933 to 1949.

    Exc.: $850   **VGood:** $700   **Good:** $500

HAMMERLI MODEL 100 FREE PISTOL

Single shot; 22 LR; 11 1/2″ octagon barrel; micrometer rear sight, interchangeable post or bead front; Euro-

pean walnut grips, forearm; adjustable set trigger; blued finish. Introduced 1933; discontinued 1949.

    Exc.: $850   **VGood:** $700   **Good:** $600
Deluxe model
    Exc.: $950   **VGood:** $800   **Good:** $700

HAMMERLI MODEL 101

Single shot; 22 LR; 11 1/2″ heavy round barrel; adjustable sights; European walnut grip, forearm; adjustable set trigger; same general specs as Model 100 Free Pistol with improved action; matte blue finish. Introduced 1956; discontinued 1960.

    Exc.: $850   **VGood:** $700   **Good:** $600

HAMMERLI MODEL 102

Single shot; 22 LR; 11 1/2″ heavy round barrel; adjustable sights; European walnut grip, forearm; adjustable set trigger; same general specs as Model 101 with high gloss blued finish. Introduced 1956; discontinued 1960.

    Exc.: $750   **VGood:** $650   **Good:** $500
Deluxe model
    Exc.: $800   **VGood:** $700   **Good:** $550

HAMMERLI MODEL 103 FREE PISTOL

Single shot; 22 LR; 11 1/2″ lightweight octagon barrel; adjustable sights; European walnut grip, forearm; adjustable set trigger; same general specs as Model 101 with high gloss blued finish. Introduced 1956; discontinued 1960.

    Exc.: $850   **VGood:** $750   **Good:** $575
With inlaid ivory carvings
    Exc.: $1000   **VGood:** $900   **Good:** $650

HAMMERLI MODEL 104 MATCH

Single shot; 22 LR; 11 1/2″ lightweight round barrel; adjustable sights; redesigned adjustable walnut grip; adjustable set trigger; action similar to Model 103; high gloss blued finish. Introduced 1961; discontinued 1965.

    Exc.: $650   **VGood:** $500   **Good:** $400

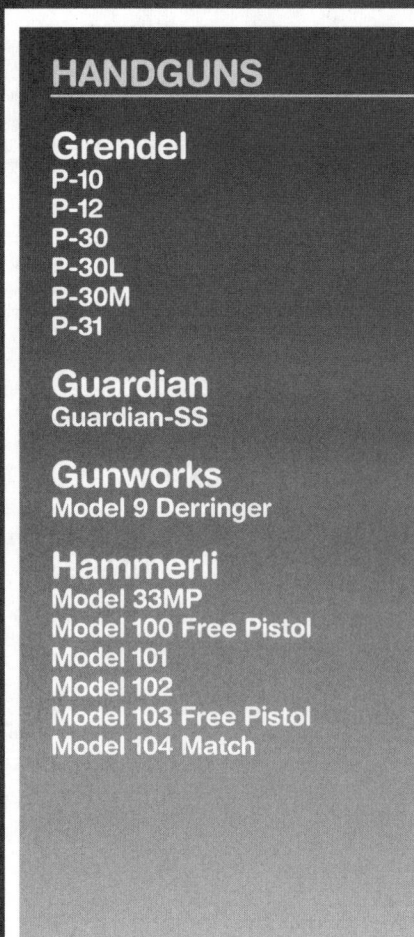

Hammerli Model 150
Free Pistol

Hammerli
Model 208
International

Hammerli
Model 208S
International

Hammerli Model 120

Hammerli Model 120

HANDGUNS

Hammerli
Model 105 Match
Model 106 Match
Model 107
Model 120
Model 120-1 Free Pistol
Model 120-2
Model 120 Heavy Barrel
Model 150 Free Pistol
Model 151 Free Pistol
Model 152 Match
Model 160 Free Pistol
Model 162 Free Pistol
Model 206 International
Model 207 International
Model 208 International
Model 208S International
Model 209 International
Model 210 International

HAMMERLI MODEL 105 MATCH
Single shot; 22 LR; 11 $1/2''$ octagon barrel; adjustable sights; redesigned adjustable walnut grip; adjustable set trigger; redesigned action; high gloss blued finish. Introduced 1962; discontinued 1965.
   Exc.: $650   **VGood:** $550   **Good:** $450

HAMMERLI MODEL 106 MATCH
Single shot; 22 LR; 11 $1/2''$ octagon barrel; adjustable sights; same general specs as Model 105 Match except redesigned trigger; replaced Model 104; matte blue finish. Introduced 1966; discontinued 1972.
   Exc.: $700   **VGood:** $600   **Good:** $500

HAMMERLI MODEL 107
Single shot; 22 LR; 11 $5/16''$ barrel; 16 $3/4''$ overall length; weighs 49 $1/2$ oz.; octagonal barrel; adjustable five-lever set trigger. Introduced 1965; discontinued 1971.
   Exc.: $750   **VGood:** $650   **Good:** $550
Deluxe model
   Exc.: $800   **VGood:** $750   **Good:** $550

HAMMERLI MODEL 120
Single shot; 22 Short; 10'' barrel; 15 $3/4''$ overall length; weighs 43 $1/4$ oz.; side-lever operation; micrometer sight; plain butt and trigger guard.
   Exc.: $350   **VGood:** $300   **Good:** $200

HAMMERLI MODEL 120-1 FREE PISTOL
Single shot; bolt action; 22 LR; 10'' barrel; 14 $3/4''$ overall length; micrometer rear sight, post front; hand-checkered walnut target grips; adjustable trigger for single- or two-stage pull; aluminum construction; blued finish. Introduced 1972; discontinued 1985.
   Exc.: $375   **VGood:** $325   **Good:** $250

Hammerli Model 120-2
Same specs as Model 120-1 Free Pistol except special contoured walnut hand rest; movable sights; blued finish. Introduced 1973; discontinued 1985.
   Exc.: $500   **VGood:** $450   **Good:** $300

Hammerli Model 120 Heavy Barrel
Same specs as Model 120-1 Free Pistol except 5 $3/4''$ bull barrel; designed to conform to existing laws governing sporting handgun sales in Great Britain. Introduced 1973; discontinued 1985.
   Exc.: $550   **VGood:** $400   **Good:** $300

HAMMERLI MODEL 150 FREE PISTOL
Single shot; 22 LR; 11 $3/8''$ barrel; 15 $3/8''$ overall length; movable front sight on collar, micrometer rear; uncheckered adjustable palm-shelf grip; Martini-type action; straight-line firing pin, no hammer; adjustable set trigger; blued finish. Introduced 1973; discontinued 1989.
   Perf.: $1750   **Exc.:** $1350   **VGood:** $1200
Left-hand
   Perf.: $1850   **Exc.:** $1450   **VGood:** $1300

HAMMERLI MODEL 151 FREE PISTOL
Single shot; 22 LR; 11 $3/8''$ barrel; 15 $3/8''$ overall length; movable front sight on collar, micrometer rear; uncheckered adjustable palm-shelf grip; Martini-type action; straight-line firing pin; no hammer; adjustable set trigger; blued finish. Replaced Model 150 Free Pistol. Introduced 1990; discontinued 1993.
   Perf.: $1400   **Exc.:** $1200   **VGood:** $1000

HAMMERLI MODEL 152 MATCH
Single shot; 22 LR; 11 $5/16''$ barrel; 16 $7/8''$ overall length; weighs 46 $5/8$ oz.; changeable post-type front sight, micrometer rear; match stocks; electronic trigger; improved Martini-style action; blued action. Introduced 1990; discontinued 1992.
   Perf.: $1500   **Exc.:** $1350   **VGood:** $1200
Left-hand
   Perf.: $1600   **Exc.:** $1450   **VGood:** $1300

HAMMERLI MODEL 160 FREE PISTOL
Single shot; 22 LR; 11 $1/4''$ barrel; 17 $1/2''$ overall length; weighs 47 oz.; walnut grip; full match-style with adjustable palm shelf, stippled surfaces; changeable blade front sight, open, fully-adjustable match rear; mechanical set trigger, fully adjustable with provision for dry firing. Imported from Switzerland by Hammerli Pistols USA. Introduced 1993; still imported. **LMSR:** $2034
   New: $1800   **Perf.:** $1700   **Exc.:** $1450

HAMMERLI MODEL 162 FREE PISTOL
Single shot; 22 LR; 11 $1/4''$ barrel; 17 $1/2''$ overall length; weighs 47 oz.; walnut grip; full match-style with adjustable palm shelf, stippled surfaces; changeable blade front sight, open, fully-adjustable match rear; electronic trigger, fully adjustable with provision for dry firing. Imported from Switzerland by Hammerli Pistols USA. Introduced 1993; still imported. **LMSR:** $2189
   New: $1900   **Perf.:** $1850   **Exc.:** $1500

HAMMERLI MODEL 206 INTERNATIONAL
Semi-automatic; single action; 22 LR, 22 Short; 8-shot magazine; 7 $1/16''$ barrel; adjustable sights; checkered walnut thumbrest grips; muzzlebrake; slide stop; adjustable trigger; blued finish. Introduced 1964; discontinued 1967.
   Exc.: $600   **VGood:** $500   **Good:** $400

HAMMERLI MODEL 207 INTERNATIONAL
Semi-automatic; single action; 22 LR, 22 Short; 8-shot magazine; 7 $1/16''$ barrel; adjustable sights; smooth walnut grips with adjustable grip plates; muzzlebrake; slide stop; adjustable trigger; blued finish. Introduced 1964; discontinued 1969.
   Exc.: $625   **VGood:** $550   **Good:** $400

HAMMERLI MODEL 208 INTERNATIONAL
Semi-automatic; single action; 22 LR; 8-shot magazine; 6'' barrel; 10'' overall length; weighs 37 $1/2$ oz.; adjustable sights; smooth walnut grips with adjustable grip plates; adjustable trigger; blued finish. Introduced 1966; discontinued 1988.
   Perf.: $1350   **Exc.:** $1200   **VGood:** $1000

Hammerli Model 208S International
Same specs as International Model 208 except interchangeable rear sight; restyled trigger guard. Introduced 1988; still in production. **LMSR:** $1768
   New: $1600   **Perf.:** $1400   **Exc.:** $1250

HAMMERLI MODEL 209 INTERNATIONAL
Semi-automatic; single action; 22 Short; 5-shot magazine; 4 $3/4''$ barrel; adjustable sights; walnut grips; adjustable muzzlebrake, barrel vents, gas-escape ports; lightweight bolt; blued finish. Introduced 1967; discontinued 1970.
   Exc.: $750   **VGood:** $650   **Good:** $500

HAMMERLI MODEL 210 INTERNATIONAL
Semi-automatic; single action; 22 Short; 5-shot magazine; 4 $3/4''$ barrel; adjustable sights; adjustable walnut grip plates; adjustable muzzlebrake, barrel vents, gas-escape ports; lightweight bolt; blued finish. Introduced 1967; discontinued 1970.
   Exc.: $750   **VGood:** $650   **Good:** $500

Hammerli
Model 211
Model 212 Hunter
Model 215
Model 230-1
Model 230-2
Model 232-1
Model 232-2
Model 280 Target
Virginian

Hammerli-Walther
American Model
(See Hammerli-Walther
Model 203)
Model 200 Olympia
Model 201
Model 202
Model 203
Model 204
Model 205
Quickfire
(See Hammerli-Walther
Model 200 Olympia)

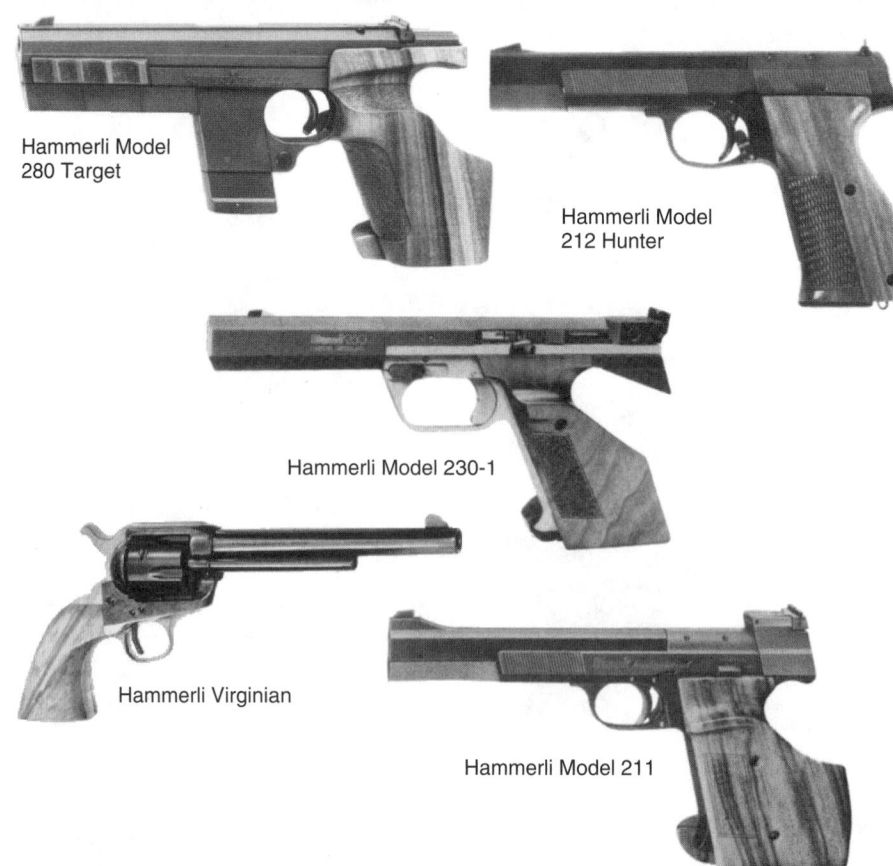

Hammerli Model
280 Target

Hammerli Model
212 Hunter

Hammerli Model 230-1

Hammerli Virginian

Hammerli Model 211

HAMMERLI MODEL 211
Semi-automatic; single action; 22 LR; 9-shot magazine; 6" barrel; 10" overall length; weighs 37 oz.; walnut stocks with adjustable palm rest; fully-adjustable match sights; interchangeable front and rear blades; fully-adjustable trigger. Imported from Switzerland by Mandall Shooting Supplies, Beeman. Introduced 1973; discontinued 1990.
Perf.: $1250   **Exc.:** $1100   **VGood:** $850

HAMMERLI MODEL 212 HUNTER
Semi-automatic; single action; 22 LR; 8-shot magazine; 5" barrel; 8 1/2" overall length; weighs 31 oz.; adjustable sights; checkered walnut stocks; adjustable trigger; blued finish. Made in Switzerland. Introduced 1984; no longer imported.
Perf.: $1300   **Exc.:** $1000   **VGood:** $950

HAMMERLI MODEL 215
Semi-automatic; single action; target model; 22 LR; 8-shot magazine; 5" barrel; adjustable sights; walnut grip with plates; adjustable trigger; blued finish. Made in Switzerland. Importation discontinued 1991.
Perf.: $1250   **Exc.:** $1050   **VGood:** $850

HAMMERLI MODEL 230-1
Semi-automatic; 22 Short; 5-shot magazine; 6 5/16" barrel; 11 5/8" overall length; micrometer rear sight, post front; uncheckered European walnut thumbrest grips; blued finish. Designed for rapid-fire International competition. Introduced 1970; discontinued 1983.
Exc.: $650   **VGood:** $550   **Good:** $400

Hammerli Model 230-2
Same specs as Model 230-1 except partially checkered stocks; adjustable heel plate. Introduced 1970; discontinued 1983.
Exc.: $700   **VGood:** $600   **Good:** $450

HAMMERLI MODEL 232-1
Semi-automatic; 22 Short; 6-shot magazine; 5" barrel with six exhaust ports; 10 3/8" overall length; weighs 44 oz.; interchangeable front, rear blades, adjustable micrometer rear sight; walnut grips; recoil-operated; adjustable trigger; blued finish. Made in Switzerland. Introduced 1984; importation discontinued 1993.
Perf.: $1250   **Exc.:** $1100   **VGood:** $900

Hammerli Model 232-2
Same specs as Model 232-1 except wrap-around grips. Introduced 1984; importation discontinued 1993.
Perf.: $1300   **Exc.:** $1150   **VGood:** $950

HAMMERLI MODEL 280 TARGET
Semi-automatic; 22 LR; 32 S&W Long WC; 6-shot magazine (22 LR), 5-shot (32 S&W); 4 1/2" barrel; 11 13/16" overall length; weighs 39 oz.; micro-adjustable match sights; stippled match-type walnut stocks; adjustable palm shelf carbon-reinforced synthetic frame and bolt/barrel housing; fully-adjustable, interchangeable trigger. Made in Switzerland. Introduced, 1990; still imported. **LMSR:** $1558
Perf.: $1250   **Exc.:** $1100   **VGood:** $900

HAMMERLI VIRGINIAN
Revolver; single action; 357 Mag., 45 Colt; 6-shot cylinder; 4 5/8", 5 1/2", 7 1/2" barrels; grooved rear sight, blade front; one-piece European walnut grips; case-hardened; chrome grip frame, triggerguard; blued frame cylinder; same general design as Colt SAA except for base pin safety feature. Manufactured in Europe, 1973 to 1976, for exclusive Interarms importation.
Exc.: $400   **VGood:** $350   **Good:** $250

HAMMERLI-WALTHER MODEL 200 OLYMPIA
Semi-automatic; 22 LR, 22 Short; 8-shot magazine; 7 1/2" barrel; micrometer rear sight, ramp front; walnut thumbrest grip; muzzle brake; adjustable barrel weights; blued finish. Also known as Quickfire. Based on 1936 Olympia with some parts interchangeable. Introduced 1950. In 1958, muzzle brake was redesigned. Discontinued 1963.
Exc.: $650   **VGood:** $550   **Good:** $500

HAMMERLI-WALTHER MODEL 201
Semi-automatic; 22 LR, 22 Short; 8-shot magazine; 9 1/2" barrel; micrometer rear sight, ramp front; adjustable custom walnut grip; muzzle brake; adjustable barrel weights; blued finish. Introduced 1955; discontinued 1957.
Exc.: $650   **VGood:** $600   **Good:** $500

HAMMERLI-WALTHER MODEL 202
Semi-automatic; 22 LR, 22 Short; 8-shot magazine; 9 1/2" barrel; micrometer rear sight, ramp front; adjustable walnut thumbrest grip; muzzle brake; adjustable barrel weights; blued finish. Introduced 1955; discontinued 1957.
Exc.: $650   **VGood:** $600   **Good:** $500

HAMMERLI-WALTHER MODEL 203
Semi-automatic; 22 LR, 22 Short; 8-shot magazine; 7 1/2" barrel; micrometer rear sight, ramp front; adjustable walnut thumbrest grip; optional muzzle brake (1958 model); slide stop; adjustable barrel weights; blued finish. Called the American Model. Introduced 1955; discontinued 1959.
Exc.: $650   **VGood:** $600   **Good:** $500
With muzzle brake
Exc.: $700   **VGood:** $650   **Good:** $550

HAMMERLI-WALTHER MODEL 204
Semi-automatic; 22 LR; 8-shot magazine; 7 1/2" barrel; micrometer rear sight, ramp front; walnut thumbrest grip; optional muzzle brake; slide stop; adjustable barrel weights; blued finish. Introduced 1956; discontinued 1963.
Exc.: $650   **VGood:** $600   **Good:** $500
With muzzle brake
Exc.: $700   **VGood:** $650   **Good:** $550

HAMMERLI-WALTHER MODEL 205
Semi-automatic; 22 LR; 8-shot magazine; 7 1/2" barrel; micrometer rear sight, ramp front; checkered French walnut thumbrest grip; detachable muzzle brake; slide stop; adjustable barrel weights; blued finish. Introduced 1956; discontinued 1963.
Exc.: $650   **VGood:** $600   **Good:** $550
With muzzle brake
Exc.: $700   **VGood:** $650   **Good:** $550

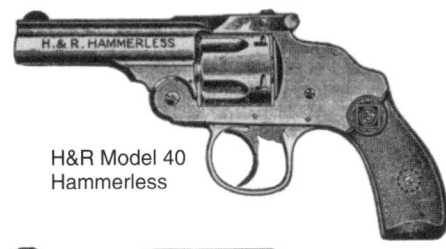

H&R Model 40
Hammerless

H&R Model 5

H&R Model 4

H&R Expert

H&R American

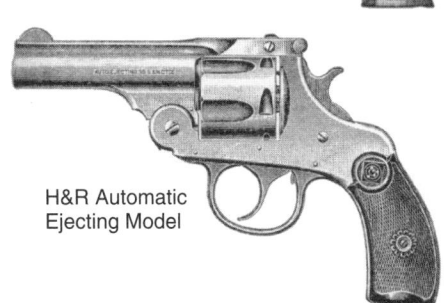

H&R Automatic
Ejecting Model

H&R 22 Special

HANDGUNS

Harrington & Richardson
22 Special
American
Automatic Ejecting Model
Bobby Model
Defender
(See H&R Model 925)
"Eureka" Sportsman
(See H&R No. 196)
Expert
Hunter Model
Model 4
Model 5
Model 6
Model 15
(See H&R Bobby Model)
Model 25 (See H&R
Model 925)
Model 40 Hammerless
Model 45
(See H&R Model 40
Hammerless)
Model 50 Hammerless
Model 55
(See H&R Model 50
Hammerless)
Model 299 (See H&R New
Defender)
Model 504
Model 532
Model 586

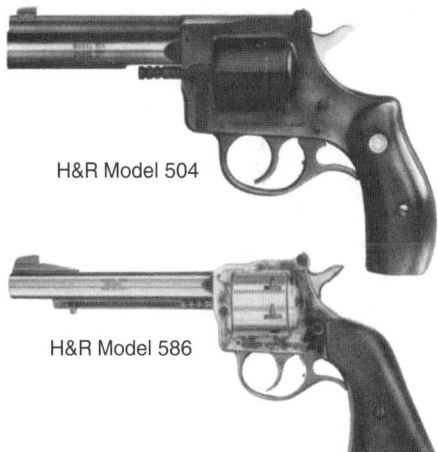

H&R Model 504

H&R Model 586

HARRINGTON & RICHARDSON 22 SPECIAL

Revolver; double action; 22 Short, 22 Long, 22 LR, 22 WMR; 9-shot cylinder; 6″ barrel; 11″ overall length; weighs 23 oz.; heavy hinged-frame; fixed notch rear sight, gold-plated front; checkered walnut grips; blued. Originally introduced as Model 944; later version with recessed cylinder for high-speed ammo was listed as Model 945. Introduced 1925; dropped 1941.

Exc.: $125　**VGood:** $110　**Good:** $85

HARRINGTON & RICHARDSON AMERICAN

Revolver; double action; 32 S&W Long, 32 S&W, 32 Colt New Police; 6-shot (32), 5-shot (38); 2 1/2″, 4 1/2″, 6″ barrel; weighs 16 oz.; solid frame; fixed sights; hard rubber grips; blued, nickel finish. Introduced 1883; dropped during WWII.

Exc.: $100　**VGood:** $75　**Good:** $50

HARRINGTON & RICHARDSON AUTOMATIC EJECTING MODEL

Revolver; double action; 32 S&W Long, 38 S&W; 6-shot cylinder (32), 5-shot (38); 3 1/4″, 4″, 5″, 6″ barrel; weighs 16 oz.; hinged-frame; fixed sights; hard rubber or checkered walnut target grips; blued, nickel finish. Introduced 1891; dropped 1941.

Exc.: $100　**VGood:** $75　**Good:** $60

HARRINGTON & RICHARDSON BOBBY MODEL

Revolver; double action; 32 S&W Long, 38 S&W; 6-shot cylinder (32), 5-shot (38); 4″ barrel; 9″ overall length; hinged-frame; fixed sights; checkered walnut grips; blued finish. Also listed as the Model 15. Designed for use by London police during WWII. Introduced 1941; dropped 1943. Collector value.

Exc.: $200　**VGood:** $150　**Good:** $125

HARRINGTON & RICHARDSON EXPERT

Revolver; double action; 22 LR, 22 WMR; 9-shot; 10″ barrel; heavy hinged frame; fixed rear sight; gold-plated front; checkered walnut grips; blued finish. Listed as the Model 955. Introduced 1929; dropped 1941.

Exc.: $135　**VGood:** $125　**Good:** $100

HARRINGTON & RICHARDSON HUNTER MODEL

Revolver; double action; 22 LR; 9-shot cylinder; 10″ octagon barrel; solid-frame; fixed sights; checkered walnut grips; blued finish. Introduced 1926; dropped 1941.

Exc.: $110　**VGood:** $100　**Good:** $85

HARRINGTON & RICHARDSON MODEL 4

Revolver; double action; 32 S&W Long, 38 S&W, 38 Colt New Police; 6-shot cylinder (32), 5-shot (38); 2 1/2″, 4 1/2″, 6″ barrels; solid frame; fixed sights; hard rubber grips; blued or nickel finish. Introduced 1905; dropped 1941.

Exc.: $85　**VGood:** $65　**Good:** $50

HARRINGTON & RICHARDSON MODEL 5

Revolver; double action; 32 S&W Short; 5-shot cylinder; 2 1/2″, 4 1/2″, 6″ barrels; light frame; fixed sights; hard rubber grips; blued or nickel finish. Introduced 1905; dropped 1941.

Exc.: $85　**VGood:** $65　**Good:** $50

HARRINGTON & RICHARDSON MODEL 6

Revolver; double action; 22 Short, 22 Long, 22 LR; 7-shot cylinder; 2 1/2″, 4 1/2″, 6″ barrels; solid frame; hard rubber grips; blued or nickel finish. Introduced 1906; dropped 1941.

Exc.: $85　**VGood:** $65　**Good:** $50

HARRINGTON & RICHARDSON MODEL 40 HAMMERLESS

Revolver; double action; 22 LR, 32 S&W Short; 7-shot cylinder (22), 5-shot (32); 2″, 3″, 4″, 5″, 6″ barrels; fixed sights; small hinged-frame; hard rubber stocks; blued or nickel finish. Also listed during late production as Model 45. Introduced 1899; dropped 1941.

Exc.: $110　**VGood:** $85　**Good:** $65

HARRINGTON & RICHARDSON MODEL 50 HAMMERLESS

Revolver; double action; 32 S&W Long, 38 S&W; 6-shot cylinder (32), 5-shot (38); 3 1/4″, 4″, 5″, 6″ barrels; fixed sights; small hinged-frame; hard rubber stocks; blued or nickel finish. Also listed during late production as Model 55. Introduced 1899; dropped 1941.

Exc.: $90　**VGood:** $60　**Good:** $40

HARRINGTON & RICHARDSON MODEL 504

Revolver; 32 H&R Mag.; 5-shot cylinder; 3″,4″, 6″ barrel; blued finish. Introduced 1984; dropped 1985.

Perf.: $135　**Exc.:** $115　**VGood:** $100

HARRINGTON & RICHARDSON MODEL 532

Revolver; 32 H&R Mag.; 5-shot cylinder; 2 1/2″, 4″ barrel; solid frame; round butt; wood grips; blued finish. Introduced 1984; dropped 1985.

Perf.: $110　**Exc.:** $100　**VGood:** $85

HARRINGTON & RICHARDSON MODEL 586

Revolver; double action; 32 H&R Mag.; 5-shot cylinder; 4 1/2″, 5 1/2″, 7 1/2″, 10″ barrel; adjustable rear sight; fixed cylinder; plastic or wood grips. Introduced 1984; dropped 1985.

Perf.: $135　**Exc.:** $115　**VGood:** $100

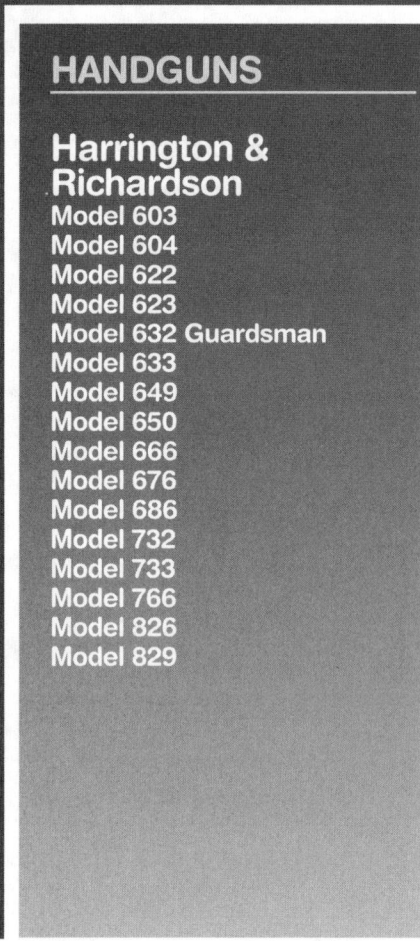

HANDGUNS

Harrington & Richardson
Model 603
Model 604
Model 622
Model 623
Model 632 Guardsman
Model 633
Model 649
Model 650
Model 666
Model 676
Model 686
Model 732
Model 733
Model 766
Model 826
Model 829

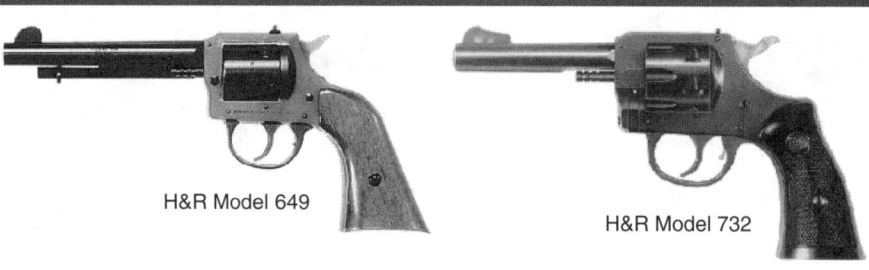

H&R Model 649

H&R Model 732

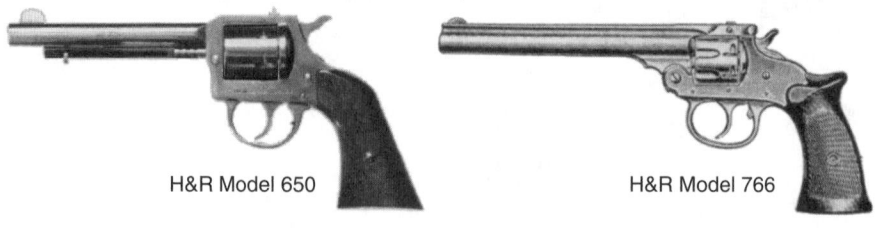

H&R Model 650

H&R Model 766

H&R Model 686

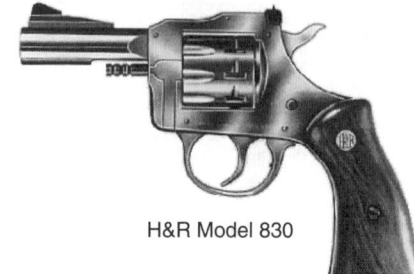

H&R Model 830

HARRINGTON & RICHARDSON
MODEL 603
Revolver; double action; 22 WMR; 6-shot cylinder; 6″ flat-sided barrel; blade front sight, fully-adjustable rear; smooth walnut grips; swing-out cylinder; coil spring construction; blued finish. Introduced 1981; dropped 1983.
Exc.: $125   **VGood:** $110   **Good:** $85

HARRINGTON & RICHARDSON
MODEL 604
Revolver; double action; 22 WMR; 6-shot cylinder; 6″ bull barrel with raised solid rib; blade front, fully-adjustable rear sight; smooth walnut grips; swing-out cylinder; coil spring construction; blued finish. Introduced 1981; dropped 1985.
Exc.: $135   **VGood:** $125   **Good:** $100

HARRINGTON & RICHARDSON
MODEL 622
Revolver; double action; 22 LR, 22 Short, 22 Long; 6-shot cylinder; 2 1/2″, 4″, 6″ barrels; solid frame; fixed sights; checkered black plastic grips; satin blued finish. Introduced 1957; dropped 1963.
Exc.: $110   **VGood:** $85   **Good:** $65

HARRINGTON & RICHARDSON
MODEL 623
Revolver; double action; 22 LR, 22 Short; 6-shot cylinder; 2 1/2″, 4″, 6″ barrels; solid frame; fixed sights; checkered plastic grips; chrome finish. Introduced 1957; dropped 1963.
Exc.: $110   **VGood:** $85   **Good:** $65

HARRINGTON & RICHARDSON
MODEL 632 GUARDSMAN
Revolver; double action; 32 S&W, 32 S&W Long; 6-shot cylinder; 2 1/2″, 4″ barrels; weighs 20 oz.; solid frame; fixed sights; checkered Tenite grips; blued finish. Introduced 1946; dropped 1957.
Exc.: $90   **VGood:** $75   **Good:** $50

HARRINGTON & RICHARDSON
MODEL 633
Revolver; double action; 32 S&W, 32 S&W Long; 6-shot cylinder; 2 1/2″, 4″ barrels; weighs 20 oz.; solid frame; fixed sights; checkered Tenite grips; nickel finish. Introduced 1946; dropped 1957.
Exc.: $100   **VGood:** $85   **Good:** $60

HARRINGTON & RICHARDSON
MODEL 649
Revolver; double action; two 6-shot cylinders; 22 LR, 22 WMR; 5 1/2″, 7 1/2″ barrel; solid frame; blade front, adjustable rear sight; one-piece walnut grip; blued finish. Introduced in 1976; dropped 1980.
Exc.: $125   **VGood:** $110   **Good:** $100

HARRINGTON & RICHARDSON
MODEL 650
Revolver; double action; 22 LR, 22 WMR; 6-shot cylinder; 5 1/2″ barrel; solid frame; blade front, adjustable rear sight; one-piece walnut grip; nickel finished. Introduced 1976; dropped 1980.
Exc.: $125   **VGood:** $110   **Good:** $100

HARRINGTON & RICHARDSON
MODEL 666
Revolver; double action; 22 LR, 22 WMR; two 6-shot cylinders; 6″ barrel; solid frame; fixed sights; plastic stocks; blued finish. Introduced 1976; dropped 1980.
Exc.: $100   **VGood:** $90   **Good:** $75

HARRINGTON & RICHARDSON
MODEL 676
Revolver; double action; 22 LR, 22 WMR; two 6-shot cylinders; 4 1/2″, 5 1/2″, 7 1/2″, 12″ barrel; solid frame; blade front, adjustable rear sight; one-piece walnut grip; color case-hardened frame; blued finish. Introduced 1976; dropped 1985.
Exc.: $110   **VGood:** $100   **Good:** $75

HARRINGTON & RICHARDSON
MODEL 686
Revolver; double action; 22 LR, 22 WMR; two 6-shot cylinders; 5 1/2″, 7 1/2″, 10″, 12″ barrel; ramp and blade front sight, fully-adjustable rear; one-piece walnut grip; color case-hardened frame; blued finish. Introduced 1981; dropped 1985.
Exc.: $135   **VGood:** $110   **Good:** $100

HARRINGTON & RICHARDSON
MODEL 732
Revolver; double action; 32 S&W, 32 S&W Long, 32 H&R Mag.; 6-shot swing-out cylinder; 2 1/2″, 4″ barrels; solid frame; windage-adjustable rear sight (4″ model), fixed on shorter barrel, ramp front; black plastic checkered grips; blued finish. Introduced 1958; dropped 1985.
Exc.: $100   **VGood:** $85   **Good:** $75
32 H&R Mag.
Exc.: $110   **VGood:** $100   **Good:** $85

HARRINGTON & RICHARDSON
MODEL 733
Revolver; double action; 32 S&W, 32 S&W Long, 32 H&R Mag.; 6-shot swing-out cylinder; 2 1/2″ barrel; solid frame; Windage-adjustable rear sight (4″ model), fixed on shorter barrel; ramp front; plastic checkered grips; same as Model 932 except nickel finish. Introduced 1958; dropped 1985.
Exc.: $125   **VGood:** $110   **Good:** $100
32 H&R Mag.
Exc.: $135   **VGood:** $125   **Good:** $110

HARRINGTON & RICHARDSON
MODEL 766
Revolver; double action; 22 LR; 7-shot cylinder; 6″ barrel; target; hinged frame; fixed sights; checkered walnut grips; blued finish. Introduced 1926; dropped 1936.
Exc.: $125   **VGood:** $110   **Good:** $100

HARRINGTON & RICHARDSON
MODEL 826
Revolver; double action; 22 LR, 22 WMR; 9-shot; 3″ bull barrel; weighs 27 oz.; ramp blade front, adjustable rear sight; uncheckered American walnut stocks; recessed muzzle; swing-out cylinder; blued finish. Introduced 1982; dropped 1984.
Exc.: $100   **VGood:** $85   **Good:** $75

HARRINGTON & RICHARDSON
MODEL 829
Revolver; double action; 22 LR; 9-shot cylinder; 3″ bull barrel; weighs 27 oz.; uncheckered American walnut stocks; ramp/blade front sight, adjustable rear; blued finish; recessed muzzle; swing-out cylinder. Introduced 1982; dropped 1984.
Exc.: $100   **VGood:** $80   **Good:** $65

H&R Model 904

H&R Model 926

H&R Model 905

H&R Model 929

H&R Model 925
(Early Model)

H&R Model 939

Harrington & Richardson
Model 830
Model 832
Model 833
Model 900
Model 901
Model 903
Model 904
Model 905
Model 922
Model 922 Bantamweight
Model 922 Camper
Model 923
Model 925
Model 926
Model 929

H&R Model 925
(1960s model)

HARRINGTON & RICHARDSON MODEL 830

Revolver; double action; 22 LR, 22 WMR; 9-shot; 3″ bull barrel; weighs 27 oz.; ramp blade front, adjustable rear sight; unchecked American walnut stocks; recessed muzzle; swing-out cylinder; nickel finish. Introduced 1982; dropped 1984.
Exc.: $100   **VGood:** $85   **Good:** $75

HARRINGTON & RICHARDSON MODEL 832

Revolver; double action; 32 S&W Long; 9-shot; 3″ bull barrel; weighs 27 oz.; ramp blade front, adjustable rear sight; unchecked American walnut stocks; recessed muzzle; swing-out cylinder; blued finish. Introduced 1982; dropped 1984.
Exc.: $90   **VGood:** $75   **Good:** $50

HARRINGTON & RICHARDSON MODEL 833

Revolver; double action; 32 S&W Long; 9-shot; 3″ bull barrel; weighs 27 oz.; ramp blade front, adjustable rear sight; unchecked American walnut stocks; recessed muzzle; swing-out cylinder; nickel finish. Introduced 1982; dropped 1984.
Exc.: $90   **VGood:** $75   **Good:** $50

HARRINGTON & RICHARDSON MODEL 900

Revolver; double action; 22 LR, 22 Short, 22 Long; 9-shot swing-out cylinder; 2 1/2″, 4″, 6″ barrels; solid frame; blade front sight; high-impact black plastic grips; blued finish. Introduced 1962; dropped 1973.
Exc.: $75   **VGood:** $60   **Good:** $40

HARRINGTON & RICHARDSON MODEL 901

Revolver; double action; 22 LR, 22 Short, 22 Long; 9-shot swing-out cylinder; 2 1/2″, 4″, 6″ barrels; solid frame; fixed sights; high-impact white plastic grips; chrome finish. Introduced 1962; dropped 1963.
Exc.: $75   **VGood:** $60   **Good:** $40

HARRINGTON & RICHARDSON MODEL 903

Revolver; double action; 22 LR; 9-shot cylinder; 6″ flat-sided barrel; blade front sight, fully-adjustable rear; smooth walnut grips; swing-out cylinder; coil spring construction; blued finish. Introduced 1981; dropped 1985.
Exc.: $100   **VGood:** $85   **Good:** $75

HARRINGTON & RICHARDSON MODEL 904

Revolver; double action; 22 LR; 9-shot cylinder; 6″ bull barrel with raised solid rib; blade front, fully-adjustable rear sight; smooth walnut grips; swing-out cylinder; coil spring construction; blued finish. Introduced 1981; dropped 1985.
Exc.: $110   **VGood:** $90   **Good:** $85

HARRINGTON & RICHARDSON MODEL 905

Revolver; double action; 22 LR; 9-shot cylinder; 4″ bull barrel with raised solid rib; blade front, fully-adjustable rear sight; smooth walnut grips; swing-out cylinder; coil spring construction; H&R Hard-Guard finish. Introduced 1981; dropped 1985.
Exc.: $110   **VGood:** $90   **Good:** $85

HARRINGTON & RICHARDSON MODEL 922

Revolver; double action; 22 Short, 22 Long, 22 LR; 9-shot cylinder; 4″, 6″, 10″ barrel; 10 1/2″ overall length (6″ barrel); weighs 21 3/4 oz.; solid frame; fixed sights; checkered walnut or Tenite grips; blued finish. Introduced in 1919. Early production had 10″ octagonal barrel, later dropped.
Exc.: $90   **VGood:** $75   **Good:** $60

Harrington & Richardson Model 922 Bantamweight

Same specs as Model 922 except 2 1/2″ barrel; weighs 20 oz.; rounded butt. Introduced 1951; no longer produced.
Exc.: $100   **VGood:** $85   **Good:** $70

Harrington & Richardson Model 922 Camper

Same specs as Model 922 except 4″ barrel; weighs 21 oz.; special small Tenite grips. Introduced 1952.
Exc.: $85   **VGood:** $70   **Good:** $50

HARRINGTON & RICHARDSON MODEL 923

Revolver; double action; 22 Short, 22 Long, 22 LR; 9-shot cylinder; 4″, 6″, 10″ barrel; 10 1/2″ overall length (6″ barrel); weighs 21 3/4 oz.; solid frame; fixed sights; checkered walnut or Tenite grips; chrome finish. Introduced in 1919. Early production had 10″ octagonal barrel, later dropped.
Exc.: $100   **VGood:** $85   **Good:** $70

HARRINGTON & RICHARDSON MODEL 925

Revolver; double action; 38 S&W; 5-shot cylinder; 2 1/2″, 4″, 6″ barrel; weighs 25 oz. (4″ barrel); hinged frame; fixed front, adjustable rear sight; one-piece smooth or checkered plastic grip; blued finish. Advertised as the Defender Model; originally introduced as Model 25. Introduced 1964; dropped 1981.
Exc.: $110   **VGood:** $90   **Good:** $75

HARRINGTON & RICHARDSON MODEL 926

Revolver; double action; 22 LR, 22 Short, 22 Long, 38 S&W; 9-shot cylinder (22), 5-shot (38 S&W); 4″ barrel; weighs 31 oz.; hinged frame; fixed front, adjustable rear sight; checkered walnut stocks; blued finish. Introduced 1968; dropped 1980.
Exc.: $110   **VGood:** $90   **Good:** $75

HARRINGTON & RICHARDSON MODEL 929

Revolver; double action; 22 LR, 22 Short, 22 Long; 9-shot cylinder; 2 1/2″, 4″, 6″ barrels; weighs 23 oz. (2 1/2″ barrel); solid frame; swing-out cylinder; auto extractor; fixed sights; checkered plastic grips; blued finish. Advertised as Sidekick Model. Introduced 1956; dropped 1985.
Exc.: $100   **VGood:** $90   **Good:** $75

HANDGUNS

Harrington & Richardson

Model 930
Model 939
Model 939 Premier
Model 940
Model 944 (See H&R 22 Special)
Model 945 (See H&R 22 Special)
Model 949
Model 949 Western
Model 950
Model 955 (See H&R Expert)
Model 976
Model 999
Model 999 Sportsman
New Defender
No. 196
No. 199
Premier Model
Self-Loading 25
Self-Loading 32

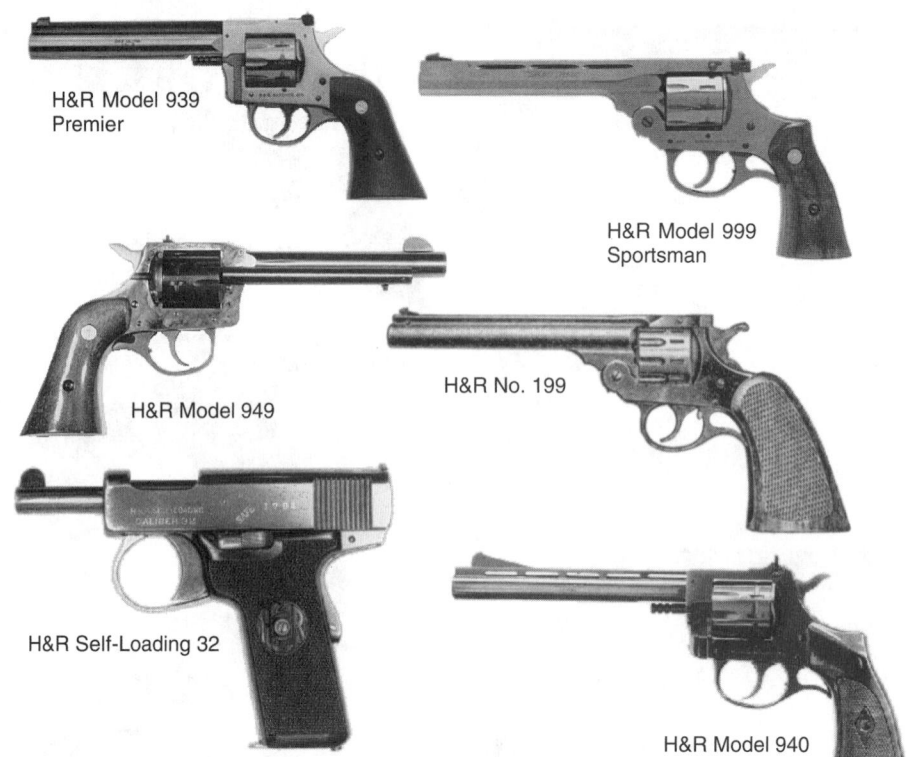

H&R Model 939 Premier

H&R Model 999 Sportsman

H&R Model 949

H&R No. 199

H&R Self-Loading 32

H&R Model 940

HARRINGTON & RICHARDSON MODEL 930

Revolver; double action; 22 LR, 22 Short, 22 Long; 9-shot cylinder; 2 1/2″, 4″, 6″ barrels; solid frame; swing-out cylinder; fixed sights; checkered plastic grips; same as Model 929 except nickel finish. Introduced 1956; dropped 1985.

Exc.: $100   **VGood:** $90   **Good:** $75

HARRINGTON & RICHARDSON MODEL 939

Revolver; double action; 22 LR; 9-shot cylinder; 6″ barrel with vent rib; solid-frame; swing-out cylinder; ramp front, adjustable rear sight; checkered walnut grips; blued finish. Advertised as the Ultra Sidekick Model. Introduced 1958; dropped 1981.

Exc.: $110   **VGood:** $100   **Good:** $85

HARRINGTON & RICHARDSON MODEL 939 PREMIER

Revolver; double action; 22 LR; 9-shot cylinder; 6″ heavy barrel; weighs 36 oz.; walnut-finished hardwood grips; blade front sight, fully-adjustable rear; swing-out cylinder; plunger-type ejection; solid barrel rib; high-polish blue finish; Western-style grip. Made in U.S. by H&R 1871, Inc. Introduced 1995; still produced. **LMSR: $185**

New: $150   **Perf.:** $135   **Exc.:** $115

HARRINGTON & RICHARDSON MODEL 940

Revolver; double action; 22 Short, 22 Long, 22 LR; 9-shot; 6″ barrel; vent rib; ramp front sight, adjustable rear; checkered hardwood grips; thumbrest; swingout cylinder; blued finish. Advertised as Ultra Sidekick. Introduced 1978; dropped 1982.

Exc.: $100   **VGood:** $85   **Good:** $75

HARRINGTON & RICHARDSON MODEL 949

Revolver; double action; 22 LR, Long, Short; 9-shot cylinder; 5 1/2″ barrel; solid-frame; side-loading, ejection; adjustable rear sight, blade front; one-piece plain walnut grip; blued, nickel finish. Advertised as Forty-Niner Model. Introduced in 1960; dropped 1985.

Exc.: $90   **VGood:** $75   **Good:** $60

HARRINGTON & RICHARDSON MODEL 949 WESTERN

Revolver; double action; 22 LR; 9-shot cylinder; 5 1/2″, 7 1/2″ barrel; weighs 36 oz.; walnut-stained hardwood grips; blade front sight, adjustable rear; color case-hardened frame and backstrap; traditional loading gate and ejector rod. Made in U.S. by Harrington & Richardson. Introduced 1994; still produced. **LMSR: $185**

New: $140   **Perf.:** $125   **Exc.:** $110

HARRINGTON & RICHARDSON MODEL 950

Revolver; double action; 22 LR, 22 Long, 22 Short; 9-shot cylinder; 5 1/2″ barrel; solid-frame; side-loading, ejection; adjustable rear sight, blade front; one-piece plain walnut grip; nickel finish. Introduced 1960; dropped 1985.

Exc.: $110   **VGood:** $90   **Good:** $80

HARRINGTON & RICHARDSON MODEL 976

Revolver; double action; 22 LR, 38 S&W; 9-shot cylinder (22 LR), 5-shot (38 S&W); 4″ barrel; hinged frame; fixed front, adjustable rear sight; checkered walnut stocks; blued finish. Introduced 1968; dropped 1981.

Exc.: $110   **VGood:** $90   **Good:** $80

HARRINGTON & RICHARDSON MODEL 999

Revolver; double action; 22 Short, 22 Long, 22 LR; 9-shot cylinder; 4″, 6″ top-break barrel; weighs 34 oz.; checkered hardwood grips; elevation-adjustable front sight, windage-adjustable rear; automatic ejection; triggerguard extension; blued finish; optional engraving. Marketed as Sportsman Model. Introduced 1936; dropped 1985.

Exc.: $135   **VGood:** $110   **Good:** $100

Engraved
Exc.: $250   **VGood:** $200   **Good:** $150

HARRINGTON & RICHARDSON MODEL 999 SPORTSMAN

Revolver; double action; 22 LR; 9-shot cylinder; 4″, 6″ barrel; 8 1/2″ overall length; weighs 30 oz.; walnut-finished hardwood grips; elevation-adjustable blade front sight, windage-adjustable rear; top-break loading; automatic shell ejection; polished blue finish. Made in U.S. by Harrington & Richardson. Reintroduced 1992; still in production. **LMSR: $280**

New: $250   **Perf.:** $200   **Exc.:** $150

HARRINGTON & RICHARDSON NEW DEFENDER

Revolver; double action; 22 LR; 2″ barrel; 6 1/4″ overall length; checkered hard rubber grips; adjustable front and rear sights; automatic ejector; triggerguard extention; blued finish. Also listed as Model 299. Introduced 1936; dropped 1941.

Exc.: $100   **VGood:** $90   **Good:** $75

HARRINGTON & RICHARDSON NO. 196

Revolver; 22 Short, 22 Long, 22 LR; 6 1/4″ barrel; 11″ overall length; weighs 32 oz.; adjustable Patridge front, adjustable rear; wide hammer; blue finish. Called the "Eureka" Sportsman.

Exc.: $175   **VGood:** $150   **Good:** $100

HARRINGTON & RICHARDSON NO. 199

Revolver; single action; 22 LR; 9-shot cylinder; 6″ barrel; 11″ overall length; weighs 30 oz.; hinged frame; adjustable target sights; checkered walnut grips; blued finish. Single action version of the double-action Model 999 Sportsman. Introduced 1933; dropped 1951.

Exc.: $125   **VGood:** $110   **Good:** $85

HARRINGTON & RICHARDSON PREMIER MODEL

Revolver; double action; hinged-frame; 22 LR, 32 S&W Short; 7-shot (22), 5-shot (32); 2″, 3″, 4″, 5″, 6″ barrel; weighs 13 oz.; fixed sights; hard rubber grips; blued nickel finish. Introduced 1895; dropped 1941.

Exc.: $90   **VGood:** $70   **Good:** $50

HARRINGTON & RICHARDSON SELF-LOADING 25

Semi-automatic; 25 ACP; 6-shot magazine; 2″ barrel; 4 1/2″ overall; hammerless checkered hard rubber stocks; fixed sights; blued finish. Variation of Webley & Scott design. Approximately 20,000 manufactured from 1912 to 1915.

Exc.: $300   **VGood:** $275   **Good:** $200

HARRINGTON & RICHARDSON SELF-LOADING 32

Semi-automatic; 32 ACP; 8-shot magazine; 3 1/2″ barrel; 6 1/2″ overall length; weighs 22 oz.; grip safety; magazine disconnector; hammerless; checkered hard rubber stocks; fixed sights; blued finish. Variation of Webley & Scott design. Approximately 40,000 manufactured from 1916 to 1939.

Exc.: $325   **VGood:** $275   **Good:** $200

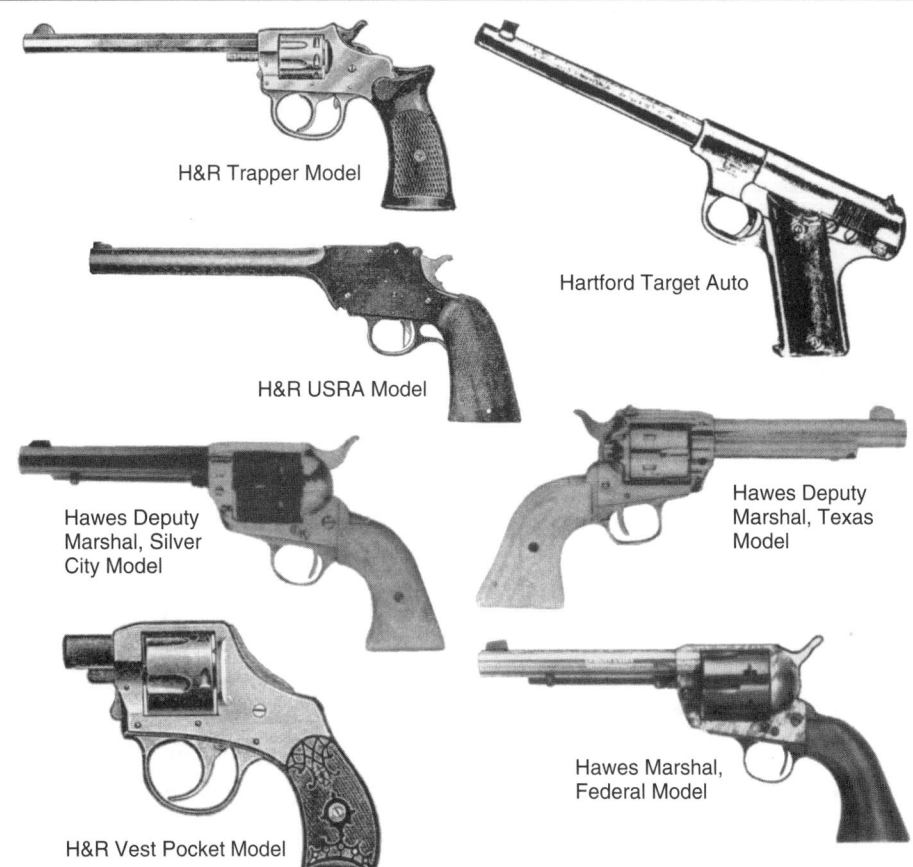

H&R Trapper Model

Hartford Target Auto

H&R USRA Model

Hawes Deputy Marshal, Silver City Model

Hawes Deputy Marshal, Texas Model

H&R Vest Pocket Model

Hawes Marshal, Federal Model

Harrington & Richardson
Sidekick (See H&R Model 929)
Sportsman (See H&R No. 199)
Trapper Model
Ultra Sidekick (See H&R Model 940)
USRA Model
Vest Pocket Model
Young America

Harris-McMillan
Signature Jr. Long Range
Wolverine

Hartford
Repeating Pistol
Target Automatic
Target Single Shot

Haskell
JS-45

Hawes
Deputy Marshal
Deputy Marshal, Denver Model
Deputy Marshal, Montana Model
Deputy Marshal, Silver City Model
Deputy Marshal, Texas Model
Favorite

HARRINGTON & RICHARDSON TRAPPER MODEL

Revolver; double action; 22 Short, 22 Long, 22 LR; 7-shot cylinder; 6″ octagonal barrel; weighs 12 1/4 oz.; solid frame; gold front sight; checkered walnut stocks; blued finish. Introduced 1924; dropped during WWII.
Exc.: $100   **VGood: $90**   **Good: $75**

HARRINGTON & RICHARDSON USRA MODEL

Single shot; 22 LR; 7″, 8″, 10″ barrel; weighs 31 oz. (10″ barrel); hinged frame target pistol; adjustable target sights; adjustable trigger pull; checkered walnut grips; blued finish. Introduced 1928; dropped 1943.
Exc.: $450   **VGood: $350**   **Good: $250**

HARRINGTON & RICHARDSON VEST POCKET MODEL

Revolver; double action; 22 Long, 32 S&W Short; 5-shot cylinder (32), 7-shot (22); 1 1/8″ barrel; weighs 8 1/2 oz.; solid frame; spurless hammer; no sights except for milled slot in top frame; hard rubber grips; blued or nickel finish. Introduced 1891; dropped during WWII.
Exc.: $100   **VGood: $85**   **Good: $75**

HARRINGTON & RICHARDSON YOUNG AMERICA

Revolver; double action; 22 Short, 22 Long, 22 LR, 32 S&W Short; 7-shot cylinder (22), 5-shot (32); 2″, 4 1/2″, 6″ barrel; solid frame; fixed sights; hard rubber grips; blued or nickel finish. Introduced 1885; dropped during WWII.
Exc.: $125   **VGood: $110**   **Good: $85**

HARRIS-McMILLAN SIGNATURE JR. LONG RANGE

Bolt action; single shot or repeater; chambered for any suitable caliber; custom barrel length to customer specs.; weighs about 5 lbs.; fiberglass stock; no sights; scope rings; right- or left-hand benchrest action of titanium or stainless steel; bipod. All parts and components made in U.S. by Harris Gunworks, Inc. Introduced 1992; still produced. **LMSR: $2400**
New: $2200   **Perf.: $1750**   **Exc.: $1500**

HARRIS-McMILLAN WOLVERINE

Semi-automatic; single action; 9mm Para., 10mm Auto, 38 Wadcutter, 38 Super, 45 Italian, 45 ACP; 6″ barrel; 9 1/2″ overall length; weighs 45 oz.; Pachmayr rubber grips; blade front sight, fully-adjustable low profile rear; integral compensator; round burr-style hammer; extended grip safety; checkered backstrap; skeletonized aluminum match trigger. Many finish options offered. Combat or Competition Match styles. All parts and components made in U.S. by Harris Gunworks, Inc. Introduced 1992. **LMSR: $1700**
New: $1600   **Perf.:1$350**   **Exc.: $1200**

HARTFORD REPEATING PISTOL

Semi-automatic; 22 LR; 10-shot magazine; 6 3/4″ barrel; 10 3/4″ overall length; hand-operated repeater; slide manually moved rearward to eject cartridge case, forward to chamber new round; same general outward design characteristics as Target Automatic; checkered, black rubber grips; target sights; blued finish.
Exc.: $300   **VGood: $250**   **Good: $200**

HARTFORD TARGET AUTOMATIC

Semi-automatic; 22 LR; 10-shot magazine; 6 3/4″ barrel; 10 3/4″ overall length; checkered black rubber grips; target sights; blued finish. Hartford was the predecessor of High Standard. Introduced 1929; dropped 1932. Has more collector than shooter value.
Exc.: $350   **VGood: $300**   **Good: $250**

HARTFORD TARGET SINGLE SHOT

Semi-automatic; single shot; 22 LR; 5 3/4″ barrel; 10 3/4″ overall length; same general outward appearance as Hartford Target Auto; black rubber or walnut grips; target sights; color case-hardened frame; blued barrel. Introduced 1929; dropped 1932.
Exc.: $325   **VGood: $275**   **Good: $200**

HASKELL JS-45

Semi-automatic; single action; 45 ACP; 7-shot magazine; 4 1/2″ barrel; 8″ overall length; weighs 44 oz.; checkered acetal resin grips; fixed low profile sights; internal drop-safe mechanism; all aluminum frame; matte black finish. From MKS Supply, Inc. Introduced 1991; still produced as Hi-Point JS-45.
LMSR: $148
Perf.:$125   **Exc.: $110**   **VGood: $100**

Brushed nickel finish
Perf.:$135   **Exc.: $125**   **VGood: $110**

HAWES DEPUTY MARSHAL

Revolver; single action; 22 LR, 22 WMR; 6-shot; 5 1/2″ barrel; 11″ overall length; blade front sight, adjustable rear; plastic or walnut grips; blued finish. Introduced 1973; dropped 1980.
Exc.: $75   **VGood: $65**   **Good: $50**

Hawes Deputy Marshal, Denver Model
Same specs as Deputy Marshal except brass frame. Introduced 1973; dropped 1980.
Exc.: $75   **VGood: $65**   **Good: $50**

Hawes Deputy Marshal, Montana Model
Same specs as Deputy Marshal except walnut grips only; brass grip frame. Introduced 1973; dropped 1980.
Exc.: $75   **VGood: $65**   **Good: $50**

Hawes Deputy Marshal, Silver City Model
Same specs as Deputy Marshal except brass grip frame; chromed frame; blued barrel and cylinder. Introduced 1973; discontinued 1980.
Exc.: $75   **VGood: $65**   **Good: $50**

Hawes Deputy Marshal, Texas Model
Same specs as Deputy Marshal except for chrome finish. Introduced 1973; discontinued 1980.
Exc.: $75   **VGood: $65**   **Good: $50**

HAWES FAVORITE

Revolver; single shot; 22 LR; 8″ barrel; 12″ overall length; tip-up action; target sights; plastic or rosewood stocks; blued barrel; chromed frame. Replica of Stevens No. 35 Target Model. Manufactured 1968 to 1976.
Exc.: $75   **VGood: $65**   **Good: $50**

Hawes
Marshal, Chief Model
Marshal, Federal Model
Marshal, Montana Model
Marshal, Silver City Model
Marshal, Texas Model
Marshal, Western Model
Medallion Model
Trophy Model

Hawes/SIG-Sauer
P220

Heckler & Koch
HK4
P7 K3
P7 M8
P7 M10
P7 M13

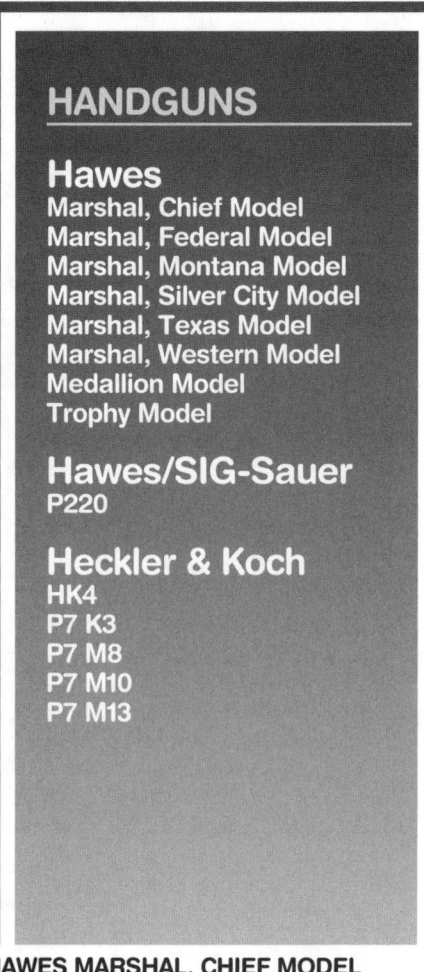

Hawes Marshal, Montana Model

H&K P9S Competition Kit

H&K P7 M13

H&K P7 M8

Hawes/Sig-Sauer P220

H&K HK4

HAWES MARSHAL, CHIEF MODEL
Revolver; single action; 357 Mag., 9mm, 44-40, 44 Mag., 45 Colt, 45 ACP; 6″ barrel; target-type front sight, adjustable rear; oversize rosewood stocks; available with interchangeable cylinder combos; blued finish. Manufactured in West Germany by Sauer & Sohn exclusively for Hawes. Introduced 1968; discontinued 1980.
Exc.: $200   **VGood:** $165   **Good:** $135

Hawes Marshal, Federal Model
Same specs as Hawes Marshal, Chief Model except color case-hardened frame; one-piece European walnut grip; brass grip frame; fixed sights. Introduced 1969; discontinued 1980.
Exc.: $200   **VGood:** $165   **Good:** $135

Hawes Marshal, Montana Model
Same specs as Hawes Marshal, Chief Model except also 22 LR, 22 WMR chamberings; 5″ barrel (rimfires); fixed sights; plastic stag (rimfires) or rosewood (centerfire) grips; brass stag frame. Introduced 1968; discontinued 1980.
Exc.: $200   **VGood:** $165   **Good:** $135

Hawes Marshal, Silver City Model
Same specs as Hawes Marshal, Chief Model except fixed sights; pearlite grips; brass frame. Introduced 1969; discontinued 1980.
Exc.: $200   **VGood:** $165   **Good:** $135

Hawes Marshal, Texas Model
Same specs as Hawes Marshal, Chief Model except also 22 LR, 22 WMR chamberings; 5″ barrel (rimfires); fixed sights; pearlite grips; nickel finish. Introduced 1969; discontinued 1980.
Exc.: $200   **VGood:** $165   **Good:** $135

Hawes Marshal, Western Model
Same specs as Hawes Marshal, Chief Model except also 22 LR, 22 WMR chamberings; 5″ barrel (rimfire); fixed sights; plastic stag (rimfire) or rosewood (centerfire) grips. Introduced 1968; discontinued 1980.
Exc.: $200   **VGood:** $165   **Good:** $135

HAWES MEDALLION MODEL
Revolver; double action; 22 LR, 38 Spl.; 6-shot; 3″, 4″, 6″ barrel; fixed sights; swing-out cylinder; blued finish. No longer in production.
Exc.: $135   **VGood:** $125   **Good:** $100

HAWES TROPHY MODEL
Revolver; double action; 22 LR, 38 Spl.; 6-shot; 6″ barrel; adjustable sights; swing-out cylinder; blued finish. No longer in production.
Exc.: $135   **VGood:** $125   **Good:** $100

HAWES/SIG-SAUER P220
Semi-automatic; 9mm Para., 38 Super, 45 ACP; 7-shot magazine (45 ACP), 9-shot (9mm, 38); 4 3/8″ barrel; 7 3/4″ overall length; checkered European walnut or black plastic stocks; windage-adjustable rear sight, blade front; square combat triggerguard. Manufactured in Germany. Introduced 1977; still in production; dropped by Hawes in 1980. Also known as Browning BDA, SIG/Sauer P220.
Exc.: $450   **VGood:** $400   **Good:** $350

HECKLER & KOCH HK4
Semi-automatic; double action; 22 LR, 25 ACP, 32 ACP, 380 ACP; 8-shot magazine; 3 1/2″ barrel; 6″ overall length; windage-adjustable rear sight, fixed front; checkered black plastic grips; blued finish. Early version available with interchangeable barrels, magazines for four calibers. Imported from Germany originally by Harrington & Richardson, then by Heckler & Koch; importation dropped 1984.
Exc.: $350   **VGood:** $275   **Good:** $200

HECKLER & KOCH P7 K3
Semi-automatic; single action; 22 LR, 380 ACP; 8-shot magazine; 3 13/16″ barrel; 26 1/2 oz.; fixed sights; black plastic grips; cocked by pressure on frontstrap; large triggerguard with heat shield; oil-filled buffer to decrease recoil; blue or nickel finish. Introduced 1988; no longer imported.
Exc.: $850   **VGood:** $700   **Good:** $600

22 LR conversion unit
Exc.: $600   **VGood:** $500   **Good:** $400
32 ACP conversion unit
Exc.: $700   **VGood:** $600   **Good:** $450

HECKLER & KOCH P7 M8
Semi-automatic; 9mm Para.; 8-shot magazine; 4 1/8″ barrel; 6 3/4″ overall length; weighs 29 oz.; blade front, adjustable rear sight with three-dot system; unique "squeeze cocker" system in frontstrap; gas-retarded action; squared combat-type triggerguard; stippled black plastic grips; blue finish. Imported from Germany by Heckler & Koch, Inc. **LMSR: $1141**
Perf.: $850   **Exc.:** $750   **VGood:** $650

HECKLER & Koch P7 M10
Semi-automatic; 40 S&W; 10-shot magazine; 4 1/8″ barrel; 7″ overall length; weighs 43 oz.; blade front, adjustable rear sight with three-dot system; unique "squeeze cocker" system in frontstrap; gas-retarded action; squared combat-type triggerguard; stippled black plastic grips; blue finish. Imported from Germany by Heckler & Koch, Inc. Introduced 1992; no longer in production.
Perf.: $1000   **Exc.:** $900   **VGood:** $750

HECKLER & KOCH P7 M13
Semi-automatic; single action; 9mm Para.; 13-shot magazine; 4 1/8″ barrel; 6 1/2″ overall length; 30 oz.; fixed sights; black plastic grips; cocked by pressure on frontstrap; large triggerguard with heat shield; ambidextrous magazine release; blue or nickel finish. Introduced 1986; importation discontinued 1994.
Perf.: $1250   **Exc.:** $1100   **VGood:** $1000
Nickel finish
Perf.: $1350   **Exc.:** $1150   **VGood:** $1000

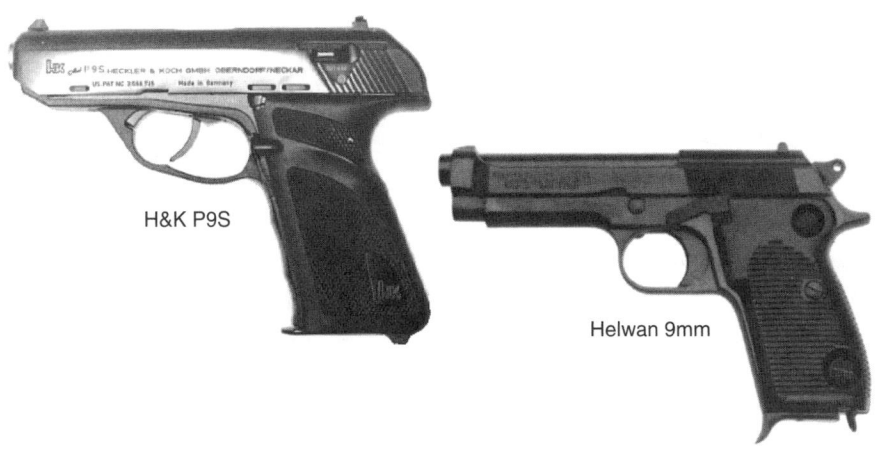

H&K P9S

Helwan 9mm

HANDGUNS

Heckler & Koch
P7 PSP
P9S
P9S Competition Kit
P9S Target
SP 89
USP 9/USP 40
USP 45
VP 70Z

Helwan
9mm

H&K USP 9mm

H&K P7 PSP

H&K VP 70Z

H&K SP 89

HECKLER & KOCH P7 PSP

Semi-automatic; single action; 9mm Para; 8-shot magazine; 4 $^1/_8$" barrel; 6 $^1/_2$" overall length; weighs 29 oz.; fixed sights; black plastic grips; cocked by pressure on frontstrap; squared combat triggerguard; blued finish. Imported from West Germany. Introduced 1982; discontinued 1986.

Perf.:$800　　**Exc.:** $700　　**VGood:** $600

HECKLER & KOCH P9S

Semi-automatic; double action; 9mm Para., 45 ACP; 9-shot magazine (9mm), 7-shot (45 ACP); 4" barrel; 5 $^7/_{16}$" overall length; fixed sights; checkered plastic grips; loaded/cocked indicators; hammer cocking lever; phosphated finish. Originally imported from Germany by Gold Rush Gun Shop in 1977, then by Heckler & Koch, Inc.; importation dropped 1984.

Exc.: $600　　**VGood:** $500　　**Good:** $400

Heckler & Koch P9S Competition Kit

Same specs as P9S except 9mm Para.; additional 5 $^1/_2$" barrel with weight; additional slide; adjustable sights, trigger; walnut stock. Introduced 1977; discontinued 1984.

Exc.: $900　　**VGood:** $750　　**Good:** $650

Heckler & Koch P9S Target

Same specs as P9S except adjustable sights, trigger. Introduced 1977; discontinued 1984.

Exc.: $700　　**VGood:** $550　　**Good:** $450

HECKLER & KOCH SP 89

Semi-automatic; single action; 9mm Para.; 15-, 30-shot magazine; 4 $^1/_2$" barrel; 12 $^7/_8$" overall length; weighs 4 $^1/_2$ lbs.; black high-impact plastic grip; post front sight, fully-adjustable diopter rear; design inspired by the HK94 rifle; has special flash-hider forend. Imported from Germany by Heckler & Koch, Inc. Introduced 1989; no longer imported.

Perf.: $3500　　**Exc.:** $3000　　**VGood:** $2500

HECKLER & KOCH USP 9/USP 40

Semi-automatic; single action/double action or double-action-only; 9mm Para., 40 S&W; 16-shot magazine (9mm), 13-shot (40 S&W); 4 $^1/_4$" barrel; 6 $^{15}/_{16}$" overall length; weighs 28 oz. (USP40); non-slip stippled black polymer grips; blade front sight, windage-adjustable rear; polymer frame; modified Browning action with recoil reduction system; single control lever; special "hostile environment" finish on all metal parts; available in SA/DA, DAO, left- and right-hand versions. Imported from Germany by Heckler & Koch, Inc. Introduced 1993; still imported. **LMSR:** $636

Right-hand
　New: $575　　**Perf.:**$500　　**Exc.:** $450
Left-hand
　New: $595　　**Perf.:**$550　　**Exc.:** $500

Heckler & Koch USP 45

Same specs as USP 9/USP 40 except 45 ACP; 10-shot magazine; 4 $^1/_8$" barrel; 7 $^7/_8$" overall length; weighs 30 oz.; adjustable three-dot sight system; available in SA/DA, DAO; left- and right-hand versions. Imported from Germany by Heckler & Koch, Inc. Introduced 1995; still imported. **LMSR: $696**

Right-hand
　New: $625　　**Perf.:**$500　　**Exc.:** $450
Left-hand
　New: $650　　**Perf.:**$550　　**Exc.:** $500

HECKLER & KOCH VP 70Z

Semi-automatic; double-action-only; 9mm Para.; 18-shot magazine; 4 $^1/_2$" barrel; 8" overall length; ramp front sight, channeled rear on slide; stippled black plastic stocks; recoil-operated; only four moving parts; double-column magazine; phosphated finish. Manufactured in West Germany. Introduced 1976; importation dropped 1989.

Perf.:$400　　**Exc.:** $350　　**VGood:** $300

HELWAN 9MM

Semi-automatic; 9mm Para.; 8-shot magazine; 4 $^1/_2$" barrel; 8 $^1/_4$" overall length; weighs 33 oz; blade front sight, drift-adjustable rear; grooved black plastic stocks; updated version of Beretta Model 951. Imported from Egypt by Steyr Daimler Puch of America. Introduced 1982; dropped 1983.

Exc.: $175　　**VGood:** $135　　**Good:** $125

Heritage
Model HA25
Rough Rider
Sentry

Hi-Point Firearms
JS-9mm
JS-9mm Compact
JS-40 S&W
JS-45

**Hi-Standard
Semi-Automatic
Pistols**
Citation (First Model 102
Series)
Citation (103 Series)
Citation (104 Series)
Citation (106 Series)
Citation (107 Series)
Citation (ML Series)
Citation (Seven Number
Series)

Heritage Sentry

Heritage Rough Rider

Hi-Point Firearms
JS-40 S&W

Hi-Point Firearms JS-45

Hi-Point Firearms
Model JS-9mm
Compact

Hi-Standard Citation
(106 Series)

HERITAGE MODEL HA25
Semi-automatic; single action; 25 ACP; 6-shot magazine; 2 1/2″ barrel; 4 5/8″ overall length; weighs 12 oz.; smooth or checkered walnut grips; fixed sights; exposed hammer; manual safety; open-top slide; polished blue or blue/gold finish. Made in U.S. by Heritage Mfg., Inc. Introduced 1993; no longer produced.

Perf.: $75	**Exc.:** $65	**VGood:** $50

Blue/gold finish

Perf.: $85	**Exc.:** $70	**VGood:** $55

HERITAGE ROUGH RIDER
Revolver; single action; 22 LR, 22 LR/22 WMR combo; 6-shot cylinder; 4 3/4″, 6 1/2″, 9″ barrel; weighs 31 to 38 oz.; Goncolo Alves grips; blade front sight, fixed rear; hammer-block safety; high polish blue or nickel finish. Made in U.S. by Heritage Mfg., Inc. Introduced 1993; still produced. **LMSR: $105**

New: $85	**Perf.:** $75	**Exc.:** $65

2″, 3″, 4″, barrel, birdshead grip

New: $95	**Perf.:** $85	**Exc.:** $75

HERITAGE SENTRY
Revolver; double action; 22 LR, 22 WMR, 9mm Para., 38 Spl.; 8-shot cylinder (22 LR), 6-shot (22 WMR, 9mm, 38 Spl.); 2″, 4″ barrel; 6 1/4″ overall length (2″ barrel); weighs 23 oz.; magnum-style round butt grips of checkered plastic; ramp front sight, fixed rear; pull-pin-type ejection; serrated hammer and trigger; polished blue or nickel finish. Made in U.S. by Heritage Mfg., Inc. Introduced 1993; still produced. **LMSR: $125**

New: $100	**Perf.:** $90	**Exc.:** $75

HI-POINT FIREARMS JS-9MM
Semi-automatic; single action; 9mm Para.; 9-shot magazine; 4 1/2″ barrel; 7 3/4″ overall length; weighs 41 oz.; textured acetal plastic grips; fixed low profile sights; scratch-resistant, non-glare blue finish. From MKS Supply, Inc. Introduced 1990; still produced. **LMSR: $140**

New: $110	**Perf.:** $100	**Exc.:** $90

Hi-Point Firearms JS-9mm Compact
Same specs as JS-9mm except 380 ACP, 9mm Para.; 8-shot magazine; 3 1/2″ barrel; 6 3/4″ overall length; weighs 35 oz.; textured acetal plastic grips; low profile, combat-style, fixed three-dot sight system; frame-mounted magazine release; scratch-resistant matte finish. From MKS Supply, Inc. Introduced 1993; still produced. **LMSR: $125**

9mm Para.

New: $100	**Perf.:** $90	**Exc.:** $75

9mm Para, polymer frame, non-slip grips

New: $120	**Perf.:** $110	**Exc.:** $95

380 ACP **LMSR: $80**

New: $70	**Perf.:** $60	**Exc.:** $50

HI-POINT FIREARMS JS-40 S&W
Semi-automatic; 40 S&W; 8-shot magazine; 4 1/2″ barrel; 7 3/4″ overall length; weighs 42 oz.; checkered acetal resin grips; fixed, low profile sights; internal drop-safe mechansim; all aluminum frame; matte black finish. From MKS Supply, Inc. Introduced 1991; still produced. **LMSR: $149**

New: $130	**Perf.:** $125	**Exc.:** $110

HI-POINT FIREARMS JS-45
Semi-automatic; single action; 45 ACP; 7-shot magazine; 4 1/2″ barrel; 8″ overall length; weighs 44 oz.; checkered acetal resin grips; fixed low profile sights; internal drop-safe mechanism; all aluminum frame. From MKS Supply, Inc. Introduced 1991; still produced. **LMSR: $149**

New: $130	**Perf.:** $125	**Exc.:** $110

HI-STANDARD SEMI-AUTOMATIC PISTOLS

HI-STANDARD CITATION (First Model 102 Series)
Semi-automatic; 22 LR; 10-shot; 6 3/4″, 8″, 10″ interchangeable tapered barrels; hammerless; push-button takedown; ramp front, click-adjustable rear sights; trigger pull and backlash adjustment screws; new rakish-looking barrels with grooves for weight attachment and knob on end for removable stabilizer; plastic grips, walnut optional; two removable weights; blued finish. Marked "SUPER-MATIC" top line; "CITATION" in smaller letters underneath on left side of frame. Introduced 1958; dropped 1963.

Two barrels

Exc.: $800	**VGood:** $700	**Good:** $600

One barrel

Exc.: $650	**VGood:** $550	**Good:** $450

Hi-Standard Citation (103 Series)
Same specs as First Model except 5 1/2″ bull barrel; tapped and notched for optional weights and stabilizer. Marked "Model 103" on right side of slide. Introduced 1962; dropped 1963.

Exc.: $700	**VGood:** $600	**Good:** $500

Hi-Standard Citation (104 Series)
Same specs as 103 Series Citation except marked "Model 104" on right side of slide. Introduced 1963; dropped 1966.

Exc.: $650	**VGood:** $550	**Good:** $500

Hi-Standard Citation (106 Series)
Same specs as 103 Series Citation except also offered in 7 1/4″ barrel; positive magazine latch; military-style grips; stippled front, backstraps; saddle-type rear sight. Introduced 1965; dropped 1967.

Exc.: $600	**VGood:** $500	**Good:** $375

Hi-Standard Citation (107 Series)
Same specs as 106 Series Citation except marked "Model 107" on right side of slide. Introduced 1968; dropped 1968.

Exc.: $600	**VGood:** $500	**Good:** $400

Hi-Standard Citation (ML Series)
Same specs as 107 Series expect "ML" prefixed serial number.

Exc.: $525	**VGood:** $450	**Good:** $375

Hi-Standard Citation (Seven Number Series)
Same specs as 107 Series except seven-digit serial number appears alone on right side of frame.

Exc.: $600	**VGood:** $525	**Good:** $425

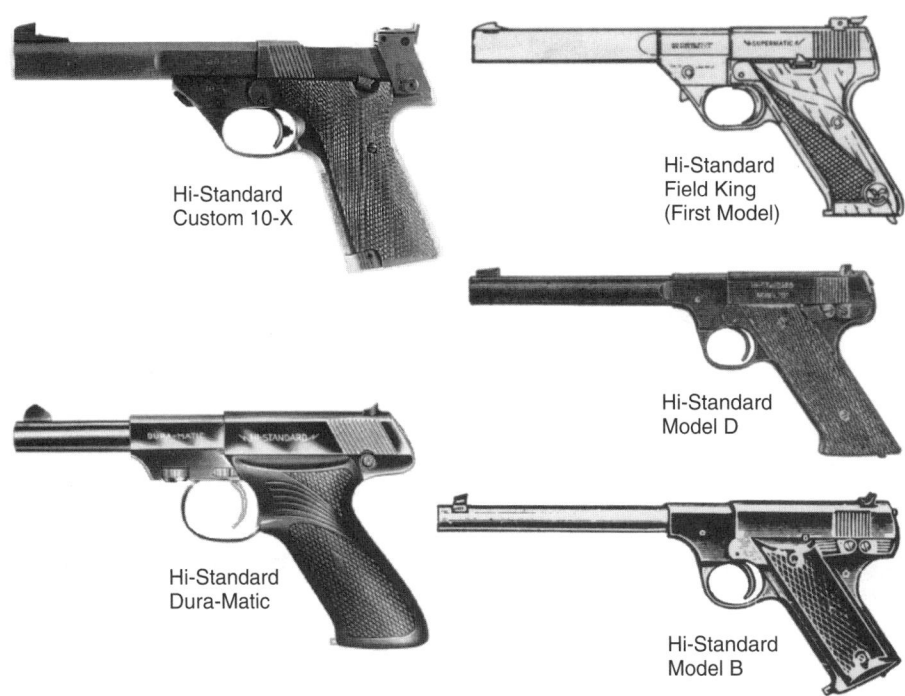

Hi-Standard
Custom 10-X

Hi-Standard
Field King
(First Model)

Hi-Standard
Model D

Hi-Standard
Dura-Matic

Hi-Standard
Model B

Hi-Standard Semi-Automatic Pistols

Citation II (SH Series)
Citation II Survival Pack
Custom 10-X
Custom 10-X Target
Dura-Matic
Field King (First Model)
Field King (101 Series)
Flite King (First Model)
Flite King (102 Series)
Flite King (103 Series)
Model A
Model B
Model C
Model D
Model E
Model G-380

HI-STANDARD CITATION II (SH Series)

Semi-automatic; 22 LR; 10-shot magazine; 5 1/2", 7 1/4" slab-sided bull barrel; weighs 45 oz.; adjustable sights; Allen-screw takedown; checkered walnut grips. Replaced Sharpshooter in High Standard line and early models inscribed "Sharpshooter" on frame with "Citation II" inscribed on barrel. Serial numbers prefixed with the letters "SH". Introduced 1982; dropped 1984.

Exc.: $450   **VGood:** $375   **Good:** $300

Hi-Standard Citation II Survival Pack

Same specs as Citation II except 5 1/2" barrel only; electroless nickel finish; padded canvas carry case; extra magazine. Introduced 1982; dropped 1984.

Exc.: $525   **VGood:** $450   **Good:** $375

HI-STANDARD CUSTOM 10-X

Semi-automatic; 22 LR; 10-shot magazine; 5 1/2" bull barrel, some with Victor barrel and vent rib with rear sight on rib; 9 3/4" overall length; weighs 44 1/2 oz.; black, ambidextrous, checkered walnut grips; undercut ramp front sight, frame-mounted adjustable rear; custom-made, fitted; fully-adjustable target trigger; stippled front, backstraps; slide lock; marketed with two extra magazines; non-reflective black/gray Parkerizing; stamped with maker's initials on frame under left grip. Approximately 600 produced. Introduced 1981; dropped 1984.

Perf.: $2000   **Exc.:** $1750   **VGood:** $1200

HI-STANDARD CUSTOM 10-X TARGET

Semi-automatic; single action; 22 LR; 10-shot magazine; 5 1/2" barrel; 9 1/2" overall length; weighs 44 oz.; ambidextrous, checkered black epoxied walnut grips; undercut ramp front sight, fully-adjustable, micrometer-click rear; hand built with select parts; adjustable trigger and sear; push-button takedown; Parkerized finish; stippled front grip and backstrap; barrel weights optional; comes with test target. Made in U.S. by High Standard Mfg. Co., Inc. Reintroduced 1994; still produced.

LMSR: $790

New: $700   **Perf.:** $600   **Exc.:** $500

HI-STANDARD DURA-MATIC

Semi-automatic; 22 LR; 4 1/2", 6 1/2" interchangeable barrels; 10 7/8" overall length (6 1/2" barrel); hammerless; screw take-down; fixed sights; modified magazine with small rectangular recession on top left side to fit clip release; plastic grips; blued finish. Marked "DURA-MATIC" on left side of frame. Introduced 1954; dropped 1970.

Exc.: $250   **VGood:** $200   **Good:** $165

HI-STANDARD FIELD KING (FIRST MODEL)

Semi-automatic; 22 LR; 10-shot magazine; 4 1/2", 6 3/4" medium weight barrels; 11 1/2" overall length (6 3/4" barrel); hammerless; lever take-down; fixed front, adjustable rear sights. Similar to Supermatic First Model except no notch groove for weight attachment; no raised rib atop barrel. Introduced 1951; dropped 1953.

Exc.: $600   **VGood:** $500   **Good:** $375

Hi-Standard Field King (101 Series)

Same specs as First Model except push-button takedown; optional slotted stabilizer (6 3/4" barrel). Marked "FK-101" on right side of slide. Introduced 1954; dropped 1957.

Exc.: $500   **VGood:** $425   **Good:** $350

HI-STANDARD FLITE KING (FIRST MODEL)

Semi-automatic; 22 Short; 10-shot magazine; 4 1/2", 6 3/4" interchangeable lightweight barrels; hammerless; push-button takedown; aluminum alloy frame and slide; fixed sights; automatic slide lock; blue finish. Marked "LW 100" or "LW 101" on right side of frame. Introduced 1954; dropped 1960.

Two barrels

Exc.: $525   **VGood:** $450   **Good:** $350

One barrel

Exc.: $450   **VGood:** $375   **Good:** $300

Hi-Standard Flite King (102 Series)

Same specs as First Model except all-steel construction; weighs 36 oz. (6 3/4" barrel); improved push-button takedown; heavier weight tapered barrels. Marked "FLITE KING" in block letters on left side of frame, "Model 102" on right side. Introduced 1958; dropped 1965.

Exc.: $425   **VGood:** $375   **Good:** $325

Hi-Standard Flite King (103 Series)

Same specs as 102 Series Flite King except marked "Model 103" on right side of frame.

Exc.: $400   **VGood:** $350   **Good:** $300

HI-STANDARD MODEL A

Semi-automatic; 22 LR; 10-shot magazine; 4 1/2", 6 3/4" barrel; 11 1/2" overall length (6 3/4" barrel); large frame; hammerless; adjustable target-type sights; new longer grip frame with checkered walnut grips. Model A had both 1B and Type II takedowns. The new grip frame style caught Colt by surprise and prompted Colt to extend grips on the 2nd Model Woodsman. Introduced 1938; dropped 1942. Approximately 7,300 produced.

Exc.: $650   **VGood:** $550   **Good:** $450

HI-STANDARD MODEL B

Semi-automatic; 22 LR; 10-shot magazine; 4 1/2", 6 3/4" barrels; 10 3/4" overall length (6 3/4" barrel); small frame; fixed sights; blued finish. Early Model B had checkered hard rubber grips. At S/N 31508 on Feb. 18, 1938, a new borderless grip with "HS" monogram in circle at center of grip was introduced. Approximately 65,000 produced. Introduced 1932; dropped 1942.

Exc.: $500   **VGood:** $400   **Good:** $250

HI-STANDARD MODEL C

Semi-automatic; 22 Short; 10-shot magazine; 4 1/2", 6 3/4" barrel; 10 3/4" overall length (6 3/4" barrel); small frame; hammerless; fixed sights; checkered hard rubber grips; blued finish. Approximately 4700 produced. Introduced 1936; dropped 1942.

Exc.: $800   **VGood:** $700   **Good:** $550

HI-STANDARD MODEL D

Semi-automatic; 22 LR only; 10-shot magazine; 4 1/2", 6 3/4" barrel; 11 1/2" overall length (6 3/4" barrel); hammerless; adjustable target-type sights; checkered walnut grips; blued finish. The Model D had both 1B and Type II takedowns with 1B type the rarest. Introduced 1938; dropped, 1942. Approximately 2500 produced.

Exc.: $450   **VGood:** $350   **Good:** $300

HI-STANDARD MODEL E

Semi-automatic; 22 LR; 10-shot; 4 1/2", 6 3/4" barrels; 11 1/2" overall length (6 3/4" barrel); hammerless; adjustable target-type sights; bull barrel and deluxe walnut grips with thumbrest (known as Roper grips); deluxe blued finish. Introduced 1937; dropped 1942. Approximately 2600 produced.

Exc.: $850   **VGood:** $700   **Good:** $550

HI-STANDARD MODEL G-380

Semi-automatic; 380 ACP; 5" barrel; visible hammer; thumb safety; first lever take-down model; fixed sights; checkered plastic grips; blued finish. Introduced 1947; dropped 1950. Approximately 7400 produced.

Exc.: $5500   **VGood:** $450   **Good:** $400

Hi-Standard Semi-Automatic Pistols
Model G-B
Model G-D
Model G-E
Model G-O (See Olympic
First Model)
Model H-A
Model H-B (1st Model)
Model H-B (2nd Model)
Model H-D
Model H-D U.S.A.
Model H-D U.S.A. Military
Model H-E
Model S-B
Olympic (First Model)
Olympic (Second Model)
Olympic (Third Model)
Olympic (102 Series)
Olympic (103 Series)

Hi-Standard Model H-A

Hi-Standard Model G-B

Hi-Standard H-D U.S.A. Military

Hi-Standard Olympic (Second Model)

Hi-Standard Model HB (Second Model)

Hi-Standard Model G-D

HI-STANDARD MODEL G-B
Semi-automatic; 22 LR; 4 1/2", 6 3/4" interchangeable barrels; 10 3/4" overall length (6 3/4" barrel); hammerless; small frame; fixed sights; brown checkered plastic grip with "HS" medallion; blued finish. Introduced 1949; dropped 1951.
Two barrels
   Exc.: $650   **VGood:** $550   **Good:** $450
One barrel
   Exc.: $550   **VGood:** $475   **Good:** $375

HI-STANDARD MODEL G-D
Semi-automatic; 22 LR; 10-shot; 4 1/2", 6 3/4" interchangeable barrels; 11 1/2" overall length (6 3/4" barrel); lever take-down; large frame; heavy barrel; adjustable target sights; checkered walnut grips; blued finish. Introduced 1949; dropped 1951. Approximately 3300 produced.
Two barrels
   Exc.: $950   **VGood:** $8000   **Good:** $450
One barrel
   Exc.: $800   **VGood:** $700   **Good:** $550

HI-STANDARD MODEL G-E
Semi-automatic; 22 LR; 10-shot; 4 1/2", 6 3/4" bull barrel; 11 1/2" overall length (6 3/4" barrel); large frame; lever take-down; adjustable sights; deluxe checkered walnut grips; high polish blued finish. Introduced 1949; dropped 1951. Approximately 3000 produced.
Two barrels
   Exc.: $1400   **VGood:** $1100   **Good:** $850
One barrel
   Exc.: $1250   **VGood:** $900   **Good:** $700

HI-STANDARD MODEL H-A
Semi-automatic; 22 LR; 10-shot magazine; 4 1/2", 6 3/4" barrels; 11 1/2" overall length (6 3/4" barrel); visible hammer; no thumb safety; small diameter barrel; checkered walnut grips. Introduced 1940; dropped 1942. Approximately 1042 produced.
   Exc.: $900   **VGood:** $700   **Good:** $500

HI-STANDARD MODEL H-B (FIRST MODEL)
Semi-automatic; 22 LR; 10-shot magazine; 4 1/2", 6 3/4" barrels; 10 3/4" overall length (6 3/4" barrel); fixed sights; checkered hard rubber grips. Same as Model B except visible hammer and no thumb safety. Marked "H-B". Introduced 1940; dropped 1942. Reintroduced 1949; dropped 1950. Approximately 2100 produced.
   Exc.: $400   **VGood:** $350   **Good:** $250

Hi-Standard Model H-B (2nd Model)
Same specs as Model H-B (1st Model) except presence of thumb safety. Marked "HB". Approximately 25,000 produced.
   Exc.: $550   **VGood:** $450   **Good:** $375

HI-STANDARD MODEL H-D
Semi-automatic; 22 LR; 10-shot magazine; 4 1/2", 6 3/4" barrel; 11 1/2" overall length (6 3/4" barrel); adjustable target-type sights; deluxe checkered walnut grips with thumbrest. Same as the Model D except visible hammer and no thumb safety. All H-D models displayed either 1B and Type II takedown. Introduced 1940; dropped 1942. Approximately 6900 produced.
   Exc.: $900   **VGood:** $700   **Good:** $550

HI-STANDARD MODEL H-D U.S.A.
Semi-automatic; 22 LR; 10-shot magazine; 4 1/2", 6 3/4" barrels; adjustable target-type sights; medium weight barrel; thumb safety; black or checkered hard rubber grips; high polish blue (early models), Parkerized finish. U.S. military training pistol. Introduced 1943; dropped 1946. Approximately 44,000 produced.
   Exc.: $650   **VGood:** $500   **Good:** $400

Hi-Standard H-D U.S.A. Military
Same specs as Model H-D U.S.A. except adjustable sights. Introduced 1946; dropped 1955. Approximately 150,000 produced.
   Exc.: $500   **VGood:** $400   **Good:** $350

HI-STANDARD MODEL H-E
Semi-automatic; 22 LR; 10-shot; 4 1/2", 6 3/4" bull barrels; 11 1/2" overall length (6 3/4" barrel); large frame; adjustable target-type sights. Same specs as the Model E except visible hammer, no thumb safety, high polished blue finish and deluxe walnut grips. Introduced 1941; dropped 1942. Approximately 1006 produced.
   Exc.: $1850   **VGood:** $1400   **Good:** $1000

HI-STANDARD MODEL S-B
Semi-automatic; 22 shot cartridge; 6 3/4" barrel; 10 3/4" overall length; smoothbore; checkered rubber grips. Introduced 1939; dropped 1940. Only 12 produced.
   Exc.: $3500   **VGood:** $3000   **Good:** $2500

HI-STANDARD OLYMPIC (FIRST MODEL)
Semi-automatic; 22 Short; 4 1/2", 6 3/4" interchangeable bull barrels; hammerless; first use of light alloy slide; adjustable rear sight; deluxe checkered walnut grips with thumbrest; lever takedown; special banana-shaped magazine. Also known as Model G-O. Introduced 1950; dropped 1951. Approximately 1200 produced.
Two barrels
   Exc.: $1350   **VGood:** $1150   **Good:** $800
One barrel
   Exc.: $100   **VGood:** $950   **Good:** $600

Hi-Standard Olympic (Second Model)
Same specs as First Model except medium barrel; interchangeable 2 oz. and 3 oz. weights attached by dovetail on bottom of barrel plus filler strip for dovetail when weights not in use; standard magazine with filler in front to fit 22 Short cartridges. Introduced 1951; dropped 1953.
Two barrels
   Exc.: $1400   **VGood:** $1200   **Good:** $850
One barrel
   Exc.: $1150   **VGood:** $900   **Good:** $650

Hi-Standard Olympic (Third Model)
Same specs as First Model except push button takedown; optional slotted stabilizer for 6 3/4" barrel. Marked "0-101" on right side of slide. Introduced 1954; dropped 1958.
   Exc.: $900   **VGood:** $700   **Good:** $500

Hi-Standard Olympic (102 Series)
Same specs as First Model except 6 3/4", 8", 10" barrel; same as Citation 102 except 22 Short with alloy slide; adjustable and removable barrel weights; removable stabilizer; aluminum slide. Marked "Model 102" on right side of slide. Introduced 1958; dropped 1965.
   Exc.: $1000   **VGood:** $800   **Good:** $600

Hi-Standard Olympic (103 Series)
Same specs as 102 Series Olympic except plastic grips standard, walnut optional. Introduced 1963; dropped 1965.
   Exc.: $1050   **VGood:** $850   **Good:** $600

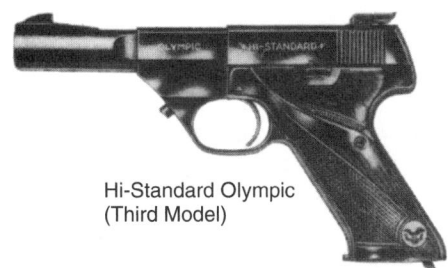

Hi-Standard Olympic
(Third Model)

Hi-Standard Sport
King (101 Series)

Hi-Standard Olympic
ISU (104 Series)

Hi-Standard
Sharpshooter

Hi-Standard Olympic
ISU Standard

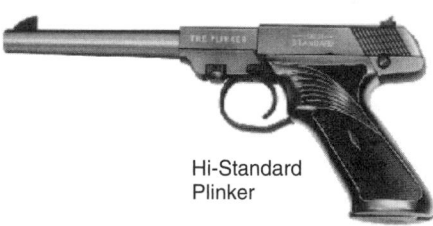

Hi-Standard
Plinker

Hi-Standard Semi-Automatic Pistols

Olympic (104 Series)
Olympic (106 Series Military)
Olympic ISU (102 Series)
Olympic ISU (103 Series)
Olympic ISU (104 Series)
Olympic ISU (106 Series Military)
Olympic ISU Military
Olympic ISU Standard
Plinker
Sharpshooter
Sport King (First Model)
Sport King (101 Series)
Sport King Lightweight (101 Series)
Sport King (102 Series)
Sport King (103 Series)
Sport King (106 Series Military Model)
Supermatic (First Model)

Hi-Standard Olympic (104 Series)
Same specs as 102/103 Series Olympic except new Allen screw backlash adjustor in trigger which carried over to all future models. Introduced 1964; dropped 1972.
Exc.: $1000   **VGood:** $750   **Good:** $550

Hi-Standard Olympic (106 Series Military)
Same specs as the 104 Series except 5 1/2" bull barrel, 6 3/4" space gun barrel; new frame with military Colt 1911-style grips; fitted magazine extension foot; new magazine release. Introduced 1965; dropped 1966.
Exc.: $1050   **VGood:** $850   **Good:** $550

HI-STANDARD OLYMPIC ISU (102 SERIES)
Semi-automatic; 22 Short; 10-shot magazine; ported 6 3/4" barrel; 11 1/2" overall length; larger push-button takedown; alloy slide; new rear target sight; high polished finish. Marked "Model 102" on right side of slide. Unlike Olympic 102 Series, the ISU has integral stabilizer and comes standard with checkered walnut grips. Introduced 1957; dropped 1960.
Exc.: $1000   **VGood:** $800   **Good:** $650

Hi-Standard Olympic ISU (103 Series)
Same specs as ISU Model 102 except 5 1/2", 6 3/4" bull barrel. Introduced 1960; dropped 1963.
Exc.: $1000   **VGood:** $800   **Good:** $650

Hi-Standard Olympic ISU (104 Series)
Same specs as ISU Model 102 except bull barrel with removable stabilizer (5 1/2" barrel); integral stabilizer (6 3/4" barrel); barrel weights and new Allen screw backlash adjuster in trigger. Introduced 1964; dropped 1964.
Exc.: $950   **VGood:** $800   **Good:** $650

Hi-Standard Olympic ISU (106 Series Military)
Same specs as the 102 Series except military Colt 1911-style grips; integral barrel stabilizer. Introduced 1965; dropped 1967.
Exc.: $1050   **VGood:** $850   **Good:** $700

Hi-Standard Olympic ISU Military
Same specs as the Olympic ISU Standard model except 5 1/2" bull barrel with removable stabilizer; high strength aluminum slide; carbon steel frame; barrel weights; adjustable trigger and sear; overall blue finish. From High Standard Mfg. Co., Inc. Reintroduced 1994; still produced. **LMSR: $504**
Perf.: $450   **Exc.:** $400   **VGood:** $300

HI-STANDARD OLYMPIC ISU STANDARD
Semi-Automatic; 22 Short; 5-shot magazine; 6 3/4" tapered barrel with intergral stabilizer; 10 1/4" overall length; weighs 45 oz.; push-button takedown; undercut ramp front sight, micro-click adjustable rear; checkered walnut grips; adjustable trigger and sear; stippled front grip and backstrap; comes with weights and brackets. Reintroduced 1994; still in production. **LMSR: $625**
Perf.: $550   **Exc.:** $450   **VGood:** $375

HI-STANDARD PLINKER
Semi-automatic; 22 LR; 10-shot magazine; interchangeable 4 1/2", 6 3/4" barrels; 9" overall length (4 1/2" barrel); hammerless; same gun as Duramatic; grooved trigger; checkered plastic target grips; fixed square-notch rear sight, ramp front; blued finish. Introduced 1962; dropped 1973.
Exc.: $300   **VGood:** $250   **Good:** $200

HI-STANDARD SHARPSHOOTER
Semi-automatic; 22 LR; 9-shot magazine; 5 1/2" bull barrel; 10 1/4" overall length; weighs 42 oz.; push-button takedown; hammerless; scored trigger; adjustable square-notch rear, ramp front sight; slidelock; checkered laminated plastic grips with medallion or walnut with medallion; blued finish. Marked "The Sharpshooter" in block letters on barrel flat. Introduced 1972; dropped 1982.
Exc.: $550   **VGood:** $450   **Good:** $300

HI-STANDARD SPORT KING (FIRST MODEL)
Semi-automatic; 22 LR; 10-shot magazine; 4 1/2", 6 3/4" lightweight interchangeable barrels; 11 1/2" overall length (6 3/4" barrel); lever takedown; fixed sights; optional adjustable sight; checkered thumbrest plastic grips; blue finish. Introduced 1950; dropped 1953.
Two barrels
Exc.: $450   **VGood:** $350   **Good:** $300
One barrel
Exc.: $400   **VGood:** $300   **Good:** $250

Hi-Standard Sport King (101 Series)
Same specs as First Model except push-button takedown; all-steel frame and slide. Marked "SK-100" or "SK-101" on right side of slide. Introduced 1954; dropped 1984.
Two barrels
Exc.: $475   **VGood:** $375   **Good:** $300
One barrel
Exc.: $400   **VGood:** $300   **Good:** $250

Hi-Standard Sport King Lightweight (101 Series)
Same specs as 101 Series Sport King except forged aluminum alloy frame; weighs 30 oz. (6 3/4" barrel). Marked "Lightweight" on left side of frame. Introduced 1954; dropped 1964.
Two barrels
Exc.: $575   **VGood:** $475   **Good:** $425
One barrel
Exc.: $500   **VGood:** $425   **Good:** $375

Hi-Standard Sport King (102 Series)
Same specs as 101 Series Sport King except brown plastic grips (black plastic grips on nickel-plated guns); new optional nickel-plated finish. Late Sport King 102s had "G" in front of serial number. Introduced 1958; dropped 1963.
Exc.: $350   **VGood:** $300   **Good:** $250

Hi-Standard Sport King (103 Series)
Same specs as 102 Series Sport King except marked "Model 103" on right side of slide.
Exc.: $350   **VGood:** $300   **Good:** $250

Hi-Standard Sport King (106 Series Military Model)
Same specs as 102 Series Sport King except military-style grips. Introduced 1977; dropped 1983.
Exc.: $300   **VGood:** $250   **Good:** $200

HI-STANDARD SUPERMATIC (FIRST MODEL)
Semi-automatic; 22 LR; 10-shot magazine; 4 1/2", 6 3/4" interchangeable barrels; 11 1/2" overall length (6 3/4" barrel); hammerless; lever takedown; adjustable ramp front, click-adjustable rear target sights; automatic slidelock; 2-, 3-oz. adjustable barrel weights; adjustable ramp front; click-adjustable rear target sights; automatic slidelock; checkered plastic thumbrest grips; blue-black finish with Parkerized slide top and barrel chamber. Introduced 1951; dropped 1958.
Two barrels
Exc.: $750   **VGood:** $550   **Good:** $375
One barrel
Exc.: $650   **VGood:** $550   **Good:** $400

Hi-Standard Semi-Automatic Pistols

Supermatic
(101 Series 2nd Model)
Supermatic Citation
Supermatic Trophy
Supermatic Tournament
Tournament (102 Series)
Tournament (103 Series)
Tournament
(106 Series Military)
Trophy (102 Series)
Trophy (103/104 Series)
Trophy (106/107 Series)
Trophy (ML Series)
Trophy (Seven Number Series)
Trophy (SH Series)
Victor (First Model 107 Series)
Victor (ML Series)
Victor (Seven Number Series)
Victor (SH Series)
Victor Target Pistol

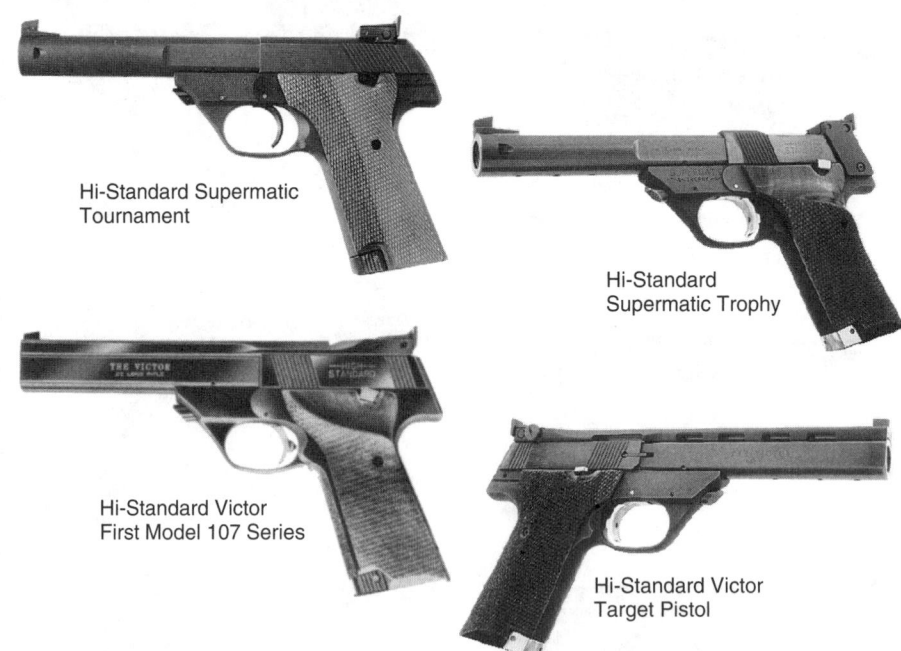

Hi-Standard Supermatic Tournament

Hi-Standard Supermatic Trophy

Hi-Standard Victor First Model 107 Series

Hi-Standard Victor Target Pistol

Hi-Standard Supermatic
(101 Series 2nd Model)
Same specs as First Model except push-button take-down; fixed front sight; integral stabilizer (6 ³/₄" barrel). Marked "S100" or "S101" on right side of slide. Introduced, 1954; dropped, 1958.
Exc.: $350   **VGood:** $275   **Good:** $225

HI-STANDARD SUPERMATIC CITATION
Semi-Automatic; 22 LR; 10-shot magazine; 5 ¹/₂", 7 ¹/₄" barrel; 9 ¹/₂" overall length; weighs 44 oz.; under-cut ramp front, micro-click adjustable rear sight; checkered walnut grip with thumbrest; nickel-plated trigger, slidelock, safety lever, magazine release; blue finish. Reintroduced 1994; still in production. **LMSR:** $425
New: $375   **Perf.:** $325   **Exc.:** $275

HI-STANDARD SUPERMATIC TROPHY
Semi-automatic; single action; 22 LR; 10-shot magazine; 5 ¹/₂", 7 ¹/₄" barrel; 9 ¹/₂" overall length; weighs 44 oz.; checkered walnut thumbrest grips; undercut ramp front sight, fully-adjustable, microm-eter-click rear; push-button takedown; removable muzzle stabilizer; gold-plated trigger, slidelock, safety-lever and magazine release; stippled front grip and backstrap; adjustable trigger and sear; drilled and tapped for scope mount; barrel weights optional. Made in U.S. by High Standard Mfg. Co., Inc. Reintroduced 1994; still produced. **LMSR:** $536
New: $400   **Perf.:** $350   **Exc.:** $275

HIGH STANDARD SUPERMATIC TOURNAMENT
Semi-Automatic; 22 LR; 10-shot magazine; 4 ¹/₂", 5 ¹/₂" barrel; 8 ¹/₂" overall length; weighs 43 oz.; push-button takedown; undercut ramp front, micro-click adjustable rear sight; black rubber ambidextrous grips; drilled and tapped for scope mount; blue finish. Reintroduced 1994. **LMSR:** $425
New: $375   **Perf.:** $350   **Exc.:** $325

HI-STANDARD TOURNAMENT
(102 SERIES)
Semi-automatic; 22 LR; 10-shot magazine; 6 ³/₄" medi-um weight, 4 ¹/₂" bull barrel; ¹/₄" high base front sight, click-adjustable rear on slide; push-button takedown; backlash adjustable screw; plastic grips; no grooving atop slide; dull black finish. Marked "SUPERMATIC" top line; "Tournament" in smaller letters underneath on left side of frame. Introduced 1958; dropped 1963.
Exc.: $600   **VGood:** $500   **Good:** $375

Hi-Standard Tournament (103 Series)
Same specs as 102 Series Tournament except 5 ¹/₂" bull barrel offering. Introduced 1958; dropped 1963.
Exc.: $550   **VGood:** $450   **Good:** $350

Hi-Standard Tournament
(106 Series Military)
Same specs as 103 Series Tournament except mili-tary-style grips. Introduced 1965; dropped 1966.
Exc.: $550   **VGood:** $450   **Good:** $375

HI-STANDARD TROPHY (102 Series)
Semi-automatic; 22 LR; 10-shot; 6 ³/₄", 8", 10" tapered "space gun" barrels; push-button takedown; 2-, 3-oz. adjustable barrel weights; detachable stabilizer; trigger pull adjustment screw at rear of frame; overtravel adjustment screw on right side of frame; adjustable sights (barrel mounted on 8", 10"); gold trigger safety; checkered walnut thumbrest grips; blue/black finish. Marked "Model 102" on right side of frame; "Supermat-ic Trophy" on left side. Introduced 1961; dropped 1966.
Exc.: $950   **VGood:** $750   **Good:** $625

Hi-Standard Trophy (103/104 Series)
Same specs as the 102 Series except 5 ¹/₂", 7 ¹/₄" bull barrels; recessed bore at muzzle. Marked "Supermatic Trophy" on left side of frame; "Model 103" or "Model 104" on right side of frame. Introduced 1960; dropped 1965.
Exc.: $900   **VGood:** $700   **Good:** $500

Hi-Standard Trophy (106/107 Series)
Same specs as the 102 Series except military-style grips; saddle-type rear sight; new slide with angled gripping ribs. Marked "Supermatic Trophy" and "HIGH STANDARD" on left side of slide; "Model 106" or "Mod-el 107" above triggerguard on right side. Introduced 1965; dropped 1967.
Exc.: $950   **VGood:** $800   **Good:** $650

Hi-Standard Trophy (ML Series)
Same specs as 106/107 Series except "ML" prefixed serial number.
Exc.: $675   **VGood:** $525   **Good:** $425

Hi-Standard Trophy
(Seven Number Series)
Same specs as 106/107 Series except seven-digit ser-ial number appears alone on right side of frame.
Exc.: $800   **VGood:** $625   **Good:** $400

Hi-Standard Trophy (SH Series)
Same specs as 106/107 Series except "SH" prefixed serial number.
Exc.: $550   **VGood:** $450   **Good:** $325

HI-STANDARD VICTOR
(FIRST MODEL 107 SERIES)
Semi-automatic; 22 LR; 10-shot magazine; 4 ¹/₂", 5 ¹/₂" barrels; 8 ³/₄" overall length (4 ¹/₂" barrel); 48 oz. (4 ¹/₂" barrel), 52 oz. (5 ¹/₂" barrel); hammerless; solid steel rib, later guns with aluminum vented rib; inter-changeable barrel feature; rib-mounted click-adjustable rear sight, undercut ramp front; checkered walnut grips with thumbrest; blued finish. Marked "The Victor" on left side of barrel. Introduced 1963; dropped 1984.
Solid rib
Exc.: $750   **VGood:** $650   **Good:** $450
Vent rib
Exc.: $800   **VGood:** $700   **Good:** $475

Hi-Standard Victor (ML Series)
Same specs as standard Victor except marked "Victor" above triggerguard. Serial number prefixed with letters "ML".
Exc.: $550   **VGood:** $450   **Good:** $350

Hi-Standard Victor
(Seven Number Series)
Same specs as Model 107 except seven-digit serial number appears alone on right side of frame.
Exc.: $525   **VGood:** $325   **Good:** $300

Hi-Standard Victor (SH Series)
Same specs as standard Victor except Allen-screw takedown; small grip thumbrest; rib cut-out for shell ejection. Serial number prefixed with letters "SH".
Exc.: $500   **VGood:** $450   **Good:** $350

HI-STANDARD VICTOR TARGET PISTOL
Semi-automatic; single action; 22 LR; 10-shot magazine; 4 ¹/₂", 5 ¹/₂" barrel; 9 ¹/₂" overall length; weighs 46 oz.; checkered walnut thumbrest grips; undercut ramp front sight, fully-adjustable, micrometer-click rear; push-button takedown; full-length aluminum vent rib (steel optional); gold-plated trigger, slidelock, safety-lever and magazine release; stippled front grip and backstrap; adjustable trig-ger and sear; comes with barrel weight; blue or Parker-ized finish. Made in U.S. by High Standard Mfg. Co., Inc. Reintroduced 1994; still produced. **LMSR:** $532
New: $500   **Perf.:** $450   **Exc.:** $400

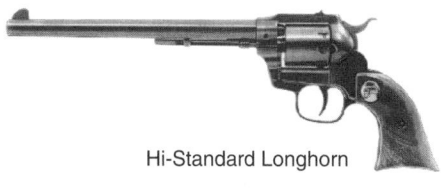

Hi-Standard Longhorn

Hi-Standard Double-
Nine Convertible

Hi-Standard High Sierra

Hi-Standard Durango

HANDGUNS

Hi-Standard Revolvers
Camp Model
Crusader
Derringer
Double-Nine
Double-Nine Convertible
Durango
High Sierra
Kit Gun
Longhorn
Longhorn Convertible
Marshall
Natchez
Posse

HI-STANDARD REVOLVERS & SINGLE SHOTS

HI-STANDARD CAMP MODEL
Revolver; double action; 22 LR, 22 WMR; 9-shot cylinder; 6" barrel; 11 1/8" overall length; weighs 28 oz.; steel frame; adjustable rear sight; target-type checkered walnut grips; blued finish. Introduced 1976; dropped 1979.
Exc.: $175   **VGood:** $135   **Good:** $110

HI-STANDARD CRUSADER
Revolver; double action; 357 Mag., 44 Mag., 45 Colt; 6-shot swing-out cylinder; 4", 6", 8" barrel (medium frame), 4", 6", 8" (large frame); weighs 38 oz. (med. frame), 48 oz. (large frame); unique internal gear mechanism which allows hammer to rest on frame when trigger is disengaged and brings hammer in direct line with firing pin when trigger is engaged; lack of transfer bar safety and trigger spring account for light weight; ramp front sight, adjustable rear; solid ribbed barrel; shrouded ejector rod; one-piece grip; blued finish. Introduced 1976; discontinued 1979. Limited production.
Exc.: $1500   **VGood:** $1200   **Good:** $1000

HI-STANDARD DERRINGER
Derringer; double action; 22 LR, 22 Short, 22 WMR; 2-shot; hammerless; over/under 3 1/2" barrels; 5 1/8" overall length; weighs 11 oz.; plastic grips; fixed sights; standard model has blue, nickel finish. Early derringers marked "Hamden Conn." on barrel; later models marked "E. Hartford" on barrel. Presentation model gold-plated, introduced in 1965; dropped 1966. Presentation model has some collector value. Standard model introduced in 1963; dropped 1984.
Early model, blue finish
Exc.: $150   **VGood:** $135   **Good:** $110
Early model, nickel finish
Exc.: $175   **VGood:** $150   **Good:** $125
Late model, blue finish
Exc.: $150   **VGood:** $135   **Good:** $110
Late model, nickel finish
Exc.: $175   **VGood:** $150   **Good:** $125
Presentation model
Exc.: $400   **VGood:** $350   **Good:** $300

HI-STANDARD DOUBLE-NINE
Revolver; double action; 22 Short, 22 LR; 9-shot swing-out cylinder; 5 1/2" barrel; 11" overall length; dummy ejection rod housing; spring-loaded ejection; rebounding hammer; movable notch rear sight, blade front; plastic simulated stag, ebony or ivory grips; blued or nickel finish. Introduced 1959; dropped 1971.
Exc.: $175   **VGood:** $150   **Good:** $125
Nickel finish
Exc.: $185   **VGood:** $160   **Good:** $135

Hi-Standard Double-Nine Convertible
Same specs as Double-Nine except two cylinders, one for 22 LR, 22 Long, 22 Short; other for 22 WMR; smooth frontier-type walnut grips; movable notched rear sight, blade front; blued, nickel finish. Introduced 1972; dropped 1984.
Exc.: $200   **VGood:** $175   **Good:** $150
Nickel finish
Exc.: $225   **VGood:** $220   **Good:** $175

HI-STANDARD DURANGO
Revolver; double action; 22 LR, 22 Long, 22 Short; 4 1/2", 5 1/2" barrels; 10" overall length (4 1/2" barrel); brass-finished triggerguard, backstrap; uncheckered walnut grips; blued only in shorter barrel length; blued, nickel (5 1/2" barrel). Introduced 1972; dropped 1975.
Exc.: $150   **VGood:** $125   **Good:** $100
Nickel finish
Exc.: $160   **VGood:** $135   **Good:** $110

HI-STANDARD HIGH SIERRA
Revolver; double action; 22 LR, 22 WMR; 9-shot; 7" octagonal barrel; 12 1/2" overall length; blade front, adjustable rear sight; smooth walnut grips; swing-out cylinder; gold-plated backstrap, triggerguard. Introduced 1978; dropped 1984.
Exc.: $275   **VGood:** $200   **Good:** $150

HI-STANDARD KIT GUN
Revolver; double action; 22 LR, 22 Long, 22 Short; 9-shot swing-out cylinder; 4" barrel; 9" overall length; micro-adjustable rear, target ramp front sight; checkered walnut grips; blued finish.
Exc.: $150   **VGood:** $125   **Good:** $100

HI-STANDARD LONGHORN
Revolver; double action; 22 LR; 4 1/2", 5 1/2" barrels; pearl-like plastic grips; 5 1/2" barrel model with plastic staghorn grips; later model with walnut grips. Introduced 1961; dropped 1966.
Exc.: $200   **VGood:** $160   **Good:** $120

Hi-Standard Longhorn Convertible
Same specs as standard Longhorn but with 9 1/2" barrel only; aluminum alloy frame; smooth walnut grips; dual cylinder to fire 22 WMR cartridge. Introduced 1971; dropped 1984.
Exc.: $225   **VGood:** $185   **Good:** $150

HI-STANDARD MARSHALL
Revolver; 22 LR; 9-shot swing-out cylinder; 5 1/2" barrel; fixed sights; Staglite western-style grips with gold medallion inset; blued finish; comes complete with leather holster. Introduced 1973; discontinued 1974. Limited production precludes pricing.

HI-STANDARD NATCHEZ
Revolver; double action; 22 LR, 22 Short; 4 1/2" barrel; 10" overall length; fluted cylinder; ivory-like plastic bird's-head grips; blued. Introduced 1961; dropped 1966.
Exc.: $175   **VGood:** $135   **Good:** $110

HI-STANDARD POSSE
Revolver; double action; 22 LR; 3 1/2" barrel; 9" overall length; no ejector rod housing; uncheckered walnut grips; brass grip frame, triggerguard; blued. Introduced 1961; dropped 1966.
Exc.: $135   **VGood:** $110   **Good:** $85

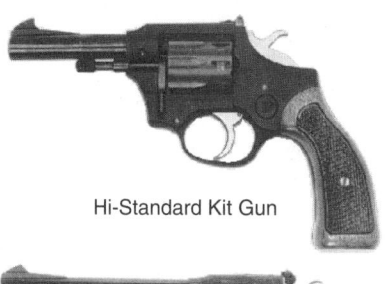

Hi-Standard Kit Gun

Hi-Standard Camp Model

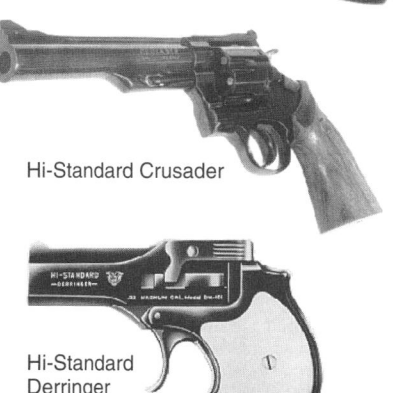

Hi-Standard Crusader

Hi-Standard Derringer

Hi-Standard Revolvers
Sentinel
Sentinel Deluxe
Sentinel Imperial
Sentinel Mark I
Sentinel Mark II
Sentinel Mark III
Sentinel Mark IV
Sentinel Snub

HJS
Antigua Derringer
Frontier Four Derringer
Lone Star Derringer

Hungarian
T-58

IAI
Automag III
Automag IV
Backup
Javelina

Iberia Firearms
JS-40 S&W

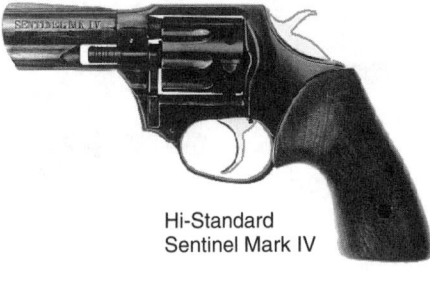

Hi-Standard
Sentinel Mark IV

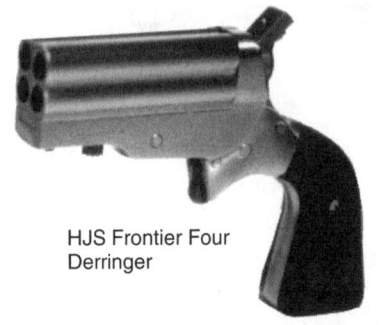

HJS Frontier Four
Derringer

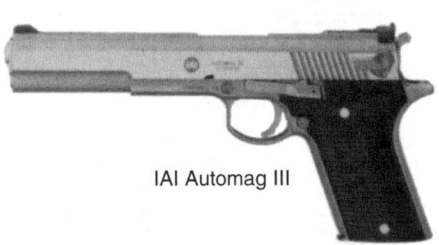

IAI Automag III

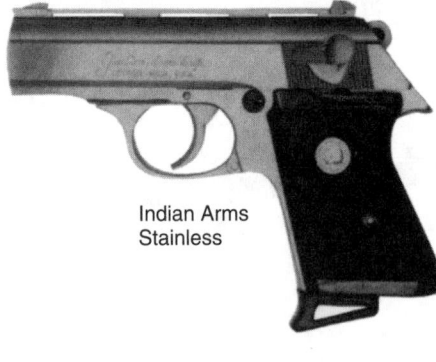

Indian Arms
Stainless

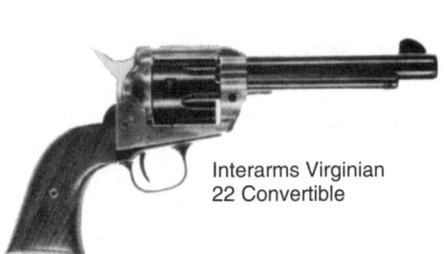

Interarms Virginian
22 Convertible

HI-STANDARD SENTINEL
Revolver; double action; 22 Short, 22 LR; 9-shot swing-out cylinder; 3″, 4″, 6″ barrels; 9″ overall length (4″ barrel); weighs 23 oz.; solid aluminum alloy frame; fixed sights; checkered plastic grips; blued or nickel finish. Introduced 1955; dropped 1974.
    Exc.: $125   **VGood:** $110   **Good:** $100
Nickel finish
    Exc.: $135   **VGood:** $125   **Good:** $110

Hi-Standard Sentinel Deluxe
Same specs as the standard Sentinel except for movable rear sight; two-piece square-butt checkered walnut grips; wide triggers; 4″, 6″ barrels only. Introduced 1957; dropped 1974.
    Exc.: $140   **VGood:** $125   **Good:** $110

Hi-Standard Sentinel Imperial
Same specs as standard Sentinel model except blade ramp front sight; two-piece checkered walnut grips; onyx-black or nickel finish. Introduced 1962; dropped 1965.
    Exc.: $150   **VGood:** $135   **Good:** $110
Nickel finish
    Exc.: $165   **VGood:** $150   **Good:** $125

HI-STANDARD SENTINEL MARK I
Revolver; double action; 22 LR; 9-shot cylinder; 2″, 3″, 4″ barrel; ramp front, fixed or adjustable rear sight; uncheckered walnut stocks; blued or nickel finish. Introduced 1974; dropped 1984.
    Exc.: $150   **VGood:** $135   **Good:** $110
Nickel finish
    Exc.: $165   **VGood:** $150   **Good:** $125

Hi-Standard Sentinel Mark II
Same specs as Sentinel Mark I except 357 Mag.; 6-shot cylinder; 2 1/2″, 4″, 6″ barrel lengths; walnut combat grips; fixed sights; blued finish. Manufactured 1973 to 1975.
    Exc.: $225   **VGood:** $200   **Good:** $175

Hi-Standard Sentinel Mark III
Same specs as Sentinel Mark II except ramp front sight, adjustable rear. Manufactured 1973 to 1975.
    Exc.: $250   **VGood:** $225   **Good:** $185

Hi-Standard Sentinel Mark IV
Same specs as Sentinel Mark I except 22 WMR. Introduced 1974; dropped 1979.
    Exc.: $160   **VGood:** $125   **Good:** $110

HI-STANDARD SENTINEL SNUB
Revolver; double action; 22 LR; 9-shot swing-out cylinder; 2 3/8″ barrel; solid aluminum alloy frame; "quick draw" sights and hammer; checkered bird's-head-type grips; blued or nickel finish. Introduced 1957; dropped 1974.
    Exc.: $150   **VGood:** $125   **Good:** $110

HJS ANTIGUA DERRINGER
Derringer; four-shot; 22 LR; 2″ barrels; 3 15/16″ overall length; weighs 5 1/2 oz.; brown plastic grips; no sights; four barrel fire with rotating firing pin; blued barrels; brass frame; brass pivot pins. Made in U.S. by HJS Arms, Inc. Introduced 1994; still produced. **LMSR: $180**
    New: $160   **Perf.:** $135   **Exc.:** $125

HJS FRONTIER FOUR DERRINGER
Derringer; four-shot; 22 LR; 2″ barrel; 3 15/16″ overall length; weighs 5 1/2 oz.; brown plastic grips; no sights; four barrels fire with rotating firing pin; stainless steel construction. Made in U.S. by HJS Arms, Inc. Introduced 1993; still produced. **LMSR: $165**
    New: $135   **Perf.:** $115   **Exc.:** $100

HJS LONE STAR DERRINGER
Derringer; single shot; 32 S&W, 380 ACP; 2″ barrel; 3 15/16″ overall length; weighs 6 oz.; brown plastic grips; groove sight; stainless steel construction; beryllium copper firing pin; button-rifled barrel. Made in U.S. by HJS Arms, Inc. Introduced 1993; still produced. **LMSR: $185**
    New: $165   **Perf.:** $135   **Exc.:** $120

HUNGARIAN T-58
Semi-automatic; single action; 7.62mm, 9mm Para.; 8-shot magazine; 4 1/2″ barrel; 7 11/16″ overall length; weighs 31 oz.; grooved composition grips; blade front sight, windage-adjustable rear; comes with both barrels and magazines; blue finish. Imported by Century International Arms; still imported. **LMSR: $187**
    Perf.: $160   **Exc.:** $145   **VGood:** $125

IAI AUTOMAG III
Semi-automatic; single action; 30 Carbine, 9mm Win. Mag.; 8-shot magazine; 6 3/8″ barrel; 10 1/2″ overall length; weighs 43 oz.; blade front sight, Millett adjustable rear; wrap-around rubber grips; hammer-drop safety; sandblasted finish; stainless steel construction. Introduced 1989; discontinued 1992.
    Perf.: $500   **Exc.:** $450   **VGood:** $400

IAI AUTOMAG IV
Semi-automatic; single-action; 45 Mag., 10mm Auto.; 7-shot magazine; 6 1/2″, 8 5/8″ barrel; 10 1/2″ overall length; weighs 46 oz; blade front sight, Millett adjustable rear; black carbon fiber grips; stainless steel construction; brushed finish. Introduced 1990; discontinued 1991.
    Perf.: $500   **Exc.:** $450   **VGood:** $400

IAI BACKUP
Semi-automatic; 380 ACP; 5-shot magazine; 2 1/2″ barrel; 4 1/2″ overall length; weighs 18 oz.; fixed sights; wrap-around rubber or walnut grips; stainless steel construction. Introduced 1988; discontinued 1989.
    Perf.: $175   **Exc.:** $150   **VGood:** $125

IAI JAVELINA
Semi-automatic; 10mm Auto; 8-shot magazine; 5″, 7″ barrel; 10 1/2″ overall length; weighs 40 oz.; blade front sight, Millett adjustable rear; wrap-around rubber stocks; brushed finish; stainless steel construction. Introduced 1990; discontinued 1991.
    Perf.: $450   **Exc.:** $400   **VGood:** $350

IBERIA FIREARMS JS-40 S&W
Semi-automatic; single action; 40 S&W; 8-shot magazine; 4 1/2″ barrel; 8″ overall length; weighs 44 oz.; checkered acetal resin grips; fixed low profile sights; internal drop-safe mechanism; all aluminum frame; matte black finish. Marketed by MKS Supply, Inc. Introduced 1991; now marketed as Hi-Point JS-40 S&W.
    Perf.: $115   **Exc.:** $100   **VGood:** $85
Brushed nickel finish
    Perf.: $125   **Exc.:** $110   **VGood:** $100

Intratec Category 9

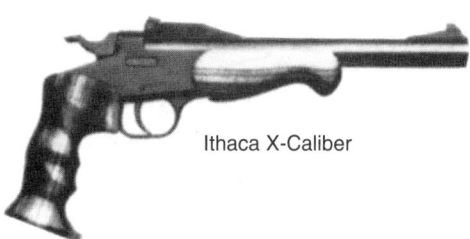

Ithaca X-Caliber

Intratec TEC-9C

INTRATEC CATEGORY 9

Semi-automatic; double action; 9mm Para.; 7-shot magazine; 3″ barrel; 5 ¹/₂″ overall length; weighs 21 oz.; textured black polymer grips; fixed channel sights; black polymer frame. Made in U.S. by Intratec. Introduced 1993; still produced. **LMSR: $225**
  New: $200   **Perf.: $175**   **Exc.: $150**

INTRATEC PROTEC-22

Semi-automatic; double-action-only; 22 LR; 10-shot magazine; 2 ¹/₂″ barrel; 5″ overall length; weighs 14 oz.; wrap-around composition grips in gray, black or driftwood color; fixed sights; choice of black, satin or Tec-Kote finishes. Made in U.S. by Intratec. Introduced 1991; still produced. **LMSR: $102**
Black finish
  Perf.: $85   **Exc.: $75**   **VGood: $65**
Satin or Tec-Kote finish
  Perf.: $90   **Exc.: $80**   **VGood: $70**

INTRATEC PROTEC-25

Semi-automatic; double-action-only; 25 ACP; 8-shot magazine; 2 ¹/₂″ barrel; 5″ overall length; weighs 14 oz.; wrap-around composition grips in gray, black or driftwood color; fixed sights; choice of black, satin or Tec-Kote finishes. Made in U.S. by Intratec. Introduced 1991; still produced. **LMSR: $102**
Black finish
  Perf.: $80   **Exc.: $70**   **VGood: $60**
Satin or Tec-Kote finish
  Perf.: $85   **Exc.: $75**   **VGood: $65**

INTRATEC TEC-9C

Semi-Automatic; 9mm Para.; 36-shot magazine; 5″ barrel; weighs 50 oz.; fixed sights; moulded composition stocks; matte blue finish. Introduced 1987; no longer in production. Produced in very limited numbers.

INTRATEC TEC-22T

Semi-automatic; single action; 22 LR; 10-shot magazine; 4″ barrel; 11 ³/₁₆″ overall length; weighs 30 oz.; moulded composition grip; protected post front, fully-adjustable front and rear sights; ambidextrous cocking knobs and safety; matte black finish; accepts any 10/22-type magazine. Made in U.S. by Intratec. Introduced 1988; still produced. **LMSR: $161**
  Perf.: $135   **Exc.: $125**   **VGood: $110**

Intratec TEC-22TK

Same specs as the TEC-22T except Tec-Kote finish. Made in U.S. by Intratec. Introduced 1988; still produced. **LMSR: $184**
  Perf.: $150   **Exc.: $135**   **VGood: $125**

INTRATEC TEC-DC9

Semi-automatic; single action; 9mm Para.; 10-shot magazine; 5″ barrel; 12 ¹/₂″ overall length; weighs 50 oz.; moulded composition grip; fixed sights; fires from closed bolt; firing pin block safety; matte blue finish. Made in U.S. by Intratec. Introduced 1985; still produced. **LMSR: $269**
  Perf.: $175   **Exc.: $150**   **VGood: $135**

Intratec TEC-DC9S

Same specs as TEC-DC9 except stainless steel construction. Made in U.S. by Intratec. Introduced 1985; still produced. **LMSR: $362**
  Perf.: $225   **Exc.: $200**   **VGood: $185**

INDIAN ARMS STAINLESS

Semi-automatic; double action; 380 ACP; 6-shot magazine; 3 ¹/₂″ barrel; 6 ¹/₁₆″ overall length; checkered walnut stocks; adjustable rear sight, blade front; made of stainless steel, but with natural or blued finish; optional safety lock. Introduced 1977; dropped 1978.
  Exc.: $350   **VGood: $300**   **Good: $250**

INTERARMS VIRGINIAN DRAGOON

Revolver; single action; 357 Mag., 44 Mag., 45 Colt; 6-shot; 6″, 7 ¹/₂″, 8 ³/₈″ barrel; 12″ buntline on special order; ramp-type Partridge front sight, micro-adjustable target rear; smooth walnut grips; spring-loaded firing pin; color case-hardened frame; blued finish. Introduced 1977; dropped 1986.
  Exc.: $300   **VGood: $225**   **Good: $175**
Buntline barrel
  Exc.: $325   **VGood: $250**   **Good: $200**

Interarms Virginian Dragoon Deputy Model

Same specs as standard model except fixed sights; 5″ barrel (357), 6″ (44 Mag.). Introduced 1983; dropped 1986.
  Exc.: $250   **VGood: $200**   **Good: $175**

Interarms Virginian Dragoon Engraved

Same specs as standard Dragoon except 44 Mag.; 6″, 7″ barrel; fluted or unfluted cylinders; stainless or blued; hand-engraved frame, cylinder, barrel; marketed in felt-lined walnut case. Introduced 1983; dropped 1986.
  Exc.: $500   **VGood: $450**   **Good: $350**

Interarms Virginian Dragoon Silhouette

Same specs as standard model except stainless steel; 357 Mag., 41 Mag., 44 Mag.; 7 ¹/₂″, 8 ³/₈″, 10 ¹/₂″ barrels; smooth walnut and Pachmayr rubber grips; undercut blade front sight, adjustable square notch rear; meets IHMSA standards. Made by Interarms. Introduced 1982; dropped 1986.
  Exc.: $350   **VGood: $300**   **Good: $250**

INTERARMS VIRGINIAN 22 CONVERTIBLE

Revolver; 22 LR, 22 WMR; 5 ¹/₂″ barrel; 10 ³/₄″ overall length; weighs 38 oz.; smaller version of standard Dragoon; comes with both cylinders; case-hardened frame, rest blued or stainless steel. Made in Italy. Introduced 1983; dropped 1987; was imported by Interarms.
  Exc.: $150   **VGood: $125**   **Good: $110**
Stainless finish
  Exc.: $165   **VGood: $135**   **Good: $125**

HANDGUNS

Indian Arms
Stainless

Interarms
Virginian Dragoon
Virginian Dragoon Deputy Model
Virginian Dragoon Engraved
Virginian Dragoon Silhouette
Virginian 22 Convertible

Intratec
Category 9
Protec-22
Protec-25
TEC-9C
TEC-22T
TEC-22TK
TEC-DC9
TEC-DC9S
TEC-DC9K
TEC-DC9M
TEC-DC9MK
TEC-DC9MS

Ithaca
Model 20
X-Caliber

Intratec TEC-DC9K

Same specs as TEC-DC9 except Tec-Kote finish. Made in U.S. by Intratec. Introduced 1985; still produced. **LMSR: $297**
  Perf.: $200   **Exc.: $175**   **VGood: $160**

Intratec TEC-DC9M

Same specs as TEC-DC9 except 3″ barrel; weighs 44 oz.; 20-shot magazine. Made in U.S. by Intratec. Introduced 1985; still produced. **LMSR: $245**
  Perf.: $175   **Exc.: $150**   **VGood: $135**

Intratec TEC-DC9MK

Same specs as the TEC-DC9 except Tec-Kote finish. Made in U.S. by Intratec. Introduced 1985; still produced. **LMSR: $277**
  Perf.: $185   **Exc.: $165**   **VGood: $140**

Intratec TEC-DC9MS

Same specs as TEC-DC9 except made of stainless steel. Made in U.S. by Intratec. Introduced 1985; still produced. **LMSR: $339**
  Perf.: $225   **Exc.: $200**   **VGood: $185**

ITHACA MODEL 20

Single shot; 22 LR, 44 Mag.; 10″, 12″ barrel; 15″ overall length (10″ barrel); weighs 3 ¹/₄ lbs.; American walnut stock with satin finish; Ithaca Gun Raybar Deerslayer sights or drilled and tapped for scope mounting; single firing pin for RF/CF use; comes with both barrels matched to one frame; matte blue finish. Made in U.S. by Ithaca Acquisition Corp. Introduced 1994; no longer produced.
  Perf.: $275   **Exc.: $250**   **VGood: $200**

ITHACA X-CALIBER

Single shot; 22 LR, 44 Mag.; 10″, 15″ barrel; 15″ overall length (10″ barrel); weighs 3 ¹/₄ lbs.; Goncalo Alves grip and forend; blade on ramp front sight, adjustable, removable target-type rear; drilled and tapped for scope mounting; dual firing pin for RF/CF use; polished blue finish. Made in U.S. by Ithaca Acquisition Corp. Made only in 1988.
  Perf.: $500   **Exc.: $450**   **VGood: $350**

Iver Johnson

9mm Double Action 1986
22 Supershot
22 Supershot Model 90
22 Supershot Sealed Eight (First Model)
22 Supershot Sealed Eight (Second Model)
22 Supershot Sealed Eight (Third Model 844)
Armsworth Model 855
Bulldog
Cadet (See Model 55S-A Cadet)
Cattleman Magnum
Cattleman Buckhorn Magnum
Cattleman Buckhorn Buntline
Cattleman Trailblazer
Champion Model 822

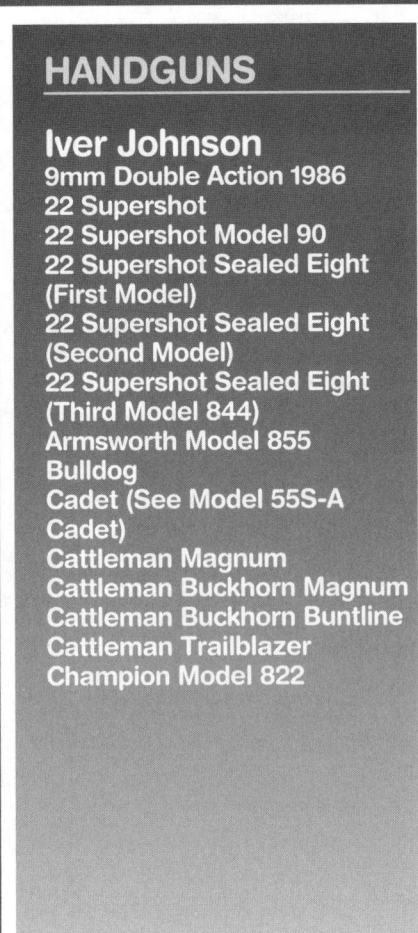

Iver Johnson 9mm Double Action 1986

Iver Johnson 22 Supershot Sealed Eight (First Model)

Iver Johnson 22 Supershot Sealed Eight (Third Model 844)

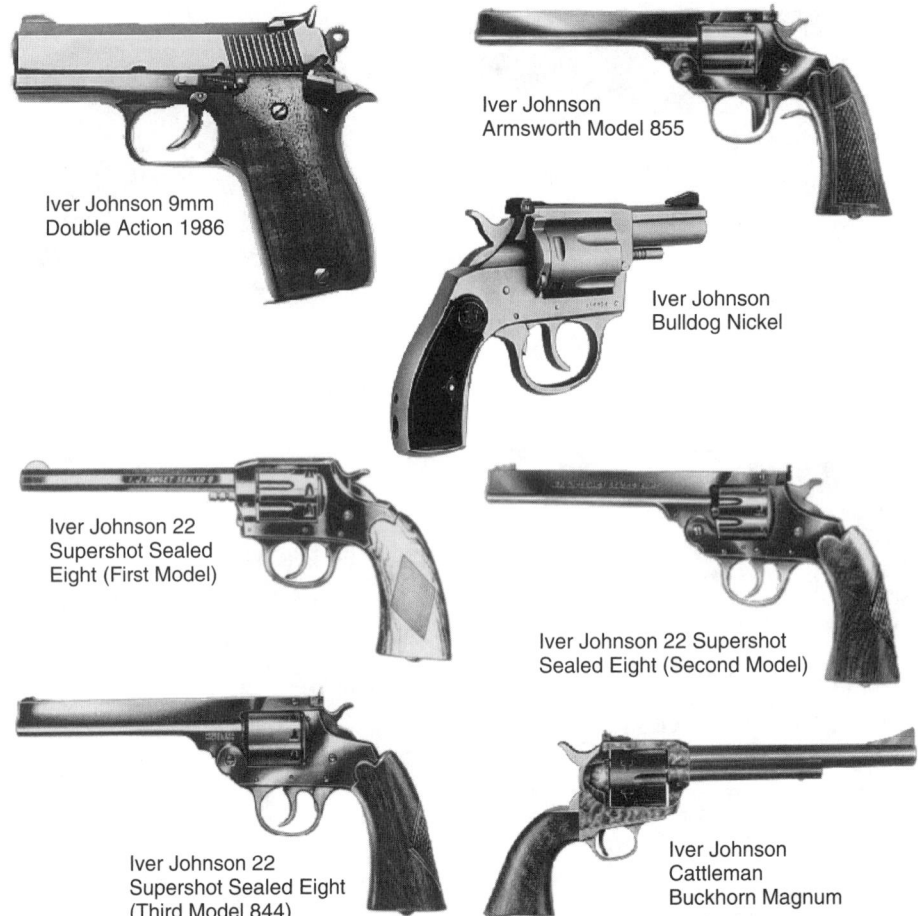

Iver Johnson Armsworth Model 855

Iver Johnson Bulldog Nickel

Iver Johnson 22 Supershot Sealed Eight (Second Model)

Iver Johnson Cattleman Buckhorn Magnum

IVER JOHNSON 9MM DOUBLE ACTION 1986

Semi-automatic; double action; 9mm; 6-shot; 3″ barrel; 6 1/4″ overall length; weighs 26 oz.; adjustable sights; smooth two-piece hardwood grips; blue finish. Manufactured in prototype only.

IVER JOHNSON 22 SUPERSHOT

Revolver; double action; 22 LR; 7-shot; 6″ barrel; 10 3/4″ overall length; weighs 24 oz.; top-break; automatic cartridge extraction; fixed sights; front sight and lettering gold-plated; one-piece saw-handle checkered walnut grips; has "Hammer the Hammer" action; blued finish. Marked "22 SUPERSHOT" in script. Introduced 1928; dropped 1931.
  VGood: $125 **Good:** $100 **Fair:** $75

IVER JOHNSON 22 SUPERSHOT MODEL 90

Revolver; double action; 22 LR; 9-shot; 6″ barrel; 10 3/4″ overall length; weighs 24 oz.; top-break; automatic cartridge extraction; fixed sights; front sight and lettering gold-plated; one-piece saw-handle shaped checkered walnut oversize grips; has "Hammer the Hammer" action; blued finish. Marked "22 SUPERSHOT" in script on left side of barrel. Pre-WWII model has adjustable finger rest. Introduced 1929; dropped 1949.
  VGood: $135 **Good:** $110 **Fair:** $85

IVER JOHNSON 22 SUPERSHOT SEALED EIGHT (FIRST MODEL)

Revolver; double action; 22 LR; 8-shot; 6″ barrel; 10 3/4″ overall length; weighs 24 oz.; large frame; top-break; recessed cylinder chambers; oversize target-type one-piece checkered walnut grips called Hi-Hold by the factory; has "Hammer the Hammer" action; blue finish. Three models: Model 88, fixed sights; Model 833, adjustable sights; Model 834, adjustable sights and adjustable finger rest. Marked "22 SUPERSHOT SEALED EIGHT" on left side of barrel. Introduced 1932; dropped 1941.
  VGood: $150 **Good:** $125 **Fair:** $100

Iver Johnson 22 Supershot Sealed Eight (Second Model)

Same specs as First Model except weighs 28 oz.; grip frame modified for ease of manufacture. Two models: Model 88, fixed sights; Model 833, adjustable sights. Marked "22 SUPERSHOT SEALED EIGHT" on left side of barrel. Introduced 1946; dropped 1954.
  VGood: $150 **Good:** $125 **Fair:** $100

Iver Johnson 22 Supershot Sealed Eight (Third Model 844)

Same specs as First Model except 4 1/2″, 6″ barrel; 9 1/4″ (4 1/4″ barrel), 10 3/4″ (6″ barrel) overall length; weighs 27 oz. (4 1/2″), 29 oz. (6″); unfluted cylinder; flat-sided barrel with full-length rib; recessed cylinder chambers with flash control front rim. Introduced 1955; dropped 1957.
  VGood: $175 **Good:** $150 **Fair:** $125

IVER JOHNSON ARMSWORTH MODEL 855

Revolver; single action; 22 LR; 6″ barrel; 10 3/4″ overall length; weighs 30 oz.; top-break; large frame; trigger pull factory set at 2 1/2 lbs.; recessed cylinder chambers with flash control front rim; flat-sided barrel with full-length rib; adjustable finger rest; adjustable sights; one-piece oversize target-type checkered walnut grip; does not have "Hammer the Hammer" action; blued finish. Limited production; rare. Introduced 1955; dropped 1957.
  VGood: $150 **Good:** $130 **Fair:** $110

IVER JOHNSON BULLDOG

Revolver; double action; 22 LR, 22 WMR, 38 Spl.; 6-shot (22), 5-shot (38); 2 1/2″, 4″ barrel; 6 1/2″ overall length (2 1/2″ barrel), 9″ overall length (4″ barrel); weighs 26 oz. (2 1/2″), 30 oz. (4″); large solid frame; pull pin cylinder release; adjustable sights; one-piece oversize molded Tenite plastic target grips (4″), two-piece pocket size plastic (2 1/2″); full-length barrel rib; does not have "Hammer the Hammer" action; blue or nickel finish. Introduced 1974; dropped 1979.
  VGood: $95 **Good:** $70 **Fair:** $45

IVER JOHNSON CATTLEMAN MAGNUM

Revolver; single action; 357 Mag., 45 LC, 44 Mag.; 6-shot; 4 3/4″, 5 1/2″, 7 1/4″ (357, 45); 4 3/4″, 6″, 7 1/4″ (44) barrel; 10 1/4″ (4 3/4″), 11″ (5 1/2″), 11 1/2″ (6″), 12 3/4″ (7 1/4″) overall length; 40 oz. (4 3/4″), 41 oz. (5 1/2″), 44 oz. (7 1/4″); large solid frame; rod ejection; automatic hammer-block safety; fixed sights; one-piece European walnut grips; brass grip frame; color case-hardened frame; blue barrel and cylinder. A limited number chambered in 44-40 caliber; add 20%. Introduced 1973; dropped 1979.
  VGood: $150 **Good:** $125 **Fair:** $100

Iver Johnson Cattleman Buckhorn Magnum

Same specs as Cattleman Magnum except 5 3/4″, 7 1/2″ (357, 45) barrel; 4 3/4″, 6″, 7 1/2″ (44) barrel; 12 3/4″ (7 1/2″ barrel), 11 1/2″ (6″ barrel), 11″ (5 3/4″ barrel), 10 3/4″ (4 3/4″ barrel) overall length; 44 oz. (7 1/2″), 42 1/2 oz. (6″), 41 oz. (5 3/4″); 41 oz. (4 3/4″). A limited number manufactured with 12″ barrel with shoulder stock lugs; add 20%. Introduced 1973; dropped 1979.
  VGood: $200 **Good:** $175 **Fair:** $135

Iver Johnson Cattleman Buckhorn Buntline

Same specs as Cattleman Magnum except 18″ barrel; 23 1/2″ overall length; weighs 56 oz.; detachable shoulder stock; adjustable sights. Introduced 1973; dropped 1979.
  VGood: $300 **Good:** $250 **Fair:** $175

Iver Johnson Cattleman Trailblazer

Same specs as Cattleman Magnum except 22 LR, 22 WMR; 5 1/2″, 6 1/2″ barrel; 10 3/4″ (5 1/2″), 11 3/4″ (6 1/2″) overall length; weighs 38 oz. (5 1/2″), 40 oz. (6 1/2″); 7/8″ size solid frame; adjustable sights. Introduced 1973; dropped 1979.
  VGood: $150 **Good:** $125 **Fair:** $100

IVER JOHNSON CHAMPION MODEL 822

Revolver; single-action; 22 LR; 6″ barrel; 8-shot; 10 3/4″ overall length; weighs 24 oz.; recessed cylin-

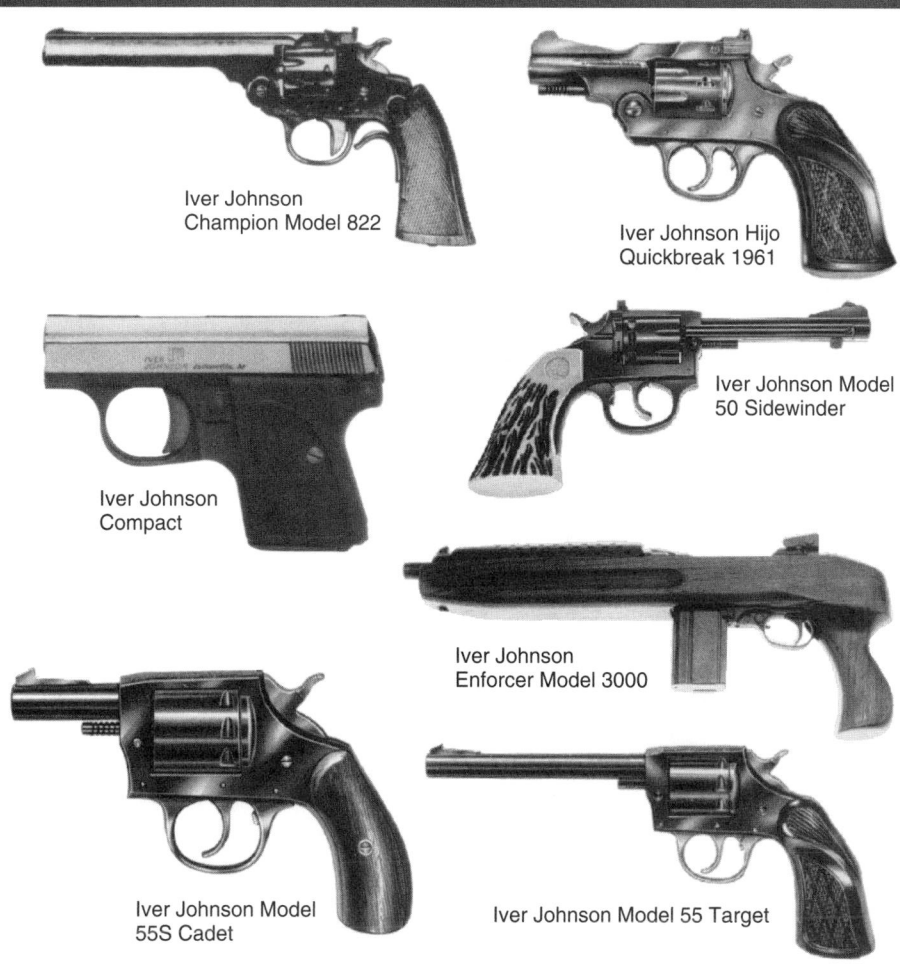

Iver Johnson
Champion Model 822

Iver Johnson Hijo
Quickbreak 1961

Iver Johnson
Compact

Iver Johnson Model
50 Sidewinder

Iver Johnson
Enforcer Model 3000

Iver Johnson Model
55S Cadet

Iver Johnson Model 55 Target

HANDGUNS

Iver Johnson
Compact
Compact Elite
Deluxe Starter Model 69
Enforcer Model 3000
Hijo Quickbreak 1961
Model 50 Sidewinder
Model 55 Target
Model 55A
Model 55S Cadet
Model 55S-A Cadet

der chambers; adjustable finger rest; adjustable target-type sights; oversize one-piece target-type checkered walnut grips; does not have "Hammer the Hammer" action; blued finish. Marked "CHAMPION SINGLE ACTION" on left side of barrel. Cataloged several years after WWII but none manufactured after 1941. Very rare; very collectible. Introduced 1939; dropped 1941.
VGood: $300 **Good:** $250 **Fair:** $200

IVER JOHNSON COMPACT
Semi-automatic; single action; 25 ACP; 6-shot; 3" barrel; 4 1/4" overall length; weighs 9 5/16" oz.; checkered composite grips; trigger block safety; matte blue frame; bright blue slide. Introduced 1991; dropped 1993.
VGood: $150 **Good:** $125 **Fair:** $100

Iver Johnson Compact Elite
Same specs as Compact except 24K gold finish; Ivorex grips. Introduced 1991; dropped 1993.
VGood: $500 **Good:** $375 **Fair:** $200

IVER JOHNSON DELUXE STARTER MODEL 69
Revolver; double action; 22, 32 blanks; 8-shot (22), 5-shot (32); 2 3/4" solid barrel; 7" overall length; weighs 22 oz.; recessed cylinder chambers; built on 66S Trailsman frame; fixed sights; two-piece small pocket-size plastic grips; does not have "Hammer the Hammer" action; matte blue finish. Introduced 1964; dropped 1974.
VGood: $65 **Good:** $45 **Fair:** $30

IVER JOHNSON ENFORCER MODEL 3000
Semi-automatic; single action; 30 Carbine; 15-, 30-shot; 9 1/2" barrel; 17" overall length; weighs 4 lbs.; gas-operated; fires from closed bolt; American walnut stock; blue finish. Pistol version of Universal M1 Carbine; Iver Johnson purchased Universal in 1985. Manufactured 1986.
VGood: $300 **Good:** $275 **Fair:** $250

IVER JOHNSON HIJO QUICKBREAK 1961
Revolver; double action; 22 LR, 32 S&W, 38 S&W; 8-shot (22), 5-shot (32, 38); 7" overall length; weighs 25 oz.; recessed cylinder chambers; rimfire model with flash control front rim; two-piece pocket-size plastic grips; adjustable sights; does not have "Hammer the Hammer" action; nickel finish only. Same as the Model 66S made to be sold by several mail-order companies in New York City area.
VGood: $100 **Good:** $75 **Fair:** $50

IVER JOHNSON MODEL 50 SIDEWINDER
Revolver; double action; 22 LR; 8-shot; 6" barrel (1961-1978/1979); 4 3/4" barrel (1974-1978/1979); 10" (4 3/4" barrel), 11 1/4" (6" barrel) overall length; 30 oz. (4 3/4"), 31 oz. (6"); large solid frame; rod ejection; recessed chambers with flash control front rim; 22 LR/22 WMR combo available after 1974; fixed sights; adjustable after 1974; Western-style walnut grips; stag plastic grips standard after 1968; several model numbers for different finish and sight combinations after 1974; matte blue and satin nickel finish. Introduced 1961; dropped 1979.
VGood: $110 **Good:** $85 **Fair:** $60

IVER JOHNSON MODEL 55 TARGET
Revolver; double action; 22 LR; 8-shot; 4 1/2", 6" barrels; 10 3/4" (4 1/4"), 9 1/4" (6") overall length; weighs 24 oz. (4 1/2"), 26 oz. (6"); large solid frame; pull pin cylinder release with flash control front rim; recessed cylinder chambers; fixed sights; oversize molded one-piece Tenite plastic target-type grips; does not have "Hammer the Hammer" action; matte blue finish. After 1958 cylinder fluted; hard chrome lined barrel. Introduced 1955; dropped 1960.
VGood: $90 **Good:** $70 **Fair:** $50

Iver Johnson Model 55A
Same specs as Model 55 except incorporation of loading gate; weighs 28 1/2 oz. (4 1/2" barrel), 30 1/2 oz. (6" barrel). After 1974/1975 this model called the "I J SPORTSMAN." Introduced 1961; dropped 1979.
VGood: $90 **Good:** $70 **Fair:** $50

Iver Johnson Model 55A

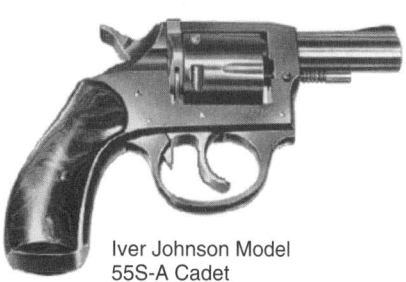

Iver Johnson Model
55S-A Cadet

Iver Johnson Model 55S Cadet
Same specs as Model 55 except 2 1/2" barrel; 7" overall length; weighs 27 oz.; two-piece small pocket-size grips; fluted cylinder after 1958. Introduced 1955; dropped 1960.
VGood: $100 **Good:** $75 **Fair:** $50

Iver Johnson Model 55S-A Cadet
Same specs as Model 55 except 22 LR, 32 S&W, 38 S&W; 8-shot (22), 5-shot (32, 38); 7" overall length; loading gate on right side of frame. After 1974/1975 this model called "I J CADET." Introduced 1961; dropped 1979.
VGood: $100 **Good:** $75 **Fair:** $50

HANDGUNS

Iver Johnson
Model 56 I J Starter Revolver
Model 56A I J Starter Revolver
Model 57 Target
Model 57A Target
Model 1900
Model 1900 Target
Model 1900 Target
Models 69 & 79
Model 1908 Petite Hammerless
Protector Sealed Eight Model 84
Rookie
Safety Automatic Hammer (Second Model)
Safety Automatic Hammer (Third Model)

Iver Johnson Model 56A
IJ Starter Revolver

Iver Johnson Model 1900

Iver Johnson
Model 57 Target

Iver Johnson Protector
Sealed Eight Model 84

Iver Johnson
Model 57A Target

Iver Johnson Safety Automatic
Hammer (Third Model)

IVER JOHNSON MODEL 56 I J STARTER REVOLVER
Revolver; double action; 22 blank; 8-shot; 2 1/2″ solid barrel; 6 3/4″ overall length; weighs 10 oz.; large solid frame; pull pin cylinder release; recessed cylinder chambers; cylinder half-length; hard rubber or walnut pocket-size grips; built on Model 1900 large frame; blue finish. Introduced 1956; dropped 1960.
VGood: $75 **Good:** $55 **Fair:** $35

Iver Johnson Model 56A I J Starter Revolver
Same specs as Model 56 except 22, 32 rimfire blank; 8-shot (22), 5-shot (32); 22 oz.; plastic grips; loading gate on right side of frame; matte blue finish; built on Model 55S-A frame. Introduced 1961; dropped 1974.
VGood: $60 **Good:** $40 **Fair:** $20

IVER JOHNSON MODEL 57 TARGET
Revolver; double action; 22 LR; 8-shot; 2 1/2″, 4 1/2″, 6″ barrel; 7 1/2″ (2 1/2″ barrel), 9 1/2″ (4 1/2″ barrel), 10 3/4″ (6″ barrel) overall length; 27 oz. (2 1/2″), 29 oz. (4 1/2″), 30 oz. (6″); large solid frame; pull pin cylinder release; unfluted cylinder; recessed cylinder chambers with flash control front rim; scored trigger; music wire springs; checkered thumbrest; half-cock safety; adjustable front and rear sight; one-piece molded Tenite plastic target-type grips; does not have "Hammer the Hammer" action; matte blue finish. Introduced 1955; dropped 1960.
VGood: $95 **Good:** $75 **Fair:** $55

Iver Johnson Model 57A Target
Same specs as Model 57 except 4 1/2″, 6″ barrel only; loading gate on right side of frame; adjustable sights. After 1974 this model called "I J TARGET DELUXE." Introduced 1961; dropped 1979.
VGood: $95 **Good:** $75 **Fair:** $55

IVER JOHNSON MODEL 1900
Revolver; double action; solid frame; pull pin cylinder release; 2 1/2″ (standard), 4 1/2″, 6″ octagon barrel; standard finish nickel, blue optional; three frame sizes. Small frame: 22 LR; 7-shot; 6″ overall length (2 1/2″ barrel); weighs 11 oz. Medium frame: 32 S&W; 5-shot; 6 1/8″ overall length (2 1/2″ barrel); weighs 12 oz. Large frame: 32 S&W Long, 38 S&W; 6-shot (32), 5-shot (38); 6 3/4″ overall length; weighs 18 oz. Small pocket-size hard rubber grips, oversize target-type hard rubber optional. Introduced 1900; dropped 1941.
VGood: $100 **Good:** $75 **Fair:** $50

Iver Johnson Model 1900 Target
Same specs as Model 1900 except blue finish only; 6″, 9″ octagon barrel; 9 1/4″ overall length (6″ barrel), 14 3/8″ (9″ barrel); weighs 12 7/8 oz. (6″ barrel), 14 3/8 oz. (9″ barrel); built on Model 1900 small frame. Two-piece oversize walnut grips. Introduced 1925; dropped 1928. Add 20% for 9″ barrel.
VGood: $135 **Good:** $110 **Fair:** $100

Iver Johnson Model 1900 Target Models 69 & 79
Same specs as Model 1900 except blue finish only; 6″ (Model 69), 10″ (Model 79) octagon barrel; 9-shot; built on Model 1900 large frame; 10 3/4″ overall length (6″ barrel), 14 3/4″ (10″ barrel); weighs 24 oz. (6″), 27 oz. (10″); one-piece oversize target-type walnut grips. Introduced 1929; dropped 1941. Add 20% for 10″ barrel.
VGood: $145 **Good:** $135 **Fair:** $110

IVER JOHNSON MODEL 1908 PETITE HAMMERLESS
Revolver; double action; 22 LR; 1 1/4″ barrel; 3 15/16″ overall length; weighs 4 7/8 oz.; solid frame; folding trigger; pull pin cylinder release; small pocket-size hard rubber grips. One of the smallest revolvers built in U.S. Less than 1000 manufactured. Manufactured 1908 only. Rare and collectible.
VGood: $300 **Good:** $200 **Fair:** $150

IVER JOHNSON PROTECTOR SEALED EIGHT MODEL 84
Revolver; double action; 22 LR; 8-shot; 2 1/2″ barrel; 10 3/4″ overall length; weighs 20 oz.; large frame; top-break; automatic extraction; recessed cylinder chambers; pocket-size one-piece checkered walnut grips; fixed sights, adjustable optional; has "Hammer the Hammer" action; blue finish. Marked "PROTECTOR SEALED EIGHT" on left side of barrel. Rare model. Introduced 1933; dropped 1941.
VGood: $200 **Good:** $175 **Fair:** $150

IVER JOHNSON ROOKIE
Revolver; double action; 38 Spl.; 5-shot; 4″ barrel; 9″ overall length; weighs 29 oz.; pull pin cylinder release; fixed sights; one-piece oversize molded Tenite plastic target-type grips; does not have "Hammer the Hammer" action; blue finish. Introduced 1974; dropped 1979.
VGood: $95 **Good:** $70 **Fair:** $45

IVER JOHNSON SAFETY AUTOMATIC HAMMER (SECOND MODEL)
Revolver; double action; hinged frame top-break; double top post barrel latch; "Hammer the Hammer" action; flat springs; two frame sizes; standard finish, nickel; blue finish optional. Small frame: 22 LR, 32 S&W; 7-shot (22), 5-shot (32); 3″(standard), 4″, 5″, 6″ barrel; 6 3/8″ overall length; weighs 12 oz. (3″ 22 LR), 12 1/4 oz. (32 S&W). Large frame: 38 S&W (blackpowder cartridges only); 3 1/4″ (standard), 2″, 4″, 5″, 6″ barrels; 7 3/8″ overall length; weighs 17 1/4 oz. Fixed sights; small pocket size hard rubber grips. Introduced 1897; dropped 1908. For rimfire models add 20%.
VGood: $100 **Good:** $75 **Fair:** $50

Iver Johnson Safety Automatic Hammer (Third Model)
Same specs as Second Model except coil springs; small frame comes in 3″ (standard), 2″, 4″, 5″, 6″ barrel; large frame model in 32 S&W Long, 38 S&W for smokeless powder. Cataloged until 1948. Introduced 1909; dropped 1941. For rimfire models add 20%.
VGood: $125 **Good:** $100 **Fair:** $75

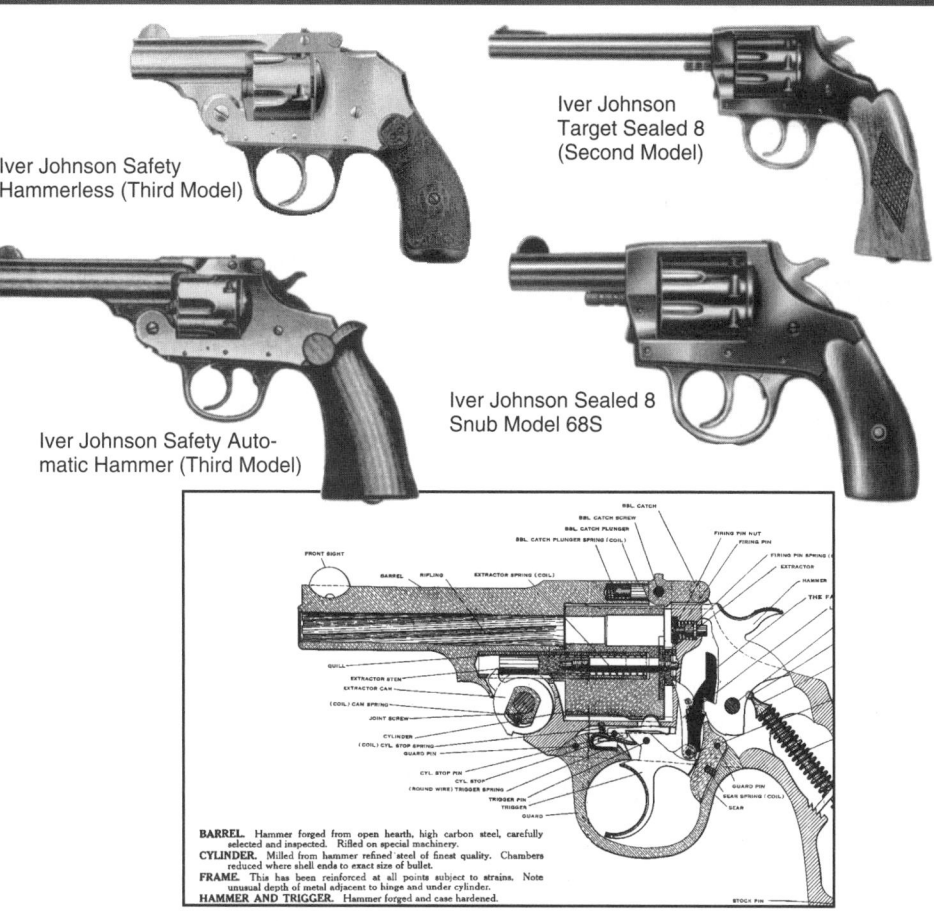

Iver Johnson Safety Hammerless (Third Model)

Iver Johnson Safety Automatic Hammer (Third Model)

Iver Johnson Target Sealed 8 (Second Model)

Iver Johnson Sealed 8 Snub Model 68S

BARREL. Hammer forged from open hearth, high carbon steel, carefully selected and inspected. Rifled on special machinery.
CYLINDER. Milled from hammer refined steel of finest quality. Chambers reduced where shell ends to exact size of bullet.
FRAME. This has been reinforced at all points subject to strains. Note unusual depth of metal adjacent to hinge and under cylinder.
HAMMER AND TRIGGER. Hammer forged and case hardened.

HANDGUNS

Iver Johnson
Safety Hammerless (Second Model)
Safety Hammerless (Third Model)
Sealed 8 Snub Model 68S
Secret Service Special Models 1917 & 1923
Sportsman (See Model 55A)
Super Enforcer Model PP30
Swingout Cylinder Model
Target Deluxe (See Model 57A Target)
Target Sealed 8 (First Model)
Target Sealed 8 (Second Model)
TP22
TP25
Trailsman

Iver Johnson Trailsman

IVER JOHNSON SAFETY HAMMERLESS (SECOND MODEL)
Revolver; double action; hinged frame top-break; double top post barrel latch; "Hammer the Hammer" action; flat springs; basic design comparable to Safety Hammer model; two frame sizes; standard finish, nickel; blue finish optional. Small frame: 22 LR, 32 S&W; 7-shot (22), 5-shot (32); 3" (standard), 2", 4", 5", 6" barrel; 6 3/8" overall length (3" barrel); weighs 12 1/2 oz. (3" 22 LR), 13 1/2 oz. (32). Large frame: 38 S&W **(blackpowder cartridges only)**; 5-shot (38); 3 1/4" (standard), 2", 4", 5", 6" barrels; 7 3/8" overall length; weighs 18 1/4 oz. (32), 19 oz. (38). Fixed sights; small pocket size hard rubber grips. Introduced 1897; dropped 1908. For rimfire models add 20%.
VGood: $100   **Good:** $75   **Fair:** $50

Iver Johnson Safety Hammerless (Third Model)
Same specs as Second Model except small frame weighs 13 1/2 oz. (3" 22 LR), 14 oz. (32); large frame weighs 19 oz. (38 S&W); 38 S&W, 32 S&W long **(smokeless cartridge)**; oversize target hard rubber or walnut grips optional. Cataloged until 1948. Introduced 1909; dropped 1941. For rimfire models add 20%.
VGood: $125   **Good:** $100   **Fair:** $75

IVER JOHNSON SEALED 8 SNUB MODEL 68S
Revolver; double action; 22 LR; 8-shot; 2 1/2" barrel; 6 3/4" overall length; weighs 20 1/2 oz.; large solid frame; pull pin cylinder release; recessed cylinder chambers; round barrel; small pocket-size two-piece walnut grips; does not have "Hammer the Hammer" action. Introduced 1952; dropped 1954. Rare and limited production.
VGood: $200   **Good:** $175   **Fair:** $150

IVER JOHNSON SECRET SERVICE SPECIAL MODELS 1917 & 1923
Revolver; double action; top break; hammer and hammerless versions; two frame sizes. Small frame: 32 S&W; 5-shot; 3" (standard), 5" overall length; weighs 12 oz. (hammer model), 13 oz. (hammerless model). Large frame: 38 S&W; 5-shot; 3 1/4" (standard), 5" barrel; 7 3/8" overall length; weighs 18 oz. (hammer model), 19 oz. (hammerless); blue finish, nickel optional; hard rub-ber pocket-size grips; oversize target grips optional. Marked on left side of barrel: "SECRET SERVICE SPECIAL" on top of barrel: "for .32 (or .38) SMITH & WESSON CARTRIDGES." Note: Not all revolvers so marked were manufactured by Iver Johnson. The name "Secret Service Special" was owned by the Fred Biffar Co. of Chicago, IL. For models with hammer-block safety add 20%.
VGood: $85   **Good:** $60   **Fair:** $40

IVER JOHNSON SUPER ENFORCER MODEL PP30
Semi-automatic; single action; 30 Carbine; 15-, 30-shot; 9 1/2" barrel; 17" overall length; weighs 4 lbs.; gas-operated; fires from closed bolt; American walnut stock; blue finish. Pistol version of Iver Johnson M1 Carbine. Introduced 1978; dropped 1992.
VGood: $325   **Good:** $300   **Fair:** $275

IVER JOHNSON SWINGOUT CYLINDER MODEL
Revolver; double action; 22 LR, 22 WMR, 32 S&W Long, 38 Spl.; 6-shot (22, 32), 5-shot (38); 2", 3", 4" plain barrel; 4", 6", 8 3/4" ventilated rib barrel; fixed sights (plain barrel), adjustable (ventilated); walnut grips; blue or nickel finish. **Manufactured in prototype only in 1977.**

IVER JOHNSON TARGET SEALED 8 (FIRST MODEL)
Revolver; double action; 22 LR; 8-shot; 6" (Model 68), 10" (Model 78) barrel; 10 3/4" (M68), 14 3/4" (M78) overall length; weighs 24 oz. (M68), 27 oz. (M78); large solid frame; pull pin cylinder release; recessed cylinder chambers; fixed sights, front blade gold-plated; one-piece oversized checkered walnut grips; octagon barrel; does not have "Hammer the Hammer" action. Introduced 1932; dropped 1941.
VGood: $125   **Good:** $100   **Fair:** $75

Iver Johnson Target Sealed 8 (Second Model)
Same specs as First Model except 4 1/2", 6" barrel; 9 1/4" (4 1/2"), 10 3/4" (6" barrel) overall length; weighs 22 1/2 oz. (4"), 24 oz. (6"); round barrel. Introduced 1946; dropped 1954.
VGood: $125   **Good:** $100   **Fair:** $75

IVER JOHNSON TP22
Semi-automatic; double action; 22 LR; 2 7/8" barrel; 5 3/8" overall length; weighs 14 1/2 oz.; hammer-block safety; no magazine safety; no slide stop; fixed sights; black plastic wrap-around grips; designed by ERMA WERKES Germany; blue and nickel finish; features patented passive firing pin block safety. Introduced 1982; dropped 1988.
VGood: $150   **Good:** $125   **Fair:** $100

IVER JOHNSON TP25
Semi-automatic; double action; 25 ACP; 2 7/8" barrel; 5 3/8" overall length; weighs 14 1/2 oz.; hammer-block safety; no magazine safety; no slide stop; fixed sights; black plastic wrap-around grips; designed by ERMA WERKES Germany; blue finish. Introduced 1981; dropped 1988.
VGood: $125   **Good:** $110   **Fair:** $95

IVER JOHNSON TRAILSMAN
Semi-automatic; single action; 22 LR; 4 1/2", 6" barrel; 8 3/4" (4 1/2" barrel) overall length; weighs 46 oz. (4 1/2"), 64 oz. (6"); checkered black plastic grips; slide hold-open latch; push button magazine release; sear block safety; fixed sights; blue finish. First 500+ actually manufactured in Argentina worth premium. Introduced 1984; dropped 1988.
VGood: $175   **Good:** $150   **Fair:** $125

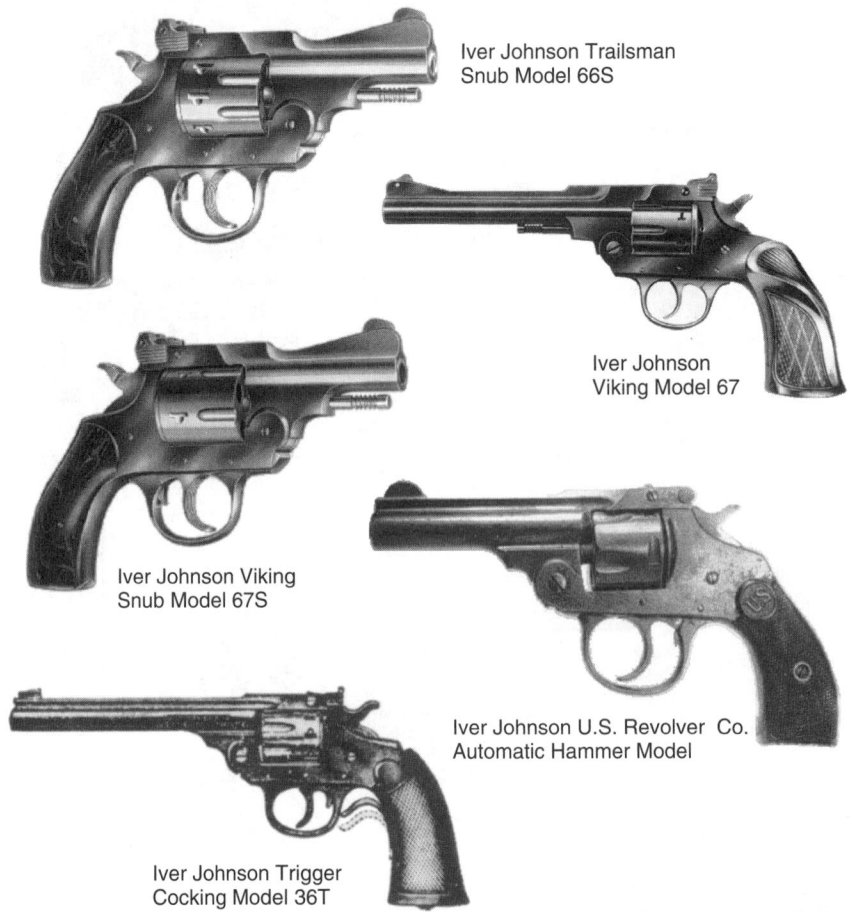

Iver Johnson Trailsman
Snub Model 66S

Iver Johnson
Viking Model 67

Iver Johnson Viking
Snub Model 67S

Iver Johnson U.S. Revolver Co.
Automatic Hammer Model

Iver Johnson Trigger
Cocking Model 36T

Iver Johnson
Trailman Model 66
Trailsman Snub Model 66S
Trigger Cocking Model 36T
U.S. Revolver Co. Automatic
Hammer Model
U.S. Revolver Co. Automatic
Hammerless Model
U.S. Revolver Co. Double
Action
Viking Model 67
Viking Snub Model 67S
X300 Pony

Iver Johnson
Trailsman Model 66

Iver Johnson X300 Pony

IVER JOHNSON TRAILSMAN MODEL 66
Revolver; double action; 22 LR; 8-shot; 4 1/2" (added in 1962; dropped 1964), 6" barrel; 9 1/2" (4 1/2" barrel), 11" (6" barrel) overall length; weighs 31 oz. (4 1/2"), 34 oz. (6"); large frame; top-break; recessed cylinder chambers with flash control front rim; manual extraction; adjustable sights; one-piece moulded Tenite plastic target-type grips; does not have "Hammer the Hammer" action; matte blue finish. Introduced 1958; dropped 1974.
VGood: $100 **Good:** $75 **Fair:** $50

Iver Johnson Trailsman Snub Model 66S
Same specs as Model 66 except 22 LR, 32 S&W, 38 S&W; 8-shot (22), 5-shot (32, 38); centerfire calibers not added until 1962; 2 3/4" barrel; 7" overall length; weighs 25 oz.; two-piece pocket-size plastic grips. Introduced 1959; dropped 1963.
VGood: $110 **Good:** $85 **Fair:** $60

IVER JOHNSON TRIGGER COCKING MODEL 36T
Revolver; single action; 22 LR; 8-shot; 6" barrel; 10 3/4" overall length; weighs 24 oz.; large frame; top-break; recessed cylinder chambers; adjustable sights; one-piece oversize target-type checkered walnut grips; adjustable finger rest; does not have "Hammer the Hammer" action; first pull of the trigger cocks the hammer; second pull releases it; blue finish. Marked "I J TRIGGER COCKING 22 SINGLE ACTION TARGET" in two lines on left side of barrel. Limited production; considered rare and collectible. Introduced 1936; dropped 1941.
VGood: $300 **Good:** $250 **Fair:** $200

IVER JOHNSON U.S. REVOLVER CO. AUTOMATIC HAMMER MODEL
Revolver; double action; top-break; automatic cartridge extraction; two frame sizes; blue finish, nickel optional. Small frame: 22 LR, 32 S&W; 7-shot (22), 5-shot (32); 3" (standard), 5" barrel; 6 3/8" overall length (3" barrel); 12 oz. (22), 12 1/2 oz. (32). Large frame: 38 S&W; 5-shot; 7 3/8" overall length; weighs 17 3/4 oz.; small hard rubber pocket-size grips, oversize grips optional. Marked on barrel top rib: "U.S. REVOLVER Co." Introduced 1910; dropped 1935.
VGood: $85 **Good:** $65 **Fair:** $45

IVER JOHNSON U.S. REVOLVER CO. AUTOMATIC HAMMERLESS MODEL
Revolver; double action; top-break; automatic cartridge extraction; two frame sizes; blue finish, nickel optional. Small frame: 22 LR, 32 S&W; 5-shot; 3" (standard), 5" barrel; 6 3/8" overall length (3" barrel); 12 1/2 oz. Large frame: 38 S&W; 5-shot; 7 3/8" overall length; weighs 18 oz.; small hard rubber pocket-size grips, two-piece oversize grips optional. Marked on barrel top rib: "U.S. REVOLVER Co." Introduced 1910; dropped 1935.
VGood: $85 **Good:** $65 **Fair:** $45

IVER JOHNSON U.S. REVOLVER CO. DOUBLE ACTION
Revolver; double action; solid frame; pull pin cylinder release; two frame sizes; round barrel; unfluted cylinder. Small frame: 22 LR, 32 S&W; 7-shot (22), 5-shot (32); 2 1/2" barrel (standard), 4 1/2" optional; 6" overall length (22), 6 1/8" (32); weighs 10 1/2 oz. (22), 12 oz. (32). Large frame: 32 S&W Long, 38 S&W; 6-shot (32), 5-shot (38); 6 3/4" overall length; weighs 17 oz.; small pocket-size hard rubber grips. Marked on frame topstrap: "U.S. REVOLVER Co." Introduced 1911; dropped 1935.
VGood: $95 **Good:** $70 **Fair:** $45

IVER JOHNSON VIKING MODEL 67
Revolver; double action; 22 LR; 8-shot; 4 1/2", 6" barrel; 9 1/2" (4 1/4" barrel), 11" (6" barrel) overall length; weighs 31 oz. (4 1/2"), 34 oz. (6"); chrome-lined barrel; large frame; top-break; recessed cylinder chambers and flash control front rim; one-piece moulded oversize target-type Tenite grip; adjustable sights; matte blue finish. Introduced 1964; dropped 1974.
VGood: $125 **Good:** $100 **Fair:** $75

Iver Johnson Viking Snub Model 67S
Same specs as Model 67 except 22 LR, 32 S&W, 38 S&W; 8-shot (22), 5-shot (32, 38); centerfire calibers not added until 1962; 2 3/4" barrel; 7" overall length; weighs 25 oz.; two-piece pocket-size plastic grips. Introduced 1964; dropped 1974.
VGood: $135 **Good:** $110 **Fair:** $85

IVER JOHNSON X300 PONY
Semi-automatic; single action; 380 ACP; 6-shot; 3" barrel; 6 1/4" overall length; weighs 20 oz.; all steel construction; inertia firing pin; loaded chamber indicator; lanyard ring; no magazine safety; adjustable rear sight; matte blue, bright blue or nickel finish. Three production periods: Middlesex, NJ 1978-1982; Jacksonville, AR 1983-1988; Jacksonville, AR by AMAC 1989-1991.
VGood: $225 **Good:** $200 **Fair:** $150

Jennings
Model J-22

Jennings Model J-25

Kareen MKII Auto Pistol

Jericho Model 941

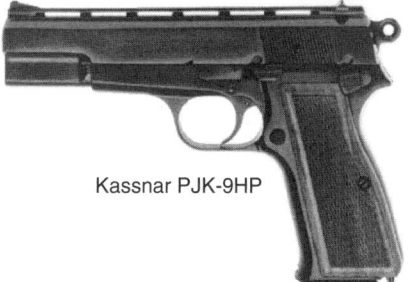
Kassnar PJK-9HP

Kahr K9

HANDGUNS

Jana
Bison

Jennings
Model J-22
Model J-25
Model M-38
Model M-48

Jericho
Model 941

JSL
Spitfire

Kahr
K9

Kareen
MKII Auto Pistol
MKII Compact

Kassnar
M-100D
Model M-100TC
PJK-9HP

JANA BISON
Revolver; single action; 22 LR, 22 WMR; 4 3/4″ barrel; fixed front sight, adjustable rear: imitation stag grips; blued finish. Introduced 1978; imported from West Germany; dropped 1980.
    Exc.: $100   **VGood:** $85   **Good:** $60

JENNINGS MODEL J-22
Semi-automatic; 22 LR; 6-shot magazine; 2 1/2″ barrel; weighs 12 oz.; 4 15/16″ overall length; fixed sights; walnut stocks. Introduced 1981; still in production. **LMSR: $80**
    Perf.: $70   **Exc.:** $55   **VGood:** $45

JENNINGS MODEL J-25
Semi-automatic; 25 ACP; 6-shot magazine; 2 1/2″ barrel; weighs 12 oz.; 4 15/16″ overall length; fixed sights; synthetic walnut or black combat-style grips. Introduced 1981; still in production. **LMSR: $80**
    Perf.: $65   **Exc.:** $50   **VGood:** $40

JENNINGS MODEL M-38
Semi-automatic; 22 LR, 32 ACP, 380 ACP; 2 13/16″ barrel; weighs 16 oz.; non-ferrous alloy construction; black grips; chrome or blue finish. No longer in production.
    Perf.: $75   **Exc.:** $60   **VGood:** $50

JENNINGS MODEL M-48
Semi-automatic; 22 LR, 32 ACP, 380 ACP; 4″ barrel; weighs 24 oz.; non-ferrous alloy construction; black grips; chrome or blue finish. No longer in production.
    Perf.: $75   **Exc.:** $60   **VGood:** $50

JERICHO MODEL 941
Semi-automatic; double action; 9mm Para.; 16-shot magazine; 4 3/8″ barrel; 8 1/8″ overall length; weighs 33 oz.; high impact black polymer grips; blade front sight, windage-adjustable rear; three tritium dots; all steel construction; polygonal rifling; ambidextrous safety. Produced in Israel by Israel Military Industries; distributed by K.B.I., Inc. Introduced 1990; dropped 1991.
    Perf.: $450   **Exc.:** $375   **VGood:** $300

JSL SPITFIRE
Semi-automatic; double action; 9mm Para.; 15-shot magazine; 4 3/8″ barrel; 8 7/8″ overall length; weighs 40 oz.; textured composition grips; blade front sight, fully-adjustable rear; stainless steel construction; ambidextrous safety. Imported from England by Rogers Ltd. International. Introduced 1992; importation dropped 1993.
    Perf.: $1200   **Exc.:** $900   **VGood:** $750

KAHR K9
Semi-automatic; double action; 9mm Para.; 7-shot magazine; 3 1/2″ barrel; 6″ overall length; weighs 25 oz.; wrap-around textured soft polymer grips; blade front sight, windage drift-adustable rear; bar-dot combat style; passive firing pin block; made of 4140 ordnance steel; matte black finish. Made in U.S. by Kahr Arms. Introduced 1994; still produced. **LMSR: $595**
    New: $500   **Perf.:** $425   **Exc.:** $350

KAREEN MKII AUTO PISTOL
Semi-automatic; single action; 9mm Para.; 13-shot magazine; 4 3/4″ barrel; 8″ overall length; weighs 32 oz.; blade front sight, windage-adjustable rear; checkered European walnut grips (early model), rubberized grips (later versions); blued finish, two-tone or matte black finish, optional. Made in Israel. Introduced 1969 by Century International Arms, Inc.; still imported by J.O. Arms. **LMSR: $425**
    Perf.: $300   **Exc.:** $275   **VGood:** $225

Kareen MKII Compact
Same specs as MKII except 3 1/4″ barrel; 6 1/2″ overall length; weighs 28 oz.; rubber grips. Introduced 1995; still in production. **LMSR: $425**
    New: $350   **Exc.:** $300   **VGood:** $275

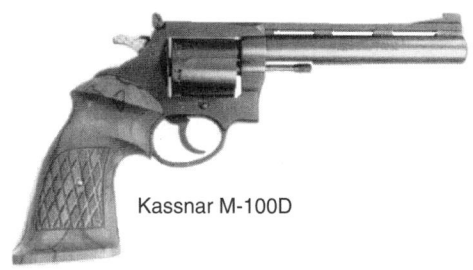
Kassnar M-100D

KASSNAR M-100D
Revolver; double action; 22 LR, 22 WMR, 38 Spl.; 6-shot cylinder; 3″, 4″, 6″ barrels; target-style checkered hardwood stocks; elevation-adjustable rear sight, ramp front; vent-rib barrel. Manufactured in the Philippines by Squires Bingham; imported by Kassnar. Introduced 1977; no longer imported.
    Exc.: $125   **VGood:** $110   **Good:** $100

KASSNAR MODEL M-100TC
Revolver; double action; 22 LR, 22 WMR, 38 Spl.; 6-shot cylinder; 3″, 4″, 6″ barrels; full ejector rod shroud; target-style checkered hardwood stocks; elevation-adjustable rear sight, ramp front; vent-rib barrel. Manufactured in the Philippines by Squires Bingham; imported by Kassnar. Introduced 1977; no longer imported.
    Exc.: $135   **VGood:** $125   **Good:** $110

KASSNAR PJK-9HP
Semi-automatic; single action; 9mm Para.; 13-shot magazine; 4 3/4″ barrel; 8″ overall length; weighs 32 oz.; adjustable rear sight, ramp front; checkered European walnut stocks; with or without full-length vent rib. Made in Hungary. Introduced 1986; no longer imported.
    Perf.: $200   **Exc.:** $175   **VGood:** $165

KBI
PSP-25

KEL-TEC
P-11

Kimber
Classic 45 Custom
Classic 45 Gold Match
Classic 45 Royal
Predator Hunter
Predator Supergrade

Kimel
AP9
AP9 Mini
AP9 Target

Kleinguenther
R-15

Korriphila
HSP 701

Korth
Semi-Automatic

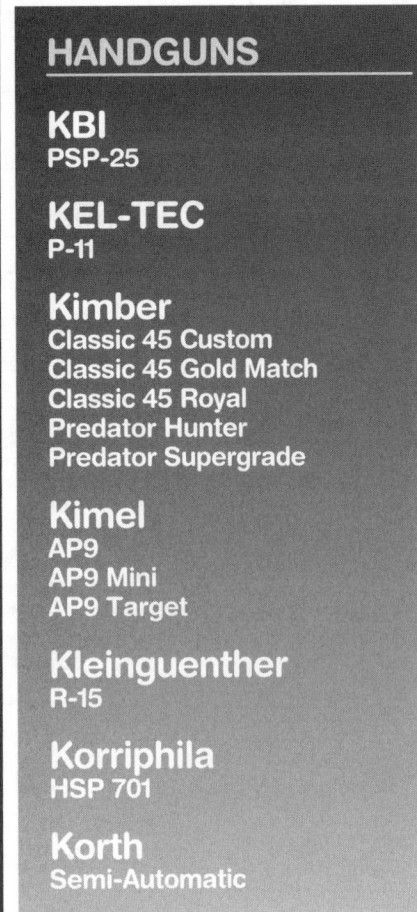

KBI PSP-25

Kimber Classic 45 Custom

Kel-Tec P-11

Korriphila HSP 701

Kimber Classic 45 Gold Match

Kimel AP9

KBI PSP-25
Semi-automatic; single action; 25 ACP; 6-shot magazine; 2 1/8" barrel; 4 1/8" overall length; weighs 9 1/2 oz.; fixed sights; checkered black plastic grips; all-steel construction; polished blue or chrome finish. Close copy of Browning Baby 25 made under F.N. license. Introduced 1990; no longer in production.
Perf.: $180   **Exc.:** $160   **VGood:** $140

KEL-TEC P-11
Semi-automatic; double-action-only; 9mm Para.; 10-shot magazine; 3 1/8" barrel; 5 9/16" overall length; weighs 14 oz.; checkered black polymer grips; blade front sight, windage-adjustable rear; ordnance steel slide, aluminum frame; blue finish. Made in U.S. by Kel-Tec CNC Industries, Inc. Introduced 1995; still produced. **LMSR:** $300
New: $250   **Perf.:** $225   **Exc.:** $200
Electroless nickel finish
New: $300   **Perf.:** $275   **Exc.:** $235
Gray finish
New: $280   **Perf.:** $250   **Exc.:** $200

KIMBER CLASSIC 45 CUSTOM
Semi-automatic; single action; 45 ACP; 8-shot magazine; 5" barrel; 8 1/2" overall length; weighs 38 oz.; checkered hard synthetic grips; McCormick dovetailed front sight, low combat rear; Chip McCormick Corp. forged frame and slide; match barrel; extended combat thumb safety; high beavertail grip safety; skeletonized lightweight composite trigger; skeletonized Commander-type hammer; elongated Commander ejector; bead-blasted black oxide finish; flat mainspring housing; short guide rod; lowered and flared ejection port; serrated front and rear of slide; relief cut under trigger guard; Wolff spring set; beveled magazine well. Made in U.S. by Kimber of America, Inc. Introduced 1995; still produced. **LMSR:** $575
New: $550   **Perf.:** $500   **Exc.:** $425
Custom stainless
New: $600   **Perf.:** $550   **Exc.:** $450

Kimber Classic 45 Gold Match
Same specs as Custom except long guide rod; polished blue finish; Bo-Mar BMCS low-mount adjustable rear sight; fancy walnut grips; tighter tolerances; comes with one 10-shot and one 8-shot magazine, factory proof target. Made in U.S. by Kimber of America, Inc. Introduced 1995; still produced.
LMSR: $925
New: $800   **Perf.:** $750   **Exc.:** $650

Kimber Classic 45 Royal
Same specs as the Custom model except has checkered diamond-pattern walnut grips; long guide rod; polished blue finish; comes with two 8-shot magazines. Made in U.S. by Kimber of America, Inc. Introduced 1995; still produced. **LMSR:** $715
New: $650   **Perf.:** $600   **Exc.:** $500

KIMBER PREDATOR HUNTER
Single-shot; 221 Fireball, 223 Rem., 6mm TCU, 6x45, 7mm TCU; 15 3/4" barrel; weighs 88 oz.; no sights; accepts Kimber scope mount system; AA Claro walnut stock; uses Kimber Model 84 mini-Mauser action. Introduced 1987; dropped 1989.
Perf.: $1400   **VGood:** $1150   **Good:** $850

Kimber Predator Supergrade
Same specs as Predator Hunter except French walnut stock; ebony forend tip; 22-line hand checkering. Introduced 1987; dropped 1989.
Perf.: $2000   **VGood:** $1600   **Good:** $1200

KIMEL AP9
Semi-automatic; single action; 9mm Para.; 20-shot magazine; 5" barrel; 11 7/8" overall length; weighs 3 1/2 lbs.; checkered plastic grip; adjustable post front sight in ring, fixed open rear; matte blue/black or nickel finish; fires from closed bolt. Made in U.S. Was available from Kimel Industries. Introduced 1988; no longer produced.
Perf.: $300   **Exc.:** $250   **VGood:** $175

Kimel AP9 Mini
Same specs as the AP9 except 3" barrel. Made in U.S. Was available from Kimel Industries. Introduced 1988; no longer produced.
Perf.: $325   **Exc.:** $275   **VGood:** $225

Kimel AP9 Target
Same specs as the AP9 except 12" barrel; grooved forend. Made in U.S. Was available from Kimel Industries. Introduced 1988; no longer produced.
Perf.: $250   **Exc.:** $200   **VGood:** $175

KLEINGUENTHER R-15
Revolver; double action; 22 LR, 22 WMR, 32 S&W; 6-shot cylinder; 6" barrel; 11" overall length; checkered thumbrest walnut stocks; adjustable rear sight, fixed front; full-length solid barrel rib; adjustable trigger; blued finish. Manufactured in Germany. Introduced 1976; dropped 1978.
Exc.: $200   **VGood:** $175   **Good:** $150

KORRIPHILA HSP 701
Semi-automatic; double action; 9mm Para., 38 Wadcutter, 38 Super, 45 ACP; 9-shot magazine (9mm), 7-shot (45 ACP); 4", 5" barrel; weighs 35 oz.; adjustable rear sight, ramp or target front; checkered walnut stocks; delayed roller lock action; limited production. Made in West Germany. Introduced 1986; importation dropped 1989.
Perf.: $1500   **Exc.:** $1350   **VGood:** $1200

KORTH SEMI-AUTOMATIC
Semi-automatic; double-action; 9mm Para., 9x21; 10-shot; 4", 5", 6" barrel; 10 1/2" overall length; weighs 35 oz.; forged, machined frame and slide; adjustable combat sights; checkered walnut stocks; matte or polished finish. Introduced 1985; dropped 1989.
Perf.: $2750   **Exc.:** $2500   **VGood:** $2000

Kimber Predator Supergrade

Korth Semi-Automatic

L.A.R. Grizzly Win. Mag. Mark I

Lahti Finnish Model L-35

Korth Revolver

Lahti Swedish Model 40

HANDGUNS

Korth
Revolver

Lahti
Finnish Model L-35
Swedish Model 40

L.A.R.
Grizzly Win. Mag. Mark I
Grizzly Win. Mag. Mark I, 8″, 10″
Grizzly Win. Mag. Mark II
Grizzly 44 Mag. Mark IV
Grizzly 50 Mark V

Laseraim Arms
Series I
Series II

KORTH REVOLVER

Revolver; double action; 22 LR, 22 Mag., 32 H&R Mag., 32 S&W Long, 357 Mag., 9mm Para.; 3″, 4″, 6″ barrel; 8″ to 11″ overall length; weighs 33 to 38 oz.; checkered walnut sport or combat grips; blade front sight, fully-adjustable rear; four interchangeable cylinders available, comes with two; high polish blue finish; presentation models have gold trim. Imported from Germany by Mandall Shooting Supplies. No longer produced.

Perf.: $2500   **Exc.:** $2200   **VGood:** $1850

LAHTI FINNISH MODEL L-35

Semi-automatic; 9mm Para.; 8-shot magazine; 4 3/4″ barrel; fixed sights; checkered plastic stocks; blued finish. Manufactured on a limited basis in Finland from 1935 to 1954.

Exc.: $1200   **VGood:** $950   **Good:** $800

LAHTI SWEDISH MODEL 40

Semi-automatic; 9mm Para.; 8-shot magazine; 4 3/4″ barrel; fixed sights; checkered plastic stocks; blued finish. Manufactured in Sweden by Husqvarna. Manufactured 1942 to 1946.

Exc.: $375   **VGood:** $300   **Good:** $250

L.A.R. GRIZZLY WIN. MAG. MARK I

Semi-automatic; single action; 30 Mauser, 357 Mag., 357/45 Grizzly Win. Mag., 10mm, 45 ACP, 45 Win. Mag.; 7-shot magazine; 5 1/2″, 6 1/2″ barrel; 10 5/8″ overall length; weighs 51 oz.; ramped blade front, adjustable rear sight; no-slip rubber combat grips; ambidextrous safeties; conversion units 45 to 357 Mag., 45 ACP, 10mm, 45 Win. Mag. available; phosphated finish. Introduced 1984; still in production. **LMSR:** $920

Perf.: $650   **Exc.:** $550   **VGood:** $475
Hard chrome or nickel finish
Perf.: $750   **Exc.:** $650   **VGood:** $600

L.A.R. Grizzly Win. Mag. Mark I, 8″, 10″

Same specs as Mark I standard except 8″, 10″ barrel; lengthened slide. Introduced 1984; still in production. **LMSR:** $1313

Perf.: $1200   **Exc.:** $1000   **VGood:** $900
10″ barrel
Perf.: $1250   **Exc.:** $1100   **VGood:** $950

L.A.R. Grizzly 50 Mark V

L.A.R. Grizzly Win. Mag. Mark II

Same specs as the Grizzly Win. Mag. Mark I except fixed sights; standard safeties. Introduced 1986; discontinued 1986.

Perf.: $550   **Exc.:** $450   **VGood:** $400

L.A.R. GRIZZLY 44 MAG. MARK IV

Semi-automatic; single action; 44 Mag.; 7-shot magazine; 5 1/2″, 6 1/2″ barrel; 10 5/8″ overall length; weighs 51 oz.; ramped blade front, adjustable rear sight; no-slip rubber combat grips; beavertail grip safety; matte blue, hard chrome, chrome or nickel finish. Introduced 1991; still in production. **LMSR:** $933

Perf.: $750   **Exc.:** $650   **VGood:** $575
Hand chrome, chrome or nickel finish
Perf.: $850   **Exc.:** $750   **VGood:** $650

L.A.R. GRIZZLY 50 MARK V

Semi-automatic; single action; 50 AE; 6-shot magazine; 5 1/2″, 6 1/2″ barrel; 10 5/8″ overall length; weighs 56 oz.; ramped blade front, adjustable rear sight; no-slip rubber combat grips; ambidextrous safeties; conversion units available; phosphated finish. Made in U.S. by L.A.R. Mfg., Inc. Introduced, 1993; still in production. **LMSR:** $1060

New: $900   **Perf.:** $850   **Exc.:** $750

LASERAIM ARMS SERIES I

Semi-automatic; single action; 10mm Auto, 45 ACP; 8-shot magazine (10mm), 7-shot (45 ACP); 4 3/8″, 6″ barrel, with compensator; 9 3/4″ overall length (6″ barrel); weighs 46 oz.; pebble-grained black composite grips; blade front sight, fully-adjustable rear; barrel compensator; stainless steel construction; ambidextrous safety-levers; extended slide release; integral mount for laser sight; matte black Teflon finish. Made in U.S. by Emerging Technologies, Inc. Introduced 1993; still produced. **LMSR:** $553

New: $400   **Perf.:** $350   **Exc.:** $300
With adjustable sight
New: $450   **Perf.:** $400   **Exc.:** $350
With fixed sight and Auto Illusion red dot sight system
New: $650   **Perf.:** $600   **Exc.:** $550
With fixed sight and Laseraim Laser with Hotdot
New: $750   **Perf.:** $700   **Exc.:** $650

LASERAIM ARMS SERIES II

Semi-automatic; single action; 10mm Auto, 45 ACP, 40 S&W; 8-shot magazine (10mm), 7-shot (45 ACP); 3 3/8″, 5″ barrel without compensator; weighs 43 oz. (5″), 37 oz. (3 3/8″); pebble-grained black composite grips; blade front sight, fixed or windage-adjustable rear; stainless steel construction; ambidextrous safety-levers; extended slide release; integral mount for laser sight; matte stainless finish. Made in U.S. by Emerging Technologies, Inc. Introduced 1993; still produced. **LMSR:** $400

New: $450   **Perf.:** $400   **Exc.:** $350
With adjustable sight (5″ barrel)
New: $475   **Perf.:** $425   **Exc.:** $375
With fixed sight and Auto Illusion red dot sight
New: $500   **Perf.:** $425   **Exc.:** $400
With fixed sight and Laseraim Laser with Hotdot
New: $550   **Perf.:** $475   **Exc.:** $450

Laseraim Arms
Series III
Series IV

Le Francais
Army Model
Pocket Model
Policeman Model

Les Incorporated
Rogak P-18

Liberty
Mustang

Lignose
Model 2 Pocket
Model 2A Einhand
Model 3 Pocket
Model 3A Einhand

Liliput
4.25
25 ACP

Ljutic
LJ II

Liberty Mustang

Les Incorporated
Rogak P-18

Le Francais
Pocket Model

Le Francais
Policeman Model

Le Francais
Army Model

Lignose Model 3A
Einhand

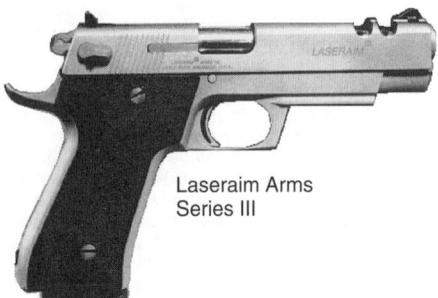

Laseraim Arms
Series III

LASERAIM ARMS SERIES III
Semi-automatic; single action; 10mm Auto, 45 ACP; 8-shot magazine (10mm), 7-shot (45 ACP); 5" barrel with dual-port compensator; 7 5/8" overall length; weighs 43 oz.; pebble-grained black composite grips; blade front sight, fixed or windage-adjustable rear; stainless steel construction; ambidextrous safety-levers; extended slide release; integral mount for laser sight; matte stainless finish. Made in U.S. by Emerging Technologies, Inc. Introduced 1994; still produced. **LMSR: $534**
With fixed sight
   New: $500   **Perf.:** $450   **Exc.:** $400
With adjustable sight
   New: $525   **Perf.:** $475   **Exc.:** $425
With fixed sight and Dream Team Laseraim laser sight
   New: $600   **Perf.:** $550   **Exc.:** $500

LASERAIM ARMS SERIES IV
Semi-Automatic; 45 ACP; 7-shot magazine; 3 3/8", 5" barrel; weighs 37 oz.; full serrated slide; diamond wood grips; stainless steel construction; blade front, fully-adjustable rear sight; ambidextrous safety levers; integral mount for laser sight. Made in U.S. by Emerging Technologies, Inc. Introduced 1996.

LE FRANCAIS ARMY MODEL
Semi-automatic; double action; 9mm Browning Long; 8-shot magazine; 5" flip up barrel; 7 3/4" overall length; fixed sights; checkered European walnut stocks; blued finish. Manufactured 1928 to 1938.
   Exc.: $1200   **VGood:** $1000   **Good:** $700

LE FRANCAIS POCKET MODEL
Semi-automatic; double action; 32 ACP; 7-shot magazine; 3 1/2" hinged barrel; 6" overall length; fixed sights; checkered hard rubber stocks; blued finish. Manufactured in France. Introduced 1950; discontinued 1965.
   Exc.: $650   **VGood:** $500   **Good:** $350

LE FRANCAIS POLICEMAN MODEL
Semi-automatic; double action; 25 ACP; 7-shot magazine; 2 1/2", 3 1/2" flip up barrel; fixed sights; hard rubber stocks; blued finish. Introduced 1914; discontinued about 1960.
   Exc.: $250   **VGood:** $225   **Good:** $185

LES INCORPORATED ROGAK P-18
Semi-automatic; double action; 9mm Para.; 18-shot magazine; 5 1/2" barrel; post front sight, V-notch rear, drift adjustable for windage; checkered resin stocks; stainless steel; matte or deluxe high gloss finishes. Introduced 1977; discontinued 1981.
   Exc.: $400   **VGood:** $350   **Good:** $300
Deluxe high gloss finish
   Exc.: $450   **VGood:** $400   **Good:** $350

LIBERTY MUSTANG
Revolver; single action; 22 LR, 22 WMR or combo; 8-shot cylinder; 5" barrel; 10 1/2" overall length; weighs 34 oz.; blade front, adjustable rear sight; smooth rosewood grips; side ejector rod; blued finish. Imported from Italy by Liberty. Introduced 1976; discontinued 1980.
   Exc.: $150   **VGood:** $135   **Good:** $110

LIGNOSE MODEL 2 POCKET
Semi-automatic; 25 ACP; 6-shot magazine; 2" barrel; 4 1/2" overall length; checkered hard rubber stocks; blued finish. Manufactured in Germany from 1920 under Bergmann name to late 1920s.
   Exc.: $200   **VGood:** $175   **Good:** $150

Lignose Model 2A Einhand
Same specs as Model 2 Pocket except trigger-type mechanism at front of triggerguard allows slide retraction with trigger finger.
   Exc.: $350   **VGood:** $250   **Good:** $200

LIGNOSE MODEL 3 POCKET
Semi-automatic; 25 ACP; 9-shot magazine; 2" barrel; 4 1/2" overall length; checkered hard rubber stocks; blued finish. Manufactured in Germany.
   Exc.: $225   **VGood:** $200   **Good:** $150

Lignose Model 3A Einhand
Same specs as Model 3 Pocket except mechanism at front of triggerguard allows slide retraction with trigger finger.
   Exc.: $350   **VGood:** $300   **Good:** $250

LILIPUT 4.25
Semi-automatic; 4.25mm; 6-shot; 1 13/16" barrel; 3 1/2" overall length; weighs 8 oz.; blue or nickel finish. Made by August Menz, Suhl, Germany. Introduced 1920.
   Exc.: $800   **VGood:** $650   **Good:** $550

LILIPUT 25 ACP
Semi-automatic; 25 ACP; 6-shot; 2" barrel; 4 1/8" overall length; weighs 10 oz.; blue or nickel finish. Made by August Menz, Suhl Germany. Introduced 1925.
   Exc.: $250   **VGood:** $200   **Good:** $150

LJUTIC LJ II
Derringer; double action; 22 WMR; 2-shot; 2 3/4" side-by-side barrels; fixed sights; checkered walnut stocks; vent rib; positive safety; stainless steel construction. Introduced 1981; dropped 1989. About 1000 produced.
   Perf.: $1500   **Exc.:** $1200   **VGood:** $1000

Llama IIIA Small Frame

Llama IX-C
New Generation
Large Model

Llama VIII

Llama XI Large
Frame Model

Liliput 4.25

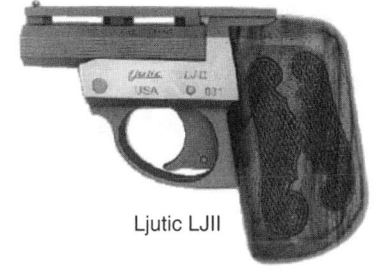
Ljutic LJII

HANDGUNS

Ljutic
Space Pistol

Llama
IIIA Small Frame
VIII
IX Large Frame
IX-A Large Frame
IX-B Compact Frame
IX-C New Generation Large Frame
IX-D New Generation Compact Frame
XA Small Frame
XI

LJUTIC SPACE PISTOL

Single shot; bolt action; 22 WMR, 357 Mag., 44 Mag., 308 Win.; 13 ¹/₂″ barrel; scope mounts; American walnut grip, forend; button trigger; Introduced 1981; prototype only.

LLAMA IIIA SMALL FRAME

Semi-automatic; single action; 380 ACP, 32 ACP (disc. 1993); 7-shot magazine; 3 ¹¹/₁₆″ barrel; 6 ¹/₂″ overall length; weighs 23 oz.; adjustable target sights; checkered thumbrest plastic grips; vent rib; grip safety; loaded chamber indicator; blued, chrome engraved, blue engraved or gold engraved finishes. Early versions were sans vent rib, had lanyard ring, no thumbrest on grips. Introduced 1951; still in production. **LMSR: $259**
Blued finish

Perf.: $250	**Exc.:** $200	**VGood:** $175

Chrome engraved

Perf.: $275	**Exc.:** $225	**VGood:** $200

Blue engraved

Perf.: $275	**Exc.:** $225	**VGood:** $200

Gold engraved

Perf.: $500	**Exc.:** $450	**VGood:** $400

LLAMA VIII

Semi-automatic; 38 Super; 9-shot magazine; 5″ barrel; 8 ¹/₂″ overall length; fixed sights; hand-checkered walnut grips; vent rib; grip safety; blued, chrome, chrome engraved or blued engraved finishes. Imported by Stoeger. Introduced 1952; no longer in production.

Perf.: $275	**Exc.:** $250	**VGood:** $225

Chrome finish

Perf.: $300	**Exc.:** $275	**VGood:** $250

Chrome engraved

Perf.: $325	**Exc.:** $300	**VGood:** $275

Blue engraved

Perf.: $300	**Exc.:** $250	**VGood:** $225

LLAMA IX LARGE FRAME

Semi-automatic; single action; 9mm, 38 Super, 45 ACP; 9-shot magazine (9mm, 38 Super), 7-shot (45 ACP); 5 ¹/₈″ barrel; 41 oz.; adjustable sights; black plastic grips; blued or satin chrome finish. Introduced 1936; no longer imported.

Perf.: $325	**Exc.:** $275	**VGood:** $225

Satin chrome finish

Perf.: $350	**Exc.:** $300	**VGood:** $250

LLAMA IX-A LARGE FRAME

Semi-automatic; 45 ACP; 7-shot magazine; 5 ¹/₈″ barrel; 8 ¹/₂″ overall length; weighs 36 oz.; hand-checkered walnut grips; vent rib; fixed sights; grip safety; loaded chamber indicator; blue, chrome, chrome engraved or blue engraved finishes. Introduced 1952; still imported as IX-C New Generation Large Frame.

Perf.: $375	**Exc.:** $325	**VGood:** $275

Chrome finish

Perf.: $400	**Exc.:** $350	**VGood:** $300

Chrome engraved

Perf.: $450	**Exc.:** $400	**VGood:** $350

Blue engraved

Perf.: $400	**Exc.:** $350	**VGood:** $300

LLAMA IX-B COMPACT FRAME

Semi-automatic; single action; 45 ACP; 7-shot magazine; 4 ¹/₄″ barrel; 7 ⁷/₈″ overall length; weighs 34 oz.; checkered polymer grips; blade front sight, fully-adjustable rear; scaled-down version of Llama Large Frame; locked breech mechanism; manual and grip safeties; blue or chrome finish. Imported from Spain by SGS Importers Int'l., Inc. Introduced 1985; still imported as IX-D model.

Perf.: $300	**Exc.:** $250	**VGood:** $225

Chrome finish

Perf.: $325	**Exc.:** $275	**VGood:** $250

LLAMA IX-C NEW GENERATION LARGE FRAME

Semi-automatic; single action; 45 ACP; 13-shot magazine; 5 ¹/₈″ barrel; 8 ¹/₂″ overall length; 41 oz.; three-dot combat sights; military-style hammer; loaded chamber indicator; anatomically designed rubber grips; non-glare matte finish. Introduced 1994; Converted to 10-shot magazine 1995; still in production. **LMSR: $400**

New: $350	**Perf.:** $300	**Exc.:** $250

LLAMA IX-D NEW GENERATION COMPACT FRAME

Semi-automatic; single action; 45 ACP; 13-shot; 4 ¹/₄″ barrel; 7 ⁷/₈″ overall length; weighs 39 oz.; three-dot combat sights; rubber grips; non-glare matte finish. Converted to 10-shot 1995; still in production. **LMSR: $400**

New: $350	**Perf.:** $300	**Exc.:** $250

LLAMA XA SMALL FRAME

Semi-automatic; 32 ACP; 7-shot magazine; 3 ¹¹/₁₆″ barrel; 6 ¹/₂″ overall length; adjustable target sights; checkered thumbrest plastic grips; grip safety; blued, chrome engraved or blue engraved finishes; successor to Llama X which had no grip safety. Imported by Stoeger. Introduced 1951; no longer in production.

Exc.: $200	**VGood:** $175	**Good:** $150

Chrome engraved

Exc.: $225	**VGood:** $200	**Good:** $150

Blue engraved

Exc.: $225	**VGood:** $200	**Good:** $175

LLAMA XI

Semi-automatic; 9mm Para.; 8-shot magazine; 5″ barrel; 8 ¹/₂″ overall length; adjustable sights; checkered thumbrest; plastic grips; vent rib; blued, chrome, chrome engraved or blued engraved finishes. Imported by Stoeger Arms. Introduced 1954; no longer in production.

Perf.: $275	**Exc.:** $250	**VGood:** $200

Chrome finish

Perf.: $300	**Exc.:** $275	**VGood:** $250

Chrome engraved

Perf.: $325	**Exc.:** $300	**VGood:** $275

Blue engraved

Perf.: $325	**Exc.:** $300	**VGood:** $275

MODERN GUN VALUES, 11TH EDITION **173**

Llama
XV Small Frame
Comanche
Comanche I
Comanche II
Comanche III
Martial
Martial Deluxe
Max-1 New Generation Large Frame
Max-1 New Generation Compact Frame
Model 82
Model 87 Competition Model

Llama Model 82

Llama Martial

Llama Martial Deluxe

Llama Medium Frame Model

Llama Comanche

Llama XV Small Frame

LLAMA XV SMALL FRAME
Semi-automatic; 22 LR; 8-shot magazine; 3 11/16″ barrel; 6 1/2″ overall length; weighs 23 oz.; adjustable target sights; checkered thumbrest plastic grips; vent rib; grip safety; blued, chrome engraved or blue engraved finishes. Introduced 1951; no longer in production.
Perf.: $225   **Exc.:** $200   **VGood:** $175
Chrome engraved
Perf.: $250   **Exc.:** $225   **VGood:** $200
Blue engraved
Perf.: $250   **Exc.:** $225   **VGood:** $200

LLAMA COMANCHE
Revolver; double action; 22 LR, 357 Mag.; 6-shot cylinder; 4″, 6″ barrel; 9 1/4″ overall length (4″ barrel); weighs 28 oz.; blade front sight, fully-adjustable rear; checkered walnut stocks; ventilated rib; wide spur hammer; blue finish. Was imported from Spain by SGS Importers International, Inc. No longer imported.
Exc.: $200   **VGood:** $175   **Good:** $150

LLAMA COMANCHE I
Revolver; double action; 22 LR; 6-shot cylinder; 6″ barrel; 11 1/4″ overall length; target-type sights; checkered walnut stocks; blued finish. Introduced in 1977, replacing Martial 22 model; discontinued 1982.
Exc.: $200   **VGood:** $175   **Good:** $150

Llama Comanche II
Same specs as Comanche I except 38 Spl.; 4″ barrel. Introduced 1977; discontinued 1982.
Exc.: $175   **VGood:** $150   **Good:** $135

Llama Comanche III
Same specs as Comanche I except 22 LR, 357 Mag.; 4″, 6″, 8 1/2″ barrel; 9 1/4″ overall length; ramp front, adjustable rear sight; blued, satin chrome or gold finish. Introduced in 1977 as "Comanche"; renamed 1977; no longer in production.
Perf.: $250   **Exc.:** $225   **VGood:** $200
Satin chrome finish
Perf.: $275   **Exc.:** $250   **VGood:** $225
Gold finish
Perf.: $750   **Exc.:** $650   **VGood:** $500

LLAMA MARTIAL
Revolver; double action; 22 LR, 38 Spl.; 6-shot cylinder; 4″ barrel (38 Spl.), 6″ (22 LR); 11 1/4″ overall length (6″ barrel); target sights; hand-checkered walnut grips; blued finish. Imported by Stoeger. Introduced 1969; discontinued 1976.
Exc.: $200   **VGood:** $175   **Good:** $150

Llama Martial Deluxe
Same specs as Martial model except choice of satin chrome, chrome engraved, blued engraved, gold engraved finishes; simulated pearl stocks.
Satin chrome finish
Exc.: $250   **VGood:** $200   **Good:** $150
Chrome engraved
Exc.: $275   **VGood:** $225   **Good:** $175
Blue engraved
Exc.: $275   **VGood:** $225   **Good:** $175
Gold engraved
Exc.: $750   **VGood:** $600   **Good:** $500

LLAMA MAX-I NEW GENERATION LARGE FRAME
Semi-automatic; single action; 9mm Para., 45 ACP; 9-shot magazine (9mm), 7-shot (45 ACP); 5 1/8″ barrel; 8 1/2″ overall length; weighs 36 oz.; blade front sight, windage-adjustable rear; black rubber grips; three-dot system; skeletonized combat-style hammer; steel frame; extended manual and grip safeties; blue or duo-tone finish. Imported from Spain by SGS Importers, International. Introduced 1995; still imported. **LMSR: $325**
New: $300   **Perf.:** $275   **Exc.:** $225
Duo-tone finish
New: $350   **Perf.:** $300   **Exc.:** $250

Llama Max-I New Generation Compact Frame
Same specs as the Max-I except 7-shot; 4 1/4″ barrel; 7 7/8″ overall length; weighs 34 oz. Imported from Spain by SGS Importers, International. Introduced 1995; still imported. **LMSR: $325**
New: $300   **Perf.:** $275   **Exc.:** $225
Duo-tone finish
New: $350   **Perf.:** $325   **Exc.:** $300

LLAMA MODEL 82
Semi-automatic; double action; 9mm Para.; 15-shot magazine; 4 1/4″ barrel; 8″ overall length; weighs 39 oz.; blade-type front sight, drift-adjustable rear; 3-dot system; matte black polymer stocks; ambidextrous safety; blued finish. Made in Spain. Introduced 1987; discontinued 1993.
Perf.: $650   **Exc.:** $600   **VGood:** $550

LLAMA MODEL 87 COMPETITION MODEL
Semi-automatic; 9mm Para.; 14-shot magazine; 6″ barrel; 9 1/2″ overall length; weighs 47 oz.; Patridge-type front sight, fully-adjustable rear; Polymer composition stocks; built-in ported compensator; oversize magazine, safety releases; fixed barrel bushing; extended triggerguard; beveled magazine well. Made in Spain. Introduced 1989; discontinued 1993.
Perf.: $900   **Exc.:** $750   **VGood:** $600

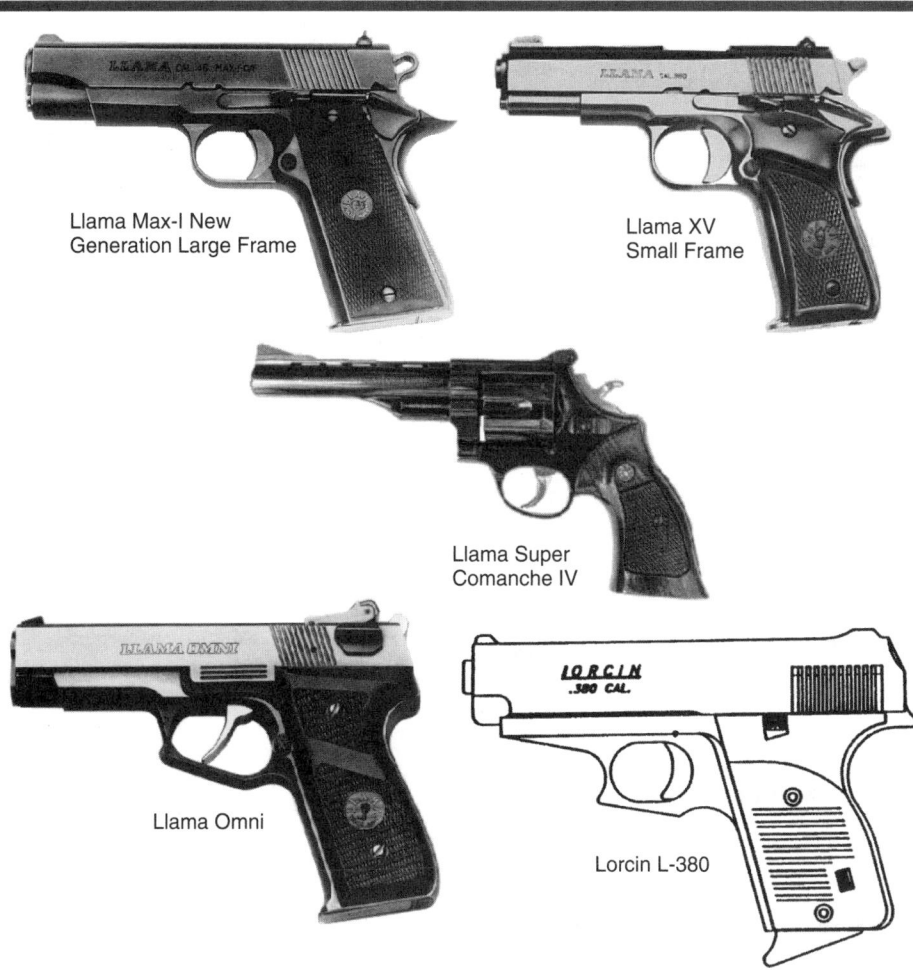

Llama Max-I New
Generation Large Frame

Llama XV
Small Frame

Llama Super
Comanche IV

Llama Omni

Lorcin L-380

HANDGUNS

Llama
Omni
Super Comanche IV
Super Comanche V

Lorcin
L-9mm
L-22
L-25
L-32
L-380
LH-380
LT-25

Lorcin L-9mm

Lorcin L-22

Lorcin L-25

LLAMA OMNI

Semi-automatic; double action; 9mm Para., 45 ACP; 13-shot magazine (9mm), 7-shot (45 ACP); 4 1/4" barrel; 7 3/4" overall length (45 ACP), 8" (9mm); weighs 40 oz.; adjustable sights; checkered plastic stocks; ball-bearing action; double sear bars; articulated firing pin; low-friction rifling; blued finish. Made in Spain. Introduced 1982; dropped 1986.

   Exc.: $350   **VGood:** $300   **Good:** $275

LLAMA SUPER COMANCHE IV

Revolver; double action; 44 Mag.; 6-shot cylinder; 6", 8 1/2" barrel; adjustable sights; oversize walnut grips; wide hammer, trigger; blued finish. No longer imported.

   Perf.: $325   **Exc.:** $275   **VGood:** $250

Llama Super Comanche V

Same specs as Super Comanche IV except 357 Mag. only; 4" barrel also available. Discontinued 1988.

   Perf.: $300   **Exc.:** $250   **VGood:** $200

LORCIN L-9MM

Semi-automatic; single action; 9mm Para.; 10-shot magazine; 4 1/2" barrel; 7 1/2" overall length; weighs 31 oz.; grooved black composition grips; fixed sights with three-dot system; matte black finish; hooked triggerguard; grip safety. Made in U.S. by Lorcin Engineering. Introduced 1994; still in production. **LMSR: $159**

   New: $140   **Perf.:** $125   **Exc.:** $110

LORCIN L-22

Semi-automatic; single action; 22 LR; 9-shot magazine; 2 1/2" barrel; 5 1/4" overall length; weighs 16 oz.; black combat, or pink or pearl grips; fixed three-dot sight system; chrome or black Teflon finish. From Lorcin Engineering. Introduced 1989; still in production. **LMSR: $89**

   Perf.: $65   **Exc.:** $50   **VGood:** $40
Pink grips
   Perf.: $50   **Exc.:** $40   **VGood:** $30

LORCIN L-25

Semi-automatic; single action; 25 ACP; 7-shot magazine; 2 5/16" barrel; 4 5/8" overall length; weighs 14 1/2 oz.; fixed sights; smooth composition stocks; black/gold, chrome/satin chrome, black finish. Introduced 1989; still in production. **LMSR: $69**

   Perf.: $50   **Exc.:** $40   **VGood:** $30

LORCIN L-32

Semi-automatic; single action; 32 ACP; 7-shot magazine; 3 1/2" barrel; 6 5/8" overall length; weighs 23 oz.; fixed sights; grooved composition grips; black Teflon or chrome finish with black grips. Made in U.S. by Lorcin Engineering. Introduced 1992; still in production. **LMSR: $89**

   New: $70   **Perf.:** $65   **Exc.:** $50

LORCIN L-380

Semi-automatic; single action; 380 ACP; 7-shot magazine; 3 1/2" barrel; 6 5/8" overall length; weighs 23 oz.; fixed sights; grooved composition grips; black Teflon or chrome finish with black grips. Made in U.S. by Lorcin Engineering. Introduced 1992; still in production. **LMSR: $100**

   New: $85   **Perf.:** $65   **Exc.:** $50

LORCIN LH-380

Semi-automatic; 380 ACP; 10-shot magazine; 4 1/2" barrel; 7 1/2" overall length; weighs 31 oz.; grooved black compostion grips; fixed sights with three-dot system; matte black finish; hooked triggerguard; grip safety. Made in U.S. by Lorcin Engineering. Introduced 1994; still in production. **LMSR: $159**

   New: $125   **Perf.:** $100   **Exc.:** $75

LORCIN LT-25

Semi-automatic; single action; 25 ACP; 7-shot magazine; 2 1/3" barrel; 4 2/3" overall length; weighs 14 1/2 oz.; fixed sights; smooth composition grips; available in chrome, black Teflon or camouflage. Made in U.S. by Lorcin Engineering. Introduced 1989; still in production. **LMSR: $79**

   New: $60   **Perf.:** $50   **Exc.:** $40

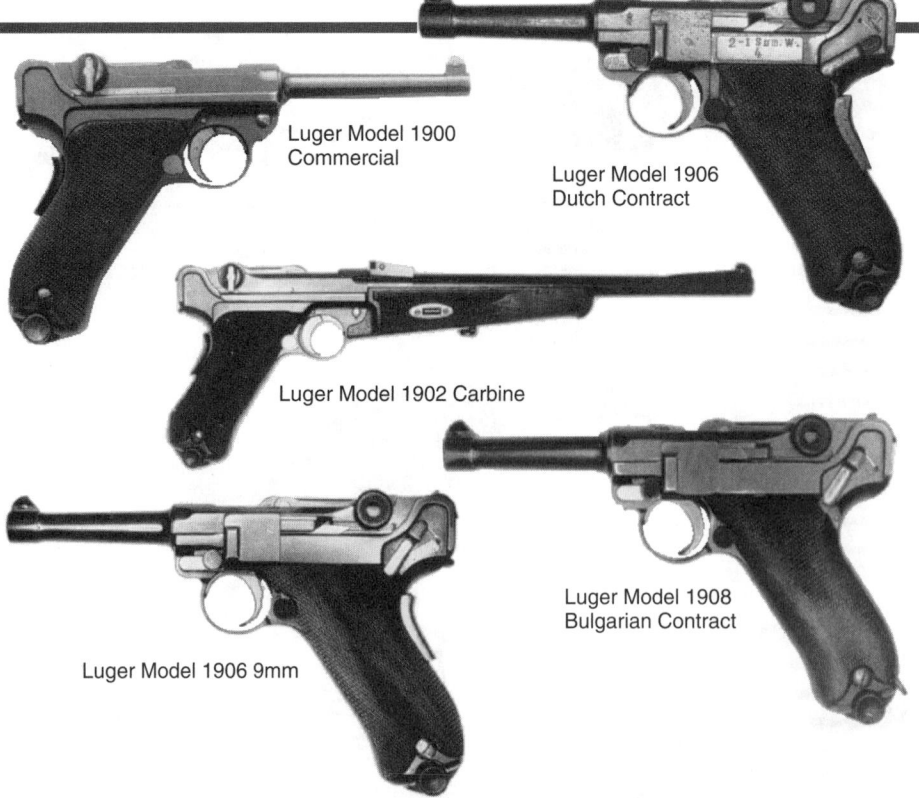

Luger Model 1900 Commercial

Luger Model 1906 Dutch Contract

Luger Model 1902 Carbine

Luger Model 1906 9mm

Luger Model 1908 Bulgarian Contract

HANDGUNS

Luger
Model 1900 Commercial
Model 1900 Swiss
Model 1900 Eagle
Model 1902
Model 1902 Carbine
Model 1906 Navy
Model 1906
Model 1908
Model 1914 Military
Model 1914 Artillery

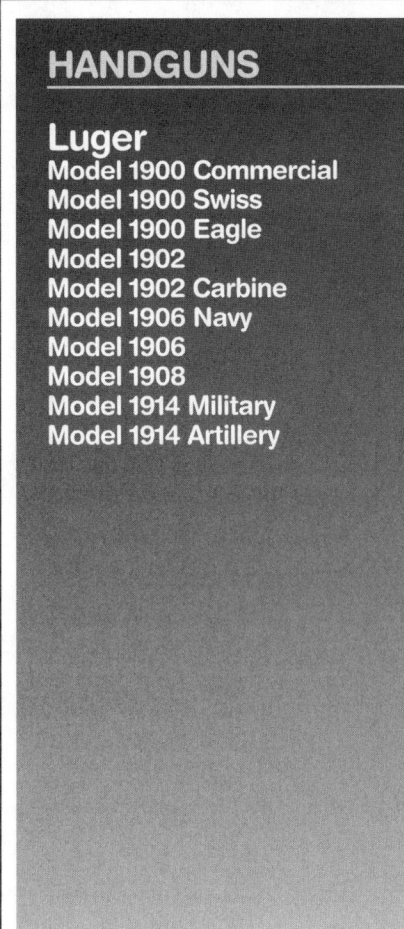

LUGER

The Luger pistol was redesigned from the 1893 Borchardt pistol by George Luger and came on the market in 1900. Its use by the German military as well as many other governments and its many variations have made it the most popular semi-automatic pistol for collectors, and it is very well regarded by shooters as well. Lugers have been widely collected and very thoroughly researched, and there are many other rare variations in addition to those listed below. To identify a Luger that does not seem to agree with those described, refer to one of the major Luger references such as *Lugers At Random* by Charles Kenyon.

Joe Schroeder

LUGER MODEL 1900 COMMERCIAL

Semi-automatic; 7.65mm Para.; 7-shot magazine; 4 3/4″ barrel; 9 1/4″ overall length; fixed sights; toggle-joint action; grip safety; checkered wood grips; blued finish. Toggle marked in script "DWM" for Deutsche Waffen und Munitionsfabrik. Distinguished by dished toggle knobs with a lock, recessed breechblock.
Exc.: $2750   **VGood:** $2250   **Good:** $1500

Luger Model 1900 Swiss

Same specs as Model 1900 Comercial except Swiss cross on chamber.
Exc.: $3750   **VGood:** $2750   **Good:** $2000

Luger Model 1900 Eagle

Same specs as Model 1900 Commercial except American eagle on chamber.
Exc.: $3500   **VGood:** $2750   **Good:** $1500
U.S. Army test guns s/n 6000-7000 distinguished by no proof marks
Exc.: $4000   **VGood:** $3250   **Good:** $2250

LUGER MODEL 1902

Semi-automatic; 9mm Para.; 7-shot magazine; 4″ fat barrel; fixed sights; toggle-joint action; grip safety; checkered walnut grips; blued finish. Made with or without American eagle. Toggle marked in script "DWM" for Deutsche Waffen und Munitionsfabrik. Distinguished by fat 9mm barrel, dished toggle knobs with a lock, recessed breechblock.
Exc.: $8500   **VGood:** $7000   **Good:** $4500

Luger Model 1902 Carbine

Same specs as Model 1902 except 7.65mm Para.; 11 3/4″ barrel with adjustable rear sight and checkered wood forend; contoured detachable wood stock.
Exc.: $14,000 **VGood:** $10,000 **Good:** $7500
Without stock
Exc.: $9500   **VGood:** $7000   **Good:** $5000

LUGER MODEL 1906 NAVY

Semi-automatic; 9mm Para.; 6″ barrel; flat checkered toggle knobs; exposed breechblock with extractor that protrudes when chamber is loaded and exposes German word "GELADEN" (LOADED); adjustable rear sight on rear toggle link; coil instead of leaf recoil spring; grip safety; stock lug on rear of frame. Usually called "Model 1906 Navy" by collectors, but was officially designated Model 1904 by the German navy. Note: Original 1904 Navies that have a toggle lock cut into right toggle knob like Model 1900 are very rare and valuable but beware of fakes.
Exc.: $3000   **VGood:** $2500   **Good:** $1500
With proper Navy stock
Exc.: $3750   **VGood:** $3250   **Good:** $2000

LUGER MODEL 1906

Semi-automatic; 7.65mm Para., 9mm Para.; 4 3/4″ (7.65), 4″ (9mm) barrel; flat checkered toggle knobs; exposed breechblock with extractor that protrudes when chamber is loaded and exposes German word "GELADEN" (LOADED); fixed rear sight on rear toggle link; coil instead of leaf recoil spring; no stock lug on rear frame. Note: 1906 Russian (crossed rifles on chamber) and Bulgarian (Bulgarian crest on chamber) very rare, very valuable.
With or without American Eagle, 7.65mm
Exc.: $2000   **VGood:** $1500   **Good:** $1000
With or without American Eagle, 9mm
Exc.: $2500   **VGood:** $1850   **Good:** $1250
With Swiss cross, 7.65mm only
Exc.: $3000   **VGood:** $2500   **Good:** $1500

Portuguese contract with M2 on chamber, 7.65mm only
Exc.: $1500   **VGood:** $1250   **Good:** $850
Brazilian contract with no chamber marking, extractor marked "CARREGADA," 7.65mm only
Exc.: $1250   **VGood:** $900   **Good:** $750
Dutch contract with safety marked "RUST" with arrow, 9mm only.
Exc.: $1750   **VGood:** $1400   **Good:** $1200
Dutch Vickers made for the Dutch during WWI by Vickers in England.
Exc.: $2750   **VGood:** $2200   **Good:** $1700

LUGER MODEL 1908

Semi-automatic; 9mm Para.; 4″ barrel; flat checkered toggle knobs; no grip safety; no hold-open device on early models, added later to many 1908 militaries; fixed rear sight on rear toggle link; coil instead of leaf recoil spring; no stock lug on rear frame.
Commercial
5-digit s/n and commercial proofs
Exc.: $1000   **VGood:** $750   **Good:** $500
Army (Military proofs, chamber date from 1910 on, "DWM" or "Erfurt" on toggle)
Exc.: $1250   **VGood:** $800   **Good:** $500
Bulgarian contract ("DWM" over chamber, Bulgarian crest on toggle, cyrillic safety markings. Rarely found in better than Good condition.)
VGood: $1500 **Good:** $950   **Fair:** $750
Navy (6″ barrel, stock lug, adjustable rear sight, no chamber date)
Exc.: $2500   **VGood:** $1750   **Good:** $1250

LUGER MODEL 1914 MILITARY

Semi-automatic; 9mm Para.; 4″ barrel; flat checkered toggle knobs; fixed rear sight; stock lug on frame; hold-open device; chamber dated 1914 to 1918.
DWM
Exc.: $850   **VGood:** $500   **Good:** $350
Erfurt
Exc.: $750   **VGood:** $400   **Good:** $300

Luger Model 1914 Artillery

Same specs as the 1914 Military model except 8″ barrel; adjustable sight on barrel. Add $300 for stock; $250 for holster (beware of fakes); $550 for 32-shot drum magazine.
DWM or Erfurt
Exc.: $1500   **VGood:** $850   **Good:** $600

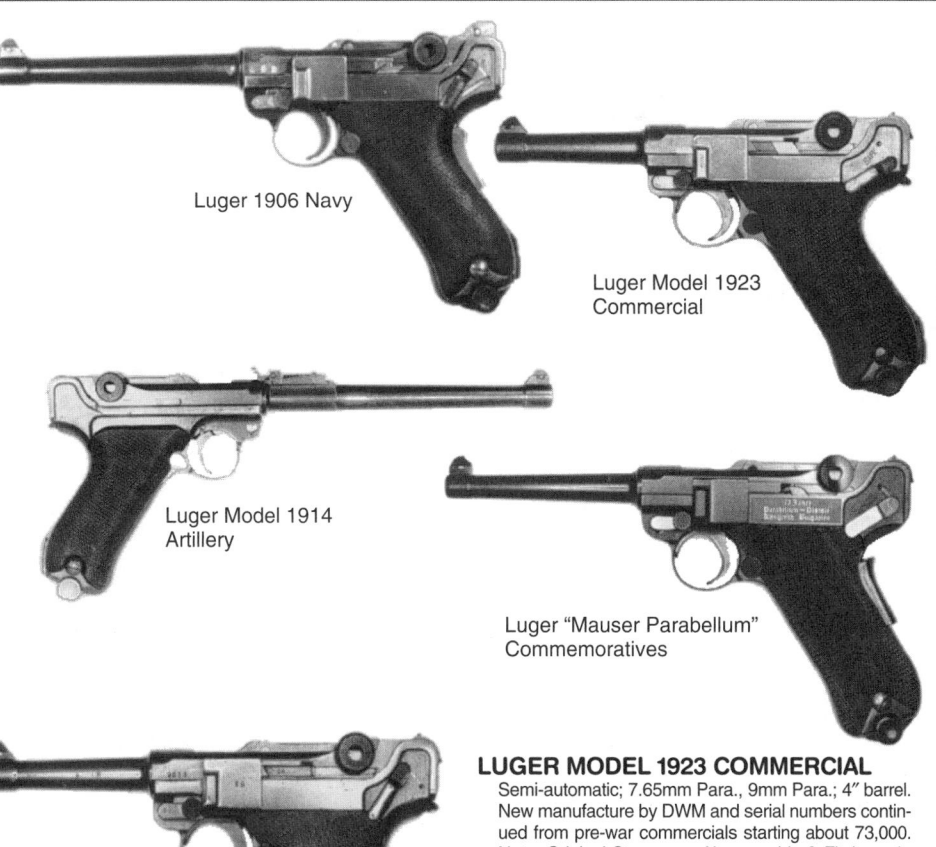

Luger 1906 Navy

Luger Model 1923 Commercial

Luger Model 1914 Artillery

Luger "Mauser Parabellum" Commemoratives

Luger Mauser Military (P.08)

Luger
Model 1914 Navy
Model 1920 Military
Model 1920 Commercial
Model 1923 Commercial
Simson & Co.
Swiss Manufacture
Mauser Commercial
Mauser Military (P.08)
Krieghoff
Mauser Post-War
Manufacture
"Mauser Parabellum"
Commemoratives

Luger Model 1914 Navy
Same specs as 1914 Military except 6″ barrel; adjustable sight on rear toggle link. DWM only.
Exc.: $2250   **VGood:** $1750   **Good:** $1200

LUGER MODEL 1920 MILITARY
Semi-automatic; 9mm Para.; 4″ barrel. The Model 1914 Military Luger dated 1920, often in addition to the original date, to indicate compliance with the Treaty of Versailles for issue to the 100,000-man post-war German army.
Exc.: $750   **VGood:** $500   **Good:** $350

Luger Model 1920 Commercial
Same specs as Military model except 7.65mm Para. or 9mm Para. Mostly reworked wartime DWM and Erfurt handguns or made up from surplus parts in every possible variation for commercial sale outside Germany. Marked "GERMANY" or "MADE IN GERMANY." Note: Original Stoeger or Abercrombie & Fitch markings will add $2000 to any variation.
7.65mm Para.
Exc.: $450   **VGood:** $300   **Good:** $225
9mm Para.
Exc.: $550   **VGood:** $400   **Good:** $300
Navy, either caliber
Exc.: $2250   **VGood:** $1750   **Good:** $1250
Artillery, either caliber
Exc.: $2000   **VGood:** $1500   **Good:** $1250
Carbine, either caliber
Exc.: $6500   **VGood:** $5000   **Good:** $3500

LUGER MODEL 1923 COMMERCIAL
Semi-automatic; 7.65mm Para., 9mm Para.; 4″ barrel. New manufacture by DWM and serial numbers continued from pre-war commercials starting about 73,000. Note: Original Stoeger or Abercrombie & Fitch markings will add $2000.
Exc.: $900   **VGood:** $600   **Good:** $450
Marked "SAFE" and "LOADED"
Exc.: $1000   **VGood:** $750   **Good:** $500

LUGER SIMSON & CO.
Semi-automatic; 9mm Para.; 4″ barrel. Made by Simson from surplus parts for military and limited commercial sale.
Undated
Exc.: $1250   **VGood:** $1000   **Good:** $750
Dated 1925 through 1928
Exc.: $2500   **VGood:** $2000   **Good:** $1500

LUGER SWISS MANUFACTURE
Semi-automatic; 7.65mm Para.; 4 3/4″ barrel. Made at Waffenfabrik Bern for Swiss army, police and commercial sale. Model 1906 (sometimes called "Model 1924") marked "Waffenfabrik Bern" with small Swiss cross on front of toggle and is mechanically same as DWM; Model 1929 has large Swiss cross in shield on front toggle; no checkering on toggle knobs; longer grip safety; other minor cosmetic differences.
Model 1906
Exc.: $2250   **VGood:** $1750   **Good:** $1250
Model 1929 Commercial ("P" before s/n)
Exc.: $2250   **VGood:** $1750   **Good:** $1250
Model 1929 Military
Exc.: $2500   **VGood:** $1850   **Good:** $1250

LUGER MAUSER COMMERCIAL
Semi-automatic; 7.65mm Para., 9mm Para. Distinguished by Mauser banner on toggle; for commercial or contract sale outside Germany, German police issue.
4″ barrel, 9mm
Exc.: $2000   **VGood:** $1500   **Good:** $900
Nazi police proofs
Exc.: $2800   **VGood:** $2150   **Good:** $1550
4″ barrel, 7.65mm
Exc.: $2250   **VGood:** $1750   **Good:** $1250
Artillery model
Exc.: $3500   **VGood:** $2500   **Good:** $1750
Luger Mauser Persian (4″ barrel)
Exc.: $4000   **VGood:** $3500   **Good:** $2500
Portuguese Contract "GNR" (7.65mm, 4 3/4″ barrel)
Exc.: $2250   **VGood:** $1750   **Good:** $1250

LUGER MAUSER MILITARY (P.08)
Semi-automatic; 9mm Para.; 4″ barrel; front toggle marked with "S/42", "42" or "byf" Nazi manufacturer's codes. Date of manufacture on chamber ("K" for 1934; "G" for 1935 or four, or later, two digits for actual year.)
Exc.: $950   **VGood:** $750   **Good:** $400
"K" date
Exc.: $1500   **VGood:** $1100   **Good:** $800
"G" date
Exc.: $1250   **VGood:** $950   **Good:** $450

LUGER KRIEGHOFF
Semi-automatic; 9mm Para.; 4″ barrel. Front toggle marked with anchor "HK" trademark over "KRIEGHOFF SUHL" and date of manufacture on chamber. A few from 1935 have "S" over chamber and only "SUHL."
Military
Exc.: $3000   **VGood:** $2250   **Good:** $1500
Commercial, "P" before s/n, usually undated chamber
Exc.: $3500   **VGood:** $2500   **Good:** $2000

LUGER MAUSER POST-WAR MANUFACTURE
In 1970 Mauser resumed limited production of the Luger using tooling purchased from the Swiss. Chambered in 7.65mm Para., 9mm Para.; 4″, 4 3/4″, 6″ barrel; front toggle marked, "Original MAUSER."
Perf.: $600   **Exc.:** $500   **VGood:** $400
With original box
New: $750   **Perf.:** $600   **Exc.:** $450

LUGER "MAUSER PARABELLUM" COMMEMORATIVES
A very limited number (250 each) of replica 1900 Swiss and Bulgarian, 1906 Russian and Navy as well as the 1902 carbine were produced by Mauser. Front toggles marked with gold inlaid "DWM" with appropriate gold inlaid inscription on side panel, cased in fitted leather case with accessories.
As new pistols
Exc.: $2500
As new carbines
Exc.: $4500

Luna
Model 200 Free Pistol
Model 300 Free Pistol

MAB
Le Chasseur (See MAB Model F)
Le Defendeur
(See MAB Model A)
Le Gendarme (See MAB
Model D)
Model A
Model B
Model C
Model D
Model E
Model F
Model PA-15
WAC Model A (See MAB
Model A)
WAC Model D (See MAB
Model D)
WAC Model E (See MAB
Model E)

Magnum Research
Baby Eagle 9mm
Baby Eagle 9mm Model F
Desert Eagle
Lone Eagle
Mountain Eagle
Mountain Eagle Compact
Edition
Mountain Eagle Target Edition
SSP-91 (See Magnum
Research Lone Eagle)

MAB Model A

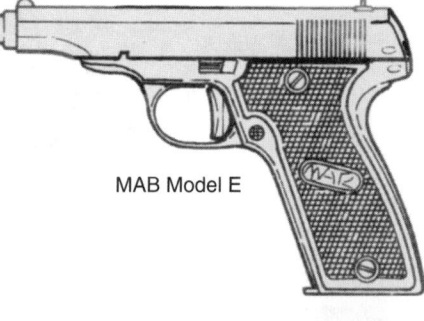

MAB Model E

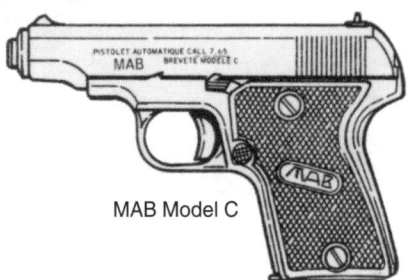

MAB Model C

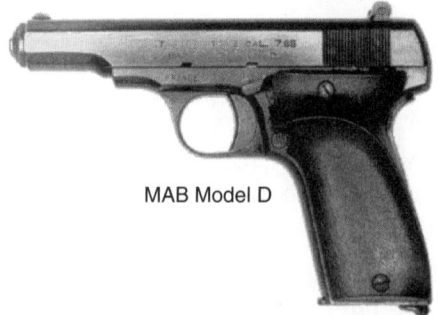

MAB Model D

LUNA MODEL 200 FREE PISTOL
Single shot; falling block; 22 LR; 11″ barrel; target sights; checkered, carved walnut stock, forearm; adjustable palm rest; set trigger; blued finish. Manufactured in Germany approximately 1929 to 1939.
Exc.: $1000   **VGood:** $850   **Good:** $500

LUNA MODEL 300 FREE PISTOL
Single shot; 22 Short; 11″ barrel; target sights; set trigger; checkered, carved walnut stock, forearm; adjustable palm rest; blued finish. Manufactured in Germany approximately 1929 to 1939.
Exc.: $850   **VGood:** $750   **Good:** $500

MAB MODEL A
Semi-automatic; 25 ACP; 6-shot magazine; 2 1/2″ barrel; 4 1/2″ overall length; no rear sight, fixed front; checkered plastic or hard rubber grips; based on Browning design; blued finish. Introduced in 1921; production suspended in 1942; production resumed in 1945 for importation into U.S. as WAC Model A or Le Defendeur; importation dropped in 1968. Manufactured by Manufacture d'Armes de Bayonne, France.
Exc.: $200   **VGood:** $160   **Good:** $125

MAB MODEL B
Semi-automatic; 25 ACP; 6-shot magazine; 2″ barrel; 4 1/2″ overall length; no rear sight, fixed front; hard rubber grips; blued finish. Introduced 1932; dropped 1949.
Exc.: $150   **VGood:** $135   **Good:** $110

MAB MODEL C
Semi-automatic; 32 ACP, 380 ACP; 7-shot magazine (32 ACP), 6-shot (380 ACP); 3 3/4″ barrel; 6″ overall length; fixed sights; black checkered hard rubber grips; push-button magazine release behind trigger; blued finish. Introduced 1933; made under German supervision during WWII. Importation discontinued 1968.
Exc.: $150   **VGood:** $135   **Good:** $100

MAB MODEL D
Semi-automatic; 32 ACP, 380 ACP; 9-shot magazine (32 ACP), 8-shot (380 ACP); 4″ barrel; 7″ overall length; fixed sights; black checkered hard rubber grips; push-button magazine release; blued finish. Introduced 1933; made under German supervision in WWII. Imported to U.S. as WAC Model D or MAB Le Gendarme. Importation discontinued 1968.
Exc.: $150   **VGood:** $135   **Good:** $100

MAB MODEL E
Semi-automatic; 25 ACP; 10-shot magazine; 3 1/4″ barrel; 6 1/8″ overall length; fixed sights; plastic grips; blued finish. Introduced 1949; imported into U.S. as WAC Model E. Importation discontinued 1968.
Exc.: $125   **VGood:** $110   **Good:** $90

MAB MODEL F
Semi-automatic; 22 LR; 10-shot magazine; 3 1/4″, 6″, 7 1/4″ barrel; 10 3/4″ overall length; windage-adjustable rear sight, ramp front; plastic thumbrest grips; blued finish. Introduced 1950; variation imported into U.S. as Le Chasseur. Importation discontinued 1968.
Exc.: $200   **VGood:** $165   **Good:** $135

MAB MODEL PA-15
Semi-automatic; 9mm Para.; 15-shot magazine; 4 1/2″ barrel; 8″ overall length; fixed sights; checkered plastic grips; blued finish. Still in production; not currently imported.
Perf.: $500   **Exc.:** $400   **VGood:** $325

MAGNUM RESEARCH BABY EAGLE 9MM
Semi-automatic; double action; 9mm Para.; 40 S&W, 41 A.E.; 4 3/8″ barrel; 8 1/16″ overall length; weighs 35 oz.; high-impact polymer grips; combat sights; polygonal rifling; ambidextrous safety; matte black or chrome finish. Made by Magnum Research. Introduced 1992; still imported. **LMSR:** $569
Perf.: $400   **Exc.:** $350   **VGood:** $300
Chrome finish
Perf.: $450   **Exc.:** $400   **VGood:** $325

Magnum Research Baby Eagle 9mm Model F
Same specs as the standard model except has frame-mounted safety on left side. Still imported by Magnum Research, Inc. **LMSR:** $569
Perf.: $425   **Exc.:** $375   **VGood:** $325

MAGNUM RESEARCH DESERT EAGLE
Semi-automatic; 357 Mag., 41 Mag., 44 Mag., 50 Mag.; 9-shot (357), 8-shot (41 Mag., 44 Mag.), 7-shot (50 Mag.); 6″, 10″, 14″ interchangeable barrels; 10 1/4″ overall length; weighs 62 oz. (357), 69 oz. (41,44), 72 oz. (50); wrap-around plastic stocks; blade on ramp

front, combat-style rear sight; rotating three-lug bolt; ambidextrous safety; combat-style trigger guard; military epoxy finish; satin, bright nickel, hard chrome, polished and blued finishes available. Imported from Israel by Magnum Research, Inc.
Perf.: $950   **Exc.:** $750   **VGood:** $575
41, 44 Mag.
Perf.: $850   **Exc.:** $700   **VGood:** $575
Stainless 41, 44 Mag.
Perf.: $950   **Exc.:** $800   **VGood:** $625
50 Mag. **LMSR:** $1249
Perf.: $1000   **Exc.:** $850   **VGood:** $750

MAGNUM RESEARCH LONE EAGLE
Single shot; 22 Hornet, 223, 22-250, 243, 7mm BR, 7mm-08, 30-30, 308, 30-06, 357 Max., 35 Rem., 358 Win., 44 Mag., 444 Marlin; 14″ interchangeable barrel; 15″ overall length; weighs 4 lbs. 3 oz. to 4 lbs. 7 oz.; composition thumbrest stock; no sights furnished; drilled and tapped for scope mounting and optional open sights; cannon-type rotating breech with spring-activated ejector; cross-bolt safety; external cocking lever on left side of gun; ordnance steel with matte blue finish. Made in U.S.; marketed by Magnum Research, Inc. Introduced 1991; still in production.
New: $350   **Perf.:** $300   **Exc.:** $250

MAGNUM RESEARCH MOUNTAIN EAGLE
Semi-automatic; single action; 22 LR; 10-shot magazine; 6 1/2″, 8″ barrel; 10 5/8″ overall length (6 1/2″ barrel); weighs 21 oz.; serrated ramp front sight with interchangeable blades, fully-adjustable rear; one-piece impact-resistant polymer grip with checkered panels; interchangeable blades; injection moulded grip frame; alloy receiver; hybrid composite barrel replicates shape of the Desert Eagle pistol; flat, smooth trigger. Made in U.S. Marketed by Magnum Research. Introduced 1992; still production.
New: $200   **Perf.:** $175   **Exc.:** $150

Magnum Research Mountain Eagle Compact Edition
Same specs as Mountain Eagle except 4 1/2″ barrel; shorter grip; windage-adjustable rear sight; weighs 19 1/4 oz. Introduced 1995; still in production.
New: $175   **Perf.:** $150   **Exc.:** $125

Magnum Research Mountain Eagle Target Edition
Same specs as the Mountain Eagle except 8″ barrel; two-stage trigger. Made in U.S. Marketed by Magnum Research. Introduced 1992; still produced. **LMSR:** $279
New: $200   **Perf.:** $175   **Exc.:** $150

MAB Model PA-15

Magnum Research
Baby Eagle 9mm

Magnum Research
Desert Eagle

Magnum Research
Lone Eagle

Magnum Research
Mountain Eagle

Manurhin Model
PPK/S

HANDGUNS

Manurhin
Model 73 Convertible
Model 73 Defense
Model 73 Gendarmerie
Model 73 Silhouette
Model 73 Sport
Model MR .32 Match
Model MR .38
Model PP
Model PPK/S

Manurhin Model PP

Manurhin Model
73 Sport

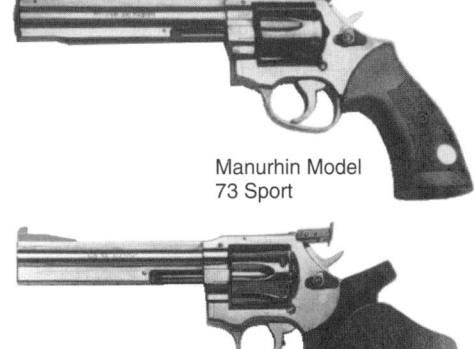

Manurhin Model
MR .32 Match

MANURHIN MODEL 73 CONVERTIBLE
Revolver; single action; 22 LR/38 Spl., 22 LR/32 ACP; 6-shot cylinder; 5 3/4" barrel (38), 6" (32); interchangeable blade front sight, adjustable micrometer rear; checkered walnut stocks; blued finish. Made in France. Introduced 1988 by Manurhin International; no longer in production.
Perf.: $1750 **Exc.:** $1500 **VGood:** $1250

MANURHIN MODEL 73 DEFENSE
Revolver; single action; 357 Mag./38 Spl.; 6-shot cylinder; 2 1/2", 3", 4" barrel; fixed sights; checkered walnut stocks; blued finish. Made in France. Introduced 1988 by Manurhin International; no longer in production.
Perf.: $1000 **Exc.:** $850 **VGood:** $700

Manurhin Model 73 Gendarmerie
Same specs as Model 73 Defense except also offered with 5 1/4", 6", 8" barrel; distinguished by prominent ramped front sight, adjustable rear with rounded edge. Used by French police forces and government agencies.
Perf.: $1500 **Exc.:** $1250 **VGood:** $850

Manurhin Model 73 Silhouette
Same specs as Model 73 Defense except also offered in 22 LR; 10", 10 3/4" heavy barrel with shroud; adjustable sights; contoured walnut stocks.
Perf.: $1750 **Exc.:** $1500 **VGood:** $1000

Manurhin Model 73 Sport
Same specs as Model 73 Defense except 6" barrel; 11" overall length; weighs 37 oz.; fully-adjustable rear sight; adjustable trigger; reduced hammer travel. Imported from France by Century International Arms. Introduced 1988; still in production. **LMSR: $1500**
Perf.: $1350 **Exc.:** $1100 **VGood:** $850

MANURHIN MODEL MR .32 MATCH
Revolver; 32 S&W Long; 6-shot cylinder; 6" barrel; 11 3/4" overall length; weighs 42 oz.; interchangeable blade front sight, adjustable micrometer rear; anatomical target shaped but unfinished grips; externally adjustable trigger; trigger shoe. Made in France. Intro-duced 1984; discontinued 1986; was imported by Manurhin International.
Perf.: $750 **Exc.:** $600 **VGood:** $500

MANURHIN MODEL MR .38
Revolver; 38 Spl.; 6-shot cylinder; 5 3/4" barrel; interchangeable blade front sight, adjustable micrometer rear; anatomical target shaped but unfinished grips; externally adjustable trigger; trigger shoe. Made in France. Introduced 1984; discontinued 1986; was imported by Manurhin International.
Perf.: $750 **Exc.:** $600 **VGood:** $500

MANURHIN MODEL PP
Semi-automatic; double action; 22 LR, 32 ACP, 380 ACP; 10-shot magazine (22 LR), 8-shot (32 ACP), 7-shot (380 ACP); 3 7/8" barrel; 7 3/4" overall length; weighs 23 oz.; white-outline front, rear sights; checkered compostion stocks; hammer drop safety; all-steel construction; supplied with two magazines; blued finish. Made in France. 22 or 380 worth 50% more than prices shown. Importation began in 1950s; dropped 1960s.
Perf.: $300 **Exc.:** $225 **VGood:** $200
Collector Model with engraving
Perf.: $350 **Exc.:** $325 **VGood:** $300
Presentation Model with ornamentation
Perf.: $450 **Exc.:** $400 **VGood:** $350

Manurhin Model PPK/S
Same specs as Model PP except 3 1/4" barrel; 6 1/8" overall length.
Perf.: $300 **Exc.:** $250 **VGood:** $225
With brushed chrome finish
Perf.: $350 **Exc.:** $300 **VGood:** $250
Collector Model with engraving
Perf.: $450 **Exc.:** $400 **VGood:** $350
Presentation Model with ornamentation
Perf.: $550 **Exc.:** $450 **VGood:** $350

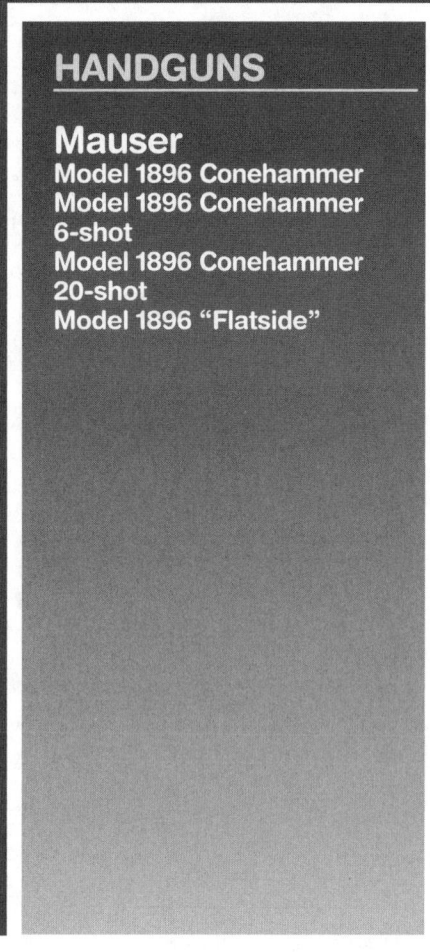

HANDGUNS

Mauser
Model 1896 Conehammer
Model 1896 Conehammer 6-shot
Model 1896 Conehammer 20-shot
Model 1896 "Flatside"

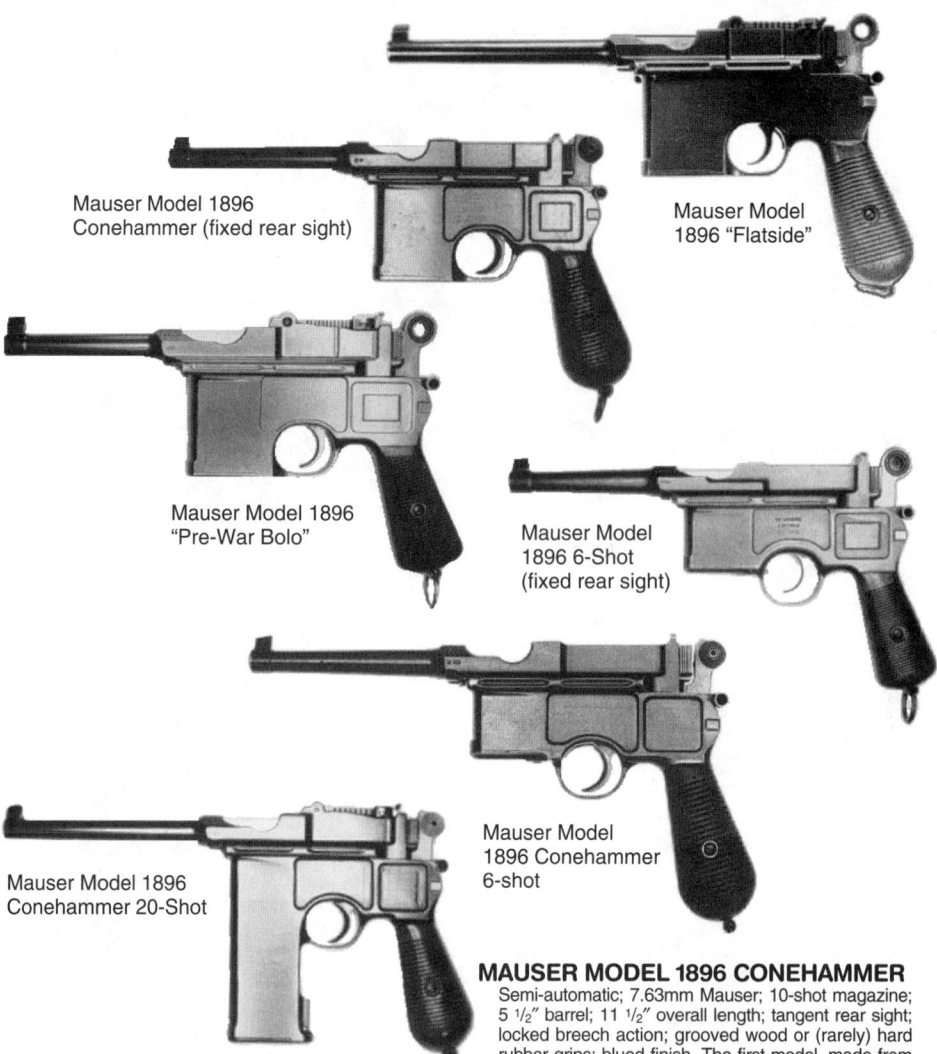

Mauser Model 1896 Conehammer (fixed rear sight)

Mauser Model 1896 "Flatside"

Mauser Model 1896 "Pre-War Bolo"

Mauser Model 1896 6-Shot (fixed rear sight)

Mauser Model 1896 Conehammer 6-shot

Mauser Model 1896 Conehammer 20-Shot

MAUSER

Mauser's Model 1896 was the first truly successful pistol, produced with relatively little change from late 1896 until just before WWII. Though some—particularly those in 9mm Parabellum—are enjoyed by shooters who also appreciate their classic status, the Model 1896 in its almost unending variations is primarily a collector's item and is usually priced accordingly.

The Model 1896 was extremely popular in China, and in recent years tens of thousands of them, including some very rare variations, have been imported from China. Unfortunately, most of these imports were in very well used to terrible condition, but many have been restored to acceptable shooter or even collector condition. A restored rarity is generally worth half to two-thirds as much as an NRA Very Good original piece, and restored shooters—usually rebored to 9mm Parabellum—sell in the $350 to $500 range. In addition, the Chinese manufactured a great many copies of the Model 1896 of wildly variable quality; these are also collectible. Since some of these copies have collector value, they are also listed at the end of the Model 1896/Model 1930 section.

The listings below present Mauser's Model 1896 variations in chronological order. Fortunately, almost all Model 1896 Mauser pistols were serial numbered in order of manufacture so most of them can be accurately identified by serial number alone. The table nearby provides a basic guide to Mauser 1896 serial numbering. However, there were often overlapping serial ranges as new changes were introduced, plus "strays" that don't conform to the standard serial pattern. In additon, some test and small contract purchased pistols will not conform to the serial ranges in the table.

For variations that don't seem to match the descriptions below, consult one of the Model 1896 reference books such as *The Broomhandle Pistol* by Pate and Erickson or *System Mauser* by Breathed and Schroeder.

Joe Schroeder

MAUSER MODEL 1896 SERIAL RANGES

Variation	Serial Range
Cone Hammer	1-15000
Italian Navy Flatside (M 1899)	1-5000*
Commercial Flatside	21000-29000
Large Ring Hammer, Panel Sides	20000-21000
	30000-35000
Large Ring Hammer 6-Shot, Bolo	29000-30000
	40000-44000
Small Ring Hammer	35000-40000
	44000-432000
"Red 9" Model 1916 Contract	1-140000*
French Gendarme	432000-447000
Post-War Bolo	447000-800000
Model 1930	800000-921000

*The Italian navy and 9mm German army pistols were still Model 1896 Mausers, though they were officially identified by their purchasers with the year of contract. Though the Italian navies were numbered in their own serial range, Mauser considered them to be Mauser's equivalent to commercial production serial numbers 15000 to 20000 and accordingly left that block unfilled. The German army contract of 1916 for 9mm Parabellum pistols was numbered in its own independent serial range.

MAUSER MODEL 1896 CONEHAMMER

Semi-automatic; 7.63mm Mauser; 10-shot magazine; 5 1/2" barrel; 11 1/2" overall length; tangent rear sight; locked breech action; grooved wood or (rarely) hard rubber grips; blued finish. The first model, made from 1896-1898 and considered antique under federal law, is distinguished by a "beehive" or cone-shaped hammer (though some have a much larger open ring hammer), wide trigger and milled frame panels. Serial numbers run up to 15000. Note: First few hundred had "SYSTEM MAUSER" engraved on top of chamber and bring a very high premium. Add $500 for matching stock holster; $100 for engraved dealer's name.

Exc.: $2000 **VGood:** $1500 **Good:** $950

Turkish contract with Turkish crest in side panel and serial/sight numbering in Cyrillic

Exc.: $4500 **VGood:** $3500 **Good:** $2000

With fixed rear sight, 4 5/8" barrel

Exc.: $2500 **VGood:** $2000 **Good:** $1500

Mauser Model 1896 Conehammer 6-Shot

Same specs as standard model except shortened 6-shot magazine; may have fixed or adjustable rear sight.

Exc.: $5000 **VGood:** $3500 **Good:** $2500

Mauser Model 1896 Conehammer 20-Shot

Same specs as standard model except fixed 20-shot magazine extending well below trigger guard; milled out or flat frame panels. Very high collector value.

Exc.: $35,000 **VGood:** $20,000 **Good:** $15,000

With original 20-shot stock holster

Exc.: $40,000 **VGood:** $25,000 **Good:** $20,000

MAUSER MODEL 1896 "FLATSIDE"

Semi-automatic; 7.63mm Mauser; 10-shot magazine; 5 1/2" barrel; 11 1/2" overall length; flat frame sides; narrow trigger; large ring hammer; tangent rear sight; locked breech action; grooved wood or (rarely) hard rubber grips; blued finish. The 6-shot flatsides are very rare and very valuable.

Exc.: $1850 **VGood:** $1350 **Good:** $900

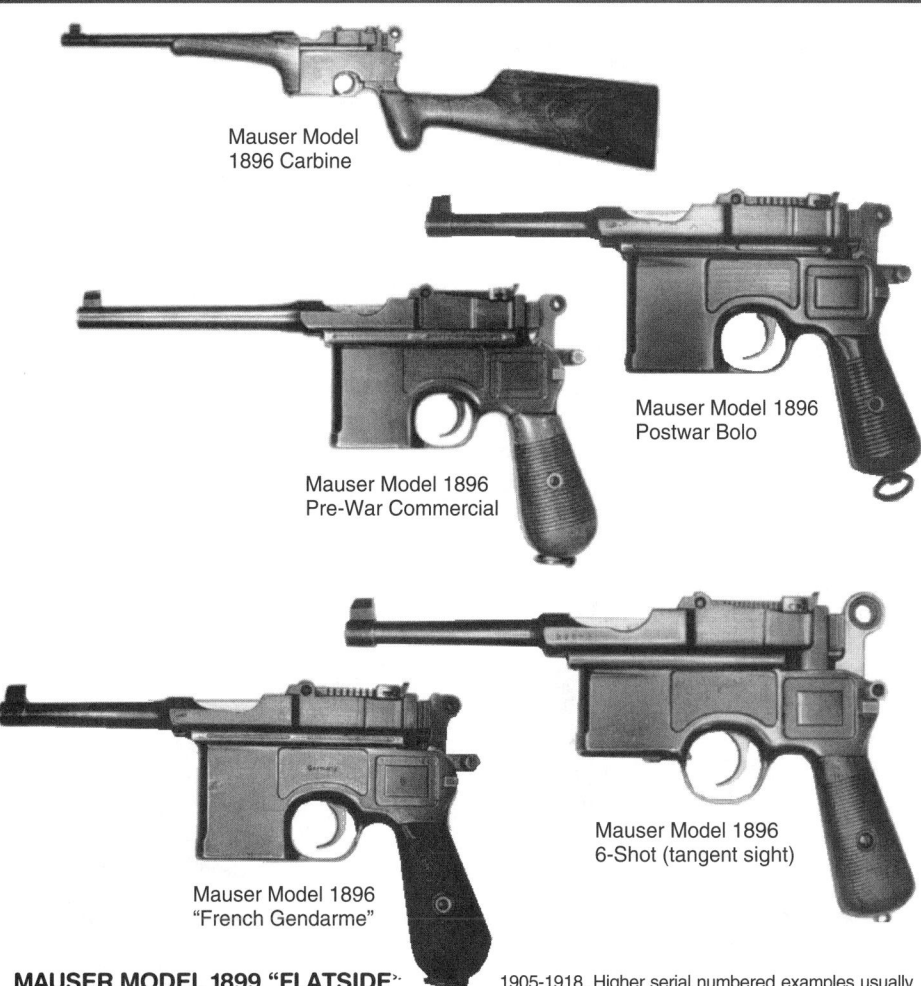

Mauser Model
1896 Carbine

Mauser Model 1896
Postwar Bolo

Mauser Model 1896
Pre-War Commercial

Mauser Model 1896
6-Shot (tangent sight)

Mauser Model 1896
"French Gendarme"

Mauser
Model 1899 "Flatside"
Italian Navy
Model 1896 Large Ring
Hammer
Model 1896 "Pre-War Bolo"
Model 1896 6-Shot
Model 1896 Pre-War
Commercial
Model 1896 9mm Export
Model 1896 Carbine
Model 1916 "Red 9"
Model 1896 Weimar Rework
Model 1896
"French Gendarme"
Model 1896 Postwar Bolo
Model 1930

MAUSER MODEL 1899 "FLATSIDE" ITALIAN NAVY
Semi-automatic; 7.63mm Mauser; 10-shot magazine; 5 1/2″ barrel; 11 1/2″ overall length; flat frame sides; narrow trigger; large ring hammer; tangent rear sight; locked breech action; grooved wood grips; blued finish. 5,000 flatsides made for Italian navy and distinguished by small "DV" on chamber side and crown over "AV" under barrel; serial numbered 1 to 5000.
   Exc.: $2000   **VGood:** $1500   **Good:** $750

MAUSER MODEL 1896 LARGE RING HAMMER
Semi-automatic; 7.63mm Mauser; 10-shot magazine; 5 1/2″ barrel; 11 1/2″ overall length; paneled frame sides like the Conehammer; narrow trigger; large ring hammer; tangent rear sight; locked breech action; grooved wood or (rarely) hard rubber grips; blued finish.
   Exc.: $1500   **VGood:** $1000   **Good:** $700

MAUSER MODEL 1896 "PRE-WAR BOLO"
Semi-automatic; 7.63mm Mauser; 10-shot magazine; 3 15/16″ barrel; 10″ overall length; paneled frame sides; narrow trigger; large or small ring hammer; tangent rear sight; locked breech action; small grip frame with wood or hard rubber grips; blued finish.
   Exc.: $2250   **VGood:** $1500   **Good:** $950
With original Bolo (short) stock-holster
   Exc.: $3500   **VGood:** $2750   **Good:** $1750

MAUSER MODEL 1896 6-SHOT
Semi-automatic; 7.63mm Mauser; 6-shot magazine 3 15/16″ barrel; large or small-ring hammer; fixed or tangent rear sight.
   Exc.: $4500   **VGood:** $3500   **Good:** $2000

MAUSER MODEL 1896 PRE-WAR COMMERCIAL
Semi-automatic; 7.63mm Mauser; 10-shot magazine; 5 1/2″ barrel; 11 1/2″ overall length; paneled frame sides like the Conehammer; narrow trigger; small ring hammer; tangent rear sight; locked breech action; grooved wood or (rarely) hard rubber grips; blued finish. Made

1905-1918. Higher serial numbered examples usually have German military proofs.
   Exc.: $1000   **VGood:** $850   **Good:** $500
With Mauser banner over chamber
   Exc.: $1250   **VGood:** $950   **Good:** $700
With Persian Lion crest on side s/n in 154000 range
   Exc.: $2500   **VGood:** $1750   **Good:** $1200

MAUSER MODEL 1896 9MM EXPORT
Semi-automatic; 9mm Mauser; 10-shot magazine; 5 1/2″ barrel; 11 1/2″ overall length; paneled frame sides like the Conehammer; narrow trigger; small ring hammer; tangent rear sight; locked breech action; grooved wood or (rarely) hard rubber grips; blued finish.
   Exc.: $1500   **VGood:** $1000   **Good:** $750

MAUSER MODEL 1896 CARBINE
Semi-automatic; 7.63mm; 10-shot; 11 3/4″, 16″ barrel; takedown sporting carbine made on Mauser pistol action. Made in limited numbers from 1896 to about 1906 with paneled or smooth sides and with all three hammer styles. Very high collector value. Note: A few very late carbines were chambered for 9mm Para. or 9mm Export and bring a premium.
   Exc.: $15,000   **VGood:** $10,000   **Good:** $8000

MAUSER MODEL 1916 "RED 9"
Semi-automatic; 9mm Para.; 10-shot; 5 1/2″ barrel. Made under contract for the German army. Called by collectors "Red 9" because most (but not all) have large figure "9" cut into the wood grips and painted red to distinguish them from 7.63mm pistols also in use by army.
   Exc.: $1350   **VGood:** $750   **Good:** $650

MAUSER MODEL 1896 WEIMAR REWORK
Many WWI Red 9 military pistols were reworked 1920-1921 to conform with the Treaty of Versailles by cutting barrels to 4″, replacing tangent sight with fixed rear sight. Usually identified by "1920" stamped on side.
   Exc.: $850   **VGood:** $600   **Good:** $500
With Weimar navy markings
   Exc.: $1250   **VGood:** $900   **Good:** $750

Mauser Model 1930

MAUSER MODEL 1896 "FRENCH GENDARME"
Semi-automatic; 7.63mm; 10-shot magazine; supposedly made for French occupation forces with 3 15/16″ Bolo barrel but full-size grip frame; checkered hard rubber grips.
   Exc.: $1000   **VGood:** $850   **Good:** $600

MAUSER MODEL 1896 POSTWAR BOLO
Semi-automatic; 7.63mm; 10-shot; 3 15/16″ barrel; small grip frame with grooved wood grips. Made about 1920 through 1930.
   Exc.: $850   **VGood:** $500   **Good:** $400

MAUSER MODEL 1930
Semi-automatic; 7.63mm; 10-shot; 5 1/4″ barrel (early) 5 1/2″ barrel. Slightly updated version of Model 1896 distinguished by small step in barrel, wide grip straps with 12-groove wood grips, high polish bright acid blue finish instead of duller rust blue. Note: A few 1930-type Mausers were made with detachable 10- or 20-shot magazines and are very rare and valuable. Beware of fakes made by welding up switch holes of a Model 1932.
   Exc.: $1500   **VGood:** $1250   **Good:** $500
With proper stock-holster with Mauser banner, no serial number
   Exc.: $2000   **VGood:** $1500   **Good:** $1100
Warning: Model 1932 (also called Model 712) pistols have a selector switch for full-auto fire as well as detachable magazine. They are illegal to own unless registered with the BATF.

Mauser
Chinese Copies of the Model 1896
Taku Naval Dockyard
Chinese Small Ring Hammer Copies
Shansei 45 Copy
Model 1910
Model 1914
Model 1910/1934
Model 1934
WTP
WTP II
HSc
HSc Post-War
HSc Super
Model 80
Model 90
Model 90 Compact

Mauser Small Ring Hammer Chinese Copy

Mauser Model 1914 Humpback

Mauser Shansei 45 Copy

Mauser Model 1910/1934

Mauser Model 1910 Early

Mauser HSc Early

CHINESE COPIES OF THE MODEL 1896

Taku Naval Dockyard
Semi-automatic; 7.63mm; 10-shot; 5 1/2″ barrel; flat frame panel sides; large ring hammer; grooved wood grips. Marked "TAKU NAVAL DOCKYARD" on top of chamber.
  Exc.: $1250 **VGood:** $850 **Good:** $500
With unmarked chamber
  Exc.: $750 **VGood:** $500 **Good:** $300

Chinese Small Ring Hammer Copies
Same specs as above except paneled frame sides; small ring hammer. Quality varies from quite good to simply awful. May be unmarked except for serial numbers with spurious Mauser markings (often misspelled) or with markings in Chinese.
  Exc.: $450 **VGood:** $300 **Good:** $200

Shansei 45 Copy
Semi-automatic; 45 ACP; 10-shot; 6 1/4″ barrel; 12 1/2″ overall length; weighs 3 lbs., 7 oz. Oversize version of the Model 1896 built in the 1930s for a Chinese warlord who liked the 45 cartridge. Note: A copy of this copy, recently made in China, has been imported and sold at $1000-$1500. If the condition of a Shansei seems too good, it is.
  Exc.: $4500 **VGood:** $3000 **Good:** $2000

MAUSER MODEL 1910
Semi-automatic; 25 ACP; 9-shot; 3 1/8″ barrel; 5 3/8″ overall length; fixed sights; checkered walnut or (rarely) hard rubber wrap-around grips. Introduced 1910; dropped 1934.
  Exc.: $250 **VGood:** $185 **Good:** $135
Early model with takedown latch above trigger (s/n below about 60000)
  Exc.: $350 **VGood:** $250 **Good:** $185

MAUSER MODEL 1914
Semi-automatic; 32 ACP; 8-shot; 3 1/4″ barrel; 6 1/8″ overall length; fixed sights; checkered walnut or (rarely) hard rubber wrap-around grip. Introduced 1914; dropped 1934.
  Exc.: $235 **VGood:** $185 **Good:** $135
Early model with hump on top of slide (s/n below about 3000)
  Exc.: $2200 **VGood:** $1800 **Good:** $1250

MAUSER MODEL 1910/1934
Semi-automatic; 25 ACP; 9-shot; 3 1/8″ barrel; 5 3/8″ overall length. Like the Model 1910 except for swept-back grip of walnut or plastic, bright high polish blue instead of rust blue finish.
  Exc.: $350 **VGood:** $250 **Good:** $175

MAUSER MODEL 1934
Semi-automatic; 32 ACP; 8-shot; 3 1/2″ barrel; 6 1/8″ overall length; fixed sights; swept-back grip of walnut or plastic; bright high polish blue instead of rust blue finish. Made from 1934 to 1940.
  Exc.: $400 **VGood:** $300 **Good:** $175
With Nazi Navy eagle over M marking
  Exc.: $600 **VGood:** $450 **Good:** $300

MAUSER WTP
Semi-automatic; 25 ACP; 6-shot; 2 1/2″ barrel; 4 1/2″ overall length; checkered wrap-around hard rubber grips; blue finish. Made from 1922 to 1938.
  Exc.: $275 **VGood:** $150 **Good:** $125

MAUSER WTP II
Semi-automatic; 25 ACP; 6-shot; 2″ barrel; 4 3/16″ overall length; checkered plastic grips; blue finish. Made from 1938 to 1944.
  Exc.: $350 **VGood:** $250 **Good:** $175

MAUSER HSc
Semi-automatic; double action; 32 ACP; 8-shot; 3 1/2″ barrel; 6 1/2″ overall length; checkered wood or plastic grips; blue finish. Made from 1940 to 1946. Last production was during French occupation.
  Exc.: $350 **VGood:** $275 **Good:** $200
With Nazi navy eagle over M marking
  Exc.: $600 **VGood:** $450 **Good:** $300
Very early model with grip screws near bottom of grips
  Exc.: $2200 **VGood:** $1800 **Good:** $1300

Mauser HSc Post-War
Same specs as above except also made in 380 ACP; 7-shot. Made from 1962 to 1976.
  Exc.: $300 **VGood:** $200 **Good:** $150

MAUSER HSc SUPER
Semi-automatic; 32 ACP, 380; 3 1/2″ barrel; 6 1/2″ overall length; 13-shot magazine; checkered wood grips; double action; blue. Made for Mauser by Gamba in Italy; imported 1968-1981.
  Exc.: $275 **VGood:** $225 **Good:** $175

MAUSER MODEL 80
Semi-automatic; single action; 9mm Para.; 13-shot magazine; 4 11/16″ barrel; 8″ overall length; weighs 32 oz.; checkered beechwood grips; blade front sight, windage-adjustable rear; uses basic Hi-Power design; polished blue finish. Made in Hungary for Mauser and imported from Germany by Precision Imports, Inc. Introduced 1992; importation dropped 1993.
  Perf.: $450 **Exc.:** $400 **VGood:** $350

MAUSER MODEL 90
Semi-automatic; double action; 9mm Para.; 14-shot magazine; 4 11/16″ barrel; 8″ overall length; weighs 35 oz.; checkered beechwood grips; blade front sight, windage-adjustable rear; uses basic Hi-Power design; polished blue finish. Made in Hungary for Mauser and imported from Germany by Precision Imports, Inc. Introduced 1992; importation dropped 1993.
  Perf.: $500 **Exc.:** $450 **VGood:** $400

Mauser Model 90 Compact
Same specs as the Model 90 except 4 1/8″ barrel; 7 1/2″ overall length; weighs 33 1/2″. Imported from Germany by Precision Imports, Inc. Introduced 1992; importation dropped 1993.
  Perf.: $500 **Exc.:** $450 **VGood:** $400

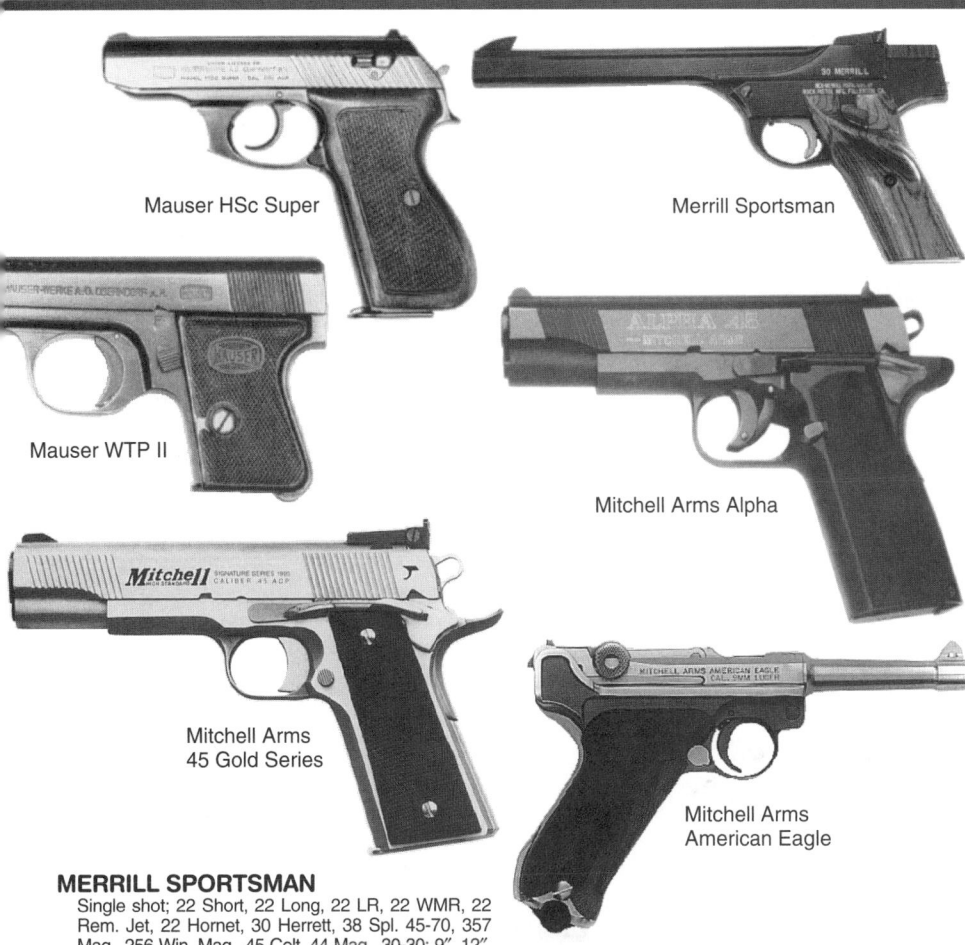

Mauser HSc Super

Mauser WTP II

Merrill Sportsman

Mitchell Arms Alpha

Mitchell Arms
45 Gold Series

Mitchell Arms
American Eagle

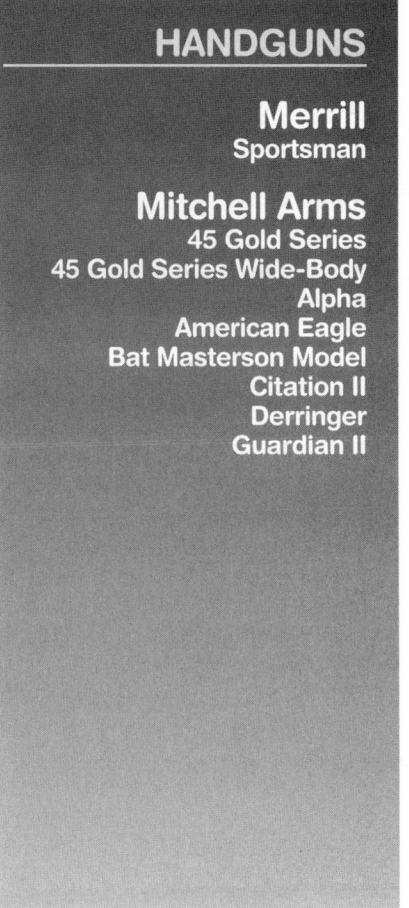

HANDGUNS

Merrill
Sportsman

Mitchell Arms
45 Gold Series
45 Gold Series Wide-Body
Alpha
American Eagle
Bat Masterson Model
Citation II
Derringer
Guardian II

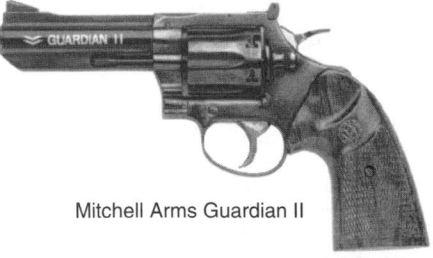

Mitchell Arms Guardian II

MERRILL SPORTSMAN
Single shot; 22 Short, 22 Long, 22 LR, 22 WMR, 22 Rem. Jet, 22 Hornet, 30 Herrett, 38 Spl. 45-70, 357 Mag., 256 Win. Mag., 45 Colt, 44 Mag., 30-30; 9″, 12″, 14″ semi-octagon hinged barrel; adjustable rear sight, fixed front; uncheckered walnut grips with thumb, heel rest; hammerless; top rib grooved for scope mounts. Introduced 1972; still in production as the RPM XL pistol.
  Perf.: $675   **Exc.:** $600   **VGood:** $450

MITCHELL ARMS 45 GOLD SERIES (HIGH STANDARD/SIGNATURE SERIES)
Semi-automatic; single action; 45 ACP; 8-, 10-shot magazine; 5″ barrel; 8 ³/₄″ overall length; weighs 39 oz.; interchangeable blade front sight, drift adjustable combat rear or fully-adjustable rear; smooth American walnut or checkered black rubber grips; bull barrel/slide lockup (no bushing design); full-length guide rod; extended ambidextrous safety; adjustable trigger; beveled magazine well; royal blue or stainless steel. Guns are marked with "High Standard" or "Signature Series" depending on date of manufacture. Made in U.S. From Mitchell Arms, Inc. Introduced 1994; still produced.
  New: $550   **Perf.:** $475   **Exc.:** $400
Stainless finish, fixed sights.
  New: $675   **Perf.:** $525   **Exc.:** $450
Blue finish, adjustable sights
  New: $575   **Perf.:** $500   **Exc.:** $450
Stainless finish, adjustable sights
  New: $650   **Perf.:** $525   **Exc.:** $475

Mitchell Arms 45 Gold Series (High-Standard/Signature Series) Wide-Body
Same specs as 45 Gold except 10-shot magazine (accepts 8- and 13-shot magazines); fixed combat sights; black rubber grips; blue or stainless steel. Guns are marked with "High Standard" or "Signature Series" depending on date of manufacture. Made in U.S. From Mitchell Arms, Inc. Introduced 1994; still produced. **LMSR: $685**
  New: $600   **Perf.:** $550   **Exc.:** $500
Stainless finish, fixed sights.
  New: $675   **Perf.:** $650   **Exc.:** $500
Blue finish, fixed sights
  New: $700   **Perf.:** $675   **Exc.:** $525
Stainless finish, adjustable sights
  New: $725   **Perf.:** $700   **Exc.:** $575

MITCHELL ARMS ALPHA
Semi-automatic; double-action-only, single-action-only or SA/DA; 45 ACP; 8-, 10-shot magazine; 5″ barrel; 8 ¹/₂″ overall length; weighs 41 oz.; interchangeable blade front sight, fully-adjustable rear or drift-adjustable rear; smooth polymer grips; accepts any single-column, 8-shot 1911-style magazine; frame-mounted decocker/safety; extended ambidextrous safety; extended slide latch; serrated combat hammer; beveled magazine well; heavy bull barrel (no bushing design); extended slide underlug; full-length recoil spring guide system. Also available as Alpha X with 14-shot magazine for police, military and export only. Made in U.S. From Mitchell Arms, Inc. Introduced 1995; still in production. **LMSR: $689**
  New: $600   **Perf.:** $550   **Exc.:** $500
Stainless finish
  New: $650   **Perf.:** $600   **Exc.:** $550
With adjustable sights
  New: $650   **Perf.:** $600   **Exc.:** $550

MITCHELL ARMS AMERICAN EAGLE
Semi-automatic; single action; 9mm Para.; 7-shot magazine; 4″ barrel; 9 ⁵/₈″ overall length; weighs 30 oz.; blade front sight, fixed rear; checkered walnut grips; recreation of the American Eagle Parabellum pistol in stainless steel. Made in U.S. Marketed by Mitchell Arms, Inc. Introduced 1992; still in production.
  New: $550   **Perf.:** $475   **Exc.:** $400

MITCHELL ARMS BAT MASTERSON MODEL
Revolver; single action; 45 Colt; 6-shot cylinder; 4 ³/₄″, 5 ¹/₂″, 7 ¹/₂″ barrel; fixed sights; one-piece walnut grip; hammer-block safety; nickel-plated. Introduced 1989; no longer imported.
  New: $350   **Perf.:** $300   **Exc.:** $275

MITCHELL ARMS CITATION II
Semi-automatic; single action; 22 LR; 10-shot magazine; 5 ¹/₂″ bull, 7 ¹/₄″ fluted barrel; weighs 44 ¹/₂ oz. (5 ¹/₂″ barrel); 9 ³/₄″ overall length; undercut ramp front sight, click-adjustable frame-mounted rear; checkered

walnut thumbrest grips; grips duplicate feel of military 45 auto; positive action magazine latch; front- and backstraps stippled; adjustable nickel-plated trigger, safety and magazine release; silver-filled roll marks; push-button barrel takedown; made of stainless steel; satin stainless or blue finish. Guns are marked with "High Standard" depending on date of manufacture. Made in U.S. Marketed by Mitchell Arms, Inc. Introduced 1992; still produced. **LMSR: $498**
  New: $400   **Perf.:** $350   **Exc.:** $300

MITCHELL ARMS DERRINGER
Derringer; over/under; 38 Spl.; 2-shot; 2 ³/₄″ barrel; 5 ¹/₄″ overall length; weighs 11 oz.; fixed sights; checkered walnut grips; has same basic design as original Remington except for ramp front sight; polished blue finish. Introduced 1981; no longer imported.
  Perf.: $100   **Exc.:** $90   **VGood:** $75
Stainless steel
  Perf.: $110   **Exc.:** $100   **VGood:** $85

MITCHELL ARMS GUARDIAN II
Revolver; double action; 38 Spl.; 6-shot cylinder; 3″, 4″ barrel; 8 ¹/₂″ overall length (3″ barrel); weighs 32 oz.; combat or target grips of checkered black rubber or walnut; blade on ramp front sight, fixed rear; target hammer; shrouded ejector rod; smooth trigger; blue finish. Made in U.S. Marketed by Mitchell Arms, Inc. Introduced 1995; still produced. **LMSR: $275**
  New: $225   **Perf.:** $200   **Exc.:** $175

Mitchell Arms
Guardian III
Model 57A
Model 70A
Model 1875 Remington
Officers Model 88A
Olympic II I.S.U.
Pistol Parabellum '08
(See Mitchell American Eagle)
Rolling Block Target
Sharpshooter II
Single Action Army
Single Action Army, U.S. Army Model
Single Action Army, Cavalry Model
Single Action Army, Cowboy Model

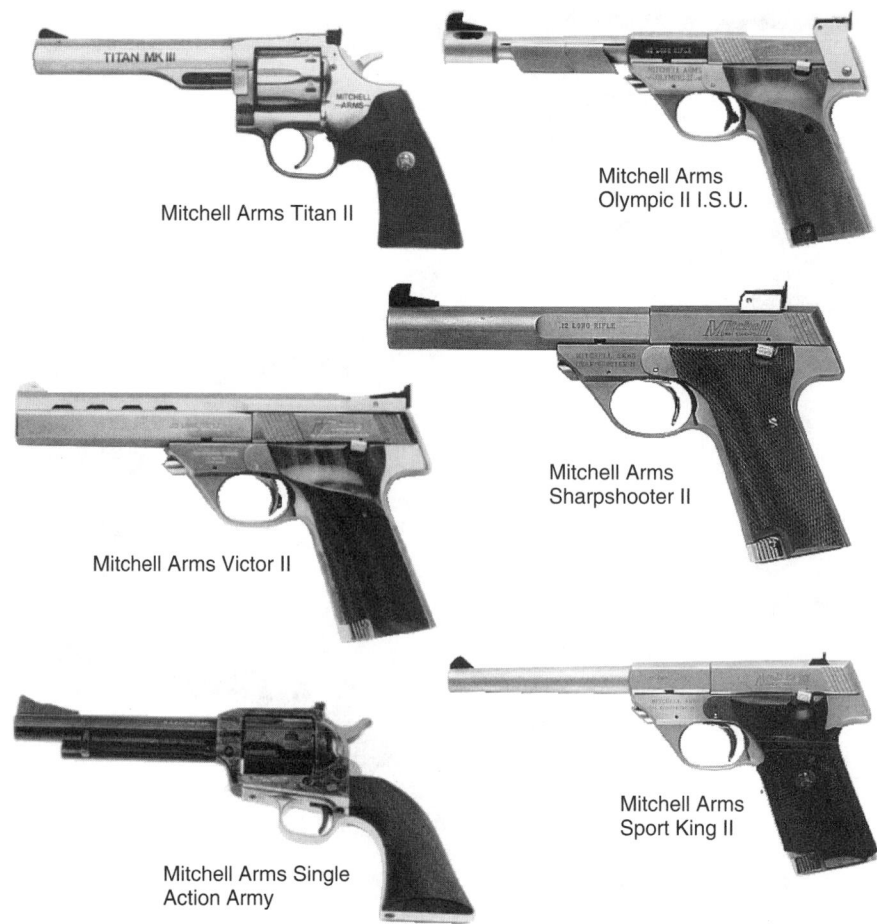

Mitchell Arms Titan II

Mitchell Arms Olympic II I.S.U.

Mitchell Arms Sharpshooter II

Mitchell Arms Victor II

Mitchell Arms Sport King II

Mitchell Arms Single Action Army

Mitchell Arms Guardian III
Same specs as Guardian II except 3″, 4″, 6″ barrel; adjustable rear sight. Made in U.S. Marketed by Mitchell Arms. Introduced 1995; still produced. **LMSR: $305**

New: $250   **Perf.:** $225   **Exc.:** $200

MITCHELL ARMS MODEL 57A
Semi-automatic; single action; 30 Mauser; 9-shot magazine; all-steel construction; magazine safety; hammer-block safety. Made in Yugoslavia. Introduced 1990; discontinued 1990.

Perf.: $175   **Exc.:** $150   **VGood:** $125

MITCHELL ARMS MODEL 70A
Semi-automatic; single action; 9mm Para.; 9-shot magazine; all-steel construction; magazine safety; hammer-block safety. Made in Yugoslavia. Introduced 1990; discontinued 1990.

Perf.: $175   **Exc.:** $135   **VGood:** $125

MITCHELL ARMS MODEL 1875 REMINGTON
Revolver; single action; 357 Mag., 45 Colt; 6-shot cylinder; fixed sights; walnut stock; color case-hardened frame; blued finish. Introduced 1990; discontinued 1991.

Perf.: $325   **Exc.:** $250   **VGood:** $225

MITCHELL ARMS OFFICERS MODEL 88A
Semi-automatic; single action; 9mm Para.; 9-shot magazine; all-steel construction; slenderized version of Model 70A; finger extension magazine. Made in Yugoslavia. Introduced 1990; discontinued 1990.

Perf.: $250   **Exc.:** $225   **VGood:** $175

MITCHELL ARMS OLYMPIC II I.S.U.
Semi-automatic; single action; 22 Short, 22 LR; 10-shot magazine; 6 ¾″ round tapered barrel with stabilizer; 11 ¼″ overall length; weighs 40 oz.; checkered walnut thumbrest grips; undercut ramp front sight, frame-mounted click-adjustable square notch rear; integral stabilizer with two removable weights; trigger adjustable for pull and over-travel; stippled front and backstraps; push-button barrel takedown; blue finish or stainless or combo. Guns are marked with "High-Standard" depending on date of manufacture. Made in U.S. Marketed by Mitchell Arms. Introduced 1992; still produced. **LMSR: $599**

New: $600   **Perf.:** $525   **Exc.:** $400

MITCHELL ARMS ROLLING BLOCK TARGET
Single shot; 22 LR, 22 WMR, 357 Mag., 45 Colt, 223 Rem.; 9 ⅞″ half-round, half-octagon barrel; 14″ overall length; weighs 44 oz.; walnut grip and forend; blade front sight, fully-adjustable rear; replica of the 1871 rolling block target pistol; brass trigger guard; color case-hardened frame; blue barrel. Imported from Italy by Mitchell Arms, Inc. Introduced 1992; dropped 1993.

Perf.: $250   **Exc.:** $200   **VGood:** $175

MITCHELL ARMS SHARPSHOOTER II
Semi-automatic; single action; 22 LR; 10-shot magazine; 5 ½″ bull barrel; 10 ¼″ overall length; weighs 45 oz.; checkered walnut thumbrest grips; ramp front sight, slide-mounted square notch fully-adjustable rear; military grip; slide lock; smooth gripstraps; push-button takedown; drilled and tapped for barrel weights; stainless steel, blue or combo. Made in U.S. Marketed by Mitchell Arms, Inc. Introduced 1992.

New: $300   **Perf.:** $250   **Exc.:** $200

MITCHELL ARMS SINGLE ACTION ARMY
Revolver; single action; 22 LR, 357 Mag., 44 Mag., 45 ACP, 45 Colt; 6-shot cylinder; 4 ¾″, 5 ½″, 6″, 7 ½″, 10″, 12″, 18″ barrel; ramp front, adjustable rear sight; one-piece walnut grips; brass grip frame; color case-hardened frame; hammer-block safety; blued finish. Introduced 1986; no longer imported.

Perf.: $250   **Exc.:** $225   **VGood:** $200

Mitchell Arms Single Action Army, U.S. Army Model
Same specs as Single Action Army except 357 Mag., 45 ACP, 45 Colt; 5 ½″ barrel; serrated ramp front sight, fixed or adjustable rear; brass or steel backstrap/trigger guard; bright nickel-plated model and dual cylinder models available. Was imported by Mitchell Arms, Inc. No longer imported.

Perf.: $250   **Exc.:** $225   **VGood:** $200
Nickel finish
Perf.: $275   **Exc.:** $250   **VGood:** $225
With 45 Colt/45 ACP dual cylinder
Perf.: $325   **Exc.:** $300   **VGood:** $275

Mitchell Arms Single Action Army, Cavalry Model
Same specs as Single Action Army except 357 Mag., 45 ACP, 45 Colt; 7 ½″ barrel; serrated ramp front sight, fixed or adjustable rear; brass or steel backstrap/trigger guard; bright nickel-plated model and dual cylinder models available. Was imported by Mitchell Arms, Inc. No longer imported.

Perf.: $250   **Exc.:** $225   **VGood:** $200
Nickel finish
Perf.: $275   **Exc.:** $250   **VGood:** $225
With 45 Colt/45 ACP dual cylinder
Perf.: $325   **Exc.:** $300   **VGood:** $275

Mitchell Arms Single Action Army, Cowboy Model
Same specs as Single Action Army except 357 Mag., 45 ACP, 45 Colt; 4 ¾″ barrel; serrated ramp front sight, fixed or adjustable rear; brass or steel backstrap/trigger guard; bright nickel-plated model and dual cylinder models available. Was imported by Mitchell Arms, Inc. No longer imported.

Perf.: $250   **Exc.:** $225   **VGood:** $200
Nickel finish
Perf.: $275   **Exc.:** $250   **VGood:** $225
With 45 Colt/45 ACP dual cylinder
Perf.: $275   **Exc.:** $250   **VGood:** $225

Mossberg Brownie

Nagant Model 1895 Training Model

MOA Maximum

MKE Kirikkale

Nagant Model 1895

Mossberg Abilene

HANDGUNS

Mitchell Arms
Skorpion
Spectre
Sport King II
Titan II
Titan III
Trophy II
Victor II
MKE
Kirikkale
MOA
Maximum
Morini
Model CM-80
Model CM-80 Super
Competition
Model 84E Free Pistol
Mossberg
Abilene
Brownie
M-S Safari Arms
(See Safari Arms)
Nagant
Model 1895
Model 1895 Training Model

MITCHELL ARMS SKORPION

Semi-automatic; single action; 32 ACP; 20-, 30-shot magazine; 4 5/8″ barrel; blued finish. Made in Yugoslavia. Introduced 1990; discontinued 1990.
Perf.: $700   **Exc.:** $600   **VGood:** $500

MITCHELL ARMS SPECTRE

Semi-automatic; single action; 9mm Para.; 30-, 50-shot magazine; 8″ barrel, shrouded; weighs 64 oz.; blued finish. Made in Yugoslavia. Introduced 1987; discontinued 1988.
Exc.: $600   **VGood:** $550   **Good:** $450

MITCHELL ARMS SPORT KING II

Semi-automatic; single action; 22 LR; 10-shot magazine; 4 1/2″, 6 3/4″ barrel; 9″ overall length (4 1/2″ barrel); weighs 39 oz.; checkered walnut or black plastic grips; blade front sight, windage-adjustable rear; military grip; standard trigger; push-button barrel takedown; stainless steel or blue. Guns are marked with "High Standard" depending on date of manufacture. Made in U.S. Marketed by Mitchell Arms, Inc. Introduced 1992; still produced. **LMSR: $325**
New: $250   **Perf.:** $225   **Exc.:** $200

MITCHELL ARMS TITAN II

Revolver; double action; 357 Mag.; 6-shot cylinder; 2″, 4″, 6″ barrel; 7 3/4″ overall length (2″ barrel); weighs 38 oz.; Pachmayr black rubber grips, combat or target; blade front, fixed rear sight; crane-mounted cylinder release; shrouded ejector rod; blue or stainless steel. Made in U.S. from Mitchell Arms, Inc. Introduced 1995; still produced. **LMSR: $339**
New: $250   **Perf.:** $225   **Exc.:** $200

Mitchell Arms Titan III

Same specs as the Titan II except adjustable rear sight. Made in U.S. Marketed by Mitchell Arms, Inc. Introduced 1995; still produced. **LMSR: $429**
New: $325   **Perf.:** $300   **Exc.:** $250

MITCHELL ARMS TROPHY II

Semi-automatic; single action; 22 LR; 10-shot magazine; 5 1/2″ bull, 7 1/4″ fluted barrel; 9 3/4″ overall length (5 1/2″ barrel); weighs 44 1/2 oz.; checkered walnut thumbrest grips; undercut ramp front sight, click-adjustable frame-mounted rear; grip duplicates feel of military 45; positive action magazine latch; front and backstraps stippled; trigger adjustable for pull, over-travel; gold-filled roll marks, gold-plated trigger, safety, magazine release; push-button barrel takedown; stainless or blue finish. Made in U.S. Marketed by Mitchell Arms, Inc. Introduced 1992; still produced. **LMSR: $498**
New: $400   **Perf.:** $350   **Exc.:** $275

MITCHELL ARMS VICTOR II

Semi-automatic; single action; 22 LR; 10-shot magazine; 4 1/2″ vent rib, 5 1/2″ vent, dovetail or Weaver rib barrels; 9 3/4″ overall length; weighs 44 oz.; military-type checkered walnut thumbrest or rubber grips; blade front sight, fully-adjustable rear mounted on rib; push-button takedown for barrel interchangeability; bright stainless steel combo or royal blue finish. Made in U.S. Marketed by Mitchell Arms. Introduced 1994; still produced. **LMSR: $595**
With 4 1/2″ vent rib barrel
  New: $500   **Perf.:** $450   **Exc.:** $400
With 5 1/2″ dovetail rib barrel
  New: $550   **Perf.:** $500   **Exc.:** $450
With 5 1/2″ Weaver rib barrel
  New: $550   **Perf.:** $500   **Exc.:** $450

MKE KIRIKKALE

Semi-automatic; double action; 32 ACP, 380 ACP; 8-shot magazine (32 ACP), 7-shot (380 ACP); 4″ barrel; 6 1/2″ overall length; adjustable notch rear sight, fixed front; checkered plastic grips; exposed hammer; safety blocks firing pin, drops hammer; chamber-loaded indicator pin; blued finish. Copy of Walther PP. Imported from Turkey by Firearms Center, Inc., then by Mandall Shooting Supplies; no longer imported.
Exc.: $250   **VGood:** $200   **Good:** $175

MOA MAXIMUM

Single shot; falling block action; 28 standard chamberings from 22 to 44 caliber; 8 3/4″, 10 3/4″, 14″ interchangeable barrel; ramp front sight, fully-adjustable open rear; drilled and tapped for scope mounts; integral grip frame/receiver; smooth walnut stocks, forend; adjustable trigger; Armaloy finish. Introduced 1983; still in production.
New: $650   **Perf.:** $550   **Exc.:** $475

MORINI MODEL CM-80

Single shot; 22 LR; 10″ free-floating barrel; 21 1/4″ overall length; weighs 30 oz.; adjustable or wrap-around stocks; adjustable match sights; adjustable grip/frame angle; adjustable barrel alignment; adjustable trigger weight, sight radius. Made in Italy. Introduced 1985; importation discontinued 1989.
Perf.: $850   **Exc.:** $750   **VGood:** $700

Morini Model CM-80 Super Competition

Same specs as Model CM-80 except deluxe finish; plexiglass front sighting system.
Perf.: $1000   **Exc.:** $850   **VGood:** $750

MORINI MODEL 84E FREE PISTOL

Single shot; 22 LR; 11 7/16″ barrel; 19 1/2″ overall length; weighs 44 oz.; adjustable match-type grip with stippled surfaces; interchangeable blade front sight, match-type fully-adjustable rear; fully-adjustable electronic trigger. Imported from Switzerland by Nygord Precision Products. Introduced 1995; still imported.
LMSR: $1495
New: $1250   **Perf.:** $1100   **Exc.:** $1000

MOSSBERG ABILENE

Revolver; single action; 357 Mag., 44 Mag., 45 Colt; 6-shot cylinder; 4 5/8″, 6″, 7 1/2″ barrel; serrated ramp front sight, click-adjustable rear; smooth walnut grips; wide hammer spur; transfer bar ignition; blued or Magnaloy finish. Introduced 1978 by United States Arms; taken over by Mossberg; not currently in production.
Exc.: $250   **VGood:** $200   **Good:** $175
Magnaloy finish
Exc.: $275   **VGood:** $225   **Good:** $200

MOSSBERG BROWNIE

Pocket pistol; top-break; double action; 22 LR, 22 Short; 4-shot; four 2 1/2″ barrels; revolving firing pin; steel extractor. Introduced in 1919; discontinued 1932.
Exc.: $325   **VGood:** $250   **Good:** $175

NAGANT MODEL 1895

Revolver; 7.62mm Nagant; 4 1/4″ barrel; 9 3/16″ overall length; 7-shot cylinder; checkered wood grips. Unique "gas-seal" action cams the cylinder forward to mate with end of barrel before firing. Russian military issue from 1895 to 1944.
Exc.: $150   **VGood:** $125   **Good:** $100

Nagant Model 1895 Training Model

Same specs as Model 1895 except 22 LR for training purposes. Conversions from 7.62mm revolver (scarce) or arsenal manufactured in 1937 on 7.62mm frames (rare).
Exc.: $600   **VGood:** $500   **Good:** $400

HANDGUNS

Nambu
Model 1902 First Type ("Grandpa")
Model 1902 Second Type ("Papa")
Model 1902 7mm ("Baby")
Type 14 (1925)
Type 26 Revolver
Type 94

Navy Arms
Frontier Model
Frontier, Buntline Model
Frontier, Target Model
Grand Prix
High Power
Luger
Model 1873
Model 1873, Economy Model
Model 1873, U.S. Cavalry Model
Model 1875 Remington

Nambu Model 1902 First Type ("Grandpa")

Nambu Type 26 Revolver

Nambu Model 1902 Second Type ("Papa")

Nambu Model 1902 7mm ("Baby")

Nambu Type 14 (1925)

NAMBU MODEL 1902 FIRST TYPE ("GRANDPA")
Semi-Automatic; 8mm Nambu; 8-shot magazine; 4 3/4" barrel; 9" overall lengh; adjustable rear (tangent) and front sights; blued finish; checkered wood grips and magazine bottom; grip safety; rear of frame slotted for shoulder-stock holster. Introduced 1903; discontinued 1906. Add $3000 for original (beware of reproductions!) shoulder-stock holster.
Exc.: $4000   **VGood:** $2500   **Good:** $1500

NAMBU MODEL 1902 SECOND TYPE ("PAPA")
Semi-automatic; same specs as first type except not slotted (except first few hundred) for or supplied with shoulder-stock holster; aluminum bottom instead of wood bottom magazine; flexible instead of fixed lanyard ring. Introduced 1906; discontinued 1928.
Exc.: $1500   **VGood:** $1000   **Good:** $700

NAMBU MODEL 1902 7MM ("BABY")
Semi-automatic; 7mm Nambu; 7-shot magazine; 3 1/4" barrel; 6 3/4" overall length; fixed rear, adjustable front sights; blued finish; checkered wood grips; grip safety. Introduced 1903; dropped 1929.
Exc.: $2500   **VGood:** $1750   **Good:** $1250

NAMBU TYPE 14 (1925)
Semi-automatic; 8mm Nambu; 8-shot magazine; 4 5/8" barrel; 9" overall length; fixed rear, adjustable front sights; blued finish; grooved wood grips. Introduced 1926; discontinued 1945. Very early production brings a small premium price.
Exc.: $400   **VGood:** $250   **Good:** $150

NAMBU TYPE 26 REVOLVER
Revolver; 9mm Japanese; 6-shot; 4 3/4" barrel; 9" overall length; fixed sights; blued finish; checkered or grooved (rare late production) wood grips. Introduced 1893; discontinued 1935.
Exc.: $325   **VGood:** $250   **Good:** $185

NAMBU TYPE 94
Semi-automatic; 8mm Nambu; 6-shot magazine; 3 3/4" barrel; 7 5/16" overall length; fixed rear, adjustable front sights; blued finish; checkered plastic or smooth wood (late war) grips. Introduced 1935; discontinued 1945.
Exc.: $375   **VGood:** $275   **Good:** $175

NAVY ARMS FRONTIER MODEL
Revolver; single action; 357 Mag., 45 Colt; 6-shot cylinder; 4 1/2", 5 1/2", 7 1/2" barrel; fixed sights; uncheckered one-piece walnut grip; brass grip frame; color case-hardened frame; blued barrel and cylinder. Manufactured in Italy. Introduced 1976; discontinued 1978.
Exc.: $300   **VGood:** $225   **Good:** $175

Navy Arms Frontier, Buntline Model
Same specs as Frontier except 16 1/2" barrel; detachable shoulder stock.
Exc.: $450   **VGood:** $375   **Good:** $300

Navy Arms Frontier, Target Model
Same specs as Frontier except ramp front, adjustable rear sight.
Exc.: $275   **VGood:** $225   **Good:** $175

NAVY ARMS GRAND PRIX
Single shot; rolling block; 44 Mag., 30-30, 7mm Spl.; 45-70; 13 3/4" barrel; weighs 64 oz.; adjustable target sights; walnut forend, thumbrest grip; adjustable aluminum barrel rib; matte blue finish. Introduced 1983; discontinued 1985.
Exc.: $300   **VGood:** $250   **Good:** $185

NAVY ARMS HIGH POWER
Semi-automatic; single action; 9mm Para.; 13-shot magazine; 4 5/8" barrel; fixed sights; black plastic grips; similar to older FN manufacture; blued finish. Introduced 1993; no longer in production.
New: $350   **Perf.:** $275   **Exc.:** $200

NAVY ARMS LUGER
Semi-automatic; 22 LR; 10-shot magazine; 4", 6", 8" barrel; 9" overall length; weighs 44 oz.; fixed sights; checkered walnut stocks; all-steel construction; blowback toggle action; blued finish. Made in U.S. Introduced 1986; discontinued 1987.
Exc.: $135   **VGood:** $110   **Good:** $100

NAVY ARMS MODEL 1873
Revolver; single action; 44-40, 45 Colt; 6-shot cylinder; 3", 4 3/4", 5 1/2", 7 1/2" barrels; 10 3/4" overall length (5 1/2" barrel); weighs 47 oz.; blade front sight, grooved rear; uncheckered walnut grips; nickel or blued finish with color case-hardened frame. Made in Italy. Introduced 1991; still imported.
New: $325   **Perf.:** $275   **Exc.:** $225
Nickel
New: $375   **Perf.:** $325   **Exc.:** $275

Navy Arms Model 1873, Economy Model
Same specs as standard Model 1873 single action except brass trigger guard, backstrap; two piece walnut grip. Introduced 1993; still in production.
New: $250   **Perf.:** $200   **Exc.:** $175

Navy Arms Model 1873, U.S. Cavalry Model
Same specs as standard Model 1873 single action except 7 1/2" barrel chambered only for 45 Colt; arsenal markings. Introduced 1991; still imported. **LMSR: $480**
New: $375   **Perf.:** $300   **Exc.:** $275

NAVY ARMS MODEL 1875 REMINGTON
Revolver; single action; 44-40, 45 Colt; 6-shot cylinder; 7 1/2" barrel; 13" overall length; weighs 41 oz.; blade front sight, grooved rear; uncheckered walnut grips; color case-hardened frame; brass trigger guard; balance blued finish. Made in Italy. Introduced 1991; discontinued 1992; resumed 1994; discontinued 1995.
New: $350   **Perf.:** $275   **Exc.:** $225

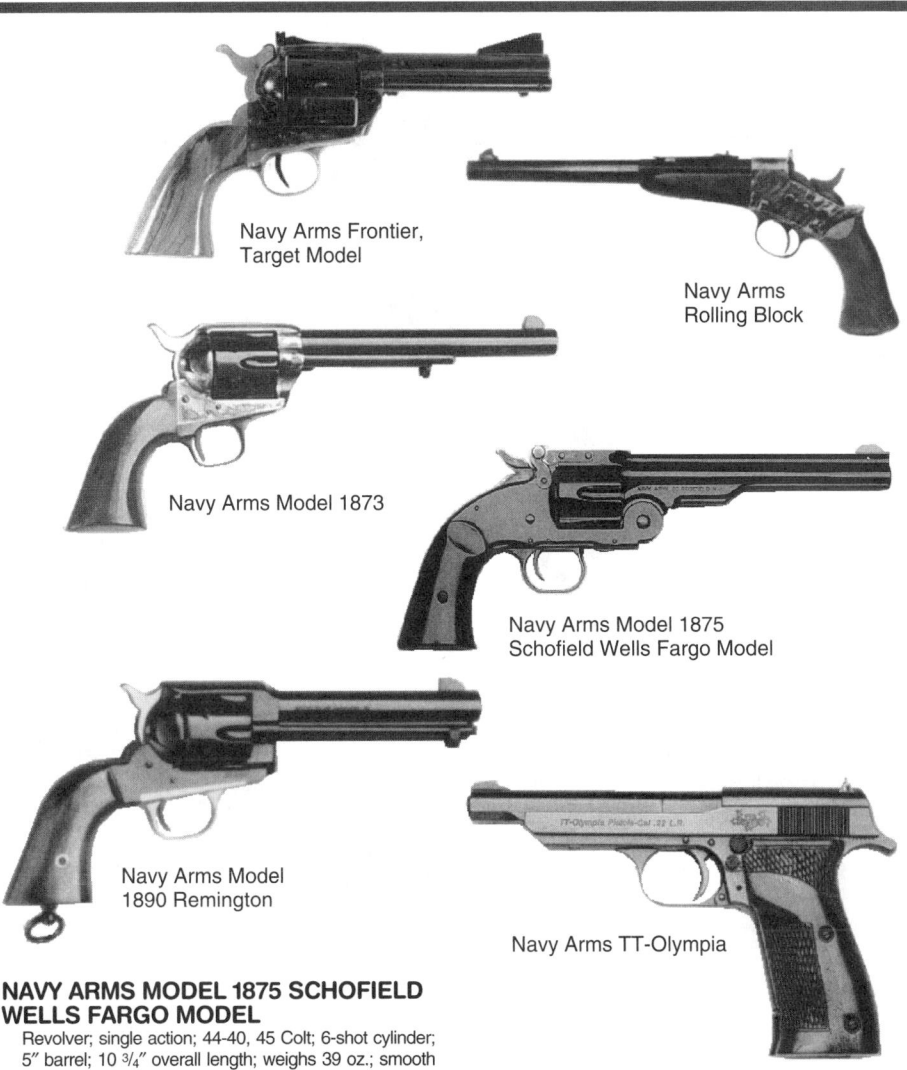

Navy Arms Frontier,
Target Model

Navy Arms
Rolling Block

Navy Arms Model 1873

Navy Arms Model 1875
Schofield Wells Fargo Model

Navy Arms Model
1890 Remington

Navy Arms TT-Olympia

Navy Arms
Model 1875 Schofield Wells
Fargo Model
Model 1875 Schofield U.S.
Cavalry Model
Model 1890 Remington
Model 1895 SAA, Artillery
Model
Rolling Block
TT-Olympia
TU-90
TU-711 Mauser
U.S. Government Model

New Advantage
Derringer

New Detonics
Combat Master
(See Detonics Comabt Master)
Compmaster
(See Detonics Servicemaster)
Jade Escort
(See New Detonics Ladies
Escort)
Ladies Escort
Midnight Escort
(See New Detonics Ladies
Escort)
Royal Escort
(See New Detonics Ladies
Escort)
Scoremaster
(See Detonics Scoremaster)
Servicemaster
(See Detonics Servicemaster)

NAVY ARMS MODEL 1875 SCHOFIELD WELLS FARGO MODEL
Revolver; single action; 44-40, 45 Colt; 6-shot cylinder; 5″ barrel; 10 3/4″ overall length; weighs 39 oz.; smooth walnut grips; blade front sight, notch rear; replica of Smith & Wesson Model 3 Schofield; top-break action with automatic ejection; polished blue finish. Imported by Navy Arms. Introduced 1994; still imported.
   New: $600   **Perf.:** $500   **Exc.:** $400

Navy Arms Model 1875 Schofield U.S. Cavalry Model
Same specs as the Wells Fargo model except 7″ barrel; original-type military markings. Imported by Navy Arms. Introduced 1994; still imported. **LMSR:** $795
   New: $600   **Perf.:** $500   **Exc.:** $400

NAVY ARMS MODEL 1890 REMINGTON
Revolver; single action; 44-40; 6-shot cylinder; 5 1/2″ barrel; 10 3/4″ overall length; weighs 39 oz.; blade front, grooved rear sight; walnut grip; brass trigger guard; lanyard loop; blued finish. Made in Italy. Introduced 1991; discontinued 1992; resumed 1994; discontinued 1995.
   New: $350   **Perf.:** $275   **Exc.:** $225

NAVY ARMS MODEL 1895 SAA, ARTILLERY MODEL
Revolver; single action; 44-40, 45 Colt; 6-shot cylinder; 5 1/2″ barrel; 10 3/4″ overall length; weighs 47 oz.; blade front, grooved rear sight; walnut grip; blued finish. Introduced 1991; still imported.
   New: $400   **Perf.:** $325   **Exc.:** $225

NAVY ARMS ROLLING BLOCK
Single shot; 22 LR, 22 Hornet, 357 Mag.; 8″ barrel; 12″ overall length; adjustable sights; uncheckered walnut stocks, forearm; color case-hardened frame; brass trigger guard; blued barrel. Manufactured in Italy. Introduced 1965; Hornet chambering discontinued 1975; 22 LR discontinued 1979. No longer imported.
   Exc.: $200   **VGood:** $150   **Good:** $135

NAVY ARMS TT-OLYMPIA
Semi-automatic; single action; 22 LR; 4 5/8″ barrel; 8″ overall length; weighs 28 oz.; checkered hardwood grips; blade front sight, windage-adjustable rear; reproduction of the Walther Olympia pistol; polished blue finish. Imported by Navy Arms. Introduced 1992; no longer imported.
   New: $200   **Perf.:** $150   **Exc.:** $125

NAVY ARMS TU-90
Semi-automatic; single action; 30 Tokarev, 9mm Para.; 8-shot magazine; 4 1/2″ barrel; weighs 30 oz.; wraparound synthetic grip; similar to TT-33 Tokarev. Made in China. Introduced 1992; no longer imported.
   New: $100   **Perf.:** $90   **Exc.:** $70

NAVY ARMS TU-711 MAUSER
Semi-automatic; 9mm Para.; 10-, 20-shot magazine; 5 1/4″ barrel; weighs 43 oz.; similar to Mauser 711. Introduced 1992; discontinued 1992.
   Perf.: $550   **Exc.:** $500   **VGood:** $450

NAVY ARMS U.S. GOVERNMENT MODEL
Semi-automatic; 45 ACP; 7-shot; 5″ barrel; fixed sights; checkered walnut grips; standard G.I. 1911 issue pistol; blue finish. Introduced 1993; no longer in production.
   New: $325   **Perf.:** $275   **Exc.:** $200

NEW ADVANTAGE DERRINGER
Double action; 22 LR, 22 WMR; 4-shot; four 2 1/2″ barrels; 4 1/2″ overall length; weighs 15 oz.; fixed sights; smooth walnut stocks; revolving firing pin; rebounding hammer; polished blue finish. Reintroduced, 1989 by New Advantage Arms Corp.; still in production.
   Perf.: $175   **Exc.:** $150   **VGood:** $135

New Advantage
Derringer

NEW DETONICS LADIES ESCORT
Semi-automatic; single action; 45 ACP; 6-shot magazine; 3 1/2″ barrel; 4 1/2″ overall length; weighs 26 oz.; checkered walnut grips; rubber mainspring housing; ramp front sight, adjustable rear; reduced grip frame size; color polymer finish. Made in U.S. by New Detonics Corp. Introduced 1990; dropped 1991.
Royal Escort, iridescent purple slide, blackened stainless frame, gold-plated hammer and trigger
   Perf.: $750   **Exc.:** $650   **VGood:** $500
Jade Escort, jade-colored slide, satin stainless frame
   Perf.: $750   **Exc.:** $650   **VGood:** $500
Midnight Escort, black slide, satin stainless frame
   Perf.: $800   **Exc.:** $700   **VGood:** $600

HANDGUNS

New England Firearms
Standard Revolvers
(See New England Firearms
Model R92, Model R73
Revolvers)
Lady Ultra
Model R73
Model R92
Single Shot
Top Break
Ultra

Norinco
Model 1911A1
Model 77B
MP-20 (See Norinco Model 77B)
Type 54-1 Tokarev
Type 59 Makarov

North American Arms
Mini-Master Black Widow
Mini-Master Target
Model 22 LR

New England
Firearms Lady Ultra

Norinco Type 54-1
Tokarev

New England
Firearms Model R92

North American Arms
Single Action Revolver

Norinco Type 59
Makarov

North American Arms
Mini-Master Target

NEW ENGLAND FIREARMS LADY ULTRA
Revolver; double action; 32 H&R Mag.; 5-shot cylinder; 3" barrel; 7 1/4" overall length; weighs 31 oz.; walnut-finished hardwood grips with NEF medallion; blade front sight, fully-adjustable rear; swing-out cylinder; polished blue finish; comes with lockable storage case. From New England Firearms Co. Introduced 1992; still produced. **LMSR: $166**
    New: $135   **Perf.:** $115   **Exc.:** $100

NEW ENGLAND FIREARMS MODEL R73
Revolver; double action; 32 H&R Mag.; 5-shot cylinder; 3" barrel; 7 1/4" overall length; weighs 31 oz.; walnut finished American hardwood grips with NEF medallion; fixed sights; blue or nickel finish. From New England Firearms Co. Introduced 1988; still in production.
    New: $120   **Perf.:** $100   **Exc.:** $75

NEW ENGLAND FIREARMS MODEL R92
Revolver; double action; 22 LR; 9-shot cylinder; 2 1/2", 4" barrel; 8 1/2" overall length (4" barrel); weighs 26 oz.; walnut-finished American hardwood grips with NEF medallion; fixed sights; blue or nickel finish. From New England Firearms Co. Introduced 1988; still produced.
    New: $110   **Perf.:** $100   **Exc.:** $85

NEW ENGLAND FIREARMS SINGLE SHOT
Single shot; rotary cannon-type action; 357 Mag., 357 Rem. Maximum, 44 Mag., 223, 30-30, 7-30 Waters, 35 Rem.; 14" barrel; 15 1/8" overall length; weighs 49 oz. (synthetic stock), 70 oz. (laminated wood); ramp front sight, adjustable rear; action cocks on opening; single stage trigger; trigger and sliding thumb safeties; matte blue finish. Made by Competitor Corp. Marketed by New England Firearms. Introduced 1995; still produced. **LMSR: $380**
    New: $350   **Perf.:** $300   **Exc.:** $275

NEW ENGLAND FIREARMS TOP BREAK
Revolver; double action; 22 LR; 9-shot cylinder; 4", 6" barrel; weighs 30 oz. (4" barrel); walnut-finished hardwood grips; elevation-adjustable front sight, windage-

adjustable rear; ventilated barrel rib; automatic ejection; blue finish. Made in U.S. by New England Firearms. Introduced 1991; no longer produced.
    Perf.: $125   **Exc.:** $115   **VGood:** $100

NEW ENGLAND FIREARMS ULTRA
Revolver; double action; 22 LR, 22 WMR; 9-shot (22 LR), 6-shot (22 WMR); 4", 6" barrel; 10 5/8" overall length (6" barrel); weighs 36 oz.; walnut-finished hardwood grips with NEF medallion; blade front sight, fully-adjustable rear; bull-style barrel with recessed muzzle; high "Lustre" blue/black finish. Made in U.S. by New England Firearms. Introduced 1989; still produced. **LMSR: $166**
    New: $135   **Perf.:** $115   **Exc.:** $100

NORINCO MODEL 1911A1
Semi-automatic; single action; 45 ACP; 7-shot magazine; 5" barrel; 8 1/2" overall length; weighs 39 oz.; checkered wood grips; blade front sight, windage-adjustable rear; matte blue finish. Comes with two magazines. Imported from China by China Sports, Inc. No longer imported.
    New: $250   **Perf.:** $200   **Exc.:** $175

NORINCO MODEL 77B
Semi-automatic; single action; 9mm Para.; 8-shot magazine; 5" barrel; 7 1/2" overall length; weighs 34 oz.; checkered wood grips; blade front sight, adjustable rear; gas-retarded recoil action; front of trigger guard able to cock the action with the trigger finger. Imported from China as the NP-20. Introduced 1989; no longer imported by China Sports, Inc.
    Perf.: $200   **Exc.:** $175   **VGood:** $150

NORINCO TYPE 54-1 TOKAREV
Semi-automatic; single action; 7.62x25mm, 9mm Para.; 8-shot magazine; 4 1/2" barrel; 7 3/4" overall length; weighs 29 oz.; grooved black plastic grips; fixed sights; matte blue finish. Imported from China. No longer imported by China Sports, Inc.
    New: $125   **Perf.:** $110   **Exc.:** $65

NORINCO TYPE 59 MAKAROV
Semi-automatic; double action; 9x18mm, 380 ACP; 8-shot magazine; 3 1/2" barrel; weighs 21 oz.; 6 3/8" overall length; checkered plastic grips; blade front sight, adjustable rear; blue finish. Direct copy of Russian-made pistol. Imported from China. Introduced 1990; no longer imported by China Sports, Inc.
    New: $185   **Perf.:** $140   **Exc.:** $125

NORTH AMERICAN ARMS MINI-MASTER BLACK WIDOW
Revolver; single action; 22 LR, 22 WMR; 5-shot cylinder; 2" heavy vent barrel; 5 7/8" overall length; weighs 8 7/8 oz.; black rubber grips; Millett Low Profile fixed sights or Millett sight adjustable for elevation only; built on the 22 WMR frame; non-fluted cylinder. Made in U.S. by North American Arms. Introduced 1989; still produced.
With adjustable sight
    New: $200   **Perf.:** $185   **Exc.:** $150
With adjustable sight, extra LR/WMR cylinder
    New: $225   **Perf.:** $210   **Exc.:** $200
With fixed sight
    New: $185   **Perf.:** $175   **Exc.:** $150
With fixed sight, extra LR/WMR cylinder
    New: $210   **Perf.:** $200   **Exc.:** $185

NORTH AMERICAN ARMS MINI-MASTER TARGET
Revolver; single action; 22 LR, 22 WMR; 5-shot; heavy vent-rib 4" barrel; 7 3/4" overall length; weighs 10 3/4 oz.; blade-type front sight, elevation-adjustable white-outline rear; checkered hard black rubber stocks. Introduced 1989; still in production.
    New: $235   **Perf.:** $210   **Exc.:** $190

NORTH AMERICAN ARMS MODEL 22 LR
Revolver; single action; 22 Short, 22 LR; 5-shot; 1 1/8", 1 5/8", 2 1/2" barrel; 3 7/8" overall length; weighs 4 1/2 oz.; fixed sights; plastic or rosewood grips; stainless steel construction. Introduced 1976; still in production.
    New: $135   **Perf.:** $115   **Exc.:** $100

ODI Viking

Ortgies Pocket Pistol

Olympic Arms Schuetzen
Pistol Works Big Deuce

Olympic Arms Schuetzen
Pistol Works Griffon

North American
Arms Model 22 LR

HANDGUNS

North American Arms
Model 22 Magnum
Model 22 Magnum
Convertible
Single Action Revolver
Viper Belt Buckle

ODI
Viking

Olympic Arms Schuetzen Pistol Works
Big Deuce
Crest Series
Griffon

Ortgies
Pocket Pistol
Vest Pocket Pistol

Pachmayr
Dominator

Pachmayr Dominator

NORTH AMERICAN ARMS MODEL 22 MAGNUM

Revolver; single action; 22 WMR; 5-shot; 1 1/8", 1 5/8", 2 1/2" barrel; 3 7/8" overall length; weighs 4 1/2 oz.; fixed sights; plastic or rosewood grips; stainless steel construction. Introduced 1976; still in production. **LMSR: $178**

New: $150   **Perf.:** $135   **Exc.:** $125

North American Arms Model 22 Magnum Convertible

Same specs as 22 Magnum, except supplied with extra 22 LR cylinder. **LMSR: $210**

New: $185   **Perf.:** $170   **Exc.:** $150

NORTH AMERICAN ARMS SINGLE ACTION REVOLVER

Revolver; single action; 45 Win. Mag., 450 Mag. Express; 5-shot; 7 1/2", 10 1/2" barrel; weighs 52 oz.; adjustable rear sight, blade front; uncheckered walnut stocks; stainless steel construction; matte finish. Introduced 1984; discontinued 1985.

Perf.: $1100   **Exc.:** $850   **VGood:** $750

High polish finish

Perf.: $1350   **Exc.:** $1000   **VGood:** $850

NORTH AMERICAN ARMS VIPER BELT BUCKLE

Revolver; single action; 22 LR; 5-shot; 1 1/8", 1 5/8" barrel; weighs 4 1/2 oz.; fixed sights; plastic or rosewood grips; belt buckle with built-in revolver; stainless steel construction. Introduced 1976; discontinued 1990; reintroduced 1993; no longer in production.

New: $175   **Perf.:** $150   **Exc.:** $135

ODI VIKING

Semi-automatic; double action; 45 ACP; 7-shot; 5" barrel; weighs 39 oz.; fixed notched rear sight, blade front; smooth teak stocks; Seecamp double-action system; spur-type hammer; stainless steel construction; brushed satin finish. Introduced 1982; discontinued 1985.

Perf.: $450   **Exc.:** $400   **VGood:** $350

OLYMPIC ARMS SCHUETZEN PISTOL WORKS BIG DEUCE

Semi-automatic; single action; 45 ACP; 7-shot magazine; 6" barrel; 9 1/2" overall length; weighs 40 oz.; smooth walnut grips; ramped blade front sight, LPA adjustable rear; stainless steel barrel; beavertail grip safety; extended thumb safety and slide release; Commander-style hammer; throated, polished and tuned; Parkerized matte black slide with satin stainless steel frame. Made in U.S. by Olympic Arms, Inc.'s specialty shop Schuetzen Pistol Works. Marked "Schuetzen Pistol Works" on the slide; "Safari Arms" on the frame. Introduced 1995; still produced. **LMSR: $1035**

New: $900   **Perf.:** $800   **Exc.:** $700

OLYMPIC ARMS SCHUETZEN PISTOL WORKS CREST SERIES

Semi-automatic; single action; 45 ACP; 6-, 7-shot magazine; 4 1/2" (4-star), 5", 5 1/2" barrel; 8 1/2" overall length; weighs 39 oz.; checkered walnut grips; ramped blade front sight, LPA adjustable rear; stainless steel barrel; right- or left-hand models available; long aluminum trigger; full-length recoil spring guide; throated, polished, tuned; satin stainless steel. Made in U.S. by Olympic Arms, Inc.'s specialty shop Schuetzen Pistol Works. Marked "Schuetzen Pistol Works" on the slide; "Safari Arms" on the frame. Introduced 1993; still produced. **LMSR: $815**

Right-hand

Perf.: $650   **Exc.:** $600   **VGood:** $550

Left-hand

Perf.: $750   **Exc.:** $700   **VGood:** $650

OLYMPIC ARMS SCHUETZEN PISTOL WORKS GRIFFON

Semi-automatic; single action; 45 ACP; 10-shot magazine; 5" barrel; 8 1/2" overall length; smooth walnut grips; ramped blade front sight, LPA adjustable rear; stainless barrel; beavertail grip safety; long aluminum trigger; full-length recoil spring guide; Commander-style hammer; throated, polished and tuned; grip size comparable to standard 1911; satin stainless steel finish. Made in U.S. by Olympic Arms, Inc.'s specialty shop Schuetzen Pistol Works. Marked "Schuetzen Pistol Works" on the slide; "Safari Arms" on the frame. Introduced 1995; still produced. **LMSR: $NA**

ORTGIES POCKET PISTOL

Semi-automatic; blowback action; 32 ACP, 380 ACP; 8-shot (32), 7-shot (380); 3 1/4" barrel; 6 1/2" overall length; fixed sights; uncheckered walnut grips; constructed without screws, uses pins and spring-loaded catches; grip safety protrudes only when firing pin is cocked; blued finish. Introduced about 1919; discontinued 1926.

Exc.: $200   **VGood:** $175   **Good:** $135

Nickel finish

Exc.: $300   **VGood:** $250   **Good:** $150

ORTGIES VEST POCKET PISTOL

Semi-automatic; blowback action; 25 ACP; 6-shot; 2 3/4" barrel; 5 3/16" overall length; fixed sights; uncheckered walnut grips; blued finish. Introduced 1920; discontinued 1926.

Exc.: $225   **VGood:** $200   **Good:** $175

Nickel finish

Exc.: $275   **VGood:** $250   **Good:** $200

PACHMAYR DOMINATOR

Single shot; bolt action on 1911A1 frame; 22 Hornet, 223, 7mm-06, 308, 35 Rem., 44 Mag.; 10 1/2" barrel in 44 Mag., 14" all other calibers; weighs 4 lbs. (14" barrel); 16" overall length; Pachmayr Signature system grips; optional adjustable sights or drilled and tapped for scope mounting. From Pachmayr. Introduced 1988; discontinued 1994.

New: $400   **Perf.:** $375   **Exc.:** $325

Para-Ordnance
P-12.45
P-13.45
P-14.40
P-14.45
P-15.40
P-16.40

Pardini
GP Rapid Fire Match
GP Rapid Fire Match
Schuman
K50 Free Pistol
Model HP Target
Model SP Ladies

Para-Ordnance
P-16.40

Para-Ordnance
P-14.45

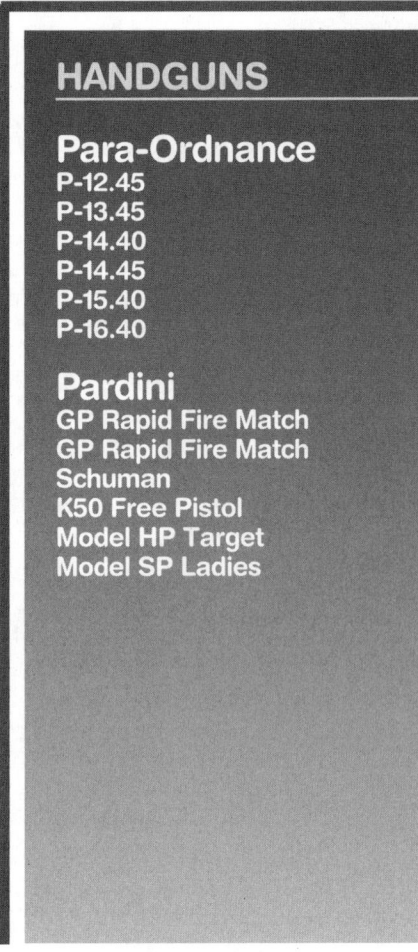

Para-Ordnance
P-13.45

Phelps Heritage I

PARA-ORDNANCE P-12.45

Semi-automatic; single action; 45 ACP; 10-shot magazine (pre-1994 guns have 12-shot); 3 1/2" barrel; 7 1/8" overall length; weighs 34 oz. (steel and stainless frame), 26 oz. (alloy frame); textured composition grips; blade front sight, adjustable rear; high visibility three-dot system; matte black finish; grooved match trigger; rounded combat-style hammer; beveled magazine well; manual thumb, grip and firing pin safeties; solid barrel bushing. Made in Canada by Para-Ordnance. Introduced 1990; still produced. **LMSR: $745**

Steel frame, matte black
New: $650	**Perf.:** $600	**Exc.:** $500

Alloy frame, matte black
New: $600	**Perf.:** $550	**Exc.:** $450

Stainless frame, stainless
New: $700	**Perf.:** $650	**Exc.:** $550

Stainless frame, Duotone
New: $685	**Perf.:** $635	**Exc.:** $575

PARA-ORDNANCE P-13.45

Semi-automatic; single action; 45 ACP; 10-shot magazine (pre-1994 guns have 13-shot); 4 1/4" barrel; 7 3/4" overall length; weighs 36 oz. (steel and stainless frame), 28 oz. (alloy frame); textured composition grips; blade front sight, adjustable rear; high visibility three-dot system; matte black finish; grooved match trigger; rounded combat-style hammer; beveled magazine well; manual thumb, grip and firing pin safeties; solid barrel bushing. Made in Canada by Para-Ordnance. Introduced 1993; still produced. **LMSR: $745**

Steel frame
New: $650	**Perf.:** $600	**Exc.:** $500

Alloy frame
New: $600	**Perf.:** $550	**Exc.:** $450

Stainless frame, stainless
New: $700	**Perf.:** $650	**Exc.:** $550

Stainless frame, Duotone
New: $685	**Perf.:** $635	**Exc.:** $575

PARA-ORDNANCE P-14.40

Semi-automatic; 40 S&W; 10-shot magazine; 3 1/2" barrel; 7 1/8" overall length; weighs 34 oz. (steel frame), 26 oz. (alloy frame); all the rest same as 14.45.

Steel frame
New: $600	**Perf.:** $550	**Exc.:** $500

Alloy frame
New: $600	**Perf.:** $550	**Exc.:** $450

Stainless frame, stainless
New: $625	**Perf.:** $575	**Exc.:** $500

Stainless frame, Duotone
New: $685	**Perf.:** $635	**Exc.:** $575

PARA-ORDNANCE P-14.45

Semi-automatic; single action; 45 ACP; 10-shot magazine (pre-1994 guns have 14-shot); 5" barrel; 8 1/2" overall length; weighs 40 oz. (steel frame), 31 oz. (alloy frame); textured composition grips; blade front sight, adjustable rear; high visibility three-dot system; matte black finish; grooved match trigger; rounded combat-style hammer; beveled magazine well; manual thumb, grip and firing pin safeties; solid barrel bushing. Made in Canada by Para-Ordnance. Introduced 1990; still produced. **LMSR: $745**

Steel frame
New: $650	**Perf.:** $600	**Exc.:** $500

Alloy frame
New: $600	**Perf.:** $550	**Exc.:** $450

Stainless frame, stainless
New: $700	**Perf.:** $650	**Exc.:** $550

Stainless frame, Duotone
New: $685	**Perf.:** $635	**Exc.:** $575

PARA-ORDNANCE P-15.40

Semi-automatic; 40 S&W; 10-shot magazine; 4 1/4" barrel; 7 3/4" overall length; weighs 36 oz. (steel frame), 28 oz. (alloy frame); blade front sight, adjustable rear; high visibility three-dot system; matte black finish; grooved match trigger; rounded combat-style hammer; beveled magazine well; manual thumb, grip and firing pin safeties; solid barrel bushing. Made in Canada by Para-Ordnance. Introduced 1990; still produced. **LMSR: $745**

Steel frame
New: $650	**Perf.:** $600	**Exc.:** $500

Alloy frame
New: $600	**Perf.:** $550	**Exc.:** $450

Stainless frame, stainless
New: $700	**Perf.:** $650	**Exc.:** $550

Stainless frame, Duotone
New: $685	**Perf.:** $635	**Exc.:** $575

PARA-ORDNANCE P-16.40

Semi-automatic; single action; 40 S&W; 10-shot magazine (pre-1994 guns have 16-shot); 5" barrel; 8 1/2" overall length; weighs 40 oz. (steel frame); textured composition grips; blade front sight, windage-adjustable rear; high visibility three-dot system; matte black finish; grooved match trigger; rounded combat-style hammer; beveled magazine well; manual thumb, grip and firing pin safeties; solid barrel bushing. Made in Canada by Para-Ordnance. Introduced 1994; still produced. **LMSR: $745**

Steel frame
New: $625	**Perf.:** $575	**Exc.:** $500

Stainless frame, stainless
New: $650	**Perf.:** $600	**Exc.:** $525

Stainless frame, Duotone
New: $685	**Perf.:** $635	**Exc.:** $575

PARDINI GP RAPID FIRE MATCH

Semi-automatic; single action; 22 Short; 5-shot magazine; 4 5/8" barrel; 11 5/8" overall length; weighs 43 oz.; wrap-around stippled walnut grips; interchangeable post front sight, fully-adjustable match rear. Imported from Italy by Nygord Precision Products. Introduced 1995; still imported. **LMSR. $995**
New: $1000	**Perf.:** $850	**Exc.:** $750

Pardini GP Rapid Fire Match Schuman

Same specs as the GP model except extended rear sight for longer sight radius. Imported from Italy by Nygord Precision Products. Introduced 1995; still imported. **LMSR $1395**
New: $1500	**Perf.:** $1350	**Exc.:** $1100

PARDINI K50 FREE PISTOL

Single shot; 22 LR; 9 7/8" barrel; 18 3/4" overall length; weighs 35 oz.; adjustable match-type wrap-around walnut grips; interchangeable post front sight, fully-adjustable match open rear; removable, adjustable match trigger; barrel weights mount above the barrel. Imported from Italy by Nygord Precision Products. Introduced 1995; still imported. **LMSR: $995**
New: $900	**Perf.:** $850	**Exc.:** $750

PARDINI MODEL HP TARGET

Semi-automatic; single action; 32 S&W; 5-shot magazine; 4 3/4" barrel; 11 5/8" overall length; weighs 40 oz.; adjustable, stippled walnut, match-type grips; interchangeable blade front sight, interchangeable, fully-adjustable rear; fully-adjustable match trigger. Imported from Italy by Nygord Precision Products. Introduced 1995; still imported. **LMSR: $1095**
New: $1000	**Perf.:** $900	**Exc.:** $800

PARDINI MODEL SP LADIES

Semi-automatic; single action; 22 LR; 5-shot magazine; 4 3/4" barrel; 11 5/8" overall length; weighs 40 oz.; adjustable, stippled walnut, small match-type grips for smaller hands; interchangeable blade front sight, interchangeable, fully-adjustable rear; fully-adjustable match trigger. Imported from Italy by Nygord Precision Products. Introduced 1986; discontinued 1990.
Perf.: $850	**Exc.:** $750	**VGood:** $650

Phoenix Arms
Model Raven

Phoenix Arms
HP22

Plainfield Model 71

HANDGUNS

Pardini
Model SP Target

Peters Stahl
PSP-07 Comabt Compensator

Phelps
Eagle I
Grizzly
Heritage I
Patriot

Phoenix Arms
HP22
HP25
Model Raven

Plainfield
Model 71
Model 72

QFI
Dark Horseman
Horseman
Model LA380
Model LA380 Tigress
Model SA25
Model SA25 Tigress
Model SO38

PARDINI MODEL SP TARGET

Semi-automatic; single action; 22 LR; 5-shot magazine; 4 3/4" barrel; 11 5/8" overall length; weighs 40 oz.; adjustable, stippled walnut, match-type grips; interchangeable blade front sight, interchangeable, fully-adjustable rear; fully-adjustable match trigger. Imported from Italy by Nygord Precision Products. Introduced 1995; still imported. **LMSR: $950**
    New: $1150   **Perf.:** $950   **Exc.:** $650

PETERS STAHL PSP-07 COMBAT COMPENSATOR

Semi-automatic; single action; 45 ACP, 10mm; 7-shot (45 ACP), 8-shot (10mm); 6" barrel; 10" overall length; weighs 45 oz.; Pachmayr Presentation rubber grips; interchangeable blade front sight, fully-adjustable Peters Stahl rear; linkless barrel with polygonal rifling and integral PS competition compensator; semi-extended PS slide stop and thumb safety; rearward extended magazine release; adjustable Videcki trigger; Wilson stainless beavertail grip safety; Pachmayr rubber mainspring housing. Imported from Germany by Federal Ordnance. Introduced 1989; no longer imported.
45 ACP
    Perf.: $2200   **Exc.:** $1800   **VGood:** $1350
10mm Auto
    Perf.: $2100   **Exc.:** $1600   **VGood:** $1100

PHELPS EAGLE I

Revolver; single action; 444 Marlin; 6-shot cylinder; 8" barrel, others to customer specifications; 15 1/2" overall length (8" barrel); weighs 5 1/2 lbs.; smooth walnut grips; ramp front sight, adjustable rear; polished blue finish; transfer bar safety. Made in U.S. by Phelps Mfg. Co. Introduced 1978; no longer produced.
    Perf.: $2000   **Exc.:** $1750   **VGood:** $1500

PHELPS GRIZZLY

Revolver; single action; 50-70; 6-shot cylinder; 8" barrel, others to customer specifications; 15 1/2" overall length (8" barrel); weighs 5 1/2 lbs.; smooth walnut grips; ramp front sight, adjustable rear; polished blue finish; transfer bar safety. Made in U.S. by Phelps Mfg. Co. Introduced 1978; no longer produced.
    Perf.: $2200   **Exc.:** $1850   **VGood:** $1600

PHELPS HERITAGE I

Revolver; single action; 45-70; 6-shot cylinder; 8" barrel, others to customer specifications; 15 1/2" overall length (8" barrel); weighs 5 1/2 lbs.; smooth walnut grips; ramp front sight, adjustable rear; polished blue finish; transfer bar safety. Made in U.S. by Phelps Mfg. Co. Introduced 1978; no longer produced.
    Perf.: $2000   **Exc.:** $1750   **VGood:** $1500

PHELPS PATRIOT

Revolver; single action; 375 Win.; 6-shot; 8" barrel, others on special order; adjustable sights; blued finish. Introduced 1993; discontinued 1995.
    New: $1850   **Perf.:** $1600   **Exc.:** $1350

PHOENIX ARMS HP22

Semi-automatic; single action; 22 LR; 10-shot magazine; 3" barrel; 5 1/2" overall length; weighs 20 oz.; checkered composition grips; blade front sight, adjustable rear; exposed hammer; manual hold-open; button magazine release; available in satin nickel, polished blue finish. Made in U.S. by Phoenix Arms. Introduced 1993; still produced.
    New: $85   **Perf.:** $75   **Exc.:** $60

PHOENIX ARMS HP25

Semi-automatic; single action; 25 ACP; 10-shot magazine; 3" barrel; 5 1/2" overall length; weighs 20 oz.; checkered composition grips; blade front sight, adjustable rear; exposed hammer; manual hold-open; button magazine release; satin nickel or polished blue finish. Made in U.S. by Phoenix Arms. Introduced 1994; still produced.
    New: $80   **Perf.:** $70   **Exc.:** $50

PHOENIX ARMS MODEL RAVEN

Semi-automatic; single action; 25 ACP; 6-shot magazine; 2 7/16" barrel; 4 3/4" overall length; weighs 15 oz.; ivory-colored or black slotted plastic grips; ramped front sight, fixed rear; available in blue, nickel or chrome finish. Made in U.S. by Phoenix Arms. Introduced 1992; still produced.
    New: $65   **Perf.:** $50   **Exc.:** $40

PLAINFIELD MODEL 71

Semi-automatic; 22 LR, 25 ACP; 10-shot (22 LR), 8-shot (25 ACP); 1" barrel; 5 1/8" overall length; fixed sights; checkered walnut stocks; stainless steel slide frame. Also made with caliber conversion kit. Introduced 1970; discontinued approximately 1980.
    Exc.: $200   **VGood:** $165   **Good:** $125
With conversion kit
    Exc.: $250   **VGood:** $225   **Good:** $175

PLAINFIELD MODEL 72

Semi-automatic; 22 LR, 25 ACP; 10-shot (22 LR), 8-shot (25 ACP); 3 1/2" barrel; 6" overall length; fixed sights; checkered walnut stocks; aluminum slide. Introduced 1970; discontinued 1978.
    Exc.: $200   **VGood:** $175   **Good:** $125
With conversion kit
    Exc.: $250   **VGood:** $200   **Good:** $150

QFI DARK HORSEMAN

Revolver; single action; 44 Mag., 45 Colt; 6 1/2", 7 1/2" barrel; adjustable sights; black composition stocks; extended grip frame; blued finish. Introduced 1991; discontinued 1992.
    New: $250   **Perf.:** $200   **Exc.:** $175

QFI HORSEMAN

Revolver; single action; 357 Mag., 44 Mag., 45 Colt; 6 1/2", 7 1/2" barrel; weighs 45 oz. (6 1/2" barrel); blade front sight, grooved rear; uncheckered walnut grips; color case-hardened frame; bright blue finish. Assembled in U.S. Introduced 1991; discontinued 1992.
    Perf.: $250   **Exc.:** $200   **VGood:** $175

QFI MODEL LA380

Semi-automatic; single action; 380 ACP; 6-shot; 3 1/8" barrel; 6 1/4" overall length; weighs 25 oz.; blade front sight, windage-adjustable rear; uncheckered European walnut grips; external hammer; magazine safety; hammer, trigger, firing pin block; blued finish. Introduced 1991; discontinued 1992.
    Perf.: $125   **Exc.:** $110   **VGood:** $100
Chrome finish
    Perf.: $125   **Exc.:** $110   **VGood:** $100
Stainless steel
    Perf.: $165   **Exc.:** $145   **VGood:** $135

QFI Model LA380 Tigress

Same specs as Model LA380 except scrimshawed white polymer grips; blue frame with gold-plated slide. Introduced 1991; discontinued 1991.
    Perf.: $200   **Exc.:** $175   **VGood:** $150

QFI MODEL SA25

Semi-automatic; single action; 25 ACP; 6-shot; 2 1/2" barrel; 4 5/8" overall length; weighs 12 oz.; fixed sights; smooth walnut or pearlite grips; external hammer; blued finish. Introduced 1991; discontinued 1991.
    Perf.: $60   **Exc.:** $45   **VGood:** $35

QFI Model SA25 Tigress

Same specs as Model SA25 except scrimshawed white polymer grips; blue frame with gold-plated slide. Introduced 1991; discontinued 1991.
    Perf.: $125   **Exc.:** $100   **VGood:** $85

QFI MODEL SO38

Revolver; double action; 38 Spl.; 6-shot; 2" solid, 4" vent-rib barrel; weighs 27 oz. (2" barrel); fixed sights; bulldog-type checkered plastic or walnut grips; one-stroke ejection. Introduced 1991; discontinued 1992.
    Perf.: $165   **Exc.:** $145   **VGood:** $120

QFI
Plains Rider
RP Series
Western Ranger

RG
14
16
17
23
25
26
30
31
38S
39
40
42

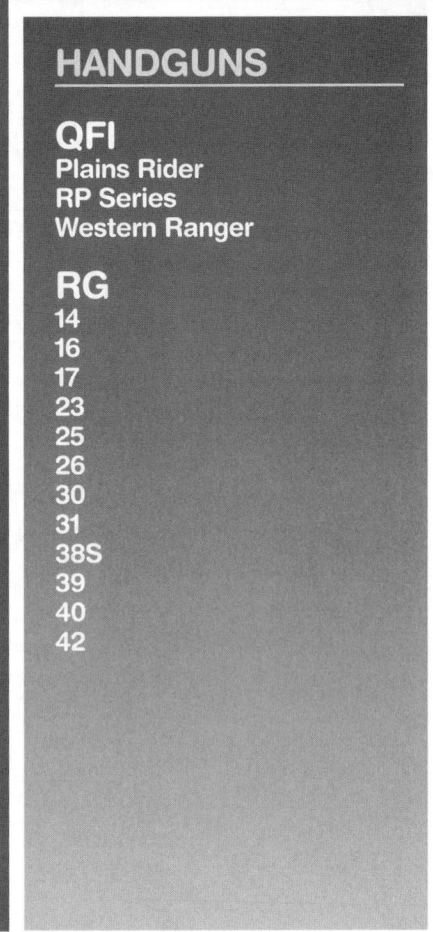

RG 14

RG 38S

RG 23

RG 39

RG 31

RG 57

RG 66

QFI PLAINS RIDER
Revolver; single action; 22 LR, 22 LR/22 WMR; 6-shot cylinder; 3″, 4 3/4″, 6 1/2″, 9″ barrel; 11″ overall length (6 1/2″ barrel); weighs 35 oz. (6 1/2″ barrel); black composition grips; blade front sight, fixed rear; blue/black finish; available with extra cylinder. From QFI. Introduced 1991; no longer produced.
22 LR, 3″, 4 3/4″, 6 1/2″ barrel
   Perf.: $85   **Exc.:** $75   **VGood:** $60
22 LR, 3″, 43/4″, 61/2″ barrel, dual cylinder
   Perf.: $100   **Exc.:** $90   **VGood:** $85
22 LR, 9″ barrel
   Perf.: $90   **Exc.:** $80   **VGood:** $65
22 LR, 9″ barrel, dual cylinder
   Perf.: $110   **Exc.:** $100   **VGood:** $90

QFI RP SERIES
Revolver; double action; 22 LR, 22 WMR, 22 LR/WMR combo, 32 S&W, 32 S&W Long, 32 H&R Mag., 38 Spl.; 6-shot cylinder; 2″, 4″ barrel; 6 1/4″ overall length (2″ barrel); weighs 23 oz.; magnum-style round butt grips of checkered plastic; ramp front sight, fixed square notch rear; one-piece solid frame; checkered hammer spur; serrated trigger; blue finish. Made in U.S. by QFI. Introduced 1991; dropped 1992.
   Perf.: $75   **Exc.:** $65   **VGood:** $50

QFI WESTERN RANGER
Revolver; single action; 22 LR, 22 LR/22 WMR combo; 3″, 4″, 43/4″, 6″, 6 1/2″, 7″, 9″ barrel; 10″ overall length (4 3/4″ barrel); weighs 31 oz.; blade front sight, notch rear; American walnut grips; blue/black finish. No longer in production.
   Perf.: $85   **Exc.:** $75   **VGood:** $60

RG 14
Revolver; double action; 22 LR; 6-shot; 1 3/4″, 3″ barrel; 5 1/2″ overall length (13/4″ barrel); fixed sights; checkered plastic grips; cylinder swings out when pin removed; blued finish. Introduced 1978; discontinued 1986.
   Exc.: $65   **VGood:** $50   **Good:** $40

RG 16
Derringer; over/under; 22 WMR; 2-shot; 3″ barrel; 5″ overall length; weighs 15 oz.; fixed sights; blue or nickel finish. Introduced 1975; no longer in production.
   Exc.: $60   **VGood:** $50   **Good:** $35

RG 17
Derringer; over/under; 38 Spl.; 2-shot; 3″ barrel; 5″ overall length; weighs 15 oz.; fixed sights; blue or nickel finish. Introduced 1975; no longer in production.
   Exc.: $60   **VGood:** $50   **Good:** $35

RG 23
Revolver; double action; 22 LR; 6-shot; 1 3/4″, 3″ barrel; 5 1/2″ overall length (13/4″ barrel); fixed sights; checkered plastic grips; central ejector system; blued finish. Introduced 1978; discontinued 1986.
   Exc.: $70   **VGood:** $60   **Good:** $45

RG 25
Semi-automatic; 25 ACP; 6-shot; 2 1/2″ barrel; 4 3/4″ overall length; weighs 12 oz.; fixed sights; thumb safety; blue or nickel finish. Introduced 1975; no longer in production.
   Exc.: $75   **VGood:** $60   **Good:** $50

RG 26
Semi-automatic; single action; 25 ACP; 6-shot; 2 1/2″ barrel; 4 3/4″ overall length; 12 oz.; fixed sights; checkered plastic stocks; thumb safety; blued finish. Introduced 1977; discontinued 1986.
   Exc.: $75   **VGood:** $60   **Good:** $50

RG 30
Revolver; double action; 22 LR, 22 WMR; 6-shot swing-out cylinder; 4″ barrel; 9″ overall length; windage-adjustable rear sight, fixed front; checkered plastic stocks; blued finish. Introduced 1977; discontinued 1986.
   Exc.: $65   **VGood:** $60   **Good:** $50
Nickel finish
   Exc.: $70   **VGood:** $65   **Good:** $55

RG 31
Revolver; double action; 32 S&W, 38 Spl.; 6-shot (32 S&W), 5-shot (38 Spl.); 2″ barrel; 6 3/4″ overall length; fixed sights; checkered plastic grips; cylinder swings out when pin removed; blued finish. Introduced 1978; discontinued 1986.
   Exc.: $110   **VGood:** $100   **Good:** $85

RG 38S
Revolver; double action; 38 Spl.; 6-shot swing-out cylinder; 3″, 4″, 6″ (38T) barrel; 9 1/4″ overall length (4″ barrel); weighs 35 oz.; windage-adjustable rear sight, fixed front; checkered plastic stocks; blued finish. Introduced 1977; discontinued 1986.
   Exc.: $125   **VGood:** $110   **Good:** $100
Nickel finish
   Exc.: $135   **VGood:** $120   **Good:** $110

RG 39
Revolver; 38 Spl., 32 S&W; 6-shot; 2″ barrel; 7″ overall length; weighs 21 oz.; fixed sights; American walnut grips; blue finish. Introduced 1981; no longer in production.
   Exc.: $125   **VGood:** $110   **Good:** $85

RG 40
Revolver; double action; 38 Spl., 32 S&W; 6-shot; 2″ barrel; 7″ overall length; fixed sights; checkered plastic grips; swing-out cylinder; spring ejector; blued finish. Introduced 1980; discontinued 1986.
   Exc.: $110   **VGood:** $100   **Good:** $75

RG 42
Semi-automatic; 25 ACP; 7-shot; 2 3/4″ barrel; 5 5/16″ overall length; fixed sights; thumb and magazine safety; blue finish.
   Exc.: $75   **VGood:** $60   **Good:** $50

RG 63

RG 88

Ram-Line
Exactor Target

Raven MP-25

Randall Raider/Service
Model-C

RG 26

HANDGUNS

RG
57
63
66
74
88

Radom
P-35

Ram-Line
Exactor Target
Ram-Tech

Randall
Curtis E. Lemay Four-Star
Model
Raider/Service Model-C
Service Model

Raven
MP-25
P-25

Record-Match
Model 200 Free Pistol

RG 57

Revolver; double action; 357 Mag., 44 Mag.; 6-shot swing-out cylinder; 4″ barrel; 9 1/2″ overall length; fixed sights; checkered plastic stocks; steel frame; blued finish. Manufactured in Germany, imported by RG Industries. Introduced 1977; discontinued 1986.
   Exc.: $125   **VGood:** $110   **Good:** $85

RG 63

Revolver; double action; 22 LR; 8-shot; 5″ barrel; 10 1/4″ overall length; fixed sights; checkered plastic stocks; Western configuration with slide ejector rod; blued finish. Introduced 1976; discontinued 1986.
   Exc.: $100   **VGood:** $85   **Good:** $65
Nickel finish
   Exc.: $110   **VGood:** $90   **Good:** $70

RG 66

Revolver; single action; 22 LR, 22 WMR; 6-shot; 4 3/4″ barrel; 10″ overall length; adjustable rear sight, fixed front; checkered plastic stocks; slide ejector rod; blued finish. Introduced 1977; discontinued 1986.
   Exc.: $90   **VGood:** $75   **Good:** $60
Nickel finish
   Exc.: $95   **VGood:** $80   **Good:** $65

RG 74

Revolver; double action; 22 LR; 6-shot; 3″ barrel; 7 3/4″ overall length; fixed sights; checkered plastic grips; swing-out cylinder with spring ejector; blued finish. Introduced 1978; discontinued 1986.
   Exc.: $65   **VGood:** $50   **Good:** $40

RG 88

Revolver; double action; 38 Spl., 357 Mag.; 6-shot swing-out cylinder; 4″ barrel; 9″ overall length; fixed sights; checkered walnut stocks; wide spur hammer, trigger; blued finish. Introduced 1977; dropped 1986.
   Exc.: $135   **VGood:** $110   **Good:** $90

RADOM P-35

Semi-automatic; 9mm Para; 8-shot; 4 3/4″ barrel; 7 3/4″ overall length; fixed sights; checkered plastic stocks; blued finish. Design based on Colt Model 1911A1. Manufactured in Poland 1936 to 1945.
Pre-War, dated 1936 to 1939
   Exc.: $1500   **VGood:** $1100   **Good:** $800
Wartime manufacture
   Exc.: $350   **VGood:** $275   **Good:** $200

RAM-LINE EXACTOR TARGET

Semi-automatic; single action; 22 LR; 15-shot magazine; 8″ barrel; 12 5/16″ overall length; weighs 23 oz.; one-piece injection moulded grips in conventional contour; checkered side panels; ridged front and backstraps; ramp front sight, adjustable rear; injection moulded grip frame, alloy receiver; hybrid composite barrel; constant force sear spring gives 2 1/2-lb. trigger pull; adapt-A-Barrel for mounting weights, flashlight; drilled and tapped receiver for scope mounting; jeweled bolt; comes with carrying case, test target. Made in U.S. by Ram-Line, Inc. Introduced 1990; no longer in production.
   Perf.: $250   **Exc.:** $200   **VGood:** $175

RAM-LINE RAM-TECH

Semi-automatic; single action; 22 LR; 15-shot magazine; 4 1/2″ barrel; weighs 19 oz.; one-piece injection moulded grips with checkered panels; ramp front sight, adjustable rear; compact frame; injection moulded grip frame, alloy receiver; hybrid composite barrel; constant force sear spring gives 3-lb. trigger pull; comes with carrying case. Made in U.S. by Ram-Line, Inc. Introduced 1994; no longer produced.
   Perf.: $175   **Exc.:** $150   **VGood:** $135

RANDALL CURTIS E. LEMAY FOUR-STAR MODEL

Semi-automatic; 9mm Para., 45 ACP; 7-shot (9mm), 6-shot (45 ACP); 4 1/4″ barrel; 7 3/4″ overall length; weighs 35 oz.; fixed or adjustable sights; checkered walnut stocks; squared trigger guard; stainless steel construction. Introduced 1984; discontinued 1984.
   Exc.: $1000   **VGood:** $850   **Good:** $700

RANDALL RAIDER/SERVICE MODEL-C

Semi-automatic; 9mm Para., 45 ACP; 6-shot; 4 1/4″ barrel; 7 3/4″ overall length; weighs 36 oz.; fixed or adjustable sights; checkered walnut stocks; squared trigger guard; extended magazine baseplate; stainless steel construction. Introduced 1983; discontinued 1984.
   Exc.: $600   **VGood:** $500   **Good:** $450

RANDALL SERVICE MODEL

Semi-automatic; 38 Super, 9mm Para., 45 ACP; 5″ barrel; 8 1/2″ overall length; weighs 38 oz.; blade front sight, fixed or adjustable rear; checkered walnut stocks; round-top or ribbed slide; all stainless steel construction. Introduced 1983; discontinued 1984.
   Exc.: $600   **VGood:** $500   **Good:** $400

RAVEN MP-25

Semi-automatic; single action; 25 ACP; 6-shot; 2 7/16″ barrel; 5 1/2″ overall length; weighs 15 oz.; fixed sights; uncheckered pearlite stocks; die-cast slide serrations; blued, nickel or satin nickel finish. Manufactured in U.S. Introduced 1984; discontinued 1992.
   Exc.: $50   **VGood:** $40   **Good:** $30

RAVEN P-25

Semi-automatic; single action; 25 ACP; 6-shot; 2 7/16″ barrel; 5 1/2″ overall length; weighs 15 oz.; fixed sights; uncheckered walnut; blued, nickel or satin nickel finish. Manufactured in U.S. Introduced 1977; discontinued 1984.
   Exc.: $50   **VGood:** $40   **Good:** $25

RECORD-MATCH MODEL 200 FREE PISTOL

Single shot; Martini action; 22 LR; 11″ barrel; micrometer rear sight, target-type front; carved, checkered European walnut stock, forearm, with adjustable hand base; set trigger, spur trigger guard; blued finish. Manufactured in Germany prior to WWII.
   Exc.: $850   **VGood:** $750   **Good:** $600

Record-Match
Model 210 Free Pistol
Model 210A Free Pistol

Reising
Target Model

Remington
Model 51
Model 95
Model 1901 Target
Model 1911 Military
XP-22R
XP-100 Hunter Pistol
XP-100 Custom HB Long
Range Pistol
XP-100 Silhouette
XP-100 Varmint Special
XP-100R KS

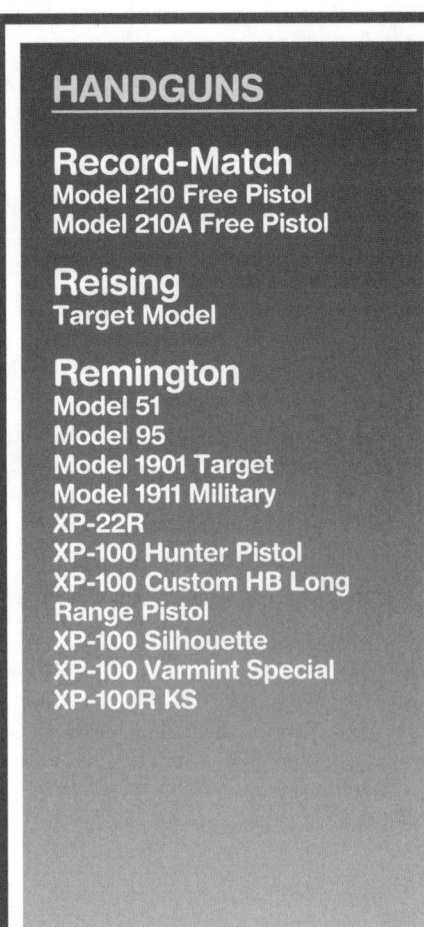

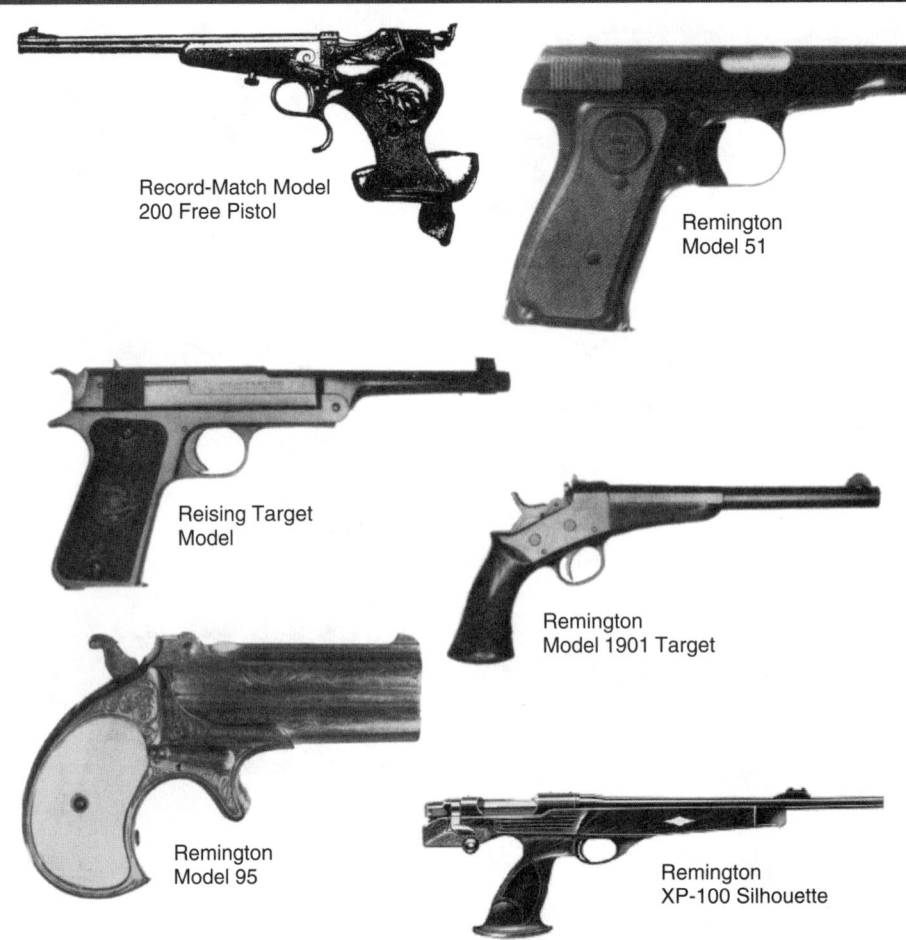

Record-Match Model
200 Free Pistol

Remington
Model 51

Reising Target
Model

Remington
Model 1901 Target

Remington
Model 95

Remington
XP-100 Silhouette

RECORD-MATCH MODEL 210 FREE PISTOL
Single shot; Martini action; 22 LR; 11" barrel; micrometer rear sight, target-type front; carved, checkered deluxe European walnut stock, forearm with adjustable hand base; button release set trigger; blued finish. Manufactured in Germany prior to WWII.
    Exc.: $1000   **VGood:** $850   **Good:** $700

Record-Match Model 210A Free Pistol
Same specs as Model 210 except alloy frame. Manufactured in Germany prior to WWII.
    Exc.: $1100   **VGood:** $900   **Good:** $700

REISING TARGET MODEL
Semi-automatic; hinged frame; 22 LR; 12-shot; 6 1/2" barrel; fixed sights; hard rubber checkered stocks; external hammers; blued finish. Manufactured 1921 to 1924.
    Exc.: $350   **VGood:** $250   **Good:** $200

REMINGTON MODEL 51
Semi-automatic; 32 ACP, 380 ACP; 8-shot; 3 1/2" barrel; 6 5/8" overall length; weighs 21 oz.; fixed sights; hard rubber grips; blued finish. Introduced 1918; discontinued 1943.
    Exc.: $300   **VGood:** $250   **Good:** $200

REMINGTON MODEL 95
Derringer; over/under; single action; 41 Short Rimfire; 3" barrels; 4 7/8" overall length. Introduced 1866; discontinued 1935. Prior to 1888 the model was stamped "E Remington & Sons"; from 1888 to 1910 the derringers were marked "Remington Arms Co."; from 1910 to 1935 guns were marked "Remington Arms-U.M.C. Co." The early styles have a two-armed extractor, long hammer spur. In later models a few have no extractor, majority have sliding extractor, short hammer spur. Available with all-blued finish, full nickel plate or blued barrels with nickel-plated frame; factory engraving; choice of checkered hard rubber, walnut, mother-of-pearl, ivory grips; fixed rear groove sight, front blade integral with top barrel.
    Exc.: $600   **VGood:** $500   **Good:** $400
Nickel finish
    Exc.: $600   **VGood:** $500   **Good:** $400
Engraved with mother-of-pearl or ivory grips
    Exc.: $950   **VGood:** $850   **Good:** $750

REMINGTON MODEL 1901 TARGET
Single shot; rolling block action; 22 Short, 22 LR, 25 Rimfire, 32 Centerfire, 44 S&W Russian; 10" barrel; 14" overall length; target sights; checkered walnut grips, forearm; blued finish. Manufactured 1901 to 1909.
    Exc.: $1500   **VGood:** $1250   **Good:** $1000

REMINGTON MODEL 1911 MILITARY
Semi-automatic; single action; 45 ACP; 8-shot; fixed sight; checkered walnut grip; Colt Model 1911 made under U.S. contract; blued finish. Introduced 1918; discontinued 1919.
    Exc.: $1000   **VGood:** $850   **Good:** $500

REMINGTON XP-22R
Single shot; bolt-action; 22 LR; 5-shot; 14 1/2" barrel; weighs 68 oz.; no sights; drilled, tapped for scope mounts or iron sights; rear-handled stock of fiberglass-reinforced Kevlar built on Remington Model 541-7 rifle action. Available on special order from Remington Custom Shop. Introduced 1991; discontinued 1992.
    Exc.: $400   **VGood:** $300   **Good:** $200

REMINGTON XP-100 HUNTER PISTOL
Single shot; bolt action; 223 Rem., 7mm BR Rem., 7mm-08 Rem., 35 Rem.; 14 1/2" barrel; 21 1/4" overall length; weighs 4 1/2 lbs.; laminated wood stock with contoured grip; no sights furnished; drilled and tapped for scope mounting; mid-handle grip design with scalloped contours for right- or left-handed use; two-position safety; matte blue finish. Introduced 1993; dropped 1995.
    New: $450   **Perf.:** $400   **Exc.:** $300

REMINGTON XP-100 CUSTOM HB LONG RANGE PISTOL
Single shot; bolt-action; 221 Rem. Fireball; 10 1/2" barrel; 16 3/4" overall length; vent rib; blade front sight, adjustable rear; receiver drilled, tapped for scope mounts; one-piece brown nylon stock; blued finish. Introduced 1963; dropped 1985.
    Exc.: $850   **VGood:** $700   **Good:** $575

REMINGTON XP-100 SILHOUETTE
Single shot; bolt action; 7mm BR Rem.; 10 1/2" barrel; 17 1/4" overall length; weighs 3 7/8 lbs.; American walnut stock; blade front sight, fully-adjustable square-notch rear; mid-handle grip with scalloped contours for left- or right-handed use; match-type trigger; two-position thumb safety; matte blue finish. Dropped 1995.
    Perf.: $500   **Exc.:** $400   **VGood:** $300

REMINGTON XP-100 VARMINT SPECIAL
Single shot; bolt action; 221 Rem. Fireball, 223 Rem.; 10 1/2" barrel (221), 14 1/2" barrel (223); weighs 70 oz.; adjustable sights; drilled, tapped for scope; one-piece nylon grip; forend weights for balance; matte blue finish. Introduced 1963; discontinued 1992.
    Perf.: $450   **Exc.:** $400   **VGood:** $325

REMINGTON XP-100R KS
Single shot; bolt action; 223 Rem., 22-250, 7mm-08 Rem., 250 Savage, 308, 350 Rem. Mag., 35 Rem.; blind magazine; 5-shot (7mm-08, 35), 6-shot (223 Rem.); 14 1/2" standard-weight barrel; weighs about 4 1/2 lbs; rear-handle, synthetic Du Pont Kevlar stock to eliminate transfer bar between forward trigger and rear trigger assembly; front and rear sling swivel studs; adjustable leaf rear sight, bead front; receiver drilled and tapped for scope mounts. From Remington Custom Shop. Introduced 1990; dropped 1995.
    New: $750   **Perf.:** $700   **Exc.:** $600

Rocky Mountain
Arms Patriot

Rossi Model
511 Sportsman

Rocky Mountain
Arms 1911A1-LH

Rossi
Model 68

Rossi Model 88
Stainless

Rossi Lady Rossi

Rossi Model 70

HANDGUNS

Rocky Mountain
Arms
1911A1-LH
Patriot

Rossi
Lady Rossi
Model 31
Model 51
Model 68
Model 68S
Model 69
Model 70
Model 84 Stainless
Model 88 Stainless
Model 89 Stainless
Model 94
Model 511 Sportsman
Model 515

ROCKY MOUNTAIN ARMS 1911A1-LH

Semi-automatic; single action; 40 S&W, 45 ACP; 7-shot magazine; 5 1/4″ barrel; 8 13/16″ overall length; weighs 37 oz.; checkered walnut grips; red insert Patridge front sight, white outline click-adjustable rear; fully left-handed pistol; slide, frame, barrel made from stainless steel; working parts coated with Teflon-S; single-stage trigger with 3 1/2-lb. pull. Made in U.S. by Rocky Mountain Arms, Inc. Introduced 1993; no longer produced.
  Perf.: $1200   **Exc.:** $950   **VGood:** $750

ROCKY MOUNTAIN ARMS PATRIOT

Semi-automatic; single action; 223 Rem.; 10-shot magazine; 7″ barrel with muzzlebrake; 20 1/2″ overall length; weighs 5 lbs.; black composition grips; no sights; milled upper receiver with enhanced Weaver base; milled lower receiver from billet plate; machined aluminum National Match handguard; finished in DuPont Teflon-S matte black or NATO green; comes with black nylon case, one magazine. Made in U.S. by Rocky Mountain Arms, Inc. Introduced 1993; still in production. **LMSR: $2500**
With A-2 handle top
  New: $2250   **Perf.:** $2000   **Exc.:** $1850
Flat top model
  New: $2400   **Perf.:** $2200   **Exc.:** $1950

ROSSI LADY ROSSI

Revolver; double action; 38 Spl.; 5-shot cylinder; 2″, 3″ barrel; 6 1/2″ overall length (2″ barrel); weighs 21 oz.; smooth rosewood grips; fixed sights; high-polish stainless steel with "Lady Rossi" engraved on frame; comes with velvet carry bag. Imported from Brazil by Interarms. Introduced 1995; still imported. **LMSR: $312**
  New: $200   **Perf.:** $160   **Exc.:** $135

ROSSI MODEL 31

Revolver; double action; 38 Spl.; 5-shot; 4″ barrel; weighs 22 oz.; ramp front sight, low-profile adjustable rear; checkered wood stocks; all-steel frame; blue or nickel finish. Introduced 1978; discontinued 1985.
  Exc.: $125   **VGood:** $110   **Good:** $100

ROSSI MODEL 51

Revolver; double action; 22 LR; 6-shot; 6″ barrel; adjustable sights; checkered wood stocks; blued finish. Discontinued 1985.
  Exc.: $125   **VGood:** $110   **Good:** $100

ROSSI MODEL 68

Revolver; double action; 38 Spl.; 5-shot; 2″, 3″ barrel; weighs 22 oz.; ramp front sight, low-profile adjustable rear; checkered rubber or wood stocks; all-steel frame; swing-out cylinder; blue or nickel finish. Made in Brazil. Introduced 1978; discontinued 1985. Reintroduced in 1993 as the Model 68S.
  Exc.: $150   **VGood:** $125   **Good:** $110

Rossi Model 68S

Same specs as Model 68 except weighs 23 oz.; fixed sights. Introduced 1993; still in production. **LMSR: $234**
  New: $175   **Perf.:** $150   **Exc.:** $125

ROSSI MODEL 69

Revolver; double action; 32 S&W; 6-shot; 3″ barrel; ramp front sight, low-profile adjustable rear; checkered wood stocks; all-steel frame; swing-out cylinder; blue or nickel finish. Made in Brazil. Introduced 1978; discontinued 1985.
  Exc.: $110   **VGood:** $90   **Good:** $80

ROSSI MODEL 70

Revolver; double action; 22 LR; 6-shot; 3″ barrel; ramp front sight, low-profile adjustable rear; checkered wood stocks; all-steel frame; swing-out cylinder; blue or nickel finish. Made in Brazil. Introduced 1978; discontinued 1985.
  Exc.: $125   **VGood:** $100   **Good:** $75

ROSSI MODEL 84 STAINLESS

Revolver; double action; 38 Spl.; 6-shot; 3″, 4″ barrel; 8″ overall length; weighs 27 1/2 oz.; fixed sights; checkered wood grips; solid raised rib; stainless steel. Made in Brazil. Introduced 1984; discontinued 1986.
  Exc.: $200   **VGood:** $175   **Good:** $150

ROSSI MODEL 88 STAINLESS

Revolver; double action; 38 Spl.; 5-shot; 2″, 3″ barrel; 8 3/4″ overall length; weighs 21 oz.; ramp front sight, drift-adjustable square-notch rear; checkered wood or rubber stocks; small frame; stainless steel construction; matte finish. Introduced 1993; still imported from Brazil by Interarms. **LMSR: $265**
  New: $200   **Perf.:** $175   **Exc.:** $150

ROSSI MODEL 89 STAINLESS

Revolver; double action; 32 S&W; 6-shot; 3″ barrel; 8 3/4″ overall length; weighs 21 oz.; ramp front sight, drift-adjustable square notch rear; checkered wood or rubber stocks; stainless steel; matte finish. Introduced 1985; discontinued 1986; reintroduced 1989; no longer imported.
  Exc.: $160   **VGood:** $150   **Good:** $125

ROSSI MODEL 94

Revolver; double action; 38 Spl.; 6-shot; 3″, 4″ barrel; 27 1/2 oz.; ramp front sight; adjustable rear; checkered wood stocks; blued finish. Introduced 1985; discontinued 1988.
  Exc.: $175   **VGood:** $150   **Good:** $135

ROSSI MODEL 511 SPORTSMAN

Revolver; double action; 22 LR; 6-shot; 4″ barrel; 9″ overall length; weighs 30 oz.; adjustable square-notch rear sight, orange-insert ramp front; checkered wood stocks; heavy barrel; integral sight rib; shrouded ejector rod; stainless steel construction. Made in Brazil. Introduced 1986; discontinued 1990.
  Exc.: $150   **VGood:** $135   **Good:** $125

ROSSI MODEL 515

Revolver; double action; 22 WMR; 6-shot cylinder; 4″ barrel; 9″ overall length; weighs 30 oz.; checkered wood and finger-groove wrap-around rubber grips; blade front sight with red insert, fully-adjustable rear; small frame; stainless steel construction; solid integral barrel rib. Imported from Brazil by Interarms. Introduced 1994; still imported.
  New: $200   **Perf.:** $175   **Exc.:** $150

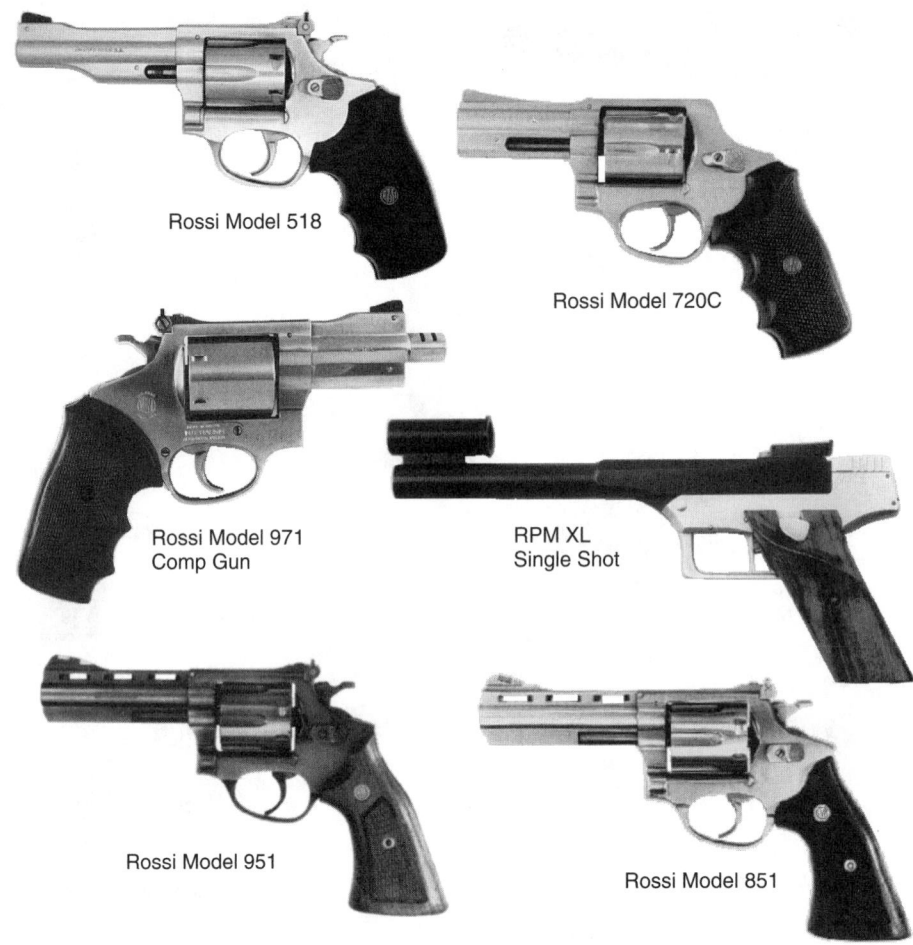

HANDGUNS

Rossi
Model 518
Model 720
Model 720C
Model 851
Model 941
Model 951
Model 971
Model 971 Comp Gun
Model 971 Stainless

Royal

RPM
XL Single Shot
XL Hunter Single Shot

Ruger
(See Sturm Ruger)

Safari Arms
Black Widow
Enforcer

Rossi Model 518

Rossi Model 720C

Rossi Model 971 Comp Gun

RPM XL Single Shot

Rossi Model 951

Rossi Model 851

ROSSI MODEL 518

Revolver; double action; 22 LR; 6-shot cylinder; 4" barrel; 9" overall length; weighs 30 oz.; checkered wood and finger-groove wrap-around rubber grips; blade front sight with red insert, fully-adjustable rear; small frame; stainless steel construction; solid integral barrel rib. Imported from Brazil by Interarms. Introduced 1994.

New: $200   **Perf.:** $185   **Exc.:** $160

ROSSI MODEL 720

Revolver; double action; 44 Spl.; 5-shot cylinder; 3" barrel; 8" overall length; weighs 27 1/2 oz.; checkered rubber, combat-style grips; red insert front sight on ramp, fully-adjustable rear; all stainless steel construction; solid barrel rib; full ejector rod shroud. Imported from Interarms. Introduced 1992.

New: $225   **Perf.:** $175   **Exc.:** $150

Rossi Model 720C

Same specs as the Model 720 except spurless hammer; double-action only. Imported from Brazil by Interarms. Introduced 1992; still imported. **LMSR: $320**

New: $225   **Perf.:** $175   **Exc.:** $150

ROSSI MODEL 851

Revolver; double action; 38 Spl.; 6-shot; 3", 4" barrel; 8" overall length (3" barrel); weighs 27 1/2 oz.; red-insert blade front sight, windage-adjustable rear; checkered Brazilian hardwood grips; medium-size frame; stainless steel construction; vent rib. Introduced 1991; still imported. **LMSR: $281**

New: $200   **Perf.:** $175   **Exc.:** $135

ROSSI MODEL 941

Revolver; double action; 38 Spl.; 6-shot; 4" solid rib barrel; 9" overall length; weighs 30 oz.; colored insert front sight, adjustable rear; checkered hardwood combat stocks; shrouded ejector rod; blued finish. Made in Brazil. Introduced 1985; discontinued 1990.

Exc.: $175   **VGood:** $150   **Good:** $125

ROSSI MODEL 951

Revolver; double action; 38 Spl.; 6-shot; 4" vent-rib barrel; 9" overall length; weighs 30 oz.; colored insert front sight, adjustable rear; checkered hardwood combat stocks; shrouded ejector rod; blued finish. Made in Brazil. Introduced 1985; discontinued 1990.

Exc.: $175   **VGood:** $150   **Good:** $125

ROSSI MODEL 971

Revolver; double action; 357 Mag.; 6-shot; 2 1/2", 4", 6" heavy barrel; 9" overall length; weighs 36 oz.; blade front sight, fully-adjustable rear; checkered Brazilian hardwood grips; matted sight rib; target trigger; wide checkered hammer spur; full-length ejector rod shroud; blued (4" only) or stainless finish. Made in Brazil. Introduced 1988.

New: $225   **Perf.:** $200   **Exc.:** $150

Rossi Model 971 Comp Gun

Same specs as Model 971 except stainless finish; 3 1/4" barrel with integral compensator; overall length 9"; weighs 32 oz.; red insert front sight, fully-adjustable rear; checkered, contoured rubber grips. Imported from Brazil by Interarms. Introduced 1993; still imported. **LMSR: $320**

New: $225   **Perf.:** $200   **Exc.:** $175

Rossi Model 971 Stainless

Same general specs as Model 971 except 2 1/2", 4", 6" barrel; rubber grips; stainless steel construction. Introduced 1989; still imported. **LMSR: $320**

New: $225   **Perf.:** $200   **Exc.:** $175

ROYAL

Semi-automatic; 7.63mm Mauser; 5 1/2" barrel; 11 3/4" overall length; 10-shot magazine; grooved wood grips. Spanish copy of Mauser Model 1896 but with different internal mechanism, round bolt. Introduced 1926; dropped 1930 when the maker replaced it with the Azul, an almost identical copy of the Mauser.

Exc.: $1500   **VGood:** $1000   **Good:** $750

RPM XL SINGLE SHOT

Single shot; 22 LR through 45-70; 8", 10 3/4", 12", 14" barrel; weighs about 60 oz.; smooth Goncalo Alves stock with thumb and heel rest; hooded front sight with interchangeable post, ISGW fully-adjustable rear; barrel tapped and drilled for scope mount; cocking indicator; spring-loaded barrel lock; positive hammer-block safety. Made in U.S. by RPM. **LMSR: $857**

New: $750   **Perf.:** $600   **Exc.:** $500

RPM XL Hunter Single Shot

Same specs as RPM XL except stainless frame; 5/16" underlug; latch lever; positive extractor. **LMSR: $1195**

New: $1000   **Perf.:** $850   **Exc.:** $750

SAFARI ARMS BLACK WIDOW

Semi-automatic; 45 ACP; 6-shot; 3 3/4" barrel; 7 11/16" overall length; weighs 28 oz.; adjustable sights; scrimshawed Micarta grip with black widow ensignia; nickel finish. Discontinued 1987.

Exc.: $500   **VGood:** $450   **Good:** $400

SAFARI ARMS ENFORCER

Semi-automatic; single action; 45 ACP; 6-shot magazine; 3 13/16" barrel; 7 5/16" overall length; weighs 36 oz.; smooth walnut grips with etched black widow spider logo; ramped blade front sight, LPA adjustable rear; stainless barrel; extended safety and slide release; Commander-style hammer; beavertail grip safety; throated, polished, tuned; Parkerized matte black or satin stainless steel finishes. In 1988 Olympic Arms bought M-S Safari Arms of Phoenix, Arizona. Some guns will be marked M-S Safari; some Safari Arms. Current production guns are still being stamped "Safari Arms." Made in U.S. by Olympic Arms, Inc.'s specialty shop Schuetzen Pistol Works. Still produced.

New: $650   **Perf.:** $600   **Exc.:** $550

Safari Arms Matchmaster

Safari Arms Model 81L

Sauer Model 38H

Safari Arms Enforcer

Sauer Model 1913

Safari Arms
Enforcer Carrycomp II
G.I. Safari
Matchmaster
Matchmaster Carrycomp I
Model 81
Model 81BP
Model 81L
Model 81NM
Model 81 Silueta
OA-93
Unlimited

Sako
Triace Match

Sauer
Model 28
Model 38H
Model 1913

Safari Arms Enforcer Carrycomp II
Same specs as the Enforcer except Wil Schueman-designed hybrid compensator system. In 1988 Olympic Arms bought M-S Safari Arms of Phoenix, Arizona. Some guns will be marked M-S Safari; some Safari Arms. Current production guns are still being stamped "Safari Arms." Made in U.S. by Olympic Arms, Inc.'s specialty shop Schuetzen Pistol Works. Introduced 1993.
New: $1100 **Perf.:** $900 **Exc.:** $750

SAFARI ARMS G.I. SAFARI
Semi-automatic; single action; 45 ACP; 7-shot magazine; 5″ barrel; 8 ¹/₂″ overall length; weighs 40 oz.; checkered walnut grips; blade front sight, fixed rear; beavertail grip safety; extended safety and slide release; Commander-style hammer; chrome-lined 4140 steel barrel; National Match 416 stainless optional; Parkerized matte black finish. Made in U.S. by Safari Arms, Inc. Introduced 1991; no longer produced.
New: $450 **Perf.:** $350 **Exc.:** $225

SAFARI ARMS MATCHMASTER
Semi-automatic; single action; 45 ACP; 7-shot magazine; 5″, 6″ barrel; 8 ¹/₂″ overall length; weighs 38 oz.; smooth walnut grips with etched scorpion logo; ramped blade front sight, LPA adjustable rear; stainless steel barrel; beavertail grip safety; extended safety; extended slide release; Commander-style hammer; throated, polished, tuned; Parkerized matte black or satin stainless steel. In 1988 Olympic Arms bought M-S Safari Arms of Phoenix, Arizona. Some guns will be marked M-S Safari; some Safari Arms. Current production guns are still being stamped "Safari Arms." Made in U.S. by Olympic Arms, Inc.'s specialty shop Schuetzen Pistol Works. Introduced 1995; still produced. **LMSR:** $770
New: $600 **Perf.:** $550 **Exc.:** $475

Safari Arms Matchmaster Carrycomp I
Same specs as Matchmaster except Wil Schueman-designed hybrid compensator system. Made in U.S. by

Olympic Arms, Inc. Introduced 1993.
New: $1000 **Perf.:** $850 **Exc.:** $750

SAFARI ARMS MODEL 81
Semi-automatic; 45 ACP, 38 Spl.; 5″ barrel; weighs 42 oz.; adjustable sights; fixed or adjustable walnut target grips; optional Aristocrat rib with extended front sight optional; nickel finish. Introduced 1983; discontinued 1987.
Exc.: $750 **VGood:** $550 **Good:** $450

Safari Arms Model 81BP
Same specs as Model 81 except contoured front grip strap; quicker slide cycle time; designed for bowling pin matches. Introduced 1983; discontinued 1987.
Exc.: $850 **VGood:** $650 **Good:** $500

Safari Arms Model 81L
Same specs as Model 81 except 6″ barrel; long slide; weighs 45 oz. Introduced 1983; discontinued 1987.
Exc.: $750 **VGood:** $550 **Good:** $400

Safari Arms Model 81NM
Same specs as Model 81 except flat front grip strap. Introduced 1983; discontinued 1987.
Exc.: $750 **VGood:** $600 **Good:** $500

Safari Arms Model 81 Silueta
Same specs as Model 81 except 38-45 wildcat, 45 ACP; 10″ barrel; designed for silhouette competition. Introduced 1983; discontinued 1987.
Exc.: $850 **VGood:** $650 **Good:** $500

SAFARI ARMS OA-93
Semi-automatic; single action; 223, 7.62x39mm; 20-, 30-shot magazine (223), 5-, 30-shot magazine (7.62x39mm); 6″, 9″, 14″ 4140 steel or 416 stainless barrel; 15 oz.; 15 ³/₄″ overall length (6″ barrel); weighs 4 lbs.; A2 stowaway pistol grip; no sights; cut-off carrying handle with attached scope rail; AR-15 receiver with special bolt carrier; short slotted aluminum handguard;

button-cut or broach-cut barrel; Vortex flash suppressor. Made in U.S. by Olympic Arms, Inc. Introduced 1993; dropped 1994.
New: $2500 **Perf.:** $2250 **Exc.:** $2000

SAFARI ARMS UNLIMITED
Single shot; bolt action; 308 or smaller; 14 ¹⁵/₁₆″ barrel; 21 ¹/₂″ overall length; weighs 72 oz.; open iron sights; fiberglass stock; electronic trigger; black finish. Introduced 1983; discontinued 1987.
Exc.: $800 **VGood:** $700 **Good:** $575

SAKO TRIACE MATCH
Semi-automatic; 22 Short, 22 LR, 32 S&W Long; 6-shot; 5 ⁷/₈″ barrel; 11″ overall length; weighs 44 to 48 oz.; adjustable trigger for weight of pull, free travel and sear engagement; marketed in case with tool/cleaning kit, two magazines. Made in Finland. Introduced 1988; discontinued 1989.
Exc.: $2200 **VGood:** $1750 **Good:** $1250

SAUER MODEL 28
Semi-automatic; 25 ACP; 6-shot; 3″ overall length; slanted serrations on slide; top ejection; checkered black rubber grips with Sauer imprint; blued finish. Introduced about 1928; discontinued 1938.
Exc.: $200 **VGood:** $165 **Good:** $135

SAUER MODEL 38H
Semi-automatic; double action; 32 ACP; 7-shot; 3 ¹/₄″ barrel; 6 ¹/₄″ overall length; fixed sights; black plastic grips; blued finish. Introduced 1938; discontinued 1944.
Exc.: $325 **VGood:** $275 **Good:** $200

SAUER MODEL 1913
Semi-automatic; 32 ACP; 7-shot; 3″ barrel; 5 ⁷/₈″ overall length; fixed sights; checkered hard rubber black grips; blued finish. Introduced 1913; discontinued 1930.
Exc.: $200 **VGood:** $150 **Good:** $125

Sauer
Model 1913 Pocket
Model 1930 (Behorden Model)
WTM

Savage
Automatic Pistol
Model 101
Model 1907
Model 1907 U.S. Army
Test Trial
Model 1915 Hammerless
Model 1917

Schuetzen Pistol Works
(See Olympic Arms)

Security Industries
Model PM357
Model PPM357
Model PSS38

Sedco
Model SP-22

Sedgley
Baby Hammerless Model

Seecamp
LWS 25 ACP Model
LWS 32

Semmerling
Semmerling LM-4
(See American Derringer
Semmerling LM-4)

Sheridan
Knockabout

Savage Model 1917

Savage Model 1907

Savage Model 101

Sauer Model 1930
(Behorden Model)

Sauer WTM

Sauer Model 1913 Pocket
Same specs as Model 1913 except smaller in size; 25 ACP; 2 1/2" barrel; 4 1/4" overall length; improved grip, safety features. Introduced 1920; discontinued 1930.
Exc.: $250   **VGood:** $200   **Good:** $150

SAUER MODEL 1930 (BEHORDEN MODEL)
Semi-automatic; 32 ACP; 7-shot; 3" barrel; fixed sights; black plastic grips; blued finish. Introduced 1930; discontinued 1938.
Exc.: $300   **VGood:** $250   **Good:** $175

SAUER WTM
Semi-automatic; 25 ACP; 6-shot; 2 1/2" barrel; 4 1/8" overall length; fixed sights; checkered hard rubber grips; top ejector; fluted slide; blued finish. Introduced 1924; discontinued 1927.
Exc.: $250   **VGood:** $200   **Good:** $150

SAVAGE AUTOMATIC PISTOL
Semi-automatic; 25 ACP; 6-shot; 2 3/8" barrel; weighs 12 oz.; hammerless; wide or narrow slide serrations. Experimental only pre-WWI. Only 50-100 produced. Extremely rare; extremely valuable.

SAVAGE MODEL 101
Single shot; single action; 22 Short, 22 LR; 5 1/2" barrel integral with chamber; 9" overall length; adjustable slotted rear sight, blade front; compressed, plastic-impreg-

nated wood grips; fake cylinder swings out for loading, ejection; manual ejection with spring-rod ejector; blued finish. Introduced 1959; discontinued 1968.
Exc.: $125   **VGood:** $110   **Good:** $100

SAVAGE MODEL 1907
Semi-automatic; 32 ACP, 380 ACP; 10-shot (32), 9-shot (380); 3 3/4" barrel, 6 1/2" overall length (32 ACP); 4 1/2" barrel, 7" overall length (380 ACP); fixed sights; checkered hard rubber stocks; exposed hammer; blued finish. The 380-caliber serial number followed by letter "B". Manufactured 1910 to 1917.
32 ACP
Exc.: $250   **VGood:** $200   **Good:** $150
380 ACP
Exc.: $350   **VGood:** $275   **Good:** $225

Savage Model 1907 U.S. Army Test Trial
Same specs as Model 1907 except enlarged model for 45 ACP only; exposed hammer; approximately 300 made for test trials from 1907 to 1911.
Exc.: $7000   **VGood:** $5500   **Good:** $3500

SAVAGE MODEL 1915 HAMMERLESS
Semi-automatic; 32 ACP, 380 ACP; 10-shot (32), 9-shot (380); 3 3/4" barrel, 6 1/2" overall length (32 ACP); 4 1/2" barrel, 7" overall (380 ACP); fixed sights; checkered hard rubber stocks; hammerless; grip safety; blued finish. Manufactured 1915 to 1917.
32 caliber
Exc.: $300   **VGood:** $250   **Good:** $200
380 caliber
Exc.: $450   **VGood:** $350   **Good:** $250

SAVAGE MODEL 1917
Semi-automatic; 32 ACP, 380 ACP; 10-shot (32), 9-shot (380); 3 3/4" barrel, 6 1/2" overall length (32 ACP); 4 1/2" barrel, 7" overall length (380 ACP); fixed sights; hard rubber grips; spurred hammer; blued finish. Marked "Savage Model 1917" on left side. Introduced 1920; discontinued 1928.
32 caliber
Exc.: $250   **VGood:** $200   **Good:** $150
380 caliber
Exc.: $400   **VGood:** $350   **Good:** $250

SECURITY INDUSTRIES MODEL PM357
Revolver; double action; 357 Mag.; 5-shot; 2 1/2" barrel; 7 1/2" overall length; fixed sights; uncheckered American walnut target stocks; stainless steel. Introduced 1973; discontinued 1977.
Exc.: $225   **VGood:** $175   **Good:** $150

SECURITY INDUSTRIES MODEL PPM357
Revolver; double action; 357 Mag.; 5-shot; 2" barrel; 6 1/8" overall length; fixed sights; checkered American

walnut target stocks; stainless steel. Introduced 1976 with spurless hammer, converted to conventional hammer 1977; discontinued 1977.
Exc.: $225   **VGood:** $175   **Good:** $150

SECURITY INDUSTRIES MODEL PSS38
Revolver; double action; 38 Spl.; 5-shot; 2" barrel; 6 1/2" overall length; fixed sights; checkered or smooth American walnut stocks; stainless steel. Introduced 1973; discontinued 1977.
Exc.: $150   **VGood:** $125   **Good:** $110

SEDCO MODEL SP-22
Semi-automatic; single action; 22 LR; 6-shot; 2 1/2" barrel; 5" overall length; weighs 11 oz.; fixed sights; simulated pearl stocks; rotary safety blocks sear, slide; chrome or black Teflon finish. Introduced 1988; discontinued 1990.
Exc.: $65   **VGood:** $50   **Good:** $40

SEDGLEY BABY HAMMERLESS MODEL
Revolver; double action; 22 LR; 6-shot; 4" overall length; solid frame; hard rubber stocks; fixed sights; folding trigger; blued or nickel finish. Manufactured 1930 to 1939.
Exc.: $300   **VGood:** $225   **Good:** $175

SEECAMP LWS 25 ACP MODEL
Semi-automatic; double action; 25 ACP; 7-shot; 2" barrel; 4 1/8" overall length; weighs 12 oz.; fixed sights; plastic grips; stainless steel; matte finish. Introduced 1982; discontinued 1985.
Exc.: $350   **VGood:** $300   **Good:** $225

SEECAMP LWS 32
Semi-automatic; double-action-only; 32 ACP Win. Silvertip; 6-shot magazine; 2" barrel; 4 1/8" overall length; weighs 10 1/2 oz.; glass-filled nylon grips; smooth, no-snag, contoured slide and barrel top serve as sights; aircraft quality 17-4 PH stainless steel; inertia-operated firing pin; magazine safety disconnector; polished stainless. Made in U.S. by L.W. Seecamp. Introduced 1985; still produced. **LMSR: $475**
New: $600   **Perf.:** $550   **Exc.:** $500

SHERIDAN KNOCKABOUT
Single shot; tip-up action; 22 Short, 22 LR; 5" barrel; 8 3/4" overall length; fixed sights; checkered plastic grips; blued finish. Introduced 1953; discontinued 1960.
Exc.: $100   **VGood:** $85   **Good:** $75

Security Industries
Model PSS38

SIG Sauer P226

Sedco Model SP-22

SIG Sauer P230

SIG
P210-1
P210-2
P210-5
P210-6
P210 22 Conversion Unit

SIG-Hammerli
Model P240

SIG Sauer
P220 "American"
P225
P226
P228
P229
P230
P230 SL Stainless

SIG Sauer P228

SIG Sauer P220
"American"

SIG P210-2
Semi-automatic; single action; 9mm Para., 7.65mm Luger; 8-shot; 4 3/4" barrel; 8 1/2" overall length; weighs 32 oz.; fixed sights; black plastic grips; standard-issue Swiss army sidearm; blued finish. Manufactured in Switzerland. Introduced 1947; still in production. **LMSR: $3500**
   Exc.: $2000   **VGood:** $1650   **Good:** $1400

SIG P210-1
Same specs as P210-2 except; walnut grips; polished blued finish. Discontinued 1986.
   Exc.: $2250   **VGood:** $2000   **Good:** $1750

SIG P210-5
Same specs as P210-2 except 6" barrel; micrometer sights; target trigger special hammer; hard rubber grips; matte finish.
   Exc.: $2250   **VGood:** $1950   **Good:** $1500

SIG P210-6
Same specs as P210-2 except fixed rear sight, target front; target trigger. Still in production. **LMSR: $2100**
   New: $2000   **Perf.:** $1750   **Exc.:** $1500

SIG P210 22 Conversion Unit
Converts P210 to 22 LR for practice; must be ordered with pistol; new 1996.

SIG-HAMMERLI MODEL P240
Semi-automatic; single action; 32 S&W Long, 38 Spl.; 5-shot; 6" barrel; 10" overall length; micrometer rear sight, post front; uncheckered European walnut stocks with target thumbrest; adjustable triggers; blued finish. Manufactured in Switzerland. Introduced 1975; discontinued 1988.
   Exc.: $1350   **VGood:** $1000   **Good:** $850

SIG SAUER P220 "AMERICAN"
Semi-automatic; double action; 38 Super, 45 ACP; 9-shot (38 Super), 7-shot (45 ACP) magazine; 4 3/8" barrel; 7 3/4" overall length; weighs 28 1/4 oz. (9mm); checkered black plastic grips; blade front sight, drift-adjustable rear; squared combat-type trigger guard; side-button magazine release. Imported from Germany by SIGARMS, Inc. Still imported.
   New: $650   **Perf.:** $600   **Exc.:** $500

With Siglite night sights
   New: $750   **Perf.:** $700   **Exc.:** $575
With K-Kote finish
   New: $675   **Perf.:** $625   **Exc.:** $525
With K-Kote finish, Siglite night sights
   New: $800   **Perf.:** $750   **Exc.:** $600

SIG SAUER P225
Semi-automatic; double action, DA-only; 9mm Para.; 8-shot; 3 7/8" barrel; 7 3/32" overall length; weighs 26 oz.; blade-type front sight, windage-adjustable rear; checkered black plastic stocks; squared combat-type trigger guard; shorter, lighter version of P220; blued finish. Made in Germany. Introduced 1985.
   New: $600   **Perf.:** $550   **Exc.:** $450
K-Kote finish
   New: $650   **Perf.:** $600   **Exc.:** $500
Nickel finish
   New: $650   **Perf.:** $600   **Exc.:** $500
Siglite night sights
   New: $700   **Perf.:** $650   **Exc.:** $525

SIG SAUER P226
Semi-automatic; double action, DA-only; 9mm Para.; 15-, 20-shot; 4 3/8" barrel; weighs 26 1/2 oz.; high contrast sights; black plastic checkered grips; blued finish. Made in Germany. Introduced 1983; still imported. **LMSR: $825**
   New: $700   **Perf.:** $600   **Exc.:** $500
K-Kote finish
   New: $750   **Perf.:** $650   **Exc.:** $550
Nickel finish
   New: $750   **Perf.:** $650   **Exc.:** $550
Siglite night sights
   New: $800   **Perf.:** $700   **Exc.:** $575

SIG SAUER P228
Semi-automatic; double action, DA-only; 9mm Para.; 10-shot; 3 7/8" barrel; 7" overall length; weighs 29 oz.; three-dot sights; black plastic grips; blued finish. Made in Germany. Introduced 1989; still imported. **LMSR: $825**
   New: $650   **Perf.:** $550   **Exc.:** $475

K-Kote finish
   New: $700   **Perf.:** $600   **Exc.:** $500
Nickel finish
   New: $700   **Perf.:** $600   **Exc.:** $500
Siglite night sights
   New: $750   **Perf.:** $650   **Exc.:** $550

SIG SAUER P229
Semi-automatic; double action, DA-only; 9mm Para., 40 S&W, 357 SIG; 12-shot; 3 7/8" barrel; 7" overall length; 30 1/2 oz.; three-dot sights; checkered black plastic grips; aluminum alloy frame; blued slide. Made in Germany. Introduced 1991; still produced. **LMSR: $875**
   New: $700   **Perf.:** $600   **Exc.:** $500
Siglite night sights
   New: $800   **Perf.:** $700   **Exc.:** $575

SIG SAUER P230
Semi-automatic; double action; 22 LR, 32 ACP, 380 ACP, 9mm Ultra; 10-shot (22 LR), 8-shot (32 ACP), 7-shot (380 ACP, 9mm); 3 3/4" barrel; 6 1/2" overall length; weighs 16 oz.; fixed sights; checkered black plastic stocks; blued finish. Introduced 1977; still in production. Manufactured in Germany.
   New: $400   **Perf.:** $350   **Exc.:** $300

SIG Sauer P230 SL Stainless
Same specs as P230 except stainless steel; 22 oz.
   New: $450   **Perf.:** $400   **Exc.:** $350

Smith & Wesson Revolvers

22 Straight Line
32 Double Action
32 Safety Hammerless
38 Double Action
38 Double Action Perfected Model
38 Safety Hammerless (New Departure)
38 Single Action Mexican Model (See S&W 38 Single Action Third Model)
38 Single Action Third Model (Model 1891)
44 Double Action First Model
44 Double Action Favorite
44 Double Action Frontier Bicycle Model (See S&W 32 Safety Hammerless)
New Model No. 3
New Model No. 3 Frontier
New Model No. 3 Target

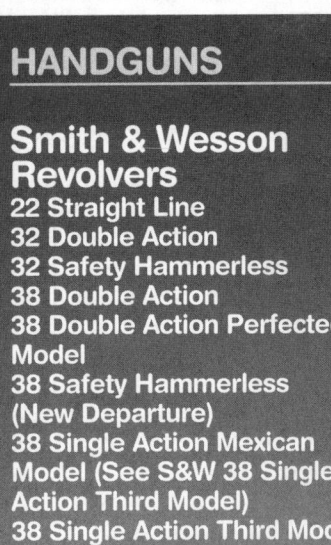

Smith & Wesson 22 Straight Line

Smith & Wesson 38 Safety Hammerless (New Departure)

Smith & Wesson 38 Double Action Perfected Model

Smith & Wesson 44 Double Action First Model

SMITH & WESSON

The section on Smith & Wesson handguns has been reorganized to make it easier to use and understand. First, it has been divided into revolvers and semi-automatics with revolvers listed first.

Second, these two major areas are delineated further with revolvers divided into four distinct groups: Early Models, Small (I and J) Frames, Medium (K and L) Frames, Large (N) Frames; and semi-automatics divided into five groups: Pre-WWII, Post-WWII, 2nd Generation, 3rd Generation and Rimfire Target.

Today's Smith & Wessons all bear a model number and serial number. However, prior to 1958, S&W handguns bore only a serial number along with the specific name of that model, e.g. 38 Military & Police. In 1958, the factory began using a model numbering system. On revolvers, the model designation is stamped in the yoke cut in the frame and covered by the cylinder arm when the cylinder is in the closed position. The official serial number can be found on the butt of the gun. For revolvers factory-issued with target stocks that cover the butt, the serial number is also stamped on the frame in the area under the yoke. Serial numbers that begin with a letter indicate post-World War II manufacture. The new triple-alpha/four-numeric system began in 1980.

Caution:

1. There is also a number stamped on the yoke which is called the work number. There is no relationship between this number and the serial number.

2. Handguns are classified by basic model numbers. The number following the model number is called the dash number and is necessary only when ordering parts.

On semi-automatics, the serial number can be found on the left side of the frame along with the model designation.

Roy Jinks

EARLY MODELS

SMITH & WESSON 22 STRAIGHT LINE

Single shot; 22 LR; 10″ barrel; pivoted frame; has the appearance of a single-action semi-automatic; smooth walnut grips; target sights; blued finish only; came in green felt-lined blued steel case with cleaning rod and screwdriver. Manufactured 1925 to 1936 with a total of 1870 produced.

Exc.: $750   **VGood:** $550   **Good:** $400
With case
Exc.: $1250   **VGood:** $750   **Good:** $500

SMITH & WESSON 32 DOUBLE ACTION

Revolver; double action; 32 S&W; 5-shot cylinder; 3″, 3 1/2″, 6″, 8″, 10″ barrels; hard rubber stocks; fixed sights; blued or nickel finish. Manufactured 1880 to 1919. Early serial numbers have great collector significance, with numbers through 50 bringing as high as $6500.

Exc.: $350   **VGood:** $225   **Good:** $150

SMITH & WESSON 32 SAFETY HAMMERLESS

Revolver; double action; 32 S&W; 5-shot; 2″, 3″, 3 1/2″, 6″ barrels; black, hard rubber or walnut grips; blue or nickel finish. Numbered in separate serial number series. The 2″ barrel models called "Bicycle Model" and demand premium prices. 249,981 produced. Manufactured from 1887 to 1937.

Exc.: $400   **VGood:** $325   **Good:** $200
2″ barrel Bicycle Model
Exc.: $450   **VGood:** $300   **Good:** $175

SMITH & WESSON 38 DOUBLE ACTION

Revolver; double action; 38 S&W; 5-shot cylinder; 4″, 4 1/2″, 5″, 6″, 8″, 10″ barrel; hinged frame; hard rubber checkered stocks; fixed sights; blued or nickel finish. Several design variations. Manufactured 1880 to 1911; some collector value.
S/N through 4000
Exc.: $700   **VGood:** $450   **Good:** $285
S/N over 4000
Exc.: $325   **VGood:** $225   **Good:** $150
8″, 10″ barrels
Exc.: $1500   **VGood:** $1250   **Good:** $800

SMITH & WESSON 38 DOUBLE ACTION PERFECTED MODEL

Revolver; double action; 38 S&W; 5-shot; 3 1/4″ to 6″ barrels; hard black rubber grips; blue or nickel finish. Distinguishing features include: manufactured on a modified 32 Hand Ejector frame; incorporates both a top barrel latch and frame-mounted side latch. The last top-break revolver introduced by S&W, the Perfected model was introduced in 1909 and produced until 1920 with a total production run of 59,400.

Exc.: $425   **VGood:** $350   **Good:** $250

SMITH & WESSON 38 SAFETY HAMMERLESS (NEW DEPARTURE)

Revolver; double action; 38 S&W; 5-shot; 3 1/4″, 4″, 5″, 6″ barrel; checkered hard rubber or walnut grips; blue or nickel finish. Numbered in separate serial number series. Total production run of 261,493. Manufactured from 1887 to 1937.

Exc.: $600   **VGood:** $375   **Good:** $225

SMITH & WESSON 38 SINGLE ACTION THIRD MODEL (MODEL 1891)

Revolver; single action; 38 S&W; 5-shot; 3 1/4″, 4″, 5″, 6″ barrel; fixed sights; checkered hard rubber or walnut grips; blued or nickel finish. A special variation was manufactured with a spur trigger assembly inserted in place of the standard trigger and trigger guard. This was called the Mexican Model and is very rare; beware of fakes.

Exc.: $2000   **VGood:** $1250   **Good:** $850

SMITH & WESSON 44 DOUBLE ACTION FIRST MODEL

Revolver; double action; 44 Russian, 38-40; 6-shot cylinder; 4″, 5″, 6″, 6 1/2″ barrel; hard rubber or walnut grips; fixed sights; blued or nickel finish. Manufactured 1881 to 1913. 53,668 manufactured.

Exc.: $1600   **VGood:** $1200   **Good:** $800

Smith & Wesson 44 Double Action Favorite

Same specs as the First Model except 5″ barrel. Approximately 1000 manufactured between 1882 to 1883.

Exc.: $5000   **VGood:** $4000   **Good:** $2500

Smith & Wesson 44 Double Action Frontier

Same specs as the First Model except 44-40; longer 1 7/16″ cylinder. Manufactured 1886 to 1910; collector interest.

Exc.: $1500   **VGood:** $1000   **Good:** $600

SMITH & WESSON NEW MODEL NO. 3

Revolver; single action; hinged frame; 44 Russian; 3 1/2″, 4″, 5″, 6″, 6 1/2″, 7″, 7 1/2″, 8″ barrel; rounded hard rubber or walnut stocks; fixed or target sights; blued or nickel finish; manufactured 1878 to 1908; broad collector value.

Exc.: $2000   **VGood:** $1600   **Good:** $800

Smith & Wesson New Model No. 3 Frontier

Same specs as New Model No. 3 except 44-40 Win.; 4″, 5″, 6 3/4″ barrel; longer 1 9/16″ cylinder. Manufactured 1885 to 1908; great collector interest.

Exc.: $3000   **VGood:** $2250   **Good:** $1600

Smith & Wesson New Model No. 3 Target

Same specs as New Model No. 3 except 32-44 S&W, 38-44 S&W; 6″, 6 1/2″ barrel; adjustable sights. Manufactured 1887 to 1910; collector interest.

Exc.: $2250   **VGood:** $1600   **Good:** $1000

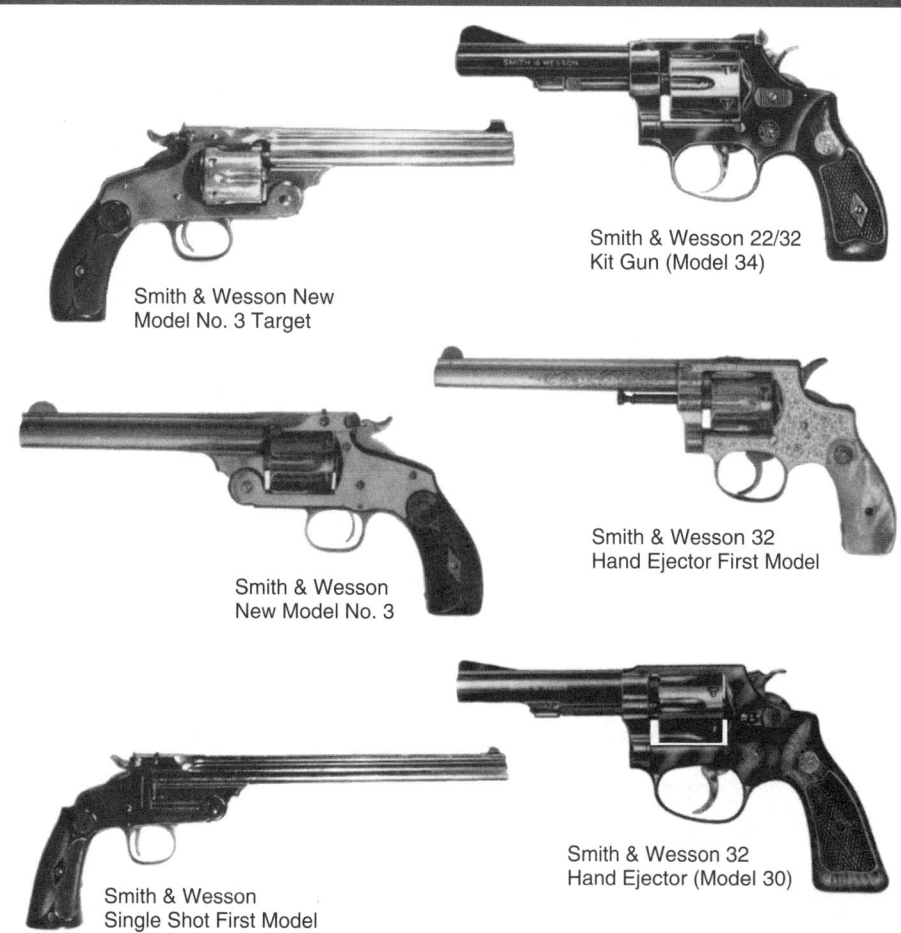

Smith & Wesson New Model No. 3 Target

Smith & Wesson 22/32 Kit Gun (Model 34)

Smith & Wesson New Model No. 3

Smith & Wesson 32 Hand Ejector First Model

Smith & Wesson Single Shot First Model

Smith & Wesson 32 Hand Ejector (Model 30)

HANDGUNS

Smith & Wesson Revolvers
New Model No. 3 Winchester
Olympic Model (See S&W
Single Shot Third Model
Perfected Single Shot)
Single Shot First Model
Single Shot Second Model
Single Shot Third Model
(Perfected Single Shot)

I and J Small Frame
Revolvers
22 M.R.F. Service Kit Gun
(Model 650)
22/32 Kit Gun
22/32 Kit Gun (Model 34)
22/32 Target
32 Hand Ejector First Model
32 Hand Ejector (Model 30)
38 Regulation Police
(Model 31)
38 Regulation Police Target
Model

Smith & Wesson New Model No. 3 Winchester
Same specs as New Model No. 3 Frontier except 38-40. Manufactured 1900-1907; extremely rare; only 74 manufactured.

Exc.: $6300   **VGood:** $4500   **Good:** $3000

SMITH & WESSON SINGLE SHOT FIRST MODEL
Single shot; 22 LR, 32 S&W, 38 S&W; 6", 8", 10" barrels; target sights; square-butt, hard rubber grips; blued or nickel finish; built on same frame as 38 Single Action Third Model. Also furnished with 38 S&W barrel, cylinder to convert to single action. 1200 manufactured between 1893 and 1905; collector interest.

22 LR
Exc.: $650   **VGood:** $450   **Good:** $300
32 S&W
Exc.: $750   **VGood:** $550   **Good:** $350
38 S&W
Exc.: $800   **VGood:** $600   **Good:** $450

Smith & Wesson Single Shot Second Model
Same specs as First Model except 22 LR only; 10" barrel; cannot be converted to revolver configuration; serial numbered in separate series.

Exc.: $600   **VGood:** $450   **Good:** $275

Smith & Wesson Single Shot Third Model (Perfected Single Shot)
Same specs as First Model except 22 LR; hinged frame; target model; redesigned to incorporate new frame design developed for the 38 Double Action Perfected Model; checkered walnut target grips. Manufactured from 1909 to 1923. A special group featuring a short chamber was manufactured in 1910 and called the Olympic Model; the 22 LR cartridge must be forced into the rifling.

Perfected Single Shot
Exc.: $600   **VGood:** $450   **Good:** $300

I and J SMALL FRAME REVOLVERS
The original Smith & Wesson small frame revolvers were called "I" frames. The "J" frame was introduced in 1950 and by 1960 became the standard small frame.

SMITH & WESSON 22 M.R.F. SERVICE KIT GUN (MODEL 650)
Revolver; double action; 22 WMR; 6-shot cylinder; 3" heavy barrel; 7 1/4" overall length; weighs 23 oz.; stainless steel construction; fixed 1/10" serrated front ramp, micro-click adjustable rear; checkered walnut round butt stocks with medallion; satin finish. Introduced 1983; dropped 1987.

Exc.: $200   **VGood:** $175   **Good:** $150

SMITH & WESSON 22/32 KIT GUN
Revolver; 22 LR; 6-shot; 4" barrel; 8" overall length; weighs 21 oz.; 1/10" Patridge or pocket revolver front sight, with rear sight adjustable for elevation and windage; checkered round-butt Circassian walnut or hard rubber stocks; (small or special oversized target square-butt stocks were offered on special order); blued or nickel finish; a compact outdoorsman's revolver based on the 22/32 Target. Introduced in 1935; replaced in 1953 by the Model 34.

Exc.: $650   **VGood:** $600   **Good:** $400

SMITH & WESSON 22/32 KIT GUN (MODEL 34)
Revolver; 22 LR; 6-shot; 2", 4" barrels; 8" overall length (4" barrel); weighs 22 1/2 oz.; fixed 1/10" serrated front ramp sight, micro-click adjustable rear; checkered walnut round or square-butt grips with medallion; blued finish. Originally introduced in 1936; pre-war variations feature round barrel and bring premium prices. Discontinued in 1991.

Exc.: $250   **VGood:** $200   **Good:** $175

SMITH & WESSON 22/32 TARGET
Revolver; 22 LR; 6-shot; 6" barrel; 10 1/2" overall length; sights, 1/10" or 1/8" Patridge front, fully-adjustable square notch rear sight; stocks, special, oversize, square-butt pattern in checkered Circassian walnut with S&W monogram; chambers countersunk at the heads around 1935 for the higher-velocity cartridges; blued finish only. Won the "Any Revolver" event at the USRA matches

several times; a forerunner of the Model 35. Introduced in 1911; superseded by the Model 35 in 1953.

Exc.: $600   **VGood:** $450   **Good:** $300

SMITH & WESSON 32 HAND EJECTOR FIRST MODEL
Revolver; double action, hand ejector; 32 S&W Long; 3 1/4", 4 1/4", 6" barrel; fixed sights; hard rubber stocks; blued or nickel finish. First S&W solid-frame revolver with swing-out cylinder. Company name and patent dates marked on cylinder rather than barrel. Total production run of 19,712. Manufactured 1896 to 1903.

Exc.: $500   **VGood:** $350   **Good:** $225

SMITH & WESSON 32 HAND EJECTOR (MODEL 30)
Revolver; double action; 32 S&W Long; will accept 32 S&W and 32 S&W Long wadcutter; 6-shot; 2", 3", 4" barrels; 6" available at one time; 8" overall length (4" barrel); weighs 18 oz.; checkered walnut grips with medallion; formerly hard rubber; serrated ramp front and square notch rear; blue or nickel finish. Introduced 1908; dropped 1972.

Exc.: $250   **VGood:** $150   **Good:** $110

SMITH & WESSON 38 REGULATION POLICE (MODEL 31)
Revolver; 32 S&W Long; accepts 32 S&W, 32 Colt New Police; 6-shot; 2", 3", 3 1/4", 4", 4 1/4", 6" barrels; 8 1/2" overall length (4" barrel); weighs 18 3/4 oz.; fixed sights with 1/10" serrated ramp front and square notch rear; checkered walnut stocks with medallion; blue or nickel finish. Introduced 1917; no longer in production.

Exc.: $300   **VGood:** $225   **Good:** $150

Smith & Wesson 38 Regulation Police Target Model
Same specs as Model 31 except target model with 6" barrel; 10 1/4" overall length; weighs 20 oz.; adjustable target sights. Introduced 1917; dropped 1941.

Exc.: $600   **VGood:** $450   **Good:** $300

HANDGUNS

Smith & Wesson
38 Regulation Police (Model 33)
1960 22/32 Kit Gun M.R.F. (Model 51)
1977 22/32 Kit Gun M.R.F. (Model 63)
Airweight Kit Gun (Model 43)
Bodyguard (Model 49)
Bodyguard (Model 649)
Bodyguard Airweight (Model 38)
Centennial (Model 40)
Centennial (Model 640)
Centennial (Model 940)
Centennial Airweight (Model 42)
Centennial Airweight (Model 442)
Centennial Airweight (Model 642)

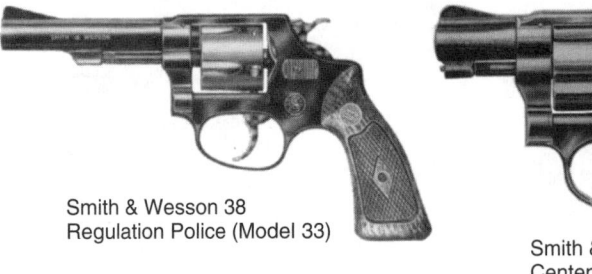

Smith & Wesson 38 Regulation Police (Model 33)

Smith & Wesson Centennial (Model 40)

Smith & Wesson 1977 22/32 Kit Gun M.R.F. (Model 63)

Smith & Wesson Centennial (Model 640)

Smith & Wesson Bodyguard (Model 649)

Smith & Wesson Centennial Airweight (Model 442)

SMITH & WESSON 38 REGULATION POLICE (MODEL 33)

Revolver; 38 S&W; accepts 38 Colt New Police; 5-shot; 4" barrel; 8 1/2" overall length; weighs 18 oz.; fixed sights, with 1/10" serrated ramp front and square notch rear; checkered walnut stocks with medallion; blue or nickel finish. Introduced in 1917; dropped 1974.

Exc.: $275   **VGood:** $175   **Good:** $125

SMITH & WESSON 1960 22/32 KIT GUN M.R.F. (MODEL 51)

Revolver; 22 WMR; 3 1/2" barrel; 8" overall length; weighs 24 oz.; all steel frame and cylinder; fixed 1/10" serrated front ramp, micro-click adjustable rear; checkered walnut, round or square-butt stocks; blue or nickel finish. Introduced 1960; dropped 1974.

Exc.: $325   **VGood:** $275   **Good:** $225
Nickel finish
Exc.: $350   **VGood:** $300   **Good:** $250

SMITH & WESSON 1977 22/32 Kit Gun M.R.F. (MODEL 63)

Revolver; 22 LR; 6-shot; 4" barrel; 8" overall length; weighs 24 1/4 oz.; stainless steel construction; fixed 1/10" serrated front ramp sight, micro-click adjustable rear; checkered walnut square-butt stocks with medallion; satin finish. Introduced 1977; still in production.

New: $375   **Perf.:** $325   **Exc.:** $250

SMITH & WESSON AIRWEIGHT KIT GUN (MODEL 43)

Revolver; 22 LR; 3 1/2" barrel; 8" overall length; weighs 14 1/4 oz.; alloy aluminum frame and cylinder; fixed 1/10" serrated front ramp, micro-click adjustable rear; checkered walnut, round or square-butt stocks; blue or nickel finish. Introduced 1955; dropped 1974.

Exc.: $350   **VGood:** $300   **Good:** $250

SMITH & WESSON BODYGUARD (MODEL 49)

Revolver; double action; 38 Spl.; 5-shot; 2" barrel; 6 5/16" overall length; steel construction; weighs 20 1/2 oz.; fixed 1/10" serrated front ramp, square notch rear; blue or nickel finish; features shrouded hammer that can be cocked manually for single-action firing. Introduced 1955; still in production.

Perf.: $250   **Exc.:** $200   **VGood:** $175

SMITH & WESSON BODYGUARD (MODEL 649)

Revolver; double action; 38 Spl.; 5-shot; 2" barrel; 6 5/16" overall length; weighs 20 1/2 oz.; stainless steel construction; fixed 1/10" serrated front ramp, square notch rear; satin finish; features shrouded hammer that can be cocked manually for single-action firing. Introduced 1955; still in production.

Perf.: $375   **Exc.:** $325   **VGood:** $250

SMITH & WESSON BODYGUARD AIRWEIGHT (MODEL 38)

Revolver; double action; 38 Spl.; 5-shot; 2" barrel; 6 5/16" overall length; weighs 14 1/2 oz.; fixed 1/10" serrated front ramp, square notch rear; blue or nickel finish; features shrouded hammer that can be cocked manually for single-action firing. Introduced 1955; still in production.

Perf.: $300   **Exc.:** $250   **VGood:** $200

SMITH & WESSON CENTENNIAL (MODEL 40)

Revolver; double action; 38 Spl; 5-shot; 2" barrel; 6 1/2" overall length; weighs 19 oz.; concealed hammer; fixed 1/10" serrated front ramp, square notch rear. Swing-out version of earlier top-break design with grip safety. Introduced 1953; dropped 1974. Collector value.

Exc.: $400   **VGood:** $350   **Good:** $275

SMITH & WESSON CENTENNIAL (MODEL 640)

Revolver; double action; 38 Spl.; 5-shot; 2", 3" barrel; 6 5/16" overall length; weighs 20 oz.; serrated ramp front, fixed notch rear sight; Goncalo Alves round-butt grips; stainless steel version of the Model 40 sans grip safety; concealed hammer; smoothed edges. Introduced 1990; still in production.

New: $375   **Perf.:** $325   **Exc.:** $275

SMITH & WESSON CENTENNIAL (MODEL 940)

Revolver; double action; 9mm Para.; 5-shot; 2", 3" barrel; 6 5/16" overall length; weighs 20 oz.; serrated ramp front, fixed notch rear sight; rubber grips; stainless steel version of the Model 40 sans grip safety; concealed hammer; smoothed edges. Introduced 1990; still in production.

New: $375   **Perf.:** $325   **Exc.:** $275

SMITH & WESSON CENTENNIAL AIRWEIGHT (MODEL 42)

Revolver; double action; 38 Spl; 5-shot; 2" barrel; 6 1/2" overall length; concealed hammer; aluminum alloy frame and cylinder; fixed 1/10" serrated front ramp, square notch rear. Lightweight version of Model 40. Introduced 1953; dropped 1974. Collector value.

Exc.: $400   **VGood:** $350   **Good:** $250

SMITH & WESSON CENTENNIAL AIRWEIGHT (MODEL 442)

Revolver; double action; 38 Spl.; 5-shot cylinder; 2" barrel; weighs 15 3/4 oz.; 6 5/8" overall length; Uncle Mike's Custom Grade Santoprene grips; serrated ramp front sight, fixed notch rear; alloy frame; carbon steel barrel and cylinder; concealed hammer. Made in U.S. by Smith & Wesson. Introduced 1993; still produced.

New: $350   **Perf.:** $300   **Exc.:** $250
Nickel finish
New: $375   **Perf.:** $325   **Exc.:** $275

SMITH & WESSON CENTENNIAL AIRWEIGHT (MODEL 642)

Revolver; double action; 38 Spl.; 5-shot; 2" carbon steel barrel; 6 5/16" overall length; anodized alloy frame; weighs 16 oz.; serrated ramp front, fixed notch rear sight; Uncle Mike's Custom Grade Santoprene grips; same as the Model 40 sans grip safety; concealed hammer; smoothed edges. Introduced 1990; no longer in production.

Perf.: $400   **Exc.:** $350   **VGood:** $250

Smith & Wesson Chief's Special (Model 36)

Smith & Wesson Model 60 3" Full Lug

Smith & Wesson Chief's Special Airweight (Model 37)

Smith & Wesson Model 651

Smith & Wesson Ladysmith (Model 60-LS)

Smith & Wesson Target Model 1953 (Model 35)

HANDGUNS

Smith & Wesson
Chief's Special (Model 36)
Chief's Special Airweight (Model 37)
Chief's Special Stainless (Model 60)
Ladysmith (Model 36-LS)
Ladysmith (Model 60-LS)
Model 60 3" Full Lug
Model 651
Target Model 1953 (Model 35)
Terrier (Model 32)

K and L Medium Frame Revolvers
32-20 Hand Ejector
38 Military & Police

SMITH & WESSON CHIEF'S SPECIAL (MODEL 36)

Revolver; double action; 38 Spl.; 5-shot; 2", 3" barrel; 7 3/8" overall length (3" barrel); weighs 21 1/2 oz.; fixed 1/10" serrated front ramp, square notch rear; all steel frame and cylinder; checkered round butt soft rubber grips most common; square-butt available; blue or nickel finish.

New: $275   **Perf.:** $225   **Exc.:** $175

SMITH & WESSON CHIEF'S SPECIAL AIRWEIGHT (MODEL 37)

Revolver; double action; 38 Spl.; 5-shot; 2" barrel; 6 1/2" overall length; weighs 19 1/2 oz.; lightweight version of Model 36, incorporating aluminum alloy frame; weighs 14 1/2 oz.; fixed 1/10" serrated front ramp, square notch rear; all steel frame and cylinder; checkered round butt grips most common; square-butt available; blue or nickel finish. Still in production.

New: $350   **Perf.:** $275   **Exc.:** $225

SMITH & WESSON CHIEF'S SPECIAL STAINLESS (MODEL 60)

Revolver; double action; 38 Spl.; 5-shot; 2" barrel; 6 1/2" overall length; weighs 19 oz.; stainless steel construction; fixed 1/10" serrated front ramp, square notch rear; checkered walnut round butt grips. Still in production.

New: $300   **Perf.:** $250   **Exc.:** $225

SMITH & WESSON LADYSMITH (MODEL 36-LS)

Revolver; double action; 38 Spl.; 5-shot; 2" barrel; 6 5/16" overall length; weighs 20 oz.; carbon steel construction; serrated front ramp, square notch rear; rosewood laminate grips; comes in fitted carry/storage case; blue finish. Introduced 1989; still in production.

New: $350   **Perf.:** $300   **Exc.:** $225

SMITH & WESSON LADYSMITH (MODEL 60-LS)

Revolver; double action; 38 Spl.; 5-shot; 2" barrel; 6 5/16" overall length; weighs 20 oz.; stainless steel construction; serrated front ramp, square notch rear; rosewood laminate grips; stainless satin finish. Introduced 1989; still in production. **LMSR: $461**

New: $350   **Perf.:** $300   **Exc.:** $250

SMITH & WESSON MODEL 60 3" FULL LUG

Revolver; double action; 38 Spl.; 5-shot; 3" full lug barrel; 7 1/2" overall length; weighs 24 1/2 oz.; stainless steel construction; fixed pinned black ramp front sight, adjustable black blade rear; rubber combat grips. Introduced 1991; still in production.

New: $350   **Perf.:** $300   **Exc.:** $250

SMITH & WESSON MODEL 651

Revolver; double action; 22 WMR; 6-shot cylinder; 4" heavy barrel; 8 11/16" overall length; weighs 24 1/2 oz.; stainless steel construction; ramp front, adjustable micrometer-click rear; semi-target hammer; combat trigger; soft rubber or checkered walnut square-butt stocks with medallion; satin finish. Introduced 1983; still in production.

New: $350   **Perf.:** $300   **Exc.:** $250

SMITH & WESSON TARGET MODEL 1953 (MODEL 35)

Revolver; 22 LR; 6-shot; 6" barrel; 10 1/2" overall length; weighs 25 oz.; ribbed barrel; micrometer rear sight; Magna-type stocks; flattened cylinder latch; blue finish. A redesign of the 22/32 Target model. Introduced 1953; dropped 1974.

Exc.: $300   **VGood:** $200   **Good:** $175

SMITH & WESSON TERRIER (MODEL 32)

Revolver; 38 S&W, 38 Colt New Police; 5-shot; 2" barrel; 6 1/4" overall length; weighs 17 oz.; fixed sights with 1/10" serrated ramp front, square notch rear; round butt checkered walnut stocks with medallion; blue or nickel finish. Introduced 1936; dropped 1974.

Exc.: $200   **VGood:** $150   **Good:** $125

K AND L MEDIUM FRAME REVOLVERS

Smith & Wesson's original medium-frame revolver was the "K" frame introduced in 1899. In 1980, S&W offered an improved medium frame revolver for the 357 Magnum cartridge and this new frame was called the "L" frame.

SMITH & WESSON 32-20 HAND EJECTOR

Revolver; 32 Winchester; 6-shot; 4", 5", 6", 6 1/2" barrels; rare; hard black rubber round or checkered walnut square-butt grips; caliber marking on early models, 32 Winchester; late models, 32 W.C.F., marked 32/20.

Exc.: $400   **VGood:** $325   **Good:** $175

SMITH & WESSON 38 MILITARY & POLICE

Revolver; double action; 38 Spl.; 6-shot; 4", 5", 6", 6 1/2" barrels; hard rubber round butt (early models) or checkered walnut square-butt grips; blue or nickel finish. The first S&W K-frame was the 38 Hand Ejector which later became known as the 38 Military & Police. This model was introduced in 1899 and has been in continuous production. During the last 96 years, S&W has produced over 5 million of these revolvers, resulting in numerous variations which affect values only slightly. The most collectable of these, the 38 Hand Ejector First Model, can be easily recognized by the lack of a locking lug located on the underside of the barrel and the serial number 1 through 20,975. Between 1899 and 1940 this model was available with both fixed sights and an adjustable target sight. The target model, referred to as the 38 Military and Police Target Model, brings a premium price and was the predecessor to all K-frame target models.

38 Hand Ejector First Model
Exc.: $450   **VGood:** $300   **Good:** $150

Standard models
Exc.: $275   **VGood:** $225   **Good:** $135

Smith & Wesson Revolvers

38 Military & Police
Victory Model
K-22 Outdoorsman
K-22 First Model
(See S&W K-22 Outdoorsman)
K-22 Second Model
K-32 First Model
Model 10 Military & Police
Model 10 Military & Police
Heavy Barrel
Model 12 Military & Police
Airweight
Model 13 H.B. Military &
Police
Model 14 K-38 Masterpiece
Model 14 K-38 Masterpiece
Full Lug
Model 14 K-38 Masterpiece
Single Action
Model 15 38 Combat
Masterpiece

Smith & Wesson Model 10
Military & Police

Smith & Wesson Model 10
Military & Police (1995 model)

Smith & Wesson Model 10
Military & Police Heavy Barrel

Smith & Wesson Model 14 K-38
Masterpiece Single Action

Smith & Wesson Model 13
H.B. Military & Police

Smith & Wesson Model 15
38 Combat Masterpiece

SMITH & WESSON 38 MILITARY & POLICE VICTORY MODEL

Revolver; 38 Spl., 38 S&W; 2", 4", 5" barrels; smooth walnut grips; midnight black finish. The Victory Model is the WWII production variation of the 38 Military & Police Model. Its name was derived from the serial number prefix "V", a good luck symbol for "Victory." The revolvers provided the U.S. government were in 38 Special; those provided to the U.S. Allied Forces in 38 S&W. At the end of WWII the "V" serial number prefix was changed to "S" which was continued to 1948 when the series again reached 1,000,000.

Exc.: $300   **VGood:** $200   **Good:** $125

SMITH & WESSON K-22 OUTDOORSMAN

Revolver; 22 LR; 6-shot; 6" round barrel; adjustable rear sight; square-butt checkered walnut grips; blue finish. Also known as the K-22 First Model, this was the first in a long line of K-frame 22 rimfire target revolvers. Introduced 1931.

Exc.: $450   **VGood:** $300   **Good:** $200

SMITH & WESSON K-22 SECOND MODEL

Revolver; 22 LR; 6-shot; 6" round barrel; checkered walnut square-butt grips; blue finish. This revolver was the first to be called the "Masterpiece." It was introduced in 1940 and, because of the factory shift to wartime production, it was manufactured for only a limited time. Total production was 1067 revolvers in serial number range 682,420 to 696,952 without a letter prefix.

Exc.: $1000   **VGood:** $850   **Good:** $500

SMITH & WESSON K-32 FIRST MODEL

Revolver; 32 S&W Long; 6" round barrel; checkered walnut square-butt grips; blue finish. The K-32 First Model was a companion model to both the 38 M&P and K-22 Outdoorsman. It was introduced in 1938 and only produced until 1940 with approximately 125 handguns manufactured. It is the rarest of the early target models.

Exc.: $3000   **VGood:** $2500   **Good:** $1800

SMITH & WESSON MODEL 10 MILITARY & POLICE

Revolver; double action; 38 Spl.; 6-shot; 2", 3", 4", 5", 6 1/2" barrel; 11 1/8" overall length with square-butt, 1/4" less for round butt; weighs 31 oz.; 1/10" service-type front sight and square notch non-adjustable rear; square-butt checkered walnut stocks, round butt pattern with choice of hard rubber or checkered walnut; blued or nickel finish. In 1957 a 4" heavy barrel became available and is the 4" barrel configuration still available. The 38 M&P has been the true workhorse of the Smith & Wesson line of revolvers. The basic frame is termed S&W's K-frame, the derivative source of the K-38, et al. It is, quite possibly, the most popular, widely accepted police duty revolver ever made. Introduced about 1902.

Perf.: $250   **Exc.:** $200   **VGood:** $135
Nickel finish
Perf.: $275   **Exc.:** $225   **VGood:** $175

Smith & Wesson Model 10 Military & Police Heavy Barrel

Same specs as standard Model 10 except 4" heavy ribbed barrel; ramp front sight, square notch rear; square-butt; blued or nickel finish. Introduced 1957; still in production.

Perf.: $250   **Exc.:** $200   **VGood:** $150

SMITH & WESSON MODEL 12 MILITARY & POLICE AIRWEIGHT

Revolver; double action; 38 Spl.; 6-shot; 2", 4" barrel; overall length, 6 7/8" (2", round butt); weighs 18 oz.; aluminum alloy frame; fixed 1/8" serrated ramp front sight, square-notch rear; checkered walnut Magna-type, round or square-butt. Introduced about 1952; no longer in production.

Exc.: $250   **VGood:** $200   **Good:** $150

SMITH & WESSON MODEL 13 H.B. MILITARY & POLICE

Revolver; double action; 357 Mag.; 6-shot cylinder; 3", 4" heavy barrel; 9 5/16" overall length (4" barrel); weighs 34 oz.; serrated ramp front sight, fixed square notch rear; square or round butt; blued finish. Introduced 1974; still in production.

New: $325   **Perf.:** $275   **Exc.:** $225

SMITH & WESSON MODEL 14 K-38 MASTERPIECE

Revolver; double action; 38 Spl.; 6-shot; 6", 8 3/8" barrel; 11 1/8" overall length (6" barrel); swing-out cylinder; micrometer rear sight, 1/8" Patridge-type front; hand-checkered service-type walnut grips; blued. Introduced 1947; dropped 1983.

Exc.: $250   **VGood:** $200   **Good:** $150

Smith & Wesson Model 14 K-38 Masterpiece Full Lug

Same specs as standard Model 14 except 6" full-lug barrel; weighs 47 oz.; pinned Patridge-type front sight, micrometer click-adjustable rear; square-butt combat grips; combat trigger, hammer; polished blue finish. Reintroduced 1991; limited production.

New: $350   **Perf.:** $275   **Exc.:** $225

Smith & Wesson Model 14 K-38 Masterpiece Single Action

Same specs as standard Model 14 except 6" barrel only; single-action only; target hammer; target trigger.

Exc.: $350   **VGood:** $300   **Good:** $250

SMITH & WESSON MODEL 15 38 COMBAT MASTERPIECE

Revolver; double action; 38 Spl.; 6-shot; 2", 4", 6", 8 3/8" barrel; 9 1/8" overall length; weighs 34 oz.; blue or nickel finish. It took some years after WWII to reestablish commercial production and begin catching up with civilian demands at S&W. By the early '50s the situation was bright enough to warrant introducing a 4" version of the K-38, which was designated the 38 Combat Masterpiece. Its only nominal companion was the 22 Combat Masterpiece and no attempt was made to match loaded weights, as in the K-series—the 38 weighing 34 oz. empty, compared to 36 1/2 oz. for the 22 version. Barrel ribs were narrower than the K-series and front sights were of the Baughman, quick-draw ramp pattern, replacing the vertical surface of the K-series Patridge-type.

New: $350   **Perf.:** $275   **Exc.:** $200
Nickel finish
New: $375   **Perf.:** $300   **Exc.:** $250

Smith & Wesson Model 16
K-32 Masterpiece Full Lug

Smith & Wesson Model
17 K-22 Masterpiece

Smith & Wesson Model
14 K-38 Masterpiece

Smith & Wesson
Model 53 22 Magnum

Smith & Wesson
Model 15 38 Combat
Masterpiece

Smith & Wesson Model 48
K-22 Masterpiece WMR

Smith & Wesson Model 19
Combat Magnum

HANDGUNS

Smith & Wesson Revolvers
Model 16 K-32 Masterpiece
Model 16 K-32 Masterpiece Full Lug
Model 17 K-22 Masterpiece
Model 17 K-22 Masterpiece Full Lug
Model 18 22 Combat Masterpiece
Model 19 Combat Magnum
Model 48 K-22 Masterpiece WMR
Model 53 22 Magnum

SMITH & WESSON MODEL 16 K-32 MASTERPIECE

Revolver; double action; 32 S&W Long; 6-shot; 6" barrel; target hammer; target trigger; red-insert front sight, white-outline rear; optional Patridge or Baughman front sights; walnut, Magna-pattern stocks with medallions standard; optional factory target stocks in exotic woods; blued finish. Originated as target version of the hand-ejector in 32 S&W Long about 1935, and dropped at the beginning of WWII. Appeared in its present form in the late '40s and designated the K-32 as a companion to the K-22 and K-38. Introduced 1947; dropped 1983.

    Exc.: $1250   **VGood:** $850   **Good:** $675

Smith & Wesson Model 16 K-32 Masterpiece Full Lug

Same specs as standard Model 16 except 32 Mag.; 4", 6", 8 3/8" full-lug barrel; Patridge-type front sight, click-adjustable micrometer rear; combat-style Goncalo Alves grips; polished blue finish. Introduced 1990; no longer in production.
Semi-target 4" barrel

    Perf.: $300   **Exc.:** $250   **VGood:** $200
Target 6" barrel
    Perf.: $325   **Exc.:** $275   **VGood:** $200
Target 8 3/8" barrel
    Perf.: $350   **Exc.:** $300   **VGood:** $250

SMITH & WESSON MODEL 17 K-22 MASTERPIECE

Revolver; double action; 22 LR; 6-shot; 6" barrel standard, 8 3/8" available; 11 1/8" overall length (6"); weighs 38 1/2 oz. (6"), 42 1/2 oz. (8 3/8") loaded; blue finish. Redesigned version of Model 16. Postwar production added the refinement of a broad barrel rib, intended to compensate for weight variations between the three Masterpiece models. Likewise added were the redesigned hammer with its broad spur and thumb tip relief notch, an adjustable anti-backlash stop for the trigger and the Magna-type grips developed in the mid-'30s to help cushion the recoil of the 357 Magnum. Introduced around 1947; no longer in production.
6" barrel

    Perf.: $350   **Exc.:** $275   **VGood:** $200
8 3/8" barrel
    Perf.: $400   **Exc.:** $325   **VGood:** $250

Smith & Wesson Model 17 K-22 Masterpiece Full Lug

Same specs as standard Model 17 except 4", 6", 8 3/8" barrel; Patridge-type front sight on two longer barrel lengths (serrated type on 4" version), S&W click-adjustable micrometer rear; grooved tang; polished blue finish. Introduced 1990; no longer in production.
4", 6" barrel

    Perf.: $325   **Exc.:** $275   **VGood:** $225
8 3/8" barrel
    Perf.: $350   **Exc.:** $300   **VGood:** $250

SMITH & WESSON MODEL 18 22 COMBAT MASTERPIECE

Revolver; double action; 22 LR, 22 Long, 22 Short; 6-shot; 4" barrel; 9 1/8" overall length; weighs 36 1/2 oz. (loaded); Baughman 1/8" quick-draw front sight on plain ramp, fully-adjustable S&W micrometer-click rear; checkered walnut, Magna-type grips with S&W medallion; options include broad-spur target hammer, wide target trigger, hand-filling target stocks, red front sight insert and white outlined rear sight notch; finish, blue only. Dropped 1985.

    Exc.: $275   **VGood:** $250   **Good:** $200

SMITH & WESSON MODEL 19 COMBAT MAGNUM

Revolver; double action; 357 Mag.; 6-shot; 2 1/2", 4", 6" barrel; available with 4" barrel; 9 1/2" overall length; weighs 35 oz.; 1/8" Baughman quick-draw front plain ramp, fully-adjustable S&W micrometer-click rear; checkered Goncalo Alves grips with S&W medallion; S&W bright blue or nickel finish; built on the lighter S&W K-frame as used on the K-38, et al., rather than on the heavier N-frame used for the Model 27 and 28. Introduced about 1956; still in production. **LMSR:** $416

    New: $350   **Perf.:** $300   **Exc.:** $225

SMITH & WESSON MODEL 48 K-22 MASTERPIECE WMR

Revolver; double action; 22 WMR; 4", 6", 8 3/8" barrel; weighs 39 oz. (6" barrel). A modification of the K-22 Model 17 without being distinctly designated as a Combat Masterpiece in 4" barrel configuration; auxiliary cylinder offered to permit the use with 22 LR cartridge with 1969 quoted price of $35.50.

    Exc.: $250   **VGood:** $200   **Good:** $175
With 22 LR and 22 WMR cylinder
    Exc.: $300   **VGood:** $250   **Good:** $200

SMITH & WESSON MODEL 53 22 Magnum

Revolver; double action; 22 Jet; 6-shot; 4", 6", 8 3/8" barrel; 11 1/4" overall length (6" barrel); weighs 40 oz.; 1/8" Baughman ramp front, adjustable S&W micrometer-click rear; checkered walnut grips with S&W medallion; blued finish only. Starting in the late '50s, there was considerable interest in converting K-22s to centerfire wildcat (i.e., nonstandard cartridge) configurations usually being chambered for a shortened version of the 22 Hornet, known as the 22 Harvey K-Chuck. With the intent of capitalizing on this interest, S&W introduced the 22 Remington CFM or centerfire magnum cartridge—also termed the 22 Jet—and the Model 53, chambered for it. The 22 Jet was a necked-down 357 case, designed to use a bullet of .222" to .223" diameter. The Model 53 was also chambered for 22 rimfire ammo by means of chamber bushings, adapting it for firing 22 rimfire ammo, by means of repositioning the striker on the hammer. Alternatively, a standard 22 LR cylinder was offered as a factory-fitted accessory at about $35.30 for interchanging with the 22 Jet cylinder. Introduced 1960; dropped 1974.

    Exc.: $650   **VGood:** $500   **Good:** $400
With chamber inserts
    Exc.: $700   **VGood:** $600   **Good:** $500
With 22 LR cylinder
    Exc.: $700   **VGood:** $600   **Good:** $500
With 8 3/8" barrel.
    Exc.: $750   **VGood:** $650   **Good:** $550

Smith & Wesson Revolvers

Model 64
Model 65
Model 65LS LadySmith
Model 66 Combat Magnum
Model 67 K-38 Combat Masterpiece
Model 547
Model 586 Distinguished Combat Magnum
Model 617 K-22 Masterpiece Full Lug
Model 648 K-22 Masterpiece WMR
Model 681
Model 686
Model 686 Midnight Black

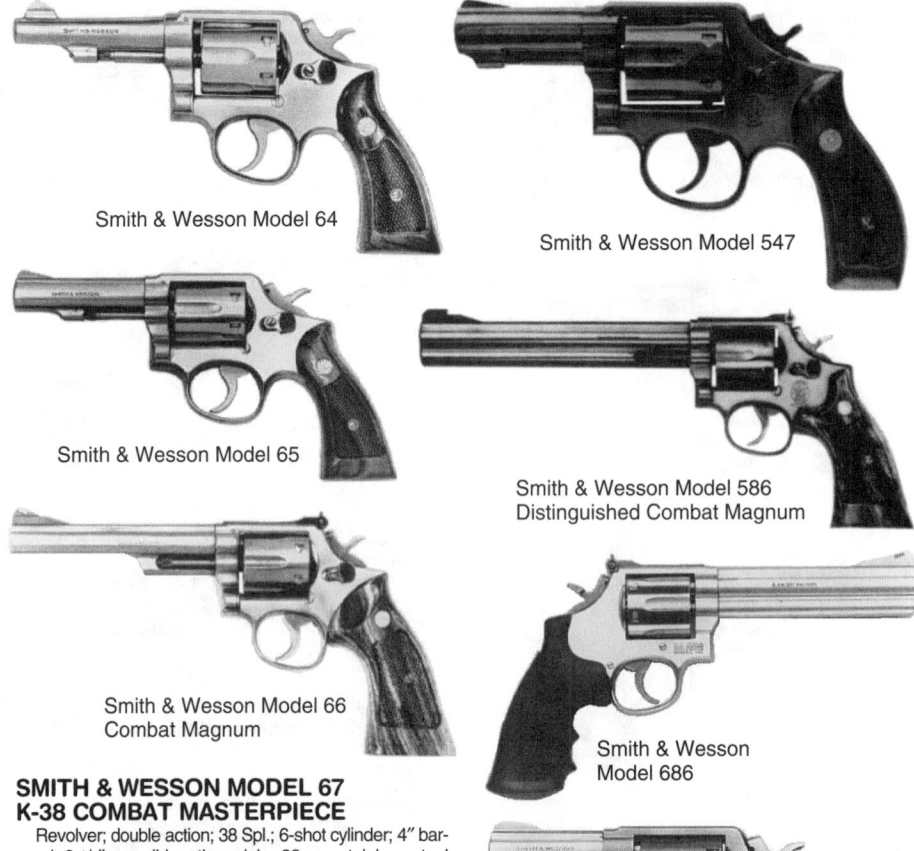

Smith & Wesson Model 64

Smith & Wesson Model 547

Smith & Wesson Model 65

Smith & Wesson Model 586 Distinguished Combat Magnum

Smith & Wesson Model 66 Combat Magnum

Smith & Wesson Model 686

Smith & Wesson Model 65LS LadySmith

SMITH & WESSON MODEL 64

Revolver; double action; 38 Spl.; 6-shot cylinder; 4″ barrel; 9 1/2″ overall length; weighs 30 1/2 oz.; Military & Police design; stainless steel construction; fixed, serrated front ramp sight, square-notch rear; service-style checkered American walnut square-butt stocks; satin finish. Introduced 1981; still in production. **LMSR: $415**
　　Perf.: $325　　**Exc.: $275**　　**VGood: $200**

SMITH & WESS0N MODEL 65

Revolver; double action; 357 Mag.; 6-shot cylinder; 3″, 4″ heavy barrel; 9 5/16″ overall length (4″ barrel); weighs 34 oz.; stainless steel construction; serrated ramp front sight, fixed square notch rear; square or round butt; blued finish. Introduced 1974; no longer in production.
　　Perf.: $350　　**Exc.: $275**　　**VGood: $225**

Smith & Wesson Model 65LS LadySmith

Same specs as the Model 65 except 3″ barrel; weighs 31 oz.; rosewood round-butt grips; stainless steel construction with frosted finish; smooth combat trigger; service hammer; shrouded ejector rod; comes with soft case. Made in U.S. by Smith & Wesson. Introduced 1992; still produced. **LMSR: $461**
　　New: $375　　**Perf.: $300**　　**Exc.: $250**

SMITH & WESSON MODEL 66 COMBAT MAGNUM

Revolver; double action; 357 Mag., 38 Spl.; 6-shot cylinder; 2 1/2″, 4″, 6″ barrel; 9 1/2″ overall length (4″ barrel); weighs 36 oz.; stainless steel construction; checkered Goncalo Alves target stocks; Baughman Quick Draw front sight on plain ramp, micro-click adjustable rear; grooved trigger, adjustable stop. Introduced 1971; still in production. **LMSR: $466**
　　Perf.: $375　　**Exc.: $325**　　**VGood: $250**

SMITH & WESSON MODEL 67 K-38 COMBAT MASTERPIECE

Revolver; double action; 38 Spl.; 6-shot cylinder; 4″ barrel; 9 1/2″ overall length; weighs 32 oz.; stainless steel construction; marketed as Combat Masterpiece; soft rubber or service-style checkered American walnut stocks; Baughman Quick Draw front sight on ramp, micro-click adjustable rear; square-butt with grooved tangs; grooved trigger, adjustable stop. Introduced 1972; reintroduced 1994; still in production. **LMSR: $467**
　　Perf.: $400　　**Exc.: $325**　　**VGood: $275**

SMITH & WESSON MODEL 547

Revolver; double action; 9mm Para.; 3″, 4″ heavy barrel; 9 1/8″ overall length (4″ barrel); half-spur hammer; special extractor system; serrated ramp front sight, fixed square notch rear; checkered square-butt Magna Service stocks with 4″ barrel; checkered round butt target stocks with 3″; blued finish. Introduced 1981; dropped 1985.
　　Exc.: $275　　**VGood: $225**　　**Good: $175**

SMITH & WESSON MODEL 586 DISTINGUISHED COMBAT MAGNUM

Revolver; double action; 357 Mag.; 4″, 6″ heavy barrel; weighs 46 oz. (6″ barrel), 41 oz. (4″ barrel); L-frame design; ejector rod shroud; combat-type trigger; semi-target hammer; Baughman red ramp front sight, micrometer-click rear; soft rubber or Goncalo Alves target stocks; blued or nickel finish. Introduced 1981; still in production.
　　New: $350　　**Perf.: $300**　　**Exc.: $275**

SMITH & WESSON MODEL 617 K-22 MASTERPIECE FULL LUG

Revolver; double action; 22 LR; 6-shot; 4″, 6″, 8 3/8″ barrel; weighs 42 oz. (4″ barrel); Patridge-type front sight on two longer barrel lengths, serrated type on 4″ version, S&W click-adjustable micrometer rear; grooved tang; polished blue finish. Has same general specs as Model 17 Full Lug version except made of stainless steel; semi-target hammer (4″ barrel); smooth combat trigger standard; smooth or serrated trigger or target hammer (6″ barrel) optional; target hammer, serrated trigger (8 3/8″ barrel). Introduced 1990; still in production.
4″ barrel
　　New: $375　　**Perf.: $300**　　**Exc.: $250**
8 3/8″ barrel with target hammer, trigger
　　Perf.: $350　　**Exc.: $300**　　**VGood: $250**

SMITH & WESSON MODEL 648 K-22 MASTERPIECE WMR

Revolver; double action; 22 WMR; 6-shot; 6″ full lug barrel; 11 1/8″ overall length; weighs 38 1/2 oz. loaded; stainless steel construction; combat trigger; semi-target hammer; square-butt combat-style grips; satin finish. Introduced 1991; no longer in production.
　　Exc.: $325　　**VGood: $250**　　**Good: $200**

SMITH & WESSON MODEL 681

Revolver; double action; 357 Mag.; 4″, 6″ heavy barrels; L-frame design; stainless steel construction; ejector rod shroud; combat-type trigger; semi-target hammer; fixed sights; Goncalo Alves target stocks; satin finish. Introduced 1981; no longer in production.
　　Perf.: $300　　**Exc.: $250**　　**VGood: $200**

SMITH & WESSON MODEL 686

Revolver; double action; 357 Mag.; 2 1/2″, 4″, 6″ heavy barrels; weighs 46 oz. (6″ barrel); L-frame design; stainless steel construction; ejector rod shroud; combat-type trigger; semi-target hammer; Baughman red ramp front sight, micrometer-click rear; soft rubber or Goncalo Alves target stocks; satin finish. Introduced 1981; still in production.
　　Perf.: $400　　**Exc.: $325**　　**VGood: $250**

Smith & Wesson Model 686 Midnight Black

Same specs as Model 686 except black finish; semi-target hammer; red ramp front sight, plain or white-outline micro rear; speedloader cut-out; full lug barrel; Goncalo Alves target stocks. Introduced 1989; no longer in production.
　　Exc.: $400　　**VGood: $350**　　**Good: $250**

Smith & Wesson 38/44
Heavy Duty (Model 20)

Smith & Wesson 41
Magnum (Model 58)

Smith & Wesson 44
Hand Ejector, 1st Model
(Triple Lock)

Smith & Wesson 44
Hand Ejector, 2nd Model

Smith & Wesson 41
Magnum (Model 57)

Smith & Wesson 44
Magnum (Model 29)

HANDGUNS

Smith & Wesson Revolvers
N Large Frame Revolvers
38/44 Heavy Duty (Model 20)
38/44 Outdoorsman (Model 23)
41 Magnum (Model 57)
41 Magnum (Model 58)
44 Hand Ejector, 1st Model
44 Hand Ejector, 2nd Model
44 Hand Ejector, 3rd Model
44 Magnum (Model 29)
44 Magnum Model 29 Classic
44 New Century
(See S&W 44 Hand Ejector, 1st Model)
Triple Lock
(See S&W 44 Hand Ejector, 1st Model)

N LARGE FRAME REVOLVERS

Smith & Wesson introduced a large frame side-swing revolver in 1908. It was the beginning of a design-style that would become most popular among collectors. The N-frame revolver was the first of the S&W handguns to feature an extractor rod shroud located on the underside of the barrel.

SMITH & WESSON 38/44 HEAVY DUTY (MODEL 20)

Revolver; 38 Spl.; 6-shot; 4″, 5″, 6 1/2″ barrel; 10 3/8″ overall length (5″ barrel); weighs 40 oz.; built on the S&W 44 frame, often termed the 38/44 designation; designed to handle high-velocity 38 Special ammunition; fixed sights, with 1/10″ service-type (semi-circle) front and square notch rear; checkered walnut Magna-type grips with S&W medallion; blued or nickel finish. Introduced 1930; discontinued 1967.

Pre-War
  Exc.: $550   **VGood:** $400   **Good:** $250
Post-War
  Exc.: $300   **VGood:** $225   **Good:** $150

SMITH & WESSON 38/44 OUTDOORSMAN (MODEL 23)

Revolver; 38 Spl.; 6-shot; 6 1/2″ barrel; 1 13/4″ overall length; weighs, 41 3/4 oz.; plain Patridge 1/8″ front sight, S&W micro-adjustable rear; blue finish. Introduced in 1930 as a companion to the Model 20; reintroduced about 1950 with ribbed barrel and Magna-type stocks. Discontinued 1967.

Pre-War
  Exc.: $800   **VGood:** $600   **Good:** $400
Post-War
  Exc.: $500   **VGood:** $400   **Good:** $300

SMITH & WESSON 41 MAGNUM (MODEL 57)

Revolver; 41 Mag.; 6-shot; 4″, 6″, 8 3/8″ barrel; 11 3/8″ overall length (6″ barrel); weighs 48 oz.; wide, grooved target trigger and broad-spur target hammer; 1/8″ red ramp front, S&W micro-adjustable rear with white-outline notch; special oversize target-type of Goncalo Alves grips with S&W medallion; bright blue or nickel finish. Introduced as a deluxe companion to the Model 58, both being chambered for a new cartridge developed especially for them at that time, carrying a bullet of .410″ diameter. The old 41 Long Colt cartridge cannot be fired in guns chambered for the 41 Magnum, nor

can any other standard cartridge. Introduced 1964; no longer in production.

4″ and 6″ barrels
  Perf.: $300   **Exc.:** $250   **VGood:** $200
8 3/8″ barrel
  Perf.: $325   **Exc.:** $275   **VGood:** $225

SMITH & WESSON 41 MAGNUM (MODEL 58)

Revolver; 41 Mag.; 6-shot; 4″ barrel; 9 1/4″ overall length; weighs 41 oz.; 1/8″ serrated ramp front, square notch rear sight; checkered Magna-type walnut grips with S&W medallion; blue or nickel finish. Also known as the 41 Military & Police, this is a fixed-sight version of the Model 57. No longer in production.

  Perf.: $450   **Exc.:** $400   **VGood:** $350
Nickel finish
  Perf.: $475   **Exc.:** $425   **VGood:** $350

SMITH & WESSON 44 HAND EJECTOR 1ST MODEL

Revolver; 44 S&W Spl., 44 S&W Russian, 44-40, 45 Colt, 45 S&W Spl., 450 Eley, 455 Mark II; 6-shot cylinder; 4″, 5″, 6 1/2″, 7 1/2″ barrel; also known as 44 Triple Lock and 44 New Century; square-butt checkered walnut grips; fixed or factory-fitted target sights. Introduced 1908; dropped 1915.

44 Russian, 44-40, 45 Colt, 45 S&W Spl., 450 Eley
  Exc.: $1500   **VGood:** $1000   **Good:** $600
44 S&W Special
  Exc.: $950   **VGood:** $600   **Good:** $400
455 Mark II
  Exc.: $800   **VGood:** $500   **Good:** $300

SMITH & WESSON 44 HAND EJECTOR 2ND MODEL

Revolver; 44 S&W Special, 38-40, 44-40, 45 Colt; 4″, 5″, 6″, 6 1/2″ barrel; internal changes; on exterior, heavy barrel lug was dropped; cylinder size and frame cut enlarged; fixed or factory target sights. Introduced 1915; dropped 1937. Collector value.

44 S&W Special
  Exc.: $450   **VGood:** $350   **Good:** $250
Other calibers
  Exc.: $1500   **VGood:** $1200   **Good:** $600

SMITH & WESSON 44 HAND EJECTOR 3RD MODEL

Revolver; 44 S&W, 44-40, 45 LC; 6-shot cylinder; 4″, 5″, 6 1/2″ barrel; also known as Model 1926; fixed or factory target sights; checkered walnut square-butt grips; blue or nickel finish. Introduced 1925; dropped 1949.

44 S&W Special
  Exc.: $500   **VGood:** $350   **Good:** $250
Other calibers
  Exc.: $1500   **VGood:** $1000   **Good:** $850

SMITH & WESSON 44 MAGNUM (MODEL 29)

Revolver; 44 Mag.; also handles 44 Spl., 44 Russian; 6-shot; 4″, 5″, 6 1/2″, 8 3/8″; 11 7/8″ overall length (6 1/2″ barrel); weighs 43 oz. (4″ barrel), 47 oz. (6 1/2″ barrel), 51 1/2 oz. (8 3/8″ barrel); 1/8″ red ramp front sight, S&W micro-adjustable rear; target-type Goncalo Alves grips with S&W medallion; broad, grooved target trigger; wide-spur target hammer; bright blue or nickel finish. As with the Model 27, the Model 29 was developed to take a new cartridge developed by lengthening the 44 Special case by 0.125″ to prevent use of 44 Magnum ammo in guns chambered for the 44 Special. The 44 Magnum is loaded to pressures approximately twice that of the 44 Special. Introduced in 1956; still in production.

LMSR: $554
  Perf.: $375   **Exc.:** $325   **VGood:** $275

Smith & Wesson 44 Magnum Model 29 Classic

Same specs as standard Model 29 except 5″, 6 1/2″, 8 3/8″ barrel; chambered cylinder front; interchangeable red ramp front sight, adjustable white-outline rear; Hogue square-butt Santoprene grips; drilled, tapped for scope mount. Introduced 1990; no longer in production.

  Perf.: $400   **Exc.:** $350   **VGood:** $300

Smith & Wesson Revolvers
44 Magnum Model 29 Classic DX
44 Magnum Model 29 Silhouette
44 Magnum Model 629
44 Magnum Model 629 Classic
44 Magnum Model 629 Classic DX
357 Magnum (Model 27)
357 Magnum Model 27 3¹/₂″
357 Magnum Model 27 5″
357 Magnum (Model 627)
455 Hand Ejector First Model
455 Hand Ejector Second Model
1950 44 Military (Model 21)
1950 Army (Model 22)
1950 44 Target (Model 24)

Smith & Wesson 44 Magnum Model 629

Smith & Wesson 357 Magnum (Model 27)

Smith & Wesson 1950 44 Military (Model 21)

Smith & Wesson 1950 Army (Model 22)

Smith & Wesson 357 Magnum (Pre-War registered)

Smith & Wesson 44 Magnum Model 29 Classic DX
Same specs as Model 29 Classic except 6 ¹/₂″, 8 ³/₈″ barrel with full lugs; Morado combat-type grips. Marketed with five different front sights, Hogue combat-style square-butt conversion grip. Introduced 1991; no longer in production.
    Exc.: $500   **VGood:** $400   **Good:** $350

Smith & Wesson 44 Magnum Model 29 Silhouette
Same specs as Model 29 except 10 ⁵/₈″ barrel; oversize target-type checkered Goncalo Alves grips; four-position click-adjustable front sight; weighs 58 oz.; 16 ¹/₈″ overall length. Introduced 1983; dropped 1991.
    Perf.: $500   **Exc.:** $425   **VGood:** $375

SMITH & WESSON 44 MAGNUM MODEL 629
Revolver; 44 Mag.; 6-shot; 4″, 6″, 8 ³/₈″ barrel; 11 ³/₈″ overall length (6″ barrel); weighs 47 oz.; ¹/₈″ red ramp front; S&W micro-adjustable rear sight; same as Model 29 Stainless except steel construction; target-type Goncalo Alves grips; broad grooved trigger; wide spur target hammer; satin finish. Still in production.
    New: $500   **Perf.:** $400   **Exc.:** $300

Smith & Wesson 44 Magnum Model 629 Classic
Same specs as Model 629 except 5″, 6 ¹/₂″, 8 ³/₈″ full-lug barrel; 10 ¹/₂″ overall length; chambered cylinder front; interchangeable red ramp front sight, adjustable white outline rear; Hogue grips. Introduced 1991; still in production. **LMSR: $629**
    New: $550   **Perf.:** $450   **Exc.:** $350

Smith & Wesson 44 Magnum Model 629 Classic DX
Same specs as Model 629 except 6 ¹/₂″, 8 ³/₈″ full-lug barrel; Hogue combat-style round butt grip; comes with five front sights. Introduced 1991; still in production.
    New: $700   **Perf.:** $600   **Exc.:** $500

SMITH & WESSON 357 MAGNUM (MODEL 27)
Revolver; 6-shot; 357 Mag., also could fire 38 Spl. (The case was lengthened to prevent its use in guns chambered for the 38 Spl. round.) Pre-WWII Model 27s offered in barrel lengths of 3 ¹/₂″, 5″, 6″, 6 ¹/₂″, 8 ³/₈″ and 8 ³/₄″; could be custom-ordered with barrels of any length up to 8 ³/₄″; 11 ³/₈″ overall length (6″ barrel); weighs 41 oz. (3 ¹/₂″ barrel), 42 ¹/₂ oz. (5″ barrel), 44 oz. (6″ barrel), 44 ¹/₂ oz. (6 ¹/₂″ barrel), 47 oz. (8 ³/₈″ barrel); could be ordered with any of S&W's standard target sights; 3 ¹/₂″ version usually furnished with Baughman quick-draw sight on a plain King ramp; finely checkered topstrap matched barrel rib, with vertically grooved front and rear grip straps and grooved trigger; S&W bright blue or nickel finishes; checkered Circassian square or Magna-type walnut stocks, with S&W medallion. Retail price at beginning of WWII was $60. Post-WWII production similar, with hammer redesigned to incorporate a wider spur and inclusion of the present pattern of S&W click-micrometer rear sight; blue or nickel finish. Post-WWII models can be identified by S or N serial number prefix. Model 27 was the first centerfire revolver with recessed cylinder; 1935-1938 guns were registered to owner with registration number stamped in yoke of frame; about 6000 made before registration stopped. The papers, themselves, have some collector value without the gun.
Pre-WWII registered
    Exc.: $1750   **VGood:** $1350   **Good:** $900
Pre-WWII not registered
    Exc.: $850   **VGood:** $600   **Good:** $450
Post-WWII model
    Perf.: $350   **Exc.:** $300   **VGood:** $200

Smith & Wesson 357 Magnum Model 27 3 ¹/₂″
Same specs as standard Model 27, except reintroduction of original barrel length; pinned black ramp front sight; oversize square-butt target stocks. Reintroduced 1991; limited production; dropped 1992.
    Perf.: $350   **Exc.:** $300   **VGood:** $250

Smith & Wesson 357 Magnum Model 27 5″
Same specs as Model 27 3 ¹/₂″ except 5″ barrel. Reintroduced 1991; limited production; dropped 1992.
    Perf.: $350   **Exc.:** $300   **VGood:** $250

SMITH & WESSON 357 MAGNUM (MODEL 627)
Revolver; 357 Mag; 6-shot; 5 ¹/₂″ barrel; stainless steel variation of Model 27; limited production with 5276 manufactured. Introduced 1989.
    Perf.: $500   **Exc.:** $400   **VGood:** $300

SMITH & WESSON 455 HAND EJECTOR FIRST MODEL
Revolver; 455 Mark II; 6 ¹/₂″ barrel; fixed sights; checkered walnut square-butt grips; blue finish. Identical in appearance to 44 Hand Ejector First Model except for caliber; designed for the British Government in 1914 in standard chambering for British armed services. Generally have British proof marks. Many released from British or Canadian service converted to 45 Colt.
    Exc.: $700   **VGood:** $600   **Good:** $475

SMITH & WESSON 455 HAND EJECTOR SECOND MODEL
Revolver; 455 Mark II; 6 ¹/₂″ barrel; fixed sights; checkered walnut square-butt grips; blue finish. A lighter weight handgun to satisfy the request of the British government. Large extractor housing under barrel eliminated. Identical to 44 Hand Ejector Second Model. Generally have British proof marks. Many released from British or Canadian service converted to 45 Colt. Produced between 1915 and 1916.
    Exc.: $350   **VGood:** $250   **Good:** $200

SMITH & WESSON 1950 44 MILITARY (MODEL 21)
Revolver; 44 Spl. (also handles 44 Russian); 5″ barrel; 10 ³/₄″ overall length; weighs 36 ¹/₄ oz.; fixed sights; blue or nickel finish; checkered walnut Magna-type grips with S&W medallion. Post-WWII version of the S&W Model 1926 with minor design refinement. Discontinued 1967.
    Exc.: $1400   **VGood:** $1150   **Good:** $900

SMITH & WESSON 1950 ARMY (MODEL 22)
Revolver; 45 ACP, 45 Auto Rim; 5 ¹/₂″ barrel; 10 ¹/₄″ overall length; weighs 36 ¹/₄ oz.; semi-circular front sight and U-shaped notch rear milled in the top of the receiver strap. Post-WWII version of the Model of 1917, with minor design refinements; remained in production until 1967. A target version was made, having adjustable rear sight.
    Exc.: $1750   **VGood:** $1350   **Good:** $800

SMITH & WESSON 1950 44 TARGET (MODEL 24)
Revolver; 44 Spl. (also handles 44 Russian); 6-shot; 6¹/₂″, 4″ barrel; 9¹/₄″ overall length; weighs 40 oz. introduced in 1950 as a refined version of the 1926 Target

Smith & Wesson Highway
Patrolman (Model 28)

Smith & Wesson
Military Model of 1917

Smith & Wesson 1950/1955
45 Target (Model 26 and 26)

Smith & Wesson
Model 624

Smith & Wesson
Model 625-2

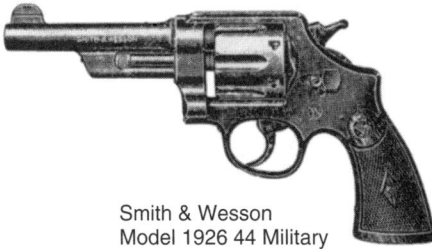

Smith & Wesson
Model 1926 44 Military

HANDGUNS

Smith & Wesson Revolvers
1983 44 Target (Model 24)
1950/1955 45 Target (Model 25 and 26)
Highway Patrolman (Model 28)
Military Model of 1917
Model 25
Model 25-5
Model 610
Model 624
Model 625-2
Model 1926 44 Military
Model 1926 44 Target

Model; Patridge-type front sight having vertical rear blade surface, S&W micro-adjustable rear; limited quantity produced in 4″-barreled version, with Baughman quick-draw front sight on serrated ramp, with the same type of rear sight; blued finish standard, although a few specimens custom-ordered in nickel. Discontinued about 1966.

    Exc.: $550   **VGood:** $450   **Good:** $300

Smith & Wesson 1983 44 Target (Model 24)
Same specs as standard Model 24 except 44 Special only; 4″, 6 1/2″ barrels; grooved topstrap; barrel rib. Limited production of 7500 guns. Made in 1983 only.

    Exc.: $375   **VGood:** $325   **Good:** $250

SMITH & WESSON 1950/1955 45 TARGET (MODEL 25 and 26)
Revolver; 45 ACP, 45 Auto Rim; 6-shot; 6 1/2″ barrel; no factory production of 4″ barrel has been reported, although some owners have cut them down to 4″ length; 11 7/8″ overall length; weighs 45 oz.; 1/8″ plain Patridge front, S&W micro-adjustable rear; checkered walnut target grips with S&W medallion; blue finish. Introduced in 1950 as a companion to the 1950 Model 24 44 Target; identical except being chambered for 45 ACP/Auto Rim. The 45 Hand Ejector model of 1950 was redesigned in 1955 to become the Model 26. The Model 1955 was called the Model 25. Modifications consisted of a heavier barrel with broad rib, similar to that of the K-38; S&W target stocks in place of the Magna-type; a target hammer and broad, target type trigger. Dropped 1983.

1955 Model 25
    Exc.: $400   **VGood:** $350   **Good:** $300
1950 Model 26
    Exc.: $600   **VGood:** $500   **Good:** $350

SMITH & WESSON HIGHWAY PATROLMAN (MODEL 28)
Revolver; 357 Mag.; 6-shot; 4″, 6″ barrel; 11 1/4″ overall length (6″ barrel); weighs 41 3/4 oz. (4″ barrel); 44 oz. (6″ barrel); 1/8″ Baughman Quick Draw front sight on plain ramp, S&W micro-adjustable rear; checkered walnut, Magna-type grips with S&W medallion; target stocks at extra cost; blued finish with sandblast stippling on barrel rib and frame edging. Introduced in 1954 as a functional version of the Model 27 minus the cost-raising frills such as the checkered topstrap.

    Exc.: $225   **VGood:** $175   **Good:** $150

SMITH & WESSON MILITARY MODEL OF 1917
Revolver; 45 ACP, 45 Auto Rim. Entry of the U.S. into WWI found facilities unable to produce sufficient quantities of the recently adopted Colt Government Model auto pistol, so approximately 175,000 Smith & Wesson revolvers were manufactured, being chambered to fire the 45 ACP cartridge by means of the two 3-shot steel clips; also fires the 45 Auto Rim round. The wartime units had a duller blued finish and smooth walnut grips, with 5 1/2″ barrel; overall length, 10 1/4″; weighs 36 1/4 oz. with lanyard ring in the butt. A commercial version remained in production after the end of WWI to the start of WWII, distinguished by a bright blue finish and checkered walnut stocks.

Military model
    Exc.: $550   **VGood:** $350   **Good:** $250
Commercial model
    Exc.: $400   **VGood:** $300   **Good:** $250

SMITH & WESSON MODEL 25
Revolver; double action; 45 Colt; 6-shot cylinder; 4″, 6″, 8 3/8″ barrel; 11 3/8″ overall length (6″ barrel); weighs about 46 oz.; target-type checkered Goncalo Alves grips; S&W red ramp front sight, S&W micrometer-click rear with white outline; available in bright blue or nickel finish; target trigger; target hammer. Made in U.S. by Smith & Wesson. Dropped 1994.

4″ or 6″ barrel, blue
    Perf.: $400   **Exc.:** $350   **VGood:** $300
8 3/8″ barrel, blue or nickel
    Perf.: $450   **Exc.:** $400   **VGood:** $350

SMITH & WESSON MODEL 25-5
Revolver; 45 Colt; 6-shot; 4″, 6″, 8 3/8″ barrel; adjustable sights; square-butt, checkered, target-type Goncalo Alves grips; blue or nickel finish. Introduced in 1978 and called the 25-5 to distinguish it from the Model 25-2 in 45 ACP. Available with presentation case. Discontinued 1987.

    Exc.: $350   **VGood:** $300   **Good:** $250

SMITH & WESSON MODEL 610
Revolver; 10mm; 5″, 6 1/2″ barrel; .500″ target hammer; .400″ smooth combat trigger; round-butt frame with smooth Goncalo Alves finger groove grips; stainless steel construction; satin finish. A total of 5000 manufactured with retail price of $510. Introduced April 1990.

    Exc.: $500   **VGood:** $450   **Good:** $400

SMITH & WESSON MODEL 624
Revolver; 44 Spl.; 6-shot cylinder; 4″, 6″ barrel; 9 1/2″ overall length (4″ barrel); weighs 41 1/2 oz.; target-type checkered Goncalo Alves stocks; black ramp front sight, adjustable micrometer-click rear; stainless steel version of Model 24. No longer in production.

    Exc.: $325   **VGood:** $275   **Good:** $225

SMITH & WESSON MODEL 625-2
Revolver; double-action; 45 ACP; 3″ full-lug barrel; same general specs as S&W Model 25, but is made of stainless steel; black pinned ramp front sight, micrometer rear; semi-target hammer; combat trigger; full lug barrel; round butt Pachmayr stocks. Introduced 1989; still in production.

    New: $500   **Perf.:** $400   **Exc.:** $350

SMITH & WESSON MODEL 1926 44 MILITARY
Revolver; primarily produced in 44 Spl., sometimes encountered in 45 Colt, 455 Webley or 455 Eley; 6-shot; 4″, 5″, 6 1/2″ barrel; 11 3/4″ overall length (6 1/2″ barrel); weighs 39 1/2 oz.; 1/10″ service-type front sight, fixed square-notch rear; checkered walnut, square or Magna- type grips with S&W medallion; blued or nickel finish. Modified version of S&W's earlier 44 Hand Ejector First Model, minus the triple-lock feature, but retaining the heavy shroud around the ejector rod. Discontinued at the start of WWII; replaced after the war by the 1950 model.

    Exc.: $600   **VGood:** $500   **Good:** $400
Nickel finish
    Exc.: $700   **VGood:** $600   **Good:** $500

Smith & Wesson Model 1926 44 Target
Same specs as 1926 44 Military except target version with rear sight adjustable for windage and elevation; produced from 1926 to the beginning of WWII; replaced after the war by the 1950 Target Model 24.

    Exc.: $1000   **VGood:** $850   **Good:** $700

LIMITED PRODUCTION SMITH & WESSON HANDGUNS

Model #	SKU #	Description	Date	Original Price	Total Production	S/N Range
10	103545	4" standard weight barrel; blued finish; square butt grips	Oct. 1988		1040	Traded to Acu-Sports 5/5/93
10	103545	Stamped "RHKP" and S/N; originally sold to Hong Kong Police in trade				
10-10	100135	3" full lug barrel; round butt grips; for Brazil	Jan. 1994			BDY2592 to BDY2733
14	100338	6" full lug barrel; 38 Spl.; Adjustable rear sight	Jan. 1991	$425		
16	100558	4" full lug barrel; 32 Mag.; blued finish	1989	368	8873	Total production all three barrel lengths
16	100560	6" full lug barrel; 32 Mag.; blued finish	1989	403	included above	
16	100562	8-3/8" full lug barrel; 32 Mag.; blued finish	1989	414	included above	
17	100542	6" full lug barrel; blued finish; Goncalo Alves Combat Stock	May 1989	415	3658	Total production complete
17	100538	4" full lug barrel; blued finish; Goncalo Alves Combat Stock	June 1989	379	1070	Total production complete
17	100544	8-3/8" full lug barrel; blued finish; Goncalo Alves Combat Stock	June 1989	427	878	Total production complete
17-7	100509	6" barrel; blued finish; stainless steel cylinder	1994		200	BRF3549-3583, BRF3586-3616, BRF3620-3671, BRC0227-0284, BRC4382-4406
19	50101	Texas Ranger Commemorative; 4" barrel; blued finish; cased	1973	250	10,000	TR1-TR10000, 8000 cased with knife
19-5	100701	2-1/2" barrel; round butt grips; glass bead blued finish	Jan. 1987			ANR7900-ANR8449 U.S. Dept. of State
19P	100727	4" barrel; 357; square butt grips; blued finish; Made for Peru	1987	358		AWF0028-AWF0508
24-3	100782	4" full lug barrel; blued finish	1984		7500	Total production includes both barrel lengths
24-3	100785	6" barrel; 44 Spl.; blued finish	1984	387	included above	
24-3	100750	3" barrel; 44 Spl.; Lew Horton Spl.	1984		1000	
24	none	6-1/2" barrel; blued finish; Target stocks	1990	350	375	BFJ4817-BFJ5788
25-3	100924	S&W's 125th Anniversary Commemorative 45 Colt	1977	439	10,000	SW0000-SW10000 Deluxe Models marked "25-4"
25-5(-7)		5" barrel; unfluted cylinder; marking lasered	April 1989		1987	BDS6812-8442
26-4	100920	5" barrel; blued finish; TH??; 45 Colt; Georgia State Police Commemorative	1988 & 89	405	802	Dispersed in BBY00354 up to BBY0434
27	101021	50th Anniversary Comemorative; 5" barrel; .357; blued finish; cased	1985	423	2500	REG0001-REG2500
27	100998	3-1/2" barrel; blued finish; Morado stocks; pinned front sight	Feb. 1991	423		
27	100996	5" barrel; blued finish; Morado stocks; pinned front sight	Feb. 1991			
29	101205	6" barrel; Patridge front sight; with Call; gold bead engraving; cased	Nov. 1984		200	S/N dispersed AEV, AEY, AFB, AFD, AFE, AFF, AFH, AFJ
29-3	101224	3" barrel; blued finish; round butt grips; Lew Horton Spl.	Nove. 1985	850	200	S/N dispersed ALA9050-9628, ALB2701-ALC0075
29-3	none	Elmer Keith Commemorative; 4" barrel etched in gold	1986	474	2500	EMK0000-EMK0100; Deluxe EMK010-EMK2500 standard
29	101230	6" full lug barrel; blued finish; unfluted cylinder; Classic Hunter	Jan. 1988	527	5271	AYE2589-2676; AYM5544-6150; AYM7110-7709
29	101233	8-3/8" barrel cut for scope mount; blued finish	Aug. 1988	498	3455	
29	101249	5" full lug barrel; unfluted cylinder; blued finish	May 1989		500	
29	101251	3" full lug barrel; unfluted cylinder; blued finish; Pachmayr grip	July 1989	460	2532	BDY7399-8898; BDZ3656-4687
29	101254	8-3/8" full lug barrel; unfluted cylinder; blued finish; Classic Hunter	Aug. 1989	508	Unknown	BDY7076-7357 known to date
29	101256	7-1/2" full lug barrel; blued finish; Hogue grips	Dec. 1989	548	750	SWN0000-0748 Made for Acu-Sport
29	101264	7-1/2" full lug barrel; Bright; blued finish; Magna Classic	Feb. 1990	999	Appx. 1500	S/N Prefix MAG
29-5	101256	7-1/2" full lug barrel; blued finish; square butt grips; Classic Mag. II	Aug. 1990		187	BFF2141-BFF2327
29	101278	6" full lug barrel; unfluted cylinder; Hogue combat grips; Classic Hunter	Jan. 1991	501		Appx. S/N Range BJA1921
29	101272	6-1/2" barrel; Classic DX same features as 629DX	Jan. 1991	686		Became production SKU
29	101272	8-3/8" barrel; Classic DX same features as 629DX	Jan. 1991	700		Became production SKU
29-5	101274	5, 6-1/2, 7-1/2 Magna Classic reject barrels	Feb. 1991	343	496	TJB34465-3961 SKU listed for 5" barrel
36	101520	2" barrel; DA only; Goncalo Alves Stocks	April 1989		255	BDW3026-3280
36	101524	2" barrel; DA only; bobbed hammer; Hogue grips	June 1989	353		
36	101522	3" barrel; DA only; bobbed hammer; Hogue grips	June 1989	353		
36	101549	3" barrel; adjustable rear sight; Hogue grips	Aug. 1989	366	615	BEA1967-2581
42	103792	2" barrel; carbon steel and aluminum; blued finish; serrated front sights	1992			BKY0334-BKY0987
60	102304	3" heavy barrel; square butt; checkered walnut grips	Dec. 1984		500	S/N dispersed AEU, AEV, AEW
60	102305	2" barrel; adjustable sights; round butt; Ashland Spl.	Sept. 1985	386	666	S/N dispersed AIV, ALA, ALU, ALV, ALW
60	102304	3" heavy barrel; Goncalo Alves combat stocks	Aug. 1988	406	282	Serial range unknown
60	102314	2" barrel; DA only; Hogue grips	April 1989	406		Made for Michigan Police Department
60	102316	2" barrel; unfluted cylinder; oiled stocks	April 1989	395		
60	102320	3" heavy barrel; DA only; Hogue grips	April 1989	406		
60	102320	2" barrel; DA only; bobbed hammer; Hogue grips	June 1989	406		
60	102298	3" full lug barrel; adjustable rear sights; Uncle Mike's grips	Dec. 1990	410		

Model #	SKU #	Description	Date	Original Price	Total Production	S/N Range
63	102405	2" barrel; adjustable rear sight; Hogue grips	May 1989	410	500	Made for Lew Horton Dist.
66	102712	3" full lug barrel; round butt; checkered walnut stocks	May 1988	409	4195	
520	none	4" barrel; "N" frame; fixed sights; 357	1979		3000	Made for New York State Police but never delivered
544	103195	Texas Wagon Train 150th Anniversary Commemorative; 44-40	1986		7801	TWT0000-TWT7800
586	103504	6" barrel; blued finish; Massachusetts State Police Commemorative	1986		631	S/N dispersed in ABT-AUC prefixes
586	103593	8 3/8" barrel; scope mount; blued finish with barrel cut for mounts	April 1989	496	999	BDY4906-4965; BDY4967-5026; mixed with 6" barrel
586	103591	4" barrel; beaded blue finish; Hogue grips	April 1989	422	1969	In same range as 4"
586	103592	6" barrel; beaded blue finish; Hogue grips	April 1989	422	1995	BDV8181-BDV8192
586	103545	6" barrel; blued finish; Uncle Mike's grips	1989			BDV8184
586	103591	4" barrel; round butt; glass bead finish; red rear; white outline	1990			BDV8185
586	103592	6" barrel; round butt; glass bead finish; red rear; white ouline	1990			BDV8185
586	103545	6" barrel; blued finish; laser smith engraved	1992			BBW3930; BAD3483-3484; BDV8186-8192
586	103497	2 1/2" full lug barrel; blued finish; round butt; 38 Spl.; red ramp front sight; Brazil	May 1994		158	BPY2805-BPY2966 to U.S. distributors
586	103498	4 1/4" barrel; 38 Spl.; Brazil	May 1994		23	BPY8039-BPY8062 to U.S. distributors 200 for Brazil
610	103578	5" full lug barrel; for 10mm cartridge	March 1990	510	4560	BFA2683-3825; BFA5483-7110; BFA7661-8148
610	103576	6 1/2" full lug barrel; for 10mm cartridge	March 1990	510	included above	BFN0558-1258; S/Ns apply to both barrel lengths
617	100555	6" standard barrel	1991		116	BEJ4194-BEJ4310
617	100563	4" full lug barrel; 22 rimfire single shot; semi-target hammer; combat trigger; Pachmayr stocks	March 1992			BHP7285-BHP7458
624	103583	3" barrel; 44 Spl.; 1st 100 engraved; Horton Spl.	July 1985		100	S/N dispersed AHS, AHT, ALU
625-2	100921	5" full lug barrel; bowling pin 1988	Feb. 1989	535	7200	BDC0000-7199
625-3	100925	4" full lug barrel; glass bead finish; Pachmayr grips	Dec. 1989			BEP6788-BEP7066
625-2	100923	3" full lug barrel; stainless glass bead finish; Pachmayr stock	Nov. 1989	535	3198	BEN7590-9664 mixed with other barrel lengths
625-2	100923	4" full lug barrel; stainless glass bead finish; Pachmayr stock	Nov. 1989	535	2980	
625-5	100430	5" full lug barrel; 45 Colt; stocking dealer direct; round butt	Oct. 1993		1550	SDS0001-SDS1550
625-5	100930	Same as above made for Acusports	Feb. 1995		49	CLS0000-0048
625-5	100930	5" stainless steel barrel; blued cylinder; round butt	Feb. 1995		48	CLS0000-0047 Made for Acusports
627	101024	5 1/2" barrel; unfluted cylinder; Goncalo Alves Combat stocks	Oct. 1989	530	5276	BEK2363-5429; BEK5630-7055
629	none	3" barrel; round butt	1986		5000	NNC prefixed serial range
629	103610	3" barrel; round butt; Goncalo Alves Smooth Combat stocks	1987	526	5000	AWF6535-AWF7156, BBF5790-BBF6714
629-1	103604	Alaska 1988 Ididarod Commemorative	Aug. 1987	502	545	AKI0001-0545; Made for Northern Consolidators
629	103615	6" full lug barrel; unfluted cylinder; Classic Hunter	July 1988	562	11952	BBF4302-5402; BBL6820-7492; BB29921-9997
629	103652	8 3/8" barrel; cut for scope mount	Aug. 1988	509	4810	BDY93093-BDZ0177; BDZ6538-7690; BEA0613-1207
629	103652	4" light barrel; chamfered cylinder; mountain gun	April 1989		8204	BNZ23738-4385 stocking dealer 1993
629		Same as above; made in several lots to meet production; may be more than above				
629	103649	7 1/2" full lug barrel; unfluted cylinder; Hogue grips	June 1989	584	750	SWN0749-1499 made for Acusports
629	103628	8 3/8" full lug barrel; unfluted cylinder; Hogue grips	Aug. 1989	525		BED5405; BDF7357-7365 incomplete
629	103650	3" full lug barrel; unfluted cylinder; Pachmayr grip; Classic Hunter	Sept. 1989	490	3200	BBW4852-4901; BED5473-6220; BEE4548-5949
629-2(3)	103632	7 1/2" full lug barrel; unfluted cylinder; polished stainless steel; Magna Classic	Feb. 1990	999	1500	MAG0000-3000 mixed with blued models
629	103632	The Magna Classic was a mix of Model 29 and 629; total production 1500 for both models				Same as above; first few odd #s blue, even #s stainless
629	103596	6" full lug barrel; adjustable front sight	March 1990	576		Made for European Arms Distr.; serial range unknown
629	103616	6" unfluted cylinder; square butt; Classic Hunter	1990			BFA8371-BFA8388
629	103644	6 1/2" full lug barrel; inter. front sight; Classic DX; Morado stocks	Jan. 1991	726		Became production SKU as Classic DX had chamfered cylinder
629	103646	8 3/8" as above; drilled and tapped for scope mount	Jan. 1991	750		Became production SKU
629-3	103368	5, 6 1/2, 7 1/2" reject barrel; Magna Classic	March 1991	528	300	TJB4001-4300 SKU for 6 1/2" barrel
629	103618	6" full lug barrel; unfluted cylinder; Hogue Combat grips; Classic Hunter	March 1991			BHE2575-2681; BHF5831-5874; BHJ6312-6347
629	130450	3" Hogue grips; called Back Packer	April 1994			
631	103664	2" barrel; 32 H&R Mag.; 6-shot; Goncalo Alves Combat stocks	March 1990	386	5474	Total probably includes both 2" & 4" serial range unknown
631	103664	4" barrel; 32 H&R Mag.; 6-shot; Goncalo Alves Combat stocks	March 1990	402	included above	same as above
631	103660	2" barrel; Lady Smith Rose stocks	March 1990	400		Serial range unknown
632	103666	2" barrel; 32 H&R Mag.; 6-shot; Uncle Mike's grips	1991	400		Serial range unknown

Model #	SKU #	Description	Date	Original Price	Total Production	S/N Range
632	103674	3" barrel; 32 H&R Mag.; 6-shot; Uncle Mike's grips	1991			Serial range unknown
637	103790	2" barrel; aluminum frame; stainless steel barrel and cylinder; 38 Spl.	1989			Serial range unknown
638	103670	2" barrel; stainless steel barrel and cylinder; clear anodized frame	Nov. 1989	405	500	Made for Ellet Bros.; serial range unknown
640	103796	2" barrel; hammerless; Goncalo Alves stocks	Jan. 1990	408	500	CEN0000-0500 first produce now a production SKU BMA1993-BMA3027, BMA5806-BMA6456
642	104790	2" barrel; stainless steel and aluminum construction; serrated fron sight; concealed hammer; combat trigger	1992			
648	103668	6" full lug barrel; 22 MRF: Goncalo ALves Combat smooth stocks	April 1990	400	unknown	Serial range unknown
649	103750	First 100 engraved; Lew Horton Spl.	Oct. 1985		100	S/N dispersed in prefix AFN, AHL, AHT, ALV
651	103902	4" barrel; 22 MRF	Dec. 1990	412		
651-1	103900	2" barrel; adjustable sights; round butt; RSR wholesale special issue	1993		300	LHS0001-0301
657	103951	3" barrel; red ramp & white outline sights; round butt	1987			ANK1020-ANK1960 not complete
657	103820	6½" full lug barrel; unfluted cylinder; Hogue combat grip; Classic Hunter	Feb. 1991	471	Approx. 1200	BFW0000-0174; BHA9261 & 9262; BPW6386-6734; BPY1739-1882; BSA4282-4401; BSB8438 & 8480; HTR0001-0363 possible out of perf. cen.??
686	104212	2½" barrel; round butt; red ramp & white ouline sights; Lew Horton Spl.	1985			Became standard product
686	104228	3" barrel; pinned black ramp front sight; U.S. Custom	Jan. 1988	410	3281	
686	104229	4" barrel; pinned black ramp front sight; U.S. Custom	Jan. 1988	410	5419	Both 4 & 6" barrels are in same serial range
686	104250	4" barrel; midnight black stainless steel finish	Dec. 1988	435	1559	BBU8080; BBU8371-9068; BBV7631-7913
686	104248	6" barrel; midnight black stainless steel finish	Dec. 1988	435	2876	BBL4145-4153; BBV3104-3304; BCA0000-0272;
686	104249	6" full lug barrel; midnight black stainless steel finish	Jan. 1989	484	7449	BBA1847; BDA2340-2559; BDR9906-9961
686	104251	8⅜" barrel; scope mount with barrel cut	May 1989	541		BDR2585-2588 for Bill Davis only partial range
686	104254	5" barrel; unfluted cylinder; dual speed loader; custom grips	June 1989	479	1059	Made for Wischo Germany BKR0393-1359
686	104218	3" barrel; satin finish; Pachmayr grips	1991			S/N intermixed cont. BKR1360-1451
686	104232	4" barrel; otherwise same as above	1991			BJC9279-BJD0179 for Wischo-Joydund Germany
686	104257	6" barrel; beaded finish; Pachmayr stocks	1992		900	
686/617	104233	Cased dual set for Zanders Sporting Goods	Dec. 1992		200 sets	686-ZZL0000 to ZZL00199
617/686	100569	Same as above both 686 & 617 in set	Dec. 1992			617-ZZK0000 to ZZK00199
686-4	104272	6" barrel with barrel port	1994			BSB9414-9634
669	104052	First 100 engraved; Lew Horton Spl.	1986		100	S/N dispersed in prefixs TAE & TAF
745	none	IPSC Commemorative				DVC0000-DVC5362
745	103721	Standard production	1987	650		DBN4960-DBN5553
745	103906	745 with adjustable rear sight	May 1990	596		Approx serial range TBK8375
1066-NS	108270	10mm with night sights	Nov. 1990	770		TFR5527; TFR5555-5577;
1076	108282	FBI Commemorative; magnum safety; night sights	Jan. 1991	825		fixed and adjustable probably same
4006	170054	4¼" barrel; Perf.?? center	1994	660		SDC0001-0207
4505	108316	Blued carbon steel; fixed sight	Jan. 1991	687	Approx. 1200	TFS4970-TFS5009; THC7340-7706; S/Ns both SKU mixed
4505	108318	Blued carbon steel; adjustable sight	Jan. 1991	559	included in above	
4516-1	103738	U.S. Marshall Commemorative	1990	740	500	USM0000-USM0499
4536	103748	3¾" barrel; 7-shot; fixed sight; decocking lever	1990			TEP3955-TEP4168
4556	108551	3¾" barrel; fixed sights; DA only	1991	735	1236	TFS1562-TFS1939
4567	108304	With night sights; blued finish; square butt grip	Jan. 1991	735		TFN8957-8974; TFR0970-1310; TFS2802-3274
4596	104490	4506-1 frame with 4516-1 slide	April 1990	600	382	TFA3952-4149; may by other S/N ranges
5903-SSV	104100	Short slide variation	March 1990	603	1500	TET7072-7535; TEU8280-8592 range ID to date
5905	108303	Blued carbon steel; fixed sight	Jan. 1991			S/N range
5905	108302	Blued carbon steel; adjustable sight	Jan. 1991	632		TFN9965-TFP1200
5906	108255	1993 Stocking dealer; fixed sight; slide high bright finish	1993			TZU5612-TZU5687; TET5000-
5906	108255	Same as above except finish wood grain grip				to TET5388
5943-SSV	108308	DA only; night sights	Nov. 1990	690		TFN7478-7622; TFP5778-6008; S/Ns to date
5967	103048	5906 frame; stainless steel with blued 3914 slide	Dec. 1989	639	3277	Made for Lew Horton Distribution

Smith & Wesson
Model 35

Smith & Wesson
Model 32 Auto

Smith & Wesson 38
Target Auto (Model 52)

Smith & Wesson
Model 59 9mm Auto

Smith & Wesson
Model 39 9mm Auto

Smith & Wesson Semi-Automatic Pistols

Pre-World War II (1913-1939)
Model 32 Auto
Model 35

Post-World War II (1955-1980)
First Generation
38 Target Auto (Model 52)
Model 39 9mm Auto
Model 59 9mm Auto

Second Generation Models (1980-1988)
Model 439
Model 459

Smith & Wesson
Model 439

SEMI-AUTOMATIC PISTOLS

Smith & Wesson semi-auto pistols are best classified by separating them into centerfire and rimfire categories. Because of the terminology used, the post-WWII models are divided into three categories called First, Second and Third generation models. These designations indicate major modifications in the pistols. First Generation models can be identified by the two-digit model number, e.g., Model 39; the Second Generation by a three-digit model number, e.g., Model 439; and the Third Generation by a four-digit model number, e.g., Model 3904.

PRE-WORLD WAR II (1913-1939)

SMITH & WESSON MODEL 32 AUTO

Semi-automatic; 32 ACP, 7.65mm; 8-shot; 4″ barrel; 7″ overall length; weighs 28 oz.; unusual grip safety just below the trigger guard; successor to S&W's original auto pistol; walnut, uncheckered grips; blue or nickel finish. Introduced 1924; discontinued 1937. Collector interest.

    Exc.: $2500   **VGood:** $2000   **Good:** $1500

SMITH & WESSON MODEL 35

Semi-automatic; 35 S&W Auto; 7-shot magazine; 3 1/2″ barrel; 6 1/2″ overall length; uncheckered walnut stocks; fixed sights; blued or nickel finish. Manufactured 1913 to 1921. Collector interest.

    Exc.: $500   **VGood:** $375   **Good:** $250

POST-WORLD WAR II (1955-1980) FIRST GENERATION

SMITH & WESSON 38 TARGET AUTO (MODEL 52)

Semi-automatic; 38 Spl.; 5-shot; 5″ barrel; 8 5/8″ overall length; weighs 41 oz.; Patridge-type front on ramp, S&W micro-adjustable rear; checkered walnut grips with S&W medallion; blue finish only. Designed to fire a mid-range loading of the 38 Special, requiring a wadcutter bullet seated flush with the case mouth; action is straight blowback, thus not suited for firing of high-velocity 38 Special ammo. Introduced 1961; no longer in production.

    Exc.: $650   **VGood:** $500   **Good:** $300

SMITH & WESSON MODEL 39 9MM AUTO

Semi-automatic; 9mm; 8-shot magazine; 4″ barrel; 7 7/16″ overall length; weighs 26 1/2 oz. *sans* magazine; 1/8″ serrated ramp front, windage-adjustable square notch rear; checkered walnut grips with S&W medallion; bright blue or nickel finish. During the first dozen years of production, a limited number were made with steel frames rather than the standard aluminum alloy and command premium price. Introduced 1954. Collector value.

    Exc.: $350   **VGood:** $250   **Good:** $175
Nickel finish
    Exc.: $400   **VGood:** $275   **Good:** $200
Steel frame model
    Exc.: $1000   **VGood:** $850   **Good:** $750

SMITH & WESSON MODEL 59 9MM AUTO

Semi-automatic; 9mm; 14-shot staggered column magazine; 4″ barrel; 7 7/16″ overall length; weighs 27 1/2 oz. *sans* magazine; 1/8″ serrated ramp front, windage-adjustable square notch rear; checkered high-impact moulded nylon; bright blue or nickel finish. Similar to the Model 39. Like the Model 39, the 59 offers the option of carrying a round in the chamber, with hammer down, available for firing via a double-action pull of the trigger. Introduced 1971; dropped 1981.

    Exc.: $375   **VGood:** $300   **Good:** $250
Nickel finish
    Exc.: $400   **VGood:** $325   **Good:** $275

SECOND GENERATION MODELS (1980-1988)

SMITH & WESSON MODEL 439

Semi-automatic; double-action; 9mm Para.; 8-shot magazine; 4″ barrel; 7 7/16″ overall length; weighs 27 1/2 oz. sans magazine; same specs as Model 39 except new trigger-actuated firing pin lock; magazine disconnector; new extractor design; 1/8″ serrated ramp front, windage-adjustable square notch rear and protective shield on both sides of blade; checkered high-impact moulded nylon; bright blue or nickel finish. Introduced 1980; dropped 1988.

    Exc.: $300   **VGood:** $250   **Good:** $200
Nickel finish
    Exc.: $325   **VGood:** $275   **Good:** $225

SMITH & WESSON MODEL 459

Semi-automatic; double-action; 9mm Para.; 14-shot magazine; 4″ barrel; 7 7/16″ overall length; weighs 27 1/2 oz. *sans* magazine; has same general specs as Model 439, except for increased magazine capacity, straighter, longer grip frame; blued or nickel finish. Introduced 1980; dropped 1989.
Blue finish
    Exc.: $350   **VGood:** $300   **Good:** $250
Nickel finish
    Exc.: $375   **VGood:** $325   **Good:** $275

HANDGUNS

Smith & Wesson Semi-Automatic Pistols

Second Generation Models (1980-1988)
Model 469
Model 539
Model 559
Model 639
Model 645
Model 659
Model 669
Model 745

Third Generation Models (1988-1995)
Model 356 TSW Limited
Model 356 TSW Compact
Model 411
Model 909
Model 910

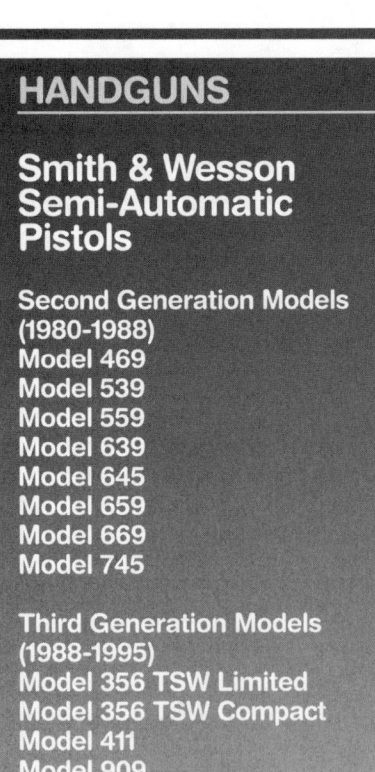

Smith & Wesson
Model 469

Smith & Wesson
Model 669

Smith & Wesson
Model 645

Smith & Wesson
Model 910

Smith & Wesson
Model 745

SMITH & WESSON MODEL 469
Semi-automatic; 9mm Para.; 3 1/2" barrel; 12-shot magazine; accepts 14-shot 459 magazine; 6 7/8" overall length; weighs 26 oz.; cut-down version of Model 459; cross-hatching on front of trigger guard, backstrap; plastic pebble-grain grips; curved finger-extension magazine; bobbed hammer; sandblasted blued finish. Introduced 1983; dropped 1988.
    Exc.: $300   **VGood:** $250   **Good:** $225

SMITH & WESSON MODEL 539
Semi-automatic; double action; 9mm Para.; 8-shot magazine; 4" barrel; 7 7/16" overall length; weighs 36 oz.; carbon steel construction; trigger-actuated firing pin lock; magazine disconnector; 1/8" serrated ramp front, windage-adjustable square notch rear and protective shield on both sides of blade; checkered high-impact moulded nylon; bright blue or nickel finish. Approximately 10,000 manufactured. Introduced 1981; dropped 1984.
    Exc.: $400   **VGood:** $350   **Good:** $300

SMITH & WESSON MODEL 559
Semi-automatic; double action; 9mm Para.; 14-shot magazine; 4" barrel; 7 7/16" overall length; weighs 40 oz. *sans* magazine; has same general specs as Model 459, except carbon steel construction; blue or nickel finish. Approximately 10,000 manufactured. Introduced 1981; dropped 1984.
    Exc.: $450   **VGood:** $400   **Good:** $350

SMITH & WESSON MODEL 639
Semi-automatic; double action; 9mm Para.; 8-shot magazine; 4" barrel; 7 7/16" overall length; weighs 27 1/2 oz. *sans* magazine; stainless steel construction; trigger-actuated firing pin lock; magazine disconnector; 1/8" serrated ramp front, windage-adjustable square notch rear with protective shield on both sides of blade; checkered high-impact moulded nylon; bright blue or nickel finish. Introduced 1981; dropped 1988.
    Exc.: $400   **VGood:** $350   **Good:** $300

SMITH & WESSON MODEL 645
Semi-automatic; double action; 45 ACP; 8-shot magazine; 5" barrel; 8 3/4" overall length; weighs 37 5/8 oz.; red ramp front sight, drift-adjustable rear; checkered nylon stocks; stainless steel construction; cross-hatch knurling on recurved front trigger guard, backstrap; beveled magazine well. Introduced 1985; dropped 1988.
    Exc.: $350   **VGood:** $300   **Good:** $250

SMITH & WESSON MODEL 659
Semi-automatic; double action; 9mm Para.; 14-shot magazine; 4" barrel; 7 7/16" overall length; weighs 27 1/2 oz. *sans* magazine; has same general specs as Model 439, except stainless steel construction. Introduced 1981; dropped 1988.
    Exc.: $350   **VGood:** $300   **Good:** $225

SMITH & WESSON MODEL 669
Semi-automatic; 9mm Para.; 3 1/2" barrel; 12-shot magazine; accepts 14-shot 459 magazine; 6 7/8" overall length; weighs 26 oz.; same specs as Model 469 except slide and barrel manufactured of stainless steel; aluminum alloy frame finished in natural finish. Introduced 1985; dropped 1988.
    Exc.: $300   **VGood:** $250   **Good:** $225

SMITH & WESSON MODEL 745
Semi-automatic; 45 ACP; 8-shot magazine; 5" barrel; 8 5/8" overall length; weighs 38 3/4 oz.; fixed Novak rear sight, serrated ramp front; blued slide, trigger, hammer, sights. Marketed with two magazines. Introduced 1987; dropped 1990.
    Exc.: $450   **VGood:** $400   **Good:** $350

THIRD GENERATION MODELS (1988-1995)

SMITH & WESSON MODEL 356 TSW LIMITED
Semi-automatic; single action; 356 TSW; 15-shot magazine; 5" barrel; weighs 44 oz.; 8 1/2" overall length; checkered black composition grips; blade front sight drift adjustable for windage, fully-adjustable Bo-Mar rear; stainless steel frame and slide; hand-fitted titanium-coated stainless steel bushing; match grade barrel; extended magazine well and oversize release; magazine pads; extended safety. Checkered front strap. Made in U.S. by Smith & Wesson; available through Lew Horton Dist. Introduced 1993; no longer in production.
    Exc.: $1200   **VGood:** $850   **Good:** $750

Smith & Wesson Model 356 TSW Compact
Same specs as the 356 TSW Limited except has 3 1/2" barrel; 12-shot magazine; Novak LoMount combat sights; 7" overall length; weighs 37 oz. Made in U.S. by Smith & Wesson; available from Lew Horton Dist. Introduced 1993; no longer in production.
    Exc.: $850   **VGood:** $725   **Good:** $550

SMITH & WESSON MODEL 411
Semi-automatic; double action; 40 S&W; 10-shot magazine; 4" barrel; weighs 27 oz.; 7 3/8" overall length; one-piece Xenoy wrap-around grips with straight backstrap; post front sight with white dot, fixed two-dot rear; alloy frame, blue carbon steel slide; slide-mounted decocking lever. Made in U.S. by Smith & Wesson. Introduced 1994; still produced.
    New: $450   **Perf.:** $400   **Exc.:** $325

SMITH & WESSON MODEL 909
Semi-automatic; double action; 9mm Para.; 9-shot magazine; 4" barrel; weighs 28 oz.; 7 3/8" overall length; one-piece Xenoy wrap-around grips with curved backstrap; post front sight with white dot, fixed two-dot rear; alloy frame, blue carbon steel slide; slide mounted decocking lever. Made in U.S. by Smith & Wesson. Introduced 1995; still produced.
    New: $350   **Perf.:** $300   **Exc.:** $250

SMITH & WESSON MODEL 910
Semi-automatic; double action; 9mm Para.; 10-shot magazine; 4" barrel; weighs 27 oz.; 7 3/8" overall length; one-piece Xenoy wrap-around grips with curved backstrap; post front sight with white dot, fixed two-dot rear; alloy frame, blue carbon steel slide; slide mounted decocking lever. Made in U.S. by Smith & Wesson. Introduced 1995; still produced.
    New: $375   **Perf.:** $325   **Exc.:** $275

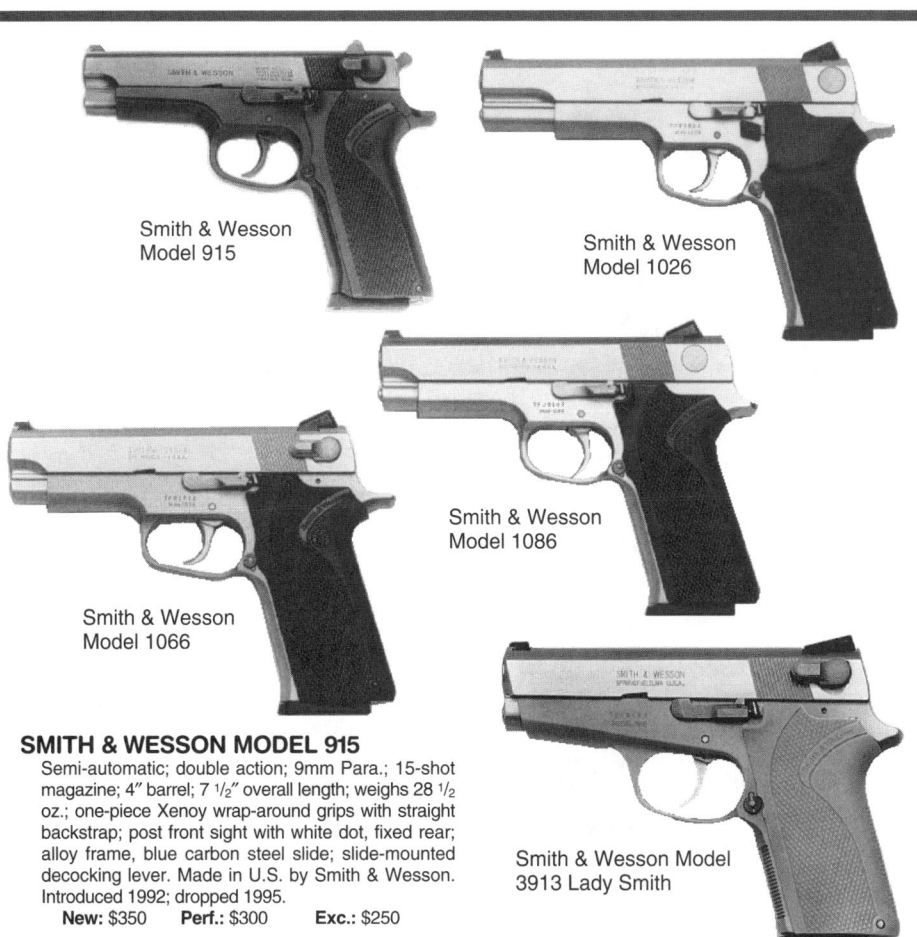

Smith & Wesson
Model 915

Smith & Wesson
Model 1026

Smith & Wesson
Model 1066

Smith & Wesson
Model 1086

Smith & Wesson Model
3913 Lady Smith

HANDGUNS

Smith & Wesson Semi-Automatic Pistols
Model 915
Model 1006
Model 1026
Model 1046
Model 1066
Model 1066-NS
Model 1076
Model 1086
Model 3904
Model 3906
Model 3913
Model 3913 LadySmith
Model 3913-NL

SMITH & WESSON MODEL 915
Semi-automatic; double action; 9mm Para.; 15-shot magazine; 4″ barrel; 7 1/2″ overall length; weighs 28 1/2 oz.; one-piece Xenoy wrap-around grips with straight backstrap; post front sight with white dot, fixed rear; alloy frame, blue carbon steel slide; slide-mounted decocking lever. Made in U.S. by Smith & Wesson. Introduced 1992; dropped 1995.
New: $350   **Perf.:** $300   **Exc.:** $250

SMITH & WESSON MODEL 1006
Semi-automatic; double action; 10mm auto; 9-shot magazine; 5″ barrel; weighs 38 oz.; one-piece Xenoy wrap-around grips with straight backstrap; rounded trigger guard; choice of Novak LoMount Carry fixed rear sight with two white dots or adjustable micro-click rear with two white dots; stainless steel construction; satin stainless finish. Introduced 1990; no longer in production.
Perf.: $550   **Exc.:** $475   **VGood:** $400

SMITH & WESSON MODEL 1026
Semi-automatic; double action; 10mm auto; 9-shot magazine; 5″ barrel; weighs 38 oz.; frame mounted decocking lever, fixed sights; one-piece Delrin grips; rounded trigger guard. Introduced 1990; dropped 1992.
Exc.: $550   **VGood:** $500   **Good:** $400

SMITH & WESSON MODEL 1046
Semi-automatic; double-action-only; 10mm auto; 9-shot magazine; 5″ barrel; weighs 38 oz.; rounded trigger guard; fixed sights; wrap-around grips with straight backstrap; stainless steel with satin finish. Introduced 1990; dropped 1992.
Exc.: $525   **VGood:** $450   **Good:** $375

SMITH & WESSON MODEL 1066
Semi-automatic; double action; 10mm auto; 9-shot magazine; 4 1/4″ barrel; fixed sights; wrap-around grips with straight backstrap; ambidextrous safety. Introduced 1990; dropped 1992.
Exc.: $500   **VGood:** $400   **Good:** $350

Smith & Wesson Model 1066-NS
Same specs as Model 1066 except Tritium night sights.
Exc.: $550   **VGood:** $450   **Good:** $400

SMITH & WESSON MODEL 1076
Semi-automatic; double action; 10mm auto; 9-shot magazine; 4 1/4″ barrel; frame mounted decocking lever; wrap-around grips with straight backstrap; fixed sights. Introduced 1990; dropped 1993.
Exc.: $500   **VGood:** $400   **Good:** $350

SMITH & WESSON MODEL 1086
Semi-automatic; double-action-only; 10mm auto; 9-shot magazine; 4 1/4″ barrel; wrap-around grips with straight backstrap; fixed sights; ambidextrous safety. Introduced 1990; no longer in production.
Perf.: $700   **Exc.:** $600   **VGood:** $450

SMITH & WESSON MODEL 3904
Semi-automatic; double action; 9mm Para.; 8-shot magazine; 4″ barrel; 7 1/2″ overall length; weighs 28 oz.; one-piece wrap-around grips; post front sight with white dot, fixed or fully-adjustable two-dot rear; blued finish; smooth trigger; serrated hammer. Introduced 1989; no longer in production.
Exc.: $400   **VGood:** $350   **Good:** $300

SMITH & WESSON MODEL 3906
Semi-automatic; double action; 9mm Para.; 8-shot magazine; 4″ barrel; 7 5/8″ overall length; weighs 28 oz.; stainless steel construction; one-piece wrap-around Delrin stocks; post front sight with white dot, fixed or fully-adjustable two-dot rear; blued finish; smooth trigger; serrated hammer. Introduced 1989; no longer in production.
Exc.: $450   **VGood:** $400   **Good:** $350

SMITH & WESSON MODEL 3913
Semi-automatic; double action; 9mm Para.; 8-shot magazine; 3 1/2″ barrel; 6 13/16″ overall length; weighs 26 oz.; post white-dot front sight, two-dot windage-adjustable Novak LoMount Carry rear; aluminum alloy frame; stainless steel slide; bobbed hammer; no half-cock notch; smooth trigger; straight backstrap. Introduced 1990; still produced.
New: $500   **Perf.:** $450   **Exc.:** $350

SMITH & WESSON MODEL 3913 LADYSMITH
Semi-automatic; double action; 9mm Para.; 8-shot magazine; 3 1/2″ barrel; 6 13/16″ overall length; weighs 26 oz.; post white-dot front sight, two-dot windage-adjustable Novak LoMount Carry rear; upswept frame

Smith & Wesson
Model 3904

Smith & Wesson
Model 3913

at front; rounded trigger guard; frosted stainless finish; gray ergonomic grips designed for smaller hands. Introduced 1990; still produced.
New: $575   **Perf.:** $500   **Exc.:** $400

Smith & Wesson Model 3913-NL
Same specs as Model 3913 except without LadySmith logo; slightly modified frame design; right-hand safety only; stainless slide in alloy frame. Introduced 1990; no longer in production.
Perf.: $450   **Exc.:** $350   **VGood:** $275

Smith & Wesson Semi-Automatic Pistols
Model 3914
Model 3914 LadySmith
Model 3914-NL
Model 3953
Model 3954
Model 4006
Model 4013
Model 4014
Model 4046
Model 4053
Model 4054
Model 4505
Model 4506
Model 4516
Model 4526

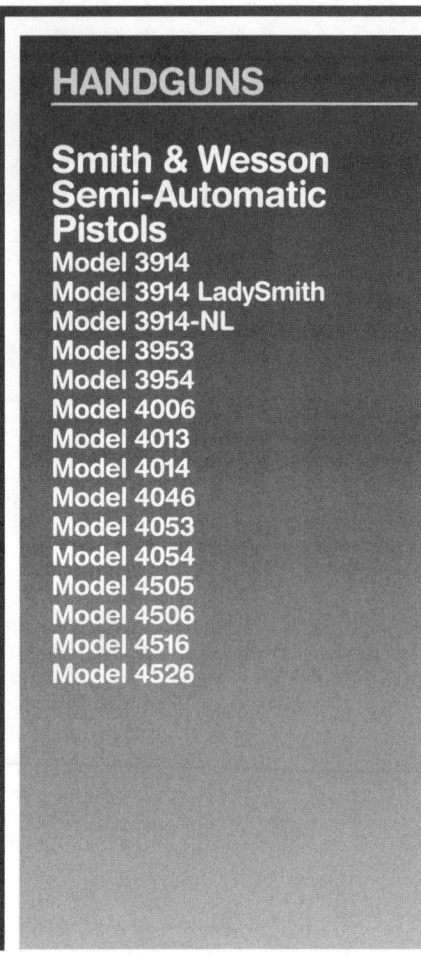

Smith & Wesson
Model 3914

Smith & Wesson
Model 4006

Smith & Wesson
Model 4053

Smith & Wesson
Model 4506

Smith & Wesson
Model 3914 LadySmith

SMITH & WESSON MODEL 3914
Semi-automatic; double action; 9mm Para.; 8-shot magazine; 3 1/2″ barrel; 6 13/16″ overall length; weighs 26 oz.; post white-dot front sight, two-dot windage-adjustable Novak LoMount Carry rear; aluminum alloy frame; blued steel slide; bobbed hammer; no half-cock notch; smooth trigger straight backstrap. Introduced 1990.
New: $400   **Perf:** $325   **Exc.:** $250

SMITH & WESSON MODEL 3914 LADYSMITH
Semi-automatic; double action; 9mm Para.; 8-shot magazine; 3 1/2″ barrel; 6 13/16″ overall length; weighs 26 oz.; post white-dot front sight, two-dot windage-adjustable Novak LoMount Carry rear; same specs as standard 3913, except slide is of blued steel. Introduced 1992.
Perf: $400   **Exc.:** $325   **VGood:** $275

Smith & Wesson Model 3914-NL
Same specs as 3914 except without LadySmith logo; slightly modified frame design; right-hand safety only; blued finish, black grips. Introduced 1990; dropped 1991.
Perf: $400   **Exc.:** $300   **VGood:** $250

SMITH & WESSON MODEL 3953
Semi-automatic; double-action-only; 9mm Para.; 8-shot magazine; 3 1/2″ barrel; 7″ overall length; weighs 25 1/2 oz.; post white-dot front sight, two-dot windage-adjustable Novak LoMount Carry rear; aluminum alloy frame; stainless steel slide; bobbed hammer; no half-cock notch; smooth trigger straight backstrap. Introduced 1990.
New: $500   **Perf:** $450   **Exc.:** $350

SMITH & WESSON MODEL 3954
Semi-automatic; double-action-only; 9mm Para.; 8-shot magazine; 3 1/2″ barrel; 7″ overall length; weighs 25 1/2 oz.; post white-dot front sight, two-dot windage-adjustable Novak LoMount Carry rear; aluminum alloy frame; blued steel slide; alloy frame; bobbed hammer; no half-cock notch; smooth trigger straight backstrap. Introduced 1990; no longer produced.
Perf: $400   **Exc.:** $350   **VGood:** $275

SMITH & WESSON MODEL 4006
Semi-automatic; 40 S&W; 11-shot magazine; 4″ barrel; 7 1/2″ overall length; weighs 38 oz.; replaceable white-dot post front sight, Novak two-dot fixed rear or two-dot micro-click adjustable rear; Xenoy wrap-around grips, checkered panels; straight backstrap; made of stainless steel; non-reflective finish. Introduced 1991.
New: $650   **Perf:** $600   **Exc.:** $500
Tritium night sights
New: $725   **Perf:** $675   **Exc.:** $575

SMITH & WESSON MODEL 4013
Semi-automatic; double action; 40 S&W; 7-shot magazine; 3 1/2″ barrel; 7″ overall length; weighs 26 oz.; white-dot post front sight, two-dot Novak LoMount Carry rear; one-piece Xenoy wrap-around grip; straight backstrap; alloy frame; stainless steel slide. Introduced 1991.
New: $550   **Perf:** $500   **Exc.:** $425

SMITH & WESSON MODEL 4014
Semi-automatic; double action; 40 S&W; 7-shot magazine; 3 1/2″ barrel; 7″ overall length; weighs 26 oz.; blued steel slide; white-dot post front sight, two-dot Novak LoMount Carry rear; one-piece Xenoy wrap-around grip; straight backstrap; alloy frame; stainless steel slide. Introduced 1991; dropped 1993
Perf: $500   **Exc.:** $450   **VGood:** $350

SMITH & WESSON MODEL 4046
Semi-automatic; double-action-only; 40 S&W; 11-shot; 4″ barrel; 7 1/2″ overall length; weighs 38 oz.; smooth trigger; semi-bobbed hammer; post white-dot front sight, Novak LoMount Carry rear. Introduced 1991.
Perf: $550   **Exc.:** $475   **VGood:** $350

SMITH & WESSON MODEL 4053
Semi-automatic; double-action-only; 40 S&W; 7-shot magazine; 3 1/2″ barrel; 7″ overall length; weighs 26 oz.; white-dot post front sight, two-dot Novak LoMount Carry rear; one-piece Xenoy wrap-around grip; straight backstrap; alloy frame; stainless steel slide. Introduced 1991.
New: $600   **Perf:** $550   **Exc.:** $450

SMITH & WESSON MODEL 4054
Semi-automatic; double-action-only; 40 S&W; 7-shot magazine; 3 1/2″ barrel; 7″ overall length; weighs 26 oz.; blued steel slide; white-dot post front sight, two-dot Novak LoMount Carry rear; one-piece Xenoy wrap-around grip; straight backstrap; alloy frame; stainless steel slide. Introduced 1991; no longer in production.
Perf: $475   **Exc.:** $400   **VGood:** $300

SMITH & WESSON MODEL 4505
Semi-automatic; 45 ACP; 8-shot magazine; 5″ barrel; one-piece wrap-around Delrin stock; arched or straight backstrap; post front sight with white dot, fixed or adjustable rear; blued finish. Introduced 1990; dropped 1991.
Perf: $550   **Exc.:** $425   **VGood:** $325

SMITH & WESSON MODEL 4506
Semi-automatic; 45 ACP; 8-shot magazine; 5″ barrel; one-piece, wrap-around Xenoy stocks; arched or straight backstrap; post front sight with white dot, fixed or adjustable Novak LoMount Carry rear; serrated hammer spur; stainless steel construction. Introduced 1989.
New: $650   **Perf:** $550   **Exc.:** $475

SMITH & WESSON MODEL 4516
Semi-automatic; 45 ACP; 7-shot magazine; 3 3/4″ barrel; one-piece wrap-around Xenoy stock; straight backstrap; post front sight with white dot; fixed Novak rear sight; bobbed hammer. Introduced 1989.
New: $600   **Perf:** $500   **Exc.:** $400

SMITH & WESSON MODEL 4526
Semi-automatic; 45 ACP; 8-shot magazine; 5″ barrel; decocking lever; one-piece wrap-around Delrin stock; arched or straight backstrap; post front sight with white dot, fixed or adjustable rear; serrated hammer spur; stainless steel construction. Introduced 1991; dropped 1993.
Perf: $575   **Exc.:** $425   **VGood:** $350

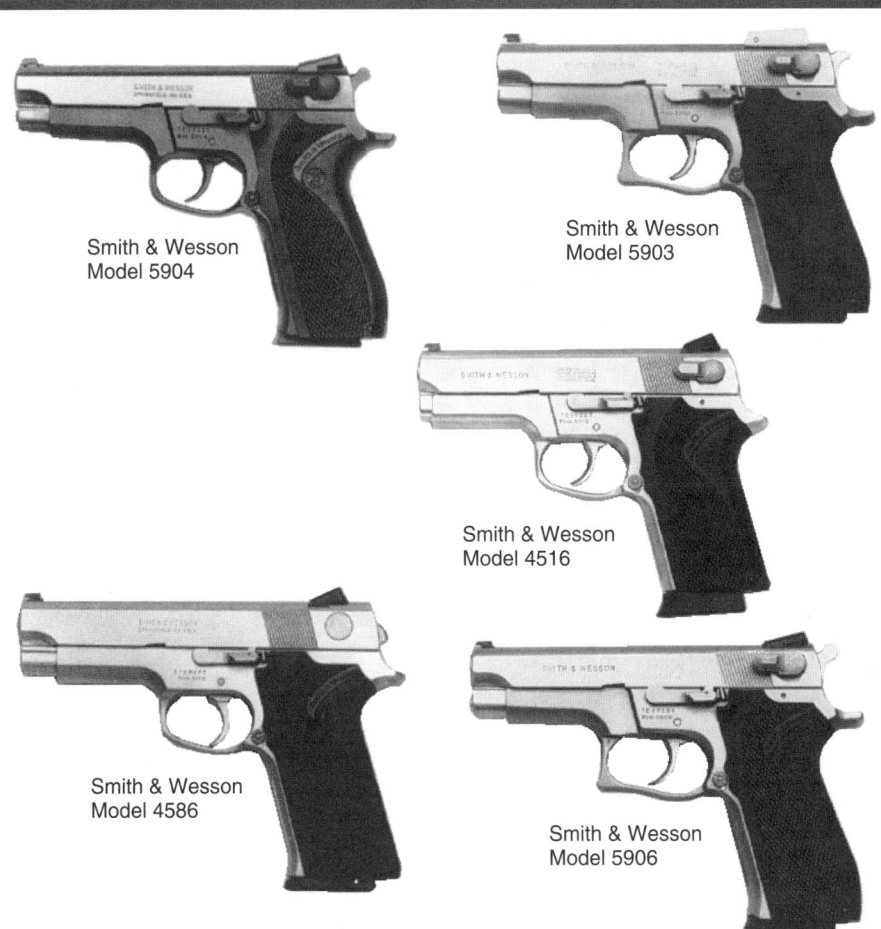

Smith & Wesson
Model 5904

Smith & Wesson
Model 5903

Smith & Wesson
Model 4516

Smith & Wesson
Model 4586

Smith & Wesson
Model 5906

HANDGUNS

Smith & Wesson
Semi-Automatic
Pistols
Model 4536
Model 4546
Model 4556
Model 4566
Model 4567-NS
Model 4576
Model 4586
Model 5903
Model 5904
Model 5905
Model 5906
Model 5924
Model 5926
Model 5943

SMITH & WESSON MODEL 4536
Semi-automatic; 45 ACP; 8-shot magazine; 3 3/4" barrel; decocking lever; one-piece wrap-around Delrin stock; arched or straight backstrap; post front sight with white dot, fixed or adjustable rear; serrated hammer spur; stainless steel construction. Introduced 1990; dropped 1991.
   Perf: $575   **Exc.:** $425   **VGood:** $350

SMITH & WESSON MODEL 4546
Semi-automatic; double-action-only; 45 ACP; 8-shot magazine; 5" barrel; decocking lever; stainless steel construction; one-piece wrap-around Delrin stock; arched or straight backstrap; post front sight with white dot, fixed or adjustable rear; serrated hammer spur; stainless steel construction. Introduced 1990; dropped 1991.
   Perf: $600   **Exc.:** $525   **VGood:** $400

SMITH & WESSON MODEL 4556
Semi-automatic; double-action-only; 45 ACP; 8-shot magazine; 3 3/4" barrel; decocking lever; stainless steel construction; one-piece wrap-around Delrin stock; arched or straight backstrap; post front sight with white dot, fixed or adjustable rear; serrated hammer spur; stainless steel construction. Introduced 1990; dropped 1991.
   Perf: $600   **Exc.:** $500   **VGood:** $425

SMITH & WESSON MODEL 4566
Semi-automatic; double-action-only; 45 ACP; 8-shot magazine; 4 1/4" barrel; ambidextrous safety; decocking lever; stainless steel construction; one-piece wrap-around Delrin stocks; arched or straight backstrap; fixed sights; serrated hammer spur; stainless steel construction. Introduced 1990.
   New: $650   **Perf:** $600   **Exc.:** $500

SMITH & WESSON MODEL 4567-NS
Semi-automatic; 45 ACP; 4 1/4" barrel; stainless steel slide; blued steel frame; bobbed hammer; one-piece wrap-around grip; straight, blued backstrap; rounded edges; Novak sights with tritium inserts. Introduced 1991; dropped 1991.
   Perf: $550   **Exc.:** $450   **VGood:** $350

SMITH & WESSON MODEL 4576
Semi-automatic; double action; 45 ACP; 8-shot magazine; 4 1/4" barrel; ambidextrous safety; decocking lever; stainless steel construction; one-piece wrap-around Delrin stocks; arched or straight backstrap; fixed sights; serrated hammer spur made of stainless steel. Introduced 1990; no longer produced.
   Perf: $600   **Exc.:** $550   **VGood:** $450

SMITH & WESSON MODEL 4586
Semi-automatic; double-action-only; 45 ACP; 8-shot magazine; 4 1/4" barrel; ambidextrous safety; decocking lever; stainless steel construction; one-piece wrap-around Xenoy stock; arched or straight backstrap; fixed sights; serrated hammer spur. Introduced 1990.
   New: $650   **Perf:** $600   **Exc.:** $500

SMITH & WESSON MODEL 5903
Semi-automatic; double action; 9mm Para.; 14-, 10-shot magazine; 4" barrel; 7 1/2" overall length; weighs 28-37 oz.; stainless steel alloy frame; ambidextrous safety; one-piece wrap-around Xenoy stocks; fixed sights; blued finish; smooth trigger; serrated hammer. Introduced 1989.
   New: $550   **Perf:** $500   **Exc.:** $425

SMITH & WESSON MODEL 5904
Semi-automatic; double action; 9mm Para.; 14-, 10-shot magazine; 4" barrel; 7 1/2" overall length; weighs 28-37 oz.; one-piece wrap-around Xenoy stocks; fixed or adjustable sights; blued finish; smooth trigger; serrated hammer. Introduced 1989.
   New: $550   **Perf:** $500   **Exc.:** $400

SMITH & WESSON MODEL 5905
Semi-automatic; double action; 9mm Para.; 14-shot magazine; 4" barrel; 7 5/8" overall length; weighs 30 oz.; blued frame, slide; ambidextrous safety; one-piece wrap-around Delrin stocks; adjustable sights; blued finish; smooth trigger; serrated hammer. Introduced 1989; dropped 1992.
   Perf: $650   **Exc.:** $600   **VGood:** $550

SMITH & WESSON MODEL 5906
Semi-automatic; double action; 9mm Para.; 14-, 10-shot magazine; 4" barrel; 7 1/2" overall length; weighs 28-38 oz.; stainless steel construction; ambidextrous safety; one-piece wrap-around Xenoy stock; fixed or adjustable sights; blued finish; smooth trigger; serrated hammer. Introduced 1989.
   New: $600   **Perf:** $525   **Exc.:** $425

SMITH & WESSON MODEL 5924
Semi-automatic; double action; 9mm Para.; 14-shot magazine; 4" barrel; 7 1/2" overall length; weighs 30 oz.; blue frame; frame mounted decocking lever; ambidextrous safety; one-piece wrap-around Delrin stocks; fixed sights; blued finish; smooth trigger; serrated hammer. Introduced 1989; dropped 1992.
   Perf: $425   **Exc.:** $350   **VGood:** $300

SMITH & WESSON MODEL 5926
Semi-automatic; double action; 9mm Para.; 14-shot magazine; 4" barrel; 7 1/2" overall length; weighs 30 oz.; stainless steel construction; frame mounted decocking lever; ambidextrous safety; one-piece wrap-around Delrin stocks; fixed sights; blued finish; smooth trigger; serrated hammer. Introduced 1989; no longer in production.
   Perf: $550   **Exc.:** $450   **VGood:** $375

SMITH & WESSON MODEL 5943
Semi-automatic; double-action-only; 9mm Para.; 14-shot magazine; 4" barrel; 7 1/2" overall length; weighs 30 oz.; stainless alloy frame; frame mounted decocking lever; ambidextrous safety; one-piece wrap-around Delrin stocks; fixed sights; blued finish; smooth trigger; serrated hammer. Introduced 1989; dropped 1992.
   Perf: $500   **Exc.:** $450   **VGood:** $350

Smith & Wesson Semi-Automatic Pistols
Model 5943-SSV
Model 5944
Model 5946
Model 6904
Model 6906
Model 6926
Model 6944
Model 6946
Sigma SW9F
Sigma SW9C
Sigma SW40F
Sigma SW40C
Sigma SW380

Smith & Wesson
Model 6946

Smith & Wesson
Model 6906

Smith & Wesson
Model 5946

Smith & Wesson
Sigma SW380

Smith & Wesson
Sigma SW40F

Smith & Wesson Model 5943-SSV
Same specs as 5943 except 3 1/2" barrel; bobbed hammer; short slide; double-action-only; alloy frame; blued slide, slide stop, magazine release, trigger, hammer; black post front sight, Novak fixed rear tritium inserts; black curved backstrap grips. Introduced 1990; dropped 1992.
Perf: $525   **Exc.:** $475   **VGood:** $350

SMITH & WESSON MODEL 5944
Semi-automatic; double-action-only; 9mm Para.; 14-shot magazine; 4" barrel; 7 1/2" overall length; weighs 30 oz.; blue alloy frame; frame mounted decocking lever; ambidextrous safety; one-piece wrap-around Delrin stocks; fixed sights; blued finish; smooth trigger; serrated hammer. Introduced 1989; dropped 1992.
Perf: $450   **Exc.:** $400   **VGood:** $325

SMITH & WESSON MODEL 5946
Semi-automatic; double-action-only; 9mm Para.; 14-shot magazine; 4" barrel; 7 5/8" overall length; weighs 30 oz.; stainless steel construction; frame mounted decocking lever; ambidextrous safety; one-piece wrap-around Delrin stocks; fixed sights; blued finish; smooth trigger; serrated hammer. Introduced 1989.
New: $600   **Perf:** $550   **Exc.:** $450

SMITH & WESSON MODEL 6904
Semi-automatic; double action; 9mm Para.; 12-, 10-shot magazine; 3 1/2" barrel; weighs 26 1/2 oz.; fixed rear sight; .260" bobbed hammer; blue finish. Introduced 1989.
New: $500   **Perf:** $450   **Exc.:** $375

SMITH & WESSON MODEL 6906
Semi-automatic; double action; 9mm Para.; 12-, 10-shot magazine; 3 1/2" barrel; weighs 26 1/2 oz.; stainless steel construction; fixed rear sight; .260" bobbed hammer. Introduced 1989.
New: $550   **Perf:** $500   **Exc.:** $400
With night sights
New: $650   **Perf:** $600   **Exc.:** $500

SMITH & WESSON MODEL 6926
Semi-automatic; double action; 9mm Para.; 12-shot magazine; 3 1/2" barrel; weighs 26 1/2 oz.; aluminum alloy frame; stainless slide; decocking lever; fixed sights; bobbed hammer. Introduced 1990; dropped 1992.
Perf: $500   **Exc.:** $450   **VGood:** $350

SMITH & WESSON MODEL 6944
Semi-automatic; double-action-only; 9mm Para.; 12-shot magazine; 3 1/2" barrel; weighs 26 1/2 oz.; aluminum alloy frame; blued steel slide; decocking lever; fixed sights; bobbed hammer. Introduced 1990; dropped 1992.
Perf: $450   **Exc.:** $400   **VGood:** $300

SMITH & WESSON MODEL 6946
Semi-automatic; double-action-only; 9mm Para.; 12-, 10-shot magazine; 3 1/2" barrel; weighs 26 1/2 oz.; aluminum alloy frame; stainless steel slide; decocking lever; fixed sights or night sights; .260" bobbed hammer. Introduced 1990.
New: $550   **Perf:** $500   **Exc.:** $400
With night sights
New: $625   **Perf:** $575   **Exc.:** $475

SMITH & WESSON SIGMA SW9F
Semi-automatic; double action; 9mm Para.; 10-shot magazine; 4 1/2" barrel; 7 1/2" overall length; weighs 26 oz.; integral polymer grips; post front sight with white dot, fixed rear with two dots; tritium night sights optional; ergonomic polymer frame; internal striker firing system; corrosion-resistant slide; Teflon-filled, electroless-nickel coated magazine. Made in U.S. by Smith & Wesson. Introduced 1994.
New: $500   **Perf:** $450   **Exc.:** $375
With fixed tritium night sights
New: $575   **Perf:** $525   **Exc.:** $450

Smith & Wesson Sigma SW9C
Same specs as the SW9F except 4" barrel; weighs 24 1/2 oz. Introduced 1995.
New: $500   **Perf:** $450   **Exc.:** $375
With fixed tritium night sights
New: $575   **Perf:** $525   **Exc.:** $450

SMITH & WESSON SIGMA SW40F
Semi-automatic; double action; 40 S&W; 10-shot magazine; 4 1/2" barrel; weighs 26 oz.; 7 1/2" overall length; integral polymer grips; post front sight with white dot, fixed rear with two dots; tritium night sights optional; ergonomic polymer frame; internal striker firing system; corrosion-resistant slide; Teflon-filled, electroless-nickel coated magazine. Made in U.S. by Smith & Wesson. Introduced 1994.
New: $450   **Perf:** $400   **Exc.:** $300
With fixed tritium night sights
New: $550   **Perf:** $500   **Exc.:** $375

Smith & Wesson Sigma SW40C
Same specs as the SW40F except 4" barrel; weighs 24 1/2 oz. Introduced 1995; still produced. **LMSR: $593**
New: $450   **Perf:** $400   **Exc.:** $325
With fixed tritium night sights
New: $550   **Perf:** $500   **Exc.:** $425

SMITH & WESSON SIGMA SW380
Semi-automatic; double-action-only; 380 ACP; 6-shot magazine; 3" barrel; 5 13/16" overall length; weighs 14 oz.; integral polymer grips; fixed groove in the slide sights; polymer frame; grooved/serrated front and rear grip straps; two passive safeties. Made in U.S. by Smith & Wesson. Introduced 1995; still produced.
New: $250   **Perf:** $200   **Exc.:** $175

Smith & Wesson
Escort Model 61

Smith & Wesson
Model 2213

Smith & Wesson
Model 2206

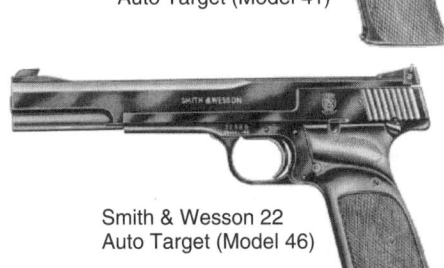

Smith & Wesson 22
Auto Match (Model 41)

Smith & Wesson 22
Auto Target (Model 41)

Smith & Wesson 22
Auto Target (Model 46)

Smith & Wesson Semi-Automatic Pistols

22 Rimfire Pistols
22 Auto Target (Model 41)
22 Auto Match (Model 41)
22 Auto Target (Model 46)
Escort Model 61
Model 422
Model 622
Model 2206
Model 2206 Target
Model 2213
Model 2214 Sportsman

22 RIMFIRE PISTOLS

SMITH & WESSON 22 AUTO TARGET (MODEL 41)
Semi-automatic; 22 LR; 10-shot clip magazine; 5″, 7 3/8″ barrel; 12″ overall length (7 3/8″ barrel); weighs 43 1/2 oz.; 3/8″ wide trigger, grooved, with adjustable stop; detachable muzzle brake supplied with 7 3/8″ barrel only (muzzle brake recently dropped); 1/8″ undercut Patridge-type front sight, fully-adjustable S&W micrometer-click rear; checkered walnut grips with modified thumbrest for right- or left-handed shooters; S&W bright blue finish. The Model 41 was at one time also available in 22 Short for international competition. Introduced about 1957; still in production.
New: $650   **Perf:** $600   **Exc.:** $500

Smith & Wesson 22 Auto Match (Model 41)
Same specs as Model 41 except 5 1/2″ heavy barrel; 9″ overall length; weighs 44 1/2 oz.; checkered walnut stocks; modified thumbrest; 1/8″ Patridge front sight on ramp base, S&W micro-click adjustable rear; grooved trigger; adjustable trigger stop; bright blued finish; matted top area. Extension front sight added 1965. Introduced 1963; dropped 1989.
Exc.: $750   **VGood:** $650   **Good:** $550

SMITH & WESSON 22 AUTO TARGET (MODEL 46)
Semi-automatic; 22 LR; 10-shot; 5″, 5 1/2″, 7″ heavy barrel; same as the Model 41 except with or without extendable front sight; less elaborate; moulded nylon stocks. Approximately 4000 maunfactured.
Exc.: $425   **VGood:** $350   **Good:** $250

SMITH & WESSON ESCORT MODEL 61
Semi-automatic; 22 LR; 5-shot magazine; 2 1/4″ barrel; 4 13/16″ overall length; hammerless; thumb safety on left side of grip; fixed sights; cocking indicator; checkered plastic grips; blued or nickel finish. Introduced in 1970; dropped 1973. Collector value.
Exc.: $225   **VGood:** $175   **Good:** $135
Nickel finish
Exc.: $250   **VGood:** $200   **Good:** $150

SMITH & WESSON MODEL 422
Semi-automatic; 22 LR; 10-shot magazine; 4 1/2″, 6″ barrel; 7 1/2″ overall length (4 1/2″ barrel); weighs 22 oz.; serrated ramp front, fixed rear field sights; serrated ramp front, adjustable rear target sights; checkered plastic stocks on field model, checkered walnut on target version; aluminum frame, steel slide; brushed blued finish; internal hammer. Introduced 1987.
Adjustable sights
New: $200   **Perf:** $175   **Exc.:** $150
Fixed sights
New: $175   **Perf:** $150   **Exc.:** $125

SMITH & WESSON MODEL 622
Semi-automatic; 22 LR; 10-shot magazine; 4 1/2″, 6″ barrel; 7 1/2″ overall length (4 1/2″ barrel); weighs 22 oz.; serrated ramp front, fixed rear field sight or serrated ramp front, adjustable rear target sights; checkered plastic stocks on field model, checkered walnut on target version; stainless steel construction; satin finish; internal hammer. Introduced 1990.
Adjustable sights
New: $250   **Perf:** $225   **Exc.:** $175
Fixed sights
New: $200   **Perf:** $150   **Exc.:** $135

SMITH & WESSON MODEL 2206
Semi-automatic; single action; 22 LR; 10-shot magazine; 4 1/2″, 6″ barrel; weighs 35 oz. (4 1/2″ barrel); fixed or adjustable sights; checkered black plastic stocks; stainless steel construction with non-reflective finish; internal hammer. Introduced 1990.
Adjustable sights
New: $350   **Perf:** $300   **Exc.:** $225
Fixed sights
New: $275   **Perf:** $225   **Exc.:** $175

Smith & Wesson Model 2006 Target
Same specs as Model 2006 except 6″ barrel; Millett Series adjustable sight system; Patridge front sight; Herett walnut target grips with thumbrest; serrated trigger with adjustable stop; bead blasted sight plane; drilled, tapped for scope mounts. Introduced 1994.
New: $375   **Perf:** $300   **Exc.:** $250

SMITH & WESSON MODEL 2213
Semi-automatic; 22 LR; 8-shot magazine; 3″ barrel 6 1/8″ overall length; weighs 18 oz.; dovetailed Patridge front sight with white dot, two-dot fixed rear; black composition grips with checkered panels; stainless steel slide and alloy frame with satin stainless finish. Comes with holster and softside carry case. Introduced 1990.
Perf: $225   **Exc.:** $175   **VGood:** $135

SMITH & WESSON MODEL 2214 SPORTSMAN
Semi-automatic; 22 LR; 8-shot magazine; 3″ barrel; 6 1/4″ overall length; weighs 18 oz.; dovetail Patridge front sight with white dot, two-dot fixed rear; black composition grips with checkered panels; matte blue finish. Introduced 1990.
Adjustable sights
New: $225   **Perf:** $175   **Exc.:** $150

Smith & Wesson
Model 422/'95

Smith & Wesson
Model 2214
Sportsman

Sokolovsky
45 Automaster

Specialized Weapons
Spectre Five

Sphinx
AT-380
AT-2000S
AT-2000C Competitor
AT-2000GM Grand Master
AT-2000H
AT-2000P
AT-2000PS

Sphinx AT-2000P

Sphinx AT-2000GM
Grand Master

Sphinx AT-2000C
Competitor

Sportarms Tokarev
Model 213

Sphinx AT-2000PS

Sokolovsky 45
Automaster

SOKOLOVSKY 45 AUTOMASTER
Semi-automatic; single action; 45 ACP; 6-shot; 6" barrel; 9 1/2" overall length; weighs 55 oz.; ramp front sight, Millett adjustable rear; smooth walnut stocks; semi-custom built with precise tolerances for target shooting; special safety trigger; primarily stainless steel. Introduced 1985; dropped 1990.

Exc.: $2500   **VGood:** $2000   **Good:** $1750

SPECIALIZED WEAPONS SPECTRE FIVE
Revolver; double action; 45 Colt/410 2", 3" shotshell; 5-shot; 2" barrel; 9" overall length; weighs 48 oz.; fixed sights; Pachmayr checkered rubber grips; ambidextrous hammer-block safety; draw bar safety; squared triggerguard; steel construction; matte blue finish. Introduced 1991; discontinued 1992. Limited production; lack of information precludes pricing.

SPHINX AT-380
Semi-automatic; double-action-only; 380 ACP; 10-shot magazine; 3 1/4" barrel; 6" overall length; weighs 25 oz.; checkered plastic grips; fixed sights;* chamber loaded indicator; ambidextrous magazine release and slide latch; blued slide, bright Palladium frame, or bright Palladium finish overall. Imported from Switzerland by Sphinx U.S.A., Inc. Introduced 1993.

New: $450   **Perf.:** $350   **Exc.:** $300

SPHINX AT-2000S
Semi-automatic; double action; 9mm Para., 9x21mm, 40 S&W; 10-shot magazine; 4 1/2" barrel; 8" overall length; weighs 36 oz.; checkered neoprene grips; blade front sight, fixed rear; three-dot system; double-action mechanism changeable to double-action-only; stainless frame, blued slide; ambidextrous safety, magazine release, slide latch. Imported from Switzerland by Sphinx U.S.A., Inc. Introduced 1993; still imported. **LMSR: $1183**
9mm, two-tone finish
New: $900   **Perf.:** $850   **Exc.:** $750
9mm, Palladium finish
New: $1000   **Perf.:** $900   **Exc.:** $800
40 S&W, two-tone finish
New: $1000   **Perf.:** $900   **Exc.:** $800
40 S&W, Palladium finish
New: $1100   **Perf.:** $1000   **Exc.:** $900

Sphinx AT-2000C Competitor
Same specs as AT-2000S except 5 5/16" barrel; 9 7/8" overall length; weighs 40 1/2 oz.; fully-adjustable Bo-Mar open sights or Tasco Pro-Point dot sight in Sphinx mount; extended magazine release; competition slide with dual-port compensated barrel; two-tone finish only. Imported from Switzerland by Sphinx U.S.A., Inc. Introduced 1993.
With Bo-Mar sights
New: $1750   **Perf.:** $1200   **Exc.:** $850
With Tasco Pro-Point and mount
New: $1900   **Perf.:** $1450   **Exc.:** $1250

Sphinx AT-2000GM Grand Master
Same specs as the AT-2000S except single-action-only trigger mechanism; squared triggerguard; extended beavertail grip, safety and magazine release; notched competition slide for easier cocking; two-tone finish; dual-port compensated barrel; fully-adjustable Bo-Mar open sights or Tasco Pro-Point and Sphinx mount. Imported from Switzerland by Sphinx U.S.A., Inc. Introduced 1993; still imported. **LMSR: $2894**
With Bo-Mar sights
New: $2500   **Perf.:** $1750   **Exc.:** $1250
With Tasco Pro-Point and mount
New: $2650   **Perf.:** $1850   **Exc.:** $1350

Sphinx AT-2000H
Same specs as the AT-2000S except shorter slide with 3 1/2" barrel; shorter frame; 7" overall length; weighs 32 oz. Imported from Switzerland by Sphinx U.S.A., Inc. Introduced 1993.
9mm, two-tone finish
New: $750   **Perf.:** $750   **Exc.:** $650
9mm, Palladium finish
New: $850   **Perf.:** $800   **Exc.:** $700
40 S&W, two-tone finish
New: $850   **Perf.:** $800   **Exc.:** $700
40 S&W, Palladium finish
New: $900   **Perf.:** $850   **Exc.:** $750

SPHINX AT-2000P
Same specs as AT-2000S except 13-shot magazine; 3 3/4" barrel; 7 1/4" overall length; weighs 34 oz.; double-action mechanism changeable to double-action-only; stainless steel frame, blued slide, or bright Paladium overall finish; ambidextrous safety magazine release and slide latch. Imported from Switzerland by Sphinx U.S.A., Inc. Introduced 1993.
9mm, two-tone finish
New: $800   **Perf.:** $750   **Exc.:** $650
9mm, Paladium finish
New: $850   **Perf.:** $800   **Exc.:** $700
40 S&W, two-tone finish
New: $850   **Perf.:** $800   **Exc.:** $700
40 S&W, Paladium finish
New: $900   **Perf.:** $850   **Exc.:** $750

Sphinx AT-2000PS
Same specs as the AT-2000S except 3 3/4" barrel; 7 1/4" overall length; full-size frame; weighs 34 oz. Imported from Switzerland by Sphinx U.S.A., Inc. Introduced 1993.
9mm, two-tone finish
New: $800   **Perf.:** $750   **Exc.:** $650
9mm, Palladium finish
New: $850   **Perf.:** $800   **Exc.:** $700
40 S&W, two-tone finish
New: $850   **Perf.:** $800   **Exc.:** $700
40 S&W, Palladium finish
New: $900   **Perf.:** $850   **Exc.:** $750

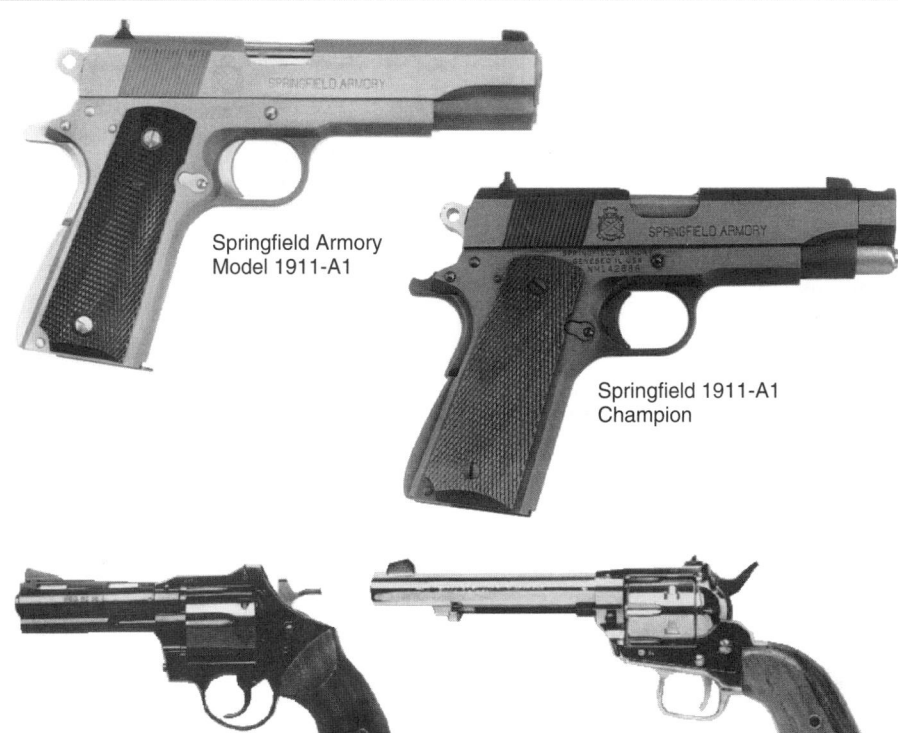

Springfield Armory
Model 1911-A1

Springfield 1911-A1
Champion

HANDGUNS

Sportarms
Model HS21S
Model HS38S
Tokarev Model 213

Springfield Armory
M6
Model 1911-A1
Model 1911-A1 Basic
Competition
Model 1911-A1 Bullseye
Wadcutter
Model 1911-A1 Wadcutter
Entry Level
Model 1911-A1 Champion
Model 1911-A1 Champion MD-1

Sportarms
Model HS38S

Sportarms
Model HS21S

Springfield Armory
1911-A1 Bullseye
Wadcutter

SPORTARMS MODEL HS21S

Revolver; single action; 22 LR, 22 LR/22 WMR combo; 6-shot cylinder; 5 1/2″ barrel; 11″ overall length; weighs 33 1/2 oz.; smooth hardwood or imitation stag grips; blade front sight, windage-adjustable rear; blue finish. Made in Germany by Herbert Schmidt; imported by Sportarms of Florida. Dropped 1994.
22 LR, blue, stag grips
Perf.: $125   Exc.: $110   VGood: $100
22 LR/22 WMR combo, blue, wood grips
Perf.: $150   Exc.: $135   VGood: $125

SPORTARMS MODEL HS38S

Revolver; double action; 38 Spl.; 6-shot cylinder; 3″, 4″ barrel; 8″ overall length (3″ barrel); weighs 31 oz.; checkered hardwood grips, round butt on 3″ model; target-style on 4″; blade front sight, adjustable rear; polished blue finish; ventilated rib on 4″ barrel. Made in Germany by Herbert Schmidt; imported by Sportarms of Florida. Dropped 1994.
Perf.: $135   Exc.: $100   VGood: $85

SPORTARMS TOKAREV MODEL 213

Semi-automatic; single action; 9mm Para.; 8-shot magazine; 4 1/2″ barrel; 7 1/2″ overall length; weighs 31 oz.; grooved plastic grips; fixed sights; blue finish; hard chrome optional; 9mm version of the famous Russian Tokarev pistol. Made in China by Norinco. Imported by Sportarms of Florida. Introduced 1988; dropped 1994.
Perf.: $125   Exc.: $100   VGood: $85

SPRINGFIELD ARMORY M6

Single shot; 22 rimfire/45 Colt; 16″ barrel; weighs 3 lbs.; rubberized wrap-around grips; blade front sight, adjustable rear; pistol version of the M6 rifle; matte blue/black finish. From Springfield Armory; now Springfield, Inc. Introduced 1991; dropped 1992.
Perf.: $175   Exc.: $135   VGood: $125

SPRINGFIELD ARMORY MODEL 1911-A1

Semi-automatic; 38 Super, 9mm Para., 10mm Auto, 45 ACP; 10-shot (38 Super, 9mm), 9-shot (10mm), 8-shot (45 ACP); 4″, 5″ barrel; 8 1/2″ overall length (5″ barrel); weighs 35 1/2 oz.; blade front, windage drift-adjustable rear; walnut grips; reproduction of original Colt model; all new forged parts; phosphated finish. Introduced 1985; replaced by '90s Edition, 1990.
Perf.: $350   Exc.: $300   VGood: $275
Blued finish
Perf.: $375   Exc.: $325   VGood: $300
Duotone finish
Perf.: $400   Exc.: $350   VGood: $325

Springfield Armory Model 1911-A1 Basic Competition

Same specs as the standard 1911-A1 except low-mounted Bo-Mar adjustable rear sight, undercut blade front; match throated barrel and bushing; polished feed ramp; lowered and flared ejection port; fitted Videcki speed trigger with tuned 3 1/2-lb. pull; fitted slide to frame; recoil buffer system; Pachmayr mainspring housing; Pachmayr grips; comes with two magazines with slam pads, plastic carrying case. From Springfield, Inc. Introduced 1992; still produced. **LMSR: $1439**
New: $1200   Perf.: $1000   Exc.: $850

SPRINGFIELD ARMORY MODEL 1911-A1 BULLSEYE WADCUTTER

Same specs as Model 1911-A1 except 45 ACP; 7-shot magazine; weighs 45 oz.; checkered walnut grips; Bo-Mar rib with undercut blade front sight, fully-adjustable rear; built for wadcutter loads only; full-length recoil spring guide rod; fitted Videcki speed trigger with 3 1/2-lb. pull; match Commander hammer and sear; beaver-tail grip safety; lowered and flared ejection port; tuned extractor; fitted slide to frame; recoil buffer system; beveled and polished magazine well; checkered

frontstrap and steel mainspring housing (flat housing standard); polished and throated National Match barrel and bushing. From Springfield, Inc. Introduced 1992.
Perf.: $1250   Exc.: $1000   VGood: $850

Springfield Armory Model 1911-A1 Wadcutter Entry Level

Same specs as the 1911-A1 Bullseye Wadcutter except low-mounted Bo-Mar adjustable rear sight, undercut blade front; match throated barrel and bushing; polished feed ramp; lowered and flared ejection port; fitted Videcki speed trigger with tuned 3 1/2-lb. pull; fitted slide to frame; Shok Buff; Pachmayr mainspring housing; Pachmayr grips; comes with two magazines with slam pads, plastic carrying case, test target. From Springfield, Inc. Introduced 1992; dropped 1993.
Perf.: $800   Exc.: $700   VGood: $600

Springfield Armory Model 1911-A1 Champion

Same specs as the standard 1911-A1 except 4 1/4″ slide; low-profile three-dot sight system; skeletonized hammer; walnut stocks; available in 45 ACP only; blue or stainless. From Springfield, Inc. Introduced 1989; still produced.
Blue finish
New: $450   Perf.: $350   Exc.: $275
Stainless steel
New: $500   Perf.: $400   Exc.: $300
Blue, with compensator
New: $700   Perf.: $600   Exc.: $500

Springfield Armory Model 1911-A1 Champion MD-1

Same specs as the 1911-A1 Champion except chambered for 380 ACP; 7-shot magazine; 3 1/2″ barrel; 7 1/8″ overall length; weighs 30 oz.; three-dot "hi-viz" sights; blued or Parkerized finish. From Springfield, Inc. Introduced 1995.
New: $450   Perf.: $350   Exc.: $250

HANDGUNS

Springfield Armory

Model 1911-A1 Commander
Model 1911-A1 Combat
Commander
Model 1911-A1 Compact
Model 1911-A1 Compact
V10 Ultra
Model 1911-A1 Competition
Model 1911-A1 Custom Carry
Model 1911-A1 Custom
Compact
Model 1911-A1 Defender
Model 1911-A1 Distinguished
Model 1911-A1 Expert Pistol
Model 1911-A1 Factory Comp
Model 1911-A1 High Capacity

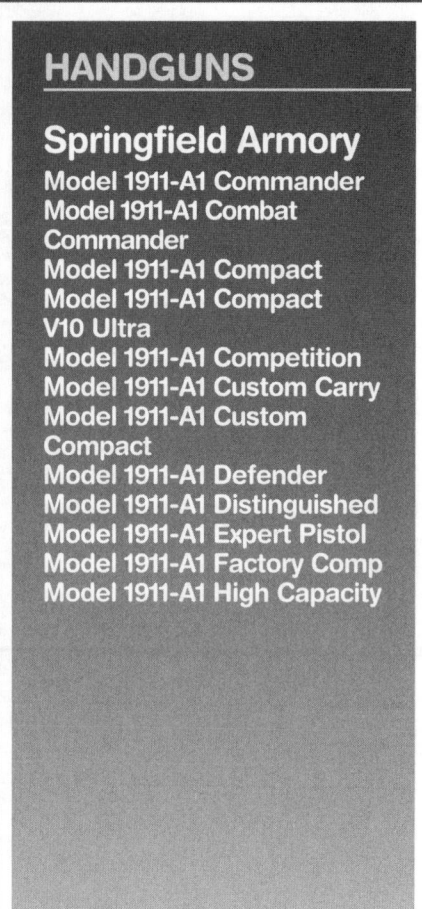

Springfield Armory Model
1911-A1 Factory Comp

Springfield Armory
Model 1911-A1
Combat Commander

Springfield Armory Model
1911-A1 High Capacity

Springfield Armory Model
1911-A1 Combat Commander

Springfield Armory Model 1911-A1 Commander

Same general specs as Model 1911-A1 except 45 ACP only; 3 5/8" barrel; shortened slide; Commander hammer; three-dot sight system; walnut stocks; phosphated finish. Introduced 1989; discontinued 1990.

Perf.: $400 **Exc.:** $350 **VGood:** $300
Blued finish
Perf.: $425 **Exc.:** $375 **VGood:** $300
Duotone finish
Perf.: $450 **Exc.:** $400 **VGood:** $350

Springfield Armory Model 1911-A1 Combat Commander

Same specs as Model 1911-A1 except 45 ACP only; 4 1/2" barrel; bobbed hammer; walnut grips. Introduced 1988; discontinued 1989.

Perf.: $425 **Exc.:** $375 **VGood:** $325
Blued finish
Perf.: $450 **Exc.:** $400 **VGood:** $350

Springfield Armory Model 1911-A1 Compact

Same general specs as Model 1911-A1 except 45 ACP only; 6-shot; 4 1/2" barrel, 7 1/4" overall length; three-dot sight system; checkered walnut stocks. Introduced 1989; discontinued 1990.

Perf.: $375 **Exc.:** $325 **VGood:** $275
Blued finish
Perf.: $400 **Exc.:** $350 **VGood:** $300
Duotone finish
Perf.: $425 **Exc.:** $375 **VGood:** $350

Springfield Armory Model 1911-A1 Compact V10 Ultra

Same specs as the 1911-A1 Compact except recoil reducing compensator built into the barrel and slide; beavertail grip safety; beveled magazine well; "hi-viz" combat sights; Videcki speed trigger; flared ejection port; stainless steel frame, blued slide; match-grade barrel; walnut grips. From Springfield, Inc. Introduced 1995; still produced.

New: $500 **Perf.:** $450 **Exc.:** $375
Without compensator
New: $450 **Perf.:** $375 **Exc.:** $300

Springfield Armory Model 1911-A1 Competition

Same specs as Model 1911-A1 except low-mounted Bo-Mar adjustable rear, brazed serrated improved ramp front sight; extended ambidextrous thumb safety; match Commander hammer and sear; serrated rear slide; Pachmayr flat mainspring housing; extended magazine release; beavertail grip safety; full-length recoil spring guide; Pachmayr wrap-around grips; 45 ACP; blue finish; comes with two magazines with slam pads, plastic carrying case. From Springfield, Inc. Introduced 1992; still produced.

New: $1250 **Perf.:** $1000 **Exc.:** $850

Springfield Armory Model 1911-A1 Custom Carry

Same specs as the standard 1911-A1 except fixed, three-dot low profile sights; Videcki speed trigger; match barrel and bushing; extended thumb and beavertail grip safeties; extended magazine well; polished feed ramp; throated barrel; match Commander hammer and sear; tuned extractor; lowered and flared ejection port; recoil buffer system; full-length spring guide rod; walnut grips; comes with two magazines with slam pads, plastic carrying case. Available in all popular calibers. From Springfield, Inc. Introduced 1992.

New: $800 **Perf.:** $650 **Exc.:** $500

Springfield Armory Model 1911-A1 Custom Compact

Same specs as Model 1911-A1 except 45 ACP; 4" barrel with compensator; Videcki speed trigger; match barrel and bushing; extended thumb and beavertail grip safeties; beveled, polished magazine well; polished feed ramp; throated barrel; match Commander hammer and sear; tuned extractor; lowered and flared ejection port; recoil buffer system; full-length spring guide rod; shortened slide; fixed three-dot sights; walnut grips; blued finish. Introduced 1991; no longer in production.

Perf.: $1400 **Exc.:** $1100 **VGood:** $850

Springfield Armory Model 1911-A1 Defender

Same specs as Model 1911-A1 except 45 ACP only; fixed combat-type sights; beveled magazine well; bobbed hammer; extended thumb safety; serrated frontstrap; walnut stocks; marketed with two stainless steel magazines; phosphated finish. Introduced 1988; discontinued 1990.

Perf.: $450 **Exc.:** $400 **VGood:** $325
Blued finish
Perf.: $475 **Exc.:** $425 **VGood:** $350

Springfield Armory Model 1911-A1 Distinguished

Same specs as the 1911-A1 except full-house pistol with match barrel with compensator; Bo-Mar low-mounted adjustable rear sight; full-length recoil spring guide rod and recoil spring retainer; beveled and polished magazine well; walnut grips; hard chrome finish; 45 ACP; comes with two magazines with slam pads, plastic carrying case. From Springfield, Inc. Introduced 1992.

New: $2200 **Perf.:** $1850 **Exc.:** $1500

Springfield Armory Model 1911-A1 Expert Pistol

Same specs as Model 1911-A1 except triple-chamber tapered cone compensator on match barrel with dovetailed front sight; lowered and flared ejection port; fully tuned for reliability; bobbed hammer; match trigger; Duotone finish; comes with two magazines, plastic carrying case. From Springfield, Inc. Introduced 1992.

New: $1600 **Perf.:** $1250 **Exc.:** $1000

Springfield Armory Model 1911-A1 Factory Comp

Same specs as the standard 1911-A1 except bushing-type dual-port compensator; adjustable rear sight; extended thumb safety; Videcki speed trigger; beveled magazine well; checkered walnut grips standard. Available in 38 Super or 45 ACP, blue only. From Springfield, Inc. Introduced 1992.

New: $750 **Perf.:** $650 **Exc.:** $500

Springfield Armory Model 1911-A1 High Capacity

Same specs as the Model 1911-A1 except 45 ACP, 9mm; 10-shot magazine; Commander-style hammer; walnut grips; ambidextrous thumb safety; beveled magazine well; plastic carrying case. From Springfield, Inc. Introduced 1993; still produced.

45 ACP
New: $600 **Perf.:** $500 **Exc.:** $350
9mm Para.
New: $500 **Perf.:** $450 **Exc.:** $375
45 ACP Factory Comp
New: $700 **Perf.:** $600 **Exc.:** $500
45 ACP Comp Lightweight, matte finish
New: $700 **Perf.:** $600 **Exc.:** $500
45 ACP Compact, blued
New: $550 **Perf.:** $475 **Exc.:** $350
45 ACP Compact, stainless steel
New: $650 **Perf.:** $550 **Exc.:** $450

Springfield Armory
Model 1911-A1 Defender

Springfield Armory Model
1911-A1 Compact V10 Ultra

Springfield Armory
Model 1911-A1 '90s
Edition

Springfield Armory
1911-A2 S.A.S.S.

HANDGUNS

Springfield Armory
Model 1911-A1 N.M. Hardball
Model 1911-A1 N.R.A. PPC
Model 1911-A1 Product
Improved Defender
Model 1911-A1 Trophy Master
Competition Pistol
Model 1911-A1 Trophy Master
Competition Expert Pistol
Model 1911-A1 Trophy Match
Model 1911-A1 '90s Edition
Model 1911-A1 '90s Edition
Champion Comp
Model 1911-A1 '90s Edition
Champion XM4 High Capacity
Model 1911-A1 '90s Edition
Combat Commander
Model 1911-A1 '90s Edition
Commander
Model 1911-A1 '90s Edition
Compact
Model 1911-A1 '90s Edition
Compact Comp
Model 1911-A1 '90s Edition
Stainless
Model 1911-A2 S.A.S.S.

Springfield Armory Model 1911-A1 N.M. Hardball

Same specs as Model 1911-A1 except Bo-Mar adjustable rear sight with undercut front blade; fitted match Videcki trigger with 4-lb. pull; fitted slide to frame; throated National Match barrel and bushing; polished feed ramp; recoil buffer system; tuned extractor; Herrett walnut grips; comes with two magazines, plastic carrying case, test target. From Springfield, Inc. Introduced 1992.

New: $1000   **Perf.:** $850   **Exc.:** $650

Springfield Armory Model 1911-A1 N.R.A. PPC

Specifically designed to comply with NRA rules for PPC competition; custom slide-to-frame fit; polished feed ramp; throated barrel; total internal honing; tuned extractor; recoil buffer system; fully checkered walnut grips; two fitted magazines; factory test target; custom carrying case. From Springfield, Inc. Introduced 1995; still produced.

New: $1350   **Perf.:** $1100   **Exc.:** $900

Springfield Armory Model 1911-A1 Product Improved Defender

Same specs as Model 1911-A1 except 4 1/4″ slide; low-profile three-dot sight system; tapered cone dual-port compensator system; rubberized grips; reverse recoil plug; full-length recoil spring guide; serrated frontstrap; extended thumb safety; skeletonized hammer with modified grip safety to match; Videcki speed trigger; Bi-Tone finish. From Springfield, Inc. Introduced 1991.

New: $800   **Perf.:** $700   **Exc.:** $600

Springfield Armory Model 1911-A1 Trophy Master Competition Pistol

Same specs as Model 1911-A1 except combat-type sights; Pachmayr wrap-around grips; ambidextrous safety; match trigger; bobbed hammer. Introduced 1988; discontinued 1990.

Perf.: $1450   **Exc.:** $1200   **VGood:** $950

Springfield Armory Model 1911-A1 Trophy Master Competition Expert Pistol

Same specs as Model 1911-A1 except match barrel with compensator; wrap-around Pachmayer grips; ambidextrous thumb safety; lowered, flared ejection port; blued finish. Introduced 1988; discontinued 1990.

Perf.: $1600   **Exc.:** $1350   **VGood:** $1000

Springfield Armory Model 1911-A1 Trophy Match

Same specs as Model 1911-A1 except factory accurized; 4- to 5 1/2-lb. trigger pull; click-adjustable rear sight; match-grade barrel and bushing; checkered walnut grips. From Springfield, Inc. Introduced 1994; still produced.
Blue finish
New: $800   **Perf.:** $700   **Exc.:** $600
Stainless steel
New: $825   **Perf.:** $725   **Exc.:** $650

SPRINGFIELD ARMORY MODEL 1911-A1 '90s EDITION

Semi-automatic; single action; 38 Super, 9mm Para., 10mm Auto, 40 S&W, 45 ACP; 10-shot (38 Super), 9-shot (9mm, 10mm), 8-shot (40 S&W, 45 ACP); 5″ barrel; 8 1/2″ overall length; weighs 36 oz.; fixed low-profile combat sights; checkered walnut grips; beveled magazine well; linkless operation; all parts forged; phosphated finish. Introduced 1990; still in production.

Perf.: $400   **Exc.:** $350   **VGood:** $275

Springfield Armory Model 1911-A1 '90s Edition Champion Comp

Same specs as Model 1911-A1 '90s Edition except 45 ACP only; 4″ barrel with compensator; three-dot sights; blued finish. Introduced 1993.

New: $750   **Perf.:** $650   **Exc.:** $550

Springfield Armory Model 1911-A1 '90s Edition Champion XM4 High Capacity

Same specs as Model 1911-A1 '90s Edition except 9mm, 45 ACP only; high-capacity variant. Introduced 1994; discontinued 1994.

Perf.: $550   **Exc.:** $500   **VGood:** $425

Springfield Armory Model 1911-A1 '90s Edition Combat Commander

Same specs as Model 1911-A1 '90s Edition except 45 ACP only; 4 1/2″ barrel; walnut grips; bobbed hammer. Introduced 1991; discontinued 1991.

Perf.: $400   **Exc.:** $350   **VGood:** $300

Springfield Armory Model 1911-A1 '90s Edition Commander

Same specs as Model 1911-A1 '90s Edition except 45 ACP only; 3 5/8″ barrel; shortened slide; three-dot sights; walnut grips; commander hammer. Introduced 1991; discontinued 1992.

Perf.: $400   **Exc.:** $350   **VGood:** $300

Springfield Armory Model 1911-A1 '90s Edition Compact

Same specs as Model 1911-A1 '90s Edition except 45 ACP only; 7-shot; 4″ barrel; shortened slide; three-dot sights; combat hammer; standard or lightweight alloy frame; phosphated finish. Introduced 1991.

New: $350   **Perf.:** $300   **Exc.:** $275
Blued finish
New: $400   **Perf.:** $350   **Exc.:** $250
Stainless steel finish
New: $475   **Perf.:** $425   **Exc.:** $375

Springfield Armory Model 1911-A1 '90s Edition Compact Comp

Same specs as Model 1911-A1 '90s Edition except 45 ACP only; 7-shot; 4″ barrel with compensator; shortened slide; three-dot sights; combat hammer; standard or lightweight alloy frame; duotone finish. Introduced 1993.

Perf.: $750   **Exc.:** $650   **VGood:** $500

Springfield Armory Model 1911-A1 '90s Edition Stainless

Same specs as Model 1911-A1 '90s Edition except 9mm Para., 45 ACP only; 8-shot; rubber grips; stainless steel construction. Introduced 1991; still in production.

New: $500   **Perf.:** $400   **Exc.:** $350

SPRINGFIELD ARMORY MODEL 1911-A2 S.A.S.S.

Single shot; break-open action; 22 LR, 223, 7mm BR, 7mm-08, 308, 357 Mag., 358 Win., 44 Mag.; 10 3/4″, 14 15/16″ barrel; 17 1/4″ overall length (14 15/16″ barrel); weighs 66 oz.; ramped blade front sight, fully adjustable open rear; drilled, tapped for scope mounts; rubberized wrap-around stock; built on standard 1911-A1 frame. Introduced 1989; discontinued 1992.

Perf.: $600   **Exc.:** $500   **VGood:** $400

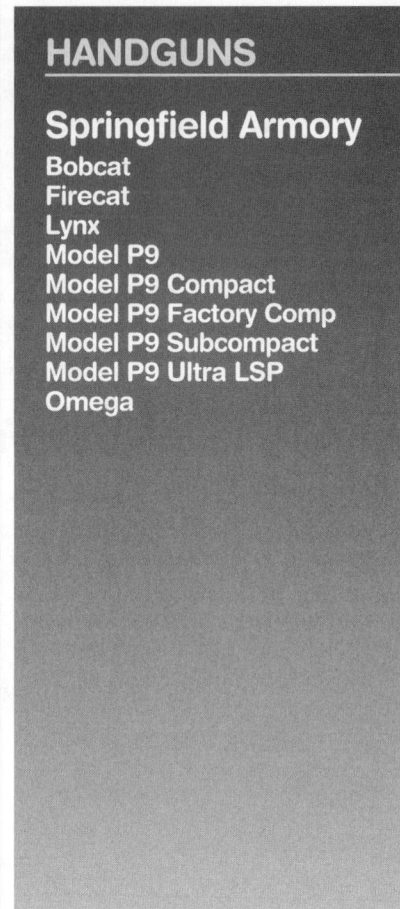

HANDGUNS

Springfield Armory

Bobcat
Firecat
Lynx
Model P9
Model P9 Compact
Model P9 Factory Comp
Model P9 Subcompact
Model P9 Ultra LSP
Omega

Springfield Armory
Panther

Springfield Armory
Omega Match

Springfield Armory
Firecat

Springfield Armory
Omega

Springfield Armory
Bobcat

Star Firestar

SPRINGFIELD ARMORY BOBCAT

Semi-automatic; 380 ACP; 13-shot magazine; 3 1/2″ barrel; 6 1/2″ overall length; weighs 22 oz.; blade front sight, windage-adjustable rear; textured composition stocks; slide-mounted ambidextrous decocker; frame-mounted slide stop; button magazine release; Commander hammer; matte blue finish. Announced 1991; never manufactured.

SPRINGFIELD ARMORY FIRECAT

Semi-automatic; single action; 9mm Para., 40 S&W; 8-shot (9mm), 7-shot (40 S&W) magazine; 3 1/2″ barrel; 6 1/2″ overall length; weighs 25 3/4 oz.; checkered walnut grips; low profile blade front sight, fixed rear; three-dot system; all-steel construction; firing pin block safety and frame-mounted thumb safety; frame-mounted slide stop; button magazine release; checkered front and rear straps; checkered and squared triggerguard; Commander hammer; matte blue only. From Springfield, Inc. Introduced 1991; dropped 1993.
9mm Para. or 40 S&W
   Perf.: $450   **Exc.:** $350   **VGood:** $300

SPRINGFIELD ARMORY LYNX

Semi-automatic; 25 ACP; 7-shot magazine; 2 1/4″ barrel; 4 3/8″ overall length; weighs 10 1/2 oz.; blade front sight, windage-adjustable rear with three-dot system; all steel construction; frame-mounted thumb safety/slide stop; Commander hammer; magazine safety; matte blue finish. Announced 1991; never manufactured.

SPRINGFIELD ARMORY MODEL P9

Semi-automatic; double action; 9mm Para., 9x21mm, 40 S&W, 45 ACP; 15-shot (9mm, 9x21), 11-shot (40 S&W, 45 ACP); 4 3/4″ barrel; 8″ overall length; weighs 35.3 oz.; blade front sight; three-dot windage-adjustable sight system; checkered walnut stocks; based upon CZ-75 design; firing pin safety block; thumb safety on frame; interchangeable magazine

catch; Commander hammer; blued finish. Introduced 1990; no longer produced.
   Perf.: $400   **Exc.:** $325   **VGood:** $225
Phosphated finish
   Perf.: $450   **Exc.:** $375   **VGood:** $250
Stainless steel
   Perf.: $475   **Exc.:** $375   **VGood:** $325

Springfield Armory Model P9 Compact

Same specs as Model P9 except 9mm Para., 40 S&W; 13-shot (9mm), 10-shot (40 S&W); 3 7/8″ barrel; 7 1/4″ overall length; weighs 32 oz.; compact slide frame; round triggerguard. Introduced 1990; discontinued 1992.
   Perf.: $350   **Exc.:** $300   **VGood:** $250

Springfield Armory P9 Factory Comp

Same specs as the standard P9 except comes with dual-port compensator system; extended sear safety; extended magazine release; fully-adjustable rear sight; extra-slim competition wood grips; stainless or bi-tone (stainless and blue) finish; overall length is 9 1/2″ with 5 1/2″ barrel; weighs 34 oz. From Springfield, Inc. Introduced 1992; dropped 1993.
9mm Para., bi-tone finish or stainless
   Perf.: $550   **Exc.:** $450   **VGood:** $350
40 S&W, bi-tone finish
   Perf.: $575   **Exc.:** $475   **VGood:** $375
40 S&W, stainless, 45 ACP bi-tone finish
   Perf.: $600   **Exc.:** $500   **VGood:** $400
45 ACP, stainless
   Perf.: $550   **Exc.:** $500   **VGood:** $450

Springfield Armory Model P9 Subcompact

Same specs as Model P9 except 9mm Para., 40 S&W; 12-shot (9mm), 9-shot (40 S&W); 3 1/2″ barrel; weighs 32 oz.; compact slide frame; squared triggerguard; extended magazine floorplate. Introduced 1990; discontinued 1992.
   Perf.: $350   **Exc.:** $300   **VGood:** $250

Springfield Armory Model P9 Ultra LSP

Same specs as Model P9 except 5″ barrel; 8 3/8″ overall length; weighs 34 1/4 oz.; long slide ported; rubber or walnut stocks; blued, Parkerized or blue/stainless finish. Introduced 1991; discontinued 1993.
   Perf.: $550   **Exc.:** $450   **VGood:** $350
Phosphated finish
   Perf.: $575   **Exc.:** $475   **VGood:** $375
Stainless steel
   Perf.: $600   **Exc.:** $500   **VGood:** $450
Stainless frame/matte black slide
   Perf.: $600   **Exc.:** $525   **VGood:** $450

SPRINGFIED ARMORY OMEGA

Semi-automatic; single action; 38 Super, 10mm, 40 S&W, 45 ACP; 5-shot (10mm, 40 S&W), 7-shot (45 ACP), 9-shot (38 Super) magazine; 5″, 6″ barrel; weighs 46 oz.; polygonal rifling; rubberized wrap-around grips; removable ramp front sight, fully-adjustable rear. Convertible between calibers; double serrated slide; built on 1911-A1 frame. Made in U.S. by Springfield Armory; now Springfield, Inc. Introduced 1987; discontinued 1990.
   Exc.: $750   **VGood:** $650   **Good:** $450

Springfield Armory
P9 Factory Comp

Star Firestar Plus

Star Model 28

Star Model 30M

Star Megastar

HANDGUNS

Springfield Armory
Omega Match
Panther

Stallard
JS-9

Star
Firestar
Firestar M45
Firestar Plus
Megastar
Model 28
Model 30M
Model 31P
Model 31PK

Springfield Armory Omega Match

Same specs as Omega except adjustable breech face; new extractor system; low profile combat sights; double serrated slide. Introduced 1991; discontinued 1992.

Perf.: $850   Exc.: $750   VGood: $600

SPRINGFIELD ARMORY PANTHER

Semi-automatic; double action; 9mm Para., 40 S&W, 45 ACP; 15-shot (9mm Para.), 11-shot (40 S&W), 9-shot (45 ACP); 40 S&W, 11-shot; 45 ACP, 9-shot magazine; 3 3/4" barrel; 7" overall length; weighs 29 oz.; narrow profile checkered walnut grips; low profile blade front sight, windage-adjustable rear; three-dot system; hammer drop and firing pin safeties; serrated front and rear straps; serrated slide top; Commander hammer; frame-mounted slide stop; button magazine release; matte blue finish. From Springfield Armory; now Springfield, Inc. Introduced 1991; dropped 1992.

Perf.: $500   Exc.: $400   VGood: $350

STALLARD JS-9

Semi-automatic; single action; 9mm Para.; 8-shot; 4 1/2" barrel; 7 3/4" overall length; weighs 48 oz.; low-profile fixed sights; textured plastic grips; non-glare blued finish. Introduced 1990; still in production.

New: $120   Perf.: $100   Exc.: $80

STAR FIRESTAR

Semi-automatic; single action; 9mm Para., 40 S&W; 7-shot (9mm), 6-shot (40 S&W); 3 3/8" barrel; 6 1/2" overall length; weighs 30 oz.; blade front sight, fully-adjustable three-dot rear; ambidextrous safety; blued or Starvel finish. Introduced 1990.

New: $275   Perf.: $250   Exc.: $200

Star Firestar M45

Same specs as Firestar except 45 ACP; 6-shot magazine; 3 1/2" barrel; 6 13/16" overall length; weighs 35 oz.; reverse-taper Acculine barrel. Imported from Spain by Interarms. Introduced 1992.
Blue finish

New: $300   Perf.: $250   Exc.: $200

Starvel finish

New: $325   Perf.: $275   Exc.: $225

Star Firestar Plus

Same specs as Firestar except 10-shot (9mm) magazine; also available in 40 S&W and 45 ACP. Imported from Spain by Interarms. Introduced 1994.
Blue finish, 9mm

New: $300   Perf.: $250   Exc.: $200

Starvel finish, 9mm

New: $325   Perf.: $275   Exc.: $225

Blue finish, 40 S&W

New: $350   Perf.: $300   Exc.: $250

Starvel finish, 40 S&W

New: $375   Perf.: $325   Exc.: $300

Blue finish, 45 ACP

New: $375   Perf.: $325   Exc.: $300

Starvel finish, 45 ACP

New: $400   Perf.: $350   Exc.: $375

STAR MEGASTAR

Semi-automatic; double action; 10mm, 45 ACP; 14-shot (10mm), 12-shot (45 ACP) magazine; 4 9/16" barrel; 8 1/2" overall length; weighs 47 1/2 oz.; checkered composition grips; blade front sight, adjustable rear; steel frame and slide; reverse-taper Acculine barrel. Imported from Spain by Interarms. Introduced 1992; no longer imported.
Blue finish, 10mm, 45 ACP

Perf.: $400   Exc.: $350   VGood: $275

Starvel finish, 10mm, 45 ACP

Perf.: $425   Exc.: $375   VGood: $300

STAR MODEL 28

Semi-automatic; double action; 9mm Para.; 15-shot; 4 1/4" barrel; 8" overall length; weighs 40 oz.; square blade front sight, square-notch click-adjustable rear; grooved triggerguard face, front and backstraps; checkered black plastic stocks; ambidextrous safety; blued finish. Introduced 1983; discontinued 1984; replaced by Model 30M.

Exc.: $400   VGood: $350   Good: $300

STAR MODEL 30M

Semi-automatic; double action; 9mm Para.; 15-shot; 4 5/16" barrel; 8" overall length; weighs 40 oz.; square blade front sight, click-adjustable rear; grooved front, backstraps, triggerguard face; checkered black plastic stocks; ambidextrous safety; steel frame; blued finish. Introduced 1985; discontinued 1991.

Perf.: $350   Exc.: $300   VGood: $250

STAR MODEL 31P

Semi-automatic; double action; 9mm Para.; 15-shot magazine; 3 7/8" barrel; 7 5/8" overall length; weighs 30 oz.; checkered black plastic grips; square blade front sight, square notch click-adjustable rear; grooved front- and backstraps and triggerguard face; ambidextrous safety cams firing pin forward; removable backstrap houses the firing mechanism. Imported from Spain by Interarms. Introduced 1984; last imported 1991.
9mm or 40 S&W, blue finish

Perf.: $350   Exc.: $300   VGood: $250

9mm or 40 S&W, Starvel finish

Perf.: $375   Exc.: $325   VGood: $275

Star Model 31PK

Same specs as the Model 31P except alloy frame with blue finish. Imported from Spain by Interarms. Introduced 1984; last imported 1991.

Perf.: $350   Exc.: $300   VGood: $250

Star Model BKM

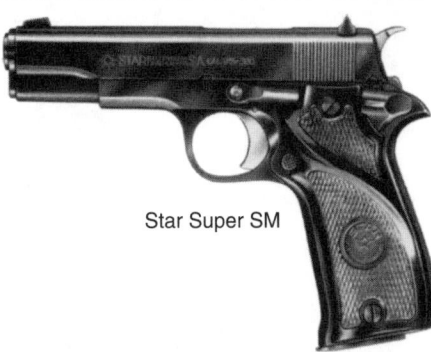
Star Super SM

Star Model A

Star Model CO Pocket

STAR MODEL A

Semi-automatic; 7.63 Mauser, 9mm Largo, 38 Super, 45 ACP; 8-shot; 5″ barrel; 8 1/2″ overall length; fixed sights; checkered walnut grips; modified version of Colt 1911; blued finish. Introduced 1934; no longer imported.
Exc.: $450   **VGood:** $300   **Good:** $150

STAR MODEL B

Semi-automatic; 9mm Para.; 9-shot; 4 1/8″, 4 3/16″, 6 5/16″ barrel; fixed sights; checkered walnut grips; blued finish. Introduced 1928; no longer imported.
Exc.: $400   **VGood:** $300   **Good:** $150

STAR MODEL BKM

Semi-automatic; single action; 9mm Para.; 8-shot; 4″ barrel; weighs 25 oz.; fixed sights; checkered walnut or plastic grips; lightweight duraluminum frame; blued finish. Introduced 1976; importation discontinued 1991.
Perf.: $275   **Exc.:** $225   **VGood:** $200

STAR MODEL BKS STARLIGHT

Semi-automatic; 9mm Para.; 8-shot; 4 1/4″ barrel; fixed sights; checkered plastic grips; magazine, manual safeties; blued, chrome finish. Introduced in 1970; discontinued 1981.
Exc.: $200   **VGood:** $175   **Good:** $150

STAR MODEL BM

Semi-automatic; single action; 9mm Para.; 8-shot; 4″ barrel; 25 oz.; fixed sights; checkered walnut or plastic grips; blued or Starvel finish. Importation discontinued 1991.
Perf.: $225   **Exc.:** $200   **VGood:** $175
Starvel or Chrome finish
Perf.: $250   **Exc.:** $225   **VGood:** $200

STAR MODEL CO POCKET

Semi-automatic; 25 ACP; 6-shot; 2 3/4″ barrel; 4 1/2″ overall length; fixed sights; checkered plastic grips; blued finish. Introduced 1934; discontinued 1957.
Exc.: $165   **VGood:** $135   **Good:** $110

STAR MODEL CU STARLET

Semi-automatic; 25 ACP; 5-shot; 2 3/8″ barrel; fixed sights; plastic grips; aluminum alloy frame; blue, chrome slide; number of color finish options. Introduced 1957; discontinued 1968.
Exc.: $200   **VGood:** $165   **Good:** $135

STAR MODEL DK

Semi-automatic; 380 ACP; 3 1/8″ barrel; 5 11/16″ overall length; fixed sights; plastic stocks; aluminum alloy frame; blue, chrome slide; number color finish options. Introduced 1957; discontinued 1968.
Exc.: $300   **VGood:** $250   **Good:** $200

STAR MODEL F

Semi-automatic; 22 LR; 10-shot; 4″ barrel; 7 1/4″ overall length; fixed sights; checkered thumbrest plastic grips; blued finish. Introduced 1942; discontinued 1969.
Exc.: $200   **VGood:** $165   **Good:** $135

Star Model F Sport

Same specs as Model F except 6″ barrel; adjustable target-type rear sight. Introduced 1942; dropped 1969.
Exc.: $225   **VGood:** $200   **Good:** $165

Star Model F Target

Same specs as Model F except adjustable target sights; 7″ barrel; weights. Introduced 1942; dropped 1969.
Exc.: $225   **VGood:** $200   **Good:** $165

STAR MODEL FM

Semi-automatic; 22 LR; 10-shot; 4 1/2″ barrel; adjustable sights; checkered plastic grips; heavy frame; triggerguard web; blued frame. Introduced 1972; no longer in production.
Perf.: $250   **Exc.:** $200   **VGood:** $165

STAR MODEL FR

Semi-automatic; 22 LR; 10-shot; 4″ barrel; adjustable sights; checkered plastic grips; slide stop; blued finish. Introduced 1967; discontinued 1972.
Exc.: $200   **VGood:** $185   **Good:** $165

STAR MODEL FRS

Semi-automatic; 22 LR; 10-shot; 6″ barrel; adjustable sights; checkered plastic grips; blued or chrome finish. Introduced 1967; no longer in production.
Perf.: $225   **Exc.:** $200   **VGood:** $175

STAR MODEL FS

Semi-automatic; 22 LR; 10-shot; 6″ barrel; adjustable sights; checkered plastic grips; blued finish. Introduced 1942; discontinued 1967.
Exc.: $200   **VGood:** $185   **Good:** $165

STAR MODEL H

Semi-automatic; 32 ACP; 7-shot; 2 3/4″ barrel; fixed sights; plastic grips; blued finish. Introduced 1934; discontinued 1941.
Exc.: $225   **VGood:** $185   **Good:** $165

STAR MODEL HK LANCER

Semi-automatic; 22 LR; 10-shot; 3 1/8″ barrel; fixed sights; plastic grips; blued finish. Introduced 1955; discontinued 1968.
Exc.: $200   **VGood:** $185   **Good:** $165

STAR MODEL HN

Semi-automatic; 380 ACP; 6-shot; 2 3/4″ barrel; fixed sights; plastic grips; blued finish. Introduced 1934; discontinued 1941.
Exc.: $250   **VGood:** $225   **Good:** $185

STAR MODEL I

Semi-automatic; 32 ACP; 9-shot; 4 3/4″ barrel; 7 1/2″ overall length; fixed sights; checkered plastic grips; blued finish. Introduced 1934; discontinued 1945.
Exc.: $200   **VGood:** $175   **Good:** $150

STAR MODEL IN

Semi-automatic; 380 ACP; 8-shot; 4 3/4″ barrel; 7 1/2″ overall length; fixed sights; checkered plastic grips; blued finish. Introduced 1934; discontinued 1945.
Exc.: $225   **VGood:** $185   **Good:** $165

STAR MODEL M

Semi-automatic; 9mm Bergman, 9mm Para., 38 Super, 45 ACP; 8-shot (9mm, 38), 7-shot (45 ACP); 5″ barrel; 8 1/2″ overall length; fixed sights; checkered walnut grips; blued finish. Not imported into U.S. Introduced 1935.
Exc.: $275   **VGood:** $225   **Good:** $150

STAR MODEL OLYMPIC

Semi-automatic; 22 Short; 9-shot; 7″ barrel; 11 1/16″ overall length; adjustable rear target sight; checkered plastic grips; alloy slide; muzzle brake; blued finish. Introduced 1942; discontinued 1967.
Exc.: $350   **VGood:** $300   **Good:** $200

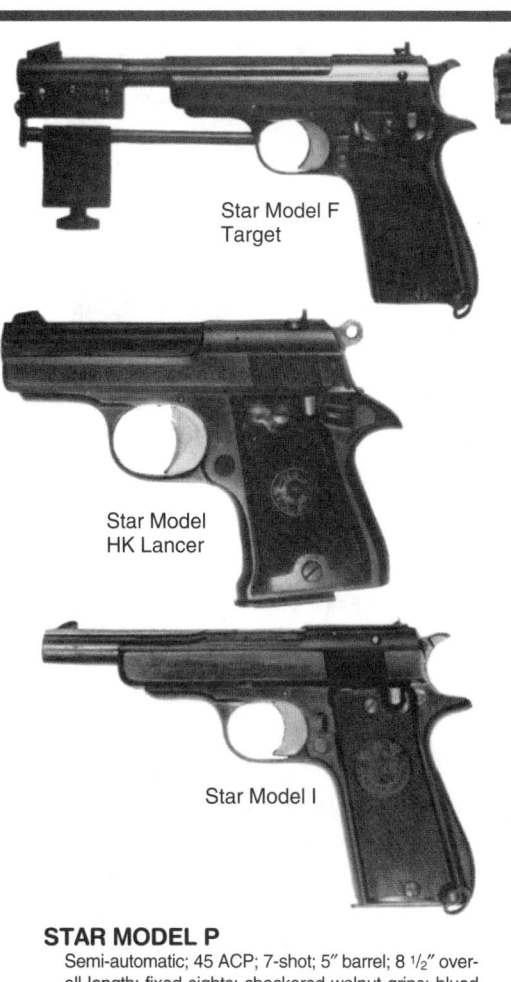

Star Model F Target

Star Model HK Lancer

Star Model I

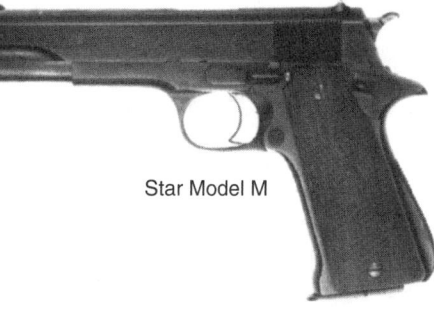

Star Model M

Star Model PD

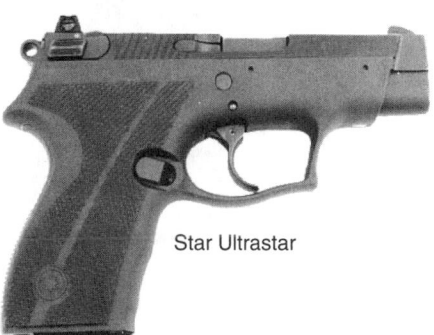

Star Ultrastar

HANDGUNS

Star
Model P
Model PD
Model S
Model SI
Model Super A
Model Super B
Model Super M
Model Super P
Model Super S
Model Super SI
Super SM
Model Super Target
Ultrastar

STAR MODEL P

Semi-automatic; 45 ACP; 7-shot; 5″ barrel; 8 1/2″ overall length; fixed sights; checkered walnut grips; blued finish. Introduced 1934; discontinued 1975.
Exc.: $350　**VGood:** $300　**Good:** $225

STAR MODEL PD

Semi-automatic; single action; 45 ACP; 6-shot; 4″ barrel; 7″ overall length; weighs 25 oz.; adjustable rear sight, ramp front; checkered walnut stocks; blued finish. Introduced 1975; discontinued 1991.
Exc.: $300　**VGood:** $250　**Good:** $200

STAR MODEL S

Semi-automatic; 380 ACP; 9-shot; 4″ barrel; 6 1/2″ overall length; fixed sights; checkered plastic grips; no grip safety; blued finish. Scaled-down modification of Colt 1911. Introduced 1941; discontinued 1965.
Exc.: $200　**VGood:** $175　**Good:** $150
Starvel finish
Exc.: $225　**VGood:** $200　**Good:** $175

STAR MODEL SI

Semi-automatic; 32 ACP; 8-shot; 4″ barrel; 6 1/2″ overall length; fixed sights; checkered plastic grips; no grip safety; blued finish. Scaled-down modification of Colt 1911. Introduced 1941; discontinued 1965.
Exc.: $175　**VGood:** $150　**Good:** $135

STAR MODEL SUPER A

Semi-automatic; 38 Super; 8-shot; 5″ barrel; luminous sights; checkered plastic grips; magazine safety; disarming bolt; loaded chamber indicator; blued finish. Introduced 1946; discontinued 1989.
Exc.: $185　**VGood:** $165　**Good:** $135

STAR MODEL SUPER B

Semi-automatic; 9mm Para.; 8-shot; 5″ barrel; luminous sights; checkered walnut grips; magazine safety; disarming bolt; loaded chamber indicator; blued finish. Discontinued 1990.
Exc.: $200　**VGood:** $175　**Good:** $150
Starvel finish
Exc.: $225　**VGood:** $200　**Good:** $175

STAR MODEL SUPER M

Semi-automatic; 9mm Bergman, 9mm Para., 38 Super, 45 ACP; 8-shot (9mm, 38), 7-shot (45); 5″ barrel; luminous sights; checkered walnut grips; magazine safety; disarming bolt; loaded chamber indicator; blued finish. Introduced 1946; discontinued 1989.
Exc.: $185　**VGood:** $165　**Good:** $135

STAR MODEL SUPER P

Semi-automatic; 45 ACP; 7-shot; 5″ barrel; luminous sights; checkered walnut grips; magazine safety; disarming bolt; loaded chamber indicator; blued finish. Introduced 1946; discontinued 1989.
Exc.: $650　**VGood:** $550　**Good:** $450

STAR MODEL SUPER S

Semi-automatic; 380 ACP; 9-shot; 4″ barrel; 6 1/2″ overall length; luminous sights; checkered plastic grips; magazine safety; disarming bolt; loaded chamber indicator; blued finish. Introduced 1946; discontinued 1972.
Exc.: $250　**VGood:** $200　**Good:** $185

STAR MODEL SUPER SI

Semi-automatic; 32 ACP; 8-shot; 4″ barrel; 6 1/2″ overall length; luminous sights; checkered plastic grips; magazine safety; disarming bolt; loaded chamber indicator; blued finish. Introduced 1946; discontinued 1972.
Exc.: $200　**VGood:** $175　**Good:** $165

STAR SUPER SM

Semi-automatic; 380 ACP; 9-shot; 4″ barrel; 6 5/8″ overall length; adjustable luminous sights; walnut grips; magazine safety; disarming bolt; loaded chamber indicator; blued finish. Introduced 1973; discontinued 1981.
Exc.: $200　**VGood:** $165　**Good:** $135

Star Model Olympic

Star Model Super P

STAR MODEL SUPER TARGET

Semi-automatic; 9mm Para., 38 Super, 45 ACP; 8-shot (9mm, 38), 7-shot (45); 5″ barrel; adjustable target sights; wood grips; blue finish. Introduced 1942; discontinued 1954.
Exc.: $350　**VGood:** $300　**Good:** $250

STAR ULTRASTAR

Semi-automatic; double action; 9mm Para.; 9-shot magazine; 3 1/2″ barrel; 7″ overall length; weighs 26 oz.; checkered black polymer grips; blade front sight, windage-adjustable rear; three-dot system; polymer frame with inside steel slide rails; ambidextrous two-position safety (Safe and Decock). Imported from Spain by Interarms. Introduced 1994.
New: $300　**Perf.:** $250　**Exc.:** $200

Sterling
Husky (See Sterling Model 285)
Husky Heavy Barrel
(See Sterling Model 295)
Model 283
Model 284
Model 285
Model 286
Model 295
Model 300
300 Target
(See Sterling Model 283)
Model 300L Target
(See Sterling Model 284)
Model 302
Model 400
Model 402
Model 450
Trapper
(See Sterling Model 286)
X-Caliber

Steel City
(See Desert Industries)

Stevens
Conlin No. 38
Diamond No. 43
Gould No. 37

Sterling Model 400

Sterling Model 302

Sterling Model 284

Sterling Model 286

Sterling Model 300

Sterling Model 450

STERLING MODEL 283
Semi-automatic; 22 LR; 10-shot; 4 1/2", 6", 8" barrel; 9" overall length (4 1/2" barrel); micrometer rear sight, blade front; checkered plastic grips; external hammer; adjustable trigger; all-steel construction; blued finish. Also designated Model 300 Target. Introduced 1970; discontinued 1972.
   Exc.: $175   **VGood:** $150   **Good:** $135

STERLING MODEL 284
Semi-automatic; 22 LR; 10-shot; 4 1/2", 6" tapered barrel; 9" overall length (4 1/2" barrel); micrometer rear sight, blade front; checkered plastic grips; adjustable trigger; all-steel construction; blued finish. Also designated Model 300L Target. Introduced 1970; discontinued 1972.
   Exc.: $185   **VGood:** $165   **Good:** $135

STERLING MODEL 285
Semi-automatic; 22 LR; 10-shot; 4 1/2" barrel; 9" overall length; fixed sights; checkered plastic grips; external hammer; adjustable trigger; all-steel construction; blued finish. Also designated Husky. Introduced 1970; discontinued 1971.
   Exc.: $165   **VGood:** $135   **Good:** $110

STERLING MODEL 286
Semi-automatic; 22 LR; 10-shot; 4 1/2", 6" tapered barrel; 9" overall length (4 1/2" barrel); fixed rear sight, serrated ramp front; checkered plastic grips; external hammer; target-type trigger; all-steel construction; blued finish. Also designated Trapper. Introduced 1970; discontinued 1972.
   Exc.: $165   **VGood:** $135   **Good:** $110

STERLING MODEL 295
Semi-automatic; 22 LR; 10-shot; 4 1/2" heavy barrel; 9" overall length; fixed rear sight, serrated ramp front; checkered plastic grips; external hammer; target-type trigger; all-steel construction; blued finish. Also designated Husky Heavy Barrel. Introduced 1970; discontinued 1972.
   Exc.: $175   **VGood:** $150   **Good:** $135

STERLING MODEL 300
Semi-automatic; blowback action; 25 ACP; 6-shot; 2 1/2" barrel; 4 1/2" overall length; no sights; black, white plastic grips; all-steel construction; blued finish. Introduced 1971; discontinued 1984.
   Exc.: $125   **VGood:** $100   **Good:** $85
Satin nickel finish
   Exc.: $135   **VGood:** $110   **Good:** $85
Stainless steel
   Exc.: $135   **VGood:** $110   **Good:** $100

STERLING MODEL 302
Semi-automatic; blowback action; 22 LR; 6-shot; 2 1/2" barrel; 4 1/2" overall length; no sights; black plastic grips; all-steel construction; blued finish. Introduced 1972; discontinued 1984.
   Exc.: $125   **VGood:** $110   **Good:** $100
Satin nickel finish
   Exc.: $135   **VGood:** $115   **Good:** $100
Stainless steel
   Exc.: $135   **VGood:** $115   **Good:** $100

STERLING MODEL 400
Semi-automatic; double action; 380 ACP; 3 1/2" barrel; 6 1/2" overall length; micrometer rear sight, fixed ramp front; checkered rosewood grips; thumb-roll safety; all-steel construction; blued finish. Introduced 1973; discontinued 1984.
   Exc.: $165   **VGood:** $135   **Good:** $110
Satin nickel finish
   Exc.: $175   **VGood:** $150   **Good:** $125
Stainless steel
   Exc.: $175   **VGood:** $165   **Good:** $135

STERLING MODEL 402
Semi-automatic; double action; 22 LR; 3 1/2" barrel; 6 1/2" overall length; micrometer rear sight, fixed ramp front; checkered rosewood grips; thumb-roll safety; all-steel construction; blued finish. Introduced 1973; discontinued 1974.
   Exc.: $125   **VGood:** $115   **Good:** $100
Satin nickel finish
   Exc.: $135   **VGood:** $125   **Good:** $115
Stainless steel
   Exc.: $150   **VGood:** $135   **Good:** $125

STERLING MODEL 450
Semi-automatic; double action; 45 ACP; 8-shot; 4" barrel; 7 1/2" overall length; adjustable rear sight; uncheckered walnut stocks; blued finish. Introduced 1977; discontinued 1984.
   Exc.: $350   **VGood:** $300   **Good:** $250

STERLING X-CALIBER
Single shot; 22 LR, 22 WMR, 357 Mag., 44 Mag.; interchangeable 8", 10" barrel; 13" overall length (8" barrel); Patridge front sight, fully-adjustable rear; drilled/tapped for scope mounts; Goncalo Alves stocks; notched hammer for easy cocking; finger-grooved grips. Introduced 1980; discontinued 1984.
   Exc.: $175   **VGood:** $150   **Good:** $135

STEVENS CONLIN NO. 38
Single shot; 22 Short, 22 LR, 22 WMR, 25 Stevens, 32 Long, 32 Short Colt; finger-rest grip; tip-down barrel; conventional trigger and guard with spur; peep rear, globe front sight. Manufactured from 1884 to 1903.
   Exc.: $750   **VGood:** $600   **Good:** $450

STEVENS DIAMOND NO. 43
Single shot; 22 Short, 22 LR; 6", 10" barrel; weighs 10 oz. (6" barrel), 12 oz. (10" barrel); round barrel; octagon breech; plain walnut grips; open or peep rear, globe front sights; spur trigger; nickel-plated finish. Manufactured from 1886 to early 1900s.
   Exc.: $250   **VGood:** $175   **Good:** $125

STEVENS GOULD NO. 37
Single shot; 22 Short, 22 LR, 22 WMR, 25 Stevens, 32 Long, 32 Short Colt; tip-down barrel; conventional trigger and guard without spur; peep rear, globe front sight. Manufactured from 1889 to 1903.
   Exc.: $750   **VGood:** $600   **Good:** $400

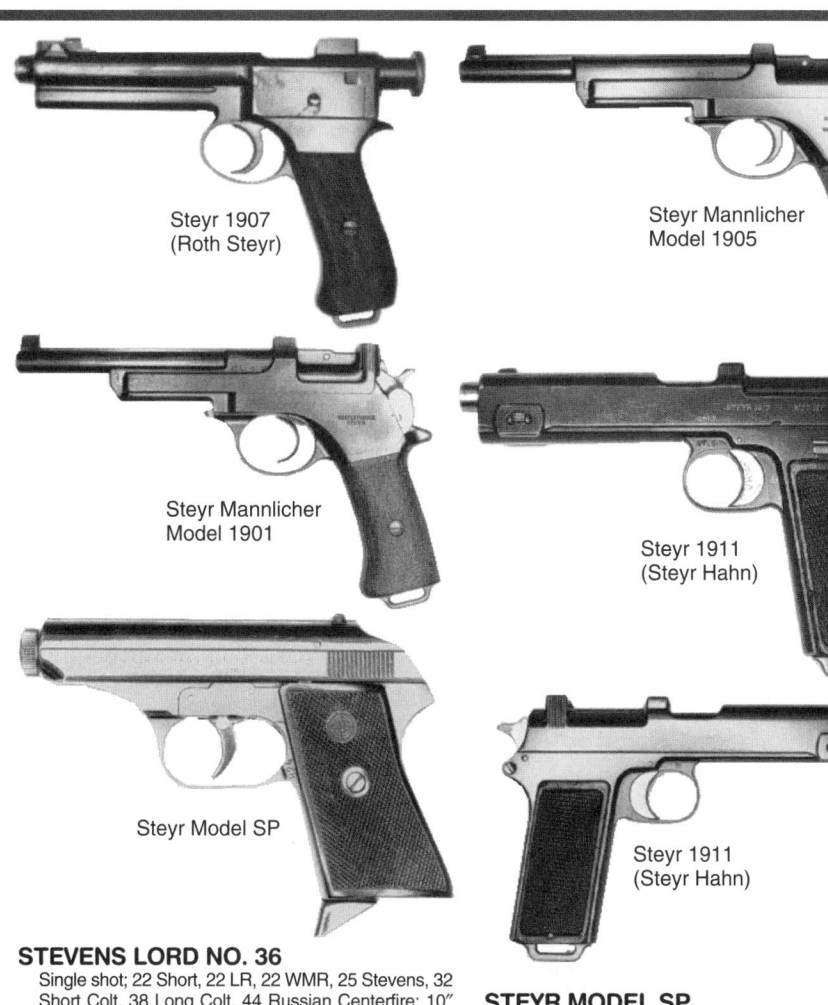

Steyr 1907
(Roth Steyr)

Steyr Mannlicher
Model 1905

Steyr Mannlicher
Model 1901

Steyr 1911
(Steyr Hahn)

Steyr Model SP

Steyr 1911
(Steyr Hahn)

HANDGUNS

Stevens
Lord No. 36
No. 10
No. 35 Off-Hand Target
No. 35 Autoshot
Tip-Up No. 41

Steyr
GB
Model SP
SSP

Steyr Mannlicher
Model 1901
Model 1905
1907 (Roth Steyr)
1911 (Steyr Hahn)

Stoeger
American Eagle Luger
American Eagle Luger Navy
Model
Luger
Luger Target

Stoeger American
Eagle Luger

STEVENS LORD NO. 36
Single shot; 22 Short, 22 LR, 22 WMR, 25 Stevens, 32 Short Colt, 38 Long Colt, 44 Russian Centerfire; 10″ heavy tip-down barrel; weighs 2 3/4 lbs. (10″ barrel); octagon breech; triggerguard with spur; sporting rear and bead front sights; checkered walnut grips; nickel-plated finish. Manufactured from 1880 to 1911.
    Exc.: $750   **VGood:** $600   **Good:** $400

STEVENS NO. 10
Single shot; tip-up action; 22 LR; 8″ barrel; 11 1/2″ overall length; target sights; hard rubber stocks; blued finish. Manufactured 1919 to 1933.
    Exc.: $200   **VGood:** $175   **Good:** $150

STEVENS NO. 35 OFF-HAND TARGET
Single shot; tip-up action; 22 LR; 6″, 8″, 10″ barrel; weighs 27 1/2 oz.; external hammer; triggerguard without spur; target sights; walnut grips; nickel-plated, case-hardened or blued finish. Introduced 1907; discontinued 1939.
    Exc.: $300   **VGood:** $250   **Good:** $150

STEVENS NO. 35 AUTOSHOT
Single shot; tip-up action; 410 shotshell; 8″, 12 1/4″ barrel; walnut grips; blued finish. Manufactured 1929 to 1934. Illegal unless registered with BATF.
    Exc.: $300   **VGood:** $250   **Good:** $150

STEVENS TIP-UP NO. 41
Single shot; 22 Short, 30 Short; 3 1/2″ barrel; weighs 7 oz.; round barrel; octagon breech; open sights; plain walnut grips; nickel-plated finish. Early manufacture with brass frame; later with iron. Manufactured from 1886 to 1916.
    Exc.: $225   **VGood:** $175   **Good:** $125

STEYR GB
Semi-automatic; double action; 9mm Para.; 18-shot; 5 1/4″ barrel; 8 3/8″ overall length; weighs 53 oz.; post front sight, fixed rear; checkered European walnut stocks; matte finish. Imported from Germany by Gun South, Inc. Introduced 1981; discontinued 1988.
    Exc.: $500   **VGood:** $400   **Good:** $300

STEYR MODEL SP
Semi-automatic; double action; 32 ACP; 7-shot; adjustable rear sight, fixed front; checkered black plastic grips; revolver-type trigger; blued finish. Very few made; collector value. Introduced 1957; discontinued 1959.
    Exc.: $800   **VGood:** $700   **Good:** $500

STEYR SSP
Semi-automatic; single action; 9mm Para.; 15- or 30-shot magazine; 5 15/16″ barrel; 12 3/4″ overall length; weighs 42 oz.; grooved synthetic grips; post elevation-adjustable front sight, open windage-adjustable rear; delayed blowback, rotating barrel operating system; synthetic upper and lower receivers; drop and crossbolt safeties; rail mount for optics. Imported from Austria by GSI, Inc. Introduced 1993; dropped 1994.
    Perf.: $700   **Exc.:** $600   **VGood:** $500

STEYR MANNLICHER MODEL 1901
Semi-automatic; 7.63mm Mannlicher; 5 1/2″ barrel; 8 3/4″ overall length; 8-shot; fine checkered wood or hard rubber (rare) grips; marked "WAFFENFABRIK STEYR" and "SYSTEM MANNLICHER" on sides.
    Exc.: $2200   **VGood:** $1750   **Good:** $1000

STEYR MANNLICHER MODEL 1905
Semi-automatic; 7.63mm Mannlicher; 6″ barrel; 9 1/2″ overall length; 10-shot; grooved wood grips; marked "WAFFENFABRIK STEYR" and "SYSTEM MANNLICHER" on sides. Also marked "Md. 1905" on side.
    Exc.: $1500   **VGood:** $850   **Good:** $400

STEYR 1907 (ROTH STEYR)
Semi-automatic; 8mm Roth Steyr; 5 1/8″ barrel; 8 13/16″ overall length; grooved wood grips.
    Exc.: $700   **VGood:** $600   **Good:** $350

STEYR 1911 (STEYR HAHN)
Semi-automatic; 9mm Steyr; 5 1/8″ barrel; 8 5/8″ overall length; checkered wood grips.
    Exc.: $300   **VGood:** $200   **Good:** $150

STOEGER AMERICAN EAGLE LUGER
Semi-automatic; single action; 9mm Para.; 7-shot magazine; 4″ barrel; 9 5/8″ overall length; weighs 32 oz.; checkered walnut grips; blade front sight, fixed rear; recreation of the American Eagle Luger pistol in stainless steel; chamber loaded indicator. Made in U.S. From Stoeger Industries. Introduced 1994; still produced.
    New: $500   **Perf.:** $450   **Exc.:** $400

Stoeger American Eagle Luger Navy Model
Same general specs as the American Eagle except has 6″ barrel. Made in U.S. From Stoeger Industries. Introduced 1994; still produced.
    New: $550   **Perf.:** $475   **Exc.:** $425

STOEGER LUGER
Semi-automatic; 22 LR; 11-shot; 4 1/2″, 5 1/2″ barrel; blade front sight, fixed rear; checkered or smooth wooden grips; blued finish. Based on original Luger design. Introduced 1970; discontinued 1978; all-steel model reintroduced 1980; discontinued 1985.
    Exc.: $165   **VGood:** $135   **Good:** $110

Stoeger Luger Target
Same general specs as Stoeger Luger, except target sights; checkered hardwood stocks. Introduced 1975; all-steel model reintroduced 1980; discontinued 1985.
    Exc.: $185   **VGood:** $155   **Good:** $130

Sturm Ruger Revolvers

Bearcat (First Issue)
Bearcat (Second Issue)
Super Bearcat
Bearcat (Third Issue)
New Super Bearcat
Bisley
Bisley Small Frame
Blackhawk
Blackhawk Convertible
Blackhawk Flat-Top
Blackhawk Flat-Top 44 Magnum
Blackhawk New Model
Blackhawk New Model 357 Maximum
Blackhawk New Model Convertible
Hawkeye

Sturm Ruger Bearcat (Third Issue) New Super Bearcat

Sturm Ruger SP101

Sturm Ruger Bisley

Sturm Ruger GP-100

Sturm Ruger Blackhawk

STURM RUGER REVOLVERS STURM RUGER BEARCAT (FIRST ISSUE)

Revolver; single action; 22 LR, 22 Long, 22 Short; 6-shot cylinder; 4″ barrel; 8 7/8″ overall length; weighs 17 oz.; non-fluted cylinder; fixed Patridge front, square notch rear sights; alloy solid frame; uncheckered walnut grips; blued. Introduced 1958; dropped 1973.

Exc.: $300   VGood: $250   Good: $175

STURM RUGER BEARCAT (SECOND ISSUE) SUPER BEARCAT

Revolver; single action; 22 LR, 22 Long, 22 Short; improved version of Bearcat except all-steel construction; weighs 22 1/2 oz.; music wire coil springs; non-fluted engraved cylinder. Introduced 1971; dropped 1975.

Exc.: $275   VGood: $225   Good: $185

STURM RUGER BEARCAT (THIRD ISSUE) NEW SUPER BEARCAT

Revolver; single action; 22 LR/22 WMR; 6-shot cylinder; 4″ barrel; weighs 23 oz.; 8 7/8″ overall length; smooth rosewood grips with Ruger medallion; blade front sight, fixed notch rear; reintroduction of the Ruger Super Bearcat with slightly lengthened frame; Ruger patented transfer bar safety system; comes with two cylinders; available in blue or stainless steel. Made in U.S. by Sturm, Ruger & Co. Introduced 1993; still produced. **LMSR: $298**

New: $250   Perf.: $200   Exc.: $165
Stainless
New: $250   Perf.: $225   Exc.: $185

STURM RUGER BISLEY

Revolver; single action; 22, 32 H&R Mag., 357, 41 Mag., 44 Mag., 45 Colt; 7 1/2″ barrel; 13″ overall length; flat-top frame; low hammer with deeply checkered wide spur; wide, smooth trigger; longer grip frame; unfluted cylinder; roll engraved with Bisley marksman, trophy. Introduced 1985; still in production. **LMSR: $415**

New: $325   Perf.: $275   Exc.: $225
22, 32 H&R
New: $275   Perf.: $225   Exc.: $185

Sturm Ruger Bisley Small Frame

Same specs as Bisley except 22 LR, 32 H&R; 6 1/2″ barrel; weighs 41 oz.; 11 1/2″ overall length; American walnut grips; front sight base accepts interchangeable square blades of various heights and fully-adjustable dovetailed rear; large oval trigger guard. Made in U.S. by Sturm, Ruger. Introduced 1985; still produced. **LMSR: $345**

New: $250   Perf.: $225   Exc.: $200

STURM RUGER BLACKHAWK

Revolver; single action; 357 Mag., 41 Mag., 45 LC; 6-shot; 4 5/8″, 6 1/2″ (357, 41 Mag.) barrel; 12″ overall length (6 1/2″ barrel); large frame; checkered hard rubber or uncheckered walnut grips; hooded adjustable rear sight, ramp front; blued. Introduced 1965; dropped 1972.

New: $300   Perf.: $250   Exc.: $200

Sturm Ruger Blackhawk Convertible

Same specs as standard Blackhawk except interchangeable cylinders for 9mm Para., 357 Mag. or 45 Colt/45 ACP. Introduced 1967; dropped 1985.

Exc.: $350   VGood: $300   Good: $225

STURM RUGER BLACKHAWK FLAT-TOP

Revolver; single action; 357 Mag.; 6-shot; 4 5/8″, 6 1/2″, 10″ barrel; 12″ overall length (6 1/2″ barrel); checkered hard rubber or uncheckered walnut grips; click-adjustable rear sight, ramp front; blued. Introduced 1955; dropped 1963.

Exc.: $450   VGood: $350   Good: $275
6 1/2″ barrel
Exc.: $600   VGood: $400   Good: $300
10″ barrel
Exc.: $1100   VGood: $850   Good: $700

Sturm Ruger Blackhawk Flat-Top 44 Magnum

Same specs as Blackhawk Flat-Top except 44 Mag.; 6 1/2″, 7 1/2″, 10″ barrel; 12 1/8″ overall length; uncheckered walnut grips; blued. Introduced 1956; dropped 1963.

Exc.: $600   VGood: $550   Good: $350
7 1/2″, 10″ barrel
Exc.: $1000   VGood: $800   Good: $600

STURM RUGER BLACKHAWK NEW MODEL

Revolver; single action; 30 Carbine, 357 Mag./38 Spl., 41 Mag., 45 LC; 6-shot cylinder; 4 5/8″, 6 1/2″ barrel; 7 1/2″ barrel (30 Carbine, 45 LC); 12 1/2″ overall length (6 1/2″ barrel); weighs 42 oz.; new Ruger interlocked mechanism; transfer bar ignition; hardened chrome-moly steel frame; wide trigger music wire springs; independent firing pin; blue, stainless or high-gloss stainless finish. Introduced 1973; still in production. **LMSR: $345**

New: $300   Perf.: $250   Exc.: $200
Stainless finish
New: $400   Perf.: $325   Exc.: $250
High-gloss stainless finish
New: $375   Perf.: $300   Exc.: $250

Sturm Ruger Blackhawk New Model 357 Maximum

Same specs as Blackhawk New Model except 357 Maximum; 7 1/2″, 10″ bull barrel; 16 7/8″ overall length (10 1/2″ barrel). Introduced 1983; dropped 1984.

Exc.: $450   VGood: $375   Good: $300

Sturm Ruger Blackhawk New Model Convertible

Same specs as Blackhawk New Model except interchangeable cylinders for 357 Mag./9mm, 44 Mag./44-40, 45 LC/45 ACP; 4 5/8″, 6 1/2″ barrel.

Exc.: $350   VGood: $275   Good: $225

STURM RUGER GP-100

Revolver; double action; 357 Mag., 38 Spl.; 6-shot cylinder; 3″, 4″, 6″ standard or heavy barrel; weighs 40 oz.; fully-adjustable rear sight, interchangeable blade front; Ruger cushioned grip with Goncalo Alves inserts; new design action, frame; full-length ejector shroud; satin blue finish or stainless steel. Introduced 1986; still in production. **LMSR: $425**

New: $350   Perf.: $300   Exc.: $250

STURM RUGER HAWKEYE

Single shot; single action; 256 Mag.; 8 1/2″ barrel; 14 1/2″ overall length; standard frame with cylinder replaced with rotating breech block; uncheckered walnut grips; click-adjustable rear sight, ramp front; barrel drilled, tapped for scope mounts; blued. Introduced 1963; dropped 1964.

Exc.: $1000   VGood: $850   Good: $700

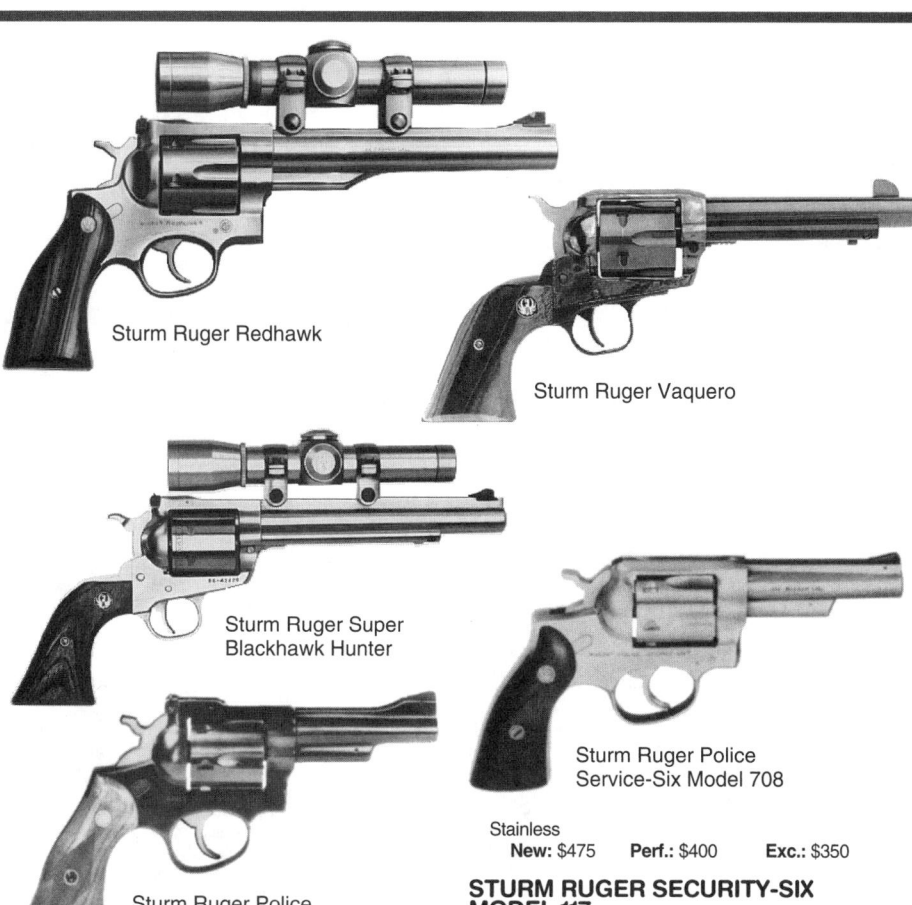

Sturm Ruger Redhawk

Sturm Ruger Vaquero

Sturm Ruger Super
Blackhawk Hunter

Sturm Ruger Police
Service-Six Model 708

Sturm Ruger Police
Service-Six Model 107

Sturm Ruger Revolvers
Police Service-Six Model 107
Police Service-Six Model 108
Police Service-Six Model 109
Police Service-Six Model 707
Police Service-Six Model 708
Redhawk
Security-Six Model 117
Security-Six Model 717
GP-100
Single-Six
Single-Six Convertible
Super Single Six
Single Six SSM
Super Blackhawk
Super Blackhawk Hunter
Super Redhawk
Vaquero

STURM RUGER POLICE SERVICE-SIX MODEL 107
Revolver; double action; 357 Mag.; 6-shot cylinder; 2 3/4", 4", 6" barrel; 6" barrel dropped in 1973; 9 1/4" overall length (4" barrel); integral ejector rod shroud and sight rib; fixed sights; semi-target-type checkered walnut grips; blued. Introduced 1972; dropped 1987.
Exc.: $225   VGood: $175   Good: $165

Sturm Ruger Police Service-Six Model 108
Same specs as Model 107 except 38 Spl.; 4" barrel. Introduced 1972; dropped 1987.
Exc.: $200   VGood: $175   Good: $150

Sturm Ruger Police Service-Six Model 109
Same specs as Model 107 except 9mm Para.; 4" barrel. Introduced 1976; dropped 1984.
Exc.: $225   VGood: $185   Good: $165

Sturm Ruger Police Service-Six Model 707
Same specs as Model 107 except 357 Mag.; stainless steel construction; 4" standard or heavy barrel. Introduced 1973; dropped 1987.
Exc.: $225   VGood: $185   Good: $165

Sturm Ruger Police Service-Six Model 708
Same specs as Model 107 except 38 Spl.; stainless steel construction; 4" standard or heavy barrel; square butt grips. Introduced 1973; dropped 1987.
Exc.: $225   VGood: $175   Good: $150

STURM RUGER REDHAWK
Revolver; double action; 41 Mag., 44 Mag.; 6-shot; 5 1/2", 7 1/2" barrel; 5 1/2" barrel added 1984; 13" overall length (7 1/2" barrel); weighs 54 oz.; stainless steel brushed satin or blued finish; Patridge-type front sight, fully-adjustable rear; square-butt Goncalo Alves grips. Introduced 1979; still in production. **LMSR: $475**
Blue
New: $450   Perf.: $375   Exc.: $300

Stainless
New: $475   Perf.: $400   Exc.: $350

STURM RUGER SECURITY-SIX MODEL 117
Revolver; double action; 357 Mag.; 6-shot cylinder; 2 3/4", 4", 6" barrel; 9 1/4" overall length (4" barrel); hand-checkered semi-target walnut grips; adjustable rear sight, Patridge type front on ramp; music wire coil springs throughout. Introduced 1974; dropped 1985.
Exc.: $225   VGood: $200   Good: $165

Sturm Ruger Security-Six Model 717
Same specs as Model 117 except 357 Mag.; all metal parts except sights of stainless steel; black alloy sights for visibility. Introduced 1974; dropped 1985.
Exc.: $275   VGood: $225   Good: $200

STURM RUGER SINGLE-SIX
Revolver; single action; 22 LR, 22 WMR; 6-shot; 4 5/8", 5 1/2", 6 1/2", 9 1/2" barrel; 10" overall length (4" barrel); checkered hard rubber grips on early production, uncheckered walnut on later versions; windage-adjustable rear sight; blued. Introduced 1953; dropped 1972.
Exc.: $200   VGood: $150   Good: $125

Sturm Ruger Single-Six Convertible
Same general specs as standard Single-Six except interchangeable cylinders, 22 LR, Long, Short/22 WRM; 5 1/2", 6 1/2", 9 1/2" barrel. Introduced 1962; dropped 1972.
Exc.: $275   VGood: $225   Good: $185

Sturm Ruger Super Single Six
Same specs as Single Six except Ruger "Interlocked" mechanism, transfer bar ignition; gate-controlled loading; hardened chrome-moly steel frame; wide trigger; music wire springs; independent firing pin. Introduced 1973; still in production. **LMSR: $298**
New: $250   Perf.: $200   Exc.: $175

Sturm Ruger Single Six SSM
Same specs as Super Single Six except 32 H&R Mag. Introduced 1985; still in production. **LMSR: $298**
New: $240   Perf.: $200   Exc.: $150

STURM RUGER SP101
Revolver; double action; 22 LR, 32 H&R Mag., 9mm Para., 38 Spl. +P, 357 Mag.; 6-shot (22 LR, 32 H&R Mag.), 5-shot (9mm, 38 Spl. +P, 357 Mag.); 2 1/4", 3 1/16", 4" barrel; weighs 25 oz. (2 1/4"), 27 oz. (3 1/16"); adjustable sights (22, 32), fixed (others); Ruger Santoprene cush-

ioned grips with Xenoy inserts; incorporates improvements and features found in the GP-100 revolvers into a compact, small frame, double-action revolver; full-length ejector shroud; stainless steel only; available with high-polish finish. Introduced 1988; still in production. **LMSR: $428**
New: $375   Perf.: $275   Exc.: $200

STURM RUGER SUPER BLACKHAWK
Revolver; 44 Mag.; 5 1/2", 7 1/2", 10 1/2" barrel; 13 3/8" overall length (7 1/2" barrel); weighs 48 oz.; 1/8" ramp front, micro-click fully-adjustable rear; American walnut grips; interlock mechanism; non-fluted cylinder; steel grip and cylinder frame; square back trigger guard; wide serrated trigger; wide spur hammer. Introduced 1985; still in production. **LMSR: $398**
New: $350   Perf.: $300   Exc.: $250
Stainless
New: $400   Perf.: $350   Exc.: $300

Sturm Ruger Blackhawk Hunter
Same specs as Super Blackhawk except stainless finish; 7 1/2"; scope rings. Introduced 1992; still in prduction. **LMSR: $498**
New: $450   Perf.: $400   Exc.: $325

STURM RUGER SUPER REDHAWK
Revolver; double action; 44 Mag.; 6-shot; 7 1/2", 9 1/2" barrel; 13" overall length (7 1/2" barrel); weighs 54 oz.; heavy extended frame with Ruger Integral Scope Mounting System; wide topstrap; ramp front with interchangeable sight blades, adjustable rear; Santoprene grips with Goncalo Alves panels; satin polished stainless steel finish. Introduced 1987; still in production. **LMSR: $574**
New: $475   Perf.: $400   Exc.: $300

STURM RUGER VAQUERO
Revolver; single action; 44-40, 44 Mag., 45 Colt; 6-shot; 4 5/8", 5 1/2", 7 1/2" barrel; 13 3/8" overall length (7 1/2" barrel); weighs 41 oz.; blade front, fixed notch rear sight; transfer bar safety system; loading gate interlock; color case-hardened frame on blue model; blue or high-polish stainless finish. Introduced 1993; still in production. **LMSR: $419**
New: $400   Perf.: $350   Exc.: $275

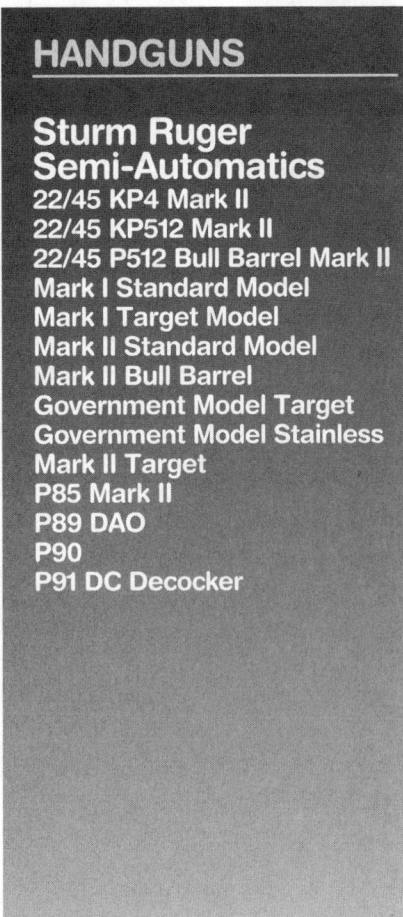

Sturm Ruger Semi-Automatics

22/45 KP4 Mark II
22/45 KP512 Mark II
22/45 P512 Bull Barrel Mark II
Mark I Standard Model
Mark I Target Model
Mark II Standard Model
Mark II Bull Barrel
Government Model Target
Government Model Stainless
Mark II Target
P85 Mark II
P89 DAO
P90
P91 DC Decocker

Sturm Ruger Mark II Government Model Target

Sturm Ruger Mark II Standard Model

Sturm Ruger 22/45 KP512 Mark II

Sturm Ruger Mark I Standard Model

Sturm Ruger P90 Decocker Model

STURM RUGER SEMI-AUTOMATICS

STURM RUGER 22/45 KP4 MARK II

Semi-automatic; single action; 22 LR; 10-shot magazine; 4 3/4" stainless steel barrel; 8 7/8" overall length; weighs 28 oz.; Zytel grip frame matches angle and magazine latch of Model 1911 45 ACP pistol; Patridge front sight, square notch windage-adjustable rear; brushed stainless steel. Made in U.S. by Sturm, Ruger & Company. Introduced 1992; still produced. **LMSR: $280**

New: $275	**Perf.:** $225	**Exc.:** $175

Sturm Ruger 22/45 KP512 Mark II

Same specs as P4 except 5 1/2" stainless steel bull barrel; 9 5/8" overall length; weighs 35 oz. Made in U.S. by Sturm, Ruger & Company. Introduced 1992; still produced. **LMSR: $330**

New: $275	**Perf.:** $250	**Exc.:** $200

Sturm Ruger 22/45 P512 Bull Barrel Mark II

Same specs as P4 except 5 1/2" steel bull barrel; 9 5/8" overall length; weighs 36 oz.; fully-adjustable rear sight; brushed satin blue finish. Made in U.S. by Sturm, Ruger & Company. Introduced 1995; still produced. **LMSR: $237**

New: $275	**Perf.:** $250	**Exc.:** $200

STURM RUGER MARK I STANDARD MODEL

Semi-automatic; 22 LR only; 9-shot magazine 4 3/4", 6" barrel (4 3/4" barrel), 8 3/4" overall length (6" barrel); checkered hard rubber grips walnut optional; windage-adjustable square-notch rear, fixed wide blade front sight; blued. Introduced 1949; dropped 1982 in favor of Mark II. Until 1951 featured red eagle insignia in grips; changed to black upon death of Alex Sturm. This type has considerable collector value.

Exc.: $150	**VGood:** $125	**Good:** $110

Red eagle model

Exc.: $350	**VGood:** $300	**Good:** $250

Sturm Ruger Mark I Target Model

Same specs as Standard Model except 5 1/2" heavy barrel, 5 3/4" tapered barrel, 6 7/8" heavy tapered barrel; adjustable rear, target front sight. Introduced 1951; dropped 1982.

Exc.: $175	**VGood:** $150	**Good:** $125

STURM RUGER MARK II STANDARD MODEL

Semi-automatic; 22 LR; 10-shot magazine; 4 3/4", 6" barrel; checkered hard rubber stocks; fixed blade front sight, square-notch windage-adjustable rear; new bolt hold-open device, magazine catch; new receiver contours; stainless steel or blued. Introduced 1982; still in production. **LMSR: $252**

Exc.: $175	**VGood:** $150	**Good:** $135

Stainless steel finish

Exc.: $225	**VGood:** $200	**Good:** $175

Sturm Ruger Mark II Bull Barrel

Same specs as standard model except 5 1/2", 10" heavy barrel. Introduced 1982; still in production. **LMSR: $310**

Perf.: $275	**Exc.:** $225	**VGood:** $200

Stainless steel finish

Exc.: $300	**VGood:** $275	**Good:** $225

Sturm Ruger Mark II Government Model Target

Same specs as Mark II Standard except 6 7/8" bull barrel; adjustable rear sight; black plastic grips; roll-stamped "Government Target Model" on right side of receiver; blued finish. Introduced 1987; still in production. **LMSR: $356**

New: $300	**Perf.:** $250	**Exc.:** $200

Sturm Ruger Mark II Government Model Stainless

Same specs as Model Target except 6 7/8" slab-sided barrel; receiver drilled, tapped for Ruger scope base adaptor; checkered walnut grip panels; right-hand thumb rest; blued open sights. Marketed with 1" stainless scope rings, integral base. Introduced 1991; still in production. **LMSR: $441**

New: $350	**Perf.:** $300	**Exc.:** $250

Sturm Ruger Mark II Target

Same specs as standard model except 5 1/4", 6 7/8" barrel; 11 1/8" overall length; weighs 42 oz.; .125" blade front sight, adjustable micro-click rear; sight radius 9 3/8". Introduced 1982; still in production. **LMSR: $310**

New: $250	**Perf.:** $200	**Exc.:** $165

Stainless steel finish

New: $300	**Perf.:** $250	**Exc.:** $200

STURM RUGER P85 MARK II

Semi-automatic; double action, DAO; 9mm Para.; 15-shot magazine; 4 1/2" barrel; 7 13/16" overall length; weighs 32 oz.; windage-adjustable square-notch rear sight, square post front; ambidextrous slide-mounted safety levers; 4140 chome-moly steel slide; aluminum alloy frame; grooved Xenoy composition stocks; ambidextrous magazine release; blue or stainless finish. Decocker version also available. Introduced 1986; no longer in production.

New: $325	**Perf.:** $300	**Exc.:** $250

Stainless finish

New: $375	**Perf.:** $325	**Exc.:** $275

STURM RUGER P89/DAO

Semi-automatic; double action, double-action-only; 9mm Para.; 15-shot magazine; 4 1/2" barrel; 7 13/16" overall length; weighs 32 oz.; improved version of P85 Mark II; windage-adjustable square-notch rear sight, square post front; internal safety; bobbed, spurless hammer; gripping grooves on rear of slide; stainless steel construction. Decocker model also available at no extra cost; blue or stainless (DAO) finish. Marketed with plastic case, extra magazine, loading tool. Introduced 1991; still in production. **LMSR: $410**

New: $325	**Perf.:** $300	**Exc.:** $250

Stainless

New: $400	**Perf.:** $350	**Exc.:** $300

STURM RUGER P90

Semi-automatic; double action; 45 ACP; 7-shot magazine; 4 1/2" barrel; 7 15/16" overall length; weighs 33 1/2 oz.; aluminum frame; stainless steel slide; ambidextrous slide-mounted safety levers; square post front, square notch windage-adjustable rear; stainless steel only. Decocker model also available at no extra cost. Introduced 1991; still in production. **LMSR: $489**

New: $400	**Perf.:** $350	**Exc.:** $300

STURM RUGER P91 DC DECOCKER

Semi-automatic; double action, double-action-only; 40 S&W; 11-shot magazine; 4 1/2" barrel; 7 15/16" overall length; weighs 33 oz.; ambidextrous slide-mounted safety levers; square post front, windage-adjustable square notch rear with white dot inserts; grooved black Xenoy composition grips; stainless finish. Marketed with plastic case, extra magazine, loading tool. Introduced 1991; no longer in production.

New: $400	**Perf.:** $350	**Exc.:** $300

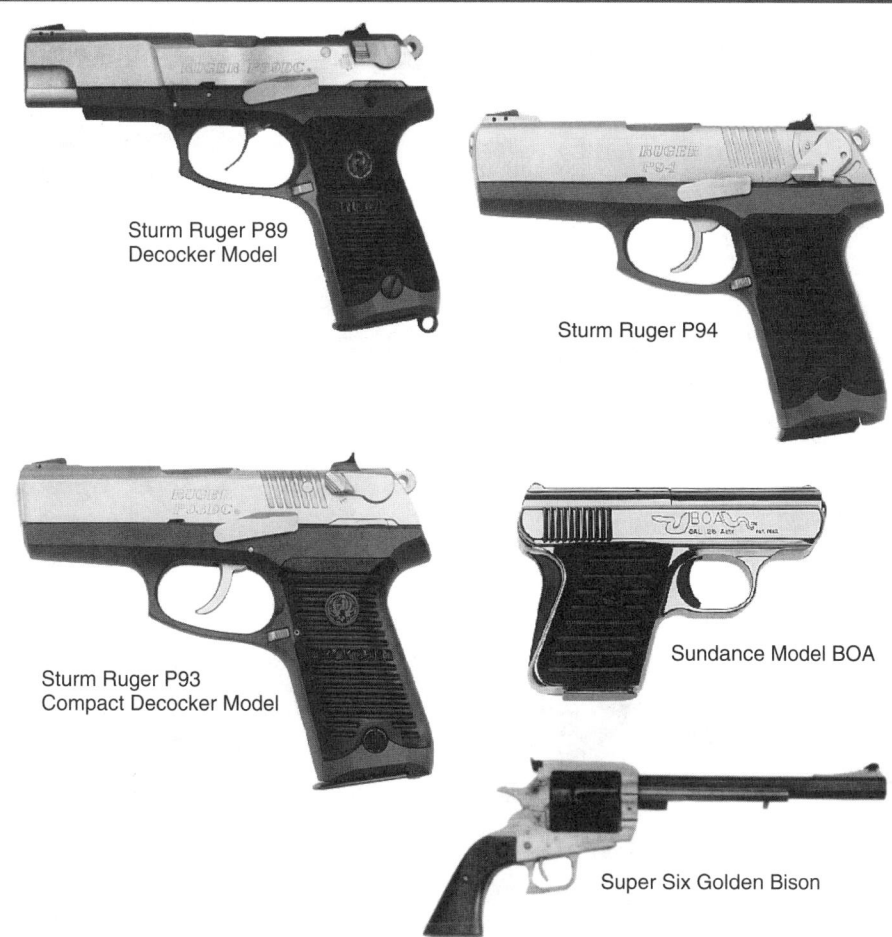

Sturm Ruger P89
Decocker Model

Sturm Ruger P94

Sturm Ruger P93
Compact Decocker Model

Sundance Model BOA

Super Six Golden Bison

HANDGUNS

Sturm Ruger Semi-Automatics
P93 Compact
P94
P94L

Sundance
Model A-25
Model BOA
Point Blank

Super Six
Golden Bison

Tanarmi
Model BTA 90
Model TA 22
Model TA 38 Over/Under Derringer
Model TA 41
Model TA 41 SS
Model TA 76

STURM RUGER P93 COMPACT

Semi-automatic; double action; 9mm Para.; 10-shot magazine; 3 15/16" barrel; 7 5/16" overall length; weighs 31 oz.; grooved black Xenoy composition grips; square post front sight, square notch windage-adjustable rear with white dot inserts; forward third of the slide tapered and polished to the muzzle; front of the slide crowned with a convex curve; seven finger grooves on slide; 400-series stainless steel slide; lightweight alloy frame. Available as decocker-only or double-action-only. Made in U.S. by Sturm, Ruger & Company. Introduced 1993; still produced. **LMSR: $520**
  New: $425　　**Perf.:** $375　　**Exc.:** $325

STURM RUGER P94

Semi-automatic; double-action-only, decock only, or manual safety; 9mm Para., 40 S&W; 10-shot magazine; 4 1/4" barrel; 7 1/2" overall length; weighs 33 oz.; grooved Xenoy grips; square post front sight, windage-adjustable rear; three-dot system; slide gripping grooves roll over top of slide; ambidextrous safety-levers; ambidextrous decocking levers; matte finish stainless slide and barrel, alloy frame; frame size between full-size P-Series and compact P93. Made in U.S. by Sturm, Ruger. Introduced 1994; still produced. **LMSR: $520**
  New: $350　　**Perf.:** $300　　**Exc.:** $275

Sturm Ruger P94L

Same specs as the KP94 except laser sight mounts in housing cast integrally with frame; Allen-head screws control windage and elevation adjustments. Made in U.S. by Sturm, Ruger. Introduced 1994; still produced for law enforcement only.

SUNDANCE MODEL A-25

Semi-automatic; 25 ACP; 7-shot; 2 1/2" barrel; 4 7/8" overall length; weighs 16 oz.; fixed sights; grooved black plastic or smooth simulated pearl stocks; rotary safety blocks rear; bright chrome or black Teflon finish. Introduced 1989; still in production. **LMSR: $80**
  New: $60　　**Perf.:** $50　　**Exc.:** $40

SUNDANCE MODEL BOA

Semi-automatic; 25 ACP; 7-shot; 2 1/2" barrel; 4 7/8" overall length; weighs 16 oz.; fixed sights; grooved ABS or simulated pearl grips; patented grip safety; manual rotary safety; button magazine release; bright chrome or black Teflon finish. Introduced 1991; still in production. **LMSR: $95**
  New: $75　　**Perf.:** $60　　**Exc.:** $50

SUNDANCE POINT BLANK

Derringer; two-shot; 22 LR; 3" barrel; 4 5/8" overall length; weighs 8 oz.; grooved composition grips; blade front sight, fixed notch rear; double action trigger; push-bar safety; automatic chamber selection; fully enclosed hammer; matte black finish. Made in U.S. by Sundance Industries. Introduced 1994; still produced. **LMSR: $99**
  New: $75　　**Perf.:** $60　　**Exc.:** $50

SUPER SIX GOLDEN BISON

Revolver; single action; 45-70; 6-shot; 8", 10 1/2" octagon barrel; 15" overall length (8" barrel); weighs 92 oz.; adjustable Millett rear sight, blaze orange blade on ramp front; walnut grips; manganese bronze used in cylinder frame, grip frame; coil springs; half-cock, cross-bolt traveling safeties; antique brown or blued finish. Marketed in fitted walnut presentation case. Discontinued 1992.
  Perf.: $1500　　**Exc.:** $1300　　**VGood:** $1000

TANARMI MODEL BTA 90

Semi-automatic; double action; 9mm Para.; 12-shot; 3 1/2" barrel; weighs 30 oz.; blade front sight, white outline rear; checkered neoprene grips; blued finish. Imported from Italy by Excam. Introduced 1987; discontinued 1990.
  Perf.: $350　　**Exc.:** $300　　**VGood:** $250
Matte chrome finish
  Perf.: $375　　**Exc.:** $325　　**VGood:** $275

TANARMI MODEL TA 22

Revolver; single action; 22 Short, 22 LR, 22 LR/22 WMR combo; 6-shot; 4 3/4" barrel; 10" overall length; blade front sight, drift-adjustable rear; checkered nylon grips, walnut optional; manual hammer-block safety frame; blued finish. Imported from Italy by Excam. Introduced 1978; discontinued 1991.
  Perf.: $75　　**Exc.:** $65　　**VGood:** $50

TANARMI MODEL TA 38 OVER/UNDER DERRINGER

Single shot; tip-up action; 38 Spl.; 3" barrel; 4 3/4" overall length; weighs 14 oz.; fixed sights; checkered white nylon stocks; blued finish. Assembled in U.S. by Excam. Discontinued 1988.
  Exc.: $75　　**VGood:** $60　　**Good:** $50

TANARMI MODEL TA 41

Semi-automatic; double action; 41 AE; 11-shot; 4 3/4" barrel; combat sights; black neoprene grips; matte blue finish. Imported from Italy by Excam; discontinued 1990.
  Exc.: $400　　**VGood:** $350　　**Good:** $275
Matte chrome finish
  Exc.: $425　　**VGood:** $375　　**Good:** $275

Tanarmi Model TA 41 SS

Same general specs as Model TA 41 except 5" compensated barrel; competition sights; blue/chrome finish. Introduced 1989; discontinued 1990.
  Exc.: $450　　**VGood:** $400　　**Good:** $350

TANARMI MODEL TA 76

Revolver; single action; 22 LR, 22 LR/22 WMR combo; 6-shot; 4 3/4" barrel; 10" overall length; weighs 32 oz.; adjustable rear sight, blade front; uncheckered walnut stocks; color case-hardened frame; brass backstrap, trigger guard; manual hammer-block safety; blued finish. Imported from Italy by Excam. Introduced 1987; discontinued 1991.
  Exc.: $65　　**VGood:** $50　　**Good:** $40

HANDGUNS

Tanarmi
Model TA 90
Model TA 90 SS

Targa
Model GT 22
Model GT 22 Target
Model GT 26
Model GT 27
Model GT 32
Model GT 32 XE
Model GT 380
Model GT 380 XE

Taurus
Model 44
Model 65
Model 66

Tanarmi Model TA 76

Taurus Model 82

Taurus Model 73 Sport

Taurus Model 66

Taurus Model 65

Taurus Model 74 Sport

TANARMI MODEL TA 90

Semi-automatic; double action; 9mm Para.; 15-shot; 4 ³/₄″ barrel; 8 ¹/₄″ overall length; weighs 38 oz.; blade front sight, white-outline rear; checkered neoprene stocks; chromed barrel, trigger; extended slide release lever; blued finish. Imported from Italy by Excam. Introduced 1987; discontinued 1990.

   Perf.: $300   **Exc.:** $250   **VGood:** $200

Tanarmi Model TA 90 SS

Same specs as Model TA 90 except 5″ compensated barrel; competition sights; blue/chrome finish. Imported from Italy by Excam. Introduced 1989; discontinued 1990.

   Perf.: $425   **Exc.:** $375   **VGood:** $325

TARGA MODEL GT 22

Semi-automatic; single action; 22 LR; 10-shot; 3 ⁷/₈″ barrel; weighs 26 oz.; blade ramp front sight, windage-adjustable rear; checkered walnut stocks; finger-rest magazine; steel frame; blued finish. Imported from Italy by Excam. Discontinued 1990.

   Perf.: $150   **Exc.:** $125   **VGood:** $110
Chrome finish
   Perf.: $165   **Exc.:** $135   **VGood:** $115

Targa Model GT 22 Target

Same specs as Model GT 22 except 6″ barrel; 12-shot; weighs 30 oz. Introduced 1988; discontinued 1990.

   Perf.: $175   **Exc.:** $135   **VGood:** $125

TARGA MODEL GT 26

Semi-automatic; single action; 25 ACP; 6-shot; 2 ¹/₂″ barrel; weighs 13 oz.; fixed sights; wooden grips; blued finish. Imported from Italy by Excam. Discontinued 1990.

   Exc.: $50   **VGood:** $40   **Good:** $30
With steel frame (GT 26S)
   Exc.: $85   **VGood:** $65   **Good:** $50

TARGA MODEL GT 27

Semi-automatic; single action; 25 ACP; 6-shot; 2 ¹/₂″ barrel; 4 ⁵/₈″ overall length; fixed sights; checkered nylon stocks; external hammer with half-cock feature; safety lever takedown; blued finish. Imported from Italy by Excam. Introduced 1977; discontinued 1990.

   Exc.: $65   **VGood:** $50   **Good:** $40

TARGA MODEL GT 32

Semi-automatic; single action; 32 ACP; 7-shot; 3 ⁷/₈″ barrel; 7 ³/₈″ overall length; weighs 26 oz.; windage-adjustable rear sight, blade front; optional checkered nylon thumbrest or walnut stocks; external hammer; blued finish. Imported from Italy by Excam. Introduced 1977; discontinued 1990.

   Exc.: $135   **VGood:** $110   **Good:** $100

Targa Model GT 32 XE

Same general specs as Model GT 32 except 12-shot; blue finish. Introduced 1980; discontinued 1985.

   Exc.: $150   **VGood:** $125   **Good:** $110

TARGA MODEL GT 380

Semi-automatic; single action; 380 ACP; 6-shot; 3 ⁷/₈″ barrel; weighs 26 oz.; windage-adjustable rear sight, blade front; optional checkered nylon thumbrest or walnut stocks; blued finish. Introduced 1977; discontinued 1990.
Blue or chrome finish
   Perf.: $150   **Exc.:** $135   **VGood:** $110
Light alloy frame
   Perf.: $125   **Exc.:** $100   **VGood:** $85

Targa Model GT 380 XE

Same specs as Model GT 380 except 12-shot; blue finish. Introduced 1980; discontinued 1990.

   Perf.: $175   **Exc.:** $150   **VGood:** $125

TAURUS MODEL 44

Revolver; double action; 44 Mag.; 6-shot cylinder; 4″, 6¹/₂″, 8 ³/₈″ barrel; weighs 44 ³/₄ oz. (4″ barrel); checkered Brazilian hardwood grips; serrated ramp front sight, micro-click fully-adjustable rear; heavy solid rib (4″), vent rib (6 ¹/₂″, 8 ³/₈″ barrel); compensated barrel; blued model with color case-hardened hammer and trigger. Imported by Taurus International. Introduced 1994; still in production. **LMSR: $439**
Blue finish, 4″ barrel
   New: $375   **Perf.:** $325   **Exc.:** $275
Blue finish, 6 ¹/₂″ or 8 ³/₈″ barrel
   New: $400   **Perf.:** $350   **Exc.:** $325
Stainless steel, 4″ barrel
   New: $450   **Perf.:** $400   **Exc.:** $350
Stainless steel, 6 ¹/₂″ or 8 ³/₈″ barrel
   New: $475   **Perf.:** $425   **Exc.:** $375

TAURUS MODEL 65

Revolver; double action; 357 Mag.; 6-shot; 2 ¹/₂″, 4″ barrel; weighs 34 oz.; fixed sights; round butt grip; checkered wood grips; blue, satin nickel or stainless steel finish. Introduced 1978; still imported. **LMSR: $299**
   New: $225   **Perf.:** $185   **Exc.:** $150
Satin nickel finish
   New: $250   **Perf.:** $200   **Exc.:** $175
Stainless steel
   New: $275   **Perf.:** $235   **Exc.:** $200

TAURUS MODEL 66

Revolver; double action; 357 Mag.; 6-shot; 2 ¹/₂″, 4″, 6″ barrel; weighs 35 oz. (4″ barrel); serrated ramp front sight, micro-click fully-adjustable rear; checkered wood grips; wide target hammer spur; floating firing pin; heavy barrel; shrouded ejector rod; blue, satin nickel or stainless steel finish. Introduced 1978; still imported. **LMSR: $329**
   New: $275   **Perf.:** $225   **Exc.:** $200
Stainless steel
   New: $325   **Perf.:** $275   **Exc.:** $225

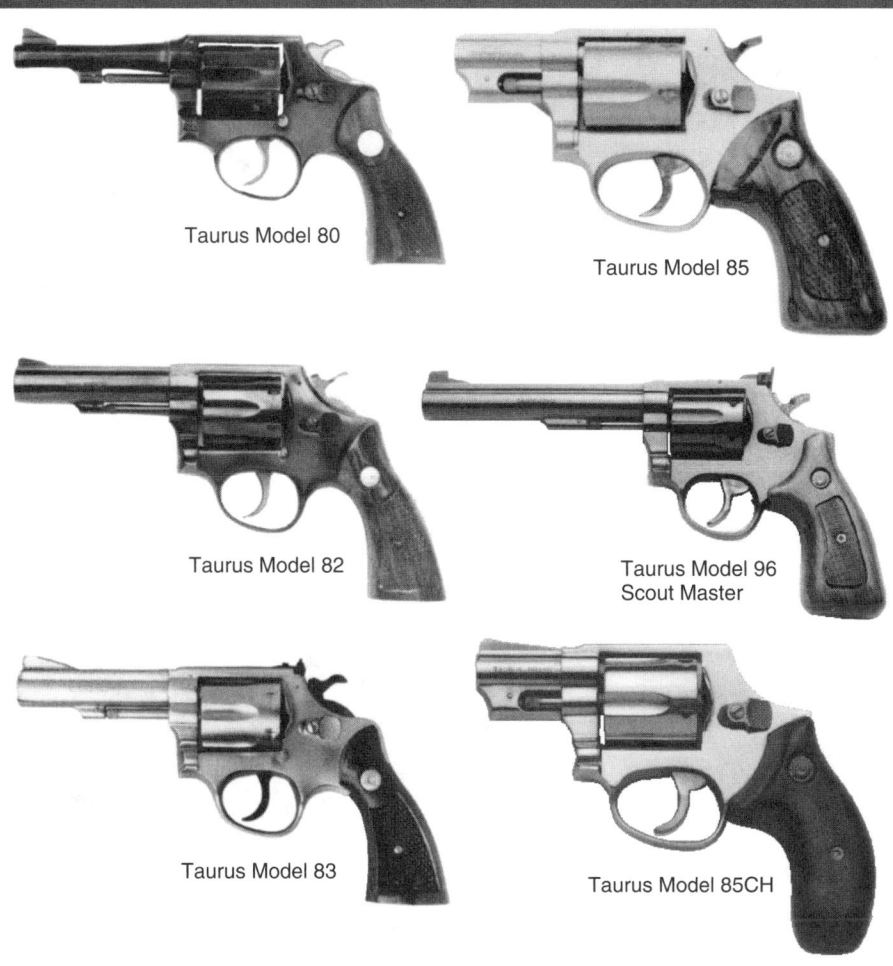

Taurus Model 80

Taurus Model 85

Taurus Model 82

Taurus Model 96
Scout Master

Taurus Model 83

Taurus Model 85CH

HANDGUNS

Taurus
Model 73 Sport
Model 74
Model 76
Model 80
Model 82
Model 83
Model 84
Model 85
Model 85CH
Model 86 Target Master
Model 94
Model 96 Scout Master

TAURUS MODEL 73 SPORT
Revolver; double action; 32 S&W Long; 6-shot; 3″ heavy barrel; 8 1/4″ overall length; weighs 20 oz.; target-type wood grips; ramp front sight, notch rear; blue or satin nickel finish. Introduced 1979; discontinued 1992.
Perf.: $175   **Exc.:** $150   **VGood:** $125
Satin nickel finish
Perf.: $200   **Exc.:** $175   **VGood:** $150

TAURUS MODEL 74 SPORT
Revolver; double action; 32 S&W Long; 6-shot; 3″ barrel; 8 1/4″ overall length; adjustable rear sight, ramp front; hand-checkered walnut stocks; blued finish. Introduced 1971; discontinued 1978.
Exc.: $135   **VGood:** $110   **Good:** $100
Nickel finish
Exc.: $150   **VGood:** $125   **Good:** $110

TAURUS MODEL 76
Revolver; double action; 32 H&R Mag.; 6-shot; 6″ heavy barrel with solid rib; weighs 34 oz.; Patridge-type front sight, fully-adjustable micro-click rear; checkered Brazilian hardwood grips; adjustable trigger; target hammer; blued finish. Introduced 1991; no longer imported.
New: $225   **Perf.:** $185   **Exc.:** $135

TAURUS MODEL 80
Revolver; double action; 38 Spl.; 6-shot; 3″, 4″ barrel; 9 1/4″ overall length; weighs 30 oz.; fixed sights; hand-checkered Brazilian hardwood stocks; blue, satin nickel or stainless steel finish. Introduced 1971; still in production. **LMSR: $260**
New: $200   **Perf.:** $150   **Exc.:** $125
Satin nickel finish
New: $225   **Perf.:** $175   **Exc.:** $135
Stainless steel
New: $250   **Perf.:** $200   **Exc.:** $175

TAURUS MODEL 82
Revolver; double action; 38 Spl.; 6-shot; 3″, 4″ heavy barrel; 9 1/4″ overall length; weighs 34 oz.; fixed sights; checkered Brazilian hardwood grips; blue, satin nickel or stainless steel finish. Introduced 1971; still in production. **LMSR: $260**
New: $200   **Perf.:** $150   **Exc.:** $125
Satin nickel finish
New: $225   **Perf.:** $165   **Exc.:** $135
Stainless steel
New: $225   **Perf.:** $175   **Exc.:** $150

TAURUS MODEL 83
Revolver; double action; 38 Spl.; 6-shot; 4″ heavy barrel; 9 1/2″ overall length; weighs 34 oz.; adjustable rear sight, ramp front; hand-checkered oversize Brazilian hardwood grips; blue or stainless steel finish. Introduced 1977; still in production. **LMSR: $274**
New: $200   **Perf.:** $150   **Exc.:** $125
Stainless steel
New: $225   **Perf.:** $185   **Exc.:** $150

TAURUS MODEL 84
Revolver; double action; 38 Spl.; 6-shot; 4″ barrel; 9 1/2″ overall length; adjustable rear sight, ramp front; checkered wood grips; blued finish. Introduced 1971; discontinued 1978.
Exc.: $135   **VGood:** $115   **Good:** $100

TAURUS MODEL 85
Revolver; double action; 38 Spl.; 5-shot; 2″, 3″ barrel; weighs 21 oz.; ramp front sight, square-notch rear; checkered Brazilian hardwood grips; blued, satin nickel or stainless finish. Introduced 1980; still imported. **LMSR: $284**
New: $225   **Perf.:** $200   **Exc.:** $175
Satin nickel finish
New: $250   **Perf.:** $225   **Exc.:** $200
Stainless steel
New: $275   **Perf.:** $250   **Exc.:** $225

Taurus Model 85CH
Same specs as Model 85 except 2″ barrel; weighs 21 oz.; smooth Brazilian hardwood grips; concealed hammer. Introduced 1992; still in production. **LMSR: $284**
Blue
New: $225   **Perf.:** $200   **Exc.:** $175
Stainless Steel
New: $275   **Perf.:** $250   **Exc.:** $225

TAURUS MODEL 86 TARGET MASTER
Revolver; double action; 38 Spl.; 6-shot cylinder; 6″ barrel; 11 1/4″ overall length; weighs 34 oz.; oversize target-type grips of checkered Brazilian hardwood; Patridge front sight, micro-click fully-adjustable rear; blue finish with non-reflective finish on barrel. Imported from Brazil by Taurus International. No longer imported.
Perf.: $275   **Exc.:** $225   **VGood:** $200

TAURUS MODEL 94
Revolver; double action; 22 LR; 9-shot; 3″, 4″ barrel; weighs 25 oz.; serrated ramp front sight, click-adjustable rear; checkered Brazilian hardwood stocks; color case-hardened hammer, trigger; floating firing pin; blued or stainless finish. Introduced 1988; still imported. **LMSR: $303**
New: $225   **Perf.:** $200   **Exc.:** $165
Stainless steel
New: $250   **Perf.:** $225   **Exc.:** $200

TAURUS MODEL 96 SCOUT MASTER
Revolver; double action; 22 LR; 6-shot; 6″ barrel; weighs 34 oz.; heavy solid barrel rib; adjustable target trigger; Patridge type front, micrometer click-adjustable rear; checkered Brazilian hardwood grips; blued finish. Introduced 1971; still in production. **LMSR: $370**
New: $275   **Perf.:** $225   **Exc.:** $175

HANDGUNS

Taurus
Model 431
Model 441
Model 605
Model 607
Model 669
Model 689
Model 741
Model 761
Model 941

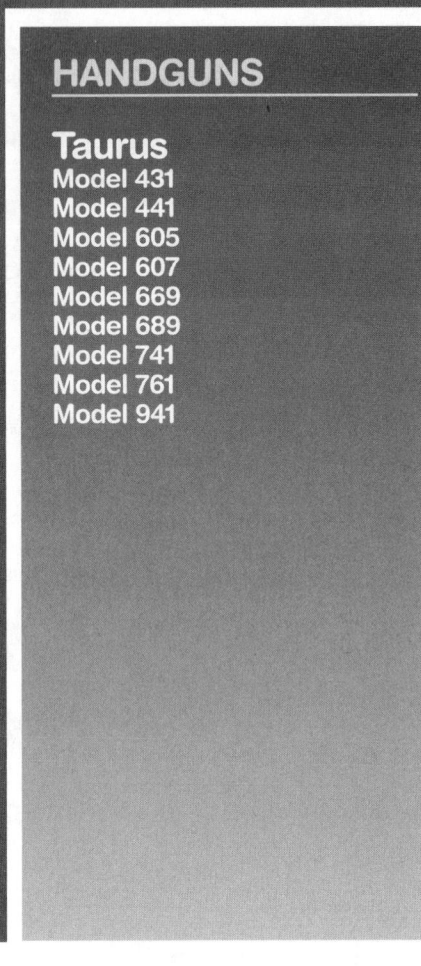

Taurus Model 431

Taurus Model 669

Taurus Model 441

Taurus Model 689

Taurus Model 607

Taurus Model 605

Taurus Model 941

TAURUS MODEL 431
Revolver; double action; 44 Spl.; 5-shot; 3″, 4″, 6″ barrel; weighs 40 1/2 oz. (6″ barrel); checkered Brazilian hardwood grips; serrated ramp front sight, fixed rear; heavy barrel with solid rib and full-length ejector shroud; blue or stainless finish. Imported by Taurus International. Introduced 1992; still imported. **LMSR: $295**
Blue finish
New: $200	**Perf.:** $175	**Exc.:** $135
Stainless steel
New: $275	**Perf.:** $225	**Exc.:** $185

TAURUS MODEL 441
Revolver; double action; 44 Spl.; 5-shot cylinder; 3″, 4″, 6″ barrel; weighs 40 1/2 oz. (6″ barrel); checkered Brazilian hardwood grips; serrated ramp front sight, micrometer click fully-adjustable rear; heavy barrel with solid rib and full-length ejector shroud; blue or stainless finish. Imported by Taurus International. Introduced 1992; still imported. **LMSR: $307**
Blue finish
New: $225	**Perf.:** $185	**Exc.:** $150
Stainless steel
New: $300	**Perf.:** $250	**Exc.:** $200

TAURUS MODEL 605
Revolver; double action; 357 Mag.; 5-shot; 2 1/4″ barrel; weighs 24 1/2 oz.; finger-groove Santoprene I grips; serrated ramp front sight, fixed notch rear; heavy, solid rib barrel; floating firing pin; blue or stainless finish. Imported by Taurus International. Introduced 1995; still imported. **LMSR: $305**
Blue finish
New: $250	**Perf.:** $200	**Exc.:** $175
Stainless steel
New: $275	**Perf.:** $225	**Exc.:** $200

TAURUS MODEL 607
Revolver; double action; 357 Mag.; 7-shot; 4″, 6 1/2″ barrel; weighs 44 oz.; Santoprene I grips with finger grooves; serrated ramp front sight, fully-adjustable rear; ventilated rib with built-in compensator (6 1/2″ barrel). Imported by Taurus International. Introduced 1995; still imported. **LMSR: $439**
Blue finish, 4″ barrel
New: $350	**Perf.:** $300	**Exc.:** $250
Blue finish, 6 1/2″ barrel
New: $350	**Perf.:** $300	**Exc.:** $250
Stainless steel, 4″ barrel
New: $400	**Perf.:** $350	**Exc.:** $300
Stainless steel, 6 1/2″ barrel
New: $425	**Perf.:** $375	**Exc.:** $325

TAURUS MODEL 669
Revolver; double action; 357 Mag.; 6-shot; 4″, 6″ barrel; weighs 37 oz. (4″ barrel); serrated ramp front sight, click-adjustable rear; checkered target-type Brazilian hardwood stocks; floating firing pin; full-length barrel shroud; target-type trigger; blued finish. Introduced 1988; still imported. **LMSR $338**
Blue
New: $250	**Perf.:** $200	**Exc.:** $175
Stainless steel
New: $350	**Perf.:** $300	**Exc.:** $250
With compensated barrel, blue
New: $275	**Perf.:** $225	**Exc.:** $200
With compensated barrel, stainless
New: $375	**Perf.:** $325	**Exc.:** $275

TAURUS MODEL 689
Revolver; double action; 357 Mag.; 6-shot cylinder; 4″, 6″ barrel; weighs 37 oz. (4″ barrel); checkered Brazilian walnut grips; serrated ramp front sight, fully-adjustable rear; wide target-type hammer; floating firing pin; full-length barrel shroud; full-length ventilated barrel rib. Imported by Taurus International. Introduced 1988; still imported. **LMSR: $352**
Blue finish
New: $275	**Perf.:** $225	**Exc.:** $175
Stainless steel
New: $325	**Perf.:** $275	**Exc.:** $200

TAURUS MODEL 741
Revolver; double action; 32 H&R Mag.; 6-shot cylinder; 3″, 4″ heavy barrel; weighs 32 oz. (4″ barrel); checkered Brazilian hardwood grips; serrated ramp; target hammer; adjustable trigger. Imported by Taurus International. Introduced 1991; dropped 1995.
Blue finish
Perf.: $200	**Exc.:** $175	**VGood:** $150
Stainless steel
Perf.: $250	**Exc.:** $225	**VGood:** $200

TAURUS MODEL 761
Revolver; double action; 32 H&R Magnum; 6-shot cylinder; 6″ heavy barrel with solid rib; weighs 34 oz.; checkered Brazilian hardwood grips; Patridge-type front sight, micro-click fully-adjustable rear; target hammer, adjustable target trigger; blue only. Imported by Taurus International. Introduced 1991; dropped 1995.
Perf.: $225	**Exc.:** $200	**VGood:** $175

TAURUS MODEL 941
Revolver; double action; 22 WMR; 8-shot; 3″, 4″ barrel; weighs 27 1/2 oz. (4″ barrel); checkered Brazilian hardwood grips; serrated ramp front sight, fully-adjustable rear; solid rib heavy barrel with full-length ejector rod shroud. Imported by Taurus International. Introduced 1992; still imported. **LMSR: $326**
Blue finish
New: $250	**Perf.:** $225	**Exc.:** $175
Stainless steel
New: $275	**Perf.:** $250	**Exc.:** $200

Taurus Model PT 92AF

Taurus Model PT 22

Taurus Model PT 58

Taurus Model PT 945

Taurus Model PT 101

Taurus Model PT 908

HANDGUNS

Taurus
Model PT 22
Model PT 25
Model PT 58
Model PT 91AF
Model PT 92AF
Model PT 92AFC
Model PT 99AF
Model PT 100
Model PT 101
Model PT 908
Model PT 945

TAURUS MODEL PT 22

Semi-automatic; double action; 22 LR; 9-shot magazine; 2 3/4" barrel; 5 1/4" overall length; weighs 12 1/2 oz.; smooth Brazilian hardwood grips; blade front sight, fixed rear; tip-up barrel for loading, cleaning. Made in U.S. by Taurus International. Introduced 1992; still produced. **LMSR: $193**

Blue finish

New: $165	**Perf.:** $135	**Exc.:** $110

Stainless steel

New: $175	**Perf.:** $150	**Exc.:** $135

TAURUS MODEL PT 25

Semi-automatic; double action; 25 ACP; 8-shot magazine; 2 3/4" barrel; 5 1/4" overall length; weighs 12 1/2 oz.; smooth Brazilian hardwood grips; blade front sight, fixed rear; tip-up barrel for loading, cleaning. Made in U.S. by Taurus International. Introduced 1992; still produced. **LMSR $193**

Blue finish

New: $150	**Perf.:** $125	**Exc.:** $110

Stainless steel

New: $175	**Perf.:** $150	**Exc.:** $135

TAURUS MODEL PT 58

Semi-automatic; double action; 380 ACP; 12-, 10-shot; 4" barrel; weighs 30 oz.; integral blade front sight, notch rear with three-dot system; Brazilian hardwood stocks; exposed hammer; inertia firing pin; blued finish. Introduced 1988; still imported. **LMSR: $462**

New: $325	**Perf.:** $300	**Exc.:** $250

Stainless steel

New: $400	**Perf.:** $350	**Exc.:** $300

TAURUS MODEL PT 91AF

Semi-automatic; double action; 41 AE; 10-shot; 5" barrel; weighs 34 oz.; fixed sights; smooth wood grips; exposed hammer; blued finish. Introduced 1990; discontinued 1991.

Perf.: $350	**Exc.:** $300	**VGood:** $250

Satin nickel finish

Perf.: $375	**Exc.:** $325	**VGood:** $275

TAURUS MODEL PT 92AF

Semi-automatic; double action; 9mm Para.; 15-, 10-shot; 5" barrel; 8 1/2" overall length; weighs 34 oz.; fixed sights; black plastic stocks; exposed hammer; chamber loaded indicator; inertia firing pin; blued, nickel or stainless finish. Introduced 1983; still imported. **LMSR: $511**

Blue, nickel finish

New: $425	**Perf.:** $375	**Exc.:** $325

Stainless steel (PT 92SS)

New: $450	**Perf.:** $425	**Exc.:** $375

Taurus Model PT 92AFC

Same specs as PT 92AF except 4" barrel; 13-, 10-shot; 7 1/2" overall length; weighs 31 oz. Introduced 1991; still imported. **LMSR: $511**

New: $350	**Perf.:** $325	**Exc.:** $275

Stainless steel

New: $400	**Perf.:** $350	**Exc.:** $300

TAURUS MODEL PT 99AF

Semi-automatic; double action; 9mm Para.; 15-, 10-shot; 5" barrel; 8 1/2" overall length; adjustable sights; uncheckered Brazilian walnut stocks; exposed hammer; chamber-loaded indicator; inertia firing pin; blue, satin nickel or stainless steel finish. Introduced 1983; still in production. **LMSR: $554**

Blue

New: $450	**Perf.:** $375	**Exc.:** $300

Stainless steel

New: $475	**Perf.:** $425	**Exc.:** $375

TAURUS MODEL PT 100

Semi-automatic; double action; 40 S&W; 11-, 10-shot; 5" barrel; weighs 34 oz.; fixed sights; three-dot combat system; uncheckered Brazilian hardwood grips; ambidextrous hammer-drop safety; exposed hammer; chamber-loaded indicator; inertia firing pin; blue or stainless steel finish. Introduced 1991; still imported. **LMSR: $522**

Blue

New: $450	**Perf.:** $400	**Exc.:** $325

Stainless steel

New: $475	**Perf.:** $425	**Exc.:** $375

TAURUS MODEL PT 101

Semi-automatic; double action; 40 S&W; 11-, 10-shot; 5" barrel; 34 oz.; adjustable sights; uncheckered Brazilian hardwood grips; ambidextrous hammer-drop safety; exposed hammer; chamber-loaded indicator; inertia firing pin; blue or stainless steel finish. Introduced 1992; still imported. **LMSR: $564**

Blue

New: $450	**Perf.:** $375	**Exc.:** $300

Stainless steel

New: $475	**Perf.:** $425	**Exc.:** $375

TAURUS MODEL PT 908

Semi-automatic; double action; 9mm Para.; 8-shot magazine; 3 13/16" barrel; 7" overall length; weighs 30 oz.; checkered black composition grips; drift-adjustable front and rear sights, three-dot combat; exposed hammer; manual ambidextrous hammer-drop; inertia firing pin; chamber loaded indicator; blue or stainless steel finish. Imported by Taurus International. Introduced 1993; still imported. **LMSR: $511**

Blue

New: $325	**Perf.:** $300	**Exc.:** $250

Stainless steel

New: $375	**Perf.:** $350	**Exc.:** $300

TAURUS MODEL PT 945

Semi-automatic; double action; 45 ACP; 8-shot magazine; 4 1/4" barrel; 7 1/2" overall length; weighs 29 1/2 oz.; Santoprene II grips; drift-adjustable front and rear sights, three-dot system; manual ambidextrous hammer drop safety; intercept notch; firing pin block; chamber loaded indicator; last-shot hold-open; blue or stainless steel finish. Imported by Taurus International. Introduced 1995; still imported. **LMSR: $570**

Blue

New: $450	**Perf.:** $400	**Exc.:** $325

Stainless steel finish

New: $475	**Perf.:** $425	**Exc.:** $375

HANDGUNS

TDE Backup
(See AMT Backup)

Texas Armory
Defender

Texas Longhorn
Grover's Improved No. 5
Jezebel
South Texas Army
Texas Border Special
West Texas Flattop Target

Thomas
45

Thompson/Center
Contender
Contender Bull Barrel Armour Alloy II
Contender Stainless
Contender Vent Rib Barrel
Contender Vent Rib Barrel Armour Alloy II
Contender Hunter
Super Contender Super 14, Super 16
Super Contender Armour Alloy II

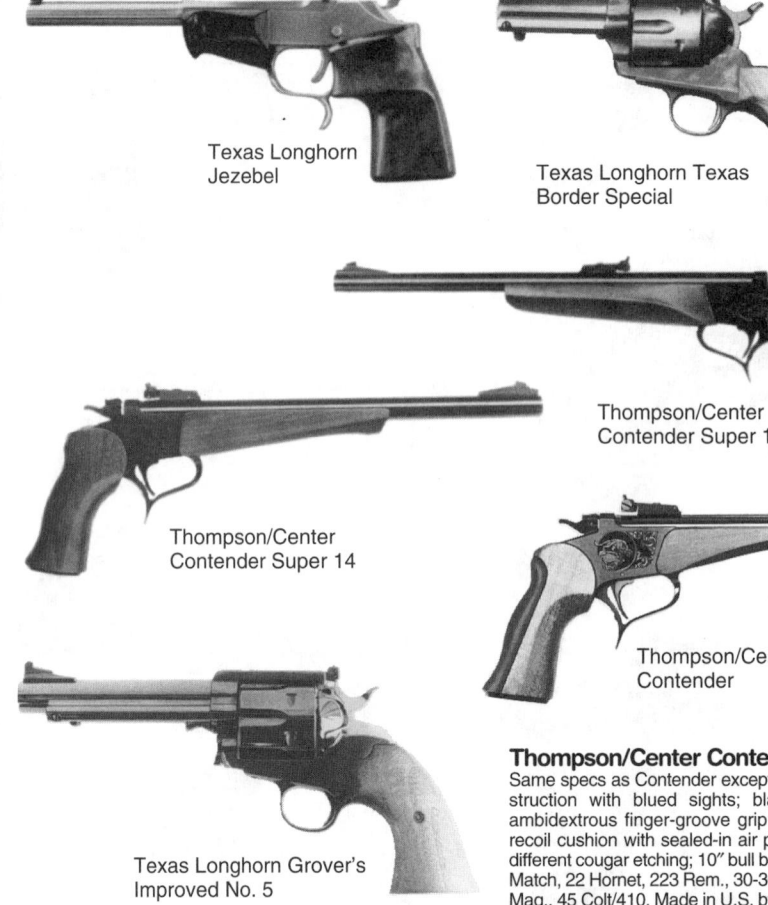

Texas Longhorn Jezebel

Texas Longhorn Texas Border Special

Thompson/Center Contender Super 16

Thompson/Center Contender Super 14

Thompson/Center Contender

Texas Longhorn Grover's Improved No. 5

TEXAS ARMORY DEFENDER
Derringer; 9mm Para., 357 Mag., 44 Mag., 45 ACP, 45 Colt/410; 2-shot; 3" barrel; 5" overall length; weighs 21 oz.; smooth wood grips; blade front sight, fixed rear; interchangeable barrels; retracting firing pins; rebounding hammer; cross-bolt safety; removable trigger guard; automatic extractor; blasted stainless steel finish. Made in U.S. by Texas Armory. Introduced 1993; still produced. **LMSR: $310**

    New: $250   **Perf.:** $200   **Exc.:** $175

TEXAS LONGHORN GROVER'S IMPROVED NO. 5
Revolver; single action; 44 Mag., 45 Colt; 6-shot; 5 1/2" barrel; weighs 44 oz.; square blade ramped front sight, fully-adjustable rear; fancy walnut stocks; double locking bolt; music wire coil springs; Elmer Keith design; polished blue finish. Introduced 1988; only 1200 made. **LMSR: $1195**

    New: $1000   **Perf.:** $850   **Exc.:** $650

TEXAS LONGHORN JEZEBEL
Single shot; top-break action; 22 LR, 22 WMR; 6" half-round, half-octagon; barrel; 8" overall length; weighs 15 oz.; automatic hammer-block safety; fixed rear sight, bead front; one-piece walnut grip; all stainless steel construction. Introduced 1986; discontinued 1992. **LMSR: $250**

    New: $225   **Perf.:** $200   **Exc.:** $165

TEXAS LONGHORN SOUTH TEXAS ARMY
Revolver; single action; 357 Mag., 44 Spl., 45 Colt; 4 3/4" barrel; blade front sight, grooved topstrap rear; one-piece fancy walnut grips; loading gate and ejector on left side of gun; cylinder rotates to left; all steel; color case-hardened frame; music wire coil springs; high-polish blue finish. Only 1000 made. Introduced 1984; still available. **LMSR: $1595**

    New: $1500   **Perf.:** $1250   **Exc.:** $1000

TEXAS LONGHORN TEXAS BORDER SPECIAL
Revolver; single action; 44 Spl., 45 Colt; 3 1/2", 4" barrel; grooved topstrap rear sight, blade front; one-piece fancy walnut bird's-head-style stocks; loading gate, ejector housing on left side; cylinder rotates to left; color case-hardened frame; music wire coil springs; high-polish blue finish. Introduced 1984; 1000 made. **LMSR: $1595**

    New: $1500   **Perf.:** $1250   **Exc.:** $1000

TEXAS LONGHORN WEST TEXAS FLATTOP TARGET
Revolver; single action; 32-20, 357 Mag., 44 Spl./44 Mag., 45 Colt; 7 1/2" barrel; standard rear sight, contoured ramp front; one-piece walnut stock; flat-top style frame; polished blue finish. Introduced 1984; 1000 made. **LMSR: $1595**

    New: $1500   **Perf.:** $1250   **Exc.:** $1000

THOMAS 45
Semi-automatic; double action; 45 ACP; 6-shot magazine; 3 1/2" barrel; 6 1/2" overall length; checkered plastic stocks; windage-adjustable rear sight, blade front; blued finish; matte sighting surface; blowback action. Introduced 1977; dropped 1978. Only about 600 ever made. Collector value.

    Exc.: $450   **VGood:** $350   **Good:** $275

THOMPSON/CENTER CONTENDER
Single shot; break-open action; chambered for 29 calibers; 8 3/4", 10", 14" round, heavy, octagon barrel; adjustable sights; checkered walnut grip, forearm; interchangeable barrels; detachable choke for shot cartridges; blued finish. Introduced 1967; still in production. **LMSR: $450**

    New: $350   **Perf.:** $300   **Exc.:** $275

Thompson/Center Contender Bull Barrel Armour Alloy II
Same specs as Contender except chambered for 7 calibers; 10" round barrel; extra-hard satin finish. Introduced 1986; discontinued 1989.

    Perf.: $300   **Exc.:** $250   **VGood:** $200

Thompson/Center Contender Stainless
Same specs as Contender except stainless steel construction with blued sights; black Rynite forend; ambidextrous finger-groove grip with built-in rubber recoil cushion with sealed-in air pocket; receiver with different cougar etching; 10" bull barrel in 22 LR, 22 LR Match, 22 Hornet, 223 Rem., 30-30 Win., 357 Mag., 44 Mag., 45 Colt/410. Made in U.S. by Thompson/Center. Introduced 1993; still produced. **LMSR: $480**

    New: $375   **Perf.:** $350   **Exc.:** $300

Thompson/Center Contender Vent Rib Barrel
Same specs as Contender except 357 Mag., 44 Mag., 45 Colt/410; 10" vent rib barrel.

    Perf.: $400   **Exc.:** $350   **VGood:** $325

Thompson/Center Contender Vent Rib Barrel Armour Alloy II
Same specs as Contender except 45 Colt/410 only; 10" vent rib barrel; extra hard satin finish. Introduced 1986; no longer in production.

    Perf.: $350   **Exc.:** $300   **VGood:** $275

THOMPSON/CENTER CONTENDER HUNTER
Single shot; break-open action; chambered for 8 calibers; 12", 14", barrel with muzzlebrake; 2 1/2x scope with illuminated reticle; walnut grips with rubber insert; interchangeable barrels; sling with studs and swivels; carrying case; blued finish. Introduced 1990; still in production. **LMSR: $765**

    New: $650   **Perf.:** $575   **Exc.:** $525

THOMPSON/CENTER SUPER CONTENDER SUPER 14, SUPER 16
Single shot; break-open action; chambered for 13 calibers; 14", 16 1/4" bull barrel; adjustable sights; walnut grips; beavertail forearm; interchangeable barrels; detachable choke for shot cartridges; blued finish. Introduced 1978; still in production. **LMSR: $465**

    New: $400   **Perf.:** $350   **Exc.:** $300

45 Colt/410

    New: $425   **Perf.:** $350   **Exc.:** $300

Thompson/Center Super Contender Armour Alloy II
Same specs as Super Contender except chambered for only 5 calibers; extra-hard satin finish. Introduced 1986; discontinued 1989.

    Perf.: $400   **Exc.:** $350   **VGood:** $300

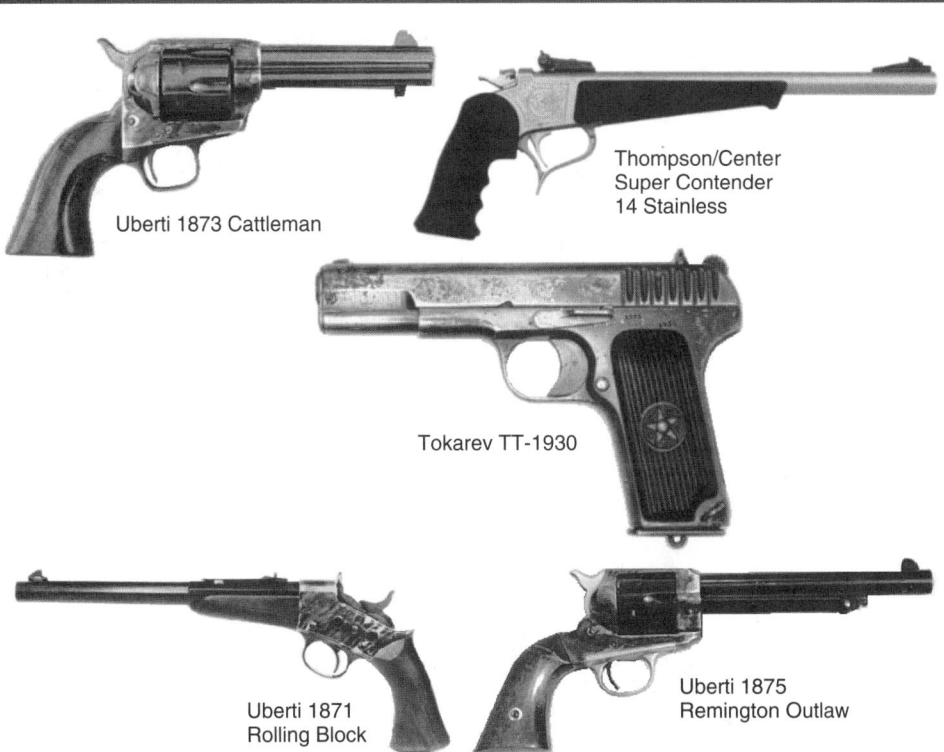

Uberti 1873 Cattleman

Thompson/Center Super Contender 14 Stainless

Tokarev TT-1930

Uberti 1871 Rolling Block

Uberti 1875 Remington Outlaw

HANDGUNS

Thompson/Center
Contender Super 14,
Super 16 Stainless
Super Contender
Vent Rib Barrel

Thunder
Five
Five T-70

TMI
Single Shot

Tokarev
TT-1930
TT-33
Post-War

Uberti
1871 Rolling Block
1873 Cattleman
1873 Cattleman, Sheriff's
Model
1873 Cattleman, Target
1873 Cattleman, Buntline
1873 Colt Stallion
1875 Remington Outlaw
1890 Remington

Thompson/Center Contender Super 14, Super 16 Stainless
Same specs as the standard Super Contender except stainless steel construction; blued sights; black Rynite forend and finger-groove; ambidextrous grip with built-in rubber recoil cushion with sealed-in air pocket; different cougar etching on receiver. Made in U.S. by Thompson/Center. Introduced 1993; still produced.
LMSR: $490
With bull barrel
New: $375 **Perf.:** $325 **Exc.:** $275
45 Colt/410, 14" barrel
New: $425 **Perf.:** $350 **Exc.:** $300

Thompson/Center Super Contender Vent Rib Barrel
Same specs as Super Contender except 45 Colt/410 only; 14", 16 1/4" vent rib barrel. **LMSR: $495**
New: $375 **Perf.:** $325 **Exc.:** $275

THUNDER FIVE
Revolver; double action; 45 Colt/410 shotshell, 2", 3"; 5-shot cylinder; 2" barrel; 9" overall length; weighs 48 oz.; Pachmayr checkered rubber grips; fixed sights; ambidextrous hammer-block safety; squared trigger-guard; internal draw bar safety; made of chrome moly steel; matte blue finish. Made in U.S. Marketed by Holston Ent., Dragun Ent. Introduced 1991; no longer produced.
Perf.: $400 **Exc.:** $350 **VGood:** $300

Thunder Five T-70
Same specs as the Thunder Five except chambered for 45-70 cartridge. Marketed by Dragun Ent.
Perf.: $450 **Exc.:** $400 **VGood:** $350

TMI SINGLE SHOT
Single shot; 22 LR, 223, 7mm TCU, 7mm Int., 30 Herrett, 357 Maximum, 41 Mag., 44 Mag., 454 Casull, 375 Super Mag., others on special order; 10 1/2", 14" barrel; smooth walnut grips with thumb-rest; ramp front sight, open adjustable rear; interchangeable barrels of blue ordnance or bright stainless steel; ventilated barrel shroud; receiver has integral scope mount. From TMI Products. Introduced 1987; no longer produced.
Exc.: $450 **VGood:** $375 **Good:** $275

TOKAREV TT-1930
Semi-automatic; 7.62mm Tokarev (same as 7.63mm Mauser); 8-shot; 4 7/16" barrel; 7 3/4" overall length; grooved plastic grips. TT-30 is rare first model with separate block in rear of frame for hammer mechanism, dated 1933, 1934 or 1935.
Exc.: $600 **VGood:** $400 **Good:** $250

Tokarev TT-33
Same as TT-30 except simplified mechanism, dates 1936-1945.
Exc.: $300 **VGood:** $250 **Good:** $175

Tokarev Post-War
Same as TT-33, except made in various countries including China. Some countries and dates have some collector value, but recent China imports have sold new for under $100.
New: $125 **Perf.:** $100 **Exc.:** $85

UBERTI 1871 ROLLING BLOCK
Single shot; 22 LR, 22 WMR, 22 Hornet, 357 Mag., 45 Colt; 9 1/2" barrel; 14" overall length; weighs 44 oz.; fully-adjustable rear sight, blade front; walnut grip, forend; brass trigger guard; color case-hardened frame; blued barrel. Introduced 1987; still imported. **LMSR: $380**
New: $300 **Perf.:** $250 **Exc.:** $200

UBERTI 1873 CATTLEMAN
Revolver; single action; 22 LR, 22 WMR, 38 Spl., 357 Mag., 38-40, 44 Spl., 44-40, 45 Colt; 6-shot; 4 3/4", 5 1/2", 7 1/2" barrel; 10 3/4" overall length (5 1/2" barrel); weighs 38 oz.; fixed or adjustable sights; wood grips; blued finish. Made in Italy, still imported. **LMSR: $365**
Brass backstrap
New: $325 **Perf.:** $275 **Exc.:** $200
Steel backstrap
New: $350 **Perf.:** $275 **Exc.:** $225

Uberti 1873 Cattleman, Sheriff's Model
Same specs as Cattleman except 44-40, 45 Colt only; 3" barrel; brass backstrap. Made in Italy, no longer imported.
Perf.: $300 **Exc.:** $250 **VGood:** $200

Uberti 1873 Cattleman, Target
Same specs as Cattleman except adjustable rear sight; brass backstrap. Discontinued 1990.
Perf.: $300 **Exc.:** $250 **VGood:** $200
Stainless steel
Perf.: $325 **Exc.:** $275 **VGood:** $225

Uberti 1873 Cattleman, Buntline
Same specs as Cattleman except 357 Mag., 44-40, 45 Colt; 18" barrel; wood grips; steel backstrap cut for shoulder stock. Made in Italy. Importation discontinued 1989; reintroduced 1993; no longer imported.
Perf.: $375 **Exc.:** $325 **VGood:** $275

UBERTI 1873 COLT STALLION
Revolver; single action; 22 LR/22 WMR combo; 4 3/4", 5 1/2", 6 1/2" barrel; 10 3/4" overall length; weighs 36 oz.; grooved or adjustable rear sight, blade front; one-piece uncheckered European walnut stocks; brass triggerguard, backstrap; blued finish. Made in Italy. Introduced 1986; discontinued 1989.
Perf.: $275 **Exc.:** $200 **VGood:** $150
Stainless steel construction
Perf.: $325 **Exc.:** $250 **VGood:** $200

UBERTI 1875 REMINGTON OUTLAW
Revolver; single action; 357 Mag., 44-40, 45 Colt, 45 ACP; 6-shot; 7 1/2" barrel; 13 3/4" overall length; weighs 44 oz; notch rear sight, blade front; uncheckered European walnut stocks; brass triggerguard, color case-hardened frame; blued finish. Replica of 1875 Remington SAA. Made in Italy. Introduced 1987; still imported. **LMSR: $405**
New: $325 **Perf.:** $250 **Exc.:** $200

UBERTI 1890 REMINGTON
Revolver; single action; 357 Mag., 44-40, 45 Colt, 45 ACP; 6-shot; 5 1/2" barrel; 12 1/2" overall length; 37 oz.; notch rear sight, blade front; uncheckered European walnut stocks; brass or steel triggerguard; blued finish. Made in Italy. Introduced 1986; discontinued 1987. **LMSR: $410**
New: $375 **Perf.:** $325 **Exc.:** $275

Uberti
Buckhorn
Inspector
Phantom Silhouette Model

Ultra Light Arms
Model 20 Hunter
Model 20 Reb

Unique
Kreigsmodell
Model 2000-U Match
Model B/Cf
Model D2
Model D6
Model DES/32U
Model DES/69U Match
Model DES/823-U Rapid Fire Match
Model L
Model Mikros Pocket
Model RR

United Sporting Arms
Seville

Uberti Inspector

Unique Model L

Unique Model DES/69U Match

Ultra Light Arms Model 20 Hunter

Voere VEC-RG Repeater

UBERTI BUCKHORN
Revolver; single action; 44-40, 44 Spl., 44 Mag.; 6-shot; 4 3/4″, 5 1/2″, 7 1/2″ barrel; wood grips; brass or steel backstrap; blued finish. Made in Italy. Importation discontinued 1989; reintroduced 1992; still imported. **LMSR: $410**
New: $350   **Perf.:** $275   **Exc.:** $200

UBERTI INSPECTOR
Revolver; double action; 32 S&W Long, 38 Spl.; 6-shot; 3″, 4″, 6″ barrel; 8″ overall length (3″ barrel); weighs 24 oz.; fixed or adjustable rear sight, blade ramp front; checkered walnut stocks; blued finish. Made in Italy. Introduced 1986; discontinued 1990.
Perf.: $350   **Exc.:** $300   **VGood:** $275
Chrome finish
Perf.: $375   **Exc.:** $325   **VGood:** $300

UBERTI PHANTOM SILHOUETTE MODEL
Single shot; 357 Mag., 44 Mag.; 10 1/2″ barrel; adjustable rear sight, blade ramp front; target-style walnut stocks; hooked trigger guard; blue finish. Made in Italy. Introduced 1986; discontinued 1989.
Perf.: $375   **Exc.:** $300   **VGood:** $250

ULTRA LIGHT ARMS MODEL 20 HUNTER
Bolt action; 22-250 through 308 Win., most silhouette calibers on request; 5-shot; 14″ barrel; weighs 64 oz.; Kevlar/graphite reinforced stock in four colors; two-position safety; Timney adjustable trigger; matte or bright stock, metal finish; left- or right-hand action. Introduced 1987; discontinued 1989.
Perf.: $1250   **Exc.:** $1000   **VGood:** $850

Ultra Light Arms Model 20 Reb
Same specs as Model 20 except includes hard case. Introduced 1994; still in production. **LMSR: $1600**
New: $1350   **Perf.:** $1250   **Exc.:** $1000

UNIQUE KREIGSMODELL
Semi-automatic; 32 ACP; 9-shot; 3 7/16″ barrel; 5 13/16″ overall length; fixed sights; grooved plastic stocks; blued finish. Manufactured in France 1940 to 1945 during WWII German occupation.
Exc.: $300   **VGood:** $250   **Good:** $175

UNIQUE MODEL 2000-U MATCH
Semi-automatic; single action; 22 Short; 5-shot; 5 15/16″ barrel; 11 3/8″ overall length; weighs 43 oz.; anatomically shaped, adjustable, stippled French walnut grips; blade front sight, fully-adjustable rear; light alloy frame, steel slide and shock absorber; five barrel vents reduce recoil, three can be blocked; trigger adjustable for position and pull weight; comes with 340-gram weight housing, 160-gram available. Imported from France by Nygord Precision Products. Introduced 1984; dropped 1993.
Perf.: $1250   **Exc.:** $1000   **VGood:** $800

UNIQUE MODEL B/Cf
Semi-automatic; 32 ACP, 380 ACP; 9-shot (32 ACP), 8-shot (380 ACP); 4″ barrel; 6 5/8″ overall length; fixed sights; thumbrest plastic stocks; blued finish. Manufactured in France. Introduced 1954; still in production, but not imported.
Exc.: $200   **VGood:** $175   **Good:** $150

UNIQUE MODEL D2
Semi-automatic; 22 LR; 10-shot; 4 1/2″ barrel; 7 1/2″ overall length; adjustable sights; thumbrest plastic stocks; blued finish. Introduced 1954; still in production, but not imported.
Perf.: $275   **Exc.:** $225   **VGood:** $185

UNIQUE MODEL D6
Semi-automatic; 22 LR; 10-shot; 6″ barrel; 9 1/4″ overall length; adjustable sights; thumbrest plastic stocks; blued finish. Introduced 1954; still in production, but not imported.
Perf.: $275   **Exc.:** $225   **VGood:** $185

UNIQUE MODEL DES/32U
Semi-automatic; single action; 32 S&W Long wadcutter; 5-, 6-shot; 5 15/16″ barrel; weighs 40 oz.; blade front sight, micrometer click rear; adjustable, stippled French walnut stocks; position-, weight-adjustable trigger; slide-stop catch; dry firing mechanism; optional sleeve weights of 120, 220, 320 grams. Made in France. Introduced 1990; still imported. **LMSR: $1350**
New: $1100   **Perf.:** $900   **Exc.:** $750

UNIQUE MODEL DES/69U MATCH
Semi-automatic; 22 LR; 5-shot; 5 7/8″ barrel; 10 1/2″ overall length; click-adjustable rear sight mounted on frame, ramp front; checkered walnut thumbrest stocks; adjustable handrest; blued finish. Introduced 1969; still in production. **LMSR: $1250**
New: $1000   **Perf.:** $850   **Exc.:** $650

UNIQUE MODEL DES/823-U RAPID FIRE MATCH
Semi-automatic; 22 Short; 5-shot; 5 7/8″ barrel; 10 7/16″ overall length; adjustable rear sight, blade front; hand-checkered walnut thumbrest stocks; adjustable handrest; adjustable trigger; blued finish. Introduced 1974; importation dropped 1988.
Perf.: $1000   **Exc.:** $850   **VGood:** $700

UNIQUE MODEL L
Semi-automatic; 22 LR, 32 ACP, 380 ACP; 10-shot (22 LR), 7-shot (32 ACP), 6-shot (380 ACP); 3 5/16″ barrel; 5 13/16″ overall length; fixed sights; checkered plastic stocks; steel or alloy frame; blued finish. Introduced 1955; still in production, but not imported.
Exc.: $175   **VGood:** $150   **Good:** $135

UNIQUE MODEL MIKROS POCKET
Semi-automatic; 22 Short, 22 LR; 6-shot; 2 1/4″ barrel; 4 7/16″ overall length; fixed sights; checkered plastic stocks; alloy or steel frame; blued finish. Introduced 1957; still in production, but not imported.
Exc.: $175   **VGood:** $150   **Good:** $125

UNIQUE MODEL RR
Semi-automatic; 32 ACP; 9-shot; 3 7/16″ barrel; 5 13/16″ overall length; fixed sights; grooved plastic stock; commercial version of Kriegsmodell with improved blued finish. No longer in production.
Exc.: $175   **VGood:** $150   **Good:** $125

UNITED SPORTING ARMS SEVILLE
Revolver; single action; 357 Mag., 41 Mag., 44 Mag., 45 Colt; 6-shot; 4 5/8″, 5 1/2″, 6 1/2″, 7 1/2″ barrel; ramp front sight with red insert, fully-adjustable rear; smooth walnut, Pachmayr or thumbrest stocks; blued finish. Introduced 1981; discontinued 1986.
Exc.: $375   **VGood:** $325   **Good:** $275

Unique Model B/Cf

Walther Model 3

Walther Model 2

Walther Model 6

HANDGUNS

United Sporting Arms
Seville Sheriff's Model
Seville Silhouette

United States Arms
Abilene (See Mossberg Abilene)

Vega
Stainless 45

Voere
VEC-95CG
VEC-RG Repeater

WAC
Model A (See MAB Model A)
Model D (See Model D)
Model E (See Model E)

Walther
Free Pistol
GSP Match
GSP-C Match
Model 1
Model 2
Model 3
Model 4
Model 5
Model 6

Walther Model 4

UNITED SPORTING ARMS SEVILLE SHERIFF'S MODEL
Revolver; single action; 357 Mag., 38 Spl., 44 Spl., 44 Mag., 45 Colt; 6-shot; 3 1/2" barrel; ramp or blade front sight, adjustable or fixed rear; square-butt or bird's-head smooth walnut grips; blued finish. Made in U.S. Introduced 1983; discontinued 1986.
   Exc.: $375   VGood: $325   Good: $275

UNITED SPORTING ARMS SEVILLE SILHOUETTE
Revolver; single action; 357 Mag., 41 Mag., 44 Mag.; 6-shot; 10 1/2" barrel; weighs 55 oz.; undercut Patridge-style front sight, adjustable rear; smooth walnut thumbrest or Pachmayr stocks; stainless steel frame with blued barrel.
   Exc.: $450   VGood: $400   Good: $325

VEGA STAINLESS 45
Semi-automatic; 45 ACP; 7-shot; 5" barrel; 8 3/8" overall length; stainless steel construction; almost exact copy of 1911-A1 Colt; fixed sights or adjustable sights; polished slide, frame flats, balance sand-blasted. Introduced 1980 by Pacific International Merchandising; dropped about 1984.
   Exc.: $325   VGood: $275   Good: $225

VOERE VEC-95CG
Bolt action; single shot; 5.56mm, 6mm UCC caseless; 12", 14" barrel; weighs 3 lbs.; black synthetic, center-grip stock; no sights furnished; fires caseless ammunition via electronic ignition; two batteries in the grip last about 500 shots; two forward locking lugs; tang safety; drilled and tapped for scope mounting. Imported from Austria by Jager-Sport, Ltd. Introduced 1995; still imported. **LMSR: $1495**
   New: $2000   Perf.: $1700   Exc.: $1350

VOERE VEC-RG REPEATER
Bolt action; 5.56mm, 6mm UCC caseless; 5-shot; 12", 14" barrel; weighs 3 lbs.; black synthetic, rear-grip stock; no sights furnished; fires caseless ammunition via electronic ignition; two batteries in the grip last about 500 shots; two forward locking lugs; tang safety; drilled and tapped for scope mounting. Imported from Austria by JagerSport, Ltd. Introduced 1995; still imported. **LMSR: $1495**
   New: $2000   Perf.: $1700   Exc.: $1350

WALTHER FREE PISTOL
Single shot; 22 LR; 11 11/16" barrel; 17 3/16" overall length; weighs 48 oz.; fully-adjustable match sights; hand-fitting walnut stocks; electronic trigger; matte blue finish. Made in Germany. Imported by Interarms; importation dropped 1991.
   Perf.: $1400   Exc.: $1000   VGood: $850

WALTHER GSP MATCH
Semi-automatic; single action; 22 LR; 5-shot; 5 3/4" barrel; 11 13/16" overall length; weighs 45 oz.; walnut, special hand-fitting design grip; fixed front sight, fully-adjustable rear; available with either 2.2-lb. (1000 gm) or 3-lb. (1360 gm) trigger; spare mag., barrel weight, tools supplied in Match Pistol Kit. Imported from Germany by Interarms. Still imported. **LMSR: $1495**
   New: $1250   Perf.: $1000   Exc.: $850

Walther GSP-C Match
Same specs as GSP except 32 S&W Long wadcutter cartridge; weighs 49 1/2 oz.; comes with case. Imported from Germany by Interarms. Still imported. **LMSR: $1595**
   New: $1500   Perf.: $1350   Exc.: $1000

WALTHER MODEL 1
Semi-automatic; 25 ACP; 6-shot magazine; 2" barrel; 4 7/16" overall length; checkered hard rubber stocks; fixed sights; blued finish. Manufactured in Germany 1908 to 1918; collector value.
   Exc.: $550   VGood: $375   Good: $300

WALTHER MODEL 2
Semi-automatic; 25 ACP; 6-shot magazine; 2" barrel; 4 7/16" overall length; checkered hard rubber stocks; fixed sights; blued finish. Same general internal design as Model 1 but has a knurled takedown nut at the muzzle. Manufactured 1909 to 1918; collector value.
   Exc.: $350   VGood: $300   Good: $250

WALTHER MODEL 3
Semi-automatic; 32 ACP; 8-shot magazine; 2 5/8" barrel; 5" overall length; checkered hard rubber stocks; fixed sights; blued finish. Manufactured 1909 to 1918. Rare; collector value.
   Exc.: $950   VGood: $750   Good: $550

WALTHER MODEL 4
Semi-automatic; 32 ACP; 8-shot magazine; 3 1/2" barrel; 5 7/8" overall length; checkered hard rubber stocks; fixed sights; blued finish. Manufactured 1910 to 1918.
   Exc.: $250   VGood: $150   Good: $100

WALTHER MODEL 5
Semi-automatic; 25 ACP; 6-shot magazine; 2" barrel; 4 7/16" overall length; checkered hard rubber stocks; fixed sights; blued finish. Same specs as Model 2 except better workmanship; improved finish. Manufactured 1913 to 1918; collector value.
   Exc.: $350   VGood: $250   Good: $200

WALTHER MODEL 6
Semi-automatic; 9mm Para.; 8-shot magazine; 4 3/4" barrel; 8 1/4" overall length; checkered hard rubber stocks; fixed sights; blued finish. Manufactured 1915 to 1917. Rare.
   Exc.: $4500   VGood: $3500   Good: $2750

Walther
Model 7
Model 8
Model 8 Lightweight
Model 9
Model HP
Model PP
Model PP Lightweight
Model PP Mark II
Model PP Post-WWII
Model PPK
Model PPK American
Model PPK Lightweight
Model PPK Mark II

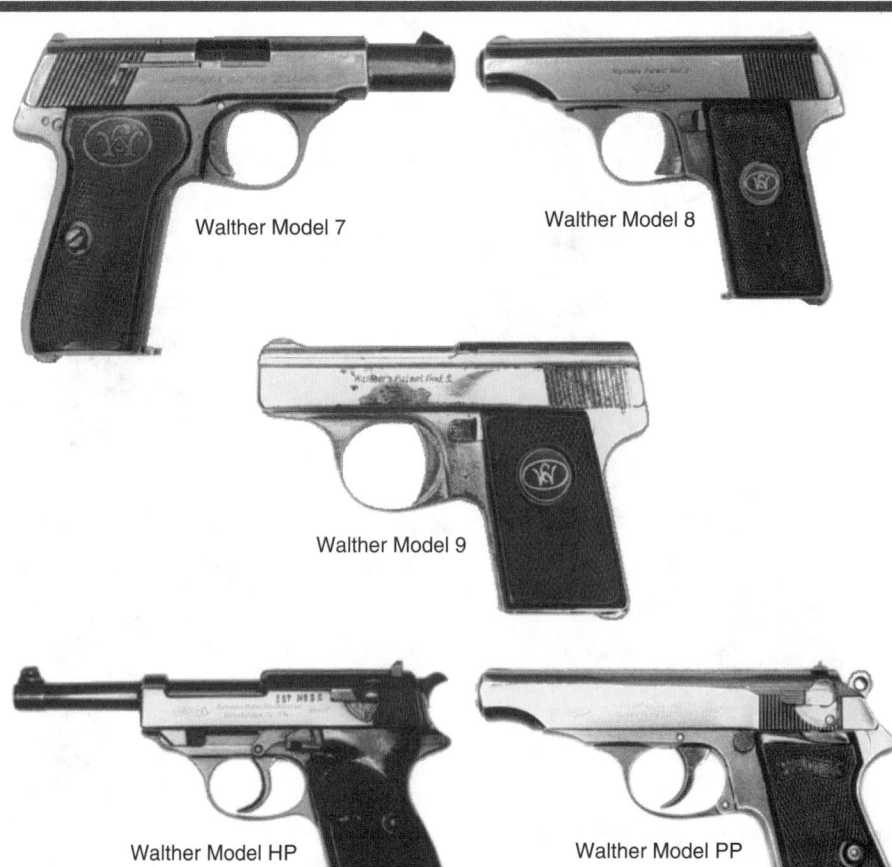

Walther Model 7

Walther Model 8

Walther Model 9

Walther Model HP

Walther Model PP

WALTHER MODEL 7
Semi-automatic; 25 ACP; 8-shot magazine; 3" barrel; 5 5/16" overall length; checkered hard rubber stocks; fixed sights; blued finish. Manufactured 1917 to 1918.
Exc.: $500   **VGood:** $425   **Good:** $350

WALTHER MODEL 8
Semi-automatic; 25 ACP; 8-shot magazine; 2 7/8" barrel; 5 1/8" overall length; fixed sights; checkered plastic grips; blued. Manufactured by Waffenfabrik Walther, Zella-Mehlis, Germany. Introduced 1920; dropped 1945.
Exc.: $400   **VGood:** $325   **Good:** $275

Walther Model 8 Lightweight
Same specs as Model 8, except aluminum alloy. Introduced 1927; dropped about 1935.
Exc.: $1000   **VGood:** $750   **Good:** $600

WALTHER MODEL 9
Semi-automatic; 25 ACP; 6-shot magazine; 2" barrel; 3 15/16" overall length; checkered plastic grips; fixed sights; blued. Introduced 1921; dropped 1945.
Exc.: $450   **VGood:** $375   **Good:** $325

WALTHER MODEL HP
Semi-automatic; 9mm, 7.65mm Para. (very rare); 10-shot magazine; 5" barrel; 8 3/8" overall length; checkered (early) or grooved plastic or checkered wood (rare) grips; fixed sights; blued. Early examples have rectangular firing pin. Introduced 1939; dropped 1945.
Rectangular firing pin
Exc.: $2250   **VGood:** $1850   **Good:** $1200
Round firing pin
Exc.: $1500   **VGood:** $950   **Good:** $700

WALTHER MODEL PP
Semi-automatic; 22 LR, 25 ACP, 32 ACP, 380 ACP; 8-shot magazine; 3 7/8" barrel; 6 5/16" overall length; designed as law-enforcement model; fixed sights; checkered plastic grips; blued. WWII production has less value because of poorer workmanship. Introduced 1929; dropped 1945. RZM, PDM and other special markings bring a premium.
Wartime models
Exc.: $350   **VGood:** $300   **Good:** $250
32 caliber
Exc.: $400   **VGood:** $300   **Good:** $225
380 caliber
Exc.: $1000   **VGood:** $750   **Good:** $500
22 caliber
Exc.: $850   **VGood:** $650   **Good:** $450
25 caliber
Exc.: $3000   **VGood:** $2500   **Good:** $2000

Walther Model PP Lightweight
Same specs as Model PP except aluminum alloy frame. Introduced 1929; dropped 1945.
32 caliber
Exc.: $750   **VGood:** $350   **Good:** $275
380 caliber
Exc.: $1200   **VGood:** $950   **Good:** $700

Walther Model PP Mark II
Same specs as pre-WWII model. Currently manufactured in France by Manufacture De Machines Du Haut-Rhin. Introduced 1953; still in production, but not imported.
Exc.: $250   **VGood:** $200   **Good:** $175

Walther Model PP Post-WWII
Same specs as pre-war except not made in 25-caliber. Currently manufactured by Carl Walther Waffenfabrik, Ulm/Donau, West Germany. Still in production. Imported by Interarms. **LMSR: $1206**
Exc.: $300   **VGood:** $250   **Good:** $200

WALTHER MODEL PPK
Semi-automatic; 22 LR, 25 ACP, 32 ACP, 380 ACP; 7-shot magazine; 3 1/4" barrel; 5 7/8" overall length; the Kurz (short) version of the PP; checkered plastic grips; fixed sights; blued finish. WWII production has less value due to poorer workmanship. Introduced 1931; dropped 1945. RZM, PDM and other special markings bring a premium.
Wartime models
Exc.: $350   **VGood:** $300   **Good:** $250
32 caliber
Exc.: $450   **VGood:** $375   **Good:** $325
380 caliber
Exc.: $1250   **VGood:** $950   **Good:** $650
22 caliber
Exc.: $900   **VGood:** $700   **Good:** $500
25 caliber
Exc.: $4500   **VGood:** $4000   **Good:** $3500

Walther Model PPK American
Same specs as PPK except 6-shot magazine; 3 7/8" barrel; 6 5/16" overall length; 380 ACP only; blued or stainless finish. Made in U.S. Introduced 1986; still marketed by Interarms. **LMSR: $651**
New: $575   **Perf.:** $500   **Exc.:** $400

Walther Model PPK Lightweight
Same specs as standard model except for incorporation of aluminum alloys. Introduced 1933; dropped 1945.
32 caliber
Exc.: $650   **VGood:** $550   **Good:** $450
380 caliber
Exc.: $2000   **VGood:** $1450   **Good:** $1000
22 caliber
Exc.: $1250   **VGood:** $950   **Good:** $800

Walther Model PPK Mark II
Same specs as pre-WWII PPK. Currently manufactured in France by Manufacture De Machines Du Haut-Rhin. Introduced 1953; no longer in production.
Exc.: $300   **VGood:** $250   **Good:** $200

Walther P-38
Military Model

Walther P-38 (banner,
round firing pin) Zero Series

Walther Olympia
Sport Model

Walther P-38
(480 code marking)

Walther Model TPH

Walther Model PPK

Walther
Model PPK Mark II
Lightweight
Model PPK Post-WWII
Model PPK/S (German)
Model PPK/S (American)
Model TPH
Model TP
Olympia Sport Model
Olympia Funkampf Model
Olympia Hunting Model
Olympia Rapid Fire Model
OSP Rapid-Fire
P-5
P-38 Military Model
P-38 Post-War
P-38IV
P-38K

Walther Model PPK Mark II Lightweight
Same specs as standard model except receiver of Dural; 22 LR, 32 ACP only. Introduced 1953; still in production, but not imported.
Exc.: $350   VGood: $300   Good: $250

Walther Model PPK Post-WWII
Same specs as pre-war model except steel or aluminum alloy construction. Currently manufactured in Germany by Carl Walther Waffenfabrik. Still in production, although U.S. importation was dropped in 1968.
32 ACP
Exc.: $350   VGood: $300   Good: $250
22 LR
Exc.: $400   VGood: $350   Good: $275
380 ACP
Exc.: $500   VGood: $400   Good: $300

WALTHER MODEL PPK/S (GERMAN)
Semi-automatic; 22 LR, 32 ACP, 380 ACP; 9-shot magazine; 3 1/4" barrel; 6 1/8" overall length; fixed sights; checkered plastic grips; blued. Same general specs as Walther PP except grip frame 1/2-inch longer and one shot larger magazine capacity to meet GCA '68 factoring system. Made in Germany. Introduced 1968; dropped 1982.
22 LR
Exc.: $450   VGood: $400   Good: $325
32 ACP, 380 ACP
Exc.: $350   VGood: $300   Good: $250

Walther Model PPK/S (American)
Same specs as German-made PPK/S except made entirely in U.S. by Interarms. Made only in 380 ACP in blue ordnance steel or stainless. Introduced 1980; still in production. **LMSR: $651**
New: $550   Perf.: $500   Exc.: $400

WALTHER MODEL TPH
Semi-automatic; 22 LR, 25 ACP; 6-shot magazine; 2 1/4" barrel; 5 3/8" overall length; weighs 14 oz.; drift-adjustable rear sight, blade front; constructed of stainless steel; scaled-down version of PP/PPK series. Made in U.S. by Interarms. Introduced 1987; still in production. **LMSR: $486**
New: $600   Perf.: $525   Exc.: $450

WALTHER MODEL TP
Semi-automatic; 22 LR, 25 ACP; 6-shot magazine; 2 1/2" barrel; 5 1/8" overall length; checkered plastic grips; fixed front sight only; blued. Updated version of Model 9 with safety lever in center of slide. Introduced 1962; dropped 1968.
Exc.: $600   VGood: $400   Good: $300

WALTHER OLYMPIA SPORT MODEL
Semi-automatic; 22 LR; 7 3/8" barrel; 10 11/16" overall length; checkered wood grips; adjustable target sights; blued. Available with set of four detachable weights. Introduced 1936; dropped during WWII.
Exc.: $1000   VGood: $850   Good: $700

Walther Olympia Funkampf Model
Same specs as Sport model except 9 5/8" barrel; 13" overall length; set of four detachable weights. Introduced 1937; dropped during WWII.
Exc.: $1350   VGood: $1000   Good: $800

Walther Olympia Hunting Model
Same specs as Sport Model except 4" barrel. Introduced 1936; dropped during WWII.
Exc.: $950   VGood: $750   Good: $600

Walther Olympia Rapid Fire Model
Same specs as Sport model except 22 Short; detachable muzzle weight. Introduced 1936; dropped during WWII.
Exc.: $1250   VGood: $950   Good: $700

WALTHER OSP RAPID-FIRE
Semi-automatic; single action; 22 Short; 5-shot; 5 3/4" barrel; 11 13/16" overall length; weighs 45 oz.; walnut, special adjustable free-style grip with hand rest; fixed front sight, fully-adjustable rear; available with either 2.2-lb. (1000 gm) or 3-lb. (1360 gm) trigger; spare mag., barrel weight, tools supplied in Match Pistol Kit. Imported from Germany by Interarms. Importation dropped 1992.
Perf.: $1350   Exc.: $1000   VGood: $800

WALTHER P-5
Semi-automatic; 9mm Para.; 3 1/2" barrel; 7" overall length; same basic double-action mechanism as P-38 but differs externally. Introduced 1978; still imported from Germany by Interarms. **LMSR: $1096**
New: $700   Perf.: $600   Exc.: $500

WALTHER MODEL P-38
Semi-automatic; 9mm Para.; 10-shot magazine; 5" barrel; 8 3/8" overall length; checkered (early) or grooved plastic grips; fixed sights; blued. Walther Model HP as adopted by German military in 1939 and produced by various makers (identified by code markings on slide) through 1945. Early Walther ("ac code") manufacture with high polish finish brings considerable premium from collectors; first few thousand had rectangular firing pin.
Walther banner, rectangular firing pin
Exc.: $4500   VGood: $3500   Good: $2200
Walther banner marked, round firing pin
Exc.: $1750   VGood: $1250   Good: $800
480 code marking
Exc.: $3500   VGood: $1500   Good: $1000
ac Code, no date, with high polish finish
Exc.: $3000   VGood: $2000   Good: $600
ac Code, dated 40, and 41, high polish finish
Exc.: $1250   VGood: $750   Good: $400
Letter codes, dated 42 to 45
Exc.: $450   VGood: $350   Good: $250

Walther P-38 Post-War
Same specs as WWII military model except improved workmanship; alloy frame. Currently manufactured by Carl Walther Waffenfabrik in Germany. Still in production. Imported by Interarms. **LMSR: $824**
New: $700   Perf.: $600   Exc.: $450

Walther P-38IV
Same specs as P-38 post-war except 4 1/2" barrel; 8" overall length. Introduced 1977; imported by Interarms.
Exc.: $475   VGood: $375   Good: $275

Walther P-38K
Same specs as P-38 except streamlined version of original with 2 1/4" barrel; 6 1/2" overall length; strengthened slide; no dust cover; windage-adjustable rear sight; hammer decocking lever; non-reflective matte finish. Imported from Germany by Interarms 1976; dropped 1980.
Exc.: $800   VGood: $650   Good: $450

Walther
P-88 Compact
PP Super
Sport Model 1926

Warner
Infallible

Weatherby
Mark IV Silhouette

Webley
Fosbery Automatic Revolver
Government Model
Mark I, II, III, IV, V Revolvers
Mark III Police Model
Mark III Police Model Target
Mark IV Police Model
Mark IV War Model
Mark IV Target Model

Webley Government Model

Walther Model P-88 Compact

Walther Sport Model 1926

Warner Infallible

Webley-Fosbery Automatic Revolver

WALTHER P-88 COMPACT
Semi-automatic; double action; 9mm Para., 14-shot; 4" barrel; 7 3/8" overall length; weighs 31 1/2 oz.; checkered black composition grips; blade front sight, fully-adjustable rear; ambidextrous decocking lever and magazine release; alloy frame; loaded chamber indicator; matte finish. Imported from Germany by Interarms. Introduced 1987; importation dropped 1994.
> **Perf.:** $800   **Exc.:** $700   **VGood:** $550

WALTHER PP SUPER
Semi-automatic; 9mm Ultra, 380 ACP; 7-shot magazine; 3 5/8" barrel; 6 15/16" overall length; weighs 29 oz. loaded; utilizes P-38 mechanism; introduced for West German police in 1972 in 9mm Ultra. Several hundred 9mm Ultra models were imported in used condition, plus 1000 new guns in 380 ACP in 1984-85 by Interarms. Some collector value.
9mm Ultra
> **Exc.:** $500   **VGood:** $400   **Good:** $300
380 ACP
> **Exc.:** $600   **VGood:** $500   **Good:** $400

WALTHER SPORT MODEL 1926
Semi-automatic; 22 LR; 6", 9" barrels; 9 7/8" overall length (6" barrel); checkered one-piece walnut grip; adjustable target sights; blued. Introduced 1926; dropped 1936.
> **Exc.:** $800   **VGood:** $650   **Good:** $500

WARNER INFALLIBLE
Semi-automatic; 32 ACP; 7-shot; 3" barrel; 6 1/2" overall length; checkered hard rubber stocks; fixed sights; blued finish. Manufactured 1917 to 1919.
> **Exc.:** $225   **VGood:** $175   **Good:** $150

WEATHERBY MARK IV SILHOUETTE
Single shot; bolt action; 22, 250, 308; 15" barrel; globe front sight with inserts, target-type rear peep; thumbhole Claro walnut stock; rosewood forend tip with grip cap; modified Mark V Varmintmaster action; drilled and tapped for scope. Introduced 1980; dropped 1981.
> **Exc.:** $3250   **VGood:** $2500   **Good:** $2000

WEBLEY
In addition to the model markings found in the illutration descriptions, Webley's also have the following markings: Webley "Flying Bullet" logo; crown and scepter proofs; BP (blackpowder) or NP (nitro-Cordite) proofs. All models officially accepted by the military are marked with a broad arrow and/or W.D. (War Department) proofs. All models are also marked with one or more of the following: "Webley Patent", "P. Webley & Son", "London & Birmingham", and "Webley & Scott."

WEBLEY-FOSBERY AUTOMATIC REVOLVER
Revolver; 455 Webley, 38 Colt Auto; 6-shot cylinder (455), 8-shot (38 Colt Auto); hinged frame; 6" barrel; 12" overall length; hand-checkered walnut stocks; fixed sights; blued finish. Recoil revolves cylinder, cocking hammer, leading to automatic terminology. Flat-top model (illustrated) was submitted for U.S. Army trials. Manufactured 1901 to 1924. Both calibers have collector value.
455 Webley
> **Exc.:** $5000   **VGood:** $4000   **Good:** $3200
38 Colt Auto
> **Exc.:** $7000   **VGood:** $5500   **Good:** $4500

WEBLEY GOVERNMENT MODEL
Revolver; 450 to 476; 6-shot; 4" to 7 1/2" barrels; hinged frame; Vulcanite or wood grips; bird's-head or square-butt grip configuration; fixed or target sights; bright blue or nickel finish. Standard Webley markings: "W.G. Army Mod.", "Webley W.G.", or "Webley W.G. Target Model." Manufactured from 1886 to 1915.
> **Exc.:** $500   **VGood:** $350   **Good:** $300

WEBLEY MARK I, II, III, IV, V REVOLVERS
Revolver; 455 Webley; 6-shot; 4" barrel; hinged frame; Vulcanite grips; brush blue finish. Standard proofs with Webley MKI through Webley MKV marks. Some models have commercial counterparts available in 450 through 476 calibers. Manufactured from 1887 to 1915.
> **Exc.:** $350   **VGood:** $275   **Good:** $225

WEBLEY MARK III POLICE MODEL
Revolver; double action; 32 S&W, 320 S&W, 38 S&W; 6-shot cylinder; 3", 4" barrel; 9 1/2" overall length; hinged frame; fixed sights; checkered Vulcanite or walnut grips; blued finish. Introduced 1897; dropped 1945.
> **Exc.:** $300   **VGood:** $250   **Good:** $200

Webley Mark III Police Model Target
Revolver; double action; 38 S&W; 6-shot cylinder; 3", 4", 6", 10" barrel; 9 1/2" overall length; hinged frame; adjustable sights; checkered Vulcanite or walnut grips; blued finish. Introduced 1897; dropped 1945.
> **Exc.:** $325   **VGood:** $250   **Good:** $200

WEBLEY MARK IV POLICE MODEL
Revolver; double action; hinged frame; 38 S&W; 6-shot cylinder; 3", 4", 5", 6" barrel; 9 1/8" overall length (5" barrel); fixed or target sights; checkered plastic grips; lanyard ring; blued finish. Introduced 1929; dropped 1968.
> **Exc.:** $250   **VGood:** $225   **Good:** $175

Webley Mark IV War Model
Same specs as the Police Model. Built during WWII, usually has poor-fitting, blue-over-unpolished surfaces. To protect the corporate reputation, most were stamped, "War Finish."
> **Exc.:** $175   **VGood:** $135   **Good:** $110

Webley Mark IV Target Model
Same specs as Police Model except 22 LR; adjustable rear sight. Built in small quantities; virtually a custom-produced handgun. Introduced 1931; dropped 1968.
> **Exc.:** $350   **VGood:** $275   **Good:** $225

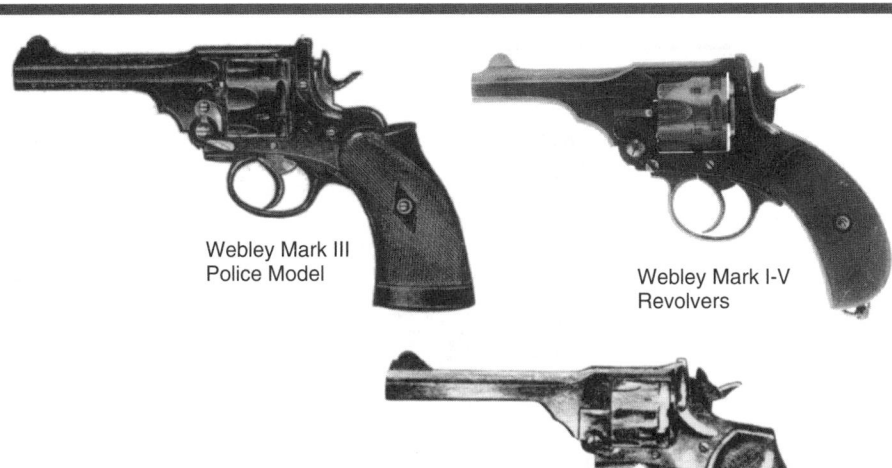

Webley Mark III
Police Model

Webley Mark I-V
Revolvers

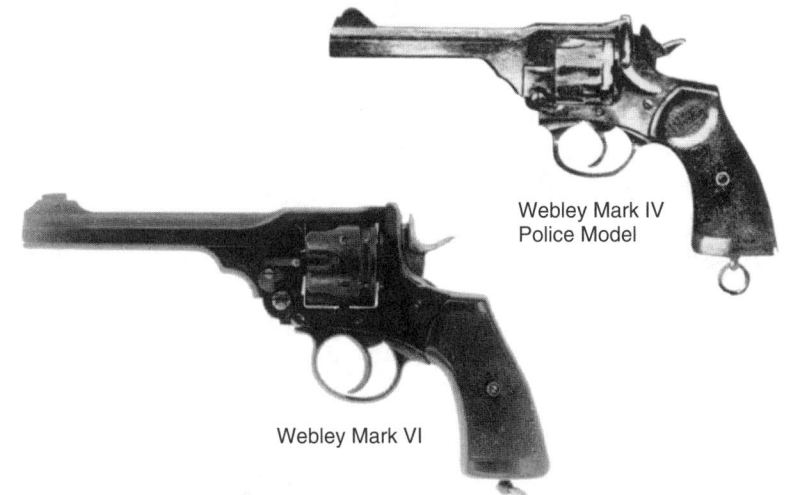

Webley Mark IV
Police Model

Webley Mark VI

HANDGUNS

Webley
Mark IV 32 Police
Mark IV Military Revolver
Model 1905 Pocket 32
Model 1907 Hammer Model
Model 1909
Model 1909 Target Pistol
Model 1910
Model 1910 High Velocity
Model 1911

Webley Model
1907 Hammer
Model

all length; 6-shot magazine; checkered wood (early) or hard rubber grips. Introduced 1907; dropped 1940.
Exc.: $275   **VGood:** $225   **Good:** $150

WEBLEY MODEL 1909
Semi-automatic; 9mm Browning Long; 8-shot magazine; 5″ barrel; 8″ overall length; external hammer; checkered hard rubber or wood grips; blue finish; only 1694 made. Introduced 1909; dropped 1914.
Exc.: $1200   **VGood:** $950   **Good:** $700

WEBLEY MODEL 1909 TARGET PISTOL
Single shot; tip-up action; 22, 32 S&W, 38 S&W; 9 7/8″ barrel; blade front sight; plastic thumbrest target grips; late versions had ballast chamber in butt to permit weight adjustment; rebounding hammer; matte-finish barrel; blued. Introduced 1909; dropped 1965.
Exc.: $350   **VGood:** $300   **Good:** $250

WEBLEY MODEL 1910
Semi-automatic; 380 ACP; 7-shot; 3 1/2″ barrel; 6 1/4″ overall length; exposed hammer; checkered hard rubber grips. Revised Model 1905 built for 380 cartridge. Introduced 1910; dropped 1932.
Exc.: $450   **VGood:** $350   **Good:** $275

WEBLEY MODEL 1910 HIGH VELOCITY
Semi-automatic; 38 ACP; 8-shot magazine; 5″ barrel; 8 1/2″ overall length; internal hammer grips. Only about 1000 made. Introduced 1910; dropped 1911.
Exc.: $1750   **VGood:** $1250   **Good:** $850

WEBLEY MODEL 1911
Semi-automatic; single shot; 22 Long; 4 1/2″, 9″ barrel; 6 1/2″ overall length (4 1/2″ barrel); external hammer; checkered hard rubber grips; blue finish. Built on revised 1905 32 ACP pistol frame for police practice; upon firing, action blows open to eject and be ready for reloading. Total production 1500. Introduced 1911; dropped 1932.
Exc.: $500   **VGood:** $400   **Good:** $250

Webley Model 1909 Target Pistol

Webley Model 1910

Webley Model 1911

Webley Mark IV 32 Police
Same specs as Police Model except 32 S&W. Introduced 1929; dropped 1968.
Exc.: $200   **VGood:** $165   **Good:** $135

WEBLEY MARK IV MILITARY REVOLVER
Revolver; 455 Webley; 6-shot; 4″, 6″ barrel; hinged frame; Vulcanite grips; brush blue finish. Manufactured at Webley from 1915 to 1923 and at Enfield Arsenal from 1921 to 1926. Standard proofs: Webley Mark IV and date or Enfield MkVI and date. Manufactured from 1915 to 1926.
Webley 6″ barrel
    Exc.: $325   **VGood:** $300   **Good:** $250
Webley 4″ barrel
    Exc.: $300   **VGood:** $250   **Good:** $200
Enfield
    Exc.: $275   **VGood:** $225   **Good:** $175
Converted to 45 ACP
    Exc.: $200   **VGood:** $175   **Good:** $150

WEBLEY 1905 POCKET 32
Semi-automatic; 32 ACP; 3 1/2″ barrel; 6 1/4″ overall length; 8-shot magazine; external hammer; checkered hard rubber grips; blue finish. Introduced 1905; extensively redesigned in 1908; dropped 1940.
    Exc.: $350   **VGood:** $300   **Good:** $250

WEBLEY MODEL 1907 HAMMER MODEL
Semi-automatic; 25 ACP only; 2 1/8″ barrel; 4 1/4″ over-

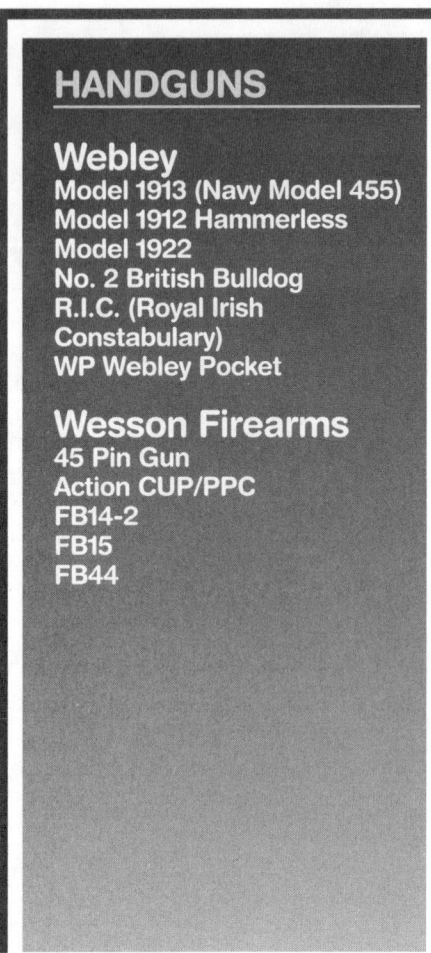

Webley
Model 1913 (Navy Model 455)
Model 1912 Hammerless
Model 1922
No. 2 British Bulldog
R.I.C. (Royal Irish Constabulary)
WP Webley Pocket

Wesson Firearms
45 Pin Gun
Action CUP/PPC
FB14-2
FB15
FB44

Webley Model 1912

Webley Model 12 Hammerless

Wesson Firearms FB15

Wesson Firearms 45 Pin Gun

Weatherby Mark V Silhouette

WEBLEY MODEL 1913 (NAVY MODEL 455)

Semi-automatic; 455 Webley Self-Loading Pistol; 5″ barrel; 8 ¹/₂″ overall length; external hammer; checkered hard rubber or wood grips; blue finish. Though usually called Model 1913 by collectors, official designation was "Pistol, self-loading, Webley & Scott .455 inch Mark IN." About 1200 made for commercial sale; 7600 for the Royal Navy, and under 500 were "Royal Horse Artillery" models cut for shoulder stock and with adjustable rear sight. Introduced 1912; dropped 1933. Note: Serial numbers of all commercial Webley auto pistols were in a single series so do not reflect quantity actually produced of any model. Military 455 autos had their own series, from 1 through 8050.
Commercial production
    Exc.: $1500   **VGood:** $1200   **Good:** $1000
Navy production
    Exc.: $1250   **VGood:** $1000   **Good:** $850
Royal Horse Artillery
    Exc.: $3500   **VGood:** $3000   **Good:** $2500

WEBLEY MODEL 1912 HAMMERLESS

Semi-automatic; 25 ACP only; 6-shot; 2 ¹/₈″ barrel; 4 ¹/₄″ overall length; striker fired; hard rubber grips. Introduced 1912; dropped 1938.
    Exc.: $250   **VGood:** $200   **Good:** $150

WEBLEY MODEL 1922

Semi-automatic; 9mm Browning Long; 8-shot; 5″ barrel; 8 ³/₈″ overall length; external hammer; checkered or plain wood grips; blued finish. Redesign of Model 1909 for possible military sale; of the 2000 made, 1000 were actually delivered to South African Defense Forces and 330 to Latvia. Introduced 1924; dropped 1932.
    Exc.: $1200   **VGood:** $950   **Good:** $700

WEBLEY NO. 2 BRITISH BULLDOG

Revolver; 320, 380, 450; 5-shot; 2″, 3″ barrel; fluted or round cylinders; solid frame; wood grips; bright blue or nickel finish. Marked "Webley No. 2" and caliber; top-strap marked "THE BRITISH BULLDOG" or with name of British retailer. Many No. 2s are marked with name of California retailer since the Bulldog was very popular in the American West. Manufactured from 1878 to 1915.
    Exc.: $300   **VGood:** $250   **Good:** $200
With American retailer or broad arrow marking
    Exc.: $350   **VGood:** $300   **Good:** $250

WEBLEY R.I.C. (ROYAL IRISH CONSTABULARY)

Revolver; 320 through 476; 5-, 6-shot cylinder; 2″-6″ barrels; solid frame; wood grips; bright blue or nickel finish. Early models had straighter grips and many had unfluted cylinders. Standard proofs with "Webley R.I.C." and caliber. Manufactured from 1868 to 1932.
    Exc.: $300   **VGood:** $250   **Good:** $200

WEBLEY WP WEBLEY POCKET

Revolver; double action; hinged frame; 32 S&W only; 6-shot; 2″, 3″ barrel; 6″ overall length (2″); checkered plastic grips; fixed sights; blued finish. Introduced 1908; dropped 1940.
    Exc.: $250   **VGood:** $200   **Good:** $165

WESSON FIREARMS 45 PIN GUN

Revolver; double action; 45 ACP; 6-shot; 5″ barrel; 12 ¹/₂″ overall length; weighs 54 oz.; 1:14″ twist; Taylor two-stage forcing cone; compensated shroud; finger-groove Hogue Monogrip grips; pin front sight, fully-adjustable rear; based on 44 Magnum frame; polished blue or brushed stainless steel; uses half-moon clips with 45 ACP, or 45 Auto Rim ammunition. Made in U.S. by Wesson Firearms Co., Inc. Introduced 1994; dropped 1995.
Blue finish, regular vent barrel shroud
    Perf.: $550   **Exc.:** $500   **VGood:** $400
Stainless steel, regular vent barrel shroud
    Perf.: $600   **Exc.:** $575   **VGood:** $500
Stainless steel vent heavy barrel shroud
    Perf.: $675   **Exc.:** $600   **VGood:** $550

WESSON FIREARMS ACTION CUP/PPC

Revolver; double action; 38 Spl., 357 Mag.; 6-shot; extra heavy 6″ barrel with bull shroud, removable underweight; weighs 4 ¹/₂ lbs.; Pachmayr Gripper grips; Tasco Pro Point II sight; competition tuned with narrow trigger; chamfered cylinder chambers; stainless steel only. Made in U.S. by Wesson Firearms. Introduced 1989; dropped 1995.
    Perf.: $700   **Exc.:** $625   **VGood:** $550

WESSON FIREARMS FB14-2

Revolver; double action; 357 Mag.; 2 ¹/₂″, 4″ fixed barrel; 8 ¹/₄″ overall length; weighs 36 oz. (2 ¹/₂″ barrel); service-style rubber grips; blade front sight, fixed rear; satin blue or stainless finish. Made in U.S. by Wesson Firearms. Introduced 1993; dropped 1995.
    New: $250   **Perf.:** $200   **Exc.:** $175

WESSON FIREARMS FB15

Revolver; double action; 357 Mag.; 6-shot; 3″, 4″, 5″, 6″ fixed barrel; 8 ³/₄″ overall length; weighs 37 oz. (3″ barrel); Hogue rubber grips; blade front sight, adjustable rear; polished blue or stainless finish. Made in U.S. by Wesson Firearms. Introduced 1993; dropped 1995.
    New: $300   **Perf.:** $275   **Exc.:** $235

WESSON FIREARMS FB44

Revolver; double action; 44 Mag.; 6-shot; 4″, 5″, 6″, 8″ barrel; 9 ³/₄″ overall length (4″ barrel); weighs 50 oz.; Hogue finger-groove rubber grips; interchangeable blade front sight, fully-adjustable rear; fixed, non-vented heavy barrel shrouds; polished blue finish. Made in U.S. by Wesson Firearms Co., Inc. Introduced 1994; dropped 1995.
    New: $350   **Perf.:** $300   **Exc.:** $275

Dan Wesson Model 12

Dan Wesson Model 32M

WESSON FIREARMS FB744
Revolver; double action; 44 Mag.; 6-shot; 4″, 5″, 6″, 8″ barrel; 9 ³/₄″ overall length (4″ barrel); weighs 50 oz.; Hogue finger-groove rubber grips; interchangeable blade front sight, fully-adjustable rear; fixed, non-vented heavy barrel shrouds; brushed stainless steel. Made in U.S. by Wesson Firearms Co., Inc. Introduced 1994; dropped 1995.
New: $400   **Perf.:** $350   **Exc.:** $325

WESSON FIREARMS HUNTER SERIES
Revolver; double action; 357 Supermag, 41 Mag., 44 Mag., 445 Supermag; 6-shot cylinder; 6″, 7 ¹/₂″ fixed barrel; 14″ overall length; weighs 64 oz.; Hogue finger-groove rubber or wood presentation grips; blade front sight, dovetailed Iron Sight Gunworks rear; 1:18.75″ twist barrel; Alan Taylor two-stage forcing cone; non-fluted cylinder; bright blue or satin stainless. Made in U.S. by Wesson Firearms Co., Inc. Introduced 1994; dropped 1995.
7 ¹/₂″ barrel, blue finish
New: $700   **Perf.:** $650   **Exc.:** $600
7 ¹/₂″ barrel, stainless steel
New: $750   **Perf.:** $700   **Exc.:** $650
6″ compensated barrel, 7″ shroud), blue finish
New: $725   **Perf.:** $675   **Exc.:** $625
6″ compensated barrel, 7″ shroud), stainless steel
New: $775   **Perf.:** $725   **Exc.:** $675
6″ compensated barrel, 7″ shroud, scope rings on shroud, blue finish
New: $775   **Perf.:** $725   **Exc.:** $625
6″ compensated barrel, 7″ shroud, scope rings on shroud, stainless steel
New: $800   **Perf.:** $750   **Exc.:** $700

WESSON FIREARMS MODEL 8-2
Revolver; double-action; 38 Spl.; 6-shot; 2 ¹/₂″, 4″, 6″ interchangeable barrels; 9 ¹/₄″ overall length (4″ barrel); weighs 30 oz.; serrated front, fixed rear sight; checkered grips; stainless or bright blue finish. Introduced 1975; dropped 1995.
Perf.: $200   **Exc.:** $175   **VGood:** $150

WESSON FIREARMS MODEL 9-2H
Revolver; double action; 38 Spl.; 2″-15″ barrel; ramp front, adjustable rear sight; heavy barrel options; blue or nickel finish. Introduced, 1975; dropped 1995.
Perf.: $250   **Exc.:** $200   **VGood:** $175

Wesson Firearms Model 9-2V
Same specs as Model 9-2H except standard vent-rib barrel. Introduced 1975; discontinued 1995.
Perf.: $275   **Exc.:** $225   **VGood:** $200

Wesson Firearms Model 9-2HV
Same specs as Model 9-2 except heavy vent-rib barrel; 38 Spl. only. Introduced 1975; discontinued 1995.
Perf.: $325   **Exc.:** $275   **VGood:** $225

WESSON FIREARMS MODEL 11
Revolver; double action; 357 Mag.; 6-shot; 2 ¹/₂″, 4″, 6″ interchangeable barrel; 9 ¹/₄″ overall length (4″ barrel); interchangeable grips; adjustable dovetail rear sight, serrated ramp front. Marketed with tools for changing barrels, grips; non-recessed barrel nut, blued. Introduced 1969; dropped 1974.
Perf.: $200   **Exc.:** $175   **VGood:** $150

WESSON FIREARMS MODEL 12
Revolver; 357 Mag.; 6-shot; 2 ¹/₂″, 3 ³/₄″, 5 ³/₄″ interchangeable barrels and grips; adjustable target-type rear sight, serrated ramp front. Marketed with tools for changing barrels, grips; blued. Introduced 1969; dropped 1974.
Perf.: $200   **Exc.:** $175   **VGood:** $150

WESSON FIREARMS MODEL 14
Revolver; double-action; 357 Mag.; 6-shot; 2 ¹/₄″, 3 ³/₄″, 5 ³/₄″ interchangeable barrel; 9″ overall length (3 ³/₄″ barrel); interchangeable walnut grips; fixed dovetail rear sight, serrated ramp front; wide trigger with adjustable over-travel stop; wide spur hammer; recessed barrel nut; blued, nickel, matte nickel finish. Introduced 1971; dropped 1975. Replaced by Model 14-2.
Perf.: $225   **Exc.:** $175   **VGood:** $150

Wesson Firearms Model 14-2
Same general specs as Model 14, except 2 ¹/₂″, 4″ Quickshift interchangeable barrels; Quickshift grips interchangeable with three other styles. Introduced 1976; dropped 1995.
New: $400   **Perf.:** $300   **Exc.:** $250

WESSON FIREARMS MODEL 15
Revolver; double action; 357 Mag.; 6-shot; 2 ¹/₄″, 3 ³/₄″, 5 ³/₄″ interchangeable barrel; fully-adjustable rear sight; wide trigger with adjustable over-travel stop; wide spur hammer; recessed barrel nut; blued, nickel, matte nickel finish. Introduced in 1971; dropped, 1975. Replaced by Model 15-2.
Perf.: $275   **Exc.:** $225   **VGood:** $175

Wesson Firearms Model 15-1
Same specs as Model 15 except reduced size of rear sight. Introduced, 1973; dropped, 1975.
Perf.: $250   **Exc.:** $200   **VGood:** $175

Wesson Firearms Model 15-2
Same specs as Model 15-1 except no barrel shroud. Introduced 1973; no longer in production.
Perf.: $300   **Exc.:** $200   **VGood:** $175

Wesson Firearms Model 15-2H
Same specs as the Model 15-2, except heavy barrel options. Introduced 1975; no longer in production.
Perf.: $300   **Exc.:** $200   **VGood:** $175

Wesson Firearms Model 15-2HV
Same specs as Model 15-2 except heavy vent-rib barrel, ranging from 2″ to 15″. Introduced 1975; no longer in production.
Perf.: $325   **Exc.:** $275   **VGood:** $225

Wesson Firearms Model 15 Gold Series
Same specs as Model 15 except smoother tuned action; 6″ or 8″ vent heavy slotted barrel shroud; blued barrel; rosewood stocks; orange dot Patridge-type front sight, white triangle on rear blade; stamped "Gold Series" with gold-filled Dan Wesson signature. Introduced 1989; dropped 1990.
Perf.: $400   **Exc.:** $325   **VGood:** $275

WESSON FIREARMS MODEL 22
Revolver; double action; 22 LR, 22 WMR; 6-shot cylinder; 2 ¹/₂″, 4″, 6″, 8″, 10″ interchangeable barrel; checkered, undercover, service or target stocks of American walnut; serrated, interchangeable front sight; adjustable white outline rear; blued or stainless steel; wide trigger; over-travel trigger adjustment; wide hammer spur. Introduced 1982; no longer in production.
Perf.: $300   **Exc.:** $200   **VGood:** $175

HANDGUNS

Wesson Firearms
FB744
Hunter Series
Model 8-2
Model 9-2H
Model 9-2V
Model 9-2HV
Model 11
Model 12
Model 14
Model 14-2
Model 15
Model 15-1
Model 15-2
Model 15-2H
Model 15-2HV
Model 15 Gold Series
Model 22
Model 22 Silhouette
Model 32M
Model 40 Silhouette
Model 41V

WESSON FIREARMS MODEL 22 SILHOUETTE
Revolver; single action; 22 LR; 6-shot; 10″ regular vent or vent heavy barrel; weighs 53 oz.; combat-style grips; Patridge-style front sight, .080″ rear. From Wesson Firearms Co., Inc. Introduced 1989; dropped 1995.
Blue finish, regular vent barrel
New: $400   **Perf.:** $350   **Exc.:** $325
Blue finish, vent heavy barrel
New: $425   **Perf.:** $375   **Exc.:** $350
Stainless steel, regular vent barrel
New: $450   **Perf.:** $400   **Exc.:** $375
Stainless steel, vent heavy barrel
New: $475   **Perf.:** $425   **Exc.:** $410

WESSON FIREARMS MODEL 32M
Revolver; 32 H&R Mag.; 6-shot; 2 ¹/₂″, 3 ³/₄″, 5 ³/₄″ barrel; adjustable rear sight, ramp front; blued or stainless steel. Introduced 1984.
Perf.: $300   **Exc.:** $200   **VGood:** $175

WESSON FIREARMS MODEL 40 SILHOUETTE
Revolver; 357 Maximum; 6-shot cylinder; 6″, 8″, 10″ barrel; 14 ⁵/₁₆″ overall length (8″ barrel); weighs 64 oz.; adjustable rear sight, serrated front; blued or stainless steel; meets IHMSA competition criteria. Introduced 1986; no longer in production.
Perf.: $400   **Exc.:** $325   **VGood:** $275

WESSON FIREARMS MODEL 41V
Revolver; double action; 41 Mag.; 6-shot cylinder; 4″, 6″, 8″, 10″ interchangeable barrel; smooth walnut stocks; serrated front sight, adjustable white-outline rear; wide trigger; adjustable trigger over-travel feature; wide hammer spur; blued or stainless steel. Introduced 1982; no longer in production.
Perf.: $350   **Exc.:** $300   **VGood:** $250

HANDGUNS

Wesson Firearms
Model 44V
Model 45V
Model 322
Model 738P
Model 7322
.455 Supermag
PPC

Whitney
Wolverine

Wichita
Classic Silhouette
International
Master
Silhouette Pistol

Wildey
Automatic Pistol

Wilkinson
Diane
Linda
Sherry

Wyoming Arms
Parker Auto

Wesson Firearms
Model 40 Silhouette

Wesson Firearms
Model 738P

Wichita International

Wildey Auto

WESSON FIREARMS MODEL 44V

Revolver; double action; 44 Mag.; 6-shot; 4″, 6″, 8″, 10″ interchangeable barrel; smooth walnut stocks; serrated front sight; adjustable white-outline rear; wide trigger with adjustable over-travel feature; wide hammer spur; blued or stainless finish. Introduced 1982; no longer in production.

Perf.: $375 **Exc.:** $325 **VGood:** $275

WESSON FIREARMS MODEL 45V

Revolver; double action; 41 Mag., 44 Mag., 45 Colt; 6-shot cylinder; 4″, 6″, 8″, 10″ interchangeable barrel; 12″ overall length (6″ barrel); weighs 48 oz. (4″ barrel); smooth wood grips; 1/8″ serrated front sight, fully-adjustable white outline rear; smooth, wide trigger with adjustable over-travel; wide hammer spur. Made in U.S. by Wesson Firearms. Introduced 1993; dropped 1995. Blue finish

New: $400 **Perf.:** $350 **Exc.:** $300
Stainless steel
New: $450 **Perf.:** $400 **Exc.:** $375

WESSON FIREARMS MODEL 322

Revolver; double action; 32-20; 6-shot cylinder; 2 1/2″, 4″, 6″, 8″ barrel, standard, vent, vent heavy; 11 1/4″ overall length; weighs 43 oz. (6″ barrel); checkered walnut grips; red ramp interchangeable front sight, fully-adjustable rear; blue finish. Made in U.S. by Wesson Firearms Co., Inc. Introduced 1991; dropped 1995.

New: $300 **Perf.:** $250 **Exc.:** $200

WESSON FIREARMS MODEL 738P

Revolver; double action; 38 Spl. +P; 5-shot cylinder; 2″ barrel; 6 1/2″ overall length; weighs 24 1/2 oz.; Pauferro wood or rubber grips; blade front sight, fixed notch rear; designed for +P ammunition; stainless steel construction. Made in U.S. by Wesson Firearms Co., Inc. Introduced 1992; dropped 1995.

New: $300 **Perf.:** $250 **Exc.:** $200

WESSON FIREARMS MODEL 7322

Revolver; double action; 32-20; 6-shot cylinder; 2 1/2″, 4″, 6″, 8″ barrel, standard, vent, vent heavy; 11 1/4″ overall length; weighs 43 oz. (6″ barrel); checkered walnut grips; red ramp interchangeable front sight, fully-adjustable rear; stainless steel. Made in U.S. by Wesson Firearms Co., Inc. Introduced 1991; dropped 1995.

New: $375 **Perf.:** $325 **Exc.:** $300

WESSON FIREARMS .455 SUPERMAG

Revolver; has same general specs as Model 40 series; chambered for the .455 Supermag cartridge; blued or stainless steel; 6″, 8″, 10″ barrel. Introduced, 1989; no longer in production.

Perf.: $325 **Exc.:** $275 **VGood:** $250

WESSON FIREARMS PPC

Revolver; double action; 38 Spl., 357 Mag.; 6-shot cylinder; 6″ extra heavy barrel with bull shroud, removable underweight; weighs 4 lbs., 7 oz.; Pachmayr Gripper grips; Aristocrat sights with three-postion rear; competition tuned with narrow trigger; chamfered cylinder chambers; bright blue finish or stainless steel. Made in U.S. by Wesson Firearms. Introduced 1989; dropped 1995.

Perf.: $550 **Exc.:** $500 **VGood:** $450

WHITNEY WOLVERINE

Semi-automatic; 22 LR; 4 1/2″ barrel; 9″ overall length; aluminuim alloy frame; windage-adjustable rear; 1/8″ Patridge-type front; top of slide serrated to reduce reflection; checkered plastic grips; blued or nickel finish. Introduced 1956; dropped 1963.

Exc.: $450 **VGood:** $325 **Good:** $250
Nickel finish
Exc.: $600 **VGood:** $500 **Good:** $400

WICHITA CLASSIC SILHOUETTE

Bolt action; all standard calibers to 2.80″; 11 1/14″ barrel; weighs 4 lbs.; hooded post front sight; open adjustable rear; AAA American walnut stock with checkered pistol grip; three locking lug bolts; three gas ports; adjustable trigger. From Wichita Arms. Introduced 1981; still in production. **LMSR: $3450**

Perf.: $3200 **Exc.:** $2400 **VGood:** $1900

WICHITA INTERNATIONAL

Single shot; 22 LR, 22 WMR, 32 H&R Mag., 357 Super Mag., 357 Mag., 7R, 7mm Super Mag., 7-30 Waters, 30-30 Win.; 10″, 10 1/2″, 14″ barrel; weighs 3 lbs. 2 oz. (10″ barrel); walnut grip and forend; Patridge front sight, adjustable rear; Wichita Multi-Range sight system optional; made of stainless steel; break-open action; grip dimensions same as Colt 45 Auto; drilled and tapped for furnished see-thru rings; extra barrels are factory fitted. Available from Wichita Arms. Introduced 1983; dropped 1994.

New: $750 **Perf.:** $650 **Exc.:** $550

WICHITA MASTER

Bolt action; 6mm BR, 7mm BR, 243, 7mm-08, 22-250, 308; 3-shot magazine; 13″, 14 7/8″ barrel; weighs 4 1/2 lbs. (13″ barrel); American walnut stock with oil finish; glass bedded; hooded post front sight, open adjustable rear; left-hand action with right-hand grip; round receiver and barrel; Wichita adjustable trigger. Made in U.S. by Wichita Arms. Introduced 1991; dropped 1994.

Perf.: $1275 **Exc.:** $850 **VGood:** $600

WICHITA SILHOUETTE PISTOL

Bolt action; 308 Win., 7mm IHSMA, 7mm-08; 14 15/16″ barrel; 21 3/8″ overall length; weighs 4 1/2 lbs.; Wichita Multi-Range sight system; American walnut glass-bedded stock; round barrel; fluted bolt with flat bolt handle; adjustable trigger. From Wichita Arms. Introduced 1979; still in production. **LMSR: $1350**

Perf.: $1500 **Exc.:** $1200 **VGood:** $950

WILDEY AUTOMATIC PISTOL

Semi-automatic; 10mm Wildey Mag., 11mm Wildey Mag., 357 Peterbuilt, 45 Win. Mag.; 7-shot; 5″, 6″, 7″, 8″, 10″, 12″, 14″ (45 Win. Mag.); 8″, 10″, 12″, 15″ (other calibers) interchangeable barrel; 11″ overall length (7″ barrel); weighs 64 oz. (5″ barrel); ramp front sight, fully-adjustable rear; hardwood grips; stainless steel construction; three lug rotary belt; double or single action; polished and matte finish. From Wildey, Inc. **LMSR: $1175**

Perf.: $1300 **Exc.:** $1150 **VGood:** $900

WILKINSON DIANE

Semi-automatic; 22 LR, 25 ACP; 8-shot magazine; 2 1/8″ barrel; 4 1/2″ overall length; internal hammer; separate ejector; checkered styrene stocks; fixed sights integral with slide; matte blued finish. Introduced 1977; no longer in production.

Exc.: $100 **VGood:** $75 **Good:** $60

WILKINSON LINDA

Semi-automatic; 9mm Para.; 31-shot magazine; 8 5/16″ barrel; 12 1/4″ overall length; weighs 5 lbs.; protected blade front sight, Wiliams adjustable rear; maple forend; checkered black plastic stocks; blowback action; cross-bolt safety. Introduced 1982; still in production. **LMSR: $533**

New: $325 **Perf.:** $275 **Exc.:** $225

WILKINSON SHERRY

Semi-automatic; 22 LR; 8-shot magazine; 2 1/8″ barrel; 4 3/4″ overall length; weighs 9 1/4 oz.; no sights; checkered black plastic stocks; crossbolt safety; blue finish or gold-plated frame; blued slide, trigger. Introduced 1985; still in production. **LMSR: $195**

New: $175 **Perf.:** $150 **Exc.:** $135

WYOMING ARMS PARKER AUTO

Semi-automatic; single action; 10mm Auto, 40 S&W, 45 ACP; 8-shot (10mm, 40 S&W), 7-shot (45 ACP); 3 3/8″, 5″, 7″ barrel; 6 3/8″ overall length (3 3/8″ barrel); weighs 29 oz. to 44 oz.; grooved composition grips; fixed or Millet adjustable sights; made of stainless steel. Made in the U.S. by Wyoming Arms Mfg. Corp. Introduced 1990; no longer produced. 3 3/8″, 5″ barrel, fixed sights

Perf.: $325 **Exc.:** $275 **VGood:** $250
7″ barrel, adjustable sights
Perf.: $400 **Exc.:** $325 **VGood:** $300

RIFLES

Directory of Manufacturers

RIFLES

A.A. Arms
AR9

AAO
Model 2000 50 Caliber

Ackley
Olympus Grade
Olympus Grade II
Standard Model
Varmint Special

Action Arms
Timberwolf

Alpha
Alaskan
Big-Five
Custom Model
Grand Slam

Ackley Olympus Grade II

Alpha Alaskan

Alpha Custom Model

Alpha Jaguar Grade I

Action Arms Timberwolf

A.A. ARMS AR9
Semi-automatic; 9mm Para.; 20-shot magazine; 16 1/4"
barrel; 33" overall length; weighs 6 1/2 lbs.; folding buttstock, checkered plastic grip; adjustable post front sight
in ring, fixed open rear; lever safety blocks trigger and sear; vented barrel shroud;
matte blue/black or nickel finish. Made in U.S. Market-
ed by Kimel Industries. Introduced 1991; no longer produced.
  Perf.: $450   **Exc.:** $400   **VGood:** $350

AAO MODEL 2000 50 CALIBER
Bolt action; 50 BMG; 5-shot magazine; 30" barrel,
1:15" twist; weighs 24 lbs.; muzzlebrake; cast alloy
stock with gray anodized finish; Kick-Ease recoil pad;
no sights; drilled and tapped for scope base; controlled feeding via rotating enclosed claw extractor;
90-degree bolt rotation; cone bolt face and barrel; trig-
ger-mounted safety blocks sear; fully-adjustable,
detachable tripod. Made in U.S. by American Arms &
Ordnance. Introduced 1994; still produced. **LMSR:**
$4000
  New: $3450   **Perf.:** $2895   **Exc.:** $1995

ACKLEY OLYMPUS GRADE
Bolt action; 22-250, 25-06, 257 Roberts, 270, 7x57,
30-06, 7mm Rem. Mag., 300 Win. Mag.; 4-shot
magazine (standard calibers), 3-shot (magnums);
24" barrel; no sights; drilled, tapped for scope
mounts; hand-checkered select American walnut
stock; rubber buttpad; swivel studs; Ackley barrel;
hinged floorplate; fully-adjustable trigger; Mark X
Mauser action. Introduced 1971; discontinued
1975.
  Perf.: $500   **Exc.:** $425   **VGood:** $350

Ackley Olympus Grade II
Same specs as Olympus Grade except finely
hand-checkered select American walnut stock; rubber recoil pad. Custom offering only; precludes
pricing.

ACKLEY STANDARD MODEL
Bolt action; 22-250, 25-06, 257 Roberts, 270, 7x57,
30-06, 7mm Rem. Mag., 300 Win. Mag.; 4-shot magazine (standard calibers), 3-shot (magnums); 24" barrel;
engraved receiver; no sights, drilled, tapped for scope
mounts; hand-checkered American walnut stock; rubber recoil pad, swivel studs; Ackley barrel; hinged floorplate; fully-adjustable trigger; Mark X Mauser action.
Introduced 1971; discontinued 1975.
  Perf.: $400   **Exc.:** $350   **VGood:** $325

ACKLEY VARMINT SPECIAL
Bolt action; 22-250, 220 Swift, 25-06, 257 Roberts, 270,
7x57, 30-06, 7mm Rem. Mag., 300 Win. Mag.; 4-shot
magazine (standard calibers), 3-shot (magnums); 26" barrel; no sights; drilled tapped for scope mounts; hand-checkered American walnut stock; varmint forend; rubber recoil
pad; swivel studs. Introduced 1971; discontinued 1975.
  Perf.: $400   **Exc.:** $350   **VGood:** $325

ACTION ARMS TIMBERWOLF
Slide action; 38 Spl./357 Mag., 44 Mag.; 10-shot maga-
zine; 18 1/2" barrel; 36 1/2" overall length; weighs 5 1/2
lbs.; blade front sight, adjustable rear; plain walnut stock;
grooved slide handle; push-button safety on trigger
guard; takedown adjustable stock; integral scope mount;
blued or chrome finish. Was imported by Action Arms
Ltd. (357 only) and Springfield Armory (44 Mag. only).
Made in Israel. Introduced 1989; no longer imported.
  Perf.: $250   **Exc.:** $230   **VGood:** $190
Chrome finish
  Perf.: $350   **Exc.:** $330   **VGood:** $300

ALPHA ALASKAN
Bolt action; 308 Win., 350 Rem. Mag., 358 Win., 458
Win. Mag.; 4-shot magazine; 20"-24" barrel; weighs 6 3/4
to 7 1/2 lbs.; no sights; drilled tapped for scope mounts;
Alphawood stock; Neidner-style grip cap; barrel band
swivel stud; right- or left-hand models; stainless steel
construction. Introduced 1985; discontinued 1987.
  Perf.: $1300   **Exc.:** $1000   **VGood:** $825

ALPHA BIG-FIVE
Bolt action; 300 H&H to 458 Win. Mag.; 4-shot
magazine; 20"-24" Douglas barrel; no sights;
drilled tapped for scope mounts; satin-finish American walnut reinforced stock; three-position safety;
honed action and trigger; lightened action; swivel
studs; decelerator recoil pad. Introduced 1987; discontinued 1987.
  Perf.: $1400   **Exc.:** $1100   **VGood:** $875

ALPHA CUSTOM MODEL
Bolt action; 17 Rem., 222, 223, 22-250, 338-284,
25-06, 35 Whelen, 257 Weatherby, 338 Win. Mag.;
other calibers available in short, medium, long
actions; 20"-24" barrel; 40"-43" overall length;
weighs 6 to 7 lbs.; no sights; drilled, tapped for
scope mounts; hand-checkered classic-style California Claro walnut stock; hand-rubbed oil finish;
ebony forend tip; custom steel grip cap; solid
buttpad; inletted swivel studs; three-lug locking
system; three-position safety; left- or right-hand
models; satin blue finish. Introduced 1984; discontinued 1987.
  Perf.: $1250   **Exc.:** $1100   **VGood:** $875

ALPHA GRAND SLAM
Bolt action; 17 Rem., 222, 223, 22-250, 338-284,
25-06, 35 Whelen, 257 Weatherby, 338 Win. Mag.;
other calibers available in short, medium, long
actions; 20"-24" barrel; 40"-43" overall length;
weighs 6 1/2 lbs.; no sights; drilled, tapped for scope
mounts; hand-checkered classic-style California
Claro walnut or laminated stock; hand-rubbed oil
finish; ebony forend tip; custom steel grip cap; solid
buttpad; inletted swivel studs; three-position safety; fluted bolt; left- or right-
hand models; matte blue finish. Introduced 1984;
discontinued 1987.
  Perf.: $1150   **Exc.:** $900   **VGood:** $750

AMAC Wagonmaster

Alpine Sporter

American Arms EXP-64

American Arms SM-64

Alpha
Jaguar Grade I
Jaguar Grade II
Jaguar Grade III
Jaguar Grade IV

Alpine
Sporter

AMAC
Targetmaster
Wagonmaster

American Arms
AKY 39
EXP-64
Mini-Max
RS Combo
SM-64

American Firearms
Stainless Model

AMT
Lightning 25/22

ALPHA JAGUAR GRADE 1

Bolt action; from 222 to 338 Win. Mag.; 4-shot magazine; 20"-24" round tapered barrel; 39 1/2" overall length; weighs 6 lbs.; no sights; drilled, tapped for scope mounts; satin-finish American walnut Monte Carlo stock; rubber buttpad; swivel studs; medium-length Mauser-type action; slide safety; cocking indicator; aluminum bedding block; Teflon-coated trigger guard, floorplate assembly. Introduced 1987; discontinued 1987.

    Perf.: $900   **Exc.: $800**   **VGood: $700**

Alpha Jaguar Grade II

Same specs as Grade I except Douglas premium barrel.

    Perf.: $1000   **Exc.: $900**   **VGood: $800**

Alpha Jaguar Grade III

Same specs as Grade I except Douglas barrel; three-position safety; honed action and trigger.

    Perf.: $1125   **Exc.: $1000**   **VGood: $900**

Alpha Jaguar Grade IV

Same specs as Grade I except Douglas barrel; three-position safety; honed action and trigger; lightened action; swivel studs.

    Perf.: $1200   **Exc.: $1050**   **VGood: $950**

ALPINE SPORTER

Bolt action; 22-250, 243 Win., 264 Win., 270, 30-06, 308, 308 Norma Mag., 7mm Rem. Mag., 8mm, 300 Win. Mag.; 5-shot magazine (standard calibers), 3-shot (magnums); 23" barrel (standard calibers), 24" (magnums); ramp front sight, open adjustable rear; checkered pistol-grip Monte Carlo stock of European walnut; recoil pad; sling swivels. Imported from England by Mandall Shooting Supplies. Introduced 1978; still imported. **LMSR: $395**

    Perf.: $350   **Exc.: $310**   **VGood: $270**

AMAC TARGETMASTER

Slide-action; 22 Short, 22 Long, 22 LR; 19-shot magazine (22 Short), 15-shot (22 Long), 12-shot (22 LR); 18 1/2" barrel; 36 1/2" overall length; weighs 5 3/4 lbs.; hooded ramp front sight, open adjustable rear; receiver grooved for scope mounts; walnut-finished hardwood stock; polished blued finish. Made in standard and youth models. Introduced 1985 by Iver Johnson; dropped 1990.

    Perf.: $190   **Exc.: $150**   **VGood: $110**

AMAC WAGONMASTER

Lever action; 22 Short, 22 Long, 22 LR, 22 WMR; 21-shot magazine (22 Short), 17-shot (22 Long), 15-shot (22 LR), 12-shot (22 WMR); 18" barrel; 36 1/2" overall length; weighs 5 3/4 lbs.; hooded ramp front sight, open adjustable rear; receiver grooved for scope mount; walnut-finished stock, forend; polished blued finish. Introduced 1985 by Iver Johnson; dropped 1990.

    Perf.: $190   **Exc.: $150**   **VGood: $110**

AMERICAN ARMS AKY 39

Semi-automatic; 7.62x39mm; 30-shot magazine; 19 1/2" barrel; 40 1/2" overall length; weighs 9 lbs.; hooded post front sight, open adjustable rear; flip-up tritium sights; teakwood stock. Manufactured in Yugoslavia. Introduced 1988; discontinued 1989.

    Perf.: $725   **Exc.: $610**   **VGood: $450**
Folding stock model

    Perf.: $750   **Exc.: $650**   **VGood: $500**

AMERICAN ARMS EXP-64

Semi-automatic; 22 LR; 10-shot magazine; 21" barrel; 40" overall length; weighs 7 lbs.; blade front sight, adjustable rear; receiver grooved for scope mounts; synthetic stock; takedown feature; crossbolt safety. Made in Italy. Introduced 1989; discontinued 1990.

    Perf.: $145   **Exc.: $125**   **VGood: $110**

AMERICAN ARMS MINI-MAX

Semi-automatic; 22 LR; 10-shot magazine; 18 3/4" barrel; 36 1/2" overall length; weighs 4 1/2 lbs.; blade front sight, adjustable open rear; black synthetic or wood stock; trigger-block safety on trigger guard; receiver grooved for scope mounts. Introduced 1990; discontinued 1990.

    Perf.: $80   **Exc.: $70**   **VGood: $55**

AMERICAN ARMS RS COMBO

Combo; over/under box lock; 222 or 308/12-ga. 3"; 24" barrel; weighs 7 lbs. 14 oz.; full-choke tube; vent rib; blade front sight, folding rear; grooved for scope mounts; European walnut Monte Carlo stock; checkered pistol grip, forend; double triggers; extractors; silver finish, engraving; barrel connectors allow for windage/elevation adjustment of rifle barrel. Made in Italy. Introduced 1989; discontinued 1989.

    Perf.: $595   **Exc.: $530**   **VGood: $450**

AMERICAN ARMS SM-64

Semi-automatic; 22 LR; 10-shot magazine; 21" barrel; 40" overall length; weighs 7 lbs.; hooded front sight, adjustable rear; checkered hardwood stock, forend; takedown feature. Made in Italy. Introduced 1989; discontinued 1990.

    Perf.: $130   **Exc.: $110**   **VGood: $90**

AMERICAN FIREARMS STAINLESS MODEL

Bolt action; 22-250, 243, 6mm Rem., 6mm Win. Mag., 25-06, 257 Win. Mag., 264 Win. Mag., 6.5mm Rem. Mag., 270 Win. Mag., 284 Win., 7x57, 7mm Rem. Mag., 7.62x39, 308 Win., 30-06, 300 Win. Mag., 338 Win. Mag., 458 Mag.; 16 1/2", 18", 20", 22", 24", 26", 28" barrel; no sights; drilled, tapped for scope mounts; hand-checkered walnut or maple stock; side safety; hinged floorplate; adjustable trigger; made of blued or satin stainless steel. Manufactured in four grades. Introduced 1972; dropped 1974.
Grade I, Presentation

    Perf.: $1495   **Exc.: $1095**   **VGood: $895**
Grade II, Deluxe

    Perf.: $1095   **Exc.: $895**   **VGood: $650**
Grade III, Standard

    Perf.: $895   **Exc.: $695**   **VGood: $495**
Grade IV, 338 & 458 Standard

    Perf.: $995   **Exc.: $795**   **VGood: $595**

AMT LIGHTNING 25/22

Semi-automatic; 22 LR; 30-shot magazine; 18" tapered barrel; 26 1/2" overall length (folded), 37" (open); weighs 6 lbs.; folding stainless stock; ramp front sight, fixed rear; made of stainless steel with matte finish; receiver dovetailed for scope mounting; extended magazine release; adjustable rear sight optionally available; youth stock available. Made in U.S. by AMT. Introduced 1984; dropped 1994.

    Perf.: $225   **Exc.: $175**   **VGood: $140**

AMT
Lightning Small-Game
Hunting Rifle II
Magnum Hunter

Anschutz
Achiever
Achiever St Super
BR-50
Kadett
Mark 525 Deluxe Auto
Mark 525 Carbine
Mark 2000
Model 54.18MS
Model 54.18MS ED
Model 54.18MS Fortner
Silhouette Rifle

AMT Magnum Hunter

Anschutz Achiever

Anschutz BR-50

Anschutz Kadett

Anschutz Mark 525 Deluxe Auto

AMT Lightning Small-Game Hunting Rifle II
Same specs as Lightning 25/22 except conventional stock of black fiberglass-filled nylon, checkered at the grip and forend, and fitted with Uncle Mike's swivel studs; removable recoil pad for ammo storage, cleaning rod and survival knife; no sights, receiver grooved for scope mounting; 22″ full-floating target weight barrel; 40 1/2″ overall length; weighs 6 3/4 lbs.; 10-shot rotary magazine. Made in U.S. by AMT. Introduced 1987; 22 WMR introduced 1992; dropped 1994.
Perf.: $225   Exc.: $175   VGood: $140

AMT MAGNUM HUNTER
Semi-automatic; 22 WMR; 10-shot magazine; 20″ barrel; 40 1/2″ overall length; weighs 6 lbs.; black fiberglass-filled nylon stock with checkered grip and forend; no sights; drilled and tapped for Weaver mount; stainless steel construction; free-floating target-weight barrel. Made in U.S. by AMT. Introduced 1995; still produced. **LMSR: $460**
Perf.: $325   Exc.: $280   VGood: $230

ANSCHUTZ ACHIEVER
Bolt action; single shot; 22 LR; 5-shot magazine; 19 1/2″ barrel; 35 1/2″ to 36 1/2″ overall length; weighs 5 lbs.; fully-adjustable open rear sight, hooded front; walnut-finished European hardwood stock; Mark 2000-type action; adjustable two-stage trigger; grooved for scope mounts. Made in Germany. Introduced 1987; still imported. **LMSR: $399**
Perf.: $340   Exc.: $265   VGood: $220

Anschutz Achiever St Super
Same specs as Achiever except 22″ barrel with 3/4″ diameter; 38 3/4″ to 39 3/4″ overall length; weighs about 6 1/2 lbs.; 13 1/2″ accessory rail on forend; designed for the advanced junior shooter with adjustable length of pull from 13 1/4″ to 14 1/4″ via removable butt spacers. Imported from Germany. Introduced 1994; still imported. **LMSR $485**
Perf.: $415   Exc.: $370   VGood: $280

ANSCHUTZ BR-50
Bolt action; 22 LR; single shot; 19 3/4″ barrel (without 11-oz. muzzle weight); 37 3/4″ to 42 1/2″ overall length; weighs about 11 lbs.; benchrest-style stock of European hardwood with stippled grip; cheekpiece vertically adjustable to 1″; stock length adjustable via spacers and buttplate; glossy blue-black paint finish; no sights; receiver grooved for mounts; barrel drilled and tapped for target mounts; uses the Anschutz 2013 target action with #5018 two-stage adjustable target trigger. Imported from Germany. Introduced 1994; still imported. **LMSR: $3312**
Perf.: $2695   Exc.: $1995   VGood: $1125

ANSCHUTZ KADETT
Bolt action; 22 LR; 5-shot clip; 22″ barrel; 40″ overall length; weighs 5 1/2 lbs.; Lyman adjustable folding leaf rear sight, hooded bead on ramp front; checkered walnut-finish European hardwood stock; Mark 2000 target action; single-stage trigger; grooved for scope mount. Made in Germany. Introduced 1987; discontinued 1988.
Perf.: $235   Exc.: $200   VGood: $160

ANSCHUTZ MARK 525 DELUXE AUTO
Semi-automatic; 22 LR; 10-shot clip magazine; 24″ barrel; 43″ overall length; weighs 6 1/2 lbs.; European hardwood stock with checkered pistol grip, Monte Carlo comb, beavertail forend; hooded ramp front sight, folding leaf rear; rotary safety; empty shell deflector; single stage trigger; receiver grooved for scope mounting. Imported from Germany. Introduced 1982; still imported. **LMSR: $547**
Perf.: $460   Exc.: $375   VGood: $290

Anschutz Mark 525 Carbine
Same specs as Mark 525 except 20″ barrel. Discontinued 1986.
Perf.: $400   Exc.: $325   VGood: $235

ANSCHUTZ MARK 2000
Bolt action; single shot; 22 LR; heavy 26″ barrel; 43″ overall length; weighs 7 1/2 lbs.; globe insert front sight, micro-click peep rear; walnut-finish hardwood stock; thumb groove in stock; pistol grip swell. Made in Germany. Introduced 1980; importation discontinued 1988.
Perf.: $340   Exc.: $290   VGood: $220

ANSCHUTZ MODEL 54.18MS
Bolt action; single shot; 22 LR; 22 1/2″ barrel; 41 3/8″ overall length; weighs 9 lbs. 4 oz.; no sights; stippled walnut stock, pistol grip, forearm; SuperMatch 54 action; two-stage trigger. Introduced 1981; still imported. **LMSR: $1579**
Right-hand action
Perf.: $1300   Exc.: $975   VGood: $800
Left-hand action
Perf.: $1350   Exc.: $1000   VGood: $825

Anschutz Model 54.18MS ED
Same specs as Model 54.18MS except 19 1/4″ barrel; 14 1/4″ extension tube; three removable muzzle weights. Introduced 1981; discontinued 1988.
Right-hand action
Perf.: $1000   Exc.: $800   VGood: $625
Left-hand action
Perf.: $1100   Exc.: $900   VGood: $700

Anschutz Model 54.18MS Fortner Silhouette Rifle
Same specs as the 54.18MS except Anschutz/Fortner system straight-pull bolt action; 21″ barrel; 3/4″-diameter; McMillan Fibergrain fiberglass silhouette stock; two-stage #5020 trigger adjustable from 3 1/2 oz. to 2 lbs.; extremely fast lock time. Imported from Germany. Introduced 1995; still imported. **LMSR: $3855**
Right-hand
Perf.: $3250   Exc.: $2795   VGood: $1995
Left-hand
Perf.: $3500   Exc.: $2995   VGood: $2125

Anschutz
Model 54.18MS REP
Model 54.18MS REP Deluxe
Model 64
Model 64MS
Model 64MS-FWT
Model 64MS Left Silhouette
Model 64S
Model 520
Model 1403B Biathlon
Model 1403D Match
Model 1407
Model 1408

Anschutz Model 54.18MS

Anschutz Model 54.18MS REP

Anschutz 54.18MS Fortner Silhouette Rifle

Anschutz Model 64MS

Anschutz Model 1403B Biathlon

Anschutz Model 54.18MS REP
Same specs as Model 54.18MS except 5-shot maga-zine; 22″-30″ barrel; weighs 7³/₄ lbs.; thumbhole wood or synthetic stock with vented forearm; Super Match 54 action. Introduced 1989; still imported. **LMSR: $2066**
   Perf.: $1600   **Exc.: $1200**   **VGood: $900**

Anschutz Model 54.18MS REP Deluxe
Same specs as Model 54.18 REP except Fibergrain McMillan thumbhole stock with stippling. Introduced 1990; still imported. **LMSR: $2450**
   Perf.: $2000   **Exc.: $1600**   **VGood: $1000**

ANSCHUTZ MODEL 64
Bolt action; single shot; 22 LR; 26″ barrel, ¹¹/₁₆″ diame-ter; 44″ overall length; weighs 7³/₄ lbs.; no sights; scope blocks receiver grooved for Anschutz sights; walnut-fin-ish hardwood stock; cheekpiece; hand-checkered pis-tol grip, beavertail forearm; adjustable buttplate; adjustable single-stage trigger; sliding side safety; for-ward sling swivel for competition sling. Marketed in U.S. as Savage/Anschutz Model 64. Introduced 1963; discontinued 1981.
Right-hand model
   Perf.: $500   **Exc.: $400**   **VGood: $300**
Left-hand model
   Perf.: $600   **Exc.: $500**   **VGood: $400**

ANSCHUTZ MODEL 64MS
Bolt action; single shot; 22 LR; 21 ¹/₂″ barrel; 39 ¹/₂″ overall length; weighs 8 lbs.; no sights; drilled, tapped for scope mounts; stippled thumbhole wood stock with pistol grip; adjustable two-stage trigger; Match 64 action. Introduced 1981; still imported. **LMSR: $987**
Right-hand model
   Perf.: $795   **Exc.: $650**   **VGood: $525**
Left-hand model
   Perf.: $795   **Exc.: $675**   **VGood: $550**

Anschutz Model 64MS-FWT
Same specs as Model 64MS except weighs 6¹/₄ lbs.; sin-gle-stage trigger. Introduced 1981; discontinued 1988.
   Perf.: $495   **Exc.: $425**   **VGood: $375**

Anschutz Model 64MS Left Silhouette
Same specs as Model 64MS except left-hand model. Imported from Germany. Introduced 1980; still import-ed. **LMSR: $1087**
   Perf.: $795   **Exc.: $675**   **VGood: $550**

ANSCHUTZ MODEL 64S
Bolt action; single shot; 22 LR; 26″ barrel; 44″ overall length; Redfield Olympic sight; walnut-finish hardwood stock; cheekpiece; hand-checkered pistol grip; beavertail forearm; adjustable buttplate; adjustable single-stage trig-ger; sliding side safety; forward sling swivel for competi-tion sling. Marketed in U.S. as Savage/Anschutz Model 64S. Introduced 1963; discontinued 1981.
Right-hand model
   Perf.: $495   **Exc.: $425**   **VGood: $350**
Left-hand model
   Perf.: $525   **Exc.: $450**   **VGood: $375**

ANSCHUTZ MODEL 520
Semi-automatic; 22 LR; 10-shot clip magazine; 24″ barrel; 43″ overall length; weighs 6 ¹/₂ lbs.; folding adjustable rear sight; hooded ramp front; receiver grooved for scope mounts; European Monte Carlo wal-nut stock, forend; hand checkering; rotary-style safety; single-stage trigger; swivel studs. Made in Germany. Introduced 1982; discontinued 1983.
   Perf.: $275   **Exc.: $200**   **VGood: $165**

ANSCHUTZ MODEL 1403B BIATHLON
Bolt action; 22 LR; 5-shot magazine; 21 ¹/₂″ barrel; 42 ¹/₂″ overall length; weighs 8 ¹/₂ lbs.; blonde-finish European hardwood stock; stippled pistol grip; globe front sight with snow cap and muzzle cover, optional micrometer peep rear with spring-hinged snow cap; uses Match 64 Target action with three-way adjustable two-stage trigger; slide safety; comes with five maga-zines; adjustable buttplate. Imported from Germany. Introduced 1991; no longer imported.
   Perf.: $850   **Exc.: $700**   **VGood: $525**

ANSCHUTZ MODEL 1403D MATCH
Bolt action; single shot; 22 LR; 26″ barrel; 44″ overall length; weighs 7 ³/₄ lbs.; no sights; receiver grooved for scope mounts; walnut-finished hardwood stock cheek-piece; checkered pistol grip, beavertail forend; adjustable buttplate; slide safety; #5053 adjustable sin-gle-stage trigger. Made in Germany. Introduced 1980; importation discontinued 1990.
Right-hand model
   Perf.: $600   **Exc.: $500**   **VGood: $400**
Left-hand model
   Perf.: $625   **Exc.: $525**   **VGood: $425**

ANSCHUTZ MODEL 1407
Bolt action; single shot; 22 LR; 26″ barrel; 44 ¹/₂″ overall length; weighs 10 lbs.; weight conforms to ISU competi-tion requirements, also suitable for NRA matches; no sights; receiver grooved for Anschutz sights; scope blocks; French walnut prone-style stock; Monte Carlo, cast-off cheekpiece; hand-stippled pistol grip, forearm; swivel rail, adjustable swivel; adjustable rubber buttplate; single-stage trigger; wing safety. Marketed in U.S. by Savage. Introduced in 1967; no longer in production.
Right-hand model
   Perf.: $375   **Exc.: $325**   **VGood: $275**
Left-hand model
   Perf.: $395   **Exc.: $340**   **VGood: $290**

ANSCHUTZ MODEL 1408
Bolt action; 22 LR; 5-, 10-shot magazine; 26″ heavy barrel; weighs 10 lbs.; receiver grooved for micrometer sights; hooded ramp front sight; European walnut stock; checkered pistol grip, forearm; adjustable cheek-piece. Marketed in the U.S. by Savage. Introduced 1967; no longer in production.
   Perf.: $375   **Exc.: $325**   **VGood: $275**

RIFLES

Anschutz
Model 1408ED
Model 1411
Model 1413
Model 1416D Classic
Model 1416D Custom
Model 1416D Deluxe
Model 1416D Fiberglass
Model 1418D
Model 1422D Classic
Model 1422D Custom
Model 1432 Classic
Model 1432 Custom
Model 1433D
Model 1449 Youth
Model 1450B Biathlon

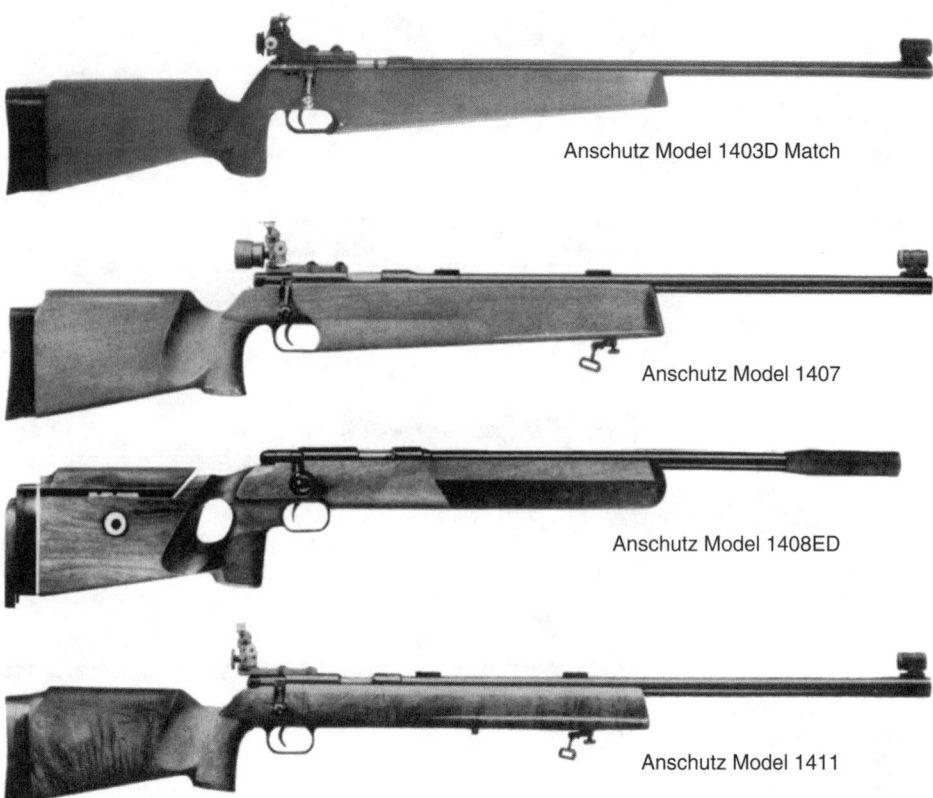

Anschutz Model 1403D Match

Anschutz Model 1407

Anschutz Model 1408ED

Anschutz Model 1411

ANSCHUTZ MODEL 1408ED
Same specs as Model 1908 except 23″ barrel; European walnut thumbhole stock; adjustable comb and buttplate; single-stage trigger; oversize bolt knob. Introduced in 1976; no longer in production.
Perf.: $550   **Exc.:** $500   **VGood:** $400

ANSCHUTZ MODEL 1411
Bolt action; single shot; 22 LR; 27 1/2″ barrel; 46″ overall length; weighs 11 lbs.; no sights; receiver grooved for Anschutz sights; scope blocks; French walnut prone-style stock; Monte Carlo, cast-off cheekpiece; hand-stippled pistol grip, forearm; swivel rail, adjustable swivel; adjustable rubber buttplate; single-stage trigger; wing safety. Marketed in U.S. by Savage. Introduced 1965; no longer in production.
Perf.: $370   **Exc.:** $325   **VGood:** $270

ANSCHUTZ MODEL 1413
Bolt action; single shot; 22 LR; 27 1/2″ barrel; 50″ overall length; weighs 15 1/2 lbs; no sights; receiver grooved for Anschutz sights; scope blocks; International-type French walnut stock; adjustable aluminum Schuetzen hook buttplate; adjustable cheekpiece; new yoke-type adjustable palm rest; single-stage trigger; wing safety. Marketed in the U.S. by Savage. Introduced 1967; no longer in production.
Perf.: $550   **Exc.:** $480   **VGood:** $400

ANSCHUTZ MODEL 1416D CLASSIC
Bolt action; 22 LR; 5-shot clip; 22 1/2″ barrel; 41″ overall length; weighs 6 lbs.; classic-style European walnut stock with straight comb, checkered pistol grip and forend; hooded ramp front sight, folding leaf rear; uses Match 64 action; adjustable single stage trigger; receiver grooved for scope mounting. Imported from Germany by Precision Sales International. Still imported.
LMSR: $785
Perf.: $610   **Exc.:** $500   **VGood:** $400
Left-hand
Perf.: $635   **Exc.:** $520   **VGood:** $420

Anschutz Model 1416D Custom
Same specs as Model 1416 Classic except European walnut stock with roll-over Monte Carlo cheekpiece; slim forend with Schnabel tip; fine cut checkering on grip and forend. Imported from Germany by Precision Sales International. Introduced 1988; still imported. **LMSR: $785**
Perf.: $650   **Exc.:** $550   **VGood:** $450

Anschutz Model 1416D Deluxe
Same specs as Model 1416 except European walnut Monte Carlo stock with cheekpiece; Schnabel forend; Model 1403 target rifle action. Imported from Germany by Precision Sales International. No longer imported.
Perf.: $650   **Exc.:** $575   **VGood:** $450

Anschutz Model 1416D Fiberglass
Same specs as Model 1416D except 40 1/4″ overall length; weighs 5 1/4 lbs.; McMillian fiberglass Monte Carlo stock with roll-over cheekpiece; Wundhammer grip swell; adjustable trigger; slide safety; 10-shot magazine and single shot adaptor optional. Imported from Germany by Precision Sales International, Inc. Introduced 1991; no longer imported.
Perf.: $725   **Exc.:** $595   **VGood:** $495

ANSCHUTZ MODEL 1418D
Bolt action; 22 LR; 5-shot clip; 19 3/4″ barrel; weighs 5 1/2 lbs.; European walnut Mannlicher-style stock with mahogany Schnabel tip, checkered pistol grip and forend; hooded ramp front sight, folding leaf rear; uses Model 1403 target rifle action; adjustable single-stage trigger; receiver grooved for scope mounting. Imported from Germany by Precision Sales International. Still imported. **LMSR: $1164**
Perf.: $975   **Exc.:** $800   **VGood:** $650

ANSCHUTZ MODEL 1422D CLASSIC
Bolt action; 22 LR; 5-shot clip; 24″ barrel; 43″ overall length; weighs 7 1/4 lbs.; hooded ramp front sight, folding leaf rear; drilled, tapped for scope mounting; select European walnut stock; checkered pistol grip, forend; adjustable single-stage trigger. Made in Germany. Introduced 1982; importation discontinued 1988.
Perf.: $995   **Exc.:** $850   **VGood:** $695

Anschutz Model 1422D Custom
Same specs as Model 1422D Classic except weighs 6 1/2 lbs.; Monte Carlo stock with roll-over cheekpiece; slim forend; Schnabel tip; palm swell on pistol grip; rosewood grip cap, diamond insert; skip-line checkering. Made in Germany. Introduced 1982; importation discontinued 1988.
Perf.: $1095   **Exc.:** $895   **VGood:** $695

ANSCHUTZ MODEL 1432 CLASSIC
Bolt action; 22 Hornet; 5-shot box magazine; 23 1/2″ barrel; 42 1/2″ overall length; weighs 7 3/4 lbs.; receiver grooved for scope mounting; folding leaf rear sight, hooded ramp front; European walnut Monte Carlo-type stock; hand-checkered pistol grip, forearm; cheekpiece. Introduced 1976; no longer in production.
Perf.: $1295   **Exc.:** $1025   **VGood:** $895

Anschutz Model 1432 Custom
Same specs as Model 1432 except 24″ barrel; 43″ overall length; weighs 6 1/2 lbs.; Monte Carlo stock with roll-over cheekpiece; palm swell on pistol grip; rosewood grip cap; diamond insert. No longer in production.
Perf.: $1095   **Exc.:** $995   **VGood:** $850

ANSCHUTZ MODEL 1433D
Bolt action; 22 Hornet; 4-shot clip magazine; 19 3/4″ barrel; 39″ overall length; weighs 6 1/4 lbs.; receiver grooved for scope mounting; folding leaf rear sight, hooded ramp front; European walnut Mannlicher-type stock; hand-checkered pistol grip, forearm; cheekpiece; Match 54 target action; single-stage or double-set trigger. Introduced 1976; discontinued 1986.
Perf.: $1100   **Exc.:** $900   **VGood:** $700

ANSCHUTZ MODEL 1449 YOUTH
Bolt-action; 22 LR; 5-shot clip; 16 1/4″ barrel; 32 1/2″ overall length; weighs 3 1/2 lbs.; hooded ramp front sight, open adjustable rear; grooved for scope mounts; walnut-finished European hardwood stock; built on Anschutz Mark 2000 action; single-shot clip adaptor and 10-shot magazine available. Made in Germany. Introduced 1990; discontinued 1991.
Perf.: $250   **Exc.:** $210   **VGood:** $185

ANSCHUTZ MODEL 1450B BIATHLON
Bolt action; single shot; 22 LR; 19 1/2″ barrel; 36″ overall length; weighs 5 lbs.; aperture sights; European hardwood stock with ventilated forend and adjustable buttplate; Mark 2000 action. Introduced 1993; discontinued 1993.
Perf.: $595   **Exc.:** $525   **VGood:** $460

Anschutz Model 1413

Anschutz Model 1416D/1516D Custom

Anschutz Model 1418D

Anschutz Model 1422D Classic

RIFLES

Anschutz
Model 1516D Classic
Model 1516D Custom
Model 1516D Deluxe
Model 1518D
Model 1518D Deluxe

ANSCHUTZ MODEL 1516D CLASSIC

Bolt action; 22 WMR; 4-shot clip; 22 ¹/₂″ barrel; 41″ overall length; weighs 6 lbs.; classic-style European walnut stock with straight comb; checkered pistol grip and forend; hooded ramp front sight, folding leaf rear; uses Match 64 action; adjustable single-stage trigger; receiver grooved for scope mounting. Imported from Germany by Precision Sales International. Still imported. **LMSR: $799**
   Perf.: $595   **Exc.:** $525   **VGood:** $460

Anschutz Model 1516D Custom

Same specs as Model 1516D Classic except European walnut stock with roll-over Monte Carlo cheekpiece; slim forend with Schnabel tip; fine cut checkering on grip and forend. Imported from Germany by Precision Sales International. Introduced 1988; still imported. **LMSR: $799**
   Perf.: $700   **Exc.:** $560   **VGood:** $450

Anschutz Model 1516D Deluxe

Same specs as Model 1516D Classic except European walnut Monte Carlo stock with cheekpiece; Schnabel forend; Model 1403 target rifle action. Imported from Germany by Precision Sales International. No longer imported.
   Perf.: $700   **Exc.:** $560   **VGood:** $450

ANSCHUTZ MODEL 1518D

Bolt action; 22 WMR; 4-shot clip; 19 ³/₄″ barrel; 38″ overall length; weighs 5 ¹/₂ lbs.; European walnut Mannlicher-style stock with mahogany Schnabel tip; checkered pistol grip and forend; hooded ramp front sight, folding leaf rear; set trigger. Introduced 1976; still in production. **LMSR: $1170**
   Perf.: $750   **Exc.:** $640   **VGood:** $500

Anschutz Model 1518D Deluxe

Same specs as Model 1518D except Model 1403 target rifle action; adjustable single stage trigger; receiver grooved for scope mounting. Imported from Germany. No longer imported.
   Perf.: $695   **Exc.:** $595   **VGood:** $470

Anschutz Model 1432 Custom

Anschutz Model 1433D

Anschutz Model 1449 Youth

Anschutz Model 1518D

Anschutz
Model 1432D (See Anschutz
Model 1700D Classic)
Model 1432D Custom
(See Anschutz Model 1700D
Custom)
Model 1522D Classic
Model 1522D Custom
Model 1532D (See Anschutz
Model 1700D Classic)
Model 1533D
Model 1700D Bavarian
Model 1700D Bavarian
Meistergrade
Model 1700D Classic
Model 1700D Classic
Meistergrade
Model 1700D Custom
Model 1700D Custom Graphite
Model 1700D Custom
Meistergrade
Model 1700D Featherweight
Model 1700D Featherweight
Deluxe
Model 1700D Graphite Hornet
Model 1733D Mannlicher
Model 1803D
Model 1803D Intermediate
Match
Model 1807 Match (See
Anschutz Model 1907 Match)
Model 1807L Match (See
Anschutz Model 1907L Match)
Model 1808D RT
Model 1810 Super Match II
(See Anschutz Model 1910
Super Match II)
Model 1811 Match
(See Anschutz Model 1911
Prone Match)
Model 1811L Match
(See Anschutz Model 1911L
Prone Match)
Model 1813 Super Match
(See Anschutz Model 1913
Super Match)

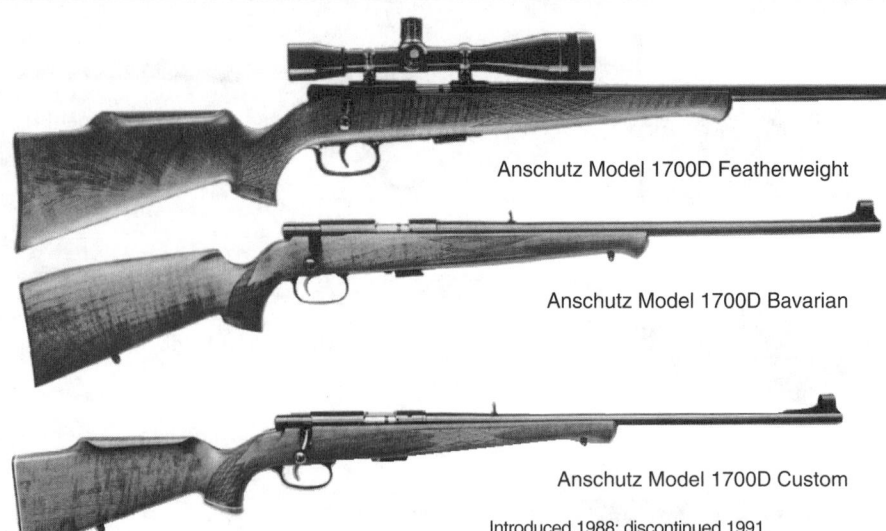

Anschutz Model 1700D Featherweight

Anschutz Model 1700D Bavarian

Anschutz Model 1700D Custom

ANSCHUTZ MODEL 1522D CLASSIC
Bolt action; 22 WMR; 5-shot clip; 24″ barrel; 43″ overall length; weighs 7 1/4 lbs.; hooded ramp front sight, folding leaf rear; drilled, tapped for scope mount; select European walnut stock; checkered pistol grip, forend; adjustable single-stage trigger. Made in Germany. Introduced 1982; importation discontinued 1988.

   Perf.: $650   **Exc.:** $575   **VGood:** $450

Anschutz Model 1522D Custom
Same specs as Model 1522D except Monte Carlo cheekpiece; slim forend; Schnabel tip; palm swell on pistol grip; rosewood grip cap, diamond insert; skip-line checkering. Introduced 1982; importation discontinued 1988.

   Perf.: $695   **Exc.:** $595   **VGood:** $470

ANSCHUTZ MODEL 1533D
Bolt action; 222 Rem.; 3-shot box magazine; 19 3/4″ barrel; receiver grooved for scope mount; folding

leaf rear sight, hooded ramp front; European walnut Mannlicher-style stock; hand-checkered pistol grip, forend; cheekpiece; single-stage or double-set trigger. Introduced 1976; no longer in production.

   Perf.: $995   **Exc.:** $850   **VGood:** $695

ANSCHUTZ MODEL 1700D BAVARIAN
Bolt action; 22 LR, 22 Hornet, 222 Rem.; 5-shot clip (22 Hornet); 24″ barrel; 43″ overall length; weighs 7 1/2 lbs.; European Monte Carlo walnut stock with Bavarian cheek rest; checkered pistol grip and forend; hooded ramp front sight, folding leaf rear; improved 1700 Match 54 action with adjustable #5096 trigger; drilled and tapped for scope mounting. Imported from Germany. Introduced 1988; still imported. **LMSR: $1364**
Rimfire

   Perf.: $910   **Exc.:** $795   **VGood:** $600
Centerfire

   Perf.: $1000   **Exc.:** $875   **VGood:** $700

Anschutz Model 1700D Bavarian Meistergrade
Same specs as Model 1700D Bavarian except select European walnut stock; gold engraved trigger guard. Introduced 1988; still imported. **LMSR: $1563**
Rimfire

   Perf.: $1000   **Exc.:** $875   **VGood:** $720
Centerfire

   Perf.: $1150   **Exc.:** $950   **VGood:** $725

Anschutz Model 1700D Classic
Same specs as Model 1700D Bavarian except 22 LR; weighs 6 3/4 lbs.; #5095 trigger; select European walnut stock with checkered pistol grip and forend. Imported from Germany. Introduced 1988; importation dropped 1994.

   Perf.: $1000   **Exc.:** $850   **VGood:** $600

Anschutz Model 1700D Classic Meistergrade
Same specs as 1700D Classic except better grade of wood; gold engraved trigger guard. Introduced 1988; importation dropped 1994.

   Perf.: $1050   **Exc.:** $900   **VGood:** $775

Anschutz Model 1700D Custom
Same specs as Model 1700D Bavarian except European walnut stock with roll-over Monte Carlo cheekpiece; slim forend; Schnabel tip; palm swell on pistol grip; skip-line checkering. Made in Germany. Introduced 1988; discontinued 1991.

   Perf.: $1100   **Exc.:** $850   **VGood:** $675

Anschutz Model 1700D Custom Graphite
Same specs as Model 1700D Custom except 22 LR; 22″ barrel; 41″ overall length; weighs 7 1/2 lbs.; McMillan graphite-reinforced stock, roll-over cheekpiece; quick-detach sling swivels; embroidered sling. Introduced 1991; still imported. **LMSR: $1364**

   Perf.: $1050   **Exc.:** $875   **VGood:** $695

Anschutz Model 1700D Custom Meistergrade
Same specs as Model 1700D Custom except select European walnut stock; gold, engraved trigger guard.

Introduced 1988; discontinued 1991.

   Perf.: $1295   **Exc.:** $995   **VGood:** $825

Anschutz Model 1700D Featherweight
Same specs as Model 1700D Bavarian except 22 LR; 22″ barrel; 41″ overall length; weighs 6 1/4 lbs.; matte black McMillan fiberglass stock; single-stage #5096 trigger. Introduced 1988; still in production. **LMSR: $1230**

   Perf.: $1000   **Exc.:** $850 **VGood:** $715

Anschutz Model 1700D Featherweight Deluxe
Same specs as Model 1700D Featherweight except Fibergrain synthetic stock with simulated wood grain and skip-line checkering. Introduced 1990; still imported. **LMSR: $1460**

   Perf.: $1050   **Exc.:** $900   **VGood:** $750

Anschutz Model 1700D Graphite Hornet
Same specs as Model 1700D Custom except 22 Hornet; 22″ barrel; 41″ overall length; weighs 7 1/2 lbs.; McMillan graphite-reinforced fiberglass stock with roll-over cheekpiece; built on Anschutz sporter action; fitted with Anschutz logo sling and quick-release swivels. Imported from Germany by Precision Sales International. Introduced 1995; still imported. **LMSR: $1299**

   Perf.: $1050   **Exc.:** $875   **VGood:** $725

ANSCHUTZ MODEL 1733D MANNLICHER
Bolt action; 22 Hornet; 5-shot clip; 19 3/4″ barrel; 39″ overall length; weighs 6 1/4 lbs.; full-length Mannlicher-style stock; hooded ramp front sight, Lyman folding rear; improved Match 54 action with #5096 single-stage trigger with 2 1/2-lb. adjustable pull weight; sling swivels. Imported from Germany. Introduced 1993; still imported. **LMSR: $1657**

   Perf.: $1250   **Exc.:** $1000   **VGood:** $795

ANSCHUTZ MODEL 1803D
Bolt action; single shot; 22 LR; 25 1/2″ heavy barrel; 43 3/4″ overall length; weighs 8 3/8 lbs.; no sights; walnut-finished European hardwood stock; adjustable cheekpiece; stippled grip, forend; built on Anschutz Match 64 action; #5091 two-stage trigger; right- or left-hand models. Made in Germany. Introduced 1987; discontinued 1993.

   Perf.: $800   **Exc.:** $725   **VGood:** $585

Anschutz Model 1803D Intermediate Match
Same specs as Model 1803D except 25″ barrel with 3/4″ diameter; weighs 9 1/2 lbs.; blonde-finished European hardwood stock with adjustable buttplate and cheekpiece; accepts Anschutz #6825 sight set. Imported from Germany. Introduced 1991; dropped 1994.

   Perf.: $695   **Exc.:** $595   **VGood:** $450

ANSCHUTZ MODEL 1808D RT
Bolt action; single shot; 22 LR; 32 1/2″ barrel; 50 1/2″ overall length; weighs 9 3/8 lbs.; no sights furnished; grooved for scope mount; European walnut stock, heavy beavertail forend; adjustable cheekpiece, buttplate; stippled grip, forend; nine-way adjustable single-stage trigger; slide safety; right- or left-hand. Made in Germany. Introduced 1991; still imported. **LMSR $2220**

   Perf.: $1500   **Exc.:** $1250   **VGood:** $1000

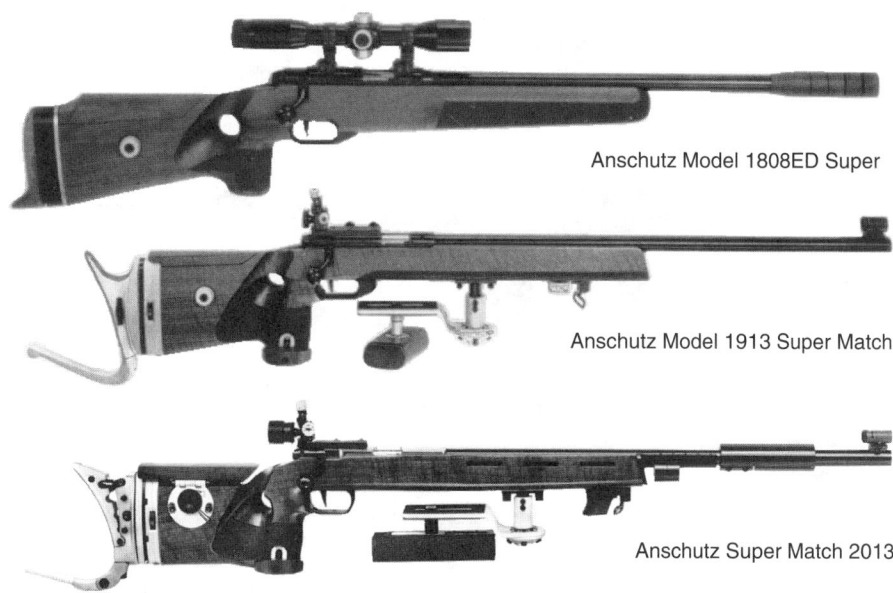

Anschutz Model 1808ED Super

Anschutz Model 1913 Super Match

Anschutz Super Match 2013

RIFLES

Anschutz
Model 1808ED Super
Model 1827B Biathlon
Model 1827BT Fortner
Biathlon
Model 1903 Match
Model 1907 Match
Model 1907-L Match
Model 1910 Super Match II
Model 1911 Prone Match
Model 1913 Super Match
Super Match 2007 ISU
Standard
Super Match 2013
Woodchucker

ANSCHUTZ MODEL 1808ED SUPER

Bolt action; single shot; 22 LR; 23 1/2" heavy barrel; 42" overall length; 9 1/4 lbs.; no sights; receiver grooved for scope mount; European hardwood stock; adjustable cheekpiece; beavertail forend; stippled pistol grip, forend; removable barrel weights; adjustable trigger; designed for Running Target. Special order only. Made in Germany. Introduced 1982; importation discontinued 1990.

Right-hand version
Perf.: $1495   **Exc.:** $1295   **VGood:** $995
Left-hand version
Perf.: $1500   **Exc.:** $1300   **VGood:** $1000

ANSCHUTZ MODEL 1827B BIATHLON

Bolt-action; 22 LR; 5-shot magazine; 21 1/2" barrel; 42 1/2" overall length; weighs 8 1/2 lbs.; special Biathlon globe front sight, micrometer rear with snow cap; adjustable trigger; adjustable wooden buttplate; adjustable hand-stop rail; Biathlon butthook. Special order only. Made in Germany. Introduced 1982; still imported. **LMSR $2457**

Right-hand version
Perf.: $1900   **Exc.:** $1500   **VGood:** $1100
Left-hand version
Perf.: $1950   **Exc.:** $1550   **VGood:** $1150

Anschutz Model 1827BT Fortner Biathlon

Same specs as the Anschutz 1827B Biathlon rifle except Anschutz/Fortner system straight-pull bolt action. Imported from Germany. Introduced 1982; still imported. **LMSR $3722**

Perf.: $2900   **Exc.:** $2100   **VGood:** $1600
Left-hand
Perf.: $2950   **Exc.:** $2150   **VGood:** $1650

ANSCHUTZ MODEL 1903 MATCH

Bolt action; single shot; 22 LR; 25" barrel, 3/4" diameter; 43 3/4" overall length; weighs 8 1/2 lbs.; walnut-finished hardwood stock with adjustable cheekpiece; stippled grip and forend; no sights; accepts #6825 sight set; Anschutz Match 64 action; #5098 two-stage trigger; designed for intermediate and advanced Junior Match competition. Imported from Germany. Introduced 1987; still imported. **LMSR $1163**

Right-hand
Perf.: $800   **Exc.:** $700   **VGood:** $525
Left-hand
Perf.: $850   **Exc.:** $750   **VGood:** $575

ANSCHUTZ MODEL 1907 MATCH

Bolt action; single shot; 22 LR; 26" barrel with 7/8" diameter barrel; 44 1/2" overall length; weighs 10 lbs.; blonde-finished European hardwood stock with vented forend; stippled pistol grip and forend; no sights; designed for ISU requirements; suitable for NRA matches. Imported from Germany. Still imported. **LMSR $1983**

Perf.: $1595   **Exc.:** $1250   **VGood:** $895

Anschutz Model 1907-L Match

Same specs as Model 1907 Match except true left-hand action and stock. Imported from Germany. Still imported. **LMSR: $2164**

Perf.: $1650   **Exc.:** $1300   **VGood:** $950

ANSCHUTZ MODEL 1910 SUPER MATCH II

Bolt action; single shot; 22 LR; 27 1/4" barrel; 46" overall length; weighs about 14 lbs.; European walnut International-type stock with adjustable cheekpiece; adjustable aluminum hook buttplate; adjustable hand stop; uses Match 54 action; sights not included. Imported from Germany. Introduced 1982; no longer imported.

Right-hand
Perf.: $2300   **Exc.:** $1900   **VGood:** $1300
Left-hand
Perf.: $2400   **Exc.:** $2000   **VGood:** $1400

ANSCHUTZ MODEL 1911 PRONE MATCH

Bolt action; 22 LR; single shot; 27 1/4" barrel; 46" overall length; weighs 12 lbs.; walnut-finished European hardwood stock of American prone style with Monte Carlo, cast-off cheekpiece; checkered pistol grip; beavertail forend with swivel rail and adjustable swivel; adjustable rubber buttplate; no sights; receiver grooved for Anschutz sights; scope blocks included; two-stage #5018 adjustable trigger. Imported from Germany. Still imported. **LMSR $2325**

Perf.: $1750   **Exc.:** $1250   **VGood:** $950

ANSCHUTZ MODEL 1913 SUPER MATCH

Bolt action; single shot; 22 LR; 27 1/4" barrel; 46" overall length; weighs 15 1/2 lbs.; international diopter sights; match-grade European walnut International-type stock with adjustable cheekpiece, aluminum hook buttplate; fully-adjustable hand stop. Introduced 1988; still in production. **LMSR $3422**

Right-hand action
Perf.: $2600   **Exc.:** $1900   **VGood:** $1100
Left-hand action
Perf.: $2700   **Exc.:** $2000   **VGood:** $1200

ANSCHUTZ SUPER MATCH 2007 ISU STANDARD

Bolt action; 22 LR; single shot; 19 3/4" barrel; 43 1/2" to 44 1/2" overall length; weighs about 11 lbs.; ISU Standard design stock of European walnut or blonde hardwood; no sights; uses improved Super Match 54 action, #5018 trigger; micro-honed barrel. Imported from Germany. Introduced 1992; still imported. **LMSR: $2961**

Right-hand
Perf.: $2200   **Exc.:** $1600   **VGood:** $1050
Left-hand
Perf.: $2300   **Exc.:** $1700   **VGood:** $1150

ANSCHUTZ SUPER MATCH 2013

Bolt action; single shot; 22 LR; 19 3/4" barrel (26" with tube installed); 43 1/2" to 45 1/2" overall length; weighs 15 1/2 lbs.; target adjustable European walnut stock; no sight; #7020/20 sight set optional; improved Super Match 54 action; #5018 trigger; micro-honed barrel; two-stage, nine-point adjustment trigger; slide safety. Imported from Germany. Introduced 1992; still imported. **LMSR: $4067**

Right-hand
Perf.: $3000   **Exc.:** $2500   **VGood:** $1700
Left-hand
Perf.: $3100   **Exc.:** $2600   **VGood:** $1800

ANSCHUTZ WOODCHUCKER

Bolt action; 22 LR; 5-shot clip; 16 1/4" barrel; 32 1/4" overall length; weighs 3 1/2 lbs.; bead front sight, U-notched rear; receiver grooved for scope mounting; walnut finished European hardwood stock; dual opposing extractors; built on Anschutz Mark 2000 action. Made in Germany. Introduced 1988; discontinued 1990.

Perf.: $230   **Exc.:** $200   **VGood:** $150

Armalite AR-7 Explorer

Armscor Model AK22

Armscor Model 14D

Armscor Model 50S

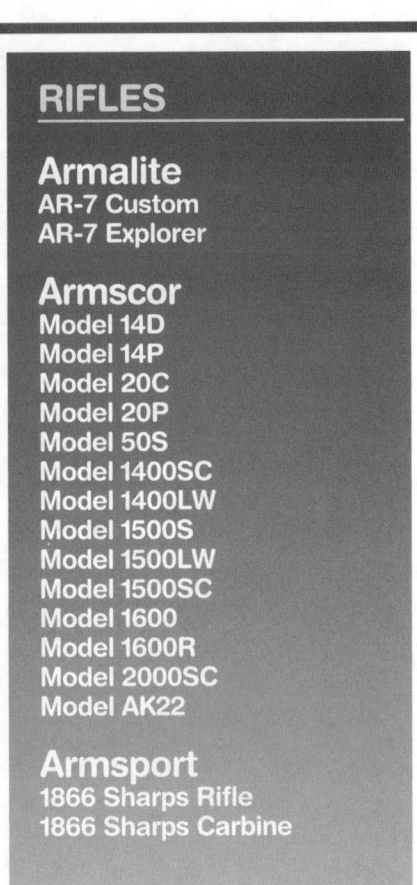

Armalite
AR-7 Custom
AR-7 Explorer

Armscor
Model 14D
Model 14P
Model 20C
Model 20P
Model 50S
Model 1400SC
Model 1400LW
Model 1500S
Model 1500LW
Model 1500SC
Model 1600
Model 1600R
Model 2000SC
Model AK22

Armsport
1866 Sharps Rifle
1866 Sharps Carbine

ARMALITE AR-7 CUSTOM
Semi-automatic; 22 LR; 8-shot box magazine; 16″ cast aluminum barrel with steel liner; 34 1/2″ overall length; peep rear sight, blade front; walnut stock; pistol grip; cheekpiece; not designed to float. Introduced 1964; discontinued 1970.
   Perf.: $170   **Exc.:** $135   **VGood:** $100

Armalite AR-7 Explorer
Same specs as AR-7 Custom except moulded Cycolac plastic stock, hollow for storing action, barrel, magazine; designed to float; snap-on rubber buttcap; modification of survival rifle designed for Air Force. Introduced in 1959 by Armalite; made by Charter Arms from 1973 to 1990; reintroduced 1992 by Survival Arms; discontinued 1995.
   Perf.: $120   **Exc.:** $100   **VGood:** $80

ARMSCOR MODEL 14D
Bolt action; 22 LR; 10-shot magazine; 23″ barrel; 41 1/2″ overall length; weighs 6 1/2 lbs.; fully-adjustable rear sight, bead ramp front; grooved for scope mounts; checkered walnut-finished mahogany stock; blued finish. Made in Philippines. Introduced 1987; still imported. **LMSR: $149**
   Perf.: $100   **Exc.:** $75   **VGood:** $55

Armscor Model 14P
Same specs as Model 14D except hooded bead ramp front sight; plain uncheckered stock. Made in the Philippines. Introduced 1987; still imported. **LMSR: $139**
   Perf.: $90   **Exc.:** $70   **VGood:** $50

ARMSCOR MODEL 20C
Semi-automatic; 22 LR; 15-shot magazine; 18 1/4″ barrel; 38″ overall length; weighs 6 1/4 lbs.; fully-adjustable rear sight, bead ramp front; receiver grooved for scope mounts; walnut finished Philippine mahogany; carbine-style stock; steel barrel band, buttplate; blued finish. Introduced 1990; still in production. **LMSR: $149**
   Perf.: $110   **Exc.:** $85   **VGood:** $60

Armscor Model 20P
Same specs as Model 20C except 20 3/4″ barrel; 40 1/2″ overall length; hooded bead ramp front sight; elevation-adjustable rear; no barrel band. Made in the Philippines. Introduced 1987; still imported. **LMSR: $129**
   Perf.: $100   **Exc.:** $70   **VGood:** $50

ARMSCOR MODEL 50S
Semi-automatic; 22 LR; 25-, 30-shot magazine; 18 1/4″ ventilated barrel shroud; 38″ overall length; weighs 6 1/2 lbs.; bead ramp front sight, open U-notch fully-adjustable rear; mahogany stock; blued finish. Introduced 1990; still imported. **LMSR: $209**
   Perf.: $150   **Exc.:** $120   **VGood:** $90

ARMSCOR MODEL 1400SC
Bolt action; 22 LR; 10-shot magazine; 23″ barrel; 41 1/4″ overall length; weighs 6 3/4 lbs.; hand-checkered American walnut Monte Carlo stock with cheekpiece; contrasting wood forend tip and grip cap; red recoil pad; engine-turned bolt; bead ramp front sight, fully-adjustable open rear; grooved for scope mount; blue finish. Imported from the Philippines by Ruko Products. Introduced 1990; dropped 1994.
   Perf.: $230   **Exc.:** $190   **VGood:** $130

Armscor Model 1400LW
Same specs as 1400SC except weighs 6 lbs.; Schnabel forend; hard rubber recoil pad; blued finish. Made in Philippines. Introduced 1990; discontinued 1992.
   Perf.: $175   **Exc.:** $145   **VGood:** $105

ARMSCOR MODEL 1500S
Bolt action; 22 WMR; 5-shot magazine; 23″ barrel; 41″ overall length; weighs 6 1/2 lbs.; fixed sights; receiver grooved for scope mounts; double-lug bolt; plain mahogany stock; blued finish. Made in the Philippines. Introduced 1987; still imported. **LMSR: $236**
   Perf.: $160   **Exc.:** $135   **VGood:** $95

Armscor Model 1500LW
Same specs as Model 1500S except lightweight European-style checkered walnut stock with recoil pad. Introduced 1990; discontinued 1992.
   Perf.: $180   **Exc.:** $145   **VGood:** $105

Armscor Model 1500SC
Same specs as Model 1500S except 41 1/4″ overall length; hand-checkered Monte Carlo stock of American walnut; contrasting wood forend tip, grip cap; red rubber recoil pad; lathe-turned bolt. Introduced 1990; no longer in production.
   Perf.: $225   **Exc.:** $175   **VGood:** $125

ARMSCOR MODEL 1600
Semi-automatic; 22 LR; 15-shot magazine; 19 1/2″ barrel; 38 1/2″ overall length; weighs 6 lbs.; peep rear sight, post front; black ebony wood stock; matte black finish; resembles Colt AR-15. Made in the Philippines. Introduced 1987; still imported. **LMSR: $199**
   Perf.: $165   **Exc.:** $135   **VGood:** $100

Armscor Model 1600R
Same specs as Model 1600 except weighs 7 lbs.; stainless steel retractable buttstock; ventilated forend. Still imported. **LMSR: $199**
   Perf.: $170   **Exc.:** $140   **VGood:** $100

ARMSCOR MODEL 2000SC
Semi-automatic; 22 LR; 15-shot magazine; 20 3/4″ barrel; 40 1/2″ overall length; weighs 6 1/4 lbs.; fully-adjustable open U-notch rear, bead ramp front sight; receiver grooved for scope mounts; checkered American walnut Monte Carlo stock with cheekpiece; rubber recoil pad; blued finish. Made in Philippines. Introduced 1987; still imported. **LMSR: $NA**
   Perf.: $110   **Exc.:** $75   **VGood:** $55

ARMSCOR MODEL AK22
Semi-automatic; 22 LR; 15-, 30-shot magazine; 18 1/4″ barrel; 36″ overall length; weighs 7 lbs.; plain mahogany stock; post front sight, open U-notch fully-adjustable rear; matte black finish; resembles the AK-47. Imported from the Philippines by Ruko Products. Introduced 1987; still imported. **LMSR: $269**
   Perf.: $190   **Exc.:** $140   **VGood:** $110
With folding steel stock
   Perf.: $210   **Exc.:** $160   **VGood:** $120

ARMSPORT 1866 SHARPS RIFLE
Single shot; 45-70; 28″ round or octagonal barrel; 46″ overall length; weighs 8 lbs.; walnut stock and forend; blade front sight, folding adjustable rear; tang sight set optionally available; replica of the 1866 Sharps; color case-hardened frame, rest blued. Still imported by Armsport. **LMSR: $860**
   Perf.: $650   **Exc.:** $495   **VGood:** $395
With octagonal barrel
   Perf.: $675   **Exc.:** $525   **VGood:** $425

Armsport 1866 Sharps Carbine
Same specs as the 1866 Rifle except 22″ round barrel. Still imported by Armsport. **LMSR: $830**
   Perf.: $650   **Exc.:** $495   **VGood:** $395

Armscor Model 1600

Auto-Ordnance Thompson M1

Armsport 1866 Sharps Carbine

A-Square Hannibal

Armsport
2785 & 2786
2800 Series
4000 Series
4500 Series
Tikka 4600 Series

A-Square
Caesar
Hamilcar
Hannibal

Auto-Ordnance
27 A-1 Thompson
1927A-3
Thompson M1

ARMSPORT 2785 & 2786
Semi-automatic; 22 LR; 10-, 15-shot magazine; adjustable rear, blade front sight; walnut stock; blued finish. Introduced 1985; discontinued 1985.
   Perf.: $140   **Exc.:** $120   **VGood:** $80

ARMSPORT 2800 SERIES
Bolt action; 30-06 (2801), 308 (2802), 270 Win. (2803), 243 Win. (2804), 7mm Rem. Mag. (2805), 300 Win. Mag. (2806); 24″ barrel; weighs 8 lbs.; open adjustable rear, ramp front sight; checkered European walnut Monte Carlo stock; blued finish, Made in Italy. Introduced 1986; discontinued 1986.
   Perf.: $595   **Exc.:** $495   **VGood:** $375

ARMSPORT 4000 SERIES
Double rifle; 243, 270, 284, 7.65mm, 308, 30-06, 7mm Rem. Mag., 9.3mm, 300 H&H, 375 H&H; interchangeable 16-, 20-ga. shotgun barrels; blade front sight with bead, windage-adjustable leaf rear; engraved receiver, sideplates; marketed in hand-fitted leather case. Introduced 1978; discontinued 1985.
Armsport 4010 (side-by-side version)
   Perf.: $15,000 **Exc.:** $11,995   **VGood:** $8995
Armsport 4011
   Perf.: $11,995 **Exc.:** $8995   **VGood:** $6995
Armsport 4012
   Perf.: $25,000 **Exc.:** $19,995   **VGood:** $15,000
Armsport 4013
   Perf.: $22,000 **Exc.:** $16,500   **VGood:** $12,000
Armsport 4020
   Perf.: $3600   **Exc.:** $2700   **VGood:** $2100
Armsport 4021
   Perf.: $3300   **Exc.:** $2400   **VGood:** $1800
Armsport 4022
   Perf.: $3350   **Exc.:** $2500   **VGood:** $1850
Armsport 4023
   Perf.: $2900   **Exc.:** $2200   **VGood:** $1750

ARMSPORT 4500 SERIES
Lever action; 20″ barrel (carbine), 24″ barrel (standard); adjustable rear, blade front sight; engraved sideplates; walnut stock; black chromed barrels; replica of Winchester 1873. Discontinued 1986.

Armsport 4500 Standard, 44-40
   Perf.: $1050   **Exc.:** $850   **VGood:** $650
Armsport 4501 Standard, 357 Mag.
   Perf.: $1000   **Exc.:** $800   **VGood:** $600
Armsport 4502 Carbine, 44-40
   Perf.: $1050   **Exc.:** $850   **VGood:** $650
Armsport 4503 Carbine, 357 Mag.
   Perf.: $1000   **Exc.:** $800   **VGood:** $600
Armsport 4504 Carbine, 357 Mag.
   Perf.: $500   **Exc.:** $410   **VGood:** $320

ARMSPORT TIKKA 4600 SERIES
Bolt action; 30-06 (4601), 308 Win. (4602), 270 Win. (4603), 243 Win. (4604), 7mm Rem. (4605), 300 Win. Mag. (4606), 222 Rem. (4607); adjustable rear, blade front sight; walnut stock; rubber recoil pad; blued finish. Discontinued 1984.
   Perf.: $550   **Exc.:** $410   **VGood:** $380

A-SQUARE CAESAR
Bolt action; numerous calibers; 20″ to 26″ barrel; weighs 8 1/2 to 11 lbs.; three-leaf express sights or scope; Claro walnut hand-rubbed oil-finish or synthetic stock; Coil-Chek recoil reducer; flush detachable swivels; double cross-bolts; claw extractor; right- or left-hand models; built on M700 receiver; matte blue finish. Introduced 1984; still in production. **LMSR: $3295**
   Perf.: $2700   **Exc.:** $2200   **VGood:** $1100

A-SQUARE HAMILCAR
Bolt action; 25-06, 6.5x55, 270 Win., 7x57, 280 Rem., 30-06, 338-06, 9.3x62, 257 Wea. Mag., 264 Win. Mag., 270 Wea. Mag., 7mm Rem. Mag., 7mm Wea. Mag., 7mm STW, 300 Win. Mag., 300 Wea. Mag.; 4- or 6-shot magazine, depending upon caliber; custom barrel lengths to customer specifications; weighs 8-8 1/2 lbs.; oil-finished Claro walnut stock with A-Square Coil-Chek, recoil pad; choice of three-leaf express, forward or normal-mount scope; matte blue finish; Mauser-style claw extractor; two-position safety; three-way target trigger. From A-Square Co., Inc. Introduced 1994; still produced. **LMSR: $3295**
   Perf.: $2700   **Exc.:** $2200   **VGood:** $1100

A-SQUARE HANNIBAL
Bolt action; numerous calibers; 20″ to 26″ barrel; weighs 9 to 11 3/4 lbs.; three-leaf express sights or scope; Claro walnut hand-rubbed oil-finish or synthetic stock; Coil-Chek recoil reducer; double cross-bolts; claw extractor; two-position safety; three-way target trigger; built on P-17 Enfield receiver. Introduced 1983; still in production. **LMSR: $3295**
   Perf.: $2700   **Exc.:** $2200   **VGood:** $1100
Note: A-Square rifles no longer are marketed by group number.

AUTO-ORDNANCE 27 A-1 THOMPSON
Semi-automatic; 45 ACP; 30-shot magazine; 16″ barrel; 42″ overall length; weighs 11 1/2 lbs.; walnut stock and vertical forend; blade front sight, open windage-adjustable rear; recreation of Thompson Model 1927; semi-auto only; finned barrel. Made in U.S. by Auto-Ordnance Corp. Still produced. **LMSR: $795**
   Perf.: $550   **Exc.:** $495   **VGood:** $420

AUTO-ORDNANCE 1927A-3
Semi-automatic; 22 LR; 10-, 30- or 50-shot magazine; 16″ finned barrel; weighs about 7 lbs; walnut stock and forend; blade front sight, open fully-adjustable rear; recreation of the Thompson Model 1927; alloy receiver. Made in U.S. by Auto-Ordnance. No longer produced.
   Perf.: $475   **Exc.:** $380   **VGood:** $300

AUTO-ORDNANCE THOMPSON M1
Semi-automatic; 45 ACP; 30-shot magazine; 16″ barrel; 42″ overall length; weighs 11 1/2 lbs.; walnut stock and horizontal forend; blade front sight, open windage-adjustable rear; side cocking knob; smooth unfinned barrel; sling swivels on butt and forend. Made in U.S. by Auto-Ordnance Corp. Introduced 1985; still produced. **LMSR: $950**
   Perf.: $650   **Exc.:** $550   **VGood:** $375

Barrett
Model 82A-1
Model 90
Model 95

Bayard
Semi-Automatic Model

Beeman
HW 60
HW 60J-ST
HW 660 Match

Benton & Brown
Model 93

Beretta
AR-70
Model 455
Model 500
Model 500 S
Model 500 DL
Model 500 DLS
Model 500 EEL
Model 500 EELLS

Bayard Semi-Automatic Model

Beeman/HW 660 Match

Beeman/HW 60J-ST

Beretta 455EELL

BARRETT MODEL 82A-1
Semi-automatic; 50 BMG; 10-shot detachable box magazine; 29″ barrel; 57″ overall length; weighs 28 1/2 lbs.; composition stock with Sorbothane recoil pad; no sights, scope optional; recoil operated with recoiling barrel; three-lug locking bolt; muzzlebrake; self-leveling bipod; fires same 50-caliber ammunition as the M2HB machinegun. From Barrett Firearms. Introduced 1985; still produced. **LMSR: $6750**
    Perf: $6200   **Exc.:** $5100   **VGood:** $3900

BARRETT MODEL 90
Bolt action; 50 BMG; 5-shot magazine; 29″ barrel; 35″ overall length; weighs 22 lbs.; Sorbothane recoil pad on stock; no sights, scope optional; bullpup design; extendable bipod legs; match-grade barrel; high efficiency muzzlebrake. Made in U.S. by Barrett Firearms Mfg., Inc. Introduced 1990; dropped 1994.
    Perf: $3850   **Exc.:** $2500   **VGood:** $2000

BARRETT MODEL 95
Bolt action; 50 BMG; 5-shot magazine; 29″ barrel; 45″ overall length; weighs 22 lbs.; Sorbothane recoil pad on butt; no sights, scope optional; updated version of the Model 90; bullpup design; extendable bipod legs; match-grade barrel; high efficiency muzzlebrake. Made in U.S. by Barrett Firearms Mfg., Inc. Introduced 1995; still produced. **LMSR: $4950**
    Perf: $3900   **Exc.:** $2800   **VGood:** $2200

BAYARD SEMI-AUTOMATIC MODEL
Single shot; 22 Short, 22 Long; 19″ barrel; 39″ overall length; weighs 3 3/4 lbs.; fixed sights; loads cartridge when release stud is activated; American walnut stock. Originally priced at $4.50. Collector value. Introduced 1908; discontinued 1914.
    Perf: $150   **Exc.:** $100   **VGood:** $80

BEEMAN/HW 60
Bolt action; single shot; 22 LR; 26 7/8″ barrel; 45 3/4″ overall length; weighs 11 lbs.; walnut stock with adjustable buttplate; stippled pistol grip and forend; accessory rail with adjustable swivel; hooded ramp front sight, match-type aperture rear; adjustable match trigger with push-button safety. Imported from Germany by Beeman. Introduced 1981; no longer imported.
Right-hand
    Perf.: $595   **Exc.:** $450   **VGood:** $375
Left-hand
    Perf.: $795   **Exc.:** $595   **VGood:** $425

BEEMAN/HW 60J-ST
Bolt action; 22 LR; 22 7/8″ barrel; 41 3/4″ overall length; weighs 6 1/2 lbs.; walnut stock with cheekpiece; cut checkered pistol grip and forend; hooded blade front, open rear sight; polished blue finish; oil-finished walnut. Imported from Germany by Beeman. Introduced 1988; dropped 1992.
    Perf.: $495   **Exc.:** $395   **VGood:** $295
222 Rem. caliber
    Perf.: $795   **Exc.:** $650   **VGood:** $495

BEEMAN/HW 660 MATCH
Bolt action; single shot; 22 LR; 26″ barrel; weighs 10 3/4 lbs.; 45 5/16″ overall length; match-type walnut stock with adjustable cheekpiece and buttplate; globe front sight, match aperture rear; adjustable match trigger; stippled pistol grip and forend; forend accessory rail. Imported from Germany by Beeman. Introduced 1988; no longer imported.
    Perf.: $695   **Exc.:** $550   **VGood:** $425

BENTON & BROWN MODEL 93
Bolt action; 243, 6mm Rem., 25-06, 270, 280 Rem., 308, 30-06 (standard calibers); 257 Wea. Mag., 264 Win. Mag., 7mm Rem. Mag., 300 Wea. Mag., 300 Win. Mag., 338 Win. Mag., 340 Wea. Mag., 375 H&H (magnum calibers); 22″ barrel (standard calibers), 24″ (magnum calibers); 41″ overall length (22″ barrel); weighs about 8 1/2 lbs.; two-piece stock design of fancy walnut or fiberglass; oil finish; 20 lpi borderless checkering; sights optional; short-throw bolt action with 60° bolt throw; takedown design; interchangeable barrels and bolt assemblies; left-hand models available; safety locks firing pin and bolt. Made in U.S. by Benton & Brown Firearms, Inc. Introduced 1995. **LMSR: $2075**
    Perf: $1750   **Exc.:** $1300   **VGood:** $1000
Fiberglass stock model
    Perf: $1550   **Exc.:** $1100   **VGood:** $800

BERETTA AR-70
Semi-automatic; 223; 8-, 30-shot magazine; 17 3/4″ barrel; weighs 8 1/4 lbs.; medium action; diopter adjustable sights; military-like epoxy resin finish. Comes with two magazines, cleaning kit and military-style carrying strap. Introduced 1984; discontinued 1989.
    Perf: $1610   **Exc.:** $1235   **VGood:** $1000

BERETTA MODEL 455
Double rifle; side-by-side; sidelock action; 375 H&H, 416 Rigby, 458 Win. Mag., 470 NE, 500 NE 3″; 23 1/2″, 25 1/2″ barrels; weighs 11 lbs.; blade front sight, folding leaf V-notch rear; hand-checkered European walnut stock; removable sideplates; double triggers; recoil pad; color case-hardened finish. Made in Italy. Introduced 1990; still imported. **LMSR: $36,000**
    Perf: $46,000   **Exc.:** $34,000   **VGood:** $22,000
Model 455EELL with select walnut, engraving
    Perf: $58,000   **Exc.:** $39,000   **VGood:** $29,000

BERETTA MODEL 500
Bolt action; 222 Rem., 223 Rem.; 5-shot; 24″ barrel; weighs 6 1/2-7 lbs.; short action; hood front sight, fully-adjustable rear; European walnut stock; oil-finished, hand-checkered; rubber buttpad. Made in Italy. Introduced 1984; importation discontinued 1990.
    Perf: $600   **Exc.:** $510   **VGood:** $420

Beretta Model 500 S
Same specs as Model 500 except with iron sights.
    Perf: $625   **Exc.:** $550   **VGood:** $460

Beretta Model 500 DL
Same specs as Model 500 except with select walnut, engraving.
    Perf: $1280   **Exc.:** $1000   **VGood:** $780

Beretta Model 500 DLS
Same specs as Model 500 except with iron sights, select walnut, engraving.
    Perf: $1190   **Exc.:** $1000   **VGood:** $810

Beretta Model 500 EEL
Same specs as Model 500 except with fine walnut, more engraving.
    Perf: $1425   **Exc.:** $1190   **VGood:** $975

Beretta Model 500 EELLS
Same specs as Model 500 except with iron sights, fine walnut, more engraving.
    Perf: $1510   **Exc.:** $1225   **VGood:** $1000

Beretta Model 500

Bernardelli Carbine

Big Horn Custom

RIFLES

Beretta
Model 501
Model 501 S
Model 501 DL
Model 501 DLS
Model 501 EELL
Model 501 EELLS
Model 502
Model 502 DL
Model 502 DLS
Model 502 EELL
Model 502 EELLS
Model 502 S
Model S689
Model S689E
Sporting Carbine
SSO Express
SSO5 Express
SSO6 Express Custom

Bernardelli
Carbine
Express VB
Model 120
Model 190
Model Combo 2000

Big Horn
Custom

BERETTA MODEL 501

Bolt action; 243, 308; 5-shot magazine; 23″ barrel; weighs 7-7 1/2 lbs.; hood front sight, fully-adjustable rear; checkered burled walnut stock; full pistol grip cap and palm swell; engraved receiver, triggerguard, magazine floorplate; rubber buttpad. Made in Italy. Introduced 1984; discontinued 1986.
Perf: $595 **Exc.:** $530 **VGood:** $450

Beretta Model 501 S
Same specs as Model 501 except with iron sights.
Perf: $625 **Exc.:** $560 **VGood:** $470

Beretta Model 501 DL
Same specs as Model 501 except with select walnut, engraving.
Perf: $1285 **Exc.:** $1070 **VGood:** $800

Beretta Model 501 DLS
Same specs as Model 501 except with iron sights, select walnut, engraving.
Perf: $1310 **Exc.:** $1040 **VGood:** $850

Beretta Model 501 EELL
Same specs as Model 501 except with fine walnut, more engraving.
Perf: $1550 **Exc.:** $1200 **VGood:** $1000

Beretta Model 501 EELLS
Same specs as Model 501 except with iron sights, fine walnut, more engraving.
Perf: $1575 **Exc.:** $1225 **VGood:** $1020

BERETTA MODEL 502

Bolt action; 6.5x55, 7mm Rem. Mag., 270, 7x64, 30-06, 300 Win. Mag., 375 H&H; 4-, 5-shot magazine; 24 1/2″ barrel; weighs 8-8 1/2 lbs.; no sights; checkered walnut stock; Schnabel forend; sling swivels; rubber buttpad. Made in Italy. Introduced, 1984; discontinued, 1986.
Perf: $640 **Exc.:** $530 **VGood:** $480

Beretta Model 502 DL
Same specs as Model 502 except with select walnut, engraving.
Perf: $1450 **Exc.:** $1175 **VGood:** $950

Beretta Model 502 DLS
Same specs as Model 502 except with iron sights, select walnut, engraving.
Perf: $1425 **Exc.:** $1195 **VGood:** $975

Beretta Model 502 EELL
Same specs as Model 502 except with fine walnut, more engraving.
Perf: $1595 **Exc.:** $1295 **VGood:** $1050

Beretta Model 502 EELLS
Same specs as Model 502 except with iron sights, fine walnut, more engraving
Perf: $1595 **Exc.:** $1295 **VGood:** $1050

Beretta Model 502 S
Same specs as Model 502 except with iron sights.
Perf: $675 **Exc.:** $550 **VGood:** $495

BERETTA MODEL S689

Double rifle; over/under boxlock; 9.3x74R; 23″ barrel; weighs 7 3/4 lbs.; open V-notch rear sight, blade front on ramp; European walnut stock; checkered grip, forend; silvered, lightly engraved receiver; double triggers; ejectors; solid buttplate. Made in Italy. Introduced 1984; importation discontinued 1990.
Perf: $3695 **Exc.:** $2995 **VGood:** $2015

Beretta Model S689E
Same specs as Model S689 except auto ejectors.
Perf: $3995 **Exc.:** $3250 **VGood:** $2250

BERETTA SPORTING CARBINE

Bolt action/semi-automatic; 22 LR; 4-, 8-, 20-shot magazine; 20 1/2″ barrel; three-leaf folding rear sight, Patridge-type front; European walnut stock; sling swivels; hand-checkered pistol-grip. When bolt handle is dropped, acts as conventional bolt-action; with bolt handle raised fires semi-auto. Produced in Italy following World War II.
Perf: $370 **Exc.:** $300 **VGood:** $230

BERETTA SSO EXPRESS

Double rifle; over/under sidelock; 9.3x74R, 375 H&H, 458 Win. Mag.; 23″ barrel; weighs 11 lbs.; blade front sight on ramp, open V-notch rear; hand-checkered European walnut stock; color case-hardened receiver, trigger, trigger guard; double triggers, ejectors; recoil pad. Made in Italy. Introduced 1984; discontinued 1990.
Perf: $12,500 **Exc.:** $10,000 **VGood:** $7500

Beretta SSO5 Express
Same specs as Model SSO except with select walnut, engraving
Perf: $13,995 **Exc.:** $11,500 **VGood:** $8995

BERETTA SSO6 EXPRESS CUSTOM

Double rifle; over/under sidelock; 9.3x74R, 375 H&H, 458 Win. Mag.; individually built to buyer's specifications; 25 1/2″ barrel; weighs 11 lbs.; blade front sight on ramp; open V-notch rear; hand-checkered European walnut stock; color case-hardened receiver; double triggers, ejectors; recoil pad. Made in Italy. Introduced 1990; still in production. **LMSR: $21,000**
Perf: $24,500 **Exc.:** $18,500 **VGood:** $12,900
With gold engraving
Perf: $29,500 **Exc.:** $23,500 **VGood:** $17,500

BERNARDELLI CARBINE

Semi-automatic; 22 LR; 5-, 10-shot magazine; 21″ barrel; 40″ overall length; weighs 5 3/8 lbs.; checkered Monte Carlo-style European walnut stock; steel receiver; hooded front sight, fully-adjustable rear; push-button safety; blued/black finish. Made in Italy. Introduced 1987; discontinued 1991.
Perf: $550 **Exc.:** $400 **VGood:** $290

BERNARDELLI EXPRESS VB

Double rifle; 9.3x74R; 25 1/2″ barrels; weighs about 8 lbs.; select walnut stock with cheekpiece; long beaver-tail-Schnabel forend; hand-checkered grip and forend; pistol grip or straight English; bead on ramp front sight, quarter-rib with leaf rear; coin-finished or color case-hardened boxlock action; double trigger; hand-cut rib. Imported from Italy by Magnum Research. Introduced 1990; no longer imported.
Perf: $5200 **Exc.:** $4400 **VGood:** $3000
With single trigger
Perf: $5400 **Exc.:** $4500 **VGood:** $3100

BERNARDELLI MODEL 120

Combination rifle; over/under boxlock; 12-ga. under 22 Hornet, 222 Rem., 5.6x50R Mag., 243, 6.5x57R, 270, 7x57R, 308, 30-06, 8x57JRS, 9.3x74R; iron sights; checkered walnut stock, forearm; double trigger; ventilated recoil pad; engraved action. Made in Italy. No longer in production.
Perf: $1995 **Exc.:** $1500 **VGood:** $995

BERNARDELLI MODEL 190

Combination rifle; over/under boxlock; 12-, 16-, 20-ga. under 243, 30-06, 308; iron sights; checkered walnut stock, forearm; double trigger; extractors. Made in Italy. Introduced 1989; discontinued 1989.
Perf: $1295 **Exc.:** $995 **VGood:** $795

BERNARDELLI MODEL COMBO 2000

Combination rifle; over/under boxlock action; 12-, 16-, 20-ga. over/under 22 Hornet, 222 Rem., 5.6x50R Mag., 243, 6.5x55, 6.5x57R, 270, 7x57R, 308, 30-06, 8x57JRS, 9.3x74R; 23 1/2″ barrels; weighs 6 3/4 lbs.; blade front sight, open rear with quarter-rib for rail-type scope mount; hand-checkered, oil-finished select European walnut stock with Bavarian-type cheekpiece; double set triggers; auto ejectors; silvered, engraved action. Made in Italy. Introduced 1990; dropped 1991.
Perf: $2000 **Exc.:** $1550 **VGood:** $1125
With extra set of barrels
Perf: $2500 **Exc.:** $2050 **VGood:** $1650

BIG HORN CUSTOM

Bolt action; chambered to customer's specs; 22-250 through all magnums; furnished with two barrels to customer specs; weighs about 6 3/4 lbs.; no sights; drilled, tapped for scope mounts; Mauser action; classic-style Claro walnut stock; Pachmayr flush swivel sockets; adjustable trigger; recoil pad. Introduced 1983; discontinued 1984.
Perf: $2000 **Exc.:** $1600 **VGood:** $1100

RIFLES

Blaser
K77A
R84
R93
R93 Safari
Ultimate Bolt Action

Bortmess
Big Horn
Classic Model
Omega

Brno
CZ 98 Hunter Classic
CZ 99 Precision 22

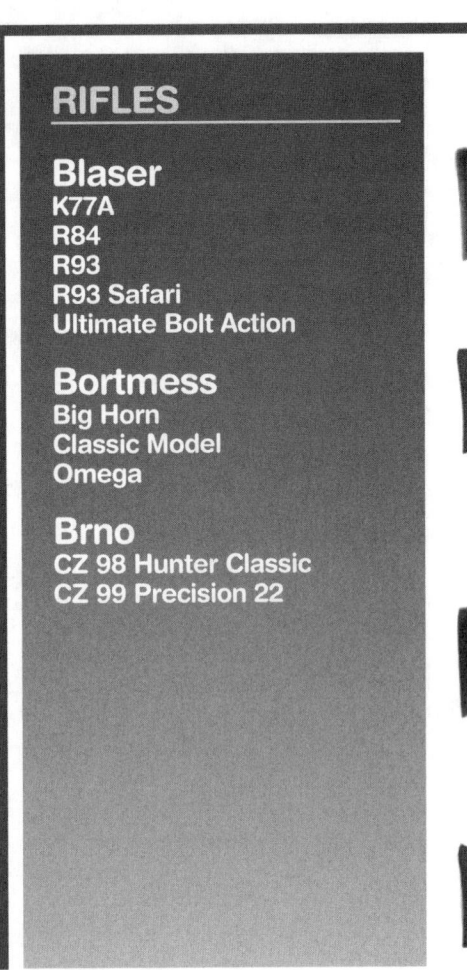

Blaser R84

Bortmess Classic Model

Bortmess Omega

Bortmess Big Horn

BLASER K77A

Single-shot; 22-250, 243, 6.5x55, 270, 280, 7x57R, 7x65R, 30-06, 7mm Rem. Mag., 300 Win. Mag., 300 Weatherby Mag.; 23", 24" barrel; 39 1/2" overall length (23" barrel); weighs 5 1/2 lbs.; break-open; no sights; marketed with Blaser scope mount; checkered two-piece Turkish walnut stock; solid buttpad; three-piece take-down; tang-mounted sliding safety; interchangeable barrels. Made in Germany. Introduced 1988; dropped 1990.

    Perf: $2000   **Exc.:** $1700   **VGood:** $1300
Magnum calibers
    Perf: $2050   **Exc.:** $1750   **VGood:** $1350
With interchangeable barrels
    Perf: $2750   **Exc.:** $2450   **VGood:** $2050

BLASER R84

Bolt action; 22-250, 243, 6mm Rem., 25-06, 270, 280, 30-06 (standard calibers); 257 Wea., 264 Win. Mag., 7mm Rem. Mag., 300 Win. Mag., 300 Wea., 338 Win. Mag., 375 H&H (magnum calibers); 23" barrel (standard calibers), 24" (magnums); 41" overall length (23" barrel); weighs 7 to 7 1/4 lbs.; two-piece Turkish walnut stock and forend; solid black buttpad; no sights; comes with low-profile Blaser scope mounts; interchangeable barrels and magnum/standard caliber bolt assemblies; left-hand models available in all calibers. No longer imported from Germany by Autumn Sales, Inc.

    Perf: $2100   **Exc.:** $1700   **VGood:** $1200
Left-hand
    Perf: $2150   **Exc.:** $1750   **VGood:** $1250

BLASER R93

Bolt action; 222, 243, 6.5x55, 270, 7x57, 308, 30-06, 7mm Rem. Mag., 300 Win. Mag., 300 Wea. Mag., 338 Win. Mag., 375 H&H, 416 Rem. Mag.; 3-shot magazine; 22" barrel (standard calibers), 24" (magnum calibers); 40" overall length (22" barrel); weighs 6 1/2 to 7 1/2 lbs.; two-piece European walnut stock; blade front sight on ramp, open rear, or no sights;

straight-pull bolt action with thumb-activated safety slide/cocking mechanism; interchangeable barrels and bolt heads. Imported from Germany by Autumn Sales, Inc. Introduced 1994; still imported. **LMSR: $2800**

    Perf: $2500   **Exc.:** $2300   **VGood:** $1400
Deluxe Grade with better wood, engraving
    Perf: $2700   **Exc.:** $2500   **VGood:** $1600
Super Deluxe Grade, best wood, gold animal inlays
    Perf: $2900   **Exc.:** $2700   **VGood:** $1800

Blaser R93 Safari

Same specs as the R93 except 375 H&H, 416 Rem. Mag.; 24" barrel; open sights; broad forend. Imported from Germany by Autumn Sales, Inc. Introduced 1994; still imported. **LMSR: $3930**

    Perf: $2995   **Exc.:** $2695   **VGood:** $1895
Safari Deluxe Grade with better wood, engraving
    Perf: $3150   **Exc.:** $2895   **VGood:** $1995
Safari Super Deluxe, best wood, gold animal inlays
    Perf: $3500   **Exc.:** $2995   **VGood:** $2250

BLASER ULTIMATE BOLT ACTION

Bolt action; 22-250, 243, 25-06, 270, 308, 30-06, 7x57, 7x64, 264 Win. Mag., 7mm Rem. Mag., 300 Win. Mag., 338 Win. Mag., 375 H&H; 22", 24" barrel; interchangeable capability; weighs 6 3/4 lbs.; select checkered walnut stock, forearm; aluminum receiver with engraving; 60°; bolt throw. Introduced 1985; discontinued 1989.

    Perf: $1300   **Exc.:** $1050   **VGood:** $850

BORTMESS BIG HORN

Bolt action; calibers from 22-250 through 458 Win.; 24", 25" barrel; no sights; drilled, tapped for scope mounts; Monte Carlo stock of American walnut; rollover cheekpiece; half-curl pistol grip; rosewood cap; tapering forend; rosewood forend cap; high-gloss finish; steel buttplate or recoil pad; built on Ranger Arms action. Introduced 1974; discontinued 1977.

    Perf: $695   **Exc.:** $595   **VGood:** $435

BORTMESS CLASSIC MODEL

Bolt action; calibers from 22-250 through 458 Win.; 24", 25" barrel; no sights; drilled, tapped for scope mounts; hand-checkered American walnut stock; plastic pistol grip cap, forend tip; built on Ranger Arms action; Pachmayr solid rubber recoil pad. Introduced 1974; discontinued 1977.

    Perf: $695   **Exc.:** $595   **VGood:** $450

BORTMESS OMEGA

Bolt action; calibers from 22-250 through 358 Norma; 24", 25" barrel; no sights; drilled, tapped for scope mounts; American walnut stock; high-gloss finish. Introduced 1974; discontinued 1977.

    Perf: $895   **Exc.:** $695   **VGood:** $550

BRNO CZ 98 HUNTER CLASSIC

Bolt action; 243, 6.5x55, 270, 7x57, 7x64, 308, 30-06, 7.92x57, 7mm Rem. Mag., 300 Win. Mag.; 24" barrel; 45" overall length; weighs 7 5/8 lbs.; walnut or synthetic stock; optional sights; integral Weaver-type base; controlled round feeding; fixed ejector; hinged floorplate; adjustable trigger; swivel studs. Imported from the Czech Republic by Springfield, Inc. Introduced 1995; still imported. **LMSR: $449**

    Perf: $400   **Exc.:** $330   **VGood:** $280
With synthetic stock
    Perf: $380   **Exc.:** $300   **VGood:** $260

BRNO CZ 99 PRECISION 22

Bolt action; 22 LR; 5-shot magazine; 20" barrel; 41" overall length; weighs 6 lbs.; hooded bead on ramp front sight, fully-adjustable rear; receiver grooved for scope mount; checkered European hardwood stock; sliding safety; polished blue finish. Made in Yugoslavia. Introduced 1990; discontinued 1991.

    Perf: $300   **Exc.:** $270   **VGood:** $220

Brno CZ 537 Sporter

Brno Model 21H

Brno CZ 99 Precision 22

Brno ZH Series 300

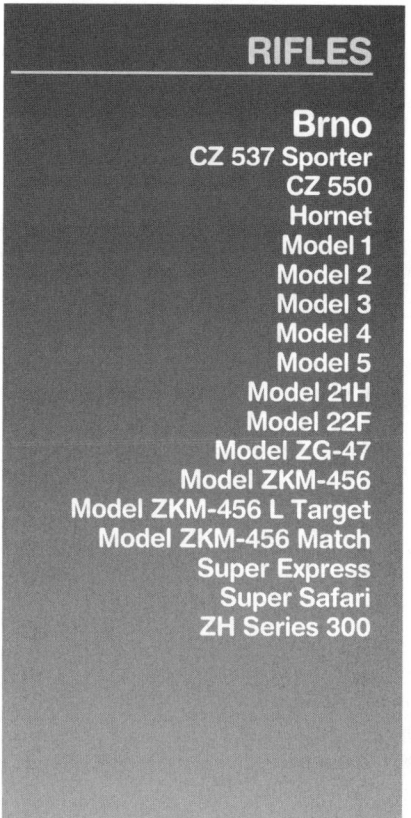

Brno
CZ 537 Sporter
CZ 550
Hornet
Model 1
Model 2
Model 3
Model 4
Model 5
Model 21H
Model 22F
Model ZG-47
Model ZKM-456
Model ZKM-456 L Target
Model ZKM-456 Match
Super Express
Super Safari
ZH Series 300

BRNO CZ 537 SPORTER
Bolt action; 270, 30-06 (internal 5-shot magazine), 243, 308 (detachable 5-shot magazine); 23 5/8″ barrel; 44 3/4″ overall length; weighs 7 lbs. 9 oz.; checkered walnut stock; hooded ramp front sight, adjustable folding leaf rear; improved standard-size Mauser-style action with non-rotating claw extractor; externally adjustable trigger; American-style safety; streamlined bolt shroud with cocking indicator. Imported from the Czech Republic by Magnum Research. Introduced 1992; still imported. **LMSR: $649**
   Perf: $550   **Exc.:** $475   **VGood:** $375
With full stock, 308, 30-06
   Perf: $650   **Exc.:** $575   **VGood:** $475

BRNO CZ 550
Bolt action; 243, 308 (4-shot detachable magazine), 7x57, 270, 30-06, 7mm Rem. Mag., 300 Win. Mag. (5-shot internal magazine); 23 5/8″ barrel; 44 3/4″ overall length; weighs 7 1/4 lbs.; walnut stock with high comb; checkered grip and forend; no sights; drilled and tapped for Remington 700-style bases; polished blue finish. Imported from the Czech Republic by Magnum Research. Introduced 1995; still imported. **LMSR: $649**
   Perf: $550   **Exc.:** $450   **VGood:** $375

BRNO HORNET
Bolt action; 22 Hornet; 5-shot box magazine; 23″ barrel; three-leaf rear sight, hooded ramp front; double-set trigger; sling swivels; hand-checkered pistol grip, forearm. Manufactured in Czechoslovakia from 1949 to 1973.
   Perf: $875   **Exc.:** $700   **VGood:** $600

BRNO MODEL 1
Bolt action; 22 LR; 5-shot magazine; 23″ barrel; weighs 6 lbs.; three-leaf rear sight, hooded ramp front; sporting stock; hand-checkered pistol grip; sling swivels. Introduced 1947; discontinued 1957.
   Perf: $550   **Exc.:** $450   **VGood:** $350

BRNO MODEL 2
Bolt action; 22 LR; 5-shot magazine; 23″ barrel; weighs 6 lbs.; three-leaf rear sight, hooded ramp front; deluxe walnut sporting stock; hand-checkered pistol grip; sling swivels. Introduced 1946; discontinued 1957.
   Perf: $550   **Exc.:** $450   **VGood:** $350

BRNO MODEL 3
Bolt action; 22 LR; 5-shot magazine; 27 1/2″ heavy barrel; weighs 9 1/2 lbs.; click-adjustable target sights; smooth target stock; sling swivels. Introduced 1949; discontinued 1956.
   Perf: $635   **Exc.:** $515   **VGood:** $430

BRNO MODEL 4
Bolt action; 22 LR; 5-shot magazine; 27 1/2″ heavy barrel; weighs 9 1/2 lbs.; click-adjustable target sights; smooth target stock; sling swivels; improved safety, trigger. Introduced 1957; discontinued 1962.
   Perf: $700   **Exc.:** $635   **VGood:** $513

BRNO MODEL 5
Bolt action; 22 LR; 5-shot magazine; 23″ barrel; weighs 6 lbs.; three-leaf rear sight, hooded ramp front; sporting stock; hand-checkered pistol grip; sling swivels. Introduced 1957; discontinued 1973.
   Perf: $650   **Exc.:** $575   **VGood:** $495

BRNO MODEL 21H
Bolt action; 6.5x55, 7x57, 7x64, 8x57mm; 5-shot box magazine; 20 1/2″, 23″ barrel; Mauser-type; two-leaf rear sight, ramp front; half-length sporting stock; hand-checkered pistol grip, forearm; sling swivels; double-set trigger. Introduced 1946; discontinued 1955.
   Perf: $825   **Exc.:** $675   **VGood:** $550

BRNO MODEL 22F
Bolt action; 6.5x55, 7x57, 7x64, 8x57mm; 5-shot box magazine; 20 1/2″, 23″ barrel; Mauser-type; two-leaf rear sight, ramp front; Mannlicher-type stock; sling swivels; double-set trigger. Introduced 1946; discontinued 1955.
   Perf: $1100   **Exc.:** $900   **VGood:** $700

BRNO MODEL ZG-47
Bolt action; 270, 30-06, 7x57, 7x64, 8x64S, 8x57, 9.3x62, 10.75x68; 23 1/2″ barrel; checkered walnut stock, pistol grip; Schnabel forend; hinged floorplate; single trigger. Made in Czechoslovakia. Introduced 1956; discontinued 1962.
   Perf: $995   **Exc.:** $875   **VGood:** $700

BRNO MODEL ZKM-456
Bolt action; 22 LR; 5-, 10-shot magazine; 24 1/2″ barrel; weighs 6 1/2 lbs.; folding rear sight, fixed front; Beech-wood stock; pistol grip; blued finish. Introduced 1992; no longer imported.
   Perf: $330   **Exc.:** $270   **VGood:** $230

Brno Model ZKM-456 L Target
Same specs as Model ZKM-456 except 27 1/2″ barrel; adjustable front and rear sights; weighs 10 lbs. Introduced 1992; no longer imported.
   Perf: $330   **Exc.:** $270   **VGood:** $230

Brno Model ZKM-456 Match
Same specs as Model ZKM-456 except single shot; 27 1/2″ barrel; aperture sights; adjustable buttplate, cheekpiece; weighs 10 lbs. Introduced 1992; no longer imported.
   Perf: $375   **Exc.:** $300   **VGood:** $250

BRNO SUPER EXPRESS
Double rifle; over/under sidelock; 7x65R, 9.3x74R, 375 H&H, 458 Win. Mag.; 23 1/2″ barrel; 40″ overall; weighs about 8 1/2 lbs.; quarter-rib with open rear sight, bead on ramp front; checkered European walnut stock; engraved sideplates; double-set triggers; selective auto ejectors; rubber recoil pad. Made in Czechoslovakia. Introduced 1986; discontinued 1992.
   Perf: $3450   **Exc.:** $2875   **VGood:** $2300

BRNO SUPER SAFARI
Double rifle; over/under sidelock; 7x64R, 375 H&H Mag., 9.3x74R; 23 1/2″ barrel; weighs 9 lbs.; open sights; checkered walnut stock, forearm; double trigger with set trigger; ventilated recoil pad. Introduced 1992; no longer imported.
   Perf: $2295   **Exc.:** $1895   **VGood:** $1350

BRNO ZH SERIES 300
Combination rifle; over/under; 5.6x52R/12-ga., 5.6x-50R Mag./12-ga., 7x57R/12-ga., 7x57R/16-ga.; 23 1/2″ barrel; 40 1/2″ overall length; weighs 8 lbs.; folding leaf rear sight, bead front; walnut stock; eight-barrel set for combo calibers and over/under shotgun barrels. Made in Czechoslovakia. Introduced 1986; discontinued 1991.
   Perf: $2995   **Exc.:** $2395   **VGood:** $1795

RIFLES

Brno
ZKB 527 Fox
ZKB 680 Fox
ZKK 600
ZKK 601
ZKK 602
ZKM 452 Standard
ZKM 452 Deluxe
ZKM 611

Bronco

Brown
Model One

Brno ZKB 527 Fox

Brno ZKB 680 Fox

Brno ZKK 600

Brno ZKK 602

Brno ZKM 452 Standard

BRNO ZKB 527 FOX

Bolt action; 22 Hornet, 222 Rem., 223 Rem.; 5-shot detachable magazine; 23 1/2″ barrel, standard or heavy; 42 1/2″ overall length; weighs 6 lbs. 1 oz.; European walnut Monte Carlo stock; hooded front sight, open adjustable rear; improved mini-Mauser action with non-rotating claw extractor; grooved receiver. Imported from the Czech Republic by Magnum Research. Still imported. **LMSR: $629**
   Perf: $595   **Exc.: $515**   **VGood: $415**

BRNO ZKB 680 FOX

Bolt action; 22 Hornet, 222 Rem.; 5-shot detachable box magazine; 23 1/2″ barrel; 42 1/2″ overall length; weighs 5 lbs. 12 oz.; Turkish walnut Monte Carlo stock; hooded front sight, open adjustable rear; adjustable double-set triggers. Imported from Czechoslovakia by T.D. Arms. No longer imported.
   Perf: $445   **Exc.: $380**   **VGood: $315**

BRNO ZKK 600

Bolt action; 30-06, 270, 7x57, 7x64; 5-shot magazine; 23 1/2″ barrel; 43″ overall length; weighs 7 1/4 lbs.; open folding adjustable rear sight, hooded ramp front; synthetic or checkered walnut stock; improved Mauser action with controlled feed, claw extractor; adjustable set trigger; easy-release sling swivels. Made in Czechoslovakia. Introduced 1986; still imported. **LMSR: $589**
   Perf: $500   **Exc.: $425**   **VGood: $370**

BRNO ZKK 601

Bolt action; 223 Rem., 243 Win., 308 Win.; 5-shot magazine; 23 1/2″ barrel; 43″ overall length; 7 1/4 lbs.; open folding adjustable rear sight, hooded ramp front; synthetic or checkered walnut stock; improved Mauser action with controlled feed, claw extractor; adjustable set trigger; easy-release sling swivels. Made in Czechoslovakia. Introduced 1986; still imported. **LMSR: $589**
   Perf: $500   **Exc.: $425**   **VGood: $370**

BRNO ZKK 602

Bolt action; 300 Win. Mag., 375 H&H, 8x68, 416 Rigby, 458 Win. Mag.; 5-shot magazine; 25″ barrel; 9 5/8 lbs.; three-leaf express rear sight, hooded ramp front; synthetic or checkered walnut stock; adjustable set trigger; easy-release sling swivels. Made in Czechoslovakia. Introduced 1986; still imported. **LMSR: $799**
   Perf: $650   **Exc.: $575**   **VGood: $495**

BRNO ZKM 452 STANDARD

Bolt action; 22 LR; 5-, 10-shot magazine; 25″ barrel; 43 1/2″ overall length; weighs 6 lbs. 9 oz.; beechwood stock; hooded bead front sight, open elevation-adjustable rear; blue finish; oiled stock; grooved receiver. Imported from Czechoslovakia by Action Arms Ltd. No longer imported.
   Perf: $285   **Exc.: $225**   **VGood: $195**

Brno ZKM 452 Deluxe

Same specs as ZKM 452 except 23 5/8″ barrel; 42 1/2″ overall length; checkered walnut stock; sling swivels. Introduced 1992; no longer imported.
   Perf: $350   **Exc.: $265**   **VGood: $210**

BRNO ZKM 611

Semi-automatic; 22 WMR; 6-shot magazine; 20″ barrel; 37″ overall length; weighs 6 lbs. 2 oz.; European walnut stock with checkered grip and forend; blade front sight, open rear; removable box magazine; pol-ished blue finish; grooved receiver for scope mounting; sling swivels; thumbscrew takedown. Imported from the Czech Republic by Magnum Research. Introduced 1995; still imported. **LMSR: $569**
   Perf: $460   **Exc.: $400**   **VGood: $330**

BRONCO

Single shot; swing-out chamber; 22 Short, 22 Long, 22 LR; 16 1/2″ barrel; weighs 3 lbs.; adjustable rear sight, blade front; early version solid construction, later, instant takedown; skeletonized crackle-finished alloy stock; cross-bolt safety. Introduced 1967; discontinued 1975.
   Perf: $140   **Exc.: $110**   **VGood: $90**

BROWN MODEL ONE

Single shot; 22 LR, 357 Mag., 44 Mag., 7-30 Waters, 30-30 Win., 375 Win., 45-70; custom chamberings from 17 Rem. through 45-caliber available; 22″ or cus-tom length barrel, bull or tapered; weighs about 6 lbs.; smooth walnut stock; custom takedown design by Woodsmith; palm swell for right- or left-hand; rubber buttpad; sights optional; drilled and tapped for scope mounting; rigid barrel/receiver; falling-block action with short lock time, automatic case ejection; air-gauged barrels by Wilson and Douglas; muzzle has 11° target crown; matte black oxide finish standard, polished and electroless nickel optional. Made in U.S. by E.A. Brown Mfg. Introduced 1988; still produced. **LMSR: $750**
   Perf: $625   **Exc.: $475**   **VGood: $350**

Browning A-Bolt 22

Browning A-Bolt Camo Stalker

Browning A-Bolt Hunter

Browning A-Bolt Stainless Stalker

Browning A-Bolt Micro Medallion

Browning Bolt-Action Rifles
A-Bolt Hunter
A-Bolt Camo Stalker
A-Bolt Composite Stalker
A-Bolt Euro-Bolt
A-Bolt Gold Medallion
A-Bolt Medallion
A-Bolt Medallion Left Hand
A-Bolt Micro Medallion
A-Bolt Stainless Stalker
A-Bolt 22
A-Bolt 22 Gold Medallion

BROWNING BOLT-ACTION RIFLES

BROWNING A-BOLT HUNTER
Bolt action; 25-06, 270, 30-06, 280, 7mm Rem. Mag., 300 Win. Mag., 338 Win. Mag., 375 H&H (long action); 22-250, 243 Win., 257 Roberts, 22-250, 7mm-08, 308 Win. (short action); 22", 24", 26" barrel; 44 ³/₄" overall length; weighs 6 ¹/₂ to 7 ¹/₂ lbs.; recessed muzzle; magnum and standard action; classic-style American walnut stock; recoil pad on magnum calibers; short throw fluted bolt; three locking lugs; plunger-type ejector; adjustable, grooved gold-plated trigger; hinged floorplate; detachable box magazine; slide tang safety; rosewood grip, forend caps. Made in Japan. Introduced 1985; no longer imported.
Perf: $415   **Exc.:** $340   **VGood:** $295

Browning A-Bolt Camo Stalker
Same specs as A-Bolt Hunter except camo-stained laminated wood stock; cut checkering; non-glare finish on metal; 270, 30-06, 7mm Rem. Mag. Made in Japan. Introduced 1987; dropped 1989.
Perf: $400   **Exc.:** $340   **VGood:** $310

Browning A-Bolt Composite Stalker
Same specs as the A-Bolt Hunter except checkered black composite stock. Imported from Japan by Browning. Introduced 1985; dropped 1993.
Perf: $395   **Exc.:** $360   **VGood:** $320

Browning A-Bolt Euro-Bolt
Same specs as the A-Bolt Hunter except satin-finished walnut stock with Continental-style cheekpiece; palm swell grip and Schnabel forend; rounded bolt shroud and Mannlicher-style flattened bolt handle; 30-06, 270 (22" barrel), 7mm Rem. Mag. (26" barrel); weighs about 6 lbs., 11 oz. Imported from Japan by Browning. Introduced 1993; no longer imported.
Perf: $575   **Exc.:** $475   **VGood:** $350

Browning A-Bolt Gold Medallion
Same specs as A-Bolt Hunter except select walnut stock; gold-filled inscription on barrel; engraved receiver; brass spacers between rubber recoil pad, rosewood grip cap, forend tip; palm swell pistol grip; Monte Carlo-style comb; double-border 22 lpi checkering; 270, 30-06, 300 Win. Mag., 7mm Rem. Mag. Made in Japan. Introduced 1988; discontinued 1993.
Perf: $650   **Exc.:** $535   **VGood:** $430

Browning A-Bolt Medallion
Same specs as A-Bolt Hunter except 375 H&H Mag.; 1" deluxe sling swivels; rosewood pistol grip, forend cap; slide tang safety; with or without sights; glossy stock finish; high polish blue. Imported from Japan by Browning. Introduced 1985; dropped 1993.
Perf: $495   **Exc.:** $395   **VGood:** $350
375 H&H, with sights
Perf: $595   **Exc.:** $495   **VGood:** $430

Browning A-Bolt Medallion Left Hand
Same specs as the Medallion except left-hand action; 270, 30-06, 7mm Rem. Mag., 375 H&H. Imported from Japan by Browning. Introduced 1987; discontinued 1993.
Perf: $500   **Exc.:** $410   **VGood:** $355

Browning A-Bolt Micro Medallion
Same specs as A-Bolt Hunter except scaled-down version with 20" barrel; shortened, 13 ⁵/₁₆" length of pull; 3-shot magazine; weighs 6 lbs., 1 oz.; 243, 308, 7mm-08, 257 Roberts, 223, 22-250; no sights. Imported from Japan by Browning. Introduced 1988; dropped 1993.
Perf: $495   **Exc.:** $450   **VGood:** $400

Browning A-Bolt Stainless Stalker
Same specs as A-Bolt Hunter except stainless steel receiver; matte silver-gray finish; graphite-fiberglass composite textured stock; no sights; 270, 30-06, 7mm Rem. Mag, 375 H&H. Made in Japan. Imported from Japan by Browning. Introduced 1987; dropped 1993.
Perf.: $550   **Exc.:** $450   **VGood:** $350
Left-hand
Perf: $570   **Exc.:** $470   **VGood:** $370
375 H&H caliber
Perf: $650   **Exc.:** $550   **VGood:** $450

BROWNING A-BOLT 22
Bolt action; 22 LR, 22 WMR; 5-, 15-shot magazine; 22" barrel, 40 ¹/₄" overall length; weighs 5 lbs. 9 oz.; walnut stock with cut checkering; rosewood grip cap and forend tip; with or without open sights; open sight model with ramp front and adjustable folding leaf rear; short 60°; bolt throw; top tang safety; grooved for 22 scope mount; drilled and tapped for full-size scope mounts; detachable magazine; gold-colored trigger. Imported from Japan by Browning. Introduced 1986; still imported. **LMSR: $405**
22 LR
Perf: $320   **Exc.:** $275   **VGood:** $220
22 WMR
Perf: $350   **Exc.:** $300   **VGood:** $240

Browning A-Bolt 22 Gold Medallion
Same specs as A-Bolt 22 except high-grade walnut stock with brass spacers between stock and rubber recoil pad, and rosewood grip cap and forend; Medallion-style engraving on receiver flats; "Gold Medallion" engraved and gold filled on right side of barrel; high-gloss stock finish; no sights. Imported from Japan by Browning. Introduced 1988; still imported. **LMSR: $540**
Perf: $400   **Exc.:** $350   **VGood:** $300

Browning Bolt-Action Rifles

A-Bolt II Hunter
A-Bolt Composite Stalker
Euro-Bolt II
A-Bolt II Gold Medallion
A-Bolt II Hunter Short Action
A-Bolt II Medallion
A-Bolt II Medallion Short Action
A-Bolt II Micro Medallion
A-Bolt II Stainless Stalker
A-Bolt II Varmint

Browning A-Bolt II Varmint

Browning Euro-Bolt II

Browning BBR

Browning Model 52

Browning T-Bolt Model T-2

BROWNING A-BOLT II HUNTER

Bolt action; 25-06, 270, 30-06, 280, 7mm Rem. Mag., 300 Win. Mag., 338 Win. Mag., 375 H&H Mag.; 4-shot (standard calibers), 3-shot (magnum calibers) detachable box magazine; 22″ medium sporter weight barrel with recessed muzzle; 26″ on magnum calibers; 44 3/4″ overall length, 41 3/4″ (short action); weighs 6 1/2 to 7 1/2 lbs.; classic-style American walnut stock; recoil pad standard on magnum calibers; short-throw (60°) fluted bolt; three locking lugs; plunger-type ejector; grooved and gold-plated adjustable trigger; hinged floorplate; slide tang safety; glossy stock finish; rosewood grip and forend cap; high polish blue; BOSS barrel vibration modulator and muzzlebrake system not available in 375 H&H. Imported from Japan by Browning. Introduced 1994; still imported. **LMSR: $590**
  Perf: $450   **Exc.:** $400   **VGood:** $350
With open sights
  Perf: $500   **Exc.:** $450   **VGood:** $380
With BOSS
  Perf: $540   **Exc.:** $480   **VGood:** $400

Browning A-Bolt II Composite Stalker

Same specs as A-Bolt II Hunter except black graphite fiberglass stock with textured finish; matte blue finish on metal; 223, 22-250, 243, 7mm-08, 308, 30-06, 270, 280, 25-06, 7mm Rem. Mag., 300 Win. Mag., 338 Win. Mag.; no sights; BOSS barrel vibration modulator and muzzlebrake system offered in all calibers. Imported from Japan by Browning. Introduced 1994; still imported. **LMSR: $595**
  Perf: $450   **Exc.:** $400   **VGood:** $350
With BOSS
  Perf: $540   **Exc.:** $480   **VGood:** $400

Browning Euro-Bolt II

Same specs as A-Bolt II Hunter except satin-finished walnut stock with Continental-style cheekpiece; palm-swell grip and Schnabel forend; rounded bolt shroud and Mannlicher-style flattened bolt handle; 30-06, 270 (22″ barrel), 7mm Rem. Mag. (26″ barrel); weighs about 6

lbs., 11 oz.; BOSS barrel vibration modulator and muzzlebrake system optional. Imported from Japan by Browning. Introduced 1993; still imported. **LMSR: $785**
  Perf: $550   **Exc.:** $500   **VGood:** $450
With BOSS
  Perf: $620   **Exc.:** $560   **VGood:** $500

Browning A-Bolt II Gold Medallion

Same specs as A-Bolt II Hunter except select walnut stock with brass spacers between rubber recoil pad, and rosewood grip cap and forend tip; gold-filled barrel inscription; palm swell pistol grip; Monte Carlo comb; 22 lpi checkering with double borders; engraved receiver flats; 270, 30-06, 7mm Rem. Mag. only. Imported from Japan by Browning. Introduced 1988; still imported. **LMSR: $905**
  Perf: $690   **Exc.:** $590   **VGood:** $500
With BOSS
  Perf: $770   **Exc.:** $665   **VGood:** $550

Browning A-Bolt II Hunter Short Action

Same specs as the standard A-Bolt II Hunter except short action for 223, 22-250, 243, 257 Roberts, 7mm-08, 284 Win., 308 chamberings; weighs 6 1/2 lbs.; BOSS barrel vibration modulator and muzzlebrake system optional. Imported from Japan by Browning. Introduced 1994; still imported. **LMSR: $577**
  Perf: $465   **Exc.:** $400   **VGood:** $340
With sights
  Perf: $495   **Exc.:** $440   **VGood:** $375
With BOSS
  Perf: $550   **Exc.:** $460   **VGood:** $400

Browning A-Bolt II Medallion

Same specs as A-Bolt II Hunter except 25-06, 270, 30-06, 280, 7mm Rem. Mag., 300 Win. Mag., 338 Win. Mag., 375 H&H Mag.; classic-style stock of American walnut; rosewood pistol grip, forend cap; slide tang safety; glossy stock finish; high polish blue. Imported from Japan by Browning. Introduced 1985; still imported. **LMSR: $780**
  Perf: $525   **Exc.:** $425   **VGood:** $400
With BOSS
  Perf: $605   **Exc.:** $550   **VGood:** $470
375 H&H
  Perf: $625   **Exc.:** $575   **VGood:** $500

Browning A-Bolt II Medallion Short Action

Same specs as the standard A-Bolt II Medallion except short action for 223, 22-250, 243, 257 Roberts, 7mm-08, 284 Win., 308; weighs 6 1/2 lbs. BOSS barrel vibration modulator and muzzlebrake system optional. Imported from Japan by Browning. Introduced 1994; still imported. **LMSR: $673**
  Perf: $525   **Exc.:** $465   **VGood:** $400

Browning A-Bolt II Micro Medallion

Same specs as A-Bolt II Medallion except scaled-down version with 20″ barrel; shortened length of pull (13 5/16″); 3-shot magazine; weighs 6 lbs. 1 oz.; 22 Hornet, 243, 308, 7mm-08, 257 Roberts, 223, 22-250; BOSS feature not available. Imported from Japan by Browning. Introduced 1994; still imported. **LMSR: $673**
  Perf: $495   **Exc.:** $440   **VGood:** $390

Browning A-Bolt II Stainless Stalker

Same specs as A-Bolt II Hunter except stainless receiver and barrel; other exposed metal surfaces in durable matte silver-gray; graphite-fiberglass composite textured stock; no sights; 223, 22-250, 243, 308, 7mm-08, 270, 30-06, 7mm Rem. Mag., 375 H&H; BOSS barrel vibration modulator and muzzlebrake system not available in 375 H&H. Imported from Japan by Browning. Introduced 1994; still imported. **LMSR: $774**
  Perf: $615   **Exc.:** $570   **VGood:** $450
With BOSS
  Perf: $695   **Exc.:** $640   **VGood:** $510
375 H&H, with sights
  Perf: $715   **Exc.:** $670   **VGood:** $550
Left-hand, with BOSS
  Perf: $695   **Exc.:** $640   **VGood:** $510

Browning A-Bolt II Varmint

Same specs as A-Bolt II Hunter except heavy varmint/target barrel; laminated wood stock with special dimensions; flat forend and palm swell grip; 223, 22-250, 308; comes with BOSS barrel vibration modulator and muzzlebrake system. Imported from Japan by Browning. Introduced 1994; still imported. **LMSR: $895**
  Perf: $695   **Exc.:** $600   **VGood:** $500

Browning Auto 22 Grade I

Browning BAR-22

Browning Auto 22 Grade III

Browning BAR Grade II

Browning BAR Grade IV

RIFLES

Browning Bolt-Action Rifles
BBR
Model 52
T-Bolt Model T-1
T-Bolt Model T-2

Browning Semi-Automatic Rifles
Auto 22 Grade I
Auto 22 Grade II
Auto 22 Grade III
Auto 22 Grade VI
BAR

BROWNING BBR

Bolt action; 25-06, 270, 30-06, 7mm Rem. Mag., 300 Win. Mag., 338 Win. Mag. (long action), 22-250, 243 Win., 257 Roberts, 7mm-08 Rem., 308 Win. (short action); 4-shot (standard calibers), 3-shot (magnum calibers); 22″, 24″ sporter barrel; recessed muzzle; American walnut Monte Carlo stock; high cheek-piece; full/checkered pistol grip, forend; recoil pad on magnums; grooved, gold-plated adjustable trigger; detachable box magazine; tang slide safety; low-profile swing swivels. Made in Japan. Introduced 1978; dropped 1984. Add $250 for 243 caliber.

Perf: $470   **Exc.:** $390   **VGood:** $320

BROWNING MODEL 52

Bolt action; 22 LR; 5-shot magazine; 24″ barrel; weighs 7 lbs.; high-grade walnut stock with oil-like finish; cut-checkered grip and forend; metal grip cap, rosewood forend tip; no sights furnished; drilled and tapped for scope mounting or iron sights; recreation of the Winchester Model 52C Sporter with minor safety improvements; duplicates the adjustable Micro-Motion trigger system; button release magazine. Only 5000 made. Imported from Japan by Browning. Introduced 1991; dropped 1992.

Perf: $550   **Exc.:** $460   **VGood:** $395

BROWNING T-BOLT MODEL T-1

Bolt action; straight-pull; 22 LR; 5-shot clip magazine; 24″ barrel; weighs 5 1/2 lbs.; unchecked walnut pistol grip stock; peep rear sight, ramp blade front; left- or right-hand models available. Introduced 1965; not regularly imported.

Perf: $375   **Exc.:** $325   **VGood:** $285

Browning T-Bolt Model T-2

Same specs as Model T-1 except figured, hand-checkered walnut stock. Not regularly imported. Last run T-2 with plain stock.

Perf: $450   **Exc.:** $400   **VGood:** $330
Late production model
Perf: $350   **Exc.:** $300   **VGood:** $250

BROWNING SEMI-AUTOMATIC RIFLES

BROWNING AUTO 22 GRADE I

Semi-automatic; 22 LR, 22 Short; 11-shot (22 LR), 16-shot (22 Short), 19 1/4″ barrel (22 LR), 22 1/4″ barrel (22 Short); weighs 4 3/4 lbs.; buttstock tube magazine; select European walnut stock; hand-checkered pistol grip, semi-beaver tail forearm; open rear sight, bead front; engraved receiver, grooved for tip-off scope mount; cross-bolt safety. Introduced in 1958. Manufactured in Belgium until 1972; production transferred to Miroku in Japan; still in production. **LMSR: $465**
Belgian manufacture
Perf: $525   **Exc.:** $400   **VGood:** $310
Japanese manufacture
Perf: $310   **Exc.:** $250   **VGood:** $200

Browning Auto 22 Grade II

Same specs as Grade I except gold-plated trigger; small game animal scenes engraved on receiver. No longer in production.
Belgian manufacture
Perf: $975   **Exc.:** $775   **VGood:** $525
Japanese manufacture
Perf: $430   **Exc.:** $350   **VGood:** $300

Browning Auto 22 Grade III

Same specs as Grade I except gold-plated trigger; extra-fancy walnut stock forearm; skip checkering on pistol grip, forearm; satin chrome-plated receiver; hand-carved, engraved scrolls, leaves, dog/game bird scenes; 22 LR only. No longer in production.
Belgian manufacture
Perf: $2000   **Exc.:** $1600   **VGood:** $900
Japanese manufacture
Perf: $800   **Exc.:** $625   **VGood:** $495

Browning Auto 22 Grade VI

Same specs as Grade I except grayed or blued receiver; extensive engraving; gold-plated animals; stock forend of high-grade walnut with double-border cut checkering. Made in Japan. Introduced 1987; still in production. **LMSR: $998**
Perf: $675   **Exc.:** $520   **VGood:** $375

BROWNING BAR

Semi-automatic; 243 Win., 308 Win., 270 Win., 280 Rem., 30-06, 7mm Rem. Mag., 300 Win. Mag., 338 Win. Mag.; 22″ barrel; 4-shot detachable box magazine (standard calibers), 3-shot (magnum calibers); adjustable folding leaf rear sight, gold bead hooded ramp front; receiver tapped for scope mounts; checkered walnut pistol-grip stock forearm. Not to be confused with military Browning selective-fire rifle. Grades I, II introduced in 1967; Grades II, III, V still in production. Grades vary according to amount of checkering, carving, engraving, inlay work. Premium for Belgian marked.
Grade I
Perf: $600   **Exc.:** $475   **VGood:** $400
Grade I Magnum
Perf: $610   **Exc.:** $485   **VGood:** $410
Grade II
Perf: $700   **Exc.:** $620   **VGood:** $500
Grade II Magnum
Perf: $725   **Exc.:** $625   **VGood:** $525
Grade III
Perf: $925   **Exc.:** $795   **VGood:** $640
Grade III Magnum
Perf: $1025   **Exc.:** $850   **VGood:** $790
Grade IV
Perf: $1500   **Exc.:** $1300   **VGood:** $900
Grade IV Magnum
Perf: $1600   **Exc.:** $1400   **VGood:** $1000
Grade V
Perf: $3000   **Exc.:** $2600   **VGood:** $1900
Grade V Magnum
Perf: $3300   **Exc.:** $2900   **VGood:** $2000

Browning BAR Mark II Safari

RIFLES

Browning Semi-Automatic Rifles
BAR-22
BAR Mark II Safari
BAR Mark II Safari Magnum

BROWNING RIFLE PRODUCTION DATES

Rifle	Introduced	Discontinued	Manufactured
Bolt Action Mauser	1959	1975	Belgium
Bolt Action Medallion & Olympian	1961	1975	Belgium
Bolt Action Sako	1963	1975	Finland
BBR (Original Version)	1978	1981	Japan
BBR (Lightning Bolt Version)	1981	1984	Japan
BBR Elk Issue High Grade	1984	Ltd Issue of 1000	Japan
A-Bolt Medallion, Hunter	1985	To date	Japan
A-Bolt Big Horn Sheep	1986	Ltd Issue of 600	Japan
A-Bolt Stainless Stalker	1987	To date	Japan
A-Bolt Camo-Stalker	1987	1989	Japan
A-Bolt Pronghorn Limited Edition	1987	Ltd Issue of 500	Japan
A-Bolt Medallion Left Hand	1988	To date	Japan
A-Bolt Gold Medallion	1988	To date	Japan
A-Bolt Micro-Medallion	1988	To date	Japan
A-Bolt Composite Stalker	1988	To date	Japan
A-Bolt Stainless Stalker LH	1990	To date	Japan
A-Bolt II Version (Replace all A-Bolt Models)	1994	To date	Japan
A-Bolt II Varmint	1994	To date	Japan
A-Bolt II 22 Hornet	1994	To date	Japan
BAR Gr I	1968	To date	Belgium
BAR Gr II	1968	1975	Belgium
BAR Gr III V	1971	1975	Belgium
BAR Gr IV	1971	1985	Belgium
BAR Gr III Reintroduced	1979	1985	Belgium
BAR Big Game	1982	Ltd Issue of 600	Belgium
BAR Mark II Safari	1993	To date	Belgium
BOSS System	1994	To date	Jap/Bel
Jonathan Borwning Mountain Rifle	1978	1983	USA
Express Rifle	1980	1985	Belgium
Centennial B-92 44 Mag.	1978	Ltd Issue of 6000	Japan
B-92 44 Mag.	1979	1987	Japan
B-92 357 Mag.	1982	1987	Japan
BLR	1969	1972	Belgium
BLR	1973	1981	Japan
Model 81 BLR	1981	To date	Japan
Model 81 BLR 284	1989	To date	Japan
Model 81 BLR Long Action	1991	To date	Japan
B-78	1973	1983	Japan
B-78 Bicentennial set, 45-70 only Belgian engraved	1976	Issue of 1000	Japan
Model 1885 Hi Wall	1985	To date	Japan
Model 1885 Low Wall	1995	To date	Japan
Model 1895 Replica 30-06 Grade I	1984	Ltd Issue of 6000	Japan
Model 1895 30-06 High Grade	1984	Ltd Issue of 1000	Japan
Model 1895 30-40 Krag Grade I	1985	Ltd Issue of 2000	Japan
Model 1895 30-40 Krag High Grade	1985	Ltd Issue of 1000	Japan
Model 1886 45-70 Grade I	1986	Ltd Issue of 7000	Japan
Model 1886 45-70 High Grade	1986	Ltd Issue of 3000	Japan
Model 1986 45-70 MT Centennial	1992	Ltd Issue of 2000	Japan
Model 1886 Carbine 45-70 Grade I	1992	Ltd Issue of 7000	Japan
Model 1886 Carbine 45-70 High Grade	1992	Ltd Issue of 3000	Japan
Model 71 Grade I 348 Win Rifle	1987	Ltd Issue of 3000	Japan
Model 71 High Grade 348 Win Rifle	1987	Ltd Issue of 3000	Japan
Model 71 Grade I 348 Win Carbine	1987	Ltd Issue of 4000	Japan
Model 71 High Grade 348 Win Carbine	1987	Ltd Issue of 3000	Japan
Model 65 218 Bee Grade I	1989	Ltd Issue of 3500	Japan
Model 65 218 Bee High Grade	1989	Ltd Issue of 1500	Japan
Model 53 32-20 Deluxe Grade	1990	Ltd Issue of 5000	Japan
Semi Auto 22 Short/Long I II III	1956	1974-75	Belgium
Semi Auto 22 Grade II III	1974	1984	Japan
Semi Auto 22 Grade I	1974	To date	Japan
Semi Auto Grade VI	1987	To date	Japan
T-Bolt I & II	1965	1976	Belgium
T-Bolt I & II Reintroduced	1979	1982	Belgium
BL-22 I & II	1969	To date	Japan
BAR 22 Grade I & II	1977	1985	Japan
(Restyled new stock and receiver in 1982)			
BPR 22	1977	1983	Japan
BPR 22 Magnum Grade I	1977	1983	Japan
BPR 22 Magnum Grade II	1980	1982	Japan
A-Bolt 22	1986	To date	Japan
A-Bolt 22 Magnum	1989	To date	Japan
Model 52C Sporter (22 Bolt Action)	1991	Ltd Issue of 5000	Japan

BROWNING BAR-22

Semi-automatic; 22 LR; 15-shot, tubular magazine; 20 1/4" barrel; weighs 5 lbs. 2 oz.; receiver grooved for scope mount; folding leaf rear sight, gold bead ramp front; French walnut stock hand-checkered pistol grip, forearm. Introduced 1977; dropped 1986. Manufactured in Japan.

Grade I
Perf: $275   **Exc.:** $225   **VGood:** $185
Grade II
Perf: $400   **Exc.:** $330   **VGood:** $230

BROWNING BAR MARK II SAFARI

Semi-automatic; 243, 270, 30-06, 308; 4-shot detachable magazine; 22" round tapered barrel; 43" overall length; weighs 7 3/8 lbs.; hand-checkered French walnut stock and forend; gold bead on hooded ramp front sight, click-adjustable rear, or no sights; updated version with new bolt release lever; removable trigger assembly with larger trigger guard; redesigned gas and buffer systems; scroll-engraved receiver tapped for scope mounting. BOSS barrel vibration modulator and muzzlebrake system available on models without sights. Imported from Belgium by Browning. Introduced 1993; still imported. **LMSR: $776**

Perf: $630   **Exc.:** $510   **VGood:** $430
Without sights
Perf: $615   **Exc.:** $495   **VGood:** $415
With BOSS
Perf: $710   **Exc.:** $590   **VGood:** $510

Browning BAR Mark II Safari Magnum

Same specs as the standard caliber model except weighs 8 3/8 lbs.; 45" overall length; 24" barrel; 3-round magazine; calibers 7mm Mag., 300 Win. Mag., 338 Win. Mag.; BOSS barrel vibration modulator and muzzlebrake system available only on models without sights. Imported from Belgium by Browning. Introduced 1993; still imported. **LMSR: $729**

With sights
Perf: $675   **Exc.:** $560   **VGood:** $460
No sights
Perf: $625   **Exc.:** $545   **VGood:** $445
With BOSS, no sights
Perf: $725   **Exc.:** $625   **VGood:** $495

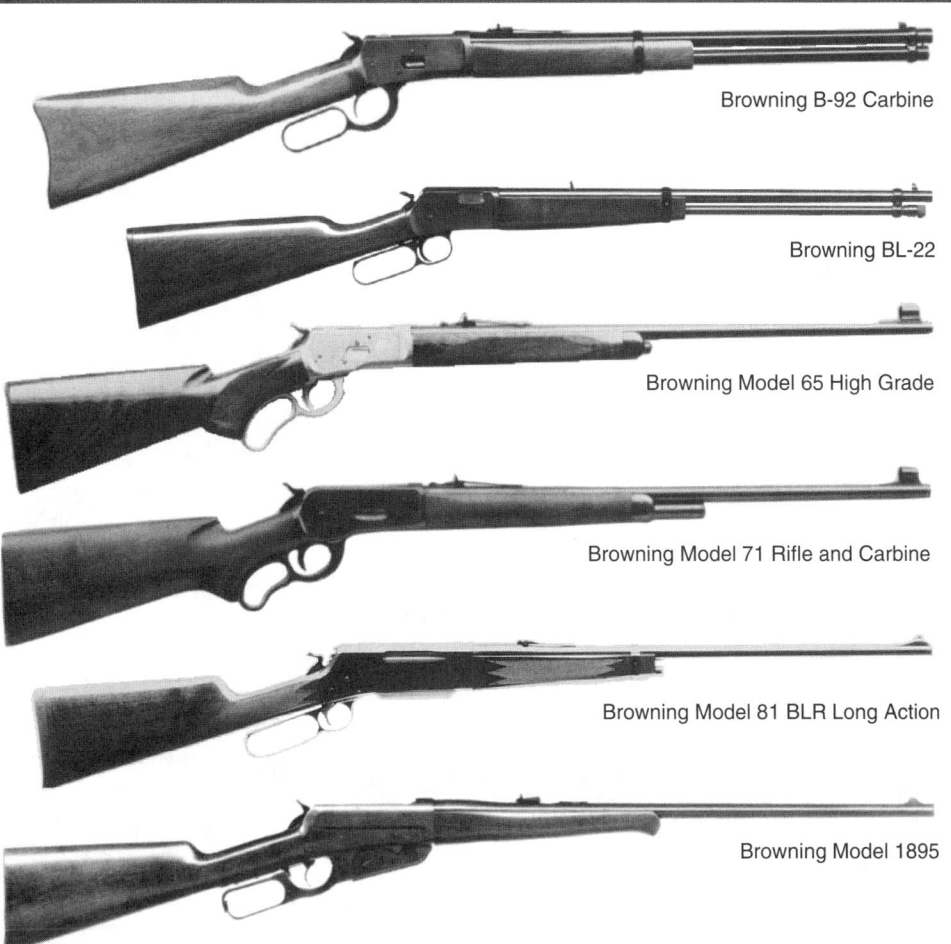

Browning B-92 Carbine

Browning BL-22

Browning Model 65 High Grade

Browning Model 71 Rifle and Carbine

Browning Model 81 BLR Long Action

Browning Model 1895

Browning Lever-Action Rifles
B-92 Carbine
BL-22
Model 53
Model 65 Grade I
Model 65 High Grade
Model 71 Rifle and Carbine
Model 81 BLR
Model 81 BLR Long Action
Model 1886 Rifle and Carbine
Model 1886 Rifle and Carbine High Grade
Model 1895

BROWNING LEVER-ACTION RIFLES
BROWNING B-92 CARBINE

Lever-action; 357 Mag., 44 Mag.; 11-shot magazine; 20″ round barrel; post front sight, classic cloverleaf rear with notched elevation ramp; straight-grip European walnut stock; high-gloss finish; steel buttplate; tubular magazine. Designed from original Model 92 lever-action. Manufactured in Japan. Introduced 1978; dropped 1988.

Perf: $450	**Exc.:** $360	**VGood:** $295

BROWNING BL-22

Lever-action; 22 LR, 22 Long, 22 Short; 22-shot magazine (22 Short), 17-shot (22 Long), 15-shot (22 LR); 20″ barrel; 36 3/4″ overall length; weighs 5 lbs.; folding leaf rear sight, bead post front; two-piece uncheckered Western-style walnut stock, forearm; barrel band; half-cock safety; receiver grooved for tip-off mounts. Grade II with engraved receiver, checkered grip, forearm. Produced by Miroku Firearms. Introduced 1970; still in production. **LMSR: $403**
Grade I

Perf: $290	**Exc.:** $230	**VGood:** $190

Grade II

Perf: $330	**Exc.:** $260	**VGood:** $220

BROWNING MODEL 53

Lever action; 32-20; 7-shot magazine; 22″ round tapered barrel; 39 1/2″ overall length; weighs 6 lbs. 8 oz.; full pistol grip stock of select walnut with semi-beavertail forend; high gloss finish; cut checkering on grip and forend; metal grip cap; post bead front sight, adjustable open rear; blue finish, including trigger. Based on the Model 92 Winchester with half-length magazine. Limited to 5000 guns. Imported from Japan by Browning. Made only in 1990.

Perf: $600	**Exc.:** $500	**VGood:** $395

BROWNING MODEL 65 GRADE I

Lever action; 218 Bee; 7-shot magazine; 24″ round tapered barrel; 41 3/4″ overall length; weighs 6 lbs.; select walnut pistol-grip stock, semi-beavertail forend; hooded ramp front sight, adjustable buckhorn-type rear; blued finish. Limited edition reproduction of Winchester

Model 65; half-length magazine. Made in Japan. Only 3500 manufactured. Introduced 1989; dropped 1990.

Perf: $550	**Exc.:** $450	**VGood:** $390

Browning Model 65 High Grade

Same specs as Grade I except better wood; cut checkering; high-gloss finish; receiver with grayed finish, scroll engraving, game scenes. Made in Japan. Only 1500 made. Introduced 1989; dropped 1990.

Perf: $820	**Exc.:** $730	**VGood:** $590

BROWNING MODEL 71 RIFLE AND CARBINE

Lever action; 348 Win.; 4-shot magazine; 20″, 24″ barrel; 45″ overall length; weighs 8 lb. 2 oz.; pistol-grip stock; classic-style forend; metal buttplate; reproduction of Winchester Model 71; half-length magazine; High Grade has blued barrel and magazine, grayed lever and receiver with scroll engraving, gold-plated game scene. Grade I production limited to 3000 rifles, 4000 carbines. Made in Japan. Introduced 1987; dropped 1989.
Grade I

Perf: $595	**Exc.:** $495	**VGood:** $420

High Grade

Perf: $795	**Exc.:** $650	**VGood:** $540

BROWNING MODEL 81 BLR

Lever action; 223, 22-250, 243, 257 Roberts, 284 Win., 7mm-08, 308 Win., 358 Win.; 4-shot detachable box magazine; 20″ round, tapered barrel; 39 3/4″ overall length; weighs 7 lbs.; gold bead front sight, hooded ramp, square-notch adjustable rear; drilled, tapped for scope mounts; oil-finished, checkered, straight-grip walnut stock; recoil pad; half-cock hammer safety; wide, grooved trigger. Made only in Japan. Introduced 1981; still imported by Browning. **LMSR: $600**
Belgian manufacture

Perf: $495	**Exc.:** $450	**VGood:** $390

Japanese manufacture

Perf: $425	**Exc.:** $390	**VGood:** $330

Browning Model 81 BLR Long Action

Same specs as Model 81 BLR except long action; 270,

30-06, 7mm Rem. Mag.; 22″, 24″ barrel; weighs 8 1/2 lbs.; six-lug fluted, rotary bolt. Made in Japan. Introduced 1991; still imported by Browning. **LMSR: $580**

Perf: $475	**Exc.:** $385	**VGood:** $315

BROWNING MODEL 1886 RIFLE AND CARBINE

Lever action; 45-70; 8-shot magazine; 22″ barrel; 40 3/4″ overall length; weighs 8 lbs. 3 oz.; satin-finished select walnut stock with metal crescent buttplate; blade front sight, open adjustable rear; full-length magazine; classic-style forend with barrel band, saddle ring; polished blue finish. Recreation. Limited to 7000 guns. Imported from Japan by Browning. Introduced 1992; dropped 1993.
Carbine

Perf: $850	**Exc.:** $700	**VGood:** $550

Rifle

Perf: $1050	**Exc.:** $850	**VGood:** $620

Browning Model 1886 Rifle and Carbine High Grade

Same specs as Model 1886 Rifle and Carbine except high grade walnut with cut-checkered grip and forend and gloss finish; grayed steel receiver and lever; scroll engraved receiver with game scenes highlighted by a special gold-plating and engraving process. Limited to 3000 guns. Introduced 1992; dropped 1993.

Perf: $1295	**Exc.:** $995	**VGood:** $795

BROWNING MODEL 1895

Lever action; 30-06, 30-40 Krag; 4-shot magazine; 24″ round barrel; 42″ overall length; weighs 8 lbs.; straight-grip walnut stock forend; buckhorn rear sight with elevator, gold bead front on ramp; replica of Browning's first box-magazine lever action; top loading magazine; half-cock hammer safety; High Grade Model with gold-plated moose, grizzly on gray engraved receiver. Made in Japan. Limited edition. Introduced 1984; dropped, 1985. 30-06 brings premium.
Grade I

Perf: $595	**Exc.:** $495	**VGood:** $395

High Grade

Perf: $995	**Exc.:** $895	**VGood:** $695

Browning Model 1885 High Wall

Browning Model 78

Browning Continental

Browning (FN) High-Power
Safari-Grade Medium-Action

Browning (FN) High-Power
Safari Olympian-Grade

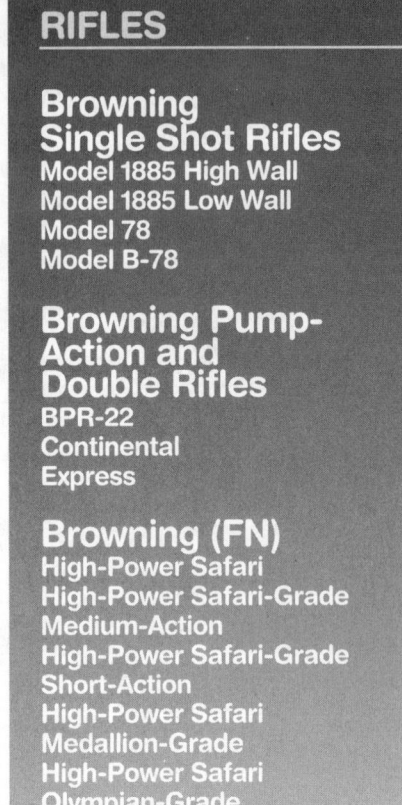

RIFLES

Browning Single Shot Rifles
Model 1885 High Wall
Model 1885 Low Wall
Model 78
Model B-78

Browning Pump-Action and Double Rifles
BPR-22
Continental
Express

Browning (FN)
High-Power Safari
High-Power Safari-Grade Medium-Action
High-Power Safari-Grade Short-Action
High-Power Safari Medallion-Grade
High-Power Safari Olympian-Grade

BROWNING SINGLE SHOT RIFLES

BROWNING MODEL 1885 HIGH WALL
Single shot; falling block; 22-250, 30-06, 270, 7mm Rem. Mag., 45-70; 28″ octagon barrel with recessed muzzle; 43 1/2″ overall length; weighs about 8 1/2 lbs.; walnut stock with straight grip; Schnabel forend; no sights furnished; drilled and tapped for scope mounting; replica of J.M. Browning's high-wall falling block rifle. Imported from Japan by Browning. Introduced 1985; still imported. **LMSR: $997**
  Perf: $690   **Exc.:** $600   **VGood:** $495

Browning Model 1885 Low Wall
Same specs as the Model 1885 High Wall except trimmer receiver; thinner 24″ octagonal barrel; forend mounted to receiver; adjustable trigger; walnut pistol grip stock; trim Schnabel forend with high-gloss finish; 22 Hornet, 223 Rem., 243 Win.; 39 1/2″ overall length; weighs 6 lbs. 4 oz.; polished blue finish. Rifling twist rates: 1:16″ (22 Hornet); 1:12″ (223); 1:10″ (243). Imported from Japan by Browning. Introduced 1995; still in production. **LMSR: $997**
  Perf: $750   **Exc.:** $600   **VGood:** $500

BROWNING MODEL 78
Single shot; 30-06, 25-06, 6mm Rem., 243, 22-250, 7mm Rem. Mag.; 26″ tapered octagon or heavy round barrel; hand-rubbed, hand-checkered select walnut stock; rubber recoil pad; no sights; furnished with scope mounts and rings; falling-block action; exposed hammer; adjustable trigger; automatic ejector. Made in Japan. Introduced 1978; dropped 1983.
  Perf: $600   **Exc.:** $520   **VGood:** $440

Browning Model B-78
Same specs as Model 78 except 45-70; 24″ heavy octagon barrel; blade front sight, step adjustable rear; drilled, tapped for scope mounts; straight-grip hand-checkered walnut stock; semi-Schnabel forend; sling swivels; curved, blued steel buttplate. Introduced 1978; dropped 1981.
  Perf: $720   **Exc.:** $600   **VGood:** $490

BROWNING PUMP-ACTION AND DOUBLE RIFLES

BROWNING BPR-22
Pump-action; 22 LR, 22 WMR; 11-shot (22 WMR), 15-shot (22 LR) tube magazine; 20 1/2″ barrel; hammerless. Grade II with engraved action; select walnut stocks. Made in Japan. Introduced in 1977; 22 LR dropped, 1982; 22 WMR dropped 1986.

22 LR
  Perf: $275   **Exc.:** $220   **VGood:** $180
22 WMR
  Perf: $300   **Exc.:** $245   **VGood:** $200
Grade II
  Perf: $375   **Exc.:** $320   **VGood:** $280

BROWNING CONTINENTAL
Double rifle; 270 Win., 30-06, 9.3x74R; interchangeable 20-ga. barrels; 24″ rifle barrels; 26 1/2″ shotgun barrel; flat face gold bead front sight on matted ramp, folding leaf rear; straight-grip, oil-finished American walnut stock hand checkered; Schnabel forend; action built on re-engineered Superposed 20-ga. frame; single selective inertia trigger; manual top tang safety barrel selector. Marketed in fitted case. Made in Belgium. Introduced 1979; dropped 1987.
  Perf: $3200   **Exc.:** $2200   **VGood:** $1750

BROWNING EXPRESS
Double rifle; 270, 30-06; 24″ barrels; 41″ overall length; weighs 7 lbs.; gold bead on ramp front sight, adjustable folding leaf rear; straight grip, oil-finished select European walnut stock; Schnabel forend; hand checkered; Superposed action; reinforced breech face; single selective trigger; selective ejectors; manual safety; hand-engraved receiver; marketed in fitted case. Made in Belgium. Introduced 1981; dropped 1987.
  Perf: $2500   **Exc.:** $2000   **VGood:** $1450

BROWNING (FN) HIGH-POWER SAFARI
Bolt-action; 243 Win., 264 Win. Mag., 270 Win., 30-06, 7mm Rem. Mag., 300 Win. Mag., 308 Norma Mag., 300 H&H, 338 Win. Mag., 375 H&H Mag., 458 Win. Mag.; 4-shot magazine (magnum cartridges), 6-shot (standard calibers); 24″ barrel (magnum calibers), 22″ (standard calibers); European walnut stock; hand-checkered pistol grip, forearm; Monte Carlo cheek-piece; recoil pad on magnum models; folding leaf rear sight, hooded ramp front; quick-detachable sling swivels. Introduced 1960; dropped 1974.

Standard calibers
  Perf: $795   **Exc.:** $675   **VGood:** $550
Magnum calibers
  Perf: $895   **Exc.:** $750   **VGood:** $650

Browning (FN) High-Power Safari-Grade Medium-Action
Same specs as standard grade Safari except action length; 22-250, 243 Win., 284 Win., 308 Win.; 22″ lightweight barrel standard, but available in 22-250, 243 with 24″ heavy barrel. Dropped 1974.
  Perf: $795   **Exc.:** $695   **VGood:** $550

Browning (FN) High-Power Safari-Grade Short-Action
Same specs as standard Safari except short action; 222 Rem., 222 Rem. Mag.; 22″ lightweight, 24″ heavy barrel. Dropped 1974.
  Perf: $795   **Exc.:** $695   **VGood:** $550

Browning (FN) High-Power Safari Medallion-Grade
Same specs as standard Safari grade except scroll-engraved receiver, barrel; engraved ram's head on floorplate; select European walnut stock with rosewood grip cap, forearm tip. Dropped 1974.
  Perf: $1400   **Exc.:** $1100   **VGood:** $750

Browning (FN) High-Power Safari Olympian-Grade
Same specs as Safari grade except engraving on barrel; chrome-plated floorplate, trigger guard, receiver, all engraved with game scenes; figured European walnut stock; 32 lpi checkering; rosewood forearm tip, grip cap; grip cap inlaid with 18-karat gold medallion. Discontinued 1974. Premium for rare cals.
  Perf: $2495   **Exc.:** $1995   **VGood:** $1495

BSA CF-2 Varminter

BSA Majestic Deluxe Featherweight

BSA Model 12/15

BSA Model 13

BSA
Centurion
CF-2 Carbine
CF-2 Heavy Barrel
CF-2 Sporter/Classic
CF-2 Stutzen
CF-2 Varminter
CFT Target
Imperial
Majestic Deluxe
Majestic Deluxe Feather-
weight
Martini International ISU
Match
Martini International Match
Martini International Match
Light
Martini International Match
Mark II
Martini International Match
Mark III
Martini International Match
Mark V
Model 12/15 Match
Model 12/15
Model 13
Model 13 Sporter

BSA CENTURION
Single shot; Martini action; 22 LR; 24" match barrel; BSA No. 30 rear sight, No. 20 front; uncheckered walnut target stock; cheekpiece; pistol grip; long semi-beavertail forend; guaranteed 1.5" groups at 100 yards. Discontinued 1932.
Perf: $495   **Exc.:** $385   **VGood:** $300

BSA CF-2 CARBINE
Bolt action; 222 Rem., 22-250, 243, 6.5x55, 7mm Mauser, 7x64, 270, 308, 30-06, 7mm Rem. Mag., 300 Win. Mag.; 20" barrel; open adjustable rear sight, hooded ramp front; roll-over Monte Carlo stock of European walnut; high-gloss finish; pistol grip, skip-line checkering; adjustable single trigger or optional double-set trigger. Manufactured in England. Discontinued 1985.
Perf: $350   **Exc.:** $300   **VGood:** $250

BSA CF-2 Heavy Barrel
Same specs as CF-2 except 222 Rem., 22-250, 243; 23"-26" heavy barrel; weighs 9 lbs.; no sights. No longer in production.
Perf: $375   **Exc.:** $325   **VGood:** $270

BSA CF-2 Sporter/Classic
Same specs as CF-2 except 23"-26" barrel; oil finished walnut stock. Manufactured in England. Introduced 1980; discontinued 1985.
Perf: $320   **Exc.:** $270   **VGood:** $240

BSA CF-2 Stutzen
Same specs as CF-2 except 222 Rem., 22-250, 243, 6.5x55, 7mm Mauser, 7x64, 270, 308, 30-06; 20 1/2" barrel; full-length Stutzen-style stock, contrasting Schnabel forend tip, grip cap; improved bolt guide. Manufactured in England. Introduced 1982; discontinued 1985.
Perf: $470   **Exc.:** $400   **VGood:** $320

BSA CF-2 Varminter
Same specs as CF-2 except 222 Rem., 22-250, 243; 23"-26" heavy barrel; matte finish; sling swivels. Introduced 1986; discontinued 1986.
Perf: $350   **Exc.:** $280   **VGood:** $255

BSA CFT TARGET
Bolt action; single shot; 7.62mm; 26 1/2" barrel; weighs 11 lbs.; globe front sight, aperture rear; European walnut stock. Discontinued 1987.
Perf: $695   **Exc.:** $595   **VGood:** $475

BSA IMPERIAL
Bolt action; 270 Win., 308 Win., 30-06 (lightweight model), 22 Hornet, 222 Rem., 243 Win., 257 Roberts, 7x57mm, 300 Savage, 30-06 (standard weight); 22" barrel; recoil reducer cut into muzzle; European walnut cheekpiece stock; hand-checkered pistol grip,

Schnabel forend; black buttplate; pistol grip cap with white spacers; drilled, tapped for scope mounts; fully-adjustable trigger. Introduced 1959; discontinued 1964.
Perf: $475   **Exc.:** $380   **VGood:** $300

BSA MAJESTIC DELUXE
Bolt action; 22 Hornet, 222 Rem., 243 Win., 308 Win., 30-06, 7x57mm; 4-shot magazine; 22" heavy barrel; weighs 6 1/4 lbs.; folding leaf rear, hooded ramp front sight; European walnut stock; checkered pistol grip, forend; cheekpiece; Schnabel forend; sling swivels. Introduced 1959; discontinued 1965.
Perf: $350   **Exc.:** $300   **VGood:** $250

BSA Majestic Deluxe Featherweight
Same specs as Majestic Deluxe except recoil pad; lightweight 22" barrel; recoil reducer; 243, 270, 308, 30-06, 458.
Perf: $340   **Exc.:** $300   **VGood:** $230

BSA MARTINI INTERNATIONAL ISU MATCH
Single shot; 22 LR; 28" barrel; 43"-44" overall length; weighs 10 3/4 lbs.; Martini action; modified PH-1 Parker-Hale tunnel front sight, PH-25 aperture rear; match-type French butt and forend; flat cheekpiece; fully-adjustable trigger; designed for ISU specs. Introduced 1968; discontinued 1986.
Perf: $600   **Exc.:** $450   **VGood:** $380

BSA MARTINI INTERNATIONAL MATCH
Single shot; 22 LR; 29" heavy barrel; Martini action; match sights; uncheckered two-piece target stock; full cheekpiece; pistol grip; broad beavertail forend; hand stop; swivels; right- or left-hand styles. Introduced 1950; discontinued 1953.
Perf: $475   **Exc.:** $375   **VGood:** $300

BSA Martini International Match Light
Same specs as Martini International Match except lightweight 26" barrel. Introduced 1950; discontinued 1953.
Perf: $475   **Exc.:** $375   **VGood:** $300

BSA Martini International Match Mark II
Same specs as Martini International Match except choice of light (26"), heavy (29") barrel; weighs 14 lbs. (29" barrel), 11 lbs. (26" barrel); redesigned stock forearm, trigger mechanism, ejection system; right- or left-hand moulds. Introduced in 1953; discontinued 1959.
Perf: $500   **Exc.:** $400   **VGood:** $320

BSA Martini International Match Mark III
Same specs as International Match Mark II heavy barrel model plus redesigned stock forend; free-floating barrel; longer frame with alloy strut to attach forend. Introduced 1959; discontinued 1967.
Perf: $550   **Exc.:** $450   **VGood:** $375

BSA Martini International Match Mark V
Same specs as International Match except 28" heavy barrel; weighs 12 1/4 lbs. Introduced 1976; no longer produced.
Perf: $650   **Exc.:** $550   **VGood:** $450

BSA MODEL 12/15 MATCH
Single shot; 22 LR; 29" barrel; Martini action; Parker-Hale PH-7A rear, PH-22 front sight; walnut target stock with high comb, cheekpiece, beavertail forend; forward sling swivel. Introduced 1938; production suspended during WWII; discontinued 1950.
Perf: $450   **Exc.:** $350   **VGood:** $300

BSA Model 12/15
Same as 12/15 Match except heavy barrel. This was designation given rifle when reintroduced following WWII. Dropped 1950.
Perf: $450   **Exc.:** $350   **VGood:** $300

BSA MODEL 13
Single shot; 22 LR; 25" barrel; Martini action; Parker-Hale No. 7 rear, hooded front sight; straight-grip walnut stock; hand-checkered forearm; lightweight version of BSA No. 12. Introduced 1913; discontinued 1929.
Perf: $400   **Exc.:** $350   **VGood:** $300

BSA Model 13 Sporter
Same specs as Model 13 except Parker-Hale Sportarget rear sight, bead front. Also available in 22 Hornet. discontinued 1932.
Perf: $375   **Exc.:** $325   **VGood:** $275
22 Hornet
Perf: $425   **Exc.:** $375   **VGood:** $325

RIFLES

BSA
Model 15
Monarch Deluxe
Monarch Deluxe Varmint
No. 12

Cabanas
Espronceda
Leyre
Master
Mini-82 Youth Pony
Phaser
R83
Taser
Varmint

Cabela's
1858 Henry Replica
1866 Winchester Replica
1873 Winchester Replica
1873 Winchester Replica
Sporting Model
Cattleman's Carbine

BSA Model 15

BSA Monarch Deluxe

Cabanas Leyre

Cabanas Phaser

Cabanas Master

BSA MODEL 15
Single shot; 22 LR; 29″ barrel; Martini action; BSA No. 30 rear, No. 20 front sight; uncheckered walnut target stock; cheekpiece, pistol grip, long semi-beavertail forend. Introduced 1915; discontinued 1932.
Perf: $395   **Exc.:** $300   **VGood:** $250

BSA MONARCH DELUXE
Bolt action; 22 Hornet, 222 Rem., 243 Win., 270 Win., 7mm Rem. Mag., 308 Win., 30-06; 22″ barrel; weighs 6 1/4 lbs.; Mauser-type; fully-adjustable trigger; bolt head encloses cartridge and in turn enclosed by barrel extension; gas-proof cocking piece; hinged floorplate; silent safety with cocking indicator; folding leaf rear sight, hooded ramp front; checkered American-style stock; hardwood forend tip, grip cap; recoil pad. Introduced 1965; discontinued 1974.
Perf: $395   **Exc.:** $340   **VGood:** $260

BSA Monarch Deluxe Varmint
Same specs as Monarch Deluxe except 222 Rem., 243 Win; 24″ heavy barrel. Introduced 1965; discontinued 1977.
Perf: $395   **Exc.:** $350   **VGood:** $300

BSA NO. 12
Single shot; 22 LR; 29″ barrel; Martini action; Parker-Hale No. 7 rear, No. 2 front sight; straight-grip walnut stock; hand-checkered forearm. Introduced 1912; discontinued 1929.
Perf: $395   **Exc.:** $300   **VGood:** $250

CABANAS ESPRONCEDA
Single shot; 177 pellet with 22 Blank cartridge; 18 3/4″ barrel; 40″ overall length; weighs 5 1/2 lbs.; open fully-adjustable rear sight, blade front; full sporter stock. Made in Mexico. Introduced 1986; still imported. **LMSR: $135**
Perf.: $125   **Exc.:** $90   **VGood:** $65

CABANAS LEYRE
Bolt action; single shot; 177 pellet with 22 Blank cartridge; 19 1/2″ barrel; 44″ overall length; weighs 8 lbs.; adjustable rear, blade front sight; sport/target walnut stock. Made in Mexico. Introduced 1984; still imported. **LMSR: $150**
Perf.: $140   **Exc.:** $100   **VGood:** $70

CABANAS MASTER
Bolt action; single shot; 177 pellet with 22 Blank cartridge; 19 1/2″ barrel; 45 1/2″ overall length; weighs 8 lbs.; adjustable rear sight, blade front; Monte Carlo target-type walnut stock. Made in Mexico. Introduced 1984; still imported. **LMSR: $190**
Perf.: $150   **Exc.:** $120   **VGood:** $80

CABANAS MINI-82 YOUTH PONY
Bolt action; single shot; 177 pellet with 22 Blank cartridge; 16 1/2″ barrel; 34″ overall length; weighs 3 1/8 lbs.; adjustable rear, blade front sight; youth-sized stock. Made in Mexico. Introduced 1984; still imported. **LMSR: $70**
Perf.: $65   **Exc.:** $50   **VGood:** $40

CABANAS PHASER
Bolt action; single shot; 177 pellet with 22 Blank cartridge; 19″ barrel; 42″ overall length; weighs 6 3/4 lbs.; open fully-adjustable rear sight, blade front; target-type thumbhole stock. Made in Mexico. Introduced 1991; no longer imported.
Perf.: $150   **Exc.:** $120   **VGood:** $80

CABANAS R83
Bolt action; single shot; 177 pellet with 22 Blank cartridge; 17″ barrel; 40″ overall length; adjustable rear, blade front sight; large youth-sized hardwood stock. Made in Mexico. Introduced 1984; still imported. **LMSR: $80**
Perf.: $65   **Exc.:** $50   **VGood:** $40

CABANAS TASER
Bolt action; single shot; 177; 19″ barrel; 42″ overall length; weighs 6 3/4 lbs.; target-type thumbhole stock; blade front sight, open fully-adjustable rear; fires round ball or pellets with 22-caliber blank cartridge. Imported from Mexico by Mandall Shooting Supplies. Still imported. **LMSR: $160**
Perf.: $150   **Exc.:** $120   **VGood:** $80

CABANAS VARMINT
Bolt action; single shot; 177 pellet with 22 Blank cartridge; 21 1/2″ barrel; 41″ overall length; weighs 4 1/2 lbs.; adjustable rear, blade front sight; varmint-type walnut stock. Made in Mexico. Introduced 1984; still imported. **LMSR: $120**
Perf.: $100   **Exc.:** $80   **VGood:** $65

CABELA'S 1858 HENRY REPLICA
Lever action; 44-40; 13-shot magazine; 24 1/4″ barrel; 43″ overall length; weighs 9 lbs.; European walnut stock; bead front sight, open adjustable rear; brass receiver and buttplate; uses original Henry loading system; faithful to the original rifle. Imported by Cabela's. Introduced 1994; still imported. **LMSR: $650**
Perf.: $495   **Exc.:** $425   **VGood:** $350

CABELA'S 1866 WINCHESTER REPLICA
Lever action; 44-40; 13-shot magazine; 24 1/4″ barrel; 43″ overall length; weighs 9 lbs.; European walnut stock; bead front sight, open adjustable rear; solid brass receiver, buttplate, forend cap; octagonal barrel. Faithful to the original Winchester '66 rifle. Imported by Cabela's. Introduced 1994; still imported. **LMSR: $500**
Perf.: $395   **Exc.:** $350   **VGood:** $295

CABELA'S 1873 WINCHESTER REPLICA
Lever action; 44-40, 45 Colt; 13-shot magazine; 24 1/4″, 30″ barrel; 43 1/4″ overall length; weighs 8 1/2 lbs.; European walnut stock; bead front sight, open adjustable rear, or globe front, tang rear; color case-hardened steel receiver. Faithful to the original Model 1873 rifle. Imported by Cabela's. Introduced 1994; still imported. **LMSR: $640**
Perf.: $450   **Exc.:** $395   **VGood:** $325
With tang sight, globe front
Perf.: $495   **Exc.:** $435   **VGood:** $395

Cabela's 1873 Winchester Replica Sporting Model
Same specs as 1873 Winchester except 30″ barrel. Imported by Cabela's. Introduced 1994; still imported. **LMSR: $600**
Perf.: $440   **Exc.:** $395   **VGood:** $350
With half-round/half-octagon barrel, half magazine
Perf.: $750   **Exc.:** $650   **VGood:** $550

CABELA'S CATTLEMAN'S CARBINE
Revolver; 44-40; 6-shot cylinder; 18″ barrel; 34″ overall length; weighs 4 lbs.; European walnut stock; blade front sight, notch rear; color case-hardened frame, rest blued. Imported by Cabela's. Introduced 1994; still imported. **LMSR: $300**
Perf.: $260   **Exc.:** $230   **VGood:** $175

Calico Liberty 50 Carbine

Century Centurion P14 Sporter

Century Custom Sporting Rifle

Century International FAL Sporter

Century Enfield Sporter #4

Cabela's
Sharps Sporting Rifle

Calico
Liberty 50 Carbine
Liberty 100 Carbine
Model M-100 Carbine
Model M-105 Sporter
Model M-900 Carbine
Model M-951 Tactical Carbine
Model M-951S Tactical Carbine

Century
Centurion P14 Sporter
Custom Sporting Rifle
Deluxe Custom Sporter
Enfield Sporter #4
International FAL Sporter
International M-14

CABELA'S SHARPS SPORTING RIFLE
Single shot; 45-70; 32″ tapered octagon barrel; 47 1/4″ overall length; weighs 9 lbs.; checkered walnut stock; blade front sight, open adjustable rear; color case-hardened receiver and hammer, rest blued. Imported by Cabela's. Introduced 1995; still imported. **LMSR: $750**
Perf.: $650   Exc.: $595   VGood: $450

CALICO LIBERTY 50 CARBINE
Semi-automatic; 9mm Para.; 50-shot magazine; 16 1/8″ barrel; 34 1/2″ overall length; weighs 7 lbs.; glass-filled, impact resistant polymer stock; adjustable post front sight, fixed notch and aperture flip rear; helical feed magazine; ambidextrous, rotating sear/striker block safety; static cocking handle; retarded blowback action; aluminum alloy receiver. Made in U.S. by Calico. Introduced 1995; still produced. **LMSR: $860**
Perf.: $550   Exc.: $460   VGood: $370

CALICO LIBERTY 100 CARBINE
Semi-automatic; 9mm Para.; 100-shot magazine; 16 1/8″ barrel; 34 1/2″ overall length; weighs 7 lbs.; glass-filled, impact resistant polymer stock; adjustable post front sight, fixed notch and aperture flip rear; helical feed magazine; ambidextrous, rotating sear/striker block safety; static cocking handle; retarded blowback action; aluminum alloy receiver. Made in U.S. by Calico. Introduced 1995; still produced. **LMSR: $860**
Perf.: $600   Exc.: $510   VGood: $425

CALICO MODEL M-100 CARBINE
Semi-automatic; 22 LR; 100-shot magazine; 16″ barrel; 35 7/8″ overall length with stock extended; weighs 5 3/4 lbs. (loaded); folding steel stock; post front sight elevation-adjustable, notch windage-adjustable rear; alloy frame and helical-feed magazine; ambidextrous safety; removable barrel assembly; pistol grip compartment; flash suppressor; bolt stop. Made in U.S. From Calico. Introduced 1986; dropped 1995.
Perf.: $295   Exc.: $230   VGood: $190

CALICO MODEL M-105 SPORTER
Semi-automatic; 22 LR; 100-shot magazine; 16″ barrel; 35 7/8″ overall length; weighs 4 3/4 lbs. (empty); hand-rubbed wood buttstock and forend; post front sight elevation-adjustable, notch windage-adjustable

rear; alloy frame and helical-feed magazine; ambidextrous safety; removable barrel assembly; pistol grip compartment; flash suppressor; bolt stop. Made in U.S. From Calico. Introduced 1987; dropped 1995.
Perf.: $285   Exc.: $230   VGood: $190

CALICO MODEL M-900 CARBINE
Semi-automatic; 9mm Para.; 50-, 100-shot magazine; 16 1/8″ barrel; weighs 28 1/2″ overall length with stock collapsed; 3 3/4 lbs. (empty) sliding steel buttstock; post front sight fully-adjustable, fixed notch rear; helical feed magazine; ambidextrous safety; static cocking handle; retarded blowback action; glass-filled polymer grip. Made in U.S. by Calico. Introduced 1989; dropped 1995.
Perf.: $560   Exc.: $475   VGood: $350

CALICO MODEL M-951 TACTICAL CARBINE
Semi-automatic; 9mm Para.; 50-, 100-shot magazine; 16 1/8″ barrel with long compensator; 28 3/4″ overall length with stock collapsed; weighs 3 3/4 lbs. (empty); sliding steel buttstock, adjustable forward grip; fully-adjustable post front sight, fixed notch rear; helical feed magazine; ambidextrous safety; static cocking handle; retarded blowback action; glass-filled polymer grip. Made in U.S. by Calico. Introduced 1989; dropped 1995.
Perf.: $495   Exc.: $410   VGood: $320

Calico Model M-951S Tactical Carbine
Same specs as the M-951 except fixed stock. Made in U.S. by Calico. Introduced 1989; dropped 1995.
Perf.: $500   Exc.: $430   VGood: $330

CENTURY CENTURION P14 SPORTER
Bolt action; 303 British, 7mm Rem. Mag., 300 Win. Mag.; 5-shot magazine; 24″ barrel; 43 5/8″ overall length; no sights; drilled, tapped for scope mounts; checkered Monte Carlo European beechwood stock; uses modified Pattern 14 Enfield action; blued finish. Introduced 1987; still available. **LMSR: $275**
Perf.: $200   Exc.: $150   VGood: $110

CENTURY CUSTOM SPORTING RIFLE
Bolt action; 308, 7.62x39mm; 5-shot magazine; 22″ barrel; 43 3/4″ overall length; weighs 6 3/4 lbs.; walnut-

finished hardwood stock; no sights; two-piece Weaver-type base; uses small ring Model 98 action; low-swing safety; blue finish. From Century International Arms. Introduced 1994; still produced. **LMSR: $275**
Perf.: $210   Exc.: $160   VGood: $120

CENTURY DELUXE CUSTOM SPORTER
Bolt action, 243, 270, 308, 30-06; 5-shot magazine; 24″ barrel; 44″ overall length; black synthetic stock; no sights, but scope base installed; Mauser 98 action; bent bolt handle for scope use; low-swing safety; matte black finish; blind magazine. From Century International Arms. Introduced 1992; still produced. **LMSR: $288**
Perf.: $230   Exc.: $180   VGood: $140

CENTURY ENFIELD SPORTER #4
Bolt action; 303 British; 10-shot magazine; 25 1/4″ barrel; 44 1/2″ overall length; adjustable aperture rear sight, blade front; checkered beechwood Monte Carlo stock; built on Lee-Enfield action; blued finish. Introduced 1987; still available. **LMSR: $156**
Perf.: $120   Exc.: $100   VGood: $80

CENTURY INTERNATIONAL FAL SPORTER
Semi-automatic; 308 Win.; 20 3/4″ barrel; 41 1/8″ overall length; weighs 9 lbs., 13 oz.; Bell & Carlson thumbhole sporter stock; protected post front sight, adjustable aperture rear; matte blue finish; rubber buttpad. From Century International Arms. Still produced. **LMSR: $625**
Perf.: $525   Exc.: $475   VGood: $400

CENTURY INTERNATIONAL M-14
Semi-automatic; 308 Win.; 10-shot magazine; 22″ barrel; 40 7/8″ overall length; weighs 8 1/4 lbs.; walnut stock with rubber recoil pad; protected blade front sight, fully-adjustable aperture rear; gas-operated; forged receiver; Parkerized finish. Imported from China by Century International Arms. Still imported. **LMSR: $469**
Perf.: $399   Exc.: $350   VGood: $295

RIFLES

Century
Mauser 98 Sporter
Swedish Sporter #38
Tiger Dragunov
Weekender

Chapuis
Agex Express Series
Boxlock Double Rifle
Ourel Exel
RGEX Express

Charter Arms
AR-7 Explorer
(See Armalite AR-7 Explorer)

Chipmunk
Single Shot

Churchill
Highlander

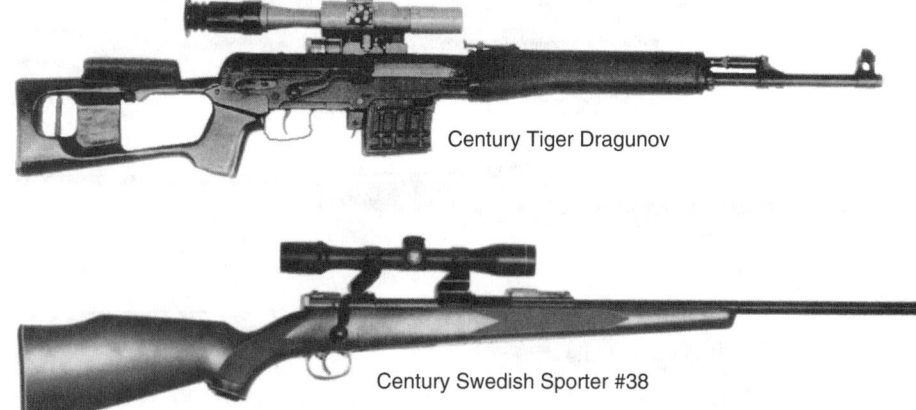

Century Tiger Dragunov

Century Swedish Sporter #38

Century Mauser 98 Sporter

Century Weekender

Chapuis Boxlock Double Rifle

CENTURY MAUSER 98 SPORTER
Bolt action; 243, 270, 308, 30-06; 24" barrel; 44" overall length; black synthetic stock; no sights; scope base installed; Mauser 98 action; bent bolt handle for scope use; low-swing safety; matte black finish; blind magazine. From Century International Arms. Introduced 1992; no longer produced.
   Perf.: $230   **Exc.:** $190   **VGood:** $140

CENTURY SWEDISH SPORTER #38
Bolt action; 6.5x55 Swede; 5-shot magazine; 24" barrel; 44" overall length; adjustable rear sight, blade front; checkered European hardwood Monte Carlo stock; marketed with Holden Ironsighter see-through scope mount; made on Model 38 Swedish Mauser action. Introduced 1987; still available. **LMSR: $237**
   Perf.: $225   **Exc.:** $180   **VGood:** $130

CENTURY TIGER DRAGUNOV
Semi-automatic; 7.62x54R; 5-shot magazine; 20 7/8" barrel; 42 15/16" overall length; weighs 8 1/2 lbs.; thumbhole stock of laminated European hardwood, black composition forend; blade front sight, open elevation-adjustable rear; 4x rangefinding scope with sunshade, lighted reticle; new manufacture of shortened version of Russian SVD sniper rifle; blued metal; quick-detachable scope mount; comes with sling, cleaning kit, gas regulator tool, case. Imported from Russia by Century International Arms. Still imported. **LMSR: $1350**
   Perf.: $1100   **Exc.:** $900   **VGood:** $750

CENTURY WEEKENDER
Bolt action; 22 LR; 5-shot magazine; 23 1/2" barrel; 42" overall length; open adjustable rear sight, hooded blade front; sling swivels; blued finish. Made in Europe. Introduced 1987; discontinued 1989.
   Perf.: $175   **Exc.:** $140   **VGood:** $100

CHAPUIS AGEX EXPRESS SERIES
Double rifle; side-by-side boxlock; 375 H&H, 416 R Chapius, 470 NE; 25 5/8" barrels; various grades of engraving; French walnut stock. Made in France. No longer imported.
Africa model
   Perf.: $17,995 **Exc.:** $15,995   **VGood:** $12,500
Brousse model
   Perf.: $9,995 **Exc.:** $7350   **VGood:** $5250
Jungle model
   Perf.: $15,950 **Exc.:** $10,995   **VGood:** $5750
Safari model
   Perf.: $28,000 **Exc.:** $24,900   **VGood:** $17,995
Savanna model
   Perf.: $36,900 **Exc.:** $31,000   **VGood:** $18,995

CHAPUIS BOXLOCK DOUBLE RIFLE
Double rifle; side-by-side; 7x65R, 8x57 JRS, 9.3x74R, 375 H&H; 23 5/8" barrels; 40 5/16" overall length; weighs 8 3/8 lbs.; ramp bead front sight, adjustable express rear on quarter rib; oil-finished French walnut pistol-grip stock; engraved, coin-finished receiver; double-hook barrels, double triggers; automatic ejectors. Made in France. Introduced 1989; no longer imported.
   Perf.: $5500   **Exc.:** $4850   **VGood:** $3650

CHAPUIS OURAL EXEL
Single shot; boxlock; 270, 300 Win. Mag., 7mm Rem. Mag.; 23 5/8" barrel; various grades of engraving; French walnut stocking. Made in France. No longer imported.
   Perf.: $4995   **Exc.:** $4450   **VGood:** $3750

CHAPUIS RGEX EXPRESS
Double rifle; 30-06, 7x65R, 8x57 JRS, 9.3x74R; 23 5/8" barrel; weighs 8-9 lbs.; deluxe walnut stock with Monte Carlo comb, oil finish; bead on ramp front sight, adjustable express rear on quarter-rib; boxlock action with long trigger guard; automatic ejectors; double hook Blitz system action with coil springs; coin metal finish; trap grip cap for extra front sight. Imported from France by Armes de Chasse. No longer imported.
   Perf.: $6100   **Exc.:** $5300   **VGood:** $4900

CHIPMUNK SINGLE SHOT
Bolt action; single-shot; 22 Short, 22 Long, 22 LR, 22 WMR; 16 1/8" barrel; 30" overall length; weighs 2 1/2 lbs.; post on ramp front sight, adjustable peep rear; drilled, tapped for scope mounting; American walnut stock. Introduced 1982; still in production. **LMSR: $195**
   Perf.: $160   **Exc.:** $125   **VGood:** $95
Deluxe model with hand checkering
   Perf.: $225   **Exc.:** $190   **VGood:** $140

CHURCHILL HIGHLANDER
Bolt action; 243, 25-06, 300 Win. Mag., 270, 308, 30-06, 7mm Rem. Mag.; 3-, 4-shot magazine, depending on caliber; 22", 24" barrel; 42 1/2" overall length (22" barrel); weighs 7 1/2 to 8 lbs.; no sights or optional fully-adjustable rear sight, gold bead on ramp front; checkered classic-style European walnut stock; oil-finish wood; swivel posts; recoil pad; positive safety. Made in Europe. Introduced 1986; importation discontinued 1989.
   Perf.: $395   **Exc.:** $350   **VGood:** $300
With optional sights
   Perf.: $425   **Exc.:** $380   **VGood:** $330

Chipmunk Single Shot

Churchill Highlander

Cimarron 1873 Winchester Replica, Long Range Model

Cimarron 1866 Winchester Replica Carbine

Churchill
Regent
Regent Combo

Cimarron
1860 Henry Replica
1866 Winchester Replica
1866 Winchester Replica
Carbine
1866 Winchester Replica
Carbine,
Indian Model
1866 Winchester Replica
Carbine,
Trapper Model
1873 Winchester Replica
Sporting Rifle
1873 Winchester Replica,
Long Range Model
1873 Winchester Replica,
Saddle Ring Carbine
1873 Winchester Replica,
Short Model
1873 Winchester Replica,
Trapper Carbine

Claridge
Hi-Tec C Carbine
Hi-Tec Model LEC-9 Carbine
Hi-Tec Model ZLEC-9
Carbine

CHURCHILL REGENT

Bolt action; 243, 25-06, 300 Win. Mag., 270, 308, 30-06, 7mm Rem. Mag.; 3-, 4-shot magazine, depending on caliber; 22″, 24″ barrel; 42 1/2″ overall length (22″ barrel); weighs 7 1/2 to 8 lbs.; no sights or optional fully-adjustable rear sight, gold bead on ramp front; Monte Carlo stock with deluxe checkering; swivel posts; recoil pad; positive safety. Made in Europe. Introduced 1986; discontinued 1988.

Perf.: $540   **Exc.:** $490   **VGood:** $320
With optional sights
Perf.: $570   **Exc.:** $500   **VGood:** $330

CHURCHILL REGENT COMBO

Combination rifle; over/under; 3″ 12-ga. over 222, 223, 243, 270, 308, 30-06; 25″ barrels; 42″ overall length; weighs 8 lbs.; open rear sight, blade on ramp front; hand-checkered, oil-finished European walnut Monte Carlo stock; dovetail scope mount; double triggers; silvered engraved receiver. Made in Europe. Introduced 1985; importation discontinued 1989.

Perf.: $800   **Exc.:** $700   **VGood:** $600

CIMARRON 1860 HENRY REPLICA

Lever action; 44-40, 44 Spl., 45 Colt; 13-shot tubular magazine; 22″ carbine, 24 1/2″ rifle barrel; 43″ overall length (24 1/2″ barrel); weighs 9 1/2 lbs.; bead front sight, open adjustable rear; original Henry loading system; brass receiver, buttplate. Made in Italy. Introduced 1991; still imported. **LMSR:** $1029

Perf.: $800   **Exc.:** $700   **VGood:** $600

CIMARRON 1866 WINCHESTER REPLICA

Lever action; 22 LR, 22 WMR, 44-40, 45 Colt; 24 1/4″ octagonal barrel; 43″ overall length; weighs 9 lbs.; bead front sight, open adjustable rear; European walnut stock; brass receiver, buttplate, forend cap. Made in Italy. Introduced 1991; still imported. **LMSR:** $839

Perf.: $700   **Exc.:** $600   **VGood:** $500

CIMARRON 1866 WINCHESTER REPLICA CARBINE

Lever action; 22 LR, 22 WMR, 38 Spl., 44-40, 45 Colt; 19″ round barrel; bead front, open adjustable rear; smooth walnut stock, forearm; brass receiver, buttplate, forend cap. Made in Italy. Introduced 1991; still imported. **LMSR:** $829

Perf.: $700   **Exc.:** $600   **VGood:** $500

Cimarron 1866 Winchester Replica Carbine, Indian Model

Same specs as 1866 Replica Carbine except engraved brass frame. Discontinued 1989.

Perf.: $595   **Exc.:** $495   **VGood:** $395

Cimarron 1866 Winchester Replica Carbine, Trapper Model

Same specs as 1866 Replica Carbine except 44-40 only; 16″ round barrel. Discontinued 1990.

Perf.: $525   **Exc.:** $435   **VGood:** $375

CIMARRON 1873 WINCHESTER REPLICA SPORTING RIFLE

Lever action; 44-40, 45 Colt; 24″ octagon barrel; 43″ overall length; 8 lbs.; adjustable semi-buckhorn rear sight, fixed front; walnut stock, forend; color case-hardened receiver. Introduced 1989; still imported. **LMSR:** $949

Perf.: $815   **Exc.:** $625   **VGood:** $525

Cimarron 1873 Winchester Replica, Long Range Model

Same specs as 1873 Winchester except 22 LR, 22 WMR, 357 Mag., 38-40, 44-40, 45 Colt; 30″ octagon barrel marked "Kings Improved"; 48″ overall length; weighs 8 1/2 lbs. Introduced 1989; still imported. **LMSR:** $999

Perf.: $850   **Exc.:** $650   **VGood:** $550

Cimarron 1873 Winchester Replica, Saddle Ring Carbine

Same specs as 1873 Winchester except 19″ round barrel; blued receiver; saddle ring. Introduced 1989; still imported. **LMSR:** $949

Perf.: $815   **Exc.:** $625   **VGood:** $495

Cimarron 1873 Winchester Replica, Short Model

Same specs as 1873 Winchester except 22 LR, 22 WMR, 357 Mag., 44-40, 45 Colt; 20″ octagon barrel; 39″ overall length; weighs 7 1/2 lbs. Introduced 1989; still imported. **LMSR:** $949

Perf.: $815   **Exc.:** $625   **VGood:** $495

Cimarron 1873 Winchester Replica, Trapper Carbine

Same specs as 1873 Winchester except 357 Mag., 44-40, 45 Colt; 16″ barrel; blued finish. Introduced 1989; discontinued 1990.

Perf.: $595   **Exc.:** $495   **VGood:** $395

CLARIDGE HI-TEC C CARBINE

Semi-automatic; 9mm Para., 40 S&W, 45 ACP; 18-shot magazine; 16 1/8″ barrel; 31 3/4″ overall length; weighs 4 lbs. 9 oz.; walnut stock; adjustable post in ring front sight, open windage-adjustable rear; aluminum or stainless frame; telescoping bolt, floating firing pin; safety locks the firing pin; sight radius of 20 1/8″; accepts same magazines as Claridge Hi-Tec pistols. Can be equipped with scope or Aimpoint sight. Made in U.S. From Claridge Hi-Tec, Inc. Introduced 1991; dropped 1993.

Perf.: $595   **Exc.:** $495   **VGood:** $395

CLARIDGE HI-TEC MODEL LEC-9 CARBINE

Semi-automatic; 9mm Para., 40 S&W, 45 ACP; 18-shot magazine; 16 1/8″ barrel; 31 3/4″ overall length; weighs 4 lbs. 9 oz.; graphite composite stock; adjustable post in ring front sight, open windage-adjustable rear; aluminum or stainless frame; telescoping bolt, floating firing pin; safety locks the firing pin; sight radius of 20 1/8″; accepts same magazines as Claridge Hi-Tec pistols. Can be equipped with scope or Aimpoint sight. Made in U.S. From Claridge Hi-Tec, Inc. Made only in 1992.

Perf.: $650   **Exc.:** $575   **VGood:** $485

Claridge Hi-Tec Model ZLEC-9 Carbine

Same specs as the LEC-9 except laser sight system. Made in U.S. by Claridge Hi-Tec, Inc. Made only in 1992.

Perf.: $995   **Exc.:** $895   **VGood:** $795

Clayco
Model 4

Clerke
Hi-Wall
Hi-Wall Deluxe

Colt
57
AR-15A2 Carbine
AR-15A2 Rifle
Colteer
Colteer 1-22
Coltsman Standard
Coltsman Custom
Coltsman Deluxe
Coltsman Long-Action
Sako Custom
Coltsman Long-Action
Sako Standard
Coltsman Medium-Action
Sako Custom
Coltsman Medium-Action
Sako Standard
Coltsman Sako
(See Colt Coltsman Standard)

Clayco Model 4

Clerke Hi-Wall

Colt Colteer

Colt Colteer 1 22

Colt Coltsman Sako Custom

CLAYCO MODEL 4
Bolt action; 22 LR; 5-shot clip; 24" barrel; 42" overall length; weighs 5 ³/₄ lbs.; adjustable open rear sight, ramp front with bead; walnut-finished hardwood stock; wing-type safety; black composition buttplate; pistol-grip cap; receiver grooved for tip-off scope. Made in China. Introduced 1983 by Clayco Sports; discontinued 1985.
   Perf.: $130   **Exc.:** $100   **VGood:** $80

CLERKE HI-WALL
Single shot; falling block; 223, 22-250, 243, 6mm Rem., 250 Savage, 257 Roberts, 25-06, 264 Win., 270, 7mm Rem. Mag., 30-30, 30-06, 300 Win., 375 H&H, 458 Win., 45-70; 26" barrel; no sights; drilled, tapped for scope mounts; walnut pistol-grip stock, forearm; no checkering; black buttplate; exposed hammer; Schnabel forearm; curved finger lever. Introduced 1970; discontinued 1975.
   Perf.: $295   **Exc.:** $240   **VGood:** $200

Clerke Hi-Wall Deluxe
Same specs as standard model except half-octagon barrel; adjustable trigger; checkered pistol grip, forearm, cheekpiece; plain or optional double-set trigger.
   Perf.: $375   **Exc.:** $320   **VGood:** $280
With double-set trigger
   Perf.: $400   **Exc.:** $350   **VGood:** $300

COLT 57
Bolt action; 243, 30-06; adjustable sights; checkered American walnut Monte Carlo stock; FN Mauser action; 5000 manufactured by Jefferson Manufacturing Co. Introduced 1957; discontinued 1957.
   Perf.: $575   **Exc.:** $485   **VGood:** $400
Deluxe stock
   Perf.: $600   **Exc.:** $500   **VGood:** $420

COLT AR-15A2 CARBINE
Semi-automatic; 223 Rem.; 5-shot magazine; 16" barrel; 5 ¹³/₁₆ lbs.; post front, adjustable rear sight; aluminum stock with collapsible butt; sling swivels. Introduced 1985; discontinued 1989.
   Perf.: $1250   **Exc.:** $1095   **VGood:** $895

Colt AR-15A2 Rifle
Same specs as AR-15A2 Carbine except 20" barrel; 39" overall length; 7 ¹/₂ lbs.; black composition stock, grip, forend; military matte black finish. Introduced 1985; discontinued 1989.
   Perf.: $1195   **Exc.:** $995   **VGood:** $895

COLT COLTEER
Semi-automatic; 22 LR; 19 ³/₈" barrel; 15-shot tube magazine; weighs 4 ³/₄ lbs.; open rear sight, hooded ramp front; uncheckered straight Western-style carbine stock; barrel band; alloy receiver. Introduced 1964; discontinued 1975.
   Perf.: $275   **Exc.:** $220   **VGood:** $170

COLT COLTEER 1-22
Bolt action; single shot; 22 Short, 22 Long, 22 LR, 22 WMR; 20", 22" barrel; open rear sight, ramp front; uncheckered walnut pistol grip; Monte Carlo stock. Introduced 1957; discontinued 1967.
   Perf.: $200   **Exc.:** $150   **VGood:** $100

COLT COLTSMAN STANDARD
Bolt action; 223, 243, 264, 308, 30-06, 300 H&H Mag.; 5-, 6-shot box magazine; 22", 24" barrel; weighs 6 ¹/₂-7 ¹/₂ lbs.; FN Mauser action; folding leaf rear sight, hooded post front; hand-checkered pistol-grip American walnut stock; Monte Carlo comb and cheekpiece; quick-detachable sling swivels. Introduced 1957; replaced in 1962 by Sako action model.
   Perf.: $495   **Exc.:** $420   **VGood:** $340

Colt Coltsman Custom
Same specs as Coltsman except 23³/₄", 23", 24¹/₂" barrel; fancy select French walnut stock; rosewood forearm, pistol grip; Monte Carlo comb, cheekpiece, engraved floorplate; recoil pad; sling swivels. Introduced 1957; replaced in 1962 by Sako action.
   Perf.: $695   **Exc.:** $595   **VGood:** $470

Colt Coltsman Deluxe
Same specs as Coltsman except better checkering, wood; adjustable rear sight, ramp front. Introduced 1957; discontinued 1962.
   Perf.: $695   **Exc.:** $595   **VGood:** $495

Colt Coltsman Long-Action Sako Custom
Same specs as Coltsman Standard except Sako action; fancy Monte Carlo stock; recoil pad; dark wood forend tip and pistol-grip cap; skip-line checkering. Introduced 1962; dropped 1965.
   Perf.: $695   **Exc.:** $595   **VGood:** $495
300, 375 H&H calibers
   Perf.: $750   **Exc.:** $650   **VGood:** $525

Colt Coltsman Long-Action Sako Standard
Same specs as Coltsman Standard except Sako action; 264 Win., 270 Win., 30-06, 300 H&H, 375 H&H; hinged floorplate; hand-checkered walnut stock; standard sling swivels; bead front sight on hooded ramp, folding leaf rear; sliding safety. Introduced 1962; dropped 1965.
   Perf.: $695   **Exc.:** $595   **VGood:** $495
300, 375 calibers
   Perf.: $750   **Exc.:** $650   **VGood:** $525

Colt Coltsman Medium-Action Sako Custom
Same specs as Coltsman Standard except Sako action; fancy Monte Carlo stock; recoil pad; dark wood forend tip; pistol-grip cap; skip-line checkering. Introduced 1962; discontinued 1965.
   Perf.: $695   **Exc.:** $595   **VGood:** $495

Colt Coltsman Medium-Action Sako Standard
Same specs as Coltsman Standard except Sako action; hinged floorplate; hand-checkered walnut stock; standard sling swivels; bead front sight on hooded ramp, folding leaf rear; sliding safety; 243 Win., 308 Win. Introduced 1962; dropped 1965.
   Perf.: $595   **Exc.:** $495   **VGood:** $395

Colt Lightning Small Frame

Colt Lightning

Colt Courier

Colt Sporter Match HBAR

Colt Sporter Lightweight

Colt
Coltsman Short-Action
Sako Custom
Coltsman Short-Action
Sako Deluxe
Courier
Lightning
Lightning Baby Carbine
Lightning Carbine
Lightning Small Frame
Sporter Target
Sporter Competition HBAR
Sporter Competition HBAR
Range
Selected
Sporter Lightweight
Sporter Match Delta HBAR
Sporter Match HBAR
Sporter Match Target
Competition HBAR II
Stagecoach

Colt Coltsman Short-Action Sako Custom
Same specs as Coltsman Standard except Sako action; 243, 308 Win. Introduced 1963; discontinued 1965.
Perf.: $595   **Exc.:** $495   **VGood:** $395

Colt Coltsman Short-Action Sako Deluxe
Same specs as Coltsman Standard except Sako action; 243, 308 Win.; action with integral scope blocks. Introduced 1963; discontinued 1965.
Perf.: $695   **Exc.:** $595   **VGood:** $495

COLT COURIER
Semi-automatic; 22 LR; 15-shot tubular magazine; 19 3/8" barrel; weighs 4 3/4 lbs.; receiver grooved for tip-off scope mount; American walnut stock forend; pistol grip. Introduced 1970; discontinued 1976.
Perf.: $275   **Exc.:** $190   **VGood:** $140

COLT LIGHTNING
Slide action; 22 LR, 32-20, 38-40, 44-40, 38-56, 50-96; 15-shot tubular magazine; 26" round or octagon barrel; open rear, blade front sight; American walnut stock, forend. Introduced 1884; discontinued 1902.
Small frame
Perf.: $2250   **Exc.:** $900   **VGood:** $650
Medium frame
Perf.: $2500   **Exc.:** $1395   **VGood:** $850
Large frame
Perf.: $4750   **Exc.:** $2795   **VGood:** $1650

Colt Lightning Baby Carbine
Same specs as Colt Lightning except 10-shot tubular magazine; 15", 16" barrel; weighs 5 1/4 lbs. Introduced 1884; discontinued 1902.
Medium frame
Perf.: $5850   **Exc.:** $3450   **VGood:** $1950
Large frame
Perf.: $10,000   **Exc.:** $6950   **VGood:** $5350

Colt Lightning Carbine
Same specs as Colt Lightning except 20", 22" barrel; weighs 6 1/4 lbs. Introduced 1884; discontinued 1902.
Medium frame
Perf.: $2700   **Exc.:** $1000   **VGood:** $600
Large frame
Perf.: $8500   **Exc.:** $6995   **VGood:** $3950

Colt Lightning Small Frame
Same specs as Colt Lightning except 22 Short, 22 Long; 24" barrel; 16-shot (22 Short), 15-shot (22 Long); bead front sight. Collector value. Manufactured 1887 to 1904.
Perf.: $1900   **Exc.:** $900   **VGood:** $650

COLT SPORTER TARGET
Semi-automatic; 223 Rem.; 5-shot detachable box magazine; 20" barrel; 39" overall length; weighs 7 1/2 lbs.; composition stock, grip, forend; post front sight, aperture fully-adjustable rear; standard-weight barrel; flash suppressor; sling swivels; forward bolt assist; military matte black finish. Made in U.S. by Colt. Introduced 1991; still produced.
Perf.: $1100   **Exc.:** $900   **VGood:** $700

Colt Sporter Competition HBAR
Same specs as Sporter Target except flat-top receiver with integral Weaver-type base for scope mounting; counter-bored muzzle; 1:9" rifling twist. Made in U.S. by Colt. Introduced 1991; still produced. **LMSR: $1073**
Perf.: $1025   **Exc.:** $950   **VGood:** $750

Colt Sporter Competition HBAR Range Selected
Same specs as Sporter Competition HBAR except range selected for accuracy; 3-9x rubber armored scope, scope mount; carrying handle with iron sights; Cordura nylon carrying case. Made in U.S. by Colt. Introduced 1992; dropped 1994.
Perf.: $1650   **Exc.:** $1450   **VGood:** $1250

Colt Sporter Lightweight
Same specs as Sporter Target except 9mm Para., 223 Rem., 7.62x39mm; 16" barrel; 34 1/2" overall length; weighs 6 3/4 lbs. (223 Rem.), 7 1/8 lbs. (9mm Para.). Made in U.S. by Colt. Introduced 1991; still produced.
Perf.: $1000   **Exc.:** $900   **VGood:** $700

Colt Sporter Match Delta HBAR
Same specs as the Sporter Target except standard stock; heavy barrel; refined and inspected by the Colt Custom Shop; 3-9x rubber armored scope; removable cheekpiece; adjustable scope mount; black leather military-style sling; cleaning kit; hard carrying case; pistol grip with Delta medallion. Made in U.S. by Colt. Introduced 1987; dropped 1991.
Perf.: $1495   **Exc.:** $1350   **VGood:** $1095

Colt Sporter Match HBAR
Same specs as the Sporter Match Delta HBAR except with heavy barrel; 800-meter M-16A2 fully-adjustable rear sight. Made in U.S. by Colt. Introduced 1991; no longer produced.
Perf.: $1195   **Exc.:** $995   **VGood:** $795

Colt Sporter Match Target Competition HBAR II
Same specs as the Sporter Match HBAR except 16 1/8" barrel; weighs 7 1/8 lbs.; 34 1/2" overall length; 1:9" twist barrel. Made in U.S. by Colt. Introduced 1995; still produced. **LMSR: $1172**
Perf.: $995   **Exc.:** $895   **VGood:** $695

COLT STAGECOACH
Semi automatic; 22 LR; 13-shot magazine; 16 1/2" barrel; 33 3/4" overall length; weighs 4 3/4 lbs.; deluxe American black walnut straight Western-style carbine stock with saddle ring; fully-adjustable rear sight; engraved receiver. Introduced 1965; discontinued 1975.
Perf.: $325   **Exc.:** $275   **VGood:** $210

Colt-Sauer
Sporter

Colt-Sharps
Falling Block

Commando
Mark 9
Mark 45

Cooper Arms
Model 21 Varmint Extreme
Model 21 Benchrest
Model 22 Pro Varmint
Extreme
Model 22 Benchrest
Model 36CF Centerfire
Sporter
Model 36RF Rimfire Sporter
Model 36RF Custom Classic
Model 36RF Featherweight
Model 38 Centerfire Sporter
Model 40 Centerfire Sporter
Model 40 Custom Classic

Colt-Sauer Sporter

Cooper Arms Model 22
Pro Varmint Extreme

Cooper Arms Model
TRP-1 ISU Standard

COLT-SAUER SPORTER
Bolt action; 22-250, 243, 308, 25-06, 270, 30-06, 7mm Rem. Mag., 300 Win. Mag., 300 Weatherby Mag.; 375 H&H Mag., 458 Win. Mag.; 3-, 4-shot detachable box magazine; 24″ barrel; 43 ¾″ overall length; weighs 8 lbs.; no sights; hand-checkered American walnut pistol-grip stock; rosewood pistol grip, forend caps; recoil pad; quick-detachable sling swivels. Introduced 1971; discontinued 1986.
Long action
    Perf.: $1095   **Exc.:** $895   **VGood:** $695
Short action
    Perf.: $1095   **Exc.:** $895   **VGood:** $695
Magnum action
    Perf.: $1150   **Exc.:** $995   **VGood:** $795
Grand Alaskan model
    Perf.: $1395   **Exc.:** $1195   **VGood:** $895
Grand African model
    Perf.: $1450   **Exc.:** $1295   **VGood:** $995

COLT-SHARPS FALLING BLOCK
Single shot; 17 Bee, 22-250, 243, 25-06, 7mm Rem. Mag., 30-06, 375 H&H Mag.; iron sights; checkered walnut stock, forearm; blued finish. Introduced 1970; discontinued 1977.
    Perf.: $2400   **Exc.:** $2000   **VGood:** $1200

COMMANDO MARK 9
Semi-automatic; 9mm Para.; 5-, 15-, 30-, 90-shot magazine; 16 ½″ barrel; weighs 8 lbs.; peep-type rear sight, blade front; walnut stock, forearm; choice of vertical or horizontal foregrip; muzzlebrake; cooling sleeve. Introduced 1969; discontinued 1976.
    Perf.: $395   **Exc.:** $330   **VGood:** $230

COMMANDO MARK 45
Semi-automatic; 45 ACP; 5-, 15-, 30-, 90-shot magazine; 16 ½″ barrel; weighs 8 lbs.; peep-type rear sight, blade front; walnut stock and forearm; available with choice of vertical or horizontal foregrip; muzzlebrake; cooling sleeve. Introduced 1969; discontinued 1976.
    Perf.: $395   **Exc.:** $340   **VGood:** $250

COOPER ARMS MODEL 21 VARMINT EXTREME
Bolt action; single shot; 17 Rem., 17 Mach IV, 221 Fireball, 222, 222 Rem. Mag.; 223, 22 PPC, 6x47; 23 ¾″ stainless steel barrel with competition step crown; free-floated; AAA Claro walnut stock with flared oval forend, ambidextrous palm swell; 22 lpi checkering; oil finish; Pachmayr buttpad; no sights; drilled and tapped for scope mounting; three mid-bolt locking lugs; adjustable trigger; glass bedded; swivel studs. Made in U.S. by Cooper Arms. Introduced 1994; still produced. **LMSR: $1495**
    Perf.: $1295   **Exc.:** $995   **VGood:** $795

Cooper Arms Model 21 Benchrest
Same specs as the Model 21 Varmint Extreme except 22 PPC, 223 Rem.; McMillan Benchrest composition stock; 24″, 1″ straight taper stainless barrel with competition step crown, match chamber; Jewell two-stage, 1-oz. fully-adjustable trigger; weighs 7 ¼ lbs. Made in U.S. by Cooper Arms. Introduced 1994; still produced.
LMSR: $1695
    Perf.: $1495   **Exc.:** $995   **VGood:** $795

COOPER ARMS MODEL 22 PRO VARMINT EXTREME
Bolt action; single shot; 22-250, 220 Swift, 243, 25-06, 6mm PPC, 308; 26″ stainless steel match-grade barrel with straight taper; free-floated; AAA Claro walnut stock with oil finish; 22 lpi wrap-around borderless ribbon checkering; beaded cheekpiece; steel grip cap; flared varminter forend; Pachmayr pad; no sights; drilled and tapped for scope mounting; three front locking lug system; available with sterling silver inlaid medallion, skeleton grip cap, and French walnut. Made in U.S. by Cooper Arms. Introduced 1995; still produced. **LMSR: $1895**
    Perf.: $1350   **Exc.:** $995   **VGood:** $750
Black Jack model with McMillan synthetic stock
    Perf.: $1450   **Exc.:** $1095   **VGood:** $795

Cooper Arms Model 22 Benchrest
Same specs as the Model 22 Pro Varmint Extreme except 6mm PPC, 243, 308; McMillan Benchrest composition stock; 1″ straight taper stainless steel barrel; competition step muzzle crown; match chamber. Made in U.S. by Cooper Arms. Introduced 1995; still produced. **LMSR: $1695**
    Perf.: $1395   **Exc.:** $995   **VGood:** $750

COOPER ARMS MODEL 36CF CENTERFIRE SPORTER
Bolt action; 17 CCM, 22 CCM, 22 Hornet; 5-shot magazine; 23″ barrel; 42 ½″ overall length; weighs 7 lbs.; AA Claro walnut stock with 22 lpi checkering, oil finish (standard grade); AAA Claro or AA French walnut (custom grade); no sights; three mid-bolt locking lugs, 45° bolt rotation; fully-adjustable trigger; swivel studs; Pachmayr buttpad. Made in U.S. by Cooper Arms. Introduced 1991; no longer made.
    Perf.: $1395   **Exc.:** $995   **VGood:** $795

COOPER ARMS MODEL 36RF RIMFIRE SPORTER
Bolt action; 22 LR; 5-shot magazine; 22 ¾″ barrel; 42 ½″ overall length; weighs 7 lbs.; hand-checkered AAA Claro walnut stock (standard grade); AA fancy French walnut or AAA Claro walnut with beaded Monte Carlo cheekpiece (Custom grade); sights optional; three front locking lugs, 45° bolt rotation; fully-adjustable single stage trigger; Wiseman/McMillan competition barrel; swivel studs; Pachmayr buttpad; oil-finished wood. Made in U.S. by Cooper Arms. Introduced 1991; still produced. **LMSR: $1495**
    Perf.: $995   **Exc.:** $695   **VGood:** $595
Custom grade
    Perf.: $1495   **Exc.:** $995   **VGood:** $795

Cooper Arms Model 36RF Custom Classic
Same specs as the Model 36RF grade except ebony forend tip; steel grip cap; Brownell No. 1 checkering pattern. Made in U.S. by Cooper Arms. Introduced 1991; no longer made.
    Perf.: $1200   **Exc.:** $900   **VGood:** $700

Cooper Arms Model 36RF Featherweight
Same specs as the Model 36RF except custom stock shape with black textured finish. Made in U.S. by Cooper Arms. Introduced 1991; no longer made.
    Perf.: $1250   **Exc.:** $950   **VGood:** $750

COOPER ARMS MODEL 38 CENTERFIRE SPORTER
Bolt action; 17 CCM, 22 CCM; 3-shot magazine; 23 ¾″ Shilen match barrel; 42 ½″ overall length; weighs 8 lbs.; AA Claro walnut stock with 22 lpi checkering, oil finish (standard grade); AAA Claro or AA French walnut, beaded Monte Carlo cheekpiece (custom grade); no sights; three front locking lugs, 45° bolt rotation; fully-adjustable single stage match trigger; swivel studs. Pachmayr buttpad. Made in U.S. by Cooper Arms. Introduced 1991; no longer produced.
    Perf.: $995   **Exc.:** $895   **VGood:** $795
Custom grade
    Perf.: $1295   **Exc.:** $1095   **VGood:** $895

COOPER ARMS MODEL 40 CENTERFIRE SPORTER
Bolt action; 17 CCM, 17 Ackley Hornet, 22 CCM, 22 Hornet, 22 K-Hornet; 5-shot magazine; 23″ barrel; 42 ½″ overall length; weighs 7 lbs.; AAA Claro walnut stock with 22 lpi borderless wrap-around ribbon checkering; oil finish; steel grip cap; Pachmayr pad; no sights; three mid-bolt locking lugs; 45° bolt rotation; fully-adjustable trigger; swivel studs. Made in U.S. by Cooper Arms. Introduced 1994; still produced. **LMSR: $1825**
    Perf.: $1400   **Exc.:** $1100   **VGood:** $800

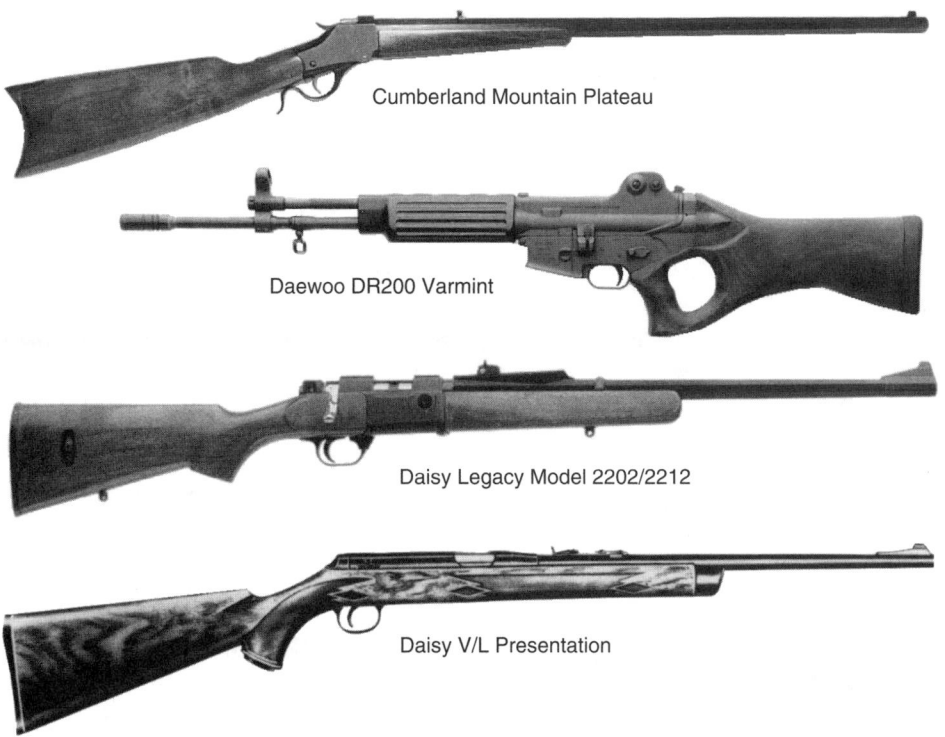

Cumberland Mountain Plateau

Daewoo DR200 Varmint

Daisy Legacy Model 2202/2212

Daisy V/L Presentation

RIFLES

Cooper Arms
Model 40 Classic Varminter
Model BR-50
Model TRP-I ISU Standard

Cumberland
Mountain Elk River
Mountain Plateau

CZ
(See Brno)

D-Max
Auto Carbine

Daewoo
DR200 Varmint

Daisy
Legacy Model 2201/2211
Legacy Model 2202/2212
Legacy Model 2203/2213
V/L
V/L Collector Kit
V/L Presentation

Cooper Arms Model 40 Custom Classic
Same specs as Model 40 except AAA Claro walnut; Monte Carlo beaded cheekpiece; oil finish. Made in U.S. by Cooper Arms. Introduced 1994; still produced. **LMSR: $1695**
    Perf.: $1595   **Exc.:** $1295   **VGood:** $995

Cooper Arms Model 40 Classic Varminter
Same specs as the Model 40 except AAA Claro walnut; wrap-around ribbon checkering; beaded cheekpiece; steel grip cap; flared varminter forend. Made in U.S. by Cooper Arms. Introduced 1994; still produced. **LMSR: $1695**
    Perf.: $1495   **Exc.:** $1195   **VGood:** $995

COOPER ARMS MODEL BR-50
Bolt action; single shot; 22 LR; 22" barrel, .860" straight; 40 1/2" overall length; weighs 6 13/16 lbs.; McMillan Benchrest stock; no sights; three mid-bolt locking lugs; fully-adjustable match-grade trigger; stainless barrel. Made in U.S. by Cooper Arms. Introduced 1994; still produced. **LMSR: $1595**
    Perf.: $1395   **Exc.:** $1095   **VGood:** $895

COOPER ARMS MODEL TRP-1 ISU STANDARD
Bolt action; single shot; 22 LR; 22" barrel; 40 1/2" overall length; weighs 10 lbs.; walnut competition-style stock with adjustable cheekpiece and buttpad; no sights; accepts Anschutz sight packages; three front locking lugs, 45° bolt rotation; fully-adjustable single stage trigger; hand-lapped match-grade Shilen stainless barrel. Made in U.S. by Cooper Arms. Introduced 1991; no longer produced.
    Perf.: $1450   **Exc.:** $1150   **VGood:** $895
With benchrest-style stock
    Perf.: $1350   **Exc.:** $1050   **VGood:** $795

CUMBERLAND MOUNTAIN ELK RIVER
Single shot; 22 through 300 Win. Mag.; 32-40, 40-65, 45-70 single shot; 24", 26", 28", round or octagon barrel; 44" overall length (28" barrel); weighs about 6 1/4 lbs.; American walnut stock; Marble's bead front sight, Marble's adjustable open rear; falling block action with underlever; blued barrel and receiver. Made in U.S. by Cumberland Mountain Arms, Inc. Introduced 1993; no longer produced.
    Perf.: $895   **Exc.:** $795   **VGood:** $695

CUMBERLAND MOUNTAIN PLATEAU
Single shot; 40-65, 45-70; barrel length up to 32", round; 48" overall length; weighs about 10 1/2 lbs. (32"

barrel); American walnut stock; Marble's bead front sight, Marble's open rear; falling block action with underlever; blued barrel and receiver; lacquer finish stock; crescent buttplate. Made in U.S. by Cumberland Mountain Arms, Inc. Introduced 1995; still produced. **LMSR: $1295**
    Perf.: $950   **Exc.:** $800   **VGood:** $675

D-MAX AUTO CARBINE
Semi-automatic; 9mm Para., 10mm Auto, 40 S&W, 45 ACP; 30-shot magazine; 16 1/4" barrel; 38 1/2" overall length; weighs about 7 3/4 lbs.; walnut butt, grip, forend; post front sight, open fully-adjustable rear; aperture rear optional; blowback operation, fires from closed bolt; trigger-block safety; side-feed magazine; integral optical sight base; Max-Coat finish. Made in U.S. by D-Max Industries. No longer produced.
    Perf.: $395   **Exc.:** $275   **VGood:** $200

DAEWOO DR200 VARMINT
Semi-automatic; 223 Rem.; 6-shot magazine; 18 3/8" barrel; 39 1/4" overall length; weighs 9 lbs.; synthetic thumbhole-style stock with rubber buttpad; post front sight in ring, aperture fully-adjustable rear; forged aluminum receiver; bolt, bolt carrier, firing pin, piston and recoil spring contained in one assembly; rotating bolt locking; uses AR-15 magazines. Imported from Korea by Kimber of America, Inc.; distributed by Nationwide Sports Dist. Introduced 1995; still imported. **LMSR: $750**
    Perf.: $650   **Exc.:** $525   **VGood:** $450

DAISY LEGACY MODEL 2201/2211
Bolt action; single shot; 22 LR; 19" barrel with octagonal shroud; weighs 6 1/2 lbs.; ramp blade front sight, fully-adjustable removable notch rear; moulded copolymer stock (Model 2201) or walnut stock (Model 2211); adjustable-length stock; removable bolt/trigger assembly; adjustable trigger; dovetailed for scope mounting; barrel interchanges with smoothbore. Introduced 1988; discontinued 1990.
Model 2201
    Perf.: $130   **Exc.:** $100   **VGood:** $70
Model 2211
    Perf.: $140   **Exc.:** $110   **VGood:** $80

DAISY LEGACY MODEL 2202/2212
Bolt action; 22 LR; 10-shot rotary magazine; 19" barrel with octagon shroud; weighs 6 1/2 lbs.; moulded copolymer stock (Model 2202) or walnut stock (Model 2212); ramp

blade front sight, fully-adjustable removable rear; receiver dovetailed for scope mounting; adjustable butt length; removable bolt/trigger assembly; barrel interchanges with smoothbore. Introduced 1988; discontinued 1990.
Model 2202
    Perf.: $140   **Exc.:** $110   **VGood:** $80
Model 2212
    Perf.: $145   **Exc.:** $115   **VGood:** $85

DAISY LEGACY MODEL 2203/2213
Semi-automatic; 22 LR; 7-shot clip; 19" barrel; 34 3/4" overall length; weighs 6 1/2 lbs.; blade on ramp front sight, fully-adjustable, removable notch rear; moulded copolymer stock (Model 2203) or American hardwood stock (Model 2213); receiver dovetailed for scope mount; removable trigger assembly. Introduced 1988; discontinued 1991.
Model 2203
    Perf.: $145   **Exc.:** $120   **VGood:** $85
Model 2213
    Perf.: $150   **Exc.:** $125   **VGood:** $90

DAISY V/L
Single shot; underlever; 22 V/L caseless cartridge; 18" barrel; adjustable open rear sight, ramp blade front; stock of wood-grained Lustran plastic. Introduced 1968; discontinued 1969; only 19,000 manufactured.
    Perf.: $130   **Exc.:** $100   **VGood:** $70

Daisy V/L Collector Kit
Same specs as V/L except American walnut stock; gold-plate engraved with owner's name, gun's serial number; gun case; brass wall hangers; 300 rounds of V/L ammo. Introduced 1968; discontinued 1969; only 1000 produced.
    Perf.: $275   **Exc.:** $220   **VGood:** $175

Daisy V/L Presentation
Same specs as V/L except American walnut stock. Introduced 1968; discontinued 1969; only 4000 manufactured.
    Perf.: $220   **Exc.:** $180   **VGood:** $110

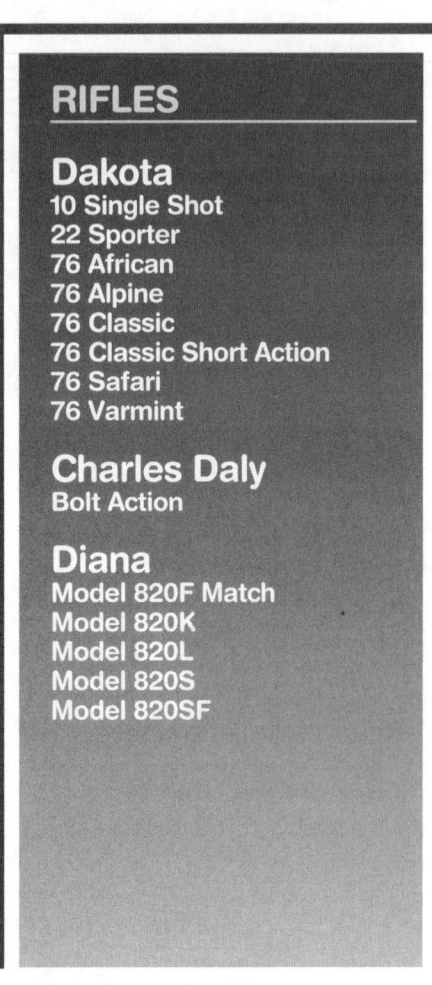

RIFLES

Dakota
10 Single Shot
22 Sporter
76 African
76 Alpine
76 Classic
76 Classic Short Action
76 Safari
76 Varmint

Charles Daly
Bolt Action

Diana
Model 820F Match
Model 820K
Model 820L
Model 820S
Model 820SF

Dakota 76 Classic

Dakota 76 African

Dakota 76 Varmint

Dakota 10 Single Shot

Dakota 76 Safari

Charles Daly Bolt Action

DAKOTA 10 SINGLE SHOT
Lever action; chambered for most rimmed and rimless commercial cartridges; 23" barrel; 39½" overall length; weighs 6 lbs.; falling block; no sights; drilled, tapped for scope mounts; medium-fancy classic-style walnut stock; checkered grip, forend; top tang safety; removable trigger plate. Made in U.S. Introduced 1990; still in production. **LMSR: $3495**
    Perf.: $2900   **Exc.:** $1800   **VGood:** $1200

DAKOTA 22 SPORTER
Bolt action; 22 LR, 22 Hornet; 5-shot magazine; 22" Premium barrel; 42½" overall length; weighs about 6½ lbs.; Claro or English walnut stock in classic design; 13½" length of pull; point panel hand checkering; swivel studs; black buttpad; no sights; comes with mount bases; combines features of Winchester 52 and Dakota 76 rifles; full-sized receiver; rear locking lugs and bolt machined from bar stock; trigger and striker-blocking safety; adjustable trigger. Made in U.S. by Dakota Arms, Inc. Introduced 1992; still produced in 22 LR only. **LMSR: $1995**
    Perf.: $1295   **Exc.:** $895   **VGood:** $695

DAKOTA 76 AFRICAN
Bolt action; 404 Jeffery, 416 Rigby, 416 Dakota, 450 Dakota; 4-shot magazine; 24" barrel; weighs 9-10 lbs.; select wood; two stock cross-bolts; ramp front sight, standing leaf rear. Made in U.S. by Dakota Arms, Inc. Introduced 1989; still produced. **LMSR: $4995**
    Perf.: $3700   **Exc.:** $2995   **VGood:** $1995

DAKOTA 76 ALPINE
Bolt action; 22-250, 243, 6mm Rem., 250-3000, 7mm-08, 308, 358; 4-shot magazine; 21", 23" barrel; weighs 6½ lbs.; slim checkered walnut stock; blind magazine. Introduced 1989; discontinued 1992.
    Perf.: $1795   **Exc.:** $1495   **VGood:** $995

DAKOTA 76 CLASSIC
Bolt action; 22-250, 243, 6mm Rem., 250-3000, 7mm-08, 308, 358, 257 Roberts, 270, 280, 30-06, 7mm Rem. Mag., 338 Win. Mag., 300 Win. Mag., 375 H&H, 458 Win. Mag.; 4-shot magazine; 21", 23" barrel; weighs 7½ lbs.; no sights; drilled, tapped for scope mounts; composite or classic-style medium-fancy walnut stock; one-piece rail trigger guard assembly; solid buttpad; steel grip cap; adjustable trigger. Based on original Winchester Model 70 design. Right- or left-handed. Introduced 1988; still in production.
    Perf.: $2820   **Exc.:** $1795   **VGood:** $1195

Dakota 76 Classic Short Action
Same specs as Model 76 Classic except 22-250, 243, 6mm Rem., 250-3000, 7mm-08, 308, 21" barrel. Introduced 1989; still in production. **LMSR: $2300**
    Perf.: $2095   **Exc.:** $1595   **VGood:** $1195

DAKOTA 76 SAFARI
Bolt action; 300 Win. Mag., 338 Win. Mag., 375 H&H, 458 Win. Mag.; 23" barrel; weighs 8½ lbs.; ramp front sight, standing leaf rear; checkered plastic composite or English walnut stock; shadowline cheekpiece; barrel band front swivel, inletted rear. Introduced 1988; still in production. **LMSR: $3300**
    Perf.: $3250   **Exc.:** $2495   **VGood:** $1995
With composite stock
    Perf.: $2750   **Exc.:** $1995   **VGood:** $1495

DAKOTA 76 VARMINT
Bolt action; single shot; 17 Rem., 22 BR, 222 Rem., 22-250, 220 Swift, 223, 6mm BR, 6mm PPC; heavy barrel contour; special stock dimensions for varmint shooting. Made in U.S. by Dakota Arms, Inc. Introduced 1994; still produced. **LMSR: $2500**
    Perf.: $2100   **Exc.:** $1695   **VGood:** $1195

CHARLES DALY BOLT ACTION
Bolt action; 22 Hornet; 5-shot box magazine; 24" barrel; miniaturized Mauser; leaf rear, ramp front sight; walnut stock; hand-checkered pistol-grip forearm; hinged floorplate. Introduced 1931; discontinued 1939.

Imported by Charles Daly; manufactured by Franz Jaeger Co. of Germany. Note: Same model was imported by A.F. Stoeger and sold as Herold Rifle.
    Perf.: $850   **Exc.:** $600   **VGood:** $500

DIANA MODEL 820F MATCH
Bolt action; single shot; 22 LR; 27⅛" barrel; 44½" overall length; weighs 15½ lbs.; walnut target stock; tunnel front sight, fully-adjustable match rear; designed for free rifle events; adjustable hook buttplate, adjustable cheekpiece, hand stop; stock stabilizer and palm rest optional; trigger adjustable from 1.4 to 8.8 oz. Imported from Germany by Dynamit Nobel-RWS, Inc. Introduced 1990; discontinued 1992.
    Perf.: $1695   **Exc.:** $1395   **VGood:** $995

Diana Model 820K
Same specs as Model 820F except 24" barrel; 41" overall length; weighs 9½ lbs.; no sights; detachable 3-piece barrel weight; designed for running boar events; European walnut stock. Made in Germany. Discontinued 1986.
    Perf.: $895   **Exc.:** $750   **VGood:** $550

Diana Model 820L
Same specs as Model 820F except 24", 26" barrel; weighs 10½ lbs; designed for three-position match shooting. Made in Germany. Introduced 1990; discontinued 1992.
    Perf.: $1195   **Exc.:** $895   **VGood:** $650

Diana Model 820S
Same specs as Model 820L except 24" barrel; 41" overall length; weighs 10⁵⁄₁₆ lbs.; Model 82 or 75 aperture rear sight. Made in Germany. Introduced 1990; discontinued 1992.
    Perf.: $995   **Exc.:** $795   **VGood:** $595

Diana Model 820SF
Same specs as Model 820F except 41" overall length; weighs 11 lbs.; heavy barrel; Model 82 or 75 aperture rear sight. Made in Germany. Introduced 1990; discontinued 1992.
    Perf.: $1095   **Exc.:** $895   **VGood:** $595

Diana Model 820L

Dixie Model 1873 Rifle

Dixie 1874 Sharps Blackpowder Silhouette

DuBiel Custom

Dumoulin

Dixie
Model 1873 Rifle
Model 1873 Carbine
1874 Sharps Blackpowder
Silhouette
1874 Sharps Lightweight
Hunter/Target
Remington Creedmore Long
Range

DuBiel
Custom

Dumoulin
Amazone
Bavaria Deluxe
Centurion
Centurion Classic
Diane
Pioneer Express
Safari
Sidelock Prestige

DIXIE MODEL 1873 RIFLE

Lever action; 44-40; 11-shot tubular magazine; 24 1/2″ octagonal barrel; leaf rear sight, blade front; walnut stock forearm, receiver engraved with scrolls, elk, buffalo; color case-hardened. Replica of Winchester '73. Introduced 1975; no longer imported.
Perf.: $795   **Exc.:** $695   **VGood:** $595

Dixie Model 1873 Carbine

Same specs as Model 1873 except no engraving or color case-hardening; 20″ barrel.
Perf.: $695   **Exc.:** $595   **VGood:** $495

DIXIE 1874 SHARPS BLACKPOWDER SILHOUETTE

Single shot; falling block; 45-70; 30″ tapered octagon barrel with 1:18″ twist; 47 1/2″ overall length; weighs 10 5/16 lbs.; oiled walnut shotgun-style butt with checkered metal buttplate; blade front sight, ladder-type hunting rear; replica of the Sharps #1 Sporter; color case-hardened receiver, hammer, lever and buttplate; blued barrel; tang drilled and tapped for tang sight; double-set triggers. Meets standards for NRA blackpowder cartridge matches. Imported from Italy by Dixie Gun Works. Introduced 1995; still imported. **LMSR:** $895
Perf.: $825   **Exc.:** $750   **VGood:** $595

Dixie 1874 Sharps Lightweight Hunter/Target

Same specs as Dixie 1874 Sharps Blackpowder Silhouette except straight-grip buttstock with military-style buttplate. Based on the 1874 military model. Imported from Italy by Dixie Gun Works. Introduced 1995; still imported. **LMSR:** $895
Perf.: $795   **Exc.:** $695   **VGood:** $575

DIXIE REMINGTON CREEDMORE LONG RANGE

Rolling block; 45-70; 30″ octagon barrel; weighs 13 lbs.; Creedmore sights; checkered wood stock; blued barrel; color case-hardening or exposed metal. No longer in production.
Perf.: $750   **Exc.:** $595   **VGood:** $495

DuBIEL CUSTOM

Bolt action; 22-250 through 458 Win. Mag., selected wildcats; barrel weights, lengths depend upon caliber; Douglas Premium barrel; no sights; integral scope mount bases; walnut, maple or laminate stocks; hand-checkered; left- or right-hand models; 5-lug locking mechanism; 36° bolt rotation; adjustable Canjar trigger; oil or epoxy stock finish; slide-type safety; floorplate release; jeweled, chromed bolt; sling swivel studs; recoil pad. Introduced 1978; discontinued 1990.
Perf.: $1995   **Exc.:** $1495   **VGood:** $1095

DUMOULIN AMAZONE

Bolt action; various calibers from 270-458; 20″ barrel; classic two-leaf rear sight, hooded blade on ramp front; full-length, hand-checkered, oil-finished European walnut stock; built on Mauser 98 action; adjustable trigger; Model 70-type safety. Made in Belgium. Introduced 1986; discontinued 1987.
Perf.: $1695   **Exc.:** $1495   **VGood:** $900

DUMOULIN BAVARIA DELUXE

Bolt action; various calibers from 222-458; 21″, 24″, 25 1/2″ octagonal barrel; weighs about 7 lbs.; classic two-leaf rear sight, hooded blade on ramp front; hand-checkered, oil-finished select European walnut stock; built on Mauser or Sako action; adjustable trigger; Model 70-type safety; quick-detachable sling swivels; solid buttpad; deluxe engraving. Made in Belgium. Imported by Midwest Gun Sport. Introduced 1986; discontinued 1987.
Perf.: $995   **Exc.:** $895   **VGood:** $795

DUMOULIN CENTURION

Bolt action; various calibers from 270-458; 21 1/2″, 24″, 25″ barrel; no sights; hand-checkered, oil-finished select European walnut stock; built on Mauser or Sako action. Made in Belgium. Introduced 1986; discontinued 1987.
Perf.: $695   **Exc.:** $595   **VGood:** $495

Dumoulin Centurion Classic

Same specs as Centurion except Mauser 98 action only; higher grade of wood stock. Introduced 1986; discontinued 1987.
Perf.: $1400   **Exc.:** $1000   **VGood:** $800

DUMOULIN DIANE

Bolt action; various calibers from 270-458; 22″ barrel; classic two-leaf rear sight, hooded blade on ramp front; hand-checkered, oil-finished European walnut stock; built on Mauser 98 action; adjustable trigger; Model 70-type safety. Made in Belguim. Introduced 1986; discontinued 1987.
Perf.: $1350   **Exc.:** $1100   **VGood:** $800

DUMOULIN PIONEER EXPRESS

Double rifle; various calibers from 22 Hornet-600 Nitro Express; 24″, 26″ barrel; 45″ overall length (24″ barrel); weighs 9 1/2 lbs.; two-leaf rear sight on quarter-rib, bead on ramp front; boxlock triple-lock system; Greener crossbolt; H&H type ejectors; articulated front trigger; various grades of engraving effect price. Made in Belgium. Introduced 1987; no longer imported.
Prices range from $6500 to $12,000.

DUMOULIN SAFARI

Bolt action; magnum calibers up to 458; 24″ to 26″ barrel; 44″ overall length (24″ barrel); weighs 8 1/2 to 9 lbs.; two-leaf rear sight, hooded front on banded ramp; built on modified Sako or Mauser Oberndorf action; classic English-style deluxe European walnut stock; oil finish; buffalo horn grip cap; rubber buttpad; Model 70-type or side safety. Custom built in Belgium. Introduced 1986; discontinued 1987.
Perf.: $3995   **Exc.:** $3250   **VGood:** $2725

DUMOULIN SIDELOCK PRESTIGE

Double rifle; sidelock; traditional chopper lump or classic Ernest Dumoulin barrel system; 22″, 24″ barrel, depending on caliber; deluxe European walnut stock; internal parts are gold-plated; Purdey lock system; ten grades, built to customer specs; differing engraving styles. Made in Belgium. Introduced 1986; no longer imported.
Perf.: $17,900 **Exc.:** $14,950 **VGood:** $9995

E.A.A.
Antonio Zoli Express Rifle
HW 660 Match
Sabatti Model 1822
Sabatti Rover 870
Weihrauch HW 60

Eagle Arms
EA-15
EA-15 Action Master
EA-15 Eagle Spirit
EA-15 Golden Eagle
EA-15A2 Post-Ban Heavy Barrel
EA-15E2 Carbine
EA-15E2 H-BAR

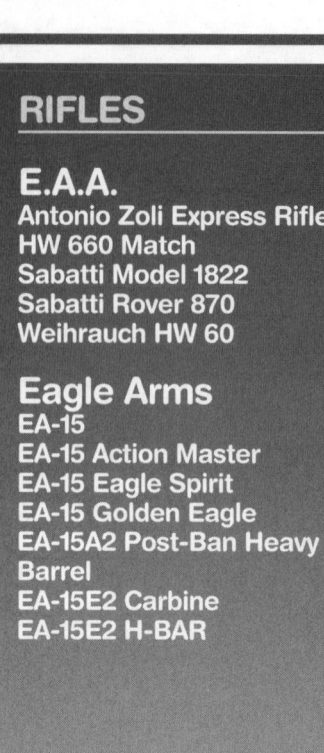

E.A.A./HW 660 Match

Eagle Arms EA-15 Golden Eagle

Eagle Arms M15A3
Post-Ban Predator

E.M.F. Sharps Carbine

E.A.A./ANTONIO ZOLI EXPRESS RIFLE
Double rifle; boxlock; 7.65R, 30-06, 9.3x74R; 25 1/2" barrels; 40 1/2" overall length; weighs about 8 lbs.; fancy European walnut stock with raised cheekpiece, fine-line checkering; bead front sight, adjustable express rear on quarter-rib; engraved coin-finish frame; double-set triggers; selective automatic ejectors; rubber recoil pad; sling swivels. Imported from Italy by European American Armory. Importation dropped 1993.
Perf.: $3695   **Exc.:** $2995   **VGood:** $2195
Express Model E, better engraving, ejectors
Perf.: $4295   **Exc.:** $3495   **VGood:** $2495
Express Model EM, elaborate engraving, extra-fancy wood, mechanical S.T.
Perf.: $4795   **Exc.:** $3995   **VGood:** $2995

E.A.A./HW 660 MATCH
Bolt action; single shot; 22 LR; 26" barrel; 45 3/8" overall length; weighs 10 3/4 lbs.; match-type walnut stock with adjustable cheekpiece and buttplate; stippled pistol grip and forend; globe front sight, match aperture rear; adjustable match trigger; forend accessory rail. Imported from Germany by European American Armory. Introduced 1988; dropped 1994.
Perf.: $750   **Exc.:** $550   **VGood:** $425

E.A.A./SABATTI MODEL 1822
Semi-automatic; 22 LR; 10-shot magazine; 18 1/2" round tapered barrel; 37" overall length; weighs 5 1/4 lbs.; stained hardwood stock; bead front sight, folding leaf elevation-adjustable rear; cross-bolt safety; blue finish. Imported from Italy by European American Armory. Introduced 1993; still imported. **LMSR:** $206
Perf.: $180   **Exc.:** $150   **VGood:** $100
Heavy model, bull barrel, no sights
Perf.: $250   **Exc.:** $200   **VGood:** $150
Thumbhole model, heavy, bull barrel, no sights, one-piece stock
Perf.: $300   **Exc.:** $250   **VGood:** $200

E.A.A./SABATTI ROVER 870
Bolt action; 22-250, 243, 25-06, 270, 30-06, 308, 7mm Rem. Mag., 300 Win. Mag., 338 Win. Mag.; 23" barrel; 42 1/2" overall length; weighs about 7 lbs.; gold bead on ramp front sight, open adjustable rear; walnut stock with straight comb, cut checkering on grip and forend; positive safety locks trigger; blue finish. Imported by E.A.A. Introduced 1986; importation dropped 1987.
Perf.: $795   **Exc.:** $695   **VGood:** $495

E.A.A./WEIHRAUCH HW 60
Bolt action; single shot; 22 LR; 26 7/8" barrel; 45 3/4" overall length; weighs 11 lbs.; walnut target stock with adjustable buttplate, stippled pistol grip and forend; forend rail with adjustable swivel; hooded ramp front sight, match-type aperture rear; adjustable match trigger with push-button safety. Imported from Germany by European American Armory. Introduced 1981; no longer imported.
Perf.: $595   **Exc.:** $450   **VGood:** $375
Left-hand model
Perf.: $795   **Exc.:** $595   **VGood:** $425

EAGLE ARMS EA-15
Semi-automatic; 223 Rem.; 30-shot magazine; 20" barrel; 39" overall length; weighs about 7 lbs.; black composition stock, trapdoor-style; post front sight, fully-adjustable rear; push-type receiver pivot pin for easy takedown; receivers hard coat anodized; E2-style forward assist mechanism; integral raised M-16A2-type fence around magazine release button. Made in U.S. by Eagle Arms, Inc. Introduced 1989; discontinued 1993.
Perf.: $750   **Exc.:** $650   **VGood:** $495

Eagle Arms EA-15 Action Master
Same specs as EA-15 except one-piece international-style upper receiver for scope mounting; no front sight; solid aluminum handguard tube; free-floating 20" premium barrel; NM trigger group; weighs about 8 3/4 lbs. Renamed the 15A3 Action Master. Made in U.S. by Eagle Arms, Inc. Introduced 1991.
Perf.: $1400   **Exc.:** $1100   **VGood:** $700

Eagle Arms EA-15 Eagle Spirit
Same specs as EA-15 rifle except 16" premium air-gauged barrel; weighs 8 lbs.; full-length tubular aluminum handguard; match accessories. Introduced 1993; still in production. **LMSR: $1475**
Perf.: $1300   **Exc.:** $1000   **VGood:** $600

Eagle Arms EA-15 Golden Eagle
Same specs as EA-15 except E2-style National Match rear sight with 1/2-MOA adjustments, elevation-adjustable NM front sight with set screw; 20" Douglas Premium extra-heavy match barrel with 1:9" twist; NM trigger group and bolt carrier group; weighs about 12 3/4 lbs. Made in U.S. by Eagle Arms, Inc. Introduced 1991.
Perf.: $1500   **Exc.:** $1200   **VGood:** $850

Eagle Arms EA-15A2 Post-Ban Heavy Barrel
Same specs as EA-15 except 10-shot magazine; elevation-adjustable front sight; A2-style forward assist mechanism. Introduced 1995; still in production. **LMSR: $895**
Perf.: $795   **Exc.:** $695   **VGood:** $595

Eagle Arms EA-15E2 Carbine
Same specs as EA-15 except collapsible carbine-type buttstock; 16" heavy carbine barrel; M177-type flash supressor; weighs about 7 1/4 lbs. Made in U.S. by Eagle Arms, Inc. Introduced 1989; still in production. **LMSR: $1525**
Perf.: $795   **Exc.:** $695   **VGood:** $595

Eagle Arms EA-15E2 H-BAR
Same specs as EA-15 except 20" slightly lighter and smaller heavy match barrel with 1:9" twist; weighs about 9 lbs. Made in U.S. by Eagle Arms, Inc. Introduced 1989; no longer in production.
Perf.: $895   **Exc.:** $795   **VGood:** $600
With NM sights
Perf.: $1195   **Exc.:** $995   **VGood:** $725

E.M.F. Sharps Rifle

E.M.F. 1866 Yellowboy Rifle

Erma EG72

Erma EG712

Eagle Arms
M15A3 Post-Ban Predator
EA-15A3 Eagle Eye
M15A3 S.P.R.

E.M.F.
1860 Henry
1866 Yellowboy Rifle
1866 Yellowboy Carbine
Model 1873 Rifle
Model 1873 Carbine
Premier Model 1873
Carbine/Rifle
Sharps Rifle
Sharps Carbine
Texas Remington Revolving
Carbine

Erma
EG72
EG73
EG712
Model EG712 L

EAGLE ARMS M15A3 POST-BAN PREDATOR

Semi-automatic; 223; 10-shot magazine; 18", 1 1/4" diameter, stainless barrel with 1:8" twist; 34 3/8" overall length; weighs 9 1/4 lbs.; sights optional; one-piece international-style upper receiver; NM trigger; free-floating barrel; receiver hard-coat anodized; black composition stock; integral raised M16A2-type fence around magazine release button; A2-style forward assist. Made in U.S. by Eagle Arms, Inc. Introduced 1995; still produced. **LMSR: $1350**
    Perf.: $1050   **Exc.:** $895   **VGood:** $650

Eagle Arms EA-15A3 Eagle Eye

Same specs as EA-15 except 24", 1 1/4" diameter, stainless, air-gauged, free-floating heavy match barrel; weighted buttstock to counterbalance barrel; match-type hand-stop on handguard; 42 1/2" overall length; weighs about 13 lbs. Match sights available on special order. Made in U.S. by Eagle Arms, Inc. Introduced 1994; still in production. **LMSR: $1495**
    Perf.: $1395   **Exc.:** $1195   **VGood:** $900

Eagle Arms M15A3 S.P.R.

Same specs as M15A3 Predator except 18" barrel with 1:9" twist and front sight housing. Made in U.S. by Eagle Arms, Inc. Introduced 1995; still produced. **LMSR: $1475**
    Perf.: $795   **Exc.:** $695   **VGood:** $595

E.M.F. 1860 HENRY

Lever action; 44-40, 44 rimfire; 24 1/4" barrel; 43 3/4" overall length; weighs about 9 lbs.; blade front sight, elevation-adjustable rear; oil-stained American walnut stock; reproduction of the original Henry rifle with brass frame and buttplate, rest blued. Imported by E.M.F. Still imported. **LMSR: $1100**
    Perf.: $695   **Exc.:** $595   **VGood:** $495

E.M.F. 1866 YELLOWBOY RIFLE

Lever action; 38 Spl., 44-40; 24" barrel; 43" overall length; weighs 9 lbs.; bead front sight, open adjustable rear; European walnut stock; solid brass frame; blued barrel, lever, hammer, buttplate. Imported from Italy by E.M.F. Still imported. **LMSR: $848**
    Perf.: $595   **Exc.:** $495   **VGood:** $395

E.M.F. 1866 Yellow Boy Carbine

Same specs as 1866 Yellow Boy Rifle except 19" barrel; 38" overall length. Still imported from Italy by E.M.F. **LMSR: $825**
    Perf.: $595   **Exc.:** $495   **VGood:** $395

E.M.F. MODEL 1873 RIFLE

Lever action; 357 Mag., 44-40, 45 Colt; 24" barrel; 43 1/4" overall length; weighs 8 lbs.; bead front sight, fully-adjustable rear; European walnut stock; color case-hardened frame. Imported by E.M.F. Still imported. **LMSR: $1050**
    Perf.: $750   **Exc.:** $650   **VGood:** $495

E.M.F. Model 1873 Carbine

Same specs as Model 1873 Rifle except 19" barrel; blued frame. Imported by E.M.F. Still imported. **LMSR: $1020**
    Perf.: $750   **Exc.:** $650   **VGood:** $495

E.M.F. PREMIER MODEL 1873 CARBINE/RIFLE

Lever action; 45 Colt; 19" barrel (carbine), 24 1/2" barrel (rifle); 43 1/4" overall length; bead front, adjustable rear sight; uncheckered walnut stock, forearm; color case-hardened frame (rifle) or blued frame (carbine). Introduced 1988; discontinued 1989.
Rifle
    Perf.: $650   **Exc.:** $495   **VGood:** $395
Carbine
    Perf.: $650   **Exc.:** $495   **VGood:** $395

E.M.F. SHARPS RIFLE

Single shot; falling block; 45-70; 28" octagonal barrel; weighs 10 3/4 lbs.; blade front sight, flip-up open rear; oiled walnut stock; color case-hardened lock; double-set trigger; blue finish. Replica of the 1874 Sharps Sporting rifle. Imported by E.M.F. Still imported. **LMSR: $950**
    Perf.: $595   **Exc.:** $495   **VGood:** $395
With browned finish
    Perf.: $625   **Exc.:** $525   **VGood:** $395

E.M.F. Sharps Carbine

Same specs as Sharps Rifle except has 22" round barrel; barrel band. Imported by E.M.F. Still imported. **LMSR: $860**
    Perf.: $595   **Exc.:** $495   **VGood:** $395

E.M.F. TEXAS REMINGTON REVOLVING CARBINE

Single action; 22 LR; 6-shot cylinder; 21" barrel; 36" overall length; blade front sight, windage-adjustable rear; smooth walnut stock; brass frame, buttplate, trigger guard; blued cylinder, barrel. Made in Italy. Introduced 1991; no longer imported.
    Perf.: $250   **Exc.:** $230   **VGood:** $190

ERMA EG72

Pump-action; 22 LR; 15-shot magazine; 18 1/2" barrel; open rear sight, hooded ramp front; receiver grooved for scope mounts; straight hardwood stock; grooved slide handle; visible hammer. Made in Germany. Introduced 1970; discontinued 1976.
    Perf.: $150   **Exc.:** $110   **VGood:** $110

ERMA EG73

Lever action; 22 WMR; 12-shot tube magazine; 18 1/2" barrel; open rear sight, hooded ramp front; receiver grooved for scope mount; European hardwood stock, forearm. Originally made in Germany. Introduced 1973; no longer in production.
    Perf.: $200   **Exc.:** $160   **VGood:** $130

ERMA EG712

Lever action; 22 Short, 22 Long, 22 LR.; 21-shot (22 Short), 17-shot (22 Long), 15-shot (22 LR) tubular magazine; 18 1/2" barrel; open rear sight, hooded ramp front; carbine-style stock forearm of European hardwood, barrel band; receiver grooved for scope mount; near replica of Model 94 Winchester. Introduced 1976; no longer in production. Made in U.S.
    Perf.: $200   **Exc.:** $175   **VGood:** $150

Erma Model EG712 L

Same specs as EG712 except European walnut stock; heavy octagon barrel; engraved, nickel silver finished receiver. Introduced 1978; no longer imported.
    Perf.: $300   **Exc.:** $250   **VGood:** $200

RIFLES

Erma
EGM1
EM1
ESG22 Carbine

Fajen
Acra M-18
Acra RA
Acra S-24
Acra Varmint

Feather
AT-9 Carbine
AT-22 Carbine
Model F2 Carbine
Model F9 Carbine

Federal Engineering
XC222
XC450

Erma EM1

Fajen Acra S-24

Feather Model F9 Carbine

Federal Engineering XC222

ERMA EGM1
Semi-automatic; 22 LR; 5-shot magazine; 18″ barrel; M-1-type sights with ramp front; receiver grooved for scope mounts; European walnut M-1 carbine-type stock; sling swivel in barrel band; patterned after U.S. Cal. 30 M-1 Carbine. Introduced 1970; no longer imported.
Perf.: $190 **Exc.:** $160 **VGood:** $120

ERMA EM1
Semi-automatic; 22 LR; 10-, 15-shot magazine; 18″ barrel; M1-type sights; receiver grooved for scope mounts; European walnut M1 Carbine-type stock; sling swivel in barrel band, oiler slot in stock; patterned after U.S. Cal. 30 M-1 Carbine. Introduced 1966; no longer imported.
Perf.: $175 **Exc.:** $140 **VGood:** $100

ERMA ESG22 CARBINE
Semi-automatic; 22 WMR; 5-, 12-shot magazine; 19″ barrel; 38″ overall length; military post front sight, adjustable peep rear; walnut stained beechwood stock; receiver grooved for scope mounts; gas-operated; styled after M-1 carbine. Made in Germany. Introduced 1978; no longer imported.
Perf.: $275 **Exc.:** $225 **VGood:** $175

FAJEN ACRA M-18
Bolt action; 270, 30-06, 243, 308, 284, 22-250, 6mm, 25-06, 7x57, 257 Roberts; 18″ barrel; no sights; full-length Mannlicher-style stock; glass-bedded stock; rosewood pistol grip cap, forend tip; Santa Barbara Mauser action; recoil pad; blued finish. Introduced 1969; discontinued 1973.
Perf.: $595 **Exc.:** $495 **VGood:** $395

FAJEN ACRA RA
Bolt action; 270, 30-06, 243, 308, 284, 22-250, 6mm, 25-06, 7x57, 257 Roberts; 24″ barrel; no sights; Monte Carlo stock of American walnut; glass-bedded stock; tenite buttplate, pistol-grip cap, forend tip; recoil pad; blued finish.
Perf.: $495 **Exc.:** $395 **VGood:** $325

FAJEN ACRA S-24
Bolt action; 270, 30-06, 243, 308, 284, 22-250, 6mm, 25-06, 7x57, 257 Roberts, 7mm Rem. Mag., 308 Norma Mag., 300 Win. Mag., 264 Win. Mag.; 24″ barrel; no sights; checkered Monte Carlo stock of American walnut; recoil pad; glass-bedded stock; Santa Barbara Mauser action; blued finish; rosewood pistol-grip cap, forend tip. Introduced 1969; discontinued 1973.
Perf.: $495 **Exc.:** $420 **VGood:** $330

FAJEN ACRA VARMINT
Bolt action; 270, 30-06, 243, 308, 284, 22-250, 6mm, 25-06, 7x57, 257 Roberts, 7mm Rem. Mag., 308 Norma Mag., 300 Win. Mag., 264 Win. Mag.; 26″ heavy barrel; no sights; checkered Monte Carlo stock of American walnut; glass-bedded stock; rosewood pistol grip cap, forend tip; recoil pad; blued finish. Introduced 1969; discontinued 1973.
Perf.: $495 **Exc.:** $395 **VGood:** $295

FEATHER AT-9 CARBINE
Semi-automatic; 9mm Para.; 17″ barrel; 35″ overall length (stock extended), 26 1/2″ (stock closed); weighs 5 lbs.; hooded post front sight, adjustable aperture rear; telescoping wire stock, composition pistol grip; matte black finish. Made in U.S. by Feather Industries. Introduced 1988; still produced. **LMSR: $500**
Perf.: $425 **Exc.:** $325 **VGood:** $265

FEATHER AT-22 CARBINE
Semi-automatic; 22 LR; 20-shot magazine; 17″ barrel; 35″ overall (stock extended), 26″ (folded); weighs 3 1/4 lbs.; protected post front sight, adjustable aperture rear; telescoping wire stock, composition pistol grip; removable barrel; matte black finish. Made in U.S. by Feather Industries. Introduced 1986; still produced. **LMSR: $250**
Perf.: $210 **Exc.:** $175 **VGood:** $130

FEATHER MODEL F2 CARBINE
Semi-automatic; 22 LR; 20-shot magazine; 17″ barrel; 35″ overall (stock extended); weighs 3 1/4 lbs.; protected post front sight, adjustable aperture rear; composition stock and pistol grip; removable barrel; matte black finish. Made in U.S. by Feather Industries. Introduced 1986; still produced. **LMSR: $280**
Perf.: $240 **Exc.:** $200 **VGood:** $150

FEATHER MODEL F9 CARBINE
Semi-automatic; 9mm Para.; 17″ barrel; 35″ overall length; weighs 5 lbs.; hooded post front sight, adjustable aperture rear; composition stock and pistol grip; matte black finish. Made in U.S. by Feather Industries. Introduced 1988; still produced. **LMSR: $535**
Perf.: $525 **Exc.:** $450 **VGood:** $325

FEDERAL ENGINEERING XC222
Semi-automatic; 22 LR; 30-shot magazine; 16 1/2″ barrel with flash hider; 34 1/2″ overall length; weighs 7 1/4 lbs.; hooded post front sight, Williams adjustable rear; sight bridge grooved for scope mounting; quick-detachable tube steel stock; quick takedown; all-steel Heli-arc welded construction; internal parts industrial hard chromed. Made in U.S. by Federal Engineering Corp. Introduced 1993; still produced. **LMSR: $459**
Perf.: $380 **Exc.:** $320 **VGood:** $250

FEDERAL ENGINEERING XC450
Semi-automatic; 45 ACP; 16-shot magazine; 16 1/2″ barrel with flash hider; 34 1/2″ overall length; weighs 8 lbs.; hooded post front sight, Williams adjustable rear; sight bridge grooved for scope mounting; quick-detachable tube steel stock; quick takedown; all-steel Heli-arc welded construction; internal parts industrial hard chromed. Made in U.S. by Federal Engineering Corp. Introduced 1984; dropped 1994.
Perf.: $950 **Exc.:** $835 **VGood:** $650

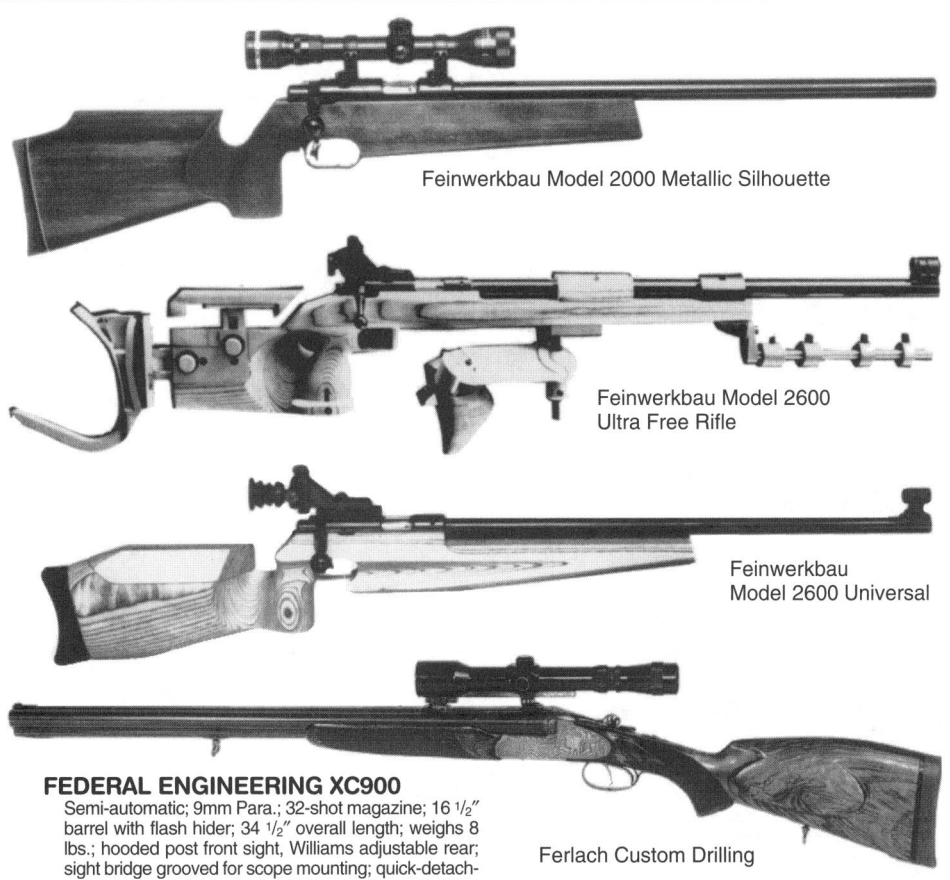

Feinwerkbau Model 2000 Metallic Silhouette

Feinwerkbau Model 2600 Ultra Free Rifle

Feinwerkbau Model 2600 Universal

Ferlach Custom Drilling

RIFLES

Federal Engineering
XC900

Federal Ordnance
M14SA Target
Model 713 Mauser Carbine
Model 713 Mauser Carbine, Field Grade Model

FEG
SA-85M

Feinwerkbau
Model 2000 Match
Model 2000 Metallic Silhouette
Model 2000 Mini-Match
Model 2000 Universal
Model 2600 Ultra Free Rifle
Model 2600 Universal

Ferlach
Custom Drilling

FEDERAL ENGINEERING XC900

Semi-automatic; 9mm Para.; 32-shot magazine; 16 1/2″ barrel with flash hider; 34 1/2″ overall length; weighs 8 lbs.; hooded post front sight, Williams adjustable rear; sight bridge grooved for scope mounting; quick-detachable tube steel stock; quick takedown; all-steel Heli-arc welded construction; internal parts industrial hard chromed. Made in U.S. by Federal Engineering Corp. Introduced 1984; dropped 1994.

Perf.: $950   **Exc.:** $835   **VGood:** $650

FEDERAL ORDNANCE M14SA TARGET

Semi-automatic; 7.62mm NATO; 20-shot magazine; 22″ barrel; 48″ overall length; weighs 9 1/2 lbs.; fully-adjustable G.I.-issue sights; fiberglass or wood stock; Parkerized metal finish. Civilian version of the M-14 service rifle. From Federal Ordnance. Introduced 1988; dropped 1992.
With fiberglass or black textured stock

Perf.: $595   **Exc.:** $525   **VGood:** $450

With walnut stock, forend

Perf.: $650   **Exc.:** $575   **VGood:** $495

FEDERAL ORDNANCE MODEL 713 MAUSER CARBINE

Semi-automatic; 7.63 Mauser, 9mm Para.; 10-, 20-shot detachable magazine; 16″ barrel; weighs 5 lbs.; adjustable sights; deluxe walnut detachable stock. Introduced 1986; discontinued 1992.

Perf.: $1495   **Exc.:** $1250   **VGood:** $995

Federal Ordnance Model 713 Mauser Carbine, Field Grade Model

Same specs as Model 713 Mauser Carbine except 10-shot fixed magazine; walnut fixed stock. Introduced, 1987; discontinued 1992.

Perf.: $750   **Exc.:** $595   **VGood:** $495

FEG SA-85M

Semi-automatic; 7.62x39mm; 6-shot magazine; 16 3/8″ barrel; 34 3/4″ overall length; weighs 7 5/8 lbs.; hardwood handguard and thumbhole buttstock; cylindrical post front sight, fully-adjustable tangent rear; matte finish; chrome-lined barrel. Imported from Hungary by K.B.I., Inc. No longer imported.

Perf.: $395   **Exc.:** $345   **VGood:** $280

FEINWERKBAU MODEL 2000 MATCH

Bolt action; single shot; 22 LR; 26 1/4″ barrel; 43 3/4″ overall length; weighs 8 3/4 lbs.; micrometer match aperture rear sight, globe front with interchangeable inserts; standard walnut match stock; stippled pistol

Ferlach Custom Drilling

grip, forend; electronic or mechanical trigger. Made in Germany. Introduced 1979; discontinued 1988.

Perf.: $995   **Exc.:** $895   **VGood:** $695

With electronic trigger

Perf.: $1250   **Exc.:** $1095   **VGood:** $895

Left-hand version

Perf.: $1095   **Exc.:** $995   **VGood:** $795

FEINWERKBAU MODEL 2000 METALLIC SILHOUETTE

Bolt action; single-shot; 22 LR; 21 13/16″ barrel; 39″ overall length; weighs 6 1/2 lbs.; no sights; grooved for standard mounts; thumbhole walnut match stock; stippled pistol grip; fully-adjustable match trigger; heavy bull barrel. Made in Germany. Introduced 1985; no longer imported.

Perf.: $695   **Exc.:** $595   **VGood:** $450

Left-hand version

Perf.: $725   **Exc.:** $615   **VGood:** $450

FEINWERKBAU MODEL 2000 MINI-MATCH

Bolt action; single shot; 22 LR; 22″ barrel; walnut-stained birchwood match stock; stippled pistol grip; micrometer match aperture rear sight, globe front with interchangeable inserts; electronic or mechanical trigger. Made in Germany. Introduced 1979; discontinued 1988.

Perf.: $1000   **Exc.:** $750   **VGood:** $595

With electronic trigger

Perf.: $1350   **Exc.:** $995   **VGood:** $695

Left-hand version

Perf.: $1150   **Exc.:** $795   **VGood:** $595

FEINWERKBAU MODEL 2000 UNIVERSAL

Bolt action; single shot; 22 LR; 26 3/8″ barrel; weighs 9 3/4 lbs.; standard walnut stock; stippled pistol grip, forend; micrometer match aperture rear sight, globe front with interchangeable inserts; electronic or mechanical trigger. Made in Germany. Introduced 1979; discontinued 1988.

Perf.: $1150   **Exc.:** $895   **VGood:** $695

With electronic trigger

Perf.: $1300   **Exc.:** $1095   **VGood:** $895

Left-hand version

Perf.: $1200   **Exc.:** $995   **VGood:** $795

FEINWERKBAU MODEL 2600 ULTRA FREE RIFLE

Bolt-action; single-shot; 22 LR; 26″ barrel; 40 7/16″ overall length; weighs 14 lbs.; micrometer match aperture rear sight, globe front with interchangeable inserts; laminated wood thumbhole stock; electronic or mechanical trigger; accessory rails for movable weights; adjustable cheekpiece; hooked buttplate; right- or left-hand styles. Made in Germany. Introduced 1983; discontinued 1988.

Perf.: $1995   **Exc.:** $1495   **VGood:** $995

With electronic trigger

Perf.: $2195   **Exc.:** $1595   **VGood:** $1095

Left-hand version

Perf.: $1995   **Exc.:** $1495   **VGood:** $995

FEINWERKBAU MODEL 2600 UNIVERSAL

Single shot; 22 LR; 26 5/16″ barrel; 43 3/4″ overall length; weighs 10 1/2 lbs.; identical smallbore companion to the Model 600 air rifle; micrometer match aperture rear sight, globe front with interchangeable inserts; laminated European hardwood stock; free-floating barrel; match trigger with fingertip weight adjustment dial. Made in Germany. Introduced 1986; no longer imported.

Perf.: $1395   **Exc.:** $995   **VGood:** $750

Left-hand version

Perf.: $1495   **Exc.:** $995   **VGood:** $795

FERLACH CUSTOM DRILLING

Combo; Blitz action; various calibers/gauges; custom-built as to barrel length, sights, stock dimensions, wood; Greener crossbolt. Manufactured in Ferlach, Austria. Introduced 1976; no longer imported by Adler Arms of Pittsburgh. Because of the variety of grades and features, it is impossible to attribute used values. Have the rifle appraised by someone experienced and aware of quality in craftsmanship.

Ferlach
Double Rifle

F.I.E.
GR-8 Black Beauty
Model 122
Model 322
Model 422

F.I.E./Franchi
Para Carbine

Finnish Lion
Champion Model
ISU Target Model
Match Model
Standard Target Model

Firearms Co.
Alpine Sporter

F.N.
Deluxe Mauser
Deluxe Mauser, Presentation Grade
Supreme Mauser
Supreme Mauser, Magnum Model

Ferlach Double Rifle

F.I.E. GR-8 Black Beauty

F.I.E. Model 122

Finnish Lion Champion Model

Finnish Lion Standard Target Model

FERLACH DOUBLE RIFLE
Side-by-side or over/under; boxlock or sidelock; various calibers; any sighting combo desired; custom stocked; auto ejection. Introduced 1980; no longer imported by Adler Arms of Pittsburgh. Because of the variety of grades and features, it is impossible to attribute used values. Have the rifle appraised by someone experienced and aware of quality in craftsmanship.

F.I.E. GR-8 BLACK BEAUTY
Semi-automatic; 22 LR; 14-shot magazine; 19 1/2" barrel; 39 1/2" overall length; weighs 4 lbs.; adjustable open rear sight, band on ramp front; moulded black nylon stock; checkered pistol grip, forend; top tang safety; receiver grooved for tip-off scope mounts. Made in Brazil. Introduced 1984; discontinued 1988.
Perf.: $90   **Exc.:** $70   **VGood:** $55

F.I.E. MODEL 122
Bolt action; 22 Short, 22 Long, 22 LR; 6- or 10-shot magazine; 21" barrel; 39" overall length; weighs 5 1/2 lbs.; fully-adjustable open rear sight, blade front; receiver grooved for scope mounts; Monte Carlo walnut stock; sliding wing safety lever; double extractors; red cocking indicator. Made in Brazil. Introduced 1986; discontinued 1988.
Perf.: $100   **Exc.:** $75   **VGood:** $55

F.I.E. MODEL 322
Bolt action; 22 Short, 22 Long, 22 LR; 6-, 10-shot magazine; 2 6 1/2" barrel; weighs 7 lbs.; adjustable sights; checkered wood stock; adjustable trigger. Introduced 1990; discontinued 1990.
Perf.: $495   **Exc.:** $375   **VGood:** $285

F.I.E. MODEL 422
Bolt action; 22 Short, 22 Long, 22 LR; 6-, 10-shot magazine; 26 1/2" heavy barrel; weighs 9 lbs.; adjustable sights; checkered wood stock; adjustable trigger. Introduced 1990; discontinued 1990.
Perf.: $495   **Exc.:** $395   **VGood:** $295

F.I.E./FRANCHI PARA CARBINE
Semi-automatic; 22 LR; 11-shot magazine; 19" barrel; 39 1/4" overall length; weighs 4 3/4 lbs.; open adjustable rear sight, hooded front; takedown; receiver grooved for scope mounts; magazine feeds through buttplate. Marketed in fitted carrying case. Made in Italy. Introduced 1986; discontinued 1988.
Perf.: $230   **Exc.:** $180   **VGood:** $130

FINNISH LION CHAMPION MODEL
Bolt action; single shot; 22 LR; 28 3/4" barrel; free rifle design; extension-type rear peep sight, aperture front; European walnut stock; full pistol grip; beavertail forearm; thumbhole, palm rest, hand stop; hooked buttplate. Introduced 1965; discontinued 1972.
Perf.: $595   **Exc.:** $450   **VGood:** $350

FINNISH LION ISU TARGET MODEL
Bolt action; single shot; 22 LR; 27 1/2" barrel; extension-type rear peep sight, aperture front; European walnut target-design stock; hand-checkered beavertail forearm; full pistol grip; adjustable buttplate; swivel. Manufactured in Finland. Introduced 1966; discontinued 1977.
Perf.: $450   **Exc.:** $325   **VGood:** $250

FINNISH LION MATCH MODEL
Bolt action; single shot; 22 LR; 28 3/4" barrel; extension-type rear peep sight, aperture front; European walnut free rifle-type stock; full pistol grip; beavertail forearm; thumbhole, palm rest, hand stop; hooked buttplate, swivel. Introduced 1937; discontinued 1972.
Perf.: $495   **Exc.:** $395   **VGood:** $295

FINNISH LION STANDARD TARGET MODEL
Bolt action; single shot; 22 LR; 27 5/8" barrel; 44 9/16" overall length; weighs 10 1/2 lbs.; aperture sights; French walnut target stock; adjustable trigger; many accessories available. Made in Finland. Introduced 1978; still imported by Mandall Shooting Supplies.
LMSR: $695
Perf.: $675   **Exc.:** $575   **VGood:** $410

FIREARMS CO. ALPINE SPORTER
Bolt action; 22-250; 243 Win., 264 Win., 270, 30-06, 308, 308 Norma Mag., 7mm Rem. Mag., 8mm, 300 Win. Mag.; 5-shot magazine, 3-shot (magnums); 23" barrel (standard calibers), 24" (magnums); ramp front, open adjustable rear sight; checkered pistol grip Monte Carlo European walnut stock; recoil pad; sling swivels. Imported from England by Mandall Shooting Supplies. Introduced 1978; still imported. **LMSR: $395**
Perf.: $350   **Exc.:** $310   **VGood:** $275

F.N. DELUXE MAUSER
Bolt action; 220 Swift, 243 Win., 244 Rem., 257 Roberts, 250-3000, 270 Win., 7mm, 300 Savage, 308 Win., 30-06, 8mm, 9.3x62, 9.5x57, 10.75x68; 5-shot box magazine; 24" barrel; Tri-range rear, hooded ramp front sight; hand-checkered European walnut stock, pistol-grip; cheekpiece; sling swivels. Manufactured in Belgium. Introduced 1947; discontinued 1963.
Perf.: $600   **Exc.:** $500   **VGood:** $400

F.N. Deluxe Mauser, Presentation Grade
Same specs as Deluxe Mauser except engraved receiver, trigger guard, barrel breech, floorplate; select walnut stock. Introduced 1947; discontinued 1963.
Perf.: $995   **Exc.:** $875   **VGood:** $650

F.N. SUPREME MAUSER
Bolt action; 243 Win., 270 Win., 7mm, 308 Win., 30-06; 5-shot magazine, 4-shot (243 and 308); 22" barrel; Tri-range rear, hooded ramp front sight; hand-checkered European walnut stock; pistol grip; Monte Carlo cheekpiece, sling swivels. Introduced 1957; discontinued 1975.
Perf.: $650   **Exc.:** $550   **VGood:** $420

F.N. Supreme Mauser, Magnum Model
Same specs as Supreme Mauser except 264 Win. Mag., 7mm Rem. Mag., 300 Win. Mag.; 3-shot magazine. Introduced 1957; discontinued 1973.
Perf.: $700   **Exc.:** $595   **VGood:** $495

F.N. Supreme Mauser

F.N. Deluxe Mauser

Fox Model FB-1

Garcia Musketeer

RIFLES

Fox
Model FB-1

Auguste Francotte
Boxlock Double
Carpathe Mountain
Rimag
Sidelock Double

Galil
Sporter

Gamba
Bayern 88
Mustang
RGZ 1000
RGZ 1000 Game
Safari Express

Garcia
Musketeer

Garrett Arms
Rolling Block

Gevarm
E-l Autoloading

FOX MODEL FB-1

Bolt action; 22 LR; 5-shot clip; 24″ barrel; 43″ overall length; hooded ramp front, adjustable open rear sight; select walnut stock; palm swell grip, roll-over cheekpiece; rosewood pistol grip cap, forend tip; receiver grooved for tipoff scope mount; also drilled, tapped; double extractors. Marketed by Savage Arms. Introduced 1981; discontinued 1983.

Perf.: $350   Exc.: $275   VGood: $200

FRANCOTTE, AUGUSTE

Francotte rifles are custom made to the customer's specifications. Because of the individuality of rifles encountered used, values will fluctuate greatly within models. Options such as engraving, inlays, sights, wood and caliber will determine each rifle's worth. An individual expert appraisal is recommended. Priced new from $15,000 to $40,000. *Used values range from 50% to 75% of new pricing.*

AUGUSTE FRANCOTTE BOXLOCK DOUBLE

Side-by-side; 243, 270, 30-06, 7x64, 7x65R, 8x57JRS, 9.3x74R, 375 H&H, 470 N.E.; 23 1/2″-26″ barrel lengths; bead front sight on ramp, quarter-rib with fixed V rear; primarily custom-built gun; oil-finished European walnut stock; checkered butt; double triggers; manual safety; floating firing pins; gas-vent safety screws; coin finish or color case-hardening; English scroll engraving; numerous options. Made in Belgium. Still imported by Armes de Chasse.

AUGUSTE FRANCOTTE CARPATHE MOUNTAIN

Single shot; 243, 270, 308, 30-06, 7x65R, 7x57R, 5.6x52R, 5.6x57R, 6.5x57R, 6.5x68R, 7mm Rem. Mag., 300 Win. Mag.; 23 1/2″-26″ barrel; no sights furnished; scope mount extra; deluxe walnut stock to customer specs; oil finish, fine checkering; boxlock or sidelock action with third fastener, extractor, manual safety; splinter forend. Many options available. Imported from Belgium by Armes de Chasse. No longer imported.

AUGUSTE FRANCOTTE RIMAG

Bolt action; 358 Norma Mag., 375 H&H, 378 Wea. Mag., 404 Jeffery, 416 Rigby, 450 Watts, 460 Wea. Mag., 458 Win. Mag., 505 Gibbs, others on request; heavy round barrel, 23 5/8″-26″; weighs 9-10 lbs.; ring-mounted front sight blade, fixed leaf rear; deluxe European walnut stock to customer specifications, oil finish, steel grip cap; A. Francotte Rimag action with three-position safety, round or square bridge. Imported in Belgium by Armes de Chasse. No longer imported.

AUGUSTE FRANCOTTE SIDELOCK DOUBLE

Side-by-side; 243, 7x64, 7x65R, 8x57JRS, 270, 30-06, 9.3x74R, 375 H&H, 470 N.E.; 23 1/2″-26″ barrel lengths; weighs 7 5/8 lbs. (medium calibers), 11 lbs. (magnums); bead front sight on ramp, leaf rear on quarter-rib; hand-checkered, oil-finshed European walnut stock; back-action sidelocks; double trigger, with hinged front trigger; auto or free safety; numerous options; largely a custom-built gun. Made in Belgium. Still imported by Armes de Chasse.

GALIL SPORTER

Semi-automatic; 223, 308; 5-shot magazine; 16 1/8″ barrel (223), 18 1/2″ (308); 40 1/2″ overall length (223); weighs 8 3/4 lbs.; hardwood thumbhole stock; hooded post front, flip-type adjustable rear sight; black-finished wood and metal. Imported by Action Arms Ltd. Introduced 1991; discontinued 1993.

Perf.: $1000   Exc.: $850   VGood: $710

GAMBA BAYERN 88

Combo; boxlock over/under; 2 3/4″ 12 ga. over 5.6x50R, 5.6x57R, 6.5x57R, 7x57R, 222 Rem. Mag., 243, 270, 30-06; 24 1/2″ barrels; modified choke; weighs 7 1/2 lbs.; folding leaf rear sight, bead front on ramp; checkered European walnut stock; double-set trigger; case-hardened receiver; game, floral scene engraving; extractors; solid barrel rib. Made in Italy. Introduced 1987; no longer imported.

Perf.: $1395   Exc.: $995   VGood: $795

GAMBA MUSTANG

Single shot; sidelock; 5.6x50, 5.6x57R, 6.5x57R, 7x65R, 222 Rem. Mag.; 243, 270, 30-06; 25 1/2″ barrel; weighs 6 1/4 lbs.; leaf rear sight, bead on ramp front; hand-checkered, oil-finished figured European walnut stock; double-set trigger; silvered receiver; Renaissance engraving signed by engraver; claw-type scope mounts. Made in Italy. Introduced 1987; no longer in production.

Perf.: $13,500   Exc.: $9950   VGood: $7500

GAMBA RGZ 1000

Bolt action; 270, 7x64, 7mm Rem. Mag., 300 Win. Mag.; 20 1/2″ barrel; weighs 7 lbs.; open notch ramp rear sight bead front; Monte Carlo-style European walnut stock; modified 98K Mauser action; single trigger. Introduced 1987; discontinued 1988.

Perf.: $1100   Exc.: $900   VGood: $720

Gamba RGZ 1000 Game

Same specs as RGZ 1000 except 23 3/4″ barrel; 7 3/4 lbs.; double-set trigger.

Perf.: $1195   Exc.: $995   VGood: $795

GAMBA SAFARI EXPRESS

Double rifle; boxlock side-by-side; 7x65R, 9.3x74R, 375 H&H Mag.; 25″ barrels; weighs about 10 lbs.; leaf rear sight, bead on ramp front; checkered select European walnut stock; cheekpiece; rubber recoil pad; triple Greener locking system; auto ejectors; double-set triggers; engraved floral, game scenes. Made in Italy. Introduced 1987; no longer imported.

Perf.: $5500   Exc.: $4250   VGood: $2995

GARCIA MUSKETEER

Bolt action; 243, 264, 270, 30-06, 308 Win., 308 Norma Mag., 7mm Rem. Mag., 300 Win. Mag.; 3-shot magazine (magnums), 5-shot (standard calibers); 44 1/2″ overall length; Williams Guide open rear, hooded ramp front sight; checkered walnut Monte Carlo stock, pistol grip; FN Mauser Supreme action; adjustable trigger; sliding thumb safety; hinged floorplate. Introduced 1970; discontinued 1972.

Perf.: $595   Exc.: $495   VGood: $375

GARRETT ARMS ROLLING BLOCK

Single shot; 45-70; 30″ octagon barrel; 48″ overall length; weighs 9 lbs.; ladder-type tang sight, open folding rear on barrel, hooded globe front; European walnut stock; reproduction of Remington rolling block long-range sporter. Made in Italy. Introduced 1987; no longer imported.

Perf.: $595   Exc.: $475   VGood: $350

GEVARM E-1 AUTOLOADING

Semi-automatic; blowback; 22 LR; 10-shot magazine; 19″ barrel; open sights; walnut stock, pistol grip; blued finish. No longer in production.

Perf.: $170   Exc.: $140   VGood: $100

Gevarm
Model A2

Gibbs
(See Parker-Hale)

Golden Eagle
Model 7000 Grade I
Model 7000 Grade I, African
Model 7000 Grade II

Greifelt
Drilling
Sport Model

Grendel
R-31 Carbine
SRT-20F Compact

Carl Gustaf
2000
Grade II
Grade II, Magnum Model
Grade III
Grade III, Magnum Model
Grand Prix
"Swede" Standard
"Swede" Deluxe

Golden Eagle Model 7000 Grade I

Grendel R-31 Carbine

Carl Gustaf "Swede" Deluxe

Carl Gustaf Varmint Target Model

Carl Gustaf Grade II

GEVARM MODEL A2
Semi-automatic; blowback; 22 LR; 8-shot clip magazine; 21 1/2" barrel; takedown; tangent rear, hooded globe front sight; uncheckered walnut stock; Schnabel forearm; no firing pin (as such) or extractor; fires from open bolt; ridge on bolt face offers twin ignition. Imported by Tradewinds, Inc. Introduced 1958; discontinued 1963.
    Perf.: $275   **Exc.:** $240   **VGood:** $170

GOLDEN EAGLE MODEL 7000 GRADE I
Bolt action; 22-250, 243 Win., 25-06, 270 Win., 270 Weatherby Mag., 7mm Rem. Mag., 30-06, 300 Weatherby Mag., 300 Win. Mag., 338 Win. Mag.; 3-shot magazine, 4-shot (22-250); 24", 26" barrel; no sights; checkered American walnut stock; contrasting grip cap; forend tip; golden eagle head inset in grip cap; recoil pad. Made in Japan. Introduced 1976; discontinued 1979.
    Perf.: $595   **Exc.:** $495   **VGood:** $385

Golden Eagle Model 7000 Grade I, African
Same specs as Model 7000 Grade I except 375 H&H Mag., 458 Win. Mag.; open sights. Introduced 1976; discontinued 1979.
    Perf.: $650   **Exc.:** $565   **VGood:** $400

GOLDEN EAGLE MODEL 7000 GRADE II
Bolt action; 22-250, 243 Win., 25-06, 270 Win., 270 Weatherby Mag., 300 Win. Mag., 338 Win. Mag.; 3-shot magazine, 4-shot (22-250); 24", 26" barrel; no sights; best grade checkered American walnut stock, contrasting grip cap; forend tip; golden eagle head inset in grip cap; recoil pad; carrying case. Made in Japan. Introduced 1976; discontinued 1979.
    Perf.: $695   **Exc.:** $595   **VGood:** $425

GREIFELT DRILLING
Over/under: 12-, 16-, 20-, 28-ga., .410 over any standard rifle caliber; 24", 26" barrels; folding rear sight; matted rib; auto or non-auto ejectors. Manufactured prior to WWII.
    Exc.: $3400   **VGood:** $2500   **Good:** $2000
Obsolete calibers
    Exc.: $2500   **VGood:** $2200   **Good:** $1800

GREIFELT SPORT MODEL
Bolt action; 22 Hornet; 5-shot box magazine; 22" barrel; two-leaf rear, ramp front sight; hand-checkered European walnut stock; pistol grip. Made in Germany prior to World War II.
    Perf.: $495   **Exc.:** $395   **VGood:** $295

GRENDEL R-31 CARBINE
Semi-automatic; 22 WMR; 30-shot magazine; 16" barrel; 23 1/2" overall length (stock collapsed); weighs 4 lbs.; fully-adjustable post front sight, aperture rear; telescoping tube stock, Zytel forend; blowback action with fluted chamber; ambidextrous safety; steel receiver; matte black finish; muzzlebrake. Scope mount optional. Made in U.S. by Grendel, Inc. Introduced 1991; discontinued 1994.
    Perf.: $295   **Exc.:** $200   **VGood:** $150

GRENDEL SRT-20F COMPACT
Bolt action; 243, 308 Win.; 9-shot magazine; 20", 24" fluted, unfluted barrel; weighs 6 5/8 lbs.; no sights; integral scope bases; muzzlebrake; folding fiberglass-reinforced Zytel stock; built on Sako A-2 action. Introduced 1987; discontinued 1990.
    Perf.: $595   **Exc.:** $465   **VGood:** $350

CARL GUSTAF 2000
Bolt action; 243, 6.5x55, 7x64, 270, 308, 30-06, 7mm Rem. Mag., 300 Win. Mag., 9.3x62; 4-shot detachable magazine; 24" barrel; 44" overall length; weighs about 7 1/2 lbs.; sights optional; receiver drilled and tapped for scope mounting; select European walnut stock with hand-rubbed oil finish; Monte Carlo cheekpiece; Wundhammar swell pistol grip; 18 lpi checkering; three-way adjustable single-stage, roller bearing trigger; three-position safety; triple front locking lugs; free-floating barrel; swivel studs. Imported from Sweden by Precision Sales International. Introduced 1991; discontinued 1993.
    Perf.: $1295   **Exc.:** $995   **VGood:** $795
Magnum calibers
    Perf.: $1495   **Exc.:** $1195   **VGood:** $995

CARL GUSTAF GRADE II
Bolt action; 22-250, 243, 25-06, 6.5x55, 270, 308, 30-06; 5-shot magazine; 23 1/2" barrel; folding leaf rear, hooded ramp front sight; walnut Monte Carlo stock with cheekpiece; rosewood forearm tip, grip cap. Available as left-hand model in 25-06, 6.5x55, 270, 30-06. Made in Sweden. Introduced 1970; discontinued 1977.
    Perf.: $495   **Exc.:** $395   **VGood:** $350

Carl Gustaf Grade II, Magnum Model
Same specs as Grade II except 7mm Rem. Mag., 300 Win. Mag.; 3-shot magazine; rubber recoil pad. Introduced 1970; discontinued 1977.
    Perf.: $525   **Exc.:** $425   **VGood:** $325

CARL GUSTAF GRADE III
Bolt action; 22-250, 243, 25-06, 6.5x55, 270, 308, 30-06; 5-shot magazine; 23 1/2" barrel; 44" overall length; weighs 7 1/8 lbs.; no sights; deluxe high-gloss French walnut stock with additional checkering, rosewood forend tip; jeweled bolt; engraved floorplate; detachable swivels. Introduced 1970; discontinued 1977.
    Perf.: $595   **Exc.:** $495   **VGood:** $395

Carl Gustaf Grade III, Magnum Model
Same specs as Grade III except 7mm Rem. Mag., 300 Win. Mag.; 3-shot magazine; recoil pad. Introduced 1970; discontinued 1977.
    Perf.: $625   **Exc.:** $525   **VGood:** $425

CARL GUSTAF GRAND PRIX
Single shot; 22 LR; 26 3/4" barrel; adjustable weight; no sights; uncheckered French walnut target-style stock; adjustable cork buttplate; single-stage adjustable trigger. Introduced 1970; no longer in production.
    Perf.: $575   **Exc.:** $450   **VGood:** $375

CARL GUSTAF "SWEDE" STANDARD
Bolt action; 22-250, 243, 25-06, 6.5x55, 270, 308, 30-06; 5-shot magazine; 23 1/2" barrel; folding leaf rear, hooded ramp front sight; Classic-style French walnut Monte Carlo or sloping comb with Schnabel forend stock; hand-checkered pistol grip, forearm; sling swivels. Available as left-hand model. Made in Sweden. Introduced 1970; discontinued 1977.
    Perf.: $395   **Exc.:** $325   **VGood:** $250

Carl Gustaf "Swede" Deluxe
Same specs as "Swede" Standard except no sights; deluxe high-gloss walnut Monte Carlo stock; rosewood Schnabel forend tip; jeweled bolt; engraved floorplate. Introduced 1970; discontinued 1977.
    Perf.: $425   **Exc.:** $375   **VGood:** $300

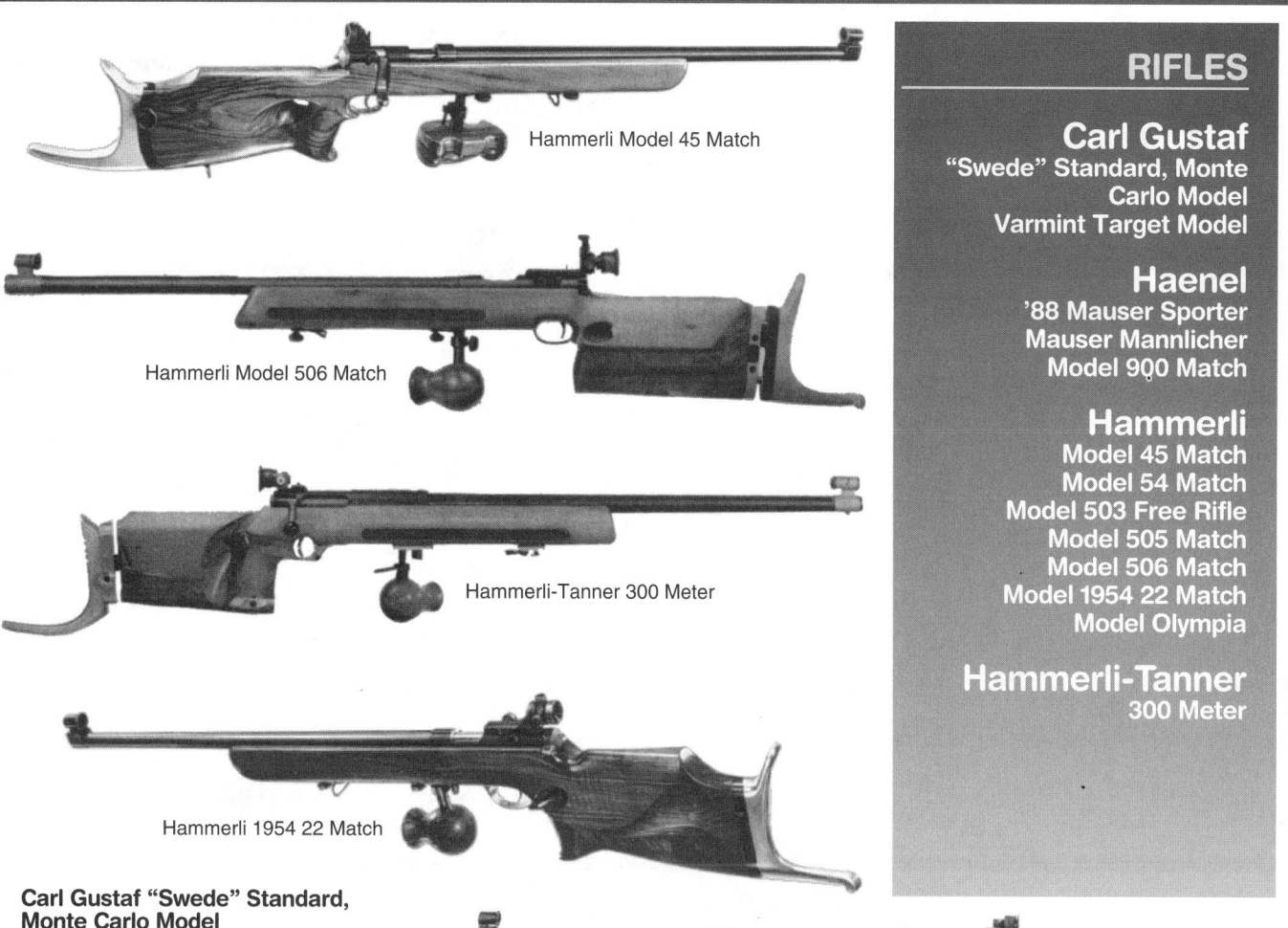

Hammerli Model 45 Match

Hammerli Model 506 Match

Hammerli-Tanner 300 Meter

Hammerli 1954 22 Match

Carl Gustaf
"Swede" Standard, Monte Carlo Model
Varmint Target Model

Haenel
'88 Mauser Sporter
Mauser Mannlicher
Model 900 Match

Hammerli
Model 45 Match
Model 54 Match
Model 503 Free Rifle
Model 505 Match
Model 506 Match
Model 1954 22 Match
Model Olympia

Hammerli-Tanner
300 Meter

Hammerli Model 503 Free Rifle

Carl Gustaf "Swede" Standard, Monte Carlo Model

Same specs as "Swede" Standard except also 7mm Rem. Mag.; Monte Carlo-style stock with cheekpiece. Introduced 1970; discontinued 1977.

Perf.: $450   **Exc.:** $395   **VGood:** $325

CARL GUSTAF VARMINT TARGET MODEL

Bolt action; 222 Rem., 22-250, 243 Win., 6.5x55mm; 26 3/4" barrel; no sights; target-style stock, French walnut; adjustable trigger. Introduced 1970; no longer in production.

Perf.: $495   **Exc.:** $395   **VGood:** $325

HAENEL '88 MAUSER SPORTER

Bolt action; 7x57, 8x57, 9x57mm; 5-shot Mauser box magazine; 22", 24" barrel, half or full octagon; leaf open rear, ramp front sight; hand-checkered European walnut sporting stock, cheekpiece, Schnabel tip, pistol grip; raised matted rib; action based on Model '88 Mauser; double-set trigger; sling swivels. Manufactured in Germany prior to WWII.

Perf.: $595   **Exc.:** $495   **VGood:** $375

HAENEL MAUSER MANNLICHER

Bolt action; 7x57, 8x57, 9x57mm; 5-shot Mannlicher-type box magazine; 22", 24" barrel, half or full octagon; leaf open rear sight, ramp front; hand-checkered European walnut sporting stock, cheekpiece, Schnabel tip, pistol grip; raised matted rib; action based on Model '88 Mauser; double-set trigger; sling swivels. Manufactured in Germany prior to WWII.

Perf.: $465   **Exc.:** $395   **VGood:** $310

HAENEL MODEL 900 MATCH

Bolt action; single shot; 22 LR; 25 15/16" barrel; 44 1/2" overall length; weighs 10 1/2 lbs.; match-type globe front sight, fully-adjustable aperture rear; sight radius adjustable from 31" to 32 1/2"; match-type beech stock with stippled grip and forend; adjustable comb and buttpad; adjustable trigger; polished blue finish. Imported from Germany by GSI, Inc. Introduced 1992; discontinued 1993.

Perf.: $695   **Exc.:** $550   **VGood:** $425

HAMMERLI MODEL 45 MATCH

Bolt action; single shot; 22 LR; 27 1/2" barrel; micrometer peep rear sight, blade front; free-rifle stock with full pistol grip, thumbhole, cheekpiece; palm rest, beavertail forearm; Swiss-type buttplate; sling swivels. Introduced 1945; discontinued 1957.

Perf.: $645   **Exc.:** $515   **VGood:** $420

HAMMERLI MODEL 54 MATCH

Bolt action; single shot; 22 LR; 27 1/2" barrel; micrometer peep rear, globe front sight; European walnut free-rifle stock; cheekpiece; adjustable hook buttplate; palm rest; thumbhole; swivel. Introduced 1954; discontinued 1957.

Perf.: $660   **Exc.:** $540   **VGood:** $450

HAMMERLI MODEL 503 FREE RIFLE

Bolt action; single shot; 22 LR; 27 1/2" barrel; micrometer rear, globe front sight; European walnut free-rifle stock; cheekpiece; adjustable hook buttplate; palm rest; thumbhole; swivel. Introduced 1957; discontinued 1962.

Perf.: $660   **Exc.:** $540   **VGood:** $450

HAMMERLI MODEL 505 MATCH

Bolt action; single shot; 22 LR; 27 1/2" barrel; aperture rear, globe front sight; European walnut free-rifle stock; cheekpiece; palm rest; thumbhole. Introduced 1957; discontinued 1962.

Perf.: $695   **Exc.:** $550   **VGood:** $450

HAMMERLI MODEL 506 MATCH

Bolt action; single shot; 22 LR; 26 3/4" barrel; micrometer peep rear, globe front sight; European walnut free-rifle stock; cheekpiece; adjustable hook buttplate; palm rest; thumbhole; swivel. Introduced 1963; discontinued 1966.

Perf.: $695   **Exc.:** $550   **VGood:** $450

HAMMERLI MODEL 1954 22 MATCH

Bolt action; single shot; 22 LR; 27 1/2" barrel; weighs 15 lbs.; double locking lugs; adjustable trigger, 5 to 16 oz.; hooded front sight, micrometer peep rear, hooded front sight; sling; palm rest; Swiss buttplate. Imported by H. Grieder.

Perf.: $695   **Exc.:** $525   **VGood:** $450

HAMMERLI MODEL OLYMPIA

Bolt action; single shot; 6.5x55 and 7.5mm (Europe), 30-06 and 300 H&H Mag. (U.S.); other calibers on special order; 29 1/2" heavy barrel; designed for 300-meter event; micrometer peep rear, hooded front sight; double-pull or double-set trigger; free-type rifle stock with full pistol grip, thumbhole; cheekpiece; palm rest; beavertail forearm; swivels; Swiss-type buttplate. Imported by H. Grieder. Introduced 1949; discontinued 1962.

Perf.: $895   **Exc.:** $750   **VGood:** $595

HAMMERLI-TANNER 300 METER

Bolt action; single shot; 7.5mm Swiss and other centerfires on request; 29 1/2" barrel; micrometer peep rear, globe front sight; unchecked European walnut free-rifle stock; cheekpiece; adjustable hook buttplate; thumbhole; palm rest; swivel. Manufactured in Switzerland. Introduced 1962; no longer in production.

Perf.: $895   **Exc.:** $795   **VGood:** $630

RIFLES

Harrington & Richardson Semi-Automatic Rifles

Leatherneck Model
(See H&R Model 165)
Model 60 Reising
Model 65
Model 150
Model 151
Model 165
Model 360
Model 361
Model 700
Model 700 Deluxe
Model 755
Model 760
Model 800
Sahara Model
(See H&R Model 755)

Harrington & Richardson Model 361

Harrington & Richardson Model 700 Deluxe

Harrington & Richardson Model 800

Harrington & Richardson
Model 158 Topper Jet

Harrington & Richardson
Model 171 Deluxe

Harrington & Richardson
Model 258 Handy Gun II

H&R SEMI-AUTOMATIC RIFLES

HARRINGTON & RICHARDSON MODEL 60 REISING
Semi-automatic; 45 ACP; 12-, 20-shot detachable box magazine; 18 1/4″ barrel; open rear, blade front sight; uncheckered hardwood pistol-grip stock. Introduced 1944; discontinued 1946. Some collector value.
   Perf.: $495   **Exc.:** $430   **VGood:** $350

HARRINGTON & RICHARDSON MODEL 65
Semi-automatic; 22 LR; 10-shot detachable box magazine; 23″ barrel; rear peep, blade front sight; uncheckered hardwood pistol-grip stock; used as training rifle by the Marine Corps in WWII; same general dimensions as M-1 Garand. Introduced 1944; discontinued 1946.
   Perf.: $295   **Exc.:** $240   **VGood:** $190

HARRINGTON & RICHARDSON MODEL 150
Semi-automatic; 22 LR; 5-shot detachable box magazine; 22″ barrel; 42″ overall length; weighs 7 1/4 lbs.; marble open rear, blade front sight on ramp; oil-finished uncheckered pistol-grip stock; cocking lever under forend; side safety; clip ejector. Introduced 1949; discontinued 1953.
   Perf.: $90   **Exc.:** $70   **VGood:** $55

HARRINGTON & RICHARDSON MODEL 151
Semi-automatic; 22 LR; 5-shot detachable box magazine; 22″ barrel; 42″ overall length; weighs 7 1/4 lbs.; Redfield No. 70 peep rear sight. Introduced 1949; discontinued 1953.
   Perf.: $100   **Exc.:** $80   **VGood:** $60

HARRINGTON & RICHARDSON MODEL 165
Semi-automatic; 22 LR; 10-shot detachable box magazine; 23″ barrel; Redfield No. 70 rear peep sight, blade front on ramp; uncheckered hardwood pistol-grip stock; sling swivels; web sling. Called the Leatherneck Model, it was a variation of Model 65 military autoloader used to train Marines in basic marksmanship during WWII. Introduced 1945; discontinued 1961.
   Perf.: $200   **Exc.:** $150   **VGood:** $100

HARRINGTON & RICHARDSON MODEL 360
Semi-automatic; gas operated; 243 Win., 308 Win.; 3-shot detachable box magazine; 22″ barrel; 43 1/2″ overall length; weighs 7 1/2 lbs.; open adjustable rear, gold bead front sight; one-piece American walnut hand-checkered stock; roll-over cheekpiece, full pistol grip; exotic wood pistol-grip cap, forearm tip; sling swivels. Introduced 1967; discontinued 1973.
   Perf.: $395   **Exc.:** $350   **VGood:** $300

HARRINGTON & RICHARDSON MODEL 361
Semi-automatic; gas operated; 243 Win., 308 Win.; 3-shot detachable box magazine; 22″ barrel; open adjustable rear, gold bead front sight; hand-checkered walnut stock; full roll-over cheekpiece, full pistol grip; exotic wood pistol-grip cap, forearm tip; sling swivels. Introduced 1970; discontinued 1973.
   Perf.: $425   **Exc.:** $370   **VGood:** $320

HARRINGTON & RICHARDSON MODEL 700
Semi-automatic; 22 WMR; 5-, 10-shot magazine; 22″ barrel; 43 1/4″ overall length; weighs 6 1/2 lbs.; folding leaf rear, blade front sight on ramp; American walnut Monte Carlo stock with full pistol grip; composition buttplate. Introduced 1977; discontinued 1985.
   Perf.: $295   **Exc.:** $250   **VGood:** $200

Harrington & Richardson Model 700 Deluxe
Same specs as Model 700 except walnut stock with cheekpiece; checkered grip, forend; rubber recoil pad; no sights; marketed with H&R Model 432 4X scope, base and rings. Introduced 1979; discontinued 1985.
   Perf.: $330   **Exc.:** $300   **VGood:** $230

HARRINGTON & RICHARDSON MODEL 755
Semi-automatic; single shot; 22 LR, 22 Long, 22 Short; 22″ barrel; open rear, military front sight; automatic ejection; hardwood Mannlicher-type stock. Advertised as the Sahara Model. Introduced 1963; discontinued 1971. Some collector interest.
   Perf.: $170   **Exc.:** $140   **VGood:** $100

HARRINGTON & RICHARDSON MODEL 760
Semi-automatic; single shot; 22 LR, 22 Long, 22 Short; 22″ barrel; open rear, military front sight; automatic ejection; hardwood sporter stock. Introduced 1965; discontinued 1970.
   Perf.: $100   **Exc.:** $90   **VGood:** $70

HARRINGTON & RICHARDSON MODEL 800
Semi-automatic; 22 LR; 5-, 10-shot clip-type magazine; 22″ barrel; open rear, bead ramp front sight; uncheckered walnut pistol-grip stock; solid frame; side thumb safety. Introduced 1958; discontinued 1960.
   Perf.: $100   **Exc.:** $80   **VGood:** $60

Harrington & Richardson Model 422

Harrington & Richardson Model 424

Harrington & Richardson
Model 5200

Harrington & Richardson Ultra Varmint

RIFLES

Harrington & Richardson Single Shot & Slide-Action Rifles

Model 058
Model 155 "Shikari"
Model 157
Model 158 Topper
Model 158 Mustang
Model 158 Topper Jet
Model 163
Model 164
Model 171
Model 171 Deluxe
Model 172
Model 173
Model 178
Model 258 Handy Gun II
Model 422
Model 424

H&R SINGLE SHOT & PUMP-ACTION RIFLES

HARRINGTON & RICHARDSON MODEL 058
Single shot; 30-30, 22 Hornet; 22" barrel; 37 1/2" overall length; weighs 6 lbs.; blade front sight, folding leaf rear; walnut finished hardwood stock; rubber buttplate.
Perf.: $145 **Exc.:** $125 **VGood:** $95

HARRINGTON & RICHARDSON MODEL 155 "SHIKARI"
Single shot; 44 Rem. Mag., 45-70; 24" (44, 45-70), 28" (45-70) barrel; 39" (24" barrel), 43" (28" barrel) overall length; weighs 7 lbs. (24"); adjustable folding leaf rear, blade front sight; uncheckered straight-grip walnut-finished hardwood stock forearm; blue-black finish with color case-hardened frame; barrel band; built on Model 158 action; brass cleaning rod. Introduced 1972; discontinued 1982.
Perf.: $150 **Exc.:** $130 **VGood:** $100

HARRINGTON & RICHARDSON MODEL 157
Single shot; 22 Hornet; 22" barrel; 37" overall length; weighs 6 1/2 lbs.; adjustable folding leaf rear, ramp-mounted blade front sight; uncheckered walnut-finished hardwood pistol-grip buttstock; full-length forearm; drilled and tapped for scope mounts; sling swivels; built on Model 158 action. Introduced 1976; discontinued 1985.
Perf.: $130 **Exc.:** $100 **VGood:** $70

HARRINGTON & RICHARDSON MODEL 158 TOPPER
Single shot; 22 Jet, 22 Hornet, 30-30, 357 Mag., 44 Mag.; 22" interchangeable barrels; 37 1/2" overall length; weighs 5 1/4 lbs.; shotgun-type action; Lyman folding adjustable open rear sight, ramp front; visible hammer; side lever; automatic ejector; uncheckered walnut pistol-grip stock, forearm; recoil pad. Introduced 1963; no longer in production.
Perf.: $120 **Exc.:** $100 **VGood:** $70

Harrington & Richardson Model 158 Mustang
Same specs as Model 158 Topper except 30-30 only; fitted with 26" 20-ga. barrel as accessory. Introduced 1968; discontinued 1985.
Perf.: $190 **Exc.:** $160 **VGood:** $130

Harrington & Richardson Model 158 Topper Jet
Same specs as Model 158 Topper except 22 Rem. Jet; interchangeable with 20-ga., 410-bore, 30-30 barrels. Introduced 1963; discontinued 1967.
Perf.: $160 **Exc.:** $140 **VGood:** $120
30-30 barrel, combo
Perf.: $200 **Exc.:** $160 **VGood:** $130
20-ga. barrel, combo
Perf.: $200 **Exc.:** $170 **VGood:** $130
410 barrel, combo
Perf.: $200 **Exc.:** $170 **VGood:** $130

HARRINGTON & RICHARDSON MODEL 163
Single shot; break open action; 22 Hornet, 30-30, 357 Mag., 44 Mag.; 22" barrel; visible hammer; side lever; automatic ejector; straight-grip stock, contoured forearm; gold-plated hammer, trigger. Introduced 1964; discontinued 1967.
Perf.: $130 **Exc.:** $120 **VGood:** $90

HARRINGTON & RICHARDSON MODEL 164
Single shot; break open action; 22 Hornet, 30-30, 357 Mag., 44 Mag.; 22" barrel; visible hammer; side lever; automatic ejector; straight-grip uncheckered walnut stock; contoured forearm; gold-plated hammer, trigger. Introduced 1964; discontinued 1967.
Perf.: $130 **Exc.:** $120 **VGood:** $100

HARRINGTON & RICHARDSON MODEL 171
Single shot; trap door action; 45-70; 22" barrel; 41" overall length; weighs 7 lbs.; open fully-adjustable rear, blade front sight; uncheckered American walnut stock. Replica of Model 1871 Springfield cavalry carbine. Introduced 1972; discontinued 1985.
Perf.: $375 **Exc.:** $325 **VGood:** $250

Harrington & Richardson Model 171 Deluxe
Same specs as Model 171 except folding leaf rear sight; engraved breechblock, side lock, hammer. Introduced 1972; discontinued 1985.
Perf.: $450 **Exc.:** $375 **VGood:** $275

HARRINGTON & RICHARDSON MODEL 172
Single shot; trapdoor action; 45-70; 22" barrel; 41" overall length; weighs 7 lbs.; tang-mounted aperture sight; fancy checkered walnut stock; silver-plated hardware. Introduced 1972; discontinued 1977.
Perf.: $695 **Exc.:** $595 **VGood:** $475

HARRINGTON & RICHARDSON MODEL 173
Single shot; trapdoor action; 45-70; 26" barrel; 44" overall length; weighs 8 lbs.; Vernier tang rear, blade front sight; engraved breechblock receiver, hammer, barrel band, lock, buttplate; checkered walnut stock; ramrod. Replica of Model 1873 Springfield Officers Model. Introduced 1972; discontinued 1977.
Perf.: $695 **Exc.:** $595 **VGood:** $495

HARRINGTON & RICHARDSON MODEL 178
Single shot; trapdoor action; 45-70; 32" barrel; 52" overall length; weighs 8 5/8 lbs.; leaf rear, blade front sight; uncheckered full-length walnut stock; barrel bands, sling swivels, ramrod. Replica of Model 1873 Springfield Infantry Rifle. Introduced 1973; discontinued 1975.
Perf.: $450 **Exc.:** $350 **VGood:** $275

HARRINGTON & RICHARDSON MODEL 258 HANDY GUN II
Single shot; 22 Hornet, 30-30, 357 Mag., 357 Maximum, 44 Mag., 20-ga. 3"; 22" barrel; 37" overall length; weighs 6 1/2 lbs.; interchangeable rifle/shotgun barrels; ramp blade front, adjustable folding leaf rear sight; bead front sight on shotgun barrel; walnut-finished American hardwood stock; electroless matte nickel finish. Introduced 1982; discontinued 1985.
Perf.: $190 **Exc.:** $160 **VGood:** $120

HARRINGTON & RICHARDSON MODEL 422
Slide-action; 22 LR, 22 Long, 22 Short; 24" barrel; 15-shot (22 LR), 17-shot (22 Long), 21-shot (22 Short) tube magazine; open rear, ramp front sight; uncheckered walnut pistol-grip stock; grooved slide handle. Introduced 1952; discontinued 1958.
Perf.: $125 **Exc.:** $100 **VGood:** $70

HARRINGTON & RICHARDSON MODEL 424
Single shot; 22 Short, 22 Long, 22 LR; 24" barrel; open rear, blade front sight; manual cocking action; uncheckered American walnut stock. Introduced 1960; discontinued 1961. Some collector value.
Perf.: $100 **Exc.:** $75 **VGood:** $45

Harrington & Richardson Single Shot & Slide-Action Rifles
Model 749
Ultra Hunter
Ultra Varmint

Harrington & Richardson Bolt-Action Rifles
Ace (See H&R Model 365)
Medalist Model (See H&R Model 451)
Model 250
Model 251
Model 265
Model 265 Targeteer Jr.
Reg'lar Model (See H&R Model 265)
Sportster Model (See H&R Model 250)
Model 300 Ultra
Model 301 Ultra
Model 317 Ultra
Model 317P Ultra
Model 330
Model 333

Harrington & Richardson Model 265

Harrington & Richardson Ultra Model 300

Harrington & Richardson Ultra Model 317

Harrington & Richardson Ultra Model 317P

Harrington & Richardson Model 330

Harrington & Richardson Model 340

HARRINGTON & RICHARDSON MODEL 749
Pump action; 22 Short, 22 Long, 22 LR; 18-shot (22 Short), 15-shot (22 Long), 13-shot (22 LR) tube magazine; 19″ barrel; 35 ¹/₂″ overall length; weighs 4 ⁵/₈ lbs.; dovetail blade front, adjustable rear sight; positive ejection; walnut-finished American hardwood stock, contoured forend. Introduced 1971; discontinued 1971.
Perf.: $150   **Exc.: $125**   **VGood: $90**

HARRINGTON & RICHARDSON ULTRA HUNTER
Single shot; 25-06, 308 Win.; 26″ (25-06), 22″ (308) barrel; weighs about 7 ¹/₂ lbs.; no sights furnished; drilled, tapped for scope mounting; cinnamon-colored laminate stock and forend; hand-checkered grip and forend; break-open action with side-lever release; positive ejection; comes with scope mount; swivel studs; blued receiver and barrel. From H&R 1971, Inc. Introduced 1995; still produced. **LMSR: $250**
Perf.: $200   **Exc.: $150**   **VGood: $110**

HARRINGTON & RICHARDSON ULTRA VARMINT
Single shot; 223 Rem.; 22″ heavy barrel; weighs about 7 ¹/₂ lbs.; no sights; drilled, tapped for scope mounting; hand-checkered laminated birch stock with Monte Carlo comb; break-open action with side-lever release; positive ejection; comes with scope mount; swivel studs; blued receiver and barrel. From H&R 1971, Inc. Introduced 1993; still produced. **LMSR: $250**
Perf.: $200   **Exc.: $150**   **VGood: $110**

H&R BOLT-ACTION RIFLES

HARRINGTON & RICHARDSON MODEL 250
Bolt action; 22 LR; 5-, 10-shot detachable box magazine; 23″ barrel; 40″ overall length; weighs 6 ¹/₂ lbs.; open rear sight, blade front on ramp; uncheckered oil-finished walnut pistol-grip stock. Advertised as the Sportster Model. Introduced 1948; discontinued 1961.
Perf.: $110   **Exc.: $90**   **VGood: $70**

HARRINGTON & RICHARDSON MODEL 251
Bolt action; 22 LR; 5-shot detachable box magazine; 22″ barrel; Lyman No. 55H rear sight, blade front on ramp; uncheckered hardwood pistol-grip stock. Introduced 1948; discontinued 1961.
Perf.: $130   **Exc.: $100**   **VGood: $80**

HARRINGTON & RICHARDSON MODEL 265
Bolt action; 22 LR; 5-, 10-shot detachable box magazine; 22″ barrel; 40″ overall length; weighs 6 ³/₄ lbs.; Lyman No. 17A rear peep sight, blade front on ramp; uncheckered hardwood pistol-grip stock; spring-type bolt release; thumb-operated safety. Introduced 1946; discontinued 1949. Called the Reg'lar Model in advertising.
Perf.: $100   **Exc.: $75**   **VGood: $50**

Harrington & Richardson Model 265 Targeteer Jr.
Same specs as Model 265 except 20″ barrel; 36 ³/₄″ overall length; weighs 7 lbs.; Redfield No. 70 rear peep sight, Lyman No. 17A front; shorter youth uncheckered walnut pistol-grip stock; sling swivels; web sling. Introduced 1948; discontinued 1951.
Perf.: $165   **Exc.: $135**   **VGood: $100**

HARRINGTON & RICHARDSON MODEL 300 ULTRA
Bolt action; 22-250, 243 Win., 270 Win., 30-06, 308 Win., 300 Win. Mag., 7mm Rem. Mag.; 3-shot magazine (magnums), 5-shot (others); 22″, 24″ barrel; FN Mauser action; with or without open rear sight, ramp front; hand-checkered American walnut stock; cheekpiece, full pistol grip; pistol-grip cap; forearm tip of contrasting exotic wood; rubber buttplate, sling swivels. Introduced 1965; discontinued 1978.
Perf.: $450   **Exc.: $395**   **VGood: $350**

HARRINGTON & RICHARDSON MODEL 301 ULTRA
Bolt action; 243 Win., 270 Win., 30-06, 308 Win., 300 Win. Mag., 7mm Rem. Mag.; 3-shot magazine (magnums), 5-shot (others); 18″ barrel; FN Mauser action; with or without open rear sight, ramp front; Mannlicher-style stock; cheekpiece; full pistol grip; pistol-grip cap; metal forearm tip; rubber buttplate, sling swivels. Introduced 1978; discontinued 1978.
Perf.: $450   **Exc.: $400**   **VGood: $350**

HARRINGTON & RICHARDSON MODEL 317 ULTRA
Bolt action; 17 Rem., 222 Rem., 223 Rem., 17/223 handloads; 6-shot magazine; 20″ tapered barrel; no sights; receiver dovetailed for scope mounts. Advertised as the Ultra Wildcat. Introduced 1966; discontinued 1976.
Perf.: $495   **Exc.: $450**   **VGood: $390**

Harrington & Richardson Model 317P Ultra
Same specs as Ultra Model 317 except better grade of walnut; basketweave checkering. Discontinued 1976.
Perf.: $595   **Exc.: $495**   **VGood: $420**

HARRINGTON & RICHARDSON MODEL 330
Bolt action; 243, 270, 30-06, 308, 7mm Rem., 300 Win.; 22″ tapered round barrel; 42 ¹/₂″ overall length; 7 ¹/₈ lbs.; gold bead on ramp front sight, fully-adjustable rear; walnut hand-checkered stock with Monte Carlo; hinged floorplate; adjustable trigger; receiver tapped for scope mounts. Introduced 1972; no longer in production.
Perf.: $495   **Exc.: $395**   **VGood: $295**

HARRINGTON & RICHARDSON MODEL 333
Bolt action; 30-06, 7mm Rem. Mag.; 22″ barrel; 42 ¹/₂″ overall length; weighs 7 ³/₄ lbs.; no sights; walnut finished hardwood stock; adjustable trigger; Sako barrel and action; sliding thumb safety.
Perf.: $400   **Exc.: $300**   **VGood: $200**

Harrington & Richardson Ultra Model 370

Harrington & Richardson Model 465

Harrington & Richardson Model 750

Harrington & Richardson Model 765

RIFLES

Harrington & Richardson Bolt-Action Rifles
Model 340
Model 365
Model 370 Ultra
Model 450
Model 451
Model 465
Model 750
Model 751
Model 765
Model 852
Model 865
Model 866
Model 5200 Sporter
Model 5200 Match
Targeteer Special
(See H&R Model 465)
Ultra Medalist
(See H&R Ultra Model 370
Ultra)
Ultra Wildcat
(See H&R Ultra Model 317
Ultra)

HARRINGTON & RICHARDSON MODEL 340
Bolt action; 243, 7x57, 308, 270, 30-06; 5-shot; 22" barrel; 43" overall length; weighs 7 1/4 lbs.; no sights; drilled and tapped for scope mounts; American walnut stock; hand-checkered pistol grip, forend; carved, beaded cheekpiece; grip cap; recoil pad; Mauser-design action; hinged steel floorplate; adjustable trigger; high-luster blued finish. Introduced 1983; discontinued 1984.

    Perf.: $395   **Exc.:** $350   **VGood:** $300

HARRINGTON & RICHARDSON MODEL 365
Bolt action; single shot; 22 LR; 22" barrel; Lyman No. 55 rear peep sight, blade front on ramp; uncheckered hardwood pistol-grip stock. Called the Ace in advertising. Introduced 1946; discontinued 1947.

    Perf.: $120   **Exc.:** $100   **VGood:** $70

HARRINGTON & RICHARDSON MODEL 370 ULTRA
Bolt action; 22-250, 243 Win., 6mm Rem.; 24" heavy target/varmint barrel; built on Sako action; no sights; tapped for open sights and/or scope mounts; uncheckered, oil-finished walnut stock; roll-over comb; adjustable trigger; recoil pad; sling swivels. Advertised as Ultra Medalist. Introduced 1967; discontinued 1974.

    Perf.: $475   **Exc.:** $425   **VGood:** $375

HARRINGTON & RICHARDSON MODEL 450
Bolt action; 22 LR; 5-, 10-shot detachable box magazine; 26" barrel; no sights; uncheckered American walnut target stock with full pistol grip, thick forend; scope bases; sling swivels, sling. Introduced 1948; discontinued 1961.

    Perf.: $150   **Exc.:** $125   **VGood:** $100

HARRINGTON & RICHARDSON MODEL 451
Bolt action; 22 LR; 5-, 10-shot detachable box magazine; 26" barrel; Lyman 524F extension rear sight, Lyman No. 77 front sight; crowned muzzle; uncheckered American walnut target stock with full pistol grip, thick forend; scope bases; sling swivels. Introduced 1948; discontinued 1961.

    Perf.: $175   **Exc.:** $150   **VGood:** $120

HARRINGTON & RICHARDSON MODEL 465
Bolt action; 22 LR; 10-shot detachable box magazine; 25" barrel; Lyman No. 57 rear peep sight, blade front on ramp; uncheckered walnut pistol-grip stock; sling swivels; web sling. Advertised as Targeteer Special. Introduced 1946; discontinued 1947.

    Perf.: $160   **Exc.:** $140   **VGood:** $100

Harrington & Richardson Model 865

HARRINGTON & RICHARDSON MODEL 750
Bolt action; single shot; 22 LR, 22 Long, 22 Short; 22" (early), 24" barrel; 39" overall length; weighs 5 lbs.; open rear, blade front sight; uncheckered walnut finished Monte Carlo hardwood stock with pistol grip; feed ramp. Introduced 1954; no longer in production.

    Perf.: $100   **Exc.:** $90   **VGood:** $65

HARRINGTON & RICHARDSON MODEL 751
Bolt action; single shot; 22 LR, 22 Long, 22 Short; 24" barrel; open rear, bead front sight; double extractors; Mannlicher stock; feed ramp. Introduced 1971; discontinued 1971.

    Perf.: $125   **Exc.:** $100   **VGood:** $70

HARRINGTON & RICHARDSON MODEL 765
Bolt action; single shot; 22 LR, 22 Long, 22 Short; 24" barrel; 41" overall length; adjustable open rear sight, Red Devil hooded bead front; uncheckered oil-finished walnut or hardwood pistol-grip stock. Introduced 1948; discontinued 1954.

    Perf.: $80   **Exc.:** $65   **VGood:** $50

HARRINGTON & RICHARDSON MODEL 852
Bolt-action; 22 LR, 22 Long, 22 Short; 24" barrel; 15-shot (22 LR), 17-shot (22 Long), 21-shot (22 Short) tube magazine; open rear, bead front sight; uncheckered pistol-grip hardwood stock. Introduced 1952; discontinued 1953.

    Perf.: $110   **Exc.:** $90   **VGood:** $70

HARRINGTON & RICHARDSON MODEL 865
Bolt action; 22 Short, 22 Long, 22 LR; 5-shot clip magazine; 24" (early), 22" round tapered barrel; 40 1/2" overall length; weighs 5 lbs.; "Cottontail" bead (early), blade front, step-adjustable open rear sight; receiver grooved for tip-off scope mounts; oil-finish walnut (early), walnut-finished American hardwood stock with Monte Carlo pistol grip; sliding side safety; cocking indicator. Introduced 1949; discontinued 1985.

    Perf.: $100   **Exc.:** $80   **VGood:** $60

HARRINGTON & RICHARDSON MODEL 866
Bolt action; 22 Short, 22 Long, 22 LR; 5-shot clip magazine; 22" round tapered barrel; 39" overall length; weighs 5 lbs.; blade front, step-adjustable open rear sight; receiver grooved for tip-off scope mounts; Mannlicher stock with Monte Carlo pistol grip; sliding side safety; cocking indicator. Introduced 1971; discontinued 1971.

    Perf.: $140   **Exc.:** $120   **VGood:** $80

HARRINGTON & RICHARDSON MODEL 5200 SPORTER
Bolt action; single shot; 22 LR; 24" barrel; 42" overall length; weighs 6 1/2 lbs.; hooded ramp front sight; fully-adjustable Lyman peep rear; classic American walnut stock with hand cut checkering; rubber buttplate; adjustable trigger; drilled and tapped for scope mounts.

    Perf.: $495   **Exc.:** $400   **VGood:** $350

Harrington & Richardson Model 5200 Match
Same specs as Model 5200 Sporter except 28" barrel; 46" overall length; weighs 11 lbs.; drilled, tapped for sights, scope; target-style American walnut stock; full-length accessory rail; rubber buttpad, palm stop; fully-adjustable trigger; dual extractors; polished blue-black metal finish. Introduced 1981; discontinued 1985.

    Perf.: $420   **Exc.:** $375   **VGood:** $315

RIFLES

Harris-McMillan

Antietam Sharps
Long Range Rifle
M-86 Sniper
M-87 Combo
M-87R
M-89 Sniper
M-92 Bullpup
M-93SN
National Match
Signature Classic Sporter
Signature Alaskan

Harris-McMillan Antietam Sharps

Harris-McMillan Long Range Rifle

Harris-McMillan M-86 Sniper

Harris-McMillan Signature Alaskan

HARRIS-McMILLAN ANTIETAM SHARPS

Falling block; single shot; 40-65, 45-75; 30″, 32″ octagon or round barrel; 47″ overall length; weighs 11 1/4 lbs.; hand-lapped stainless or chrome-moly barrel; Montana Vintage Arms #111 Low Profile Spirit Level front sight, #108 mid-range tang rear with windage adjustments; choice of straight grip, pistol grip or Creedmoor-style stock; Schnabel forend; pewter tip optional; standard A Fancy wood, higher grades available; recreation of the 1874 Sharps sidehammer; action color case-hardened; barrel satin black; chrome-moly barrel optionally blued; optional sights include #112 Spirit Level Globe front with windage, #107 Long Range rear with windage. All parts and components made in U.S. by Harris Gunworks, Inc. Introduced 1994; still produced. **LMSR: $2000**
   Perf.: $1850 **Exc.:** $1400   **VGood:** $1050

HARRIS-McMILLAN LONG RANGE RIFLE

Bolt action; single shot; 300 Win. Mag., 7mm Rem. Mag., 300 Phoenix, 338 Lapua; 26″ stainless steel match-grade barrel; 46 1/2″ overall length; weighs 14 lbs.; barrel band and Tompkins front sight; no rear sight furnished; fiberglass stock with adjustable buttplate and cheekpiece; adjustable for length of pull, drop, cant and cast-off; solid bottom single shot action and Canjar trigger; barrel twist of 1:12″. All parts and components made in U.S. by Harris Gunworks, Inc. Introduced 1989; still produced. **LMSR: $2600**
   Perf.: $1595 **Exc.:** $1295   **VGood:** $995

HARRIS-McMILLAN M-86 SNIPER

Bolt action; 308, 30-06, 300 Win. Mag.; 4-shot (308, 30-06), 3-shot (300 Win. Mag.) magazine; 24″ McMillan match-grade barrel in heavy contour; 43 1/2″ overall length; weighs 11 1/4 lbs. (308), 11 1/2 lbs. (30-06, 300); no sights; specially designed fiberglass stock with textured grip and forend; recoil pad; repeating action; comes with bipod; matte black finish; sling swivels. All parts and components made in U.S. by Harris Gun-

works, Inc. Introduced 1989; still produced. **LMSR: $2700**
   Perf.: $2000 **Exc.:** $1700   **VGood:** $1150
Takedown model
   Perf.: $2200 **Exc.:** $1900   **VGood:** $1350

HARRIS-McMILLAN M-87 COMBO

Bolt action; single shot; 50 BMG; 29″ barrel, with muzzlebrake; 53″ overall length; weighs 21 1/2 lbs.; no sights; fiberglass stock; stainless steel receiver; chrome-moly barrel with 1:15″ twist. All parts and components made in U.S. by Harris Gunworks, Inc. Introduced 1987; still in production. **LMSR: $3885**
   Perf.: $3250 **Exc.:** $2800   **VGood:** $2000

Harris-McMillan M-87R

Same specs as Combo M-87 except 5-shot repeater. All parts and components made in U.S. by Harris Gunworks, Inc. Introduced 1990; still produced. **LMSR: $4000**
   Perf.: $3795 **Exc.:** $3100   **VGood:** $2230

HARRIS-McMILLAN M-89 SNIPER

Bolt action; 308 Win.; 5-shot magazine; 28″ barrel with suppressor; weighs 15 1/4 lbs.; no sights; fiberglass stock adjustable for length; recoil pad; drilled and tapped for scope mounting; repeating action; comes with bipod. All parts and components made in U.S. by Harris Gunworks, Inc. Introduced 1990; still produced. **LMSR: $3200**
   Perf.: $1995 **Exc.:** $1950   **VGood:** $1225

HARRIS-McMILLAN M-92 BULLPUP

Bolt action; single shot; 50 BMG; bullpup barrel, with muzzlebrake; no sights; fiberglass bullpup stock; stainless steel receiver; chrome-moly barrel with 1:15″ twist. All parts and components made in U.S. by Harris Gunworks, Inc. Introduced 1993; still in production. **LMSR: $4770**
   Perf.: $4000 **Exc.:** $3200   **VGood:** $2100

HARRIS-McMILLAN M-93SN

Bolt action; 50 BMG; 5-, 10-shot magazine; 29″ barrel, with muzzlebrake; 53″ overall length; weighs about 21 1/2 lbs.; no sights; folding fiberglass stock; right-handed stainless steel receiver; chrome-moly barrel

with 1:15″ twist. All parts and components made in U.S. by Harris Gunworks, Inc. Introduced 1987; still produced. **LMSR: $4300**
   Perf.: $3895 **Exc.:** $3250   **VGood:** $2350

HARRIS-McMILLAN NATIONAL MATCH

Bolt action; 7mm-08, 308; 5-shot magazine; 24″ stainless steel barrel; 43″ overall length; weighs about 11 lbs.; barrel band and Tompkins front sight, no rear sight; modified ISU fiberglass stock with adjustable buttplate; McMillan repeating action with clip slot, Canjar trigger; match-grade barrel. Fibergrain stock, sight installation, special machining and triggers optional. All parts and components made in U.S. by Harris Gunworks, Inc. Introduced 1989; still produced. **LMSR: $2600**
   Perf.: $2295 **Exc.:** $1795   **VGood:** $1095

HARRIS-McMILLAN SIGNATURE CLASSIC SPORTER

Bolt action; 22-250, 243, 6mm Rem., 7mm-08, 284, 308 (short action); 25-06, 270, 280 Rem., 30-06, 7mm Rem. Mag., 300 Win. Mag., 300 Wea. (long action); 338 Win. Mag., 340 Wea., 375 H&H (magnum); 4-shot magazine (standard calibers), 3-shot (magnums); 22″, 24″, 26″ barrel; weighs about 7 lbs. (short action); no sights; fiberglass stock in green, beige, brown or black; recoil pad and 1″ swivels installed; length of pull up to 14 1/4″; comes with 1″ rings and bases; right- or left-hand action with matte black finish; aluminum floorplate. Fibergrain and wood stocks optional. All parts and components made in U.S. by Harris Gunworks, Inc. Introduced 1987; still produced. **LMSR: $2700**
   Perf.: $2200 **Exc.:** $1750   **VGood:** $1200

Harris-McMillan Signature Alaskan

Same specs as Classic Sporter except 270, 280 Rem., 30-06, 7mm Rem. Mag., 300 Win. Mag., 300 Wea., 358 Win., 340 Wea., 375 H&H; match-grade barrel with single leaf rear sight, barrel band front; 1″ detachable rings and mounts; steel floorplate; electroless nickel finish; wood Monte Carlo stock with cheekpiece; palm swell grip; solid buttpad. All parts and components made in U.S. by Harris Gunworks, Inc. Introduced 1989; still produced. **LMSR: $3800**
   Perf.: $3225 **Exc.:** $1975   **VGood:** $1150

Heckler & Koch HK300

Heckler & Koch HK91 A-2

Heckler & Koch HK93 A-2

Heckler & Koch HK94 A-2

RIFLES

Harris-McMillan
Signature Classic Stainless
Sporter
Signature Super Varminter
Signature Titanium Mountain
Rifle
Talon Safari
Talon Sporter

Heckler & Koch
HK91 A-2
HK91 A-3
HK93 A-2
HK93 A-3
HK94 A-2
HK94 A-3
HK270
HK300
HK630

Harris-McMillan Signature Classic Stainless Sporter

Same specs as Classic Sporter except addition of 416 Rem. Mag.; barrel and action made of stainless steel; fiberglass stock; right- or left-hand action in natural stainless, glass bead or black chrome sulfide finishes. All parts and components made in U.S. by Harris Gunworks, Inc. Introduced 1990; still produced. **LMSR: $2500**
   Perf.: $2195   **Exc.:** $1895   **VGood:** $1195

Harris-McMillan Signature Super Varminter

Same specs as Classic Sporter except 223, 22-250, 220 Swift, 243, 6mm Rem., 25-06, 7mm-08, 7mm BR, 308, 350 Rem. Mag.; heavy-contour barrel; adjustable trigger; field bipod and special hand-bedded fiberglass stock; comes with 1″ rings and bases. Fibergrain optional. All parts and components made in U.S. by Harris Gunworks, Inc. Introduced 1989; still produced. **LMSR: $2700**
   Perf.: $2195   **Exc.:** $1895   **VGood:** $1095

Harris-McMillan Signature Titanium Mountain Rifle

Same specs as Classic Sporter except 270, 280 Rem., 30-06, 7mm Rem. Mag., 300 Win. Mag.; weighs 5 1/2 lbs.; action of titanium alloy; barrel of chrome-moly steel; graphite reinforced fiberglass stock. Fibergrain stock optional. All parts and components made in U.S. by Harris Gunworks, Inc. Introduced 1989; still produced. **LMSR: $3000**
   Perf.: $2695   **Exc.:** $2250   **VGood:** $1795

HARRIS-McMILLAN TALON SAFARI

Bolt action; 300 Win. Mag., 300 Wea. Mag., 300 Phoenix, 338 Win. Mag., 30-378, 338 Lapua, 300 H&H, 340 Wea. Mag., 375 H&H, 404 Jeffery, 416 Rem. Mag., 458 Win. Mag. (Safari Magnum); 378 Wea. Mag., 416 Rigby, 416 Wea. Mag., 460 Wea. Mag. (Safari Super Magnum); 24″ match grade barrel; 43″ overall length; weighs about 9-10 lbs.; barrel band front ramp sight, multi-leaf express rear; McMillan Safari action; fiberglass Safari stock; quick-detachable 1″ scope mounts; positive locking steel floorplate; barrel band sling swivel; matte black finish standard. All parts and components made in U.S. by Harris Gunworks, Inc. Introduced 1989; still produced. **LMSR: $3600**
   Perf.: $2995   **Exc.:** $2395   **VGood:** $1995
Talon Safari Super Magnum
   Perf.: $3395   **Exc.:** $2795   **VGood:** $2195

HARRIS-McMILLAN TALON SPORTER

Bolt action; 22-250, 243, 6mm Rem., 6mm BR, 7mm BR, 7mm-08, 25-06, 270, 280 Rem., 284, 308, 30-06, 350 Rem. Mag. (Long Action) 7mm Rem. Mag., 7mm STW, 300 Win. Mag., 300 Wea. Mag., 300 H&H, 338 Win. Mag., 340 Wea. Mag., 375 H&H, 416 Rem. Mag.; 24″ barrel; no sights; weighs about 7 1/2 lbs.; choice of walnut or fiberglass stock; comes with rings and bases; uses pre-'64 Model 70-type action with cone breech, controlled feed, claw extractor and three-position safety; barrel and action of stainless steel; open sights optional and chrome-moly optional. All parts and components made in U.S. by Harris Gunworks, Inc. Introduced 1991; still produced. **LMSR: $2900**
   Perf.: $2400   **Exc.:** $1995   **VGood:** $1210

HECKLER & KOCH HK91 A-2

Semi-automatic; 308 Win.; 5-, 20-shot detachable box magazine; 17 3/4″ barrel; weighs 9 3/4 lbs.; V and aperture rear sight, post front; plastic buttstock forearm; delayed roller-lock blowback action. Introduced 1976; discontinued 1989. Imported for law enforcement sales only.
   Perf.: $2000   **Exc.:** $1600   **VGood:** $1300

Heckler & Koch HK91 A-3

Same specs as HK91 A-2 except collapsible metal buttstock. Introduced 1976; discontinued 1989. Imported for law enforcement sales only.
   Perf.: $2300   **Exc.:** $1800   **VGood:** $1500

HECKLER & KOCH HK93 A-2

Semi-automatic; 223 Rem.; 25-shot magazine; 16 1/8″ barrel; weighs 8 lbs.; V and aperture rear sight, post front; plastic buttstock, forearm; delayed roller-lock blowback action. Introduced 1976; discontinued 1989. Imported for law enforcement sales only.
   Perf.: $1950   **Exc.:** $1400   **VGood:** $1100

Heckler & Koch HK93 A-3

Same specs as HK93 A-2 except collapsible metal buttstock. Introduced 1976; discontinued 1989. Imported for law enforcement sales only.
   Perf.: $2000   **Exc.:** $1650   **VGood:** $1250

HECKLER & KOCH HK94 A-2

Semi-automatic; 9mm Para.; 15-shot magazine; 16″ barrel; 34 3/4″ overall length; weighs 6 1/2 lbs. (fixed stock); hooded post front sight, fully-adjustable aperture rear; high-impact plastic butt and forend or retractable metal stock; delayed roller-locked action; accepts H&K quick-detachable scope mount. Imported from Germany by Heckler & Koch, Inc. Introduced 1983; discontinued 1994.
   Perf.: $2500   **Exc.:** $2000   **VGood:** $1500

Heckler & Koch HK94 A-3

Same specs as HK94 A-2 except retractable metal stock. Imported from Germany by Heckler & Koch, Inc. Introduced 1983; discontinued 1994.
   Perf.: $3000   **Exc.:** $2500   **VGood:** $2000

HECKLER & KOCH HK270

Semi-automatic; 22 LR; 5-, 20-shot magazine; 19 3/4″ barrel; 38 1/4″ overall length; weighs 5 3/4 lbs.; post front sight, fully-adjustable diopter rear; unchecked European walnut cheekpiece stock; intregal H&K scope mounts. Manufactured in Germany. Introduced 1977; discontinued 1985.
   Perf.: $470   **Exc.:** $380   **VGood:** $310

HECKLER & KOCH HK300

Semi-automatic; 22 WMR; 5-, 15-shot detachable box magazine; 19 3/4″ barrel; post windage-adjustable front, adjustable V-notch rear; hand-checkered European walnut Monte Carlo stock with cheekpiece; checkered pistol grip and Schnabel forend integral H&K scope mounts. Manufactured in Germany. Introduced 1977; discontinued 1989.
   Perf.: $675   **Exc.:** $525   **VGood:** $450

HECKLER & KOCH HK630

Semi-automatic; 223; 4-, 10-shot magazine; 17 3/4″ barrel; weighs 7 lbs.; V-notch rear, ramp front sight; European walnut stock with checkered forend, pistol grip; magazine catch at front of trigger guard; receiver dovetailed to accept clamp-type scope. Discontinued 1986.
   Perf.: $900   **Exc.:** $750   **VGood:** $600

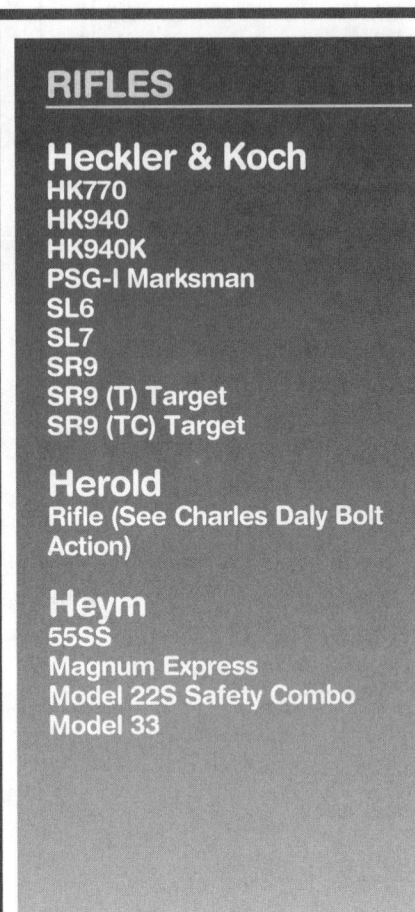

RIFLES

Heckler & Koch
HK770
HK940
HK940K
PSG-I Marksman
SL6
SL7
SR9
SR9 (T) Target
SR9 (TC) Target

Herold
Rifle (See Charles Daly Bolt Action)

Heym
55SS
Magnum Express
Model 22S Safety Combo
Model 33

Heckler & Koch HK770

Heckler & Koch PSG-1 Marksman

Heckler & Koch SL7

Heckler & Koch SR9

HECKLER & KOCH HK770
Semi-automatic; 308 Win.; 3-, 10-shot magazine; 19 1/2″ barrel; weighs 8 lbs.; vertically-adjustable blade front, open fold-down windage-adjustable rear sight; checkered European walnut pistol-grip stock; polygonal rifling; delayed roller-locked bolt system; receiver top dovetailed for clamp-type scope mount. Imported from Germany. Introduced 1976; discontinued 1986.
Perf.: $950   **Exc.:** $800   **VGood:** $600

HECKLER & KOCH HK940
Semi-automatic; 30-06; 3-, 10-shot magazine; 21 1/2″ barrel; weighs 8 3/8 lbs.; vertically-adjustable blade front, open fold-down windage-adjustable rear sight; checkered European walnut pistol-grip stock; polygonal rifling; delayed roller-locked bolt system; receiver top dovetailed for clamp-type scope mount. Imported from Germany. Discontinued 1986.
Perf.: $1100   **Exc.:** $800   **VGood:** $600

Heckler & Koch HK940K
Same specs as HK940 except 16″ barrel; fuller cheekpiece. Introduced 1984; discontinued 1984.
Perf.: $1200   **Exc.:** $900   **VGood:** $700

HECKLER & KOCH PSG-1 MARKSMAN
Semi-automatic; 308 Win.; 5-, 20-shot magazines; 25 1/2″ heavy barrel; 47 1/2″ overall length; weighs 17 1/2 lbs.; no iron sights; 6x42 Hendsoldt scope; matte black high-impact plastic stock adjustable for length; pivoting buttcap; adjustable cheekpiece; target-type pistol grip, palm shelf; built on H&K 91 action; T-way rail for tripod, sling swivel. Made in Germany. Introduced 1986; still imported. **LMSR: $10,100**
Perf.: $8995   **Exc.:** $6995   **VGood:** $4250

HECKLER & KOCH SL6
Semi-automatic; 223 Rem.; 4-shot magazine; 17 1/2″ barrel; weighs 8 3/8 lbs.; adjustable aperture rear, hooded post-front sight; oil-finished European walnut stock; polygonal rifling; delayed roller-locked action; dovetailed for quick-detachable scope mount. Made in Germany. Introduced 1983; discontinued 1986.
Perf.: $800   **Exc.:** $650   **VGood:** $510

HECKLER & KOCH SL7
Semi-automatic; 308 Win.; 3-shot magazine; 17 1/2″ barrel; 39 3/4″ overall length; weighs 8 3/8 lbs.; adjustable aperture rear, hooded post-front sight; oil-finished European walnut stock; polygon rifling; delayed roller-locked action; dovetailed for H&K quick-detachable scope mount. Made in Germany. Introduced 1983; no longer imported by H&K.
Perf.: $800   **Exc.:** $650   **VGood:** $510

HECKLER & KOCH SR9
Semi-automatic; 308 Win.; 5-shot magazine; 19 3/4″ bull barrel with polygonal rifling; 42 7/16″ overall length; weighs 11 lbs.; post-front sight, fully-adjustable aperture rear; Kevlar reinforced fiberglass thumbhole stock with woodgrain finish; redesigned version of the HK91 rifle. Imported from Germany by Heckler & Koch, Inc. Introduced 1990; discontinued 1994.
Perf.: $1500   **Exc.:** $1200   **VGood:** $900

Heckler & Koch SR9 (T) Target
Same specs as SR9 except MSG90 adjustable buttstock; trigger group from the PSG1 Marksman's Rifle; and the PSG1 contoured pistol grip with palm shelf. Imported from Germany by Heckler & Koch, Inc. Introduced 1992; discontinued 1994.
Perf.: $1600   **Exc.:** $1300   **VGood:** $1100

Heckler & Koch SR9 (TC) Target
Same specs as SR9 except PSG1 adjustable buttstock; target/competition version of the SR9 rifle; trigger group and contoured grip. Imported from Germany by Heckler & Koch, Inc. Introduced 1993; discontinued 1994.
Perf.: $2000   **Exc.:** $1700   **VGood:** $1500

HEYM 55SS
Double rifle; over/under; 7x65R, 308, 30-06, 8x57JRS, 9.3x74R; 25″ barrels; 42″ overall length; weighs about 8 lbs.; silver bead ramp front, open V-type rear sight; dark European walnut stock with hand-checkered pistol grip and forend, oil finish; boxlock or full sidelock action; Kersten double cross-bolt; cocking indicators; hand-engraved hunting scenes. Options include interchangeable barrels, Zeiss scopes in claw mounts, deluxe engravings and stock carving, etc. and increase values. Imported from Germany by Heckler & Koch, Inc. No longer imported.
Perf.: $4500   **Exc.:** $3895   **VGood:** $2995

HEYM MAGNUM EXPRESS
Bolt action; 338 Lapua Mag., 375 H&H, 378 Wea. Mag., 416 Rigby, 500 Nitro Express 3″ 460 Wea. Mag., 500 A-Square, 450 Ackley, 600 N.E.; 5-shot magazine (416 Rigby); 24″ barrel; 45 1/4″ overall length; weighs about 10 lbs.; adjustable post front sight on ramp, three-leaf express rear; classic English stock of AAA-grade European walnut with cheekpiece; solid rubber buttpad; steel grip cap; modified magnum Mauser action with double square bridge; Timney single trigger; special hinged floorplate; barrel-mounted quick detachable swivel, quick detachable rear; vertical double recoil lug in rear of stock; three-position safety. Imported from Germany by JagerSport, Ltd. Introduced 1989; no longer imported.
Perf.: $5750   **Exc.:** $4495   **VGood:** $3250
600 Nitro Express caliber
Perf.: $9995   **Exc.:** $7995   **VGood:** $5150

HEYM MODEL 22S SAFETY COMBO
Combination gun; 16- or 20-ga. (2 3/4″, 3″), 12-ga. (2 3/4″) over 22 Hornet, 22 WMR, 222 Rem., 223, 243 Win., 5.6x50R, 5.6x52R, 6.5x55, 6.5x57R, 7x57R, 8x57 JRS; 24″ barrel with solid rib; weighs about 5 1/2 lbs.; silver bead ramp front, folding leaf rear sight; dark European walnut stock with hand-checkered pistol grip and forend; oil finish; tang-mounted cocking slide; floating rifle barrel; single-set trigger; base supplied for quick-detachable scope mounts; patented rocker-weight system automatically uncocks gun if accidentally dropped or bumped hard. Imported from Germany by Heckler & Koch, Inc. No longer imported.
Perf.: $3500   **Exc.:** $2895   **VGood:** $1995

HEYM MODEL 33
Drilling; 5.6x50R Mag., 5.6x52R, 6.5x55, 6.5x57R, 7x57R, 7x65R, 8x57JRS, 9.3x74R, 243, 308, 30-06; 16x16 (2 3/4″), 20x20 (3″); 25″ (Full & Mod.) barrels; 42″ overall length; weighs about 6 1/2 lbs.; silver bead front, folding leaf rear sight; dark European walnut stock with checkered pistol grip and forend; oil finish; automatic sight positioner; boxlock action with Greener-type cross-bolt and safety, double under lugs; double-set triggers; plastic or steel trigger guard. Imported from Germany by Heckler and Koch, Inc. No longer imported.
Perf.: $7500   **Exc.:** $5500   **VGood:** $4350

Heym Magnum Express

Heym Model 22S Safety Combo

Heym Model 33

Heym Model 55B

Heym Model 88B Safari

RIFLES

Heym
Model 33 Deluxe
Model 37B Double Rifle Drilling
Model 37B Deluxe
Model 37B Sidelock Drilling
Model 55B
Model 55BF O/U Combo Gun
Model 55BW
Model 55FW O/U Combo
Model 77B
Model 77BF
Model 88B
Model 88B Safari
Model 88BSS

Heym Model 33 Deluxe
Same specs as Model 33 except extensive hunting scene engraving. Imported from Germany by Heckler & Koch, Inc. No longer imported.
Perf.: $7695 **Exc.:** $6150 **VGood:** $4595

HEYM MODEL 37B DOUBLE RIFLE DRILLING
Drilling; 7x65R, 30-06, 8x57JRS, 9.3x74R; 20-ga. (3″); 25″ barrels choked Full or Mod.; 42″ overall length; weighs about 8 1/2 lbs.; silver bead front, folding leaf rear sight; dark European walnut stock with hand-checkered pistol grip and forend; oil finish; full sidelock construction; Greener-type crossbolt; double under lugs; cocking indicators. Imported from Germany by Heckler & Koch, Inc. No longer imported.
Perf.: $12,995 **Exc.:** $10,000 **VGood:** $7995

Heym Model 37B Deluxe
Same specs as Model 37B except extensive hunting scene engraving. Imported from Germany by Heckler & Koch, Inc. No longer imported.
Perf.: $15,500 **Exc.:** $12,000 **VGood:** $8995

Heym Model 37B Sidelock Drilling
Same specs as Model 37B except 12x12, 16x16 or 20x20 over 5.6x50R Mag., 5.6x52R, 6.5x55, 6.5x57R, 7x57R, 7x65R, 8x57JRS, 9.3x74R, 243, 308, 30-06; rifle barrel manually cocked and uncocked. Imported from Germany by Heckler & Koch, Inc. No longer imported.
Perf.: $11,995 **Exc.:** $8995 **VGood:** $6995

HEYM MODEL 55B
Double rifle; over/under; 7x65R, 308, 30-06, 8x57JRS, 8x75 RS, 9.3x74R, 375 H&H, 458 Win. Mag., 470 N.E.; 25″ barrels; 42″ overall length; weighs about 8 lbs., depending upon caliber; silver bead ramp front, open V-type rear sight; dark European walnut stock with hand-checkered pistol grip and forend; oil finish; boxlock or full sidelock action;

Kersten double cross-bolt; cocking indicators; hand-engraved hunting scenes. Options include interchangeable barrels, Swarovski scopes in claw mounts, deluxe engravings and stock carving, etc. Imported from Germany by JagerSport, Ltd. No longer imported.
Perf.: $8995 **Exc.:** $7500 **VGood:** $5250

Heym Model 55BF O/U Combo Gun
Same specs as Model 55B O/U rifle except 12-, 16-, 20-ga. (2 3/4″ or 3″) over 5.6x50R, 222 Rem., 223 Rem., 5.6x52R, 243, 6.5x57R, 270, 7x57R, 7x65R, 308, 30-06, 8x57JRS, 9.3x74R; solid rib barrel. Available with interchangeable shotgun and rifle barrels. No longer imported.
Perf.: $6500 **Exc.:** $5350 **VGood:** $3995

Heym Model 55BW
Same specs as Model 55B except solid rib barrel. Available with interchangeable shotgun and rifle barrels. No longer imported.
Perf.: $9295 **Exc.:** $7725 **VGood:** $5495

Heym Model 55FW O/U Combo
Same specs as Model 55B O/U rifle except 12-, 16-, 20-ga. (2 3/4″ or 3″) over 7x65R, 308, 30-06, 8x57JRS, 8x75 RS, 9.3x74R, 375 H&H, 458 Win. Mag., 470 N.E.; solid rib barrel. Available with interchangeable shotgun and rifle barrels. No longer imported.
Perf.: $10,995 **Exc.:** $8995 **VGood:** $6995

HEYM MODEL 77B
Double rifle; over/under boxlock; 9.3x74R, 375 H&H, 458 Win. Mag.; 25″, 28″ barrels; weighs 8 1/2 lbs.; silver bead front, folding leaf rear sight; dark oil-finished European walnut stock; hand-checkered pistol grip, forend; fine engraving; Greener-type crossbolt; double under lugs. Introduced 1980; discontinued 1986.
Perf.: $5250 **Exc.:** $4695 **VGood:** $3395

HEYM MODEL 77BF
Same specs as Model 77B except over/under combo; boxlock or sidelock; 12-, 16-, 20-ga. over 5.6x50R, 222 Rem., 5.6x57R, 243, 6.5x57R, 270, 7x57R, 7x65R, 308, 30-06, 8x57 JRS, 9.3x74R, 375 H&H; solid rib barrel. Introduced 1986.
Boxlock (77BF)
Perf.: $5495 **Exc.:** $4895 **VGood:** $3695
Sidelock (77BFSS)
Perf.: $9500 **Exc.:** $8250 **VGood:** $5250

HEYM MODEL 88B
Side-by-side double rifle; 30-06, 8x57JRS, 9.3x74R, 375 H&H; 25″ Krupp steel barrels; 42″ overall length; weighs 7 3/4 lbs. (standard calibers), 8 1/2 lbs. (magnums); silver bead post on ramp front sight, fixed or three-leaf express rear; fancy French walnut stock of classic North American design; Monte Carlo cheekpiece; hand-checkered forend; pistol grip; action with complete coverage hunting scene engraving; Anson-type boxlock action; Greener-type cross-bolt; double under-locking lugs; recoil pad. Imported from Germany by JagerSport, Ltd. No longer imported.
Perf.: $10,895 **Exc.:** $8995 **VGood:** $6995

Heym Model 88B Safari
Same specs as Model 88B except 375 H&H, 458 Win. Mag., 470 NE, 500 NE; weighs about 10 lbs.; large frame; ejectors. Imported from Germany by Jager-Sport, Ltd. No longer imported.
Perf.: $13,995 **Exc.:** $10,995 **VGood:** $8995

Heym Model 88BSS
Same specs as Model 88B except quick detachable sidelocks. Imported from Germany by JagerSport, Ltd. No longer imported.
Perf.: $13,995 **Exc.:** $10,195 **VGood:** $8995

Heym
Model SR20 Trophy
Model SR20 Alpine
Model SR20 Classic Safari
Model SR20 Classic Sportsman
Model SR40

Heym-Ruger
Model HR30
Model HR30G
Model HR38
Model HR38G

Hi-Standard
Flite King
Hi-Power Field
Hi-Power Field Deluxe
Sport King Field
Sport King Carbine
Sport King Deluxe
Sport King Special

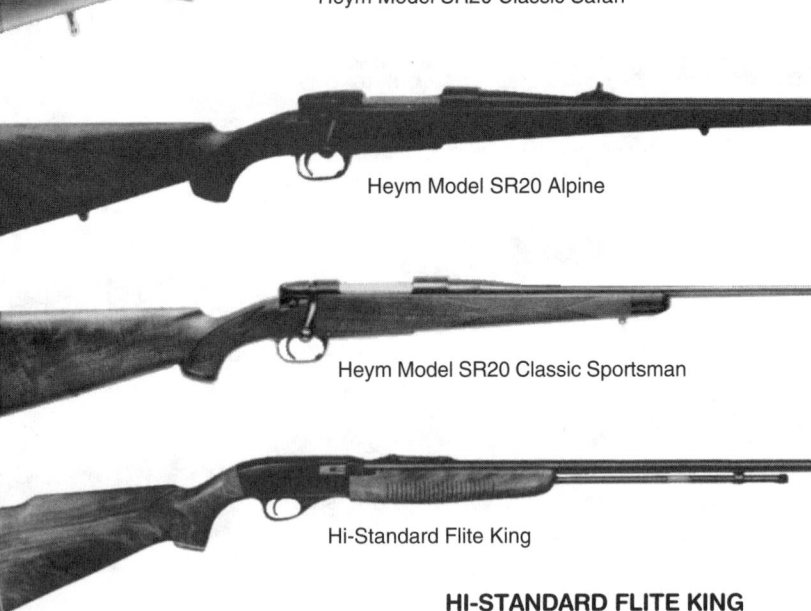

Heym Model SR20 Classic Safari

Heym Model SR20 Alpine

Heym Model SR20 Classic Sportsman

Hi-Standard Flite King

HEYM MODEL SR20 TROPHY
Bolt action; 243, 7x57, 270, 308, 30-06, 7mm Rem. Mag., 338 Win. Mag., 375 H&H, other calibers on request; 22" barrel (standard calibers), 24" (magnums); weighs about 7 lbs.; German silver bead ramp front sight, open rear on quarter-rib; drilled and tapped for scope mounting; AAA-grade European walnut stock with cheekpiece; solid rubber buttpad; checkered grip and forend; oil finish; rosewood grip cap; octagonal barrel; barrel-mounted quick detachable swivel, standard quick detachable rear swivel. Imported from Germany by Heckler & Koch, Inc. No longer imported.
Perf.: $2495   **Exc.:** $1995   **VGood:** $1195

Heym Model SR20 Alpine
Same specs as SR20 Trophy except 243, 270, 7x57, 308, 30-06, 6.5x55, 7x64, 8x57JS; 20" barrel; open sights; full-length "Mountain Rifle" stock with steel forend cap, steel grip cap. Imported from Germany by Heckler & Koch, Inc. Introduced 1989; no longer imported.
Perf.: $1895   **Exc.:** $1650   **VGood:** $995

Heym Model SR20 Classic Safari
Same specs as SR20 Trophy except 404 Jeffery, 425 Express, 458 Win. Mag.; 24" barrel; large post front, three-leaf express rear sight; barrel-mounted ring-type front quick detachable swivel, quick detachable rear; double-lug recoil bolt in stock. Imported from Germany by Heckler & Koch, Inc. Introduced 1989; no longer imported.
Perf.: $2175   **Exc.:** $1695   **VGood:** $1295

Heym Model SR20 Classic Sportsman
Same specs as SR20 Trophy except round barrel without sights. Imported from Germany by Heckler & Koch, Inc. Introduced 1989; no longer imported.
Perf.: $1695   **Exc.:** $1395   **VGood:** $995
Magnum calibers
Perf.: $1795   **Exc.:** $1495   **VGood:** $1095

HEYM MODEL SR40
Bolt action; 222, 223, 5.6x50 Mag.; 24" barrel; 44" overall length; weighs 6 1/4 lbs.; silver bead ramp front, adjustable folding leaf rear sight; carbine-length Mannlicher-style stock; recoil pad; hinged floorplate; three-position safety. Made in Germany. Introduced 1984; discontinued 1988.
Perf.: $895   **Exc.:** $695   **VGood:** $495
Left-hand model
Perf.: $995   **Exc.:** $795   **VGood:** $595

HEYM-RUGER MODEL HR30
Single shot; 243, 6.5x57R, 7x64, 7x65R, 270, 308, 30-06; 24" round barrel; weighs 6 3/8 lbs.; bead ramp front, leaf rear sight; hand-checkered European walnut Mannlicher or sporter stock; oil finish; recoil pad; Ruger No. 1 action. Custom-made gun, many options available. Introduced 1978; discontinued 1987.
Sporter stock
Perf.: $1895   **Exc.:** $1595   **VGood:** $1195
Mannlicher stock
Perf.: $1995   **Exc.:** $1695   **VGood:** $1295

Heym-Ruger Model HR30G
Same specs as HR30 except 6.5x68R, 300 Win. Mag., 8x68S, 9.3x74R; 25" barrel.
Sporter stock
Perf.: $1995   **Exc.:** $1695   **VGood:** $1295
Mannlicher stock
Perf.: $2095   **Exc.:** $1795   **VGood:** $1395

HEYM-RUGER MODEL HR38
Single shot; 243, 6.5x57R, 7x64, 7x65R, 270, 308, 30-06; 24" octagon barrel; weighs 6 3/8 lbs.; bead ramp front, leaf rear sight; hand-checkered European walnut Mannlicher or sporter stock; oil finished; recoil pad; Ruger No. 1 action. Custom-made gun, many options available. Introduced 1978; discontinued 1987.
Sporter stock
Perf.: $1895   **Exc.:** $1595   **VGood:** $1195
Mannlicher stock
Perf.: $1995   **Exc.:** $1695   **VGood:** $1295

Heym-Ruger Model HR38G
Same specs as HR38 except 6.5x68R, 300 Win. Mag., 8x68S.
Sporter stock
Perf.: $1995   **Exc.:** $1695   **VGood:** $1295
Mannlicher stock
Perf.: $2195   **Exc.:** $1795   **VGood:** $1395

HI-STANDARD FLITE KING
Pump-action; hammerless; 22 LR, 22 Long, 22 Short; 17-shot (22 LR), 19-shot (22 Long), 24-shot (22 Short) tubular magazine; 24" barrel; 41 3/4" overall length; weighs 5 1/2 lbs.; Patridge-type rear sight, bead front; checkered hardwood Monte Carlo pistol-grip stock; grooved slide handle. Introduced 1962; discontinued 1975.
Perf.: $120   **Exc.:** $95   **VGood:** $80

HI-STANDARD HI-POWER FIELD
Bolt action; 270, 30-06; 4-shot magazine; 22" barrel; 42 3/4" overall length; weighs 7 lbs.; built on Mauser-type action; folding leaf open rear sight, ramp front; uncheckered walnut field-style pistol-grip stock; sliding safety; quick-detachable sling swivels. Introduced 1962; discontinued 1966.
Perf.: $350   **Exc.:** $300   **VGood:** $240

Hi-Standard Hi-Power Field Deluxe
Same specs as Hi-Power Field except impressed checkering on Monte Carlo stock; sling swivels. Introduced 1962; discontinued 1966.
Perf.: $375   **Exc.:** $325   **VGood:** $265

HI-STANDARD SPORT KING FIELD
Semi-automatic; 22 LR, 22 Long, 22 Short; 22 1/4" tapered barrel; 15-shot (22 LR), 17-shot (22 Long), 21-shot (22 Short) tubular magazine; open rear, bead post front sight; uncheckered pistol-grip stock. Introduced 1960; dropped 1966.
Perf.: $100   **Exc.:** $90   **VGood:** $75

Hi-Standard Sport King Carbine
Same specs as Sport King except 12-shot (22 LR), 14-shot (22 Long), 17-shot (22 Short) tubular magazine; 18" barrel; 38 1/2" overall length; weighs 5 1/2 lbs.; open rear, bead post front sight; receiver grooved for scope mounts; straight-grip stock; brass buttplate; sling swivels; golden trigger guard, trigger, safety. Introduced 1964; discontinued 1973.
Perf.: $150   **Exc.:** $130   **VGood:** $100

Hi-Standard Sport King Deluxe
Same specs as Sport King Field except impressed checkering on stock. Introduced 1966; dropped 1975.
Perf.: $170   **Exc.:** $150   **VGood:** $120

Hi-Standard Sport King Special
Same specs as Sport King Field except Monte Carlo stock; semi-beavertail forearm. Introduced 1960; discontinued 1966.
Perf.: $130   **Exc.:** $110   **VGood:** $90

Hi-Standard Hi-Power Field

Hi-Standard Sport King Field

Howa Lightning

Holland & Holland Best Quality

Holland & Holland Royal

Holland & Holland
Best Quality
Deluxe
No. 2 Model
Royal
Royal Deluxe

Howa
Heavy Barrel Varmint
Lightning
Lightning Woodgrain
Model 1500 Hunter
Model 1500 Lightning
Model 1500 Trophy
Realtree Camo

HOLLAND & HOLLAND BEST QUALITY
Bolt action; 240 Apex, 300 H&H Mag., 375 H&H Mag.; 4-shot box magazine; 24″ barrel; built on Mauser or Enfield action; folding leaf rear, hooded ramp front; cheekpiece stock of European walnut; hand-checkered pistol grip, forearm; sling swivels, recoil pad. Still in production; not imported to U.S.
    Perf.: $15,500 **Exc.:** $12,950   **VGood:** $9995

HOLLAND & HOLLAND DELUXE
Bolt action; 240 Apex, 300 H&H Mag., 375 H&H Mag.; 4-shot box magazine; 24″ barrel; built on Mauser or Enfield action; folding leaf rear, hooded ramp front sight; exhibition-grade European walnut cheekpiece stock; hand-checkered pistol grip, forearm; sling swivels; recoil pad. Introduced following World War II; still in production; not imported to U.S. **Prices vary based on features.**

HOLLAND & HOLLAND NO. 2 MODEL
Double rifle; sidelock; side-by-side; 240 Apex, 7mm H&H Mag., 375 H&H Mag., 458 Win. Mag., 465 H&H Mag., 577 H&H Mag.; 24″, 26″, 28″ barrels; folding leaf rear, ramp front sight; two-piece European stock; hand-checkered pistol grip, forearm; swivels. Still in production; not imported to U.S.
    Perf.: $15,000 **Exc.:** $12,500   **VGood:** $9950

HOLLAND & HOLLAND ROYAL
Double rifle; hammerless sidelock; 240 Apex, 7mm H&H Mag., 300 H&H Mag., 375 H&H Mag., 458 Win., 465 H&H; 24″, 26″, 28″ barrels; folding leaf rear, ramp front sight; two-piece choice European stock; hand-checkered pistol grip, forearm; swivels; custom-engraved receiver. Special-order rifle in the realm of semi-production, with original buyer's options available. Still in production; not imported to U.S. **Prices range from $20,000 to $100,000 based on features.**

Holland & Holland Royal Deluxe
Same specs as Royal Model except more ornate engraving; better grade European walnut stock; better fitting. Still in production; not imported to U.S. **Prices vary based on features.**

HOWA HEAVY BARREL VARMINT
Bolt action; 223, 22-250, 308; 24″ heavy barrel; 42″ overall length; weighs 7 1/2-7 3/4 lbs.; no sights; drilled and tapped for scope mounts; American walnut stock with Monte Carlo comb and cheekpiece; 18 lpi checkering on pistol grip and forend; single unit trigger guard and magazine box with hinged floorplate; Parkerized finish; quick detachable swivel studs; composition non-slip buttplate with white spacer. Imported from Japan by Interarms. Introduced 1989; discontinued 1992.
    Perf.: $425   **Exc.:** $375   **VGood:** $300

HOWA LIGHTNING
Bolt action; 223, 22-250, 243, 270, 308, 30-06, 7mm Rem. Mag., 300 Win. Mag., 338 Win. Mag.; 22″ barrel (standard calibers), 24″ (magnums); 42″ overall length (22″ barrel); weighs 7 1/2 lbs.; no sights; drilled and tapped for scope mounting; black Bell & Carlson Carbelite composite stock with Monte Carlo comb; checkered grip and forend; sliding thumb safety; hinged floorplate; polished blue/black finish. From Interarms. Introduced 1993; still imported.
  LMSR: $469
    Perf.: $375   **Exc.:** $335   **VGood:** $300

Howa Lightning Woodgrain
Same specs as Lightning except 243, 270, 30-06, 308, 7mm Rem. Mag.; 5-shot magazine; lightweight Carbelite synthetic stock with simulated woodgrain. Introduced 1994; no longer in production.
    Perf.: $395   **Exc.:** $350   **VGood:** $320

HOWA MODEL 1500 HUNTER
Bolt action; 22-250, 223, 243, 270, 308, 30-06, 300 Win. Mag., 7mm Rem. Mag.; 3-, 5-shot magazine; 22″, 24″ barrel; weighs 7 1/2 lbs.; ramped gold bead front, adjustable open rear sight; Monte-Carlo-type walnut stock; checkered pistol grip, forend; swivels; adjustable trigger. Made in Japan. Introduced 1988; discontinued 1988.
    Perf.: $375   **Exc.:** $325   **VGood:** $290

Howa Model 1500 Lightning
Same specs as Model 1500 Hunter except 270, 30-06, 300 Win. Mag., 7mm Rem. Mag.; lightweight Carbelite synthetic stock. Made in Japan. Introduced 1988; discontinued 1991.
    Perf.: $395   **Exc.:** $325   **VGood:** $300

HOWA MODEL 1500 TROPHY
Bolt action; 223, 22-250, 243, 270, 30-06, 308, 7mm Rem. Mag., 300 Win. Mag., 338 Win. Mag.; 22″ barrel (standard calibers), 24″ (magnums); 42″ overall length; weighs 7 1/2-7 3/4 lbs.; hooded ramp gold bead front sight, open round-notch fully-adjustable rear; drilled and tapped for scope mounts; American walnut stock with Monte Carlo comb and cheekpiece; 18 lpi checkering on pistol grip and forend; single unit trigger guard and magazine box with hinged floorplate; quick detachable swivel studs; composition non-slip buttplate with white spacer; magnum models with rubber recoil pad. Imported from Japan by Interarms. Introduced 1979; discontinued 1992.
    Perf.: $495   **Exc.:** $395   **VGood:** $280

HOWA REALTREE CAMO
Bolt action; 270, 30-06; 5-shot magazine; 22″ barrel; 42 1/4″ overall length; weighs 8 lbs.; no sights; drilled and tapped for scope mounting; Bell & Carlson Carbelite composite stock with straight comb, checkered grip and forend; completely covered with Realtree camo finish, except bolt; sliding thumb safety; hinged floorplate; sling swivel studs; recoil pad. Imported from Japan by Interarms. Introduced 1993; discontinued 1994.
    Perf.: $495   **Exc.:** $395   **VGood:** $320

Husqvarna
Hi-Power
Model 456
Model 1950
Model 1951
Model 8000
Model 9000
Series 1000
Series 1100
Series 3000
Series 3100
Series 4000
Series 4100
Series 6000
Series 7000
Series P-3000

Husqvarna Hi-Power

Husqvarna Model 456

Husqvarna Series 1100

Husqvarna Series 3000

Husqvarna Series 3100

HUSQVARNA HI-POWER
Bolt action; 220 Swift, 270 Win., 30-06, 6.5x55, 8x57, 9.3x57; 23 3/4″ barrel; 5-shot box magazine; Mauser-type action; open rear, hooded ramp front sight; hand-checkered pistol-grip beech stock; sling swivels. Introduced 1946; discontinued 1959.
   Perf.: $395   **Exc.:** $350   **VGood:** $320

HUSQVARNA MODEL 456
Bolt action; 243, 270, 30-06, 308, 7mm Rem. Mag.; 5-shot box magazine; Husqvarna improved Mauser action; open adjustable rear, hooded ramp front sight; full-length European walnut stock; slope-away cheekpiece; metal forearm cap; sling swivels. Introduced 1959; discontinued 1970.
   Perf.: $495   **Exc.:** $430   **VGood:** $370

HUSQVARNA MODEL 1950
Bolt action; 220 Swift, 270 Win., 30-06; 5-shot box magazine; 23 3/4″ barrel; Mauser-type action; open rear, hooded ramp front sight; hand-checkered pistol-grip beech stock; sling swivels. Introduced 1950; discontinued 1952.
   Perf.: $425   **Exc.:** $395   **VGood:** $330

HUSQVARNA MODEL 1951
Bolt action; 220 Swift, 30-06; 5-shot box magazine; 23 3/4″ barrel; Mauser-type action; open rear, hooded ramp front sight; hand-checkered pistol-grip beech stock with high-comb, low safety. Introduced 1951; discontinued 1951.
   Perf.: $425   **Exc.:** $385   **VGood:** $330

HUSQVARNA MODEL 8000
Bolt action; 270 Win., 30-06, 300 Win. Mag., 7mm Rem. Mag.; 5-shot box magazine; 23 3/4″ barrel; improved Husqvarna action; no sights; hand-checkered deluxe French walnut stock; Monte Carlo cheekpiece; rosewood forearm tip; pistol-grip cap; adjustable trigger; jeweled bolt; hinged engraved floorplate. Introduced 1971; discontinued 1972.
   Perf.: $695   **Exc.:** $595   **VGood:** $475

HUSQVARNA MODEL 9000
Bolt action; 270 Win., 30-06, 300 Win. Mag., 7mm Rem. Mag.; 5-shot box magazine; 23 3/4″ barrel; improved Husqvarna action; folding leaf rear, hooded ramp front sight; Monte Carlo cheekpiece stock; adjustable trigger. Introduced 1971; discontinued 1972.
   Perf.: $495   **Exc.:** $430   **VGood:** $360

HUSQVARNA SERIES 1000
Bolt action; 220 Swift, 270 Win., 30-06; 5-shot box magazine; 23 3/4″ barrel; Mauser-type action; open rear, hooded ramp front sight; European walnut stock, with cheekpiece; Monte Carlo comb. Introduced 1952; discontinued 1956.
   Perf.: $450   **Exc.:** $395   **VGood:** 320

HUSQVARNA SERIES 1100
Bolt action; 220 Swift, 270, 30-06, 6.5x55, 8x57, 9.3x57; 5-shot box magazine; 23 1/2″ barrel; Mauser-type action; open rear, hooded ramp front sight; European walnut Monte Carlo stock; jeweled bolt. Introduced 1952; discontinued 1956.
   Perf.: $475   **Exc.:** $425   **VGood:** $340

HUSQVARNA SERIES 3000
Bolt action; 243, 270, 30-06, 308, 7mm Rem. Mag.; 5-shot box magazine; 23 3/4″ barrel; Husqvarna improved Mauser action; open rear, hooded ramp front sight; European walnut Monte Carlo-style stock; sling swivels. Introduced 1954; discontinued 1976.
   Perf.: $450   **Exc.:** $395   **VGood:** $320

HUSQVARNA SERIES 3100
Bolt action; 243 Win., 270 Win., 7mm Rem. Mag., 30-06, 308 Win.; 5-shot box magazine; 23 3/4″ barrel; Husqvarna improved Mauser action; open rear, hooded ramp front sight; hand-checkered European walnut pistol-grip stock; cheekpiece; black forearm tip; pistol-grip cap; sling swivels. Introduced 1954; discontinued 1976.
   Perf.: $495   **Exc.:** $430   **VGood:** $360

HUSQVARNA SERIES 4000
Bolt action; 243 Win., 270 Win., 30-06, 308 Win., 7mm Rem. Mag.; 5-shot box magazine; 20 1/2″ barrel; Husqvarna improved Mauser action; no rear sight, hooded ramp front; drilled, tapped for scope mounts; European walnut Monte Carlo stock; hand-checkered pistol grip, forearm; sling swivels. Introduced 1954; discontinued 1976.
   Perf.: $495   **Exc.:** $430   **VGood:** $360

HUSQVARNA SERIES 4100
Bolt action; 243 Win., 270 Win., 30-06, 308 Win., 7mm Rem. Mag.; 5-shot box magazine; 20 1/2″ barrel; Husqvarna improved action; adjustable open rear, hooded ramp front sight; lightweight European walnut stock with cheekpiece; sling swivels. Introduced 1954; discontinued 1976.
   Perf.: $475   **Exc.:** $395   **VGood:** $350

HUSQVARNA SERIES 6000
Bolt action; 243 Win., 270 Win., 30-06, 308 Win., 7mm Rem. Mag.; 5-shot box magazine; 23 3/4″ barrel; three-leaf folding rear, hooded ramp front sight; fancy walnut stock; adjustable trigger; sling swivels. Introduced 1968; discontinued 1970.
   Perf.: $595   **Exc.:** $495   **VGood:** $395

HUSQVARNA SERIES 7000
Bolt action; 243 Win., 270 Win., 30-06, 308 Win.; 5-shot box magazine; 20 1/2″ barrel; three-leaf folding rear, hooded ramp front sight; lightweight fancy walnut stock; adjustable trigger; sling swivels. Introduced 1968; discontinued 1970.
   Perf.: $595   **Exc.:** $495   **VGood:** $395

HUSQVARNA SERIES P-3000
Bolt action; 243 Win., 270 Win., 30-06, 7mm Rem. Mag.; 5-shot box magazine; 23 3/4″ barrel; Husqvarna improved action; open rear, hooded ramp front sight; top-grade walnut stock; engraved action; adjustable trigger; sling swivels. Introduced 1968; discontinued 1970.
   Perf.: $795   **Exc.:** $650   **VGood:** $495

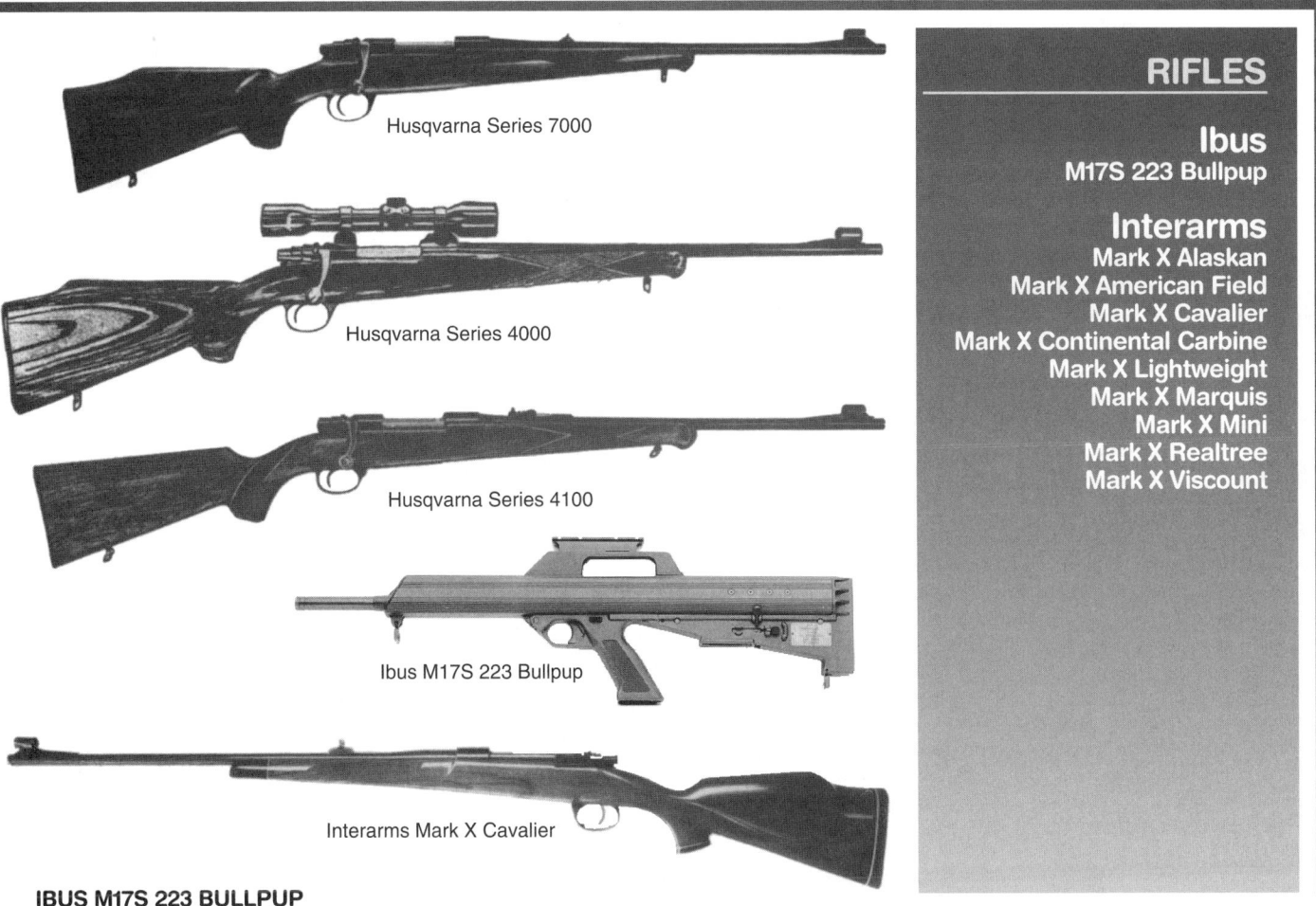

Husqvarna Series 7000

Husqvarna Series 4000

Husqvarna Series 4100

Ibus M17S 223 Bullpup

Interarms Mark X Cavalier

Ibus
M17S 223 Bullpup

Interarms
Mark X Alaskan
Mark X American Field
Mark X Cavalier
Mark X Continental Carbine
Mark X Lightweight
Mark X Marquis
Mark X Mini
Mark X Realtree
Mark X Viscount

Interarms Mark X Lightweight

IBUS M17S 223 BULLPUP
Semi-automatic; 223 Rem.; 10-shot magazine; 21 $1/2''$ barrel; 30″ overall length; weighs 8 $1/4$ lbs.; no sights; comes with scope mount for Weaver-type rings; Zytel glass-filled bullpup-style nylon stock; gas-operated, short-stroke piston system; ambidextrous magazine release. Made in U.S. by Quality Parts Co./Bushmaster Firearms. Introduced 1993; still produced. **LMSR: $975**
   Perf.: $895   Exc.: $750   VGood: $595

INTERARMS MARK X ALASKAN
Bolt action; 375 H&H Mag., 458 Win. Mag.; 3-shot magazine; 24″ barrel; 44 $3/4''$ overall length; weighs 8 $1/4$ lbs.; adjustable folding leaf rear, ramp front sight with removable hood; select walnut Monte Carlo-style stock with crossbolt; hand-checkered pistol grip, forend; sling swivels; heavy-duty recoil pad. Introduced 1976; discontinued 1985.
   Perf.: $495   Exc.: $395   VGood: $295

INTERARMS MARK X AMERICAN FIELD
Bolt action; 22-250, 243, 25-06, 270, 7x57, 7mm Rem. Mag., 308 Win., 30-06, 300 Win. Mag.; 24″ barrel; 45″ overall length; weighs 7 lbs.; Mauser-system action; ramp front sight with removable hood, open fully-adjustable rear; drilled and tapped for scope mounts and receiver sight; hand-checkered walnut stock; 1″ sling swivels; one-piece trigger guard with hinged floorplate; adjustable trigger; hammer-forged chrome vanadium steel barrel. Imported from Yugoslavia by Interarms. Discontinued 1994.
   Perf.: $395   Exc.: $350   VGood: $295

INTERARMS MARK X CAVALIER
Bolt action; 22-250, 243, 25-06, 270, 7x57, 7mm Rem. Mag., 30-06, 300 Win. Mag., 308 Win.; 3-, 5-shot magazine; 24″ barrel; 44″ overall length; weighs 7 $1/2$ lbs.; adjustable folding leaf rear, ramp front sight with removable hood; drilled, tapped for receiver sights, scope mounts; checkered walnut stock with roll-over cheekpiece; rosewood grip cap and forend tip; recoil pad; adjustable trigger. Introduced 1974; discontinued 1983.
   Perf.: $390   Exc.: $340   VGood: $310

INTERARMS MARK X CONTINENTAL CARBINE
Bolt action; 243, 270, 308, 30-06, 7x57; 3-, 5-shot magazine; 20″ barrel; 40″ overall length; weighs 7 $1/2$ lbs.; adjustable folding leaf rear, ramp front sight with removable hood; hand-checkered European walnut full-length stock with roll-over cheekpiece; recoil pad; double-set trigger; button-release hinged floorplate. Introduced 1976; discontinued 1983.
   Perf.: $495   Exc.: $395   VGood: $310

INTERARMS MARK X LIGHTWEIGHT
Bolt action; 22-250, 270, 30-06, 7mm Rem. Mag.; 3-, 5-shot magazine; 20″ barrel; adjustable folding leaf rear, ramp front sight with removable hood; lightweight Carbolite synthetic stock; adjustable trigger. Introduced 1988; discontinued 1990; reintroduced 1994; no longer in production.
   Perf.: $350   Exc.: $290   VGood: $245

INTERARMS MARK X MARQUIS
Bolt action; 243, 270 Win., 308, 30-06, 7x57mm; 3-, 5-shot magazine; 20″ barrel; 40″ overall length; weighs 7 $1/2$ lbs.; adjustable folding leaf rear, ramp front sight with removable hood; hand-checkered European walnut full-length stock; adjustable trigger; quick-detachable sling swivels; blue steel forend cap; Mark X action; white line spacers buttplate, pistol grip cap. Discontinued 1984.
   Perf.: $475   Exc.: $400   VGood: $330

INTERARMS MARK X MINI
Bolt action; 223 Rem., 7.62x39mm; 20″ barrel; 39 $3/4''$ overall length; weighs 6 $3/8$ lbs.; blade on ramp front sight, open adjustable rear; drilled and tapped for scope mounting; checkered European hardwood stock; adjustable trigger; miniature M98 Mauser action. Imported from Yugoslavia by Interarms. Introduced 1987; importation dropped 1994.
   Perf.: $350   Exc.: $310   VGood: $270

INTERARMS MARK X REALTREE
Bolt action; 270, 30-06; 3-, 5-shot magazine; 22″ barrel; 42 $1/4''$ overall length; weighs 8 lbs.; adjustable folding leaf rear, ramp front sight with removable hood; Realtree camo stock; adjustable trigger; sliding thumb safety; hinged floorplate; mono-block receiver. Introduced 1994; no longer imported.
   Perf.: $450   Exc.: $350   VGood: $290

INTERARMS MARK X VISCOUNT
Bolt action; 22-250, 243, 25-06, 270, 7x57, 308, 30-06, 7mm Rem. Mag., 300 Win. Mag.; 24″ barrel; 44″ overall length; weighs about 7 lbs.; blade on ramp front sight, open fully-adjustable rear; drilled and tapped for scope mounting; European hardwood stock with Monte Carlo comb; checkered grip and forend; polished blue finish; uses Mauser system action with sliding thumb safety, hinged floorplate, adjustable trigger. Imported from Yugoslavia by Interarms. Reintroduced 1987; discontinued 1994.
   Perf.: $395   Exc.: $330   VGood: $290

Interarms
Mark X Whitworth
Whitworth Express
Whitworth Mannlicher-Style
Carbine

Ithaca
BSA CF-2 (See BSA CF-2)
BSA CF-2 Stutzen
(See BSA CF-2 Stutzen)
Model LSA-55
Model LSA-55 Deluxe
Model LSA-55 Heavy Barrel
Model LSA-55 Turkey Gun
Model LSA-65
Model LSA-65 Deluxe
Model 37 $1000 Grade (See
Ithaca Model 37 $3000
Grade)
Model 49 Saddlegun
Model 49 Deluxe
Model 49 Presentation
Model 49 Saddlegun
Repeater
Model 49 Saddlegun Youth
Model 72 Saddlegun
Model 72 Saddlegun Deluxe
Model X5-C
Model X5-T
Model X15

Ithaca Model LSA-55

Ithaca Model 49 Presentation

Ithaca Model 49 Saddlegun Repeater

Ithaca Model 49 Saddlegun

Ithaca Model X5-T

INTERARMS MARK X WHITWORTH
Bolt action; 22-250, 243, 25-06, 270, 7x57, 308, 30-06, 7mm Rem. Mag., 300 Win. Mag.; 5-shot magazine, 3-shot (300 Win. Mag.); 24″ barrel; 44″ overall length; weighs 7 lbs.; hooded blade on ramp front sight, open fully-adjustable rear; select grade European walnut stock with checkered grip and forend, straight comb; Mauser system action with sliding thumb safety, hinged floorplate, adjustable trigger; polished blue finish; swivel studs. Imported from Yugoslavia by Interarms. Introduced 1984; discontinued 1994.
Perf.: $475   **Exc.:** $395   **VGood:** $310

INTERARMS WHITWORTH EXPRESS
Bolt action; 375 H&H, 458 Win. Mag.; 24″ barrel; 44 3/4″ overall length; weighs about 8 1/4-8 1/2 lbs.; ramp front sight with removable hood, three-leaf open rear calibrated for 100, 200, 300 yards, on 1/4-rib; solid rubber recoil pad; classic English Express rifle design of hand-checkered, select European walnut; barrel-mounted sling swivel; adjustable trigger; hinged floorplate; solid steel recoil cross-bolt. From Interarms. Introduced 1974; no longer produced.
Perf.: $595   **Exc.:** $495   **VGood:** $400

INTERARMS WHITWORTH MANNLICHER-STYLE CARBINE
Bolt action; 243, 270, 308, 30-06, 7x57; 5-shot magazine; 20″ barrel; 40″ overall length; weighs 7 lbs.; hooded ramp front sight, fully-adjustable rear; full-length checkered European walnut stock with cheekpiece; sling swivels; rubber buttplate. Introduced 1984; discontinued 1987.
Perf.: $565   **Exc.:** $485   **VGood:** $350

ITHACA MODEL LSA-55
Bolt action; 22-250, 222 Rem., 6mm Rem., 243, 25-06, 270 Win., 30-06; 3-shot detachable box magazine; 23″ free-floating barrel; 41 1/2″ overall length; weighs 6 1/2 lbs.; adjustable rear, hooded ramp front sight; drilled, tapped for scope mounts; checkered European walnut pistol-grip Monte Carlo stock; adjustable trigger. Introduced 1972; discontinued 1976.
Perf.: $395   **Exc.:** $350   **VGood:** $300

Ithaca Model LSA-55 Deluxe
Same specs as LSA-55 except no sights; scope mounts; roll-over cheekpiece; rosewood forearm tip, grip cap; white spacers; sling swivel.
Perf.: $475   **Exc.:** $390   **VGood:** $320

Ithaca Model LSA-55 Heavy Barrel
Same specs as LSA-55 except 22-250, 22 Rem.; 23″ heavy barrel; no sights; redesigned stock; beavertail forearm. Introduced 1974; discontinued 1976.
Perf.: $450   **Exc.:** $380   **VGood:** $300

Ithaca Model LSA-55 Turkey Gun
Same specs as Model LSA-55, except over/under combo; 12-ga./222 Rem.; 24 1/2″ ribbed barrel; folding leaf rear sight, bead front; plain extractor; single trigger; exposed hammer. Imported from Finland. Introduced 1970; discontinued 1979.
Perf.: $595   **Exc.:** $525   **VGood:** $395

ITHACA MODEL LSA-65
Bolt action; 25-06, 270, 30-06; 4-shot magazine; 23″ barrel; 41 1/2″ overall length; weighs 6 1/2 lbs.; adjustable rear, hooded ramp front sight; drilled, tapped for scope mounts; checkered European walnut pistol-grip Monte Carlo stock; adjustable trigger. Introduced 1969; discontinued 1976.
Perf.: $430   **Exc.:** $400   **VGood:** $360

Ithaca Model LSA-65 Deluxe
Same specs as LSA-65 except no sights; scope mounts; roll-over cheekpiece; rosewood grip cap, forend tip. Introduced 1969; discontinued 1976.
Perf.: $475   **Exc.:** $395   **VGood:** $380

ITHACA MODEL 49 SADDLEGUN
Lever action; single shot; 22 Short, 22 Long, 22 LR, 22 WMR; 18″ barrel; 34 1/2″ overall length; weighs 5 1/2 lbs.; blank tube magazine for appearance only; open adjustable rear, bead post front sight; straight uncheckered two-piece Western-style carbine stock; barrel band on forearm; Martini-type action. Introduced 1961; discontinued 1976.
Perf.: $130   **Exc.:** $100   **VGood:** $80

Ithaca Model 49 Deluxe
Same specs as Model 49 Saddlegun except figured walnut stock; gold-plated hammer, trigger, sling swivels. Introduced 1962; discontinued 1975.
Perf.: $170   **Exc.:** $140   **VGood:** $100

Ithaca Model 49 Presentation
Same specs as Model 49 Saddlegun except fancy figured walnut stock; gold nameplate inlay; gold trigger, hammer; engraved receiver. Introduced 1962; discontinued 1974.
Perf.: $220   **Exc.:** $160   **VGood:** $120

Ithaca Model 49 Saddlegun Repeater
Same specs as Model 49 Saddlegun except tubular magazine; 20″ barrel; open rear, bead front sight; checkered grip. Introduced 1968; discontinued 1971.
Perf.: $210   **Exc.:** $170   **VGood:** $120

Ithaca Model 49 Saddlegun Youth
Same specs as Model 49 Saddlegun except shorter stock. Introduced 1961; discontinued 1976.
Perf.: $130   **Exc.:** $100   **VGood:** $80

ITHACA MODEL 72 SADDLEGUN
Lever action; 22 Short, 22 Long, 22 WMR; 11-, 15-shot tube magazine; 18 1/2″ barrel; weighs 5 lbs.; step-adjustable open rear sight, hooded ramp front; grooved for scope mounts; uncheckered Western-style straight American walnut stock; barrel band on forearm; half-cock safety. Introduced 1972; discontinued 1977.
Perf.: $220   **Exc.:** $180   **VGood:** $130

Ithaca Model 72 Saddlegun Deluxe
Same specs as Model 72 Saddlegun except engraved, silver-finished receiver; octagon barrel; semi-fancy European walnut stock, forearm. Introduced 1974; discontinued 1976.
Perf.: $275   **Exc.:** $190   **VGood:** $140

ITHACA MODEL X5-C
Semi-automatic; 22 LR; 7-shot clip magazine; 22″ barrel; 40 1/2″ overall length; weighs 6 lbs.; open rear, Ray-bar front sight; uncheckered hardwood pistol-grip stock; grooved forearm. Introduced 1958; discontinued 1964.
Perf.: $150   **Exc.:** $120   **VGood:** $100

Ithaca Model X5-T
Same specs as Model X5-C except 16-shot tube magazine; smooth forearm. Introduced 1959; discontinued 1963.
Perf.: $160   **Exc.:** $130   **VGood:** $110

ITHACA MODEL X15
Semi-automatic; 22 LR; 7-shot clip magazine; 22″ barrel; 40 1/2″ overall length; weighs 6 lbs.; open rear, Ray-bar front sight; uncheckered walnut pistol-grip stock; beavertail forend. Introduced 1964; discontinued 1967.
Perf.: $180   **Exc.:** $150   **VGood:** $110

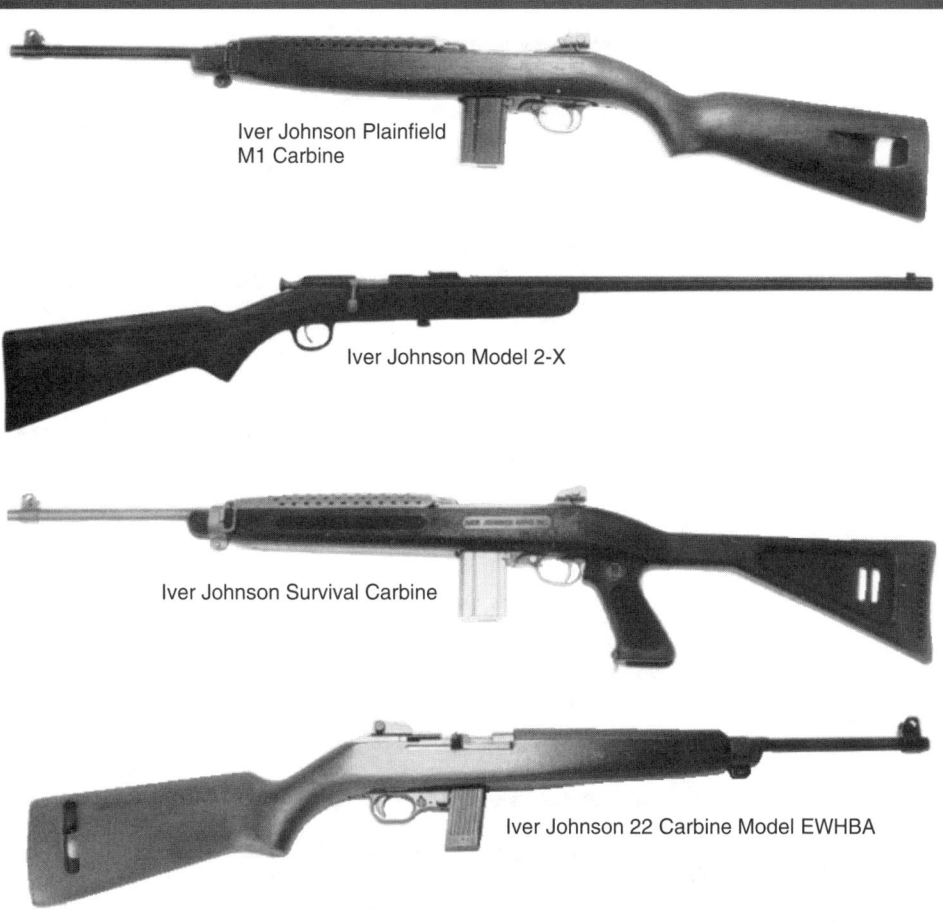

Iver Johnson Plainfield M1 Carbine

Iver Johnson Model 2-X

Iver Johnson Survival Carbine

Iver Johnson 22 Carbine Model EWHBA

RIFLES

Ithaca-Perazzi
(See Perazzi)

Ithaca-SKB
(See SKB)

Iver Johnson
22 Carbine Model EWHBA
Li'l Champ
Model 2-X
Model 2-XA
Model 1003 Universal Carbine
Model 1256 Universal Carbine
Model 5100 Special Application Rifle
Model X (Early Model)
Model X (Late Model)
Plainfield Model M1 Carbine
Survival Carbine
Targetmaster EW22HBP
Trailblazer Model IJ22HBA

IVER JOHNSON 22 CARBINE MODEL EWHBA
Semi-automatic; 22 LR, 22 WMR; 15-shot; 18 1/2″ barrel; 38″ overall length; weighs 5 3/4 lbs.; 22 WMR gas-operated; military-style front sight protected by wings, fully-adjustable peep-style rear; sling swivels; walnut finished hardwood stock and front hand guard; blue finish. Imported from Germany. Introduced 1985; dropped 1988.
Perf.: $175   **Exc.:** $140   **VGood:** $110

IVER JOHNSON LI'L CHAMP
Bolt action; single shot; 22 Short, 22 Long, 22 LR; 16 1/4″ barrel; 32 1/2″ overall length; weighs 3 1/8 lbs.; front ramp sight, step-adjustable rear; nickeled bolt; blue receiver and barrel; stock designed for young shooters. Introduced 1986; dropped 1988.
Perf.: $80   **Exc.:** $60   **VGood:** $40

IVER JOHNSON MODEL 2-X
Bolt action; single shot; 22 Short, 22 Long, 22 LR; 24″ round tapered barrel; 41″ overall length; weighs 4 1/2 lbs.; thumb-screw takedown; drop at comb 1 3/8″; drop at heel 2 3/8″; length of pull 14″; patented automatic safety; adjustable rear sight, blade front; steel buttplate; checkered full pistol grip; grooves in forend; chromium-plated bolt and trigger. Introduced 1930; dropped 1957.
Perf.: $100   **Exc.:** $90   **VGood:** $50

Iver Johnson Model 2-XA
Same specs as 2-X except Lyman #55 adjustable rear sight, Lyman #3 ivory 1/16″ bead front; sling swivels; leather sling strap. Introduced 1930; dropped 1941.
Perf.: $120   **Exc.:** $100   **VGood:** $70

IVER JOHNSON MODEL 1003 UNIVERSAL CARBINE
Semi-automatic; 30 Carbine; 5-, 15-, 30-shot detachable magazine; 18″ barrel; 35 3/4″ overall length; weighs 5 1/2 lbs.; twin guide springs; drilled and tapped for scope mounting; hardwood stock with sling swivels; stainless or blue finish. Not made to military specifications. Manufactured 1986 only.
Perf.: $240   **Exc.:** $200   **VGood:** $150

IVER JOHNSON MODEL 1256 UNIVERSAL CARBINE
Semi-automatic; 256 Win. Mag.; 5-shot detachable magazine; 18″ barrel; 35 3/4″ overall length; weighs 5 1/2 lbs.; twin guide springs; drilled and tapped for scope mounting; hardwood stock with sling swivels; stainless or blue finish. Not made to military specifications. Manufactured 1986 only.
Perf.: $330   **Exc.:** $275   **VGood:** $200

IVER JOHNSON MODEL 5100 SPECIAL APPLICATION RIFLE
Bolt action; single shot; 338/416, 50 BMG; 29″ free-floating fluted barrel; 51 1/2″ overall length; weighs 36 lbs.; sniper rifle; marketed with Leupold Ultra M1 20x scope; adjustable composition stock; adjustable trigger; limited production in both calibers.
Perf.: $4350   **Exc.:** $3500   **VGood:** $2695

IVER JOHNSON MODEL X (EARLY MODEL)
Bolt action; single shot; 22 Short, 22 Long, 22 LR; 22″ round tapered barrel; 39 1/4″ overall length; weighs 4 lbs.; thumb-screw takedown; drop at comb 1 3/8″; drop at heel 2 5/8″; length of pull 13 1/4″; patented automatic safety; fixed open rear sight, blade front; steel buttplate; stock with large knob at front of forend; solid walnut stock. Introduced 1928; dropped 1932.
Perf.: $100   **Exc.:** $75   **VGood:** $50

Iver Johnson Model X (Late Model)
Same specs as Early Model except 24″ barrel; weighs 4 1/2 lbs.; larger, heavier stock; drop at heel 2 3/8″; length of pull 14″; overall length 41″; finger grooves in forend. Model X-A with adjustable sights. Introduced 1933; dropped 1941.
Perf.: $110   **Exc.:** $80   **VGood:** $60

IVER JOHNSON PLAINFIELD MODEL M1 CARBINE
Semi-automatic; 30 Carbine, 5.7mmJ; 15-shot detachable magazine; 18″ barrel; 35 1/2″ overall length; weighs 6 1/2 lbs.; exact copy of WWII U.S. military carbine and manufactured to military specifications; American walnut or hardwood stock; blue or stainless finish; model names and numbers changed several times; M-2 full-automatic model available in eary 1980s; Paratrooper Model with folding stock also available; 5.7mmJ caliber dropped early; 9mm caliber added in 1986. Introduced 1978; dropped 1992.
Perf.: $200   **Exc.:** $170   **VGood:** $150

IVER JOHNSON SURVIVAL CARBINE
Semi-automatic; 30 Carbine, 5.7mmJ; 15-shot detachable magazine; 18 1/2″ barrel; 35 1/2″ overall length; weighs 6 1/2 lbs.; exact copy of U.S. WWII military carbine and manufactured to military specifications; Zytel plastic stock with pistol grip standard; folding stock available; stainless steel or blue finish. Introduced 1983; dropped 1985.
Perf.: $200   **Exc.:** $150   **VGood:** $120

IVER JOHNSON TARGETMASTER EW22HBP
Pump action; 22 Short, 22 Long, 22 LR; 19-shot (22 Short), 15-shot (22 Long), 12-shot (22 LR); 18 1/2″ barrel; 36 1/2″ overall length; weighs 5 3/4 lbs.; tubular magazine under barrel; hooded front sight, step-adjustable rear; walnut finished hardwood buttstock and forearm; plastic buttplate; blue finish. Imported from Germany. Introduced 1985; dropped 1988.
Perf.: $170   **Exc.:** $125   **VGood:** $115

IVER JOHNSON TRAILBLAZER MODEL IJ22HBA
Semi-automatic; 22 LR; 10-shot magazine; 18 1/2″ barrel; 38″ overall length; weighs 5 3/4 lbs.; blade front sight, rear step-adjustable for elevation; Monte Carlo-styled checkered hardwood stock; plastic buttplate; blue finish. Imported from Canada. Introduced 1984; dropped 1986.
Perf.: $120   **Exc.:** $90   **VGood:** $70

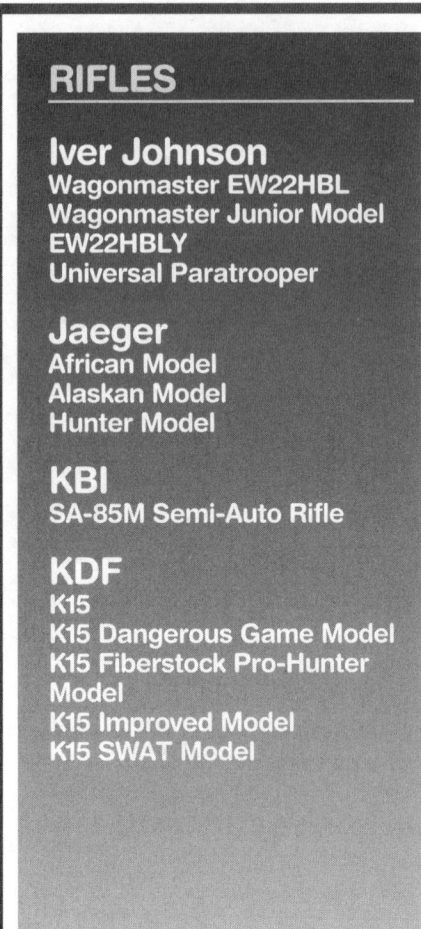

RIFLES

Iver Johnson
Wagonmaster EW22HBL
Wagonmaster Junior Model EW22HBLY
Universal Paratrooper

Jaeger
African Model
Alaskan Model
Hunter Model

KBI
SA-85M Semi-Auto Rifle

KDF
K15
K15 Dangerous Game Model
K15 Fiberstock Pro-Hunter Model
K15 Improved Model
K15 SWAT Model

Jaeger Hunter Model

Jaeger Alaskan Model

KDF K22

KDF K15 Improved Model

KDF K15

IVER JOHNSON WAGONMASTER EW22HBL

Lever action; 22 LR, 22 Short, 22 Long, 22 WMR; 21-shot (22 Short), 17-shot (22 Long), 15-shot (22 LR), 12-shot (22 WMR); 18 1/2" barrel; 36 1/2" overall length; weighs 5 3/4 lbs.; tubular magazine under barrel; hooded front sight, step-adjustable rear; walnut finished hardwood buttstock and forearm; plastic buttplate; blue finish. Imported from Germany. Introduced 1985; dropped 1988.

Perf.: $170   **Exc.:** $140   **VGood:** $110

Iver Johnson Wagonmaster Junior Model EW22HBLY

Same specs as Wagonmaster except 22 Short, 22 Long, 22 LR; 18-shot (22 Short), 15-shot (22 Long); 14-shot (22 LR); 16 1/4" barrel; 33" overall length; weighs 5 1/4 lbs. Imported from Germany. Introduced 1985; dropped 1988.

Perf.: $170   **Exc.:** $140   **VGood:** $120

IVER JOHNSON UNIVERSAL PARATROOPER

Semi-automatic; 30 Carbine; 5-, 15-, 30-shot detachable magazine; 18" barrel; 36" overall length (stock open), 27" overall length (stock folded); twin action guide springs; drilled and tapped for scope mounting; folding Schmeisser-type stock. Not manufactured to military specifications. Manufactured 1986 only.

Perf.: $240   **Exc.:** $200   **VGood:** $150

JAEGER AFRICAN MODEL

Bolt action; 375 H&H Mag., 416 Taylor, 458 Win. Mag.; 3-shot magazine; 22", 24" barrel; weighs 9 lbs.; V-notch rear, hooded ramp front sight; deluxe black fiberglass stock with graphite-reinforcing; swivel studs; recoil pad; single-stage adjustable trigger; hinged floorplate; Mauser-type action; blue/black finish. Introduced 1989; discontinued 1990.

Perf.: $795   **Exc.:** $675   **VGood:** $495

JAEGER ALASKAN MODEL

Bolt action; 7mm Rem. Mag., 300 Win. Mag., 338 Win. Mag.; 4-shot magazine; 22", 24" barrel; weighs 8 lbs.; Williams open rear, silver bead ramp front sight; black fiberglass stock; integral sling; rubber recoil pad; checkered pistol-grip, forend; single-stage trigger; hinged floorplate; Mauser-type action; black wrinkle finish. Introduced 1989; discontinued 1990.

Perf.: $595   **Exc.:** $495   **VGood:** $395

JAEGER HUNTER MODEL

Bolt action; 243, 257 Roberts, 25-06, 7x57, 7mm-08, 280, 308, 30-06; 4-shot magazine; 22", 24" barrel; weighs 7 lbs.; no sights; drilled, tapped for scope mounts; black fiberglass stock; integral sling; rubber recoil pad; checkered pistol grip, forend; single-stage adjustable trigger; hinged floorplate; Mauser-type action; black wrinkle finish. Introduced 1989; discontinued 1990.

Perf.: $595   **Exc.:** $495   **VGood:** $395
Laminated wood stock
Perf.: $550   **Exc.:** $495   **VGood:** $375

KBI SA-85M SEMI-AUTO RIFLE

Semi-automatic; 7.62x39mm; 6-shot magazine; 16 3/8" barrel; 34 3/4" overall length; weighs 7 1/2 lbs.; European hardwood thumbhole stock; post front sight, open adjustable rear; BATF-approved version of the gas-operated Kalashnikov rifle; black phosphate finish; comes with one magazine, cleaning rod, cleaning/tool kit. Imported from Hungary by K.B.I., Inc. Introduced 1995; dropped 1995.

Perf.: $495   **Exc.:** $420   **VGood:** $360

KDF K15

Bolt action; 25-06, 257 Wea. Mag., 270, 270 Wea. Mag., 7mm Rem. Mag., 30-06, 300 Win. Mag., 300 Wea. Mag., 338 Win. Mag., 340 Wea. Mag., 375 H&H, 411 KDF Mag., 416 Rem. Mag., 458 Win. Mag.; 4-shot magazine (standard calibers), 3-shot (Magnums); 22" barrel (standard calibers), 24" optional; 44" overall (24" barrel); weighs about 8 lbs.; sights optional; drilled and tapped for scope mounting; laminated stock standard; Kevlar composite or AAA walnut in Monte Carlo, classic, Schnabel or thumbhole styles optional; three-lug locking design with 60° bolt lift; ultra-fast lock time; fully-adjustable trigger. Imported from Germany by KDF, Inc. Introduced 1976; no longer imported.

Perf.: $1695   **Exc.:** $1395   **VGood:** $950

KDF K15 Dangerous Game Model

Same specs as K15 except 411 KDF Mag.; choice of iron sights or scope mounts; oil-finished deluxe walnut stock; gloss blue, matte blue, Parkerized or electroless nickel finish; hinged floorplate. Introduced 1986; discontinued 1988.

Perf.: $1795   **Exc.:** $1495   **VGood:** $995

KDF K15 Fiberstock Pro-Hunter Model

Same specs as K15 except Brown Precision fiberglass stock in black, green, brown or camo with wrinkle finish; Parkerized, matte blue or electroless finish; recoil arrestor. Introduced 1986; discontinued 1988.

Perf.: $1750   **Exc.:** $1450   **VGood:** $975

KDF K15 Improved Model

Same specs as K15 except 243, 25-06, 270, 7x57, 308, 30-06, 257 Weatherby, 270 Weatherby, 308 Norma Mag., 375 H&H Mag.; hand-checkered, oil-finished featherweight European walnut stock with Schnabel or Monte Carlo style. Introduced 1987; discontinued 1988.

Perf.: $1795   **Exc.:** $1495   **VGood:** $995

KDF K15 SWAT Model

Same specs as K15 except 308; 24", 26" barrel; weighs 10 lbs.; oil-finished walnut target stock; Parkerized metal. Introduced 1986; discontinued 1988.

Perf.: $1650   **Exc.:** $1395   **VGood:** $895

Kimber Model 82, Varminter Model

Kimber Model 82

Kimber Model 82, Government Target Model

Kimber Model 82C Classic

Kimber Model 82C SuperAmerica

RIFLES

KDF
K22
K22 Deluxe

Kimber
Model 82
Model 82, All-American Match
Model 82, Continental
Model 82, Government
Target
Model 82, Hunter
Model 82, Mini Classic
Model 82, Sporter
Model 82, SuperAmerica
Model 82, Varminter
Model 82C Classic
Model 82C Custom Match
Model 82C SuperAmerica
Model 82C SuperAmerica
Custom
Model 82C SuperClassic

KDF K22
Bolt action; 22 LR, 22 WMR; 5-, 6-shot magazine; 21 1/2″ barrel; 40″ overall length; weighs 6 1/2 lbs.; no sights; receiver grooved for scope mounts; hand-checkered, oil-finished European walnut Monte Carlo stock; front-locking lugs on bolt; pillar bedding system. Made in Germany. Introduced 1984; discontinued 1988.
Perf.: $325 **Exc.:** $250 **VGood:** $200

KDF K22 Deluxe
Same specs as K22 except 22 LR only; quick-detachable swivels; rosewood forend tip; rubber recoil pad. Introduced 1984; discontinued 1988.
Perf.: $395 **Exc.:** $340 **VGood:** $270

KIMBER MODEL 82
Bolt action; 22 Short, 22 Long, 22 LR, 22 WMR, 22 Hornet; 5-shot magazine (22 Short, 22 Long, 22 LR), 4-shot (22 WMR), 3-shot (22 Hornet); 22″, 24″ barrel; weighs 6 1/2 lbs.; blade front sight on ramp, open adjustable rear; receiver grooved for special Kimber scope mounts; classic-style or Cascade-design select walnut stock; hand-checkered pistol grip, forend; rocker-type silent safety; checkered steel buttplate; steel grip cap; all-steel construction; blued finish. Introduced 1980; discontinued 1988.
Perf.: $850 **Exc.:** $695 **VGood:** $595

Kimber Model 82, All-American Match
Same specs as Model 82 except 22 LR; 25″ target barrel; weighs 9 lbs.; fully-adjustable stock; palm swell, thumb dent pistol grip; step-crowned .9″ diameter free-floating barrel; forend inletted for weights; adjustable trigger. Introduced 1990; discontinued 1991.
Perf.: $750 **Exc.:** $635 **VGood:** $495

Kimber Model 82, Continental
Same specs as Model 82 except deluxe full-length walnut stock. Introduced 1987; discontinued 1988.
Perf.: $1395 **Exc.:** $995 **VGood:** $695

Kimber Model 82, Government Target
Same specs as Model 82 except 22 LR; 25″ heavy target barrel; 43 1/2″ overall length; weighs 10 3/4 lbs.; oversize target-type Claro walnut stock; single-stage adjustable trigger; designed as U.S. Army trainers. Introduced 1987; discontinued 1991.
Perf.: $695 **Exc.:** $595 **VGood:** $495

Kimber Model 82, Hunter
Same specs as Model 82 except 22 LR; 22″ barrel; laminated stock; rubber recoil pad; matte finish. Introduced 1990; discontinued 1990.
Perf.: $750 **Exc.:** $635 **VGood:** $495

Kimber Model 82, Mini Classic
Same specs as Model 82 except 22 LR; 18″ barrel; sling swivels. Introduced 1988; discontinued 1988.
Perf.: $595 **Exc.:** $495 **VGood:** $395

Kimber Model 82, Sporter
Same specs as Model 82 except 22 LR; 22″ barrel; 4-shot magazine; round-top receiver. Introduced 1991; discontinued 1991.
Perf.: $895 **Exc.:** $795 **VGood:** $595

Kimber Model 82, SuperAmerica
Same specs as Model 82 except checkered top quality California Claro walnut stock; beaded cheekpiece; ebony forend tip; detachable scope mounts. Introduced 1982; discontinued 1988; reintroduced 1990; discontinued 1991.
Perf.: $1095 **Exc.:** $950 **VGood:** $795

Kimber Model 82, Varminter
Same specs as Model 82 except 22 LR; 5-, 10-shot magazine; 25″ heavy barrel; weighs 8 1/4 lbs.; laminated stock; rubber recoil pad. Introduced 1990; discontinued 1991.
Perf.: $795 **Exc.:** $695 **VGood:** $595

KIMBER MODEL 82C CLASSIC
Bolt action; 22 LR; 4-shot magazine; 21″ premium air-gauged barrel; 40 1/2″ overall length; weighs 6 1/2 lbs.; no sights; drilled and tapped for Warne scope mounts optionally available from factory; classic-style stock of Claro walnut; 13 1/2″ length of pull; hand-checkered; red rubber buttpad; polished steel grip cap; action with aluminum pillar bedding for consistent accuracy; single-set fully-adjustable trigger with 2 1/2-lb. pull. Made in U.S. by Kimber of America, Inc. Reintroduced 1994; still produced. **LMSR: $785**
Perf.: $695 **Exc.:** $595 **VGood:** $495

Kimber Model 82C Custom Match
Same specs as Model 82C Classic except high-grade stock French walnut stock with black ebony forend tip; full coverage 22 lpi borderless checkering; steel Neidner (uncheckered) buttplate; satin rust blue finish. Made in U.S. by Kimber of America, Inc. Reintroduced 1995; still produced. **LMSR: $1850**
Perf.: $1650 **Exc.:** $1295 **VGood:** $995

Kimber Model 82C SuperAmerica
Same specs as Model 82C except AAA fancy grade Claro walnut with beaded cheekpiece; ebony forend cap; hand-checkered 22 lpi patterns with wrap-around coverage; black rubber buttpad. Made in U.S. by Kimber of America, Inc. Reintroduced 1994; still produced. **LMSR: $1175**
Perf.: $995 **Exc.:** $895 **VGood:** $695

Kimber Model 82C SuperAmerica Custom
Same specs as SuperAmerica except Neidner-style buttplate. Available options include: steel skeleton grip cap and buttplate; quarter-rib and open express sights; jewelled bolt; checkered bolt knob; special length of pull; rust blue finish. Made in U.S. by Kimber of America, Inc. Reintroduced 1994; still produced. **LMSR: $1250**
Perf.: $1195 **Exc.:** $995 **VGood:** $795

Kimber Model 82C SuperClassic
Same specs as Model 82C except AAA Claro walnut stock with black rubber buttpad, as used on the SuperAmerica. Made in U.S. by Kimber of America, Inc. Introduced 1995; still produced. **LMSR: $1090**
Perf.: $995 **Exc.:** $895 **VGood:** $695

Kimber

Model 82C Varmint
Model 84
Model 84, Continental Model
Model 84, Custom Classic Model
Model 84, Hunter Model
Model 84, Sporter Model
Model 84, SuperAmerica
Model 84, Super Grade
Model 84, Ultra Varminter
Model 89
Model 89 African
Model 89, Featherweight Deluxe
Model 89, Heavyweight Deluxe
Model 89, Hunter
Model 89, Medium Weight Deluxe
Model 89, Super Grade

Kimber Model 84, SuperAmerica

Kimber Model 89, Medium Weight Deluxe

Kimber Model 89 African

Kleinguenther K-15

Kleinguenther K-22

Kimber Model 82C Varmint
Same specs as Model 82C except slightly larger forend to accommodate the medium/heavy barrel profile; fluted, stainless steel match-grade barrel; weighs about 7 1/2 lbs. Made in U.S. by Kimber of America, Inc. Introduced 1995; still produced.
LMSR: $885
Perf.: $695   **Exc.:** $595   **VGood:** $495

KIMBER MODEL 84
Bolt action; 17 Rem., 17 Mach IV, 6x45, 6x47, 5.6x50, 221 Fireball, 221 Rem., 222 Rem., 223 Rem.; 5-shot magazine; 22″ barrel; 40 1/2″ overall length; weighs 6 1/4 lbs.; hooded ramp front sight with bead, folding leaf rear are optional; grooved for scope mounts; Claro walnut plain straight comb stock with hand-cut borderless checkering; steel grip cap; checkered steel buttplate; new Mauser-type head locking action; steel trigger guard; three-position safety (new in '87); hinged floorplate; Mauser-type extractor; fully-adjustable trigger. Introduced 1984; discontinued 1989.
Perf.: $950   **Exc.:** $795   **VGood:** $595

Kimber Model 84, Continental Model
Same specs as Model 84 except 221 Fireball, 222 Rem., 223 Rem.; 22″ barrel; full-length deluxe checkered walnut stock. Introduced 1987; discontinued 1988.
Perf.: $1200   **Exc.:** $1095   **VGood:** $895

Kimber Model 84, Custom Classic Model
Same specs as Model 84 except select-grade Claro walnut stock with ebony forend tip, Neider-style buttplate. Introduced 1984; discontinued 1988.
Perf.: $995   **Exc.:** $895   **VGood:** $695

Kimber Model 84, Hunter Model
Same specs as Model 84 except 17 Rem., 222 Rem., 223 Rem.; 22″ barrel; laminated stock; matte finish. Introduced 1990; discontinued 1990.
Perf.: $825   **Exc.:** $750   **VGood:** $595

Kimber Model 84, Sporter Model
Same specs as Model 84 except 17 Rem., 22 Hornet, 222 Rem., 22-250, 223 Rem., 250 Savage, 35 Rem.; 4-shot magazine; hand-checkered Claro walnut. Introduced 1991; discontinued 1991.
Perf.: $950   **Exc.:** $795   **VGood:** $695

Kimber Model 84, SuperAmerica
Same specs as Model 84 except 17 Rem., 221 Rem., 22 Hornet, 222 Rem., 223 Rem., 250 Savage, 35 Rem.; 4-shot magazine; 22″ barrel; detachable scope mounts; top-quality stock; right- or left-hand versions. Introduced 1985; discontinued 1988; reintroduced 1990; discontinued 1991.
Perf.: $1425   **Exc.:** $1095   **VGood:** $895

Kimber Model 84, Super Grade
Same specs as Model 84 except 22 LR, 17 Rem., 222 Rem., 223 Rem.; Mauser bolt action; 22″ barrel; grade AAA walnut stock. Introduced 1989; discontinued 1989.
Perf.: $1300   **Exc.:** $1095   **VGood:** $895

Kimber Model 84, Ultra Varminter
Same specs as Model 84 except 17 Rem., 22 Hornet, 221 Rem., 222 Rem., 22-250, 223 Rem.; 24″ barrel; weighs 7 3/4 lbs.; laminated birchwood stock. Introduced 1989; discontinued 1991.
Perf.: $1295   **Exc.:** $995   **VGood:** $795

KIMBER MODEL 89
Bolt action; 257 Roberts, 25-06, 7x57, 270, 280, 7mm Rem. Mag., 30-06, 300 Win. Mag., 300 H&H, 35 Whelen, 338 Win. Mag., 375 H&H Mag.; 3-, 5-shot magazine; 22″, 24″ barrel; 42″ overall length (22″ barrel); weighs 7 3/4 lbs.; no sights; Claro or English walnut stock in Classic, Custom Classic design; Model 70-type trigger, ejector design; Mauser-type extractor; three-position safety. Introduced 1988; discontinued 1989.
Perf.: $795   **Exc.:** $695   **VGood:** $595

Kimber Model 89 African
Same specs as Model 89 except 375 H&H Mag., 404 Jeffery, 416 Rigby, 460 Weatherby Mag., 505 Gibbs; 5-shot magazine (375 H&H Mag.), 4-shot (404 Jeffery, 416 Rigby), 3-shot (460 Weatherby, 505 Gibbs); 26″ heavy barrel; 47″ overall length; weighs 10 1/2 lbs.; ramped blade front, express rear sight; English walnut stock with cheekpiece; rubber buttpad; double crossbolts. Introduced 1990; discontinued 1991.
Perf.: $4100   **Exc.:** $3695   **VGood:** $2595

Kimber Model 89, Featherweight Deluxe
Same specs as Model 89 except 257 Roberts, 25-06, 7x57, 270, 280, 30-06; 5-shot magazine; 22″ light barrel; weighs 7 1/2 lbs.; deluxe walnut stock. Introduced 1989; discontinued 1990.
Perf.: $1495   **Exc.:** $1095   **VGood:** $895

Kimber Model 89, Heavyweight Deluxe
Same specs as Model 89 except 375 H&H Mag., 458 Win. Mag.; 3-shot magazine; 24″ heavy barrel; weighs 9 lbs. Introduced 1989; discontinued 1990.
Perf.: $1695   **Exc.:** $1295   **VGood:** $995

Kimber Model 89, Hunter
Same specs as Model 89 except 270, 30-06, 300 Win. Mag., 338 Win. Mag., 7mm Rem. Mag.; 22″ barrel; laminated stock; matte finish. Introduced 1990; discontinued 1991.
Perf.: $1195   **Exc.:** $995   **VGood:** $750

Kimber Model 89, Medium Weight Deluxe
Same specs as Model 89 except 300 Win. Mag., 300 H&H Mag., 300 Weatherby, 338 Win. Mag., 35 Whelen, 7mm Rem. Mag.; 3-shot magazine; 24″ barrel; weighs 8 lbs. Introduced 1989; discontinued 1990.
Perf.: $1595   **Exc.:** $1095   **VGood:** $795

Kimber Model 89, Super Grade
Same specs as Model 89 except 22″ barrel; grade AAA walnut stock; square receiver. Introduced 1989; discontinued 1989.
Perf.: $1495   **Exc.:** $1195   **VGood:** $895

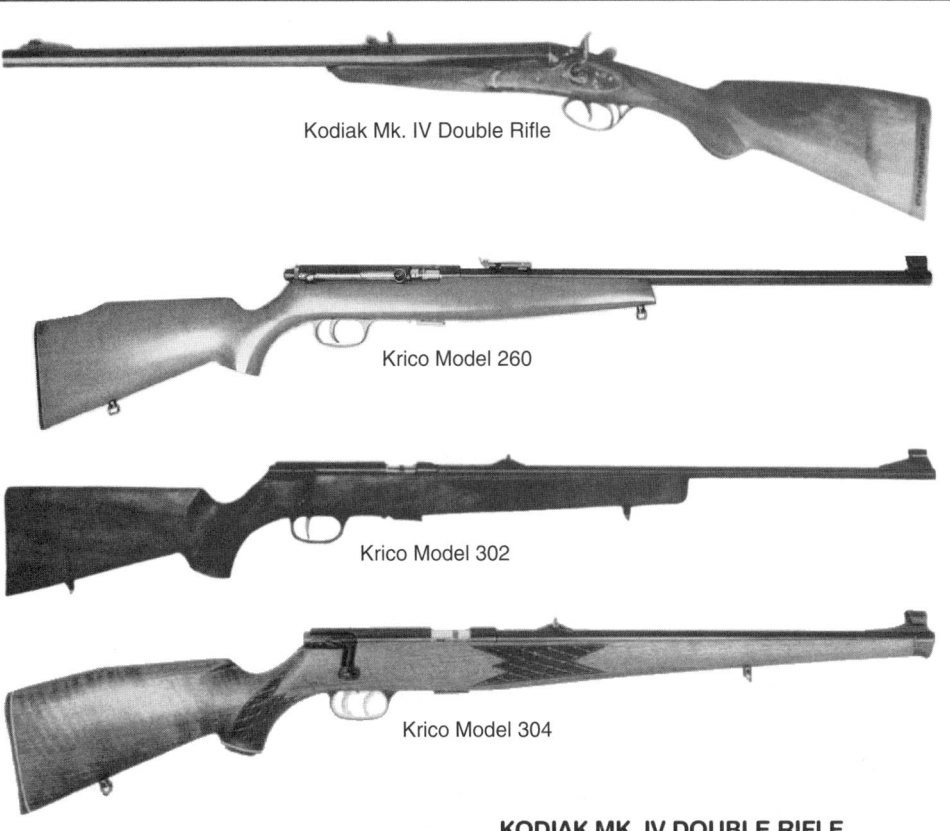

Kodiak Mk. IV Double Rifle

Krico Model 260

Krico Model 302

Krico Model 304

RIFLES

Kimel
AR9

Kintrek
Model KBP-1

Kleinguenther
K-15
K-22

Kodiak
Mk. IV Double Rifle

Krico
Carbine
Model 120M
Model 260
Model 300
Model 300 Deluxe
Model 300 SA
Model 300 Stutzen
Model 302
Model 302DR C
Model 302E
Model 304

KIMEL AR9

Semi-automatic; 9mm Para.; 20-shot magazine; 16 1/4″ barrel; 33″ overall length; weighs 6 1/2 lbs.; adjustable post front sight in ring, fixed open rear; folding butt-stock, checkered plastic grip; fires from closed bolt; lever safety blocks trigger and sear; vented barrel shroud; matte blue/black or nickel finish. Made in U.S. Marketed by Kimel Industries. Introduced 1991; no longer produced.
Perf.: $410   **Exc.:** $350   **VGood:** $300

KINTREK MODEL KBP-1

Semi-automatic; 22 LR; 17-shot magazine; 25″ barrel; 31 1/2″ overall length; weighs 5 1/2 lbs.; post front sight, fully-adjustable aperture rear; drilled and tapped for scope mount; solid black synthetic bullpup stock with smooth pebble finish; grip safety and trigger-blocking safeties; ejects empties out through bottom of stock; bolt hold-open operated by grip safety; matte black finish. Made in U.S. by Kintrek, Inc. Introduced 1991; discontinued 1992.
Perf.: $175   **Exc.:** $150   **VGood:** $130

KLEINGUENTHER K-15

Bolt action; 243, 25-06, 270, 30-06, 308 Win., 7x57, 308 Norma Mag., 300 Weatherby Mag., 7mm Rem. Mag., 375 H&H, 257 Winchester, 270 Weatherby Mag., 300 Weatherby Mag.; 24″ barrel (standard calibers); 26″ (Magnums); no sights; drilled, tapped for scope mounts; hand-checkered Monte Carlo stock of European walnut; rosewood grip cap; high-luster or satin finish; recoil pad; many optional features available on special order. Manufactured in Germany. Imported and assembled by Kleinguenther; discontinued 1985.
Perf.: $1295   **Exc.:** $995   **VGood:** $750

KLEINGUENTHER K-22

Bolt action; 22 LR, 22 WMR; 5-shot magazine; 21 1/2″ barrel; 40″ overall length; weighs 6 1/2 lbs.; no sights; drilled, tapped for scope mounts; walnut-stained beech-wood Monte Carlo stock; hand-cut checkering; sling swivels; no sights; two forward locking lugs; adjustable trigger; optional set trigger; silent safety. Made in Germany. Introduced 1984; discontinued 1985.
Perf.: $350   **Exc.:** $290   **VGood:** $200
With double-set trigger.
Perf.: $395   **Exc.:** $320   **VGood:** $220

KODIAK MK. IV DOUBLE RIFLE

Double rifle; 45-70; 24″ barrels; 42 1/2″ overall length; weighs 10 lbs.; ramp front sight with bead, adjustable two-leaf rear; European walnut stock with semi-pistol grip; exposed hammers; color case-hardened locks; rubber recoil pad. Imported from Italy by Trail Guns Armory. Introduced 1988; discontinued 1994.
Perf.: $995   **Exc.:** $895   **VGood:** $750

KRICO CARBINE

Bolt action; 22 Hornet, 222 Rem; 4-shot clip magazine; 20″, 22″ barrel; open rear, hooded ramp front sight; hand-checkered European walnut full-length stock; miniature Mauser-type action; single- or double-set trigger; sling swivels. Made in Germany. Introduced 1956; discontinued 1962.
Perf.: $695   **Exc.:** $595   **VGood:** $450

KRICO MODEL 120M

Bolt action; 22 LR; 5-shot magazine; 19 1/2″ barrel; 46″ overall length; weighs 6 lbs.; military trainer; tangent elevation-adjustable rear; hooded blade front; receiver grooved for scope mounts; European hardwood stock; blued finish; adjustable trigger. Made in Germany. Introduced No longer imported.
Perf.: $350   **Exc.:** $290   **VGood:** $220

KRICO MODEL 260

Semi-automatic; 22 LR; 5-shot magazine; 19 1/2″ barrel; 39″ overall length; weighs 6 3/8 lbs.; ramp blade front sight, open adjustable rear; receiver grooved for scope mounting; beechwood stock; sliding safety. Made in Germany. Introduced 1989; still imported. **LMSR:** $700
Perf.: $495   **Exc.:** $395   **VGood:** $285

KRICO MODEL 300

Bolt action; 22 LR, 22 WMR, 22 Hornet; 19 1/2″, 23 1/2″ barrel; weighs 6 lbs.; ramp blade front sight, open adjustable rear; walnut-stained beech stock; double triggers; sliding safety; checkered grip, forend. Made in Germany. Introduced 1989; still imported. **LMSR:** $700
Perf.: $550   **Exc.:** $435   **VGood:** $325

Krico Model 300 Deluxe

Same specs as Model 300 except European walnut stock; better checkering. Introduced 1989; still imported. **LMSR:** $795
Perf.: $625   **Exc.:** $495   **VGood:** $395

Krico Model 300 SA

Same specs as Model 300 except Monte Carlo-style walnut stock. Introduced 1989; still imported. **LMSR:** $750
Perf.: $575   **Exc.:** $450   **VGood:** $350

Krico Model 300 Stutzen

Same specs as Model 300 except full-length Stutzen-type walnut stock. Introduced 1989; still imported. **LMSR:** $825
Perf.: $595   **Exc.:** $495   **VGood:** $395

KRICO MODEL 302

Bolt action; 22 LR; 5-, 10-shot magazine; 24″ barrel; 43″ overall length; weighs 6 1/2 lbs.; hooded post front, windage-adjustable rear sight; European walnut stock; checkered pistol grip, forend; single- or double-set trigger. Made in Germany. Introduced 1982; discontinued 1984.
Perf.: $575   **Exc.:** $450   **VGood:** $350
With double-set trigger.
Perf.: $595   **Exc.:** $495   **VGood:** $395

Krico Model 302DR C

Same specs as Model 302 except 22 LR, 22 WMR; classic-style European walnut stock. Introduced 1983; discontinued 1984.
Perf.: $595   **Exc.:** $495   **VGood:** $395

Krico Model 302E

Same specs as Model 302 except straight forend; walnut-finished hardwood stock; single trigger. Introduced 1982; discontinued 1983.
Perf.: $595   **Exc.:** $495   **VGood:** $395

KRICO MODEL 304

Bolt action; 22 LR; 5-, 10-shot magazine; 20″ barrel; weighs 6 1/4 lbs.; hooded ramp front, windage-adjustable rear sight; full-length Mannlicher-type stock; single- or double-set trigger. Made in Germany. Introduced 1982; discontinued 1984.
Perf.: $595   **Exc.:** $495   **VGood:** $395

RIFLES

Krico
Model 311
Model 320
Model 330S Match
Model 340
Model 340 Kricotronic
Model 340 Mini-Sniper
Model 352E
Model 360S Biathlon
Model 360S2 Biathlon
Model 400
Model 400E
Model 400L
Model 400 Match
Model 420L
Model 500 Kricotronic Match

Krico Model 320

Krico Model 340 Mini-Sniper

Krico Model 360S Biathlon

Krico Model 400

Krico Model 340 Match

KRICO MODEL 311
Bolt action; 22 LR; 5-, 10-shot clip magazine; 22″ barrel; open rear sight, hooded ramp front; hand-checkered European walnut stock, pistol grip, cheekpiece; single- or double-set trigger; sling swivels. Also available with 2 1/2X scope. Introduced 1958; discontinued 1962.
Perf.: $350   **Exc.:** $300   **VGood:** $225
With scope.
Perf.: $440   **Exc.:** $280   **VGood:** $290

KRICO MODEL 320
Bolt action; 22 LR; 5-shot magazine; 19 1/2″ barrel; 38 1/2″ overall length; weighs 6 lbs.; windage-adjustable open rear sight, blade front on ramp; Mannlicher-style select European walnut stock; cut-checkered grip, forend; blued steel forend cap; detachable box magazine; single- or double-set triggers. Made in Germany. Introduced 1986; discontinued 1988.
Perf.: $625   **Exc.:** $550   **VGood:** $450

KRICO MODEL 330S MATCH
Single shot; 22 LR; 25 1/2″ heavy barrel; hooded front sight with interchangeable inserts, diopter match rear, rubber eye-cup; walnut-finished beechwood match stock; built-in hand stop; adjustable recoil pad; factory-set match trigger; stippled pistol grip. Introduced 1981; no longer imported.
Perf.: $695   **Exc.:** $595   **VGood:** $450

KRICO MODEL 340
Bolt action; 22 LR; 5-shot clip magazine; 21″ match barrel; 39 1/2″ overall length; weighs 7 1/2 lbs.; no sights; receiver grooved for tip-off mounts; European walnut match-style stock; stippled grip, forend; free-floating barrel; adjustable two-stage match trigger or double-set trigger; meets NRA MS rules. Made in Germany. Introduced 1983; discontinued 1988.
Perf.: $695   **Exc.:** $595   **VGood:** $450

Krico Model 340 Kricotronic
Same specs as Model 340 except electronic ignition system, replacing firing pin; fast lock time. Made in Germany. Introduced 1985; discontinued 1988.
Perf.: $1200   **Exc.:** $900   **VGood:** $750

Krico Model 340 Mini-Sniper
Same specs as Model 340 except weighs 8 lbs.; 40″ overall length; high comb; palm swell, stippled ventilated forend; receiver grooved for scope mounts; free-floating bull barrel; muzzlebrake; large bolt knob; match-quality single trigger; sandblasted barrel, receiver. Made in Germany. Introduced 1984; discontinued 1986.
Perf.: $995   **Exc.:** $795   **VGood:** $575

KRICO MODEL 352E
Bolt action; 22 WMR; 24″ barrel; 43″ overall length; weighs 6 1/2 lbs.; hooded post front, windage-adjustable rear sight; walnut finished hard stock; wood straight forend; no white line spaces. Introduced 1982; discontinued 1984.
Perf.: $375   **Exc.:** $325   **VGood:** $235

KRICO MODEL 360S BIATHLON
Bolt action; 22 LR; 5-shot magazine; 21 1/4″ barrel; 40 1/2″ overall length; weighs 9 1/4 lbs.; globe front, fully-adjustable Diana 62 match peep rear sight; high-comb European walnut stock; adjustable buttplate; muzzle/sight snow cap; straight-pull action; match trigger. Marketed with five magazines, four stored in stock. Made in Germany. Introduced 1991; still imported.
LMSR: $1695
Perf.: $1295   **Exc.:** $995   **VGood:** $695

Krico Model 360S2 Biathlon
Same specs as Model 360S except weighs 9 lbs.; biathlon stock design of black epoxy-finished walnut with pistol grip; pistol-grip-activated action. Imported from Germany by Mandall Shooting Supplies. Introduced 1991; no longer imported.
Perf.: $1250   **Exc.:** $895   **VGood:** $650

KRICO MODEL 400
Bolt action; 22 LR, 22 Hornet; 5-shot detachable magazine; 23 1/2″ barrel; 41″ overall length; weighs 8 1/2 lbs.; hooded post front, windage-adjustable open rear sight; hand-checkered deluxe European walnut stock; Schnabel forend; solid rubber recoil pad; rear locking lugs; twin extractors; sling swivels; single- or double-set triggers. Introduced 1983; discontinued 1989.
Perf.: $795   **Exc.:** $695   **VGood:** $595

Krico Model 400E
Same specs as Model 400 except straight forend; walnut-finished beech stock; no forend tip; hard rubber buttplate; single- or double-set trigger. Introduced 1982; discontinued 1983.
Perf.: $595   **Exc.:** $495   **VGood:** $375

Krico Model 400L
Same specs as Model 400 except select French walnut stock; no Schnabel. Introduced 1982; discontinued 1983.
Perf.: $795   **Exc.:** $695   **VGood:** $585

Krico Model 400 Match
Same specs as Model 400 except single shot; 22 LR; 23 1/2″ heavy barrel; no sights; match-type stock. Introduced Still in production. **LMSR: $950**
Perf.: $825   **Exc.:** $695   **VGood:** $495

KRICO MODEL 420L
Bolt action; 22 Hornet; 5-shot magazine; 20″ barrel; weighs 6 1/2 lbs.; hooded post front, windage-adjustable open rear sight; full-length walnut stock; solid rubber buttpad; rear locking lugs; twin extractors; sling swivels; single- or double-set trigger. Introduced 1982; discontinued 1983.
Perf.: $850   **Exc.:** $725   **VGood:** $525

KRICO MODEL 500 KRICOTRONIC MATCH
Bolt action; single shot; 22 LR; 23 1/2″ barrel; 45 1/5″ overall length; weighs 9 1/5 lbs.; globe front, match-type rear sight; match-type European walnut stock; adjustable butt; electronic ignition system; adjustable trigger. Made in Germany. Still imported. **LMSR: $3950**
Perf.: $3700   **Exc.:** $2695   **VGood:** $1695

Krico Model 600 EAC

Krico Model 620

Krico Model 640S

Krico Model 650S

Krico Model 700

KRICO MODEL 600

Bolt action; 17 Rem., 22-250, 222, 222 Rem. Mag., 223, 243, 308, 5.6x50 Mag.; 3-, 4-shot magazine; 23 1/2″ barrel; 43 3/4″ overall length; weighs 8 lbs.; hooded ramp front, fixed rear sight; tangent rear sight optional; hand-checkered European walnut Monte Carlo stock; Schnabel forend; classic American-style stock also available; adjustable single- or double-set trigger, silent safety; double front locking lugs. Introduced 1983; still imported. **LMSR: $1295**
 Perf.: $995 **Exc.:** $750 **VGood:** $650

Krico Model 600 EAC

Same specs as Model 600 except American-style classic stock; sling swivels. Introduced 1983; discontinued 1984.
 Perf.: $995 **Exc.:** $795 **VGood:** $595

Krico Model 600 Deluxe

Same specs as Model 600 except traditional European-style select fancy walnut stock; rosewood Schnabel forend tip; Bavarian cheekpiece; fine checkering; front sling swivel attaches to barrel; butterknife bolt handle; gold-plated single-set trigger. Introduced 1983; discontinued 1984.
 Perf.: $1095 **Exc.:** $795 **VGood:** $625

Krico Model 600 Match

Same specs as Model 600 except no sights; drilled, tapped for scope mounts; cheekpiece match stock; vented forend; rubber recoil pad. Introduced 1983; still imported. **LMSR: $1250**
 Perf.: $995 **Exc.:** $895 **VGood:** $750

Krico Model 600 Single Shot

Same specs as Model 600 except match barrel; no sights; drilled, tapped for scope mounts; adjustable buttplate; match-type stock; flash hider; Parkerized finish; large bolt knob; wide trigger shoe. Made in Germany. Introduced 1983; still imported. **LMSR: $1295**
 Perf.: $995 **Exc.:** $895 **VGood:** $750

Krico Model 600 Sniper

Same specs as Model 600 except match barrel; no sights; drilled, tapped for scope mounts; adjustable buttplate; flash hider; Parkerized finish; large bolt knob; wide trigger shoe. Made in Germany. Introduced 1983; still imported. **LMSR: $2645**
 Perf.: $1395 **Exc.:** $995 **VGood:** $795

KRICO MODEL 620

Bolt action; 222 Rem., 223 Rem., 22-250, 243, 308, 5.6x50 Mag.; 3-shot magazine; 20 3/4″ barrel; weighs 6 3/4 lbs.; hooded ramp front, fixed rear sight; tangent rear sight optional; drilled, tapped for scope mounts; full-length walnut stock; double-set trigger. Introduced 1988; discontinued 1988.
 Perf.: $1165 **Exc.:** $850 **VGood:** $710

KRICO MODEL 640S

Bolt action; 222, 223, 22-250, 308; 5-shot magazine; 20″ semi-bull barrel; no sights; pistol-grip stock of French walnut; ventilated forend; single- or double-set triggers. Introduced 1981; discontinued 1988.
 Perf.: $1295 **Exc.:** $995 **VGood:** $750

KRICO MODEL 640 VARMINT

Bolt action; 22-250, 222 Rem., 223 Rem.; 4-shot magazine; 23 3/4″ barrel; weighs 9 3/8 lbs.; no sights; drilled, tapped for scope mounts; cut-checkered select European walnut stock; high Monte Carlo comb; Wundhammer palm swell; rosewood forend tip. Made in Germany. Importation discontinued 1988.
 Perf.: $825 **Exc.:** $695 **VGood:** $495

KRICO MODEL 650S

Bolt action; 222, 223, 308; 3-shot magazine; 23″ bull barrel; muzzlebrake/flash hider; no sights; drilled, tapped for scope mounts; oil-finished select European walnut stock; adjustable cheekpiece, recoil pad; match trigger; single- or double-set trigger available; all metal; matte blue finish. Introduced 1981; discontinued 1988.
 Perf.: $1395 **Exc.:** $995 **VGood:** $795

Krico Model 650S2

Same specs as Model 650 except special benchrest stock of French walnut. Introduced 1981; discontinued 1983.
 Perf.: $1395 **Exc.:** $1095 **VGood:** $825

Krico Model 650 Super Sniper

Same specs as Model 650 except 223, 308; 26″ bull barrel. Made in Germany. Introduced 1981; discontinued 1988.
 Perf.: $1495 **Exc.:** $1095 **VGood:** $825

RIFLES

Krico
Model 600
Model 600 EAC
Model 600 Deluxe
Model 600 Match
Model 600 Single Shot
Model 600 Sniper
Model 620
Model 640S
Model 640 Varmint
Model 650S
Model 650S2
Model 650 Super Sniper
Model 700
Model 700 Deluxe
Model 700 Stutzen
Model 720

KRICO MODEL 700

Bolt action; 17 Rem., 222, 222 Rem. Mag., 223, 5.6x50 Mag., 243, 308, 5.6x57 RWS, 22-250, 6.5x55, 6.5x57, 7x57, 7x64, 30-06, 9.3x62, 6.5x68, 7mm Rem. Mag., 300 Win. Mag., 8x68S, 7.5 Swiss, 9.3x64, 6x62 Freres; 23 1/2″ barrel (standard calibers), 25 1/2″ (magnums); 43 3/8″ overall length (23 1/2″ barrel); weighs about 7 lbs.; blade on ramp front sight, open adjustable rear; drilled, tapped for scope mounting; European walnut stock with Bavarian cheekpiece; removable box magazine; sliding safety. Imported from Germany by Mandall Shooting Supplies. Still imported. **LMSR: $1249**
 Perf.: $925 **Exc.:** $795 **VGood:** $595

Krico Model 700 Deluxe

Same specs as Model 700 except better grade of wood, checkering. Imported from Germany by Mandall Shooting Supplies. Still imported. **LMSR: $1379**
 Perf.: $1050 **Exc.:** $875 **VGood:** $725

Krico Model 700 Stutzen

Same specs as Model 700 except full-length stock. Imported from Germany by Mandall Shooting Supplies. Still imported. **LMSR: $1450**
 Perf.: $1195 **Exc.:** $895 **VGood:** $695

KRICO MODEL 720

Bolt action; 17 Rem., 222, 222 Rem. Mag., 223, 5.6x50 Mag., 243, 308, 5.6x57 RWS, 22-250, 6.5x55, 6.5x57, 7x57, 270, 7x64, 30-06, 9.3x62, 6.5x68, 7mm Rem. Mag., 300 Win. Mag., 8x68S, 7.5 Swiss, 9.3x64, 6x62 Freres; 3-shot magazine; 20 3/4″ barrel; weighs 6 3/4 lbs.; hooded ramp front, fixed rear sight; tangent rear sight optional; drilled, tapped for scope mounting; full-length walnut stock; double-set trigger. Introduced 1983; discontinued 1990.
 Perf.: $1095 **Exc.:** $925 **VGood:** $695

RIFLES

Krico
Sporter
Varmint Model

Krieghoff
Neptun
Teck Combo
Teck Double Rifle
Trumpf
Ulm Combo
Ulm Double Rifle
Ulm-Primus

Lakefield Arms
Mark I
Mark II

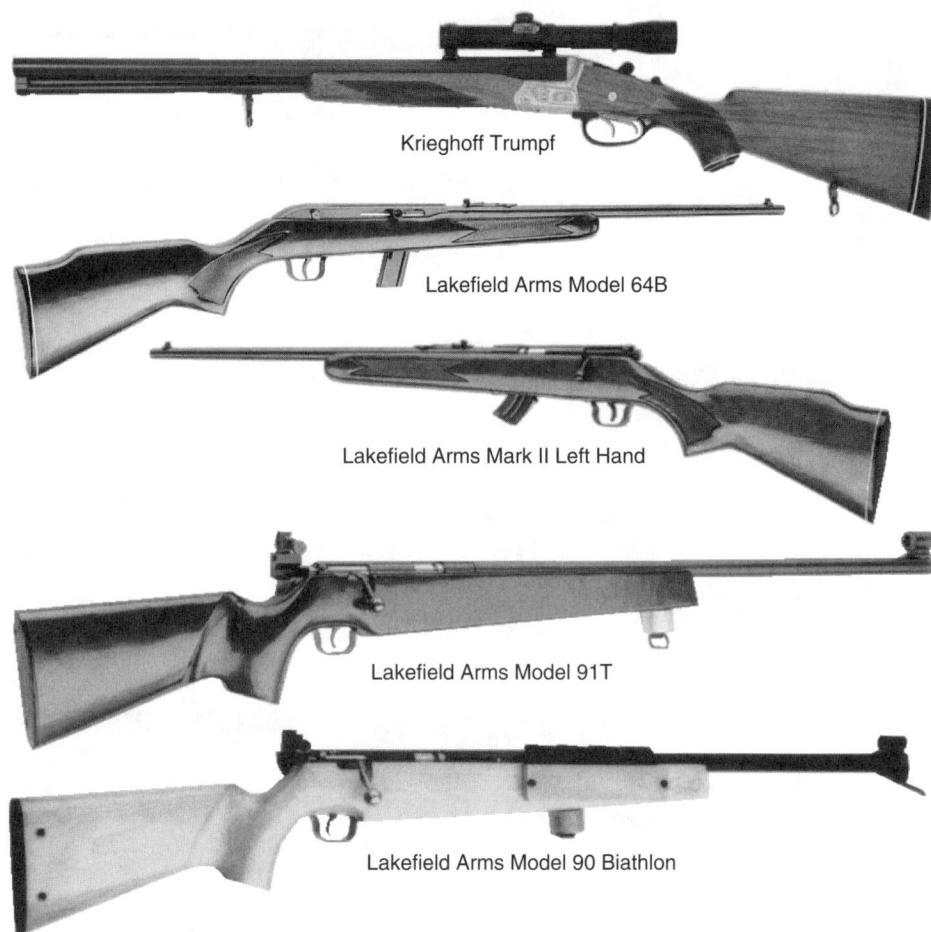

Krieghoff Trumpf

Lakefield Arms Model 64B

Lakefield Arms Mark II Left Hand

Lakefield Arms Model 91T

Lakefield Arms Model 90 Biathlon

KRICO SPORTER
Bolt action; 22 Hornet, 222 Rem.; 4-shot clip magazine; 22″, 24″, 26″ barrel; open rear, hooded ramp front sight; hand-checkered European walnut stock, cheekpiece, pistol grip, forend tip; miniature Mauser-type action; single- or double-set trigger; sling swivels. Made in Germany. Introduced 1956; discontinued 1962.
Perf.: $595 **Exc.:** $495 **VGood:** $375

KRICO VARMINT MODEL
Bolt action; 222 Rem.; 4-shot magazine; 22″, 24″, 26″ heavy barrel; no sights. Made in Germany. Introduced 1956; discontinued 1962.
Perf.: $625 **Exc.:** $525 **VGood:** $395

KRIEGHOFF NEPTUN
Drilling; 12-, 16-, 20-ga./22 Hornet, 222 Rem., 243, 270, 30-06, 308; standard European calibers also available; 25″ barrels; shot barrels choked Imp. Mod. & Full; optional free-floating rifle barrel available; weighs about 7 1/2 lbs.; bead front sight, automatic pop-up open rear; hand-checkered European walnut stock with German-style grip and cheekpiece, oil finish; full sidelock action with double or optional single trigger, top tang shotgun safety; fine, light scroll engraving. Imported from Germany by Krieghoff International, Inc. No longer imported.
Perf.: $12,900 **Exc.:** $9995 **VGood:** $6500

KRIEGHOFF TECK COMBO
Combination gun; over/under; 12-, 16-, 20-ga./22 Hornet, 222, 243, 270, 30-06, 308 and standard European calibers; O/U rifle also available in 458 Win. on special order; 25″ barrels on double rifle combo, 28″ on O/U shotgun; optional free-floating rifle barrel; weighs about 7-7 1/2 lbs.; white bead front sight on shotgun, open or folding on rifle or combo; hand-checkered European walnut stock with German-style grip and cheekpiece; boxlock action with non-selective single trigger or optional single/double trigger; Greener cross-bolt; ejectors standard on all but O/U rifle; top tang safety; light

scroll engraving. Imported from Germany by Krieghoff International, Inc. No longer imported.
Perf.: $6295 **Exc.:** $5000 **VGood:** $2995

KRIEGHOFF TECK DOUBLE RIFLE
Double rifle; boxlock; over/under; 7x57R, 7x64, 7x65R, 308 Win., 30-06, 300 Win. Mag., 8x57JRS, 9.3x74R, 375 H&H Mag., 458 Win. Mag.; 25″ barrel; Express rear, ramp front sight; hand-checkered European walnut stock, forearm; double crossbolt; double underlugs. Introduced 1967; No longer imported.
Perf.: $7995 **Exc.:** $6250 **VGood:** $3695

KRIEGHOFF TRUMPF
Drilling; 12-, 16-, 20-ga./22 Hornet, 222 Rem., 243, 270, 30-06, 308; standard European calibers also available; 25″ barrels; shot barrels choked Imp. Mod. & Full; optional free-floating rifle barrel; weighs about 7 1/2 lbs.; bead front sight, automatic pop-up open rear; hand-checkered European walnut stock with German-style grip and cheekpiece; oil finish; boxlock action with double or optional single trigger; top tang shotgun safety; fine, light scroll engraving. Imported from Germany by Krieghoff International, Inc. No longer imported.
Perf.: $8100 **Exc.:** $6250 **VGood:** $3695

KRIEGHOFF ULM COMBO
Combination gun; over/under; 12-, 16-, 20-ga./22 Hornet, 222, 243, 270, 30-06, 308 and standard European calibers; O/U rifle also available in 458 Win. on special order; 25″ barrels (double rifle combo), 28″ (O/U shotgun); optional free-floating rifle barrel; weighs about 7-7 1/2 lbs.; white bead front sight on shotgun, open or folding on rifle or combo; hand-checkered European walnut stock with German-style grip and cheekpiece; full sidelock action with non-selective single trigger or optional single/double trigger; Greener cross-bolt; ejectors standard on all but O/U rifle; top tang safety; light scroll engraving. Imported from Germany by Krieghoff International, Inc. No longer imported.
Perf.: $14,000 **Exc.:** $8500 **VGood:** $5995

KRIEGHOFF ULM DOUBLE RIFLE
Double rifle; boxlock; over/under; 7x57R, 7x64, 7x65R, 308 Win., 30-06, 300 Win. Mag., 8x57JRS, 9.3x74R, 375 H&H Mag., 458 Win. Mag.; 25″ barrel; Express rear, ramp front sight; hand-checkered European walnut stock, forearm; double cross-bolt; double underlugs; leaf arabesque engraving. Introduced 1963; still imported.
Perf.: $12,995 **Exc.:** $9950 **VGood:** $6995

Krieghoff Ulm-Primus
Same specs as Ulm except deluxe modifications including higher grade of engraving, wood; detachable side locks. Introduced 1963; discontinued 1983.
Perf.: $18,500 **Exc.:** $12,500 **VGood:** $7250

LAKEFIELD ARMS MARK I
Bolt action; single shot; 22 LR; 20 1/2″ barrel; 39 1/2″ overall length; weighs 5 1/2 lbs.; bead front, open adjustable rear sight; receiver grooved for scope mounts; checkered Monte Carlo-type stock of walnut-finished hardwood; thumb-activated rotating safety; right-, left-hand variations; youth model available with smaller dimensions; blued finish. Made in Canada. Introduced 1990; still in production. **LMSR:** $135
Perf.: $100 **Exc.:** $75 **VGood:** $60

LAKEFIELD ARMS MARK II
Bolt-action; 22 LR; 10-shot magazine; 20 1/2″ barrel; 39 1/2″ overall length; weighs 5 1/2 lbs.; bead front, open adjustable rear sight; receiver grooved for scope mounts; checkered Monte Carlo-type stock of walnut-finished hardwood; thumb-activated rotary safety; right-, left-hand variations; youth model available with smaller dimensions; blued finish. Made in Canada. Introduced 1990; still in production. **LMSR:** $140
Perf.: $110 **Exc.:** $80 **VGood:** $65

Lakefield Arms Model 92S Silhouette

Lakefield Arms Model 93M

L.A.R. Grizzly 50 Big Boar

Lebeau-Courally Sidelock

Magtech Model 122.2R

RIFLES

Lakefield Arms
Model 64B
Model 90 Biathlon
Model 91T
Model 91TR
Model 92S Silhouette
Model 93M

L.A.R.
Grizzly 50 Big Boar

Lebeau-Courally
Boxlock
Sidelock

Ljutic
Space Gun

Magtech
Model 122.2R
Model 122.2S
Model 122.2T
Model MT-22C
Model MT-22T

LAKEFIELD ARMS MODEL 64B
Semi-automatic; 22 LR; 10-shot magazine; 20″ barrel; 40″ overall length; weighs 5 ¹/₂ lbs.; bead front, open adjustable rear sight; checkered Monte Carlo-type walnut stained hardwood stock; side ejection; bolt hold-open device; thumb-activated rotary safety; blued finish. Made in Canada. Introduced 1990; still in production. **LMSR: $143**
   Perf.: $110   **Exc.:** $90   **VGood:** $80

LAKEFIELD ARMS MODEL 90 BIATHLON
Bolt action; 22 LR; 5-shot magazine; 21″ barrel; 39 ¹/₂″ overall length; weighs 8 ¹/₄ lbs.; target front sight with inserts, adjustable peep rear; natural-finish hardwood stock; clip holder; carrying, shooting rails; hand stop; butthook; snow-cap muzzle protector; right-, left-hand versions. Marketed with five magazines. Made in Canada. Introduced 1991; discontinued 1995. **LMSR: $570**
   Perf.: $400   **Exc.:** $360   **VGood:** $290

LAKEFIELD ARMS MODEL 91T
Bolt action; single shot; 22 LR; 25″ barrel; 43 ¹/₂″ overall length; weighs 8 lbs.; target front sight with inserts; click-adjustable peep rear; target-type walnut-finished hardwood stock; shooting rail; hand stop. Made in Canada. Introduced 1991; discontinued 1995. **LMSR: $455**
   Perf.: $330   **Exc.:** $290   **VGood:** $220

Lakefield Arms Model 91TR
Same specs as Model 91T except repeater with 5-shot magazine. Made in Canada. Introduced 1991; discontinued 1995. **LMSR: $485**
   Perf.: $350   **Exc.:** $300   **VGood:** $230

LAKEFIELD ARMS MODEL 92S SILHOUETTE
Bolt action; 22 LR; 5-shot magazine; 21″ barrel; 39 ¹/₂″ overall length; weighs 8 lbs.; no sights; receiver drilled, tapped for scope base; high-comb target-type stock of walnut-finished hardwood; clip holder; carrying, shooting rails; hand stop; butthook; snow-cap muzzle protector; right-, left-hand versions. Made in Canada by Lakefield Arms. Introduced 1992; discontinued 1995. **LMSR: $388**
   Perf.: $300   **Exc.:** $270   **VGood:** $200
Left-hand model.
   Perf.: $320   **Exc.:** $280   **VGood:** $210

LAKEFIELD ARMS MODEL 93M
Bolt action; 22 WMR; 5-shot magazine; 20 ³/₄″ barrel; 39 ¹/₂″ overall length; weighs 5 ³/₄ lbs.; bead front sight, adjustable open rear; receiver grooved for scope mount; walnut-finished hardwood stock with Monte Carlo-type comb; checkered grip and forend; thumb-operated rotary safety; blue finish. Made in Canada by Lakefield Arms Ltd. Introduced 1994; still produced. **LMSR: $168**
   Perf.: $130   **Exc.:** $100   **VGood:** $90

L.A.R. GRIZZLY 50 BIG BOAR
Bolt action; single shot; 50 BMG; 36″ barrel; 45 ¹/₂″ overall length; weighs 28 ¹/₂ lbs.; no sights; scope mount; integral stock; ventilated rubber recoil pad; bullpup design; thumb safety; all-steel construction. Made in U.S. by L.A.R. Mfg., Inc. Introduced 1994; still produced. **LMSR: $2400**
   Perf.: $2200   **Exc.:** $1695   **VGood:** $1195

LEBEAU-COURALLY BOXLOCK
Double rifle; 8x57JRS, 9.3x74R; 23¹/₂″-26″ chopper lump barrels; made to customer specs; weighs 8 lbs.; Express rear sight on quarter-rib, bead on ramp front; French walnut stock; pistol grip with cheekpiece, splinter or beavertail forend; steel grip cap; ejectors; several engraving patterns. Made in Belgium. Introduced 1987; no longer imported.
   Perf.: $16,995   **Exc.:** $14,950   **VGood:** $9995

LEBEAU-COURALLY SIDELOCK
Double rifle; 8x57 JRS, 9.3x74R, 375 H&H, 458 Win. Mag., 470 NE, 577 NE; 23 ¹/₂″-26″ chopper lump barrels; weighs 8 lbs.; made to customer specs; Express rear sight on quarter-rib, bead on ramp front; French walnut stock; pistol grip with cheekpiece, splinter or beavertail forend; steel grip cap; ejectors; reinforced action; several engraving patterns. Made in Belgium. Introduced 1987; no longer imported.
   Perf.: $34,000   **Exc.:** $28,000   **VGood:** $19,000

LJUTIC SPACE GUN
Single shot; 22-250, 30-30, 30-06, 308; 24″ barrel; 44″ overall length; iron sights or scope mounts; American walnut stock grip, forend; anti-recoil mechanism; twist-bolt action. Introduced 1981; discontinued 1988.
   Perf.: $2495   **Exc.:** $1995   **VGood:** $1495

MAGTECH MODEL 122.2R
Bolt action; 22 Short, 22 Long, 22 LR; 6-, 10-shot magazines; 24″ six-groove barrel; 43″ overall length; weighs 6 ¹/₂ lbs.; blade front sight, open fully-adjustable rear; receiver grooved for scope mount; Brazilian hardwood stock; sliding safety; double extractors. Imported from Brazil by Magtech Recreational Products, Inc. Introduced 1994; still imported. **LMSR: $150**
   Perf.: $100   **Exc.:** $80   **VGood:** $60

Magtech Model 122.2S
Same specs as Model 122.2R except no sights; receiver grooved for scope mounting. Imported from Brazil by Magtech Recreational Products, Inc. Introduced 1994; still imported. **LMSR: $140**
   Perf.: $90   **Exc.:** $70   **VGood:** $50

Magtech Model 122.2T
Same specs as Model 122.2R except ramp front sight, micrometer-adjustable open rear; receiver grooved for scope mounting. Imported from Brazil by Magtech Recreational Products, Inc. Introduced 1994; still imported. **LMSR: $170**
   Perf.: $120   **Exc.:** $100   **VGood:** $80

MAGTECH MODEL MT-22C
Bolt action; 22 Short, 22 Long, 22 LR; 6-, 10-shot magazines; 21″ six-groove barrel; 39″ overall length; weighs 5 ³/₄ lbs.; blade front sight, fully-adjustable open rear; receiver grooved for scope mount; Brazilian hardwood stock; sliding wing-type safety; double extractors; red cocking indicator. Imported from Brazil by Magtech Recreational Products, Inc. Introduced 1991; discontinued 1993.
   Perf.: $100   **Exc.:** $80   **VGood:** $60

Magtech Model MT-22T
Same specs as MT-22C except 15-shot tubular magazine. Imported from Brazil by Magtech Recreational Products, Inc. Introduced 1991; discontinued 1993.
   Perf.: $150   **Exc.:** $120   **VGood:** $90

Magtech
Model MT-52-T Target
Model MT-66

Mannlicher-Schoe-nauer
High Velocity
Model 1903 Carbine
Model 1905 Carbine
Model 1908 Carbine
Model 1910 Carbine
Model 1924 Carbine
Model 1950
Model 1950 Carbine
Model 1950 Carbine 6.5
Model 1952 Improved

Magtech Model AT-66

Magtech Model MT-52-T Target

Mannlicher-Schoenauer Model 1950 Carbine 6.5

FREE FLOATING BARREL

Section A-B

A !

Mannlicher-Schoenauer
Model 1952 Improved Carbine

MAGTECH MODEL MT-52-T TARGET
Bolt action; 22 LR; 6-, 10-shot magazine; 26 1/4" eight-groove barrel; 43 3/8" overall length; weighs about 9 lbs.; globe front sight with inserts, match-type aperture rear; receiver grooved for scope mounting; Brazilian hardwood stock with stippled grip; fully-adjustable rubber buttplate; aluminum forend rail with sling support; fully-adjustable trigger; free-floating barrel. Imported from Brazil by Magtech Recreational Products, Inc. Introduced 1991; discontinued 1993.
Perf.: $150   **Exc.:** $120   **VGood:** $90

MAGTECH MODEL MT-66
Semi-automatic; 22 LR; 14-shot tubular magazine; 19 5/8" six-groove barrel; 38 1/2" overall length; weighs 4 1/2 lbs.; blade front sight, open adjustable rear; receiver grooved for scope mounts; moulded black nylon stock with checkered pistol grip and forend; tube magazine loads through buttplate; top tang safety. Imported from Brazil by Magtech Recreational Products, Inc. Introduced 1991; discontinued 1993.
Perf.: $90   **Exc.:** $80   **VGood:** $70

MANNLICHER-SCHOENAUER HIGH VELOCITY
Bolt action; 30-06, 7x64 Brenneke, 8x60 Mag., 9.3x62, 10.75x57mm; 5-shot rotary magazine; 23 1/2" barrel; British-type three-leaf open rear sight, ramp front; hand-checkered traditional sporting stock of European walnut; cheekpiece, pistol grip; sling swivels. Introduced 1922; discontinued 1937.
Perf.: $1100   **Exc.:** $895   **VGood:** $725

MANNLICHER-SCHOENAUER MODEL 1903 CARBINE
Bolt action; 6.5x53mm; 5-shot rotary magazine; 17 1/2" barrel; two-leaf rear, ramp front sight; full-length uncheckered European walnut stock; metal forearm cap; pistol grip; cartridge trap in buttplate; double-set trigger; flat bolt handle; sling swivels. Introduced 1903; discontinued 1937.
Perf.: $1195   **Exc.:** $995   **VGood:** $775

MANNLICHER-SCHOENAUER MODEL 1905 CARBINE
Bolt action; 9x56mm; 5-shot rotary magazine; 19 1/2" barrel; two-leaf rear, ramp front sight; full-length uncheckered European walnut stock; metal forearm cap; pistol grip; cartridge trap in buttplate; double-set trigger; flat bolt handle; sling swivels. Introduced 1905; discontinued 1937.
Perf.: $895   **Exc.:** $775   **VGood:** $650

MANNLICHER-SCHOENAUER MODEL 1908 CARBINE
Bolt action; 7x57mm, 8x56mm; 5-shot rotary magazine; 19 1/2" barrel; two-leaf rear, ramp front sight; full-length uncheckered European walnut stock; metal forearm cap; pistol grip; cartridge trap in buttplate; double-set trigger; flat bolt handle; sling swivels. Introduced 1908; discontinued 1947.
Perf.: $895   **Exc.:** $795   **VGood:** $675

MANNLICHER-SCHOENAUER MODEL 1910 CARBINE
Bolt action; 9.5x56, 9.5x57, 375 Express; 5-shot rotary magazine; 19 1/2" barrel; two-leaf rear, ramp front sight; full-length uncheckered European walnut stock; metal forearm cap; pistol grip; cartridge trap in buttplate; double-set trigger; flat bolt handle; sling swivels. Introduced 1910; discontinued 1937.
Perf.: $1050   **Exc.:** $875   **VGood:** $775

MANNLICHER-SCHOENAUER MODEL 1924 CARBINE
Bolt action; 30-06; 5-shot rotary magazine; 19 1/2" barrel; two-leaf rear, ramp front sight; full-length uncheckered European walnut stock; metal forearm cap; pistol grip; cartridge trap in buttplate; double-set trigger; flat bolt handle; sling swivels. Introduced 1924; discontinued 1937.
Perf.: $995   **Exc.:** $825   **VGood:** $725

MANNLICHER-SCHOENAUER MODEL 1950
Bolt action; 244, 257 Roberts, 270, 280, 30-06, 358 Win. Mag., 6.5x54, 7x57, 8x57, 9.3x62; 5-shot rotary magazine; 23 1/2" barrel; weighs 7 1/2 lbs.; folding leaf open rear sight, hooded ramp front; standard hand-checkered European walnut half-stock; pistol grip, cheekpiece, ebony forend tip; single- or double-set trigger; flat bolt handle; shotgun-type safety; sling swivels. Original price: $179.60. Introduced 1950; discontinued 1952.
Perf.: $850   **Exc.:** $775   **VGood:** $600

Mannlicher-Schoenauer Model 1950 Carbine
Same specs as Model 1950 rifle except weighs 7 lbs.; full-length walnut stock; metal forend cap; 20" barrel. Introduced 1950; discontinued 1952.
Perf.: $850   **Exc.:** $775   **VGood:** $600

Mannlicher-Schoenauer Model 1950 Carbine 6.5
Same specs as Model 1950 except 6.5x53; 18 1/4" barrel; full-length walnut stock. Introduced 1950; discontinued 1952.
Perf.: $975   **Exc.:** $750   **VGood:** $625

MANNLICHER-SCHOENAUER MODEL 1952 IMPROVED
Bolt action; 244, 257 Roberts, 270, 280, 30-06, 358 Win. Mag., 6.5x54, 7x57, 8x57, 9.3x62; 5-shot rotary magazine; 24" barrel; weighs 7 1/4 lbs.; folding leaf open rear sight, hooded ramp front; improved hand-checkered European walnut half-stock; pistol grip, deluxe hand-carved cheekpiece; ebony forend tip; single- or double-set trigger; swept-back bolt handle; shotgun-type safety; sling swivels. Introduced 1952; discontinued 1956.
Perf.: $825   **Exc.:** $725   **VGood:** $575

Mannlicher-Schoenauer Model 1956

Mannlicher-Schoenauer 1961MCA Carbine

Marathon Centerfire

Marathon Super Shot 22

RIFLES

Mannlicher-Schoenauer
Model 1952 Improved Carbine
Model 1952 Improved Carbine 6.5
Model 1956
Model 1956 Carbine
Model 1956 Magnum
Model 1961MCA
Model 1961MCA Carbine

Mapiz
Zanardini Oxford

Marathon
Centerfire
Super Shot 22

Marcel Thys
King Royal
Liege Royal Lux

Mannlicher-Schoenauer Model 1952 Improved Carbine

Same specs as Model 1952 Improved except 257 Roberts, 270, 30-06; 20″, 24″ barrel; weighs 7 lbs.; improved full-length walnut stock design; swept back bolt handle. Introduced 1952; discontinued 1956.

Perf.: $850   **Exc.:** $675   **VGood:** $600

Mannlicher-Schoenauer Model 1952 Improved Carbine 6.5

Same specs as Model 1952 Improved except 6.5x53; 18 1/4″ barrel; weighs 6 3/4 lbs.; improved full-length walnut stock. Introduced 1952; discontinued 1956.

Perf.: $975   **Exc.:** $750   **VGood:** $650

MANNLICHER-SCHOENAUER MODEL 1956

Bolt action; 243 Win., 30-06; 5-shot rotary magazine; 22″ barrel; folding leaf open rear sight, hooded ramp front; high-comb improved walnut half-stock; pistol grip; cheekpiece; ebony forend tip; single- or double-set trigger; swept-back bolt handle; shotgun-type safety; sling swivels. Introduced 1956; discontinued 1960.

Perf.: $695   **Exc.:** $560   **VGood:** $495

Mannlicher-Schoenauer Model 1956 Carbine

Same specs as Model 1956 except 243, 257 Roberts, 270, 30-06, 308, 6.5x53, 7mm; 20″ barrel; redesigned high-comb walnut full-length stock. Introduced 1956; discontinued 1960.

Perf.: $775   **Exc.:** $650   **VGood:** $595

Mannlicher-Schoenauer Model 1956 Magnum

Same specs as Model 1956 except 257 Weatherby Mag., 6.5x68, 8x68, 458 Win. Mag. Introduced 1956; discontinued 1960.

Perf.: $950   **Exc.:** $795   **VGood:** $675

MANNLICHER-SCH0ENAUER MODEL 1961MCA

Bolt action; 243 Win., 270 Win., 30-06; 5-shot rotary magazine; 22″ barrel; folding leaf open rear sight, hooded ramp front; Monte Carlo walnut half-stock; pistol grip; cheekpiece; ebony forend tip; single- or double-set trigger; swept-back bolt handle; shotgun-type safety; sling swivels. Introduced 1961; discontinued 1971.

Perf.: $850   **Exc.:** $735   **VGood:** $650

Mannlicher-Schoenauer 1961MCA Carbine

Same specs as Model 1956 except 243 Win., 270 Win., 30-06, 308 Win., 6.5x53; 20″ barrel; Monte Carlo walnut full-length stock. Introduced 1961; discontinued 1971.

Perf.: $895   **Exc.:** $725   **VGood:** $595

MAPIZ ZANARDINI OXFORD

Double rifle; boxlock side-by-side; 444 Marlin, 7x65R, 6.5x57R, 9.3x74R, 375 H&H Mag., 458 Win. Mag., 465 N.E., 470 N.E., 577 N.E., 600 N.E., 375 Flanged N.E.; 24″, 25″, 25 1/2″ barrel; oil-finished European walnut stock; rubber or hard buttplate; automatic ejectors; double triggers; heavy engraving. Made in Italy. Introduced 1989; discontinued 1990.

Perf.: $5595   **Exc.:** $4995   **VGood:** $3995

MARATHON CENTERFIRE

Bolt action; 243, 308, 7x57, 30-06, 270, 7mm Rem. Mag., 300 Win. Mag.; 5-shot magazine; 24″ barrel; 45″ overall length; weighs 8 lbs.; open adjustable rear sight, bead front on ramp; select walnut Monte Carlo stock; rubber recoil pad; Santa Barbara Mauser action; triple thumb locking safety; blued finish. Made in Spain. Introduced 1984; discontinued 1986.

Perf.: $310   **Exc.:** $270   **VGood:** $200

MARATHON SUPER SHOT 22

Bolt action; single shot; 22 LR; 24″ barrel; 41 1/2″ overall length; weighs 5 lbs.; step-adjustable open rear sight, bead front; receiver grooved for scope mounts; select hardwood stock; blued finish. Made in Spain. Introduced 1984; discontinued 1986.

Perf.: $60   **Exc.:** $50   **VGood:** $40

MARCEL THYS KING ROYAL

Double rifle; sidelock side-by-side; 22 LR, 22 Hornet, 30-06, 375 H&H Mag., 450 #2, 458 Win. Mag., 470 Nitro, 500 3″, 577, 600 Nitro; 24″ - 27″ chopper lump barrels; weighs 6-14 lbs. depending on caliber, accessories; bead on ramp front sight, one standing leaf, two folding leaves on quarter-rib; oil-finished European walnut stock; hand-detachable locks; reinforced top tang extension; reinforced Holland-type frame; full-coverage game scene engraving with gold inlay; coin finish or color case-hardened action. Made in Belgium. Introduced 1990; discontinued 1992.

Perf.: $16,000 **Exc.:** $13,900 **VGood:** $11,500

MARCEL THYS LIEGE ROYAL LUX

Double rifle; sidelock side-by-side; 22 LR, 22 Hornet, 30-06, 375 H&H Mag., 450 #2, 458 Win. Mag., 470 Nitro, 500 3″, 577, 600 Nitro; 24″-27″ chopper lump barrels; weighs 6-14 lbs. depending on caliber, accessories; bead on ramp front sight, one standing leaf, two folding leaves on quarter-rib; oil-finished dark European walnut stock; hand-detachable locks; reinforced top tang extension; reinforced Holland-type frame; full-coverage game scene engraving; coin finish or color case-hardened action. Made in Belgium. Introduced 1990; discontinued 1992.

Perf.: $14,500 **Exc.:** $12,500 **VGood:** $9950

Marlin Semi-Automatic Rifles

Model 9 Camp Carbine
Model 9N
Model 45 Camp Carbine
Model 49
Model 49DL
Model 50
Model 50E
Model 60
Model 60SS
Model 70HC
Model 70P Papoose
Model 70PSS
Model 75C
Model 88C
Model 88DL
Model 89C

Marlin Model 9 Camp Carbine

Marlin Model 45 Camp Carbine

Marlin Model 49

Marlin Model 49DL

Marlin Model 50

MARLIN SEMI-AUTOMATIC RIFLES

MARLIN MODEL 9 CAMP CARBINE

Semi-automatic; 9mm Para.; 12-shot magazine; 16 1/2″ barrel; 35 1/2″ overall length; weighs 6 3/4 lbs.; adjustable open rear sight, ramp front with bead, hood; drilled, tapped for scope mount; wood stock; rubber buttpad; manual hold-open; Garand-type safety; magazine safety; loaded chamber indicator. Introduced 1985; still in production. **LMSR: $443**
   Perf.: $310   **Exc.:** $260   **VGood:** $200

Marlin Model 9N

Same specs as Model 9 except nickel finish. Introduced 1991.
   Perf.: $330   **Exc.:** $290   **VGood:** $250

MARLIN MODEL 45 CAMP CARBINE

Semi-automatic; 45 ACP; 7-shot magazine; 16 1/2″ barrel; 35 1/2″ overall length; weighs 6 3/4 lbs.; ramp front, open adjustable rear sight; walnut-finished hardwood stock; rubber buttpad; sling swivels. Introduced 1986; still in production. **LMSR: $443**
   Perf.: $320   **Exc.:** $280   **VGood:** $245

MARLIN MODEL 49

Semi-automatic; 22 LR; 19-shot tubular magazine; 22″ barrel; 40 1/4″ overall length; weighs 5 1/2 lbs.; open rear, hooded ramp front sight; two-piece checkered or uncheckered hardwood stock; same as Model 99C except solid top receiver; manual bolt hold-open. Introduced 1968; discontinued 1971.
   Perf.: $100   **Exc.:** $80   **VGood:** $63

Marlin Model 49DL

Same specs as Model 49 except capped pistol grip; checkered forearm and pistol grip; gold trigger; scroll-engraved receiver; grooved for tip-off scope mounts. Introduced 1970; no longer produced.
   Perf.: $110   **Exc.:** $90   **VGood:** $70

MARLIN MODEL 50

Semi-automatic; 22 LR; 6-shot detachable box magazine; 22″ barrel; take-down action; open adjustable rear sight, bead front; uncheckered pistol-grip walnut stock, grooved forearm. Introduced 1931; discontinued 1934.
   Perf.: $125   **Exc.:** $100   **VGood:** $80

Marlin Model 50E

Same specs as Model 50 except hooded front sight, peep rear.
   Perf.: $135   **Exc.:** $110   **VGood:** $90

MARLIN MODEL 60

Semi-automatic; 22 LR; 14-shot tubular magazine; 22″ round tapered barrel; 40 1/2″ overall length; weighs 5 1/2 lbs.; ramp front sight, open adjustable rear; matted receiver; grooved for tip-off mounts; walnut-finished Monte Carlo hardwood stock; auto last-shot bolt hold-open. Originally marketed as Marlin Glenfield Model 60. Introduced 1982; still in production. **LMSR: $158**
   Perf.: $100   **Exc.:** $85   **VGood:** $70

Marlin Model 60SS

Same specs as Model 60 except breech bolt, barrel and outer magazine tube of stainless steel construction; most other parts either nickel-plated or coated to match the stainless finish; black/gray Maine birch laminate Monte Carlo stock; nickel-plated swivel studs; rubber buttpad. Made in U.S. by Marlin. Introduced 1993; still produced. **LMSR: $273**
   Perf.: $190   **Exc.:** $150   **VGood:** $130

MARLIN MODEL 70HC

Semi-automatic; 22 LR; 7-, 15-shot clip magazine; 18″ barrel; 36 3/4″ overall length; weighs 5 1/2 lbs.; ramp front sight, adjustable open rear; receiver grooved for scope mount; Monte Carlo-type walnut-finished hardwood stock; cross-bolt safety; manual bolt hold-open. Introduced 1988; still in production. **LMSR: $167**
   Perf.: $160   **Exc.:** $140   **VGood:** $100

Marlin Model 70P Papoose

Same specs as Model 70HC except takedown model with easily removable barrel; 16 1/4″ Micro-Groove® barrel; walnut-finished hardwood stock; 35 1/4″ overall length; weighs 3 1/4 lbs. Comes with zippered case. Made in U.S. by Marlin. Introduced 1986; discontinued 1995.
   Perf.: $150   **Exc.:** $120   **VGood:** $90

Marlin Model 70PSS

Same specs as Model 70P except fiberglass filled synthetic stock; take-down model. Introduced 1986; still in production. **LMSR: $278**
   Perf.: $170   **Exc.:** $140   **VGood:** $110

MARLIN MODEL 75C

Semi-automatic; 22 LR; 13-shot tubular magazine; 18″ barrel; 36 1/2″ overall length; weighs 5 lbs.; ramp front, adjustable open rear sight; receiver grooved for scope mounts; walnut-finished Monte Carlo hardwood stock; cross-bolt safety; auto last-shot bolt hold-open. Introduced 1982; no longer in production.
   Perf.: $120   **Exc.:** $100   **VGood:** $90

MARLIN MODEL 88C

Semi-automatic; 22 LR; 15-shot; 24″ barrel; 45″ overall length; weighs 6 3/4 lbs.; tube magazine in buttstock; open rear, hooded front sight; chrome-plated cocking handle; safety button; positive sear lock safety; black walnut uncheckered pistol-grip stock. Introduced 1948; discontinued 1956.
   Perf.: $90   **Exc.:** $80   **VGood:** $70

Marlin Model 88DL

Same specs as Model 88C except hand-checkered stock; sling swivels; receiver peep sight in rear. Introduced 1953; discontinued 1956.
   Perf.: $100   **Exc.:** $90   **VGood:** $80

MARLIN MODEL 89C

Semi-automatic; 22 LR; 5-shot clip magazine (early model), 12-shot (late model); 24″ barrel; 45″ overall length; weighs 6 1/2 lbs.; hooded front, open rear sight; uncheckered pistol-grip stock. Introduced 1948; discontinued 1961.
   Perf.: $90   **Exc.:** $80   **VGood:** $70

Marlin Model 60

Marlin Model 70HC

Marlin Model 70P Papoose

Marlin Model 75C

Marlin Model 70PSS

Marlin Semi-Automatic Rifles
Model 89DL
Model 98
Model 99
Model 99C
Model 99DL
Model 99G
Model 99M1 Carbine

Marlin Model 89DL
Same specs as Model 89C except sling swivels; receiver peep sight.

Perf.: $110   **Exc.:** $90   **VGood:** $70

MARLIN MODEL 98
Semi-automatic; 22 LR; 15-shot buttstock tube magazine; 22″ barrel; 42″ overall length; weighs 6 1/2 lbs.; fully-adjustable peep rear, hooded ramp front sight; solid frame; Bishop-style uncheckered Monte Carlo walnut stock, cheekpiece. Introduced 1957; discontinued 1959.

Perf.: $100   **Exc.:** $90   **VGood:** $70

MARLIN MODEL 99
Semi-automatic; 22 LR; 18-shot tube magazine; 22″ barrel; 42″ overall length; weighs 5 1/2 lbs.; open rear, hooded ramp front sight; uncheckered walnut pistol-grip stock. Introduced 1959; discontinued 1961.

Perf.: $100   **Exc.:** $90   **VGood:** $70

Marlin Model 99C
Same specs as Model 99 except uncheckered walnut pistol-grip stock with fluted comb; grooved receiver for tip-off scope mounts; gold-plated trigger. Later production features checkering on pistol grip, forearm. Introduced 1962; discontinued 1978.

Perf.: $110   **Exc.:** $90   **VGood:** $80

Marlin Model 99DL
Same specs as Model 99 except for uncheckered black walnut Monte Carlo stock; jeweled bolt; gold-plated trigger; sling, sling swivels. Introduced 1960; discontinued 1965.

Perf.: $120   **Exc.:** $100   **VGood:** $80

Marlin Model 99G
Same specs as Model 99C except plainer stock; bead front sight. Introduced 1960; discontinued 1965.

Perf.: $100   **Exc.:** $90   **VGood:** $70

Marlin Model 99M1 Carbine
Same specs as Model 99C except designed after U.S. 30M1 carbine; 18″ barrel; 37″ overall length; weighs 4 3/4 lbs.; 10-shot tube magazine; uncheckered pistol grip carbine stock; hand guard with barrel band; open rear, military-type ramp front sight; drilled, tapped for receiver sights; grooved for tip-off mount; gold-plated trigger; sling swivels. Introduced 1964; discontinued 1978.

Perf.: $170   **Exc.:** $150   **VGood:** $100

Marlin Model 88C

Marlin Model 89C

Marlin Model 98

Marlin Model 99DL

Marlin Model 99M1 Carbine

Marlin Semi-Automatic Rifles
Model 922 Magnum
Model 989M2 Carbine
Model 990
Model 990L
Model 995
Model 995SS
Model A-1
Model A-1C
Model A-1DL
Model A-1E

Marlin Bolt-Action Rifles
Model 15
Model 15Y "Little Buckaroo"
Model 15YN "Little Buckaroo"
Model 25B
Model 25M
Model 25MB Midget Magnum

Marlin Model 922 Magnum

Marlin Model 989M2 Carbine

Marlin Model 995

Marlin Model A-1

Marlin Model 15Y "Little Buckaroo"

Marlin Model 15YN "Little Buckaroo"

MARLIN MODEL 922 MAGNUM
Semi-automatic; 22 WMR; 7-shot magazine; 20 ¹/₂″ barrel; 39 ³/₄″ overall length; weighs 6 ¹/₂ lbs.; ramp front sight with bead and removable Wide-Scan hood, adjustable folding semi-buckhorn rear; action based on the centerfire Model 9 Carbine; receiver drilled and tapped for scope mounting; checkered American black walnut stock with Monte Carlo comb; swivel studs; rubber buttpad; automatic last-shot bolt hold-open; magazine safety. Made in U.S. by Marlin Firearms Co. Introduced 1993; still in production. **LMSR: $441**
Perf.: $300   **Exc.:** $270   **VGood:** $200

MARLIN MODEL 989M2 CARBINE
Semi-automatic; 22 LR; 7-shot detachable clip magazine; 18″ barrel; 37″ overall length; weighs 4 ³/₄ lbs.; open rear sight, military-type ramp front; uncheckered pistol-grip carbine walnut stock and handguard; barrel band; sling swivels; gold-plated trigger. Introduced 1965; discontinued 1978.
Perf.: $120   **Exc.:** $100   **VGood:** $80

MARLIN MODEL 990
Semi-automatic; 22 LR; 18-shot tubular magazine; 22″ barrel; 40 ³/₄″ overall length; weighs 5 ¹/₂ lbs.; ramp bead hooded front, adjustable folding semi-buckhorn rear sight; American walnut Monte Carlo stock; checkered fluted comb; full pistol grip; cross-bolt safety; receiver grooved for tip-off mount. Introduced 1979; discontinued 1988.
Perf.: $120   **Exc.:** $100   **VGood:** $80

Marlin Model 990L
Same specs as Model 990 except 14-shot magazine; 40 ¹/₂″ overall length; weighs about 5 ³/₄ lbs.; laminated hardwood stock with black rubber rifle buttpad and swivel studs; gold-plated trigger; manual bolt hold-open; automatic last-shot bolt hold-open. Made in U.S. by Marlin Firearms Co. Introduced 1992; dropped 1994.
Perf.: $165   **Exc.:** $140   **VGood:** $110

MARLIN MODEL 995
Semi-automatic; 22 LR; 7-shot clip magazine; 18″ Micro-Groove barrel; 36 ³/₄″ overall length; weighs 5 lbs.; ramp bead front sight with Wide-Scan hood, adjustable folding semi-buckhorn rear; receiver grooved for scope mount; American black walnut stock with Monte Carlo; full pistol grip; checkered grip and forend; white buttplate spacer; Mar-Shield finish; bolt hold-open device; cross-bolt safety. Introduced 1979; dropped 1994.
Perf.: $140   **Exc.:** $120   **VGood:** $100

Marlin Model 995SS
Same specs as Model 995 except stainless steel 18″ barrel; ramp front sight with orange post and cut-away Wide-Scan hood, screw-adjustable open rear; black fiberglass-filled synthetic stock with nickel-plated swivel studs, moulded-in checkering; stainless steel breech-bolt. Introduced 1979; still produced. **LMSR: $231**
Perf.: $160   **Exc.:** $140   **VGood:** $110

MARLIN MODEL A-1
Semi-automatic; 22 LR; 6-shot detachable box magazine; 24″ barrel; open rear sight, bead front; uncheckered walnut pistol-grip stock. Introduced 1935; discontinued 1946.
Perf.: $135   **Exc.:** $90   **VGood:** $70

Marlin Model A-1C
Same specs as Model A-1 except off/on safety; military-style one-piece buttstock with fluted comb and semi-beavertail forearm. Introduced 1940; discontinued 1946.
Perf.: $135   **Exc.:** $90   **VGood:** $70

Marlin Model A-1DL
Same specs as Model A-1C except sling swivels; hooded front sight, peep rear.
Perf.: $140   **Exc.:** $100   **VGood:** $80

Marlin Model A-1E
Same specs as Model A-1 except hooded front sight, peep rear.
Perf.: $135   **Exc.:** $90   **VGood:** $70

MARLIN BOLT-ACTION RIFLES

MARLIN MODEL 15
Bolt Action; 22 Short, 22 Long, 22 LR; 22″ barrel; 41″ overall length; weighs 5 ¹/₂ lbs.; ramp front sight, adjustable open rear; walnut finished hardwood Monte Carlo stock with full pistol grip; receiver grooved for tip-off scope mount; thumb safety; cocking indicator.
Perf.: $100   **Exc.:** $80   **VGood:** $65

MARLIN MODEL 15Y "LITTLE BUCKAROO"
Bolt action; single shot; 22 Short, 22 Long, 22 LR; 16 ¹/₄″ barrel; 33 ¹/₄″ overall length; weighs 4 ¹/₄ lbs.; adjustable open rear sight, ramp front; marketed with 4x15 scope, mount; walnut-finished hardwood stock. Introduced 1984; discontinued 1986.
Perf.: $120   **Exc.:** $100   **VGood:** $80

Marlin Model 15YN "Little Buckaroo"
Same specs as Model 15Y except without scope. Reintroduced 1989; still produced. **LMSR: $163**
Perf.: $110   **Exc.:** $90   **VGood:** $70

MARLIN MODEL 25B
Bolt action; 22 Short, 22 Long, 22 LR; 7-shot clip magazine; 22″ barrel; 41″ overall length; weighs 5 lbs.; ramp front, open rear sight; walnut-finished hardwood pistol-grip stock. Introduced 1982; no longer produced.
Perf.: $90   **Exc.:** $75   **VGood:** $60

Marlin Model 25M
Same specs as Model 25 except 22 WMR.
Perf.: $100   **Exc.:** $80   **VGood:** $65

Marlin Model 25MB Midget Magnum
Same specs as Model 25 except 22 WMR; 16 ¹/₄″ barrel; 35 ¹/₄″ overall length; weighs 4 ³/₄ lbs.; take-down; walnut-finished hardwood stock; grooved for tip-off scope; marketed with iron sights, 4x scope, zippered nylon case. Introduced 1987; discontinued 1988.
Perf.: $145   **Exc.:** $115   **VGood:** $95

Marlin Model 100

Marlin Model 25

Marlin Model 25M

Marlin Model 80C

Marlin Model 101

Marlin Bolt-Action Rifles
Model 25MN
Model 25N
Model 65
Model 65E
Model 80
Model 80C
Model 80DL
Model 80E
Model 81
Model 81C
Model 81DL
Model 81E
Model 100
Model 100S
Model 100SB
Model 101
Model 101DL

Marlin Model 25MN
Same specs as Model 25 except 22 WMR; weighs 6 lbs.; walnut-finished Maine birch pistol-grip Monte Carlo stock; ramp front sight, adjustable open rear. Introduced 1989; still in production. **LMSR: $223**
   Perf.: $130   **Exc.:** $100   **VGood:** $85

Marlin Model 25N
Same specs as Model 25 except 22 LR; adjustable open rear sight; receiver grooved for scope mounting. Introduced 1989; still in production. **LMSR: $168**
   Perf.: $95   **Exc.:** $80   **VGood:** $65

MARLIN MODEL 65
Bolt action; single shot; 22 Short, 22 Long, 22 LR; 24″ barrel; open rear, bead front sight; uncheckered walnut pistol-grip stock, grooved forearm. Introduced 1935; discontinued 1937.
   Perf.: $80   **Exc.:** $60   **VGood:** $45

Marlin Model 65E
Same specs as Model 65 except equipped with hooded front sight, peep rear.
   Perf.: $90   **Exc.:** $70   **VGood:** $50

MARLIN MODEL 80
Bolt action; 22 Short, 22 Long, 22 LR; 8-shot detachable box magazine; 24″ barrel; open rear, bead front sight; uncheckered walnut pistol-grip stock; black buttplate. Introduced 1934; discontinued 1939.
   Perf.: $90   **Exc.:** $80   **VGood:** $70

Marlin Model 80C
Same specs as Model 80 except one-piece military-style buttstock with fluted comb and semi-beavertail forearm; off/on safety. Model 80-C replaced standard Model 80 in 1940; discontinued 1970.
   Perf.: $90   **Exc.:** $80   **VGood:** $70

Marlin Model 80DL
Same specs as Model 80C except peep rear, hooded front sight; sling swivels. Introduced 1940; discontinued 1965.
   Perf.: $100   **Exc.:** $90   **VGood:** $70

Marlin Model 80E
Same specs as Model 80 except peep rear sight, hooded front.
   Perf.: $110   **Exc.:** $100   **VGood:** $80

MARLIN MODEL 81
Bolt action; 22 Short, 22 Long, 22 LR; 18-shot tubular magazine (22 LR), 20-shot (22 Long), 25-shot (22 Short); 24″ round tapered barrel; 42 1/2″ overall length; weighs 6 1/2 lbs.; open rear, silver bead front sight; uncheckered pistol-grip stock. Introduced 1937; discontinued 1965.
   Perf.: $80   **Exc.:** $65   **VGood:** $50

Marlin Model 81C
Same specs as Model 81 except improved military-type one-piece buttstock with fluted comb and semi-beavertail forearm. Introduced 1940; discontinued 1971.
   Perf.: $80   **Exc.:** $70   **VGood:** $60

Marlin Model 81DL
Same specs as Model 81C except hooded front, peep rear sight; sling swivels. Introduced 1940; discontinued 1965.
   Perf.: $90   **Exc.:** $80   **VGood:** $70

Marlin Model 81E
Same specs as Model 81 except hooded front, peep rear sight.
   Perf.: $90   **Exc.:** $80   **VGood:** $70

MARLIN MODEL 100
Bolt action; take-down single shot; 22 Short, 22 Long, 22 LR; 24″ barrel; weighs 4 1/2 lbs.; open adjustable rear sight, bead front; uncheckered walnut pistol-grip stock. Introduced 1936; discontinued 1941.
   Perf.: $70   **Exc.:** $60   **VGood:** $50

Marlin Model 100S
Same specs as Model 100 except sling; peep rear, hooded front sight. It is known as the Tom Mix Special, allegedly because Tom Mix used such a rifle in his vaudeville act in the '30s. Manufactured 1937 only. Value is based largely upon rarity.
   Perf.: $300   **Exc.:** $270   **VGood:** $220

Marlin Model 100SB
Same specs as Model 100 except smoothbore for use with 22 shot cartridge; shotgun sight. Actually, this probably is the version used by Tom Mix in his act, as he used shot cartridges for breaking glass balls and other on-stage targets. Introduced 1936; discontinued 1941.
   Perf.: $130   **Exc.:** $100   **VGood:** $80

MARLIN MODEL 101
Bolt action; single shot; 22 Short, 22 Long, 22 LR; 22″ barrel; 40″ overall length; weighs 4 1/2 lbs.; fully-adjustable semi-buckhorn rear sight, hooded Wide-Scan front; black plastic trigger guard; ring or T-shaped cocking piece; uncheckered walnut stock with beavertail forearm; receiver grooved for tip-off scope mount. Improved version of Model 100 with improved bolt, redesigned stock. Introduced 1941; discontinued 1976.
   Perf.: $80   **Exc.:** $70   **VGood:** $50

Marlin Model 101DL
Same specs as Model 101 except 24″ barrel; 42 1/2″ overall length; weighs 5 lbs.; peep rear, hooded front sight; sling swivels. Introduced 1941; discontinued 1945.
   Perf.: $90   **Exc.:** $75   **VGood:** $50

Marlin Bolt-Action Rifles
Model 122
Model 322
Model 455
Model 780
Model 781
Model 782
Model 783
Model 880
Model 880SS
Model 881
Model 882
Model 882L
Model 882SS
Model 883

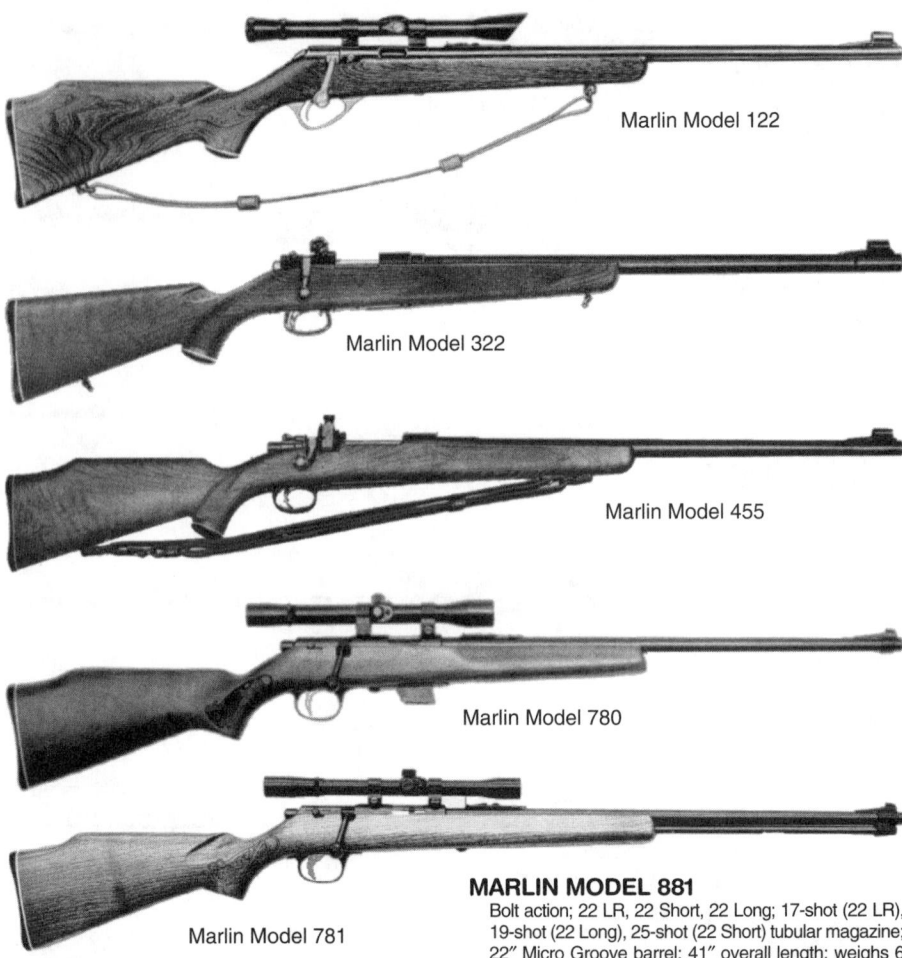

Marlin Model 122

Marlin Model 322

Marlin Model 455

Marlin Model 780

Marlin Model 781

MARLIN MODEL 122
Bolt action; single shot; 22 Short, 22 Long, 22 LR; 22″ barrel; 40″ overall length; weighs 5 lbs.; open rear, hooded ramp front sight; drilled, tapped for receiver sights, tip-off scope mount; junior target model with shortened uncheckered stock; walnut Monte Carlo pistol-grip stock. Introduced 1962; discontinued 1965.
Perf.: $80   **Exc.:** $70   **VGood:** $50

MARLIN MODEL 322
Bolt action; 222 Rem.; 3-shot clip-type magazine; 24″ medium weight barrel; 42″ overall length; weighs 7 1/2 lbs.; short Sako Mauser-type action; two-position peep sight rear, hooded ramp front with hunting bead; checkered walnut stock, forearm; sling swivels. Introduced 1954; discontinued 1958.
Perf.: $380   **Exc.:** $330   **VGood:** $270

MARLIN MODEL 455
Bolt action; 270, 308, 30-06; 5-shot box magazine; 24″ medium-weight stainless steel barrel; 42 1/2″ overall length; weighs 7 3/4 lbs.; Lyman No. 48 receiver sight, hooded ramp front; FN Mauser action; Sako trigger; checkered walnut Monte Carlo stock with cheekpiece; checkered forearm; receiver drilled, tapped for scope mounts. Introduced 1956; discontinued 1959.
Perf.: $450   **Exc.:** $390   **VGood:** $310

MARLIN MODEL 780
Bolt action; 22 Short, 22 Long, 22 LR; 7-shot clip magazine; 22″ barrel; 40 1/2″ overall length; weighs 5 1/2 lbs.; open rear, ramp front sight; receiver grooved for scope mounts; Monte Carlo American walnut stock; checkered pistol grip, forearm; gold-plated trigger. Introduced 1971; discontinued 1988.
Perf.: $100   **Exc.:** $80   **VGood:** $65

MARLIN MODEL 781
Bolt action; 22 Short, 22 Long, 22 LR; 17-shot tubular magazine; 22″ barrel; weighs 6 lbs.; open rear, ramp front sight; receiver grooved for scope mounts; American walnut Monte Carlo stock; checkered pistol-grip, forearm. Introduced 1971; discontinued 1988.
Perf.: $100   **Exc.:** $80   **VGood:** $70

MARLIN MODEL 782
Bolt action; 22 WMR; 7-shot clip magazine; 22″ barrel; 41″ overall length; weighs 6 lbs.; open rear, ramp front sight; receiver grooved for scope mounts; American walnut Monte Carlo stock; checkered pistol grip, forearm; sling swivels; leather sling. Introduced 1971; discontinued 1988.
Perf.: $130   **Exc.:** $100   **VGood:** $80

MARLIN MODEL 783
Bolt action; 22 WMR; 13-shot tubular magazine; 22″ barrel; 41″ overall length; weighs 6 lbs.; open rear, ramp front sight; receiver grooved for scope mounts; select American walnut Monte Carlo stock; checkered pistol grip, forearm; white line spacer; gold-plated trigger; sling swivels; leather sling. Introduced 1971; discontinued 1988.
Perf.: $130   **Exc.:** $100   **VGood:** $80

MARLIN MODEL 880
Bolt action; 22 LR; 7-shot clip magazine; 22″ Micro Groove barrel; 41″ overall length; weighs 5 1/2 lbs.; ramp front, adjustable semi-buckhorn rear sight; receiver grooved for tip-off scope mount; black walnut Monte Carlo stock; checkered pistol grip, forend; swivel studs; rubber buttpad. Mar-Shield finish. Replaces Model 780. Introduced 1989; still in production. **LMSR:** $251
Perf.: $170   **Exc.:** $140   **VGood:** $120

Marlin Model 880SS
Same specs as Model 880 except barrel, receiver, front breech bolt, striker knob, trigger stud, cartridge lifter stud and outer magazine tube made of stainless steel; most other parts nickel-plated to match stainless finish; black fiberglass-filled AKZO synthetic stock with moulded-in checkering; stainless steel swivel studs. Made in U.S. by Marlin Firearms Co. Introduced 1994; still in production. **LMSR:** $289
Perf.: $220   **Exc.:** $170   **VGood:** $120

MARLIN MODEL 881
Bolt action; 22 LR, 22 Short, 22 Long; 17-shot (22 LR), 19-shot (22 Long), 25-shot (22 Short) tubular magazine; 22″ Micro Groove barrel; 41″ overall length; weighs 6 lbs.; ramp front, adjustable semi-buckhorn rear sight; receiver grooved for tip-off scope mount; black walnut Monte Carlo stock; checkered pistol grip, forend; swivel studs; rubber buttpad. Mar-Shield finish. Replaces Model 781. Introduced 1989; still in production. **LMSR:** $261
Perf.: $180   **Exc.:** $150   **VGood:** $125

MARLIN MODEL 882
Bolt action; 22 WMR; 7-shot clip magazine; 22″ Micro Groove barrel; 41″ overall length; weighs 5 1/2 lbs.; ramp front, adjustable semi-buckhorn rear sight; receiver grooved for tip-off scope mount; black walnut Monte Carlo stock; checkered pistol grip, forend; swivel studs; rubber buttpad. Mar-Shield finish. Replaces Model 782. Introduced 1989; still in production. **LMSR:** $296
Perf.: $190   **Exc.:** $150   **VGood:** $100

Marlin Model 882L
Same specs as Model 882 except laminated hardwood stock. Still in production. **LMSR:** $272
Perf.: $200   **Exc.:** $180   **VGood:** $140

Marlin Model 882SS
Same specs as Model 882 except stainless steel front breech bolt, barrel, receiver and bolt knob; other parts either stainless steel or nickel-plated; black Monte Carlo stock of fiberglass-filled polycarbonate with moulded-in checkering; nickel-plated swivel studs. Made in U.S. by Marlin Firearms Co. Introduced 1995; still produced. **LMSR:** $275
Perf.: $200   **Exc.:** $170   **VGood:** $120

MARLIN MODEL 883
Bolt action; 22 WMR; 12-shot tubular magazine; 22″ Micro Groove barrel; 41″ overall length; weighs 5 1/2 lbs.; ramp front, adjustable semi-buckhorn rear sight; receiver grooved for tip-off scope mount; black walnut Monte Carlo stock; checkered pistol grip, forend; swivel studs; rubber buttpad. Mar-Shield finish. Replaces Model 783. Introduced 1989; still in production. **LMSR:** $308
Perf.: $190   **Exc.:** $150   **VGood:** $100

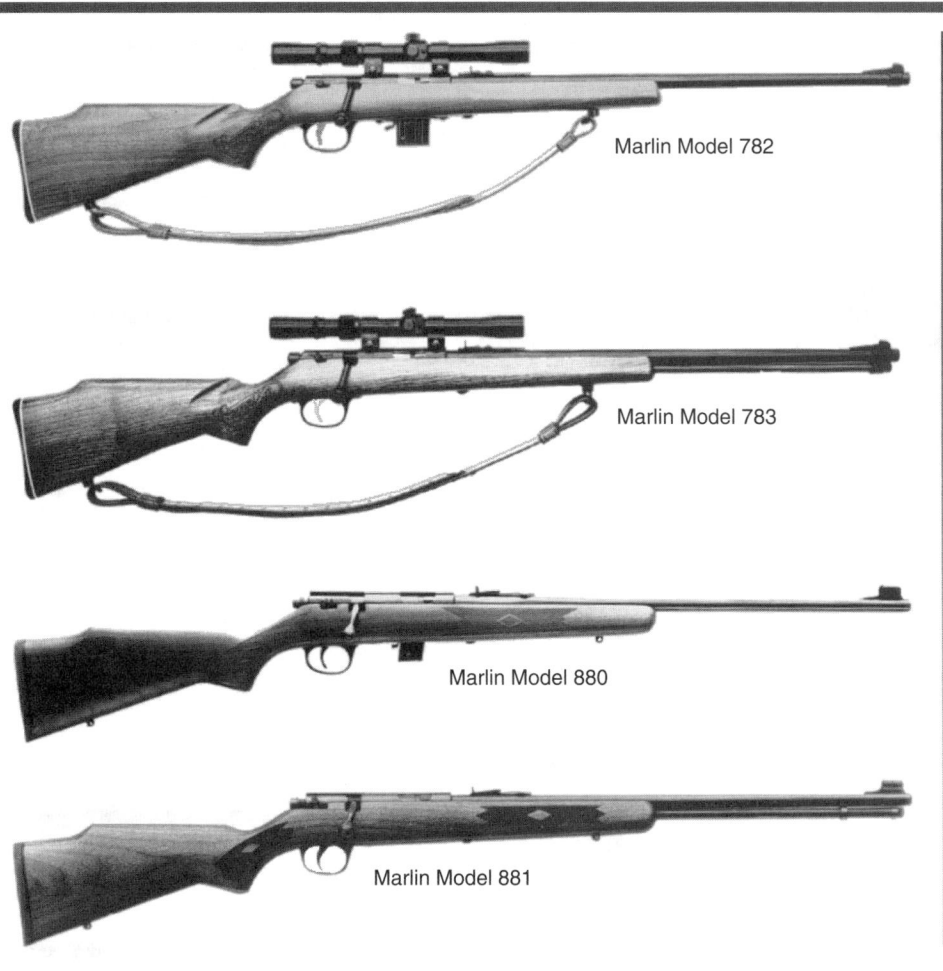

Marlin Model 782

Marlin Model 783

Marlin Model 880

Marlin Model 881

Marlin Bolt-Action Rifles
Model 883SS
Model 980
Model 2000 Target
Model 2000A

Marlin Model 882SS

Marlin Model 882

Marlin Model 883

Marlin Model 980

Marlin Model 2000 Target

Marlin Model 883SS

Same specs as Model 883 except front breech bolt, striker knob, trigger stud, cartridge lifter stud, outer magazine tube of stainless steel; other parts nickel-plated; two-tone brown laminated Monte Carlo stock with swivel studs; rubber buttpad. Made in U.S. by Marlin Firearms Co. Introduced 1993; still in production.
LMSR: $326
    Perf.: $220   **Exc.:** $150   **VGood:** $110

MARLIN MODEL 980

Bolt action; 22 WMR; 8-shot clip-type magazine; 24" barrel; 43" overall length; weighs 6 lbs.; open adjustable rear sight, hooded ramp front; uncheckered walnut Monte Carlo-style stock with white spacers at pistol grip and buttplate; sling; sling swivels. Introduced 1966; discontinued 1971.
    Perf.: $150   **Exc.:** $130   **VGood:** $100

MARLIN MODEL 2000 TARGET

Bolt action; single shot; 22 LR; 22" heavy barrel; 41" overall length; weighs 8 lbs.; hooded Lyman front sight, fully-adjustable Lyman target peep rear; high-comb, blue-enameled fiberglass/Kevlar stock; stippled grid forend; adjustable buttplate; aluminum forend rail; quick-detachable swivel; two-stage target trigger; red cocking indicator. Marketed with seven front sight inserts. Current models with Williams sights. Introduced 1991; still produced. **LMSR: $602**
    Perf.: $475   **Exc.:** $360   **VGood:** $250

Marlin Model 2000A

Same specs as Model 2000 except adjustable comb; ambidextrous pistol grip; Marlin logo moulded into the blue Carbelite stock. Weighs 8 1/2 lbs. Made in U.S. by Marlin Firearms Co. Introduced 1994; discontinued 1995.
    Perf.: $500   **Exc.:** $425   **VGood:** $320

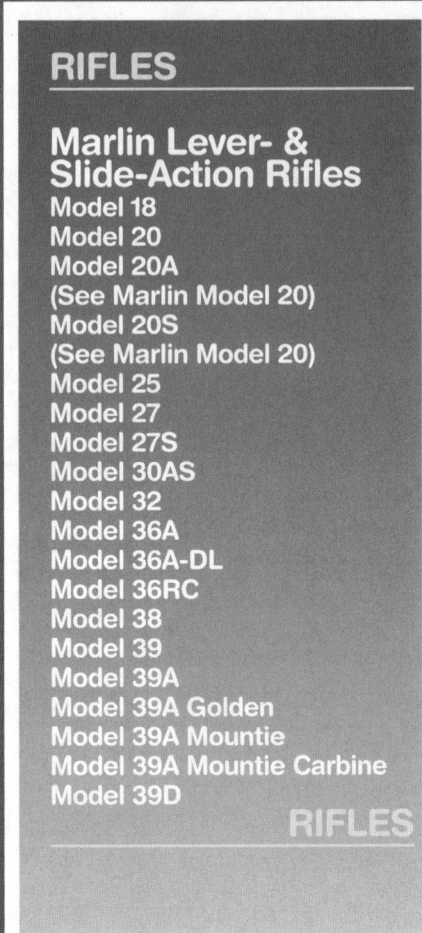

Marlin Model 20

Marlin Model 27

Marlin Model 38

Marlin Model 36A

Marlin Model 39

Marlin Model 39A

MARLIN LEVER & SLIDE ACTION RIFLES

MARLIN MODEL 18

Slide action; 22 Short, 22 Long, 22 LR; tube magazine; 20″ round or octagonal barrel; exposed hammer; open rear sight, bead front; solid frame; uncheckered straight stock, slide handle. Deluxe special-order models available. Introduced 1906; discontinued 1909. Some collector value.

Perf.: $350   **Exc.:** $290   **VGood:** $200

MARLIN MODEL 20

Slide action; 22 Short, 22 Long, 22 LR; 18-shot (22 LR) full-length tubular magazine, 10-shot (22 LR) half-length magazine; 24″ octagonal barrel; open rear sight, bead front; uncheckered straight grip stock. Later versions designated as 20S and 20A. Introduced 1907; discontinued 1922. Some collector value.

Perf.: $295   **Exc.:** $250   **VGood:** $170

MARLIN MODEL 25

Slide action; 22 Short; 15-shot tubular magazine; 23″ barrel; exposed hammer; open rear sight, bead front; uncheckered straight grip stock, slide handle. Introduced 1909; discontinued 1910. Some collector value.

Perf.: $350   **Exc.:** $300   **VGood:** $200

MARLIN MODEL 27

Slide action; 25-20, 32-20; 7-shot tubular magazine; 24″ octagonal barrel; open rear, bead front sight; uncheckered straight-grip stock; grooved slide handle. Introduced 1910; discontinued 1916. Some collector value.

Perf.: $330   **Exc.:** $280   **VGood:** $200

Marlin Model 27S

Same specs as Model 27 except 25 Stevens RF; round barrel. Introduced 1920; discontinued 1932.

Perf.: $320   **Exc.:** $250   **VGood:** $180

MARLIN MODEL 30AS

Lever action; 30-30: 6-shot tube magazine; 20″ barrel; 38 1/2″ overall length; weighs 7 lbs.; adjustable rear sight, bead front; walnut-finished Maine birch pistol grip stock; hammer-block safety. Introduced 1985; still in production as Model 336AS. **LMSR: $337**

Perf.: $280   **Exc.:** $230   **VGood:** $150

MARLIN MODEL 32

Slide action; 22 Short, 22 Long, 22 LR; 10-shot short tubular magazine, 18-shot full-length magazine; 24″ octagonal barrel; open rear, bead front sight; uncheckered pistol-grip stock, grooved slide handle. Some collector value. Introduced 1914; discontinued 1915.

Perf.: $390   **Exc.:** $320   **VGood:** $260

MARLIN MODEL 36-DL DELUXE RIFLE

Lever action; 32 Spl., 30-30; 6-shot tubular magazine; 24″ barrel; 42″ overall length; weighs 6 3/4 lbs.; "Huntsmen" hooded ramp front sight with silver bead, flat-top rear; checkered American black walnut pistol-grip stock; semi-beavertail forearm; case-hardened receiver, lever; visible hammer; drilled, tapped for tang peep sight; detachable swivels; 1″ leather sling. Introduced 1938; discontinued 1947.

Perf.: $290   **Exc.:** $250   **VGood:** $200

Marlin Model 36-DL Carbine

Same specs as Model 36A except 20″ barrel; 38″ overall length; no checkering, swivels, sling.

Perf.: $320   **Exc.:** $280   **VGood:** $240

Marlin Model 36RC

Same specs as Model 36 Carbine except "Huntsman" ramp front sight with silver bead and hood. Introduced 1938; discontinued 1947.

Perf.: $310   **Exc.:** $270   **VGood:** $200

MARLIN MODEL 38

Slide action; 22 Short, 22 Long, 22 LR; 10-shot tube magazine (22 LR), 12-shot (22 Long), 15-shot (22 Short); 24″ octagon barrel; hammerless takedown; open rear, bead front sight; uncheckered pistol-grip walnut stock; grooved slide handle. Introduced 1920; discontinued 1930.

Perf.: $330   **Exc.:** $290   **VGood:** $190

MARLIN MODEL 39

Lever action; 22 Short, 22 Long, 22 LR; 18-shot (22 LR), 20-shot (22 Long), 25-shot (22 Short); 24″ octagon barrel with tube magazine; open rear, bead front sight; uncheckered pistol-grip walnut stock, forearm. Introduced 1922; discontinued 1938.

Perf.: $1295   **Exc.:** $995   **VGood:** $450

Marlin Model 39A

Same specs as Model 39 except heavier stock; semi-beavertail forearm; semi-heavy round barrel. Introduced 1939; discontinued 1987.

Perf.: $470   **Exc.:** $400   **VGood:** $300

Marlin Model 39A Golden

Same specs as Model 39 except gold-plated trigger, other refinements. Introduced 1960.

Perf.: $250   **Exc.:** $220   **VGood:** $190

Marlin Model 39A Mountie

Same specs as Model 39A except 15-shot tubular magazine (22 LR), 16-shot (22 Long), 21-shot (22 Short); American black walnut straight-grip stock; slim forearm; 20″ barrel; 36″ overall length; weighs 6 lbs. Designation changed in 1972 to Model 39M. Introduced 1953; discontinued 1969.

Perf.: $250   **Exc.:** $220   **VGood:** $190

Marlin Model 39A Mountie Carbine

Same specs as Model 39A Mountie except weighs 5 1/4 lbs.; shorter tube magazine; 18-shot (22 Short, 14-shot (22 Long), 12-shot (22 LR).

Perf.: $250   **Exc.:** $220   **VGood:** $190

Marlin Model 39D

Same specs as Model 39 except 20 1/2″ barrel; 36 1/2″ overall length; weighs 5 3/4 lbs.; 21-shot (22 Short), 16-shot (22 Long), 15-shot (22 LR); adjustable rear, target front sight; grip cap; white line spacers; offset hammer spur. Introduced 1971; discontinued 1973.

Perf.: $250   **Exc.:** $200   **VGood:** $150

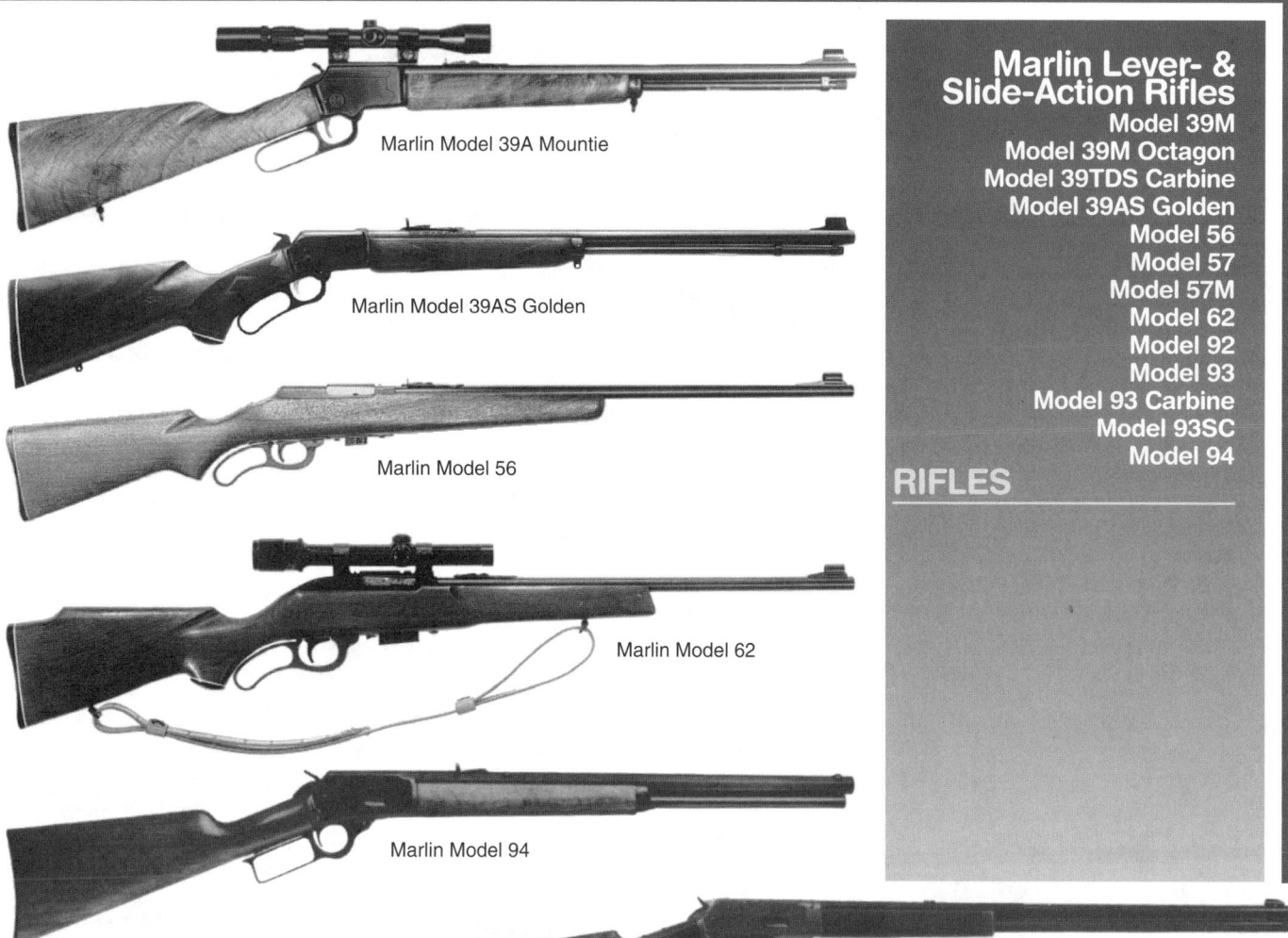

Marlin Model 39A Mountie

Marlin Model 39AS Golden

Marlin Model 56

Marlin Model 62

Marlin Model 94

Marlin Lever- & Slide-Action Rifles
Model 39M
Model 39M Octagon
Model 39TDS Carbine
Model 39AS Golden
Model 56
Model 57
Model 57M
Model 62
Model 92
Model 93
Model 93 Carbine
Model 93SC
Model 94

RIFLES

Marlin Model 39M
Same specs as Model 39 except 20″ barrel; 36″ overall length; weighs 6 lbs.; 21-shot (22 Short), 16-shot (22 Long), 15-shot (22 LR); straight-grip stock. Introduced 1963; discontinued 1987.
  Perf.: $250   **Exc.:** $220   **VGood** $190

Marlin Model 39M Octagon
Same specs as Model 39M except octagon barrel; bead front sight; hard rubber buttplate; squared lever. Introduced 1973.
  Perf.: $650   **Exc.:** $500   **VGood** $400

Marlin Model 39TDS Carbine
Same specs as Model 39A except 16-shot magazine (22 Short), 12-shot (22 Long), 10-shot (22 LR); 16 1/2″ Micro-Groove barrel; 32 5/8″ overall length; weighs 5 1/4 lbs.; hooded ramp front sight, adjustable semi-buckhorn rear; straight-grip American black walnut stock; hammer-block safety; rebounding hammer; gold-plated trigger; take-down style; blued finish. Marketed with case. Introduced 1988; discontinued 1996. **LMSR: $443**
  Perf.: $340   **Exc.:** $250   **VGood** $190

MARLIN MODEL 39AS GOLDEN
Lever action; 22 Short, 22 Long, 22 LR; 26-shot tubular magazine (22 Short), 21-shot (22 Long), 19-shot (22 LR); 24″ Micro-Groove barrel; 40″ overall length; weighs 6 1/2 lbs.; bead on ramp front sight with detachable Wide-Scan hood, folding fully-adjustable semi-buckhorn rear; receiver tapped for scope mount (supplied); checkered American black walnut stock with white line spacers at pistol-grip cap and buttplate; Mar-Shield finish; swivel studs; rubber buttpad; hammer-block safety; rebounding hammer; offset hammer spur; gold-plated steel trigger. Made in U.S. by Marlin. Introduced 1988; still in production. **LMSR: $509**
  Perf.: $310   **Exc.:** $230   **VGood** $175

Marlin Model 93

MARLIN MODEL 56
Lever action; 22 Short, 22 Long, 22 LR; 7-shot clip magazine; 22″ barrel; 41″ overall length; weighs 6 1/4 lbs.; open rear, ramp front sight; one-piece Bishop-style uncheckered walnut Monte Carlo-type pistol-grip stock. Introduced 1955; discontinued 1964.
  Perf.: $190   **Exc.:** $160   **VGood** $130

MARLIN MODEL 57
Lever action; 22 Short, 22 Long, 22 LR; 25-shot tube magazine (22 Short), 20-shot (22 Long), 18-shot (22 LR); 22″ barrel; 41″ overall length; weighs 6 1/4 lbs.; open rear, hooded ramp front; one-piece Bishop-style uncheckered walnut Monte Carlo-type pistol-grip stock. Introduced 1959; discontinued 1965.
  Perf.: $200   **Exc.:** $170   **VGood** $130

Marlin Model 57M
Same specs as Model 57 except 24″ barrel; 43″ overall length; weighs 6 1/2 lbs.; 22 WMR; 15-shot magazine; 25° lever throw. Introduced 1959; discontinued 1969.
  Perf.: $230   **Exc.:** $200   **VGood** $150

MARLIN MODEL 62
Lever action; 256 Mag., 22 Jet, 30 Carbine; 4-shot clip magazine; 24″ barrel; 43″ overall length; weighs 7 lbs.; open windage-adjustable rear sight, hooded ramp front; swivels; sling; uncheckered American black walnut Monte Carlo-type pistol-grip stock. The 256 Mag. was introduced in 1963; discontinued 1969.
256 Magnum
  Perf.: $350   **Exc.:** $300   **VGood** $250
30 Carbine
  Perf.: $300   **Exc.:** $260   **VGood** $210

MARLIN MODEL 92
Lever action; 22 Short, 22 Long, 22 LR, 32 Short, 32 LR, 32 Centerfire (with interchangeable firing pin); 16″, 24″, 26″, 28″ barrel; tubular magazine; open rear, blade front sight; uncheckered straight-grip stock, forearm. Originally marketed as Model 1892. Introduced 1892; discontinued 1915. Collector value.
22 Caliber
  Perf.: $1350   **Exc.:** $1095   **VGood** $895
32 Caliber
  Perf.: $1225   **Exc.:** $995   **VGood** $695

MARLIN MODEL 93
Lever action; 25-36 Marlin, 30-30, 32 Spl., 32-40, 38-55; 10-shot tubular magazine; 26″, 28″, 30″, 32″ round or octagonal barrel; solid frame or take-down; open rear, bead front sight; uncheckered straight-grip stock, forearm. Originally marketed as Model 1893. Introduced 1893; discontinued 1935. Collector value.
  Perf.: $1095   **Exc.:** $895   **VGood** $595

Marlin Model 93 Carbine
Same specs as Model 93 except 20″ barrel; 7-shot magazine; 30-30, 32 Spl.; carbine sights. Collector value.
  Perf.: $1295   **Exc.:** $995   **VGood** $800

Marlin Model 93SC
Same specs as Model 93 Carbine except shorter 5-shot magazine. Designated Sports Carbine. Collector value.
  Perf.: $1350   **Exc.:** $1095   **VGood** $895

MARLIN MODEL 94
Lever action; 25-20, 32-20, 38-40, 44-40; 10-shot magazine; 24″ (standard), 15″, 20″, 26″, 28″, 30″, 32″ round or octagonal barrel; solid frame or takedown; open rear, bead front sight; uncheckered pistol grip or straight stock, forearm. Originally marketed as Model 1894. Introduced 1894; discontinued 1933. Collector value.
  Perf.: $1200   **Exc.:** $895   **VGood** $695

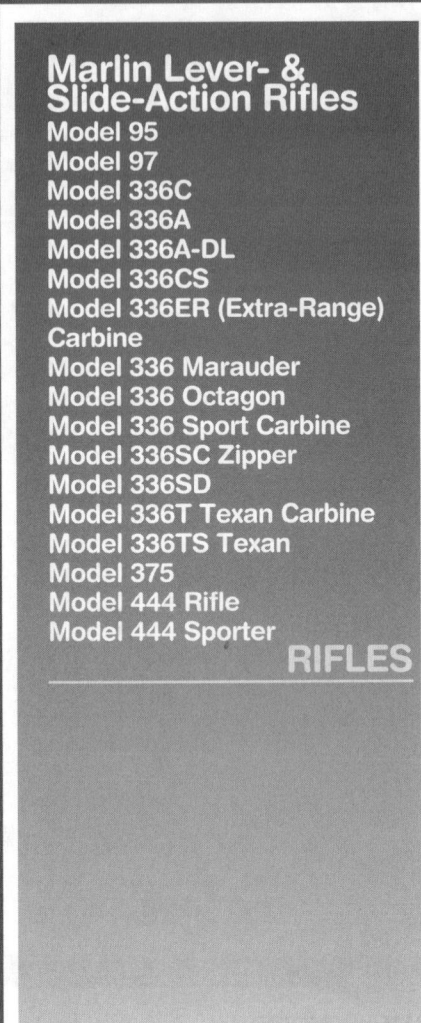

Marlin Lever- & Slide-Action Rifles

Model 95
Model 97
Model 336C
Model 336A
Model 336A-DL
Model 336CS
Model 336ER (Extra-Range) Carbine
Model 336 Marauder
Model 336 Octagon
Model 336 Sport Carbine
Model 336SC Zipper
Model 336SD
Model 336T Texan Carbine
Model 336TS Texan
Model 375
Model 444 Rifle
Model 444 Sporter

RIFLES

Marlin Model 336A

Marlin Model 336C

Marlin Model 336CS

Marlin Model 336T Texan Carbine

Marlin Model 375

MARLIN MODEL 95

Lever action; 33 WCF, 38-56, 40-65, 40-70, 40-82, 45-70; 9-shot tubular magazine; 24″ round or octagonal barrel, other lengths on special order; open rear, bead front sight; uncheckered straight or pistol-grip stock, forearm. Originally marketed as Model 1895. Introduced 1895; discontinued 1915. Collector value.
   Perf.: $1895   **Exc.:** $1495   **VGood:** $895

MARLIN MODEL 97

Lever action; 22 Short, 22 Long, 22 LR; tubular magazine; 16″, 24″, 26″, 28″ barrel; open rear, bead front sight; uncheckered straight or pistol grip stock, forearm. Originally marketed as Model 1897. Introduced 1897; discontinued 1916. Collector value.
   Perf.: $1300   **Exc.:** $895   **VGood:** $695

MARLIN MODEL 336C

Lever-action; 30-30, 35 Rem., 32 Spl.; 6-shot tubular magazine; 20″ barrel; 38 1/2″ overall length; weighs 7 lbs.; updated version of Model 36 carbine; semi-buck-horn adjustable folding rear sight, ramp front with Wide-Scan hood; receiver tapped for scope mounts; round breech bolt; gold-plated trigger; offset hammer spur; top of receiver sandblasted to reduce glare. Introduced 1948; discontinued 1983.
   Perf.: $250   **Exc.:** $230   **VGood:** $170
Pre-1962
   Perf.: $290   **Exc.:** $250   **VGood:** $200

Marlin Model 336A

Same specs as Model 336C except 24″ round barrel; two-thirds magazine tube; 6-shot; blued forearm cap; sling swivels. Introduced 1950; temporarily discontinued 1962. Reintroduced 1973; discontinued 1980.
   Perf.: $290   **Exc.:** $250   **VGood:** $200

Marlin Model 336A-DL

Same specs as Model 336A except sling; swivels; hand-checkered stock, forearm. Discontinued 1962.
   Perf.: $470   **Exc.:** $370   **VGood:** $290

Marlin Model 336CS

Same specs as Model 336A except 30-30, 35 Rem.; 6-shot two-thirds tube magazine; 20″ barrel; 38 1/2″ overall length; weighs 7 lbs.; select black walnut stock; capped pistol grip; hammer-block safety; tapped for scope mounts; offset hammer spur. Introduced 1984; still in production. **LMSR: $493**
   Perf.: $310   **Exc.:** $240   **VGood:** $190

Marlin Model 336 ER (Extra-Range) Carbine

Same specs as Model 336A except 356 Win., 307 Win.; 4-shot tubular magazine; new hammer-block safety; rubber buttpad; marketed with detachable sling swivels, branded leather sling. Introduced 1983; discontinued 1986.
   Perf.: $400   **Exc.:** $350   **VGood:** $280

Marlin Model 336 Marauder

Same specs as Model 336A except 30-30, 35 Rem.; 16 1/4″ barrel. Introduced 1963; discontinued 1964.
   Perf.: $410   **Exc.:** $360   **VGood:** $290

Marlin Model 336 Octagon

Same specs as Model 336A except octagon barrel. Introduced 1973.
   Perf.: $430   **Exc.:** $390   **VGood:** $310

Marlin Model 336 Sport Carbine

Same specs as Model 336A except 20″ barrel; weighs 7 lbs.; two-thirds 6-shot magazine. Introduced 1948; discontinued 1963.
   Perf.: $300   **Exc.:** $250   **VGood:** $200

Marlin Model 336SC Zipper

Same specs as Model 336 Sport Carbine except weighs 7 3/4 lbs.; only 219 Zipper. Introduced 1955; discontinued 1960.
   Perf.: $550   **Exc.:** $500   **VGood:** $410

Marlin Model 336SD

Same specs as Model 336A except 20″ barrel; checkered pistol grip, forend; 38 1/2″ overall length; weighs 7 1/2 lbs.; leather sling; deluxe sports carbine. Introduced 1954; discontinued 1962.
   Perf.: $400   **Exc.:** $295   **VGood:** $220

Marlin Model 336T Texan Carbine

Same specs as Model 336A except 18 1/2″ (1953), 20″ barrel; straight-grip uncheckered walnut stock; squared lever. Called the Texan Model. Introduced in 1953; discontinued 1983. Originally chambered in 30-30, 35 Rem.; chambered in 44 Mag. from 1963 to 1967. Deluxe 336TDL version manufactured 1962-1963. No longer in production.
   Perf.: $300   **Exc.:** $230   **VGood:** $175
Deluxe 336TDL
   Perf.: $400   **Exc.:** $330   **VGood:** $240

Marlin Model 336TS Texan

Same specs as Model 336CS except 30-30; 18 1/2″ barrel; straight grip-stock; cross-bolt safety. Introduced 1984; discontinued 1987.
   Perf.: $270   **Exc.:** $195   **VGood:** $155

MARLIN MODEL 375

Lever action; 375 Win.; 5-shot tube magazine; 20″ barrel; 38 1/2″ overall length; weighs 6 3/4 lbs.; hooded ramp front, adjustable semi-buckhorn folding rear sight; checkered American black walnut stocks; quick-detachable swivels; leather sling. Introduced 1980; discontinued 1983.
   Perf.: $330   **Exc.:** $270   **VGood:** $220

MARLIN MODEL 444 RIFLE

Lever action; 444 Marlin; 5-shot tube magazine; 24″ barrel; 42 1/2″ overall length; weighs 7 1/2 lbs.; open elevation-adjustable rear sight, hooded ramp front; action is strengthened version of Model 336; straight-grip Monte Carlo stock of uncheckered American walnut; recoil pad; carbine forearm; barrel band; sling; sling swivels. Introduced 1965; discontinued 1972.
   Perf.: $325   **Exc.:** $250   **VGood:** $190

Marlin Model 444 Sporter

Same specs as Model 444 Rifle except 22″ barrel, 40 1/2″ overall length; folding rear, brass bead front sight; Model 336A-type pistol-grip stock, forearm; recoil pad; detachable swivels; sling. Introduced 1972; discontinued 1983.
   Perf.: $285   **Exc.:** $230   **VGood:** $200

Marlin Model 444 Sporter

Marlin Model 1894C

Marlin Model 1894CL Classic

Marlin Model 1894CS Carbine

Marlin Model 1895

Marlin Model 1895

Marlin Lever- & Slide-Action Rifles
Model 444SS
Model 1892 (First Model)
(See Marlin Model 92)
Model 1893 (First Model)
(See Marlin Model 93)
Model 1894 (First Model)
(See Marlin Model 94)
Model 1894
Model 1894C
Model 1894CL Classic
Model 1894CS Carbine
Model 1894M
Model 1894 Octagon
Model 1894 Sporter
Model 1894S
Model 1895 (First Model)
(See Marlin Model 95)
Model 1895
Model 1895S
Model 1895SS
Model 1897 (First Model) (See
Marlin Model 97)
Model 1936 Rifle/Carbine
Sports Carbine
(See Marlin Model 93SC)
Tom Mix Special
(See Marlin Model 100S)

RIFLES

Marlin Model 444SS

Same specs as Model 444 Sporter except Micro Groove barrel; fully-adjustable folding semi-buckhorn rear sight; hammer-block safety; offset hammer spur; Mar-Shield finish. Introduced 1984; still in production. **LMSR: $582**

Perf.: $390   **Exc.: $330**   **VGood: $230**

MARLIN MODEL 1894

Lever action; 41 Mag., 44 Mag., 45 Colt; 10-shot tube magazine; 20″ carbine barrel; 37 1/2″ overall length; weighs 6 lbs.; hooded ramp front sight, semi-buckhorn adjustable rear; uncheckered straight grip black walnut stock, forearm; gold-plated trigger; receiver tapped for scope mount; offset hammer spur; solid top receiver sand blasted to reduce glare. 41 Mag. produced 1985-1988; 45 Colt produced 1989-1990. Introduced 1969; still in production as Model 1894S.

Perf.: $400   **Exc.: $300**   **VGood: $250**

Marlin Model 1894C

Same specs as Model 1894 except 357 Mag.; 9-shot tube magazine; 18 1/2″ barrel; 35 1/2″ overall length; weighs 6 lbs. Introduced 1979; discontinued 1985.

Perf.: $380   **Exc.: $290**   **VGood: $225**

Marlin Model 1894CL Classic

Same specs as 1894CL except 218 Bee, 25-20, 32-30; brass bead front sight, adjustable semi-buckhorn folding rear; hammer-block safety; rubber rifle buttpad; swivel studs. Introduced 1988; discontinued 1994.

Perf.: $390   **Exc.: $300**   **VGood: $250**

Marlin Model 1894CS Carbine

Same specs Model 1894 except 38 Spl., 357 Mag.; 18 1/2″ barrel; 9-shot magazine; brass bead front sight; hammer-block safety. Introduced 1985; still in production. **LMSR: $510**

Perf.: $385   **Exc.: $290**   **VGood: $240**

Marlin Model 1894M

Same specs as Model 1894 except 22 WMR; 11-shot magazine; no gold-plated trigger; receiver tapped for scope mount or receiver sight. Introduced 1982; discontinued 1986.

Perf.: $270   **Exc.: $230**   **VGood: $190**

Marlin Model 1894 Octagon

Same specs as Model 1894CS Carbine except octagon barrel; bead front sight. Introduced 1973.

Perf.: $390   **Exc.: $300**   **VGood: $250**

Marlin Model 1894 Sporter

Same specs as Model 1894CS Carbine except 22″ barrel, 6-shot magazine. Introduced 1973.

Perf.: $390   **Exc.: $300**   **VGood: $250**

Marlin Model 1894S

Same specs as Model 1894 except 41 Mag., 44 Spl., 44 Mag.; buttpad; swivel studs; hammer-block safety; offset hammer spur. Still in production. **LMSR: $510**

Perf.: $370   **Exc.: $300**   **VGood: $230**

MARLIN MODEL 1895

Lever action; 45-70; 4-shot tube magazine; 22″ round barrel; 40 1/2″ overall length; weighs 7 lbs.; offset hammer spur; adjustable semi-buckhorn folding rear sight, bead front; solid receiver tapped for scope mounts, receiver sights; two-piece uncheckered straight grip American walnut stock, forearm of black walnut; rubber buttplate; blued steel forend cap. Meant to be a recreation of the original Model 1895 discontinued in 1915. Actually built on action of Marlin Model 444. Introduced 1972; discontinued 1979.

Perf.: $320   **Exc.: $280**   **VGood: $230**

Marlin Model 1895S

Same specs as Model 1895 except full pistol-grip stock; straight buttpad. Introduced 1980; discontinued 1983.

Perf.: $300   **Exc.: $250**   **VGood: $200**

Marlin Model 1895SS

Same specs as Model 1895 except full pistol-grip stock; hammer-block safety. Introduced 1984; still in production. **LMSR: $582**

Perf.: $350   **Exc.: $290**   **VGood: $230**

MARLIN MODEL 1936 RIFLE/CARBINE

Lever action; 32 Spl., 30-30; 6-shot tube magazine; 20″, 24″ barrel; open rear sight, bead front; uncheckered walnut pistol-grip stock; semi-beavertail forearm; carbine barrel band. Introduced 1936; discontinued 1937.

Perf.: $500   **Exc.: $380**   **VGood: $275**

Marlin-Glenfield
Model 15
Model 20
Model 30
Model 30A
Model 30GT
Model 60
Model 70
Model 75C
Model 81G
Model 101G
Model 989G

Mauser
Model 10 Varminter
Model 66
Model 66 Safari
Model 66 Stutzen
Model 66SP Match

RIFLES

Mauser Model 10 Varminter

Mauser Model 66 Magnum

Mauser Model 66SP Match

Mauser Model 77 Big Game

MARLIN-GLENFIELD MODEL 15
Bolt action; single shot; 22 Short, 22 Long, 22 LR; 22″ barrel; 41″ overall length; weighs 5 1/2 lbs.; ramp front, adjustable open rear sight; receiver grooved for tip-off scope mount; Monte Carlo walnut-finished hardwood stock; checkered full pistol grip; red cocking indicator; thumb safety. Introduced 1979; discontinued 1983.
Perf.: $70   **Exc.:** $60   **VGood:** $50

MARLIN-GLENFIELD MODEL 20
Bolt action; 22 Short, 22 Long, 22 LR; 8-shot clip magazine; 22″ barrel; bead front sight; walnut finished stock.
Perf.: $80   **Exc.:** $70   **VGood:** $60

MARLIN-GLENFIELD MODEL 30
Lever action; 30-30; 4-shot tubular magazine; 20″ barrel; 38 1/2″ overall length; weighs 7 lbs.; semi-buckhorn adjustable folding rear sight, ramp front with Wide-Scan hood; receiver tapped for scope mounts; walnut finished hardwood stock with semi-beavertail forend; round breech bolt; gold-plated trigger; offset hammer spur; top of receiver sandblasted to reduce glare; blued forearm cap; sling swivels. Introduced 1966; discontinued 1972.
Perf.: $180   **Exc.:** $155   **VGood:** $140

Marlin-Glenfield Model 30A
Same specs as Model 30 except impressed checkering on stock, forearm. Introduced 1969; still in production as Marlin Model 30AS.
Perf.: $180   **Exc.:** $155   **VGood:** $130

Marlin-Glenfield Model 30GT
Same specs as Marlin-Glenfield Model 30 except 18 1/2″ barrel; brass bead front, adjustable rear sight; straight grip walnut-finished hardwood stock; receiver sandblasted. Introduced 1979; discontinued 1981.
Perf.: $180   **Exc.:** $155   **VGood:** $130

MARLIN-GLENFIELD MODEL 60
Semi-automatic; 22 LR; 14-, 18-shot tubular magazine; 22″ tapered, round barrel; 40 1/2″ overall length; weighs

5 1/2 lbs.; ramp front, fully-adjustable rear sight; press-checkered, walnut-finished Maine birch stock with Monte Carlo full pistol grip; manual bolt hold-open; Mar-Shield finish. Introduced 1966; still in production as Marlin Model 60. **LMSR: $158**
Perf.: $100   **Exc.:** $80   **VGood:** $70

MARLIN-GLENFIELD MODEL 70
Semi-automatic; 22 LR; 7-shot detachable clip magazine; 18″ barrel; 36 1/2″ overall length; weighs 4 1/2 lbs.; open adjustable rear, ramp front sight; chrome-plated trigger; cross-bolt safety; bolt hold-open; chrome-plated magazine; sling swivels; walnut-finished hardwood stock. Introduced 1967; discontinued 1994.
Perf.: $110   **Exc.:** $90   **VGood:** $70

MARLIN-GLENFIELD MODEL 75C
Semi-automatic; 22 LR; 14-shot magazine; 18″ barrel; 36 3/4″ overall length; weighs 5 lbs.; ramp front, open adjustable rear sight; walnut finished hardwood Monte Carlo stock with full pistol grip; bolt hold-open; cross-bolt safety; receiver grooved for scope mounts. Introduced 1980.
Perf.: $115   **Exc.:** $100   **VGood:** $80

MARLIN-GLENFIELD MODEL 81G
Bolt action; 22 Short, 22 Long, 22 LR; 18-shot (22 LR), 20-shot (22 Long), 24-shot (22 Short) tube magazine; open rear, bead front sight; standard uncheckered pistol-grip stock; chrome-plated trigger. Introduced 1937; discontinued 1971.
Perf.: $70   **Exc.:** $60   **VGood:** $50

MARLIN-GLENFIELD MODEL 101G
Bolt action; single shot; 22 Short, 22 Long, 22 LR; 22″ barrel; semi-buckhorn fully-adjustable rear sight, hooded wide-scan front; black plastic triggerguard; T-shaped cocking piece; uncheckered hardwood stock with beavertail forearm; receiver grooved for tip-off scope mount. Introduced 1960; discontinued 1965.
Perf.: $60   **Exc.:** $50   **VGood:** $45

MARLIN-GLENFIELD MODEL 989G
Semi-automatic; 22 LR; 7-shot clip detachable magazine; 18″ barrel; bead front, open rear sight; plain stock. Same as Marlin Model 989M2. Introduced 1962; discontinued 1964.
Perf.: $100   **Exc.:** $85   **VGood:** $70

MAUSER MODEL 10 VARMINTER
Bolt action; 22-250; 5-shot box magazine; 24″ heavy barrel; no sights; drilled, tapped for scope mounts; hand-checkered European walnut Monte Carlo pistol-grip stock; externally adjustable trigger; hammer-forged Krupp Special Ordnance steel barrel. Introduced 1973; discontinued 1975.
Perf.: $495   **Exc.:** $440   **VGood:** $400

MAUSER MODEL 66
Bolt action; 243, 270, 308, 30-06, 5.6x57, 6.5x57, 7x64, 9.3x62, 7mm Rem. Mag., 300 Wea. Mag., 300 Win. Mag., 6.5x68, 8x68S, 9.3x64; 3-shot magazine; 24″ barrel (standard calibers), 26″ (magnum caliber); 39″ overall length (standard calibers); weighs 7 1/2 to 9 3/8 lbs.; blade front sight on ramp, fully-adjustable open rear; hand-checkered European walnut stock with Monte Carlo comb; hand-rubbed oil finish; rosewood forend and grip caps; telescoping short-stroke action; interchangeable, free-floated, medium-heavy barrels; mini-claw extractor; adjustable single-stage trigger; internal magazine. Imported from Germany by Precision Imports, Inc. Introduced 1989; discontinued 1994.
Perf.: $2195   **Exc.:** $1295   **VGood:** $795

Mauser Model 66 Safari
Same specs as Model 66 except 375 H&H, 458 Win. Mag.; weighs 9 3/8 lbs. Imported from Germany by Precision Imports, Inc. Introduced 1989; discontinued 1994.
Perf.: $2995   **Exc.:** $1895   **VGood:** $995

Mauser Model 66 Stutzen
Same specs as Model 66 except 21″ barrel; full-length Stutzen stock. Imported from Germany by Precision Imports, Inc. Introduced 1989; discontinued 1994.
Perf.: $2395   **Exc.:** $1495   **VGood:** $895

MAUSER MODEL 66SP MATCH
Bolt action; 308 Win.; 3-shot magazine; 27 1/2″ barrel, muzzle brake; no sights; scope supplied; match-design European walnut stock; thumbhole pistol grip; spring-loaded cheekpiece; Morgan adjustable recoil pad; adjustable match trigger. Introduced 1976; discontinued 1990.
Perf.: $3995   **Exc.:** $3495   **VGood:** $2495

Mauser Model 77 Sportsman

Mauser Model 98 Standard

Mauser Model 99 Monte Carlo Stock

Mauser Model 99 Classic Stock

Mauser
Model 77
Model 77 Big Game
Model 77 Magnum
Model 77 Mannlicher
Model 77 Sportsman
Model 77 Ultra
Model 83
Model 83 Single Shot
Model 83 Match UIT
Model 86-SR Specialty Rifle
Model 98 Standard
Model 99
Model 107

RIFLES

MAUSER MODEL 77

Bolt action; 243, 270, 6.5x57, 7x64, 308, 30-06; 3-shot detachable magazine; 24" barrel; weighs 7 1/4 lbs.; ramp front sight, open adjustable rear; oil-finished European walnut stock; rosewood grip cap; forend tip; palm-swell pistol grip; Bavarian cheekpiece; recoil pad; interchangeable double-set or single trigger. Introduced 1981; no longer imported.
 Perf.: $1050   **Exc.:** $910   **VGood:** $750

Mauser Model 77 Big Game
Same specs as Model 77 except 375 H&H, 458 Win. Mag.; 26" barrel; weighs 8 1/2 lbs. Introduced 1981; no longer imported.
 Perf.: $1195   **Exc.:** $895   **VGood:** $695

Mauser Model 77 Magnum
Same specs as Model 77 except 7mm Rem. Mag., 6.5x68, 300 Win. Mag., 300 Weatherby Mag., 9.3x62, 9.3x64, 8x68S; 26" barrel; weighs 8 1/8 lbs. Introduced 1981; no longer imported.
 Perf.: $1095   **Exc.:** $895   **VGood:** $695

Mauser Model 77 Mannlicher
Same specs as Model 77 except 20" barrel; 7 3/4 lbs.; European walnut full-stock; set trigger. Introduced 1981; no longer imported.
 Perf.: $1195   **Exc.:** $995   **VGood:** $750

Mauser Model 77 Sportsman
Same specs as Model 77 except 243, 308; weighs 9 lbs.; no sights; set trigger. No longer imported.
 Perf.: $1395   **Exc.:** $995   **VGood:** $750

Mauser Model 77 Ultra
Same specs as Model 77 except 6.5x57, 7x64, 30-06; 20" barrel; weighs 7 3/4 lbs. Introduced 1981; no longer imported.
 Perf.: $995   **Exc.:** $795   **VGood:** $625

MAUSER MODEL 83

Bolt action; 308; 10-shot magazine; 26" barrel; European walnut match stock; adjustable buttplate comb; match trigger. No longer imported.
 Perf.: $2195   **Exc.:** $1695   **VGood:** $1345

Mauser Model 83 Single Shot
Same specs as Model 83 except single shot action. No longer imported.
 Perf.: $1995   **Exc.:** $1495   **VGood:** $995

Mauser Model 83 Match UIT
Same specs as Model 83 except single shot action; free rifle designed for UIT competition. No longer imported.
 Perf.: $2250   **Exc.:** $1795   **VGood:** $1375

MAUSER MODEL 86-SR SPECIALTY RIFLE

Bolt action; 308 Win.; 9-shot detachable magazine; 25 5/8" fluted barrel with 1:12" twist; 47 3/4" overall length; weighs about 11 lbs.; no sights; competition metallic sights or scope mount optional; laminated wood, fiberglass, or special match thumbhole wood stock with rail in forend and adjustable recoil pad; match barrel with muzzle brake; action with two front bolt locking lugs; action bedded in stock with free-floated barrel; match trigger adjustable as single or two-stage, fully-adjustable for weight, slack, and position; silent safety locks bolt, firing pin. Imported from Germany by Precision Imports, Inc. Introduced 1989; discontinued 1994.
With fiberglass stock
 Perf.: $4395   **Exc.:** $3500   **VGood:** $2750
With match thumbhole stock
 Perf.: $4195   **Exc.:** $3295   **VGood:** $2500

MAUSER MODEL 98 STANDARD

Bolt action; 7mm Mauser, 7.9mm Mauser; 5-shot box magazine; 23 1/2" barrel; adjustable rear, blade front sight; military-style uncheckered European walnut stock; straight military-type bolt handle; Mauser trademark on receiver ring; commercial version of German military rifle. Introduced post-WWI; discontinued 1938.
 Perf.: $695   **Exc.:** $495   **VGood:** $350

MAUSER MODEL 99

Bolt action; 243, 25-06, 270, 308, 30-06, 5.6x57, 6.5x57, 7x57, 7x64 (standard calibers); 7mm Rem. Mag., 257 Wea. Mag., 270 Wea. Mag., 300 Wea. Mag., 300 Win. Mag., 338 Win. Mag., 375 H&H, 8x68S, 9.3x64 (magnum calibers); removable 4-shot magazine (standard calibers), 3-shot (magnums); 24" barrel (standard calibers), 26" (magnums); 44" overall length (standard calibers); weighs about 8 lbs.; no sights; drilled and tapped for scope mounting; hand-checkered European walnut stock with rosewood grip cap; accuracy bedding with free-floated barrel, three front-locking bolt lugs, 60° bolt throw; adjustable single-stage trigger; silent safety locks bolt, sear, trigger. Imported from Germany by Precision Imports, Inc. Introduced 1989; discontinued 1994.
Classic stock, oil finish
 Perf.: $1175   **Exc.:** $975   **VGood:** $695
Classic stock, high luster finish
 Perf.: $1150   **Exc.:** $925   **VGood:** $695
Monte Carlo stock, oil finish
 Perf.: $1195   **Exc.:** $995   **VGood:** $695
Monte Carlo stock, high luster finish
 Perf.: $1175   **Exc.:** $975   **VGood:** $695

MAUSER MODEL 107

Bolt action; 22 LR; 5-shot magazine; 21 1/2" barrel; 40" overall length; weighs about 5 lbs.; hooded blade front sight, adjustable open rear; grooved receiver for scope mounting; walnut-stained beechwood stock with Monte Carlo comb; checkered grip and forend; sling swivels; dual extractors; 60° bolt throw; steel triggerguard and floorplate; satin blue finish. Imported from Germany by Precision Imports, Inc. Introduced 1992; discontinued 1994.
 Perf.: $300   **Exc.:** $260   **VGood:** $220

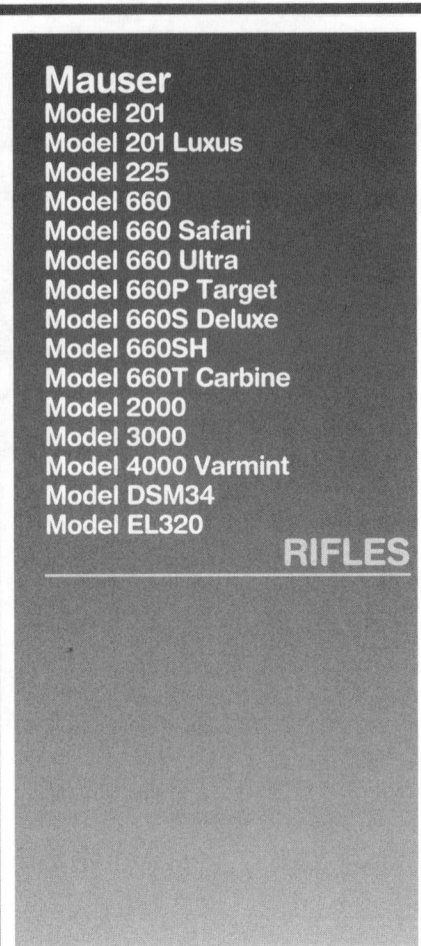

Mauser
Model 201
Model 201 Luxus
Model 225
Model 660
Model 660 Safari
Model 660 Ultra
Model 660P Target
Model 660S Deluxe
Model 660SH
Model 660T Carbine
Model 2000
Model 3000
Model 4000 Varmint
Model DSM34
Model EL320

RIFLES

Mauser Model 201

Mauser Model 225

Mauser Model 660 Safari

Mauser Model 660

Mauser Model 3000

MAUSER MODEL 201

Bolt action; 22 LR, 22 WMR; 5-shot magazine; 21" barrel; 40" overall length; weighs about 6 1/2 lbs.; sold with or without sights; receiver accepts rail mounts and drilled, tapped for scope mounting; walnut-stained beechwood stock with Monte Carlo comb and cheekpiece; checkered grip and forend; hammer forged medium-heavy, free-floated barrel; bolt with two front locking lugs, dual extractors; adjustable trigger; safety locks bolt, sear and trigger. Imported from Germany by Precision Imports, Inc. Introduced 1989; discontinued 1994.

22 LR
| **Perf.:** $595 | **Exc.:** $495 | **VGood:** $395 |

22 WMR
| **Perf.:** $650 | **Exc.:** $550 | **VGood:** $450 |

Mauser Model 201 Luxus
Same specs as Model 201 except walnut stock with rosewood forend tip. Imported from Germany by Precision Imports, Inc. Introduced 1989; discontinued 1994.

22 LR
| **Perf.:** $695 | **Exc.:** $595 | **VGood:** $495 |

22 WMR
| **Perf.:** $730 | **Exc.:** $630 | **VGood:** $495 |

MAUSER MODEL 225

Bolt action; 243, 25-06, 270, 7x57, 308, 30-06, 257 Weatherby, 270 Weatherby, 7mm Rem. Mag., 300 Win. Mag., 300 Weatherby, 308 Norma Mag., 375 H&H Mag.; 3-shot magazine (magnums), 4-shot (standard); 24" barrel (standard), 26" (magnums); 44 1/2" overall length with longer barrel; weighs 8 lbs.; no sights; drilled, tapped for scope mounts; hand-checkered, oil-finished European Monte Carlo stock; recoil pad; swivel studs. Made in Germany. Introduced 1988; discontinued 1989.

Standard calibers
| **Perf.:** $1050 | **Exc.:** $950 | **VGood:** $750 |

Magnum calibers
| **Perf.:** $1100 | **Exc.:** $1000 | **VGood:** $800 |

MAUSER MODEL 660

Bolt action; 243 Win., 25-06, 270 Win., 308 Win., 30-06, 7x57, 7mm Rem. Mag.; 24" barrel; no sights; drilled, tapped for scope mounts; checkered European Monte Carlo walnut stock; white line pistol-grip cap, recoil pad; adjustable single-stage trigger; push-button safety. Introduced 1973; discontinued 1975.
| **Perf.:** $895 | **Exc.:** $795 | **VGood:** $595 |

Mauser Model 660 Safari
Same specs as Model 660 except only 458 Win. Mag., 375 H&H Mag., 338 Win. Mag., 7mm Rem. Mag.; 28" barrel; express rear sight, fixed ramp front. Introduced 1973; discontinued 1975.
| **Perf.:** $995 | **Exc.:** $795 | **VGood:** $595 |

Mauser Model 660 Ultra
Same specs Model 660 except 21" barrel. Introduced 1965; discontinued 1975.
| **Perf.:** $795 | **Exc.:** $695 | **VGood:** $495 |

Mauser Model 660P Target
Same specs as Model 660 except 308 Win., other calibers on special order; 3-shot magazine; 26 1/2" heavy barrel, muzzle brake; dovetail rib for scope mounting; European walnut target stock with full pistol-grip, thumbhole, adjustable cheekpiece, adjustable rubber buttplate. Not imported.
| **Perf.:** $1295 | **Exc.:** $895 | **VGood:** $695 |

Mauser Model 660S Deluxe
Same specs as Model 660 except 21" barrel; carbine version; engraving; gold/silver inlay work; heavy carving on select walnut stock. Special order only; no longer imported. Prices vary depending on special order features. **$2500+**

Mauser Model 660SH
Same specs as Model 660 except 6.5x68, 7mm Rem. Mag., 7mm SE v. Hoff, 300 Win. Mag., 8x68S, 9.3x64. Introduced 1965; discontinued 1975.
| **Perf.:** $995 | **Exc.:** $795 | **VGood:** $595 |

Mauser Model 660T Carbine

Mauser Model 660T Carbine
Same specs as Model 660 except 21" barrel; full-length stock. Introduced 1965; discontinued 1975.
| **Perf.:** $1050 | **Exc.:** $875 | **VGood:** $695 |

MAUSER MODEL 2000

Bolt action; 270 Win., 308 Win., 30-06; 5-shot magazine; 24" barrel; folding leaf rear sight, hooded ramp front; hand-checkered European walnut stock; Monte Carlo comb; cheekpiece; forend tip; sling swivels. Introduced 1969; discontinued 1971.
| **Perf.:** $450 | **Exc.:** $400 | **VGood:** $370 |

MAUSER MODEL 3000

Bolt action; 243 Win., 270 Win., 308 Win., 30-06, 375 H&H Mag., 7mm Rem. Mag.; 3-shot magazine (magnum), 4-shot (standard); 22" barrel (standard calibers), 26" (magnums); no sights; drilled, tapped for scope mounts; hand-checkered European walnut Monte Carlo stock; white line spacer on pistol-grip cap; recoil pad; sliding safety; fully-adjustable trigger; detachable sling swivels. Left-hand action available. Introduced 1971; discontinued 1975.
| **Perf.:** $495 | **Exc.:** $450 | **VGood:** $400 |

Left-hand model
| **Perf.:** $550 | **Exc.:** $500 | **VGood:** $460 |

Magnum calibers
| **Perf.:** $595 | **Exc.:** $530 | **VGood:** $500 |

Left-hand model
| **Perf.:** $625 | **Exc.:** $575 | **VGood:** $500 |

MAUSER MODEL 4000 VARMINT

Bolt action; 222 Rem., 223 Rem.; 4-shot magazine; 22" barrel; folding leaf rear sight, hooded ramp front; hand-checkered European walnut Monte Carlo stock; rubber buttplate. Introduced 1971; discontinued 1975.
| **Perf.:** $495 | **Exc.:** $430 | **VGood:** $380 |

MAUSER MODEL DSM34

Bolt action; single shot; 22 LR; 26" barrel; tangent curve open rear sight, barleycorn front; Model 98 Mauser military-type stock; no checkering; sling swivels. Introduced 1934; discontinued 1939.
| **Perf.:** $450 | **Exc.:** $370 | **VGood:** $300 |

MAUSER MODEL EL320

Bolt action; single shot; 22 LR; 23 1/2" barrel; adjustable open rear sight, bead front; sporting-style European walnut stock; hand-checkered pistol grip, forearm; grip cap; sling swivels. Introduced 1927; discontinued 1935.
| **Perf.:** $475 | **Exc.:** $400 | **VGood:** $300 |

Mauser Model 4000 Varmint

Mauser Model MS420B

Mauser Sporter

Mauser Special British Model Type A

Masuer Special British Model Type A Magnum

Mauser Type B

Mauser
Model EN310
Model ES340
Model ES340B
Model ES350
Model ES350B
Model KKW
Model MM410
Model MM410B
Model MS350B
Model MS420
Model MS420B
Sporter
Sporter Carbine
Sporter, Military Model
Sporter, Short Model
Special British Model Type A
Special British Model Type A
Magnum
Special British Model Type A
Short
Type B

RIFLES

MAUSER MODEL EN310
Bolt action; single shot; 22 LR; 19 3/4″ barrel; fixed open rear sight, blade front; uncheckered European walnut pistol-grip stock. Introduced post-WWI; discontinued 1935.
Perf.: $450   Exc.: $360   VGood: $300

MAUSER MODEL ES340
Bolt action; single shot; 22 LR; 25 1/2″ barrel; tangent curve rear sight, ramp front; European walnut sporting stock; hand-checkered pistol grip; grooved forearm; sling swivels. Introduced 1923; discontinued 1935.
Perf.: $425   Exc.: $350   VGood: $290

Mauser Model ES340B
Same specs as Model ES340 except 26 3/4″ barrel; uncheckered pistol-grip stock. Introduced 1935; discontinued 1939.
Perf.: $425   Exc.: $350   VGood: $290

MAUSER MODEL ES350
Bolt action; repeater; 22 LR; 27 1/2″ barrel; open micrometer rear sight, ramp front; target-style walnut stock; hand-checkered pistol grip, forend; grip cap. Introduced 1925; discontinued 1935.
Perf.: $650   Exc.: $575   VGood: $450

Mauser Model ES350B
Same specs as Model ES350 except single shot; 26 3/4″ barrel; receiver grooved for scope mount. Introduced 1935; discontinued 1938.
Perf.: $525   Exc.: $450   VGood: $400

MAUSER MODEL KKW
Bolt action; single shot; 22 LR; 26″ barrel; tangent curve open rear sight, barleycorn front; Model 98 Mauser military-type walnut stock; no checkering; sling swivels; used in training German troops in WWII. Introduced 1935; discontinued 1939.
Perf.: $495   Exc.: $425   VGood: $375

MAUSER MODEL MM410
Bolt action; 22 LR; 5-shot detachable box magazine; 23 1/2″ barrel; tangent curve open rear sight, ramp front; European walnut sporting stock; hand-checkered pistol grip; sling swivels. Introduced 1926; discontinued 1935.
Perf.: $795   Exc.: $595   VGood: $475

Mauser Model MM410B
Same specs as MM410 except lightweight sporting stock. Introduced 1935; discontinued 1939.
Perf.: $595   Exc.: $495   VGood: $395

MAUSER MODEL MS350B
Bolt action; 22 LR; 5-shot detachable box magazine; 26 3/4″ barrel; micrometer open rear sight, ramp front; barrel grooved for detachable rear sight/scope; target stock of European walnut; hand-checkered pistol grip, forearm; sling swivels. Replaced Model ES350. Introduced 1935; discontinued 1939.
Perf.: $695   Exc.: $520   VGood: $405

MAUSER MODEL MS420
Bolt action; 22 LR; 5-shot detachable box magazine; 25 1/2″ barrel; tangent curve open rear sight, ramp front; European walnut sporting stock; hand-checkered pistol grip; grooved forearm; sling swivels. Introduced 1925; discontinued 1935.
Perf.: $495   Exc.: $420   VGood: $390

Mauser Model MS420B
Same specs as Model MS420 except 25 3/4″ barrel; better stock wood. Introduced 1935; discontinued 1939.
Perf.: $525   Exc.: $440   VGood: $400

MAUSER SPORTER
Bolt action; 6.5x55, 6.5x58, 7x57, 9x57, 9.3x62, 10.75x68; 5-shot box magazine; 23 1/2″ barrel; tangent-curve rear sight, ramp front; uncheckered European walnut stock, pistol grip; Schnabel-tipped forearm; sling swivels; double-set trigger. Made in Germany prior to World War I.
Perf.: $1250   Exc.: $995   VGood: $700

Mauser Sporter Carbine
Same specs Sporter except only 6.5x54, 6.5x58, 7x57, 8x57, 9x57; 19 3/4″ barrel. Manufactured prior to WWI.
Perf.: $1395   Exc.: $1095   VGood: $750

Mauser Sporter, Military Model
Same specs as Sporter except 7x57, 8x57, 9x57; military front sight; Model 98-type barrel; double-pull trigger. Manufactured prior to WWI.
Perf.: $795   Exc.: $695   VGood: $495

Mauser Sporter, Short Model
Same specs as Sporter except only 6.5x54, 8x58; 19 3/4″ barrel. Manufactured prior to WWI.
Perf.: $2495   Exc.: $1995   VGood: $1195

MAUSER SPECIAL BRITISH MODEL TYPE A
Bolt action; 7x57, 8x60, 9x57, 9.3x62mm, 30-06; 5-shot box magazine; 23 1/2″ barrel; express rear sight, hooded ramp front; hand-checkered pistol grip; Circassian walnut sporting stock; buffalo horn grip cap, forearm tip; military-type trigger; detachable sling swivels. Introduced before WWI; discontinued 1938.
Perf.: $2995   Exc.: $2395   VGood: $1495

Mauser Special British Model Type A Magnum
Same specs as Type A except magnum action; 10.75x68mm, 280 Ross, 318 Westley Richards Express, 404 Jeffery. Introduced before WWI; discontinued 1939.
Perf.: $3695   Exc.: $2995   VGood: $1995

Mauser Special British Model Type A Short
Same specs as Type A except short action; 6.5x54, 8x51mm, 250-3000; 21 1/2″ barrel. Introduced before WWI; discontinued 1938.
Perf.: $1995   Exc.: $1395   VGood: $995

MAUSER TYPE B
Bolt action; 7x57, 8x57, 8x60, 9x57, 9.3x62, 10.75x68mm, 30-06; 5-shot box magazine; 23 1/2″ barrel; three-leaf rear sight, ramp front; hand-checkered walnut pistol-grip stock; Schnabel forearm tip; grip cap; double-set triggers; sling swivels. Introduced before WWI; discontinued 1938.
Perf.: $1995   Exc.: $1395   VGood: $995

McMillan

300 Phoenix Long Range
Rifle
M-87 Combo
M-87R
M-40 Sniper Rifle
M-86 Sniper
M-89 Sniper
M-92 Bullpup
M-93SN
National Match
Signature Alaskan
Signature Classic Sporter
Signature Classic Sporter
Stainless
Signature Titanium Mountain
Signature Varminter
Standard Sporter

RIFLES

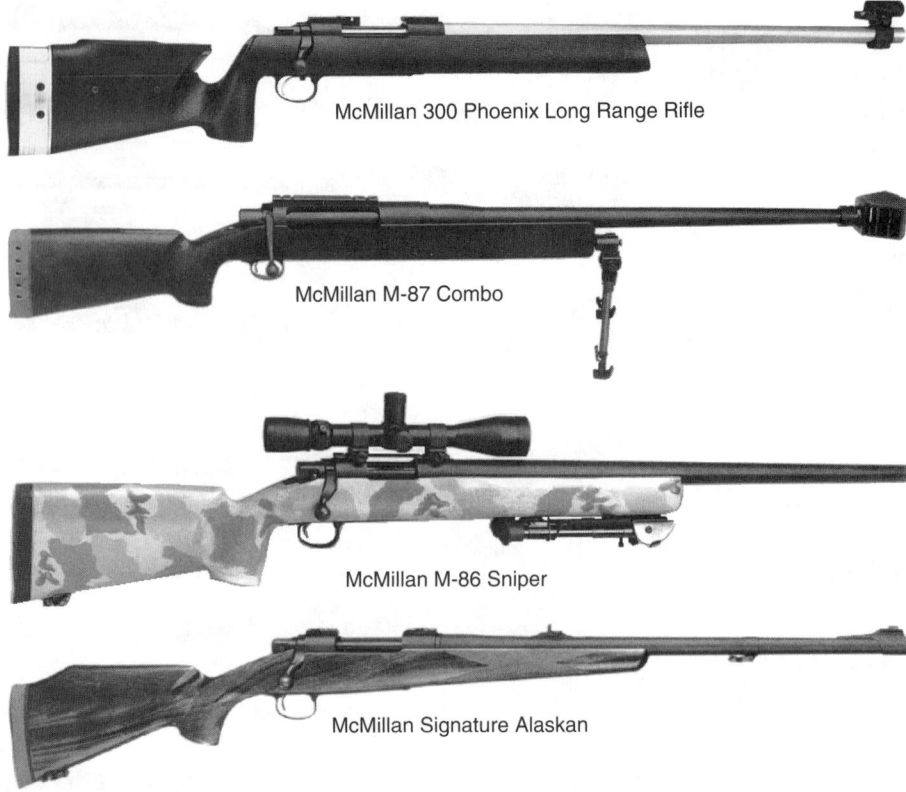

McMillan 300 Phoenix Long Range Rifle

McMillan M-87 Combo

McMillan M-86 Sniper

McMillan Signature Alaskan

McMILLAN 300 PHOENIX LONG RANGE RIFLE

Bolt action; 300 Phoenix; 28″ barrel; weighs 12 1/2 lbs.; no sights furnished; fiberglass stock with adjustable cheekpiece, adjustable buttplate; comes with rings and bases; matte black finish; textured stock. All parts and components made in U.S. by Harris Gunworks. Introduced 1992; still produced. **LMSR: $3000**

 Perf.: $3600 **Exc.: $2100** **VGood: $1300**

McMILLAN M-87 COMBO

Bolt action; 50 BMG; single shot; 29″ barrel, with muzzle brake; 53″ overall length; weighs about 21 1/2 lbs.; no sights furnished; McMillan fiberglass stock; right-handed McMillan stainless steel receiver, chrome-moly barrel with 1:15″ twist. All parts and components made in U.S. by Harris Gunworks. Introduced 1987; still produced. **LMSR: $4300**

 Perf.: $3200 **Exc.: $2400** **VGood: $1900**

McMILLAN M-87R

Same specs as M-87 except 5-shot repeater. All parts and components made in U.S. by Harris Gunworks. Introduced 1987; still produced. **LMSR: $4000**

 Perf.: $3495 **Exc.: $2495** **VGood: $1995**

McMILLAN M-40 SNIPER MODEL

Bolt action; 308 Win.; 4-shot magazine; 24″, 26″ match-grade heavy contour barrel; weighs 9 lbs.; no sights; fiberglass stock; recoil pad. Introduced 1990; still in production. **LMSR: $1900**

 Perf.: $1595 **Exc.: $1195** **VGood: $995**

McMILLAN M-86 SNIPER

Bolt action; 308, 30-06, 300 Win. Mag.; 4-shot (308, 30-06), 3-shot (300 Win. Mag.); 24″ heavy match-grade barrel; 43 1/2″ overall length; weighs 11 1/4 lbs. (308), 11 1/2 lbs. (30-06, 300); McHale fiberglass stock with textured grip and forend; recoil pad; no sights; bipod; sling swivels; matte black finish. Introduced 1989; no longer in production.

 Perf.: $1695 **Exc.: $1195** **VGood: $995**

McMILLAN M-89 SNIPER

Bolt action; 308 Win.; 5-shot magazine; 28″ barrel with optional suppressor; weighs 15 1/4 lbs.; drilled, tapped for scope mounts; length-adjustable McMillan fiberglass stock; bipod. Introduced 1990; still in production. **LMSR: $2300**

 Perf.: $1995 **Exc.: $1395** **VGood: $995**

McMILLAN M-92 BULLPUP

Bolt action; single shot; 50 BMG; 24″, 26″ barrel; no sights; fiberglass bullpup-style stock; recoil pad. Introduced 1995; still in production. **LMSR: $4000**

 Perf.: $3710 **Exc.: $2995** **VGood: $2195**

McMILLAN M-93SN

Bolt action; 50 BMG; 5-, 10-shot detachable magazine; 29″ chrome-moly with muzzle brake; 53″ overall length; 21 1/2 lbs.; no sights; folding fiberglass stock. Introduced 1995; still in production. **LMSR: $4300**

 Perf.: $3850 **Exc.: $2995** **VGood: $2395**

McMILLAN NATIONAL MATCH

Bolt action; 308; 5-shot magazine; 24″ stainless steel barrel; 43″ overall length; weighs 11 lbs.; barrel band-type Tomkins front sight, no rear sight; modified ISU fiberglass stock; adjustable buttplate; Canjar trigger; right-hand only. Options include Fibergrain stock, sights. Introduced 1989; still in production. **LMSR: $2600**

 Perf.: $2300 **Exc.: $1795** **VGood: $1295**

McMILLAN SIGNATURE ALASKAN

Bolt action; 270, 280 Rem., 30-06, 7mm Rem. Mag., 300 Win. Mag., 300 Weatherby, 358 Win. Mag., 340 Weatherby, 375 H&H Mag.; 3-shot magazine (magnums), 4-shot (standard calibers); 22″, 24″, 26″ match-grade barrel; single-leaf rear sight; barrel band front; Monte Carlo walnut stock, cheekpiece, palm-swell grip; steel floorplate; Teflon or electroless nickel finish. Introduced 1989; still in production. **LMSR: $3300**

 Perf.: $2900 **Exc.: $2195** **VGood: $1800**

McMILLAN SIGNATURE CLASSIC SPORTER

Bolt action; 22-250, 243, 6mm Rem., 7mm-08, 284, 308, 25-06, 270, 280 Rem., 30-06, 7mm Rem. Mag., 300 Win. Mag., 300 Weatherby, 338 Win. Mag., 340 Weatherby Mag., 375 H&H Mag.; 3-shot magazine (magnum), 4-shot (standard); 22″, 24″, 26″ barrel; weighs 7 lbs. (magnums); no sights; 1″ rings, bases furnished; McMillan fiberglass stock; recoil pad; sling swivels; matte-black McMillan right- or left-hand action; aluminum floorplate; Fibergrain, wood stocks optional. Introduced 1987; still in production. **LMSR: $2400**

 Perf.: $2100 **Exc.: $1495** **VGood: $1095**

McMILLAN Signature Classic Sporter Stainless

Same specs as Classic Sporter except barrel and action made of stainless steel; same calibers in addition to 416 Rem. Mag.; fiberglass stock; right- or left-hand action; natural stainless, glass bead or black chrome sulfide finishes. All parts and components made in U.S. by Harris Gunworks. Introduced 1990; still produced. **LMSR: $2500**

 Perf.: $1995 **Exc.: $1395** **VGood: $995**

McMILLAN SIGNATURE TITANIUM MOUNTAIN

Bolt action; 270, 280, 30-06, 300 Win. Mag., 338 Win. Mag., 7mm Rem. Mag; 3-shot magazine (magnum), 4-shot (standard); 22″, 24″, 26″ barrel; chrome-moly steel or titanium match-grade barrel; weighs 6 1/2 lbs.; graphite-reinforced fiberglass stock; titanium-alloy action. Introduced 1989; still in production. **LMSR: $3000**
Steel barrel
 Perf.: $2495 **Exc.: $1895** **VGood: $1295**
Titatium barrel
 Perf.: $2995 **Exc.: $2195** **VGood: $1695**

McMILLAN SIGNATURE VARMINTER

Bolt action; 223, 22-250, 220 Swift, 243, 6mm Rem., 25-06, 7mm-08, 308; 4-shot magazine; 22″, 24″, 26″ barrel; weighs 7 lbs.; heavy contoured barrel; hand-bedded fiberglass or Fibergrain stock; adjustable trigger; field bipod; marketed with rings, bases. Introduced 1989; still in production. **LMSR: $2400**

 Perf.: $2200 **Exc.: $1495** **VGood: $995**

McMILLAN STANDARD SPORTER

Bolt action; 6.5x55, 7mm-08, 308, 270, 280 Rem., 30-06, 7mm Rem. Mag., 300 Win. Mag., 338 Win. Mag.; 24″ barrel; weighs 7 1/2 lbs.; no sights; drilled, tapped for scope mounts; American walnut or painted fiberglass stock; chrome-moly action; button rifled barrel; Shilen trigger; hinged floorplate. Introduced 1991; discontinued 1992.

 Perf.: $995 **Exc.: $795** **VGood: $695**

McMillan Signature Classic Sporter

McMillan Classic Sporter Stainless

McMillan Talon Safari

Merkel Model 210E Combination Gun

McMillan
Talon Safari
Talon Safari, Super Mag
Talon Sporter

Merkel
Model 90S Drilling
Model 90K Drilling
Model 95S Drilling
Model 95K Drilling
Model 140-1 Double Rifle
Model 140-1.1 Double Rifle
Model 150-1 Double Rifle
Model 150-1.1 Double Rifle
Model 160 Double Rifle
Model 210E Combination Gun
Model 211E Combination Gun

RIFLES

McMILLAN TALON SAFARI

Bolt action; 300 Win. Mag., 300 Weatherby, 338 Win. Mag., 340 Weatherby, 375 Weatherby, 375 H&H Mag., 378 Weatherby, 416 Taylor, 416 Rigby, 458 Win. Mag.; 4-shot magazine; 24″ stainless barrel; 43″ overall length; weighs 10 lbs.; barrel-band ramp front sight, multi-leaf express rear; fiberglass stock; barrel-band sling swivels; 1″ scope mounts; steel floorplate; matte black finish. Introduced 1989; still in production. **LMSR: $3600**
   Perf.: $3200   **Exc.:** $2195   **VGood:** $1695

McMillan Talon Safari Super Mag

Same specs as Talon Safari except 338 Lapua, 378 Wea. Mag., 416 Rigby, 460 Weatherby, 416 Weatherby, 300 Phoenix, 30-378. Still in production. **LMSR: $4200**
   Perf.: $3495   **Exc.:** $2850   **VGood:** $1195

McMILLAN TALON SPORTER

Bolt action; 25-06, 270, 280 Rem., 30-06, 7mm Rem. Mag., 300 Win. Mag., 300 Weatherby Mag., 300 H&H, 338 Win. Mag., 340 Weatherby Mag., 375 H&H Mag., 416 Rem. Mag.; 24″ barrel; weighs 7 1/2 lbs.; no sights; drilled, tapped for scope mounts; American walnut or McMillan fiberglass stock; built on pre-'64 Model 70-type action; cone breech; controlled feed; claw extractor; three-position safety; stainless steel barrel, action; chrome-moly optional. Introduced 1991. **LMSR: $2600**
   Perf.: $2400   **Exc.:** $1795   **VGood:** $1195

MERKEL MODEL 90S DRILLING

Drilling; 12-, 20-ga., 3″ chambers, 16-ga., 2 3/4″ chambers; 22 Hornet, 5.6x50R Mag., 5.6x52R, 222 Rem., 243 Win., 6.5x55, 6.5x57R, 7x57R, 7x65R, 308, 30-06, 8x57JRS, 9.3x74R, 375 H&H; 25 5/8″ barrel; weighs 8 to 8 1/2 lbs. depending upon caliber; blade front sight, fixed rear; oil-finished walnut stock with pistol-grip; cheekpiece on 12-, 16-gauge; selective sear safety; double barrel locking lug with Greener cross-bolt; scroll-engraved, case-hardened receiver; automatic trigger safety; Blitz action; double triggers. Imported from Germany by GSI. Still imported. **LMSR: $5995**
   Perf.: $5500   **Exc.:** $4995   **VGood:** $3495

Merkel Model 90K Drilling

Same specs as Model 90S Drilling except manually cocked rifle system. Imported from Germany by GSI. Still imported. **LMSR: $6495**
   Perf.: $5995   **Exc.:** $5250   **VGood:** $3995

MERKEL MODEL 95S DRILLING

Drilling; 12, 20, 3″ chambers, 16, 2 3/4″ chambers; 22 Hornet, 5.6x50R Mag., 5.6x52R, 222 Rem., 243 Win., 6.5x55, 6.5x57R, 7x57R, 7x65R, 308, 30-06, 8x57JRS, 9.3x74R, 375 H&H; 25 5/8″ barrels; weighs 8 to 8 1/2 lbs., depending upon caliber; blade front sight, fixed rear; oil-finished walnut stock with pistol grip; cheekpiece on 12-, 16-gauge; double barrel locking lug with Greener cross-bolt; scroll-engraved, case-hardened receiver; selective sear safety; Blitz action; double triggers. Imported from Germany by GSI. **LMSR: $7195**
   Perf.: $6995   **Exc.:** $5995   **VGood:** $4995

Merkel Model 95K Drilling

Same specs as Model 95S Drilling except manually cocked rifle system. Imported from Germany by GSI. Still imported. **LMSR: $7695**
   Perf.: $7250   **Exc.:** $6650   **VGood:** $5995

MERKEL MODEL 140-1 DOUBLE RIFLE

Double rifle; side-by-side; 22 Hornet, 5.6x50R Mag., 5.6x52R, 222 Rem., 243 Win., 6.5x55, 6.5x57R, 7x57R, 7x65R, 308, 30-06, 8x57JRS, 9.3x74R, 375 H&H; 25 5/8″ barrel; weighs about 7 3/4 lbs., depending upon caliber; blade front sight on ramp, fixed rear; oil-finished walnut stock with pistol grip, cheekpiece; Anson & Deeley boxlock action with cocking indicators, double triggers; engraved color case-hardened receiver. Imported from Germany by GSI. Still imported. **LMSR: $5995**
   Perf.: $3995   **Exc.:** $3450   **VGood:** $2795

Merkel Model 140-1.1 Double Rifle

Same as Model 140-1 Double Rifle except engraved; silver-gray receiver. Imported from Germany by GSI. Still imported. **LMSR: $5595**
   Perf.: $4800   **Exc.:** $3795   **VGood:** $2995

MERKEL MODEL 150-1 DOUBLE RIFLE

Double rifle; side-by-side; 22 Hornet, 5.6x50R Mag., 5.6x52R, 222 Rem., 243 Win., 6.5x55, 6.5x57R, 7x57R, 7x65R, 308, 30-06, 8x57JRS, 9.3x74R, 375 H&H; 25 5/8″ barrels; weighs about 7 3/4 lbs., depending upon caliber; blade front sight on ramp, fixed rear; oil-finished walnut stock with pistol grip, cheekpiece; Anson & Deeley boxlock action with cocking indicators, double triggers, false sideplates; silver-gray receiver with Arabesque engraving. Imported from Germany by GSI. Still imported. **LMSR: $7495**
   Perf.: $5800   **Exc.:** $4695   **VGood:** $3495

Merkel Model 150-1.1 Double Rifle

Same specs as Model 150-1 Double Rifle except English Arabesque engraving. Imported from Germany by GSI. Still imported. **LMSR: $8995**
   Perf.: $6500   **Exc.:** $5250   **VGood:** $4295

MERKEL MODEL 160 DOUBLE RIFLE

Double rifle; side-by-side; 22 Hornet, 5.6x50R Mag., 5.6x52R, 222 Rem., 243 Win., 6.5x55, 6.5x57R, 7x57R, 7x65R, 308, 30-06, 8x57JRS, 9.3x74R, 375 H&H; 25 5/8″ barrels; weighs about 7 3/4 lbs., depending upon caliber; blade front sight on ramp, fixed rear; oil-finished walnut stock with pistol grip, cheekpiece; sidelock action; double barrel locking lug with Greener cross-bolt; fine engraved hunting scenes on sideplates; Holland & Holland ejectors; double triggers. Imported from Germany by GSI. Still imported. **LMSR: $10,995**
   Perf.: $9875   **Exc.:** $7995   **VGood:** $5995

MERKEL MODEL 210E COMBINATION GUN

Combo; over/under; 12-, 16-, 20-ga. (2 3/4″ chamber) over 22 Hornet, 5.6x50R, 5.6x52R, 222 Rem., 243 Win., 6.5x55, 6.5x57R, 7x57R, 7x65R, 308, 30-06, 8x57JRS, 9.3x74R, 375 H&H; 25 5/8″ barrel; weighs about 7 1/2 lbs.; bead front sight, fixed rear; oil-finished walnut stock with pistol grip, cheekpiece; Kersten double cross-bolt lock; scroll-engraved, color case-hardened receiver; Blitz action; double triggers. Imported from Germany by GSI. Still imported. **LMSR: $6195**
   Perf.: $5700   **Exc.:** $4395   **VGood:** $3250

MERKEL MODEL 211E COMBINATION GUN

Combo; over/under; 12-, 16-, 20-ga. (2 3/4″ chamber) over 22 Hornet, 5.6x50R, 5.6x52R, 222 Rem., 243 Win., 6.5x55, 6.5x57R, 7x57R, 7x65R, 308 Win., 30-06, 8x57JRS, 9.3x74R, 375 H&H; 25 5/8″ barrel; weighs about 7 1/2 lbs.; bead front sight, fixed rear; oil-finished walnut stock with pistol grip, cheekpiece; Kersten double cross-bolt lock; silver-grayed receiver with fine hunting scene engraving; Blitz action; double triggers. Imported from Germany by GSI. Still imported. **LMSR: $7495**
   Perf.: $6200   **Exc.:** $4895   **VGood:** $3995

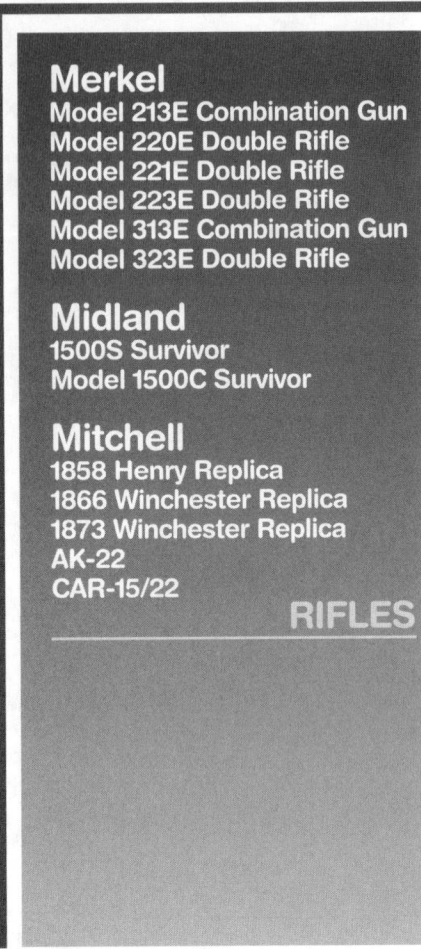

Merkel
Model 213E Combination Gun
Model 220E Double Rifle
Model 221E Double Rifle
Model 223E Double Rifle
Model 313E Combination Gun
Model 323E Double Rifle

Midland
1500S Survivor
Model 1500C Survivor

Mitchell
1858 Henry Replica
1866 Winchester Replica
1873 Winchester Replica
AK-22
CAR-15/22

RIFLES

Mitchell 1858 Henry Replica

Mitchell 1873 Winchester Replica

Mitchell AK-22

Mitchell CAR-15/22

Mitchell Galil/22

MERKEL MODEL 213E COMBINATION GUN
Combo; over/under; 12-, 16-, 20-ga. (2 ³/₄" chamber) over 22 Hornet, 5.6x50R, 5.6x52R, 222 Rem., 243 Win., 6.5x55, 6.5x57R, 7x57R, 7x65R, 308 Win., 30-06, 8x57JRS, 9.3x74R, 375 H&H; 25 ⁵/₈" barrel; weighs about 7 ¹/₂ lbs.; bead front sight, fixed rear; oil-finished walnut stock with pistol grip, cheekpiece; Kersten double cross-bolt lock; English-style, large scroll Arabesque engraving; sidelock action; double triggers. Imported from Germany by GSI. Still imported. **LMSR: $14,795**
   Perf.: $11,995 Exc.: $9995   VGood: $6995

MERKEL MODEL 220E DOUBLE RIFLE
Double rifle; over/under; boxlock; 22 Hornet, 5.6x50R Mag., 5.6x52R, 222 Rem., 243 Win., 6.5x55, 6.5x57R, 7x57R, 7x65R, 308, 30-06, 8x57JRS, 9.3x74R; 25 ⁵/₈" barrel; weighs about 7 ³/₄ lbs., depending upon caliber; blade front sight, fixed rear; oil-finished walnut stock with pistol grip, cheekpiece; Kersten double cross-bolt lock; scroll-engraved, case-hardened receiver; Blitz action with double triggers. Imported from Germany by GSI. Still imported. **LMSR: $10,795**
   Perf.: $9495 Exc.: $7995   VGood: $5995

MERKEL MODEL 221E DOUBLE RIFLE
Over/under; double rifle; boxlock; 22 Hornet, 5.6x50R Mag., 5.6x52R, 222 Rem., 243 Win., 6.5x55, 6.5x57R, 7x57R, 7x65R, 308, 30-06, 8x57JRS, 9.3x74R; 25 ⁵/₈" barrel; weighs about 7 ³/₄ lbs, depending upon caliber; blade front sight, fixed rear; oil-finished walnut stock with pistol grip, cheekpiece; Kersten double cross-bolt lock; silver-grayed receiver with hunting scene engraving; Blitz action with double triggers. Imported from Germany by GSI. Still imported. **LMSR: $10,895**
   Perf.: $9500 Exc.: $7995   VGood: $5995

MERKEL MODEL 223E DOUBLE RIFLE
Double rifle; over/under; sidelock; 22 Hornet, 5.6x50R Mag., 5.6x52R, 222 Rem., 243 Win., 6.5x55, 6.5x57R, 7x57R, 7x65R, 308, 30-06, 8x57JRS, 9.3x74R; 25 ⁵/₈" barrel; weighs about 7 ³/₄ lbs, depending upon caliber;

blade front sight, fixed rear; oil-finished walnut stock with pistol grip, cheekpiece; sidelock action; English-style large-scroll Arabesque engraving; Blitz action with double triggers. Imported from Germany by GSI. No longer imported.
   Perf.: $14,995 Exc.: $12,500   VGood: $9500

MERKEL MODEL 313E COMBINATION GUN
Combo; over/under; 12-, 16-, 20-ga. (2 ³/₄" chamber) over 22 Hornet, 5.6x50R, 5.6x52R, 222 Rem., 243 Win., 6.5x55, 6.5x57R, 7x57R, 7x65R, 308 Win., 30-06, 8x57JRS, 9.3x74R, 375 H&H; 25 ⁵/₈" barrel; weighs about 7 ¹/₂ lbs.; bead front sight, fixed rear; oil-finished walnut stock with pistol grip, cheekpiece; Kersten double cross-bolt lock; medium-scroll engraving on receiver; sidelock action; double triggers. Imported from Germany by GSI. Still imported. **LMSR: $20,695**
   Perf.: $16,950 Exc.: $13,995   VGood: $9750

MERKEL MODEL 323E DOUBLE RIFLE
Double rifle; over/under; sidelock; 22 Hornet, 5.6x50R Mag., 5.6x52R, 222 Rem., 243 Win., 6.5x55, 6.5x57R, 7x57R, 7x65R, 308, 30-06, 8x57JRS, 9.3x74R; 25 ⁵/₈" barrel; weighs about 7 ³/₄ lbs., depending upon caliber; blade front sight, fixed rear; oil-finished walnut stock with pistol grip, cheekpiece; medium-scroll engraving on action; Blitz action with double triggers. Imported from Germany by GSI. Still imported. **LMSR: $27,195**
   Perf.: $20,000 Exc.: $17,500   VGood: $13,500

MIDLAND 1500S SURVIVOR
Bolt action; 308 Win.; 5-shot magazine; 22" barrel; 43" overall length; weighs 7 lbs.; hooded ramp front sight, open adjustable rear; black composite stock with recoil pad; Monte Carlo cheekpiece; stainless steel barreled action with satin chromed bolt. Made by Gibbs Rifle Co. Introduced 1993; discontinued 1994.
   Perf.: $695 Exc.: $595   VGood: $495

Midland Model 1500C Survivor
Same specs as Model 1500S Survivor except detachable clip magazine. Made by Gibbs Rifle Co. Introduced 1993; discontinued 1994.
   Perf.: $695 Exc.: $595   VGood: $495

MITCHELL 1858 HENRY REPLICA
Lever action; 44-40; 13-shot magazine; 24 ¹/₄" barrel; 43" overall length; weighs 9 ¹/₂ lbs.; bead front sight, open adjustable rear; brass receiver and buttplate; uses original Henry loading system; faithful to the original rifle. Imported by Mitchell Arms, Inc. Introduced 1990; discontinued 1993.
   Perf.: $800 Exc.: $630   VGood: $500

MITCHELL 1866 WINCHESTER REPLICA
Lever action; 44-40; 13-shot magazine; 24 ¹/₄" barrel; 43" overall length; weighs 9 lbs.; bead front sight, open adjustable rear; European walnut stock; solid brass receiver, buttplate, forend cap; octagonal barrel; faithful to the original Winchester '66 rifle. Imported by Mitchell Arms, Inc. Introduced 1990; discontinued 1993.
   Perf.: $695 Exc.: $530   VGood: $430

MITCHELL 1873 WINCHESTER REPLICA
Lever action; 45 Colt; 13-shot magazine; 24 ¹/₄" barrel; 43" overall length; weighs 9 ¹/₂ lbs.; bead front sight, open adjustable rear; European walnut stock; color case-hardened steel receiver; faithful to the original Model 1873 rifle. Imported by Mitchell Arms, Inc. Introduced 1990; discontinued 1993.
   Perf.: $750 Exc.: $625   VGood: $495

MITCHELL AK-22
Semi-automatic; 22 LR, 22 WMR; 20-shot magazine (22 LR), 10-shot (22 WMR); 18" barrel; 36" overall length; weighs 6 ¹/₂ lbs.; post front sight, open adjustable rear; European walnut stock; replica of the AK-47 rifle; wide magazine to maintain appearance. Imported from Italy by Mitchell Arms, Inc. Introduced 1985; discontinued 1993.
   Perf.: $230 Exc.: $200   VGood: $150

MITCHELL CAR-15/22
Semi-automatic; 22 LR; 15-shot magazine; 16 ³/₄" barrel; telescoping butt; 32" overall length when collapsed; adjustable post front sight, adjustable aperture rear. Scope mount available. Replica of the CAR-15 rifle. Imported by Mitchell Arms, Inc. Introduced 1990; discontinued 1993.
   Perf.: $230 Exc.: $200   VGood: $150

Mitchell High Standard 20/22

Mitchell High Standard 9302

Mitchell MAS/22

Mitchell High Standard 16/22

Mitchell PPS/5

Mitchell
Galil/22
High Standard 15/22
High Standard 15/22D
High Standard 20/22
High Standard 20/22D
High Standard 20/22 Special
High Standard 9301
High Standard 9302
High Standard 9303
High Standard 9304
High Standard 9305
M-16A-1
MAS/22
PPS/50

MITCHELL GALIL/22

Semi-automatic; 22 LR, 22 WMR; 20-shot magazine (22 LR), 10-shot (22 WMR); 18″ barrel; 36″ overall length; weighs 6 1/2 lbs.; elevation-adjustable post front sight, windage-adjustable rear; European walnut grip and forend with metal folding or wood fixed stock; replica of the Israeli Galil rifle. Imported by Mitchell Arms, Inc. Introduced 1987; discontinued 1993.

   Perf.: $270   **Exc.:** $230   **VGood:** $180

MITCHELL HIGH STANDARD 15/22

Semi-automatic; 22 LR; 15-shot magazine, 30-shot available; 20 1/2″ barrel; 37 1/2″ overall length; weighs 6 1/4 lbs.; blade on ramp front sight, open adjustable rear; American walnut stock; polished blue finish; barrel band on forend. Imported from Philippines by. Mitchell Arms, Inc. Introduced 1994; discontinued 1995.

   Perf.: $140   **Exc.:** $120   **VGood:** $90

Mitchell High Standard 15/22D

Same specs as Mitchell High Standard 15/22 except fancy walnut stock with checkering; rosewood grip and forend caps. Imported from the Philippines by Mitchell Arms, Inc. Introduced 1994; discontinued 1995.

   Perf.: $150   **Exc.:** $130   **VGood:** $100

MITCHELL HIGH STANDARD 20/22

Semi-automatic; 22 LR; 10-shot magazine; 20 1/2″ barrel; 37 1/2″ overall length; weighs 6 1/4 lbs.; blade on ramp front sight, open adjustable rear; American walnut stock; polished blue finish; barrel band on forend. Imported from Philippines by Mitchell Arms, Inc. Introduced 1994; discontinued 1995.

   Perf.: $100   **Exc.:** $80   **VGood:** $70

Mitchell High Standard 20/22D

Same specs as Mitchell High Standard 20/22 except fancy walnut stock with checkering; rosewood grip and forend caps. Imported from the Philippines by Mitchell Arms, Inc. Introduced 1994; discontinued 1995.

   Perf.: $120   **Exc.:** $100   **VGood:** $80

Mitchell High Standard 20/22 Special

Same specs as Mitchell High Standard 20/22 except heavy barrel. Imported from the Philippines by Mitchell Arms, Inc. Introduced 1994; discontinued 1995.

   Perf.: $110   **Exc.:** $90   **VGood:** $80

MITCHELL HIGH STANDARD 9301

Bolt action; 22 LR; 10-shot magazine; 22 1/2″ barrel; 40 3/4″ overall length; weighs about 6 1/2 lbs.; bead on ramp front sight, open adjustable rear; American walnut with rosewood grip and forend caps; checkering; polished blue finish. Imported from the Philippines by Mitchell Arms, Inc. Introduced 1994; discontinued 1995.

   Perf.: $250   **Exc.:** $200   **VGood:** $140

MITCHELL HIGH STANDARD 9302

Bolt action; 22 WMR; 5-shot magazine; 22 1/2″ barrel; 40 3/4″ overall length; weighs about 6 1/2 lbs.; bead on ramp front sight, open adjustable rear; American walnut stock with rosewood grip and forend caps; checkering; polished blue finish. Imported from the Philippines by Mitchell Arms, Inc. Introduced 1994; discontinued 1995.

   Perf.: $270   **Exc.:** $210   **VGood:** $160

MITCHELL HIGH STANDARD 9303

Bolt action; 22 LR; 10-shot magazine; 22 1/2″ barrel; 40 3/4″ overall length; weighs about 6 1/2 lbs.; bead on ramp front sight, open adjustable rear; American walnut stock; polished blue finish. Imported from the Philippines by Mitchell Arms, Inc. Introduced 1994; discontinued 1995.

   Perf.: $210   **Exc.:** $160   **VGood:** $120

MITCHELL HIGH STANDARD 9304

Bolt action; 22 WMR; 5-shot magazine; 22 1/2″ barrel; 40 3/4″ overall length; weighs about 6 1/2 lbs.; bead on ramp front sight, open adjustable rear; American walnut stock; rosewood grip and forend caps; checkering; polished blue finish. Imported from the Philippines by Mitchell Arms, Inc. Introduced 1994; discontinued 1995.

   Perf.: $230   **Exc.:** $175   **VGood:** $130

MITCHELL HIGH STANDARD 9305

Bolt action; 22 LR; 10-shot magazine; 22 1/2″ heavy barrel; 40 3/4″ overall length; weighs about 7 lbs.; bead on ramp front sight, open adjustable rear; American walnut stock; polished blue finish. Imported from the Philippines by Mitchell Arms, Inc. Introduced 1994; discontinued 1995.

   Perf.: $270   **Exc.:** $200   **VGood:** $150

MITCHELL M-16A-1

Semi-automatic; 22 LR; 15-shot magazine; 20 1/2″ barrel; 38 1/2″ overall length; weighs 7 lbs.; adjustable post front sight, adjustable aperture rear; full width magazine; black composition stock; comes with military-type sling. Replica of the AR-15 rifle. Imported by Mitchell Arms, Inc. Introduced 1990; discontinued 1994.

   Perf.: $240   **Exc.:** $190   **VGood:** $150

MITCHELL MAS/22

Semi-automatic; 22 LR; 20-shot magazine; 18″ barrel; 28 1/2″ overall length; weighs 7 1/2 lbs.; adjustable post front sight, flip-type aperture rear; walnut butt, grip and forend; top cocking lever; flash hider; bullpup design resembles French armed forces rifle. Imported by Mitchell Arms, Inc. Introduced 1987; discontinued 1993.

   Perf.: $270   **Exc.:** $200   **VGood:** $160

MITCHELL PPS/50

Semi-automatic; 22 LR; 20- or 50-shot magazine; 16 1/2″ barrel; 33 1/2″ overall length; weighs 5 1/2 lbs.; blade front sight, adjustable rear; walnut stock; full-length perforated barrel shroud; matte finish. Imported by Mitchell Arms, Inc. Introduced 1989; discontinued 1993.

20-shot
   Perf.: $240   **Exc.:** $190   **VGood:** $150
50-shot
   Perf.: $330   **Exc.:** $250   **VGood:** $200

Mossberg Bolt-Action Rifles

Boy Scout Target Model
(See Mossberg Model 320B)
Classic Hunter
(See Mossberg Model 1700LS)
Model 10
Model 14
Model 20
Model 21
Model 25
Model 25A
Model 26B
Model 26C
Model 26T
Model 30
Model 34
Model 35
Model 35A
Model 35A-LS
Model 40
Model 42
Model 42A
Model L42A
Model 42B

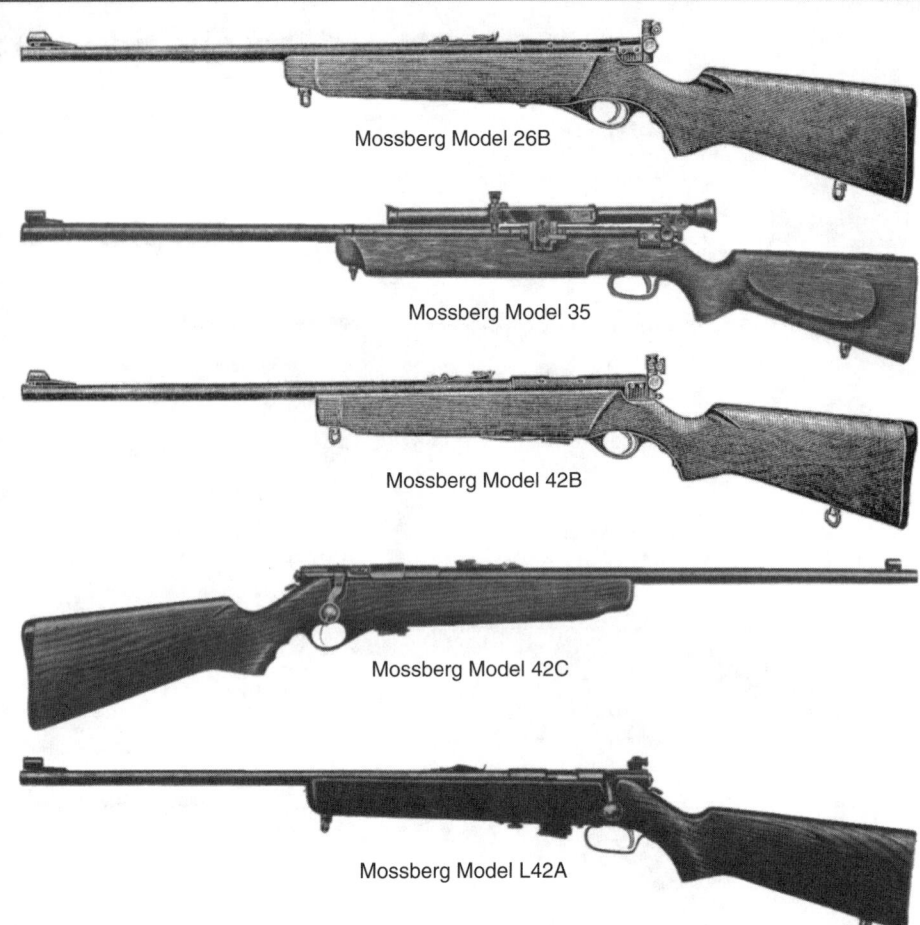

Mossberg Model 26B

Mossberg Model 35

Mossberg Model 42B

Mossberg Model 42C

Mossberg Model L42A

MOSSBERG BOLT-ACTION RIFLES

MOSSBERG MODEL 10
Bolt action; single shot; 22 Short, 22 Long, 22 LR; 22" barrel; take-down; open rear sight, bead front; uncheckered walnut pistol-grip stock; swivels, sling. Introduced 1933; discontinued 1935.
Perf.: $160   **Exc.:** $130   **VGood:** $90

MOSSBERG MODEL 14
Bolt action; single shot; 22 Short, 22 Long, 22 LR; 24" barrel; take down; rear peep sight, hooded ramp front; uncheckered pistol-grip stock; semi-beavertail forearm; sling swivels. Introduced 1934; discontinued 1935.
Perf.: $200   **Exc.:** $150   **VGood:** $100

MOSSBERG MODEL 20
Bolt action; single shot; 22 Short, 22 Long, 22 LR; 24" barrel; take down; open rear sight, bead front; uncheckered pistol-grip stock; forearm with finger grooves. Introduced 1933; discontinued 1935.
Perf.: $170   **Exc.:** $130   **VGood:** $90

MOSSBERG MODEL 21
Bolt action; single shot; 22 Short, 22 Long, 22 LR; 24" barrel; take down. Similar to Model 20. Introduced 1934; discontinued 1935.
Perf.: $170   **Exc.:** $130   **VGood:** $90

MOSSBERG MODEL 25
Bolt action; single shot; 22 Short, 22 Long, 22 LR; 24" barrel; take down; rear peep sight, hooded ramp front; uncheckered pistol-grip stock; semi-beavertail forearm; sling swivels. Introduced 1935; discontinued 1936.
Perf.: $190   **Exc.:** $140   **VGood:** $95

Mossberg Model 25A
Same specs as Model 25 except with master action; better wood and finish. Introduced 1937; discontinued 1938.
Perf.: $250   **Exc.:** $180   **VGood:** $120

MOSSBERG MODEL 26B
Bolt action; single shot; 22 Short, 22 Long, 22 LR; 26" tapered barrel; 41 3/4" overall length; weighs 5 1/2 lbs.; No. 2A open sporting or No. 4; micrometer peep rear sight, hooded ramp front; uncheckered pistol-grip stock; sling swivels. Manufactured 1938 to 1941.
Perf.: $190   **Exc.:** $140   **VGood:** $110

Mossberg Model 26C
Same specs as Model 26B except iron sights; no sling swivels.
Perf.: $170   **Exc.:** $120   **VGood:** $90

MOSSBERG MODEL 26T
Bolt action; single shot; 22 smoothbore with smoothbore and rifled screw-on adapters. Less than 1000 manufactured. Introduced 1940; discontinued 1942.
Perf.: $625   **Exc.:** $450   **VGood:** $300

MOSSBERG MODEL 30
Bolt action; single shot; 22 Short, 22 Long, 22 LR; 24" barrel; takedown; rear peep sight, hooded ramp bead front; uncheckered pistol-grip stock; grooved forearm. Introduced 1933; discontinued 1935.
Perf.: $130   **Exc.:** $90   **VGood:** $70

MOSSBERG MODEL 34
Bolt action; single shot; 22 Short, 22 Long, 22 LR; 24" barrel; takedown; rear peep sight, hooded ramp bead front; uncheckered pistol-grip stock with semi-beavertail forearm. Introduced 1934; discontinued 1935.
Perf.: $160   **Exc.:** $120   **VGood:** $90

MOSSBERG MODEL 35
Bolt action; single shot; 22 LR; 26" barrel, target grade; micrometer rear peep sight; hooded ramp front; target stock with full pistol grip, beavertail forearm; sling swivels. Introduced 1935; discontinued 1937.
Perf.: $275   **Exc.:** $220   **VGood:** $150

Mossberg Model 35A
Same specs as Model 35 except heavy barrel; new "master action"; uncheckered target stock with cheekpiece, full pistol grip. Introduced 1937; discontinued 1938.
Perf.: $295   **Exc.:** $230   **VGood:** $150

Mossberg Model 35A-LS
Same specs as Model 35A except Lyman 17A front, Lyman No. 57 rear sight.
Perf.: $345   **Exc.:** $250   **VGood:** $160

MOSSBERG MODEL 40
Bolt action; 22 Short, 22 Long, 22 LR; 16-shot tube magazine (22 LR), 18-shot (22 Long), 22-shot (22 Short); takedown repeater. Introduced 1933; discontinued 1935.
Perf.: $140   **Exc.:** $110   **VGood:** $90

MOSSBERG MODEL 42
Bolt action; 22 Short, 22 Long, 22 LR; 7-shot detachable box magazine; 24" barrel; takedown; receiver peep sight, open rear sight, hooded ramp front; uncheckered walnut pistol-grip stock; sling swivels. Introduced 1935; discontinued 1937.
Perf.: $125   **Exc.:** $100   **VGood:** $80

Mossberg Model 42A
Same specs as Model 42 with minor upgrading. Introduced 1937 to replace dropped Model 42; discontinued 1938.
Perf.: $150   **Exc.:** $120   **VGood:** $90

Mossberg Model L42A
Same specs as Model 42A except left-handed action. Introduced 1938; discontinued 1941.
Perf.: $250   **Exc.:** $190   **VGood:** $130

Mossberg Model 42B
Same specs as Model 42A except minor design improvements; replaced the Model 42A; micrometer peep sight, with open rear; 5-shot detachable box magazine. Introduced 1938; discontinued 1941.
Perf.: $150   **Exc.:** $120   **VGood:** $90

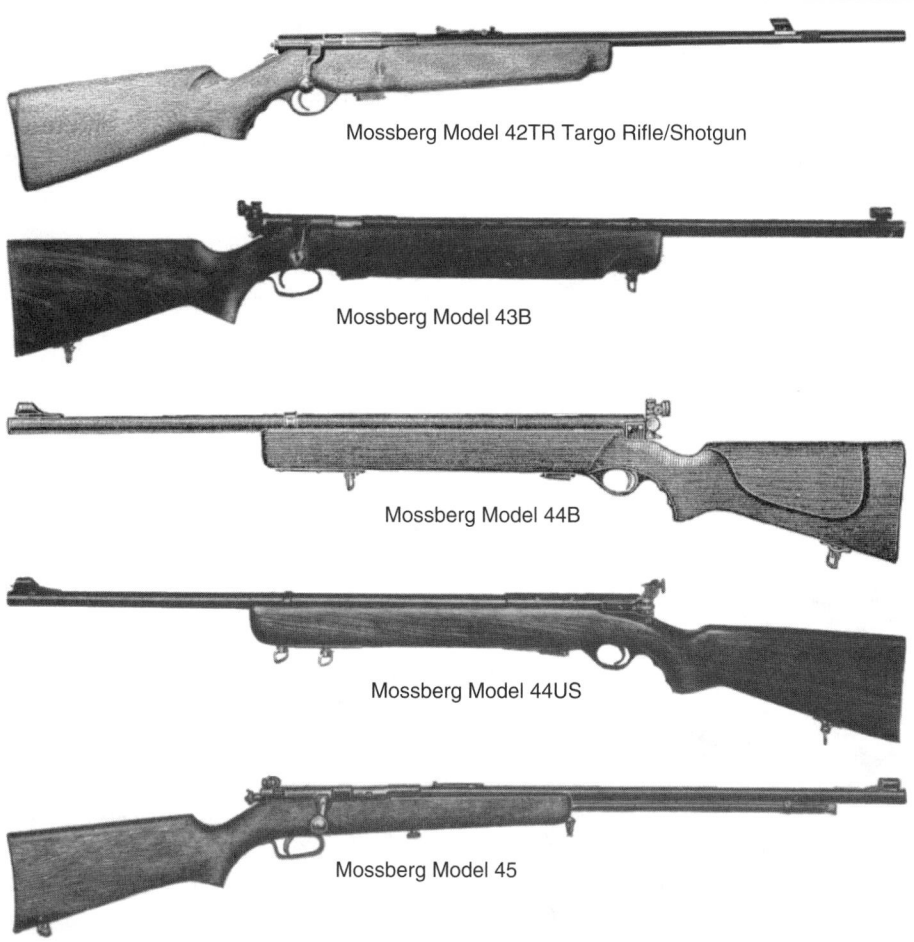

Mossberg Model 42TR Targo Rifle/Shotgun

Mossberg Model 43B

Mossberg Model 44B

Mossberg Model 44US

Mossberg Model 45

Mossberg Model 42C
Same specs as Model 42B except no rear peep sight.
Perf.: $110 **Exc.:** $90 **VGood:** $70

Mossberg Model 42M
Same specs as Model 42 except 23″ tapered barrel; 40″ overall length; weighs 6 3/4 lbs.; two-piece Mannlicher-type stock with pistol grip, cheekpiece; micrometer receiver peep sight, open rear sight, hooded ramp front. Introduced 1940; discontinued 1950.
Perf.: $175 **Exc.:** $145 **VGood:** $110

Mossberg Model 42M (a), 42M (b), 42M (c)
Same specs as Model 42M except minor changes in extractors and sights. No military markings. Introduced 1944; discontinued 1950.
Perf.: $175 **Exc.:** $145 **VGood:** $110

Mossberg Model 42MB
Same same specs as Model 42M except no cheekpiece; approximately 50,000 made for U.S. and Great Britain; U.S. Property and British proofmarks. Produced only during World War II. Some collector value. U.S. Property marked
Perf.: $250 **Exc.:** $225 **VGood:** $190
British proof marks
Perf.: $325 **Exc.:** $275 **VGood:** $250

MOSSBERG MODEL 42TR TARGO RIFLE/SHOTGUN
Bolt action; 22 Short; Long, Long Rifle, 22 Long Rifle Shot, 7-shot box magazine; smoothbore and rifled screw-on adapters; no sling swivels. Introduced commercially 1940; discontinued 1942.
Perf.: $450 **Exc.:** $375 **VGood:** $295

Mossberg Model 42T Targo Rifle/Shotgun
Same specs as Model 42TR except not provided with rifled adapter; front sight hood or rear sight. Introduced commercially 1940; discontinued 1942.
Perf.: $425 **Exc.:** $350 **VGood:** $275

MOSSBERG MODEL 43
Bolt action; 22 LR; 7-shot detachable box magazine; 26″ barrel; Lyman No. 57 rear sight, selective aperture front; adjustable trigger, speed lock; target stock with cheekpiece, full pistol grip; beavertail forend; adjustable front swivel. Introduced 1937; discontinued 1938.
Perf.: $275 **Exc.:** $225 **VGood:** $175

Mossberg Model L43
Same specs as Model 43 except left-hand action.
Perf.: $375 **Exc.:** $325 **VGood:** $275

MOSSBERG MODEL 43B
Bolt action; 22 LR; 7-shot detachable magazine; 26″ heavy barrel; Lyman No. 57 receiver sight, Lyman No. 17A front sight; walnut target stock with cheekpiece, full pistol-grip; beavertail forearm; not styled after standard Model 43. Introduced 1938; discontinued 1939.
Perf.: $275 **Exc.:** $225 **VGood:** $175

MOSSBERG MODEL 44B
Bolt action; target configuration; 22 LR; 7-shot clip magazine; 26″ heavy barrel; 43″ overall length; weighs 8 1/2 lbs.; micrometer receiver peep sight, hooded front; speed lock; thumb safety; grooved adjustable trigger; walnut target stock with cheekpiece, full pistol grip; beavertail forearm; adjustable sling swivels. Introduced 1938; discontinued 1943.
Perf.: $275 **Exc.:** $225 **VGood:** $175

MOSSBERG MODEL 44US
Bolt action; 22 LR; 7-shot detachable box magazine; 26″ heavy barrel; 43″ overall length; weighs 8 1/2 lbs.; same specs as Model 44B except designed primarily for teaching marksmanship to Armed Forces during World War II; drilled, tapped for Mossberg side-mounting scope; uncheckered walnut target stock; sling swivels. Introduced 1943; discontinued 1948. Collector value.
Perf.: $200 **Exc.:** $170 **VGood:** $120
U.S. Property marked
Perf.: $210 **Exc.:** $180 **VGood:** $130

Mossberg Model 44US(a), 44US(b), 44US(c), 44US(d)
Same specs as Model 44US except minor changes in sights and extractors.
Perf.: $210 **Exc.:** $180 **VGood:** $130

MOSSBERG MODEL 45
Bolt action; 22 Short, 22 Long, 22 LR; 15-shot tube magazine (22 LR), 18-shot (22 Long), 22-shot (22 Short); 24″ heavy barrel; 42 1/2″ overall length; weighs 6 3/4 lbs.; take-down, repeater; receiver peep sight, open rear, hooded ramp front; uncheckered pistol-grip stock; sling swivels. Introduced 1935; discontinued 1937.
Perf.: $150 **Exc.:** $100 **VGood:** $80

Mossberg Model 45A
Same specs as Model 45 except improved version of discontinued Model 45 with minor design variations. Introduced 1937; discontinued 1938.
Perf.: $150 **Exc.:** $100 **VGood:** $80

Mossberg Model 45AC
Same specs as Model 45A except without receiver peep sight.
Perf.: $125 **Exc.:** $90 **VGood:** $70

Mossberg Model 45B
Same specs as Model 45A except open rear sight. Introduced 1938; discontinued 1940.
Perf.: $125 **Exc.:** $90 **VGood:** $70

Mossberg Model 45C
Same specs as Model 45 except no sights; designed for use only with scope sight.
Perf.: $125 **Exc.:** $90 **VGood:** $80

Mossberg Model L45A
Same specs as Model 45A except left-hand action. Introduced 1937; discontinued 1938.
Perf.: $225 **Exc.:** $175 **VGood:** $115

RIFLES

Mossberg Bolt-Action Rifles
Model 46
Model 46A
Model 46AC
Model 46A-LS
Model 46B
Model 46BT
Model 46C
Model 46M
Model 46M (a), 46M (b)
Model 46T
Model L46A-LS
Model 140B
Model 140K
Model 142
Model 142K
Model 144
Model 144LS
Model 144LS-A
Model 144LS-B (Target)

Mossberg Model 46

Mossberg Model 46B

Mossberg Model 140B

Mossberg Model 140K

Mossberg Model 144

Mossberg Model 144LS

MOSSBERG MODEL 46
Bolt action; 22 Short, 22 Long, 22 LR; 15-shot (22 LR), 18-shot (22 Long), 22-shot (22 Short) tube magazine; 26″ barrel; 44 1/2″ overall length; weighs 7 1/2 lbs.; takedown repeater; micrometer rear peep sight; hooded ramp front; uncheckered walnut pistol-grip stock with cheekpiece; beavertail forearm; sling swivels. Introduced 1935; discontinued 1937.
Perf.: $150   **Exc.:** $100   **VGood** $90

Mossberg Model 46A
Same specs as discontinued Model 46 with minor design improvements; detachable sling swivels. Introduced 1937; discontinued 1938.
Perf.: $150   **Exc.:** $100   **VGood** $90

Mossberg Model 46AC
Same specs as Model 46A except open rear sight instead of micrometer peep sight.
Perf.: $140   **Exc.:** $90   **VGood** $80

Mossberg Model 46A-LS
Same specs as Model 46A except equipped with factory-supplied Lyman No. 57 receiver sight.
Perf.: $225   **Exc.:** $150   **VGood** $100

Mossberg Model 46B
Same specs as Model 46A except receiver peep and open rear sights. Introduced 1938; discontinued 1940.
Perf.: $160   **Exc.:** $120   **VGood** $100

Mossberg Model 46BT
Same specs as Model 46B except heavier barrel; target-styled stock. Introduced 1938; discontinued 1939.
Perf.: $225   **Exc.:** $150   **VGood** $110

Mossberg Model 46C
Same specs as Model 46 except heavier barrel.
Perf.: $250   **Exc.:** $165   **VGood** $120

Mossberg Model 46M
Same specs as Model 46 except 23″ barrel; 40″ overall length; weighs 7 lbs.; micrometer receiver peep sight, open rear, hooded ramp front; two-piece Mannlicher-type stock with pistol grip, cheekpiece; sling swivels. Introduced 1940; discontinued 1952.
Perf.: $190   **Exc.:** $140   **VGood** $110

Mossberg Model 46M(a), 46M(b)
Same specs as Model 46M except minor sight changes.
Perf.: $190   **Exc.:** $140   **VGood** $110

Mossberg Model 46T
Same specs as Model 46 except heavier barrel and stock; target version. Introduced 1936; discontinued 1937.
Perf.: $225   **Exc.:** $150   **VGood** $110

Mossberg Model L46A-LS
Same specs as Model 46A-LS except left-hand action.
Perf.: $350   **Exc.:** $260   **VGood** $190

MOSSBERG MODEL 140B
Bolt action; 22 Short, 22 Long, 22 LR; 7-shot clip magazine; 24 1/2″ barrel; 42″ overall length; weighs 5 3/4 lbs.; target/sporter version of Model 140K; peep rear sight, ramp front. Introduced 1957; discontinued 1958.
Perf.: $165   **Exc.:** $130   **VGood** $100

Mossberg Model 140K
Same specs as Model 140B except uncheckered walnut pistol-grip stock; Monte Carlo with cheekpiece; open rear sight, bead front; sling swivels. Introduced 1955; discontinued 1958.
Perf.: $140   **Exc.:** $110   **VGood** $90

MOSSBERG MODEL 142
Bolt-action; 22 Short, 22 Long, 22 LR; 7-shot removable clip magazine; 18″ tapered barrel; 36″ overall length; weighs 5 lbs.; rear fully-adjustable peep sight, military-style ramp front; walnut Monte Carlo pistol-grip stock; as with Model 152, sling swivels, sling mount on left side of stock; forearm hinges down to act as handgrip. Some barrels marked 142A. Introduced 1949; discontinued 1957.
Perf.: $175   **Exc.:** $140   **VGood** $110

Mossberg Model 142K
Same specs as Model 142 except peep sight is replaced by open rear sight. Introduced 1953; discontinued 1957.
Perf.: $150   **Exc.:** $120   **VGood** $100

MOSSBERG MODEL 144
Bolt action; 22 LR; 26″ heavy barrel; 7-shot detachable clip magazine; 43″ overall length; weighs 8 lbs.; target model; micrometer receiver peep sight, hooded front; pistol-grip target stock; grooved trigger; grooved receiver for Mossberg No. 4M4 scope; thumb safety; T-shaped bolt handle; adjustable hand stop; beavertail forearm; sling swivels. Introduced 1949; discontinued 1954.
Perf.: $220   **Exc.:** $160   **VGood** $120

Mossberg Model 144LS
Same specs as Model 144 except Lyman No. 57MS receiver peep sight, Lyman No. 17A front. Introduced 1954; discontinued 1960.
Perf.: $250   **Exc.:** $210   **VGood** $140

Mossberg Model 144LS-A
Same specs as Model 144LS except Mossberg 330 rear peep sight. Introduced 1960; discontinued 1978.
Perf.: $220   **Exc.:** $160   **VGood** $120

Mossberg Model 144LS-B (Target)
Same specs as Model 144 except 27″ round barrel; Lyman 17A hooded front sight with inserts, Mossberg 5331 receiver peep sight; target-style American walnut stock; adjustable forend hand stop. Introduced 1978; discontinued 1985.
Perf.: $290   **Exc.:** $230   **VGood** $190

Mossberg Model 146B

Mossberg Model 320K

Mossberg Model 340TR Targo Rifle/Shotgun

Mossberg Model 340B

Mossberg Model 342

Mossberg Bolt-Action Rifles
Model 146B
Model 146B-A
Model 320B
Model 320K
Model 320K-A
Model 320TR Targo Rifle/Shotgun
Model 321B
Model 321K
Model 340B, 340B-A
Model 340K, 340K-A
Model 340M
Model 340TR Targo Rifle/Shotgun
Model 341
Model 342
Model 342K, 342K-A
Model 344
Model 344K
Model 346B
Model 346K, 346K-A
Model 620K, 620K-A

MOSSBERG MODEL 146B
Bolt-action; 22 Short, 22 Long, 22 LR; 20-shot (22 LR), 23-shot (22 Long), 30-shot (22 Short) tube magazine; 26" barrel; 43 1/4" overall length; weighs 7 lbs.; take-down repeater; micrometer fully-adjustable peep rear sight on receiver, hooded front sight, open fully-adjustable rear; uncheckered walnut pistol-grip Monte Carlo-type stock with cheekpiece; adjustable trigger; moulded finger-groove trigger guard; Schnabel forearm; sling swivels. Introduced 1949; discontinued 1954.
Perf.: $165   **Exc.:** $130   **VGood:** $100

Mossberg Model 146B-A
Same specs as Model 146B except different sights. Introduced 1954; discontinued 1958.
Perf.: $165   **Exc.:** $130   **VGood:** $100

MOSSBERG MODEL 320B
Bolt action; single shot; 22 LR, 22 Short, 22 Long; 24" tapered barrel; 43 1/2" overall length; weighs 5 3/4 lbs.; hooded ramp front, rear target precision peep sight; automatic safety; hammerless; walnut pistol-grip Monte Carlo stock; sling swivels. Designated by manufacturer as a Boy Scout target model. Introduced 1960; discontinued 1971.
Perf.: $150   **Exc.:** $120   **VGood:** $90

Mossberg Model 320K
Same specs as Model 320B except open sights; no sling swivels; drop-in loading platform; automatic safety. Introduced 1958; discontinued 1960.
Perf.: $100   **Exc.:** $80   **VGood:** $60

Mossberg Model 320K-A
Same specs as Model 320K.
Perf.: $100   **Exc.:** $80   **VGood:** $60

MOSSBERG MODEL 320TR TARGO RIFLE/SHOTGUN
Bolt action; single shot; 22 Short, 22 Long, 22 LR; smoothbore with rifled and smoothbore barrel adjustable adapters; fully-adjustable notch rear sight, blade front; comes with Model 1A hand trap thrower. Introduced 1961; discontinued 1962.
Perf.: $340   **Exc.:** $290   **VGood:** $190

MOSSBERG MODEL 321B
Bolt action; single shot; 22 Short, 22 Long, 22 LR; 24" barrel; 43 1/2" overall length; weighs 6 1/2 lbs.; S330 peep sight with 1/4-minute click adjustments; hardwood stock with walnut finish; cheekpiece; checkered pistol grip, forearm; hammerless bolt-action with drop-in load-ing platform; automatic safety. Discontinued 1976.
Perf.: $100   **Exc.:** $80   **VGood:** $60

Mossberg Model 321K
Same specs as Model 321B except adjustable open rear sight, ramp front. Introduced 1972; no longer in production.
Perf.: $90   **Exc.:** $70   **VGood:** $50

MOSSBERG MODEL 340B, 340B-A
Bolt action; 22 Short, 22 Long, 22 LR; 7-shot clip magazine; 24" barrel; 43 1/2" overall length; weighs 6 1/2 lbs.; hammerless; rear precision target peep sight, hooded ramp front; walnut Monte Carlo stock with cheekpiece, pistol-grip; buttplate white line spacer. Introduced 1958; no longer in production.
Perf.: $150   **Exc.:** $120   **VGood:** $90

Mossberg Model 340K, 340K-A
Same specs as Model 340B except for open rear sight, bead front. Introduced 1958; discontinued 1971.
Perf.: $140   **Exc.:** $110   **VGood:** $80

Mossberg Model 340M
Same specs as Model 340K except for 18 1/2" barrel; 38 1/2" overall length; weighs 5 1/4 lbs.; one-piece Mannlicher-style Monte Carlo pistol-grip stock; sling; sling swivels. Introduced 1970; discontinued 1971.
Perf.: $240   **Exc.:** $190   **VGood:** $140

MOSSBERG MODEL 340TR TARGO RIFLE/SHOTGUN
Bolt action; 22 Short, 22 Long, 22 LR; 7-shot; smoothbore; rifled and choke adapters screw on muzzle for shooting bullets or shot. Smoothbore was designed for trap shooting with hand trap. Special device fitted barrel to allow shooter to spring trap. Introduced 1961; discontinued 1962.
Perf.: $340   **Exc.:** $290   **VGood:** $190

MOSSBERG MODEL 341
Bolt action; 22 Short, 22 Long, 22 LR; 7-shot clip magazine; 24" barrel; 43 1/2" overall length; weighs 6 1/2 lbs.; fully-adjustable open rear sight, bead post ramp front; walnut Monte Carlo stock with cheekpiece; checkered pistol grip, forearm; plastic buttplate with white line spacer; sliding side safety; sling swivels. Introduced 1972; discontinued 1976.
Perf.: $110   **Exc.:** $90   **VGood:** $80

MOSSBERG MODEL 342
Bolt action; 22 Short, 22 Long, 22 LR; 7-shot clip magazine; 18" tapered barrel; hammerless action; peep rear sight, bead front; receiver grooved for scope mounts; uncheckered walnut Monte Carlo pistol-grip stock; two-position forearm that folds down for rest or handgrip; thumb safety; sling swivels; web sling. Intro-

duced 1957; discontinued 1959.
Perf.: $150   **Exc.:** $120   **VGood:** $90

Mossberg Model 342K, 342K-A
Same specs as Model 342 except with open rear sight. Introduced 1958; discontinued 1971.
Perf.: $140   **Exc.:** $110   **VGood:** $80

MOSSBERG MODEL 344
Bolt action; 22 LR; 7-shot clip magazine; walnut finished, checkered stock. Introduced 1985.
Perf.: $155   **Exc.:** $120   **VGood:** $100

Mossberg Model 344K
Same specs as Model 344 except carbine length.
Perf.: $155   **Exc.:** $120   **VGood:** $100

MOSSBERG MODEL 346B
Bolt action; 22 Short, 22 Long, 22 LR; 18-shot tube magazine (22 LR), 20-shot (22 Long), 25-shot (22 Short); 24" barrel; hammerless action; triple sight combination; hooded ramp front, receiver peep with open rear sight; uncheckered walnut stock with pistol grip; Monte Carlo comb; cheekpiece; quick-detachable sling swivels. Introduced 1958; discontinued 1967.
Perf.: $150   **Exc.:** $120   **VGood:** $90

Mossberg Model 346K, 346K-A
Same specs as Model 346B except open rear sight, bead front. Introduced 1958; discontinued 1971.
Perf.: $140   **Exc.:** $110   **VGood:** $80

MOSSBERG MODEL 620K, 620K-A
Bolt action; single shot; 22 WMR; 24" barrel; 44 3/4" overall length; weighs 6 lbs.; fully-adjustable open rear sight, bead front; extra heavy duty receiver and bolt; grooved trigger; thumb safety; receiver grooved for scope mounts; hammerless action; walnut Monte Carlo pistol-grip stock with cheekpiece; sling swivels; impressed checkering on pistol-grip, forearm. Introduced 1958; discontinued 1974.
Perf.: $140   **Exc.:** $120   **VGood:** $100

Mossberg Bolt-Action Rifles
Model 640K
Model 640KS
Model 640M
Model 642K
Model 800
Model 800D Super Grade
Model 800M
Model 800SM
Model 800VT
Model 810, 810A
Model 810ASM
Model 810B
Model 810BSM
Model 810C
Model 810D
Model 1500 Mountaineer
Model 1500 Mountaineer Deluxe
Model 1500 Mountaineer Varmint Deluxe

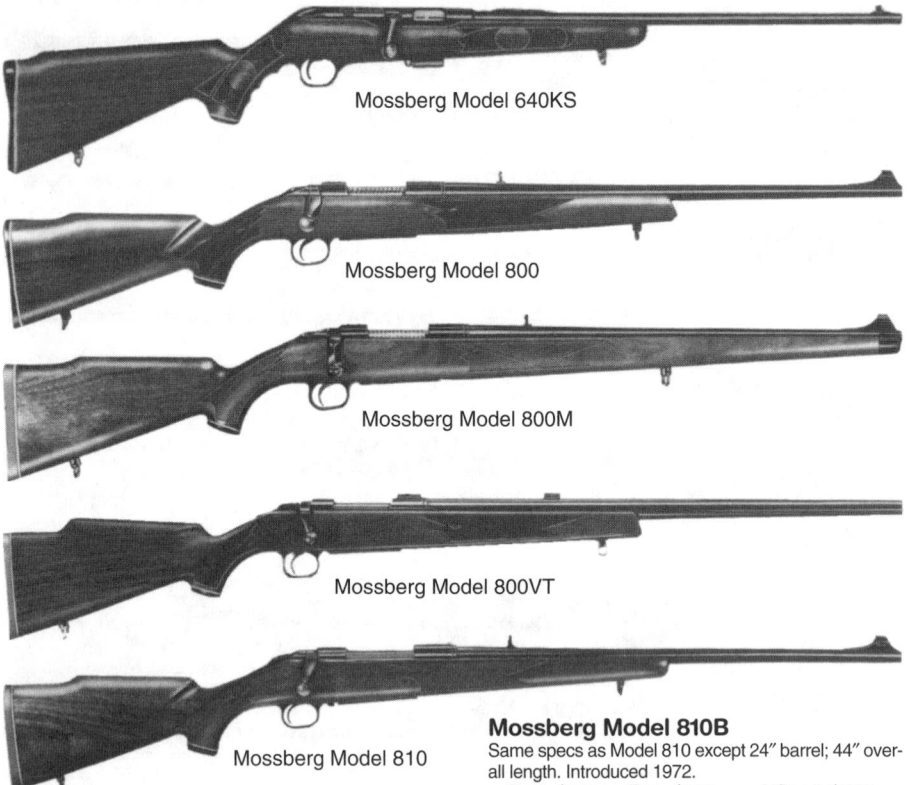

Mossberg Model 640KS

Mossberg Model 800

Mossberg Model 800M

Mossberg Model 800VT

Mossberg Model 810

MOSSBERG MODEL 640K
Bolt action; 22 WRM; 5-shot detachable clip magazine; 24″ barrel; 44 ³/₄″ overall length; weighs 6 lbs.; hammerless action; fully-adjustable open rear sight, bead front; receiver grooved for scope mounts; extra heavy duty receiver and bolt; grooved trigger; thumb safety; walnut Monte Carlo pistol-grip stock with cheekpiece; open rear sight, bead front; sling swivels; impressed checkering on pistol grip, forearm. Called the "Chuckster". Introduced 1959; no longer in production.
   Perf.: $190   **Exc.:** $140   **VGood:** $110

Mossberg Model 640KS
Same specs as Model 640K except select walnut stock, checkered pistol grip, forearm; gold-plated front sight, rear sight elevator and trigger. Discontinued 1974.
   Perf.: $215   **Exc.:** $180   **VGood:** $140

Mossberg Model 640M
Same specs as Model 640K except full-length checkered Monte Carlo Mannlicher stock; 20″ barrel; heavy receiver; jeweled bolt. Manufactured 1971.
   Perf.: $240   **Exc.:** $210   **VGood:** $160

MOSSBERG MODEL 642K
Bolt action; 22 WMR; 5-shot clip magazine; 18″ barrel; 38 ¼″ overall length; weighs 5 lbs.; bead front sight, fully-adjustable rear sight, fold-down rear; walnut Monte Carlo pistol-grip stock; grip cap and buttplate white line spacers; web sling. Introduced 1960; discontinued 1968.
   Perf.: $240   **Exc.:** $190   **VGood:** $140

MOSSBERG MODEL 800
Bolt action; 222 Rem., 22-250, 243 Win., 308 Win.; 4-shot magazine, 3-shot (22-250); 22″ barrel; 42″ overall length; weighs 6 ½ lbs.; folding-leaf rear sight, ramp front; checkered pistol-grip Monte Carlo stock, forearm; recessed bolt head with six locking lugs; top tang safety; hinged floorplate with button release; cheekpiece; sling swivels. Manufactured 1967; discontinued 1980.
   Perf.: $230   **Exc.:** $190   **VGood:** $160

Mossberg Model 800D Super Grade
Same specs as Model 800 except 22-250, 243 Win., 308 Win.; Monte Carlo stock with roll-over comb, cheekpiece; rosewood forend tip, pistol-grip cap; hand-Damascened bolt. Manufactured 1970 to 1973.
   Perf.: $330   **Exc.:** $280   **VGood:** $250

Mossberg Model 800M
Same specs as Model 800 except 20″ barrel; 40″ overall length; 22-250, 243 Win., 308 Win.; Mannlicher-type stock; flat bolt handle. Manufactured 1969-1972.
   Perf.: $320   **Exc.:** $270   **VGood:** $240

Mossberg Model 800SM
Same specs as Model 800 except weighs 7 ½ lbs.; equipped with Mossberg Model 84 4x scope.
   Perf.: $340   **Exc.:** $300   **VGood:** $270

Mossberg Model 800VT
Same specs as Model 800 except varmint/target model; 222 Rem., 22-250, 243 Win.; 24″ heavy barrel; weighs 9 ½ lbs.; 44″ overall length; no sights; Marksman-style stock; scope mount blocks. Manufactured 1968 to 1980.
   Perf.: $260   **Exc.:** $200   **VGood:** $180

MOSSBERG MODEL 810, 810A
Bolt action; 270 Win., 30-06, 7mm Rem. Mag., 338 Win. Mag.; 4-shot (standard calibers), 3-shot (magnum calibers) detachable or internal magazine; 22″ barrel (standard calibers), 24″ (magnum calibers); 42″ overall length; weighs 7 ½-8 lbs.; folding leaf rear sight, ramp front with gold bead, fully-adjustable folding leaf middle sight; receiver drilled, tapped for peep or scope; hinged floorplate; checkered pistol-grip walnut stock, forearm; Monte Carlo comb, cheekpiece; pistol-grip cap; recoil pad with white liners; sling swivels. Manufactured 1970 to 1980.
Standard calibers
   Perf.: $300   **Exc.:** $280   **VGood:** $250
Magnum calibers
   Perf.: $340   **Exc.:** $290   **VGood:** $260

Mossberg Model 810ASM
Same Specs as Model 810 except 30-06; no sights; comes fitted with Mossberg Model 84 4x scope. Introduced 1972.
   Perf.: $300   **Exc.:** $270   **VGood:** $250

Mossberg Model 810B
Same specs as Model 810 except 24″ barrel; 44″ overall length. Introduced 1972.
   Perf.: $330   **Exc.:** $300   **VGood:** $270

Mossberg Model 810BSM
Same specs as Model 810B except 7mm Rem. Mag.; no sights; comes fitted with Mossberg Model 84 4x scope. Introduced 1972.
   Perf.: $370   **Exc.:** $325   **VGood:** $295

Mossberg Model 810C
Same specs as Model 810 except 270 Win. Introduced 1973.
   Perf.: $360   **Exc.:** $315   **VGood:** $285

Mossberg Model 810D
Same specs as Model 810 except 338 Win. Mag. Introduced 1973.
   Perf.: $375   **Exc.:** $330   **VGood:** $300

MOSSBERG MODEL 1500 MOUNTAINEER
Bolt action; 222, 223, 22-250, 243, 25-06, 270, 30-06, 308, 7mm Rem. Mag., 300 Win. Mag., 338 Win. Mag., 22″, 24″ barrel; weighs 7 ½ lbs.; open round-notch adjustable rear sight, hooded ramp gold bead front; American walnut stock; Monte Carlo comb, cheekpiece; cut checkering; single-unit triggerguard/magazine box; hinged floorplate; swivel studs; composition buttplate; magnums with rubber recoil pad. Made in Japan. Originally marketed by S&W. Introduced 1979; discontinued 1987.
Standard calibers
   Perf.: $330   **Exc.:** $300   **VGood:** $250
Magnum calibers
   Perf.: $350   **Exc.:** $310   **VGood:** $270

Mossberg Model 1500 Mountaineer Deluxe
Same specs as Model 1500 except sans sights; engine-turned bolt; decorated floorplate with scrollwork; skip-line checkering; sling, swivels, swivel posts included; magnum models with vent, recoil pad.
Standard calibers
   Perf.: $350   **Exc.:** $300   **VGood:** $250
Magnum calibers
   Perf.: $370   **Exc.:** $320   **VGood:** $270

Mossberg Model 1500 Mountaineer Varmint Deluxe
Same specs as Model 1500 except 222, 22-250, 223; 22″ heavy barrel; fully-adjustable trigger; skip-line checkering; quick-detachable swivels. Originally marketed by S&W. Introduced 1982; discontinued by Mossberg, 1987.
   Perf.: $370   **Exc.:** $310   **VGood:** $280

Mossberg Model 1500 Mountaineer Deluxe

Mossberg Model RM-7A

Mossberg Model Model 50

Mossberg Model 51

Mossberg Model 151M

Mossberg Model 51M

MOSSBERG MODEL 1550
Bolt action; 243, 270, 30-06; removable box magazine; 22", 24" barrel; weighs 7 1/2 lbs.; with or without sights; American walnut stock; Monte Carlo comb, cheekpiece; cut checkering; single-unit triggerguard/magazine box; hinged floorplate; swivel studs; composition buttplate; magnums with rubber recoil pad. Made in Japan. Introduced 1986; discontinued 1987.
    Perf.: $330   **Exc.:** $280   **VGood:** $250

MOSSBERG MODEL 1700LS
Bolt action; 243, 270, 30-06; 4-shot magazine; classic-style stock; tapered forend; Schnabel tip; hand-checkering; black rubber buttpad; flush-mounted sling swivels; jeweled bolt body. Marketed as Classic Hunter. Made in Japan. Originally marketed by S&W. Introduced 1983; discontinued 1987.
    Perf.: $400   **Exc.:** $350   **VGood:** $275

MOSSBERG MODEL B
Bolt aciton; single shot; 22 Short, 22 Long, 22 LR; 22" barrel; take-down; uncheckered pistol-grip walnut stock. Introduced 1930; discontinued 1932.
    Perf.: $175   **Exc.:** $120   **VGood:** $70

MOSSBERG MODEL C
Bolt action; single shot; 22 Short, 22 Long, 22 LR; 24" barrel; ivory bead front sight, open sporting rear. Introduced 1931; discontinued 1932.
    Perf.: $175   **Exc.:** $120   **VGood:** $70

Mossberg Model C-1
Same specs as Model C except Lyman front and rear sights; leather sling, swivels.
    Perf.: $225   **Exc.:** $150   **VGood:** $100

MOSSBERG MODEL L
Bolt action; single shot; 22 Short, 22 Long, 22 LR; 24" barrel; take-down; Martini-design falling-block lever-action; open rear sight, bead front; uncheckered walnut pistol-grip stock, forearm. Introduced 1929; discontinued 1932.
    Perf.: $450   **Exc.:** $325   **VGood:** $200

Mossberg Model L-1
Same specs as Model L except target version with Lyman 2A tang sight; factory sling. Introduced about 1930; discontinued about 1931.
    Perf.: $500   **Exc.:** $375   **VGood:** $250

MOSSBERG MODEL R
Bolt action; 22 Short, 22 Long, 22 LR; 24" barrel; tube magazine; take-down; open rear sight, bead front; uncheckered pistol-grip stock. Introduced 1930; discontinued 1932.
    Perf.: $165   **Exc.:** $125   **VGood:** $90

MOSSBERG MODEL RM-7A
Bolt action; 30-06; 22" barrel; 5-shot magazine; gold bead front sight on ramp, adjustable folding leaf rear; drilled, tapped for scope mounts; classic-style checkered pistol-grip stock of American walnut; rotary magazine; three-position bolt safety; sling swivel studs. Introduced 1978; discontinued 1981. Very rare.
    Perf.: $500   **Exc.:** $400   **VGood:** $300

Mossberg Model RM-7B
Same specs as Model RM-7A except 7mm Rem. Mag.; 4-shot magazine; 24" barrel; 45 1/4" overall length. Introduced 1978; discontinued 1981. Very rare.
    Perf.: $650   **Exc.:** $550   **VGood:** $450

MOSSBERG SEMI-AUTOMATIC RIFLES MOSSBERG MODEL 50
Semi-automatic; 22 LR; 24" barrel; 15-shot tube magazine in buttstock; 43 3/4" overall length; weighs 6 3/4 lbs.; takedown; open rear sight, hooded ramp front; uncheckered walnut pistol-grip stock; finger grooves in grip; no sling swivels. Introduced 1939; discontinued 1942.
    Perf.: $170   **Exc.:** $130   **VGood:** $100

MOSSBERG MODEL 51
Semi-automatic; 22 LR; 15-shot tube magazine; 24" barrel; 43 3/4" overall length; weighs 7 1/4 lbs.; receiver peep sight; sling swivels; cheekpiece stock; heavy beavertail forearm. Made only in 1939.
    Perf.: $190   **Exc.:** $140   **VGood:** $110

Mossberg Model 51M
Same specs as Model 51 except 20" round tapered barrel; 40" overall length; weighs 7 lbs.; two-piece Mannlicher-style stock. Introduced 1939; discontinued 1946.
    Perf.: $200   **Exc.:** $150   **VGood:** $120

MOSSBERG MODEL 151M
Semi-automatic; 22 LR; 15-shot tube magazine; 20" barrel; 40" overall length; weighs 7 lbs.; No. S101 hooded ramp front, No. S107 open sporting rear or No. S130 micro-click adjustable rear; two-piece Mannlicher stock; sling swivels; steel buttplate. Introduced 1946; discontinued 1958.
    Perf.: $175   **Exc.:** $140   **VGood:** $120

Mossberg Model 151M(a), 151M(b), 151M(c)
Same specs as Model 151M except minor changes in sights and buttplate. Introduced 1947; discontinued 1958.
    Perf.: $175   **Exc.:** $140   **VGood:** $120

Mossberg Model 151K
Same specs as Model 151M except 24" tapered barrel; 44" overall length; weighs 6 lbs.; no peep sight; sling swivels; uncheckered sporting stock; Monte Carlo comb; cheekpiece. Introduced 1950; discontinued 1951.
    Perf.: $150   **Exc.:** $130   **VGood:** $100

Mossberg Semi-Automatic Rifles
Model 152
Model 152K
Model 333
Model 350K, 350K-A
Model 351K, 351K-A
Model 351C
Model 352
Model 352K, 352K-A, 352K-B
Model 353
Model 377
Model 380, 380S
Model 430
Model 432
Plinkster (See Mossberg Model 377)

Mossberg Model 152

Mossberg Model 352K

Mossberg Model 350K

Mossberg Model 351K

Mossberg Model 353

Mossberg Model 377

Mossberg Model 380

MOSSBERG MODEL 152
Semi-automatic; 22 LR; 7-shot removable clip magazine; 18″ barrel; 38″ overall length; weighs 5 lbs.; rear peep sight, military-type ramp front; walnut Monte Carlo pistol-grip stock; hinged forearm swings down to act as forward handgrip; sling swivels and adjustable sling mounted on left side. Introduced 1948; discontinued 1957.
Perf.: $200   **Exc.:** $150   **VGood:** $120

Mossberg Model 152K
Same specs as Model 152 except open rear sight. Introduced 1950; discontinued 1957.
Perf.: $180   **Exc.:** $140   **VGood:** $110

MOSSBERG MODEL 333
Semi-automatic; 22 LR; 15-shot tubular magazine; 20″ barrel; 39 1/2″ overall length; weighs 6 1/4 lbs.; fully-adjustable open rear sight, beaded ramp front; checkered pistol-grip Monte Carlo American walnut stock, forearm; buttplate, pistol grip white line spacers; automatic bolt hold-open; tang safety; gold-plated, grooved trigger; Damascened bolt; receiver grooved for scope mounts; barrel band; sling swivels. Manufactured 1972 to 1973.
Perf.: $150   **Exc.:** $130   **VGood:** $100

MOSSBERG MODEL 350K, 350K-A
Semi-automatic; 22 high-speed Short, 22 Long, 22 LR; 7-shot clip magazine; 23 1/2″ barrel; 43 1/2″ overall length; weighs 6 lbs.; fully-adjustable open rear sight with "U" notch, bead front; receiver grooved for scope mount; walnut Monte Carlo pistol-grip stock. Introduced 1958; discontinued 1971.
Perf.: $130   **Exc.:** $100   **VGood:** $90

MOSSBERG MODEL 351K, 351K-A
Semi-automatic; 22 LR; 15-shot tube magazine in buttstock; 24″ barrel; 43″ overall length; weighs 6 lbs.; fully-adjustable open rear sight with "U" notch, bead front; receiver grooved for scope mount; walnut Monte Carlo pistol-grip stock. Introduced 1960; discontinued 1971.
Perf.: $130   **Exc.:** $100   **VGood:** $90

Mossberg Model 351C
Same specs as Model 351K except 18 1/2″ barrel; 38 1/2″ overall length; weighs 5 1/2 lbs.; straight Western-type carbine stock with barrel band; sling swivels. Called the "Jack Rabbit Special". Introduced 1965; discontinued 1971.
Perf.: $140   **Exc.:** $120   **VGood:** $100

MOSSBERG MODEL 352
Semi-automatic; 22 Short, 22 Long, 22 LR; 7-shot clip-fed magazine; 18″ barrel; 38″ overall length; weighs 5 lbs.; peep rear sight, bead front; uncheckered walnut Monte Carlo pistol-grip stock; two-position Tenite forearm extension folds down for rest or hand grip; sling swivels; web sling. Introduced 1958; discontinued 1971.
Perf.: $150   **Exc.:** $125   **VGood:** $110

Mossberg Model 352K, 352K-A, 352K-B
Same specs as Model 352 except with fully-adjustable open rear sight. Introduced 1960; discontinued 1971.
Perf.: $140   **Exc.:** $115   **VGood:** $100

MOSSBERG MODEL 353
Semi-automatic; 22 LR; 7-shot clip magazine; 18″ barrel; 38″ overall length; weighs 5 lbs.; fully-adjustable open rear sight, ramp front with bead; American walnut Monte Carlo Pistol-grip stock; black Tenite fold down forend extension; checkered pistol grip, forend; receiver grooved for scope mount; sling swivels, web strap on left side of stock. Introduced 1972.
Perf.: $150   **Exc.:** $125   **VGood:** $110

MOSSBERG MODEL 377
Semi-automatic; 22 LR; 15-shot tubular magazine; 20″ barrel; 40″ overall length; weighs 6 1/4 lbs.; no open sights; 4x scope mounted; moulded polystyrene straight-line thumbhole stock; roll-over cheekpiece; Monte Carlo comb; checkered forearm. Advertised as the Plinkster. Introduced 1977; discontinued 1985.
Perf.: $175   **Exc.:** $140   **VGood:** $120

MOSSBERG MODEL 380, 380S
Semi-automatic; 22 LR; 15-shot tubular magazine in stock; 20″ tapered barrel; weighs about 5 1/2 lbs.; bead front sight, adjustable open rear; receiver grooved for scope mounting; walnut-finished hardwood stock; black plastic buttplate. Introduced 1981; discontinued 1985.
Perf.: $150   **Exc.:** $120   **VGood:** $100

MOSSBERG MODEL 430
Semi-automatic; 22 LR; 18-shot tube magazine; 24″ barrel; 43 1/2″ overall length; weighs 6 1/4 lbs.; open rear sight, gold bead front; walnut Monte Carlo stock; checkered pistol grip, forearm; gold-plated, grooved trigger; Damascened bolt; buttplate, pistol grip cap white line spacers; top tang safety; hammerless action. Introduced 1970; discontinued 1971.
Perf.: $140   **Exc.:** $110   **VGood:** $100

MOSSBERG MODEL 432
Semi-automatic; 22 LR; 15-shot magazine; 20″ barrel; 39 1/2″ overall length; weighs 6 lbs.; gold-plated, grooved trigger; Damascened bolt; buttplate, pistol grip white line spacers; top tang safety; hammerless action; carbine; uncheckered straight-grip carbine stock, forearm; barrel band; sling swivels. Introduced 1970; discontinued 1971.
Perf.: $140   **Exc.:** $110   **VGood:** $90

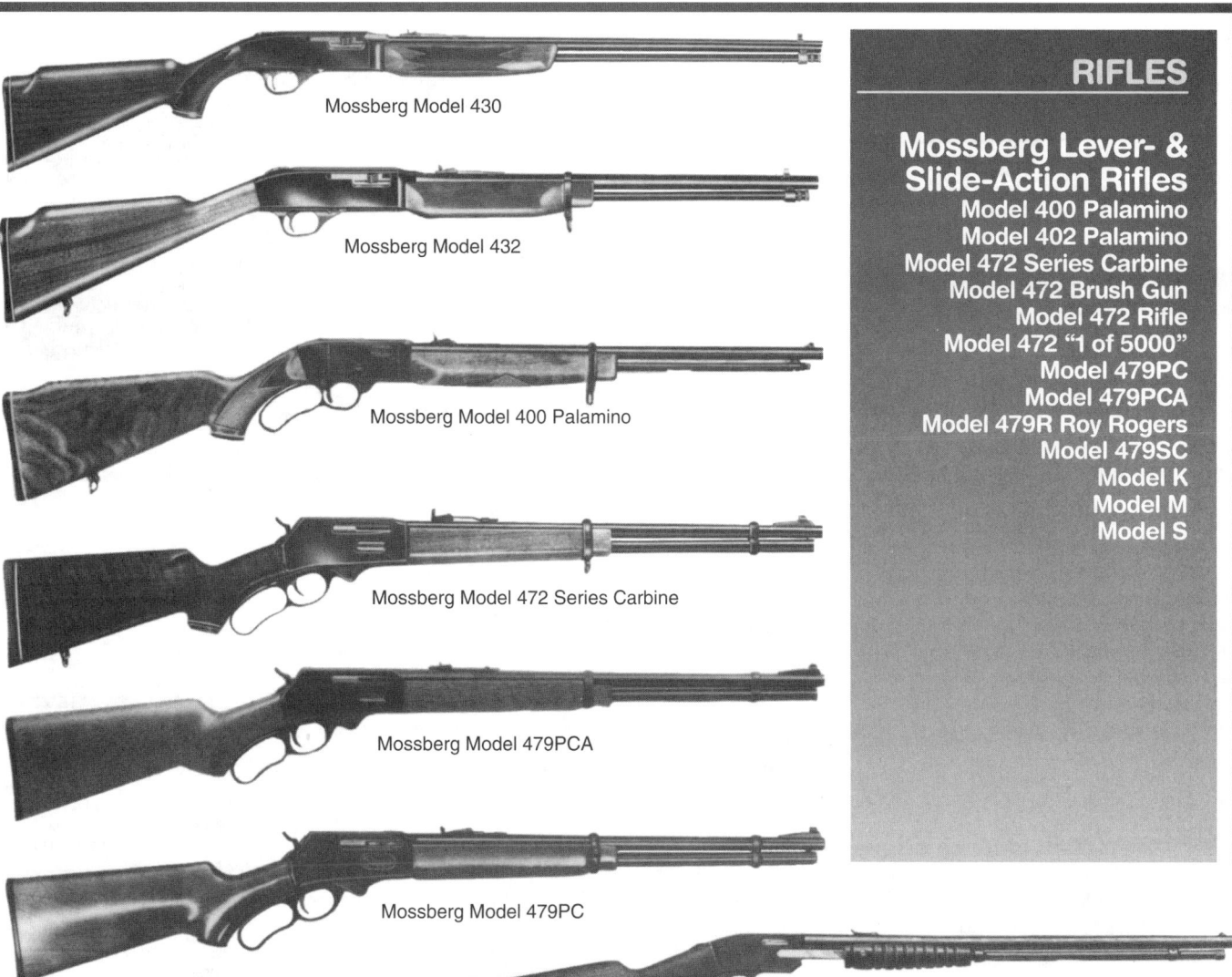

Mossberg Model 430

Mossberg Model 432

Mossberg Model 400 Palamino

Mossberg Model 472 Series Carbine

Mossberg Model 479PCA

Mossberg Model 479PC

Mossberg Model K

RIFLES

Mossberg Lever- & Slide-Action Rifles
Model 400 Palamino
Model 402 Palamino
Model 472 Series Carbine
Model 472 Brush Gun
Model 472 Rifle
Model 472 "1 of 5000"
Model 479PC
Model 479PCA
Model 479R Roy Rogers
Model 479SC
Model K
Model M
Model S

MOSSBERG LEVER- & SLIDE-ACTION RIFLES

MOSSBERG MODEL 400 PALAMINO
Lever action; 22 Short, 22 Long, 22 LR; 15-shot (22 LR), 17-shot (22 Long), 20-shot (22 Short) tube magazine; 24" barrel; 41" overall length; weighs 5 1/2 lbs.; fully-adjustable open notch rear sight, bead front; walnut Monte Carlo stock; moulded pistol-grip cap and buttplate; beavertail forearm; hammerless action; lightweight high-tensile alloy receiver; removable sideplate; grooved trigger; cross-bolt safety; blued finish; receiver grooved for scope. Introduced 1959; discontinued 1963.
Perf.: $225   **Exc.:** $180   **VGood:** $130

Mossberg Model 402 Palamino
Same specs as Model 400 except 18 1/2", 20" barrel; 36 1/2" overall length; weighs 4 3/4 lbs.; barrel band on forearm; magazine holds two less rounds in 22 Long, 22 LR lengths. Manufactured 1961 to 1971.
Perf.: $230   **Exc.:** $190   **VGood:** $130

MOSSBERG MODEL 472 SERIES CARBINE
Lever action; 30-30, 35 Rem.; 6-shot tubular magazine; 20" barrel; adjustable buckhorn rear sight, ramp front; with gold bead; barrel tapped for scope mounting; offset hammer spur; side-ejecting; pistol-grip or straight stock; barrel band on forearm; sling swivels on pistol-grip style, removable saddle ring on straight-stock model. Manufactured 1972 to 1980.
Perf.: $200   **Exc.:** $170   **VGood:** $130

Mossberg Model 472 Brush Gun
Same specs as Model 472 Carbine except 18" barrel; 30-30; 5-shot magazine; straight stock. Manufactured 1974 to 1976.
Perf.: $220   **Exc.:** $190   **VGood:** $150

Mossberg Model 472 Rifle
Same specs as Model 472 Carbine except 24" barrel; 5-shot magazine; pistol-grip stock only. Manufactured 1974 to 1976.
Perf.: $220   **Exc.:** $190   **VGood:** $150

Mossberg Model 472 "1 of 5000"
Same specs as Model 472 Brush Gun except etched Native American frontier scenes on receiver; gold-plated trigger; brass saddle ring, buttplate, barrel bands; select walnut stock forearm; limited edition, numbered 1 through 5000. Manufactured 1974.
Perf.: $430   **Exc.:** $255   **VGood:** $200

MOSSBERG MODEL 479PC
Lever action; 30-30, 35 Rem.; 6-shot tube magazine; 20" barrel; 38 1/2" overall length; weighs 6 3/4-7 lbs.; ramp front sight, elevation-adjustable rear; American walnut stock fluted comb; composition buttplate; pistol-grip cap; hammer-block safety; side ejection. Introduced 1978; discontinued 1981.
Perf.: $220   **Exc.:** $190   **VGood:** $150

Mossberg Model 479PCA
Same specs as Model 479PC except 30-30; bead on ramp front sight, adjustable open rear; walnut-finished hardwood stock; rebounding hammer; trigger built into cocking lever; blued finish. Reintroduced 1983; no longer in production.
Perf.: $210   **Exc.:** $180   **VGood:** $150

Mossberg Model 479RR Roy Rogers
Same specs as Model 479PCA except 5-shot magazine; 18" barrel; 36 1/2" overall length; gold bead on ramp front sight, adjustable semi-buckhorn rear; American walnut stock; gold-finished trigger; barrel bands; Rogers' signature, American eagle, stars and stripes etched in receiver. 5000 guns produced only in 1983.
Perf.: $375   **Exc.:** $275   **VGood:** $225

Mossberg Model 479SC
Same specs as Model 479PC except straight grip stock. Introduced 1978; discontinued 1981.
Perf.: $220   **Exc.:** $190   **VGood:** $150

MOSSBERG MODEL K
Slide action; 22 Short, 22 Long, 22 LR; 14-shot (22 LR), 16-shot (22 Long), 20-shot (22 Short) tube magazine; 22" barrel; take-down; hammerless; open rear sight, bead front; unchecked straight-grip walnut stock; grooved slide handle. Introduced 1922; discontinued 1931.
Perf.: $295   **Exc.:** $225   **VGood:** $190

MOSSBERG MODEL M
Slide action; 22 Short, 22 Long, 22 LR; 14-shot (22 LR), 16-shot (22 Long), 20-shot (22 Short) tube magazine; 24" octagonal barrel; take-down; hammerless; open rear sight, bead front; unchecked pistol-grip stock; grooved slide handle. Introduced 1928; discontinued 1931.
Perf.: $330   **Exc.:** $270   **VGood:** $195

MOSSBERG MODEL S
Slide action; 22 Short, 22 Long, 22 LR; 19 3/4" barrel; similar to Model K except shorter magazine tube. Introduced 1927; dropped 1931.
Exc.: $250   **VGood:** $200   **Good:** $145

Mossberg/New Haven
Model 220K
Model 240K
Model 246K
Model 250K
Model 250KB

M-S Safari Arms
1000-Yard Match Model
Silhouette Model
Varmint Model

Musgrave
Premier NR5
RSA Single Shot NR1
Valiant NR6

Navy Arms
45-70 Mauser Rifle
45-70 Mauser Carbine
1873 Winchester-Style Rifle
1873 Winchester-Style Sporting Rifle
1873 Winchester-Style Carbine

M-S Safari Arms 1000-Yard Match Model

M-S Safari Arms Silhouette Model

M-S Safari Arms Varmint Model

Musgrave Premier NR5

Musgrave RSA Single Shot NR1

MOSSBERG/NEW HAVEN MODEL 220K
Bolt action; single shot; 22 Short, 22 Long, 22 LR; 24″ barrel with crowned muzzzle; weighs 5 ³/₄ lbs.; open adjustable rear sight, bead front; walnut finished Monte Carlo pistol-grip stock; hand-rubbed oil finish; automatic safety. Introduced 1961.
Perf.: $100   **Exc.:** $80   **VGood:** $65

MOSSBERG/NEW HAVEN MODEL 240K
Bolt action; 22 Short, 22 Long, 22 LR; 24″ barrel with crowned muzzle; fully-adjustable open rear sight, bead front; walnut finished Monte Carlo pistol-grip stock; hammerless; takedown; grooved trigger; thumb safety. Lower-priced version of the Mossberg Model 340K. Introduced 1961.
Perf.: $105   **Exc.:** $85   **VGood:** $65

MOSSBERG/NEW HAVEN MODEL 246K
Bolt action; 22 Short, 22 Long, 22 LR; 25-shot (22 Short), 20-shot (22 Long), 18-shot (22 LR); 24″ round tapered barrel with crowned muzzle; 43 ¹/₄ overall length; weighs 6 ¹/₄ lbs.; fully-adjustable open rear sight, sporting front; receiver grooved for scope mounts; tapped and drilled for Mossberg No. S-330 peep; hammerless; takedown; grooved trigger; thumb safety. A lower-priced version of the Mossberg Model 346K. Introduced 1961.
Perf.: $100   **Exc.:** $80   **VGood:** $65

MOSSBERG/NEW HAVEN MODEL 250K
Semi-automatic; 22 LR; 7-shot clip magazine; 24″ round tapered barrel; 44″ overall length; weighs 5 ¹/₂ lbs.; open adjustable rear sight, bead front; American walnut Monte Carlo stock with tapered forend; hand-rubbed oil finish; receiver grooved for scope mounting. A lower-priced version of the Mossberg Model 350K. Introduced 1960.
Perf.: $120   **Exc.:** $110   **VGood:** $80

Mossberg/New Haven Model 250KB
Same specs as Model 250K except 22 Long, 22 LR; weighs 6 lbs.; Ac-Kro-Gruv rifling; fully-adjustable rear sight. Introduced 1961.
Perf.: $120   **Exc.:** $100   **VGood:** $80

M-S SAFARI ARMS 1000-YARD MATCH MODEL
Bolt action; single shot; 28″ heavy barrel; no sights; drilled/tapped for scope mount; sleeved stainless steel action; electronic trigger; custom-painted fully-adjustable fiberglass stock; custom-built to buyer's specs. Introduced 1982; discontinued 1985.
Perf.: $1695   **Exc.:** $1395   **VGood:** $995

M-S SAFARI ARMS SILHOUETTE MODEL
Bolt action; single shot; 22 LR, all standard centerfires; 23″ barrel (22 LR), 24″ (centerfires); no sights; drilled/tapped for scope mounts; custom-painted fiberglass silhouette stock; stainless steel action; electronic trigger; custom built to buyer's specs. Introduced 1982; discontinued 1985.
Perf.: $995   **Exc.:** $795   **VGood:** $695

M-S SAFARI ARMS VARMINT MODEL
Bolt action; single shot; any standard centerfire chambering; 24″ stainless steel barrel; no sights; drilled, tapped for scope mounts; custom-painted thumbhole or pistol-grip fiberglass stock; stainless steel action; electronic trigger; custom built to buyer's specs. Introduced 1983; discontinued 1985.
Perf.: $1295   **Exc.:** $995   **VGood:** $795

MUSGRAVE PREMIER NR5
Bolt action; 243 Win., 270 Win., 30-06, 308 Win., 7mm Rem. Mag.; 5-shot magazine; 25 ¹/₂″ barrel; no sights; European walnut Monte Carlo stock; cheekpiece; hand-checkered pistol-grip, forearm; pistol-grip cap, forend tip; recoil pad; sling swivel studs. Introduced 1972; discontinued 1973.
Perf.: $450   **Exc.:** $400   **VGood:** $340

MUSGRAVE RSA SINGLE SHOT NR1
Bolt action; 308 Win.; 26″ barrel; aperture rear sight, tunnel front; European walnut target stock; barrel band; rubber buttplate; sling swivels. Introduced 1972; discontinued 1973.
Perf.: $450   **Exc.:** $400   **VGood:** $340

MUSGRAVE VALIANT NR6
Bolt action; 243 Win., 270 Win., 30-06, 308 Win., 7mm Rem. Mag.; 5-shot magazine; 24″ barrel; removable leaf rear sight, hooded ramp front; straight-comb stock; skip-line checkering; sans grip cap, forend tip. Introduced 1972; discontinued.
Perf.: $425   **Exc.:** $375   **VGood:** $310

NAVY ARMS 45-70 MAUSER RIFLE
Bolt action; 45-70 Government; 3-shot magazine; 24″, 26″ barrel; open rear sight, ramp front; hand-checkered Monte Carlo stock; built on Siamese Mauser action. Introduced, 1973; no longer in production.
Perf.: $375   **Exc.:** $330   **VGood:** $290

Navy Arms 45-70 Mauser Carbine
Same specs as 45-70 Mauser except 18″ barrel; straight walnut stock.
Perf.: $375   **Exc.:** $330   **VGood:** $290

NAVY ARMS 1873 WINCHESTER-STYLE RIFLE
Lever action; 44-40, 45 Colt; 12-shot magazine; 24″ barrel; 43″ overall length; weighs 8 ¹/₄ lbs.; blade front sight, buckhorn rear; European walnut stock and forend; color case-hardened frame, rest blued; full-octagon barrel. Imported by Navy Arms. Introduced 1991; still imported. **LMSR:** $790
Perf.: $695   **Exc.:** $595   **VGood:** $495

Navy Arms 1873 Winchester-Style Sporting Rifle
Same specs as 1873 Winchester-Style rifle except checkered pistol-grip stock; 30″ octagonal barrel (24″ available). Imported by Navy Arms. Introduced 1992; still imported. **LMSR:** $930
Perf.: $750   **Exc.:** $675   **VGood:** $595

Navy Arms 1873 Winchester-Style Carbine
Same specs as 1873 Winchester-Style Rifle except 19″ barrel. Imported by Navy Arms. Introduced 1991; still imported. **LMSR:** $780
Perf.: $695   **Exc.:** $595   **VGood:** $500

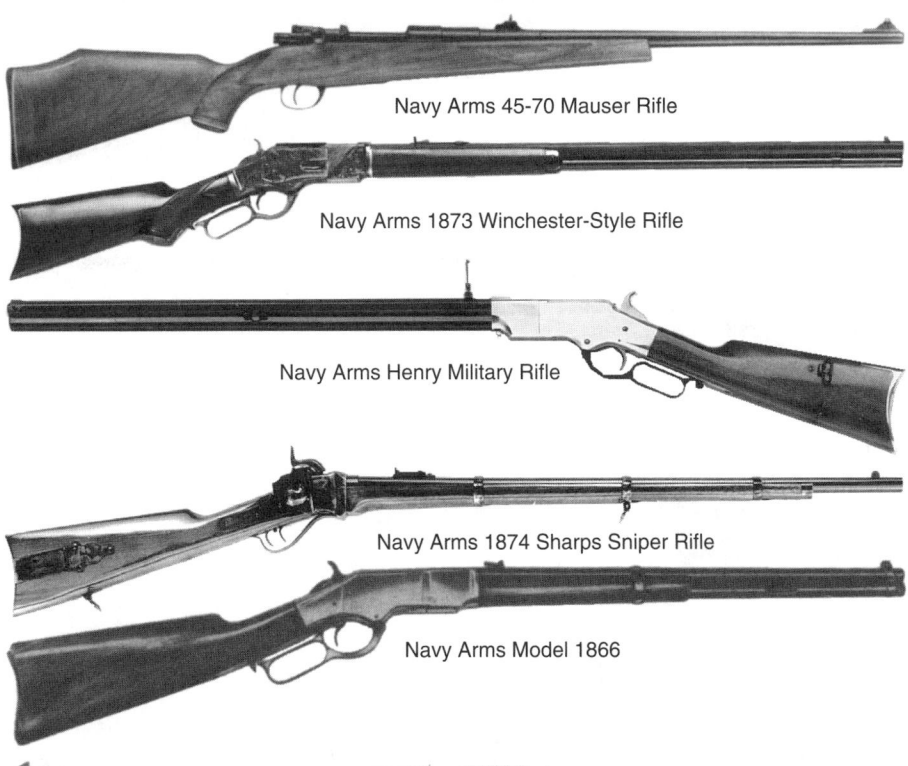

Navy Arms 45-70 Mauser Rifle

Navy Arms 1873 Winchester-Style Rifle

Navy Arms Henry Military Rifle

Navy Arms 1874 Sharps Sniper Rifle

Navy Arms Model 1866

Navy Arms Revolving Carbine

Navy Arms
1874 Sharps Cavalry Carbine
1874 Sharps Infantry Rifle
1874 Sharps Sniper Rifle
EM-331
Henry Military Rifle
D. Henry Carbine
Henry Iron Frame
Henry Trapper Model
Martini
Model 1866
Model 1866 Carbine
Model 1866 Trapper
Model 1866 Yellowboy
Model 1866 Yellowboy Carbine
Model 1866 Yellowboy Trapper
Revolving Carbine
Rolling Block Baby Carbine

NAVY ARMS 1874 SHARPS CAVALRY CARBINE

Falling block; single shot; 45-70; 22" barrel; 39" overall length; weighs 7 3/4 lbs.; blade front sight, military ladder-type rear; walnut stock; color case-hardened receiver and furniture. Replica of the 1874 Sharps military carbine. Imported by Navy Arms. Introduced 1991; still imported. **LMSR: $935**

    Perf.: $695   **Exc.:** $595   **VGood:** $495

Navy Arms 1874 Sharps Infantry Rifle

Same specs as 1874 Sharps Calvary Carbine except 30" barrel; double-set triggers; weighs 8 1/2 lbs.; 46 3/4" overall length; three-band model. Imported by Navy Arms. Introduced 1984; still imported. **LMSR: $1060**

    Perf.: $850   **Exc.:** $695   **VGood:** $525

Navy Arms 1874 Sharps Sniper Rifle

Same specs as Navy Arms 1874 Sharps Carbine except 30" barrel; double-set triggers; weighs 8 1/2 lbs.; 46 3/4" overall length. Imported by Navy Arms. Introduced 1984; still imported. **LMSR: $1055**

    Perf.: $895   **Exc.:** $750   **VGood:** $575

NAVY ARMS EM-331

Bolt action; 7.62x39mm; 5-shot detachable magazine; 21" barrel; 42" overall length; weighs 7 1/2 lbs.; hooded ramp front sight, open adjustable rear; receiver dovetailed for scope rings; Monte Carlo style stock with sling swivels, rubber recoil pad, contrasting forend tip; polished blue finish. Imported by Navy Arms. Introduced 1994; discontinued 1994.

    Perf.: $275   **Exc.:** $240   **VGood:** $200

NAVY ARMS HENRY MILITARY RIFLE

Lever action; 44-40, 44 Rem.; 13-shot magazine; 24 1/4" barrel; weighs 9 1/4 lbs.; blade front sight, elevation-adjustable rear; oil-stained American walnut stock; brass frame, buttplate; sling swivels; no forend; blued barrel; optional engraving. Recreation of the model used by cavalry units in the Civil War. Introduced 1985; still in production. **LMSR: $895**

    Perf.: $695   **Exc.:** $595   **VGood:** $495

Navy Arms D. Henry Carbine

Same specs as Henry except 22" barrel; 41" overall length; weighs 8 3/4 lbs.; no sling swivels. Imported from Italy by Navy Arms. Introduced 1992; no longer imported. **LMSR: $875**

    Perf.: $695   **Exc.:** $595   **VGood:** $495

Navy Arms Henry Iron Frame

Same specs as Henry except receiver of blued or color case-hardened steel. Made in Italy. Introduced 1991; still imported. **LMSR: $945**

    Perf.: $725   **Exc.:** $650   **VGood:** $550

Navy Arms Henry Trapper Model

Same specs as Henry except 8-shot magazine; 16 1/2" barrel; weighs 7 1/2 lbs.; brass frame, buttplate; blued barrel. Made in Italy. Introduced, 1991; still imported. **LMSR: $875**

    Perf.: $675   **Exc.:** $575   **VGood:** $475

NAVY ARMS MARTINI

Single shot; 444 Marlin, 45-70; 26", 30" half or full octagon barrel; Creedmore tang peep sight, open middle sight, blade front; checkered pistol-grip stock; Schnabel forend; cheekpiece. Introduced 1972; discontinued 1984.

    Perf.: $495   **Exc.:** $420   **VGood:** $350

NAVY ARMS MODEL 1866

Lever action; 38 Spl., 44-40; full-length tube magazine; 24" octagon barrel; 39 1/2" overall length; weighs 9 1/4 lbs.; open leaf rear sight, fixed blade front; walnut straight grip stock, forearm; barrel band; polished brass frame, buttplate; other parts blued. Introduced in 1966; no longer imported.

    Perf.: $595   **Exc.:** $475   **VGood:** $385

Navy Arms Model 1866 Carbine

Same specs as Model 1866 except 10-shot magazine; 19" barrel; carbine forearm, barrel band. Introduced 1967; no longer in production.

    Perf.: $595   **Exc.:** $495   **VGood:** $395

Navy Arms Model 1866 Trapper

Same specs as Model 1866 except 22 LR; 8-shot magazine; 16 1/2" barrel.

    Perf.: $595   **Exc.:** $495   **VGood:** $395

Navy Arms Model 1866 Yellowboy

Same specs as Model 1866 except 44-40; 12-shot tubular magazine; 24" full-octagon barrel; 42 1/2" overall length; weighs 8 1/2 lbs.; ladder-type adjustable rear sight; European walnut stock; brass forend tip; blued barrel, hammer, lever. Made in Italy. Introduced 1991; still imported. **LMSR: $675**

    Perf.: $595   **Exc.:** $475   **VGood:** $385

Navy Arms Model 1866 Yellowboy Carbine

Same specs as Model 1866 Yellowboy except 10-shot magazine; 19" barrel. Made in Italy. Introduced 1991; still imported. **LMSR: $670**

    Perf.: $595   **Exc.:** $495   **VGood:** $395

Navy Arms Model 1866 Yellowboy Trapper

Same specs as Model 1866 Yellowboy except 44-40; 8-shot magazine; 16 1/2" barrel. Made in Italy. Introduced 1991; no longer imported.

    Perf.: $595   **Exc.:** $495   **VGood:** $395

NAVY ARMS REVOLVING CARBINE

Carbine; 357 Mag., 44-40, 45 Colt; 6-shot cylinder; 20" barrel; 38" overall length; weighs 5 lbs.; open rear sight, blade front; straight-grip stock; brass buttplate, triggerguard. Action based on Remington Model 1874 revolver. Manufactured in Italy. Introduced, 1968; discontinued 1984.

    Perf.: $575   **Exc.:** $450   **VGood:** $360

NAVY ARMS ROLLING BLOCK BABY CARBINE

Single shot; 22 LR, 22 Hornet, 357 Mag., 44-40; 20" octagon barrel, 22" round barrel; open rear sight; blade front; unchecked straight-grip stock forearm; case-hardened frame; brass triggerguard; brass buttplate. Introduced, 1968; discontinued 1984.

    Perf.: $225   **Exc.:** $190   **VGood:** $130

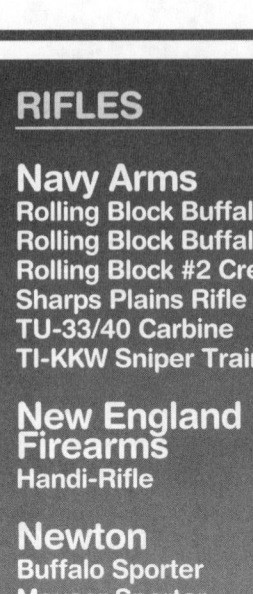

RIFLES

Navy Arms
Rolling Block Buffalo
Rolling Block Buffalo Carbine
Rolling Block #2 Creedmore
Sharps Plains Rifle
TU-33/40 Carbine
TI-KKW Sniper Trainer

New England Firearms
Handi-Rifle

Newton
Buffalo Sporter
Mauser Sporter
Standard Sporter (Type I)
Standard Sporter (Type II)

Noble
Model 10
Model 20
Model 33
Model 33A

Navy Arms Rolling Block Buffalo

Navy Arms Rolling Block #2 Creedmore

Navy Arms TU-KKW Sniper Trainer

New England Firearms Handi-Rifle

Noble Model 10

Navy Arms Rolling Block Buffalo
Same specs as Rolling Block Baby Carbine except 45-70; 26″, 30″ half or full octagonal barrel; brass barrel band. Introduced 1971; still in production. **LMSR: $765**
Perf.: $450   Exc.: $395   VGood: $325

Navy Arms Rolling Block Buffalo Carbine
Same specs as Rolling Block Buffalo except 18″ barrel. Introduced 1971; discontinued 1985.
Perf.: $395   Exc.: $350   VGood: $320

NAVY ARMS ROLLING BLOCK #2 CREEDMORE
Single shot; 45-70, 45-90; 26″, 28″, 30″ heavy half or full octagon barrel; Creedmore tang peep sight; checkered walnut stock; color case-hardened receiver. Still in production. **LMSR: $930**
Perf.: $675   Exc.: $595   VGood: $495

NAVY ARMS SHARPS PLAINS RIFLE
Single shot; falling block; 45-70; 28 1/2″ barrel; 45 3/4″ overall length; weighs 8 lbs., 10 oz.; blade front sight, open windage-adjustable rear; checkered walnut butt and forend; color case-hardened action, rest blued. Imported by Navy Arms. Introduced 1991; no longer imported.
Perf.: $695   Exc.: $595   VGood: $420

NAVY ARMS TU-33/40 CARBINE
Bolt action; 7.62x39mm; 4-shot magazine; 20 3/4″ barrel; 38″ overall length; weighs 6 1/2 lbs.; hooded barleycorn front sight, military V-notch adjustable rear; miniature Mauser-style action; hardwood stock; comes with leather sling. Based on Mauser G.33/40 carbine. Imported by Navy Arms. Introduced 1992; discontinued 1994.
Perf.: $200   Exc.: $175   VGood: $140

NAVY ARMS TU-KKW SNIPER TRAINER
Bolt action; 7.62x39mm; 4-shot magazine; 20 3/4″ barrel; 38″ overall length; weighs 7 3/8 lbs.; hooded barleycorn front sight, military V-notch adjustable rear; Type 89 2 3/4x scope with quick detachable mount system; hardwood stock; miniature Mauser-style action; comes with leather sling; cleaning rod. Based on Mauser G.33/40 carbine. Imported by Navy Arms. Introduced 1992; discontinued 1994.
Perf.: $240   Exc.: $210   VGood: $180

NEW ENGLAND FIREARMS HANDI-RIFLE
Break-open; single shot; 22 Hornet, 223, 30-30, 45-70; 22″ barrel; weighs 7 lbs.; ramp front sight, folding rear; drilled, tapped for scope mounts; walnut-finished hardwood stock; break-open action; side-lever release; blued finish. Introduced 1989; still in production. **LMSR: $220**
Perf.: $180   Exc.: $150   VGood: $110

NEWTON BUFFALO SPORTER
Bolt action; 256, 30, 35 Newton, 30-06; 5-shot box magazine; 24″ barrel; open rear sight, ramp front; hand-checkered pistol-grip stock; reversed set-trigger; Enfield-designed bolt handle. Introduced 1922; discontinued 1924. Collector value.
Perf.: $1000   Exc.: $850   VGood: $650

NEWTON-MAUSER SPORTER
Bolt action; 256 Newton; 5-shot box magazine; 24″ barrel; open rear sight, ramp front; checkered American walnut pistol-grip stock; double-set triggers; hinged floorplate. Introduced 1914; discontinued 1915.
Perf.: $795   Exc.: $695   VGood: $525

NEWTON STANDARD SPORTER (TYPE I)
Bolt action; 22, 256, 280, 30, 33, 35 Newton, 30-06; 24″ barrel; open rear sight, ramp front; hand-checkered pistol-grip stock; double-set triggers. Introduced 1916; discontinued 1918. Collector value.
Perf.: $1200   Exc.: $950   VGood: $750

Newton Standard Sporter (Type II)
Same specs as Sporter Type I except improved bolt-action design; 256, 30, 35 Newton, 30-06; 5-shot box magazine; reversed set-trigger; Enfield-design bolt handle. Collector value. Introduced 1921; discontinued 1922.
Perf.: $1000   Exc.: $850   VGood: $650

NOBLE MODEL 10
Bolt action; single shot; 22 Short, 22 Long, 22 LR; 24″ barrel; open rear sight, bead front; uncheckered hardwood pistol-grip stock. Introduced 1955; discontinued 1958.
Perf.: $65   Exc.: $50   VGood: $35

NOBLE MODEL 20
Bolt action; single shot; 22 Short, 22 Long, 22 LR; 22″ barrel; weighs 5 lbs.; open rear sight, bead front; uncheckered walnut pistol-grip stock. Introduced 1958; discontinued 1963.
Perf.: $65   Exc.: $50   VGood: $35

NOBLE MODEL 33
Slide action; 22 Short, 22 Long, 22 LR; 21-shot (22 Short), 17-shot (22 Long), 15-shot (22 LR) tube magazine; 24″ barrel; 41 3/4″ overall length; weighs 6 lbs.; hammerless; open rear sight, bead front; Tenite stock; grooved wood slide handle. Introduced 1949; discontinued 1953.
Perf.: $80   Exc.: $60   VGood: $40

Noble Model 33A
Same specs as Model 33 except hardwood stock and slide handle. Introduced 1953, as replacement for Model 33; discontinued 1955.
Perf.: $100   Exc.: $70   VGood: $50

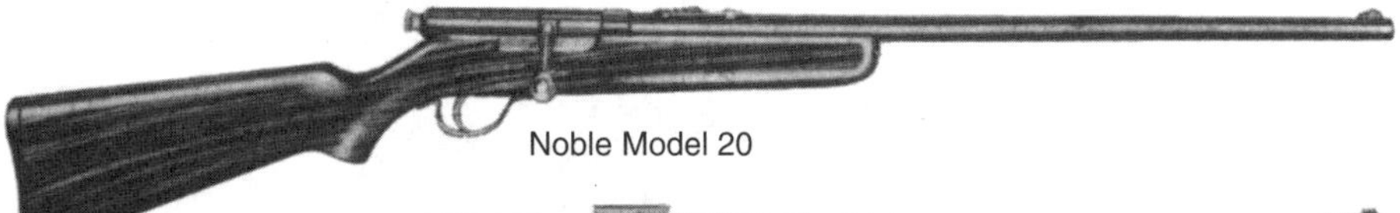

RIFLES

Noble
Model 222
Model 235
Model 275
Model 285

Norinco
MAK 90
Model ATD
Model EM-321
Model JW-15
Model JW-27
Type EM-332

Norrahammar
Model N-900

NS Firearms
Model 522

Olympic Arms
AR-15 Match

Noble Model 20

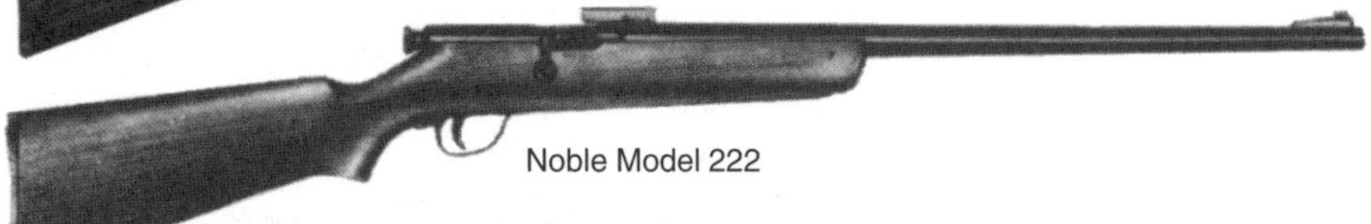

Noble Model 222

Noble Model 235

Norinco Model JW-27

Norinco Model ATD

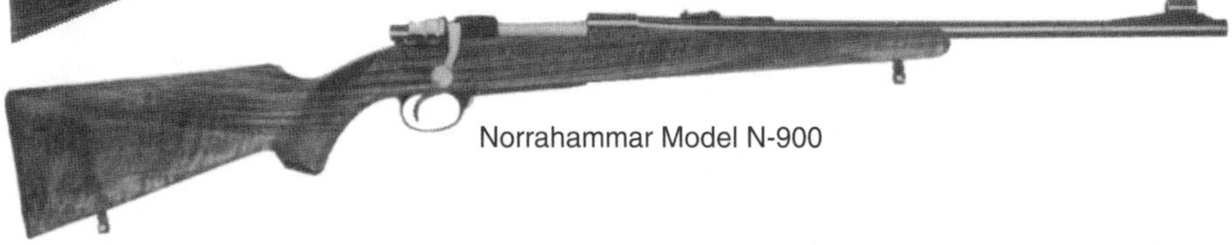

Norrahammar Model N-900

NOBLE MODEL 222

Bolt action; single shot; 22 Short, 22 Long, 22 LR; 22" barrel; 38" overall length; weighs 5 lbs.; barrel, receiver milled as integral unit; interchangeable peep and V-notch rear sight, ramp front; scope mounting base; uncheckered hardwood pistol-grip stock. Introduced 1958; discontinued 1971.

    Perf.: $65   **Exc.:** $50   **VGood:** $35

NOBLE MODEL 235

Slide-action; 22 Short, 22 Long, 22 LR; 21-shot (22 Short), 17-shot (22 Long), 15-shot (22 LR) tube magazine; 24" barrel; hammerless; open rear sight, ramp front; hardwood pistol-grip stock; grooved wood slide handle. Introduced 1951; discontinued 1973.

    Perf.: $100   **Exc.:** $70   **VGood:** $50

NOBLE MODEL 275

Lever action; 22 Short, 22 Long, 22 LR; 21-shot (22 Short), 17-shot (22 Long), 15-shot (22 LR) tube magazine; 24" barrel; 42" overall length; weighs 5 1/2 lbs.; hammerless; open rear sight, ramp front; receiver grooved for tip-off scope mount; uncheckered one-piece hardwood full pistol-grip stock. Introduced 1958; discontinued 1971.

    Perf.: $100   **Exc.:** $70   **VGood:** $50

NOBLE MODEL 285

Semi-automatic; 22 LR; 15-shot tube magazine; 22" barrel; 40" overall length; weighs 5 1/2 lbs.; ramp front sight, adjustable open rear; top thumb safety; receiver grooved for tip-off mount; walnut finished pistol-grip stock.

    Perf.: $115   **Exc.:** $95   **VGood:** $65

NORINCO MAK 90

Semi-automatic; 7.62x39, 223; 5-shot magazine; 16 1/4" barrel; 35 1/2" overall length; weighs 8 lbs.; adjustable post front sight, open adjustable rear; walnut-finished thumbhole stock with recoil pad; chrome-lined barrel; forged receiver; black oxide finish. Comes with extra magazine, oil bottle, cleaning kit, sling. Imported from China by Century International Arms. Still imported. **LMSR:** $312

    Perf.: $450   **Exc.:** $400   **VGood:** $350

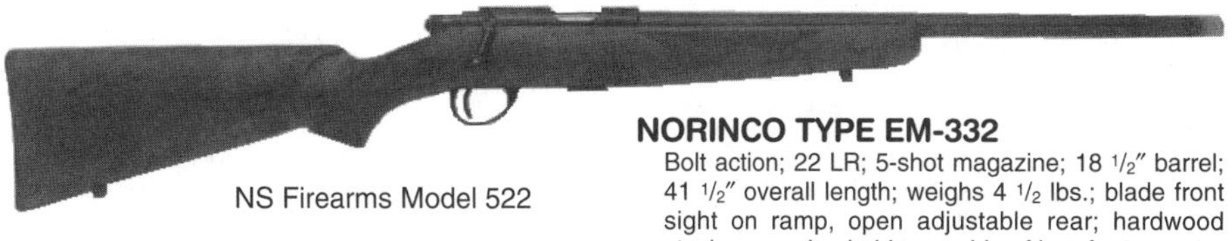

NS Firearms Model 522

NORINCO MODEL ATD

Semi-automatic; 22 LR; 11-shot magazine; 19 3/8" barrel; 36 5/8" overall length; weighs 4 3/8 lbs.; blade front sight, open adjustable rear; checkered hardwood stock; Browning-type take-down action; tube magazine loads through buttplate; cross-bolt safety; blued finish; engraved receiver. Made in China. Introduced 1987; still imported. **LMSR:** $166

    Perf.: $150   **Exc.:** $130   **VGood:** $100

NORINCO MODEL EM-321

Slide action; 22 LR; 9-shot magazine; 19 1/2" barrel; 37" overall length; weighs 6 lbs.; blade front sight, open folding rear; hardwood stock; blue finish; grooved slide handle. Imported from China by China Sports, Inc. No longer imported.

    Perf.: $125   **Exc.:** $100   **VGood:** $80

NORINCO MODEL JW-15

Bolt action; 22 LR; 5-shot detachable magazine; 24" barrel; 41 3/4" overall length; weighs 5 3/4 lbs.; hooded blade front sight, windage-adjustable open rear; walnut-stained hardwood stock; sling swivels; wing-type safety. Made in China. Introduced 1991; still imported. **LMSR:** $118

    Perf.: $90   **Exc.:** $80   **VGood:** $70

NORINCO MODEL JW-27

Bolt action; 22 LR; 5-shot magazine; 22 3/4" barrel; 41 3/4" overall length; weighs 5 3/4 lbs.; dovetailed bead on blade front sight, fully-adjustable rear; receiver grooved for scope mounting; walnut-finished hardwood stock with checkered grip and forend; blued finish. Imported from China by Century International Arms. Introduced 1992; still imported. **LMSR:** $107

    Perf.: $85   **Exc.:** $75   **VGood:** $65

NORINCO TYPE EM-332

Bolt action; 22 LR; 5-shot magazine; 18 1/2" barrel; 41 1/2" overall length; weighs 4 1/2 lbs.; blade front sight on ramp, open adjustable rear; hardwood stock; magazine holder on side of butt for two extra magazines; blue finish. Imported from China by China Sports, Inc. Introduced 1990; no longer imported.

    Perf.: $240   **Exc.:** $200   **VGood:** $160

NORRAHAMMAR MODEL N-900

Bolt action; 243 Win., 270 Win., 308 Win., 30-06; 20 1/4" barrel; hooded front sight, adjustable rear; hand-checkered European walnut pistol-grip stock, forearm; single-stage trigger; ebony grip cap; buttplate; side safety; hinged floorplate; sling swivels. Made in Sweden. Introduced 1957; discontinued 1967.

    Perf.: $425   **Exc.:** $370   **VGood:** $310

NS FIREARMS MODEL 522

Bolt action; 22 LR; 5-shot magazine; 21" barrel; 39 1/2" overall length; weighs 7 3/4 lbs.; no sights; receiver grooved for scope mount; walnut stock with cut-checkered grip and forend; satin finish; free-floated hammer-forged heavy barrel; forged receiver and bolt with two locking lugs, dual extractors; safety locks bolt and trigger. Imported by Keng's Firearms Specialty. Introduced 1993; discontinued 1994.

    Perf.: $270   **Exc.:** $220   **VGood:** $170

OLYMPIC ARMS AR-15 MATCH

Semi-automatic; 223; 20- or 30-shot magazine; 20" stainless steel barrel; 39 1/2" overall length; weighs 8 3/4 lbs.; post front sight, fully-adjustable aperture rear; cut-off carrying handle with scope rail attached; button-rifled 4140 ordnance steel or 416 stainless barrel with 1:9" twist standard, 1:7", 1:12", 1:14" twists optional; weighs 8 1/2 lbs. Made in U.S. by Olympic Arms, Inc. Introduced 1993; discontinued 1994.

    Perf.: $695   **Exc.:** $595   **VGood:** $495

Olympic Arms
Intercontinental Match
International Match
K-4 AR-15
Multimatch
PCR-1
PCR-2
PCR-3
PCR-4
PCR-5
PCR-6
PCR-9
PCR-15
PCR-40
PCR-45

Olympic Arms International Match

Olympic Arms Multimatch

Olympic Arms PCR-1

Olympic Arms PCR-2

Olympic Arms PCR-5

OLYMPIC ARMS INTERCONTINENTAL MATCH

Semi-automatic; 223; 5-shot magazine; 20″ stainless steel barrel; 39 1/2″ overall length; weighs about 10 lbs.; no sights; cut-off carrying handle with scope rail attached; synthetic woodgrain thumbhole buttstock; broach-cut, free-floating barrel with 1:10″ or 1:8.5″ twist; magazine well floorplate; fluting optional. Made in U.S. by Olympic Arms, Inc. Introduced 1992; discontinued 1994.

Perf.: $1595   **Exc.:** $1195   **VGood:** $950

OLYMPIC ARMS INTERNATIONAL MATCH

Semi-automatic; 223; 20- or 30-shot magazine; 20″ or 24″ stainless steel barrel; 39 1/2″ overall length (20″ barrel); weighs 10 lbs.; standard AR-15-type post front sight, adjustable target peep rear; cut-off carrying handle with scope rail attached; black composition A2 stow-away butt and grip; broach-cut, free-floating barrel with 1:10″ or 1:8.5″ twist; fluting optional. Based on the AR-15 rifle. Made in U.S. by Olympic Arms, Inc. Introduced 1985; no longer made.

Perf.: $1395   **Exc.:** $995   **VGood:** $825

OLYMPIC ARMS K-4 AR-15

Semi-automatic; 223; 30-shot magazine; 20″ barrel; 39″ overall length; weighs about 8 lbs.; post elevation-adjustable front sight, peep-style windage-adjustable rear; E-2-style sight system optionally available; full-length M-16-style black composition stock; heavy match-grade barrel; trapdoor in buttstock; A-2 stow-away pistol grip. Made in U.S. by Olympic Arms, Inc. Introduced 1975; no longer produced.

Perf.: $795   **Exc.:** $695   **VGood:** $595

OLYMPIC ARMS MULTIMATCH

Semi-automatic; 223; 30-shot magazine; 16″ free-floating, broach-cut, stainless steel barrel with 1:10″ or 1:8 1/2″ twist; 36″ overall length; weighs 7 1/2 lbs.; E2 fully adjustable or International Match upper receiver for scope mounting; telescoping stock or collapsible aluminum buttstock; cut-off carrying handle with scope rail attached (ML2); based on the AR-15 rifle. Made in U.S. by Olympic Arms, Inc. Introduced 1991; no longer produced.

Perf.: $995   **Exc.:** $895   **VGood:** $750

OLYMPIC ARMS PCR-1

Semi-automatic; 223; 10-shot magazine; 20″ barrel of 416 stainless steel; 38 1/4″ overall length; weighs about 10 lbs.; no sights; flat-top upper receiver, cut-down front sight base; black composition A2 stowaway grip and trapdoor butt; broach-cut, free-floating barrel with 1:8.5″ or 1:10″ twist; no bayonet lug; crowned barrel; fluting available. Based on the AR-15 rifle. Made in U.S. by Olympic Arms, Inc. Introduced 1994; discontinued 1995.

Perf.: $1095   **Exc.:** $950   **VGood:** $800

Olympic Arms PCR-2

Same specs as PCR-1 except 16″ barrel; weighs about 8 lbs.; post front sight, fully-adjustable aperture rear. Made in U.S. by Olympic Arms, Inc. Introduced 1994; discontinued 1995.

Perf.: $995   **Exc.:** $895   **VGood:** $750

Olympic Arms PCR-3

Same specs as PCR-1 except flat-top upper receiver, cut-down front sight base. Made in U.S. by Olympic Arms, Inc. Introduced 1994; dropped 1995.

Perf.: $995   **Exc.:** $895   **VGood:** $750

Olympic Arms PCR-4

Same specs as PCR-1 except weighs 8 1/2 lbs.; post front sight, A1 windage-adjustable rear; button-rifled 1:7″, 1:9″, 1:12″ or 1:14″ twist barrel. Made in U.S. by Olympic Arms, Inc. Introduced 1994; discontinued 1995.

Perf.: $850   **Exc.:** $775   **VGood:** $600

Olympic Arms PCR-5

Same specs as PCR-1 except 9mm Para., 40 S&W, 45 ACP, 223; 16″ barrel; 34 3/4″ overall length; weighs 7 lbs.; post front sight, A1 windage-adjustable rear. Based on the CAR-15. Made in U.S. by Olympic Arms, Inc. Introduced 1994; discontinued 1995, reintroduced.

Perf.: $750   **Exc.:** $650   **VGood:** $495

Olympic Arms PCR-6

Same specs as PCR-1 except 7.62x39mm; 16″ barrel; 34 3/4″ overall length; weighs 7 lbs.; post front sight, A1 windage-adjustable rear. Based on the CAR-15. Made in U.S. by Olympic Arms, Inc. Introduced 1994; discontinued 1995.

Perf.: $795   **Exc.:** $650   **VGood:** $475

Olympic Arms PCR-9

Same specs as PCR-1 except 9mm Para; 16″ barrel; 34″ overall length (stock extended); weighs 7 lbs.; post elevation-adjustable front sight, windage-adjustable rear; telescoping buttstock. Made in U.S. by Olympic Arms, Inc. Introduced 1982; no longer made.

Perf.: $795   **Exc.:** $695   **VGood:** $550

Olympic Arms PCR-15

Same specs as PCR-1 except 16″ barrel; 34″ overall length (stock extended); weighs 7 lbs.; post elevation-adjustable front sight, windage-adjustable rear; telescoping buttstock. Based on the AR-15 rifle. Made in U.S. by Olympic Arms, Inc. Introduced 1982; no longer made.

Perf.: $1095   **Exc.:** $995   **VGood:** $895

Olympic Arms PCR-40

Same specs as PCR-1 except 40 S&W; 16″ barrel; 34″ overall length (stock extended); weighs 7 lbs.; post elevation-adjustable front sight, windage-adjustable rear; telescoping buttstock. Made in U.S. by Olympic Arms, Inc. Introduced 1982; no longer made.

Perf.: $995   **Exc.:** $895   **VGood:** $795

Olympic Arms PCR-45

Same specs as PCR-1 except 45 ACP; 16″ barrel; 34″ overall length (stock extended); weighs 7 lbs.; post elevation-adjustable front sight, windage-adjustable rear; telescoping buttstock. Made in U.S. by Olympic Arms, Inc. Introduced 1982; no longer made.

Perf.: $995   **Exc.:** $895   **VGood:** $795

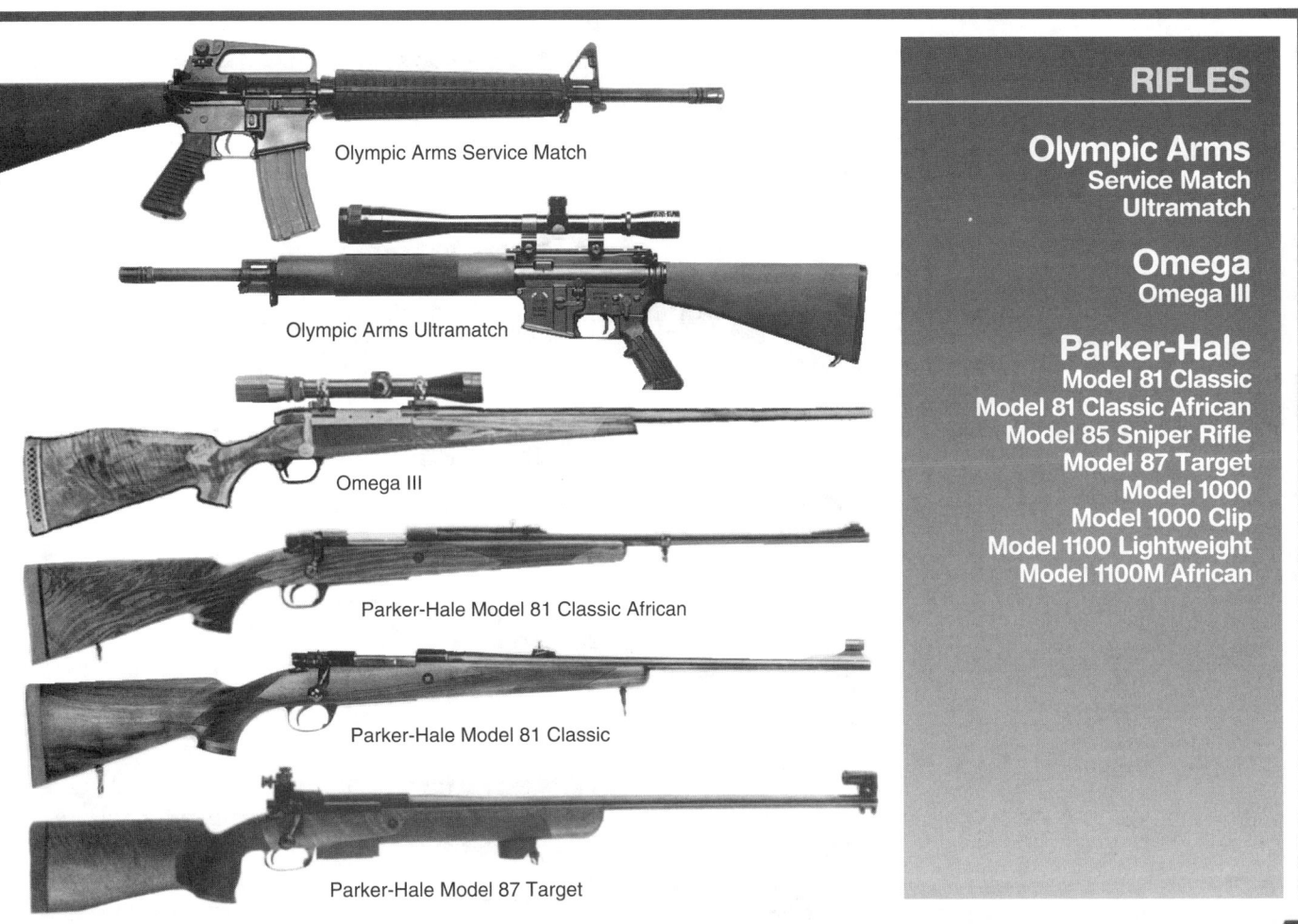

Olympic Arms Service Match

Olympic Arms Ultramatch

Omega III

Parker-Hale Model 81 Classic African

Parker-Hale Model 81 Classic

Parker-Hale Model 87 Target

RIFLES

Olympic Arms
Service Match
Ultramatch

Omega
Omega III

Parker-Hale
Model 81 Classic
Model 81 Classic African
Model 85 Sniper Rifle
Model 87 Target
Model 1000
Model 1000 Clip
Model 1100 Lightweight
Model 1100M African

OLYMPIC ARMS SERVICE MATCH

Semi-automatic; 223; 20-, 30-shot magazine; 20″ stainless steel barrel; 39 1/2″ overall length; weighs 8 3/4 lbs.; post front sight, fully-adjustable aperture rear; black composition A2 standard stock; barrel broach-cut and free-floating with 1:10″ or 1:8.5″ twist; fluting optional. Based on the AR-15 rifle. Conforms to all DCM standards. Made in U.S. by Olympic Arms, Inc. Introduced 1989; no longer made.
Perf.: $1495   Exc.: $1195   VGood: $950

OLYMPIC ARMS ULTRAMATCH

Semi-automatic; 223; 20-, 30-shot magazine; 20″ or 24″ stainless steel barrel; 39 1/2″ overall length (20″ barrel); weighs about 10 lbs.; no sights; cut-off carrying handle with scope rail attached; black composition A2 stowaway butt and grip; broach-cut, free-floating barrel with 1:10″ or 1:8.5″ twist; fluting optional. Based on the AR-15 rifle. Made in U.S. by Olympic Arms, Inc. Introduced 1985; no longer made.
Perf.: $1550   Exc.: $1250   VGood: $925

OMEGA III

Bolt action; 25-06, 270, 30-06, 7mm Rem. Mag., 300 Win. Mag., 338 Win. Mag., 358 Norma Mag.; 4-shot (belted), 5-shot (standard) rotary magazine; 22″, 24″ barrel; no sights; choice of Monte Carlo, Classic or thumbhole varminter; Claro walnut English laminated or laminated walnut/maple; right- or left-hand action; octagonal bolt; square locking system, enclosed bolt face; fully-adjustable trigger. Introduced 1973; discontinued 1976.
Perf.: $795   Exc.: $650   VGood: $495

PARKER-HALE MODEL 81 CLASSIC

Bolt action; 22-250, 243, 6mm Rem., 270, 6.5x55, 7x57, 7x64, 308, 30-06, 300 Win. Mag., 7mm Rem. Mag.; 4-shot magazine; 24″ barrel; 44 1/2″ overall length; weighs 7 3/4 lbs.; Mauser-style action; one-piece steel Oberndorf-style triggerguard with hinged floorplate; drilled and tapped for open sights and scope mounting; scope bases included; rubber buttpad; quick-detachable sling swivels. Introduced 1985; no longer produced.
Perf.: $750   Exc.: $600   VGood: $485

Parker-Hale Model 1000

Parker-Hale Model 81 Classic African

Same specs as Model 81 Classic except 375 H&H, 9.3x62; African express rear sight; adjustable trigger; barrel band front swivel; engraved receiver; classic-style stock with solid buttpad; checkered pistol grip and forend. Made by Gibbs Rifle Co. Introduced 1986; no longer produced.
Perf.: $895   Exc.: $750   VGood: $595

PARKER-HALE MODEL 85 SNIPER RIFLE

Bolt action; 308 Win.; 10-shot magazine; 24 1/4″ barrel; 45″ overall length; weighs 12 1/2 lbs. (with scope); post front windage-adjustable sight, fold-down elevation-adjustable rear; McMillan fiberglass stock with several color patterns available; quick-detachable bipod; palm stop with rail; sling swivels; matte finish. Made by Gibbs Rifle Co. No longer made.
Perf.: $1695   Exc.: $1395   VGood: $995

PARKER-HALE MODEL 87 TARGET

Bolt action; 308 Win., 243, 6.5x55, 308, 30-06, 300 Win. Mag. (other calibers on request); 5-shot detachable box magazine; 26″ heavy barrel; 45″ overall length; weighs about 10 lbs.; no sights; walnut target-style stock adjustable for length of pull; solid buttpad; accessory rail with hand-stop; deeply stippled grip and forend; receiver dovetailed for Parker-Hale "Roll-Off" scope mounts; Mauser-style action with large bolt knob; Parkerized finish. Made by Gibbs Rifle Co. Introduced 1987; no longer made.
Perf.: $1200   Exc.: $950   VGood: $750

PARKER-HALE MODEL 1000

Bolt action; 22-250, 243, 6mm, 270, 6.5x55, 7x57, 7x64, 308, 30-06, 7mm Rem. Mag.; 4-shot magazine; 22″ barrel, 24″ (22-250); 42 1/2″ overall length; weighs 7 1/4 lbs.; no sights; drilled and tapped for scope mounting; walnut Monte Carlo stock with checkered palm swell pistol grip and forend; rosewood grip cap; Mauser-style action; one-piece steel Oberndorf-style trigger guard with hinged floorplate. Made by Gibbs Rifle Co. Introduced 1992; no longer made.
Perf.: $395   Exc.: $340   VGood: $300

Parker-Hale Model 1000 Clip

Same specs as Model 1000 except detachable box magazine. Made by Gibbs Rifle Co. Introduced 1992; no longer made.
Perf.: $395   Exc.: $340   VGood: $300

PARKER-HALE MODEL 1100 LIGHTWEIGHT

Bolt action; 22-250, 243, 6mm Rem., 270, 6.5x55, 7x57, 7x64, 308, 30-06, 300 Win. Mag., 7mm Rem. Mag.; 4-shot magazine; 24″ barrel; 44 1/2″ overall length; weighs 7 3/4 lbs.; no sights; drilled, tapped for open sights or scope mounting; classic-style European walnut stock with hand-cut checkering; palm swell; rosewood grip cap; Mauser-style action; Oberndort-style triggerguard; buttpad; sling swivels. Imported from England by Precision Sport, Inc. Introduced 1984; no longer imported.
Perf.: $450   Exc.: $350   VGood: $300

Parker-Hale Model 1100M African

Same specs as Model 1100 except 375 H&H Mag., 458 Win. Mag; 46″ overall length; weighs 9 1/2 lbs.; hooded post front sight; V-notch rear; heavily reinforced, glass-beaded, weighted stock; vent recoil pad. Introduced 1984; no longer imported.
Perf.: $625   Exc.: $490   VGood: $400

Parker-Hale
Model 1200 Super
Model 1200P Presentation
Model 1200V Varminter
Model 1300C Scout
Model 2100 Midland
Model 2600 Midland
Model 2700 Lightweight Midland
Model 2800 Midland

Pedersen
Model 3000
Model 3500
Model 4700

Perugini-Visini
Boxlock Express

Parker-Hale Model 1100 Lightweight

Parker-Hale Model 1100M African Magnum

Parker-Hale Model 1200 Super

Parker-Hale Model 2100 Midland

Pedersen Model 3000

PARKER-HALE MODEL 1200 SUPER

Bolt action; 22-250, 243, 6mm, 270, 6.5x55, 7x57, 7x64, 308, 30-06, 300 Win. Mag., 7mm Rem. Mag.; 22″ barrel; 43″ overall length; weighs about 7 lbs.; Mauser-type action with twin locking lugs; hinged floorplate; adjustable single stage trigger; silent side safety; hooded post front, flip-up rear sight; European walnut stock; cut-checkered pistol grip and forend; sling swivels. Introduced 1984; no longer produced.

Perf.: $525   **Exc.:** $450   **VGood:** $375

Parker-Hale Model 1200P Presentation

Same specs as Model 1200 Super except 243, 30-06; no sights; engraved action, floorplate, triggerguard; detachable sling swivels. Introduced 1969; discontinued 1975.

Perf.: $495   **Exc.:** $450   **VGood:** $395

Parker-Hale Model 1200V Varminter

Same specs as Model 1200 Super except 22-250, 6mm Rem., 25-06, 243; 24″ barrel; no sights. Introduced 1969; discontinued 1983.

Perf.: $510   **Exc.:** $450   **VGood:** $380

PARKER-HALE MODEL 1300C SCOUT

Bolt action; 243, 308; 10-shot magazine; 20″ barrel; 41″ overall length; weighs 8 ¹/₂ lbs.; no sights; drilled and tapped for scope mounting; checkered laminated birch stock; detachable magazine; muzzle brake; polished blue finish. Made by Gibbs Rifle Co. Introduced 1992; no longer made.

Perf.: $650   **Exc.:** $540   **VGood:** $400

PARKER-HALE MODEL 2100 MIDLAND

Bolt action; 22-250, 243, 6mm, 270, 6.5x55, 7x57, 7x64, 308, 30-06; 22″ barrel; 43″ overall length; weighs about 7 lbs.; hooded post front, flip-up open rear; Mauser-type action has twin front locking lugs; rear safety lug, claw extractor; hinged floorpate; adjustable single stage trigger; silent side safety. Imported from England by Precision Sports, Inc. Introduced 1984; discontinued 1989.

Perf.: $300   **Exc.:** $250   **VGood:** $200

PARKER-HALE MODEL 2600 MIDLAND

Bolt action; 22-250, 243, 6mm, 270, 6.5x55, 7x57, 7x64, 308, 30-06, 300 Win. Mag., 7mm Rem. Mag.; 22″ barrel; 43″ overall length; weighs about 7 lbs.; hooded post front sight, flip-up open rear; European hardwood stock with cut-checkered pistol grip and forend; sling swivels; Mauser-type action with twin front locking lugs, rear safety lug, and claw extractor; hinged floorplate; adjustable single-stage trigger; silent side safety. Made by Gibbs Rifle Co. Introduced 1984; no longer produced.

Perf.: $295   **Exc.:** $250   **VGood:** $220

PARKER-HALE MODEL 2700 LIGHTWEIGHT MIDLAND

Bolt action; 22-250, 243, 6mm, 270, 6.5x55, 7x57, 7x64, 308, 30-06, 7mm Rem. Mag.; 22″ light barrel; 43″ overall length; weighs about 6 ¹/₂ lbs.; hooded post front sight, flip-up open rear; receiver drilled and tapped for scope mounting; lightweight European walnut stock with checkered pistol grip and forend; sling swivels. Made by Gibbs Rifle Co. Introduced 1992; no longer made.

Perf.: $350   **Exc.:** $300   **VGood:** $240

PARKER-HALE MODEL 2800 MIDLAND

Bolt action; 22-250, 243, 6mm, 270, 6.5x55, 7x57, 7x64, 308, 300 Win. Mag., 7mm Rem. Mag.; 22″ barrel; 43″ overall length; weighs about 7 lbs.; hooded post front sight, flip-up open rear; laminated birch Monte Carlo stock with cut-checkered pistol grip and forend; sling swivels; Mauser-type action with twin front locking lugs, rear safety lug, and claw extractor; hinged floorplate; adjustable single-stage trigger; silent side safety. Made by Gibbs Rifle Co. No longer produced.

Perf.: $325   **Exc.:** $275   **VGood:** $230

PEDERSEN MODEL 3000

Bolt action; 270, 30-06, 7mm Rem. Mag., 338 Win. Mag.; 3-shot magazine; 22″ barrel (270, 30-06), 24″ (magnums); no sights; drilled, tapped for scope mount; American walnut stock, roll-over cheekpiece; hand-checkered pistol-grip, forearm; Mossberg Model 800 action; adjustable trigger; sling swivels. Grades differ in amount of engraving and quality of stock wood. Introduced 1972; discontinued 1975.

Grade I
  Perf.: $950   **Exc.:** $750   **VGood:** $550
Grade II
  Perf.: $650   **Exc.:** $550   **VGood:** $450
Grade III
  Perf.: $550   **Exc.:** $450   **VGood:** $375

PEDERSEN MODEL 3500

Bolt action; 270 Win., 30-06, 7mm Rem. Mag.; 22″ barrel (standard calibers), 24″ (7mm Mag.); 3-shot magazine; drilled, tapped for scope mounts; hand-checkered black walnut stock, forearm; rosewood pistol-grip cap, forearm tip; hinged steel floorplate; Damascened bolt; adjustable trigger. Introduced 1973; discontinued 1975.

Perf.: $550   **Exc.:** $480   **VGood:** $420

PEDERSEN MODEL 4700

Lever action; 30-30, 35 Rem.; 5-shot tubular magazine; 24″ barrel; open rear sight, hooded ramp front; black walnut stock beavertail forearm; Mossberg Model 472 action; barrel band swivels. Introduced 1975; discontinued 1975.

Perf.: $220   **Exc.:** $230   **VGood:** $190

PERUGINI-VISINI BOXLOCK EXPRESS

Double rifle; side-by-side; 9.3x74R, 444 Marlin; 25″ barrel; 41 ¹/₂″ overall length; bead on ramp front sight, express rear on rib; oil-finished European walnut stock; hand-checkered pistol grip, forend; cheekpiece; non-ejector action; double triggers; color case-hardened receiver; rubber recoil pad; marketed in trunk-type case. Made in Italy. Introduced 1983; discontinued 1989.

Perf.: $2995   **Exc.:** $2695   **VGood:** $1995

Perugini-Visini Boxlock Express

Perugini-Visini Sidelock Express

Plainfield M-1 Carbine

Plainfield M-1 Carbine, Deluxe Sporter

Plainfield M-1 Carbine, Commando Model

Perugini-Visini
Boxlock Magnum
Eagle
Model Selous
Professional
Professional Deluxe
Sidelock Express
Victoria

Plainfield
M-1 Carbine
M-1 Carbine, Commando Model
M-1 Carbine, Deluxe Sporter
M-1 Carbine, Military Sporter

Purdey
Big Game Sporter
Double Rifle

Quality Parts Bushmaster
V Match

PERUGINI-VISINI BOXLOCK MAGNUM
Double rifle; over/under; 7mm Rem. Mag., 7x65R, 9.3x74R, 270 Win., 338 Win. Mag., 375 H&H Mag., 458 Win. Mag.; 24″ barrels; 40 ½″ overall length; bead on ramp front sight, express rear on rib; oil-finished European walnut stock; cheekpiece; hand-checkered pistol grip, forend; ejectors; silvered receiver, other metal parts blued; marketed in trunk-type case. Made in Italy. Introduced 1983; discontinued 1989.
Perf.: $5495   Exc.: $4895   VGood: $3695

PERUGINI-VISINI EAGLE
Single shot; boxlock; 17 Rem., 222, 22-250, 243, 270, 30-06, 7mm Rem. Mag., 300 Win. Mag., 5.6x50R, 5.6x57R, 6.5x68R, 7x57R, 9.3x74R, 10.3x60R; 24″, 26″ barrel; no sights; claw-type scope mounts; oil-finished European walnut stock; adjustable set trigger ejector; engraved, case-hardened receiver. Made in Italy. Introduced 1986; discontinued 1989.
Perf.: $4995   Exc.: $4250   VGood: $2995

PERUGINI-VISINI MODEL SELOUS
Double rifle; 30-06, 7mm Rem. Mag., 7x65R, 9.3x74R, 270 Win., 300 H&H, 338 Win., 375 H&H, 458 Win. Mag., 470 Nitro; 22″-26″ barrels; 41″ overall length (24″ barrels); weighs 7 ¼ to 10 ½ lbs.; bead on ramp front sight, express rear on quarter-rib; oil-finished walnut stock with checkered grip and forend, cheekpiece; true sidelock action with ejectors; sideplates hand detachable; comes with leather trunk case. Imported from Italy by Wm. Larkin Moore. Introduced 1983; discontinued 1992.
Perf.: $21,995 Exc.: $17,995 VGood: $10,995

PERUGINI-VISINI PROFESSIONAL
Bolt action; 222 Rem. through 458 Win. Mag.; 24″, 26″ barrel; no sights; oil-finished European walnut stock; modified Mauser 98k bolt action; single or double-set triggers. Made in Italy. Introduced 1986; discontinued 1987.
Perf.: $3495   Exc.: $2995   VGood: $1995

Perugini-Visini Professional Deluxe
Same specs as Professional except open sights;

checkered oil-finished European walnut stock; knurled bolt handle. Made in Italy. Introduced 1986; discontinued 1987.
Perf.: $3495   Exc.: $2995   VGood: $1995

PERUGINI-VISINI SIDELOCK EXPRESS
Double rifle; side-by-side; various up to 470 Nitro Express; 22″, 26″ barrels; bead on ramp front sight, express rear on rib; oil-finished walnut stock; hand-checkered grip, forend; cheekpiece; ejectors; hand-detachable side plates; marketed in trunk-type case. Made in Italy. Introduced 1983; discontinued 1989.
Perf.: $8995   Exc.: $6995   VGood: $5495

PERUGINI-VISINI VICTORIA
Double rifle; 30-06, 7x65R, 9.3x74R, 375 H&H, 458 Win. Mag., 470 Nitro; 22″-26″ barrels; 41″ overall length (24″ barrels); weighs 7 ¼ to 10 ½ lbs.; bead on ramp front sight, express rear on quarter-rib; oil-finished walnut stock with checkered grip and forend, cheekpiece; true sidelock action with automatic ejectors; sideplates hand detachable; double triggers; comes with leather trunk case. Imported from Italy by Wm. Larkin Moore. Introduced 1983; discontinued 1992.
Perf.: $6995   Exc.: $5495   VGood: $3700

PLAINFIELD M-1 CARBINE
Semi-automatic; 5.7mm, 30 Carbine, 256 Ferret; 15-shot magazine; 18″ barrel; weighs 5 ½ lbs.; open sights; military-style wood stock; early models with standard military fittings, some from surplus parts; later models with ventilated metal hand guard, no bayonet lug. Introduced 1960; discontinued 1976.
30 Carbine
Perf.: $200   Exc.: $180   VGood: $160
256 Ferret
Perf.: $330   Exc.: $270   VGood: $220

Plainfield M-1 Carbine, Commando Model
Same specs as M-1 Carbine except telescoping wire stock; pistol-grips front and rear.
Perf.: $250   Exc.: $200   VGood: $170

Plainfield M-1 Carbine, Deluxe Sporter
Same specs as M-1 Carbine except Monte Carlo sporting stock. Introduced 1960; discontinued 1973.
Perf.: $250   Exc.: $200   VGood: $170

Plainfield M-1 Carbine, Military Sporter
Same specs as M-1 Carbine except unslotted buttstock; wood handguard.
Perf.: $230   Exc.: $190   VGood: $160

PURDEY BIG GAME SPORTER
Bolt action; 7x57, 300 H&H Mag., 375 H&H Mag., 404 Jeffery, 10.75x73; 3-shot magazine; 24″ barrel; Mauser-design; folding leaf rear sight, hooded ramp front; hand-checkered European walnut pistol-grip stock. Introduced post World War I; still in production; not imported.
Perf.: $6995   Exc.: $5695   VGood: $4295

PURDEY DOUBLE RIFLE
Double rifle; sidelock; 300 H&H Mag., 375 H&H Mag., 375 Flanged Nitro Express, 500-465 Nitro Express, 470 Nitro Express, 577 Nitro Express, 600 Nitro Express; 25 ½″ barrel; hammerless; folding leaf rear sight, hooded ramp front; hand-checkered European cheekpiece stock; ejectors; sling swivels; recoil pad. Introduced prior to World War I; still in production; not imported. **Prices range from $25,000 to $75,000.**

QUALITY PARTS/BUSHMASTER V MATCH
Semi-automatic; 223; 10-shot magazine; 20″, 24″, 26″ barrel with 1:9″ twist; 38 ¼″ overall length (20″ barrel); weighs 8 ¼ lbs.; .950″ barrel diameter with counter-bored crown, integral flash suppressor; no sights; comes with scope mount base installed; black composition stock; hand-built match gun with E2 lower receiver, push-pin-style takedown; upper receiver with brass deflector; free-floating steel handguard accepts laser sight, flashlight, bipod. Made in U.S. by Quality Parts Co./Bushmaster Firearms Co. No longer produced.
Perf.: $995   Exc.: $850   VGood: $675

Quality Parts/ Bushmaster
XM-15 E-2 Shorty Carbine
XM-15 E-2 Dissipator
XM-15 E-2 Target

Rahn
Deer Series
Elk Series
Himalayan Series
Safari Series

Ranger Arms
Standard Model
Texas Magnum
Texas Maverick

Red Willow Armory
Ballard No. 1½ Hunting Rifle
Ballard No. 4½ Target Rifle
Ballard No. 5 Pacific
Ballard No. 8 Union Hill
Ballard No. 8 Union Hill Deluxe Rifle

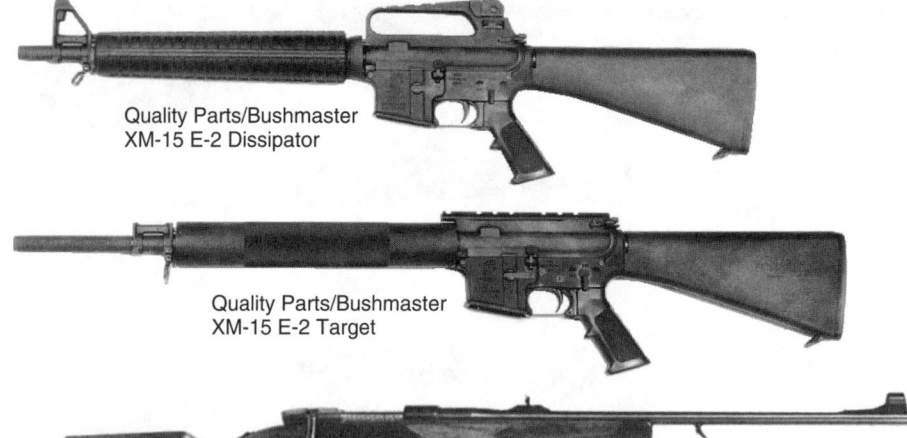

Quality Parts/Bushmaster
XM-15 E-2 Dissipator

Quality Parts/Bushmaster
XM-15 E-2 Target

Rahn Elk Series

QUALITY PARTS/BUSHMASTER XM-15 E-2 SHORTY CARBINE
Semi-automatic; 223; 10-shot magazine; 16" barrel; 34 ½" overall length; weighs 7 ¼ lbs.; adjustable post front sight, adjustable aperture rear; fixed black composition stock; chrome-lined barrel with manganese phosphate finish; "Shorty" handguards; E-2 lower receiver with push-pin takedown. Patterned after Colt M-16A2. Made in U.S. by Quality Parts Co./Bushmaster Firearms. No longer produced.
Perf.: $950   **Exc.:** $800   **VGood:** $650

Quality Parts/Bushmaster XM-15 E-2 Dissipator
Same specs as Shorty XM-15 E-2 Carbine except "Dissipator" handguard and fixed stock. Made in U.S. by Quality Parts Co./Bushmaster Firearms. No longer produced.
Perf.: $920   **Exc.:** $830   **VGood:** $650

Quality Parts/Bushmaster XM-15 E2 Target
Same specs as XM-15 Carbine except 20", 24", 26" heavy barrel with 1:7" or 1:9" twist; 38" overall length (20" barrel); weighs 8 ⅜ lbs. Made in U.S. by Quality Parts Co./ Bushmaster Firearms Co. No longer produced.
Perf.: $950   **Exc.:** $850   **VGood:** $700

RAHN DEER SERIES
Bolt action; 25-06, 308, 270; 24" barrel; open adjustable rear sight, bead front; drilled, tapped for scope mounts; Circassian walnut stock with rosewood forend, grip caps or Monte Carlo-style with semi-Schnabel forend; hand-checkered; one-piece triggerguard; hinged, engraved floorplate; rubber recoil pad; 22 rimfire conversion insert available. Introduced 1986; discontinued 1993.
Perf.: $695   **Exc.:** $595   **VGood:** $495

Rahn Elk Series
Same specs as Rahn Deer Series except 6x56, 30-06, 7mm Rem. Mag.; elk head engraved on floorplate. Introduced 1986; discontinued 1993.
Perf.: $695   **Exc.:** $595   **VGood:** $495

Rahn Himalayan Series
Same specs as Deer Series except 5.6x57, 6.5x68S; short stock of walnut or fiberglass; yak engraved on floorplate with scroll border. Introduced 1986; discontinued 1993.
Perf.: $725   **Exc.:** $625   **VGood:** $525

Rahn Safari Series
Same specs as Deer Series except 308 Norma Mag., 300 Win. Mag., 8x68S, 9x64; choice of Cape buffalo, rhino, elephant engraving; gold oval nameplate with three initials. Introduced 1986; discontinued 1993.
Perf.: $795   **Exc.:** $695   **VGood:** $595

RANGER ARMS STANDARD MODEL
Bolt action; various from 22-250 through 458 Win. Mag.; various barrel lengths, contour to order; no sights; drilled, tapped for scope mounting; hand-checkered roll-over cheekpiece, thumbhole, Mannlicher stock in variety of woods; rosewood pistol-grip cap, forend tip; recoil pad; push-button safety; left- or right-hand models. Introduced 1969; discontinued 1974.
Perf.: $530   **Exc.:** $450   **VGood:** $380

RANGER ARMS TEXAS MAGNUM
Bolt action; 270 Win., 30-06, 7mm Rem. Mag., 300 Win. Mag., 338 Win. Mag., 358 Norma Mag.; 24", 25" barrel; no sights; drilled, tapped for scope mounts; hand-checkered American walnut or Claro stock; hand-rubbed to high polish; epoxy finish; recoil pad; swivel studs; jeweled, chromed bolt body; rosewood pistol-grip cap; forend tip. Introduced 1969; discontinued 1974.
Perf.: $550   **Exc.:** $470   **VGood:** $400

RANGER ARMS TEXAS MAVERICK
Bolt action; 22-250, 243, 6mm Rem., 308 Win., 6.5mm, 350 Rem. Mag.; 22", 24" round tapered barrels; no sights; drilled, tapped for scope mounts; Monte Carlo stock of English or Claro walnut; skip-line checkering; cheekpiece; rosewood pistol-grip cap, forend tip; recoil pad; adjustable trigger; push-button safety; right- and left-hand models. Introduced 1969; discontinued 1974.
Perf.: $510   **Exc.:** $430   **VGood:** $380
With thumbhole stock
Perf.: $550   **Exc.:** $470   **VGood:** $400
With full-length stock
Perf.: $575   **Exc.:** $475   **VGood:** $410

RED WILLOW ARMORY BALLARD NO. 1½ HUNTING RIFLE
Single shot; dropping block; 32-40 Win., 38-55, 40-65 Win., 40-70 Ballard, 40-85 Ballard, 45-70; other calibers on special order; 30" medium-heavy tapered round barrel; weighs 9 lbs.; blade front sight, buckhorn rear; oil-finished American walnut stock with crescent butt, Schnabel forend; exact recreation of the Ballard No. 1 ½; single trigger; under-barrel wiping rod; drilled and tapped for tang sight; ring lever; mid- and long-range tang sights; Swiss tube with bead or Lyman globe front sight; fancy wood, double-set triggers optionally available. Made in U.S. by Red Willow Tool & Armory, Inc. Introduced 1992; discontinued 1993.
Perf.: $1300   **Exc.:** $1200   **VGood:** $1000

RED WILLOW ARMORY BALLARD NO. 4½ TARGET RIFLE
Single shot; dropping block; 32-40 Win., 38-55, 40-65 Win., 40-85, 45-70; 30" part-round, part-octagon medium-heavy barrel; weighs 10-11 ½ lbs.; Swiss bead front sight, buckhorn rear; oil-finished American walnut stock with pistol grip; checkered steel shotgun-style buttplate; double-set triggers; under-barrel wiping rod; drilled and tapped for tang sight; full loop lever. Made in U.S. by Red Willow Tool & Armory, Inc. Introduced 1992; discontinued 1993.
Perf.: $2300   **Exc.:** $2200   **VGood:** $2000

RED WILLOW ARMORY BALLARD NO. 5 PACIFIC
Single shot; dropping block; 32-40 Win., 38-55, 40-65 Win., 40-70 Ballard, 40-85 Ballard, 45-70; other calibers on special order; 30" tapered octagon barrel; weighs 10-11 ½ lbs.; blade front sight, buckhorn rear; oil-finished American walnut stock with crescent butt, Schnabel forend; exact recreation of the Ballard No. 5 Pacific; double-set triggers; under-barrel wiping rod; drilled and tapped for tang sight; ring lever; mid- and long-range tang sights; fancy wood, single trigger optionally available. Made in U.S. by Red Willow Tool & Armory, Inc. Introduced 1992; discontinued 1993.
Perf.: $1850   **Exc.:** $1700   **VGood:** $1400

RED WILLOW ARMORY BALLARD NO. 8 UNION HILL
Single shot; dropping block; 32-40 Win., 38-55, 40-65 Win., 40-70 Ballard, 40-85 Ballard, 45-70; other calibers on special order; 30" part round-part octagon barrel; weighs 10 lbs.; Swiss tube sight with bead front, drilled and tapped for tang sight; oil-finished American walnut stock with cheek rest, pistol grip, nickeled off-hand-style buttplate; exact recreation of the original model, with double-set triggers; mid- and long range tang sights, Lyman globe front fancy wood, single trigger, buckhorn rear sight optionally available. Made in U.S. by Red Willow Tool & Armory, Inc. Introduced 1992; discontinued 1993.
Perf.: $2500   **Exc.:** $2400   **VGood:** $2000

Red Willow Armory Ballard No. 8 Union Hill Deluxe Rifle
Same specs as Ballard No. 8 Union Hill except mid-range tang sight; fancy wood and checkering. Made in U.S. by Red Willow Tool & Armory, Inc. Introduced 1992; discontinued 1993.
Perf.: $3000   **Exc.:** $2800   **VGood:** $2500

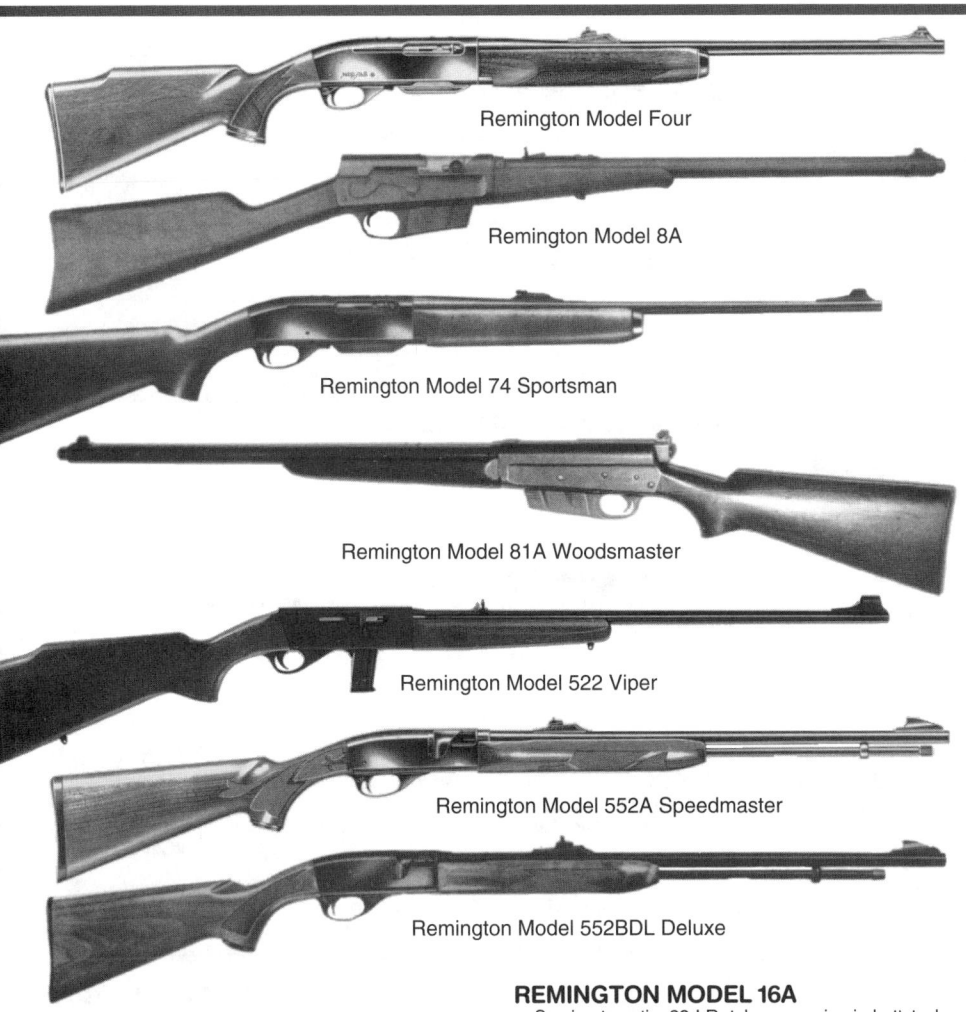

Remington Model Four

Remington Model 8A

Remington Model 74 Sportsman

Remington Model 81A Woodsmaster

Remington Model 522 Viper

Remington Model 552A Speedmaster

Remington Model 552BDL Deluxe

RIFLES

Remington Semi-Automatic Rifles
Gallery Special
(See Remington Model 550GS)
Model Four
Model Four Collectors Edition
Model 8A
Model 74 Sportsman
Model 16A
Model 24A
Model 81A Woodsmaster
Model 522 Viper
Model 550A
Model 550GS
Model 550P
Model 552A Speedmaster
Model 552BDL Deluxe
Model 552C
Model 552GS Gallery Special

REMINGTON SEMI-AUTOMATIC RIFLES REMINGTON MODEL FOUR

Semi-automatic; 243 Win., 6mm Rem., 270, 7mm Express Rem., 280 Rem., 308 Win., 308 Accelerator, 30-06, 30-06 Accelerator; 4-shot clip magazine; 22" round tapered barrel; 42" overall length; weighs 7 1/2 lbs.; gold bead ramp front sight, windage-adjustable sliding ramp rear; checkered American walnut Monte Carlo pistol-grip stock; cartridge head inset on receiver bottom denotes caliber; positive cross-bolt safety; tapped for scope mount. Redesign of Model 742. Introduced 1981; discontinued 1988.
Perf.: $395 **Exc.:** $350 **VGood:** $320
D Peerless Grade
Perf.: $1995 **Exc.:** $1695 **VGood:** $995
F Premier Grade
Perf.: $4195 **Exc.:** $3795 **VGood:** $2695
F Premier Gold Grade
Perf.: $6495 **Exc.:** $4995 **VGood:** $3995

Remington Model Four Collectors Edition

Same specs as Model Four except 30-06; etched receiver; 24K gold inlays; all metal parts with high-luster finish. Only 1500 made in 1982.
Perf.: $1000 **Exc.:** $795 **VGood:** $595

REMINGTON MODEL 8A

Semi-automatic; 25, 30, 32, 35 Rem.; 5-shot detachable box magazine; 22" barrel; bead front sight, open rear; plain walnut straight stock forearm. Introduced 1906; discontinued 1936.
Exc.: $425 **VGood:** $350 **Good:** $250

REMINGTON MODEL 74 SPORTSMAN

Semi-automatic; 30-06; 4-shot magazine; 22" barrel; weighs 7 1/2 lbs.; open adjustable sights; walnut-finished hardwood stock, forend; cartridge head inset on receiver bottom denotes caliber; positive cross-bolt safety; tapped for scope mounts. Introduced 1984; discontinued 1987.
Perf.: $295 **Exc.:** $260 **VGood:** $210

REMINGTON MODEL 16A

Semi-automatic; 22 LR; tube magazine in buttstock; 22" barrel; takedown; bead front sight, open rear; uncheckered straight walnut stock. Introduced 1914; discontinued 1928.
Exc.: $345 **VGood:** $300 **Good:** $225

REMINGTON MODEL 24A

Semi-automatic; 22 Short, 22 LR; 15-shot (22 Short), 10-shot (22 LR) tube magazine in buttstock; 21" barrel; takedown; bead front sight, open rear; uncheckered walnut stock, forearm. Introduced 1922; discontinued 1935.
Exc.: $380 **VGood:** $290 **Good:** $220

REMINGTON MODEL 81A WOODSMASTER

Semi-automatic; 30, 32, 35 Rem., 300 Savage; 5-shot non-detachable box magazine; 22" barrel; 41 1/2" overall length; weighs 8 lbs.; white metal bead front sight, step-adjustable rear; takedown; hammerless; solid breech; uncheckered American walnut pistol-grip stock, forearm; shotgun-style buttplate. Introduced 1936; discontinued 1950.
Exc.: $425 **VGood:** $350 **Good:** $275

REMINGTON MODEL 522 VIPER

Semi-automatic; 22 LR; 10-shot magazine; 20" barrel; 40" overall length; weighs 4 5/8 lbs.; bead on ramp front sight, fully-adjustable open rear; integral grooved rail on receiver for scope mounting; black synthetic stock with positive checkering; beavertail forend; synthetic receiver; magazine safety; cocking indicator; manual and last-shot hold-open; trigger mechanism with primary and secondary sears; integral ejection port shield. Made in U.S. by Remington. Introduced 1993; still produced. **LMSR: $165**
Perf.: $140 **Exc.:** $115 **VGood:** $95

REMINGTON MODEL 550A

Semi-automatic; 22 Short, 22 Long, 22 LR; 22-shot (22 Short), 17-shot (22 Long), 15-shot (22 LR) tube magazine; 24" barrel; 43 1/2" overall length; weighs 6 1/4 lbs.; white metal bead front sight, step-adjustable rear; uncheckered American walnut one-piece pistol-grip stock; semi-beaver-tail forend; thumb safety; shotgun-style Bakelite checkered buttplate. Introduced 1941; discontinued 1971.
Perf.: $170 **Exc.:** $150 **VGood:** $110

Remington Model 550GS

Same specs as Model 550A except 22" barrel; fired shell deflector; screw eye for counter chain; originally listed as Gallery Special.
Perf.: $190 **Exc.:** $160 **VGood:** $120

Remington Model 550P

Same specs as Model 550A except Patridge-type ramp blade front sight on ramp, peep-type rear.
Perf.: $180 **Exc.:** $150 **VGood:** $115

REMINGTON MODEL 552A SPEEDMASTER

Semi-automatic; 22 Short, 22 Long, 22 LR; 20-shot (22 Short), 17-shot (22 Long), 15-shot (22 LR) 23", 25" barrel; 42" overall length; weighs 5 1/2 lbs.; white metal bead front sight, step-adjustable rear; uncheckered walnut-finished hardwood half-pistol-grip stock; semi-beavertail forearm; composition checkered buttplate; cross-bolt safety. Introduced 1958; no longer in production.
Perf.: $160 **Exc.:** $135 **VGood:** $110

Remington Model 552BDL Deluxe

Same specs as Model 552A except 21" barrel; 41" overall length; weighs 5 3/4 lbs.; Du Pont RKW-finished walnut stock; checkered forend; capped pistol-grip stock; semi-beavertail forearm. Introduced 1958; still in production. **LMSR: $319**
Perf.: $250 **Exc.:** $200 **VGood:** $150

Remington Model 552C

Same specs as Model 552A except 21" barrel; 41" overall length. Introduced 1961; discontinued 1977.
Perf.: $170 **Exc.:** $150 **VGood:** $120

Remington Model 552GS Gallery Special

Same specs as Model 552A except 22 Short. No longer produced.
Perf.: $200 **Exc.:** $150 **VGood:** $120

RIFLES

Remington Semi-Automatic Rifles

Model 740A
Model 740ADL
Model 740BDL
Model 742A
Model 742ADL
Model 742BDL
Model 742C Carbine
Model 742CDL Carbine
Model 742D Peerless
Model 742F Premier
Model 742F Premier Gold
Model 7400
Model 7400 Carbine
Model 7400D Peerless
Model 7400F Premier
Model 7400F Premier Gold
Model 7400 Special Purpose
Nylon 66
Nylon 77
Nylon 77 Apache

Remington Model 740A

Remington Model 742BDL

Remington Model 7400

Remington Nylon 66

REMINGTON MODEL 740A
Semi-automatic; 244, 280, 308 Win., 30-06; 4-shot detachable box magazine; 22″ barrel; 42 1/2″ overall length; weighs 7 1/2 lbs.; white metal bead ramp front sight, step-adjustable, semi-buckhorn rear; hammerless; solid frame; side ejection; uncheckered American walnut pistol-grip stock; grooved semi-beavertail forearm. Introduced 1955; discontinued 1960.
Perf.: $295   **Exc.:** $260   **VGood:** $210

Remington Model 740ADL
Same specs as Model 740A except deluxe checkered walnut stock, forend; grip cap; sling swivels; standard or high comb.
Perf.: $350   **Exc.:** $310   **VGood:** $240

Remington Model 740BDL
Same specs as Model 740A except select checkered walnut stock, forend; grip cap; sling swivels; standard or high comb.
Perf.: $380   **Exc.:** $330   **VGood:** $280

REMINGTON MODEL 742A
Semi-automatic; 6mm Rem., 243 Win., 280 Rem., 308 Win., 30-06; 4-shot box magazine; 22″ barrel; 42″ overall length; weighs 7 1/2 lbs.; gold bead front sight on ramp, step-adjustable rear; versions in 1960s with impressed checkering on stock, forearm; later versions, cut checkering. Introduced 1960; discontinued 1980.
Perf.: $330   **Exc.:** $300   **VGood:** $250

Remington Model 742ADL
Same specs as Model 742A except deluxe checkered walnut stock; engraved receiver; sling swivels; grip cap. Introduced 1960.
Perf.: $350   **Exc.:** $300   **VGood:** $260

Remington Model 742BDL
Same specs as Model 742A except 308 Win., 30-06; Monte Carlo cheekpiece; black tip on forearm; basketweave checkering.
Perf.: $370   **Exc.:** $320   **VGood:** $280
Left-hand model
Perf.: $370   **Exc.:** $320   **VGood:** $290

Remington Model 742C Carbine
Same specs as Model 742A except 280, 308 Win., 30-06; 18 1/2″ barrel; 38 1/2″ overall length; weighs 6 1/2 lbs.
Perf.: $350   **Exc.:** $300   **VGood:** $260

Remington Model 742CDL Carbine
Same specs as Model 742C except checkered stock, forend; decorated receiver.
Perf.: $390   **Exc.:** $325   **VGood:** $290

Remington Model 742D Peerless
Same specs as Model 742A except fancy wood; scroll engraving.
Perf.: $2000   **Exc.:** $1600   **VGood:** $1100

Remington Model 742F Premier
Same specs as Model 742A except figured American walnut best-grade stock, cheekpiece; extensive engraved receiver, barrel.
Perf.: $3995   **Exc.:** $3495   **VGood:** $2750

Remington Model 742F Premier Gold
Same specs as Model 742A except best-grade wood; extensive engraving; gold inlays.
Perf.: $5995   **Exc.:** $4995   **VGood:** $3985

REMINGTON MODEL 7400
Semi-automatic; 6mm Rem., 243, 270, 280, 30-06, 30-06 Accelerator, 308, 308 Accelerator, 35 Whelen; 4-shot detachable magazine; 22″ barrel; 42″ overall length; weighs 7 1/2 lbs.; gold bead ramp front sight, windage-adjustable sliding ramp rear; tapped for scope mounts; impressed checkered American walnut pistol-grip stock; positive cross-bolt safety. Introduced 1981; still in production. **LMSR: $612**
Perf.: $485   **Exc.:** $370   **VGood:** $310

Remington Model 7400 Carbine
Same specs as Model 7400 except 30-06; 18 1/2″ barrel; weighs 7 1/4 lbs. Introduced 1988; still in production.
LMSR: $612
Exc.: $370   **VGood:** $310
Perf.: $485

Remington Model 7400D Peerless
Same specs as Model 7400 except fine checkered fancy walnut stock; engraving.
Perf.: $2495   **Exc.:** $1995   **VGood:** $1395

Remington Model 7400F Premier
Same specs as Model 7400 except best-grade fine checkered walnut stock; extensive engraving.
Perf.: $5295   **Exc.:** $4495   **VGood:** $2695

Remington Model 7400F Premier Gold
Same specs as Model 7400 except best-grade fine checkered walnut stock; extensive engraving with gold inlays.
Perf.: $7995   **Exc.:** $6450   **VGood:** $2995

Remington Model 7400 Special Purpose
Same specs as Model 7400 except 270, 30-06; non-glare finish on the American walnut stock; all exposed metal with non-reflective matte black finish; quick-detachable sling swivels; camo-pattern Cordura carrying sling. Made in U.S. by Remington. Introduced 1993; discontinued 1994.
Perf.: $430   **Exc.:** $350   **VGood:** $290

REMINGTON NYLON 66
Semi-automatic; 22 LR; 14-shot tube magazine in buttstock; 19 1/2″ barrel; 38 1/2″ overall length; weighs 4 lbs.; blade front sight, fully-adjustable open rear; receiver grooved for tip-off mounts; moulded nylon stock with checkered pistol grip, forearm; white diamond forend inlay; black pistol-grip cap; available in three stock colors: Seneca green, Mohawk brown and Apache black; latter has chrome-plated receiver cover. Introduced 1959; discontinued 1988.
Perf.: $140   **Exc.:** $120   **VGood:** $100

REMINGTON NYLON 77
Semi-automatic; 22 LR; 5-shot clip magazine; 19 1/2″ barrel; 38 1/2″ overall length; weighs 4 lbs.; blade front sight, open rear; receiver grooved for tip-off mounts; moulded nylon stock with checkered pistol grip, forearm. Available in same colors as Nylon 66. Introduced 1970; discontinued 1971.
Perf.: $140   **Exc.:** $120   **VGood:** $100

Remington Nylon 77 Apache
Same specs as Nylon 77 except 10-shot clip magazine; green nylon stock; made for K-Mart. Introduced 1987; discontinued 1987.
Perf.: $120   **Exc.:** $90   **VGood:** $75

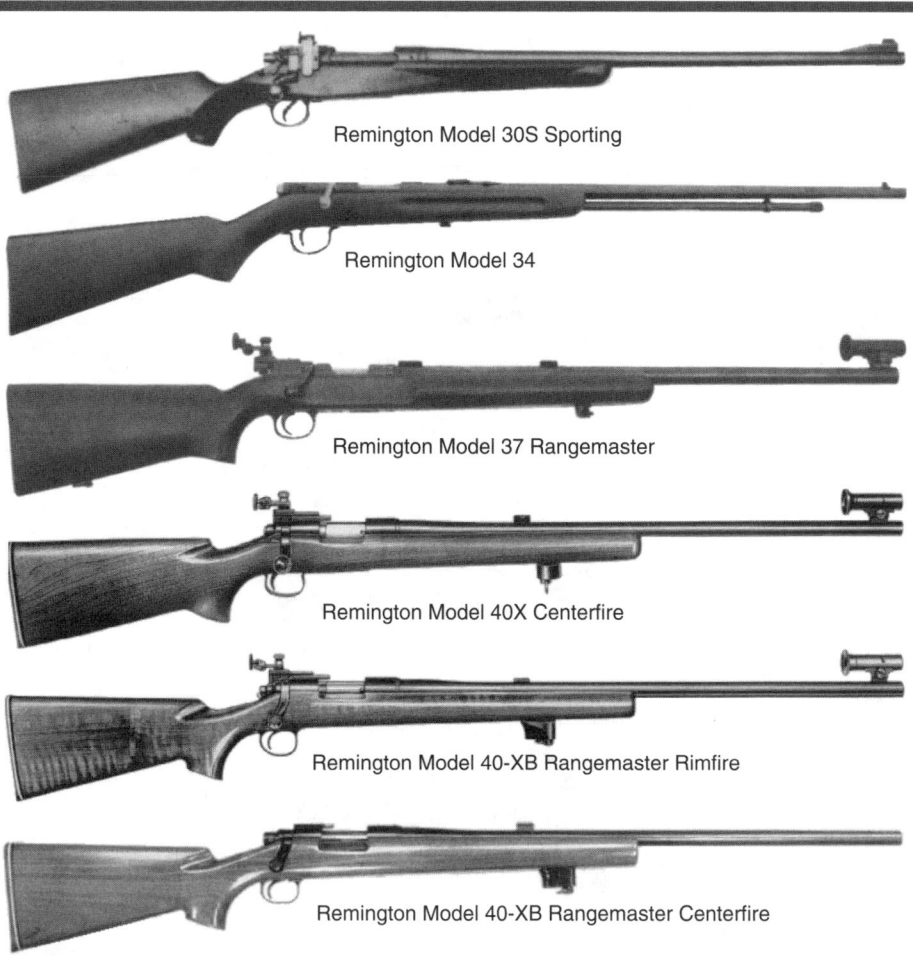

Remington Model 30S Sporting

Remington Model 34

Remington Model 37 Rangemaster

Remington Model 40X Centerfire

Remington Model 40-XB Rangemaster Rimfire

Remington Model 40-XB Rangemaster Centerfire

Remington Bolt-Action Rifles
Model 30A
Model 30R Carbine
Model 30S Sporting
Model 33
Model 33 NRA Junior
Model 34
Model 34 NRA
Model 37 Rangemaster
Model 37 of 1940
Model 40X
Model 40X Centerfire
Model 40X Standard
Model 40-XB Rangemaster Rimfire
Model 40-XB Rangemaster Centerfire

REMINGTON BOLT-ACTION RIFLES

REMINGTON MODEL 30A

Bolt action; 25, 30, 32, 35 Rem., 7mm Mauser, 30-06; 5-shot box magazine; early models with 24" barrel; military-type double-stage trigger; Schnabel forearm tip; later versions with 22" barrel, checkered walnut stock forearm; uncheckered on earlier version with finger groove in forearm; pistol grip; bead front sight, open rear; modified commercial version of 1917 Enfield action. Introduced 1921; discontinued 1940.

Exc.: $450   **VGood:** $390   **Good:** $300

Remington Model 30R Carbine

Same specs as Model 30A except 20" barrel; plain stock.

Exc.: $550   **VGood:** $430   **Good:** $330

Remington Model 30S Sporting

Same specs as Model 30A except 25 Rem., 257 Roberts, 7mm Mauser, 30-06; 5-shot box magazine, 24" barrel; bead front sight, No. 48 Lyman receiver sight; long full forearm, high-comb checkered stock. Introduced 1930; discontinued 1940.

Exc.: $595   **VGood:** $495   **Good:** $400

REMINGTON MODEL 33

Bolt action; single shot; 22 Short, 22 Long, 22 LR; 24" barrel; takedown; bead front sight, open rear; uncheckered pistol-grip stock; grooved forearm. Introduced 1931; discontinued 1936.

Exc.: $180   **VGood:** $140   **Good:** $90

Remington Model 33 NRA Junior

Same specs as Model 33 except Patridge front sight, Lyman peep-style rear; 7/8" sling; swivels.

Exc.: $270   **VGood:** $200   **Good:** $130

REMINGTON MODEL 34

Bolt action; 22 Short, 22 Long, 22 LR; 22-shot (22 Short), 17-shot (22 Long), 15-shot (22 LR) tube magazine; 24" barrel; takedown; bead front sight, open rear; uncheckered hardwood pistol-grip stock; grooved forearm. Introduced 1932; discontinued 1936.

Exc.: $190   **VGood:** $130   **Good:** $90

Remington Model 34 NRA

Same specs as Model 34 except Patridge front sight, Lyman peep rear; swivels; 7/8" sling.

Exc.: $270   **VGood:** $200   **Good:** $130

REMINGTON MODEL 37 RANGEMASTER

Bolt action; 22 LR; 5-shot box magazine; single shot adapter supplied as standard; 28" heavy barrel; 46 1/2" overall length; weighs 12 lbs.; Remington peep rear sight, hooded front; models available with Wittek Vaver or Marble-Goss receiver sight; scope bases; uncheckered target stock with or without sling; swivels. When introduced, barrel band held forward section of stock to barrel; with modification of forearm, barrel band eliminated in 1938. Introduced 1937; discontinued 1940.

Exc.: $550   **VGood:** $470   **Good:** $400

Remington Model 37 of 1940

Same specs as Model 37 except Miracle trigger mechanism; high-comb stock; beavertail forearm. Introduced 1940; discontinued 1954.

Exc.: $650   **VGood:** $550   **Good:** $460

REMINGTON MODEL 40X

Bolt action; single shot; 22 LR; 28" standard or heavy barrel; 46 3/4" overall length; weighs 12 3/4 lbs. (heavy barrel); Redfield Olympic sights optional; scope bases; American walnut, high comb target stock; built-in adjustable bedding device; adjustable swivel; rubber buttplate; action similar to Model 722; adjustable trigger. Introduced 1955; discontinued 1964.

Perf.: $575   **Exc.:** $495   **VGood:** $395
With sights
Perf.: $625   **Exc.:** $530   **VGood:** $450

Remington Model 40X Centerfire

Same specs as Model 40X except only 222 Rem., 222 Rem. Mag., 30-06, 308, other calibers on special request at additional cost. Introduced 1961; discontinued 1964.

Perf.: $550   **Exc.:** $450   **VGood:** $350
With sights
Perf.: $600   **Exc.:** $530   **VGood:** $430

Remington Model 40X Standard

Same specs as Model 40X except lighter standard-weight 28" barrel; weighs 10 3/4 lbs.

Perf.: $550   **Exc.:** $450   **VGood:** $350
With sights
Perf.: $600   **Exc.:** $530   **VGood:** $430

REMINGTON MODEL 40-XB RANGEMASTER RIMFIRE

Bolt action; single shot; 22 LR; 28" barrel, standard or heavyweight; 47" overall length; weighs 10 3/4 lbs. (standard barrel); no sights; American walnut target stock, with palm rest; adjustable swivel block on guide rail; adjustable trigger; rubber buttplate. Replaced Model 40X. Introduced 1964; discontinued 1974.

Perf.: $600   **Exc.:** $530   **VGood:** $420

Remington Model 40-XB Rangemaster Centerfire

Same specs as Model 40-XB Rangemaster except 222 Rem., 222 Rem. Mag., 223 Rem., 6mm, 7.62mm, 220 Swift, 243, 25-06, 7mm BR, 7mm Rem. Mag., 30-338, 300 Win. Mag., 30-06, 6mmx47, 6.5x55, 22-250, 244, 30-06, 308; stainless steel 27 1/4" barrel, standard or heavyweight; 47" overall length; weighs 11 1/4 lbs.; no sights; target stock; adjustable swivel block on guide rail; rubber buttplate. Introduced 1964; still in production.

LMSR: $1565
Perf.: $1100   **Exc.:** $825   **VGood:** $650

RIFLES

Remington Bolt-Action Rifles
Model 40-XB Rangemaster Repeater
Model 40-XB Rangemaster Varmint Special KS
Model 40-XB-BR
Model 40-XB-BR KS
Model 40-XC National Match
Model 40-XC KS National Match
Model 40-XR
Model 40-XR Custom Sporter
Model 40-XR KS
Model 40-XR KS Sporter
Model 41A Targetmaster
Model 41AS
Model 41P
Model 41SB
Model 78 Sportsman
Model 241A Speedmaster
Model 341A
Model 341P
Model 341SB

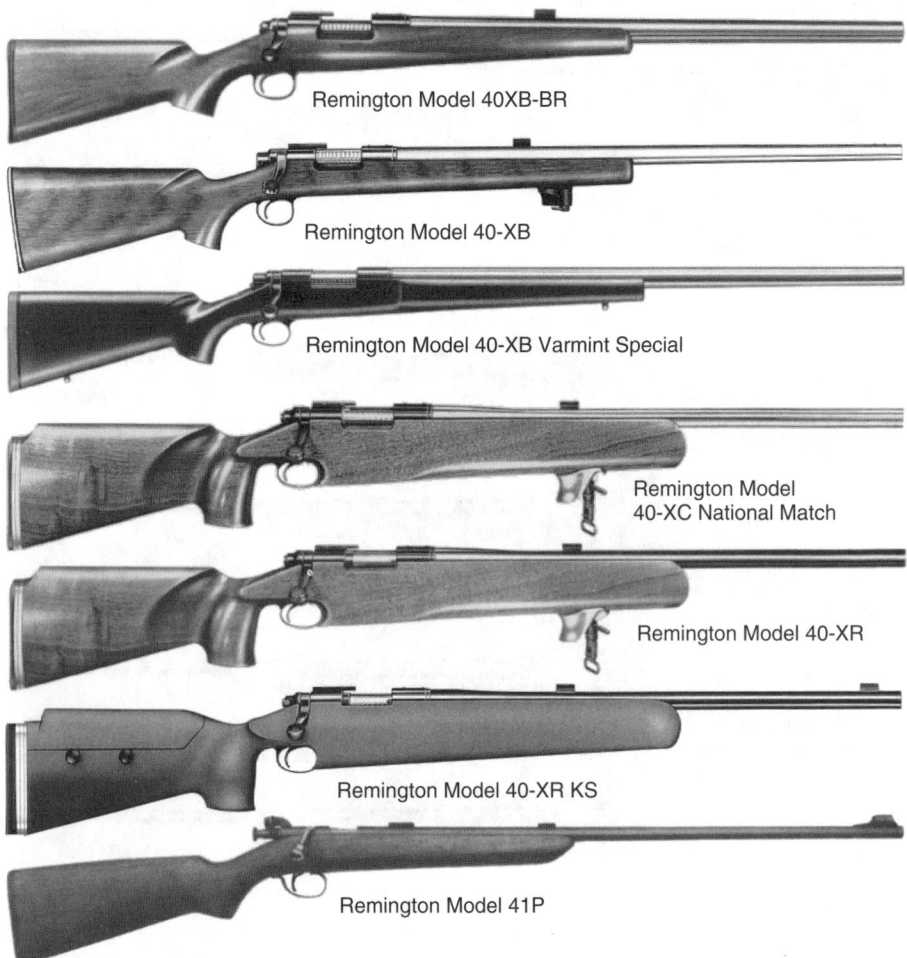

Remington Model 40XB-BR

Remington Model 40-XB

Remington Model 40-XB Varmint Special

Remington Model 40-XC National Match

Remington Model 40-XR

Remington Model 40-XR KS

Remington Model 41P

Remington Model 40-XB Rangemaster Repeater
Same specs as Model 40-XB Rangemaster except repeater with 5-shot magazine. No longer in production.
Perf.: $895   **Exc.:** $750   **VGood:** $640

Remington Model 40-XB Rangemaster Varmint Special KS
Same specs as Model 40-XB except single shot or repeater; 220 Swift; 27 1/4" stainless steel barrel; 45 3/4" overall length; weighs 9 3/4 lbs.; no sights; Kevlar aramid fiber stock; straight comb; cheekpiece; palm swell grip; black recoil pad; removable swivel studs; custom-built to order. Introduced 1987; discontinued 1990.
Perf.: $1195   **Exc.:** $895   **VGood:** $730

REMINGTON MODEL 40-XB-BR
Bolt action; single shot; 22 BR Rem., 222, 222 Rem. Mag., 223, 6mmx47, 6mm BR Rem., 7.62 NATO; stainless steel 20" light varmint, 24" light varmint, 26" heavy varmint barrel; 38" overall length (20" barrel); weighs 9 1/4 lbs.; no sights; supplied with scope blocks; select walnut stock; wide squared-off forend; adjustable trigger. Introduced 1978; discontinued 1989.
Perf.: $995   **Exc.:** $795   **VGood:** $650

Remington Model 40-XB-BR KS
Same specs as Model 40XB-BR except Kevlar stock. Introduced 1989; still in production. **LMSR: $1432**
Perf.: $1195   **Exc.:** $895   **VGood:** $730

REMINGT0N MODEL 40-XC NATIONAL MATCH
Bolt action; 7.62 NATO; 5-shot top-loading clip-slot magazine; 24" stainless steel barrel; 43 1/2" overall length; weighs 11 lbs.; no sights; position-style American walnut stock; palm swell; adjustable buttplate; adjustable trigger to meet all ISU Army Rifle specs. Introduced 1978; still in production with Kevlar stock as 40-XC KS.
Perf.: $1195   **Exc.:** $895   **VGood:** $730

Remington Model 40-XC KS National Match
Same specs as Model 40-XC except Kevlar stock. Introduced 1992.
Perf.: $1250   **Exc.:** $995   **VGood:** $795

REMINGTON MODEL 40-XR
Bolt-action; single-shot; 22 LR; 24" heavy target barrel; 42 1/2" overall length; weighs 9 1/4 lbs.; no sights; drilled, tapped, furnished with scope blocks; American walnut position-style stock, with front swivel block, forend guide rail; adjustable buttplate; adjustable trigger. Meets ISU specs. Introduced 1975; discontinued 1991.
Perf.: $995   **Exc.:** $850   **VGood:** $695

Remington Model 40-XR Custom Sporter
Same specs as Model 40-XR except Custom Shop model; available in four grades. Duplicates Model 700 Centerfire rifle. Introduced 1986; discontinued 1991.
Perf.: $1100   **Exc.:** $925   **VGood:** $795

Remington Model 40-XR KS
Same specs as Model 40-XR except 43 1/2" overall length; weighs 10 1/2 lbs.; Kevlar stock. Introduced 1989; still in production. **LMSR: $1728**
Perf.: $1150   **Exc.:** $895   **VGood:** $695

Remington Model 40-XR KS Sporter
Same specs as Model 40-XR except match chamber; Kevlar stock; special order only. Introduced 1994; still in production.
Perf.: $1225   **Exc.:** $980   **VGood:** $735

REMINGTON MODEL 41A TARGETMASTER
Bolt action; single shot; 22 Short, 22 Long, 22 LR; 27" barrel; takedown; bead front sight, open rear; uncheckered pistol-grip stock. Introduced 1936; discontinued 1940.
Perf.: $160   **Exc.:** $130   **VGood:** $90

Remington Model 41AS
Same specs as Model 41A Targetmaster except 22 Remington Special (22WRF) cartridge.
Perf.: $200   **Exc.:** $150   **VGood:** $120

Remington Model 41P
Same specs as Model 41A Targetmaster except hooded front sight, peep-type rear.
Perf.: $180   **Exc.:** $140   **VGood:** $100

Remington Model 41SB
Same specs as Model 41A Targetmaster except smoothbore barrel for 22 shot cartridges.
Perf.: $230   **Exc.:** $160   **VGood:** $120

REMINGTON MODEL 78 SPORTSMAN
Bolt action; 223, 243, 308, 270, 30-06; 4-shot magazine; 22" barrel; 42 1/2" overall length; weighs 7 1/2 lbs.; open adjustable sights; straight-comb walnut-finished hardwood stock; cartridge head inset on receiver bottom denotes caliber; positive cross-bolt safety; tapped for scope mount. Introduced 1984; discontinued 1989.
Perf.: $285   **Exc.:** $235   **VGood:** $195

REMINGTON MODEL 241A SPEEDMASTER
Semi-automatic; 22 Short, 22 LR; 15-shot (22 Short), 10-shot (22 LR) tube magazine; 24" barrel; 41 1/2" overall length; weighs 6 lbs.; takedown; white metal bead front sight, step-adjustable sporting rear; uncheckered walnut stock, semi-beavertail forearm, shotgun-style checkered buttplate. Introduced 1935; discontinued 1951.
Exc.: $375   **VGood:** $300   **Good:** $240

REMINGTON MODEL 341A
Bolt action; 22 Short, 22 Long, 22 LR; 22-shot (22 Short), 17-shot (22 Long), 15-shot (22 LR) tube magazine; 27" barrel; takedown; bead front sight, open rear; uncheckered hardwood pistol-grip stock. Introduced 1936; discontinued 1940.
Exc.: $160   **VGood:** $120   **Good:** $90

Remington Model 341P
Same specs as Model 341A except hooded front sight, peep rear.
Exc.: $190   **VGood:** $140   **Good:** $100

Remington Model 341SB
Same specs as Model 341A except smoothbore barrel for 22 shot cartridges.
Exc.: $250   **VGood:** $200   **Good:** $150

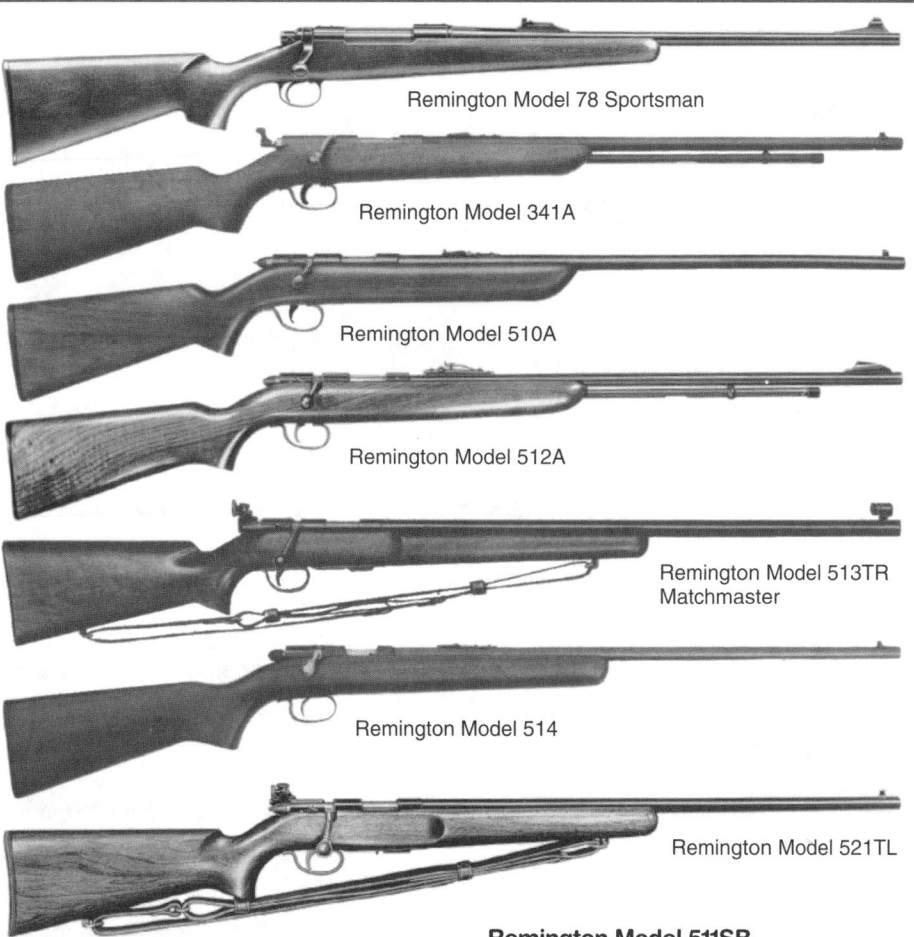

Remington Model 78 Sportsman

Remington Model 341A

Remington Model 510A

Remington Model 512A

Remington Model 513TR Matchmaster

Remington Model 514

Remington Model 521TL

REMINGTON MODEL 510A
Bolt action; single shot; 22 Short, 22 Long, 22 LR; 25″ barrel; 43″ overall length; 5 ¾ lbs.; takedown; bead front sight, open rear; uncheckered American walnut pistol-grip stock; shotgun-style checkered buttplate; semi-beavertail forend; grooved trigger; double locking lugs, extractors; thumb safety; auto ejector. Introduced 1939; discontinued 1963.
    Exc.: $120   **VGood:** $90   **Good:** $70

Remington Model 510P
Same specs as Model 510A except Patridge-design front sight on ramp, "Point-crometer" peep rear.
    Exc.: $140   **VGood:** $120   **Good:** $80

Remington Model 510SB
Same specs as Model 510A except smoothbore barrel for 22 shot cartridge; shotgun bead front sight, no rear sight.
    Exc.: $175   **VGood:** $140   **Good:** $100

Remington Model 510X
Same specs as Model 510A except fully-adjustable rear sights. Introduced 1964; discontinued 1966.
    Exc.: $120   **VGood:** $90   **Good:** $70

Remington 510X-SB
Same specs as Model 510X except smoothbore barrel for use with 22-shot cartridge.
    Exc.: $175   **VGood:** $140   **Good:** $100

REMINGTON MODEL 511A
Bolt action; 22 Short, 22 Long, 22 LR; 6-shot detachable clip magazine; 25″ barrel; 43″ overall length; weighs 5 ¾ lbs.; takedown; white metal bead front sight, step-adjustable rear; uncheckered pistol-grip stock; grooved trigger; thumb safety. Introduced 1939; discontinued 1962.
    Exc.: $130   **VGood:** $100   **Good:** $85

Remington Model 511P
Same specs as Model 511A except "Point-crometer" peep sight on rear of receiver; Patridge-type ramp blade front.
    Exc.: $140   **VGood:** $110   **Good:** $95

Remington Model 511SB
Same specs as Model 511A except smoothbore for use with 22-shot cartridge; open sights.
    Exc.: $175   **VGood:** $140   **Good:** $110

Remington Model 511X
Same specs as Model 511A except box or clip-type magazine; improved sights. Introduced 1965; discontinued 1966.
    Exc.: $130   **VGood:** $100   **Good:** $85

REMINGTON MODEL 512A
Bolt action; 22 Short, 22 Long, 22 LR; 22-shot (22 Short), 17-shot (22 Long), 15-shot (22 LR) tube magazine; 25″ barrel; 43″ overall length; weighs 6 lbs.; takedown; double extractors, locking lugs; automatic ejector; thumb safety; white metal bead front sight, open rear; uncheckered pistol-grip stock; semi-beavertail forearm. Introduced 1940; discontinued 1963.
    Exc.: $140   **VGood:** $110   **Good:** $90

Remington Model 512P
Same specs as Model 512A except Patridge-type blade front sight on ramp, "Point-crometer" peep rear.
    Exc.: $150   **VGood:** $120   **Good:** $100

Remington Model 512SB
Same specs as Model 512 except smoothbore for use with 22-shot cartridge; open sights.
    Exc.: $170   **VGood:** $135   **Good:** $110

Remington Model 512X
Same specs as Model 512A except one cartridge shorter magazine; improved sights. Introduced 1964; discontinued 1966.
    Exc.: $130   **VGood:** $100   **Good:** $80

REMINGTON MODEL 513S
Bolt action; 22 LR; 6-shot detachable box magazine; 27″ barrel; 45″ overall length; weighs 6 ¾ lbs.; Patridge-type ramp front sight, Marble open rear; receiver drilled, tapped for Redfield No. 75 micrometer sight; adjustable trigger stop; double extractors, locking

lugs; side-lever safety; checkered American walnut sporter-style stock; checkered steel buttplate; sling swivels. Introduced 1940; discontinued 1956.
    Exc.: $550   **VGood:** $450   **Good:** $350

Remington Model 513TR Matchmaster
Same specs as Model 513TR except weighs 9 lbs.; Redfield globe front sight, Redfield No. 75 peep-type rear; uncheckered target-style stock; 1 ¼″ government-style leather sling; adjustable trigger stop; double extractors, locking lugs; side-lever safety. Introduced 1940; discontinued 1969.
    Exc.: $370   **VGood:** $310   **Good:** $210

REMINGTON MODEL 514
Bolt action; single shot; 22 Short, 22 Long, 22 LR; 24 ¾″ barrel; 42″ overall length; weighs 5 ¼ lbs.; flat integral bead front sight, step-adjustable rear; uncheckered pistol-grip stock; black plastic buttplate. Introduced 1948; discontinued 1971.
    Exc.: $100   **VGood:** $90   **Good:** $70

Remington Model 514P
Same specs as Model 514 except for ramp front sight, peep-type rear.
    Exc.: $110   **VGood:** $100   **Good:** $80

Remington Model 514SB
Same specs as Model 514 except smoothbore barrel for 22 shot cartridge.
    Perf.: $150   **Exc.:** $120   **VGood:** $100

REMINGTON MODEL 521TL
Bolt action; 22 LR; 6-shot detachable box magazine; 25″ barrel; 43″ overall length; weighs 7 lbs.; takedown; Patridge-type blade front sight, Lyman 57RS peep-type rear; double locking lugs, extractors; firing indicator; side-lever safety; uncheckered American walnut target stock; black Bakelite buttplate; sling; swivels. Introduced 1947; discontinued 1969.
    Exc.: $290   **VGood:** $230   **Good:** $175

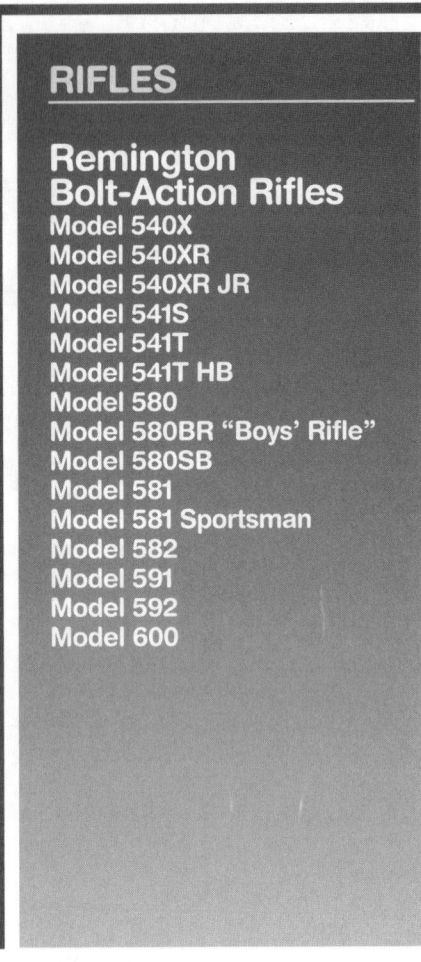

RIFLES

Remington Bolt-Action Rifles
Model 540X
Model 540XR
Model 540XR JR
Model 541S
Model 541T
Model 541T HB
Model 580
Model 580BR "Boys' Rifle"
Model 580SB
Model 581
Model 581 Sportsman
Model 582
Model 591
Model 592
Model 600

Remington Model 541S

Remington Model 541T

Remington Model 580

Remington Model 581

Remington Model 581 Sportsman

Remington Model 582

REMINGTON MODEL 540X
Bolt action; single shot; 22 LR; 26″ heavy barrel; 43 1/2″-47″ overall length; weighs 8 lbs.; no sights; drilled, tapped for scope blocks; fitted with front sight base; target-style stock with Monte Carlo comb and thumb rest groove; full-length guide rail; adjustable buttplate. Introduced 1969; discontinued 1974.
Perf.: $350   **Exc.:** $275   **VGood:** $220

REMINGTON MODEL 540XR
Bolt action; single shot; 22 LR; 26″ medium-weight barrel; 43 1/2″-46 3/4″ overall length; weighs 8 7/8 lbs.; no sights; drilled, tapped for scope blocks; fitted with front sight base; Monte Carlo position-style stock; thumb groove, cheekpiece; adjustable buttplate; full-length guide rail; adjustable match trigger. Introduced 1974; discontinued 1983.
Perf.: $375   **Exc.:** $300   **VGood:** $250

Remington Model 540XR JR
Same specs as Model 540XR except 1 3/4″ shorter stock to fit junior shooter; adjustable length of pull. Introduced 1974; discontinued 1983.
Perf.: $360   **Exc.:** $300   **VGood:** $250

REMINGTON MODEL 541S
Bolt action; 22 Short, 22 Long, 22 LR; 5-shot clip magazine; 24″ barrel; 42 5/8″ overall length; weighs 5 1/2 lbs.; no sights; drilled, tapped for scope mounts, receiver sights; American walnut Monte Carlo stock; checkered pistol grip, forearm; rosewood colored forend tip, pistol grip cap; checkered buttplate; thumb safety; engraved receiver, trigger guard. Introduced 1972; discontinued 1983.
Perf.: $475   **Exc.:** $400   **VGood:** $350

Remington Model 541T
Same specs as Model 541S except drilled, tapped for scope mounts only; satin finish; no engraving. Introduced 1986; still in production. **LMSR: $425**
Perf.: $350   **Exc.:** $275   **VGood:** $240

Remington Model 541T HB
Same specs as Model 541T except weighs about 6 1/2 lbs; heavy target-type barrel without sights; receiver drilled and tapped for scope mounting; American walnut stock with straight comb, satin finish, cut checkering, black checkered buttplate, black grip cap and forend tip. Made in U.S. by Remington. Introduced 1993; still produced. **LMSR: $542**
Perf.: $375   **Exc.:** $320   **VGood:** $260

REMINGTON MODEL 580
Bolt action; single shot; 22 Short, 22 Long, 22 LR; 24″ tapered barrel; 42 3/8″ overall length; weighs 5 lbs.; screw-lock adjustable rear sight, bead post front; hardwood stock with Monte Carlo comb, pistol grip; black composition buttplate; side safety; integral loading platform; receiver grooved for tipoff mounts. Introduced 1957; discontinued 1976.
Perf.: $135   **Exc.:** $100   **VGood:** $80

Remington Model 580BR "Boys' Rifle"
Same specs as Model 580 except youth model with stock 1″ shorter. Introduced 1971; discontinued 1978.
Perf.: $140   **Exc.:** $120   **VGood:** $90

Remington Model 580SB
Same specs as Model 580 except smoothbore barrel for 22 shot cartridges.
Perf.: $175   **Exc.:** $140   **VGood:** $120

REMINGTON MODEL 581
Bolt action; 22 Short, 22 Long, 22 LR; 5-shot clip magazine; 24″ round barrel; 42 3/8″ overall length; weighs 5 1/4 lbs.; screw-adjustable open rear sight, bead on post front; hardwood Monte Carlo pistol-grip stock; side safety; wide trigger; receiver grooved for tip-off mounts. Introduced 1967; discontinued 1984.
Perf.: $180   **Exc.:** $150   **VGood:** $120

Remington Model 581 Sportsman
Same specs as Model 581 except with single-shot adapter. Reintroduced 1986; still in production. **LMSR: $225**
Perf.: $190   **Exc.:** $160   **VGood:** $120

REMINGTON MODEL 582
Bolt action; 22 Short, 22 Long, 22 LR; 20-shot (22 Short), 15-shot (22 Long), 14-shot (22 LR) tube magazine; 24″ round barrel; 42 3/8″ overall length; weighs 5 1/4 lbs.; screw-adjustable open rear sight, bead on post front; hardwood Monte Carlo pistol-grip stock; side safety; wide trigger; receiver grooved for tip-off mounts. Introduced 1967; discontinued 1983.
Perf.: $150   **Exc.:** $125   **VGood:** $110

REMINGTON MODEL 591
Bolt action; 5mm Rem. Mag.; 4-shot clip magazine; 24″ barrel; 42 3/8″ overall length; weighs 5 lbs.; screw-adjustable open rear sight, bead post front; uncheckered hardwood stock, with Monte Carlo comb; black composition pistol-grip cap, buttplate; side safety; wide trigger; receiver grooved for tip-off scope mounts. Introduced 1970; discontinued 1974.
Perf.: $180   **Exc.:** $130   **VGood:** $90

REMINGTON MODEL 592
Bolt action; 5mm Rem. Rimfire; 10-shot tube magazine; 24″ barrel; 42 3/8″ overall length; weighs 5 1/2 lbs.; screw-adjustable open rear sight, bead post front; uncheckered hardwood stock, with Monte Carlo comb; black composition pistol-grip cap, buttplate; side safety; wide trigger; receiver grooved for tip-off scope mounts. Introduced 1970; discontinued 1974.
Perf.: $180   **Exc.:** $135   **VGood:** $95

REMINGTON MODEL 600
Bolt action; 222 Rem., 6mm Rem., 243, 308, 35 Rem.; 5-, 6-shot (222 Rem. only) magazine; 18 1/2″ round barrel with ventilated nylon rib; 37 1/4″ overall length; weighs 5 1/2 lbs.; blade ramp front sight, fully-adjustable open rear; drilled, tapped for scope mounts; checkered walnut Monte Carlo pistol-grip stock. Introduced 1964; discontinued 1967.
Perf.: $400   **Exc.:** $320   **VGood:** $275
223 Rem. Caliber Rare.
Perf.: $950   **Exc.:** $800   **VGood:** $600

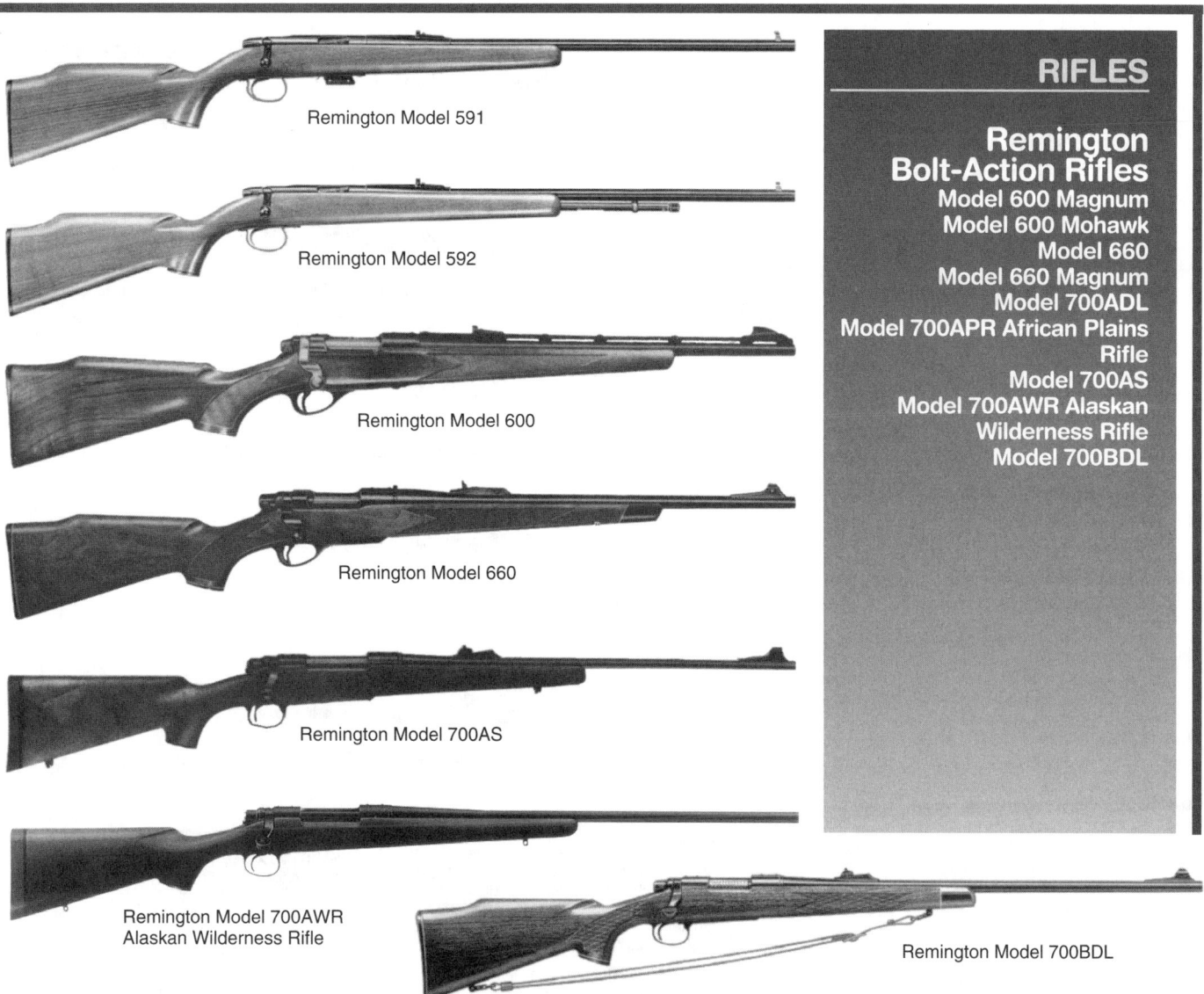

Remington Model 591

Remington Model 592

Remington Model 600

Remington Model 660

Remington Model 700AS

Remington Model 700AWR
Alaskan Wilderness Rifle

Remington Model 700BDL

RIFLES

Remington Bolt-Action Rifles
Model 600 Magnum
Model 600 Mohawk
Model 660
Model 660 Magnum
Model 700ADL
Model 700APR African Plains Rifle
Model 700AS
Model 700AWR Alaskan Wilderness Rifle
Model 700BDL

Remington Model 600 Magnum

Same specs as Model 600 except 6.5mm Rem. Mag., 350 Rem. Mag.; 4-shot box magazine; heavy Magnum-type barrel; weighs 6 1/2 lbs.; laminated walnut/beech stock; recoil pad; swivels; sling. Introduced 1965; discontinued 1967.

Perf.: $800   **Exc.:** $700   **VGood:** $550

Remington Model 600 Mohawk

Same specs as Model 600 except 222 Rem., 6mm Rem., 243, 308; 18 1/2" barrel with no rib; promotional model. Introduced 1971; discontinued 1980.

Perf.: $350   **Exc.:** $275   **VGood:** $225

REMINGTON MODEL 660

Bolt action; 222 Rem., 6mm Rem., 243, 308; 5-, 6-shot (222 Rem. only) magazine; 20" barrel; 38 3/4" overall length; weighs 6 1/2 lbs.; brass bead front sight on ramp, open adjustable rear; checkered Monte Carlo stock; black pistol-grip cap, forearm tip. Introduced 1968; discontinued 1971.

Perf.: $470   **Exc.:** $425   **VGood:** $350

Remington Model 660 Magnum

Same specs as Model 660 except 6.5mm Rem. Mag., 350 Rem. Mag.; 4-shot magazine; laminated walnut/beech stock for added strength; quick-detachable sling swivels, sling; recoil pad. Introduced 1968; discontinued 1971.

Perf.: $700   **Exc.:** $550   **VGood:** $450

REMINGTON MODEL 700ADL

Bolt action; 222, 222 Rem. Mag., 22-250, 243, 25-06, 264 Win. Mag., 6mm Rem., 270, 280, 7mm Rem. Mag., 308, 30-06; 4-shot (264, 7mm), 6-shot (222, 222 Rem. Mag.), 5-shot (others); 20", 22", 24" round tapered barrel; 39 1/2" overall length (222, 222 Rem. Mag., 243 Win.); removable, adjustable rear sight with windage screw, gold bead ramp front; tapped for scope mounts; walnut Monte Carlo stock, with pistol grip; originally introduced with hand-checkered pistol grip, forearm; made for several years with RKW finish, impressed checkering; more recent models with computerized cut checkering. Introduced 1962; still in production. **LMSR: $452**

Perf.: $400   **Exc.:** $320   **VGood:** $285

With optional laminated stock.

Perf.: $400   **Exc.:** $350   **VGood:** $300

REMINGTON MODEL 700APR AFRICAN PLAINS RIFLE

Bolt action; 7mm Rem. Mag., 300 Win. Mag., 300 Wea. Mag., 338 Win. Mag., 375 H&H; 26" Custom Shop barrel with satin finish; Magnum receiver; laminated wood stock with raised cheekpiece; satin finish; black buttpad; 20 lpi cut checkering. Made in U.S. by Remington. Introduced 1994; still produced. **LMSR: $1466**

Perf.: $1250   **Exc.:** $1050   **VGood:** $850

REMINGTON MODEL 700AS

Bolt action; 22-250, 243, 270, 280, 30-06, 308, 7mm Rem. Mag., 300 Weatherby Mag.; 4-shot magazine, 3-shot (7mm, 300 Wea. Mag.); 22", 24" (22-250, 7mm Rem. Mag., 300 Wea. Mag.) barrel; weighs 6 1/2-7 lbs.; no sights; Arylon thermoplastic resin stock; black, light-ly textured finish; hinged floorplate; solid buttpad; right-hand only. Introduced 1989; discontinued 1991.

Perf.: $450   **Exc.:** $400   **VGood:** $330

REMINGTON MODEL 700AWR ALASKAN WILDERNESS RIFLE

Bolt action; 7mm Rem. Mag., 300 Win. Mag., 300 Weatherby Mag., 338 Win. Mag., 375 H&H; 24" Custom Shop barrel profile; stainless barreled action with satin blue finish; matte gray stock of fiberglass and graphite, reinforced with DuPont Kevlar; straight comb with raised cheekpiece; Magnum-grade black rubber recoil pad. Made in U.S. by Remington. Introduced 1994; still produced. **LMSR: $1256**

Perf.: $1100   **Exc.:** $950   **VGood:** $750

REMINGTON MODEL 700BDL

Bolt action; 17 Rem., 22-250, 222 Rem., 222 Rem. Mag., 243, 25-06, 264 Win. Mag., 270, 280, 300 Savage, 30-06, 308, 35 Whelen, 6mm Rem., 7mm Rem. Mag., 7mm-08, 300 Win. Mag., 338 Win. Mag., 8mm Mag.; 4-shot magazine; 20", 22", 24" barrel; stainless barrel (7mm Rem. Mag., 264, 300 Win. Mag.); with or without sights; select walnut, hand-checkered Monte Carlo stock; black forearm tip, pistol-grip cap, skip-line fleur-de-lis checkering; matted receiver top; quick-release floorplate; quick-detachable swivels, sling. Still in production. **LMSR: $576**

Perf.: $465   **Exc.:** $400   **VGood:** $325

Left-hand model

Perf.: $525   **Exc.:** $450   **VGood:** $350

Remington Bolt-Action Rifles
Model 700BDL DM
Model 700BDL European
Model 700BDL Limited Classic
Model 700BDL Mountain
Model 700BDL Mountain Custom KS Wood Grained
Model 700BDL Mountain Stainless
Model 700BDL Safari
Model 700BDL Safari Custom KS
Model 700BDL Safari Custom KS Stainless
Model 700BDL SS
Model 700BDL SS DM
Model 700BDL Varmint Special

Remington Model 700BDL DM

Remington Model 700BDL SS DM

Remington Model 700BDL Varmint Special

Remington Model 700 Classic

Remington Model 700 Custom Mountain KS

Remington Model 700BDL DM
Same specs as Model 700BDL except right-hand action calibers: 6mm, 243, 25-06, 270, 280, 7mm-08, 30-06, 308, 7mm Rem. Mag., 300 Win. Mag., 338 Win. Mag.; left-hand calibers: 243, 270, 7mm-08, 30-06, 7mm Rem. Mag., 300 Win. Mag.; 3-shot (Magnums), 4-shot (standard) detachable box magazine; glossy stock finish; open sights; recoil pad; sling swivels. Made in U.S. by Remington. Introduced 1995; still produced. **LMSR: $603**
Right-hand, standard calibers.
   Perf.: $500   **Exc.:** $450   **VGood:** $380
Left-hand, standard calibers, right-hand Magnums.
   Perf.: $525   **Exc.:** $475   **VGood:** $400
Left-hand, Magnum calibers.
   Perf.: $550   **Exc.:** $485   **VGood:** $425

Remington Model 700BDL European
Same specs as Model 700BDL except 243, 270, 7mm-08, 280 Rem., 30-06 (22″ barrel), 7mm Rem. Mag. (24″ barrel); oil-finished walnut stock. Made in U.S. by Remington. Introduced 1993; discontinued 1995.
Standard calibers.
   Perf.: $440   **Exc.:** $385   **VGood:** $325
7mm Rem. Mag.
   Perf.: $465   **Exc.:** $425   **VGood:** $375

Remington Model 700BDL Limited Classic
Same specs as Model 700BDL except 220 Swift, 222 Rem., 22-250, 250-3000, 6.5x55 Swedish, 6mm, 7x57, 243, 25-06, 257 Roberts, 264 Win. Mag., 270, 300 Weatherby Mag., 30-06, 7mm Weatherby Mag., 338 Win. Mag., 350 Rem. Mag., 35 Whelen, 300 H&H Mag., 375 H&H Mag.; classic-style, hand-checkered walnut straight stock; black forearm tip, pistol-grip cap; high gloss blued finish. New caliber announced annually. Introduced 1981; no longer in production.
   Perf.: $510   **Exc.:** $450   **VGood:** $385

Remington Model 700BDL Mountain
Same specs as Model 700BDL except 243, 25-06, 257 Roberts, 270, 7mm-08, 280, 30-06, 308, 7x57; 22″ tapered barrel; checkered American walnut stock with cheekpiece; ebony forend tip. Introduced 1986; no longer in production.
   Perf.: $435   **Exc.:** $375   **VGood:** $325

Remington Model 700BDL Mountain Custom KS Wood Grained
Same specs as Model 700BDL Mountain except 375 H&H Mag., 416 Rem. Mag., 458 Win. Mag.; wood-grained Kevlar stock. Introduced 1992; discontinued 1993.
   Perf.: $1000   **Exc.:** $875   **VGood:** $625

Remington Model 700BDL Mountain Stainless
Same specs as Model 700BDL Mountain except 25-06, 270, 280, 30-06; 22″ stainless barrel; weighs 7 1/4 lbs.; black synthetic stock with checkering. Introduced 1993; discontinued 1993.
   Perf.: $460   **Exc.:** $400   **VGood:** $375

Remington Model 700BDL Safari
Same specs as Model 700BDL except 375 H&H Mag., 8mm Rem. Mag., 416 Rem. Mag., 458 Win. Mag.; 3-shot magazine; 24″, 26″ barrel; 46 3/8″ overall length; weighs 9 lbs.; no sights; Classic or Monte Carlo hand-checkered walnut stock; recoil pad. Introduced 1962; still in production. **LMSR: $1041**
   Perf.: $875   **Exc.:** $695   **VGood:** $530
Left-hand model.
   Perf.: $925   **Exc.:** $750   **VGood:** $565

Remington Model 700BDL Safari Custom KS
Same specs as Model 700BDL Safari except 8mm Rem. Mag., 375 H&H Mag., 416 Rem. Mag., 458 Win. Mag.; lightweight Kevlar straight-line classic stock; comb; no cheekpiece. Introduced 1989; still in production. **LMSR: $1198**
   Perf.: $1100   **Exc.:** $925   **VGood:** $700
Left-hand model.
   Perf.: $1150   **Exc.:** $950   **VGood:** $725

Remington Model 700BDL Safari Custom KS Stainless
Same specs as Model 700BDL Safari except 375 H&H Mag., 416 Rem. Mag., 458 Win. Mag.; Kevlar stock; stainless barrel and action. Introduced 1993; still in production. **LMSR: $1338**
   Perf.: $1200   **Exc.:** $950   **VGood:** $750

Remington Model 700BDL SS
Same specs as Model 700BDL Safari Custom KS Stainless rifle except 223, 243, 6mm Rem., 7mm-08 Rem., 308 (short action), 25-06, 270, 280 Rem., 30-06, 7mm Rem. Mag., 7mm Weatherby Mag., 300 Win. Mag., 300 Weatherby Mag., 338 Win. Mag. (standard long action); 24″ barrel; Magnum-contour barrel (Magnums); weighs 6 3/4-7 lbs.; corrosion-resistant follower and fire control; stainless #416 BDL-style barreled action with fine matte finish; synthetic stock with straight comb and cheekpiece, textured finish, positive checkering, plated swivel studs. Made in U.S. by Remington. Introduced 1993; still produced. **LMSR: $603**
Standard calibers.
   Perf.: $500   **Exc.:** $450   **VGood:** $400
Magnum calibers.
   Perf.: $535   **Exc.:** $465   **VGood:** $410

Remington Model 700BDL SS DM
Same specs as Model 700BDL SS except 6mm, 243, 25-06, 270, 280, 7mm-08, 7mm Rem. Mag., 300 Win. Mag., 300 Weatherby Mag.; detachable box magazine; black synthetic stock. Made in U.S. by Remington. Introduced 1995; still produced. **LMSR: $656**
Standard calibers.
   Perf.: $520   **Exc.:** $450   **VGood:** $400
Magnum calibers.
   Perf.: $545   **Exc.:** $475   **VGood:** $425

Remington Model 700BDL Varmint Special
Same specs as Model 700BDL except 24″ heavy barrel; 43 1/2″ overall length; weighs 9 lbs.; no sights; 222, 223, 22-250, 25-06, 6mm Rem., 243, 7mm-08 Rem., 308; 5-, 6-shot magazine. Introduced 1967.
   Perf.: $470   **Exc.:** $400   **VGood:** $325

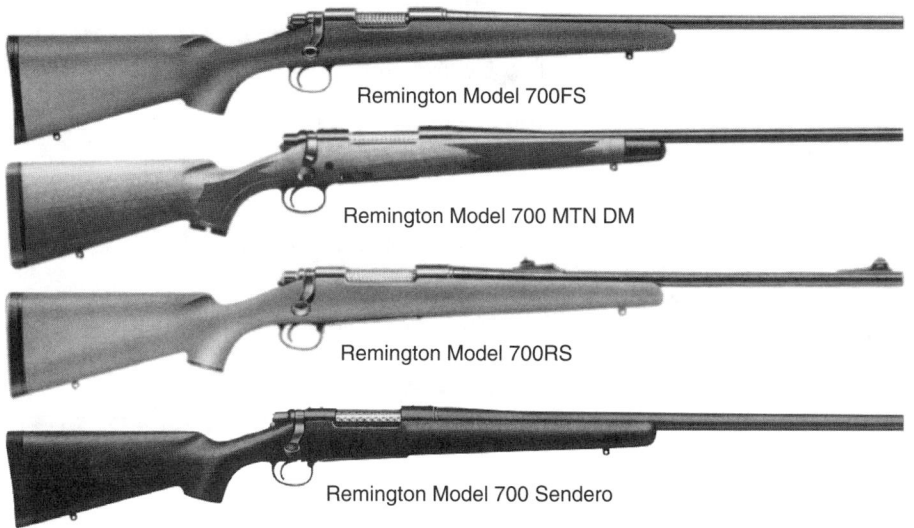

Remington Model 700FS

Remington Model 700 MTN DM

Remington Model 700RS

Remington Model 700 Sendero

Remington Model 700BDL Varmint Synthetic
Same specs as Model 700BDL Varmint except 220 Swift, 22-250, 223 Rem., 308; black/gray synthetic stock; aluminum bedding block; matte metal finish. Introduced 1992; still in production. **LMSR: $665**
   Perf.: $540   **Exc.: $420**   **VGood: $350**

REMINGTON MODEL 700C
Bolt action; 17 Rem., 22-250, 222 Rem., 222 Rem. Mag., 223 Rem., 243, 25-06, 264 Mag., 270, 280, 300 Savage, 30-06, 308, 35 Whelen, 6mm Rem., 6.5mm Rem. Mag., 7mm Rem. Mag., 7mm-08, 300 Win. Mag., 338 Win. Mag., 350 Rem. Mag., 8mm Mag.; 4-shot magazine; 20″, 22″, 24″ barrel; with or without sights; select walnut, hand-checkered stock; rosewood forearm tip, grip cap; hand-lapped barrel; hinged floorplate; Custom Shop rifle. Introduced 1964; discontinued 1983.
   Perf.: $895   **Exc.: $795**   **VGood: $650**

Remington Model 700D Peerless
Same specs as Model 700C except best-grade wood; scroll engraving.
   Perf.: $1695   **Exc.: $1295**   **VGood: $995**

Remington Model 700F Premier
Same specs as Model C except best-grade wood; extensive engraving.
   Perf.: $3125   **Exc.: $2625**   **VGood: $2100**

REMINGTON MODEL 700 CAMO SYNTHETIC RIFLE
Bolt action; 22-250, 243, 7mm-08, 270, 280, 30-06, 308, 7mm Rem. Mag., 300 Weatherby Mag.; 4-shot, 3-shot (7mm, 300 Wea. Mag.) magazine; open adjustable sights; synthetic stock, stock and metal (except bolt and sights) fully camouflaged in Mossy Oak Bottomland camo; swivel studs. Made in U.S. by Remington. Introduced 1992; discontinued 1994.
   Perf.: $450   **Exc.: $375**   **VGood: $330**

REMINGTON MODEL 700 CLASSIC
Bolt action; 22-250, 6mm Rem., 243, 270, 7mm Mauser, 30-06, 30-06 Accelerator, 7mm Rem. Mag., 350 Rem. Mag., 375 H&H, 300 Win. Mag.; 24″ barrel; 44 1/2″ overall length; weighs 7 3/4 lbs.; no sights; drilled and tapped for scope mounting; American walnut straight-comb stock with 20 lpi checkered pistol grip, forend; rubber recoil pad; sling swivel studs; hinged floorplate. Introduced 1978; still in production. **LMSR: $576**
   Perf.: $500   **Exc.: $450**   **VGood: $400**

Remington Model 700 Classic Limited
Same specs as Model 700 Classic except produced in limited quantities of a single but different caliber each year.
   Perf.: $550   **Exc.: $475**   **VGood: $400**

REMINGTON MODEL 700 CUSTOM MOUNTAIN KS
Bolt action; 270, 280, 300 Win. Mag., 300 Weatherby Mag., 30-06, 338 Win. Mag., 35 Whelen, 7mm Rem. Mag., 375 H&H Mag.; 4-shot magazine; 22″ barrel; 42 1/2″ overall length; weighs 6 3/8 lbs.; no sights; Kevlar reinforced resin stock; left- and right-hand versions; Custom Shop rifle. Introduced 1986; still in production. **LMSR: $1037**
   Perf.: $995   **Exc.: $895**   **VGood: $650**
Left-hand model
   Perf.: $1025   **Exc.: $925**   **VGood: $675**

Remington Model 700 Mountain
Same specs as Model 700 Custom Mountain KS except walnut stock with satin finish, fine-line checkering. Introduced 1986; no longer in production.
   Perf.: $450   **Exc.: $375**   **VGood: $325**

REMINGTON MODEL 700FS
Bolt action; 243, 270, 30-06, 308, 7mm Rem. Mag.; 4-shot magazine; 22″, 24″ (7mm) barrel; weighs 6 1/4-6 3/4 lbs.; open sights; classic-style gray or gray camo fiberglass stock reinforced with Kevlar; black Old English-type rubber recoil pad; right- or left-hand actions. Introduced 1987; discontinued 1988.
   Perf.: $525   **Exc.: $465**   **VGood: $400**

REMINGTON MODEL 700LS
Bolt action; 243, 270, 30-06, 7mm Rem. Mag.; 5-shot, 4-shot (7mm); 22″, 24″ barrel (7mm); weighs 7 1/4 lbs.; open sights; laminated light/dark wood Monte Carlo pistol-grip stock with cut checkering; sling swivels. Introduced 1990; no longer in production.
   Perf.: $450   **Exc.: $385**   **VGood: $310**

Remington Model 700MTN DM
Bolt action; 243, 270 Win., 7mm-08, 25-06, 280 Rem., 30-06; 4-shot; 22″ barrel; 42 1/2″ overall length; weighs 6 3/4 lbs.; no sights; drilled, tapped for scope mounting; redesigned pistol grip; straight comb; contoured cheekpiece; hand-rubbed oil satin finish; hinged floorplate, magazine follower; two position thumb safety. Introduced 1986; still in production. **LMSR: $603**
   Perf.: $550   **Exc.: $430**   **VGood: $375**

REMINGTON MODEL 700MTRSS
Bolt action; 25-06, 270, 280 Rem., 30-06; weighs 6 3/4 lbs.; 22″ barrel; stainless steel barreled action; textured black synthetic stock with positive checkering; straight comb and cheekpiece. Made in U.S. by Remington. Introduced 1993; discontinued 1994.
   Perf.: $510   **Exc.: $420**   **VGood: $360**

REMINGTON MODEL 700RS
Bolt action; 270, 280, 30-06; 4-shot magazine; 22″ barrel; weighs 7 1/4 lbs.; open sights; glass-reinforced Rynite thermoplastic resin stock; textured gray or camouflage finish; solid buttpad; hinged floorplate. Introduced 1987; discontinued 1989.
   Perf.: $500   **Exc.: $425**   **VGood: $380**

REMINGTON MODEL 700 SENDERO
Bolt action; 25-06, 270, 7mm Rem. Mag., 300 Win. Mag.; 26″ barrel with spherical concave crown; long action for Magnum calibers; composite stock. Made in U.S. by Remington. Introduced 1994; still produced. **LMSR: $665**
   Perf.: $585   **Exc.: $475**   **VGood: $400**

REMINGTON MODEL 700VLS VARMINT LAMINATED STOCK
Bolt action; 222 Rem., 223 Rem., 22-250, 243, 308; 26″ heavy barrel; no sights; brown laminated stock with forend tip, grip cap, rubber buttpad; polished blue finish. Made in U.S. by Remington. Introduced 1995; still produced. **LMSR: $585**
   Perf.: $510   **Exc.: $440**   **VGood: $380**

REMINGTON MODEL 700VS SF
Bolt action; 223, 220 Swift, 22-250, 308; 26″ fluted barrel; satin-finish stainless barreled action; spherical concave muzzle crown. Made in U.S. by Remington. Introduced 1994; still produced. **LMSR: $798**
   Perf.: $725   **Exc.: $565**   **VGood: $450**

REMINGTON MODEL 720A
Bolt action; 257 Roberts, 270, 30-06; 5-shot detachable box magazine; 22″ barrel; 42 1/2″ overall length; weighs 8 lbs.; bead front sight on ramp, open rear; models available with Redfield No. 70RST micrometer or Lyman No. 48 receiver sights; checkered American walnut pistol-grip stock; modified Model 1917 Enfield action. Made only in 1941 as factory facilities were converted to wartime production. Note: Most 720s were made in 30-06; few in 270; and only a handful in 257 Roberts. Add premium.
   Perf.: $1250   **Exc.: $1000**   **VGood: $700**

Remington Model 720R
Same specs as Model 720A except 20″ barrel; models available with Redfield No. 70RST micrometer or Lyman No. 48 receiver sights.
   Perf.: $1250   **Exc.: $1000**   **VGood: $700**

Remington Model 720S
Same specs as Model 720A except 24″ barrel; models available with Redfield No. 70RST micrometer or Lyman No. 48 receiver sights.
   Perf.: $1250   **Exc.: $1000**   **VGood: $700**

Remington Model 721A Magnum

Remington Model 725ADL

Remington Model 788

Remington International Match Free Rifle

Remington Model Seven

Remington Bolt-Action Rifles
Model 721A
Model 721A Magnum
Model 721ADL
Model 721ADL Magnum
Model 721BDL
Model 721BDL Magnum
Model 722A
Model 722ADL
Model 722BDL
Model 725ADL
Model 725ADL Kodiak Magnum
Model 788
Model 788 Carbine
Model 788 Left-Hand
Model International Free Rifle
Model International Match Free Rifle
Model Seven

REMINGTON MODEL 721A
Bolt action; 270, 280, 30-06; 4-shot box magazine; 24″ barrel; 44 1/4″ overall length; weighs 7 1/4 lbs.; white metal bead front sight on ramp, step-adjustable sporting rear; unchecked American walnut sporter stock with pistol grip; semi-beavertail forend; checkered shotgun-style buttplate; thumb safety; receiver drilled, tapped for scope mounts, micrometer sights. Introduced 1948; discontinued 1962.
Perf.: $360   Exc.: $285   VGood: $250

Remington Model 721A Magnum
Same specs as Model 721A except only 264 Win. Mag., 300 H&H Mag.; 3-shot magazine; heavy 26″ barrel; 46 1/4″ overall length; weighs 8 1/4 lbs.; recoil pad.
Perf.: $500   Exc.: $400   VGood: $340

Remington Model 721ADL
Same specs as Model 721A except deluxe checkered walnut sporter stock, forearm. Introduced 1948; discontinued 1962.
Perf.: $425   Exc.: $375   VGood: $325

Remington Model 721ADL Magnum
Same specs as Model 721ADL except 264 Win. Mag., 300 H&H Mag.; 3-shot magazine; 26″ barrel; recoil pad.
Perf.: $550   Exc.: $420   VGood: $375

REMINGTON MODEL 721BDL
Bolt action; 270, 280, 30-06; 4-shot magazine; 24″ barrel; bead front sight on ramp, open rear; select checkered walnut sporter stock, forearm. Introduced 1948; discontinued 1962.
Perf.: $575   Exc.: $495   VGood: $410

Remington Model 721BDL Magnum
Same specs as Model 721BDL except 264 Win. Mag., 300 H&H Mag.; 3-shot magazine; 26″ barrel; recoil pad.
Perf.: $650   Exc.: $525   VGood: $475

REMINGTON MODEL 722A
Bolt action; 222 Rem., 222 Rem. Mag., 243 Win., 244 Rem., 257 Roberts, 300 Savage, 308; 4-, 5-shot (222 Rem.) magazine; 24″, 26″ (222 Rem., 244 Rem.) barrel; 43 1/4″ overall length; weighs 7 lbs.; shorter action than Model 721A; white metal bead front sight on ramp, step-adjustable sporting rear; receiver drilled, tapped for scope mounts or micrometer sights; uncheckered walnut sporter stock. Introduced 1948; discontinued 1962.
Perf.: $360   Exc.: $310   VGood: $275

Remington Model 722ADL
Same specs as Model 722A except deluxe checkered walnut sporter stock.
Perf.: $430   Exc.: $370   VGood: $320

Remington Model 722BDL
Same specs as Model 722A except select checkered walnut sporter stock.
Perf.: $575   Exc.: $475   VGood: $375

REMINGTON MODEL 725ADL
Bolt action; 222 Rem., 243, 244, 270, 280, 30-06; 4-, 5-shot (222 Rem.) box magazine; 22″, 24″ (222 Rem.) barrel; 42 1/2″, 43 1/2″ (222) overall length; weighs 7 lbs.; removable hood ramp front sight, adjustable open rear; Monte Carlo comb, walnut stock; hand-checkered pistol grip, forearm; hinged floorplate; swivels. Some calibers bring premium. Introduced 1958; discontinued 1961.
Perf.: $595   Exc.: $490   VGood: $400

Remington Model 725ADL Kodiak Magnum
Same specs as Model 725ADL except 375 H&H Mag., 458 Win. Mag.; 3-shot box magazine; 26″ barrel; reinforced deluxe Monte Carlo stock; recoil pad; special recoil reduction device built into barrel; black forearm tip; sling, swivels. Made only in 1962.
Perf.: $3000   Exc.: $2600   VGood: $2100

REMINGTON MODEL 788
Bolt action; 222, 22-250, 223, 6mm Rem., 243, 7mm-08, 308, 30-30, 44 Mag.; 3-, 4-shot (222) detachable box magazine; 22″, 24″ (222, 22-250) barrel; 41″-43 5/8″ overall length; weighs 7-7 1/2 lbs.; open fully-adjustable rear sight, blade ramp front; receiver tapped for scope mounts; American walnut or walnut-finished hardwood pistol-grip stock, uncheckered with Monte Carlo comb; thumb safety; artillery-type bolt. Introduced 1967; discontinued 1983.
Perf.: $330   Exc.: $275   VGood: $240
44 Mag., 7mm-08.
Perf.: $395   Exc.: $330   VGood: $275

Remington Model 788 Carbine
Same specs as Model 788 except 18 1/2″ barrel.
Perf.: $350   Exc.: $290   VGood: $250

Remington Model 788 Left-Hand
Same specs as Model 788 except 6mm, 308; left-hand version. Introduced 1969.
Perf.: $375   Exc.: $310   VGood: $260

REMINGTON MODEL INTERNATIONAL FREE RIFLE
Bolt action; single shot; 22 LR; 222 Rem., 222 Rem. Mag., 223 Rem., 7.62 NATO, 30-06; 27 1/4″ barrel; 47″ overall length; weighs 15 lbs.; no sights; hand-finished stock; adjustable buttplate and hook; movable front sling swivel; adjustable palm rest; 2-oz. adjustable trigger. Introduced 1964; discontinued 1974.
Perf.: $1900   Exc.: $1600   VGood: $1300

REMINGTON MODEL INTERNATIONAL MATCH FREE RIFLE
Bolt action; single shot; 22 LR, 222 Rem., 222 Rem. Mag., 30-06, 7.62mm NATO, others on special order; 28″ barrel; no sights; freestyle hand-finished stock with thumbhole; interchangeable, adjustable rubber buttplate and hook-type buttplate; adjustable palm rest; adjustable sling swivel; 2-oz. adjustable trigger. Introduced 1961; discontinued 1964.
Perf.: $2000   Exc.: $1700   VGood: $1400

REMINGTON MODEL SEVEN
Bolt action; 17 Rem., 222 Rem., 223 Rem., 243, 7mm-08, 6mm, 308; 4-shot magazine, 5-shot (222 only); 18 1/2″ barrel; 37 1/2″ overall length; weighs 6 1/4 lbs.; ramp front sight, adjustable open rear; American walnut stock; modified Schnabel forend; machine-cut checkering; short action; silent side safety; free-floated barrel. Introduced 1983; still in production. **LMSR: $549**
Perf.: $450   Exc.: $370   VGood: $320

Remington Model Seven SS

Remington Model Seven Custom KS

Remington Model Seven FS

Remington Nylon 11

Remington Bolt-Action Rifles
Model Seven Custom KS
Model Seven Custom MS
Model Seven FS
Model Seven SS
Model Seven Youth
Nylon 10
Nylon 10-C
Nylon 11
Nylon 12
Lee Sporter
Lee Sporter Deluxe

Remington Model Seven Custom KS
Same specs as Model Seven except 223 Rem., 7mm BR, 7mm-08, 308, 35 Rem., 350 Rem. Mag.; 20" barrel; weighs 5 3/4 lbs.; Kevlar aramid fiber stock; iron sights; drilled, tapped for scope mounts. Special order through Remington Custom Shop. Introduced 1987; still in production. **LMSR: $1037**
   Perf.: $950   **Exc.:** $750   **VGood:** $600

Remington Model Seven Custom MS
Same specs as Model Seven except 222 Rem., 223, 22-250, 243, 6mm Rem., 7mm-08 Rem., 308, 350 Rem. Mag.; 20" barrel; weighs 6 3/4 lbs.; full-length Mannlicher-style stock of laminated wood with straight comb, solid black recoil pad, black steel forend tip, cut checkering, gloss finish; polished blue finish; calibers 250 Savage, 257 Roberts, 35 Rem. available on special order. From Remington Custom Shop. Introduced 1993; still produced. **LMSR: $1041**
   Perf.: $910   **Exc.:** $710   **VGood:** $650

Remington Model Seven FS
Same specs as Model Seven except 243, 7mm-08, 308; weighs 5 1/4 lbs.; fiberglass/Kevlar classic-style stock gray or camo; rubber buttpad. Introduced 1987; discontinued 1989.
   Perf.: $550   **Exc.:** $475   **VGood:** $400

Remington Model Seven SS
Same specs as Model Seven except 243, 7mm-08, 308; 20" barrel; stainless steel barreled action; black synthetic stock. Made in U.S. by Remington. Introduced 1994; still produced. **LMSR: $603**
   Perf.: $475   **Exc.:** $430   **VGood:** $390

Remington Model Seven Youth
Same specs as Model Seven except 6mm Rem., 243, 7mm-08; hardwood stock with 12 3/16" length of pull. Made in U.S. by Remington. Introduced 1993; still produced. **LMSR: $452**
   Perf.: $360   **Exc.:** $320   **VGood:** $280

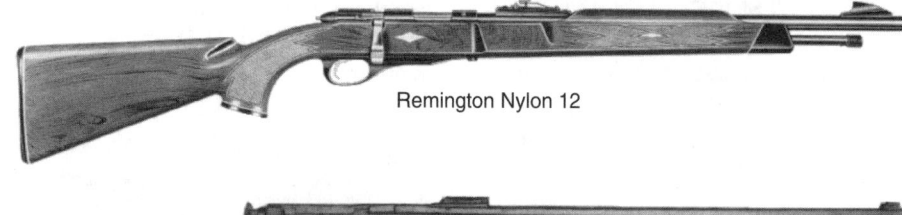

Remington Nylon 12

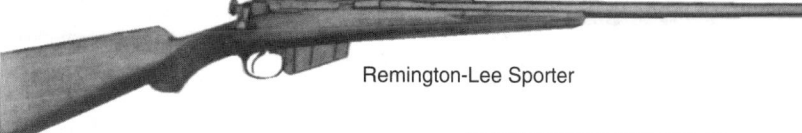

Remington-Lee Sporter

REMINGTON NYLON 10
Bolt action; single shot; 22 Short, 22 Long, 22 LR; 19 5/8" barrel; 38 1/2" overall length; weighs 4 1/2 lbs.; blade front sight, adjustable open rear; Mohawk brown nylon stock with white diamond inlay; checkered pistol grip, forearm. Introduced 1963; discontinued 1964.
   Perf.: $150   **Exc.:** $110   **VGood:** $90

Remington Nylon 10-C
Same specs as Nylon 10 except semi-automatic with 10-shot magazine. Introduced 1971; discontinued 1978.
   Perf.: $120   **Exc.:** $100   **VGood:** $80

REMINGTON NYLON 11
Bolt action; 22 Short, 22 Long, 22 LR; 6-, 10-shot clip-type magazines; 19 5/8" barrel; 38 1/2" overall length; weighs 4 1/2 lbs.; blade front sight, thumb-screw adjustable rear; Mohawk brown nylon stock with diamond inlay; checkered pistol grip, forearm grip cap, buttplate with white line spacers. Introduced 1962; discontinued 1964.
   Perf.: $160   **Exc.:** $135   **VGood:** $110

REMINGTON NYLON 12
Bolt action; 22 Short, 22 Long, 22 LR; 22-shot (22 Short), 17-shot (22 Long), 15-shot (22 LR) tube magazine; 19 5/8" barrel; 38 1/2" overall length; weighs 4 1/2 lbs.; blade front sight, thumb-screw adjustable rear; Mohawk brown nylon stock with diamond inlay; checkered pistol grip, forearm grip cap, buttplate with white line spacers. Introduced 1962; discontinued 1964.
   Perf.: $170   **Exc.:** $140   **VGood:** $120

REMINGTON-LEE SPORTER
Bolt-action; 6mm USN, 30-30, 30-40, 303 British, 7mm Mauser, 7.65 Mauser, 32 Spl., 32-40, 35 Spl., 38-55, 38-72, 405 Win., 43 Spanish, 45-70, 45-90; 5-shot detachable box magazine; 24", 26" barrel; open rear sight, bead front; hand-checkered American walnut pistol-grip stock. Collector value. Introduced 1886; discontinued 1906.
   Exc.: $1050   **VGood:** $795   **Good:** $595

Remington-Lee Sporter Deluxe
Same specs as Remington-Lee Sporter except half-octagon barrel; Lyman sights; deluxe walnut stock. Collector value. Introduced 1886; discontinued 1906.
   Exc.: $1395   **VGood:** $995   **Good:** $695

Remington Slide- & Lever-Action Rifles

Model Six
Model 12A
Model 12B
Model 12C
Model 12CS
Model 14A
Model 14R Carbine
Model 14½
Model 14½ Carbine
Model 25A
Model 25R Carbine
Model 76 Sportsman
Model 121A Fieldmaster
Model 121S
Model 121SB
Model 141A Gamemaster
Model 141R Carbine

Remington Model Six

Remington Model 12A

Remington Model 14½

Remington Model 25A

Remington Model 121A Fieldmaster

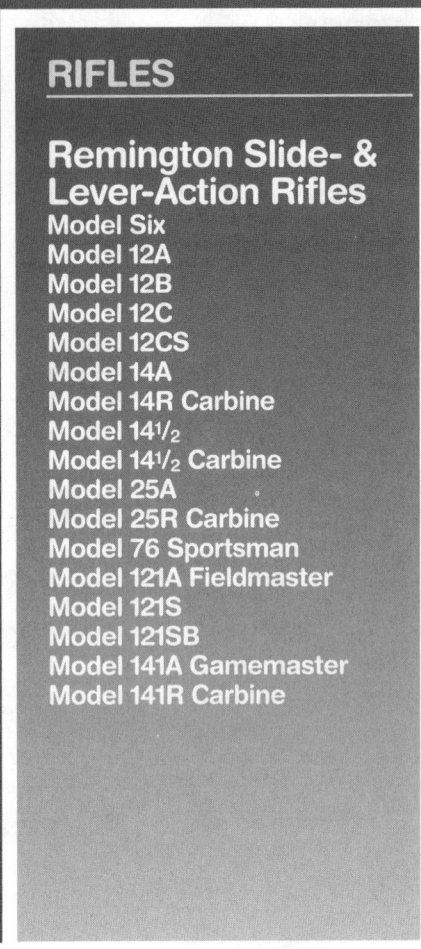

Remington Model 141A Gamemaster

REMINGTON SLIDE- & LEVER-ACTION RIFLES

REMINGTON MODEL SIX

Slide action; 6mm Rem., 243, 270, 308 Win., 308 Accelerator, 30-06, 30-06 Accelerator; 4-shot detachable clip magazine; 22″ round tapered barrel; 42″ overall length; weighs 7 1/2 lbs.; gold bead front sight on matted ramp, open adjustable sliding ramp rear; cut-checkered Monte Carlo walnut stock with full cheekpiece; cross-bolt safety; tapped for scope mount; cartridge head medallion on receiver bottom to denote caliber. Improved version of Model 760. Introduced 1981; discontinued 1988.

   Perf.: $395   **Exc.:** $350   **VGood:** $300
D Peerless Grade.
   Perf.: $1995   **Exc.:** $1795   **VGood:** $1095
F Premier Grade.
   Perf.: $4195   **Exc.:** $3795   **VGood:** $2695
F Premier Gold Grade.
   Perf.: $6495   **Exc.:** $5195   **VGood:** $3995

REMINGTON MODEL 12A

Slide action; 22 Short, 22 Long, 22 LR; 15-shot (22 Short), 12-shot (22 Long), 10-shot (22 LR) tube magazine; 22″ barrel; hammerless takedown; bead front sight, open rear; uncheckered straight stock; grooved slide handle. Introduced 1909; discontinued 1936.

   Exc.: $425   **VGood:** $350   **Good:** $300

Remington Model 12B

Same specs as Model 12A except 22 Short; octagonal barrel.

   Exc.: $440   **VGood:** $350   **Good:** $300

Remington Model 12C

Same specs as Model 12A except 24″ octagonal barrel; pistol-grip stock.

   Exc.: $460   **VGood:** $385   **Good:** $315

Remington Model 12CS

Same specs as Model 12A except 22 Rem. Spl. (22 WRF); 12-shot magazine; 14″ octagonal barrel; pistol-grip stock.

   Exc.: $450   **VGood:** $360   **Good:** $310

REMINGTON MODEL 14A

Slide action; 25, 30, 32, 35 Rem.; 5-shot tube magazine; 22″ barrel; hammerless; takedown; bead front sight, open rear; uncheckered walnut straight stock; grooved slide handle. Introduced 1912; discontinued 1935.

   Exc.: $370   **VGood:** $280   **Good:** $220

Remington Model 14R Carbine

Same specs as Model 14A except 18″ barrel.

   Exc.: $400   **VGood:** $330   **Good:** $285

REMINGTON MODEL 14½

Slide action; 38-40, 44-40; 11-shot tube magazine; 22 1/2″ barrel; hammerless; takedown; bead front sight, open rear; uncheckered walnut straight stock; grooved slide handle. Introduced 1912; discontinued 1925.

   Exc.: $630   **VGood:** $560   **Good:** $470

Remington Model 14½ Carbine

Same specs as Model 14 1/2 except 9-shot full-length tube magazine; 18 1/2″ barrel. Introduced 1912; discontinued 1925.

   Exc.: $660   **VGood:** $580   **Good:** $490

REMINGTON MODEL 25A

Slide action; 25-20, 32-20; 10-shot tube magazine; 24″ barrel; hammerless; takedown; blade front sight, open rear; uncheckered walnut pistol-grip stock; grooved slide handle. Introduced 1923; discontinued 1936.

   Exc.: $410   **VGood:** $345   **Good:** $300

Remington Model 25R Carbine

Same specs as Model 25A except 6-shot magazine; 18″ barrel; straight stock.

   Exc.: $495   **VGood:** $420   **Good:** $350

REMINGTON MODEL 76 SPORTSMAN

Slide action; 30-06; 4-shot magazine; 22″ barrel; 42″ overall length; weighs 7 1/2 lbs.; open adjustable sights; walnut-finished hardwood stock, forend; cartridge head inset on receiver bottom denotes caliber; cross-bolt safety; tapped for scope mounts. Introduced 1984; discontinued 1987.

   Perf.: $300   **Exc.:** $270   **VGood:** $225

REMINGTON MODEL 121A FIELDMASTER

Slide action; 22 Short, 22 Long, 22 LR; 20-shot (22 Short), 15-shot (22 Long), 14-shot (22 LR); 24″ round barrel; weighs 6 lbs.; hammerless; takedown; white metal bead ramp front sights, step-adjustable rear; uncheckered pistol-grip stock; grooved semi-beavertail slide handle. Introduced 1936; discontinued 1954.

   Exc.: $425   **VGood:** $325   **Good:** $250

Remington Model 121S

Same specs as Model 121A except 22 Rem. Spl.; 12-shot magazine.

   Exc.: $520   **VGood:** $430   **Good:** $335

Remington Model 121SB

Same specs as Model 121A except smoothbore barrel for 22 shot cartridge.

   Exc.: $620   **VGood:** $500   **Good:** $400

REMINGTON MODEL 141A GAMEMASTER

Slide action; 30, 32, 35 Rem.; 5-shot tube magazine; 24″ barrel; 42 3/4″ overall length; weighs 7 3/4 lbs.; hammerless; takedown; white metal bead ramp front sight, step-adjustable rear; uncheckered American walnut half-pistol-grip stock; grooved slide handle. Introduced 1936; discontinued 1950.

   Exc.: $400   **VGood:** $325   **Good:** $270

Remington Model 141R Carbine

Same specs as Model 141 except 30, 32 Rem., 18 1/2″ barrel.

   Exc.: $425   **VGood:** $350   **Good:** $285

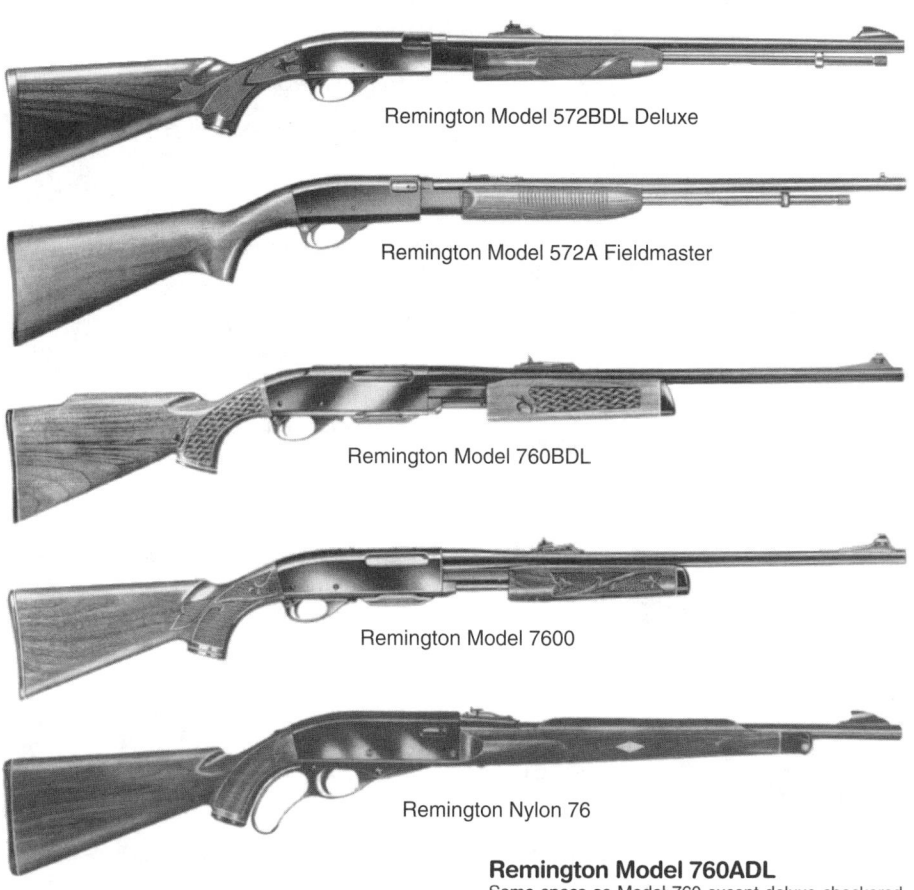

Remington Model 572BDL Deluxe

Remington Model 572A Fieldmaster

Remington Model 760BDL

Remington Model 7600

Remington Nylon 76

RIFLES

**Remington Slide- &
Lever-Action Rifles**
Model 572A Fieldmaster
Model 572 Fieldmaster
Model 572BDL Deluxe
Model 572SB
Model 760
Model 760ADL
Model 760BDL
Model 760 Carbine
Model 760CDL
Model 760D Peerless
Model 760F Premier
Model 760F Premier Gold
Model 7600
Model 7600 Carbine
Model 7600 Special Purpose
Model 7600D Peerless
Model 7600F Premier
Model 7600F Premier Gold
Nylon 76

REMINGTON MODEL 572A FIELDMASTER

Slide action; 22 Short, 22 Long, 22 LR; 20-shot (22 Short), 17-shot (22 Long), 15-shot (22 LR) tube magazine; 25″ barrel; weighs 5 ¹/₂ lbs.; hammerless; ramp front sight, open rear; uncheckered hardwood pistol-grip stock, grooved forearm. Introduced 1955; discontinued 1988.

Perf.: $160   **Exc.:** $150   **VGood:** $120

Remington Model 572 Fieldmaster

Same specs as Model 572A except 23″ barrel; weighs 4 lbs.; "Sun-Grain" stock, forend in Buckskin tan or Crow-Wing black colors; chrome-plated magazine tube, trigger, floorplate.

Perf.: $225   **Exc.:** $175   **VGood:** $125

Remington Model 572BDL Deluxe

Same specs as Model 572A except pistol-grip cap; RKW wood finish; checkered grip, slide handle; adjustable rear sight, ramp front. Still in production.
LMSR: $332
Perf.: $250   **Exc.:** $200   **VGood:** $150

Remington Model 572SB

Same specs as Model 572A except smoothbore barrel for 22 shot cartridge. No longer produced.
Perf.: $370   **Exc.:** $300   **VGood:** $220

REMINGTON MODEL 760

Slide action; 222 Rem., 223 Rem., 6mm Rem., 243 Win., 244 Rem., 257 Roberts, 270 Win., 280 Rem., 30-06, 300 Savage, 308 Win., 35 Rem.; 4-shot magazine; 22″ barrel; 42 ¹/₄″ overall length; weighs 7 ¹/₂ lbs.; hammerless; white metal bead front sight on ramp, step-adjustable, semi-buckhorn rear. Models from mid-'60s to early '70s with impressed checkering; others hand-checkered on pistol grip, slide handle; early versions with grooved slide handle, no checkering on stock. Introduced 1952; discontinued 1980.

Perf.: $375   **Exc.:** $320   **VGood:** $250
222 Rem., 223 Rem.
Perf.: $1295   **Exc.:** $995   **VGood:** $695
257 Roberts.
Perf.: $810   **Exc.:** $670   **VGood:** $475

Remington Model 760ADL

Same specs as Model 760 except deluxe checkered stock, forend; grip cap; sling swivels; standard or high comb. Introduced 1953; discontinued 1963.
Perf.: $395   **Exc.:** $320   **VGood:** $270

Remington Model 760BDL

Same specs as Model 760 except 270, 30-06, 308; basketweave checkering on pistol grip, forearm; black forearm tip; early versions in right- or left-hand styles.
Perf.: $400   **Exc.:** $350   **VGood:** $300

Remington Model 760 Carbine

Same specs as Model 760 except 270, 280, 30-06, 308, 35 Rem.; 18 ¹/₂″ barrel; 38 ¹/₂″ overall length; weighs 6 ¹/₂ lbs.
Perf.: $430   **Exc.:** $370   **VGood:** $300

Remington Model 760CDL

Same specs as Model 760 Carbine except checkered stock, forend; decorated receiver.
Perf.: $435   **Exc.:** $375   **VGood:** $310

Remington Model 760D Peerless

Same specs as Model 760 except fine checkered fancy stock; engraving.
Perf.: $1000   **Exc.:** $850   **VGood:** $650

Remington Model 760F Premier

Same specs as Model 760 except figured American walnut best-grade stock; fine checkered stock, forend; extensive engraved game scenes.
Perf.: $2420   **Exc.:** $1910   **VGood:** $1500

Remington Model 760F Premier Gold

Same specs as Model 760 except best-grade fine checkered fancy stock; extensive engraving with gold inlay.
Perf.: $5000   **Exc.:** $4000   **VGood:** $3000

REMINGTON MODEL 7600

Slide action; 6mm Rem., 243, 270, 280, 30-06, 30-06 Accelerator, 308, 308 Accelerator, 35 Whelen; 4-shot detachable magazine; 22″ barrel; 42″ overall length; weighs 7 ¹/₂ lbs.; removable gold bead ramp front, windage-adjustable sliding ramp rear sight; tapped for scope mount; impressed checkered American walnut

pistol-grip stock; positive cross-bolt safety. Introduced 1981; still in production. **LMSR: $513**
Perf.: $425   **Exc.:** $350   **VGood:** $275

Remington Model 7600 Carbine

Same specs as Model 7600 except 30-06; 18 ¹/₂″ barrel; weighs 7¹/₄ lbs.
Perf.: $425   **Exc.:** $350   **VGood:** $275

Remington 7600 Special Purpose

Same specs as Model 7600 except 270, 30-06; non-glare finished American walnut stock; all exposed metal with non-reflective matte black finish; quick-detachable sling swivels; camo-pattern Cordura carrying sling. Made in U.S. by Remington. Introduced 1993; discontinued 1994.
Perf.: $400   **Exc.:** $360   **VGood:** $280

Remington 7600D Peerless

Same specs as Model 7600 except fine checkered fancy walnut stock; engraving.
Perf.: $2100   **Exc.:** $1600   **VGood:** $1200

Remington 7600F Premier

Same specs as Model 7600 except best-grade fine checkered walnut stock; extensive engraving.
Perf.: $4600   **Exc.:** $4000   **VGood:** $2600

Remington 7600F Premier Gold

Same specs as Model 7600 except best-grade fine checkered walnut stock; extensive engraving with gold inlays.
Perf.: $7000   **Exc.:** $5000   **VGood:** $3200

REMINGTON NYLON 76

Lever-action; 22 LR; 14-shot tube magazine in butt-stock; 19 ¹/₂″ barrel; weighs 4 ¹/₂ lbs.; blade front sight, open rear; receiver grooved for tip-off mounts; moulded nylon stock with checkered pistol grip, forearm; available in two stock colors: Mohawk brown and Apache black; latter has chrome-plated receiver. Introduced 1962; discontinued 1964.
Perf.: $200   **Exc.:** $150   **VGood:** $120

RIFLES

Remington Single Shot Rifles

Baby Carbine
Model 1902 (See Remington
No. 5 Military)
No. 2
No. 3
No. 3 High Power
No. 4
No. 4-S Military
No. 5
No. 5 Military
No. 6
No. 7

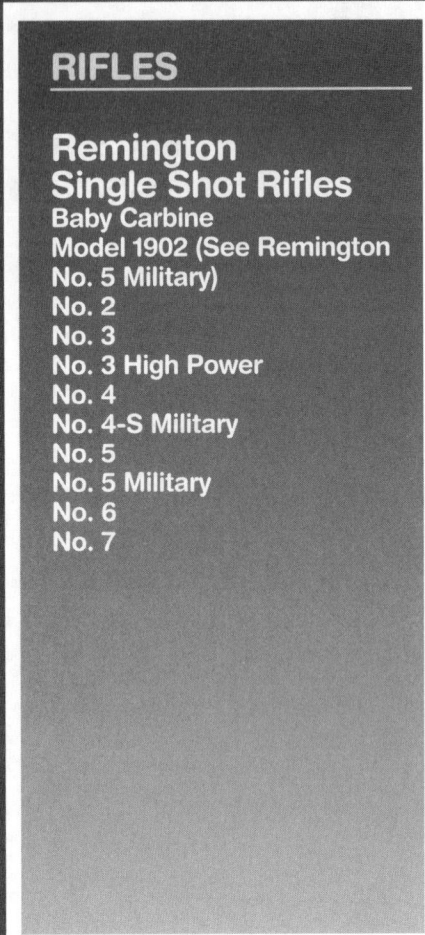

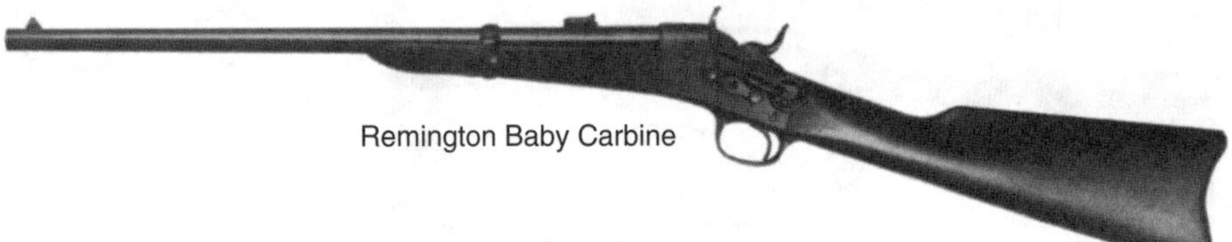

Remington Baby Carbine

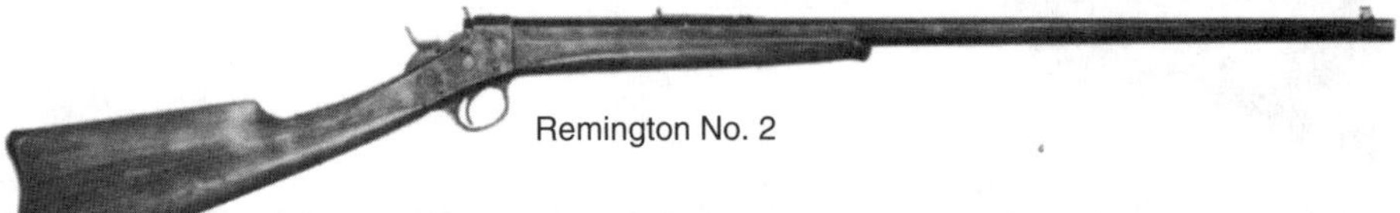

Remington No. 2

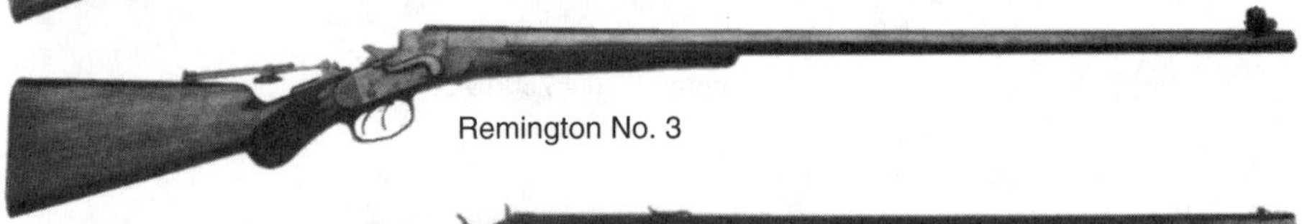

Remington No. 3

Remington No. 4

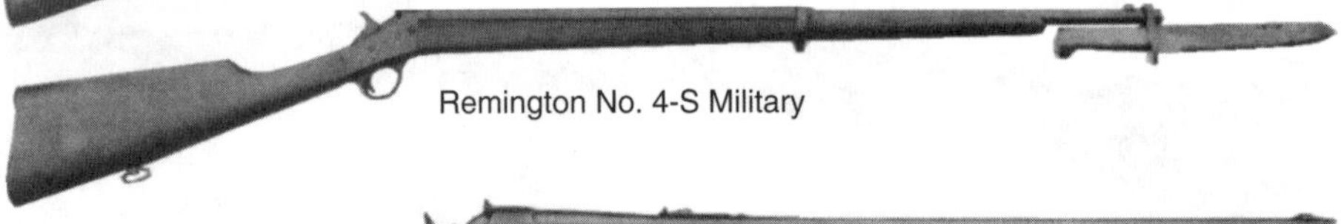

Remington No. 4-S Military

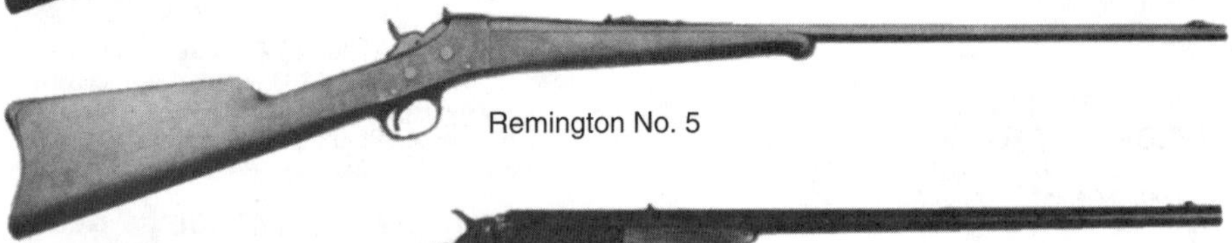

Remington No. 5

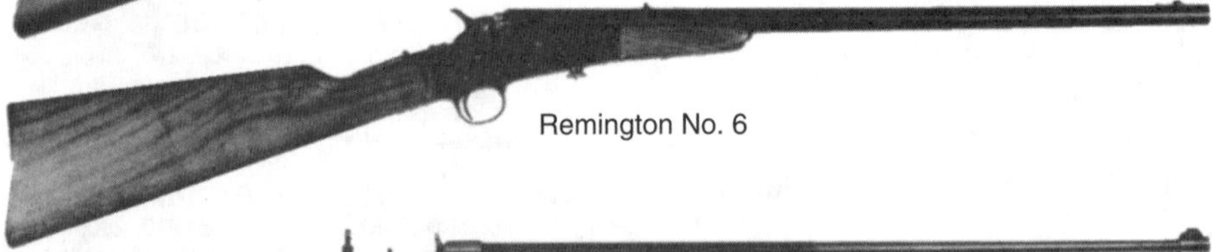

Remington No. 6

Remington No. 7

REMINGTON SINGLE SHOT RIFLES

REMINGTON BABY CARBINE

Single shot; rolling block; 44 Win.; 20″ barrel; open rear sight, blade front; uncheckered American walnut carbine-style straight stock, forearm; barrel band. Introduced 1902; discontinued 1902.

Perf.: $3000   **Exc.:** $2500   **VGood:** $1700

REMINGTON NO. 2

Single shot; rolling block; 22, 25, 32, 38, 44 rimfire or centerfire; 24″, 26″, 28″, 30″ barrel; open rear sight, bead front; American walnut straight-grip stock, Schnabel forearm. Collector value. Introduced 1873; discontinued 1910.

Exc.: $795   **VGood:** $650   **Good:** $400

REMINGTON NO. 3

Single shot; falling block side-lever; 22 WCF, 22 Extra Long, 25-20 Stevens, 25-21 Stevens, 25-25 Stevens, 32 WCF, 32 Ballard & Marlin, 32-40 Rem., 38 WCF, 38-40 Rem., 38-50 Rem., 38-55 Ballard & Marlin, 40-60 Ballard & Marlin, 40-60 WCF, 45-60 Rem., 40-82 WCF, 45-70 Govt., 45-90 WCF; 28″, 30″ half- or full-octagon barrel; open rear sight, blade front; hand-checkered pistol-grip stock. Collector value. Introduced 1880; discontinued 1911.

Exc.: $1300   **VGood:** $1100   **Good:** $800

Remington No. 3 High Power

Same specs as Model No. 3 except 30-30, 30-40, 32 Spl., 32-40, 38-55, 38-72; open rear sight, bead front; hand-checkered pistol-grip stock, forearm. Introduced 1893; discontinued 1907. Collector value.

Exc.: $1350   **VGood:** $1150   **Good:** $850

REMINGTON NO. 4

Single shot; rolling block; 22 Short, 22 Long, 22 LR, 25 Stevens rimfire, 32 Short, 32 Long, 32 Rimfire; 22 ½″, 24″ (32 Rimfire) octagonal barrel; blade front sight, open rear; plain walnut stock, forearm; solid-frame and takedown models. 1890 to 1901, solid frame; 1901 to 1926 first takedown with lever on right side; 1926 to 1933, second takedown with large screw head. Introduced 1890; discontinued 1933.

Exc.: $495   **VGood:** $325   **Good:** $200

Remington No. 4-S Military

Same specs as No. 4 except 22 Short, 22 Long, 22 LR; 26″ barrel; military stock, including stacking swivel, sling; bayonet stud on barrel; bayonet, scabbard originally included. Early production was marked "American Boy Scout"; later production was marked "Military Model." This rifle was never the officially adopted arm of the Boy Scouts of America; the reasons behind the early production markings are still unknown. Collector value. Introduced 1913; discontinued 1923.

Exc.: $1050   **VGood:** $895   **Good:** $575

REMINGTON NO. 5

Single shot; rolling block; 7mm Mauser, 30-30, 30-40 Krag, 303 British, 32-40, 32 Spl., 38-55; 24″, 26″, 28″ barrel; open sporting rear sight, blade front; uncheckered straight-grip stock. Some collector value. Introduced 1897; discontinued 1905.

Exc.: $850   **VGood:** $595   **Good:** $395

Remington No. 5 Military

Same specs as No. 5 except only 8mm Lebel, 7mm Mauser, 7,62 Russian, 30 U.S.; 30″ barrel. Also known as Model 1902.

Exc.: $850   **VGood:** $595   **Good:** $395

REMINGTON NO. 6

Single shot; rolling block; 22 Short, 22 Long, 22 LR, 32 Rimfire Short, 32 Rimfire Long; 20″ barrel; takedown; open front, rear sights; tang peep sight optional; plain straight stock, forearm. Introduced 1901; discontinued 1933.

Exc.: $375   **VGood:** $325   **Good:** $250

REMINGTON NO. 7

Single shot; 22 Short, 22 Long, 22 LR, 22 Stevens RF; other calibers on special order; 24″, 26″, 28″ half-octagon barrel; Lyman combination rear sight, Beach combination front; hand-checkered American walnut stock, forearm. Collector value. Introduced 1903; discontinued 1911.

Exc.: $2695   **VGood:** $2195   **Good:** $1795

Rigby Second Quality

Rigby 350 Magnum

Rigby Standard Bolt Action

Rigby Best Quality

Rossi Model 62 SAC Carbine

Rossi Model 62 SA

RIFLES

Rigby
350 Magnum
416 Magnum
Best Quality
Big Game Bolt Action
Second Quality
Standard Bolt Action
Standard Lightweight
Third Quality

Ross
Model 1905 Mark II
Model 1910 Mark III

Rossi
Model 59 SA
Model 62 SA
Model 62 SAC Carbine
Model 62 SA Stainless
Model 65 SRC

REMINGTON-STYLE ROLLING BLOCK CARBINE
Single shot; rolling block; 45-70; 30" octagonal barrel; 46 1/2" overall length; weighs 11 3/4 lbs.; blade front sight, adjustable rear; walnut stock and forend; color case-hardened receiver, brass trigger guard, buttplate and barrel band, blued barrel. Imported from Italy by E.M.F. No longer imported.
Perf.: $495   **Exc.:** $395   **VGood:** $250

RIGBY NOTE:
Rifles should be individually appraised as caliber and options can affect values.

RIGBY 350 MAGNUM
Bolt action; 350 Mag.; 24" barrel; 5-shot box magazine; high-quality walnut stock with checkered full pistol grip, forearm; folding leaf rear sight, bead front. No longer in production.
Perf.: $3995   **Exc.:** $3400   **VGood:** $2600

RIGBY 416 MAGNUM
Bolt action; 375 H&H, 404, 416 Rigby, 416 Rem. Mag., 458 Win. Mag., 505; 24" barrel; 4-shot box magazine; walnut sporting stock with checkered pistol grip, forearm; folding leaf rear sight, bead front. Still in production.
Perf.: $9250   **Exc.:** $7995   **VGood:** $5495

RIGBY BEST QUALITY
Double rifle; sidelock; 22 LR, 275 Mag., 350 Mag., 416 Rigby, 465, 470 NE, 577 NE, 600 NE; 24"-28" barrel; hammerless ejector; folding leaf rear sight, bead front; hand-checkered walnut stock; pistol-grip forearm; engraved receiver. Still in production.
Perf.: $47,000   **Exc.:** $32,500   **VGood:** $25,000

RIGBY BIG GAME BOLT ACTION
Bolt action; 350 Mag., 375 H&H Mag., 404, 416 Rigby, 416 Rem. Mag., 458 Win. Mag., 505; 4-, 5-shot magazine; 21"-24" barrel; folding leaf rear sight, bead front; high-quality walnut sporting stock with checkered pistol grip, forearm. Still in production.
Perf.: $8500   **Exc.:** $7600   **VGood:** $4900

RIGBY SECOND QUALITY
Double rifle; boxlock; 22 LR, 275 Mag., 350 Mag., 416 Rigby, 458 Win. Mag., 465, 470 NE, 577 NE, 600 NE; 22"-26" barrel; hammerless ejector; folding leaf rear sight, bead front; hand-checkered walnut stock; pistol grip, forearm; engraved receiver. Still in production.
Perf.: $16,500   **Exc.:** $12,995   **VGood:** $9995

RIGBY STANDARD BOLT ACTION
Bolt action; 243, 270, 275 Rigby, 7x57, 30-06, 300 H&H Mag., 300 Win. Mag., 308, 7mm Rem. Mag., 375 H&H Mag., 404 Rigby, 458 Win. Mag.; 3-, 4-, 5-shot magazine; 20" barrel; folding leaf rear sight, bead front; walnut sporting stock with hand-checkered half pistol grip, forearm. Still in production.
Perf.: $9995   **Exc.:** $7500   **VGood:** $5250
Magnum calibers.
Perf.: $10,500   **Exc.:** $7995   **VGood:** $5650

Rigby Standard Lightweight
Same specs as Standard Bolt Action except 24" barrel; lighter weight. Still in production.
Perf.: $9995   **Exc.:** $7500   **VGood:** $5250
Magnum calibers.
Perf.: $10,500   **Exc.:** $7995   **VGood:** $5650

RIGBY THIRD QUALITY
Double rifle; boxlock; 22 LR, 275 Mag., 350 Mag., 416 Rigby, 458 Win. Mag., 465, 470 NE, 577 NE, 600 NE; 22"-26" barrel; hammerless ejector; folding leaf rear sight, bead front; checkered walnut stock; engraved receiver. No longer in production.
Perf.: $12,000   **Exc.:** $10,000   **VGood:** $7500

ROSS MODEL 1905 MARK II
Bolt action; 303 British; 4-shot magazine; 28" barrel; two-leaf open rear sight, bead front; hand-checkered American walnut sporting stock; straight-pull action. Introduced prior to WWII; no longer in production.
Perf.: $375   **Exc.:** $295   **VGood:** $225

ROSS MODEL 1910 MARK III
Bolt action; 280 Ross, 303 British; 4- or 5-shot magazine; 22", 24", 26" barrel; two-leaf open rear sight, bead front; hand-checkered American walnut sporting stock. Made in Canada. Some collector value. May be unsafe to fire. Introduced 1910; discontinued 1918.
Perf.: $375   **Exc.:** $325   **VGood:** $275

ROSSI MODEL 59 SA
Slide action; 22 WMR; 23" round or octagonal barrel; 39 1/4" overall length; weighs 5 3/4 lbs.; takedown; blade front sight, adjustable rear; walnut stock with straight grip, grooved forend; blue finish. Imported from Brazil by Interarms. Introduced by Interarms 1978; no longer imported.
Perf.: $220   **Exc.:** $170   **VGood:** $125

ROSSI MODEL 62 SA
Slide action; 22 Short, 22 Long, 22 LR; 20-shot (22 Short), 16-shot (22 Long), 14-shot (22 LR) tube magazine; 23" round or octagon barrel; 39 1/4" overall length; weighs 5 3/4 lbs.; takedown; blade front sight, adjustable rear; walnut stock with straight grip, grooved forend; blued or nickel finish. Imported from Brazil. Introduced by Interarms 1978; still imported.
LMSR: $226
Perf.: $180   **Exc.:** $150   **VGood:** $120
Nickel.
Perf.: $200   **Exc.:** $170   **VGood:** $130
Blue, with octagonal barrel.
Perf.: $200   **Exc.:** $170   **VGood:** $130

Rossi Model 62 SAC Carbine
Same specs as Model 62 SA except 16 1/4" barrel. Introduced 1975; still in production. **LMSR:** $226
Perf.: $180   **Exc.:** $150   **VGood:** $120

Rossi Model 62 SA Stainless
Same specs as Model 62 SA except stainless steel barrel and action. Introduced 1986; discontinued 1986.
Perf.: $180   **Exc.:** $150   **VGood:** $120

ROSSI MODEL 65 SRC
Lever action; 44 Spl./44 Mag., 44-40; 10-shot tube magazine; 20" barrel; weighs 5 3/4 lbs.; blade front sight, buckhorn rear; uncheckered walnut stock; saddle ring. Introduced 1989; still in production. **LMSR:** $364
Perf.: $290   **Exc.:** $250   **VGood:** $200

Rossi
Model 92 SRC
Model 92 SRC Puma Short
Carbine

Ruger
(See Sturm Ruger)

Sako
Classic
Fiberglass
Fiberglass Carbine
Finnbear Sporter
Finnfire
Finnsport Model 2700
Finnwolf
Forester Sporter
Forester Sporter Carbine
Forester Sporter Heavy Barrel
Hunter
Hunter Carbine
Hunter LS

Rossi Model 92 SRC

Sako Fiberglass

Sako Finnsport Model 2700

Sako Classic

Sako Finnfire

Sako Finnbear Sporter

Rossi Model 92 SRS Puma
Short Carbine

ROSSI MODEL 92 SRC
Lever action; 38 Spl., 357 Mag., 44-40, 44 Mag.; 9-shot (38 Spl.), 8-shot (other calibers); 20″ barrel; blade front sight, buckhorn rear; uncheckered walnut stock. Made in Brazil. Introduced 1978; still imported. **LMSR: $347**
Perf.: $280 **Exc.:** $225 **VGood:** $180
Engraved model.
Perf.: $350 **Exc.:** $300 **VGood:** $250

Rossi Model 92 SRS Puma Short Carbine
Same specs as Model 92 except 38-357; 16″ barrel; 33″ overall length. Early production with Puma medallion on side of receiver. Imported from Brazil by Interarms. Introduced 1986; still imported. **LMSR: $347**
Perf.: $300 **Exc.:** $260 **VGood:** $200

SAKO CLASSIC
Bolt action; 243, 270, 30-06, 7mm Rem. Mag.; 21 3/4″ barrel; weighs 6 lbs.; classic-style stock with straight comb; matte finish wood. Imported from Finland by Stoeger. Introduced 1993; still imported. **LMSR: $1030**
Perf.: $800 **Exc.:** $680 **VGood:** $530

SAKO FIBERGLASS
Bolt action; 22-250, 243, 308, 7mm-08, 25-06, 270, 30-06, 7mm Rem. Mag., 300 Win. Mag., 338 Win. Mag., 375 H&H Mag.; 4-shot (Magnums), 5-shot magazine (standard calibers); 23″ barrel; no sights; marketed with scope mounts; black fiberglass stock; wrinkle finish; rubber buttpad. Made in Finland. Introduced 1985; still in production. **LMSR: $1360**
Perf.: $1050 **Exc.:** $800 **VGood:** $600
Left-hand model.
Perf.: $1100 **Exc.:** $1000 **VGood:** $800

SAKO Fiberglass Carbine
Same specs as Fiberglass except 243, 308, 25-06, 270, 30-06, 7mm Rem. Mag., 300 Win. Mag., 375 H&H Mag.; 18 1/2″ barrel. Introduced 1986; discontinued 1991.
Perf.: $1050 **Exc.:** $800 **VGood:** $600

SAKO FINNBEAR SPORTER
Bolt action; 25-06, 264 Win. Mag., 270, 30-06, 300 Win. Mag., 338 Mag., 7mm Rem. Mag., 375 H&H Mag.; 4-shot (Magnums), 5-shot magazine (standard calibers); 24″ barrel; no rear sight, hooded ramp front; drilled, tapped for scope mounts; hand-checkered European walnut, Monte Carlo, pistol-grip stock; recoil pad; sling swivels. Introduced 1961; discontinued 1971.
Perf.: $995 **Exc.:** $795 **VGood:** $650

SAKO FINNFIRE
Bolt action; 22 LR; 5-shot magazine; 22″ barrel; 40″ overall length; weighs 5 1/4 lbs.; hooded blade front sight, open adjustable rear; European walnut stock with checkered grip and forend; adjustable single-stage trigger; 50-degree bolt lift. Imported from Finland by Stoeger Industries. Introduced 1994; still imported. **LMSR: $685**
Perf.: $700 **Exc.:** $600 **VGood:** $450

SAKO FINNSPORT MODEL 2700
Bolt action; 270, 300 Win. Mag.; 4-shot magazine; 21″, 22″ barrel; no sights; scope mounts; Monte Carlo stock design. Introduced 1983; no longer in production.
Perf.: $795 **Exc.:** $675 **VGood:** $550

SAKO FINNWOLF
Lever action; 243 Win., 308 Win.; 3-, 4-shot magazine (early models); 23″ barrel; hammerless; no rear sight, hooded ramp front; drilled, tapped for scope mounts; hand-checkered European walnut Monte Carlo stock in left- or right-hand styling; sling swivels. Introduced 1964; discontinued 1972.
Perf.: $800 **Exc.:** $700 **VGood:** $500

SAKO FORESTER SPORTER
Bolt action; 22-250, 243 Win., 308 Win.; 5-shot magazine; 23″ barrel; no rear sight, hooded ramp front; drilled, tapped for scope mounts; hand-checkered walnut Monte Carlo pistol-grip stock; sling swivel studs. Introduced 1957; discontinued 1971.
Perf.: $950 **Exc.:** $775 **VGood:** $600

Rossi Model 92 SRS Puma Short Carbine

SAKO Forester Sporter Carbine
Same specs as Forester Sporter except 20″ barrel; Mannlicher-type stock.
Perf.: $1195 **Exc.:** $925 **VGood:** $695

SAKO Forester Sporter Heavy Barrel
Same specs as Forester Sporter except 24″ heavy barrel.
Perf.: $895 **Exc.:** $695 **VGood:** $595

SAKO HUNTER
Bolt action; 17 Rem., 222 Rem., 223 Rem., 22-250, 243, 308, 7mm-08, 25-06, 270, 280, 30-06, 7mm Rem. Mag., 300 Win. Mag., 300 Weatherby Mag., 338 Win. Mag., 375 H&H Mag., 416 Rem. Mag.; 4-shot (Magnums), 5-shot magazine (standard calibers); 21 1/4″, 21 3/4″, 22″ barrel; no sights; scope mounts; checkered classic-style French walnut stock; sling swivels. Made in Finland. Introduced 1986; still in production. **LMSR: $1030**
Perf.: $795 **Exc.:** $675 **VGood:** $495
Left-hand model.
Perf.: $850 **Exc.:** $750 **VGood:** $575

SAKO Hunter Carbine
Same specs as Hunter except 22-250, 243, 7mm-08, 308, 300 Win. Mag., 25-06, 6.5x55, 270, 7x64, 30-06, 7mm Rem. Mag., 375 H&H Mag.; 18 1/2″ barrel; oil-finished stock. Introduced 1986; discontinued 1991.
Perf.: $775 **Exc.:** $650 **VGood:** $450

SAKO Hunter LS
Same specs as Hunter except laminated stock with dull finish. Introduced 1987; still in production. **LMSR: $1110**
Perf.: $975 **Exc.:** $795 **VGood:** $595
Left-hand model.
Perf.: $995 **Exc.:** $825 **VGood:** $595

Sako Forester Sporter

Sako Finnwolf

Sako Model 74

Sako Hunter LS

Sako Lightweight Deluxe

Sako Mauser

RIFLES

Sako
Lightweight Deluxe
Mannlicher Carbine
Mauser
Mauser Magnum
Model 72
Model 73
Model 74
Model 74 Carbine
Model 78
PPC Model
Safari Grade
Standard Sporter

SAKO LIGHTWEIGHT DELUXE

Bolt action; 17 Rem., 222, 223, 22-250, 243, 308, 7mm-08, 25-06, 270, 280 Rem., 30-06, 7mm Rem. Mag., 300 Win. Mag., 338 Win. Mag., 300 Wea., 375 H&H, 416 Rem. Mag.; 21", 22" barrel; select wood with skip-line checkering; rosewood grip cap and forend tip; ventilated recoil pad; fine checkering on top surfaces of integral dovetail bases, bolt sleeve, bolt handle root and bolt knob; mirror finish bluing. Imported from Finland by Stoeger Industries. Still imported. **LMSR: $1445**
Standard calibers

   Perf.: $1125   **Exc.:** $925   **VGood:** $675
Magnum calibers

   Perf.: $1175   **Exc.:** $950   **VGood:** $695

SAKO MANNLICHER CARBINE

Bolt action; 17 Rem., 222, 223, 243, 25-06, 270, 308, 30-06, 7mm Rem. Mag., 300 Win. Mag., 375 H&H Mag.; 4-shot (Magnums), 5-shot magazine (standard calibers); 18 1/2" barrel; open sights; full Mannlicher-style stock. Made in Finland. Introduced 1977; still in production. **LMSR: $1245**

   Perf.: $995   **Exc.:** $795   **VGood:** $595
Magnum calibers.

   Perf.: $1050   **Exc.:** $850   **VGood:** $650

SAKO MAUSER

Bolt action; 270, 30-06; 5-shot magazine; 24" barrel; open leaf rear sight, Patridge front; hand-checkered European walnut, Monte Carlo cheekpiece stock; sling swivel studs. Introduced 1946; discontinued 1961.

   Perf.: $695   **Exc.:** $550   **VGood:** $450

SAKO Mauser Magnum

Same specs as Mauser except 8x60S, 8,2x57, 300 H&H Mag., 375 H&H Mag.; recoil pad. Discontinued 1961.

   Perf.: $775   **Exc.:** $650   **VGood:** $560

SAKO MODEL 72

Bolt action; 222 Rem., 223 Rem., 22-250, 243 Win., 25-06, 270 Win., 30-06, 308, 7mm Rem. Mag., 300 Win. Mag., 338 Win. Mag., 375 H&H Mag.; 23", 24" barrel; adjustable rear sight, hooded front; hand-checkered European walnut stock; adjustable trigger; hinged floorplate. Introduced 1973; discontinued 1976.

   Perf.: $595   **Exc.:** $530   **VGood:** $470

SAKO MODEL 73

Lever action; 243 Win., 308 Win.; 3-shot clip magazine; 23" barrel; no rear sight, hooded ramp front; drilled, tapped for scope mounts; hand-checkered European walnut Monte Carlo stock; sling swivels; flush floorplate. Introduced 1973; discontinued 1975.

   Perf.: $595   **Exc.:** $495   **VGood:** $420

SAKO MODEL 74

Bolt action; 222 Rem., 223 Rem., 220 Swift, 22-250, 243, 25-06, 270, 7mm Rem. Mag., 30-06, 300 Win. Mag., 338 Win. Mag., 375 H&H Mag.; 4-shot (Magnums), 5-shot (standard calibers) magazine; 23 1/2", 24" standard or heavy barrel; no sights; hand-checkered European walnut Monte Carlo stock; Mauser-type action; detachable sling swivels. Introduced 1974; no longer in production.

   Perf.: $650   **Exc.:** $575   **VGood:** $450

SAKO Model 74 Carbine

Same specs as Model 74 except only 30-06; 20" barrel; Mannlicher-design full-length stock. Introduced 1974; no longer in production.

   Perf.: $725   **Exc.:** $595   **VGood:** $495

SAKO MODEL 78

Bolt action; 22 WMR, 22 LR, 22 Hornet; 4-shot (22 Hornet), 5-shot magazine; 22 1/2" barrel; no sights; hand-checkered European walnut Monte Carlo stock. Introduced 1977; no longer in production.
Rimfire.

   Perf.: $495   **Exc.:** $420   **VGood:** $350
22 Hornet.

   Perf.: $550   **Exc.:** $480   **VGood:** $420

SAKO PPC MODEL

Bolt action; single shot; 22 PPC, 6mm PPC; 5-shot magazine; 21 3/4≤, 23 3/4≤ (Benchrest) barrel; weighs 8 1/2 lbs. (heavy barrel); checkered walnut stock; beavertail forend; rosewood pistol grip, forearm (Deluxe). Introduced 1989; no longer in production.
Hunter model.

   Perf.: $1100   **Exc.:** $900   **VGood:** $650
Deluxe model.

   Perf.: $1350   **Exc.:** $1050   **VGood:** $900
Benchrest model.

   Perf.: $995   **Exc.:** $895   **VGood:** $695

SAKO SAFARI GRADE

Bolt action; 338 Win. Mag., 375 H&H Mag., 416 Rem. Mag.; 4-shot magazine; 22≤ barrel; quarter-rib "express" rear sight, hooded ramp front; French walnut stock; checkered 20 lpi, solid rubber buttpad; grip cap and forend tip; front sling swivel band-mounted on barrel. Imported from Finland by Stoeger Industries. Still imported.
LMSR: $2715

   Perf.: $2100   **Exc.:** $1750   **VGood:** $1100

SAKO STANDARD SPORTER

Bolt action; 17 Rem., 222, 223 (short action); 22-250, 220 Swift, 243, 308 (medium action); 25-06, 270, 30-06, 7mm Rem. Mag., 300 Win. Mag., 338 Win. Mag., 375 Mag., 375 H&H Mag. (long action); 23", 24" barrel; no sights; hand-checkered European walnut pistol-grip stock; hinged floorplate; adjustable trigger. Imported from Finland. Introduced 1978; no longer imported.
Short action.

   Perf.: $750   **Exc.:** $650   **VGood:** $450
Medium action.

   Perf.: $750   **Exc.:** $650   **VGood:** $450
Long action.

   Perf.: $750   **Exc.:** $650   **VGood:** $450

Sako

Standard Sporter Carbine
Standard Sporter Classic
Standard Sporter Deluxe
Standard Sporter Heavy Barrel
Standard Sporter Safari
Standard Sporter Super Deluxe
Standard Sporter Varmint Heavy Barrel
TRG-21
TRG-S
Vixen Sporter
Vixen Sporter Carbine
Vixen Sporter Heavy Barrel

Sauer

202 TR Target
Drilling

Sako Standard Sporter

Sako Standard Sporter Safari

Sako Standard Sporter Deluxe

Sako TRG-S

Sako TRG-21

SAKO Standard Sporter Carbine
Same specs as Standard Sporter except 222, 243, 270, 30-06; 20″ barrel; full Mannlicher-type stock. Introduced 1977; no longer imported.
Perf.: $795   **Exc.:** $625   **VGood:** $525

SAKO Standard Sporter Classic
Same specs as Standard Sporter except 243, 270, 30-06, 7mm Rem. Mag.; receiver drilled, tapped for scope mounts; straight-comb stock with oil finish; solid rubber recoil pad; recoil lug. Introduced 1980; no longer imported.
Perf.: $695   **Exc.:** $575   **VGood:** $495

SAKO Standard Sporter Deluxe
Same specs as Standard Sporter except select wood; rosewood pistol-grip cap, forend tip; metal checkering on dovetail bases, bolt sleeve, bolt handle; ventilated recoil pad; skip-line checkering. No longer imported.
Perf.: $795   **Exc.:** $625   **VGood:** $525

SAKO Standard Sporter Heavy Barrel
Same specs as Standard Sporter except beavertail forend; made with short, medium actions only.
Perf.: $695   **Exc.:** $595   **VGood:** $495

SAKO Standard Sporter Safari
Same specs as Standard Sporter except long action; 7mm Rem. Mag., 300 Win. Mag., 338 Win Mag., 375 H&H Mag.; quarter-rib express rear sights, hooded ramp front; hand-checkered European walnut stock, solid rubber recoil pad; pistol-grip cap, forend tip; front sling swivel mounted on barrel. No longer imported.
Perf.: $895   **Exc.:** $695   **VGood:** $525

SAKO Standard Sporter Super Deluxe
Same specs as Standard Sporter Deluxe except select European walnut stock with deep-cut oak leaf carving; rosewood pistol-grip cap, forend tip; high-gloss finish; metal checkering on dovetail bases, bolt sleeve, bolt handle; metal has super-high polish. Still imported. **LMSR: $3030**
Perf.: $2500   **Exc.:** $2000   **VGood:** $1500

SAKO Standard Sporter Varmint Heavy Barrel
Same specs as Standard Sporter Super Deluxe except 17 Rem., 222, 223 (short action), 22-250, 243, 308, 7mm-08 (medium action); 5-shot magazine; weighs from 8 1/4-8 1/2 lbs.; beavertail forend. Imported from Finland by Stoeger Industries. **LMSR: $1215**
Perf.: $1000   **Exc.:** $800   **VGood:** $650

SAKO TRG-21
Bolt action; 308 Win.; 10-shot magazine; free-floating 25 3/4″ heavy stainless barrel; 46 1/2″ overall length; weighs 10 1/2 lbs.; no sights; optional quick-detachable, one-piece scope mount base, 1″ or 30mm rings; reinforced polyurethane stock with fully-adjustable cheekpiece and buttplate; resistance-free bolt; 60-degree bolt lift; two-stage trigger adjustable for length, pull, horizontal or vertical pitch. Imported from Finland by Stoeger. Introduced 1993; **LMSR: $4185**
Perf.: $3200   **Exc.:** $2600   **VGood:** $1800

SAKO TRG-S
Bolt action; 243, 7mm-08, 270, 30-06, 7mm Rem. Mag., 300 Win. Mag., 338 Win. Mag., 375 H&H, 416 Rem. Mag.; 5-shot magazine (4-shot for 375 H&H); 22″, 24″ (Magnum calibers) barrel; 45 1/2″ overall length; weighs 7 3/4 lbs.; no sights; reinforced polyurethane stock with Monte Carlo comb; resistance-free bolt with 60-degree lift; recoil pad adjustable for length; free-floating barrel; detachable magazine; fully-adjustable trigger; matte blue metal. Imported from Finland by Stoeger. Introduced 1993; still imported. **LMSR: $775**
Perf.: $700   **Exc.:** $550   **VGood:** $475

SAKO VIXEN SPORTER
Bolt action; 218 Bee, 22 Hornet, 222 Rem., 222 Rem. Mag., 223; 5-shot magazine; no rear sight, hooded ramp front; drilled, tapped for scope mounts; checkered European walnut, Monte Carlo, pistol-grip stock; cheekpiece; sling swivels. Introduced 1946; discontinued 1976.
Perf.: $995   **Exc.:** $795   **VGood:** $650

SAKO Vixen Sporter Carbine
Same specs as Vixen Sporter except 20″ barrel; Mannlicher-type stock.
Perf.: $1095   **Exc.:** $895   **VGood:** $695

SAKO Vixen Sporter Heavy Barrel
Same specs as Vixen Sporter except only 222 Rem., 222 Rem. Mag., 223; heavy barrel; target-style stock; beavertail forearm.
Perf.: $895   **Exc.:** $775   **VGood:** $595

SAUER 202 TR TARGET
Bolt action; 6.5x55mm, 308 Win.; 5-shot magazine; 26″, 28 1/2″ heavy match target barrel; 44 1/2″ overall length; weighs 12 lbs.; globe front sight, Sauer-Busk 200-600m diopter rear; drilled, tapped for scope mounting; one-piece true target-type stock of laminated beechwood/ epoxy; adjustable buttplate and cheekpiece; interchangeable free-floating, hammer-forged barrel; two-stage adjustable trigger; vertical slide safety; 3 millisecond lock time; rail for swivel, bipod; right- or left-hand; converts to 22 Rimfire. Was imported from Germany by Paul Co. Introduced 1994; **LMSR: $1900**
Perf.: $1500   **Exc.:** $1100   **VGood:** $800

SAUER DRILLING
Drilling; side-by-side; 12-gauge, 2 3/4″ chambers/243, 6.5x57R, 7x57R, 7x65R, 30-06, 9.3x74R; 16-gauge, 2 3/4″ chambers/6.5x57R, 7x57R, 7x65R, 30-06, 25″ barrel; 46″ overall length; weighs 7 1/2 lbs.; bead front sight, automatic pop-up rifle rear; fancy French walnut stock with checkered grip and forend, hog-back comb, sculptured cheekpiece, hand-rubbed oil finish; Greener boxlock cross-bolt action with double underlugs; Greener side safety; separate rifle cartridge extractor; nitride-coated, hand-engraved receiver available with English Arabesque or relief game animal scene engraving; Lux Model with profuse relief-engraved game scenes, extra-fancy stump wood. Still imported from Germany by Paul Co. **LMSR: $4600**
Standard.
Perf.: $3995   **Exc.:** $3250   **VGood:** $2695
Lux.
Perf.: $5495   **Exc.:** $4695   **VGood:** $3950

Sako Vixen Sporter Carbine

Sako Vixen Sporter Heavy Barrel

Sauer Model 90

Sauer Model 200

RIFLES

Sauer
Mauser Sporter
Model 90
Model 90 Lux
Model 90 Safari
Model 90 Safari Lux
Model 90 Stutzen
Model 90 Stutzen Lux
Model 90 Supreme
Model 200
Model 200 Carbon Fiber
Model 200 Lux
Model 202
Model 202 Alaska
Model 202 Hunter Match
Model SSG 2000

SAUER MAUSER SPORTER

Bolt action; 7x57, 8x57, 30-06; other calibers on special order; 5-shot box magazine; 22″, 24″ half-octagon barrels, matted raised rib; three-leaf open rear sight, ramp front; hand-checkered pistol-grip European walnut stock; raised side panels; Schnabel tip; double-set triggers; sling swivels; also manufactured with full-length stock, 20″ barrel. Manufactured prior to WWII.

Perf.: $695   **Exc.:** $595   **VGood:** $495
Full-length stock
Perf.: $795   **Exc.:** $695   **VGood:** $550

SAUER MODEL 90

Bolt action; 222, 22-250, 243, 308, 25-06, 270, 30-06, 7mm Rem. Mag., 300 Win. Mag., 300 Weatherby Mag., 338 Win. Mag., 375 H&H Mag., 458 Win. Mag.; 3-shot (Magnums), 4-shot detachable box magazine (standard calibers); 22 1/2″, 26″ barrel; windage-adjustable open rear sight, post front on ramp; oil-finished European walnut stock; recoil pad; rear bolt-locking lugs; front sling swivel on barrel band. Made in Germany. Introduced 1986; discontinued 1989.

Perf.: $795   **Exc.:** $695   **VGood:** $550

Sauer Model 90 Lux

Same specs as Model 90 except only 300 Win. Mag., 300 Weatherby Mag., 338 Win. Mag., 375 H&H Mag.; deluxe oil-finished European walnut stock; rosewood forearm tip, pistol-grip cap; gold trigger. Still in production.

Perf.: $1295   **Exc.:** $995   **VGood:** $795

Sauer Model 90 Safari

Same specs as Model 90 except 458 Win. Mag.; 23 2/3″ barrel. Introduced 1986; discontinued 1988.

Perf.: $1295   **Exc.:** $995   **VGood:** $795

Sauer Model 90 Safari Lux

Same specs as Model 90 except 458 Win. Mag.; 23 2/3″ barrel; Williams sights; rosewood forearm, pistol-grip cap; gold trigger. **LMSR: $1995**

Perf.: $1400   **Exc.:** $1100   **VGood:** $900

Sauer Model 90 Stutzen

Same specs as Model 90 except 222, 22-250, 243, 308, 25-06, 270, 30-06; Mannlicher-style full stock. Introduced 1986; discontinued 1989.

Perf.: $800   **Exc.:** $700   **VGood:** $600

Sauer Model 90 Stutzen Lux

Same specs as Model 90 Lux except Mannlicher-style stock.

Perf.: $1195   **Exc.:** $895   **VGood:** $795

Sauer Model 90 Supreme

Same specs as Model 90 except 25-06, 270, 30-06, 300 Win. Mag., 300 Weatherby Mag., 7mm Rem. Mag., 338 Win. Mag., 375 H&H Mag.; high-gloss European walnut stock; jeweled bolt; gold trigger. Introduced 1987; still in production. **LMSR: $1495**

Grade I engraving.
Perf.: $1295   **Exc.:** $995   **VGood:** $875
Grade II engraving.
Perf.: $2595   **Exc.:** $1995   **VGood:** $1195
Grade III engraving.
Perf.: $2895   **Exc.:** $2295   **VGood:** $1395
Grade IV engraving.
Perf.: $3495   **Exc.:** $2995   **VGood:** $1995

SAUER MODEL 200

Bolt action; various calibers from 243 to 375 H&H Mag.; 3-shot (Magnums), 4-shot detachable box magazine (standard calibers); 24″ interchangeable barrels; 44″ overall length; no sights; drilled, tapped for iron sights or scope mounts; checkered European walnut pistol-grip stock; left-hand model available; steel or alloy versions. Made in Germany. Introduced 1986; discontinued 1993.

Perf.: $1195   **Exc.:** $895   **VGood:** $695
Left-hand model.
Perf.: $1250   **Exc.:** $950   **VGood:** $750

Sauer Model 200 Carbon Fiber

Same specs as Model 200 except carbon fiber stock. Introduced 1986; discontinued 1988. **LMSR: $1495**

Perf.: $800   **Exc.:** $700   **VGood:** $625

Sauer Model 200 Lux

Same specs as Model 200 except deluxe walnut stock; rosewood forearm tip, pistol-grip cap; gold trigger. Introduced 1986; discontinued 1988.

Perf.: $1295   **Exc.:** $995   **VGood:** $795
Left-hand model.
Perf.: $1425   **Exc.:** $1050   **VGood:** $850

SAUER MODEL 202

Bolt action; 22 LR, 243, 6.5x55, 6.5x57, 25-06, 270, 280, 7x64, 308, 30-06, 9,3x62 (standard calibers); 6.5x68, 8x68S, 7mm Rem. Mag., 300 Win. Mag., 300 Wea. Mag., 338 Win. Mag., 375 H&H (Magnum calibers); interchangeable 24″ barrel (standard calibers), 26″ (Magnum calibers); weighs 7 1/2 lbs. (steel), 6 1/2 lbs. (alloy); sights optional; drilled, tapped for scope mounting; modular receiver accepts interchangeable barrels; steel or alloy receiver; standard stock of fancy Claro walnut, two-piece, with Monte Carlo comb, palm swell grip, semi-Schnabel forend tip; Super Grade with extra-fancy Claro walnut with rosewood grip cap, forend tip, high-gloss epoxy finish; French walnut Euro-stock with oil finish available; right- or left-hand bolt; tang safety; fully-adjustable trigger; cocking indicator; detachable magazine. Was imported from Germany by Paul Co. Introduced 1994; **LMSR: $899**

Perf.: $795   **Exc.:** $695   **VGood:** $575
Super Grade.
Perf.: $895   **Exc.:** $795   **VGood:** $675
375 H&H Magnum.
Perf.: $950   **Exc.:** $795   **VGood:** $695
375 H&H Magnum Super Grade.
Perf.: $1050   **Exc.:** $895   **VGood:** $750

Sauer Model 202 Alaska

Same specs as Model 202 except 300 Wea. Mag. or 300 Win. Mag.; 26″ barrel; weighs 8 1/4 lbs.; laminated brown stock; metal coated with "Ilaflon" for protection. Accepts any Model 202 Magnum barrel. Was imported from Germany by Paul Co. Introduced 1994; still imported. **LMSR: $1335**

Perf.: $1095   **Exc.:** $925   **VGood:** $750

Sauer Model 202 Hunter Match

Same specs as Model 202 except 6.5x55, 308; 28″ match barrel; French walnut sporter stock; matte black finish. Introduced 1994; still in production. **LMSR: $1495**

Perf.: $1295   **Exc.:** $995   **VGood:** $795

SAUER MODEL SSG 2000

Bolt action; 223, 308, 7.5 Swiss, 300 Weatherby Mag.; 4-shot detachable box magazine; 24″, 26″ barrel; weighs 13 1/5 lbs.; no sights; scope mounts; thumbhole walnut stock; adjustable comb, buttplate, forend rail; stippled grip, forend; right- or left-hand models; flash hider/muzzlebrake; double-set triggers; sliding safety. Made in Germany. Introduced 1985; discontinued 1986.

Perf.: $3495   **Exc.:** $2595   **VGood:** $1750

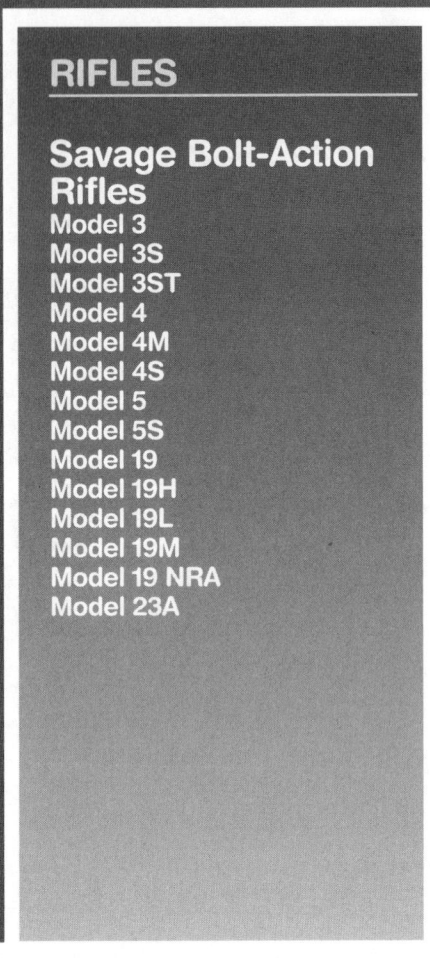

RIFLES

Savage Bolt-Action Rifles
Model 3
Model 3S
Model 3ST
Model 4
Model 4M
Model 4S
Model 5
Model 5S
Model 19
Model 19H
Model 19L
Model 19M
Model 19 NRA
Model 23A

Savage Model 3

Savage Model 4

Savage Model 5

Savage Model 19

Savage Model 23AA

Savage Model 23D

Savage Model 40

SAVAGE BOLT-ACTION RIFLES

SAVAGE MODEL 3
Bolt action; single shot; 22 Short, 22 Long, 22 LR; 26″ barrel (pre-WWII), 24″ (post-WWII); 43″ overall length; weighs 5 lbs.; takedown; chromium-plated bolt, trigger; adjustable flat-top rear sight, gold bead front; receiver drilled. tapped for scope mounting; uncheckered American walnut one-piece pistol-grip stock; forearm with finger grooves; checkered hard rubber or steel buttplate. Introduced 1933; discontinued 1952.
Perf.: $90   **Exc.:** $80   **VGood:** $60

Savage Model 3S
Same specs as Model 3 except rear peep sight, hooded front sight with three inserts. Introduced 1933; discontinued 1942.
Perf.: $100   **Exc.:** $90   **VGood:** $70

Savage Model 3ST
Same specs as Model 3S except rear peep sight, hooded front sight; swivels, sling. Introduced 1933; discontinued 1942.
Perf.: $110   **Exc.:** $90   **VGood:** $80

SAVAGE MODEL 4
Bolt action; 22 Short, 22 Long, 22 LR; 5-shot detachable box magazine; 24″ barrel; weighs 5 1/2 lbs.; takedown; chromium-plated bolt, trigger; pre-WWII version with open rear sight, bead front; checkered pistol grip; American walnut stock, grooved forearm; post-WWII model with uncheckered stock; rubber buttplate. Introduced 1933; discontinued 1965. Pre-WWII model.
Perf.: $110   **Exc.:** $95   **VGood:** $80
Post-WWII model.
Perf.: $115   **Exc.:** $100   **VGood:** $90

Savage Model 4M
Same specs as post-war Model 4 except 22 WMR; gold-plated trigger. Introduced 1961; discontinued 1965.
Perf.: $110   **Exc.:** $95   **VGood:** $80

Savage Model 4S
Same specs as Model 4 except fully-adjustable rear peep sight, hooded front with three inserts. Introduced 1933; discontinued 1942.
Perf.: $120   **Exc.:** $100   **VGood:** $90

SAVAGE MODEL 5
Bolt action; 22 Short, 22 Long, 22 LR; 21-shot (22 Short), 17-shot (22 Long), 15-shot (22 LR) tube magazine; 24″ tapered barrel with crowned muzzle; weighs 6 lbs.; take-down; chromium-plated bolt, trigger; fully-adjustable sporting rear sight, gold bead front; redesigned American walnut stock, bolt handle, trigger guard; hard rubber buttplate. Introduced 1936; discontinued 1961.
Perf.: $120   **Exc.:** $105   **VGood:** $90

Savage Model 5S
Same specs as Model 5 except No. 15 Micro peep rear sight, hooded No. 150 front. Introduced 1936; discontinued 1942.
Perf.: $125   **Exc.:** $110   **VGood:** $90

SAVAGE MODEL 19
Bolt action; 22 LR; 5-shot detachable box magazine; 25″ barrel; weighs 8 lbs.; early model with adjustable rear peep sight, blade front; later model with hooded front sight, fully-adjustable aperture extension rear; target-type uncheckered one-piece walnut stock; full pistol grip, 2-inch wide beavertail forearm; checkered buttplate; receiver drilled for scope blocks; 1 1/4″ sling; sling swivels. Introduced 1933; discontinued 1946.
Perf.: $230   **Exc.:** $200   **VGood:** $150

Savage Model 19H
Same specs as Model 19 except 22 Hornet. Introduced 1933; discontinued 1942.
Perf.: $495   **Exc.:** $440   **VGood:** $300

Savage Model 19L
Same specs as Model 19 except Lyman 48Y receiver sight, No. 17A front sight. Introduced 1933; discontinued 1942.
Perf.: $325   **Exc.:** $265   **VGood:** $200

Savage Model 19M
Same specs as Model 19 except heavy 28″ barrel; scope bases. Introduced 1933; discontinued 1942.
Perf.: $350   **Exc.:** $290   **VGood:** $220

Savage Model 19 NRA
Same specs as Model 19 except adjustable rear sight, blade front; American walnut military-type full stock, pistol grip; uncheckered. Introduced 1919; discontinued 1933.
Perf.: $230   **Exc.:** $190   **VGood:** $150

SAVAGE MODEL 23A
Bolt action; 22 LR; 5-shot detachable box magazine; 23″ barrel; open rear sight, blade or bead front; uncheckered American walnut pistol-grip stock, thin forearm, Schnabel tip. Introduced 1923; discontinued 1933.
Perf.: $220   **Exc.:** $180   **VGood:** $130

Savage Model 63

Savage Model 65

Savage Model 23D

Savage Model 40

Savage Model 110B

Savage Bolt-Action Rifles
Model 23AA
Model 23B
Model 23C
Model 23D
Model 34
Model 34M
Model 35
Model 35M
Model 36
Model 40
Model 45
Model 46
Model 63
Model 63M
Model 65
Model 65M
Model 110
Model 110B

Savage Model 23AA
Same specs as Model 23A except 22 Short, 22 Long, 22 LR; checkered stock; swivel studs; speed lock; receiver tapped for No. 10 Savage aperture rear sight. Introduced 1933; discontinued 1942.
Perf.: $270   **Exc.:** $220   **VGood:** $170

Savage Model 23B
Same specs as Model 23A except 25-20; 25″ barrel; weighs 6 ½ lbs.; swivel studs; no Schnabel tip. Introduced 1923; discontinued 1942.
Perf.: $250   **Exc.:** $200   **VGood:** $150

Savage Model 23C
Same specs as Model 23B except 32-20; 25″ barrel; swivel studs; no Schnabel tip. Introduced 1923; discontinued 1942.
Perf.: $250   **Exc.:** $200   **VGood:** $150

Savage Model 23D
Same specs as Model 23B except 22 Hornet; 25″ barrel; weighs 6 ½ lbs.; swivel studs; no Schnabel tip; flattop adjustable rear sight, gold bead front. Introduced 1923; discontinued 1947.
Perf.: $310   **Exc.:** $270   **VGood:** $220

SAVAGE MODEL 34
Bolt action; 22 Short, 22 Long, 22 LR; 5-shot magazine; 20″ barrel; open rear sight, blade front; uncheckered hardwood pistol grip or checkered Monte Carlo stock. Introduced 1969; discontinued 1973.
Perf.: $100   **Exc.:** $90   **VGood:** $80

Savage Model 34M
Same specs as Model 34 except 22 WMR. Introduced 1969; discontinued 1973.
Perf.: $110   **Exc.:** $100   **VGood:** $90

SAVAGE MODEL 35
Bolt action; 22 LR; 5-shot magazine; 22″ barrel; 41″ overall length; ramp front sight, step-adjustable open rear; walnut-finished hardwood Monte Carlo stock; checkered pistol grip, forend; receiver grooved for scope mount. Introduced 1982; discontinued 1985.
Perf.: $100   **Exc.:** $90   **VGood:** $80

Savage Model 35M
Same specs as Model 35 except 22 WMR. Introduced 1982; discontinued 1985.
Perf.: $120   **Exc.:** $100   **VGood:** $90

SAVAGE MODEL 36
Bolt action; single shot; 22 LR; 22″ barrel; 41″ overall length; ramp front sight, step-adjustable rear; walnut-finished hardwood Monte Carlo stock; checkered pistol grip, forend; receiver grooved for scope mount. Introduced 1982; discontinued 1985.
Perf.: $90   **Exc.:** $80   **VGood:** $65

SAVAGE MODEL 40
Bolt action; 250-3000, 30-30, 300 Savage, 30-06; 4-shot detachable box magazine; 22″, 24″ barrel (300 Savage, 30-06); weighs 7 ½ lbs.; semi-buckhorn rear sight, white metal bead front on ramp; uncheckered American walnut pistol-grip stock; tapered forearm; Schnabel tip; steel buttplate. Introduced 1928; discontinued 1940.
Perf.: $350   **Exc.:** $300   **VGood:** $250

SAVAGE MODEL 45
Bolt action; 250-3000, 30-30, 300 Savage, 30-06; 4-shot detachable box magazine; 22″ (250-3000, 30-30), 24″ barrel; Lyman receiver sight, bead front on ramp; hand-checkered pistol-grip stock, forearm, Schnabel tip. Introduced 1928; discontinued 1940.
Perf.: $400   **Exc.:** $350   **VGood:** $300

SAVAGE MODEL 46
Bolt action; 22 Short, 22 Long, 22 LR; 20-shot (22 Short), 17-shot (22 Long), 15-shot (22 LR) tubular magazine; open rear sight, blade front; uncheckered hardwood pistol grip or checkered Monte Carlo stock. Introduced 1969; discontinued 1973.
Perf.: $100   **Exc.:** $80   **VGood:** $70

SAVAGE MODEL 63
Bolt action; single shot; 22 Short, 22 Long, 22 LR; 18″ barrel; 36″ overall length; weighs 4 lbs.; open rear sight, hooded ramp front; one-piece full-length walnut-finished hardwood pistol-grip stock; sling swivels; fur-nished with key to lock trigger. Introduced 1970; discontinued 1972.
Perf.: $95   **Exc.:** $85   **VGood:** $70

Savage Model 63M
Same specs as Model 63K except only 22 WMR. Introduced 1970; discontinued 1972.
Perf.: $100   **Exc.:** $90   **VGood:** $80

SAVAGE MODEL 65
Bolt action; 22 LR; 5-shot detachable clip magazine; 20″ lightweight tapered barrel; 39″ overall length; weighs 4 ¾ lbs.; gold bead ramp front sight, open rear with elevator; receiver grooved for scope mounting; select walnut Monte Carlo pistol-grip stock; buttplate with white line spacer; raised fluted comb; thumb safety; double extractors. Introduced 1965; discontinued 1974.
Perf.: $140   **Exc.:** $120   **VGood:** $100

Savage Model 65M
Same specs as Model 65 except 22 WMR. Introduced 1966; discontinued 1981.
Perf.: $150   **Exc.:** $130   **VGood:** $110

SAVAGE MODEL 110
Bolt action; 243, 270, 308, 30-06; 4-shot box magazine; 22″ medium-weight barrel; 43″ overall length; weighs 6 ¾ lbs.; step-adjustable rear sight, gold bead ramp front; hand-checkered American walnut pistol-grip stock; aluminum buttplate; pistol-grip cap. Introduced 1958; discontinued 1963.
Perf.: $250   **Exc.:** $200   **VGood:** $175

Savage Model 110B
Same specs as Model 110 except 243, 270, 30-06; 4-shot internal magazine; checkered American walnut Monte Carlo pistol-grip stock, forearm. Introduced 1977; discontinued 1979.
Perf.: $300   **Exc.:** $275   **VGood:** $200
Left-hand model.
Perf.: $340   **Exc.:** $300   **VGood:** $270

Savage Bolt-Action Rifles

Model 110B Laminate
Model 110C
Model 110CY Ladies/Youth
Model 110D
Model 110E
Model 110ES
Model 110F
Model 110FNS
Model 110FP Police
Model 110FCXP3
Model 110FXP3
Model 110G
Model 110GC
Model 110GCXP3

Savage Model 110C

Savage Model 110C Left-Hand

Savage Model 110CY Ladies/Youth

Savage Model 110D

Savage Model 110D Left-Hand

Savage Model 110E

Savage Model 110B Laminate
Same specs as Model 110 except 300 Win. Mag., 338 Win. Mag.; 3-shot internal magazine; 24″ barrel; brown laminated hardwood Monte Carlo pistol-grip stock, forearm. Introduced 1989; discontinued 1991.
Perf.: $310   **Exc.:** $270   **VGood:** $240

Savage Model 110C
Same specs as Model 110 except 22-250, 243 Win., 25-06, 270 Win., 30-06, 308 Win., 7mm Rem. Mag., 300 Win. Mag.; 3-shot (Magnums), 4-shot detachable box magazine; 22″, 24″ (22-250, Magnums) barrel; open folding leaf rear sight, gold bead ramp front; hand-checkered Monte Carlo American walnut pistol-grip stock; recoil pad (Magnums). Introduced 1966.
Perf.: $300   **Exc.:** $265   **VGood:** $230
Left-hand model.
Perf.: $310   **Exc.:** $270   **VGood:** $240

Savage Model 110CY Ladies/Youth
Same specs as Model 110 except 223, 243, 270, 300 Sav., 308; 5-shot magazine; ramp front sight, fully-adjustable rear; drilled and tapped for scope mounting; walnut-stained hardwood stock with high comb, cut checkering; 12 1/2″ length of pull; red rubber buttpad. Made in U.S. by Savage. Introduced 1991; still produced. **LMSR: $362**
Perf.: $325   **Exc.:** $275   **VGood:** $225

Savage Model 110D
Same specs as Model 110 except 22-250, 223 Rem., 243, 25-06, 270, 308, 30-06, 7mm Rem. Mag., 264, 300 Win. Mag., 338 Win. Mag.; 3-shot (Magnums), 4-shot magazine; 22″, 24″ (22-250) barrel; 42 1/2-45″ overall length; weighs 6 3/4-8 lbs. (22-250); semi-buck-horn step-adjustable folding rear sight; hinged floor-plate; aluminum buttplate or hard rubber (22-250) recoil pad. Introduced 1966; discontinued 1988.
Perf.: $340   **Exc.:** $290   **VGood:** $260
Left-hand model.
Perf.: $350   **Exc.:** $300   **VGood:** $270

Savage Model 110E
Same specs as Model 110 except 22-250, 223, 243 Win., 30-06, 270, 7mm Rem. Mag., 308; 3-shot (Magnum), 4-shot box magazine; 20″, 24″ (7mm, stainless) barrel; 40 1/2″ overall length (20″ barrel), 45 1/2″ (24″); weighs 6 3/4 lbs.; uncheckered Monte Carlo stock (early versions); checkered pistol grip, forearm (later models); recoil pad (Magnum). Introduced 1963; discontinued 1989.
Perf.: $300   **Exc.:** $250   **VGood:** $200
Left-hand model.
Perf.: $310   **Exc.:** $260   **VGood:** $210

Savage Model 110ES
Same specs as Model 110E except 30-06, 243, 308 Win.; comes with 4x scope and mount. Introduced 1983.
Perf.: $340   **Exc.:** $300   **VGood:** $250

Savage 110F
Same as Model 110 except 22-250, 223 Rem., 243 Win., 250 Savage, 25-06, 308 Win., 30-06, 270 Win., 7mm Rem Mag., 300 Savage, 300 Win. Mag., 7mm-08, 338 Win. Mag.; removable open sights; black Du Pont Rynite® stock with black buttpad, swivel studs; right-hand only. Made in U.S. by Savage. Introduced 1988; discontinued 1994.
Perf.: $310   **Exc.:** $260   **VGood:** $210

Savage 110FNS
Same as Model 110F except no sights; black composite stock. Made in U.S. by Savage. No longer made.
Perf.: $300   **Exc.:** $240   **VGood:** $190

Savage Model 110FP Police
Same specs as Model 110F except 223, 308, 300 Win. Mag., 7mm Rem. Mag., 25-06; 4-shot internal magazine; 24″ heavy barrel; 45 1/2″ overall length; weighs 8 1/2 lbs.; no sights; drilled, tapped for scope mounts; black granite/fiberglass composition stock; double swivel studs on forend for swivels or bipod; matte finish on all metal parts. Introduced 1990; still in production. **LMSR: $425**
Perf.: $350   **Exc.:** $300   **VGood:** $230

Savage 110FCXP3
Same as Savage 110F except 223, 22-250, 308, 243, 30-06, 270, 7mm Rem. Mag., 300 Win. Mag.; detachable box magazine; black composite stock; 3-9x32 scope, Kwik-Site rings and bases; Savage/Pathfinder leather sling; Uncle Mike's swivels, gun lock, ear plugs, shooting glasses and sight-in target. Made in U.S. by Savage. Introduced 1991; still produced. **LMSR: $447**
Perf.: $400   **Exc.:** $330   **VGood:** $290

Savage Model 110FXP3
Same as Savage 110FCXP3 except non-detachable magazine. Introduced 1991; still in production. **LMSR: $488**
Perf.: $410   **Exc.:** $340   **VGood:** $300

Savage Model 110G
Same specs as Model 110 except 22-250, 223 Rem., 250 Savage, 25-06, 270 Win., 300 Savage, 308 Win., 30-06, 243 Win., 7mm-08, 7mm Rem. Mag., 300 Win. Mag.; 5-shot (standard calibers), 4-shot (Magnums); 22″ 24″ round tapered barrels; 42″ overall length (22″ barrel); checkered walnut-finished Monte Carlo hardwood stock; hard rubber buttplate; ramp front sight, step-adjustable rear; top tang safety; tapped for scope mounts; floating barrel; adjustable trigger. Introduced, 1989; still in production.
Perf.: $325   **Exc.:** $285   **VGood:** $250
Left-hand model (110GL).
Perf.: $330   **Exc.:** $300   **VGood:** $260

Savage 110GC
Same specs as Model 110G except 30-06, 270, 7mm Rem. Mag., 300 Win. Mag.; removable box magazine. Made in U.S. by Savage. No longer made.
Perf.: $340   **Exc.:** $300   **VGood:** $250

Savage Model 110GCXP3
Same specs as Model 110G except 270, 30-06, 7mm Rem. Mag., 300 Win. Mag.; detachable box magazine; factory-mounted and bore-sighted 3-9x32 scope, rings and bases; quick-detachable swivels; sling; Monte Carlo-style hardwood stock with walnut finish; left-hand models available in all calibers. Made in U.S. by Savage Arms, Inc. Introduced 1994; **LMSR: $480**
Perf.: $410   **Exc.:** $370   **VGood:** $320

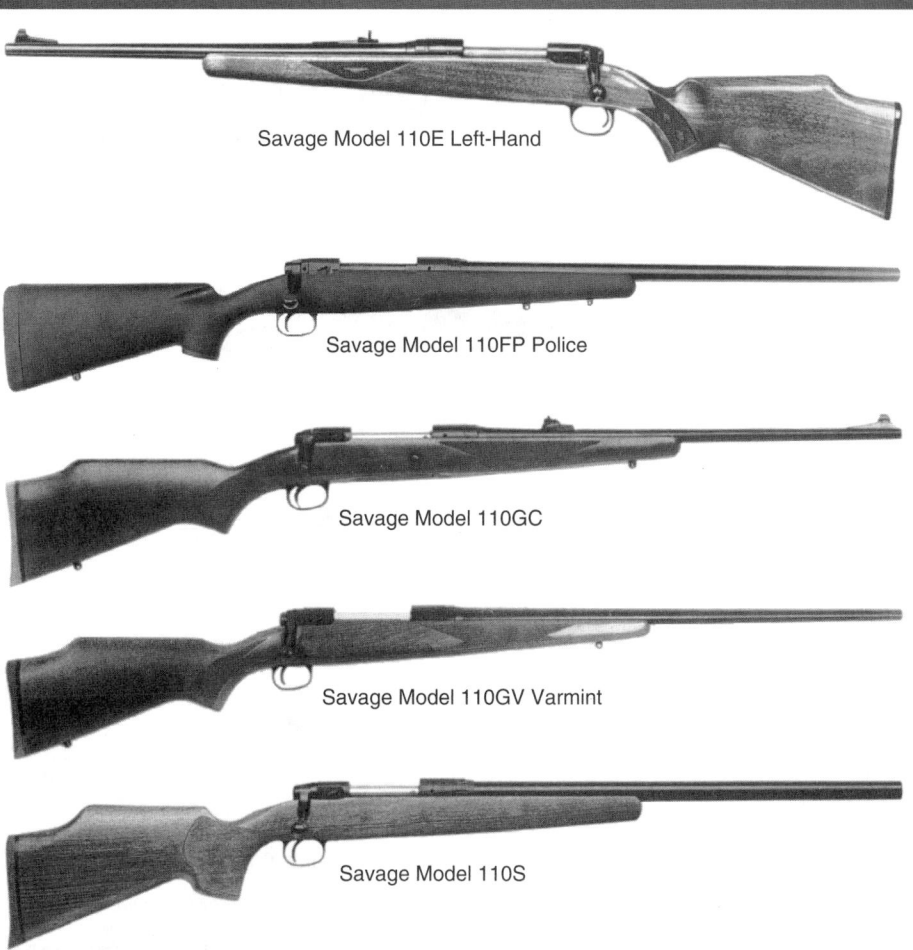

Savage Model 110E Left-Hand

Savage Model 110FP Police

Savage Model 110GC

Savage Model 110GV Varmint

Savage Model 110S

Savage Bolt-Action Rifles
Model 110GLNS
Model 110GNS
Model 110GV Varmint
Model 110GXP3
Model 110K
Model 110M
Model 110MC
Model 110P Premier Grade
Model 110PE Presentation Grade
Model 110S
Model 110V
Model 110WLE One of One Thousand Limited Edition
Model 111 Chieftan

Savage 110GLNS
Same specs as Model 110G except 30-06, 270, 7mm Rem. Mag.; true left-hand action; no sights. Made in U.S. by Savage. No longer made.
Perf.: $320   **Exc.:** $260   **VGood:** $220

Savage 110GNS
Same specs as Model 110G except no sights. Made in U.S. by Savage. No longer made.
Perf.: $300   **Exc.:** $270   **VGood:** $230

Savage 110GV Varmint
Same as Model 110G except 22-250, 223; no sights; receiver drilled and tapped for scope mounting; medium-weight varmint barrel. Made in U.S. by Savage. Introduced 1989; discontinued 1993.
Perf.: $330   **Exc.:** $290   **VGood:** $240

Savage Model 110GXP3
Same specs as Model 110G except 223, 22-250, 243, 250 Savage, 25-06, 270, 300 Sav., 30-06, 308, 7mm Rem. Mag., 7mm-08, 300 Win. Mag.; factory mounted bore-sighted 3-9x32 scope, rings and bases; quick-detachable swivels; sling; Monte Carlo-style hardwood stock with walnut finish; left-hand models available in all calibers. Made in U.S. by Savage Arms, Inc. Introduced 1991; still produced. **LMSR: $418**
Perf.: $385   **Exc.:** $330   **VGood:** $280

Savage Model 110K
Same specs as Model 110 except 243, 270, 30-06, 7mm Rem. Mag., 338 Win. Mag.; laminated camouflage stock. Introduced 1986; discontinued 1988.
Perf.: $330   **Exc.:** $290   **VGood:** $240

Savage Model 110M
Same specs as Model 110 except 264 Win. Mag., 300 Win. Mag., 7mm Rem. Mag., 338 Win. Mag.; 3-shot magazine; 24″ stainless steel barrel; 45″ overall length; weighs 8 lbs. (338); hand-checkered American walnut pistol-grip Monte Carlo stock; recoil pad. Introduced 1963; discontinued 1969.
Perf.: $340   **Exc.:** $290   **VGood:** $240
Left-hand model (110ML).
Perf.: $360   **Exc.:** $320   **VGood:** $260

Savage Model 110MC
Same specs as Model 110 except 22-250; 24″ barrel; hand-checkered American walnut pistol-grip Monte Carlo stock. Introduced 1959; discontinued 1969.
Perf.: $370   **Exc.:** $330   **VGood:** $270
Left-hand model (110MC-L)
Perf.: $390   **Exc.:** $340   **VGood:** $300

Savage Model 110P Premier Grade
Same specs as Model 110 except 243 Win., 30-06, 7mm Rem. Mag.; 3-shot (Magnum), 4-shot magazine; 22″, 24″ (7mm, stainless) barrel; 43″-45″ overall length; weighs 7-7 3/4 lbs.; recessed bolt head; double front locking lugs; top tang safety; aluminum buttplate; skip-checkered Monte Carlo select French walnut stock; hand-carved roll-over cheekpiece; rosewood pistol-grip cap, forearm tip with white inlay; recoil pad (Magnum); left- or right-hand action. Introduced 1964; discontinued 1970.
Perf.: $440   **Exc.:** $330   **VGood:** $310
7mm Rem. Mag. model.
Perf.: $460   **Exc.:** $350   **VGood:** $330

Savage Model 110PE Presentation Grade
Same specs as Model 110 except 243 Win., 30-06, 7mm Rem. Mag.; 3-shot (Magnum), 4-shot magazine; 22″, 24″ (Magnum) barrel; skip-checkered Monte Carlo fancy French walnut stock; rosewood pistol-grip cap, forearm tip; recoil pad (Magnum); right-, or left-hand action; engraved receiver, trigger guard, floorplate. Introduced 1958; discontinued 1970.
Perf.: $695   **Exc.:** $595   **VGood:** $495
7mm Rem. Mag. model
Perf.: $725   **Exc.:** $625   **VGood:** $525

Savage Model 110S
Same specs as Model 110 except 308, 7mm-08; 5-shot magazine; 22″ heavy tapered barrel; weighs 8 1/2 lbs.; no sights; drilled, tapped for scope mounts; special silhouette stock with Wundhammer swell stippled pistol grip, forend; rubber recoil pad. Introduced 1978; discontinued 1985.
Perf.: $350   **Exc.:** $300   **VGood:** $275

Savage Model 110V
Same specs as Model 110 except 22-250, 223; 5-shot magazine; 26″ heavy barrel; no sights; checkered American walnut varmint stock. Introduced 1983; discontinued 1989.
Perf.: $350   **Exc.:** $300   **VGood:** $250

Savage 110WLE One of One Thousand Limited Edition
Same specs as Model 110 except 7x57mm Mauser, 250-3000 Savage, 300 Savage; high-luster #2 fancy-grade American walnut stock with cut checkering; swivel studs, and recoil pad; highly polished barrel; the bolt with laser-etched Savage logo. Comes with gun lock, ear plugs, sight-in target and shooting glasses. Made in U.S. by Savage. Introduced 1992; no longer made.
Perf.: $460   **Exc.:** $400   **VGood:** $360

SAVAGE MODEL 111 CHIEFTAN
Bolt action; 243 Win., 270 Win., 7x57, 7mm Rem. Mag., 30-06; 3-shot (Magnum), 4-shot detachable clip magazine; 22″, 24″ (Magnum, stainless) barrel; 43″-45″ overall length; 7 1/2-8 1/4 lbs.; leaf rear sight, hooded ramp front; select checkered American walnut pistol-grip stock; Monte Carlo comb; cheekpiece; pistol-grip cap; hard rubber or recoil pad (Magnum) with white line spacer; detachable swivels; sling. Introduced 1974; discontinued 1978.
Perf.: $370   **Exc.:** $330   **VGood:** $280

Savage Bolt-Action Rifles

Model 111F Classic Hunter
Model 111FCXP3
Model 111FXP3
Model 111G Classic Hunter
Model 112BV
Model 112BT Competition Grade
Model 112BVSS
Model 112FV Varmint Rifle
Model 112FVSS
Model 112FVSS-S
Model 112R
Model 112V

Savage Model 111FCXP3

Savage Model 111G Classic Hunter

Savage Model 112V

Savage Model 116FSS Weather Warrior

Savage Model 111F Classic Hunter

Same specs as Model 111 except 338 Win. Mag.; top-loading magazine; graphite/fiberglass filled classic-style stock with non-glare finish; positive checkering; black recoil pad; swivel studs. Right- or left-hand action. Made in U.S. by Savage Arms, Inc. Model number reintroduced 1994; still produced. **LMSR: $376**
With sights.
  Perf.: $350   **Exc.:** $300   **VGood:** $250
Without sights as Model 111FNS.
  Perf.: $290   **Exc.:** $250   **VGood:** $210
With detachable magazine as Model 111FC.
  Perf.: $300   **Exc.:** $270   **VGood:** $230

Savage Model 111FCXP3

Same specs as Model 111 except 223, 22-250, 243, 270, 30-06, 308, 7mm-08, 7mm Rem. Mag., 300 Win. Mag., 338 Win. Mag.; detachable box magazine; factory-mounted and bore-sighted 3-9x32 scope; lightweight, black graphite/fiberglass composite stock with non-glare finish, positive checkering; quick-detachable swivels, sling. Right- or left-hand action. Made in U.S. by Savage Arms, Inc. Introduced 1994; still produced. **LMSR: $488**
  Perf.: $465   **Exc.:** $400   **VGood:** $350

Savage Model 111FXP3

Same specs as Model 111FCXP3 except non-detachable magazine. Made in U.S. by Savage Arms, Inc. Introduced 1994; still produced. **LMSR: $447**
  Perf.: $400   **Exc.:** $350   **VGood:** $290

Savage Model 111G Classic Hunter

Same specs as Model 111 except 223, 22-250, 243, 250 Sav., 25-06, 270, 300 Sav., 30-06, 308, 7mm Rem. Mag., 7mm-08, 300 Win. Mag.; weighs about 6 1/2 lbs.; ramp front sight, open fully-adjustable rear (or no sights); receiver drilled, tapped for scope mounting; classic-style, walnut-finished hardwood stock with straight comb, ventilated red rubber recoil pad; three-position top tang safety; double front locking lugs; free-floated button-rifled barrel; trigger lock, target, ear puffs. Right- or left-hand action. Made in U.S. by Savage Arms, Inc. Model number reintroduced 1994; still produced. **LMSR: $362**
With sights.
  Perf.: $325   **Exc.:** $300   **VGood:** $250
Without sights as Model 111GNS.
  Perf.: $300   **Exc.:** $275   **VGood:** $225
With detachable magazine as Model 111GC.
  Perf.: $300   **Exc.:** $270   **VGood:** $230

SAVAGE MODEL 112BV

Bolt action; 22-250, 223 Rem.; 4-shot magazine; 26" barrel; no sights; drilled, tapped for scope mounts; brown laminate stock, Wundhammer pistol grip; alloy steel construction. Introduced 1993; discontinued 1993.
  Perf.: $475   **Exc.:** $430   **VGood:** $365

Savage Model 112BT Competition Grade

Same specs as Model 112BV except 223, 300 Win. Mag., 308 Win.; 5-shot (223, 308 Win.) magazine; single shot (300 Win.); stainless barrel with black finish; laminated wood stock with straight comb, adjustable cheek rest, Wundhammer palm swell, ventilated forend; recoil pad adjustable for length of pull; matte black alloy receiver; bolt with black titanium nitride coating, large handle ball; alloy accessory rail on forend. Comes with safety gun lock, target and ear puffs. Made in U.S. by Savage Arms, Inc. Introduced 1994; still produced. **LMSR: $1000**
  Perf.: $910   **Exc.:** $800   **VGood:** $675

Savage Model 112BVSS

Same specs as Model 112BV except 223, 22-250 (4-shot magazine), 220 Swift (single shot); weighs 10 1/2 lbs; fluted stainless steel barrel, bolt handle, trigger guard; uncheckered heavy-prone laminated stock with high comb, Wundhammer swell. Made in U.S. by Savage Arms, Inc. Introduced 1991; no longer produced.
  Perf.: $475   **Exc.:** $410   **VGood:** $350

Savage Model 112FV Varmint Rifle

Same specs as Model 112BV except 220 Swift (single shot), 22-250, 223 (4-shot magazine); weighs 8 1/2 lbs.; blued, heavy 26" barrel with recessed target-style muzzle; black graphite/fiberglass filled composite stock with positive checkering; double front swivel studs for attaching bipod. Made in U.S. by Savage Arms, Inc. Introduced 1991; no longer produced.
  Perf.: $360   **Exc.:** $300   **VGood:** $240

Savage Model 112FVSS

Same specs as Model 112BV except 223, 22-250, 25-06, 7mm Rem. Mag., 300 Win. Mag.; heavy 26" barrel with recessed target-style muzzle; stainless fluted steel barrel, bolt handle, trigger guard; black graphite/fiberglass filled composite stock with positive checkering; double front swivel studs for attaching bipod. Made in U.S. by Savage Arms, Inc. Introduced 1991; no longer produced.
  Perf.: $470   **Exc.:** $400   **VGood:** $345

Savage Model 112FVSS-S

Same specs as Model 112BV except 223, 22-250, 220 Swift; solid bottom single shot action; heavy 26" barrel with recessed target-style muzzle; fluted stainless steel barrel, bolt handle, trigger guard; black graphite/fiberglass filled composite stock with positive checkering; double front swivel studs for attaching bipod. Made in U.S. by Savage Arms, Inc. Introduced 1991; no longer produced.
  Perf.: $450   **Exc.:** $380   **VGood:** $320

Savage Model 112R

Same specs as Model 112BV except 22-250, 243, 25-06; 5-shot magazine; 26" tapered free-floating barrel; no sights; drilled, tapped for scope mounts; American walnut stock; fluted comb, Wundhammer swell at pistol grip; top tang safety. Introduced 1979; discontinued 1980.
  Perf.: $330   **Exc.:** $280   **VGood:** $240

Savage Model 112V

Same specs as Model 112BV except 220 Swift, 222 Rem., 223 Rem., 225 Win., 22-250 Rem., 243 Win., 25-06 Rem; 26" heavy barrel; 47" overall length; weighs 9 1/4 lbs.; scope bases; select American walnut varminter stock; high comb; checkered pistol grip; detachable sling swivels; recoil pad with white line spacer. Introduced 1975; discontinued 1979.
  Perf.: $340   **Exc.:** $290   **VGood:** $250

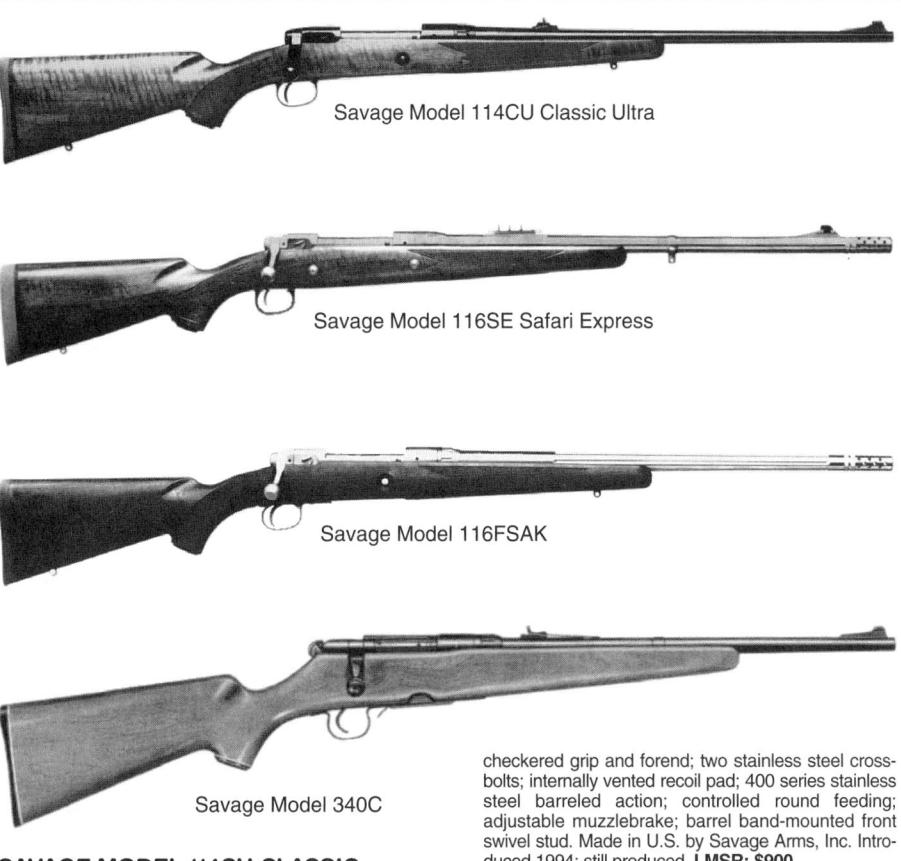

Savage Model 114CU Classic Ultra

Savage Model 116SE Safari Express

Savage Model 116FSAK

Savage Model 340C

Savage Bolt-Action Rifles
Model 114CU Classic Ultra
Model 116FSS Weather Warrior
Model 116FSAK
Model 116FSK
Model 116SE Safari Express
Model 116US Ultra Stainless Rifle
Model 340
Model 340C
Mdoel 340S Deluxe
Model 340V
Model 342
Model 342S Deluxe
Model 1904
Model 1905
Model 1920
Model 1920-1926

SAVAGE MODEL 114CU CLASSIC ULTRA
Bolt action; 270, 30-06, 7mm Rem. Mag., 300 Win. Mag.; 4-shot (Magnums), 5-shot removable box magazine; 22", 24" (Magnums) barrel; weighs 7 1/8 lbs.; ramp front sight, step-adjustable rear; tapped for scope mounts; checkered, high-gloss straight American walnut stock; grip cap; recoil pad; cut checkering; tang safety. Introduced 1991; still in production. **LMSR: $525**
    Perf.: $460   **Exc.:** $410   **VGood:** $350

SAVAGE MODEL 116FSS WEATHER WARRIOR
Bolt action; 223, 243, 270, 30-06, 308, 7mm Rem. Mag., 300 Win. Mag., 338 Win. Mag.; free-floated 22" barrel (standard calibers), 24" (Magnums); weighs 6 3/4 lbs.; no sights; drilled, tapped for scope mounting; graphite/ fiberglass filled black composite stock; stainless steel barreled action; right- or left-hand action. Made in U.S. by Savage Arms, Inc. Introduced 1991; still produced. **LMSR: $489**
    Perf.: $420   **Exc.:** $380   **VGood:** $320
With detachable magazine as Model 116FCS.
    Perf.: $430   **Exc.:** $385   **VGood:** $340

Savage Model 116FSAK
Same specs as Model 116FSS except 270, 30-06, 7mm Rem. Mag.; Savage "Adjustable Muzzle Brake" with fluted barrel. Made in U.S. by Savage Arms, Inc. Introduced 1994; still produced. **LMSR: $581**
    Perf.: $510   **Exc.:** $445   **VGood:** $380
With detachable magazine as Model 116FCSAK.
    Perf.: $500   **Exc.:** $440   **VGood:** $370

Savage Model 116FSK
Same specs as Model 116FSS except 270, 30-06, 7mm-Rem. Mag., 338 Win. Mag.; 22" "Shock-Suppressor" barrel; fixed muzzle brake. Made in U.S. by Savage Arms, Inc. Introduced 1993; still produced. **LMSR: $552**
    Perf.: $495   **Exc.:** $430   **VGood:** $370

SAVAGE MODEL 116SE SAFARI EXPRESS
Bolt Action; 300 Win. Mag., 338 Win. Mag., 425 Express, 458 Win. Mag.; 3-shot magazine; 24" barrel; 45 1/2" overall length; weighs 8 1/2 lbs.; bead on ramp front sight, three-leaf rear; select grade classic-style American walnut stock with ebony forend tip; cut-checkered grip and forend; two stainless steel crossbolts; internally vented recoil pad; 400 series stainless steel barreled action; controlled round feeding; adjustable muzzlebrake; barrel band-mounted front swivel stud. Made in U.S. by Savage Arms, Inc. Introduced 1994; still produced. **LMSR: $900**
    Perf.: $765   **Exc.:** $675   **VGood:** $540

Savage Model 116US Ultra Stainless Rifle
Same specs as Model 116SE except 270, 30-06, 7mm Rem. Mag., 300 Win. Mag.; stock has high-gloss finish; no open sights; stainless steel barreled action with satin finish. Made in U.S. by Savage Arms, Inc. Introduced 1995; still produced. **LMSR: $700**
    Perf.: $600   **Exc.:** $500   **VGood:** $400

SAVAGE MODEL 340
Bolt action; 22 Hornet, 222 Rem., 30-30; 3-, 4-shot magazine; 20", 22", 24" medium-weight barrel; 40" overall length (20" barrel); weighs 6 3/4 lbs.; click-adjustable middle sight, ramp front; uncheckered one-piece American walnut pistol-grip stock; thumb safety at right rear of receiver. Introduced 1950; discontinued 1986.
    Perf.: $230   **Exc.:** $200   **VGood:** $170

Savage Model 340C
Same specs as Model 340 except 30-30; 3-shot magazine; 18 1/2" medium-weight barrel; 36 1/2" overall length; weighs 6 1/4 lbs.; checkered American walnut pistol-grip stock; pistol grip, buttplate white line spacer; sling swivels. Introduced 1962; discontinued 1965.
    Perf.: $235   **Exc.:** $205   **VGood:** $180

Savage Model 340S Deluxe
Same specs as Model 340 Savage No. 175 except peep rear sight, hooded gold bead front; 3-shot magazine (30-30) hand-checkered American walnut pistol-grip stock; swivel studs. Introduced 1952; discontinued 1960.
    Perf.: $260   **Exc.:** $225   **VGood:** $205

Savage Model 340V
Same specs as Model 340 except 225 Win.; 24" barrel; American walnut varmint-style stock. Introduced 1967; no longer in production.
    Perf.: $295   **Exc.:** $270   **VGood:** $230

SAVAGE MODEL 342
Bolt action; 22 Hornet; 4-shot clip magazine; 22", 24" barrel; 40" overall length; weighs 6 3/4 lbs.; click-adjustable middle sight, ramp front; tapped for Weaver scope side mounts; uncheckered one-piece American walnut pistol-grip stock; after 1953 was incorporated into Model 340 line. Introduced 1950; discontinued 1953.
    Perf.: $250   **Exc.:** $220   **VGood:** $200

Savage Model 342S Deluxe
Same specs as Model 340 except Savage No. 175 peep rear sight, hooded ramp front with gold bead; hand-checkered American walnut pistol-grip stock; swivel studs; after 1953 was incorporated into Model 340 line.
    Perf.: $260   **Exc.:** $230   **VGood:** $210

SAVAGE MODEL 1904
Bolt action; single shot; 22 Short, 22 Long, 22 LR; 18" barrel; takedown; open rear sight, bead front; uncheckered one-piece straight stock. Introduced 1904; discontinued 1917.
    Perf.: $150   **Exc.:** $100   **VGood:** $75

SAVAGE MODEL 1905
Bolt action; single shot; 22 Short, 22 Long, 22 LR; 22" barrel; takedown; open rear sight, bead front; uncheckered one-piece straight stock. Introduced 1905; discontinued 1918.
    Perf.: $150   **Exc.:** $100   **VGood:** $75

SAVAGE MODEL 1920
Bolt action; 250-3000, 300 Savage; 5-shot box magazine; 22" (250-3000), 24" (300 Savage) barrel; Mauser-type action; open rear sight, bead front; hand-checkered American walnut pistol-grip stock; slender Schnabel forearm. Introduced 1920; discontinued 1926.
    Perf.: $350   **Exc.:** $310   **VGood:** $250

SAVAGE MODEL 1920-1926
Bolt action; 250-3000, 300 Savage; 5-shot box magazine; 24" barrel; Mauser-type action; Lyman No. 54 rear peep sight; redesigned stock. Introduced 1926; discontinued 1929.
    Perf.: $350   **Exc.:** $310   **VGood:** $250

RIFLES

Savage Slide- & Lever-Action Rifles

Model 25
Model 29
Model 29-G
Model 99
Model 99-358
Model 99-375
Model 99A (Early Model)
Model 99A (Late Model)
Model 99B
Model 99C
Model 99CD
Model 99DE Citation
Model 99DL
Model 99E (Early Model)
Model 99E (Late Model)
Model 99EG (Early Model)

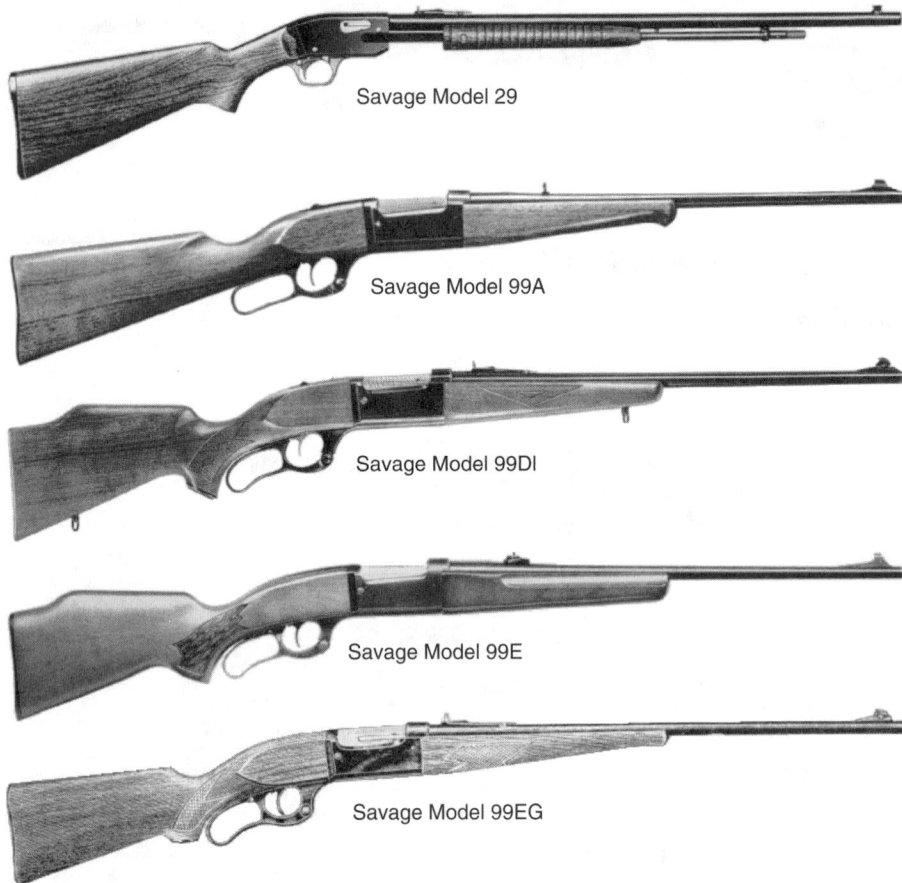

Savage Model 29

Savage Model 99A

Savage Model 99DI

Savage Model 99E

Savage Model 99EG

SAVAGE SLIDE- & LEVER-ACTION RIFLES

SAVAGE MODEL 25
Slide action; 22 Short, 22 Long, 22 LR; 20-shot (22 Short), 17-shot (22 Long), 15-shot (22 LR) tube magazine; 24" octagon barrel; hammerless; takedown; open rear sight, blade front; uncheckered American walnut pistol-grip stock; grooved slide handle. Introduced 1925; discontinued 1929.
Perf.: $270   **Exc.:** $240   **VGood:** $200

SAVAGE MODEL 29
Slide action; 22 Short, 22 Long, 22 LR; 20-shot (22 Short), 17-shot (22 Long), 15-shot (22 LR) tube magazine; hammerless; takedown; 21" overall length; weighs 5 3/4 lbs.; push-button safety rear of trigger guard; flat-top sporting rear sight, gold bead front; open rear sight, blade front; pre-WWII model with 24" octagon barrel, hand-checkered walnut pistol-grip stock, slide handle; post-WWII model with 24" round barrel, uncheckered stock. Introduced 1929; discontinued 1967.
Pre-WWII model
Perf.: $295   **Exc.:** $250   **VGood:** $200
Post-WWII model
Perf.: $235   **Exc.:** $200   **VGood:** $160

Savage Model 29-G
Same specs as Model 29 post-WWII except shorter magazine, 16-shot (22 Long), 14-shot (22 LR); short 1 1/4" slide action. Introduced 1955.
Perf.: $235   **Exc.:** $200   **VGood:** $160

SAVAGE MODEL 99
Lever action; 25-35, 30-30, 303 Savage, 32-40, 38-55,; 5-shot rotary magazine; 22" (round), 26" (half, full octagon) barrel; hammerless; open rear sight, bead front; uncheckered walnut straight-grip stock, tapered forearm. Introduced, 1899; discontinued 1922.
Perf.: $695   **Exc.:** $595   **VGood:** $495

Savage Model 99-358
Same specs as Model 99 except only 358 Win.; grooved forend; recoil pad. Introduced 1977; discontinued 1980.
Perf.: $495   **Exc.:** $400   **VGood:** $310

Savage Model 99-375
Same specs as Model 99 except only 375 Win.; recoil pad. Introduced 1980; discontinued 1980.
Perf.: $510   **Exc.:** $405   **VGood:** $315

Savage Model 99A (Early Model)
Same specs as Model 99 except 30-30, 300 Savage, 303 Savage; 24" barrel. Introduced 1920; discontinued 1936.
Perf.: $575   **Exc.:** $500   **VGood:** $400

Savage Model 99A (Late Model)
Same specs as Model 99 except 243, 250 Savage, 300 Savage, 308, 375 Win.; 20", 22" barrel; 39 3/4" (20" barrel) overall length; weighs 7 lbs.; grooved trigger; straight walnut stock with Schnabel forend. Introduced 1971; discontinued 1981.
Perf.: $395   **Exc.:** $350   **VGood:** $300

Savage Model 99B
Same specs as Model 99 except takedown design. Introduced 1922; discontinued 1936.
Perf.: $695   **Exc.:** $595   **VGood:** $495

Savage Model 99C
Same specs as Model 99 except clip magazine replaces rotary type; 22-250, 243 Win., 284 Win., 308 Win. (284 dropped in 1974); 4-shot detachable magazine, 3-shot (284); 22" barrel; 41 3/4" overall length; weighs 6 3/4 lbs.; hammerless; solid breech; Damascened bolt; case-hardened lever; blue receiver; gold-plated trigger; top tang safety; hooded gold bead ramp front, adjustable folding ramp rear sight; receiver tapped for scope mounts; walnut stock with checkered pistol grip, forend; Monte Carlo comb; swivel studs. Introduced in 1965; still in production.
Perf.: $530   **Exc.:** $430   **VGood:** $370

Savage Model 99CD
Same specs as Model 99C except 22-250, 243, 284 Win., 7mm-08, 308; white line recoil pad, pistol-grip cap; quick-detachable swivels, sling. Introduced 1976; discontinued 1980.
Perf.: $530   **Exc.:** $430   **VGood:** $370

Savage Model 99DE Citation
Same specs as Model 99 except 243, 284 Win., 308; 22" barrel; checkered American walnut Monte Carlo pistol-grip stock, forearm; engraved, receiver, tang, lever; not as fancy as Model 110PE; quick detachable swivels. Introduced 1968; discontinued 1970.
Perf.: $895   **Exc.:** $795   **VGood:** $650

Savage Model 99DL
Same specs as Model 99 except 243, 250-3000, 284, 300 Savage, 308, 358 Win.; 24", 22" barrel; 43 3/4" overall length; 6 3/4 lbs.; hammerless; solid breach; Damascened bolt; case-hardened laver; blued receiver; gold-plated trigger; top tang safety; high comb, checkered; walnut Monte Carlo pistol-grip stock; step adjustable, rear sight, gold bead front; semi-buckhorn; sling swivels; deluxe model. Introduced 1960; discontinued 1974.
Perf.: $395   **Exc.:** $340   **VGood:** $290

Savage Model 99E (Early Model)
Same specs as Model 99 except 22 Hi-Power, 250-3000, 30-30, 300 Savage, 303 Savage; 22" 24" (300 Savage) barrel; solid frame. Introduced 1920; discontinued 1936.
Perf.: $800   **Exc.:** $700   **VGood:** $550

Savage Model 99E (Late Model)
Same specs as Model 99 except 243, 250 Savage, 300 Savage, 308; 20", 22", 24" medium-weight barrel; weighs 7 1/4 lbs.; checkered hardwood pistol-grip stock; Schnabel forend; trigger guard safety. Introduced 1960; discontinued 1982.
Perf.: $395   **Exc.:** $320   **VGood:** $270

Savage Model 99EG (Early Model)
Same specs as Model 99 except 22 Hi-Power, 250-3000, 30-30, 300 Savage, 303 Savage; 22", 24" (300 Savage) barrel; solid frame model; no checkering. Introduced 1936; discontinued 1941.
Perf.: $695   **Exc.:** $595   **VGood:** $495

Savage Model 99PE

Savage Model 1903

Savage Model 1914

Savage Model 170

Savage Model 170C Carbine

Savage Slide- & Lever-Action Rifles
Model 99EG (Late Model)
Model 99F Featherweight (Early Model)
Model 99F Featherweight (Late Model)
Model 99G
Model 99H
Model 99K
Model 99M (See Savage Models 99DE, 99F Featherweight [Late Model] and 99PE)
Model 99PE
Model 99R (Early Model)
Model 99R (Late Model)
Model 99RS (Early Model)
Model 99RS (Late Model)
Model 99T Featherweight
Model 170
Model 170C Carbine
Model 1899 (See Savage Model 99)
Model 1903
Model 1909
Model 1914

Savage Model 99EG (Late Model)
Same specs as Model 99 except 243, 250-3000, 250 Savage, 300 Savage, 308, 358 Win.; 24″ barrel; 43 1/4 overall length; weighs 7 1/4 lbs.; checkered pistol-grip stock. Introduced 1946; discontinued 1960.
Perf.: $450   **Exc.:** $390   **VGood:** $330

Savage Model 99F Featherweight (Early Model)
Same specs as Model 99 except 22 Hi-Power, 250-3000, 30-30, 300 Savage, 303 Savage; 22″, 24″ (300 Savage) barrel; lightweight takedown model. Introduced 1920; discontinued 1942.
Perf.: $595   **Exc.:** $495   **VGood:** $375

Savage Model 99F Featherweight (Late Model)
Same specs as Model 99 except 243, 250-3000, 284 Win., 300 Savage, 308, 358 Win.; 22″ lightweight barrel; 41 1/4″ overall length; weighs 6 1/2 lbs.; solid breech; polished Damascened finish bolt; case-hardened lever; blued receiver; grooved trigger; checkered walnut pistol-grip stock; featherweight solid frame model. Introduced 1955; discontinued 1972.
Perf.: $495   **Exc.:** $395   **VGood:** $320

Savage Model 99G
Same specs as Model 99 except 22 Hi-Power, 250-3000, 30-30, 300 Savage, 308, 303 Savage; 5-shot rotary box-type magazine; 22″, 24″ (300 Savage) medium-weight barrel; weighs 7 3/4 lbs.; takedown; hand-checkered full pistol grip, tapered forearm; shotgun butt. Introduced 1920; discontinued 1942.
Perf.: $695   **Exc.:** $595   **VGood:** $475

Savage Model 99H
Same specs as Model 99 except 250-3000, 30-30, 303 Savage; 20″ barrel; solid frame; carbine stock with barrel band. Introduced 1931; discontinued 1942.
Perf.: $795   **Exc.:** $595   **VGood:** $375

Savage Model 99K
Same specs as Model 99 except 22 Hi-Power, 250-3000, 30-30, 300 Savage, 303 Savage; 22″ 24″ (300 Savage) barrel; takedown; Lyman peep rear sight, folding middle sight; fancy grade checkered walnut pistol-grip stock, forearm; engraved receiver, barrel. Introduced 1931; discontinued 1942.
Perf.: $2195   **Exc.:** $1895   **VGood:** $1195

Savage Model 99PE
Same specs as Model 99 except 243, 284 Win., 308; 22″ barrel; 41 3/4″ overall length; weighs 6 3/4 lbs.; Damascened bolt; gold-plated trigger; case-hardened lever; hand-checkered fancy American walnut Monte Carlo pistol-grip stock, forearm; game scene engraved on receiver sides; engraved tang, lever; quick-detachable swivels. Introduced 1966; discontinued 1970.
Perf.: $1350   **Exc.:** $995   **VGood:** $795

Savage Model 99R (Early Model)
Same specs as Model 99 except 250-3000, 300 Savage; 22″ (250-3000), 24″ (300 Savage) tapered medium-weight barrel; weighs 7 1/4 lbs.; hand-checkered American walnut oversize pistol-grip stock, forearm; solid frame model; semi-buckhorn rear sight, raised ramp gold bead front. Introduced 1936; discontinued 1942.
Perf.: $595   **Exc.:** $495   **VGood:** $450

Savage Model 99R (Late Model)
Same specs as Model 99 except 243, 250-3000, 300 Savage, 308, 358 Win.; 24″ barrel; 43 3/4″ overall length; weighs 7 1/2 lbs.; tapped for Weaver scope mount; hand-checkered American walnut oversize pistol-grip stock, semi-beavertail forearm; sling swivel studs. Introduced 1946; discontinued 1960.
Perf.: $495   **Exc.:** $440   **VGood:** $390

Savage Model 99RS (Early Model)
Same specs as Model 99 except 250-3000, 300 Savage; 22″, 24″ (300 Savage) tapered medium-weight barrel; weighs 7 1/2 lbs.; Lyman rear peep sight, gold bead front; hand-checkered American walnut oversize pistol-grip stock, forearm; sling, swivels; solid frame model. Introduced 1936; discontinued 1942.
Perf.: $695   **Exc.:** $595   **VGood:** $495

Savage Model 99RS (Late Model)
Same specs as Model 99 except 243, 250 Savage, 300 Savage, 308, 358 Win.; 24″ barrel; weighs 7 1/2 lbs.; Redfield 70LH adjustable receiver sight; milled slot for a middle sight; No. 34 Lyman gold bead front; hand-checkered American walnut oversize pistol-grip stock, forearm; quick-release sling swivel; leather sling. Introduced 1946; discontinued 1958.
Perf.: $495   **Exc.:** $450   **VGood:** $400

Savage Model 99T Featherweight
Same specs as Model 99 except 22 Hi-Power, 30-30, 300 Savage, 303 Savage; 20″ 22″ (300 Savage) barrel; hand-checkered walnut pistol-grip stock; beavertail forend; solid frame model. Introduced 1936; discontinued 1942.
Perf.: $650   **Exc.:** $575   **VGood:** $475

SAVAGE MODEL 170
Slide action; 30-30, 35 Rem.; 3-shot tubular magazine; 22″ barrel; 41 1/2″ overall length; weighs 6 3/4 lbs.; folding leaf rear sight, gold bead ramp front; receiver drilled, tapped for scope mount; select checkered American walnut pistol-grip stock; Monte Carlo comb, grooved slide handle; hard rubber buttplate. Introduced 1970; discontinued 1981.
Perf.: $200   **Exc.:** $170   **VGood:** $140

Savage Model 170C Carbine
Same specs as Model 170 except 30-30; 18 1/2″ barrel; weighs 6 lbs.; straight-comb stock. Introduced 1974; discontinued 1981.
Perf.: $210   **Exc.:** $180   **VGood:** $150

SAVAGE MODEL 1903
Slide action; 22 Short, 22 Long, 22 LR; hammerless; takedown; detachable box magazine; 24″ octagonal barrel; open rear sight, bead front; checkered one-piece pistol grip or straight stock. Introduced 1903; discontinued 1921.
Perf.: $270   **Exc.:** $220   **VGood:** $160

SAVAGE MODEL 1909
Slide action; 22 Short, 22 Long, 22 LR; detachable box magazine; 20″ round barrel; open rear sight, bead front; uncheckered one-piece pistol grip or straight stock. Introduced 1909; discontinued 1915.
Perf.: $200   **Exc.:** $165   **VGood:** $100

SAVAGE MODEL 1914
Slide action; 22 Short, 22 Long, 22 LR; 17-shot (22 LR) tubular magazine; 24″ octagonal barrel; hammerless; takedown; open rear sight, bead front; uncheckered pistol-grip stock, grooved slide handle. Introduced 1914; discontinued 1924.
Perf.: $200   **Exc.:** $170   **VGood:** $130

Savage Semi Automatic Rifles & Combos
Model 6
Model 6P
Model 6S
Model 7
Model 7S
Model 24
Model 24C
Model 24D
Model 24DL
Model 24F
Model 24F-12T Turkey Gun
Model 24FG
Model 24 Field
Model 24MDL
Model 24MS
Model 24S
Model 24V

Savage Model 6

Savage Model 6S

Savage Model 7

Savage Model 7S

Savage Model 24F

Savage Model 60

SAVAGE SEMI-AUTOMATIC RIFLES & COMBOS

SAVAGE MODEL 6
Semi-automatic; 22 Short, 22 Long, 22 LR; 21-shot (22 Short), 17-shot (22 Long), 15-shot (22 LR) tube magazine; 24″ tapered barrel with crowned muzzle; weighs 6 lbs.; takedown; open fully-adjustable sporting rear sight, gold bead front; receiver tapped for Weaver scope; pre-WWII version has checkered pistol-grip walnut stock; post-WWII model has uncheckered stock; receiver, triggerguard with silver-gray finish. Introduced 1938; discontinued 1968.
Pre-WWII model
  Perf.: $150   **Exc.:** $130   **VGood:** $110
Post-WWII model
  Perf.: $140   **Exc.:** $120   **VGood:** $100

Savage Model 6P
Same specs as Model 6 except checkered pistol grip, forend; rosewood forend tip come with 0420 scope and mount. Introduced 1965.
  Perf.: $175   **Exc.:** $150   **VGood:** $110

Savage Model 6S
Same specs as Model 6 except fully-adjustable peep rear sight with two discs, hooded front with three inserts. Introduced 1938; discontinued 1942.
  Perf.: $160   **Exc.:** $140   **VGood:** $120

SAVAGE MODEL 7
Semi-automatic; 22 Short, 22 Long, 22 LR; 5-shot detachable clip magazine; 24″ tapered barrel with crowned muzzle; weighs 6 lbs.; takedown; fully-adjustable sporting rear sight, gold bead front; takedown; pre-WWII model with checkered pistol-grip walnut stock; post-WWII model with uncheckered stock; hard rubber buttplate. Introduced 1939; discontinued 1951.
Pre-WWII model
  Perf.: $140 `   **Exc.:** $120   **VGood:** $100
Post-WWII model
  Perf.: $150   **Exc.:** $130   **VGood:** $100

Savage Model 7S
Same specs as Model 7 except fully-adjustable peep rear sight with two discs, hooded front with three inserts. Introduced 1938; discontinued 1942.
  Perf.: $150   **Exc.:** $130   **VGood:** $100

SAVAGE MODEL 24
Over/under combo; 22 LR, 22 Short, 22 Long/410 3″ or 2 ½″; 24″ separated barrels; 40 ½″ overall length; weighs 7 lbs.; takedown; barrel selection slide button on right side of receiver; full choke shotgun barrel; top-lever break-open; ramp front sight, elevation-adjustable rear; uncheckered walnut pistol-grip stock; visible hammer; single trigger; barrel selector on hammer. Introduced 1950; discontinued 1965.
  Perf.: $220   **Exc.:** $200   **VGood:** $140

Savage Model 24C
Same specs as Model 24 except 22 LR/20-ga.; 20″ barrel; Cylinder bore; weighs 5 ¾ lbs.; straight walnut stock; storage in buttstock for ten 22 LR cartridges and one 20-ga. shell; nickel finish. Introduced 1972; discontinued 1988.
  Perf.: $220   **Exc.:** $175   **VGood:** $150

Savage Model 24D
Same specs as Model 24 except 22 LR, 22 WMR/20-ga., 410; black or color case-hardened frame with engraving; checkered walnut pistol-grip stock. Introduced 1981; no longer in production.
  Perf.: $250   **Exc.:** $230   **VGood:** $180

Savage Model 24DL
Same specs as Model 24 except 22/20-ga., 410; checkered walnut Monte Carlo pistol-grip stock; checkered forend; top lever; scope dovetails; satin chrome frame with game scenes; pistol grip cap; pistol grip, buttplate white line spacer. Introduced 1962; discontinued 1969.
  Perf.: $250   **Exc.:** $220   **VGood:** $170

Savage Model 24F
Same specs as Model 24 except 22 LR, 22 Hornet, 223, 30-30/12-, 20-ga.; black Rynite composition or walnut stock; removable buttcap for storage; removable grip cap with integral compass, screwdriver. Introduced 1989; still in production. **LMSR: $449**
  Perf.: $350   **Exc.:** $275   **VGood:** $220

Savage Model 24F-12T Turkey Gun
Same specs as Model 24F except 22 Hornet, 223 over 12-gauge with 3″ chamber; camouflage Rynite stock and Full, Imp. Cyl., Mod. choke tubes. Made in U.S. by Savage. Introduced 1989; no longer in production.
  Perf.: $375   **Exc.:** $320   **VGood:** $250

Savage Model 24FG
Same specs as Model 24 except 22 LR/20-ga., 410; top lever; scope dovetails; color case-hardened frame; walnut finish hardwood straight stock; no checkering. Introduced 1972; no longer in production.
  Perf.: $240   **Exc.:** $210   **VGood:** $160

Savage Model 24 Field
Same specs as Model 24 except 22 LR, 22 WMR/20-ga., 410; weighs 6 ¾ lbs.; lightweight field model. Discontinued 1989.
  Perf.: $250   **Exc.:** $220   **VGood:** $170

Savage Model 24MDL
Same specs as Model 24 except 22 WMR/20-ga., 410; checkered walnut pistol-grip stock; top lever; scope dovetails; satin chrome frame. Introduced 1965; discontinued 1969.
  Perf.: $260   **Exc.:** $230   **VGood:** $190

Savage Model 24MS
Same specs as Model 24 except 22 WMR/20-ga., 410; sidelever; scope dovetails. Introduced 1962; discontinued 1971.
  Perf.: $200   **Exc.:** $170   **VGood:** $140

Savage Model 24S
Same specs as Model 24 except 22 LR/20-ga., 410; sidelever; scope dovetails. Introduced 1965; discontinued 1971.
  Perf.: $200   **Exc.:** $170   **VGood:** $140

Savage Model 24V
Same specs as Model 24 except 22 Hornet, 222, 223, 30-30, 357 Mag., 357 Max./20-ga. 3″; stronger receiver; color case-hardened frame; scope dovetails; walnut finished hardwood stock. Introduced 1971; discontinued 1989.
  Perf.: $270   **Exc.:** $240   **VGood:** $200

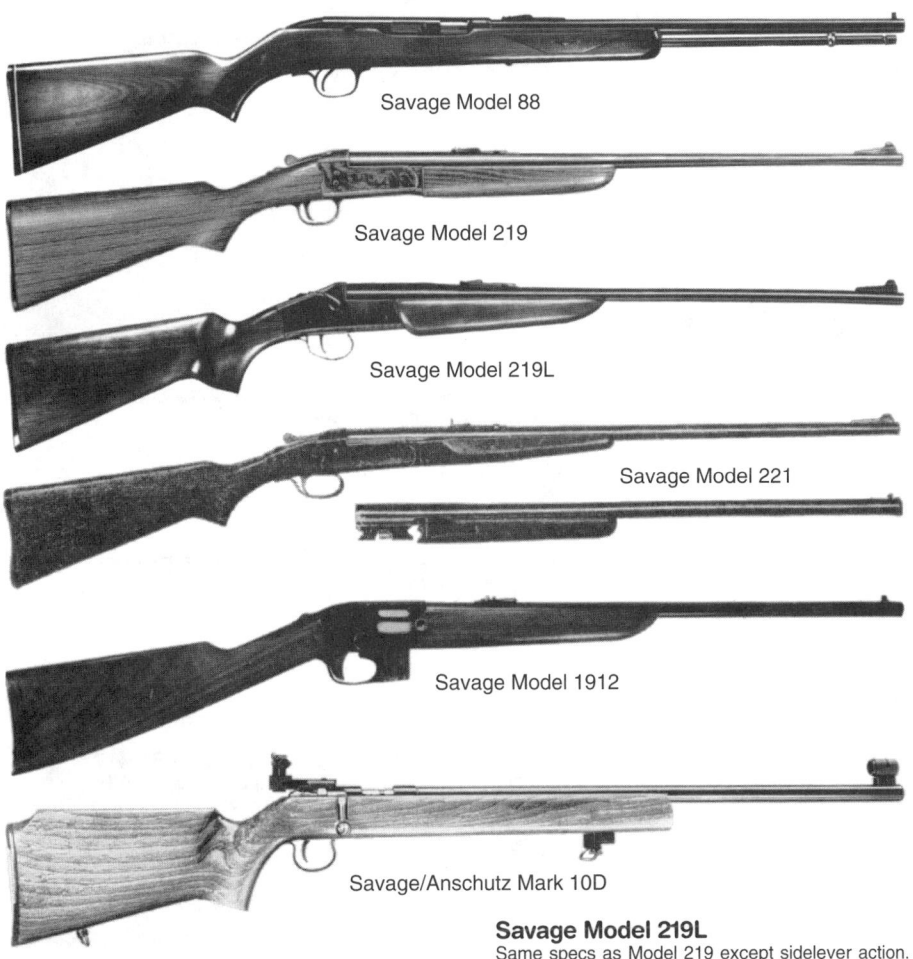

Savage Model 88

Savage Model 219

Savage Model 219L

Savage Model 221

Savage Model 1912

Savage/Anschutz Mark 10D

Savage Semi-Automatic Rifles & Combos
Model 24VS
Model 60
Model 88
Model 90 Carbine
Model 219
Model 219L
Model 221
Model 222
Model 223
Model 227
Model 228
Model 229
Model 389
Model 980DL
Model 1912

Savage-Anschutz
Mark 10
Mark 10D

Savage Model 24VS
Same specs as Model 24 except 357 Mag./20-ga.; 20″ barrel; cylinder bore; pistol-grip walnut stock; nickel finish. No longer in production.
Perf.: $270   **Exc.:** $240   **VGood:** $200

SAVAGE MODEL 60
Semi-automatic; 22 LR; 15-shot tubular magazine; 20″ barrel; 40 1/2″ overall length; weighs 6 lbs.; open rear sight, gold bead ramp front; tang slide safety; American walnut Monte Carlo stock; checkered pistol-grip, forearm; Wundhammer swell pistol grip; beavertail forend; buttplate white line spacer. Introduced 1969; discontinued 1972.
Perf.: $110   **Exc.:** $95   **VGood:** $80

SAVAGE MODEL 88
Semi-automatic; 22 LR; 15-shot tubular magazine; 20″ barrel; open rear sight, bead front; checkered American walnut pistol-grip stock; side safety. Introduced 1969; discontinued 1972.
Perf.: $100   **Exc.:** $90   **VGood:** $75

SAVAGE MODEL 90 CARBINE
Semi-automatic; 22 LR; 10-shot tubular magazine; 16 1/2″ barrel; 37 1/2″ overall length; weighs 5 3/4 lbs.; satin blued receiver; tang slide safety; folding leaf rear sight, gold bead front; uncheckered walnut carbine stock; barrel band; sling swivels. Introduced 1969; discontinued 1972.
Perf.: $100   **Exc.:** $90   **VGood:** $75

SAVAGE MODEL 219
Single shot; 22 Hornet, 25-20, 30-30, 32-20; 26″ tapered medium-weight barrel; weighs 6 lbs.; hammerless; takedown; break-open; top tang safety; adjustable flat-top rear sight, bead front on ramp; uncheckered walnut pistol-grip stock, forearm; top lever. Introduced 1938; discontinued 1965.
Perf.: $220   **Exc.:** $170   **VGood:** $120

Savage Model 219L
Same specs as Model 219 except sidelever action. Introduced 1965; discontinued 1967.
Perf.: $200   **Exc.:** $160   **VGood:** $120

SAVAGE MODEL 221
Single shot; 30-30; 26″ barrel; takedown; hammerless; break-open; open rear sight, bead front on ramp; uncheckered walnut pistol-grip stock, forearm; top lever; interchangeable 30″ 12-ga. shotgun barrel. Introduced 1939; discontinued 1960.
Perf.: $180   **Exc.:** $150   **VGood:** $120

SAVAGE MODEL 222
Single shot; 30-30; 26″ barrel; takedown; hammerless; break-open; open rear sight, bead front on ramp; uncheckered walnut pistol-grip stock, forearm; top lever; interchangeable 28″ 16-ga. shotgun barrel.
Perf.: $180   **Exc.:** $150   **VGood:** $120

SAVAGE MODEL 223
Single shot; 30-30; 26″ barrel; takedown; hammerless; break-open; open rear sight, bead front on ramp; uncheckered walnut pistol-grip stock, forearm; top lever; interchangeable 28″ 20-ga. shotgun barrel.
Perf.: $180   **Exc.:** $150   **VGood:** $120

SAVAGE MODEL 227
Single shot; 22 Hornet; 26″ barrel; takedown; hammerless; break-open; open rear sight, bead front on ramp; uncheckered walnut pistol-grip stock, forearm; top lever; interchangeable 30″ 12-ga. shotgun barrel.
Perf.: $190   **Exc.:** $150   **VGood:** $120

SAVAGE MODEL 228
Single shot; 22 Hornet; 26″ barrel; takedown; hammerless; break-open; open rear sight, bead front on ramp; uncheckered walnut pistol-grip stock, forearm; top lever; interchangeable 28″ 16-ga. shotgun barrel.
Perf.: $190   **Exc.:** $150   **VGood:** $120

SAVAGE MODEL 229
Single shot; 22 Hornet; 26″ barrel; takedown; hammerless; break-open; open rear sight, bead front on ramp; uncheckered walnut pistol-grip stock, forearm; top lever; interchangeable 28″ 20-ga. shotgun barrel.
Perf.: $200   **Exc.:** $160   **VGood:** $130

SAVAGE MODEL 389
Over/under combo; 12-ga./222, 308; 25 3/4″ barrels; hammerless; floating front mount for windage, elevation adjustment; blade front sight, folding leaf rear; oil-finished walnut stock; recoil pad; checkered grip, forend; matte finish; double triggers; swivel studs. Introduced 1988; discontinued 1990.
Perf.: $650   **Exc.:** $550   **VGood:** $450

SAVAGE MODEL 980DL
Semi-automatic; 22 LR; 15-shot tube magazine; 20″ barrel; 40 1/2″ overall length; weighs 6 lbs.; solid steel receiver; receiver grooved for scope mounting; hooded ramp front sight, folding leaf adjustable rear; checkered walnut Monte Carlo stock, pistol grip, forend. Introduced 1981.
Perf.: $120   **Exc.:** $90   **VGood:** $80

SAVAGE MODEL 1912
Semi-automatic; 22 LR; 7-shot detachable box magazine; takedown; open rear sight, bead front; uncheckered straight stock. Introduced 1912; discontinued 1916.
Perf.: $210   **Exc.:** $195   **VGood:** $145

SAVAGE-ANSCHUTZ MARK 10
Bolt action; single shot; 22 LR; 26″ barrel; 44″ overall length; weighs 8 1/2 lbs.; micrometer click-adjustable rear sight, globe front; European walnut pistol-grip target stock; cheekpiece; adjustable hand stop; adjustable single-stage trigger; sling swivels. Imported from Germany. Introduced 1967; discontinued 1972.
Perf.: $370   **Exc.:** $320   **VGood:** $270

Savage-Anschutz Mark 10D
Same specs as Mark 10 except different rear sight; Monte Carlo stock. Imported 1972 only.
Perf.: $380   **Exc.:** $330   **VGood:** $280

RIFLES

Savage-Anschutz
Mark 12
Model 54
Model 54M
Model 64
(See Anschutz Model 64)
Model 64S (See Anschutz
Model 64S)
Model 153
Model 164
Model 164M
Model 184
Model 1410
Model 1418 Sporter
Model 1518 Sporter

Savage-Stevens
Model 35
Model 35M
Model 36
Model 72 Crackshot
Model 73
Model 73Y
Model 74
Model 80

Savage/Anschutz Mark 12

Savage/Anschutz Model 54

Savage/Anschutz Model 164M

Savage/Anschutz Model 184

SAVAGE-ANSCHUTZ MARK 12
Bolt action; single shot; 22 LR; 26" heavy barrel; 43" overall length; weighs 8 lbs.; globe front sight, micro-click peep rear; walnut-finished hardwood stock; thumb groove, Wundhammer pistol-grip swell, adjustable hand stop; sling swivels. Imported from Germany. Introduced 1978; discontinued 1981.
Perf.: $400   **Exc.:** $350   **VGood:** $300

SAVAGE-ANSCHUTZ MODEL 54
Bolt action; 22 LR; 5-shot clip magazine; 22 1/2" barrel; 42" overall length; weighs 6 3/4 lbs.; folding leaf sight, hooded ramp gold bead front; receiver grooved for tip-off mount, tapped for scope blocks; French walnut stock with Monte Carlo roll-over comb; Schnabel forearm tip; hand-checkered pistol grip, forearm; adjustable single-stage trigger; wing safety. Introduced 1966; discontinued 1981.
Perf.: $695   **Exc.:** $595   **VGood:** $495

Savage-Anschutz Model 54M
Same specs as Model 54 except 22 WMR. Introduced 1973; discontinued 1981.
Perf.: $725   **Exc.:** $625   **VGood:** $525

SAVAGE-ANSCHUTZ MODEL 153
Bolt action; 222 Rem.; 24" barrel; 43" overall length; weighs 6 3/4 lbs.; recessed bolt head with double locking lugs; wing-type safety; adjustable single-stage trigger; folding leaf open rear sight, hooded ramp front; skip-checkered high grade French walnut stock; hand-carved roll-over cheekpiece; rosewood grip cap, forearm tip with white inlay; sling swivels. Introduced 1964; discontinued 1967.
Perf.: $645   **Exc.:** $540   **VGood:** $430

SAVAGE-ANSCHUTZ MODEL 164
Bolt action; 22 LR; 5-shot detachable clip magazine; 24" barrel; 40 3/4" overall length; receiver grooved for tip-off mount; hooded ramp gold bead front, folding leaf rear; European walnut stock; hand-

checkered pistol grip, forearm; Monte Carlo comb, cheekpiece; sliding side safety; Schnabel forearm; fully-adjustable single-stage trigger. Introduced 1966; discontinued 1981.
Perf.: $395   **Exc.:** $350   **VGood:** $300

Savage-Anschutz Model 164M
Same specs as Model 164 except 22 WMR; 4-shot magazine.
Perf.: $430   **Exc.:** $380   **VGood:** $330

SAVAGE-ANSCHUTZ MODEL 184
Bolt action; 22 LR; 5-shot detachable clip magazine; 21 1/2" barrel; folding leaf rear sight, hooded ramp front; receiver grooved for scope mounts; European walnut stock with Monte Carlo comb, Schnabel forearm; hand-checkered pistol grip, forearm. Introduced 1966; discontinued 1974.
Perf.: $395   **Exc.:** $350   **VGood:** $300

SAVAGE-ANSCHUTZ MODEL 1410
Bolt action; single shot; 22 LR; 27 1/2" barrel; 50" overall length; weighs 15 1/2 lbs.; no sights; receiver grooved for Anschutz sights; scope blocks; International-type stock with thumb hole, aluminum hook buttplate; vertically adjustable buttplate. Introduced 1968.
Perf.: $400   **Exc.:** $350   **VGood:** $300

SAVAGE-ANSCHUTZ MODEL 1418 SPORTER
Bolt action; 22 LR; 5-shot detachable clip magazine; 24" barrel; receiver grooved for tip-off mount; European Mannlicher stock, inlays, skip-line hand-checkering; double-set or single-stage trigger. Made in Germany. Introduced 1981; discontinued 1981.
Perf.: $695   **Exc.:** $595   **VGood:** $495

SAVAGE-ANSCHUTZ MODEL 1518 SPORTER
Bolt action; 22 WMR; 5-shot detachable clip magazine; 24" barrel; receiver grooved for tip-off mount; European Mannlicher stock, inlays, skip-line hand-checkering; double-set or single-stage trigger. Made in Germany. Introduced 1981; discontinued 1981.
Perf.: $755   **Exc.:** $645   **VGood:** $535

SAVAGE-STEVENS MODEL 35
Bolt action; 22 LR; 5-shot clip magazine; 22" barrel; 41" overall length; weighs 4 3/4 lbs.; ramp front sight, step-adjustable rear; walnut finished hardwood stock; checkered pistol grip, forend; receiver grooved for scope mounts. Introduced 1982; discontinued 1985.
Perf.: $90   **Exc.:** $80   **VGood:** $55

Savage-Stevens Model 35M
Same specs as Model 35 except 22 WMR.
Perf.: $100   **Exc.:** $90   **VGood:** $70

SAVAGE-STEVENS MODEL 36
Bolt action; single shot; 22 LR; 22" barrel; 41" overall length; 4 3/4 lbs.; ramp front sight, step-adjustable open rear; walnut-finished hardwood stock; checkered pistol grip, forend; receiver grooved for scope mounts. Introduced 1983.
Perf.: $90   **Exc.:** $80   **VGood:** $60

SAVAGE-STEVENS MODEL 72 CRACKSHOT
Lever action; single shot; 22 Short, 22 Long, 22 LR; 22" octagonal barrel; falling block; open rear sight, bead front; uncheckered American walnut straight-grip stock, forearm; color case-hardened frame. Introduced 1972; discontinued 1987.
Perf.: $170   **Exc.:** $140   **VGood:** $110

SAVAGE-STEVENS MODEL 73
Bolt action; single shot; 22 Short, 22 Long, 22 LR; 20" barrel; 38" overall length; weighs 4 1/2 lbs.; plated trigger; takedown screw; cocking knob; open rear sight, blade front; receiver grooved for scope mounting; one-piece uncheckered hardwood pistol grip stock. Introduced 1965; discontinued 1980.
Perf.: $80   **Exc.:** $70   **VGood:** $60

Savage-Stevens Model 73Y
Same specs as Model 73 except 18" barrel; 35" overall length; weighs 4 lbs.; shorter buttstock. Introduced 1965; discontinued 1980.
Perf.: $80   **Exc.:** $70   **VGood:** $60

SAVAGE-STEVENS MODEL 74
Lever action; single shot; 22 Short, 22 Long, 22 LR; 22" round barrel; hardwood stock; black-finished frame. Introduced 1972; discontinued 1974.
Perf.: $165   **Exc.:** $135   **VGood:** $110

SAVAGE-STEVENS MODEL 80
Semi-automatic; 22 LR; 15-shot tube magazine; 20" barrel; 40" overall length; weighs 6 lbs.; blade front sight, fully-adjustable open rear; select walnut Monte Carlo stock with checkered forens, pistol grip; buttplate, pistol grip white line spacers; receiver grooved for scope mounting. Introduced 1977; no longer in production.
Perf.: $100   **Exc.:** $90   **VGood:** $80

Savage-Stevens Model 80

Savage-Stevens Model 125

Savage-Stevens Model 88

Sedgely Springfield Sporter

Serrifle Schuetzen Model

RIFLES

Savage-Stevens
Model 88
Model 89 Crackshot
Model 120
Model 125
Model 141

Schultz & Larsen
Model 47
Model 54
Model 54J
Model 61
Model 62
Model 68

Sedgely Springfield
Sporter
Sporter Carbine

Serrifle
Schuetzen Model

C. Sharps Arms
Model 1874 Sporting

SAVAGE-STEVENS MODEL 88
Semi-automatic; 22 LR; 15-shot tube magazine; 20″ barrel; 40 1/2″ overall length; weighs 5 3/4 lbs.; open rear sight, ramp gold bead front; checkered hardwood straight stock with pistol grip; slide safety on tang. Introduced 1969; discontinued 1972.

    Perf.: $120   **Exc.:** $100   **VGood** $80

SAVAGE-STEVENS MODEL 89 CRACKSHOT
Lever action; single shot; 22 LR; 18 1/2″ barrel; 35″ overall length; weighs 5 lbs.; sporting front sight, open rear; walnut finished hardwood straight stock; hammer cocked by hand; automatic ejection; hard rubber buttplate black satin finish. Introduced 1977; discontinued 1985.

    Perf.: $170   **Exc.:** $140   **VGood:** $105

SAVAGE-STEVENS MODEL 120
Bolt action; single shot; 22 Short, 22 Long, 22 LR; tubular magazine; 20″ round, tapered barrel; action cocks on opening of bolt; sporting front/rear sights; grooved for scope mounts; walnut-finished hardwood stock; pistol-grip; thumb safety; double extractors; recessed bolt face. Introduced 1979; discontinued 1983.

    Perf.: $70   **Exc.:** $60   **VGood:** $50

SAVAGE-STEVENS MODEL 125
Bolt action; single shot; 22″ barrel; 39″ overall length; sporting front sight, open rear with elevator; walnut-finished stock; blued finish. Introduced 1981; no longer made.

    Perf.: $80   **Exc.:** $70   **VGood:** $60

SAVAGE-STEVENS MODEL 141
Bolt action; 22 LR; 5-shot clip magazine; 22″ barrel; 40 3/4″ overall length; weighs 5 1/2 lbs.; right side slide safety; adjustable single-stage trigger; gold bead hooded ramp front sight, folding leaf rear; receiver grooved for scope mount, tapped for aperture sight; French walnut Monte Carlo stock with cheekpiece; skip-line hand-checkered pistol grip, forend; rosewood forend tip with white inlay; buttplate with white-line spacer; sling swivel studs. Introduced 1966.

    Perf.: $150   **Exc.:** $125   **VGood:** $100

SCHULTZ & LARSEN MODEL 47
Bolt action; single shot; 22 LR; 28 1/2″ barrel; micrometer receiver sight, globe front; free-rifle style European walnut stock; cheekpiece; thumbhole; buttplate; palm rest; sling swivels; set trigger. Made in Germany. No longer in production.

    Perf.: $695   **Exc.:** $595   **VGood:** $495

SCHULTZ & LARSEN MODEL 54
Bolt action; single shot; 6.5x55mm, American calibers available on special order; 27 1/2″ barrel; micrometer receiver sight, globe front; free-rifle style European walnut stock; adjustable buttplate; cheekpiece; palm rest; sling swivels. No longer in production.

    Perf.: $895   **Exc.:** $795   **VGood:** $625

SCHULTZ & LARSEN MODEL 54J
Bolt action; 270 Win., 30-06, 7x61 Sharpe & Hart; 3-shot magazine; 24″, 26″ barrels; no sights; hand-checkered European walnut sporter stock; Monte Carlo comb, cheekpiece.

    Perf.: $695   **Exc.:** $595   **VGood:** $495

SCHULTZ & LARSEN MODEL 61
Bolt action; single shot; 22 LR; 28 1/2″ barrel; micrometer receiver sight, globe front; free-rifle style European walnut stock; cheekpiece; buttplate; palm rest; sling swivels; set trigger. No longer in production.

    Perf.: $895   **Exc.:** $795   **VGood:** $625

SCHULTZ & LARSEN MODEL 62
Bolt action; single shot; various calibers; 28 1/2″ barrel; micrometer receiver sight, globe front; free-rifle style European walnut stock; cheekpiece; buttplate; palm rest; sling swivels; set trigger. No longer in production.

    Perf.: $995   **Exc.:** $895   **VGood:** $710

SCHULTZ & LARSEN MODEL 68
Bolt action; 22-250, 243, 6mm Rem., 264 Win. Mag., 270, 30-06, 308, 7x61 Sharpe & Hart, 7mm Rem. Mag., 8x57JS, 300 Win. Mag., 308 Norma Mag., 338 Win. Mag., 358 Norma Mag., 458 Win. Mag.; 24″ barrel; no sights; select French walnut sporting stock; adjustable trigger. No longer in production.

    Perf.: $795   **Exc.:** $695   **VGood:** $595

SEDGELY SPRINGFIELD SPORTER
Bolt action; 218 Bee, 22-3000, 220 Swift, 22-4000, 22 Hornet, 25-35, 250-3000, 257 Roberts, 270 Win., 30-06, 7mm; 24″ barrel; Lyman No. 48 rear sight, bead front on matted ramp; hand-checkered walnut stock; sling swivels, grip cap. Introduced 1928; discontinued 1941.

    Perf.: $1295   **Exc.:** $1125   **VGood:** $895

Left-hand model

    Perf.: $1495   **Exc.:** $1295   **VGood:** $995

Sedgely Springfield Sporter Carbine
Same specs as Springfield Sporter except 20″ barrel; full-length walnut stock.

    Perf.: $1495   **Exc.:** $1295   **VGood:** $995

SERRIFLE SCHUETZEN MODEL
Single shot; 22, 32, 38, 41, 44, 45; octagon, half-octagon or round barrel to 32″ at customer's preference; no sights; furnished with scope blocks; fancy Helm-pattern walnut stock; Niedner-type firing pin; coil spring striker based on Winchester Hi-Wall action. Introduced 1984; discontinued 1986.

    Perf.: $1295   **Exc.:** $1050   **VGood:** $895

C. SHARPS ARMS MODEL 1874 SPORTING
Single shot; 40, 45, 50; 30″ octagon barrel; weighs 9 1/2 lbs.; Lawrence-type open rear sight, blade front; American walnut stock; color case-hardened receiver, buttplate, barrel bands; blued barrel; recreation of original Sharps models in six variations. Introduced 1985; no longer in production.

Sporting Rifle No. 1

    Perf.: $795   **Exc.:** $695   **VGood:** $595

Sporting Rifle No. 3

    Perf.: $750   **Exc.:** $695   **VGood:** $595

Long Range Express Rifle

    Perf.: $895   **Exc.:** $795   **VGood:** $695

Military Rifle

    Perf.: $895   **Exc.:** $795   **VGood:** $695

Carbine

    Perf.: $795   **Exc.:** $695   **VGood:** $595

Business Rifle

    Perf.: $750   **Exc.:** $695   **VGood:** $595

RIFLES

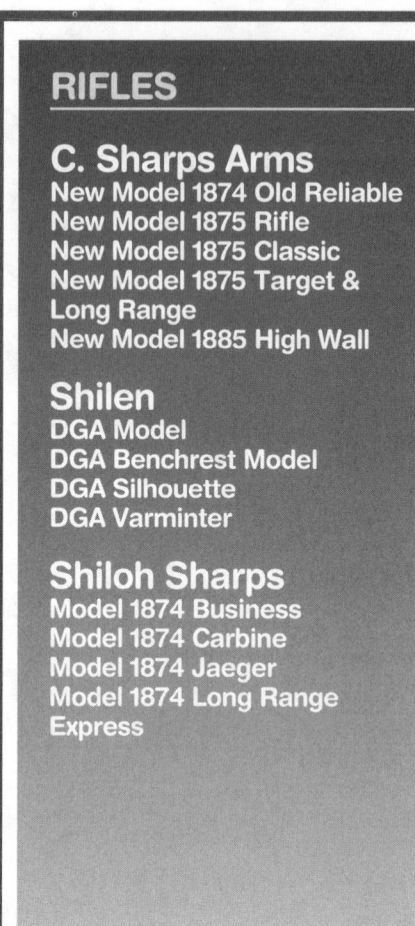

C. Sharps Arms
New Model 1874 Old Reliable
New Model 1875 Rifle
New Model 1875 Classic
New Model 1875 Target &
Long Range
New Model 1885 High Wall

Shilen
DGA Model
DGA Benchrest Model
DGA Silhouette
DGA Varminter

Shiloh Sharps
Model 1874 Business
Model 1874 Carbine
Model 1874 Jaeger
Model 1874 Long Range
Express

C. Sharps Arms New Model 1875 Classic

C. Sharps Arms New Model 1875 Rifle

C. Sharps Arms New Model 1875 Rifle

Shilen DGA Varminter

Shiloh Sharps Model 1874 Jaeger

Shiloh Sharps Model 1874 Long Range Express

C. SHARPS ARMS NEW MODEL 1874 OLD RELIABLE

Single shot; 40-50, 40-70, 40-90, 45-70, 45-90, 45-100, 45-110, 45-120, 50-70, 50-90, 50-140; 26", 28", 30" tapered, octagon barrel; weighs 10 lbs.; blade front sight, buckhorn rear; drilled, tapped for tang sight installation; straight-grip American black walnut stock; shotgun butt; heavy forend with Schnabel tip; recreation of original C. Sharps design; double-set triggers. Manufactured by Shiloh Rifle Mfg. Co. Reintroduced 1991; still in production.
LMSR: $995
  Perf.: $895   **Exc.: $795**   **VGood: $695**

C. SHARPS ARMS NEW MODEL 1875 RIFLE

Single shot; 22 LR, 32-40 Ballard, 38-55 Ballard, 38-56 WCF, 40-65 WCF, 40-90 3 1/4" 40-90 2 5/8" 40-70 2 1/10" 40-70 2 1/4" 40-70 2 1/2" 40-50 1 11/16" 40-50 1 7/8" 45-90, 40-70, 45-100, 45-110, 45-120, 50-70, 50-90, 50-140; 24", 26", 28", 30", 32", 34" barrel; improved Lawrence-type buckhorn rear sight, blade front; straight-grip walnut stock; case-colored receiver; reproduction of 1875 Sharps rifle. Manufactured by Shiloh Rifle Mfg. Co. Introduced 1986; still in production. **LMSR: $825**
Carbine model
  Perf.: $725   **Exc.: $610**   **VGood: $495**
Saddle Ring model
  Perf.: $830   **Exc.: $720**   **VGood: $610**
Sporting model
  Perf.: $820   **Exc.: $710**   **VGood: $595**
Business model
  Perf.: $720   **Exc.: $610**   **VGood: $495**

C. Sharps Arms New Model 1875 Classic

Same specs as New Model 1875 except 30" full octagon barrel; weighs 10 lbs.; Rocky Mountain buckhorn rear sight; crescent buttplate with toe plate; Hartford-style forend with German silver nose cap. Introduced 1987; still in production. **LMSR: $1075**
  Perf.: $895   **Exc.: $795**   **VGood: $695**

C. Sharps Arms New Model 1875 Target & Long Range

Same specs as New Model 1875 except available in all listed calibers except 22 LR; 34" tapered octagon barrel; globe with post front sight, Long Range Vernier tang sight with windage adjustments; pistol-grip stock with cheek rest; checkered steel buttplate. Made in U.S. by C. Sharps Arms Co., distributed by Montana Armory, Inc. Introduced 1991; still produced. **LMSR: $1165**
  Perf.: $995   **Exc.: $895**   **VGood: $795**

C. SHARPS ARMS NEW MODEL 1885 HIGH WALL

Single shot; 22 LR, 22 Hornet, 219 Zipper, 30-40 Krag, 32-40, 38-55, 40-65 WCF, 45-70; 26", 28", 30" Douglas Premium tapered octagon barrels; Marble's ivory bead front sight, #66 long-blade, flat-top rear with reversible notch, elevator; American black walnut stock; recreation of original octagon-top High Wall; coil-spring action; tang drilled for tang sight; color case-hardened finish on most external parts. Introduced 1991; no longer in production.
  Perf.: $1095   **Exc.: $995**   **VGood: $775**

SHILEN DGA MODEL

Bolt action; 17 Rem., 222, 223, 22-250, 220 Swift, 6mm Rem., 243 Win., 250 Savage, 257 Roberts, 284 Win., 308 Win., 358 Win.; 3-shot magazine; 24" barrel; no sights; uncheckered Monte Carlo walnut stock; cheekpiece; pistol-grip; sling swivels. Introduced 1976; discontinued 1987.
  Perf.: $1295   **Exc.: $995**   **VGood: $795**

Shilen DGA Benchrest Model

Same specs as DGA Model except single shot; 26" medium or heavy barrel; classic or thumbhole walnut or fiberglass stock. Introduced 1977; no longer in production.
  Perf.: $1295   **Exc.: $995**   **VGood: $795**

Shilen DGA Silhouette

Same specs as DGA Model except 308 Win.; 25" heavy barrel. No longer in production.
  Perf.: $1250   **Exc.: $995**   **VGood: $750**

Shilen DGA Varminter

Same specs as DGA Model except 25" heavy barrel. Introduced 1976; discontinued 1987.
  Perf.: $1250   **Exc.: $995**   **VGood: $750**

SHILOH SHARPS MODEL 1874 BUSINESS

Single shot; 40-50 BN, 40-70 BN, 40-90 BN, 45-70 ST, 45-90 ST, 50-70 ST, 50-100 ST, 32-40, 38-55; 28" heavy round barrel; blade front sight, buckhorn rear; straight-grip steel buttplate; double-set trigger. Introduced 1986; still in production. **LMSR: $1010**
  Perf.: $950   **Exc.: $850**   **VGood: $695**

Shiloh Sharps Model 1874 Carbine

Same specs as Model 1874 Business except 24" round barrel; single trigger. Introduced 1986; no longer in production.
  Perf.: $950   **Exc.: $850**   **VGood: $695**

Shiloh Sharps Model 1874 Jaeger

Same specs as Model 1874 Business except 30-40, 30-30, 307 Win., 45-70; 26" half-octagon lightweight barrel; standard supreme black walnut stock; shotgun, pistol grip or military-style butt. Introduced 1986; no longer in production.
  Perf.: $895   **Exc.: $795**   **VGood: $650**

Shiloh Sharps Model 1874 Long Range Express

Same specs as Model 1874 Business except 34" tapered octagon barrel; sporting tang rear sight, globe front; semi-fancy, oil-finished pistol-grip walnut stock; shotgun-type buttplate; Schnabel forend. Introduced 1985; still in production. **LMSR: $1134**
  Perf.: $995   **Exc.: $895**   **VGood: $795**

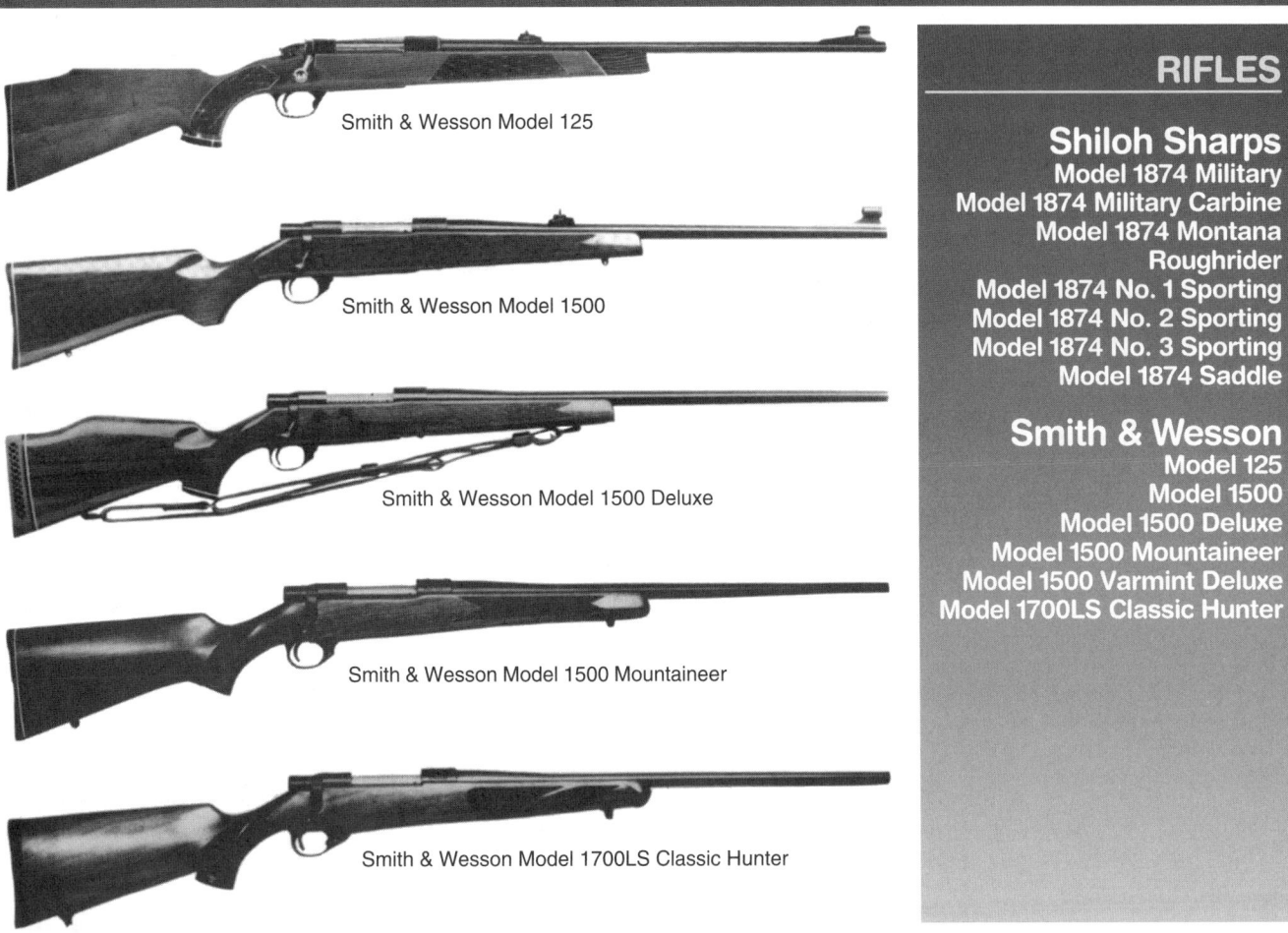

Smith & Wesson Model 125

Smith & Wesson Model 1500

Smith & Wesson Model 1500 Deluxe

Smith & Wesson Model 1500 Mountaineer

Smith & Wesson Model 1700LS Classic Hunter

RIFLES

Shiloh Sharps
Model 1874 Military
Model 1874 Military Carbine
Model 1874 Montana
Roughrider
Model 1874 No. 1 Sporting
Model 1874 No. 2 Sporting
Model 1874 No. 3 Sporting
Model 1874 Saddle

Smith & Wesson
Model 125
Model 1500
Model 1500 Deluxe
Model 1500 Mountaineer
Model 1500 Varmint Deluxe
Model 1700LS Classic Hunter

Shiloh Sharps Model 1874 Military

Same specs as Model 1874 Business excepty 40-50x1 $^{11}/_{16}$″ BN, 40-70x2 $^{1}/_{10}$″ BN, 40-90 BN, 45-70x2 $^{1}/_{10}$″ ST, 50-70 ST; 30″ round barrel; Lawrence-style rear ladder sight, iron block front; semi-fancy, oil-finished pistol-grip walnut stock; military butt; Schnabel forend; patchbox; three barrel bands; single trigger. Introduced 1985; no longer in production.

Perf.: $810   **Exc.:** $740   **VGood:** $650

Shiloh Sharps Model 1874 Military Carbine

Same specs as Model 1874 Business except 40-70 BN, 45-70, 50-70; 22″ round barrel; blade-type front sight, full buckhorn ladder-type rear; semi-fancy, oil-finished pistol-grip walnut stock; Schnabel forend; steel buttplate, saddle bar and ring; military-type buttstock; patchbox; single trigger; barrel band on forend. Introduced 1989; no longer in production.

Perf.: $810   **Exc.:** $745   **VGood:** $660

Shiloh Sharps Model 1874 Montana Roughrider

Same specs as Model 1874 Business except 30-40, 30-30, 40-50x1 $^{11}/_{16}$ BN, 40-70x2 $^{1}/_{10}$, 45-70x2 $^{1}/_{10}$ ST; 24″, 26″, 28″, 30″, 34″ half-octagon or full-octagon barrel; globe front and tang sight optional; standard supreme or semi-fancy oil-finished pistol-grip walnut stock; shotgun, pistol grip or military-style butt; Schnabel forend. Introduced 1985; still in production. **LMSR: $904**

Perf.: $825   **Exc.:** $725   **VGood:** $595

Shiloh Sharps Model 1874 No. 1 Sporting

Same specs as Model 1874 Business except 40-50 BN, 40-90 BN, 45-70 ST, 45-110 ST, 50-70 ST, 50-90 ST, 50-110 ST, 32-40, 38-55, 40-70 ST, 40-90 ST; 30″ octagon barrel; semi-fancy, oil-finished pistol-grip walnut stock; shotgun-type buttplate; Schnabel forend. Introduced 1985; still in production. **LMSR: $1108**

Perf.: $950   **Exc.:** $850   **VGood:** $695

Shiloh Sharps Model 1874 No. 2 Sporting

Same specs as Model 1874 Business except 45-70, 45-90, 45-120, 50-90, 50-140; 30″ octagon barrel; semi-fancy, oil-finished pistol-grip walnut stock; shotgun-type buttplate; Schnabel forend. No longer in production.

Perf.: $995   **Exc.:** $895   **VGood:** $795

Shiloh Sharps Model 1874 No. 3 Sporting

Same specs as Model 1874 Business except 40-50 BN, 40-90 BN, 45-70 ST, 45-110 ST, 50-70 ST, 50-110 ST, 32-40, 38-55, 40-70 ST, 40-90 ST; 30″ octagon barrel; oil-finished straight-grip stock standard wood; shotgun-type buttplate. Introduced 1985; still in production. **LMSR: $1004**

Perf.: $950   **Exc.:** $810   **VGood:** $650

Shiloh Sharps Model 1874 Saddle

Same specs as Model 1874 Business except 26″ octagon barrel; semi-fancy shotgun-type butt. Introduced 1986; still in production. **LMSR: $1062**

Perf.: $950   **Exc.:** $850   **VGood:** $695

SMITH & WESSON MODEL 125

Bolt action; 270 Win., 30-06; 5-shot magazine; 24″ barrel; step-adjustable rear sight, hooded ramp front; action drilled, tapped for scope mounts; thumb safety; standard grade with hand-checkered stock of European walnut; deluxe grade adds rosewood forearm tip, pistol-grip cap. Introduced 1973; discontinued 1973.

Standard grade
Perf.: $395   **Exc.:** $350   **VGood:** $300
Deluxe grade
Perf.: $450   **Exc.:** $400   **VGood:** $350

SMITH & WESSON MODEL 1500

Bolt action; 22-250, 222, 223, 25-06, 243, 270, 30-06, 308, 7mm Rem. Mag., 300 Win. Mag., 338 Win. Mag.; 5-shot box magazine; 22″, 24″ (magnum)

barrel; hooded gold bead front sight, open round-notch fully-adjustable rear; checkered American walnut stock; single-set trigger; one-piece trigger guard; hinged floorplate; quick-detachable swivel studs; composition non-slip buttplate; rubber recoil pad (magnums). Introduced 1979; discontinued 1984.

Perf.: $375   **Exc.:** $330   **VGood:** $300

Smith & Wesson Model 1500 Deluxe

Same specs as Model 1500 except no sights; decorative scrollwork on floorplate; skip-line checkering; pistol-grip cap with S&W seal; sling, swivels. Introduced 1980; discontinued 1984.

Perf.: $375   **Exc.:** $330   **VGood:** $290

Smith & Wesson Model 1500 Mountaineer

Same specs as Model 1500 except 22″, 24″ barrel; no sights; drilled, tapped for scope mounts; satin-finished, checkered American walnut stock; recoil pad (magnum). Introduced 1983; discontinued 1984.

Perf.: $350   **Exc.:** $330   **VGood:** $280

Smith & Wesson Model 1500 Varmint Deluxe

Same specs as Model 1500 except 222, 22-250, 223; 22″ heavy barrel; oil-finished stock; quick-detachable swivels; fully-adjustable trigger; blued or Parkerized finish. Introduced 1982; discontinued 1984.

Perf.: $370   **Exc.:** $340   **VGood:** $300

SMITH & WESSON MODEL 1700LS CLASSIC HUNTER

Bolt action; 243, 270, 30-06; 5-shot removable magazine; hooded globe bead front sight, open round-notch fully-adjustable rear; checkered American classic walnut stock, tapered forend, Schnabel tip; ribbon checkering pattern; black rubber buttpad; flush-mounted sling swivels; jeweled bolt body; knurled bolt knob. Introduced 1983; discontinued 1984.

Perf.: $400   **Exc.:** $350   **VGood:** $310

Smith & Wesson
Model A
Model B
Model C
Model D
Model E

Sovereign
TD 22

Springfield Armory, Geneseo, IL
(See Springfield, Inc.)

Springfield, Inc.
BM-59 Italian Model
BM-59 Alpine Paratrooper
BM-59 Nigerian Model
M1A
Match M1A
Super Match M1A
Tactical M1A/M-21
M-6 Scout Rifle/Shotgun

Smith & Wesson Model A

Smith & Wesson Model B

Smith & Wesson Model C

Smith & Wesson Model D

Smith & Wesson Model E

SMITH & WESSON MODEL A
Bolt action; 22-250, 243 Win., 270 Win., 308 Win., 30-06, 7mm Rem. Mag., 300 Win. Mag.; 3-shot (magnum), 5-shot magazine; 23 3/4" barrel; folding leaf rear sight, hooded ramp front; hand-checkered European walnut stock; Monte Carlo, rosewood pistol-grip cap, forend tip; sling swivels. Manufactured in Sweden by Husqvarna. Introduced 1969; discontinued 1972.
Perf.: $400   **Exc.:** $350   **VGood:** $310

Smith & Wesson Model B
Same specs as Model A except 243 Win., 270 Win., 30-06, 308; 23 3/4" light barrel; three-leaf folding rear sight, hooded German silver bead front; Monte Carlo stock; Schnabel forend tip. Introduced 1969; discontinued 1972.
Perf.: $400   **Exc.:** $350   **VGood:** $290

Smith & Wesson Model C
Same specs as Model A except 243 Win., 270 Win., 30-06, 308; 23 3/4" light barrel; open sporting rear sight, hooded German silver bead front; straight-comb stock; cheekpiece, Schnabel forend tip. Introduced 1969; discontinued 1972.
Perf.: $400   **Exc.:** $350   **VGood:** $310

Smith & Wesson Model D
Same specs as Model A except 243 Win., 270 Win., 30-06, 308; 20 3/4" light barrel; straight comb full-length Mannlicher-type stock; cheekpiece; open sporting rear sight, hooded German silver bead front. Introduced 1969; discontinued 1972.
Perf.: $475   **Exc.:** $400   **VGood:** $365

Smith & Wesson Model E
Same specs as Model A except 243 Win., 270 Win., 30-06, 308; 20 3/4" light barrel; full-length Mannlicher-type Monte Carlo stock; Schnabel forend tip; open sporting rear sight, hooded German silver bead front. Introduced 1969; discontinued 1972.
Perf.: $475   **Exc.:** $400   **VGood:** $355

SOVEREIGN TD 22
Semi-automatic; 22 LR; 10-shot clip; 21" barrel; 41" overall length; weighs 6 1/2 lbs.; takedown; fully-adjustable open rear sight, hooded ramp front; walnut-finished hardwood stock; blued finish. Introduced 1986; no longer in production.
Perf.: $90   **Exc.:** $80   **VGood:** $60

SPRINGFIELD INC. BM-59 ITALIAN MODEL
Semi-automatic; 7.62mm NATO (308 Win.); 20-shot box magazine; 19 3/8" barrel; 43 3/4" overall length; weighs 9 1/4 lbs.; military square-blade front sight, click-adjustable peep rear; walnut stock with trapped rubber buttpad; full military-dress Italian service rifle with winter trigger, grenade launcher and sights, tri-compensator, bipod. Refined version of the M-1 Garand. Made in Italy, assembled by Springfield Armory, Inc. Introduced 1981; no longer available.
Perf.: $1795   **Exc.:** $1495   **VGood:** $1195

Springfield Inc. BM-59 Alpine Paratrooper
Same specs as BM-59 Italian Model except Beretta-made folding metal buttstock with pistol grip. Made in Italy, assembled in U.S. by Springfield Armory, Inc. Introduced 1981; no longer available.
Perf.: $1995   **Exc.:** $1695   **VGood:** $1295

Springfield Inc. BM-59 Nigerian Model
Same specs as BM-59 Italian Model except Beretta-made vertical pistol grip stock. Made in Italy, assembled in U.S. by Springfield Armory, Inc. Introduced 1981; no longer available.
Perf.: $2050   **Exc.:** $1695   **VGood:** $1350

SPRINGFIELD INC. M1A
Semi automatic; gas-operated; 308 Win. (7.62 NATO), 243 Win., 7mm-08; 5-, 10-, 20-shot box magazine; 22" barrel; flash suppressor; adjustable aperture rear sight, blade front; walnut, birch or fiberglass stock; fiberglass hand guard; sling swivels, sling. Same general specs as military M14 sans full-auto capability.

Maker is private Illinois firm, not a U.S. government facility. Introduced 1974; still in production. **LMSR:** $1329
Perf.: $995   **Exc.:** $895   **VGood:** $795
Fiberglass stock
Perf.: $950   **Exc.:** $850   **VGood:** $750

Springfield Inc. Match M1A
Same specs as M1A except National Match grade barrel, sights; glass-bedded American walnut or fiberglass stock; better trigger pull; modified gas system, mainspring guide. Still in production. **LMSR:** $1995
Perf.: $1410   **Exc.:** $1100   **VGood:** $955

Springfield Inc. Super Match M1A
Same specs as M1A except heavier, premium-grade barrel; National Match sights; glass-bedded oversize American walnut or fiberglass stock; better trigger pull; modified gas system, mainspring guide. Still in production. **LMSR:** $2449
Perf.: $1620   **Exc.:** $1295   **VGood:** $995

Springfield Inc. Tactical M1A/M-21
Same specs as M1A Super Match except Douglas Premium match barrel and special sniper stock with adjustable cheekpiece; weighs 11 1/4 lbs. From Springfield, Inc. **LMSR:** $2975
Perf.: $2125   **Exc.:** $1675   **VGood:** $1450

SPRINGFIELD INC. M-6 SCOUT RIFLE/SHOTGUN
Combination gun; 22 LR, 22 Hornet over 410-bore; 18 1/4" barrel; 32" overall length; weighs 4 1/2 lbs.; blade front sight, military aperture for 22, V-notch for 410; folding steel stock with storage for fifteen 22 LR, four 410 shells; all-metal construction; designed for quick disassembly and minimum maintenance; folds for compact storage. Early examples made in U.S. by Springfield, Inc. Current guns imported from the Czech Republic. Introduced 1982; reintroduced 1995. **LMSR:** $185
Perf.: $125   **Exc.:** $100   **VGood:** $85

Springfield, Inc. M1A/M-21

Springfield, Inc. M-6 Scout Rifle/Shotgun

Springfield, Inc. M1A

Springfield, Inc. SAR-4800

Springfield, Inc. Super Match M1A

Squires Bingham Model 14D

SPRINGFIELD INC. M-21 LAW ENFORCEMENT
Semi-automatic; 243, 7mm-08, 308 Win.; 20-shot magazine; air-gauged 22″ Douglas heavy barrel; 44 1/4″ overall length; weighs 11 7/8 lbs.; National Match front and rear sights; glass-bedded heavy walnut stock with adjustable comb, ventilated recoil pad. Refinement of the standard M1A rifle with specially knurled shoulder for figure-eight operating rod guide. From Springfield Armory, Inc. Introduced 1987; no longer made.
   Perf.: $1695   Exc.: $1395   VGood: $1195

SPRINGFIELD INC. M-21 SNIPER
Semi-automatic; 308 Win.; 20-shot box magazine; 22″ heavy barrel; 44 1/4″ overall length; weighs 15 1/4 lbs. with bipod, scope mount; National Match sights; heavy American walnut stock with adjustable comb, ventilated recoil pad; based on M1A rifle; folding, removable bipod; leather military sling. Introduced 1987; discontinued 1992.
   Perf.: $1695   Exc.: $1395   VGood: $1195

SPRINGFIELD INC. MODEL 700 BASR
Bolt action; 308 Win.; 5-shot magazine; 26″ heavy Douglas Premium barrel; 46 1/4″ overall length; weighs 13 1/2 lbs.; no sights; synthetic fiber stock; rubber recoil pad; marketed with military leather sling; adjustable, folding Parker-Hale bipod. Introduced 1987; discontinued 1988.
   Perf.: $795   Exc.: $695   VGood: $595

SPRINGFIELD INC. SAR-8 SPORTER
Semi-automatic; 308 Win.; 20-shot magazine; 18″ barrel; 40 3/8″ overall length; weighs 8 3/4 lbs.; protected post front sight, rotary-style adjustble rear; black or green composition forend; wood thumbhole butt; delayed roller-lock action; fluted chamber; matte black finish. From Springfield Inc. Introduced 1990; discontinued; reintroduced 1995. **LMSR: $1204**
   Perf.: $995   Exc.: $895   VGood: $795

SPRINGFIELD INC. SAR-4800
Semi-automatic; 5.56, 7.62 NATO (308 Win.); 10-shot magazine; 21″ barrel; 43 3/8″ overall length; weighs 9 1/2 lbs.; protected post front sight, adjustable peep rear; fiberglass forend; wood thumbhole butt. New production. From Springfield Inc. Reintroduced 1995; still available. **LMSR: $1249**
   Perf.: $1060   Exc.: $915   VGood: $795

Springfield Inc. SAR-4800 22
Same specs as SAR-4800 except 22 LR. No longer in production.
   Perf.: $750   Exc.: $650   VGood: $495

Springfield Inc. SAR-4800 Bush
Same specs as SAR-4800 except 18″ barrel. No longer in production.
   Perf.: $1080   Exc.: $935   VGood: $810

SQUIRES BINGHAM MODEL 14D
Bolt action; 22 LR; 5-shot box magazine; 24″ barrel; V-notch rear sight, hooded ramp front; grooved receiver for scope mounts; exotic hand-checkered wood stock, contrasting forend tip, pistol-grip cap. Made in the Philippines. No longer in production.
   Perf.: $80   Exc.: $70   VGood: $50

SQUIRES BINGHAM MODEL 15
Bolt action; 22 WMR; 5-shot box magazine; 24″ barrel; V-notch rear sight, hooded ramp front; grooved receiver for scope mounts; exotic hand-checkered wood stock, contrasting forend tip; pistol-grip cap. Made in the Philippines. No longer in production.
   Perf.: $90   Exc.: $80   VGood: $60

SQUIRES BINGHAM MODEL M16
Semi-automatic; 22 LR; 15-shot detachable box magazine; 19 1/2″ barrel; muzzlebrake/flash hider; integral rear sight, ramped post front; black painted mahogany buttstock, forearm; similar to military M16. No longer imported.
   Perf.: $140   Exc.: $100   VGood: $80

SQUIRES BINGHAM MODEL M20D
Semi-automatic; 22 LR; 15-shot detachable box magazine; 19 1/2″ barrel; muzzlebrake/flash hider; V-notch rear sight, blade front; grooved receiver for scope mount; hand-checkered exotic wood stock; contrasting pistol-grip cap, forend tip. No longer imported.
   Perf.: $100   Exc.: $90   VGood: $70

STANDARD MODEL G
Semi-automatic; 25 Rem., 30 Rem., 35 Rem.; 4-shot (35 Rem.), 5-shot magazine; 22 3/8″ barrel; gas operated; hammerless; takedown; open rear sight, ivory bead front; American walnut buttstock; slide handle-type forearm; gas port, when closed, required manual slide action. Introduced 1912. Some collector value.
   Exc.: $495   VGood: $395   Good: $295

STANDARD MODEL M
Slide action; 25 Rem., 30 Rem., 35 Rem.; 4-shot (35 Rem.), 5-shot magazine; 22 3/8″ barrel; hammerless; takedown; open rear sight, ivory bead front; American walnut buttstock, slide handle-type forearm. Introduced 1912. Some collector value.
   Exc.: $430   VGood: $370   Good: $290

RIFLES

Stevens Bolt-Action & Semi-Automatic Rifles
Buckhorn Model 053
Buckhorn Model 53
Buckhorn Model 056
Buckhorn Model 56
Buckhorn Model 066
Buckhorn Model 66
Buckhorn No. 057
Buckhorn No. 57
Buckhorn No. 076
Buckhorn No. 76
Model 15
Model 15Y
Model 84
Model 84S
Model 85
Model 85K
Model 85S

Stevens Buckhorn Model 053

Stevens Buckhorn Model 056

Stevens Buckhorn Model 066

Stevens Buckhorn Model 66

Stevens Buckhorn No. 076

Stevens Model 84

STEVENS BOLT-ACTION & SEMI-AUTOMATIC RIFLES

STEVENS BUCKHORN MODEL 053
Bolt action; single shot; 25 Stevens, 22 Short, 22 Long, 22 LR, 22 WMR; 24″ barrel; 41 1/4″ overall length; weighs 5 1/2 lbs.; takedown; receiver peep sight, open folding middle sight, hooded ramp front with three inserts, removable hood; uncheckered walnut stock; pistol grip; black forearm tip. Introduced 1935; discontinued 1948.
Exc.: $120   **VGood:** $90   **Good:** $70

Stevens Buckhorn Model 53
Same specs as Buckhorn Model 053 except open rear sight, gold bead front.
Exc.: $90   **VGood:** $80   **Good:** $60

STEVENS BUCKHORN MODEL 056
Bolt action; 22 Short, 22 Long, 22 LR; 5-shot detachable box magazine; 24″ barrel; 43 1/2″ overall length; weighs 6 lbs.; takedown; receiver peep sight, open middle sight, hooded front; uncheckered walnut sporter-type stock; pistol grip, black forearm tip; rubber buttplate. Introduced 1935; discontinued 1948.
Exc.: $120   **VGood:** $90   **Good:** $70

Stevens Buckhorn Model 56
Same specs as Buckhorn Model 056 except open rear sight, gold bead front.
Exc.: $90   **VGood:** $80   **Good:** $60

STEVENS BUCKHORN MODEL 066
Bolt action; 22 Short, 22 Long, 22 LR; 21-shot (22 Short), 17-shot (22 Long), 15-shot (22 LR) tube magazine; 24″ barrel; 43 1/2″ overall length; weighs 6 lbs.; receiver peep sight, folding sporting middle sight, hooded front; uncheckered walnut sporting stock; pistol grip, black forearm tip; rubber buttplate. Introduced 1935; discontinued 1948.
Exc.: $120   **VGood:** $90   **Good:** $70

Stevens Buckhorn Model 66
Same specs as Model 066 except fully-adjustable sporting rear sight, gold bead front.
Exc.: $90   **VGood:** $80   **Good:** $60

STEVENS BUCKHORN NO. 057
Semi-automatic; 22 LR; 5-shot box magazine; 24″ barrel; takedown; peep receiver sight, hooded front; uncheckered sporter-style stock; black forearm tip. Introduced 1939; discontinued 1948.
Exc.: $90   **VGood:** $80   **Good:** $60

Stevens Buckhorn No. 57
Same specs as Buckhorn No. 057 except open rear sight, plain bead front.
Exc.: $90   **VGood:** $80   **Good:** $60

STEVENS BUCKHORN NO. 076
Semi-automatic; 22 LR; 15-shot tube magazine; 24″ barrel; weighs 6 lbs.; takedown; cross-bolt locks for use as single shot with 22 Short, 22 Long, 22 LR; peep rear receiver sight, folding flat-top middle sight, hooded front; uncheckered sporter-style stock; black forearm tip; rubber buttplate. Introduced 1938; discontinued 1948.
Exc.: $110   **VGood:** $100   **Good:** $90

Stevens Buckhorn No. 76
Same specs as Buckhorn No. 076 except open rear sight; gold bead front.
Exc.: $110   **VGood:** $100   **Good:** $90

STEVENS MODEL 15
Bolt action; single shot; 22 Short, 22 Long, 22 LR; 22″, 24″ barrel; 43″ overall length; weighs 4 lbs.; elevation-adjustable open rear sight, gold bead front; uncheckered walnut pistol-grip stock. Introduced 1948; discontinued 1965.
Exc.: $70   **VGood:** $60   **Good:** $40

Stevens Model 15Y
Same specs as Model 15 except 21″ barrel; 35 1/4″ overall length; shorter youth-size buttstock. Introduced 1958; discontinued 1965.
Exc.: $70   **VGood:** $60   **Good:** $40

STEVENS MODEL 84
Bolt action; 22 Short, 22 Long, 22 LR; 15-shot (22 LR), 17-shot (22 Long), 21-shot (22 Short) tube magazine; 24″ barrel; 43″ overall length; weighs 6 lbs.; cross-bolt locks for single shot operation; takedown; elevation-adjustable micro peep rear sight, gold bead front; uncheckered walnut oval militry-style full pistol-grip stock; plated bolt, trigger. Introduced 1949; discontinued 1965.
Exc.: $90   **VGood:** $80   **Good:** $60

Stevens Model 84S
Same specs as Model 84 except No. 150 micro peep rear sight, No. 150 hooded front. Introduced 1948; discontinued 1952.
Exc.: $100   **VGood:** $80   **Good:** $60

STEVENS MODEL 85
Semi-automatic; 22 LR; 5-shot detachable clip magazine; 24″ barrel; 44″ overall length; weighs 6 lbs.; takedown; cross-bolt locks for single shot operation; open rear sight, bead front; uncheckered American walnut full pistol-grip stock; black forearm tip; hard rubber buttplate. Introduced 1948; discontinued 1976.
Exc.: $100   **VGood:** $80   **Good:** $70

Stevens Model 85K
Same specs as Model 85 except 20″ barrel; 40″ overall length; weighs 5 3/4 lbs. Introduced 1959.
Exc.: $100   **VGood:** $80   **Good:** $70

Stevens Model 85S
Same specs as Model 85 except peep rear sight, hooded front. Introduced 1948; discontinued 1976.
Exc.: $100   **VGood:** $80   **Good:** $70

Stevens Model 84S

Stevens Model 15Y

Stevens Model 86S

Stevens Model 87

Stevens Model 110ES

Stevens Model 325

STEVENS MODEL 86

Bolt action; 22 Short, 22 Long, 22 LR; 21-shot (22 Short), 17-shot (22 Long), 15-shot (22 LR) tube magazine; 24" barrel; 43 1/2" overall length; weighs 6 lbs.; takedown; sporting elevation-adjustable rear sight, gold bead front; uncheckered walnut oval military-style full pistol-grip stock; plated bolt, trigger. Introduced 1935; discontinued 1948.
  Exc.: $100   **VGood:** $80   **Good:** $70

Stevens Model 86S

Same specs as Model 86 except No. 150 peep rear sight, hooded front. Introduced 1948; discontinued 1952.
  Exc.: $100   **VGood:** $80   **Good:** $70

STEVENS MODEL 87

Semi-automatic; 22 LR; 15-shot tube magazine; 24" barrel (until late '60s), 20" barrel (later version); weighs 6 lbs.; takedown; cross-bolt locks for single shot operation; open rear sight, bead front; uncheckered American walnut full pistol-grip stock; black forearm tip; hard rubber buttplate. Marketed as Springfield Model 87 from 1938 to 1948, when trade name dropped. Introduced as the Stevens Model 87 in 1948; discontinued 1976.
  Exc.: $100   **VGood:** $90   **Good:** $70

Stevens Model 87K

Same specs as Model 87 except 20" barrel; 40" overall length; weighs 5 3/4 lbs. Introduced 1959.
  Exc.: $110   **VGood:** $90   **Good:** $70

Stevens Model 87S

Same specs as Model 87 except peep rear sight, hooded front. Introduced with Stevens name in 1948; discontinued 1953.
  Exc.: $105   **VGood:** $95   **Good:** $75

STEVENS MODEL 110E

Bolt action; 308, 30-06, 243 ; 4-shot magazine; 22" round tapered barrel; removable bead ramp front sight, step-adjustable rear; drilled, tapped for peep sight or scope mounts; walnut-finished hardwood Monte Carlo stock; checkered pistol grip, forend; hard rubber buttplate; top tang safety. Introduced 1978; discontinued 1984.
  Exc.: $295   **VGood:** $250   **Good:** $200

Stevens Model 110ES

Same specs as Model 110E except 5-shot magazine; free-floating barrel; marketed with 4x scope, mounts. Introduced 1981; discontinued 1985.
  Exc.: $320   **VGood:** $270   **Good:** $210

STEVENS MODEL 120

Bolt action; single shot; 22 Short, 22 Long, 22 LR; 20" round tapered barrel; sporting front/rear sights; grooved for scope mounts; walnut-finished hardwood stock; pistol grip; action cocks on opening of bolt; thumb safety; double extractors; recessed bolt face. Introduced 1979; discontinued 1983.
  Exc.: $80   **VGood:** $70   **Good:** $50

STEVENS MODEL 125

Bolt action; single shot; 22 Short, 22 Long, 22 LR; 22" barrel; 39" overall length; sporting front sight, open rear with elevator; walnut-finished stock; manual cocking; blued finish. Introduced 1981; no longer made.
  Exc.: $80   **VGood:** $70   **Good:** $50

STEVENS MODEL 322

Bolt action; 22 Hornet; 4-shot detachable box magazine; 21" barrel; open rear sight, ramp front; uncheckered walnut pistol grip stock. Introduced 1947; discontinued 1950.
  Exc.: $220   **VGood:** $180   **Good:** $150

Stevens Model 322S

Same specs as Model 322 except peep rear sight. Introduced 1947; discontinued 1950.
  Exc.: $230   **VGood:** $190   **Good:** $160

STEVENS MODEL 325

Bolt action; 30-30; 4-shot detachable clip magazine; 21" barrel; open rear sight, bead front; uncheckered pistol-grip stock. Introduced 1947; discontinued 1950.
  Exc.: $220   **VGood:** $180   **Good:** $150

Stevens Model 110ES

Stevens No. 419 Junior Target

Stevens Model 325S

Same specs as Model 325 except peep rear sight. Introduced 1947; discontinued 1950.
  Exc.: $230   **VGood:** $190   **Good:** $160

STEVENS MODEL 416-2

Bolt action; 22 LR; 5-shot detachable clip magazine; 26" heavy barrel; weighs 9 1/2 lbs.; No. 106 receiver peep sight, No. 25 hooded front; uncheckered American walnut target-type stock; checkered steel buttplate; sling swivels, 1 1/4" sling. Introduced 1937; discontinued 1949.
  Exc.: $150   **VGood:** $130   **Good:** $110
  With U.S. Property markings
  Exc.: $295   **VGood:** $250   **Good:** $200

STEVENS NO. 419 JUNIOR TARGET

Bolt action; single shot; 22 LR; 26" barrel; takedown; Lyman No. 55 peep rear sight, blade front; uncheckered walnut junior target stock; pistol grip; sling; swivels. Introduced 1932; discontinued 1936.
  Exc.: $170   **VGood:** $140   **Good:** $100

STEVENS NO. 66

Bolt action; 22 Short, 22 Long, 22 LR; 19-shot (22 Short), 15-shot (22 Long), 13-shot (22 LR) tube magazine; 24" barrel; takedown; open rear sight, bead front; uncheckered walnut pistol-grip stock; grooved forearm. Introduced 1931; discontinued 1935.
  Exc.: $75   **VGood:** $60   **Good:** $50

STEVENS NO. 76

Semi-automatic or single shot; 22 LR (semi-auto), 22 Short, 22 Long, 22 LR (single shot); 15-shot tubular magazine; 24" tapered barrel with crowned muzzle; weighs 6 lbs.; hammer release mechanism for single shot operation; open rear sight, gold bead front; American walnut full pistol-grip stock; hard rubber buttplate. Introduced 1947.
  Exc.: $85   **VGood:** $75   **Good:** $65

Stevens Lever- & Slide-Action Rifles

Stevens Favorite No. 17

Stevens Ideal No. 44

Stevens No. 70 "Visible Loader"

Stevens No. 414 Armory

Stevens No. 417-0 Walnut Hill

STEVENS LEVER- & SLIDE-ACTION RIFLES

STEVENS FAVORITE NO. 17
Lever action; single shot; 22 LR, 25 rimfire, 32 rimfire; 24" barrel, other lengths available on special order; takedown; open rear sight, Rocky Mountain front; uncheckered walnut straight-grip stock; tapered forearm. Introduced 1894; discontinued 1935. Collector value.
Exc.: $200   VGood: $170   Good: $145

STEVENS FAVORITE NO. 18
Lever action; single shot; 22 LR, 25 rimfire, 32 rimfire; 24" barrel, other lengths available on special order; takedown; Vernier peep rear sight, Beach combination front, leaf middle sight; uncheckered walnut straight-grip stock, tapered forearm. Introduced 1894; discontinued 1917. Collector value.
Exc.: $220   VGood: $190   Good: $160

STEVENS FAVORITE NO. 19
Lever action; single shot; 22 LR, 25 rimfire, 32 rimfire; 24" barrel, other lengths available on special order; takedown; Lyman combination rear sight, Lyman front, leaf middle sight; uncheckered walnut straight-grip stock, tapered forearm. Introduced 1895; discontinued 1917. Collector value.
Exc.: $220   VGood: $190   Good: $160

STEVENS FAVORITE NO. 20
Lever action; single shot; 22 rimfire, 32 rimfire only; smoothbore barrel; takedown; open rear sight, Rocky Mountain front; uncheckered walnut straight-grip stock; tapered forearm. Introduced 1895; discontinued 1935. Collector value.
Exc.: $225   VGood: $200   Good: $170

STEVENS FAVORITE NO. 26 CRACKSHOT
Lever action; single shot; 22 LR, 32 rimfire; 18", 22" barrel; takedown; open rear sight, blade front; uncheckered straight-grip walnut stock, tapered forearm. Introduced 1913; discontinued 1939. Collector value.
Exc.: $200   VGood: $170   Good: $125

Stevens Favorite No. 26 ½ Crackshot
Same specs as No. 26 Crackshot except smoothbore barrel for 22, 32 rimfire shot cartridges. Introduced 1914; discontinued 1939. Collector value.
Exc.: $225   VGood: $175   Good: $135

STEVENS FAVORITE NO. 27
Lever action; single shot; 22 LR, 25 rimfire, 32 rimfire; 24" octagon barrel, other lengths available on special order; takedown; open rear sight, Rocky Mountain front; uncheckered walnut straight-grip stock; tapered forearm. Introduced 1912; discontinued 1939. Collector value.
Exc.: $200   VGood: $170   Good: $140

STEVENS FAVORITE NO. 28
Lever action; single shot; 22 LR, 25 rimfire, 32 rimfire; 24" octagon barrel, other lengths available on special order; takedown; Vernier peep sight, Beach combination front, leaf middle sight; uncheckered walnut straight-grip stock, tapered forearm. Introduced 1896; discontinued 1935. Collector value.
Exc.: $220   VGood: $190   Good: $160

STEVENS FAVORITE NO. 29
Lever action; single shot; 22 LR, 25 rimfire, 32 rimfire; 24" octagon barrel, other lengths available on special order; takedown; Lyman combination rear sight, Lyman front, leaf middle sight; uncheckered walnut straight-grip stock, tapered forearm. Introduced 1896; discontinued 1935. Collector value.
Exc.: $220   VGood: $190   Good: $160

STEVENS IDEAL NO. 44
Lever action; single shot; 22 LR, 25 rimfire, 32 rimfire, 25-20, 32-20, 32-40, 38-40, 38-55, 44-40; 24", 26" round, full- or half-octagon barrel; rolling block; takedown; open rear sight, Rocky Mountain front; uncheck-ered straight-grip walnut stock, forearm. Introduced 1894; discontinued 1932. Primarily of collector interest.
Exc.: $575   VGood: $500   Good: $425

Stevens Ideal No. 44 ½
Same specs as Ideal No. 44 except falling block, lever action. Introduced 1903; discontinued 1916. Collector interest.
Exc.: $820   VGood: $740   Good: $575

STEVENS IDEAL NOS. 45 THROUGH 56
Lever actions; rolling and falling block; 22 LR, 25 rimfire, 32 rimfire, 25-20, 32-20, 32-40, 38-40, 38-55, 44-40; 24", 26" round, full-, half-octagon barrel; other lengths available on special order; various sights available; various walnut stock options; similar to Ideal No. 44 with numerous updated improvements. Introduced 1896; discontinued 1916. Because of special order features, triggers, barrel makers, stocks and engraving, these models require separate evaluation. Higher grades and Scheutzens are very collectible and desireable.

STEVENS NO. 70 "Visible Loader"
Slide action; 22 Short, 22 Long, 22 LR; 15-shot (22 Short), 13-shot (22 Long), 11-shot (22 LR) tube magazine; 22" barrel; open rear sight, bead front; uncheckered straight-grip stock; grooved slide handle. Introduced 1907; discontinued 1934. Collector value.
Exc.: $225   VGood: $175   Good: $125

STEVENS NO. 414 ARMORY
Lever action; single shot; 22 Short, 22 LR; 26" barrel; rolling block; Lyman receiver peep sight, blade front; checkered straight-grip walnut stock; military-style forearm; sling swivels. Introduced 1912; discontinued 1932.
Exc.: $450   VGood: $375   Good: $300

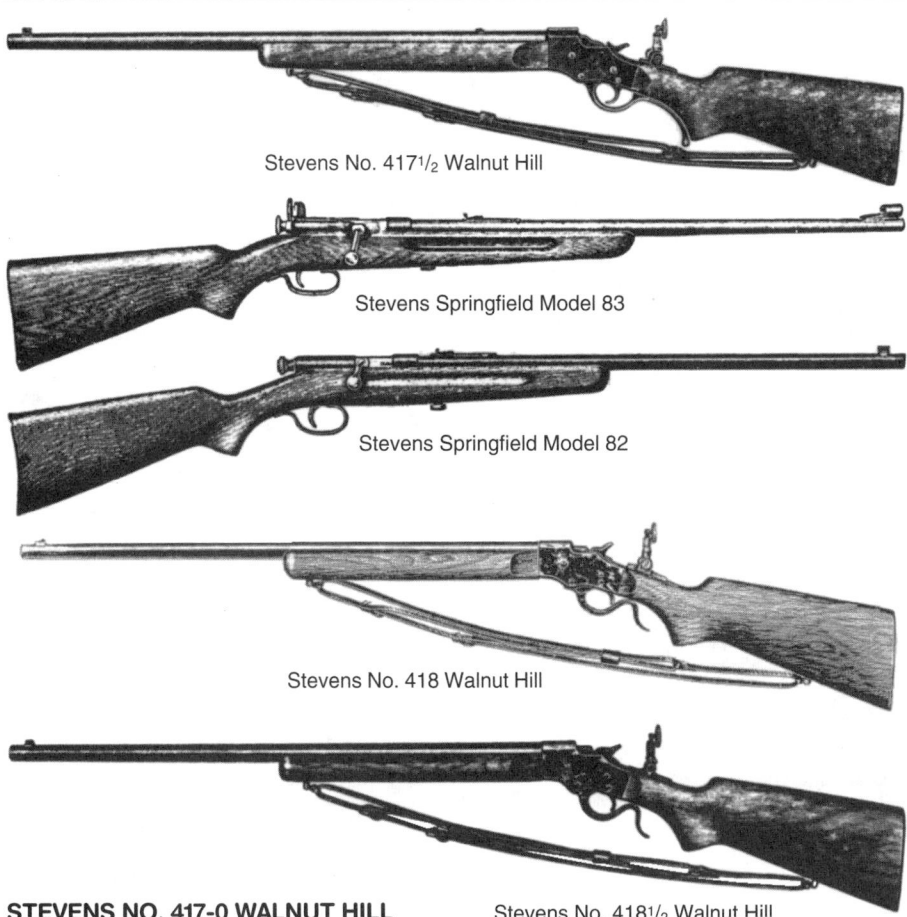

Stevens No. 417½ Walnut Hill

Stevens Springfield Model 83

Stevens Springfield Model 82

Stevens No. 418 Walnut Hill

Stevens Lever- & Slide- Action Rifles
No. 417-0 Walnut Hill
No. 417-1 Walnut Hill
No. 417-2 Walnut Hill
No. 417-3 Walnut Hill
No. 417½ Walnut Hill
No. 418 Walnut Hill
No. 418½ Walnut Hill

Stevens Springfield
Model 15
Model 82
Model 83
Model 084
Model 84
Model 085
Model 85
Model 086
Model 86
Model 087
Model 87

Stevens No. 418½ Walnut Hill

STEVENS NO. 417-0 WALNUT HILL
Lever action; single shot; 22 Short, 22 LR, 22 Hornet; 28" heavy, 29" extra heavy barrel; Lyman No. 52L extension rear sight, No. 17A front; uncheckered walnut target stock; full pistol grip, beavertail forearm; sling swivels; barrel band; sling. Introduced 1932; discontinued 1947.
    Exc.: $895   **VGood:** $750   **Good:** $395

Stevens No. 417-1 Walnut Hill
Same specs as No. 417-0 Walnut Hill except Lyman No. 48L receiver sight. Introduced 1932; discontinued 1947.
    Exc.: $900   **VGood:** $795   **Good:** $425

Stevens No. 417-2 Walnut Hill
Same specs as No. 417-0 Walnut Hill except Lyman No. 1441 tang sight. Introduced 1932; discontinued 1947.
    Exc.: $950   **VGood:** $795   **Good:** $425

Stevens No. 417-3 Walnut Hill
Same specs as No. 417-0 Walnut Hill except no sights. Introduced 1932; discontinued 1947.
    Exc.: $895   **VGood:** $750   **Good:** $395

Stevens No. 417½ Walnut Hill
Same specs as No. 417-0 Walnut Hill except 22 Hornet, 25 rimfire, 22 WRF, 22 LR; 28" barrel; Lyman No. 144 tang peep sight, folding middle sight, bead front; uncheckered walnut sporting-style stock; pistol grip; semi-beavertail forearm. Introduced 1932; discontinued 1940.
    Exc.: $895   **VGood:** $750   **Good:** $470

STEVENS NO. 418 WALNUT HILL
Lever action; single shot; 22 Short, 22 LR; 26" barrel; takedown; Lyman No. 144 tang peep sight, bead front; uncheckered walnut stock; pistol grip, semi-beavertail forearm; sling swivels, sling. Introduced 1932; discontinued 1940.
    Exc.: $650   **VGood:** $495   **Good:** $300

Stevens No. 418 ½ Walnut Hill
Same specs as No. 418 Walnut Hill except 22 Short, 22 LR, 22 WRF, 25 Stevens; Lyman No. 2A tang peep sight, bead front. Introduced 1932; discontinued 1940.
    Exc.: $695   **VGood:** $495   **Good:** $300

STEVENS SPRINGFIELD MODEL 15
Bolt action; single shot; 22 Short, 22 Long, 22 LR; 22" barrel; 37" overall length; weighs 4 lbs.; takedown; elevation-adjustable sporting rear sight, gold bead front; uncheckered walnut-finished pistol-grip stock. Introduced 1937; discontinued 1948.
    Exc.: $65   **VGood:** $50   **Good:** $40
NOTE: Springfield was used as a brand name from 1935 until 1948, with the designation being dropped at that time. It should not be confused with Springfield Armory, although the name probably was registered with such mistaken identity in mind. After the Springfield brand name was dropped, rifles were known strictly by the Stevens name.

STEVENS SPRINGFIELD MODEL 82
Bolt action; single shot; 22 Short, 22 Long, 22 LR; 22" barrel; takedown; open rear sight, bead front; uncheckered walnut pistol-grip stock; grooved forearm. Introduced 1935; discontinued 1939.
    Exc.: $65   **VGood:** $50   **Good:** $40

STEVENS SPRINGFIELD MODEL 83
Bolt action; 25 Stevens, 22 WRF, 22 Short, 22 Long, 22 LR; 24" barrel; takedown; open rear sight, bead front; uncheckered walnut pistol-grip stock; grooved forearm. Introduced 1935; discontinued 1939.
    Exc.: $80   **VGood:** $65   **Good:** $50

STEVENS SPRINGFIELD MODEL 084
Bolt action; 22 Short, 22 Long, 22 LR; 5-shot detachable clip magazine; 24" barrel; 43 ½" overall length; weighs 6 lbs.; takedown; rear peep sight, hooded front; uncheckered walnut pistol-grip stock; black forearm tip; plated bolt, trigger. Introduced 1935; discontinued 1948.
    Exc.: $90   **VGood:** $70   **Good:** $60

Stevens Springfield Model 84
Same specs Model 084 except open sights. Introduced 1935; discontinued 1948.
    Exc.: $80   **VGood:** $65   **Good:** $50

STEVENS SPRINGFIELD MODEL 085
Semi-automatic; 22 LR; 5-shot detachable box magazine; 24" barrel; takedown; peep rear sight, hooded front; uncheckered pistol-grip stock; black forend tip. Introduced 1939; discontinued 1948.
    Exc.: $90   **VGood:** $70   **Good:** $50

Stevens Springfield Model 85
Same specs as Model 085 except open sights. Introduced 1939; discontinued 1948.
    Exc.: $85   **VGood:** $65   **Good:** $45

STEVENS SPRINGFIELD MODEL 086
Bolt action; 22 Short, 22 Long, 22 LR; 21-shot (22 Short), 17-shot (22 Long), 15-shot (22 LR) tube magazine; 24" barrel; 41 ½" overall length; weighs 5 ½ lbs.; chromium-plated bolt, trigger; takedown; peep rear sight, folding sporting middle, hooded front; uncheckered walnut pistol-grip stock; black forearm tip; plated bolt, trigger. Introduced 1935; discontinued 1948.
    Exc.: $90   **VGood:** $70   **Good:** $50

Stevens Springfield Model 86
Same specs as Model 086 except open sights. Introduced 1935; discontinued 1948.
    Exc.: $85   **VGood:** $65   **Good:** $45

STEVENS SPRINGFIELD MODEL 087
Semi-automatic; 22 LR; 24" barrel; takedown; peep rear sight, hooded front; uncheckered pistol-grip stock; black forearm tip. Introduced 1938; discontinued 1948.
    Exc.: $90   **VGood:** $70   **Good:** $50

Stevens Springfield Model 87
Same specs as Model 087 except open rear sight, bead front. Introduced 1938; discontinued 1948.
    Exc.: $85   **VGood:** $65   **Good:** $50

RIFLES

Steyr
A.U.G.-SA
Model 90 Carbine
Model 95

Steyr-Mannlicher
Jagd Match
Match SPG-UIT
Model L
Model L Carbine
Model L Luxus
Model L Luxus Carbine
Model L Varmint
Model M
Model M Carbine
Model M Luxus
Model M Luxus Carbine
Model M Professional
ML 79 Luxus (See Steyr-Mannlicher Models L and M Luxus)

Steyr-Mannlicher Model M Luxus

Steyr-Mannlicher ML79 Luxus

Steyr Model 90 Carbine

Steyr-Mannlicher Model L Varmint

Steyr-Mannlicher Model M

STEYR A.U.G.-SA
Semi-automatic; 223 Rem.; 30- or 42-shot magazine; 20″ barrel; 31″ overall length; weighs 8 1/2 lbs.; synthetic green stock; one-piece moulding houses receiver group, hammer mechanism and magazine; integral 1.5x scope; scope and mount form the carrying handle; gas-operated action; conversion to suit right- or left-handed shooters, including ejection port; folding vertical front grip. Imported from Austria by Gun South, Inc. Introduced 1983; no longer imported. Values unpredictable due to import ban. Can fluctuate greatly.
Perf.: $3500   **Exc.:** $2700   **VGood:** $2400

STEYR MODEL 90 CARBINE
Bolt action; 22 LR; 5-shot detachable box magazine; 19″ barrel; leaf rear sight, hooded bead front; hand-checkered European walnut Mannlicher-type stock; sling swivels. Manufactured in Austria. Introduced 1953; discontinued 1967.
Perf.: $1195   **Exc.:** $995   **VGood:** $725

STEYR MODEL 95
Bolt action; 8x50R Mannlicher; 30″ barrel; straight pull; adjustable leaf rear sight, hooded bead front; military walnut Mannlicher-style stock; sling swivels. No longer produced.
Perf.: $165   **Exc.:** $115   **VGood:** $85

STEYR-MANNLICHER JAGD MATCH
Bolt action; 222 Rem., 243 Win., 308 Win.; 23 1/2″ heavy barrel; Mannlicher sights; checkered laminate half-stock. Still in production. **LMSR:** $2275
Perf.: $1595   **Exc.:** $1495   **VGood:** $1095

STEYR-MANNLICHER MATCH SPG-UIT
Bolt action; 308 Win.; 25 1/2″ barrel; 44″ overall length; weighs 10 lbs.; Steyr globe front sight, Steyr peep rear; laminated and ventilated stock of special UIT Match design; double-pull trigger adjustable for let-off point, slack, weight of first-stage pull, release force and length; buttplate adjustable for height and length.

Meets UIT specifications. Imported from Austria by GSI, Inc. Introduced 1992; still imported.
Perf.: $3495   **Exc.:** $2995   **VGood:** $2195

STEYR-MANNLICHER MODEL L
Bolt action; 22-250, 5.6x57mm, 243 Win., 6mm Rem., 308 Win.; 5-shot detachable rotary magazine; 23 5/8″ barrel; 42 1/2″ overall length; open rear sight, hooded ramp front; European walnut half-stock; Monte Carlo comb, cheekpiece; skip-checkered pistol grip, forearm; double-set triggers; detachable sling-swivels; buttpad. Introduced, 1968; **LMSR:** $2250
Perf.: $1720   **Exc.:** $1370   **VGood:** $810

Steyr-Mannlicher Model L Carbine
Same specs as Model L except 20″ barrel; 39″ overall length; full-length checkered European walnut stock. Introduced 1968; **LMSR:** $2450
Perf.: $1790   **Exc.:** $1410   **VGood:** $900

Steyr-Mannlicher Model L Luxus
Same specs as Model L except only 5.6x57, 6mm Rem., 22-250, 243, 308; 3-shot detachable steel magazine; single set trigger; rear tang slide safety; European walnut full- or half-stock. **LMSR:** $2950
Perf.: $2100   **Exc.:** $1495   **VGood:** $1195
With full-stock
Perf.: $2400   **Exc.:** $1595   **VGood:** $1295

Steyr-Mannlicher Model L Luxus Carbine
Same specs as Model L except 5.6x57, 6mm Rem., 22-250, 243, 308; 3-shot magazine; 20″ barrel; European walnut full-stock. **LMSR:** $3150
Perf.: $2450   **Exc.:** $1595   **VGood:** $1295

Steyr-Mannlicher Model L Varmint
Same specs as Model L except 22-250, 243 Win.; 26″ heavy barrel; no sights; ventilated square forearm. Introduced 1969; **LMSR:** $2450
Perf.: $1775   **Exc.:** $1100   **VGood:** $840

STEYR-MANNLICHER MODEL M
Bolt action; 6.5x57mm, 7x57mm, 7x64mm, 25-06, 270 Win., 30-06, 8x57JS, 9.3x62; 5-shot detachable rotary magazine; 23 1/2″ barrel; open rear sight, hooded ramp front; European walnut full- or half-stock; forend tip, recoil pad; detachable sling swivels. Also left-hand model, with 6.5x55, 7.5mm Swiss as additional calibers. Standard version introduced, 1969; left-hand version, 1977; **LMSR:** $2250
Perf.: $1700   **Exc.:** $1295   **VGood:** $895
Left-hand model
Perf.: $1900   **Exc.:** $1495   **VGood:** $995

Steyr-Mannlicher Model M Carbine
Same specs as Model M except 20″ barrel; full-length stock. Introduced 1977; **LMSR:** $2250
Perf.: $1850   **Exc.:** $1350   **VGood:** $995
Left-hand model
Perf.: $2000   **Exc.:** $1550   **VGood:** $1050

Steyr-Mannlicher Model M Luxus
Same specs as Model M except 6.5x57, 270, 7x64, 30-06, 9.3x62, 7.5 Swiss; 3-shot detachable magazine; single set trigger; rear tang slide safety; full or half-stock. Imported from Austria by GSI, Inc. **LMSR:** $2950
Perf.: $2000   **Exc.:** $1295   **VGood:** $895
With full-stock
Perf.: $2200   **Exc.:** $1495   **VGood:** $995

Steyr-Mannlicher Model M Luxus Carbine
Same specs as Model M except only 6.5x55, 6.5x57, 7x64, 7.5 Swiss, 270, 30-06; 3-shot magazine; 20″ barrel; European walnut full-stock. **LMSR:** $3150
Perf.: $2200   **Exc.:** $1495   **VGood:** $995

Steyr-Mannlicher Model M Professional
Same specs as Model M except 270, 7x57, 7x64, 30-06, 9.3x62; 20″, 23 2/3″ barrel; Cycolac synthetic stock; Parkerized finish. Introduced 1977; discontinued 1993.
Perf.: $1475   **Exc.:** $995   **VGood:** $795

Steyr-Mannlicher Model SSG Match

Steyr-Mannlicher Model S

Steyr-Mannlicher Model SL

Steyr-Mannlicher SSG P-IV

Steyr-Mannlicher SSG Match UIT

RIFLES

Steyr-Mannlicher
Model MII Professional
Model M72 L/M
Model M72 Model L/M Carbine
Model S
Model SL
Model SL Carbine
Model SL Varmint
Model SSG
Model SSG Match
Model SSG Match UIT
SSG P-II
SSG P-IIK
SSG P-III
SSG P-IV
Model S/T
Model S Luxus

Stoeger
Herold
(See Charles Daly Bolt Action)

Steyr-Mannlicher Model MIII Professional
Same specs as the Model M except 6.5x57, 270, 30-06, 7x64, 9.3x62; 23 1/2″ barrel; weighs about 7 1/2 lbs.; no sights; black ABS Cycolac or stippled, checkered European wood half-stock; single or optional double-set trigger. Imported from Austria by GSI, Inc. Reintroduced 1994; **LMSR: $995**
   Perf.: $795   **Exc.:** $695   **VGood:** $595

STEYR-MANNLICHER MODEL M72 L/M
Bolt action; 22-250, 5.6x57mm, 6mm Rem., 243 Win., 6.5x57mm, 270 Win., 7x57mm, 7x64mm, 308 Win., 30-06; 23 2/3″ barrel; 5-shot rotary magazine; open rear sight, hooded ramp front; European walnut half-stock; hand-checkered pistol-grip forearm, rosewood forend tip; recoil pad; interchangeable single- or double-set trigger; detachable sling swivels; front locking bolt. Introduced 1972; no longer in production.
   Perf.: $795   **Exc.:** $695   **VGood:** $595

Steyr-Mannlicher Model M72 Model L/M Carbine
Same specs as Model M72 L/M except 20″ barrel; full-length stock. Introduced 1972; no longer in production.
   Perf.: $895   **Exc.:** $750   **VGood:** $650

STEYR-MANNLICHER MODEL S
Bolt action; 6.5x68mm, 257 Weatherby Mag., 264 Win. Mag., 7mm Rem. Mag., 300 Win. Mag., 300 H&H Mag., 308 Norma Mag., 8x68S, 338 Win. Mag., 9.3x64mm, 375 H&H Mag., 458 Win. Mag.; 5-shot detachable rotary magazine; 26″ barrel; open rear sight, hooded ramp front; European walnut half-stock; Monte Carlo comb, cheekpiece; skip-checkered pistol grip, forearm; forend tip; recoil pad. Introduced 1969; **LMSR: $2179**
   Perf.: $1725   **Exc.:** $1195   **VGood:** $895

STEYR-MANNLICHER MODEL SL
Bolt action; 5.6x50, 22-250, 222 Rem., 222 Rem. Mag., 223 Rem.; 5-shot detachable rotary Makrolon magazine; 23 5/8″ barrel; 42 1/2″ overall length; open rear sight, hooded ramp front with .080 gold bead; European walnut half-stock; Monte Carlo comb, cheekpiece; skip-checkered pistol-grip, forearm; interchangeable single- or double-set triggers; detachable sling swivels; buttpad. Introduced 1968; **LMSR: $2023**
   Perf.: $1495   **Exc.:** $1195   **VGood:** $895

Steyr-Mannlicher Model SL Carbine
Same specs as Model SL except 20″ barrel; 39″ overall length; full-length stock. Introduced 1968; **LMSR: $2179**
   Perf.: $1595   **Exc.:** $1295   **VGood:** $995

Steyr-Mannlicher Model SL Varmint
Same specs as Model SL except only 22-250, 222 Rem., 222 Rem. Mag., 223; 26″ barrel; no sights; vent. square forearm. Introduced 1969; **LMSR: $2179**
   Perf.: $1695   **Exc.:** $1295   **VGood:** $995

STERY-MANNLICHER MODEL SSG
Bolt action; 308 Win.; 5- or 10-shot magazine; 25 1/2″ heavy barrel; 44 1/2″ overall length; weighs 8 3/8 lbs.; micrometer peep rear sight, globe front; European walnut or synthetic target stock; full pistol grip; wide forearm; swivel rail; adjustable buttplate; single trigger; single shot plug; also made with high-impact ABS Cycolac plastic stock. Introduced 1969; **LMSR: $1995**
   Perf.: $1695   **Exc.:** $1295   **VGood:** $995
With plastic stock
   Perf.: $1495   **Exc.:** $995   **VGood:** $795

Steyr-Mannlicher Model SSG Match
Same specs as Model SSG except 26″ heavy barrel; weighs 11 lbs.; Walther target peep sights; brown synthetic or walnut stock; adjustable rail in forend; match bolt. Introduced 1981; discontinued 1991.
Synthetic stock
   Perf.: $1895   **Exc.:** $1495   **VGood:** $1095
Walnut stock
   Perf.: $1995   **Exc.:** $1695   **VGood:** $1195

Steyr-Mannlicher Model SSG Match UIT
Same specs as Model SSG except 243, 308; weighs 11 lbs.; 10-shot magazine; Walther rear peep sight, globe front; UIT Match walnut stock; stippled grip, forend; adjustable double-pull trigger; adjustable buttplate. Introduced 1984; still in production. **LMSR: $3995**
   Perf.: $3375   **Exc.:** $2150   **VGood:** $1495

Steyr-Mannlicher SSG P-II
Same specs as Model SSG except 243, 308; large modified bolt handle; no sights; forend rail. Still in production. **LMSR: $1699**
   Perf.: $1595   **Exc.:** $1350   **VGood:** $995

Steyr-Mannlicher SSG P-IIK
Same specs as Model SSG P-II except 20″ heavy barrel. **LMSR: $1995**
   Perf.: $1595   **Exc.:** $1350   **VGood:** $995

Steyr-Mannlicher SSG P-III
Same specs as Model SSG except 26″ heavy barrel; diopter match sight bases; H-S Precision Pro-Series stock (black only). Imported from Austria by GSI, Inc. Introduced 1992; discontinued 1993.
   Perf.: $2025   **Exc.:** $1295   **VGood:** $995

Steyr-Mannlicher SSG P-IV
Same specs as Model SSG except 16 3/4″ heavy barrel with flash hider; ABS Cycolac synthetic stock in green or black. Imported from Austria by GSI, Inc. Introduced 1992; **LMSR: $2660**
   Perf.: $2200   **Exc.:** $1495   **VGood:** $995

STEYR-MANNLICHER MODEL S/T
Bolt action; 9.3x64, 375 H&H Mag., 458 Win. Mag.; 5-shot detachable rotary magazine; 26″ heavy barrel; open rear sight, hooded ramp front; European walnut half-stock; Monte Carlo comb, cheekpiece; skip-checkered pistol grip, forearm; forend tip; recoil pad. Introduced 1975; **LMSR: $2850**
   Perf.: $2050   **Exc.:** $1395   **VGood:** $995

Steyr-Mannlicher Model S Luxus
Same specs as the Model S/T except 6.5x68, 7mm Rem. Mag., 300 Win. Mag., 8x68S; 3-shot detachable steel, in-line magazine; single set trigger; rear tang slide safety; half-stock only. Imported from Austria by GSI, Inc. **LMSR: $3250**
   Perf.: $2300   **Exc.:** $1395   **VGood:** $995

Stoner
SR-25 Match
SR-25 Match Lightweight
SR-25 Sporter
SR-25 Carbine

Sturm Ruger
Mini-14
Mini-14 Ranch Rifle
Mini-14 Stainless
Mini-14/5
Mini Thirty Rifle
Model 10/22 Carbine
Model 10/22 Deluxe Sporter
Model 10/22 International

Stoner SR-25 Match

Stoner SR-25 Sporter

Sturm Ruger Mini-14

Sturm Ruger Mini-14/5

Sturm Ruger Mini-14 Folding Stock

STONER SR-25 MATCH

Semi-automatic; 7.62 NATO; 10-shot steel maga-zine, 5-shot optional; 24″ heavy match barrel with 1:11.25″ twist; 44″ overall length; weighs 10 3/4 lbs.; no sights; black synthetic stock of AR-15A2 design; full floating forend of Mil-spec synthetic attaches to upper receiver at a single point; integral Weaver-style rail; rings and iron sights optional; improved AR-15 trigger; AR-15-style seven-lug rotating bolt; gas block rail mounts detachable front sight. Made in U.S. by Knight's Mfg. Co. Introduced 1993; **LMSR: $2995**

  Perf.: $2495   **Exc.:** $1995   **VGood:** $1295

Stoner SR-25 Match Lightweight

Same specs as SR-25 Match except 20″ medium match target contour barrel; 40″ overall length; weighs 9 1/2 lbs. **LMSR: $2995**

  Perf.: $2495   **Exc.:** $1995   **VGood:** $1295

Stoner SR-25 Sporter

Same specs as SR-25 Match except 20″ barrel; 40″ overall length; weighs 8 3/4 lbs.; AR-15A2-style elevation-adjustable front sight, detachable windage-adjustable rear; upper and lower receivers made of lightweight aircraft aluminum alloy; quick-detachable carrying handle/rear sight assembly; two-stage target trigger, shell deflector, bore guide, scope rings optional. Made in U.S. by Knight's Mfg. Co. Introduced 1993; **LMSR: $2995**

  Perf.: $2395   **Exc.:** $1895   **VGood:** $1195

Stoner SR-25 Carbine

Same as the SR-25 Sporter except 16″ light/hunt-ing contour barrel; 36″ overall length; weighs 7 3/4 lbs.; no sights; integral Weaver-style rail; scope rings; iron sights optional. Made in U.S. by Knight's Mfg. Co. Introduced 1995; **LMSR: $2995**

  Perf.: $2395 **Exc.:** $1895 **VGood:** $1195

NOTE: In 1976 Sturm Ruger marked virtually all then-current rifles, except the Mini-14 Stainless, all Mini-30s and Model 77-22s, with the words "Made in the 200th yr. of American Liberty." Rifles so marked command prices 30% above standard versions. This applies to new, unfired rifles in the box as well.

STURM RUGER MINI-14

Semi-automatic; 223 Rem., 222 Rem.; 5-shot detach-able box magazine; 18 1/2″ barrel; weighs 6 1/2 lbs.; gas-operated, fixed-piston carbine; uncheckered, reinforced American hardwood carbine-type stock (early versions with walnut); positive primary extrac-tion; fully-adjustable rear sight, gold bead front. Intro-duced 1973.

  Perf.: $400   **Exc.:** $370   **VGood:** $330

Sturm Ruger Mini-14 Ranch Rifle

Same specs as Mini-14 except 222 (1984 only), 223; steel-reinforced hardwood stock; ramp-type front sight; marketed with Sturm Ruger S100R scope rings; 20-shot magazine available; blued or stainless steel.

  Perf.: $420   **Exc.:** $390   **VGood:** $340

Sturm Ruger Mini-14 Stainless

Same specs as Mini-14 except barrel, action built of stainless steel. Introduced 1978; no longer in pro-duction.

  Perf.: $430   **Exc.:** $400   **VGood:** $355

Sturm Ruger Mini-14/5

Same specs as Mini-14 Ranch Rifle except metal fold-ing stock; checkered plastic vertical pistol grip; 47 3/4″ overall length with stock extended, 27 1/2″ stock closed; weighs 7 3/4 lbs.; blued or stainless steel. Still in pro-duction. **LMSR: $556**

  Perf.: $475 **Exc.:** $400 **VGood:** $355

STURM RUGER MINI THIRTY RIFLE

Semi-automatic; 7.62x39mm Russian; 5-shot detachable staggered box magazine; 18 1/2″ barrel; weighs 7 1/4 lbs.; six-groove barrel; Sturm Ruger integral scope mount bases; folding peep sight; blued or stainless finish. Introduced 1987; still in pro-duction. **LMSR: $649**

  Perf.: $460   **Exc.:** $390   **VGood:** $340
Stainless finish
  Perf.: $480   **Exc.:** $400   **VGood:** $355

STURM RUGER MODEL 10/22 CARBINE

Semi-automatic; 22 LR; 10-shot detachable rotary magazine; 18 1/2″ barrel; uncheckered walnut carbine stock on early versions; as of 1980, standard models with birch stocks; barrel band; receiver tapped for scope blocks or tip-off mount; adjustable folding leaf rear sight, gold bead front; blue or stainless finish. Intro-duced 1964; still in production. **LMSR: $230**

  Perf.: $150   **Exc.:** $120   **VGood:** $100
Uncheckered stock walnut
  Perf.: $175   **Exc.:** $135   **VGood:** $120

Sturm Ruger Model 10/22 Deluxe Sporter

Same specs as Model 10/22 except checkered walnut stock; flat buttplate; sling swivels; no barrel band. Intro-duced 1971; still in production. **LMSR: $274**

  Perf.: $190   **Exc.:** $165   **VGood:** $130

Sturm Ruger Model 10/22 International

Same specs as Model 10/22 except full-length Mannlicher-type walnut stock (early manufacture); birch full-length Mannlicher stock (1994); sling swivels. Introduced 1964; discontinued 1971. Reintroduced 1994; still in production. **LMSR: $262**

  Perf.: $200   **Exc.:** $170   **VGood:** $140
Checkered stock
  Perf.: $300   **Exc.:** $250   **VGood:** $200

Sturm Ruger Model 10/22 Deluxe Sporter

Sturm Ruger Model 10/22 International

Sturm Ruger Model 77RS Tropical

Sturm Ruger Model 44

RIFLES

Sturm Ruger
Model 10/22 Sporter
Model 44
Model 44 International
Model 44RS
Model 44 Sporter
Model 77R
Model 77RL Ultra Light
Model 77RS
Model 77RS Tropical
Model 77 Round Top Magnum

Sturm Ruger Model 10/22 Sporter
Same specs as Model 10/22 except Monte Carlo stock with grooved forearm; grip cap; sling swivels. Dropped in 1971, but reintroduced in 1973, with hand-checkered walnut pistol-grip stock. Dropped 1975.
    Perf.: $200   **Exc.:** $170   **VGood:** $140

STURM RUGER MODEL 44
Semi-automatic; 44 Mag.; 4-shot tube magazine; 18 1/2" barrel; magazine release button incorporated in 1967; uncheckered walnut pistol-grip carbine stock; barrel band; receiver tapped for scope mount; folding leaf rear sight, gold bead front. Introduced 1961; dropped 1985.
    Perf.: $395   **Exc.:** $350   **VGood:** $300

Sturm Ruger Model 44 International
Same specs as Model 44 except full-length Mannlicher-type walnut stock; sling swivels. Dropped 1971.
    Perf.: $795   **Exc.:** $625   **VGood:** $395

Sturm Ruger Model 44RS
Same specs as Model 44 except sling swivels; built-in peep sight.
    Perf.: $450   **Exc.:** $390   **VGood:** $330

Sturm Ruger Model 44 Sporter
Same specs as Model 44 except sling swivels; Monte Carlo sporter stock; grooved forearm; grip cap; flat buttplate. Dropped 1971.
    Perf.: $525   **Exc.:** $430   **VGood:** $380

STURM RUGER MODEL 77R
Bolt-action; 22-250, 220 Swift, 243 Win., 7mm-08, 6.5 Rem. Mag., 280 Rem., 284 Win., 308 Win., 300 Win. Mag., 338 Win. Mag., 350 Rem. Mag., 25-06, 257 Roberts, 250-3000, 6mm Rem., 270 Win., 7x57mm, 7mm Rem. Mag., 30-06; 3-, 5-shot magazine, depending upon caliber; 22" tapered barrel; hinged floorplate; adjustable trigger; hand-checkered American walnut stock; pistol-grip cap; sling swivel studs; recoil pad; integral scope mount base; optional folding leaf adjustable rear sight, gold bead ramp front. Introduced in 1968; no longer in production. Replaced by the Model M77 Mark II.
    Perf.: $400   **Exc.:** $330   **VGood:** $300
350 Rem. Mag. caliber
    Perf.: $450   **Exc.:** $390   **VGood:** $330

Sturm Ruger Model 77R

Sturm Ruger Model 77 Round Top Magnum

Sturm Ruger Model 77RL Ultra Light

Sturm Ruger Model 77RL Ultra Light
Same specs as Model 77R except 243, 308, 270, 30-06, 257, 22-250, 250-3000; 20" light barrel; Sturm Ruger 1" scope rings. Introduced 1983; **LMSR:** $592
    Perf.: $420   **Exc.:** $350   **VGood:** $300

Sturm Ruger Model 77RS
Same specs as Model 77R except magnum-size action; 257 Roberts, 25-06, 270 Win., 30-06, 7mm Rem. Mag., 300 Win. Mag., 338 Win. Mag.; 3-, 5-shot, depending upon caliber; 22" barrel (270, 30-06, 7x57, 280 Rem.); 24" barrel (all others).
    Perf.: $410   **Exc.:** $350   **VGood:** $300

Sturm Ruger Model 77RS Tropical
Same specs as Model 77RS except 458 Win. Mag.; 24" barrel; steel triggerguard, floorplate; weighs 8 3/4 lbs.; open sights; 1" scope rings. Introduced 1990; dropped 1991.
    Perf.: $550   **Exc.:** $480   **VGood:** $400

Sturm Ruger Model 77 Round Top Magnum
Same specs as Model 77R except round top action; drilled, tapped for standard scope mounts, open sights; 25-06, 270 Win., 7mm Rem. Mag., 30-06, 300 Win. Mag., 338 Win. Mag. Introduced 1971; dropped 1985.
    Perf.: $390   **Exc.:** $350   **VGood:** $300

Sturm Ruger
Model 77V Varmint
Model 77 RSI International
Model 77 RLS Ultra Light Carbine
Model 77R Mark II
Model 77R Mark II Express
Model 77R Mark II Magnum
Model 77 Mark II All-Weather Stainless
Model 77/22
Model 77/22 Hornet
Model 77/22MP
Model 77/22RM
Model 77/22 RMP
Model 77/22RP

Sturm Ruger Model 77V Varmint

Sturm Ruger Model 77R Mark II

Sturm Ruger Model 77R Mark II Magnum

Sturm Ruger Model 77/22

Sturm Ruger Model 77/22 Stainless

Sturm Ruger Model 77/22 Varmint

Sturm Ruger Model 77V Varmint
Same specs as Model 77R except 22-250, 220 Swift, 243 Win., 25-06, 308; 24" heavy straight tapered barrel, 26" (220 Swift); drilled, tapped for target scope mounts or integral scope mount bases on receiver; checkered American walnut stock. Introduced 1970; dropped 1992. Replaced by Model 77V Mark II in 1992.
Perf.: $410   **Exc.:** $360   **VGood:** $320

Sturm Ruger Model 77 RSI International
Same specs as Model 77R except 18 1/2" barrel; full-length Mannlicher-style stock; steel forend cap; loop-type sling swivels; open sights; Sturm Ruger steel scope rings; improved front sight; 22-250, 250-3000, 243, 308, 270, 30-06; weighs 7 lbs; 38 3/8" overall length. Introduced 1986; still in production. **LMSR: $713**
Perf.: $475   **Exc.:** $390   **VGood:** $330

Sturm Ruger Model 77 RLS Ultra Light Carbine
Same specs as Model 77R except 18 1/2" barrel; Sturm Ruger scope mounting system; iron sights; hinged floorplate; 270, 30-06, 243, 308; 38 7/8" overall length; weighs 6 lbs. Introduced 1987; no longer in production.
Perf.: $430   **Exc.:** $370   **VGood:** $310

STURM RUGER MODEL 77R Mark II
Bolt-action; 223, 22-250, 243 Win., 25-06, 257 Roberts, 270 Roberts, 270 Win., 280 Rem., 6mm Rem., 6.5x55, 7x57, 30-06, 308 Win. Mag., 7mm Rem. Mag., 300 Win. Mag., 338 Win. Mag.; 4-shot magazine; 20" barrel; 39 1/2" overall length; weighs 6 1/2 lbs.; American walnut stock; no sights; Ruger integral scope mount base; marketed with 1" scope rings; three-position safety; redesigned trigger; short action. Introduced 1989; still in production. **LMSR: $634**
Perf.: $390   **Exc.:** $350   **VGood:** $300

Sturm Ruger Model 77R Mark II Express
Same specs as Model 77R Mark II except 270, 30-06, 7mm Rem. Mag. 300 Win. Mag., 338 Win. Mag.; 4-shot magazine (3-shot for magnums); 22" barrel with integral sight rib; checkered French walnut stock; live-rubber recoil pad; steel grip cap; rear swivel stud, barrel-mounted front swivel stud; blade front sight, adjustable V-notch folding-leaf express rear sight with one standing leaf, one folding; three-position safety; Mauser-type extractor; controlled feeding. Made in U.S. by Sturm, Ruger & Co. Introduced 1991; still in production. **LMSR: $1625**
Perf.: $1250   **Exc.:** $995   **VGood:** $795

Sturm Ruger Model 77R Mark II Magnum
Same specs as Model 77R Mark II except 7/16" longer receiver with increased locking area and lengthened front; 375 H&H, 404 Jeffery, 416 Rigby; 4-shot magazine (375, 404), 3-shot (416) magazine; 22" barrel with integral sight rib; ramp front, three-leaf express sight; high-grade Circassian walnut stock with checkered grip and forend; black live-rubber recoil pad; steel floorplate and triggerguard. Made in U.S. by Sturm Ruger & Co. Introduced 1989; **LMSR: $1550**
Perf.: $1195   **Exc.:** $995   **VGood:** $795

Sturm Ruger Model 77 Mark II All-Weather Stainless
Same specs as Model 77R Mark II except all metal parts of stainless steel; fiberglass-reinforced Zytel stock; 223, 243, 270, 308, 30-06, 7mm Rem. Mag.; fixed blade-type ejector; new triggerguard; patented floorplate latch; three-position safety; integral scope base, 1" Sturm Ruger scope rings; built-in swivel loops. Introduced 1990; **LMSR: $574**
Perf.: $495   **Exc.:** $450   **VGood:** $400

STURM RUGER MODEL 77/22
Bolt-action; 22 LR; 10-shot rotary magazine; 20" barrel; 39 3/4" overall length; weighs 5 3/4 lb.; checkered American walnut or nylon-reinforced Zytel stock; gold bead front sight, adjustable folding leaf rear; 1" Sturm Ruger rings optional; Mauser-type action; three-position safety simplified bolt stop; 10/22 dual screw barrel attachment system. Blued model introduced 1983; stainless steel and blued models with synthetic stock introduced 1989; still in production. **LMSR: $458**
Perf.: $320   **Exc.:** $290   **VGood:** $250
With synthetic stock
Perf.: $300   **Exc.:** $260   **VGood:** $220

Sturm Ruger Model 77/22 Hornet
Same specs as Model 77/22 except 22 Hornet; 6-shot rotary magazine; weighs about 6 lbs.; checkered American walnut stock; black rubber buttpad; brass bead front sight, open adjustable rear; slightly lengthened receiver; comes with 1" Sturm Ruger scope rings. Made in U.S. by Sturm Ruger & Co. Introduced 1994; still produced. **LMSR: $525**
Perf.: $380   **Exc.:** $320   **VGood:** $290

Sturm Ruger Model 77/22MP
Same specs as Model 77/22M except 22 WMR; stainless steel; synthetic stock. Introduced 1989; **LMSR: $431**
Perf.: $320   **Exc.:** $290   **VGood:** $250

Sturm Ruger Model 77/22RM
Same specs as Model 77/22 except 22 WMR. Introduced 1989; still produced. **LMSR: $498**
Perf.: $330   **Exc.:** $270   **VGood:** $230

Sturm Ruger Model 77/22 RMP
Same specs as Ruger Model 77/22 RM except stainless finish; synthetic stock. Still in production. **LMSR: $431**
Perf.: $300   **Exc.:** $270   **VGood:** $230

Sturm Ruger Model 77/22RP
Same specs as Model 77/22 except synthetic stock; no iron sights; comes with rings. Made in U.S. by Sturm, Ruger & Co. Introduced 1989; still in production. **LMSR: $498**
Perf.: $345   **Exc.:** $290   **VGood:** $250

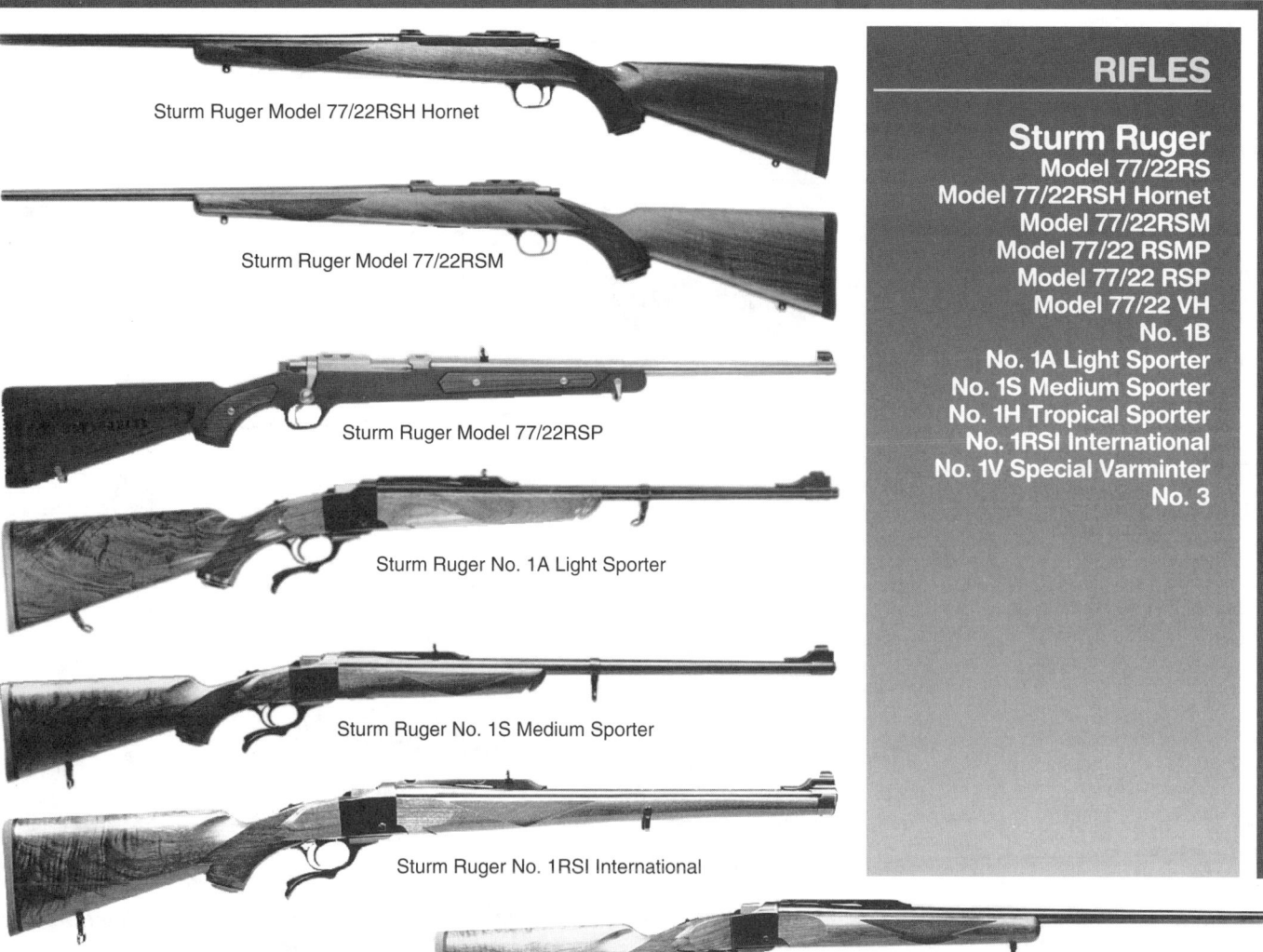

Sturm Ruger Model 77/22RSH Hornet

Sturm Ruger Model 77/22RSM

Sturm Ruger Model 77/22RSP

Sturm Ruger No. 1A Light Sporter

Sturm Ruger No. 1S Medium Sporter

Sturm Ruger No. 1RSI International

Sturm Ruger No. 1B

Sturm Ruger No. 3

RIFLES

Sturm Ruger
Model 77/22RS
Model 77/22RSH Hornet
Model 77/22RSM
Model 77/22 RSMP
Model 77/22 RSP
Model 77/22 VH
No. 1B
No. 1A Light Sporter
No. 1S Medium Sporter
No. 1H Tropical Sporter
No. 1RSI International
No. 1V Special Varminter
No. 3

Sturm Ruger Model 77/22RS
Same specs as the Model 77/22 except blade front sight, fully-adjustable open rear; comes with scope rings. Made in U.S. by Sturm, Ruger & Co. Still produced. **LMSR: $498**
Perf.: $3500   **Exc.:** $250   **VGood:** $230

Sturm Ruger Model 77/22RSH Hornet
Same as the 77/22 except 22 Hornet; blade front sight, fully-adjustable open rear adjustable. Made in U.S. by Sturm, Ruger & Co. Introduced 1994; still produced. **LMSR: $498**
Perf.: $385   **Exc.:** $320   **VGood:** $290

Sturm Ruger Model 77/22RSM
Same specs as the Model 77/22R except 22 WMR; blade front sight, fully-adjustable open rear; blue finish; checkered walnut stock. Made in U.S. by Sturm Ruger & Co. Still produced. **LMSR: $498**
Perf.: $350   **Exc.:** $260   **VGood:** $230

Sturm Ruger Model 77/22 RSMP
Same specs as Model 77/22 RSM except stainless finish; synthetic stock. Still in production. **LMSR: $539**
Perf.: $360   **Exc.:** $260   **VGood:** $230

Sturm Ruger Model 77/22 RSP
Same specs as Model 77/22 except stainless finish; synthetic stock. Still in production. **LMSR: $498**
Perf.: $340   **Exc.:** $260   **VGood:** $230

Sturm Ruger Model 77/22 VH
Same specs as Model 77/22 except 22 Hornet; laminated stock. Still in production. **LMSR: $575**
Perf.: $380   **Exc.:** $325   **VGood:** $270

STURM RUGER NO. 1B
Single-shot; 22-250, 220 Swift, 243 Win., 223, 257 Roberts, 280, 6mm Rem., 25-06, 270 Win., 30-06, 7mm Rem. Mag., 300 Win. Mag., 338 Mag., 270 Weatherby, 300 Weatherby; 26″ barrel with quarter rib; American walnut, two-piece stock; hand-checkered pistol grip, forearm; open sights or integral scope mounts; hammerless falling-block design; automatic ejector; top-tang safety. Introduced 1967; still in production. **LMSR: $774**
Perf.: $540   **Exc.:** $420   **VGood:** $365

Sturm Ruger No. 1A Light Sporter
Same specs as No. 1B except 22″ barrel; Alex Henry-style forearm; iron sights; 243 Win., 270 Win., 30-06, 7x57mm. Introduced 1968; still in production. **LMSR: $774**
Perf.: $565   **Exc.:** $420   **VGood:** $365

Sturm Ruger No. 1S Medium Sporter
Same specs as No. 1A Light Sporter except 7mm Rem. Mag., 300 Win. Mag., 338 Win. Mag., 45-70; 26″ barrel, 22″ (45-70) barrel. Introduced 1968; still in production. **LMSR: $774**
Perf.: $570   **Exc.:** $430   **VGood:** $370

Sturm Ruger No. 1H Tropical Model
Same specs as No. 1S Medium Sporter except 375 H&H Mag., 458 Win. Mag.; 24″ heavy barrel; open sights. Introduced 1968; still in production. **LMSR: $774**
Perf.: $570   **Exc.:** $430   **VGood:** $370

Sturm Ruger No. 1RSI International
Same specs as No. 1B, except full-length Mannlicher-style stock of American walnut; 243, 30-06, 7x57, 270. Introduced 1983; still in production. **LMSR: $794**
Perf.: $590   **Exc.:** $440   **VGood:** $390

Sturm Ruger No. 1V Special Varminter
Same specs as No. 1B except 24″ heavy barrel; 22-250, 220 Swift, 223, 25-06, 6mm; supplied with target scope bases. Introduced 1970; still in production. **LMSR: $774**
Perf.: $560   **Exc.:** $420   **VGood:** $365

STURM RUGER NO. 3
Single-shot; 22 Hornet, 223, 30-40 Krag, 375 Win., 44 Mag., 45-70; 22″ barrel; same action as Sturm Ruger No. 1, except for different lever; uncheckered American walnut, two-piece carbine-type stock; folding leaf rear sight, gold bead front; adjustable trigger; barrel band on forearm; automatic ejector. Introduced 1969; 30-40 chambering dropped 1978; model discontinued 1986.
Perf.: $375   **Exc.:** $290   **VGood:** $230
223, 44 Mag. calibers
Perf.: $425   **Exc.:** $350   **VGood:** $285

Survival Arms
AR-7 Explorer
AR-7 Sporter
AR-7 Wildcat

Swiss
K-31 Target Model

Tanner
50-Meter Free Rifle
300-Meter Free Rifle
300-Meter UIT Standard Free Rifle

Thompson/Center
Contender Carbine
Contender Carbine Rynite
Contender Carbine Stainless
Contender Carbine Survival System
Contender Carbine Youth Model
Custom Shop TCR '87
TCR '83 Hunter

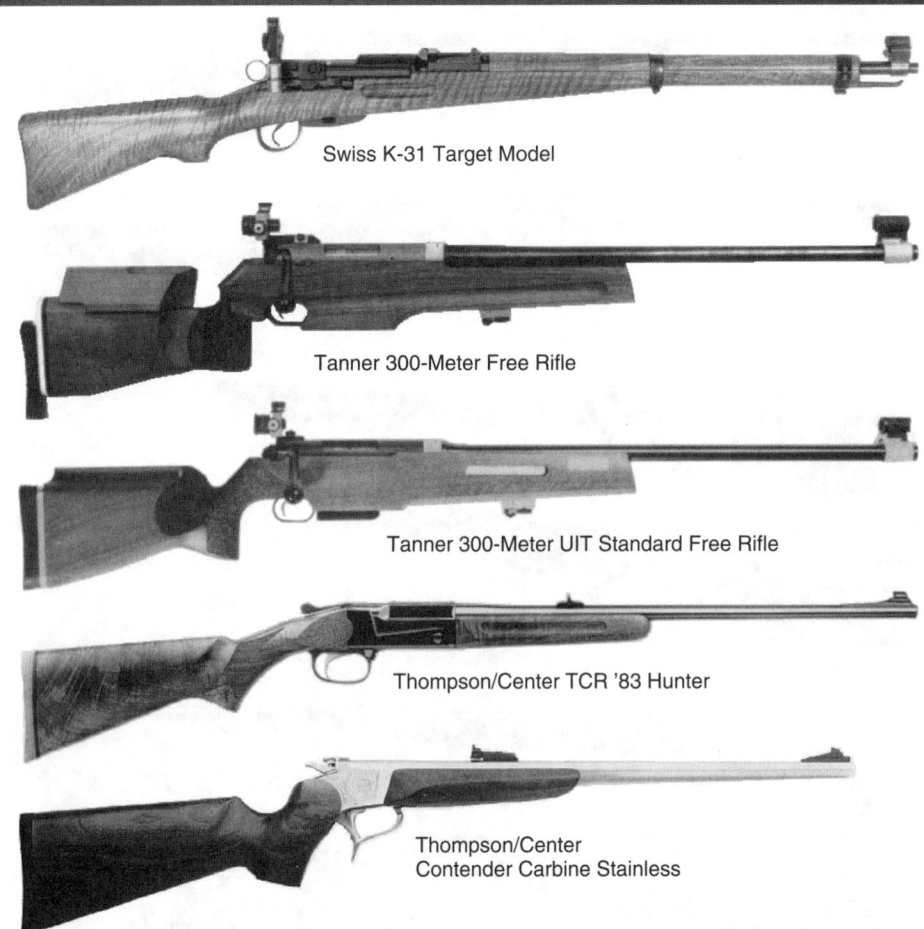

Swiss K-31 Target Model

Tanner 300-Meter Free Rifle

Tanner 300-Meter UIT Standard Free Rifle

Thompson/Center TCR '83 Hunter

Thompson/Center
Contender Carbine Stainless

SURVIVAL ARMS AR-7 EXPLORER
Semi-automatic; 22 LR; 8-shot magazine; 16″ barrel; 34 1/2″ overall length with barrel mounted, 16 1/2″ stowed; weighs 2 1/2 lbs.; square blade front sight, aperture elevation-adjustable rear; moulded Cycolac stock with snap-on rubber buttcap; takedown design stores barrel and action in hollow stock; black, silvertone or camouflage finish. Light enough to float. Made in U.S. by Survival Arms, Inc. Reintroduced 1992; discontinued 1995.
Perf.: $130   **Exc.:** $115   **VGood:** $90

Survival Arms AR-7 Sporter
Same specs as AR-7 Explorer except 25-shot magazine; black finish with telescoping stock. Made in U.S. by Survival Arms, Inc. Introduced 1992; discontinued 1995.
Perf.: $180   **Exc.:** $150   **VGood:** $125

Survival Arms AR-7 Wildcat
Same specs as AR-7 Explorer except black finish and wood stock. Made in U.S. by Survival Arms, Inc. Introduced 1992; discontinued 1995.
Perf.: $165   **Exc.:** $130   **VGood:** $110

SWISS K-31 TARGET MODEL
Bolt action; 308; 6-shot magazine; 26″ barrel; 44″ overall length; straight pull; protected blade front sight, ladder-type adjustable rear; European walnut stock; sling, muzzle cap; based on straight-pull Schmidt-Rubin design. Made in Switzerland. Introduced 1982; no longer in production.
Perf.: $595   **Exc.:** $450   **VGood:** $300

TANNER 50-METER FREE RIFLE
Bolt action; single shot; 22 LR; 27 2/3″ barrel; 43 2/5″ overall length; weighs 14 lbs.; micrometer-diopter rear sight, globe front with interchangeable inserts; nutwood stock with palm rest; accessory rail; adjustable hook buttplate; adjustable set trigger. Made in Switzerland. Introduced 1984; dropped 1988.
Perf.: $2795   **Exc.:** $2195   **VGood:** $1695

TANNER 300-METER FREE RIFLE
Bolt action; single shot; 308, 7.5 Swiss; 28 5/8″ barrel; 45 5/16″ overall length; weighs 15 lbs.; Tanner micrometer-diopter rear sight, globe front with interchangeable inserts; thumbhole walnut stock; accessory rail; palm rest; three-lug revolving bolt design; adjustable set trigger. Made in Switzerland. Introduced 1984; discontinued 1988.
Perf.: $3750   **Exc.:** $3100   **VGood:** $2600

Tanner 300-Meter UIT Standard Free Rifle
Same specs as 300-Meter Free Rifle except repeater with 10-shot magazine; 24 13/16″ barrel; 40 5/8″ overall length; weighs 10 1/2 lbs.; no palm rest; no adjustable buttplate. Made in Switzerland. Introduced 1984; discontinued 1988.
Perf.: $3800   **Exc.:** $3100   **VGood:** $2200

THOMPSON/CENTER CONTENDER CARBINE
Single shot; 22 LR, 22 Hornet, 223 Rem., 7mm TCU, 7-30 Waters, 30-30, 357 Rem. Max., 35 Rem., 44 Mag., 410; 21″ barrel; 35″ overall length; weighs 5 1/8 lbs.; open adjustable rear sight, blade front; drilled, tapped for scope mounts; checkered American walnut stock; rubber buttpad; built on T/C Contender action; interchangeable barrels. Introduced 1985; still in production. **LMSR: $500**
Perf.: $400   **Exc.:** $325   **VGood:** $270

Thompson/Center Contender Carbine Rynite
Same specs as Contender Carbine except Rynite stock, forend. Introduced 1990; discontinued 1993.
Perf.: $330   **Exc.:** $280   **VGood:** $230

Thompson/Center Contender Carbine Stainless
Same specs as Contender Carbine except stainless steel with blued sights; walnut or Rynite stock, forend. Made in U.S. by Thompson/Center. Introduced 1993; still produced. **LMSR: $495**
Perf.: $390   **Exc.:** $320   **VGood:** $250

Thompson/Center Contender Carbine Survival System
Same specs as Contender Carbine except Rynite stock; two 16 1/4″ barrels, 223 Rem., 45 Colt/410. Marketed in Cordura nylon case. Introduced 1991; no longer in production.
Perf.: $470   **Exc.:** $410   **VGood:** $340

Thompson/Center Contender Carbine Youth Model
Same specs as Contender Carbine except 22 LR, 22 WMR, 223, 7-30 Waters, 30-30, 35 Rem., 44 Mag.; 16 1/4″ barrel; 29″ overall length; fully-adjustable open sights; shorter buttstock; also available with rifled vent rib barrel chambered for 45/410. Introduced 1987; still in production. **LMSR: $465**
Perf.: $350   **Exc.:** $300   **VGood:** $250

THOMPSON/CENTER CUSTOM SHOP TCR '87
Single shot; 22 Hornet, 222 Rem., 22-250, 243 Win., 270, 308, 7mm-08, 30-06, 32-40 Win., 12-ga. slug, 10-ga. and 12-ga. field barrels; 23″ (standard), 25 7/8″ (heavy) barrel; 39 1/2″ overall length; weighs about 6 3/4 lbs.; break-open; no sights; checkered American black walnut stock, forend; interchangeable barrels; single-stage trigger; cross-bolt safety. Made in U.S. by T/C. Available only through the T/C custom shop. Introduced 1987; no longer offered.
Perf.: $500   **Exc.:** $410   **VGood:** $355

THOMPSON/CENTER TCR '83 HUNTER
Single shot; 223, 22-250, 243 Win., 7mm Rem. Mag., 30-06; 23″ barrel; 39 1/2″ overall length; weighs 6 7/8 lbs.; break-open; blade on ramp front sight, windage-adjustable open rear; American black walnut stock; cut-checkering on pistol-grip, forend; interchangeable barrels; cross-bolt safety; single-stage or double-set trigger. Introduced 1983; discontinued 1986.
Perf.: $420   **Exc.:** $380   **VGood:** $310

Tikka Model 55

Tikka Model 55 Sporter

Tikka Model 65

Tikka Model 512S Double Rifle

Tikka Premium Grade

Tikka Standard Model

RIFLES

Thompson/Center
TCR '87 Hunter

Tikka
Model 07
Model 55
Model 55 Deluxe
Model 55 Sporter
Model 65
Model 65 Deluxe
Model 65 Target
Model 77K
Model 412S Double Rifle
Model 512S Combination Gun
Model 512S Double Rifle
Premium Grade
Standard

THOMPSON/CENTER TCR '87 HUNTER

Single shot; 22 Hornet, 222 Rem., 223 Rem., 22-250, 243 Win., 270, 308, 7mm-08, 30-06, 32-40 Win., 375 H&H, 416 Rem. Mag.; 12-gauge slug; 23″ (standard), 25 7/8″ heavy barrel; break-open; no sights; checkered American black walnut stock; interchangeable barrels; single-stage trigger; cross-bolt safety. Introduced 1987; discontinued 1992.

Perf.: $420   Exc.: $380   VGood: $310

TIKKA MODEL 07

Over/under combo; 222, 5.6x52R, 5.6x50R Mag., beneath 12-ga. barrel; 22 3/4″ barrel; bead front sight, open windage-adjustable rear; Monte Carlo stock of European walnut; palm swell with pistol grip; exposed hammer, sling swivels; vent rib, rosewood pistol-grip cap. Introduced 1979; discontinued 1982.

Perf.: $595   Exc.: $495   VGood: $395

TIKKA MODEL 55

Bolt action; 17 Rem., 222, 22-250, 6mm Rem., 243, 308; 3-shot detachable magazine; 23″ barrel; bead ramped front sight, fully-adjustable rear; drilled, tapped for scope mounts; hand-checkered Monte Carlo stock of European walnut; palm swell on pistol grip. Introduced 1979; discontinued 1981.

Perf.: $470   Exc.: $400   VGood: $350

Tikka Model 55 Deluxe

Same specs as Model 55 except roll-over cheekpiece; forend tip; grip cap of rosewood. Introduced 1979; discontinued 1981.

Perf.: $500   Exc.: $430   VGood: $370

Tikka Model 55 Sporter

Same specs as Model 55 except 222, 22-250, 243, 308; 5-, 10-shot magazine; 23″ heavy barrel; no sights; varmint-type stock; oil-finish. Introduced 1979; discontinued 1981.

Perf.: $450   Exc.: $390   VGood: $330

TIKKA MODEL 65

Bolt action; 25-06, 6.5x55, 7x57, 7x64, 270, 308, 30-06, 7mm Rem. Mag., 300 Win. Mag.; 5-shot magazine; 22″ barrel; bead ramped front sight, fully-adjustable rear; drilled, tapped for scope mounts; hand-checkered Monte Carlo stock of European walnut; palm swell on pistol grip; adjustable trigger. Introduced 1979; discontinued 1981.

Perf.: $480   Exc.: $420   VGood: $370

Tikka Model 65 Deluxe

Same specs as Model 65 except roll-over cheekpiece; forend tip; rosewood grip cap. Introduced 1979; discontinued 1981.

Perf.: $500   Exc.: $430   VGood: $370

Tikka Model 65 Target

Same specs as Model 65 except 25-06, 6.5x55, 270, 308, 30-06; 22″ heavy barrel; no sights; target-type walnut stock; stock designed to meet ISU requirements, with stippled forend, palm swell. Introduced 1969; discontinued 1981.

Perf.: $570   Exc.: $500   VGood: $430

TIKKA MODEL 77K

Combo; over/under; 222, 5.6x52R, 6.5x55, 7x57R, 7x65R, 308 beneath 12-ga. barrel; 22 3/4″ barrel; hammerless; bead front sight, open windage-adjustable rear; Monte Carlo stock of European walnut; palm swell with pistol-grip; sling swivels; ventilated rib; rosewood pistol-grip cap. Introduced 1979; discontinued 1982.

Perf.: $625   Exc.: $495   VGood: $395

TIKKA MODEL 412S DOUBLE RIFLE

Double rifle; 9.3x74R; 24″ barrel; weighs 8 5/8 lbs.; ramp front sight, adjustable open rear; American walnut stock with Monte Carlo-style comb; barrel selector mounted in trigger; cocking indicators in tang; recoil pad; ejectors. Imported from Italy by Stoeger. Introduced 1980; still imported as Model 512S.

Perf.: $1195   Exc.: $895   VGood: $695

TIKKA MODEL 512S COMBINATION GUN

Combo; over/under; 12 gauge over 222, 308; 24″ barrel choked Imp. Mod.; weighs 7 5/8 lbs.; blade front sight, flip-up-type open rear; American walnut stock with recoil pad, Monte Carlo comb; standard measurements 14″x1 3/5″x 2″x2 3/5″; barrel selector on trigger; hand-checkered stock and forend; barrels are screw-adjustable to change point of bullet impact; interchangeable barrels. Imported from Italy by Stoeger. Introduced 1980 as Model 412S; still imported. **LMSR: $1350**

Perf.: $1100   Exc.: $920   VGood: $710

TIKKA MODEL 512S DOUBLE RIFLE

Double rifle; 9.3x74R; 24″ barrel; weighs 8 5/8 lbs.; ramp front sight, adjustable open rear; American walnut stock with Monte Carlo-style comb; barrel selector mounted in trigger; cocking indicators in tang; recoil pad; ejectors. Imported from Italy by Stoeger. Introduced 1980; still imported. **LMSR: $1525**

Perf.: $1300   Exc.: $1100   VGood: $810

TIKKA PREMIUM GRADE

Bolt action; 22-250, 223, 243, 270, 30-06, 7mm Rem. Mag., 300 Win. Mag., 338 Win. Mag.; 3-, 5-shot magazine; 22″ (standard), 24″ (magnum) barrel; 43″ overall length; weighs 7 1/4 lbs.; no sights; Monte Carlo select walnut stock with roll-over cheekpiece; rosewood grip, forend caps; hand-checkered grip, forend; polished, blued barrel. Made in Finland. Introduced 1990; no longer in production.

Perf.: $800   Exc.: $680   VGood: $520

TIKKA STANDARD

Bolt action; 22-250, 223, 243, 270, 30-06, 7mm Rem. Mag., 300 Win. Mag., 338 Win. Mag.; 3-, 5-shot magazine; 22″ barrel (standard), 24″ (magnums); 43″ overall length; weighs 7 1/4 lbs.; no sights; Monte Carlo European walnut stock; checkered grip, forend; rubber buttplate; receiver dovetailed for scope mounts. Made in Finland. Introduced 1988; no longer in production.

Perf.: $675   Exc.: $595   VGood: $475

Tikka
Varmint/Continental
Whitetail/Battue

Tradewinds
Husky Model 5000
Model 260-A
Model 311-A

Tyrol
Custom Crafted Model

Uberti
Henry Rifle
Henry Carbine
Henry Trapper
Model 1866 Sporter Rifle
Model 1866 Yellowboy
Carbine
Model 1866 Trapper Carbine
Model 1873 Buckhorn
Model 1873 Cattleman
Model 1873 Sporting Rifle

Tradewinds Model 260-A

Tradewinds Husky Model 5000

Tyrol Custom Crafter Model

Uberti Rolling Block Baby Carbine

Uberti Model 1873 Sporting Rifle

Ultra-Hi Model 2200

TIKKA VARMINT/CONTINENTAL
Bolt-action; 22-250, 223, 243, 308; 3-, 5-shot magazine; 24" heavy barrel; no sights; Monte Carlo European walnut stock, extra-wide forend; rubber buttplate; receiver dovetailed for scope mounts. Made in Finland. Introduced 1991; no longer in production.
Perf.: $675   **Exc.:** $550   **VGood:** $460

TIKKA WHITETAIL/BATTUE
Bolt action; 270, 308, 30-06, 7mm Rem. Mag., 300 Win. Mag., 338 Win. Mag.; 3-, 5-shot magazine; 20 1/2" barrel; raised quarter-rib; V-shaped rear sight, blade front; Monte Carlo European walnut stock; checkered pistol grip, forend; rubber buttplate; receiver dovetailed for scope mounts. Made in Finland. Introduced 1991; no longer in production.
Perf.: $500   **Exc.:** $440   **VGood:** $370

TRADEWINDS HUSKY MODEL 5000
Bolt action; 22-250, 243, 270, 30-06, 308; removable magazine; 23 3/4" barrel; fixed hooded front sight, adjustable rear; hand-checkered Monte Carlo stock of European walnut; white-line spacers on pistol-grip cap, forend tip, buttplate; recessed bolt head; adjustable trigger. Imported from Europe. Introduced 1973; discontinued 1983.
Perf.: $350   **Exc.:** $320   **VGood:** $290

TRADEWINDS MODEL 260-A
Semi-automatic; 22 LR; 5-shot magazine; 22 1/2" barrel; 41 1/2" overall length; hooded ramp front sight, three-leaf folding rear; walnut stock; hand-checkered pistol grip, forend; double extractors; sliding safety; sling swivels; receiver grooved for scope mount. Made in Japan. Introduced 1975; discontinued 1989.
Perf.: $200   **Exc.:** $170   **VGood:** $140

TRADEWINDS MODEL 311-A
Bolt action; 22 LR; 5-shot magazine; 22 1/2" barrel; 41 1/4" overall length; hooded ramp front sight, folding leaf rear; receiver grooved for scope mount; Monte Carlo walnut stock; hand-checkered pistol grip, forend; sliding safety. Made in Europe. Introduced 1976; no longer in production.
Perf.: $180   **Exc.:** $150   **VGood:** $130

TYROL CUSTOM CRAFTER MODEL
Bolt action; 243, 25-06, 30-06, 308, 7mm, 300 Win.; 23 3/4" barrel; hooded ramp front sight, adjustable rear; drilled, tapped for scope mounts; hand-checkered Monte Carlo stock of European walnut; adjustable trigger; shotgun-type tang safety. Manufactured in Austria. Introduced 1973; discontinued 1975.
Perf.: $450   **Exc.:** $400   **VGood:** $330

UBERTI HENRY RIFLE
Lever action; 44-40; 24 1/2" half-octagon barrel; 43 3/4" overall length; weighs 9 lbs.; elevation-adjustable rear sight, blade front; uncheckered walnut stock; brass buttplate, frame, elevator, magazine follower; rest charcoal blued or of polished steel. Made in Italy. Introduced 1987; still in production. **LMSR: $940**
Perf.: $750   **Exc.:** $520   **VGood:** $510

Uberti Henry Carbine
Same specs as Henry Rifle except 22 1/4" barrel. Imported by Uberti USA; still in production. **LMSR: $940**
Perf.: $750   **Exc.:** $620   **VGood:** $510

Uberti Henry Trapper
Same specs as Henry Rifle except 16", 18" barrel. Imported by Uberti USA; still in production. **LMSR: $950**
Perf.: $750   **Exc.:** $620   **VGood:** $510

UBERTI MODEL 1866 SPORTER RIFLE
Lever action; 22 LR, 22 WMR, 38 Spl., 44-40; 24 1/2" octagonal barrel; 43 1/4" overall length; weighs 8 lbs.; elevation-adjustable rear sight, windage-adjustable blade front; polished brass frame, buttplate, forend cap; rest charcoal blued. Made in Italy. Introduced 1987; still in production. **LMSR: $839**
Perf.: $690   **Exc.:** $570   **VGood:** $440

Uberti Model 1866 Yellowboy Carbine
Same specs as Model 1866 except 19" round barrel; 38 1/4" overall length; weighs 7 7/16 lbs. Introduced 1987; still in production. **LMSR: $760**
Perf.: $690   **Exc.:** $570   **VGood:** $440

Uberti Model 1866 Trapper Carbine
Same specs as Model 1866 except 22 LR, 38 Spl., 44-40; 16 1/8" barrel. Introduced 1987; discontinued 1989.
Perf.: $690   **Exc.:** $570   **VGood:** $440

UBERTI MODEL 1873 BUCKHORN
Revolving carbine; 44 Mag., 44 Spl., 44-40; 6-shot cylinder; 18" barrel; 34" overall length; weighs 4 1/2 lbs.; grooved or target rear sight, blade front; carbine version of single-action revolver; color case-hardened frame; blued cylinder, barrel; brass buttplate. Introduced 1987; discontinued 1989.
Perf.: $410   **Exc.:** $350   **VGood:** $290
With target sights
Perf.: $440   **Exc.:** $380   **VGood:** $320

UBERTI MODEL 1873 CATTLEMAN
Revolving carbine; 22 LR/22 WMR, 38 Spl., 357 Mag., 44-40, 45 Colt; 6-shot cylinder; 18" barrel; 34" overall length; weighs 4 1/2 lbs.; grooved or target rear sight, blade front; carbine version of single-action revolver; case hardened frame; blued cylinder, barrel; brass buttplate. Made in Italy. Introduced 1987; discontinued 1989.
Perf.: $400   **Exc.:** $330   **VGood:** $270
With target sights
Perf.: $425   **Exc.:** $350   **VGood:** $280

UBERTI MODEL 1873 SPORTING RIFLE
Lever action; 22 LR, 22 WMR, 38 Spl., 357 Mag., 44-40, 45 Colt; 24 1/4", 30" octagonal barrel; 43 1/4" overall length; weighs 8 lbs.; elevation-adjustable rear sight, windage-adjustable front sight; uncheckered walnut stock; brass elevator; color case-hardened frame; blued barrel, hammer, lever, buttplate. Made in Italy. Introduced 1987; still in production. **LMSR: $973**
Perf.: $800   **Exc.:** $675   **VGood:** $540

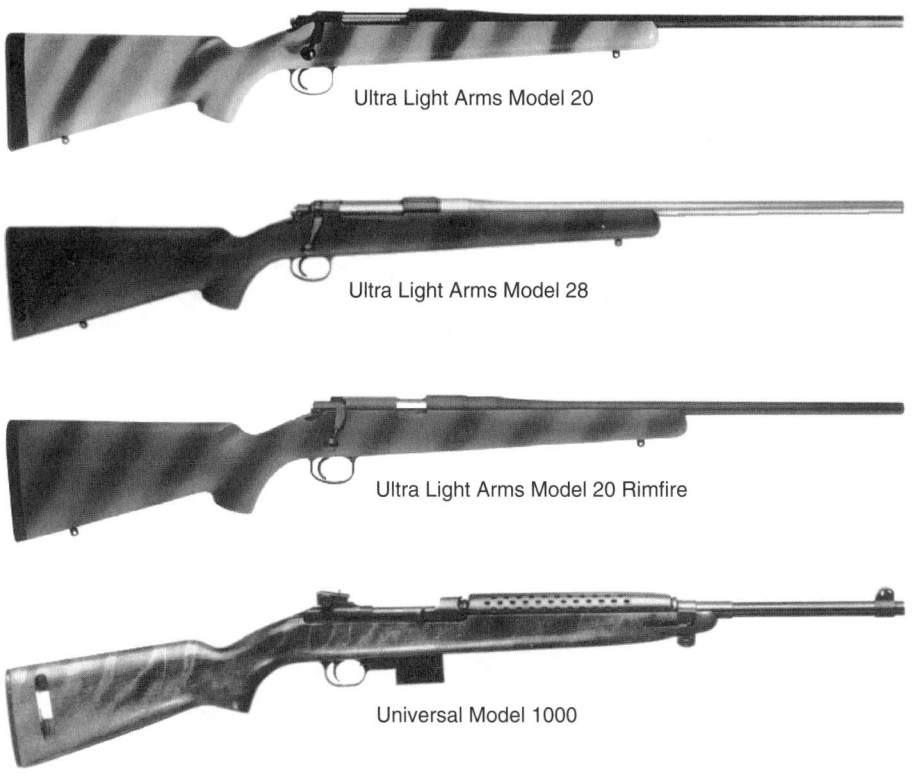

Ultra Light Arms Model 20

Ultra Light Arms Model 28

Ultra Light Arms Model 20 Rimfire

Universal Model 1000

RIFLES

Uberti
Model 1873 Sporting Carbine
Model 1873 Sporting Carbine, Trapper Model
Model 1875 Army Target
Rolling Block Baby Carbine

Ultimate Accuracy
Model 5100A1 Long-Range Rifle
Model 5100A1 Improved Model

Ultra-Hi
Model 2200

Ultra Light Arms
Model 20
Model 20 Rimfire
Model 24
Model 28
Model 40

Universal
Model 1000
Model 1002

Uberti Model 1873 Sporting Carbine
Same specs as Model 1873 Sporting except 19" round barrel; 38 1/4" overall length; weighs 7 1/4 lbs.; available with nickel plating. Introduced 1987; still in production.
LMSR: $910
  Perf.: $780 **Exc.:** $625 **VGood:** $500
Nickel finish
  Perf.: $800 **Exc.:** $645 **VGood:** $525

Uberti Model 1873 Sporting Carbine, Trapper Model
Same specs as Model 1873 Sporting except 357 Mag., 44-40, 45 Colt; 16" barrel. Discontinued 1990.
  Perf.: $700 **Exc.:** $590 **VGood:** $500

UBERTI MODEL 1875 ARMY TARGET
Revolving carbine; 357 Mag., 44-40, 45 Colt; 6-shot cylinder; 18" barrel; 37" overall length; weighs 4 1/2 lbs.; carbine version of 1875 single-action revolver; elevation-adjustable rear sight; ramp front; unchecked walnut stock; polished brass buttplate, triggerguard; case-hardened frame; blued or nickel-plated cylinder, barrel. Made in Italy. Introduced 1987; discontinued 1989.
Blued finish
  Perf.: $400 **Exc.:** $325 **VGood:** $260
Nickel finish
  Perf.: $430 **Exc.:** $350 **VGood:** $285

UBERTI ROLLING BLOCK BABY CARBINE
Single shot; 22 LR, 22 WMR, 22 Hornet, 357 Mag.; 22" barrel; 35 1/2" overall length; weighs 4 13/16 lbs.; copy of Remington New Model No. 4 carbine; brass buttplate, triggerguard; blued barrel; color case-hardened frame. Introduced 1986; still imported. **LMSR: $460**
  Perf.: $345 **Exc.:** $300 **VGood:** $270

ULTIMATE ACCURACY MODEL 5100A1 LONG-RANGE RIFLE
Bolt action; 50 BMG; 5-shot magazine; 29" fully fluted, free-floating barrel; 51 1/2" overall length; weighs 36 lbs.; no sights; optional Leupold Ultra M1 16x scope; composition buttstock with comb adjustment for drop; adjustable trigger; breaks down for transport, storage. Made in U.S. by Ultimate Accuracy. Introduced 1994; no longer in production.
  Perf.: $2995 **Exc.:** $2495 **VGood:** $1995

Ultimate Accuracy Model 5100A1 Improved Model
Same specs as Model 5100A1 except receiver is drilled and tapped for scope mount; manual safety; one-piece muzzle brake. Made in U.S. by Ultimate Accuracy. No longer in production.
  Perf.: $2995 **Exc.:** $2495 **VGood:** $1995

ULTRA-HI MODEL 2200
Bolt action; single shot; 22 Short, 22 Long, 22 LR; 23" barrel; weighs 5 lbs.; blade front sight, open rear; pistol-grip Monte Carlo hardwood stock. Made in Japan. Introduced 1977; no longer in production.
  Perf.: $60 **Exc.:** $50 **VGood:** $40

ULTRA LIGHT ARMS MODEL 20
Bolt action; 17 Rem., 222, 223, 22 Hornet, 22-250, 243, 6mm Rem., 257 Roberts, 7x57, 7x57 Ackley, 7mm-08, 284, 308, 358 Win.; other calibers on request; 22" barrel; 41 1/2" overall length; weighs 4 1/2 lbs.; no sights; marketed with scope mount; Kevlar/graphite stock in green, black brown or camo; two-position, three-function safety; Timney trigger; bright or matte finish; marketed in hard case; left- or right-hand action. Introduced 1986; still in production. **LMSR: $2600**
  Perf.: $2100 **Exc.:** $1750 **VGood:** $1250
Left-hand model
  Perf.: $2150 **Exc.:** $1800 **VGood:** $1300

ULTRA LIGHT ARMS MODEL 20 RIMFIRE
Bolt action; 22 LR; single shot or 5-shot repeater with removable magazine; 22" Douglas Premium, #1 contour barrel; 41 1/2" overall length; weighs 5 1/4 lbs.; no sights; drilled, tapped for scope mount; scope mounts; composite Kevlar, graphite reinforced stock with Du Pont Imron paint; 13 1/2" length of pull. Made in U.S. by Ultra Light Arms, Inc. Introduced 1993; still produced. **LMSR: $800**
  Perf.: $700 **Exc.:** $600 **VGood:** $500

ULTRA LIGHT MODEL 24
Bolt action; 25-06, 270, 280, 30-06, 7mm Express; 22" barrel; weighs 5 1/4 lbs.; no sights; marketed with scope mount; Kevlar/graphite stock; two-position, three-function safety; Timney trigger; bright or matte finish; left- or right-hand aciton. Introduced 1986; still in production. **LMSR: $2600**
  Perf.: $2200 **Exc.:** $1900 **VGood:** $1300
Left-hand model
  Perf.: $2250 **Exc.:** $1950 **VGood:** $1350

ULTRA LIGHT ARMS MODEL 28
Bolt action; 264, 7mm Rem. Mag., 300 Win. Mag., 338 Win. Mag.; 24" Douglas Premium No. 2 contour barrel; 45" overall length; weighs 5 1/2 lbs.; Timney adjustable trigger; two-position, three-function safety; benchrest-quality action; matte or gloss metal and stock finish; KDF or ULA recoil arrestor built in. Right- or left-hand action. Made in U.S. by Ultra Light Arms, Inc. Introduced 1993; still produced.
LMSR: $2900
  Perf.: $2500 **Exc.:** $2000 **VGood:** $1400
Left-hand model
  Perf.: $2525 **Exc.:** $2025 **VGood:** $1425

ULTRA LIGHT ARMS MODEL 40
Bolt action; 300 Wea. Mag., 416 Rigby; 24" Douglas Premium No. 2 contour barrel; 45" overall length; weighs 5 1/2 lbs.; Timney adjustable trigger; two-position, three-function safety; benchrest-quality action; matte or gloss metal and stock finish; KDF or ULA recoil arrestor built in; right- or left-hand action. Made in U.S. by Ultra Light Arms, Inc. Introduced 1993; still produced. **LMSR: $3000**
  Perf.: $2650 **Exc.:** $2100 **VGood:** $1500
Left-hand model
  Perf.: $2700 **Exc.:** $2150 **VGood:** $1550

UNIVERSAL MODEL 1000
Semi-automatic; 30 carbine; 5-shot magazine; 18" barrel; 35 1/4" overall length; weighs 6 lbs.; military open sights; receiver tapped for scope mount; birchwood stock; G.I. version of M-1 Carbine with bayonet lug. Two other versions of the Model 1000 were offered in 256 one of which was the Model 1256 Ferret. Introduced about 1965. No longer in production.
  Perf.: $250 **Exc.:** $200 **VGood:** $170

Universal Model 1002
Same specs as Model 1003 except Military type with metal handguard.
  Perf.: $250 **Exc.:** $200 **VGood:** $150

Unique Model T.66 Match

Valmet Hunter Model

Valmet 412 Double Rifle

Varner Favorite Hunter

Vickers Empire Model

UNIVERSAL MODEL 1003
Semi-automatic; 30 carbine; 5-shot magazine; 18" barrel; 35 1/2" overall length; weighs 5 1/2 lbs.; military open sights; receiver tapped for scope mount; birchwood stock; similar to 30 M1 Carbine without bayonet lug. No longer in production.
   Perf.: $250   **Exc.:** $200   **VGood:** $150

Universal Model 1005
Same specs as Model 1003 except super-mirrored blue finish; Monte Carlo walnut stock; weighs 6 1/2 lbs. No longer in production.
   Perf.: $310   **Exc.:** $260   **VGood:** $200

Universal Model 1010
Same specs as Model 1005 except nickel-plated.
   Perf.: $280   **Exc.:** $230   **VGood:** $180

Universal Model 1015
Same specs as Model 1005 except gold-plated finish.
   Perf.: $300   **Exc.:** $250   **VGood:** $200

Universal Model 1020
Same specs as Model 1005 except black Du Pont teflon coated.
   Perf.: $290   **Exc.:** $250   **VGood:** $200

Universal Model 1025
Same specs as Model 1005 except camouflage olive Du Pont teflon coated.
   Perf.: $280   **Exc.:** $240   **VGood:** $190

UNIVERSAL MODEL 1256 FERRET
Semi-automatic; 256 Win. Mag.; 5-shot magazine; 18" barrel; no sights; 4x Universal scope; birchwood stock; M1 action; satin blue finish. No longer in production.
   Perf.: $300   **Exc.:** $250   **VGood:** $210

UNIQUE MODEL G.21
Semi-automatic; 22 LR; 5-, 10-shot magazine; 18" barrel; 33 1/2" overall length; weighs 6 lbs.; adjustable rear sight; French walnut stock; magazine and manual safeties. Was imported by Nygord Precision Products; no longer imported.
   Perf.: $300   **Exc.:** $280   **VGood:** $250

UNIQUE MODEL T AUDAX
Bolt action; 22 LR, 22 Short; 5-shot (22 Short), 10-shot magazine; 21 1/2" barrel; 39 1/4" overall length; weighs 6 1/2 lbs.; adjustable rear sight; dovetailed grooves on receiver for scope or target sight; lateral safety; French walnut stock. Was imported by Nygord Precision Products; no longer imported.
   Perf.: $545   **Exc.:** $500   **VGood:** $400

UNIQUE MODEL T DIOPTRA
Bolt action; 22 LR, 22 WMR; 5-shot (22 WMR), 10-shot (22 LR) magazine; 23 1/2" barrel; 41" overall length; weighs 6 1/2 lbs.; adjustable rear sight; dovetailed grooves on receiver for scope or target sight; lateral safety; French Monte Carlo walnut stock. Was imported by Nygord Precision Products; no longer imported.
   Perf.: $750   **Exc.:** $620   **VGood:** $510

UNIQUE MODEL T/SM
Bolt action; 22 LR, 22 WMR; 5-shot (22 WMR), 10-shot (22 LR) magazine; 20 1/2" barrel; 38 1/2" overall length; weighs 6 1/2 lbs.; adjustable rear sight; adjustable trigger; dovetailed grooves on receiver for scope or target sight; French Monte Carlo walnut stock; left-handed stock available. Was imported by Nygord Precision Products; no longer imported.
   Perf.: $770   **Exc.:** $630   **VGood:** $525

Unique Model T/SM Biathalon Junior
Same specs as Model T/SM except metallic grooves on stock for sling and harness mounting. Was imported by Nygord Precision Products; no longer imported.
   Perf.: $725   **Exc.:** $595   **VGood:** $495

Unique Model T/SM Match Junior
Same specs as Model T/SM except adjustable buttplate; metallic grooves on stock for sling and harness mounting. Was imported by Nygord Precision Products; no longer imported.
   Perf.: $725   **Exc.:** $595   **VGood:** $495

UNIQUE MODEL T.66 MATCH
Bolt action; single shot; 22 LR; 25 1/2" barrel; 44 1/8" overall length; weighs 10 7/8 lbs.; interchangeable globe front sight, Micro-Match rear; French walnut stock; stippled forend, pistol-grip; left-hand model available; meets NRA, UIT standards. Imported from France. Introduced 1980; no longer in production.
   Perf.: $450   **Exc.:** $410   **VGood:** $360

UNIQUE MODEL T.791 BIATHLON
Bolt action; 22 LR; 5-shot magazine; 22 1/2" barrel; 40 1/4" overall length; weighs 10 lbs.; interchangeable globe front sight, MIcro-Match rear; French walnut stock; stippled forend, pistol grip; left-hand model available; meets NRA, UIT standards. Imported from France. Introduced 1980; no longer in production.
   Perf.: $750   **Exc.:** $695   **VGood:** $475

UNIQUE MODEL X51 BIS
Semi-automatic; 22 LR; 5-, 10-shot magazine; 23 1/2" barrel; 40 1/2" overall length; weighs 6 lbs.; adjustable rear sight; French walnut stock; magazine and manual safeties. Was imported by Nygord Precision Products; no longer imported.
   Perf.: $350   **Exc.:** $290   **VGood:** $250

VALMET HUNTER MODEL
Semi-automatic; 223, 243, 308; 5-, 9-, 20-shot magazine; 20 1/2" barrel; 42" overall length; weighs 8 lbs.; open rear sight, blade front; American walnut butt, forend; checkered palm swell pistol grip, forend; Kalashnikov-type action. Made in Finland. Introduced 1986; discontinued 1989.
   Perf.: $950   **Exc.:** $800   **VGood:** $700

Vickers Jubilee Model

Voere Model 1007

Voere Model 2107

Voere Model 2155

RIFLES

Valmet
Model 412 Double Rifle
Model 412S Combo Gun

Varner
Favorite Hunter
Favorite Scheutzen

Vickers
Empire Model
Jubliee Model

Voere
Model 1007
Model 1013
Model 2107
Model 2107 Deluxe
Model 2114S
Model 2115
Model 2150
Model 2155
Model 2165

VALMET MODEL 412 DOUBLE RIFLE
Double rifle; 243, 375 Win., 308, 30-06; 24" barrels; weighs 8 5/8 lbs.; ramp front sight, adjustable open rear; barrel selector mounted on trigger; hand-checkered Monte Carlo stock of American walnut; recoil pad. Importation began 1980; no longer imported from Finland by Valmet.
Perf.: $1100 **Exc.:** $900 **VGood:** $700

VALMET MODEL 412S COMBO GUN
Combo; 12 ga. over 222, 223, 243, 308, 30-06; 24" barrel (Imp. & Mod.); weighs 7 5/8 lbs.; blade front, flip-up-type open rear; hand-checkered American walnut stock, forend; recoil pad. Importation began 1980; no longer imported from Finland by Valmet.
Perf.: $1000 **Exc.:** $875 **VGood:** $685

VARNER FAVORITE HUNTER
Single shot; 22 LR; 21 1/2" half-round/half-octagon barrel; weighs 5 lbs.; takedwon; blade front sight, open step-adjustable rear and peep; checkered American walnut stock; recreation of Stevens Favorite rifle with takedown barrel. Introduced 1988; discontinued 1990.
Perf.: $375 **Exc.:** $300 **VGood:** $220

Varner Favorite Scheutzen
Same specs as Favorite Hunter except 24" half-round/half-octagon target-grade barrel; ladder-type tang-mounted peep sight, globe-type front with six inserts; checkered AAA fancy walnut pistol-grip perch-belly stock, extended forend; scroll-engraved color case-hardened frame, lever; recreation of the Stevens Ladies Favorite Scheutzen, originally produced 1910 to 1916. Reintroduced 1989; discontinued 1990.
Perf.: $995 **Exc.:** $825 **VGood:** $695

VICKERS EMPIRE MODEL
Single shot; 22 LR; 27", 30" barrel; Martini-type action; perfection rear peep sight, Parker-Hale No. 2 front; straight-grip walnut stock. Manufactured prior to WWII.
Perf.: $425 **Exc.:** $350 **VGood:** $275

VICKERS JUBILEE MODEL
Single-shot; 22 LR; 28" heavy barrel; Martini-type action; Perfection rear peep sight, Parker-Hale No. 2 front; one-piece European walnut target stock; full pistol grip; forearm. Manufactured prior to World War II.
Perf.: $450 **Exc.:** $380 **VGood:** $310

VOERE MODEL 1007
Bolt action; 22 LR; 18" barrel; weighs 5 1/2 lbs.; open adjustable rear sight, hooded front; military-type oil-finished beech stock; single-stage trigger; sling swivels; convertible to single-shot; Biathlon model. Made in Austria. Introduced 1984; discontinued 1991.
Perf.: $230 **Exc.:** $190 **VGood:** $150

VOERE MODEL 1013
Bolt action; 22 WMR; 5-, 8-shot magazine; 18" barrel; weighs 5 1/2 lbs.; open adjustable rear sight, hooded front; military-type oil-finished stock; double-set trigger; sling swivels. Made in Austria. Introduced 1984; discontinued 1991.
Perf.: $280 **Exc.:** $240 **VGood:** $190

VOERE MODEL 2107
Bolt action; 22 LR; 5-, 8-shot magazine; 19 1/2" barrel; 41" overall length; weighs 6 lbs.; fully adjustable open rear sight, hooded front; European hardwood Monte Carlo stock; swivel studs; buttpad. Made in Germany. Introduced 1986; discontinued 1988.
Perf.: $325 **Exc.:** $300 **VGood:** $240

Voere Model 2107 Deluxe
Same specs as Model 2107 except checkered stock; raised cheekpiece. Introduced 1986; discontinued 1988.
Perf.: $350 **Exc.:** $280 **VGood:** $240

VOERE MODEL 2114S
Semi-automatic; 22 LR; 8-, 15-shot clip magazine; 18" barrel; 37 2/3" overall length; weighs 5 3/4 lbs.; leaf rear sight, hooded ramp front; walnut-finished beechwood stock, pistol grip, forend; single-stage trigger; wing-type safety. Made in Austria. Introduced 1984; no longer in production.
Perf.: $265 **Exc.:** $210 **VGood:** $160

VOERE MODEL 2115
Semi-automatic; 22 LR; 8-, 10-, 15-shot clip magazine; 18" barrel; 37 3/4" overall length; weighs 5 3/4 lbs.; leaf rear sight, hooded ramp front; walnut-finished beechwood stock; checkered pistol grip, forend; cheekpiece; single-stage trigger wing type safety. Made in Austria. Introduced 1984; discontinued 1995.
Perf.: $300 **Exc.:** $260 **VGood:** $230

VOERE MODEL 2150
Bolt action; 22-250, 243, 270, 7x57, 7x64, 308, 30-06 (standard), 7mm Rem. Mag., 300 Win. Mag., 9.3x64, 338 Win. Mag. (375 H&H, 458 Win. Mag. on special order); 22" barrel, 24" (magnums); weighs about 8 lbs.; ramp front sight, adjustable open rear; walnut stock with hand-rubbed oil finish, hand-checkered grip and forend; barrel-mounted front sling swivel; K-98 Mauser action with hinged floorplate; Was imported from Austria; importation dropped 1995.
Perf.: $800 **Exc.:** $675 **VGood:** $520

VOERE MODEL 2155
Bolt action; 270, 243, 30-06 (standard), 308, 22-250, 7x64, 5.6x57, 6.5x55, 8x57 JRS, 7mm Rem. Mag., 300 Win. Mag., 8x68S, 9.3x62, 9.3x64, 6.5x68; 5-shot non-detachable box magazine; 22" (standard), 24" barrel; Mauser-type action; no sights; drilled, tapped for scope mount; European walnut stock; checkered pistol grip, forend; single- or double-set trigger. Made in Austria. Introduced 1984; importation dropped 1995.
Perf.: $800 **Exc.:** $650 **VGood:** $530

VOERE MODEL 2165
Bolt action; 22-250, 243, 270, 7x57, 7x64, 308, 30-06 (standard), 7mm Rem. Mag., 300 Win. Mag., 9.3x64 (magnum); 5-shot magazine (standard calibers), 3-shot (magnums); 22" barrel (standard calibers), 24" (magnums); 44 1/2" overall length (22" barrel); weighs 7 to 7 1/2 lbs.; ramp front sight, open adjustable rear; European walnut stock with Bavarian cheekpiece, Schnabel forend tip, rosewood grip cap; built on Mauser 98-type action; tang safety; detachable box magazine; comes with extra magazine. No longer imported.
Perf.: $850 **Exc.:** $700 **VGood:** $550

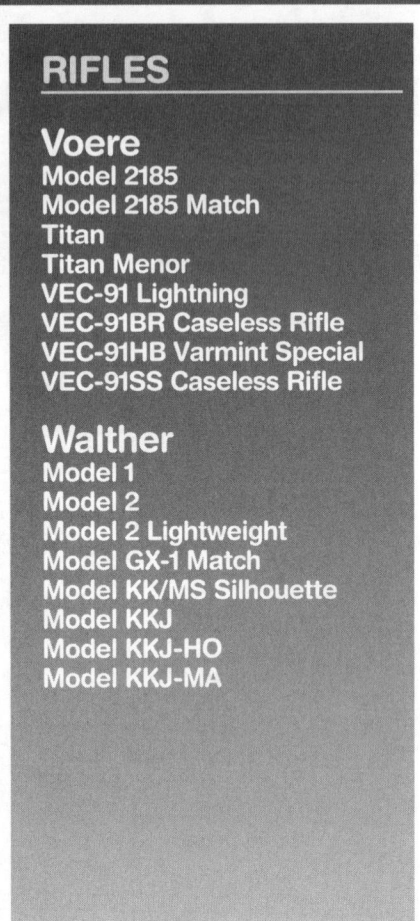

RIFLES

Voere
Model 2185
Model 2185 Match
Titan
Titan Menor
VEC-91 Lightning
VEC-91BR Caseless Rifle
VEC-91HB Varmint Special
VEC-91SS Caseless Rifle

Walther
Model 1
Model 2
Model 2 Lightweight
Model GX-1 Match
Model KK/MS Silhouette
Model KKJ
Model KKJ-HO
Model KKJ-MA

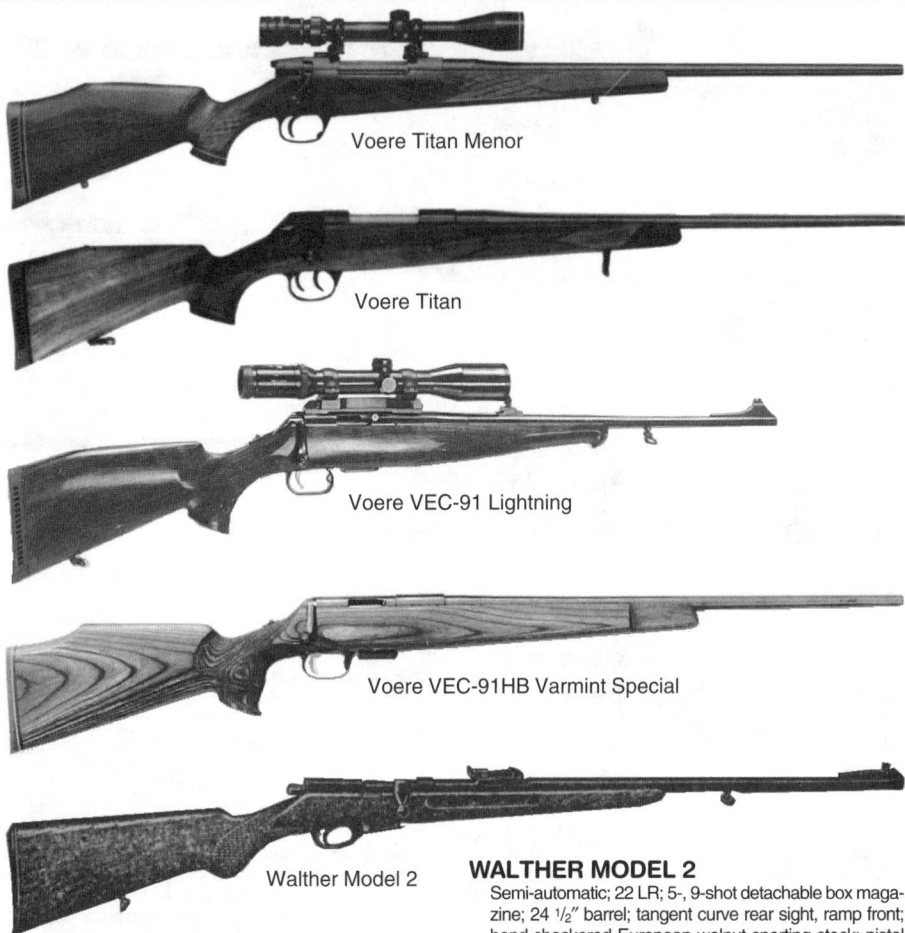

Voere Titan Menor

Voere Titan

Voere VEC-91 Lightning

Voere VEC-91HB Varmint Special

Walther Model 2

VOERE MODEL 2185
Semi-automatic; 7x64, 308, 30-06; 2-shot detachable magazine; 20″ barrel; 43 1/2″ overall length; weighs 7 3/4 lbs.; blade on ramp front sight, open adjustable rear; receiver drilled, tapped for scope mounts; European walnut stock with checkered grip and forend; ventilated rubber recoil pad; oil finish; gas-operated action with three forward locking lugs; hammer-forged free-floating barrel; two-stage trigger; cocking indicator inside trigger guard. Importation dropped 1994.
   Perf.: $1700   **Exc.:** $1350   **VGood:** $1100
With Mannlicher-style full stock
   Perf.: $1750   **Exc.:** $1400   **VGood:** $1200

Voere Model 2185 Match
Same specs as Model 2185 except 5-shot magazine; hooded post front sight, aperture rear; laminated match-type stock with glass bedding; adjustable cheekpiece and buttplate. Introduced 1992; discontinued 1994.
   Perf.: $2900   **Exc.:** $2400   **VGood:** $1900

VOERE TITAN
Bolt action; 243, 25-06, 270, 7x57, 308, 30-06, 257 Weatherby, 270 Weatherby, 7mm Rem. Mag., 300 Win. Mag., 300 Weatherby, 308 Norma Mag., 375 H&H; 3-, 4-shot magazine, depending on caliber; 24″, 26″ barrel; 44 1/2″ overall length (24″ barrel); weighs 8 lbs.; no sights; drilled, tapped for scope mounts; hand-checkered, oil-finished European walnut Monte Carlo stock; recoil pad; swivel studs; three-lug, front-locking action. Made in Austria. Introduced 1986; discontinued 1988.
   Perf.: $800   **Exc.:** $675   **VGood:** $575

VOERE TITAN MENOR
Bolt action; 222 Rem., 223 Rem.; 3-shot magazine; 23 1/2″ barrel; 42″ overall length; weighs 6 lbs.; no sights; drilled, tapped for scope mounts; hand-checkered, oil-finished European walnut Monte Carlo stock; rosewood grip cap, forend tip. Made in Austria. Introduced 1986; discontinued 1988.
   Perf.: $600   **Exc.:** $500   **VGood:** $400

VOERE VEC-91 LIGHTNING
Bolt action; 5.56 UCC (223-cal.), 6mm UCC caseless; 5-shot magazine; 20″ barrel; 39″ overall length; weighs 6 lbs.; blade on ramp front sight, open adjustable rear; drilled, tapped for scope mounts; European walnut stock with cheekpiece; checkered grip and Schnabel forend; top tang safety; fires caseless ammunition via electric ignition; two batteries housed in pistol grip last about 5000 shots; trigger adjustable from 5 oz. to 7 lbs. Introduced 1991; still imported. **LMSR: $1995**
   Perf.: $1700   **Exc.:** $1400   **VGood:** $900

Voere VEC-91BR Caseless Rifle
Same specs as VEC-91 except single shot; heavy 20″ barrel; synthetic benchrest stock. Imported from Austria. Introduced 1995; still imported. **LMSR: $1995**
   Perf.: $1700   **Exc.:** $1400   **VGood:** $900

Voere VEC-91HB Varmint Special
Same specs as VEC-91 except 22″ heavy sporter barrel; black synthetic or laminated wood stock. Imported from Austria. Introduced 1995; still imported. **LMSR: $1695**
   Perf.: $1400   **Exc.:** $1100   **VGood:** $800

Voere VEC-91SS Caseless Rifle
Same specs as VEC-91 except no sights; synthetic stock with straight comb, matte-finished metal. Imported from Austria. Introduced 1995; still imported. **LMSR: $1495**
   Perf.: $1200   **Exc.:** $950   **VGood:** $750

WALTHER MODEL 1
Semi-automatic; 22 LR; 5-, 9-shot detachable box magazine; 20″ barrel; tangent curve rear sight, ramp front; hand-checkered European walnut sporting stock; pistol grip; grooved forend; sling swivels; with bolt-action feature makes it possible to fire as semi-automatic, single-shot or as bolt-operated repeater. Manufactured prior to WWII.
   Perf.: $470   **Exc.:** $400   **VGood:** $350

WALTHER MODEL 2
Semi-automatic; 22 LR; 5-, 9-shot detachable box magazine; 24 1/2″ barrel; tangent curve rear sight, ramp front; hand-checkered European walnut sporting stock; pistol grip; grooved forend; sling swivels; bolt-action feature makes it possible to fire as semi-automatic, single-shot or as bolt-operated repeater. Manufactured prior to WWII.
   Perf.: $590   **Exc.:** $490   **VGood:** $365

Walther Model 2 Lightweight
Same specs as Model 2 except 20″ barrel; lightweight stock.
   Perf.: $590   **Exc.:** $490   **VGood:** $365

WALTHER MODEL GX-1 MATCH
Bolt action; single shot; 22 LR; 25 1/2″ heavy barrel; micrometer aperture rear sight, globe front; European walnut thumbhole stock; adjustable cheekpiece, buttplate; removable butthook; accessory rail; hand stop; palm rest; counterweight; sling swivels; free rifle design. No longer in production.
   Perf.: $1875   **Exc.:** $1375   **VGood:** $1000

WALTHER MODEL KK/MS SILHOUETTE
Bolt action; single shot; 22 LR; 25 1/2″ barrel; 44 3/4″ overall length; weighs 8 3/4 lbs.; no sights; receiver grooved for scope mounts; thumbhole European walnut stock; stippled grip, forend; adjustable trigger; over-size bolt knob; rubber buttpad. Made in Germany. Introduced 1989; discontinued 1991.
   Perf.: $895   **Exc.:** $795   **VGood:** $595

WALTHER MODEL KKJ
Bolt action; 22 LR; 5-shot clip magazine; 22 1/2″ medium-heavy target barrel; open rear sight, ramp front; checkered European walnut pistol-grip stock forearm; high tapered comb; sling swivels. Introduced in 1957; no longer imported.
   Perf.: $700   **Exc.:** $575   **VGood:** $450

Walther Model KKJ-HO
Same specs as Model KKJ except 22 Hornet.
   Perf.: $800   **Exc.:** $680   **VGood:** $550

Walther Model KKJ-MA
Same specs as Model KKJ except 22 WMR.
   Perf.: $700   **Exc.:** $600   **VGood:** $420

Walther Model KKJ

Walther Model KKM Match

Walther MovingTarget Model

Walther Running Boar Model 500

Walther Model SSV

Walther
Model KKM Match
Model KKM-S
Model KKW
Model Prone 400
Model SSV
Model V
Model V Meisterbusche
Moving Target Model
Olympic Model
Running Boar Model 500
U.I.T. BV Universal
U.I.T. Match Model
U.I.T.-E Match Model

Walther U.I.T. BV Universal

Walther U.I.T. Match Model

WALTHER MODEL KKM MATCH
Bolt action; single shot; 22 LR; 26″ barrel; micrometer rear sight, Olympic front with post, aperture inserts; European walnut stock with adjustable hook buttplate, hand shelf, ball-type offset yoke palm rest; fully-adjustable match trigger. Imported from Germany. Introduced 1957; no longer in production.
    Perf.: $900   **Exc.:** $800   **VGood:** $600

Walther Model KKM-S
Same specs as Model KKM except adjustable cheekpiece.
    Perf.: $950   **Exc.:** $825   **VGood:** $650

WALTHER MODEL KKW
Bolt action; single shot; 22 LR; tangent curve rear sight, ramp front; walnut military-style stock. Manufactured prior to WWII.
    Perf.: $675   **Exc.:** $525   **VGood:** $375

WALTHER MODEL PRONE 400
Bolt action; single shot; 22 LR; 25 1/2″ barrel; micrometer rear sight, interchangeable post or aperture front; European walnut stock with adjustable length, drop; cheekpiece; forearm guide rail for sling or palm rest; fully-adjustable trigger; especially designed for prone shooting. Introduced 1972; no longer in production.
    Perf.: $750   **Exc.:** $630   **VGood:** $510

WALTHER MODEL SSV
Bolt action; single shot; 22 LR, 22 Hornet; 25 1/2″ barrel; no sights; European walnut Monte Carlo stock; high cheekpiece; full pistol grip, forend. No longer in production.
    Perf.: $600   **Exc.:** $520   **VGood:** $420
22 Hornet model
    Perf.: $700   **Exc.:** $620   **VGood:** $520

WALTHER MODEL V
Bolt action; single shot; 22 LR; 26″ barrel; open rear sight, ramp front; uncheckered European walnut pistol-grip sporting stock, grooved forend; sling swivels. Manufactured prior to World War II.
    Perf.: $400   **Exc.:** $375   **VGood:** $300

Walther Model V Meisterbusche
Same specs as Model V except micrometer open rear sight; checkered pistol grip.
    Perf.: $480   **Exc.:** $440   **VGood:** $340

WALTHER MOVING TARGET MODEL
Bolt action; single shot; 22 LR; 23 1/2″ barrel; micrometer rear sight, globe front; receiver grooved for dovetail scope mounts; European walnut thumbhole stock; stippled forearm, pistol grip; adjustable cheekpiece, buttplate; especially designed for running boar competition. Imported from Germany. Introduced 1972; no longer in production.
    Perf.: $1295   **Exc.:** $1095   **VGood:** $850

WALTHER OLYMPIC MODEL
Bolt action; single shot; 22 LR; 26″ heavy barrel; extension micrometer rear sight, interchangeable front; hand-checkered pistol-grip target stock of European walnut; full rubber-covered beavertail forend; thumbhole; palm rest; adjustable buttplate; sling swivels. Manufactured in Germany prior to WWII.
    Perf.: $950   **Exc.:** $850   **VGood:** $695

WALTHER RUNNING BOAR MODEL 500
Bolt action; single shot; 22 LR; 23 1/2″ barrel; no sights; receiver grooved for dovetail scope mounts; European walnut thumbhole stock; stippled pistol grip; forend; adjustable cheekpiece, buttplate; left-hand stock available. Introduced 1975; discontinued 1990.
    Perf.: $1200   **Exc.:** $1000   **VGood:** $800

WALTHER U.I.T. BV UNIVERSAL
Bolt action; single shot; 22 LR; 25 1/2″ barrel; 44 3/4″ overall length; weighs about 10 lbs.; globe-type front sight, fully-adjustable aperture rear; European walnut stock, adjustable for length, drop; forend rail; meets NRA, UIT requirements; fully-adjustable trigger. Made in Germany. Introduced 1988; discontinued 1990.
    Perf.: $1400   **Exc.:** $1180   **VGood:** $950

Walther U.I.T. Match Model
Same specs as U.I.T. BV except new tapered forend profile; scope mount bases; fully stippled forend. Imported from Germany by Interarms. Introduced 1966; no longer imported.
    Perf.: $1200   **Exc.:** $1000   **VGood:** $775

Walther U.I.T.-E Match Model
Same specs as U.I.T. BV except state-of-the-art electronic trigger. Introduced 1984; no longer imported.
    Perf.: $1325   **Exc.:** $1050   **VGood:** $875

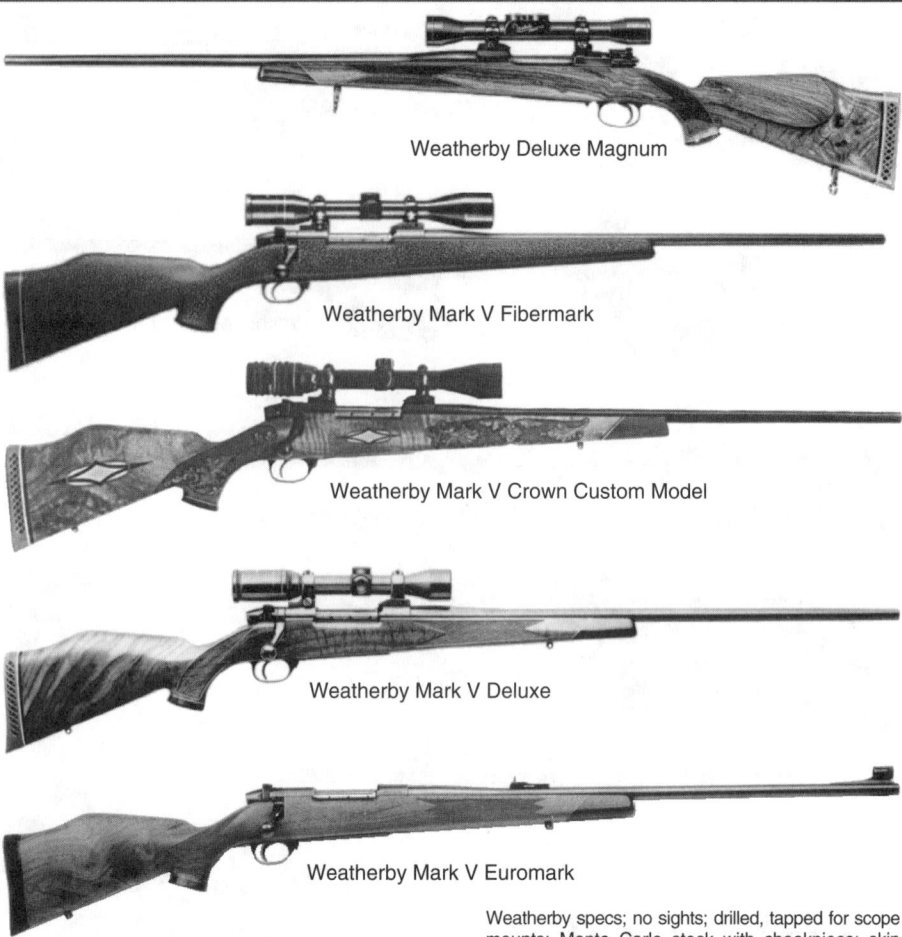

RIFLES

Weatherby
Classicmark I
Classicmark II
Classicmark II Safari Classic
Deluxe
Deluxe Magnum
Mark V Crown Custom Model
Mark V Custom Model
Mark V Deluxe
Mark V Euromark
Mark V Eurosport

Weatherby Deluxe Magnum

Weatherby Mark V Fibermark

Weatherby Mark V Crown Custom Model

Weatherby Mark V Deluxe

Weatherby Mark V Euromark

WEATHERBY CLASSICMARK I
Bolt action; 240 through 460 Wea. Mag.; 22″, 26″ barrel; straight comb stock of hand-selected American Claro walnut with oil finish; 18 lpi panel point checkering; 1″ Presentation recoil pad; all metal satin finished; Mark V action. Available in right- or left-hand versions. Imported from Japan by Weatherby. Introduced 1992; discontinued 1993.

Perf.: $1000   Exc.: $825   VGood: $700
Calibers 300, 340 Wea. Mag., 378 Wea. Mag., 416 Wea. Mag., 460 Wea. Mag.
Perf.: $1125   Exc.: $895   VGood: $695

Weatherby Classicmark II
Same as the Classicmark I except stock of deluxe hand-selected American walnut with shadow-line cheekpiece, rounded forend; 22 lpi wrap-around point checkering, oil finish; steel grip cap; Old English recoil pad; satin-finished metal; uses Mark V action. Available in right-hand version only. Imported from Japan by Weatherby. Made only in 1992.

Perf.: $1400   Exc.: $1175   VGood: $975
Calibers 300, 340 Wea. Mag., 378 Wea. Mag.
Perf.: $1500   Exc.: $1275   VGood: $1075
Calibers 416 Wea. Mag., 460 Wea. Mag.
Perf.: $1800   Exc.: $1475   VGood: $1275

Weatherby Classicmark II Safari Classic
Same specs as Classicmark II except 375 H&H; 24″ barrel; engraved floorplate; ramp front sight, quarter-rib express rear; barrel band front sling swivel. Made in 1992 only.

Perf.: $2200   Exc.: $1850   VGood: $1550

WEATHERBY DELUXE
Bolt action; 270, 30-06; 24″ barrel; no sights; Monte Carlo stock with cheekpiece; hand-checkered pistol-grip, forearm; black grip cap, forearm tip; quick-detachable sling swivels; Mauser action built to Weatherby specs by FN; some Springfield actions were used. Introduced 1948; discontinued 1955.

Perf.: $1495   Exc.: $1295   VGood: $895

Weatherby Deluxe Magnum
Same specs as Deluxe except 220 Rocket, 257 Weatherby Mag., 7mm Weatherby Mag., 300 Weatherby Mag., 375 Weatherby Mag., 378 Weatherby Mag.; 24″, 26″ barrel. Introduced 1948; discontinued 1955.

Perf.: $1695   Exc.: $1395   VGood: $995

WEATHERBY MARK V CROWN CUSTOM MODEL
Bolt action; 240 Weatherby Mag., 257 Weatherby Mag., 7mm Weatherby Mag., 300 Weatherby Mag., 30-06, 340 Weatherby Mag.; 3-, 5-shot magazine; 24″, 26″ barrel; optional sights; super-fancy walnut stock with carving, inlays; utilizes hand-honed, engraved Mark V barreled action; fully checkered bolt knob; engraved floorplate; damascened bolt, follower; gold monogramed name or initials; right-hand only. Introduced 1989; still in production. **LMSR: $NA**

Perf.: $4495   Exc.: $3995   VGood: $2995

WEATHERBY MARK V CUSTOM MODEL
Bolt-action; 22-250, 30-06, 224 Weatherby Varmintmaster (standard), 240, 257, 270, 7mm, 300, 340, 378, 460 (magnum); 2-, 5-shot box magazine, depending on caliber; 24″, 26″ barrel; available in right- or left-hand model; scope mounts; super deluxe stock with buttstock inlay, two forend inlays; gold monogram inlay; engraved with name or initials; stock carving with stained background; fully checkered bolt knob; damascened bolt and follower; hand-honed action; engraved floorplate. **LMSR: $3553**

Perf.: $2995   Exc.: $2495   VGood: $1795

WEATHERBY MARK V DELUXE
Bolt-action; 22-250, 30-06, 224 Weatherby Varmintmaster (standard), 240, 257, 270, 7mm, 300, 340, 378, 460 (magnum); 2-, 5-shot box magazine, depending on caliber; 24″, 26″ barrel; available in right- or left-hand model; some actions made by Sauer in Germany to Weatherby specs; no sights; drilled, tapped for scope mounts; Monte Carlo stock with cheekpiece; skip checkering on pistol grip, forearm; forearm tip, pistol grip cap, recoil pad; quick-detachable sling swivels. Introduced in 1958; made in U.S. from 1958 to 1960, then production transferred to Germany; still in production, except made in Japan. **LMSR: $1599**
Standard calibers, German-made
Perf.: $1295   Exc.: $1095   VGood: $820
Standard calibers, Japan-made
Perf.: $995   Exc.: $895   VGood: $750
Magnum calibers, German-made
Perf.: $1495   Exc.: $1195   VGood: $995
Magnum calibers, Japanese-made
Perf.: $1095   Exc.: $995   VGood: $850

WEATHERBY MARK V EUROMARK
Bolt action; chambered for all Weatherby calibers except 224, 22-250; 24″, 26″ round tapered barrel; 44 1/4″ overall length (24″ barrel); weighs 6 1/2 to 10 1/2 lbs.; open sights optional; walnut Monte Carlo stock with extended tail, fine-line hand checkering; satin oil finish; ebony forend tip and grip cap with maple diamond; solid buttpad; cocking indicator; adjustable trigger; hinged floorplate; thumb safety; quick-detachable sling swivels. Uses Mark V action. Introduced 1986; still produced. **LMSR: $1699**
Perf.: $1050   Exc.: $900   VGood: $750
Calibers 378 Wea. Mag., 416 Wea. Mag., 460 Wea. Mag.
Perf.: $1500   Exc.: $1100   VGood: $925

WEATHERBY MARK V EUROSPORT
Bolt-action; 22-250, 30-06, 224 Weatherby Varmintmaster (standard), 240, 257, 270, 7mm, 300, 340, 378, 460 (magnum); 2-, 5-shot box magazine, depending on caliber; 24″ barrel; Mark V action; available in right-hand only; no sights; drilled, tapped for scope mounts; raised-comb Monte Carlo stock with hand-rubbed satin oil finish; no grip cap, forend tip; recoil pad; quick-detachable sling swivels. Introduced 1995; still in production. **LMSR: $1049**
Perf.: $775   Exc.: $660   VGood: $580
375 caliber
Perf.: $825   Exc.: $690   VGood: $610

Weatherby Mark V Sporter

Weatherby Mark V Synthetic

Weatherby Mark V Safari Grade Custom

Weatherby Mark V Lazermark

Weatherby Ultramark

RIFLES

Weatherby
Mark V Fibermark
Mark V Fibermark Alaskan
Mark V Lazermark
Mark V Safari Grade Custom
Mark V Sporter
Mark V Stainless
Mark V Synthetic
Mark XXII Clip
Mark XXII Tubular
Ultramark

WEATHERBY MARK V FIBERMARK

Bolt-action; 240 WM through 340 WM; 24″, 26″ (left-hand in 300, 340 Wea.) round tapered barrel; 44 1/4″ overall length (24″ barrel); weighs 6 1/2 to 10 1/2 lbs.; fiberglass stock with black wrinkle, non-glare finish; black recoil pad; receiver, floorplate sandblasted, blued; fluted bolt. Introduced 1983; discontinued 1992.

Perf.: $1100   Exc.: $900   VGood: $725

Weatherby Mark V Fibermark Alaskan

Same specs as Fibermark except 270 Weatherby Mag., 7mm Weatherby Mag., 300 Weatherby Mag., 340 Weatherby Mag.; all metal parts plated with electroless nickel for corrosion protection. Introduced 1991; discontinued 1994.

Perf.: $1150   Exc.: $995   VGood: $895

WEATHERBY MARK V LAZERMARK

Bolt action; 240, 257, 270, 7mm Wea. Mag., 300, 340, 378 WM, 416 WM, 460 WM, 300 WM; 3-, 5-shot magazine; 24″, 26″ round tapered barrel; checkered pistol-grip American walnut stock; rosewood forearm, pistol grip; lazer carved stock, forend. Introduced 1981; still in production. (Add $250 for 416 WM & 460 WM calibers.) **LMSR: $1799**

Perf.: $1200   Exc.: $1100   VGood: $875

WEATHERBY MARK V SAFARI GRADE CUSTOM

Bolt action; 300 WM, 340 WM, 378 WM, 416 WM, 460 WM; 24″ barrel; satin oil-finished European walnut stock; ebony forend tip, pistol grip cap; black recoil pad; fine-line checkering; matte blue finish; barrel band front sling swivel; quarter-rib rear sight, hooded ramp front with brass bead; engraved floorplate; custom-built rifle; allow 8 to 10 months for delivery. Introduced, 1989; still produced. (Add $200 for 378 WM and larger calibers.) **LMSR: $NA**

Perf.: $2695   Exc.: $1995   VGood: $1395

WEATHERBY MARK V SPORTER

Bolt action; 257, 270, 7mm, 300, 340 Wea. Mag., 7mm Rem Mag., 300 Win Mag., 338 Win. Mag., 375 H&H; 24″, 26″ barrel; Mark V action; low-luster blue metal; Monte Claro walnut stock with high-gloss epoxy finish; Monte Carlo comb, recoil pad. Introduced 1993; still produced. **LMSR: $878**

Perf.: $695   Exc.: $595   VGood: $530

Weatherby Mark XXII Clip

Weatherby Mark XXII Tubular

WEATHERBY MARK V STAINLESS

Bolt-action; 22-250, 30-06, 224 Weatherby Varmintmaster (standard), 240, 257, 270, 7mm, 300, 340, 378, 460 (magnum); 2-, 5-shot box magazine, depending on caliber; 24″, 26″ barrel; available in right-hand only; 400 series stainless steel construction; lightweight injection-moulded synthetic stock with raised Monte Carlo comb; checkered grip and forend; custom floorplate release. Made in U.S. From Weatherby. Introduced 1995; still in production. **LMSR: $999**

Perf.: $855   Exc.: $695   VGood: $585
375 H&H
Perf.: $895   Exc.: $795   VGood: $700

WEATHERBY MARK V SYNTHETIC

Bolt action; 257, 270, 7mm, 300, 340 Wea. Magnums, 7mm Rem Mag., 300 Win. Mag., 338 Win. Mag, 375 H&H; 24″, 26″ barrel; weighs 7 1/2 lbs.; synthetic stock with raised Monte Carlo comb, dual-taper checkered forend; low-luster blued metal; uses Mark V action. Right-hand only. Made in U.S. From Weatherby. Introduced 1995; still produced. **LMSR: $799**

Perf.: $650   Exc.: $540   VGood: $475
375 H&H
Perf.: $750   Exc.: $630   VGood: $575

WEATHERBY MARK XXII CLIP

Semi-automatic; 22 LR; 5-, 10-shot clip magazine; 24″ barrel; folding leaf open rear sight, ramp front; Monte Carlo stock cheekpiece; skip checkering on pistol grip, forearm; forearm tip, grip cap, quick-detachable sling swivels. Introduced 1963; discontinued 1989.

Perf.: $450   Exc.: $325   VGood: $295

Weatherby Mark XXII Tubular

Same specs as Mark XXII Clip except 15-shot tube magazine. Introduced 1973; discontinued 1989.

Perf.: $450   Exc.: $325   VGood: $295

WEATHERBY ULTRAMARK

Bolt action; 240 Wea. Mag., 257 Wea. Mag., 270 Wea. Mag., 30-06, 7mm Wea. Mag., 300 Wea. Mag., 378 Wea. Mag., 416 Wea. Mag.; 3-, 5-shot magazine; 24″, 26″ barrel; fancy American walnut stock, basketweave checkering; hand-honed jeweled action; engraved floorplate; right- or left-hand models. Introduced 1989; discontinued 1990.

Perf.: $995   Exc.: $895   VGood: $795
378 Wea. Mag.
Perf.: $1250   Exc.: $1100   VGood: $950
416 Wea. Mag.
Perf.: $1350   Exc.: $1200   VGood: $1050

Weatherby
Vanguard
Vanguard Alaskan
Vanguard Classic I
Vanguard Classic II
Vanguard Fiberguard
Vanguard VGL
Vanguard VGS
Vanguard VGX
Vanguard Weatherguard
Weathermark
Weathermark Alaskan

Weihrauch
Model HW 60
Model HW 60J
Model HW 66
Model HW 660 Match

Wesson & Harrington
Buffalo Classic

Weatherby Vanguard Classic I

Weatherby Vanguard VGL

Weihrauch Model HW 60

Westley Richards Magazine Model

Westley Richards Double

WEATHERBY VANGUARD
Bolt action; 25-06, 243, 270, 30-06, 308, 264, 7mm Rem. Mag.; 300 Win. Mag.; 3-, 5-shot magazine, depending on caliber; 24″ hammer-forged barrel; 44 1/2″ overall length; weighs 7 7/8 lbs.; no sights; receiver drilled, tapped for scope mounts; American walnut stock; pistol-grip cap; forearm tip; hand checkered forearm, pistol grip; adjustable trigger; hinged floorplate. Introduced 1970; discontinued 1983.
Perf.: $450   **Exc.:** $375   **VGood:** $325

Weatherby Vanguard Alaskan
Same as the Vanguard except 223, 243, 7mm-08, 270 Win., 7mm Rem. Mag., 308, 30-06; forest green or black wrinkle-finished synthetic stock; all metal finished with electroless nickel; right-hand only. Introduced 1992; discontinued 1993.
Perf.: $600   **Exc.:** $530   **VGood:** $465

Weatherby Vanguard Classic I
Same specs as Vanguard except 223, 243, 270, 7mm-08, 7mm Rem. Mag., 30-06, 308; 24″ barrel; classic-style stock without Monte Carlo comb, no forend tip; distinctive Weatherby grip cap; satin finish on stock. Introduced 1989; discontinued 1993.
Perf.: $475   **Exc.:** $400   **VGood:** $325

Weatherby Vanguard Classic II
Same specs as Vanguard except 22-250, 243, 270, 7mm Rem. Mag., 30-06, 300 Win. Mag., 338 Win. Mag., 270 Wea. Mag., 300 Wea. Mag.; rounded forend with black tip, black grip cap with walnut diamond inlay, 20 lpi checkering; solid black recoil pad; oil-finished stock. Introduced 1989; discontinued 1993.
Perf.: $625   **Exc.:** $510   **VGood:** $410

Weatherby Vanguard Fiberguard
Same specs as Vanguard except 223, 243, 308, 270, 7mm Rem. Mag., 30-06; 20″ barrel; weighs 6 1/2 lbs.; forest green wrinkle-finished fiberglass stock; matte blued metal. Made in Japan. Introduced 1985; discontinued 1988.
Perf.: $465   **Exc.:** $395   **VGood:** $350

Weatherby Vanguard VGL
Same specs as Vanguard except 223, 243, 270, 30-06, 7mm Rem. Mag., 308; 5-, 6-shot magazine; 20″ barrel; non-glare blued finish; satin-finished stock; hand checkering; black buttpad, spacer. Made in Japan. Introduced 1984; discontinued 1988.
Perf.: $415   **Exc.:** $360   **VGood:** $300

Weatherby Vanguard VGS
Same specs as except 222-250, 25-06, 243, 270, 30-06, 7 mm Rem. Mag., 300 Win. Mag.; side safety. Made in Japan. Introduced 1984; discontinued 1988.
Perf.: $415   **Exc.:** $360   **VGood:** $300

Weatherby Vanguard VGX
Same specs as Vanguard except 22-250, 243, 270 Wea. Mag., 7mm Rem. Mag., 30-06, 300 Win. Mag., 300 Wea. Mag., 338 Win. Mag.; 24″ No. 2 contour barrel; walnut stock with high luster finish; rosewood grip cap and forend tip. Imported from Japan by Weatherby. Introduced 1984; discontinued 1993.
Perf.: $515   **Exc.:** $435   **VGood:** $365

Weatherby Vanguard Weatherguard
Same specs as Vanguard except 223, 243, 308, 270, 7mm-08, 7mm Rem. Mag., 30-06; 24″ barrel; 44 1/2″ overall length; weighs 7 1/2 lbs.; accepts same scope mount bases as Mark V action; forest green or black wrinkle-finished synthetic stock; all metal is matte blue. Right-hand only. Introduced 1989; discontinued 1993.
Perf.: $400   **Exc.:** $350   **VGood:** $300

WEATHERBY WEATHERMARK
Bolt action; 240, 257, 270, 7mm, 300, 340 Wea. Mags, 7mm Rem. Mag., 270 Win., 30-06; 22″, 24″ barrel; weighs 7 1/2 lbs.; impregnated-color black composite stock with raised point checkering; Mark V action; right-hand only. Introduced 1992; no longer made.
Perf.: $600   **Exc.:** $520   **VGood:** $465
375 H&H
Perf.: $695   **Exc.:** $620   **VGood:** $570

Weatherby Weathermark Alaskan
Same specs as Weathermark except all metal plated with electroless nickel. Available in right-hand only. Introduced 1992; no longer produced.
Perf.: $750   **Exc.:** $655   **VGood:** $525
375 H&H
Perf.: $895   **Exc.:** $820   **VGood:** $695

WEIHRAUCH MODEL HW 60
Single shot; 22 LR; 26 3/4″ barrel; hooded ramp front sight, match-type aperture rear; European walnut stock with stippled pistol-grip, forend; adjustable buttplate; rail with adjustable swivel; adjustable trigger; push-button safety. Introduced 1981; no longer imported from Germany.
Perf.: $650   **Exc.:** $530   **VGood:** $410

Weihrauch Model HW 60J
Same specs as Model HW 60 except 22 LR, 222; checkered walnut sporter-style stock. Discontinued 1992.
Perf.: $525   **Exc.:** $410   **VGood:** $375
222 caliber
Perf.: $725   **Exc.:** $625   **VGood:** $595

WEIHRAUCH MODEL HW 66
Bolt action; 22 Hornet, 222; 22 3/4″ barrel; 41 3/4″ overall length; weighs 6 1/2 lbs.; hooded blade ramp front sight, open rear; oil-finished walnut stock with cheekpiece; checkered pistol grip, forend. Introduced 1988; discontinued 1990.
Perf.: $605   **Exc.:** $495   **VGood:** $390
Double-set trigger
Perf.: $655   **Exc.:** $550   **VGood:** $450

WEIHRAUCH MODEL HW 660 MATCH
Bolt action; 22 LR; 26″ barrel; 45 1/3″ overall length; weighs 10 3/4 lbs.; globe front sight, match aperture rear; match-type walnut stock with adjustable cheekpiece, buttplate; checkered pistol grip, forend; forend accessory rail; adjustable match trigger. Introduced 1990; no longer in production.
Perf.: $725   **Exc.:** $650   **VGood:** $550

WESSON & HARRINGTON BUFFALO CLASSIC
Single shot; 45-70; 32″ heavy barrel; 52″ overall length; weighs 9 lbs.; no sights; drilled and tapped for peep sight; American black walnut stock and forend; barrel dovetailed for front sight; color case-hardened Handi-Rifle action with exposed hammer; color case-hardened crescent buttplate; 19th century checkering pattern. Made in U.S. by H&R 1871, Inc. Introduced 1995; still produced. **LMSR: $350**
Perf.: $270   **Exc.:** $240   **VGood:** $195

Whitworth Express

Wichita Silhouette Model

Wichita Stainless Magnum

Wichita Classic Model

Wickliffe Stinger

RIFLES

Westley Richards
Double
Magazine Model

Whitworth
Express
Express Carbine
Safari Express

Wichita
Classic Model
Silhouette Model
Stainless Magnum
Varmint Model

Wickliffe
Model 76
Model 76 Deluxe
Stinger
Stinger Deluxe
Traditionalist

Wilkinson
Terry Carbine

WESTLEY RICHARDS DOUBLE

Double rifle; boxlock; 30-06, 318 Accelerated Express, 375 Mag., 425 Mag. Express, 465 Nitro Express, 470 Nitro Express; 25″ barrels; hammerless; leaf rear sight, hooded front; French walnut stock; hand-checkered pistol-grip, forearm; cheekpiece; horn forearm tip; sling swivels; ejectors. Favored for African big game. Values vary according to caliber and engraving; range $16,000 to $45,000.

WESTLEY RICHARDS MAGAZINE MODEL

Bolt action; 30-06, 7mm high velocity, 318 Accelerated Express, 375 Mag., 404 Nitro Express, 425 Mag.; 22″ barrel (7mm High Velocity), 25″ (425 Mag.) 24″ other calibers; leaf rear sight, hooded front; sporting stock of French walnut; hand-checkered pistol grip, forearm; cheekpiece; horn forearm tip; sling swivels.
    Perf.: $9500   **Exc.:** $7900   **VGood:** $5800

WHITWORTH EXPRESS

Bolt-action; 22-250, 243, 25-06, 270, 7x57, 308, 30-06, 300 Win. Mag., 7mm Rem. Mag., 375 H&H, 458 Win. Mag.; 24″ barrel; 44″ overall length; 3-leaf open sight (magnums); open sights (others); classic European walnut English Express stock; hand checkering; adjustable trigger; hinged floorplate; steel recoil crossbolt; solid recoil pad; barrel-mounted sling swivel. Made originally in England; barreled actions later produced in Yugoslavia. Introduced 1974; no longer imported.
    Perf.: $470   **Exc.:** $410   **VGood:** $350

Whitworth Express Carbine

Same specs as Safari Express except 243, 270, 308, 7x57, 30-06; 20″ barrel; Mannlicher-style stock. Introduced 1986; discontinued 1988.
    Perf.: $500   **Exc.:** $430   **VGood:** $400

WHITWORTH SAFARI EXPRESS

Bolt action; 375 H&H, 458 Win. Mag.; 24″ barrel; 44″ overall length; weighs 7 1/2-8 lbs.; ramp front sight with removable hood, three-leaf open rear calibrated for 100, 200, 300 yards on 1/4-rib; classic English Express rifle stock design of hand checkered, select European walnut; solid rubber recoil pad; barrel-mounted sling swivel; adjustable trigger; hinged floorplate; solid steel recoil crossbolt. From Interarms. No longer imported.
    Perf.: $540   **Exc.:** $500   **VGood:** $425

WICHITA CLASSIC MODEL

Bolt action; single shot; calibers 17 through 308 Win.; 21 1/8″ octagon barrel; no sights; drilled, tapped for scope mounts; hand-checkered American walnut pistol-grip stock; steel grip cap; Pachmayr rubber recoil pad; right-, left-hand action; Canjar trigger; checkered bolt handle; jeweled bolt; non-glare blue finish. Introduced 1978; still in production. **LMSR: $3495**
    Perf.: $3000   **Exc.:** $2250   **VGood:** $1800
Left-hand model
    Perf.: $3200   **Exc.:** $2450   **VGood:** $2000

WICHITA SILHOUETTE MODEL

Bolt action; single shot; all standard calibers; 24″ free-floated barrel; no sights; drilled/tapped for scope mounts; metallic gray fiberthane stock; vent rubber recoil pad; 2-oz. Canjar trigger; fluted bolt; left- or right-hand; marketed in hard case. Introduced 1983; no longer in production.
    Perf.: $2250   **Exc.:** $1800   **VGood:** $1100
Left-hand model
    Perf.: $2450   **Exc.:** $2000   **VGood:** $1300

WICHITA STAINLESS MAGNUM

Bolt action; calibers 270 Win. through 458 Win. Mag.; single shot or with blind magazine; 22″, 24″ target-grade barrel; no sights; drilled, tapped for Burris scope mounts; hand-inletted, glass-bedded fancy American walnut stock; hand-checkered; steel pistol-grip cap; Pachmayr rubber recoil pad; fully-adjustable trigger; stainless steel barrel, action. Introduced 1980; no longer in production.
    Perf.: $1725   **Exc.:** $1350   **VGood:** $1100

WICHITA VARMINT MODEL

Bolt action; calibers 17 Rem. through 308 Win.; 3-shot magazine; 21 1/8″ Atkinson chrome-moly barrel; no sights; drilled, tapped for scope mounts; hand inletted, hand-rubbed, hand-checkered American walnut pistol-grip stock; steel grip cap; Pachmayr rubber recoil pad; checkered bolt handle; jeweled bolt; non-glare blued finish. Introduced 1978; still in production. **LMSR: $2695**
    Perf.: $2200   **Exc.:** $1700   **VGood:** $1100

WICKLIFFE MODEL 76

Single shot; 22 Hornet, 223 Rem., 22-250, 243 Win., 25-06, 308 Win., 30-06, 45-70; 22″ lightweight, 26″ heavy sporter barrel; falling block; no sights; American walnut Monte Carlo stock; cheekpiece, pistol grip, semi-beavertail forend. Introduced 1976; discontinued 1979.
    Perf.: $400   **Exc.:** $350   **VGood:** $300

Wickliffe Model 76 Deluxe

Same specs as Model 76 except 30-06; 22″ barrel; high-luster blue finish; nickel-silver grip cap, better wood. Introduced 1976; discontinued 1979.
    Perf.: $490   **Exc.:** $420   **VGood:** $370

WICKLIFFE STINGER

Single shot; 22 Hornet, 223 Rem.; 22″ barrel; falling block; no sights; American walnut Monte Carlo stock, Continental-type forend; etched receiver logo; quick-detachable sling swivels. Introduced 1979; discontinued 1980.
    Perf.: $400   **Exc.:** $350   **VGood:** $300

Wickliffe Stinger Deluxe

Same specs as Stinger except 22″ lightweight barrel; high-luster blue finish; nickel-silver grip cap; better wood. Introduced 1979; discontinued 1980.
    Perf.: $490   **Exc.:** $420   **VGood:** $370

WICKLIFFE TRADITIONALIST

Single shot; 30-06, 45-70; 24″ chrome-moly barrel; falling block; open sights; American walnut classic-style buttstock; hand-cut checkering; sling; sling swivels. Introduced 1979; discontinued 1980.
    Perf.: $400   **Exc.:** $350   **VGood:** $300

WILKINSON TERRY CARBINE

Semi-automatic; 9mm Para.; 30-shot magazine; 16 3/16″ barrel; 30″ overall length; weighs about 6 lbs.; protected post front sight, aperture rear; dovetailed receiver for scope mounting; uncheckered maple or black synthetic stock forend; bolt-type safety; ejection port has automatic trap door; blowback action; fires from closed bolt. Introduced 1975; still in production. **LMSR: $636**
    Perf.: $425   **Exc.:** $365   **VGood:** $295

Winchester Bolt-Action Rifles

Lee Musket
Lee Sporter
Model 43
Model 43 Special Grade
Model 47
Model 52
Model 52 Sporter
Model 52A
Model 52A Heavy Barrel
Model 52A Sporter
Model 52B
Model 52B Bull Gun
Model 52B Heavy Barrel
Model 52B Sporter
Model 52B Sporter (1993)
Model 52C

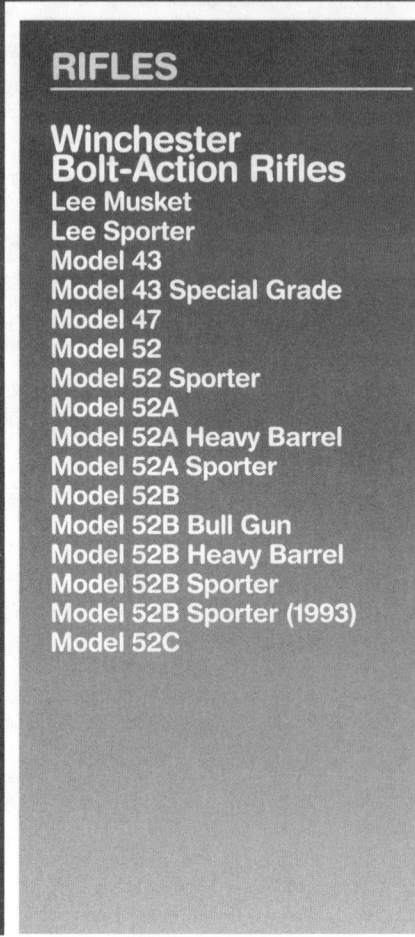

Winchester Lee Sporter

Winchester Model 43

Winchester Model 43 Special Grade

Winchester Model 47

Winchester Model 52A

Winchester Model 52A Heavy Barrel

WINCHESTER BOLT-ACTION RIFLES

WINCHESTER LEE MUSKET

Bolt action; 6mm (236 USN); 5-shot box magazine; 28″ barrel; folding leaf rear sight, post front; military semi-pistol-grip stock; blued finish. Commercial version of Lee Navy Model 1895 rifle. Introduced 1897; discontinued 1905. Collector value.
Exc.: $1300   **VGood:** $1100   **Good:** $770

Winchester Lee Sporter

Same specs as Lee Musket except 24″ barrel; open rear sight, bead front; sporter-style walnut stock. Introduced 1897; discontinued 1905. Collector value.
Exc.: $1350   **VGood:** $1050   **Good:** $825

WINCHESTER MODEL 43

Bolt action; 218 Bee, 22 Hornet, 25-20, 32-20; 3-shot box magazine, 2-shot (32-20); 24″ barrel; 42 1/2″ overall length; weighs 6 lbs.; open rear sight, No. 103 bead front on hooded ramp; uncheckered American walnut pistol-grip stock; swivels. Introduced 1950; discontinued 1957.
Exc.: $600   **VGood:** $500   **Good:** $400

Winchester Model 43 Special Grade

Same specs as Model 43 except for grip cap; checkered pistol grip, forearm; Lyman 59A micrometer rear sight.
Exc.: $650   **VGood:** $550   **Good:** $475

WINCHESTER MODEL 47

Bolt action; single shot; 22 Short, 22 Long, 22 LR; 25″ barrel; peep or open rear sight, No. 97 bead front on ramp with detachable sight-cover or No. 95 bead; uncheckered American walnut pistol grip stock. Introduced 1949; discontinued 1954.
Exc.: $300   **VGood:** $220   **Good:** $150

WINCHESTER MODEL 52

Bolt action; 22 Short, 22 LR; 5-shot box magazine; 28″ barrel; folding leaf peep rear sight, blade front; scope bases; semi-military stock; pistol grip; grooves on forearm. Later versions had higher comb, semi-beavertail forearm; slow lock model was replaced in 1929 by speed lock. Last arms of model bore serial number followed by letter "A". Introduced 1919; discontinued 1937.
Exc.: $475   **VGood:** $415   **Good:** $375

Winchester Model 52 Sporter

Same specs as Model 52 except for 24″ lightweight barrel; weighs 7 1/4 lbs.; Lyman No. 48F receiver sight, Redfield gold bead on hooded ramp at front; deluxe walnut sporting stock with cheekpiece; checkered forend, pistol grip; black forend tip; leather sling. Beware of fakes. Introduced 1934; discontinued 1958.
Exc.: $2500   **VGood:** $2000   **Good:** $1500

WINCHESTER MODEL 52A

Bolt action; 22 Short, 22 LR; 5-shot box magazine; 28″ barrel; folding leaf peep rear sight, blade front; scope bases; semi-military stock; pistol grip; grooves on forearm; speedlock action. Introduced 1929; discontinued 1939.
Exc.: $475   **VGood:** $425   **Good:** $375

Winchester Model 52A Heavy Barrel

Same specs as Model 52A except 28″ heavy barrel; Lyman 17G front sight. Discontinued 1939.
Exc.: $700   **VGood:** $600   **Good:** $500

Winchester Model 52A Sporter

Same specs as Model 52A except 24″ lightweight barrel; Lyman No. 48 receiver sight, gold bead on hooded ramp front; deluxe walnut sporting stock, checkered; black forend tip; cheekpiece. Introduced 1937; discontinued 1939.
Exc.: $2500   **VGood:** $2000   **Good:** $1500

WINCHESTER MODEL 52B

Bolt action; 22 Short, 22 LR; 5-shot box magazine; 28″ barrel; various sighting options; target or Marksman pistol-grip stock with high comb, beavertail forearm; round-top receiver; redesigned Model 52A action. Introduced 1940; discontinued 1947.
Exc.: $600   **VGood:** $520   **Good:** $450

Winchester Model 52B Bull Gun

Same specs as Model 52B except 28″ extra-heavy barrel; Marksman pistol-grip stock; Vaver No. 35 Mielt extension receiver sight, Vaver WIIAT front.
Exc.: $695   **VGood:** $620   **Good:** $495

Winchester Model 52B Heavy Barrel

Same specs as Model 52B except 28″ heavy barrel; Lyman No. 48FH rear sight, Lyman No. 77 front.
Exc.: $665   **VGood:** $600   **Good:** $500

Winchester Model 52B Sporter

Same specs as Model 52 except 5-shot detachable box magazine; 24″ lightweight barrel; Lyman No. 48 receiver sight, gold bead on hooded ramp front; deluxe walnut sporting stock, checkered; black forend tip; cheekpiece; sling swivels; single shot adapter.
Exc.: $2500   **VGood:** $1900   **Good:** $1495

Winchester Model 52B Sporter (1993)

Same specs as Model 52B except no sights; drilled, tapped for scope mounting; Model 52C mechanism with stock configuration of the Model 52B; Micro-Motion trigger. Production limited to 6000 rifles. Introduced 1993; discontinued 1993.
Perf.: $500   **Exc.:** $425   **VGood:** $360

WINCHESTER MODEL 52C

Bolt action; 22 Short, 22 LR; 5-, 10-shot box magazine; 28″ barrel; various sighting systems; Marksman pistol-grip stock with high comb, beavertail forearm; Micro-Motion trigger; single shot adapter. Introduced 1947; discontinued 1961.
Perf.: $725   **Exc.:** $620   **VGood:** $550

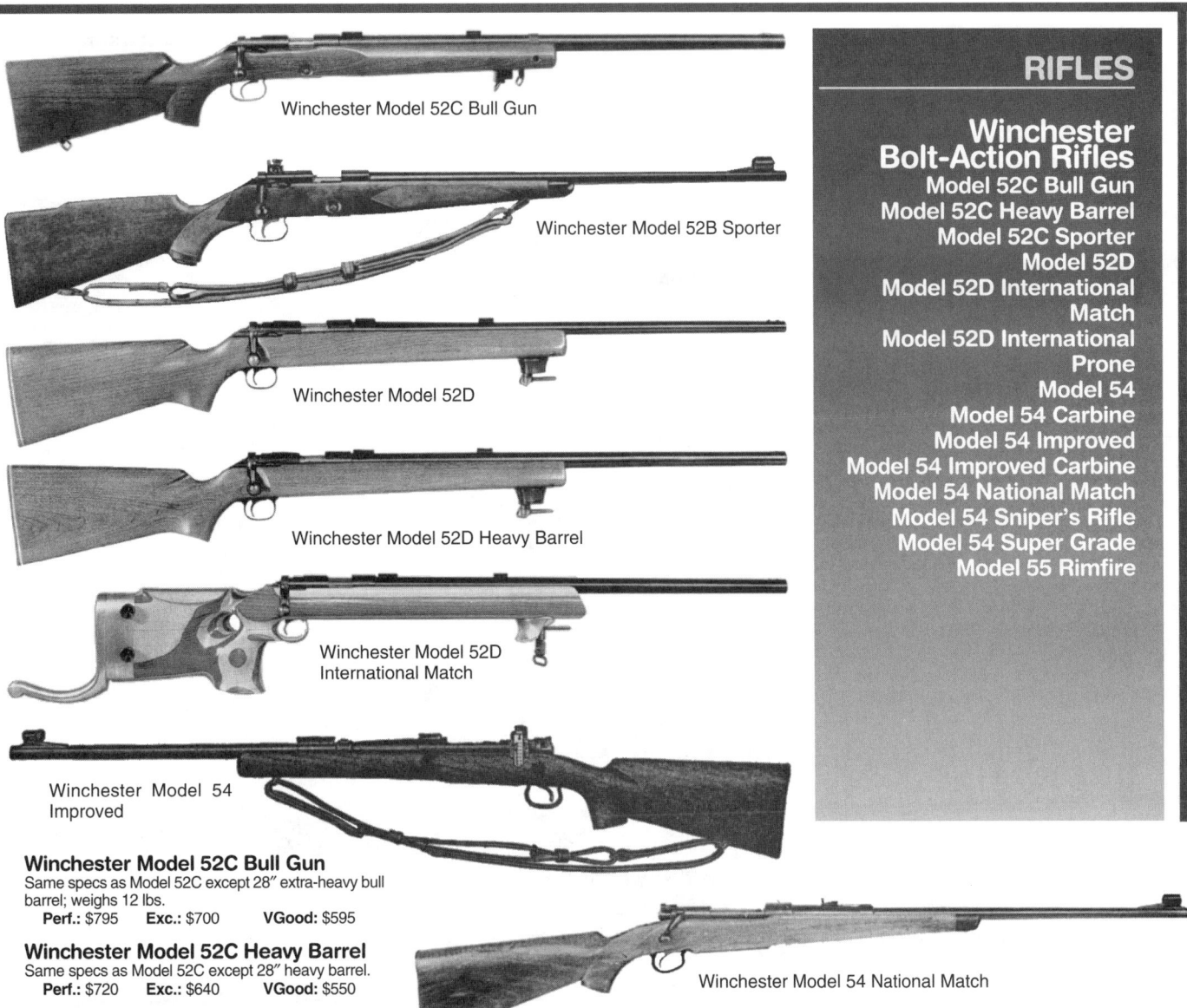

Winchester Model 52C Bull Gun

Winchester Model 52B Sporter

Winchester Model 52D

Winchester Model 52D Heavy Barrel

Winchester Model 52D International Match

Winchester Model 54 Improved

Winchester Model 54 National Match

RIFLES

Winchester Bolt-Action Rifles
Model 52C Bull Gun
Model 52C Heavy Barrel
Model 52C Sporter
Model 52D
Model 52D International Match
Model 52D International Prone
Model 54
Model 54 Carbine
Model 54 Improved
Model 54 Improved Carbine
Model 54 National Match
Model 54 Sniper's Rifle
Model 54 Super Grade
Model 55 Rimfire

Winchester Model 52C Bull Gun
Same specs as Model 52C except 28″ extra-heavy bull barrel; weighs 12 lbs.
Perf.: $795   **Exc.:** $700   **VGood:** $595

Winchester Model 52C Heavy Barrel
Same specs as Model 52C except 28″ heavy barrel.
Perf.: $720   **Exc.:** $640   **VGood:** $550

Winchester Model 52C Sporter
Same specs as Model 52C except 24″ lightweight barrel; Lyman No. 48 receiver sight, gold bead on hooded ramp front; deluxe walnut sporting stock, checkered; black forend tip; cheekpiece; sling swivels; single shot adapter; Micro-Motion trigger.
Perf.: $3100   **Exc.:** $2700   **VGood:** $2000

WINCHESTER MODEL 52D
Bolt action; single shot; 22 LR; 28″ free-floating standard or heavy barrel; 46″ overall length; weighs 9 3/4 lbs. (standard barrel), 11 lbs. (heavy barrel); no sights; scope blocks for standard target scopes; redesigned Marksman stock; rubber buttplate; accessory channel in stock with forend stop. Introduced 1961; discontinued 1980.
Perf.: $700   **Exc.:** $620   **VGood:** $550

Winchester Model 52D International Match
Same specs as Model 52D except laminated international-style stock with aluminum forend assembly; adjustable palm rest; ISU or Kenyon trigger. Introduced 1969; discontinued 1980.
ISU trigger
Perf.: $860   **Exc.:** $750   **VGood:** $640
Kenyon trigger
Perf.: $900   **Exc.:** $800   **VGood:** $700

Winchester Model 52D International Prone
Same specs as Model 52D International Match except oil-finished stock with removable roll-over cheekpiece for easy bore cleaning. Introduced 1975; discontinued 1980.
Perf.: $700   **Exc.:** $600   **VGood:** $500

WINCHESTER MODEL 54
Bolt action; 270, 7x57mm, 30-30, 30-06, 7.65x53mm, 9x57mm; 5-shot box magazine; 24″ barrel; open rear sight, bead front; checkered pistol-grip stock; two-piece firing pin; steel buttplate (checkered from 1930 on). Introduced 1925; discontinued 1930.
Exc.: $675   **VGood:** $575   **Good:** $475

Winchester Model 54 Carbine
Same specs as Model 54 except 20″ barrel; plain pistol-grip stock; grooves on forearm. Introduced 1927; discontinued 1930.
Exc.: $750   **VGood:** $675   **Good:** $450

Winchester Model 54 Improved
Same specs as Model 54 except 22 Hornet, 220 Swift, 250-3000, 257 Roberts, 270, 7x57mm, 30-06; 24″, 26″ (220 Swift) barrel; NRA-type stock; checkered pistol-grip, forearm; one-piece firing pin; speed lock. Introduced 1930; discontinued 1936.
Exc.: $775   **VGood:** $700   **Good:** $475

Winchester Model 54 Improved Carbine
Same specs as Model 54 except 22 Hornet, 220 Swift, 250-3000, 257 Roberts, 270, 7x57mm, 30-06; 20″ barrel; lightweight or NRA-type stock; checkered pistol-grip, forearm; one-piece firing pin; speed lock. Introduced 1930; discontinued 1936.
Exc.: $800   **VGood:** $750   **Good:** $525

Winchester Model 54 National Match
Same specs as Model 54 except Lyman sights; Marksman target stock; scope bases. Introduced 1935; no longer in production.
Exc.: $895   **VGood:** $795   **Good:** $595

Winchester Model 54 Sniper's Rifle
Same specs as Model 54 except 30-06; heavy 26″ barrel; Lyman No. 48 rear peep sight, blade front sight; semi-military type stock.
Exc.: $895   **VGood:** $795   **Good:** $595

Winchester Model 54 Super Grade
Same specs as Model 54 except deluxe stock with pistol grip cap, cheekpiece, black forend tip; quick-detachable sling swivels; 1″ leather sling. Introduced 1934; discontinued 1935.
Exc.: $900   **VGood:** $750   **Good:** $600

WINCHESTER MODEL 55 RIMFIRE
Single shot; 22 Short, 22 Long, 22 LR; 22″ barrel; weighs 5 1/2 lbs.; top loading; open rear sight, bead front; one-piece uncheckered walnut stock. Introduced 1958; discontinued 1961.
Exc.: $160   **VGood:** $130   **Good:** $100

Winchester Bolt-Action Rifles
Model 56
Model 57
Model 58
Model 59
Model 60
Model 60A

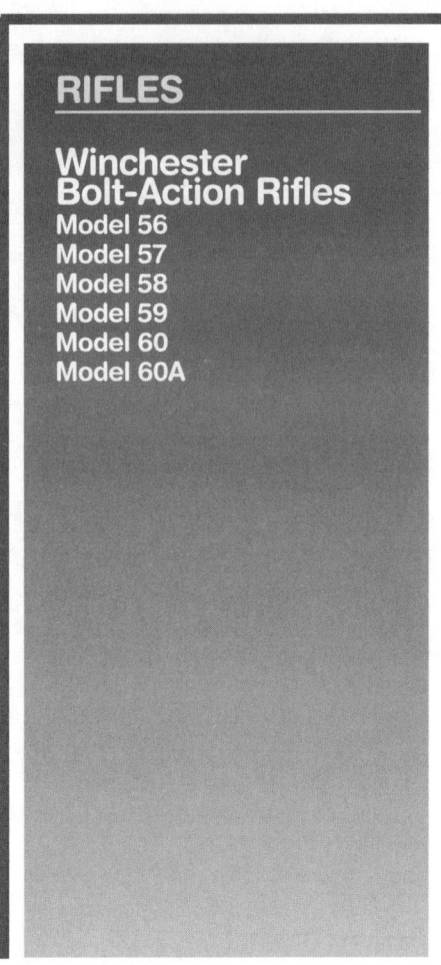

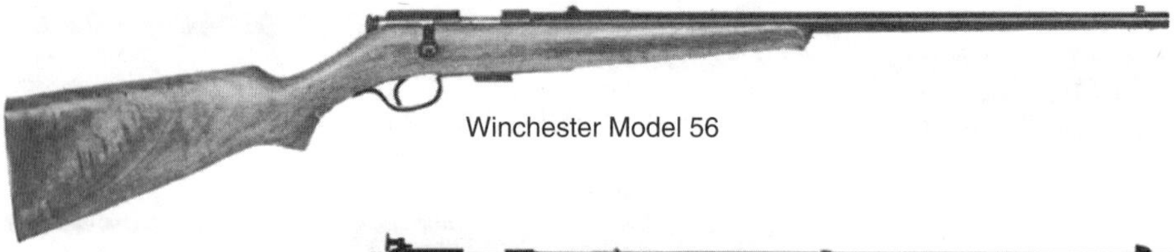

Winchester Model 56

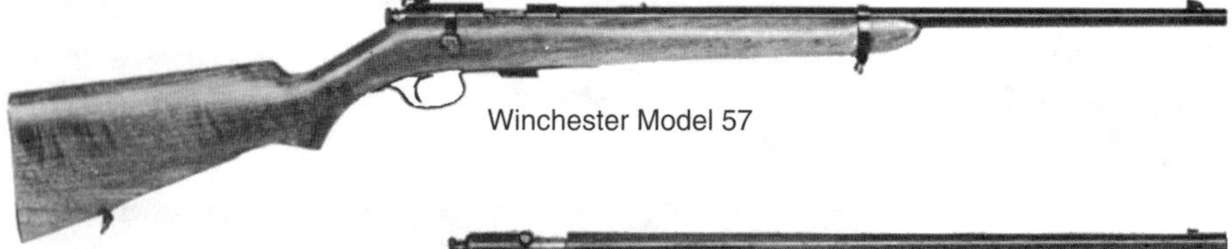

Winchester Model 57

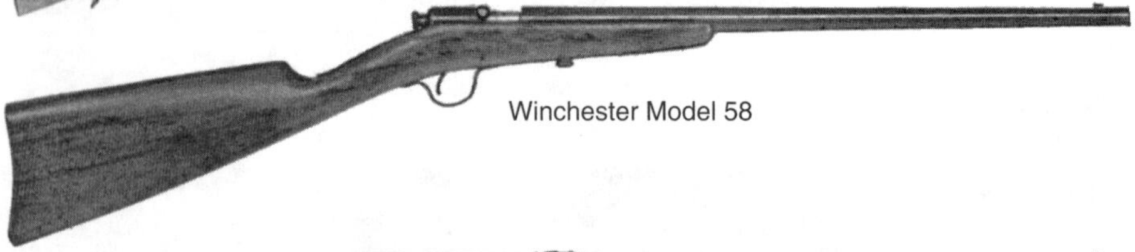

Winchester Model 58

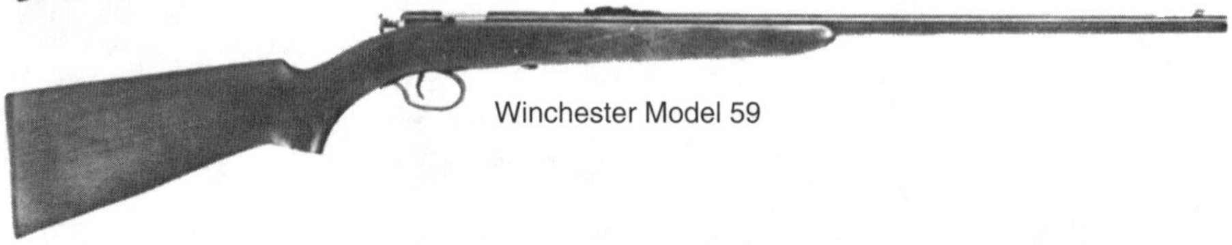

Winchester Model 59

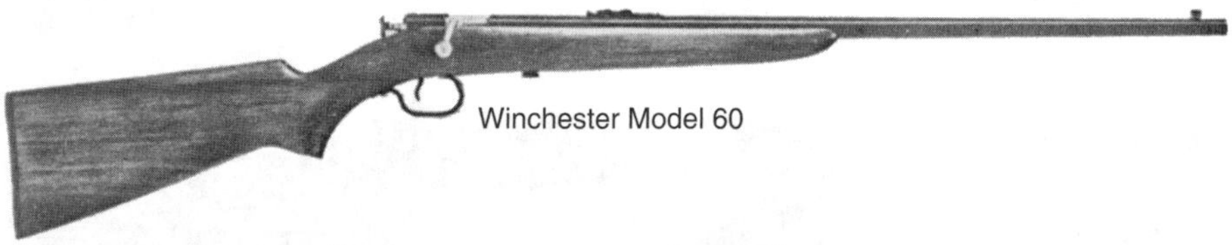

Winchester Model 60

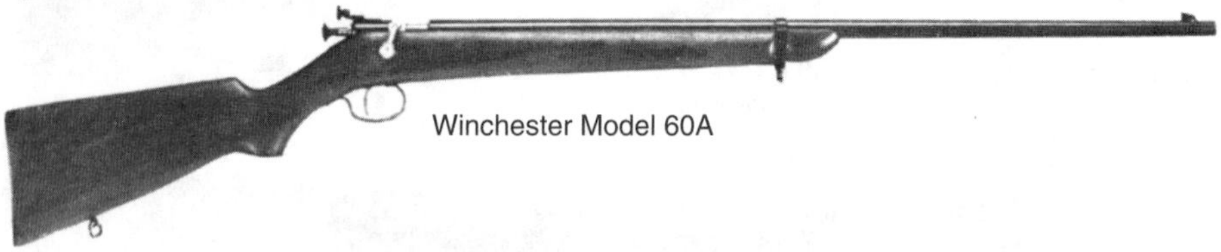

Winchester Model 60A

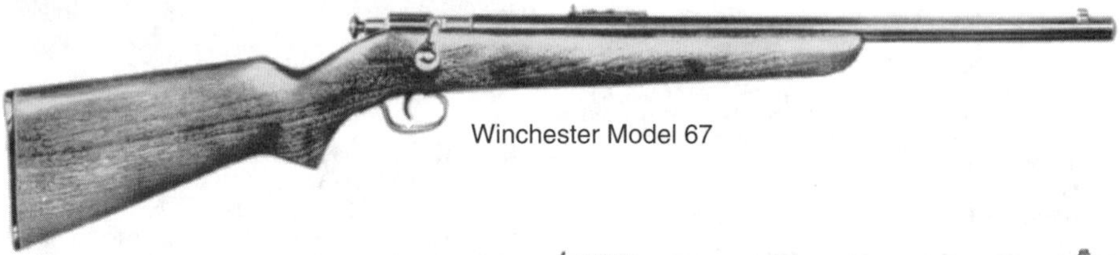

Winchester Model 67

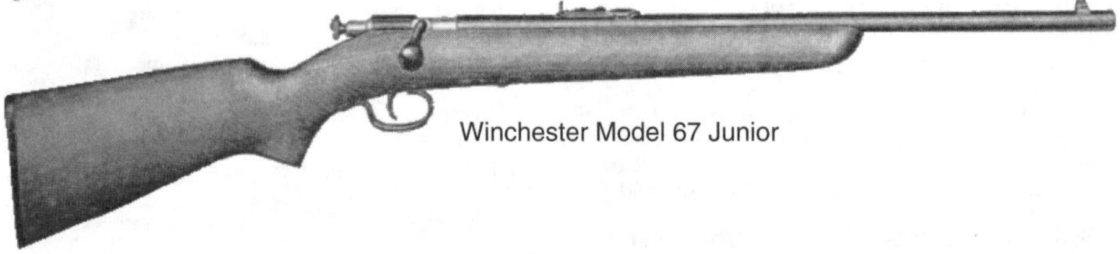

Winchester Model 67 Junior

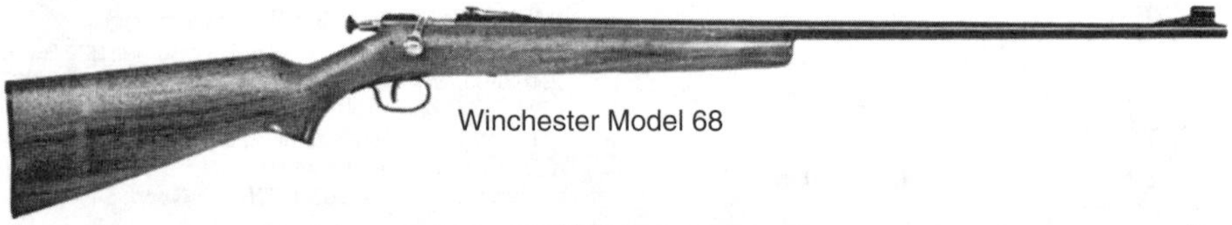

Winchester Model 68

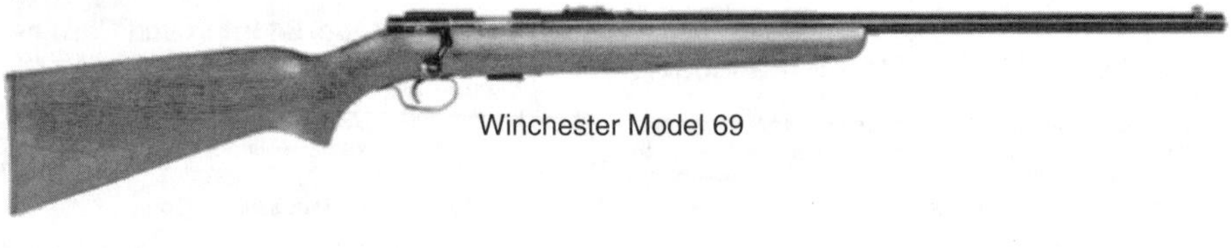

Winchester Model 69

WINCHESTER MODEL 56
Bolt action; 22 Short, 22 LR; 5-, 10-shot magazine; 22" barrel; open rear sight, bead front; uncheckered pistol-grip stock; Schnabel-type forearm. Introduced 1926; discontinued 1929.
    Exc.: $495   **VGood:** $395   **Good:** $250
Deluxe model with checkering
    Exc.: $695   **VGood:** $595   **Good:** $395

WINCHESTER MODEL 57
Bolt action; 22 Short, 22 LR; 5-, 10-shot magazine; 22" barrel; Lyman peep sight, blade front; drilled, tapped receiver; semi-military-type pistol-grip target stock; swivels; web sling; barrel band. Introduced 1927; discontinued 1936.
    Exc.: $495   **VGood:** $395   **Good:** $275

WINCHESTER MODEL 58
Bolt action; single shot; 22 Short, 22 Long, 22 LR; 18" barrel; open rear sight, blade front sight; uncheckered straight-grip hardwood stock; takedown; enlarged trigger guard. Introduced 1928; discontinued 1931.
    Exc.: $300   **VGood:** $200   **Good:** $150

WINCHESTER MODEL 59
Bolt action; single shot; 22 Short, 22 Long, 22 LR; 23" barrel; open rear sight, blade front; pistol grip, uncheckered one-piece hardwood stock; buttplate; takedown; enlarged trigger guard. Introduced 1930; discontinued 1931.
    Exc.: $400   **VGood:** $300   **Good:** $250

WINCHESTER MODEL 60
Bolt action; single shot; 22 Short, 22 Long, 22 LR; 23" barrel (until 1933), 27" thereafter; open rear sight, blade front; uncheckered pistol-grip hardwood stock; buttplate; takedown. Introduced 1931; discontinued 1934.
    Exc.: $315   **VGood:** $200   **Good:** $160

Winchester Model 60A
Same specs as Model 60 except 27" barrel; Lyman rear peep sight, square-top front; heavy semi-military target stock; web sling. Introduced 1933; discontinued 1939.
    Exc.: $495   **VGood:** $295   **Good:** $200

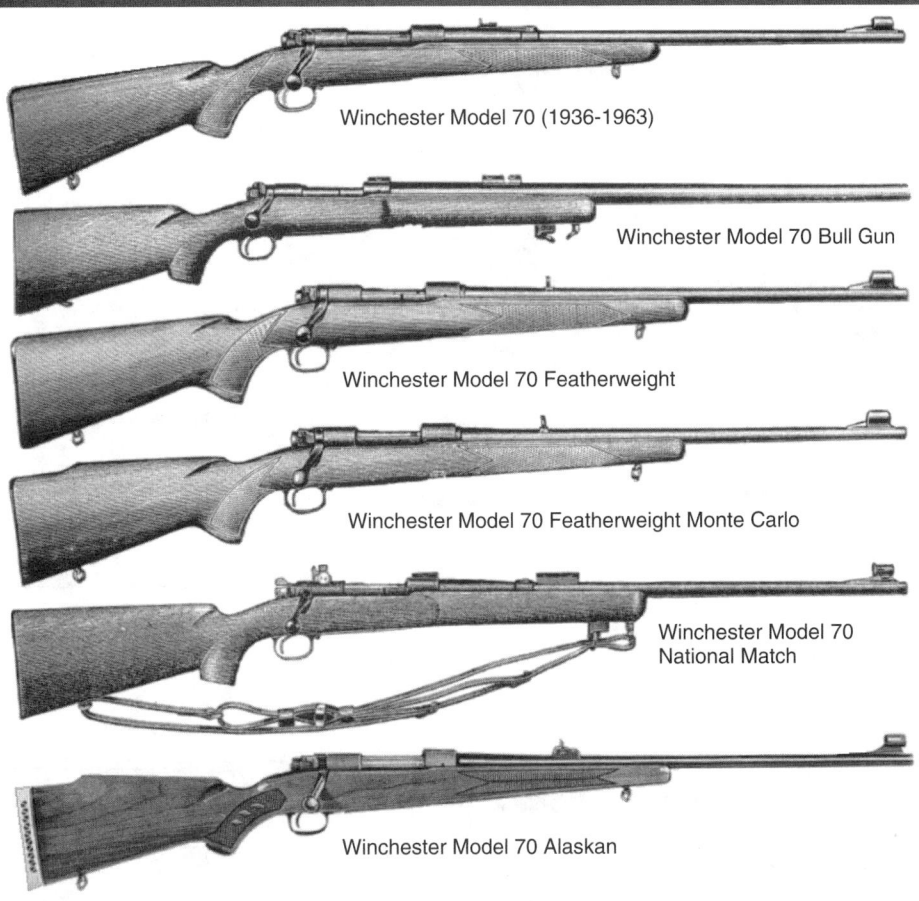

Winchester Model 70 (1936-1963)

Winchester Model 70 Bull Gun

Winchester Model 70 Featherweight

Winchester Model 70 Featherweight Monte Carlo

Winchester Model 70 National Match

Winchester Model 70 Alaskan

Winchester Bolt-Action Rifle
Model 67
Model 67 Junior
Model 68
Model 69
Model 69 Match
Model 69 Target
Model 70
Model 70 (1936-1963)
Model 70 Alaskan
Model 70 Bull Gun
Model 70 Carbine
Model 70 Featherweight
Model 70 National Match

WINCHESTER MODEL 67
Bolt action; single shot; 22 Short, 22 Long, 22 LR, 22 WMR, 22-shot cartridge; 24" rifled, 27" rifled or smoothbore barrel; takedown; open rear sight, bead front; unchecked pistol-grip stock; grooved forearm (early models). Introduced 1934; discontinued 1963.
Exc.: $190   **VGood:** $150   **Good:** $100

Winchester Model 67 Junior
Same specs as Model 67 except 20" barrel; shorter stock.
Exc.: $250   **VGood:** $200   **Good:** $140

WINCHESTER MODEL 68
Bolt action; single shot; 22 Short, 22 Long, 22 LR, 22 WMR; 24", 27"; takedown; adjustable rear peep sight, ramp front; unchecked walnut pistol-grip stock; grooved forearm (early models). Introduced 1934; discontinued 1946.
Exc.: $215   **VGood:** $175   **Good:** $125

WINCHESTER MODEL 69
Bolt action; 22 Short, 22 Long, 22 LR; 5-, 10-shot detachable box magazine; 25" barrel; takedown; Winchester No. 80A peep or open rear sight, bead ramp front; unchecked pistol-grip stock. Introduced 1935; discontinued 1963.
Exc.: $295   **VGood:** $220   **Good:** $150

Winchester Model 69 Match
Same specs as Model 69 except Lyman No. 57EW receiver sight, Winchester No. 101 front; Army-type leather sling.
Exc.: $315   **VGood:** $235   **Good:** $170

Winchester Model 69 Target
Same specs as Model 69 except rear peep sight, Winchester No. 93 blade front sight; sling swivels; Army-type leather sling.
Exc.: $335   **VGood:** $250   **Good:** $190

WINCHESTER MODEL 70
This bolt action, centerfire repeating rifle is a versatile longarm, having been produced in more variations and configurations than any other of the manufacturer's firearms. The rifle is divided into roughly three historical

categories, the original variations having been made from 1936 to 1963; at that time, the rifle was redesigned to a degree, actually downgraded in an effort to meet rising costs, but to hold the retail price. This series of variations was produced from 1964 until 1972, at which time the rifle was upgraded and the retail price increased. Additional changes have been made in years since.

WINCHESTER MODEL 70 (1936-1963)
Bolt action; 22 Hornet, 220 Swift, 243, 250-3000, 7mm, 257 Roberts, 264 Win. Mag., 270, 7x57mm, 300 Savage, 300 H&H Mag., 300 Win. Mag., 30-06, 308, 338 Win. Mag., 35 Rem., 358 Win., 375 H&H Mag., (other calibers on special order such as 9x57mm and 7.65mm); 4-shot box magazine (magnums); 5-shot (other calibers); 20", 24", 25", 26" barrel; claw extractor; hooded ramp front sight, open rear; hand-checkered walnut pistol-grip stock; Monte Carlo comb on later productions. Introduced 1936; discontinued 1963.
Note: 300 Savage, 35 Rem., 7.65mm, 9x57mm very rare-see appraiser.
Pre-WWII (1936-1945)
Standard calibers
    Exc.: $995   **VGood:** $750   **Good:** $550
220 Swift, 257 Roberts, 300 H&H Mag.
    Exc.: $1195   **VGood:** $795   **Good:** $595
375 H&H Mag., 7x57mm, 250-3000
    Exc.: $1495   **VGood:** $995   **Good:** $795
Post-WWII (1946-1963)
Standard calibers
    Exc.: $900   **VGood:** $800   **Good:** $600
220 Swift, 243, 257 Roberts
    Exc.: $1100   **VGood:** $900   **Good:** $650
22 Hornet, 300 H&H Mag.,
    Exc.: $1600   **VGood:** $1300   **Good:** $1100
300 Win. Mag., 338 Win. Mag., 375 H&H Mag.
    Exc.: $1500   **VGood:** $1200   **Good:** $1000
Note: 250-3000, 300 Savage, 308, 35 Rem., 7x57mm Very rare-see appraiser.

Winchester Model 70 Alaskan
Same specs as Model 70 (1936-1963) except 300 Win. Mag., 338 Win. Mag., 375 H&H Mag.; 3-, 4-shot (375 H&H Mag.) magazine; 24" barrel (300 Win. Mag.), 25"

barrel; 45 5/8" overall length; weighs 8-8 3/4 lbs.; bead front, sight, folding leaf rear; tapped for scope mounts, receiver sights; Monte Carlo stock; recoil pad. Introduced 1960; discontinued 1963.
Exc.: $1695   **VGood:** $1295   **Good:** $995

Winchester Model 70 Bull Gun
Same specs as Model 70 (1936-1963) except 30-06, 300 H&H Mag.; 28" extra heavy barrel; scope bases; walnut target Marksman stock; Lyman No. 77 front sight, Lyman No. 48WH rear; Army-type leather sling strap. Discontinued 1963.
Exc.: $1995   **VGood:** $1595   **Good:** $1295

Winchester Model 70 Carbine
Same specs as Model 70 (1936-1963) except 22 Hornet, 250-3000, 257 Roberts, 270, 7x57mm, 30-06; 20" barrel. The 250-3000, 7x57mm rare, see appraiser. Introduced 1936; discontinued 1946.
Calibers 270, 30-06
    Exc.: $1495   **VGood:** $1195   **Good:** $895
22 Hornet, 257 Roberts
    Exc.: $2495   **VGood:** $1995   **Good:** $1295

Winchester Model 70 Featherweight
Same specs as Model 70 (1936-1963) except 243, 264 Win. Mag., 270, 308, 30-06, 358 Win.; 22" lightweight barrel, 24" (special order); weighs 6 1/2 lbs.; lightweight American walnut Monte Carlo or straight comb stock; checkered pistol grip, forend; aluminum triggerguard, checkered buttplate, floorplate; 1" swivels. Introduced 1952; discontinued 1963.
Calibers 243, 270, 30-06, 308
    Exc.: $900   **VGood:** $800   **Good:** $650
264 Win. Mag., 358 Win.
    Exc.: $1400   **VGood:** $1300   **Good:** $1000

Winchester Model 70 National Match
Same specs as Model 70 (1936-1963) except 30-06; 24" barrel; Lyman No. 77 front ramp sight, Lyman No. 48WH rear; Army-type leather sling strap; scope bases; checkered walnut Marksman target stock. Discontinued 1960.
Exc.: $1475   **VGood:** $1200   **Good:** $850

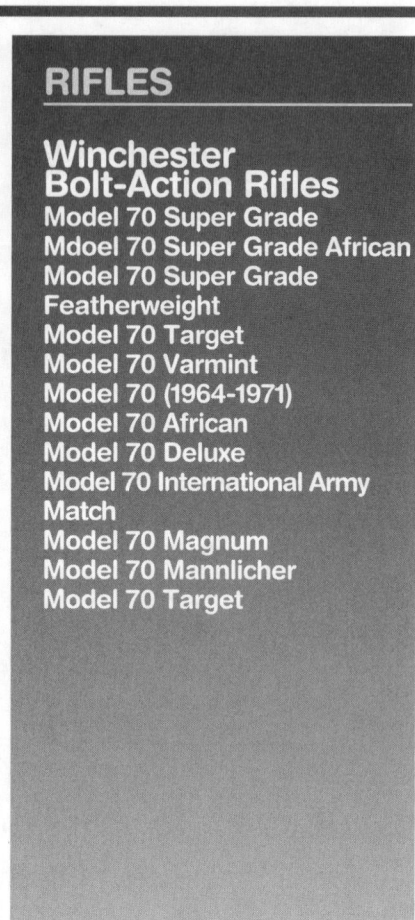

RIFLES

Winchester Bolt-Action Rifles
Model 70 Super Grade
Mdoel 70 Super Grade African
Model 70 Super Grade Featherweight
Model 70 Target
Model 70 Varmint
Model 70 (1964-1971)
Model 70 African
Model 70 Deluxe
Model 70 International Army Match
Model 70 Magnum
Model 70 Mannlicher
Model 70 Target

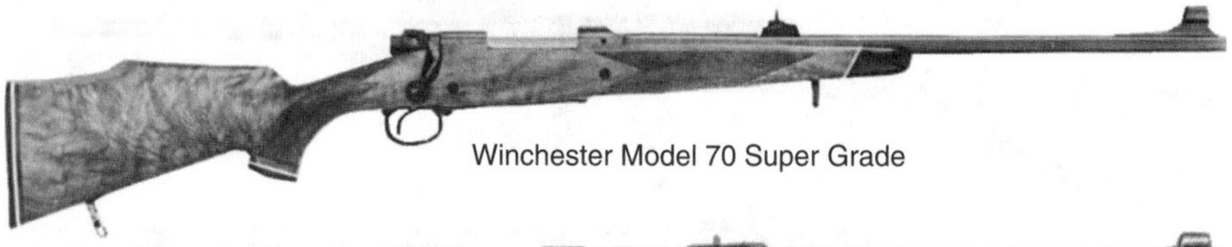

Winchester Model 70 Super Grade

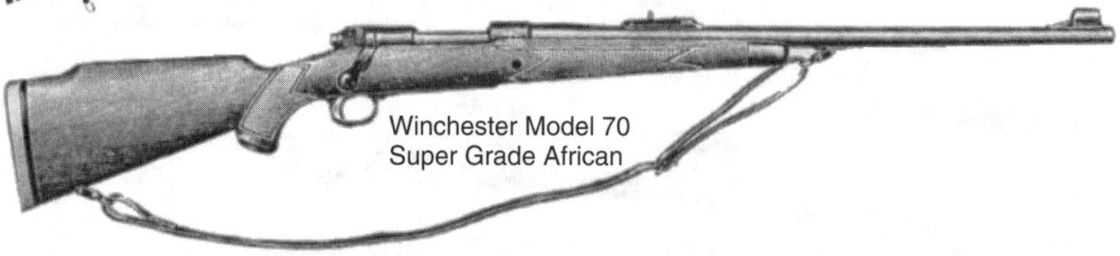

Winchester Model 70 Super Grade African

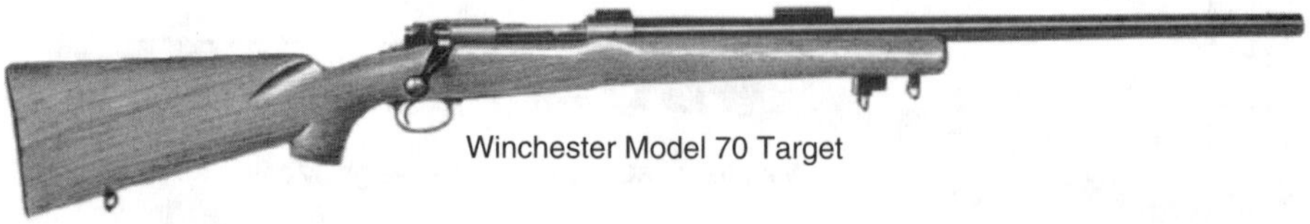

Winchester Model 70 Target

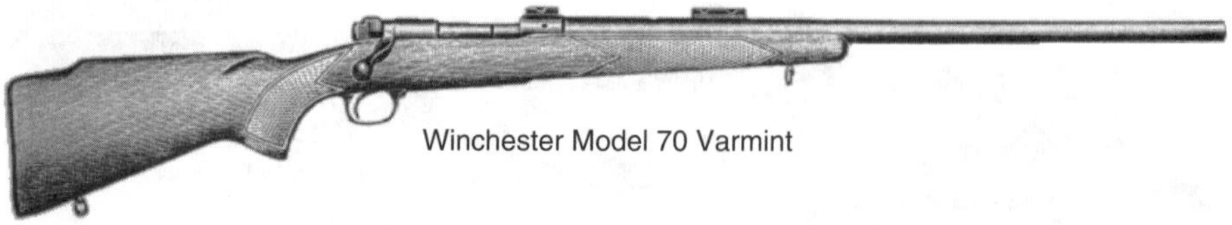

Winchester Model 70 Varmint

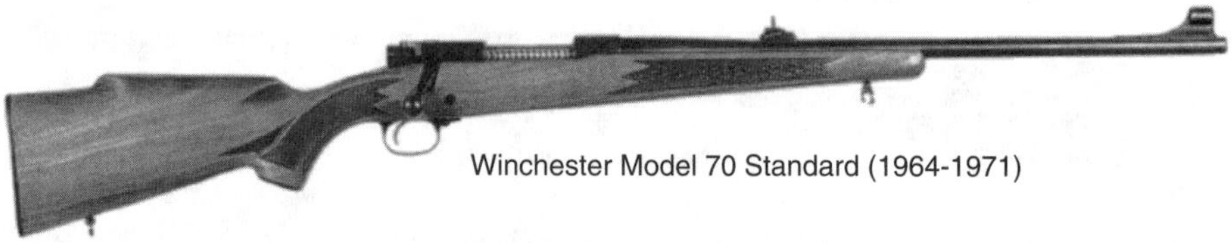

Winchester Model 70 Standard (1964-1971)

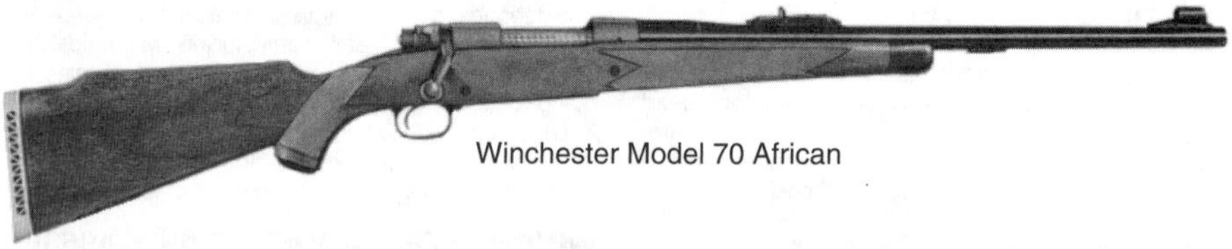

Winchester Model 70 African

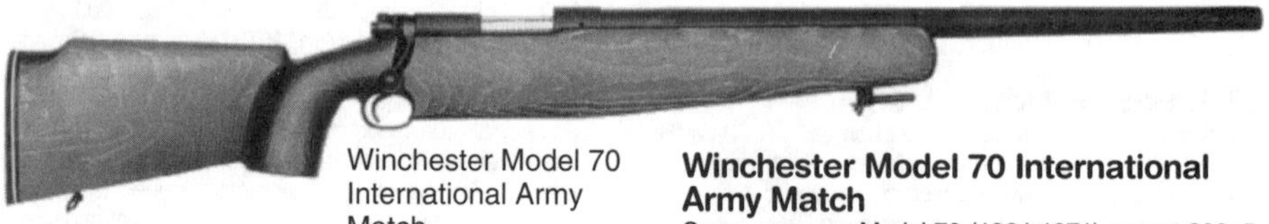

Winchester Model 70 International Army Match

Winchester Model 70 Super Grade
Same specs as Model 70 (1936-1963) except 24", 25", 26" barrel; Winchester 22G open sporting rear sight; deluxe stock, cheekpiece; black forearm tip; sling; quick-detachable sling swivels; grip cap. Introduced 1936; discontinued 1960.
Exc.: $1650　**VGood:** $1300　**Good:** $1000
375 H&H Mag.
Exc.: $2700　**VGood:** $2300　**Good:** $1600

Winchester Model 70 Super Grade African
Same specs as Model 70 (1936-1963) except 458 Win. Mag.; 3-shot magazine; 25" barrel; weighs 9 1/2 lbs.; Monte Carlo checkered walnut pistol-grip stock; cheekpiece; recoil pad; crossbolt. Introduced 1956; discontinued 1963.
Exc.: $3500　**VGood:** $3000　**Good:** $2300

Winchester Model 70 Super Grade Featherweight
Same specs as Model 70 (1936-1963) except 243, 270, 30-06, 308; 22" barrel; lightweight deluxe Monte Carlo stock, cheekpiece; black pistol-grip cap, forearm tip; aluminum buttplate, triggerguard, floorplate; sling; quick-detachable swivels. Discontinued 1960.
Exc.: $2895　**VGood:** $1995　**Good:** $1095
308 caliber
Exc.: $1895　**VGood:** $1495　**Good:** $995

Winchester Model 70 Target
Same specs as Model 70 (1936-1963) except 24", 26" medium-weight barrel; scope bases; walnut target Marksman stock; Lyman No. 77 front sight, Lyman No. 48WH rear, Army-type leather sling strap.
Exc.: $1895　**VGood:** $995　**Good:** $795

Winchester Model 70 Varmint
Same specs as Model 70 (1936-1963) except 220 Swift, 243 Win.; 26" stainless heavy barrel; weighs 9 3/4 lbs. scope bases; checkered walnut varminter stock. Introduced 1956; discontinued 1963.
Exc.: $1095　**VGood:** $895　**Good:** $750

WINCHESTER MODEL 70 (1964-1971)
Bolt action; 22-250, 222, 225, 243, 270, 308, 30-06; 5-shot box magazine; 22" heavy barrel; 42 1/2" overall length; weighs 7 lbs.; plunger-type extractor; hooded ramp front sight, adjustable open rear; checkered walnut Monte Carlo stock, cheekpiece; sling swivels. Introduced 1964; discontinued 1971.
Exc.: $395　**VGood:** $325　**Good:** $300

Winchester Model 70 African
Same specs as Model 70 (1964-1971) except 375 H&H Mag., 458 Win. Mag.; 22" barrel (375 H&H Mag.), 24" (458 Win. Mag.); 42 1/2" overall length; weighs 8 1/2 lbs.; special sights; hand-checkered Monte Carlo stock; ebony forearm tip; recoil pad; quick-detachable swivels; sling; twin crossbolts. Introduced 1964; discontinued 1971.
Perf.: $695　**Exc.:** $625　**VGood:** $550

Winchester Model 70 Deluxe
Same specs as Model 70 (1964-1971) except 243, 270 Win., 30-06, 300 Win. Mag.; 3-shot magazine (magnum), 5-shot (other calibers); 22", 24" barrel (magnum); hand-checkered walnut Monte Carlo stock, forearm; ebony forearm tip; recoil pad (magnum). Introduced 1964; discontinued 1971.
Perf.: $450　**Exc.:** $395　**VGood:** $300

Winchester Model 70 International Army Match
Same specs as Model 70 (1964-1971) except 308; 5-shot box magazine; 24" heavy barrel; optional sights; International Shooting Union stock; forearm rail for accessories; adjustable buttplate; externally adjustable trigger. Introduced 1971; discontinued 1971.
Perf.: $750　**Exc.:** $650　**VGood:** $600

Winchester Model 70 Magnum
Same specs as Model 70 (1964-1971) except 264 Win. Mag., 7mm Rem. Mag., 300 H&H Mag., 300 Win. Mag., 338 Win. Mag., 375 H&H Mag.; 3-shot magazine; 24" barrel; 44 1/2" overall length; weighs 7 1/4 lbs.
Perf.: $450　**Exc.:** $395　**VGood:** $330

Winchester Model 70 Mannlicher
Same specs as Model 70 (1964-1971) except 243, 220, 30-06, 308; 19" barrel; checkered Mannlicher stock with Monte Carlo comb, cheekpiece; steel forearm cap. Introduced 1969; discontinued 1972.
Perf.: $695　**Exc.:** $595　**VGood:** $495

Winchester Model 70 Target
Same specs as Model 70 (1964-1971) except 30-06, 308; 5-shot box magazine; 24" heavy barrel; 44 1/2" overall length; weighs 10 1/4 lbs.; no sights; target scope blocks; checkered heavy high-comb Marksman stock; aluminum hand stop. Introduced 1964; discontinued 1971.
Perf.: $600　**Exc.:** $500　**VGood:** $400

Winchester Model 70 Mannlicher

Winchester Model 70 Deluxe

Winchester Model 70 Magnum

Winchester Model 70 Target

Winchester Model 70 Varmint

RIFLES

Winchester Bolt-Action Rifles
Model 70 (1972 to present)
Model 70 Classic Custom Grade
Model 70 Classic Custom Grade Featherweight
Model 70 Classic Custom Grade Sharpshooter
Model 70 Classic Custom Grade Sporting Sharpshooter
Model 70 Classic DBM
Model 70 Classic DBM-S
Model 70 Classic SM
Model 70 Classic Sporter

WINCHESTER MODEL 70 (1972 to present)

Bolt action; 22-250, 222, 25-06, 243, 270, 308, 30-06; 5-shot box magazine; 22″ swaged, floating barrel; removable hooded ramp bead front sight, open rear; tapped for scope mounts; walnut Monte Carlo stock; cut checkering on pistol grip, forearm; forend tip; hinged floorplate; steel grip cap; sling swivels. Introduced 1972; discontinued 1980.

Perf.: $360   **Exc.:** $320   **VGood:** $290

Winchester Model 70 Classic Custom Grade

Same specs as Model 70 (1972 to present) except 270, 30-06, 7mm Rem. Mag., 300 Win. Mag., 338 Win. Mag.; 3-shot (magnum), 5-shot magazine; 24″, 26″ barrel; fancy satin-finished walnut stock; hand-honed and fitted parts. Introduced 1990; discontinued 1994.

Perf.: $1295   **Exc.:** $995   **VGood:** $695

Winchester Model 70 Classic Custom Grade Featherweight

Same specs as Model 70 (1972 to present) except 22-250, 223, 243, 270, 280, 30-06, 308, 7mm-08 Rem.; no sights; checkered, satin-finished, high-grade American walnut stock, Schnabel forend; rubber buttpad; high polish blued finish; controlled round feeding. Introduced 1992; no longer in production.

Perf.: $1295   **Exc.:** $995   **VGood:** $695

Winchester Model 70 Classic Custom Grade Sharpshooter

Same specs as Model 70 (1972 to present) except 223, 22-250, 308 Win., 300 Win. Mag.; 24″ (308), 26″ (223, 22-250, 300 Win. Mag.) barrel; 44 1/2″ overall length (24″ barrel); weighs 11 lbs.; no sights; scope bases and rings; glass-bedded McMillan A-2 target style stock with recoil pad, swivel studs; controlled round feeding; hand-honed and fitted action; Schneider barrel; matte blue finish. Introduced 1992; discontinued 1994.

Perf.: $1495   **Exc.:** $1095   **VGood:** $795

Winchester Model 70 Classic Custom Grade Sporting Sharpshooter

Same specs as Model 70 (1972 to present) except 220 Swift, 270, 7mm STW, 300 Win. Mag.; 24″ (270), 26″ stainless steel Schneider barrel, glass-bedded with natural finish; 44 1/2″ overall length (24″ barrel); no sights; scope bases and rings; McMillan sporter-style, gray-finished composite stock; blued receiver; pre-'64-style action with controlled round feeding. Introduced 1994; discontinued 1994.

Perf.: $1495   **Exc.:** $1195   **VGood:** $895

Winchester Model 70 Classic DBM

Same specs as Model 70 (1972 to present) except 22-250, 243, 308, 284 Win., 270, 30-06, 7mm Rem. Mag., 300 Win. Mag.; 3-shot detachable box magazine; 24″, 26″ barrel; with or without sights; scope bases and rings; pre-'64-type action with controlled round feeding. Introduced 1994; discontinued 1995.

Perf.: $475   **Exc.:** $395   **VGood:** $330

Winchester Model 70 Classic DBM-S

Same specs as Model 70 (1972 to present) except 270, 30-06, 7mm Rem. Mag., 300 Win. Mag.; 3-shot detachable box magazine; 24″ barrel; with or without sights; scope bases and rings; black fiberglass/graphite composite stock; pre-'64-type controlled round feeding. Introduced 1994; discontinued 1994.

Perf.: $460   **Exc.:** $390   **VGood:** $330

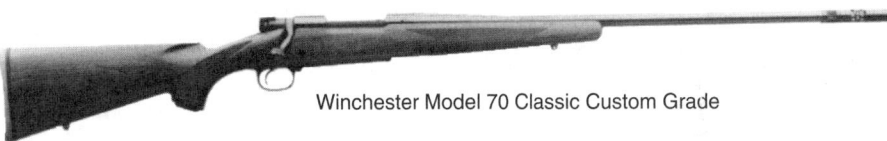

Winchester Model 70 Classic Custom Grade

Winchester Model 70 Classic SM

Same specs as Model 70 (1972 to present) except 264 Win. Mag., 270, 30-06, 7mm Rem. Mag., 300 Win. Mag., 300 Weatherby Mag., 338 Win. Mag., 375 H&H Mag.; 3-shot magazine (magnum), 5-shot (others); 24″, 26″ barrel; weighs 7 3/4 lbs.; with or without sights; scope bases and rings; black composite, graphite-impregnated stock; matte-finished metal; pre-'64-type action with controlled feeding; BOSS barrel vibration modulator and muzzle brake system optional. Introduced 1994; still in production.

LMSR: $602

Perf.: $475   **Exc.:** $410   **VGood:** $360
375 H&H

Perf.: $550   **Exc.:** $460   **VGood:** $390
With BOSS (270, 30-06, 7mm Rem. Mag., 300 Win. Mag., 338 Win. Mag.)

Perf.: $600   **Exc.:** $510   **VGood:** $440

Winchester Model 70 Classic Sporter

Same specs as Model 70 (1972 to present) except 25-06, 264 Win. Mag., 270, 270 Win. Mag., 30-06, 300 Win. Mag., 300 Weatherby Mag., 338 Win. Mag., 7mm Rem. Mag.; 3-shot magazine (magnum), 5-shot (others); 24″, 26″ barrel; controlled round feeding. Introduced 1994; still in production. **LMSR:** $632

Perf.: $495   **Exc.:** $410   **VGood:** $360

RIFLES

Winchester Bolt-Action Rifles

Model 70 Classic Stainless
Model 70 Classic Super Grade
Model 70 Classic Express Magnum
Model 70 Classic Super Express Magnum
Model 70 Custom Sharpshooter
Model 70 DBM
Model 70 Featherweight Classic
Model 70 Featherweight WinTuff
Model 70 Heavy Barrel Varmint
Model 70 Heavy Varmint
Model 70 Lightweight
Model 70 Lightweight Carbine

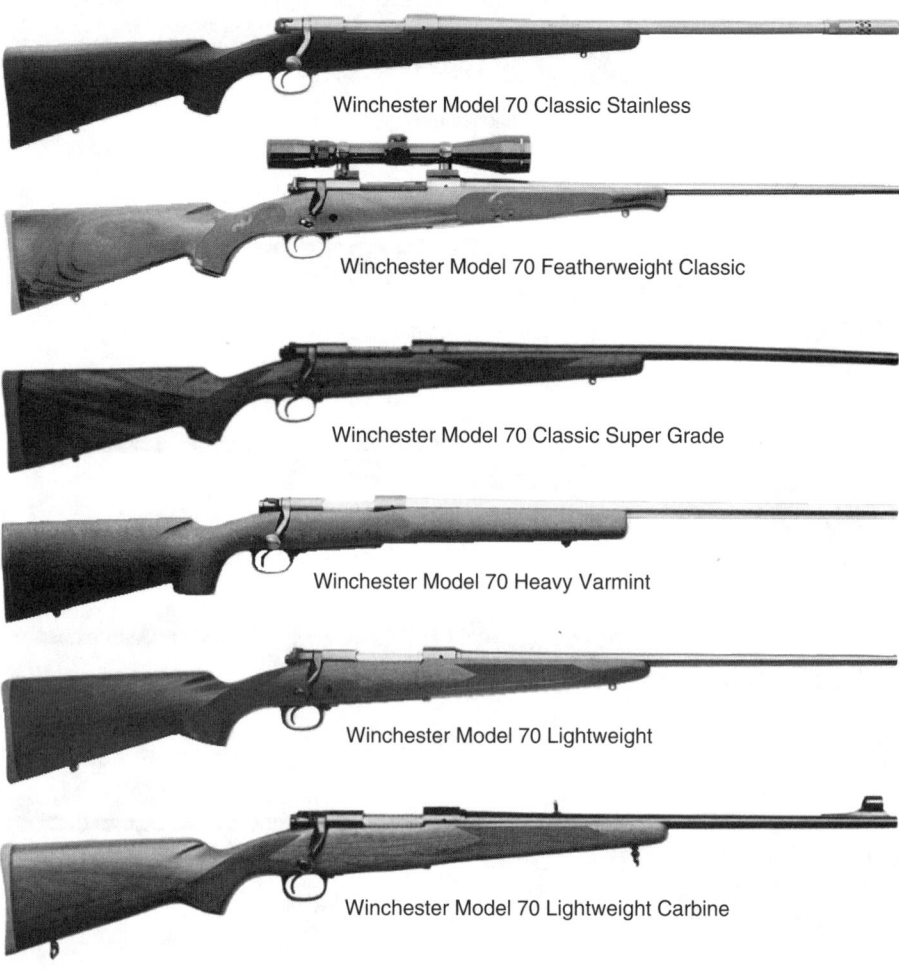

Winchester Model 70 Classic Stainless

Winchester Model 70 Featherweight Classic

Winchester Model 70 Classic Super Grade

Winchester Model 70 Heavy Varmint

Winchester Model 70 Lightweight

Winchester Model 70 Lightweight Carbine

Winchester Model 70 Classic Stainless
Same specs as Model 70 (1972 to present) except 22-250, 243, 308, 270, 30-06, 7mm Rem. Mag., 300 Win. Mag., 300 Weatherby Mag., 338 Win. Mag., 375 H&H Mag.; 3-shot magazine (magnum), 5-shot (others); 22", 24", 26" stainless steel barrel; weighs 6 3/4 lbs.; no sights; drilled, tapped for scope mounts; black fiberglass/graphite composite stock; matte gray finish; stainless steel pre-'64-type action with controlled feeding; BOSS barrel vibration modulator and muzzlebrake system optional. Introduced 1994; still in production. **LMSR: $737**
   Perf.: $525   **Exc.:** $420   **VGood:** $360
375 H&H Mag., with sights
   Perf.: $575   **Exc.:** $470   **VGood:** $410
With BOSS (all except 300 Wea. Mag., 375 H&H Mag.)
   Perf.: $635   **Exc.:** $535   **VGood:** $430

Winchester Model 70 Classic Super Grade
Same specs as Model 70 (1972 to present) except 270, 30-06, 7mm Rem. Mag., 300 Win. Mag., 338 Win. Mag.; 3-shot magazine (magnum), 5-shot (others); 24" barrel; no sights; scope bases and rings; straight-comb walnut stock with sculptured cheekpiece, tapered forend; wraparound cut checkering; solid buttplate; stainless steel claw extractor, magazine follower, bolt guide rail; three-position safety; hinged floorplate; pre-'64-type action with controlled round feeding; BOSS barrel vibration modulator and muzzlebrake system optional. Introduced 1994; still in production. **LMSR: $816**
   Perf.: $640   **Exc.:** $550   **VGood:** $470
With BOSS system
   Perf.: $670   **Exc.:** $595   **VGood:** $495

Winchester Model 70 Classic Super Express Magnum
Same specs as Model 70 (1972 to present) except 375 H&H, 375 JRS, 7mm STW, 416 Rem. Mag., 458 Win. Mag., 470 Capstick; 3-shot magazine; 22" barrel (458 Win. Mag.), 24" (others); weighs 8 1/2 lbs.; three-leaf express rear sight; deluxe checkered, satin-finished walnut stock; controlled round feed action; two steel crossbolts; barrel mounted front swivel stud. Introduced 1990; still in production. **LMSR: $865**
   Perf.: $725   **Exc.:** $595   **VGood:** $520

Winchester Model 70 Custom Sharpshooter
Same specs as Model 70 (1972 to present) except 223, 22-250, 308 Win., 300 Win. Mag.; 24" barrel (308), 26" (others); weighs 11 lbs.; no sights; scope bases and rings; McMillan A-2 target-style stock, glass bedded; hand-honed and fitted action; Schneider barrel; matte blue finish. Introduced 1992; discontinued 1994.
   Perf.: $1495   **Exc.:** $1195   **VGood:** $895

Winchester Model 70 DBM
Same specs as Model 70 (1972 to present) except 22-250, 223, 243, 270, 30-06, 308, 7mm Rem. Mag., 300 Win. Mag.; 3-shot detachable box magazine; 24", 26" barrel; with or without sights; scope bases and rings. Introduced 1992; discontinued 1994.
   Perf.: $450   **Exc.:** $395   **VGood:** $325

Winchester Model 70 Featherweight Classic
Same specs as Model 70 (1972 to present) except 22-250, 223, 243, 270, 280, 30-06, 308, 7mm-08; no sights; scope bases and rings; standard-grade walnut stock; claw extractor; controlled-round feeding system. Introduced 1992; still in production. **LMSR: $680**
   Perf.: $470   **Exc.:** $410   **VGood:** $330

Winchester Model 70 Featherweight WinTuff
Same specs as Model 70 (1972 to present) except 22-250, 223, 243, 270, 30-06, 308; no sights; scope bases and rings; brown laminated checkered pistol-grip stock; Schnabel forend. Introduced 1988; discontinued 1990.
   Perf.: $425   **Exc.:** $375   **VGood:** $310

Winchester Model 70 Heavy Barrel Varmint
Same specs as Model 70 (1972 to present) except 22-250, 223, 243, 308; heavy 26" barrel with counter-bored muzzle; 46" overall length; weighs 9 lbs.; no sights; drilled, tapped for scope mounts; sporter-style stock; rubber buttpad. Introduced 1989; discontinued 1994.
   Perf.: $450   **Exc.:** $400   **VGood:** $320

Winchester Model 70 Heavy Varmint
Same specs as Model 70 (1972 to present) except 220 Swift, 22-250, 223, 243, 308; weighs 10 3/4 lbs.; heavy 26" stainless barrel with counter-bored muzzle; no sights; black synthetic stock with full-length Pillar Plus AccuBlock, beavertail forend; rubber buttpad. Introduced 1989; discontinued 1994.
   Perf.: $595   **Exc.:** $520   **VGood:** $450

Winchester Model 70 Lightweight
Same specs as Model 70 (1972 to present) except 22-250, 223 Rem., 243 Win., 270 Win., 280, 308 Win., 30-06; 5-shot magazine, 6-shot (223 Rem.); 22" barrel; no sights; drilled, tapped for scope mounts; satin-finished American walnut stock; machine-cut checkering; three-position safety; stainless steel magazine follower; hinged floorplate. Introduced 1984; still in production. **LMSR: $513**
   Perf.: $400   **Exc.:** $330   **VGood:** $300

Winchester Model 70 Lightweight Carbine
Same specs as Model 70 (1972 to present) except 22-250, 222, 223, 243, 250 Savage, 270, 308, 30-06; 5-shot magazine, 6-shot (223); 20" barrel; no sights; drilled, tapped for scope mounts; satin-finished American walnut stock; cut checkering; stainless steel magazine follower; three-position safety; hinged floorplate. Introduced 1984; discontinued 1986.
   Perf.: $350   **Exc.:** $310   **VGood:** $280

Winchester Model 70 Winlite

Winchester Model 70 XTR

Winchester Model 70 XTR Featherweight

Winchester Model 70 XTR Super Express Magnum

Winchester Model 70 Varmint

Winchester Model 70A

RIFLES

Winchester Bolt-Action Rifles
Model 70 SHB
(Synthetic Heavy Barrel)
Model 70 Sporter SSM
Model 70 Sporter WinTuff
Model 70 Synthetic Heavy Varmint
Model 70 Target
Model 70 Varmint
Model 70 Winlite
Model 70 XTR
Model 70 XTR Featherweight
Model 70 XTR Featherweight European
Model 70 XTR Sporter
Model 70 XTR Super Express Magnum

Winchester Model 70 SHB (Synthetic Heavy Barrel)
Same specs as Model 70 (1972 to present) except 308; heavy 26″ barrel with counter-bored muzzle; weighs 9 lbs.; black synthetic stock with checkering; matte blue finish. Introduced 1992; discontinued 1992.
Perf.: $460   **Exc.:** $400   **VGood:** $340

Winchester Model 70 Sporter SSM
Same specs as Model 70 (1972 to present) except 223, 22-250, 243, 270, 308, 30-06, 7mm Rem. Mag., 300 Win. Mag., 338 Win. Mag., 375 H&H Mag.; 3-shot magazine (magnum), 5-shot (others); 24″, 26″ barrel; weighs 7 ¾ lbs.; no sights; scope bases and rings; black composite, graphite-impregnated stock; matte finish metal. Introduced 1992; discontinued 1993.
Perf.: $430   **Exc.:** $360   **VGood:** $310

Winchester Model 70 Sporter WinTuff
Same specs as Model 70 (1972 to present) except 270, 30-06, 7mm Rem. Mag., 300 Win. Mag., 300 Weatherby Mag., 338 Win. Mag.; 3-shot magazine (magnum), 5-shot (others); 24″ barrel; bases and rings for scope mounts; checkered brown laminated stock; rubber recoil pad. Introduced 1992; discontinued 1992.
Perf.: $440   **Exc.:** $370   **VGood:** $320

Winchester Model 70 Synthetic Heavy Varmint
Same specs as Model 70 (1972 to present) except 223, 22-250, 243, 308; 26″ heavy stainless steel barrel; weighs 10 ¾ lbs.; fiberglass/graphite stock; full-length Pillar Plus AccuBlock stock bedding system. Introduced 1993; still in production. **LMSR:** $742
Perf.: $595   **Exc.:** $495   **VGood:** $410

Winchester Model 70 Target
Same specs as Model 70 (1972 to present) except 30-06, 308; 26″ heavy barrel; contoured aluminum hand stop for either left- or right-handed shooter; high-comb target stock; tapped for micrometer sights. No longer in production.
Perf.: $495   **Exc.:** $420   **VGood:** $380

Winchester Model 70 Varmint
Same specs as Model 70 (1972 to present) except 22-250, 223, 225, 243, 308; 24″ heavy barrel; no sights; black serrated buttplate; black forend tip; high-luster finish on walnut stock. Introduced 1972; designated XTR 1978; designation dropped 1989; discontinued 1993.
Perf.: $460   **Exc.:** $400   **VGood:** $340

Winchester Model 70 Winlite
Same specs as Model 70 (1972 to present) except 25-06, 270, 280, 30-06, 7mm Rem. Mag., 300 Win. Mag., 300 Weatherby Mag., 338 Win. Mag.; 3-, 4-shot magazine; 22″, 24″ (magnum) barrel; McMillan brown fiberglass stock. Introduced 1986; discontinued 1990.
Perf.: $450   **Exc.:** $400   **VGood:** $350

Winchester Model 70 XTR
Same specs as Model 70 (1972 to present) except 222, 22-250, 25-06, 243, 270, 308, 30-06, 264 Win. Mag., 7mm Rem. Mag., 300 Win. Mag., 338 Win. Mag.; 3-shot magazine (magnum), 5-shot (others); 22″, 24″ (magnum) barrel; satin-finished stock. Introduced 1978; discontinued 1989.
Perf.: $395   **Exc.:** $340   **VGood:** $300

Winchester Model 70 XTR Featherweight
Same specs as Model 70 (1972 to present) except 22-250, 223, 243, 25-06, 257 Roberts, 270, 280, 7x57, 6.5x55 Swedish, 30-06, 7mm-08, 7mm Rem. Mag., 300 Win. Mag.; 22″, 24″ (300 Win. Mag.) tapered barrel; no sights; checkered satin-finished American walnut stock, Schnabel forend; red rubber buttpad; high-polish blued finish on metal; optional blade front sight, adjustable folding rear. Introduced 1981; XTR designation dropped 1989; no longer in production.
Perf.: $410   **Exc.:** $350   **VGood:** $300
With sights
Perf.: $430   **Exc.:** $360   **VGood:** $300

Winchester Model 70 XTR Featherweight European
Same specs as Model 70 (1972 to present) except 6.5x55 Swedish Mauser; weighs 6 ¾ lbs. Introduced 1986; discontinued 1986.
Perf.: $470   **Exc.:** $430   **VGood:** $390

Winchester Model 70 XTR Sporter
Same specs as Model 70 (1972 to present) except 22-250, 223, 243, 25-06, 264 Win. Mag., 270, 270 Weatherby Mag., 30-06, 300 Win. Mag., 300 Weatherby Mag., 300 H&H, 308, 7mm Rem. Mag., 338 Win. Mag.; 3-shot magazine (magnum), 5-shot (others); 24″ barrel; three-position safety; stainless steel magazine follower; rubber buttpad. Introduced 1980; XTR designation dropped 1989; no longer in production.
Perf.: $395   **Exc.:** $330   **VGood:** $300

Winchester Model 70 XTR Super Express Magnum
Same specs as Model 70 (1972 to present) except 375 H&H Mag., 458 Win Mag.; 3-shot magazine; 22″, 24″ barrel; steel crossbolts; contoured rubber buttpad. Introduced 1981; discontinued 1989.
Perf.: $695   **Exc.:** $595   **VGood:** $450

WINCHESTER MODEL 70A
Bolt action; 222 Rem., 22-250, 243 Win., 25-06, 264 Win. Mag., 270 Win., 30-06, 308 Win., 7mm Rem. Mag., 300 Win. Mag.; 22″, 24″ barrel; adjustable leaf rear sight with white diamond for quick sighting, hooded ramp front; dark American walnut with high-comb Monte Carlo, undercut cheekpiece; three-position safety. Introduced 1972 as replacement for Model 770, more closely following style of Model 70; discontinued 1978.
Perf.: $330   **Exc.:** $290   **VGood:** $240

RIFLES

Winchester Bolt-Action Rifles
Model 72
Model 72 Gallery
Model 72 Target
Model 75 Sporter
Model 75 Target
Model 99 Thumb Trigger
Model 121
Model 131
Model 141
Model 310
Model 320
Model 670

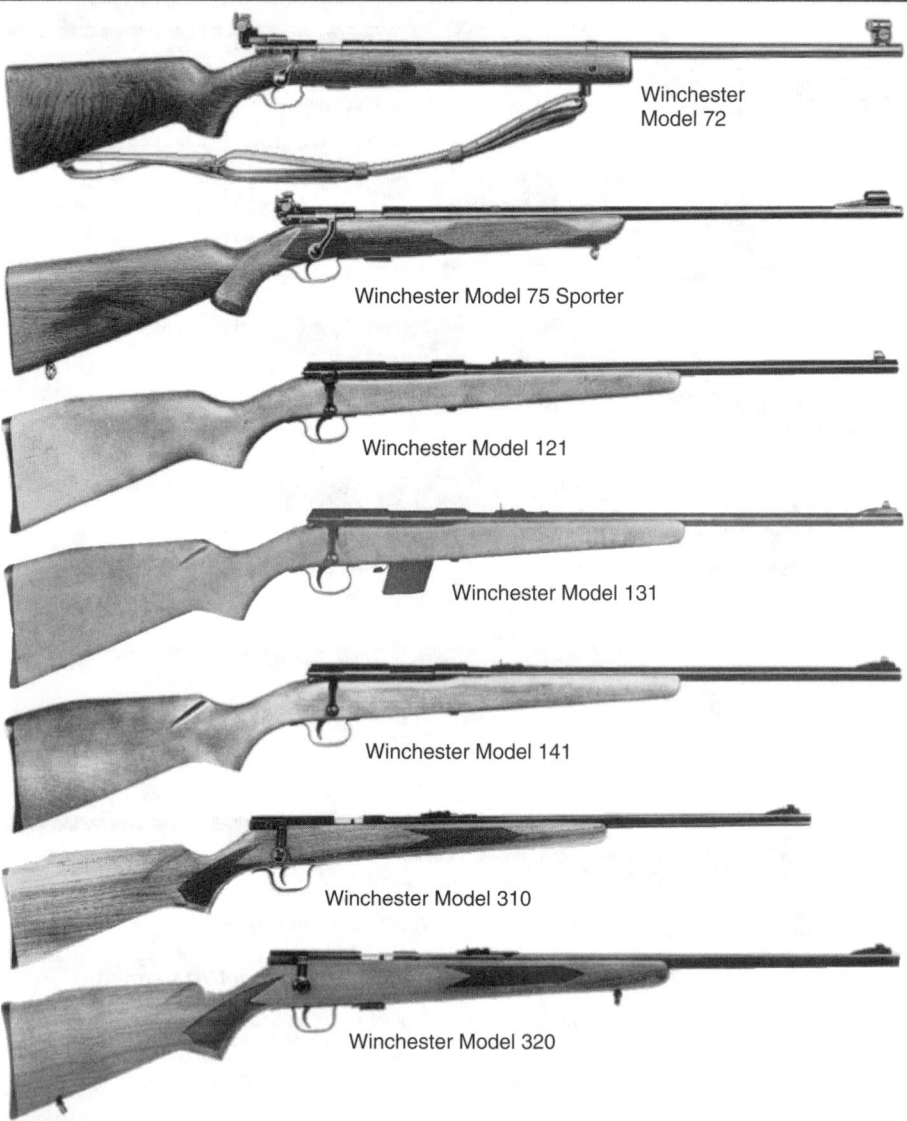

Winchester Model 72

Winchester Model 75 Sporter

Winchester Model 121

Winchester Model 131

Winchester Model 141

Winchester Model 310

Winchester Model 320

WINCHESTER MODEL 72
Bolt action; 22 Short, 22 Long, 22 LR; 22-shot tubular magazine (22 Short), 16-shot (22 Long), 15-shot (22 LR); 25″ barrel; takedown; peep or open rear sight, bead front; uncheckered pistol-grip stock. Introduced 1938; discontinued 1959.
Perf.: $350   **Exc.:** $250   **VGood:** $185

Winchester Model 72 Gallery
Same specs as Model 72 except 22 Short. Introduced 1939; discontinued 1942.
Perf.: $465   **Exc.:** $295   **VGood:** $200

Winchester Model 72 Target
Same specs as Model 72 except rear peep sight, blade front; sling swivels.
Perf.: $375   **Exc.:** $270   **VGood:** $190

WINCHESTER MODEL 75 SPORTER
Bolt action; 22 LR; 5-, 10-shot clip magazine; 24″ barrel; weighs 5 ½ lbs.; open rear sight, hooded ramp blade front; checkered select walnut stock, pistol grip; hard rubber grip cap; swivels; checkered steel buttplate; cocked with opening movement of bolt. Introduced 1939; discontinued 1958.
Perf.: $750   **Exc.:** $625   **VGood:** $525

Winchester Model 75 Target
Same specs as Model 75 except 28″ barrel; 44 ³/₄″ overall length; weighs 8 ⁵/₈ lbs.; target scope or variety of sights; uncheckered walnut stock, semi-beavertail forearm, pistol grip; 1″ Army-type leather sling. Introduced 1938; discontinued 1958.
Perf.: $470   **Exc.:** $340   **VGood:** $290

WINCHESTER MODEL 99 THUMB TRIGGER
Bolt action; single shot; 22 Short, 22 Long, 22 Extra Long; 18″ round barrel; takedown; open rear sight, blade front; straight-grip one-piece American walnut stock; button at rear of cocking piece serves as trigger. Introduced 1904; discontinued 1923. Some collector interest.
Exc.: $1995   **VGood:** $995   **Good:** $695

WINCHESTER MODEL 121
Bolt action; single shot; 22 Short, 22 Long, 22 LR; 20 ³/₄″ barrel; weighs 5 lbs.; standard post bead front sight, adjustable V rear; one-piece American hardwood pistol-grip stock with modified Monte Carlo profile; grooved for tip-off scope mounts. Introduced 1967; discontinued 1973.
Perf.: $120   **Exc.:** $90   **VGood:** $80
Winchester Model 121Y Youth with shortened stock
Perf.: $125   **Exc.:** $100   **VGood:** $90
Winchester Model 121 Deluxe with ramp front sight, swivels
Perf.: $125   **Exc.:** $100   **VGood:** $90

WINCHESTER MODEL 131
Bolt action; 22 Short, 22 Long, 22 LR; 7-shot clip magazine; 20″ barrel; weighs 5 lbs.; ramped bead post front sight, adjustable rear; receiver grooved for telescopic sight mounts; one-piece American hardwood stock with fluted comb, modified Monte Carlo profile; red safety; red cocking indicator. Introduced 1967; discontinued 1973.
Perf.: $150   **Exc.:** $120   **VGood:** $100

WINCHESTER MODEL 141
Bolt action; 22 Short, 22 Long, 22 LR; 19-shot tube magazine (22 Short), 15-shot (22 Long), 13-shot (22 LR); 20 ³/₄″ barrel; weighs 5 lbs.; ramped bead post front, adjustable rear sight; American hardwood stock with fluted comb, modified Monte Carlo; red cocking indicator; red-marked safety. Introduced 1967; discontinued 1973.
Perf.: $150   **Exc.:** $120   **VGood:** $110

WINCHESTER MODEL 310
Bolt action; single shot; 22 Short, 22 Long, 22 LR; 22″ barrel; 39″ overall length; weighs 5 ⁵/₈ lbs.; ramped bead post front sight, adjustable rear; grooved for scope sight; drilled, tapped for micrometer rear sight; checkered American walnut Monte Carlo pistol-grip stock, forearm; serrated trigger; positive safety lever; swivels. Introduced 1971; discontinued 1974.
Perf.: $200   **Exc.:** $170   **VGood:** $130

WINCHESTER MODEL 320
Bolt action; 22 Short, 22 Long, 22 LR; 5-shot magazine; 22″ barrel; 30 ¹/₂″ overall length; weighs 5 ⁵/₈ lbs.; ramped bead post front sight, adjustable rear; grooved for scope mounts; drilled, tapped for micrometer rear sight; American walnut Monte Carlo pistol-grip stock, forearm; sling swivels; serrated trigger; positive safety. Introduced 1971; discontinued 1974.
Perf.: $330   **Exc.:** $250   **VGood:** $200

WINCHESTER MODEL 670
Bolt action; 225, 243, 270, 308, 30-06; 4-shot magazine; 19″, 24″ barrel; ramp bead front sight, adjustable rear; sights easily detached for scope mounting; hardwood Monte Carlo pistol-grip stock. Introduced 1967; discontinued 1973.
Perf.: $300   **Exc.:** $250   **VGood:** $200

Winchester Model 670 Magnum
Same specs as Model 670 except 264 Win. Mag., 7mm Rem. Mag., 300 Win. Mag.; 3-shot magazine. Introduced 1967; discontinued 1970.
Perf.: $330   **Exc.:** $270   **VGood:** $220

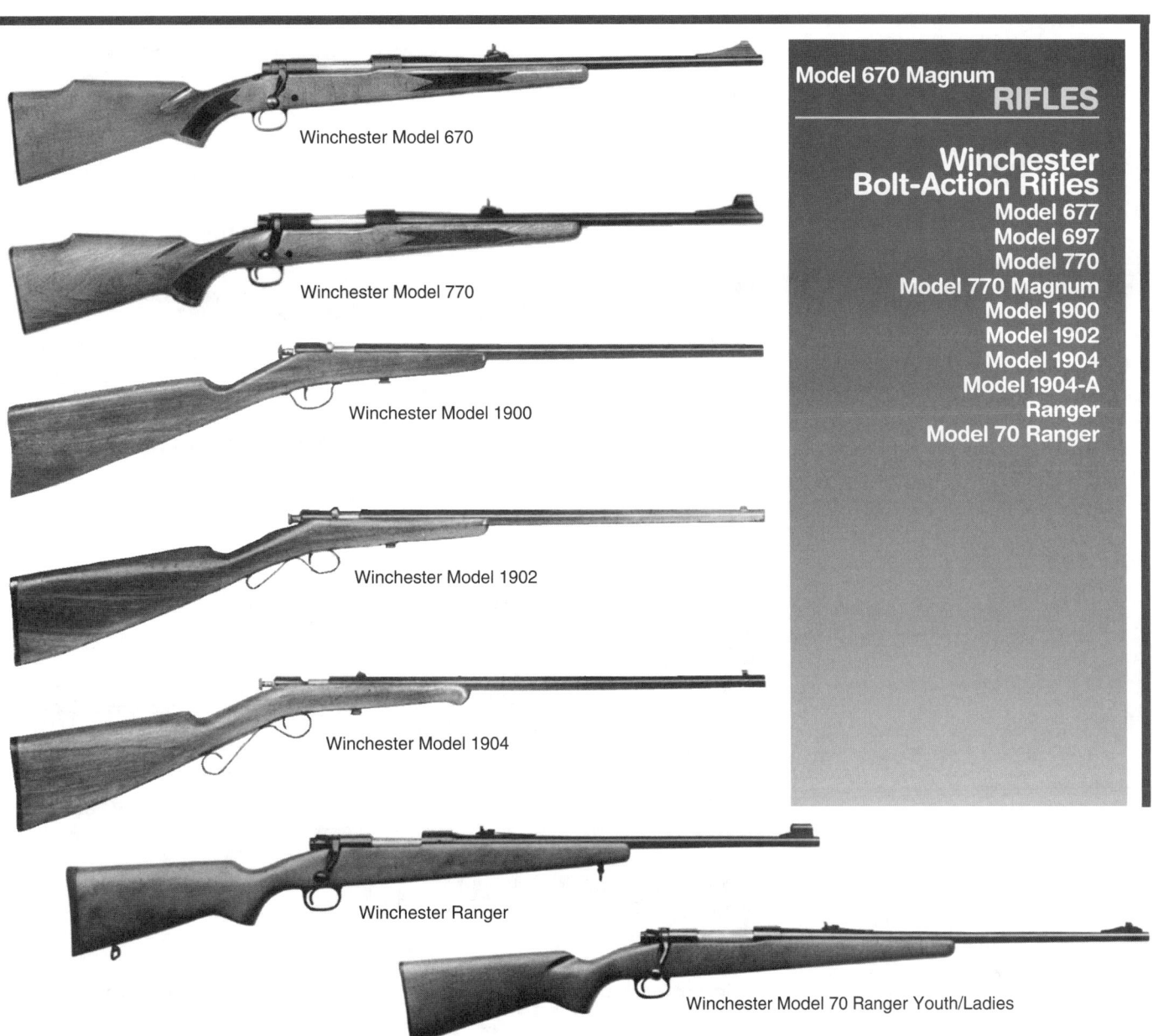

Winchester Model 670

Winchester Model 770

Winchester Model 1900

Winchester Model 1902

Winchester Model 1904

Winchester Ranger

Winchester Model 70 Ranger Youth/Ladies

Model 670 Magnum

RIFLES

Winchester Bolt-Action Rifles
Model 677
Model 697
Model 770
Model 770 Magnum
Model 1900
Model 1902
Model 1904
Model 1904-A
Ranger
Model 70 Ranger

WINCHESTER MODEL 677

Bolt action; single shot; 22 Short, 22 Long, 22 LR, 22 WMR, 22-shot cartridge; 24″ rifled, 27″ rifled or smootherbore barrel; takedown; no sights; scope mounts mounted on barrel; uncheckered pistol-grip stock, grooved forearm. Enjoyed little success due to poor scope-mounting system. Introduced 1937; discontinued 1939. Only 2239 produced. Some collector value.

Exc.: $695   VGood: $495   Good: $295

WINCHESTER MODEL 697

Bolt action; 22 Short, 22 Long, 22 LR, 22 WMR; 5-, 10-shot detachable box magazine; 25″ barrel; no sights; scope bases attached to barrel; choice of 2 ³/₄x or 5x scope. Introduced 1937; discontinued 1941. Collector interest.

Exc.: $795   VGood: $595   Good: $365
22 WMR
Exc.: $895   VGood: $695   Good: $475

WINCHESTER MODEL 770

Bolt action; 222 Rem., 22-250, 243 Win., 270 Win., 308 Win., 30-06; 22″ barrel; weighs 7 ¹/₂ lbs.; hooded ramp front sight, adjustable rear; checkered walnut Monte Carlo pistol-grip stock, cheekpiece, forearm; composition buttplate; red cocking indicator. This rifle was designed as a lower-echelon Model 70 but failed to meet acceptance, thus was dropped after only four years and replaced by Model 70A. Introduced 1969; discontinued 1972.

Perf.: $330   Exc.: $300   VGood: $250

Winchester Model 770 Magnum

Same specs as Model 770 except 264 Win. Mag., 7mm Rem. Mag., 300 Win. Mag.; 3-shot magazine; 24″ barrel; rubber recoil pad. Introduced 1969; discontinued 1972. Replaced by Model 70A.

Perf.: $350   Exc.: $320   VGood: $270

WINCHESTER MODEL 1900

Bolt action; single shot; 22 Short, 22 Long; 18″ round barrel; takedown; open rear sight, blade front; straight-grip one-piece American walnut stock. Introduced 1899; discontinued 1902. Some collector interest.

Exc.: $400   VGood: $325   Good: $220

WINCHESTER MODEL 1902

Bolt action; single shot; 22 Short, 22 Long, 22 Extra Long; 18″ round barrel; takedown; open rear sight, blade front; straight-grip one-piece American walnut stock; enlarged trigger guard. Introduced 1902; discontinued 1931.

Exc.: $275   VGood: $175   Good: $125

WINCHESTER MODEL 1904

Bolt action; single shot; 22 Short, 22 Long, 22 Extra Long, 22 LR; 21″ round barrel; takedown; open rear sight, blade front; straight-grip one-piece American walnut stock; enlarged trigger guard. Introduced 1904; discontinued 1931.

Exc.: $300   VGood: $200   Good: $160

Winchester Model 1904-A

Same specs as Model 1904 except 22 LR. Introduced 1927; discontinued 1931.

Exc.: $300   VGood: $200   Good: $160

WINCHESTER RANGER

Bolt action; 223, 243, 270, 30-06, 7mm Rem. Mag.; 3-shot magazine (7mm Rem. Mag.), 4-shot (others); 22″, 24″ barrel; weighs 7 ¹/₈ lbs.; American hardwood stock; no checkering; composition buttplate; matte blue finish. Introduced 1985; **LMSR: $503**

Perf.: $370   Exc.: $320   VGood: $235

Winchester Model 70 Ranger Youth/Ladies

Same specs as Ranger 243, 308; 4-shot magazine; 20″, 22″ barrel; weighs 5 ³/₄ lbs.; scaled-down American hardwood stock; no checkering; composition buttplate; matte blue finish. Introduced 1985; still in production. **LMSR: $468**

Perf.: $370   Exc.: $320   VGood: $220

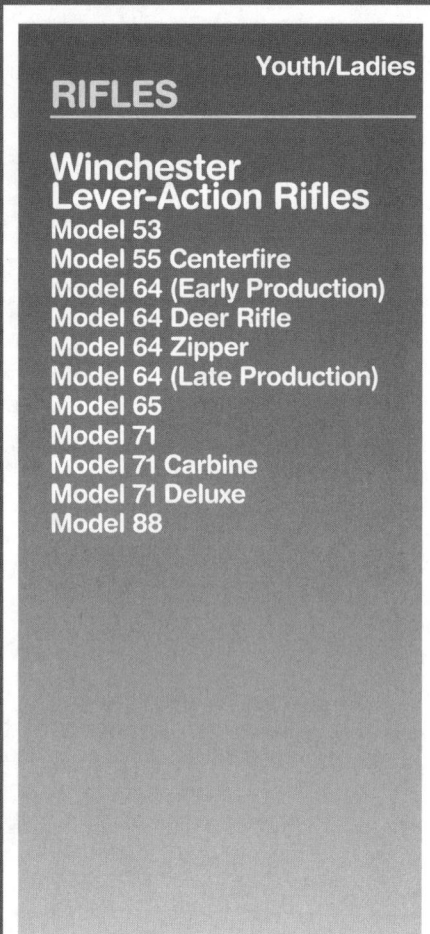

Winchester Lever-Action Rifles

Model 53
Model 55 Centerfire
Model 64 (Early Production)
Model 64 Deer Rifle
Model 64 Zipper
Model 64 (Late Production)
Model 65
Model 71
Model 71 Carbine
Model 71 Deluxe
Model 88

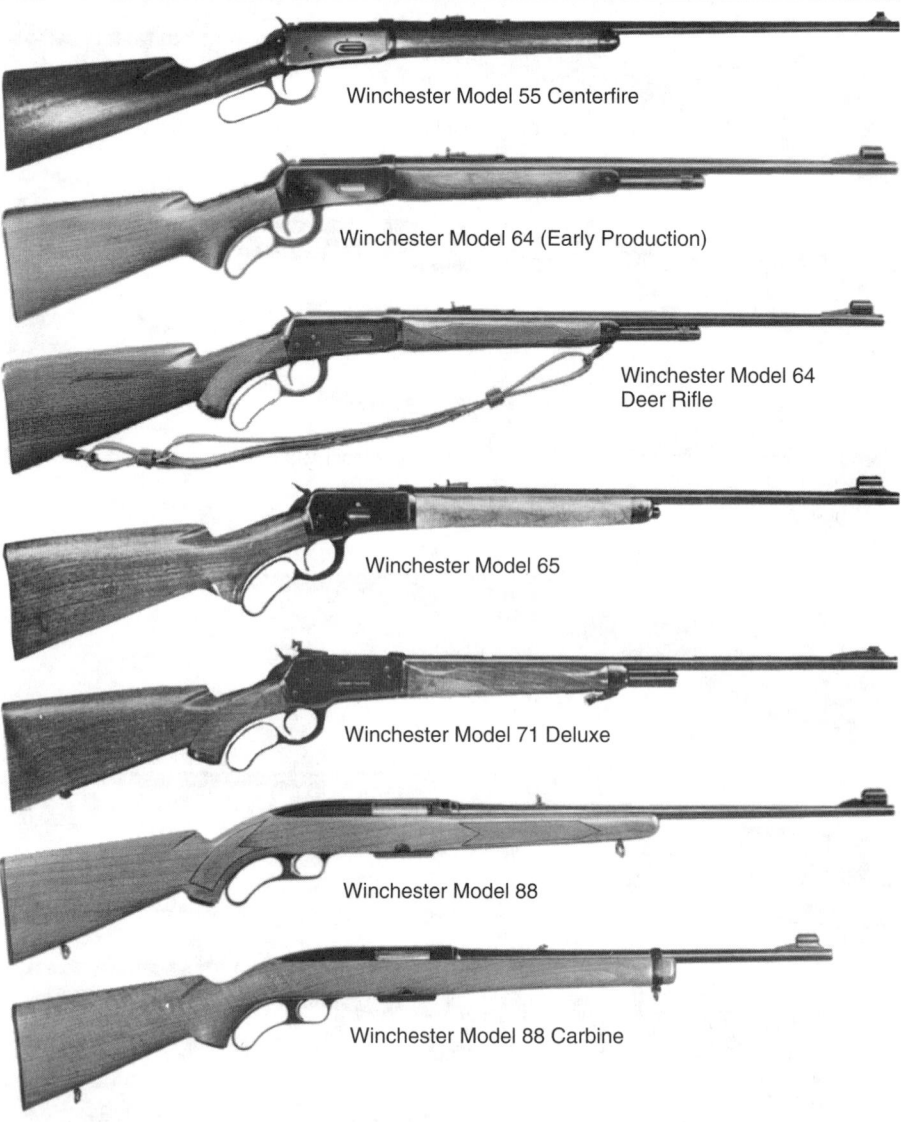

Winchester Model 55 Centerfire

Winchester Model 64 (Early Production)

Winchester Model 64 Deer Rifle

Winchester Model 65

Winchester Model 71 Deluxe

Winchester Model 88

Winchester Model 88 Carbine

WINCHESTER LEVER-ACTION RIFLES

WINCHESTER MODEL 53

Lever action; 25-20, 32-20, 44-40; 6-shot half-magazine; 22″ barrel; open rear sight, bead front; walnut pistol-grip or straight-grip stock; blued finish. Introduced 1924; discontinued 1932.

   Exc.: $2295   **VGood:** $1295   **Good:** $795
Takedown model
   Exc.: $2495   **VGood:** $1495   **Good:** $995
(Premium for .44-40 Cal)

WINCHESTER MODEL 55 CENTERFIRE

Lever action; 25-35, 30-30, 32 Spl.; 3-shot tubular magazine; 24″ round barrel; open rear sight, bead front; uncheckered American walnut straight grip stock, forend. Based on Model 94 design. Collector value. Introduced 1924; discontinued 1932.

   Exc.: $1495   **VGood:** $895   **Good:** $595
Takedown model
   Exc.: $1595   **VGood:** $995   **Good:** $650

WINCHESTER MODEL 64 (EARLY PRODUCTION)

Lever action; 219 Zipper, 25-35, 30-30, 32 Spl.; 20″, 24″ barrel; weighs 7 lbs.; Winchester No. 22H open sporting or Lyman No. 56 (20″ barrel) open rear sight, hooded ramp bead front; uncheckered American walnut pistol-grip stock, forend. Originally manufactured 1933 to 1957; additional production with 30-30 chambering only and 24″ barrel from 1972 to 1973. Collector value on original production.

   Exc.: $1295   **VGood:** $895   **Good:** $540

Winchester Model 64 Deer Rifle

Same specs as Model 64 except 30-30, 32 Spl.; weighs 7 ¾ lbs.; hand-checkered pistol-grip stock, forend; 1″ sling swivels; sling; checkered steel buttplate. Introduced 1933; discontinued 1956. Collector value.

   Exc.: $1295   **VGood:** $950   **Good:** $650

Winchester Model 64 Zipper

Same specs as Model 64 except 219 Zipper; 26″ barrel; weighs 7 lbs.; Winchester No. 98A peep rear sight. Introduced 1938; discontinued 1941. Collector value.

   Exc.: $1695   **VGood:** $1295   **Good:** $995

WINCHESTER MODEL 64 (LATE PRODUCTION)

Lever action; 30-30; 5-shot tube magazine; 24″ barrel; open rear sight, ramped bead front; uncheckered American walnut pistol-grip stock, forend. Introduced 1972; discontinued 1973.

   Exc.: $495   **VGood:** $400   **Good:** $300

WINCHESTER MODEL 65

Lever action; 218 Bee, 25-20, 32-20; 7-shot tube half-magazine; 22″, 24″ (218 Bee) barrel; weighs 6 ½ lbs.; open or peep rear sight, Lyman gold bead front on ramp base; plain pistol-grip stock, forearm; shotgun-type butt with checkered steel buttplate. Introduced 1933; discontinued 1947.

   Exc.: $2395   **VGood:** $1695   **Good:** $1150

WINCHESTER MODEL 71

Lever-action; 348 Win.; 4-shot tube magazine; 24″ barrel; weighs 8 lbs.; open rear sight, bead front on ramp with hood; plain walnut pistol-grip stock, beavertail forend; blued finish. Introduced 1936; discontinued 1957.

   Exc.: $995   **VGood:** $795   **Good:** $625

Winchester Model 71 Carbine

Same specs as Model 71 except 20″ barrel. Introduced 1936; discontinued 1938.

   Exc.: $2995   **VGood:** $2395   **Good:** $1650

Winchester Model 71 Deluxe

Same specs as Model 71 except No. 98A rear peep sight; checkered stock, forearm, grip cap; quick-detachable sling swivels; leather sling.

   Exc.: $1295   **VGood:** $995   **Good:** $750

WINCHESTER MODEL 88

Lever action; 243 Win., 284 Win., 308 Win., 358 Win.; 5-shot detachable box magazine; 22″ barrel; 39 ½″ overall length; weighs 6 ½ lbs.; hammerless; hooded white metal bead front sight, Lyman folding leaf middle sight; one-piece checkered walnut stock with steel-capped pistol grip, fluted comb, sling swivels; three-lug bolt; crossbolt safety; side ejection. Introduced 1955; discontinued 1974.

308 Win.
   Perf.: $495   **Exc.:** $450   **VGood:** $380
243 Win. (pre-1964)
   Perf.: $595   **Exc.:** $500   **VGood:** $400
243 Win. (post-1964)
   Perf.: $495   **Exc.:** $450   **VGood:** $380
284 Win. (pre-1964)
   Perf.: $825   **Exc.:** $650   **VGood:** $540
284 Win. (post-1964)
   Perf.: $620   **Exc.:** $495   **VGood:** $420
358 Win.
   Perf.: $1200   **Exc.:** $920   **VGood:** $730

Winchester Model 88 Carbine

Same specs as Model 88 except 243, 284, 308; 19″ barrel; barrel band. Introduced 1968; discontinued 1973.

   Perf.: $700   **Exc.:** $590   **VGood:** $510

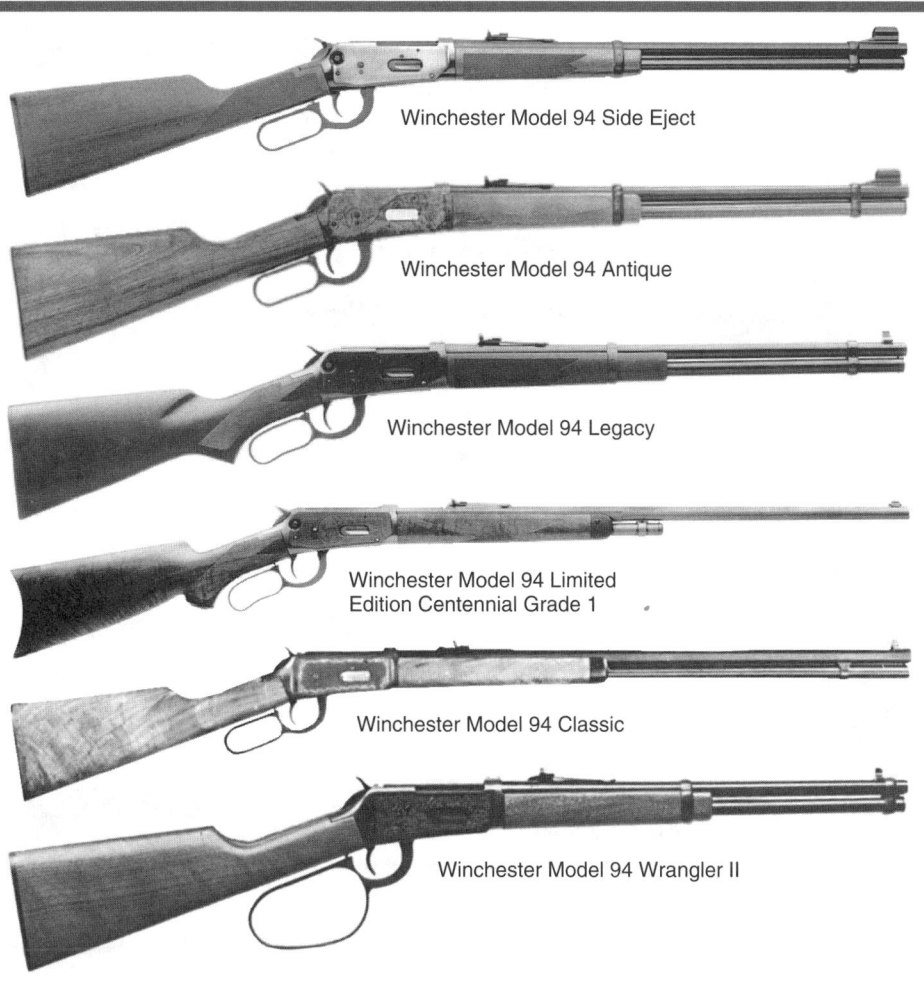

Winchester Model 94 Side Eject

Winchester Model 94 Antique

Winchester Model 94 Legacy

Winchester Model 94 Limited Edition Centennial Grade 1

Winchester Model 94 Classic

Winchester Model 94 Wrangler II

Winchester Lever-Action Rifles
Model 94
Model 94 Antique
Model 94 Classic
Model 94 Deluxe
Model 94 Limited Edition Centennial Grade I
Model 94 Limited Edition Centennial High Grade Rifle
Model 94 Limited Edition Centennial High Grade Custom
Model 94 Legacy
Model 94 Ranger
Model 94 Saddle Ring Carbine
Model 94 Trapper
Model 94 Win-Tuff
Model 94 Wrangler
Model 94 Wrangler II
Model 94 Wrangler Large

WINCHESTER MODEL 94
Lever action; 30-30, 25-35, 32 Spl., 7-30 Waters, 44 Mag.; 6-tube magazine; 20″, 24″ barrel; weighs 6 1/4 lbs.; open rear sight, ramp front; plain American walnut straight-grip stock; barrel band on forearm; saddle ring; side or angle ejection (post-1982); blued finish. Introduced 1964; still in production. **LMSR: $352**
    Perf.: $270   **Exc.:** $240   **VGood:** $170
Side or Angle Eject Model
    Perf.: $280   **Exc.:** $250   **VGood:** $185

Winchester Model 94 Antique
Same specs as Model 94 except 20″ barrel; case-hardened receiver with scrollwork; gold-plated saddle ring. Introduced 1964; discontinued 1983.
    Perf.: $260   **Exc.:** $230   **VGood:** $200

Winchester Model 94 Classic
Same specs as Model 94 except 30-30; 20″, 26″ octagonal barrel; semi-fancy American walnut stock, forearm; steel buttplate; scrollwork on receiver. Introduced 1967; discontinued 1970.
    Perf.: $350   **Exc.:** $310   **VGood:** $250

Winchester Model 94 Deluxe
Same specs as Model 94 except 30-30; checkered walnut stock, forearm. Introduced 1988; still in production as Model 94 Traditional - CW. **LMSR: $381**
    Perf.: $290   **Exc.:** $250   **VGood:** $205

Winchester Model 94 Limited Edition Centennial Grade I
Same specs as Model 94 except 30-30; 26″ half-octagon barrel; tube half-magazine; drilled, tapped for tang sight; close reproduction of the turn-of-the-century #9 factory engraving; diamond-style "H" checkering pattern on the pistol-grip stock, forend; engraving on receiver sides shows whitetail deer profiles. Introduced 1994; discontinued 1994.
    Perf.: $725   **Exc.:** $550   **VGood:** $395

Winchester Model 94 Limited Edition Centennial High Grade Rifle
Same specs as Model 94 except 30-30; 26″ half-octagon barrel; tube half-magazine; Lyman No. 2 tang sight; spade-pattern checkering on pistol-grip stock; blued receiver with gold inlays and #6-style engraving pattern showing a gold deer on the right, gold mountain sheep on the left. Only 3,000 produced. Introduced 1994; discontinued 1994.
    Perf.: $1150   **Exc.:** $850   **VGood:** $595

Winchester Model 94 Limited Edition Centennial High Grade Custom
Same specs as Model 94 except 30-30; 26″ half-octagon barrel; tube half-magazine; Lyman No. 2 tang sight; spade-pattern checkering on pistol-grip stock; top ejection; hand-engraved, grayed action with inking in #5 style with gold inlaid ovals containing a pair of pronghorns on the right, caribou on the left; crescent buttplate. Only 94 produced. Introduced 1994; discontinued 1994.
    Perf.: $4295   **Exc.:** $3495   **VGood:** $1995

Winchester Model 94 Legacy
Same specs as Model 94 except 30-30; half-pistol-grip walnut stock; checkered grip and forend. Introduced 1995; still in production. **LMSR: $446**
    Perf.: $300   **Exc.:** $250   **VGood:** $200

Winchester Model 94 Ranger
Same specs as Model 94 except 30-30; 5-shot magazine; 20″ barrel; uncheckered hardwood stock, forearm. Introduced 1985; still in production. **LMSR: $347**
    Perf.: $230   **Exc.:** $200   **VGood:** $170

Winchester Model 94 Saddle Ring Carbine
Same specs as Model 94 except 44 Mag.; 20″ barrel; saddle ring; top ejection. Introduced 1967; discontinued 1972.
    Perf.: $330   **Exc.:** $270   **VGood:** $200

Winchester Model 94 Trapper
Same specs as Model 94 except 30-30 (standard), 357 Mag., 44 Mag./44 Spl., 45 Colt; 5-shot tube magazine in 30-30, 9 shots in other calibers; 16″ barrel; side ejection. Introduced 1985; still in production. **LMSR: $352**
    Perf.: $280   **Exc.:** $240   **VGood:** $200

Winchester Model 94 Win-Tuff
Same specs as Model 94 except 30-30; drilled, tapped for scope mounts; checkered hardwood stock, forearm. Introduced 1987; **LMSR: $404**
    Perf.: $300   **Exc.:** $260   **VGood:** $220

Winchester Model 94 Wrangler
Same specs as Model 94 except 32 Spl.; 16″ barrel; top ejection; roll-engraved Western scene on receiver; hoop finger lever. Introduced 1983; no longer in production.
    Perf.: $340   **Exc.:** $300   **VGood:** $230

Winchester Model 94 Wrangler II
Same specs as Model 94 except 32 Spl., 38-55; 16″ barrel; angle ejection; loop-type finger lever; roll-engraved Western scenes on receiver. Introduced 1983; discontinued 1986.
    Perf.: $290   **Exc.:** $240   **VGood:** $200

Winchester Model 94 Wrangler Large Loop
Same specs as Model 94 except 30-30, 44 Mag.; 16″ barrel; extra-large loop lever. Introduced 1992; **LMSR: $400**
    Perf.: $285   **Exc.:** $240   **VGood:** $200
Side or Angle Eject Model
    Perf.: $290   **Exc.:** $240   **VGood:** $200

RIFLES

Winchester Lever-Action Rifles

Model 94 XTR
Model 94 XTR Big Bore
Model 94 XTR Deluxe
Model 150
Model 250
Model 250 Deluxe
Model 255
Model 255 Deluxe
Model 1873
Model 1873 Carbine
Model 1873 Musket
Model 1873 Rimfire
Model 1873 Special
Model 1876
Model 1876 Carbine

Winchester Model 94 XTR Angle Eject

Winchester Model 94 XTR Big Bore

Winchester Model 150

Winchester Model 250 Deluxe

Winchester Model 255

Winchester Model 94 XTR
Same specs as Model 94 except 30-30, 7-30 Waters; 20″, 24″ barrel (7-30 Waters); hooded front sight; select checkered walnut straight-grip stock. Introduced 1978; discontinued 1988.
Perf.: $270   **Exc.:** $230   **VGood:** $200
7-30 Waters
Perf.: $300   **Exc.:** $250   **VGood:** $230

Winchester Model 94 XTR Big Bore
Same specs as Model 94 except 307 Win., 356 Win., 375 Win.; 6-shot tube magazine; 20″ barrel; cut-checkered satin-finish American walnut stock, rubber recoil pad. XTR designation dropped 1989. Introduced 1978;
LMSR: $393
Perf.: $295   **Exc.:** $250   **VGood:** $220
Side or Angle Eject model
Perf.: $300   **Exc.:** $250   **VGood:** $230

Winchester Model 94 XTR Deluxe
Same specs as Model 94 except 30-30; 20″ barrel; deluxe checkered walnut straight-grip stock, long forearm; rubber buttplate. Introduced 1987; discontinued 1988.
Perf.: $370   **Exc.:** $300   **VGood:** $250

WINCHESTER MODEL 150
Lever action; 22 Short, 22 Long, 22 LR; 21-shot tube magazine (22 Short), 17-shot (22 Long), 15-shot (22 LR); 20 1/2″ barrel; weighs 5 lbs.; aluminum alloy receiver grooved for scope sight; walnut-finished American hardwood stock; forearm has frontier-style barrel band; straight grip; no checkering. Introduced 1967; discontinued 1974.
Perf.: $130   **Exc.:** $110   **VGood:** $100

WINCHESTER MODEL 250
Lever action; 22 Short, 22 Long, 22 LR; 21-shot tube magazine (22 Short), 17-shot (22 Long), 15-shot (22 LR); 20 1/2″ barrel; 39″ overall length; weighs 5 lbs.; hammerless; ramped square post front sight, adjustable square notch rear; aluminum alloy receiver grooved for tip-off scope mounts; walnut-finished hardwood stock; crossbolt safety on front of trigger guard. Introduced 1963; discontinued 1973.
Perf.: $140   **Exc.:** $120   **VGood:** $90

Winchester Model 250 Deluxe
Same specs as Model 250 except select walnut stock fluted comb, cheekpiece; basketweave checkering; white spacer between buttplate and stock; sling swivels. Introduced 1965; discontinued 1971.
Perf.: $180   **Exc.:** $130   **VGood:** $110

WINCHESTER MODEL 255
Lever action; 22 WMR; 11-shot tube magazine; 20 1/2″ barrel; 39″ overall length; weighs 5 lbs.; hammerless; ramped square post front sight, adjustable square notch rear; aluminum alloy receiver grooved for scope mounts; walnut-finish hardwood pistol-grip stock; crossbolt safety. Introduced 1964; discontinued 1970.
Perf.: $180   **Exc.:** $140   **VGood:** $110

Winchester Model 255 Deluxe
Same specs as Model 255 except high-gloss Monte Carlo select walnut stock, fluted comb, cheekpiece; basketweave checkering; white spacer between buttplate and stock; sling swivels. Introduced 1965; discontinued 1971.
Perf.: $200   **Exc.:** $160   **VGood:** $130

WINCHESTER MODEL 1873
Lever action; 32-20, 38-40, 44-40; 6-, 12-, 15- or 17-shot tubular magazine; 24″ barrel, others available; open rear sight, bead or blade front; uncheckered American walnut straight-grip stock, forend; blued finish. Introduced 1873; discontinued 1924. 1 of 1000, and 1 of 100 are extremely rare. Appraisal recommended. Beware of fakes. Collector value.
Exc.: $4000-$9000
VGood: $2500-$4900
Good: $900-$2500

Winchester Model 1873 Carbine
Same specs as Model 1873 except 20″ barrel; 12-shot magazine; two barrel bands. Collector value.
Exc.: $5000-$12000
VGood: $2900-$9000
Good: $1000-$3000

Winchester Model 1873 Musket
Same specs as Model 1873 except 30″ barrel; extra capacity tube magazine; three barrel bands. Values from $1000 to $7000.

Winchester Model 1873 Rimfire
Same specs as Model 1873 except 22 Short, 22 Long, 22 Extra Long. Introduced 1884; discontinued 1904. Collector value.
Exc.: $7000   **VGood:** $3750   **Good:** $2500

Winchester Model 1873 Special
Same specs as Model 1873 except octagon barrel only; color case-hardened receiver; select American walnut pistol-grip stock. Case-hardened receivers brings substantial premium. Need appraisal. Collector value.

WINCHESTER MODEL 1876
Lever action; 40-60, 45-60, 45-75, 50-95; tube magazine; 26″, 28″ barrel; open rear sight, bead or blade front; uncheckered American walnut straight-grip stock, forend; blued finish. Model 1876 values to $6000. 1 of 1000 and 1 of 100 Model values to $75,000, an appraiser is needed. Introduced 1876; discontinued 1897.

Winchester Model 1876 Carbine
Same specs as Model 1876 except 22″ barrel; full-length forearm; one barrel band. Values to $7500.

Winchester Model 1876 Musket
Same specs as Model 1876 except 32″ barrel; carbine-like forend tip; one barrel band. Values to $8000.

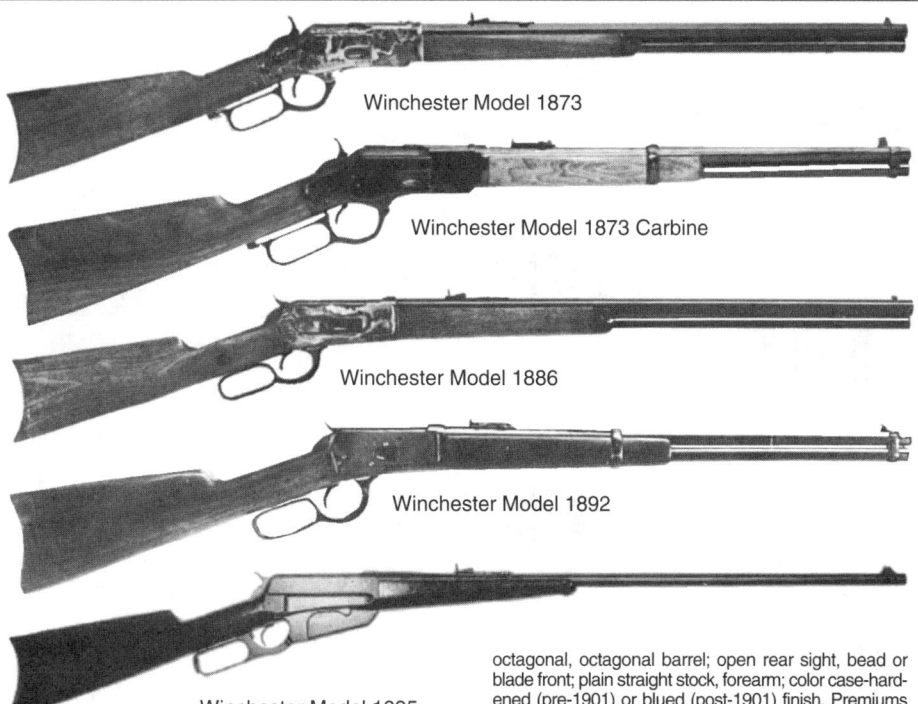

Winchester Model 1873

Winchester Model 1873 Carbine

Winchester Model 1886

Winchester Model 1892

Winchester Model 1895

Winchester Lever-Action Rifles
Model 1885 Highwall
Model 1885 Highwall Deluxe Grade
Model 1885 Highwall Schuetzen
Model 1885 Low Wall Sporter
Model 1885 Low Wall Winder Musket
Model 1886
Model 1886 Carbine
Model 1886 Lightweight
Model 1886 Musket
Model 1886 Takedown
Model 1892
Model 1892 Carbine
Model 1892 Musket
Model 1892 Trapper's Carbine
Model 1894
Model 1894 Carbine
Model 1894 Saddle Ring Carbine
Model 1894 Trapper Carbine
Model 1895

Note: All Model 1885 Single Shot rifles were available in many calibers and barrel contours. Each example merits expert appraisal.

WINCHESTER MODEL 1885 HIGHWALL
Lever action; single shot; 22 RF through 50; 30″ barrel of various weights; open rear sight, blade front; uncheckered American walnut straight-grip stock, forend; solid frame. Introduced 1885; discontinued 1920. Collector value.
Color case-hardened (pre-1901)
   Exc.: $3495   **VGood:** $2995   **Good:** $1995
Blued finish (post-1901)
   Exc.: $2695   **VGood:** $2195   **Good:** $1650

Winchester Model 1885 HighWall Deluxe Grade
Same specs as Model 1885 HighWall except better wood; hand-checkered stock, forend. Collector value.
   Exc.: $4995   **VGood:** $4250   **Good:** $3295

Winchester Model 1885 HighWall Schuetzen
Same specs as Model 1885 HighWall except 30″ octagonal barrel; Vernier rear tang peep sight wind-gauge front; European walnut Schuetzen-type stock; hand-checkered forend, pistol grip; Schuetzen buttplate, adjustable palm rest; spur finger lever; double-set trigger. Collector value.
   Exc.: $8995   **VGood:** $6995   **Good:** $4250
Takedown model
   Exc.: $9995   **VGood:** $7995   **Good:** $5250

WINCHESTER MODEL 1885 LOW WALL SPORTER
Lever action; single shot; 22 RF; 28″ round, octagonal barrel; solid frame; open rear sight, blade front; uncheckered American walnut straight stock, forend. Introduced 1885; discontinued 1920. Collector value.
   Exc.: $1295   **VGood:** $995   **Good:** $795

Winchester Model 1885 Low Wall Winder Musket
Same specs as Model 1885 Low Wall except 22 Short, 22 LR; 28″ round barrel; solid frame or takedown; musket-type rear sight, blade front; straight-grip military-type stock, forend. Collector value.
   Exc.: $1895   **VGood:** $995   **Good:** $695

WINCHESTER MODEL 1886
Lever action; 33 WCF, 45-70, 45-90, 40-82, 40-65, 38-56, 40-70, 38-70, 50-100-450, 50-110 Express; 8-shot tube magazine or 4-shot half-magazine; 26″ round, half-octagonal, octagonal barrel; open rear sight, bead or blade front; plain straight stock, forearm; color case-hardened (pre-1901) or blued (post-1901) finish. Premiums for many features and calibers. Appraisal necessary. Introduced 1886; discontinued 1935. Collector value.
   Exc.: $8000   **VGood:** $5000   **Good:** $2500
With deluxe checkered stock
   Exc.: $12000   **VGood:** $8995   **Good:** $4250

Winchester Model 1886 Carbine
Same specs as Model 1886 except 22″ round barrel. Rare.
   Exc.: $10950   **VGood:** $7450   **Good:** $3995
With full-length stock
   Exc.: $13995   **VGood:** $9995   **Good:** $5250

Winchester Model 1886 Lightweight
Same specs as Model 1886 except 33 WCF, 45-70; 22″, 24″ round barrel; rubber buttplate; blued finish.
   Exc.: $3795   **VGood:** $2850   **Good:** $1495

Winchester Model 1886 Musket
Same specs as Model 1886 except 30″ round barrel; one-barrel band.
   Exc.: $12995   **VGood:** $9995   **Good:** $5700

Winchester Model 1886 Takedown
Same specs as Model 1886 except 24″ round barrel; takedown feature.
   Exc.: $7495   **VGood:** $4500   **Good:** $2450

WINCHESTER MODEL 1892
Lever action; 218 Bee, 25-20, 32-20, 38-40, 44-40; 13-shot tube magazine, 7-shot half-magazine, 10-shot two-thirds magazine; 24″ round or octagonal barrel; open rear sight, bead front; plain, straight stock, forearm; blued finish. Introduced 1892; discontinued 1941.
   Exc.: $2495   **VGood:** $1495   **Good:** $875
With takedown feature
   Exc.: $2995   **VGood:** $1995   **Good:** $1250

Winchester Model 1892 Carbine
Same specs as Model 1892 except 5-, 11-shot magazine; 20″ barrel; two barrel bands.
   Exc.: $2995   **VGood:** $1995   **Good:** $1250

Winchester Model 1892 Musket
Same specs as Model 1892 except 30″ barrel; three-barrel bands; buttplate.
   Exc.: $8995   **VGood:** $5350   **Good:** $3450

Winchester Model 1892 Trapper's Carbine
Same specs as Model 1892 except 12″, 14″, 15″, 16″, 18″ barrel; shortened tube magazine; two barrel bands; browned finish.
   Exc.: $7995   **VGood:** $4995   **Good:** $3500

WINCHESTER MODEL 1894
Lever action; 25-35, 30-30, 32-40, 32 Spl., 38-55; 4-, 6-shot tube magazine; 22″, 26″ barrel; open rear sight, blade or ramp (post-1931) front; plain American walnut straight-grip stock; blued finish. Introduced 1894; discontinued 1936.
   Exc.: $1995   **VGood:** $1295   **Good:** $695
Takedown model
   Exc.: $2495   **VGood:** $1895   **Good:** $895

Winchester Model 1894 Carbine
Same specs as Model 1894 except 20″ barrel. Introduced 1940; discontinued 1964.
   Exc.: $495   **VGood:** $370   **Good:** $300

Winchester Model 1894 Saddle Ring Carbine
Same specs as Model 1894 except 20″ barrel; saddle ring. Introduced 1894; discontinued 1940.
   Exc.: $1495   **VGood:** $875   **Good:** $695

Winchester Model 1894 Trapper Carbine
Same specs as Model 1894 except 14″, 15″, 16″, 17″, 18″ barrel. No longer in production.
   Exc.: $3500   **VGood:** $2500   **Good:** $1500

WINCHESTER MODEL 1895
Lever action; 30-40 Krag, 30 Govt. (03), 30 Govt. (06), 303 British, 7.62mm Russian, 35 Win., 38-72, 405 Win., 40-72; 4-, 5-shot (30-40, 303 British) box magazine; 24″, 28″ barrel; open rear sight, bead or blade front; uncheckered American walnut straight-grip stock, forend; blued finish. Introduced 1895; discontinued 1931. Collector value.
   Exc.: $3000   **VGood:** $2300   **Good:** $1400
Takedown model
   Exc.: $3300   **VGood:** $2500   **Good:** $1600

Winchester Model 1895 Carbine
Same specs as Model 1895 except 30-40 Krag, 30 Gov't (03), 30 Gov't (06), 303 British; 22″ barrel; one-barrel band; carbine stock; solid frame. Collector value.
   Exc.: $2900   **VGood:** $2000   **Good:** $1000
With U.S. Government markings
   Exc.: $4000   **VGood:** $3300   **Good:** $1800

RIFLES

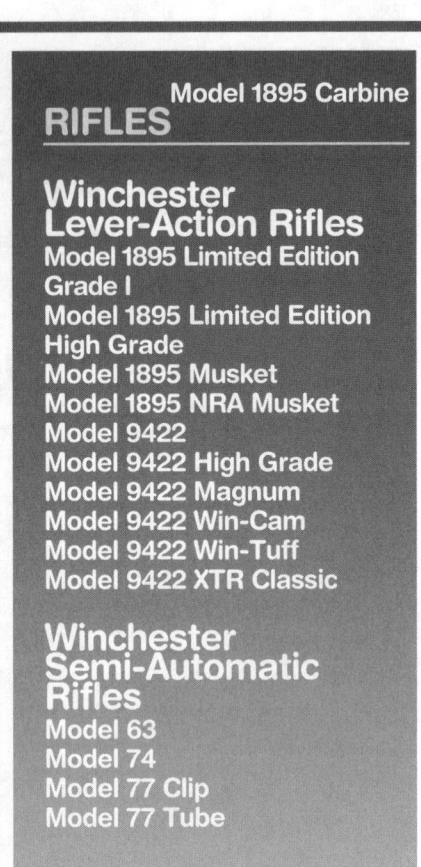

Winchester Lever-Action Rifles
Model 1895 Limited Edition Grade I
Model 1895 Limited Edition High Grade
Model 1895 Musket
Model 1895 NRA Musket
Model 9422
Model 9422 High Grade
Model 9422 Magnum
Model 9422 Win-Cam
Model 9422 Win-Tuff
Model 9422 XTR Classic

Winchester Semi-Automatic Rifles
Model 63
Model 74
Model 77 Clip
Model 77 Tube

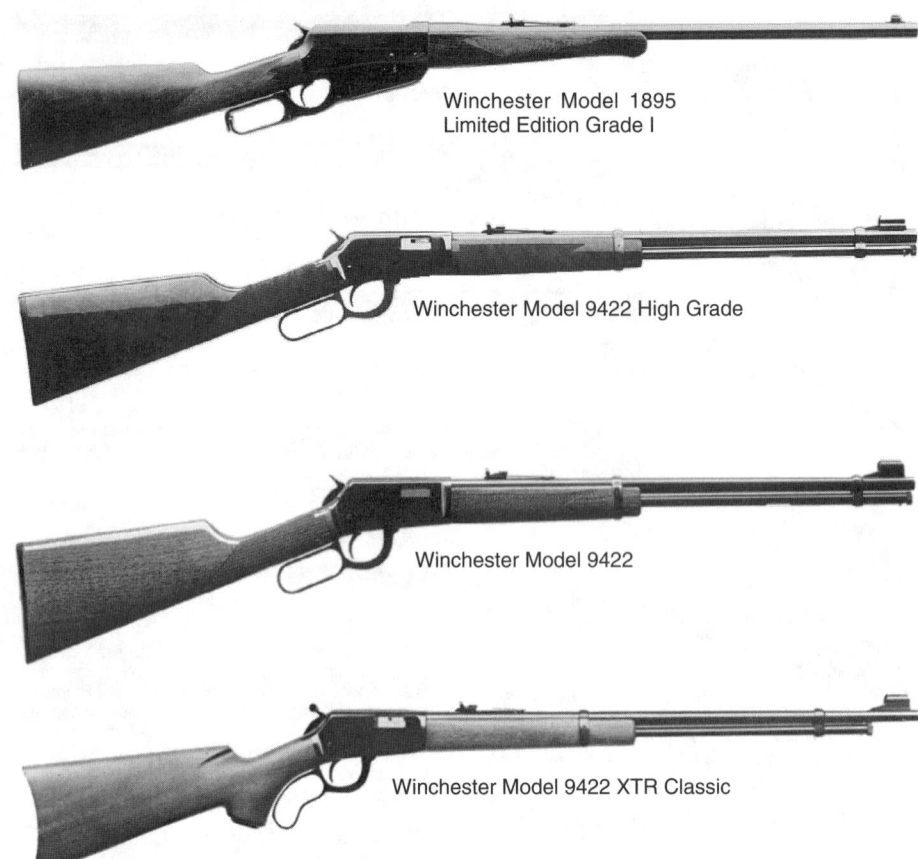

Winchester Model 1895 Limited Edition Grade I

Winchester Model 9422 High Grade

Winchester Model 9422

Winchester Model 9422 XTR Classic

Winchester Model 1895 Limited Edition Grade I
Same specs as Model 1895 except 30-06; 4-shot magazine; 24″ round barrel; 42″ overall length; weighs 8 lbs.; gold bead front sight, buckhorn rear adjustable for elevation; polished blue finish with Nimschke-style scroll engraving on receiver; scalloped receiver; two-piece cocking lever. Only 4000 rifles produced. Introduced 1995; still in production.
LMSR: $853
  Perf.: $750   **Exc.:** $640   **VGood:** $510

Winchester Model 1895 Limited Edition High Grade
Same specs as Model 1895 except 30-06; silvered receiver with extensive engraving showing two scenes portraying large bighorn sheep on right side, bull elk and cow elk on left side; gold borders accent the scenes; engraved magazine, cocking lever; classic Winchester H-style checkering pattern on fancy grade American walnut. Only 4000 rifles made. Introduced 1994; **LMSR: $1540**
  Perf.: $1295   **Exc.:** $995   **VGood:** $795

Winchester Model 1895 Musket
Same specs as Model 1895 except 30-40 Krag, 30 Gov't (03), 30 Gov't (06); 28″, 30″ barrel; two-barrel bands; handguard on barrel.
  Exc.: $1895   **VGood:** $1495   **Good:** $1195
With U.S. Government markings
  Exc.: $2195   **VGood:** $1650   **Good:** $1295

Winchester Model 1895 NRA Musket
Same specs as Model 1895 except 24″ barrel; meets NRA specs for competition.
  Exc.: $2500   **VGood:** $1900   **Good:** $1400

WINCHESTER MODEL 9422
Lever action; 22 Short, 22 Long, 22 LR; 21-shot tube magazine (22 Short), 17-shot (22 Long), 15-shot (22 LR); 20 1/2″ barrel; weighs 6 1/2 lbs.; hooded ramp bead front sight, adjustable semi-buckhorn rear; grooved for scope mounts; checkered American walnut straight-grip stock, forearm. Introduced 1972; designated XTR 1978; designation dropped 1989; still in production.
LMSR: $437
  Perf.: $300   **Exc.:** $270   **VGood:** $215

Winchester Model 9422 High Grade
Same specs as Model 9422 except 22 LR; high grade walnut with gloss finish; blued and engraved receiver with coonhound on right side, racoon profile on left, both framed with detailed Nimschke-style scrollwork. Introduced 1995; **LMSR: $475**
  Perf.: $470   **Exc.:** $395   **VGood:** $295

Winchester Model 9422 Magnum
Same specs as Model 9422 except 22 WMR; 11-shot tube magazine. Introduced 1972; still in production.
LMSR: $457
  Perf.: $300   **Exc.:** $270   **VGood:** $230

Winchester Model 9422 Win-Cam
Same specs as Model 9422 except 22 WMR; 11-shot tube magazine; checkered green laminated hardwood stock, forearm. Introduced 1987; **LMSR: $457**
  Perf.: $300   **Exc.:** $270   **VGood:** $225

Winchester Model 9422 Win-Tuff
Same specs as Model 9422 except 22 LR, 22 WMR; checkered brown laminated hardwood stock, forearm. Introduced 1988; still in production. **LMSR: $423**
  Perf.: $300   **Exc.:** $270   **VGood:** $220

Winchester Model 9422 XTR Classic
Same specs as Model 9422 except 22 1/2″ barrel; 39 1/8″ overall length; satin-finished walnut stock fluted comb; no checkering; crescent steel buttplate; curved finger lever; capped pistol grip. Introduced 1985; discontinued 1988.
  Perf.: $400   **Exc.:** $350   **VGood:** $310

WINCHESTER SEMI-AUTOMATIC RIFLES

WINCHESTER MODEL 63
Semi-automatic; 22 LR, 22 LR Super Speed; 10-shot tube magazine in buttstock; 20″ barrel (early series), 23″ (later series); takedown; open rear sight, bead front; plain pistol-grip stock forearm. Introduced 1933; discontinued 1958.
  20″ barrel
    Exc.: $1350   **VGood:** $1000   **Good:** $775
  23″ barrel
    Exc.: $700   **VGood:** $620   **Good:** $530

WINCHESTER MODEL 74
Semi automatic; 22 Short, 22 LR; 20-shot tube magazine in buttstock (22 Short), 14-shot (22 LR); 24″ barrel; takedown; open rear sight, bead front; uncheckered one-piece pistol-grip stock; square-top receiver. Introduced 1939; discontinued 1955.
  Exc.: $295   **VGood:** $190   **Good:** $160

WINCHESTER MODEL 77 CLIP
Semi automatic; 22 LR; 8-shot clip magazine; 22″ barrel; weighs 5 1/2 lbs.; #75C bead front sight, #32B open rear; plain, American walnut one-piece stock; pistol grip; beavertail forend; rotary thumb safety; black composition buttplate. Introduced 1955; discontinued 1963.
  Exc.: $230   **VGood:** $190   **Good:** $160

Winchester Model 77 Tube
Same specs as Model 77 Clip except for 15-shot tube magazine.
  Exc.: $280   **VGood:** $240   **Good:** $200

WINCHESTER MODEL 100
Semi-automatic; gas-operated; 243 Win., 284 Win., 308 Win.; 3-shot magazine (284 Win.), 4-shot (others); 22″ barrel; 42 1/2″ overall length; weighs 7 1/2 lbs.; hooded bead front sight, folding leaf rear; tapped for receiver sights, scope mounts; one-piece walnut with checkered pistol-grip stock, forearm; sling swivels. Introduced 1960; discontinued 1974.
  Pre-1964 Model
    Perf.: $525   **Exc.:** $490   **VGood:** $380
  Post-1964 Model
    Perf.: $475   **Exc.:** $450   **VGood:** $325

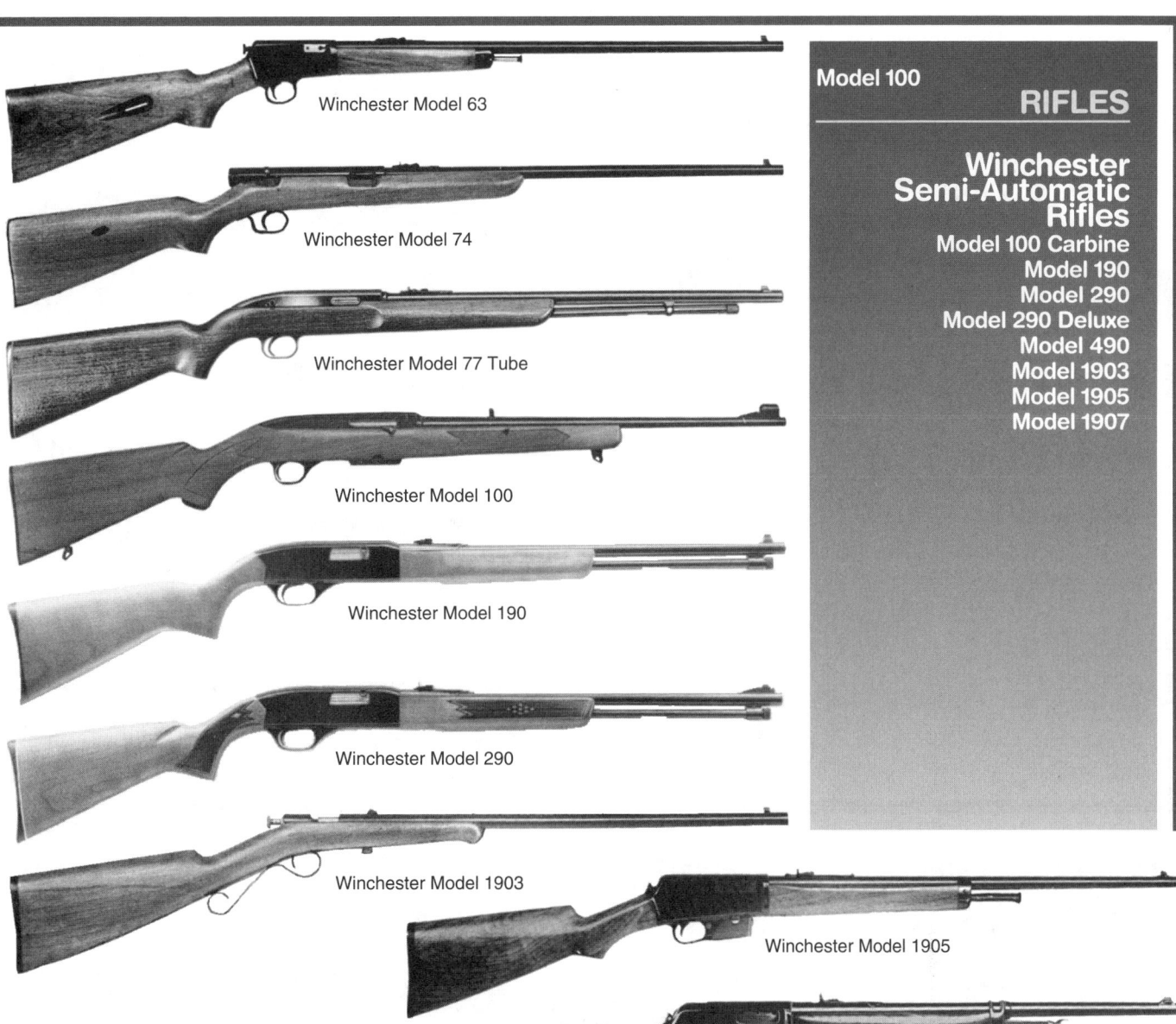

Winchester Model 63

Winchester Model 74

Winchester Model 77 Tube

Winchester Model 100

Winchester Model 190

Winchester Model 290

Winchester Model 1903

Winchester Model 1905

Winchester Model 1907

Winchester Model 1910

Model 100

RIFLES

Winchester Semi-Automatic Rifles
Model 100 Carbine
Model 190
Model 290
Model 290 Deluxe
Model 490
Model 1903
Model 1905
Model 1907

Winchester Model 100 Carbine
Same specs as Model 100 except 19" barrel; no checkering; barrel band. Introduced 1967; discontinued 1973.
Perf.: $595   **Exc.:** $530   **VGood:** $470

WINCHESTER MODEL 190
Semi-automatic; 22 Short, 22 Long, 22 LR; 21-shot tube magazine (22 Short), 17-shot (22 Long), 15-shot (22 LR); 20 1/2", 24" barrel with 1:16" twist; 39" overall length; weighs 5 lbs; bead post front sight, adjustable V rear; aluminum alloy receiver grooved for scope mounts; American hardwood stock with plain, uncapped pistol-grip, forearm encircled with barrel band; sling swivels. Introduced 1967; discontinued 1980.
Perf.: $130   **Exc.:** $110   **VGood:** $100

WINCHESTER MODEL 290
Semi-automatic; 22 Short, 22 Long, 22 LR; 21-shot tube magazine (22 Short), 17-shot (22 Long), 15-shot (22 LR); 20 1/2" barrel; 39" overall length; weighs 5 lbs.; open rear sight, ramp front; uncheckered or impress-checkered hardwood pistol-grip stock, forend. Introduced 1963; discontinued 1977.
Perf.: $120   **Exc.:** $100   **VGood:** $90

Winchester Model 290 Deluxe
Same specs as Model 290 except fancy American walnut Monte Carlo stock, forend. Introduced 1965; discontinued 1977.
Perf.: $150   **Exc.:** $120   **VGood:** $100

WINCHESTER MODEL 490
Semi-automatic; 22 LR; 10-, 15-shot clip magazine; 22" barrel; folding leaf rear sight, hooded ramp front; impress-checkered American walnut pistol-grip stock, forend. Introduced 1975; discontinued 1978.
Perf.: $275   **Exc.:** $220   **VGood:** $185

WINCHESTER MODEL 1903
Semi-automatic; 22 Win. Auto RF; 10-shot tube magazine in butt; 20" barrel; takedown; open rear sight, bead front; uncheckered straight-grip stock, forend. Introduced 1903; discontinued 1932. Collector value.
Exc.: $765   **VGood:** $515   **Good:** $345

WINCHESTER MODEL 1905
Semi-automatic; 32 Win. Self-Loading, 35 Win. Self-loading; 5-, 10-shot box magazine; 22" barrel; takedown; open rear sight, bead front; uncheckered American walnut pistol-grip stock, forend. Introduced 1905; discontinued 1920. Collector value.
Exc.: $595   **VGood:** $495   **Good:** $275

WINCHESTER MODEL 1907
Semi-automatic; 351 Win. Self-Loading; 5-, 10-shot box magazine; 20" barrel; takedown; open rear sight, bead front; uncheckered walnut pistol-grip stock, forend. Introduced 1907; discontinued 1957. Collector value.
Exc.: $495   **VGood:** $430   **Good:** $285

WINCHESTER MODEL 1910
Semi-automatic; 401 Win. Self-Loading; 4-shot box magazine; 20" barrel; takedown; open rear sight, bead front; uncheckered walnut pistol-grip stock, forend. Introduced 1910; discontinued 1936. Collector value.
Exc.: $595   **VGood:** $495   **Good:** $295

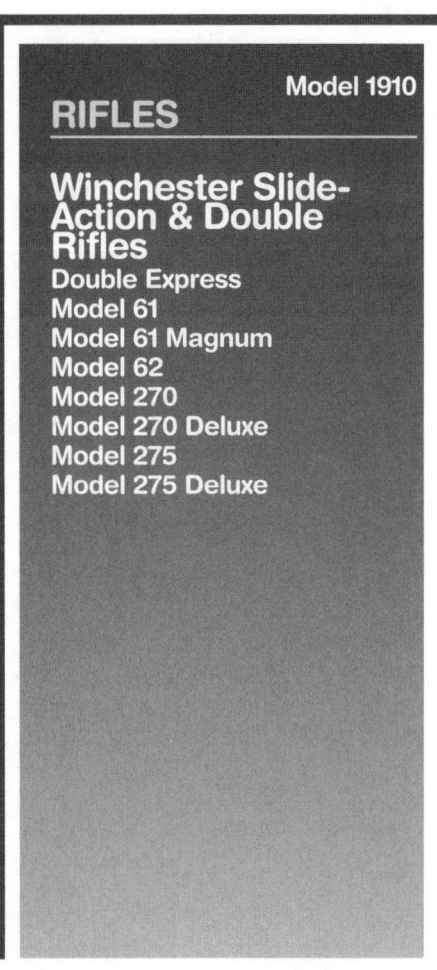

RIFLES

Model 1910

Winchester Slide-Action & Double Rifles

Double Express
Model 61
Model 61 Magnum
Model 62
Model 270
Model 270 Deluxe
Model 275
Model 275 Deluxe

Winchester Double Express

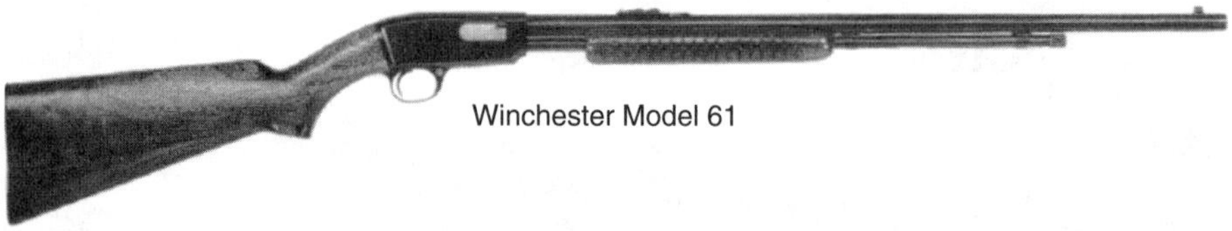

Winchester Model 61

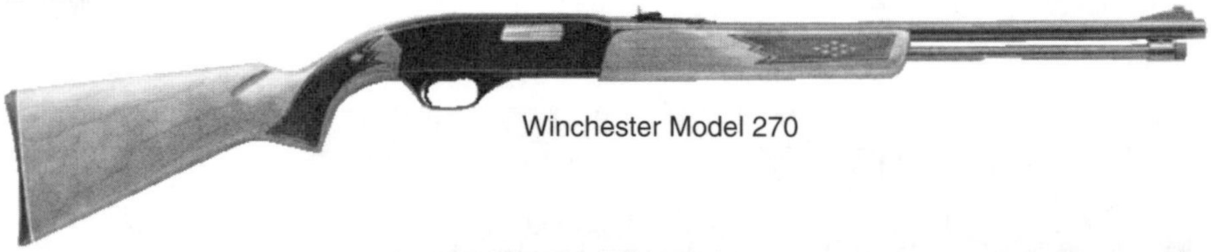

Winchester Model 270

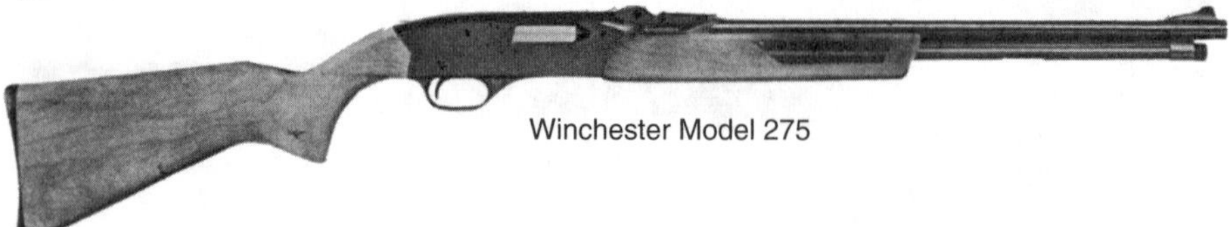

Winchester Model 275

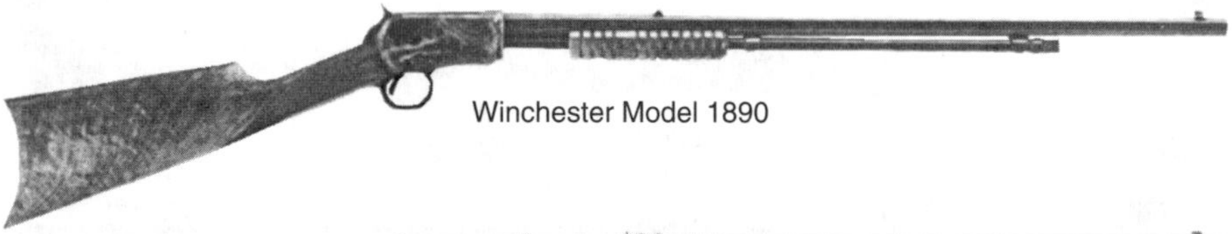

Winchester Model 1890

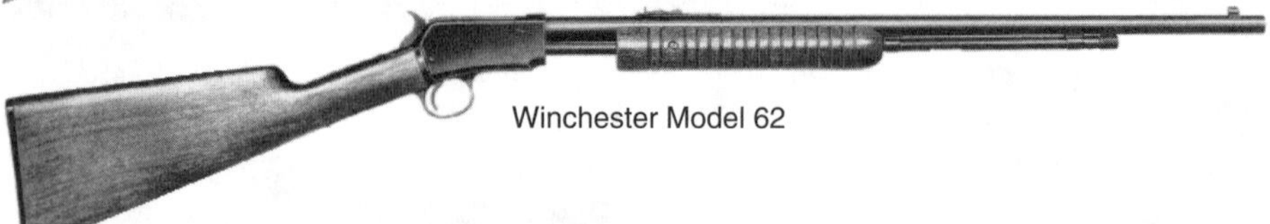

Winchester Model 62

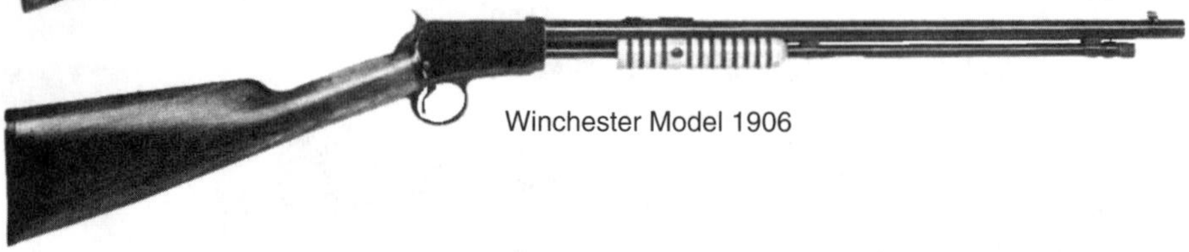

Winchester Model 1906

WINCHESTER SLIDE-ACTION & DOUBLE RIFLES

WINCHESTER DOUBLE EXPRESS

Double rifle; over/under; 30-06, 9.3x74R, 6.5mm, 270, 7.65R, 257 Roberts; 23 1/2″ barrel; 39 5/8 ″ overall length; bead on ramp front sight, folding leaf rear on quarter-rib; integral scope bases; fancy American walnut stock; hand-checkered pistol grip, forend; uses Model 101 shotgun action; silvered, engraved receiver; blued barrels; quick-detachable sling swivels; marketed in hard case. Made in Japan. Introduced 1982; discontinued 1985.

 Perf.: $1695   **Exc.:** $1495   **VGood:** $1195

WINCHESTER MODEL 61

Slide action; 22 Short, 22 Long, 22 LR; 20-shot tube magazine (22 Short), 16-shot (22 Long), 14-shot (22 LR); 24″ round or octagon barrel; hammerless; takedown; open rear sight, bead front; uncheckered pistol-grip stock; grooved semi-beavertail slide handle. Introduced 1932; discontinued 1963.

 Exc.: $650   **VGood:** $500   **Good:** $345
Octagon barrel model
 Exc.: $1095   **VGood:** $795   **Good:** $495

Winchester Model 61 Magnum

Same specs as Model 61 except 22 WMR; 12-shot tube magazine. Introduced 1960; discontinued 1963.
 Exc.: $750   **VGood:** $605   **Good:** $465

WINCHESTER MODEL 62

Slide action; 22 Short, 22 Long, 22 LR; 20-shot tube magazine (20 Short), 16-shot (22 Long), 14-shot (22 LR); 23″ barrel; visible hammer; bead front sight, open rear; plain, straight-grip stock; grooved semi-beavertail slide handle; also available in gallery model in 22 Short only. Introduced 1932; discontinued 1959.
 Exc.: $595   **VGood:** $465   **Good:** $350

WINCHESTER MODEL 270

Slide action; 22 Short, 22 Long, 22 LR; 21-shot tube magazine (22 Short), 17-shot (22 Long), 15-shot (22 LR); 20 1/2″ barrel; 39″ overall length; ramped square post front sight, adjustable square notch rear; aluminum alloy receiver grooved for tip-off scope mounts; walnut-finished hardwood or black or brown cycolac plastic pistol-grip stock; crossbolt safety. Introduced 1963; discontinued 1973.
Walnut stock
 Perf.: $140   **Exc.:** $120   **VGood:** $90
Plastic stock
 Perf.: $90   **Exc.:** $80   **VGood:** $70

Winchester Model 270 Deluxe

Same specs as Model 270 except high-gloss Monte Carlo select walnut stock, fluted comb, cheekpiece; basketweave checkering; white buttspacer. Introduced 1965; discontinued 1973.
 Perf.: $170   **Exc.:** $130   **VGood:** $95

WINCHESTER MODEL 275

Slide action; 22 WMR; 11-shot tube magazine; 20 1/2″ barrel; 39″ overall length; weighs 5 lbs.; ramped square post front sight, adjustable square notch rear; alu-

minum alloy receiver grooved for tip-off scope mounts; walnut-finished hardwood pistol-grip stock; crossbolt safety. Introduced 1964; discontinued 1971.
 Perf.: $180   **Exc.:** $140   **VGood:** $110

Winchester Model 275 Deluxe

Same specs as Model 275 except Monte Carlo select American walnut stock, fluted comb, cheekpiece; basketweave checkering; white buttspacer. Introduced 1965; discontinued 1971.
 Perf.: $200   **Exc.:** $160   **VGood:** $120

WINCHESTER MODEL 1890

Slide action; 22 Short, 22 Long, 22 LR, 22 WRF; 15-shot tube magazine (22 Short), 12-shot (22 Long), 11-shot (22 LR), 10-shot (22 WRF); 24″ octagonal barrel; open rear, bead front sight; plain, straight stock, grooved slide handle; visible hammer; originally solid-frame design; after serial No. 15,552, all were takedowns; color case-hardened (pre-1901) or blued (post-1901) finish. Introduced 1890; discontinued 1934.
Color case-hardened model
 Exc.: $4500   **VGood:** $3600   **Good:** $2100
Blued finish
 Exc.: $1295   **VGood:** $695   **Good:** $465

73

Fancy Finished Winchester Rifles.

FANCY
RIFLES

The ornamenting of Winchester rifles affords excellent opportunities for displaying both fine engraving and artistic carving. The ornamental work shown below was done at our armory by our own employes, under our own immediate supervision.

The prices named cover only the engraving of the metal parts of the gun, and the carving or checking of the wood. These prices are to be added to the list price of the rifle to ascertain the price of the completed gun. For instance: The standard Model 1886, Octagon Barrel rifle, with plain stock, is listed at $21.00; to this add for pistol grip stock and forearm of fancy walnut, $10.00; for hand carving stock and forearm, $60.00; for engraving and inlaying, $250.00, making total price of the 1886 gun, shown on this page, $341.00. Prices for special engraving, carving, etc., are subject to the same trade discounts as the gun it is put upon.

An Elaborately Ornamented Model 1886 Winchester Rifle.

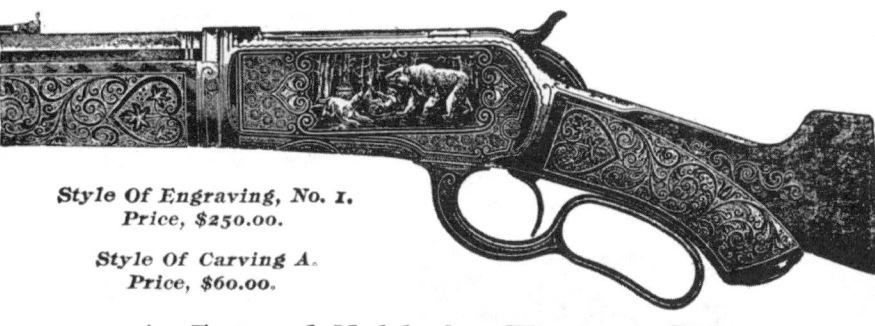

Style Of Engraving, No. 1.
Price, $250.00.

Style Of Carving A.
Price, $60.00.

An Engraved Model 1894 Winchester Rifle.

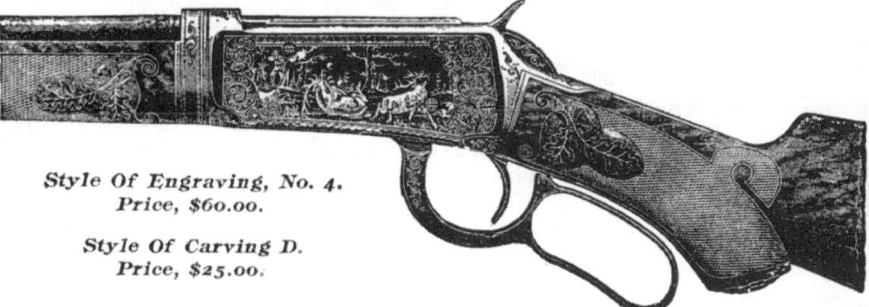

Style Of Engraving, No. 4.
Price, $60.00.

Style Of Carving D.
Price, $25.00.

An Inexpensively Engraved Model 1892 Winchester Rifle.

Style Of Engraving, No. 6.
Price, $25.00.

Style Of Carving H.
Price, $5.00.

For further description of Fancy Finished Winchester Rifles, send 10 cents for 28 page illustrated catalogue, showing many different styles of engraving, carving, etc.

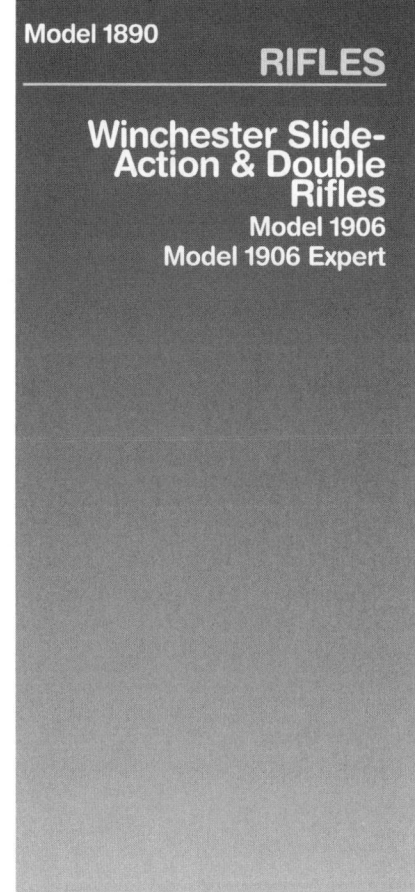

Model 1890

RIFLES

Winchester Slide-Action & Double Rifles
Model 1906
Model 1906 Expert

WINCHESTER MODEL 1906
Slide action; 22 Short, 22 Long, 22 LR; 15-shot tube magazine (22 Short), 12-shot (22 Long), 11-shot (22 LR); takedown; open rear sight, bead front; straight-grip uncheckered American walnut stock, grooved forend; visible hammer; shotgun buttplate. Introduced 1906; discontinued 1932. Collector value.
  Exc.: $995   **VGood:** $695   **Good:** $345

Winchester Model 1906 Expert
Same specs as Model 1906 except pistol-grip stock, redesigned forend; blue/nickel or nickel finish. Introduced 1918; discontinued 1924.
  Exc.: $1695   **VGood:** $950   **Good:** $620

WINCHESTER SUPER GRADE COMBO
Over/under; 30-06/12-ga.; 25″ Winchoke shot barrel; 41 1/4″ overall length; bead front sight, folding leaf rear; fancy American walnut stock; hand-checkered pistol grip, forend; full-length top barrel rib; silvered, engraved receiver; blued barrels; integral scope bases; single selective mechanical trigger. Made in Japan. Introduced 1982; discontinued 1985.
  Perf.: $1695 **Exc.:** $1395 **VGood:** $995

Winchester Super Grade Combo

MODERN GUN VALUES, 10TH EDITION **423**

RIFLES

Winslow
Bolt-Action Sporter

A. Zoli
Combinato
Model AZ-1900 Classic
Model AZ-1900M

WINSLOW BOLT-ACTION SPORTER

Bolt action; all popular standard and magnum centerfire calibers; 3-, 4-shot magazine; 24" barrel, 26" (magnums); two stock styles: slim pistol grip with beavertail forearm or hooked pistol grip with wide flat forearm; Monte Carlo stock with cheekpiece in walnut, maple or myrtle; rosewood pistol grip, forend tip; custom rifle in eight grades. Introduced 1963; discontinued 1978.

Commander Grade
Perf.: $495   **Exc.:** $430   **VGood:** $350

Regal Grade
Perf.: $595   **Exc.:** $530   **VGood:** $450

Regent Grade
Perf.: $750   **Exc.:** $650   **VGood:** $550

Regimental Grade
Perf.: $950   **Exc.:** $825   **VGood:** $695

Crown Grade
Perf.: $1395   **Exc.:** $1195   **VGood:** $930

Royal Grade
Perf.: $1595   **Exc.:** $1295   **VGood:** $1100

Imperial Grade
Perf.: $3495   **Exc.:** $2895   **VGood:** $2000

Emperor Grade
Perf.: $6100   **Exc.:** $4695   **VGood:** $4000

A. ZOLI COMBINATO

Combo; over/under; boxlock; 222, 243/12.ga.; 24" barrels; blade front sight, flip-up rear; checkered European walnut stock; double triggers. Introduced 1980; discontinued 1993.

Perf.: $1695   **Exc.:** $1295   **VGood:** $995

ZOLI MODEL AZ-1900 CLASSIC

Bolt action; 243, 6.5x55, 270, 308, 30-06, 7mm Rem. Mag, 300 Win. Mag.; 21" barrel (24" on 7mm Rem. Mag., 300 Win. Mag.); 41 3/4" overall length with 21" barrel; weighs 7 1/4 lbs.; open sights supplied with gun but not mounted; drilled, tapped for scope mounts; checkered Turkish circassian walnut stock; polished blue finish; oil-finished stock; engine-turned bolt. Imported from Italy by European American Armory. Introduced 1989; discontinued 1993.

Perf.: $800   **Exc.:** $670   **VGood:** $500

Zoli Model AZ-1900M

Same specs as Model AZ-1900 except Bell & Carlson composite stock. Imported from Italy by European American Armory. Imported only in 1991.

Perf.: $695   **Exc.:** $495   **VGood:** $425

ZOLI SAVANA DOUBLE RIFLE

Double rifle; 7x65R, 30-06, 9.3x74R; 25 1/2" barrels; gold bead front sight, fixed V-notch rear in quarter-rib; premium grade French walnut stock with full pistol grip, cheekpiece; Anson & Deeley boxlock action with choice of single or double triggers; bushed firing pins; cocking indicators; silvered, engraved frame. Imported from Italy by European American Armory. Introduced 1989; discontinued 1991.

Perf.: $5995   **Exc.:** $4895   **VGood:** $3295

Winslow Bolt-Action Sporter

A. Zoli AZ-1900 Classic

SHOTGUNS

Directory of Manufacturers

SHOTGUNS

American Arms
Bristol
Brittany
Derby
Excelsior
FS200 Competition
FS300 Competition
FS400 Competition
FS500 Competition
Gentry
Grulla #2
Lince
Royal
Shogun

American Arms Bristol

American Arms Brittany

American Arms Derby

American Arms Gentry

American Arms Grulla #2

AMERICAN ARMS BRISTOL
Over/under; boxlock; 12-, 20-ga.; 3" chambers; 24", 26", 28" barrels; standard choke combos or choke tubes; front bead sight; glossy hand-checkered walnut stock; ejectors; single set trigger; engraved dummy sideplates; silver-finished frame; gold trigger. Made in Spain. Introduced 1987; discontinued 1988.
Fixed chokes
   Perf.: $625   **Exc.:** $500   **VGood:** $425
Choke tubes
   Perf.: $685   **Exc.:** $550   **VGood:** $470

AMERICAN ARMS BRITTANY
Side-by-side; boxlock ejectors; 12-, 20-ga; 3" chambers; 27" barrel (12-ga.), 25" barrel (20-ga.); weighs 6 1/2 lbs.; Imp. Cyl., Mod., Full choke tubes; straight English-style hand-checkered European walnut stock; semi-beavertail forend; engraving; single selective trigger; automatic safety; rubber recoil pad; case-color finish. Made in Spain. Introduced 1989; still in production.
LMSR: $849
   Perf.: $615   **Exc.:** $490   **VGood:** $425

AMERICAN ARMS DERBY
Side-by-side; sidelock; 12-, 20-, 28-ga., 410; 3" chambers, 2 3/4" (28-ga.); 26", 28" barrel; weighs 6 1/2 to 6 3/4 lbs.; standard chokes; chromed bores; metal bead front sight; hand-checkered European walnut straight-grip stock, splinter forend; single selective trigger; hand rubbed oil finish; chromed receiver. Made in Spain. Introduced 1987; no longer imported.
   Perf.: $800   **Exc.:** $635   **VGood:** $550

AMERICAN ARMS EXCELSIOR
Over/under; sidelock; 12-, 20-ga.; 3" chambers; 26", 28" barrels; standard choke combos or choke tubes; metal middle bead sight, glow worm front; hand-checkered European walnut stock, forend; auto selective ejectors; single non-selective trigger; manual safety; raised relief hand-engraved, gold-plated hunting scenes on sideplates. Made in Spain. Introduced 1987; discontinued 1987.

Fixed chokes
   Perf.: $1625   **Exc.:** $1300   **VGood:** $1175
Choke tubes
   Perf.: $1685   **Exc.:** $1350   **VGood:** $1150

AMERICAN ARMS FS200 COMPETITION
Over/under; boxlock; 12-ga.; 2 3/4" chambers; 26", 32" barrels; Skeet/Skeet, Improved/Full; weighs 7 3/4 lbs.; hand-checkered, oil-finished European walnut stock; palm swell; ventilated recoil pad; selective auto ejectors; single selective trigger; black or satin chrome-finished frame. Made in Spain. Introduced 1987; discontinued 1987.
   Perf.: $685   **Exc.:** $550   **VGood:** $470

American Arms FS300 Competition
Same specs as FS200 Competition except 26", 30", 32" barrels; engraved false sideplates; satin chrome-finished receiver. Introduced 1986; discontinued 1986.
   Perf.: $785   **Exc.:** $625   **VGood:** $530

American Arms FS400 Competition
Same specs as FS200 Competition except sidelock; 26", 30", 32" barrels; single trigger; engraved chrome-finished receiver. Introduced 1986; discontinued 1986.
   Perf.: $1125   **Exc.:** $900   **VGood:** $765

American Arms FS500 Competition
Same specs as FS200 Competition except sidelock; 26", 30", 32" barrels; single trigger; engraved chrome-finished receiver. Introduced 1985; discontinued 1985.
   Perf.: $1125   **Exc.:** $895   **VGood:** $775

AMERICAN ARMS GENTRY
Side-by-side; boxlock; 12-, 20-, 28-ga., 410; 3" chambers, 2 3/4" (28-ga.); 26", 28" barrel; weighs about 6 1/4 to 6 3/4 lbs.; standard chokes; chromed bores; metal bead front sight; hand-checkered European walnut stock with semi-gloss finish; English-type scroll engraving; extractors; floating firing pin; silver finish on receiver. Introduced 1987; still in production.
LMSR: $725
   Perf.: $475   **Exc.:** $375   **VGood:** $325

AMERICAN ARMS GRULLA #2
Side-by-side; sidelock; 12-, 20-, 28-ga., 410; 26", 28" barrel; weighs about 6 lbs.; standard choke combos; select European walnut straight English stock; splinter forend; hand-rubbed oil finish; checkered grip, forend, butt; double triggers; detachable locks; automatic selective ejectors; cocking indicators; gas escape valves; English-style concave rib; color case-hardened receiver; scroll engraving. Made in Spain. Introduced 1989; still imported. **LMSR: Special Order**
   Perf.: $2575   **Exc.:** $2050   **VGood:** $1750

AMERICAN ARMS LINCE
Over/under; boxlock ejector; 12-, 20-ga.; 3" chambers; weighs 6 3/4 lbs.; metal bead front sight; standard choke combinations; checkered European walnut stock; single selective trigger; manual safety; rubber recoil pad; chrome-lined barrels; scroll engraving; blue or silver finish. Introduced 1986; discontinued 1986.
   Perf.: $495   **Exc.:** $390   **VGood:** $350
410
   Perf.: $535   **Exc.:** $430   **VGood:** $390

AMERICAN ARMS ROYAL
Over/under; sidelock; 12-, 20-ga.; 3" chambers; 26", 28" barrels; weighs 7 1/4 lbs.; standard choke combos or choke tubes; metal middle bead sight, glow worm front; hand-checkered European walnut stock, forend; auto selective ejectors; single non-selective trigger; manual safety; silvered sideplates; English-style hand-engraved scrollwork. Made in Spain. Introduced 1987; discontinued 1987.
Fixed chokes
   Perf.: $1275   **Exc.:** $1000   **VGood:** $925
Choke tubes
   Perf.: $1350   **Exc.:** $1030   **VGood:** $950

AMERICAN ARMS SHOGUN
Side-by-side; boxlock ejector; 10-ga.; 3 1/2" chambers; double triggers; English-style hand-engraved scrollwork; chrome-finished receiver. Introduced 1986; discontinued 1986.
   Perf.: $450   **Exc.:** $335   **VGood:** $295

American Arms Silver I

American Arms Silver II

American Arms Silver WS/OU 12

American Arms Silver Sporting

American Arms WS/SS 10

SHOTGUNS

American Arms
Silver I
Silver II
Silver II Upland Light
Silver Sporting
Silver Sporting Skeet
Silver Sporting Trap
Silver WS/OU 12
Silver WT/OU 10
TS/OU 12
TS/SS 12 Double
TS/SS 10 Double
WS/SS 10
York

AMERICAN ARMS SILVER I

Over/under; boxlock; 12-, 20-, 28-ga., 410; 2 3/4" chamber (28-ga.), 3" chamber (others); 26", 28" barrels; standard choke combinations; weighs 6 3/4 lbs.; checkered European walnut stock with cast-off; single selective trigger; extractors; manual safety; rubber recoil pad; chrome-lined barrels; scroll engraving; metal bead front sight; silver finish. Made in Europe. Introduced 1987; still imported. **LMSR: $599**
   Perf.: $450   **Exc.:** $365   **VGood:** $325

American Arms Silver II

Same specs as Silver I except 26", 28" barrels in 12-ga., 26" (others); choke tubes. Also available in two-barrel set in 28/410. Introduced 1987; still in production. **LMSR: $699**
   Perf.: $595   **Exc.:** $485   **VGood:** $395

American Arms Silver II Upland Light

Same specs as Silver II except 12-, 20-ga.; 26" barrels; Franchoke tubes; weighs 6 1/4 lbs. (12-ga.), 5 3/4 lbs. (20-ga.); ejectors; ventilated rib; engraved frame with antique silver finish. Made in Spain. Introduced 1994; still in production. **LMSR: $899**
   Perf.: $800   **Exc.:** $625   **VGood:** $535

AMERICAN ARMS SILVER SPORTING

Over/under; boxlock; 12-, 20-ga.; 2 3/4" (12-ga.), 3" (20-ga.) chambers; 28" (20-ga.), 30" barrels; weighs 7 3/8 lbs.; wide vent rib; Franchoke tubes; figured walnut stock with cut checkering; elongated forcing cones; ported barrels; pistol-grip stock with palm swell; radiused recoil pad; tapered target rib; target bead sights; mechanical single selective trigger; nickel finish. Introduced 1990; still in production. **LMSR: $899**
   Perf.: $675   **Exc.:** $525   **VGood:** $475

American Arms Silver Sporting Skeet

Same specs as Silver Sporting except 26" ported barrels with elongated forcing cones; weighs 7 3/8 lbs.; stock dimensions of 14 3/8" x 1 3/8" x 2 3/8"; target-type ventilated rib with two bead sights; Skeet, Skeet, Imp. Cyl., Mod. choke tubes. Introduced 1992; no longer in production.
   Perf.: $725   **Exc.:** $560   **VGood:** $475

American Arms Silver Sporting Trap

Same specs as Silver Sporting except 30" ported barrels with elongated forcing cones; weighs 7 3/4 lbs.; stock dimensions of 14 3/8" x 1 1/2" x 1 5/8"; Mod., Mod., Full, Full choke tubes; target-type ventilated rib with two sight beads. Introduced 1992; no longer produced.
   Perf.: $725   **Exc.:** $560   **VGood:** $475

AMERICAN ARMS SILVER WS/OU 12

Over/under; boxlock ejector; 12-ga.; 3 1/2" chambers; 28" barrels; 46" overall length; weighs 7 lbs., 2 oz.; cut-checkered European walnut stock; black vented recoil pad; single selective trigger; chromed bores; matte metal finish. Made in Italy. Introduced 1988; still in production. **LMSR: $765**
   Perf.: $675   **Exc.:** $365   **VGood:** $300

AMERICAN ARMS SILVER WT/OU 10

Over/under; boxlock; 10-ga.; 3 1/2" chambers; 26" barrels; Full/Full choke tubes; weighs 9 3/8 lbs.; extractors; chromed bores; non-reflective wood stock; dull metal finish; recoil pad; single trigger; ventilated rib; top tang safety. Introduced 1988; still in production. **LMSR: $1029**
   Perf.: $775   **Exc.:** $750   **VGood:** $650

AMERICAN ARMS TS/OU 12

Over/under; boxlock ejector; 12-ga.; 3 1/2" chambers; 24" barrels; 46" overall length; weighs 6 lbs., 14 oz.; cut-checkered European walnut stock; black vented recoil pad; single selective trigger; chromed bores; matte metal finish. Made in Italy. Introduced 1988; still in production. **LMSR: $765**
   Perf.: $685   **Exc.:** $545   **VGood:** $475

AMERICAN ARMS TS/SS 12 DOUBLE

Side-by-side; boxlock; 12-ga.; 3 1/2" chambers; 26" barrels; choke tubes; extractors; single selective trigger; camouflage sling, sling swivels; recoil pad; chromed bores; raised matted rib; wood and metal matte finish. Made in Spain. Introduced 1988; still in production. **LMSR: $750**
   Perf.: $625   **Exc.:** $435   **VGood:** $375

American Arms TS/SS 10 Double

Same specs as TS/SS 12 except 10-ga.; weighs about 11 lbs.; double triggers; AAI choke tubes Full/Full. Made in Spain. Introduced 1988; still in production.
   Perf.: $625   **Exc.:** $435   **VGood:** $375

AMERICAN ARMS WS/SS 10

Side-by-side; boxlock; 10-ga.; 3 1/2" chambers; 32" barrels choked Full & Full; weighs 11 1/4 lbs.; stock dimensions of 14 5/16" x 1 3/8" x 2 3/8"; flat rib; hand-checkered European walnut stock with beavertail forend; full pistol grip; dull finish; rubber recoil pad; double triggers; extractors; camouflaged sling; sling swivels. All metal with Parkerized finish. Introduced 1987; no longer produced.
   Perf.: $550   **Exc.:** $425   **VGood:** $385

AMERICAN ARMS YORK

Side-by-side; boxlock; 12-, 20-, 28-ga., 410; 3" chambers; 25", 28" barrels; standard choke combos; weighs 7 1/4 lbs.; gold bead front sight; gloss-finished, hand-checkered European walnut stock; beavertail forend; pistol grip; double triggers; extractors; manual safety; independent floating firing pins. Made in Spain. Introduced 1987; discontinued 1988.
   Perf.: $475   **Exc.:** $375   **VGood:** $320
Optional single selective trigger
   Perf.: $500   **Exc.:** $400   **VGood:** $345

SHOTGUNS

American Arms/Franchi
Black Magic 48/AL
(See Franchi Black Magic 48/AL)
Falconet 2000
(See Franchi Falconet 2000)
Sporting 2000
(See Franchi Sporting 2000)

Arizaga
Model 31

Armalite
AR-17 Golden Gun

Armscor
Model 30D
Model 30D/IC
Model 30DG
Model 30FS
Model 30K
Model 30P
Model 30R
Model 30RP

Armsport
Model 1032 Goose Gun
Model 1040
Model 1041
Model 1042
Model 1043
Model 1050 Series

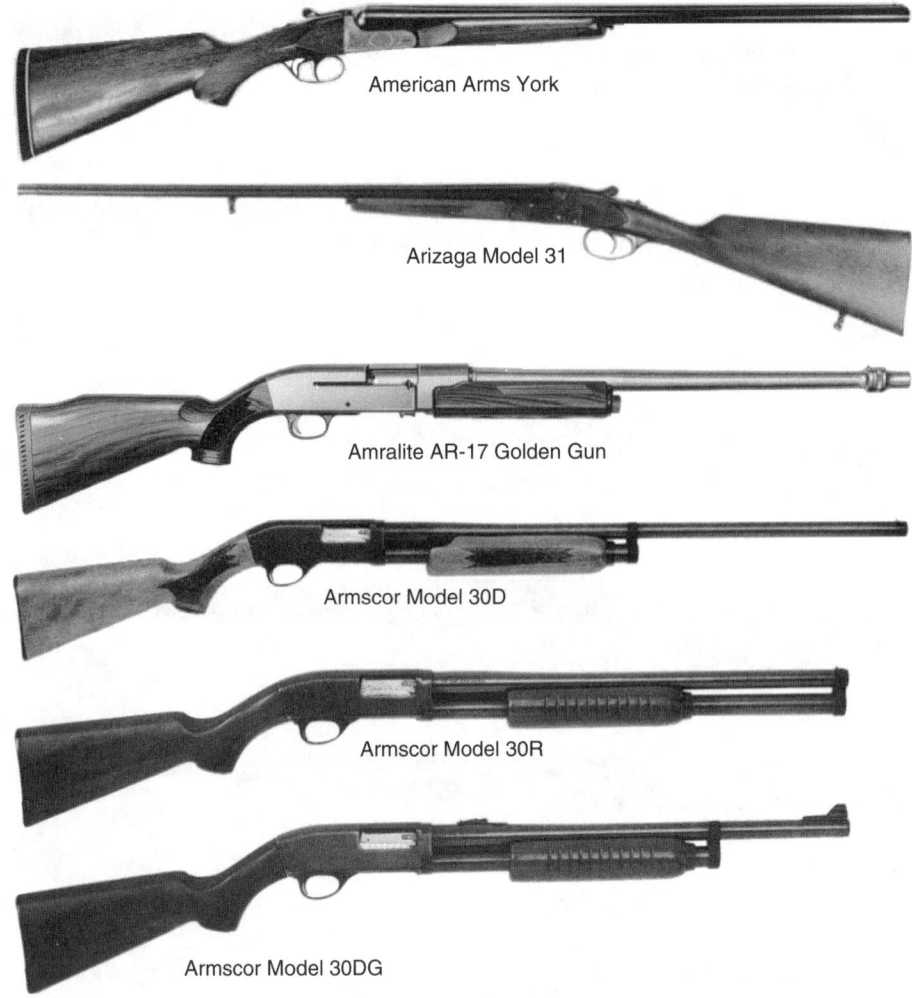

American Arms York

Arizaga Model 31

Amralite AR-17 Golden Gun

Armscor Model 30D

Armscor Model 30R

Armscor Model 30DG

ARIZAGA MODEL 31

Side-by-side; boxlock; 12-, 16-. 20-, 28-ga., 410; 26″, 28″ barrels; standard choke combos; 45″ overall length; weighs 6 5/8 lbs; European walnut; English-style straight or pistol-grip stock; double triggers; engraved receiver; blued finish. Made in Spain. Introduced 1986; still in production. **LMSR: $550**
  Perf.: $450   **Exc.:** $365   **VGood:** $300

ARMALITE AR-17 GOLDEN GUN

Semi-auto; 12-ga.; 2-shot; 28″ barrel; recoil operated; interchangeable choke tubes for Improved, Modified, Full chokes; polycarbonate stock, forearm; barrel, receiver housing of aluminum alloy; recoil pad; gold-anodized finish; also with black anodized finish. Introduced 1964; discontinued 1965. Only 2000 made.
  Exc.: $435   **VGood:** $375   **Good:** $300

ARMSCOR MODEL 30D

Slide-action; 12-ga.; 5-shot magazine; 26″ Modified, 30″ Full barrel; 47″ overall length; weighs 7 1/4 lbs.; metal bead front sight; Philippine plain mahogany stock; checkered double slidebars; blued finish. Made in the Philippines. Introduced 1990; discontinued 1991.
  Perf.: $195   **Exc.:** $160   **VGood:** $135

Armscor Model 30D/IC

Same specs as Model 30D except interchangeable choke tubes; checkered stock. No longer imported.
  Perf.: $220   **Exc.:** $185   **VGood:** $160

Armscor Model 30DG

Same specs as Model 30D except 20″ plain barrel; rifle sights; 6-, 8-shot; plain pistol grip and grooved forend. Introduced 1987; discontinued 1991.
  Perf.: $185   **Exc.:** $150   **VGood:** $130

Armscor Model 30FS

Same specs as Model 30D except black folding stock, pistol grip. Introduced 1990; discontinued 1991.
  Perf.: $240   **Exc.:** $215   **VGood:** $190

Armscor Model 30K

Same specs as Model 30D except 7-shot magazine; 21″ barrel; olive green butt, forend. Introduced 1990; discontinued 1991.
  Perf.: $215   **Exc.:** $190   **VGood:** $165

Armscor Model 30P

Same specs as Model 30D except 11 1/2″ barrel; 3-, 5-shot; "bolo-type" short stock; 22 1/2″ overall length; weighs 5 1/2 lbs. For law enforcement. No longer imported.
  Perf.: $250   **Exc.:** $225   **VGood:** $200

Armscor Model 30R

Same specs as Model 30D except 6-, 8-shot; 20″ Cylinder-bore barrel; 39″ overall length; weighs 6 3/4 lbs.; bead front sight; smooth pistol-grip stock and grooved forend. No longer imported.
  Perf.: $175   **Exc.:** $150   **VGood:** $125

Armscor Model 30RP

Same specs as Model 30D except 18″ barrel; 27″ overall length; weighs 6 1/8 lbs.; black wood butt, forend. For law enforcement. Introduced 1987; discontinued 1991.
  Perf.: $195   **Exc.:** $170   **VGood:** $145

ARMSPORT MODEL 1032 GOOSE GUN

Side-by-side; 10-ga.; 3 1/2″ chambers: 32″ barrels; weighs 11 lbs.; solid matted rib; engraved; checkered pistol-grip European walnut stock, forend; double triggers; vent rubber recoil pad with white spacer. Made in Spain. Introduced 1979; no longer imported.
  Perf.: $360   **Exc.:** $300   **VGood:** $265

ARMSPORT MODEL 1040

Side-by-side; 12-ga.; 26″ barrel; Imp./Mod.; oil-finished walnut stock and forend; hand-engraved receiver. No longer imported.
  Perf.: $370   **Exc.:** $325   **VGood:** $295

ARMSPORT MODEL 1041

Side-by-side; 12-ga.; 28″ barrel; Mod./Full.; oil-finished walnut stock and forend; hand-engraved receiver. No longer imported.
  Perf.: $370   **Exc.:** $325   **VGood:** $295

ARMSPORT MODEL 1042

Side-by-side; 20-ga.; 26″ barrel; Imp./Mod.; oil-finished walnut stock and forend; hand-engraved receiver. No longer imported.
  Perf.: $370   **Exc.:** $325   **VGood:** $295

ARMSPORT MODEL 1043

Side-by-side; 20-ga.; 28″ barrel; Mod./Full.; oil-finished walnut stock and forend; hand-engraved receiver. No longer imported.
  Perf.: $370   **Exc.:** $325   **VGood:** $295

ARMSPORT MODEL 1050 SERIES

Side-by-side; boxlock; 12, 20, 28, 410; 3″ chambers; 26″, 28″ barrels; weighs 6 3/4 lbs.; chrome-lined barrels; European walnut stock; double triggers; extractors; silvered, engraved receiver. Made in Italy. Introduced 1986; still in production. **LMSR: $785**
Model 1050 (12-ga.)
  Perf.: $595   **Exc.:** $475   **VGood:** $395
Model 1053 (20-ga.)
  Perf.: $600   **Exc.:** $480   **VGood:** $400
Model 1054 (410)
  Perf.: $625   **Exc.:** $500   **VGood:** $425
Model 1055 (28-ga.)
  Perf.: $625   **Exc.:** $500   **VGood:** $425

Amsport Model 2700 Goose Gun

Armsport Model 1050 Series

Armsport Model 2500 Series

Armsport Model 2700 Series

SHOTGUNS

Armsport
Model 1212 Western
Model 1626
Model 1628
Model 1726
Model 1728
Model 1810
Model 2526
Model 2528
Model 2626
Model 2628
Model 2697
Model 2698
Model 2699
Model 2700 Goose Gun
Model 2700 Series
Model 2701, 2702
Model 2703, 2704
Model 2705
Model 2706 Commander
Model 2707
Model 2708 Slug Gun
Model 2716, 2717
Model 2718, 2719

ARMSPORT MODEL 1212 WESTERN
Side-by-side; 12-ga.; 3″ chambers; 20″ barrels; weighs 6 ½ lbs.; checkered pistol-grip stock of European walnut, beavertail forend; metal front bead on matted solid rib; exposed hammers. Made in Spain. Introduced 1979; no longer imported.
Perf.: $350   **Exc.:** $275   **VGood:** $240

ARMSPORT MODEL 1626
Over/under; boxlock; 12-ga.; 3″ chambers; 26″ barrels; Improved/Modified; non-ejector; single selective trigger; vent rib; hand-checkered European walnut stock, forend; recoil pad; engraved. Made in Italy. No longer imported.
Perf.: $560   **Exc.:** $360   **VGood:** $300

ARMSPORT MODEL 1628
Over/under; boxlock; 12-ga.; 3″ chambers; 28″ barrels; Modified/Full; non-ejector; single selective trigger; vent rib; hand-checkered European walnut stock, forend; recoil pad; engraved. Made in Italy. No longer imported.
Perf.: $560   **Exc.:** $360   **VGood:** $300

ARMSPORT MODEL 1726
Over/under; boxlock; 20-ga.; 3″ chambers; 26″ barrels; Improved/Modified; non-ejector; single selective trigger; vent rib; hand-checkered European walnut stock, forend; recoil pad; engraved. Made in Italy. No longer imported.
Perf.: $560   **Exc.:** $360   **VGood:** $300

ARMSPORT MODEL 1728
Over/under; boxlock; 20-ga.; 3″ chambers; 28″ barrels; Modified/Full; non-ejector; single selective trigger; vent rib; hand-checkered European walnut stock, forend; recoil pad; engraved. Made in Italy. No longer imported.
Perf.: $570   **Exc.:** $370   **VGood:** $300

ARMSPORT MODEL 1810
Over/under; 410-bore; 3″ chambers; 26″ barrels; Full/Full chokes; blued receiver and barrels. No longer imported.
Perf.: $575   **Exc.:** $375   **VGood:** $300

ARMSPORT MODEL 2526
Over/under; 12-ga.; 26″ barrels; Imp./Mod.; vent rib; hand-checkered European walnut pistol-grip stock; single selective trigger; auto ejectors; engraved receiver. Manufactured in Europe. Introduced 1979; no longer imported.
Perf.: $585   **Exc.:** $375   **VGood:** $300

ARMSPORT MODEL 2528
Over/under; 12-ga.; 28″ barrels; Modified/Full; vent rib; hand-checkered European walnut pistol-grip stock; single selective trigger; auto ejectors; engraved receiver. Manufactured in Europe. Introduced 1979; no longer imported.
Perf.: $585   **Exc.:** $375   **VGood:** $300

ARMSPORT MODEL 2626
Over/under; boxlock; 20-ga.; 26″ barrels; Imp./Mod.; automatic ejectors; vent rib; single selective trigger; walnut stock and forend; recoil pad; engraved. No longer imported.
Perf.: $585   **Exc.:** $375   **VGood:** $300

ARMSPORT MODEL 2628
Over/under; boxlock; 20-ga.; 28″ barrels; Mod./Full; automatic ejectors; vent rib; single selective trigger; walnut stock and forend; recoil pad; engraved. No longer imported.
Perf.: $585   **Exc.:** $375   **VGood:** $300

ARMSPORT MODEL 2697
Over/under; boxlock; 10-ga. Magnum; 3 ½″ chambers; 28″ chrome-lined barrels; choke tubes; engraved silver finished receiver; wide vent rib; hand-checkered walnut forend and stock. No longer imported.
Perf.: $925   **Exc.:** $700   **VGood:** $600

ARMSPORT MODEL 2698
Over/under; boxlock; 10-ga. Magnum; 3 ½″ chambers; 32″ chrome-lined barrels; choke tubes; engraved silver finished receiver; wide vent rib; hand-checkered walnut forend and stock. No longer imported.
Perf.: $925   **Exc.:** $700   **VGood:** $600

ARMSPORT MODEL 2699
Over/under; boxlock; 10-ga. Magnum; 3 ½″ chambers; 28″ chrome-lined barrels; Full/Improved-Modified chokes; engraved silver finished receiver; wide vent rib; hand-checkered walnut forend and stock. No longer imported.
Perf.: $995   **Exc.:** $605   **VGood:** $565

ARMSPORT MODEL 2700 GOOSE GUN
Over/under; 10-ga; 3 ½″ chambers; 28″ Full/Imp. Mod., 32″ Full/Full barrels; weighs 9 ½ lbs.; Boss-type action; European walnut stock; double triggers; extractors. Made in Italy. Introduced 1986; still in production. **LMSR:** $1190
Perf.: $875   **Exc.:** $605   **VGood:** $565

ARMSPORT MODEL 2700 SERIES
Over/under; boxlock; 10-ga. Mag.; 3 ½″ chambers; 32″ chrome-lined barrels; Full/Full chokes; engraved silver finished receiver; extractors; wide vent rib; hand-checkered walnut forend and stock. No longer imported.
Perf.: $875   **Exc.:** $605   **VGood:** $565

Armsport Model 2701, 2702
Same specs as 2700 except 12-ga.; 3″ Mag. chambers; 28″ barrels; Modified/Full. No longer imported by Armsport.
Perf.: $450   **Exc.:** $350   **VGood:** $290

Armsport Model 2703, 2704
Same specs as 2700 except 20-ga.; 3″ Mag. chambers; Improved/Modified chokes. No longer imported by Armsport.
Perf.: $475   **Exc.:** $350   **VGood:** $290

Armsport Model 2705
Same specs as 2700 except 410; 26″ barrels; Modified/Full chokes. Made in Italy. No longer imported.
Perf.: $625   **Exc.:** $425   **VGood:** $355

Armsport Model 2706 Commander
Same specs as 2700 except 12-ga.; 20″ barrels; double triggers; extractors. No longer imported by Armsport.
Perf.: $300   **Exc.:** $240   **VGood:** $225

Armsport Model 2707
Same specs as 2700 except 28-ga.; Improved/Modified; double trigger. No longer imported by Armsport.
Perf.: $625   **Exc.:** $425   **VGood:** $355

Armsport Model 2708 Slug Gun
Same specs as 2700 except 12-ga.; 20″ or 23″ barrels; double or single trigger. No longer imported by Armsport.
Perf.: $575   **Exc.:** $450   **VGood:** $390

Armsport Model 2716, 2717
Same specs as 2700 except 12-ga.; 3″ Mag. chambers; 28″ barrels; Modified/Full chokes; single selective trigger; extractors. No longer imported.
Perf.: $600   **Exc.:** $475   **VGood:** $390

Armsport Model 2718, 2719
Same specs as 2700 except 20-ga.; 3″ Mag. chambers; 26″ barrels; Improved/Modified chokes; single selective trigger; extractors. No longer imported.
Perf.: $600   **Exc.:** $475   **VGood:** $390

SHOTGUNS

Armsport

Model 2720
Model 2725
Model 2730
Model 2731
Model 2733
Model 2734
Model 2735
Model 2736
Model 2741
Model 2742 Sporting Clays
Model 2744 Sporting Clays
Model 2746
Model 2747
Model 2750 Sporting Clays
Model 2751 Sporting Clays
Model 2751
Model 2755
Model 2900 Tri-Barrel
Model 2901 Tri-Barrel
Model 3030 Deluxe Grade Trap
Model 3032 Deluxe Grade Trap
Model 3101 Deluxe Mono Trap
Model 3102 Deluxe Mono Trap Set
Model 3103 Deluxe Mono Trap
Model 3104 Deluxe Mono Trap
Model 4000 Emperor Over/Under
Model 4010 Emperor Side-By-Side
Model 4031

Armsport Model 2720
Same specs as 2700 except 410; 26" barrels; Modified/Full chokes; single selective trigger; extractors. No longer imported.
Perf.: $640   **Exc.:** $515   **VGood:** $430

Armsport Model 2725
Same specs as 2700 except 28-ga.; 26" barrels; Improved/Modified choke tubes. No longer imported.
Perf.: $640   **Exc.:** $515   **VGood:** $430

Armsport Model 2730
Same specs as 2700 except 12-ga.; 28" barrels; Skeet set with choke tubes; single trigger; ejectors. No longer imported.
Perf.: $740   **Exc.:** $525   **VGood:** $475

Armsport Model 2731
Same specs as 2700 except 20-ga.; 26" barrels; Skeet set with choke tubes; single trigger; ejectors. No longer imported.
Perf.: $740   **Exc.:** $525   **VGood:** $475

Armsport Model 2733
Same specs as 2700 except 12-ga.; 28" barrels; Modified/Full; "Boss-type" action; extractors; single selective trigger. No longer imported.
Perf.: $685   **Exc.:** $460   **VGood:** $370

Armsport Model 2734
Same specs as 2700 except 12-ga.; 28" barrels; choke tubes; extractors; single selective trigger. No longer imported.
Perf.: $720   **Exc.:** $505   **VGood:** $455

Armsport Model 2735
Same specs as 2700 except 20-ga.; 26" barrels; Improved/Modified; "Boss-type" action; extractors; single selective trigger. No longer imported.
Perf.: $685   **Exc.:** $460   **VGood:** $370

Armsport Model 2736
Same specs as 2700 except 20-ga.; 26" barrels; choke tubes; "Boss-type" action; extractors; single selective trigger. No longer imported.
Perf.: $600   **Exc.:** $475   **VGood:** $425

Armsport Model 2741
Same specs as 2700 except 12-ga.; 28" barrels; Modified/Full; "Boss-type" action; ejectors; single selective trigger. No longer imported.
Perf.: $640   **Exc.:** $440   **VGood:** $375

Armsport Model 2742 Sporting Clays
Same specs as 2700 except 12-ga.; 28" barrels; choke tubes; single selective trigger. No longer imported.
Perf.: $600   **Exc.:** $450   **VGood:** $390

Armsport Model 2744 Sporting Clays
Same specs as 2700 except 20-ga.; 26" barrels; choke tubes; ejectors; single selective trigger. No longer imported.
Perf.: $625   **Exc.:** $450   **VGood:** $390

Armsport Model 2746
Same specs as 2700 except 12-ga. Mag.; 3½" chambers; 28" barrels; choke tubes. No longer imported.
Perf.: $695   **Exc.:** $425   **VGood:** $390

Armsport Model 2747
Same specs as 2700 except 12-ga. Mag.; 3½" chambers; 32" barrels; choke tubes. No longer imported.
Perf.: $695   **Exc.:** $425   **VGood:** $390

Armsport Model 2750 Sporting Clays
Same specs as 2700 except 12-ga.; 28" barrels; choke tubes; single selective trigger; ejectors; equipped with decorative sideplates. No longer imported.
Perf.: $725   **Exc.:** $575   **VGood:** $525

Armsport Model 2751 Sporting Clays
Same specs as 2700 except 20-ga.; 26" barrels; choke tubes; single selective trigger; ejectors; equipped with decorative sideplates. No longer imported.
Perf.: $925   **Exc.:** $575   **VGood:** $525

ARMSPORT MODEL 2751
Semi-automatic; 12-ga.; 3" chamber; 28" Modified, 30" Full (2751A) choke barrels; weighs 7 lbs.; choke tube version available; European walnut stock; rubber recoil pad; blued or silvered receiver with engraving. Made in Italy. Introduced 1986; no longer imported.
Perf.: $400   **Exc.:** $320   **VGood:** $275
With choke tubes
Perf.: $420   **Exc.:** $340   **VGood:** $285
Silvered receiver
Perf.: $420   **Exc.:** $340   **VGood:** $285

ARMSPORT MODEL 2755
Pump-action; 12-ga.; 3" chamber; 28" Modified, 30" Full (2755A) barrel; weighs 7 lbs.; choke tubes available; vent rib; European walnut stock; rubber recoil pad; blued finish. Made in Italy. Introduced 1986; no longer imported.
Perf.: $310   **Exc.:** $220   **VGood:** $165
With choke tubes
Perf.: $330   **Exc.:** $240   **VGood:** $185

ARMSPORT MODEL 2900 TRI-BARREL
Over/under; 12-ga.; 3" chambers; 28" barrels; weighs 7¾ lbs.; choked Improved/Modified/Full; standard choke combos; European walnut stock; double triggers; top-tang barrel selector; silvered, engraved frame. Made in Italy. Introduced 1986; no longer imported.
Perf.: $2675   **Exc.:** $2150   **VGood:** $1700

Armsport MODEL 2901 Tri-Barrel
Same specs as 2900 Tri-Barrel except 12-ga.; 3" Mag. chambers; choked Modified/Full/Full. No longer imported.
Perf.: $2675   **Exc.:** $2150   **VGood:** $1700

ARMSPORT MODEL 3030 DELUXE GRADE TRAP
Over/under; 12-ga.; 32" barrels; wide ventilated rib; ventilated lateral rib; single selective trigger; automatic ejectors; fluorescent bead front sight; checkered walnut stock; beavertail forend; Greener-type cross-bolt; engraved sideplates. No longer imported.
Perf.: $625   **Exc.:** $475   **VGood:** $375

ARMSPORT MODEL 3032 DELUXE GRADE TRAP
Over/under; 12-ga.; 30" barrels; wide ventilated rib; ventilated lateral rib; single selective trigger; automatic ejectors; fluorescent bead front sight; checkered walnut stock; beavertail forend; Greener-type cross-bolt; engraved sideplates. No longer imported.
Perf.: $645   **Exc.:** $495   **VGood:** $385

ARMSPORT MODEL 3101 DELUXE MONO TRAP
Single or over/under; 12-ga.; 30" (o/u), 32" (single) barrels; single selective trigger; automatic ejectors; highly grained walnut stock, forend with high lustre finish; choice of mono trap or as a set combo set *should be 30% higher with over/under barrels. No longer imported.
Perf.: $650   **Exc.:** $500   **VGood:** $400

ARMSPORT MODEL 3102 DELUXE MONO TRAP SET
Single or over/under; 12-ga.; 32" (o/u), 34" (single) barrels; single selective trigger; automatic ejectors; highly grained walnut stock, forend with high lustre finish; choice of mono trap or as a set * combo set should be 30% higher with over/under barrels. No longer imported.
Perf.: $1100   **Exc.:** $895   **VGood:** $700

ARMSPORT MODEL 3103 DELUXE MONO TRAP
Single shot; 12-ga.; 32" barrels; single selective trigger; automatic ejectors; highly grained walnut stock, forend with high lustre finish; choice of mono trap or as a set * combo set should be 30% higher with over/under barrels. No longer imported.
Perf.: $700   **Exc.:** $575   **VGood:** $400

ARMSPORT MODEL 3104 DELUXE MONO TRAP
Single shot; 12-ga.; 34" barrels; single selective trigger; automatic ejectors; highly grained walnut stock, forend with high lustre finish; choice of mono trap or as a set* combo set should be 30% higher with over/under barrels. No longer imported.
Perf.: $700   **Exc.:** $575   **VGood:** $400

ARMSPORT MODEL 4000 EMPEROR OVER/UNDER
Over/under; 12-, 16-, 20-ga.; any barrel length; any choke; handcrafted barrels, action, root walnut stock, forend; hand engraved receiver, sideplate; custom made to order; luggage-type case. No longer imported.
Perf.: $1200   **Exc.:** $975   **VGood:** $725

ARMSPORT MODEL 4010 EMPEROR SIDE-BY-SIDE
Side-by-side; sidelock; *If sidelock prices should be 12-16-, 20-ga.; any barrel length; any choke; handcrafted barrels, action, root walnut stock, forend; engraved receiver, sideplate; luggage-type case; custom made. No longer imported.
Perf.: $1750   **Exc.:** $1425   **VGood:** $1200

ARMSPORT MODEL 4031
Side-by-side; sidelock; 12-, 20-ga. 26", 28" barrels; Improved/Modified or Modified/Full chokes; English-style straight walnut stock; hand detachable locks; silver finish hand-engraved receiver; highly polished steel barrels. Made in Italy. No longer imported.
Perf.: $3800   **Exc.:** $2500   **VGood:** $1950

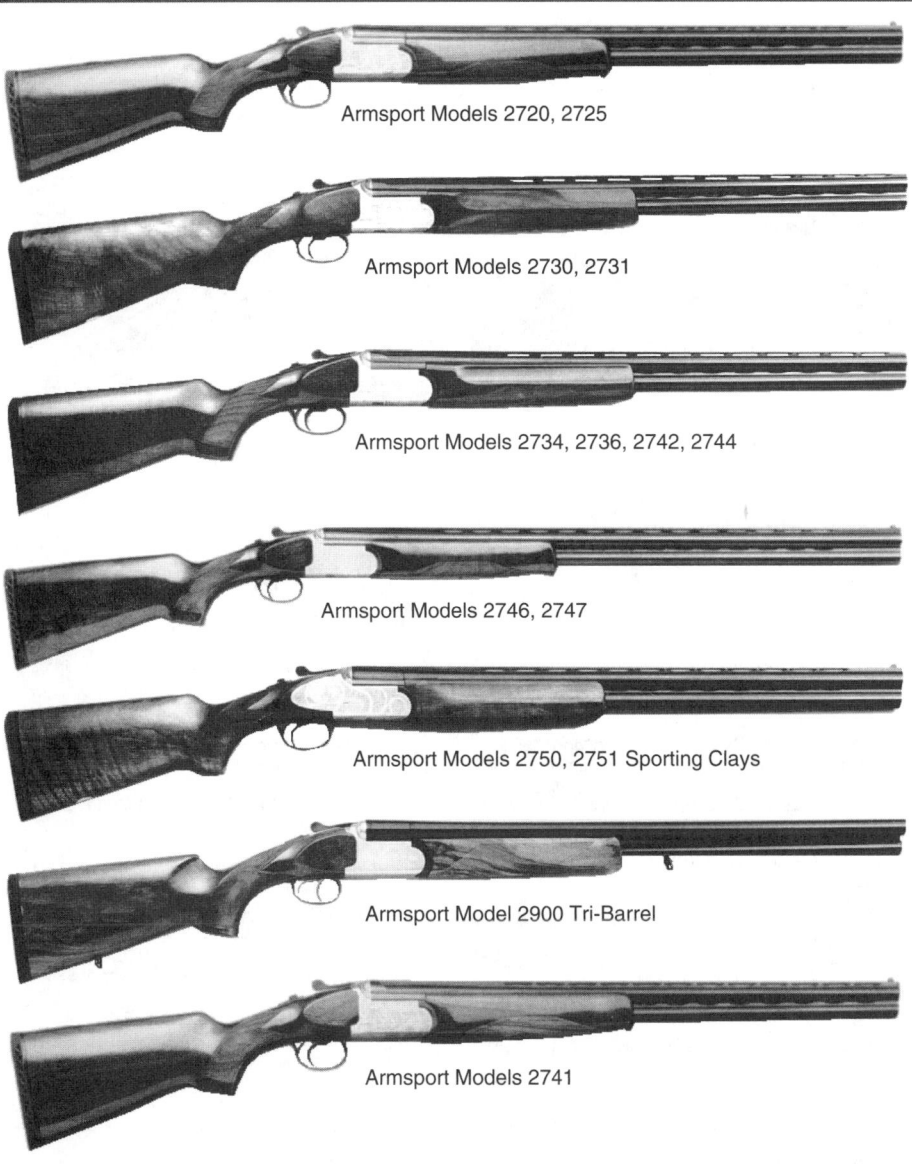

Armsport Models 2720, 2725

Armsport Models 2730, 2731

Armsport Models 2734, 2736, 2742, 2744

Armsport Models 2746, 2747

Armsport Models 2750, 2751 Sporting Clays

Armsport Model 2900 Tri-Barrel

Armsport Models 2741

SHOTGUNS

Armsport
Model 4032 Mono Trap
Model 4033 Mono Trap
Model 4034 Mono Trap Set
Model 4034 O/U Trap Set
Model 4035 Mono Trap Set
Model 4035 O/U Trap Set
Model 4040 Premier Slug
Model 4041 Premier Slug
Model 4046 Mono Trap
Model 4047 Mono Trap
Model 4048 Mono Trap Set

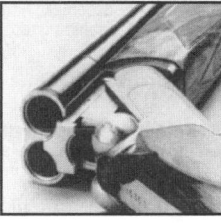

ARMSPORT MODEL 4032 MONO TRAP
Single shot; 12-ga.; 32″ barrel; vent rib; single trigger; walnut stock with pistol grip; recoil pad; automatic ejector; premier grade; hand engraved. No longer imported.

    Perf.: $1750   **Exc.:** $1300   **VGood:** $1100

ARMSPORT MODEL 4033 MONO TRAP
Single shot; 12-ga.; 34″ barrel; vent rib; single trigger; walnut stock with pistol grip; recoil pad; automatic ejector; premier grade; hand engraved. No longer imported.

    Perf.: $1750   **Exc.:** $1300   **VGood:** $1100

ARMSPORT MODEL 4034 MONO TRAP SET
Single shot; 12-ga.; 32″ barrel; vent rib; single trigger; walnut stock with pistol grip; recoil pad; automatic ejector; premier grade; hand engraved. Additional set of 30″ O/U barrels. No longer imported.

    Perf.: $2500   **Exc.:** $1900   **VGood:** $1000

ARMSPORT MODEL 4034 O/U TRAP SET
Single shot; 12-ga.; 30″ barrels; vent rib; single trigger; walnut stock with pistol grip; recoil pad; automatic ejector; premier grade; hand engraved. Additional set of 30″ O/U barrels. No longer imported.

    Perf.: $2500   **Exc.:** $1900   **VGood:** $1650

ARMSPORT MODEL 4035 MONO TRAP SET
Single shot; 12-ga.; 34″ barrel; vent rib; single trigger; walnut stock with pistol grip; recoil pad; ejector; premier grade; hand engraved. Additional set of 32″ O/U barrels. No longer imported.

    Perf.: $2500   **Exc.:** $1900   **VGood:** $1650

ARMSPORT MODEL 4035 O/U TRAP SET
Single shot; 12-ga.; 32″ barrels; vent rib; single trigger; walnut stock with pistol grip; recoil pad; automatic ejectors; premier grade; hand engraved. Additional set of 32″ O/U barrels. No longer imported.

    Perf.: $2500   **Exc.:** $1900   **VGood:** $1650

ARMSPORT MODEL 4040 PREMIER SLUG
Side-by-side; 12-ga.; 23″ barrels; Cyl./Imp. Cyl.; rib with leaf rear sight, blade front; hand-checkered walnut stock with half pistol grip and beavertail forend; hand relief engraved receiver. No longer imported.

    Perf.: $1250   **Exc.:** $800   **VGood:** $700

ARMSPORT MODEL 4041 PREMIER SLUG
Side-by-side; 12-ga.; 23″, 28″ barrels; Cyl./Imp. Cyl. (23″), Mod./Full (28″); rib with leaf rear sight, blade front; hand-checkered walnut stock with half pistol grip and beavertail forend; hand relief engraved receiver. No longer imported.

    Perf.: $1250   **Exc.:** $800   **VGood:** $700

ARMSPORT MODEL 4046 MONO TRAP
Single shot; 12-ga.; 34″ barrel; specially reinforced receiver carved from a solid block of gun steel; unique coil spring trigger assembly; handcrafted Italian leather fitted case. No longer imported.

    Perf.: $2750   **Exc.:** $2150   **VGood:** $1800

ARMSPORT MODEL 4047 MONO TRAP
Single shot; 12-ga.; 32″ barrel; specially reinforced receiver carved from a solid block of gun steel; unique coil spring trigger assembly; handcrafted Italian leather fitted case. No longer imported.

    Perf.: $2750   **Exc.:** $2150   **VGood:** $1800

ARMSPORT MODEL 4048 MONO TRAP SET
Single shot; 12-ga.; 32″ (over/under), 34″ (mono) barrel; specially reinforced receiver carved from a solid block of gun steel; unique coil spring trigger assembly; comes with extra trigger assembly; handcrafted Italian leather fitted case. No longer imported.

    Perf.: $1100   **Exc.:** $875   **VGood:** $700

Armsport
Model 4049 Mono Trap Set
Single

Arrieta
Sidelock Double Shotguns

Astra
Model 650
Model 750

AyA
Augusta
Bolero
Coral
Cosmos
Iberia
Iberia/E
Iberia II
Matador
Matador II

AyA Cosmos

AyA Coral Grade A

AyA Matador II

AyA Model 37 Super Grade A

AyA Iberia II

ARMSPORT MODEL 4049 MONO TRAP SET
Single shot; 12-ga.; 30" (over/under), 32" (mono) barrel; specially reinforced receiver carved from a solid block of gun steel; unique coil spring trigger assembly; comes with extra trigger assembly; handcrafted Italian leather fitted case. No longer imported.
Perf.: $1100 **Exc.:** $875 **VGood:** $700

ARMSPORT SINGLE
Single-shot; 12-, 20-ga.; 3" chamber; 26", 28" barrel; weighs 6 1/2 lbs.; oil-finished hardwood stock; chrome-lined barrel; cocking indicator opening lever behind trigger guard; manual safety. Made in Europe. Introduced 1987; still imported. **LMSR: $100**
Perf.: $75 **Exc.:** $50 **VGood:** $40
Model 1125 (12-ga., Modified)
Perf.: $75 **Exc.:** $50 **VGood:** $40
Model 1126 (12-ga., Full)
Perf.: $75 **Exc.:** $50 **VGood:** $40
Model 1127 (20-ga., Modified)
Perf.: $75 **Exc.:** $50 **VGood:** $40

ARRIETA SIDELOCK DOUBLE SHOTGUNS
Side-by-side; sidelock; 12-, 16-, 20-, 28-ga., 410; barrel length and chokes to customer specs.; select European walnut stock with oil finish, straight English grip, checkered butt (standard), or pistol-grip; essentially a custom gun with myriad options; Holland & Holland-pattern hand-detachable sidelocks, selective automatic ejectors, double triggers (hinged front) standard. Some have self-opening action. Finish and engraving to customer specs. Imported from Spain by Wingshooting Adventures.
Model 557, detachable engraved sidelocks, auto ejectors
Perf.: $2650 **Exc.:** $1650 **VGood:** $1300
Model 570, non-detachable sidelocks, auto ejectors
Perf.: $2900 **Exc.:** $1800 **VGood:** $1400
Model 578, scrollwork engraved, auto ejectors
Perf.: $3450 **Exc.:** $1925 **VGood:** $1700
Model 600 Imperial, lightly engraved, self-opening
Perf.: $4450 **Exc.:** $3250 **VGood:** $3100
Model 601 Imperial Tiro, nickel-plated action, self-opening

Perf.: $6600 **Exc.:** $4250 **VGood:** $3500
Model 801, detachable sidelocks, fine engraving, self-opening
Perf.: $7200 **Exc.:** $6500 **VGood:** $5500
Model 802, non-detachable sidelocks, finely engraved
Perf.: $7100 **Exc.:** $6300 **VGood:** $5300
Model 803, detachable sidelocks, finely engraved, self-opening
Perf.: $5000 **Exc.:** $3750 **VGood:** $3300
Model 871, auto ejectors, scroll engraved, rounded frame action
Perf.: $4000 **Exc.:** $2675 **VGood:** $2200
Model 872, self-opening, rounded frame action, finely engraved
Perf.: $8500 **Exc.:** $7700 **VGood:** $6000
Model 873, self-opening, automatic ejectors, engraved with game scene
Perf.: $6150 **Exc.:** $4200 **VGood:** $3950
Model 874, self-opening, gold engraved
Perf.: $7500 **Exc.:** $4500 **VGood:** $3700
Model 875, self-opening, custom-made only
Perf.: $11,500 **Exc.:** $9000 **VGood:** $7775

ASTRA MODEL 650
Over/under; 12-ga; 28", 30" barrels; standard chokes; vent rib; hand-checkered European walnut stock; double triggers; ejectors or extractors; scroll-engraved receiver. Was imported from Spain. Introduced 1980; discontinued 1987.
Perf.: $450 **Exc.:** $375 **VGood:** $300
Extractors
Perf.: $410 **Exc.:** $335 **VGood:** $260
Trap and Skeet models
Perf.: $460 **Exc.:** $375 **VGood:** $300

ASTRA MODEL 750
Over/under; 12-ga.; 28", 30" barrels; vent rib; hand-checkered European walnut stock; single selective trigger; ejectors or extractors; scroll-engraved receiver. Introduced 1980; discontinued 1987.
Perf.: $440 **Exc.:** $365 **VGood:** $290
Extractors only
Perf.: $410 **Exc.:** $335 **VGood:** $275

AYA AUGUSTA
Over/under; sidelock; 12-ga.; vent rib; single or double trigger; high-grade scroll engraving; fancy walnut; Kersten cross bolts. A deluxe gun modeled after the Merkel. May still be imported by Armes de Chasse. **LMSR: $28,000**
Perf.: $20,000 **Exc.:** $5600 **VGood:** $4500

AYA BOLERO
Side-by-side; boxlock; extractors; 12-, 16-, 20-, 28-ga.,

410; single non-selective trigger. Imported as Model 400 by Firearms International. Introduced 1955; discontinued 1963.
Perf.: $340 **Exc.:** $275 **VGood:** $225

AYA CORAL
Over/under; boxlock; 12-, 16-ga.; 26", 28", 30" barrels; 2 3/4" chambers; any choke desired; Kersten crossbolt; ejectors; pistol-grip stock; single or double trigger; two grades of engraving. No longer imported.
Perf.: $1300 **Exc.:** $1000 **VGood:** $800

AYA COSMOS
Single barrel; single shot; 12 Mag., 12-, 16-, 20-, 24-, 28-ga., 410; ejector; hammer; top lever; pistol-grip walnut stock; buttpad; sling swivels; interchangeable barrels. No longer imported.
Perf.: $345 **Exc.:** $295 **VGood:** $250

AYA IBERIA
Side-by-side; boxlock; 12-ga.; 2 3/4" chambers; 28" barrels; straight, half or full pistol-grip stock; hand checkered walnut; minimal engraving; extractors; double trigger. No longer imported.
Perf.: $525 **Exc.:** $375 **VGood:** $325

AyA Iberia/E
Same specs as Iberia except ejectors. No longer imported.
Perf.: $575 **Exc.:** $425 **VGood:** $375

AyA Iberia II
Same specs as Iberia except 12-, 16-ga.; no engraving; earlier version had straight receiver-to-wood joint; newer Iberia has a single scallop. No longer imported.
Perf.: $500 **Exc.:** $350 **VGood:** $300

AYA MATADOR
Side-by-side; boxlock; ejectors; 10-, 12-, 16-, 20-, 28-ga., 410; 2 3/4", 3" (20-ga.), 3 1/2" (10-ga.) chambers; 26", 28", 30" barrels; any standard choke combo; European walnut stock; checkered pistol grip; beavertail forearm; single-selective trigger. Imported as Model 400E by Firearms International. Introduced 1955; discontinued 1963.
Exc.: $350 **VGood:** $300 **Good:** $200

AyA Matador II
Same specs as Matador except 12-, 20-ga.; better wood. Manufactured 1964 to 1969.
Exc.: $340 **VGood:** $310 **Good:** $250

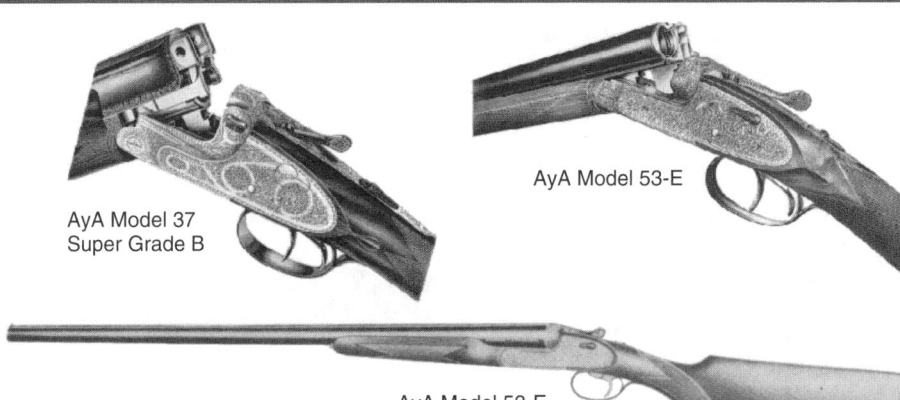

AyA Model 37
Super Grade B

AyA Model 53-E

AyA Model 53-E

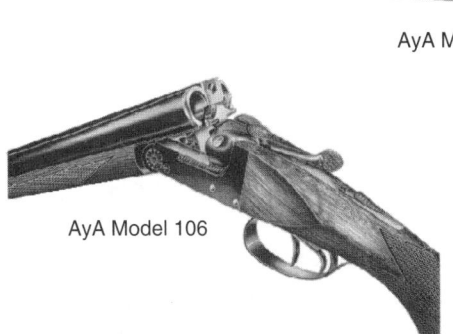

AyA Model 106

AyA Model 56

AyA Model 107

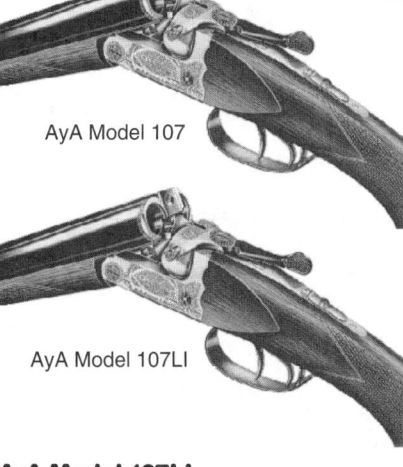

AyA Model 107LI

SHOTGUNS

AyA
Model 4 Deluxe
Model 37 Super
Model 53-E
Model 56
Model 106
Model 107
Model 107LI
Model 110
Model 112
Model 116
Model 117

AYA MODEL 4 DELUXE

Side-by-side; boxlock; 12-, 16-, 20-, 28-ga., 410; 26″, 27″, 28″ barrels, depending upon gauge; weighs 5 to 7 lbs.; select European walnut stock and forend; Anson & Deeley system with double locking lugs; chopper lump barrels; bushed firing pins; automatic safety and ejectors; articulated front trigger. Still imported. **LMSR: $3000**
 Perf.: $2700 **Exc.:** $1200 **VGood:** $800

AYA MODEL 37 SUPER

Over/under; sidelock; 12-, 16-, 20-ga.; 26″, 26″, 30″ barrels; any standard choke combo; vent rib; single selective or double trigger(s); automatic ejectors; hand-checkered stock, forearm; heavily engraved pistol-grip or straight stock. Introduced 1963; no longer in production.
 Perf.: $2000 **Exc.:** $1600 **VGood:** $1350

AYA MODEL 53-E

Side-by-side; sidelock ejector; 12-, 16-, 20-ga.; 2 ³/₄″ chambers, 3″ Mag. (available in 20-ga.); 26″, 28″, 30″ barrels; any choke; finely hand-engraved with full coverage; hand-detachable locks; side clips; third locking bite; gas vents; single or double trigger; straight grip; half or full pistol grip on request; stock made to measure on request; plain or vent rib. Still imported by Armes de Chasse. **LMSR: $4660**
 Perf.: $3700 **Exc.:** $1800 **VGood:** $1200

AYA MODEL 56

Side-by-side; sidelock ejectors; triple-bolting; 12-, 16-, 20-ga.; barrel length, chokes to customers specs; European walnut stock to customer's specs; semi-custom manufacture; matted rib, vent rib available; automatic safety; cocking indicators; gas escape valves; folding front trigger; highly engraved receiver. Introduced 1972. 12-, 20-ga. still imported by Armes de Chasse. **LMSR: $7560**
 Perf.: $6500 **Exc.:** $2500 **VGood:** $2250

AYA MODEL 106

Side-by-side; boxlock; 12-, 16-, 20-ga.; 2 ³/₄″ chambers; 28″ barrels; half or full pistol-grip with or without cheekpiece; almost no engraving; Greener-type cross bolt. No longer imported.
 Perf.: $505 **Exc.:** $410 **VGood:** $350

AYA MODEL 107

Side-by-side; boxlock; 12-, 16-, 20-ga.; 2 ³/₄″ chambers; 26″, 28″, 30″ barrels; Modified/Full; cross bolt; brightly-finished receiver or color-case hardened; extractors; double trigger; checkered wood. No longer imported.
 Perf.: $550 **Exc.:** $440 **VGood:** $390

AyA Model 107LI

Same specs as Model 107 except 12-, 16-ga.; Half/Full choke; side clips; scalloped receiver; modest engraving. No longer imported.
 Perf.: $650 **Exc.:** $540 **VGood:** $490

AYA MODEL 110

Side-by-side; hammer gun; sidelock; 12-, 16-ga.; 2 3/4″ chambers; 28″, 30″ barrels; cross bolt and double underlugs; hand-checkered walnut; extractors. No longer imported.
 Perf.: $725 **Exc.:** $525 **VGood:** $440

AYA MODEL 112

Side-by-side; sidelock; 12-, 16-, 20-ga.; 2 ³/₄″ chambers; 28″, 30″ barrels; hand-checkered walnut stock and forend; sideclips; cross bolt; extractors; modest engraving; double triggers; straight, semi- or full pistol grip. No longer imported.
 Perf.: $690 **Exc.:** $475 **VGood:** $400

AYA MODEL 116

Side-by-side; sidelock; 12-, 16-, 20-ga.; 2 ³/₄″ chambers; 27″, 28″, 30″ barrels; hand-detachable locks; double trigger; third bite; hand-checkered walnut stock, forend; extractors. No longer imported.
 Perf.: $925 **Exc.:** $750 **VGood:** $695

AYA MODEL 117

Side-by-side; sidelock ejector; 12-, 16-, 20-ga.; 2 ³/₄″ chambers; 26″, 27″, 28″, 30″ barrels; standard choke

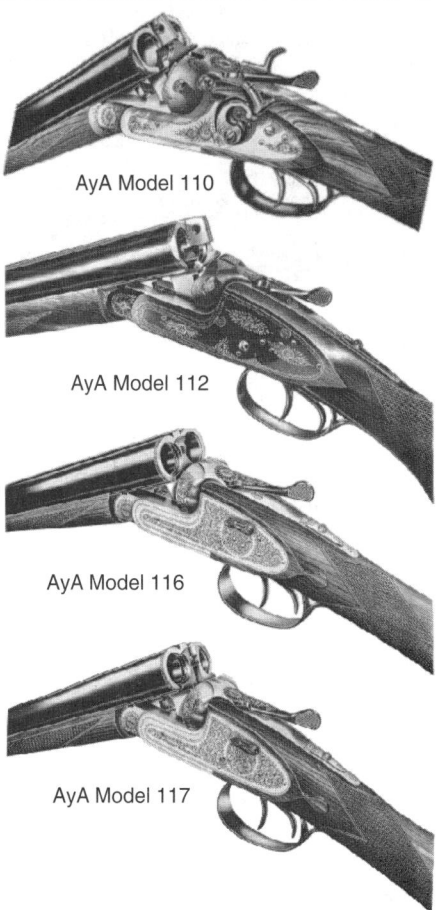

AyA Model 110

AyA Model 112

AyA Model 116

AyA Model 117

SHOTGUNS

AyA
Model 400 (See AyA Bolero)
Model 400E (See AyA Matador)
Model 931
Model XXV/SL
Model XXV/BL
No. 1
No. 2
No. 3-A
No. 4
Yeoman

Baikal
66 Coach Gun
IJ-18
IJ-18EM

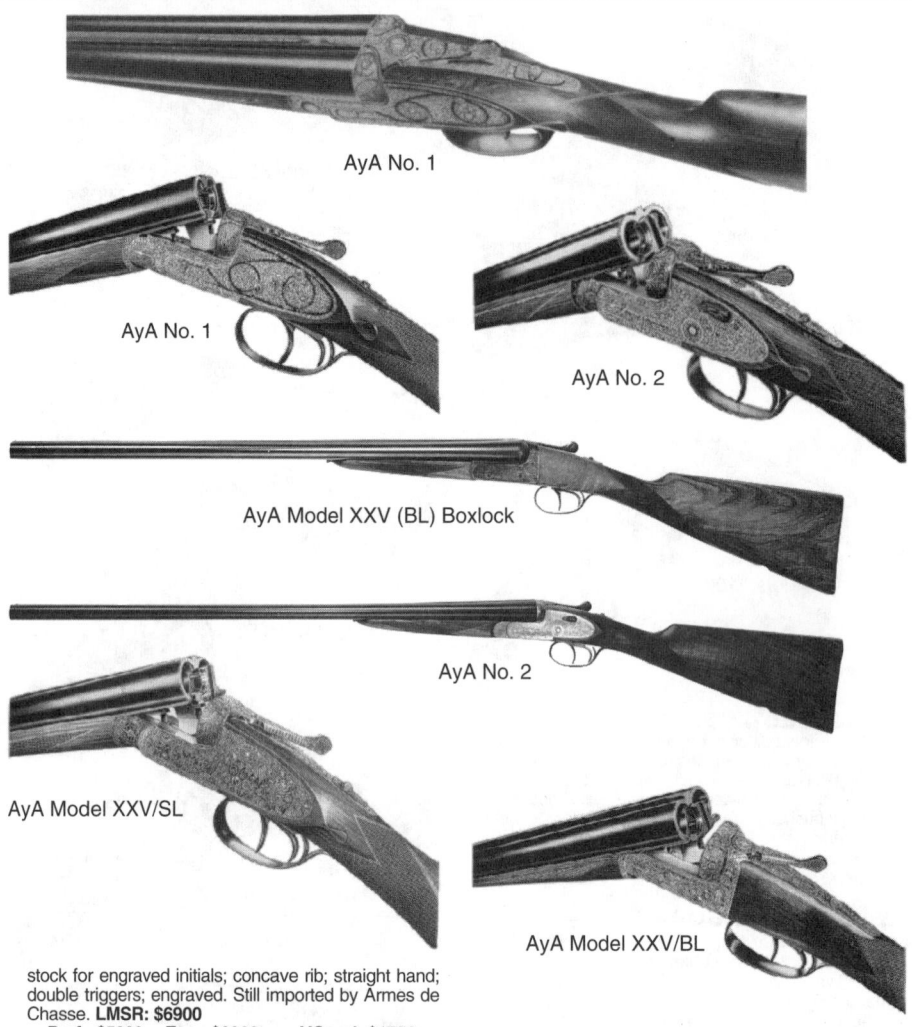

AyA No. 1

AyA No. 1

AyA No. 2

AyA Model XXV (BL) Boxlock

AyA No. 2

AyA Model XXV/SL

AyA Model XXV/BL

combos; hand-checkered European walnut stock; beavertail forend; straight, half or full pistol grip; Holland & Holland design hand-detachable locks; double or selective single trigger. Was imported by Precision Sports. No longer in production.
Perf.: $850   Exc.: $700   VGood: $625

AYA MODEL 931
Over/under; boxlock; 12-, 16-, 20-, 28-ga., 410; 26", 27", 28" barrels; weighs 5 to 7 lbs; self-opening; Anson & Deeley system with double locking lugs; chopper lump barrels; bushed firing pins; automatic safety, ejectors; articulated front trigger. Imported by Armes de Chasse. Still in production. **LMSR: $14,500**
Perf.: $1000   Exc.: $750   VGood: $525

AYA MODEL XXV/SL
Side-by-side; boxlock; 12-, 20-ga.; 26", 27", 28", 29" barrels, depending upon gauge; weighs 5 to 7 lbs.; European walnut stock and forend; Holland & Holland system with double locking lugs; chopper lump barrels; bushed firing pins; automatic safety and ejectors; articulated front trigger. Still imported. **LMSR: $3900**
Perf.: $2750   Exc.: $1500   VGood: $1000

AYA MODEL XXV/BL
Side-by-side; boxlock ejector; 12- or 20-ga.; 2 1/2" chambers; 2 3/4" chambers on request; 25" barrels; standard chokes; straight-grip European walnut stock; half or full pistol grip on request; checkered butt; fine hand engraving; double triggers; Churchill-type narrow rib; color case-hardened or coin-finished receiver. Introduced 1981; still imported by Armes de Chasse. **LMSR: $2800**
Perf.: $2000   Exc.: $1150   VGood: $900

AYA NO. 1
Side-by-side; sidelock; 12-, 20-ga.; 26", 27", 28", 29" barrels; weighs 5-7 lbs.; double triggers; articulated front trigger; automatic safety; selective automatic ejectors; cocking indicators; bushed firing pins; replaceable hinge pins and chopper lump barrels; hand-cut checkered walnut stocks with oil finish; metal oval on butt-

stock for engraved initials; concave rib; straight hand; double triggers; engraved. Still imported by Armes de Chasse. **LMSR: $6900**
Perf.: $5900   Exc.: $2300   VGood: $1750

AYA NO. 2
Side-by-side; sidelock; 12-, 16-, 20-, 28-ga., .410; 3" chambers; 28" barrels; standard choke combos; weighs 7 lbs.; silver bead front sight; English-style oil-finished walnut stock, splinter forend; checkered butt; single selective or double triggers; engraved sideplates; auto safety; gas escape valves. Made in Spain. Introduced 1987; discontinued 1988. Now imported by Armes de Chasse. **LMSR: $3370**
12-, 20-ga., double triggers
Perf.: $2775   Exc.: $1050   VGood: $800
12-, 20-ga., single trigger
Perf.: $2825   Exc.: $1100   VGood: $850
28-ga., 410, double triggers
Perf.: $2850   Exc.: $1150   VGood: $900
28-ga., 410, single trigger
Perf.: $2875   Exc.: $1175   VGood: $925

AYA NO. 3-A
Side-by-side; boxlock; 12-, 16-, 20-, 28-ga., 410; 26", 28", 30" barrels; 3" chambers (12-ga.); standard chokes; straight, half or full pistol grip; double triggers; border-engraved only; Spanish walnut stock; hand-checkered; chopper lump barrels; double underlug lock only. Discontinued 1985.
Perf.: $625   Exc.: $500   VGood: $400

AYA NO. 4
Side-by-side; boxlock; 12-, 16-, 20-, 28-ga., 410; 26", 28" 30" barrels; standard chokes; hand-checkered straight-grip European walnut stock; other stock types made to order; checkered butt; classic forend; automatic ejectors, safety; double or single triggers; color case-hardened receiver. Made in Spain. Introduced 1981; still imported by Armes de Chasse. **LMSR: $1825**
12-, 16ga.
Perf.: $1400   Exc.: $525   VGood: $425
20-ga.
Perf.: $1400   Exc.: $525   VGood: $425

28-ga., 410
Perf.: $1450   Exc.: $575   VGood: $475

AYA YEOMAN
Side-by-side; boxlock; 12-ga.; ejectors; very plain gun with no engraving; color-hardened receiver; hand-checkered walnut stock. No longer imported.
Perf.: $475   Exc.: $395   VGood: $325

BAIKAL 66 COACH GUN
Side-by-side; exposed hammers; 12-ga.; 2 3/4" chambers; 20" Improved/Modified, 28" Modified/Full barrels; weighs 6 1/4 lbs. (20"), 7 1/4 lbs. (28"); hand-checkered European hardwood pistol-grip stock, beavertail forearm; chromed bores, chambers; hand engraved receiver; extractors. Also known as the TOZ-66/54. Made in Russia. No longer imported.
Perf.: $175   Exc.: $140   VGood: $125

BAIKAL IJ-18
Single shot; single barrel; exposed hammer; 12-, 16-, 20-ga., .410; 2 3/4" chamber; 26" Modified, 28" Modified, 30" Full choke barrel; 44 1/2" overall length; weighs 6 lbs.; hand-checkered European walnut pistol-grip stock, forearm; white spacers at pistol grip, plastic buttplate; chrome-lined bore, chamber; extractor; cocking indicator; crossbolt safety. Made in Russia. Introduced 1973; still in production. **LMSR: $95**
Perf.: $75   Exc.: $60   VGood: $50

Baikal IJ-18EM
Same specs as IJ-18 except automatic ejector. Imported from Russia by Century International Arms. Still imported. **LMSR: $108**
Perf.: $80   Exc.: $65   VGood: $55

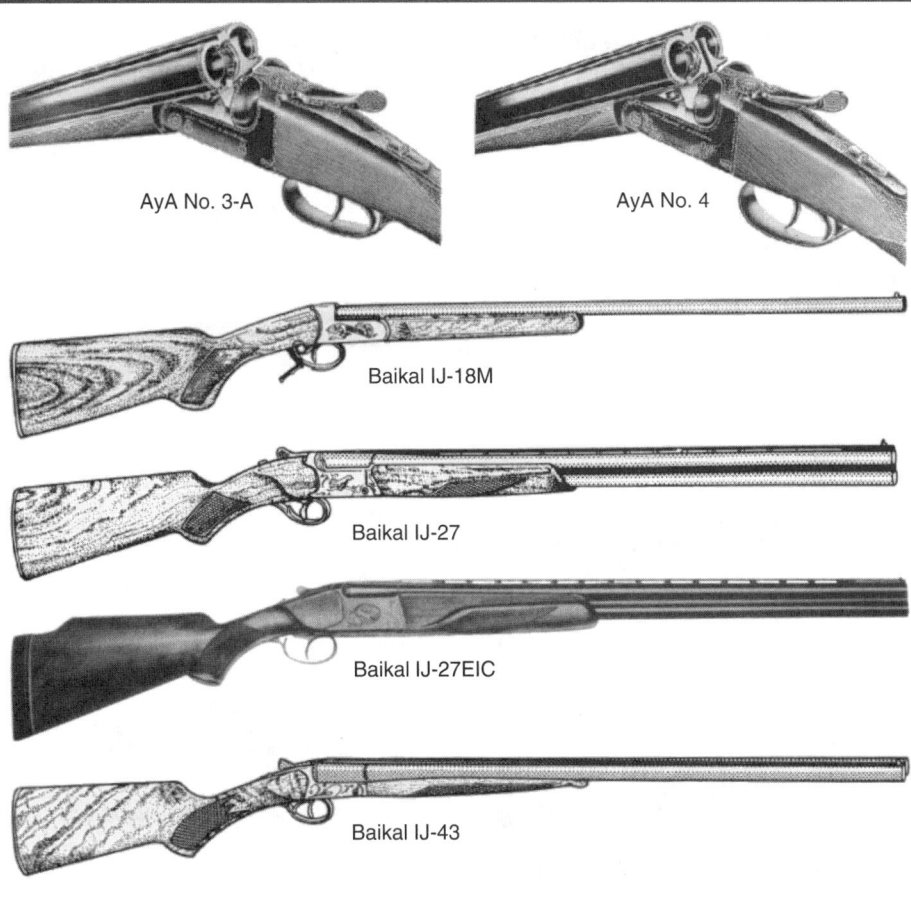

AyA No. 3-A

AyA No. 4

Baikal IJ-18M

Baikal IJ-27

Baikal IJ-27EIC

Baikal IJ-43

SHOTGUNS

Baikal
IJ-18M
IJ-25
IJ-27
IJ-27EIC
IJ-27EM
IJ-27M
IJ-43
IJ-43EM
IJ-43M
IJ-58M
MC-5
MC-6
MC-7
MC-8

Baikal IJ-18M

Same specs as IJ-18 except 12-, 16-, 20-ga., .410; 2 3/4" chamber (12, 16), 3" chamber (20, .410); 26", 28" barrel; choked Improved Cylinder, Full, Modified; trigger block safety; engraved, blued receiver. Imported from Russia by K.B.I., Inc. Reintroduced 1994; still imported. **LMSR: $69**

Perf.: $60   **Exc.:** $50   **VGood:** $40

BAIKAL IJ-25

Over/under; 12-ga; 2 3/4" chambers; 26" Skeet/Skeet, 28" Modified/Full, 30" Improved/Full barrels; vent rib; hand-checkered European walnut pistol-grip stock, ventilated forearm; white spacers at pistol-grip cap, recoil pad; single non-selective trigger; chrome-lined barrels, chambers, internal parts; hand-engraved, silver inlaid receiver, forearm latch, triggerguard. Made in Russia. Introduced 1973; discontinued 1978.

Perf.: $305   **Exc.:** $225   **VGood:** $175

BAIKAL IJ-27

Over/under; 12- (2 3/4"), 20-ga. (3"); 26" Skeet/Skeet, 28" Modified/Full barrels; hand-fitted ventilated rib; hand-checkered European walnut pistol-grip stock, ventilated forend; white spacers at pistol-grip cap, rubber recoil pad; double triggers; automatic safety; extractors; chrome-lined barrels, chambers, internal parts; hand-engraved, silver inlaid receiver, forearm latch, triggerguard. Made in Russia. Introduced 1973; still in production. **LMSR: $299**

Perf.: $260   **Exc.:** $225   **VGood:** $175

Baikal IJ-27EIC

Same specs as IJ-27 except automatic ejectors; single trigger. Still in production. **LMSR: $339**

Perf.: $325   **Exc.:** $275   **VGood:** $210

Baikal IJ-27EM

Same specs as IJ-27 except 12-ga.; 2 3/4" chambers; 28 1/2" chrome-lined barrels; weighs 7 1/2 lbs.; European hardwood stock; double triggers; selective automatic ejectors. Imported from Russia by Century International Arms. Still imported. **LMSR: $365**

Perf.: $330   **Exc.:** $260   **VGood:** $215

Baikal IJ-27M

Same specs as IJ-27 except 12-ga.; 2 3/4" chambers; 28 1/2" chrome-lined barrels; European hardwood stock; double triggers; extractors; sling swivels. Imported from Russia by Century International Arms; still imported. **LMSR: $340**

Perf.: $310   **Exc.:** $250   **VGood:** $210

BAIKAL IJ-43

Side-by-side; boxlock; 12-ga.; 2 3/4" chambers; 20" Cylinder/Cylinder, 28" Modified/Full barrels; weighs 6 3/4 lbs.; checkered walnut stock, forend; double triggers; extractors; blued, engraved receiver. Imported from Russia by K.B.I., Inc. Reintroduced 1994; still imported. **LMSR: $249**

Perf.: $165   **Exc.:** $115   **VGood:** $75

Baikal IJ-43EM

Same specs as IJ-43 except 28 1/2" chrome-lined barrels; European hardwood stock; automatic ejectors. Imported from Russia by Century International Arms. Still imported. **LMSR: $270**

Perf.: $250   **Exc.:** $190   **VGood:** $140

Baikal IJ-43M

Same specs as IJ-43 except 28 1/2" chrome-lined barrels; European hardwood stock; automatic safety. Imported from Russia by Century International Arms. Still imported. **LMSR: $255**

Perf.: $240   **Exc.:** $190   **VGood:** $160

BAIKAL IJ-58M

Side-by-side; hammerless; 12-ga.; 2 3/4" chambers; 26" Improved/Modified, 28" Modified/Full barrels; hand-checkered European walnut pistol-grip stock, beavertail forend; hinged front double trigger; chrome-lined barrels, chambers; hand-engraved receiver; hammer interceptors; recoil pad; extractors. Made in Russia. Introduced 1973; discontinued 1982.

Perf.: $225   **Exc.:** $175   **VGood:** $125

BAIKAL MC-5

Over/under; 12-, 20-ga.; 2 3/4" chambers; 26" Improved/Modified, Skeet/Skeet barrels; weighs 5 3/4 lbs.; hand-fitted solid rib; fancy hand-checkered walnut stock; straight or pistol grip, with or without cheekpiece;

non-removable forend; double triggers; extractors; hammer interceptors; silver-finished, fully engraved receiver; chrome-lined barrels, chambers, internal parts; marketed with case. Made in Russia. Introduced 1973; discontinued 1982.

Perf.: $325   **Exc.:** $350   **VGood:** $295

BAIKAL MC-6

Over/under; 12-ga.; 2 3/4" chambers; 26" Skeet/Skeet, Improved Cylinder/Modified barrels; hand-fitted solid raised rib; fancy hand-checkered walnut stock; pistol grip or straight stock, with or without cheekpiece; non-removable forend; single non-selective or double triggers, extractors; hammer interceptors; silver-finished, fully engraved receiver; chrome-lined barrels, chambers, internal parts; marketed with case. Made in Russia. Introduced 1973; discontinued 1982.

Perf.: $495   **Exc.:** $395   **VGood:** $300

BAIKAL MC-7

Over/under; 12-, 20-ga.; 2 3/4" chambers; 26" Improved/Modified, 28" Modified/Full barrels; solid raised rib; hand-checkered fancy walnut straight or pistol-grip stock, beavertail forend; ejectors; double triggers; optional single selective trigger; hand-chiseled, engraved receiver; chrome-lined barrels, chambers, internal parts; marketed with case. Made in Russia. Introduced 1973; discontinued 1982.

Perf.: $900   **Exc.:** $750   **VGood:** $625

BAIKAL MC-8

Over/under; 12-ga.; 2 3/4" chambers; 2-barrel set of 26" Skeet/Skeet, 28" Modified/Full; hand-fitted vent rib; weighs 7 3/4 lbs.; hand-checkered fancy European walnut Monte Carlo pistol-grip stock, non-removable beavertail forend; double triggers; extractors; optional single selective trigger, selective ejectors; hand-engraved, blued receiver; chrome-lined barrels, chambers, internal parts; marketed with case. Made in Russia. Introduced 1973; discontinued 1982.

Perf.: $875   **Exc.:** $700   **VGood:** $595

With single selective trigger, selective ejectors

Perf.: $925   **Exc.:** $750   **VGood:** $625

Baikal MC-110

Baikal TOZ-34P

SHOTGUNS

Baikal
MC-10
MC-21
MC-109
MC-110
MC-111
TOZ-34P
TOZ-66
TOZ-66/54
(See Baikal 66 Coach Gun)

Baker
Batavia Ejector
Batavia Leader
Batavia Special
Black Beauty Special
Grade R
Grade S

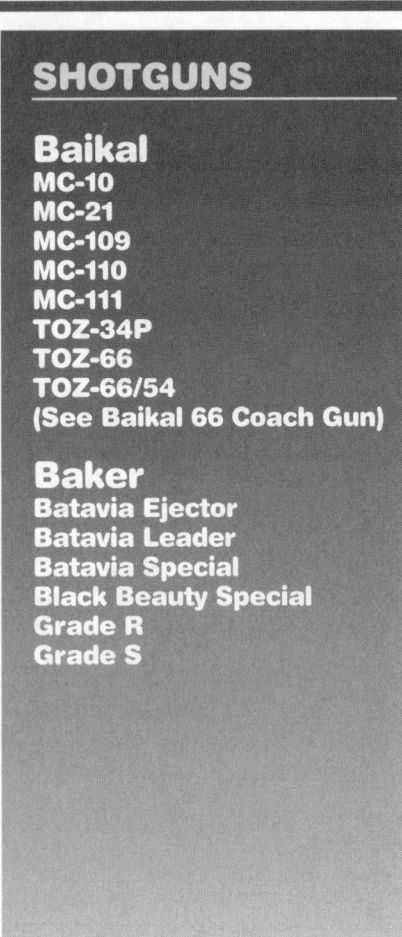

Baker Grade R

Baker Batavia Special

Baker Batavia Leader

BAKER MC-10 / BAIKAL MC-10

Side-by-side; hammerless; 12-, 20-ga.; 2 3/4″ chambers; 26″ Improved/Modified, 28″ Modified/Full barrels; raised solid rib; hand-checkered fancy European walnut stock, semi-beavertail forearm; straight or pistol grip; chrome-lined barrels, chambers, internal parts; double triggers; auto safety; extractors or selective ejectors; receiver engraved with animal, bird scenes; engraved trigger guard, tang. Made in Russia. Introduced 1973; discontinued 1978.
Perf.: $375   Exc.: $310   VGood: $275

BAIKAL MC-21
Autoloader; takedown; 12-ga; 5-shot magazine; 26″ Improved, 28″ Modified, 30″ Full choke barrel; vent rib; hand-rubbed, hand-checkered European walnut cheekpiece stock, grooved forearm; white spacers at pistol grip, buttplate; chrome-lined barrel, chamber; reversible safety; target-grade trigger. Made in Russia. Introduced 1973; discontinued 1978.
Perf.: $275   Exc.: $200   VGood: $175

BAIKAL MC-109
Over/under; hand-detachable sidelock; 12-ga.; 2 3/4″ chambers; 28″ barrels; raised rib; weighs 7 1/4 lbs.; hand-carved, checkered fancy walnut stock, beavertail forend; straight or pistol grip with or without cheekpiece; single selective trigger; chromed barrels, chambers, internal parts; hand-chiseled scenes on receiver to customer specs; marketed with case. Made in Russia. Introduced 1975; discontinued 1982.
Perf.: $925   Exc.: $700   VGood: $595

BAIKAL MC-110
Side-by-side; 12-, 20-ga.; 2 3/4″ chambers; 26″ Improved Cylinder/Modified (20-ga.), 28″ Modified/Full (12-ga.) barrels; raised solid rib; weighs 6 lbs. (20-ga.), 6 3/4 lbs. (12-ga.); hand-checkered fan-

cy walnut straight or pistol-grip stock, semi-beavertail forend; double triggers; chromed barrels, chambers, internal parts; hammer interceptors; extractors; auto safety; fully engraved receiver, triggerguard, tang; marketed in case. Made in Russia. Introduced 1975; discontinued 1982.
Perf.: $500   Exc.: $400   VGood: $300

BAIKAL MC-111
Side-by-side; hand-detachable sidelock; 12-ga.; 2 3/4″ chambers; barrel lengths, chokes to customer's specifications; hand-checkered European walnut straight or pistol-grip stock, semi-beavertail forend; single selective trigger; hammer interceptors; cocking indicators; gold, silver inlays in butt; hand-chiseled bird, animal scenes on receiver; chromed barrels, chambers, internal parts; marketed with case. Made in Russia. Introduced 1975; discontinued 1982.
Perf.: $975   Exc.: $750   VGood: $625

BAIKAL TOZ-34P
Over/under; boxlock; 12-ga.; 2 3/2″ chambers; 28″ Full/Improved Cylinder barrels; vent rib; 44″ overall length; weighs 7 1/2 lbs.; European walnut stock, forend; extractors; optional ejectors; cocking indicator; double triggers; engraved, blued receiver; ventilated rubber buttpad. Imported from Russia by Century International Arms. Still imported. **LMSR: $405**
Perf.: $350   Exc.: $280   VGood: $210
With ejectors
Perf.: $375   Exc.: $300   VGood: $235

BAIKAL TOZ-66
Side-by-side; exposed hammers; 12-ga.; 2 3/4″ chambers; 20″ Improved/Modified, 28″ Modified/Full barrels; hand-checkered European hardwood pistol-grip stock, beavertail forearm; chrome-lined barrels, chambers; hand-engraved receiver; extractors. Made in Russia. Introduced 1973; discontinued 1978.
Perf.: $225   Exc.: $150   VGood: $100

BAKER BATAVIA EJECTOR
Side-by-side; sidelock; hammerless; 12-, 16-, 20-ga.; 26″ to 32″ Damascus or forged steel barrels; any standard choke combo; best-grade hand-checkered American walnut pistol-grip stock, forearm; auto ejectors. Introduced 1921; discontinued 1930.
Exc.: $760   VGood: $700   Good: $625
Damascus
Exc.: $500   VGood: $400   Good: $300

Baker Batavia Leader
Same specs as Batavia Ejector except forged steel barrels; extractors or auto ejectors; less expensive finish. Discontinued around 1930.
Extractors
Exc.: $350   VGood: $310   Good: $250
Auto ejectors
Exc.: $420   VGood: $380   Good: $310

Baker Batavia Special
Same specs as Batavia Ejector except forged steel barrels; extractors; less expensive finish. Discontinued around 1930.
Exc.: $295   VGood: $260   Good: $210

BAKER BLACK BEAUTY SPECIAL
Side-by-side; sidelock; hammerless; 12-, 16-, 20-ga.; 26″ to 32″ barrels; any standard choke combination; ejectors or extractors; hand-checkered American walnut pistol-grip stock, forearm; engraved. Introduced 1921; discontinued 1930.
Extractors
Exc.: $625   VGood: $570   Good: $500
Auto ejectors
Exc.: $725   VGood: $670   Good: $600

BAKER GRADE R
Side-by-side; sidelock; hammerless; 12-, 16-, 20-ga.; 26″, 28″, 30″, 32″ Damascus or Krupp steel barrels; select hand-checkered European walnut pistol-grip stock, forend; extractors or auto ejectors; extensive engraving with game scenes. Introduced 1915; discontinued 1933.
Extractors
Exc.: $1000   VGood: $925   Good: $800
Auto ejectors
Exc.: $1200   VGood: $1125   Good: $1000
Paragon Grade, extractors
Exc.: $1300   VGood: $1125   Good: $1000
Paragon Grade, auto ejectors
Exc.: $1500   VGood: $1275   Good: $1000
Expert Grade, ejectors, select finish
Exc.: $1975   VGood: $1350   Good: $1100
Deluxe Grade, best-grade finish
Exc.: $3250   VGood: $2650   Good: $2000

BAKER GRADE S
Side-by-side; sidelock; hammerless; 12-, 16-, 20-ga.; 26″, 28″, 30″, 32″ fluid-tempered steel barrels; select hand-checkered European walnut pistol-grip stock, forend; scroll, line engraving; extractors or auto ejectors. Introduced 1915; discontinued 1933.
Extractors
Exc.: $800   VGood: $710   Good: $650
Auto ejectors
Exc.: $1000   VGood: $900   Good: $850

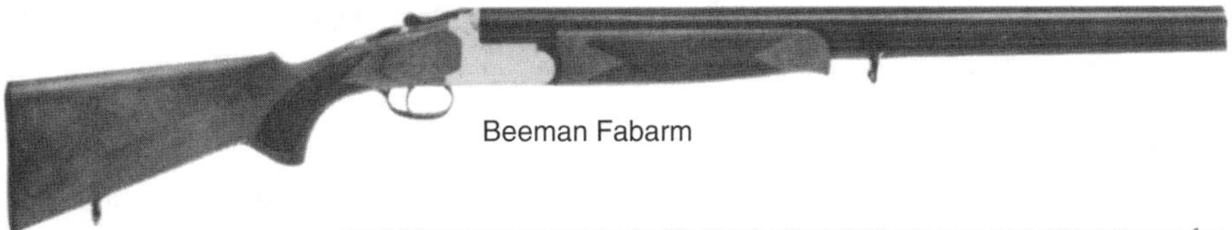

Beeman Fabarm

Benelli Black Eagle Competition

Benelli Black Eagle Super

Benelli Black Eagle Super Slug

Benelli Executive Series Type III

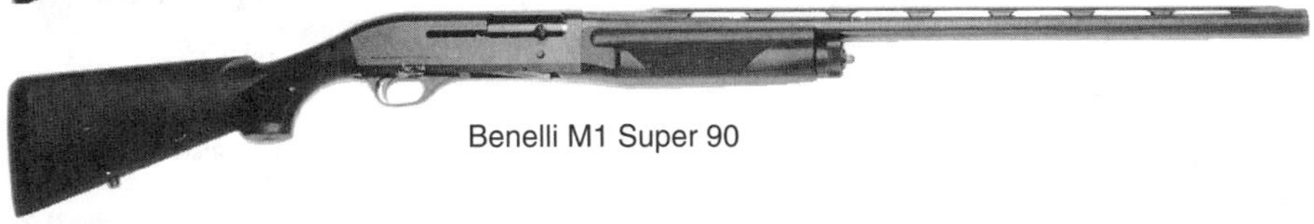

Benelli M1 Super 90

Beeman
Fabarm

Benelli
Black Eagle
Black Eagle Competition
Black Eagle Competition
"Limited Edition"
Black Eagle Slug Gun
Black Eagle Super
Black Eagle Super Slug
Executive Series
Extraluxe
M1 Super 90
M1 Super 90 Defense

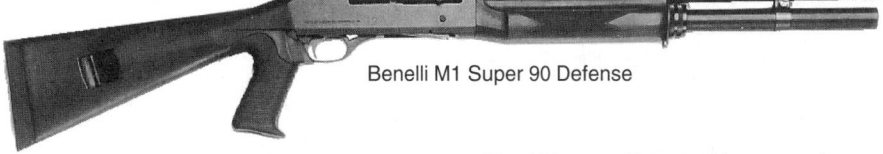

Benelli M1 Super 90 Defense

BEEMAN FABARM

Over/under; boxlock; 12-ga.; 2 3/4" chambers; 26 1/2" Skeet/Skeet, 29" Full/Modified barrels; weighs about 7 1/2 lbs.; red bead front sight, white bead middle; select cut-checkered walnut stock, Schnabel forend; single selective trigger; auto ejectors; chrome-lined bores, chambers; silvered, engraved receiver; Skeet/Trap combo with interchangeable barrels cased. Made in Italy. Introduced 1984; discontinued 1984.

Field Grade
Perf.: $650   **Exc.:** $525   **VGood:** $410
Trap/Skeet
Perf.: $750   **Exc.:** $625   **VGood:** $500
Combo
Perf.: $1100   **Exc.:** $860   **VGood:** $750

BENELLI BLACK EAGLE

Autoloader; Montefeltro recoil action; 12-ga.; 3" chambers; 5-shot magazine; 21", 24", 26", 28" vent-rib barrels; choke tubes; weighs 7 3/8 lbs.; high-gloss European walnut stock; marketed with drop adjustment kit; two-piece steel/aluminum receiver; silver finish on lower receiver, blued upper. Made in Italy. Introduced 1989; discontinued 1991.

Perf.: $720   **Exc.:** $495   **VGood:** $360

Benelli Black Eagle Competition

Same specs as Black Eagle except 26", 28" barrel; mid-bead sight; competition stock; two-piece etched aluminum receiver. Made in Italy. Introduced 1989; still imported. **LMSR:** $1156

Perf.: $975   **Exc.:** $650   **VGood:** $510

Benelli Black Eagle Competition "Limited Edition"

Same specs as Black Eagle except gold inlays; fancy wood; special serial numbers. **LMSR:** $2000

Perf.: $1850   **Exc.:** $1280   **VGood:** $1050

Benelli Black Eagle Slug Gun

Same specs as Black Eagle except 24" rifled barrel; no rib; weighs 7 1/8 lbs.; two-piece steel/alloy receiver;

drilled, tapped for scope mount; marketed with scope mount. Made in Italy. Introduced 1989; discontinued 1991.

Perf.: $750   **Exc.:** $495   **VGood:** $410

Benelli Black Eagle Super

Same specs as Black Eagle except 3 1/2" chamber; 24", 26", 28" barrels; 49 3/8" overall length (28" barrel); weighs 7 1/2 lbs.; bead front sight; European walnut stock adjustable for drop; satin or gloss finish. Introduced 1991; still imported by Heckler & Koch. **LMSR:** $1144

Perf.: $900   **Exc.:** $620   **VGood:** $500

Benelli Black Eagle Super Slug

Same specs as Benelli Black Eagle except 3 1/2" chamber; 24" E.R. Shaw Custom rifled barrel; 45 1/2" overall length; weighs 7 1/2 lbs.; scope mount base; matte-finish receiver; wood or polymer stocks available. Imported from Italy by Heckler & Koch, Inc. Introduced 1992; still imported. **LMSR:** $1171

Perf.: $950   **Exc.:** $620   **VGood:** $500

BENELLI EXECUTIVE SERIES

Autoloader; Montefeltro recoil action; 12-ga.; 3" chamber; 5-shot magazine; 21", 24", 26", 28" barrel; choke tubes; weighs 7 3/8 lbs.; highest grade of walnut stock with drop adjustment kit; two-piece steel/aluminum receiver; grayed steel lower receiver; hand-engraved and gold inlaid (Type III). Special order only. Made in Italy. Introduced 1995; still in production. **LMSR:** $4375

Type I (about two-thirds engraving coverage)
Perf.: $4000   **Exc.:** $3200   **VGood:** $2800
Type II (full coverage engraving)
Perf.: $4300   **Exc.:** $3400   **VGood:** $3000
Type III (full coverage, gold inlays)
Perf.: $5200   **Exc.:** $4500   **VGood:** $3800

BENELLI EXTRALUXE

Autoloader; inertial action; 12-ga.; 2 3/4" chamber; 5-shot magazine; 26", 28" barrel; vent rib; weighs 6 3/4 lbs.; hand-checkered walnut stock; crossbolt safety; interchangeable barrels; similar to SL-121 but with extra-fancy wood; fine hand engraving; gold trigger; carrying case. No longer imported.

Perf.: $295   **Exc.:** $250   **VGood:** $200

BENELLI M1 SUPER 90

Autoloader; rotating Montefeltro bolt system; 12-ga; 3" chamber; 3-shot magazine; extended magazine available with 26", 28" barrels; 21", 24", 26", 28" barrels; choke tubes; vent rib; weighs 7 1/2 lbs.; metal bead front sight; wood or polymer stock, forearm; blued finish. Made in Italy. Introduced 1987; still imported. **LMSR:** $848

Perf.: $700   **Exc.:** $390   **VGood:** $350

Benelli M1 Super 90 Defense

Same specs as M1 Super 90 except 5-shot magazine; 18 1/2" Cylinder barrel; open or aperture (ghost ring) rifle-type sights; polymer pistol-grip stock; matte black finish. Introduced 1993; still imported. **LMSR:** $816

Perf.: $680   **Exc.:** $370   **VGood:** $320

SHOTGUNS

Benelli

**M1 Super 90 Entry
M1 Super 90 Montefeltro
M1 Super 90 Montefeltro
20-Gauge Limited Edition
M1 Super 90 Montefeltro,
Left-Hand
M1 Super 90 Montefeltro
Uplander
M1 Super 90 Montefeltro
Slug
M1 Super 90 Slug
M1 Super 90 Sporting
Special
M1 Super 90 Tactical
M3 Super 90
SL-121
SL-123
SL-201**

Benelli M1 Super 90 Montefeltro

Benelli M1 Super 90
Tactical

Benelli M3 Super 90

Benelli SL-121

Benelli SL-121 Slug

Benelli SL-123

Benelli M1 Super 90 Entry
Same specs as M1 Super 90 except 5-shot magazine; 14" Cylinder barrel; open or aperture (ghost ring) rifle-type sights; polymer straight or pistol-grip stock; matte black finish. Introduced 1992; still imported. **LMSR: $850**

> **Perf.:** $700   **Exc.:** $390   **VGood:** $350

Benelli M1 Super 90 Montefeltro
Same specs as M1 Super 90 except 12- or 20-ga.; 4-shot magazine; checkered European walnut drop-adjustable stock with satin finish; matte blue metal finish. Made in Italy. Introduced 1987; still imported. **LMSR: $868**

> **Perf.:** $730   **Exc.:** $420   **VGood:** $365

Benelli M1 Super 90 Montefeltro 20-Gauge Limited Edition
Same specs as M1 Super 90 except 20-ga.; 3" chamber; 4-shot magazine; 26" barrel; weighs 5 3/4 lbs.; checkered European walnut drop-adjustable stock with satin finish; special nickel-plated and engraved receiver inlaid with gold. Made in Italy. Introduced 1993; still imported. **LMSR: $2000**

> **Perf.:** $1700   **Exc.:** $1250   **VGood:** $900

Benelli M1 Super 90 Montefeltro, Left-Hand
Same specs as M1 Super 90 except left-hand action; 12- or 20-ga.; 4-shot magazine; checkered European walnut drop-adjustable stock with satin finish; matte blue metal finish. Made in Italy. Introduced 1987; still imported. **LMSR: $887**

> **Perf.:** $750   **Exc.:** $470   **VGood:** $410

Benelli M1 Super 90 Montefeltro Uplander
Same specs as M1 Super 90 except 12-, 20-ga.; 4-shot magazine; 21", 24" barrels; checkered European walnut drop-adjustable stock with satin finish. Made in Italy. Introduced 1987; discontinued 1992.

> **Perf.:** $600   **Exc.:** $350   **VGood:** $300

Benelli M1 Super 90 Montefeltro Slug
Same specs as M1 Super 90 except 12-ga.; 3" chamber; 4-shot magazine; 24" barrel; Cylinder choke; 45 1/2" overall length; weighs 7 lbs.; checkered European walnut drop-adjustable stock with satin finish; matte blue metal finish. Made in Italy. Introduced, 1988; discontinued, 1992.

> **Perf.:** $590   **Exc.:** $340   **VGood:** $290

Benelli M1 Super 90 Slug
Same specs as M1 Super 90 except 5-shot magazine; 18 1/2" Cylinder barrel; open or aperture (ghost ring) rifle-type sights; polymer straight stock; matte black finish. Introduced 1986; still imported. **LMSR: $785**

> **Perf.:** $615   **Exc.:** $310   **VGood:** $280

Benelli M1 Super 90 Sporting Special
Same specs as M1 Super 90 except 18 1/2" barrel; Improved Cylinder, Modified, Full choke tubes; 39 3/4" overall length; weighs 6 1/2 lbs.; ghost ring sight; sporting-style polymer stock with drop adjustment; matte black finish receiver. Made in Italy. Introduced 1993; still imported. **LMSR: $868**

> **Perf.:** $680   **Exc.:** $410   **VGood:** $375

Benelli M1 Super 90 Tactical
Same specs as M1 Super 90 except 5-shot magazine; 18 1/2" Improved Cylinder, Modified, Full barrel; open or aperture (ghost ring) rifle-type sights; polymer straight or pistol-grip stock; matte black finish. Numerous law enforcement and military options. Introduced 1993; still imported. **LMSR: $850**

> **Perf.:** $660   **Exc.:** $385   **VGood:** $320

BENELLI M3 SUPER 90
Convertible autoloader/slide-action; 12-ga.; 3" chamber; 7-shot magazine; 19 3/4" Cylinder bore barrel; 41" overall length; weighs 7 1/2 lbs.; post front sight; windage-adjustable buckhorn rear; optional ghost-ring sight; high-impact polymer stock with sling loop; rubberized pistol grip on optional SWAT stock; alloy receiver; matte finish; auto shell release lever. May be illegal for civilian use in some jurisdictions; designed for law enforcement. Made in Italy. Introduced 1989; still imported. **LMSR: $975**

> **Perf.:** $800   **Exc.:** $440   **VGood:** $375

With ghost-ring sight
> **Perf.:** $840   **Exc.:** $480   **VGood:** $410

With folding stock
> **Perf.:** $900   **Exc.:** $540   **VGood:** $475

With pistol-grip stock
> **Perf.:** $815   **Exc.:** $455   **VGood:** $390

BENELLI SL-121
Autoloader; 12-ga.; 2 3/4" chamber; 5-shot magazine; 26", 28" barrel; plain or vent rib (SL-121V); 6 3/4 lbs.; metal bead front sight; hand-checkered European walnut pistol-grip stock; crossbolt safety; quick-interchangeable barrels; optional engraving; slug model available. Made in Italy. Introduced 1977; no longer imported.

> **Perf.:** $350   **Exc.:** $250   **VGood:** $200

Engraved model
> **Perf.:** $390   **Exc.:** $290   **VGood:** $240

Slug model
> **Perf.:** $400   **Exc.:** $300   **VGood:** $250

BENELLI SL-123
Autoloader; 12-ga; 2 3/4" chamber; 5-shot magazine; 26", 28" barrels; plain or vent rib (SL-123V); weighs 6 3/4 lbs.; metal bead front sight; fancy hand-checkered European walnut pistol-grip stock; crossbolt safety; quick-interchangeable barrels; optional engraving; slug model available. Made in Italy. Introduced 1977; no longer imported.

> **Perf.:** $350   **Exc.:** $250   **VGood:** $200

Engraved model
> **Perf.:** $390   **Exc.:** $290   **VGood:** $240

Slug model
> **Perf.:** $400   **Exc.:** $300   **VGood:** $250

BENELLI SL-201
Autoloader; 20-ga.; 4-shot magazine; 26" barrel; vent rib; weighs 5 3/4 lbs.; hand-checkered walnut stock; crossbolt safety; quick-interchangeable barrels; similar to SL-121. No longer imported.

> **Perf.:** $350   **Exc.:** $250   **VGood:** $200

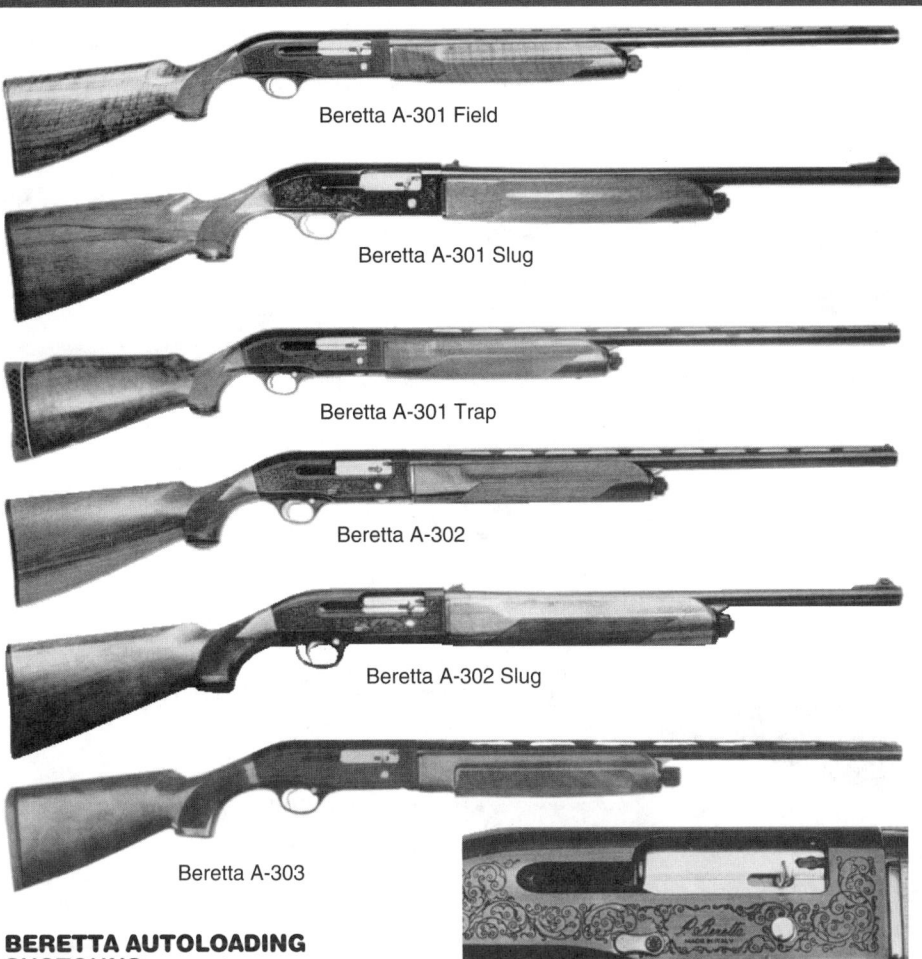

Beretta A-301 Field

Beretta A-301 Slug

Beretta A-301 Trap

Beretta A-302

Beretta A-302 Slug

Beretta A-303

SHOTGUNS

Beretta Autoloading Shotguns
A-301 Field
A-301 Magnum
A-301 Skeet
A-301 Slug
A-301 Trap
A-302
A-303
A-303 Matte Finish
A-303 Sporting Clays
A-303 Upland Model
A-303 Youth
AL-2

Beretta Engraving

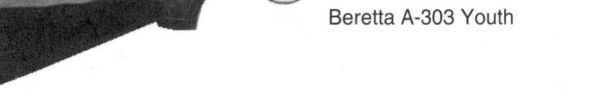

Beretta A-303 Youth

BERETTA AUTOLOADING SHOTGUNS

BERETTA A-301 FIELD
Gas-operated autoloader; 12- (2 3/4"), 20-ga. (3"); 3-shot magazine; 26" Improved, 28" Full, Modified barrels; vent rib; checkered European walnut pistol-grip stock, forend. Introduced 1977; discontinued 1982.

Perf.: $380   **Exc.:** $310   **VGood:** $240

Beretta A-301 Magnum
Same specs as A-301 Field except 12-ga.; 3" chamber; 30" Full choke barrel; recoil pad. Introduced 1977; discontinued 1982.

Perf.: $410   **Exc.:** $300   **VGood:** $250

Beretta A-301 Skeet
Same specs as A-301 Field except 26" Skeet barrel; checkered European walnut Skeet-style stock; gold-plated trigger. Introduced 1977; discontinued 1982.

Perf.: $380   **Exc.:** $300   **VGood:** $250

Beretta A-301 Slug
Same specs as A-301 Field except 22" barrel; Slug choke; no rib; rifle sights. Introduced 1977; discontinued 1982.

Perf.: $380   **Exc.:** $310   **VGood:** $260

Beretta A-301 Trap
Same specs as A-310 Field except 30" Full choke barrel; checkered European walnut Monte Carlo stock; recoil pad; gold-plated trigger. Introduced 1977; discontinued 1982.

Perf.: $390   **Exc.:** $310   **VGood:** $265

BERETTA A-302
Gas-operated autoloader; 12-, 20-ga.; 2 3/4", 3" chamber; 3-shot magazine; 22" Slug, 26" Improved or Skeet, 28" Modified or Full, 30" (12-ga. only) Full or Full Trap barrel; optional Mobilchokes; cut-checkered European walnut pistol-grip stock, forend; scroll-engraved alloy receiver; push-button safety. Made in Italy. Introduced 1983; discontinued 1987.

Standard chokes
Perf.: $400   **Exc.:** $350   **VGood:** $290
Mobilchokes
Perf.: $420   **Exc.:** $370   **VGood:** $310

BERETTA A-303
Gas-operated autoloader; 12-, 20-ga.; 2 3/4", 3" chamber; 3-shot magazine; 22", 24", 26", 28", 30", 32" barrel; optional Mobilchokes; weighs 6 lbs. (20-ga.), 7 lbs. (12-ga.); hand-checkered American walnut stock; alloy receiver; magazine cut-off; push-button safety. Made in Italy. Introduced 1983; 20-ga. still imported. **LMSR: $799**
Standard chokes
Perf.: $600   **Exc.:** $340   **VGood:** $300
Mobilchokes
Perf.: $650   **Exc.:** $390   **VGood:** $340

Beretta A-303 Matte Finish
Same specs as A-303 except 3" chamber; 24", 26", 28", 30" barrel; Mobilchoke tubes; non-reflective finish on all wood, metal surfaces; studs, sling swivels. Introduced 1991; discontinued 1992.

Perf.: $650   **Exc.:** $390   **VGood:** $340

Beretta A-303 Sporting Clays
Same specs as A-303 except 12-, 20-ga.; 2 3/4" chamber; 28" barrel; Mobilchoke tubes; wide competition-style vent rib; slightly different stock dimensions; special rubber buttpad. Made in Italy. Introduced 1989; still imported. **LMSR: $822**

Perf.: $670   **Exc.:** $450   **VGood:** $380

Beretta A-303 Upland Model
Same specs as A-303 except 12-, 20-ga.; 2 3/4", 3" chamber; 24" vent rib barrel; Mobilchoke tubes; straight English-style stock. Made in Italy. Introduced 1989; still imported. **LMSR: $772**

Perf.: $600   **Exc.:** $360   **VGood:** $330

Beretta A-303 Youth
Same specs as A-303 except shorter length of pull; weighs 6 lbs. Introduced 1991; still in production. **LMSR: $736**

Perf.: $500   **Exc.:** $360   **VGood:** $300

BERETTA AL-2
Gas-operated autoloader; 12-, 20-ga.; 3-shot magazine; 2 3/4", 3" chamber; 26", 28", 30" interchangeable barrel; Full, Modified, Improved Cylinder, Skeet chokes; vent rib; medium front bead sight; diamond-point hand-checkered European walnut pistol-grip stock, forend. Introduced 1969; discontinued 1975.
Standard grade
Perf.: $300   **Exc.:** $250   **VGood:** $210
Trap/Skeet model
Perf.: $380   **Exc.:** $320   **VGood:** $260
Magnum model
Perf.: $400   **Exc.:** $320   **VGood:** $250

Beretta Autoloading Shotguns

AL-3
Gold Lark
Model 390 Field Grade
Model 390 Field Grade Gold Mallard
Model 390 Field Grade Silver Mallard
Model 309 Field Grade Silver Slug
Model 390 Competition Skeet Sport
Model 390 Competition Skeet Sport Super
Model 390 Competition Sporting
Model 390 Competition Trap Sport
Model 390 COmpetition Trap Super
Model 1200F
Model 1200FP

Beretta Model 390 Field Grade

Beretta Model 390 Competition Skeet Sport

Beretta Model 390 Competition Trap Sport

Beretta Model 1200F

Beretta Model 1201FP3

Beretta Silver Lark

BERETTA AL-3
Gas-operated autoloader; 12-, 20-ga.; 3-shot magazine; 2 3/4", 3" chamber; 26", 28", 30" interchangeable barrel; Full, Modified, Improved Cylinder, Skeet chokes; vent rib; medium front bead sight; diamond-point hand-checkered select European walnut pistol-grip stock, forend; hand-engraved receiver. Introduced 1969; discontinued 1975.
Standard grade
  Perf.: $380   **Exc.:** $320   **VGood:** $240
Trap/Skeet model
  Perf.: $390   **Exc.:** $310   **VGood:** $250
Magnum model
  Perf.: $410   **Exc.:** $310   **VGood:** $260

BERETTA GOLD LARK
Gas-operated autoloader; takedown; 12-ga.; 26", 28", 30", 32" vent-rib barrels; Improved Cylinder, Modified, Full choke; hand-checkered European walnut pistol-grip stock, beavertail forearm; push-button safety in triggerguard; all parts hand polished; fine engraving. Introduced 1960; discontinued 1967.
  Exc.: $300   **VGood:** $210   **Good:** $190

BERETTA MODEL 390 FIELD GRADE
Gas-operated autoloader; 12-ga.; 3" chamber; 3-shot magazine; 22", 24", 26", 28", 30" barrel; Mobilchoke tubes; floating vent rib; weighs 7 lbs.; select walnut stock adjustable for drop and cast; self-compensating valve allows action to shoot all loads without adjustment; alloy receiver; reversible safety; chrome-plated bore. Made in Italy. Introduced 1992; still imported. **LMSR:** $779
  Perf.: $680   **Exc.:** $420   **VGood:** $360

Beretta Model 390 Field Grade Gold Mallard
Same specs as Model 390 Field Grade except highly select walnut stock; receiver with gold engraved animals. Introduced 1992; still imported. **LMSR:** $936
  Perf.: $790   **Exc.:** $475   **VGood:** $400

Beretta Model 390 Field Grade Silver Mallard
Same specs as Model 390 Field Grade except select walnut stock with gloss or matte finish; medium or thin rubber buttpad. Introduced 1992; still imported. **LMSR:** $779
  Perf.: $680   **Exc.:** $420   **VGood:** $360

Beretta Model 390 Field Grade Silver Slug
Same specs as Model 390 Field Grade except 22" slug barrel; no rib; weighs 7 1/4 lbs.; grooved receiver for scope mount; blade front sight; adjustable rear; swivels. Introduced 1992; still imported. **LMSR:** $779
  Perf.: $680   **Exc.:** $420   **VGood:** $360

BERETTA MODEL 390 COMPETITION SKEET SPORT
Gas-operated autoloader; 12-ga.; 3" chamber; 3-shot magazine; 26", 28" Skeet-choke barrel; vent rib; select walnut Skeet-style stock, forearm; self-compensating valve allows action to shoot all loads without adjustment; reversible safety; chrome-plated bore; lower-contour, rounded alloy receiver; recoil pad; ported barrel available. Made in Italy. Introduced 1995; still imported. **LMSR:** $779
  Perf.: $680   **Exc.:** $420   **VGood:** $360
With ported barrel
  Perf.: $750   **Exc.:** $520   **VGood:** $460

Beretta Model 390 Competition Skeet Sport Super
Same specs as Model 390 Competition Skeet Sport except 28" Skeet-choke ported barrel; wide ventilated rib with orange front sight; select walnut Skeet-style stock; adjustable comb and length of pull. Made in Italy. Introduced 1993; still imported. **LMSR:** $1203
  Perf.: $900   **Exc.:** $600   **VGood:** $500

Beretta Model 390 Competition Sporting
Same specs as Model 390 Competition Skeet Sport except 28", 30" vent-rib barrel; Full, Modified, Improved Cylinder, Skeet choke tubes; select walnut Monte Carlo stock. Made in Italy. Introduced 1995; still imported. **LMSR:** $807
  Perf.: $650   **Exc.:** $380   **VGood:** $300
With ported barrel
  Perf.: $680   **Exc.:** $410   **VGood:** $330

Beretta Model 390 Competition Trap Sport
Same specs as Model 390 Competition Skeet Sport except 30", 32" step-tapered vent-rib barrel; Full, Improved Modified, Modified choke tubes; white front bead; select walnut trap-style stock. Introduced 1995; still imported. **LMSR:** $821
  Perf.: $680   **Exc.:** $375   **VGood:** $340
With ported barrel
  Exc.: $475   **VGood:** $440

Beretta Model 390 Competition Trap Super
Same specs as 390 Competition Skeet Sport except 30", 32" ported barrel; Mobilchoke tubes; wide ventilated rib with orange front sight; select walnut trap-style stock; adjustable comb and length of pull. Made in Italy. Introduced 1993; still imported. **LMSR:** $1256
  Perf.: $1000   **Exc.:** $620   **VGood:** $580

BERETTA MODEL 1200F
Recoil-operated autoloader; 12-ga.; 2 3/4" chamber; 4-shot magazine; 26", 28" vent-rib barrel; Mobilchokes; weighs 7 1/8 lbs.; checkered European walnut or matte-black finished polymer stock, forend; recoil pad. Left-handed model available. Made in Italy. Introduced 1984; discontinued 1990.
  Perf.: $500   **Exc.:** $360   **VGood:** $300

Beretta Model 1200FP
Same specs as Model 1200F except 6-shot magazine; 20" Cylinder barrel; no vent rib; adjustable rifle-type sights; designed for law enforcement. Made in Italy. Introduced 1988; discontinued 1990.
  Perf.: $500   **Exc.:** $320   **VGood:** $280

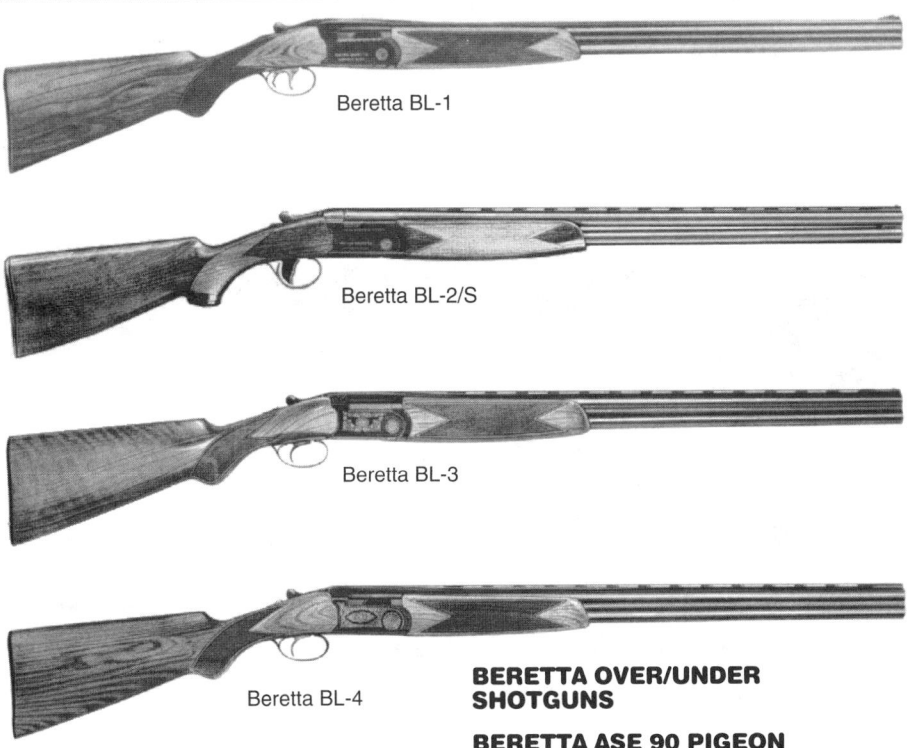

Beretta BL-1

Beretta BL-2/S

Beretta BL-3

Beretta BL-4

Beretta Autoloading Shotguns
Model 1201F
Model 1201FP
Model 1201FP3
Pintail
Ruby Lark
Silver Lark
Vittoria (See Beretta Pintail)

Beretta Over/Under Shotguns
ASE 90 Gold Pigeon
ASE 90 Gold Skeet
ASE 90 Gold Sporting Clays
ASE 90 Gold Trap
ASE 90 Gold Trap Combo
ASEL
BL-1
BL-2
BL-2/S
BL-3
BL-4

BERETTA MODEL 1201F

Recoil-operated autoloader; 12-ga.; 3″ chamber; 4-shot magazine; 24″, 26″, 28″ vent-rib barrel; Mobilchokes; weighs 7 1/4 lbs.; matte-black finished polymer stock, forend; adjustable butt, recoil pad. Made in Italy. Introduced 1988; discontinued 1994.
    Perf.: $520   **Exc.:** $360   **VGood:** $280

Beretta Model 1201FP

Same specs as Model 1201F except 6-shot magazine; 20″ Cylinder barrel; no vent rib; adjustable rifle-type sights; designed for law enforcement. Made in Italy. Introduced 1991; still imported. **LMSR:** $715
    Perf.: $500   **Exc.:** $320   **VGood:** $280

Beretta Model 1201FP3

Same specs as Model 1201F except 20″ Cylinder barrel; fixed rifle type sight; 6-shot magazine. Introduced 1988; still in production. **LMSR:** $683
    Perf.: $475   **Exc.:** $300   **VGood:** $225

BERETTA PINTAIL

Recoil-operated autoloader; 12-ga.; 3″ chamber; 24″ rifled Slug choke, 24″, 26″ Mobilchoke barrels; weighs 7 lbs.; metal bead front sight; select checkered hardwood stock, forend with matte finish; recoil pad; crossbolt safety; matte-black receiver. Slug version has rifle sights, sling swivels. Imported from Italy by Beretta U.S.A. Introduced 1993; still imported. **LMSR:** $743
    Perf.: $645   **Exc.:** $400   **VGood:** $350

BERETTA RUBY LARK

Gas-operated autoloader; takedown; 12-ga.; 26″, 28″, 30″, 32″ vent-rib stainless barrels; Improved Cylinder, Modified, Full choke; hand-checkered European walnut pistol-grip stock, beavertail forearm; push-button safety in triggerguard; all parts hand polished; extensive engraving. Introduced 1960; discontinued 1967.
    Exc.: $485   **VGood:** $350   **Good:** $250

BERETTA SILVER LARK

Gas-operated autoloader; takedown; 12-ga.; 26″, 28″, 30″, 32″ barrels; Improved Cylinder, Modified, Full choke; hand-checkered European walnut pistol-grip stock, beavertail forearm; push-button safety in trigger guard; all parts hand polished. Introduced 1960; discontinued 1967.
    Exc.: $250   **VGood:** $210   **Good:** $170

BERETTA OVER/UNDER SHOTGUNS

BERETTA ASE 90 PIGEON

Over/Under; 12-ga.; 2 3/4″ chambers; 3-shot magazine; 28″ Improved Modified/Full barrels; vent rib; weighs about 8 lbs.; high-grade checkered walnut stock, forend; drop-out trigger assembly; recoil pad; silver-finished or blued receiver; gold etching; hard-chrome bores; comes with hard case. Made in Italy. Introduced 1992; discontinued 1994.
    Perf.: $7100   **Exc.:** $5000   **VGood:** $4000

Beretta ASE 90 Gold Skeet

Same specs as ASE 90 Gold Pigeon except 28″ Skeet/Skeet barrels; high-grade checkered walnut Skeet-style stock, forend. Made in Italy. Introduced 1992; still imported. **LMSR:** $8314
    Perf.: $7500   **Exc.:** $5000   **VGood:** $4000

Beretta ASE 90 Gold Sporting Clays

Same specs as ASE 90 Gold Pigeon except 28″, 30″ vent-rib barrel; Mobilchoke tubes; high-grade checkered walnut Monte Carlo stock, forend. Introduced 1992; still imported. **LMSR:** $8387
    Perf.: $7550   **Exc.:** $5000   **VGood:** $4000

Beretta ASE 90 Gold Trap

Same specs as ASE 90 Gold Pigeon except 30″ vent-rib barrel; fixed chokes or Mobilchokes; high-grade checkered walnut trap-style stock, forend; extra trigger group. Introduced 1992; still imported. **LMSR:** $8387
    Perf.: $7550   **Exc.:** $5000   **VGood:** $4000

Beretta ASE 90 Gold Trap Combo

Same specs as ASE 90 Gold Pigeon except two-barrel set with 30″ vent-rib over/under and 32″ or 34″ single over barrel; fixed chokes or Mobilchokes; high-grade checkered walnut trap-style stock, forend; extra trigger group. Introduced 1992; still imported. **LMSR:** $9763
    Perf.: $9500   **Exc.:** $7000   **VGood:** $5000

BERETTA ASEL

Over/under; boxlock; 12-, 20-ga.; 26″, 28″, 30″ barrels; Improved/Modified, Modified/Full chokes; hand-checkered pistol-grip stock, forend; single nonselective trigger; selective automatic ejectors. Introduced 1947; discontinued 1964.
12-ga.
    Exc.: $1250   **VGood:** $800   **Good:** $700
20-ga.
    Exc.: $1800   **VGood:** $1200   **Good:** $800

BERETTA BL-1

Over/under; boxlock; 12-ga.; 2 3/4″ chambers; 26″, 28″, 30″ chrome moly steel barrels; Improved/Modified, Modi-fied/Full chokes; ramp front sight with fluorescent inserts; hand-checkered European walnut pistol-grip stock, forearm; double triggers; extractors; automatic safety; Monoblock design. Introduced 1969; discontinued 1972.
    Exc.: $300   **VGood:** $200   **Good:** $170

BERETTA BL-2

Over/under; boxlock; 12-ga.; 2 3/4″ chambers; 26″, 28″, 30″ chrome moly steel barrels; Improved/Modified, Modified/Full chokes; ramp front sight with fluorescent inserts; hand-checkered European walnut pistol-grip stock, forearm; single selective triggers; extractors; automatic safety; Monoblock design; engraving. Introduced 1969; discontinued 1972.
    Exc.: $385   **VGood:** $250   **Good:** $175

Beretta BL-2/S

Same specs as BL-2 except 2 3/4″, 3″ chambers; vent rib. Introduced 1974; discontinued 1976.
    Perf.: $420   **Exc.:** $310   **VGood:** $290

BERETTA BL-3

Over/under; boxlock; 12-, 20-ga; 2 3/4″, 3″ chambers; 26″, 28″, 30″ chrome-moly steel barrels; vent rib; Improved/Modified, Modified/Full chokes; ramp front sight with fluorescent inserts; hand-checkered European walnut pistol-grip stock, forearm; single selective triggers; ejectors; automatic safety; Monoblock design; engraved receiver. Introduced 1968; discontinued 1976.
    Perf.: $580   **Exc.:** $500   **VGood:** $460

BERETTA BL-4

Over/under; boxlock; 12-, 20-, 28-ga; 2 3/4″, 3″ chambers; 26″, 28″, 30″ chrome-moly steel barrels; vent rib; Improved/Modified, Modified/Full chokes; ramp front sight with fluorescent inserts; hand-checkered select European walnut pistol-grip stock, forearm; single selective triggers; ejectors; automatic safety; Monoblock design; heavily engraved receiver. Introduced 1968; discontinued 1976.
    Perf.: $770   **Exc.:** $600   **VGood:** $580

SHOTGUNS

Beretta Over/Under Shotguns

BL-5
BL-6
Golden Snipe
Golden Snipe Deluxe
Grade 100
Grade 200
Model 57E
Model 682 Competition Gold Skeet
Model 682 Competition Gold Skeet Super
Model 682 Competition Gold Sporting
Model 682 Competition Gold Sporting Ported
Model 682 Competition Gold Sporting Super
Model 682 Competition Gold Trap
Model 682 Competition Gold Trap Live Bird

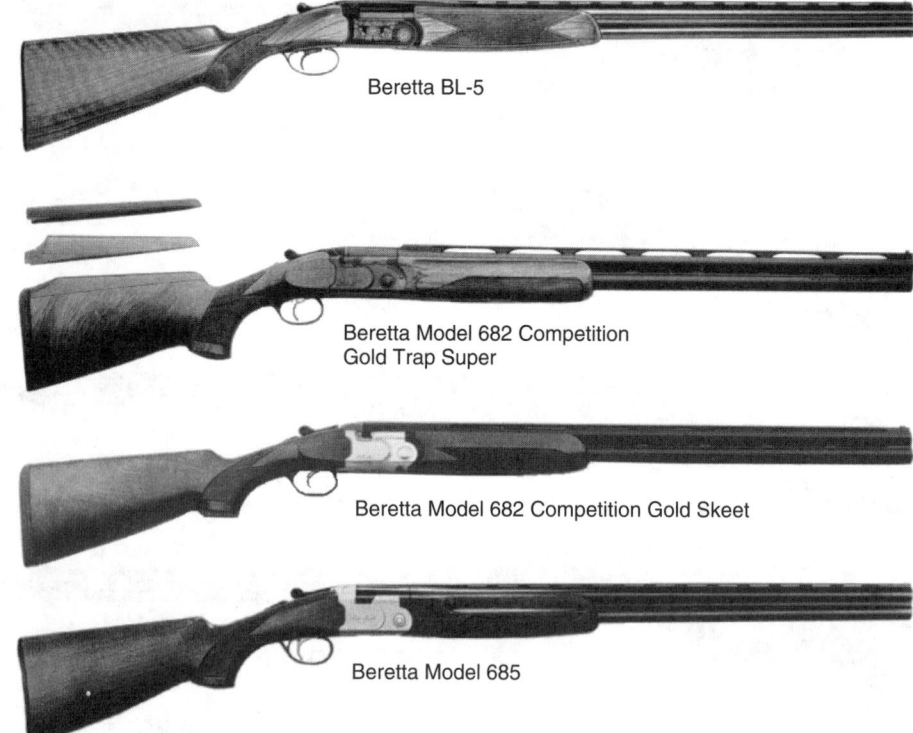

Beretta BL-5

Beretta Model 682 Competition Gold Trap Super

Beretta Model 682 Competition Gold Skeet

Beretta Model 685

BERETTA BL-5

Over/under; boxlock; 12-, 20-ga; 2 3/4", 3" chambers; 26", 28", 30" chrome-moly steel barrels; vent rib; Improved/Modified, Modified/Full chokes; ramp front sight with fluorescent inserts; hand-checkered fancy European walnut pistol-grip stock, forearm; single selective triggers; ejectors; automatic safety; Monoblock design; heavily engraved receiver. Introduced 1968; discontinued 1976.

Perf.: $1000 **Exc.:** $820 **VGood:** $750

BERETTA BL-6

Over/under; boxlock; 12-, 20-ga; 2 3/4", 3" chambers; 26", 28", 30" chrome-moly steel barrels; vent rib; Improved/Modified, Modified/Full chokes; ramp front sight with fluorescent inserts; best-grade hand-checkered fancy European walnut pistol-grip stock, forearm; single selective triggers; ejectors; automatic safety; Monoblock design; false sideplates; extensive engraving. Introduced 1973; discontinued 1976.

Perf.: $1400 **Exc.:** $1100 **VGood:** $1000

BERETTA GOLDEN SNIPE

Over/under; boxlock; 12-, 20-ga.; 2 3/4", 3" chambers; 26", 28", 30" barrels; Improved/Modified, Modified/Full, Full/Full chokes; vent rib; hand-checkered European walnut pistol-grip stock, forearm; non-selective, selective trigger; auto ejectors; nickel steel receiver. Introduced 1955; discontinued 1975.
Nonselective trigger
Exc.: $540 **VGood:** $460 **Good:** $400
Selective trigger
Exc.: $580 **VGood:** $500 **Good:** $440

Beretta Golden Snipe Deluxe

Same specs as Golden Snipe except finer engraving; better walnut. Available in Skeet, Trap models. Introduced 1955; discontinued 1975.

Exc.: $640 **VGood:** $520 **Good:** $460

BERETTA GRADE 100

Over/under; sidelock; 12-ga.; 26", 28", 30" barrel; any standard choke combination; hand-checkered straight or pistol-grip European walnut stock, forend; double triggers; automatic ejectors. Introduced post-World War II; discontinued 1960s.

Exc.: $1500 **VGood:** $1000 **Good:** $775

BERETTA GRADE 200

Over/under; sidelock; 12-ga.; 26", 28", 30" barrel; any standard choke combination; hand-checkered select European walnut straight or pistol-grip stock, forend; double triggers; automatic ejectors; chrome-plated bores, action; engraving. Introduced post-World War II; discontinued 1960s.

Exc.: $1750 **VGood:** $1400 **Good:** $1000

BERETTA MODEL 57E

Over/under; boxlock; 12-, 20-ga.; 2 3/4", 3" chambers; 26", 28", 30" vent-rib barrel; hand-checkered European walnut pistol-grip stock, beavertail forend; automatic ejectors; single non-selective trigger; optional single selective trigger; similar to Golden Snipe model but overall quality is higher. Introduced 1955; discontinued 1967.

Exc.: $675 **VGood:** $520 **Good:** $450
Single selective trigger
Exc.: $700 **VGood:** $560 **Good:** $490

BERETTA MODEL 682 COMPETITION GOLD SKEET

Over/under; low-profile, improved boxlock; 12-, 20-, 28-ga., 410; 26", 28" Skeet/Skeet barrels; weighs 7 1/2 lbs.; white bead front sight; highly select hand-checkered European walnut Skeet-type stock, beavertail forearm with matte finish; manual safety; single selective trigger with adjustable length of pull; recoil pad; gray-finished receiver. Introduced 1991; still in production. **LMSR:** $2597

Perf.: $2200 **Exc.:** $1350 **VGood:** $1000
Four-barrel set in all four gauges
Perf.: $4895 **Exc.:** $3000 **VGood:** $2700

Beretta Model 682 Competition Gold Skeet Super

Same specs as Model 682 Competition Gold Skeet except adjustable comb and buttpads; ported barrel. Introduced 1991; still in production. **LMSR:** $3006

Perf.: $2650 **Exc.:** $1720 **VGood:** $1500

Beretta Model 682 Competition Gold Sporting

Same specs as Model 682 Competition Gold Skeet except 12-ga.; 2 3/4" chamber; 28", 30", 32" barrels; Mobilchokes; 1/2" wide ventilated rib; weighs 7 3/8 lbs.; highly select hand-checkered European walnut Sporting stock, Schnabel forend with matte finish. Made in Italy. Still imported. **LMSR:** $2915

Perf.: $2750 **Exc.:** $2500 **VGood:** $2300

Beretta Model 682 Competition Gold Sporting Ported

Same specs as Model 682 Competition Gold Skeet except 12-ga.; 2 3/4" chamber; 28", 30" ported barrels; Mobilchokes; 1/2" wide ventilated rib; weighs 7 3/8 lbs.; highly select hand-checkered European walnut Sporting stock, Schnabel forend with matte finish. Made in Italy. Still imported. **LMSR:** $2649

Perf.: $2300 **Exc.:** $1350 **VGood:** $1000

Beretta Model 682 Competition Gold Sporting Super

Same specs as Model 682 Competition Gold Skeet except 12-ga.; 2 3/4" chamber; 28", 30" ported barrels; Mobilchokes; 1/2" wide ventilated tapered rib; weighs 7 3/8 lbs.; adjustable highly select hand-checkered European walnut Sporting stock, Schnabel forend with matte finish. Made in Italy. Still imported. **LMSR:** $3017

Perf.: $2600 **Exc.:** $1750 **VGood:** $1000

Beretta Model 682 Competition Gold Trap

Same specs as Model 682 Competition Gold Skeet except 12-ga.; 2 3/4" chamber; 30", 32" barrel; Mobilchokes; stepped, tapered rib; highly select hand-checkered European walnut Monte Carlo stock. Made in Italy. Introduced 1987; still imported. **LMSR:** $2789

Perf.: $2500 **Exc.:** $1700 **VGood:** $1200
Combo with extra single 32", 34" over barrel
Perf.: $3000 **Exc.:** $2100 **VGood:** $1500

Beretta Model 682 Competition Gold Trap Live Bird

Same specs as Model 682 Competition Gold Skeet except 12-ga.; 2 3/4" chamber; 30" barrel; Mobilchokes; Strada or flat tapered rib; highly select hand-checkered European walnut tap-style stock, beavertail forend with matte finish. Still in production. **LMSR:** $2605

Perf.: $2200 **Exc.:** $1500 **VGood:** $1000

Beretta Model 686 Field

Beretta Model 686 Field EL Gold Perdiz

Beretta Model 686 Field Essential

Beretta Model 686 Competition Sporting Onyx

Beretta Model 686 Competition Sporting Silver Perdiz

Beretta Over/Under Shotguns
Model 682 Competition Gold Trap Mono
Model 682 Competition Gold Trap Mono Super
Model 682 Competition Gold Trap Super
Model 682 Competition Sporting Continental Course
Model 685
Model 686 Competition Skeet Silver Perdiz
Model 686 Competition Sporting Collection Sport
Model 686 Competition Sporting Onyx
Model 686 Competition Sporting Onyx English Course
Model 686 Competition Sporting Silver Perdiz
Model 686 Competition Sporting Silver Pigeon
Model 686 Competition Trap International
Model 686 Field EL Gold Perdiz
Model 686 Field Essential

Beretta Model 682 Competition Gold Trap Mono
Same specs as Model 682 Competition Gold Skeet except 12-ga.; 2 3/4" chambers; single 32", 34" over barrel; Multichokes; vent rib; highly select hand-checkered European walnut Monte Carlo stock. Made in Italy. Introduced 1987; still imported. **LMSR $2734**
   Perf.: $2300 **Exc.:** $1500 **VGood:** $1000

Beretta Model 682 Competition Gold Trap Mono Super
Same specs as Model 682 Competition Skeet except 12-ga.; 2 3/4" chambers; single 32", 34" over barrel; fixed or Mobilchokes; vent rib; highly select hand-checkered European walnut Monte Carlo stock. Made in Italy. Introduced 1991; still imported. **LMSR $3083**
   Perf.: $2500 **Exc.:** $1650 **VGood:** $1000

Beretta Model 682 Competition Gold Trap Super
Same specs as Model 682 Competition Gold Skeet except 12-ga.; 2 3/4" chamber; 30", 32" ported barrels; Mobilchokes; stepped, tapered rib; adjustable highly select hand-checkered European walnut Monte Carlo stock with matte finish; silver sideplates with hand engraved game scenes. Introduced 1991; still imported. **LMSR $3083**
   Perf.: $2500 **Exc.:** $1650 **VGood:** $1000
Combo with extra single 32", 34" over barrel
   Perf.: $3000 **Exc.:** $2000 **VGood:** $1400

Beretta Model 682 Competition Sporting Continental Course
Same specs as Model 682 Competition Gold Skeet except 12-ga.; 2 3/4" chambers; 28", 30" Mobilchoke barrel; reverse tapered rib; Highly select hand-checkered European walnut stock, Schnabel forend with matte finish; palm swell; black receiver; made for English Sporting Clays competition. Introduced 1994; still in production. **LMSR: $2796**
   Perf.: $2400 **Exc.:** $1500 **VGood:** $1000

BERETTA MODEL 685
Over/under; boxlock; 12-, 20-ga.; 2 3/4" or 3" chambers; 26", 28", 30" barrels; standard chokes; hand-checkered walnut stock; steel receiver; single trigger; extractors; light engraving. Discontinued 1986.
   Perf.: $620 **Exc.:** $490 **VGood:** $400

BERETTA MODEL 686 COMPETITION SKEET SILVER PERDIZ
Over/Under; low-profile, improved boxlock; 12-ga.; 2 3/4" chambers; 28" Skeet/Skeet barrels; tapered rib; weighs 7 5/8 lbs.; white front bead sight; select checkered walnut stock, beavertail forend with matte finish; manual safety; single trigger; recoil pad; silver receiver with scroll engraving. Made in Italy. Introduced 1994; still imported. **LMSR: $1434**
   Perf.: $1150 **Exc.:** $800 **VGood:** $600

Beretta Model 686 Competition Sporting Collection Sport
Same specs as Model 686 Competition Skeet Silver Perdiz except 12-ga.; 3" chambers; 28" Strada-rib barrel; Mobilchokes; select checkered multi-colored walnut stock, beavertail forend with matte finish; single selective trigger; multi-colored receiver. Made in Italy. Introduced 1996; still imported. **LMSR $1499**
   Perf.: $1200 **Exc.:** $850 **VGood:** $700

Beretta Model 686 Competition Sporting Onyx
Same specs as Model 686 Competition Skeet Silver Perdiz except 12-ga.; 3" chambers; 28", 30" Strada-rib barrels; Mobilchokes; single selective trigger; semi-matte black receiver with gold lettering. Made in Italy. Introduced 1994; still imported. **LMSR: $1427**
   Perf.: $1150 **Exc.:** $900 **VGood:** $550

Beretta Model 686 Competition Sporting Onyx English Course
Same specs as Model 686 Competition Skeet Silver Perdiz except 12-ga.; 3" chamber; 28" vent-rib barrels; Mobilchokes; semi-matte black receiver; designed for English Sporting Clays competition. Made in Italy. Introduced 1991; discontinued 1992.
   Perf.: $1700 **Exc.:** $1100 **VGood:** $950

Beretta Model 686 Competition Sporting Silver Perdiz
Same specs as Model 686 Competition Skeet Silver Perdiz except 12-, 20-ga.; 3" chamber; 28", 30" barrels; Mobilchokes; Strada rib (12-ga.), flat rib (20-ga.); weighs 7 3/4 lbs.; checkered select walnut stock, semi-beavertail forend with matte finish. Made in Italy. Introduced 1994; discontinued 1996.
   Perf.: $1200 **Exc.:** $775 **VGood:** $500
Combo with two-barrel set
   Perf.: $1800 **Exc.:** $1300 **VGood:** $1100

Beretta Model 686 Competition Sporting Silver Pigeon
Same specs as Model 686 Competition Skeet Silver Perdiz except 12-, 20-ga.; 3" chamber; 28", 30" barrels; Mobilchokes; Strada rib (12-ga.), flat rib (20-ga.); weighs 7 3/4 lbs.; checkered select walnut stock, Schnabel forend with matte finish; single selective trigger. Made in Italy. Introduced 1996; still imported. **LMSR $1573**
   Perf.: $1250 **Exc.:** $900 **VGood:** $600

Beretta Model 686 Competition Trap International
Same specs as Model 686 Competition Skeet Silver Perdiz except 30" Improved Modified/Full barrels; vent rib; select checkered walnut trap-style stock, forearm. Made in Italy. Introduced, 1994; discontinued 1995.
   Perf.: $1000 **Exc.:** $600 **VGood:** $525

BERETTA MODEL 686 FIELD EL GOLD PERDIZ
Over/under; low-profile, improved boxlock; 12-, 20-ga.; 3" chamber; 26", 28" vent-rib barrels; Mobilchokes; weighs 6 7/8 lbs.; highly select checkered walnut stock, beavertail forend with gloss finish; auto safety; single selective trigger; rubber recoil pad; silver receiver with full sideplates and floral scroll engraving. Made in Italy. Introduced 1994; still imported. **LMSR: $2266**
   Perf.: $1800 **Exc.:** $1120 **VGood:** $910

Beretta Model 686 Field Essential
Same specs as Model 686 Field EL Gold Perdiz except 3" chamber; weighs 6 3/4 lbs.; matte black plain receiver. Made in Italy. Introduced 1994; still imported. **LMSR: $1186**
   Perf.: $910 **Exc.:** $590 **VGood:** $510

SHOTGUNS

Beretta Over/Under Shotguns

Model 686 Field Onyx
Model 686 Field Onyx Magnum
Model 686 Field Silver Perdiz
Model 686 Field Silver Pigeon
Model 686 Field Ultralight
Model 687 Competition Skeet EELL Diamond Pigeon
Model 687 Competition Sporting EELL Diamond Pigeon
Model 687 Competition Sporting EL Gold Pigeon
Model 687 Competition Sporting L Silver Pigeon
Model 687 Competition Sporting Silver Perdiz English Course
Model 687 Competition Trap EELL Diamond Pigeon
Model 687 Competition Trap EELL Diamond Pigeon Combo
Model 687 Competition Trap EELL Diamond Pigeon Mono
Model 687 Field EELL Diamond Pigeon
Model 687 Field EL Gold Pigeon
Model 687 Field EL Onyx
Model 687 Field Golden Onyx
Model 687 Field L Onyx
Model 687 Field L Silver Pigeon
S-55B
S-56E
S-58 Skeet

Beretta Model 686 Field Onyx

Same specs as Model 686 Field EL Gold Perdiz except 12-, 20-ga.; 3", 3 1/2" chambers; weighs 6 7/8 lbs.; checkered select walnut straight English-style or pistol-grip stock, semi-beavertail forend (Schnabel forend for English stock) with matte finish; semi-matte black receiver; ventilated rib. Imported from Italy by Beretta U.S.A. Introduced 1988; still imported. **LMSR:** $1399

Perf.: $1100   **Exc.:** $650   **VGood:** $540

Beretta Model 686 Field Onyx Magnum

Same specs as Model 686 Field EL Gold Perdiz except 12-ga.; 3 1/2" chambers; checkered select walnut straight English-style or pistol-grip stock, semi-beavertail forend (Schnabel forend for English stock) with matte finish; matte black plain receiver. Made in Italy. Introduced 1990; discontinued 1993.

Perf.: $1175   **Exc.:** $700   **VGood:** $600

Beretta Model 686 Field Silver Perdiz

Same specs as Model 686 Field EL Gold Perdiz except 12-, 20-, 28-ga.; 2 3/4", 3" chambers; checkered

select walnut straight English-style or pistol-grip stock, semi-beavertail forend (Schnabel forend for English stock) with gloss finish; silver receiver with floral scroll engraving. Made in Italy. Introduced 1994; discontinued 1996.

Perf.: $1100   **Exc.:** $750   **VGood:** $650

Combo with two-barrel set for 20-, 28-ga.

Perf.: $1475   **Exc.:** $1250   **VGood:** $1100

Beretta Model 686 Field Silver Pigeon

Same specs as Model 686 Field EL Gold Perdiz except 12-, 20-, 28-ga.; 2 3/4", 3" chambers; highly select checkered walnut stock, Schnabel forend with gloss finish; silver receiver with floral scroll engraving. Made in Italy. Introduced 1996; still imported. **LMSR:** $1544

Perf.: $1225   **Exc.:** $825   **VGood:** $575

Beretta Model 686 Field Ultralight

Same specs as Model 686 Field EL Gold Perdiz except 12-ga.; 2 3/4" chamber; weighs 5 3/4 lbs.; checkered select walnut stock, Schnabel forend with matte finish; aluminum alloy receiver with titanium plate; game scene engraving. Made in Italy. Introduced 1994; still imported. **LMSR:** $1716

Perf.: $1450   **Exc.:** $1000   **VGood:** $750

BERETTA MODEL 687 COMPETITION SKEET EELL DIAMOND PIGEON

Over/under; low-profile, improved boxlock; 12-ga.; 2 3/4" chambers; 28" Skeet/Skeet barrels; tapered rib; weighs 7 1/2 lbs.; white front bead sight; highly select checkered walnut stock, beavertail forend with matte finish; rubber recoil pad; manual safety; single selective trigger; silver receiver with full sideplates; hand-engraved game scenes; comes with a case. Made in Italy. Introduced 1994; still imported. **LMSR:** $4590

Perf.: $3950   **Exc.:** $2600   **VGood:** $2200

Four-barrel Set in 12-, 20-, 28-ga., .410

Perf.: $6750   **Exc.:** $4900   **VGood:** $4000

Beretta Model 687 Competition Sporting EELL Diamond Pigeon

Same specs as Model 687 Competition Skeet EELL Diamond Pigeon except 12-ga.; 2 3/4", 3" chambers; 28", 30" Strada-rib barrels; Mobilchokes; highly select checkered walnut stock, Schnabel forend with matte finish; stock oval plate. Made in Italy. Introduced 1994; still imported. **LMSR:** $5098

Perf.: $4350   **Exc.:** $2700   **VGood:** $2000

Beretta Model 687 Competition Sporting EL Gold Pigeon

Same specs as Model 687 Competition Skeet EELL Diamond Pigeon except 12-ga.; 3" chambers; 28", 30" Strada-rib barrels; Mobilchokes; weighs 7 1/4 lbs.; highly select checkered walnut stock, Schnabel forend with matte finish; stock oval plate; silver receiver with full sideplates; gold engraved animals. Made in Italy. Introduced 1994; still imported. **LMSR:** $3489

Perf.: $2800   **Exc.:** $1500   **VGood:** $1100

Beretta Model 687 Competition Sporting L Silver Pigeon

Same specs as Model 687 Competition Skeet EELL Diamond Pigeon except 12-, 20-ga.; 3" chambers; 28", 30" barrel; Mobilchokes; Strada rib (12-ga.), flat rib (20-ga.); weighs 7 1/4 lbs.; select checkered walnut stock, Schnabel forend with matte finish; stock oval plate. Made in Italy. Introduced 1994; still imported. **LMSR:** $2474

Perf.: $2150   **Exc.:** $1600   **VGood:** $1200

Beretta Model 687 Competition Sporting Silver Perdiz English Course

Same specs as Model 687 Competition Skeet EELL Diamond Pigeon except 28" vent-rib barrels; Mobilchokes; designed for English Sporting Clays competition. Made in Italy. Introduced 1991; discontinued 1992.

Perf.: $2350   **Exc.:** $1400   **VGood:** $1125

Beretta Model 687 Competition Trap EELL Diamond Pigeon

Same specs as Model 687 Competition Skeet EELL Diamond Pigeon except 30" barrel; Mobilchokes; stepped, tapered rib; weighs 8 7/8 lbs.; highly select checkered walnut Monte Carlo stock, beavertail forend with matte finish; single adjustable trigger. Made in Italy. Introduced 1994; still imported. **LMSR:** $4991

Perf.: $4250   **Exc.:** $2700   **VGood:** $2200

Beretta Model 687 Competition Trap EELL Diamond Pigeon Combo

Same specs as Model 687 Competition Skeet EELL Diamond Pigeon except two-barrel set with 32" over/under and single 34" over; Mobilchokes; stepped, tapered rib; weighs 8 7/8 lbs.; highly select checkered walnut Monte Carlo stock, beavertail forend with matte finish; single adjustable trigger. Made in Italy. Introduced 1994; still imported. **LMSR:** $6292

Perf.: $5000   **Exc.:** $3500   **VGood:** $3000

Beretta Model 687 Competition Trap EELL Diamond Pigeon Mono

Same specs as Model 687 Competition Skeet EELL Diamond Pigeon except 32", 34" barrel; Full or Mobilchokes; stepped, tapered rib; weighs 8 1/8 lbs.; highly select checkered walnut Monte Carlo stock, beavertail forend with gloss finish; single adjustable trigger. Made in Italy. Introduced 1994; still imported. **LMSR:** $5241

Perf.: $3250   **Exc.:** $2600   **VGood:** $2300

BERETTA MODEL 687 FIELD EELL DIAMOND PIGEON

Over/under; low-profile, improved boxlock; 12-, 20-, 28-ga.; 2 3/4", 3" chambers; 26", 28" vent-rib barrels; Mobilchokes; weighs 6 7/8 lbs.; metal front bead sight; highly select checkered walnut stock, Schnabel forend with gloss finish; rubber recoil pad; stock oval plate; auto safety; single selective trigger; silver receiver with full sideplates; hand-engraved game scenes; comes with case. Made in Italy. Introduced 1994; still imported. **LMSR:** $4850

Perf.: $4175   **Exc.:** $2500   **VGood:** $2100

Beretta Model 687 Field EL Gold Pigeon

Same specs as Model 687 Field EELL Diamond Pigeon except 12-, 20-, 28-ga., .410; 2 3/4", 3" chambers; silver receiver with full sideplates; gold engraved animals. Made in Italy. Introduced 1994; still imported. **LMSR:** $3446

Perf.: $2925   **Exc.:** $1975   **VGood:** $1700

Beretta Model 687 Field EL Onyx

Same specs as Model 687 Field EELL Diamond Pigeon except 12-, 20-ga.; 3" chambers; black receiver with dummy sideplates; engraving. Made in Italy. Introduced 1990; discontinued 1991.

Perf.: $2350   **Exc.:** $1775   **VGood:** $1520

Beretta Model 687 Field Golden Onyx

Same specs as Model 687 Field EELL Diamond Pigeon except 12-, 20-ga.; 3" chambers; black receiver; gold engraving. Made in Italy. Introduced 1988; discontinued 1989.

Perf.: $1400   **Exc.:** $900   **VGood:** $795

Beretta Model 687 Field L Onyx

Same specs as Model 687 Field EELL Diamond Pigeon except 12-, 20-ga.; 3" chambers; black receiver; engraved game scenes. Made in Italy. Introduced 1990; discontinued 1991.

Perf.: $1300   **Exc.:** $860   **VGood:** $720

Beretta Model 687 Field L Silver Pigeon

Same specs as Model 687 Field EELL Diamond Pigeon except 12-, 20-ga.; 3" chambers; silver receiver with game scenes. Made in Italy. Introduced 1994; still imported. **LMSR:** $2031

Perf.: $1625   **Exc.:** $950   **VGood:** $800

BERETTA S-55B

Over/under; boxlock; 12-, 20-ga.; 2 3/4", 3" chambers; 26", 28", 30" vent-rib barrels; standard choke combos; checkered European walnut pistol-grip stock, forend; selective single trigger; extractors. Introduced 1977; no longer in production.

Perf.: $565   **Exc.:** $455   **VGood:** $390

BERETTA S-56E

Over/under; boxlock; 12-, 20-ga.; 2 3/4", 3" chambers; 26", 28", 30" vent-rib barrels; standard choke combos; checkered European walnut pistol-grip stock, forend; selective single trigger; ejectors; scroll engraved. Introduced 1977; no longer in production.

Perf.: $620   **Exc.:** $525   **VGood:** $480

BERETTA S-58 SKEET

Over/under; boxlock; 12-, 20-ga.; 2 3/4", 3" chambers; 26" Skeet/Skeet barrels; vent rib; checkered European walnut Skeet-style stock, forend; recoil pad; selective single trigger; ejectors; scroll engraved. Introduced

Beretta Model 687 Field EL

Beretta Model 687 Field Golden Onyx

Beretta Silver Snipe

Beretta SO-2 Presentation

Beretta SO-3

Beretta SO-4

Beretta SO-5

Beretta SO-6 EELL

SHOTGUNS

Beretta Over/Under Shotguns

S-58 Trap
Silver Snipe
SO-2 Presentation
SO-3
SO-4
SO-5
SO-6
SO-6 EELL
SO-7
SO-9

1977; no longer in production.
Perf.: $750　　**Exc.:** $620　　**VGood:** $500

Beretta S-58 Trap
Same specs as S-58 Skeet except 30″ Improved Modified/Full; checkered European walnut Monte Carlo stock, forend. Introduced 1977; no longer in production.
Perf.: $680　　**Exc.:** $520　　**VGood:** $450

BERETTA SILVER SNIPE
Over/under; boxlock; 12-, 20-ga.; 2 3/4″, 3″ chambers; 26″, 28″, 30″ barrels; Improved/Modified, Modified/Full, Full/Full chokes; plain or vent rib; hand-checkered European walnut pistol-grip stock, forearm; nonselective, selective trigger; extractors; nickel steel receiver. Introduced 1955; discontinued 1967.
Nonselective trigger
　Exc.: $350　　**VGood:** $280　　**Good:** $240
Selective trigger
　Exc.: $430　　**VGood:** $350　　**Good:** $310

BERETTA SO-2 PRESENTATION
Over/under; sidelock; 12-ga.; 26″ Improved/Modified, 28″ Modified/Full Boehler anti-rust barrels; vent rib; hand-checkered European walnut straight or pistol-grip stock; chrome-nickel receiver; all interior parts chromed; checkered trigger, safety, top lever; scroll engraving; silver pigeon inlaid in top lever. Made in Field, Skeet, Trap models. Introduced 1965; no longer in production.
Perf.: $3000　　**Exc.:** $2500　　**VGood:** $2200

BERETTA SO-3
Over/under; sidelock; 12-ga.; 26″ Improved/Modified, 28″ Modified/Full Boehler anti-rust barrels; vent rib; fancy hand-checkered European walnut straight or pistol-grip stock; chrome-nickel receiver; all interior parts chromed; checkered trigger, safety, top lever; profuse scroll and relief engraving; silver pigeon inlaid in top lever. Made in Field, Skeet, Trap models. Made in Italy. Introduced 1965; no longer in production.
Perf.: $6500　　**Exc.:** $4800　　**VGood:** $4300

BERETTA SO-4
Over/under; hand-detachable sidelock; 12-ga.; 26″ Improved/Modified, 28″ Modified/Full Boehler anti-rust barrels; vent rib; fancy hand-checkered European full-grain walnut straight or pistol-grip stock; chrome-nickel receiver; all interior parts chromed; checkered trigger, safety, top lever; elaborate scroll and relief engraving; silver pigeon inlaid in top lever. Made in Field, Skeet, Trap models. Made in Italy. Introduced 1965; no longer in production.
Perf.: $7500　　**Exc.:** $6000　　**VGood:** $5500

BERETTA SO-5
Over/under; hand-detachable sidelock; 12-ga.; hand-made gun built to customer specs; fancy hand-checkered European full-grain walnut straight or pistol-grip stock; chrome-nickel receiver; all interior parts chromed; checkered trigger, safety, top lever; elaborate scroll and relief engraving; Crown Grade symbol inlaid in top lever. Made in Field, Skeet, Trap models. Made in Italy. Still in production on special-order basis. **LMSR: $13,000**
Perf.: $9500　　**Exc.:** $7750　　**VGood:** $6500

BERETTA SO-6
Over/under; hand-detachable sidelock; 12-ga.; hand-made gun built to customer specs; fancy hand-checkered European full-grain walnut straight or pistol-grip stock; chrome-nickel receiver; all interior parts chromed; checkered trigger, safety, top lever; elaborate scroll and relief engraving; Crown Grade symbol inlaid in top lever. Made in Skeet, Trap, Sporting models. Made in Italy. Still in production on special-order basis.
LMSR: $17,500
Perf.: $15,500　**Exc.:** $10,000　**VGood:** $8000

Beretta SO-6 EELL
Same specs as SO-6 except produced on custom order to buyer's specs; special engraving and inlays. Made in Italy. Introduced 1989; still imported. **LMSR: $28,000**
Perf.: $25,000　**Exc.:** $17,000　**VGood:** $12,500

BERETTA SO-7
Over/under; sidelock; 12-ga.; 26″ Improved/Modified, 28″ Modified/Full Boehler anti-rust barrels; vent rib; fancy hand-checkered European full-grain walnut straight or pistol-grip stock; chrome-nickel receiver; all interior parts chromed; checkered trigger, safety, top lever; elaborate scroll and relief engraving; silver pigeon inlaid in top lever. Made in Field model. Made in Italy. Introduced 1948; no longer in production.
Perf.: $9000　　**Exc.:** $8300　　**VGood:** $7500

BERETTA SO-9
Over/under; hand-detachable sidelock; 12-, 20-, 28-ga., .410; handmade gun built to customer specs; Boehler anti-rust steel barrels; fixed or screw-in Mobilchoke system; fancy hand-checkered European full-grain walnut straight or pistol-grip stock; chrome-nickel receiver; all interior parts chromed; checkered trigger, safety, top lever; highly decorated, engraved. Made in Skeet, Trap, Sporting models. Made in Italy. Introduced 1989; still imported. **LMSR: $31,000**
Perf.: $27,000　**Exc.:** $18,000　**VGood:** $14,000

Beretta Side-By-Side Shotguns
GR-2
GR-3
GR-4
Model 409PB
Model 410
Model 410E
Model 411E
Model 424
Model 426
Model 452
Model 452 EELL
Model 626
Model 626 Onyx
Model 627 EL

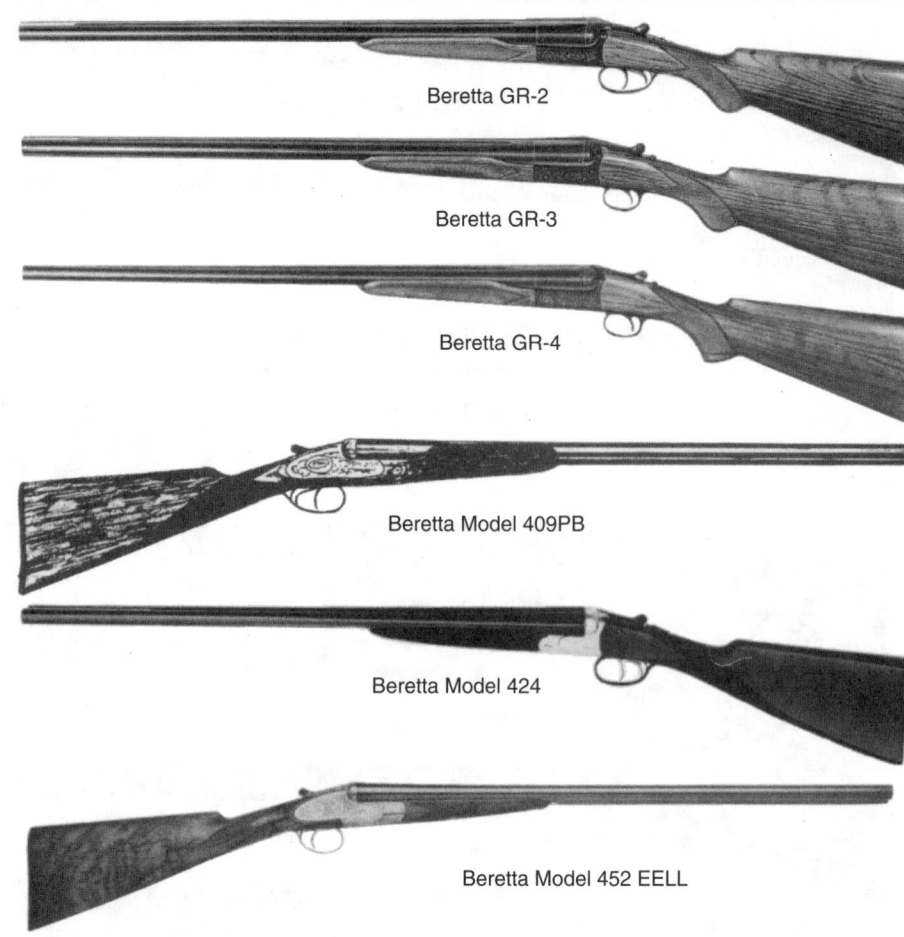

Beretta GR-2

Beretta GR-3

Beretta GR-4

Beretta Model 409PB

Beretta Model 424

Beretta Model 452 EELL

BERETTA SIDE-BY-SIDE SHOTGUNS

BERETTA GR-2
Side-by-side; boxlock; 12- (2 3/4"), 20-ga. (3"); 26", 28" 30" barrels; standard choke combos; checkered European walnut pistol-grip stock, forend; plain extractors; double triggers. Introduced 1968; discontinued 1976.
Perf.: $670 **Exc.:** $560 **VGood:** $520

BERETTA GR-3
Side-by-side; boxlock; 12-, 20-ga.; 3" chambers; 26", 28" 30" barrels; standard choke combos; checkered European walnut pistol-grip stock, forend; plain extractors; single selective trigger; recoil pad. Introduced 1968; discontinued 1976.
Perf.: $780 **Exc.:** $660 **VGood:** $575

BERETTA GR-4
Side-by-side; boxlock; 12-ga.; 2 3/4" chambers; 26", 28" 30" barrel; standard choke combos; checkered select European walnut pistol-grip stock, forend; auto ejectors; single selective trigger; engraving. Introduced 1968; discontinued 1976.
Perf.: $900 **Exc.:** $780 **VGood:** $650

BERETTA MODEL 409PB
Side-by-side; boxlock; hammerless; 12-, 16-, 20-, 28-ga.; 27 1/2", 28 1/2", 30" barrels; Improved/Modified, Modified/Full chokes; hand-checkered European walnut straight or pistol-grip stock, beavertail forearm; double triggers; plain extractors; engraved action. Introduced 1934; discontinued 1964.
Exc.: $650 **VGood:** $500 **Good:** $420

BERETTA MODEL 410
Side-by-side; boxlock; hammerless; 10-ga.; 3 1/2" chambers; 27 1/2", 28 1/2", 30" barrels; Improved/Modified, Modified/Full chokes; hand-checkered European walnut straight stock; recoil pad; plain extractors; double triggers. Introduced 1934; discontinued 1963.
Exc.: $900 **VGood:** $720 **Good:** $600

Beretta Model 410E
Same specs as Model 410 except 12-, 16-, 20-, 28-ga.; hand-checkered European walnut straight or pistol-grip stock, beavertail forend; double triggers; automatic ejectors; engraved action. Introduced 1934; discontinued 1964.
Exc.: $725 **VGood:** $590 **Good:** $525

BERETTA MODEL 411E
Side-by-side; boxlock; hammerless; 12-, 16-, 20-, 28-ga.; 27 1/2", 28 1/2", 30" barrels; Improved/Modified, Modified/Full chokes; hand-checkered European walnut straight or pistol-grip stock; beavertail forend; sideplates; engraved action; automatic ejectors; double triggers. Introduced 1934; discontinued 1964.
Exc.: $1000 **VGood:** $830 **Good:** $780

BERETTA MODEL 424
Side-by-side; boxlock; hammerless; 12-, 20-ga.; 26" Improved/Modified, 28" Modified/Full barrels; checkered European walnut straight-grip stock, forend; plain extractors; border engraving on action. Introduced 1977; discontinued 1984. Redesignated as Model 625.
Perf.: $850 **Exc.:** $620 **VGood:** $590

BERETTA MODEL 426
Side-by-side; boxlock; hammerless; 12-, 20-ga.; 26" Improved/Modified, 28" Modified/Full barrels; high-grade checkered European walnut straight-grip stock, forend; ejectors; finely engraved action; silver pigeon inlaid on top lever. Introduced 1977; discontinued 1984. Redesignated as Model 626.
Perf.: $850 **Exc.:** $620 **VGood:** $590

BERETTA MODEL 452
Side-by-side; detachable sidelock; 12-ga; 26", 28", 30" barrels; choked to customer specs; weighs 6 3/4 lbs.; highly figured European walnut stock; double bolting; ejectors; double triggers; optional single non-selective trigger; manual safety; coin-finished frame. Essentially custom built. Made in Italy. Introduced 1990; still in production. **LMSR: $22,500**
Perf.: $20,000 **Exc.:** $12,000 **VGood:** $9500

Beretta Model 452 EELL
Same specs as Model 452 except walnut briar stock; full engraving. Introduced 1992; still imported. **LMSR: $31,000**
Perf.: $27,000 **Exc.:** $18,000 **VGood:** $13,500

BERETTA MODEL 626
Side-by-side; boxlock; 12-, 20-ga.; 2 3/4" chambers; 26", 28" barrels; standard choke combos; concave matted rib; hand-checkered European walnut straight or pistol-grip stock, forearm; coil springs; double underlugs; bolts; double triggers; auto safety; ejectors. Made in Italy. Introduced 1985; discontinued 1988. Formerly designated as Beretta 426.
Perf.: $850 **Exc.:** $620 **VGood:** $590

Beretta Model 626 Onyx
Same specs as Model 626 except 12-, 20-ga.; 3 1/2", 3" chambers; choke tubes; select hand-checkered European walnut stock, forearm. Introduced 1988; discontinued 1993.
Perf.: $1350 **Exc.:** $800 **VGood:** $650

BERETTA MODEL 627 EL
Side-by-side; boxlock; 12-, 20-ga.; 2 3/4", 3" chambers; 26", 28" barrels; fixed or Mobilchokes; front, center bead sights; concave matted barrel rib; fine-quality hand-checkered European walnut straight or pistol-grip stock, forend; coil springs; double underlugs, bolts; single trigger; auto safety ejectors; fine engraving; gold inlays on dummy sideplates; comes with case. Made in Italy. Introduced 1985; discontinued 1993.
Perf.: $2700 **Exc.:** $1600 **VGood:** $1400

Beretta Model 626

Beretta Model SO-7 SXS

Beretta FS-1

Beretta Mark II

Beretta Model 410E

Beretta Side-By-Side Shotguns
Model 627 EELL
Silver Hawk Featherweight
Silver Hawk Magnum
SO-6 SXS
SO-7 SXS

Beretta Single Shot & Slide-Action Shotguns
FS-1
Gold Pigeon
Mark II
Model 680 Competition Mono Trap
Model 680 Competition Trap
Model 680 Competition Trap Combo
Model 680 Competition Skeet
Ruby Pigeon
Silver Pigeon
SL-2
TR-1
TR-2

Beretta Model 627 EELL
Same specs as 627 EL except profuse engraving on dummy sideplates. Made in Italy. Introduced 1985; discontinued 1993.
Perf.: $4500 **Exc.:** $3000 **VGood:** $2700

BERETTA SILVER HAWK FEATHERWEIGHT
Side-by-side; boxlock; 12-, 16-, 20-, 28-ga.; 26" to 32" barrels; all standard choke combos; matted rib; hand-checkered European walnut stock, beavertail forearm; plain extractors; double or non-selective single trigger. Introduced 1954; discontinued 1967.
Single trigger
Exc.: $400 **VGood:** $310 **Good:** $270
Double triggers
Exc.: $350 **VGood:** $260 **Good:** $220

Beretta Silver Hawk Magnum
Same specs as Silver Hawk Featherweight except 10- (3 1/2"), 12-ga. (3"); 30", 32" barrels; chrome-plated bores; raised rib; recoil pad. Introduced 1954; discontinued 1967.
Single trigger
Exc.: $500 **VGood:** $410 **Good:** $370
Double triggers
Exc.: $450 **VGood:** $360 **Good:** $300

BERETTA SO-6 SXS
Side-by-side; sidelock; 12-ga.; 26", 30" barrels; any choke; walnut stock; ejectors; single selective trigger; entire gun made-to-order. No longer in production.
Perf.: $6200 **Exc.:** $5200 **VGood:** $4800

BERETTA SO-7 SXS
Side-by-side; sidelock; 12-ga.; 26", 30" barrels; any choke; select walnut stock; ejectors; single selective trigger; entire gun made-to-order. No longer in production.
Perf.: $8500 **Exc.:** $7000 **VGood:** $6200

BERETTA SINGLE SHOT & SLIDE-ACTION SHOTGUNS

BERETTA FS-1
Single barrel; single shot; hammerless under-lever; 12-, 16-, 20-, 28-ga., 410; 26" (28-ga., 410), 28" (16-, 20-ga.), 30" barrel (12-ga.); Full choke; hand-checkered pistol-grip stock, forearm; folds to length of barrel; barrel release ahead of trigger guard. Advertised as the Companion. Introduced 1959; no longer in production.
Exc.: $110 **VGood:** $90 **Good:** $70

BERETTA GOLD PIGEON
Slide action; 12-ga.; 5-shot magazine; 26", 30", 32" vent-rib barrels; standard chokes; hand-checkered walnut pistol-grip stock, beavertail forearm; hand-polished, engine-turned bolt; gold trigger; moderate engraving; inlaid gold pigeon. Introduced 1959; discontinued 1966.
Exc.: $300 **VGood:** $230 **Good:** $190

BERETTA MARK II
Single shot; boxlock; 12-ga.; 32", 34" barrel; wide vent rib; weighs about 8 3/8 lbs.; checkered European walnut Monte Carlo pistol-grip stock, beavertail forend; ejector; recoil pad. Introduced 1972; discontinued 1976.
Perf.: $480 **Exc.:** $390 **VGood:** $340

BERETTA MODEL 680 COMPETITION MONO TRAP
Single shot; boxlock; 12-ga.; 2 3/4" chamber; 32", 34" barrel; ventilated rib; weighs about 8 lbs.; luminous front sight, center bead; hand-checkered European walnut Monte Carlo stock, forearm; automatic ejector; single trigger; deluxe trap recoil pad; silver finish. Introduced 1979; no longer in production.
Perf.: $1050 **Exc.:** $700 **VGood:** $650

Beretta Model 680 Competition Trap
Same specs as Model 680 Competition Mono Trap except over/under; 29 1/2" Improved Modified/Full barrels. No longer imported.
Perf.: $1150 **Exc.:** $800 **VGood:** $710

Beretta Model 680 Competition Trap Combo
Same specs as Model 680 Competition Mono Trap except over/under; two-barrel set with 29 1/2" Improved Modified/Full over/under and 32" or 34" single over barrels. No longer in production.
Perf.: $1150 **Exc.:** $800 **VGood:** $700

Beretta Model 680 Competition Skeet
Same specs as Model 680 Competition Mono Trap except over/under; 28" Skeet/Skeet barrels; smooth recoil pad. No longer in production.
Perf.: $1150 **Exc.:** $800 **VGood:** $700

BERETTA RUBY PIGEON
Slide action; 12-ga.; 5-shot magazine; 26", 30", 32" vent-rib barrels; standard chokes; hand-checkered walnut pistol-grip stock, beavertail forearm; hand-polished, engine-turned bolt; gold trigger; elaborate engraving; inlaid gold pigeon with ruby eye. Introduced 1959; discontinued 1966.
Exc.: $400 **VGood:** $300 **Good:** $250

BERETTA SILVER PIGEON
Slide action; 12-ga.; 5-shot magazine; 26", 30", 32" barrels; standard chokes; hand-checkered walnut pistol-grip stock, beavertail forearm; hand-polished, engine-turned bolt; chromed trigger; light engraving; inlaid silver pigeon. Introduced 1959; discontinued 1966.
Exc.: $160 **VGood:** $140 **Good:** $120

BERETTA SL-2
Slide action; 12-ga.; 3-shot magazine; 26" Improved, 28" Full or Modified, 30" Full barrels; vent rib; checkered European walnut stock, forend. Introduced 1968; discontinued 1971.
Exc.: $225 **VGood:** $175 **Good:** $150

BERETTA TR-1
Single shot; underlever; 12-ga.; 32" vent-rib barrel; checkered European walnut Monte Carlo pistol-grip stock, beavertail forend; engraved frame; recoil pad. Introduced 1968; discontinued 1971.
Exc.: $250 **VGood:** $165 **Good:** $110

BERETTA TR-2
Single shot; underlever; 12-ga.; 32" extended vent-rib barrel; checkered European walnut Monte Carlo pistol-grip stock, beavertail forend; engraved frame; recoil pad. Introduced 1968; discontinued 1972.
Exc.: $250 **VGood:** $165 **Good:** $110

SHOTGUNS

Bernardelli Over/Under Shotguns
Model 115
Model 115E
Model 115L
Model 115S
Model 115 Target
Model 115E Target
Model 115L Target
Model 115S Target
Model 115S Target Monotrap
Model 190
Model 190MC
Model 190 Special
Model 190 Special MS
Model 192
Model 192MS MC
Model 192MS MC WF
Model 220MS

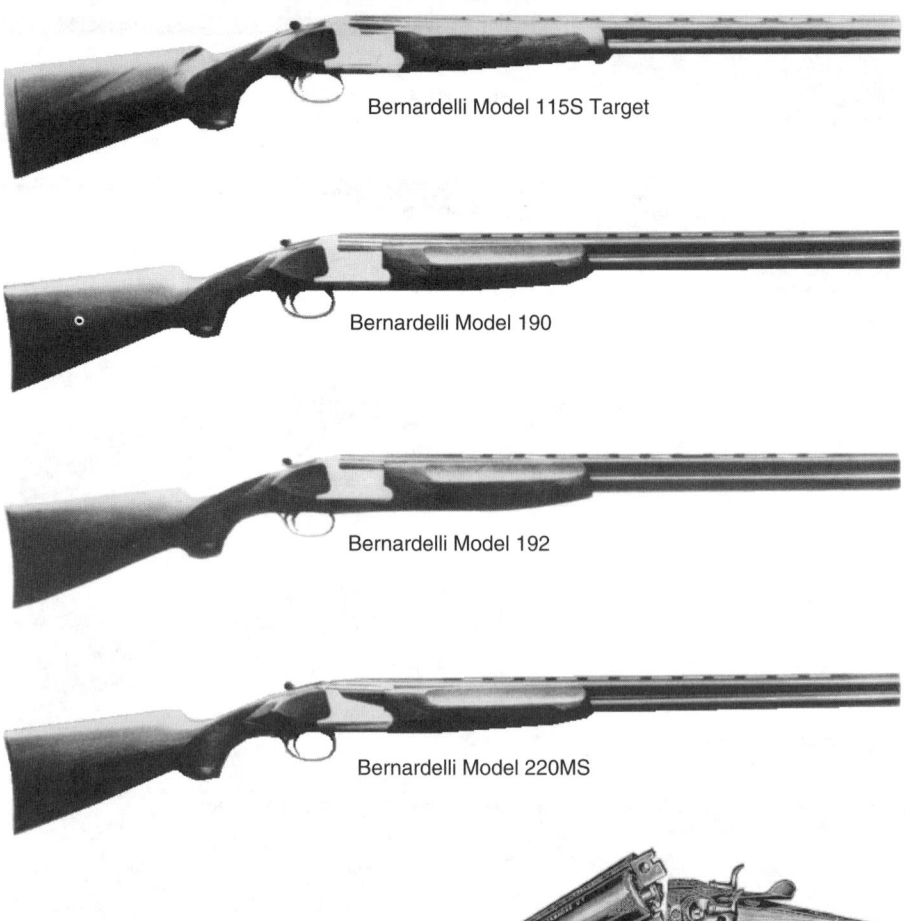

Bernardelli Model 115S Target

Bernardelli Model 190

Bernardelli Model 192

Bernardelli Model 220MS

Bernardelli Italia

Bernardelli Brescia

BERNARDELLI OVER/UNDER SHOTGUNS

BERNARDELLI MODEL 115
Over/under; boxlock; 12-ga.; 2 3/4″ chambers; 25 1/2″, 26 3/4″, 29 1/2″ barrels; Multichokes or standard chokes; concave top rib, vented middle rib; specially designed anatomical grip stock for competition; Schnabel forend; leather-faced recoil pad; inclined-plane locking; ejectors; selective or non-selective trigger. Made in Italy. Introduced 1989; discontinued 1991.
  Perf.: $1650   **Exc.:** $1175   **VGood:** $950

Bernardelli Model 115E
Same specs as Model 115 except sideplates; engraving; silver finish.
  Perf.: $4500   **Exc.:** $3525   **VGood:** $3000

Bernardelli Model 115L
Same specs as Model 115 except engraving; silver finish.
  Perf.: $2500   **Exc.:** $2000   **VGood:** $1650

Bernardelli Model 115S
Same specs as Model 115 except engraving.
  Perf.: $2000   **Exc.:** $1650   **VGood:** $1400

Bernardelli Model 115 Target
Same specs as Model 115 except configured for trap shooting. Introduced 1989; discontinued 1991.
  Perf.: $1750   **Exc.:** $1300   **VGood:** $1100

Bernardelli Model 115E Target
Same specs as Model 115 Target except sideplates; engraving; silver finish.
  Perf.: $5800   **Exc.:** $4000   **VGood:** $3450

Bernardelli Model 115L Target
Same specs as Model 115 Target except engraving; silver finish.
  Perf.: $3550   **Exc.:** $2550   **VGood:** $2250

Bernardelli Model 115S Target
Same specs as Model 115 Target except engraving.
  Perf.: $3100   **Exc.:** $1700   **VGood:** $1400

Bernardelli Model 115S Target Monotrap
Same specs as Model 115 Target except single barrel.
  Perf.: $1550   **Exc.:** $1100   **VGood:** $900

BERNARDELLI MODEL 190
Over/under; 12-ga.; various barrel lengths; checkered walnut English straight or pistol-grip stock, forearm; ejectors; single selective trigger; silver finished receiver; engraving. Introduced 1986; discontinued 1989.
  Perf.: $1400   **Exc.:** $900   **VGood:** $750

Bernardelli Model 190MC
Same specs as Model 190 except Monte Carlo stock.
  Perf.: $950   **Exc.:** $650   **VGood:** $500

Bernardelli Model 190 Special
Same specs as Model 190 except select walnut; elaborate engraving.
  Perf.: $1250   **Exc.:** $800   **VGood:** $700

Bernardelli Model 190 Special MS
Same specs as Model 190 except select walnut stock; single trigger; elaborate engraving.
  Perf.: $1300   **Exc.:** $850   **VGood:** $750

BERNARDELLI MODEL 192
Over/under; boxlock; 12-ga.; 2 3/4″ chambers; 25 1/2″, 26 3/4″, 28″, 29 1/2″ barrels; standard choke combos or multichoke tubes; hand-checkered European walnut English straight or pistol-grip stock; ejectors; single selective trigger silvered, engraved action. Made in Italy. Introduced 1989; still imported. **LMSR:** $1340
  Perf.: $1150   **Exc.:** $700   **VGood:** $600

Bernardelli Model 192MS MC
Same specs as Model 192 except 3″ chambers; five choke tubes. Introduced 1989; still imported. **LMSR:** $2140
  Perf.: $1750   **Exc.:** $1200   **VGood:** $900

Bernardelli Model 192MS MC WF
Same specs as Model 192 except 3 1/2″ chambers; three choke tubes. Introduced 1989; still imported. **LMSR:** $1460
  Perf.: $1200   **Exc.:** $800   **VGood:** $650

BERNARDELLI MODEL 220MS
Over/under; boxlock; 12-, 20-ga.; 25 1/2″, 26 3/4″, 28″, 29 1/2″ barrels; standard choke combos or multichoke tubes; hand-checkered walnut English straight or pistol-grip stock; ejectors; double triggers; silver-gray receiver with engraving. Made in Italy. Introduced 1989; still in production. **LMSR:** $1490
  Perf.: $1350   **Exc.:** $810   **VGood:** $700
With selective trigger
  Perf.: $1350   **Exc.:** $860   **VGood:** $750

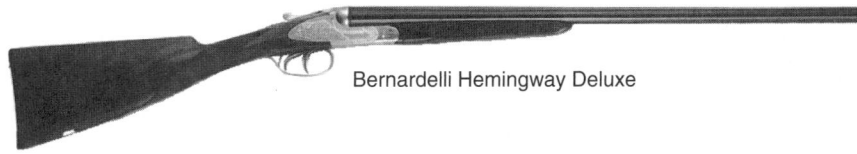

Bernardelli Hemingway Deluxe

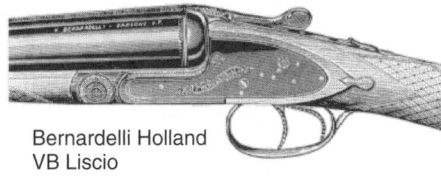

Bernardelli Holland
VB Liscio

Bernardelli Holland
VB Deluxe

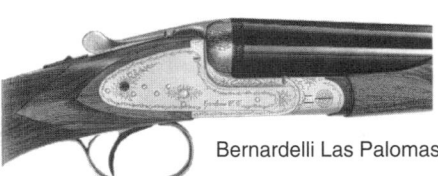

Bernardelli Las Palomas

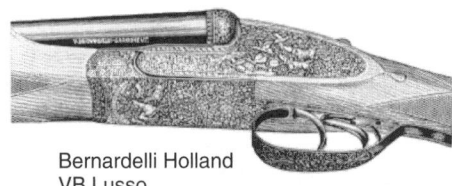

Bernardelli Holland
VB Lusso

Bernardelli Game Cock

SHOTGUNS

Bernardelli Autoloading & Side-By-Side Shotguns
9mm Flobert
Brescia
Elio
Gamecock
Hemingway
Hemingway Deluxe
Holland VB
Italia
Italia Extra
Las Palomas
Model 112

BERNARDELLI AUTOLOADING & SIDE-BY-SIDE SHOTGUNS

BERNARDELLI 9MM FLOBERT

Autoloader; rimfire 9mm Flobert shot cartridge; 3-shot magazine; 24 $^2/_5$″ barrel; uncheckered walnut stock, forearm. Introduced 1987; discontinued 1991.

Perf.: $300   **Exc.:** $180   **VGood:** $150

BERNARDELLI BRESCIA

Side-by-side; sidelock; exposed hammer; 12-, 16-, 20-ga.; 2 $^3/_4$″, 3″ chambers; 25 $^1/_2$″ Cylinder/Modified or Improved Cylinder/Improved Modified, 26 $^3/_4$″ Improved Cylinder/Improved Modified or Modified/Full, 29 $^1/_2$″ Improved Modified/Full barrels; weighs about 7 lbs.; checkered European walnut English straight-grip stock; extractors; double triggers; color case-hardened action. Made in Italy. Introduced 1960; no longer imported.

Perf.: $1850   **Exc.:** $700   **VGood:** $500

BERNARDELLI ELIO

Side-by-side; boxlock; hammerless; 12-ga.; 25″ Improved/Modified, 28″ Modified/Full barrels; hand-checkered European walnut straight stock; double triggers; extractors; silver-finished reciver; English-style scroll engraving; lightweight game model. Introduced 1970; no longer imported.

Perf.: $1200   **Exc.:** $900   **VGood:** $750
Elio E with ejectors
Perf.: $1250   **Exc.:** $950   **VGood:** $800
Elio M with single trigger
Perf.: $1250   **Exc.:** $950   **VGood:** $800
Elio EM
Perf.: $1300   **Exc.:** $1000   **VGood:** $850

BERNARDELLI GAMECOCK

Side-by-side; boxlock; hammerless; 12-, 16-, 20-, 28-ga.; 25″ Improved/Modified, 28″ Modified/Full barrels; hand-checkered European walnut straight stock; extractors; double triggers. Introduced 1970; no longer imported.

Perf.: $900   **Exc.:** $620   **VGood:** $580
With ejectors
Perf.: $1050   **Exc.:** $770   **VGood:** $620
With Deluxe engraving
Perf.: $1250   **Exc.:** $950   **VGood:** $800

BERNARDELLI HEMINGWAY

Side-by-side; boxlock; 12-, 16-, 20-, 28-ga.; 2 $^3/_4$″, 3″ chambers; 23 $^1/_2$ to 28″ barrels; weighs 6 $^1/_4$ lbs.; European walnut stock; ejectors; front trigger folds on double-trigger; silvered, engraved action. Introduced 1990; still in production. **LMSR: $1750**

Perf.: $1600   **Exc.:** $1000   **VGood:** $850
With single trigger
Perf.: $1680   **Exc.:** $1080   **VGood:** $900
With single selective trigger
Perf.: $1750   **Exc.:** $1150   **VGood:** $1000

Bernardelli Hemingway Deluxe

Same specs as Hemingway except sideplates; select wood; elaborate engraving.
Perf.: $1800   **Exc.:** $1100   **VGood:** $900
With single trigger
Perf.: $1900   **Exc.:** $1200   **VGood:** $1000

BERNARDELLI HOLLAND VB

Side-by-side; Holland & Holland-design sidelock; 12-ga.; 26″ to 32″ barrels; all standard choke combos; checkered European walnut straight or pistol-grip stock, forearm; double triggers; automatic ejectors. Introduced 1946; discontinued 1992.
Liscio Model
Perf.: $8000   **Exc.:** $4200   **VGood:** $3500
Inciso Model with engraving
Perf.: $10,000   **Exc.:** $6000   **VGood:** $4500
Lusso Model with engraving, select walnut
Perf.: $9000   **Exc.:** $6800   **VGood:** $5000
Extra Model with deluxe engraving, select wood
Perf.: $13,000   **Exc.:** $7000   **VGood:** $5800
Gold Model
Perf.: $40,000   **Exc.:** $25,000   **VGood:** $17,000

BERNARDELLI ITALIA

Side-by-side; sidelock; exposed hammer; 12-, 16-, 20-ga.; 2 $^3/_4$″, 3″ chambers; 25 $^1/_2$″ Cylinder/Modified or Improved Cylinder/Improved Modified, 26 $^3/_4$″ Improved Cylinder/Improved Modified or Modified/Full, 29 $^1/_2$″ Improved Modified/Full barrels; weighs about 7 lbs.; checkered high-grade European walnut English straight-grip stock; profuse engraving. Made in Italy. Introduced 1990; no longer imported.

Perf.: $2100   **Exc.:** $850   **VGood:** $700

Bernardelli Italia Extra

Same specs as Italia except highest grade of engraving and wood. Made in Italy. Introduced 1990; no longer imported.

Perf.: $6200   **Exc.:** $2000   **VGood:** $1550

BERNARDELLI LAS PALOMAS

Side-by-side; Anson & Deeley boxlock; 12-ga.; 2 $^3/_4$″ chambers; 28″ barrels bored for pigeon shooting; select European walnut stock; Australian pistol grip with shaped heel, palm swell; double trigger; special trigger guard; manual safety; auto safety optional; vent recoil pad; optional features. Made in Italy. Introduced 1984; still in production. **LMSR: $3700**

Perf.: $3000   **Exc.:** $1850   **VGood:** $1600
Las Palomas M with single trigger
Perf.: $3600   **Exc.:** $2400   **VGood:** $2150

BERNARDELLI MODEL 112

Side-by-side; 12-ga.; Anson & Deeley boxlock; 12-ga.; 2 $^3/_4$″ chambers; 25 $^5/_8$″, 26 $^3/_4$″, 28″, 29 $^1/_8$″ barrels; Modified/Full; European walnut stock; extractors; double triggers. Made in Italy. Introduced 1989; discontinued 1989.

Perf.: $800   **Exc.:** $625   **VGood:** $580
Model 112E with engraving
Perf.: $975   **Exc.:** $750   **VGood:** $650
Model 112M with single trigger
Perf.: $850   **Exc.:** $675   **VGood:** $625

Bernardelli Autoloading & Side-By-Side Shotguns
Model 112EM
Model 112EM MC
Model 112EM MC WF
Roma 3
Roma 3E
Roma 3EM
Roma 3M
Roma 4
Roma 4E
Roma 4EM
Roma 4M
Roma 6
Roma 6E
Roma 6EM
Roma 6M
Roma 7
Roma 8
Roma 9
S. Uberto 1
S. Uberto 1E
S. Uberto 1EM
S. Uberto 1M

Bernardelli Roma 3

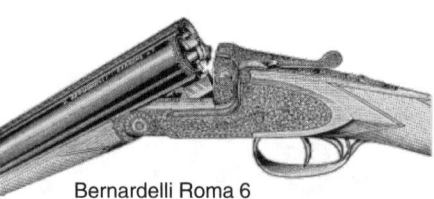

Bernardelli Roma 6

Bernardelli Roma 4

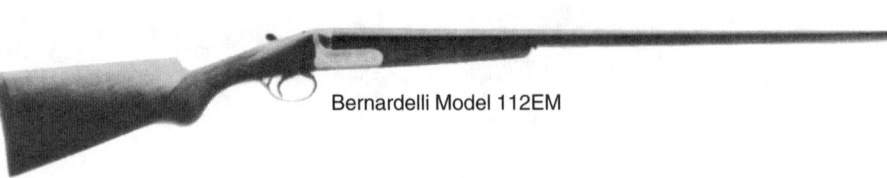

Bernardelli Model 112EM

Bernardelli Model 112EM
Same specs as Model 112 except engraving; single trigger. Introduced 1989; still in production. **LMSR:** $1385

Perf.: $1200 **Exc.:** $800 **VGood:** $650

Bernardelli Model 112EM MC
Same specs as Model 112EM except 3″ chamber; 5 choke tubes; engraving; single trigger. Introduced 1990; discontinued 1992.

Perf.: $1550 **Exc.:** $810 **VGood:** $740

Bernardelli Model 112EM MC WF
Same specs as Model 112EM except 3 1/2″ chamber; 3 choke tubes; engraving; single trigger. Introduced 1990; discontinued 1991.

Perf.: $1200 **Exc.:** $800 **VGood:** $725

BERNARDELLI ROMA 3
Side-by-side; Anson & Deeley boxlock; false sideplates; hammerless; 12-, 16-, 20-, 28-ga.; 27 1/2″, 29 1/2″ barrels; Modified/Full chokes; hand-checkered European walnut straight or pistol-grip stock, forearm; extractors; double triggers; color case-hardened receiver. Introduced 1946; still in production. **LMSR:** $1470

Perf.: $1400 **Exc.:** $845 **VGood:** $725

Bernardelli Roma 3E
Same specs as Roma 3 except ejectors. Still in production. **LMSR:** $1615

Perf.: $1600 **Exc.:** $1100 **VGood:** $900

Bernardelli Roma 3EM
Same specs as Roma 3 except ejectors; single trigger. No longer in production.

Perf.: $1700 **Exc.:** $1200 **VGood:** $1000

Bernardelli Roma 3M
Same specs as Roma 3 except single trigger. No longer in production.

Perf.: $1450 **Exc.:** $900 **VGood:** $775

BERNARDELLI ROMA 4
Side-by-side; Anson & Deeley boxlock; false sideplates; hammerless; 12-, 16-, 20-, 28-ga.; 27 1/2″, 29 1/2″ barrels; Modified/Full chokes; hand-checkered European walnut straight or pistol-grip stock, forearm; extractors; double triggers; scroll engraving; silver finish. Still in production. **LMSR:** $1800

Perf.: $1550 **Exc.:** $900 **VGood:** $695

Bernardelli Roma 4E
Same specs as Roma 4 except ejectors. Still in production. **LMSR:** $2000

Perf.: $1750 **Exc.:** $1100 **VGood:** $875

Bernardelli Roma 4EM
Same specs as Roma 4 except ejectors; single trigger. No longer in production.

Perf.: $1850 **Exc.:** $1200 **VGood:** $975

Bernardelli Roma 4M
Same specs as Roma 4 except single trigger. Still in production. **LMSR:** $1975

Perf.: $1650 **Exc.:** $1000 **VGood:** $800

BERNARDELLI ROMA 6
Side-by-side; Anson & Deeley boxlock; hammerless; 12-, 16-, 20-, 28-ga.; 27 1/2″, 29 1/2″ barrels; Modified/Full chokes; select hand-checkered European walnut straight or pistol-grip stock, forearm; extractors; double triggers; elaborate engraved sideplates; silver finish. Still in production. **LMSR:** $1970

Perf.: $1700 **Exc.:** $1100 **VGood:** $900

Bernardelli Roma 6E
Same specs as Roma 6 except ejectors. Still in production. **LMSR:** $2175

Perf.: $1850 **Exc.:** $1300 **VGood:** $1050

Bernardelli Roma 6EM
Same specs as Roma 6 except ejectors; single trigger. No longer in production.

Perf.: $1900 **Exc.:** $1350 **VGood:** $1100

Bernardelli Roma 6M
Same specs as Roma 6 except single trigger. Still in production. **LMSR:** $2140

Perf.: $1850 **Exc.:** $1250 **VGood:** $1050

BERNARDELLI ROMA 7
Side-by-side; Anson & Deeley boxlock; hammerless; 12-ga.; 27 1/2″, 29 1/2″ barrels; Modified/Full chokes; fancy hand-checkered European walnut straight or pistol-grip stock, forearm; ejectors; double triggers; elaborately engraved, silver-finished sideplates. Introduced 1994; still in production. **LMSR:** $2750

Perf.: $2500 **Exc.:** $1500 **VGood:** $1200

BERNARDELLI ROMA 8
Side-by-side; Anson & Deeley boxlock; hammerless; 12-ga.; 27 1/2″, 29 1/2″ barrels; Modified/Full chokes; fancy select hand-checkered European walnut straight or pistol-grip stock, forearm; ejectors; double triggers; elaborately engraved, silver-finished sideplates. Introduced 1994; still in production. **LMSR:** $3250

Perf.: $2900 **Exc.:** $1800 **VGood:** $1550

BERNARDELLI ROMA 9
Side-by-side; Anson & Deeley boxlock; hammerless; 12-ga.; 27 1/2″, 29 1/2″ barrels; Modified/Full chokes; best-grade hand-checkered European walnut straight or pistol-grip stock, forearm; ejectors; double triggers; very elaborately engraved, silver-finished sideplates. Introduced 1994; still in production. **LMSR:** $3850

Perf.: $3400 **Exc.:** $2350 **VGood:** $1800

BERNARDELLI S. UBERTO 1
Side-by-side; Anson & Deeley-style boxlock; Purdey locks; 12-, 16-, 20-, 28-ga.; 2 3/4″, 3″ chambers; 25 5/8″, 26 3/4″, 28″, 29 1/8″ barrels; Modified/Full; select hand-checkered European walnut stock; extractors; double triggers; color case-hardened receiver. Made in Italy. Introduced 1946; discontinued 1990.

Perf.: $1000 **Exc.:** $750 **VGood:** $600

Bernardelli S Uberto 1E
Same specs as S. Uberto 1 except ejectors. No longer in production.

Perf.: $1100 **Exc.:** $850 **VGood:** $700

Bernardelli S. Uberto 1EM
Same specs as S. Uberto 1 except ejectors; single trigger. No longer in production.

Perf.: $1150 **Exc.:** $900 **VGood:** $750

Bernardelli S. Uberto 1M
Same specs as S. Uberto 1 except single trigger. No longer in production.

Perf.: $1050 **Exc.:** $800 **VGood:** $650

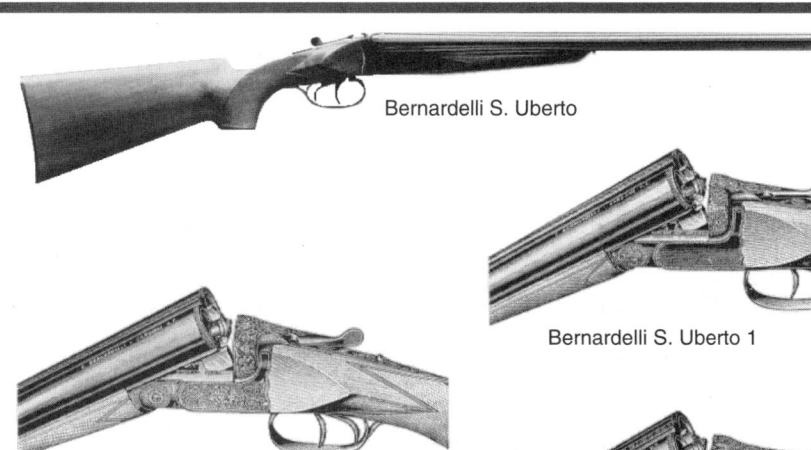

Bernardelli S. Uberto

Bernardelli S. Uberto 1

Bernardelli S. Uberto 2

Bernardelli S. Uberto FS

SHOTGUNS

**Bernardelli
Autoloading
& Side-By-Side
Shotguns**
S. Uberto 2
S. Uberto 2E
S. Uberto 2EM
S. Uberto 2M
S. Uberto FS
S. Uberto FSE
S. Uberto FSEM
S. Uberto FSM
Slug Gun
Slug Lusso
Slug Lusso M
Slug M Gun
XXVSL

Boss
Side-By-Side
Over/Under

Breda
Autoloader

Boss Over/Under

Beda Autoloader

BERNARDELLI S. UBERTO 2

Side-by-side; Anson & Deeley-style boxlock; Purdey locks; 12-, 16-, 20-, 28-ga.; 2 3/4", 3" chambers; 25 5/8", 26 3/4", 28", 29 1/8" barrels; Modified/Full; select hand-checkered European walnut stock; extractors; double triggers; engraved, silver-finished receiver. Made in Italy. Introduced 1946; still in production. **LMSR: $1435**

Perf.: $1300 **Exc.:** $800 **VGood:** $650

Bernardelli S Uberto 2E

Same specs as S. Uberto 2 except ejectors. Still in production. **LMSR: $1555**

Perf.: $1400 **Exc.:** $900 **VGood:** $700

Bernardelli S. Uberto 2EM

Same specs as S. Uberto 2 except ejectors; single trigger. No longer in production.

Perf.: $1450 **Exc.:** $950 **VGood:** $750

Bernardelli S. Uberto 2M

Same specs as S. Uberto 2 except single trigger. No longer in production.

Perf.: $1350 **Exc.:** $850 **VGood:** $700

BERNARDELLI S. UBERTO FS

Side-by-side; Anson & Deeley-style boxlock; Purdey locks; 12-, 16-, 20-, 28-ga.; 2 3/4", 3" chambers; 25 5/8", 26 3/4", 28", 29 1/8" barrels; Modified/Full; select hand-checkered European walnut stock; extractors; double triggers; elaborately engraved, silver-finished receiver. Made in Italy. Introduced 1987; no longer in production.

Perf.: $1600 **Exc.:** $1050 **VGood:** $800

Bernardelli S Uberto FSE

Same specs as S. Uberto FS except ejectors. Still in production. **LMSR: $1750**

Perf.: $1400 **Exc.:** $1000 **VGood:** $875

Bernardelli S. Uberto FSEM

Same specs as S. Uberto FS except ejectors; single trigger. No longer in production.

Perf.: $1500 **Exc.:** $1100 **VGood:** $975

Bernardelli S. Uberto FSM

Same specs as S. Uberto FS except single trigger. No longer in production.

Perf.: $1650 **Exc.:** $1100 **VGood:** $850

BERNARDELLI SLUG GUN

Side-by-side; Anson & Deeley boxlock; 12-ga.; 23 3/4" barrels; slug bores; walnut English straight or pistol-grip stock, forearm; extractors; double trigger; silver finished receiver; engraving. Discontinued 1990.

Perf.: $1250 **Exc.:** $800 **VGood:** $710

Bernardelli Slug Lusso

Same specs as Slug Gun except ejectors; sideplates; elaborate engraving. Discontinued 1992.

Perf.: $2200 **Exc.:** $1100 **VGood:** $900

Bernardelli Slug Lusso M

Same specs as Slug Gun single trigger; ejectors; sideplates; elaborate engraving. No longer in production.

Perf.: $2300 **Exc.:** $1200 **VGood:** $1000

Bernardelli Slug M Gun

Same specs as Slug Gun except single trigger. No longer inproduction.

Perf.: $1300 **Exc.:** $850 **VGood:** $760

BERNARDELLI XXVSL

Side-by-side; Holland & Holland-type sidelock; 12-ga.; 25" demi-block barrels; standard chokes; custom-fitted European walnut stock, classic or beavertail forend; selective ejectors; manual or auto safety; double sears; fitted luggage case. Made in Italy. Introduced 1982; discontinued 1984.

Perf.: $1700 **Exc.:** $1200 **VGood:** $800

BOSS SIDE-BY-SIDE

Side-by-side; sidelock; 12-, 16-, 20-, 28-ga., 410; 26", 28", 30", 32" barrels; any desired choke combo; hand-checkered European walnut stock, forearm; straight or pistol-grip stock; automatic ejectors; double or non-selective single trigger; selective single trigger extra; round or flat action sides. Made to order. Introduced 1880; still in production. **LMSR: $43,195**
12-ga.

Perf.: $40,000 **Exc.:** $30,000 **VGood:** $20,000
Smaller gauges

Perf.: $45,000 **Exc.:** $35,000 **VGood:** $25,000

BOSS OVER/UNDER

Over/under; sidelock; 12-, 16-, 20-, 28-ga., 410; 26", 28", 30", 34" barrels; any desired choke combo; matted or vent rib; hand-checkered European walnut stock, forearm; recoil pad; automatic ejectors; double or non-selective single trigger; selective single trigger extra. Made to order. Introduced 1909; still in production. **LMSR: $61,710**
12-ga.

Perf.: $60,000 **Exc.:** $40,000 **VGood:** $30,000
Smaller gauges

Perf.: $65,000 **Exc.:** $45,000 **VGood:** $35,000

BREDA AUTOLOADER

Autoloader; takedown; 12-, 20-ga.; 4-shot tube magazine; 25 1/2", 27 1/2" barrels; chrome bore; plain or with matted rib; hand-checkered European walnut straight or pistol-grip stock, forearm; available in 3 grades with chromed receivers, engraving. Grade, value depends upon amount of engraving, quality of wood. Made in Italy. Introduced 1946; no longer imported.
Standard

Exc.: $240 **VGood:** $200 **Good:** $180
Grade I

Exc.: $450 **VGood:** $400 **Good:** $350
Grade II

Exc.: $500 **VGood:** $400 **Good:** $350
Grade III

Exc.: $650 **VGood:** $600 **Good:** $500

SHOTGUNS

Breda
Autoloader Magnum 12

Bretton
Baby
Baby Deluxe

BRI
Special Model

Brno
CZ-581
Model 500
Super
ZH-301 Series
ZP-49
ZP-149
ZP-349

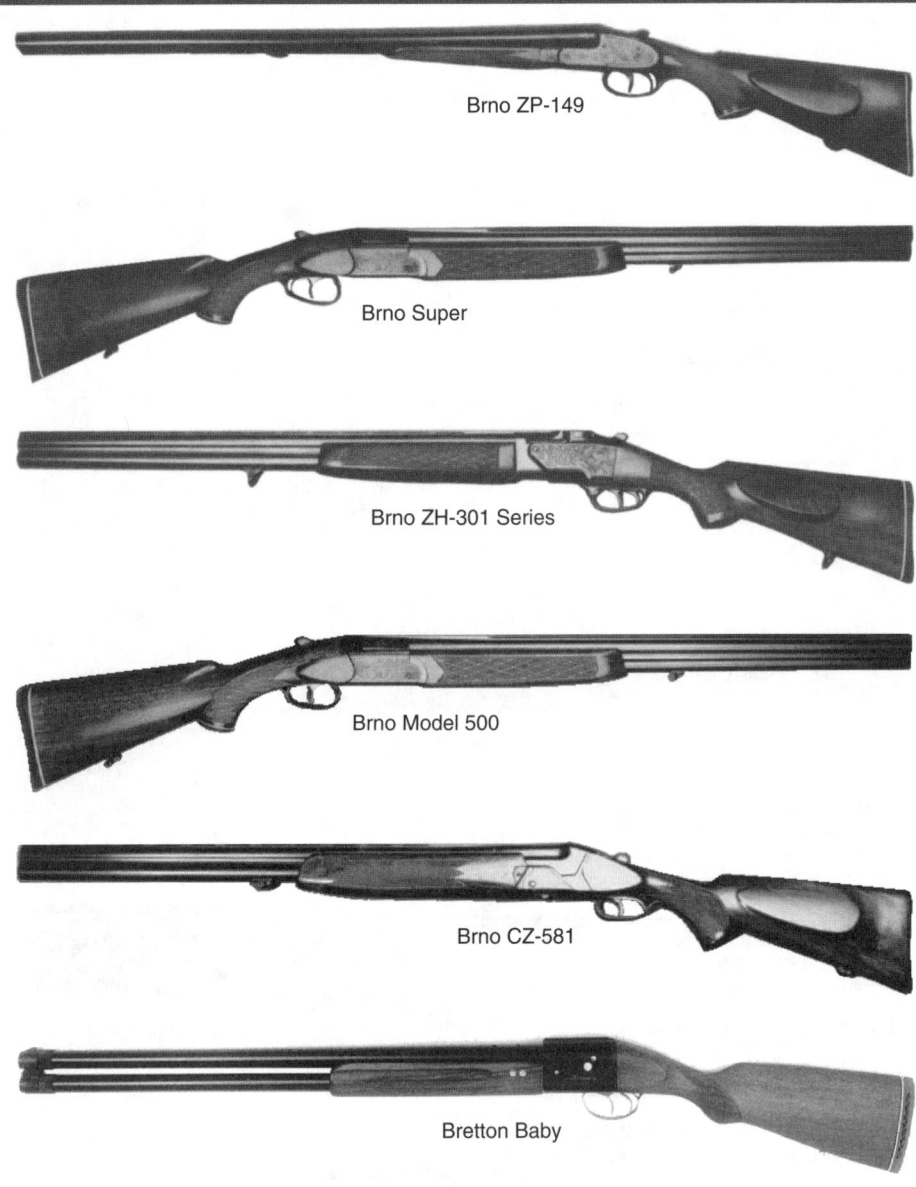

Brno ZP-149

Brno Super

Brno ZH-301 Series

Brno Model 500

Brno CZ-581

Bretton Baby

Breda Autoloader Magnum 12
Same specs as Autoloader except 12-ga.; 3″ chamber. Introduced 1950; no longer imported.
Plain barrel
  Exc.: $350   **VGood:** $300   **Good:** $250
Matte rib
  Exc.: $370   **VGood:** $320   **Good:** $270

BRETTON BABY
Over/under; 12-, 20-ga.; 2 3/4″ chambers; 27 1/2″ barrels; standard choke tubes; weighs 5 lbs.; oil-finished checkered European walnut stock; receiver slides open on guide rods, locks with thumblever; extractors. Made in France. Introduced 1986; still in production.
LMSR: $995
  Perf.: $850   **Exc.:** $600   **VGood:** $550

Bretton Baby Deluxe
Same specs as Bretton Baby except 12-, 16-, 20-ga.; double triggers; silvered engraved receiver. Discontinued 1994.
  Perf.: $825   **Exc.:** $600   **VGood:** $500

BRI SPECIAL MODEL
Slide action; 12-ga.; 3″ chamber; 24″ rifled Cylinder barrel; 44″ overall length; weighs 7 1/2 lbs.; no sights; scope mount on barrel; high, straight comb walnut stock; rubber recoil pad; built on Mossberg Model 500 Trophy Slugster action. Made by Ballistic Research Industries. Introduced, 1988; discontinued, 1990.
  Perf.: $300   **Exc.:** $225   **VGood:** $180

BRNO CZ-581
Over/under; boxlock; 12-ga.; 2 3/4″ chambers; 28″ barrels; vent rib; 45 1/2″ overall length; weighs 7 1/3 lbs.; Turkish walnut stock; auto safety; sling swivels; selective ejectors; double triggers. Made in Czechoslovakia. Introduced 1986; discontinued 1991.
  Perf.: $550   **Exc.:** $425   **VGood:** $350

BRNO MODEL 500
Over/under; boxlock; 12-ga.; 2 3/4″ chambers; 27″ Full/Modified barrels; weighs 7 lbs.; European walnut raised cheekpiece stock; ejectors; double triggers; acid-etched engraving. Made in Czechoslovakia. Introduced 1987; discontinued 1991.
  Perf.: $700   **Exc.:** $500   **VGood:** $400

BRNO SUPER
Over/under; sidelock; 12-ga.; 2 3/4″ chambers; 27 1/2″ Full/Modified barrels; weighs 7 1/2 lbs.; European walnut raised cheekpiece stock; double safety interceptor sears; selective ejectors; double or single triggers; engraved sideplates. Made in Czechoslovakia. Introduced 1987; discontinued 1991.
  Perf.: $750   **Exc.:** $620   **VGood:** $550

BRNO ZH-301 SERIES
Over/under; boxlock; 12-ga.; 2 3/4″ chambers; 27 1/2″ barrels; Modified/Full; weighs 7 lbs.; European walnut stock; double triggers; acid-etched engraving. Made in Czechoslovakia. Introduced 1987; discontinued 1991.
Brno ZH-301 Field
  Perf.: $575   **Exc.:** $325   **VGood:** $250
Brno ZH-302 Skeet with 26″ Skeet/Skeet barrels
  Perf.: $600   **Exc.:** $350   **VGood:** $250
Brno ZH-303 Trap with 30″ barrels
  Perf.: $600   **Exc.:** $350   **VGood:** $250

BRNO ZP-49
Side-by-side; sidelock; 12-ga.; 2 3/4″ chambers; 28 1/2″ Full/Modified barrels; Turkish or Yugoslavian walnut straight or pistol-grip stock; barrel indicators; ejectors; double triggers; auto safety; sling swivels. Introduced 1986; discontinued 1994.
  Perf.: $500   **Exc.:** $410   **VGood:** $350

BRNO ZP-149
Side-by-side; sidelock; 12-ga.; 23/4″ chambers; 28 1/2″ Full/Modified barrels; weighs 7 1/4 lbs.; Turkish or Yugoslavian walnut straight or pistol-grip stock; raised cheekpiece; barrel indicators; ejectors; double triggers; auto safety; engraving available. Made in Czechoslovakia. Introduced 1986; discontinued 1991.
  Perf.: $500   **Exc.:** $400   **VGood:** $300
With engraving
  Perf.: $550   **Exc.:** $450   **VGood:** $350

BRNO ZP-349
Side-by-side; sidelock; 12-ga.; 2 3/4″ chambers; 28 1/2″ Full/Modified barrels; weighs 7 1/4 lbs.; Turkish or Yugoslavian walnut straight or pistol-grip stock, beavertail forearm; raised cheekpiece; barrel indicators; extractors; double triggers; auto safety; engraving available. Made in Czechoslovakia. Introduced 1986; discontinued 1991.
  Perf.: $425   **Exc.:** $325   **VGood:** $295
With engraving
  Perf.: $470   **Exc.:** $375   **VGood:** $340

452 SHOTGUNS

Browning A-500

Browning A-500G

Browning Auto-5 American Grade I (1940-1949)

Browning Auto-5 American Utility Field Model (1940-1949)

Browning Auto-5 Belgium Standard (1947-1953)

SHOTGUNS

Browning Autoloading Shotguns
A-500
A-500G
A-500G Buck Special
A-500G Sporting Clays
A-500R
A-500R Buck Special
Auto-5 (1900-1940)
Auto-5 (1900-1940) Grade III
Auto-5 (1900-1940) Grade IV
Auto-5 American Grade (1900-1949)
Auto-5 American Special (1940-1949)
Auto-5 American Special Skeet (1940-1949)
Auto-5 American Utility Field Model (1940-1949)
Auto-5 Belgium Standard (1947-1953)

BROWNING AUTOLOADING SHOTGUNS

NOTE: Some models are made in Japan, some In Belgium and some in both countries. Collector interest has increased in Belgian-made models and prices are approximately 25% higher than for Japanese production.

BROWNING A-500

Short-recoil autoloader; 12-ga.; 3″ chamber; 26″, 28″, 30″ barrel; Invector choke tubes; 49 1/2″ overall length (30″ barrel); weighs 7 1/2 lbs.; metal bead front sight; gloss-finished select checkered walnut pistol-grip stock, forend; black vent recoil pad; four-lug rotary bolt; coil spring buffering system; magazine cut-off. Redesignated as A-500R in 1989. Also available as A-500R Buck Special. Made in Belgium. Introduced 1987; discontinued 1993.

   Perf.: $450   **Exc.:** $350   **VGood:** $300

BROWNING A-500G

Gas-operated autoloader; 12-ga.; 3″ chamber; 26″, 28″, 30″ interchangeable barrel with Invector choke tubes; ventilated rib; weighs 7 3/4 lbs.; 47 1/2″ overall length; select walnut rounded pistol-grip stock with gloss finish; recoil pad; four-lug rotary bolt; crossbolt safety; patented gas metering system to handle all loads; built-in buffering system to absorb recoil, reduce stress on internal parts; high-polish blue finish with light engraving on receiver and "A-500G" in gold color. Made in Japan. Introduced 1990; discontinued 1993.

   Perf.: $520   **Exc.:** $340   **VGood:** $300

Browning A-500G Buck Special

Same specs as A-500G except 24″ barrel; 45 1/2″ overall length; weighs 7 3/4 lbs.; screw adjustable rear sight, countoured ramp front with gold bead. Made in Japan. Introduced 1990; discontinued 1992.

   Perf.: $540   **Exc.:** $360   **VGood:** $320

Browning A-500G Sporting Clays

Same specs as A-500G except 28″, 30″ Invector choke barrel; receiver has semi-gloss finish with "Sporting Clays" in gold lettering. Made in Japan. Introduced 1992; discontinued 1993.

   Perf.: $520   **Exc.:** $340   **VGood:** $300

BROWNING A-500R

Short-recoil autoloader; 12-ga.; 3″ chamber; 26″, 28″, 30″ barrel; Invector choke tubes; weighs about 7 1/2 lbs.; select walnut pistol-grip stock, forend with gloss finish; black ventilated recoil pad; four-lug rotary bolt; composite and coil spring buffering system; shoots all loads without adjustment; magazine cut-off. Made in Belgium. Introduced 1987; discontinued 1993.

   Perf.: $450   **Exc.:** $350   **VGood:** $300

Browning A-500R Buck Special

Same specs as A-500R except 24″ barrel; ramp front sight, open adjustable rear. Made in Belgium. Introduced 1987; discontinued 1993.

   Perf.: $470   **Exc.:** $370   **VGood:** $330

BROWNING AUTO-5 (1900-1940)

Recoil-operated autoloader; takedown; 12-, 16-ga.; 3- (pre-WWII), 4-shot magazine; 26″ to 32″ barrels; various chokes; plain barrel, vent or raised matted rib; hand-checkered European walnut pistol-grip stock, forearm. Made by FN. Introduced 1900; has been made in a wide variety of configurations grades and gauges; redesignated as Grade I in 1940.

With plain barrel
   Exc.: $310   **VGood:** $240   **Good:** $190
With raised rib
   Exc.: $340   **VGood:** $260   **Good:** $220
With ventilated rib
   Exc.: $380   **VGood:** $300   **Good:** $260

Browning Auto-5 (1900-1940) Grade III

Same specs as Auto-5 (1900-1940) except better wood, checkering and more engraving. Discontinued 1940.

   Exc.: $1800   **VGood:** $1500   **Good:** $1000
With raised matted rib
   Exc.: $1950   **VGood:** $1650   **Good:** $1150
With ventilated rib
   Exc.: $2100   **VGood:** $1800   **Good:** $1300

Browning Auto-5 (1900-1940) Grade IV

Same specs as Auto-5 (1900-1940) except better wood, checkering and engraving; profuse inlays of green, yellow gold. Sometimes called Midas Grade. Discontinued 1940.

   Exc.: $3400   **VGood:** $2700   **Good:** $2000
With raised matted rib
   Exc.: $3500   **VGood:** $2800   **Good:** $2100

BROWNING AUTO-5 AMERICAN GRADE I (1940-1949)

Autoloader; 12-, 16-, 20-ga.; 2 3/4″ chamber; 3-, 5-shot tube magazine; 26″, 28″, 30″, 32″ barrel; standard chokes; no rib; weighs 6 7/8 lbs. (20-ga.), 7 1/4 lbs. (16-ga.), 8 lbs. (12-ga.); hand-checkered American walnut pistol-grip stock, forend; identical to the Remington Model 11A; manufactured by Remington for Browning.

Introduced 1940; discontinued 1949.
   Exc.: $250   **VGood:** $190   **Good:** $170

Browning Auto-5 American Special (1940-1949)

Same specs as Auto-5 American Grade I (1940-1949) except either ventilated rib or matted raised rib. Introduced 1940; discontinued 1949.

With ventilated rib
   Exc.: $275   **VGood:** $220   **Good:** $190
With raised rib
   Exc.: $260   **VGood:** $200   **Good:** $180

Browning Auto-5 American Special Skeet (1940-1949)

Same specs as Auto-5 American Grade I (1940-1949) except 26″ barrel; ventilated rib; Cutts Compensator. Introduced 1940; discontinued 1949.

   Exc.: $275   **VGood:** $225   **Good:** $195

Browning Auto-5 American Utility Field Model (1940-1949)

Same specs as Auto-5 American Grade I (1940-1949) except 28″ barrel; PolyChoke. Introduced 1940; discontinued 1949.

   Exc.: $265   **VGood:** $215   **Good:** $190

BROWNING AUTO-5 BELGIUM STANDARD (1947-1953)

Autoloader; 12-, 16-ga.; 3-, 5-shot magazine; 2 3/4″ chambers; 26″, 28″, 30″, 32″ barrel; standard choke tubes; plain, matted or ventilated rib; weighs 8 lbs.; hand-checkered French walnut stock with high-gloss finish; hand-engraved receiver. Introduced 1947; discontinued 1953.

With plain barrel
   Exc.: $275   **VGood:** $225   **Good:** $200
With hollow matted rib
   Exc.: $280   **VGood:** $235   **Good:** $215
With ventilated rib
   Exc.: $295   **VGood:** $245   **Good:** $215

Browning Autoloading Shotguns

Auto-5 Classic
Auto-5 Classic Gold
Auto-5 Light (FN)
Auto-5 Light (Miroku)
Auto-5 Light Buck Special (Miroku)
Auto-5 Light Skeet (FN)
Auto-5 Light Skeet (Miroku)
Auto-5 Light Stalker
Auto-5 Light Sweet 16 (FN)
Auto-5 Light Sweet 16 (Miroku)
Auto-5 Light Trap (FN)
Auto-5 Magnum (FN)
Auto-5 Magnum (Miroku)
Auto-5 Magnum Stalker
B-80

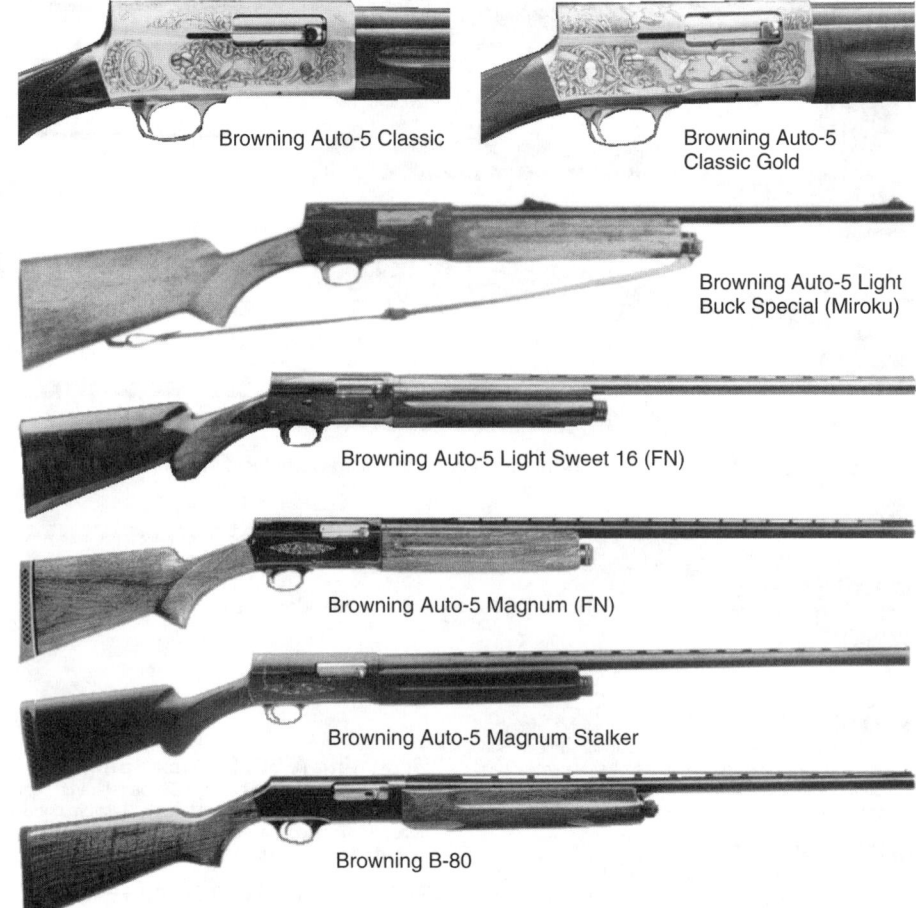

Browning Auto-5 Classic

Browning Auto-5 Classic Gold

Browning Auto-5 Light Buck Special (Miroku)

Browning Auto-5 Light Sweet 16 (FN)

Browning Auto-5 Magnum (FN)

Browning Auto-5 Magnum Stalker

Browning B-80

BROWNING AUTO-5 CLASSIC
Autoloader; 12-ga.; 2 3/4″ chamber; 5-shot magazine with 3-shot plug furnished; 28″ Modified barrel; select figured walnut stock; special checkering; hunting, wildlife scenes engraved in satin gray receiver; engraved portrait of John M. Browning; engraved legend, "Browning Classic. One of Five Thousand". Only 5000 manufactured. Introduced 1984; discontinued 1984.
Exc.: $1000   **VGood:** $650   **Good:** $400

Browning Auto-5 Classic Gold
Same specs as Auto-5 Classic except engraved scenes are inlaid with gold animals portrait; engraved with "Browning Gold Classic One of Five Hundred". Only 500 made. Introduced 1984; discontinued 1984.
Perf.: $3650   **Exc.:** $2000   **VGood:** $1500

BROWNING AUTO-5 LIGHT (FN)
Autoloader; 12-, 20-ga.; 2 3/4″ chamber; 5-shot magazine with 3-shot plug furnished; 26″, 28″, 30″ barrel; standard chokes; checkered walnut pistol-grip stock, forend; hand-engraved receiver; gold-plated trigger; double extractors; interchangeable barrels; with or without ventilated rib. Made by FN Belgium. Introduced 1952; discontinued 1976.
Without rib
Perf.: $425   **Exc.:** $350   **VGood:** $300
With ventilated rib
Perf.: $550   **Exc.:** $400   **VGood:** $340

Browning Auto-5 Light (Miroku)
Same specs as Auto-5 Light (FN) except ventilated rib; Invector choke tubes. Made by Miroku of Japan. Introduced 1976; still in production. **LMSR: $800**
Perf.: $650   **Exc.:** $400   **VGood:** $325

Browning Auto-5 Light Buck Special (Miroku)
Same specs as Auto-5 Light (FN) except 24″ barrel; choked for slugs; vent rib; adjustable rear sight, gold bead front on contoured ramp; detachable swivels, sling optional on current model. Made by Miroku of Japan. Introduced 1976; discontinued 1985; reintro-

duced 1989; still in production. **LMSR: $790**
Perf.: $600   **Exc.:** $400   **VGood:** $350

Browning Auto-5 Light Skeet (FN)
Same specs as the Auto-5 Light (FN) except for 26″, 28″ Skeet/Skeet barrel; plain barrel or vent rib (late models). Made by FN Belgium. Introduced 1952; discontinued 1976.
Perf.: $475   **Exc.:** $375   **VGood:** $300

Browning Auto-5 Light Skeet (Miroku)
Same specs as the Auto-5 Light (FN) except for 26″, 28″ Skeet/Skeet barrel; vent rib. Made by Miroku of Japan. Introduced 1976; discontinued 1983.
Perf.: $440   **Exc.:** $340   **VGood:** $300

Browning Auto-5 Light Stalker
Same specs as Auto-5 Light (FN) except 12-ga.; 2 3/4″ chamber; 26″, 28″ vent-rib barrel; Invector choke tubes; weighs 8 lbs. (26″); black graphite-fiberglass stock, forend with checkered panels; matte blue metal finish. Made by Miroku of Japan. Introduced 1992; still imported. **LMSR: $799.95**
Perf.: $600   **Exc.:** $395   **VGood:** $325

Browning Auto-5 Sweet 16 (FN)
Same specs as Auto-5 (FN) except 16-ga.; 2 9/16″, 2 3/4″ chamber; weighs 6 3/4 lbs. (without rib); gold-plated trigger, safety, safety latch. Introduced 1937; discontinued 1976.
Without rib
Exc.: $360   **VGood:** $260   **Good:** $200
With hollow matted rib
Exc.: $450   **VGood:** $300   **Good:** $225
With ventilated rib
Exc.: $580   **VGood:** $380   **Good:** $300

Browning Auto-5 Sweet 16 (Miroku)
Same specs as Auto-5 (FN) except 16-ga.; 2 3/4″ chamber; Invector chokes; vent rib; weighs 7 1/4 lbs. Made by Miroku of Japan. Introduced 1987; discontinued 1992.
Perf.: $550   **Exc.:** $400   **VGood:** $350

Browning Auto-5 Trap (FN)
Same specs as Auto-5 Light (FN) except 12-ga.; 30″ Full barrel; vent rib; weighs 8 1/4 lbs.; trap stock. Made by FN Belgium. Discontinued 1971.
Exc.: $460   **VGood:** $300   **Good:** $280

BROWNING AUTO-5 MAGNUM (FN)
Autoloader; 12-, 20-ga.; 3″ chamber; 26″, 28″, 30″, 32″ barrels; Full, Modified, Improved Cylinder chokes; plain or vent-rib; checkered walnut pistol-grip stock, forend; recoil pad. Made by FN Belgium. Introduced 1958; discontinued 1976.
With plain barrel
Perf.: $525   **Exc.:** $390   **VGood:** $340
With ventilated rib
Perf.: $650   **Exc.:** $500   **VGood:** $450

Browning Auto-5 Magnum (Miroku)
Same specs as Auto-5 Magnum (FN) except 26″, 28″ barrel; Invector choke tubes. Made by Miroku of Japan. Introduced 1976; still in production. **LMSR: $743**
Perf.: $600   **Exc.:** $400   **VGood:** $350

Browning Auto-5 Magnum Stalker
Same specs as the Auto-5 Magnum (FN) except 26″, 28″ back-bored vent-rib barrel; Invector choke tubes; weighs 8 3/4 lbs. (28″); black graphite-fiberglass stock, forend with checkered panels; matte blue metal finish. Made by Miroku of Japan. Introduced 1992; still imported. **LMSR: $825**
Perf.: $650   **Exc.:** $450   **VGood:** $400

BROWNING B-80
Autoloader; 12-, 20-ga.; 2 3/4″, 3″ chamber; 5-shot magazine; 22″ slug barrel, 26″ Improved Cylinder, Cylinder, Cylinder Skeet, Full, Modified, 28″ Full, Modified, 30″ Full, 32″ Full choke barrels; ventilated rib; hand-checkered French walnut stock; solid black recoil pad; steel receiver; interchangeable barrels. Made in Belgium. Introduced 1981; discontinued 1989.
Perf.: $425   **Exc.:** $300   **VGood:** $275

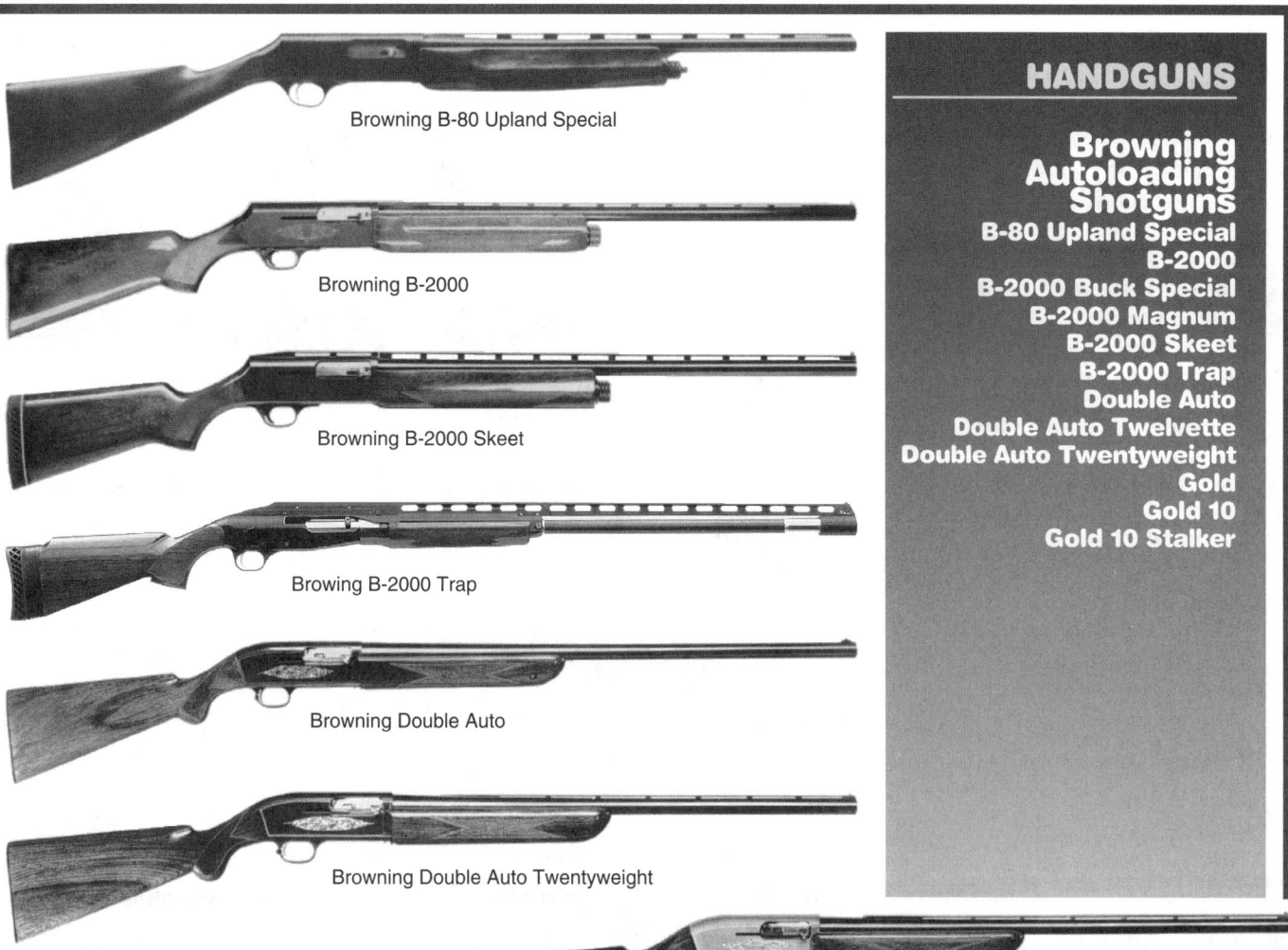

Browning B-80 Upland Special

Browning B-2000

Browning B-2000 Skeet

Browing B-2000 Trap

Browning Double Auto

Browning Double Auto Twentyweight

Browning Double Auto Twelvette

Browning Gold

Browning B-80 Upland Special
Same specs as B-80 except 2³/₄″ chamber; 22″ barrel; hand-checkered French walnut straight-grip stock. Made in Belgium. Introduced 1986; discontinued 1989.
Perf.: $450   **Exc.:** $320   **VGood:** $290

BROWNING B-2000
Autoloader; 12-, 20-ga.; 2 ³/₄″ chamber; 4-shot magazine; 26″, 28″, 30″ barrel; choice of standard chokes; vent rib or plain matted barrel; checkered pistol-grip stock of European walnut. Manufactured in Belgium, assembled in Portugal. Introduced 1974; discontinued 1982.
Perf.: $345   **Exc.:** $280   **VGood:** $240

Browning B-2000 Buck Special
Same specs as B-2000 except 12- (2 ³/₄″), 20-ga. (3″); 24″ slug barrel; no rib; open rear sight, front ramp. Introduced 1974; discontinued 1982.
Perf.: $350   **Exc.:** $280   **VGood:** $250

Browning B-2000 Magnum
Same specs as B-2000 except 3″ chamber; 3-shot magazine; 26″, 28″, 30″, 32″ barrel; vent rib. Introduced 1974; discontinued 1982.
Perf.: $350   **Exc.:** $300   **VGood:** $275

Browning B-2000 Skeet
Same specs as B-2000 except 26″ Skeet-choke barrel; Skeet stock; recoil pad. Introduced 1974; discontinued 1982.
Perf.: $350   **Exc.:** $290   **VGood:** $260

Browning B-2000 Trap
Same specs as B-2000 except 12-ga.; 2 ³/₄″ chamber; 30″, 32″ barrel; high post vent rib; Modified, Improved, Full chokes; Monte Carlo-style stock; recoil pad. Introduced 1974; discontinued 1982.
Perf.: $350   **Exc.:** $290   **VGood:** $260

BROWNING DOUBLE AUTO
Short-recoil autoloader; takedown; 12-ga.; 2-shot magazine; 26″, 28″, 30″ barrel; any standard choke; plain or recessed rib barrel; weighs about 7 ³/₄ lbs; hand-checkered European walnut pistol-grip stock, forearm; steel receiver; conservative engraving. Introduced 1955; discontinued 1961.
Exc.: $340   **VGood:** $240   **Good:** $195
With ventilated rib
Exc.: $400   **VGood:** $300   **Good:** $250

Browning Double Auto Twelvette
Same specs as Double Auto except aluminum receiver; barrel with plain matted top or vent rib; black anodized receiver with gold-wiped engraving; some receivers anodized in brown, gray or green with silver-wiped engraving; weighs about a pound less than standard model. Introduced 1955; discontinued 1971.
Exc.: $330   **VGood:** $230   **Good:** $200
With ventilated rib
Exc.: $400   **VGood:** $320   **Good:** $250

Browning Double Auto Twentyweight
Same specs as Double Auto except 26 ¹/₂″ barrel; weighs almost 2 lbs. less, largely due to thinner stock. Introduced 1956; discontinued 1971.
Exc.: $350   **VGood:** $210   **Good:** $190
With ventilated rib
Exc.: $500   **VGood:** $350   **Good:** $280

BROWNING GOLD
Gas-operated autoloader; 12-, 20-ga; 3″ chamber; 4-shot magazine; 26″, 28″, 30″ back-bored (12-ga.) or 26″, 30″ (20-ga.) barrel; Invector choke tubes; weighs about 7 ⁵/₈ lbs. (12-ga.); cut-checkered select walnut stock, palm swell grip with gloss finish; self-regulating, self-cleaning gas system shoots all loads; lightweight receiver with special non-glare deep black finish; large reversible safety button; large rounded trigger guard, gold trigger; crossbolt safety; recoil pad. Made in Japan. Introduced 1994; still imported. **LMSR: $700**
Perf.: $560   **Exc.:** $340   **VGood:** $300

Browning Gold 10
Same specs as Gold except 10-ga.; 3 ¹/₂″ chamber; 26″, 28″, 30″ barrel; Invector tubes choked Improved Cylinder, Modified, Full; weighs about 10 ¹/₂ lbs.; forged steel receiver with polished blue finish. Made in Japan. Introduced 1993; still imported. **LMSR: $960**
Perf.: $820   **Exc.:** $550   **VGood:** $480

Browning Gold 10 Stalker
Same specs as Gold except 10-ga.; 3 ¹/₂″ chamber; 26″, 28″, 30″ barrel; Invector tubes choked Improved Cylinder, Modified, Full; weighs about 10¹/₂ lbs.; checkered black graphite-fiberglass composite stock with dull finish; non-glare metal finish. Made in Japan. Introduced 1993; still imported. **LMSR: $960**
Perf.: $820   **Exc.:** $550   **VGood:** $480

Browning Side-By-Side & Over/Under Shotguns

BSS
BSS Sporter
BSS Sidelock
Citori Hunting
Citori Hunting Lightning
Citori Hunting Lightning Gran
Citori Hunting Lightning Micro
Citori Hunting Sporter
Citori Hunting Superlight
Citori Skeet

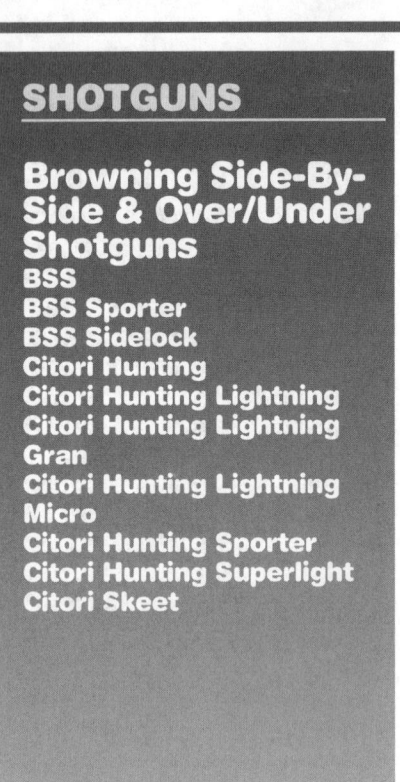

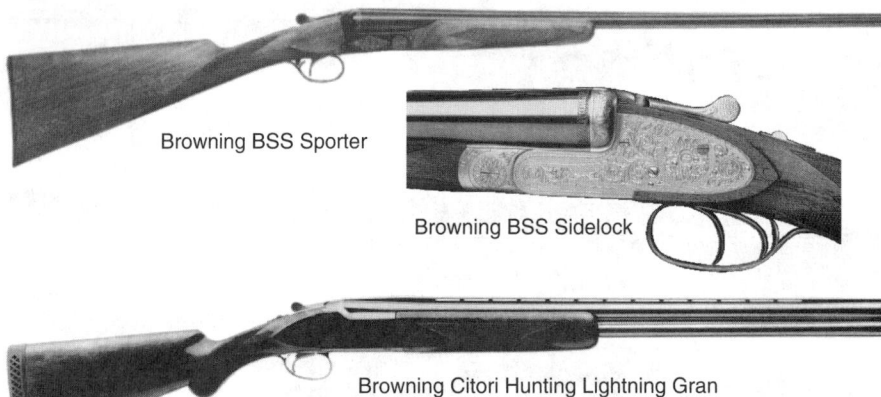

Browning BSS Sporter

Browning BSS Sidelock

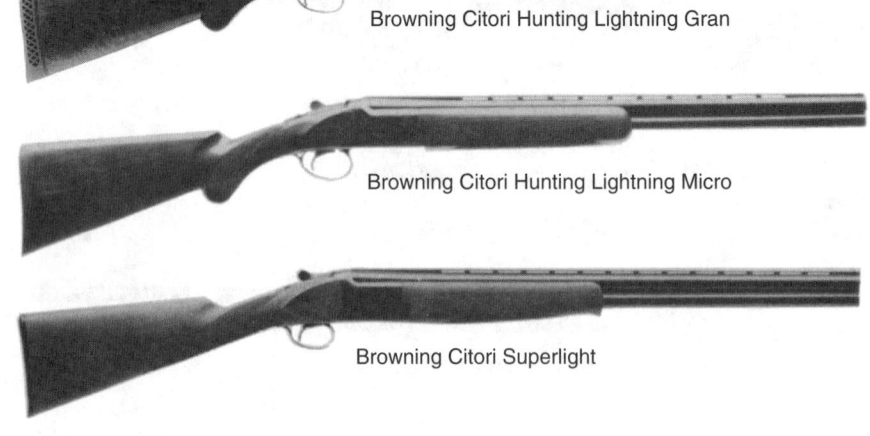

Browning Citori Hunting Lightning Gran

Browning Citori Hunting Lightning Micro

Browning Citori Superlight

BROWNING SIDE-BY-SIDE & OVER/UNDER SHOTGUNS

BROWNING BSS
Side-by-side; boxlock; hammerless; 12-, 20-ga.; 3″ chambers; 26″, 28″, 30″ barrel; standard choke combos; checkered walnut pistol grip stock, beavertail forend; non-selective single trigger. Made in Japan. Introduced 1972; discontinued 1987.
Perf.: $575   **Exc.:** $425   **VGood:** $375

Browning BSS Sporter
Same specs as BSS except 20-ga.; 26″, 28″ barrel; walnut straight stock, forend; selective trigger. Made in Japan. Introduced 1977; discontinued 1987.
Perf.: $650   **Exc.:** $500   **VGood:** $450

BROWING BSS SIDELOCK
Side-by-side; sidelock; 12-, 20-ga.; 26″, 28″ barrel; straight-grip French walnut stock; checkered butt; double triggers; auto safety; cocking indicator; receiver, forend iron, trigger guard, top lever, tang are satin gray with rosettes, scrollwork. Made in Japan. Introduced 1984; discontinued 1987.
12-Gauge
Perf.: $1825   **Exc.:** $1100   **VGood:** $900
20-Gauge
Perf.: $2250   **Exc.:** $1250   **VGood:** $1000

BROWNING CITORI HUNTING
Over/under; boxlock; 12-, 16-, 20-, 28-ga., .410; 3″, 3 1/2″ chambers; 26″, 28″, 30″ barrel; Improved/Modified, Modified/Full, Full/Full chokes or Invector choke tubes; checkered pistol-grip stock, semi-beavertail forearm; automatic ejectors; selective single trigger; recoil pad. Various grades available with price dependent on type of engraving. Made in Japan. Introduced 1973; still in production. **LMSR: $1270**
Grade I
Perf.: $720   **Exc.:** $540   **VGood:** $500
Grade I, 12-ga., 3 1/2″
Perf.: $950   **Exc.:** $720   **VGood:** $600

Grade II
Perf.: $950   **Exc.:** $700   **VGood:** $650
Grade III, grayed steel receiver, light engraving
Perf.: $1350   **Exc.:** $800   **VGood:** $700
Grade V, extensive engraving
Perf.: $1450   **Exc.:** $1000   **VGood:** $950
Grade VI, blue or grayed receiver, extensive engraving and inlays
Perf.: $2000   **Exc.:** $1200   **VGood:** $1000

Browning Citori Hunting Lightning
Same specs as Citori Hunting except Invector choke tubes; hand-checkered French walnut pistol-grip stock, slim forearm; ivory bead sights. Introduced 1988; still in production. **LMSR: $1350**
Grade I
Perf.: $950   **Exc.:** $520   **VGood:** $480
Grade III, grayed steel receiver, light engraving
Perf.: $1350   **Exc.:** $800   **VGood:** $700
Grade VI, blue or grayed receiver, extensive engraving and inlays
Perf.: $2000   **Exc.:** $1100   **VGood:** $1000

Browning Citori Hunting Lightning Gran
Same specs as Citori Hunting except 3″ chambers; 26″, 28″ barrels; Invector choke tubes; high-grade walnut stock; satin-oil finish. Introduced 1990; still in production. **LMSR: $1780**
Perf.: $1400   **Exc.:** $800   **VGood:** $650

Browning Citori Hunting Lightning Micro
Same specs as Citori Hunting except scaled down for smaller shooters; 24″ barrels; Invector choke system; weighs 6 1/8 lbs. Introduced 1991; still imported. **LMSR: $1360**
Grade I
Perf.: $1000   **Exc.:** $600   **VGood:** $500
Grade III, grayed steel receiver, light engraving
Perf.: $1350   **Exc.:** $800   **VGood:** $650
Grade VI, blue or grayed receiver, extensive engraving and inlays
Perf.: $1900   **Exc.:** $1150   **VGood:** $950

Browning Citori Hunting Sporter
Same specs as Citori Hunting except 3″ chamber; 26″ barrel; walnut straight-grip stock, Schnabel forend with satin oil finish. Introduced 1978; no longer in production.
Grade I
Perf.: $800   **Exc.:** $550   **VGood:** $500

Grade II
Perf.: $1150   **Exc.:** $980   **VGood:** $850
Grade V
Perf.: $1450   **Exc.:** $1150   **VGood:** $1000

Browning Citori Hunting Superlight
Same specs as Citori Hunting except 12-, 20-, 28-ga., .410; 2 3/4″ chambers; 24″, 26″, 28″ barrels; straight-grip stock; Schnabel forend. Made in Japan. Introduced 1982; still imported. **LMSR: $1370**
Grade I
Perf.: $950   **Exc.:** $575   **VGood:** $480
Grade III, grayed steel receiver, light engraving
Perf.: $1400   **Exc.:** $800   **VGood:** $700
Grade VI, blue or grayed receiver, extensive engraving and inlays
Perf.: $1950   **Exc.:** $1175   **VGood:** $1000

BROWING CITORI SKEET
Over/under; boxlock; 12-, 20-, 28-ga., .410; 26″, 28″ barrels; Skeet/Skeet fixed choke of Invector choke tubes; ventilated rib or target high-post wide rib; checkered walnut pistol-grip Skeet-style stock, forearm; single selective trigger; ejectors; recoil pad. Various grades available with price dependent on type of engraving. Made in Japan. Introduced 1974; still in production. **LMSR: $1510**
Grade I
Perf.: $1150   **Exc.:** $700   **VGood:** $550
Grade I with 3-barrel set
Perf.: $2400   **Exc.:** $1750   **VGood:** $1350
Grade III
Perf.: $1500   **Exc.:** $820   **VGood:** $700
Grade III with 4-barrel set
Perf.: $4500   **Exc.:** $2650   **VGood:** $2100
Grade III with 3-barrel set
Perf.: $3000   **Exc.:** $2000   **VGood:** $1400
Grade VI
Perf.: $1950   **Exc.:** $1500   **VGood:** $1100
Grade VI with 3-barrel set
Perf.: $3100   **Exc.:** $1850   **VGood:** $1650
Grade VI with 4-barrel set
Perf.: $4450   **Exc.:** $2650   **VGood:** $2100
Golden Clays
Perf.: $2300   **Exc.:** $1500   **VGood:** $1100
Golden Clays with 3-barrel set
Perf.: $4000   **Exc.:** $2750   **VGood:** $2000
Golden Clays with 4-barrel set
Perf.: $5400   **Exc.:** $3250   **VGood:** $2500

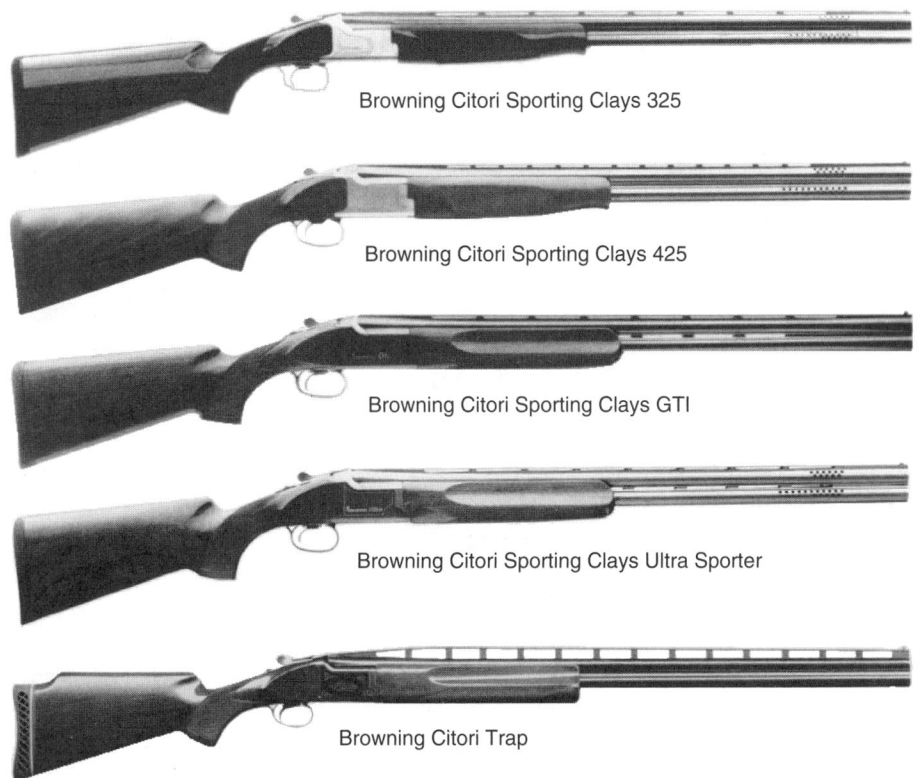

Browning Citori Sporting Clays 325

Browning Citori Sporting Clays 425

Browning Citori Sporting Clays GTI

Browning Citori Sporting Clays Ultra Sporter

Browning Citori Trap

Browning Side-By-Side & Over/Under Shotguns
Citori Sporting Clays 325
Citori Sporting Clays 325 Golden Clays
Citori Sporting Clays 425
Citori Sporting Clays 425 Golden Clays
Citori Sporting Clays 425 WSSF
Citori Sporting Clays GTI
Citori Sporting Clays Lightning
Citori Sporting Clays Special
Citori Sporting Clays Ultra Sporter
Citori Trap

BROWNING CITORI SPORTING CLAYS 325
Over/under; boxlock; 12-, 20-ga.; 2 3/4″ chambers; back-bored 28″, 30″, 32″ vent-rib barrels; barrels are ported on 12-gauge guns; Invector Plus choke tubes; weighs about 7 3/4 lbs.; select cut-checkered walnut stock, Schnabel forend with gloss finish; three interchangeable trigger shoes to adjust length of pull; grayed receiver with engraving; blued barrels. Made in Japan. Introduced 1993; discontinued 1994.
    Perf.: $1250 **Exc.:** $800 **VGood:** $700

Browning Citori Sporting Clays 325 Golden Clays
Same specs as Citori Sporting Clays 325 except high-grade walnut; grayed receiver highlighted with 24-karat gold. Made in Japan. Introduced 1993; discontinued 1994.
    Perf.: $2250 **Exc.:** $1350 **VGood:** $1050

BROWNING CITORI SPORTING CLAYS 425
Over/under; 12-, 20-ga.; 2 3/4″ chambers; back-bored 28″, 30″, 32″ vent-rib barrels; barrels are ported on 12-ga. guns; Invector Plus choke tubes; weighs about 7 3/4 lbs.; select cut-checkered walnut stock, Schnabel forend with gloss finish; three interchangeable trigger shoes to adjust length of pull; grayed receiver with engraving; blued barrels. Made in Japan. Introduced 1993; still imported. **LMSR: $1690**
    Perf.: $1380 **Exc.:** $800 **VGood:** $700

Browning Citori Sporting Clays 425 Golden Clays
Same specs as Sporting Clays 425 except high-grade walnut; grayed receiver highlighted with 24-karat gold. Made in Japan. Introduced 1993; still imported. **LMSR: $3150**
    Perf.: $2500 **Exc.:** $1350 **VGood:** $1000

Browning Citori Sporting Clays 425 WSSF
Same specs as Sporting Clays 425 except 12-ga.; stock dimensions specifically tailored to women shooters; top lever and takedown lever are easier to operate; stock and forend have teal-colored finish with WSSF logo. Made in Japan. Introduced 1995; still in production. **LMSR: $1690**
    Perf.: $1400 **Exc.:** $800 **VGood:** $700

BROWNING CITORI SPORTING CLAYS GTI
Over/under; boxlock; 12-ga.; 28″, 30″ vent-rib barrels (ported or non-ported); Invector Plus choke tubes; ventilated side ribs; checkered walnut semi-pistol grip stock, grooved semi-beavertail forend with satin finish; radiused rubber buttpad; three interchangeable trigger shoes for three length of pull adjustments. Made in Japan. Introduced 1989; discontinued 1994.
Grade I, non-ported barrels
    Perf.: $1100 **Exc.:** $550 **VGood:** $450
Grade I, ported barrels
    Perf.: $1200 **Exc.:** $650 **VGood:** $550
Golden Clays, ported barrels
    Perf.: $2200 **Exc.:** $1350 **VGood:** $1000

BROWNING CITORI SPORTING CLAYS LIGHTNING
Over/under; boxlock; 12-ga.; back-bored 30″ Invector Plus barrels (ported or non-ported); high-post tapered rib or hunting-type rib; checkered walnut pistol-grip stock, forend with gloss stock finish; adjustable comb, recoil pad; engraved, gold filled receiver. Made in Japan. Introduced, 1989; still imported. **LMSR: $1425**
Grade I, non-ported
    Perf.: $1100 **Exc.:** $625 **VGood:** $490
Grade I, ported
    Perf.: $1150 **Exc.:** $700 **VGood:** $550
Pigeon Grade, ported
    Perf.: $1250 **Exc.:** $800 **VGood:** $650
Golden Clays, ported
    Perf.: $2200 **Exc.:** $1350 **VGood:** $1000

BROWNING CITORI SPORTING CLAYS SPECIAL
Over/under; boxlock; 12-ga.; back-bored 28″, 30″, 32″ barrels (ported or non-ported); Invector Plus choke tubes; high-post tapered rib; checkered walnut pistol-grip stock, forend with gloss finish; palm swell. Also available as 28″, 30″ two-barrel set. Made in Japan. Introduced 1989; still imported. **LMSR: $1490**
Grade I, non-ported
    Perf.: $1050 **Exc.:** $900 **VGood:** $700
Grade I, ported
    Perf.: $1100 **Exc.:** $950 **VGood:** $750
Grade I, two-barrel set
    Perf.: $1900 **Exc.:** $1700 **VGood:** $1500
Golden Clays, ported
    Perf.: $3100 **Exc.:** $2800 **VGood:** $2200

BROWNING CITORI SPORTING CLAYS ULTRA SPORTER
Over/under; boxlock; 12-ga.; back-bored 28″ or 30″ barrels (ported or non-ported); Invector Plus choke tubes; ventilated side ribs; checkered walnut pistol-grip stock, grooved semi-beavertail forend with satin finish; radiused rubber buttpad; three interchangeable trigger shoes for three length of pull adjustments. Made in Japan. Introduced 1989; still imported. **LMSR: $1640**
Grade I, non-ported
    Perf.: $1500 **Exc.:** $1250 **VGood:** $1000
Grade I, ported
    Perf.: $1595 **Exc.:** $1300 **VGood:** $1100
Golden Clays
    Perf.: $3300 **Exc.:** $2850 **VGood:** $2200

BROWING CITORI TRAP
Over/under; boxlock; 12-ga.; 30″, 32″ ported or non-ported barrels; Full/Full, Improved Modified/Full, Modified/Full, or Invector Plus choke tubes; vent rib or high-post target wide vent rib; fitted with trap-style recoil pad; checkered walnut Monte Carlo cheekpiece stock, beavertail forearm. Introduced 1974; still in production. **LMSR: $1510**
Grade I, Invector Plus, ported
    Perf.: $1500 **Exc.:** $1250 **VGood:** $1000
Grade I Combo, with extra 34″ single barrel, Invector Plus, ported
    Perf.: $1800 **Exc.:** $1500 **VGood:** $1300
Grade III, Invector Plus, ported
    Perf.: $1500 **Exc.:** $1250 **VGood:** $1000
Grade V, Invector Plus, ported
    Perf.: $2650 **Exc.:** $2200 **VGood:** $1950
Grade VI, Invector Plus, ported
    Perf.: $2100 **Exc.:** $1800 **VGood:** $1650
Golden Clays, satin-gray receiver with engraving, inlays
    Perf.: $3000 **Exc.:** $2750 **VGood:** $2400
Pigeon Grade, deluxe wood, inlays
    Perf.: $3200 **Exc.:** $2950 **VGood:** $2650
Signature Painted, red and black painted stock
    Perf.: $1475 **Exc.:** $1175 **VGood:** $950

SHOTGUNS

Browning Side-By-Side & Over/Under Shotguns

Citori Trap Plus
Liege
ST-100
Superposed
Superposed Broadway Trap
Superposed Diana Grade
Superposed Grade I
Lightning Model
Superposed Magnum
Superposed Midas Grade
Superposed Pigeon Grade
Superposed Skeet
Superposed Presentation Super-Light
Superposed Presentation Super-Light Broadway Trap

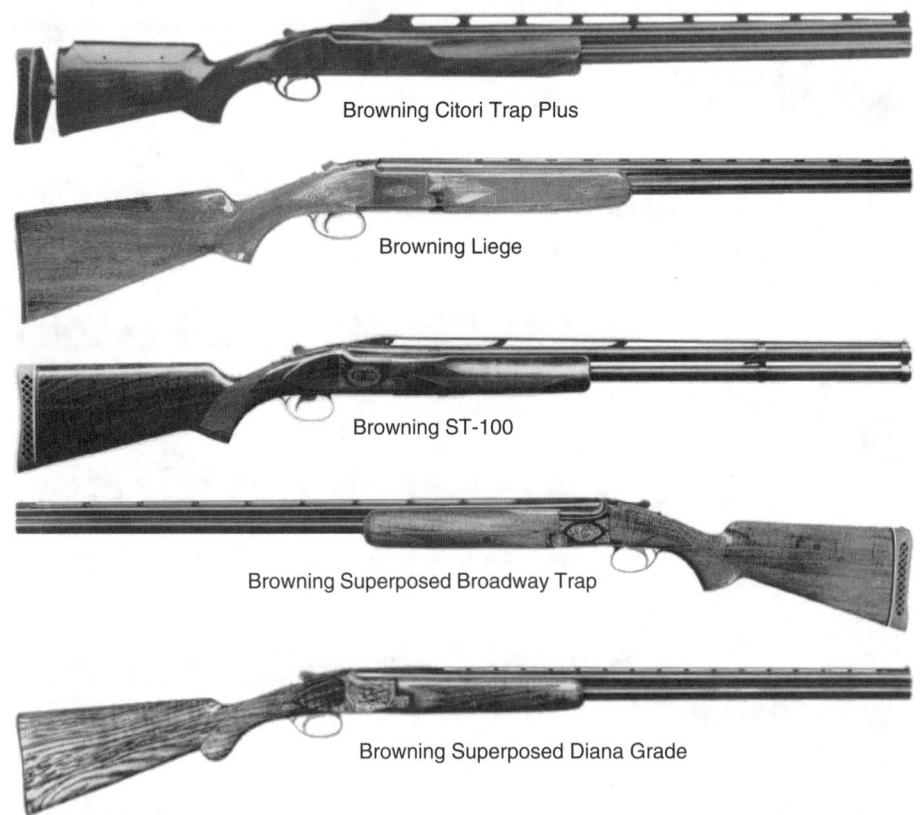

Browning Citori Trap Plus

Browning Liege

Browning ST-100

Browning Superposed Broadway Trap

Browning Superposed Diana Grade

Browning Citori Trap Plus

Same specs as Citori Trap except back-bored 30″, 32″ barrels; .745 over-bore; Invector Plus choke tubes with Full, Improved Modified and Modified; high post, ventilated, tapered, target rib adjustable for impact from 3″ to 12″ above point of aim; with or without ported barrels; select walnut Monte Carlo stock, modified beavertail forend with high-gloss finish; fully adjustable for length of pull, drop at comb, drop at Monte Carlo; Browning Recoil Reduction System. Made in Japan. Introduced 1989; no longer imported.
LMSR: $2075
Grade I, Invector Plus, ported
   Perf.: $1500   **Exc.:** $900   **VGood:** $750
Grade I Combo with extra 34″ single barrel, Invector Plus, ported
   Perf.: $2600   **Exc.:** $1900   **VGood:** $1700
Golden Clays, satin-gray receiver with engraving, inlays
   Perf.: $4300   **Exc.:** $2650   **VGood:** $2100
Pigeon Grade, deluxe wood, inlays
   Perf.: $1650   **Exc.:** $850   **VGood:** $700

BROWNING LIEGE

Over/under; boxlock; 12-ga.; 2 3/4″, 3″ chamber; 26 1/2″, 28″, 30″ barrels; standard choke combos; vent rib; checkered walnut pistol-grip stock, forearm; automatic ejectors; non-selective single trigger. Introduced 1973; discontinued 1975.
   Perf.: $720   **Exc.:** $550   **VGood:** $490

BROWNING ST-100

Over/under; 12-ga.; 30″ barrels; floating Broadway rib; five-position impact adjustment; hand-checkered select walnut stock, semi-beavertail forend with high-gloss finish; selective auto ejectors; single selective mechanical trigger; manual top tang safety. Introduced 1979; discontinued 1982.
   Perf.: $2200   **Exc.:** $1650   **VGood:** $1400

BROWNING SUPERPOSED

Over/under; 12-, 20-, 28-ga., .410; 26 1/2″, 28″, 30″, 32″ (discontinued WWII) barrels; gun is made in a wide spectrum of variations, grades, choke combinations; early models and pre-WWII had double triggers, twin single triggers or non-selective single trigger (pre-WWII production ended when Belgium was invaded); hand-checkered French walnut pistol-grip or straight stock, forearm. Standard Grade is listed as Grade I, with raised matted rib or vent rib. Introduced 1928; no longer in production.
   Exc.: $950   **VGood:** $800   **Good:** $575

Browning Superposed Broadway Trap

Same specs as Superposed except 30″, 32″ barrel; 5/8″-wide Broadway vent rib. No longer in production.
   Exc.: $850   **VGood:** $720   **Good:** $500

Browning Superposed Diana Grade

Same specs as Superposed except raised matted rib or vent rib before WWII; post-war models have only vent rib; better wood, more extensive engraving; redesignated as Grade V after WWII; improved general quality. No longer in production.
   Perf.: $4700   **Exc.:** $2700   **VGood:** $2100

Browning Superposed Grade I Lightning Model

Same specs as Superposed except matted barrel; no rib before WWII; post-war models have vent rib. Other specs generally the same as Standard Superposed model. No longer in production.
   Perf.: $1550   **Exc.:** $925   **VGood:** $800

Browning Superposed Magnum

Same specs as Superposed except 12-ga.; 3″ chamber; 30″ vent-rib barrel; choked Full/Full or Full/Mod.; recoil pad. No longer in production.
   Perf.: $1350   **Exc.:** $875   **VGood:** $800

Browning Superposed Midas Grade

Same specs as Superposed except heavily engraved, gold inlaid; pre-war versions with raised matted rib or vent rib; later versions have only wide vent rib. No longer in production.
   Perf.: $6500   **Exc.:** $3750   **VGood:** $3000

Browning Superposed Pigeon Grade

Same specs as Superposed except better wood, finer checkering, more engraving than Standard Superposed model; raised matted rib or vent rib. Was redesignated as Grade II after WWII. No longer in production.
   Perf.: $3300   **Exc.:** $2000   **VGood:** $1850

Browning Superposed Skeet

Same specs as Superposed except 12-, 20-, 28-ga., .410; 26 1/2″, 28″ barrel; choked Skeet/Skeet. No longer in production.
   Exc.: $900   **VGood:** $750   **Good:** $560

BROWNING SUPERPOSED PRESENTATION SUPERLIGHT

Over/under; boxlock; 12-, 20-ga; 26 1/2″ barrel; solid or vent rib; hand-checkered straight-grip stock of select walnut; top lever single selective trigger barrel selector combined with manual tang safety. Options too numerous to mention; available in 4 grades. Introduced 1977; no longer imported.
Grade I
   Perf.: $2800   **Exc.:** $1900   **VGood:** $1680
Grade II
   Perf.: $3300   **Exc.:** $2200   **VGood:** $2000
Grade III
   Perf.: $5500   **Exc.:** $4000   **VGood:** $3500
Grade IV
   Perf.: $6500   **Exc.:** $4500   **VGood:** $3900

Browning Superposed Presentation Superlight Broadway Trap

Same specs as Superposed Presentation Superlight except 30″, 32″ barrel; wide vent rib; different stock measurements. Discontinued 1984.
Grade I
   Perf.: $2500   **Exc.:** $1650   **VGood:** $1500
Grade II
   Perf.: $3000   **Exc.:** $1900   **VGood:** $1750
Grade III
   Perf.: $5000   **Exc.:** $3600   **VGood:** $3200
Grade IV
   Perf.: $6000   **Exc.:** $4000   **VGood:** $3500

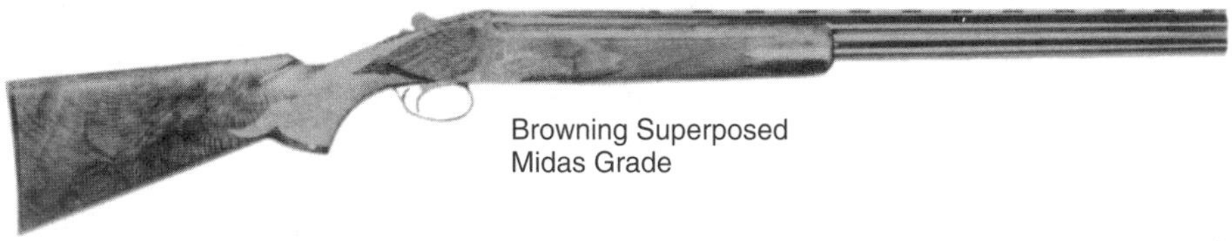

Browning Superposed
Midas Grade

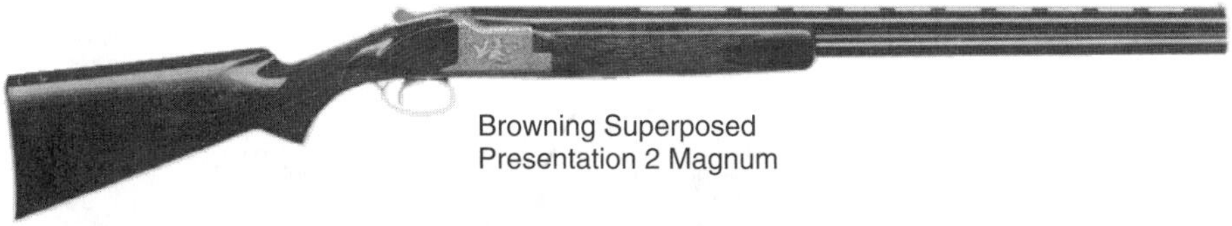

Browning Superposed
Presentation 2 Magnum

Browning Superposed
Super-Light Field Grade

Browning Superposed Presentation
Superlight Lightning Trap

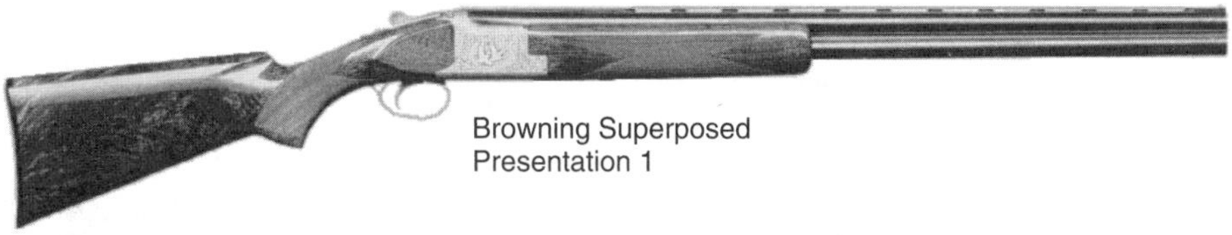

Browning Superposed
Presentation 1

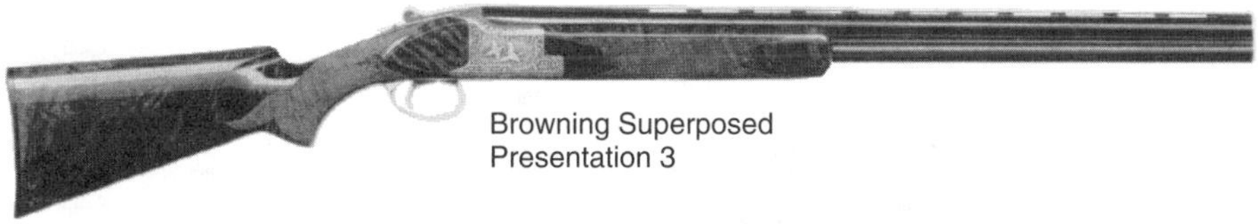

Browning Superposed
Presentation 3

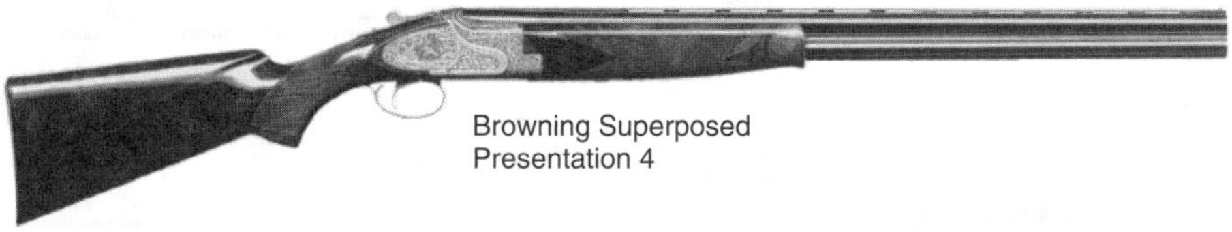

Browning Superposed
Presentation 4

Browning Side-By-Side & Over/Under Shotguns
Superposed Presentation
Superlight Lightning Skeet
Superposesd Presentation
Superlight Lightning Trap
Superposed Presentation
Superlight Magnum
Superposed Superlight

Browning Superposed
Presentation 2 Magnum
Left Side

Browning Superposed
Presentation 2 Magnum
Right Side

Browning Superposed
Presentation 2 Magnum
Bottom View

Browning Superposed Presentation Superlight Lightning Skeet

Same specs as Superposed Presentation Superlight except 12-, 20-, 28-ga., .410; center and front ivory bead sights; special Skeet stock, forend. Introduced 1977; discontinued 1982.

Grade I
  Perf.: $2500   **Exc.:** $1650   **VGood:** $1500
Grade II
  Perf.: $3000   **Exc.:** $1900   **VGood:** $1750
Grade III
  Perf.: $5000   **Exc.:** $3600   **VGood:** $3200
Grade IV
  Perf.: $6000   **Exc.:** $4000   **VGood:** $3500

Browning Superposed Presentation Superlight Lightning Trap

Same specs as Superposed Presentation Superlight except 30″ barrels; trap stock, semi-beavertail forend. Discontinued 1984.

Grade I
  Perf.: $2500   **Exc.:** $1650   **VGood:** $1500
Grade II
  Perf.: $3000   **Exc.:** $1900   **VGood:** $1750
Grade III
  Perf.: $5000   **Exc.:** $3600   **VGood:** $3200
Grade IV
  Perf.: $6000   **Exc.:** $4000   **VGood:** $3500

Browning Superposed Presentation Superlight Magnum

Same specs as Superposed Presentation Superlight except 3″ chambers; 30″ barrels; factory fitted recoil pad. Discontinued 1984.

Grade I
  Perf.: $2500   **Exc.:** $1650   **VGood:** $1500
Grade II
  Perf.: $3000   **Exc.:** $1900   **VGood:** $1750
Grade III
  Perf.: $5000   **Exc.:** $3600   **VGood:** $3200
Grade IV
  Perf.: $6000   **Exc.:** $4000   **VGood:** $3500

BROWNING SUPERPOSED SUPERLIGHT

Over/under; boxlock; 12-, 20-ga.; 2 ³/₄″ chambers; 26 ¹/₂″ vent-rib barrels; choked Modified/Full or Improved/Modified; hand-checkered select walnut straight-grip stock, forearm; top lever; barrel selector combined with manual tang safety; single selective trigger; engraved receiver. Introduced 1967; discontinued 1976.

12-ga.
  Perf.: $1650   **Exc.:** $1150   **VGood:** $1000
20-ga.
  Perf.: $2300   **Exc.:** $1400   **VGood:** $1200

Browning Bolt- & Slide-Action Shotguns
A-Bolt Hunter Shotgun
A-Bolt Stalker Shotgun
BPS

BROWNING SHOTGUN PRODUCTION DATES

Shotgun	Introduced	Discontinued	Manufactured
Superposed 12 Ga Grade I	1931	1976	Belgium
Superposed 20 Ga Grade I	1949	1976	Belgium
Superposed 28 Ga & 410	1960	1976	Belgium
Superposed Bicentennial	1976	Ltd Issue of 51	Belgium
(One for each state & one for District of Columbia)			
Presentation Series Superposed	1977	1986	Belgium
Superposed Cent. Continental	1978	Ltd Issue of 500	Belgium
Superposed Continental	1980	1986	Belgium
Mallard Duck Issue Superposed	1981	Ltd Issue of 500	Belgium
Pintail Duck Issue Superposed	1982	Ltd Issue of 500	Belgium
Black Duck Issue Superposed	1983	Ltd Issue of 500	Belgium
ST-100 Superposed Trap	1979	1981	Belgium
Grade I Superposed Reintroduced	1983	1985	Belgium
Super Pigeon, Pointer, Diana, Midas	1985	To date	Belgium
B-125	1988	To date	Belgium
Over/Under Classic	1986	Ltd Issue of 5000	Japan
Over/Under Gold Classic	1986	Ltd Issue of 500	Belgium
Auto-5	1903	1976	Belgium
Auto-5	1976	To date	Japan
Auto 5 Classic	1984	Ltd Issue of 5000	Japan
Auto 5 Gold Classic	1984	Ltd Issue of 500	Belgium
Auto 5 Invector	1984	To date	Japan
Auto 5 Sweet 16	1936	1976	Belgium
Auto 5 Sweet 16 Reintroduced	1988	1992	Belgium
Auto 5 Stalker	1992	To date	Japan
Auto 5 Invector Plus	1993	To date	Japan
Double Automatic	1954	1972	Belgium
(only shotgun offered in colors. Also, "Twelvette & Twentyweight")			
B-SS Grade I (Selective Trigger)	1978	1987	Japan
B-SS Grade II (Selective Trigger)	1978	1984	Japan
B-SS Sidelock 12 Gauge	1983	1987	Japan
B-SS Sidelock 20 Gauge	1984	1987	Japan
B-2000 All Versions	1974	1981	Bel/Port
B-80	1981	1988	Portugal
B-80 Superlight	1982	1984	Portugal
B-80 Invector	1984	1988	Portugal
B-80 Upland Special	1986	1988	Portugal
Citori	1973	To date	Japan
Citori Sporter (3" chambers)	1978	1983	Japan
Citori Grade II	1978	1983	Japan
Citori Grade V	1978	1984	Japan
Citori Grade V Sideplate 20 Ga	1981	1984	Japan
Citori All Gauge Skeet Set	1981	To date	Japan
Citori Superlight	1982	To date	Japan
Citori Grade VI	1983	To date	Japan
Citori Grade III	1985	To date	Japan
Citori Trap Combo Set	1978	1983	Japan
Citori Upland Special	1984	To date	Japan
Citori Invector	1983	To date	Japan
Citori 16 Gauge All Grades	1987	1989	Japan

BROWNING BOLT- AND SLIDE-ACTION SHOTGUNS

BROWNING A-BOLT HUNTER SHOTGUN

Bolt action; 12-ga.; 3" chamber; 2-shot detachable magazine; 22" fully rifled barrel or 23" barrel with 5" Invector choke tube; weighs 7 1/8 lbs.; 44 3/4" overall length; blade front sight with red insert, open adjustable rear, or no sights; drilled and tapped for scope mount; walnut stock with satin finish; A-Bolt rifle action with 60° bolt throw; front-locking bolt with claw extractor; hinged floorplate; swivel studs; matte finish on barrel, receiver. Imported by Browning. Introduced 1995; still imported. **LMSR: $790**
Rifled barrel
   Perf.: $650   **Exc.:** $470   **VGood:** $400
Invector barrel
   Perf.: $600   **Exc.:** $420   **VGood:** $350

Browning A-Bolt Stalker Shotgun

Same specs as A-Bolt Hunter except black, non-glare composite pistol-grip stock, forend. Imported by Browning. Introduced 1995; still imported. **LMSR: $710**
Rifled barrel
   Perf.: $600   **Exc.:** $450   **VGood:** $400
Invector barrel
   Perf.: $550   **Exc.:** $400   **VGood:** $350

BROWNING BPS

Slide action; 10-, 12-, 20-, 28-ga.; 3", 3 1/2" chamber; 4-, 5-shot magazine; 24", 26", 28", 30" barrel; Invector chokes; high-post vent rib; 48 3/4" overall length; weighs 7 1/2 lbs.; select walnut pistol-grip stock, semi-beavertail forend; bottom feeding, ejection; receiver top safety. Made in Japan. Introduced 1977; still imported by Browning. **LMSR: $510**
   Perf.: $400   **Exc.:** $325   **VGood:** $290

Browning A-Bolt Stalker Shotgun

Browning BPS

Browning BPS Waterfowl 10-Gauge

Browning BPS Game Gun Deer Special

BROWNING SHOTGUN PRODUCTION DATES

Shotgun	Introduced	Discontinued	Manufactured
Citori Skeet (3 Barrel Set)	1987	To date	Japan
Citori Plus 3 1/2″ 12 Gauge	1989	To date	Japan
Citori Plus Trap	1989	1994	Japan
Citori Lightning & Secial Sporting Clay	1989	To date	Japan
Citori Gran Lightning	1990	To date	Japan
Invector Plus (Back-bored)	1989	To date	Japan
Grade I Micro Lightning 20 Gauge	1991	To date	Japan
Plus Trap Combo Inv Plus	1992	To date	Japan
325 Sporting Clays	1993	1994	Japan
425 Sporting Clays	1995	To date	Japan
GTI Sporting Clay	1989	1994	Japan
Ultra Sporting Clays	1995	To date	Japan
Special Trap	1995	To date	Japan
Special Skeet	1995	To date	Japan
High Grade Micro Lightning	1993	To date	Japan
Signature Painted Sporting Clays	1993	To date	Japan
Pigeon Painted Sporting Clays	1993	To date	Japan
BT-99 Trap	1968	1977	Japan
BT-99 Competition Trap	1978	1994	Japan
BT-99 Pigeon Grade	1978	1984	Japan
BT-99 Invector	1983	1994	Japan
BT-99 Plus	1989	1994	Japan
BT-99 Plus Micro	1991	1994	Japan
BT-99 Stainless	1993	1994	Japan
BT-99 Invector Plus Competition	1992	1992	Japan
BT-100	1995	To date	Japan
BT-99 Max	1995	To date	Japan
BPS 12 & 20	1977	1984	Japan
BPS Trap	1978	1985	Japan
BPS Invector	1983	To date	Japan
BPS Upland Special	1984	To date	Japan
BPS Youth and Ladies 20 Ga	1986	To date	Japan
BPS Stalker 10 & 12	1987	To date	Japan
BPS 10 & 3 1/2″ 12 Ga	1989	To date	Japan
BPS With Engraving	1991	To date	Japan
BPS Pigeon Grade	1992	To date	Japan
BPS Game Guns	1992	To date	Japan
A-500 & A500R (R stamping 1990)	1987	1993	Bel/Port
A-500G	1990	1993	Bel/Port
A-500G Sporting Clays	1992	1993	Bel/Port
Gold 10 Gauge Semi Auto	1993	To date	Japan
Gold 12 & 20 Gauge Semi Auto	1994	To date	Bel/Port
Model 12 20 Gauge Grade I	1988	Ltd Issue of 8000	Japan
Model 12 20 Gauge Grade V	1988	Ltd Issue of 4000	Japan
Model 12 28 Gauge Grade I	1990	Ltd Issue of 8000	Japan
Model 12 28 Gauge Grade V	1990	Ltd Issue of 4000	Japan
Model 42 Grade I	1991	Ltd Issue of 7000	Japan
Model 42 Grade V	1991	Ltd Issue of 5000	Japan
Recoilless Trap	1993	To date	Japan
A-Bolt Shotgun	1995	To date	Japan

Browning Bolt- & Slide-Action Shotguns
BPS Buck Special
BPS Game Gun Deer Special
BPS Game Gun Turkey Special
BPS Ladies & Youth Model
BPS Pigeon Grade
BPS Stalker
BPS Upland Special
BPS Waterfowl 10-Gauge
BSA 10 (See Browning Gold 10)
BSA 10 Stalker (See Browning Gold 10 Stalker)
Model 12
Model 42
Recoilless Trap
Recoilless Trap Micro

high-grade walnut; crossbolt safety in trigger guard; polished blue finish. Made in Japan. Reproduction of the Winchester Model 12. The 20-ga. was limited to 8000 guns; introduced 1988, discontinued 1990. The 28-ga. was limited to 8500 Grade I and 4000 Grade V guns; introduced 1991, discontinued 1992.

Grade I
| Perf.: $450 | Exc.: $320 | VGood: $280 |

Grade V
| Perf.: $795 | Exc.: $600 | VGood: $450 |

BROWNING MODEL 42
Slide action; 410; 3″ chamber; 26″ Full barrel; high post floating rib with grooved sighting plane; weighs about 6 3/4 lbs.; cut-checkered select walnut stock, forearm with semi-gloss finish; Grade V has high-grade walnut; crossbolt safety in trigger guard; polished blue finish. Made in Japan. Reproduction of the Winchester Model 42. Limited to 6000 Grade I and 6000 Grade V guns. Introduced 1991; discontinued 1993.

Grade I
| Perf.: $500 | Exc.: $370 | VGood: $300 |

Grade V
| Perf.: $795 | Exc.: $625 | VGood: $500 |

BROWNING RECOILLESS TRAP
Bolt action; single shot; 12-ga.; 2 3/4″ chamber; back-bored 30″ barrel; Invector Plus tubes; ventilated rib adjusts to move point of impact; weighs about 9 lbs.; cut-checkered select walnut stock, forearm with high gloss finish; adjustable for drop at comb and length of pull; bolt action eliminates up to 72 percent of recoil; forend is used to cock action when the action is forward. Made in Japan. Introduced 1993; still imported. **LMSR: $1900**
| Perf.: $825 | Exc.: $790 | VGood: $680 |

Browning Recoilless Trap Micro
Same specs as Recoilless Trap except 27″ barrel; weighs 8 5/8 lbs.; stock length of pull is adjustable from 13″ to 13 3/4″. Made in Japan. Introduced 1993; still imported. **LMSR: $1900**
| Perf.: $825 | Exc.: $790 | VGood: $680 |

Browning BPS Buck Special
Same specs as BPS except 24″ Cylinder barrel; no Invector choke tubes. Introduced 1977; reintroduced 1989; still in production. **LMSR: $495**
| Perf.: $375 | Exc.: $240 | VGood: $200 |
10-ga.
| Perf.: $525 | Exc.: $375 | VGood: $340 |

Browning BPS Game Gun Deer Special
Same specs as BPS except 12-ga.; 3″ chamber; heavy 20 1/2″ barrel with rifled choke tube; rifle-type sights with adjustable rear; solid receiver scope mount; "rifle" stock dimensions for scope or open sights; gloss or matte finished wood with checkering; newly designed receiver/magazine tube/barrel mounting system to eliminate play; sling swivel studs; polished blue metal. Made by Miroku in Japan. Introduced 1992; still imported. **LMSR: $575**
| Perf.: $430 | Exc.: $240 | VGood: $200 |

Browning BPS Game Gun Turkey Special
Same specs as BPS except 12-ga.; 3″ chamber; light 20 1/2″ barrel; Extra-Full Invector choke tube; drilled and tapped for scope mounting; rifle-style satin-finished walnut stock; swivel studs; dull-finished barrel, receiver. Made in Japan by Miroku. Introduced 1992; still imported. **LMSR: $545**
| Perf.: $400 | Exc.: $225 | VGood: $190 |

Browning BPS Ladies & Youth Model
Same specs as BPS except 20-ga.; 22″ barrel; shortened walnut pistol-grip stock; recoil pad. Made in Japan. Introduced 1986; still imported. **LMSR: $510**
| Perf.: $380 | Exc.: $220 | VGood: $180 |

Browning BPS Pigeon Grade
Same specs as BPS except 12-ga.; 3″ chamber; 26″, 28″ vent-rib barrels; select high grade walnut stock, forend; gold-trimmed receiver. Made in Japan. Introduced 1992; still imported. **LMSR: $680**
| Perf.: $450 | Exc.: $400 | VGood: $325 |

Browning BPS Stalker
Same specs as BPS except 10-, 12-ga.; 3″, 3 1/2″ chamber; 24″, 26″, 28″, 30″ barrel; black-finished synthetic stock, forearm; matte blued finish on metal; black recoil pad. Made in Japan. Introduced 1987; still imported. **LMSR: $510**
| Perf.: $375 | Exc.: $240 | VGood: $200 |

Browning BPS Upland Special
Same specs as BPS except 12-, 20-ga.; 22″ barrel; walnut straight-grip stock, forearm. Made in Japan. Introduced 1989; still imported. **LMSR: $510**
| Perf.: $375 | Exc.: $240 | VGood: $200 |

Browning BPS Waterfowl 10-Gauge
Same specs as BPS except 10-ga.; 3 1/2″ chamber; 28″, 30″ vent-rib barrel; high grade stock; gold-trimmed receiver. Made in Japan. Introduced 1993; still imported. **LMSR: $820**
| Perf.: $625 | Exc.: $430 | VGood: $375 |

BROWNING MODEL 12
Slide action; 20-, 28-ga.; 2 3/4″ chamber; 26″ Modified barrel; high post floating rib with grooved sighting plane; weighs about 7 lbs.; cut-checkered select walnut stock, forearm with semi-gloss finish; Grade V has

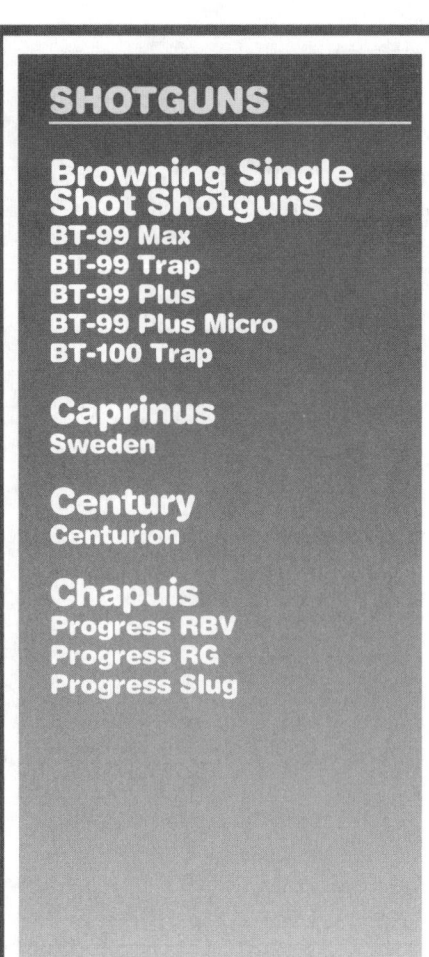

SHOTGUNS

Browning Single Shot Shotguns
BT-99 Max
BT-99 Trap
BT-99 Plus
BT-99 Plus Micro
BT-100 Trap

Caprinus
Sweden

Century
Centurion

Chapuis
Progress RBV
Progress RG
Progress Slug

Browning BT-99 Grade I

Browning BT-99 Plus Trap

Browning BT-100 Trap

Century Centurion

Chapuis S/S Model

BROWNING SINGLE SHOT SHOTGUNS

BROWNING BT-99 MAX
Single shot; boxlock; 12-ga.; 2 3/4" chamber; 32", 34" vent-rib barrel; Invector Plus choke tubes; weighs 8 5/8 lbs.; ivory front sight, middle bead; hand-checkered walnut pistol-grip Monte Carlo or standard stock, forend; gold-plated trigger; automatic ejector; recoil pad. Made in Japan. Introduced 1995; still in production. **LMSR: $1496**
    Perf.: $1200 **Exc.:** $710   **VGood:** $580
Stainless model with black ported barrel
    Perf.: $1450 **Exc.:** $825   **VGood:** $650

BROWNING BT-99 TRAP
Same specs as BT-99 Max except Improved, Full fixed choke or Invector Plus chokes (standard 1992); no gold plating; also available in stainless version with black barrel. Made in Japan. Introduced 1969; discontinued 1994.
    Perf.: $800 **Exc.:** $500   **VGood:** $420
Stainless model with black barrel
    Perf.: $1275 **Exc.:** $850   **VGood:** $650
Pigeon Grade high grade stock
    Perf.: $1150 **Exc.:** $700   **VGood:** $600
Signature Painted with red, black stock
    Perf.: $950 **Exc.:** $500   **VGood:** $450
Golden Clays with gold inlays
    Perf.: $1950 **Exc.:** $1400   **VGood:** $1150

Browning BT-99 Plus
Same specs as BT-99 Max except high, wide rib, Monte Carlo-style stock. Introduced 1976; Competition name dropped 1978; discontinued 1994.
    Perf.: $1200 **Exc.:** $900   **VGood:** $750
Stainless model with black barrel
    Perf.: $1475 **Exc.:** $900   **VGood:** $750
Pigeon Grade high grade stock
    Perf.: $1425 **Exc.:** $900   **VGood:** $750
Signature Painted with red, black stock
    Perf.: $1400 **Exc.:** $900   **VGood:** $780
Golden Clays with gold inlays
    Perf.: $2475 **Exc.:** $2000   **VGood:** $1775

Browning BT-99 Plus Micro
Same specs as BT-99 Max except 28", 30" ported barrel with adjustable rib system; Invector Plus choke system and back-bored barrel; buttstock with adjustable length of pull range of 13 1/2" to 14"; Browning's recoil reducer system; scaled down for smaller shooters; marketed in Travel Vault case. Made in Japan. Introduced 1991; discontinued 1994.
    Perf.: $1325 **Exc.:** $900   **VGood:** $750
Stainless model with black barrel
    Perf.: $1625 **Exc.:** $1400   **VGood:** $1200
Pigeon Grade high grade stock
    Perf.: $1475 **Exc.:** $1250   **VGood:** $1000
Signature Painted with red, black stock
    Perf.: $1475 **Exc.:** $1250   **VGood:** $1000
Golden Clays with gold inlays
    Perf.: $2625 **Exc.:** $2000   **VGood:** $1600

BROWNING BT-100 TRAP
Single shot; 12-ga.; 2 3/4" chamber; back-bored 32", 34" barrel with Invector Plus, or fixed Full choke; weighs 8 5/8 lbs.; 48 1/2" overall length (32" barrel); cut-checkered walnut Monte Carlo stock, wedge-shaped forend with finger groove; high gloss finish; drop-out trigger adjustable for weight of pull from 3 1/2 to 5 1/2 lbs., and for three length positions; Ejector-Selector allows ejection or extraction of shells; stainless steel or blue finish; optional adjustable comb stock with thumbhole. Made in Japan. Introduced 1995; still imported. **LMSR: $1900**
Blue finish
    Perf.: $1500 **Exc.:** $910   **VGood:** $740
Stainless steel
    Perf.: $1900 **Exc.:** $1200   **VGood:** $1000
Thumbhole stock, blue or stainless
    Perf.: $1850 **Exc.:** $1160   **VGood:** $990

CAPRINUS SWEDEN
Over/under; 12-ga.; 2 3/4" chamber; 28", 30" barrel; interchangeable choke tubes; high-grade European walnut, optional Monte Carlo stock; oil finish; checkered butt; single selective trigger ejectors; double safety system; stainless steel construction. Made in Sweden. Introduced 1982; discontinued 1985.
Skeet Special
    Perf.: $3600 **Exc.:** $2700   **VGood:** $2350
Skeet Game
    Perf.: $3650 **Exc.:** $2750   **VGood:** $2400
Game
    Perf.: $3700 **Exc.:** $2800   **VGood:** $2400
Trap
    Perf.: $3600 **Exc.:** $2700   **VGood:** $2350

CENTURY CENTURION
Over/under; boxlock; 12-ga.; 2 3/4" chamber; 28" Modified/Full barrel; weighs 7 1/4 lbs.; European walnut stock; double triggers; polished blue finish. Introduced 1993; still imported. **LMSR: $380**
    Perf.: $320 **Exc.:** $220   **VGood:** $200

CHAPUIS PROGRESS RBV
Side-by-side; boxlock; sideplates; 12-, 20-ga.; 26 1/2", 27 1/2" barrels; ventilated rib; weighs about 6 1/4 lbs.; fine-checkered oil-finished French or American walnut straight or pistol-grip stock; auto ejectors; double triggers; chromed bores; scroll engraving. Made in France. Introduced 1979; discontinued 1982.
    Perf.: $1800 **Exc.:** $1400   **VGood:** $1200

Chapuis Progress RG
Same specs as Progress RBV except boxlock. Introduced 1979; discontinued 1982; reintroduced 1989; no longer imported.
    Perf.: $1200 **Exc.:** $900   **VGood:** $750

Chapuis Progress Slug
Same specs as Progress RBV except boxlock; right barrel is rifled for slugs. No longer imported.
    Perf.: $1200 **Exc.:** $900   **VGood:** $750

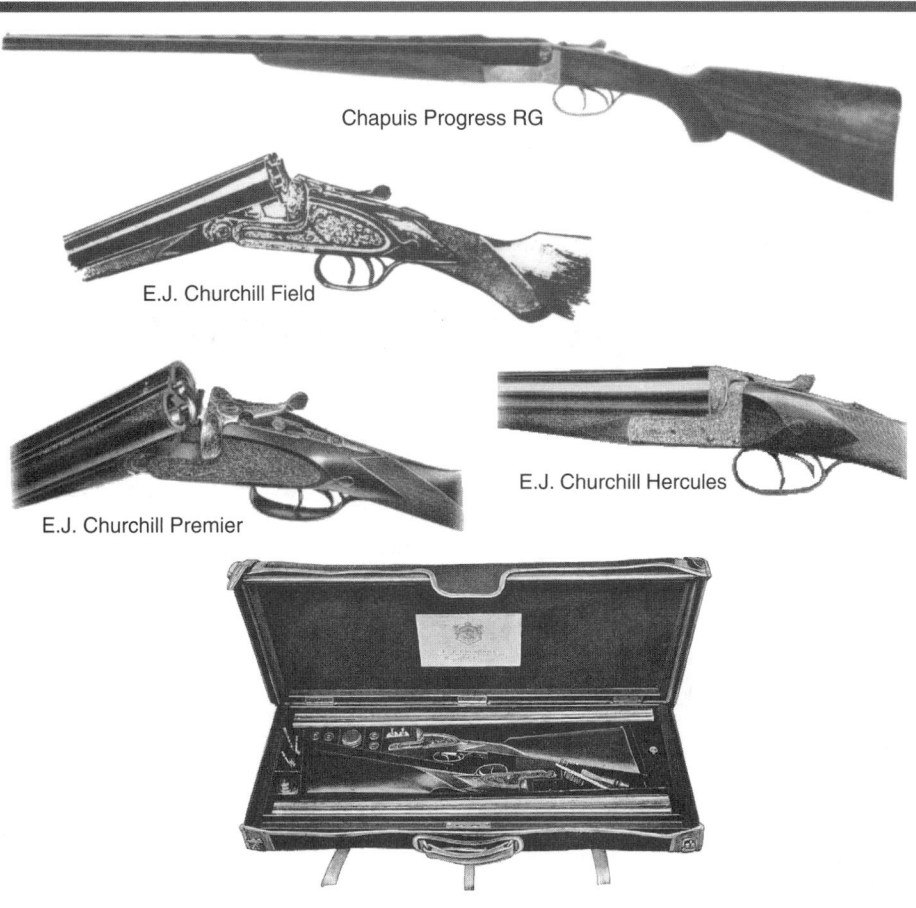

Chapuis Progress RG

E.J. Churchill Field

E.J. Churchill Premier

E.J. Churchill Hercules

SHOTGUNS

Chapuis
O/U

E.J. Churchill
Crown
Field
Hercules
Imperial
Premier
Premier Quality
Regal

CHAPUIS O/U

Over/under; sidelock; 12-, 16-, 20-ga.; 22″, 23 1/2″, 26 2/3″, 27 1/2″, 31 1/2″ barrel; choked to customer specs; choice of rib styles; French walnut straight English or pistol-grip stock; auto ejectors or extractors; long trigger guard. Made in France. Introduced 1989; no longer imported.

    Perf.: $3450 **Exc.:** $2250 **VGood:** $1900

NOTE: 20-, 28-gauges bring 25% to 35% more than prices shown. The .410 is as much as 50% higher.

CHURCHILL, E.J., CROWN

Side-by-side; boxlock; hammerless; 12-, 16-, 20-, 28-ga., 410; 25″, 28″, 30″, 32″ barrel; any desired choke combo; third-grade, hand-checkered European walnut stock, forearm; pistol grip or straight English-style; ejectors; double triggers or single selective trigger. Introduced 1900; discontinued 1982. Made in England.

    Perf.: $4750 **Exc.:** $3200 **VGood:** $2600
With single selective trigger
    Perf.: $5200 **Exc.:** $3700 **VGood:** $3000
20-ga.
    Perf.: $5500 **Exc.:** $3800 **VGood:** $3000
28-ga.
    Perf.: $6500 **Exc.:** $4500 **VGood:** $3500
410-bore
    Perf.: $7500 **Exc.:** $5000 **VGood:** $4200

CHURCHILL, E.J., FIELD

Side-by-side; sidelock; hammerless; 12-, 16-, 20-, 28-ga.; 25″, 28″, 30″, 32″ barrel; any desired choke combo; third grade, hand-checkered European walnut stock, forearm; pistol grip or straight English-style; ejectors; double triggers or single selective trigger; engraving. Introduced 1900; discontinued 1982. Made in England.

    Perf.: $9500 **Exc.:** $7500 **VGood:** $6500
With single selective trigger
    Perf.: $10,500 **Exc.:** $8500 **VGood:** $7500
20-ga.
    Perf.: $12,500 **Exc.:** $9000 **VGood:** $8000
28-ga.
    Perf.: $14,500 **Exc.:** $10,500 **VGood:** $9000

CHURCHILL, E.J., HERCULES

Side-by-side; boxlock; hammerless; 12-, 16-, 20-, 28-ga.; 25″, 28″, 30″, 32″ barrel; any desired choke combo; top-grade, hand-checkered European walnut stock, forearm; pistol grip or straight English-style; ejectors; double triggers or single selective trigger; engraving. Introduced 1900; discontinued 1982. Made in England.

    Perf.: $9000 **Exc.:** $7000 **VGood:** $6000
With single selective trigger
    Perf.: $10,000 **Exc.:** $8000 **VGood:** $7000
20-ga.
    Perf.: $12,000 **Exc.:** $8500 **VGood:** $7500
28-ga.
    Perf.: $14,000 **Exc.:** $9500 **VGood:** $8000

CHURCHILL, E.J., IMPERIAL

Side-by-side; boxlock; hammerless; 12-, 16-, 20-, 28-ga.; 25″, 28″, 30″, 32″ barrel; any desired choke combo; second-grade, hand-checkered European walnut stock, forearm; pistol grip or straight English-style; ejectors; double triggers or single selective trigger; engraving. Introduced 1900; discontinued 1982. Made in England.

    Perf.: $19,500 **Exc.:** $13,500 **VGood:** $11,500
With single selective trigger
    Perf.: $20,000 **Exc.:** $14,000 **VGood:** $12,000
20-ga.
    Perf.: $24,000 **Exc.:** $16,000 **VGood:** $13,500
28-ga.
    Perf.: $28,000 **Exc.:** $19,000 **VGood:** $16,000

CHURCHILL, E.J., PREMIER

Over/under; sidelock; hammerless; 12-, 16-, 20-, 28-ga.; 25″, 28″, 30″, 32″ barrel; hand-checkered European walnut stock, forearm; pistol grip or straight English-style; ejectors; double triggers or single selective trigger; English engraving. Introduced 1925; discontinued 1955. Made in England.

    Perf.: $18,000 **Exc.:** $12,000 **VGood:** $9000
With single selective trigger
    Perf.: $19,000 **Exc.:** $13,000 **VGood:** $10,000

20-ga.
    Perf.: $22,000 **Exc.:** $14,500 **VGood:** $11,000
28-ga.
    Perf.: $25,000 **Exc.:** $16,500 **VGood:** $13,000

CHURCHILL, E.J., PREMIER QUALITY

Side-by-side; sidelock; hammerless; 12-, 16-, 20-, 28-ga.; 25″, 28″, 30″, 32″ barrel; any desired choke combo; top-grade, hand-checkered European walnut stock, forearm; pistol grip or straight English-style; ejectors; double triggers or single selective trigger; English engraving. Introduced 1900; discontinued 1982. Made in England.

    Perf.: $17,000 **Exc.:** $11,000 **VGood:** $8000
With single selective trigger
    Perf.: $18,000 **Exc.:** $12,000 **VGood:** $9000
20-ga.
    Perf.: $21,000 **Exc.:** $13,500 **VGood:** $10,000
28-ga.
    Perf.: $24,000 **Exc.:** $15,500 **VGood:** $12,000

CHURCHILL, E.J., REGAL

Side-by-side; boxlock; hammerless; 12-, 16-, 20-, 28-ga., .410; 25″, 28″, 30″, 32″ barrel; any desired choke combo; second-grade, hand-checkered European walnut stock, forearm; pistol grip or straight English-style; ejectors; double triggers or single selective trigger. Introduced during WWII; discontinued 1982. Made in England.

    Perf.: $5500 **Exc.:** $3800 **VGood:** $3200
With single selective trigger
    Perf.: $6000 **Exc.:** $4300 **VGood:** $3700
20-ga.
    Perf.: $6700 **Exc.:** $4500 **VGood:** $3850
28-ga.
    Perf.: $7500 **Exc.:** $4500 **VGood:** $3800
.410-bore
    Perf.: $8500 **Exc.:** $6000 **VGood:** $5000

SHOTGUNS

E.J. Churchill
Utility

Churchill
Deerfield Semi-Auto
Regent IV S/S Boxlock
Regent VI S/S Sidelock
Regent V O/U
Regent O/U Competition
Regent O/U Flyweight
Regent O/U Trap/Skeet
Regent VII O/U
Regent Semi-Auto
Royal
Turkey Automatic Shotgun
Windsor III O/U

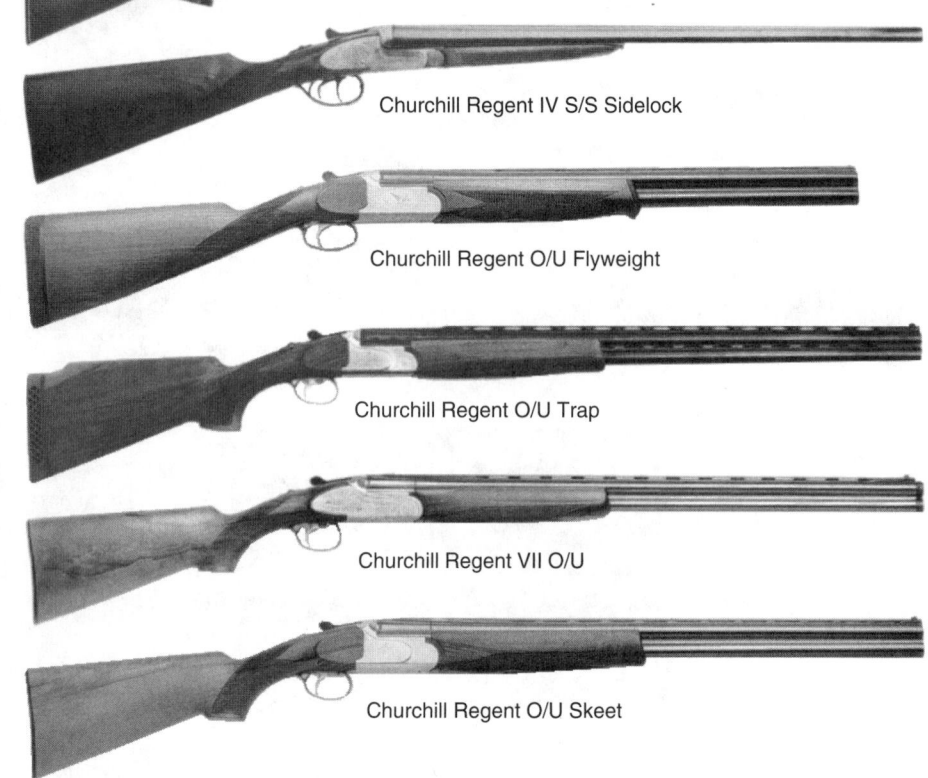

Churchill Regent Semi-Auto

Churchill Regent IV S/S Sidelock

Churchill Regent O/U Flyweight

Churchill Regent O/U Trap

Churchill Regent VII O/U

Churchill Regent O/U Skeet

CHURCHILL, E.J., UTILITY
Side-by-side; boxlock; hammerless; 12-, 16-, 20-, 28-ga., .410; 25″, 28″, 30″, 32″ barrel; any desired choke combo; second-grade, hand-checkered European walnut stock, forearm; pistol grip or straight English-style; ejectors; double triggers or single selective trigger. Introduced 1900; discontinued WWII. Made in England.
  Exc.: $3600 **VGood:** $2600 **Good:** $2000
With single selective trigger
  Exc.: $4000 **VGood:** $3000 **Good:** $2400
20-ga.
  Exc.: $4300 **VGood:** $3500 **Good:** $2800
28-ga.
  Exc.: $5000 **VGood:** $4000 **Good:** $3000
.410-bore
  Exc.: $5500 **VGood:** $4500 **Good:** $3500

CHURCHILL DEERFIELD SEMI-AUTO
Autoloader; 12-ga.; 2 3/4″, 3″ chamber; 20″ slug barrel; standard chokes or ICT choke tubes; weighs 7 1/2 lbs.; checkered select Claro walnut pistol-grip stock, forearm; high gloss or matte finish wood; rosewood grip cap; stainless steel gas piston; crossbolt safety; aluminum alloy receiver. Made in Japan. Introduced 1984; discontinued 1986.
  Perf.: $450 **Exc.:** $350 **VGood:** $300

CHURCHILL REGENT IV S/S BOXLOCK
Side-by-side; boxlock; 12-, 20-ga.; 2 3/4″ chamber; 25″, 28″ barrel; tapered Churchill rib; hand-checkered European walnut straight stock; double triggers. Made in Spain. Introduced 1984; discontinued 1987.
  Perf.: $525 **Exc.:** $400 **VGood:** $325
Left-hand version
  Perf.: $575 **Exc.:** $450 **VGood:** $375

Churchill Regent VI S/S Sidelock
Same specs as Regent IV S/S except sidelock. No longer in production.
  Perf.: $800 **Exc.:** $600 **VGood:** $500

CHURCHILL REGENT V O/U
Over/under; boxlock; 12-, 20-ga.; 27″ barrel; interchangeable choke tubes; ventilated rib; ejectors; SST; pewter finish. Made in Italy. Introduced 1984; discontinued 1991.
  Perf.: $850 **Exc.:** $690 **VGood:** $600

Churchill Regent O/U Competition
Same specs as Regent V except available with trap and Skeet models. 12-ga.; 2 3/4″ chamber; 26″ Skeet/Skeet, 30″ Improved Modified/Full (trap) barrel; checkered European walnut stock, Schnabel forend; oil finish; single selective trigger; automatic ejectors; silvered, engraved receiver. Introduced 1991; discontinued 1992.
Trap version
  Perf.: $800 **Exc.:** $575 **VGood:** $550
Skeet version
  Perf.: $750 **Exc.:** $525 **VGood:** $500

Churchill Regent O/U Flyweight
Same specs as Regent V except 23″ barrel; choke tubes. Made in Japan. Introduced 1984; discontinued 1986.
  Perf.: $850 **Exc.:** $790 **VGood:** $700

Churchill Regent O/U Trap/Skeet
Same specs as Regent V except sideplates; 26″, 30″ barrel; checkered Monte Carlo stock (trap model); oil-finished wood; vent recoil pad; chrome bores; silvered, engraved receivers. Introduced 1984; discontinued 1991.
  Perf.: $800 **Exc.:** $600 **VGood:** $500

Churchill Regent VII O/U
Same specs as Regent V except sideplates. No longer imported.
  Perf.: $850 **Exc.:** $690 **VGood:** $600

CHURCHILL REGENT SEMI-AUTO
Autoloader; 12-ga.; 2 3/4″, 3″ chamber; 26″, 28″, 30″ barrel; standard chokes or ICT choke tubes; weighs 7 1/2 lbs.; checkered select Monte Claro walnut pis-

tol-grip stock, forend; high gloss or matte finish wood; rosewood grip cap; stainless steel gas piston; crossbolt safety; etched, polished aluminum receiver. Made in Japan. Introduced 1984; discontinued 1986.
  Perf.: $400 **Exc.:** $300 **VGood:** $280

CHURCHILL ROYAL
Side-by-side; boxlock; 10-, 12-, 20-, 28-ga., .410; 3″, 3 1/2″ chamber; 26″, 28″ barrel; standard choke combos; concave rib; checkered European walnut straight-grip stock; double triggers; extractors; chromed bores; color case-hardened finish. Made in Spain. Introduced 1988; discontinued 1991.
10-ga.
  Perf.: $450 **Exc.:** $350 **VGood:** $300
12-, 20-ga.
  Perf.: $450 **Exc.:** $350 **VGood:** $300
28-ga.
  Perf.: $480 **Exc.:** $380 **VGood:** $330
.410
  Perf.: $525 **Exc.:** $425 **VGood:** $375

CHURCHILL TURKEY AUTOMATIC SHOTGUN
Gas-operated autoloader; 12-ga.; 3″ chamber; 5-shot magazine; 25″ barrel; standard Modified, Full, Extra Full choke tubes; weighs 7 lbs.; hand-checkered walnut stock with satin finish; magazine cut-off; non-glare metal finish; gold-colored trigger. Introduced 1990; still imported.
  LMSR: $570
  Perf.: $480 **Exc.:** $360 **VGood:** $300

CHURCHILL WINDSOR III O/U
Over/under; boxlock; 12-, 20-ga., .410; 3″ chamber; 26″, 28″, 30″ barrel; standard choke combos or interchangeable choke tubes; checkered European walnut pistol-grip stock, forend; extractors; single selective trigger; silvered, engraved finish. Made in Italy. Introduced 1984; no longer in production.
  Perf.: $500 **Exc.:** $400 **VGood:** $340

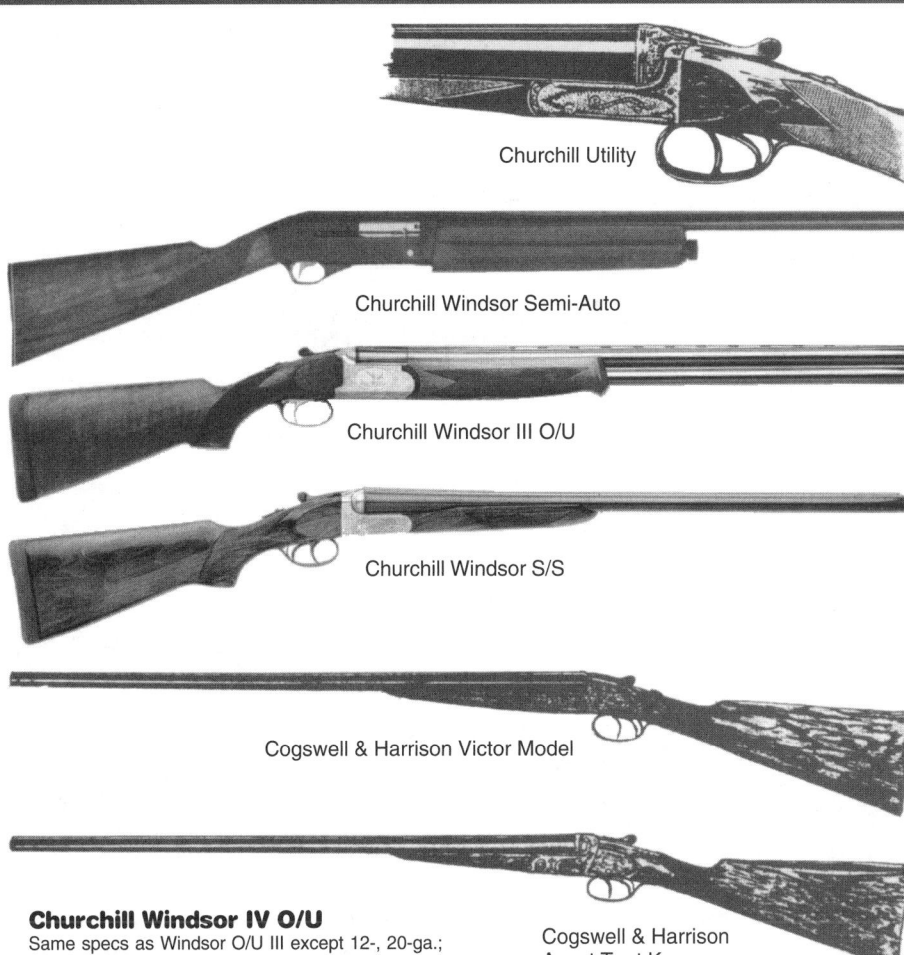

Churchill Utility

Churchill Windsor Semi-Auto

Churchill Windsor III O/U

Churchill Windsor S/S

Cogswell & Harrison Victor Model

Cogswell & Harrison
Avant Tout Konor

Churchill
Windsor IV O/U
Windsor O/U Flyweight
**Windsor O/U Sporting
Clays**
Windsor S/S
Windsor S/S II
Windsor Semi-Auto

Clayco
Model 6

Cogswell &
Harrison
Ambassador
Avant Tout Konor
Avant Tout Rex
Avant Tout Sandhurst
Huntic Model
Markor Model
Primac Model
Victor Model

Churchill Windsor IV O/U
Same specs as Windsor O/U III except 12-, 20-ga.; ejectors. Introduced 1984; still imported. **LMSR: $932**
Perf.: **$750**   Exc.: **$525**   VGood: **$440**

Churchill Windsor O/U Flyweight
Same specs as Windsor O/U except 23" barrel; ICT choke tubes; checkered walnut straight-grip stock. No longer imported.
Perf.: **$600**   Exc.: **$500**   VGood: **$450**

Churchill Windsor O/U Sporting Clays
Same specs as Windsor IV except 12-ga.; 28", 30" ported, back-bored barrels; choke tubes; tapered ventilated rib, ventilated side rib; select walnut stock, sporting-style forend with finger grooves; palm swell grip; lengthened forcing cones. Introduced 1995; still imported. **LMSR: $1125**
Perf.: **$900**   Exc.: **$750**   VGood: **$600**

CHURCHILL WINDSOR S/S
Side-by-side; boxlock; 10-, 12-, 16-, 20-, 28-ga., .410; 2 3/4" (16-ga.), 3" chamber; 24", 26", 28", 30", 32" barrel; hand-checkered European walnut stock, beavertail forend; rubber butt pad; auto safety; double triggers; extractors; silvered, engraved finish. Made in Spain. Introduced 1984; discontinued 1991.
Perf.: **$500**   Exc.: **$410**   VGood: **$320**
10-ga.
Perf.: **$495**   Exc.: **$380**   VGood: **$300**

Churchill Windsor S/S II
Same specs as Windsor S/S except ejectors. No longer in production.
Perf.: **$550**   Exc.: **$460**   VGood: **$370**

CHURCHILL WINDSOR SEMI-AUTO
Autoloader; 12-ga.; 2 3/4", 3" chamber; 26" to 30" barrel; standard chokes or ICT choke tubes; weighs 7 1/2 lbs.; checkered select Monte Claro walnut pistol-grip stock, forend; high gloss or matte finish wood; rosewood grip cap; stainless steel gas piston; crossbolt safety; anodized alloy receiver. Made in Japan. Introduced 1984; discontinued 1991.
Perf.: **$350**   Exc.: **$275**   VGood: **$210**

CLAYCO MODEL 6
Over/under; 12-ga.; 2 3/4" chamber; 26", 28" barrel; vent top rib; checkered walnut-finished hardwood pistol-grip stock, forend; single non-selective trigger; auto safety; vent rubber recoil pad; blued scroll-engraved receiver. Made in China. Introduced 1983; discontinued 1985.
Perf.: **$200**   Exc.: **$150**   VGood: **$100**

COGSWELL & HARRISON AMBASSADOR
Side-by-side; boxlock; 12-, 16-, 20-ga.; 26", 28", 30" barrel; any choke combo; hand-checkered European walnut straight-grip stock, forearm; ejectors; double triggers; sideplates feature game engraving or rose and scroll engraving. Introduced 1970; no longer in production. Made in England.
Perf.: **$3650**   Exc.: **$3000**   VGood: **$2600**

COGSWELL & HARRISON AVANT TOUT KONOR
Side-by-side; boxlock; hammerless; 12-, 16-, 20-ga.; 25", 27 1/2", 30" barrels; any desired choke combo; hand-checkered European walnut straight-grip stock, forearm; pistol-grip stock available on special order; sideplates; double, single or single selective triggers; engraving. Introduced 1920s; no longer in production. Made in England.
Exc.: **$2500**   VGood: **$2200**   Good: **$1800**

Cogswell & Harrison Avant Tout Rex
Same specs as Avant Tout Konor except no sideplates; lower grade wood, checkering, engraving, overall finish. No longer in production. Made in England.
Exc.: **$1650**   VGood: **$1300**   Good: **$1000**

Cogswell & Harrison Avant Tout Sandhurst
Same specs as Avant Tout Konor except less-intricate engraving, checkering; lower grade of wood, overall workmanship. No longer in production. Made in England.
Exc.: **$2300**   VGood: **$1850**   Good: **$1500**

COGSWELL & HARRISON HUNTIC MODEL
Side-by-side; sidelock; 12-, 16-, 20-ga.; 25", 27 1/2", 30" barrel; any desired choke combo; hand-checkered European walnut straight-grip stock, forearm; pistol-grip stock on special order; ejectors; double, single or single selective triggers. Introduced in late 1920s; no longer in production. Made in England.
Exc.: **$3250**   VGood: **$2600**   Good: **$2200**

COGSWELL & HARRISON MARKOR MODEL
Side-by-side; boxlock; 12-, 16-, 20-ga.; 27 1/2", 30" barrels; any standard choke combo; hand-checkered European walnut straight-grip stock, forearm; double triggers; ejectors or extractors. Introduced in late 1920s; no longer in production. Made in England.
Extractors
Exc.: **$1300**   VGood: **$1000**   Good: **$895**
Ejectors
Exc.: **$1450**   VGood: **$1150**   Good: **$1000**

COGSWELL & HARRISON PRIMAC MODEL
Side-by-side; hand-detachable sidelock; 12-, 16-, 20-ga.; 25", 26", 27 1/2", 30" barrels; any choke combo; hand-checkered European walnut straight-grip stock, forearm; ejectors; double triggers, single or single selective triggers; English-style engraving. Introduced in 1920s; no longer in production. Made in England.
Exc.: **$4950**   VGood: **$3750**   Good: **$2400**

COGSWELL & HARRISON VICTOR MODEL
Side-by-side; hand-detachable sidelock; 12-, 16-, 20-ga.; 25", 26", 27 1/2", 30" barrels; any choke combo; hand-checkered European walnut straight-grip stock, forearm; ejectors; double triggers, single or single selective triggers; high quality English-style engraving. Introduced in 1920s; no longer in production. Made in England.
Exc.: **$6000**   VGood: **$4000**   Good: **$3000**

Colt
Armsmear Over/Under
Coltsman Pump
Coltsman Pump Custom
Custom Double Barrel
Magnum Auto
Magnum Auto Custom
Sauer Drilling
Ultra Light Auto
Ultra Light Auto Custom

Connecticut Valley Classics
Classic Sporter
Classic Field Waterfowler

Cosmi
Semi-Automatic

Crucelegui Hermanos
Model 150

Charles Daly
Auto

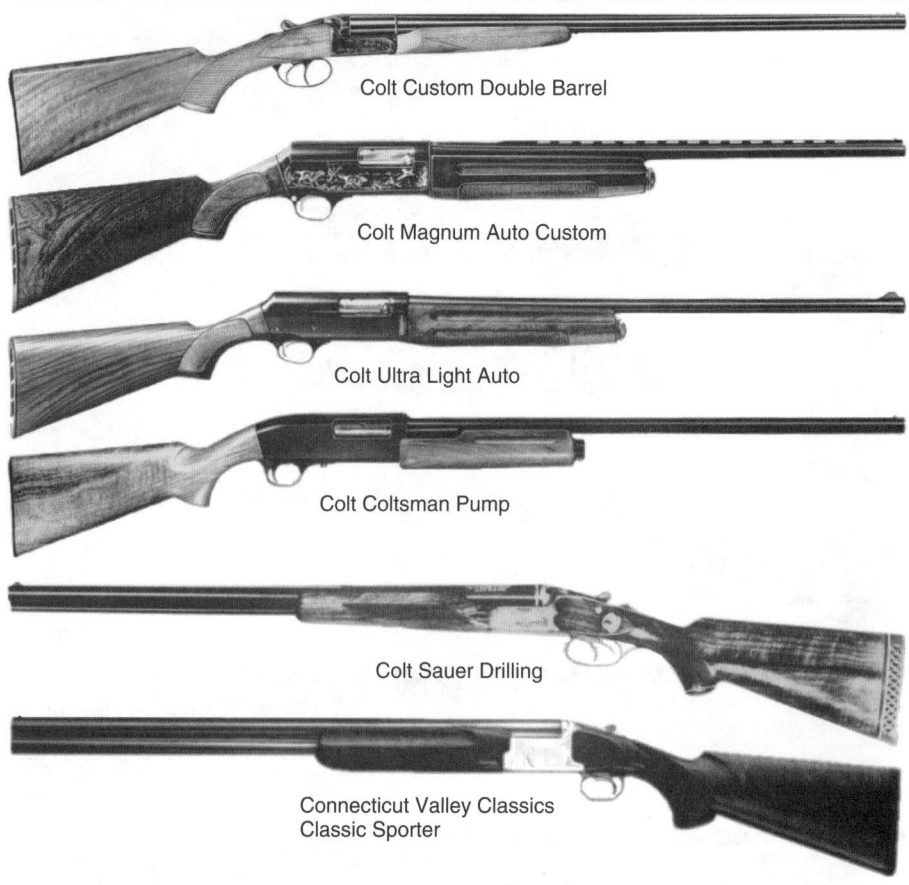

Colt Custom Double Barrel

Colt Magnum Auto Custom

Colt Ultra Light Auto

Colt Coltsman Pump

Colt Sauer Drilling

Connecticut Valley Classics
Classic Sporter

COLT ARMSMEAR OVER/UNDER
Though advertised in 1995, overseas production problems prevented this gun from ever reaching the market and the project was shelved.

COLT COLTSMAN PUMP
Slide action; takedown; 12-, 16-, 20-ga.; 4-shot magazine; 26″ Improved, 28″ Modified, 30″ Full choke barrel; weighs about 6 lbs.; uncheckered European walnut stock. Introduced 1961; discontinued 1965.
    Exc.: $250   **VGood:** $200   **Good:** $150

Colt Coltsman Pump Custom
Same specs as Coltsman Pump except vent rib; hand-checkered pistol grip, forearm. Introduced 1961; discontinued 1963.
    Exc.: $300   **VGood:** $250   **Good:** $200

COLT CUSTOM DOUBLE BARREL
Boxlock; 12-, 16-ga.; 2 ³/₄″, 3″ chamber (12-ga.); 26″ Improved/Modified, 28″ Modified/Full, 30″ Full/Full choke barrels; hand-checkered European walnut pistol-grip stock, beavertail forearm. Introduced 1961; discontinued 1961.
    Exc.: $600   **VGood:** $450   **Good:** $225

COLT MAGNUM AUTO
Autoloader; takedown; 12-, 20-ga.; 3″ chamber; 28″ (20-ga.), 32″ (12-ga.) interchangeable chrome-lined barrels; Full choke; plain, solid or vent rib; weighs 8 ¹/₄ lbs. (12-ga.) or 6 lbs. (20-ga.); walnut stock; steel receiver. Introduced 1964; discontinued 1966.
    Exc.: $350   **VGood:** $290   **Good:** $230

Colt Magnum Auto Custom
Same specs as Magnum Auto except solid or vent rib; select walnut stock, forearm; engraved receiver. Introduced 1964; discontinued 1966.
    Exc.: $395   **VGood:** $330   **Good:** $270

COLT SAUER DRILLING
Side-by-side; boxlock; top lever; 30-06 or 243 Winchester under 12-ga.; 25″ barrels; shotgun barrels choked Modified/Full; folding leaf rear sight, blade front with brass bead; hand-checkered oil-finished American walnut pistol-grip stock, forearm; black pistol-grip cap; recoil pad; cocking indicators; tang barrel selector; automatic sight positioner; set rifle trigger; side safety; crossbolt safety. Made for Colt by Sauer in Germany. No longer imported by Colt.
    Perf.: $3200   **Exc.:** $2300   **VGood:** $1900

COLT ULTRA LIGHT AUTO
Autoloader; takedown; 12-, 20-ga.; 2 ³/₄″ chamber; 4-shot magazine; 26″ Improved or Modified, 28″ Modified or Full, 30″ Full chrome-lined interchangeable barrels; plain, solid or vent rib; weighs 6 ¹/₄ lbs. (12-ga.), 5 lbs. (20-ga.); walnut stock; crossbolt safety; aluminum alloy receiver. Introduced 1964; discontinued 1966.
    Perf.: $180   **Exc.:** $150   **VGood:** $120

Colt Ultra Light Auto Custom
Same specs as Ultra Light Auto except solid or vent rib; select walnut stock, forearm; engraved with bird-scenes. Introduced 1964; discontinued 1966.
    Perf.: $225   **Exc.:** $175   **VGood:** $140

CONNECTICUT VALLEY CLASSICS CLASSIC SPORTER
Over/under; boxlock; 12-ga.; 3″ chambers; 28″, 30″, 32″ barrels; Skeet, Improved Cylinder, Modified, Full CV choke tubes; weighs 7 ³/₄ lbs.; hand-checkered American black walnut pistol-grip stock, forend; elongated forcing cones; stainless receiver with fine engraving; chrome-lined bores and chambers suitable for steel shot; receiver duplicates Classic Doubles M101 specifications. Introduced 1993; still in production. **LMSR: $2995**
    Perf.: $2450   **Exc.:** $2000   **VGood:** $1500

CONNECTICUT VALLEY CLASSICS CLASSIC FIELD WATERFOWLER
Over/under; boxlock; 12-ga.; 3″ chambers; 30″ barrels; Skeet, Improved Cylinder, Modified, Full CV choke tubes; weighs 7 ³/₄ lbs.; hand-checkered American black walnut pistol-grip stock, forend; elongated forcing cones; blued, non-reflective receiver with fine engraving; chrome-lined bores and chambers suitable for steel shot; receiver duplicates Classic Doubles M101 specifications. Introduced 1995; still in production. **LMSR: $2495**
    Perf.: $2200   **Exc.:** $1300   **VGood:** $1000

COSMI SEMI-AUTOMATIC
Recoil-operated autoloader; breakopen; 12-, 20-ga.; 2 ³/₄″, 3″ chamber; 3-, 8-shot magazine in buttstock; 22″ to 34″ barrel to customers specs; choke tubes; weighs 6 ¹/₄ lbs.; hand-checkered exhibition-grade Circassian walnut stock; double ejectors; double safety system; stainless steel construction; various grades of engraving can double price; marketed in fitted leather case; essentially a hand-made custom gun. Made in Italy. Introduced 1985; still imported. **LSMR: $7400**
    Perf.: $5750   **Exc.:** $5000   **VGood:** $4500

CRUCELEGUI HERMANOS MODEL 150
Side-by-side; Greener triple crossbolt action; exposed hammers; 12-, 20-ga.; 20″, 26″, 28″, 30″, 32″ barrels; hand-checkered European walnut stock, beavertail forend; double triggers; color case-hardened receiver; chromed bores; sling swivels. Made in Spain. Introduced 1979; still in production. **LMSR: $450**
    Perf.: $350   **Exc.:** $290   **VGood:** $240
NOTE: On this model, .410 is worth 50% more than prices shown, 28-gauge worth 40% more, 20-gauge, 25% more.

CHARLES DALY AUTO
Recoil-operated autoloader; 12-ga.; 5-shot magazine; 3-shot plug furnished; 26″ Improved, 28″ Modified or Full, 30″ Full barrels; vent rib; hand-checkered walnut pistol-grip stock, forearm; button safety; copy of early Browning patents. Made in Japan. Introduced 1973; discontinued 1975.
    Perf.: $300   **Exc.:** $250   **VGood:** $210

Charles Daly Diamond Grade Trap (1971)

Charles Daly Diamond Grade (1984)

Charles Daly Empire Double

Charles Daly Empire Quality Trap

Charles Daly Field Grade

Charles Daly Field III

SHOTGUNS

Charles Daly
Diamond Grade
Diamond-Regent Grade
Diamond Grade (1984)
Diamond Skeet (1984)
Diamond Trap (1984)
Empire Double
Empire Quality Trap
Field Auto
Field Grade
Field Grade (1989)
Field Grade Deluxe (1989)
Field III
Hammerless Drilling

CHARLES DALY DIAMOND GRADE

Over/under; boxlock; 12-, 20-, 28-ga., .410; 26″, 28″, 30″ barrels; choked Skeet/Skeet, Improved/Modified, Modified/Full, Full/Full; vent rib; extensive hand-checkered French walnut pistol-grip stock, beavertail forearm; selective single trigger; selective ejection system; safety/barrel selector; engraved receiver, trigger guard. Made in Japan. Introduced 1967; discontinued 1973.
Field model
  Perf.: $1000 **Exc.:** $900 **VGood:** $800
Skeet model
  Perf.: $1050 **Exc.:** $900 **VGood:** $800
Trap model
  Perf.: $900 **Exc.:** $780 **VGood:** $700

Charles Daly Diamond-Regent Grade

Same specs as Diamond Grade except highly figured French walnut stock; profuse engraving; hunting scenes inlaid in gold, silver; firing pins removable through breech face. Made in Japan. Introduced 1967; discontinued 1973.
  Perf.: $1900 **Exc.:** $1350 **VGood:** $1150

CHARLES DALY DIAMOND GRADE (1984)

Over/under; boxlock; 12-, 20-ga.; 27″ barrels; three choke tubes; hand-checkered oil-finished European walnut stock; single selective competition trigger; selective auto ejectors; silvered, engraved receiver. Made in Italy. Introduced 1984; discontinued 1987.
  Perf.: $650 **Exc.:** $500 **VGood:** $480

Charles Daly Diamond Skeet (1984)

Same specs as Diamond Grade (1984) except 12-ga.; 26″ Skeet/Skeet barrels; competition vent rib; Skeet stock; target trigger. Introduced 1984; discontinued 1988.
  Perf.: $750 **Exc.:** $450 **VGood:** $400

Charles Daly Diamond Trap (1984)

Same specs as Diamond Grade (1984) except 12-ga.; 30″ barrels; competition vent top and middle ribs; Monte Carlo stock; target trigger. Introduced 1984; discontinued 1988.
  Perf.: $800 **Exc.:** $500 **VGood:** $450

CHARLES DALY EMPIRE DOUBLE

Side-by-side; boxlock; hammerless; 12-, 16-, 20-ga; 2 3/4″ (16-ga.), 3″ (12-, 20-ga.) chamber; 26″, 28″, 30″ barrels; standard choke combos; checkered walnut pistol-grip stock, beavertail forearm; plain extractors; non-selective single trigger. Made in Japan. Introduced 1968; discontinued 1971.
  Exc.: $500 **VGood:** $380 **Good:** $300

CHARLES DALY EMPIRE QUALITY TRAP

Single barrel; Anson & Deeley-type boxlock; 12-ga.; 30″, 32″, 34″ vent-rib barrel; hand-checkered European walnut pistol-grip stock, forearm; automatic ejector. Made in Suhl, Germany. Introduced 1920; discontinued 1933.
  Exc.: $1200 **VGood:** $900 **Good:** $700

CHARLES DALY FIELD AUTO

Autoloader; 12-ga.; 2 3/4″, 3″ chamber; 27″, 30″ barrel; standard chokes or Invector choke tubes; weighs 7 1/4 lbs.; checkered walnut pistol-grip stock, forend; cross-bolt safety; alloy receiver; chromed bolt; stainless steel gas piston. Made in Japan. Introduced 1984; discontinued 1987.
  Perf.: $300 **Exc.:** $250 **VGood:** $210

CHARLES DALY FIELD GRADE

Over/under; boxlock; 12-, 20-, 28-ga., .410; 2 3/4″, 3″ chambers (12-, 20-ga.); 26″, 28″, 30″ barrels; choked Skeet/Skeet, Improved/Modified, Modified/Full, Full/Full; vent rib; hand-checkered walnut pistol-grip stock, fluted forearm; recoil pad (12-ga. Magnum); selective single trigger; automatic selective ejectors; safety/barrel selector; engraved receiver. Made in Japan. Introduced 1963 (12-, 20-ga.); introduced 1965 (28-ga., .410); discontinued 1976.
  Perf.: $600 **Exc.:** $500 **VGood:** $450

CHARLES DALY FIELD GRADE (1989)

Over/under; boxlock; 12-, 20-ga.; 3″ chambers; 26″, 28″ barrels; standard fixed choke combos; cut-checkered walnut pistol-grip stock, forend with semi-gloss finish; black vent rubber recoil pad; manual safety extractors; single-stage selective trigger; color case-hardened, engraved receiver. Made in Europe. Introduced 1989; still in production. **LMSR: $545**
  Perf.: $450 **Exc.:** $350 **VGood:** $280

Charles Daly Field Grade Deluxe (1989)

Same specs as Field Grade (1989) except 12-, 20-ga.; Improved Cylinder, Modified and Full choke tubes; automatic selective ejectors; antique silver finish on frame. Introduced 1989; still imported. **LMSR: $770**
  Perf.: $670 **Exc.:** $450 **VGood:** $400

CHARLES DALY FIELD III

Over/under; boxlock; 12-, 20-ga.; 26″, 28″, 30″ chrome-lined barrels; standard choke combos; vent rib; checkered European walnut stock, forend; single selective trigger extractors; blued, engraved frame. Made in Italy. Introduced 1984; discontinued 1988.
  Perf.: $375 **Exc.:** $320 **VGood:** $295

CHARLES DALY HAMMERLESS DRILLING

Anson & Deeley-type boxlock; 12-, 16-, 20-ga.; rifle barrel chambered for 25-20, 25-35, 30-30; automatic rear sight; hand-checkered European walnut pistol-grip stock, forearm; plain extractors. Made in three grades that differ only in checkering, amount of engraving, wood, overall quality. Made in Suhl, Germany. H.A. Linder models exhibit extensive higher grade engraving, and inlays which affect value. Introduced 1921; discontinued 1933.
Superior Quality
  Exc.: $2200 **VGood:** $1650 **Good:** $1300
Diamond Quality
  Exc.: $4000 **VGood:** $3000 **Good:** $2500
Regent Diamond Quality (Linder)
  Exc.: $7500 **VGood:** $6850 **Good:** $6000
Regent Diamond Quality (Sauer)
  Exc.: $6500 **VGood:** $4500 **Good:** $3800

SHOTGUNS

Charles Daly
Hammerless Over/Under
Hammerless Side-By-Side
Lux Over/Under
Lux Side-By-Side
Model 100 Commander
Model 200 Commander
Model DSS
Multi-XII
Novamatic Lightweight
Novamatic Lightweight
Magnum
Novamatic Lightweight
Trap
Novamatic Lightweight
Super

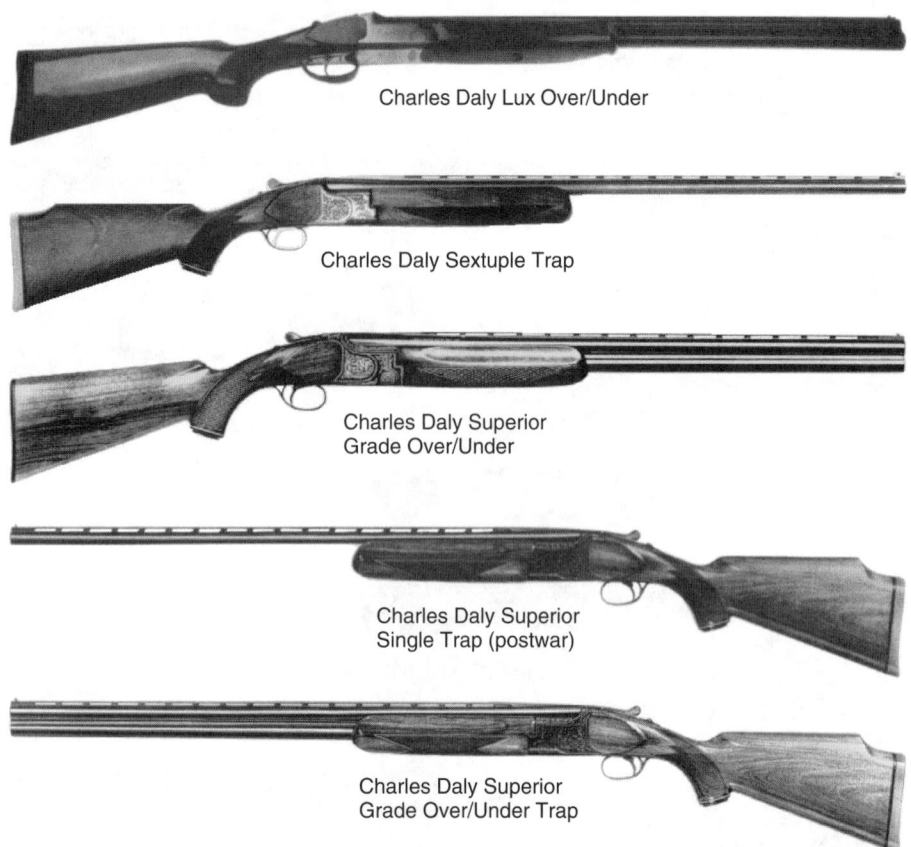

Charles Daly Lux Over/Under

Charles Daly Sextuple Trap

Charles Daly Superior
Grade Over/Under

Charles Daly Superior
Single Trap (postwar)

Charles Daly Superior
Grade Over/Under Trap

CHARLES DALY HAMMERLESS OVER/UNDER
Over/under; Anson & Deeley-type boxlock; 12-, 16-, 20-ga.; 26″ to 30″ barrels; any standard choke combo; European walnut stock hand-checkered pistol grip, forearm; double triggers; automatic ejectors. Made in two grades that differ in grade of wood, checkering, amount of engraving, general overall quality. Made in Suhl, Germany. Introduced 1920; discontinued 1933.
Empire Quality
    Exc.: $2500   **VGood:** $2150   **Good:** $1750
Diamond Quality
    Exc.: $3600   **VGood:** $2800   **Good:** $2200

CHARLES DALY HAMMERLESS SIDE-BY-SIDE
Side-by-side; Anson & Deeley-type boxlock; 10-, 12-, 16-, 20-, 28-ga.; .410; 26″ to 28″ barrels; any choke combo; checkered walnut pistol-grip stock, forearm; made in four grades; all have automatic ejectors, except Superior Quality; grades differ in grade of wood, checkering, amount of engraving, general overall quality. Made in Suhl, Germany. Shotguns made by H.A. Linder exhibit extensive higher grade engraving, inlays and other custom features which affect value. Introduced 1920; discontinued 1933.
Superior Quality
    Exc.: $700   **VGood:** $500   **Good:** $400
Empire Quality (Linder)
    Exc.: $4000   **VGood:** $3000   **Good:** $2000
Diamond Quality (Linder)
    Exc.: $9500   **VGood:** $7000   **Good:** $5000
Regent Diamond Quality (Linder)
    Exc.: $13,000   **VGood:** $9000   **Good:** $7000
Empire Quality (Sauer)
    Exc.: $3000   **VGood:** $2000   **Good:** $1500
Diamond Quality (Sauer)
    Exc.: $5000   **VGood:** $3500   **Good:** $2200
Regent Diamond Quality (Sauer)
    Exc.: $6500   **VGood:** $4800   **Good:** $3600

CHARLES DALY LUX OVER/UNDER
Over/under; boxlock; 12-, 20-, 28-ga., .410; 26″, 28″ vent rib, chrome-lined barrels; three internal choke tubes; cut-checkered walnut pistol-grip stock, forend with semi-gloss finish; single selective trigger; auto selective ejectors; antique silver-finished frame. Made in Europe. Introduced 1989; no longer in production.
    Perf.: $600   **Exc.:** $440   **VGood:** $390

CHARLES DALY LUX SIDE-BY-SIDE
Side-by-side; boxlock; 12-, 20-ga.; 26″ barrel; choke tubes; cut-checkered walnut pistol-grip stock, semi-beavertail forend; single selective trigger; ejectors; recoil pad. Introduced 1990; discontinued 1994.
    Perf.: $520   **Exc.:** $370   **VGood:** $300

CHARLES DALY MODEL 100 COMMANDER
Over/under; Anson & Deeley-type boxlock; 12-, 16-, 20-, 28-ga., .410; 26″, 28″, 30″ barrels; Improved/Modified, Modified/Full chokes; hand-checkered European walnut straight or pistol-grip stock, forend; automatic ejectors; double triggers or Miller single selective trigger; engraved receiver. Made in Liege, Belgium. Introduced 1933; discontinued at beginning of WWII.
    Exc.: $425   **VGood:** $350   **Good:** $275
With single selective trigger
    Exc.: $450   **VGood:** $370   **Good:** $300

Charles Daly Model 200 Commander
Over/under; Anson & Deeley-type boxlock; 12-, 16-, 20-, 28-ga., .410; 26″, 28″, 30″ barrels; Improved/Modified, Modified/Full chokes; hand-checkered select European walnut straight or pistol-grip stock, forearm; automatic ejectors; double triggers or Miller single selective trigger; elaborate engraved receiver. Made in Liege, Belgium. Introduced 1933; discontinued at beginning of WWII.
    Exc.: $495   **VGood:** $350   **Good:** $290
With single selective trigger
    Exc.: $525   **VGood:** $425   **Good:** $360

CHARLES DALY MODEL DSS
Side-by-side; boxlock; 12-, 20-ga.; 3″ chambers; 26″ barrels; choke tubes; weighs about 6 3/4 lbs.; cut-checkered figured walnut pistol-grip stock, semi-beavertail forend; black rubber recoil pad; automatic selective ejectors; automatic safety; gold single trigger; engraved, silvered frame. Introduced 1990; no longer imported.
    Exc.: $550   **VGood:** $400   **Good:** $325

CHARLES DALY MULTI-XII
Gas-operated autoloader; 12-ga.; 3″ chamber; 27″ vent-rib barrel; Invector choke system; checkered walnut pistol-grip stock, forend; rosewood grip cap; brown vent recoil pad; shoots all loads without adjustment; scroll engraved receiver. Made in Japan. Introduced 1986; discontinued 1987.
    Perf.: $400   **Exc.:** $300   **VGood:** $220

CHARLES DALY NOVAMATIC LIGHTWEIGHT
Autoloader; 12-ga.; 2 3/4″ chamber; 4-shot tubular magazine; 26″, 28″ barrel; two-tube Quick-Choke system; plain or vent-rib barrel; checkered walnut pistol-grip stock, forearm. Made in Italy by Breda. Introduced 1968; discontinued 1968.
With plain barrel
    Exc.: $265   **VGood:** $200   **Good:** $150
With vent rib
    Exc.: $295   **VGood:** $230   **Good:** $180

Charles Daly Novamatic Lightweight Magnum
Same specs as Novamatic Lightweight except 12-, 20-ga.; 3″ chambers; 3-shot magazine; 28″ vent-rib barrel; Full choke. Introduced 1968; discontinued 1968.
    Exc.: $285   **VGood:** $250   **Good:** $190

Charles Daly Novamatic Lightweight Trap
Same specs as Novamatic Lightweight except 30″ vent-rib barrel; Full choke; Monte Carlo trap stock; recoil pad. Introduced 1968; discontinued 1968.
    Exc.: $300   **VGood:** $240   **Good:** $190

Charles Daly Novamatic Lightweight Super
Same specs as Novamatic Lightweight except 12-, 20-ga.; 26″, 28″ (only 12-ga.) barrel; Skeet choke available; standard vent rib (only 12-ga.); lighter weight. Introduced 1968; discontinued 1968.
    Exc.: $280   **VGood:** $240   **Good:** $180

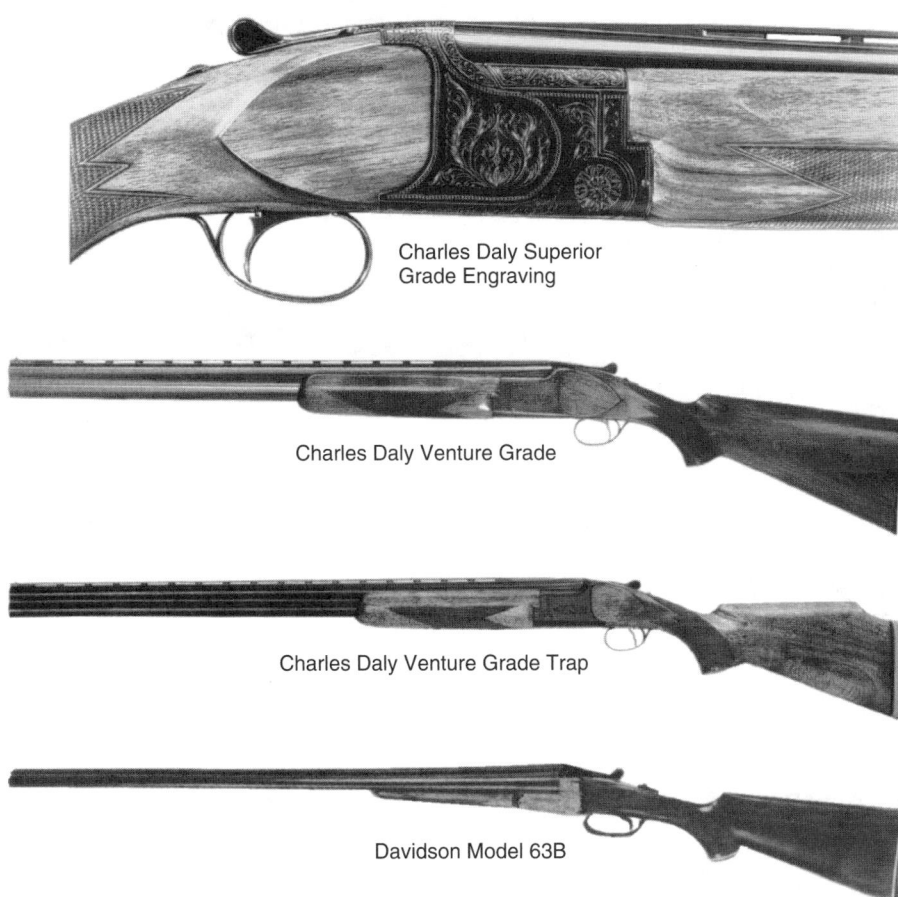

Charles Daly Superior Grade Engraving

Charles Daly Venture Grade

Charles Daly Venture Grade Trap

Davidson Model 63B

SHOTGUNS

Charles Daly
Presentation Grade
Sextuple Trap
Superior Grade
Over/Under
Superior Grade
Side-By-Side
Superior II
Superior Single Trap
Venture Grade

Darne
Model R11 Bird Hunter
Model R15 Pheasant
Hunter
Model V19 Quail Hunter
Supreme

Davidson
Model 63B
Model 63B Magnum
Model 69SL

CHARLES DALY PRESENTATION GRADE

Over/under; boxlock; 12-, 20-ga.; 27" barrel; three choke tubes; hand-checkered, oil-finished European walnut stock; single selective competition trigger; selective auto ejectors; silvered, engraved receiver, sideplates. Made in Italy. Introduced 1984; discontinued 1986.

    Perf.: $1000   **Exc.:** $800   **VGood:** $695

CHARLES DALY SEXTUPLE TRAP

Single barrel; Anson & Deeley-type boxlock; 12-ga.; 30", 32", 34" vent-rib barrel; hand-checkered European walnut pistol-grip stock, forearm; six locking bolts; automatic ejector. Made in two grades that differ only in checkering, amount of engraving, grade of wood, improved overall quality. Made in Suhl, Germany. Introduced 1920; discontinued 1933.

Empire Quality
    Exc.: $1850   **VGood:** $1300   **Good:** $1050
Regent Diamond Quality
    Exc.: $2500   **VGood:** $1750   **Good:** $1100

CHARLES DALY SUPERIOR GRADE OVER/UNDER

Over/under; boxlock; 12-, 20-, 28-ga.; .410; 26", 28", 30" barrels; choked Skeet/Skeet, Improved/Modified, Modified/Full, Full/Full; vent-rib; hand-checkered walnut pistol-grip stock, beavertail forearm; selective single trigger; selective ejection system; safety/barrel selector; engraved receiver. Made in Japan. Introduced 1963; discontinued 1976.

Standard model
    Perf.: $700   **Exc.:** $550   **VGood:** $425
Trap model
    Perf.: $750   **Exc.:** $600   **VGood:** $460

CHARLES DALY SUPERIOR GRADE SIDE-BY-SIDE

Side-by-side; boxlock; 12-, 20-ga.; 26", 28", 30" chrome-lined barrels; standard choke combos; vent-rib; checkered European walnut stock, forend; single selective trigger. Introduced 1984; discontinued 1985.

    Perf.: $510   **Exc.:** $395   **VGood:** $320

CHARLES DALY SUPERIOR II

Over/under; boxlock; 12-, 20-ga.; 2 3/4", 3" chambers (12-ga.); 26", 28", 30" chrome-line barrels; vent-rib; standard choke combos; checkered European walnut stock, forend; single selective trigger; automatic ejectors; silvered receiver; engraving. Introduced 1984; discontinued 1988.

    Perf.: $650   **Exc.:** $440   **VGood:** $390
12-ga., 3" chambers
    Perf.: $650   **Exc.:** $440   **VGood:** $390

CHARLES DALY SUPERIOR SINGLE TRAP

Single shot; boxlock; hammerless; 12-ga.; 32", 34" barrels; Full choke; vent-rib; checkered walnut Monte Carlo pistol-grip stock, forearm; recoil pad; auto ejector. Made in Japan. Introduced 1968; discontinued 1976.

    Perf.: $575   **Exc.:** $500   **VGood:** $400

CHARLES DALY VENTURE GRADE

Over/under; boxlock; 12-, 20-ga.; 26" Skeet/Skeet or Improved/Modified, 28" Modified/Full, 30" Improved/Full barrels; vent rib; checkered walnut pistol-grip stock, forearm; manual safety; automatic ejectors; single selective trigger. Made in three variations. Made in Japan. Introduced 1973; discontinued 1976.

Standard model
    Perf.: $575   **Exc.:** $500   **VGood:** $400
Trap model, Monte Carlo stock, 30" barrel
    Perf.: $550   **Exc.:** $480   **VGood:** $390
Skeet model, 26" barrel
    Perf.: $600   **Exc.:** $520   **VGood:** $490

DARNE MODEL R11 BIRD HUNTER

Side-by-side; sliding breech action; 12-, 20-ga.; 25 1/2" Improved/Modified barrels; raised rib; deluxe hand-checkered walnut stock, forearm; double triggers; automatic selective ejection; case-hardened receiver. Made in France. Discontinued 1979.

    Perf.: $950   **Exc.:** $620   **VGood:** $505

DARNE MODEL R15 PHEASANT HUNTER

Side-by-side; sliding breech action; 12-ga.; 27 1/2" Modified/Full barrels; raised rib; fancy hand-checkered walnut stock, forearm; double triggers; automatic selective ejection system; case-hardened engraved receiver. Discontinued 1979.

    Perf.: $2200   **Exc.:** $1690   **VGood:** $1550

DARNE MODEL V19 QUAIL HUNTER SUPREME

Side-by-side; sliding breech action; 20-, 28-ga.; 25 1/2" Improved/Modified barrels; raised rib; extra-fancy hand-checkered walnut stock, forearm; double triggers; automatic selective ejection; case-hardened receiver; elaborate engraving. Discontinued 1979.

    Perf.: $3400   **Exc.:** $2500   **VGood:** $2100

DAVIDSON MODEL 63B

Double barrel; Anson & Deeley boxlock; 12-, 16-, 20-, 28-ga., .410; 25" (.410), 26", 28" 30" (12-ga.) barrels; Improved Cylinder/Modified, Modified/Full, Full/Full chokes; hand-checkered walnut pistol-grip stock; plain extractors; automatic safety; engraved, nickel-plated frame. Made in Spain. Introduced 1963; no longer in production.

    Exc.: $210   **VGood:** $160   **Good:** $140

Davidson Model 63B Magnum

Same specs as Model 63B except 10-, 12-, 20-ga.; 3" (12-, 20-ga.), 3 1/2" (10-ga.) chambers; 32" Full/Full barrels (10-ga. only). Made in Spain. No longer in production.

12-, 20-ga. Magnum
    Exc.: $300   **VGood:** $210   **Good:** $165
10-ga. Magnum
    Exc.: $350   **VGood:** $240   **Good:** $200

DAVIDSON MODEL 69SL

Double barrel; sidelock; 12-, 20-ga.; 26", 28" barrels; Modified/Improved (26"), Modified/Full (28") choke; hand-checkered European walnut stock; detachable locks; nickel-plated, engraved action. Introduced 1963; discontinued 1979.

    Perf.: $405   **Exc.:** $360   **VGood:** $310

Davidson
Model 73

Desert Industries
Big Twenty

Diarm
(See American Arms)

Dumoulin
Liege

E.A.A./Sabatti
Falcon-Mon
Saba-Mon Double
Sporting Clays Pro-Gold

ERA
Bird Hunter
Full Limit
Winner

Erbi
Model 76AJ
Model 76ST
Model 80

Exel
Model 101
Model 102
Model 103
Model 104
Model 105

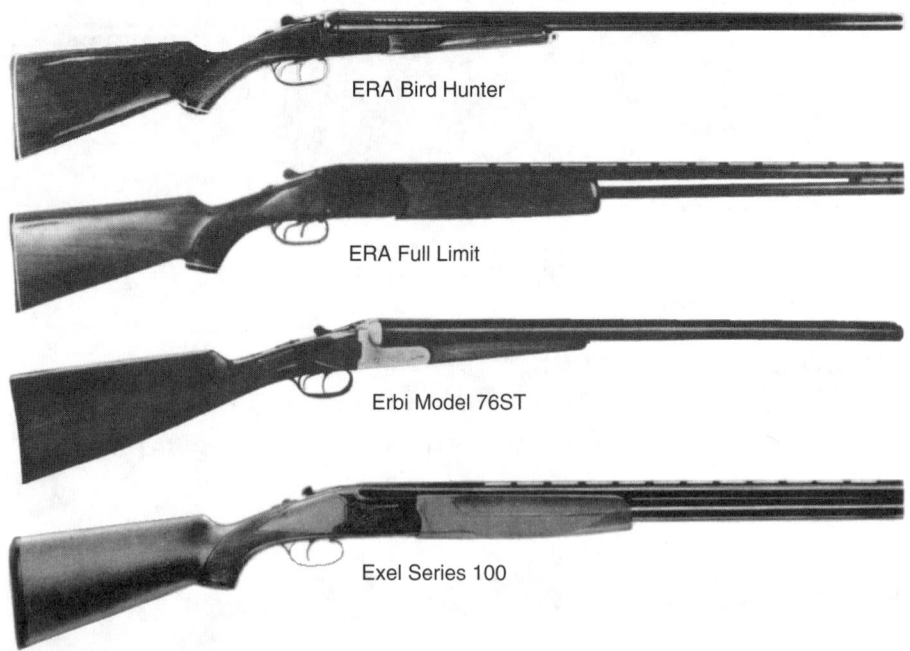

ERA Bird Hunter

ERA Full Limit

Erbi Model 76ST

Exel Series 100

DAVIDSON MODEL 73
Side-by-side; sidelock; exposed hammer; 12-, 20-ga.; 3" chambers; 20" barrels; checkered European walnut pistol-grip stock, forearm; detachable sideplates; plain extractors; double triggers. Made in Spain. Introduced 1976; no longer in production.
Perf.: $260   **Exc.:** $210   **VGood:** $190

DESERT INDUSTRIES BIG TWENTY
Single shot; 20-ga.; 19" Cylinder bore barrel; 31 3/4" overall length; weighs 4 3/4 lbs.; bead front sight; fixed formed-wire stock with buttplate; walnut forend, grip; all-steel construction; blued finish. Introduced 1990; no longer in production.
Perf.: $180   **Exc.:** $155   **VGood:** $125

DUMOULIN LIEGE
Side-by-side; Anson & Deeley boxlock or sidelock; 12-, 16-, 20-, 28-ga.; 2 3/4", 3" chambers; 26" to 32" barrels choked to customers specs; weighs 6 1/4 lbs. Custom-built in Belgium. Introduced 1986; discontinued 1988.
Luxe Model
Perf.: $5500   **Exc.:** $4100   **VGood:** $3450
Grand Luxe
Perf.: $7200   **Exc.:** $5200   **VGood:** $4400

E.A.A./SABATTI FALCON-MON
Over/under; boxlock; 12-, 20-, 28-ga., .410; 3" chambers; 26", 28" barrels; standard chokes; ventilated rib; weighs about 7 lbs.; cut-checkered select walnut pistol-grip stock, beavertail forend with gloss finish; gold-plated single selective trigger; extractors; engraved, blued receiver. Made in Italy. Introduced 1993; discontinued 1994.

12- or 20-ga.
Perf.: $600   **Exc.:** $420   **VGood:** $350
28-ga. or .410
Perf.: $650   **Exc.:** $470   **VGood:** $400

E.A.A./SABATTI SABA-MON DOUBLE
Side-by-side; boxlock; 12-, 20-, 28-ga., .410; 3" chambers; 26", 28" barrels; standard chokes; European walnut straight English or pistol-grip stock; single selective trigger; automatic selective ejectors; blue finish. Made in Italy. Introduced 1993; discontinued 1994.
Perf.: $920   **Exc.:** $680   **VGood:** $500

E.A.A./SABATTI SPORTING CLAYS PRO-GOLD
Over/under; boxlock; 12-ga.; 3" chambers; 28", 30" barrels; six choke tubes; weighs 7 3/4 lbs.; target-style flourescent bar front sight; European walnut pistol-grip stock, forend with gloss finish; special Sporting Clays recoil pad; automatic ejectors; gold-plated single selective trigger; engraved, blued receiver with gold inlays. Comes with lockable hard shell plastic case. Made in Italy. Introduced 1993; no longer imported.
Perf.: $800   **Exc.:** $600   **VGood:** $500

ERA BIRD HUNTER
Side-by-side; boxlock; 12-, 16-, 20-ga., .410; 26", 28", 30" barrels; raised matted rib; hand-checkered walnut stock, beavertail forend; extractors; auto disconnector; double triggers; engraved receiver. Made in Brazil. Introduced 1979; no longer in production.
Perf.: $150   **Exc.:** $125   **VGood:** $100

ERA FULL LIMIT
Over/under; 12-, 20-ga.; 28" barrels; vent, top and middle ribs; hand-checkered walnut-finished hardwood stock; Monte Carlo or straight styles; auto safety, extractors; double triggers; engraved receiver. Made in Brazil. Introduced 1980; no longer in production.
Perf.: $250   **Exc.:** $200   **VGood:** $180

ERA WINNER
Single barrel; exposed hammer; 12-, 16-, 20-ga., .410; 28" barrel; metal bead front sight; walnut-stained hardwood stock, beavertail forend; triggerguard button opens action; auto ejectors. Made in Brazil. Introduced 1980; no longer in production.
Perf.: $125   **Exc.:** $100   **VGood:** $75

ERBI MODEL 76AJ
Side-by-side; 10-, 12-, 20-, 28-ga.; 26", 28" Modified/Full barrels; medium bead front sight; hand-checkered European walnut straight-grip stock; double triggers; automatic ejectors; engraved, silvered receiver. Made in Spain. Introduced 1982 by Toledo Armas; discontinued 1984.
Perf.: $550   **Exc.:** $460   **VGood:** $400

Erbi Model 76ST
Same specs as Model 76AJ except automatic extractors. Introduced 1982 by Toledo Armas; discontinued 1984.
Perf.: $500   **Exc.:** $400   **VGood:** $350

ERBI MODEL 80
Side-by-side; sidelock; 10-, 12-, 20-, 28-ga.; 26", 28" Modified/Full barrels; medium bead front sight; hand-checkered European walnut straight-grip stock; double triggers; extractors; engraved silvered receiver. Introduced 1982 by Toledo Armas; discontinued 1984.
Perf.: $850   **Exc.:** $675   **VGood:** $550

EXEL MODEL 101
Over/under; boxlock; 12-ga.; 2 3/4", 3" chambers; 26" Improved Cylinder/Modified barrels; vent-rib; checkered European walnut stock; single selective trigger; extractors. Made in Spain. Introduced 1984; discontinued 1987.
Perf.: $350   **Exc.:** $300   **VGood:** $260

EXEL MODEL 102
Over/under; boxlock; 12-ga.; 2 3/4", 3" chambers; 27 5/8" Improved Cylinder/Modified barrels; vent-rib; checkered European walnut stock; single selective trigger; extractors. Made in Spain. Introduced 1984; discontinued 1987.
Perf.: $360   **Exc.:** $300   **VGood:** $260

EXEL MODEL 103
Over/under; boxlock; 12-ga.; 2 3/4", 3" chambers; 29 1/2" Modified/Full barrels; vent-rib; checkered European walnut stock; single selective trigger; extractors. Made in Spain. Introduced 1984; discontinued 1987.
Perf.: $360   **Exc.:** $300   **VGood:** $260

EXEL MODEL 104
Over/under; boxlock; 12-ga.; 2 3/4", 3" chambers; 27 5/8" Improved Cylinder/Improved Modified barrels; vent-rib; checkered European walnut stock; single selective trigger; selective auto ejectors. Made in Spain. Introduced 1984; discontinued 1987.
Perf.: $360   **Exc.:** $300   **VGood:** $260

EXEL MODEL 105
Over/under; boxlock; 12-ga.; 2 3/4", 3" chambers; 27 5/8" barrel with screw-in choke tubes; vent-rib; deluxe checkered European walnut stock; single selective trigger; selective auto ejectors; engraved satin finish. Made in Spain. Introduced 1984; discontinued 1987.
Perf.: $525   **Exc.:** $410   **VGood:** $380

Exel Model 201

Exel Series 300

Exel Model 251

SHOTGUNS

Exel
Model 106
Model 107
Model 201
Model 202
Model 203
Model 204
Model 205
Model 206
Model 207
Model 208
Model 209
Model 210
Model 240
Model 251
Model 281
Series 300

EXEL MODEL 106

Over/under; boxlock; 12-ga.; 2 3/4", 3" chambers; 27 5/8" barrel with screw-in choke tubes; vent-rib; deluxe checkered European walnut stock; single selective trigger; selective auto ejectors; engraved blue finish. Made in Spain. Introduced 1984; discontinued 1987.
Perf.: $700   **Exc.:** $500   **VGood:** $430

EXEL MODEL 107

Over/under; boxlock; 12-ga.; 2 3/4", 3" chambers; 29 1/2" barrel with Full/interchangeable tubes; vent-rib; deluxe checkered European walnut stock; single selective trigger; selective auto ejectors; engraved blue finish. Made in Spain. Introduced 1984; discontinued 1987.
Perf.: $700   **Exc.:** $500   **VGood:** $430

EXEL MODEL 201

Side-by-side; boxlock; 12-, 20-ga.; 2 3/4" chambers; 28" barrels; high matted rib; metal bead front sight; hand-checkered European walnut straight or pistol-grip stock; double triggers; extractors; color case-hardened finish. Made in Spain. Introduced 1984; discontinued 1987.
Perf.: $350   **Exc.:** $220   **VGood:** $190

EXEL MODEL 202

Side-by-side; boxlock; 12-, 20-ga.; 2 3/4" chambers; 26" barrels; high matted rib; metal bead front sight; hand-checkered European walnut straight or pistol-grip stock; double triggers; extractors; color case-hardened finish. Made in Spain. Introduced 1984; discontinued 1987.
Perf.: $350   **Exc.:** $220   **VGood:** $190

EXEL MODEL 203

Side-by-side; boxlock; 12-, 20-ga.; 3" chambers; 27" barrels; high matted rib; metal bead front sight; hand-checkered European walnut straight or pistol-grip stock; double triggers; extractors; color case-hardened finish. Made in Spain. Introduced 1984; discontinued 1987.
Perf.: $350   **Exc.:** $220   **VGood:** $190

EXEL MODEL 204

Side-by-side; boxlock; 12-, 20-ga.; 2 3/4" chambers; 28" barrels; high matted rib; metal bead front sight; hand-checkered European walnut straight or pistol-grip stock; single or double triggers; automatic selective ejectors; silvered, engraved receiver. Made in Spain. Introduced 1984; discontinued 1987.
Perf.: $525   **Exc.:** $400   **VGood:** $340

EXEL MODEL 205

Side-by-side; boxlock; 12-, 20-ga.; 2 3/4" chambers; 26" barrels; high matted rib; metal bead front sight; hand-checkered European walnut straight or pistol-grip stock; single or double triggers; automatic selective ejectors; silvered, engraved receiver. Made in Spain. Introduced 1984; discontinued 1987.
Perf.: $525   **Exc.:** $400   **VGood:** $340

EXEL MODEL 206

Side-by-side; boxlock; 12-, 20-ga.; 2 3/4" chambers; 27" barrels; high matted rib; metal bead front sight; hand-checkered European walnut stock; straight or full pistol-grip; single or double triggers; automatic selective ejectors; silvered, engraved receiver. Made in Spain. Introduced 1984; discontinued 1987.
Perf.: $525   **Exc.:** $400   **VGood:** $340

EXEL MODEL 207

Side-by-side; sidelock; 12-ga.; 2 3/4" chambers; 28" barrels; high matted rib; metal bead front sight; hand-checkered European walnut straight or pistol-grip stock; double triggers; extractors or ejectors; color case-hardened finish. Made in Spain. Introduced 1984; discontinued 1987.
Perf.: $600   **Exc.:** $490   **VGood:** $400

EXEL MODEL 208

Side-by-side; sidelock; 12-ga.; 2 3/4" chambers; 28" barrels; high matted rib; metal bead front sight; deluxe hand-checkered European walnut straight or pistol-grip stock; double triggers; extractors or ejectors; color case-hardened finish; engraved receiver. Made in Spain. Introduced 1984; discontinued 1987.
Perf.: $675   **Exc.:** $525   **VGood:** $500

EXEL MODEL 209

Side-by-side; sidelock; 20-ga.; 2 3/4" chambers; 26" barrel; high matted rib; metal bead front sight; deluxe hand-checkered European walnut straight or pistol-grip stock; double triggers; extractors or ejectors; color case-hardened finish; engraved receiver. Made in Spain. Introduced 1984; discontinued 1987.
Perf.: $550   **Exc.:** $465   **VGood:** $400

EXEL MODEL 210

Side-by-side; sidelock; 20-ga.; 2 3/4" chambers; 27" barrel; high matted rib; metal bead front sight; deluxe hand-checkered European walnut straight or pistol-grip stock; double triggers; extractors or ejectors; color case-hardened finish; engraved receiver. Made in Spain. Introduced 1984; discontinued 1987.
Perf.: $550   **Exc.:** $465   **VGood:** $400

EXEL MODEL 240

Side-by-side; boxlock; .410; 2 3/4" chambers; 28" barrel; high matted rib; metal bead front sight; hand-checkered European walnut straight or pistol-grip stock; double triggers; extractors; color case-hardened finish. Made in Spain. Introduced 1984; discontinued 1987.
Perf.: $425   **Exc.:** $280   **VGood:** $210

EXEL MODEL 251

Single shot; exposed hammer; .410; 3" chamber; folding stock; splinter forend; non-ejector; case-hardened frame. Made in Spain. Introduced 1985; discontinued 1987.
Perf.: $160   **Exc.:** $125   **VGood:** $100

EXEL MODEL 281

Side-by-side; boxlock; 28-ga.; 2 3/4" chambers; 28" barrel; high matted rib; metal bead front sight; hand-checkered European walnut straight or pistol-grip stock; double triggers; extractors; color case-hardened finish. Made in Spain. Introduced 1984; discontinued 1987.
Perf.: $400   **Exc.:** $295   **VGood:** $250

EXEL SERIES 300

Over/under; boxlock; 12-ga.; 2 3/4" chambers; 28", 29" barrels; vent-rib; checkered European walnut pistol-grip field or Monte Carlo stock, forend; auto selective ejectors; silvered, engraved finish. Offered in ten variations with differing degrees of ornamentation. Made in Spain. Introduced 1984; discontinued 1987.
Models 301 & 302
Perf.: $450   **Exc.:** $360   **VGood:** $300
Models 303 & 304
Perf.: $520   **Exc.:** $430   **VGood:** $390
Models 305 & 306
Perf.: $600   **Exc.:** $500   **VGood:** $450
Models 307 & 308
Perf.: $550   **Exc.:** $450   **VGood:** $400
Models 309 & 310
Perf.: $600   **Exc.:** $500   **VGood:** $450

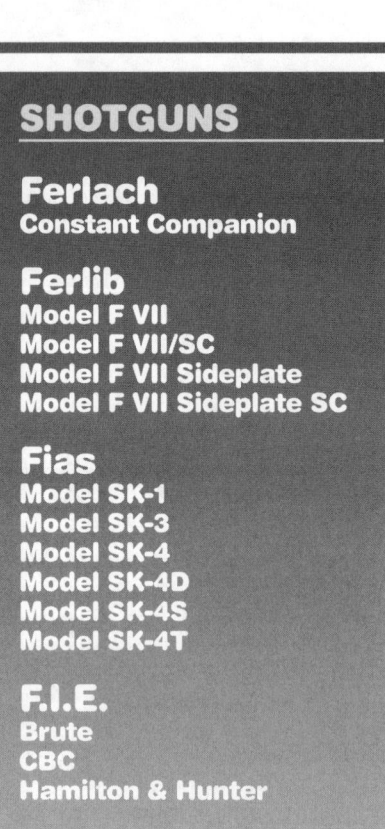

SHOTGUNS

Ferlach
Constant Companion

Ferlib
Model F VII
Model F VII/SC
Model F VII Sideplate
Model F VII Sideplate SC

Fias
Model SK-1
Model SK-3
Model SK-4
Model SK-4D
Model SK-4S
Model SK-4T

F.I.E.
Brute
CBC
Hamilton & Hunter

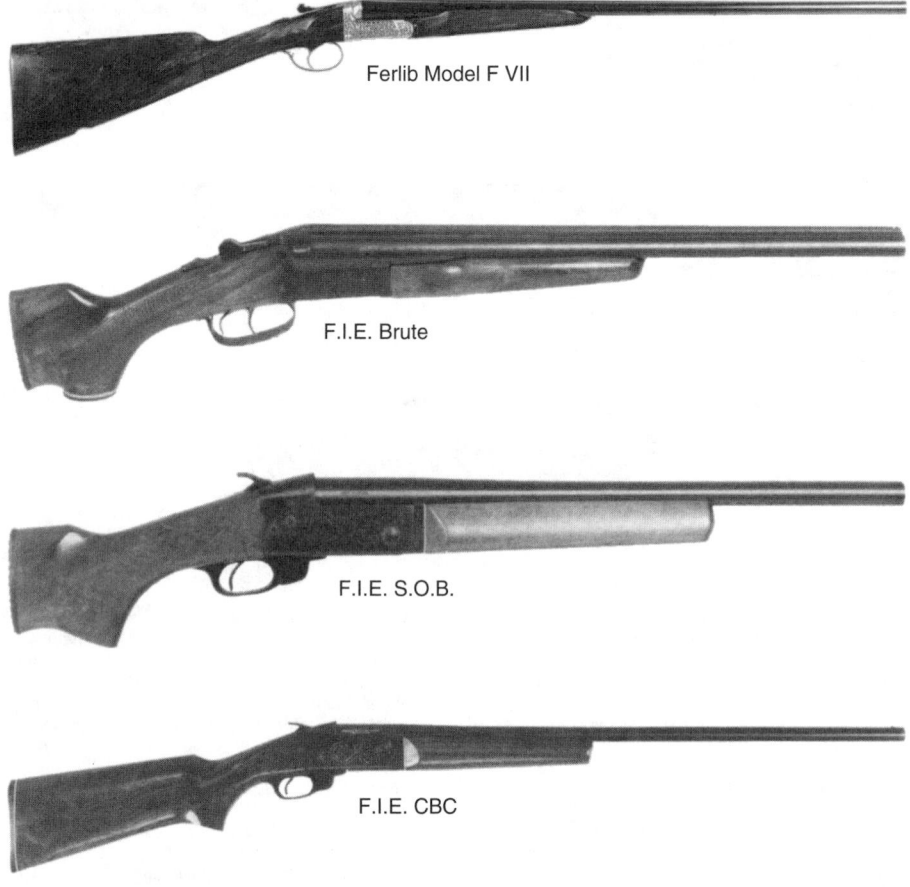

Ferlib Model F VII

F.I.E. Brute

F.I.E. S.O.B.

F.I.E. CBC

FERLACH CONSTANT COMPANION
Side-by-side; Anson & Deeley-type action; 12-, 16-, 20-ga.; 28″, 30″ barrels; tapered boring; hand-checkered black walnut pistol-grip stock, cheekpiece; quadruple Greener bolt; auto safety; ejectors; double triggers; engraved receiver. Made in Austria. Introduced 1956; discontinued 1958.
Perf.: $875   **Exc.:** $775   **VGood:** $700

FERLIB MODEL F VII
Side-by-side; boxlock; 12-, 16-, 20-, 28-ga., .410; 25″, 26″, 27″, 28″ barrels; standard chokes; oil-finished checkered European walnut straight-grip stock, forend; beavertail forend optional; single or double triggers; silvered, scroll-engraved receiver. Made in Italy. Introduced 1983; still in production. **LSMR: $7500**
12-, 16-, 20-ga.
Perf.: $6300   **Exc.:** $5100   **VGood:** $4000
28-ga., .410
Perf.: $7000   **Exc.:** $5600   **VGood:** $4500

Ferlib Model F VII/SC
Same specs as Model F VII except elaborate engraving with gold inlays. Made in Italy. Introduced 1983; still in production. **LMSR: $9000**
12-, 16-, 20-ga.
Perf.: $7500   **Exc.:** $6000   **VGood:** $5000
28-ga., .410
Perf.: $8100   **Exc.:** $6500   **VGood:** $5400

Ferlib Model F VII Sideplate
Same specs as Model F VII except sideplates; single trigger; elaborate engraving. Made in Italy. Introduced 1983; still in production. **LMSR: $9600**
12-, 16-, 20-ga.
Perf.: $8200   **Exc.:** $6300   **VGood:** $5200
28-ga., .410
Perf.: $8800   **Exc.:** $6900   **VGood:** $5700

Ferlib Model F VII Sideplate SC
Same specs as Model F VII except sideplates; single trigger; elaborate engraving with gold inlays. Made in Italy. Introduced 1983; still in production. **LMSR: $13,000**
12-, 16-, 20-ga.
Perf.: $10,000 **Exc.:** $8500   **VGood:** $7000
28-ga., .410
Perf.: $11,000 **Exc.:** $9000   **VGood:** $7500

FIAS MODEL SK-1
Over/under; boxlock; 12-, 20-ga.; 3″ chambers; 26″ Improved Cylinder/Modified, 28″ Modified/Full, 30″ Modified/Full or Full/Full, 32″ Full/Full barrels; vent-rib; hand-checkered select European walnut pistol-grip stock, forearm; top lever break; Greener crossbolt; double triggers; extractors. Made in Europe. Introduced 1974; discontinued 1984.
Perf.: $275   **Exc.:** $230   **VGood:** $190

FIAS MODEL SK-3
Over/under; boxlock; 12-, 20-ga.; 3″ chambers; 26″ Improved Cylinder/Modified, 28″ Modified/Full, 30″ Modified/Full or Full/Full, 32″ Full/Full barrels; vent-rib; hand-checkered select European walnut pistol-grip stock, forearm; top lever break; Greener crossbolt; selective single trigger; extractors. Made in Europe. Introduced 1974; discontinued 1984.
Perf.: $290   **Exc.:** $240   **VGood:** $200

FIAS MODEL SK-4
Over/under; boxlock; 12-, 20-ga.; 3″ chambers; 26″ Improved Cylinder/Modified, 28″ Modified/Full, 30″ Modified/Full or Full/Full, 32″ Full/Full barrels; vent-rib; hand-checkered select European walnut pistol-grip stock, beavertail forearm; top lever break; Greener crossbolt; selective single trigger; selective auto ejectors. Made in Europe. Introduced 1974; discontinued 1984.
Perf.: $350   **Exc.:** $280   **VGood:** $210

Fias Model SK-4D
Same specs as SK-4 except sideplates; better grade wood; deluxe engraving. Introduced 1974; discontinued 1984.
Perf.: $480   **Exc.:** $320   **VGood:** $270

Fias Model SK-4S
Same specs as SK-4 except 26″ Skeet 1/Skeet 2 barrels. Introduced 1974; discontinued 1984.
Perf.: $320   **Exc.:** $280   **VGood:** $210

Fias Model SK-4T
Same specs as Model SK-4 except 12-ga.; 2 ¾″ chambers; 30″ Improved Modified/Full, 32″ Improved Modified/Full barrels; middle sight; select European walnut Monte Carlo pistol-grip stock, beavertail forearm. Introduced 1974; discontinued 1984.
Perf.: $350   **Exc.:** $280   **VGood:** $210

F.I.E. BRUTE
Side-by-side; boxlock; 12-, 20-ga., .410; 19″ barrels; 30″ overall length; short hand-checkered walnut stock, beavertail forend. Made in Brazil. Introduced 1979; no longer in production.
Perf.: $160   **Exc.:** $120   **VGood:** $100

F.I.E. CBC
Single shot; takedown; exposed hammer; 12-, 16-, 20-ga., .410; 28″ Full-choke barrel; metal bead front sight; walnut-stained hardwood stock, beavertail forend; triggerguard button breaks action; auto ejector. Made in Brazil. Introduced 1982; discontinued 1984.
Perf.: $90   **Exc.:** $75   **VGood:** $50

F.I.E. HAMILTON & HUNTER
Single shot; takedown; exposed hammer; 12-, 20-ga., .410; 3″ chamber; 28″ barrel; weighs 6 ½ lbs.; metal bead front sight; walnut-stained hardwood stock, beavertail forend; break-button on triggerguard; auto ejector. Made in Brazil. Introduced 1986; discontinued 1990.
Perf.: $100   **Exc.:** $80   **VGood:** $60

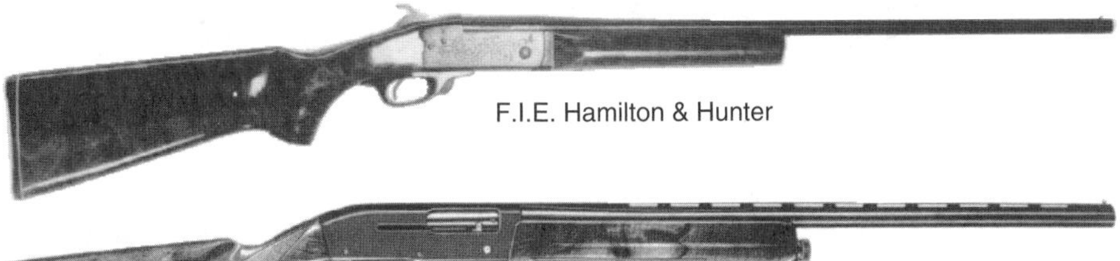

F.I.E. Hamilton & Hunter

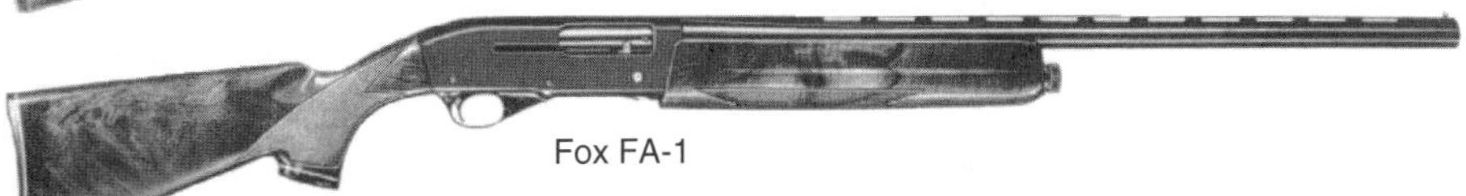

Fox FA-1

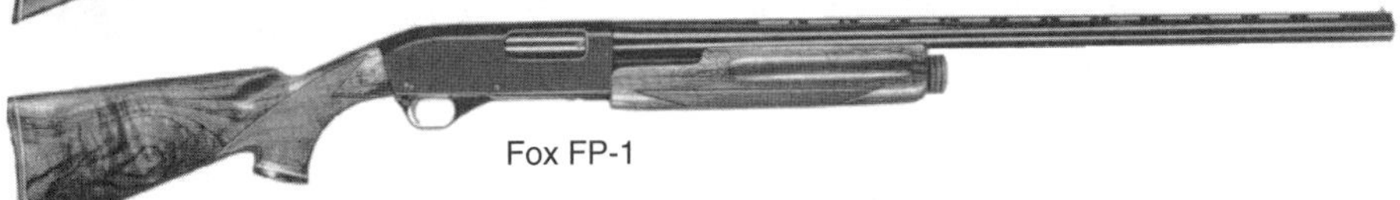

Fox FP-1

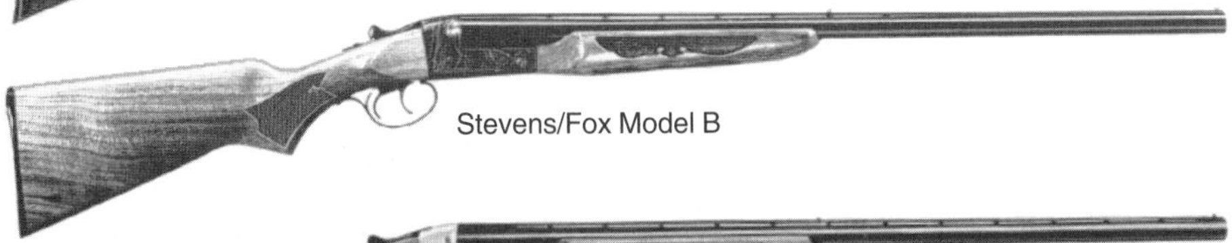

Stevens/Fox Model B

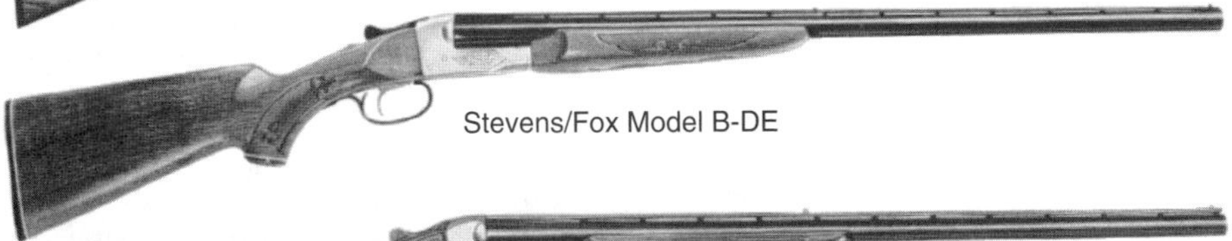

Stevens/Fox Model B-DE

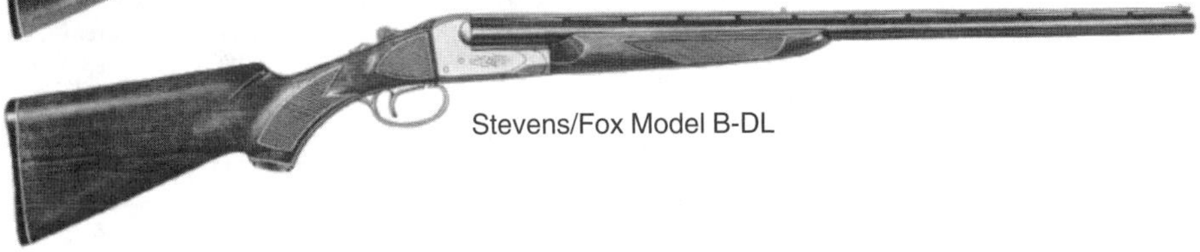

Stevens/Fox Model B-DL

SHOTGUNS

F.I.E.
S.O.B.
S.S.S. Model
Sturdy
Stury Deluxe Priti
Sturdy Model 12 Deluxe

Fox
FA-1
FP-1

Stevens/Fox
Model B
Model B-DE
Model B-DL
Model B-SE
Model B-ST

F.I.E. S.O.B.

Single barrel; exposed hammer; 12-, 20-ga., .410; 18 1/2″ barrel; metal bead front sight; short walnut-finished hardwood stock, beavertail forend; auto ejector. Made in Brazil. Introduced 1980; discontinued 1984.
    Perf.: $110   **Exc.:** $80   **VGood:** $60

F.I.E. S.S.S. MODEL

Single barrel; exposed hammer; 12-, 20-ga., .410; 3″ chamber; 18 1/2″ Cylinder barrel; weighs 6 1/2 lbs.; walnut-finished hardwood stock, full beavertail forend; auto ejectors; break-button on triggerguard. Made in Brazil. Introduced 1986; discontinued 1990.
    Perf.: $90   **Exc.:** $75   **VGood:** $50

F.I.E. STURDY

Over/under; 12-, 20-ga.; 3″ chamber; 28″ vent-rib barrel; walnut stock; double triggers; extractors; engraved silver finished receiver. Introduced 1985; discontinued 1988.
    Perf.: $280   **Exc.:** $240   **VGood:** $220

F.I.E. Sturdy Deluxe Priti

Same specs as Sturdy except deluxe walnut stock. Introduced 1985; discontinued 1988.
    Perf.: $310   **Exc.:** $240   **VGood:** $200
With optional single selective trigger; ejectors; choke tubes
    Perf.: $360   **Exc.:** $290   **VGood:** $250

F.I.E. Sturdy Model 12 Deluxe

Same specs as Sturdy except only 12-ga.; choke tubes; select walnut stock; single selective trigger; ejectors. Introduced 1988; discontinued 1988.
    Perf.: $300   **Exc.:** $240   **VGood:** $200

FOX FA-1

Autoloader; 12-ga.; 28″ Modified, 30″ Full choke barrels; metal bead front sight; walnut pistol-grip stock; rosewood grip cap with inlay; self-compensating gas system; crossbolt safety; chrome-moly barrel, polished

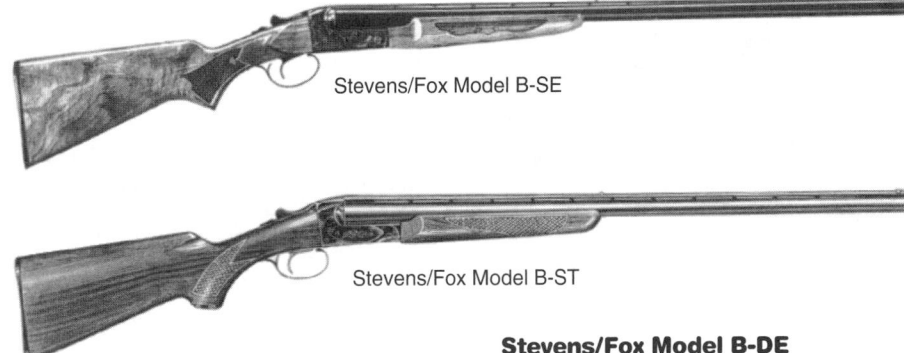

Stevens/Fox Model B-SE

Stevens/Fox Model B-ST

receiver Made in Japan. Introduced 1981 by Savage Arms; discontinued 1982.
    Perf.: $300   **Exc.:** $260   **VGood:** $225

FOX FP-1

Slide action; 12-ga.; 28″, 30″ barrels; vent. rib; metal bead front sight; checkered pistol-grip walnut stock; rosewood pistol-grip cap; crossbolt safety; dual action bars. Made in Japan. Introduced 1981 by Savage Arms; discontinued 1982.
    Perf.: $290   **Exc.:** $250   **VGood:** $210

STEVENS/FOX MODEL B

Side-by-side; boxlock; hammerless; 12-, 16-, 20-ga., .410; 26″, 28″, 30″ barrels; Modified/Full, Improved/Modified, Full/Full chokes; hand-checkered American walnut pistol-grip stock, forearm; double triggers; plain extractors; case-hardened frame. Introduced 1940; no longer in production.
    Exc.: $220   **VGood:** $170   **Good:** $125

Stevens/Fox Model B-DE

Same specs as Model B except for checkered select walnut pistol-grip stock, beavertail forearm; non-selective single trigger; satin chrome-finished frame; replaced Model B-DL. Introduced 1965; discontinued 1966.
    Exc.: $240   **VGood:** $190   **Good:** $145

Stevens/Fox Model B-DL

Same specs as Model B except for checkered select walnut pistol-grip stock, beavertail forearm; non-selective single trigger; satin chrome-finished frame; sideplates. Introduced 1962; discontinued 1965.
    Exc.: $230   **VGood:** $180   **Good:** $140

Stevens/Fox Model B-SE

Same specs as Model B except single trigger; selective ejectors. Introduced 1966; discontinued 1987.
    Perf.: $400   **Exc.:** $310   **VGood:** $260

Stevens/Fox Model B-ST

Same specs as Model B except non-selective single trigger. Introduced 1955; discontinued 1966.
    Exc.: $275   **VGood:** $200   **Good:** $120

SHOTGUNS

Fox, A.H.

Hammerless Double Barrel (1902-1905)
Hammerless Double Barrel (1905-1946)
Hammerless Double Barrel A Grade
Hammerless Double Barrel AE Grade
Hammerless Double Barrel B Grade
Hammerless Double Barrel BE Grade
Hammerless Double Barrel C Grade
Hammerless Double Barrel CE Grade
Hammerless Double Barrel D Grade
Hammerless Double Barrel DE Grade
Hammerless Double Barrel F Grade
Hammerless Double Barrel FE Grade
Hammerless Double Barrel GE Grade
Hammerless Double Barrel HE Grade (See Fox Hammerless Double Barrel Super-Fox)
Hammerless Double Barrel Skeeter
Hammerless Double Barrel Trap Grade
Hammerless Double Barrel XE Grade

A.H. Fox Hammerless
Double Barrel CE Grade

A.H. Fox Hammerless
Double Barrel XE Grade

A.H. Fox Sterlingworth Standard

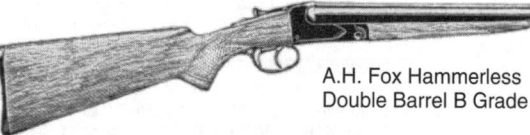

A.H. Fox Hammerless
Double Barrel B Grade

A.H. FOX HAMMERLESS DOUBLE BARREL (1902-1905)

Side-by-side; boxlock; 12-ga.; 28", 30", 32" Krupp fluid steel barrels; Damascus barrels optional in grades C and higher; Whitworth fluid steel barrels optional in Grade H; English walnut stocks; pistol grip in Grade A, straight or pistol grip in others; made by Philadelphia Gun Company, Philadelphia. Introduced 1902; discontinued 1905.

Grade A
  Perf.: $1250   Exc.: $875   VGood: $700
Grade B
  Perf.: $2000   Exc.: $1400   VGood: $1100
Grade C
  Perf.: $2900   Exc.: $2000   VGood: $1600
Grade D
  Perf.: $6500   Exc.: $4900   VGood: $3600
Grade E
  Perf.: $10,000   Exc.: $7500   VGood: $5500
Grade F
  Perf.: $13,500   Exc.: $10,000   VGood: $7400
Grade H
  Perf.: $15,500   Exc.: $12,000   VGood: $8600

A.H. FOX HAMMERLESS DOUBLE BARREL (1905-1946)

Side-by-side; boxlock; 12-, 16-, 20-ga.; 2 9/16" (until mid-1930s), 2 3/4", 3" chambers; Krupp fluid steel (until about 1910), Chromox steel barrels (used exclusively after the early 1920s); walnut half-pistol-grip stock standard, straight or pistol grip optional; earliest graded smallbores have snap-on forend latch; standard engraving patterns for all grades except CE were changed 1913-1914; all graded 12-ga. and graded smallbores built after 1913 have Deeley finger-lever forend latch; Fox-Kautzky single trigger available in all grades after 1914; other options include beavertail forend, vent rib, skeleton steel butt, recoil pad, extra sets of barrels, custom stocks. 12-ga. introduced 1905; 16-, 20-ga. introduced 1912 and built on the same, scaled-down frame. Made by A.H. Fox Gun Company, Philadelphia, 1905-1929; taken over by Savage Arms Company in late 1929 and production moved to Utica, New York in 1930; production of the original A.H. Fox gun discontinued in 1946. Approximate production totals: graded 12-ga., 35,000; graded 16-ga., 3521; graded 20-ga., 3434.

A.H. Fox Hammerless Double Barrel A Grade
Same specs as Hammerless Double Barrel (1905-1946) except 26" to 32" barrels; American walnut stocks; double or single trigger; extractors. Introduced 1905; discontinued 1942.
  Exc.: $1400   VGood: $950   Good: $800

A.H. Fox Hammerless Double Barrel AE Grade
Same specs as Hammerless Double Barrel (1905-1946) except 26" to 32" barrels; American walnut stocks; double or single trigger; ejectors. Introduced 1905; discontinued 1946.
  Exc.: $1600   VGood: $1150   Good: $950

A.H. Fox Hammerless Double Barrel B Grade
Same specs as Hammerless Double Barrel (1905-1946) except 26" to 32" barrels; English walnut stocks; double or single trigger; extractors. Introduced 1905; discontinued 1918.
  Exc.: $2300   VGood: $1700   Good: $1450

A.H. Fox Hammerless Double Barrel BE Grade
Same specs as Hammerless Double Barrel (1905-1946) except 26" to 32" barrels; English walnut stocks; double or single trigger; ejectors. Introduced 1905; discontinued 1918.
  Exc.: $2500   VGood: $1900   Good: $1650

A.H. Fox Hammerless Double Barrel C Grade
Same specs as Hammerless Double Barrel (1905-1946) except 26" to 32" barrels; English walnut stocks; double or single trigger; extractors. Introduced 1905; discontinued 1913.
  Exc.: $3400   VGood: $2900   Good: $2400

A.H. Fox Hammerless Double Barrel CE Grade
Same specs as Hammerless Double Barrel (1905-1946) except 26" to 32" barrels; English walnut stocks; double or single trigger; ejectors. Introduced 1905; discontinued 1946.
  Exc.: $3800   VGood: $3100   Good: $2600

A.H. Fox Hammerless Double Barrel D Grade
Same specs as Hammerless Double Barrel (1905-1946) except 26" to 32" barrels; Circassian walnut stocks; double or single trigger; extractors. Introduced 1905; discontinued 1913.
  Exc.: $7700   VGood: $5700   Good: $4300

A.H. Fox Hammerless Double Barrel DE Grade
Same specs as Hammerless Double Barrel (1905-1946) except 26" to 32" barrels; Circassian walnut stocks; double or single trigger; ejectors. Introduced 1905; discontinued 1945.
  Exc.: $8000   VGood: $6000   Good: $4500

A.H. Fox Hammerless Double Barrel F Grade
Same specs as Hammerless Double Barrel (1905-1946) except 26" to 32" barrels; Circassian walnut stocks; extractors; made to order with any available option at no extra charge. Introduced 1906; discontinued 1913.
  Exc.: $16,000   VGood: $11,000   Good: $9500

A.H. Fox Hammerless Double Barrel FE Grade
Same specs as Hammerless Double Barrel (1905-1946) except 26" to 32" barrels; Circassian walnut stocks; ejectors; made to order with any available option at no extra charge. Introduced 1906; discontinued 1940.
  Exc.: $16,500   VGood: $12,000   Good: $10,000

A.H. Fox Hammerless Double Barrel GE Grade
Same specs as Hammerless Double Barrel (1905-1946) except reputedly the highest-grade A.H. Fox gun; included in price lists from 1922 through 1930, but never shown in catalogs.

A.H. Fox Hammerless Double Barrel Skeeter
Same specs as Hammerless Double Barrel (1905-1946) except 12-, 20-ga.; 28" barrels; vent rib; ivory beads; walnut pistol-grip stock, beavertail forend; recoil pad; ejectors; double triggers; single trigger optional. Introduced 1931; discontinued 1931.
  Exc.: $2500   VGood: $1800   Good: $1400

A.H. Fox Hammerless Double Barrel Trap Grade
Same specs as Hammerless Double Barrel (1905-1946) except 26" to 32" barrels; vent rib; ivory beads; walnut pistol-grip stock, beavertail forend; recoil pad; double triggers; single trigger optional. Introduced 1932; discontinued 1936.
  Exc.: $2500   VGood: $1800   Good: $1400

A.H. Fox Hammerless Double Barrel XE Grade
Same specs as Hammerless Double Barrel (1905-1946) except 26" to 32" barrels; Circassian walnut stocks; Monte Carlo comb optional; double or single trigger; ejectors. Introduced 1914; discontinued 1945.
  Exc.: $5750   VGood: $4800   Good: $3800

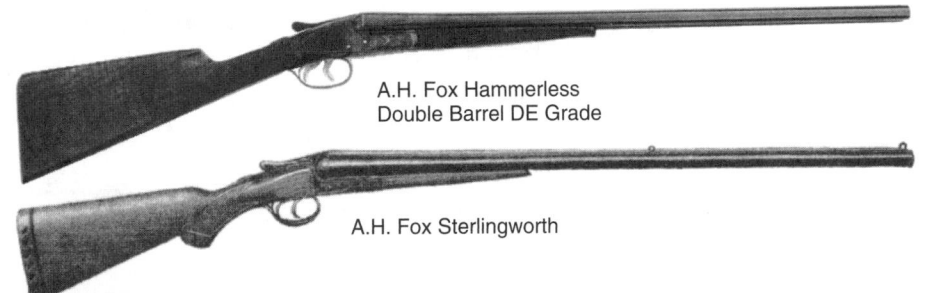
A.H. Fox Hammerless Double Barrel DE Grade

A.H. Fox Sterlingworth

Fox
Hammerless Double Barrel (Current Production)
Model B Lightweight
Single-Barrel Trap
SP Grade
SPE Grade
SPE Skeet Grade
SPE Skeet & Upland Game Gun
SPR Grade
Sterlingworth
Sterlingworth Brush
Sterlingworth Deluxe
Sterlingworth Deluxe Ejector
Sterlingworth Ejector
Sterlingworth Field
Sterlingworth Skeet
Sterlingworth Skeet Ejector
Sterlingworth Standard
Sterlingworth Trap
Sterlingworth Wildfowl
Super-Fox

A.H. FOX HAMMERLESS DOUBLE BARREL (Current Production)

Side-by-side; boxlock; 20-ga.; 26″, 28″, 30″ barrels; American walnut stock, splinter forend; stocks made to order; double triggers; optional beavertail forend, skeleton steel butt, recoil pad, checkered butt, single trigger; grades CE, DE, FE, Exhibition. Introduced 1993 by Connecticut Shotgun Manufacturing Company, New Britain, Connecticut; still in production. **LMSR: $7200**
CE Grade
   Perf.: $6500   **Exc.:** $4900   **VGood:** $4100
DE Grade
   Perf.: $11,000   **Exc.:** $8200   **VGood:** $7000
FE Grade
   Perf.: $16,000   **Exc.:** $12,500   **VGood:** $9000
Exhibition Grade
   Perf.: $22,500   **Exc.:** $17,500   **VGood:** $12,500

A.H. FOX MODEL B LIGHTWEIGHT

Side-by-side; boxlock; hammerless; 12-, 20-ga., .410; 24″ Improved/Modified (12-, 20-ga. only), 26″ (.410 only), 26″ Improved/Modified, 28″ Modified/Full, 30″ Modified/Full (12-ga. only) barrels; vent rib; checkered select walnut pistol-grip stock, beavertail forearm; double triggers; color case-hardened frame. Introduced 1973; no longer in production.
   Perf.: $150   **Exc.:** $120   **VGood:** $100

A.H. FOX SINGLE-BARREL TRAP

Single shot; boxlock; 12-ga.; 30″, 32″ barrel; vent rib; American walnut straight, half- or pistol-grip stock; Monte Carlo comb optional for early guns, standard after 1931; ejector; available in four grades. Introduced 1919; discontinued 1939. Approximate production totals: J Grade, 460; K Grade, 77; L Grade, 25; M Grade, 9.
J Grade
   Exc.: $1900   **VGood:** $1750   **Good:** $1400
K Grade
   Exc.: $2950   **VGood:** $2500   **Good:** $2200
L Grade
   Exc.: $3500   **VGood:** $3250   **Good:** $3000
M Grade
   Exc.: $8000   **VGood:** $7250   **Good:** $5000

A.H. FOX SP GRADE

Side-by-side; boxlock; 12-, 16-, 20-ga.; 26″, 28″, 30″, 32″ barrels; American walnut pistol-grip stock; double triggers; extractors; optional single trigger, beavertail forend, recoil pad, ivory beads; a distinct model mechanically identical to standard Fox guns but with a squarer, smooth-sided frame and no cheek panels on the stock; originally called Fox Special Grade, changed to SP Grade. Introduced 1932; discontinued 1946.
   Perf.: $2000   **Exc.:** $1400   **VGood:** $1100

A.H. Fox SPE Grade

Same specs as SP Grade except ejectors. Introduced 1932; discontinued 1946.
   Perf.: $2100   **Exc.:** $1500   **VGood:** $1200

A.H. Fox SPE Skeet Grade

Same specs as SP Grade except ejectors; double or single trigger. Introduced 1934; discontinued 1934.
   Perf.: $2100   **Exc.:** $1500   **VGood:** $1200

A.H. Fox SPE Skeet and Upland Game Gun

Same specs as SP Grade except 26″, 28″ Cylinder/Quarter barrels; American walnut straight or pistol-grip stock, beavertail forend; single trigger; ejectors; optional double triggers, vent rib, recoil pad. Introduced 1935; discontinued 1939.
   Perf.: $2100   **Exc.:** $1500   **VGood:** $1200

A.H. Fox SPR Grade

Same specs as SP Grade except vent rib; ejectors. Introduced 1936; discontinued 1936.
   Perf.: $2300   **Exc.:** $1700   **VGood:** $1400

A.H. FOX STERLINGWORTH

Side-by-side; boxlock; 12-, 16-, 20-ga.; 2 9/16″ (until mid-1930s), 2 3/4″, 3″ chambers; 26″ to 32″ Krupp fluid steel (until about 1910), Chromox steel barrels (used exclusively after the early 1920s); checkered American walnut half-pistol-grip stock, snap-on forend standard; straight or pistol grip optional; snap-on forend latch; double triggers; optional Fox-Kautzky single trigger available in all grades after 1914; extractors; standard engraving patterns for all grades except CE were changed 1913-1914; other options include beavertail forend, vent rib, skeleton steel butt, recoil pad, extra sets of barrels, custom stocks; mechanically same as other A.H. Fox doubles, but purely production-line guns without much handwork. Introduced 1911; discontinued 1946. Made by A.H. Fox Gun Company, Philadelphia, 1905-1929; taken over by Savage Arms Company in late 1929 and production moved to Utica, New York in 1930; production of the original A.H. Fox gun discontinued in 1946. Approximate production totals: 12-ga. Sterlingworth, 110,000; Sterlingworth 16-ga., 28,000; 20-ga. Sterlingworth, 21,000.
   Exc.: $1000   **VGood:** $700   **Good:** $450

A.H. Fox Sterlingworth Brush

Same specs as Fox Sterlingworth except 26″ Cylinder/Modified barrels. Introduced 1911; discontinued 1930.
   Exc.: $1000   **VGood:** $700   **Good:** $450

A.H. Fox Sterlingworth DeLuxe

Same specs as Fox Sterlingworth except Lyman ivory beads; recoil pad. Introduced 1930; discontinued 1945.
   Exc.: $1250   **VGood:** $1000   **Good:** $800

A.H. Fox Sterlingworth DeLuxe Ejector

Same specs as Fox Sterlingworth except Lyman ivory beads; ejectors; recoil pad. Introduced 1931; discontinued 1945.
   Exc.: $1400   **VGood:** $1150   **Good:** $950

A.H. Fox Sterlingworth Ejector

Same specs as Fox Sterlingworth except ejectors. Introduced 1911; discontinued 1946.
   Exc.: $1200   **VGood:** $900   **Good:** $650

A.H. Fox Sterlingworth Field

Same specs as Fox Sterlingworth except 28″ Modified/Full barrels. Introduced 1911; discontinued 1930.
   Exc.: $1000   **VGood:** $700   **Good:** $450

A.H. Fox Sterlingworth Skeet

Same specs as Fox Sterlingworth except 26″, 28″ barrels; checkered American walnut straight-grip stock; extractors; double triggers; optional single trigger, beavertail forend, recoil pad. Introduced 1935; discontinued 1945.
   Perf.: $2200   **Exc.:** $1600   **VGood:** $1200

A.H. Fox Sterlingworth Skeet Ejector

Same specs as Fox Sterlingworth except 26″, 28″ barrels; checkered American walnut straight-grip stock; ejectors; double triggers; optional single trigger, beavertail forend, recoil pad. Introduced 1935; discontinued 1945.
   Perf.: $2250   **Exc.:** $1650   **VGood:** $1250

A.H. Fox Sterlingworth Standard

Same specs as Fox Sterlingworth except 30″ Full/Full barrels. Introduced 1911; discontinued 1930.
   Exc.: $1000   **VGood:** $700   **Good:** $450

A.H. Fox Sterlingworth Trap

Same specs as Fox Sterlingworth except 32″ Full/Full barrels. Introduced 1911; discontinued 1930.
   Exc.: $1100   **VGood:** $750   **Good:** $500

A.H. Fox Sterlingworth Wildfowl

Same specs as Fox Sterlingworth except 12-ga.; 30″, 32″ barrels; ejectors; not a true Sterlingworth but rather a Super-Fox engraved and stamped "Fox Sterlingworth." Introduced 1934; discontinued 1940.
   Exc.: $1300   **VGood:** $850   **Good:** $700

A.H. FOX SUPER-FOX

Over/Under; boxlock; 12-, 20-ga.; over-bored 30″, 32″ Chromox steel barrels; checkered American walnut half-pistol-grip stock, snap-on forend standard; straight or pistol grip optional; snap-on forend latch; double triggers; optional Fox-Kautzky single trigger available; ejectors; standard engraving patterns; built on special, oversized frames; other options include beavertail forend, vent rib, ivory beads; skeleton steel butt, recoil pad, extra sets of barrels, custom stocks; special Fox model designed for long-range shooting; although designated HE Grade, it was also available in any standard Fox grade on special order; high grades rare. A few very early Super-Foxes are stamped "Barrels Guaranteed - See Tag"; most later ones are stamped "Barrels Not Guaranteed - See Tag"; both stamps refer to pattern density, not quality of barrels. 12-ga. introduced 1923; 20-ga. introduced 1925; 20-ga. discontinued 1931; 12-ga. discontinued 1942. Produced in very small numbers, about 300 guns in 12-ga. and only about 60 in 20-ga.
   Exc.: $1600   **VGood:** $1000   **Good:** $900

Franchi Semi-Automatic Shotguns
48/AL (See Franchi Standard Semi-Automatic (48/AL))
Black Magic 48/AL
Black Magic Game
Black Magic Skeet
Black Magic Trap
Crown Grade Diamond (See Franchi Alcione)
Diamond Grade
Dynamic 12 (See Franchi Standard Autoloader (48/AL))
Dynamic 12 Slug (See Franchi Slug Gun)
Dynamic 12 Skeet (See Franchi Skeet Gun)
Eldorado
Elite
Falcon (See Franchi Falconet)
Hunter Semi-Automatic
Hunter Magnum
Imperial Grade
LAW-12
Model 500
Model 520 Deluxe
Model 520 Eldorado Gold
Model 530

Franchi Black Magic 48/AL

Franchi Black Magic Game

Franchi Black Magic Skeet

Franchi Eldorado

Franchi Crown Grade

FRANCHI SEMI-AUTOMATIC SHOTGUNS

FRANCHI BLACK MAGIC 48/AL
Recoil-operated autoloader; 12-, 20-ga.; 2 3/4" chamber; 5-shot magazine; 24" rifled, 24", 26", 28" barrel; Improved Cylinder, Modified, Full Franchoke tubes; vent rib; weighs 5 1/4 lbs.; checkered walnut pistol-grip stock, forend; chrome-lined bore; crossbolt safety. Made in Italy. Introduced 1950; still in production.
LMSR: $625
Perf.: $550   Exc.: $400   VGood: $300

FRANCHI BLACK MAGIC GAME
Autoloader; 12-, 20-ga.; 3" chamber; 24", 26", 28" barrel; vent rib; choke tubes; bead front sight; checkered walnut stock, forearm; crossbolt safety; magazine cut-off; recoil pad. Made in Italy. Introduced 1989; discontinued 1991.
Perf.: $500   Exc.: $350   VGood: $300

FRANCHI BLACK MAGIC SKEET
Autoloader; 12-, 20-ga.; 2 3/4" chamber; 28" barrel; vent rib; choke tubes; bead front sight; checkered walnut stock, forearm; crossbolt safety; magazine cut-off; recoil pad. Made in Italy. Introduced 1989; discontinued 1991.
Perf.: $540   Exc.: $405   VGood: $320

FRANCHI BLACK MAGIC TRAP
Autoloader; 12-ga.; 2 3/4" chamber; 30" barrel; vent rib; choke tubes; bead front sight; checkered European walnut trap-style stock, forearm; crossbolt safety; magazine cut-off; recoil pad. Made in Italy. Introduced 1989; discontinued 1991.
Perf.: $600   Exc.: $410   VGood: $340

FRANCHI CROWN GRADE
Recoil-operated autoloader; 12-, 20-ga.; 5-shot magazine; 24", 26", 28", 30" interchangeable vent-rib barrel; Improved, Modified, Full chokes; hand-checkered European walnut pistol-grip stock, forearm; alloy receiver; chrome-lined barrels; simplified takedown; engraved hunting scene. Introduced 1954; discontinued 1975.
Perf.: $1350   Exc.: $1200   VGood: $1000

FRANCHI DIAMOND GRADE
Recoil-operated autoloader; 12-, 20-ga.; 5-shot magazine; 24", 26", 28", 30" interchangeable vent-rib barrel; Improved, Modified, Full chokes; hand-checkered European walnut pistol-grip stock, forearm; alloy receiver; chrome-lined barrels; simplified takedown; scroll engraved with silver inlay. Introduced 1954; discontinued 1975.
Perf.: $1725   Exc.: $1400   VGood: $1250

FRANCHI ELDORADO
Recoil-operated autoloader; 12-, 20-ga.; 5-shot magazine; 24", 26", 28", 30" interchangeable vent-rib barrel; Improved, Modified, Full chokes; select hand-checkered European walnut pistol-grip stock, forearm; alloy receiver; chrome-lined barrels; simplified takedown; scroll engraving covering 75% of receiver surfaces; gold-plated trigger; chrome-plated breech bolt. Introduced 1955; discontinued 1975.
Perf.: $425   Exc.: $375   VGood: $320

FRANCHI ELITE
Gas-operated autoloader; 12-ga.; 2 3/4", 3" chamber; 24", 26", 28", 30", 32"; 7mm vent rib; all standard chokes; weighs 7 3/8 lbs.; oil-finished, checkered European walnut stock; blued receiver; stainless steel gas piston; engraved receiver. Made in Italy. Introduced 1985; discontinued 1989.
Perf.: $550   Exc.: $410   VGood: $325

FRANCHI HUNTER SEMI-AUTOMATIC
Recoil-operated autoloader; 12-, 20-ga.; 5-shot magazine; 24", 26", 28", 30" interchangeable vent-rib barrel; Improved, Modified, Full chokes; hand-checkered select European walnut pistol-grip stock, forearm; alloy receiver; chrome-lined barrels; simplified takedown; etched receiver. Introduced 1950; discontinued 1990.
Perf.: $400   Exc.: $290   VGood: $240

Franchi Hunter Magnum
Same specs as Hunter Semi-Automatic except 12-, 20-ga.; 3" chamber; 32" (12-ga.), 28" (20-ga.) vent-rib, Full-choke barrel; recoil pad. Introduced 1955; discontinued 1973.
Perf.: $390   Exc.: $350   VGood: $280

FRANCHI IMPERIAL GRADE
Recoil-operated autoloader; 12-, 20-ga.; 5-shot magazine; 24", 26", 28", 30" interchangeable vent-rib barrels; Improved, Modified, Full chokes; hand-checkered European walnut pistol-grip stock, forearm; alloy receiver; chrome-lined barrels; simplified takedown; elaborate engraving with gold inlay. Introduced 1954; discontinued 1975.
Perf.: $2250   Exc.: $2000   VGood: $1800

FRANCHI LAW-12
Autoloader; 12-ga.; 2 3/4" chamber; 5-, 8-shot tube magazine; 21 1/2" barrel; synthetic pistol-grip stock; alloy receiver. Introduced 1988; discontinued 1994.
Perf.: $505   Exc.: $390   VGood: $300

FRANCHI MODEL 500
Gas-operated autoloader; 12-ga.; 4-shot magazine; 26", 28" barrel; vent rib; any standard choke; checkered European walnut pistol-grip stock. Introduced 1976; no longer in production.
Perf.: $310   Exc.: $290   VGood: $240

FRANCHI MODEL 520 DELUXE
Gas-operated autoloader; 12-ga.; 4-shot magazine; 26", 28" barrel; vent rib; any standard choke; checkered select European walnut pistol-grip stock; engraved receiver. Introduced 1976; no longer in production.
Perf.: $350   Exc.: $300   VGood: $270

Franchi Model 520 Eldorado Gold
Same specs as Model 520 Deluxe except high quality of wood; better engraving; gold inlays in receiver. Introduced 1977; no longer in production.
Perf.: $850   Exc.: $700   VGood: $650

FRANCHI MODEL 530
Autoloader; 12-, 20-, 28-ga; 2 3/4", 3" chamber; 5-shot magazine; 30" barrel; interchangeable choke tubes; vent rib; oil finished, hand-checkered French walnut straight or Monte Carlo stock; interchangeable stock drilled, tapped for recoil reducer; specially tuned gas system; target-grade trigger; chrome-lined bore, chamber; matte blue finish. Introduced 1978; discontinued 1982.
Perf.: $600   Exc.: $500   VGood: $400

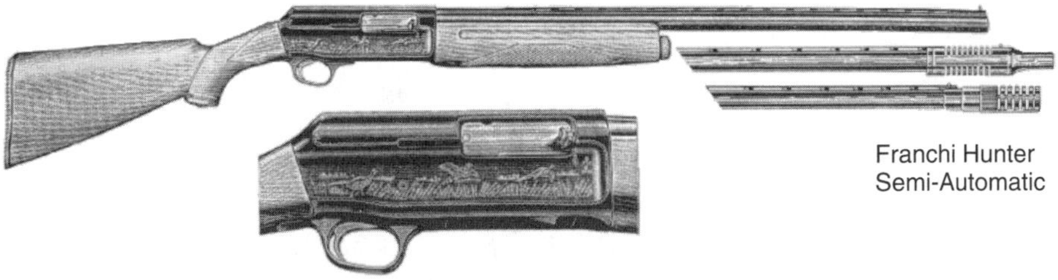

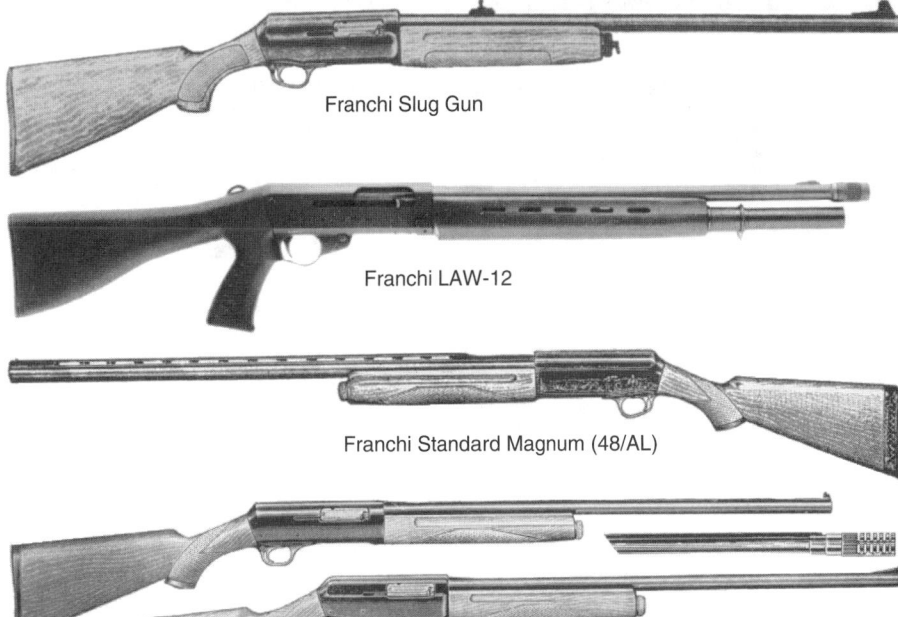

Franchi Hunter
Semi-Automatic

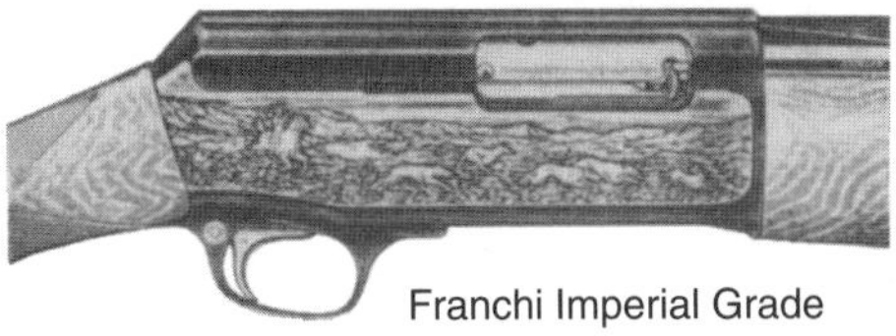

Franchi Imperial Extra

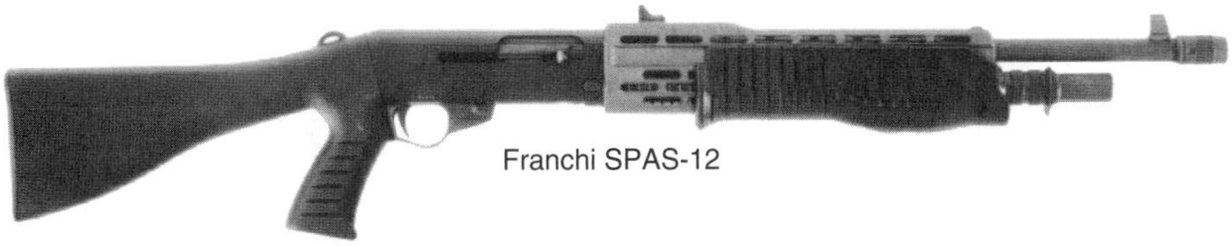

Franchi Imperial Grade

Franchi SPAS-12

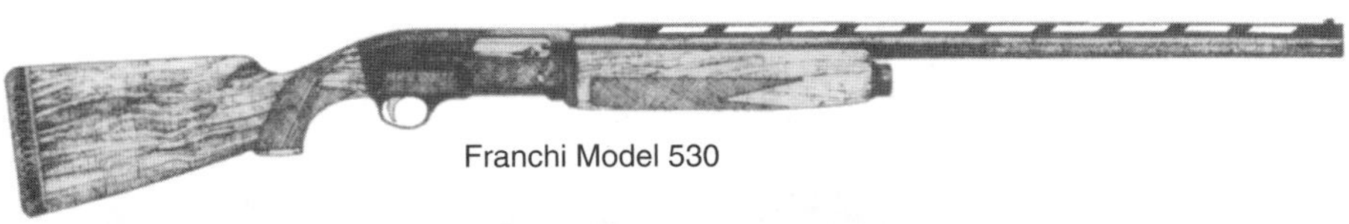

Franchi Model 530

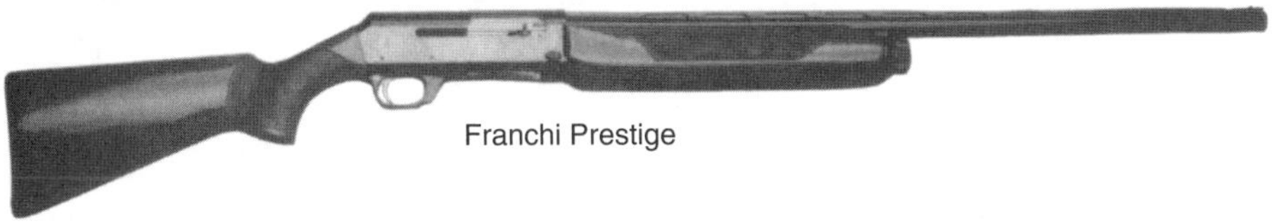

Franchi Prestige

Franchi Slug Gun

Franchi LAW-12

Franchi Standard Magnum (48/AL)

Franchi Standard Semi-Automatic (48/AL)

SHOTGUNS

Franchi Semi-Automatic Shotguns
Prestige
Skeet Gun
Slug Gun
SPAS-12
Standard Semi-Automatic (48/AL)
Standard Magnum (48/AL)
Turkey Gun

FRANCHI PRESTIGE
Gas-operated autoloader; 12-ga.; 2 3/4", 3" chamber; 24", 26", 28", 30", 32" barrel; 7mm vent-rib; all standard chokes; weighs 7 3/8 lbs.; oil-finished, checkered European walnut stock; blued receiver; stainless steel gas piston. Made in Italy. Introduced 1985; discontinued 1989.
Perf.: $600   **Exc.:** $390   **VGood:** $290

FRANCHI SKEET GUN
Recoil-operated autoloader; 12-, 20-ga.; 5-shot magazine; 26" vent-rib barrel; Skeet choke; hand-checkered fancy European walnut pistol-grip stock, forearm; alloy receiver; chrome-lined barrels; simplified takedown. Introduced 1972; discontinued 1974.
Perf.: $375   **Exc.:** $310   **VGood:** $290

FRANCHI SLUG GUN
Recoil-operated autoloader; 12-, 20-ga.; 5-shot magazine; 22" Cylinder-bore barrel; raised gold bead front sight, Lyman folding leaf open rear; hand-checkered European walnut pistol-grip stock, forearm; sling swivels. Introduced 1955; no longer in production.
Exc.: $300   **VGood:** $250   **Good:** $200

FRANCHI SPAS-12
Autoloader/slide action; 12-ga.; 2 3/4" chamber; 5-, 8-shot tube magazine; 21 1/2" barrel; synthetic pistol-grip stock; alloy receiver; button-activated switch to change action type. Introduced 1988; discontinued 1994.
Perf.: $575   **Exc.:** $475   **VGood:** $340

FRANCHI STANDARD SEMI-AUTOMATIC (48/AL)
Recoil-operated autoloader; 12-, 20-ga.; 5-shot magazine; 24", 26", 28", 30" interchangeable vent-rib barrel; Improved, Modified, Full chokes; hand-checkered European walnut pistol-grip stock, forearm; alloy receiver; chrome-lined barrels; simplified takedown. Introduced 1950; still in production.
LMSR: $625
Perf.: $480   **Exc.:** $290   **VGood:** $240

Franchi Standard Magnum (48/AL)
Same specs as Standard Semi-automatic (48/AL) except 12-, 20-ga.; 3" chamber; 28" (20-ga.), 32" (12-ga.) plain or vent-rib barrel; recoil pad. Introduced 1952; discontinued 1990.
Perf.: $400   **Exc.:** $300   **VGood:** $225

FRANCHI TURKEY GUN
Recoil-operated autoloader; 12-ga.; 3" chamber; 5-shot magazine; 32" vent-rib barrel; Extra Full choke; hand-checkered European walnut pistol-grip stock, forearm; alloy receiver; chrome-lined barrels; simplified takedown; engraved turkey scene on receiver. Introduced 1963; discontinued 1965.
Perf.: $360   **Exc.:** $300   **VGood:** $265

Franchi Over/Under Shotguns
Alcione
Alcione SL
Aristocrat
Aristocrat Deluxe
Aristocrat Imperial
Aristocrat Magnum
Aristocrat Monte Carlo
Aristocrat Silver-King
Aristocrat Skeet
Aristocrat Supreme
Aristocrat Trap
Black Magic Lightweight
Hunter
Black Magic Sporting
Hunter
Falconet
Falconet International
Skeet
Falconet International
Trap
Falconet Skeet
Falconet Super

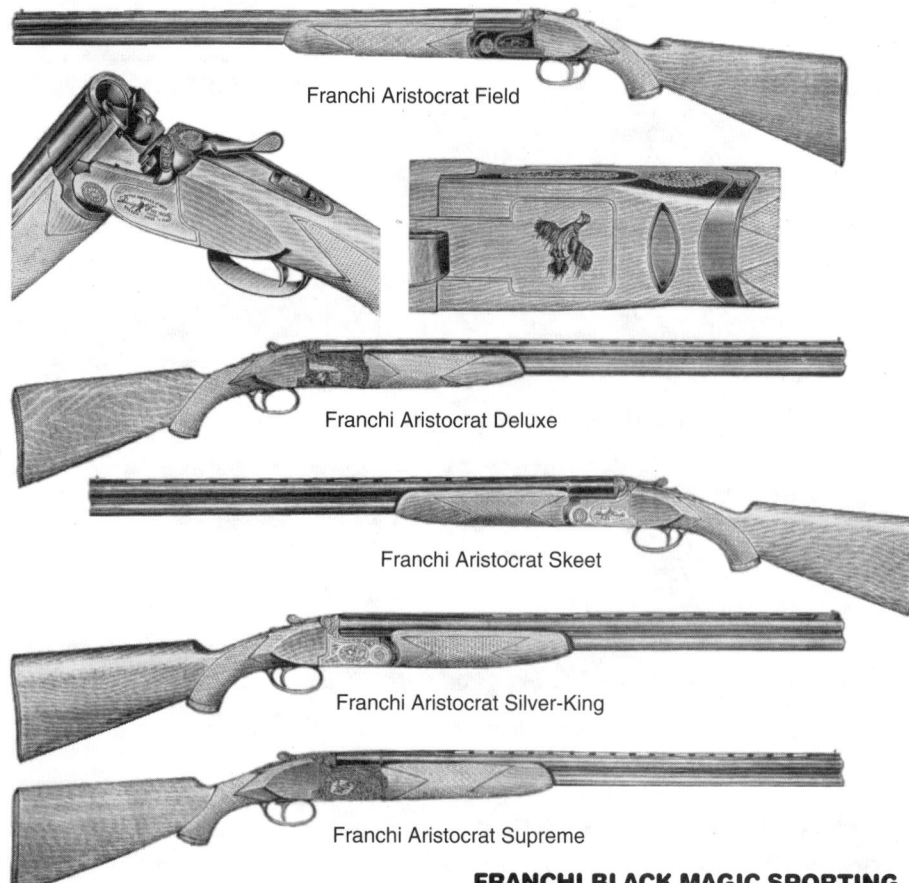

Franchi Aristocrat Field

Franchi Aristocrat Deluxe

Franchi Aristocrat Skeet

Franchi Aristocrat Silver-King

Franchi Aristocrat Supreme

FRANCHI OVER/UNDER SHOTGUNS
FRANCHI ALCIONE
Over/under; 12-ga.; 2 ¾" chamber; 28" Modified/Full barrel; cut-checkered French walnut pistol-grip stock, forend; top tang safety; single selective trigger; auto ejectors; chrome-plated bores; silvered receiver; decorative scroll engraving. Made in Italy. Introduced 1982; discontinued 1989.
Perf.: $650   Exc.: $440   VGood: $400

Franchi Alcione SL
Same specs as Alcione except 27" Improved/Modified, 28" Modified/Full barrels; ivory bead front sight; better engraving; silvered receiver; gold-plated trigger; gold-inlaid receiver; marketed in fitted case. Made in Italy. Introduced 1982; discontinued 1986.
Perf.: $1100   Exc.: $840   VGood: $760

FRANCHI ARISTOCRAT
Over/under; 12-ga.; 24" Cylinder/Improved Cylinder, 26" Improved/Modified, 28" Modified/Full, 30" Modified/Full vent-rib barrel; hand-checkered Italian walnut pistol-grip stock, forearm; selective automatic ejectors; automatic safety; selective single trigger; English scroll engraving on receiver; blue-black finish. Introduced 1960; discontinued 1968.
Exc.: $400   VGood: $360   Good: $300

Franchi Aristocrat Deluxe
Same specs as Aristocrat except select walnut stock forearm; heavy relief engraving on receiver, tang and trigger guard. Introduced 1960; discontinued 1965.
Exc.: $800   VGood: $700   Good: $640

Franchi Aristocrat Imperial
Same specs as Aristocrat except top-quality European walnut stock; exquisite engraving. Introduced 1967 on customer-order basis; discontinued 1969.
Exc.: $2000   VGood: $1750   Good: $1400

Franchi Aristocrat Magnum
Same specs as Aristocrat except 3" chamber; 32" Full choke barrels; recoil pad. Introduced 1962; discontinued 1965.
Exc.: $400   VGood: $350   Good: $300

Franchi Aristocrat Monte Carlo
Same specs as Aristocrat except top-of-the-line European walnut stock; top-of-the-line engraving. Introduced 1967 on customer-order basis; discontinued 1969.
Exc.: $3000   VGood: $2500   Good: $1900

Franchi Aristocrat Silver-King
Same specs as Aristocrat except selected ultra-deluxe walnut stock forearm, cut from the same blank for match of grain, color, finish; hand-checkered pistol grip, forearm; engraved, bright-finish receiver. Introduced 1965; discontinued 1968.
Exc.: $525   VGood: $450   Good: $400

Franchi Aristocrat Skeet
Same specs as Aristocrat except 26" Skeet/Skeet vent-rib barrel; grooved beavertail forearm. Introduced 1962; discontinued 1968.
Exc.: $425   VGood: $390   Good: $350

Franchi Aristocrat Supreme
Same specs as Aristocrat except select walnut stock, forearm; heavy relief engraving on receiver, tang and trigger guard; gold-inlaid game bird figures on receiver. Introduced 1960; discontinued 1965.
Exc.: $1050   VGood: $900   Good: $800

Franchi Aristocrat Trap
Same specs as Aristocrat except 30" Modified/Full vent-rib barrels; hand-checkered select deluxe-grade European walnut pistol-grip stock, beavertail forend; Monte Carlo comb; color case-hardened receiver. Introduced 1962; discontinued 1968.
Exc.: $505   VGood: $450   Good: $390

FRANCHI BLACK MAGIC LIGHTWEIGHT HUNTER
Over/under; boxlock; 12-, 20-ga.; 2 ¾" chambers; 25" barrel; choke tubes; weighs 6 lbs.; cut-checkered figured walnut stock, Schnabel forend with semi-gloss finish; solid black recoil pad; single selective trigger auto safety; auto selective ejectors; polished blue finish; gold accents. Made in Italy. Introduced 1989; discontinued 1991.
Perf.: $920   Exc.: $700   VGood: $620

FRANCHI BLACK MAGIC SPORTING HUNTER
Over/under; boxlock; 12-ga.; 3" chambers; 28" barrel; choke tubes; weighs 7 lbs.; cut-checkered figured walnut stock, Schnabel forend with semi-gloss finish; solid black recoil pad; single selective trigger auto safety; auto selective ejectors; polished blue finish; gold accents. Made in Italy. Introduced 1989; discontinued 1991.
Perf.: $945   Exc.: $730   VGood: $690

FRANCHI FALCONET
Over/under; 12-, 16-, 20-, 28-ga., .410; 24", 26", 28", 30" (12-ga. only) barrels; standard choke combos; checkered walnut pistol-grip stock, forearm; epoxy finish; selective single trigger; barrel selector; automatic safety; selective automatic ejectors; alloy receiver, fully engraved. Introduced 1969 as Falcon; discontinued 1985.
Perf.: $530   Exc.: $450   VGood: $410

Franchi Falconet International Skeet
Same specs as Falconet except 26" Skeet-choked barrels; wide vent rib; select Skeet-style stock, forearm; engraved color case-hardened receiver. Introduced 1970; discontinued 1974.
Perf.: $925   Exc.: $840   VGood: $780

Franchi Falconet International Trap
Same specs as Falconet except 12-ga.; 30" Modified/Full choke barrels; wide vent rib; select checkered trap-style Monte Carlo stock, forearm; engraved color case-hardened receiver; recoil pad. Introduced 1970; discontinued 1974.
Perf.: $925   Exc.: $840   VGood: $780

Franchi Falconet Skeet
Same specs as Falconet except 26" Skeet-choked barrels; wide vent rib; Skeet-style stock, forearm; color case-hardened receiver. Introduced 1970; discontinued 1974.
Perf.: $910   Exc.: $800   VGood: $750

Franchi Falconet Super
Same specs as Falconet except 12-ga.; 27", 28" barrels; translucent front sight; engraved silver finish lightweight alloy receiver; rubber buttplate. Made in Italy. Introduced 1982; discontinued 1986.
Perf.: $825   Exc.: $610   VGood: $525

Franchi Aristocrat Trap

Franchi Black Magic Lightweight Hunter

Franchi Falconet 2000

Franchi Falconet International Skeet

Franchi Falconet International Trap

Franchi Black Magic Lightweight Hunter

Franchi Model 2000 Sporting

Franchi Over/Under Shotguns
Falconet Trap
Falconet 2000
Model 2000 Sporting
Model 2003 Trap
Model 2004 Trap
Model 2005 Combination Trap
Model 2005/3 Combination Trap
Model 3000/2
Peregrin Model 400
Peregrin Model 451

Franchi Falconet Trap
Same specs as Falconet except 12-ga.; 30″ Modified/Full choke barrel; wide vent rib; checkered trap-style Monte Carlo stock, forearm; color case-hardened receiver; recoil pad. Introduced 1970; discontinued 1974.
Perf.: $900   **Exc.:** $810   **VGood:** $780

FRANCHI FALCONET 2000
Over/under; boxlock; 12-ga.; 2 3/4″ chambers; 26″ barrels; Improved Cylinder, Modified, Full Franchoke tubes; weighs 6 lbs.; white flourescent bead front sight; checkered walnut stock; silvered action with gold-plated game scene; single selective trigger; automatic selective ejectors. Made in Italy. Reintroduced 1992; no longer imported.
Perf.: $1075   **Exc.:** $825   **VGood:** $700

FRANCHI MODEL 2000 SPORTING
Over/under; boxlock; 12-ga.; 2 3/4″ chambers; 28″ ported barrels; Skeet, Improved Cylinder, Modified, Full Franchoke tubes; weighs 7 3/4 lbs.; white flourescent bead front sight; checkered walnut stock; single selective mechanical trigger; automatic selective ejectors; blued action. Made in Italy. Introduced 1992; no longer imported.
Perf.: $1175   **Exc.:** $900   **VGood:** $800

FRANCHI MODEL 2003 TRAP
Over/under; boxlock; 12-ga.; 30″, 32″, Improved Modified/Full, Full/Full barrels; high vent rib; checkered European walnut straight or Monte Carlo stock, beavertail forearm; single selective trigger; auto ejectors; recoil pad; marketed with carrying case. Introduced 1976; discontinued 1982.
Perf.: $1150   **Exc.:** $1000   **VGood:** $900

FRANCHI MODEL 2004 TRAP
Single barrel; boxlock; 12-ga.; 32″, 34″, Full choke barrel; high vent rib; checkered European walnut straight or Monte Carlo stock, beavertail forearm; single selective trigger; auto ejectors; recoil pad; marketed with carrying case. Introduced 1976; discontinued 1982.
Perf.: $1100   **Exc.:** $950   **VGood:** $825

FRANCHI MODEL 2005 COMBINATION TRAP
Over/under; boxlock; 12-ga.; incorporates barrels of Model 2003, single barrel of Model 2004 as interchangeable set; high vent rib; checkered European walnut straight or Monte Carlo stock, beavertail forearm; single selective trigger; auto ejectors; recoil pad. Introduced 1976; discontinued 1982.
Perf.: $1675   **Exc.:** $1500   **VGood:** $1250

Franchi Model 2005/3 Combination Trap
Same specs as Model 2005 Combination Trap except three-barrel set incorporating over/under barrels of Model 2003, single barrel of Model 2004. Introduced 1976; discontinued 1982.
Perf.: $2250   **Exc.:** $1900   **VGood:** $1650

FRANCHI MODEL 3000/2
Over/under; 12-ga.; 30″, 32″ over and 32″, 34″ under barrel; 3 interchangeable choke tubes; high vent rib; checkered European walnut straight or Monte Carlo stock, beavertail forearm; single selective trigger; auto ejectors; recoil pad. Marketed with fitted case. No longer in production.
Perf.: $2600   **Exc.:** $2200   **VGood:** $1950

FRANCHI PEREGRIN MODEL 400
Over/under; 12-ga.; 26 1/2″, 28″ vent-rib barrels; standard choke combos; checkered European walnut pistol-grip stock, forearm; steel receiver; auto ejectors; selective single trigger. Introduced 1975; no longer in production.
Perf.: $620   **Exc.:** $510   **VGood:** $450

FRANCHI PEREGRIN MODEL 451
Over/under; 12-ga.; 26 1/2″, 28″ vent-rib barrels; standard choke combos; checkered European walnut pistol-grip stock, forearm; alloy receiver; auto ejectors; selective single trigger. Introduced 1975; no longer in production.
Perf.: $600   **Exc.:** $525   **VGood:** $475

Franchi Side-By-Side Shotguns
Airone
Astore
Astore S
Astore II
Custom Sidelock
SAS-12

Francotte
Boxlock
Jubilee
Knockabout
Model 9/40E/38321

Franchi Airone

Franchi Astore S

Franchi Custom Sidelock Imperial Monte Carlo Extra

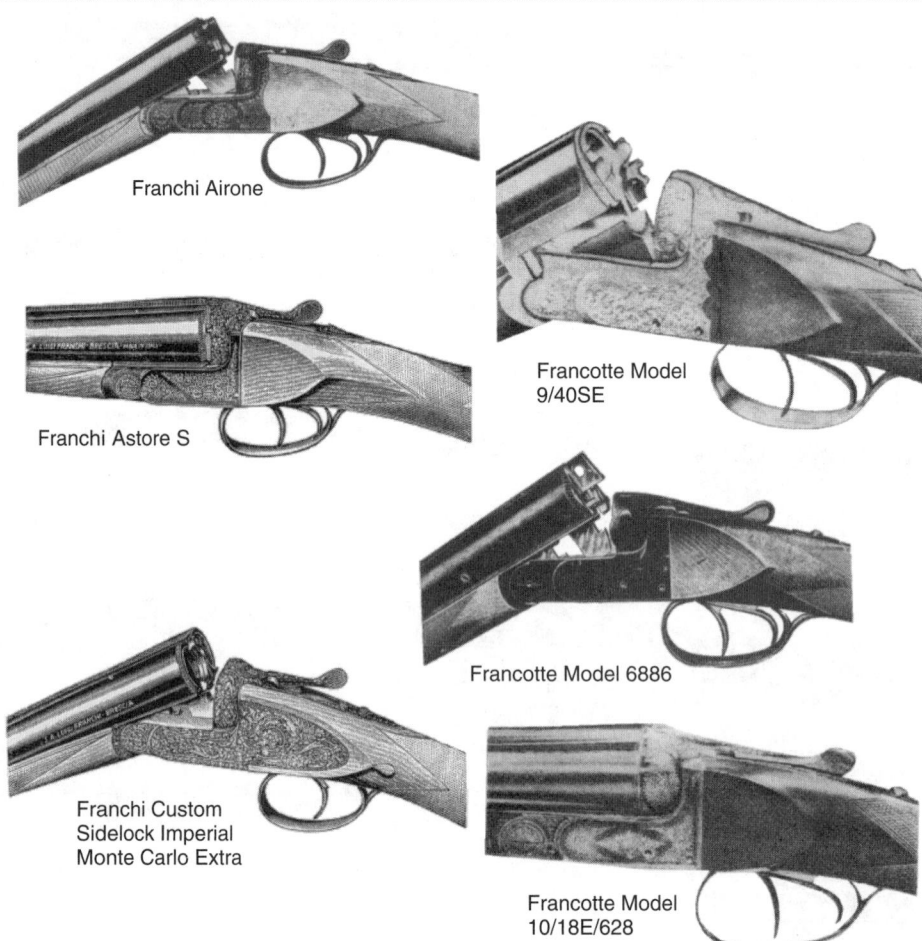

Francotte Model 9/40SE

Francotte Model 6886

Francotte Model 10/18E/628

FRANCHI SIDE-BY-SIDE SHOTGUNS

FRANCHI AIRONE
Side-by-side; boxlock; 12-ga.; all standard barrel lengths, choke combinations; hand-checkered European walnut straight-grip stock, forearm; automatic ejectors; double triggers. Introduced 1956; special order only; discontinued 1968.
Exc.: $900   **VGood:** $700   **Good:** $600

FRANCHI ASTORE
Side-by-side; boxlock; 12-ga.; all standard barrel lengths, choke combinations; hand-checkered European walnut straight-grip stock, forearm; plain extractors; double triggers. Introduced 1956; special order only; discontinued 1968.
Exc.: $700   **VGood:** $600   **Good:** $500

Franchi Astore S
Same specs as Astore except high-grade hand-checkered European walnut straight-grip stock, forearm; auto ejectors; fine engraving on frame. No longer in production.
Exc.: $1700   **VGood:** $1400   **Good:** $1250

Franchi Astore II
Same specs as Astore except 27", 28" barrels; standard chokes; European walnut pistol-grip stock; plain extractors or auto ejectors; engraving. No longer in production.
Exc.: $900   **VGood:** $750   **Good:** $600

FRANCHI CUSTOM SIDELOCK
Side-by-side; 12-, 16-, 20-, 28-ga.; barrel lengths, chokes custom-ordered; hand-checkered European walnut straight or pistol-grip stock, forearm; automatic ejectors; hand-detachable locks; self-opening action; single trigger optional. Made in six grades; variations depending upon quality of wood, amount of engraving, checkering, overall workmanship. For self-opening action, add $75; for single trigger, add $200. Introduced 1956, special order only; discontinued 1968.
Condor
  Exc.: $6500   **VGood:** $5700   **Good:** $4000
Imperial
  Exc.: $9000   **VGood:** $7500   **Good:** $6000
Imperial
  Exc.: $9200   **VGood:** $7900   **Good:** $6500
No. 5 Imperial Monte Carlo
  Exc.: $12,000 **VGood:** $9500   **Good:** $7500
No. 11 Imperial Monte Carlo
  Exc.: $12,500 **VGood:** $10,000 **Good:** $9000
Imperial Monte Carlo Extra
  Exc.: $15,000 **VGood:** $13,000 **Good:** $10,000

FRANCHI SAS-12
Slide action; 12-ga.; 3" chamber; 8-shot tube magazine; 21 1/2" barrel; synthetic pistol-grip stock; alloy receiver. Introduced 1988; discontinued 1990.
  Perf.: $405   **Exc.:** $350   **VGood:** $295

FRANCOTTE BOXLOCK
Side-by-side; boxlock; 12-, 16-, 20-, 28-ga., .410; 2 3/4", 3" chambers; 26" to 29" barrels; bead front sight; custom deluxe oil-finished European walnut straight or pistol-grip stock, splinter or beavertail forend; checkered butt; double locks, triggers; manual or auto safety; ejectors; English scroll engraving, coin finish or color case-hardening; many custom options. Made in Belgium. Introduced 1989; still imported by Armes de Chasse. **LMSR:** $16,500
  Perf.: $15,000 **Exc.:** $9000   **VGood:** $7000

FRANCOTTE JUBILEE
Side-by-side; boxlock with sideplates; 12-, 16-, 20- 28-ga., .410; 2 3/4", 3" chambers; customer-specified chokes, triggers and stock style; ejectors; imported in numbered grades 14, 18, 20, 25, 30 and No. 45 Eagle Grade.
No. 14
  Exc.: $1300   **VGood:** $1000   **Good:** $750
No. 18
  Exc.: $1850   **VGood:** $1600   **Good:** $1400
No. 20
  Exc.: $2400   **VGood:** $2025   **Good:** $1550
No. 25
  Exc.: $2550   **VGood:** $2000   **Good:** $1800
No. 30
  Exc.: $4200   **VGood:** $2700   **Good:** $2200
No. 45 Eagle Grade
  Exc.: $5000   **VGood:** $4000   **Good:** $2700

FRANCOTTE KNOCKABOUT
Side-by-side; boxlock; 12-, 16-, 20-, 28-ga., .410; 26" to 32" (12-ga.), 26" to 28" barrels in other gauges; any desired choke combo; crossbolt safety; double triggers; ejectors; series was made in several grades for distribution exclusively in this country by Abercrombie & Fitch; grades varied only in overall quality, grade of wood, checkering and engraving. Introduced prior to WWII; discontinued in 1960s. A&F applied the Knockabout name to a different gun imported from Italy.
  Exc.: $900   **VGood:** $750   **Good:** $600

FRANCOTTE MODEL 9/40E/38321
Side-by-side; Anson & Deeley boxlock; sideplates; 12-, 16-, 20-, 28-ga., .410; all standard barrel lengths, choke combos; hand-checkered European walnut stock, straight or pistol-grip design; reinforced frame, side clips; Purdey-type bolt; automatic ejectors; fine English engraving. Introduced about 1950; no longer in production.
  Perf.: $4200   **Exc.:** $3000   **VGood:** $2200

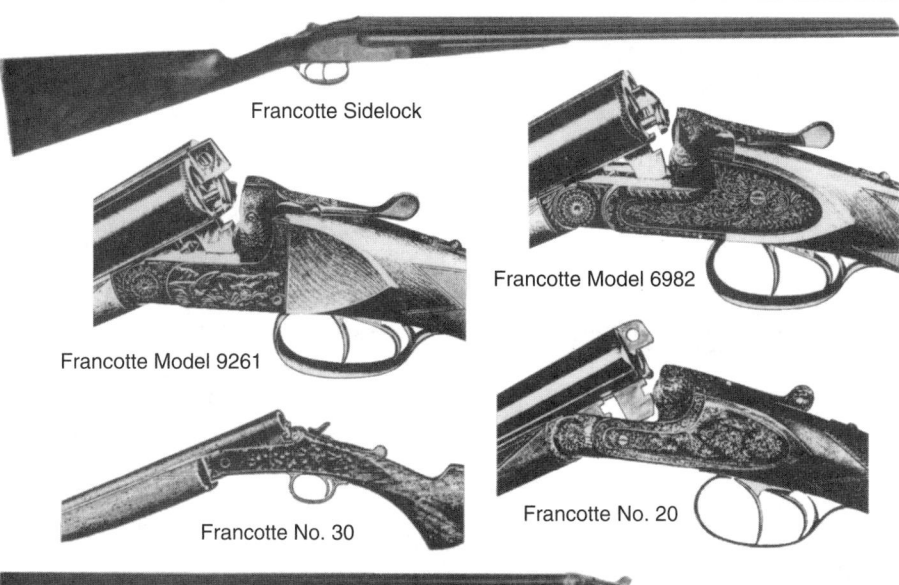

Francotte Sidelock

Francotte Model 9261

Francotte Model 6982

Francotte No. 30

Francotte No. 20

Francotte Boxlock

Francotte Model 8446

SHOTGUNS

Francotte
Model 9/40SE
Model 10/18E/628
Model 11/18E
Model 120.HE/328
Model 4996
Model 6886
Model 6930
Model 6982
Model 8446
Model 8455
Model 8457
Model 9261
Model 10594
Model SOB.E/11082
Sidelock

FRANCOTTE MODEL 9/40SE
Over/under; Anson & Deeley boxlock; 12-, 16-, 20-, 28-ga., .410; all standard barrel lengths choke combos; hand-checkered European walnut straight or pistol-grip stock, forearm; automatic ejectors; double triggers; elaborate engraving. Introduced about 1950; no longer in production.
Perf.: $8700 **Exc.:** $6000 **VGood:** $4800

FRANCOTTE MODEL 10/18E/628
Side-by-side; Anson & Deeley boxlock; 12-, 16-, 20-, 28-ga., .410; all standard barrel lengths, choke combos; hand-checkered European walnut straight or pistol-grip stock, forearm; Purdey-type bolt; Greener crossbolt; side clips; automatic ejectors; double triggers. Manufactured before World War II.
Perf.: $3000 **Exc.:** $2100 **VGood:** $1650

FRANCOTTE MODEL 11/18E
Side-by-side; Anson & Deeley boxlock; 12-, 16-, 20-, 28-ga., .410; all standard barrel lengths, choke combos; hand-checkered European walnut straight or pistol-grip stock, forearm; Purdey-type bolt; Greener crossbolt; side clips; automatic ejectors; double triggers. Manufactured before World War II.
Perf.: $3000 **Exc.:** $2100 **VGood:** $1650

FRANCOTTE MODEL 120.HE/328
Side-by-side; sidelock; made in all standard gauges; chokes, barrel lengths to customer's order; hand-checkered European walnut straight or pistol-grip stock, forearm; automatic ejectors; double triggers. Introduced about 1950; no longer in production.
Perf.: $8100 **Exc.:** $5700 **VGood:** $4450

FRANCOTTE MODEL 4996
Side-by-side; Anson & Deeley boxlock; 12-, 16-, 20-, 28-ga., .410; all standard barrel lengths, choke combinations available; hand-checkered European walnut straight or pistol-grip stock, forearm; Greener crossbolt; side clips; automatic ejectors; double triggers. Manufactured before World War II.
Perf.: $2750 **Exc.:** $1900 **VGood:** $1500

FRANCOTTE MODEL 6886
Side-by-side; boxlock; 12-, 16-, 20-, 28-ga., .410; all standard barrel lengths, choke combinations available; hand-checkered European walnut straight or pistol-grip stock, forearm; automatic ejectors; double triggers. Manufactured before World War II.
Perf.: $2250 **Exc.:** $1600 **VGood:** $1250

FRANCOTTE MODEL 6930
Side-by-side; boxlock; 12-, 16-, 20-, 28-ga., .410; all standard barrel lengths, choke combos; hand-checkered European walnut straight or pistol-grip stock, forearm; square crossbolt; automatic ejectors; double triggers. Manufactured before World War II.
Perf.: $2350 **Exc.:** $1850 **VGood:** $1300

FRANCOTTE MODEL 6982
Side-by-side; Anson & Deeley boxlock; sideplates; 12-, 16-, 20-, 28-ga., .410; all standard barrel lengths, choke combos; hand-checkered European walnut stock, straight or pistol-grip design; reinforced frame, side clips; Purdey-type bolt; automatic ejectors. Introduced about 1950; no longer in production.
Perf.: $3125 **Exc.:** $2200 **VGood:** $1700

FRANCOTTE MODEL 8446
Side-by-side; boxlock; 12-, 16-, 20-, 28-ga., .410; all standard barrel lengths, choke combinations available; hand-checkered European walnut straight or pistol-grip stock, forearm; Greener crossbolt; side clips; automatic ejectors; double triggers. Manufactured before World War II.
Perf.: $2700 **Exc.:** $1900 **VGood:** $1500

FRANCOTTE MODEL 8455
Side-by-side; Anson & Deeley boxlock; sideplates; 12-, 16-, 20-, 28-ga., .410; all standard barrel lengths, choke combos; hand-checkered European walnut straight or pistol-grip design; reinforced frame, side clips; Greener crossbolt; Purdey-type bolt; automatic ejectors; engraving. Introduced about 1950; no longer in production.
Perf.: $3100 **Exc.:** $2200 **VGood:** $1700

FRANCOTTE MODEL 8457
Side-by-side; Anson & Deeley boxlock; 12-, 16-, 20-, 28-ga., .410; all standard barrel lengths, choke combos; hand-checkered European walnut straight or pistol-grip stock, forearm; Greener-Scott crossbolt; side clips; automatic ejectors; double triggers. Manufactured before World War II.
Perf.: $3000 **Exc.:** $2100 **VGood:** $1700

FRANCOTTE MODEL 9261
Side-by-side; Anson & Deeley boxlock; 12-, 16-, 20-, 28-ga., .410; all standard barrel lengths, choke combos; hand-checkered European walnut straight or pistol-grip stock, forearm; Greener crossbolt; side clips; automatic ejectors; double triggers. Manufactured before World War II.
Perf.: $2350 **Exc.:** $1650 **VGood:** $1300

FRANCOTTE MODEL 10594
Side-by-side; Anson & Deeley boxlock; sideplates; 12-, 16-, 20-, 28-ga., .410; all standard barrel lengths, choke combos; hand-checkered European walnut stock, straight or pistol-grip design; reinforced frame, side clips; Purdey-type bolt; automatic ejectors. Introduced about 1950; no longer in production.
Perf.: $3000 **Exc.:** $2100 **VGood:** $1650

FRANCOTTE MODEL SOB.E/11082
Over/under; Anson & Deeley boxlock; 12-, 16-, 20-, 28-ga., .410; all standard barrel lengths, choke combos; hand-checkered European walnut straight or pistol-grip stock, forearm; automatic ejectors; double triggers; engraving. Introduced about 1950; no longer in production.
Perf.: $6300 **Exc.:** $4400 **VGood:** $3500

FRANCOTTE SIDELOCK
Side-by-side; sidelock; 12-, 16-, 20-, 28-ga., .410; 2 3/4", 3" chamber; 26", 29" barrel; custom chokes; bead front sight; deluxe oil-finished European walnut straight or pistol-grip stock, splinter or beavertail forend; checkered butt; double intercepting sears; double triggers; ejectors; English scroll engraving, coin finish or color case-hardening; numerous custom options. Made in Belgium. Introduced 1989; still imported by Armes de Chasse. **LMSR: $20,000-25,000**
Perf.: $24,000 **Exc.:** $20,000 **VGood:** $17,500

Galef
Companion
Monte Carlo Trap

Galef/Zabala

Galef/Zoli
Golden Snipe
Silver Hawk
Silver Snipe
Silver Snipe Trap
Silver Snipe Skeet

Gamba
2100
Ambassador
Ambassador Executive
Daytona
Daytona SL
Daytona SL HH
Daytona Competition

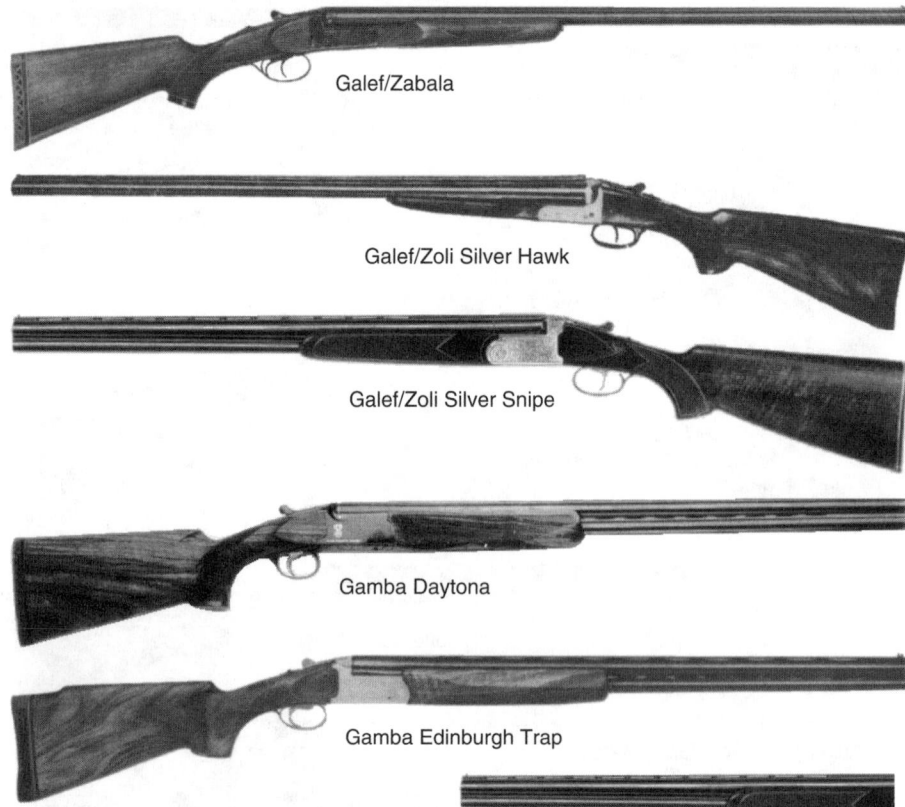

Galef/Zabala

Galef/Zoli Silver Hawk

Galef/Zoli Silver Snipe

Gamba Daytona

Gamba Edinburgh Trap

Galef Companion

GALEF COMPANION
Single shot; 12-, 16-, 20-, 28-ga., .410; 26″, 28″, 30″ barrel; plain or vent rib; European walnut pistol-grip stock, forearm; special folding feature. Made in Italy. Introduced 1968; no longer imported.
Exc.: $100   **VGood:** $75   **Good:** $50

GALEF MONTE CARLO TRAP
Single shot; underlever; 12-ga.; 32″ barrel; Full choke; vent rib; checkered European walnut pistol-grip stock, beavertail forearm; extractor; recoil pad. Made in Italy. Introduced 1968; no longer imported.
Exc.: $180   **VGood:** $140   **Good:** $110

GALEF/ZABALA
Side-by side; boxlock; 10-, 12-, 16-, 20-, 28-ga., .410; 22″ (12-ga.) Improved/Modified, 26″ (12-, 20-, 28-ga.) Improved/Modified or (.410) Modified/Full, 28″ (10-, 12-, 16-, 20-, 28-ga.) Modified/Full, 30″ (12-ga.) Modified/Full, 32″ (10-, 12-ga.) Full/Full barrels; hand-checkered European walnut stock, beavertail forearm; recoil pad; automatic safety; extractors. Made in Spain; no longer in production.
Standard gauges
Exc.: $160   **VGood:** $130   **Good:** $100
.410-bore
Exc.: $170   **VGood:** $140   **Good:** $110

GALEF/ZOLI GOLDEN SNIPE
Over/under; boxlock; 12-, 20-ga.; 3″ chambers; 26″ Improved/Modified, 28″ Modified/Full, 30″ Modified/Full (12-ga. only) barrels; vent rib; hand-checkered European walnut pistol-grip stock, forearm; crossbolt safety; automatic safety; automatic selective ejectors; single trigger; chrome-lined barrels. Made in Italy. No longer imported.
Field model
Exc.: $400   **VGood:** $300   **Good:** $225
Trap/Skeet models
Exc.: $380   **VGood:** $280   **Good:** $210

GALEF/ZOLI SILVER HAWK
Side-by-side; boxlock; 12-, 20-ga.; 3″ chambers; 26″, 28″, 30″ barrels; Improved/Modified, Modified/Full chokes; hand-checkered European walnut pistol-grip stock. Made in Italy. Introduced 1967; discontinued 1972.
Exc.: $365   **VGood:** $250   **Good:** $195

GALEF/ZOLI SILVER SNIPE
Over/under; boxlock; 12-, 20-ga.; 3″ chambers; 26″ Improved/Modified, 28″ Modified/Full, 30″ Modified/Full (12-ga. only) barrels; vent rib; hand-checkered European walnut pistol-grip stock, forearm; crossbolt safety; automatic safety; single trigger; chrome-lined barrels. Made in Italy. No longer imported.
Exc.: $320   **VGood:** $260   **Good:** $200

Galef/Zoli Silver Snipe Trap
Same specs as Silver Snipe except 30″ Full/Full barrel; non-automatic safety. No longer imported.
Exc.: $375   **VGood:** $295   **Good:** $200

Galef/Zoli Silver Snipe Skeet
Same specs as Silver Snipe except 26″ Skeet/Skeet barrels; non-automatic safety. No longer imported.
Exc.: $375   **VGood:** $295   **Good:** $200

GAMBA 2100
Slide action; 12-ga.; 7-shot magazine; 20″ barrel; weighs 7 lbs.; European walnut stock; flat black finish; designed for law enforcement. No longer imported.
Perf.: $590   **Exc.:** $375   **VGood:** $300

GAMBA AMBASSADOR
Side-by-side; sidelock; 12-, 20-ga.; 2 3/4″ chamber; 27 1/2″ barrels; choked to customer's specs; checkered straight English-style root walnut stock; double or single trigger; ejectors; inlaid gold shield in forend; engraved, gold-inlaid blued barrels. Made in Italy. Introduced 1987; discontinued 1987.
Perf.: $24,000 **Exc.:** $17,500 **VGood:** $14,000

Gamba Ambassador Executive
Same specs as Ambassador except custom built to customer's specs; has better wood; extensive signed engraving. Made in Italy. Introduced 1987; discontinued 1987.
Perf.: $30,000 **Exc.:** $20,000 **VGood:** $16,500

GAMBA DAYTONA
Over/under; boxlock; 12-, 20-ga.; 2 3/4″ chambers; 26 3/4″, 28″, 29″, 32″ barrel; weighs 5 1/2-8 1/2 lbs.; European walnut stock in Monte Carlo, traditional, Skeet/Sporting Clays or hunting configuration; interchangeable Schnabel, half or full beavertail forend; ejectors; single selective, adjustable or release trigger available. Made in Italy. Introduced 1990; no longer imported.
Perf.: $4750   **Exc.:** $3250   **VGood:** $2700

Gamba Daytona SL
Same specs as Daytona except sidelock; cased. No longer imported.
Perf.: $10,000 **Exc.:** $7500   **VGood:** $6000

Gamba Daytona SL HH
Same specs as Daytona except hand-detachable sidelocks; cased. No longer imported.
Perf.: $27,500 **Exc.:** $20,000 **VGood:** $17,500

GAMBA DAYTONA COMPETITION
Over/under; boxlock; 12-, 20-ga.; 2 3/4″, 3″ chambers; 26 3/4″ Skeet/Skeet, 28″ Skeet/Skeet, 28″ Modified/Full (SC model), 30″ Improved Modified/Full (SC model), 30″ Improved Modified/Full (trap), 32″ Modified/Full (trap), 32″ Full (monotrap) barrels; hand-checkered select walnut stock with oil finish; detachable trigger mechanism; Boss-type locking system; automatic ejectors; anatomical single trigger; optional single selective, release or adjustable trigger; black or chrome frame. Made in Italy. Introduced 1990; still imported. **LMSR: $5650**
Trap, 12- or 20-ga.
Perf.: $5200   **Exc.:** $3200   **VGood:** $2750
Skeet, 12-ga.
Perf.: $5250   **Exc.:** $3750   **VGood:** $3300
Pigeon, 12-ga.
Perf.: $5500   **Exc.:** $3500   **VGood:** $3250
Sporting, 12-ga.
Perf.: $5800   **Exc.:** $3700   **VGood:** $3200
American Trap, 12-ga.
Perf.: $5950   **Exc.:** $4400   **VGood:** $3850

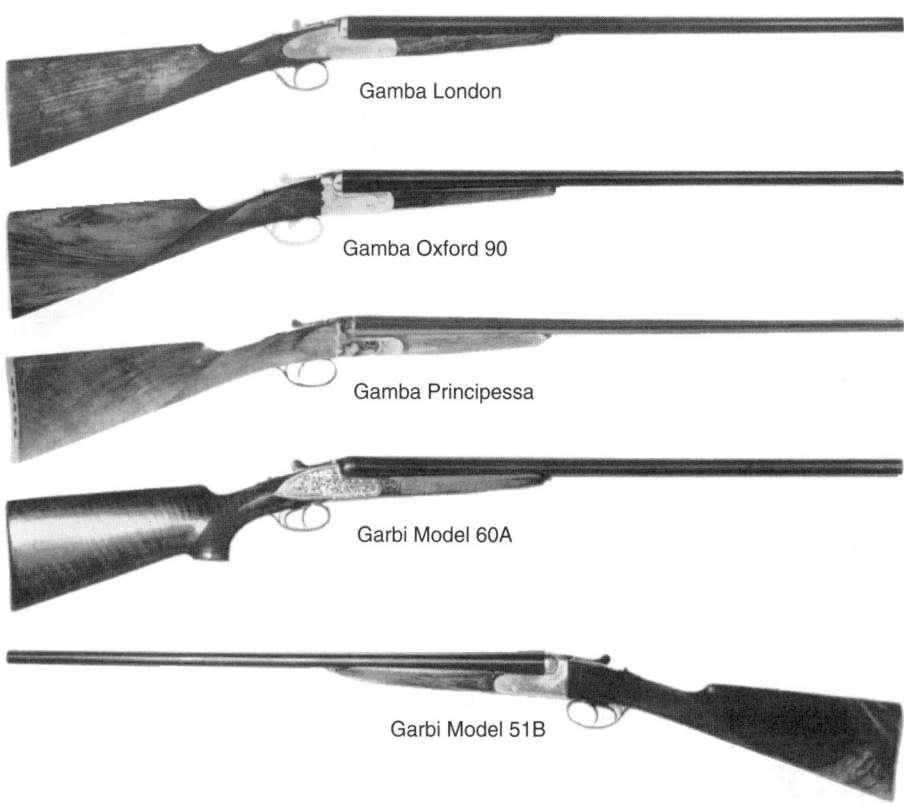

Gamba London

Gamba Oxford 90

Gamba Principessa

Garbi Model 60A

Garbi Model 51B

SHOTGUNS

Gamba
Edinburgh Skeet
Edinburgh Trap
Europa 2000
Grifone
Grifone Skeet
Grifone Trap
Hunter Super
London
LS2000
Milano
Oxford 90
Principessa
Victory Skeet
Victory Trap

Garbi
Model 51A
Model 51B
Model 60A
Model 60B

GAMBA EDINBURGH SKEET
Over/under; 12-ga.; 2 3/4″ chambers; 26 1/2″ Skeet/Skeet barrels; European walnut stock with high-gloss lacquer finish; recoil pad; chrome-lined barrels; double vent ribs; single trigger; auto ejectors; scroll border engraving; silver-finished receiver. Made in Italy. Introduced 1980; no longer imported.
    Perf.: $1450   **Exc.:** $1050   **VGood:** $950

Gamba Edinburgh Trap
Same specs as Edinburgh except 30″, 32″, 34″ barrels; oil-finished walnut Monte Carlo stock. No longer imported.
    Perf.: $1450   **Exc.:** $1050   **VGood:** $950

GAMBA EUROPA 2000
Over/under; boxlock; 12-ga.; 2 3/4″ chamber; 27″ Improved/Modified or Modified/Full barrels; oil-finished, select walnut stock; dummy sideplates; ejectors; anti-recoil buttplate; floral, English engraving patterns. Made in Italy. Introduced 1987; discontinued 1987.
    Perf.: $1150   **Exc.:** $800   **VGood:** $790

GAMBA GRIFONE
Over/under; boxlock; 12-, 20-ga.; 2 3/4″ chambers; 27″ Improved/Modified, Modified/Full barrels; weighs 7 lbs.; checkered European walnut stock; ejectors; double or single triggers; silvered receiver engraved game, floral scenes. Made in Italy. Introduced 1987; discontinued 1987.
    Perf.: $750   **Exc.:** $600   **VGood:** $500

GAMBA GRINTA SKEET
Over/under; boxlock; 12-ga.; 2 3/4″ chambers; 27″ Skeet/Skeet barrels; weighs 8 lbs.; checkered European walnut Skeet-style stock; ejectors; single selective trigger. No longer imported.
    Perf.: $1400   **Exc.:** $900   **VGood:** $750

Gamba Grinta Trap
Same specs as Grinta Skeet except 29″ Modified/Full barrels; checkered European walnut trap-style stock. No longer imported.
    Perf.: $1400   **Exc.:** $900   **VGood:** $750

GAMBA HUNTER SUPER
Side-by-side; boxlock; 12-ga.; 2 3/4″ chambers; 26 3/4″ Modified/Full, 28″ Improved/Modified barrels; weighs 6 3/4 lbs.; checkered European walnut stock; double triggers; extractors; silvered, engraved receiver. Made in Italy. Introduced 1987; importation by Armes de Chasse discontinued 1987.
    Perf.: $1300   **Exc.:** $840   **VGood:** $700

GAMBA LONDON
Side-by-side; sidelock; 12-, 20-ga.; 2 3/4″ chambers; 26 3/4″, 27 1/2″ barrels; weighs 6 3/4 lbs.; checkered straight English-style root walnut stock; double sear safety; ejectors; double triggers; silvered receiver; fine English-type scroll, rosette engraving. Made in Italy. Originally marketed by Renato Gamba. Introduced 1981; discontinued 1987.
    Perf.: $8800   **Exc.:** $8500   **VGood:** $7000

GAMBA LS2000
Over/under; boxlock; 12-, 20-, 28- ga.; .410; 27 1/2″ Modified/Full barrels; weighs 6 1/4 lbs.; European walnut stock; single trigger extractors; game scene engraving; folding design. Made in Italy. Introduced 1987; discontinued 1987.
    Perf.: $500   **Exc.:** $300   **VGood:** $250

GAMBA MILANO
Single shot; 12-, 16-, 20-, 28-, 32-ga.; .410; 27 1/2″ Full choke barrel; vent rib; weighs 5 3/4 lbs.; checkered European hardwood stock; chromed, photo-engraved receiver; folding design. Made in Italy. Introduced 1987; discontinued 1987.
    Perf.: $300   **Exc.:** $210   **VGood:** $175

GAMBA OXFORD 90
Side-by-side; boxlock with sideplates; 12-, 20-ga.; 26 3/4″, 27 1/2″ barrels; straight English or pistol-grip root walnut stock; double or single trigger auto ejectors; better grade of engraving. Made in Italy. Introduced 1981; discontinued 1987; reintroduced by Heckler & Koch 1990; discontinued 1991.
    Perf.: $3650   **Exc.:** $2700   **VGood:** $2300

GAMBA PRINCIPESSA
Side-by-side; 12-, 20-ga.; 2 3/4″ chambers; straight English-style walnut stock; double or single trigger; ejectors; fine English-style scroll engraving; silvered receiver. Made in Italy. Introduced 1981; discontinued 1987; reintroduced by Heckler & Koch 1990; discontinued 1991.
    Perf.: $1650   **Exc.:** $1350   **VGood:** $1100

GAMBA VICTORY SKEET
Over/under; boxlock; 12-ga.; 2 3/4″ chambers; 27″ Skeet/Skeet barrels; recessed chokes; weighs 7 3/4 lbs.; stippled European walnut Skeet-style stock; removable rubber recoil pad; ejectors; single trigger; blued receiver; light engraving; gold nameplate. Made in Italy. Introduced 1987; discontinued 1987.
    Perf.: $1450   **Exc.:** $1050   **VGood:** $900

Gamba Victory Trap
Same specs as Victory Skeet except 29″ Modified/Full barrels; stippled European walnut trap-stlye stock. Made in Italy. Introduced 1987; discontinued 1987.
    Perf.: $1450   **Exc.:** $1050   **VGood:** $900

GARBI MODEL 51A
Side-by-side; boxlock; 12-ga.; 2 3/4″ chambers; 28″ Modified/Full barrels; hand-checkered European walnut pistol-grip stock, forend; double triggers; extractors; hand-engraved receiver. Made in Spain. Introduced 1980; no longer imported.
    Perf.: $475   **Exc.:** $340   **VGood:** $310

Garbi Model 51B
Same specs as Model 51A except 12-, 16-, 20-ga.; automatic ejectors. Introduced 1980; discontinued 1988.
    Perf.: $800   **Exc.:** $540   **VGood:** $500

GARBI MODEL 60A
Side-by-side; sidelock; 12-ga.; 2 3/4″ chambers; 26″, 28″, 30″ barrel; custom chokes; hand-checkered select European walnut custom pistol-grip stock, forend; double triggers; extractors; scroll-engraved receiver. Made in Spain. Introduced 1981; discontinued 1988.
    Perf.: $735   **Exc.:** $550   **VGood:** $495

Garbi Model 60B
Same specs as 60A except 12-, 16-, 20-ga.; ejectors; demi-block barrels. Introduced 1981; no longer imported.
    Perf.: $1150   **Exc.:** $800   **VGood:** $700

Garbi
Model 62A
Model 62B
Model 71
Model 100
Model 101
Model 102
Model 103A
Model 103B
Model 110
Model 200
Model 300
Special Model

Garcia
Bronco
Bronco 22/410

GIB
Magnum (See Ugartecha 10-Gauge Magnum)

Golden Eagle
Model 5000 Grade I Field
Model 5000 Grade II Field

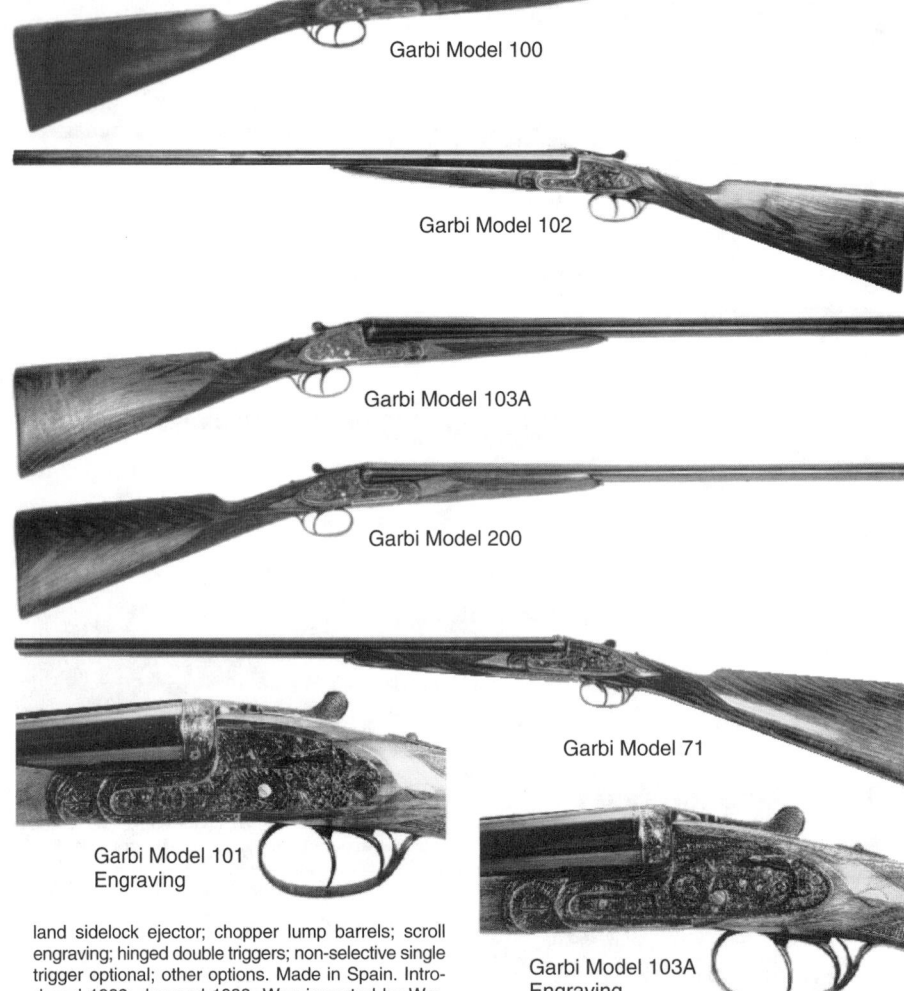

Garbi Model 100

Garbi Model 102

Garbi Model 103A

Garbi Model 200

Garbi Model 71

Garbi Model 101 Engraving

Garbi Model 103A Engraving

GARBI MODEL 62A
Side-by-side; sidelock; 12-ga.; 2 3/4″ chamber; 26″-30″ barrels; Modified/Full chokes; demi-block barrels; gas exhaust valves; extractors; jointed double triggers; plain receiver with engraved border. Made in Spain. Introduced 1981; discontinued 1988.
Perf.: $695   **Exc.:** $550   **VGood:** $490

Garbi Model 62B
Same specs as 62A except 12-, 16-, 20-ga.; ejectors. Introduced 1982; discontinued 1988.
Perf.: $1100   **Exc.:** $740   **VGood:** $700

GARBI MODEL 71
Side-by-side; hand-detachable sidelock; 12-, 16-, 20-ga.; 26″, 28″ barrels; oil-finished checkered European walnut straight-grip stock, classic forend; ejectors; double triggers; color case-hardened action. Made in Spain. Introduced 1980; discontinued 1988.
Perf.: $2100   **Exc.:** $1450   **VGood:** $1300

GARBI MODEL 100
Side-by-side; hand-detachable sidelock; 12-, 16-, 20-ga.; 26″, 28″ custom-choked barrels; checkered European walnut straight-grip stock, classic forend; ejectors; double triggers; forged barrel lumps; color case-hardened action; numerous options. Made in Spain. Introduced 1986; still imported. **LMSR: $4100**
Perf.: $3500   **Exc.:** $2200   **VGood:** $1700

GARBI MODEL 101
Side-by-side; hand-detachable sidelock; 12-, 16-, 20-ga.; file-cut Churchill or vent top rib; select European walnut stock; ejectors; Continental-style floral, scroll engraving; 12-ga. pigeon or wildfowl designs available; better quality than Model 71. **LMSR: $5250**
Perf.: $4700   **Exc.:** $2900   **VGood:** $2400

GARBI MODEL 102
Side-by-side; 12-, 16-, 20-, 28-ga.; 25″ to 30″ barrels; standard chokes; select European walnut stock; Holland sidelock ejector; chopper lump barrels; scroll engraving; hinged double triggers; non-selective single trigger optional; other options. Made in Spain. Introduced 1982; dropped 1988. Was imported by Wm. Larkin Moore, L. Joseph Rahn.
Perf.: $4500   **Exc.:** $4000   **VGood:** $3000

GARBI MODEL 103A
Side-by-side; 12-, 16-, 20-ga.; 25″ to 30″ barrels; chopper lump barrels; select European walnut stock; ejectors; hinged double triggers; non-selective single trigger optional; Purdey-type scroll, rosette engraving. Made in Spain. Introduced 1982; still imported. **LMSR: $6550**
Perf.: $5600   **Exc.:** $3300   **VGood:** $2800

Garbi Model 103B
Same specs as 103 except Holland & Holland-type assisted-opening mechanism; nickel-chrome steel barrels. Introduced 1982; still imported. **LMSR: $9200**
Perf.: $7800   **Exc.:** $5000   **VGood:** $4000

GARBI MODEL 110
Side-by-side; sidelock; 12-, 20-, 28-ga.; barrel lengths, chokes, stocks to custom order. Made in Spain. Introduced 1982 by Toledo Armas; discontinued 1984.
Perf.: $7300   **Exc.:** $4500   **VGood:** $4000

GARBI MODEL 200
Side-by-side; sidelock; 12-, 16-, 20-, 28-ga.; 26″, 28″ chopper-lump, nickel-chrome steel barrels; finely figured European walnut stock; ejectors; double triggers; heavy-duty locks; Continental-style floral, scroll engraving. Made in Spain. Introduced 1982; still imported.
LMSR: $8800
Perf.: $7400   **Exc.:** $4500   **VGood:** $4000

GARBI MODEL 300
Side-by-side; sidelock; 12-, 20-, 28-ga.; barrel lengths, chokes, stocks to custom order; engraving. Made in Spain. No longer imported.
Perf.: $8900   **Exc.:** $6500   **VGood:** $5000

GARBI SPECIAL MODEL
Side-by-side; sidelock; 12-, 16-, 20-, 28-ga.; 26″, 28″ barrels; fancy European walnut stock; ejectors; game scene engraved in receiver some with gold inlays; top quality metalwork, wood. Made in Spain. Introduced 1982; discontinued 1988.
Perf.: $7600   **Exc.:** $5000   **VGood:** $4100

GARCIA BRONCO
Single shot; swing-out action; takedown; .410; 18 1/2″ barrel; weighs 3 1/2 lbs.; one-piece metal frame skeletonized stock, receiver; crinkle finish. Made in Italy. Introduced 1968; discontinued 1978.
Perf.: $110   **Exc.:** $90   **VGood:** $70

GARCIA BRONCO 22/410
Over/under combo; swing-out action; takedown; 22 LR over, .410 under 18 1/2″ barrel; one-piece metal frame stock, receiver; crinkle finish. Made in Italy. Introduced 1976; discontinued 1978.
Perf.: $125   **Exc.:** $100   **VGood:** $80

GOLDEN EAGLE MODEL 5000 GRADE I FIELD
Over/under; boxlock; 12-, 20-ga.; 2 3/4″, 3″ chambers; 26″, 28″, 30″ barrels; Improved/Modified, Modified/Full chokes; vent rib; checkered walnut pistol-grip stock, semi-beavertail forearm; selective single trigger; auto ejectors; engraved receiver; gold eagle head inlaid in frame. Made in Japan. Introduced 1975; discontinued 1980.
Perf.: $810   **Exc.:** $685   **VGood:** $600

Golden Eagle Model 5000 Grade II Field
Same specs as Model 5000 Grade I Field except more elaborate engraving; spread-wing eagle inlaid in gold; fancier wood. Introduced 1975; discontinued 1980.
Perf.: $920   **Exc.:** $780   **VGood:** $700

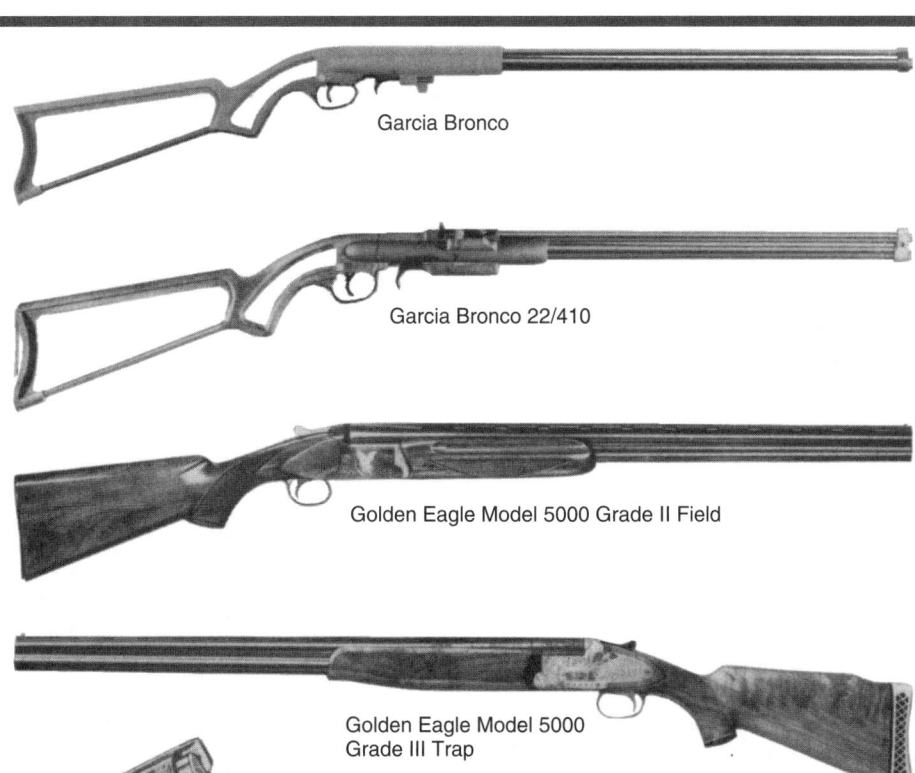

Garcia Bronco

Garcia Bronco 22/410

Golden Eagle Model 5000 Grade II Field

Golden Eagle Model 5000 Grade III Trap

Greener Crown Grade DH75

SHOTGUNS

Golden Eagle
Model 5000 Grade I Trap
Model 5000 Grade II Trap
Model 5000 Grade I Skeet
Model 5000 Grade II Skeet
Model 5000 Grade III

Gorosabel
501
502
503
Blackpoint
Silverpoint

Greener
Crown Grade DH75
Empire
Empire Deluxe

Golden Eagle Model 5000 Grade I Trap
Same specs as Model 5000 Grade I Field except 30″, 32″ barrels; Modified/Full, Improved Modified/Full, Full/Full choke; wide vent rib; trap-style stock; recoil pad. Introduced 1975; discontinued 1980.
Perf.: $825   **Exc.:** $700   **VGood:** $600

Golden Eagle Model 5000 Grade II Trap
Same specs as Model 5000 Grade I Field except 30″, 32″ barrels; Modified/Full, Improved Modified/Full, Full/Full choke; wide vent rib; vent side ribs; trap-style stock; recoil pad; more elaborate engraving; spread-wing eagle gold inlay; inertia trigger; fancier wood. Introduced 1975; discontinued 1980.
Perf.: $940   **Exc.:** $800   **VGood:** $700

Golden Eagle Model 5000 Grade I Skeet
Same specs as Model 5000 Grade I Field except 26″, 28″ Skeet/Skeet barrels; vent rib. Introduced 1975; discontinued 1980.
Perf.: $840   **Exc.:** $700   **VGood:** $640

Golden Eagle Model 5000 Grade II Skeet
Same specs as Model 5000 Grade I Field except 26″, 28″ Skeet/Skeet barrels; vent rib; vent side ribs; more elaborate engraving; spread-wing eagle gold inlay; inertia trigger; fancier wood. Introduced 1975; discontinued 1980.
Perf.: $925   **Exc.:** $800   **VGood:** $700

Golden Eagle Model 5000 Grade III
Same specs as Grade I Field except game scene engraving; scroll-engraved frame, barrels, sideplates; fancy wood; available in Field, Trap and Skeet versions; Trap model has Monte Carlo comb, full pistol grip, recoil pad. Introduced 1976; discontinued 1980.

Field version
Perf.: $2250   **Exc.:** $1900   **VGood:** $1600
Trap/Skeet versions
Perf.: $2200   **Exc.:** $1800   **VGood:** $1500

GOROSABEL 501
Side-by-side; boxlock; 12-, 16-, 20-, 28-ga., .410; 26″, 27″, 28″ barrels; standard choke combos; hand-checkered select walnut English or pistol-grip stock, splinter or beavertail forend; ejectors. Introduced 1986; discontinued 1988.
Perf.: $700   **Exc.:** $590   **VGood:** $500

GOROSABEL 502
Side-by-side; boxlock; 12-, 16-, 20-, 28-ga., .410; 26″, 27″, 28″ barrels; standard choke combos; hand-checkered English or pistol-grip select walnut stock, splinter or beavertail forend; ejectors; engraving. Introduced 1986; discontinued 1988.
Perf.: $890   **Exc.:** $790   **VGood:** $700

GOROSABEL 503
Side-by-side; boxlock; 12-, 16-, 20-, 28-ga., 410; 26″, 27″, 28″ barrels; standard choke combos; hand-checkered fancy walnut English or pistol-grip stock; splinter or beavertail forend; ejectors; scalloped frame; scroll engraving. Introduced 1986; discontinued 1988.
Perf.: $1000   **Exc.:** $850   **VGood:** $750

GOROSABEL BLACKPOINT
Side-by-side; sidelock; 12-, 20-ga.; 2 ¾″, 3″ chambers; 26″, 27″, 28″ barrels; standard choke combos; hand-checkered select European walnut stock, splinter or beavertail forend; Purdey-style scroll, rose engraving; numerous options. Made in Italy. Introduced 1986, discontinued 1988.
Perf.: $1550   **Exc.:** $1200   **VGood:** $1000

GOROSABEL SILVERPOINT
Side-by-side; sidelock; 12-, 20-ga.; 2 ¾″, 3″ chambers;

Greener Empire

26″, 27″, 28″ barrels; standard choke combos; hand-checkered European walnut stock, splinter or beaver-tail forend; large scroll engraving. Made in Italy. Introduced 1986, discontinued 1988.
Perf.: $1000   **Exc.:** $800   **VGood:** $700
Note: On all variations, add $175 for non-selective single trigger; $350 for selective trigger.

GREENER CROWN GRADE DH75
Side-by-side; boxlock; 12-, 16-, 20-ga.; 26″, 28″, 30″ barrels; any choke combo; top-quality hand-checkered European walnut straight or pistol-grip stock, forearm; Greener crossbolt; ejectors; double triggers; non-selective or selective single trigger at added cost; best-grade engraved action. Introduced 1875; discontinued 1965.
Exc.: $4200   **VGood:** $3975   **Good:** $3350

GREENER EMPIRE
Side-by-side; boxlock; 12-ga.; 28″, 30″, 32″ barrels; any choke combo; hand-checkered European walnut straight or half-pistol-grip stock, forearm; non-ejector; ejectors at additional cost. Introduced about 1893; discontinued 1962.
Exc.: $1600   **VGood:** $1100   **Good:** $900

Greener Empire Deluxe
Same specs as Empire except deluxe finish and better craftsmanship.
Exc.: $1800   **VGood:** $1300   **Good:** $1000

Greener
Far-Killer
General Purpose
Jubilee Grade DH35
Sovereign Grade DH40

Greifelt
Grade No. 1
Grade No. 3
Model 22
Model 22E
Model 103
Model 103E
Model 143E

Bill Hanus
Birdgun

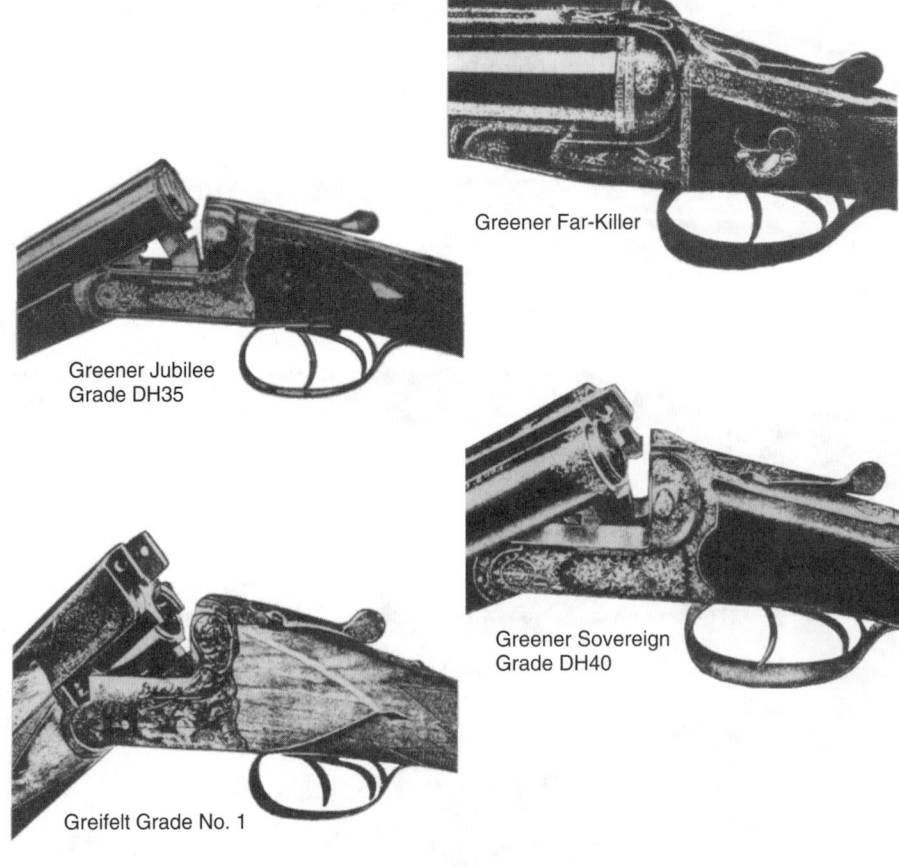

Greener Far-Killer

Greener Jubilee
Grade DH35

Greener Sovereign
Grade DH40

Greifelt Grade No. 1

Bill Hanus Birdgun

GREENER FAR-KILLER
Side-by-side; boxlock; 12-, 10-, 8-ga.; 28″, 30″, 32″ barrels; any desired choke combination; hand-checkered European walnut straight or half-pistol-grip stock; nonejectors; automatic ejectors at added cost. Introduced about 1895; discontinued 1962.

12-ga.
    Exc.: $2300   **VGood:** $1900   **Good:** $1650

10-, 8-ga.
    Exc.: $2500   **VGood:** $2100   **Good:** $1790

GREENER GENERAL PURPOSE
Single shot; Martini-type action; takedown; 12-ga.; 26″, 30″, 32″ barrel; Modified, Full choke; hand-checkered European walnut straight-grip stock, forearm; ejector. Introduced 1910; discontinued 1964.
    Exc.: $300   **VGood:** $225   **Good:** $200

GREENER JUBILEE GRADE DH35
Side-by-side; boxlock; 12-, 16-, 20-ga.; 26″, 28″, 30″ barrels; any choke combo; hand-checkered European walnut straight or pistol-grip stock, forearm; Greener crossbolt; ejectors; double triggers; non-selective or selective single trigger at added cost; engraved action. Introduced 1875; discontinued 1965.
    Exc.: $2400   **VGood:** $1700   **Good:** $1500

GREENER SOVEREIGN GRADE DH40
Side-by-side; boxlock; 12-, 16-, 20-ga.; 26″, 28″, 30″ barrels; any choke combo; hand-checkered select European walnut straight or pistol-grip stock, forearm; Greener crossbolt; ejectors; double triggers; non-selective or selective single trigger at added cost; engraved action. Introduced 1875; discontinued 1965.
    Exc.: $2500   **VGood:** $2000   **Good:** $1800

GREIFELT GRADE NO. 1
Over/under; boxlock; 12-, 16-, 20-, 28-ga., .410; 26″, 28″, 30″, 32″ barrels; any desired choke combo; solid matted rib standard; vent rib at added cost; hand-checkered European walnut straight or pistol-grip stock; ejectors; double triggers; single trigger at added cost; elaborate engraving. Made in Germany prior to World War II.

Solid rib
    Exc.: $2750   **VGood:** $2000   **Good:** $1650

Vent rib
    Exc.: $3000   **VGood:** $2250   **Good:** $1900

.410-bore
    Exc.: $3650   **VGood:** $2650   **Good:** $2400

GREIFELT GRADE NO. 3
Over/under; boxlock; 12-, 16-, 20-, 28-ga., .410; 26″, 28″, 30″, 32″ barrels; any desired choke combo; solid matted rib; vent rib at added cost; hand-checkered European walnut straight or pistol-grip stock; ejectors; double triggers; single trigger at added cost; engraving. Made in Germany prior to World War II.

Solid rib
    Exc.: $2150   **VGood:** $1600   **Good:** $1200

Vent rib
    Exc.: $2400   **VGood:** $1850   **Good:** $1450

.410-bore
    Exc.: $2500   **VGood:** $1900   **Good:** $1500

GREIFELT MODEL 22
Side-by-side; boxlock; 12-, 20-ga.; 28″, 30″ barrels; Modified/Full choke; hand-checkered straight English or pistol-grip stock, forend; cheekpiece; plain extractors; double triggers; sideplates. Introduced about 1950; no longer in production.
    Exc.: $1500   **VGood:** $1050   **Good:** $925

Greifelt Model 22E
Same specs as Model 22 except ejectors. Introduced about 1950; no longer in production.
    Exc.: $1900   **VGood:** $1450   **Good:** $1300

GREIFELT MODEL 103
Side-by-side; boxlock; 12-, 16-ga.; 28″, 30″ barrels; Modified/Full chokes; hand-checkered European walnut straight English or pistol-grip stock, forearm; extractors; double triggers. Introduced about 1950; no longer in production.
    Exc.: $1500   **VGood:** $1000   **Good:** $750

Greifelt Model 103E
Same specs as Model 103 except ejectors. Introduced about 1950; no longer in production.
    Exc.: $1600   **VGood:** $1050   **Good:** $800

GREIFELT MODEL 143E
Over/under; boxlock; 12-, 16-, 20-ga.; 26″, 28″, 30″ barrels; any desired choke; solid or vent rib; hand-checkered European walnut straight or pistol-grip stock; single or double trigger; engraved; post-WWII version of the Grade I but of lower quality. Introduced about 1950. No longer imported.
    Exc.: $1750   **VGood:** $1300   **Good:** $1100

BILL HANUS BIRDGUN
Side-by-side; boxlock; 16-, 20-, 28-ga.; 26″ Skeet/Skeet barrels; weighs 6 1/4 lbs. (12-ga.); hand-checkered straight-grip walnut stock, semi-beavertail forend; single non-selective trigger; raised Churchill rib; auto safety; ejectors; color case-hardened action. Made in Spain. Introduced 1991; still imported Bill Hanus Birdguns, LLC.
LMSR: $1200
    Perf.: $1000   **Exc.:** $850   **VGood:** $750

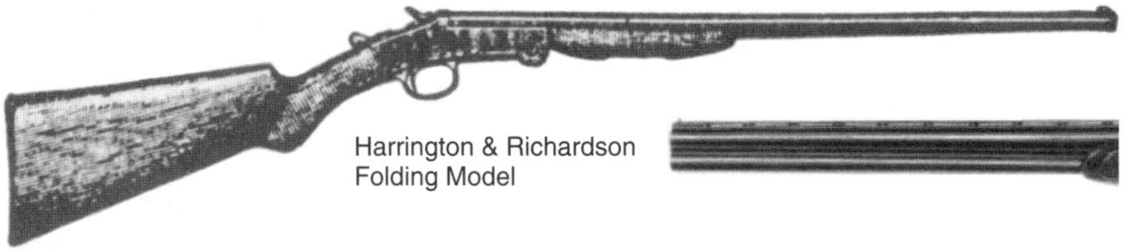

Harrington & Richardson
Folding Model

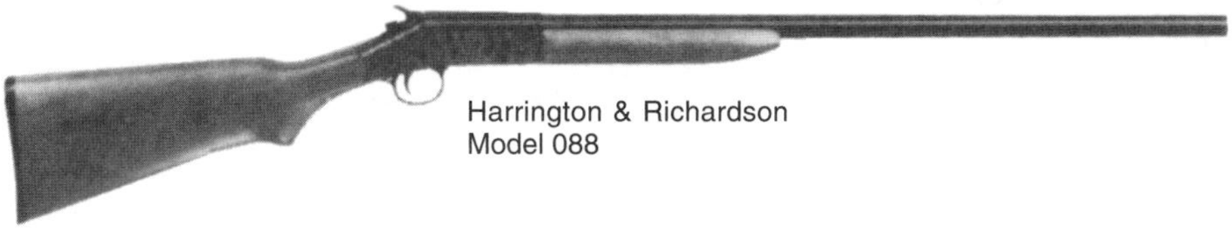

Harrington & Richardson
Model 088

Harrington & Richardson
Model 099 Deluxe

Harrington & Richardson
Model 162 Slug Gun

Harrington & Richardson
Model 176

SHOTGUNS

Harrington & Richardson
**Folding Model
Model 088
Model 088 Junior
Model 099 Deluxe
Model 159 Golden Squire
Model 162 Slug Gun
Model 176
Model 176 Slug Gun
Model 348 Gamester
Model 349 Gamester Deluxe**

HARRINGTON & RICHARDSON FOLDING MODEL

Single barrel; 12-, 16-, 20-, 28-ga. or 410 with 26″ barrel (Heavy Frame model); 28-ga., .410 with 22″ barrel (Light Frame model); Full choke; bead front sight; checkered pistol-grip stock, forearm; gun is hinged at front of frame so barrel folds against stock for storage, transport. Introduced about 1910; discontinued 1942.
Exc.: $150   **VGood:** $100   **Good:** $75

HARRINGTON & RICHARDSON MODEL 088

Single shot; sidelever; external hammer; takedown; 12-, 16-, 20-, 28-ga., .410; 26″, 28″, 30″, 32″, 36″ barrels; walnut-finished hardwood semi-pistol-grip stock; auto ejector; case-hardened frame. No longer in production.
Exc.: $75   **VGood:** $60   **Good:** $30

Harrington & Richardson Model 088 Junior

Same specs as Model 088 except 20-ga., .410; 25″ Modified (20-ga.) or Full (.410) choke barrel; stock dimensions for smaller shooters.
Exc.: $75   **VGood:** $60   **Good:** $30

HARRINGTON & RICHARDSON MODEL 099 DELUXE

Single shot; 12-, 16-, 20-ga., .410; 2 3/4″, 3″ chamber; 25″, 26″, 28″ barrel; bead front sight; walnut-finished hardwood semi-pistol-grip stock, semi-beavertail forend; electroless matte nickel finish. Introduced 1982; discontinued 1985.
Perf.: $85   **Exc.:** $60   **VGood:** $50

HARRINGTON & RICHARDSON MODEL 159 GOLDEN SQUIRE

Single barrel; exposed hammer; 12-, 20-ga.; 28″ (20-ga.), 30″ (12-ga.) barrel; Full choke; bead front sight; uncheckered hardwood straight-grip stock, Schnabel forearm. Introduced 1964; discontinued 1966.
Exc.: $75   **VGood:** $65   **Good:** $55

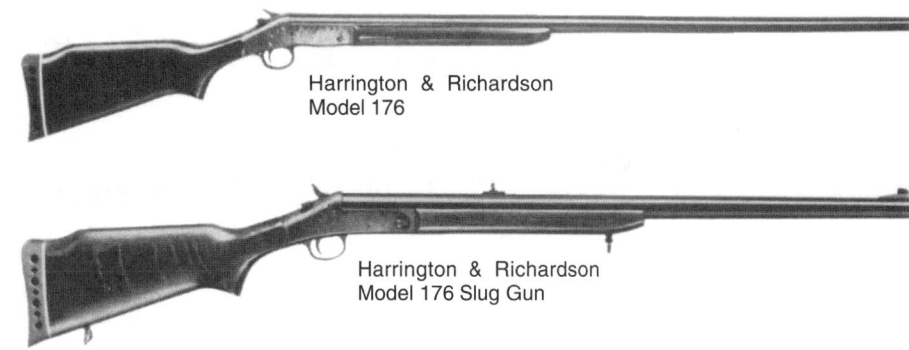

Harrington & Richardson
Model 176

Harrington & Richardson
Model 176 Slug Gun

HARRINGTON & RICHARDSON MODEL 162 SLUG GUN

Single barrel; sidelever breakopen; exposed hammer; takedown; 12-, 20-ga.; 3″ chamber; 24″ Cylinder choke barrel; rifle sights; uncheckered hardwood pistol-grip stock, forend; recoil pad. Introduced 1968; discontinued 1985.
Perf.: $105   **Exc.:** $85   **VGood:** $75

HARRINGTON & RICHARDSON MODEL 176

Single shot; sidelever breakopen; exposed hammer; takedown; 10-, 12-ga.; 3″, 3 1/2″ chamber; 32″, 36″ heavy barrel; Full choke; weighs 9 1/4 lbs.; bead front sight; uncheckered hardwood pistol-grip stock, special long forend; recoil pad. Introduced 1978; discontinued 1985.
Perf.: $110   **Exc.:** $85   **VGood:** $65

Harrington & Richardson Model 176 Slug Gun

Same specs as Model 176 except 10-ga. rifled slugs; 3

1/2″ chamber; 28″ Cylinder barrel; weighs 9 1/4 lbs.; ramp front sight, folding leaf rear; uncheckered hardwood pistol-grip stock, magnum-type forend; recoil pad; sling swivels. Introduced 1982; discontinued 1985.
Perf.: $125   **Exc.:** $95   **VGood:** $80

HARRINGTON & RICHARDSON MODEL 348 GAMESTER

Bolt action; takedown; 12-, 16-ga.; 2-shot tube magazine; 28″ barrel; Full choke; uncheckered hardwood pistol-grip stock, forearm. Introduced 1949; discontinued 1954.
Exc.: $95   **VGood:** $75   **Good:** $50

HARRINGTON & RICHARDSON MODEL 349 GAMESTER DELUXE

Bolt action; takedown; 12-, 16-ga.; 2-shot tube magazine; 26″ barrel; adjustable choke device; uncheckered hardwood pistol-grip stock, forearm; recoil pad. Introduced 1953; discontinued 1955.
Exc.: $100   **VGood:** $75   **Good:** $50

Harrington & Richardson

Model 351 Huntsman
Model 400
Model 401
Model 402
Model 403
Model 404
Model 404C
Model 440
Model 442
Model 459 Golden Squire Junior
Model 1212
Model 1212 Waterfowler
No. 1 Harrich
No. 3
No. 5
No. 6

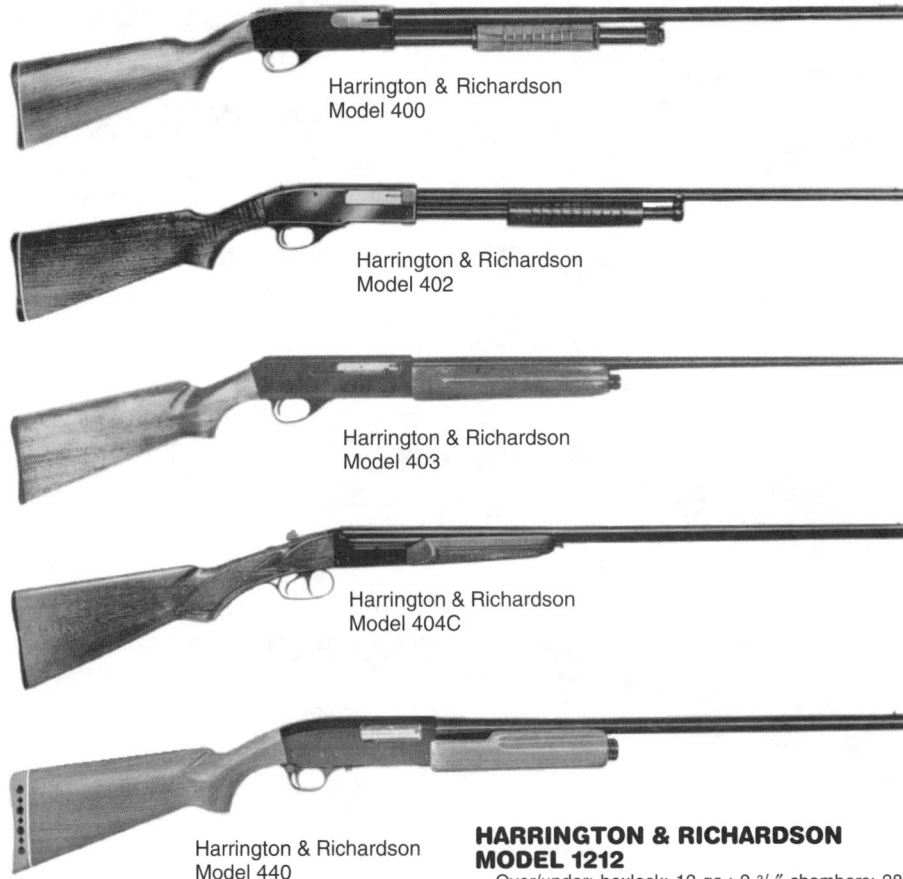

Harrington & Richardson
Model 400

Harrington & Richardson
Model 402

Harrington & Richardson
Model 403

Harrington & Richardson
Model 404C

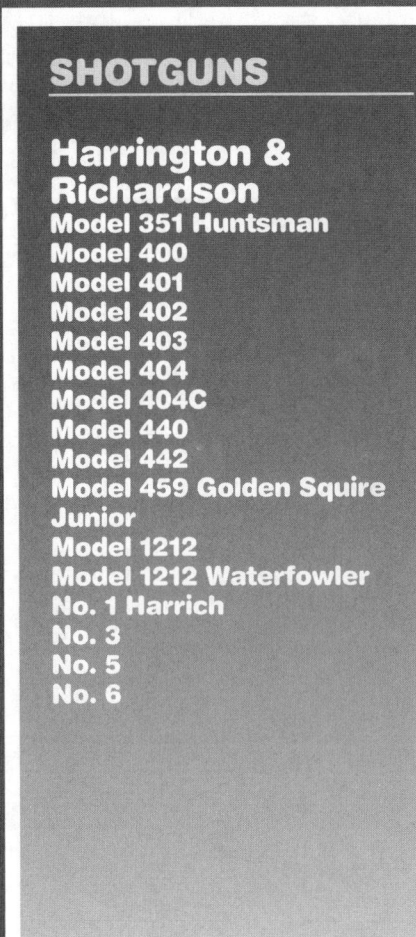

Harrington & Richardson
Model 440

HARRINGTON & RICHARDSON MODEL 351 HUNTSMAN

Bolt action; takedown; 12-, 16-ga.; 2-shot tube magazine; 26" barrel; adjustable choke device; uncheckered American hardwood Monte Carlo stock; recoil pad; push-button safety. Introduced 1956; discontinued 1958.

Exc.: $95 **VGood:** $75 **Good:** $45

HARRINGTON & RICHARDSON MODEL 400

Slide action; hammerless; 12-, 16-, 20-ga.; 5-shot tube magazine; 28" Full choke barrel; uncheckered pistol-grip stock, grooved slide handle; recoil pad (12-, 16-ga. only). Introduced 1955; discontinued 1967.

Exc.: $120 **VGood:** $85 **Good:** $70

HARRINGTON & RICHARDSON MODEL 401

Slide action; hammerless; 12-, 16-, 20-ga.; 5-shot tube magazine; 28" barrel; adjustable choke device; uncheckered pistol-grip stock, grooved slide handle; recoil pad (12-, 16-ga. only). Introduced 1956; discontinued 1963.

Exc.: $135 **VGood:** $95 **Good:** $70

HARRINGTON & RICHARDSON MODEL 402

Slide action; hammerless; .410; 5-shot tube magazine; 28" Full choke barrel; uncheckered pistol-grip stock, grooved slide handle; recoil pad (12-, 16-ga. only). Introduced 1959; discontinued 1963.

Exc.: $140 **VGood:** $100 **Good:** $90

HARRINGTON & RICHARDSON MODEL 403

Autoloader; takedown; .410; 4-shot tube magazine; 26" barrel; Full choke; uncheckered walnut pistol-grip stock, forearm. Introduced 1964; discontinued 1964.

Exc.: $175 **VGood:** $150 **Good:** $90

HARRINGTON & RICHARDSON MODEL 404

Side-by-side; boxlock; 12-, 20-ga., .410; 25" Full/Full (.410), 26" Modified/Improved (20-ga.), 28" Modified/Full (12-ga.) barrels; checkered hardwood pistol-grip stock, forearm; double triggers; plain extractors. Made to H&R specifications in Brazil. Introduced 1968; discontinued 1972.

Exc.: $160 **VGood:** $125 **Good:** $100

Harrington & Richardson Model 404C

Same specs as Model 404 except Monte Carlo stock. No longer in production.

Exc.: $165 **VGood:** $125 **Good:** $100

HARRINGTON & RICHARDSON MODEL 440

Slide action; hammerless; 12-, 16-, 20-ga.; 4-shot magazine; 24", 26", 28" barrel; standard chokes; uncheckered American walnut pistol-grip stock, forend; recoil pad; slide ejection. Introduced 1972; discontinued 1975.

Exc.: $150 **VGood:** $130 **Good:** $110

HARRINGTON & RICHARDSON MODEL 442

Slide action; hammerless; 12-, 16-, 20-ga.; 4-shot magazine; 24", 26", 28" barrel; standard chokes; vent rib; checkered American walnut pistol-grip stock, forend; recoil pad; slide ejection. No longer in production.

Exc.: $150 **VGood:** $130 **Good:** $100

HARRINGTON & RICHARDSON MODEL 459 GOLDEN SQUIRE JUNIOR

Single barrel; exposed hammer; 20-ga., .410; 26" barrel; Full choke; bead front sight; youth-sized uncheckered hardwood straight-grip stock, Schnabel forearm. Introduced 1964; discontinued 1964.

Exc.: $75 **VGood:** $65 **Good:** $55

HARRINGTON & RICHARDSON MODEL 1212

Over/under; boxlock; 12-ga.; 2 3/4" chambers; 28" barrels; Improved Cylinder/Improved Modified chokes; vent rib; checkered European walnut pistol-grip stock, fluted forearm; selective single trigger; plain extractors. Made in Spain. Introduced 1976; discontinued 1981.

Perf.: $300 **Exc.:** $270 **VGood:** $240

Harrington & Richardson Model 1212 Waterfowler

Same specs as Model 1212 except 3" chamber; 30" barrels; Modified/Full choke; recoil pad. Introduced 1976; discontinued 1981.

Perf.: $310 **Exc.:** $280 **VGood:** $250

HARRINGTON & RICHARDSON NO. 1 HARRICH

Single shot; 12-ga.; 32", 34" barrel; Full choke; high vent rib; hand-checkered European walnut Monte Carlo pistol-grip stock, beavertail forearm; Kersten top locks, double under lugs; recoil pad; trap model. Made in Austria. Introduced 1971; discontinued 1974.

Perf.: $1700 **Exc.:** $1550 **VGood:** $1400

HARRINGTON & RICHARDSON NO. 3

Single barrel; toplever breakopen; hammerless; takedown; 12-, 16-, 20-, 24-, 28-ga., .410; 26", 28", 30", 32" barrels; Full choke; bead front sight; uncheckered American walnut pistol-grip stock, forearm; automatic ejector. Introduced 1908; discontinued 1942.

Exc.: $75 **VGood:** $50 **Good:** $40

HARRINGTON & RICHARDSON NO. 5

Single barrel; toplever breakopen; exposed hammer; takedown; 28-ga., .410; 26", 28" barrel; Full choke; bead front sight; uncheckered American walnut pistol-grip stock, forearm; automatic ejector; lightweight configuration. Introduced 1908; discontinued 1942.

Exc.: $80 **VGood:** $60 **Good:** $45

HARRINGTON & RICHARDSON NO. 6

Single barrel; top-lever breakopen; exposed hammer; takedown; 10-, 12-, 16-, 20-ga.; 28", 30", 32", 34", 36" barrels; bead front sight; uncheckered American walnut pistol-grip stock, forearm; heavy breech; automatic ejector. Introduced 1908; discontinued 1942.

Exc.: $85 **VGood:** $60 **Good:** $50

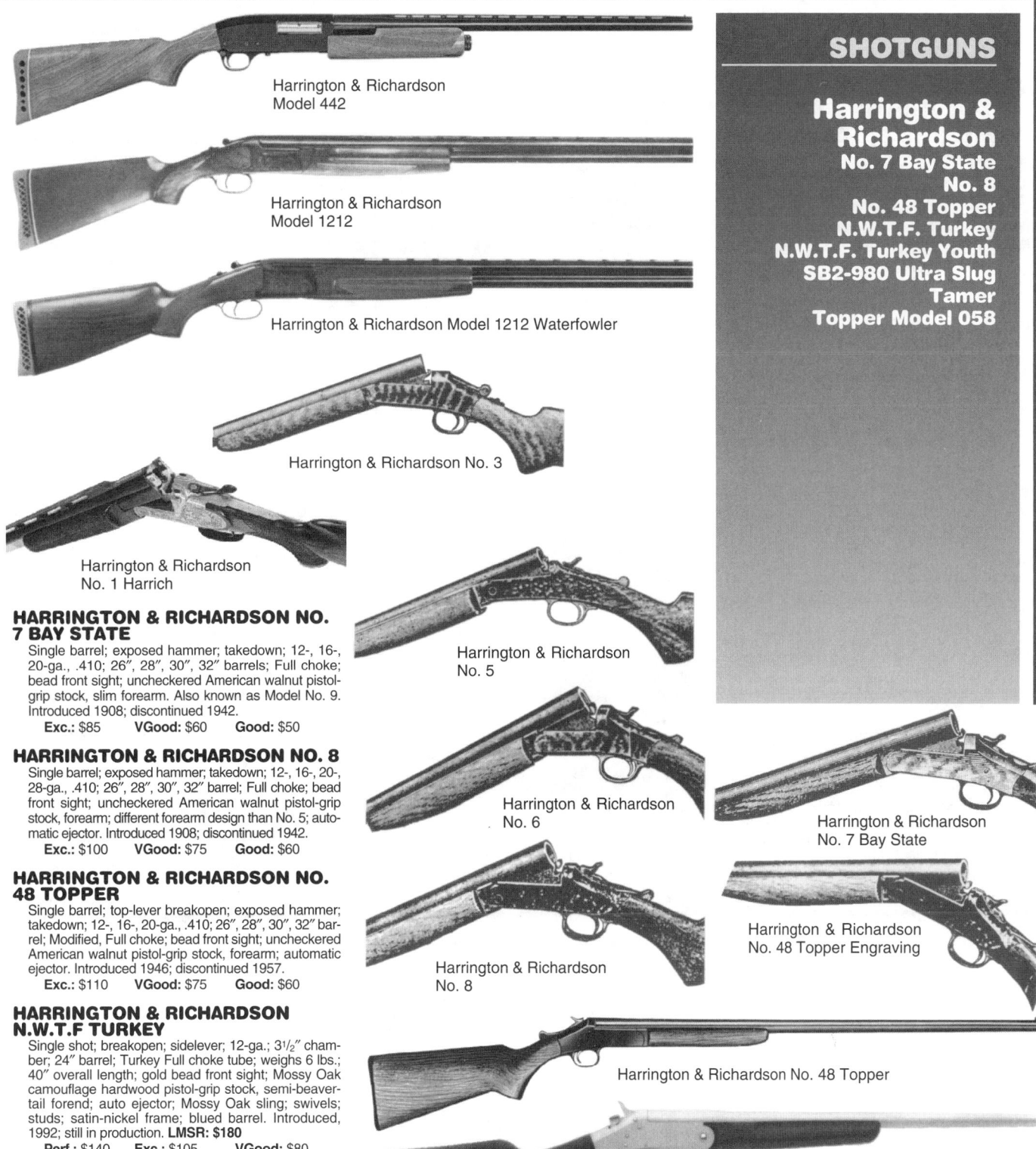

Harrington & Richardson
Model 442

Harrington & Richardson
Model 1212

Harrington & Richardson Model 1212 Waterfowler

Harrington & Richardson No. 3

Harrington & Richardson
No. 1 Harrich

Harrington & Richardson
No. 5

Harrington & Richardson
No. 6

Harrington & Richardson
No. 8

Harrington & Richardson
No. 7 Bay State

Harrington & Richardson
No. 48 Topper Engraving

Harrington & Richardson No. 48 Topper

Harrington & Richardson Tamer

SHOTGUNS

Harrington & Richardson
**No. 7 Bay State
No. 8
No. 48 Topper
N.W.T.F. Turkey
N.W.T.F. Turkey Youth
SB2-980 Ultra Slug
Tamer
Topper Model 058**

HARRINGTON & RICHARDSON NO. 7 BAY STATE

Single barrel; exposed hammer; takedown; 12-, 16-, 20-ga., .410; 26″, 28″, 30″, 32″ barrels; Full choke; bead front sight; uncheckered American walnut pistol-grip stock, slim forearm. Also known as Model No. 9. Introduced 1908; discontinued 1942.

    Exc.: $85   **VGood:** $60   **Good:** $50

HARRINGTON & RICHARDSON NO. 8

Single barrel; exposed hammer; takedown; 12-, 16-, 20-, 28-ga., .410; 26″, 28″, 30″, 32″ barrel; Full choke; bead front sight; uncheckered American walnut pistol-grip stock, forearm; different forearm design than No. 5; automatic ejector. Introduced 1908; discontinued 1942.

    Exc.: $100   **VGood:** $75   **Good:** $60

HARRINGTON & RICHARDSON NO. 48 TOPPER

Single barrel; top-lever breakopen; exposed hammer; takedown; 12-, 16-, 20-ga., .410; 26″, 28″, 30″, 32″ barrel; Modified, Full choke; bead front sight; uncheckered American walnut pistol-grip stock, forearm; automatic ejector. Introduced 1946; discontinued 1957.

    Exc.: $110   **VGood:** $75   **Good:** $60

HARRINGTON & RICHARDSON N.W.T.F TURKEY

Single shot; breakopen; sidelever; 12-ga.; 3½″ chamber; 24″ barrel; Turkey Full choke tube; weighs 6 lbs.; 40″ overall length; gold bead front sight; Mossy Oak camouflage hardwood pistol-grip stock, semi-beavertail forend; auto ejector; Mossy Oak sling; swivels; studs; satin-nickel frame; blued barrel. Introduced, 1992; still in production. **LMSR: $180**

    Perf.: $140   **Exc.:** $105   **VGood:** $80

Harrington & Richardson N.W.T.F. Turkey Youth

Same specs as N.W.T.F. Turkey except 20-ga.; 3″ chamber; 22″ Full choke barrel; recoil pad. Introduced 1992; still in production. **LMSR: $160**

    Perf.: $135   **Exc.:** $100   **VGood:** $75

HARRINGTON & RICHARDSON SB2-980 ULTRA SLUG

Single shot; 12-ga.; 3″ chamber; 24″ fully rifled barrel; weighs 9 lbs.; scope mount; walnut-stained hardwood Monte Carlo stock; sling swivels; black nylon sling; H&R 10-gauge action with heavy-wall barrel. Introduced 1995; still in production. **LMSR: $210**

    Perf.: $180   **Exc.:** $140   **VGood:** $110

HARRINGTON & RICHARDSON TAMER

Single shot; .410; 3″ chamber; 19½″ barrel; Full choke; weighs 5-6 lbs.; 33″ overall length; high-density black-polymer thumbhole stock holds four spare shotshells; H&R Topper action with matte electroless nickel finish. Introduced 1994; still in production. **LMSR: $125**

    Perf.: $110   **Exc.:** $90   **VGood:** $75

HARRINGTON & RICHARDSON TOPPER MODEL 058

Single shot; sidelever; exposed hammer; takedown; 12-, 16-, 20-, 28-ga., .410; 26″, 28″, 30″ barrel; walnut-finished hardwood pistol-grip stock; auto ejector; case-hardened frame. Introduced 1978; discontinued 1981.

    Perf.: $80   **Exc.:** $65   **VGood:** $50

SHOTGUNS

Harrington & Richardson

Topper Model 098
Topper Model 098 Classic Youth
Topper Model 098 Deluxe
Topper Model 098 Deluxe Rifled Slug Gun
Topper Model 098 Junior
Topper Model 148
Topper Model 158
Topper Model 188 Deluxe
Topper Model 198 Deluxe
Topper Model 480 Junior
Topper Model 488 Deluxe
Topper Model 490 Junior

Harrington & Richardson
Topper Model 098

Harrington & Richardson
Topper Model 098 Deluxe

Harrington & Richardson
Topper Model 098 Junior

Harrington & Richardson
Topper Model 158

Harrington & Richardson
Topper Model 198 Deluxe

Harrington & Richardson
Topper Model 490 Greenwing

Harrington & Richardson
Topper Model 590 Junior

HARRINGTON & RICHARDSON TOPPER MODEL 098

Single shot; breakopen; sidelever; 12-, 16- (2 ³/₄"), 20-, 28-ga.; .410; 3" chamber; 26", 28" barrel; Modified, Full choke; gold bead front sight; black-finish hardwood pistol-grip stock, semi-beavertail forend; auto ejector; satin nickel frame; blued barrel. Introduced, 1992; still in production. **LMSR: $110**
 Perf.: $90   Exc.: $75   VGood: $60

Harrington & Richardson Topper Model 098 Classic Youth

Same specs as Topper Model 098 except 20-, 28-ga. (2 ³/₄"), .410; 3" chamber; Modified, Full choke; 22" barrel; cut-checkered American black walnut pistol-grip stock, forend; ventilated rubber recoil pad with white line spacers; blued barrel, blued frame. Introduced 1992; still produced. **LMSR: $140**
 Perf.: $115   Exc.: $90   VGood: $65

Harrington & Richardson Topper Model 098 Deluxe

Same specs as Topper Model 098 except 12-ga.; 3 ¹/₂" chamber; 28" barrel; Modified choke tube. Introduced 1992; still in production. **LMSR: $130**
 Perf.: $105   Exc.: $85   VGood: $60

Harrington & Richardson Topper Model 098 Deluxe Rifled Slug Gun

Same specs as Topper Model 098 except 12-ga.; 3" chamber; 28" fully rifled and ported barrel; ramp front sight, fully adjustable rear; black-finish hardwood pistol-grip stock, semi-beavertail forend; barrel twist is 1:35"; nickel-plated frame; blued barrel. Introduced 1995; still produced. **LMSR: $170**
 Perf.: $150   Exc.: $125   VGood: $100

Harrington & Richardson Topper Model 098 Junior

Same specs as Topper Model 098 except 20-ga., .410; 22" barrel; Modified, Full choke; stock dimensions for

smaller shooters. Introduced 1992; still in production. **LMSR: $115**
 Perf.: $90   Exc.: $75   VGood: $60

HARRINGTON & RICHARDSON TOPPER MODEL 148

Single barrel; sidelever; exposed hammer; takedown; 12-, 16-, 20-ga., .410; 28", 30", 32", 36" barrel; Full choke; uncheckered American walnut pistol-grip stock, forearm; recoil pad. Introduced 1958; discontinued 1961.
 Exc.: $100   VGood: $75   Good: $60

HARRINGTON & RICHARDSON TOPPER MODEL 158

Single barrel; sidelever; exposed hammer; takedown; 12-, 16-, 20-ga., .410; 28", 30", 32", 36" barrel; Full, Modified choke; bead front sight; uncheckered hardwood pistol-grip stock, forearm; recoil pad; improved version of Topper Model 148. Introduced 1962; discontinued 1985.
 Perf.: $90   Exc.: $75   VGood: $60

HARRINGTON & RICHARDSON TOPPER MODEL 188 DELUXE

Single barrel; sidelever; exposed hammer; takedown; .410; 28", 30", 32", 36" barrels; Full choke; uncheckered American walnut pistol-grip stock, forearm finished with black, red, yellow, blue, green, pink or purple lacquer; recoil pad; chrome-plated frame. Introduced 1958; discontinued 1961.
 Exc.: $125   VGood: $100   Good: $75

HARRINGTON & RICHARDSON TOPPER MODEL 198 DELUXE

Single barrel; sidelever; exposed hammer; takedown; 20-ga., .410; 28" barrel; Full, Modified choke; uncheckered American walnut pistol-grip stock, forearm finished with black lacquer; chrome-plated frame. Introduced 1962; discontinued 1985.
 Perf.: $90   Exc.: $75   VGood: $60

HARRINGTON & RICHARDSON TOPPER MODEL 480 JUNIOR

Single barrel; top-lever breakopen; exposed hammer; takedown; .410; 26" barrel; Modified, Full choke; bead front sight; youth-sized uncheckered American walnut pistol-grip stock, forearm; automatic ejector. Introduced 1958; discontinued 1960.
 Exc.: $70   VGood: $60   Good: $50

HARRINGTON & RICHARDSON TOPPER MODEL 488 DELUXE

Single barrel; top-lever breakopen; exposed hammer; takedown; 12-, 16-, 20-ga., .410; 26", 28", 30", 32" barrel; Modified, Full choke; bead front sight; uncheckered American walnut pistol-grip stock, forearm with black-lacquer finish; recoil pad; automatic ejector; chrome-plated frame. Introduced 1946; discontinued 1957.
 Exc.: $75   VGood: $65   Good: $55

HARRINGTON & RICHARDSON TOPPER MODEL 490 JUNIOR

Single shot; sidelever breakopen; takedown; 20-, 28-ga., .410; 26" barrel; Modified (20-, 28-ga.), Full (.410) choke; weighs 5 lbs.; bead front sight; youth-sized uncheckered hardwood pistol-grip stock, forend; recoil pad. Introduced 1962; discontinued 1985.
 Perf.: $90   Exc.: $75   VGood: $60

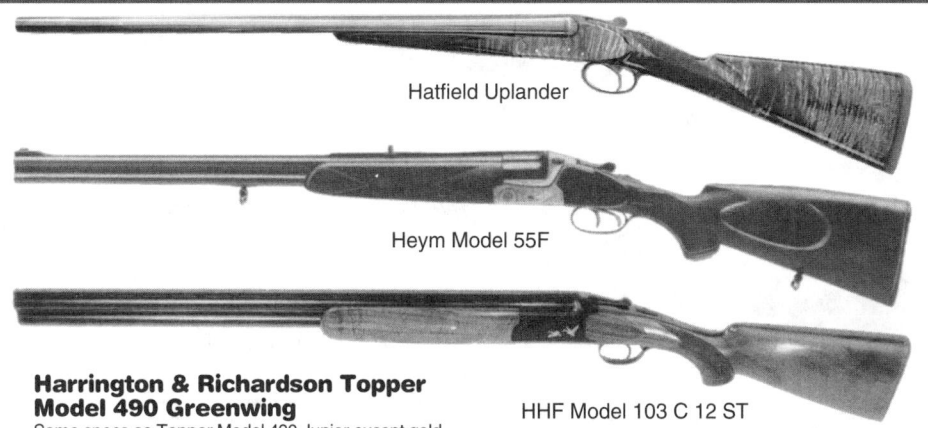

Hatfield Uplander

Heym Model 55F

HHF Model 103 C 12 ST

SHOTGUNS

Harrington & Richardson
Topper Model 490 Greenwing
Topper Model 580 Junior
Topper Model 590 Junior

Hatfield
Uplander

Heym
Model 55 F
Model 55 SS

HHF
Model 101 B 12 AT-DT
Model 101 B 12 ST Trap
Model 103 B 12 ST
Model 103 C 12 ST
Model 103 D 12 ST
Model 103 F 12 ST
Model 104 A 12 ST
Model 200 A 12 ST Double
Model 202 A 12 ST

Harrington & Richardson Topper Model 490 Greenwing

Same specs as Topper Model 490 Junior except gold-finished trigger; inscription on frame; polished blued finish. Discontinued 1984.

Perf.: $100   **Exc.:** $85   **VGood:** $75

HARRINGTON & RICHARDSON TOPPER MODEL 580 JUNIOR

Single barrel; top-lever breakopen; exposed hammer; takedown; 12-, 16-, 20-ga.; .410; 26", 28", 30", 32" barrel; Modified, Full choke; bead front sight; unchecked American walnut pistol-grip stock, forearm with black, red, yellow, blue, green, pink or purple lacquer finish; automatic ejector. Introduced 1958; discontinued 1961.

Exc.: $120   **VGood:** $90   **Good:** $70

HARRINGTON & RICHARDSON TOPPER MODEL 590 JUNIOR

Single shot; sidelever breakopen; exposed hammer; takedown; 20-, 28-ga.; .410; 26" barrel; Modified (20-, 28-ga.), Full (410) choke; weighs 5 lbs.; bead front sight; youth-sized unchecked hardwood pistol-grip stock, forearm with black lacquer finish; chrome-plated frame; recoil pad. Introduced 1962; discontinued 1963.

Exc.: $75   **VGood:** $60   **Good:** $50

HATFIELD UPLANDER

Side-by-side; boxlock; 20-ga.; 3" chambers; 26" Improved Cylinder/Modified barrel; weighs 5 3/4 lbs.; fancy maple straight English-style stock, splinter forend; double locking underlug; single non-selective trigger; color case-hardened frame; half-coverage hand-engraved action with French gray finish; comes with English-style oxblood leather luggage case with billiard felt interior. Made in eight grades with various levels of engraving, inlays and fancy wood. Introduced 1988; Grades I and II still in production; Grades III through VIII no longer in production. **Grade I LMSR: $2249. Grade II LMSR: $2995.**

Grade I
  Perf.: $1500   **Exc.:** $1300   **VGood:** $1000
Grade II
  Perf.: $2000   **Exc.:** $1750   **VGood:** $1550
Grade III
  Perf.: $2300   **Exc.:** $1875   **VGood:** $1575
Grade IV
  Perf.: $3500   **Exc.:** $3000   **VGood:** $2500
Grade V
  Perf.: $4700   **Exc.:** $3500   **VGood:** $2900
Grade VI
  Perf.: $5000   **Exc.:** $3850   **VGood:** $3100
Grade VII
  Perf.: $5200   **Exc.:** $3900   **VGood:** $3250
Grade VIII
  Perf.: $10,000 **Exc.:** $9000   **VGood:** $7500

HEYM MODEL 55 F

Over/under; boxlock; 12-, 16-, 20-ga.; 28" barrel; hand-checkered European walnut stock; Kersten double crossbolt; double underlugs; Arabesque or hunting engraving; numerous options at added cost. Introduced 1979; discontinued 1992.

Perf.: $5000   **Exc.:** $3750   **VGood:** $3400

Heym Model 55 SS

Same specs as Model 55 F except sidelock action; large engraving scenes. No longer in production.

Perf.: $4800   **Exc.:** $3500   **VGood:** $2900

HHF MODEL 101 B 12 AT-DT

Over/under; boxlock; 12-ga.; 3" chambers; combo set with 30" and 32" barrels; fixed chokes or choke tubes; 16mm rib; weighs about 8 lbs.; Circassian walnut Monte Carlo stock to trap dimensions; palm swell grip; recoil pad; single selective trigger; manual safety; automatic ejectors or extractors; silvered frame with 50-percent engraving coverage. Many custom features available. Made in Turkey. Introduced 1995; still imported. **LMSR: $2295**

With extractors
  Perf.: $2000   **Exc.:** $1750   **VGood:** $1500
With ejectors
  Perf.: $2100   **Exc.:** $1850   **VGood:** $1600

HHF Model 101 B 12 ST Trap

Same specs as Model 101 B 12 AT-DT except 30" over/under barrels. Made in Turkey. Introduced 1995; still imported. **LMSR: $1050**

With extractors
  Perf.: $950   **Exc.:** $750   **VGood:** $600
With ejectors
  Perf.: $1025   **Exc.:** $825   **VGood:** $675

HHF MODEL 103 B 12 ST

Over/under boxlock; 12-, 20-, 28-ga.; .410; 3" chambers; 28" barrels; choke tubes or fixed chokes; weighs about 7 1/2 lbs.; Circassian walnut stock; dummy sideplates; double triggers; manual safety; extractors; 80-percent engraving coverage, inlaid animals on blackened sideplates. Made in Turkey. Introduced 1995; still imported. **LMSR: $995**

12-, 20-ga., fixed chokes
  Perf.: $900   **Exc.:** $725   **VGood:** $550
12-, 20-ga., choke tubes
  Perf.: $950   **Exc.:** $775   **VGood:** $600
28-ga., .410, fixed chokes
  Perf.: $950   **Exc.:** $775   **VGood:** $600

HHF Model 103 C 12 ST

Same specs as Model 103 B 12 ST except 12-, 20-ga.; 3" chambers; extractors or ejectors; black receiver with 50-percent engraving coverage. Made in Turkey. Introduced 1995; still imported. **LMSR: $1050**

With extractors
  Perf.: $950   **Exc.:** $750   **VGood:** $600
With ejectors
  Perf.: $1025   **Exc.:** $825   **VGood:** $675

HHF Model 103 D 12 ST

Same specs as Model 103 B 12 ST except standard boxlock; 12-, 20-ga.; 3" chambers; extractors or ejectors; 80-percent engraving coverage. Made in Turkey. Introduced 1995; still imported. **LMSR: $1050**

With extractors
  Perf.: $950   **Exc.:** $750   **VGood:** $600
With ejectors
  Perf.: $1025   **Exc.:** $825   **VGood:** $675

HHF Model 103 F 12 ST

Same specs as Model 103 B 12 ST except 12-, 20-ga.; 3" chambers; extractors or ejectors; 100-percent engraving coverage. Made in Turkey. Introduced 1995; still imported. **LMSR: $1120**

With extractors
  Perf.: $1000   **Exc.:** $800   **VGood:** $650
With ejectors
  Perf.: $1075   **Exc.:** $875   **VGood:** $725

HHF MODEL 104 A 12 ST

Over/under; boxlock; 12-, 20-, 28-ga.; .410; 3" chambers; 28" barrels; fixed chokes or choke tubes; weighs about 7 1/2 lbs.; Circassian walnut stock with field dimensions; manual safety; extractors or ejectors; double triggers; silvered, engraved receiver with 15-percent engraving coverage. Made in Turkey. Introduced 1995; still imported. **LMSR: $925**

12-, 20-ga., fixed chokes, extractors
  Perf.: $850   **Exc.:** $675   **VGood:** $500
12-, 20-ga., choke tubes, ejectors
  Perf.: $900   **Exc.:** $725   **VGood:** $550
28-ga., 410, fixed chokes, extractors
  Perf.: $900   **Exc.:** $725   **VGood:** $550

HHF MODEL 200 A 12 ST DOUBLE

Side-by-side; boxlock; 12-, 20-, 28-ga.; .410; 3" chambers; 28" barrels; fixed chokes or choke tubes; weighs about 7 1/2 lbs.; Circassian walnut stock; single selective trigger; extractors; manual safety; silvered receiver with 15-percent engraving coverage. Made in Turkey. Introduced 1995; still imported. **LMSR: $1050**

12-, 20-ga.
  Perf.: $950   **Exc.:** $750   **VGood:** $600
28-ga., .410
  Perf.: $1025   **Exc.:** $825   **VGood:** $675

HHF MODEL 202 A 12 ST

Side-by-side; boxlock; 12-, 20-, 28-ga.; .410; 3" chambers; 28" barrels; fixed chokes or choke tubes; weighs about 7 1/2 lbs.; Circassian walnut stock; double triggers; extractors; manual safety; silvered receiver with 30-percent engraving coverage. Made in Turkey. Introduced 1995; still imported. **LMSR: $1025**

12-, 20-ga.
  Perf.: $925   **Exc.:** $725   **VGood:** $575
28-ga., .410
  Perf.: $1000   **Exc.:** $800   **VGood:** $700

SHOTGUNS

Hi-Standard
Flite-King Brush
Flite-King Brush Deluxe
Flite-King Brush (1966)
Flite-King Brush Deluxe (1966)
Flite-King Deluxe Rib
Flite-King Deluxe (1966)
Flite-King Deluxe Rib (1966)
Flite-King Deluxe Skeet (1966)
Flite-King Deluxe Trap (1966)
Flite-King Field
Flite-King Skeet
Flite-King Special

Hi-Standard Flite-King Brush

Hi-Standard Flite-King Brush Deluxe

Hi-Standard Flite-King Brush Deluxe (1966)

Hi-Standard Flite-King Deluxe Rib (1966)

Hi-Standard Flite-King Deluxe (1966)

Hi-Standard Flite-King Deluxe 28 (1966)

Hi-Standard Flite-King Deluxe Rib 12

HI-STANDARD FLITE-KING BRUSH
Slide-action; 12-ga.; 2 ¾″ chamber; 4-shot magazine; 18″, 20″ Cylinder bore barrel; rifle sights; uncheckered walnut pistol-grip stock, slide handle. Introduced 1962; discontinued 1964.
  Exc.: $150   **VGood:** $140   **Good:** $125

Hi-Standard Flite-King Brush Deluxe
Same specs as Flite-King Brush except 20″ Cylinder barrel; adjustable peep rear sight; checkered pistol grip, fluted slide handle; recoil pad; sling swivels; sling. Introduced 1964; discontinued 1966.
  Exc.: $190   **VGood:** $150   **Good:** $130

HI-STANDARD FLITE-KING BRUSH (1966)
Slide action; 12-ga.; 2 ¾″ chamber; 5-shot magazine; 20″ Cylinder barrel; rifle sights; checkered American walnut pistol-grip stock, slide handle; recoil pad. General specs follow Flite-King series dropped in 1966, except for damascened bolt, new checkering design. Introduced 1966; discontinued 1975.
  Perf.: $195   **Exc.:** $170   **VGood:** $150

Hi-Standard Flite-King Brush Deluxe (1966)
Same specs as Flite-King Brush (1966) except adjustable peep rear sight; sling swivels; sling. Introduced 1966; discontinued 1975.
  Perf.: $270   **Exc.:** $200   **VGood:** $180

HI-STANDARD FLITE-KING DELUXE RIB
Slide action; 12-, 16-, 20-, 28-ga., .410; 2 ¾″ (12-ga.), 3″ chamber; 3- (20-ga.), 4-shot magazine; 28″ Full or Modified, 30″ Full barrel; vent rib; checkered walnut pistol-grip stock, slide handle. Introduced 1961; discontinued 1966.
  Exc.: $175   **VGood:** $160   **Good:** $130

HI-STANDARD FLITE-KING DELUXE (1966)
Slide action; 12-, 20-, 28-ga., .410; 2 ¾″ (12-ga.), 3″ chamber; 5-shot magazine; 26″ Improved Cylinder, 27″ adjustable choke, 28″ Modified or Full, 30″ Full choke barrel; checkered American walnut pistol-grip stock, slide handle; recoil pad. General specs follow Flite-King series dropped in 1966, except for damascened bolt, new checkering design. Introduced 1966; discontinued 1975.
Fixed choke
  Perf.: $195   **Exc.:** $175   **VGood:** $140
Adjustable choke
  Perf.: $200   **Exc.:** $180   **VGood:** $150

Hi-Standard Flite-King Deluxe Rib (1966)
Same specs as Flite-King Deluxe (1966) except 27″ adjustable choke, 28″ Modified or Full, 30″ Full choke barrel; vent rib. General specs follow Flite-King series dropped in 1966, except for damascened bolt, new checkering design. Introduced 1966; discontinued 1975.
Fixed choke
  Perf.: $210   **Exc.:** $180   **VGood:** $150
Adjustable choke
  Perf.: $220   **Exc.:** $190   **VGood:** $160

Hi-Standard Flite-King Deluxe Skeet (1966)
Same specs as Flite-King Deluxe (1966) except 26″ vent-rib barrel; Skeet choke; optional recoil pad. General specs follow Flite-King series dropped 1966, except for damascened bolt, new checkering design. Introduced 1966; discontinued 1975.
  Perf.: $270   **Exc.:** $200   **VGood:** $170

Hi-Standard Flite-King Deluxe Trap (1966)
Same specs as Flite-King Deluxe (1966) except 12-ga.; 2 ¾″ chamber; 30″ Full choke barrel; vent rib; trap-style checkered walnut pistol-grip stock, slide handle. General specs follow Flite-King series dropped in 1966, except for damascened bolt, new checkering design. Introduced 1966; discontinued 1975.
  Perf.: $270   **Exc.:** $200   **VGood:** $170

HI-STANDARD FLITE-KING FIELD
Slide action; 12-, 16-, 20-, 28-ga., .410; 2 ¾″ (12-ga.), 3″ chamber; 3- (20-ga.), 4-shot magazine; 26″ Improved Cylinder, 28″ Modified or Full, 30″ Full barrel; uncheckered walnut pistol-grip stock, grooved slide handle. Introduced 1960; discontinued 1966.
  Exc.: $135   **VGood:** $115   **Good:** $100

Hi-Standard Flite-King Skeet
Same specs as Flite-King Field except 12-, 20-, 28-ga., .410; 2 ¾″ (12-ga.), 3″ chamber; 26″ Skeet-choked barrel; vent rib; checkered walnut pistol-grip stock, slide handle. Introduced 1962; discontinued 1966.
  Exc.: $190   **VGood:** $150   **Good:** $135

Hi-Standard Flite-King Special
Same specs as Flite-King Field except 12-, 16-, 20-, 28-ga.; 2 ¾″ (12-ga.), 3″ chamber; 27″ adjustable choke barrel. Introduced 1960; discontinued 1966.
  Exc.: $110   **VGood:** $90   **Good:** $80

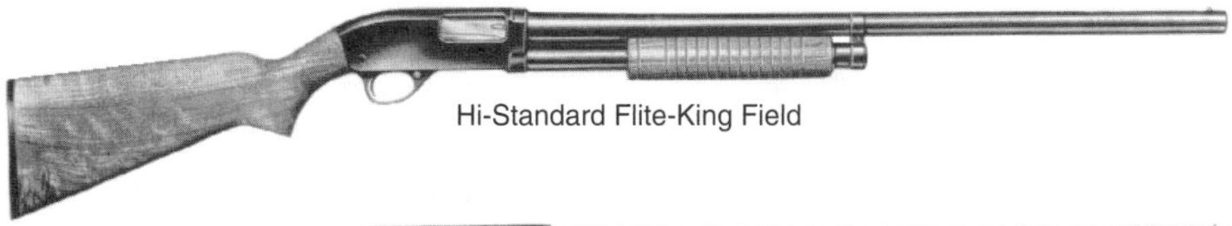

Hi-Standard Flite-King Field

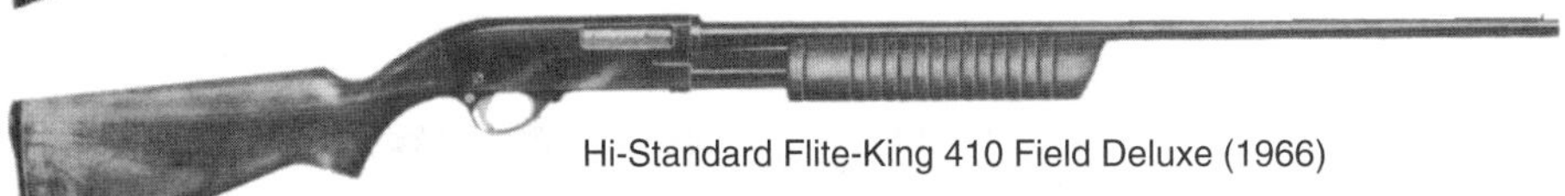

Hi-Standard Flite-King 410 Field Deluxe (1966)

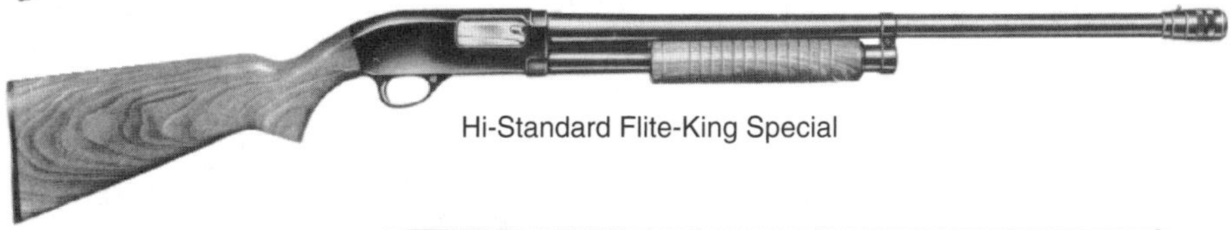

Hi-Standard Flite-King Special

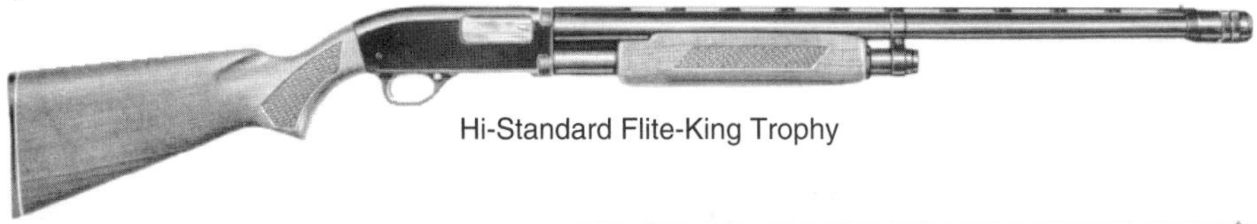

Hi-Standard Flite-King Trophy

Hi-Standard Shadow Auto

Hi-Standard Shadow Indy

SHOTGUNS

Hi-Standard
Flite-King Trap
Flite-King Trophy
Shadow Auto
Shadow Indy
Shadow Seven
Supermatic Deer
Supermatic Deluxe
Supermatic Deluxe Rib
Supermatic Duck
Supermatic Duck Deluxe

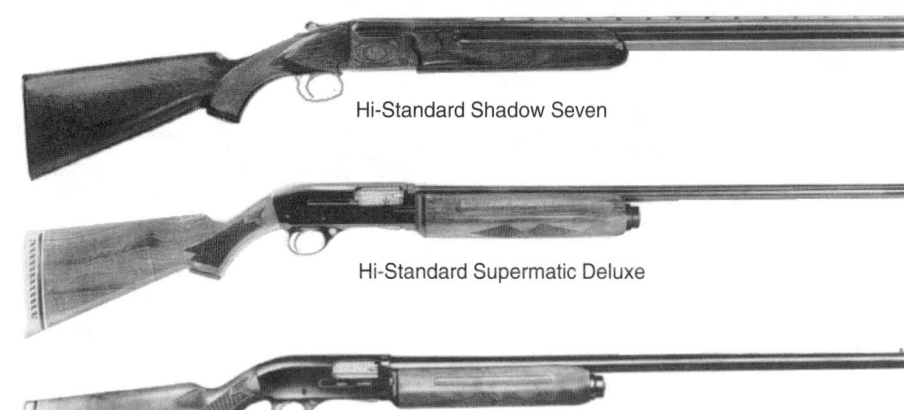

Hi-Standard Shadow Seven

Hi-Standard Supermatic Deluxe

Hi-Standard Supermatic Duck

Hi-Standard Flite-King Trap
Same specs as Flite-King Field except 12-ga.; 2 3/4" chamber; 4-shot magazine; 30" barrel; Full choke; vent rib; trap-style checkered walnut pistol-grip stock, slide handle; recoil pad. Introduced 1962; discontinued 1966.
Exc.: $190   **VGood:** $150   **Good:** $135

Hi-Standard Flite-King Trophy
Same specs as Flite-King Field except Slide-action; 12-, 16-, 20-, 28-ga.; 2 3/4" (12-ga.), 3" chamber; 3- (20-ga.), 4-shot magazine; 27" adjustable choke, 28" Modified or Full barrel (20-ga.); vent rib; checkered walnut pistol-grip stock, slide handle. Introduced 1960; discontinued 1966.
Exc.: $170   **VGood:** $150   **Good:** $130

HI-STANDARD SHADOW AUTO
Gas-operated autoloader; 12-, 20-ga.; 2 3/4", 3" chamber; 3- (3"), 4-shot magazine; 26" Improved Cylinder or Skeet, 28" Modified, Improved Modified or Full, 30" Trap or Full barrel; magnum available only with 30" barrel; checkered walnut pistol-grip stock, forearm. Made in Japan. Introduced 1974; discontinued 1975.
Perf.: $360   **Exc.:** $280   **VGood:** $240

HI-STANDARD SHADOW INDY
Over/under; boxlock; 12-ga.; 2 3/4" chamber; 27 1/2" Skeet, 29 3/4" Improved Modified/Full or Full/Full barrels; Airflow vent rib; skip-checkered walnut pistol-grip stock, ventilated forearm; selective auto ejectors; selective single trigger; engraved receiver; recoil pad. Made in Japan. Introduced 1974; discontinued 1975.
Perf.: $800   **Exc.:** $750   **VGood:** $680

Hi-Standard Shadow Seven
Same specs as Shadow Indy except 27 1/2" Improved Cylinder/Modified or Modified/Full barrels; vent rib; skip-checkered walnut pistol-grip stock, forearm; no recoil pad. Made in Japan. Introduced 1974; discontinued 1975.
Perf.: $700   **Exc.:** $600   **VGood:** $500

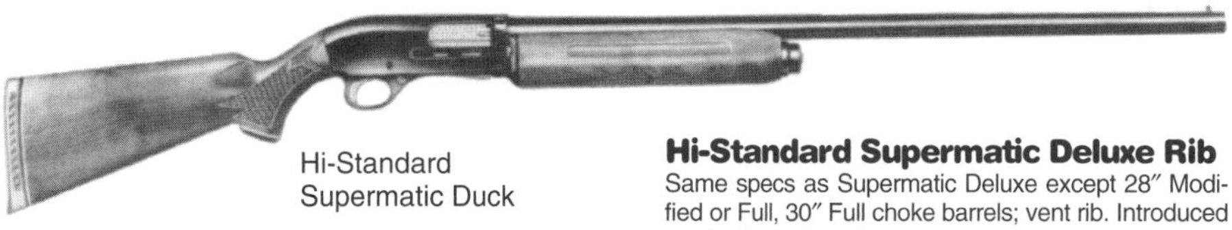

HI-STANDARD SUPERMATIC DEER
Autoloader; 12-ga.; 4-shot magazine; 22" Cylinder barrel; rifle sights; checkered American walnut stock, forearm; recoil pad. Introduced 1965; discontinued 1965.
Exc.: $210   **VGood:** $170   **Good:** $150

HI-STANDARD SUPERMATIC DELUXE
Autoloader; 12-, 20-ga.; 2 3/4" (12-ga.), 3" chamber; 3- (20-ga.), 4-shot magazine; 27" adjustable-choke barrel (discontinued 1970), 26" Improved, 28" Modified or Full, 30" Full choke barrels; checkered American walnut pistol-grip stock, forearm; recoil pad. Differs from original Supermatic; has new checkering, damascened bolt. Introduced 1966; discontinued 1975.
Fixed choke
Perf.: $225   **Exc.:** $150   **VGood:** $140
Adjustable choke
Perf.: $225   **Exc.:** $165   **VGood:** $150

Hi-Standard Supermatic Deluxe Rib
Same specs as Supermatic Deluxe except 28" Modified or Full, 30" Full choke barrels; vent rib. Introduced 1966; discontinued 1975.
Fixed choke
Perf.: $240   **Exc.:** $175   **VGood:** $140
Adjustable choke
Perf.: $250   **Exc.:** $175   **VGood:** $150

HI-STANDARD SUPERMATIC DUCK
Autoloader; 12-ga.; 3" chamber; 3-shot magazine; 30" Full-choke barrel; uncheckered American walnut pistol-grip stock, forearm; recoil pad. Introduced 1961; discontinued 1966.
Exc.: $200   **VGood:** $175   **Good:** $150

Hi-Standard Supermatic Duck Deluxe
Same specs as Supermatic Duck except checkered American walnut pistol-grip stock, forearm. Introduced 1966; discontinued 1975.
Perf.: $250   **Exc.:** $190   **VGood:** $165

SHOTGUNS

Hi-Standard
Supermatic Duck Rib
Supermatic Duck Rib Deluxe
Supermatic Field
Supermatic Skeet
Supermatic Skeet Deluxe
Supermatic Special
Supermatic Trap
Supermatic Trap Deluxe
Supermatic Trophy

Holland & Holland
Badminton
Badminton Game Gun

Hi-Standard Supermatic Skeet

Hi-Standard Supermatic Trap Deluxe

Hi-Standard Supermatic Duck Rib

Hi-Standard Supermatic Special

Hi-Standard Supermatic Trap

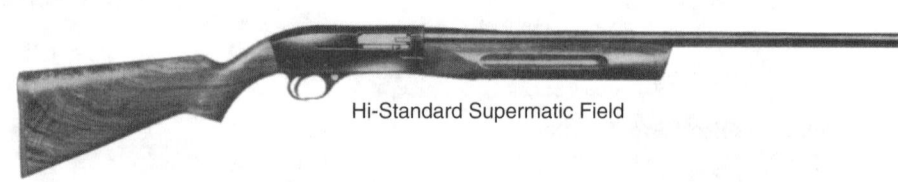

Hi-Standard Supermatic Field

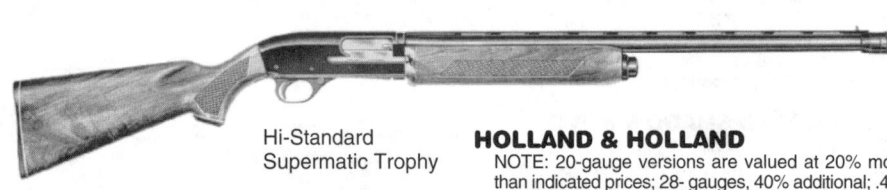

Hi-Standard Supermatic Trophy

Hi-Standard Supermatic Duck Rib
Same specs as Supermatic Duck except vent rib. Introduced 1961; discontinued 1966.
Exc.: $225   **VGood:** $200   **Good:** $175

Hi-Standard Supermatic Duck Rib Deluxe
Same specs as Supermatic Duck Rib except vent rib; checkered American walnut pistol-grip stock, forearm. Introduced 1966; discontinued 1975.
Perf.: $290   **Exc.:** $225   **VGood:** $200

HI-STANDARD SUPERMATIC FIELD
Gas-operated autoloader; 12-, 20-ga.; 2 3/4" (12-ga.), 3" chamber; 3- (20-ga.), 4-shot magazine; 26" Improved, 28" Modified or Full, 30" Full barrel; uncheckered American walnut pistol-grip stock, forearm. Introduced 1960; discontinued 1966.
Exc.: $190   **VGood:** $160   **Good:** $120

HI-STANDARD SUPERMATIC SKEET
Autoloader; 12-, 20-ga.; 2 3/4" (12-ga.), 3" chamber; 3- (20-ga.), 4-shot magazine; 26" vent-rib barrel; Skeet choke; uncheckered Skeet-style American walnut pistol-grip stock, forearm; recoil pad. Introduced 1962; discontinued 1966.
Exc.: $230   **VGood:** $180   **Good:** $155

Hi-Standard Supermatic Skeet Deluxe
Same specs as Supermatic Skeet except checkered Skeet-style American walnut pistol-grip stock, forearm. Introduced 1966; discontinued 1975.
Perf.: $310   **Exc.:** $240   **VGood:** $190

HI-STANDARD SUPERMATIC SPECIAL
Autoloader; 12-, 20-ga.; 2 3/4" (12-ga.), 3" chamber; 3- (20-ga.), 4-shot magazine; 27" barrel; adjustable choke; uncheckered American walnut pistol-grip stock, forearm. Introduced 1960; discontinued 1966.
Exc.: $180   **VGood:** $145   **Good:** $125

HI-STANDARD SUPERMATIC TRAP
Autoloader; 12-ga.; 4-shot magazine; 30" Full barrel; vent rib; trap-style uncheckered American walnut pistol-grip stock, forearm; recoil pad. Introduced 1962; discontinued 1966.
Exc.: $240   **VGood:** $200   **Good:** $150

Hi-Standard Supermatic Trap Deluxe
Same specs as Supermatic Trap except trap-style checkered American walnut pistol-grip stock, forearm. Introduced 1966; discontinued 1975.
Perf.: $250   **Exc.:** $235   **VGood:** $215

HI-STANDARD SUPERMATIC TROPHY
Autoloader; 12-, 20-ga.; 2 3/4" (12-ga.), 3" chamber; 3- (20-ga.), 4-shot magazine; 27" adjustable choke barrel; vent rib; checkered American walnut pistol-grip stock, forearm; recoil pad. Introduced 1961; discontinued 1966.
Exc.: $225   **VGood:** $195   **Good:** $175

HOLLAND & HOLLAND
NOTE: 20-gauge versions are valued at 20% more than indicated prices; 28- gauges, 40% additional; .410 at 60% higher.

HOLLAND & HOLLAND BADMINTON
Side-by-side; sidelock; hammerless; 12-, 16-, 20-, 28-ga.; barrel lengths, chokes to customer specs; hand-checkered straight-grip or pistol-grip stock, forearm; non-self-opening action; auto ejectors; double triggers or single trigger; engraving. Introduced 1902; no longer in production.
Double triggers
Perf.: $10,000 **Exc.:** $8000   **VGood:** $7000
Single trigger
Perf.: $11,000 **Exc.:** $9000   **VGood:** $8000

Holland & Holland Badminton Game Gun
Same specs as Badminton except only 12-, 20-ga. Discontinued 1988.
Double trigger
Perf.: $20,000 **Exc.:** $15,000   **VGood:** $12,000
Single trigger
Perf.: $22,000 **Exc.:** $17,000   **VGood:** $14,000

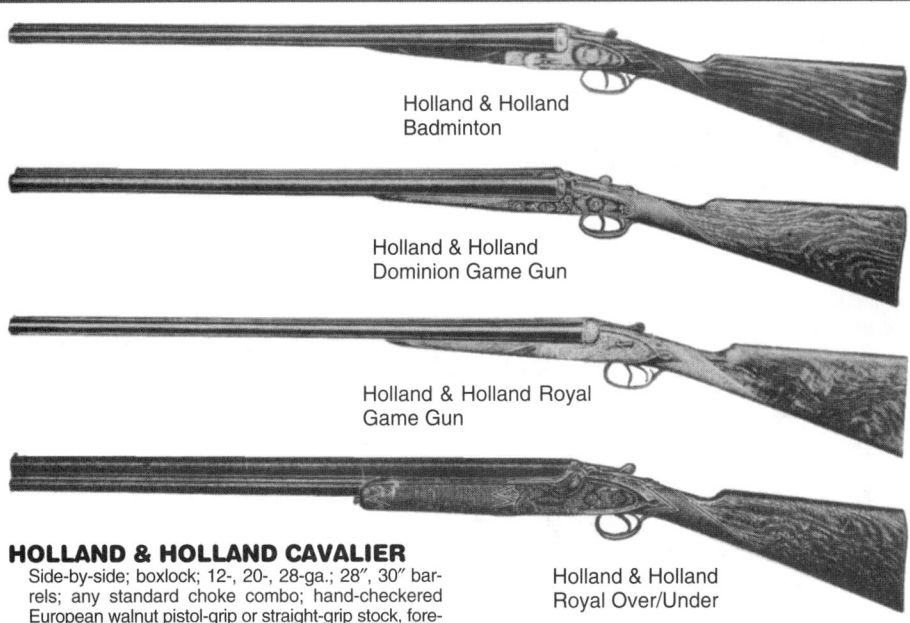

Holland & Holland
Badminton

Holland & Holland
Dominion Game Gun

Holland & Holland Royal
Game Gun

Holland & Holland
Royal Over/Under

HOLLAND & HOLLAND CAVALIER
Side-by-side; boxlock; 12-, 20-, 28-ga.; 28", 30" barrels; any standard choke combo; hand-checkered European walnut pistol-grip or straight-grip stock, forearm; double trigger; auto ejectors; engraving; color case-hardened receiver. Discontinued 1993.
Perf.: $9000 **Exc.:** $7000 **VGood:** $6000

Holland & Holland Cavalier Deluxe
Same specs as Cavalier except better engraving and select wood. Discontinued 1993.
Perf.: $10,000 **Exc.:** $8000 **VGood:** $7000

HOLLAND & HOLLAND DOMINION GAME GUN
Side-by-side; sidelock; hammerless; 12-, 16-, 20-ga.; 25", 28", 30" barrels; choked to customer's specs; hand-checkered European walnut straight-grip stock, forearm; double triggers; automatic ejectors. Introduced 1935; discontinued 1989.
Perf.: $6500 **Exc.:** $5000 **VGood:** $4000

HOLLAND & HOLLAND NORTHWOOD
Side-by-side; boxlock; 12-, 16-, 20-, 28-ga.; 28", 30" barrels; any standard choke combo; hand-checkered European walnut pistol-grip stock or straight-grip stock, forearm; double triggers; auto ejectors; engraving; color case-hardened receiver. Discontinued 1993.
Perf.: $6000 **Exc.:** $4800 **VGood:** $4000

Holland & Holland Northwood Deluxe
Same specs as Northwood except better engraving and select wood. Discontinued 1993.
Perf.: $7000 **Exc.:** $5500 **VGood:** $4500

HOLLAND & HOLLAND NORTHWOOD GAME MODEL
Side-by-side; Anson & Deeley boxlock; 12-, 16-, 20-, 28-ga.; 28" barrels; any standard choke combo; hand-checkered European walnut straight-grip or pistol-grip stock, forearm; double triggers, automatic ejectors. Introduced prior to World War II; discontinued late 1960s.
Perf.: $7000 **Exc.:** $5500 **VGood:** $4500

Holland & Holland Northwood Pigeon Model
Same specs as Northwood Game Model except 12-, 16-, 20-ga.; more engraving. No longer in production.
Perf.: $8000 **Exc.:** $6500 **VGood:** $5000

Holland & Holland Northwood Wildfowl Model
Same specs as Northwood Game Model except 12-ga.; 3" chamber; 30" barrels. No longer in production.
Perf.: $8000 **Exc.:** $6500 **VGood:** $5000

HOLLAND & HOLLAND RIVIERA
Side-by-side; sidelock; hammerless; 12-, 16-, 20-, 28-ga.; barrel lengths, chokes to customer specs; hand-checkered straight-grip or pistol-grip stock, forearm; non-self-opening action; auto ejectors; double triggers; two interchangeable barrels; engraving. Introduced 1945; discontinued 1967.
Exc.: $10,000 **VGood:** $7300 **Good:** $6000

HOLLAND & HOLLAND ROYAL
Side-by-side; sidelock; hammerless; 12-, 16-, 20-, 28-ga.; barrel lengths, chokes to customer's specs; hand-checkered straight-grip or pistol-grip stock, forearm; non-self-opening action; automatic ejectors; double triggers or single non-selective trigger; English engraving. Introduced 1885; no longer in production.
Double triggers
Exc.: $20,000 **VGood:** $17,500 **Good:** $7000
Single trigger
Exc.: $21,500 **VGood:** $18,500 **Good:** $8000

HOLLAND & HOLLAND ROYAL GAME GUN
Side-by-side; sidelock; hammerless; 12-, 16-, 20-, 28-ga., .410; barrel lengths, chokes to customer's specs; hand-checkered straight-grip or pistol-grip stock, forearm; self-opening action; double triggers or single non-selective trigger; ornate engraving. Introduced 1922; still in production. **LMSR: $48,050**
Double triggers
Perf.: $48,000 **Exc.:** $27,000 **VGood:** $19,000
Single trigger
Perf.: $48,000 **Exc.:** $30,000 **VGood:** $22,000

Holland & Holland Royal Game Gun Deluxe
Same specs as Royal Game Gun except top-of-the-line model with better engraving and wood. Still in production. **LMSR: $56,900**
Double triggers
Perf.: $56,900 **Exc.:** $35,000 **VGood:** $20,000
Single trigger
Perf.: $56,900 **Exc.:** $39,000 **VGood:** $24,000

HOLLAND & HOLLAND ROYAL OVER/UNDER
Over/under; hand-detachable sidelocks; hammerless; 12-ga.; barrel lengths, chokes to customer's specs; hand-checkered European walnut straight-grip stock, forearm; automatic ejectors; double triggers or single trigger. Introduced 1925; discontinued 1950.

Holland & Holland Royal Over/Under (New Model)
Same specs as Royal Over/Under except narrower, improved action. Introduced 1951; discontinued 1965.

HOLLAND & HOLLAND ROYAL OVER/UNDER GAME GUN
Over/under; hand-detachable sidelocks; hammerless; 12-, 20-ga.; 2 3/4" chambers; 25" to 30" barrels; any standard choke combo; select hand-checkered walnut straight-grip or pistol-grip stock, forearm; automatic ejectors; single or double triggers; engraved color case-hardened receiver. Introduced 1992; still in production. **LMSR: $60,375**
Double triggers
Perf.: $60,375 **Exc.:** $35,000 **VGood:** $24,000
Single trigger
Perf.: $60,375 **Exc.:** $39,000 **VGood:** $28,000

HOLLAND & HOLLAND SINGLE BARREL TRAP
Single barrel; boxlock; 12-ga.; 30", 32" barrel; Full choke; vent rib; European walnut pistol-grip stock, forearm; auto ejector; recoil pad. Discontinued 1992.
Perf.: $14,500 **Exc.:** $12,000 **VGood:** $10,000

HOLLAND & HOLLAND SPORTING OVER/UNDER
Blitz action; 12-, 20-ga.; 2 3/4" chambers; 28" to 32" barrels; screw-in choke tubes; hand-checkered European walnut straight-grip or pistol-grip stock, forearm; single selective trigger; auto ejectors. Introduced 1993; still in production. **LMSR: $29,380**
Perf.: $25,000 **Exc.:** $22,000 **VGood:** $16,000

Holland & Holland Sporting Over/Under Deluxe
Same specs as Sporting Over/Under except better engraving and select wood. Introduced 1993; still in production. **LMSR: $34,775**
Perf.: $34,775 **Exc.:** $23,000 **VGood:** $18,000

HOLLAND & HOLLAND SUPER TRAP
Single barrel; Anson & Deeley boxlock; 12-ga.; 30", 32" barrel; Extra-Full choke; European walnut Monte Carlo stock, full beavertail forearm; automatic ejector; no safety; recoil pad. Made in three grades with varying types of engraving and wood. Introduced prior to World War II; discontinued late 1960s.
Standard Grade
Perf.: $4500 **Exc.:** $4000 **VGood:** $3500
Deluxe Grade
Perf.: $7500 **Exc.:** $6000 **VGood:** $4500
Exhibition Grade
Perf.: $10,250 **Exc.:** $8500 **VGood:** $7000

Hunter
Fulton
Special

IGA
Coach Model
Condor O/U
Condor II
Condor Supreme
ERA 2000
Reuna Single Barrel
Uplander

Industrias Danok
Red Prince

Ithaca Semi-Automatic Shotguns
Mag-10
Mag-10 Deluxe
Mag-10 Supreme
Model 51
Model 51A
Model 51A Deerslayer
Model 51A Magnum

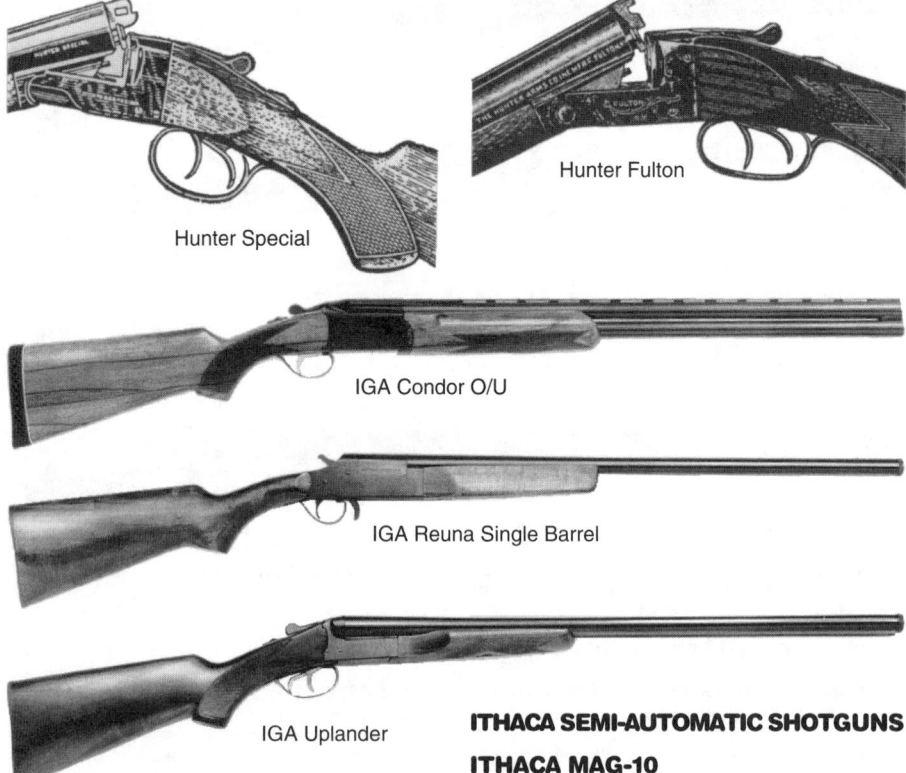

Hunter Special

Hunter Fulton

IGA Condor O/U

IGA Reuna Single Barrel

IGA Uplander

HUNTER FULTON
Side-by-side; boxlock; hammerless; 12-, 16-, 20-ga.; .410; 26″ to 32″ barrels; standard choke combos; walnut stock; hand-checkered pistol grip, forearm; double triggers or non-selective single trigger. Introduced 1920; discontinued 1948. Made by Hunter Arms Company, Fulton, New York.
Double triggers
   Exc.: $180   **VGood:** $150   **Good:** $130
Single trigger
   Exc.: $190   **VGood:** $160   **Good:** $140

HUNTER SPECIAL
Side-by-side; boxlock; hammerless; 12-, 16-, 20-ga.; 26″ to 30″ barrels; standard choke combos; hand-checkered walnut pistol-grip stock, forearm; extractors. Introduced 1920; discontinued 1948.
Double triggers
   Exc.: $210   **VGood:** $180   **Good:** $150
Single trigger
   Exc.: $220   **VGood:** $190   **Good:** $160

IGA COACH MODEL
Side-by-side; 12-, 20-ga., .410; 2 3/4″ (12-, 20-ga.), 3″ (.410) chambers; 20″ barrels; Improved Cylinder/Modified chokes; solid matted rib; oil-finished hand-checkered Brazilian hardwood pistol grip stock, forend; extractors; auto safety; double triggers. Made in Brazil. Introduced 1983; still imported by Stoeger. **LMSR: $382**
   Perf.: $240   **Exc.:** $170   **VGood:** $130

IGA CONDOR O/U
Over/under; boxlock; 12-, 20-ga.; 3″ chambers; 26″. 28″ barrels; Improved Cylinder/Modified, Modified/Full choke tubes; vent top rib; weighs 7 lbs.; oil-finished checkered Brazilian hardwood pistol-grip stock, forend; manual safety; single trigger; extractors. Made in Brazil. Introduced 1983; still imported by Stoeger. **LMSR: $500**
   Perf.: $350   **Exc.:** $280   **VGood:** $200

IGA Condor II
Same specs as the Condor except double triggers, moulded buttplate. Made in Brazil. Introduced 1983; still imported by Stoeger Industries. **LMSR: $375**
   Perf.: $300   **Exc.:** $240   **VGood:** $190

IGA Condor Supreme
Same specs as the Condor except automatic ejectors. Made in Brazil. Introduced 1983; still imported by Stoeger Industries. **LMSR: $689**
   Perf.: $495   **Exc.:** $400   **VGood:** $360

IGA ERA 2000
Over/under; boxlock; 12-ga.; 3″ chambers; 26″, 28″ barrels; Full, Modified, Improved Cylinder choke tubes; oil-finished hand-checkered Brazilian hardwood pistol grip stock, forend; single trigger; blue finish. Made in Brazil. Introduced 1992; still imported by Stoeger Industries. **LMSR: $665**
   Perf.: $425   **Exc.:** $325   **VGood:** $250

IGA REUNA SINGLE BARREL
Single shot; exposed hammer; 12-, 20-ga., .410; 2 3/4″ (12-ga), 3″ (20-ga., 410) chamber; 26″, 28″ barrel; Full choke tube; weighs 5 1/4 lbs.; metal bead front sight; Brazilian hardwood stock; half-cock safety; extractor; blued finish. Made in Brazil. Introduced 1987; still imported by Stoeger. **LMSR: $120**
   Perf.: $90   **Exc.:** $60   **VGood:** $45

IGA UPLANDER
Side-by-side; boxlock; 12-, 20-, 28-ga., .410; 2 3/4″ (12, 20-, 28-ga.), 3″ (.410) chambers; 26″. 28″ barrels; Improved Cylinder/Modified, Modified/Full, Full/Full (.410) chokes; solid matted rib; oil-finished hand-checkered hardwood pistol-grip stock, forend; extractors; auto safety; double triggers. Made in Brazil. Introduced 1983; still imported by Stoeger. **LMSR: $398**
   Perf.: $280   **Exc.:** $180   **VGood:** $120

INDUSTRIAS DANOK RED PRINCE
Side-by-side; 12-ga.; 2 3/4″ chambers; Modified/Full barrels; medium bead front sight; checkered European walnut straight-grip stock, forend; double triggers; auto ejectors; hand-engraved action. Made in Spain. Introduced 1982 by Toledo Armas; discontinued 1984.
   Perf.: $900   **Exc.:** $700   **VGood:** $550

ITHACA SEMI-AUTOMATIC SHOTGUNS

ITHACA MAG-10
Gas-operated autoloader; 10-ga.; 3 1/2″ chamber; 3-shot magazine; various barrel lengths, chokes; plain or vent rib; walnut stock, forearm; recoil pad. Introduced 1977; discontinued 1986.
With plain barrel
   Perf.: $650   **Exc.:** $520   **VGood:** $440
With vent rib
   Perf.: $700   **Exc.:** $570   **VGood:** $490

Ithaca Mag-10 Deluxe
Same specs as Mag-10 except vent rib; checkered semi-fancy walnut stock, forearm; sling swivels. Introduced 1977; discontinued 1986.
   Perf.: $725   **Exc.:** $580   **VGood:** $525

Ithaca Mag-10 Supreme
Same specs as Mag-10 except vent rib; checkered fancy walnut stock, forearm; sling swivels. Introduced 1974; discontinued 1986.
   Perf.: $850   **Exc.:** $740   **VGood:** $625

ITHACA MODEL 51
Gas-operated autoloader; takedown; 12-, 20-ga.; 2 3/4″ chamber; 3-shot tube magazine; 26″ Improved or Skeet, 28″ Full or Modified or Skeet, 30″ Full barrel; optional vent rib; Raybar front sight; hand-checkered American walnut pistol stock; white spacer on pistol grip; reversible safety; engraved receiver. Introduced 1970; discontinued 1982.
Plain barrel
   Perf.: $260   **Exc.:** $210   **VGood:** $180
Vent rib
   Perf.: $290   **Exc.:** $240   **VGood:** $210

Ithaca Model 51A
Same specs as Model 51 except cosmetic improvements. Introduced 1982; discontinued 1986.
   Perf.: $260   **Exc.:** $210   **VGood:** $180

Ithaca Model 51A Deerslayer
Same specs as Model 51 except special 24″ slug barrel; Raybar front sight, open adjustable rear; sight base grooved for scope mounts. Introduced 1983; discontinued 1984.
   Perf.: $295   **Exc.:** $250   **VGood:** $210

Ithaca Model 51A Magnum
Same specs as Model 51 except 12-ga.; 3″ chamber. Introduced 1984; discontinued 1985.
   Perf.: $275   **Exc.:** $220   **VGood:** $170

Ithaca Mag-10

Ithaca Mag-10 Supreme

Ithaca Model 51

Ithaca Model 51 Deerslayer

Ithaca Model 300

Ithaca Model XL 900

Ithaca Semi-Automatic Shotguns
Model 51A Magnum Waterfowler
Model 51A Presentation
Model 51A Supreme Skeet
Model 51A Supreme Trap
Model 51 Deerslayer
Model 51 Featherlight
Model 51 Featherlight Deluxe Trap
Model 51 Magnum
Model 51 Skeet
Model 51 Trap
Model 51 Turkey
Model 300
Model 900 Deluxe
Model 900 Deluxe Slug
Model XL 300
Model XL 900
Model XL 900 Slug Gun

Ithaca Model 51A Magnum Waterfowler
Same specs as Model 51 except 12-ga.; 3″ chamber; vent rib; matte-finished metal; sling; swivels. Introduced 1985; discontinued 1986.
Perf.: $325 **Exc.:** $290 **VGood:** $210

Ithaca Model 51A Presentation
Same specs as Model 51 except 12-ga.; deluxe checkered walnut pistol-grip stock, forearm; elaborate engraving. Introduced 1984; discontinued 1986.
Perf.: $1050 **Exc.:** $800 **VGood:** $700

Ithaca Model 51A Supreme Skeet
Same specs as Model 51 except 12-, 20-ga.; 26″ Skeet barrel; fancy American walnut Skeet stock. Introduced 1983; discontinued 1986.
Perf.: $425 **Exc.:** $350 **VGood:** $290

Ithaca Model 51A Supreme Trap
Same specs as Model 51 except 30″ Full-choke barrel; fancy American walnut trap stock; Monte Carlo stock optional; recoil pad. Introduced 1983; discontinued 1986.
Perf.: $425 **Exc.:** $350 **VGood:** $290

Ithaca Model 51 Deerslayer
Same specs as Model 51 except 12-, 20-ga.; special 24″ slug barrel; Raybar front sight, open adjustable rear; sight base grooved for scope. Introduced 1972; discontinued 1983.
Perf.: $330 **Exc.:** $270 **VGood:** $220

Ithaca Model 51 Featherlight
Same specs as Model 51 except hand-fitted; engraved receiver; weighs 7 1/2 lbs. Introduced 1978; discontinued 1982.
Perf.: $375 **Exc.:** $300 **VGood:** $275

Ithaca Model 51 Featherlight Deluxe Trap
Same specs as Model 51 except 28″ Full or Improved Modified, 30″ Full or Improved Cylinder barrel; weighs 7 1/2 lbs.; fancy walnut trap-style stock; hand-fitted, engraved receiver. No longer produced.
Perf.: $400 **Exc.:** $315 **VGood:** $270

Ithaca Model 51 Magnum
Same specs as Model 51 except 12-ga.; 3″ chamber. Introduced 1970; discontinued 1985.
Plain barrel
Perf.: $250 **Exc.:** $200 **VGood:** $175
Vent rib
Perf.: $280 **Exc.:** $230 **VGood:** $200

Ithaca Model 51 Skeet
Same specs as Model 51 except 12-, 20-ga.; 26″ Skeet barrel; vent rib; select Skeet-style stock; recoil pad. Introduced 1970; discontinued 1986.
Perf.: $360 **Exc.:** $310 **VGood:** $260

Ithaca Model 51 Trap
Same specs as Model 51 except 30″. 32″ barrel; Full choke; vent rib; select trap-style stock; recoil pad. Introduced 1970; discontinued 1986.
Perf.: $360 **Exc.:** $310 **VGood:** $260

Ithaca Model 51 Turkey
Same specs as Model 51 except 12-ga.; 3″ chamber; 26″ barrel; sling; swivels; matte finish. Introduced 1984; discontinued 1986.
Perf.: $350 **Exc.:** $270 **VGood:** $240

ITHACA MODEL 300
Recoil-operated autoloader; takedown; 12-ga.; 26″ Improved Cylinder, 28″ Full or Modified, 30″ Full-choke barrel; optional vent rib; checkered American walnut pistol-grip stock, fluted forearm; crossbolt safety; automatic magazine cutoff allows changing loads without unloading magazine. Introduced 1969; discontinued 1973.
Plain barrel
Perf.: $275 **Exc.:** $200 **VGood:** $150
Vent rib
Perf.: $295 **Exc.:** $220 **VGood:** $170

ITHACA MODEL 900 DELUXE
Recoil-operated autoloader; takedown; 12-, 20-ga.; 25″ Improved Cylinder, 28″ Full or Modified, 30″ Full-choke (12-ga.) vent-rib barrel; hand-checkered American walnut pistol-grip stock, forearm; white spacers on grip cap, buttplate; interchangeable barrels; crossbolt safety; gold-filled engraving on receiver; gold-plated trigger; nameplate inlaid in stock. Introduced 1969; discontinued 1973.
Perf.: $420 **Exc.:** $340 **VGood:** $270

Ithaca Model 900 Deluxe Slug
Same specs as Model 900 Deluxe except 24″ barrel; rifle sights. Introduced 1969; discontinued 1973.
Perf.: $420 **Exc.:** $340 **VGood:** $270

ITHACA MODEL XL 300
Gas-operated autoloader; 12-, 20-ga.; 26″ Improved Cylinder or Skeet, 28″ Full or Modified, 30″ Full or Modified barrel; optional vent rib; checkered American walnut pistol-grip stock, fluted forearm; self-compensating gas system; reversible safety. Introduced 1973; discontinued 1976.
Perf.: $325 **Exc.:** $250 **VGood:** $200

ITHACA MODEL XL 900
Gas-operated autoloader; 12-, 20-ga.; 5-shot tube magazine; 26″ Improved Cylinder or Skeet, 28″ Full or Modified, 30″ Full or Improved Cylinder barrels; Bradley-type front sight on target-grade guns; Raybar front sight on vent-rib field guns; checkered walnut-finished stock; self-compensating gas system; reversible safety; action release button. Introduced 1973; discontinued 1978.
Perf.: $350 **Exc.:** $270 **VGood:** $200
Vent rib
Perf.: $370 **Exc.:** $290 **VGood:** $220
Skeet grade
Perf.: $385 **Exc.:** $300 **VGood:** $250
Trap grade
Perf.: $385 **Exc.:** $300 **VGood:** $250

Ithaca Model XL 900 Slug Gun
Same specs as Model XL 900 except 24″ slug barrel; rifle sights. Introduced 1973; discontinued 1978.
Perf.: $335 **Exc.:** $270 **VGood:** $240

Ithaca Slide- & Lever-Action Shotguns

Model 37
Model 37 $3000 Grade
Model 37 Basic Featherlight
Model 37D Featherlight
Model 37 Deerslayer
Model 37 Deluxe Featherlight
Model 37DV Featherweight
Model 37 English Ultralight
Model 37 Featherlight
Model 37 Featherlight Presentation
Model 37 Field Grade
Model 37 Field Grade Magnum
Model 37R
Model 37R Deluxe
Model 37S Skeet Grade
Model 37 Super Deerslayer
Model 37 Supreme Grade

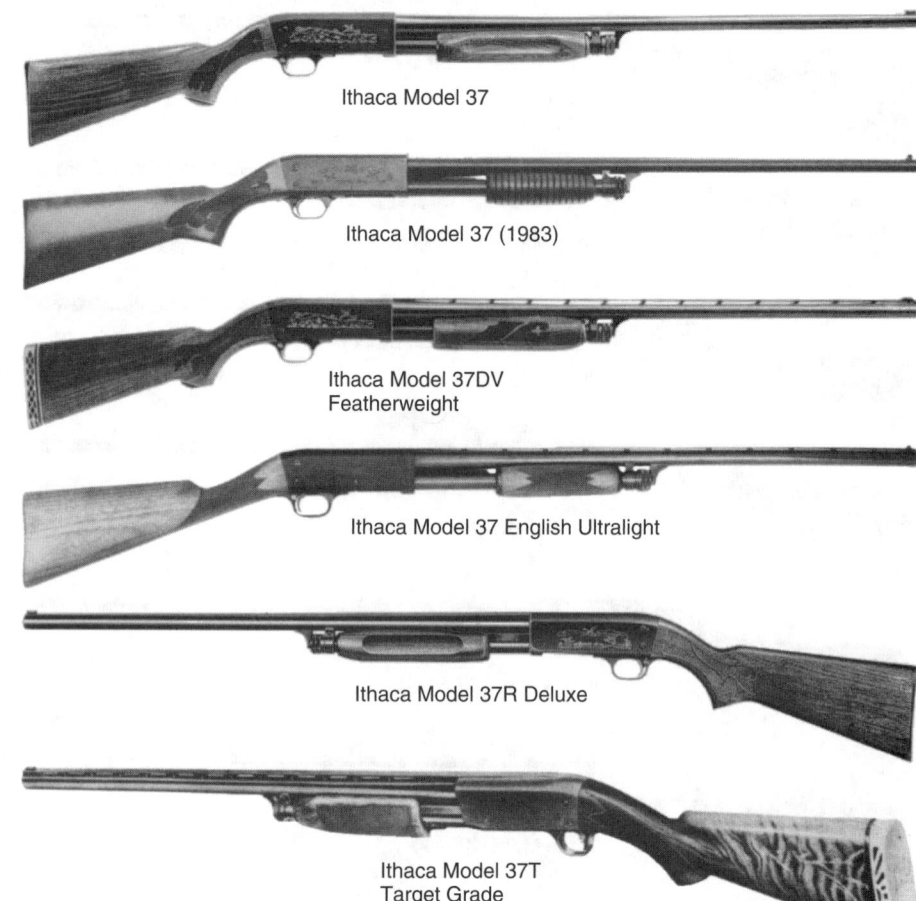

Ithaca Model 37

Ithaca Model 37 (1983)

Ithaca Model 37DV Featherweight

Ithaca Model 37 English Ultralight

Ithaca Model 37R Deluxe

Ithaca Model 37T Target Grade

ITHACA SLIDE- & LEVER-ACTION SHOTGUNS

ITHACA MODEL 37

Slide action; hammerless; takedown; 12-, 16-, 20-ga.; 4-shot tube magazine; 26″, 28″, 30″ barrel; hand-checkered American walnut pistol-grip stock, slide handle, or uncheckered stock, grooved slide handle. Introduced 1937; discontinued 1986.
Checkered stock
   Perf.: $250   **Exc.:** $200   **VGood:** $160
Uncheckered stock
   Perf.: $240   **Exc.:** $190   **VGood:** $150

Ithaca Model 37 $3000 Grade

Same specs as Model 37 except custom-built; hand-finished parts; hand-checkered pistol-grip stock, slide handle of select figured walnut; gold-inlaid engraving; recoil pad. Was listed as $1000 Grade prior to World War II. Introduced 1937; discontinued 1967.
   Exc.: $4000   **VGood:** $3500   **Good:** $2800

Ithaca Model 37 Basic Featherlight

Same specs as Model 37 except plain or vent rib; tung-oil-finished walnut stock, traditional ringtail forend; matte finish on all metal. Introduced 1980; discontinued 1983.
   Perf.: $250   **Exc.:** $190   **VGood:** $150

Ithaca Model 37D Featherlight

Same specs as Model 37 except 12-, 20-ga.; 5-shot magazine; Ithaca Raybar front sight; checkered American walnut pistol-grip stock, beavertail forend; decorated receiver; crossbolt safety; recoil pad. Introduced 1954; discontinued 1981.
   Perf.: $280   **Exc.:** $225   **VGood:** $190

Ithaca Model 37 Deerslayer

Same specs as Model 37 except 20″, 26″ barrel bored for rifled slugs; open rifle-type rear sight, ramp front. Introduced 1969; discontinued 1986.
   Perf.: $290   **Exc.:** $200   **VGood:** $175

Ithaca Model 37 Deluxe Featherlight

Same specs as Model 37 except vent rib; deluxe oil-finished cut-checkered straight-grip stock, slide handle. Introduced 1981; discontinued 1986.
   Perf.: $280   **Exc.:** $225   **VGood:** $195

Ithaca Model 37DV Featherweight

Same specs as Model 37 except 12-, 20-ga.; 5-shot magazine; Ithaca Raybar front sight; vent rib; checkered American walnut pistol-grip stock, beavertail forend; decorated receiver; crossbolt safety; recoil pad. Introduced 1962; discontinued 1981.
   Perf.: $325   **Exc.:** $240   **VGood:** $200

Ithaca Model 37 English Ultralight

Same specs as Model 37 except 12-, 20-ga.; 25″ vent-rib barrel; oil-finished cut-checkered straight-grip stock, slide handle. Introduced 1981; discontinued 1986.
   Perf.: $360   **Exc.:** $280   **VGood:** $240

Ithaca Model 37 Featherlight

Same specs as Model 37 except 12-, 20-ga.; 5-shot magazine; Ithaca Raybar front sight; checkered American walnut pistol-grip stock; decorated receiver; crossbolt safety. Introduced 1980; discontinued 1986.
   Perf.: $250   **Exc.:** $220   **VGood:** $180

Ithaca Model 37 Featherlight Presentation

Same specs as Model 37 except 12-ga.; 5-shot magazine; Ithaca Raybar front sight; fancy checkered American walnut pistol-grip stock; gold decorated receiver; engraving; crossbolt safety. Introduced 1981; discontinued 1986.
   Perf.: $1175   **Exc.:** $1000   **VGood:** $890

Ithaca Model 37 Field Grade

Same specs as Model 37 except 12-, 20-ga.; 2 ³/₄″ chamber; standard or vent rib barrel; Raybar front sight; American hardwood stock, ring-tail forend. Introduced 1983; discontinued 1986.
   Perf.: $250   **Exc.:** $200   **VGood:** $175

Ithaca Model 37 Field Grade Magnum

Same specs as Model 37 except 12-, 20-ga.; 3″ chamber; 28″, 30″ vent-rib barrel; Raybar front sight; American hardwood stock, ringtail forend; elongated receiver; grip cap has flying mallard; recoil pad. Introduced 1978; discontinued 1986.
   Perf.: $310   **Exc.:** $200   **VGood:** $175

Ithaca Model 37R

Same general specs as Model 37 except raised solid rib. Introduced 1937; discontinued 1967.
Checkered stock
   Exc.: $225   **VGood:** $180   **Good:** $140
Uncheckered stock
   Exc.: $200   **VGood:** $170   **Good:** $130

Ithaca Model 37R Deluxe

Same specs as Model 37 except raised solid rib; hand-checkered fancy walnut stock, slide handle. Introduced 1955; discontinued 1961.
   Exc.: $280   **VGood:** $225   **Good:** $195

Ithaca Model 37S Skeet Grade

Same specs as Model 37 except vent rib; checkered stock, extension slide handle. Introduced 1937; discontinued 1955.
   Exc.: $380   **VGood:** $300   **Good:** $250

Ithaca Model 37 Super Deerslayer

Same specs as Model 37 except 20″, 26″ barrel bored for rifled slugs; open rifle-type rear sight, ramp front; improved wood in stock, slide handle. Introduced 1962; discontinued 1979.
   Exc.: $310   **VGood:** $240   **Good:** $180

Ithaca Model 37 Supreme Grade

Same specs as Model 37 except hand-checkered fancy walnut stock, slide handle; available in Skeet or trap configurations. Introduced 1967; discontinued 1979.
   Perf.: $400   **Exc.:** $300   **VGood:** $250

Ithaca Model 37
Deerslayer

Ithaca Model 37
$3000 Grade

Ithaca Model 37
Super Deerslayer

Ithaca Model 66

Ithaca Model 66RS
Buck Buster

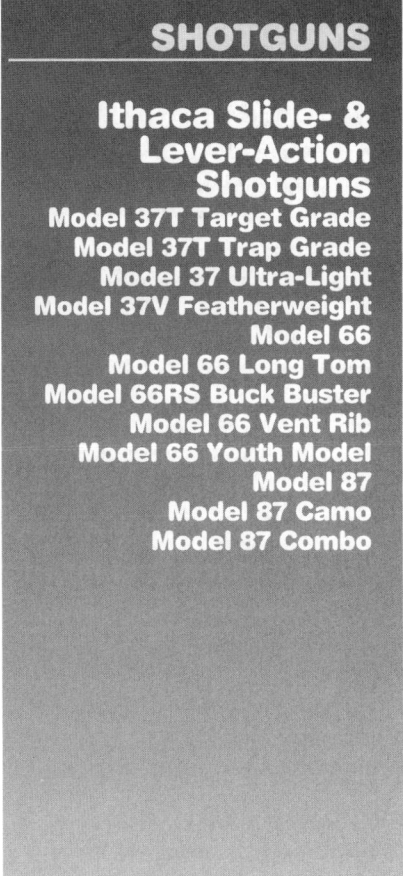

Ithaca Slide- & Lever-Action Shotguns
Model 37T Target Grade
Model 37T Trap Grade
Model 37 Ultra-Light
Model 37V Featherweight
Model 66
Model 66 Long Tom
Model 66RS Buck Buster
Model 66 Vent Rib
Model 66 Youth Model
Model 87
Model 87 Camo
Model 87 Combo

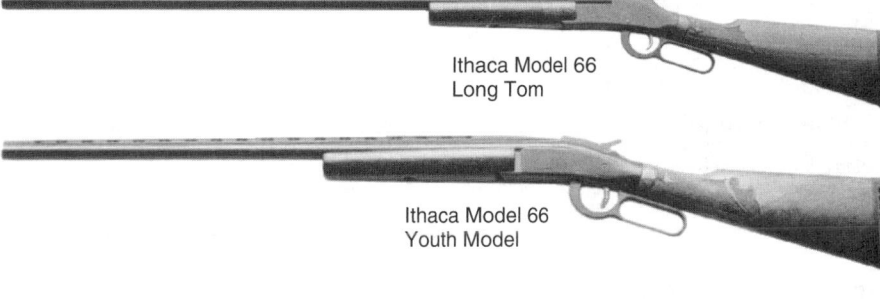

Ithaca Model 66
Long Tom

Ithaca Model 66
Youth Model

Ithaca Model 87
Supreme

Ithaca Model 37T Target Grade
Same specs as Model 37 except vent rib; hand-checkered fancy walnut stock, slide handle; choice of Skeet or Trap stock. Replaced Model 37S Skeet Grade and Model 37T Trap Grade. Introduced 1955; discontinued 1961.
   Perf.: $400   **Exc.:** $300   **VGood:** $250

Ithaca Model 37T Trap Grade
Same specs as Model 37 except vent rib; checkered select walnut straight trap-style stock, extension slide-handle; recoil pad. Introduced 1937; discontinued 1955.
   Perf.: $400   **Exc.:** $300   **VGood:** $250

Ithaca Model 37 Ultra-light
Same specs as Model 37 except 12-, 20-ga.; 5-shot magazine; 25″ vent-rib barrel; checkered American walnut pistol-grip stock; recoil pad; gold-plated trigger; Sid Bell-designed grip cap. No longer in production.
   Perf.: $390   **Exc.:** $275   **VGood:** $220

Ithaca Model 37V Featherweight
Same specs as Model 37 except 12-, 20-ga.; 5-shot magazine; vent rib; Ithaca Raybar front sight; checkered American walnut pistol-grip stock; decorated receiver; crossbolt safety. No longer in production.
   Perf.: $250   **Exc.:** $195   **VGood:** $170

ITHACA MODEL 66
Lever action; single shot; manually cocked hammer; 12-, 20-ga., .410; 26″ Full (.410), 28″ Full or Modified (12-, 20-ga.), 30″ Full (12-ga.) barrel; checkered straight stock, uncheckered forearm. Introduced 1963; discontinued 1979.
   Exc.: $155   **VGood:** $80   **Good:** $60

Ithaca Model 66 Long Tom
Same specs as Model 66 except 12-ga.; 36″ Full-choke barrel; recoil pad. Introduced 1969; discontinued 1974.
   Exc.: $175   **VGood:** $100   **Good:** $75

Ithaca Model 66RS Buck Buster
Same specs as Model 66 except 12-, 20-ga.; 22″ Cylinder barrel; rifle sights. Introduced 1967; 12-ga. discontinued 1970; 20-ga. discontinued 1979.
   Perf.: $140   **Exc.:** $100   **VGood:** $80

Ithaca Model 66 Vent Rib
Same specs as Model 66 except vent-rib barrel. Introduced 1969; discontinued 1974.
   Perf.: $150   **Exc.:** $110   **VGood:** $90

Ithaca Model 66 Youth Model
Same specs as Model 66 except 20-ga., .410; 25″ barrel; shorter stock; recoil pad. Introduced 1965; discontinued 1979.
   Exc.: $125   **VGood:** $80   **Good:** $60

ITHACA MODEL 87
Slide action; 12-, 20-ga.; 3″ chamber; 5-shot magazine; 26″, 28″, 30″ barrel; choke tubes; Raybar front sight; checkered walnut pistol-grip stock, forend; bottom ejection; crossbolt safety. Introduced 1988; still in production. **LMSR: $477**
   Perf.: $350   **Exc.:** $220   **VGood:** $180

Ithaca Model 87 Camo
Same specs as Model 87 except 12-ga.; 24″, 26″, 28″ vent-rib barrel; Modified choke tube; camouflage finish. Introduced 1988; still in production. **LMSR: $547**
   Perf.: $410   **Exc.:** $240   **VGood:** $180

Ithaca Model 87 Combo
Same specs as Model 87 except 28″ barrel with choke tubes, 20″, 25″ rifled barrel; no checkering. No longer in production.
   Perf.: $380   **Exc.:** $250   **VGood:** $200

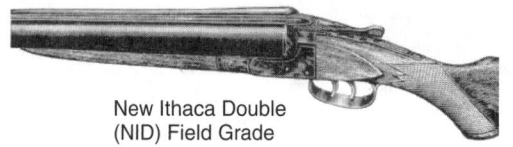
New Ithaca Double
(NID) Field Grade

New Ithaca Double (NID)
Skeet Model Field Grade

Ithaca Slide- & Lever-Action Shotguns
Model 87 Deerslayer
Model 87 Deerslayer II
Model 87 Deerslayer II Fast Twist
Model 87 Deerslayer Deluxe
Model 87 Deerslayer Deluxe Combo
Model 87 Deluxe
Model 87 Deluxe Combo
Model 87 Deluxe Magnum
Model 87 English
Model 87 M&P
Model 87 Supreme
Model 87 Turkey Gun
Model 87 Ultra Deerslayer
Model 87 Ultralight
Model 87 Ultralight Deluxe

Ithaca Side-By-Side & Over/Under Shotguns
Crass Model No. 1
Crass Model No. 1 Special
Crass Model No. 1½
Crass Model No. 2
Crass Model No. 3
Crass Model No. 4
Crass Model No. 5
Crass Model No. 6
Crass Model No. 7
Double (NID) Field Grade

Ithaca Model 87 Deerslayer
Same specs as Model 87 except 20″, 25″ slug or rifled barrel; Raybar blade front sight on ramp, rear adjustable for windage, elevation; grooved for scope mounting. Introduced 1988; discontinued 1993.
Perf.: $325   **Exc.:** $210   **VGood:** $180

Ithaca Model 87 Deerslayer II
Same specs as Model 87 except rifled barrel; Raybar blade front sight on ramp, rear adjustable for windage and elevation; grooved for scope mounting; high-gloss checkered American walnut Monte Carlo stock, forend; solid-frame construction. Introduced 1988; still in production. **LMSR: $567**
Perf.: $425   **Exc.:** $275   **VGood:** $210

Ithaca Model 87 Deerslayer II Fast Twist
Same specs as Model 87 except 12-ga.; 25″ non-removable rifled barrel; Raybar blade front sight on ramp, rear ajustable for windage and elevation; grooved for scope mounting; high-gloss checkered American walnut Monte Carlo stock, forend; solid-frame construction. Introduced 1992; discontinued 1993.
Perf.: $420   **Exc.:** $300   **VGood:** $250

Ithaca Model 87 Deerslayer Deluxe
Same specs as Model 87 except 20″, 25″ slug or rifled barrel; Raybar blade front sight on ramp, rear adjustable for windage and elevation; grooved for scope mounting; high-gloss checkered walnut pistol-grip stock, forend; gold trigger. Introduced 1989; still in production. **LMSR: $465**
Perf.: $375   **Exc.:** $240   **VGood:** $195

Ithaca Model 87 Deerslayer Deluxe Combo
Same specs as Model 87 except 20, 25″ slug or rifled barrel; extra 28″ barrel; Raybar blade front sight on ramp, rear adjustable for windage and elevation; grooved for scope mounting; high-gloss checkered walnut pistol-grip stock, forend; gold trigger. Introduced 1989; still in production. **LMSR: $585**
Perf.: $475   **Exc.:** $340   **VGood:** $295

Ithaca Model 87 Deluxe
Same specs as Model 87 except vent-rib barrel; choke tubes; high-gloss walnut stock, forend; gold trigger. Introduced 1988; still in production. **LMSR: $533**
Perf.: $400   **Exc.:** $300   **VGood:** $240

Ithaca Model 87 Deluxe Combo
Same specs as Model 87 except extra 28″ barrel; vent rib; choke tubes; high-gloss walnut stock, forend; gold trigger. No longer in production.
Perf.: $450   **Exc.:** $350   **VGood:** $290

Ithaca Model 87 Deluxe Magnum
Same specs as Model 87 except vent-rib barrel; choke tubes; high-gloss walnut stock, forend; gold trigger. Introduced 1981; discontinued 1988.
Perf.: $310   **Exc.:** $240   **VGood:** $190

Ithaca Model 87 English
Same specs as Model 87 except 20-ga.; 3″ chamber; 24″, 26″ vent-rib barrel; choke tubes; recoil pad. Introduced 1991; still in production. **LMSR: $545**
Perf.: $410   **Exc.:** $275   **VGood:** $225

Ithaca Model 87 M&P
Same specs as Model 87 except 5-, 8-shot magazine; 18½″, 20″ Cylinder bore barrel; weighs 7 lbs.; black polymer pistol-grip stock, forearm; Parkerized finish. Introduced 1988; discontinued 1993.
Perf.: $250   **Exc.:** $200   **VGood:** $170

Ithaca Model 87 Supreme
Same specs as Model 87 except vent-rib; fancy-grade walnut stock, forend; engraving. Introduced 1988; still in production. **LMSR: $809.**
Perf.: $650   **Exc.:** $400   **VGood:** $300

Ithaca Model 87 Turkey Gun
Same specs as Model 87 except 12-ga.; 24″ barrel; fixed Full choke or Full choke tube; matte blue or Camoseal finish. Introduced 1988; still in production. **LMSR: $466**
Perf.: $375   **Exc.:** $280   **VGood:** $200

Ithaca Model 87 Ultra Deerslayer
Same specs as Model 87 except 20″, 25″ slug or rifled barrel; Raybar blade front sight on ramp, rear adjustable for windage and elevation; grooved for scope mounting; high-gloss checkered walnut pistol-grip stock, forend; gold trigger; aluminum construction. Introduced 1989; discontinued 1990.
Perf.: $325   **Exc.:** $225   **VGood:** $180

Ithaca Model 87 Ultralight
Same specs as Model 87 except 12-, 20-ga.; 3″ chamber; 20″, 24″, 25″, 26″ barrel; fixed or choke tubes; aluminum construction. Introduced 1988; discontinued 1990.
Perf.: $375   **Exc.:** $260   **VGood:** $200

Ithaca Model 87 Ultralight Deluxe
Same specs as Model 87 except 12-, 20-ga.; 3″ chamber; 20″, 24″, 25″, 26″ barrel; fixed or choke tubes; high-gloss select-grade checkered walnut stock, forend; gold trigger; aluminum construction. Introduced 1989; discontinued 1991.
Perf.: $410   **Exc.:** $300   **VGood:** $235

ITHACA SIDE-BY-SIDE & OVER/UNDER SHOTGUNS

ITHACA CRASS MODEL NO. 1
Side-by-side; boxlock; hammerless; (also available with exposed hammers); 10-, 12-, 16-ga. (introduced about 1898); twist barrels; American walnut pistol-grip stock; double triggers; extractors (ejectors optional after 1897). Ithaca's first hammerless model; original grades, No. 1 through No. 7; No. 1 Special and No. 1½ added about 1898. Introduced, about 1888; discontinued, 1901; serial number range: 6981-49104.
Exc.: $350   **VGood:** $300   **Good:** $240

Ithaca Crass Model No. 1 Special
Same specs as Crass Model No. 1 except steel barrels.
Exc.: $400   **VGood:** $350   **Good:** $295

Ithaca Crass Model No. 1½
Same specs as Crass Model No. 1.
Exc.: $450   **VGood:** $400   **Good:** $325

Ithaca Crass Model No. 2
Same specs as Crass Model No. 1 except English walnut pistol-grip stock.
Exc.: $570   **VGood:** $425   **Good:** $350

Ithaca Crass Model No. 3
Same specs as Crass Model No. 1 except twist or steel barrels; English walnut pistol-grip stock.
Exc.: $660   **VGood:** $500   **Good:** $400

Ithaca Crass Model No. 4
Same specs as Crass Model No. 1 except twist or steel barrels; French walnut pistol-grip stock.
Exc.: $1350   **VGood:** $1000   **Good:** $800

Ithaca Crass Model No. 5
Same specs as Crass Model No. 1 except twist or Krupp steel barrels; French walnut pistol-grip stock.
Exc.: $1800   **VGood:** $1450   **Good:** $1100

Ithaca Crass Model No. 6
Same specs as Crass Model No. 1 except twist or Krupp steel barrels; French walnut pistol-grip stock.
Exc.: $2200   **VGood:** $1850   **Good:** $1500

Ithaca Crass Model No. 7
Same specs as Crass Model No. 1 except twist or Whitworth steel barrels; French walnut pistol-grip stock.
Exc.: $2500   **VGood:** $2150   **Good:** $1800

NEW ITHACA DOUBLE (NID) FIELD GRADE
Side-by-side; boxlock; hammerless; 10- (3″. 3½″ chambers), 12- (2¾″. 3″ chambers), 16-, 20-ga., .410; 26″ (12-, 16-, 20-ga., .410), 28″ (12-, 16-, 20-ga., .410), 30″ (12-, 16-, 20-ga.), 32″ (10-ga.), 34″ (10-ga.) fluid-steel barrels; American walnut pistol-grip stock; extractors (ejectors optional); double triggers (single trigger optional); optional recoil pad, ivory sight. Produced in nine grades. Magnum 10-ga. available in any grade, on special-order only, 1932-1942; total production about 887. Magnum 12-ga. available on special-order only, after 1937; total production about 87. Early NIDs (1926-1936) have snail-ear cocking indicators at the top of the frame. Introduced 1926; discontinued 1948; serial number range: 425000-470099.
Exc.: $590   **VGood:** $480   **Good:** $400

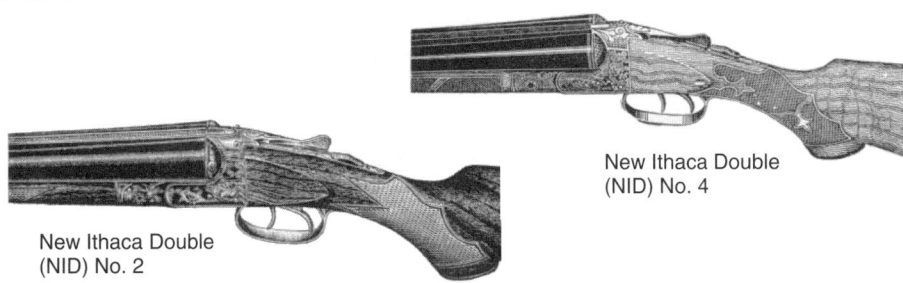

New Ithaca Double
(NID) No. 4

New Ithaca Double
(NID) No. 2

New Ithaca Double (NID) Skeet Model Field Grade
Same specs as New Ithaca Double (NID) Field Grade except ivory sight; American walnut pistol-grip stock, beavertail forend; ejectors; recoil pad. Introduced 1935; discontinued 1948.
    Exc.: $750   **VGood: $600**   **Good: $475**

New Ithaca Double (NID) No. 2
Same specs as New Ithaca Double (NID) Field Grade.
    Exc.: $950   **VGood: $550**   **Good: $525**

New Ithaca Double (NID) No. 3
Same specs as New Ithaca Double (NID) Field Grade; ejectors standard after 1935.
    Exc.: $1100   **VGood: $700**   **Good: $600**

New Ithaca Double (NID) No. 4
Same specs as New Ithaca Double (NID) Field Grade except extractors (ejectors optional until 1935, then standard); double triggers (single trigger optional); optional beavertail forend, vent rib, recoil pad, ivory sight.
    Exc.: $2250   **VGood: $1600**   **Good: $1200**

New Ithaca Double (NID) No. 5
Same specs as New Ithaca Double (NID) Field Grade except extractors (ejectors optional until 1926, then standard); double triggers (single trigger optional); optional beavertail forend, vent rib, recoil pad, ivory sight.
    Exc.: $3000   **VGood: $2650**   **Good: $1900**

New Ithaca Double (NID) No. 7
Same specs as New Ithaca Double (NID) Field Grade except extractors (ejectors optional until 1926, then standard); double triggers (single trigger optional); optional beavertail forend, vent rib, recoil pad, ivory sight.
    Exc.: $6000   **VGood: $5000**   **Good: $4500**

New Ithaca Double (NID) Sousa Special
Same specs as New Ithaca Double (NID) Field Grade except extractors (ejectors optional until 1926, then standard); double triggers (single trigger optional); optional beavertail forend, vent rib, recoil pad, ivory sight. Extreme rarity precludes pricing.

ITHACA FLUES MODEL FIELD GRADE
Side-by-side; boxlock; hammerless (also available with exposed hammers until about 1915); 10-, 12-, 16-, 20-ga.; 24″ (20-ga.), 26″ (12-, 16-, 20-ga.), 28″ (12-, 16-, 20-ga.), 30″ (10-, 12-, 16-ga.), 32″ (10-, 12-ga.) fluid-steel barrels; American walnut half-pistol-grip stock; extractors (ejectors optional); double triggers (Infallible single trigger optional, 1915-1916); produced in eleven grades. Introduced 1908; discontinued 1926; serial number range: 175000-398365.
    Exc.: $750   **VGood: $500**   **Good: $350**

Ithaca Flues Model No. 1
Same specs as Flues Model Field Grade except also 28-ga. with 24″, 26″ barrels; twist barrels; American walnut pistol-grip stock.
    Exc.: $800   **VGood: $650**   **Good: $400**

Ithaca Flues Model No. 1 Special
Same specs as Flues Model Field Grade except also 28-ga. with 24″, 26″ barrels; American walnut pistol-grip stock.
    Exc.: $850   **VGood: $700**   **Good: $500**

Ithaca Flues Model No. 1 ¹/₂
Same specs as Flues Model Field Grade except also 28-ga. with 24″, 26″ barrels; twist barrels; American walnut pistol-grip stock.
    Perf.: $750   **Exc.: $500**   **VGood: $350**

Ithaca Flues Model No. 2
Same specs as Flues Model Field Grade except also 28-ga. with 24″, 26″ barrels; twist barrels; American walnut pistol-grip stock.
    Exc.: $750   **VGood: $500**   **Good: $350**

Ithaca Flues Model No. 2 Krupp Pigeon Gun
Same specs as Flues Model Field Grade except also 28-ga. with 24″, 26″ barrels; Krupp steel barrels; European walnut stock; pistol, half-pistol or straight grip.
    Exc.: $800   **VGood: $550**   **Good: $400**

Ithaca Flues Model No. 3
Same specs as Flues Model Field Grade except also 28-ga. with 24″, 26″ barrels; twist or Krupp steel barrels; European walnut stock; pistol, half-pistol or straight grip.
Krupp steel
    Exc.: $825   **VGood: $550**   **Good: $400**

Ithaca Flues Model No. 4
Same specs as Flues Model Field Grade except also 28-ga. with 24″, 26″ barrels; twist or Krupp steel barrels; European walnut stock; pistol, half-pistol or straight grip.
Krupp steel
    Exc.: $1350   **VGood: $1000**   **Good: $800**

Ithaca Flues Model No. 5
Same specs as Flues Model Field Grade except also 28-ga. with 24″, 26″ barrels; twist or Krupp steel barrels; European walnut stock; pistol, half-pistol or straight grip.
Krupp steel
    Exc.: $1800   **VGood: $1450**   **Good: $1100**

Ithaca Flues Model No. 6
Same specs as Flues Model Field Grade, except also 28-ga. with 24″, 26″ barrels; twist or Krupp steel barrels; English walnut stock; pistol, half-pistol or straight grip. Extreme rarity precludes pricing.

Ithaca Flues Model No. 7
Same specs as Flues Model Field Grade, except also 28-ga. with 24″, 26″ barrels; Whitworth steel barrels; English walnut stock; pistol, half-pistol or straight grip. Extreme rarity precludes pricing.

ITHACA FLUES ONE-BARREL TRAP GUN NO. 4
Side-by-side; boxlock; hammerless; 12-ga.; 30″, 32″, 34″ Krupp steel barrels; vent rib; American walnut stock; pistol, half-pistol or straight grip; ejector; decoration similar to equivalent grades of Flues Model double guns. Introduced 1914; discontinued 1922; serial number range same as Flues Model doubles; total production about 4700.
    Exc.: $1050   **VGood: $800**   **Good: $650**

Ithaca Flues One-Barrel Trap Gun No. 5
Same specs as Flues One-Barrel Trap Gun No. 4.
    Exc.: $1350   **VGood: $1000**   **Good: $800**

Ithaca Flues One-Barrel Trap Gun No. 6
Same specs as Flues One-Barrel Trap Gun No. 4. Extreme rarity precludes pricing.

SHOTGUNS

Ithaca Side-By-Side & Over/Under Shotguns

New Ithaca Double (NID) Skeet Model Field Grade
New Ithaca Double (NID) No. 2
New Ithaca Double (NID) No. 3
New Ithaca Double (NID) No. 4
New Ithaca Double (NID) No. 5
New Ithaca Double (NID) No. 7
New Ithaca Double (NID) Sousa Special
Flues Model Field Grade
Flues Model No. 1
Flues Model No. 1 Special
Flues Model No. 1 ¹/₂
Flues Model No. 2
Flues Model No. 2 Krupp Pigeon Gun
Flues Model No. 3
Flues Model No. 4
Flues Model No. 5
Flues Model No. 6
Flues Model No. 7
Flues One-Barrel Trap Gun No. 4
Flues One-Barrel Trap Gun No. 5
Flues One-Barrel Trap Gun No. 6
Flues One-Barrel Trap Gun No. 7
Flues One-Barrel Gun Sousa Special
Flues One-Barrel Trap Gun Victory Grade
Knickerbocker One-Barrel Trap Gun No. 4
Knickerbocker One-Barrel Trap No. 5

Ithaca Flues One-Barrel Trap Gun No. 7
Same specs as Flues One-Barrel Trap Gun No. 4.
    Exc.: $2500   **VGood: $2200**   **Good: $1900**

Ithaca Flues One-Barrel Trap Gun Sousa Special
Same specs as Flues One-Barrel Trap Gun No. 4. Introduced 1918; discontinued 1922. Extreme rarity precludes pricing.

Ithaca Flues One-Barrel Trap Gun Victory Grade
Same specs as Flues One-Barrel Trap Gun No. 4 except extractor; straight-grip stock. Introduced 1919; discontinued 1922.
    Exc.: $800   **VGood: $600**   **Good: $450**

ITHACA KNICKERBOCKER ONE-BARREL TRAP GUN NO. 4
Boxlock; hammerless; 12-ga.; 30″, 32″, 34″ fluid-steel barrel; vent rib; American walnut straight or pistol-grip stock; recoil pad; ejector; decoration similar to equivalent grades of New Ithaca Double guns; named for its designer, Frank Knickerbocker, as it was widely known among shooters as the Knick. Introduced 1922; discontinued 1977, except on special order; serial number range: 400000-405739.
    Exc.: $1495   **VGood: $995**   **Good: $750**

Ithaca Knickerbocker One-Barrel Trap No. 5
Same specs as Ithaca Knickerbocker One-Barrel Trap Gun No. 4. Introduced 1922; discontinued 1982.
    Exc.: $3500   **VGood: $3000**   **Good: $2500**

Ithaca Side-By Side & Over/Under Shotguns

Knickerbocker One-Barrel Trap No. 6
Knickerbocker One-Barrel Trap No. 7
Knickerbocker One-Barrel Trap Sousa Special
Knickerbocker One-Barrel Trap Victory Grade
Knickerbocker One-Barrel Trap $1000 Grade
Minier Model Field Grade
Minier Model No. 1
Minier Model No. 1 Special
Minier Model No. 1½
Minier Model No. 2
Minier Model No. 2 Krupp Pigeon Gun
Minier Model No. 3
Minier Model No. 4
Minier Model No. 5
Minier Model No. 6
Minier Model No. 7
Lewis Model No. 1
Lewis Model No. 1 Special
Lewis Model No. 1½
Lewis Model No. 2
Lewis Model No. 3
Lewis Model No. 4
Lewis Model No. 5
Lewis Model No. 6
Lewis Model No. 7
LSA-55 Turkey Gun
Single Barrel 5E Custom Trap

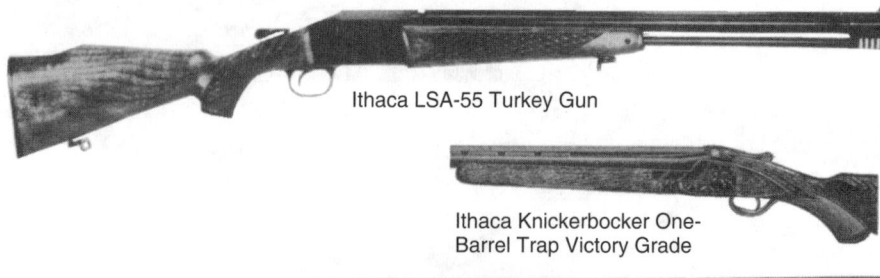

Ithaca LSA-55 Turkey Gun

Ithaca Knickerbocker One-Barrel Trap Victory Grade

Ithaca Single Barrel 5E Custom Trap

Ithaca Knickerbocker One-Barrel Trap No. 6
Same specs as Ithaca Knickerbocker One-Barrel Trap Gun No. 4. Though never a catalogue item, a few Knicks in grade No. 6 were built in 1922-1923.
Exc.: $3800   **VGood:** $2900   **Good:** $2200

Ithaca Knickerbocker One-Barrel Trap No. 7
Same specs as Ithaca Knickerbocker One-Barrel Trap Gun No. 4. Introduced 1922; discontinued early 1960s.
Exc.: $4200   **VGood:** $3200   **Good:** $2500

Ithaca Knickerbocker One-Barrel Trap Sousa Special
Same specs as Ithaca Knickerbocker One-Barrel Trap Gun No. 4. Introduced 1922; discontinued 1937. Extreme rarity precludes pricing.

Ithaca Knickerbocker One-Barrel Trap Victory Grade
Same specs as Ithaca Knickerbocker One-Barrel Trap Gun No. 4 except 32″ barrel; straight-grip stock. Introduced 1922; discontinued 1937.
Exc.: $795   **VGood:** $600   **Good:** $450

Ithaca Knickerbocker One-Barrel Trap $1000 Grade
Same specs as Ithaca Knickerbocker One-Barrel Trap Gun No. 4. Introduced in 1937 to replace Sousa Special Grade; the name was changed periodically to

reflect price increases: $1500 Grade, $2000 Grade, $6500 Grade; after 1980, it was simply Dollar Grade; discontinued 1981.
Perf.: $6000   **Exc.:** $4800   **VGood:** $4000

ITHACA MINIER MODEL FIELD GRADE
Side-by-side; boxlock; hammerless (also available with exposed hammers); 10-, 12-, 16-ga.; 26″. 28″. 30″. 32″ fluid-steel barrels; American walnut half-pistol-grip stock, forearm; extractors (ejectors optional except as noted); double triggers; mechanically quite different from the Lewis Model; produced in eleven grades. Introduced 1906; discontinued 1908; serial number range: 130000-151770.
Exc.: $400   **VGood:** $350   **Good:** $290

Ithaca Minier Model No. 1
Same specs as Minier Model Field Grade except 10-, 12-, 16-, 20-ga.; twist barrels; American walnut pistol-grip stock.
Exc.: $400   **VGood:** $350   **Good:** $290

Ithaca Minier Model No. 1 Special
Same specs as Minier Model Field Grade except American walnut pistol-grip stock.
Exc.: $425   **VGood:** $375   **Good:** $300

Ithaca Minier Model No. 1 1/2
Same specs as Minier Model Field Grade except 10-, 12-, 16-, 20-ga.; twist barrels; American walnut pistol-grip stock.
Perf.: $400   **Exc.:** $350   **VGood:** $290

Ithaca Minier Model No. 2
Same specs as Minier Model Field Grade except 10-, 12-, 16-, 20-ga.; twist barrels; American walnut pistol-grip stock.
Exc.: $470   **VGood:** $400   **Good:** $350

Ithaca Minier Model No. 2 Krupp Pigeon Gun
Same specs as Minier Model Field Grade except 10-, 12-, 16-, 20-ga.; Krupp steel barrels; American walnut pistol-grip stock.
Exc.: $750   **VGood:** $550   **Good:** $400

Ithaca Minier Model No. 3
Same specs as Minier Model Field Grade except 10-, 12-, 16-, 20-ga.; twist or Krupp steel barrels; American walnut pistol-grip stock.
Exc.: $850   **VGood:** $625   **Good:** $450

Ithaca Minier Model No. 4
Same specs as Minier Model Field Grade except 10-, 12-, 16-, 20-ga.; twist or Krupp steel barrels; American walnut straight or pistol-grip stock.
Exc.: $1400   **VGood:** $1100   **Good:** $900

Ithaca Minier Model No. 5
Same specs as Minier Model Field Grade except 10-, 12-, 16-, 20-ga.; twist or Krupp steel barrels; American walnut straight or pistol-grip stock.
Exc.: $2000   **VGood:** $1650   **Good:** $1400

Ithaca Minier Model No. 6
Same specs as Minier Model Field Grade except 10-, 12-, 16-, 20-ga.; twist or Krupp steel barrels; American

walnut straight or pistol-grip stock.
Exc.: $3500   **VGood:** $3000   **Good:** $1500

Ithaca Minier Model No. 7
Same specs as Minier Model Field Grade, except 10-, 12-, 16-, 20-ga.; twist or Whitworth steel barrels; American walnut straight or pistol-grip stock.
Exc.: $3600   **VGood:** $3200   **Good:** $1700

ITHACA LEWIS MODEL NO. 1
Side-by-side; boxlock; hammerless; 10-, 12-, 16-, 20-ga. (intro. 1904); twist barrels; American walnut pistol-grip stock; extractors (ejectors optional); same grades as and slightly redesigned version of Crass Model. Introduced 1901; discontinued 1906; serial number range: 94109-123677.
Exc.: $500   **VGood:** $300   **Good:** $240

Ithaca Lewis Model No. 1 Special
Same specs as Lewis Model No. 1 except steel barrels.
Exc.: $800   **VGood:** $525   **Good:** $240

Ithaca Lewis Model No. 1 1/2
Same specs as Lewis Model No. 1.
Exc.: $800   **VGood:** $550   **Good:** $240

Ithaca Lewis Model No. 2
Same specs as Lewis Model No. 1 except English walnut pistol-grip stock.
Exc.: $825   **VGood:** $600   **Good:** $350

Ithaca Lewis Model No. 3
Same specs as Lewis Model No. 1 except twist or steel barrels; English walnut pistol-grip stock.
Exc.: $1400   **VGood:** $750   **Good:** $400

Ithaca Lewis Model No. 4
Same specs as Lewis Model No. 1 except twist or steel barrels; French walnut pistol-grip stock.
Exc.: $1900   **VGood:** $1700   **Good:** $800

Ithaca Lewis Model No. 5
Same specs as Lewis Model No. 1 except twist or Krupp steel barrels; French walnut pistol-grip stock.
Exc.: $1800   **VGood:** $1450   **Good:** $1100

Ithaca Lewis Model No. 6
Same specs as Lewis Model No. 1 except twist or Krupp steel barrels; French walnut pistol-grip stock.
Exc.: $4250   **VGood:** $3000   **Good:** $2400

Ithaca Lewis Model No. 7
Same specs as Lewis Model No. 1 except twist or Whitworth steel barrels; French walnut pistol-grip stock.
Exc.: $4250   **VGood:** $3200   **Good:** $2400

ITHACA LSA-55 TURKEY GUN
Over/under combo; exposed hammers; 12-ga./.222 Rem.; 24 1/2″ vent-rib barrels; folding leaf rear sight, bead front; checkered walnut Monte Carlo stock, forearm; plain extractor; single trigger. Made in Finland. Introduced 1970; discontinued 1979.
Perf.: $600   **Exc.:** $480   **VGood:** $400

ITHACA SINGLE BARREL 5E CUSTOM TRAP
Single barrel; boxlock; 12-ga.; 2 3/4″ chamber; 32″, 34″ Full choke barrel; weighs about 8 1/2 lbs.; white front bead, brass middle bead sights; checkered fancy American walnut stock, forend; extensively engraved and gold inlaid frame, top lever, trigger guard. Reintroduced 1988; discontinued 1991.
Perf.: $3000   **Exc.:** $2400   **VGood:** $2000

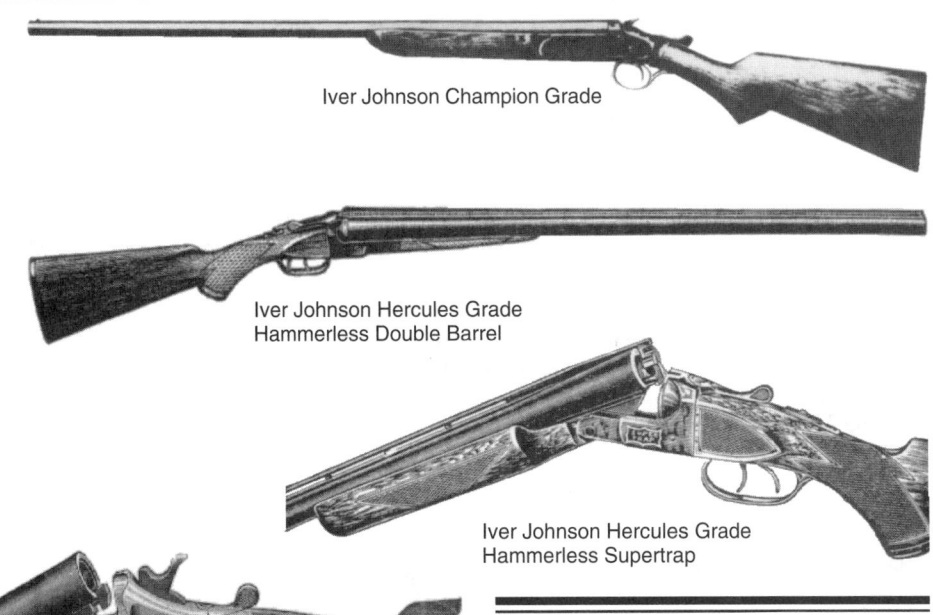

Iver Johnson Champion Grade

Iver Johnson Hercules Grade
Hammerless Double Barrel

Iver Johnson Hercules Grade
Hammerless Supertrap

Iver Johnson Champion
Top Snap

SHOTGUNS

Iver Johnson
Champion Matted Rib Model
Champion Model 36
Champion Model 36 Junior
Champion Model 39
Champion Model 39 Junior
Champion Top Snap
Champion Top Snap New Model
Champion Top Snap New Model Junior
Ejector Single
Ejector Single Junior
Hammerless Double Barrel
Hercules Grade Hammerless Double Barrel
Hercules Grade Hammerless Supertrap

Iver Johnson
Hercules Grade Hammerless

Ga.	Barrel (ins.)	Optional Barrels	Choke Lt./Rt.	Wgt./lbs. Extractor	Wgt./lbs. Ejector
12	30	28, 32	Full/Mod.	7	7 1/4
16	28	30, 32	Full/Mod.	6 3/4	7
20	28	None	Full/Mod.	6 1/2	6 3/4
410	26	None	Full/Full	5 3/4	6
	32		Full/Full		

IVER JOHNSON CHAMPION MATTED RIB MODEL

Single shot; top-lever breaking; 12-, 16-, 20-ga., .410 (intro. 1928); 28″, 30″, 32″ barrel; Full choke; solid raised full-length rib finely matted; 45″ overall length (30″ barrel); weighs 7 1/4 lbs. (12-ga.), 7 lbs. (16-, 20-ga.); full-checkered pistol-grip stock; automatic extractor; optional automatic ejector; hard rubber pistol-grip cap, buttplate; large knob on forend. Introduced 1913; discontinued 1941.
   Exc.: $100   **VGood:** $75   **Good:** $50

IVER JOHNSON CHAMPION MODEL 36

Single shot; top lever breaking; 12-, 16-, 20-, 24-, 28-ga., .410; 28″, 30″, 32″ barrel; Full choke; 45″ overall length (30″ barrel); weighs 6 3/4 lbs.; American black walnut stock, forend; rebounding hammer; barrel and lug forged in one; extractor; optional automatic ejector; case-hardened frame, nickel optional; browned barrel; Introduced 1909; name changed to Champion Single Barrel Shotgun in 1913; .410, 24-, 28-ga. introduced 1913; 24-ga. discontinued 1928; 16-, 20-, 28-ga., .410 discontinued 1941; 12-ga. discontinued 1957.
   Exc.: $115   **VGood:** $100   **Good:** $75

Iver Johnson Champion Model 36 Junior

Same specs as Champion Model 36 except 26″ barrel; shortened stock. Introduced 1909; dropped 1941.
   Exc.: $115   **VGood:** $100   **Good:** $75

IVER JOHNSON CHAMPION MODEL 39

Single shot; top-lever breaking; 24-, 28-ga.; 28″. 30″ plain barrel; Full choke; 45″ overall length (30″ barrel); weighs 5 3/4 lbs.; American black walnut stock, forend; hard rubber pistol-grip cap, buttplate; barrel and lug forged as one; rebounding hammer; extractor; optional automatic ejector; case-hardened frame, nickel optional; brown barrel. Introduced 1909; discontinued 1913.
   Exc.: $135   **VGood:** $120   **Good:** $100

Iver Johnson Champion Model 39 Junior

Same specs as Champion Model 39 except 26″. 28″ barrel; shortened stock. Introduced 1909; dropped 1913.
   Exc.: $135   **VGood:** $120   **Good:** $100

IVER JOHNSON CHAMPION TOP SNAP

Single shot; top lever breaking; 12-ga.; 30″ plain twist steel barrel; 45 3/4″ overall length; weighs 6 1/2 lbs.; American black walnut stock, forend; rebounding hammer; double locking bolt; extractor; nickeled frame; browned barrel imported from Belgium. Introduced 1880; discontinued 1905.
   Exc.: $75   **VGood:** $60   **Good:** $40

IVER JOHNSON CHAMPION TOP SNAP NEW MODEL

Single shot; top lever breaking; 12-, 16-ga.; 28″, 30″, 32″ steel barrel; Full choke; 45″ overall length (30″ barrel); weighs 6 3/4 lbs.; American black walnut stock, forend; hard rubber pistol grip, buttplate cap; rebounding hammer; automatic ejector; case-hardened frame; browned barrel. Introduced 1900; discontinued 1908.
   Exc.: $90   **VGood:** $65   **Good:** $50

Iver Johnson Champion Top Snap New Model Junior

Same specs as Champion Top Snap New Model except shortened stock. Introduced 1900; dropped 1908.
   Exc.: $90   **VGood:** $65   **Good:** $50

IVER JOHNSON EJECTOR SINGLE

Single shot; ring trigger breakopen; semi-hammerless; 12-, 16-ga.; 28″, 30″, 32″ plain twist steel barrel; 43 3/4″ overall length (28″ barrel); weighs 7 lbs.; American black walnut stock, forend; bobbed rebounding hammer; automatic ejector; case-hardened frame; browned barrel; after 1900 ring trigger model available with full hammer. Introduced 1885; 16-ga. introduced 1900; discontinued 1905.
   Exc.: $85   **VGood:** $60   **Good:** $40

Iver Johnson Ejector Single Junior

Same specs as Ejector Single except shortened stock. Introduced 1900; dropped 1905.
   Exc.: $90   **VGood:** $65   **Good:** $50

IVER JOHNSON HAMMERLESS DOUBLE BARREL

Side-by-side; boxlock; 12-ga.; 28″, 30″, 32″ barrel; 28″,

30″ Full (left barrel) and Modified (right), 32″ Full choke; 46 3/8″ overall length; weighs 7 1/2 lbs.; American black walnut stock, forend; hard rubber buttplate, pistol-grip cap; breech sleeve area is solid drop-forged piece; hammer can be lowered without snapping on empty chambers; automatic safety; automatic ejectors; double triggers. Introduced 1913; discontinued 1923.
   Exc.: $300   **VGood:** $275   **Good:** $250

IVER JOHNSON HERCULES GRADE HAMMERLESS DOUBLE BARREL

Side-by-side; boxlock; 12-, 16-, 20-ga., .410; 26″, 28″, 30″ barrels; 46 3/8″ overall length (30″ barrels); hand-checkered American black walnut stock with 14″ pull; .410 has straight stock with 13 1/2″ pull; stock drop at comb, 1 3/4″; drop at heel, 2 3/4″; plain extractor model with slim forend; automatic ejector model with beavertail forend with D&E fastener; hard rubber buttplate and pistol grip cap; barrels, lugs forged as one then joined together; automatic safety; double triggers. Introduced 1924; discontinued 1941.
   Exc.: $550   **VGood:** $450   **Good:** $300

IVER JOHNSON HERCULES GRADE HAMMERLESS SUPERTRAP

Side-by-side; boxlock; 12-ga.; 32″ barrel; raised full-length ventilated rib; 48 3/8″ overall length; 8 1/2 lbs.; two Lyman ivory bead sights; hand-checkered American black walnut stock, wide beavertail forend with Deeley & Edge fastener; 14 1/2″ length of stock; 1 1/2″ drop at comb; 2″ drop at heel; hard rubber buttplate and pistol-grip cap; each barrel and lug forged in one, proof-tested, then joined together; automatic selective ejectors; double triggers; anti-flinch recoil pad; automatic safety. Introduced 1924; discontinued 1941.
   Exc.: $900   **VGood:** $800   **Good:** $700

Iver Johnson
Semi-Octagon Barrel
Silver Shadow
SKEET-ER
Special Trap Model
Trigger Action Ejector Single
Trigger Action Ejector
Single Junior

Kassnar
Fox
Grade I
Grade II
Omega O/U
Omega Pump
Omega SxS

Iver Johnson Matted Rib Grade

Iver Johnson SKEET-ER

Iver Johnson Silver Shadow

Iver Johnson Special Trap Model

Iver Johnson Special Trap Model

IVER JOHNSON SEMI-OCTAGON BARREL

Single shot; 12-, 16-ga.; 28″, 30″, 32″ barrel; top half of barrel 5″ from breech semi-octagon; top surface of barrel matted; Full choke; 45″ overall length; weighs 6 3/4 lbs. (12-ga.), 6 1/2 lbs. (16-ga.); American black walnut stock, plain-finished forend with large knob; hard rubber buttplate and pistol-grip cap; extractor; optional automatic ejector; case-hardened frame; browned barrel. Introduced 1913; discontinued 1927.

Exc.: $95   **VGood:** $85   **Good:** $75

IVER JOHNSON SILVER SHADOW

Over/under; boxlock; 12-ga.; 3″ chambers; 26″ Improved Cylinder/Modified, 28″ Modified/Full, 30″ Full/Full barrels; full-length ventilated rib; checkered European walnut pistol-grip stock, forend; plastic pistol-grip cap; plain extractors; double triggers; optional non-selective single trigger. Made in Italy. Introduced 1973; discontinued 1978.

Perf.: $400   **Exc.:** $350   **VGood:** $300

IVER JOHNSON SKEET-ER

Side-by-side; boxlock; 12-, 16- 20-, 28-ga., .410; 2 3/4″, 3″ (.410) chamber; optional 2 1/2″ chamber; 26″, 28″ barrel; 8 1/2 lbs. (12-, 16-, 20-ga.), 7 1/2 lbs. (.410); Skeet choke, 75% at 30 yds. (right barrel), 75% at 20 yds. (left); hand-checkered, lacquer-finished select fancy figured American black walnut straight (.410) or pistol-grip stock, large beavertail forend with Deeley & Edge fastener; 14 1/2″ length; drop at heel 2 5/8″; hard rubber grip cap; each barrel, lug forged in one, proofed then joined together; automatic extractors or automatic ejectors; automatic or manual safety. Introduced 1927; discontinued 1941.

Exc.: $1500   **VGood:** $1250   **Good:** $1000

IVER JOHNSON SPECIAL TRAP MODEL

Single shot; top-lever breaking; 12-ga.; 32″ barrel; Full choke; raised full-length ventilated rib, finely matted; 47″ overall length; weighs 7 lbs.; two Lyman ivory sight beads; hand-checkered American black walnut stock, large trap-style forend; hard rubber buttplate, pistol-grip cap; length of stock, 14 1/2″; drop at comb, 1 1/2″; drop at heel, 2″; automatic ejectors; optional rubber recoil pad. Introduced 1927; discontinued 1941.

Exc.: $250   **VGood:** $225   **Good:** $200

IVER JOHNSON TRIGGER ACTION EJECTOR SINGLE

Single shot; 12-, 16-ga.; 28″, 30″, 32″ barrel; Full choke; 45″ overall length (30″ barrel); weighs 6 lbs.; American black walnut stock, forend; hard rubber pistol-grip cap, buttplate; automatic ejectors; trigger operates locking bolt; fires by small trigger inset in main trigger; case-hardened frame; browned barrel. Introduced 1900; discontinued 1908.

Exc.: $100   **VGood:** $80   **Good:** $60

Iver Johnson Trigger Action Ejector Single Junior

Same specs as Trigger Action Ejector Single except shortened stock; weighs 5 1/2 lbs.; 42″ overall length. Introduced 1900; dropped 1908.

Exc.: $110   **VGood:** $90   **Good:** $70

KASSNAR FOX

Autoloader; 12-ga.; 26″, 28″, 30″ barrels; standard choke combos; vent rib; metal bead front sight; American walnut stock; crossbolt safety; interchangeable barrel. Imported by Kassnar. Introduced 1979; discontinued 1983.

Perf.: $275   **Exc.:** $240   **VGood:** $200

KASSNAR GRADE I

Over/under; boxlock; 12-, 20-, 28-ga., 410; 3″ chambers; 26″, 28″ barrels; fixed or choke tubes; vent rib; checkered European walnut stock, forend; extractors; gold-plated single selective trigger; blued, engraved receiver. Introduced 1990; still imported by K.B.I. Inc.

LMSR: $599

With fixed chokes, extractors
Perf.: $500   **Exc.:** $400   **VGood:** $350
With choke tubes, extractors
Perf.: $550   **Exc.:** $450   **VGood:** $400
With choke tubes; automatic ejectors
Perf.: $600   **Exc.:** $500   **VGood:** $450

KASSNAR GRADE II

Side-by-side; boxlock; 10-, 12-, 16-, 20-, 28-ga., .410; 3″, 3 1/2″ chambers; 26″, 28″, 32″ barrels; concave rib; checkered European walnut stock; double triggers; auto top tang safety; extractors; color case-hardened action. Made in Europe. Introduced 1989; discontinued 1990.

Perf.: $500   **Exc.:** $395   **VGood:** $300

KASSNAR OMEGA O/U

Over/under; .410; 3″ chambers; 24″ Full/Full barrels; vent rib; checkered European walnut stock; auto safety; single trigger; folds for storage, transport. Made in Italy. Introduced 1984; discontinued 1985.

Perf.: $280   **Exc.:** $220   **VGood:** $190

KASSNAR OMEGA PUMP

Slide action; 12-ga.; 2 3/4″ chambers; 5-, 8-shot magazine; 20″, 26″, 28″, 30″ barrels; bead front sight; rifle-type on slug gun; stained hardwood or Philippine mahogany stock; damascened bolt; crossbolt safety. Made in the Philippines. Introduced 1984; discontinued 1986.

Perf.: $240   **Exc.:** $190   **VGood:** $150

KASSNAR OMEGA SxS

Side-by-side; .410; 3″ chambers; 24″ barrels; checkered beech (standard) or walnut (deluxe) semi-pistol-grip stock; top tang safety; blued barrels, receiver; folds for storage, transport. Made in Spain. Introduced 1984; discontinued 1985.
Standard
Perf.: $210   **Exc.:** $185   **VGood:** $150
Deluxe
Perf.: $230   **Exc.:** $205   **VGood:** $170

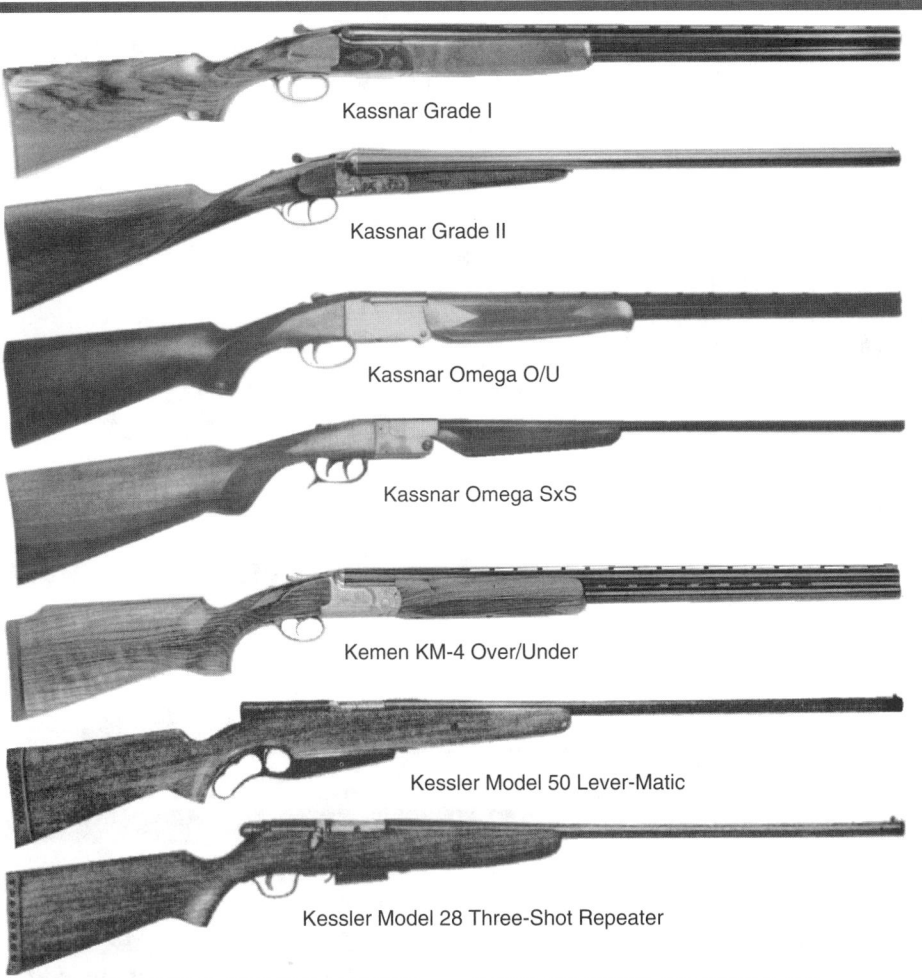

Kassnar Grade I

Kassnar Grade II

Kassnar Omega O/U

Kassnar Omega SxS

Kemen KM-4 Over/Under

Kessler Model 50 Lever-Matic

Kessler Model 28 Three-Shot Repeater

SHOTGUNS

Kassnar
Omega Single Barrel

Kawaguchiya
FG
M-250
M-250 Deluxe
OT-Skeet E1
OT-Skeet E2
OT-Trap E1
OT-Trap E2

Kemen
KM-4 Over/Under

Kessler
Model 28 Three-Shot
Repeater
Model 50 Lever-Matic

KDF
Brescia

KASSNAR OMEGA SINGLE BARREL

Single barrel; 12-, 16-, 20-, 28-ga.; .410; 2 3/4", 3" chambers; 26", 28", 30" Full choke barrel; checkered hardwood (standard) or walnut (deluxe) stock, forearm; chromed (standard) or blued (deluxe) receiver; bottom lever (standard) or top lever (deluxe) break; folds for storage, transport. Introduced 1984; discontinued 1985.
Standard
Perf.: $100   **Exc.:** $80   **VGood:** $70
Deluxe
Perf.: $120   **Exc.:** $100   **VGood:** $80

KAWAGUCHIYA FG

Over/under; boxlock; 12-ga.; 2 3/4" chambers; 26", 28" barrels; vent rib; sterling silver front bead sight; high-grade French walnut stock; selective single trigger; selective auto ejectors; chrome-lined bore; chromed trigger. Made in Japan. Introduced 1981 by La Paloma Marketing; discontinued 1985.
Perf.: $625   **Exc.:** $500   **VGood:** $450

KAWAGUCHIYA M-250

Gas-operated autoloader; 12-ga.; 2 3/4" chamber; 26", 28", 30" barrel; standard fixed chokes or Tru-Choke System; vent rib; checkered French walnut pistol-grip stock; reversible crossbolt safety. Made in Japan. Introduced 1980 by La Paloma Marketing; discontinued 1986.
With fixed chokes
Perf.: $325   **Exc.:** $245   **VGood:** $200
With Tru-Choke System
Perf.: $375   **Exc.:** $300   **VGood:** $250

Kawaguchiya M-250 Deluxe

Same specs as M-250 except silvered, etched receiver. Introduced 1980; discontinued 1986.
With fixed chokes
Perf.: $410   **Exc.:** $300   **VGood:** $260
With Tru-Choke System
Perf.: $460   **Exc.:** $350   **VGood:** $310

KAWAGUCHIYA OT-SKEET E1

Over/under; boxlock; 12-ga.; 2 3/4" chambers; 26", 28" Skeet/Skeet barrels; 13mm vent rib; middle and front bead sights; high-grade French walnut stock; gold-colored wide trigger; plastic buttplate; push-button forend release. Introduced 1981 by La Paloma Marketing; discontinued 1985.
Perf.: $900   **Exc.:** $750   **VGood:** $690

Kawaguchiya OT-Skeet E2

Same specs as OT-Skeet E1 except super-deluxe French Walnut stock, forend; chromed receiver; better scroll engraving. Introduced 1981 by La Paloma Marketing; discontinued 1985.
Perf.: $1350   **Exc.:** $1000   **VGood:** $900

KAWAGUCHIYA OT-TRAP E1

Over/under; boxlock; 12-ga.; 2 3/4" chamber; 30" barrels; 13mm vent rib; bone white middle, front bead sights; high-grade oil-finished French walnut stock; rubber recoil pad; blued scroll-engraved receiver. Introduced 1981 by LaPaloma Marketing; discontinued 1985.
Perf.: $910   **Exc.:** $725   **VGood:** $650

Kawaguchiya OT-Trap E2

Same specs as OT-Trap E1 except super-deluxe French walnut stock, forend; chromed receiver; better scroll engraving. Introduced 1981 by LaPaloma Marketing; discontinued 1985.
Perf.: $1295   **Exc.:** $1000   **VGood:** $900

KEMEN KM-4 OVER/UNDER

Over/under; boxlock; 12-ga,; 2 3/4", 3" chambers; 27 5/8" barrels choked Hunting, Pigeon, Sporting Clays, Skeet; 30", 32" barrels choked Sporting Clays, Trap; ventilated flat or step top rib; ventilated, solid or no side ribs; weighs 7 1/4 to 8 1/2 lbs.; high-grade walnut stock made to customer dimensions; drop-out trigger assembly; low-profile receiver with black finish on Standard model, antique silver on sideplate models and all engraved, gold inlaid models; barrels, forend, trigger parts interchangeable with Perazzi. Comes with hard case, accessory tools, spares. Imported from Spain by USA Sporting Clays. Introduced 1989; still imported.
LMSR: $6179
KM-4 Standard
   Perf.: $5900   **Exc.:** $5000   **VGood:** $4600
KM-4 Luxe-A (engraved scroll), Luxe-B (game scenes)
   Perf.: $9000   **Exc.:** $7000   **VGood:** $5000
KM-4 Super Luxe (engraved game scene)
   Perf.: $11,500 **Exc.:** $9500   **VGood:** $8000
KM-4 Extra Luxe-A (scroll engraved sideplates)
   Perf.: $14,000 **Exc.:** $10,500   **VGood:** $9000
KM-4 Extra Luxe-B (game scene sideplates)
   Perf.: $16,000 **Exc.:** $12,500   **VGood:** $10,000
KM-4 Extra Gold (inlays, game scene)
   Perf.: $18,500 **Exc.:** $15,000   **VGood:** $12,000

KESSLER MODEL 28 THREE-SHOT REPEATER

Bolt action; takedown; 12-, 16-, 20-ga.; 3-shot detachable box magazine; 26" (20-ga.), 28" (12-,16-ga.) barrel; Full choke; uncheckered one-piece pistol-grip stock; recoil pad. Introduced 1951; discontinued 1953.
Exc.: $95   **VGood:** $75   **Good:** $30

KESSLER MODEL 50 LEVER-MATIC

Lever action; takedown; 12-, 16-, 20-ga.; 3-shot magazine; 26", 28", 30" barrels; Full choke; uncheckered pistol-grip stock; recoil pad. Introduced 1951; discontinued 1953.
Exc.: $100   **VGood:** $70   **Good:** $50

KDF BRESCIA

Side-by-side; 12-, 20-ga.; Anson & Deeley-type action; 28" Full/Modified or Improved/Modified chrome-lined barrels; hand-checkered European walnut pistol-grip stock, forearm; recoil pad; double triggers; engraved action. Made in Italy. No longer imported.
Perf.: $300   **Exc.:** $260   **VGood:** $200

KDF
Condor
Condor Trap

Krieghoff
K-80 Live Bird
K-80 Live Bird Lightweight
K-80 Skeet
K-80 Skeet Four-Barrel Set
K-80 Skeet International
K-80 Skeet Lightweight
K-80 Skeet Lightweight
Two-Barrel Set
K-80 Skeet Special
K-80 Sporting Clays

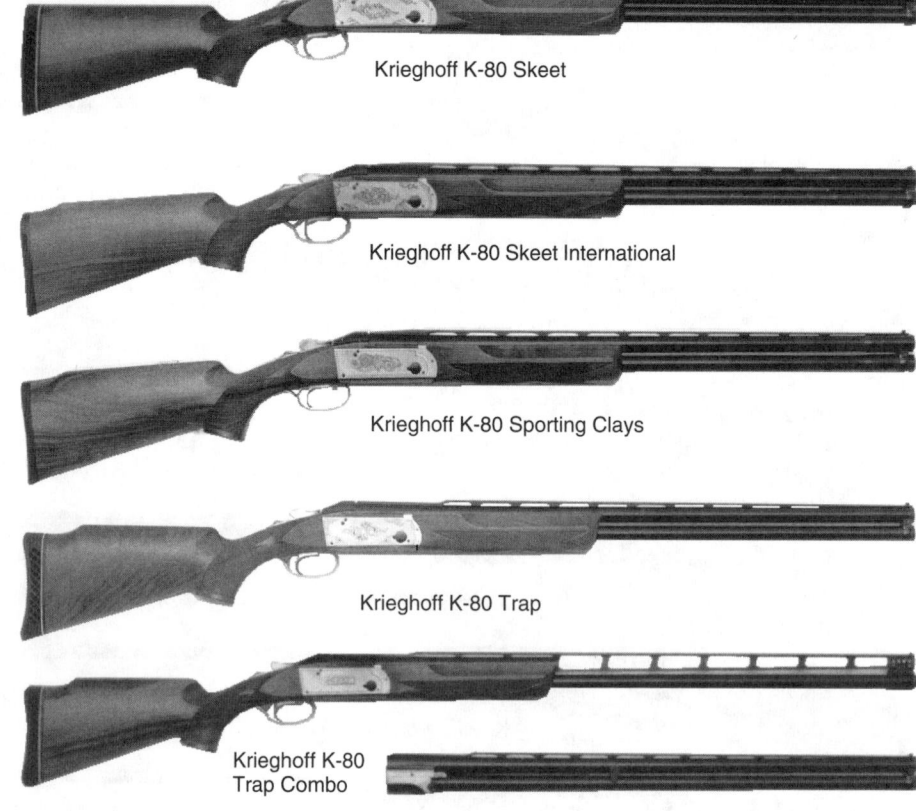

Krieghoff K-80 Skeet

Krieghoff K-80 Skeet International

Krieghoff K-80 Sporting Clays

Krieghoff K-80 Trap

Krieghoff K-80 Trap Combo

KDF CONDOR
Over/under; boxlock; 26″ Improved/Modified or Skeet/Skeet; 28″ Full/Modified or Modified/Modified, 30″ (12-ga.) Full/Modified or Full/Full barrels; vent rib; Skeet model has extra-wide rib; hand-checkered European walnut pistol-grip stock, forearm; single selective trigger; automatic ejectors. Made in Italy. No longer imported.
Field Grade
    Perf.: $600   **Exc.:** $570   **VGood:** $525
Skeet Grade
    Perf.: $625   **Exc.:** $595   **VGood:** $550

KDF Condor Trap
Same specs as Condor except 12-ga.; 28″ Full/Modified, 30″, 32″ Modified/Full or Full/Full barrels; wide vent rib; Monte Carlo stock. No longer imported.
    Perf.: $625   **Exc.:** $595   **VGood:** $550

KRIEGHOFF K-80 LIVE BIRD
Over/under; 12-ga.; 2 ¾″ chambers; 28″, 29″, 30″ barrels; Improved Modified/ Extra-Full chokes; weighs 8 lbs.; checkered European walnut stock; free-floating barrels; steel receiver satin gray finish; engraving; marketed in aluminum case made in four grades with differing wood and engraving options. Introduced 1980; still in production. **LMSR:** $5950
    Perf.: $5100   **Exc.:** $3500   **VGood:** $2800
Bavaria Model
    Perf.: $9200   **Exc.:** $6000   **VGood:** $4800
Danube Model
    Perf.: $15,000 **Exc.:** $10,000 **VGood:** $8000
Gold Target Model
    Perf.: $18,500 **Exc.:** $12,000 **VGood:** $9000

Krieghoff K-80 Live Bird Lightweight
Same specs as K-80 Live Bird except lightweight barrels and action. Introduced 1980; still in production.
LMSR: $5950
    Perf.: $5100   **Exc.:** $3500   **VGood:** $2800
Bavaria Model
    Perf.: $9200   **Exc.:** $6000   **VGood:** $4800
Danube Model
    Perf.: $15,000 **Exc.:** $10,000 **VGood:** $8000
Gold Medal Target
    Perf.: $18,500 **Exc.:** $12,000 **VGood:** $9000

KRIEGHOFF K-80 SKEET
Over/under; 12-ga.; 2 ½″ chambers; 28″ Skeet/Skeet barrels; optional Tula or choke tubes; vent rib; Skeet-style checkered European walnut stock; selective adjustable trigger; satin-gray finish receiver; made in four grades with differing wood and engraving options. Made in Germany. Introduced 1980; still in production.
LMSR: $6290
    Perf.: $5700   **Exc.:** $3500   **VGood:** $2900
Bavaria Model
    Perf.: $9500   **Exc.:** $5300   **VGood:** $4800
Danube Model
    Perf.: $15,500 **Exc.:** $10,000 **VGood:** $8500
Gold Medal Target
    Perf.: $19,000 **Exc.:** $12,000 **VGood:** $9000

Krieghoff K-80 Skeet Four-Barrel Set
Same specs as K-80 Skeet except comes with four 28″ barrels for 12-, 20-, 28-ga., .410; marketed in aluminum case; made in four grades with differing wood and engraving options. Introduced 1980; still in production.
LMSR: $14,200
    Perf.: $12,000 **Exc.:** $5500   **VGood:** $4800
Bavaria Model
    Perf.: $16,000 **Exc.:** $10,000 **VGood:** $8500
Danube Model
    Perf.: $22,000 **Exc.:** $15,000 **VGood:** $11,000
Gold Medal Target
    Perf.: $25,000 **Exc.:** $16,000 **VGood:** $12,000

Krieghoff K-80 Skeet International
Same specs as K-80 Skeet except Tula chokes with gas release ports; ½″ vent Broadway-type rib; International Skeet stock; marketed in fitted aluminum case. Introduced 1980; still in production. **LMSR:** $6995
    Perf.: $6200   **Exc.:** $4000   **VGood:** $3400

Krieghoff K-80 Skeet Lightweight
Same specs as K-80 Skeet except weighs 7 lbs.; made in three grades with differing wood and engraving options. Made in Germany. Introduced 1980; still in production. **LMSR:** $6590
    Perf.: $6000   **Exc.:** $3800   **VGood:** $3200
Bavaria Model
    Perf.: $9700   **Exc.:** $5500   **VGood:** $5000

Danube Model
    Perf.: $16,000 **Exc.:** $10,500 **VGood:** $9000

Krieghoff K-80 Skeet Lightweight Two-Barrel Set
Same specs as K-80 Skeet except set of two 28″, 30″ barrels; weighs 7 lbs.; made in four grades with differing wood and engraving options. Made in Germany. Introduced 1988; still in production. **LMSR:** $10,935
    Perf.: $8800   **Exc.:** $5000   **VGood:** $3900
Bavaria Model
    Perf.: $13,000 **Exc.:** $8200   **VGood:** $7000
Danube Model
    Perf.: $19,000 **Exc.:** $11,500 **VGood:** $9100
Gold Medal Target
    Perf.: $22,500 **Exc.:** $14,500 **VGood:** $11,500

Krieghoff K-80 Skeet Special
Same specs as K-80 Skeet except Skeet/Skeet choke tubes; tapered flat rib. Introduced 1980; still in production. **LMSR:** $6895
    Perf.: $6100   **Exc.:** $3900   **VGood:** $3300
Bavaria Model
    Perf.: $10,000 **Exc.:** $7000   **VGood:** $5900
Danube Model
    Perf.: $16,000 **Exc.:** $11,500 **VGood:** $9000
Gold Medal Target
    Perf.: $19,000 **Exc.:** $12,500 **VGood:** $10,000

KRIEGHOFF K-80 SPORTING CLAYS
Over/under; 12-ga.; 2 ¾″ chambers; 28″, 30″, 32″ barrels; choke tubes; tapered flat or step rib styles; checkered European walnut sporting stock for gun-down shooting style; free-floating barrels; standard or lightweight receiver; satin nickel finish; scroll engraving; marketed in aluminum case; made in four grades with differing wood and engraving options. Made in Germany. Introduced 1988; still in production. **LMSR:** $7350
    Perf.: $6400   **Exc.:** $4200   **VGood:** $3500
Bavaria Model
    Perf.: $10,000 **Exc.:** $6400   **VGood:** $5000
Danube Model
    Perf.: $15,500 **Exc.:** $10,000 **VGood:** $8000
Gold Medal Target
    Perf.: $20,000 **Exc.:** $11,500 **VGood:** $9200

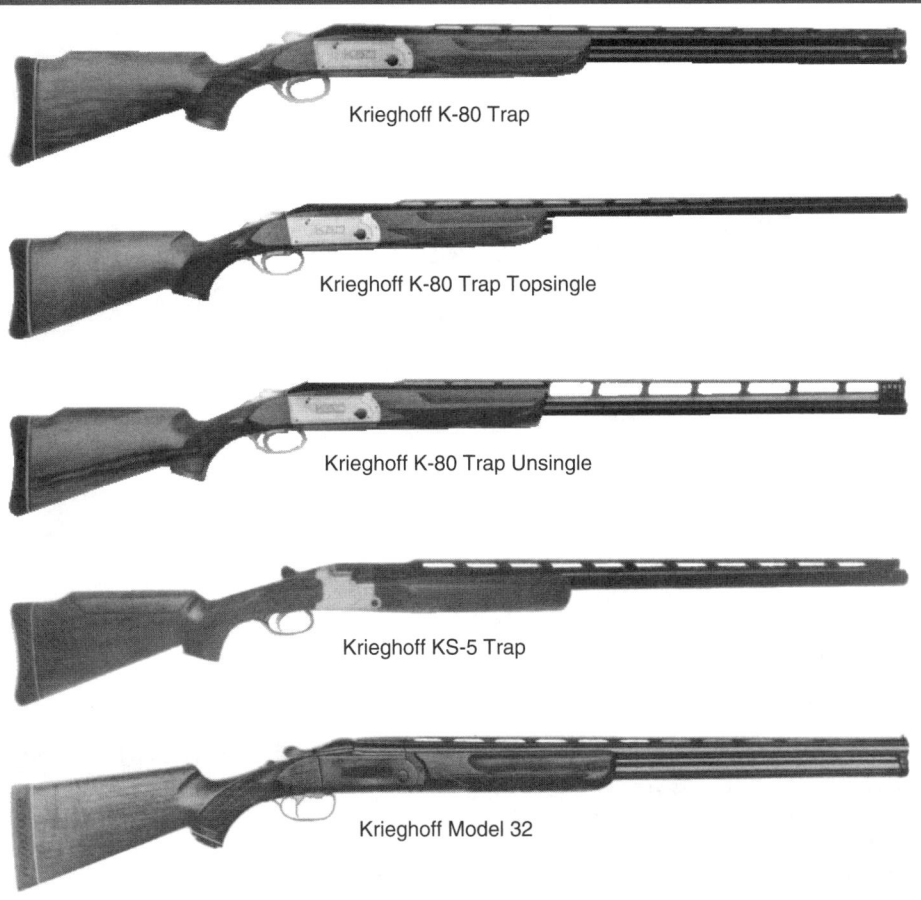

Krieghoff K-80 Trap

Krieghoff K-80 Trap Topsingle

Krieghoff K-80 Trap Unsingle

Krieghoff KS-5 Trap

Krieghoff Model 32

SHOTGUNS

Krieghoff
**K-80 Trap
K-80 Trap Combo
K-80 Trap Topsingle
K-80 Trap Unsingle
KS-5 Trap
KS-5 Special
Model 32
Model 32 Four-Barrel
Skeet Set
Model 32 Single-Barrel Trap
Neptun**

KRIEGHOFF K-80 TRAP

Over/under; 12-ga.; 2 3/4" chambers; 30", 32" barrels; standard choke combos or choke tubes; vent step rib; weighs 8 1/2 lbs.; checkered European walnut stock; palm swell grip; four stock dimensions or adjustable stock; selective adjustable trigger; satin nickel receiver; made in four grades with differing wood and engraving options. Made in Germany. Introduced 1980; still in production. **LMSR: $6695**
    Perf.: $6000 **Exc.:** $4000   **VGood:** $3000
Bavaria Model
    Perf.: $10,000 **Exc.:** $6200   **VGood:** $5000
Danube Model
    Perf.: $15,500 **Exc.:** $9500   **VGood:** $8500
Gold Medal Target
    Perf.: $19,500 **Exc.:** $12,000   **VGood:** $9000

Krieghoff K-80 Trap Combo

Same specs as K-80 Trap except two-barrel set cased with standard frame; made in four grades with differing wood and engraving options. Introduced 1980; still in production. **LMSR: $9375**
    Perf.: $9000 **Exc.:** $6000   **VGood:** $4000
Bavaria Model
    Perf.: $12,000 **Exc.:** $8200   **VGood:** $7000
Danube Model
    Perf.: $20,500 **Exc.:** $15,000   **VGood:** $11,000
Gold Medal Target
    Perf.: $24,000 **Exc.:** $17,500   **VGood:** $14,000

Krieghoff K-80 Trap Topsingle

Same specs as K-80 Trap except single top 34" barrel; Full choke; made in four grades with differing wood and engraving options. Introduced 1980; still in production. **LMSR: special order**
    Perf.: $6600 **Exc.:** $4500   **VGood:** $3500
Bavaria Model
    Perf.: $11,000 **Exc.:** $6900   **VGood:** $6000
Danube Model
    Perf.: $19,000 **Exc.:** $12,500   **VGood:** $10,000
Gold Medal Target
    Perf.: $21,000 **Exc.:** $14,000   **VGood:** $11,000

Krieghoff K-80 Trap Unsingle

Same specs as K-80 Trap except single lower 32", 34" barrel; Full choke; made in four grades with differing wood and engraving options. Introduced 1980; still in production. **LMSR: $7300**
    Perf.: $6400 **Exc.:** $4200   **VGood:** $3500
Bavaria Model
    Perf.: $10,000 **Exc.:** $6400   **VGood:** $5000
Danube Model
    Perf.: $15,500 **Exc.:** $10,000   **VGood:** $8000
Gold Medal Target
    Perf.: $20,000 **Exc.:** $11,500   **VGood:** $9200

KRIEGHOFF KS-5 TRAP

Single shot; 12-ga.; 2 3/4" chamber; 32", 34" barrels; Full choke or choke tubes; vent tapered step rib; weighs 8 1/2 lbs.; high Monte Carlo, low Monte Carlo or factory-adjustable stock of European walnut; adjustable trigger; blued or nickel receiver; marketed in fitted aluminum case. Made in Germany. Introduced 1988; still in production. **LMSR: $3950**
With fixed chokes
    Perf.: $3000 **Exc.:** $2000   **VGood:** $1600
With choke tubes
    Perf.: $3300 **Exc.:** $2300   **VGood:** $1900

Krieghoff KS-5 Special

Same specs as KS-5 Trap except fully adjustable rib; adjustable stock. Made in Germany. Introduced 1990; still in production. **LMSR: $4450**
With fixed chokes
    Perf.: $3300 **Exc.:** $2200   **VGood:** $1750
With choke tubes
    Perf.: $3600 **Exc.:** $2500   **VGood:** $2000

KRIEGHOFF MODEL 32

Over/under; boxlock; 12-, 20-, 28-ga., .410; 26 1/2", 28", 30", 32" barrels; any choke combo; checkered European walnut pistol-grip stock, forend; single trigger; auto ejectors; manufactured in Field, Skeet, Trap configurations; patterned after Remington Model 32. Introduced 1958; discontinued 1980.
    Perf.: $2200 **Exc.:** $1850   **VGood:** $1500
Low-rib two-barrel trap
    Perf.: $3200 **Exc.:** $2500   **VGood:** $2000

High-rib Vandalia two-barrel trap
    Perf.: $3200 **Exc.:** $2500   **VGood:** $2000

Krieghoff Model 32 Four-Barrel Skeet Set

Same specs as Model 32 except four sets of matched barrels (12-, 20-, 28-ga., 410); Skeet chokes, stock design; marketed in fitted case; made in seven grades depending upon quality of wood, amount and quality of engraving. No longer in production.
Standard Grade
    Perf.: $11,000 **Exc.:** $8400   **VGood:** $7300
München Grade
    Perf.: $12,500 **Exc.:** $9200   **VGood:** $8000
San Remo Grade
    Perf.: $15,800 **Exc.:** $10,000   **VGood:** $8500
Monte Carlo Grade
    Perf.: $18,800 **Exc.:** $13,500   **VGood:** $10,500
Crown Grade
    Perf.: $23,000 **Exc.:** $16,500   **VGood:** $13,500
Super Crown Grade
    Perf.: $28,000 **Exc.:** $19,500   **VGood:** $16,000
Exhibition Grade
    Perf.: $32,000 **Exc.:** $23,000   **VGood:** $18,500

Krieghoff Model 32 Single-Barrel Trap

Same specs as Model 32 except single shot; 12-ga.; 32", 34" barrel; Modified, Improved Modified, Full choke; high vent rib; checkered Monte Carlo stock, beavertail forend; recoil pad. Introduced 1959; no longer in production.
    Perf.: $1700 **Exc.:** $1100   **VGood:** $900

KRIEGHOFF NEPTUN

Drilling; sidelock; 12-, 20-ga. (2 3/4", 3" chamber); 22 Hornet, 222 Remington, 243 Winchester, 270 Winchester, 6.5x57R, 7x57R, 7x65R, 30-06, other calibers available on special order; 25" barrels; solid rib; folding leaf rear sight, post or bead front; checkered European walnut pistol-grip stock, forend; cheekpiece; steel or dural receiver; slit extractor or ejector for shotgun barrels; double triggers; sling swivels; engraving. Introduced 1960; still in production. **LMSR: $14,950**
    Perf.: $13,500 **Exc.:** $9500   **VGood:** $7500

Krieghoff
Neptun Primus
Plus
Teck
Teck Combination
Trumpf
Ulm
Ulm Combination
Ulm-Primus
Ulm-Primus Combination
Ultra Combination
Ultra-B Combination
Vandalia Trap

Lanber
82
87 Deluxe

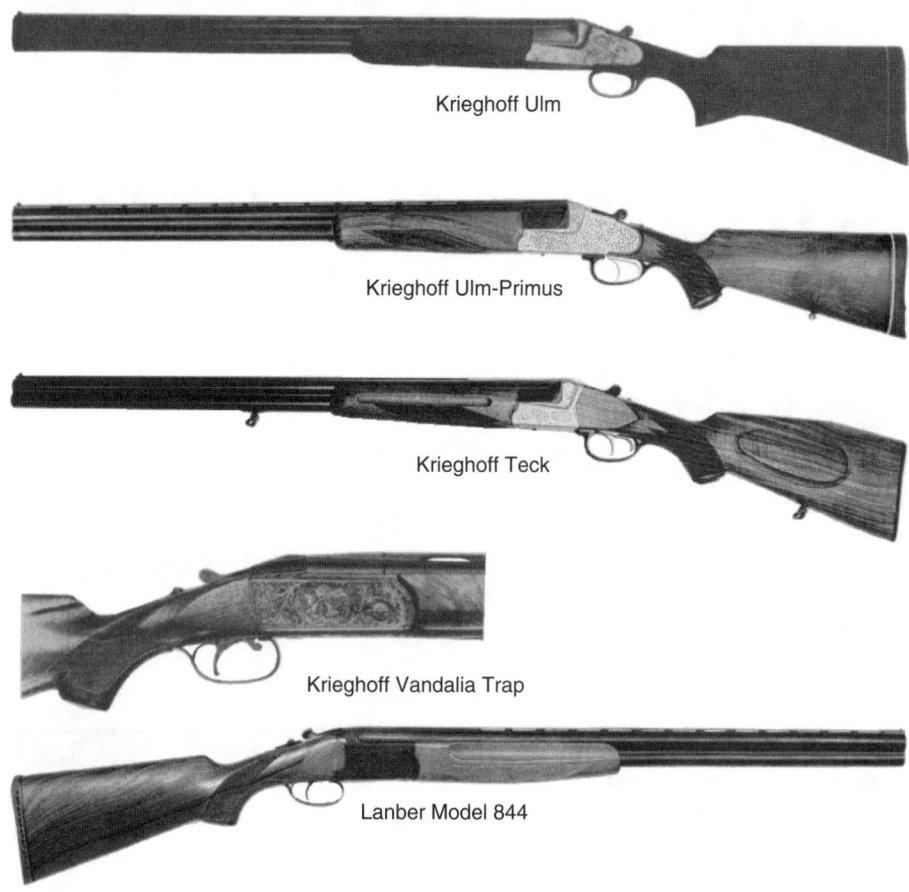

Krieghoff Ulm

Krieghoff Ulm-Primus

Krieghoff Teck

Krieghoff Vandalia Trap

Lanber Model 844

Krieghoff Neptun Primus
Same specs as Neptun except detachable sidelocks; fancier figured walnut; higher grade of engraving. Introduced 1962; still in production. **LMSR: $19,950**
   Perf.: $18,000 **Exc.:** $12,500   **VGood:** $9000

KRIEGHOFF PLUS
Drilling; boxlock; 12-, 20-ga. (2 ¾", 3" chamber); 222 Remington, 243 Winchester, 270 Winchester, 6.5x57R, 7x65R, 30-06, other calibers on special order; 25" barrels; solid rib; folding leaf rear sight, post or bead front; checkered European walnut pistol-grip stock, forend; cheekpiece; slit extractor or ejector for shotgun barrels; double triggers. Introduced 1988; still in production. **LMSR: $5245**
   Perf.: $4600   **Exc.:** $2800   **VGood:** $2000

KRIEGHOFF TECK
Over/under; boxlock; 12-, 16-, 20-ga.; 28" vent-rib barrel; Modified/Full choke; checkered European walnut pistol-grip stock, forend; Kersten double crossbolt; auto ejectors; double or single triggers; steel or dural receiver. Made in Germany. Introduced 1967; still in production. **LMSR: $7400**
   Perf.: $6500   **Exc.:** $4400   **VGood:** $3750

Krieghoff Teck Combination
Same specs as Teck except rifle barrel in 22 Hornet, 222 Remington, 222 Remington Magnum, 7x57R, 7x64, 7x65R, 30-30, 300 Winchester Magnum, 30-06, 308, 9.3x-74R; solid rib; folding leaf rear sight, post or bead front; cheekpiece; semi-beavertail forend; sling swivels. Introduced 1967; still in production. **LMSR: $7400**
   Perf.: $6500   **Exc.:** $4400   **VGood:** $3750

KRIEGHOFF TRUMPF
Drilling; boxlock; 12-, 16-, 20-ga. (2 ¾", 3" chamber); 22 Hornet, 222 Remington, 243 Winchester, 270 Winchester, 6.5x57R, 7x57R, 7x65R, 30-06, other calibers

available on special order; 25" barrels; solid rib; folding leaf rear sight, post or bead front; checkered European walnut pistol-grip stock, forend; cheekpiece; steel or Dural receiver; slit extractor or ejector for shotgun barrels; double triggers; sling swivels. Introduced 1953; still in production. **LMSR: $9450**
   Perf.: $8000   **Exc.:** $5200   **VGood:** $4400

KRIEGHOFF ULM
Over/under; sidelock; 12-, 16-, 20-ga.; 28" vent-rib barrel; Modified/Full choke; checkered European walnut pistol-grip stock, forend; Kersten double crossbolt; auto ejectors; double or single triggers; steel or Dural receiver; Arabesque engraving. Introduced 1958; still in production. **LMSR: $12,900**
   Perf.: $10,500 **Exc.:** $7200   **VGood:** $6000

Krieghoff Ulm Combination
Same specs as Ulm except rifle barrel in 22 Hornet, 222 Remington, 222 Remington Magnum, 7x57R, 7x64, 7x65R, 30-30, 300 Winchester Magnum, 30-06, 308, 9.3x74R; solid rib; folding leaf rear sight, post or bead front; cheekpiece; semi-beavertail forend; sling swivels. Introduced 1963; still in production. **LMSR: $12,900**
   Perf.: $10,500 **Exc.:** $7200   **VGood:** $6000

KRIEGHOFF ULM-PRIMUS
Over/under; detachable sidelock; 28" vent-rib barrel; Modified/Full choke; fancy checkered European walnut pistol-grip stock, forend; Kersten double crossbolt; auto ejectors; double or single triggers; steel or dural receiver; high-grade engraving. Introduced 1958; still in production. **LMSR: $18,790**
   Perf.: $15,500 **Exc.:** $8500   **VGood:** $7000

Krieghoff Ulm-Primus Combination
Same specs as Ulm-Primus except rifle barrel in 22 Hornet, 222 Remington, 222 Remington Magnum, 7x57R, 7x64, 7x65R, 30-30, 300 Winchester Magnum, 30-06, 308, 9.3x74R; solid rib; folding leaf rear sight, post or bead front; cheekpiece; semi-beavertail forend;

sling swivels. Introduced 1963; still in production. **LMSR: $18,790**
   Perf.: $15,500 **Exc.:** $8500   **VGood:** $7000

KRIEGHOFF ULTRA COMBINATION
Over/under; boxlock; 12-ga. over 7x57R, 7x64, 7x65R, 30-06, 308; 25" vent-rib barrel; folding leaf rear sight, post or bead front; checkered European walnut pistol-grip stock, forend; Kersten double crossbolt; auto ejectors; double triggers; thumb-cocking mechanism; satin finish receiver. Introduced 1985; still in production. **LMSR: $4450**
   Perf.: $3500   **Exc.:** $1750   **VGood:** $1400

Krieghoff Ultra-B Combination
Same specs as Ultra Combo except barrel selector for front set trigger. Introduced 1985; still in production. **LMSR: $4990**
   Perf.: $4200   **Exc.:** $1900   **VGood:** $1500

KRIEGHOFF VANDALIA TRAP
Single barrel or over/under; boxlock; 12-ga.; 30", 32", 34" barrels; vent rib; hand-checkered European walnut pistol-grip stock, beavertail forearm; three-way safety; selective single trigger; ejectors; optional silver, gold inlays, relief engraving, fancier wood. Made in Germany. Introduced 1973; discontinued 1976.
   Perf.: $2200   **Exc.:** $1850   **VGood:** $1500

LANBER 82
Over/under; boxlock; 12-, 20-ga.; 3" chambers; 26" Improved Cylinder/Modified, 28" Modified/Full barrels; European walnut stock, forend; double triggers; silvered, engraved receiver. Imported from Spain by Eagle Imports, Inc. Introduced 1994; still imported. **LMSR: $585**
   Perf.: $480   **Exc.:** $375   **VGood:** $325

LANBER 87 DELUXE
Over/under; boxlock; 12-, 20-ga.; 3" chambers; 26", 28" barrels; choke tubes; single selective trigger; silvered, engraved receiver. Imported from Spain by Eagle Imports, Inc. Introduced 1994; still imported. **LMSR: $915**
   Perf.: $820   **Exc.:** $550   **VGood:** $480

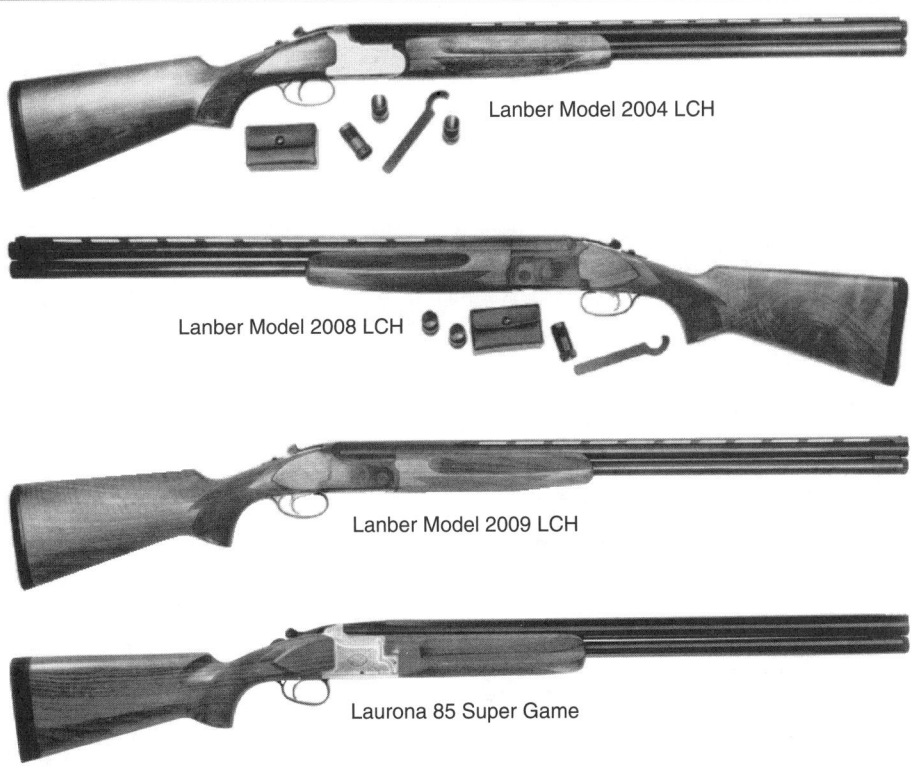

Lanber Model 2004 LCH

Lanber Model 2008 LCH

Lanber Model 2009 LCH

Laurona 85 Super Game

SHOTGUNS

Lanber
97 Sporting Clays
Model 844
Model 844 EST
Model 844 MST
Model 2004 LCH
Model 2004 LCH Skeet
Model 2004 LCH Trap
Model 2008 LCH
Model 2009 LCH

Miguel Larranaga
Traditional

Laurona
82 Super Game
82 Trap Competition
83 Super Game
84 Super Game
84 Super Trap
85 Super Game
85 Super Skeet
85 Super Trap

LANBER 97 SPORTING CLAYS

Over/under; boxlock; 12-ga.; 2 3/4″ chambers; 28″ barrels; choke tubes; European walnut stock, forend; single selective trigger; silvered, engraved receiver. Imported from Spain by Eagle Imports, Inc. Introduced 1994; still imported. **LMSR: $965**

Perf.: $870　**Exc.: $575**　**VGood: $500**

LANBER MODEL 844

Over/under; boxlock; 12-ga.; 2 3/4″ chambers; 28″ Improved Cylinder/Improved Modified barrels; checkered European walnut pistol-grip stock, forend; single trigger; extractors. Made in Spain. Introduced 1981; discontinued 1986.

Exc.: $375　**VGood: $300**　**Good: $275**

Lanber Model 844 EST

Same specs as Model 844, except ejectors. Introduced 1981; discontinued 1986.

Exc.: $395　**VGood: $320**　**Good: $295**

Lanber Model 844 MST

Same specs as Model 844, except 12-ga.; 3″ chambers; 30″ Modified/Full barrels. Introduced 1981; discontinued 1986.

Exc.: $375　**VGood: $300**　**Good: $275**

LANBER MODEL 2004 LCH

Over/under; boxlock; 12-ga.; 2 3/4″ chambers; 28″ barrels; interchangeable LanberChoke system; middle rib between barrels; checkered European walnut pistol-grip stock, forend; single selective trigger; ejectors; satin finish. Made in Spain. Introduced 1981; discontinued 1986.

Perf.: $650　**Exc.: $500**　**VGood: $390**

Lanber Model 2004 LCH Skeet

Same specs as Model 2004 LCH, except select walnut stock; engraving; blued finish. Made in Spain. Introduced 1981; discontinued 1986

Perf.: $710　**Exc.: $550**　**VGood: $500**

Lanber Model 2004 LCH Trap

Same specs as Model 2004 LCH, except 30″ barrels; trap-style walnut stock; blued finish. Introduced 1981; discontinued 1986.

Perf.: $650　**Exc.: $540**　**VGood: $500**

LANBER MODEL 2008 LCH

Over/under; boxlock; 12-ga.; 2 3/4″ chambers; 28″ barrels; interchangeable LanberChoke system; checkered European walnut pistol-grip stock, forend; single selective trigger; ejectors; blued finish. Introduced 1981; discontinued 1986.

Perf.: $750　**Exc.: $625**　**VGood: $410**

LANBER MODEL 2009 LCH

Over/under; boxlock; 12-ga.; 2 3/4″ chambers; 30″ barrels; interchangeable LanberChoke system; checkered European walnut pistol-grip stock, forend; single selective trigger; ejectors; blued finish. Introduced 1981; discontinued 1986.

Perf.: $570　**Exc.: $460**　**VGood: $410**

MIGUEL LARRANAGA TRADITIONAL

Side-by-side; exposed hammers; 12-, 20-ga.; 2 3/4″ chambers; 28″ Modified/Full barrels; medium bead front sight; hand-checkered European walnut straight-grip stock; checkered butt; hand-engraved locks. Made in Spain. Introduced 1982 by Toledo Armas; discontinued 1984.

Perf.: $395　**Exc.: $325**　**VGood: $295**

LAURONA 82 SUPER GAME

Over/under; boxlock; 12-, 20-ga.; 2 3/4″ chambers; 28″ Full/Modified, Improved Cylinder/Improved Modified barrels; vent rib; checkered European walnut stock, forend; twin single triggers; auto selective ejectors; silvered, engraved frame; black chrome barrels. Made in Spain. Introduced 1986; discontinued 1989

Perf.: $925　**Exc.: $750**　**VGood: $650**

Laurona 82 Trap Competition

Same specs as 82 Super Game except 29″ barrels; checkered European walnut Monte Carlo stock, forend; single trigger; rubber recoil pad. Introduced 1986; discontinued 1986.

Perf.: $600　**Exc.: $475**　**VGood: $375**

LAURONA 83 SUPER GAME

Over/under; boxlock; 12-, 20-ga.; 2 3/4″, 3″ chambers; 28″ vent-rib barrels; Multichokes; checkered European walnut stock, forend; twin single triggers; auto selective ejectors; silvered, engraved frame; black chrome barrels. Made in Spain. Introduced 1986; no longer in production.

Perf.: $1150　**Exc.: $850**　**VGood: $750**

LAURONA 84 SUPER GAME

Over/under; boxlock; 12-, 20-ga.; 2 3/4″, 3″ chambers; 28″ Full/Modified, Improved Cylinder/Improved Modified barrels; vent rib; checkered European walnut stock, forend; single selective trigger; auto selective ejectors; silvered, engraved frame; black chrome barrels. Made in Spain. Introduce 1986; discontinued 1989.

Perf.: $900　**Exc.: $740**　**VGood: $650**

Laurona 84 Super Trap

Same specs as 84 Super Game, except 29″ Improved Modified/Full, Modified/Full barrels; checkered European walnut pistol-grip stock, beavertail forend; rubber recoil pad. Made in Spain. Introduced 1986; no longer in production.

Perf.: $1250　**Exc.: $925**　**VGood: $830**

LAURONA 85 SUPER GAME

Over/under; boxlock; 12-, 20-ga.; 2 3/4″, 3″ chambers; 28″ vent-rib barrels; Multichokes; checkered European walnut stock, forend; single selective trigger; auto selective ejectors; silvered, engraved frame; black chrome barrels. Made in Spain. Introduced 1986; no longer in production.

Perf.: $1150　**Exc.: $825**　**VGood: $750**

Laurona 85 Super Skeet

Same specs as 85 Super Game, except only 12-ga.; 2 3/4″ chambers; mechanical triggers. Made in Spain. Introduced 1986; no longer in production.

Perf.: $1450　**Exc.: $925**　**VGood: $825**

Laurona 85 Super Trap

Same specs as 85 Super Game, except 29″ vent-rib barrels; Full/Multichokes; checkered European walnut pistol-grip stock, beavertail forend; rubber recoil pad. Made in Spain. Introduced 1986; no longer in production.

Perf.: $1575　**Exc.: $950**　**VGood: $850**

Laurona
Grand Trap GTO Combo
Grand Trap GTU Combo
Silhouette 300 Sporting Clays
Silhouette 300 Trap
Silhouette Single Barrel Trap
Silhouette Ultra-Magnum

Lebeau-Courally
Boxlock
Model 1225
Sidelock

Lefever Arms
Automatic Hammerless Double
Automatic Hammerless Double Grade I
Automatic Hammerless Double Grade A
Automatic Hammerless Double Grade AA
Automatic Hammerless Double Grade B
Automatic Hammerless Double Grade BE
Automatic Hammerless Double Grade C

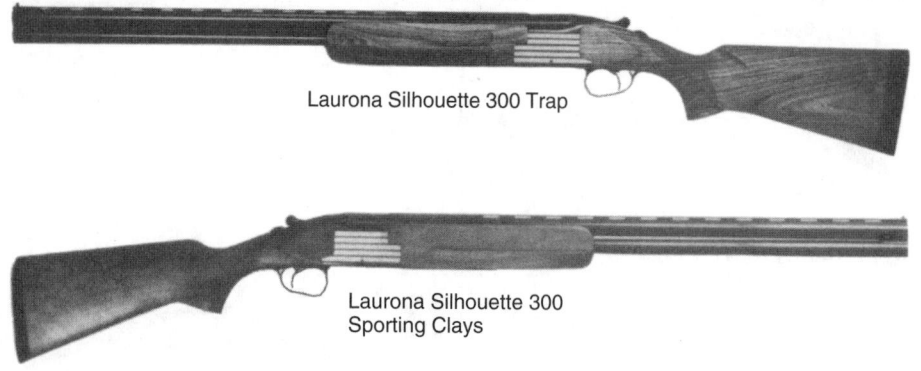

Laurona Silhouette 300 Trap

Laurona Silhouette 300 Sporting Clays

LAURONA GRAND TRAP GTO COMBO
Single shot or over/under; 12-ga.; 2″ chambers; 34″ single top barrel; 29″ over/under barrels; Multichokes; 10mm steel rib; European walnut Monte Carlo stock, full beavertail, finger-grooved forend; curved trap recoil pad; elongated forcing cone; bottom chamber area fitted with buffered recoil system. Made in Spain. Introduced 1990; discontinued 1991.
Perf.: $2000   Exc.: $1250   VGood: $1000

Laurona Grand Trap GTU Combo
Same specs as Grand Trap GTO Combo, except single barrel is on bottom; floating steel rib; teardrop forend. Introduced 1990; discontinued 1991.
Perf.: $2100   Exc.: $1350   VGood: $1100

LAURONA SILHOUETTE 300 SPORTING CLAYS
Over/under; 12-ga.; 3″ chambers; 28″ vent-rib barrels; Multichokes; weighs 7 1/4 lbs.; European walnut pistol-grip stock, beavertail forend; selective single trigger; auto selective ejectors; rubber buttpad; blued finish. Made in Spain. Introduced 1988; no longer in production.
Perf.: $1350   Exc.: $900   VGood: $800

Laurona Silhouette 300 Trap
Same specs as Silhouette 300 Sporting Clays, except 29″ barrels; weighs about 8 lbs.; trap-style European walnut pistol-grip stock, forend. Introduced 1988; no longer in production.
Perf.: $1500   Exc.: $950   VGood: $850

LAURONA SILHOUETTE SINGLE BARREL TRAP
Single shot; 12-ga.; 2 3/4″ chamber; top single or bottom single 34″ vent-rib barrel; Multichokes; European walnut pistol-grip stock, beavertail forend; rubber buttpad. Made in Spain. Introuded 1991; no longer in production.
Perf.: $1000   Exc.: $850   VGood: $700

LAURONA SILHOUETTE ULTRA-MAGNUM
Over/under; boxlock; 12-ga.; 3 1/2″ chamber; 28″ barrels; Multichokes; checkered European walnut pistol-grip stock, beavertail forend; single selective trigger; auto selective ejectors; rubber buttpad. Made in Spain. Introduced 1990; no longer in production.
Perf.: $1350   Exc.: $925   VGood: $825

LEBEAU-COURALLY BOXLOCK
Side-by-side; Anson & Deeley boxlock; 12-, 16-, 20-, 28-ga.; 26″ to 30″ barrels; choked to custom specs; hand-rubbed, oil-finished French walnut stock, forend; double triggers; ejectors; optional sideplates; Purdey-type fastener; choice of rib style, engraving. Made in Belgium to custom order. Introduced 1987; still in production. **LMSR: $14,650**
Perf.: $13,000   Exc.: $9000   VGood: $7500

LEBEAU-COURALLY MODEL 1225
Side-by-side; Holland & Holland sidelock; 12-, 20-, 28-ga.; 2 3/4″, 3″ chambers; 26″ to 28″ custom-choked barrels; weighs 6 3/8 lbs.; custom Grand Luxe walnut stock; double triggers; auto ejectors; color case-hardened frame; English engraving. Made to order in Belgium. Introduced 1987; discontinued 1988.
Perf.: $25,000   Exc.: $17,500   VGood: $12,000

LEBEAU-COURALLY SIDELOCK
Side-by-side; Holland & Holland sidelock; 12-, 16-, 20-, 28-ga., .410; 26″ to 30″ custom-choked barrels; best-quality checkered French walnut custom pistol-grip stock, splinter forend; ejectors; chopper lump barrels; choice of rib type, engraving pattern; some have H&H-type self-opening mechanism. Made to custom order in Belgium. Introduced 1987; no longer in production.
Perf.: $32,000   Exc.: $22,500   VGood: $19,000

LEFEVER ARMS AUTOMATIC HAMMERLESS DOUBLES
Side-by-side; sidelock; hammerless; 8-, 10-, 12-, 16-, 20-ga.; 26″ (only 12-, 16-, 20-ga.), 28″ (only 10-, 12-, 16-, 20-ga.), 30″, 32″, 34″ (only 8-ga.) twist and fluid-steel barrels; ejectors optional in some grades, standard in others; double triggers standard; single trigger catalogue option after 1913; earliest guns semi-boxlock, later ones fully boxlock, both with decorative sideplates. Originally called the Automatic Hammerless, the Lefever Arms gun has an unusual ball-and-socket hinge joint and numerous adjustable features in the action, fastening system, triggers, safety and ejectors. Originally offered in eight grades; several more were added after 1890; grades differ in quality of finish, amount and quality of engraving and checkering, quality of wood, and in some cases simply the addition of ejectors. Introduced 1885; discontinued 1915. Lefever Arms was purchased by Ithaca Gun Company in 1915; all subsequent guns bearing the Lefever name—Nitro Special, Long Range, Grade A, and Single Barrel Trap—were manufactured by Ithaca. None of these were built on the same mechanical system as either the Lefever Arms or D.M. Lefever guns.

Lefever Arms Automatic Hammerless Double Grade I
Same specs as Automatic Hammerless Double except 12-, 16-, 20-ga.; steel barrels; walnut half-pistol-grip stock; double triggers; extractors. The Grade I never appeared in Lefever Arms catalogues and may have been made on contract for Schoverling, Daly & Gales sporting-goods company of New York. It is the only grade Lefever Arms ever discontinued. Introduced 1899; discontinued 1907.
Exc.: $1000   VGood: $830   Good: $700

Lefever Arms Automatic Hammerless Double Grade A
Same specs as Automatic Hammerless Double except 8-, 10-, 12-, 16-, 20-ga.; twist or Krupp steel barrels; Circassian or French walnut stock; pistol, half-pistol or straight grip; horn or skeleton-steel buttplate; ejectors. 8-gauge extremely rare, precludes pricing.
Other gauges
Exc.: $14,500   VGood: $10,000   Good: $9000

Lefever Arms Automatic Hammerless Double Grade AA
Same specs as Automatic Hammerless Double except 8-, 10-, 12-, 16-, 20-ga.; twist or Krupp steel barrels; Circassian or French walnut stock; pistol, half-pistol or straight grip; horn or skeleton-steel buttplate; ejectors. 8-gauge extremely rare, precludes pricing.
Other gauges
Exc.: $20,000   VGood: $15,000   Good: $11,000

Lefever Arms Automatic Hammerless Double Grade B
Same specs as Automatic Hammerless Double except 8-, 10-, 12-, 16-, 20-ga.; twist or Krupp steel barrels; English walnut stock; pistol, half-pistol or straight grip; extractors. 8-gauge extremely rare, precludes pricing.
Other gauges
Exc.: $4700   VGood: $4000   Good: $3000

Lefever Arms Automatic Hammerless Double Grade BE
Same specs as Automatic Hammerless Double except 8-, 10-, 12-, 16-, 20-ga.; twist or Krupp steel barrels; English walnut stock; pistol, half-pistol or straight grip; ejectors. 8-gauge extremly rare, precludes pricing.
Other gauges
Exc.: $8000   VGood: $6500   Good: $5500

Lefever Arms Automatic Hammerless Double Grade C
Same specs as Automatic Hammerless Double except 8-, 10-, 12-, 16-, 20-ga.; twist or Krupp steel barrels; English walnut stock; pistol, half-pistol or straight grip; extractors. 8-gauge extremely rare, precludes pricing.
Other gauges
Exc.: $3850   VGood: $2650   Good: $2400

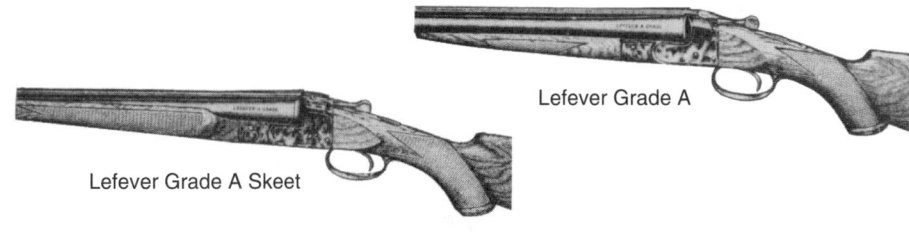

Lefever Grade A

Lefever Grade A Skeet

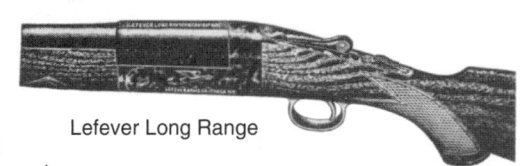

Lefever Long Range

SHOTGUNS

Lefever Arms
Automatic Hammerless
Double Grade CE
Automatic Hammerless
Double Grade D
Automatic Hammerless
Double Grade DE
Automatic Hammerless
Double Grade DS
Automatic Hammerless
Double Grade DSE
Automatic Hammerless
Double Grade E
Automatic Hammerless
Double Grade EE
Automatic Hammerless
Double Grade F
Automatic Hammerless
Double Grade FE
Automatic Hammerless
Double Grade G
Automatic Hammerless
Double Grade GE
Automatic Hammerless
Double Grade H
Automatic Hammerless
Double Grade HE
Automatic Hammerless
Double Optimus
Automatic Hammerless-
Double Thousand Dollar
Grade

Lefever
Grade A
Grade A Skeet
Long Range

Lefever Arms Automatic Hammerless Double Grade CE

Same specs as Automatic Hammerless Double except 8-, 10-, 12-, 16-, 20-ga.; twist or Krupp steel barrels; English walnut stock; pistol, half-pistol or straight grip; ejectors. 8-gauges extremly rare, precludes pricing. Other gauges

Exc.: $6500 **VGood:** $4300 **Good:** $3900

Lefever Arms Automatic Hammerless Double Grade D

Same specs as Automatic Hammerless Double except 8-, 10-, 12-, 16-, 20-ga.; twist or Krupp steel barrels; English walnut stock; pistol, half-pistol or straight grip; extractors. 8-gauge extremly rare, precludes pricing. Other gauges

Exc.: $2400 **VGood:** $1900 **Good:** $1750

Lefever Arms Automatic Hammerless Double Grade DE

Same specs as Automatic Hammerless Double except 8-, 10-, 12-, 16-, 20-ga.; twist or Krupp steel barrels; English walnut stock; pistol, half-pistol or straight grip; ejectors. 8-gauge extremly rare, precludes pricing. Other gauges

Exc.: $3900 **VGood:** $3000 **Good:** $2500

Lefever Arms Automatic Hammerless Double Grade DS

Same specs as Automatic Hammerless Double except 12-, 16-, 20-ga.; steel barrels; walnut half-pistol-grip stock; double triggers; extractors; grade stamp on water table.

Exc.: $1000 **VGood:** $800 **Good:** $700

Lefever Arms Automatic Hammerless Double Grade DSE

Same specs as Automatic Hammerless Double except 12-, 16-, 20-ga.; steel barrels; walnut half-pistol-grip stock; double triggers; ejectors; grade stamp on water table.

Exc.: $1150 **VGood:** $900 **Good:** $790

Lefever Arms Automatic Hammerless Double Grade E

Same specs as Automatic Hammerless Double except 8-, 10-, 12-, 16-, 20-ga.; twist or Krupp steel barrels; English walnut stock; pistol, half-pistol or straight grip; extractors. 8-gauge extremly rare, precludes pricing. Other gauges

Exc.: $2200 **VGood:** $1800 **Good:** $1700

Lefever Arms Automatic Hammerless Double Grade EE

Same specs as Automatic Hammerless Double except 8-, 10-, 12-, 16-, 20-ga.; twist or Krupp steel barrels; English walnut stock; pistol, half-pistol or straight grip; ejectors. 8-gauge extremly rare, precludes pricing. Other gauges

Exc.: $3000 **VGood:** $2500 **Good:** $2200

Lefever Arms Automatic Hammerless Double Grade F

Same specs as Automatic Hammerless Double except 10-, 12-, 16-, 20-ga.; twist or steel barrels; English walnut stock; pistol, half-pistol or straight grip; double triggers; extractors.

Exc.: $1650 **VGood:** $1450 **Good:** $1300

Lefever Arms Automatic Hammerless Double Grade FE

Same specs as Automatic Hammerless Double except 10-, 12-, 16-, 20-ga.; twist or steel barrels; English walnut stock; pistol, half-pistol or straight grip; double triggers; ejectors.

Exc.: $2200 **VGood:** $1650 **Good:** $1400

Lefever Arms Automatic Hammerless Double Grade G

Same specs as Automatic Hammerless Double except 10-, 12-, 16-, 20-ga.; twist or steel barrels; English walnut stock; pistol, half-pistol or straight grip; double triggers; extractors.

Exc.: $1500 **VGood:** $1350 **Good:** $1200

Lefever Arms Automatic Hammerless Double Grade GE

Same specs as Automatic Hammerless Double except 10-, 12-, 16-, 20-ga.; twist or steel barrels; English walnut stock; pistol, half-pistol or straight grip; double triggers; ejectors.

Exc.: $2100 **VGood:** $1600 **Good:** $1350

Lefever Arms Automatic Hammerless Double Grade H

Same specs as Automatic Hammerless Double except 12-, 16-, 20-ga.; twist (12-, 16-ga.) or steel (12-, 16-, 20-ga.) barrels; English walnut stock; pistol, half-pistol or straight grip; double triggers; extractors.

Exc.: $1250 **VGood:** $1000 **Good:** $900

Lefever Arms Automatic Hammerless Double Grade HE

Same specs as Automatic Hammerless Double except 12-, 16-, 20-ga.; twist (12-, 16-ga.) or steel (12-, 16-, 20-ga.) barrels; English walnut stock; pistol, half-pistol or straight grip; double triggers; ejectors.

Exc.: $1700 **VGood:** $1300 **Good:** $1000

Lefever Arms Automatic Hammerless Double Optimus

Same specs as Automatic Hammerless Double except 8-, 10-, 12-, 16-, 20-ga.; twist or Whitworth steel barrels; Circassian or French walnut stock; pistol, half-pistol or straight grip; horn or skeleton-steel buttplate. 8-ga.

Exc.: $5000 **VGood:** $3700 **Good:** $3000

Other gauges

Exc.: $4000 **VGood:** $3000 **Good:** $2000

Lefever Arms Automatic Hammerless Double Thousand Dollar Grade

Same specs as Automatic Hammerless Double except built to order. 8-ga.

Exc.: $10,000 **VGood:** $8000 **Good:** $6500

Other gauges

Exc.: $9000 **VGood:** $7000 **Good:** $5500

LEFEVER GRADE A

Side-by-side; boxlock; hammerless; 12-, 16-, 20-ga., .410; 26″, 28″, 30″, 32″ barrels; standard choke combos; hand-checkered American walnut pistol-grip stock, forend; plain extractors or auto ejectors; double, single triggers. Made by Ithaca. Introduced 1934; discontinued 1942.

With extractors, double triggers

Exc.: $700 **VGood:** $500 **Good:** $390

With ejectors, double triggers

Exc.: $900 **VGood:** $700 **Good:** $550

With extractors, single trigger

Exc.: $760 **VGood:** $560 **Good:** $450

With ejectors, single trigger

Exc.: $900 **VGood:** $700 **Good:** $600

NOTE: 16-gauge 20% higher; 20-gauge 50% higher; 410 200% higher.

Lefever Grade A Skeet

Same specs as Grade A, except 26″ barrels, Skeet chokes; beavertail forend; integral auto ejector; single trigger. Introduced 1934; discontinued 1942.

Exc.: $950 **VGood:** $800 **Good:** $700

NOTE: 16-gauge 50% higher; 20-gauge 100% higher; 410 200% higher.

LEFEVER LONG RANGE

Single shot; boxlock; hammerless; 12-, 16-, 20-ga., .410; 26″, 28″, 30″, 32″ barrels; standard chokes; vent-rib; bead front sight; hand-checkered American walnut pistol-grip stock, forearm; ejector. Introduced 1923; discontinued 1942.

Exc.: $225 **VGood:** $150 **Good:** $100

Lefever
Nitro Special
Single Barrel Trap

D.M. LeFever
Hammerless Double
Hammerless Double Grade
No. 4, AA
Hammerless Double Grade
No. 5, B
Hammerless Double Grade
No. 6, C
Hammerless Double Grade
No. 8, E
Hammerless Double Grade
No. 9, F
Hammerless Double Grade
O Excelsior
Hammerless Double Uncle
Dan Grade
Single Barrel Trap

Ljutic
Bi-Gun
Bi-Gun Combo
Bi-Gun Four Barrel Skeet
Bi-Matic
Dyna Trap
LM-6
LM-6 Combo
LM-6 Four-Barrel Skeet Set
Model X-73

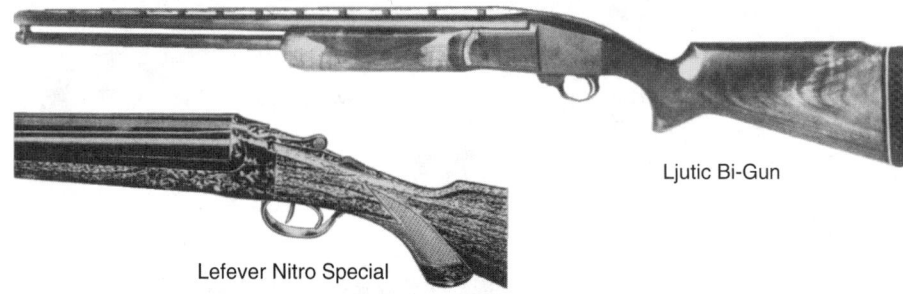

Ljutic Bi-Gun

Lefever Nitro Special

LEFEVER NITRO SPECIAL
Side-by-side; boxlock; 12-, 16-, 20-ga., .410; 26", 28", 30", 32" barrels; standard choke combos; hand-checkered American walnut pistol-grip stock, forearm; single nonselective trigger; optional double triggers; plain extractors. Introduced 1921; discontinued 1948.
With double triggers
  Exc.: $340   **VGood:** $210   **Good:** $190
With single trigger
  Exc.: $400   **VGood:** $250   **Good:** $210
NOTE: 16-gauge 20% higher; 20-gauge 50% higher than prices shown; .410, 200% higher.

LEFEVER SINGLE BARREL TRAP
Single shot; boxlock; hammerless; 12-ga.; 30", 32" barrels; Full choke; vent rib; bead front sight; hand-checkered American walnut pistol-grip stock, forearm; ejector; recoil pad. Introduced 1923; discontinued 1942.
  Exc.: $350   **VGood:** $240   **Good:** $200
NOTE: 16-gauge 20% higher; 20-gauge 50% higher; .410 200% higher.

D.M. LEFEVER HAMMERLESS DOUBLE
Side-by-side; boxlock; hammerless; 12-, 16-, 20-ga.; 26", 28", 30" 32" twist or fluid-steel barrels; double triggers; optional single trigger; ejectors standard in most grades; same ball-and-socket hinge joint as Lefever Arms and most of the same adjustable features; quite rare, only about 1000 were built. Introduced 1901; discontinued 1906. Extreme rarity precludes pricing.

D.M. Lefever Hammerless Double Grade No. 4, AA
Same specs as Hammerless Double except twist, Krupp steel or Whitworth steel barrels; French walnut stock; pistol, half-pistol or straight grip; ejectors; double triggers; optional single trigger; skeleton-steel or Lefever buttplate.
With double triggers
  Exc.: $9000   **VGood:** $8000   **Good:** $6000

With single trigger
  Exc.: $10,500   **VGood:** $9000   **Good:** $7000

D.M. Lefever Hammerless Double Grade No. 5, B
Same specs as Hammerless Double except twist or Krupp steel barrels; English walnut stock; pistol, half-pistol or straight grip; skeleton-steel or Lefever buttplate; ejectors; double triggers; optional single trigger.
With double triggers
  Exc.: $7500   **VGood:** $6000   **Good:** $5000
With single trigger
  Exc.: $8000   **VGood:** $7000   **Good:** $6000

D.M. Lefever Hammerless Double Grade No. 6, C
Same specs as Hammerless Double except twist or Krupp steel barrels; English walnut stock; pistol, half-pistol or straight grip; skeleton-steel or Lefever buttplate; ejectors; double triggers; optional single trigger.
With double triggers
  Exc.: $5000   **VGood:** $4000   **Good:** $3000
With single trigger
  Exc.: $6000   **VGood:** $5000   **Good:** $4000

D.M. Lefever Hammerless Double Grade No. 8, E
Same specs as Hammerless Double except twist, steel, or Krupp steel barrels; English walnut stock; pistol, half-pistol or straight grip; ejectors; double triggers; optional single trigger.
With double triggers
  Exc.: $4000   **VGood:** $3500   **Good:** $2500
With single trigger
  Exc.: $5200   **VGood:** $4500   **Good:** $3500

D.M. Lefever Hammerless Double Grade No. 9, F
Same specs as Hammerless Double except twist, steel or Krupp steel barrels; American walnut stock; pistol, half-pistol or straight grip; ejectors; double triggers; optional single trigger.
With double triggers
  Exc.: $3000   **VGood:** $2750   **Good:** $2000
With single trigger
  Exc.: $3750   **VGood:** $3400   **Good:** $3000

D.M. Lefever Hammerless Double Grade O Excelsior
Same specs as Hammerless Double except twist or steel barrels; American walnut stock; pistol, half-pistol or straight grip; extractors; optional ejectors; double triggers only.
With extractors
  Exc.: $2500   **VGood:** $2000   **Good:** $1500
With ejectors
  Exc.: $3100   **VGood:** $2800   **Good:** $2300

D.M. Lefever Hammerless Double Uncle Dan Grade
Same specs as Hammerless Double except made to order.
  Exc.: $7000   **VGood:** $6200   **Good:** $5000

D.M. LEFEVER SINGLE BARREL TRAP
Single shot; boxlock; 12-ga.; 26", 28", 30", 32" steel barrel; Full choke; checkered American walnut pistol-grip stock, forend; ejector; presumably available in same grades as double guns; extremely rare, total production less than 50; have the gun appraised by an expert. Introduced 1904; discontinued 1906.
  Exc.: $1000   **VGood:** $850   **Good:** $700

LJUTIC BI-GUN
Over/under; 12-ga.; 28", 33" barrels; choked to customer specs; hollow-milled rib; oil-finished, hand-checkered American walnut stock; choice of pull or release trigger; push-button opener in front of trigger guard; custom-made gun. Introduced 1978; no longer in production.
  Perf.: $9000   **Exc.:** $8400   **VGood:** $7000

Ljutic Bi-Gun Combo
Same specs as Bi-Gun except interchangeable single barrel; two trigger assemblies (one for single trigger, one for double). No longer in production.
  Perf.: $15,000   **Exc.:** $12,500   **VGood:** $8500

Ljutic Bi-Gun Four Barrel Skeet
Same specs as Bi-Gun except comes with matched 28" barrels in 12-, 20-, 28-ga., .410; custom checkered fancy American or French walnut stock; Ljutic Paternator chokes integral to barrels. No longer in production.
  Perf.: $19,000   **Exc.:** $17,500   **VGood:** $15,500

LJUTIC BI-MATIC
Autoloader; 12-ga.; 2 ³/₄" chamber; 2-shot magazine; 26", 28", 30", 32" barrel; choked to customer's specs; oil-finished, hand-checkered American walnut stock; designed for trap, Skeet; left- or right-hand ejection; many options. Introduced 1983; still in production.
LMSR: $5995
  Perf.: $2500   **Exc.:** $2000   **VGood:** $1500

LJUTIC DYNA TRAP
Single barrel; 12-ga.; 33" Full barrel; trap-style walnut stock; extractor; push-button opening. Introduced 1981; discontinued 1984.
  Perf.: $2000   **Exc.:** $1500   **VGood:** $1250

LJUTIC LM-6
Over/under; 12-ga.; 2 ³/₄" chambers; 28" to 32" barrels; choked to customer's specs; hollow-milled rib; custom oil-finished, hand-checkered walnut stock; pull or release trigger push-button opener. Introduced 1988; still in production. **LMSR: $14,995**
  Perf.: $14,000   **Exc.:** $9000   **VGood:** $7800

Ljutic LM-6 Combo
Same specs as LM-6 except interchangeable single barrel; two trigger guards (one for single trigger, one for double triggers). Introduced 1988; still in production.
LMSR: $21,995
  Perf.: $20,000   **Exc.:** $15,000   **VGood:** $12,000

Ljutic LM-6 Four-Barrel Skeet Set
Same specs as LM-6, except matched set of 28" barrels in 12-, 20-, 28-ga. and .410; integral Ljutic Paternator chokes; custom-checkered American or French walnut stock. Introduced 1988; still in production to order. **LMSR: $29,995**
  Perf.: $28,500   **Exc.:** $20,000   **VGood:** $18,000

LJUTIC MODEL X-73
Single barrel; 12-ga.; 33" Full barrel; Monte Carlo walnut stock; extractor; push-button opening. Introduced 1981; no longer in production.
  Perf.: $2400   **Exc.:** $1900   **VGood:** $1800

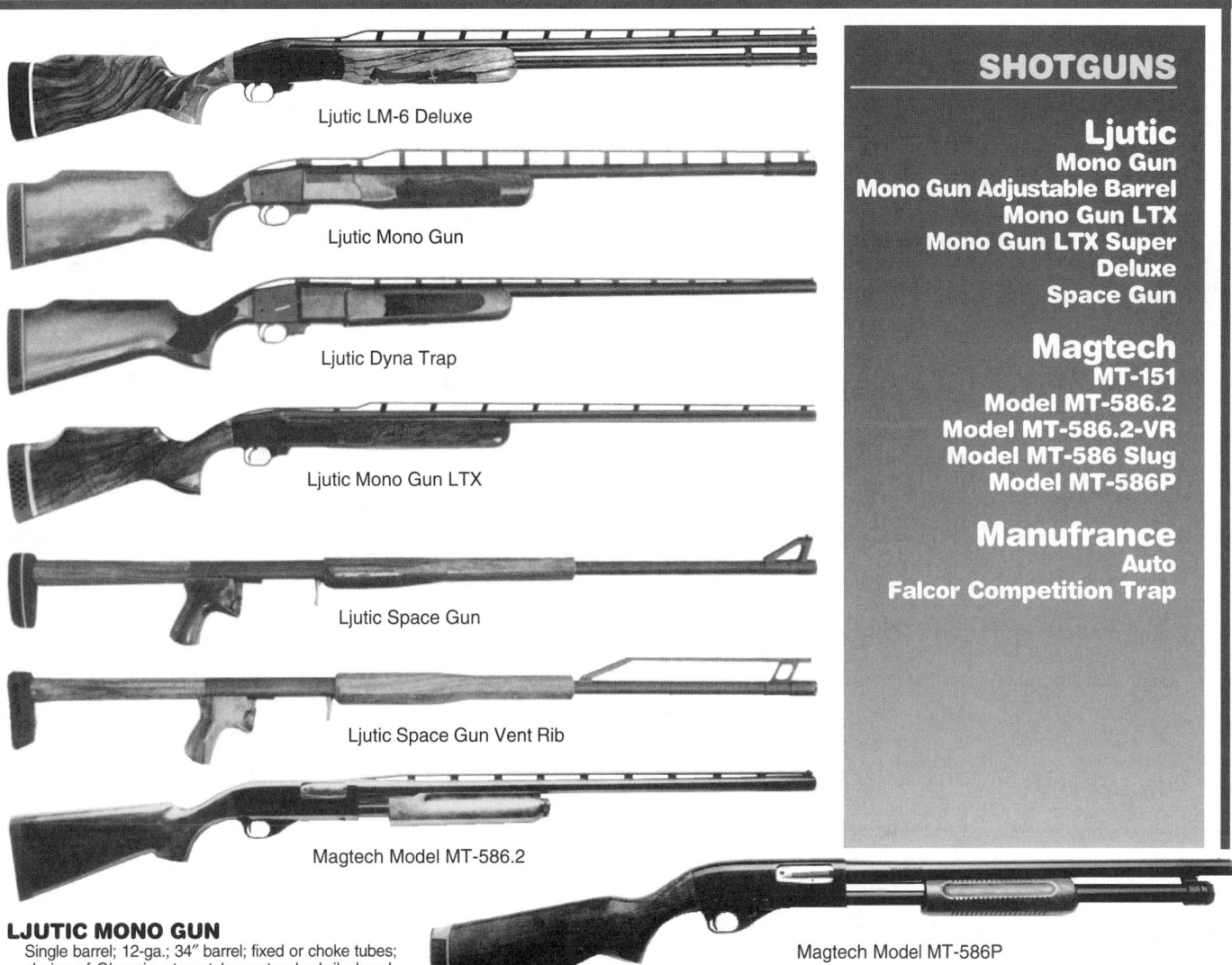

Ljutic LM-6 Deluxe

Ljutic Mono Gun

Ljutic Dyna Trap

Ljutic Mono Gun LTX

Ljutic Space Gun

Ljutic Space Gun Vent Rib

Magtech Model MT-586.2

SHOTGUNS

Ljutic
Mono Gun
Mono Gun Adjustable Barrel
Mono Gun LTX
Mono Gun LTX Super
Deluxe
Space Gun

Magtech
MT-151
Model MT-586.2
Model MT-586.2-VR
Model MT-586 Slug
Model MT-586P

Manufrance
Auto
Falcor Competition Trap

Magtech Model MT-586P

LJUTIC MONO GUN
Single barrel; 12-ga.; 34″ barrel; fixed or choke tubes; choice of Olympic, step-style or standard rib; hand-checkered, oil-finished fancy walnut stock; pull or release trigger; removable trigger guard; push-button opening. Introduced 1962; still in production. **LMSR: $4595**
With fixed choke
    Perf.: $3500   **Exc.:** $3000   **VGood:** $2500
With choke tubes
    Perf.: $4500   **Exc.:** $3100   **VGood:** $2600

Ljutic Mono Gun Adjustable Barrel
Same specs as Mono Gun, except adjustable choke to adjust pattern. Introduced 1978; discontinued 1983.
    Perf.: $4600   **Exc.:** $3200   **VGood:** $2700

Ljutic Mono Gun LTX
Same specs as Mono Gun, except 33″ extra-light barrel; double recessed choking; Olympic rib; exhibition-quality wood; extra-fancy checkering. Introduced 1985; still in production. **LMSR: $5795**
With fixed choke
    Perf.: $5000   **Exc.:** $3600   **VGood:** $3000
With choke tubes
    Perf.: $5650   **Exc.:** $3750   **VGood:** $3150

Ljutic Mono Gun LTX Super Deluxe
Same specs as Mono Gun, except 33″ extra-light barrel; double recessed choking; medium rib; exhibition-quality wood; extra-fancy checkering; release trigger. Introduced 1984; still in production. **LMSR: $6195**
    Perf.: $6000   **Exc.:** $5000   **VGood:** $3900
With choke tubes
    Perf.: $6200   **Exc.:** $5200   **VGood:** $4100

LJUTIC SPACE GUN
Single barrel; 12-ga.; 30″ barrel; front sight on vent rib; fancy American walnut stock with medium or large pistol grip, universal comb; pull or release button trigger;

anti-recoil device. Introduced 1981; still in production. **LMSR: $5995**
    Perf.: $4500   **Exc.:** $4000   **VGood:** $3000

MAGTECH MT-151
Single shot; exposed hammer; three-piece takedown; 12-, 16-, 20-ga., .410; 2 ¾″ (16-ga.), 3″ chamber; 25″ (.410 only), 26″, 28″, 30″ barrel; weighs about 5 ¾ lbs.; Brazilian hardwood stock, beavertail forend; trigger guard opener button. Made in Brazil. Introduced 1991; dropped 1992.
    Perf.: $100   **Exc.:** $80   **VGood:** $60

MAGTECH MODEL MT-586.2
Slide action; 12-ga.; 2 ¾″, 3″ chamber; 28″ Full, Modified or Improved Modified barrel; Brazilian Embuia hardwood stock, grooved slide handle; crossbolt safety; dual-action slide bars; blued finish. Made in Brazil. Introduced 1991; discontinued 1995.
    Perf.: $180   **Exc.:** $150   **VGood:** $135

Magtech Model MT-586.2-VR
Same specs as Model MT-586.2 except 12-ga.; 3″ chamber; 26″, 28″ barrel; choke tubes; weighs 7 ¼ lbs.; 46 ½″ overall length (26″ barrel); ventilated rib with bead front sight. Made in Brazil. Introduced 1995; still imported. **LMSR: $255**
    Perf.: $230   **Exc.:** $180   **VGood:** $150

Magtech Model MT-586 Slug
Same specs as Model MT-586.2 except 24″ Cylinder barrel; rifle sights; Brazilian Embuia hardwood Monte Carlo stock, slide handle; matte finish. Made in Brazil. Introduced 1991; discontinued 1995.
    Perf.: $180   **Exc.:** $150   **VGood:** $130

MAGTECH MODEL MT-586P
Slide action; 12-ga.; 3″ chamber; 7-shot magazine; 19″ Cylinder bore barrel; weighs 7 ¼ lbs.; bead front sight; Brazilian hardwood stock, pump handle; crossbolt safety dual-action slide bars; blued finish. Made in Brazil. Introduced 1991; still in production. **LMSR: $219**
    Perf.: $180   **Exc.:** $150   **VGood:** $130

MANUFRANCE AUTO
Gas-operated autoloader; 12-ga.; 2 ¾″, 3″ chamber; 3-shot magazine; 26″ Improved Cylinder, 28″ Modified, 30″ Full choke barrels; vent rib; hand-checkered French walnut pistol-grip stock; black matte finish; quick takedown; interchangeable barrels. Made in France. Introduced 1978; no longer in production.
    Perf.: $325   **Exc.:** $260   **VGood:** $210

MANUFRANCE FALCOR COMPETITION TRAP
Over/under; boxlock; 12-ga.; 2 ¾″ chambers; 30″ chrome-lined barrels; alloy high-post rib; 48″ overall length; ivory bead front sight on metal sleeve, middle ivory bead; hand-checkered French walnut stock, smooth beavertail forend; hand-rubbed oil finish; inertia-type trigger; top tang safety/barrel selector; auto ejectors. Made in France. Introduced 1984; discontinued 1985.
    Perf.: $700   **Exc.:** $600   **VGood:** $540

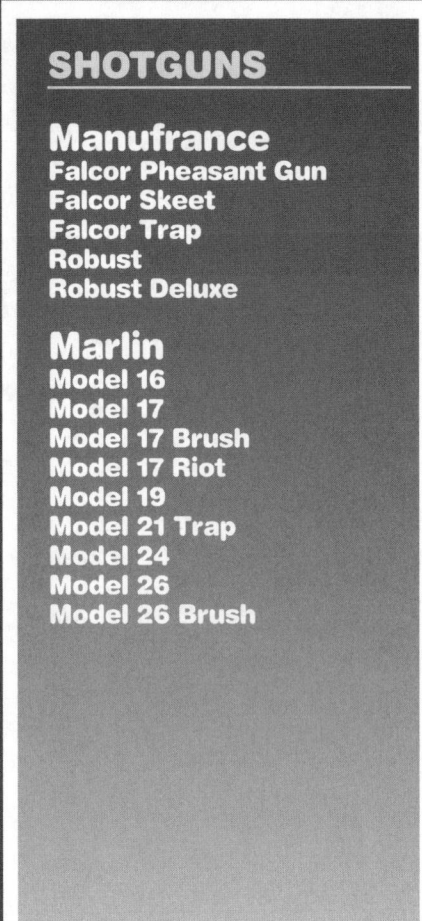

SHOTGUNS

Manufrance
Falcor Pheasant Gun
Falcor Skeet
Falcor Trap
Robust
Robust Deluxe

Marlin
Model 16
Model 17
Model 17 Brush
Model 17 Riot
Model 19
Model 21 Trap
Model 24
Model 26
Model 26 Brush

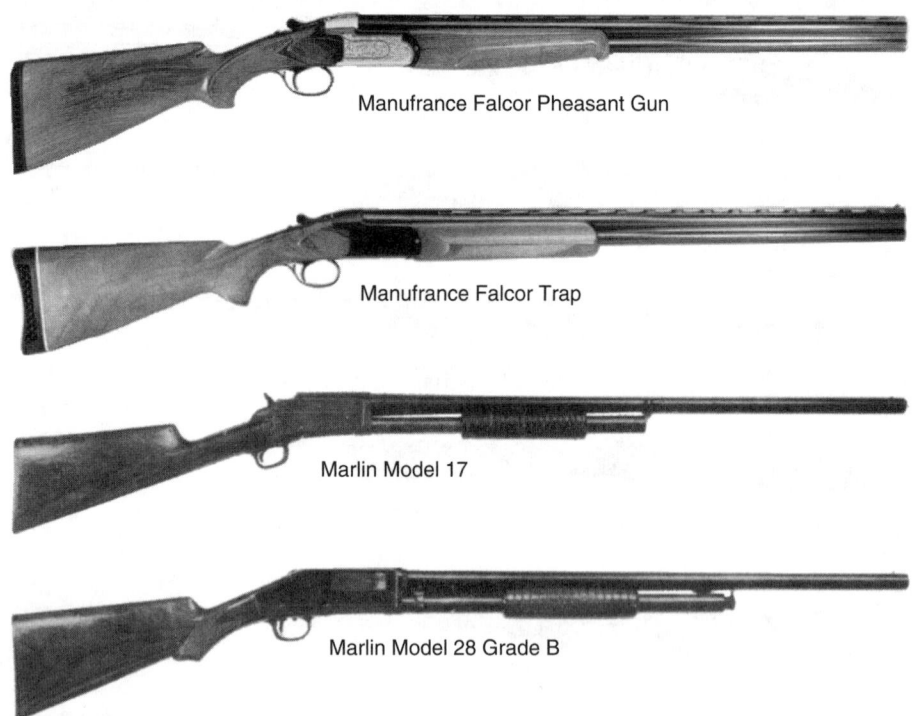

Manufrance Falcor Pheasant Gun

Manufrance Falcor Trap

Marlin Model 17

Marlin Model 28 Grade B

Manufrance Falcor Pheasant Gun
Same specs as Falcor Competition Trap, except 27 1/2″ chrome-lined barrels; metal bead front sight; checkered forend; single trigger; silver-gray finish, scroll engraving on receiver top lever, trigger guard. Made in France. Introduced 1984; discontinued 1985.
Perf.: $800   **Exc.:** $700   **VGood:** $640

Manufrance Falcor Skeet
Same specs as Falcor Competition Trap except 27 1/2″ Skeet barrels; vent top and middle rib; luminous yellow front bead in metal sleeve; checkered forend; smooth Skeet buttpad. Introduced 1984; discontinued 1985.
Perf.: $700   **Exc.:** $600   **VGood:** $540

Manufrance Falcor Trap
Same specs as Falcor Competition Trap, except vent top and middle rib; luminous yellow front bead in metal sleeve; checkered forend. Made in France. Introduced 1984; discontinued 1985.
Perf.: $700   **Exc.:** $600   **VGood:** $540

MANUFRANCE ROBUST
Side-by-side; boxlock; 12-ga.; 2 3/4″ chambers; 20 1/2″, 27 1/2″ barrels; hand-checkered French walnut stock, beavertail forend; double triggers; extractors; top tang safety; color case-hardened receiver. Introduced 1984; discontinued 1985.
Perf.: $500   **Exc.:** $400   **VGood:** $340

Manufrance Robust Deluxe
Same specs as Robust except 27 1/2″ chrome-lined barrels; auto ejectors; silver-gray finish, scroll engraving on receiver, top lever, trigger guard; optional retractable sling in butt. Made in France. Introduced 1984; discontinued 1985.
Perf.: $550   **Exc.:** $450   **VGood:** $390

MARLIN MODEL 16
Slide action; visible hammer; takedown; 16-ga.; 5-shot tubular magazine; 26″, 28″ barrel; standard chokes; American walnut pistol-grip stock, grooved slide handle; made in four grades, differing in quality of wood, engraving and checkering on Grades C, D. Introduced 1903; discontinued 1910.
Grade A
    Exc.: $200   **VGood:** $150   **Good:** $100
Grade B
    Exc.: $340   **VGood:** $300   **Good:** $225
Grade C
    Exc.: $450   **VGood:** $400   **Good:** $320
Grade D
    Exc.: $900   **VGood:** $700   **Good:** $600

MARLIN MODEL 17
Slide action; visible hammer; solid frame; 12-ga.; 5-shot tubular magazine; 30″, 32″ barrel; Full choke; unchecked-ered American walnut straight-grip stock, grooved slide handle. Introduced 1906; discontinued 1908.
    Exc.: $300   **VGood:** $210   **Good:** $150

Marlin Model 17 Brush
Same specs as Model 17, except for 26″ Cylinder barrel. Introduced 1906; discontinued 1908.
    Exc.: $275   **VGood:** $150   **Good:** $130

Marlin Model 17 Riot
Same specs as Model 17, except for 20″ Cylinder barrel. Introduced 1906; discontinued 1908.
    Exc.: $260   **VGood:** $150   **Good:** $130

MARLIN MODEL 19
Slide action; visible hammer; 12-ga.; 5-shot tubular magazine; 26″, 28″, 30″, 32″ barrel; standard chokes; American walnut pistol-grip stock, grooved slide handle; checkering on higher grades; two extractors; matted sighting groove on top of lightweight receiver; made in four grades, differing in quality of workmanship, wood and amount of engraving. Introduced 1906; discontinued 1907.
Grade A
    Exc.: $325   **VGood:** $240   **Good:** $150
Grade B
    Exc.: $425   **VGood:** $325   **Good:** $240
Grade C
    Exc.: $550   **VGood:** $475   **Good:** $325
Grade D
    Exc.: $1000   **VGood:** $750   **Good:** $650

MARLIN MODEL 21 TRAP
Slide action; visible hammer; 12-ga.; 5-shot tubular magazine; 26″, 28″, 30″, 32″ barrel; standard chokes; American walnut straight-grip stock, grooved slide handle; checkering on higher grades; two extractors; matted sighting groove on top of lightweight receiver; made in four grades, differing in workmanship, quality of wood and engraving. Introduced 1907; discontinued 1908.
Grade A
    Exc.: $375   **VGood:** $225   **Good:** $120
Grade B
    Exc.: $425   **VGood:** $325   **Good:** $225
Grade C
    Exc.: $550   **VGood:** $450   **Good:** $300
Grade D
    Exc.: $1175   **VGood:** $1050   **Good:** $600

MARLIN MODEL 24
Slide action; visible hammer; improved takedown; 12-ga.; 5-shot tubular magazine; 26″, 28″, 30″, 32″ barrel; standard chokes; solid matted rib attached to frame; American walnut pistol-grip stock, grooved slide handle; checkering or higher grades; two extractors; matted sighting groove on top of lightweight receiver; automatic recoil safety lock; made in four grades. Introduced 1908; discontinued 1917.
Grade A
    Exc.: $325   **VGood:** $275   **Good:** $150
Grade B
    Exc.: $450   **VGood:** $400   **Good:** $280
Grade C
    Exc.: $575   **VGood:** $500   **Good:** $350
Grade D
    Exc.: $1100   **VGood:** $800   **Good:** $600

MARLIN MODEL 26
Slide action; visible hammer; solid frame; 12-ga.; 5-shot tubular magazine; 30″, 32″ Full-choke barrel; solid matted rib attached to frame; American walnut straight-grip stock, grooved slide handle; two extractors; matted sighting groove on top of lightweight receiver; automatic recoil safety lock. Introduced 1909; discontinued 1916.
    Exc.: $250   **VGood:** $200   **Good:** $120

Marlin Model 26 Brush
Same specs as Model 26, except 26″ Cylinder barrel. Introduced 1909; discontinued 1916.
    Exc.: $250   **VGood:** $180   **Good:** $120

Marlin Model 30 Grade D

Marlin Model 42A

Marlin Model 43A

Marlin Model 43T

Marlin Model 50

Marlin
Model 26 Riot
Model 28
Model 28T
Model 28TS
Model 30
Model 30 Field
Model 31
Model 31 Field
Model 42A
Model 43A
Model 43T
Model 43TS
Model 44A
Model 44S
Model 49
Model 50

Marlin Model 26 Riot
Same specs as Model 26, except 20″ Cylinder barrel. Introduced 1909; discontinued 1915.
   Exc.: $250   **VGood:** $175   **Good:** $110

MARLIN MODEL 28
Slide action; hammerless; takedown; 12-ga.; 5-shot tubular magazine; 26″, 28″, 30″, 32″ barrel; standard chokes; American walnut pistol-grip stock, grooved slide handle; matted barrel top or solid matted rib (28D); made in four grades, differing in quality of wood, amount of engraving on 28C, 28D. Introduced 1913; 28B, 28C, 28D discontinued 1915; 28A discontinued 1922.
Grade A
   Exc.: $275   **VGood:** $225   **Good:** $175
Grade B
   Exc.: $475   **VGood:** $325   **Good:** $250
Grade C
   Exc.: $600   **VGood:** $500   **Good:** $395
Grade D
   Exc.: $1200   **VGood:** $875   **Good:** $625

Marlin Model 28T
Same specs as Model 28, except 30″ Full-choke barrel; matted rib; hand-checkered American walnut straight-grip stock; high fluted comb; fancier wood. Introduced 1915; discontinued 1915.
   Exc.: $500   **VGood:** $420   **Good:** $350

Marlin Model 28TS
Same specs as Model 28, except 30″ Full-choke barrel; American walnut straight-grip stock; matted barrel top. Introduced 1915; discontinued 1915.
   Exc.: $400   **VGood:** $350   **Good:** $250

MARLIN MODEL 30
Slide action; visible hammer; improved takedown; 16-ga.; 5-shot tubular magazine; 26″, 28″ barrel; standard chokes; solid matted rib on frame; American walnut pistol-grip stock, grooved slide handle; auto recoil safety lock; made in four grades depending upon quality of wood, amount and quality of engraving. Introduced 1911; discontinued 1917.
Grade A
   Exc.: $325   **VGood:** $275   **Good:** $225
Grade B
   Exc.: $475   **VGood:** $375   **Good:** $300
Grade C
   Exc.: $575   **VGood:** $495   **Good:** $375
Grade D
   Exc.: $1200   **VGood:** $975   **Good:** $695

Marlin Model 30 Field
Same specs as Model 30, except 25″ barrel; Modified choke; American walnut straight-grip stock. Introduced 1911; discontinued 1917.
   Exc.: $300   **VGood:** $200   **Good:** $100

MARLIN MODEL 31
Slide action; hammerless; 16-, 20-ga.; 5-shot tubular magazine; 25″, 26″, 28″ barrel; standard chokes; American walnut pistol-grip stock, grooved slide handle; matted barrel top; scaled-down version of Model 28; made in four grades, differing in quality wood, checkering and engraving. Introduced 1914; discontinued 1917.
Grade A
   Exc.: $350   **VGood:** $250   **Good:** $140
Grade B
   Exc.: $400   **VGood:** $300   **Good:** $260
Grade C
   Exc.: $550   **VGood:** $475   **Good:** $350
Grade D
   Exc.: $1100   **VGood:** $800   **Good:** $500

Marlin Model 31 Field
Same specs as Model 31, except 25″ barrel; Modified choke; American walnut straight- or pistol-grip stock. Introduced 1914; discontinued 1917.
   Exc.: $305   **VGood:** $250   **Good:** $200

MARLIN MODEL 42A
Slide action; visible hammer; takedown; 12-ga.; 5-shot tubular magazine; 26″ Cylinder, 28″ Modified, 30″ or 32″ Full barrel; bead front sight; uncheckered American walnut pistol-grip stock, grooved slide handle. Introduced 1922; discontinued 1933.
   Exc.: $260   **VGood:** $225   **Good:** $175

MARLIN MODEL 43A
Slide action; hammerless; takedown; 12-ga.; 5-shot tubular magazine; 26″ 28″, 30″, 32″ barrel; standard chokes; American walnut pistol-grip stock, grooved slide handle; matted barrel top. Introduced 1922; discontinued 1930.
   Exc.: $275   **VGood:** $250   **Good:** $200

Marliln Model 43T
Same specs as Model 43A, except 30″ Full-choke barrel; matted rib; American walnut straight-grip stock. Introduced 1922; discontinued 1930.
   Exc.: $350   **VGood:** $300   **Good:** $250

Marliln Model 43TS
Same specs as Model 43, except 30″ Full-choke barrel; American walnut straight-grip stock. Introduced 1922; discontinued 1930.
   Exc.: $400   **VGood:** $340   **Good:** $250

MARLIN MODEL 44A
Slide action; hammerless; 20-ga.; 5-shot tubular magazine; 25″, 26″, 28″ barrel; standard chokes; American walnut pistol-grip stock, grooved slide handle; matted barrel top. Introduced 1922; discontinued 1933.
   Exc.: $325   **VGood:** $275   **Good:** $225

Marlin Model 44S
Same specs as Model 44A, except select hand-checkered American walnut stock, slide handle. Introduced 1922; discontinued 1933.
   Exc.: $350   **VGood:** $300   **Good:** $250

MARLIN MODEL 49
Slide action; visible hammer; takedown; 12-ga.; 5-shot tubular magazine; 26″ Cylinder, 28″ Modified, 30″ or 32″ Full barrel; bead front sight; uncheckered American walnut pistol-grip stock, groooved slide handle. Used by Marlin as a premium, with purchase of four shares of corporate stock; less than 3000 made. Introduced 1928; discontinued 1930.
   Exc.: $300   **VGood:** $225   **Good:** $160

MARLIN MODEL 50
Bolt action; 12-, 20-ga.; 2-shot clip magazine; 28″ (12-ga.), 26″ (20-ga.) barrel; Full or adjustable choke; uncheckered American walnut one-piece pistol-grip stock; thumb safety; sling swivels; leather sling; double extractors; tapped for receiver sights. Introduced 1967; discontinued 1975.
With fixed choke
   Perf.: $125   **Exc.:** $80   **VGood:** $60
With adjustable choke
   Perf.: $140   **Exc.:** $90   **VGood:** $70

SHOTGUNS

Marlin
Model 53
Model 55
Model 55 Goose Gun
Model 55 Swamp Gun
Model 55S Slug
Model 59 Olympic
Model 59 Olympic Junior
Model 60
Model 63A
Model 63T
Model 63TS Trap Special
Model 90
Model 90-DT
Model 90-ST
Model 120 Magnum

Marlin Model 53

Marlin Model 55

Marlin Model 55 Fixed Choke

Marlin Model 55 Adjustable Choke

Marlin Model 55 Swamp Gun

Marlin Model 55 Goose Gun

MARLIN MODEL 53
Slide action; hammerless; takedown; 12-ga.; 5-shot tubular magazine; 26″, 28″, 30″, 32″ barrel; standard chokes; American walnut pistol-grip stock, grooved slide handle; matted barrel top. Introduced 1929; discontinued 1931.
Exc.: $300 **VGood:** $250 **Good:** $200

MARLIN MODEL 55
Bolt action; takedown; 12-, 16-, 20-ga.; 28″ (12-, 16-ga.); 26″ (20-ga.) barrel; Full or adjustable choke; uncheckered American walnut one-piece pistol-grip stock; recoil pad (12-ga.). Introduced 1954; discontinued 1965.
With fixed choke
Exc.: $100 **VGood:** $60 **Good:** $40
With adjustable choke
Exc.: $110 **VGood:** $70 **Good:** $50

Marlin Model 55 Goose Gun
Same general specs as Model 55 except 12-ga.; 2-shot clip magazine; 36″ barrel; Full choke; thumb safety; sling swivels; leather carrying strap; double extractors; tapped for receiver sights. Introduced 1962; still in production. **LMSR: $299**
Perf.: $220 **Exc.:** $160 **VGood:** $120

Marlin Model 55 Swamp Gun
Same specs as Model 55, except 12-ga., 3″ chamber; 20 1/2″ barrel; adjustable choke; sling swivels. Introduced 1963; discontinued 1965.
Exc.: $100 **VGood:** $85 **Good:** $40

Marlin Model 55S Slug
Same specs as Model 55, except 12-ga.; 2-shot clip magazine; 24″ Cylinder barrel; rifle sights; tapped for receiver sights; thumb safety; sling swivels; leather carrying strap; double extractors. Introduced 1973; discontinued 1979.
Perf.: $135 **Exc.:** $100 **VGood:** $80

MARLIN MODEL 59 OLYMPIC
Single shot; bolt action; .410; 2 1/2″, 3″ chamber; 24″ Full barrel; bead front sight; uncheckered one-piece walnut pistol-grip stock, also available with Junior stock with 12″ length of pull; self-cocking bolt; automatic thumb safety. Introduced 1959; discontinued 1965.
Exc.: $85 **VGood:** $60 **Good:** $50

Marlin Model 59 Olympic Junior
Same specs as Model 59 Olympic except smaller sized stock with 12″ length of pull. No longer in production.
Exc.: $85 **VGood:** $60 **Good:** $50

MARLIN MODEL 60
Single shot; boxlock; visible hammer; takedown; 12-ga.; 30″, 32″ Full barrel; walnut pistol-grip stock, beavertail forend; automatic ejector. Limited production of 600 to 3000. Introduced 1923; discontinued 1923.
Exc.: $145 **VGood:** $120 **Good:** $100

MARLIN MODEL 63A
Slide action; hammerless; takedown; 12-ga.; 5-shot magazine; 26″, 28″, 30″, 32″ barrel; standard chokes; American walnut pistol-grip stock, grooved slide handle; matted barrel top. Replaced model 43A. Introduced 1931; discontinued 1933.
Exc.: $300 **VGood:** $225 **Good:** $120

Marlin Model 63T
Same specs as Model 63A, except 30″ Full-choke barrel; matted rib; American walnut straight-grip stock. Introduced 1931; discontinued 1933.
Exc.: $325 **VGood:** $275 **Good:** $225

Marlin Model 63TS Trap Special
Same specs as Model 63A, except 30″ Full-choke barrel; matted rib; American walnut trap-style stock with dimensions to special order. Introduced 1931; discontinued 1933.
Exc.: $375 **VGood:** $325 **Good:** $275

MARLIN MODEL 90
Over/under; boxlock; 12-, 16-, 20-ga., .410; 26″, 28″, 30″ barrels; Improved/Modified, Modified/Full chokes; full-length rib between barrels; bead front sight; hand-checkered American walnut pistol-grip stock, forearm; double triggers; optional single non-selective trigger; recoil pad. Introduced 1937; discontinued during WWII.
With double triggers
Exc.: $340 **VGood:** $280 **Good:** $200
With single trigger
Exc.: $450 **VGood:** $380 **Good:** $300
.410 model, double triggers
Exc.: $400 **VGood:** $320 **Good:** $260
.410 model, single trigger
Exc.: $500 **VGood:** $420 **Good:** $350

Marlin Model 90-DT
Same specs as Model 90 except post-WWII version with double triggers; no rib between barrels; no recoil pad. Introduced 1949; discontinued 1958.
Exc.: $340 **VGood:** $280 **Good:** $200
.410 model
Exc.: $400 **VGood:** $320 **Good:** $250

Marlin Model 90-ST
Same specs as Model 90, except post-WWII version with single non-selective trigger; no rib between barrels; no recoil pad. Introduced 1949; discontinued 1958.
Exc.: $500 **VGood:** $380 **Good:** $300
.410 model
Exc.: $550 **VGood:** $420 **Good:** $350

MARLIN MODEL 120 MAGNUM
Slide action; hammerless; 12-ga.; 2 3/4″, 3″ chamber; 26″ Improved Cylinder, 28″ Modified, 30″ Full barrel; vent rib; checkered walnut pistol-grip stock, semi-beavertail forearm; slide release button; crossbolt safety; interchangeable barrels; side ejection. Introduced 1971; discontinued 1984.
Exc.: $280 **VGood:** $200 **Good:** $155

Marlin Model 59 Olympic

Marlin Model 90-DT

Marlin Model 120 Magnum

Marlin Model 410

Marlin Model 512 Slugmaster

Marlin Model 5510

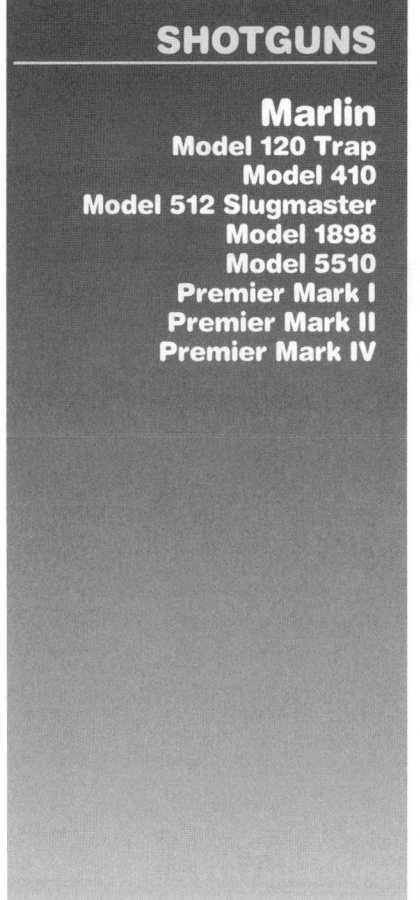

SHOTGUNS

Marlin
Model 120 Trap
Model 410
Model 512 Slugmaster
Model 1898
Model 5510
Premier Mark I
Premier Mark II
Premier Mark IV

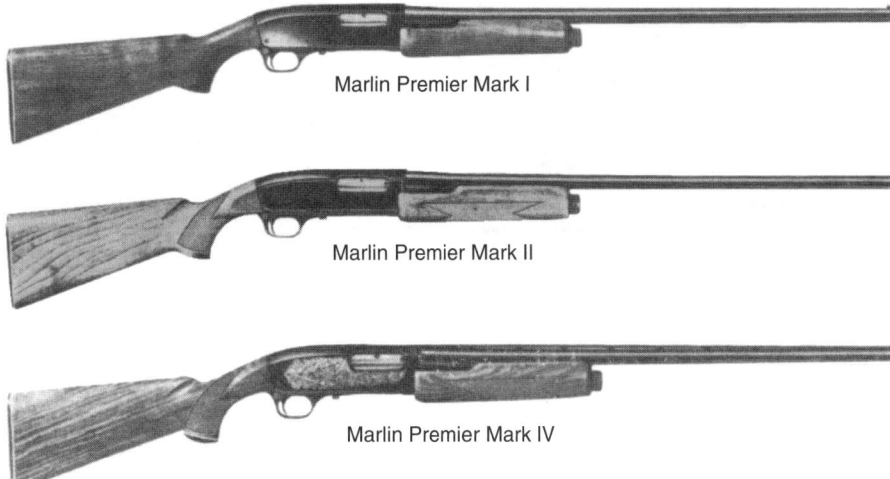

Marlin Premier Mark I

Marlin Premier Mark II

Marlin Premier Mark IV

Marlin Model 120 Trap
Same specs as Model 120 Magnum, except 30″ Full or Modified barrel; hand-checkered Monte Carlo stock, full forearm. Introduced 1973; discontinued 1975.
> **Exc.:** $300   **VGood:** $210   **Good:** $160

MARLIN MODEL 410
Lever action; visible hammer; solid frame; .410, '2 1/2″ shell; 4-shot tubular magazine; 22″, 26″ barrel; Full choke; uncheckered American walnut pistol-grip stock, grooved beavertail forend. Used by Marlin as a promotional firearm for shareholders. Introduced 1929; discontinued 1932.
> **Exc.:** $1000   **VGood:** $850   **Good:** $675

MARLIN MODEL 512 SLUGMASTER
Bolt action; 12-ga.; 3″ chamber; 2-shot detachable box magazine; 21″ rifled barrel with 1:28″ twist; weighs 8 lbs.; 44 3/4″ overall length; ramp front sight with brass bead and removable hood, adjustable folding semi-buckhorn rear; drilled and tapped for scope mounting; press-checkered, walnut-finished Maine birch stock; ventilated recoil pad. Uses Model 55 action with thumb safety. Designed for shooting saboted slugs. Introduced 1994; still in production. **LMSR: $353**
> **Perf.:** $275   **Exc.:** $210   **VGood:** $180

MARLIN MODEL 1898
Slide action; visible hammer; takedown; 12-ga.; 5-shot tubular magazine; 26″, 28″, 30″, 32″ barrel; standard chokes; American walnut pistol-grip stock, grooved slide handle; checkering on higher grades; grades differ in quality of woods, amount of engraving; Marlin's first shotgun. Introduced 1898; discontinued 1905.
Grade A
> **Exc.:** $325   **VGood:** $250   **Good:** $190

Grade B
> **Exc.:** $420   **VGood:** $320   **Good:** $275

Grade C
> **Exc.:** $600   **VGood:** $500   **Good:** $420

Grade D
> **Exc.:** $1250   **VGood:** $1000   **Good:** $850

MARLIN MODEL 5510
Bolt action; takedown; 10-ga.; 3 1/2″ chamber; 2-shot clip magazine; 34″ heavy barrel; Full choke; uncheckered American walnut one-piece pistol-grip stock; recoil pad; sling swivels; leather sling. Introduced 1976; discontinued 1985.
> **Perf.:** $205   **Exc.:** $170   **VGood:** $150

MARLIN PREMIER MARK I
Slide-action; hammerless; takedown; 12-ga.; 3-shot magazine; 26″ Improved or Skeet, 28″ Modified, 30″ Full barrel; bead front sight; French walnut pistol-grip stock, forearm; bead front sight; side ejection; crossbolt safety. Made in France. Introduced 1961; discontinued 1963.
> **Exc.:** $200   **VGood:** $175   **Good:** $100

Marlin Premier Mark II
Same specs as Premier Mark I, except checkered pistol grip, forearm; scroll-engraved receiver. Made in France. Introduced 1961; discontinued 1963.
> **Exc.:** $210   **VGood:** $190   **Good:** $110

Marlin Premier Mark IV
Same specs as Premier Mark I except optional vent rib; fine checkering; better wood; pistol-grip cap; full-coverage engraved receiver; engraved trigger guard. Made in France. Introduced 1961; discontinued 1963.
> **Exc.:** $275   **VGood:** $225   **Good:** $175

With vent rib
> **Exc.:** $305   **VGood:** $250   **Good:** $190

Marlin-Glenfield
Model 55G
Model 60G
Model 60G Junior
Model 778

Marocchi
America
Avanza
Avanza Sporting Clays
Conquista Skeet
Conquista Sporting Clays
Conquista Trap
Lady Sport

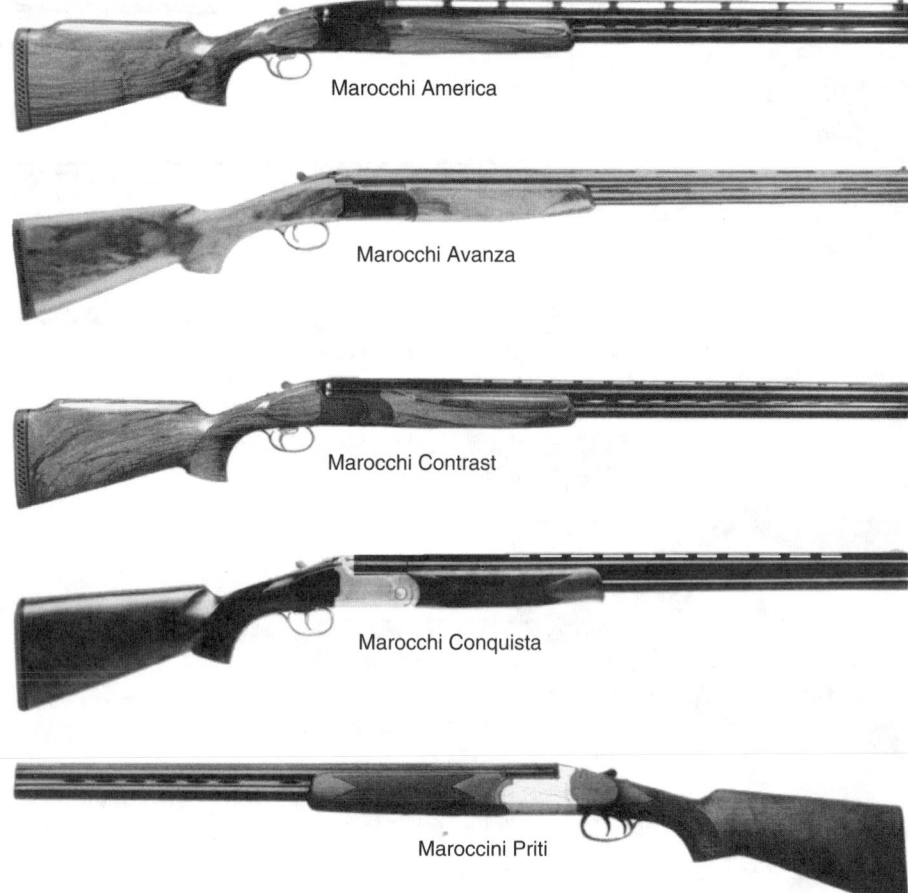

Marocchi America

Marocchi Avanza

Marocchi Contrast

Marocchi Conquista

Maroccini Priti

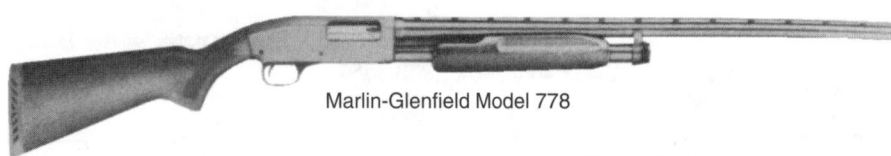

Marlin-Glenfield Model 778

MARLIN-GLENFIELD MODEL 55G

Bolt action; takedown; 12-, 16-, 20-ga.; 28" (12-, 16-ga.); 26" (20-ga.) barrel; Full or adjustable choke; uncheckered walnut-finished hardwood one-piece pistol-grip stock; recoil pad (12-ga.). Introduced 1954; discontinued 1965.
With fixed choke
   Exc.: $85   **VGood:** $60   **Good:** $40
With adjustable choke
   Exc.: $95   **VGood:** $70   **Good:** $50

MARLIN-GLENFIELD MODEL 60G

Single shot; bolt action; .410; 2 ½", 3" chamber; 24" Full barrel; bead front sight; uncheckered walnut-finished hardwood one-piece pistol-grip stock; also available with junior stock with 12" length of pull; self-cocking bolt; automatic thumb safety. Introduced 1960; discontinued 1964.
   Exc.: $75   **VGood:** $60   **Good:** $40

Marlin-Glenfield Model 60G Junior

Same specs as Model 60G except smaller sized stock with 12" length of pull. No longer in production.
   Exc.: $75   **VGood:** $60   **Good:** $40

MARLIN-GLENFIELD MODEL 778

Slide action; 12-ga.; 2 ¾", 3" chambers; 5-shot magazine; 20" slug barrel with sights, 26" Improved, 28" Modified, 30" Full barrel; plain or vent rib; walnut-finished hardwood stock, semi-beavertail forend; steel receiver; engine-turned bolt; double action bars; vent recoil pad. Introduced 1979; discontinued 1984.
With plain barrel
   Perf.: $230   **Exc.:** $180   **VGood:** $125
With vent rib
   Perf.: $250   **Exc.:** $200   **VGood:** $150

MAROCCHI AMERICA

Over/under; 12-, 20-ga.; 2 ¾" chambers; 26" to 29" Skeet, 27" to 32" trap, barrels, 30", 32" single barrel; hand-checkered select European walnut stock, Schnabel or beavertail forend; left- or right-hand palm swell; medium engraving coverage on frame; custom engraving, inlays offered. Marketed in fitted hard case. Made in Italy. Introduced 1983; discontinued 1986.
   Perf.: $1800   **Exc.:** $1550   **VGood:** $1200
With deluxe engraving
   Perf.: $1950   **Exc.:** $1700   **VGood:** $1350

MAROCCHI AVANZA

Over/under; 12-, 20-ga.; 2 ¾" chambers; 26", 28" barrels; fixed or Interchokes; vent top, middle ribs; weighs 6 ¾ lbs.; cut-checkered select walnut stock; recoil pad; auto mechanical barrel cycling; auto selective ejectors; auto safety; single selective trigger. Made in Italy. Introduced 1990; discontinued 1993.
   Perf.: $700   **Exc.:** $575   **VGood:** $480

Marocchi Avanza Sporting Clays

Same specs as Avanza except 12-ga.; 28" vent-rib barrels; Interchokes; trigger adjustable for length. Made in Italy. Introduced 1990; still in production. **LMSR: $889**
   Perf.: $800   **Exc.:** $550   **VGood:** $490

MAROCCHI CONQUISTA SKEET

Over/under; boxlock; 12-ga.; 2 ¾" chambers; 28" Skeet/Skeet barrels; 10mm concave vent rib; checkered American walnut pistol-grip stock, forend; ergonomically shaped trigger adjustable for pull length and weight; automatic selective ejectors; coin-finished receiver, blued barrels; hard case; stock wrench. Also available as true left-hand model—opening lever operates from left to right; stock has left-hand cast. Made in Italy. Introduced 1994; still imported. **LMSR: $1895**
Grade I
   Perf.: $1700   **Exc.:** $1150   **VGood:** $1050
Grade II
   Perf.: $1800   **Exc.:** $1400   **VGood:** $1200
Grade III
   Perf.: $2800   **Exc.:** $2000   **VGood:** $1800

Marocchi Conquista Sporting Clays

Same specs as Conquista Skeet except 28", 30", 32" barrels; Contrechoke tubes; Sporting Clays buttpad; no left-hand model. Made in Italy. Introduced 1994; still imported. **LMSR: $1895**
Grade I
   Perf.: $1700   **Exc.:** $1150   **VGood:** $1050
Grade II
   Perf.: $1800   **Exc.:** $1400   **VGood:** $1200
Grade III
   Perf.: $2800   **Exc.:** $2000   **VGood:** $1800

Marocchi Conquista Trap

Same specs as the Conquista Skeet except 30", 32" barrels; Full/Full; weighs about 8 ¼ lbs.; trap stock dimensions; no left-hand model. Made in Italy. Introduced 1994; still imported. **LMSR: $1895**
Grade I
   Perf.: $1700   **Exc.:** $1150   **VGood:** $1050
Grade II
   Perf.: $1800   **Exc.:** $1400   **VGood:** $1200
Grade III
   Perf.: $2800   **Exc.:** $2000   **VGood:** $1800

MAROCCHI LADY SPORT

Over/under; boxlock; 12-ga.; 2 ¾" chambers; 28", 30" barrels; five Contrechoke tubes; 10mm concave vent rib; weighs about 7 ½ lbs.; checkered American walnut pistol-grip stock, forend; buttpad; ergonomically shaped trigger adjustable for pull length and weight; automatic selective ejectors; coin-finished receiver, blued barrels; hard case; stock wrench; ergonomically designed specifically for women shooters; available with colored graphics finish on frame and opening

Mauser-Bauer Model 71E

Mauser-Bauer Model 72E

Mauser-Bauer Model 496

Mauser-Bauer Model 496 Competition Grade

Mauser-Bauer Model 610 Phantom

Mauser-Bauer Model 620

SHOTGUNS

Marocchi
Contrast
SM-28 SXS

Maroccini
Priti

Mauser
Model Contest

Mauser-Bauer
Model 71E
Model 72E
Model 496
Model 496 Competition Grade
Model 580 St. Vincent
Model 610 Phantom
Model 610 Skeet
Model 620

lever. Also available as true left-hand model—opening lever operates from left to right; stock has left-hand cast. Made in Italy. Introduced 1995; still imported.
LMSR: $1945
Grade I
Perf.: $1750 **Exc.:** $1200 **VGood:** $1100
Grade II
Perf.: $1850 **Exc.:** $1450 **VGood:** $1250
Grade III
Perf.: $2850 **Exc.:** $2050 **VGood:** $1850

MAROCCHI CONTRAST
Over/under; 12-, 20-ga.; 2 3/4″ chambers; 26″ to 29″ Skeet, 27″ to 32″ trap barrels; select European walnut stock; hand-checkered pistol grip, Schnabel or beavertail forend; hand-rubbed wax finish; left or right-hand palm swell; light engraving on standard grade; custom engraving, inlays at added cost. Marketed in fitted hard case. Made in Italy. Introduced 1983; discontinued 1986.
Perf.: $1800 **Exc.:** $1450 **VGood:** $1100
With deluxe engraving
Perf.: $1950 **Exc.:** $1600 **VGood:** $1250

MAROCCHI SM-28 SXS
Side-by-side; 12-, 20-ga.; 2 3/4″ chamber; totally handmade to customers specs and dimensions. Supplied with fitted leather luggage case. Introduced 1983; discontinued 1986.
Perf.: $10,000 **Exc.:** $8500 **VGood:** $7500

MAROCCINI PRITI
Over/under; 12-, 20-ga.; 3″ chambers; 28″ barrels; vent top and middle ribs; hand-checkered walnut stock; auto safety; extractors; double triggers; engraved antique silver receiver. Made in Italy. Introduced 1984; discontinued 1988.
Perf.: $324 **Exc.:** $285 **VGood:** $250

MAUSER MODEL CONTEST
Over/under; 12-ga.; 27 1/2″ chrome-lined barrels; select European walnut stock; receiver engraved with hunting scenes; dummy sideplates; auto ejectors; single selective trigger. Made in Germany. Introduced 1981; no longer in production.
Trap Model
Perf.: $1100 **Exc.:** $950 **VGood:** $800
Skeet Model
Perf.: $1500 **Exc.:** $1200 **VGood:** $900

MAUSER-BAUER MODEL 71E
Over/under; Greener crossbolt boxlock; 12-ga.; 28″ Modified/Full, Improved/Modified barrels; vent rib; hand-checkered European walnut pistol-grip stock, beavertail forearm; double triggers; automatic ejectors. Made in Italy. Introduced 1972; discontinued 1973.
Perf.: $375 **Exc.:** $250 **VGood:** $200

MAUSER-BAUER MODEL 72E
Over/under; boxlock; 12-ga.; 28″ Full/Modified, 30″ Full/Full barrels; wide vent rib; hand-checkered European walnut pistol-grip stock, beavertail forearm; Greener crossbolt; single trigger; automatic ejectors; engraved receiver. Made in Italy. Introduced 1972; discontinued 1973.
Perf.: $300 **Exc.:** $260 **VGood:** $225

MAUSER-BAUER MODEL 496
Single barrel; single shot; boxlock; 12-ga.; 32″ Modified, 34″ Full-choke barrel; matted vent rib; hand-checkered European walnut Monte Carlo stock, forearm; recoil pad; automatic ejector; auto safety; Greener crossbolt; double underlocking blocks; color case-hardened action; scroll engraving. Introduced 1972; discontinued 1974.
Perf.: $495 **Exc.:** $385 **VGood:** $295

Mauser-Bauer Model 496 Competition Grade
Same specs as Model 496 except high ramp rib; front, middle sight bead; hand finishing on wood and metal parts. Introduced 1973; discontinued 1974.
Perf.: $600 **Exc.:** $435 **VGood:** $325

MAUSER-BAUER MODEL 580 ST. VINCENT
Side-by-side; Holland & Holland sidelock; 12-ga.; 28″, 30″, 32″ barrels; standard choke combos; hand-checkered European walnut straight stock, forearm; split sear levers; coil hammer springs; single or double triggers; scroll-engraved receiver. Introduced 1973; discontinued 1974.
Perf.: $900 **Exc.:** $795 **VGood:** $695

MAUSER-BAUER MODEL 610 PHANTOM
Over/under; 12-ga.; 30″, 32″ barrels; standard choke combos; raised rib; vent rib between barrels for heat reduction; hand-checkered European walnut stock, forearm; recoil pad; coil springs throughout working parts; color case-hardened action. Introduced 1973; discontinued 1974.
Perf.: $695 **Exc.:** $580 **VGood:** $500

Mauser-Bauer Model 610 Skeet
Same specs as Model 610 except three Purbaugh tubes to convert gun for all-gauge competition (20-, 28-ga., .410). Introduced 1973; discontinued 1974.
Perf.: $1100 **Exc.:** $850 **VGood:** $700

MAUSER-BAUER MODEL 620
Over/under; boxlock; 12-ga.; 28″ Modified/Full, Improved/Modified, Skeet/Skeet, 30″ Full/Modified barrels; vent rib; hand-checkered European walnut pistol-grip stock, beavertail forearm; Greener crossbolt; single nonselective adjustable trigger; optional selective or double triggers; automatic ejectors; recoil pad. Made in Italy. Introduced 1972; discontinued 1974.
With double triggers
Perf.: $750 **Exc.:** $650 **VGood:** $575
With single selective trigger
Perf.: $795 **Exc.:** $680 **VGood:** $600

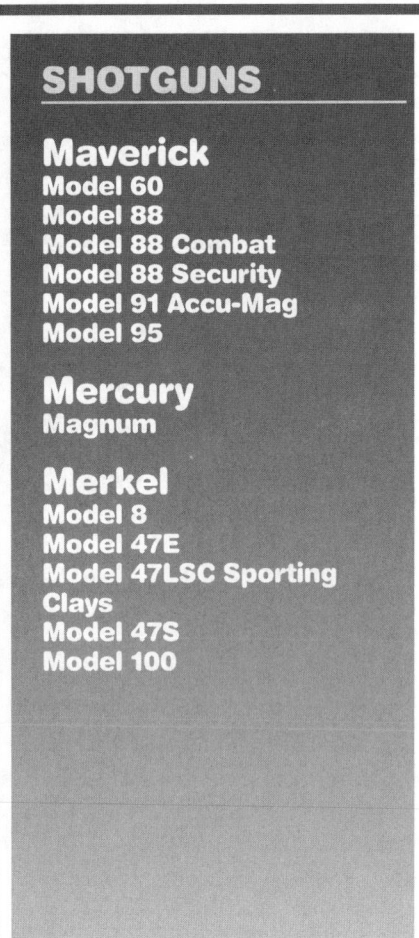

SHOTGUNS

Maverick
Model 60
Model 88
Model 88 Combat
Model 88 Security
Model 91 Accu-Mag
Model 95

Mercury
Magnum

Merkel
Model 8
Model 47E
Model 47LSC Sporting Clays
Model 47S
Model 100

Maverick Model 88

Maverick Model 88 Security

Maverick Model 91 Accu-Mag

Maverick Model 95

Merkel Model 147E

Merkel Model 47S

MAVERICK MODEL 60

Autoloader; 12-ga.; 2 3/4", 3" chamber; 5-shot magazine; 18 1/2" (2 3/4" chamber only) Cylinder, 24" Full and rifled choke tubes, 28" vent-rib Modified Accu-Choke barrel; weighs 7 1/4 lbs.; black synthetic stock and forend; designated barrels for magnum and non-magnum loads; action release button; blued receiver. Introduced 1993; no longer in production.

   Perf.: $175   **Exc.:** $150   **VGood:** $125

MAVERICK MODEL 88

Slide action; 12-ga.; 3" chamber; 6-shot magazine; 24" with rifle sights, 28", 32" barrels; fixed or Accu-chokes; plain or vent rib; 48" overall length with 28" barrel; weighs 7 1/4 lb.; bead front sight; black synthetic stock, ribbed forend; alloy receiver; interchangeable barrels; crossbolt safety; rubber recoil pad; blue finish. Marketed with Mossberg Cablelock. Introduced 1988; still in production. **LSMR: $230**

With plain barrel
   Perf.: $195   **Exc.:** $150   **VGood:** $125
With vent rib
   Perf.: $200   **Exc.:** $160   **VGood:** $140
With 24" barrel with rifle sights
   Perf.: $205   **Exc.:** $145   **VGood:** $115

Maverick Model 88 Combat

Same specs as Model 88 except 18 1/2" Cylinder barrel with vented shroud; black synthetic pistol-grip stock; built-in carrying handle. Introduced 1990; discontinued 1992.

   Perf.: $230   **Exc.:** $200   **VGood:** $170

Maverick Model 88 Security

Same specs as Model 88 except 6-, 8-shot magazine; 18 1/2" Cylinder barrel; synthetic straight- or pistol-grip stock, unribbed forend. Introduced 1993; still in production. **LSMR: $228**

   Perf.: $195   **Exc.:** $130   **VGood:** $100

MAVERICK MODEL 91 ACCU-MAG

Slide action; 12-ga.; 3 1/2" chamber; 28" vent-rib barrel; ACCU-MAG Modified tube; weighs about 7 3/4 lbs.; black synthetic stock, slide handle; dual slide bars; crossbolt safety; rubber recoil pad. Accessories interchangeable with Mossberg Model 835. Introduced 1993; still in production. **LMSR: $271**

   Perf.: $250   **Exc.:** $205   **VGood:** $180

MAVERICK MODEL 95

Bolt action; 12-ga.; 3" chamber; 2-shot magazine; 25" Modified barrel; weighs 6 1/2 lbs.; full-length textured black synthetic stock with integral magazine; ambidextrous rotating safety; twin extractors; rubber recoil pad; blue finish. Introduced 1995; still in production. **LMSR: $176**

   Perf.: $160   **Exc.:** $100   **VGood:** $90

MERCURY MAGNUM

Side-by-side; boxlock; 10-, 12-, 20-ga.; 3", 3 1/2" chambers; 28", 32" barrels; checkered European walnut pistol-grip stock; double triggers; auto safety; extractors; safety gas ports; engraved frame. Made in Spain and imported by Tradewinds, Inc. Also called Model G1032. Introduced 1970; discontinued 1989.

10-ga.
   Perf.: $390   **Exc.:** $310   **VGood:** $280
12-, 20-ga.
   Perf.: $300   **Exc.:** $260   **VGood:** $210

MERKEL MODEL 8

Side-by-side; boxlock; 12-, 16-, 20-ga.; 2 3/4" chambers; 26", 26 3/4", 28" barrels; weighs 6+ lbs.; checkered European walnut English straight-grip or pistol-grip stock, forend; double triggers or single selective trigger; double lugs; Greener crossbolt locking system; automatic ejectors; sling swivels; built to handle steel shot; engraved, color case-hardened receiver. Made in Germany. Discontinued 1994.

   Perf.: $1050   **Exc.:** $800   **VGood:** $700

MERKEL MODEL 47E

Side-by-side; boxlock; hammerless; 12-, 20-ga.; 2 3/4" chambers; standard barrel lengths, choke combos; hand-checkered European walnut stock, forearm; pistol grip and cheekpiece or straight English style; double hook bolting; Greener crossbolt; double triggers or single selective trigger; automatic ejectors; cocking indicators; engraving; sling swivels; built to handle steel shot. Still in production. **LMSR: $1795**

   Perf.: $1650   **Exc.:** $1100   **VGood:** $900

Merkel Model 47LSC Sporting Clays

Same specs as Model 47E except 12-ga.; 3" chambers; 28" barrel; Briley choke tubes; weighs about 7 1/4 lbs.; fancy figured walnut pistol-grip stock, beavertail forend; recoil pad; single selective trigger adjustable for length of pull; H&H-type ejectors; manual safety; cocking indicators; lengthened forcing cones; color case-hardened receiver with Arabesque engraving. Comes with fitted leather luggage case. Made in Germany. Introduced 1993; no longer imported.

   Perf.: $2600   **Exc.:** $1950   **VGood:** $1700

Merkel Model 47S

Same specs as Model 47E except sidelock; 12-, 16-, 20-, 28-ga., .410, 2 3/4", 3" chambers; double triggers or single trigger. Introduced prior to WWII; still in production. **LMSR: $4495**

   Perf.: $3950   **Exc.:** $2800   **VGood:** $2400

MERKEL MODEL 100

Over/under; boxlock; hammerless; 12-, 16-, 20-ga.; standard barrel lengths, choke combos; plain or optional ribbed barrel; hand-checkered European walnut stock, forearm; pistol grip and cheekpiece or straight English style; Greener crossbolt safety; double triggers; plain extractors. Made in Germany prior to World War II.

With plain barrel
   Exc.: $950   **VGood:** $800   **Good:** $600
With ribbed barrel
   Exc.: $1050   **VGood:** $900   **Good:** $700

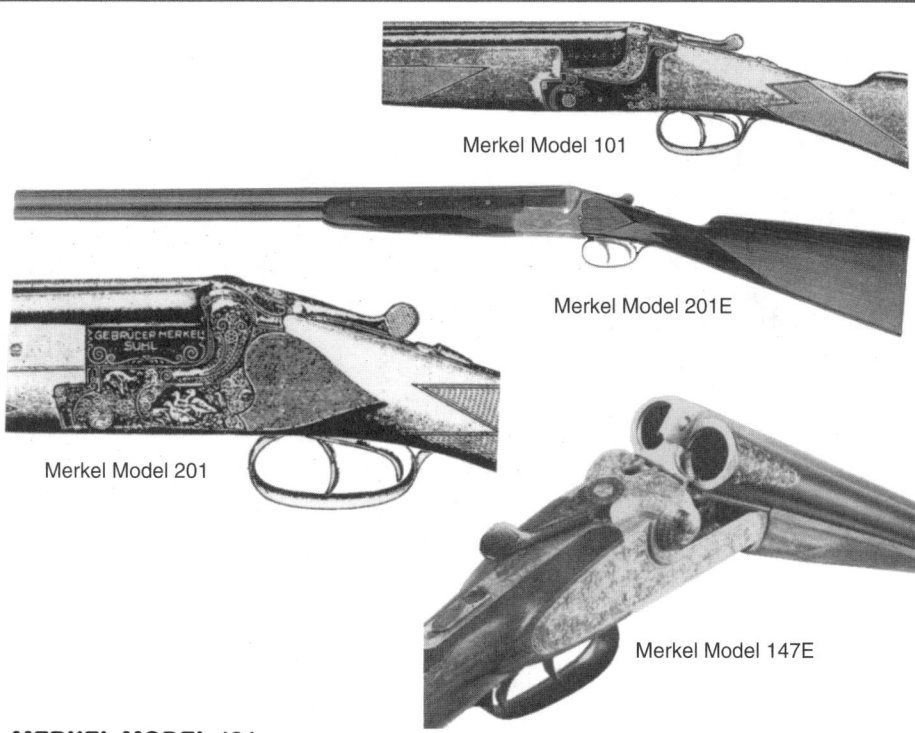

Merkel Model 101

Merkel Model 201E

Merkel Model 201

Merkel Model 147E

SHOTGUNS

Merkel
Model 101
Model 101E
Model 122
Model 127
Model 130
Model 147
Model 147E
Model 147S
Model 200
Model 200E
Model 200E Skeet
Model 200ET Trap
Model 200SC Sporting Clays
Model 201
Model 201E
Model 201E (Recent)
Model 201ES

MERKEL MODEL 101

Over/under; boxlock; hammerless; 12-, 16-, 20-ga.; standard barrel lengths, choke combos; ribbed barrel; hand-checkered European walnut stock, forearm; pistol grip and cheekpiece or straight English style; Greener crossbolt safety; double triggers; separate extractors; engraving. Made in Germany prior to World War II.
Exc.: $1050   **VGood:** $900   **Good:** $790

Merkel Model 101E

Same specs as Model 101 except ejectors. Made in Germany prior to World War II.
Exc.: $1200   **VGood:** $950   **Good:** $700

MERKEL MODEL 122

Side-by-side; Anson & Deely boxlock; 12-, 16-, 20-ga.; 3″ chambers; 26 ¾″, 28″ barrels; any standard choke combo; hand-checkered straight-grip or pistol-grip stock, forend; automatic ejectors; double triggers or single selective trigger; dummy sideplates; Greener crossbolt. Made in Germany. Introduced 1993; still in production. **LMSR:** $3795
Perf.: $3200   **Exc.:** $2000   **VGood:** $1650

MERKEL MODEL 127

Side-by-side; Holland & Holland-type hand-detachable sidelocks; hammerless; 12-, 16-, 20-, 28-ga., .410; standard barrel lengths, choke combos; hand-checkered European walnut stock, forearm; pistol-grip and cheekpiece or straight English style; double triggers; automatic ejectors; elaborately engraved with Arabesque or hunting scene. Manufactured in Germany prior to WWII.
Perf.: $13,500   **Exc.:** $9000   **VGood:** $7000

MERKEL MODEL 130

Side-by-side; Anson & Deeley-type boxlock; hammerless; 12-, 16-, 20-, 28-ga., .410; standard barrel lengths, choke combos; hand-checkered European walnut stock, forearm; pistol-grip and cheekpiece or straight English style; dummy sideplates; double triggers; automatic ejectors; elaborate Arabesque or hunting-scene engraving. Made in Germany prior to World War II.
Perf.: $7000   **Exc.:** $5000   **VGood:** $4000

MERKEL MODEL 147

Side-by-side; Anson & Deeley boxlock; 12-, 16-, 20-ga.; 26 ¾″, 28″ barrels; other lengths on special order; any standard choke combo; hand-checkered straight-grip or pistol-grip stock, forend; extractors; double triggers or single selective trigger. Made in Germany since World War II; still in production. **LMSR:** $1895
Perf.: $1600   **Exc.:** $1100   **VGood:** $900

Merkel Model 147E

Same specs as Model 147 except 12-, 16-, 20-, 28-ga.; automatic ejectors. Made in Germany since World War II; still in production. **LMSR:** $2295
Perf.: $2150   **Exc.:** $1850   **VGood:** $1600
28-ga.
Perf.: $2250   **Exc.:** $1950   **VGood:** $1700

Merkel Model 147S

Same specs as Model 147 except sidelock; 12-, 16-, 20-, 28-ga., .410; 25 ½″, 26″, 26 ¾″, 28″ barrels; automatic ejectors; engraved hunting scene on action. Made in Germany since World War II; still in production. **LMSR:** $5595
Perf.: $5300   **Exc.:** $3600   **VGood:** $3100

MERKEL MODEL 200

Over/under; boxlock; hammerless; 12-, 16-, 20-, 24-, 28-, 32-ga.; standard barrel lengths, choke combos; ribbed barrels; hand-checkered European walnut stock, forearm; pistol grip and cheekpiece or straight English style; Kersten double crossbolt; separate extractors; double triggers; scalloped frame; Arabesque engraving. Made in Germany prior to World War II.
Exc.: $1000   **VGood:** $800   **Good:** $700

Merkel Model 200E

Same specs as the Model 200 except ejectors; double, single or single selective trigger. Introduced prior to WWII; 24-, 28-, 32-ga. dropped during WWII; still in production. **LMSR:** $3395
Perf.: $3100   **Exc.:** $2250   **VGood:** $1900

Merkel Model 200E Skeet

Same specs as the Model 200 except 12-ga.; 2 ¾″ chambers; 26″ Skeet/Skeet barrels; tapered ventilated rib; competition pistol-grip stock; single selective trigger; half-coverage Arabesque engraving on silver-grayed receiver. Made in Germany. Introduced 1993; still imported. **LMSR:** $4895
Perf.: $4650   **Exc.:** $3700   **VGood:** $3200

Merkel Model 200ET Trap

Same specs as the Model 200 except 12-ga.; 2 ¾″ chambers; 30″ Full/Full barrels; tapered ventilated rib; competition pistol-grip stock; single selective trigger; half-coverage Arabesque engraving on silver-grayed receiver. Made in Germany. Introduced 1993; still imported. **LMSR:** $4895
Perf.: $4650   **Exc.:** $3700   **VGood:** $3200

Merkel Model 200SC Sporting Clays

Same specs as the Model 200 except Blitz action; 30″ barrels with lengthened forcing cones; fixed or five Bri-

ley choke tubes; tapered vent rib select-grade stock; single selective trigger adjustable for length of pull; Kersten double crossbolt lock; competition recoil pad; color case-hardened receiver; fitted luggage case. Made in Germany. Introduced 1995; still imported. **LMSR:** $6995
With fixed chokes
Perf.: $6700   **Exc.:** $3900   **VGood:** $3200
With choke tubes
Perf.: $7100   **Exc.:** $4300   **VGood:** $3600

MERKEL MODEL 201

Over/under; boxlock; hammerless; 12-, 16-, 20-ga.; standard barrel lengths, choke combos; ribbed barrels; hand-checkered European walnut stock, forearm; pistol grip and cheekpiece or straight English style; separate extractors; double triggers; Greener crossbolt; scalloped frame; engraving. Made in Germany prior to World War II. No longer in production.
Exc.: $1400   **VGood:** $900   **Good:** $780

Merkel Model 201E

Same specs as Model 201 except ejectors. Made in Germany prior to World War II.
Exc.: $1550   **VGood:** $1150   **Good:** $950

Merkel Model 201E (Recent)

Same specs as Model 201 except 12-, 16-, 20-, 28-ga.; ejectors; double, single or single selective trigger. Made in Germany. Still in production. **LMSR:** $4895
Perf.: $4200   **Exc.:** $2750   **VGood:** $2200
28-ga. model
Perf.: $5000   **Exc.:** $3250   **VGood:** $2700

Merkel Model 201ES

Same specs as Model 201 except 12-ga.; 2 ¾″ chamber; 26 ¾″ Skeet/Skeet barrels; full-coverage engraving. Made in Germany. Still in production. **LMSR:** $7495
Perf.: $6800   **Exc.:** $4350   **VGood:** $3200

SHOTGUNS

Merkel
Model 201ET
Model 202
Model 202E
Model 202E (Recent)
Model 203
Model 203E
Model 203ES
Model 203ET
Model 204E
Model 210
Model 247S
Model 300
Model 300E
Model 301
Model 301E
Model 302
Model 303E

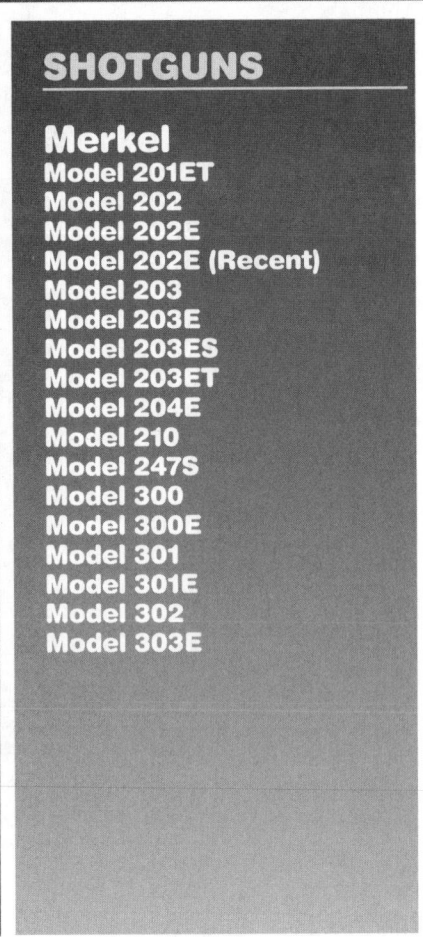
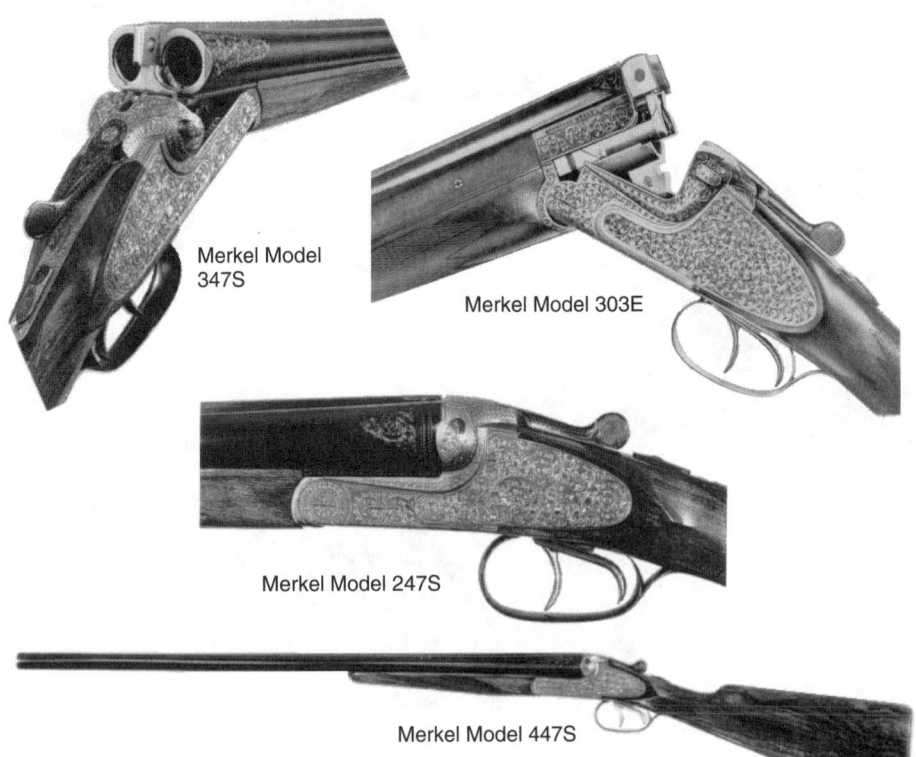

Merkel Model 347S

Merkel Model 303E

Merkel Model 247S

Merkel Model 447S

Merkel Model 201ET
Same specs as Model 201 except 12-ga.; 2 3/4" chamber; 30" Full/Full barrels; full-coverage engraving. Made in Germany. Still in production. **LMSR: $7495**
   Perf.: $3000   **Exc.:** $2800   **VGood:** $2600

MERKEL MODEL 202
Over/under; boxlock; hammerless; 12-, 16-, 20-ga.; standard barrel lengths, choke combos; ribbed barrels; selective hand-checkered European walnut stock, forearm; pistol grip and cheekpiece or straight English style; Greener separate extractors; double triggers; dummy sideplates. Made in Germany prior to World War II.
   Exc.: $2950   **VGood:** $2500   **Good:** $2300

Merkel Model 202E
Same specs as Model 202 except ejectors. Made in Germany prior to World War II.
   Exc.: $3200   **VGood:** $2750   **Good:** $2400

Merkel Model 202E (Recent)
Same specs as Model 202, except automatic ejectors; full-coverage engraving. Made in Germany. Introduced 1993; still in production. **LMSR: $8895**
   Perf.: $8200   **Exc.:** $5500   **VGood:** $4000

MERKEL MODEL 203
Over/under; hand-detachable sidelocks; hammerless; 12-, 16-, 20-ga.; ribbed barrels in standard lengths, choke combos; hand-checkered European walnut stock, forearm; pistol grip and cheekpiece or straight English style; Kersten double crossbolt; automatic ejectors; double triggers; optional single selective or non-selective trigger; Arabesque or hunting scene engraving. Made in Germany prior to World War II.
With double triggers
   Perf.: $5000   **Exc.:** $3500   **VGood:** $2800
With single trigger
   Perf.: $5500   **Exc.:** $4000   **VGood:** $3000
With single selective trigger
   Perf.: $6000   **Exc.:** $4500   **VGood:** $3500

Merkel Model 203E
Same specs as Model 203 except standard sidelocks. Made in Germany. Still in production. **LMSR: $10,695**
   Perf.: $9500   **Exc.:** $4900   **VGood:** $3800

Merkel Model 203ES
Same specs as Model 203 except standard sidelocks; 12-ga.; 2 3/4" chamber; 26 3/4" Skeet/Skeet barrels; full coverage engraving. Made in Germany. Still in production. **LMSR: $12,950**
   Perf.: $11,500 **Exc.:** $7500   **VGood:** $6500

Merkel Model 203ET
Same specs as Model 203 except standard sidelocks; 12-ga.; 2 3/4" chamber; 30" Full/Full barrels; full coverage engraving. Made in Germany. Still in production. **LMSR: $12,950**
   Perf.: $11,500 **Exc.:** $7500   **VGood:** $6500

MERKEL MODEL 204E
Over/under; Merkel sidelocks; hammerless; 12-, 16-, 20-ga.; ribbed barrels in standard lengths, choke combos; hand-checkered European walnut stock, forearm; pistol grip and cheekpiece or straight English style; Kersten double crossbolt; automatic ejectors; double triggers; fine English-style engraving. Made in Germany prior to World War II.
   Exc.: $4000   **VGood:** $3200   **Good:** $2400

MERKEL MODEL 210
Over/under; boxlock; hammerless; 12-, 16-, 20-, 24-, 28-, 32-ga.; ribbed barrels in standard lengths, choke combos; select hand-checkered European walnut stock, forearm; pistol grip and cheekpiece or straight English style; Kersten double crossbolt; scalloped frame; elaborate engraving; separate extractors; double triggers. Made in Germany prior to World War II.
   Exc.: $1250   **VGood:** $900   **Good:** $800

MERKEL MODEL 247S
Side-by-side; sidelock; 12-, 16-, 20-ga.; 25 1/2", 26", 26 3/4" barrels; standard choke combos; hand-checkered straight-grip or pistol-grip stock, forend; automatic ejectors; double triggers or single selective trigger; engraving. Made in Germany. Still in production. **LMSR: $6895**
   Perf.: $6500   **Exc.:** $3800   **VGood:** $3200

MERKEL MODEL 300
Over/under; Merkel-Anson boxlock; hammerless; 12-, 16-, 20-, 24-, 28-, 32-ga.; ribbed barrels in standard lengths, choke combos; hand-checkered European walnut stock; pistol grip and cheekpiece or straight English style; Kersten double crossbolt; two underlugs; scalloped frame; Arabesque or hunting scene engraving; separate extractors. Made in Germany prior to World War II.
   Exc.: $1650   **VGood:** $1400   **Good:** $1200

Merkel Model 300E
Same general specs as Model 300 except automatic ejectors. Made in Germany prior to World War II.
   Exc.: $1850   **VGood:** $1550   **Good:** $1350

MERKEL MODEL 301
Over/under; Merkel-Anson boxlock; hammerless; 12-, 16-, 20-, 24-, 28-, 32-ga.; ribbed barrels in standard lengths, choke combos; select hand-checkered European walnut stock; pistol grip and cheekpiece or straight English style; Kersten double crossbolt; two underlugs; scalloped frame; elaborate engraving; separate extractors. Made in Germany prior to World War II.
   Exc.: $4000   **VGood:** $3000   **Good:** $2400

Merkel Model 301E
Same specs as Model 301 except automatic ejectors. Made in Germany prior to World War II.
   Exc.: $4200   **VGood:** $3200   **Good:** $2600

MERKEL MODEL 302
Over/under; hammerless; 12-, 16-, 20-, 24-, 38-, 32-ga.; ribbed barrels in standard lengths, choke combos; select hand-checkered European walnut stock; pistol grip and cheekpiece or straight English style; Kersten double crossbolt; two underlugs; scalloped frame; elaborate engraving; automatic ejectors; dummy sideplates. Made in Germany prior to World War II.
   Exc.: $7000   **VGood:** $5000   **Good:** $4000

MERKEL MODEL 303E
Over/under; Holland & Holland-type hand-detachable sidelocks; hammerless; ribbed barrels in standard lengths, choke combos; select hand-checkered European walnut stock, forearm; pistol grip and cheekpiece or straight English style; Kersten crossbolt; double underlugs; automatic ejectors. Introduced prior to World War II; still in production. **LMSR: $19,995**
   Perf.: $18,000 **Exc.:** $13,000 **VGood:** $10,000

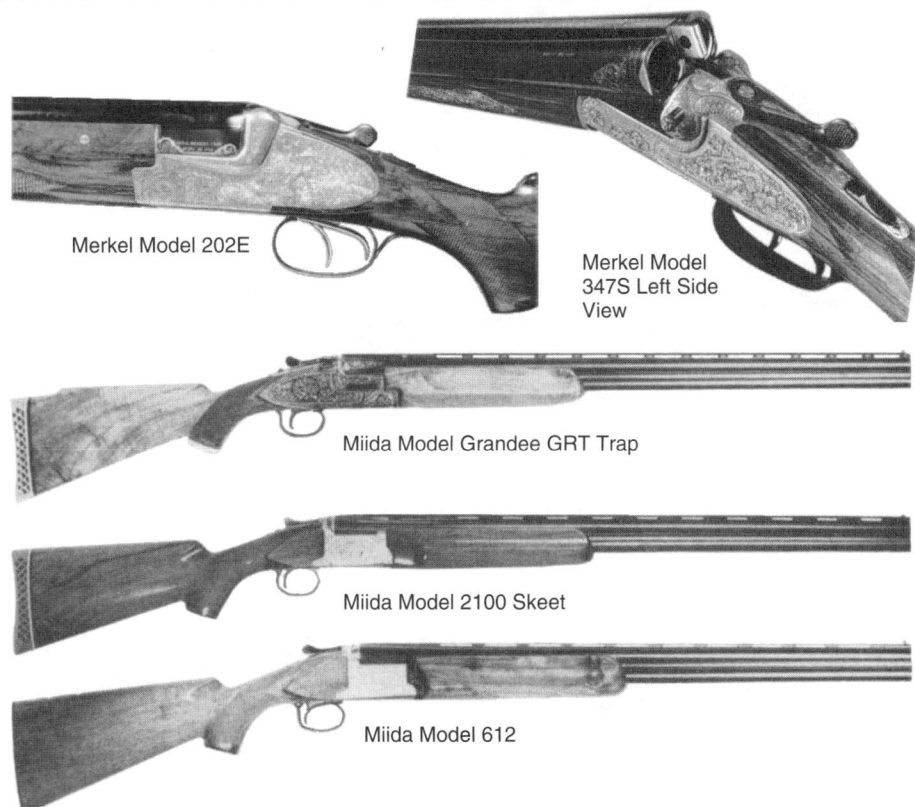

Merkel Model 202E

Merkel Model 347S Left Side View

Miida Model Grandee GRT Trap

Miida Model 2100 Skeet

Miida Model 612

SHOTGUNS

Merkel
Model 304E
Model 310E
Model 347S
Model 400
Model 400E
Model 410
Model 410E
Model 447S

Miida
Model 612
Model 2100 Skeet
Model 2200S Skeet
Model 2200T Trap
Model 2300S Skeet
Model 2300T Trap
Model Grandee GRS Skeet
Model Grandee GRT Trap

MERKEL MODEL 304E

Over/under; Holland & Holland-type hand-detachable sidelocks; hammerless; 12-, 16-, 20-, 24-, 28-, 32-ga.; ribbed barrels in standard lengths, choke combos; fancy hand-checkered European walnut stock; pistol grip and cheekpiece or straight English style; Kersten double crossbolt; two underlugs; automatic ejectors; scalloped frame; elaborate engraving. Made in Germany prior to World War II. Optional extras can greatly affect price.
Perf.: $20,000 **Exc.:** $15,000 **VGood:** $10,000

MERKEL MODEL 310E

Over/under; Merkel-Anson boxlock; hammerless; 12-, 16-, 20-, 24-, 28-, 32-ga.; ribbed barrels in standard lengths, choke combos; hand-checkered European walnut stock; pistol grip and cheekpiece or straight English style; Kersten double crossbolt; two underlugs; automatic ejectors; scalloped frame; Arabesque or hunting scene engraving. Made in Germany prior to World War II.
Exc.: $5000 **VGood:** $3900 **Good:** $3000

MERKEL MODEL 347S

Side-by-side; sidelock; 12-,16-, 20-ga.; 25 1/2", 26", 26 3/4" barrels; any standard choke combo; select hand-checkered straight-grip or pistol-grip stock, forend; automatic ejectors; double triggers or single selective trigger; elaborate engraving. Made in Germany. Still in production. **LMSR: $7895**
Perf.: $7300 **Exc.:** $4200 **VGood:** $3500

MERKEL MODEL 400

Over/under; boxlock; hammerless; 12-, 16-, 20-ga.; ribbed barrels in standard lengths, choke combos; hand-checkered European walnut stock, forearm; pistol grip and cheekpiece or straight English style; Kersten double crossbolt; double triggers; separate extractors; Arabesque engraving on receiver. Made in Germany prior to World War II.
Exc.: $1350 **VGood:** $1050 **Good:** $800

Merkel Model 400E

Same specs as Model 400 except for ejectors. Made in Germany prior to World War II.
Exc.: $1400 **VGood:** $1100 **Good:** $850

MERKEL MODEL 410

Over/under; boxlock; hammerless; 12-, 16-, 20-ga.; ribbed barrels in standard lengths, choke combos; select hand-checkered European walnut stock, forearm; pistol grip and cheekpiece or straight English style; Kersten double crossbolt; double triggers; separate extractors; elaborate engraving. Made in Germany prior to World War II.
Exc.: $1100 **VGood:** $900 **Good:** $750

Merkel Model 410E

Same specs as Model 410 except ejectors. Made in Germany prior to World War II.
Exc.: $1300 **VGood:** $1100 **Good:** $950

MERKEL MODEL 447S

Side-by-side; sidelock; 12-, 16-, 20-ga.; 25 1/2", 26", 26 3/4" barrels; standard choke combos; select hand-checkered straight-grip or pistol-grip stock, forend; automatic ejectors; double triggers or single selective trigger; fancy engraving. Made in Germany. Still in production. **LMSR: $8995**
Perf.: $8500 **Exc.:** $4400 **VGood:** $3200

MIIDA MODEL 612

Over/under; boxlock; 12-ga.; 26" Improved Cylinder/Modified; 28" Modified/Full barrels; vent rib; checkered walnut pistol-grip stock, forearm; auto ejectors; single selective trigger; engraving. Made in Japan; imported by Marubeni America Corp. Introduced 1972; discontinued 1974.
Perf.: $725 **Exc.:** $600 **VGood:** $500

MIIDA MODEL 2100 SKEET

Over/under; boxlock; 12-ga.; 27" vent-rib barrels; 12-ga.; 27" Skeet/Skeet barrels; vent rib; select checkered walnut pistol-grip stock, forearm; selective single trigger; auto ejectors; 50% engraving coverage. Made in Japan; imported by Marubeni America Corp. Introduced 1972; discontinued 1974.
Perf.: $800 **Exc.:** $680 **VGood:** $600

MIIDA MODEL 2200S SKEET

Over/under; boxlock; 12-ga.; 27" Skeet/Skeet barrels; vent rib; checkered fancy walnut Skeet stock, semi-beavertail forearm; 60% engraving coverage. Made in Japan; imported by Marubeni America Corp. Introduced 1972; discontinued 1974.
Perf.: $850 **Exc.:** $700 **VGood:** $600

Miida Model 2200T Trap

Same specs as 2200S, except 29 3/4" Improved Modified/Full barrels; trap stock; recoil pad. Made in Japan; imported by Marubeni America Corp. Introduced 1972; discontinued 1974.
Perf.: $850 **Exc.:** $700 **VGood:** $600

MIIDA MODEL 2300S SKEET

Over/under; boxlock; 12-ga.; 27" Skeet/Skeet barrels; vent rib; checkered fancy walnut Skeet stock, semi-beavertail forearm; 70% engraving coverage. Made in Japan; imported by Marubeni America Corp. Introduced 1972; discontinued 1974.
Perf.: $900 **Exc.:** $750 **VGood:** $650

Miida Model 2300T Trap

Same specs as Model 2300S Skeet, except 29 3/4" Improved Modified/Full barrels; trap stock; recoil pad. Made in Japan; imported by Marubeni America Corp. Introduced 1972; discontinued 1974.
Perf.: $900 **Exc.:** $750 **VGood:** $650

MIIDA MODEL GRANDEE GRS SKEET

Over/under; boxlock; 12-ga.; 27" Skeet/Skeet barrels; wide vent rib; extra fancy walnut Skeet stock, semi-beavertail forearm; single selective trigger; auto ejectors; sideplates; gold-inlaid, fully engraved frame, breech ends of barrels, locking lever, trigger guard. Made in Japan; imported by Marubeni America Corp. Introduced 1972; discontiued 1974.
Perf.: $2250 **Exc.:** $1800 **VGood:** $1500

Miida Model Grandee GRT Trap

Same specs as Model Grandee GRS Skeet, except 29 3/4" Full/Full barrels; trap stock; recoil pad. Made in Japan; imported by Marubeni America Corp. Introduced 1972; discontinued 1974.
Perf.: $2250 **Exc.:** $1800 **VGood:** $1800

SHOTGUNS

Mitchell High Standard
Model 9104
Model 9104-CT
Model 9105
Model 9108-B
Model 9108-BL
Model 9108-CT
Model 9108-HG-B
Model 9108-HG-BL
Model 9109
Model 9111-B
Model 9111-BL
Model 9111-CT
Model 9113-B
Model 9114-PG
Model 9115
Model 9115-CT

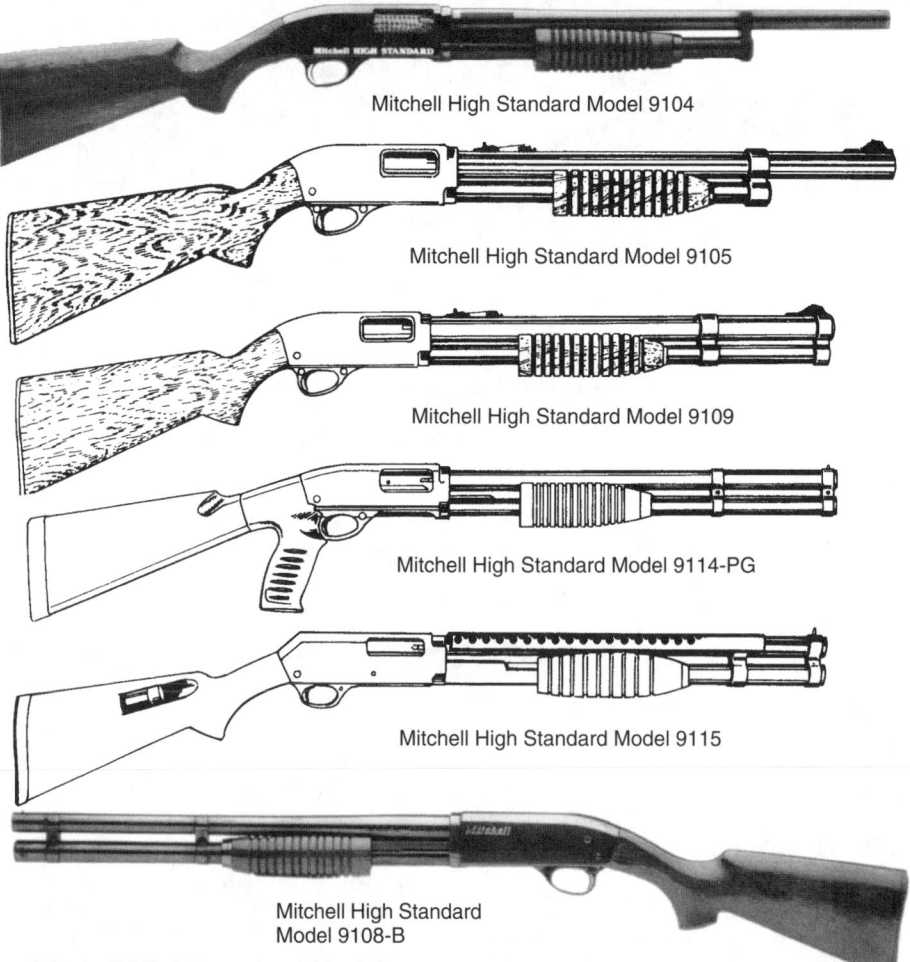

Mitchell High Standard Model 9104

Mitchell High Standard Model 9105

Mitchell High Standard Model 9109

Mitchell High Standard Model 9114-PG

Mitchell High Standard Model 9115

Mitchell High Standard Model 9108-B

MITCHELL HIGH STANDARD MODEL 9104
Slide action; 12-ga.; 2 3/4″ chamber; 6-shot magazine; 20″ Cylinder barrel; bead sight; walnut-finished hardwood stock, slide handle; fixed barrel with steel receiver; polished blue finish. Made in the Philippines. Introduced 1994; still imported. **LMSR: $279**
Perf.: $230   **Exc.:** $180   **VGood:** $140

Mitchell High Standard Model 9104-CT
Same specs as Model 9104 except Improved Cylinder, Modified, Full choke tubes. Introduced 1994; still imported. **LMSR: $299**
Perf.: $250   **Exc.:** $200   **VGood:** $160

MITCHELL HIGH STANDARD MODEL 9105
Slide action; 12-ga.; 2 3/4″ chamber; 6-shot magazine; 20″ Cylinder barrel; rifle sights; walnut-finished hardwood stock, slide handle; fixed barrel with steel receiver; polished blue finish. Made in the Philippines. Introduced 1994; still imported. **LMSR: $299**
Perf.: $250   **Exc.:** $200   **VGood:** $160

MITCHELL HIGH STANDARD MODEL 9108-B
Slide action; 12-ga.; 2 3/4″ chamber; 8-shot magazine; 20″ Cylinder barrel; bead sight; brown-finished hardwood stock, slide handle; fixed barrel with steel receiver; polished blue finish. Made in the Philippines. Introduced 1994; still imported. **LMSR: $279**
Perf.: $230   **Exc.:** $180   **VGood:** $140

Mitchell High Standard Model 9108-BL
Same specs as Model 9108-B except black-finished hardwood stock, slide handle. Introduced 1994; still imported. **LMSR: $279**
Perf.: $230   **Exc.:** $180   **VGood:** $140

Mitchell High Standard Model 9108-CT
Same specs as Model 9108-B except Improved Cylinder, Modified, Full choke tubes. Introduced 1994; still imported. **LMSR: $299**
Perf.: $250   **Exc.:** $200   **VGood:** $160

Mitchell High Standard Model 9108-HG-B
Same specs as Model 9108-B except brown-finished hardwood hand-grip stock, slide handle. Introduced 1994; still imported. **LMSR: $299**
Perf.: $250   **Exc.:** $200   **VGood:** $160

Mitchell High Standard Model 9108-HG-BL
Same specs as Model 9108-B except black-finished hardwood hand-grip stock, slide handle. Introduced 1994; still imported. **LMSR: $299**
Perf.: $250   **Exc.:** $200   **VGood:** $150

MITCHELL HIGH STANDARD MODEL 9109
Slide action; 12-ga.; 2 3/4″ chamber; 8-shot magazine; 20″ Cylinder barrel; rifle sights; walnut-finished hardwood stock, slide handle; fixed barrel with steel receiver; polished blue finish. Made in the Philippines. Introduced 1994; still imported. **LMSR: $299**
Perf.: $250   **Exc.:** $200   **VGood:** $150

MITCHELL HIGH STANDARD MODEL 9111-B
Slide action; 12-ga.; 2 3/4″ chamber; 7-shot magazine; 18 1/2″ Cylinder barrel; bead sight; brown-finished hardwood hand-grip stock, slide handle; fixed barrel with steel receiver; polished blue finish. Made in the Philippines. Introduced 1994; still imported. **LMSR: $299**
Perf.: $250   **Exc.:** $200   **VGood:** $150

Mitchell High Standard Model 9111-BL
Same specs as Model 9108-B except black-finished hardwood hand-grip stock, slide handle. Introduced 1994; still in production. **LMSR: $299**
Perf.: $250   **Exc.:** $200   **VGood:** $150

Mitchell High Standard Model 9111-CT
Same specs as Model 9108-B except Improved Cylinder, Modified, Full choke tubes. Introduced 1994; still imported. **LMSR: $299**
Perf.: $250   **Exc.:** $200   **VGood:** $150

MITCHELL HIGH STANDARD MODEL 9113-B
Slide action; 12-ga.; 2 3/4″ chamber; 7-shot magazine; 18 1/2″ Cylinder barrel; rifle sights; brown-finished hardwood hand-grip stock, slide handle; fixed barrel with steel receiver; polished blue finish. Made in the Philippines. Introduced 1994; still imported. **LMSR: $299**
Perf.: $250   **Exc.:** $200   **VGood:** $150

MITCHELL HIGH STANDARD MODEL 9114-PG
Slide action; 12-ga.; 2 3/4″ chamber; 7-shot magazine; 18 1/2″ Cylinder barrel; rifle sights; pistol-grip stock; slide handle; fixed barrel with steel receiver; polished blue finish. Introduced 1994; still in production. **LMSR: $349**
Perf.: $290   **Exc.:** $220   **VGood:** $180

MITCHELL HIGH STANDARD MODEL 9115
Slide action; 12-ga.; 2 3/4″ chamber; 7-shot magazine; 18 1/2″ Cylinder barrel with ventilated heat shield; bead sight; walnut-finished hardwood SAS-style stock, slide handle; stock stores four shells; fixed barrel with steel receiver; Parkerized finish. Made in the Philippines. Introduced 1994; still imported. **LMSR: $349**
Perf.: $290   **Exc.:** $220   **VGood:** $180

Mitchell High Standard Model 9115-CT
Same specs as Model 9115 except Improved Cylinder, Modified, Full choke tubes. Introduced 1994; still imported. **LMSR: $369**
Perf.: $310   **Exc.:** $230   **VGood:** $190

Mossberg Model 1000 Junior

Mossberg Model 5500 MkII

Mossberg Model 5500

Mossberg Model 9200 Crown Grade

Mossberg Model 9200 Crown Grade Camo

SHOTGUNS

Mitchell High Standard
Model 9115-HG

Monte Carlo
Single

Mossberg Semi-Automatic Shotguns
Model 1000
Model 1000 Junior
Model 1000 Skeet
Model 1000 Slug
Model 1000 Super
Model 1000 Super Skeet
Model 1000 Super Trap
Model 1000 Super Waterfowler
Model 5500
Model 5500 Mk II
Model 5500 Mark II Camo
Model 6000
Model 9200 Crown Grade
Model 9200 Crown Grade Camo
Model 9200 NWTF Edition

Mitchell High Standard Model 9115-HG
Same specs as Model 9115 except walnut finished hardwood SAS-style hand-grip stock, slide handle. Introduced 1994; still imported. **LMSR: $349**
Perf.: $290   **Exc.:** $220   **VGood:** $180

MONTE CARLO SINGLE
Single barrel; 12-ga.; 32″ trap barrel; hand-checkered European walnut pistol-grip stock, beavertail forend; recoil pad; auto ejector; slide safety; gold-plated trigger. Made in Italy. Introduced 1968; no longer in production.
Exc.: $275   **VGood:** $240   **Good:** $180

MOSSBERG SEMI-AUTOMATIC SHOTGUNS

MOSSBERG MODEL 1000
Gas-operated autoloader; 12-, 20-ga.; 2 3/4″ chamber; 4-shot tube magazine; 26″, 28″ vent-rib barrel; standard chokes; optional screw-in choke tubes; front, middle bead sights; checkered American walnut stock; interchangeable crossbolt safety; pressure compensator, floating piston for recoil control; engraved alloy receiver. Made in Japan. Formerly marketed by Smith & Wesson; discontinued 1987.
Perf.: $390   **Exc.:** $320   **VGood:** $280

Mossberg Model 1000 Junior
Same specs as Model 1000 except 20-ga.; 22″ Multi-Choke barrel; vent rib; junior-sized stock. Introduced 1986; discontinued 1987.
Perf.: $410   **Exc.:** $330   **VGood:** $290

Mossberg Model 1000 Skeet
Same specs as Model 1000 except 26″ Skeet barrel; vent rib; Skeet-style stock; steel receiver. Introduced 1986; discontinued 1986.
Perf.: $375   **Exc.:** $300   **VGood:** $250

Mossberg Model 1000 Slug
Same specs as Model 1000 except 22″ barrel; rifle sights; recoil pad. Introduced 1986; discontinued 1987.
Perf.: $400   **Exc.:** $320   **VGood:** $260

Mossberg Model 1000 Super
Same specs as Model 1000 except gas-metering system for 2 3/4″, 3″ shells; 26″, 28″, 30″ Multi-Choke barrel; recoil pad. Introduced 1986; discontinued 1987.
Perf.: $480   **Exc.:** $375   **VGood:** $315

Mossberg Model 1000 Super Skeet
Same specs as Model 1000 except gas-metering system for 2 3/4″, 3″ shells; 25″ barrel; recessed-type Skeet choke; oil-finished, select-grade walnut stock, forend; palm swell; recoil-reduction compensator system; contoured trigger; forend cap weights for changing balance. Introduced 1986; discontinued 1987.
Perf.: $550   **Exc.:** $450   **VGood:** $400

Mossberg Model 1000 Super Trap
Same specs as Model 1000 except 12-ga.; gas-metering system for 2 3/4″, 3″ shells; 30″ Multi-choke barrel; vent rib; walnut-finished hardwood Monte Carlo stock; recoil pad. Introduced 1986; discontinued 1987.
Perf.: $465   **Exc.:** $370   **VGood:** $300

Mossberg Model 1000 Super Waterfowler
Same specs as Model 1000 except 12-ga.; gas-metering system for 2 3/4″, 3″ shells; 28″ Multi-choke barrel; sling swivels; camo sling; Parkerized finish. Introduced 1986; discontinued 1987.
Perf.: $500   **Exc.:** $400   **VGood:** $360

MOSSBERG MODEL 5500
Autoloader; 12-ga.; 2 3/4″, 3″ chamber; interchangeable 18 1/2″ Cylinder, 24″ Slugster, 26″ Improved, 28″ Modified, 30″ Full barrel; Accu-Choke tubes; bead front sight; walnut-finished hardwood stock. Introduced 1983; discontinued 1985.
Perf.: $235   **Exc.:** $200   **VGood:** $170
Slugster model
Perf.: $260   **Exc.:** $215   **VGood:** $190

MOSSBERG MODEL 5500 MKII
Same specs as Model 5500 except 24″ rifled, 24″, 26″, 28″ vent-rib barrel; Accu-II choke tubes. Also marketed as two-barrel set with both 26″ (2 3/4″) and 28″ (3″) barrels. Introduced 1988; discontinued 1992.
Perf.: $250   **Exc.:** $200   **VGood:** $175
Combo set
Perf.: $300   **Exc.:** $250   **VGood:** $225

Mossberg Model 5500 Mark II Camo
Same specs as Model 5500 except 28″ barrel; Accu-II choke tubes; synthetic stock; camouflage finish. Introduced 1988; discontinued 1992.
Perf.: $280   **Exc.:** $230   **VGood:** $200

MOSSBERG MODEL 6000
Autoloader; 12-ga.; 3″ chamber; 28″ barrel; Modified Accu-Choke tube; weighs about 7 1/2 lbs.; walnut-finished stock with high-gloss finish; alloy receiver; shoots all 2 3/4″ or 3″ loads without adjustment. Introduced 1993; discontinued 1993.
Perf.: $275   **Exc.:** $200   **VGood:** $180

MOSSBERG MODEL 9200 CROWN GRADE
Autoloader; 12-ga.; 3″ chamber; 24″ rifled, 24″, 28″ vent-rib barrel; Accu-Choke tubes; weighs about 7 1/2 lbs.; cut-checkered walnut stock with high-gloss finish; ambidextrous safety; alloy receiver; shoots all 2 3/4″ or 3″ loads without adjustment. Introduced 1992; still in production. **LMSR: $478**
With standard barrel
Perf.: $425   **Exc.:** $350   **VGood:** $290
With 24″ rifled barrel; scope base; Dual-Comb stock
Perf.: $445   **Exc.:** $370   **VGood:** $310

Mossberg Model 9200 Crown Grade Camo
Same specs as Model 9200 Crown Grade except 24″ vent-rib barrel; Accu-Choke tubes; synthetic stock, forend; completely covered with Mossy Oak Tree Stand or OFM camouflage finish. Introduced 1993; still in production. **LMSR: $463**
With Mossy Oak finish
Perf.: $400   **Exc.:** $325   **VGood:** $265
With OFM camo finish
Perf.: $400   **Exc.:** $325   **VGood:** $265

Mossberg Model 9200 NWTF Edition
Same specs as Model 9200 Crown Grade except black matte receiver with turkey scenes etched on both sides; Realtree camo finish on rest of gun. Introduced 1994; still in production. **LMSR: $562**
Perf.: $490   **Exc.:** $395   **VGood:** $350

Mossberg Semi-Automatic Shotguns
Model 9200 Regal
Model 9200 USST

Mossberg Slide-Action Shotguns
Home Security 410
Model 200D
Model 200K
Model 500
Model 500AGVD
Model 500AHT (Hi-Rib Trap)
Model 500AHTD (Hi-Rib Trap)
Model 500ALD
Model 500ALMR (Heavy Duck)
Model 500ALS Slugster
Model 500ASG Slugster
Model 500APR Pigeon Grade Trap
Model 500 Bantam

Mossberg Home Security 410

Mossberg Model 500

Mossberg Model 500AGVD

Mossberg Model 500AHT (Hi-Rib Trap)

Mossberg Model 500ALS Slugster

Mossberg Model 500ASG Slugster

Mossberg Model 9200 Regal
Same specs as Model 9200 Crown Grade except 24" rifled, 28" vent-rib barrel; Accu-Choke tubes; less fancy wood. Introduced 1992; discontinued 1993.
Perf.: $425   Exc.: $350   VGood: $290

Mossberg Model 9200 USST
Same specs as Model 9200 Crown Grade except 26" vent-rib barrel; Accu-Choke tubes (including Skeet); cut-checkered walnut-finished stock, forend; "United States Shooting Team" custom engraved receiver. Introduced 1993; still in production. **LMSR: $478**
Perf.: $415   Exc.: $330   VGood: $280

MOSSBERG SLIDE-ACTION SHOTGUNS

MOSSBERG HOME SECURITY 410
Slide action; .410; 3" chamber; 18 1/2" barrel; spreader choke; 37 1/2" overall length; weighs 6 1/4 lbs; synthetic pistol-grip stock, forend; recoil pad; blued finish; also available with laser installed in forend. Marketed with Mossberg Cablelock. Introduced 1990; still in production. **LMSR: $293**
Perf.: $260   Exc.: $195   VGood: $165
With laser
Perf.: $375   Exc.: $300   VGood: $275

MOSSBERG MODEL 200D
Slide action; 12-ga.; 3-shot detachable box magazine; 28" barrel; interchangeable choke tubes; uncheckered, walnut-finished hardwood pistol-grip stock, grooved black nylon slide handle; recoil pad. Introduced 1955; discontinued 1959.
Exc.: $135   VGood: $95   Good: $50

Mossberg Model 200K
Same specs as Model 200D except variable C-Lect-Choke instead of interchangeable choke tubes. Introduced 1955; discontinued 1959.
Exc.: $140   VGood: $100   Good: $50

MOSSBERG MODEL 500
Slide action; hammerless; takedown; 12-, 20-ga.; .410; 6-shot tube magazine; 3-shot plug furnished; 24" Slugster Cylinder with rifle sights (12-ga. only), 26" Improved, 28" Modified or Full, 30" Full (12-ga. only) barrel; adjustable C-Lect-Choke, Accu-Choke, Accu-II, Accu-Steel tubes; white bead front sight, brass mid-bead; checkered or uncheckered American walnut pistol-grip stock, grooved slide handle; recoil pad; twin extractors; ambidextrous safety; disconnecting safety dual action bars. Marketed with Cablelock. Introduced 1961; still in production. **LMSR: $281**
With Slugster barrel
Perf.: $240   Exc.: $170   VGood: $140
With fixed-choke barrel
Perf.: $220   Exc.: $150   VGood: $130
With adjustable-choke barrel
Perf.: $240   Exc.: $170   VGood: $140
Combo with extra barrel
Perf.: $280   Exc.: $210   VGood: $180

Mossberg Model 500AGVD
Same specs as Model 500 except 12-, 20-ga.; 3" chamber; 28" vent-rib barrel with Accu-Choke system for Improved, Modified, Full. No longer in production.
Perf.: $240   Exc.: $170   VGood: $140

Mossberg Model 500AHT (Hi-Rib Trap)
Same specs as Model 500 except 12-ga.; 28", 30" Full-choke barrel; Simmons Olympic-style free-floating rib; built-up Monte Carlo trap stock. Introduced 1978; discontinued 1986.
Perf.: $275   Exc.: $240   VGood: $200

Mossberg Model 500AHTD (Hi-Rib Trap)
Same specs as Model 500 except 12-ga.; 28", 30" barrel; three choke tubes; Simmons Olympic-style free-floating rib; built-up Monte Carlo trap stock. Introduced 1978; discontinued 1986.
Perf.: $305   Exc.: $255   VGood: $215

Mossberg Model 500ALD
Same specs as Model 500 except 12-ga.; 2 3/4", 3" chamber; 28" vent-rib barrel; Accu-Choke or three interchangeable choke tubes; game scene etched in receiver. Introduced 1977; no longer in production.
Perf.: $245   Exc.: $180   VGood: $150

Mossberg Model 500 ALMR (Heavy Duck)
Same specs as Model 500 except 12-ga.; 3" chamber; 30", 32" Full-choke barrel; vent rib. Introduced 1977; no longer in production.
Perf.: $250   Exc.: $185   VGood: $155

Mossberg Model 500ALS Slugster
Same specs as Model 500 except 12-ga.; 2 3/4", 3" chamber; 18 1/2", 24" Cylinder-bore barrel; rifle sights; game scene etched in receiver. Introduced 1977; no longer in production.
Perf.: $230   Exc.: $160   VGood: $130

Mossberg Model 500ASG Slugster
Same specs as Model 500 except 12-, 20-ga.; 2 3/4", 3" chamber; 18 1/2", 24" slug barrel; ramp front sight, open adjustable folding leaf rear; running deer etched on receiver. No longer in production.
Perf.: $225   Exc.: $180   VGood: $150

Mossberg Model 500APR Pigeon Grade Trap
Same specs as Model 500 except for 30" Full-choke barrel; vent rib; checkered walnut Monte Carlo trap-style stock, beavertail slide handle. Introduced 1968; discontinued 1975.
Perf.: $425   Exc.: $310   VGood: $240

Mossberg Model 500 Bantam
Same specs as Model 500 except 20-ga.; 22" vent rib barrel; Modified Accu-Choke tube; 1" shorter stock with reduced length from pistol grip to trigger, reduced forend reach. Introduced 1992; still in production. **LMSR: $281**
Perf.: $245   Exc.: $200   VGood: $160

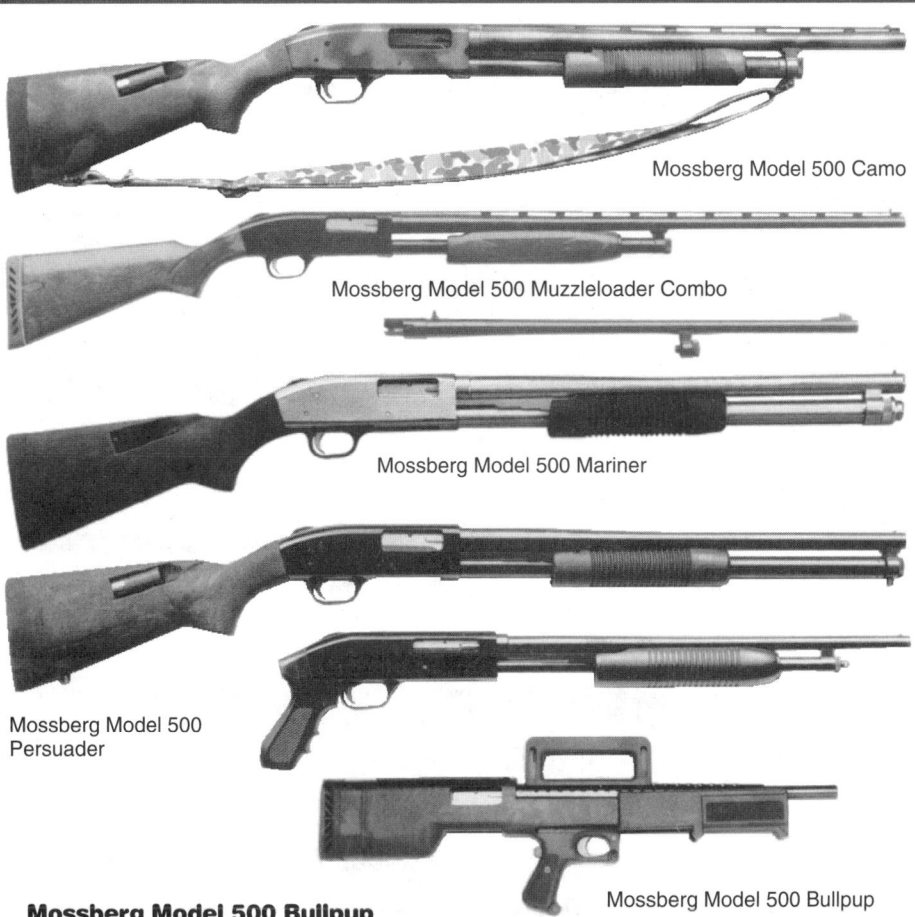

Mossberg Model 500 Camo

Mossberg Model 500 Muzzleloader Combo

Mossberg Model 500 Mariner

Mossberg Model 500 Persuader

Mossberg Model 500 Bullpup

Mossberg Slide-Action Shotguns
Model 500 Bullpup
Model 500 Camo
Model 500 Camper
Model 500CLD
Model 500CLDR
Model 500CLS Slugster
Model 500E
Model 500EGV
Model 500EL
Model 500ELR
Model 500 Ghost Ring
Model 500 Intimidator
Model 500 Mariner
Model 500 Medallion
Model 500 Muzzleloader Combo
Model 500 Persuader

Mossberg Model 500 Bullpup
Same specs as Model 500 except 12-ga.; 3″ chamber; 6-, 9-shot magazine; 18 1/2″, 20″ Cylinder-bore barrel; weighs 9 1/2 lbs.; fixed sights mounted in carrying handle; plastic bullpup-design stock; crossbolt, grip safeties; marketed with Cablelock. Designed for law enforcement; may be illegal in some jurisdictions. Introduced 1986; discontinued 1990.
6-shot model
Perf.: $330   **Exc.:** $240   **VGood:** $200
8-shot model
Perf.: $350   **Exc.:** $260   **VGood:** $220

Mossberg Model 500 Camo
Same specs as Model 500 except 12-ga.; receiver drilled, tapped for scope mount; synthetic field or Speedfeed stock; sling swivels; camo sling; entire surface covered with camo finish; marketed with Mossberg Cablelock. Introduced 1986; still in production. **LMSR: $296**
Perf.: $250   **Exc.:** $150   **VGood:** $100
Camo Combo with extra barrel
Perf.: $340   **Exc.:** $220   **VGood:** $170

Mossberg Model 500 Camper
Same specs as Model 500 except .410 bore only; 18 1/2″ Cylinder-bore barrel; synthetic pistol-grip stock; carry case. Introduced 1988; dropped 1988.
Perf.: $325   **Exc.:** $280   **VGood:** $250

Mossberg Model 500CLD
Same specs as Model 500 except 20-ga.; 2 3/4″ chamber; 28″ barrel; Accu-Choke or three interchangeable choke tubes; game scene etched on receiver. Introduced 1977; no longer in production.
Perf.: $250   **Exc.:** $175   **VGood:** $130

Mossberg Model 500CLDR
Same specs as Model 500 except 20-ga.; 2 3/4″ chamber; 28″ vent-rib barrel; Accu-Choke or three interchangeable choke tubes; game scene etched on receiver. Introduced 1977; no longer in production.
Perf.: $270   **Exc.:** $195   **VGood:** $150

Mossberg Model 500CLS Slugster
Same specs as Model 500 except 20-ga.; 2 3/4″, 3″ chamber; 24″ Cylinder-bore barrel; rifle sights; game scene etched in receiver. Introduced 1977; no longer in production.
Perf.: $235   **Exc.:** $170   **VGood:** $130

Mossberg Model 500E
Same specs as Model 500 except .410; 26″ barrel; Full, Modified, Improved chokes; fluted comb. No longer in production.
Perf.: $245   **Exc.:** $175   **VGood:** $145
Skeet barrel with vent rib, checkering
Perf.: $275   **Exc.:** $195   **VGood:** $175

Mossberg Model 500EGV
Same specs as Model 500 except .410; 26″ Full-choke barrel; vent rib; checkered pistol grip, forend; fluted comb; recoil pad. No longer in production.
Perf.: $275   **Exc.:** $195   **VGood:** $175

Mossberg Model 500EL
Same specs as Model 500 except .410; 2 1/2″, 3″ chamber; 26″ Full-choke barrel; game scene etched in receiver. Introduced 1977; no longer in production.
Perf.: $275   **Exc.:** $195   **VGood:** $175

Mossberg Model 500ELR
Same specs as Model 500 except .410; 2 1/2″, 3″ chamber; 26″ Full-choke barrel; vent rib; game scene etched in receiver. Introduced 1977; no longer in production.
Perf.: $275   **Exc.:** $195   **VGood:** $175

Mossberg Model 500 Ghost Ring
Same specs as Model 500 except 12-ga.; 3″ chamber; 6-, 9-shot magazine; 18 1/2″ Cylinder or Accu-Choke barrel; adjustable blade front sight, adjustable Ghost Ring rear with protective ears; synthetic or Speedfeed stock; blue or Parkerized finish. Introduced 1990; still in production. **LMSR: $331**
With blue finish
Perf.: $270   **Exc.:** $190   **VGood:** $150
With Parkerized finish
Perf.: $300   **Exc.:** $210   **VGood:** $180
With Parkerized finish, Accu-Choke barrel
Perf.: $350   **Exc.:** $260   **VGood:** $225

Mossberg Model 500 Intimidator
Same specs as Model 500 except 12-ga.; 3″ chamber; 6-, 9-shot magazine; 18 1/2″ Cylinder-bore barrel; integral laser sight incorporated in synthetic forend; blued or Parkerized finish; marketed with Mossberg Cablelock. Introduced 1990; discontinued 1993.
With blue finish
Perf.: $450   **Exc.:** $360   **VGood:** $300

With Parkerized finish
Perf.: $460   **Exc.:** $370   **VGood:** $310

Mossberg Model 500 Mariner
Same specs as Model 500 except 12-ga.; 3″ chamber; 6-, 9-shot magazine; 18 1/2″, 20″ Cylinder-bore barrel; ghost ring or metal bead front sight; walnut-finished hardwood, synthetic or Speedfeed stock; all metal parts are finished with Marinecoat to resist rust, corrosion; marketed with Cablelock. Still in production.
LMSR: $403
Perf.: $350   **Exc.:** $220   **VGood:** $180
Mini Combo with extra barrel, handguard, pistol grip
Perf.: $450   **Exc.:** $320   **VGood:** $280

Mossberg Model 500 Medallion
Same specs as Model 500 except 12-, 20-ga.; 28″ vent-rib barrel; Accu-Choke tubes; game bird medallion inset in receiver: pheasant or duck (12-ga.), grouse or quail (20-ga.); only 5000 made in each category. Introduced 1983; discontinued 1984.
Perf.: $260   **Exc.:** $190   **VGood:** $175

Mossberg Model 500 Muzzleloader Combo
Same specs as Model 500 except two-barrel set with 24″ rifled Slugster barrel with rifle sights and 24″ fully rifled 50-caliber muzzle-loading barrel with ramrod; Uses #209 standard primer. Introduced 1992; still in production. **LMSR: $457**
Perf.: $390   **Exc.:** $280   **VGood:** $220

Mossberg Model 500 Persuader
Same specs as Model 500 except 12-, 20-ga.; 6-, 8-shot magazine; 18 1/2″, 20″ barrel; Cylinder bore or Accu-Choke tubes; shotgun or rifle sights; uncheckered wood or synthetic pistol-grip stock; grooved slide handle; blue or Parkerized finish; optional bayonet lug. Designed specifically for law enforcement. Introduced 1977; still in production.
LMSR: $281
Perf.: $245   **Exc.:** $175   **VGood:** $150
With bayonet lug
Perf.: $260   **Exc.:** $190   **VGood:** $165

Mossberg Slide-Action Shotguns

Model 500 Security Series
Model 500 Super Grade
Model 500 Trophy Slugster
Model 500 Turkey
Model 590
Model 590 Bullpup
Model 590 Ghost Ring
Model 590 Intimidator
Model 590 Mariner
Model 835 American Field
Model 835 Ulti-Mag
Model 835 Ulti-Mag Crown Grade

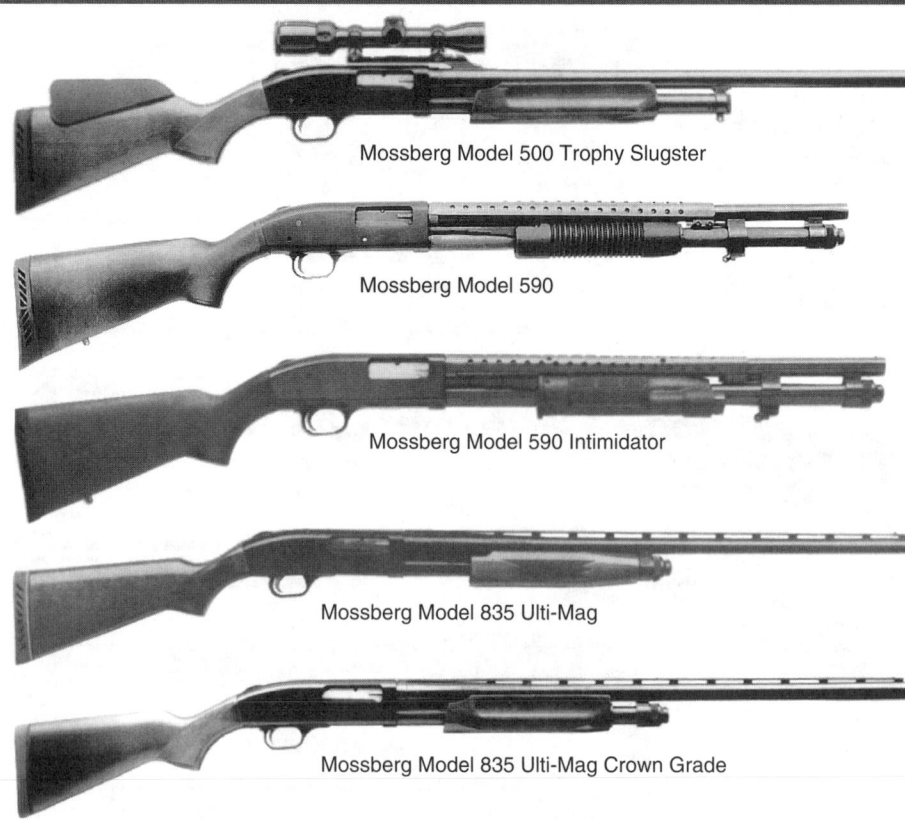

Mossberg Model 500 Trophy Slugster

Mossberg Model 590

Mossberg Model 590 Intimidator

Mossberg Model 835 Ulti-Mag

Mossberg Model 835 Ulti-Mag Crown Grade

Mossberg Model 500 Security Series

Same specs as Model 500 except 12-ga.; 3″ chamber; 6-, 8-shot magazine; 18 1/2″, 20″ Cylinder bore barrel; weighs 7 lbs.; metal bead front sight; walnut-finished hardwood, synthetic or Speedfeed stock; blue, Parkerized or nickel finish.

With blue or nickel finish
Perf.: $240   **Exc.:** $170   **VGood:** $150
With Parkerized finish
Perf.: $260   **Exc.:** $190   **VGood:** $170
Mini Combo with extra barrel
Perf.: $290   **Exc.:** $220   **VGood:** $190

Mossberg Model 500 Super Grade

Same specs as Model 500 except vent-rib barrel; checkered pistol grip, slide handle. Introduced 1961; discontinued 1976.

With fixed-choke barrel
Perf.: $245   **Exc.:** $190   **VGood:** $150
With adjustable-choke barrel
Perf.: $265   **Exc.:** $210   **VGood:** $170

Mossberg Model 500 Trophy Slugster

Same specs as Model 500 except 12-ga.; 3″ chamber; 24″ smooth or rifled bore; 44″ overall length; weighs 7 1/4 lbs.; scope mount; walnut-stained hardwood stock; swivel studs; recoil pad; Introduced 1988; still in production. **LMSR: $354**
Perf.: $285   **Exc.:** $200   **VGood:** $180

Mossberg Model 500 Turkey

Same specs as Model 500 except 24″ Accu-Choke barrel with Improved Cylinder, Modified, Full, Extra-Full lead shot choke tubes; Ghost-Ring sights; swivel studs; camo sling; overall OFM camo finish. Introduced 1992. still in production. **LMSR: $384**
Perf.: $300   **Exc.:** $240   **VGood:** $200

MOSSBERG MODEL 590

Slide action; 12-ga.; 3″ chamber; 9-shot magazine; 20″ Cylinder-bore barrel; synthetic field or Speedfeed stock with heat shield; bayonet lug; swivel studs; rubber recoil pad; blued or Parkerized finish; marketed with Cablelock. Introduced 1990; still in production. **LMSR: $329**
With blued finish, field stock
Perf.: $290   **Exc.:** $220   **VGood:** $190
With blued finish, Speedfeed stock
Perf.: $310   **Exc.:** $240   **VGood:** $210
With Parkerized finish, field stock
Perf.: $340   **Exc.:** $270   **VGood:** $240
With Parkerized finish, Speedfeed stock
Perf.: $360   **Exc.:** $290   **VGood:** $260

Mossberg Model 590 Bullpup

Same specs as Model 590 except shrouded 20″ Cylinder-bore barrel; fixed sights mounted in carrying handle; bullpup-configuration plastic stock; crossbolt, grip safties. Designed for law enforcement. Introduced 1989; discontinued 1990.
Perf.: $410   **Exc.:** $310   **VGood:** $260

Mossberg Model 590 Ghost Ring

Same specs as Model 590 except ghost-ring sights. Introduced 1990; still in production. **LMSR: $379**
With blued finish, field stock
Perf.: $310   **Exc.:** $240   **VGood:** $210
With blued finish, Speedfeed stock
Perf.: $330   **Exc.:** $260   **VGood:** $230
With Parkerized finish, field stock
Perf.: $360   **Exc.:** $290   **VGood:** $260
With Parkerized finish, Speedfeed stock
Perf.: $380   **Exc.:** $310   **VGood:** $280

Mossberg Model 590 Intimidator

Same specs as Model 590 except has integral laser sight built into synthetic field stock, forend; marketed with Mossberg Cablelock. Introduced 1990; discontinued 1993.
With blued finish
Perf.: $425   **Exc.:** $330   **VGood:** $280
With Parkerized finish
Perf.: $450   **Exc.:** $365   **VGood:** $300

Mossberg Model 590 Mariner

Same specs as Model 590 except all parts coated with Marinecoat. Introduced 1989; discontinued 1993.
With field stock
Perf.: $300   **Exc.:** $210   **VGood:** $170
With Speedfeed stock
Perf.: $320   **Exc.:** $230   **VGood:** $190

MOSSBERG MODEL 835 AMERICAN FIELD

Slide action; 12-ga.; 3 1/2″ chamber handles all size shells; backbored 28″ Accu-Mag Modified choke tube for steel or lead shot; weighs 7 3/4 lbs.; cut-checkered, walnut-stained hardwood stock; ambidextrous thumb safety; twin extractors; dual slide bars. Introduced 1988; still in production. **LMSR: $310**
Perf.: $225   **Exc.:** $180   **VGood:** $150

Mossberg Model 835 Ulti-Mag

Same specs as Model 835 American Field except 24″ rifled bore with rifle sights, 24″, 28″ backbored vent-rib barrel; four Accu-Mag choke tubes for steel or lead shot; cut-checkered, walnut-stained hardwood or camo synthetic stock. Introduced 1988; discontinued 1991.
Perf.: $380   **Exc.:** $275   **VGood:** $200

Mossberg Model 835 Ulti-Mag Crown Grade

Same specs as Model 835 American Field except 24″ rifled bore with rifle sights, 24″, 28″ backbored vent-rib barrel; four Accu-Mag choke tubes for steel or lead shot; cut-checkered walnut standard or Dual-Comb, or camo synthetic stock. Introduced 1988; still in production. **LMSR: $404**
With standard barrel; standard walnut stock
Perf.: $380   **Exc.:** $310   **VGood:** $270
With standard barrel; Dual-Comb walnut stock
Perf.: $385   **Exc.:** $315   **VGood:** $275
With rifled barrel; scope base; Dual-Comb walnut stock
Perf.: $400   **Exc.:** $320   **VGood:** $290
Combo with 24″ rifled, 28″ barrels; Dual-Comb walnut stock
Perf.: $420   **Exc.:** $340   **VGood:** $300
Realtree or Mossy Oak Camo with standard barrel; synthetic stock
Perf.: $430   **Exc.:** $350   **VGood:** $310
Realtree Camo Combo with 24″ rifled, 24″ barrels; synthetic stock, hard case
Perf.: $525   **Exc.:** $450   **VGood:** $400
OFM Camo with standard barrel; synthetic stock
Perf.: $410   **Exc.:** $330   **VGood:** $300
OFM Camo Combo with 24″ rifled, 28″ barrels; synthetic stock
Perf.: $460   **Exc.:** $400   **VGood:** $360

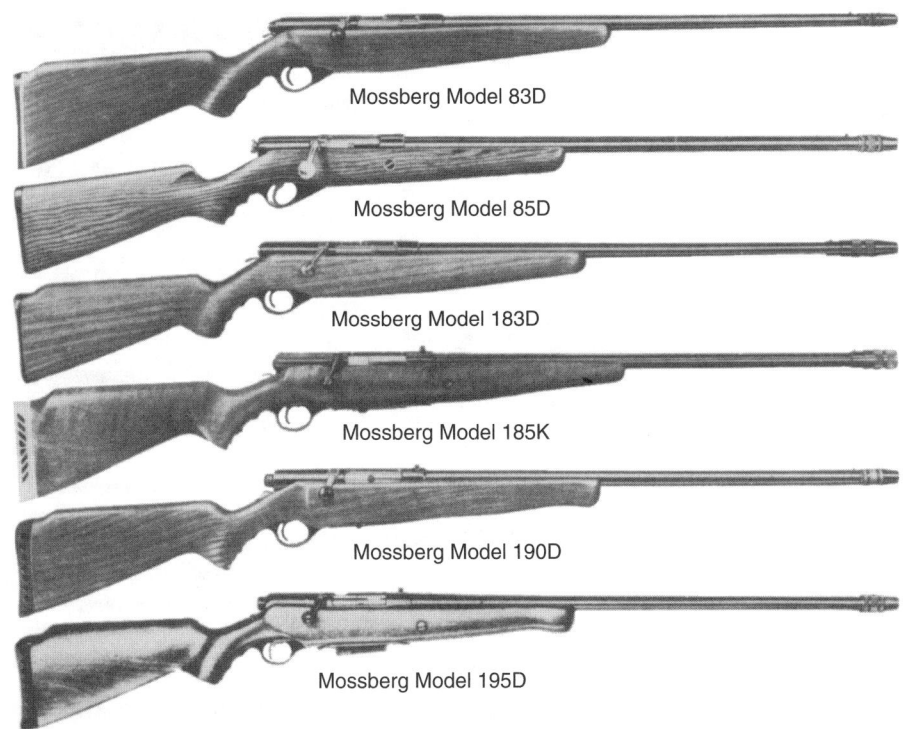

Mossberg Model 83D
Mossberg Model 85D
Mossberg Model 183D
Mossberg Model 185K
Mossberg Model 190D
Mossberg Model 195D

MOSSBERG BOLT-ACTION SHOTGUNS

MOSSBERG MODEL 70
Bolt action; single shot; takedown; .410; 24″ barrel; interchangeable Modified, Full choke tubes; unchecked- ered one-piece, finger-grooved, pistol-grip stock. Intro- duced 1933; discontinued 1935.
Exc.: $100   **VGood:** $75   **Good:** $50

MOSSBERG MODEL 73
Bolt action; single shot; takedown; .410; 24″ barrel; interchangeable Modified, Full choke tubes; unchecked- ered one-piece, finger-grooved, pistol-grip stock. Intro- duced 1936; discontinued 1939.
Exc.: $125   **VGood:** $95   **Good:** $75

MOSSBERG MODEL 75
Bolt action; single shot; takedown; 20-ga.; 26″ barrel; interchangeable Modified, Full choke tubes; unchecked- ered one-piece, finger-grooved, pistol-grip stock. Intro- duced 1934; discontinued 1937.
Exc.: $125   **VGood:** $95   **Good:** $75

MOSSBERG MODEL 80
Bolt action; takedown; .410; 2-shot fixed top-loading magazine; 24″ barrel; interchangeable Modified, Full choke tubes; uncheckered one-piece, finger-grooved, pistol-grip stock. Introduced 1934; discontinued 1935.
Exc.: $100   **VGood:** $80   **Good:** $60

MOSSBERG MODEL 83D
Bolt action; takedown; .410; 2-shot fixed top-loading magazine; 23″ barrel; interchangeable Modified, Full choke tubes; uncheckered one-piece, finger-grooved, pistol grip stock. Introduced 1940; discontinued 1947 replaced by Model 183D.
Exc.: $100   **VGood:** $80   **Good:** $60

MOSSBERG MODEL 85
Bolt action; takedown; 20-ga.; 2-shot fixed top-loading magazine; 26″ barrel; interchangeable Modified, Full choke tubes; uncheckered one-piece, finger-grooved, pistol-grip stock. Introduced 1934; discontinued 1939.
Exc.: $90   **VGood:** $60   **Good:** $40

Mossberg Model 85D
Same specs as Model 85 except 2-shot detachable box magazine; 25″ barrel; interchangeable Full, Modified, Improved Cylinder choke tubes; black plastic buttplate. Intro- duced 1940; discontinued, 1947 replaced by Model 185D.
Exc.: $80   **VGood:** $60   **Good:** $40

MOSSBERG MODEL 173
Bolt action; single shot; takedown; .410; 2 1/2″, 3″ chamber; 24″ Full-choke barrel; 43 1/2″ overall length; weighs 5 1/2 lbs.; uncheckered one-piece, finger- grooved, pistol-grip stock; plastic buttplate. Introduced 1957; discontinued 1973.
Exc.: $80   **VGood:** $60   **Good:** $40

Mossberg Model 173Y
Same specs as Model 173 except 22″ Full-choke bar- rel; youth-size uncheckered one-piece, finger-grooved, pistol-grip stock. Introduced 1957; discontinued 1973.
Exc.: $80   **VGood:** $60   **Good:** $40

MOSSBERG MODEL 183D
Bolt action; .410; 2-shot fixed top-loading magazine; 24″ barrel; interchangeable Modified, Full choke tubes; 43 1/2″ overall length; weighs 5 1/2 lbs.; uncheckered one-piece, finger-grooved, pistol-grip stock. Introduced 1947; discontinued 1971.
Exc.: $90   **VGood:** $60   **Good:** $45

Mossberg Model 183K
Same specs as Model 183D, except variable C-Lect- Choke instead of interchangeable tubes; 44 1/2″ overall length. Introduced 1953; discontinued 1985.
Exc.: $95   **VGood:** $60   **Good:** $45

MOSSBERG MODEL 185D
Bolt action; takedown; 20-ga.; 2 3/4″ chamber; 2-shot detachable box magazine; 26″ barrel; Full, Improved Cylinder choke tubes; 44 1/2″ overall length; weighs 6 1/4 lbs.; uncheckered one-piece, finger-grooved, pistol- grip stock; black plastic buttplate. Introduced 1947; dis- continued 1971.
Exc.: $95   **VGood:** $60   **Good:** $45

Mossberg Model 185K
Same specs as Model 185D except variable C-Lect- Choke instead of interchangeable tubes; 45 1/2″ overall length. Introduced 1951; discontinued 1963.
Exc.: $95   **VGood:** $60   **Good:** $45

MOSSBERG MODEL 190D
Bolt action; takedown; 16-ga.; 2 3/4″ chamber; 2-shot detachable box magazine; 26″ barrel; Full, Improved Cylinder choke tubes; 44 1/2″ overall length; weighs 6 3/4 lbs.; uncheckered one-piece, finger-grooved, pistol- grip stock; black plastic buttplate. Introduced 1955; dis- continued 1971.
Exc.: $100   **VGood:** $65   **Good:** $50

Mossberg Model 190K
Same specs as Model 190D except variable C-Lect- Choke instead of interchangeable choke tubes; 45 1/2″ overall length. Introduced 1956; discontinued 1963.
Exc.: $90   **VGood:** $60   **Good:** $45

MOSSBERG MODEL 195D
Bolt action; takedown; 12-ga.; 2 3/4″ chamber; 2-shot detachable box magazine; 26″ barrel; interchangeable Full, Improved Cylinder choke tubes; 45 1/2″ overall length; weighs 7 1/2 lbs.; uncheckered one-piece, fin- ger-grooved, pistol-grip stock; black plastic buttplate. Introduced 1955; discontinued 1971.
Exc.: $90   **VGood:** $60   **Good:** $45

Mossberg Model 195K
Same specs as the Model 195D except variable C-Lect- Choke instead of interchangeable choke tubes; 46 1/2″ overall length. Introduced 1956; discontinued 1963.
Exc.: $95   **VGood:** $65   **Good:** $50

MOSSBERG MODEL 385K
Bolt action; 20-ga.; 3″ chamber; 2-shot detachable clip magazine; 26″ barrel; C-Lect-Choke; walnut-finished hardwood Monte Carlo stock; recoil pad. Introduced 1963; no longer in production.
Exc.: $95   **VGood:** $70   **Good:** $50

MOSSBERG MODEL 390K
Bolt action; takedown; 16-ga.; 3″ chamber; 2-shot detachable clip magazine; 28″ barrel; C-Lect-Choke; walnut-finished hardwood Monte Carlo stock; recoil pad. Introduced 1963; no longer in production.
Exc.: $95   **VGood:** $70   **Good:** $50

MOSSBERG MODEL 395K
Bolt action; takedown; 12-ga.; 3″ chamber; 2-shot detachable clip magazine; 28″ barrel; C-Lect-Choke; walnut-finished hardwood Monte Carlo stock; recoil pad. Introduced 1963; no longer in production.
Exc.: $95   **VGood:** $70   **Good:** $50

SHOTGUNS

Navy Arms
Model 83 Bird Hunter
Model 84 Bird Hunter
(See Navy Arms Model 93)
Model 93
Model 95
Model 96
Model 100 SxS
Model 100 O/U
Model 105
Model 150
Model 600

New England Firearms
10-Gauge (See New England Firearms Turkey/Goose Gun)
Handi-Gun
Pardner
Pardner Deluxe
Pardner Special Purpose
Pardner Youth
Survivor

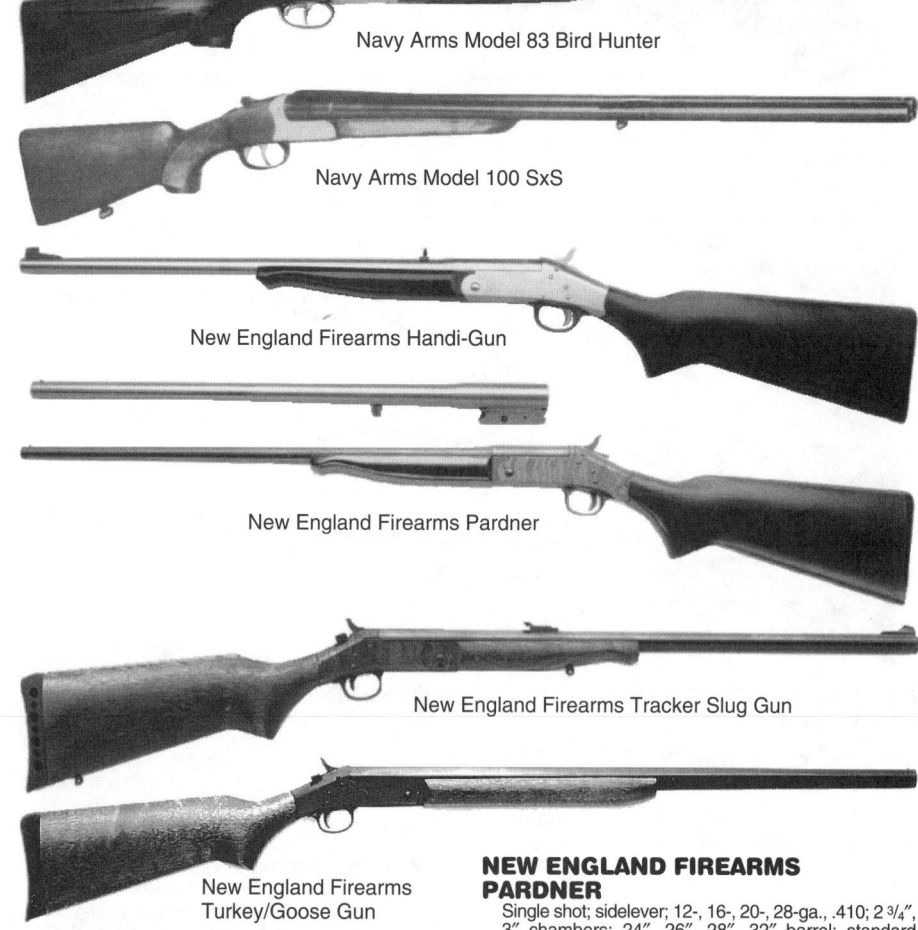

Navy Arms Model 83 Bird Hunter

Navy Arms Model 100 SxS

New England Firearms Handi-Gun

New England Firearms Pardner

New England Firearms Tracker Slug Gun

New England Firearms Turkey/Goose Gun

NAVY ARMS MODEL 83 BIRD HUNTER
Over/under; boxlock; 12-, 20-ga.; 3″ chambers; 28″ barrels; standard choke combos; vent top, middle ribs; weighs 7 ¹/₂ lbs.; metal bead front sight; checkered European walnut stock; double triggers; extractors; silvered, engraved receiver. Made in Italy. Introduced 1984; discontinued 1990.
Perf.: $260 **Exc.:** $200 **VGood:** $180

NAVY ARMS MODEL 93
Over/under; boxlock; 12-, 20-ga.; 3″ chambers; 28″ barrels; standard choke tubes; vent top, middle ribs; metal bead front sight; checkered European walnut stock, forend; double triggers; ejectors; silvered engraved receiver. Made in Italy. Introduced 1985; discontinued 1990.
Perf.: $315 **Exc.:** $240 **VGood:** $200

NAVY ARMS MODEL 95
Over/under; boxlock; 12-, 20-ga.; 3″ chambers; 28″ barrels; five choke tubes; vent top, middle ribs; weighs 7 ¹/₂ lbs.; metal bead front sight; checkered European walnut stock; single trigger; extractors; silvered, engraved receiver. Made in Italy. Introduced 1985; discontinued 1990.
Perf.: $365 **Exc.:** $300 **VGood:** $270

NAVY ARMS MODEL 96
Over/under; boxlock; 12-, 20-ga.; 3″ chambers; 28″ barrels; five choke tubes; vent top, middle ribs; weighs 7 ¹/₂ lbs.; metal bead front sight; checkered European walnut stock; single trigger; ejectors; gold-plated, engraved receiver. Made in Italy. Introduced 1986; discontinued 1990.
Perf.: $450 **Exc.:** $360 **VGood:** $300

NAVY ARMS MODEL 100 SxS
Side-by-side; 12-, 20-ga.; 3″ chambers; 28″ Improved Cylinder/Modified, Modified/Full chrome-lined barrels; weighs 7 lbs.; checkered European walnut stock; extractors; gold-plated double triggers; engraved, hard-chromed receiver. Made in Italy. Introduced 1985; discontinued 1987.
Perf.: $300 **Exc.:** $290 **VGood:** $225

NAVY ARMS MODEL 100 O/U
Over/under; 12-, 20-, 28-ga., .410; 3″ chambers; 26″ vent-rib, chrome-lined barrels; checkered walnut stock; single trigger; extractors; engraved, hard-chrome finished receiver. Introduced 1985; discontinued 1988.
Perf.: $375 **Exc.:** $290 **VGood:** $260

NAVY ARMS MODEL 105
Single shot; top lever; hammerless; 12-, 20-ga., .410; 3″ chamber; 26″, 28″ Full-choke barrel; metal bead front sight; checkered walnut-stained hardwood stock; folds for storage, transport; engraved chrome receiver. Made in Italy. Introduced 1987; discontinued 1990.
Perf.: $80 **Exc.:** $60 **VGood:** $40

NAVY ARMS MODEL 150
Side-by-side; 12-, 20-ga.; 3″ chambers; 28″ Improved Cylinder/Modified, Modified/Full chrome-lined barrels; weighs 7 lbs.; checkered European walnut stock; ejectors; gold-plated double triggers; engraved, hard-chromed receiver. Introduced 1985; discontinued 1987.
Perf.: $465 **Exc.:** $375 **VGood:** $315

NAVY ARMS MODEL 600
Single shot; top lever; hammerless; 12-, 20-ga., .410; 26″, 28″ chrome-lined barrel; checkered beech stock; engraved hard-chromed receiver; folds for storage, transport. Made in Italy. Introduced 1986; discontinued 1987.
Perf.: $80 **Exc.:** $60 **VGood:** $40

NEW ENGLAND FIREARMS HANDI-GUN
Single shot; breakopen; interchangeable rifle, shotgun barrels; 22 Hornet, 223, 243, 30-30, 30-06, 45-70; 12-, 20-ga. with 3″ chamber; 37″ overall length; weighs 6 ¹/₂ lbs.; American hardwood stock; rifle barrels have ramp front sight, open rear; drilled, tapped for scope mounts; matte electroless nickel or blued finish. Introduced 1987; discontinued 1990.
Perf.: $200 **Exc.:** $160 **VGood:** $130

NEW ENGLAND FIREARMS PARDNER
Single shot; sidelever; 12-, 16-, 20-, 28-ga., .410; 2 ³/₄″, 3″ chambers; 24″, 26″, 28″, 32″ barrel; standard chokes; weighs 5 ¹/₂ lbs.; walnut-finished hardwood stock; transfer-bar ignition; blued finish. Introduced 1987; still in production. **LMSR: $100**
Perf.: $90 **Exc.:** $65 **VGood:** $50

New England Firearms Pardner Deluxe
Same specs as Pardner except Double Back-up buttstock holding two extra shotshells. Introduced 1987; discontinued 1991.
Perf.: $100 **Exc.:** $75 **VGood:** $60

New England Firearms Pardner Special Purpose
Same specs as Pardner except 10-ga.; 3 ¹/₂″ chamber; 28″ barrel; weighs 9 ¹/₂ lbs.; recoil pad; blued or camo finish. Introduced 1988; still in production. **LMSR: $160**
Perf.: $125 **Exc.:** $85 **VGood:** $70

New England Firearms Pardner Youth
Same specs as Pardner except 20-, 28-ga., .410; 22″ barrel; shorter stock. Introduced 1989; still in production. **LMSR: $110**
Perf.: $90 **Exc.:** $65 **VGood:** $50

NEW ENGLAND FIREARMS SURVIVOR
Single shot; 12-, 20-ga., .410/45 Colt; 3″ chamber; 20″ rifled (.410/45 Colt), 22″ smoothbore with Modified choke tube; weighs 6 lbs.; 36″ overall length; black polymer thumbhole/pistol-grip stock, beavertail forend; sling swivels; buttplate removes for extra ammunition storage; forend also holds extra ammunition; black or nickel finish. Introduced 1993; still in production. **LMSR: $121**
With black finish
Perf.: $100 **Exc.:** $75 **VGood:** $50
With nickel finish
Perf.: $110 **Exc.:** $85 **VGood:** $60
With black finish, .410/45 Colt
Perf.: $135 **Exc.:** $95 **VGood:** $75
With nickel finish, .410/45 Colt
Perf.: $140 **Exc.:** $100 **VGood:** $80

New England Firearms Survivor

New England Firearms
Turkey/Goose N.W.T.F.

New Haven Model 273

New Haven Model 285

New Haven Model 495

New Haven Model
600AST Slugster

New England Firearms
Tracker Slug Gun
Tracker II Slug Gun
Turkey/Goose Gun
Turkey/Goose N.W.T.F.
Turkey/Goose Special

New Haven
Model 273
Model 283
Model 285
Model 290
Model 295
Model 495
Model 600
Model 600AST Slugster
Model 600ETV
Model 600K

NEW ENGLAND FIREARMS TRACKER SLUG GUN
Single shot; breakopen; sidelever; 12-, 20-ga.; 3″ chamber; 24″ barrel; Cylinder choke; weighs 6 lbs.; 40″ overall length; blade front sight, fully adjustable rifle-type rear; walnut-finished hardwood pistol-grip stock; recoil pad; blued barrel, color case-hardened frame. Introduced 1992; still in production. **LMSR: $125**
　　Perf.: $100　　**Exc.:** $60　　**VGood:** $50

New England Firearms Tracker II Slug Gun
Same specs as the Tracker Slug Gun except fully rifled bore. Introduced 1992; still in production. **LMSR: $130**
　　Perf.: $125　　**Exc.:** $95　　**VGood:** $55

NEW ENGLAND FIREARMS TURKEY/GOOSE GUN
Single shot; breakopen; sidelever; 10-ga.; 3 1/2″ chamber; 28″ barrel; Full choke; weighs 9 1/2 lbs.; 44″ overall length; ventilated rubber recoil pad; ejector; matte finish on metal. Introduced 1992; still in production. **LMSR: $160**
　　Perf.: $140　　**Exc.:** $90　　**VGood:** $70

New England Firearms Turkey/Goose N.W.T.F.
Same specs as Turkey/Goose Gun except 10-, 20-ga.; 24″ barrel; interchangeable choke tubes (comes with Turkey Full, others optional); drilled, tapped for long-eye relief scope mount; completely covered with Mossy Oak camouflage finish; Mossy Oak sling included. Introduced 1992; still in production. **LMSR: $150/$230**
10-ga. model
　　Perf.: $200　　**Exc.:** $145　　**VGood:** $100

20-ga. model
　　Perf.: $125　　**Exc.:** $70　　**VGood:** $50

New England Firearms Turkey/Goose Special
Same specs as the Turkey/Goose Gun except 12-ga.; 3″ chamber; 24″ (fixed Full Turkey choke); weighs 5-6 lbs.; 40″ overall length; modified pistol-grip stock; recoil pad; swivel studs; full coverage Realtree camouflage. Introduced 1994; still in production. **LMSR: $115**
　　Perf.: $95　　**Exc.:** $75　　**VGood:** $50

NEW HAVEN MODEL 273
Bolt action; single shot; top-loading; .410; 24″ tapered barrel; Full choke; oil-finished American walnut Monte Carlo-style, pistol-grip stock; thumb safety. Made by Mossberg. Introduced 1960; discontinued 1965.
　　Exc.: $75　　**VGood:** $50　　**Good:** $40

NEW HAVEN MODEL 283
Bolt action; takedown; .410; 3″ chamber; 2-shot clip detachable clip magazine; 25″ barrel; detachable Full-choke tube; other chokes available at added cost; oil-finished American walnut Monte Carlo-style, pistol-grip stock; thumb safety. Made by Mossberg. Introduced 1960; discontinued 1965.
　　Exc.: $80　　**VGood:** $55　　**Good:** $45

NEW HAVEN MODEL 285
Bolt action; takedown; 20-ga.; 3″ chamber; 2-shot clip detachable clip magazine; 25″ barrel; detachable Full-choke tube; other chokes available at added cost; oil-finished American walnut Monte Carlo-style, pistol-grip stock; thumb safety. Made by Mossberg. Introduced 1960; discontinued 1965.
　　Exc.: $80　　**VGood:** $55　　**Good:** $45

NEW HAVEN MODEL 290
Bolt action; takedown; 16-ga.; 2-shot detachable clip magazine; 28″ barrel; detachable Full-choke tube; other choke tubes available at added cost; oil-finished American walnut Monte Carlo-style, pistol-grip stock; thumb safety. Made by Mossberg. Introduced 1960; discontinued 1965.
　　Exc.: $80　　**VGood:** $55　　**Good:** $45

NEW HAVEN MODEL 295
Bolt action; takedown; 12-ga.; 2-shot detachable clip magazine; 28″ barrel; detachable Full-choke tube; other choke tubes available at added cost; oil-finished American walnut Monte Carlo-style, pistol-grip stock; thumb safety. Made by Mossberg. Introduced 1960; discontinued 1965.
　　Exc.: $80　　**VGood:** $55　　**Good:** $45

NEW HAVEN MODEL 495
Bolt action; takedown; 12-ga.; 2-shot detachable clip magazine; 28″ Full-choke barrel; uncheckered walnut-finished hardwood Monte Carlo-style, pistol-grip stock; thumb safety. Introduced 1964; discontinued 1965.
　　Exc.: $80　　**VGood:** $55　　**Good:** $45

NEW HAVEN MODEL 600
Slide action; takedown; 12-ga.; 2 3/4″, 3″ chamber; 6-shot magazine; 26″ Improved Cylinder, 28″ Full or Modified, 30″ Full-choke barrel; choice of standard or magnum barrel; uncheckered walnut pistol-grip stock, extension slide handle; safety on top of receiver. Same general design as Mossberg Model 500. Introduced 1962; discontinued 1965.
　　Exc.: $140　　**VGood:** $110　　**Good:** $90

New Haven Model 600AST Slugster
Same specs as Model 600 except 12-, 20-ga.; 18 1/2″, 24″ Slugster barrel; ramp front sight, open adjustable folding leaf rear; running deer scene etched on receiver. Introduced 1978; no longer in production.
　　Exc.: $140　　**VGood:** $110　　**Good:** $90

New Haven Model 600ETV
Same specs as Model 600 except .410; 26″ vent-rib barrels; checkered walnut-finished, pistol-grip stock, forend; fluted comb; recoil pad. No longer in production.
　　Exc.: $150　　**VGood:** $120　　**Good:** $100

New Haven Model 600K
Same specs as Model 600 except C-Lect-Choke feature. No longer in production.
　　Exc.: $150　　**VGood:** $120　　**Good:** $100

SHOTGUNS

Noble

Model 40
Model 50
Model 60
Model 60ACP
Model 60AF
Model 65
Model 66CLP
Model 66RCLP
Model 66RLP
Model 66XL
Model 70
Model 70CLP
Model 70RCLP
Model 70RLP
Model 70X
Model 70XL
Model 80
Model 160 Deergun
Model 166L Deergun

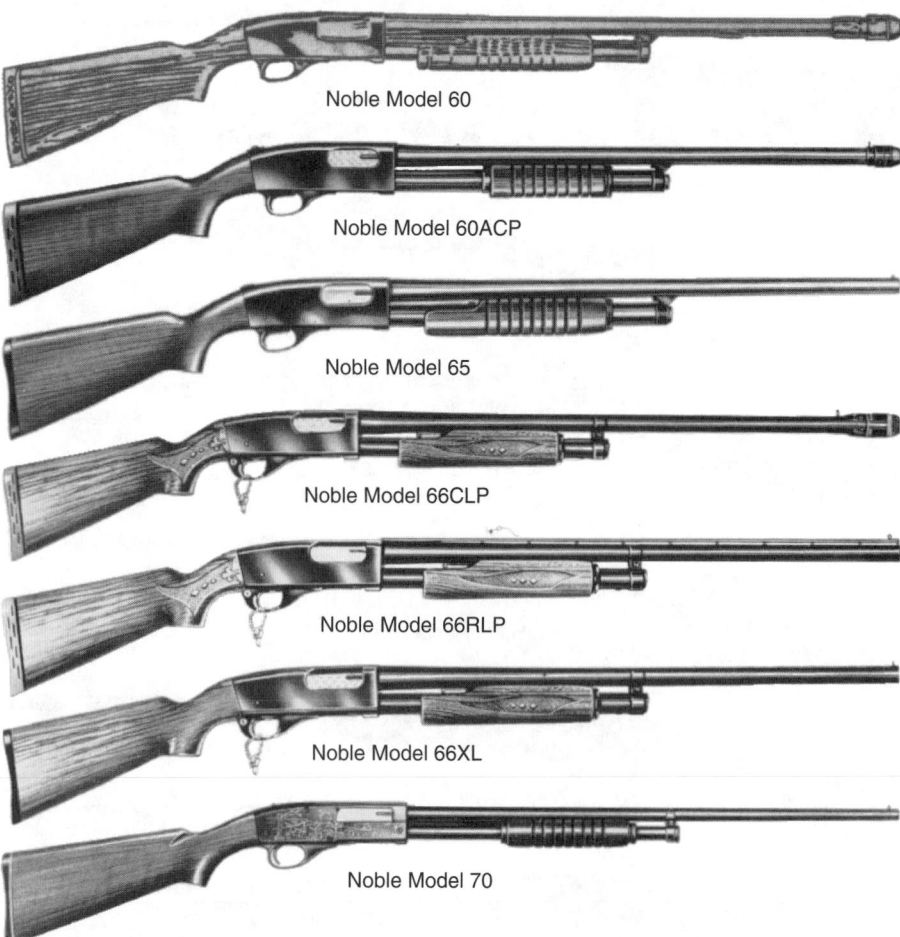

Noble Model 60

Noble Model 60ACP

Noble Model 65

Noble Model 66CLP

Noble Model 66RLP

Noble Model 66XL

Noble Model 70

NOBLE MODEL 40

Slide action; solid frame; 12-, 16-ga; 6-shot magazine; 28″ barrel; Multi-choke; uncheckered American walnut pistol-grip stock, grooved forearm; recoil pad; push-button safety. Introduced 1952; discontinued 1956.
Exc.: $150 **VGood:** $110 **Good:** $75

NOBLE MODEL 50

Slide action; solid frame; 12-, 16-ga.; 6-shot magazine; 28″ barrel; uncheckered American walnut pistol-grip stock, grooved forearm; push-button safety. Introduced 1954; discontinued 1956.
Exc.: $150 **VGood:** $110 **Good:** $75

NOBLE MODEL 60

Slide action; solid frame; 12-, 16-ga.; 6-shot magazine; 28″ barrel; Master choke; uncheckered American walnut pistol-grip stock, grooved slide handle; crossbolt safety; recoil pad. Introduced 1957; discontinued 1969.
Exc.: $150 **VGood:** $110 **Good:** $75

Noble Model 60ACP

Same specs as Model 60 except receiver is machined from single block of steel, all lock surfaces are hardened. Replaced Model 60 and Model 60AF. Introduced 1967; discontinued 1971.
Exc.: $160 **VGood:** $120 **Good:** $85

Noble Model 60AF

Same specs as Model 60 except selected steel barrel; damascened bolt; select walnut stock with fluted comb. Introduced 1965; discontinued 1966.
Exc.: $150 **VGood:** $110 **Good:** $75

NOBLE MODEL 65

Slide action; solid frame; 12-, 16-ga.; 6-shot magazine; 28″ barrel; Full or Modified choke; uncheckered American walnut pistol-grip stock, grooved slide handle; crossbolt safety. Introduced 1967; discontinued 1969.
Exc.: $150 **VGood:** $110 **Good:** $75

NOBLE MODEL 66CLP

Slide action; solid frame; 12-, 16-ga.; 3″ (12-ga.) chamber; 6-shot tubular magazine; 28″ barrel; adjustable choke; checkered American walnut pistol-grip stock, slide handle; keylock safety mechanism; recoil pad. Introduced 1967; discontinued 1979.
Exc.: $155 **VGood:** $115 **Good:** $80

Noble Model 66RCLP

Same specs as Model 66CLP except vent-rib barrel. Introduced 1967; discontinued 1970.
Exc.: $160 **VGood:** $120 **Good:** $85

Noble Model 66RLP

Same specs as Model 66CLP except Modified or Full choke. Introduced 1967; discontinued 1970.
Exc.: $160 **VGood:** $120 **Good:** $85

Nobel Model 66XL

Same specs as Model 66CLP except Modified or Full choke; checkered slide handle; no recoil pad. Introduced 1967; discontinued 1970.
Exc.: $155 **VGood:** $115 **Good:** $80

NOBLE MODEL 70

Slide action; solid frame; .410; 3″ chamber; 6-shot magazine; 26″ barrel; Full choke; uncheckered walnut pistol-grip stock, grooved forearm; top safety. Introduced 1959; discontinued 1967.
Exc.: $150 **VGood:** $110 **Good:** $75

Noble Model 70CLP

Same specs as Model 70 except adjustable choke. Introduced 1958; discontinued 1970.
Exc.: $150 **VGood:** $110 **Good:** $75

Noble Model 70RCLP

Same specs as Model 70 except adjustable choke; vent rib. Introduced 1967; discontinued 1970.
Exc.: $155 **VGood:** $115 **Good:** $89

Noble Model 70RLP

Same specs as Model 70 except vent rib. Introduced 1967; discontinued 1970.
Exc.: $155 **VGood:** $115 **Good:** $80

Noble Model 70X

Same specs as Model 70 except side ejection; damascened bolt. Replaced Model 70. Introduced 1967; discontinued 1971.
Exc.: $160 **VGood:** $120 **Good:** $85

Noble Model 70XL

Same specs as Model 70 except checkered buttstock. Introduced 1958; discontinued 1970.
Exc.: $160 **VGood:** $120 **Good:** $85

NOBLE MODEL 80

Recoil-operated autoloader; .410; 2 1/2″, 3″ chamber; 6-shot magazine; 26″ barrel; Full choke; uncheckered American walnut pistol-grip stock, grooved forearm; fluted comb; action release button; push-button safety. Introduced 1965; discontinued 1967.
Exc.: $225 **VGood:** $200 **Good:** $140

NOBLE MODEL 160 DEERGUN

Slide action; 12-, 16-ga.; 6-shot magazine; 24″ barrel; Lyman adjustable peep rear sight, ramp post front; tapped for scope; uncheckered American walnut pistol-grip stock, grooved slide handle; hard rubber buttplate; sling swivels; detachable carrying strap. Introduced 1965; discontinued 1966.
Exc.: $140 **VGood:** $90 **Good:** $75

NOBLE MODEL 166L DEERGUN

Slide action; 12-, 16-ga.; 6-shot magazine; 24″ barrel; Lyman adjustable peep rear sight, ramp post front; tapped for scope; uncheckered American walnut pistol-grip stock, grooved slide handle; hard rubber buttplate; sling swivels; detachable carrying strap. Replaced Model 160 Deergun. Introduced 1967; discontinued 1971.
Exc.: $185 **VGood:** $150 **Good:** $120

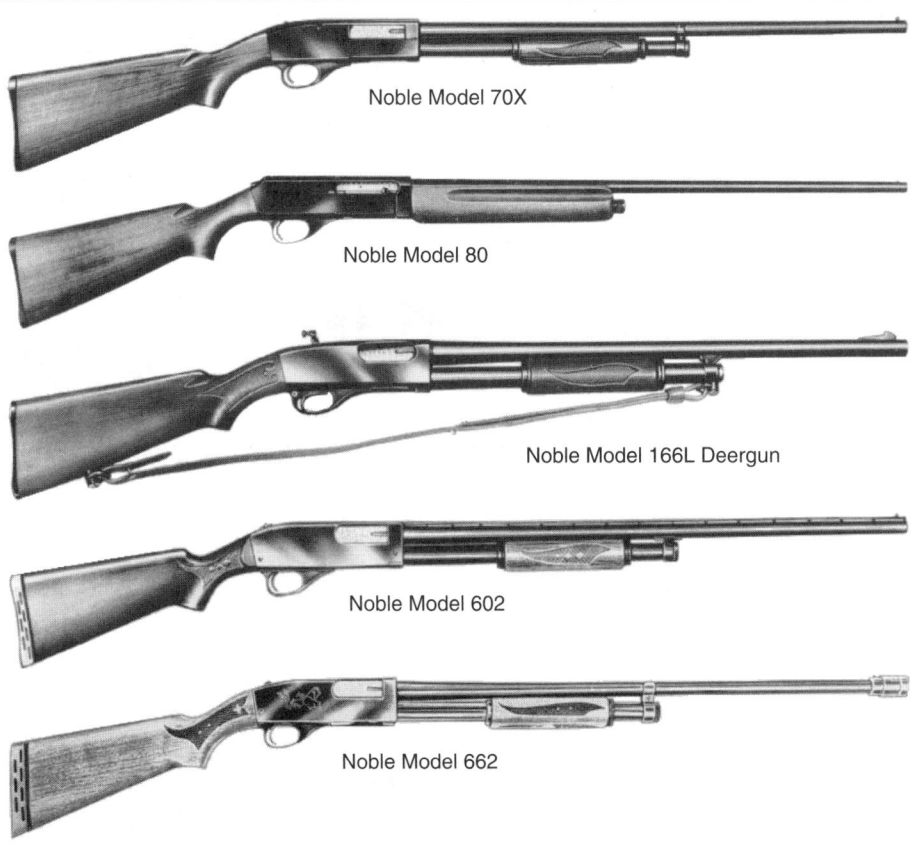

Noble Model 70X

Noble Model 80

Noble Model 166L Deergun

Noble Model 602

Noble Model 662

SHOTGUNS

Noble
Model 390 Deergun
Model 420
Model 420EK
Model 450E
Model 520
Model 550
Model 602
Model 602CLP
Model 602RCLP
Model 602RLP
Model 602XL
Model 662
Model 757
Series 200
Series 300

NOBLE MODEL 390 DEERGUN

Slide action; solid frame; 12-ga.; 3″ chamber; 6-shot magazine; 24″ rifled slug barrel; Lyman adjustable peep rear sight, ramp post front; impressed-checkered American walnut stock, slide handle; tang safety; sling swivels; detachable carrying strap. Introduced 1972; discontinued 1972.

Exc.: $185   **VGood:** $150   **Good:** $120

NOBLE MODEL 420

Double barrel; hammerless; top lever; 12-, 16-, 20-ga.; 28″ barrels; Full/Modified chokes; matted rib; checkered pistol-grip stock, forearm; double triggers; automatic safety. Introduced 1959; discontinued 1971.

Exc.: $275   **VGood:** $225   **Good:** $90

Noble Model 420EK

Same specs as Model 420 except demi-block with triple lock; front, middle bead sights; hand-checkered Circassian walnut pistol-grip stock, beavertail forearm; recoil pad; automatic selective ejectors; hand-engraved action; gold inlay on top lever. Introduced 1968; discontinued 1968.

Exc.: $290   **VGood:** $240   **Good:** $120

NOBLE MODEL 450E

Double barrel; demi-block with triple lock; 12-, 16-, 20-ga.; 28″ barrel; Modified/Full chokes; front, middle bead sights; hand-checkered Circassian walnut pistol-grip stock, beavertail forend; double triggers; engraving; gold inlay; recoil pad. Replaced Model 420EK. Introduced 1969; discontinued 1971.

Exc.: $290   **VGood:** $240   **Good:** $120

NOBLE MODEL 520

Side-by-side; hammerless; 12-, 20-, 16-, 28-ga. .410; 2 ³/₄″ chamber; 26″, 28″ barrels; standard choke combos; matted top rib; hand-checkered Circassian walnut stock, forend; double triggers; Holland-design extractors; hand-engraved frame. Introduced 1970; discontinued 1972.

Exc.: $250   **VGood:** $150   **Good:** $120

NOBLE MODEL 550

Side-by-side; hammerless; 12-, 20-ga.; 2 ³/₄″ chambers; 28″ barrels; Modified/Full chokes; front, middle bead sights; hand-checkered Circassian walnut pistol-grip stock, beavertail forend; grip cap; double triggers; double

auto selective ejectors; tang safety; rubber recoil pad; custom hand-engraved frame; knight's head medallion inlaid on top snap. Introduced 1970; discontinued 1972.

Exc.: $270   **VGood:** $160   **Good:** $130

NOBLE MODEL 602

Slide action; solid frame; 20-ga.; 3″ chamber; 6-shot magazine; 28″ barrel; adjustable choke; uncheckered American walnut pistol-grip stock, grooved slide handle; top safety; side ejection; recoil pad. Introduced 1963; discontinued 1971.

Exc.: $175   **VGood:** $140   **Good:** $80

Noble Model 602CLP

Same specs as Model 602 except keylock safety mechanism; checkered pistol-grip stock, slide handle. Introduced 1958; discontinued 1970.

Exc.: $175   **VGood:** $140   **Good:** $80

Noble Model 602RCLP

Same specs as Model 602 except keylock safety mechanism; vent rib; checkered pistol-grip stock, slide handle. Introduced 1967; discontinued 1970.

Exc.: $175   **VGood:** $140   **Good:** $85

Noble Model 602RLP

Same specs as Model 602 except keylock safety mechanism; vent rib; Full or Modified choke; checkered pistol-grip stock, slide handle. Introduced 1967; discontinued 1970.

Exc.: $175   **VGood:** $140   **Good:** $85

Noble Model 602XL

Same specs as Model 602 except keylock safety mechanism; Full or Modified choke; checkered slide handle; no recoil pad. Introduced 1958; discontinued 1970.

Exc.: $165   **VGood:** $125   **Good:** $80

NOBLE MODEL 662

Slide action; solid frame; 20-ga.; 6-shot magazine; 26″ barrel; Full choke; uncheckered walnut pistol-grip stock, grooved slide forearm; top safety; aluminum alloy barrel, receiver. Introduced 1966; discontinued 1970.

Exc.: $170   **VGood:** $140   **Good:** $100

NOBLE MODEL 757

Slide action; solid frame; 20-ga.; 2 ³/₄″ chamber; 5-shot magazine; 28″ aircraft alloy barrel; adjustable choke; impressed-checkered American walnut pistol-grip stock, slide handle; tang safety; barrel, receiver black anodized; decorated receiver. Introduced 1972; discontinued 1972.

Exc.: $170   **VGood:** $120   **Good:** $100

NOBLE SERIES 200

Slide action; solid frame; 20-ga.; 3″ chamber; 6-shot magazine; 28″ barrel; optional vent rib; adjustable choke or Modified, Full choke; impressed-checkered American walnut stock, slide handle; tang safety; side ejection; keylock safety mechanism; recoil pad. Introduced 1972; discontinued 1972.

With fixed choke
Exc.: $150   **VGood:** $100   **Good:** $80
With fixed choke, vent rib
Exc.: $155   **VGood:** $105   **Good:** $85
With adjustable choke
Exc.: $155   **VGood:** $105   **Good:** $85
With adjustable choke, vent rib
Exc.: $160   **VGood:** $110   **Good:** $90

NOBLE SERIES 300

Slide action; solid frame; 12-, 16-ga.; 3″ (12-ga.) chamber; 6-shot magazine, 3-shot plug furnished; 28″ barrel; adjustable choke or Modified, Full choke; optional vent rib; impressed-checkered American walnut stock, slide handle; tang safety; side ejection; keylock safety mechanism; damascened bolt. Introduced 1972; discontinued 1972.

With fixed choke
Exc.: $150   **VGood:** $100   **Good:** $80
With fixed choke, vent rib
Exc.: $155   **VGood:** $105   **Good:** $85
With adjustable choke
Exc.: $155   **VGood:** $105   **Good:** $85
With adjustable choke, vent rib
Exc.: $160   **VGood:** $110   **Good:** $90

SHOTGUNS

Noble
Series 400

Omega
Folding Over/Under
Folding Over/Under Deluxe
Folding Side-By-Side
Folding Side-By-Side Deluxe
Folding Single Barrel
Folding Single Barrel Deluxe

Parker
Hammerless Double

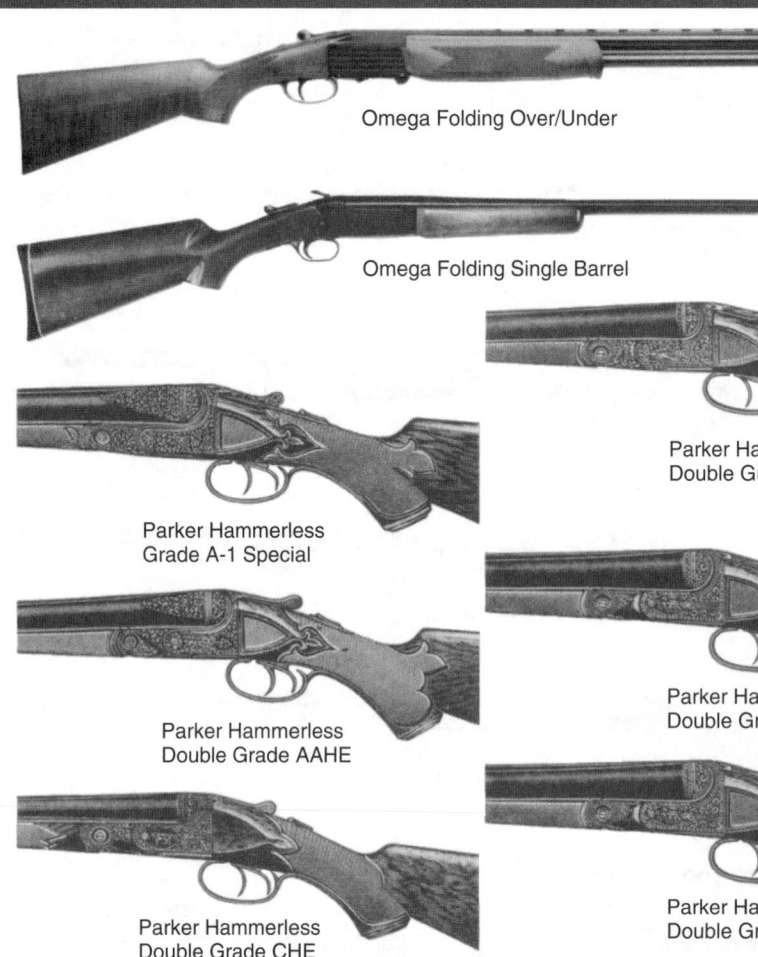

Omega Folding Over/Under

Omega Folding Single Barrel

Parker Hammerless Grade A-1 Special

Parker Hammerless Double Grade AAHE

Parker Hammerless Double Grade CHE

Parker Hammerless Double Grade AHE

Parker Hammerless Double Grade BHE

Parker Hammerless Double Grade DHE

Parker Hammerless Double Grade GHE

Noble Model 450E

NOBLE SERIES 400
Slide action; solid frame; .410; 3″ chamber; 6-shot magazine; 25″ barrel; adjustable choke or Modified, Full choke; optional vent rib; impressed-checkered American walnut pistol-grip stock, slide handle; tang safety; side ejection; keylock safety mechanism; damascened bolt. Introduced 1972; discontinued 1972.
With fixed choke
   Exc.: $155   **VGood:** $105   **Good:** $85
With fixed choke, vent rib
   Exc.: $160   **VGood:** $110   **Good:** $90
With adjustable choke
   Exc.: $160   **VGood:** $110   **Good:** $90
With adjustable choke, vent rib
   Exc.: $165   **VGood:** $115   **Good:** $95

OMEGA FOLDING OVER/UNDER
Over/under; boxlock; 12-, 20-, 28-ga., .410; 3″ chambers; 26″, 28″ barrels; standard choke combos; vent rib; weighs 5 ½ lbs.; checkered European walnut stock; auto safety; single trigger; extractors; folding takedown design. Made in Italy. Introduced 1986; discontinued 1994.
   Perf.: $330   **Exc.:** $250   **VGood:** $200

Omega Folding Over/Under Deluxe
Same specs as Folding Over/Under except walnut stock. Introduced 1986; discontinued 1990.
   Perf.: $400   **Exc.:** $310   **VGood:** $265

OMEGA FOLDING SIDE-BY-SIDE
Side-by-side; boxlock; 10-, 28-ga., .410; 26″ barrels; standard choke combos; vent rib; weighs 5 ½ lbs.; checkered beechwood stock, forearm; auto safety; double triggers; extractors; folding takedown design. Introduced 1986; discontinued 1989.
   Perf.: $180   **Exc.:** $130   **VGood:** $100

Omega Folding Side-By-Side Deluxe
Same specs as Folding Side-By-Side except select walnut stock. Introduced 1986; discontinued 1990.
   Perf.: $210   **Exc.:** $180   **VGood:** $130

OMEGA FOLDING SINGLE BARREL
Single shot; 12-, 16-, 20-, 28-ga., .410; 2 ¾″, 3″ chamber; 26″, 28″, 30″ barrel; Full choke; metal bead front sight; checkered beech stock; top opening lever; matte chromed receiver; folds for storage, transport. Made in Italy. Introduced 1984; discontinued 1988.
   Perf.: $145   **Exc.:** $100   **VGood:** $80

Omega Folding Single Barrel Deluxe
Same specs as Folding Single Barrel except vent rib; checkered walnut stock; blued receiver. Introduced 1984; discontinued 1988.
   Perf.: $180   **Exc.:** $125   **VGood:** $100

PARKER HAMMERLESS DOUBLE
Side-by-side; boxlock; 10-, 12-, 16-, 20-, 28-ga., .410; 26″ to 32″ barrels; any standard choke combo; hand-checkered select walnut stock, forearm; choice of straight, half- or pistol-grip stock; automatic ejectors; double or selective single trigger. After Parker Brothers was absorbed by Remington Arms in 1934, shotgun was designated as Remington Parker Model 920.
Because of the wide variations in styles and extras, as well as the number of grades—differing in engraving, checkering and general workmanship—there is a wide range of values. The selective trigger was introduced in 1922, with the raised vent rib; the beavertail forend was introduced in 1923; all add to used value. Some guns were put together from available parts and stocks by Remington until 1942. Grades are in descending values, with the A-designated model being worth several times that of the V model. Non-ejector models (pre-1934) are worth about 30% less than value shown for ejector models; if gun has interchangeable barrels, it is worth 30 to 35% more than shown. Those in 20-ga. are 35% higher, 28-ga. are 75% higher and .410 are 100% higher than values shown. Prices shown are for 12-, 16-ga. configurations. For single trigger, add $250; vent rib, add $300; single selective trigger, add $200 to $300; raised vent rib, add $325 to $350; beavertail forearm addition in grades VHE, GHE, DHE, CHE, add $200 to $250 to base price; for grades BHE, AHE, AAHE, add $450 to $500 to base; for A1 Special, add $500 to $750.

Grade A-1 Special
   Exc.: $65,000 **VGood:** $50,000 **Good:** $42,000
Grade AAHE
   Exc.: $30,000 **VGood:** $22,500 **Good:** $18,500
Grade AHE
   Exc.: $17,500 **VGood:** $10,000 **Good:** $8000
Grade BHE
   Exc.: $7500   **VGood:** $5000   **Good:** $4000
Grade CHE
   Exc.: $5000   **VGood:** $3500   **Good:** $2500
Grade DHE
   Exc.: $4000   **VGood:** $2500   **Good:** $2000
Grade GHE
   Exc.: $3000   **VGood:** $2000   **Good:** $1500
Grade VHE
   Exc.: $2100   **VGood:** $1250   **Good:** $1000

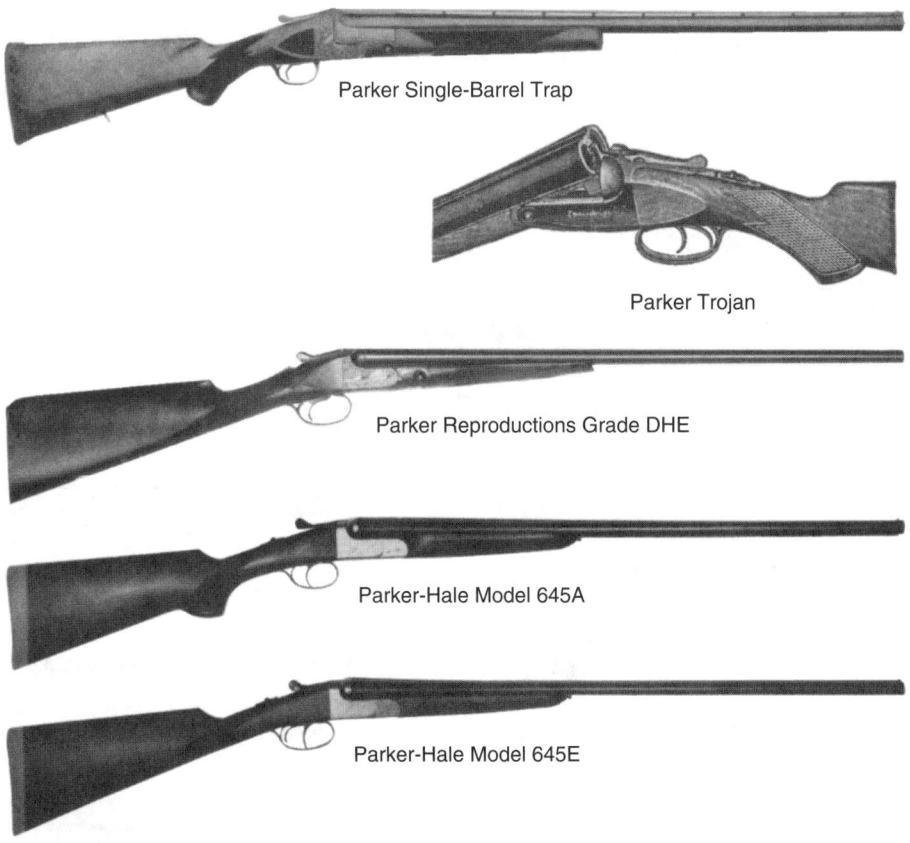

Parker Single-Barrel Trap

Parker Trojan

Parker Reproductions Grade DHE

Parker-Hale Model 645A

Parker-Hale Model 645E

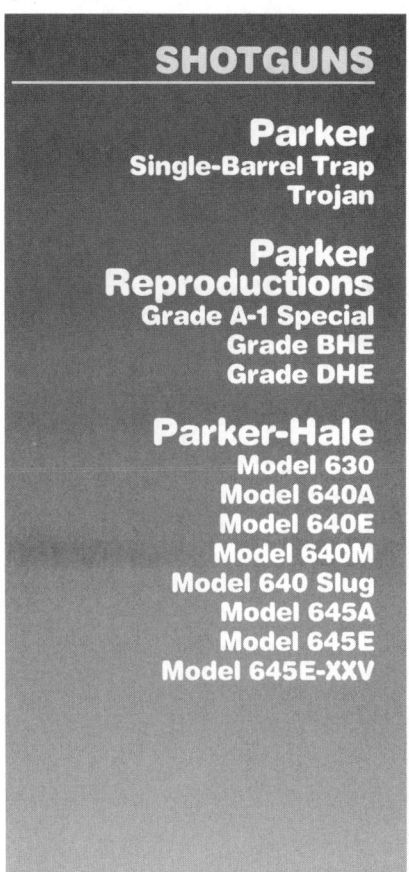

SHOTGUNS

Parker
Single-Barrel Trap
Trojan

Parker Reproductions
Grade A-1 Special
Grade BHE
Grade DHE

Parker-Hale
Model 630
Model 640A
Model 640E
Model 640M
Model 640 Slug
Model 645A
Model 645E
Model 645E-XXV

PARKER SINGLE-BARREL TRAP

Single barrel; boxlock; hammerless; 12-ga.; 30″, 32″, 34″ barrel; any designated choke; vent rib; hand-checkered select walnut pistol grip stock, forearm; choice of straight, half- or pistol grip; ejector. Various grades differ with amount of workmanship, checkering, engraving, etc. General specs are the same for all variations. After absorption of Parker by Remington, model was listed as Remington Parker Model 930. Introduced 1917; discontinued 1942.

Grade SA-1 Special
Exc.: $12,500 **VGood:** $9500 **Good:** $8000
Grade SAA
Exc.: $5200 **VGood:** $4000 **Good:** $3000
Grade SA
Exc.: $4000 **VGood:** $3500 **Good:** $2500
Grade SB
Exc.: $3500 **VGood:** $3000 **Good:** $2600
Grade SC
Exc.: $3000 **VGood:** $2100 **Good:** $1700

PARKER TROJAN

Double barrel; boxlock; hammerless; 12-, 16-, 20-ga.; 26″, 28″ Modified/Full, 30″ Full barrels; hand-checkered American walnut pistol grip stock, forearm; plain extractors; double or single triggers. Introduced 1915; discontinued 1939.

12-, 16-ga. model with double triggers
Exc.: $1600 **VGood:** $1000 **Good:** $800
12-, 16-ga. model with single trigger
Exc.: $1900 **VGood:** $1200 **Good:** $950
20-ga. model with double triggers
Exc.: $2400 **VGood:** $1650 **Good:** $1400
20-ga. model with single trigger
Exc.: $2900 **VGood:** $2000 **Good:** $1700

PARKER REPRODUCTIONS GRADE A-1 SPECIAL

Side-by-side; boxlock; 12-, 20-, 28-ga.; 2 ³/₄″, 3″ chambers; 26″, 28″ barrels; metal bead front sight; checkered select American walnut straight or pistol-grip stock, forend; double or single trigger; skeleton or hard rubber buttplate; two-barrel set; engraving; reproduction of original; all parts interchange with original. Introduced 1988; no longer in production.

Perf.: $10,000 **Exc.:** $7500 **VGood:** $6000

PARKER REPRODUCTIONS GRADE BHE

Side-by-side; 12-, 20-, 28-ga.; 2 ³/₄″, 3″ chambers; checkered fancy American walnut straight or pistol-grip stock, splinter or beavertail forend; checkered butt; double or single selective triggers; selective ejectors; hand engraving; case-hardened frame; marketed in fitted leather trunk-type case; reproduction of original; all parts interchange with original. Made in Japan. Introduced 1984; discontinued 1989.

Perf.: $3500 **Exc.:** $2800 **VGood:** $2100

PARKER REPRODUCTIONS GRADE DHE

Side-by-side; boxlock; 12-, 20-, 28-ga.; 2 ³/₄″, 3″ chambers; 26″, 28″ barrels; metal bead front sight; checkered American walnut straight or pistol-grip stock, forend; double or single trigger; skeleton or hard rubber buttplate; reproduction of original; all parts interchange with original. Made in Japan. Introduced 1984; no longer in production.

Perf.: $2800 **Exc.:** $2100 **VGood:** $1750

PARKER-HALE MODEL 630

Side-by-side; boxlock; 12-ga.; 3″ chambers; 26″, 28″ barrels; standard choke combos; checkered straight-grip English-style stock, forend; auto safety; extractors; double triggers; color case-hardened action. Made in Spain. Introduced 1993; discontinued 1993.

Perf.: $600 **Exc.:** $425 **VGood:** $380

PARKER-HALE MODEL 640A

Side-by-side; boxlock; 12-, 16-, 20-, 28-ga., .410; 2 ³/₄″, 3″ chambers; 25″, 26″, 27″, 28″ barrels; raised rib; hand-checkered, oil-finished walnut pistol-grip stock, beavertail forend; auto safety; extractors; single trigger; buttplate; silvered, engraved action. Made in Spain. Introduced 1986; discontinued 1993.

Perf.: $800 **Exc.:** $600 **VGood:** $540

Parker-Hale Model 640E

Same specs as Model 640A except concave rib; hand-checkered, oil-finished straight-grip English-style walnut stock, splinter forend; checkered butt; double triggers. Available in a wide variety of styles. Made in Spain. Introduced 1986; discontinued 1993. Name changed to Precision Sports Model 800 Series in 1990.

Perf.: $760 **Exc.:** $520 **VGood:** $450

Parker-Hale Model 640M

Same specs as Model 640A except 10-ga.; 3 ¹/₂″ chambers; 26″, 30″, 32″ Full/Full barrels; checkered straight-grip English-style stock, forend; auto safety; extractors; double triggers; recoil pad. Made in Spain. Introduced 1989; discontinued 1993.

Perf.: $825 **Exc.:** $625 **VGood:** $550

Parker-Hale Model 640 Slug

Same specs as Model 640A except 12-ga.; 25″ Improved Cylinder/Improved Cylinder barrels. Made in Spain. Introduced 1991; discontinued 1993.

Perf.: $900 **Exc.:** $700 **VGood:** $600

PARKER-HALE MODEL 645A

Side-by-side; boxlock; 12-, 16-, 20-, 28-ga., .410; 3″ chambers; 26″, 28″ barrels; raised rib; hand-checkered, oil-finished walnut pistol-grip stock, beavertail forend; auto safety; ejectors; single trigger; buttplate; silvered, engraved action. Made in Spain. Introduced 1986; discontinued 1993.

Perf.: $1000 **Exc.:** $725 **VGood:** $600

Parker-Hale Model 645E

Same specs as Model 645A except concave rib; hand-checkered, oil-finished straight-grip English-style walnut stock, splinter forend; checkered butt; double triggers. Made in Spain. Introduced 1986; discontinued 1993.

Perf.: $910 **Exc.:** $600 **VGood:** $550

Parker-Hale Model 645E-XXV

Same specs as Model 645E except 25″ barrels; Churchill rib; hand-checkered, oil-finished straight-grip English-style walnut stock, splinter forend; checkered butt; double triggers. Made in Spain. Introduced 1986; discontinued 1993.

Perf.: $900 **Exc.:** $600 **VGood:** $550

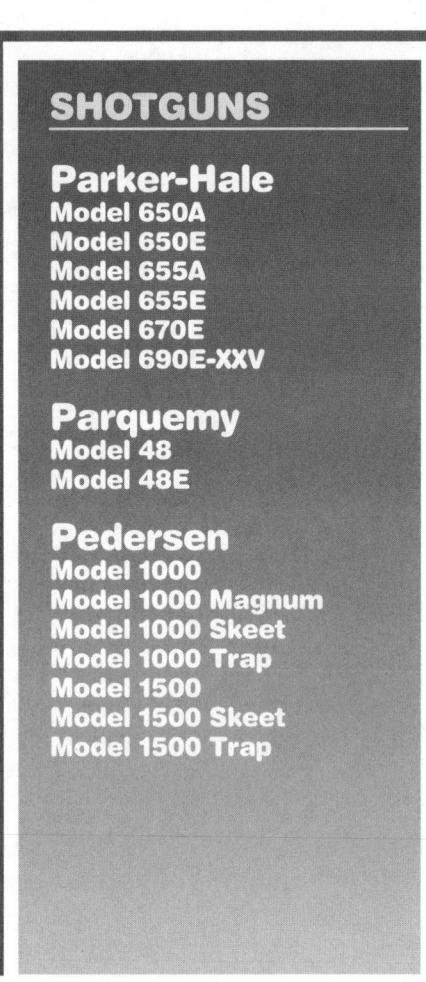

Parker-Hale
Model 650A
Model 650E
Model 655A
Model 655E
Model 670E
Model 690E-XXV

Parquemy
Model 48
Model 48E

Pedersen
Model 1000
Model 1000 Magnum
Model 1000 Skeet
Model 1000 Trap
Model 1500
Model 1500 Skeet
Model 1500 Trap

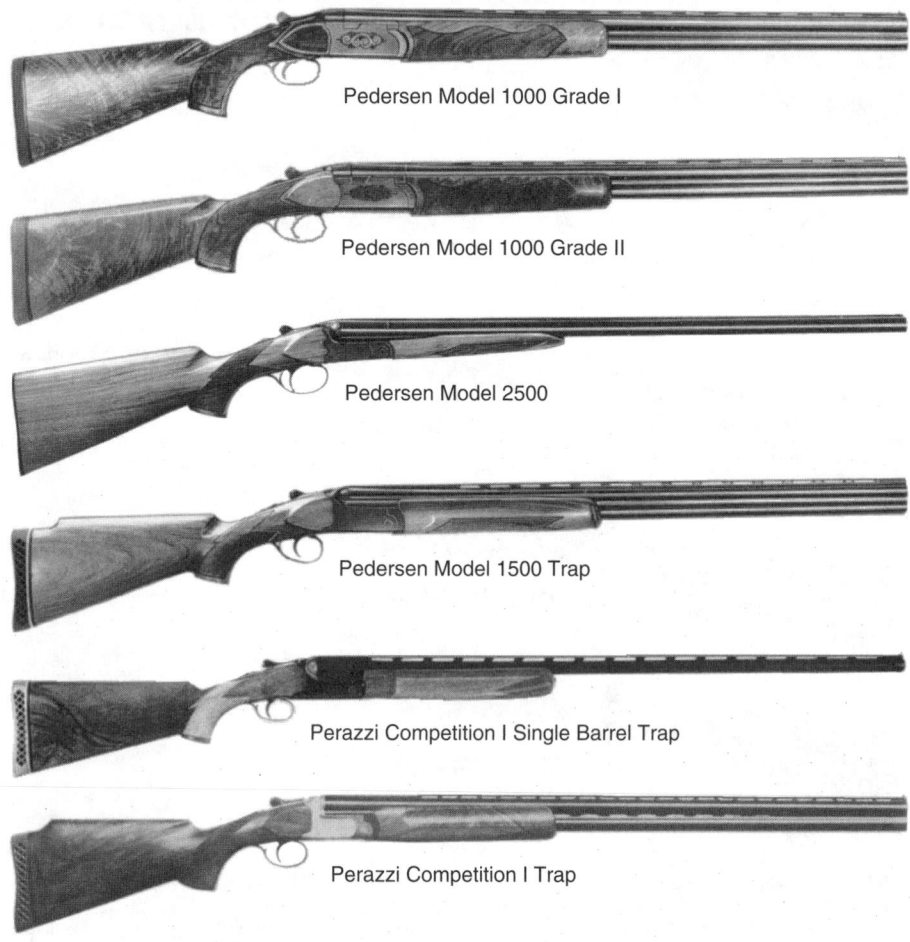

Pedersen Model 1000 Grade I

Pedersen Model 1000 Grade II

Pedersen Model 2500

Pedersen Model 1500 Trap

Perazzi Competition I Single Barrel Trap

Perazzi Competition I Trap

PARKER-HALE MODEL 650A
Side-by-side; boxlock; 12-ga.; 28″ barrels; choke tubes; raised rib; checkered pistol-grip walnut stock, beavertail forend; auto safety; extractors; single trigger; buttplate; silvered action. Made in Spain. Introduced 1992; discontinued 1993.
Perf.: $850   **Exc.:** $650   **VGood:** $600

Parker-Hale Model 650E
Same specs as Model 650A except concave rib; straight-grip English-style walnut stock, splinter forend; checkered butt; double triggers. Made in Spain. Introduced 1992; discontinued 1993.
Perf.: $800   **Exc.:** $600   **VGood:** $550

PARKER-HALE MODEL 655A
Side-by-side; boxlock; 12-ga.; 28″ barrels; choke tubes; raised rib; checkered pistol-grip walnut stock, beavertail forend; auto safety; ejectors; single trigger; buttplate; silvered action. Made in Spain. Introduced 1992; discontinued 1993.
Perf.: $975   **Exc.:** $775   **VGood:** $700

Parker-Hale Model 655E
Same specs as Model 655A except concave rib; straight-grip English-style walnut stock, splinter forend; checkered butt; double triggers. Made in Spain. Introduced 1992; discontinued 1993.
Perf.: $900   **Exc.:** $700   **VGood:** $625

PARKER-HALE MODEL 670E
Side-by-side; sidelock; 12-, 16-, 20-ga.; 2 ¾″, 3″ chambers; 26″, 27″, 28″ barrels; concave rib; hand-checkered, oil-finished straight-grip English-style walnut stock, splinter forend; auto safety; ejectors; double triggers; checkered butt; silvered, engraved action. Made in Spain. Introduced 1986; discontinued 1993.
Perf.: $3500   **Exc.:** $2500   **VGood:** $1900

PARKER-HALE MODEL 680E-XXV
Side-by-side; sidelock; 12-, 16-, 20-ga.; 2 ¾″, 3″ chambers; 25″ barrels; Churchill rib; hand-checkered, oil-finished straight-grip English-style walnut stock, splinter forend; auto safety; ejectors; double triggers; checkered butt; color case-hardened action. Made in Spain. Introduced 1986; discontinued 1993.
Perf.: $3300   **Exc.:** $2300   **VGood:** $1700

PARQUEMY MODEL 48
Side-by-side; .410; 3″ chambers; Modified/Full barrels; medium bead front sight; hand-checkered straight-grip European walnut stock; double triggers; checkered butt; extractors; hand-engraved locks. Made in Spain. Introduced 1983 by Toledo Armas; discontinued 1984.
Perf.: $525   **Exc.:** $400   **VGood:** $350

Parquemy Model 48E
Same specs as Model 48 except 12-, 20-, 28-ga.; 2 ¾″ chambers; ejectors. Introduced 1982 by Toledo Armas; discontinued 1984.
Perf.: $650   **Exc.:** $525   **VGood:** $450

PEDERSEN MODEL 1000
Over/under; boxlock; 12-, 20-ga.; 2 ¾″ chambers; 26″, 28″, 30″ barrels; vent rib; hand-checkered American walnut pistol-grip stock, forearm; rubber recoil pad; automatic ejectors; single selective trigger; hand-engraved, gold-filled receiver; silver inlays. Introduced 1973; discontinued 1975.
Grade I
Perf.: $2000   **Exc.:** $1700   **VGood:** $1600
Grade II
Perf.: $1700   **Exc.:** $1400   **VGood:** $1200
@ MGV-HD2:Pedersen Model 1000 Magnum
Same specs as Model 1000 except 12-ga.; 3″ chamber; 30″ barrels; Improved Modified/Full chokes. Introduced 1973; discontinued 1975.
Grade I
Perf.: $2000   **Exc.:** $1700   **VGood:** $1600
Grade II
Perf.: $1700   **Exc.:** $1400   **VGood:** $1300

Pedersen Model 1000 Skeet
Same specs as Model 1000 except 12-ga.; 2 ¾″ chambers; 26″, 28″ barrels; Skeet chokes; Skeet-style walnut stock. Introduced 1973; discontinued 1975.
Grade I
Perf.: $2100   **Exc.:** $1950   **VGood:** $1700
Grade II
Perf.: $1900   **Exc.:** $1450   **VGood:** $1300

Pedersen Model 1000 Trap
Same specs as Model 1000 except made in 12-ga.; 2 ¾″ chambers; 30″, 32″ barrels; Modified/Full or Improved Modified/Full chokes; Monte Carlo trap-style walnut stock. Introduced 1973; discontinued 1975.
Grade I
Perf.: $2000   **Exc.:** $1600   **VGood:** $1400
Grade II
Perf.: $1550   **Exc.:** $1300   **VGood:** $1100

PEDERSEN MODEL 1500
Over/under; boxlock; 12-ga.; 2 ¾″, 3″ chambers; 26″, 28″, 30″, 32″ barrels; standard choke combos; vent rib; hand-checkered European walnut pistol-grip stock, forearm; rubber recoil pad; automatic selective ejectors; choice of sights. Introduced 1973; discontinued 1975.
Perf.: $650   **Exc.:** $450   **VGood:** $350

Pedersen Model 1500 Skeet
Same specs as Model 1500 except 27″ barrels; Skeet choke; Skeet-style walnut stock. Introduced 1973; discontinued 1975.
Perf.: $700   **Exc.:** $530   **VGood:** $450

Pedersen Model 1500 Trap
Same specs as Model 1500 except 30″, 32″ barrels; Modified/Full or Improved Modified/Full chokes; Monte Carlo trap-style walnut stock. Introduced 1973; discontinued 1975.
Perf.: $750   **Exc.:** $500   **VGood:** $400

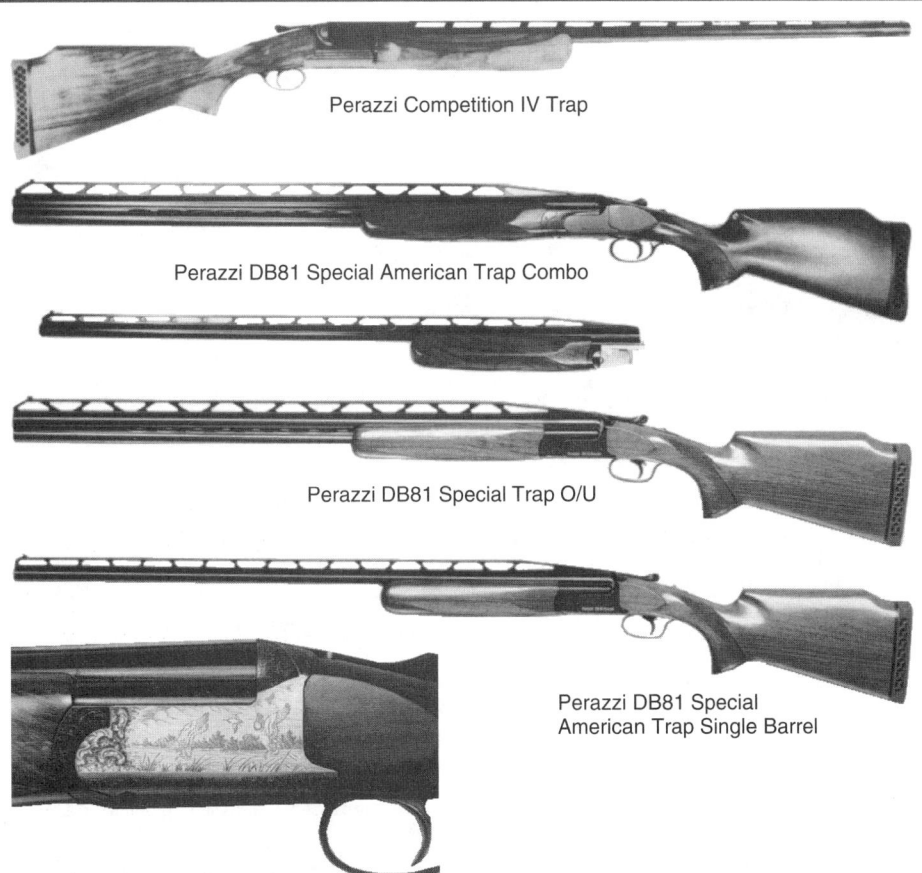

Perazzi Competition IV Trap

Perazzi DB81 Special American Trap Combo

Perazzi DB81 Special Trap O/U

Perazzi DB81 Special American Trap Single Barrel

Pedersen
Model 2000
Model 2500
Model 4000
Model 4000 Trap
Model 4500
Model 4500 Trap

Perazzi
Competition I Single Barrel Trap
Competition I Skeet
Competition I Trap
Competition IV Trap
DB81 Special American Trap Combo
DB81 Special American Trap Single Barrel
DB81 Special Trap O/U

PEDERSEN MODEL 2000
Side-by-side; boxlock; 12-, 20-ga.; 26″, 28″, 30″ barrels; hand-checkered American walnut pistol-grip stock forearm; automatic selective ejectors; barrel selector/safety; single selective trigger; automatic safety; gold-filled, hand-engraved receiver. Introduced 1973; discontinued 1975. Grade I
   Perf.: $2500 **Exc.:** $2100 **VGood:** $1800
Grade II
   Perf.: $2250 **Exc.:** $1900 **VGood:** $1650

PEDERSEN MODEL 2500
Side-by-side; boxlock; 12-, 20-ga.; hand-checkered European walnut pistol-grip stock, beavertail forearm; automatic selective ejectors; barrel selector/safety; single selective trigger; automatic safety; no receiver engraving. Introduced 1973; discontinued 1975.
   Perf.: $450 **Exc.:** $375 **VGood:** $300

PEDERSEN MODEL 4000
Slide action; 12-, 20-ga., .410; 3″ chambers; 26″, 28″, 30″ vent-rib barrel; standard chokes; checkered select American walnut stock, slide handle; full-coverage engraving on receiver. Based upon Mossberg Model 500. Introduced 1973; discontinued 1973.
   Perf.: $450 **Exc.:** $375 **VGood:** $300

Pedersen Model 4000 Trap
Same specs as Model 4000 except 12-ga.; 30″ Full-choke barrel; Monte Carlo trap-style walnut stock; recoil pad. Introduced 1975; discontinued 1975.
   Perf.: $475 **Exc.:** $365 **VGood:** $325

PEDERSEN MODEL 4500
Slide action; 12-, 20-ga., .410; 3″ chamber; 26″, 28″, 30″ vent-rib barrel; standard chokes; checkered select American walnut stock, slide handle; some engraving on receiver. Introduced 1975; discontinued 1975.
   Perf.: $400 **Exc.:** $320 **VGood:** $280

Pedersen Model 4500 Trap
Same specs as Model 4500 except 12-ga.; 30″ Full-choke barrel; Monte Carlo trap-style walnut stock; recoil pad. Introduced 1975; discontinued 1975.
   Perf.: $420 **Exc.:** $330 **VGood:** $290

PERAZZI COMPETITION I SINGLE BARREL TRAP
Single shot; boxlock; 12-ga.; 32″, 34″ vent-rib barrel; Full choke; checkered Monte Carlo stock, beavertail forend; auto ejectors; single trigger; recoil pad. Introduced by Ithaca 1973; discontinued 1978.
   Perf.: $2650 **Exc.:** $1950 **VGood:** $1700

Perazzi Competition I Skeet
Same specs as Competition I Single Barrel Trap except over/under; 26 3/4″ barrels; integral muzzlebrakes; Skeet chokes; Skeet-style checkered European walnut pistol-grip stock, forend. Introduced by Ithaca 1969; discontinued 1974.
   Perf.: $4400 **Exc.:** $3300 **VGood:** $2800

Perazzi Competition I Trap
Same specs as Competition I Single Barrel Trap except over/under; 30″, 32″ barrels; Improved Modified/Full chokes; checkered European walnut pistol-grip stock, forend. Introduced by Ithaca 1969; discontinued 1974.
   Perf.: $4000 **Exc.:** $3200 **VGood:** $2700

PERAZZI COMPETITION IV TRAP
Single barrel; boxlock; 12-ga.; 32″, 34″ barrel; four interchangeable choke tubes; high, wide vent rib; checkered European walnut stock, beavertail forend; auto ejectors; single selective trigger; recoil pad; marketed in fitted case. Introduced by Ithaca 1977; discontinued 1978.
   Perf.: $2000 **Exc.:** $1400 **VGood:** $1100

PERAZZI DB81 SPECIAL AMERICAN TRAP COMBO
Over/under; boxlock; 12-ga.; two-barrel set of 29 1/2″ or 31 1/2″ O/U and 32″ or 34″ single; fixed or choke tubes; high vent rib; checkered European walnut Monte Carlo stock, beavertail forend; removable trigger group; external selector. Made in Italy. Introduced 1988; still in production. **LMSR:** $11,600
Standard Grade
   Perf.: $10,000 **Exc.:** $7000 **VGood:** $5000
SC3 Grade
   Perf.: $16,000 **Exc.:** $11,200 **VGood:** $8000
SC0 Grade
   Perf.: $27,500 **Exc.:** $19,250 **VGood:** $13,750

SC0 Gold Grade
   Perf.: $30,000 **Exc.:** $20,000 **VGood:** $15,000
SC0 Sideplates Grade
   Perf.: $40,000 **Exc.:** $28,000 **VGood:** $20,000
SC0 Gold Sideplates Grade
   Perf.: $45,000 **Exc.:** $31,500 **VGood:** $22,500

Perazzi DB81 Special American Trap Single Barrel
Same specs as DB81 Special American Trap Combo except only single shot; 32″, 34″ barrel. Made in Italy. Introduced 1988; discontinued 1994.
Standard Grade
   Perf.: $5800 **Exc.:** $3650 **VGood:** $3100
SC3 Grade
   Perf.: $9500 **Exc.:** $6100 **VGood:** $5000
SC0 Grade
   Perf.: $16,000 **Exc.:** $10,000 **VGood:** $8500
SC0 Gold Grade
   Perf.: $17,500 **Exc.:** $11,000 **VGood:** $9000
SC0 Sideplates Grade
   Perf.: $23,000 **Exc.:** $15,500 **VGood:** $12,500
SC0 Gold Sideplates Grade
   Perf.: $26,500 **Exc.:** $20,000 **VGood:** $16,000

Perazzi DB81 Special Trap O/U
Same specs as DB81 Special American Trap Combo except only 29 1/2″, 31 1/2″ O/U barrels. Made in Italy. Introduced 1988; still in production. **LMSR:** $8550
Standard Grade
   Perf.: $8000 **Exc.:** $5600 **VGood:** $4100
SC3 Grade
   Perf.: $12,500 **Exc.:** $8500 **VGood:** $6500
SC0 Grade
   Perf.: $22,000 **Exc.:** $14,500 **VGood:** $11,500
SC0 Gold Grade
   Perf.: $25,500 **Exc.:** $17,500 **VGood:** $13,250
SC0 Sideplates Grade
   Perf.: $34,000 **Exc.:** $23,500 **VGood:** $17,500
SC0 Gold Sideplates Grade
   Perf.: $40,000 **Exc.:** $28,000 **VGood:** $21,000

SHOTGUNS

Perazzi
Grand American 88 Special American Trap Combo
Grand American 88 Special American Trap Single Barrel
Grand American 88 Special Trap O/U
Light Game Model
Mirage Classic Sporting O/U
Mirage Live Bird (Early Mfg.)
Mirage Skeet (Early Mfg.)
Mirage Trap (Early Mfg.)
Mirage Pigeon-Electrocibles O/U
Mirage Skeet O/U
Mirage Trap O/U

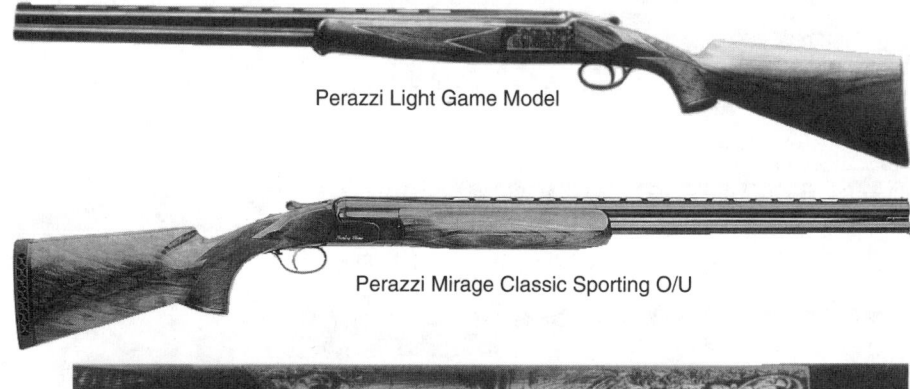

Perazzi Light Game Model

Perazzi Mirage Classic Sporting O/U

PERAZZI GRAND AMERICAN 88 SPECIAL AMERICAN TRAP COMBO
Over/under; boxlock; 12-ga.; two-barrel set with 29 1/2" or 31 1/2" O/U and 32" or 34" single; fixed or choke tubes; vent rib; checkered European walnut Monte Carlo stock, beavertail forend; removable, adjustable trigger group; single selective trigger; external selector. Made in Italy. Discontinued 1992.
Standard Grade
    Perf.: $9500 **Exc.:** $6150   **VGood:** $4950
SC3 Grade
    Perf.: $15,000 **Exc.:** $9750   **VGood:** $7800
SC0 Grade
    Perf.: $23,000 **Exc.:** $15,000 **VGood:** $11,950
SC0 Gold Grade
    Perf.: $26,000 **Exc.:** $16,900 **VGood:** $13,500
SC0 Sideplates Grade
    Perf.: $33,500 **Exc.:** $21,950 **VGood:** $17,500
SC0 Gold Sideplates Grade
    Perf.: $38,500 **Exc.:** $25,000 **VGood:** $20,000

Perazzi Grand American 88 Special American Trap Single Barrel
Same specs as Grand American 88 Special American Trap Combo except 32", 34" single barrel only. Made in Italy. Discontinued 1992.
Standard Grade
    Perf.: $6500 **Exc.:** $3800   **VGood:** $3000
SC3 Grade
    Perf.: $12,000 **Exc.:** $7000   **VGood:** $5500
SC0 Grade
    Perf.: $20,000 **Exc.:** $11,500 **VGood:** $9000
SC0 Gold Grade
    Perf.: $22,000 **Exc.:** $13,500 **VGood:** $10,000
SC0 Sideplates Grade
    Perf.: $29,500 **Exc.:** $18,000 **VGood:** $13,000
SC0 Gold Sideplates Grade
    Perf.: $34,500 **Exc.:** $21,000 **VGood:** $17,000

Perazzi Grand American 88 Special Trap O/U
Same specs as Grand American 88 Special American Trap Combo except 29 1/2", 30 3/4", 31 1/2" O/U barrels; checkered European walnut pistol-grip stock, beavertail forend. Made in Italy. Discontinued 1992.
Standard Grade
    Perf.: $6900 **Exc.:** $4500   **VGood:** $3600
SC3 Grade
    Perf.: $12,500 **Exc.:** $8150   **VGood:** $6500
SC0 Grade
    Perf.: $20,500 **Exc.:** $13,300 **VGood:** $10,500
SC0 Gold Grade
    Perf.: $23,000 **Exc.:** $14,500 **VGood:** $12,000
SC0 Sideplates Grade
    Perf.: $31,000 **Exc.:** $20,000 **VGood:** $16,100
SC0 Gold Sideplates Grade
    Perf.: $35,500 **Exc.:** $23,000 **VGood:** $18,500

PERAZZI LIGHT GAME MODEL
Over/under; boxlock; 12-ga.; 27 1/2" barrels; Modified/Full, Improved Cylinder/Modified chokes; checkered European walnut field stock, forend; single trigger; auto ejectors. Introduced by Ithaca 1972; discontinued 1974.
    Perf.: $4300 **Exc.:** $3100   **VGood:** $2750

PERAZZI MIRAGE CLASSIC SPORTING O/U
Over/under; boxlock; 12-ga.; 28 3/8", 29 1/2", 31 1/2" barrels; fixed or choke tubes; vent rib; checkered European walnut pistol-grip stock, beavertail forend; single selective trigger; removable trigger group; external selector; engraving. Made in Italy. Still in production. **LMSR: $9900**
    Perf.: $9000 **Exc.:** $6300   **VGood:** $4900

PERAZZI MIRAGE LIVE BIRD (EARLY MFG.)
Over/under; boxlock; 12-ga.; 28" Modified/Extra-Full barrels; vent rib; checkered European walnut pistol-grip stock, forend; selective ejectors; non-selective single trigger; recoil pad. Introduced by Ithaca 1973; discontinued 1978.
    Perf.: $2500 **Exc.:** $1700   **VGood:** $1300

Perazzi Mirage Skeet (Early Mfg.)
Same specs as Mirage Live Bird (Early Mfg.) except 28" barrels; integral muzzle brakes; Skeet chokes; Skeet-style checkered European walnut pistol-grip stock, forend. Introduced by Ithaca 1973; discontinued 1978.
    Perf.: $2500 **Exc.:** $1700   **VGood:** $1300

Perazzi Mirage Trap (Early Mfg.)
Same specs as Mirage Live Bird (Early Mfg.) except 30", 32" barrels; Improved Modified/Full chokes; tapered vent rib; checkered European walnut Monte Carlo stock, forend. Introduced by Ithaca 1973; discontinued 1978.
    Perf.: $2500 **Exc.:** $1700   **VGood:** $1300

PERAZZI MIRAGE PIGEON-ELECTROCIBLES O/U
Over/under; boxlock; 12-ga.; 27 1/2", 28 3/8", 29 1/2", 31 1/2" barrels; fixed or choke tubes; vent rib; checkered European walnut pistol-grip stock, Schnabel beavertail forend; removable trigger group. Made in Italy. Introduced 1995; still in production. **LMSR: $7850**
Standard Grade
    Perf.: $7000 **Exc.:** $4500   **VGood:** $3600
SC3 Grade
    Perf.: $12,500 **Exc.:** $8000   **VGood:** $6500
SC0 Grade
    Perf.: $21,500 **Exc.:** $14,000 **VGood:** $11,000
SC0 Gold Grade
    Perf.: $24,000 **Exc.:** $15,500 **VGood:** $12,500
SC0 Sideplates Grade
    Perf.: $32,500 **Exc.:** $21,000 **VGood:** $17,000
SC0 Gold Sideplates Grade
    Perf.: $38,500 **Exc.:** $25,000 **VGood:** $20,000

Perazzi Mirage Skeet O/U
Same specs as Mirage Pigeon-Electrocibles O/U except 26 3/4", 27 1/2" barrels; beavertail forend. Made in Italy. Introduced 1993; still in production. **LMSR: $7850**
Standard Grade
    Perf.: $7000 **Exc.:** $4500   **VGood:** $3600
SC3 Grade
    Perf.: $12,500 **Exc.:** $8000   **VGood:** $6500
SC0 Grade
    Perf.: $21,500 **Exc.:** $14,000 **VGood:** $11,000
SC0 Gold Grade
    Perf.: $24,000 **Exc.:** $15,500 **VGood:** $12,500
SC0 Sideplates Grade
    Perf.: $32,500 **Exc.:** $21,000 **VGood:** $17,000
SC0 Gold Sideplates Grade
    Perf.: $38,500 **Exc.:** $25,000 **VGood:** $20,000

Perazzi Mirage Trap O/U
Same specs as Mirage Pigeon-Electrocibles O/U except 29 1/2", 30 3/4", 31 1/2" barrels; beavertail forend. Made in Italy. Still in production. **LMSR: $7850**
Standard Grade
    Perf.: $7000 **Exc.:** $4500   **VGood:** $3600
SC3 Grade
    Perf.: $12,500 **Exc.:** $8000   **VGood:** $6500
SC0 Grade
    Perf.: $21,500 **Exc.:** $14,000 **VGood:** $11,000
SC0 Gold Grade
    Perf.: $24,000 **Exc.:** $15,500 **VGood:** $12,500
SC0 Sideplates Grade
    Perf.: $32,500 **Exc.:** $21,000 **VGood:** $17,000
SC0 Gold Sideplates Grade
    Perf.: $38,500 **Exc.:** $25,000 **VGood:** $20,000

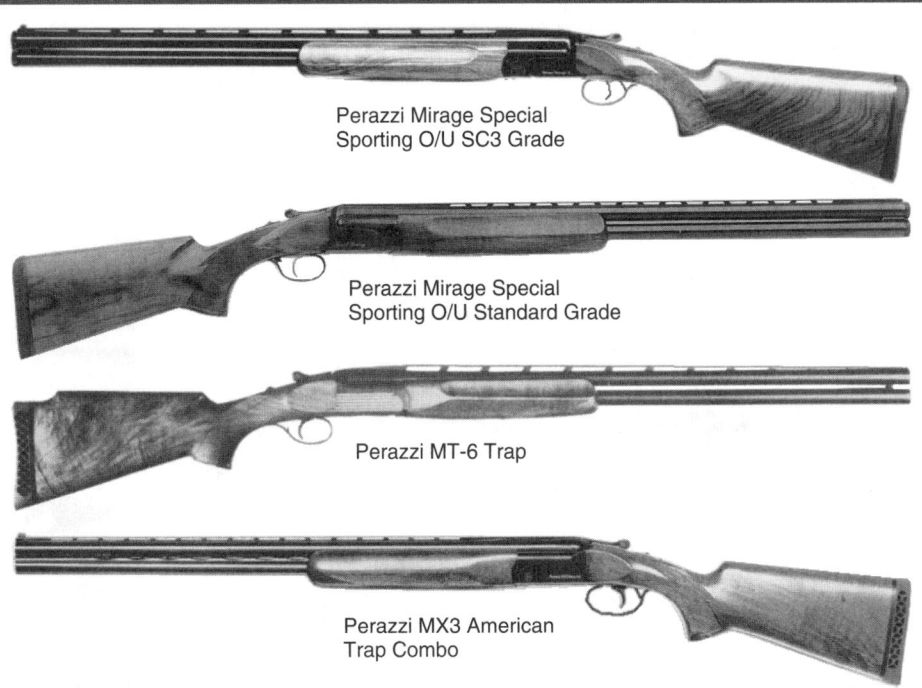

Perazzi Mirage Special
Sporting O/U SC3 Grade

Perazzi Mirage Special
Sporting O/U Standard Grade

Perazzi MT-6 Trap

Perazzi MX3 American
Trap Combo

Perazzi
**Mirage Special Pigeon-Electrocibles O/U
Mirage Special Skeet O/U
Mirage Special Sporting O/U
Mirage Special Trap O/U
MT-6 Skeet
MT-6 Trap
MT-6 Trap Combo
MX1B Pigeon-Electrocibles O/U
MX1B Sporting O/U
MX3 Special American Trap Combo**

PERAZZI MIRAGE SPECIAL PIGEON-ELECTROCIBLES O/U

Over/under; boxlock; 12-ga.; 28 3/8″, 29 1/2″, 31 1/2″ barrels; fixed or choke tubes; vent rib; checkered European walnut pistol-grip stock; Schnabel beavertail forend; removable, adjustable trigger. Made in Italy. Introduced 1995; still in production. **LMSR: $8320**
Standard Grade
    Perf.: $7500 **Exc.:** $5000   **VGood:** $3900
SC3 Grade
    Perf.: $12,500 **Exc.:** $8000   **VGood:** $6500
SC0 Grade
    Perf.: $22,000 **Exc.:** $14,000 **VGood:** $11,500
SC0 Gold Grade
    Perf.: $25,000 **Exc.:** $16,250 **VGood:** $13,000
SC0 Sideplates Grade
    Perf.: $33,000 **Exc.:** $21,500 **VGood:** $17,000
SC0 Gold Sideplates Grade
    Perf.: $38,000 **Exc.:** $25,000 **VGood:** $19,750

Perazzi Mirage Special Skeet O/U

Same specs as Mirage Special Pigeon-Electrocibles O/U except 26 3/4″, 27 1/2″ barrels; beavertail forend. Made in Italy. Still in production. **LMSR: $8320**
Standard Grade
    Perf.: $7500 **Exc.:** $5000   **VGood:** $3900
SC3 Grade
    Perf.: $12,500 **Exc.:** $8000   **VGood:** $6500
SC0 Grade
    Perf.: $22,000 **Exc.:** $14,000 **VGood:** $11,500
SC0 Gold Grade
    Perf.: $25,000 **Exc.:** $16,250 **VGood:** $13,000
SC0 Sideplates Grade
    Perf.: $33,000 **Exc.:** $21,500 **VGood:** $17,000
SC0 Gold Sideplates Grade
    Perf.: $38,000 **Exc.:** $25,000 **VGood:** $19,750

Perazzi Mirage Special Sporting O/U

Same specs as Mirage Special Pigeon-Electrocibles O/U except 28 3/8″, 29 1/2″, 31 1/2″ barrels; beavertail forend; single selective trigger; external selector. Made in Italy. Still in production. **LMSR: $8320**
Standard Grade
    Perf.: $7500 **Exc.:** $5000   **VGood:** $3900
SC3 Grade
    Perf.: $12,500 **Exc.:** $8000   **VGood:** $6500
SC0 Grade
    Perf.: $22,000 **Exc.:** $14,000 **VGood:** $11,500
SC0 Gold Grade
    Perf.: $25,000 **Exc.:** $16,250 **VGood:** $13,000
SC0 Sideplates Grade
    Perf.: $33,000 **Exc.:** $21,500 **VGood:** $17,000
SC0 Gold Sideplates Grade
    Perf.: $38,000 **Exc.:** $25,000 **VGood:** $19,750

Perazzi Mirage Special Trap O/U

Same specs as Mirage Special Pigeon-Electrocibles O/U except 29 1/2″, 30 3/4″, 31 1/2″ barrels; beavertail forend. Made in Italy. Still in production. **LMSR: $8320**
Standard Grade
    Perf.: $7500 **Exc.:** $5000   **VGood:** $3900
SC3 Grade
    Perf.: $12,500 **Exc.:** $8000   **VGood:** $6500
SC0 Grade
    Perf.: $22,000 **Exc.:** $14,000 **VGood:** $11,500
SC0 Gold Grade
    Perf.: $25,000 **Exc.:** $16,250 **VGood:** $13,000
SC0 Sideplates Grade
    Perf.: $33,000 **Exc.:** $21,500 **VGood:** $17,000
SC0 Gold Sideplates Grade
    Perf.: $38,000 **Exc.:** $25,000 **VGood:** $19,750

PERAZZI MT-6 SKEET

Over/under; boxlock; 12-ga.; 28″ separated barrels; five interchangeable choke tubes; wide vent rib; Skeet-style checkered European walnut pistol-grip stock, forend; non-selective single trigger; auto selective ejectors; recoil pad; marketed in fitted case. Introduced by Ithaca 1976; discontinued 1978.
    Perf.: $4000   **Exc.:** $3100   **VGood:** $2750

Perazzi MT-6 Trap

Same specs as MT-6 Skeet except 30″, 32″ barrels; trap-style checkered European walnut pistol-grip stock, forend. Introduced by Ithaca 1976; discontinued 1978.
    Perf.: $2500   **Exc.:** $1700   **VGood:** $1300

Perazzi MT-6 Trap Combo

Same specs as MT-6 Skeet except two-barrel set with 30″ or 32″ O/U and 32″ or 34″ single under barrel; seven interchangeable choke tubes; high aluminum vent rib; trap-style checkered European walnut pistol-grip stock, forend; marketed in fitted case. Introduced by Ithaca 1977; discontinued 1978.
    Perf.: $4200   **Exc.:** $3000   **VGood:** $2200

PERAZZI MX1B PIGEON-ELECTROCIBLES O/U

Over/Under; boxlock; 12-ga.; 27 1/2″ barrels; fixed or choke tubes; vent rib; checkered European pistol-grip walnut stock, Schnabel forend; removable trigger group. Made in Italy. Introduced 1995; still in production. **LMSR: $7850**
Standard Grade
    Perf.: $7000 **Exc.:** $4500   **VGood:** $3600
SC3 Grade
    Perf.: $12,500 **Exc.:** $8000   **VGood:** $6500
SC0 Grade
    Perf.: $21,500 **Exc.:** $14,000 **VGood:** $11,000

SC0 Gold Grade
    Perf.: $24,000 **Exc.:** $15,000 **VGood:** $12,500
SC0 Sideplates Grade
    Perf.: $32,500 **Exc.:** $21,000 **VGood:** $17,000
SC0 Gold Sideplates Grade
    Perf.: $38,500 **Exc.:** $25,000 **VGood:** $20,000

Perazzi MX1B Sporting O/U

Same specs as MX1B Pigeon-Electrocibles O/U except 28 3/8″, 29 1/2″, 31 1/2″ barrels; choke tubes; single selective trigger; external selector, Made in Italy. No longer in production.
Standard Grade
    Perf.: $7000 **Exc.:** $4500   **VGood:** $3600
SC3 Grade
    Perf.: $12,500 **Exc.:** $8000   **VGood:** $6500
SC0 Grade
    Perf.: $21,500 **Exc.:** $14,000 **VGood:** $11,000
SC0 Gold Grade
    Perf.: $24,000 **Exc.:** $15,000 **VGood:** $12,500
SC0 Sideplates Grade
    Perf.: $32,500 **Exc.:** $21,000 **VGood:** $17,000
SC0 Gold Sideplates Grade
    Perf.: $38,500 **Exc.:** $25,000 **VGood:** $20,000

PERAZZI MX3 SPECIAL AMERICAN TRAP COMBO

Over/under; boxlock; 12-ga.; two-barrel set with 29 1/2″ or 31 1/2″ and 32″ O/U or 34″ single; fixed or choke tubes; vent rib; checkered European walnut Monte Carlo stock, beavertail forend; removable, adjustable trigger group; single selective trigger; external selector. Made in Italy. Discontinued 1992.
    Perf.: $7500 **Exc.:** $4900   **VGood:** $3900
SC3 Grade
    Perf.: $13,000 **Exc.:** $8500   **VGood:** $6750
SC0 Grade
    Perf.: $21,500 **Exc.:** $14,000 **VGood:** $11,000
SC0 Gold Grade
    Perf.: $24,000 **Exc.:** $15,500 **VGood:** $12,500

Perazzi
MX3 Special American Trap Single Barrel
MX3 Special Skeet O/U
MX3 Special Sporting O/U
MX3 Special Trap O/U
MX6 American Trap Single Barrel
MX6 Skeet O/U
MX6 Sporting O/U
MX6 Trap O/U
MX7 American Trap Single Barrel
MX7 Skeet O/U
MX7 Sporting O/U
MX7 Trap O/U
MX8 Game O/U

Perazzi MX3 Special Skeet O/U

Perazzi MX3 Special American Trap Single Barrel

Perazzi MX3 Special Trap O/U

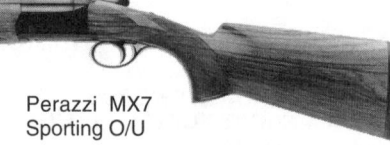
Perazzi MX7 Sporting O/U

Perazzi MX3 Special American Trap Single Barrel
Same specs as MX3 Special American Trap Combo except 32″, 34″ single barrel. Made in Italy. Discontinued 1992.
Standard Grade
  Perf.: $5500   **Exc.:** $3500   **VGood:** $2850
SC3 Grade
  Perf.: $9500   **Exc.:** $6200   **VGood:** $5000
SC0 Grade
  Perf.: $17,500 **Exc.:** $11,500 **VGood:** $9100
SC0 Gold Grade
  Perf.: $18,000 **Exc.:** $11,750 **VGood:** $9500

Perazzi MX3 Special Skeet O/U
Same specs as MX3 Special American Trap Combo except 26 ³/₄″, 27 ¹/₂″ barrels; checkered European walnut pistol-grip stock, beavertail forend. Made in Italy. Discontinued 1992.
Standard Grade
  Perf.: $6000   **Exc.:** $3900   **VGood:** $3100
SC3 Grade
  Perf.: $10,000 **Exc.:** $6500   **VGood:** $5200
SC0 Grade
  Perf.: $17,500 **Exc.:** $11,500 **VGood:** $9100
SC0 Gold Grade
  Perf.: $20,000 **Exc.:** $13,000 **VGood:** $10,500

Perazzi MX3 Special Sporting O/U
Same specs as MX3 Special American Trap Combo except 28 ³/₈″, 29 ¹/₂″, 31 ¹/₂″ barrels; checkered European walnut pistol-grip stock, beavertail forend. Made in Italy. Discontinued 1992.
Standard Grade
  Perf.: $6200   **Exc.:** $4000   **VGood:** $3200
SC3 Grade
  Perf.: $10,500 **Exc.:** $6800   **VGood:** $5500
SC0 Grade
  Perf.: $18,000 **Exc.:** $11,700 **VGood:** $9400
SC0 Gold Grade
  Perf.: $20,500 **Exc.:** $13,500 **VGood:** $10,500

Perazzi MX3 Special Trap O/U
Same specs as MX3 Special American Trap Combo except 29 ¹/₂″, 30 ³/₄″, 31 ¹/₂″ barrels; checkered European walnut pistol-grip stock, beavertail forend. Made in Italy. Discontinued 1992.
Standard Grade
  Perf.: $6000   **Exc.:** $3900   **VGood:** $3100
SC3 Grade
  Perf.: $10,000 **Exc.:** $6500   **VGood:** $5200
SC0 Grade
  Perf.: $17,500 **Exc.:** $11,500 **VGood:** $9100
SC0 Gold Grade
  Perf.: $20,000 **Exc.:** $13,000 **VGood:** $10,500

PERAZZI MX6 AMERICAN TRAP SINGLE BARREL
Single shot; 12-ga.; 32″, 34″ barrel; fixed or choke tubes; raised vent rib; checkered European walnut Monte Carlo stock, beavertail forend; removable trigger group. Made in Italy. Introduced 1995; still in production. **LMSR:** $4900
  Perf.: $4200   **Exc.:** $2900   **VGood:** $2200
Combo model with extra O/U barrels
  Perf.: $6800   **Exc.:** $4500   **VGood:** $3600

Perazzi MX6 Skeet O/U
Same specs as MX6 American Trap Single Barrel except over/under; boxlock; 26 ³/₄″, 27 ¹/₂″ barrels. Made in Italy. Introduced 1995; still in production. **LMSR:** $5700
  Perf.: $5000   **Exc.:** $3500   **VGood:** $2650

Perazzi MX6 Sporting O/U
Same specs as MX6 American Trap Single Barrel except over/under; boxlock; 28 ³/₈″, 29 ¹/₂″, 31 ¹/₂″ barrels; single selective trigger; external selector. Made in Italy. Introduced 1995; still in production. **LMSR:** $6740
  Perf.: $6000   **Exc.:** $4000   **VGood:** $3100

Perazzi MX6 Trap O/U
Same specs as MX6 American Trap Single Barrel except over/under; boxlock; 29 ¹/₂″, 30 ³/₄″, 31 ¹/₂″ barrels. Made in Italy. Introduced 1995; still in production. **LMSR:** $5700
  Perf.: $5000   **Exc.:** $3500   **VGood:** $2650

PERAZZI MX7 AMERICAN TRAP SINGLE BARREL
Single shot; 12-ga.; 32″, 34″ barrel; fixed or choke tubes; raised vent rib; checkered European walnut Monte Carlo stock, beavertail forend; fixed trigger group. Made in Italy. Introduced 1995; still in production. **LMSR:** $5650
  Perf.: $4900   **Exc.:** $3500   **VGood:** $2600

Combo model with extra O/U barrels
  Perf.: $7000   **Exc.:** $4900   **VGood:** $3700

Perazzi MX7 Skeet O/U
Same specs as MX7 American Trap Single Barrel except over/under; boxlock; 27 ¹/₂″ barrels; checkered European walnut pistol-grip stock, beavertail forend; single selective trigger; external selector. Made in Italy. Introduced 1993; still in production. **LMSR:** $6100
  Perf.: $5200   **Exc.:** $3700   **VGood:** $2800

Perazzi MX7 Sporting O/U
Same specs as MX7 American Trap Single Barrel except over/under; boxlock; 28 ³/₈″, 29 ¹/₂″, 31 ¹/₂″ barrels; checkered European walnut pistol-grip stock, beavertail forend; single selective trigger; external selector. Made in Italy. Introduced 1992; still in production. **LMSR:** $6670
  Perf.: $5800   **Exc.:** $4000   **VGood:** $3000

Perazzi MX7 Trap O/U
Same specs as MX7 American Trap Single Barrel except over/under; boxlock; 29 ¹/₂″, 30 ³/₄″, 31 ¹/₂″ barrels; checkered European walnut pistol-grip stock, beavertail forend; single selective trigger; external selector. Made in Italy. Introduced 1993; still in production. **LMSR:** $6100
  Perf.: $5200   **Exc.:** $3700   **VGood:** $2800

PERAZZI MX8 GAME O/U
Over/under; boxlock; 12-, 20-ga.; 26 ³/₄″, 27 ¹/₂″ barrels; fixed or choke tubes; vent rib; checkered European walnut pistol-grip (12-ga.) or straight-grip (20-ga.) stock; Schnabel or tapered forend; removable trigger group; single selective trigger; external selector. Made in Italy. Introduced 1993; still in production. **LMSR:** $7850
Standard Grade
  Perf.: $7000   **Exc.:** $4500   **VGood:** $3600
SC3 Grade
  Perf.: $12,500 **Exc.:** $8000   **VGood:** $6500
SC0 Grade
  Perf.: $21,500 **Exc.:** $14,000 **VGood:** $11,000
SC0 Gold Grade
  Perf.: $24,000 **Exc.:** $15,500 **VGood:** $12,500
SC0 Sideplates Grade
  Perf.: $32,500 **Exc.:** $21,000 **VGood:** $17,000
SC0 Gold Sideplates Grade
  Perf.: $38,500 **Exc.:** $25,000 **VGood:** $20,000

Perazzi MX3 SC0
Grade Engraving

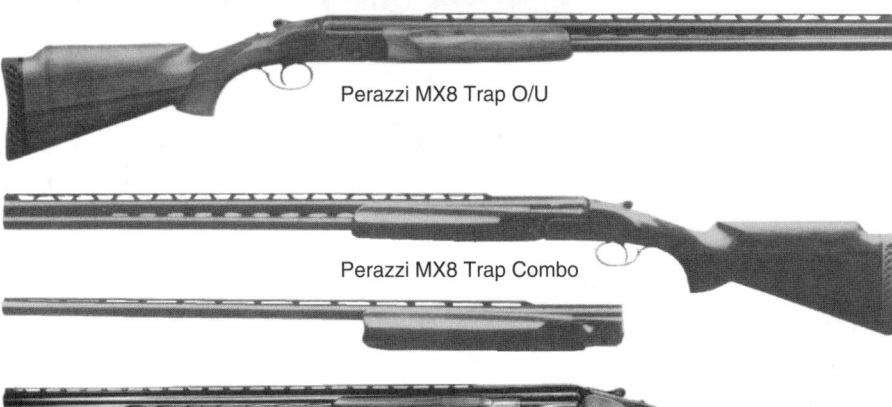

Perazzi MX8 Trap O/U

Perazzi MX8 Trap Combo

Perazzi MX8 Special
Skeet O/U

SHOTGUNS

Perazzi
MX8 Skeet O/U
MX8 Trap O/U
MX8 Trap Combo
MX8 Special American Trap Combo
MX8 Special American Trap Combo Single Barrel
MX8 Special Skeet O/U
MX8 Special Sporting O/U
MX8 Special Trap O/U
MX9 American Trap Combo

Perazzi MX8 Skeet O/U

Same specs as MX8 Game O/U except 26 3/4", 27 1/2" barrels; checkered European walnut pistol-grip stock, beavertail forend. Made in Italy. Introduced 1993; still in production. **LMSR: $7850**
Standard Grade
  Perf.: $7000 **Exc.:** $4500   **VGood:** $3600
SC3 Grade
  Perf.: $12,500 **Exc.:** $8000   **VGood:** $6500
SC0 Grade
  Perf.: $21,500 **Exc.:** $14,000   **VGood:** $11,000
SC0 Gold Grade
  Perf.: $24,000 **Exc.:** $15,500   **VGood:** $12,500
SC0 Sideplates Grade
  Perf.: $32,500 **Exc.:** $21,000   **VGood:** $17,000
SC0 Gold Sideplates Grade
  Perf.: $38,500 **Exc.:** $25,000   **VGood:** $20,000

Perazzi MX8 Trap O/U

Same specs as MX8 Game O/U except 29 1/2", 30 3/4", 31 1/2" barrels; checkered European walnut pistol-grip stock, beavertail forend. Made in Italy. Introduced by Ithaca 1969; still in production by Perazzi. **LMSR: $7850**
Standard Grade
  Perf.: $7000 **Exc.:** $4500   **VGood:** $3600
SC3 Grade
  Perf.: $12,500 **Exc.:** $8000   **VGood:** $6500
SC0 Grade
  Perf.: $21,500 **Exc.:** $14,000   **VGood:** $11,000
SC0 Gold Grade
  Perf.: $24,000 **Exc.:** $15,500   **VGood:** $12,500
SC0 Sideplates Grade
  Perf.: $32,500 **Exc.:** $21,000   **VGood:** $17,000
SC0 Gold Sideplates Grade
  Perf.: $38,500 **Exc.:** $25,000   **VGood:** $20,000

Perazzi MX8 Trap Combo

Same specs as MX8 Game O/U except two-barrel set with 29 1/2" or 31 1/2" O/U and 32" or 34" single; checkered European walnut pistol-grip stock, forend; two trigger groups. Introduced by Ithaca 1973; discontinued 1978.
  Perf.: $3500 **Exc.:** $2300   **VGood:** $1800

PERAZZI MX8 SPECIAL AMERICAN TRAP COMBO

Over/under; boxlock; 12-ga.; two-barrel set of 29 1/2" or 31 1/2" O/U and 32" or 34" single; fixed or choke tubes; raised vent rib; checkered European walnut Monte Carlo stock, beavertail forend; adjustable, removable trigger group; external selector. Made in Italy. Introduced 1988; still in production. **LMSR: $10,950**
Standard Grade
  Perf.: $9200 **Exc.:** $6000   **VGood:** $4800

Perazzi MX8 Special American Trap Single Barrel

Same specs as MX8 Special American Trap Combo except 32", 34" single barrel. Made in Italy. Introduced 1988; discontinued 1994.
Standard Grade
  Perf.: $6000 **Exc.:** $3900   **VGood:** $3100
SC3 Grade
  Perf.: $11,000 **Exc.:** $7100   **VGood:** $5700
SC0 Grade
  Perf.: $19,000 **Exc.:** $12,300   **VGood:** $9900
SC0 Gold Grade
  Perf.: $21,000 **Exc.:** $13,500   **VGood:** $11,000
SC0 Sideplates Grade
  Perf.: $30,000 **Exc.:** $17,500   **VGood:** $15,500
SC0 Gold Sideplates Grade
  Perf.: $35,000 **Exc.:** $22,750   **VGood:** $18,000

Perazzi MX8 Special Skeet O/U

Same specs as MX8 Special American Trap Combo except 26 3/4", 27 1/2" O/U barrels; checkered European walnut pistol-grip stock, beavertail forend. Made in Italy. Still in production. **LMSR: $8320**
Standard Grade
  Perf.: $7500 **Exc.:** $5000   **VGood:** $3900
SC3 Grade
  Perf.: $12,500 **Exc.:** $8000   **VGood:** $6500
SC0 Grade
  Perf.: $22,000 **Exc.:** $14,000   **VGood:** $11,500
SC0 Gold Grade
  Perf.: $25,000 **Exc.:** $16,250   **VGood:** $13,000
SC0 Sideplates Grade
  Perf.: $33,000 **Exc.:** $21,500   **VGood:** $17,000
SC0 Gold Sideplates Grade
  Perf.: $38,000 **Exc.:** $25,000   **VGood:** $19,750

Perazzi MX8 Special Sporting O/U

Same specs as MX8 Special American Trap Combo except 12-, 20-ga.; 28 3/8", 29 1/2", 31 1/2" O/U barrels; checkered European walnut pistol-grip stock, beavertail forend. Made in Italy. Still in production. **LMSR: $8890**
Standard Grade
  Perf.: $7800 **Exc.:** $5000   **VGood:** $4000
SC3 Grade
  Perf.: $13,000 **Exc.:** $8500   **VGood:** $6750
SC0 Grade
  Perf.: $22,500 **Exc.:** $14,600   **VGood:** $11,700
SC0 Gold Grade
  Perf.: $25,500 **Exc.:** $16,500   **VGood:** $13,250

Perazzi MX8 Special Trap O/U

Same specs as MX8 Special American Trap Combo except 29 1/2", 30 3/4", 31 1/2" O/U barrels; checkered European walnut pistol-grip stock, beavertail forend. Made in Italy. Still in production. **LMSR: $8320**
Standard Grade
  Perf.: $7500 **Exc.:** $5000   **VGood:** $3900
SC3 Grade
  Perf.: $12,500 **Exc.:** $8000   **VGood:** $6500
SC0 Grade
  Perf.: $22,000 **Exc.:** $14,000   **VGood:** $11,500
SC0 Gold Grade
  Perf.: $25,000 **Exc.:** $16,250   **VGood:** $13,000
SC0 Sideplates Grade
  Perf.: $33,000 **Exc.:** $21,500   **VGood:** $17,000
SC0 Gold Sideplates Grade
  Perf.: $38,000 **Exc.:** $25,000   **VGood:** $19,750

PERAZZI MX9 AMERICAN TRAP COMBO

Over/under; boxlock; 12-ga.; two-barrel set of 29 1/2" or 31 1/2" O/U and 32" or 34" single; fixed or choke tubes; adjustable rib to change point of impact; checkered European walnut Monte Carlo adjustable stock, beavertail forend; removable trigger. Made in Italy. Introduced 1993; discontinued 1994.
Standard Grade
  Perf.: $11,500 **Exc.:** $7500   **VGood:** $6000
SC3 Grade
  Perf.: $17,000 **Exc.:** $11,000   **VGood:** $8800
SC0 Grade
  Perf.: $26,500 **Exc.:** $17,000   **VGood:** $13,800
SC0 Gold Grade
  Perf.: $29,000 **Exc.:** $19,000   **VGood:** $15,000
SC0 Sideplates Grade
  Perf.: $37,500 **Exc.:** $24,000   **VGood:** $19,500
SC0 Gold Sideplates Grade
  Perf.: $42,000 **Exc.:** $27,000   **VGood:** $21,800

SHOTGUNS

Perazzi
MX9 American Trap
Single Barrel
MX9 Trap O/U
MX10 American
Trap Combo
MX10 American Trap
Single Barrel
MX10 Pigeon-Electrocibles
O/U
MX10 Skeet O/U
MX10 Sporting O/U
MX10 Trap O/U
MX11 American Trap Combo
MX11 American Trap
Single Barrel
MX11 Pigeon-Electrocibles
O/U
MX11 Skeet O/U
MX11 Sporting O/U
MX11 Trap O/U

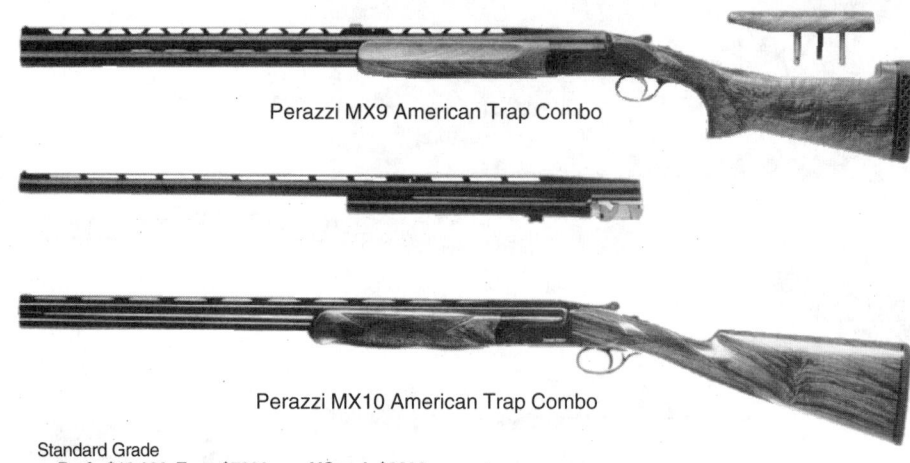

Perazzi MX9 American Trap Combo

Perazzi MX10 American Trap Combo

Perazzi MX9 American Trap Single Barrel
Same specs as MX9 American Trap Combo except 32″, 34″ single barrel. Made in Italy. Introduced 1993; discontinued 1994.
Standard Grade
  Perf.: $8500   **Exc.:** $5500   **VGood:** $4400
SC3 Grade
  Perf.: $13,000 **Exc.:** $8500   **VGood:** $6700
SC0 Grade
  Perf.: $21,500 **Exc.:** $14,000   **VGood:** $11,000
SC0 Gold Grade
  Perf.: $23,500 **Exc.:** $15,000   **VGood:** $12,000
SC0 Sideplates Grade
  Perf.: $32,500 **Exc.:** $21,000   **VGood:** $17,000
SC0 Gold Sideplates Grade
  Perf.: $37,500 **Exc.:** $24,000   **VGood:** $19,500

Perazzi MX9 Trap O/U
Same specs as MX9 American Trap Combo except 29 1/2″, 30 3/4″, 31 1/2″ O/U barrels; checkered European walnut pistol-grip adjustable stock, beavertail forend. Made in Italy. Introduced 1993; discontinued 1994.
Standard Grade
  Perf.: $8800   **Exc.:** $5700   **VGood:** $4500
SC3 Grade
  Perf.: $13,400 **Exc.:** $8700   **VGood:** $7000
SC0 Grade
  Perf.: $22,000 **Exc.:** $14,000   **VGood:** $11,500
SC0 Gold Grade
  Perf.: $24,100 **Exc.:** $15,600   **VGood:** $12,500
SC0 Sideplates Grade
  Perf.: $33,200 **Exc.:** $21,600   **VGood:** $17,000
SC0 Gold Sideplates Grade
  Perf.: $38,300 **Exc.:** $25,000   **VGood:** $20,000

PERAZZI MX10 AMERICAN TRAP COMBO
Over/under; boxlock; 12-ga.; two-barrel set of 29 1/2″ or 31 1/2″ O/U and 32″ or 34″ single; fixed or choke tubes; checkered European walnut adjustable stock, beavertail forend; removable trigger group; external selector. Made in Italy. Introduced 1993; still in production. **LMSR: $13,700**

Standard Grade
  Perf.: $12,000 **Exc.:** $7800   **VGood:** $6200
SC3 Grade
  Perf.: $18,000 **Exc.:** $11,700   **VGood:** $9400
SC0 Grade
  Perf.: $28,000 **Exc.:** $18,200   **VGood:** $14,500
SC0 Gold Grade
  Perf.: $31,000 **Exc.:** $20,000   **VGood:** $16,000

Perazzi MX10 American Trap Single Barrel
Same specs as MX10 American Trap Combo except 32″, 34″ single barrel. Made in Italy. Introduced 1993; discontinued 1993.
Standard Grade
  Perf.: $8500   **Exc.:** $5500   **VGood:** $4400
SC3 Grade
  Perf.: $13,000 **Exc.:** $8500   **VGood:** $6700
SC0 Grade
  Perf.: $21,000 **Exc.:** $13,600   **VGood:** $11,000
SC0 Gold Grade
  Perf.: $24,000 **Exc.:** $15,600   **VGood:** $12,500
SC0 Sideplates Grade
  Perf.: $32,000 **Exc.:** $20,800   **VGood:** $16,500
SC0 Gold Sideplates Grade
  Perf.: $38,000 **Exc.:** $25,000   **VGood:** $20,000

Perazzi MX10 Pigeon-Electrocibles O/U
Same specs as MX10 American Trap Combo except 27 1/2″, 29 1/2″ O/U barrels. Made in Italy. Introduced 1995; still in production. **LMSR: $10,300**
Standard Grade
  Perf.: $9000   **Exc.:** $5800   **VGood:** $4700
SC3 Grade
  Perf.: $14,000 **Exc.:** $9100   **VGood:** $7300
SC0 Grade
  Perf.: $23,500 **Exc.:** $15,000   **VGood:** $12,000
SC0 Gold Grade
  Perf.: $26,000 **Exc.:** $17,000   **VGood:** $13,500

Perazzi MX10 Skeet O/U
Same specs as MX10 American Trap Combo except 27 1/2″, 29 1/2″ O/U barrels. Made in Italy. Introduced 1995; still in production. **LMSR: $10,300**
Standard Grade
  Perf.: $9000   **Exc.:** $5800   **VGood:** $4700
SC3 Grade
  Perf.: $14,000 **Exc.:** $9100   **VGood:** $7300
SC0 Grade
  Perf.: $23,500 **Exc.:** $15,000   **VGood:** $12,000
SC0 Gold Grade
  Perf.: $26,000 **Exc.:** $17,000   **VGood:** $13,500

Perazzi MX10 Sporting O/U
Same specs as MX10 American Trap Combo except 28 3/8″, 29 1/2″, 31 1/2″ O/U barrel. Made in Italy. Introduced 1993; still in production. **LMSR: $11,340**
Standard Grade
  Perf.: $10,000 **Exc.:** $6500   **VGood:** $5200
SC3 Grade
  Perf.: $15,000 **Exc.:** $9750   **VGood:** $7800
SC0 Grade
  Perf.: $24,500 **Exc.:** $16,000   **VGood:** $13,000
SC0 Gold Grade
  Perf.: $27,500 **Exc.:** $18,000   **VGood:** $14,300

Perazzi MX10 Trap O/U
Same specs as MX10 American Trap Combo except 12-, 20-ga.; 29 1/2″, 30 3/4″, 31 1/2″ O/U barrels. Made in Germany. Introduced 1993; still in production. **LMSR: $10,300**
Standard Grade
  Perf.: $9000   **Exc.:** $5800   **VGood:** $4700
SC3 Grade
  Perf.: $14,000 **Exc.:** $9100   **VGood:** $7300
SC0 Grade
  Perf.: $23,500 **Exc.:** $15,000   **VGood:** $12,000
SC0 Gold Grade
  Perf.: $26,000 **Exc.:** $17,000   **VGood:** $13,500

PERAZZI MX11 AMERICAN TRAP COMBO
Over/under; boxlock; 12-ga.; two-barrel set of 29 1/2″ or 31 1/2″ O/U and 32″ or 34″ single; fixed or choke tubes; vent rib; checkered European walnut Monte Carlo adjustable stock, beavertail forend; removable trigger group; single selective trigger; external selector. Made in Italy. Introduced 1995; still in production. **LMSR: $9350**
  Perf.: $8200   **Exc.:** $5800   **VGood:** $4300

Perazzi MX11 American Trap Single Barrel
Same specs as MX11 American Trap Combo except 32″, 34″ single barrel. Made in Italy. Introduced 1995; still in production. **LMSR: $7030**
  Perf.: $6200   **Exc.:** $4300   **VGood:** $3300

Perazzi MX11 Pigeon-Electrocibles O/U
Same specs as MX11 American Trap Combo except 27 1/2″ O/U barrels; checkered European walnut pistol grip adjustable stock, beavertail forend. Made in Italy. Introduced 1995; still in production. **LMSR: $7400**
  Perf.: $6500   **Exc.:** $4500   **VGood:** $3400

Perazzi MX11 Skeet O/U
Same specs as MX11 American Trap Combo except 26 3/4″, 27 1/2″ O/U barrels; checkered European walnut pistol-grip adjustable stock, beavertail forend. Made in Italy. Introduced 1995; still in production. **LMSR: $7400**
  Perf.: $6500   **Exc.:** $4500   **VGood:** $3400

Perazzi MX11 Sporting O/U
Same specs as MX11 American Trap Combo except 28 3/8″, 29 1/2″, 31 1/2″ O/U barrels; checkered European walnut pistol-grip adjustable stock, beavertail forend. Made in Italy. Introduced 1995; still in production. **LMSR: $8440**
  Perf.: $7500   **Exc.:** $5000   **VGood:** $4000

Perazzi MX11 Trap O/U
Same specs as MX11 American Trap Combo except 29 1/2″, 30 3/4″, 31 1/2″ O/U barrels; checkered European walnut pistol-grip adjustable stock, beavertail forend. Made in Italy. Introduced 1995; still in production. **LMSR: $7400**
  Perf.: $6500   **Exc.:** $4500   **VGood:** $3400

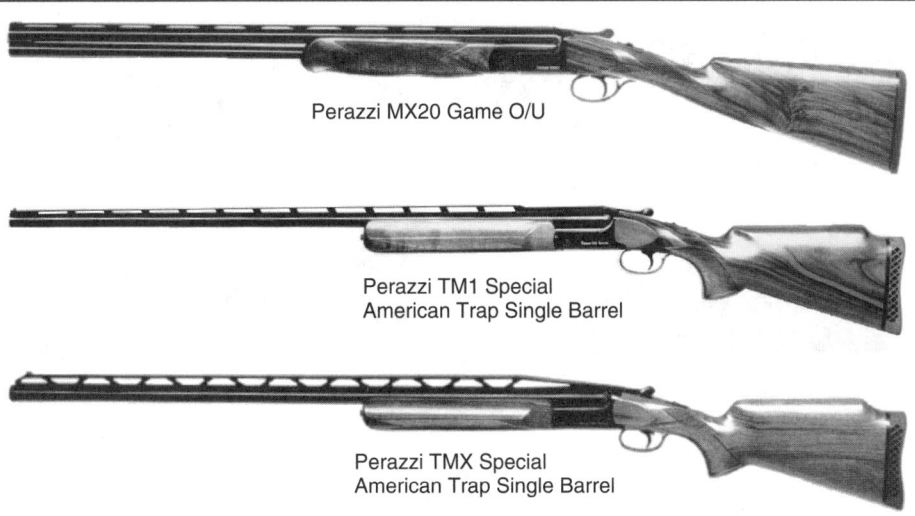

Perazzi MX20 Game O/U

Perazzi TM1 Special
American Trap Single Barrel

Perazzi TMX Special
American Trap Single Barrel

SHOTGUNS

Perazzi
MX12 Game O/U
MX14 American Trap
Single Barrel
MX20 Game O/U
MX28 Game O/U
MX410 Game O/U
Single Barrel Trap
TM1 Special American
Trap Single Barrel
TMX Special American
Trap Single Barrel

Perugini-Visini
Classic Double
Liberty Double

Piotti
Boss Over/Under
King No. 1

PERAZZI MX12 GAME O/U

Over/under; boxlock; 12-ga.; 26 3/4", 27 1/2" barrels; fixed or choke tubes; vent rib; checkered European walnut straight-grip stock, Schnabel or tapered forend; single selective trigger; non-removable trigger group; engraving. Made in Italy. Still in production. **LMSR: $7850**

Standard Grade
   Perf.: $6200   **Exc.:** $4400   **VGood:** $3500
SC3 Grade
   Perf.: $12,500 **Exc.:** $8000   **VGood:** $6500
SC0 Grade
   Perf.: $21,000 **Exc.:** $13,500 **VGood:** $11,000
SC0 Gold Grade
   Perf.: $24,000 **Exc.:** $15,500 **VGood:** $12,500
SC0 Sideplates Grade
   Perf.: $33,000 **Exc.:** $21,500 **VGood:** $17,000
SC0 Gold Sideplates Grade
   Perf.: $38,500 **Exc.:** $25,000 **VGood:** $20,000

PERAZZI MX14 AMERICAN TRAP SINGLE BARREL

Single shot; 12-ga.; 34" barrel; fixed or choke tubes; vent rib; checkered European walnut Monte Carlo adjustable stock, beavertail forend; removable trigger group; unsingle configuration. Made in Italy. Introduced 1995; still in production. **LMSR: $7030**
   Perf.: $6200   **Exc.:** $4300   **VGood:** $3300
Combo model with extra O/U barrels
   Perf.: $8500   **Exc.:** $5500   **VGood:** $4400

PERAZZI MX20 GAME O/U

Over/under; boxlock; 20-ga.; 26", 26 3/4", 27 1/2" barrels; fixed or choke tubes; vent rib; checkered European walnut straight-grip stock, Schnabel or tapered forend; single selective trigger; non-removable trigger group; engraving. Made in Italy. Still in production. **LMSR: $7850**

Standard Grade
   Perf.: $6700   **Exc.:** $4400   **VGood:** $3500
SC3 Grade
   Perf.: $12,500 **Exc.:** $8000   **VGood:** $6500
SC0 Grade
   Perf.: $21,000 **Exc.:** $13,500 **VGood:** $11,000
SC0 Gold Grade
   Perf.: $24,000 **Exc.:** $15,500 **VGood:** $12,500
SC0 Sideplates Grade
   Perf.: $33,000 **Exc.:** $21,500 **VGood:** $17,000
SC0 Gold Sideplates Grade
   Perf.: $38,500 **Exc.:** $25,000 **VGood:** $20,000

PERAZZI MX28 GAME O/U

Over/under; boxlock; 28-ga.; 26", 26 3/4", 27 1/2" barrels; fixed or choke tubes; vent rib; checkered European walnut straight-grip stock, tapered forend; non-removable trigger group; engraving. Made in Italy. Introduced 1993; still in production. **LMSR: $15,700**
Standard Grade
   Perf.: $11,500 **Exc.:** $9700   **VGood:** $7300
SC3 Grade
   Perf.: $20,000 **Exc.:** $13,000 **VGood:** $10,500

SC0 Grade
   Perf.: $29,000 **Exc.:** $19,000 **VGood:** $15,000
SC0 Gold Grade
   Perf.: $31,500 **Exc.:** $20,500 **VGood:** $16,000
SC0 Sideplates Grade
   Perf.: $40,000 **Exc.:** $26,000 **VGood:** $21,000
SC0 Gold Sideplates Grade
   Perf.: $46,000 **Exc.:** $30,000 **VGood:** $24,000

PERAZZI MX410 GAME O/U

Over/under; boxlock; .410; 26", 26 3/4", 27 1/2" barrels; fixed chokes; vent rib; checkered European walnut straight-grip stock, tapered forend; non-removable trigger group; engraving. Made in Italy. Introduced 1993; still in production. **LMSR: $15,700**
Standard Grade
   Perf.: $11,500 **Exc.:** $9700   **VGood:** $7300
SC3 Grade
   Perf.: $20,000 **Exc.:** $13,000 **VGood:** $10,500
SC0 Grade
   Perf.: $29,000 **Exc.:** $19,000 **VGood:** $15,000
SC0 Gold Grade
   Perf.: $31,500 **Exc.:** $20,500 **VGood:** $16,000
SC0 Sideplates Grade
   Perf.: $40,000 **Exc.:** $26,000 **VGood:** $21,000
SC0 Gold Sideplates Grade
   Perf.: $46,000 **Exc.:** $30,000 **VGood:** $24,000

PERAZZI SINGLE BARREL TRAP

Single shot; boxlock; 12-ga.; 34" barrel; Full choke; vent rib; checkered European walnut pistol-grip stock, forend; auto ejector; recoil pad. Introduced by Ithaca 1971; discontinued 1972.
   Perf.: $2200   **Exc.:** $1700   **VGood:** $1500

PERAZZI TM1 SPECIAL AMERICAN TRAP SINGLE BARREL

Single shot; 12-ga.; 32", 34" barrel; fixed or choke tubes; raised vent rib; checkered European walnut Monte Carlo stock, beavertail forend; removable, adjustable four-position trigger. Made in Italy. Introduced 1988; still in production. **LMSR: $6150**
Standard Grade
   Perf.: $5200   **Exc.:** $3400   **VGood:** $2700
SC0 Grade
   Perf.: $16,000 **Exc.:** $10,500 **VGood:** $8300
SC0 Gold Grade
   Perf.: $18,500 **Exc.:** $12,000 **VGood:** $9600

PERAZZI TMX SPECIAL AMERICAN TRAP SINGLE BARREL

Single shot; 12-ga.; 32", 34" barrel; fixed or choke tubes; high vent rib; checkered European walnut Monte Carlo stock, beavertail forend; removable, adjustable four-position trigger. Made in Italy. Introduced 1988; still in production. **LMSR: $6400**
Standard Grade
   Perf.: $5500   **Exc.:** $3600   **VGood:** $2900
SC0 Grade
   Perf.: $17,000 **Exc.:** $11,000 **VGood:** $8900
SC0 Gold Grade
   Perf.: $19,500 **Exc.:** $12,500 **VGood:** $10,000

PERUGINI-VISINI CLASSIC DOUBLE

Side-by-side; sidelock; 12-, 20-ga.; 2 3/4", 3" chambers; various barrel lengths, chokes; high-grade oil-finished straight English briar walnut stock; H&H-type hand-detachable sidelocks; internal parts gold-plated; single or double triggers; auto ejectors; numerous options. Made in Italy. Introduced, 1986; discontinued 1989.
   Perf.: $10,000 **Exc.:** $7000   **VGood:** $6500

PERUGINI-VISINI LIBERTY DOUBLE

Side-by-side; boxlock; 12-, 20-, 28-ga.; 410; 2 3/4", 3" chambers; various barrel lengths, chokes; high-grade oil-finished straight English briar walnut stock; internal parts gold-plated; single or double trigger; auto ejectors; numerous options. Made in Italy. Introduced 1986; discontinued 1989.
   Perf.: $5000   **Exc.:** $3700   **VGood:** $3200

PIOTTI BOSS OVER/UNDER

Over/under; sidelock; 12-, 20-ga.; 26" to 32" barrels; standard chokes; stock dimensions to customer specs; best quality figured walnut. Essentially a custom-made gun with many options. Made in Italy. Introduced 1993; still imported. **LMSR: $33,000**
   Perf.: $26,000 **Exc.:** $17,000 **VGood:** $14,000

PIOTTI KING NO. 1

Side-by-side; H&H sidelock; 12-, 16-, 20-, 28-ga.; .410; 25" to 30" (12-ga.), 25" to 28" (other gauges) barrels; file-cut or concave vent rib; fine-figured straight-grip European walnut stock, split or beavertail forend; oil or satin luster finish; double triggers; optional single non-selective trigger; coin finish or color case-hardening; full-coverage engraving; gold crown on top lever; name in gold; gold crest on forend. Made in Italy to customer's requirements. Introduced 1983; still in production. **LMSR: $18,600**
   Perf.: $16,000 **Exc.:** $10,000 **VGood:** $8500
With single trigger
   Perf.: $17,000 **Exc.:** $11,000 **VGood:** $9500

SHOTGUNS

Piotti
King No. 1 EELL
King No. 1 Extra
Lunik
Monte Carlo
Piuma

Precision Sports
Model 600 Series
(See Parker-Hale 640E)
Model 640A
Model 640E
Model 640M "Big Ten"
Model 640 Slug Gun
Model 645A
Model 645E
Model 645E-XXV
Model 650A

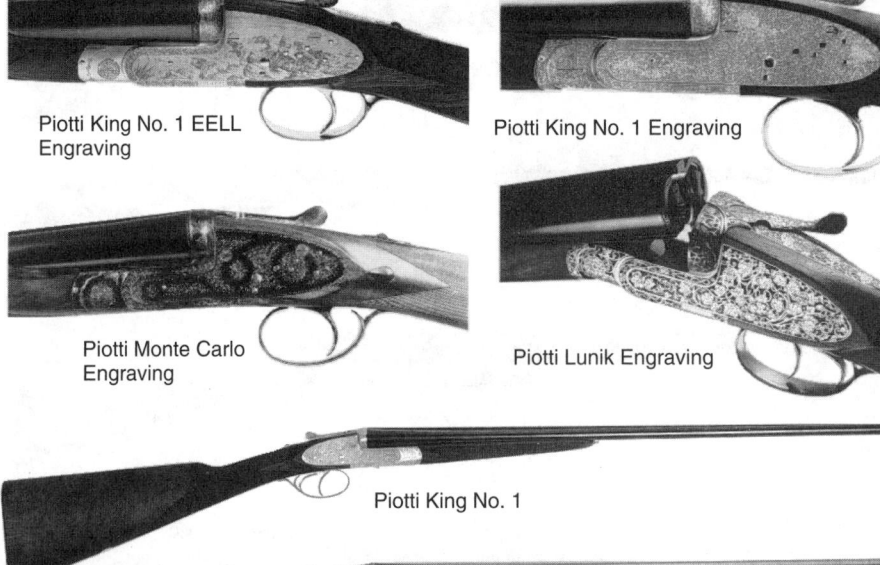

Piotti King No. 1 EELL
Engraving

Piotti King No. 1 Engraving

Piotti Monte Carlo
Engraving

Piotti Lunik Engraving

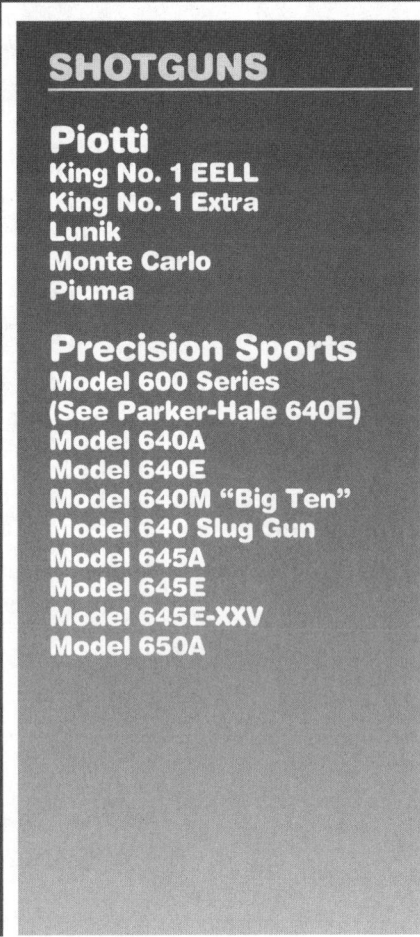

Piotti King No. 1

Piotti Piuma

Piotti Lunik

Piotti King No. 1 EELL
Same specs as King No. 1 except highest quality wood, metal work; engraved scenes, gold inlays; signed by master engraver. Made in Italy. Introduced 1983; no longer imported. Extremely rare, precludes pricing.

Piotti King No. 1 Extra
Same specs as King No. 1 except exhibition grade wood; better metal work; Bulino game scene engraving or game scene engraving with gold inlay work signed by master engraver. Made in Italy. Still in production. **LMSR: $22,500**
Perf.: $20,000 **Exc.:** $12,500 **VGood:** $10,500
With single trigger
Perf.: $21,000 **Exc.:** $13,500 **VGood:** $11,500

PIOTTI LUNIK
Side-by-side; H&H sidelock; 12-, 16-, 20- 28-ga., .410; 25" to 30" (12-ga.), 25" to 28" (other gauges) barrels; fine-figured straight-grip European walnut stock, split or beavertail forend; oil or satin luster finish; double triggers; optional single non-selective trigger; Renaissance-style scroll engraving, gold crown on top lever, gold name and gold crest in forend; demi-block barrels. Made in Italy. Introduced 1983; still in production.
LMSR: $20,000
Perf.: $18,000 **Exc.:** $11,000 **VGood:** $9500
With single trigger
Perf.: $19,000 **Exc.:** $12,000 **VGood:** $10,500

PIOTTI MONTE CARLO
Side-by-side; H&H sidelock; 12-, 16-, 20- 28-ga., .410; 25" to 30" (12-ga.), 25" to 28" (other gauges) barrels; fine-figured straight-grip European walnut stock, split or beavertail forend; oil or satin luster finish; double triggers; optional single non-selective trigger; ejectors; Purdey-type engraving. Made in Italy. Introduced 1983; discontinued 1989.
Perf.: $9500 **Exc.:** $8000 **VGood:** $6800
With single trigger
Perf.: $10,500 **Exc.:** $9000 **VGood:** $7800

PIOTTI PIUMA
Side-by-side; Anson & Deeley boxlock; 12-, 16-, 20-, 28-ga., .410; 25" to 30" (12-ga.), 25" to 28" (other gauges) barrels; file-cut rib; oil-finished, straight-grip European walnut stock, splinter type or beavertail forend; satin luster finish optional; ejectors; chopper lump barrels; double triggers, hinged front; optional single non-selective trigger; coin finish or color case-hardening; scroll, rosette, scallop engraving. Made in Italy. Introduced 1983; still in production. **LMSR: $10,800**
Perf.: $9000 **Exc.:** $7000 **VGood:** $6000
With single trigger
Perf.: $10,000 **Exc.:** $8000 **VGood:** $7000

PRECISION SPORTS MODEL 640A
Side-by-side; boxlock; 12-, 16-, 20-, 28-ga., .410; 2 3/4" chambers; 26", 28" barrels; Improved Cylinder/Modified or Modified/Full chokes; raised matte rib; hand-checkered walnut pistol-grip stock, beavertail forend with oil finish; checkered butt; buttplate; automatic safety; ejectors or extractors; single non-selective trigger; silvered, engraved action. Made in Spain by Ugartechea. Originally imported as Parker-Hale brand. Introduced 1986; discontinued 1993.
Exc.: $800 **VGood:** $650 **Good:** $500
28-ga., .410, ejectors
Perf.: $900 **Exc.:** $750 **VGood:** $600

Precision Sports Model 640E
Same specs as Model 640A except concave rib; hand-checkered walnut straight-grip stock, splinter forend with oil finish; double triggers. Made in Spain by Ugartechea. Originally imported as Parker-Hale brand. Introduced 1986; discontinued 1993.
Perf.: $700 **Exc.:** $500 **VGood:** $400
28-ga., .410, ejectors
Perf.: $800 **Exc.:** $600 **VGood:** $500

Precision Sports
Model 640M "Big Ten"
Same specs as Model 640A except chambered for 10-ga.; 26", 30", 32" barrels; Full/Full. Originally imported as Parker-Hale brand. Introduced 1986; discontinued 1993.
Perf.: $800 **Exc.:** $650 **VGood:** $550

Precision Sports
Model 640 Slug Gun
Same specs as Model 640A except 12-ga. only; 25" barrels; Improved Cylinder/Improved Cylinder. Originally imported as Parker-Hale brand. Introduced 1986; discontinued 1993.
Perf.: $950 **Exc.:** $700 **VGood:** $600

PRECISION SPORTS MODEL 645A
Side-by-side; boxlock; 12-, 16-, 20-, 28-ga., .410; 2 3/4" chambers; 26", 28" barrels; Improved Cylinder/Modified or Modified/Full; raised matte rib; hand-checkered walnut pistol-grip stock, beavertail forend with oil finish; checkered butt; buttplate; automatic safety; ejectors; single non-selective trigger; silvered, engraved action. Made in Spain by Ugartechea. Originally imported as Parker-Hale brand. Introduced 1986; discontinued 1993.
Perf.: $1100 **Exc.:** $750 **VGood:**600
28-ga., .410, ejectors
Perf.: $1200 **Exc.:** $850 **VGood:** $700

Precision Sports Model 645E
Same specs as Model 645A except concave rib; hand-checkered walnut straight-grip stock, splinter forend with oil finish; double triggers. Made in Spain by Ugartechea. Originally imported as Parker-Hale brand. Introduced 1986; discontinued 1993.
Perf.: $900 **Exc.:** $650 **VGood:** $550
28-ga., .410, ejectors
Perf.: $980 **Exc.:** $730 **VGood:** $630

Precision Sports Model 645E-XXV
Same specs as Model 645A except 25" barrels; Churchill-type rib. Originally imported as Parker-Hale brand. Introduced 1986; discontinued 1993.
Perf.: $900 **Exc.:** $650 **VGood:** $550
28-ga., .410, ejectors
Perf.: $980 **Exc.:** $730 **VGood:** $630

PRECISION SPORTS MODEL 650A
Side-by-side; boxlock; 12-ga.; 2 3/4" chambers; 26", 28" barrels; Improved Cylinder/Modified choke tubes; raised matte rib; hand-checkered walnut pistol-grip stock, beavertail forend with oil finish; checkered butt; buttplate; automatic safety; extractors; single non-selective trigger; silvered, engraved action. Made in Spain by Ugartechea. Originally imported as Parker-Hale brand. Introduced 1986; discontinued 1993.
Perf.: $800 **Exc.:** $650 **VGood:** $550

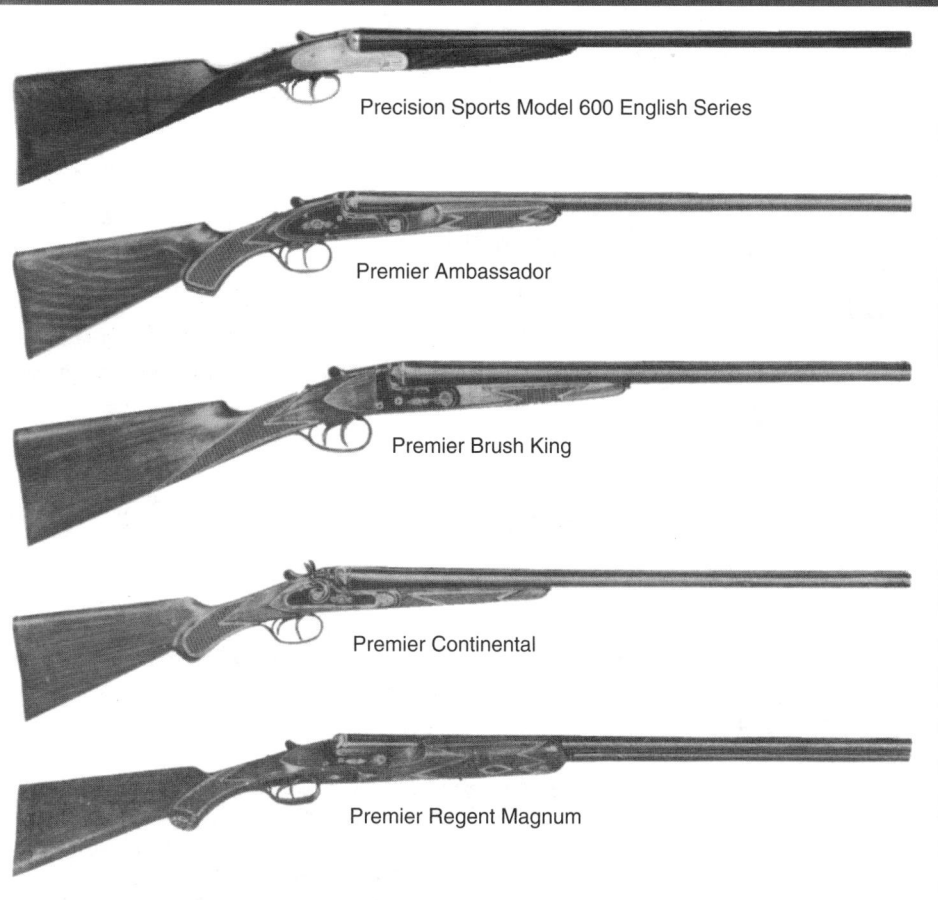

Precision Sports Model 600 English Series

Premier Ambassador

Premier Brush King

Premier Continental

Premier Regent Magnum

SHOTGUNS

Precision Sports
Model 650E
Model 655A
Model 655E
Model 800 American Series

Premier
Ambassador
Brush King
Continental
Monarch Supreme
Presentation Custom Grade
Regent
Regent Magnum

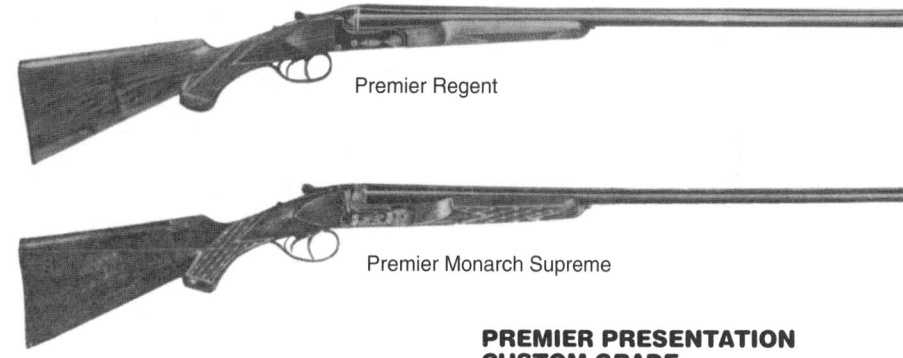

Premier Regent

Premier Monarch Supreme

Precision Sports Model 650E
Same specs as Model 650A except concave rib; hand-checkered walnut straight-grip stock, splinter forend with oil finish; double triggers. Made in Spain by Ugartechea. Originally imported as Parker-Hale brand. Introduced 1986; discontinued 1993.
   Perf.: $800   **Exc.:** $650   **VGood:** $550

PRECISION SPORTS MODEL 655A
Side-by-side; boxlock; 12-ga.; 2 3/4″ chambers; 28″ barrels; Improved Cylinder/Modified choke tubes; concave rib; hand-checkered walnut pistol-grip stock, beavertail forend with oil finish; automatic safety; ejectors; checkered butt; buttplate; single non-selective trigger; silvered, engraved action. Made in Spain by Ugartechea. Originally imported as Parker-Hale brand. Introduced 1986; discontinued 1993.
   Perf.: $900   **Exc.:** $700   **VGood:** $600

Precision Sports Model 655E
Same specs as Model 655A except concave rib; hand-checkered walnut straight-grip stock, splinter forend with oil finish; double triggers. Made in Spain by Ugartechea. Originally imported as Parker-Hale brand. Introduced 1986; discontinued 1993.
   Perf.: $900   **Exc.:** $700   **VGood:** $600

PRECISION SPORTS MODEL 800 AMERICAN SERIES
Side by side; boxlock; 10-, 12-, 16-, 20-, 28-ga., .410; 2 3/4″, 3″, 3 1/2″ chambers depending upon gauge; 25″, 26″, 27″, 28″ barrels; raised matte rib; hand-checkered European walnut pistol-grip stock, beavertail forend; buttplate; auto safety; ejectors or extractors; single non-selective trigger; engraved, silvered frame. Made in Spain. Introduced 1986; discontinued 1993.
   Perf.: $1000   **Exc.:** $750   **VGood:** $650

PREMIER AMBASSADOR
Side-by-side; 12-, 16-, 20-ga., .410; 22″ (except .410), 26″ barrels; Modified/Full chokes; European walnut stock; hand-checkered pistol-grip stock, forearm; double triggers; cocking indicators; automatic safety. Made in Europe. Introduced 1957; discontinued 1981.
   Perf.: $350   **Exc.:** $300   **VGood:** $275

PREMIER BRUSH KING
Side-by-side; 12-, 20-ga.; 22″ Improved/Modified barrels; matted tapered rib; hand-checkered European walnut pistol-grip stock, forearm; double triggers; automatic safety. Introduced 1959; discontinued 1981.
   Perf.: $250   **Exc.:** $200   **VGood:** $180

PREMIER CONTINENTAL
Side-by-side; exposed hammers; 12-, 16-, 20-ga.; 22″, 26″ barrels; Modified/Full chokes; hand-checkered European walnut pistol-grip stock, forearm; double triggers; cocking indicators; automatic safety. Made in Europe. No longer in production.
   Perf.: $225   **Exc.:** $190   **VGood:** $170

PREMIER MONARCH SUPREME
Side-by-side; boxlock; 12-, 20-ga.; 2 3/4″ (12-ga.), 3″ (20-ga.) chambers; 26″ Improved Cylinder/Modified, 28″ Modified/Full barrels; checkered fancy European walnut pistol-grip stock, beavertail forearm; double triggers; auto ejectors. Introduced 1959; discontinued 1981.
   Perf.: $400   **Exc.:** $355   **VGood:** $300

PREMIER PRESENTATION CUSTOM GRADE
Side-by-side; boxlock; 12-, 20-ga.; 2 3/4″ (12-ga.), 3″ (20-ga.) chambers; 26″ Improved Cylinder/Modified, 28″ Modified/Full barrels; hand-checkered high-grade European walnut pistol-grip stock, beavertail forearm; engraved hunting scene; gold and silver inlays; custom-order gun. Introduced 1959; discontinued 1981.
   Perf.: $1050   **Exc.:** $800   **VGood:** $725

PREMIER REGENT
Side-by-side; 12-, 16-, 20-, 28-ga., .410; 26″ Improved/Modified, Modified/Full, 28″ Modified/Full barrels; matted tapered rib; hand-checkered European walnut pistol-grip stock, forearm; double triggers; automatic safety. Introduced 1955; discontinued 1981.
   Perf.: $260   **Exc.:** $210   **VGood:** $180

Premier Regent Magnum
Same specs as Regent except 10-, 12-ga.; 3″, 3 1/2″ chambers; 30″, 32″ Full/Full barrels; hand-checkered European walnut pistol-grip stock, beavertail forearm; recoil pad. Discontinued 1981.
10-ga.
   Perf.: $300   **Exc.:** $280   **VGood:** $250
12-ga.
   Perf.: $290   **Exc.:** $260   **VGood:** $230

SHOTGUNS

Purdey
Side-By-Side
Over/Under
Single Barrel Trap

Remington Semi-Automatic Shotguns
Model 11A
Model 11B Special
Model 11D Tournament
Model 11E Expert
Model 11F Premier
Model 11R Riot Gun
Model 11-48A
Model 11-48A Riot
Model 11-48B Special
Model 11-48D Tournament
Model 11-48F Premier
Model 11-48RSS Slug Gun
Model 11-48SA Skeet

Purdey Over/Under

Purdey Single Barrel Trap

Remington Model 11A

Remington Model 11-48A

Remington Model 11-48RSS Slug

Remington Model 11-48SA Skeet

PURDEY SIDE-BY-SIDE
Side-by-side; sidelock; 12-, 16-, 20-, 28-ga., .410-ga.; 26", 27", 28", 30" barrels; 2 1/2, 2 3/4, or 3"; any choke combo; choice of rib style; hand-checkered European walnut straight-grip stock, forearm; pistol-grip stock on special order; double triggers or single trigger; automatic ejectors; made in several variations including Game Model, Featherweight Game, Pigeon Gun, with side clips. Prices are identical for all. Introduced 1880; still in production. **LMSR: $44,275**
With double triggers
  Perf.: $44,000 **Exc.:** $35,000 **VGood:** $30,000
With single trigger
  Perf.: $45,000 **Exc.:** $36,000 **VGood:** $31,000

PURDEY OVER/UNDER
Over/under; sidelock; pre-WWII guns are built on Purdey action, post-war versions on Woodward action; 12-, 16-, 20-, 28-ga., .410-ga.; 26", 27", 28" barrels; any choke combo; any rib style to customer's preference; hand-checkered European walnut straight or pistol-grip stock, forearm; double or single trigger; engraved receiver. Introduced 1925; still in production. **LMSR: $56,210**
With Purdey action, double triggers
  Perf.: $50,000 **Exc.:** $36,000 **VGood:** $33,000
With Purdey action, single trigger
  Perf.: $52,000 **Exc.:** $38,000 **VGood:** $35,000
With Woodward action, double trigger
  Perf.: $56,000 **Exc.:** $40,000 **VGood:** $37,000
With Woodward action, single trigger
  Perf.: $60,000 **Exc.:** $44,000 **VGood:** $40,000

PURDEY SINGLE BARREL TRAP
Single shot; Purdey action; 12-ga.; barrel length, choke to customer's specs; vent rib; hand-checkered European walnut straight or pistol-grip stock, forearm; engraved receiver. Introduced 1917; discontinued prior to WWII.
  Perf.: $12,500 **Exc.:** $9000 **VGood:** $8200

REMINGTON SEMI-AUTOMATIC SHOTGUNS

REMINGTON MODEL 11A
Autoloader; hammerless; takedown; 12-, 16-, 20-ga.; 5-shot tube magazine; 26", 28", 30", 32" barrel; Full, Modified, Improved Cylinder, Skeet chokes; plain, solid or vent rib; checkered pistol-grip stock, forearm. Introduced 1905; discontinued 1949.
Plain barrel
  Exc.: $200 **VGood:** $170 **Good:** $140
Solid rib
  Exc.: $210 **VGood:** $180 **Good:** $150
Vent rib
  Exc.: $300 **VGood:** $250 **Good:** $210

Remington Model 11B Special
Same specs as Model 11A except higher grade of walnut, checkering and engraving. Discontinued 1948.
  Exc.: $350 **VGood:** $295 **Good:** $250

Remington Model 11D Tournament
Same specs as Model 11A except select grade of walnut, checkering and engraving. Discontinued 1948.
  Exc.: $700 **VGood:** $450 **Good:** $400

Remington Model 11E Expert
Same specs as Model 11A except fine grade of walnut, checkering and engraving. Discontinued 1948.
  Exc.: $800 **VGood:** $600 **Good:** $500

Remington Model 11F Premier
Same specs as Model 11A except best grade of walnut, checkering and engraving. Discontinued 1948.
  Exc.: $1200 **VGood:** $900 **Good:** $700

Remington Model 11R Riot Gun
Same specs as Model 11A except 12-ga.; special 20" barrel; sling swivels. Introduced 1921; discontinued 1948.
  Exc.: $300 **VGood:** $250 **Good:** $210

REMINGTON MODEL 11-48A
Autoloader; hammerless; takedown; 12-, 16-, 20-, 28-ga., .410; 4-, 5-shot tube magazine; 26" Improved Cylinder, 28" Modified or Full, 30" Full barrels; plain matted or vent-rib barrel; hand-checkered half-pistol-grip stock, forend; redesigned version of Model 11. Introduced 1949; discontinued 1969.
Plain barrel
  Exc.: $300 **VGood:** $240 **Good:** $195
Vent-rib barrel
  Exc.: $450 **VGood:** $350 **Good:** $300

Remington Model 11-48A Riot
Same specs as Model 11-48A except 12-ga.; 20" plain barrel. Introduced 1949; discontinued 1969.
  Exc.: $300 **VGood:** $240 **Good:** $195

Remington Model 11-48B Special
Same specs as Model 11-48A except higher grade of wood, checkering and engraving. Introduced 1949; discontinued 1969.
  Exc.: $325 **VGood:** $265 **Good:** $220

Remington Model 11-48D Tournament
Same specs as Model 11-48A except select grade of wood, checkering and engraving. Introduced 1949; discontinued 1969.
  Exc.: $375 **VGood:** $300 **Good:** $260

Remington Model 11-48F Premier
Same specs as Model 11-48A except best grade of wood, checkering and engraving. Introduced 1949; discontinued 1969.
  Exc.: $400 **VGood:** $320 **Good:** $280

Remington Model 11-48 RSS Slug Gun
Same specs as Model 11-48 except 12-ga. slug; 26" plain barrel; adjustable rifle-type gold bead front sight, step-adjustable rear. Introduced 1959; no longer in production.
  Exc.: $350 **VGood:** $265 **Good:** $200

Remington Model 11-48SA Skeet
Same specs as Model 11-48A except 25" vent-rib barrel; Skeet choke; Skeet-style walnut stock, forend. Introduced 1952; discontinued 1969.
  Exc.: $550 **VGood:** $480 **Good:** $400

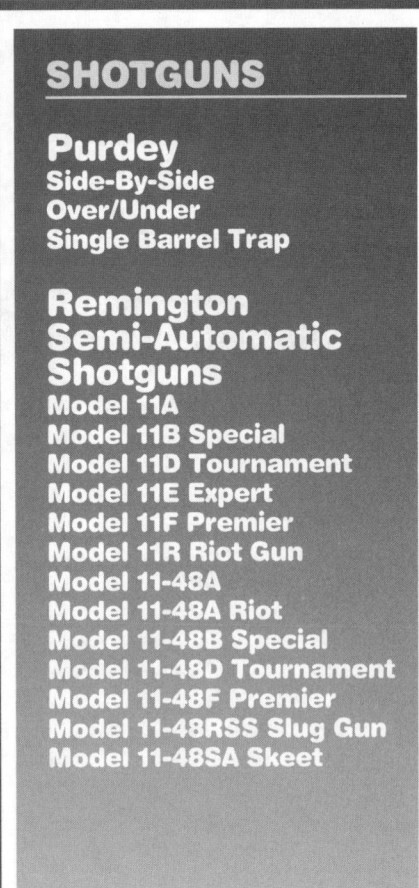

Remington Model 11-87 Premier

Remington Model 11-87 Premier Special Purpose Magnum

Remingtom Model 11-87 Premier Special Purpose Deer Gun

Remington Model 11-87 Premier SPS-BG-Camo Deer/Turkey

Remington Model 11-87 Premier SPS-Deer

Remington Model 11-87 Premier SPS-T Camo

SHOTGUNS

Remington Semi-Automatic Shotguns
Model 11-87 Premier
Model 11-87 Premier 175th Anniversary
Model 11-87 Premier N.W.T.F
Model 11-87 Premier Skeet
Model 11-87 Premier Special Purpose Deer Gun
Model 11-87 Premier Special Purpose Magnum
Model 11-87 Premier Special Purpose Synthetic Camo
Model 11-87 Premier Sporting Clays
Model 11-87 Premier SPS-BG-Camo Deer/Turkey
Model 11-87 Premier SPS Cantilever
Model 11-87 Premier SPS-Deer
Model 11-87 Premier SPS-T Camo

REMINGTON MODEL 11-87 PREMIER
Gas-operated autoloader; 12-ga.; 3″ chamber; 26″, 28″, 30″ vent-rib barrel; Rem-Choke tubes; weighs 8 1/4 lb.; metal bead middle sight, Bradley-type white-faced front; high-gloss or satin-finished, cut-checkered walnut stock, pinned forend; brown buttpad; pressure compensating system handles 2 3/4″ or 3″ shells; stainless steel magazine tube; barrel support ring on operating bars; left- or right-hand versions. Introduced, 1987; still in production.
LMSR: $644
Right-hand model
    Perf.: $500   **Exc.:** $350   **VGood:** $260
Left-hand model
    Perf.: $550   **Exc.:** $400   **VGood:** $310

Remington Model 11-87 Premier 175th Anniversary
Same specs as Model 11-87 Premier except 28″ vent-rib barrel; walnut wood with high-gloss finish; receiver engraved with Remington's 175th anniversary scroll design with an American eagle. Introduced 1991; discontinued 1991.
    Perf.: $500   **Exc.:** $350   **VGood:** $260

Remington Model 11-87 Premier N.W.T.F.
Same specs as Model 11-87 Premier except 21″ vent-rib barrel; Rem-Choke Improved Cylinder and Turkey Extra-Full tubes; synthetic brown Trebark camouflage-finished stock; camo sling; swivels. Introduced 1992; discontinued 1993.
    Perf.: $525   **Exc.:** $375   **VGood:** $285

Remington Model 11-87 Premier Skeet
Same specs as Model 11-87 Premier except 12-ga.; 2 3/4″ chamber; 26″ vent-rib barrel; cut-checkered, deluxe walnut skeet-style stock with satin finish; two-piece buttplate, Introduced 1987; still in production. **LMSR: $700**
    Perf.: $600   **Exc.:** $420   **VGood:** $320

Remington Model 11-87 Premier Special Purpose Deer Gun
Same specs as Model 11-87 Premier except 21″ barrel; rifled and Improved Cylinder choke tubes; rifle sights; cantilever scope mount, rings; gas system handles all 2 3/4″ and 3″ slug, buckshot, high-velocity field and magnum loads; not designed to function with light 2 3/4″ field loads; dull stock finish, Parkerized exposed metal surfaces; bolt and carrier have blackened color. Introduced 1987; discontinued 1995.
    Perf.: $500   **Exc.:** $350   **VGood:** $260

Remington Model 11-87 Premier Special Purpose Magnum
Same specs as Model 11-87 Premier except black synthetic or satin wood stock; Parkerized metal finish; blackened bolt, carrier; quick-detachable sling swivels; camo padded nylon sling. Introduced 1987; still in production. **LMSR: $625**
    Perf.: $500   **Exc.:** $350   **VGood:** $260

Remington Model 11-87 Premier Special Purpose Synthetic Camo
Same specs as Model 11-87 Premier except 26″ vent-rib barrel; synthetic stock; camo sling; swivels; all surfaces (except bolt and trigger guard) covered with Mossy Oak Bottomland camo finish. Introduced 1992; still in production. **LMSR: $705**
    Perf.: $600   **Exc.:** $420   **VGood:** $320

Remington Model 11-87 Premier Sporting Clays
Same specs as Model 11-87 Premier except 12-ga.; 2 3/4″ chamber; 26″, 28″ medium-height vent-rib Light Contour barrel; Skeet, Improved Cylinder, Modified, Full long Rem-Choke tubes; ivory bead front sight; special stock dimensions; shortened magazine tube and forend; Sporting Clays buttpad; lengthened forcing cone; competition trigger; top of receiver, barrel and rib matte finished; comes in two-barrel fitted hard case. Introduced 1992; still in production. **LMSR: $725**
    Perf.: $625   **Exc.:** $450   **VGood:** $350

Remington Model 11-87 Premier SPS-BG-Camo Deer/Turkey
Same specs as Model 11-87 Premier except 21″ barrel; Improved Cylinder, Super-Full Turkey (.665″ diameter with knurled extension) Rem-Choke tubes and rifled choke tube insert; rifle sights; synthetic stock, forend; quick-detachable swivels; camo Cordura carrying sling; all surfaces Mossy Oak Bottomland camouflage-finished. Introduced 1993; discontinued 1995.
    Perf.: $525   **Exc.:** $365   **VGood:** $300

Remington Model 11-87 Premier SPS Cantilever
Same specs as Model 11-87 Premier except 20″ barrel; Improved Cylinder, 3 1/2″ rifled Rem-Choke tubes; cantilever scope mount; synthetic Monte Carlo stock; sling; swivels. Introduced 1994; still in production.
LMSR: $706
    Perf.: $600   **Exc.:** $420   **VGood:** $320

Remington Model 11-87 Premier SPS-Deer
Same specs as Model 11-87 Premier except fully-rifled 21″ barrel; rifle sights; black non-reflective synthetic stock, forend; black carrying sling. Introduced 1993; still in production. **LMSR: $647**
    Perf.: $525   **Exc.:** $375   **VGood:** $290

Remington Model 11-87 Premier SPS-T Camo
Same specs as Model 11-87 Premier except 21″ vent-rib barrel; Improved Cylinder, Super-Full Turkey (.665″ diameter with knurled extension) Rem-Choke tubes; synthetic stock; all surfaces Mossy Oak Green Leaf camouflage-finished; non-reflective black bolt body, trigger guard and recoil pad. Introduced 1993; still in production. **LMSR: $718**
    Perf.: $620   **Exc.:** $440   **VGood:** $340

SHOTGUNS

Remington Semi-Automatic Shotguns
Model 11-87 Premier Trap
Model 878A Automaster
Model 1100D Tournament
Model 1100 Deer
Model 1100
Deer Special Purpose
Model 1100
Deer Cantilever 20-Gauge
Model 1100F Premier
Model 1100F Premier Gold
Model 1100 Field
Model 1100 Field
Collectors Edition
Model 1100 Special Field
Model 1100
Lightweight Magnum
Model 1100 LT-20

Remington Model 878A Automaster

Remington Model 1100D Tournament

Remington Model 1100 Deer

Remington Model 1100F Premier

Remington Model 1100 Field

Remington Model 1100 Special Field

Remington Model 11-87 Premier Trap
Same specs as Model 11-87 Premier except 12-ga.; 2 ³/₄" chamber; 30" vent-rib barrel; checkered deluxe walnut trap-style stock with satin-finish; straight or Monte Carlo comb; right- or left-hand models. Introduced 1987; still in production.
LMSR: $725
Right-hand model
   Perf.: $625   **Exc.:** $450   **VGood:** $345
Left-hand model
   Perf.: $675   **Exc.:** $500   **VGood:** $395

REMINGTON MODEL 878A AUTOMASTER
Gas-operated autoloader; 12-ga.; 3-shot tube magazine; plain or vent rib; 26" Improved Cylinder; 28" Modified, 30" Full choke barrels; uncheckered pistol-grip stock, forend. Introduced 1959; discontinued 1962.
Plain barrel
   Exc.: $190   **VGood:** $160   **Good:** $130
Vent-rib barrel
   Exc.: $215   **VGood:** $185   **Good:** $155

REMINGTON MODEL 1100D TOURNAMENT
Gas-operated autoloader; hammerless; takedown; 12-, 16-, 20-, 28-ga., .410; 5-shot magazine; 26", 28", 30" vent-rib barrel; fixed or choke tubes; high-grade checkered walnut pistol-grip stock, forend; buttplate; engraving; custom-order gun. Introduced 1963; still in production. **LMSR: $2610**
   Perf.: $2000   **Exc.:** $1100   **VGood:** $800

REMINGTON MODEL 1100 DEER
Gas-operated autoloader; hammerless; takedown; 12-, 20-ga.; 5-shot magazine; 20", 21", 22" Improved Cylinder barrel; rifle sights; checkered walnut pistol-grip stock, forend; recoil pad. Still in production.
LMSR: $565
   Perf.: $400   **Exc.:** $290   **VGood:** $235

Left-hand model
   Perf.: $450   **Exc.:** $340   **VGood:** $285

Remington Model 1100 Deer Special Purpose
Same specs as Model 1100 Deer except matte finish on checkered walnut pistol-grip stock, forend; matte metal finish. Introduced 1986; discontinued 1986.
   Perf.: $325   **Exc.:** $280   **VGood:** $240

Remington Model 1100 Deer Cantilever 20-Gauge
Same specs as Model 1100 Deer except 20-ga.; fully rifled 21" slug barrel; cantilever scope mount; sling; swivels. Introduced 1994; still in production. **LMSR: $682**
   Perf.: $500   **Exc.:** $350   **VGood:** $260

REMINGTON MODEL 1100F PREMIER
Gas-operated autoloader; hammerless; takedown; 12-, 16-, 20-, 28-ga., .410; 5-shot magazine; 26", 28", 30" vent-rib barrel; fixed or choke tubes; best-grade checkered walnut pistol-grip stock, forend; buttplate; engraving; custom-order gun. Introduced 1963; still in production. **LMSR: $5377**
   Perf.: $4200   **Exc.:** $2650   **VGood:** $2000

Remington Model 1100F Premier Gold
Same specs as Model 110F Premier except gold inlays. Introduced 1963; still in production. **LMSR: $8062**
   Perf.: $6800   **Exc.:** $3200   **VGood:** $2600

REMINGTON MODEL 1100 FIELD
Gas-operated autoloader; hammerless; takedown; 12-, 16-, 20-ga.; 5-shot magazine; 26", 28", 30" barrel; fixed or choke tubes; plain or vent rib; checkered walnut pistol-grip stock, forend; buttplate. Introduced 1963; discontinued 1988.

Plain barrel
   Perf.: $260   **Exc.:** $190   **VGood:** $150
Vent-rib barrel
   Perf.: $310   **Exc.:** $240   **VGood:** $200

Remington Model 1100 Field Collectors Edition
Same specs as Model 1100 Field except positive cut-checkered, richly figured walnut stock; deep-relief etching; gold highlights; marketed with certificate showing serial number. Only 3000 made. Introduced 1981; discontinued 1981.
   Perf.: $1000   **Exc.:** $650   **VGood:** $550

Remington Model 1100 Special Field
Same specs as Model 1100 Field except 12-, 20-ga., .410; 23" vent-rib barrel; fixed or choke tubes; checkered walnut straight-grip stock, shortened forend; matte-finished receiver; no engraving. Introduced 1983; still in production. **LMSR: $605**
   Perf.: $500   **Exc.:** $340   **VGood:** $260

Remington Model 1100 Lightweight Magnum
Same specs as Model 1100 Field except 20-ga.; 3" chamber; lightweight frame. Introduced 1977; no longer in production.
Plain barrel
   Perf.: $395   **Exc.:** $260   **VGood:** $220
Vent rib
   Perf.: $425   **Exc.:** $290   **VGood:** $250

Remington Model 1100 LT-20
Same specs as Model 1100 Field except 20-ga.; 2 ³/₄" chamber; 26", 28" plain barrel; 6¹/₂ lbs.; satin or high-gloss mahogany pistol-grip stock, forend; lightweight receiver. Still in production. **LMSR: $605**
   Perf.: $500   **Exc.:** $340   **VGood:** $260

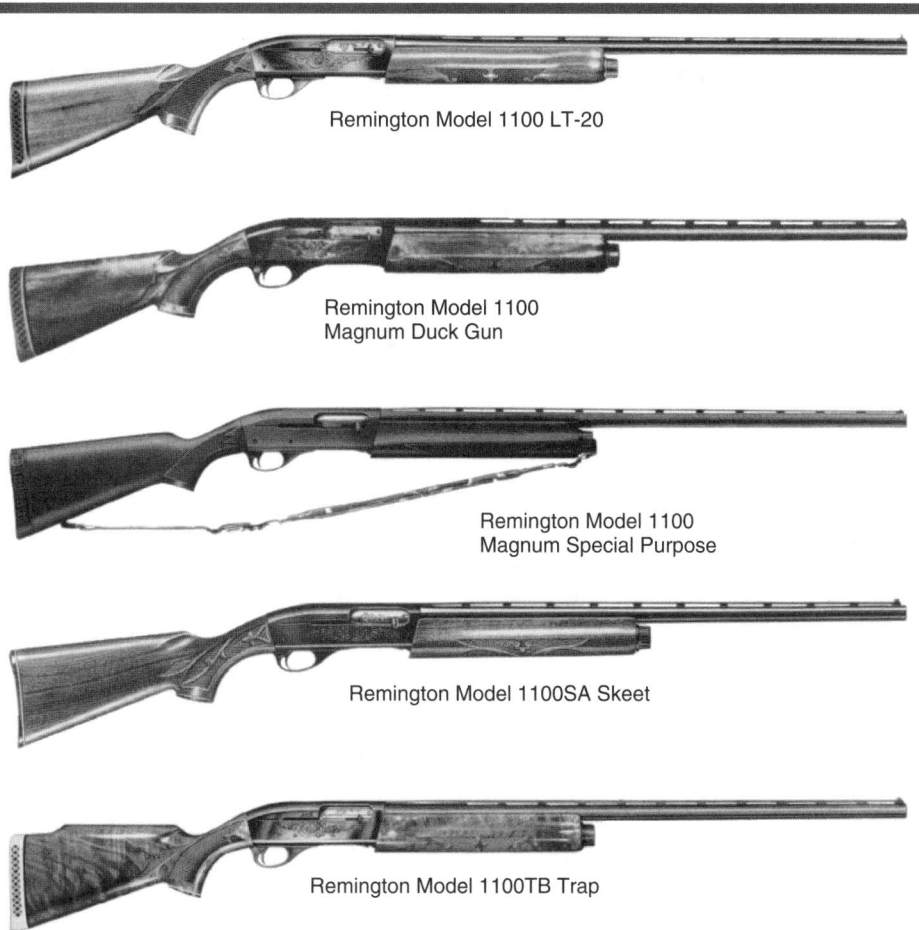

Remington Model 1100 LT-20

Remington Model 1100
Magnum Duck Gun

Remington Model 1100
Magnum Special Purpose

Remington Model 1100SA Skeet

Remington Model 1100TB Trap

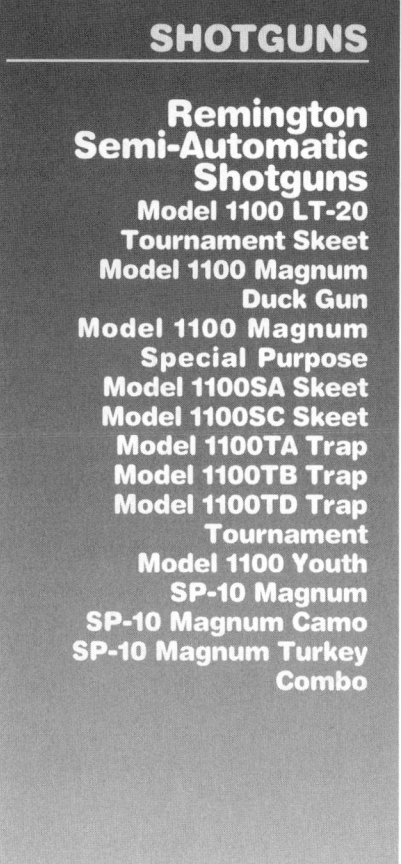

SHOTGUNS

Remington Semi-Automatic Shotguns
Model 1100 LT-20 Tournament Skeet
Model 1100 Magnum Duck Gun
Model 1100 Magnum Special Purpose
Model 1100SA Skeet
Model 1100SC Skeet
Model 1100TA Trap
Model 1100TB Trap
Model 1100TD Trap Tournament
Model 1100 Youth
SP-10 Magnum
SP-10 Magnum Camo
SP-10 Magnum Turkey Combo

Remington Model 1100 LT-20 Tournament Skeet

Same specs as Model 1100 except 20-, 28-ga., .410; 26" plain barrel; special Skeet choke; vent rib; ivory bead front, metal bead middle sights. No longer in production.

Perf.: $500   **Exc.: $380**   **VGood: $300**

Remington Model 1100 Magnum Duck Gun

Same specs as Model 1100 Field except 12-, 20-ga.; 3" chamber; 28" (20-ga.), 30" (12-ga.) barrel; recoil pad. Introduced 1963; discontinued 1988.
Plain barrel

Perf.: $365   **Exc.: $240**   **VGood: $200**
Vent-rib barrel

Perf.: $395   **Exc.: $270**   **VGood: $225**

Remington Model 1100 Magnum Special Purpose

Same specs as Model 1100 Field except 12-ga.; 3" chamber; 26", 30" vent-rib barrel; oil-finished wood; dark recoil pad; quick-detachable swivels; padded sling; chrome-lined bore; all exposed metal finished in nonreflective black. Introduced 1985; discontinued 1986.

Perf.: $350   **Exc.: $290**   **VGood: $225**

Remington Model 1100SA Skeet

Same specs as Model 1100 Field except Skeet boring; vent rib; ivory bead front sight, metal bead middle; cut checkering; new receiver scroll pattern. Introduced 1963; no longer made.

Perf.: $460   **Exc.: $375**   **VGood: $300**

Remington Model 1100SC Skeet

Same specs as Model 1100 Field except Skeet boring; vent rib; ivory bead front sight, metal bead middle; new receiver scroll pattern; selected wood. No longer made.

Perf.: $460   **Exc.: $375**   **VGood: $300**

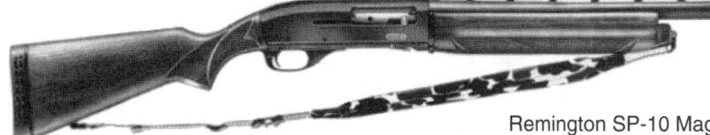

Remington SP-10 Magnum

Remington Model 1100TA Trap

Same specs as Model 1100 Field except 12-ga.; 30" Full-choke barrel; recoil pad; right- or left-hand models. Introduced 1979; discontinued 1986.

Perf.: $400   **Exc.: $320**   **VGood: $280**

Remington Model 1100TB Trap

Same specs as Model 1100 Field except 12-ga.; 30" Full-choke barrel; vent rib; trap-style stock; Monte Carlo or straight comb; recoil pad. Introduced 1963; discontinued 1981.
Straight-comb stock

Perf.: $440   **Exc.: $350**   **VGood: $290**
Monte Carlo stock

Perf.: $460   **Exc.: $370**   **VGood: $300**

Remington Model 1100TD Trap Tournament

Same specs as Model 1100 Field except 12-ga.; 30" Full-choke barrel; vent rib; select-grade trap-style stock; Monte Carlo or straight comb; recoil pad. Introduced 1979; discontinued 1986.
Straight-comb stock

Perf.: $550   **Exc.: $400**   **VGood: $350**
Monte Carlo stock

Perf.: $570   **Exc.: $420**   **VGood: $370**

Remington Model 1100 Youth

Same specs as Model 1100 Field except 20-ga.; 2 3/4" chamber; 21" barrel; youth-sized mahogany stock, forend; lightweight receiver. Still in production. **LMSR: $605**

Perf.: $475   **Exc.: $340**   **VGood: $260**

REMINGTON SP-10 MAGNUM

Autoloader; 10-ga.; 3 1/2" chamber; 3-shot magazine; 26", 30 1/2" barrel; Full, Modified Rem-Choke tubes; 47 1/2" overall length with shorter barrel; weighs about 11 lbs.; metal bead front sight; checkered satin-finished American walnut stock; brown recoil pad; padded Cordura sling; stainless steel gas system that lessens recoil; matte-finished receiver, barrel. Introduced 1989; still in production. **LMSR: $993**

Perf.: $750   **Exc.: $525**   **VGood: $400**

Remington SP-10 Magnum Camo

Same specs as Model SP-10 Magnum except 23" vent-rib barrel; Extra-Full Turkey Rem-Choke tube; mid-rib bead, Bradley-style front sight; swivel studs; quick-detachable swivels; non-slip camo Cordura carrying sling; Mossy Oak Bottomland camo-finished buttstock, forend, receiver, barrel and magazine cap; matte black bolt body and trigger guard. Introduced 1993; still in production. **LMSR: $1078**

Perf.: $850   **Exc.: $600**   **VGood: $500**

Remington SP-10 Magnum Turkey Combo

Same specs as Model SP-10 except two-barrel set with 26" or 30" vent-rib barrel and extra 22" rifle-sighted barrel; Modified, Full, Extra-Full Turkey Rem-Choke tubes; camo sling; swivels. Introduced 1991; discontinued 1995.

Perf.: $900   **Exc.: $750**   **VGood: $650**

Remington Semi-Automatic Shotguns

Sportsman Auto
Sportsman Model 48A
Sportsman Model 48B Special
Sportsman Model 48D Tournament
Sportsman Model 48F Premier
Sportsman Model 48SA Skeet
Sportsman Model 48SC Skeet
Sportsman Model 48SF Skeet Premier
Sportsman Model 58ADL
Sportsman Model 58BDL Deluxe Special
Sportsman Model 58D Tournament
Sportsman Model 58F Premier
Sportsman Model 58SA Skeet
Sportsman Model 58SC Skeet Target
Sportsman Model 58SD Skeet Tournament
Sportsman Model 58SF Skeet Premier

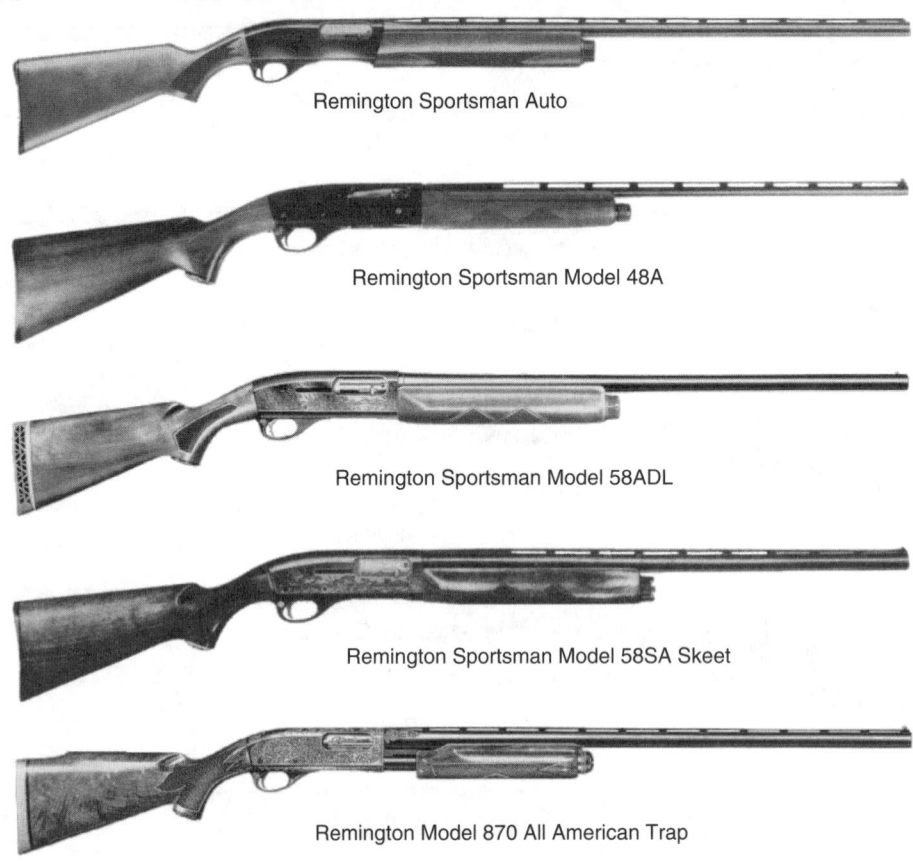

Remington Sportsman Auto

Remington Sportsman Model 48A

Remington Sportsman Model 58ADL

Remington Sportsman Model 58SA Skeet

Remington Model 870 All American Trap

REMINGTON SPORTSMAN AUTO

Autoloader; 12-ga.; 2 ¾" chamber; 5-shot tube magazine; 28", 30" vent-rib barrel; fixed or choke tubes; 7 ¾" lbs.; walnut-stained hardwood pistol-grip stock, forend; buttplate. Introduced 1985; discontinued 1986.
  Perf.: $290   **Exc.:** $240   **VGood:** $200

REMINGTON SPORTSMAN MODEL 48A

Autoloader; hammerless; takedown; 12-, 16-, 20-ga.; 3-shot tube magazine; 26" Improved Cylinder, 28" Modified or Full, 30" Full-choke barrel; plain matted or vent-rib barrel available; hand-checkered American walnut pistol-grip stock, grooved forend; streamlined receiver. Introduced 1949; discontinued 1959.
Plain barrel
  Exc.: $210   **VGood:** $165   **Good:** $130
Vent-rib barrel
  Exc.: $240   **VGood:** $195   **Good:** $160

Remington Sportsman Model 48B Special

Same specs as Model 48A except high-quality wood, checkering; engraved receiver. Introduced 1949; discontinued 1959.
  Exc.: $300   **VGood:** $250   **Good:** $210

Remington Sportsman Model 48D Tournament

Same specs as 48A except improved wood; finer checkering; more engraving. Introduced 1949; discontinued 1959.
  Exc.: $650   **VGood:** $550   **Good:** $490

Remington Sportsman Model 48F Premier

Same specs as 48A except top-quality wood; fully engraved receiver. Introduced 1949; discontinued 1959.
  Exc.: $1600   **VGood:** $1000   **Good:** $800

REMINGTON SPORTSMAN MODEL 48SA SKEET

Autoloader; hammerless; takedown; 12-, 16-, 20-ga.; 3-shot magazine; 26" Skeet-choke barrel; plain matted surface or vent rib; ivory bead front sight, metal bead rear; checkered American walnut pistol-grip stock, forend. Introduced 1949; discontinued 1960.
Plain barrel
  Exc.: $240   **VGood:** $200   **Good:** $175
Vent-rib barrel
  Exc.: $280   **VGood:** $240   **Good:** $205

Remington Sportsman Model 48SC Skeet Target

Same specs as Model 48SA Skeet except higher grade of wood, checkering and engraving. Introduced 1949; discontinued 1960.
  Exc.: $300   **VGood:** $260   **Good:** $200

Remington Model 48SF Skeet Premier

Same specs as Model 48SA Skeet except best grade of wood, checkering and engraving. Introduced 1949; discontinued 1960.
  Exc.: $1500   **VGood:** $1000   **Good:** $800

REMINGTON SPORTSMAN MODEL 58ADL

Gas-operated autoloader; 12-ga.; 3-shot tube magazine; 26", 28", 30" barrels; plain or vent rib; Improved Cylinder, Modified, Full choke, Remington Skeet boring; checkered pistol-grip stock, forend. Introduced 1956; discontinued 1964.
Plain barrel
  Exc.: $200   **VGood:** $150   **Good:** $120
Vent-rib barrel
  Exc.: $250   **VGood:** $200   **Good:** $170

Remington Sportsman Model 58BDL Deluxe Special

Same specs as Sportsman Model 58ADL except higher grade of walnut wood. Introduced 1956; discontinued 1964.
  Exc.: $230   **VGood:** $200   **Good:** $180

Remington Sportsman Model 58D Tournament

Same specs as Sportsman Model 58ADL except vent rib; select grade of wood, checkering and engraving. Introduced 1956; discontinued 1964.
  Exc.: $600   **VGood:** $500   **Good:** $430

Remington Sportsman Model 58F Premier

Same specs as Sportsman Model 58ADL except vent rib; best grade of wood, checkering and engraving. Introduced 1956; discontinued 1964.
  Exc.: $1100   **VGood:** $950   **Good:** $800

REMINGTON SPORTSMAN MODEL 58SA SKEET

Gas-operated autoloader; 12-ga.; 3-shot tube magazine; 26" vent-rib barrel; Skeet choke; special Skeet-style stock, forend. Introduced 1956; discontinued 1964.
  Exc.: $295   **VGood:** $240   **Good:** $190

Remington Sportsman Model 58SC Skeet Target

Same specs as Sportsman Model 58SA Skeet except higher grade of wood, checkering and engraving. Introduced 1956; discontinued 1964.
  Exc.: $400   **VGood:** $325   **Good:** $260

Remington Sportsman Model 58SD Skeet Tournament

Same specs as Sportsman Model 58SA Skeet except select grade of wood, checkering and engraving. Introduced 1956; discontinued 1964.
  Exc.: $605   **VGood:** $500   **Good:** $430

Remington Sportsman Model 58SF Skeet Premier

Same specs as Sportsman Model 58SA Skeet except best grade of wood, checkering and engraving. Introduced 1956; discontinued 1964.
  Exc.: $1125   **VGood:** $1025   **Good:** $950

Remington Model 10A

Remington Model 17A

Remington Model 31A

Remington Model 31S Trap Special

REMINGTON SLIDE-ACTION

REMINGTON MODEL 10A
Slide action; hammerless; takedown; 12-ga.; 6-shot tube magazine; 26", 28", 30", 32" barrel; Full, Modified, Cylinder choke; unchecked pistol-grip stock, grooved slide handle. Introduced 1907; discontinued 1929.
  Exc.: $230   **VGood:** $175   **Good:** $140

REMINGTON MODEL 17A
Slide action; hammerless; takedown; 26", 28", 30", 32" barrel; plain barrel or solid rib; Modified, Full, Cylinder choke; unchecked stock with pistol grip, grooved slide handle; Browning design. Introduced 1921; discontinued 1933.
Plain barrel
  Exc.: $210   **VGood:** $175   **Good:** $130
Solid rib
  Exc.: $300   **VGood:** $260   **Good:** $200

REMINGTON MODEL 29A
Slide action; hammerless; takedown; 12-ga.; 5-shot tubular magazine; 26", 28", 30", 32" barrel; plain barrel, solid or vent rib; Full, Modified, Cylinder choke; hand-checkered pistol-grip stock, slide handle. Introduced 1929; discontinued 1933.
Plain barrel
  Exc.: $220   **VGood:** $140   **Good:** $100
Solid rib
  Exc.: $250   **VGood:** $165   **Good:** $110
Vent rib
  Exc.: $270   **VGood:** $195   **Good:** $130

Remington Model 29S Trap Special
Same specs as Model 29A except ventilated rib; on trap-style straight-grip stock, longer slide handle. No longer in production.
  Exc.: $400   **VGood:** $325   **Good:** $290

REMINGTON MODEL 31A
Slide action; hammerless; takedown 12-, 16-, 20-ga.; 3-, 5-shot magazine; 26", 28", 30", 32" barrel; plain barrel, solid or vent rib; Full, Modified, Improved Cylinder, Skeet choke; early models had checkered pistol-grip stock, slide handle; later styles had plain stock, grooved slide handle. Introduced 1931; discontinued 1949.
Plain barrel
  Exc.: $295   **VGood:** $230   **Good:** $190
Solid rib
  Exc.: $350   **VGood:** $280   **Good:** $200
Vent rib
  Exc.: $375   **VGood:** $300   **Good:** $225

Remington Model 31B Special
Same specs as Model 31A except higher grade of wood, checkering and engraving. Introduced 1931; discontinued 1949.
  Exc.: $400   **VGood:** $350   **Good:** $300

Remington Model 31D Tournament
Same specs as Model 31A except select grade of wood, checkering and engraving. Introduced 1931; discontinued 1949.
  Exc.: $700   **VGood:** $550   **Good:** $490

Remington Model 31E Expert
Same specs as Model 31A except fine grade of wood, checkering and engraving. Introduced 1931; discontinued 1949.
  Exc.: $900   **VGood:** $700   **Good:** $600

Remington Model 31F Premier
Same specs as Model 31A except best grade of wood, checkering and engraving. Introduced 1931; discontinued 1949.
  Exc.: $1550   **VGood:** $1200   **Good:** $900

Remington Model 31H Hunter's Special
Same specs as Model 31A except 12-ga.; 30", 32" solid-rib barrel; Full choke; sporting uncheckered walnut pistol-grip stock, forend. Introduced 1931; discontinued 1949.
  Exc.: $350   **VGood:** $250   **Good:** $200

Remington Model 31R Riot Gun
Same specs as Model 31A except 12-ga.; 20" barrel. Introduced 1931; discontinued 1949.
  Exc.: $200   **VGood:** $165   **Good:** $120

Remington Model 31S Trap Special
Same specs as Model 31A except 12-ga.; 30", 32" solid-rib barrel; Full choke; trap-style uncheckered walnut half-pistol-grip stock, forend. Introduced 1931; discontinued 1949.
  Exc.: $360   **VGood:** $260   **Good:** $220

Remington Model 31 Skeet
Same specs as Model 31A except 26" barrel; solid or vent rib; Skeet choke; beavertail forend. Introduced 1931; discontinued 1949.
Solid rib
  Exc.: $375   **VGood:** $280   **Good:** $200
Vent rib
  Exc.: $450   **VGood:** $370   **Good:** $300

Remington Model 31TC Trap
Same as Model 31A except 12-ga.; 30", 32" vent-rib barrel; Full choke; trap-style checkered walnut pistol-grip stock, beavertail forend; recoil pad. Introduced 1931; discontinued 1948.
  Exc.: $475   **VGood:** $400   **Good:** $320

SHOTGUNS

Remington Slide-Action Shotguns
Model 10A
Model 17A
Model 29A
Model 29S Trap Special
Model 31A
Model 31B Special
Model 31D Tournament
Model 31E Expert
Model 31F Premier
Model 31H Hunter's Special
Model 31R Riot Gun
Model 31S Trap Special
Model 31 Skeet
Model 31TC Trap
Model 870 ADL Wingmaster
Model 870 All American Trap
Model 870AP Wingmaster
Model 870AP Wingmaster Magnum

REMINGTON MODEL 870ADL WINGMASTER
Slide action; hammerless; takedown; 12-, 16-, 20-ga.; 5-shot tube magazine; 3-shot plug furnished with gun; 26" Improved Cylinder, 28" Modified or Full, 30" Full barrel; choice of plain, matted top surface or vent rib; fine-checkered pistol-grip stock, beavertail forend. Introduced 1950; discontinued 1963.
Plain barrel
  Exc.: $170   **VGood:** $125   **Good:** $100
Vent rib
  Exc.: $200   **VGood:** $150   **Good:** $125

REMINGTON MODEL 870 ALL AMERICAN TRAP
Slide action; hammerless; takedown; 12-ga.; 5-shot tube magazine; 30" Full-choke barrel; fancy American walnut straight-comb or Monte Carlo stock, forend; custom-grade engraved receiver, trigger guard, barrel. Introduced 1972; discontinued 1976.
  Exc.: $600   **VGood:** $500   **Good:** $425

REMINGTON MODEL 870AP WINGMASTER
Slide action; hammerless; takedown; 12-, 16-, 20-ga.; 5-shot tube magazine; 3-shot plug furnished with gun; 26" Improved Cylinder, 28" Modified or Full, 30" Full barrel; choice of plain, matted top surface or vent rib; walnut stock, grooved slide handle. Introduced 1950; discontinued 1963.
Plain barrel
  Exc.: $160   **VGood:** $130   **Good:** $100
Vent rib
  Exc.: $190   **VGood:** $160   **Good:** $130

Remington Model 870AP Wingmaster Magnum
Same specs as Model 870AP Wingmaster except 12-ga.; 3" chamber; 30" Full-choke barrel; recoil pad. Introduced 1955; discontinued 1963.
Plain barrel
  Exc.: $170   **VGood:** $140   **Good:** $110
Vent rib
  Exc.: $200   **VGood:** $170   **Good:** $140

Remington Slide-Action Shotguns

Model 870AP Wingmaster Magnum Deluxe
Model 870BDL Wingmaster
Model 870 Brushmaster Deluxe
Model 870 Competition Trap
Model 870 Express
Model 870 Express Deer
Model 870 Express Deer Cantilever
Model 870 Express HD
Model 870 Express Synthetic
Model 870 Express Turkey
Model 870 Express Youth
Model 870 Field Wingmaster
Model 870 Lightweight
Model 870 Lightweight Magnum

Remington Model 870 Brushmaster Deluxe

Remington Model 870 Competition Trap

Remington Model 870 Express

Remington Model 870 Express Deer

Remington Model 870 Express HD

Remington Model 870 Express Turkey

Remington Model 870AP Wingmaster Magnum Deluxe

Same specs as Model 870AP Wingmaster except 12-ga.; 3″ chamber; 30″ Full-choke barrel; checkered stock, extension beavertail slide handle; matted top surface barrel.

Exc.: $180   **VGood:** $150   **Good:** $120

Remington Model 870BDL Wingmaster

Same specs as Model 870AP Wingmaster except select American walnut pistol-grip stock, forend. Introduced 1950; discontinued 1963.

Plain barrel
Exc.: $190   **VGood:** $155   **Good:** $125
Vent rib
Exc.: $240   **VGood:** $190   **Good:** $165

Remington Model 870 Brushmaster Deluxe

Same specs as Model 870AP Wingmaster except 12-, 20-ga.; 2 ³/₄″, 3″ chamber; 20″ barrel; Improved Cylinder or choke tube; adjustable rear sight; ramp front; satin-finished checkered wood stock, forearm; recoil pad; right- or left-hand model. Discontinued 1994.

Right-hand model
Perf.: $340   **Exc.:** $240   **VGood:** $190
Left-hand model
Perf.: $380   **Exc.:** $270   **VGood:** $220

Remington Model 870 Competition Trap

Same specs as Model 870AP Wingmaster except single shot; 12-ga.; 30″ vent-rib barrel; Full choke; select walnut trap-style stock, forend; gas reduction system to lessen recoil. Introduced 1980; discontinued 1986.
Perf.: $480   **Exc.:** $390   **VGood:** $300

REMINGTON MODEL 870 EXPRESS

Slide action; hammerless; takedown; 12-, 20-, 28-ga.; .410; 5-shot tube magazine; 3-shot plug furnished with gun; 20″, 25″, 26″, 28″ vent-rib barrel; Modified Rem-Choke tube; press-checkered, walnut-finished hardwood stock, forearm; solid black recoil pad; Parkerized metal surfaces. Introduced 1987; still in production. **LMSR: $292**
Perf.: $250   **Exc.:** $175   **VGood:** $150

Remington Model 870 Express Deer

Same specs as Model 870 Express except 12-ga.; 20″ smooth bore or rifled barrel; fixed Improved Cylinder choke; open iron sights; Monte Carlo stock. Introduced 1991; still in production. **LMSR: $287**

Smooth bore barrel
Perf.: $240   **Exc.:** $175   **VGood:** $140
Rifled barrel
Perf.: $270   **Exc.:** $205   **VGood:** $170

Remington Model 870 Express Deer Cantilever

Same specs as Model 870 Express except 12-ga.; 20″ smooth bore or rifled barrel; fixed Improved Cylinder Rem-Chokes for slugs or buckshot; barrel-mounted cantilever scope mount; scope rings; Monte Carlo stock; sling; swivels; Introduced 1991; discontinued 1992.

Smooth bore barrel
Perf.: $300   **Exc.:** $235   **VGood:** $200
Rifled barrel
Perf.: $330   **Exc.:** $265   **VGood:** $230

Remington Model 870 Express HD

Same specs as Model 870 Express except 12-ga.; 18″ Cylinder barrel; bead front sight; positive-checkered synthetic stock, forend with non-reflective black finish. Introduced 1995; still in production. **LMSR: $292**
Perf.: $250   **Exc.:** $180   **VGood:** $160

Remington Model 870 Express Synthetic

Same specs as Model 870 Express except 12-, 20-ga.; 26″, 28″ barrel; synthetic stock, forend. Made in U.S. by Remington. Introduced 1994; still in production. **LMSR: $299**
Perf.: $250   **Exc.:** $175   **VGood:** $140

Remington Model 870 Express Turkey

Same specs as Model 870 Express except 12-ga.; 3-chamber; 21″ vent-rib barrel; Extra-Full Rem-Choke turkey tube. Introduced 1991; still in production. **LMSR: $305**
Perf.: $255   **Exc.:** $180   **VGood:** $145

Remington Model 870 Express Youth

Same specs as Model 870 Express except 20-ga.; 21″ vent-rib barrel; Modified Rem-Choke; low-luster hardwood stock; 12″ length of pull. Introduced 1991; still in production. **LMSR: $292**
Perf.: $250   **Exc.:** $175   **VGood:** $140

REMINGTON MODEL 870 FIELD WINGMASTER

Slide action; hammerless; takedown; 12-, 16-, 20-, 28-ga., .410; 5-shot tube magazine; 25″ Full or Modified, 26″ Improved Cylinder, 28″ Modified or Full, 30″ Full barrel; fixed or Rem-Choke tubes; choice of plain, matted top surface or vent rib; checkered walnut stock, slide handle with high-gloss or satin finish. Introduced 1965; still in production. **LMSR: $473**

Plain barrel
Perf.: $300   **Exc.:** $200   **VGood:** $140
Vent rib
Perf.: $380   **Exc.:** $260   **VGood:** $200

REMINGTON MODEL 870 LIGHTWEIGHT

Slide action; hammerless; takedown; 20-ga.; 5-shot magazine; 23″ barrel; plain or vent rib; 5 lbs.; lightweight mahogany stock, forearm. Introduced 1972; discontinued 1983.

Plain barrel
Perf.: $265   **Exc.:** $200   **VGood:** $175
Vent rib
Perf.: $300   **Exc.:** $235   **VGood:** $210

Remington Model 870 Lightweight Magnum

Same specs as Model 870 Lightweight except 3″ chamber; 26″, 28″ barrel; fixed or choke tubes; plain or vent rib; 6 lbs.; lightweight mahogany stock, forearm. Introduced 1972; discontinued 1994.

Plain barrel
Perf.: $300   **Exc.:** $210   **VGood:** $175
Vent rib
Perf.: $350   **Exc.:** $250   **VGood:** $200

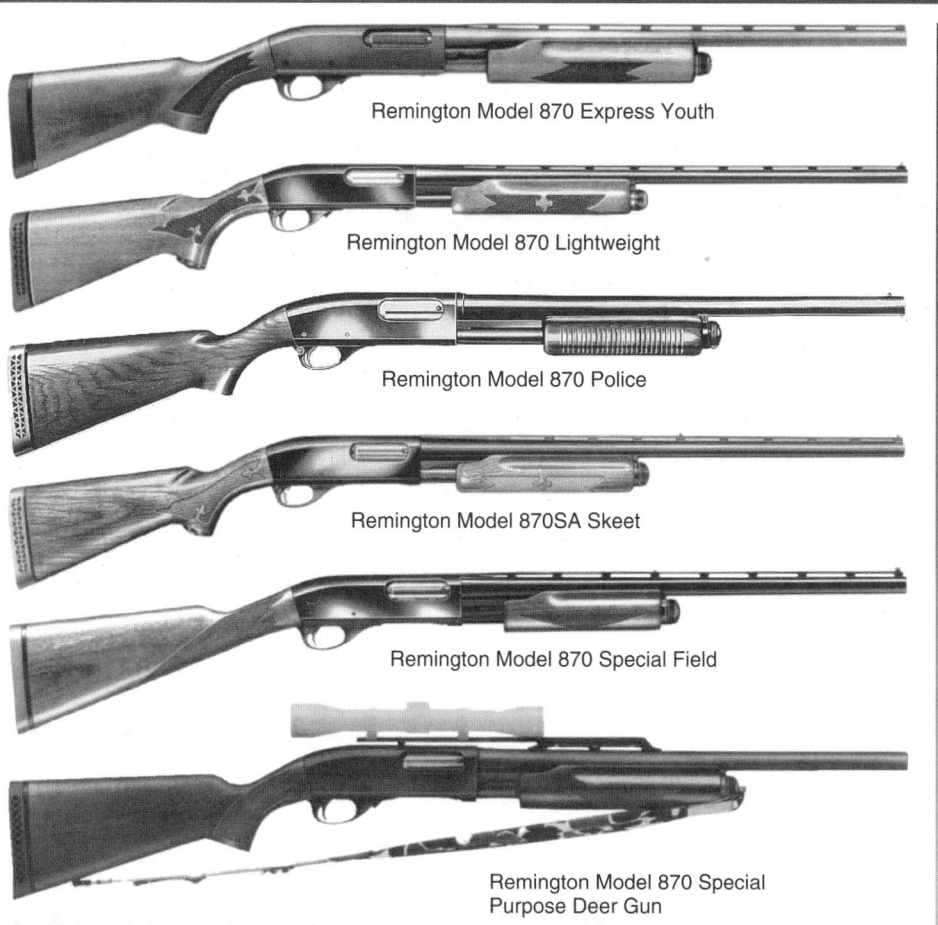

Remington Model 870 Express Youth

Remington Model 870 Lightweight

Remington Model 870 Police

Remington Model 870SA Skeet

Remington Model 870 Special Field

Remington Model 870 Special
Purpose Deer Gun

SHOTGUNS

Remington Slide-Action Shotguns
Model 870 Magnum Duck Gun
Model 870 Marine Magnum
Model 870 Police
Model 870ASA Skeet
Model 870SC Skeet Target
Model 870SD Skeet Tournament
Model 870SF Skeet Premier
Model 870 Special Field
Model 870 Special Purpose
Model 870 Special Purpose Deer Gun
Model 870 Special Purpose Synthetic Camo
Model 870 SPS-BG-Camo Deer/Turkey

Remington Model 870 Special
Purpose Synthetic Camo

REMINGTON MODEL 870 MAGNUM DUCK GUN
Slide action; hammerless; takedown; 12-, 20-ga.; 3″ chamber; 26″, 28″, 30″ vent-rib barrel; Full or Modified choke; checkered walnut stock, slide handle; recoil pad. Introduced 1964; still in production. **LMSR: $479**
Perf.: $385   **Exc.:** $265   **VGood:** $210

REMINGTON MODEL 870 MARINE MAGNUM
Slide action; hammerless; takedown; 12-ga.; 3″ chamber; 7-shot magazine; 18″ plain barrel; Cylinder choke; bead front sight; black synthetic stock, forend; electroless nickel-plated metal surfaces. Introduced 1992; still in production. **LMSR: $489**
Perf.: $400   **Exc.:** $280   **VGood:** $210

REMINGTON MODEL 870 POLICE
Slide action; hammerless; takedown; 12-ga.; 3″ chamber; 18″, 20″ Police Cylinder or 20″ Improved Cylinder; weighs 7 lbs.; metal bead front or rifle sights; lacquered hardwood stock; steel receiver; double slide bars; blued or Parkerized finish. Introduced 1994; still in production. **LMSR: $399**
Perf.: $325   **Exc.:** $250   **VGood:** $200

REMINGTON MODEL 870SA SKEET
Slide action; hammerless; takedown; 12-, 20-, 28-ga., .410; 5-shot tube magazine; 26″ vent-rib barrel; Skeet choke; ivory bead front sight, metal bead in rear; checkered pistol-grip stock, extension beavertail slide handle. Introduced 1950; discontinued 1982.
Perf.: $300   **Exc.:** $250   **VGood:** $200

Remington Model 870SC Skeet Target
Same specs as Model 870SA Skeet except higher grade of wood, checkering and engraving. Introduced 1950; discontinued 1982.
Perf.: $350   **Exc.:** $300   **VGood:** $250

Remington Model 870SD Skeet Tournament
Same specs as Model 870SA Skeet except select grade of wood, checkering and engraving. Introduced 1950; discontinued 1982.
Perf.: $375   **Exc.:** $325   **VGood:** $275

Remington Model 870SF Skeet Premier
Same specs as Model 870SA Skeet except best grade of wood, checkering and engraving. Introduced 1950; discontinued 1982.
Perf.: $400   **Exc.:** $350   **VGood:** $300

REMINGTON MODEL 870 SPECIAL FIELD
Slide action; hammerless; takedown; 12-, 20-ga.; 3″ chamber; 5-shot tube magazine; 21″, 23″ vent-rib barrel; fixed or choke tubes; checkered straight-grip stock, short forend. Introduced 1984; still in production. **LMSR: $473**
Perf.: $300   **Exc.:** $265   **VGood:** $210

REMINGTON MODEL 870 SPECIAL PURPOSE
Slide action; hammerless; takedown; 12-ga.; 3″ chamber; 26″, 28″ vent-rib barrel; Rem-Choke tubes; oil-finished wood or black synthetic stock, forend; quick detachable sling swivels; padded sling; black metal finish; chrome-lined bore. Introduced 1985; still in production. **LMSR: $399**
Synthetic stock
Perf.: $325   **Exc.:** $250   **VGood:** $200
Wood stock
Perf.: $385   **Exc.:** $310   **VGood:** $260

Remington Model 870 Special Purpose Deer Gun
Same specs as Model 870 Special Purpose except 20″ barrel; rifled and Improved Cylinder choke tubes; rifle sights or cantilever scope mount with rings; walnut Monte Carlo stock with satin finish; recoil pad; detachable camo Cordura nylon sling; black, non-glare metal finish. Introduced 1989; still in production. **LMSR: $432**
With rifle sights
Perf.: $350   **Exc.:** $250   **VGood:** $200
With scope mount and rings
Perf.: $400   **Exc.:** $300   **VGood:** $250

Remington Model 870 Special Purpose Synthetic Camo
Same specs as Model 870 Special Purpose except 26″ vent-rib barrel; Rem-Choke tubes; synthetic stock; all surfaces (except bolt and trigger guard) Mossy Oak Bottomland camo-finished; camo sling; swivels. Introduced 1992; still in production. **LMSR: $465**
Perf.: $380   **Exc.:** $260   **VGood:** $205

Remington Model 870 SPS-BG-Camo Deer/Turkey
Same specs as Model 870 Special Purpose except 20″ barrel; Improved Cylinder, Super-Full Turkey (.665″ diameter with knurled extension) Rem-Choke tubes and rifled choke tube insert; rifle sights, synthetic stock, forend; quick-detachable swivels; camo Cordura carrying sling; all surfaces Mossy Oak Bottomland camouflage-finished. Introduced 1993; discontinued 1995.
Perf.: $340   **Exc.:** $240   **VGood:** $200

SHOTGUNS

Remington Slide-Action Shotguns

Model 870 SPS Cantilever
Model 870 SPS Deer
Model 870 SPS Special Purpose Magnum
Model 870 SPS-T Camo
Model 870 SPS-T Special Purpose Magnum (See Remington Model 870 SPS Special Purpose Magnum)
Model 870TA Trap
Model 870TB Trap
Model 870TC Trap
Sportsman
Sportman Pump

Remington Single Shot Shotguns

90-T Super Single
Rider No. 3
Rider No. 9

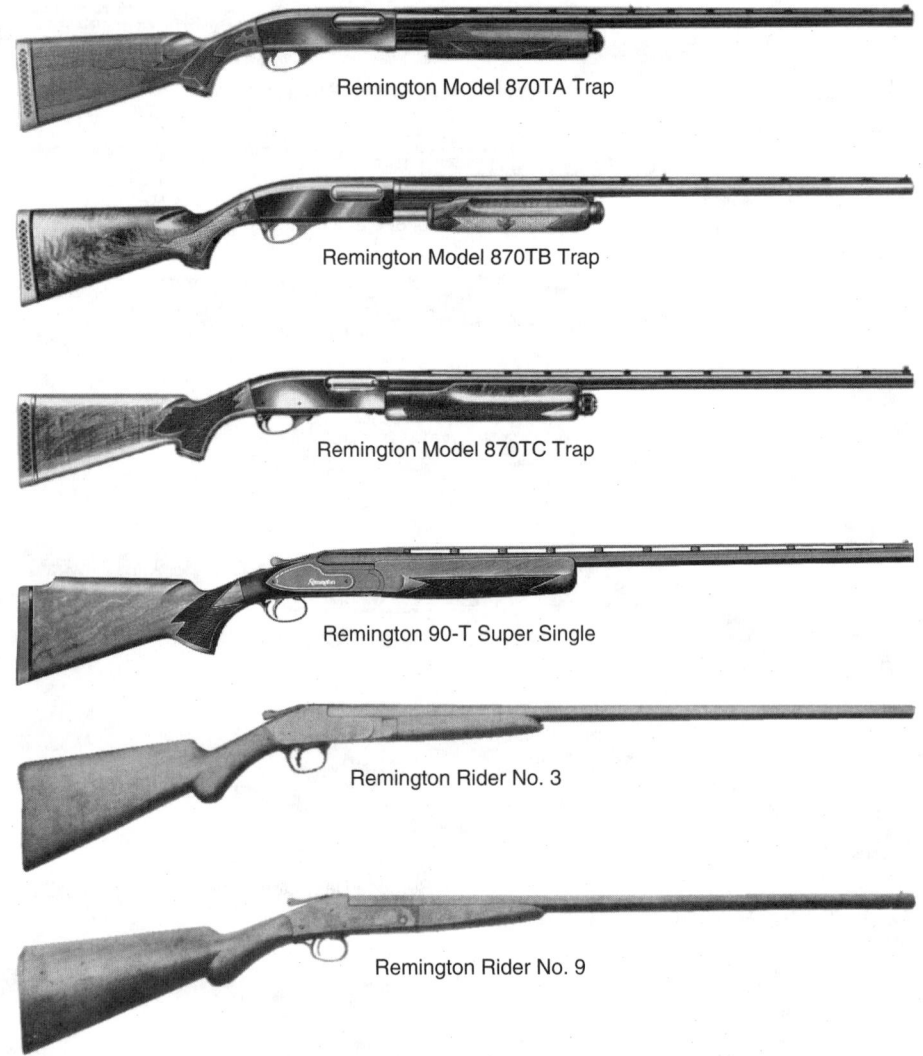

Remington Model 870TA Trap

Remington Model 870TB Trap

Remington Model 870TC Trap

Remington 90-T Super Single

Remington Rider No. 3

Remington Rider No. 9

Remington Model 870 SPS Cantilever

Same specs as Model 870 Special Purpose except 20″ smooth bore barrel; optional rifled barrel; Improved Cylinder and 3 1/2″ rifled Rem-Choke tubes; cantilever scope mount; synthetic Monte Carlo stock; sling; swivels. Introduced 1994; still in production. **LMSR:** $425

Perf.: $350	**Exc.:** $250	**VGood:** $200

With fully rifled barrel

Perf.: $400	**Exc.:** $300	**VGood:** $250

Remington Model 870 SPS Deer

Same specs as Model 870 Special Purpose except fully-rifled 20″ barrel; rifle sights; black non-reflective synthetic stock, forend; black carrying sling. Introduced 1993; still in production. **LMSR:** $407

Perf.: $340	**Exc.:** $240	**VGood:** $190

Remington Model 870 SPS Special Purpose Magnum

Same specs as Model 870 Special Purpose except 26″, 28″ barrel; Rem-Choke tubes; black synthetic stock, forend; dark recoil pad; padded Cordura 2″ wide sling; quick-detachable swivels; dull, non-reflective black metal finish. Introduced 1985; still in production as SPS-T Special Purpose Magnum. **LMSR:** $367

Perf.: $310	**Exc.:** $230	**VGood:** $190

Remington Model 870 SPS-T Camo

Same specs as Model 870 Special Purpose except 21″ vent-rib barrel; Improved Cylinder and Super-Full Turkey (.665″ diameter with knurled extension) Rem-Choke tubes; synthetic stock; Mossy Oak Green Leaf camouflage-finished; non-reflective black bolt body, trigger guard and recoil pad. Introduced 1993; still in production. **LMSR:** $479

Perf.: $390	**Exc.:** $270	**VGood:** $215

REMINGTON MODEL 870TA TRAP

Slide action; hammerless; takedown; 12-ga. 5-shot tube magazine; 30″ Modified, Full barrel; ivory front bead, white metal middle; deluxe checkered walnut trap-style stock, forend; hand-fitted action, parts; recoil pad; special hammer, sear, trigger assembly. Discontinued 1986.

Perf.: $350	**Exc.:** $260	**VGood:** $210

Remington Model 870TB Trap

Same specs as Model 870TA Trap except 28″, 30″ vent-rib barrel; Full choke; metal bead front; no rear sight. Introduced 1950; Discontinued 1981.

Perf.: $400	**Exc.:** $290	**VGood:** $225

Remington Model 870TC Trap

Same specs as Model 870TA Trap except 30″ vent-rib barrel; Rem-Choke; vent rib; optional Monte Carlo stock.

Perf.: $550	**Exc.:** $355	**VGood:** $295

With Monte Carlo stock

Perf.: $570	**Exc.:** $375	**VGood:** $310

REMINGTON SPORTSMAN

Slide action; 12-, 16-, 20-ga; 5-shot tube magazine; 20″, 26″ barrel; various choke options; walnut pistol-grip stock, forearm; made in Field, Riot and Skeet models. Introduced 1931; discontinued 1948.

Plain barrel

Exc.: $265	**VGood:** $180	**Good:** $140

Solid rib

Exc.: $310	**VGood:** $225	**Good:** $170

Vent rib

Exc.: $375	**VGood:** $290	**Good:** $235

REMINGTON SPORTSMAN PUMP

Slide action; 12-ga.; 3″ chamber; 28″, 30″ vent-rib barrel; 7 1/2 lbs.; checkered walnut-stained hardwood stock, forend; recoil pad; similar to Model 870. Introduced 1984; discontinued 1986.

Perf.: $210	**Exc.:** $175	**VGood:** $150

REMINGTON SINGLE SHOT SHOTGUNS

REMINGTON 90-T SUPER SINGLE

Single shot; 12-ga.; 2 3/4″ chamber; 30″, 32″, 34″ barrel; fixed choke or Rem-Chokes; ported or non-ported barrel; weighs 8 3/4 lb.; figured, checkered American walnut stock; cavities in forend, buttstock for added weights; black vented rubber recoil pad; elongated forcing cones; drop-out trigger unit. Introduced 1990; still in production. **LMSR:** $3199

Perf.: $2000	**Exc.:** $1800	**VGood:** $1600

With adjustable rib

Perf.: $2300	**Exc.:** $2100	**VGood:** $1850

REMINGTON RIDER NO. 3

Single shot; breakopen; hammerless; 10-, 12-, 16-, 20-, 24-, 28-ga.; 30″, 32″ barrel; uncheckered American walnut pistol-grip stock, forearm. Introduced 1893; discontinued 1903.

Exc.: $250	**VGood:** $200	**Good:** $160

REMINGTON RIDER NO. 9

Single shot; breakopen; hammerless; 10-, 12-, 16-, 20-, 24-, 28-ga.; 30″, 32″ barrel; uncheckered American walnut pistol-grip stock, forearm; automatic ejector. Introduced 1902; discontinued 1910.

Exc.: $275	**VGood:** $225	**Good:** $180

Remington Model 32 Expert

Remington Model 32A

Remington Model 3200 Skeet

Remington Model 3200 Competition Skeet

Remington Model 3200 Trap

Remington Model 3200 Competition Trap

Remington Over/Under & Side-By-Side Shotguns
Model 32A
Model 32D Tournament
Model 32 Expert
Model 32F Premier
Model 32 Skeet Grade
Model 32TC Trap
Model 3200
Model 3200 Competition Skeet
Model 3200 Competition Trap
Model 3200 Magnum
Model 3200 Skeet
Model 3200 Special Trap
Model 3200 Trap

REMINGTON OVER/UNDER & SIDE-BY-SIDE SHOTGUNS

REMINGTON MODEL 32A
Over/under; double lock; hammerless; takedown; 12-ga.; 26", 28", 30" barrels; various fixed chokes; plain barrels, solid or vent rib; walnut checkered pistol-grip stock, forend; double or single trigger; auto ejectors. Introduced 1931; discontinued 1947.
Single trigger, plain barrels
  Exc.: $1500 **VGood:** $1100 **Good:** $900
Single trigger, solid or vent rib
  Exc.: $1600 **VGood:** $1200 **Good:** $1000
Double triggers, plain barrels
  Exc.: $1450 **VGood:** $1050 **Good:** $800
Double triggers, solid or vent rib
  Exc.: $1500 **VGood:** $1100 **Good:** $900

Remington Model 32D Tournament
Same specs as Model 32A except higher grade of wood, checkering and engraving. Introduced 1931; discontinued 1947.
Single trigger, plain barrels
  Exc.: $2700 **VGood:** $2350 **Good:** $1400
Single trigger, solid or vent rib
  Exc.: $3000 **VGood:** $2475 **Good:** $1700
Double triggers, plain barrels
  Exc.: $2500 **VGood:** $2000 **Good:** $1200
Double triggers, solid or vent rib
  Exc.: $2700 **VGood:** $2200 **Good:** $1400

Remington Model 32 Expert
Same specs as Model 32A except select grade of wood, checkering and engraving. Introduced 1931; discontinued 1947.
Single trigger, plain barrels
  Exc.: $4300 **VGood:** $3200 **Good:** $2600
Single trigger, solid or vent rib
  Exc.: $4700 **VGood:** $3600 **Good:** $3000
Double triggers, plain barrels
  Exc.: $3900 **VGood:** $2800 **Good:** $2200
Double triggers, solid or vent rib
  Exc.: $4300 **VGood:** $3200 **Good:** $2600

Remington Model 32F Premier
Same specs as Model 32A except best grade of wood, checkering and engraving. Introduced 1931; discontinued 1947.
Single trigger, plain barrels
  Exc.: $5300 **VGood:** $3900 **Good:** $3000
Single trigger, solid or vent rib
  Exc.: $5800 **VGood:** $4400 **Good:** $3500
Double triggers, plain barrels
  Exc.: $4800 **VGood:** $3400 **Good:** $2500
Double triggers, solid or vent rib
  Exc.: $5300 **VGood:** $3900 **Good:** $3000

Remington Model 32 Skeet Grade
Same specs as Model 32A except 26", 28" barrels; skeet boring; single selective trigger, beavertail forend. Introduced 1932; discontinued 1942.
Plain barrels
  Exc.: $1400 **VGood:** $1000 **Good:** $750
Solid or vent rib
  Exc.: $1500 **VGood:** $1100 **Good:** $850

Remington Model 32TC Trap
Same specs as Model 32A except 30", 32" vent-rib barrel; Full choke; checkered trap-style pistol-grip stock, beavertail forend.
Double triggers
  Exc.: $2200 **VGood:** $1500 **Good:** $1100
Single trigger
  Exc.: $2400 **VGood:** $1700 **Good:** $1200

REMINGTON MODEL 3200
Over/under; boxlock; 12-ga.; 2³/₄" chamber; 26" Improved Cylinder/Modified, 28" Modified/Full, 30" Modified/Full barrels; vent rib; checkered American walnut pistol-grip stock, forearm; auto ejectors; selective single trigger. Similar to earlier Model 32. Introduced 1973; discontinued 1984.
  Perf.: $1000 **Exc.:** $875 **VGood:** $800

Remington Model 3200 Competition Skeet
Same specs as Model 3200 except 26", 28" Skeet-choke barrels; select walnut Skeet-style stock, full beavertail forearm; engraved forend latch plate, trigger guard; gilt scrollwork on frame. Introduced 1973; discontinued 1984.
  Perf.: $1550 **Exc.:** $1100 **VGood:** $900

Remington Model 3200 Competition Trap
Same specs as Model 3200 except 30", 32" Improved Modified/Full or Full/Full barrels; select fancy walnut straight comb or Monte Carlo stock, beavertail forend; engraved forend latch plate, trigger guard; gilt scrollwork on frame. Introduced 1973; discontinued 1984.
  Perf.: $2000 **Exc.:** $1300 **VGood:** $1100

Remington Model 3200 Magnum
Same specs as Model 3200 except 12-ga.; 3" chamber; 30" Modified/Full or Full/Full barrels. Introduced 1973; discontinued 1984.
  Perf.: $1050 **Exc.:** $900 **VGood:** $800

Remington Model 3200 Skeet
Same specs as Model 3200 except for 26", 28" Skeet-choke barrels; skeet-style stock, full beavertail forearm. Introduced 1973; discontinued 1984.
  Perf.: $1250 **Exc.:** $900 **VGood:** $800

Remington Model 3200 Special Trap
Same specs as Model 3200 except 30", 32" Improved Modified/Full or Full/Full barrels; higher grade select walnut straight comb or Monte Carlo trap-style stock, beavertail forearm. Introduced 1973; discontinued 1984.
  Perf.: $1100 **Exc.:** $950 **VGood:** $850

Remington Model 3200 Trap
Same specs as Model 3200 except 30", 32" Improved Modified/Full or Full/Full barrels; straight comb or Monte Carlo trap-style stock, beavertail forearm. Introduced 1973; discontinued 1984.
  Perf.: $900 **Exc.:** $850 **VGood:** $775

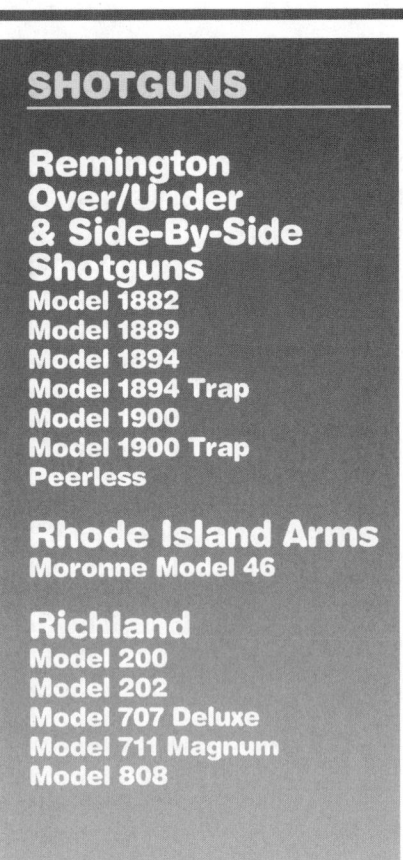

Remington Over/Under & Side-By-Side Shotguns
Model 1882
Model 1889
Model 1894
Model 1894 Trap
Model 1900
Model 1900 Trap
Peerless

Rhode Island Arms
Moronne Model 46

Richland
Model 200
Model 202
Model 707 Deluxe
Model 711 Magnum
Model 808

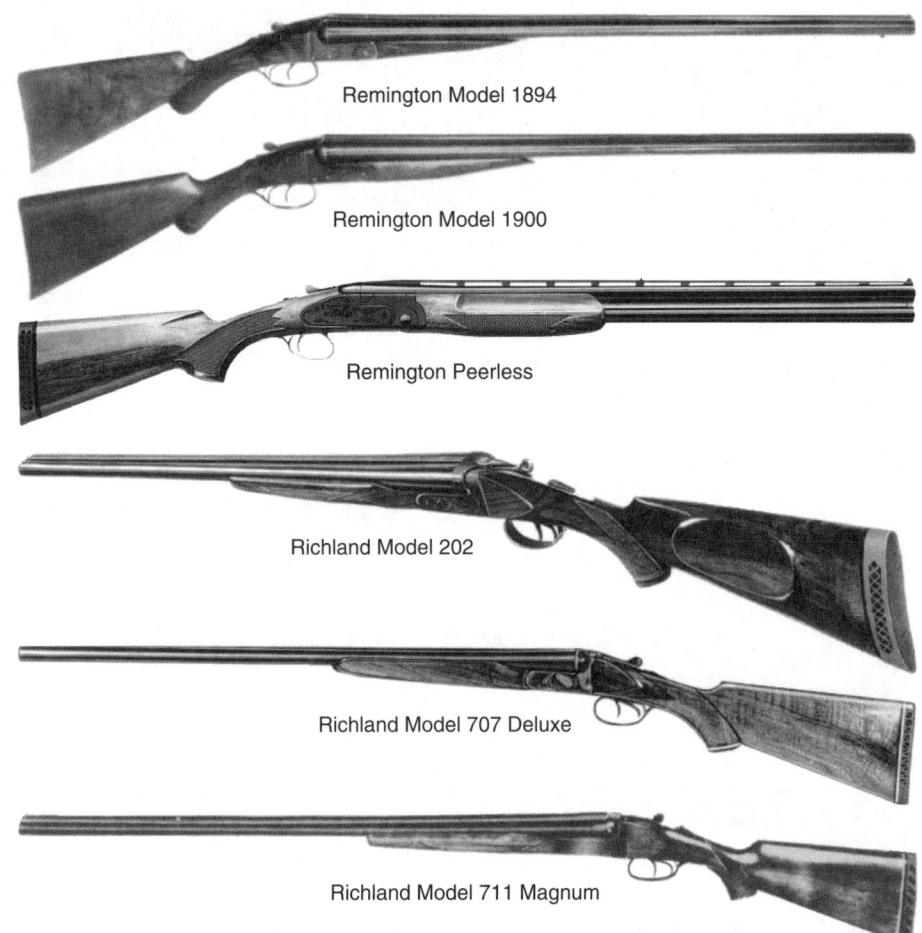

Remington Model 1894

Remington Model 1900

Remington Peerless

Richland Model 202

Richland Model 707 Deluxe

Richland Model 711 Magnum

REMINGTON MODEL 1882
Side-by-side; visible hammers; 10-, 12-, 16-ga.; 28″, 30″, 32″ steel or Damascus barrels; hand-checkered American walnut pistol-grip stock, forearm; double triggers. Introduced 1882; discontinued 1910.
Damascus barrels
 Exc.: $1000   **VGood:** $700   **Good:** $575
Steel barrels
 Exc.: $1000   **VGood:** $900   **Good:** $800

REMINGTON MODEL 1889
Side-by-side; visible hammers; 10-, 12-, 16-ga.; 28″, 30″, 32″ steel or Damascus barrels; double triggers; hand-checkered American walnut stock, slim forend. Introduced 1889; discontinued 1908. Prices shown are for lowest grade.
Damascus barrels
 Exc.: $1100   **VGood:** $800   **Good:** $700
Steel barrels
 Exc.: $1250   **VGood:** $950   **Good:** $800

REMINGTON MODEL 1894
Side-by-side; boxlock; hammerless; 10-, 12-, 16-ga.; 26″, 28″, 30″, 32″ barrels; hand-checkered American walnut straight-grip stock, forearm; double triggers; auto ejectors. Introduced 1894; discontinued 1910.
 Exc.: $750   **VGood:** $625   **Good:** $525

Remington Model 1894 Trap
Same specs as Model 1894 except 32″ Full-Choke barrels; trap-style walnut stock. Introduced 1894; discontinued 1910.
 Exc.: $750   **VGood:** $625   **Good:** $525

REMINGTON MODEL 1900
Side-by-side; boxlock; hammerless; 10-, 12-, 16-ga.; 28″, 30″ barrels; hand-checkered American walnut straight-grip stock, heavy forearm; double triggers; auto ejectors. Introduced 1900; discontinued 1910.
 Exc.: $600   **VGood:** $425   **Good:** $390

Remington Model 1900 Trap
Same specs as Model 1900 except 32″ Full-Choke barrels; select trap-style walnut stock. Introduced 1900; discontinued 1910.
 Exc.: $700   **VGood:** $425   **Good:** $390

REMINGTON PEERLESS
Over/under; boxlock; 12-ga.; 3″ chamber; 26″, 28″, 30″ barrels; Improved, Modified, Full Rem-Choke tubes; weighs 7 1/4 lbs.; cut-checkered American walnut pistol-grip stock, forend with Imron gloss finish; black venti-lated recoil pad; gold-plated, single selective trigger; automatic safety; automatic ejectors; polished blue fin-ish with light scrollwork on removable sideplates; Rem-ington logo on bottom of receiver. Introduced 1993; still in production. **LMSR: $1225**
 Perf.: $925   **Exc.:** $700   **VGood:** $600

RHODE ISLAND ARMS MORONNE MODEL 46
Over/under; boxlock; 12-, 20-ga.; 26″ Improved Cylin-der/Modified, 28″ Modified/Full barrels; plain barrel or vent rib; checkered straight or pistol-grip stock; non-selective single trigger; plain extractors. Fewer than 500 made. Introduced 1949; discontinued 1953.
12-ga., plain barrel
 Exc.: $750   **VGood:** $575   **Good:** $500
12-ga., vent rib
 Exc.: $900   **VGood:** $725   **Good:** $650
20-ga., plain barrel
 Exc.: $900   **VGood:** $725   **Good:** $650
20-ga., vent rib
 Exc.: $1050   **VGood:** $875   **Good:** $800

RICHLAND MODEL 200
Side-by-side; Anson & Deeley-type boxlock; hammer-less; 12-, 16-, 20-, 28-ga., .410; 26″ Improved/Modified or Modified/Full, 28″ Modified/Full barrels; hand-checkered European walnut pistol-grip stock, beavertail forearm; cheekpiece; double triggers; plain extractors; recoil pad. Made in Spain. Introduced 1963; discontinued 1985.
 Perf.: $300   **Exc.:** $250   **VGood:** $210

RICHLAND MODEL 202
Side-by-side; Anson & Deeley boxlock; hammerless; 12-, 20-ga.; comes in two-barrel set for a single gauge; 26″ Improved/Modified and 30″ Full/Full (12-ga.) or 22″ Improved/Modified and 26″ Modified/Full (20-ga.) bar-rels; hand-checkered European walnut pistol-grip stock, beavertail forearm; cheekpiece; double triggers; plain extractors; recoil pad. Made in Spain. Introduced 1963; discontinued 1971.
 Exc.: $250   **VGood:** $210   **Good:** $150

RICHLAND MODEL 707 DELUXE
Side-by-side; boxlock; hammerless; 12-, 20-ga.; 26″ Improved/Modified, 28″ Modified/Full, 30″ Full/Full barrels; hand-checkered European walnut stock, forearm; triple bolting system; double triggers; plain extractors; recoil pad. Made in Spain. Introduced 1963; discontinued 1972.
 Exc.: $280   **VGood:** $240   **Good:** $200

RICHLAND MODEL 711 MAGNUM
Side-by-side; hammerless; Anson & Deeley-type boxlock; Purdey-type triple lock; 10- (3 1/2″), 12- (3″), 20-ga.; 30″, 32″ Full/Full barrels; hand-checkered Euro-pean walnut pistol-grip stock, forearm; double triggers; plain extractors; automatic safety; recoil pad. Adver-tised as Long Range Waterfowl Magnum. Made in Spain. Introduced 1963; discontinued 1985.
10-ga.
 Perf.: $390   **Exc.:** $280   **VGood:** $200
12-ga.
 Perf.: $320   **Exc.:** $240   **VGood:** $200
20-ga.
 Perf.: $425   **Exc.:** $300   **VGood:** $250

RICHLAND MODEL 808
Over/under; boxlock; 12-ga.; 26″ Improved/Modified, 28″ Modified/Full, 30″ Full/Full barrels; vent rib; hand-checkered European walnut stock, forearm; plain extractors; non-selective single trigger. Made in Italy. Introduced 1963; discontinued 1968.
 Exc.: $300   **VGood:** $280   **Good:** $240

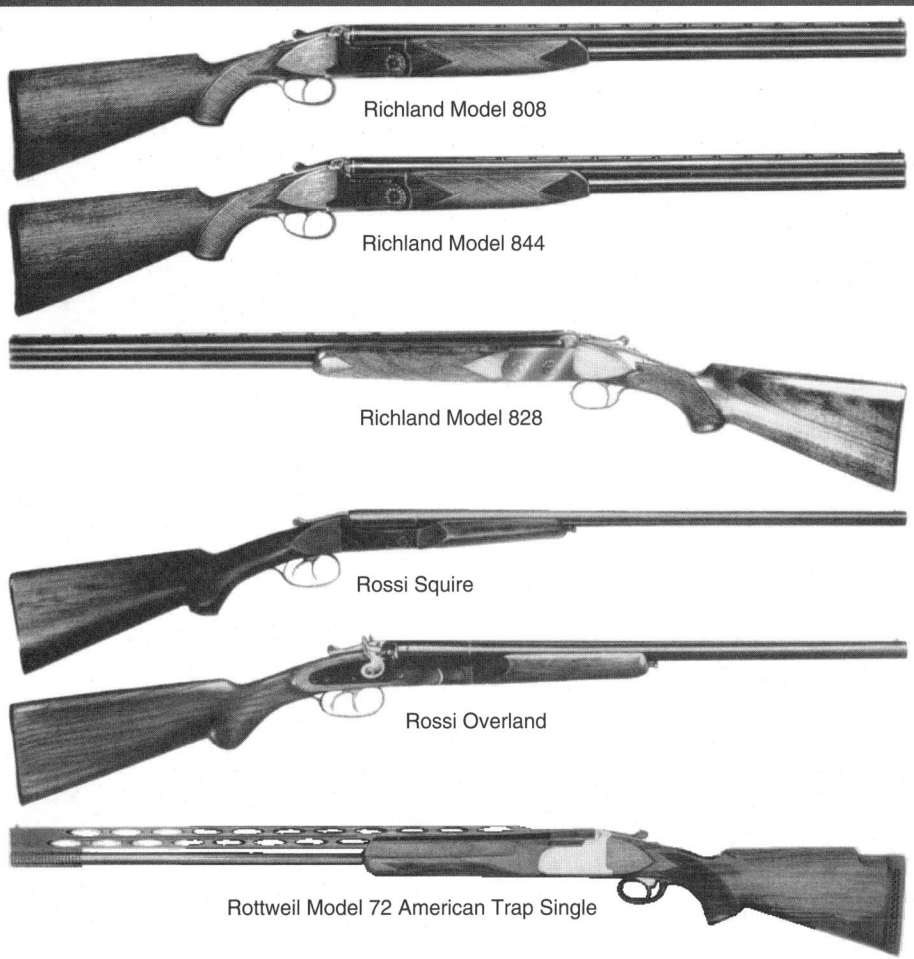

Richland Model 808

Richland Model 844

Richland Model 828

Rossi Squire

Rossi Overland

Rottweil Model 72 American Trap Single

SHOTGUNS

Richland
Model 828
Model 844

Rigby
Best Quality Sidelock
Sidelock Double
Boxlock Double

Rizzini
Boxlock
Sidelock

Rossi
Squire
Overland

Rottweil
Model 72 American Skeet
Model 72 American Trap
Single
Model 72 American Trap
Combo

RICHLAND MODEL 828

Over/under; boxlock; 28-ga.; 26″ Improved/Modified, 28″ Full/Modified barrels; vent rib; hand-checkered European walnut stock, quick-detachable forearm; sliding crossbolt lock; non-automatic safety; plain extractors; single selective trigger; color case-hardened receiver; rosette engraving. Made in Italy. Introduced 1971; no longer in production. Available on special order only.

Perf.: $500   Exc.: $400   VGood: $320

RICHLAND MODEL 844

Over/under; boxlock; 12-ga.; 2 3/4″, 3″ chambers; 26″ Improved/Modified, 28″ Modified/Full, 30″ Full/Full barrels; hand-checkered European walnut pistol-grip stock, forearm; plain extractors; non-selective single trigger. Made in Italy. Introduced 1971; no longer in production.

Perf.: $325   Exc.: $280   VGood: $250

RIGBY BEST QUALITY SIDELOCK

Side-by-side; 12-, 20-ga.; various barrel lengths, chokes to customer's specs; hand-checkered European walnut straight-grip stock, forearm; automatic ejectors; double triggers; English engraving. Still in production. **Perf.: $35,000 Exc.: $28,000 VGood: $22,000**

RIGBY SIDELOCK DOUBLE

Side-by-side; all gauges; various barrel lengths, chokes to customer's specs; hand-checkered European walnut straight-grip stock, forearm; automatic ejectors; double triggers; English engraving; made in two grades, differing in overall quality, amount of engraving. Introduced 1885; discontinued 1955.
Sandringham Grade
Exc.: $8500   VGood: $6000   Good: $5000
Regal Grade
Exc.: $12,000 VGood: $9000   Good: $7500

RIGBY BOXLOCK DOUBLE

Side-by-side; all gauges; various barrel lengths, chokes to customer's specs; hand-checkered European walnut straight-grip stock, forearm; automatic ejectors; double triggers; English engraving; made in two grades, differ-

ing in amount and nature of engraving, overall quality. Introduced 1900; discontinued 1955.
Chatsworth Grade
Exc.: $4000   VGood: $3000   Good: $2500
Sackville Grade
Exc.: $5500   VGood: $4500   Good: $4000

RIZZINI BOXLOCK

Side-by-side; Anson & Deeley boxlock: 12-, 16-, 20-, 28-ga., .410; 25″ to 30″ barrels; file-cut rib; weighs 5 1/2 to 6 1/4 lbs., depending upon gauge; hand-rubbed oil finish stock; ejectors; chopper lump barrels; double triggers, single non-selective optional; scroll, rosette engraving on scalloped frame; coin finish; optional color case-hardening; primarily a custom gun with stock, choke tubes to customer specs. Made in Italy. Introduced 1989; still in production. **LMSR: $25,000**
12-, 16-, 20-ga.
Perf.: $22,000 Exc.: $11,000 VGood: $9000
28-ga., .410
Perf.: $25,000 Exc.: $12,000 VGood: $10,000

RIZZINI SIDELOCK

Side-by-side; Holland & Holland sidelock; 12-, 16-, 20-, 28-ga., .410; 25″ to 30″ barrels; optional rib styles; weighs 6 1/2 to 8 lbs., depending upon gauge; rubbed oil finish or optional satin luster finish stock; auto ejectors; double triggers; single selective trigger optional; coin finish or optional color case-hardening; full coverage scroll engraving, with floral bouquets, gold crown on top lever; customer's name in gold; gold crest inset in forend. Primarily a custom gun with stock, choke tubes to customer specs. Made in Italy. Introduced 1989; still in production. **LMSR: $41,000**
12-, 16-, 20-ga.
Perf.: $35,000 Exc.: $17,500 VGood: $15,000
28-ga., .410
Perf.: $40,000 Exc.: $21,000 VGood: $16,500

ROSSI SQUIRE

Side-by-side; hammerless; 12-, 20-ga., .410; 20″, 28″ barrels; raised matted rib; walnut-finished hardwood

pistol-grip stock, beavertail forearm; double triggers; twin underlugs; synchronized sliding bolts. Made in Brazil. Introduced 1978; discontinued 1990.
Perf.: $250   Exc.: $200   VGood: $150

ROSSI OVERLAND

Side-by-side; sidelock; 12-, 20-ga., .410; 20″, 26″, 28″ barrels; solid raised matted rib; European walnut pistol-grip stock, beavertail forend; double triggers; external hammers; Greener crossbolt. Made in Brazil. Introduced 1978; discontinued 1988.
Perf.: $240   Exc.: $165   VGood: $140

ROTTWEIL MODEL 72 AMERICAN SKEET

Over/under; boxlock; 12-ga.; 26 3/4″ barrels; vent rib; metal bead front sight; hand-checkered French walnut pistol-grip stock, forend; ejectors; single trigger; engraved. Introduced 1977; discontinued 1986.
Perf.: $1800   Exc.: $1350   VGood: $1100

Rottweil Model 72 American Trap Single

Same specs as Model 72 American Skeet except 34″ single barrel; hand-honed chokes; high vent rib; center bead sight, plastic front in metal sleeve; hand-checkered, oil-finished French walnut Monte Carlo stock; muzzle collar changes point of impact; double vent recoil pad; single selective or double triggers. Introduced 1977; discontinued 1986.
Perf.: $850   Exc.: $750   VGood: $600

Rottweil Model 72 American Trap Combo

Same specs as Model 72 American Skeet except two-barrel set with 32″ separated over/under barrels, 34″ single-barrel; hand-honed chokes; high vent rib; center bead sight, plastic front in metal sleeve; hand-checkered, rubbed European walnut Monte Carlo stock; muzzle collar changes point of impact; double vent recoil pad; single selective or double triggers. Introduced 1977; discontinued 1986.
Perf.: $1750   Exc.: $1600   VGood: $1500

SHOTGUNS

Rottweil
Model 72 American Trap Double
Model 72 Field Supreme
Model 72 International Skeet
Model 72 International Trap

Royal Arms
Model 87SC
Model 87T Trap

Sabatti
(See E.A.A.)

SAE
Model 66C
Model 70
Model 70 Multichoke
Model 209E
Model 210S
Model 340X

San Marco
10-Gauge
Field Special
Wildfowler

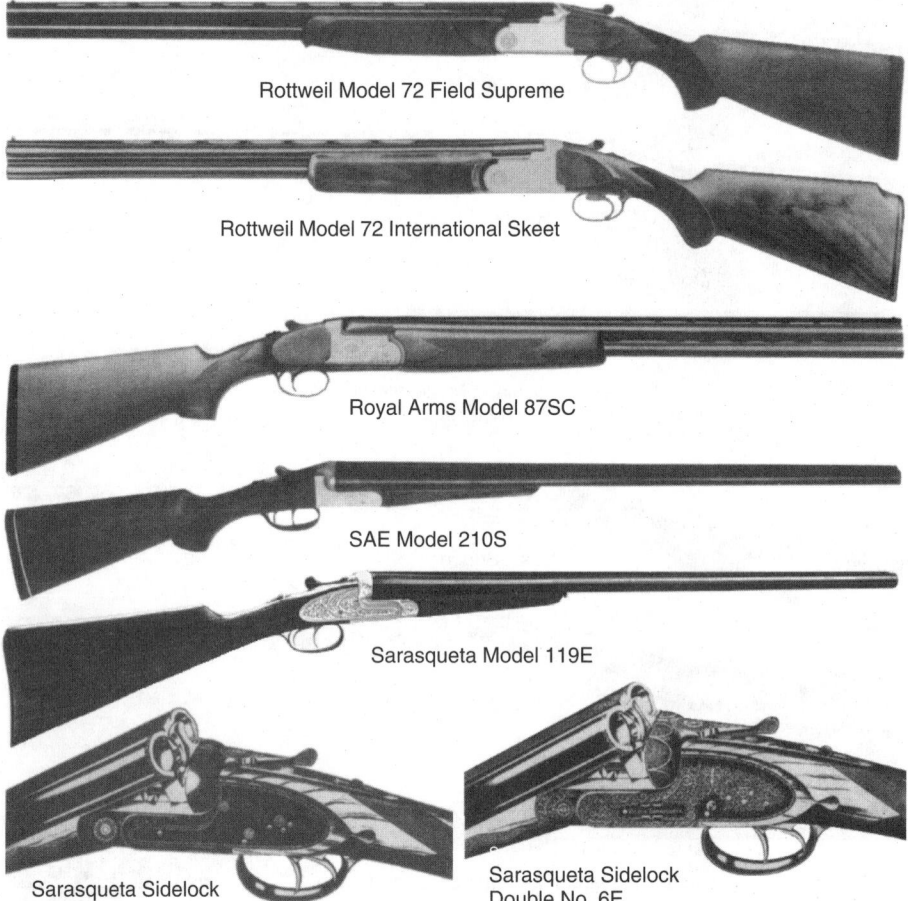

Rottweil Model 72 Field Supreme

Rottweil Model 72 International Skeet

Royal Arms Model 87SC

SAE Model 210S

Sarasqueta Model 119E

Sarasqueta Sidelock Double No. 4

Sarasqueta Sidelock Double No. 6E

Rottweil Model 72 American Trap Double
Same specs as Model 72 American Skeet except 32" barrels; hand-honed chokes; high vent rib; center bead sight, plastic front in metal sleeve; hand-checkered, oil-finished Monte Carlo French walnut stock; muzzle collar changes point of impact; double vent recoil pad; single selective or double triggers. Introduced 1977; discontinued 1986.
    Perf.: $1650   **Exc.:** $1300   **VGood:** $1100

Rottweil Model 72 Field Supreme
Same specs as Model 72 American Skeet except 28" barrels; plastic buttplate; removable interchangeable single trigger assembly; engraved action. Made in Germany. Introduced 1976; discontinued 1986.
    Perf.: $1750   **Exc.:** $1300   **VGood:** $1100

Rottweil Model 72 International Skeet
Same specs as Model 72 American Skeet except 26 3/4" chrome-lined barrels; flared chokes; hand-checkered, oil-finished French walnut stock, modified beavertail forend; retracting spring-mounted firing pins; selective single inertia-type trigger. Introduced 1976; discontinued 1986.
    Perf.: $1750   **Exc.:** $1300   **VGood:** $1100

Rottweil Model 72 International Trap
Same specs as Model 72 American Skeet except 30" barrels; raised vent rib. Introduced 1976; discontinued 1986.
    Perf.: $1750   **Exc.:** $1300   **VGood:** $1100

ROYAL ARMS MODEL 87SC
Over/under; boxlock; 12-ga.; 2 3/4" chambers; 27 5/8" barrels; choke tubes; tapered raised rib; weighs 7 3/8 lb.; select walnut stock; auto ejectors; auto safety; single selective trigger; coin-finish receiver; arabesque scroll engraving. Made in Italy. Introduced 1987; discontinued 1987.
    Perf.: $575   **Exc.:** $475   **VGood:** $400

Royal Arms Model 87T Trap
Same specs as Model 87SC except trap-style stock; other trap configurations. Introduced 1987; discontinued 1987.
    Perf.: $600   **Exc.:** $440   **VGood:** $340

SAE MODEL 66C
Over/under; 12-ga.; 26" Skeet/Skeet, 28" Modified/Full barrels; oil-finished Monte Carlo walnut stock, beavertail forearm; selective auto ejectors; auto safety; single mechanical trigger; dummy sideplates; gold inlays; extensive engraving. Made in Spain. Introduced 1987; discontinued 1988.
    Perf.: $900   **Exc.:** $675   **VGood:** $550

SAE MODEL 70
Over/under; boxlock; 12-, 20-ga.; 3" chambers; 26" Modified/Full, 28" Improved/Modified barrels; weighs 6 3/4 lbs.; European walnut stock; selective auto ejectors; auto safety; single mechanical trigger; engraved receiver; blued finish. Made in Spain. Introduced 1987; discontinued 1988.
    Perf.: $390   **Exc.:** $250   **VGood:** $200

SAE Model 70 Multichoke
Same specs as Model 70 except 12-ga.; 27" barrels; choke tubes; silvered receiver. Made in Spain. Introduced 1987; discontinued 1988.
    Perf.: $425   **Exc.:** $350   **VGood:** $300

SAE MODEL 209E
Side-by-side; sidelock; 12-, 20-ga., .410; 2 3/4" chambers; 26" Modified/Full, 28" Modified/Improved Cylinder barrels; fancy oil-finished checkered select walnut stock, forend; double triggers; selective ejectors; engraved, coin-finished receiver. Made in Spain. Introduced 1987; discontinued 1988.
    Perf.: $900   **Exc.:** $675   **VGood:** $550

SAE MODEL 210S
Side-by-side; boxlock; 12-, 20-ga., .410; 3" chambers; 25" Modified/Full, 28" Modified Improved barrels; weighs 7 lbs.; checkered pistol-grip European walnut stock, splinter forend; auto safety; extractors; double triggers; silver-finished, engraved action. Made in Spain. Introduced 1987; discontinued 1988.
    Perf.: $400   **Exc.:** $300   **VGood:** $210

SAE MODEL 340X
Side-by-side; sidelock; 12-, 20-ga.; 2 3/4" chambers; 26" Modified/Full, 28" Modified/Improved Cylinder barrels; weighs about 7 lbs.; high-gloss straight-grip select walnut stock, forearm; double triggers; selective ejectors; color case-hardened receiver, engraving. Made in Spain. Introduced 1987; discontinued 1988.
    Perf.: $650   **Exc.:** $480   **VGood:** $400

SAN MARCO 10-GAUGE
Over/under; boxlock; 10-ga.; 3 1/2" chambers; 28" Modified/Modified, 32" Modified/Full barrels; chrome-lined bores; solid 3/8" barrel rib; weighs 9 to 9 1/2 lbs.; walnut stock with waterproof finish; long forcing cones; double triggers; extractors; Deluxe grade has automatic ejectors; engraved receiver with game scenes, matte finish. Made in Italy. Introduced 1990; discontinued 1994.
    Perf.: $700   **Exc.:** $500   **VGood:** $370
Deluxe grade
    Perf.: $800   **Exc.:** $600   **VGood:** $470

SAN MARCO FIELD SPECIAL
Over/under; boxlock; 12-, 20-, 28-ga.; 3" chambers; 26" Improved Cylinder/Modified, 28" Full/Modified barrels; vented top and middle ribs; weighs 5 1/2 to 6 lbs.; engraved, silvered receiver; single trigger. Made in Italy. Introduced 1990; discontinued 1994.
    Perf.: $600   **Exc.:** $430   **VGood:** $330

SAN MARCO WILDFOWLER
Over/under; boxlock; 12-ga.; 3 1/2" chambers; 28" Modified/Modified, Full/Modified barrels; chrome-lined bores; vented top and middle ribs; weighs 7 3/4 lbs.; checkered waterproof walnut pistol-grip stock, forend; long forcing cones; single non-selective trigger; extractors on Standard, automatic ejectors on Deluxe; silvered, engraved action. Made in Italy. Introduced 1990; discontinued 1994.
    Perf.: $525   **Exc.:** $380   **VGood:** $300
Deluxe grade
    Perf.: $600   **Exc.:** $430   **VGood:** $330

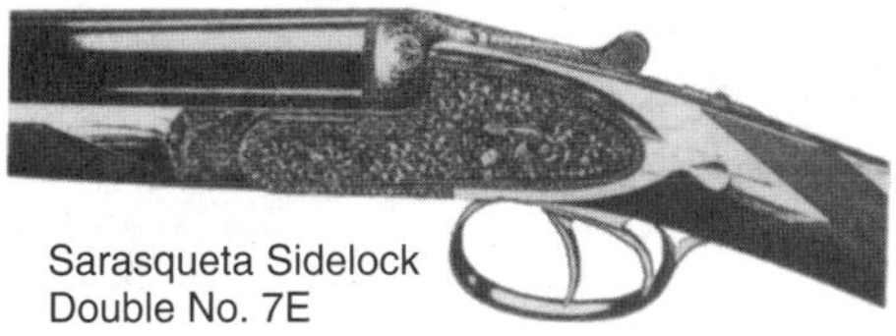

Sarasqueta
Boxlock Double
Model 119E
Sidelock Double
Super Deluxe

Sarriugarte
Model 101DS
Model 101E
Model 101E DS
Model 200 Trap
Model 501E Special
Model 501E Special
Excelsior
Model 501E Special Niger

Sarasqueta Sidelock
Double No. 7E

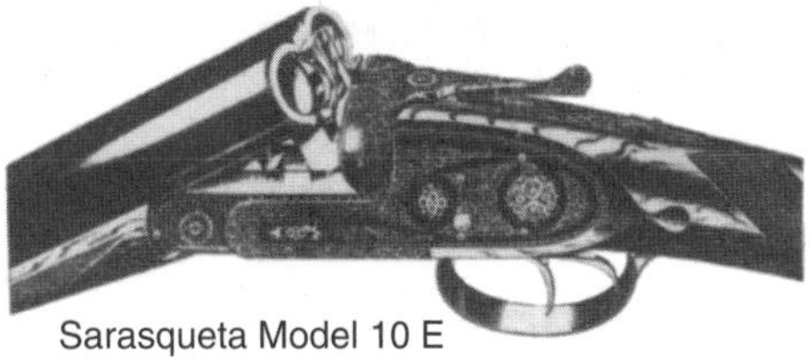

Sarasqueta Model 10 E

Sarasqueta Sidelock
Double Model 11E

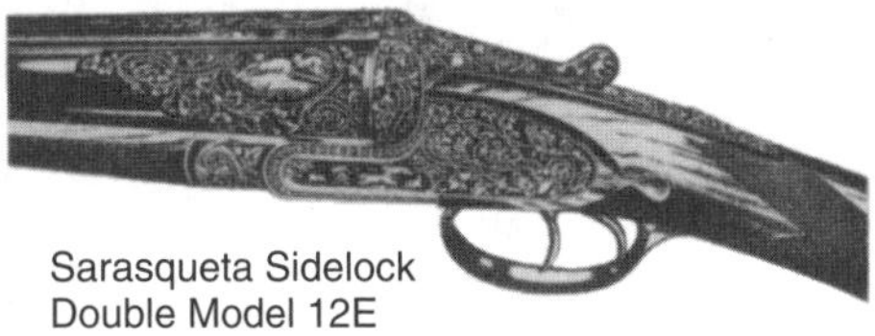

Sarasqueta Sidelock
Double Model 12E

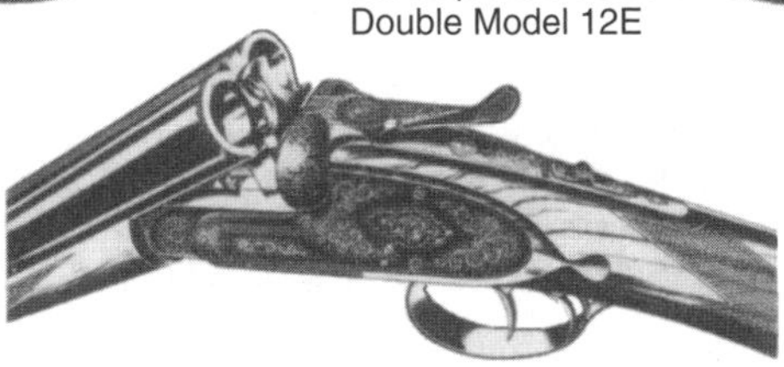

Sarasqueta Super
Deluxe

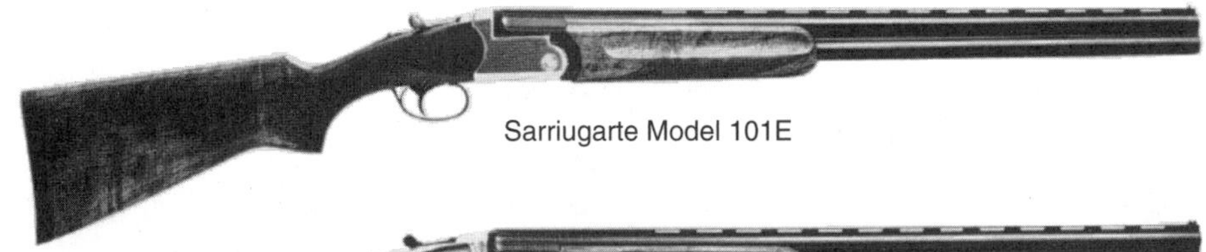

Sarriugarte Model 101E

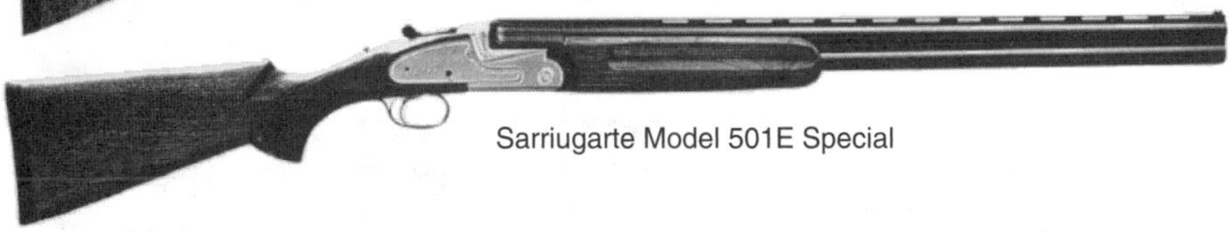

Sarriugarte Model 501E Special

SARASQUETA BOXLOCK DOUBLE

Side-by-side; boxlock; hammerless; 12-, 16-, 20-, 28-ga.; various standard barrel lengths, choke combinations; hand-checkered European walnut straight-grip stock, forearm; plain extractors; double triggers; Greener crossbolt (No.2); optional ejectors; engraved. Introduced mid-1930s; no longer in production.
No. 2
   Perf.: $325   **Exc.:** $250   **VGood:** $180
No. 2E
   Perf.: $375   **Exc.:** $280   **VGood:** $230
No. 3
   Perf.: $375   **Exc.:** $280   **VGood:** $230
No. 3E
   Perf.: $425   **Exc.:** $330   **VGood:** $280

SARASQUETA MODEL 119E

Side-by-side; 12-ga.; 2³/₄″ chambers; 28″ Modified/Full, Improved/Modified barrels; medium bead front sight; European walnut straight-grip stock; auto ejectors; double triggers; hand-engraved locks. Made in Spain. Introduced 1982 by Toledo Armas; discontinued 1984.
   Perf.: $450   **Exc.:** $350   **VGood:** $300

SARASQUETA SIDELOCK DOUBLE

Side-by-side; hammerless; 12-, 16-, 20-, 28-ga., .410; barrel lengths, choke combinations to customer's order; hand-checkered European walnut straight-grip stock, forearm; double triggers; optional automatic ejectors; made in 18 grades, differing in quality of wood, checkering and engraving. Made in Spain.
No. 4
   Perf.: $600   **Exc.:** $500   **VGood:** $450
No. 4E
   Perf.: $650   **Exc.:** $550   **VGood:** $500
No. 6E
   Perf.: $780   **Exc.:** $680   **VGood:** $600
No. 7E
   Perf.: $850   **Exc.:** $750   **VGood:** $700
No. 10E
   Perf.: $1650   **Exc.:** $1400   **VGood:** $1300

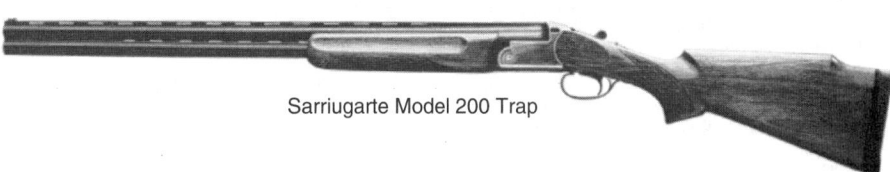

Sarriugarte Model 200 Trap

No. 11E
   Perf.: $1700   **Exc.:** $1500   **VGood:** $1400
No. 12E
   Perf.: $2000   **Exc.:** $1800   **VGood:** $1650

SARASQUETA SUPER DELUXE

Side-by-side; sidelock; hammerless; 12-ga.; barrel lengths, choke combos to customer's order; hand-checkered European walnut pistol-grip stock, forearm; automatic ejectors; double triggers; engraved action. Introduced 1930s; no longer in production.
   Perf.: $1800   **Exc.:** $1600   **VGood:** $1200

SARRIUGARTE MODEL 101DS

Over/under; 12-ga.; 2³/₄″ chambers; 26″ Improved/Modified, 28″ Modified/Full barrels; medium bead front sight; hand-checkered European walnut pistol-grip stock; extractors; selective trigger; border engraving on receiver. Made in Spain. Introduced 1982 by Toledo Armas; discontinued 1984.
   Perf.: $425   **Exc.:** $340   **VGood:** $295

Sarriugarte Model 101E

Same specs as Model 101DS except single trigger; auto ejectors. Introduced 1982; discontinued 1984.
   Perf.: $500   **Exc.:** $375   **VGood:** $320

Model 101E DS

Same specs as Model 101DS except ejectors. Introduced 1982; discontinued 1984.
   Perf.: $550   **Exc.:** $425   **VGood:** $350

SARRIUGARTE MODEL 200 TRAP

Over/under; 12-ga.; 2³/₄″ chambers; 30″ Full/Full barrels; vent middle rib; European walnut Monte Carlo stock; single trigger; auto ejectors. Made in Spain. Introduced 1982 by Toledo Armas; discontinued 1984.
   Perf.: $700   **Exc.:** $570   **VGood:** $500

SARRIUGARTE MODEL 501E SPECIAL

Over/under; 12-ga.; 2³/₄″ chambers; 30″ Full/Full barrels; vent middle rib; select European walnut Monte Carlo stock; single trigger; auto ejectors; elaborate engraving. Introduced 1982; discontinued 1984.
   Perf.: $1500   **Exc.:** $1150   **VGood:** $900

Sarriugarte Model 501E Special Excelsior

Same specs as Model 501E Special except top of the line with best wood available; best grade of engraving. Introduced 1982; discontinued 1984.
   Perf.: $2000   **Exc.:** $1550   **VGood:** $1300

Sarriugarte Model 501E Special Niger

Same specs as Model 501E Special except higher grade wood; better engraving. Introduced 1982; discontinued 1984.
   Perf.: $1900   **Exc.:** $1400   **VGood:** $1100

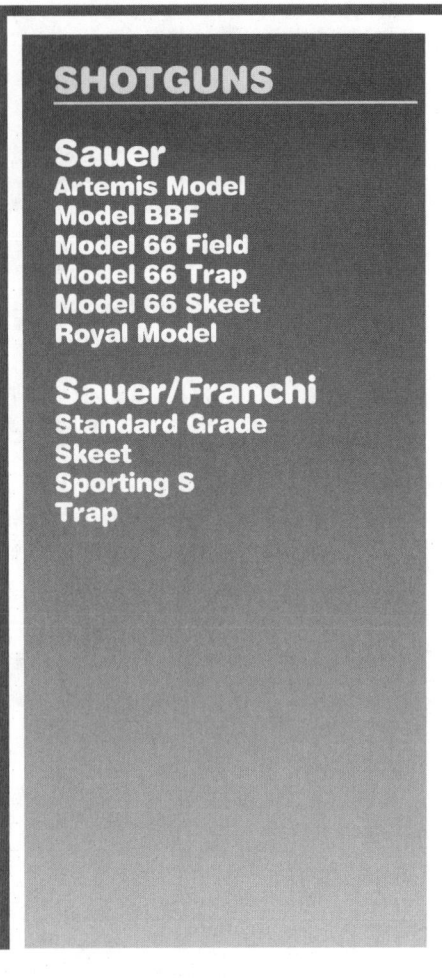

Sauer
Artemis Model
Model BBF
Model 66 Field
Model 66 Trap
Model 66 Skeet
Royal Model

Sauer/Franchi
Standard Grade
Skeet
Sporting S
Trap

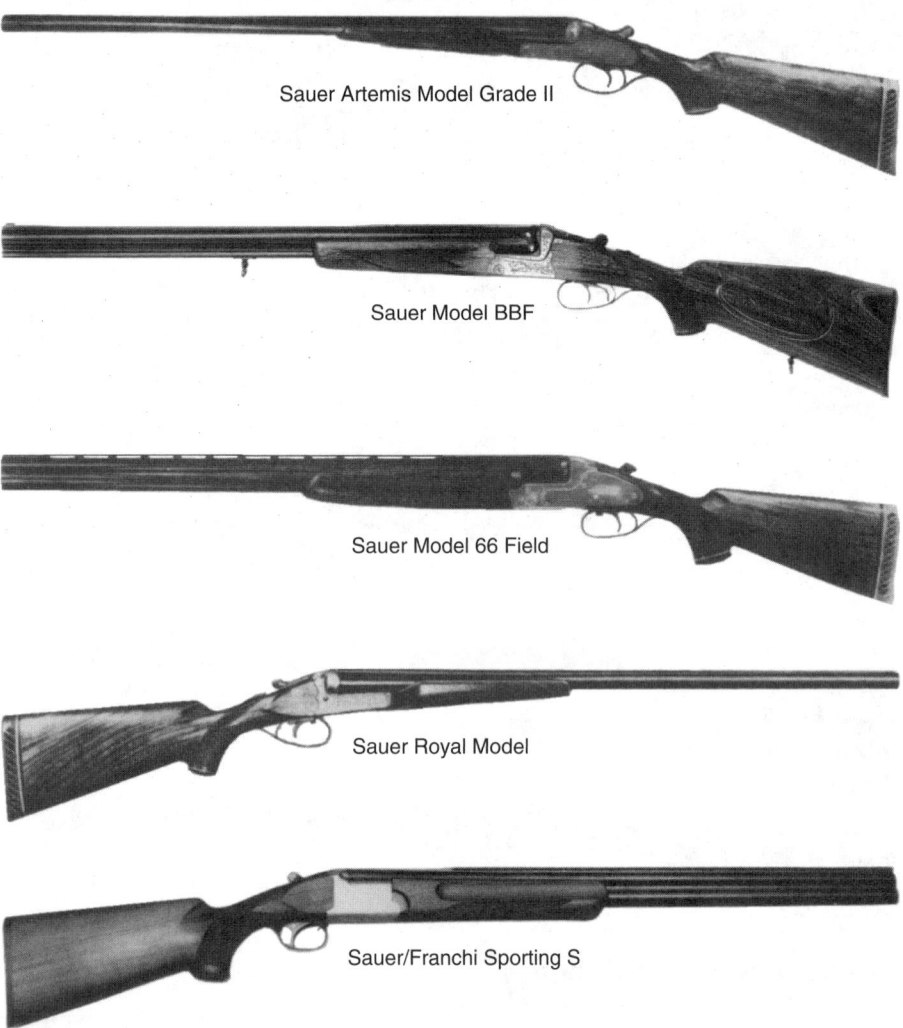

Sauer Artemis Model Grade II

Sauer Model BBF

Sauer Model 66 Field

Sauer Royal Model

Sauer/Franchi Sporting S

SAUER ARTEMIS MODEL

Side-by-side; Holland & Holland-type sidelock; 12-ga.; 28″ Modified/Full barrels; hand-checkered European walnut pistol-grip stock, beavertail forearm; Greener-type crossbolt; double underlugs; double sear safeties; automatic ejectors; single selective trigger; recoil pad; Grade I has fine-line engraving; Grade II has English arabesque motif. Introduced 1960s; discontinued 1977.

Grade I
Perf.: $5000   **Exc.:** $4000   **VGood:** $3600
Grade II
Perf.: $6000   **Exc.:** $5000   **VGood:** $4500

SAUER MODEL BBF

Over/under combo; blitz action; Kersten lock; 16-ga. top barrel; choice of 30-30, 30-06, 7x65R rifle barrel; 25″ barrels; shotgun barrel Full choke; folding leaf rear sight; hand-checkered European walnut pistol-grip stock, forearm; modified Monte Carlo comb, cheek-piece; front set trigger activates rifle barrel; sling swivels; arabesque engraving pattern. Introduced 1950s; discontinued 1985.

Standard Grade
Perf.: $2150   **Exc.:** $1600   **VGood:** $1250
Deluxe Grade
Perf.: $2350   **Exc.:** $1800   **VGood:** $1450

SAUER MODEL 66 FIELD

Over/under; Holland & Holland sideplates; 12-ga.; 28″ Modified/Full barrels; hand-checkered European walnut stock, forearm; automatic safety; selective automatic ejectors; single selective trigger; recoil pad; three grades of engraving. Introduced 1960s; discontinued 1975.

Grade I
Perf.: $2000   **Exc.:** $1650   **VGood:** $1300
Grade II
Perf.: $2900   **Exc.:** $2000   **VGood:** $1700
Grade III
Perf.: $3750   **Exc.:** $3000   **VGood:** $2600

Sauer Model 66 Trap

Same specs as Model 66 Field except 30″ Full/Full or Modified/Full barrels; wide vent rib; trap-style checkered European walnut stock, ventilated beavertail forearm; non-automatic safety. Introduced 1960s; discontinued 1975.

Grade I
Perf.: $1850   **Exc.:** $1400   **VGood:** $1250
Grade II
Perf.: $2800   **Exc.:** $1800   **VGood:** $1400
Grade III
Perf.: $3600   **Exc.:** $2500   **VGood:** $2300

Sauer Model 66 Skeet

Same specs as Model 66 Field except 26″ Skeet/Skeet barrels; vent rib; Skeet-style checkered European walnut stock, ventilated beavertail forearm. Introduced 1966; discontinued 1975.

Grade I
Perf.: $2000   **Exc.:** $1650   **VGood:** $1300
Grade II
Perf.: $2900   **Exc.:** $2000   **VGood:** $1700
Grade III
Perf.: $3750   **Exc.:** $3000   **VGood:** $2600

SAUER ROYAL MODEL

Side-by-side; Anson & Deeley-type boxlock; 12-, 16-, 20-ga.; 26″ Improved/Modified, 28″ Modified/Full, 30″ Full/Full barrels; hand-checkered European walnut pistol-grip stock, beavertail forearm; Greener crossbolt; single selective trigger; automatic ejectors; automatic safety; double underlugs; scalloped frame; recoil pad; arabesque engraving. Introduced 1955; discontinued 1977.

Perf.: $1500   **Exc.:** $1100   **VGood:** $900

SAUER/FRANCHI STANDARD GRADE

Over/under; 12-ga.; 28″, 29″ barrels; vent rib; standard choke combos; weighs 7 lbs.; checkered European walnut stock; single selective trigger; selective auto ejectors; blued finish; made in four grades with differing wood, finish and engraving. Made in Germany. Introduced 1986; discontinued 1988.

Standard Grade
Perf.: $350   **Exc.:** $300   **VGood:** $280
Regent Grade
Perf.: $450   **Exc.:** $380   **VGood:** $325
Favorite Grade
Perf.: $525   **Exc.:** $450   **VGood:** $400
Diplomat Grade
Perf.: $850   **Exc.:** $700   **VGood:** $600

SAUER/FRANCHI SKEET

Over/under; 12-ga.; 28″ Skeet/Skeet barrels; vent rib; checkered select European walnut stock, forearm; ejectors; single selective trigger; silvered receiver. Introduced 1986; discontinued 1988.

Perf.: $900   **Exc.:** $690   **VGood:** $600

SAUER/FRANCHI SPORTING S

Over/under; 12-ga.; 28″ barrels; standard choke combos; vent rib; checkered select European walnut stock, forearm; ejectors; single selective trigger; silvered receiver. Introduced 1986; discontinued 1988.

Perf.: $800   **Exc.:** $700   **VGood:** $600

SAUER/FRANCHI TRAP

Over/under; 12-ga.; 29″ barrels; standard choke combos; vent rib; checkered select European walnut trap-style stock, forearm; ejectors; single selective trigger; silvered receiver. Introduced 1986; discontinued 1988.

Perf.: $850   **Exc.:** $700   **VGood:** $600

Savage Model 24

Savage Model 30

Savage Model 30AC

Savage Model 30D

Savage Model 30 Slug Gun

Savage Model 220

SHOTGUNS

Savage
Model 24
Model 24DL
Model 24M
Model 24MDL
Model 28A
Model 28B
Model 28D
Model 30
Model 30AC
Model 30ACL
Model 30D
Model 30FG
Model 30 Slug Gun
Model 30T
Model 220
Model 220AC
Model 220L
Model 220P
Model 242

SAVAGE MODEL 24

Over/under combo; breakopen; 22 Short, Long, LR upper barrel; .410; 3", Full-choke lower barrel; 24" barrels; open rear sight, ramp front rifle sight; uncheckered walnut pistol-grip stock; sliding button selector; single trigger. Introduced 1950; discontinued 1965.

Exc.: $100   **VGood:** $80   **Good:** $70

Savage Model 24DL

Same specs as Model 24 except top lever; 20-ga., .410 lower barrel; checkered Monte Carlo stock, beavertail forearm; satin chrome-finished receiver, trigger guard. No longer in production.

Exc.: $110   **VGood:** $95   **Good:** $80

Savage Model 24M

Same specs as Model 24 except 22 WMR upper barrel. Introduced 1965; discontinued 1971.

Exc.: $115   **VGood:** $95   **Good:** $80

Savage Model 24MDL

Same specs as Model 24L except top lever; 20-ga., .410 lower barrel; 22 WMR upper barrel; checkered Monte Carlo stock, beavertail forearm; satin chrome-finished receiver, trigger guard. Introduced 1965; discontinued 1969.

Exc.: $120   **VGood:** $100   **Good:** $90

SAVAGE MODEL 28A

Slide action; hammerless; takedown; 12-, 16-, 20-ga.; 5-shot tube magazine; 26", 28", 30", 32" plain barrel; Modified, Cylinder, Full choke; uncheckered American walnut pistol-grip stock, grooved slide handle; black plastic buttplate. Introduced 1920s; discontinued 1940.

Exc.: $225   **VGood:** $175   **Good:** $150

Savage Model 28B

Same specs as Model 28A except raised matted rib. No longer in production.

Exc.: $235   **VGood:** $185   **Good:** $160

Savage Model 28D

Same specs as Model 28A except Full-choke barrel; matted rib; hand-checkered pistol-grip trap-style stock, slide handle. No longer in production.

Exc.: $235   **VGood:** $185   **Good:** $160

SAVAGE MODEL 30

Slide action; hammerless; solid frame; 12-, 16-, 20-ga., .410; 4- (.410), 5-shot magazine; 26", 28", 30" barrel; Improved Modified, Full chokes; vent rib; uncheckered American walnut stock, grooved slide handle; hard rubber buttplate. Introduced 1958; discontinued 1979.

Exc.: $160   **VGood:** $130   **Good:** $100

Savage Model 30AC

Same specs as Model 30 except 12-ga.; 26" barrel; adjustable choke; checkered stock. Introduced 1959; discontinued 1975.

Exc.: $180   **VGood:** $150   **Good:** $120

Savage Model 30ACL

Same specs as Model 30 except 12-ga.; 26" barrel; adjustable choke; checkered stock; ejection port, safety on left side. Introduced 1960; discontinued 1964.

Exc.: $180   **VGood:** $150   **Good:** $120

Savage Model 30D

Same specs as Model 30 except takedown; 12-, 20-ga., .410; 3" chamber; vent rib; checkered pistol-grip stock, fluted extension slide handle; recoil pad; alloy receiver; etched pattern on receiver. Introduced 1972; discontinued 1979.

Exc.: $170   **VGood:** $140   **Good:** $110

Savage Model 30FG

Same specs as Model 30 except takedown; 12-, 20-ga., .410; 3" chamber; checkered pistol-grip stock, fluted extension slide handle; recoil pad; alloy receiver. Introduced 1970; discontinued 1975.

Perf.: $180   **Exc.:** $130   **VGood:** $100

Savage Model 30 Slug Gun

Same specs as Model 30 except 12-, 20-ga.; 22" slug barrel; rifle sights. Introduced 1964; discontinued 1982.

Perf.: $180   **Exc.:** $150   **VGood:** $130

Savage Model 30T

Same specs as Model 30 except 12-ga.; 30" Full-choke barrel; Monte Carlo trap-style stock; recoil pad. Introduced 1964; discontinued 1975.

Perf.: $180   **Exc.:** $150   **VGood:** $130

SAVAGE MODEL 220

Single barrel; single shot; hammerless; takedown; 12-, 16-, 20-ga., .410; 28" to 32" barrels; Full choke; uncheckered American walnut pistol-grip stock, forearm; automatic ejector. Introduced 1947; discontinued 1965.

Exc.: $125   **VGood:** $70   **Good:** $50

Savage Model 220AC

Same specs as Model 220 except Savage adjustable choke. No longer in production.

Exc.: $120   **VGood:** $70   **Good:** $50

Savage Model 220L

Same specs as Model 220 except side lever. Introduced 1965; discontinued 1972.

Exc.: $100   **VGood:** $60   **Good:** $40

Savage Model 220P

Same specs as Model 220 except 12-, 16-, 20-, 28-ga.; Poly Choke; recoil pad. No longer in production.

Exc.: $125   **VGood:** $60   **Good:** $40

SAVAGE MODEL 242

Over/under; .410; 24" barrels; Full choke; open rear, ramp front sight; single exposed hammer; uncheckered walnut pistol-grip stock; lever barrel selector; single trigger. Introduced 1977; discontinued 1979.

Perf.: $325   **Exc.:** $275   **VGood:** $225

SHOTGUNS

Savage
Model 312 Field
Model 312SC Sporting Clays
Savage Model 312T Trap
Model 320
Model 330
Model 333
Model 333T Trap
Model 420
Model 430
Model 440
Model 440T Trap
Model 444 Deluxe
Model 550
Model 720

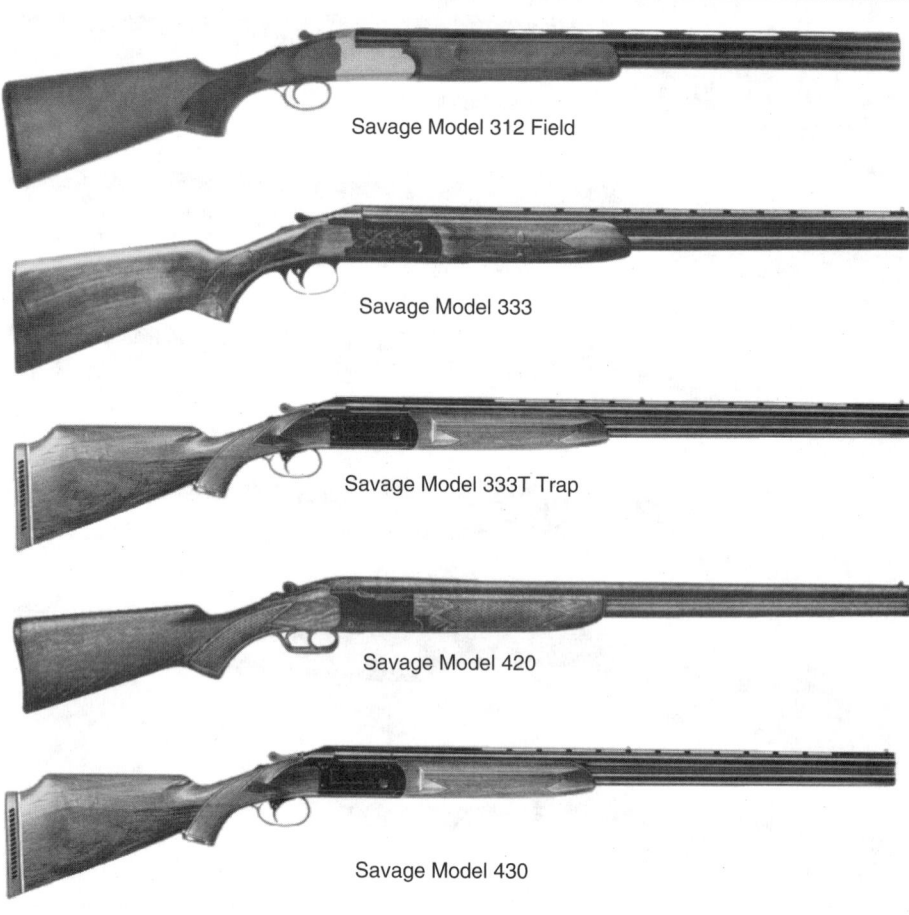

Savage Model 312 Field

Savage Model 333

Savage Model 333T Trap

Savage Model 420

Savage Model 430

SAVAGE MODEL 312 FIELD
Over/under; boxlock; 12-ga.; 3″ chambers; 26″, 28″ barrels; Improved Cylinder, Modified, Full choke tubes; ventilated top, middle ribs; weighs 7 lbs.; checkered walnut stock; ventilated recoil pad; single trigger; satin chrome-finished frame. Introduced 1990; discontinued 1993.

    Perf.: $575   **Exc.:** $460   **VGood:** $400

Savage Model 312SC Sporting Clays
Same as the Model 312 Field except 28″ barrels; Skeet 1, Skeet 2, Improved Cylinder, Modified, Full choke tubes; curved target-type recoil pad; receiver marked "Sporting Clays" on each side. Introduced 1990; discontinued 1993.

    Perf.: $575   **Exc.:** $460   **VGood:** $400

Savage Model 312T Trap
Same as the Model 312 Field except 30″ barrels; Full, Modified choke tubes; weighs 7 1/4 lbs.; checkered walnut Monte Carlo stock; rubber recoil pad. Introduced 1990; discontinued 1993.

    Perf.: $600   **Exc.:** $485   **VGood:** $430

SAVAGE MODEL 320
Over/under; 20-ga.; 3″ chambers; 26″ barrels; choke tubes; vent top, middle ribs; weighs 6 3/4 lbs.; high-gloss, uncheckered stock, forend; recoil pad; single trigger; marketed with gun lock, shooting glasses, ear protection. Introduced 1991; discontinued 1992.

    Perf.: $500   **Exc.:** $425   **VGood:** $325

SAVAGE MODEL 330
Over/under; boxlock; 12- (2 3/4″), 20-ga. (3″); 26″ Improved Cylinder/Modified, 28″ Modified/Full, 30″ Modified/Full barrels; checkered European walnut pistol-grip stock, forearm; selective single trigger; plain extractors. Made in Finland by Valmet. Introduced 1969; discontinued 1979.

    Perf.: $500   **Exc.:** $400   **VGood:** $325

SAVAGE MODEL 333
Over/under; boxlock; 12- (2 3/4″), 20-ga. (3″); 26″ Improved Cylinder/Modified or Skeet/Skeet, 28″ Modified/Full, 30″ Modified/Full barrels; vent rib; checkered European walnut pistol-grip stock, broad forearm; single selective triggers; auto ejectors. Made in Finland by Valmet.

    Perf.: $575   **Exc.:** $475   **VGood:** $425

Savage Model 333T Trap
Same specs as Model 333 except 12-ga.; 30″ Improved Modified/Full barrels; checkered European walnut Monte Carlo pistol-grip trap stock; extractors; recoil pad. Introduced 1972; discontinued 1979.

    Perf.: $525   **Exc.:** $425   **VGood:** $375

SAVAGE MODEL 420
Over/under; boxlock; hammerless; takedown; 12-, 16-, 20-ga.; 26″, 28″, 30″ barrels; Modified/Full or Cylinder/Improved; uncheckered American walnut pistol-grip stock, forearm; double triggers; optional single non-selective trigger; extractors; automatic safety. Introduced 1930s; discontinued 1942.
Double triggers

    Exc.: $300   **VGood:** $200   **Good:** $175
Single trigger

    Exc.: $340   **VGood:** $235   **Good:** $200

SAVAGE MODEL 430
Over/under; boxlock; hammerless; takedown; 12-, 16-, 20-ga.; 26″, 28″, 30″ barrels; Modified/Full, Cylinder/Improved chokes; matted top barrel; hand-checkered American walnut stock, forearm; double triggers; optional single non-selective trigger; automatic safety; recoil pad. Introduced 1930s; discontinued 1942.
Double triggers

    Exc.: $350   **VGood:** $250   **Good:** $200
Single trigger

    Exc.: $395   **VGood:** $300   **Good:** $250

SAVAGE MODEL 440
Over/under; boxlock; 12- (2 3/4″), 20-ga. (3″); 26″ Improved Cylinder/Modified or Skeet/Skeet, 28″ Modified/Full, 30″ Modified/Full barrels; checkered American walnut pistol-grip stock, forearm; single selective trigger; plain extractors. Introduced 1968; discontinued 1972.

    Perf.: $480   **Exc.:** $400   **VGood:** $375

Savage Model 440T Trap
Same specs as Model 440, except 12-ga.; 30″ Improved Modified/Full barrels; wide vent rib; checkered select American walnut Monte Carlo trap stock, semi-beavertail forearm; recoil pad. Introduced 1969; discontinued 1972.

    Perf.: $500   **Exc.:** $450   **VGood:** $400

SAVAGE MODEL 444 DELUXE
Over/under; boxlock; 12- (2 3/4″), 20-ga. (3″); 26″ Improved Cylinder/Modified or Skeet/Skeet, 28″ Modified/Full, 30″ Modified/Full barrels; select checkered American walnut pistol-grip stock, semi-beavertail forearm; single selective trigger; auto ejectors. Introduced 1969; discontinued 1972.

    Perf.: $500   **Exc.:** $450   **VGood:** $400

SAVAGE MODEL 550
Side-by-side; boxlock; hammerless; 12- (2 3/4″), 20-ga. (3″); 26″ Improved Cylinder/Modified, 28″ Modified/Full, 30″ Modified/Full barrels; checkered American walnut pistol-grip stock, semi-beavertail forearm; non-selective single trigger; auto ejectors. Introduced 1971; discontinued 1973.

    Perf.: $325   **Exc.:** $275   **VGood:** $175

SAVAGE MODEL 720
Autoloader; Browning design; takedown; 12-, 16-ga.; 4-shot tube magazine; 26″, 28″, 30″, 32″ barrels; Cylinder, Modified, Full chokes; hand-checkered American walnut pistol-grip stock forearm; black plastic buttplate. Introduced 1930; discontinued 1949.

    Exc.: $250   **VGood:** $200   **Good:** $120

Savage Model 440

Savage Model 440T Trap

Savage Model 550

Savage Model 720

Savage Model 726

Savage Model 755

SHOTGUNS

Savage
Model 726
Model 740C
Model 745 Lightweight
Model 750
Model 750AC
Model 750SC
Model 755
Model 755SC
Model 775 Lightweight
Model 775SC Lightweight
Model 2400 Combo

Savage-Fox
Model B-SE (See Fox Model B-SE)

Savage-Stevens
Model 311 (See Stevens Model 311)

SAVAGE MODEL 726
Autoloader; takedown; 12-, 16-ga.; 2-shot tube magazine; 26″, 28″, 30″, 32″ barrels; Cylinder, Modified, Full chokes; hand-checkered American walnut pistol-grip stock, forearm; black plastic buttplate; engraved receiver. Introduced 1930; discontinued 1949.
Exc.: $250   **VGood:** $200   **Good:** $120

SAVAGE MODEL 740C
Autoloader; takedown; 12-, 16-ga.; 2-shot tube magazine; 24 1/2″ barrel; Skeet choke; hand-checkered American walnut pistol-grip Skeet-style stock, beavertail forearm; Cutts Compensator; black plastic buttplate. Introduced 1939; discontinued 1949.
Exc.: $250   **VGood:** $200   **Good:** $175

SAVAGE MODEL 745 LIGHTWEIGHT
Autoloader; takedown; 12-ga.; 3-, 5-shot tube magazine; 28″ barrel; Cylinder, Modified, Full chokes; hand-checkered American walnut pistol-grip stock, forearm; black plastic buttplate; alloy receiver. Introduced 1946; discontinued 1949.
Exc.: $225   **VGood:** $150   **Good:** $120

SAVAGE MODEL 750
Autoloader; Browning design; takedown; 12-ga. 4-shot tube magazine; 26″ Improved, 28″ Full or Modified barrel; checkered American walnut pistol-grip stock, grooved forearm. Introduced 1960; discontinued 1963.
Exc.: $225   **VGood:** $150   **Good:** $120

Savage Model 750AC
Same specs as Model 750 except 26″ barrel; adjustable choke. Introduced 1964; discontinued 1967.
Exc.: $225   **VGood:** $150   **Good:** $110

Savage Model 750SC
Same specs as Model 750 except 26″ barrel; Savage Super Choke. Introduced 1962; discontinued 1963.
Exc.: $225   **VGood:** $150   **Good:** $120

Savage Model 775 Lightweight

Savage Model 775SC Lightweight

Savage Model 2400 Combo

SAVAGE MODEL 755
Autoloader; takedown; 12-, 16-ga.; 2-, 4-shot tube magazine; 26″ Improved Cylinder, 28″ Full or Modified, 30″ Full barrel; hand-checkered American walnut pistol-grip stock, forearm; rounded receiver. Introduced 1949; discontinued 1958.
Exc.: $210   **VGood:** $160   **Good:** $120

Savage Model 755SC
Same specs as Model 755 except 25″ barrel; Savage Super Choke. No longer in production.
Exc.: $210   **VGood:** $160   **Good:** $120

SAVAGE MODEL 775 LIGHTWEIGHT
Autoloader; takedown; 12-, 16-ga.; 2-, 4-shot magazine; 26″ Improved Cylinder, 28″ Full or Modified, 30″ Full barrel; hand-checkered American walnut pistol-grip stock, forearm; alloy receiver. Introduced 1953; discontinued 1960.
Exc.: $210   **VGood:** $150   **Good:** $125

Savage Model 775SC Lightweight
Same specs as Model 775 Lightweight, except 26″ barrel; Savage Super Choke. No longer in production.
Exc.: $225   **VGood:** $165   **Good:** $120

SAVAGE MODEL 2400 COMBO
Over/under; boxlock; 12-ga.; Full-choke upper barrel, 308 Win. or 222 Rem. lower barrel; 23 1/2″ barrels; solid matted rib; blade front sight, folding leaf rear; dovetail for scope mounting; checkered European walnut Monte Carlo pistol-grip stock, semi-beavertail forearm; recoil pad. Made by Valmet in Finland. Introduced 1975; discontinued 1979.
Perf.: $575   **Exc.:** $500   **VGood:** $450

SHOTGUNS

Scattergun Technologies
Tactical Response TR-870

W&C Scott (Gunmakers) Ltd.
Blenheim
Bowood Deluxe
Chatsworth Grandeluxe
Crown
Kinmount
Texan

W&C Scott & Son
Monte Carlo B

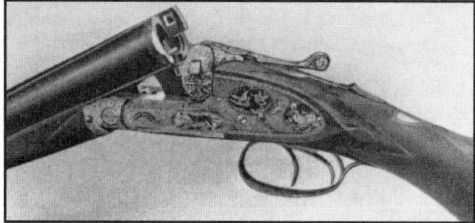
W&C Scott & Son
Premier Hammerless

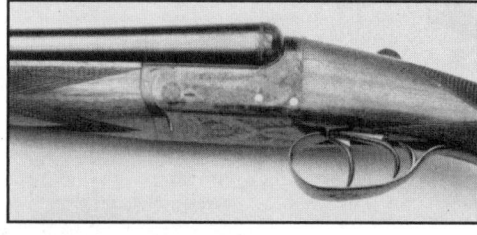
W&C Scott (Gunmakers)
Ltd. Kinmount

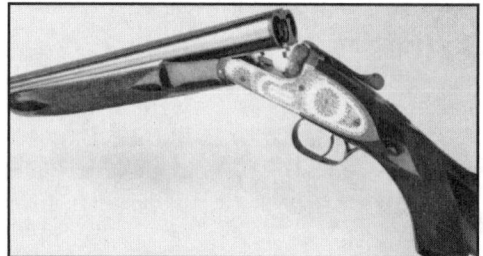
W&C Scott (Gunmakers)
Ltd. Texan

W&C Scott (Gunmakers)
Ltd. Crown

Scattergun Technologies
Tactical Response TR-870

SCATTERGUN TECHNOLOGIES TACTICAL RESPONSE TR-870

Slide action; 12-ga.; 3″ chamber; 7-shot magazine; 18″ Cylinder-choked barrel; ramp front sight with tritium insert, adjustable Ghost Ring rear; Davis Speed Feed II synthetic stock; adjustable nylon sling; Jumbo Head safety; 6-shot ammo holder on left side of receiver; modified Remington 870P; additions include recoil-absorbing buttplate; nylon forend with flashlight. Made in variety of law enforcement versions. Introduced 1991; still in production. **LMSR: $735**

Perf.: $650   **Exc.:** $400   **VGood:** $300

W&C SCOTT (GUNMAKERS) LTD. BLENHEIM

Side-by-side; bar-action sidelock; 12-ga.; 25″ to 30″ barrels; choking to order; concave, Churchill or flat rib; figured walnut with 28 lpi checkering; gold oval on belly at stock; fine scroll and floral engraving; optional full or half pistol grip stock, semi-beavertail forend. Introduced 1983; discontinued 1991. A best quality gun. Made in England.

Perf.: $25,000 **Exc.:** $18,500 **VGood:** $15,000
Deluxe, upgraded wood and metal work
Perf.: $30,000 **Exc.:** $23,000 **VGood:** $19,000

W&C SCOTT (GUNMAKERS) LTD. BOWOOD DELUXE

Side-by-side; boxlock; 12-, 16-, 20-, 28-ga.; 25″ to 30″ barrels; Improved/Modified choking or to order; concave, Churchill or flat rib; hand-checkered custom French walnut stock; scroll engraving. Made in England. Introduced 1980; discontinued 1991.
12-, 16-ga.
Perf.: $7500   **Exc.:** $6000   **VGood:** $5000
20-, 28-ga.
Perf.: $8500   **Exc.:** $7000   **VGood:** $6000

W&C SCOTT (GUNMAKERS) LTD. CHATSWORTH GRANDELUXE

Side-by-side; boxlock; 12-, 16-, 20-, 28-ga.; 25″ to 30″ barrels; Improved/Modified choking or to order; concave, Churchill or flat rib; fine hand-checkered custom French walnut stock; hand-fitted; extensive scroll engraving. Made in England. Introduced 1980; discontinued 1991.
12-, 16-ga.
Perf.: $9500   **Exc.:** $7000   **VGood:** $6000
20-, 28-ga.
Perf.: $10,500 **Exc.:** $8000   **VGood:** $7000

W&C SCOTT (GUNMAKERS) LTD. CROWN

Side-by-side; hammerless boxlock; 12-, 16-, 20-, 28-ga.; 25″ to 30″ barrels; Improved/Modified choking or to order; concave, Churchill or flat rib; checkered French walnut stock with side panels and drop points; engraved case-colored action. Introduced 1982; discontinued 1991. Made in England.

Perf.: $n/a   **Exc.:** $n/a   **VGood:** $n/a
Good: $n/a

W&C SCOTT (GUNMAKERS) LTD. KINMOUNT

Side-by-side; boxlock; 12-, 16-, 20-, 28-ga.; 25″ to 30″ barrels; Improved/Modified choking or to order; con-cave, Churchill or flat rib; checkered French walnut stock; engraving. Introduced 1981; discontinued 1991. Made in England.
12-, 16-ga.
Perf.: $7000   **Exc.:** $5500   **VGood:** $4500
20-, 28-ga.
Perf.: $8000   **Exc.:** $6500   **VGood:** $5500

W&C SCOTT (GUNMAKERS) LTD. TEXAN

Side-by-side; hammerless boxlock with false side-plates; 12-, 16-, 20-, 28-ga.; 25″ to 30″ barrels; standard chokes; flat rib; single trigger; ejectors; fancy walnut; pistol grip stock, beavertail forend. Made for the American market. Introduced 1981; discontinued 1991. Made in England.

Perf.: $7000   **Exc.:** $5500   **VGood:** $4500

W&C SCOTT & SON MONTE CARLO B

Side-by-side; hammerless; bar-action sidelock; 10-, 12-, 16-, 20-ga.; 28″ to 32″ Damascus or fluid steel barrels; choking to order; solid flat rib; figured walnut stock and forend; extractors or ejectors; double or single trigger; early examples have crystal (cocking) indicators on sideplate; two grades of engraving; case-colored action. This was the company's most popular sidelock. Introduced 1898; discontinued 1935. Made in England.

Perf.: $5000   **Exc.:** $3500   **VGood:** $2500

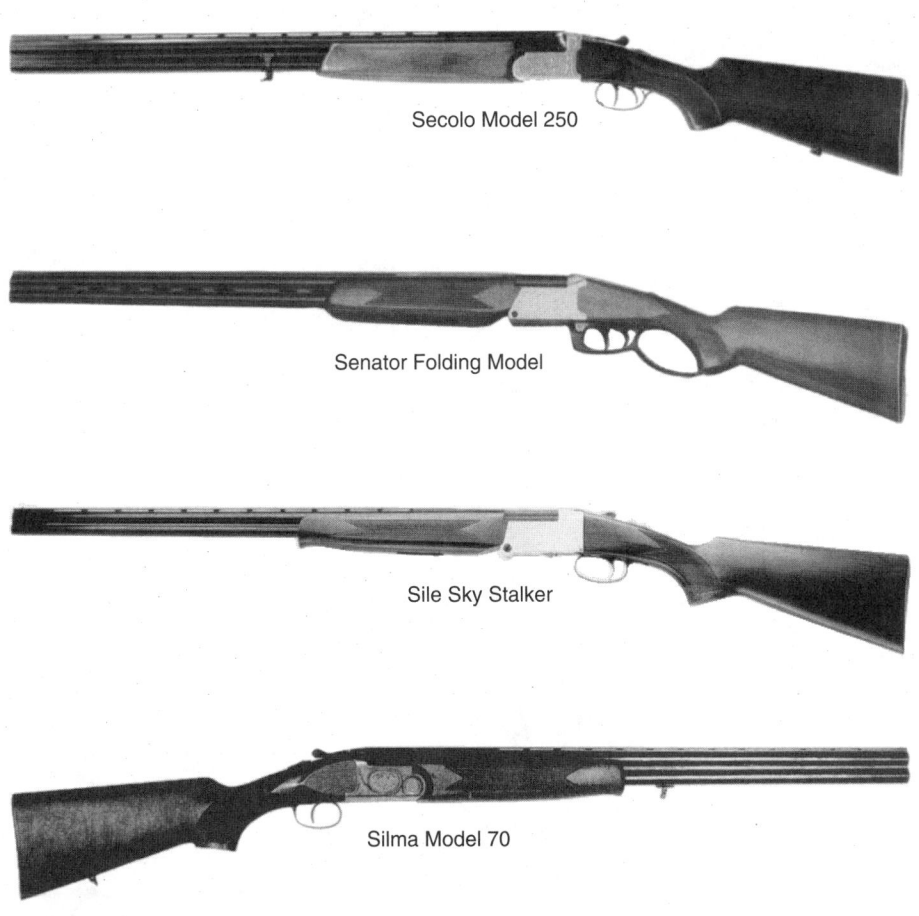

Secolo Model 250

Senator Folding Model

Sile Sky Stalker

Silma Model 70

W&C Scott & Son
Pigeon Club Gun
Premier Hammerless
Imperial Premier
Hammerless
Premier Quality
Hammer Gun
Reliance/Continental
Single Barrel Trap

Secolo
Model 250
Model 530
Model 540 Mono Trap
Model 550 Trap
Model 560 Skeet

Senator
Folding Model

Sile
Sky Stalker

Silma
Model 70

W&C SCOTT & SON PIGEON CLUB GUN

Side-by-side; bar-action sidelock; exposed hammers; 10-, 12-ga.; 27″ to 32″ Damascus or fluid steel barrel; choking to order; solid flat rib; figured walnut stock and forend; fully engraved case-colored action; made in grades A, B, C, but not marked. Made in England. Introduced 1879; discontinued 1924.

Perf.: $4500   **Exc.:** $3500   **VGood:** $2850

W&C SCOTT & SON PREMIER HAMMERLESS

Side-by-side; bar-action sidelock; hammerless; 10-, 12-, 16-, 20-ga.; 28″ to 32″ Damascus or fluid steel barrel; choking to order; solid flat rib; figured walnut stock and forend; ejectors; square cross bolt third lock; sideclips; fully engraved case-colored action. Made in England. This was the company's best gun. Introduced 1897; discontinued 1927.

Perf.: $9000   **Exc.:** $7500   **VGood:** $6000

W&C Scott & Son Imperial Premier Hammerless

Same specs as the Premier Hammerless except with extensive high-quality engraving; finest Italian walnut with fleur-de-lis checkering. Introduced 1897; discontinued 1932.

Perf.: $11,500 **Exc.:** $9000   **VGood:** $7000

W&C SCOTT & SON PREMIER QUALITY HAMMER

Side-by-side; bar-action sidelock; exposed hammers; 4-, 8-, 10-, 12-, 16-ga.; 27″ to 32″ Damascus or fluid steel barrels; choking to order; solid flat rib; figured walnut stock and forend; fully engraved case-colored action. Made in England. Introduced 1873; discontinued 1921.

Perf.: $2800   **Exc.:** $2000   **VGood:** $1750

W&C SCOTT & SON RELIANCE/CONTINENTAL

Side-by-side; hammerless boxlock; 10-, 12-, 16-, 20-ga.; 28″ to 32″ Damascus or fluid steel barrels; standard choking; solid flat rib; figured walnut; ejectors; double trigger; Greener cross bolt; sideclips fully engraved, case-colored scalloped action. This was the company's best boxlock shotgun. Made in England. Discontinued 1933.

Perf.: $3000   **Exc.:** $2400   **VGood:** $1900

W&C SCOTT & SON SINGLE BARREL TRAP

Single shot; bar-action sidelock; 12-ga.; 27″ to 36″ fluid steel barrel; vent-rib with twin ivory sight beads; pistol grip stock, beavertail forend; recoil pad; engraved action; Italian walnut. Sold for $400 in 1928; 70 guns made. Made in England. Introduced 1914; discontinued 1929. Extreme rarity precludes pricing.

SECOLO MODEL 250

Over/under; 12-ga.; 2 3/4″ chambers; 28″ Modified/Full barrels; hand-checkered European walnut pistol-grip stock, forend; extractors; single or double triggers; sling swivels; silvered frame; light engraving. Made in Spain. Introduced 1983; discontinued 1984.

Perf.: $375   **Exc.:** $295   **VGood:** $225

SECOLO MODEL 530

Single barrel; 12-ga.; 2 3/4″ chamber; 30″, 32″ lower barrel; 5 interchangeable choke tubes; vent rib; European walnut Monte Carlo stock; silvered or case-hardened receiver. Made in Spain. Introduced 1983; discontinued 1984.

Perf.: $900   **Exc.:** $775   **VGood:** $675

SECOLO MODEL 540 MONO TRAP

Single barrel; 12-ga.; 2 3/4″ chamber; 30″, 32″ upper barrel; 5 interchangeable choke tubes; vent rib; European walnut Monte Carlo stock; silvered or case-hardened receiver. Made in Spain. Introduced 1983; discontinued 1984.

Perf.: $900   **Exc.:** $775   **VGood:** $675

SECOLO MODEL 550 TRAP

Over/under; 12-ga.; 2 3/4″ chambers, 30″, 32″ barrels; 5 interchangeable choke tubes; vent rib; European walnut Monte Carlo stock; silvered or case-hardened receiver. Made in Spain. Introduced 1983; discontinued 1984.

Perf.: $800   **Exc.:** $625   **VGood:** $540

SECOLO MODEL 560 SKEET

Over/under; 12-ga.; 2 3/4″ chambers; 30″, 32″ Skeet/Skeet barrels; vent rib; Skeet-style European walnut stock; silvered or case-hardened receiver. Made in Spain. Introduced 1983; discontinued 1984.

Perf.: $825   **Exc.:** $650   **VGood:** $575

SENATOR FOLDING MODEL

Over/under; boxlock; 12-, 20-ga., .410; 3″ chambers; 26″, 28″ barrels; vent top, middle ribs; weighs 7 lbs.; European walnut stock; engraved; underlever cocking/opening system. Made in Italy. Introduced 1986; discontinued 1988.

Perf.: $210   **Exc.:** $160   **VGood:** $110

SILE SKY STALKER

Over/under; .410; 3″ chambers; 26″ barrels; checkered walnut stock, Schnabel forend; folds for storage, carrying; chrome-lined bores; matted hard chrome finish on receiver. Made in Italy. Introduced 1984; discontinued 1990.

Perf.: $275   **Exc.:** $230   **VGood:** $185

SILMA MODEL 70

Over/under; boxlock; 12-ga.; 3″ chambers; 27 1/2″ barrels; Modified/Improved Cylinder; weighs 7 lbs.; European walnut pistol-grip stock; single trigger; sling swivels; engraved, blued action. Made in Italy. Introduced 1995; still in production. **LMSR:** $540

Perf.: $480   **Exc.:** $360   **VGood:** $300

SHOTGUNS

SKB Semi-Automatic Shotguns

Model 300
Model 900 Deluxe
Model 900 Deluxe Slug
Model 1300 Upland Mag
Model 1900 Field
Model 1900 Trap
Model XL 100
Model XL 300
Model XL 900
Model XL 900 Skeet
Model XL 900 Slug Gun
Model XL 900 Trap

SKB Model 900 Deluxe Slug

SKB Model 900 Deluxe

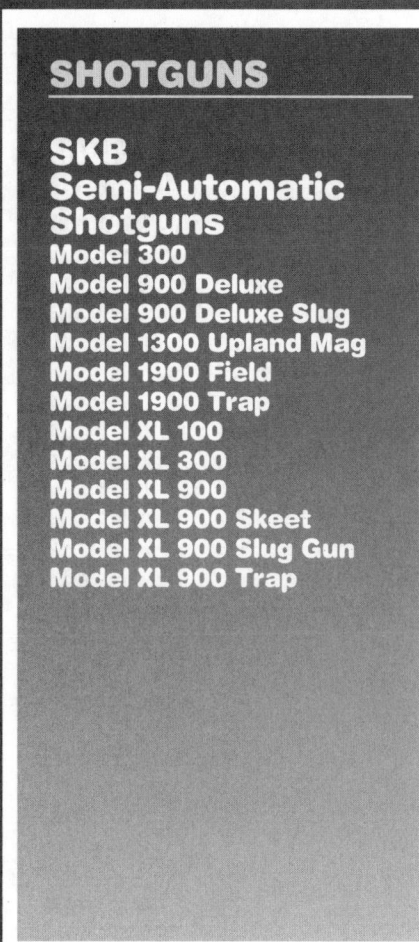

SKB Model 1900 Field

SKB Model 1300 Upland Mag

SKB Model 500

SKB SEMI-AUTOMATIC SHOTGUNS

SKB MODEL 300

Recoil-operated autoloader; takedown; 12-, 20-ga.; 5-shot magazine; 26″ Improved Cylinder, 28″ Full or Modified, 30″ Full barrel; optional vent rib; checkered American walnut pistol-grip stock, fluted forearm; crossbolt safety; automatic magazine cutoff allows changing loads without unloading magazine. Introduced 1969; discontinued 1973.
With plain barrel
Perf.: $300   **Exc.:** $200   **VGood:** $150
With vent rib
Perf.: $325   **Exc.:** $225   **VGood:** $175

SKB MODEL 900 DELUXE

Recoil-operated autoloader; takedown; 12-, 20-ga; 5-shot magazine; 25″ Improved Cylinder, 28″ Full or Modified, 30″ Full barrel; vent rib; hand-checkered American walnut pistol-grip stock, forearm; white spacers on grip cap, buttplate; interchangeable barrels; crossbolt safety; gold-filled engraving on receiver; gold-plated trigger; nameplate inlaid in stock. Introduced 1969; discontinued 1973.
Perf.: $300   **Exc.:** $250   **VGood:** $200

SKB Model 900 Deluxe Slug

Same specs as Model 900 Deluxe except 24″ slug barrel; rifle sights. Introduced 1969; discontinued 1973.
Perf.: $275   **Exc.:** $220   **VGood:** $200

SKB MODEL 1300 UPLAND MAG

Gas-operated autoloader; 12-, 20-ga.; 2 3/4″, 3″ chamber; 22″ Slug barrel, 26″, 28″ vent-rib barrel; Inter-Choke tubes; hand-checkered walnut pistol-grip stock, forend; magazine cut-off system; blued receiver. Made in Japan. Introduced 1988; discontinued 1992.
Perf.: $400   **Exc.:** $325   **VGood:** $290

SKB MODEL 1900 FIELD

Gas-operated autoloader; 12-, 20-ga.; 2 3/4″, 3″ chambers; 22″ Slug barrel, 26″, 28″ vent-rib barrel; Inter-Choke tubes; hand-checkered walnut pistol-grip stock, forend; magazine cut-off system; gold-plated trigger; engraved bright-finish receiver. Made in Japan. Introduced 1988; discontinued 1992.
Perf.: $450   **Exc.:** $375   **VGood:** $300

SKB Model 1900 Trap

Same specs as Model 1900 Field except 12-ga.; 2 3/4″ chamber; 30″ barrel; Inter-Choke tubes; 9.5mm wide raised rib. Made in Japan. Introduced 1988; discontinued 1992.
Perf.: $450   **Exc.:** $375   **VGood:** $300

SKB MODEL XL 100

Gas-operated autoloader; slug gun; 12-ga.; 2 3/4″ chamber; 4-shot magazine; 20″ Cylinder barrel; 40 3/8″ overall length; red ramp front sight with Ray-type blade, adjustable rifle rear sight; French walnut stock forend; reversible crossbolt safety; marketed with sling swivels; aluminum alloy receiver black anodized finish. Made in Japan. Introduced 1978; discontinued 1980.
Perf.: $225   **Exc.:** $175   **VGood:** $150

SKB MODEL XL 300

Gas-operated autoloader; 12-, 20-ga.; 5-shot magazine; 26″ Improved Cylinder or Skeet, 28″ Full or Modified, 30″ Full barrel; optional vent rib; checkered American walnut pistol-grip stock, fluted forearm; self-compensating gas system; reversible safety. Introduced 1973; discontinued 1976.
With plain barrel
Perf.: $275   **Exc.:** $225   **VGood:** $190
With vent rib
Perf.: $300   **Exc.:** $250   **VGood:** $210

SKB MODEL XL 900

Gas-operated autoloader; 12-, 20-ga; 5-shot tube magazine; 26″ Improved Cylinder; 28″ Full or Modified, 30″ Full barrel; vent rib; Raybar front sight; checkered walnut-finished stock; self-compensating gas system; reversible safety; action release button. Introduced 1973; discontinued 1978.
Perf.: $300   **Exc.:** $250   **VGood:** $210

SKB Model XL 900 Skeet

Same specs as Model XL 900 except 26″ Skeet barrel; Bradley front sight. Introduced 1973; discontinued 1978.
Perf.: $325   **Exc.:** $275   **VGood:** $235

SKB Model XL 900 Slug Gun

Same specs as Model XL 900 except 24″ slug barrel; rifle sights. Introduced 1973; discontinued 1978.
Perf.: $300   **Exc.:** $250   **VGood:** $200

SKB Model XL 900 Trap

Same specs as Model XL 900 except 12-ga.; 30″ Full or Improved Cylinder barrel; Bradley front sight. Introduced 1973; discontinued 1978.
Perf.: $325   **Exc.:** $275   **VGood:** $235

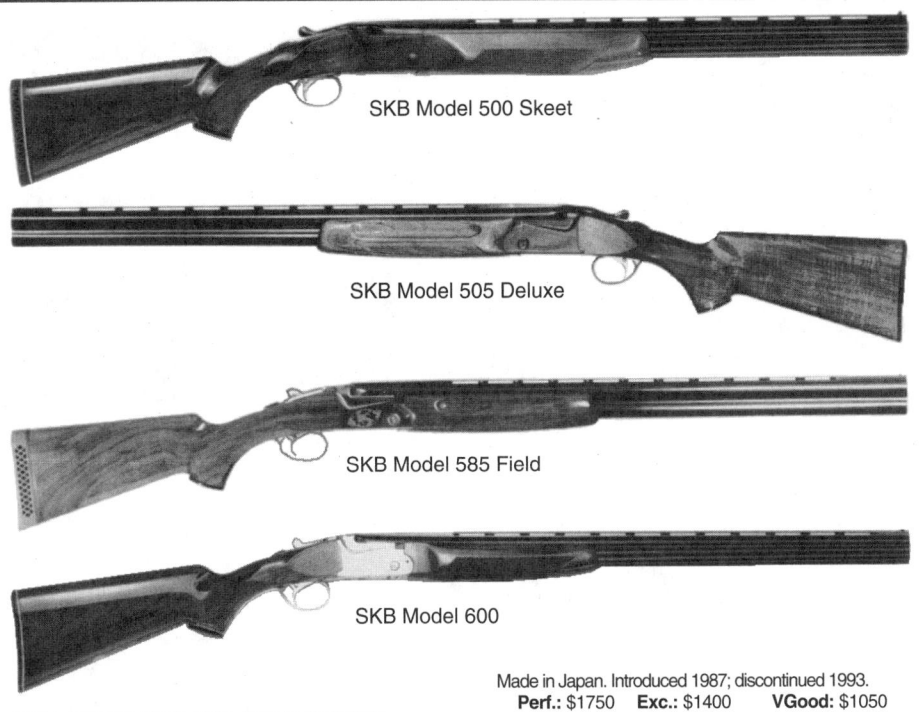

SKB Model 500 Skeet

SKB Model 505 Deluxe

SKB Model 585 Field

SKB Model 600

SHOTGUNS

SKB Over/Under Shotguns
Model 500
Model 500 Magnum
Model 500 Skeet
Model 505 Deluxe
Model 505 Skeet
Model 505 Skeet Set
Model 505 Trap
Model 505 Trap Combo
Model 585 Field
Model 585 Skeet
Model 585 Sporting Clays
Model 585 Trap
Model 585 Waterfowler
Model 585 Youth
Model 600

SKB OVER/UNDER SHOTGUNS

SKB MODEL 500

Over/under; boxlock; hammerless; 12-, 20-, 28-ga., .410; 26″ Improved/Modified, 28″ Improved/Modified or Modified/Full, 30″ Modified/Full barrels; Raybar front sight; hand-checkered walnut pistol-grip stock, forearm; pistol-grip cap; fluted comb; gold-plated single selective trigger; automatic ejectors; non-automatic safety; chrome-lined barrels, action; scroll-engraved border on receiver. Introduced 1967; discontinued 1979.

Perf.: $500   **Exc.:** $400   **VGood:** $350

SKB Model 500 Magnum

Same specs as Model 500 except 12-ga.; 3″ chambers. Introduced 1967; discontinued 1979.

Perf.: $550   **Exc.:** $450   **VGood:** $400

SKB Model 500 Skeet

Same specs as Model 500 except 12-, 20-ga.; fixed Skeet/Skeet chokes; recoil pad with white line spacer; white front sight with middle bead. Introduced 1979; no longer in production.

Perf.: $500   **Exc.:** $400   **VGood:** $350

SKB MODEL 505 DELUXE

Over/under; boxlock; 12-, 20-, 28-ga. (2 ³/₄″), .410; 3″ chambers; 26″, 28″, 30″, 32″, 34″ barrels; all 12-ga. barrels are back-bored, have lengthened forcing cones and longer choke tube system; Inter-Choke tubes or fixed (.410 only) Improved Cylinder/Modified, Modified/Full chokes; ventilated side ribs; weighs 6 ¹/₂ to 8 ¹/₂ lbs.; hand-checkered walnut stock with high-gloss finish; manual safety; automatic ejectors; single selective trigger; silver nitride-finish receiver with Field or Target pattern engraving, gold inlay. Made in Japan. Introduced 1987; discontinued 1993.
Field

Perf.: $750   **Exc.:** $600   **VGood:** $500
Two-barrel Field Set (12- & 20-ga., 20- & 28-ga. or 28-ga. & .410)

Perf.: $1050   **Exc.:** $900   **VGood:** $800

SKB Model 505 Skeet

Same specs as the Model 505 Deluxe except 12-, 20-, 28-ga., .410; 26″ 28″ barrels; Skeet/Skeet chokes; Skeet stock dimensions. Made in Japan. Introduced 1987; discontinued 1993.

Perf.: $750   **Exc.:** $600   **VGood:** $500

SKB Model 505 Skeet Set

Same specs as the Model 505 Deluxe except 12-ga. gun comes with 20-, 28-ga. and .410 barrel sets in 26″ or 28″ lengths; fixed Skeet/Skeet chokes; Skeet stock dimensions.

Made in Japan. Introduced 1987; discontinued 1993.

Perf.: $1750   **Exc.:** $1400   **VGood:** $1050

SKB Model 505 Trap

Same specs as the Model 505 Deluxe except 12-ga.; 2 ³/₄″ chambers; 30″ barrels; fixed Improved Modified/Full chokes; trap stock dimensions. Made in Japan. Introduced 1987; discontinued 1993.

Perf.: $850   **Exc.:** $600   **VGood:** $500

SKB Model 505 Trap Combo

Same specs as the Model 505 Deluxe except comes with 30″ or 32″ over/under barrels and 32″ or 34″ top single barrel; standard or Monte Carlo stock. Made in Japan. Introduced 1987; discontinued 1993.

Perf.: $1150   **Exc.:** $900   **VGood:** $800

SKB MODEL 585 FIELD

Over/under; boxlock; 12-, 20-, 28-ga. (2 ³/₄″), .410; 3″ chambers; 26″, 28″, 30″, 32″, 34″ barrels; all 12-gauge barrels are back-bored, have lengthened forcing cones and longer choke tube system; Inter-Choke tubes or fixed (.410 only) Improved Cylinder/Modified or Modified/Full chokes; ventilated side ribs; hand checkered walnut stock with high-gloss finish; stock dimensions of 14 ¹/₈″x1 ¹/₂″x2 ³/₁₆″; target stocks available in standard and Monte Carlo; manual safety, automatic ejectors, single selective trigger; silver nitride-finished receiver with Field or Target pattern engraving. Made in Japan. Introduced 1992; still in production. **LMSR:** $1179

Perf.: $950   **Exc.:** $750   **VGood:** $650
Two-barrel Field Set (12- & 20-ga., 20- & 28-ga. or 28-ga. & .410)

Perf.: $1650   **Exc.:** $1400   **VGood:** $1200

SKB Model 585 Skeet

Same specs as Model 585 Field except Competition series choke tubes; Skeet-dimensioned stock, radiused recoil pad; oversize bores and lengthened forcing cones (12-ga. models); chrome-plated bores, chambers, ejectors. Made in Japan. Introduced 1992; still in production. **LMSR:** $1279

Perf.: $1050   **Exc.:** $800   **VGood:** $750
Four-barrel Skeet set

Perf.: $2400   **Exc.:** $1725   **VGood:** $1200

SKB Model 585 Sporting Clays

Same specs as Model 585 Field except 12-, 20-, 28-ga.; 28″, 30″ 32″ barrels; Competition series choke tubes (12-ga.), standard Inter-Choke tubes (20-, 28-ga.); traditional ³/₈″ narrow stepped rib or semi-wide channeled ¹⁵/₁₆″ stepped rib; nickel center bead, white front; special stock dimensions. Made in Japan. Introduced 1992; still in production. **LMSR:** $1379

Perf.: $1100   **Exc.:** $850   **VGood:** $750

SKB Model 585 Trap

Same specs as Model 585 Field except 12-ga.; 30″, 32″ barrels; Competition series choke tubes; wide step rib; nickel center bead, white front; standard trap stock or Monte Carlo; oversize bores; lengthened forcing cones. Made in Japan. Introduced 1992; still in production. **LMSR:** $1279

Perf.: $1050   **Exc.:** $800   **VGood:** $750
Two-barrel combo

Perf.: $1550   **Exc.:** $1300   **VGood:** $1150

SKB Model 585 Waterfowler

Same specs as Model 585 Field except 12-ga.; 28″, 30″ barrels; Improved Cylinder, Skeet 1, Modified Inter-Choke tubes; oil-finished stock, forend; bead-blasted receiver with silver nitride finish; bead-blasted, blued barrels. Made in Japan. Introduced 1995; still in production. **LMSR:** $1329

Perf.: $1100   **Exc.:** $850   **VGood:** $750

SKB Model 585 Youth

Same specs as Model 585 Field except 12-, 20-ga.; 26″, 28″ barrels; 13 ¹/₂″ length of pull; .755″ bores, lengthened forcing cones and Competition series choke tubes. Made in Japan. Introduced 1994; still in production. **LMSR:** $1179

Perf.: $950   **Exc.:** $750   **VGood:** $650

SKB MODEL 600

Over/under; boxlock; hammerless; 12-, 20-, 28-ga., .410; 26″ Improved/Modified, 28″ Improved/Modified or Modified/Full, 30″ Modified/Full barrel; Raybar front sight; select hand-checkered walnut pistol-grip stock, forearm; pistol-grip cap; fluted comb; gold-plated single selective trigger; automatic ejectors; non-automatic safety; chrome-lined barrels; silver-finished, scroll-engraved receiver. Introduced 1967; discontinued 1979.

Perf.: $650   **Exc.:** $475   **VGood:** $400

SKB Over/Under Shotguns

Model 600 Magnum
Model 600 Skeet Grade
Model 600 Small Bore
Model 600 Trap Grade
Model 605 Field
Model 605 Skeet
Model 605 Sporting Clays
Model 605 Trap
Model 680 English
Model 685 Field
Model 685 Sporting Clays
Model 685 Trap
Model 685 Skeet
Model 700 Skeet
Model 700 Trap

SKB Model 600 Field

SKB Model 600 Magnum

SKB Model 600 Trap Grade

SKB Model 605 Trap

SKB Model 600 Magnum
Same specs as Model 600 except 12-ga.; 3″ chamber. Introduced 1969; discontinued 1972.
- **Perf.:** $680 **Exc.:** $500 **VGood:** $430

SKB Model 600 Skeet Grade
Same specs as Model 600 except 26″, 28″ Skeet/Skeet barrels; recoil pad. Introduced 1967; discontinued 1979.
- **Perf.:** $650 **Exc.:** $500 **VGood:** $425

SKB Model 600 Small Bore
Same specs as Model 600 except 20-, 28-ga., .410; 28″ Skeet/Skeet three-barrel set; recoil pad. Introduced 1979; discontinued 1979.
- **Perf.:** $750 **Exc.:** $600 **VGood:** $525

SKB Model 600 Trap Grade
Same specs as Model 600, except 12-ga; 30″, 32″ barrels; Full/Full or Full/Improved; walnut straight or Monte Carlo pistol-grip stock; recoil pad. Introduced 1967; discontinued 1979.
- **Perf.:** $650 **Exc.:** $500 **VGood:** $425

SKB MODEL 605 FIELD
Over/under; boxlock; 12-, 20-, 28-ga. (2 3/4″), .410; 3″ chambers; 26″, 28″, 30″, 32″, 34″ barrels; all 12-gauge barrels are back-bored, have lengthened forcing cones and longer choke tube system; Inter-Choke tubes or fixed Improved Cylinder/Modified, Modified/Full chokes; ventilated side ribs; weighs 6 1/2 to 8 1/2 lbs.; hand-checkered semi-fancy American walnut stock with high-gloss finish; manual safety; automatic ejectors; single selective trigger; silver nitride-finished receiver with Field or Target pattern engraving, gold inlay. Made in Japan. Introduced 1987; discontinued 1992.
Field
- **Perf.:** $975 **Exc.:** $800 **VGood:** $700
Two-barrel Field Set (12- & 20-ga. or 28-ga. & .410)
- **Perf.:** $1275 **Exc.:** $1100 **VGood:** $1050

SKB Model 605 Skeet
Same specs as Model 605 Field except 12-, 20-, 28-ga., .410; 28″ barrels; Skeet/Skeet chokes; Skeet stock dimensions. Made in Japan. Introduced 1987; discontinued 1992.
- **Perf.:** $1000 **Exc.:** $750 **VGood:** $650
Four-barrel Skeet set
- **Perf.:** $1850 **Exc.:** $1500 **VGood:** $1100

SKB Model 605 Sporting Clays
Same specs as Model 605 Field except 12-ga.; 2 3/4″ chambers; 28″, 30″ barrels; Inter-Choke tubes; Sporting Clays stock dimensions. Made in Japan. Introduced 1987; discontinued 1993.
- **Perf.:** $1050 **Exc.:** $800 **VGood:** $700

SKB Model 605 Trap
Same specs as Model 605 Field except 12-ga.; 2 3/4″ chambers; 30″, 32″ barrels; Improved Modified/Full choke tubes; trap stock dimensions. Made in Japan. Introduced 1987; discontinued 1992.
- **Perf.:** $1000 **Exc.:** $750 **VGood:** $650
Two-barrel trap combo
- **Perf.:** $1300 **Exc.:** $1050 **VGood:** $950

SKB MODEL 680 ENGLISH
Over/under; boxlock; hammerless; 12-, 20-, 28-ga., .410; 26″, 28″ chrome-lined barrels; Full/Modified or Modified/Improved chokes; vent rib; Bradley sights; checkered select walnut English-style straight-grip stock; single selective trigger; automatic selective ejectors; black chromed exterior surfaces. Introduced 1973; discontinued 1979.
- **Perf.:** $700 **Exc.:** $575 **VGood:** $500

SKB MODEL 685 FIELD
Over/under; boxlock; 12- 20-, 28-ga. (2 3/4″), .410; 3″ chambers; 26″, 28″, 30″, 32″, 34″ barrels; all 12-gauge barrels are back-bored, have lengthened forcing cones and longer choke tube system; Inter-Choke tubes or fixed (.410 only) Improved Cylinder/Modified or Modified/Full chokes; hand-checkered walnut stock with high-gloss finish; stock dimensions of 14 1/8″x1 1/2″x2 3/16″; target stocks available in standard and Monte Carlo; manual safety; automatic ejectors; single selective trigger; jeweled barrel block; silver-finished receiver with fine engraving, gold inlay. Made in Japan. No longer imported.
- **Perf.:** $1250 **Exc.:** $925 **VGood:** $750
Two-barrel Field set (12- & 20-ga. or 28-ga. & .410)
- **Perf.:** $1650 **Exc.:** $1350 **VGood:** $1200

SKB Model 685 Sporting Clays
Same specs as Model 685 Field except 12-, 20-, 28-ga.; 28″, 30″ 32″ barrels; choke tubes; 3/8″ stepped target-style rib; nickel center bead, white front; special stock dimensions; matte finish receiver. Made in Japan. No longer imported.
- **Perf.:** $1250 **Exc.:** $850 **VGood:** $750
Two-barrel Sporting Clays set (12-, 20-ga.)
- **Perf.:** $1650 **Exc.:** $1250 **VGood:** $1150

SKB Model 685 Trap
Same specs as Model 685 Field except 12-ga.; 30″, 32″ barrels; wide step rib; nickel center bead, white front; Competition series choke tubes; standard trap or Monte Carlo stock; oversize bores and lengthened forcing cones. Made in Japan. No longer imported.
- **Perf.:** $1250 **Exc.:** $850 **VGood:** $750
Two-barrel combo
- **Perf.:** $1650 **Exc.:** $1250 **VGood:** $1150

SKB Model 685 Skeet
Same specs as Model 685 Field except Competition series choke tubes; Skeet-dimensioned stock, radiused recoil pad; all 12-ga. models have oversize bores and lengthened forcing cones; chrome-plated bores, chambers, ejectors. Made in Japan. No longer imported.
- **Perf.:** $1250 **Exc.:** $850 **VGood:** $750
Four-barrel Skeet set
- **Perf.:** $1750 **Exc.:** $1350 **VGood:** $1200

SKB MODEL 700 SKEET
Over/under; boxlock; hammerless; 12-, 20-, 28-ga., .410; 26″, 28″ Skeet/Skeet chrome-lined barrels; Raybar front sight; select hand-checkered oil-finished walnut pistol-grip stock, forearm; pistol-grip cap; fluted comb; gold-plated single selective trigger; automatic ejectors; non-automatic safety; recoil pad; silver-finished engraved receiver. Introduced 1967; discontinued 1979.
- **Perf.:** $800 **Exc.:** $700 **VGood:** $625

SKB Model 700 Trap
Same specs as Model 700 Skeet except 12-, 20-ga.; 30″, 32″ barrels; Full/Full or Full/Improved chokes; straight or Monte Carlo walnut pistol-grip stock. Introduced 1967; discontinued 1979.
- **Perf.:** $750 **Exc.:** $695 **VGood:** $625

SKB Model 680 English

SKB Model 700 Trap

SKB Model 785 Field

SKB Model 885 Field

SKB Over/Under Shotguns
Model 785 Field
Model 785 Skeet
Model 785 Sporting Clays
Model 785 Trap
Model 800 Field
Model 800 Skeet
Model 800 Trap
Model 880 Crown Grade
Model 885 Field
Model 885 Sporting Clays
Model 885 Skeet
Model 885 Trap
Model 5600
Model 5700
Model 5800

SKB MODEL 785 FIELD
Over/under; boxlock; 12-, 20-, 28-ga. (2 3/4"), .410; 3" chambers; 26", 28", 30", 32" barrels; chrome-plated, oversize, back-bored barrels with lengthened forcing cones; Inter-Choke tubes; hand-checkered American black walnut stock, semi-beavertail forend with high-gloss finish; single selective chrome-plated trigger, chrome-plated selective ejectors; manual safety. Made in Japan. Introduced 1995; still in production. **LMSR: $1949**

Perf.: $1750 **Exc.:** $1400 **VGood:** $1200
Field set (12- and 20-ga.)
Perf.: $2200 **Exc.:** $1800 **VGood:** $1500

SKB Model 785 Skeet
Same specs as Model 785 Field except Competition series choke tubes; Skeet-dimensioned stock, radiused recoil pad; all 12-ga. barrels back-bored with lengthened forcing cones. Made in Japan. Introduced 1995; still in production. **LMSR: $1949**

Perf.: $1750 **Exc.:** $1400 **VGood:** $1200
Three-barrel Skeet set (20-, 28-ga., .410)
Perf.: $2350 **Exc.:** $1800 **VGood:** $1500

SKB Model 785 Sporting Clays
Same specs as Model 785 Field except 12-, 20-, 28-ga.; 28", 30", 32" barrels; target-style ventilated rib; Sporting Clays stock dimensions; radiused recoil pad. Made in Japan. Introduced 1995; still in production. **LMSR: $2029**

Perf.: $1800 **Exc.:** $1400 **VGood:** $1150
Sporting Clays set (12- and 20-ga.)
Perf.: $2400 **Exc.:** $2000 **VGood:** $1550

SKB Model 785 Trap
Same specs as Model 785 Field except 12-ga.; 30", 32" barrels; back-bored barrels with lengthened forcing cones, chrome-plated chambers; Competition series choke tubes; standard trap or Monte Carlo stock. Made in Japan. Introduced 1995; still in production. **LMSR: $1949**

Perf.: $1750 **Exc.:** $1400 **VGood:** $1200
Two-barrel combo
Perf.: $2200 **Exc.:** $1800 **VGood:** $1500

SKB MODEL 800 FIELD
Over/under; boxlock; hammerless; 12-, 20-, 28-ga., .410; 26" Improved/Modified, 28" Improved/Modified or Modified/Full, 30" Modified/Full barrels; Raybar front sight; select hand-checkered walnut pistol-grip stock, forearm; pistol-grip cap; fluted comb; gold-plated single selective trigger; automatic ejectors; non-automatic safety; chrome-lined barrels, action; fine scroll-engraved border on receiver. Introduced 1979; discontinued 1980.

Perf.: $1100 **Exc.:** $800 **VGood:** $700

SKB Model 800 Skeet
Same specs as Model 800 Field except 12-, 20-ga.; 26", 28" Skeet/Skeet chrome-lined barrels; recoil pad. Introduced 1969; discontinued 1979.

Perf.: $1150 **Exc.:** $800 **VGood:** $700

SKB Model 800 Trap
Same specs as Model 800 Field except 12-ga.; 30", 32" barrels; Full/Full or Full/Improved; straight or Monte Carlo walnut pistol-grip stock. Introduced 1967; discontinued 1979.

Perf.: $1150 **Exc.:** $800 **VGood:** $700

SKB MODEL 880 CROWN GRADE
Over/under; boxlock; sideplates; 12-, 20-, 28-ga., .410; 26" Skeet/Skeet, 28" Skeet/Skeet, 30" Full/Improved, 32" Full/Improved barrels; Bradley-type front sight; trap or Skeet hand-checkered fancy French walnut pistol-grip stock, forearm; hand-honed action; engraved receiver; gold-inlaid crown on bottom of frame. Introduced 1973; discontinued 1976.

Perf.: $1500 **Exc.:** $1000 **VGood:** $800

SKB MODEL 885 FIELD
Over/under; boxlock; dummy sideplates; 12-, 20-, 28-ga. (2 3/4"), .410; 3" chambers; 26", 28", 30", 32", 34" barrels; all 12-gauge barrels are back-bored, have lengthened forcing cones and longer choke tube system; Inter-Choke tubes or fixed (.410 only) Improved Cylinder/Modified or Modified/Full chokes; hand-checkered select-grade walnut stock with high-gloss finish; stock dimensions of 14 1/8"x1 1/2"x2 3/16"; target stocks available in standard and Monte Carlo; manual safety; automatic ejectors; single selective trigger; jeweled barrel block; silver-finished receiver with fine engraving, gold inlay. Made in Japan. No longer imported.

Perf.: $1450 **Exc.:** $800 **VGood:** $700
Two-barrel Field set (12- & 20-ga. or 28-ga. & .410)
Perf.: $2000 **Exc.:** $1400 **VGood:** $1200

SKB Model 885 Sporting Clays
Same specs as Model 885 Field except 12-, 20-, 28-ga.; 28", 30" 32" barrels; choke tubes; 3/8" stepped target-style rib optional; nickel center bead, white front; special stock dimensions; matte finish receiver. Made in Japan. No longer imported.

Perf.: $1450 **Exc.:** $800 **VGood:** $700

SKB Model 885 Skeet
Same specs as Model 885 Field except Competition series choke tubes; Skeet-dimensioned stock, radiused recoil pad; all 12-ga. barrels oversize with lengthened forcing cones; chrome-plated bores, chambers, ejectors. Made in Japan. No longer imported.

Perf.: $1450 **Exc.:** $800 **VGood:** $700
Four-barrel Skeet set
Perf.: $2800 **Exc.:** $2000 **VGood:** $1800

SKB Model 885 Trap
Same specs as Model 885 Field except 12-ga.; 30", 32" barrels; Competition series choke tubes; wide step rib; nickel center bead, white front; standard trap stock or Monte Carlo; oversize bores and lengthened forcing cones. Made in Japan. No longer imported.

Perf.: $1450 **Exc.:** $800 **VGood:** $700
Two-barrel combo
Perf.: $2000 **Exc.:** $1400 **VGood:** $1200

SKB MODEL 5600
Over/under; 12-ga.; 26", 28", 30" barrels; vent rib; checkered walnut pistol-grip stock; auto selective ejectors; mechanical single trigger; hand-polished, blued frame, barrels; made in trap and Skeet configurations. Made in Japan. Introduced 1979; discontinued 1980.

Perf.: $525 **Exc.:** $450 **VGood:** $400

SKB MODEL 5700
Over/under; 12-ga.; 26", 28", 30" barrels; vent rib; high-grade checkered walnut pistol-grip stock; auto selective ejectors; mechanical single trigger; hand-polished, blued frame, barrels; engraving; made in trap and Skeet configurations. Introduced 1979; discontinued 1980.

Perf.: $675 **Exc.:** $500 **VGood:** $425

SKB MODEL 5800
Over/under; 12-ga.; 26", 28", 30" barrels; vent rib; best-grade checkered walnut pistol-grip stock; auto selective ejectors; mechanical single trigger; hand-polished, blued frame, barrels; elaborate engraving; made in trap and Skeet configurations. Introduced 1979; discontinued 1980.

Perf.: $900 **Exc.:** $650 **VGood:** $575

SHOTGUNS

SKB Side-By-Side Shotguns
Model 100
Model 150
Model 200E
Model 200E Skeet Grade
Model 200 Magnum
Model 280
Model 400E
Model 400E Skeet Grade
Model 480

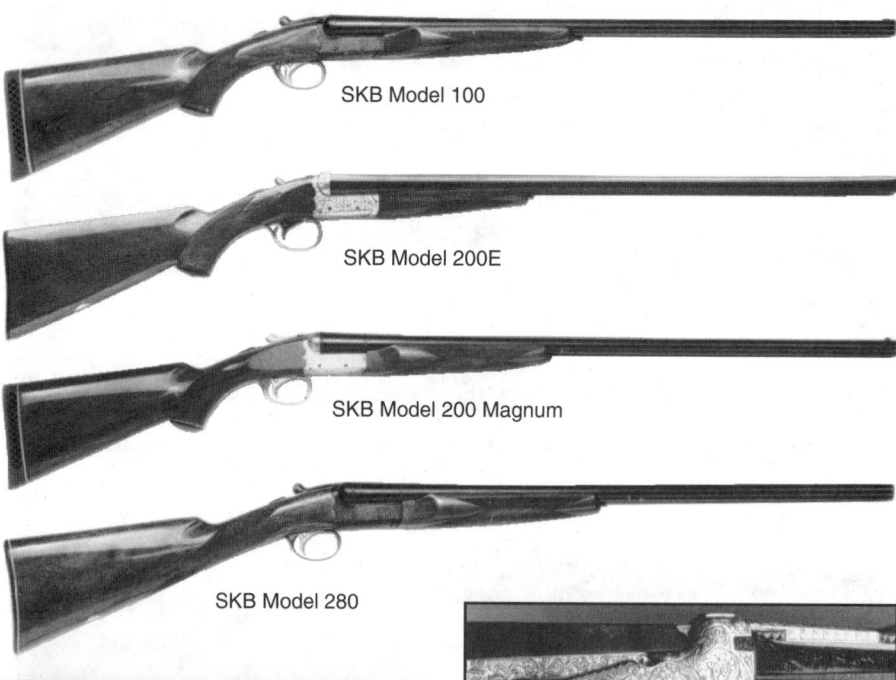

SKB Model 100

SKB Model 200E

SKB Model 200 Magnum

SKB Model 280

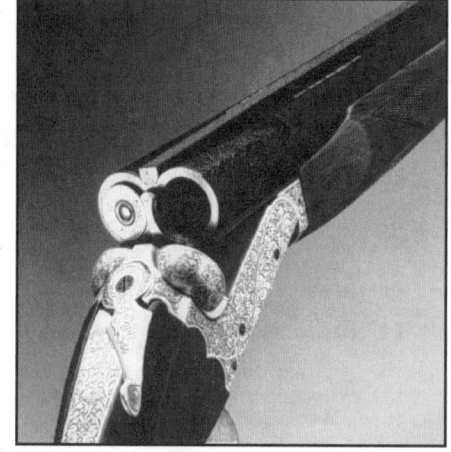

SKB SIDE-BY-SIDE SHOTGUNS

SKB MODEL 100
Side-by side; boxlock; 12-, 20-ga.; 26″ Improved/Modified, 28″ Full/Modified, 30″ Full/Full barrels; hand-checkered pistol grip stock, forearm; single selective trigger; plain extractors; automatic safety. Made in Japan. Introduced 1967; discontinued 1979.
Perf.: $450   **Exc.:** $375   **VGood:** $300

SKB MODEL 150
Side-by-side; boxlock; 12-, 20-ga.; 26″ Improved/Modified, 28″ Full/Modified, 30″ Full/Full barrels; hand-checkered pistol-grip stock, beavertail forearm; single selective trigger; plain extractors; automatic safety. Introduced 1972; discontinued 1974.
Perf.: $475   **Exc.:** $395   **VGood:** $325

SKB MODEL 200E
Side-by-side; boxlock; 12-, 20-ga.; 2 3/4″ chambers; 26″ Improved/Modified, 28″ Full/Modified, 30″ Full/Full barrels; hand-checkered pistol-grip stock, beavertail forearm; recoil pad; automatic selective ejectors; engraved, silver-plated frame; gold-plated nameplate, trigger. Introduced 1967; discontinued 1979.
Perf.: $600   **Exc.:** $500   **VGood:** $450

SKB Model 200E Skeet Grade
Same specs as Model 200E except 25″ Skeet/Skeet barrels; straight-grip English-style stock; non-automatic safety; recoil pad. Introduced 1967; discontinued 1979.
Perf.: $600   **Exc.:** $500   **VGood:** $450

SKB Model 200 Magnum
Same specs as Model 200E except 3″ chambers; 26″, 28″, 30″ barrels; white line spacer recoil pad. Introduced 1979; no longer in production.
Perf.: $625   **Exc.:** $550   **VGood:** $500

SKB MODEL 280
Side-by-side; boxlock; 12-, 20-ga.; 2 3/4″ chambers; 25″, 26″, 28″ barrels; standard choke combos; hand-checkered straight-grip stock, forearm; automatic selective ejectors; single selective trigger; gold-plated nameplate, trigger; game scene engraving. Introduced 1971; discontinued 1979.
Perf.: $700   **Exc.:** $600   **VGood:** $550

SKB MODEL 400E
Side-by-side; boxlock; 12-, 20-ga.; 3″ chambers; 26″ Improved/Modified, 28″ Full/Modified, 30″ Full/Full barrels; hand-checkered pistol-grip stock, beavertail forearm; recoil pad; automatic selective ejectors; engraved, silver-plated frame, sideplates; gold-plated nameplate, trigger. No longer in production.
Perf.: $650   **Exc.:** $500   **VGood:** $450

SKB Model 400E Skeet Grade
Same specs as Model 400E except 25″ Skeet/Skeet barrel; straight-grip English-style stock; non-automatic safety; recoil pad. Introduced 1967; discontinued 1979.
Perf.: $650   **Exc.:** $500   **VGood:** $450

SKB MODEL 480
Side-by-side; boxlock; 12-, 20-ga.; 3″ chambers; 25″, 26″, 28″ barrels; standard choke combos; hand-checkered straight-grip stock, forearm; automatic selective ejectors; single selective trigger; gold-plated nameplate, trigger; game scene engraving. Introduced 1971; discontinued 1979.
Perf.: $1000   **Exc.:** $750   **VGood:** $600

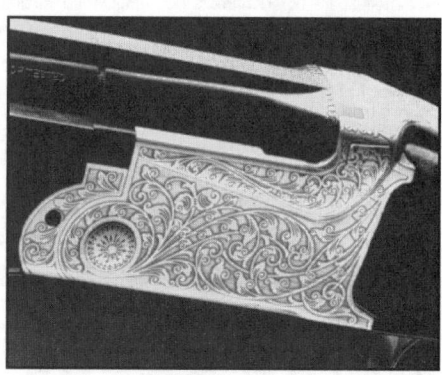

SKB Model 7300

SKB Model 7900 Target
Grade, Trap

SKB Model 7900 Target
Grade, Skeet

SHOTGUNS

SKB Single Barrel & Slide-Action Shotguns
Century
Century II
Model 7300
Model 7900 Target Grade

S&M
10-Gauge

L.C. Smith
Hammerless Double Barrel
(1890-1912)
Hammerless Double Barrel
No. 00 (1890-1912)
Hammerless Double Barrel
No. 0 (1890-1912)
Hammerless Double Barrel
No. 1 (1890-1912)
Hammerless Double Barrel
Pigeon Gun (1890-1912)
Hammerless Double Barrel
No. 2 (1890-1912)
Hammerless Double Barrel
No. 3 (1890-1912)
Hammerless Double Barrel
No. 4 (1890-1912)

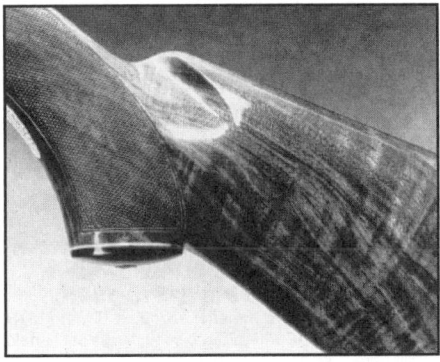

SKB SINGLE BARREL & SLIDE-ACTION SHOTGUNS

SKB CENTURY
Single-barrel trap; boxlock; 12-ga.; 30″, 32″ barrel; vent rib; full choke; checkered pistol-grip walnut stock, beavertail forearm; straight or Monte Carlo comb; recoil pad; auto ejector. Introduced 1973; discontinued 1974.
Perf.: $500　**Exc.:** $450　**VGood:** $400

SKB Century II
Same specs as Century except redesigned locking iron; Monte Carlo stock, reverse-taper forearm. Introduced 1975; discontinued 1979.
Perf.: $575　**Exc.:** $475　**VGood:** $425

SKB MODEL 7300
Slide action; 12-, 20-ga.; 2 ³/₄″, 3″ chambers; 24″, 26″, 28″, 30″ barrels; vent rib; Ray-type front sights; hand-checkered French walnut stock, beavertail forend; recoil pad; white line spacer; double action bars. Made in Japan. Introduced 1979; discontinued 1980.
Perf.: $275　**Exc.:** $200　**VGood:** $180

SKB MODEL 7900 TARGET GRADE
Slide action; 12-, 20-ga.; 2 ³/₄″, 3″ chambers; 26″ Skeet/Skeet, 30″ trap-choked barrels; white front sight, middle bead; hand-checkered French walnut stock, beavertail forend; recoil pad (trap) or composition buttplate (Skeet); blued scroll-etched receiver; made in trap and Skeet configurations. Introduced 1979; discontinued 1980.
Perf.: $325　**Exc.:** $275　**VGood:** $225

S&M 10-GAUGE
Over/under; boxlock; 10-ga.; 3 ¹/₂″ chambers; 28″, 32 ¹/₂″ Full/Full barrels; weighs 9 lbs.; checkered walnut stock; double triggers; extractors; matte-finished metal. Made in Europe. Introduced by Ballistic Products 1986; discontinued 1988.
Perf.: $475　**Exc.:** $375　**VGood:** $300

L.C. SMITH HAMMERLESS DOUBLE BARRELS (1890-1912)
Side-by-side; sidelock; hammerless; 8-, 10-, 12-, 16-, 20-ga.; 26″, 28″, 30″, 32″ twist or fluid steel barrels; lowest grades stocked in American walnut, high grades in European walnut; splinter or beavertail forend; straight, half-, or pistol grip; double triggers; selective single trigger optional after 1904; ejectors optional in lower grades, standard in high grades; Featherweight frame was available in certain grades after 1909; only 30 made in 8-ga., 1895-1897; made in 12 grades until revised in 1913. Manufactured by Hunter Arms Co., Fulton, New York. Introduced 1890; discontinued 1912.

L.C. Smith Hammerless Double Barrel No. 00 (1890-1912)
Same specs as Hammerless Double Barrel (1890-1912), except 12-, 16-, 20-ga.; fluid steel barrels; double trigger, extractors standard; single trigger, ejectors optional; total production, 57,795; with ejectors, 5,874. Introduced 1889; discontinued 1912.
Exc.: $1150　**VGood:** $700　**Good:** $550

L.C. Smith Hammerless Double Barrel No. 0 (1890-1912)
Same specs as Hammerless Double Barrel (1890-1912), except twist (10-, 12-, 16-ga.) or fluid steel (10-, 12-, 16-, 20-ga.) barrels; double triggers, extractors standard; ejectors optional after 1898; Featherweight frame available after 1907; total production, 29,360; with ejectors, 6607. Introduced 1895; discontinued 1912.
Exc.: $1350　**VGood:** $850　**Good:** $650

L.C. Smith Hammerless Double Barrel No. 1 (1890-1912)
Same specs as Hammerless Double Barrel (1890-1912), except twist (10-, 12-, 16-ga.) or fluid steel (10-, 12-, 16-, 20-ga.) barrels; fluid steel barrels available after 1907; double triggers, extractors standard; single trigger, ejectors optional; Featherweight frame available after 1907; in decoration, corresponds to the earlier Quality 2; total production, 10,221; with ejectors, 1640. Introduced 1890; discontinued 1912.
Exc.: $1800　**VGood:** $1100　**Good:** $900

L.C. Smith Hammerless Double Barrel Pigeon Gun (1890-1912)
Same specs as Hammerless Double Barrel (1890-1912), except twist (10-, 12-, 16-ga.) or fluid steel (12-, 16-, 20-ga.) barrels; French walnut straight or pistol grip stock; Monte Carlo comb optional; double triggers, ejectors standard; single trigger optional; lightweight version available in 12- and 16-ga.; total production, 1214. Introduced 1895; discontinued 1912.
Exc.: $1850　**VGood:** $1300　**Good:** $1000

L.C. Smith Hammerless Double Barrel No. 2 (1890-1912)
Same specs as Hammerless Double Barrel (1890-1912), except 28″ to 32″ fluid steel barrels standard; twist barrels optional; European walnut pistol-grip stock; double triggers, extractors standard; ejectors optional; Featherweight frame available; total production, 12,887; with ejectors, 5044. Introduced 1892; discontinued 1912.
Exc.: $2600　**VGood:** $1800　**Good:** $1400

L.C. Smith Hammerless Double Barrel No. 3 (1890-1912)
Same specs as Hammerless Double Barrel (1890-1912), except fluid steel (12-, 16-, 20-ga.) or twist (10-, 12-, 16-, 20-ga.) barrels; English walnut pistol-grip stock; double triggers, extractors standard; single trigger, ejectors optional; total production, 3790; with ejectors, 2093. Introduced 1892; discontinued 1912.
Exc.: $2800　**VGood:** $2000　**Good:** $1500

L.C. Smith Hammerless Double Barrel No. 4 (1890-1912)
Same specs as Hammerless Double Barrel (1890-1912), except twist (10-, 12-, 16-, 20-ga.) or fluid steel (12-, 16-, 20-ga.) barrels; French or English walnut straight or pistol-grip stock; Monte Carlo comb optional; double triggers, extractors standard; ejectors optional; total production, 455; with ejectors, 321. Introduced 1892; discontinued 1913.
Exc.: $7500　**VGood:** $5900　**Good:** $3500

SHOTGUNS

L.C. Smith

Hammerless Double Barrel A1 Grade (1890-1912)
Hammerless Double Barrel No. 5 (1890-1912)
Hammerless Double Barrel Monogram (1890-1912)
Hammerless Double Barrel A2 Grade (1890-1912)
Hammerless Double Barrel A3 Grade (1890-1912)
Hammerless Double Barrels (1913-1945)
Hammerless Double Barrel Field Grade (1913-1945)
Hammerless Double Barrel Ideal Grade (1913-1945)
Hammerless Double Barrel Trap Grade (1913-1945)
Hammerless Double Barrel Skeet Special (1913-1945)

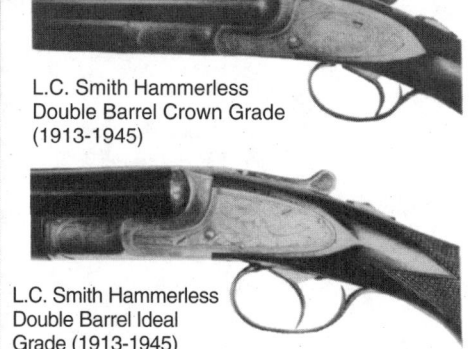

L.C. Smith Hammerless Double Barrel Crown Grade (1913-1945)

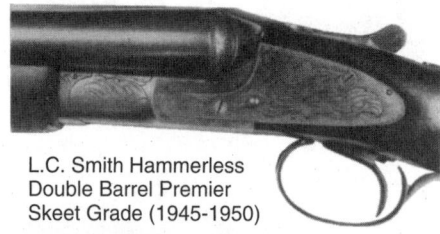

L.C. Smith Hammerless Double Barrel Ideal Grade (1913-1945)

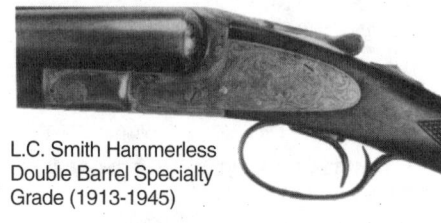

L.C. Smith Hammerless Double Barrel Premier Skeet Grade (1945-1950)

L.C. Smith Hammerless Double Barrel Specialty Grade (1913-1945)

L.C. Smith Hammerless Double Barrel Eagle Grade (1913-1945)

L.C. Smith Hammerless Double Barrel Monogram Grade (1913-1945)

L.C. Smith Hammerless Double Barrel Premier Grade (1913-1945)

L.C. Smith Hammerless Double Barrel Skeet Special (1913-1945)

L.C. Smith Hammerless Double Barrel A1 Grade (1890-1912)
Same specs as Hammerless Double Barrel (1890-1912), except 10-, 12-, 16-ga.; 28″ to 32″ twist barrels; English walnut half-, three-quarter- or pistol-grip stock; double triggers, ejectors standard; total production, 739. Introduced 1890; discontinued 1898.
  Exc.: $4500   **VGood:** $3100   **Good:** $2500

L.C. Smith Hammerless Double Barrel No. 5 (1890-1912)
Same specs as Hammerless Double Barrel (1890-1912), except twist (10-, 12-, 16-ga.) or fluid steel (12-, 16-, 20-ga.) barrels; English or French walnut stocks; double triggers, ejectors standard; extractors, single trigger optional; in 1913, No. 5 was renamed Crown Grade; total production before 1913, 484; with ejectors, 373. Introduced 1892; discontinued 1912.
  Exc.: $7000   **VGood:** $4200   **Good:** $3300

L.C. Smith Hammerless Double Barrel Monogram (1890-1912)
Same specs as Hammerless Double Barrel (1890-1912), except twist barrels; Whitworth fluid steel barrels available after 1896; English, French or Circassian walnut straight or pistol-grip stock; Monte Carlo comb optional; ejectors standard after 1896; double triggers standard; single trigger optional; Featherweight frame, with Whitworth steel barrels, available after 1907; Monogram was the only grade not changed in 1913; total production (1895-1912), 102. Introduced 1895; discontinued 1945.
  Exc.: $9000   **VGood:** $6300   **Good:** $5000

L.C. Smith Hammerless Double Barrel A2 Grade (1890-1912)
Same specs as Hammerless Double Barrel (1890-1912), except twist or Whitworth fluid steel barrels; French, English or Circassian walnut straight or pistol-grip stock; Monte Carlo comb optional; ejectors, double triggers standard; single trigger optional; lightweight version with Krupp steel barrels available; total production, 100. Introduced 1892; discontinued 1912.
  Exc.: $12,000   **VGood:** $8500   **Good:** $6600

L.C. Smith Hammerless Double Barrel A3 Grade (1890-1912)
Same specs as Hammerless Double Barrel (1890-1912), except Whitworth steel barrels; Krupp steel barrels optional; ejectors, double triggers standard; single trigger optional; Circassian walnut straight or pistol-grip stock; Monte Carlo comb optional; lightweight version available; total production, 17. Introduced 1896; discontinued 1915. Extremely rare, precludes pricing.

L.C. SMITH HAMMERLESS DOUBLE BARRELS (1913-1945)
Side-by-side; sidelock; hammerless; 10-, 12-, 16-, 20-ga., .410; 26″, 28″, 30″, 32″ twist or fluid steel barrels; lowest grades stocked in American walnut, high grades in European walnut; straight-, half- or pistol-grip; splinter or beavertail forend. L.C. Smith guns built after 1912 follow the same basic specifications in gauges, barrel lengths, and other features as those made earlier. Their appearance is somewhat different, however, in that the frames and lockplates lack many of the complex, graceful curves and planes that characterize the older guns; this simpler, more straightforward shaping was an attempt by Hunter Arms to reduce the amount of milling and handwork that went into the guns, and thereby reduce the cost of manufacture. Beavertail forends became available in 1920; .410-bore guns in 1926; 10-ga. phased out in early 1920s; Featherweight frame available in some grades up to 1927, in all grades after 1927; special, extra-high solid rib available in all grades after 1939; manufactured by Hunter Arms Co., Fulton, New York. Introduced 1913; discontinued 1945.

L.C. Smith Hammerless Double Barrel Field Grade (1913-1945)
Same specs as Hammerless Double Barrel (1913-1945), except 12-, 16-, 20-ga., .410; fluid steel barrels; American walnut pistol-grip stock; double triggers, extractors standard; single trigger, ejectors optional; a version featuring ivory sights and recoil pad was introduced in 1939 as the Field Special; total production, 141,844. Introduced 1913; discontinued 1945.
  Exc.: $900   **VGood:** $600   **Good:** $500

L.C. Smith Hammerless Double Barrel Ideal Grade (1913-1945)
Same specs as Hammerless Double Barrel (1913-1945), except twist or fluid steel barrels; fluid steel barrels only, after 1917; 28″ barrels only in .410; American walnut pistol-grip stock; straight or half-pistol grip optional; double triggers, extractors standard; single trigger, ejectors optional; total production, 21,862. Introduced 1913; discontinued 1945.
  Exc.: $1300   **VGood:** $900   **Good:** $700

L.C. Smith Hammerless Double Barrel Trap Grade (1913-1945)
Same specs as Hammerless Double Barrel (1913-1945), except fluid steel barrels standard; twist barrels optional until 1917; American walnut pistol-grip stock standard; straight or half-pistol grip optional; ejectors and single trigger standard; double triggers optional; total production 3335. Introduced 1913; discontinued 1939.
  Exc.: $1400   **VGood:** $1000   **Good:** $800

L.C. Smith Hammerless Double Barrel Skeet Special (1913-1945)
Same specs as Hammerless Double Barrel (1913-1945), except 12-, 16-, 20-ga., .410; 26″, 27″, 28″ fluid steel barrels; American walnut straight-grip stock, beavertail forend standard; half- and pistol grip optional; single trigger; ejectors; checkered butt standard; Featherweight frame standard; total production 771. Introduced 1929; discontinued 1944.
  Exc.: $1850   **VGood:** $1300   **Good:** $1000

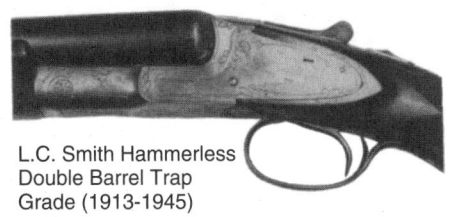

L.C. Smith Hammerless Double Barrel Trap Grade (1913-1945)

L.C. Smith Hammerless Double Barrel Olympic Grade (1913-1945)

L.C. Smith Hammerless Double Barrel Olympic Grade (1913-1945)

Same specs as Hammerless Double Barrel (1913-1945), except 12-, 16-, 20-ga.; 28″ to 32″ fluid steel barrels; vent rib; ivory bead sight; American walnut pistol-grip stock, beavertail forend standard; straight or half-pistol grip optional; single trigger; recoil pad; total production, 26. Introduced 1932; discontinued 1938. Extemely rare; precludes pricing.

L.C. Smith Hammerless Double Barrel Specialty Grade (1913-1945)

Same specs as Hammerless Double Barrel (1913-1945), except fluid steel barrels standard; twist barrels optional until 1917; vent rib optional; American walnut straight-, half- or pistol-grip stock; double triggers, extractors standard; single trigger, ejectors optional; total production, 6,565. Introduced 1913; discontinued 1945.

Exc.: $2000   VGood: $1400   Good: $1100

L.C. Smith Hammerless Double Barrel Eagle Grade (1913-1945)

Same specs as Hammerless Double Barrel (1913-1945), except 10-, 12-, 16-, 20-ga.; fluid steel barrels standard; twist barrels optional until 1917; European walnut straight-, half- or pistol-grip stock; double triggers, extractors standard; single trigger, ejectors optional; total production, 580. Introduced 1913; discontinued 1932.

Exc.: $4000   VGood: $2800   Good: $2200

L.C. Smith Hammerless Double Barrel Crown Grade (1913-1945)

Same specs as Hammerless Double Barrel (1913-1945), except fluid steel barrels standard; twist barrels optional until 1917; made largely to order, with all options: vent rib; European walnut straight-, half- or pistol-grip stock, beavertail forend; single trigger; ejectors; total production, 842. Introduced 1913; discontinued 1945.

Exc.: $5000   VGood: $3500   Good: $2750

L.C. Smith Hammerless Double Barrel Monogram Grade (1913-1945)

Same specs as Hammerless Double Barrel (1913-1945), except twist or Whitworth fluid steel barrels; English, French or Circassian walnut straight or pistol-grip stock; Monte Carlo comb optional; ejectors; double triggers standard; single trigger optional; Featherweight frame, with Whitworth steel barrels, available; same specifications as pre-1913 Monogram Grade; total production (1913-1945), 164. Introduced 1895; discontinued 1945.

Exc.: $9000   VGood: $6300   Good: $5000

L.C. Smith Hammerless Double Barrel Premier Grade (1913-1945)

Same specs as Hammerless Double Barrel (1913-1945), except 12-, 16-, 20-ga.; Whitworth fluid steel barrels; made to order; all options available; total production, 28. Introduced 1913; discontinued 1941. Extremely rare; precludes pricing.

L.C. Smith Hammerless Double Barrel DeLuxe Grade (1913-1945)

Same specs as Hammerless Double Barrel (1913-1945), except 12-, 16-, 20-ga.; Whitworth fluid steel barrels; made to order; all options available; total production, 30. Introduced 1913; discontinued 1945. Extreme rarity precludes pricing.

L.C. Smith Hammerless Double Barrel Long Range Wild Fowl Gun (1913-1945)

Same specs as Hammerless Double Barrel (1913-1945), except 12-ga.; 3″ chambers; 30″ or 32″ barrels; European walnut pistol-grip stock; straight- or New: half-pistol-grip optional; extractors or ejectors; double or single trigger; available in any standard grade. Introduced 1924; discontinued 1945.

Exc.: $900   VGood: $600   Good: $500

L.C. SMITH HAMMERLESS DOUBLE BARRELS (1945-1950)

Side-by-side; sidelock; hammerless; 10-, 12-, 16-, 20-ga., .410; 26″, 28″, 30″, 32″ twist or fluid steel barrels; lowest grades stocked in American walnut, high grades in European walnut; straight-, half- or pistol-grip; splinter or beavertail forend. Marlin Firearms Company purchased Hunter Arms Company in 1945; guns built after 1945 were stamped L.C. SMITH GUN COMPANY; all grades of double gun except Field, Ideal, Specialty and Crown were discontinued; specifications for these grades remained same as before; manufactured by Marlin Firearms, Co. Introduced 1945; discontinued 1950.

L.C. Smith Hammerless Double Barrel Field Grade (1945-1950)

Same specs as Hammerless Double Barrel (1945-1950), except 12-, 16-, 20-ga., .410; fluid steel barrels; American walnut pistol-grip stock; double triggers, extractors standard; single trigger, ejectors optional; total production, 43,312. Introduced 1945; discontinued 1950.

Exc.: $900   VGood: $600   Good: $500

L.C. Smith Hammerless Double Barrel Ideal Grade (1945-1950)

Same specs as Hammerless Double Barrel (1945-1950), except fluid steel barrels; 26″, 28″ barrels only in .410; American walnut pistol-grip stock; straight or half-pistol grip optional; double triggers, extractors standard; single trigger, ejectors optional; total production, 3950. Introduced 1945; discontinued 1950.

Exc.: $1300   VGood: $900   Good: $700

L.C. Smith Hammerless Double Barrel Specialty Grade (1945-1950)

Same specs as Hammerless Double Barrel (1945-1950), except fluid steel barrels standard; vent rib optional; American walnut straight-, half- or pistol-grip stock; double triggers, extractors standard; single trigger, ejectors optional; total production, 109. Introduced 1945; discontinued 1950.

Exc.: $2000   VGood: $1400   Good: $1100

L.C. Smith Hammerless Double Barrel Crown Grade (1945-1950)

Same specs as Hammerless Double Barrel (1945-1950), except fluid steel barrels standard; made largely to order, with all option;: vent rib; European walnut straight-, half- or pistol-grip stock, beavertail forend; single trigger; ejectors; total production, 48. Introduced 1945; discontinued 1950.

Exc.: $5000   VGood: $3500   Good: $2750

L.C. Smith Hammerless Double Barrel Premier Skeet Grade (1945-1950)

Same specs as Hammerless Double Barrel (1945-1950), except 12-, 20-ga.; 26″ or 28″ barrels; high solid rib; European walnut straight-grip stock, beavertail forend; ejectors; single trigger; checkered butt; manufactured 1949-1950; total production, 507. Introduced 1949; discontinued 1950.

Exc.: $1850   VGood: $1300   Good: $1000

L.C. SMITH HAMMERLESS DOUBLE BARREL FIELD GRADE (1967-1971)

Side-by-side; sidelock; hammerless; 12-ga.; 28″ barrels; vent rib; American walnut pistol-grip stock, splinter forend; double triggers; extractors. Marlin Firearms Co. briefly reintroduced to L.C. Smith double gun; total production, 2351. Introduced 1967; discontinued 1971.

Exc.: $700   VGood: $550   Good: $450

L.C. Smith Hammerless Double Barrel Field Grade Deluxe

Same specs as Hammerless Double Barrel Field Grade (1967-1971), except Simmons vent rib; select American walnut pistol-grip stock, beavertail forend; total production, 188. Introduced 1971; discontinued 1971.

Exc.: $950   VGood: $725   Good: $550

L.C. SMITH SINGLE-BARREL TRAP GUN

Single shot; boxlock; hammerless; 12-ga.; 30″ to 34″ barrels (Olympic Grade 32″ only); vent rib; hand-checkered American walnut pistol-grip stock, beavertail forend; ejectors; recoil pad; made in 10 grades with differing quality of workmanship, engraving and wood. Specialty, Eagle, Crown and Monogram grades introduced 1917; other grades introduced later (Olympic in 1928); Eagle Grade discontinued 1932; all others (except Olympic and Specialty) discontinued 1945; Olympic and Specialty grades discontinued 1950. Production totals as follows: Field Grade, 1; Ideal Grade, 5; Olympic Grade, 622; Specialty Grade, 1861; Trap Grade, 1; Eagle Grade, 56; Crown Grade, 88; Monogram Grade, 15; Premier Grade, 2; DeLuxe Grade, 3. The rarity of some grades precludes pricing.

Olympic Grade
Exc.: $1900   VGood: $1400   Good: $1000
Specialty Grade
Exc.: $2500   VGood: $1800   Good: $1400
Crown Grade
Exc.: $4500   VGood: $3000   Good: $2500
Monogram Grade
Exc.: $9000   VGood: $6000   Good: $5000

SHOTGUNS

Smith & Wesson
Model 916
Model 916T
Model 1000
Model 1000 Magnum
Model 1000S
Model 1000 Trap
Model 1000 Waterfowler
Model 3000
Model 3000 Waterfowler

Sporting Arms
Snake Charmer
Snake Charmer II
Snake Charmer II New Generation

Squires Bingham
Model 30D (See Kassnar Omega Pump)

Stevens
Model 22-410
Model 58

Smith & Wesson Model 916

Smith & Wesson Model 1000 Trap

Smith & Wesson Model 1000 Waterfowler

Smith & Wesson Model 3000

Smith & Wesson Model 3000 Waterfowler

SMITH & WESSON MODEL 916
Slide action; 12-, 16-, 20-ga.; 6-shot magazine; 20″ Cylinder, 26″ Improved Cylinder, 28″ Modified, Full or adjustable choke, 30″ Full barrel; vent rib (26″, 28″ barrel); uncheckered walnut stock, fluted comb, grooved slide handle; optional recoil pad; satin-finished steel receiver, no-glare top. Introduced 1973; discontinued 1980.
With plain barrel
   Perf.: $180   **Exc.:** $150   **VGood:** $130
With plain barrel, recoil pad
   Perf.: $180   **Exc.:** $150   **VGood:** $130
With vent rib
   Perf.: $200   **Exc.:** $170   **VGood:** $150
With vent rib, recoil pad
   Perf.: $200   **Exc.:** $170   **VGood:** $150

Smith & Wesson Model 916T
Same specs as Model 916 except takedown; 12-ga.; no 20″ barrel. Introduced 1976; discontinued 1980.
With plain barrel
   Perf.: $200   **Exc.:** $150   **VGood:** $130
With plain barrel, recoil pad
   Perf.: $200   **Exc.:** $150   **VGood:** $130
With vent rib
   Perf.: $225   **Exc.:** $175   **VGood:** $155
With vent rib, recoil pad
   Perf.: $225   **Exc.:** $175   **VGood:** $155

SMITH & WESSON MODEL 1000
Gas-operated autoloader; 12-, 20-ga.; 2 ¾″ chamber; 4-shot magazine; 26″ Skeet or Improved Cylinder, 28″ Improved, Modified or Full barrel; vent rib; front, middle beads; walnut checkered pistol-grip stock, forearm; crossbolt safety; pressure compensator; engraved alloy receiver. Made in Japan. Introduced 1973; discontinued 1984.
   Perf.: $325   **Exc.:** $280   **VGood:** $250

Smith & Wesson Model 1000 Magnum
Same specs as Model 1000 except 12-, 20-ga.; 3″ chamber; 28″, 30″ Modified or Full choke barrel; recoil pad. Introduced 1977; discontinued 1984.
   Perf.: $330   **Exc.:** $285   **VGood:** $255

Smith & Wesson Model 1000S
Same specs as Model 1000 except recessed-type Skeet choke with compensator to soften recoil, reduce muzzle jump; fluorescent red front bead; oil-finished select walnut stock with palm swell. Introduced 1979; discontinued 1984.
   Perf.: $375   **Exc.:** $325   **VGood:** $280

Smith & Wesson Model 1000 Trap
Same specs as Model 1000 except 30″ Multi-Choke barrel; stepped rib; white middle bead, Bradley front; Monte Carlo trap-style stock; shell catcher; steel receiver. Introduced 1983; discontinued 1984.
   Perf.: $550   **Exc.:** $400   **VGood:** $350

Smith & Wesson Model 1000 Waterfowler
Same specs as Model 1000 except 3″ chamber; 30″ Full-choke barrel; dull oil stock finish; quick-detachable swivels; padded camouflage sling; Parkerized finish; black oxidized bolt. Introduced 1982; discontinued 1984.
   Perf.: $350   **Exc.:** $300   **VGood:** $260

SMITH & WESSON MODEL 3000
Slide action; 12-, 20-ga.; 3″ chamber; 22″ slug with rifle sights, 26″ Improved, 28″ Modified, 30″ Full plain or vent-rib barrel; Multi-Choke available; American walnut stock; crossbolt reversible safety for left-handers; dual action bars; chrome-plated bolt; steel receiver; rubber recoil pad. Introduced 1980; discontinued 1984.
   Perf.: $325   **Exc.:** $265   **VGood:** $200
With slug barrel
   Perf.: $290   **Exc.:** $230   **VGood:** $170

Smith & Wesson Model 3000 Waterfowler
Same specs as Model 3000 except 3″ chamber; 30″ Full-Choke barrel; dull oil-finished stock; quick-detachable sling swivels; padded camo sling; Parkerized finish; black oxidized bolt. Introduced 1982; discontinued 1984.
   Perf.: $325   **Exc.:** $265   **VGood:** $200

SPORTING ARMS SNAKE CHARMER
Single barrel; breakopen; .410; 3″ chamber; 18 ⅛″ barrel; no sights; plastic thumbhole stock; storage compartment in buttstock for spare ammo; all stainless steel construction. Introduced 1978; discontinued 1988.
   Perf.: $100   **Exc.:** $75   **VGood:** $60

Sporting Arms Snake Charmer II
Same specs as Snake Charmer. Reintroduced 1989; still in production. **LMSR:** $149
   Perf.: $110   **Exc.:** $85   **VGood:** $70

Sporting Arms Snake Charmer II New Generation
Same specs as Snake Charmer I except black carbon steel barrel. Introduced 1989; still in production. **LMSR:** $139
   Perf.: $90   **Exc.:** $70   **VGood:** $55

STEVENS MODEL 22-410
Over/under combo; exposed hammer; takedown; 22 Short, 22 Long, 22 LR barrel over .410 shotgun barrel; 24″ barrels; Full choke; open rear sight, rifle-type ramp front; original models had uncheckered American walnut pistol-grip stock, forearm; later production had Tenite plastic stock, forearm; single trigger. Introduced 1938; discontinued 1950. Still in production by Savage Arms as Model 24, with variations.
With walnut stock
   Exc.: $100   **VGood:** $80   **Good:** $60
With Tenite stock
   Exc.: $125   **VGood:** $90   **Good:** $70

STEVENS MODEL 58
Bolt action; takedown; .410; 3-shot detachable box magazine; 24″ barrel; Full choke; uncheckered, one-piece, walnut-finished hardwood pistol-grip stock; late models had checkering; plastic buttplate. Introduced 1937; discontinued 1945.
   Exc.: $75   **VGood:** $60   **Good:** $50

Stevens
Model 59
Model 67
Model 67 Slug
Model 67 VRT-K Camo
Model 67 VRT-Y
Model 69 RXL
Model 77
Model 77SC
Model 79-VR
Model 94C
Model 94Y
Model 107

Sporting Arms Snake Charmer

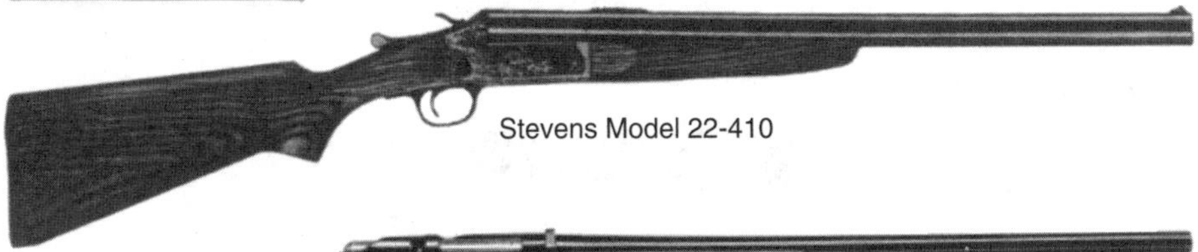

Stevens Model 22-410

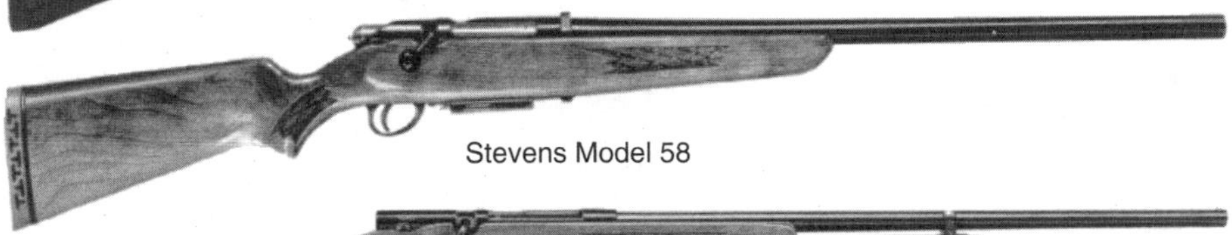

Stevens Model 58

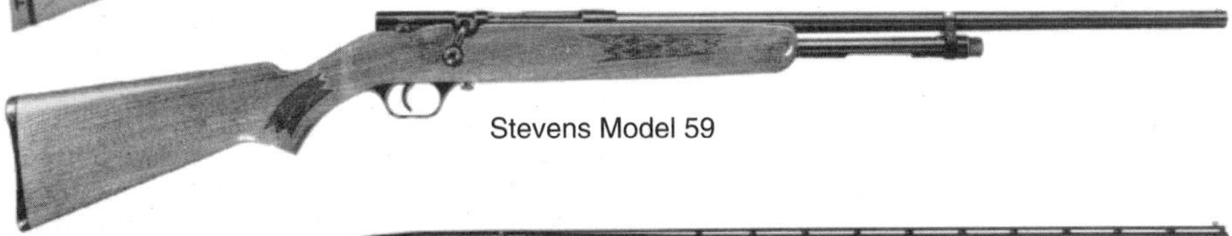

Stevens Model 59

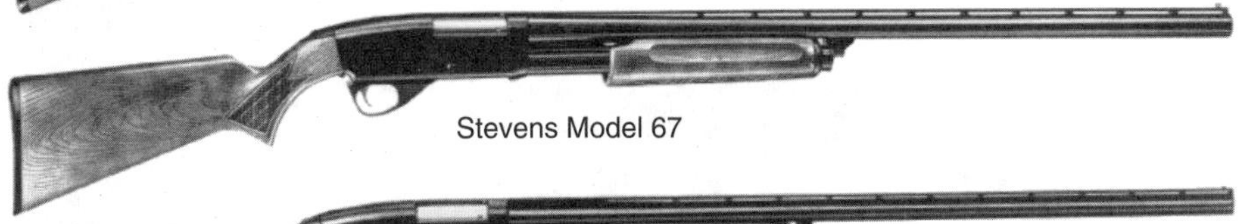

Stevens Model 67

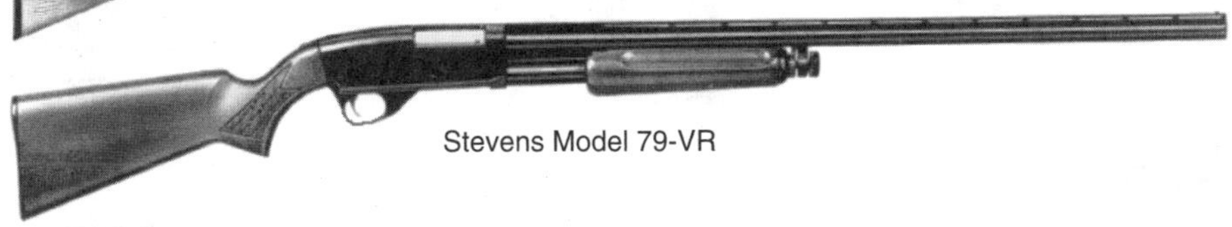

Stevens Model 79-VR

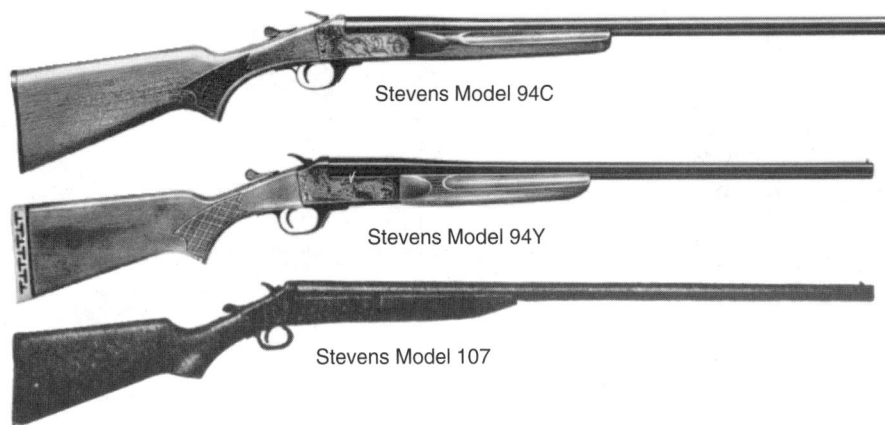

Stevens Model 94C

Stevens Model 94Y

Stevens Model 107

STEVENS MODEL 59

Bolt action; takedown; .410; 5-shot tube magazine; 24″ barrel; Full choke; uncheckered one-piece walnut-finished hardwood pistol-grip stock; plastic buttplate. Introduced 1934; discontinued 1973.

Exc.: $80   **VGood:** $65   **Good:** $55

STEVENS MODEL 67

Slide action; 12-, 20-ga., .410; 3″ chamber; 5-shot tube magazine; 26″, 28″, 30″ barrel; optional vent rib; fixed or choke tubes; metal bead front sight; checkered walnut-finished hardwood pistol-grip stock, tapered slide handle; top tang safety; steel receiver. Introduced 1981; discontinued 1988.

Perf.: $175   **Exc.:** $150   **VGood:** $125
With vent rib
Perf.: $185   **Exc.:** $160   **VGood:** $135

Stevens Model 67 Slug

Same specs as Model 67 except 12-ga.; 21″ slug barrel; rifle sights. Introduced 1986; discontinued 1989.

Perf.: $175   **Exc.:** $150   **VGood:** $125

Stevens Model 67 VRT-K Camo

Same specs as Model 67 except 12-, 20-ga.; 28″ vent-rib barrel; choke tubes; laminated hardwood camouflage stock, slide handle. Introduced 1986; discontinued 1988.

Perf.: $225   **Exc.:** $180   **VGood:** $160

Stevens Model 67 VRT-Y

Same specs as Model 67 except 20-ga.; 22″ vent-rib barrel; choke tubes; youth-sized checkered walnut-finish hardwood stock, forearm. Introduced 1987; discontinued 1988.

Perf.: $200   **Exc.:** $150   **VGood:** $125

STEVENS MODEL 69 RXL

Slide action; 12-ga.; 5-shot tube magazine; 18 1/4″ Cylinder barrel; checkered walnut-finished hardwood stock, slide handle; recoil pad. Introduced 1981; discontinued 1989.

Perf.: $200   **Exc.:** $150   **VGood:** $125

STEVENS MODEL 77

Slide-action; hammerless; solid frame; 12, 16-, 20-ga.; 5-shot tube magazine; 26″ Improved; 28″ Modified or Full barrel; uncheckered walnut-finished hardwood stock, grooved slide handle. Introduced 1954; discontinued 1971.

Exc.: $130   **VGood:** $90   **Good:** $70

Stevens Model 77SC

Same specs as Model 77 except Savage Super Choke. No longer in production.

Exc.: $130   **VGood:** $90   **Good:** $70

STEVENS MODEL 79-VR

Slide action; 12-, 20-ga.; 26″, 28″, 30″ vent-rib barrel; metal bead front sight; checkered walnut-finished hardwood pistol-grip stock, tapered slide handle; top tang safety; interchangeable barrels. Introduced 1981; discontinued 1983.

Perf.: $200   **Exc.:** $150   **VGood:** $125

STEVENS MODEL 94C

Single barrel; single shot; exposed hammer; breakopen; early models side lever breaking; 12-, 16-, 20-ga., .410; 28″, 30″, 32″, 36″ barrel; Full choke; checkered walnut finished hardwood pistol-grip stock, forearm; automatic ejector; color case-hardened frame. Introduced 1937; discontinued 1984.

Perf.: $100   **Exc.:** $80   **VGood:** $60

Stevens Model 94Y

Same specs as Model 94C except top lever breaking; 20-ga., .410; 26″ barrel; youth stock; recoil pad. Discontinued 1984.

Perf.: $100   **Exc.:** $80   **VGood:** $60

STEVENS MODEL 107

Single barrel; exposed hammer; takedown; 12-, 16-, 20-ga., .410; 26″, 28″, 30″ barrel; Full choke; uncheckered walnut-finished hardwood pistol-grip stock, forearm automatic ejector. Introduced 1937; discontinued 1953.

Exc.: $75   **VGood:** $50   **Good:** $40

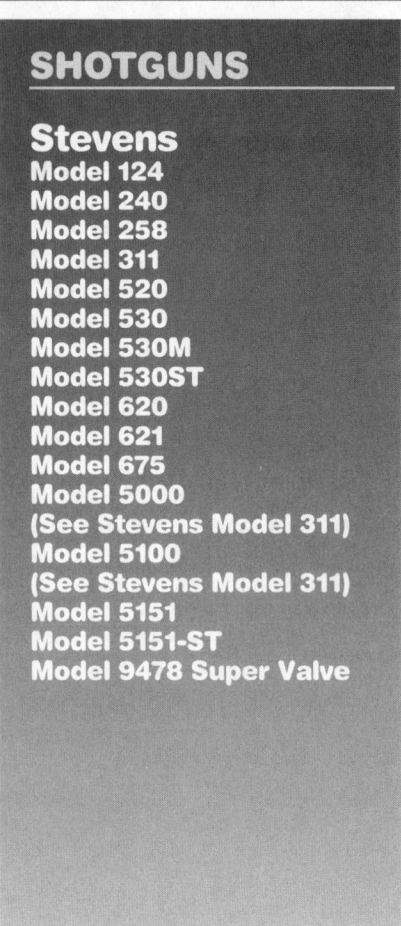

SHOTGUNS

Stevens
Model 124
Model 240
Model 258
Model 311
Model 520
Model 530
Model 530M
Model 530ST
Model 620
Model 621
Model 675
Model 5000
(See Stevens Model 311)
Model 5100
(See Stevens Model 311)
Model 5151
Model 5151-ST
Model 9478 Super Valve

Stevens Model 124

Stevens Model 240

Stevens Model 258

Stevens Model 311

Stevens Model 520

Stevens Model 530

Stevens Model 620

Stevens Model 9478 Super Value

STEVENS MODEL 124
Straight-pull bolt action; solid frame, hammerless; 12-ga.; 2-shot tube magazine; 28" barrel; Improved, Modified, Full chokes; checkered Tenite plastic stock, forearm. Introduced 1947; discontinued 1952.
   Exc.: $175   **VGood:** $140   **Good:** $125

STEVENS MODEL 240
Over/under; takedown; .410; 26" barrels; Full choke; early models had uncheckered American walnut pistol-grip stock, forearm; later versions had Tenite plastic stock forearm; double triggers. Introduced 1939; discontinued 1942.
With walnut stock
   Exc.: $225   **VGood:** $180   **Good:** $160
With Tenite stock
   Exc.: $250   **VGood:** $200   **Good:** $180

STEVENS MODEL 258
Bolt action; takedown; 20-ga.; 3-shot detachable box magazine; 25" barrel; Full choke; uncheckered hardwood, one-piece, pistol-grip stock; black plastic forearm cap, buttplate. Introduced 1937; discontinued 1965.
   Exc.: $100   **VGood:** $75   **Good:** $50

STEVENS MODEL 311
Side-by-side; boxlock; hammerless; 12-, 16-, 20-ga., .410; 3" chambers; 26", 28", 30", 32" barrels; standard choke combos; early models had uncheckered American walnut pistol-grip stock, forearm; later models had walnut-finished hardwood pistol-grip stock, fluted comb; double triggers, auto top tang safety; plastic buttplate; case-hardened finish on frame. Originally introduced 1931 as Model 311, but stamped Model 5000; in 1941 stamp was changed to Model 5100, but still listed in catalog as Model 311A; in 1950 gun marked Model 311; discontinued 1988.
   Perf.: $225   **Exc.:** $175   **VGood:** $150

STEVENS MODEL 520
Slide action; 12-, 16-, 20-ga.; 5-shot magazine; 30" barrel; weighs 7 1/2 lbs.; checkered American walnut stock, forend; safety inside trigger guard; made exclusively for Sears, Roebuck; marketed as Sears Ranger. Introduced 1915; discontinued about 1930.
   Exc.: $150   **VGood:** $110   **Good:** $80

STEVENS MODEL 530
Side-by-side; boxlock; hammerless; 12-, 16-, 20-ga., .410; 26", 28", 30", 32" barrels; Modified/Full, Cylinder/Modified, Full/Full chokes; hand-checkered American walnut pistol-grip stock, forearm; double triggers; early models have recoil pads. Introduced 1936; discontinued 1954.
   Exc.: $200   **VGood:** $140   **Good:** $110

Stevens Model 530M
Same specs as Model 530 except Tenite plastic stock, forearm. Introduced before WWII; discontinued 1947.
   Exc.: $350   **VGood:** $200   **Good:** $120

Stevens Model 530ST
Same specs as Model 530 except single selective trigger. Introduced 1947; discontinued 1954.
   Exc.: $375   **VGood:** $225   **Good:** $135

STEVENS MODEL 620
Slide action; hammerless; takedown; 12-, 16-, 20-ga.; 5-shot tube magazine; 26", 28", 30", 32" barrel; Cylinder, Improved, Modified, Full choke; hand-checkered American walnut pistol-grip stock, slide handle; black plastic buttplate. Introduced 1927; discontinued 1953.
   Exc.: $250   **VGood:** $185   **Good:** $125

STEVENS MODEL 621
Slide action; hammerless; takedown; 12-, 16-, 20-ga.; 5-shot tube magazine; 26", 28", 30", 32" barrel; Cylinder, Improved, Modified Full choke; raised solid matted rib; hand-checkered American walnut pistol-grip stock, slide handle; black plastic buttplate. Introduced 1927; discontinued 1953.
   Exc.: $250   **VGood:** $185   **Good:** $125

STEVENS MODEL 675
Slide action; 12-ga.; 5-shot tube magazine; 24" vent-rib barrel; choke tubes; rifle iron sights; checkered walnut-finished hardwood pistol-grip stock, forearm; recoil pad. Introduced 1987; discontinued 1988.
   Perf.: $240   **Exc.:** $185   **VGood:** $160

STEVENS MODEL 5151
Side-by-side; boxlock; hammerless; 12-, 16-, 20-ga., .410; 3" chambers; 26", 28", 30", 32" barrels; Ivoroid sights; hand-checkered American walnut pistol-grip stock, forearm; recoil pad. Introduced 1936; discontinued 1942.
   Exc.: $200   **VGood:** $140   **Good:** $110

Stevens Model 5151-ST
Same specs as Model 5151 except non-selective single trigger. No longer in production.
   Exc.: $220   **VGood:** $160   **Good:** $120

STEVENS MODEL 9478 SUPER VALUE
Single barrel; exposed hammer; 10-, 12-, 20-ga., .410; 26", 28", 30", 36" barrel; walnut-finished hardwood stock; bottom action opening button; auto ejection; color case-hardened frame. Introduced 1979; discontinued 1986.
   Perf.: $110   **Exc.:** $80   **VGood:** $60

Sturm, Ruger Red Label

Sturm, Ruger Woodside

Tar-Hunt RSG-12

Techni-Mec Model SPL 640

Techni-Mec Model 610

Stoeger/IGA
(See IGA)

Sturm, Ruger
Red Label
Red Label Sporting Clays
Red Label Sporting Clays
20-Gauge
Woodside

Tar-Hunt
RSG-12 Professional
RSG-12 Professional
Matchless Model
RSG-12 Professional Peer-
less Model
RSG-12 Turkey Model

Techni-Mec
Model 610
Model SPL 640
Model SPL 642
Model SR 690 Skeet
Model SR 690 Trap
Model SR 692 EM

STURM, RUGER RED LABEL

Over/under; boxlock; 12-, 20-ga.; 3" chambers; 26", 28" vent-rib barrels; Full/Modified, Improved Cylinder/Modified, Skeet choke tubes; checkered American walnut pistol-grip stock, forearm; optional straight-grip stock; single selective trigger; auto ejectors. Introduced (20-ga.) 1977; (12-ga.) 1982; still in production. **LMSR: $1215**

    Perf.: $950   **Exc.:** $700   **VGood:** $600
With straight-grip stock
    Perf.: $950   **Exc.:** $700   **VGood:** $600

Sturm, Ruger Red Label Sporting Clays

Same specs as Red Label except 12-ga.; 3" chambers; 30" barrels backbored to .744" diameter; two Skeet, one Improved Cylinder, one Modified stainless steel choke tubes; optional Full and Extra-Full tubes available; free-floating serrated vent rib; brass front, mid-rib beads; weighs 7 3/4 lbs.; overall length 47"; stock dimensions of 14 1/8"x1 1/2"x2 1/2"; no barrel side ribs. Introduced 1992; still in production. **LMSR: $1349**

    Perf.: $1100   **Exc.:** $800   **VGood:** $700

Sturm, Ruger Red Label Sporting Clays 20-Gauge

Same specs as the Red Label except 20-ga.; 3" chambers; 30" barrels backbored to .631"-.635" diameter; four special 2" long interchangeable, screw-in choke tubes: two Skeet, one Modified, one Improved Cylinder; optional Full and Extra-Full tubes available; no barrel side spacers. Introduced 1994; still in production. **LMSR: $1349**

    Perf.: $1100   **Exc.:** $800   **VGood:** $700

STURM, RUGER WOODSIDE

Over/under; boxlock; 12-, 20-ga.; 3" chambers; 26", 28", 30" backbored barrels; Full, Modified, Improved Cylinder and two Skeet stainless steel choke tubes; serrated free-floating rib; weighs 7 1/2 to 8 lbs.; select Circassian walnut pistol-grip or straight-grip stock; buttstock extends forward into action as two sidepanels; stock dimensions of 14 1/8"x1 1/2"x2 1/2"; newly patented Ruger cocking mechanism for easier, smoother opening; single selective mechanical trigger; selective automatic ejectors; blued barrels; stainless steel receiver; optional engraved action. Introduced 1995; still in production. **LMSR: $1675**

    Perf.: $1500   **Exc.:** $1100   **VGood:** $850

TAR-HUNT RSG-12 PROFESSIONAL

Bolt action; 12-, 20-ga.; 2 3/4" chamber; 21 1/2" fully rifled barrel with muzzlebrake; weighs 7 3/4 lbs.; Weaver-style scope mounting bases and Burris Zee steel rings; matte black McMillan fiberglass stock; Pachmayr Decelerator pad; rifle-style action with two locking lugs; two-position safety; single-stage, adjustable rifle trigger; many options available; right- and left-hand models at same prices. Introduced 1991; still in production. **LMSR: $1395**

    Perf.: $1200   **Exc.:** $850   **VGood:** $750

Tar-Hunt RSG-12 Professional Matchless Model

Same specs as RSG-12 Professional except McMillan Fibergrain or camouflage stock; 400-grit gloss metal finish. Introduced 1991; still in production. **LMSR: $1783**

    Perf.: $1500   **Exc.:** $1100   **VGood:** $950

Tar-Hunt RSG-12 Professional Peerless Model

Same specs as RSG-12 Professional except McMillan Fibergrain fiberglass stock; NP-3 nickel/teflon metal finish. Introduced 1991; still in production. **LMSR: $1973**

    Perf.: $1750   **Exc.:** $1200   **VGood:** $1000

Tar-Hunt RSG-12 Turkey Model

Same specs as RSG-12 Professional except smoothbore barrel; Remington Rem-Choke thread system. Introduced 1991; still in production. **LMSR: $1439**

    Perf.: $1200   **Exc.:** $850   **VGood:** $750

TECHNI-MEC MODEL 610

Over/under; boxlock; 10-ga.; 3 1/2" chambers; 32" barrels; Improved Modified/Full chokes; hand-checkered walnut stock, forend; rubber recoil pad; single selective trigger; silvered and engraved frame, blued barrels. Made in Italy. Introduced 1991; no longer imported.

    Perf.: $875   **Exc.:** $700   **VGood:** $625

TECHNI-MEC MODEL SPL 640

Over/under; boxlock; 12-, 16-, 20-, 28-ga.; 2 3/4" chambers; 26" chrome-lined barrel; Modified/Full chokes; ventilated rib; weighs 5 1/2 lbs.; European walnut stock; single or double triggers; folds in half for storage, transportation; photo-engraved silvered

receiver. Made in Italy. Introduced 1984; no longer imported.

    Perf.: $440   **Exc.:** $300   **VGood:** $250

TECHNI-MEC MODEL SPL 642

Folding over/under; 12-, 16-, 20-, 28-ga., .410; 26" Modified/Full barrels; vent rib; checkered European walnut stock, forearm; single or double triggers; chrome-lined barrels; photo-engraved, silvered receiver. No longer in production.
With single trigger
    Perf.: $440   **Exc.:** $300   **VGood:** $250
With double triggers
    Perf.: $420   **Exc.:** $280   **VGood:** $230

TECHNI-MEC MODEL SR 690 SKEET

Over/under; boxlock; 12-ga.; 2 3/4" chambers; 25", 28" barrels; Skeet/Skeet chokes; Ray-type sights; select European walnut Monte Carlo stock with Skeet dimensions; single selective trigger; automatic ejectors; antique silver finish on receiver. Made in Italy. Introduced 1984; no longer imported.

    Perf.: $525   **Exc.:** $375   **VGood:** $300

Techni-Mec Model SR 690 Trap

Same specs as Model SR 690 Skeet except 30" barrels; Full/Full chokes; select European walnut Monte Carlo stock with trap dimensions. Made in Italy. Introduced 1984; no longer imported.

    Perf.: $525   **Exc.:** $375   **VGood:** $300

TECHNI-MEC MODEL SR 692 EM

Over/under; boxlock; 12-, 16-, 20-ga.; 2 3/4", 3" chambers; 26", 28", 30" barrels; Modified, Full, Improved Cylinder, Cylinder chokes; checkered European walnut pistol-grip stock, forend; single selective trigger; automatic ejectors; dummy sideplates with fine game scene engraving. Made in Italy. Introduced 1984; no longer imported.

    Perf.: $440   **Exc.:** $350   **VGood:** $290

Thompson/Center
Custom Shop TCR '87 Hunter

Tikka
Model 77
Model 412S Field Grade
Model 412S Sporting Clays
Model 412ST Skeet
Model 412ST Skeet Grade II
Model 412ST Trap
Model 412ST Trap Grade II
Model 512S (See Tikka Model 412S)

Toledo Armas
Valezquez

Tradewinds
H-170
Model G1032 (See Mercury Magnum)

Trident
Supertrap II

Ugartechea
10-Gauge Magnum

Union Armera
Luxe
Winner

Thompson/Center Custom Shop TCR '87 Hunter

Tikka Model 77

Tikka Model 412S Field Grade

Tikka Model 412S Sporting Clays

Tradewinds H-170

THOMPSON/CENTER CUSTOM SHOP TCR '87 HUNTER
Single shot; boxlock; 10-, 12-ga.; 3, 3 1/2" chamber; 25" barrel; Full choke designed for steel shot; weighs 8 lbs.; uncheckered walnut stock; stock has extra 7/16" drop at heel; same receiver as TCR '87 rifle models. Introduced through T/C custom shop 1989; discontinued 1994.
   Perf.: $525   **Exc.:** $395   **VGood:** $325

TIKKA MODEL 77
Over/under; 12-ga.; 27", 30" vent-rib barrels; skip-line-checkered European walnut Monte Carlo pistol-grip stock with roll-over cheekpiece, forend; ejectors; barrel selector; single trigger. Made in Finland. Introduced 1979; no longer imported.
   Perf.: $675   **Exc.:** $525   **VGood:** $475

TIKKA MODEL 412S FIELD GRADE
Over/under; boxlock; 12-, 20-ga.; 3" chambers; 24", 26", 28", 30" barrels; Improved Cylinder, Modified, Improved Modified, Full stainless steel screw-in chokes; weighs about 7 1/4 lbs.; checkered American walnut pistol-grip stock, forend; barrel selector in trigger; automatic top tang safety; barrel cocking indicators; system allows free interchangeability of barrels, stocks and forends into double rifle model, combination gun, etc. Name changed to Model 512S in 1993. Made in Finland and Italy. Introduced 1980; still imported as Tikka Model 512S. **LMSR: $1290**
   Perf.: $1000   **Exc.:** $700   **VGood:** $525

Tikka Model 412S Sporting Clays
Same as the Model 412S except 12-ga.; 28", 30" barrels; five choke tubes; manual safety. Made in Finland and Italy. Introduced 1992; still imported as Model 512S. **LMSR: $1270**
   Perf.: $1050   **Exc.:** $750   **VGood:** $600

TIKKA MODEL 412ST SKEET
Over/under; 12-, 20-ga.; 28" stepped-rib barrels; choke tubes; European walnut Skeet-style stock; mechanical single trigger; auto ejectors; elongated forcing cone; cocking indicators; hand-honed action. Made in Finland. Introduced 1980; discontinued 1990.
   Perf.: $1000   **Exc.:** $750   **VGood:** $650

Tikka Model 412ST Skeet Grade II
Same specs as Model 412ST Skeet except checkered semi-fancy European walnut Skeet-style stock; drilled for recoil-reducing unit; matte nickel receiver; matte blue locking bolt, lever; gold trigger. Made in Finland. Introduced 1989; discontinued 1990.
   Perf.: $1200   **Exc.:** $850   **VGood:** $775

Tikka Model 412ST Trap
Same specs as Model 412ST Skeet except 12-ga.; 30", 32" barrels; European walnut stock, palm swell. Made in Finland. Introduced 1980; discontinued 1990.
   Perf.: $1000   **Exc.:** $750   **VGood:** $650

Tikka Model 412ST Trap Grade II
Same specs as Model 412ST Skeet except 12-ga.; 30", 32" barrels; checkered semi-fancy European walnut stock, palm swell; drilled for recoil-reducing unit; matte nickel receiver; matte blue locking bolt, lever; gold trigger. Made in Finland. Introduced 1989; discontinued 1990.
   Perf.: $1200   **Exc.:** $850   **VGood:** $775

TOLEDO ARMAS VALEZQUEZ
Side-by-side; 12-ga. 2 3/4" chambers; custom barrel lengths, chokes; custom exhibition-grade European walnut stock; auto ejectors; hand-engraved action; many options. Made in Spain. Introduced 1982; discontinued 1984.
   Perf.: $2600   **Exc.:** $2000   **VGood:** $1700

TRADEWINDS H-170
Recoil-operated autoloader; 12-ga.; 2 3/4" chambers; 5-shot tube magazine; 26" Modified, 28" Full barrel; vent rib; hand-checkered select European walnut pistol-grip stock; light alloy receiver. Made in Japan. Introduced 1970; no longer in production.
   Perf.: $250   **Exc.:** $200   **VGood:** $175

TRIDENT SUPERTRAP II
Single shot; 12-ga.; 32", 34" ported barrel; Multi-Choke tubes; weighs 8 1/2 lbs.; white front bead, brass middle bead; checkered American walnut stock, forend; vent rubber recoil pad; pull/release-convertible trigger; long forcing cone. Introduced 1990; discontinued 1992.
   Perf.: $1800   **Exc.:** $1400   **VGood:** $1200

UGARTECHEA 10-GAUGE MAGNUM
Side-by-side; boxlock; 10-ga.; 3 1/2" chambers; 32" barrels; matted rib; weighs 11 lbs.; front, center metal beads; checkered European walnut stock; vent rubber recoil pad; Purdey-type forend release; double triggers; color case-hardened action; rest is blued. Made in Spain. Introduced 1990; still in production. **LMSR: $700**
   Perf.: $625   **Exc.:** $545   **VGood:** $500

UNION ARMERA LUXE
Side-by-side; 12-, 20-ga.; 2 3/4" chambers; custom built to customer's specs; top-grade European walnut stock; auto ejectors; hand-engraved action; numerous options. Made in Spain. Introduced 1982; discontinued 1984.
   Perf.: $5200   **Exc.:** $3750   **VGood:** $3200

UNION ARMERA WINNER
Side-by-side; 12-, 20-ga.; 2 3/4" chambers; custom built to customer's specs; top-grade European walnut stock; auto ejectors; hand-engraved action; numerous options. Made in Spain. Introduced 1982; discontinued 1984.
   Perf.: $2700   **Exc.:** $2250   **VGood:** $1900

Ugartechea 10-Gauge Magnum

Union Armera Winner

Universal Firearms Auto Wing

Universal Firearms Double Wing

Universal Firearms Duck Wing

Universal Firearms Single Wing

Universal Firearms Over Wing

Urbiola Model 160E

UNIVERSAL FIREARMS AUTO WING
Recoil-operated autoloader; takedown; 12-ga.; 2 3/4″ chamber; 5-shot magazine, 3-shot plug furnished; 25″, 28″, 30″ barrel; Improved, Modified, Full chokes; vent rib; ivory bead front, middle sights; checkered European walnut pistol-grip stock, grooved forearm; crossbolt safety; interchangeable barrels. Introduced 1970; discontinued 1974.
    Perf.: $150   **Exc.:** $125   **VGood:** $100

UNIVERSAL FIREARMS DOUBLE WING
Side-by-side; boxlock; 10-, 12-, 20-ga.; 26″ Improved/Modified, 28″ or 30″ Modified/Full barrels; checkered European walnut pistol-grip stock, beavertail forearm; double triggers; recoil pad. Introduced 1970; discontinued 1974.
    Perf.: $350   **Exc.:** $295   **VGood:** $250

UNIVERSAL FIREARMS DUCK WING
Recoil-operated autoloader; takedown; 12-ga.; 2 3/4″ chamber; 5-shot magazine, 3-shot plug furnished; 28″, 30″ barrel; Full choke; ivory bead front, middle sights; checkered European walnut pistol-grip stock, grooved forearm; crossbolt safety; interchangeable barrels; exposed metal parts coated with olive green Teflon-S. Introduced 1970; discontinued 1972.
    Perf.: $150   **Exc.:** $125   **VGood:** $95

UNIVERSAL FIREARMS MODEL 101
Single shot; top break; external hammer; takedown; 12-ga.; 3″ chamber; 28″, 30″ Full-choke barrel; uncheckered pistol-grip stock, beavertail forearm. Introduced 1967; discontinued 1969. Replaced by Single Wing model.
    Perf.: $75   **Exc.:** $45   **VGood:** $30

UNIVERSAL FIREARMS MODEL 202
Side-by-side; boxlock; 12-, 20-ga.; 3″ chambers; 26″ Improved/Modified, 28″ Modified/Full barrels; hand-checkered European walnut pistol-grip stock, European-style forearm; double triggers. Introduced 1967; discontinued 1969. Replaced by Double Wing model.
    Perf.: $130   **Exc.:** $110   **VGood:** $85

UNIVERSAL FIREARMS MODEL 203
Side-by-side; boxlock; 10-ga.; 3 1/2″ chambers; 32″ Full/Full barrels; hand-checkered European walnut pistol-grip stock, European-style forearm; double triggers. Introduced 1967; discontinued 1969.
    Perf.: $140   **Exc.:** $120   **VGood:** $95

UNIVERSAL FIREARMS MODEL 2030
Side-by-side; boxlock; 10-ga.; 3″ chambers; 32″ Full/Full barrels; checkered European walnut pistol-grip stock, beavertail forearm; double triggers; recoil pad. Introduced 1970; discontinued 1974.
    Perf.: $160   **Exc.:** $140   **VGood:** $110

UNIVERSAL FIREARMS OVER WING
Over/under; boxlock; hammerless; 12-, 20-ga.; 3″ chambers; 26″ Improved/Modified, 28″ or 30″ Modified/Full barrels; vent rib; front, middle sights; checkered European walnut pistol-grip stock, forearm; double triggers; single-trigger model with engraved receiver at added cost. Introduced 1970; discontinued 1974.

With double triggers
    Perf.: $225   **Exc.:** $195   **VGood:** $150
With single trigger, engraved receiver
    Perf.: $250   **Exc.:** $220   **VGood:** $175

UNIVERSAL FIREARMS SINGLE WING
Single-shot; top break; external hammer; takedown; 12-ga.; 3″ chamber; 28″ Full or Modified barrel; uncheckered European walnut pistol-grip stock, beavertail forearm; automatic ejector. Introduced 1970; discontinued 1974.
    Perf.: $75   **Exc.:** $45   **VGood:** $30

URBIOLA MODEL 160E
Side-by-side; 12-, 20-ga.; 2 3/4″ chambers; 26″ Improved/Full barrels; hand-checkered European walnut straight-grip stock; checkered butt; automatic ejectors; double triggers; hand-engraved locks. Made in Spain. Introduced 1982 by Toledo Armas; discontinued 1984.
    Perf.: $625   **Exc.:** $500   **VGood:** $425

Valmet
Lion
Model 412K
Mdoel 412KE
Model 412S American
Model 412ST Skeet
Model 412ST Skeet
Grade II
Model 412ST Trap
Model 412ST Trap Grade II

Ventura
Avanti Small Gauge
Avanti Small Gauge Extra
Lusso
Contento
Contento Extra Lusso
Contento Lusso Grade
Model 51
Model 53
Model 62
Model 64

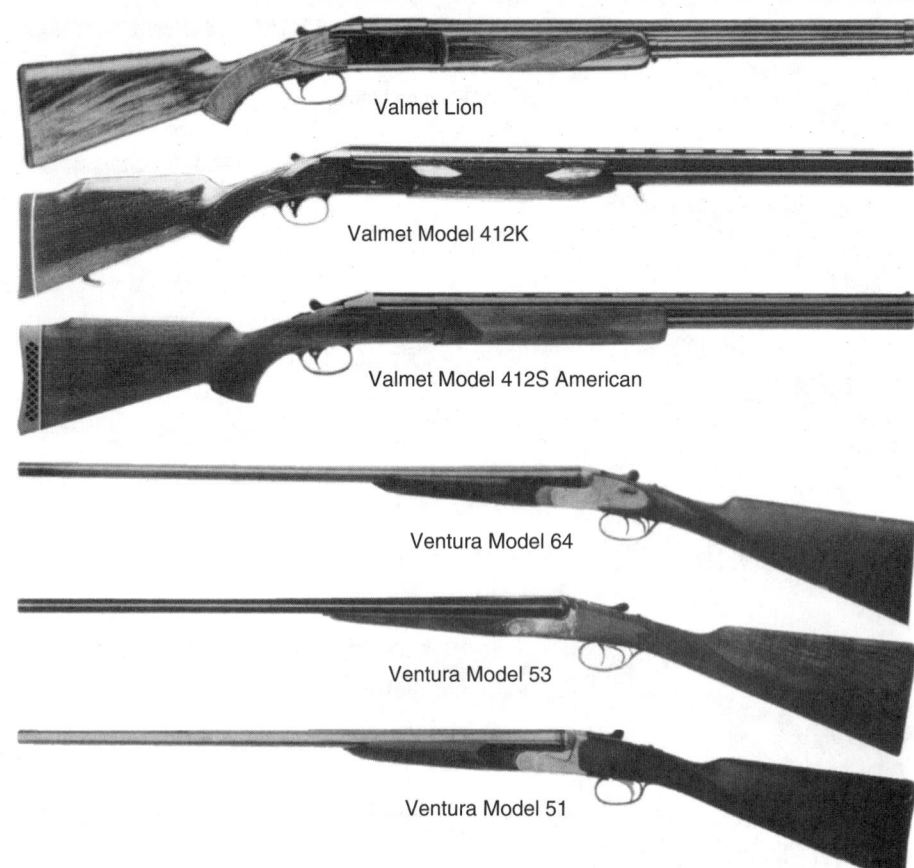

Valmet Lion

Valmet Model 412K

Valmet Model 412S American

Ventura Model 64

Ventura Model 53

Ventura Model 51

VALMET LION
Over/under; boxlock; 12-ga.; 26″ Improved/Modified, 28″ Modified/Full, 30″ Modified/Full or Full/Full barrels; hand-checkered walnut stock; single selective trigger; plain extractors. Made in Finland. Introduced 1951; discontinued 1967.
Exc.: $360 **VGood:** $300 **Good:** $250

VALMET MODEL 412K
Over/under; 12-, 20-ga.; 26″, 28″, 30″ barrels; vent rib; checkered American walnut stock; extractors; interchangeable barrels, stocks, forends; barrel selector on trigger; auto top tang safety; barrel cocking indicators; optional double triggers. Made in Finland. Introduced 1980; discontinued 1990.
Perf.: $700 **Exc.:** $500 **VGood:** $400

Valmet Model 412KE
Same specs as Model 412K except auto ejectors; non-auto safety. Introduced 1980; discontinued 1983.
Perf.: $775 **Exc.:** $575 **VGood:** $475

Valmet Model 412S American
Same specs as Model 412K except luminous sights; better wood, checkering; palm swell on pistol grip; new forend latch spring mechanism; improved firing pin; made in trap, Skeet, field versions. Made in Finland. Introduced 1980; discontinued 1984.
Perf.: $1000 **Exc.:** $750 **VGood:** $600

VALMET MODEL 412ST SKEET
Over/under; 12-, 20-ga.; 28″ stepped-rib barrels; choke tubes; European walnut Skeet-style stock; mechanical single trigger; auto ejectors; elongated forcing cone; cocking indicators; hand-honed action. Made in Finland. Introduced 1980; discontinued 1990.
Perf.: $950 **Exc.:** $650 **VGood:** $550

Valmet Model 412ST Skeet Grade II
Same specs as Model 412ST Skeet except checkered semi-fancy European walnut Skeet-style stock drilled for recoil-reducing unit; matte nickel receiver; matte blue locking bolt, lever; gold trigger. Made in Finland. Introduced 1989; discontinued 1990.
Perf.: $1200 **Exc.:** $800 **VGood:** $700

Valmet Model 412ST Trap
Same specs as Model 412ST Skeet except 12-ga.; 30″, 32″ barrels; European walnut stock, palm swell. Made in Finland. Introduced 1980; discontinued 1990.
Perf.: $950 **Exc.:** $650 **VGood:** $550

Valmet Model 412ST Trap Grade II
Same specs as Model 412ST Skeet except 12-ga.; 30″, 32″ barrels; checkered semi-fancy European walnut stock, palm swell; stock drilled for recoil-reducing unit; matte nickel receiver; matte blue locking bolt, lever; gold trigger. Made in Finland. Introduced 1989; discontinued 1990.
Perf.: $1200 **Exc.:** $800 **VGood:** $700

VENTURA AVANTI SMALL GAUGE
Over/under; boxlock; 28-ga. (2 3/4″), .410 (3″); 26″ barrels; vent top, side ribs; weighs 5 3/4 lbs.; straight English-type French walnut stock; single selective trigger; auto ejectors; fully engraved. Made in Italy. Introduced 1987; discontinued 1988.
Perf.: $750 **Exc.:** $600 **VGood:** $525

Ventura Avanti Small Gauge Extra Lusso
Same specs as Avanti Small Gauge except highly figured French walnut stock; more ornate engraving. Introduced 1987; discontinued 1988.
Perf.: $1050 **Exc.:** $850 **VGood:** $750

VENTURA CONTENTO
Over/under; boxlock; 12-ga.; 26″, 28″, 29 1/2″, 32″ barrels; high post rib, vent side ribs; hand-checkered European walnut Monte Carlo stock; Woodward side lugs; double internal bolts; selective single trigger; auto ejectors. Introduced 1975, discontinued 1982.
Perf.: $1800 **Exc.:** $1250 **VGood:** $1000

Ventura Contento Extra Lusso
Same specs as Contento except best-grade fancy walnut stock; extensive Florentine engraving. No longer in production.
Perf.: $2000 **Exc.:** $1475 **VGood:** $1200

Ventura Contento Lusso Grade
Same specs as Contento except better wood; engraved action. No longer in production.
Perf.: $1900 **Exc.:** $1350 **VGood:** $1100

VENTURA MODEL 51
Side-by-side; 12-, 20-ga.; 27 1/2″, 30″ barrels; hand-checkered select European walnut straight or pistol-grip stock, slender beavertail forend; single selective trigger; auto ejectors; hand-engraved action. Made in Spain. Introduced 1980; discontinued 1985.
Perf.: $600 **Exc.:** $450 **VGood:** $375

VENTURA MODEL 53
Side-by-side; 12-, 20-, 28-ga., .410; 25″, 27 1/2″, 30″ barrels; hand-checkered select European walnut straight or pistol-grip stock, slender beavertail forend; single selective or double triggers; auto ejectors; hand-engraved frame. Made in Spain. Introduced 1980; discontinued 1985.
Perf.: $640 **Exc.:** $490 **VGood:** $410

VENTURA MODEL 62
Side-by-side; H&H sidelock; 12-, 20-, 28-ga.; 25″, 27 1/2″, 30″ barrels; select figured English walnut straight or pistol-grip stock, slender beavertail forend; single selective or double triggers; auto ejectors; cocking indicator; gas escape valve; intercepting safety; double underbolts; Purdey-style engraving. Made in Spain. Introduced 1980; discontinued 1982.
Perf.: $1125 **Exc.:** $850 **VGood:** $750

VENTURA MODEL 64
Side-by-side; H&H sidelock; 12-, 20-, 28-ga.; 25″, 27 1/2″, 30″ barrels; select figured English walnut straight or pistol-grip stock, slender beavertail forend; single selective or double triggers; auto ejectors; cocking indicator; gas escape valve; intercepting safety; Florentine engraving. Introduced 1978; no longer in production.
Perf.: $1125 **Exc.:** $850 **VGood:** $750

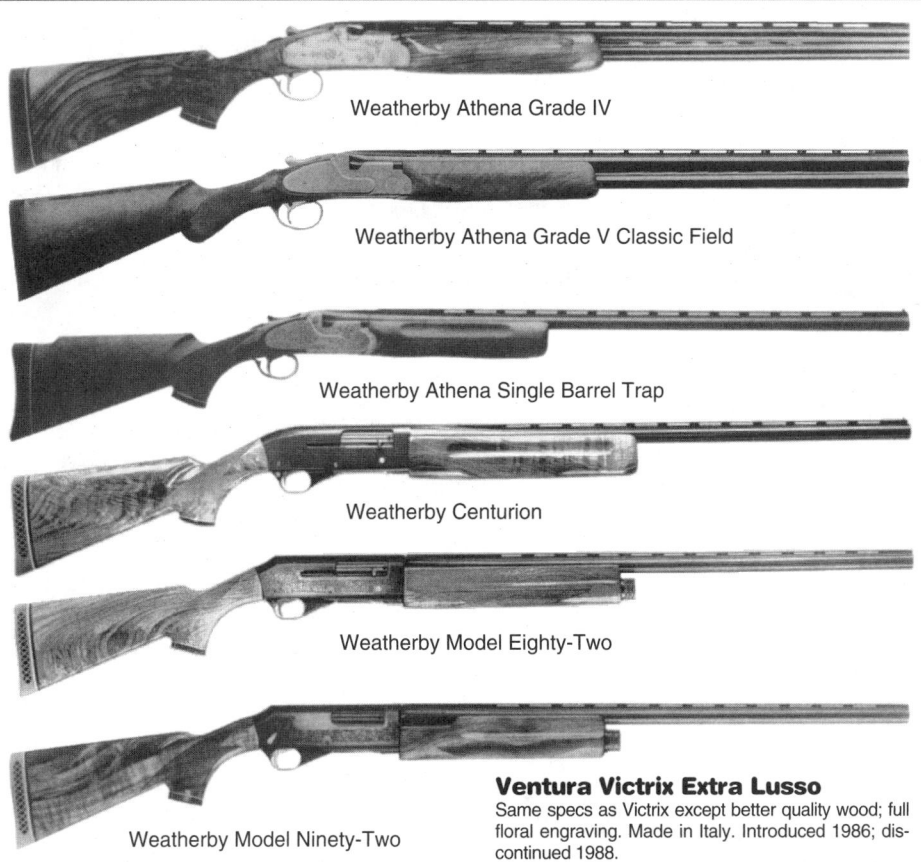

Weatherby Athena Grade IV

Weatherby Athena Grade V Classic Field

Weatherby Athena Single Barrel Trap

Weatherby Centurion

Weatherby Model Eighty-Two

Weatherby Model Ninety-Two

Ventura
Model 66
Model XXV
Regis Model
Victrix
Victrix Extra Lusso

Weatherby
Athena Grade IV
Athena Grade V
Classic Field
Athena Single Barrel
Trap
Centurion
Centurion Deluxe
Centurion Trap
Model Eighty-Two
Model Eighty-Two
Buckmaster
Model Eighty-Two Trap
Model Ninety-Two

VENTURA MODEL 66

Side-by-side; H&H sidelock; 12-, 20-, 28-ga.; 25″, 27 1/2″, 30″ barrels; select figured English walnut straight or pistol-grip stock, slender beavertail forend; single selective or double triggers; auto ejectors; cocking indicator; gas escape valve; intercepting safety; treble bolting; side clips; Florentine engraving. Introduced 1980; no longer in production.

   Perf.: $1125 **Exc.:** $850   **VGood:** $750

VENTURA MODEL XXV

Side-by-side; 12-, 20-, 28-ga.; .410; 25″ barrels; Churchill rib; hand-checkered select European walnut straight or pistol-grip stock, slender beavertail forend; single selective or double triggers; auto ejectors; hand-engraved frame. Made in Spain. Introduced 1980; no longer in production.

   Perf.: $850 **Exc.:** $625   **VGood:** $550

VENTURA REGIS MODEL

Side-by-side; H&H sidelock; 12-, 20-, 28-ga., .410; 2 3/4″, 3″ chambers; 26″, 28″ barrels; weighs 6 1/2 lbs.; hand-checkered select figured French walnut stock, sliver beavertail forend; intercepting safeties; triple locks; auto ejectors; single selective, double triggers; floral engraving; several options. Made in Italy. Introduced 1986; discontinued 1988.

   Perf.: $1500 **Exc.:** $1350   **VGood:** $1100

VENTURA VICTRIX

Side-by-side; Anson & Deeley boxlock; 12-, 20-, 28-ga., .410; 2 3/4″, 3″ chambers; 26″, 28″ barrels; fixed chokes; optional screw-in chokes; weighs 6 1/2 lbs.; hand-checkered French walnut stock, beavertail forend; triple locks; auto ejectors; double or single selective trigger; marketed in leather trunk-type case. Made in Italy. Introduced 1986; discontinued 1988.

With fixed chokes
   Perf.: $800 **Exc.:** $600   **VGood:** $500
With choke tubes
   Perf.: $850 **Exc.:** $650   **VGood:** $550

Ventura Victrix Extra Lusso

Same specs as Victrix except better quality wood; full floral engraving. Made in Italy. Introduced 1986; discontinued 1988.

With fixed chokes
   Perf.: $1125 **Exc.:** $850   **VGood:** $750
With choke tubes
   Perf.: $1175 **Exc.:** $900   **VGood:** $800

WEATHERBY ATHENA GRADE IV

Over/under; boxlock; dummy sideplates; 12-, 20-ga.; 3″ chambers; 26″, 28″ barrels; three IMC Multi-Choke tubes; checkered American walnut pistol-grip stock, forend; mechanically operated single selective trigger (selector inside trigger guard); selective auto ejectors; top tang safety, Greener crossbolt; fully engraved receiver. Made in Japan. Introduced 1982; still in production. **LMSR: $2200**

   Perf.: $1750 **Exc.:** $1000   **VGood:** $800

Weatherby Athena Grade V Classic Field

Same specs as the Athena Grade IV except 26″, 28″, 30″ barrels; oil-finished, fine-line-checkered Claro walnut rounded-pistol-grip stock, slender forend; Old English recoil pad; sideplate receiver rose and scroll engraved. Made in Japan. Introduced 1993; still imported. **LMSR: $2527**

   Perf.: $2000 **Exc.:** $1200   **VGood:** $1000

Weatherby Athena Single Barrel Trap

Same specs as the Athena Grade IV except 12-ga.; 2 3/4″ chambers; 32″, 34″ single top barrel; Full, Modified, Improved Modified Multi-Choke tubes; white front sight, brass middle bead; trap stock dimensions. Made in Japan. Introduced 1988; discontinued 1992.

   Perf.: $1550 **Exc.:** $950   **VGood:** $850
Combo with extra O/U barrel set
   Perf.: $2000 **Exc.:** $1500   **VGood:** $1200

WEATHERBY CENTURION

Autoloader; 12-ga.; 2 3/4″ chamber; 26″ Skeet or Improved Cylinder, 28″ Improved or Modified or Full, 30″ Full barrel; vent rib; front, middle bead sights; hand-checkered American walnut pistol-grip stock, forearm; pressure compensator; engraved alloy receiver. Made in Japan. Introduced 1970; discontinued 1982; replaced by Model Eighty-Two.

   Perf.: $325 **Exc.:** $275   **VGood:** $225

Weatherby Centurion Deluxe

Same specs as Centurion except fancy-grade wood; etched receiver. Introduced 1972; discontinued 1982.

   Perf.: $395 **Exc.:** $325   **VGood:** $250

Weatherby Centurion Trap

Same specs as Centurion except 30″ Full-choke barrel; trap-style walnut stock. Introduced 1972; discontinued 1982.

   Perf.: $340 **Exc.:** $290   **VGood:** $200

WEATHERBY MODEL EIGHTY-TWO

Gas-operated autoloader; 12-ga.; 2 3/4″, 3″ chamber; 26″, 28″, 30″ barrels; 22″ slug barrel with rifle sights available; fixed or interchangeable choke tubes; hand-checkered pistol-grip stock, forend; rubber recoil pad; floating piston; fluted bolt; crossbolt safety; gold-plated trigger. Made in Japan. Introduced 1982; discontinued 1988.

With fixed choke
   Perf.: $350 **Exc.:** $275   **VGood:** $250
With choke tubes
   Perf.: $375 **Exc.:** $300   **VGood:** $275

Weatherby Model Eighty-Two Buckmaster

Same specs as Model Eighty-Two except 22″ Skeet barrel; rifle sights. Introduced 1982; discontinued 1987.

   Perf.: $375 **Exc.:** $300   **VGood:** $275

Weatherby Model Eighty-Two Trap

Same specs as Model Eighty-Two except 30″ Full-choke barrel; trap-style walnut stock. Introduced 1982; discontinued 1984.

With fixed choke
   Perf.: $350 **Exc.:** $275   **VGood:** $250
With choke tubes
   Perf.: $375 **Exc.:** $300   **VGood:** $275

WEATHERBY MODEL NINETY-TWO

Slide action; 12-ga.; 2 3/4″, 3″ chamber; 26″ Modified or Improved or Skeet, 28″ Full or Modified, 30″ Full barrel; vent rib; hand-checkered American walnut pistol-grip stock, forend; grip cap; recoil pad; crossbolt safety; engraved black alloy receiver. Made in Japan. Introduced 1982; discontinued 1988.

With fixed choke
   Perf.: $300 **Exc.:** $250   **VGood:** $190
With Multi-chokes
   Perf.: $325 **Exc.:** $275   **VGood:** $200

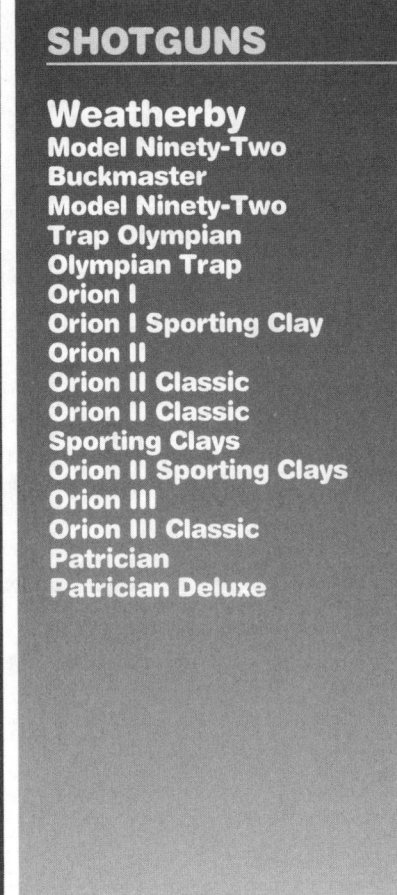

SHOTGUNS

Weatherby

Model Ninety-Two
Buckmaster
Model Ninety-Two
Trap Olympian
Olympian Trap
Orion I
Orion I Sporting Clay
Orion II
Orion II Classic
Orion II Classic
Sporting Clays
Orion II Sporting Clays
Orion III
Orion III Classic
Patrician
Patrician Deluxe

Weatherby Orion I

Weatherby Orion II Classic Sporting Clays

Weatherby Olympian

Weatherby Patrician

Weatherby Patrician Deluxe

Weatherby Model Ninety-Two Buckmaster

Same specs as Model Ninety-two except 22″ Skeet barrel; rifle sights. Introduced 1982; discontinued 1987.

Perf.: $325   **Exc.:** $275   **VGood:** $225

Weatherby Model Ninety-Two Trap

Same specs as Model Ninety-Two except 30″ Full-choke barrel; trap-style walnut stock. Introduced 1982; discontinued 1984.

With fixed choke
Perf.: $300   **Exc.:** $220   **VGood:** $190
With Multi-chokes
Perf.: $325   **Exc.:** $240   **VGood:** $200

WEATHERBY OLYMPIAN

Over/under; boxlock; 12-, 20-ga.; 26″, 28″, 30″ barrels; checkered American walnut pistol-grip stock; selective auto ejectors; single selective trigger; top tang safety; Greener crossbolt. Made in Italy. Introduced 1978; discontinued 1982; replaced by Athena.

Perf.: $875   **Exc.:** $750   **VGood:** $650

Weatherby Olympian Trap

Same specs as Olympian except 12-ga.; 30″, 32″ barrels; vent rib; trap-style walnut stock. Introduced 1978; discontinued 1982.

Perf.: $875   **Exc.:** $750   **VGood:** $650

WEATHERBY ORION I

Over/under; boxlock; 12-, 20-ga.; 3″ chambers; 26″, 28″ Full/Modified or Modified/Improved or Skeet/Skeet, 30″ Full/Modified barrels; Multi-chokes; checkered American walnut pistol-grip stock, forend; rubber recoil pad; selective auto ejectors; single selective trigger; top tang safety; Greener-type crossbolt. Made in Japan. Introduced 1982; still in production. **LMSR: $1225**

Perf.: $1000   **Exc.:** $750   **VGood:** $600

Weatherby Orion I Sporting Clays

Same specs as Orion I except 12-ga.; 28″ barrels; IMC choke tubes; raised rib; center and front beads; special Sporting Clays stock dimensions; rounded buttpad; elongated forcing cones; blued receiver with two 24-Karat gold clay targets on the bottom. Made in Japan. Introduced 1991; discontinued 1992. **LMSR: $1436**

Perf.: $1200   **Exc.:** $950   **VGood:** $800

WEATHERBY ORION II

Over/under; boxlock; 12-, 20-, 28-ga., .410; 3″ chambers; 26″, 28″, 30″ barrels; fixed or choke tubes; high-gloss checkered American walnut pistol-grip stock, forend; rubber recoil pad; selective auto ejectors; single selective trigger; top tang safety; Greener crossbolt; engraving. Discontinued 1993.

Perf.: $1100   **Exc.:** $800   **VGood:** $675

Weatherby Orion II Classic

Same specs as Orion II except 12-, 20-, 28-ga.; 3″ chambers; fine-line-checkered Claro walnut rounded-pistol-grip, slender forend with high-gloss finish; Old English recoil pad; silver-gray nitride receiver with engraved waterfowl and upland scenes. Made in Japan. Introduced 1993; still in production. **LMSR: $1365**

Perf.: $1100   **Exc.:** $800   **VGood:** $675

Weatherby Orion II Classic Sporting Clays

Same specs as Orion II except 12-ga.; 28″, 30″ barrels; IMC choke tubes; stepped Broadway-style competition vent top, side rib; Sporting Clays rounded-pistol-grip stock, slender forend with high-gloss finish; rounded buttpad; elongated forcing cones; silver-gray nitride receiver; scroll engraving with clay pigeon monogram in gold-plate overlay. Made in Japan. Introduced 1993; still in production. **LMSR: $1460**

Perf.: $1250   **Exc.:** $1000   **VGood:** $850

Weathery Orion II Sporting Clays

Same specs as Orion II except 12-ga.; 2 3/4″ chambers; 28″, 30″ barrels; Improved Cylinder, Modified, Full chokes; competition center vent rib; mid-barrel and enlarged front beads; Sporting Clays rounded-pistol-grip stock, slender forend with high-gloss finish; rounded recoil pad; lengthened forcing cones; silver nitride receiver with acid-etched, gold-plate clay pigeon monogram. Made in Japan. Introduced 1992; still in production. **LMSR: $1460**

Perf.: $1250   **Exc.:** $1000   **VGood:** $850

WEATHERBY ORION III

Over/under; boxlock; 12-, 20-ga.; 3″ chambers; 26″, 28″, 30″ barrels; Multi-chokes; high-gloss checkered American walnut pistol-grip stock, forend; rubber recoil pad; selective auto ejectors; single selective trigger; top tang safety; Greener crossbolt; silvered, engraved receiver. Introduced 1989; still in production. **LMSR: $1545**

Perf.: $1250   **Exc.:** $850   **VGood:** $700

Weatherby Orion III Classic

Same specs as Orion III except fine-line-checkered Claro walnut rounded-pistol-grip stock, slender forend with oil finish; Old English recoil pad; silver-gray nitride receiver with engraved waterfowl and upland scenes and gold-plate overlay. Made in Japan. Introduced 1993; still in production. **LMSR: $1626**

Perf.: $1300   **Exc.:** $900   **VGood:** $750

WEATHERBY PATRICIAN

Slide action; 12-ga.; 2 3/4″ chamber; 26″ Modified or Improved or Skeet, 28″ Full or Modified, 30″ Full barrels; hand-checkered stock, pistol-grip forearm; recoil pad; hidden magazine cap; crossbolt safety. Made in Japan. Introduced 1970; discontinued 1982; replaced by Model Ninety-Two.

Perf.: $280   **Exc.:** $200   **VGood:** $170

Weatherby Patrician Deluxe

Same specs as Patrician except fancy-grade wood; etched receiver. Introduced 1972; discontinued 1982.

Perf.: $310   **Exc.:** $240   **VGood:** $200

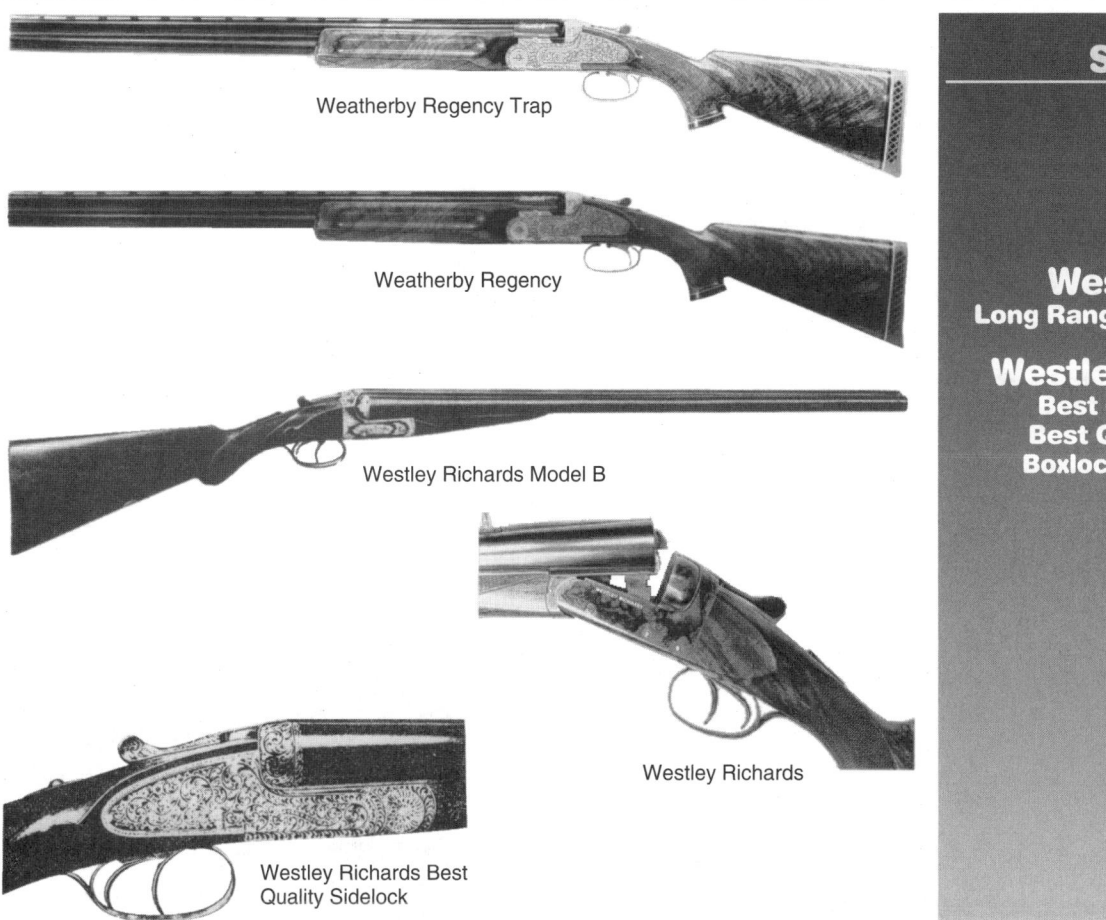

Weatherby Regency Trap

Weatherby Regency

Westley Richards Model B

Westley Richards

Westley Richards Best Quality Sidelock

Weatherby
Patrician Trap
Regency
Regency Trap

Western Arms
Long Range Hammerless

Westley Richards
Best Quality Boxlock
Best Quality Sidelock
Boxlock Model Deluxe
Quality
Model B
Ovundo

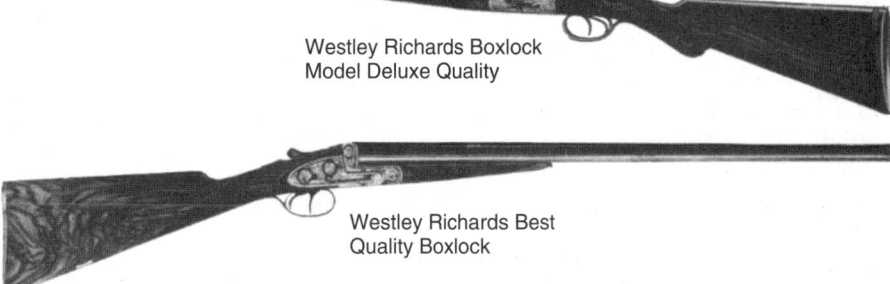

Westley Richards Boxlock
Model Deluxe Quality

Westley Richards Best
Quality Boxlock

Weatherby Patrician Trap
Same specs as Patrician except 30″ Full-choke barrel; trap-style walnut stock. Introduced 1972; discontinued 1982.
   Perf.: $280   **Exc.:** $200   **VGood:** $170

WEATHERBY REGENCY
Over/under; 12-, 20-ga.; boxlock; 28″ Full/Modified, Modified/Improved, Skeet/Skeet barrels; vent rib; bead front sight; hand-checkered American walnut pistol-grip stock, forearm; recoil pad; selective automatic ejectors; single selective trigger; simulated sidelocks; fully engraved receiver. Made first in Italy; then Japan. Introduced 1972; discontinued 1982; replaced by Orion.
   Perf.: $850   **Exc.:** $700   **VGood:** $625

Weatherby Regency Trap
Same specs as Regency except 30″, 32″ barrels; Modified/Full, Improved Modified/Full, Full/Full chokes; vent side, top ribs; walnut Monte Carlo or straight trap-style stock. Introduced 1965; discontinued 1982.
   Perf.: $850   **Exc.:** $700   **VGood:** $625

WESTERN ARMS LONG RANGE HAMMERLESS
Side-by-side; boxlock; 12-, 16-, 20-ga., .410; 26″ to 32″ Modified/Full barrels; uncheckered walnut stock, forearm; double or single trigger; plain extractors. Introduced 1924; discontinued 1942. Made by Western Arms Corp., later absorbed by Ithaca Gun Co.
With double triggers
   Exc.: $250   **VGood:** $175   **Good:** $150
With single trigger
   Exc.: $275   **VGood:** $200   **Good:** $175

WESTLEY RICHARDS BEST QUALITY BOXLOCK
Side-by-side; hand-detachable boxlock; hammerless; 12-, 16-, 20-, 28-ga., .410; barrel lengths, chokes to order; hand-checkered walnut straight or half-pistol-grip stock, forearm; hinged lockplate; selective ejectors; double or single selective trigger. Still in production.
LMSR: $15,900

With double triggers
   Perf.: $15,500   **Exc.:** $9000   **VGood:** $7000
With single trigger
   Perf.: $16,500   **Exc.:** $10,000   **VGood:** $8000

WESTLEY RICHARDS BEST QUALITY SIDELOCK
Side-by-side; hand-detachable sidelocks; hammerless; 12-, 16-, 20-, 28-ga., .410; barrel lengths, chokes to order; hand-checkered walnut straight or half-pistol-grip stock, forearm; selective ejectors; double or single selective trigger. Introduced 1910; still in production.
LMSR: $25,000
With double triggers
   Perf.: $25,000   **Exc.:** $18,000   **VGood:** $12,000
With single trigger
   Perf.: $26,000   **Exc.:** $19,000   **VGood:** $13,000

WESTLEY RICHARDS BOXLOCK MODEL DELUXE QUALITY
Side-by-side; hand-detachable boxlock; hammerless; 12-, 16-, 20-, 28-, .410-ga.; barrel lengths, chokes to order; hand-checkered European stock, forearm; straight or half-pistol grip available in Pigeon or Wildfowl Model at same price; triple-bite leverwork; selective ejectors; double triggers or single selective trigger.

Introduced 1890; still in production. **LMSR: $12,500**
With double triggers
   Perf.: $11,500   **Exc.:** $8500   **VGood:** $7000
With single trigger
   Perf.: $12,500   **Exc.:** $9500   **VGood:** $8000

WESTLEY RICHARDS MODEL B
Side-by-side; Anson & Deeley boxlock; hammerless; 12-, 16-, 20-ga.; barrel lengths, choking to order; hand-checkered European walnut straight or half-pistol-grip stock, forearm; Pigeon or Wildfowl Model available at same price; selective ejectors or extractors; double triggers. Introduced late 1920s; still in production. **LMSR: $6000**
With ejectors
   Perf.: $5500   **Exc.:** $4000   **VGood:** $3000
With extractors
   Perf.: $5000   **Exc.:** $3500   **VGood:** $2500

WESTLEY RICHARDS OVUNDO
Over/under; hand-detachable boxlock; hammerless; 12-ga.; barrel lengths, chokes to order; hand-checkered European walnut straight or half-pistol-grip stock, forearm; dummy sideplates; single selective trigger. Introduced 1920; still in production. **LMSR: $18,000**
   Perf.: $15,500   **Exc.:** $12,500   **VGood:** $10,000

SHOTGUNS

Winchester Semi-Automatic Shotguns
Model 40
Model 40 Skeet
Model 50
Model 50 Featherweight
Model 50 Pigeon
Model 50 Skeet
Model 50 Trap
Model 59
Model 1400
Model 1400 Deer Gun
Model 1400 Mark II
Model 1400 Skeet

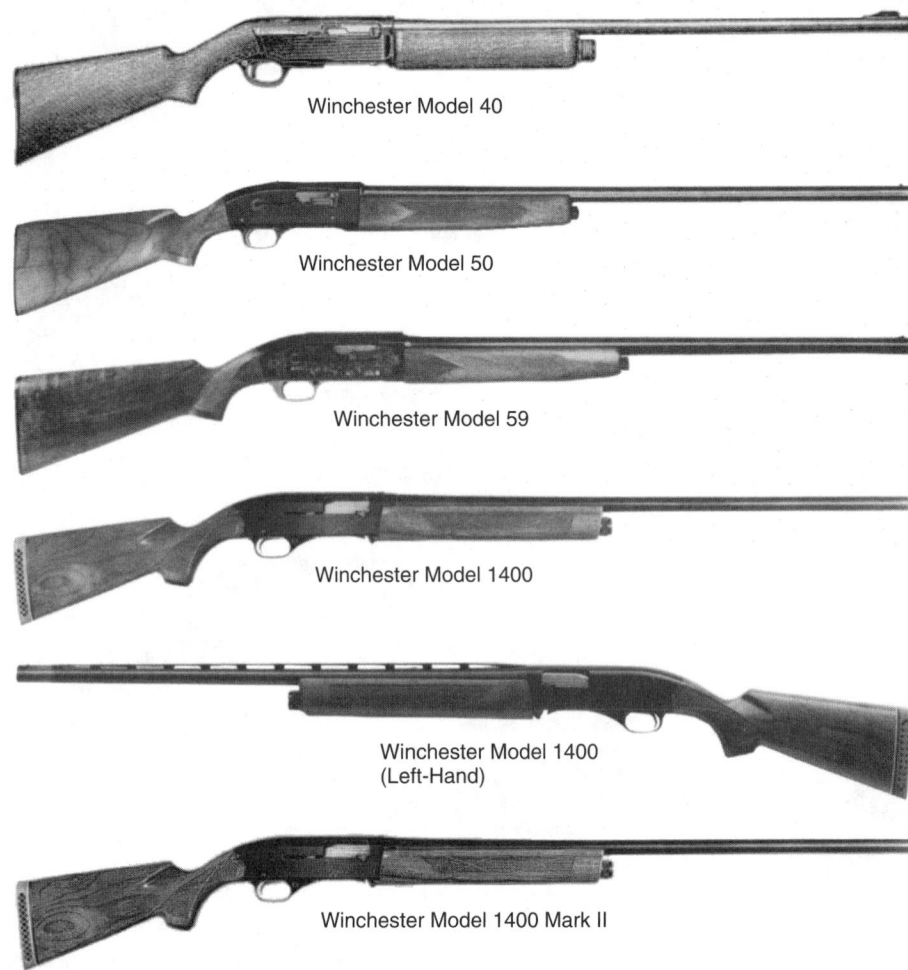

Winchester Model 40

Winchester Model 50

Winchester Model 59

Winchester Model 1400

Winchester Model 1400 (Left-Hand)

Winchester Model 1400 Mark II

WINCHESTER SEMI-AUTOMATIC

WINCHESTER MODEL 40
Autoloader; hammerless; 12-ga.; 4-shot tube magazine; 28", 30" barrel; Modified, Full choke; ramp bead front sight; unchecked pistol-grip stock, semi-beavertail forend; streamlined receiver. Introduced 1940; discontinued 1941.
Exc.: $390   VGood: $275   Good: $200

Winchester Model 40 Skeet
Same specs as Model 40 except 24" barrel; Cutts Compensator; checkered pistol-grip stock, forearm; grip cap. No longer in production.
Exc.: $370   VGood: $250   Good: $180

WINCHESTER MODEL 50
Autoloader; 12-, 20-ga.; 2-shot magazine; 28", 30" barrel; Improved, Modified, Full chokes; optional vent rib; 7 1/4 lbs.; bead-front sight; hand-checkered American walnut stock; fluted comb; composition buttplate; side ejection; short recoil action; interchangeable barrels. Introduced 1954; discontinued 1961.
Exc.: $290   VGood: $225   Good: $180
With vent rib
Exc.: $340   VGood: $275   Good: $220

Winchester Model 50 Featherweight
Same specs as Model 50 except 12-ga. weighs 7 lbs.; 20-ga. weighs 5 3/4 lbs.; alloy construction. Introduced 1958; discontinued 1961.
Exc.: $325   VGood: $265   Good: $210

Winchester Model 50 Pigeon
Same specs as Model 50 except best-grade wood, carving and engraving. Introduced 1954; discontinued 1961.
Standard model
Exc.: $1000   VGood: $825   Good: $650

Featherweight model
Exc.: $1100   VGood: $800   Good: $600
Skeet model
Exc.: $1100   VGood: $900   Good: $750
Trap model
Exc.: $1100   VGood: $900   Good: $750

Winchester Model 50 Skeet
Same specs as Model 50 except 12-ga.; 26" vent-rib barrel; Skeet choke; hand-checkered American walnut Skeet-style stock. Introduced 1954; discontinued 1961.
Exc.: $400   VGood: $275   Good: $220

Winchester Model 50 Trap
Same specs as Model 50 except 12-ga.; 30" vent-rib barrel; Full choke; hand-checkered American walnut Monte Carlo stock. Introduced 1954; discontinued 1961.
Exc.: $400   VGood: $275   Good: $220

WINCHESTER MODEL 59
Autoloader; 12-ga.; 3-shot magazine; 26" Improved Cylinder, 28" Modified or Full, 30" Full choke barrel; special-order 26" fiberglass-wrapped steel barrel with Versalite choke system of cylinder tubes (introduced in 1961) to allow any choke variation; checkered walnut stock, forearm. (Winchester also made a Model 59 rimfire rifle in 1930—don't be confused by the model numbers.) Introduced 1959; discontinued 1965.
Perf.: $375   Exc.: $250   VGood: $200

WINCHESTER MODEL 1400
Gas-operated autoloader; takedown; 12-, 16-, 20-ga.; 2 3/4" chamber; 2-shot magazine; 26", 28", 30" barrel; fixed Improved Cylinder, Modified, Full chokes or Win-choke tubes; plain or vent rib; checkered walnut stock, forearm; optional Cycolac stock and recoil pad with recoil-reduction system. Introduced 1964; discontinued 1968; replaced by Model 1400 Mark II.

With plain barrel, walnut stock
Exc.: $210   VGood: $165   Good: $150
With vent rib, walnut stock
Exc.: $240   VGood: $195   Good: $175
With plain barrel, Cycolac stock
Exc.: $275   VGood: $210   Good: $180
With vent rib, Cycolac stock
Exc.: $325   VGood: $245   Good: $200

Winchester Model 1400 Deer Gun
Same specs as Model 1400 except 12-ga.; 22" barrel for slugs or buckshot; rifle sight; walnut stock, forearm. Introduced 1965; discontinued 1974.
Perf.: $250   Exc.: $200   VGood: $180

Winchester Model 1400 Mark II
Same specs as Model 1400 except restyled walnut stock, forearm; push-button carrier release; front-locking, rotating bolt locking into barrel extension; self-compensating gas system for standard and magnum 2 3/4" loads; aluminum receiver; engine turned bolt; push-button action release; crossbolt safety. Made in right- or left-hand versions. Introduced 1968; replaced Model 1400; discontinued 1980.
With plain barrel
Perf.: $275   Exc.: $225   VGood: $200
With vent rib
Perf.: $305   Exc.: $255   VGood: $230

Winchester Model 1400 Skeet
Same specs as Model 1400 except 12-, 20-ga.; 26" vent-rib barrel; Skeet choke; semi-fancy walnut Skeet-style stock, forearm; optional Cycolac stock with recoil reduction system. Introduced 1965; discontinued 1968.
With walnut stock
Exc.: $300   VGood: $230   Good: $190
With Cycolac stock with recoil reduction
Exc.: $350   VGood: $280   Good: $240

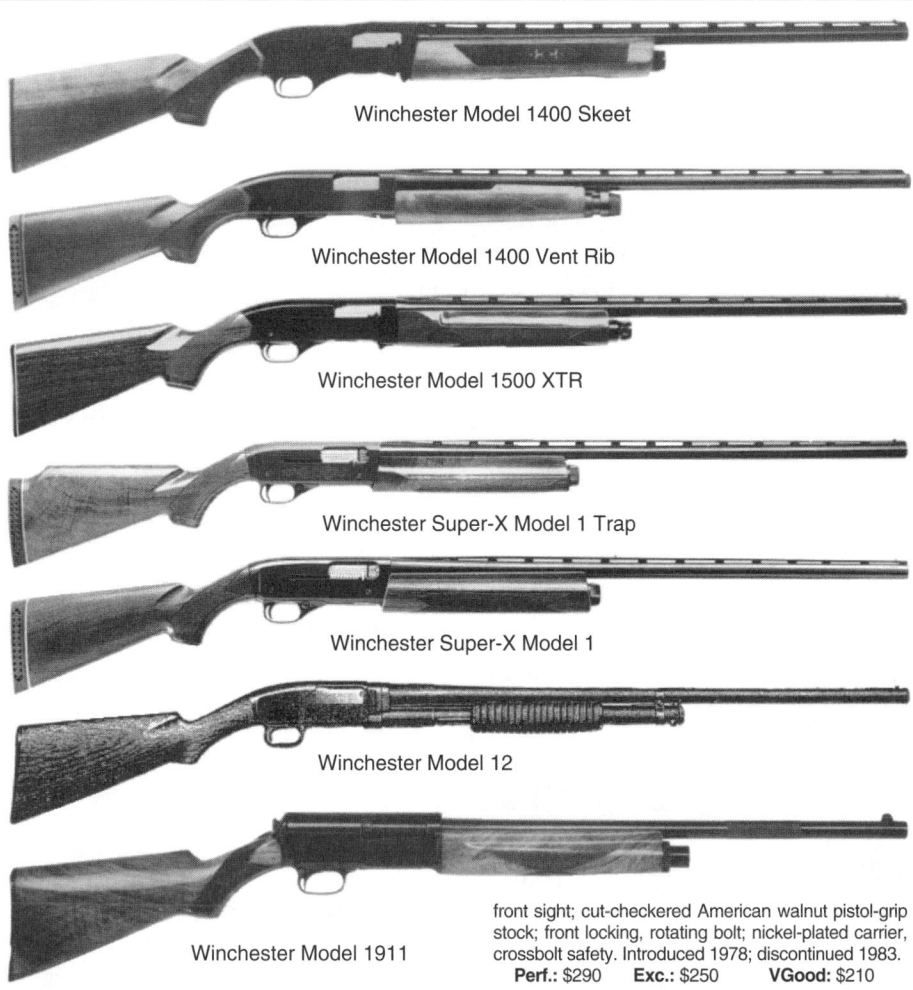

Winchester Model 1400 Skeet

Winchester Model 1400 Vent Rib

Winchester Model 1500 XTR

Winchester Super-X Model 1 Trap

Winchester Super-X Model 1

Winchester Model 12

Winchester Model 1911

SHOTGUNS

Winchester Semi-Automatic Shotguns
Model 1400 Trap
Model 1400 (Recent)
Model 1400 Custom (Recent)
Model 1400 Slug Hunter (Recent)
Model 1500 XTR
Model 1911
Model 1911 Fancy Finished
Model 1911 Pigeon
Model 1911 Trap
Super-X Model 1
Super-X Model 1 Skeet
Super-X Model 1 Trap

Winchester Slide-Action Shotguns
Model 12
Model 12 Featherweight
Model 12 Heavy Duck

Winchester Model 1400 Trap
Same specs as Model 1400 except 12-ga.; 30″ vent-rib barrel; Full choke; semi-fancy walnut Monte Carlo stock, forearm; optional Cycolac stock with recoil reduction system. Introduced 1965; discontinued 1973. With walnut stock
   Exc.: $300 **VGood:** $230 **Good:** $190
With Cycolac stock with recoil reduction
   Exc.: $350 **VGood:** $280 **Good:** $240

WINCHESTER MODEL 1400 (RECENT)
Gas-operated autoloader; 12-, 20-ga.; 2 3/4″ chamber; 3-shot magazine; 22″, 26″, 28″ vent-rib barrel; Win-Choke tubes; cut-checkered American walnut stock, forearm. Introduced 1989; discontinued 1994.
   Perf.: $340 **Exc.:** $270 **VGood:** $230

Winchester Model 1400 Custom (Recent)
Same specs as Model 1400 (Recent) except 12-ga.; 28″ vent-rib barrel; semi-fancy walnut stock, forend; hand-engraved receiver; made in Winchester custom shop. Introduced 1991; discontinued 1992.
   Perf.: $1100 **Exc.:** $800 **VGood:** $700

Winchester Model 1400 Slug Hunter (Recent)
Same specs as Model 1400 (Recent) except 12-ga.; 22″ smoothbore barrel; Improved Cylinder, Sabot Winchoke tubes; adjustable open sights; drill, tapped for scope mounts. Introduced 1990; no longer in production.
   Perf.: $345 **Exc.:** $275 **VGood:** $235

WINCHESTER MODEL 1500 XTR
Gas-operated autoloader; 12-, 20-ga.; 26″, 28″, 30″ barrel; plain or vent rib; WinChoke tubes; metal bead front sight; cut-checkered American walnut pistol-grip stock; front locking, rotating bolt; nickel-plated carrier, crossbolt safety. Introduced 1978; discontinued 1983.
   Perf.: $290 **Exc.:** $250 **VGood:** $210

WINCHESTER MODEL 1911
Autoloading; hammerless; takedown; 12-ga.; 26″, 28″ barrel standard chokes; uncheckered laminated birch stock, forend. Introduced 1911; discontinued 1925.
   Exc.: $375 **VGood:** $275 **Good:** $210

Winchester Model 1911 Fancy Finished
Same specs as Model 1911 except checkered walnut stock, forend. Introduced 1911; discontinued 1918.
   Exc.: $400 **VGood:** $325 **Good:** $260

Winchester Model 1911 Pigeon
Same specs as Model 1911 except matted barrel; checkered fancy walnut straight-grip stock, forend; pistol grip optional; elaborate engraving. Introduced 1913; discontinued 1926.
   Exc.: $500 **VGood:** $400 **Good:** $350

Winchester Model 1911 Trap
Same specs as Model 1911 except matted barrel; checkered fancy walnut straight-grip stock, forend; pistol grip optional. Introduced 1913; discontinued 1926.
   Exc.: $400 **VGood:** $305 **Good:** $260

WINCHESTER SUPER-X MODEL 1
Gas-operated autoloader; takedown; 12-ga.; 2 3/4″ chamber; 4-shot magazine; 26″ Improved Cylinder, 28″ Modified or Full, 30″ Full-choke barrel; vent rib; checkered American walnut pistol-grip stock, forearm. Introduced 1974; discontinued 1981.
   Perf.: $400 **Exc.:** $350 **VGood:** $300

Winchester Super-X Model 1 Skeet
Same specs as Super-X Model 1 except 26″ Skeet-choke barrel; select American walnut Skeet-style stock. Introduced 1974; discontinued 1981.
   Exc.: $625 **VGood:** $450 **Good:** $400

Winchester Super-X Model 1 Trap
Same specs as Super-X Model 1 except 30″ barrel; Improved Modified or Full choke; select American walnut straight or Monte Carlo trap-style stock, forearm; recoil pad. Introduced 1974; discontinued 1981.
   Exc.: $525 **VGood:** $400 **Good:** $375

WINCHESTER SLIDE-ACTION SHOTGUNS

WINCHESTER MODEL 12
Slide action; hammerless; 12-, 16-, 20-, 28-ga.; 2 3/4″, 3″ chamber; 5-shot magazine; 26″, 28″, 30″, 32″ barrel; Improved Cylinder, Modified, Full choke; optional matted or vent rib; uncheckered walnut pistol-grip stock, forearm; blued. Introduced 1912; discontinued 1963.
With plain barrel
   Exc.: $375 **VGood:** $280 **Good:** $225
With matted rib
   Exc.: $450 **VGood:** $350 **Good:** $300
With vent rib
   Exc.: $500 **VGood:** $400 **Good:** $350

Winchester Model 12 Featherweight
Same specs as Model 12 except 12-ga.; 26″ Improved Cylinder, 28″ Modified or Full, 30″ Full barrel; alloy guard. Introduced 1959; discontinued 1962.
   Exc.: $375 **VGood:** $295 **Good:** $200

Winchester Model 12 Heavy Duck
Same specs as Model 12 except 12-ga.; 3″ chamber; 3-shot magazine; 30″, 32″ Full barrel; plain, matted or vent rib; recoil pad. Introduced 1937; discontinued 1963.
With plain barrel
   Exc.: $500 **VGood:** $350 **Good:** $300
With matted rib
   Exc.: $525 **VGood:** $425 **Good:** $375
With vent rib
   Exc.: $1000 **VGood:** $800 **Good:** $650

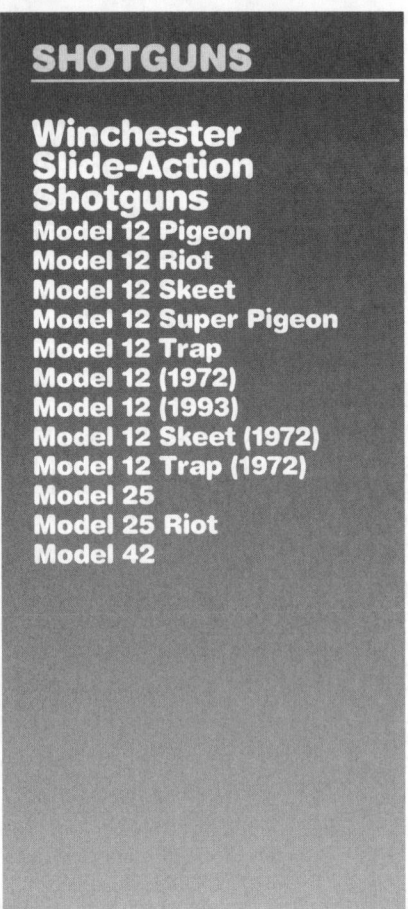

Winchester Slide-Action Shotguns

Model 12 Pigeon
Model 12 Riot
Model 12 Skeet
Model 12 Super Pigeon
Model 12 Trap
Model 12 (1972)
Model 12 (1993)
Model 12 Skeet (1972)
Model 12 Trap (1972)
Model 25
Model 25 Riot
Model 42

Winchester Model 12 Skeet

Winchester Model 12 (1972)

Winchester Model 12 Trap (1972)

Winchester Model 12 (1993)

Winchester Model 25

Winchester Model 12 Pigeon

Same specs as Model 12 except hand-checkered fancy walnut stock, forearm; engine-turned bolt, carrier; hand-worked action; optional carving and engraving. Introduced 1912; discontinued 1963.
Standard model with plain barrel
   Exc.: $1250   **VGood:** $800   **Good:** $700
Standard model with vent rib
   Exc.: $1375   **VGood:** $950   **Good:** $800
Skeet model with plain barrel
   Exc.: $1250   **VGood:** $800   **Good:** $700
Skeet model with plain barrel, Cutts Compensator
   Exc.: $1000   **VGood:** $600   **Good:** $500
Skeet model with vent rib
   Exc.: $1250   **VGood:** $950   **Good:** $800
Trap model with matted rib
   Exc.: $1200   **VGood:** $900   **Good:** $800
Trap model with vent rib
   Exc.: $1250   **VGood:** $950   **Good:** $850

Winchester Model 12 Riot

Same specs as Model 12 except 12-ga.; 20″ Cylinder barrel; vent handguard. Introduced 1918; discontinued 1963.
Military Version
   Exc.: $850   **VGood:** $600   **Good:** $500
Civilian Version
   Exc.: $450   **VGood:** $375   **Good:** $275

Winchester Model 12 Skeet

Same specs as Model 12 except 26″ Skeet barrel; plain barrel or vent rib; red or ivory bead front sight; 94B middle; checkered walnut pistol-grip stock, extension slide handle; recoil pad. Introduced 1937; discontinued 1963.
With plain barrel
   Exc.: $500   **VGood:** $475   **Good:** $400
With plain barrel, Cutts Compensator
   Exc.: $475   **VGood:** $375   **Good:** $300
With vent rib
   Exc.: $625   **VGood:** $575   **Good:** $500

Winchester Model 12 Super Pigeon

Same specs as Model 12 except 12-ga.; 26″, 28″, 30″ vent-rib barrel; hand-checkered fancy walnut stock, forearm; hand-honed and fitted action; engine-turned bolt, carrier; carving; engraving; top-of-the-line custom gun. Introduced 1965; discontinued 1975.
Standard model with plain barrel
   Exc.: $1750   **VGood:** $1300   **Good:** $1200
Standard model with vent rib
   Exc.: $1950   **VGood:** $1600   **Good:** $1350
Skeet model with plain barrel
   Exc.: $1750   **VGood:** $1500   **Good:** $1200
Skeet model with plain barrel, Cutts Compensator
   Exc.: $1500   **VGood:** $1000   **Good:** $850
Skeet model with vent rib
   Exc.: $1950   **VGood:** $1600   **Good:** $1350
Trap model with matted rib
   Exc.: $1800   **VGood:** $1500   **Good:** $1350
Trap model with vent rib
   Exc.: $1950   **VGood:** $1600   **Good:** $1450

Winchester Model 12 Trap

Same specs as Model 12 except 12-ga.; 30″ Full-choke barrel; vent or matted rib; checkered walnut pistol-grip stock, extension slide handle; optional Monte Carlo stock; recoil pad. Introduced 1937; discontinued 1963.
With matted rib
   Exc.: $700   **VGood:** $500   **Good:** $425
With vent rib
   Exc.: $750   **VGood:** $550   **Good:** $475
With vent rib, Monte Carlo stock
   Exc.: $775   **VGood:** $575   **Good:** $500

Winchester Model 12 (1972)

Same specs as Model 12 except 12-ga.; 26″, 28″, 30″ barrels; standard chokes; vent rib; hand-checkered American walnut stock, slide handle; engine-turned bolt, carrier. Introduced 1972; discontinued 1975.
   Perf.: $575   **Exc.:** $525   **VGood:** $425

Winchester Model 12 (1993)

Same specs as Model 12 except limited production. Introduced 1993, still in production.
   Perf.: $950   **Exc.:** $800   **VGood:** $650

Winchester Model 12 Skeet (1972)

Same specs as Model 12 except 26″ Skeet-choke barrel; hand-checkered American walnut Skeet-style stock, slide handle; engine-turned bolt, carrier; recoil pad. Introduced 1972; discontinued 1975.
   Perf.: $900   **Exc.:** $650   **VGood:** $525

Winchester Model 12 Trap (1972)

Same specs as Model 12 except 30″ Full-choke barrel; vent rib; checkered American walnut straight or Monte Carlo trap-style stock; engine-turned bolt, carrier; recoil pad. Introduced 1972; discontinued 1975.
   Perf.: $700   **Exc.:** $575   **VGood:** $475

WINCHESTER MODEL 25

Slide action; hammerless; solid frame; 12-ga.; 4-shot tubular magazine; 26″, 28″ plain barrel; Improved Cylinder, Modified or Full choke; metal bead front sight; uncheckered walnut pistol grip stock, grooved slide handle. Introduced 1950; discontinued 1954.
   Exc.: $250   **VGood:** $190   **Good:** $140

Winchester Model 25 Riot

Same specs as Model 25 except 20″ Cylinder barrel. Introduced 1949; discontinued 1955.
   Exc.: $240   **VGood:** $180   **Good:** $130

WINCHESTER MODEL 42

Slide action; hammerless; .410; 2 1/2″, 3″ chamber; 5-(3″), 6-shot (2 1/2″) magazine; 26″, 28″ barrel; Full, Modified, Cylinder choke; plain, matted or vent rib; weighs 6 lbs.; uncheckered walnut stock, grooved slide handle. Introduced 1933; discontinued 1963.
With plain barrel
   Exc.: $750   **VGood:** $600   **Good:** $500
With matted rib
   Exc.: $1150   **VGood:** $1000   **Good:** $850
With vent rib
   Exc.: $1350   **VGood:** $1200   **Good:** $950

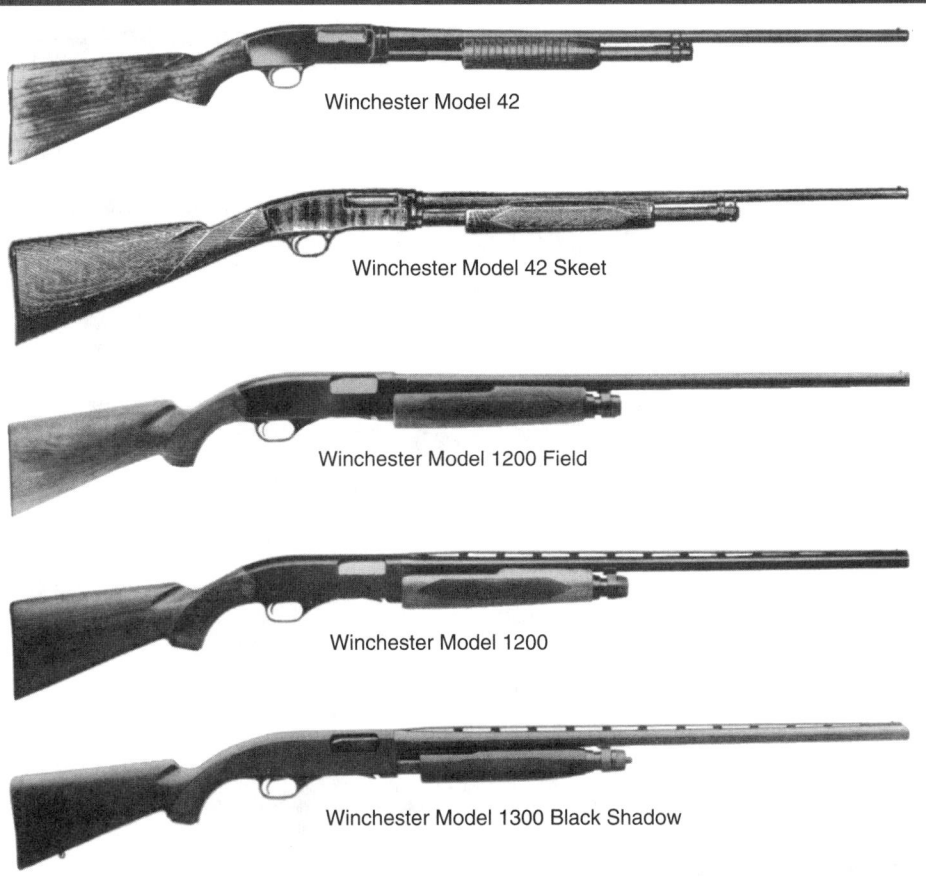

Winchester Model 42

Winchester Model 42 Skeet

Winchester Model 1200 Field

Winchester Model 1200

Winchester Model 1300 Black Shadow

SHOTGUNS

Winchester Slide-Action Shotguns
Model 42 Deluxe
Model 42 Skeet
Model 42 Trap
Model 42 High Grade (1993)
Model 97 (See Stoeger 1942)
Model 1200
Model 1200 Deer
Model 1200 Magnum
Model 1200 Skeet
Model 1200 Trap
Model 1300 Black Shadow
Model 1300 Black Shadow Deer
Model 1300 Black Shadow Turkey
Model 1300 N.W.T.F. Series I

Winchester Model 42 Deluxe
Same specs as Model 42 except top-of-the-line model with best-grade wood and finish. Introduced 1933; discontinued 1963.
Exc.: $2200   **VGood:** $1750   **Good:** $1300

Winchester Model 42 Skeet
Same specs as Model 42 except Skeet choke; matted rib; checkered fancy walnut Skeet-style stock, forearm. Introduced 1933; discontinued 1963.
Exc.: $1850   **VGood:** $1350   **Good:** $1200

Winchester Model 42 Trap
Same specs as Model 42 except matted rib; checkered fancy walnut trap-style stock, forearm. Introduced 1933; discontinued 1940.
Exc.: $1850   **VGood:** $1350   **Good:** $1200

WINCHESTER MODEL 42 HIGH GRADE (1993)
Slide action; hammerless; .410; 2 3/4" chamber; 26" vent-rib barrel; Full choke; weighs 7 lbs.; checkered high-grade walnut pistol-grip stock, forend; engraved receiver with gold inlays. Only 850 guns made. Made in Japan. Introduced 1993; discontinued 1993. **LMSR:** $1617
Perf.: $1250   **Exc.:** $900   **VGood:** $750

WINCHESTER MODEL 1200
Slide action; takedown; 12-, 16-, 20-ga.; 2 3/4" chambers; 4-shot magazine; 26", 28", 30" barrel; Improved Cylinder, Modified, Full fixed choke or interchangeable WinChoke tubes for Cylinder, Modified, Full; plain or vent rib; press-checkered walnut stock, slide handle; optional Cycolac stock with recoil-reduction system; front-lock rotary bolt; recoil pad. Introduced 1964; discontinued 1981.
With plain barrel, walnut stock
Perf.: $200   **Exc.:** $170   **VGood:** $150
With vent-rib barrel, walnut stock
Perf.: $240   **Exc.:** $210   **VGood:** $190
With plain barrel, Cycolac stock
Perf.: $260   **Exc.:** $230   **VGood:** $200
With vent-rib barrel, Cycolac stock
Perf.: $300   **Exc.:** $270   **VGood:** $240

Winchester Model 1200 Deer
Same specs as Model 1200 except 12-ga.; 22" barrel for rifled slugs or buckshot; rifle-type sights; walnut stock only; sling swivels. Introduced 1965; discontinued 1974.
Perf.: $210   **Exc.:** $150   **VGood:** $140

Winchester Model 1200 Magnum
Same specs as Model 1200 except 12-, 20-, 28-ga.; 3" chamber; 28", 30" Full-choke barrel; plain or vent rib. Introduced 1964; discontinued 1981.
With plain barrel, walnut stock
Perf.: $220   **Exc.:** $165   **VGood:** $150
With vent-rib barrel, walnut stock
Perf.: $270   **Exc.:** $210   **VGood:** $190
With plain barrel, Cycolac stock
Perf.: $280   **Exc.:** $240   **VGood:** $200
With vent-rib barrel, Cycolac stock
Perf.: $325   **Exc.:** $260   **VGood:** $230

Winchester Model 1200 Skeet
Same specs as Model 1200 except 12-, 20- ga.; 2-shot magazine; 26" vent-rib barrel; Skeet choke; semi-fancy walnut stock, forend; optional Cycolac stock with recoil-reduction system; tuned trigger. Introduced 1965; discontinued 1974.
With walnut stock
Perf.: $295   **Exc.:** $240   **VGood:** $200
With Cycolac stock
Perf.: $350   **Exc.:** $300   **VGood:** $250

Winchester Model 1200 Trap
Same specs as Model 1200 except 12-ga.; 2-shot magazine; 28" WinChoke, 30" Full-choke barrel; vent rib; semi-fancy walnut standard or Monte Carlo trap-style stock, forearm; optional Cycolac stock with recoil-reduction system. Introduced 1965; discontinued 1974.
With standard stock
Perf.: $325   **Exc.:** $275   **VGood:** $200
With Monte Carlo stock
Perf.: $345   **Exc.:** $300   **VGood:** $220
With Cycolac stock
Perf.: $360   **Exc.:** $325   **VGood:** $250

WINCHESTER MODEL 1300 BLACK SHADOW
Slide action; hammerless; 12-, 20-ga.; 3" chamber; 5-shot magazine; 26", 28" vent-rib barrel; Modified WinChoke tubes; metal bead front sight; black composite pistol-grip stock, forend; black rubber recoil pad; cross-bolt safety; twin action slide bars; front-locking rotating bolt; matte black finish. Introduced 1995; still in production. **LMSR:** $296
Perf.: $270   **Exc.:** $210   **VGood:** $180

Winchester Model 1300 Black Shadow Deer
Same specs as Model 1300 Black Shadow except 12-ga.; 3" chamber; 22" vent-rib barrel; Improved Cylinder WinChoke tube; rampr-type front sight, fully adjustable rear; receiver drilled and tapped for scope mounts. Introduced 1994; still in production. **LMSR:** $296
Perf.: $270   **Exc.:** $210   **VGood:** $180

Winchester Model 1300 Black Shadow Turkey
Same specs as Model 1300 Black Shadow except 12-ga.; 3" chamber; 22" vent-rib barrel; Full WinChoke tube; receiver drilled and tapped for scope mounts. Introduced 1994; still in production. **LMSR:** $296
Perf.: $270   **Exc.:** $210   **VGood:** $180

WINCHESTER MODEL 1300 N.W.T.F. SERIES I
Slide action; hammerless; 12-ga.; 3" chamber; 5-shot magazine; 22" vent-rib barrel; Extra Full, Full, Modified WinChoke tubes; checkered green Win-Cam laminated stock, forend; black rubber recoil pad; crossbolt safety; twin action slide bars; front-locking rotating bolt; camo sling and swivels; receiver roll-engraved with turkey scenes; right side has National Wild Turkey Federation name in script; matte finish metal. Introduced 1989; no longer in production.
Perf.: $350   **Exc.:** $265   **VGood:** $210

Winchester Slide-Action Shotguns

Model 1300 N.W.T.F. Series II
Model 1300 N.W.T.F. Series III
Model 1300 N.W.T.F. Series IV
Model 1300 Ranger
Model 1300 Ranger Combo
Model 1300 Ranger Deer
Model 1300 Ranger Ladies/Youth
Model 1300 Realtree® Turkey Gun
Model 1300 Slug Hunter
Model 1300 Stainless Marine
Model 1300 Turkey
Model 1300 XTR
Model 1300 XTR Deer Gun
Model 1300 XTR Defender

Winchester Model 1300 Ranger

Winchester Model 1300 Ranger Combo

Winchester Model 1300 Ranger Ladies/Youth

Winchester Model 1300 Slug Hunter

Winchester Model 1300 Turkey

Winchester Model 1300 N.W.T.F. Series II
Same specs as the Model 1300 N.W.T.F. Series I except 12-, 20-ga.; WinCam green laminated stock, ribbed forend; ladies/youth-sized stock available. Introduced 1990; discontinued 1991.

Perf.: $350   **Exc.:** $265   **VGood:** $210

Winchester Model 1300 N.W.T.F. Series III
Same specs as the Model 1300 N.W.T.F. Series I except rifle-type adjustable sights; WinCam green laminated stock, ribbed forend. Introduced 1992; discontinued 1993.

Perf.: $350   **Exc.:** $265   **VGood:** $210

Winchester Model 1300 N.W.T.F. Series IV
Same specs as the Model 1300 N.W.T.F. Series I except WinTuff black laminated stock, ribbed forend. Introduced 1993; discontinued 1994.

Perf.: $350   **Exc.:** $265   **VGood:** $210

WINCHESTER MODEL 1300 RANGER
Slide action; hammerless; 12-, 20-ga.; 3″ chamber; 5-shot magazine; 26″, 28″ vent-rib barrel; Full, Modified, Improved Cylinder WinChoke tubes; checkered walnut-finished hardwood pistol-grip stock, forend; early guns had ribbed forend; black rubber recoil pad; crossbolt safety; twin action slide bars; front-locking rotating bolt. Introduced 1983; still in production. **LMSR: $309**

Perf.: $280   **Exc.:** $220   **VGood:** $185

Winchester Model 1300 Ranger Combo
Same specs as Model 1300 Ranger except two-barrel set with 22″ Cylinder smoothbore or rifled deer barrel with rifle-type sights and an interchangeable 28″ vent-rib barrel with Full, Modified, Improved Cylinder WinChoke tubes; drilled and tapped for scope mounts; rings and bases. Introduced 1983; still in production.

LMSR: $379
Perf.: $330   **Exc.:** $270   **VGood:** $230

Winchester Model 1300 Ranger Deer
Same specs as Model 1300 Ranger except 22″ Cylinder smoothbore or rifled deer barrel with rifle-type sights; drilled and tapped; for scope mounts; rings and bases. Introduced 1983; still in production. **LMSR: $343**

Perf.: $300   **Exc.:** $245   **VGood:** $200

Winchester Model 1300 Ranger Ladies/Youth
Same specs as Model 1300 Ranger except 20-ga.; 3″ chamber; 22″ vent-rib barrel; Full, Modified, Improved Cylinder WinChoke tubes; weighs $6 \frac{1}{2}$ lbs.; overall length 41 $5/8$″; hardwood stock (walnut available) with 13″ pull length. Introduced 1983; still in production. **LMSR: $309**

With hardwood stock
Perf.: $280   **Exc.:** $220   **VGood:** $185
With walnut stock
Perf.: $300   **Exc.:** $240   **VGood:** $200

Winchester Model 1300 Realtree® Turkey Gun
Same specs as Model 1300 Ranger except 12-ga.; 3″ chamber; 22″ barrel; Extra Full, Full, Modified WinChoke tubes; drilled and tapped for scope mounts; synthetic Realtree® camo stock, forend; padded, adjustable sling; matte finished barrel, receiver. Introduced 1994; still in production. **LMSR: $370**

Perf.: $325   **Exc.:** $285   **VGood:** $225

Winchester Model 1300 Slug Hunter
Same specs as Model 1300 Ranger except 12-ga.; 22″ smoothbore or rifled barrel; Improved Cylinder, Sabot choke tubes (smoothbore only); adjustable open sights; receiver drilled and tapped for scope mounts; scope bases; cut-checkered walnut stock, forend. Introduced 1990; smoothbore dropped 1992; still in production as Model 1300 Slug Hunter Deer. **LMSR: $404**

Perf.: $350   **Exc.:** $265   **VGood:** $200

Winchester Model 1300 Stainless Marine
Same specs as Model 1300 Ranger except 5-, 8-shot magazine; 18″ Cylinder stainless barrel; rifle-type sights; bright chrome finish; phosphate-coated receiver for corrosion resistance. Introduced 1989; still in production. **LMSR: $460**

Perf.: $395   **Exc.:** $275   **VGood:** $210

Winchester Model 1300 Turkey
Same specs as Model 1300 Ranger except 12-ga.; 22″ barrel; Modified, Full, Extra Full WinChoke tubes; WinCam green camo laminated stock; recoil pad, Cordura sling; swivels; matte finish wood and metal. Introduced 1985; discontinued 1988.

Perf.: $300   **Exc.:** $240   **VGood:** $185

WINCHESTER MODEL 1300 XTR
Slide action; 12-, 20-ga.; 3″ chamber; 4-shot magazine; 26, 28″ plain or vent-rib barrel; metal bead front sight; cut-checkered American walnut stock, grooved forearm; twin action bars; crossbolt safety; engine-turned bolt; alloy receiver, trigger guard. Introduced 1978; XTR designation dropped 1989; discontinued 1993.

With plain barrel
Perf.: $260   **Exc.:** $200   **VGood:** $160
With vent-rib barrel
Perf.: $295   **Exc.:** $240   **VGood:** $185

Winchester Model 1300 XTR Deer Gun
Same specs as Model 1300 XTR except 12-ga.; 24 $1/8$″ barrel; rifle-type sights. No longer in production.

Perf.: $295   **Exc.:** $240   **VGood:** $185

Winchester Model 1300 XTR Defender
Same specs as Model 1300 XTR except 5-, 8-shot magazine; 18″, 24″ Cylinder barrel; 38 $3/5$″ overall length; weighs 6 $3/4$ lb.; metal bead front sight; walnut-finished hardwood or synthetic pistol-grip stock, ribbed forend; crossbolt safety; front-locking rotary bolt; twin slide bars; black rubber buttpad. Introduced 1984; XTR designation dropped 1989; still in production. **LMSR: $290**

Perf.: $240   **Exc.:** $165   **VGood:** $120
Defender Combo with extra 28″ vent-rib barrel
Perf.: $320   **Exc.:** $240   **VGood:** $180

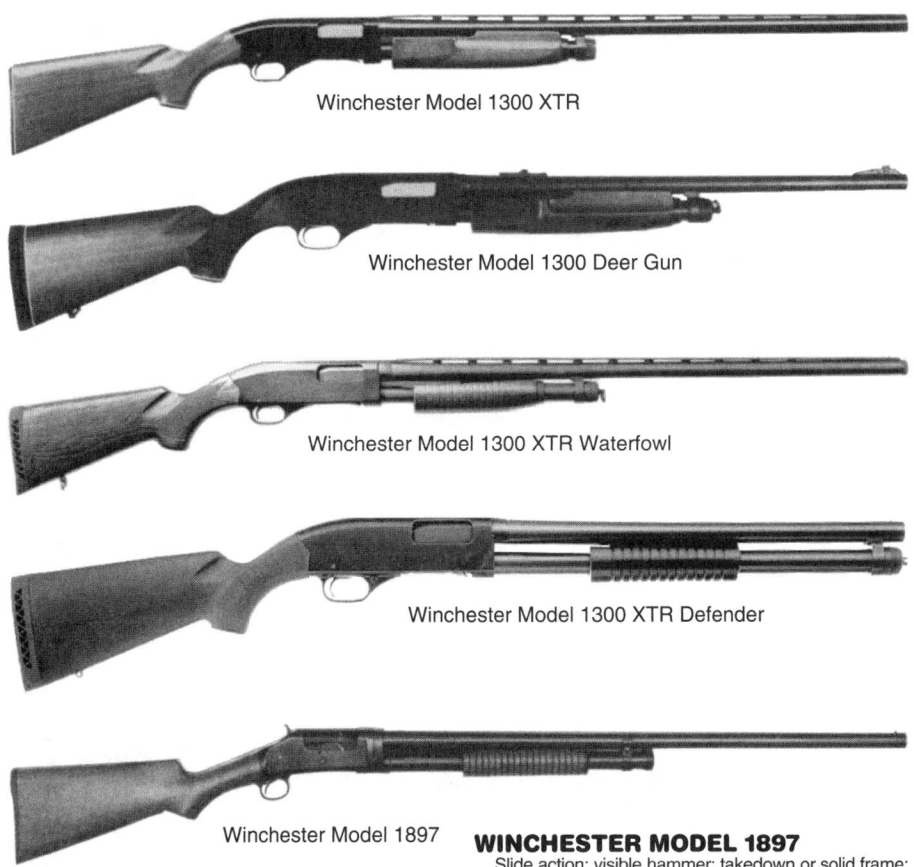

Winchester Model 1300 XTR

Winchester Model 1300 Deer Gun

Winchester Model 1300 XTR Waterfowl

Winchester Model 1300 XTR Defender

Winchester Model 1897

SHOTGUNS

Winchester Slide-Action Shotguns
Model 1300 XTR Featherweight
Model 1300 XTR Ladies/Youth
Model 1300 XTR Walnut
Model 1300 XTR Waterfowl
Model 1893
Model 1897
Model 1897 Brush
Model 1897 Pigeon
Model 1897 Special Trap
Model 1897 Standard Trap
Model 1897 Tournament
Model 1897 Trap
Model 1897 Trench

Winchester Over/Under Shotguns
Model 101
Model 101 Magnum
Model 101 Pigeon Grade
Model 101 Skeet

Winchester Model 1300 XTR Featherweight
Same specs as Model 1300 XTR except 22″ vent-rib barrel; crossbolt safety with red indicator; roll-engraved alloy receiver. Introduced 1984; XTR designation dropped 1989; discontinued 1993.
Perf.: $320   **Exc.:** $240   **VGood:** $200

Winchester Model 1300 XTR Ladies/Youth
Same specs as Model 1300 XTR except 22″ vent-rib barrel; youth-sized walnut stock with high luster finish; crossbolt safety with red safety indicator; highly polished blue finish. Introduced 1983; XTR designation dropped 1989; still in production.
LMSR: $309
Perf.: $250   **Exc.:** $220   **VGood:** $185

Winchester Model 1300 XTR Walnut
Same specs as Model 1300 XTR 26″, 28″ vent-rib barrel; walnut stock with high luster finish; crossbolt safety with red safety indicator; highly polished blued finish. Introduced, 1984; XTR designation dropped 1989; still in production. **LMSR:** $340
Perf.: $300   **Exc.:** $245   **VGood:** $200

Winchester Model 1300 XTR Waterfowl
Same specs as Model 1300 XTR except 12-ga.; 28″, 30″ vent-rib barrel; matte walnut or brown WinTuff stock, forearm; recoil pad; camo sling; swivels. Introduced 1984; XTR designation dropped 1989; discontinued 1991.
Perf.: $280   **Exc.:** $230   **VGood:** $200

WINCHESTER MODEL 1893
Slide action; exposed hammer; solid frame; 12-ga.; 2 5/8″ chamber; 5-shot tubular magazine; 30″, 32″ steel barrel; Damascus barrel optional; Full choke standard; Modified or Cylinder bore optional; uncheckered American walnut half-pistol-grip stock; fancy wood, checkering optional. Introduced 1893; discontinued 1897.
Exc.: $800   **VGood:** $600   **Good:** $500
With fancy wood, checkering
Exc.: $900   **VGood:** $675   **Good:** $550

WINCHESTER MODEL 1897
Slide action; visible hammer; takedown or solid frame; 12-, 16-ga.; 2 3/4″ chamber; 5-shot tube magazine; 26″, 28″, 30″, 32″ (12-ga. only) barrel; Full, Modified, Cylinder bore; intermediate chokes added 1931; uncheckered walnut half-pistol-grip stock, grooved forend. Introduced 1897; discontinued 1957.
Exc.: $500   **VGood:** $300   **Good:** $235

Winchester Model 1897 Brush
Same specs as Model 1897 except 26″ Cylinder bore barrel. Introduced 1897; discontinued 1931.
Exc.: $425   **VGood:** $300   **Good:** $235

Winchester Model 1897 Pigeon
Same specs as Model 1897 except checkered fancy walnut stock, standard or beavertail forearm; engraving. Introduced 1897; discontinued 1939.
Exc.: $3000   **VGood:** $1750   **Good:** $1300

Winchester Model 1897 Special Trap
Same specs as Model 1897 except checkered fancy walnut trap-style stock, standard or beavertail forearm; engraving. Introduced 1931; discontinued 1939.
Exc.: $1500   **VGood:** $1000   **Good:** $800

Winchester Model 1897 Standard Trap
Same specs as Model 1897 except trap-style stock; replaced Model 1897 Trap. Introduced 1931; discontinued 1939.
Exc.: $700   **VGood:** $500   **Good:** $400

Winchester Model 1897 Tournament
Same specs as Model 1897 except checkered fancy walnut stock, standard or beavertail forearm; engraving. Introduced 1910; discontinued 1931.
Exc.: $1200   **VGood:** $750   **Good:** $600

Winchester Model 1897 Trap
Same specs as Model 1897 except trap-style stock. Introduced 1897; discontinued 1931.
Exc.: $600   **VGood:** $475   **Good:** $400

Winchester Model 1897 Trench
Same specs as Model 1897 except ventilated steel handguard; bayonet stud; built for U.S. Army during WWI. Introduced 1920; discontinued 1935.
Exc.: $2500   **VGood:** $1500   **Good:** $1300

WINCHESTER OVER/UNDER SHOTGUNS

WINCHESTER MODEL 101
Over/under; boxlock; 12- (2 3/4″), 20- (3″), 28-ga. (2 3/4″), .410 (3″); 26 1/2″ Improved/Modified, 28″ Modified/Full, 30″ Modified/Full (12-ga. only) barrels; vent rib; checkered french walnut stock, forearm; auto ejectors; single selective trigger; combo barrel selector, safety; engraving. Made in Japan by Olin Kodensha. Introduced 1963; discontinued 1978.
12-, 20-ga. model
Perf.: $875   **Exc.:** $700   **VGood:** $600
28-ga., .410 model
Perf.: $1050   **Exc.:** $875   **VGood:** $750

Winchester Model 101 Magnum
Same specs as Model 101 except 12-, 20-ga.; 3″ chambers; 30″ Full/Full, Modified/Full barrels; recoil pad. Introduced 1966; discontinued 1981.
Perf.: $850   **Exc.:** $650   **VGood:** $600

Winchester Model 101 Pigeon Grade
Same specs as Model 101 except best-grade wood, checkering and engraving. Introduced 1963; discontinued 1984.
Field model
Perf.: $1300   **Exc.:** $900   **VGood:** $800
Skeet model
Perf.: $1300   **Exc.:** $900   **VGood:** $800
Trap model
Perf.: $1400   **Exc.:** $1000   **VGood:** $900

Winchester Model 101 Skeet
Same specs as Model 101 except 26″ (12-ga.), 26 1/2″ (20-ga.), 28″ (28-ga., .410) barrels; Skeet chokes; Skeet-style stock, forearm. Introduced 1966; discontinued 1984.
Perf.: $950   **Exc.:** $750   **VGood:** $695

SHOTGUNS

Winchester Over/Under Shotguns
Model 101 Trap
Model 101 Trap Single Barrel
Model 101 Waterfowl
Model 101 Diamond Grade Skeet
Model 101 Diamond Grade Trap
Model 101 Diamond Grade Trap Combo
Model 101 Diamond Grade Trap Single
Model 101 XTR
Model 101 XTR American Flyer Live Bird
Model 101 XTR American Flyer Live Bird Combo Set
Model 101 XTR Pigeon Grade Featherweight
Model 101 XTR Pigeon Grade Lightweight
Model 101 XTR Pigeon Grade Skeet
Model 101 XTR Pigeon Grade Trap

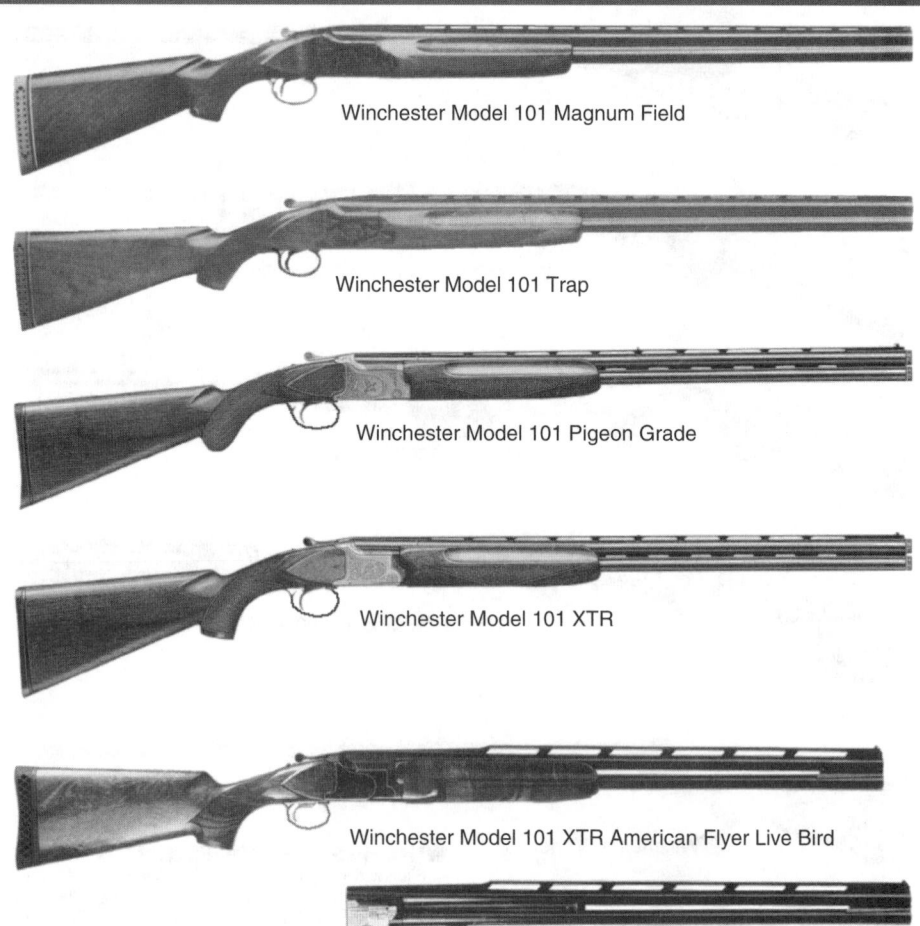

Winchester Model 101 Magnum Field

Winchester Model 101 Trap

Winchester Model 101 Pigeon Grade

Winchester Model 101 XTR

Winchester Model 101 XTR American Flyer Live Bird

Winchester Model 101 Trap
Same specs as Model 101 except 12-ga.; 30″, 32″ Improved Modified/Full, Full/Full barrels; trap-style stock; Monte Carlo or straight comb. Introduced 1966; discontinued 1984.
With Monte Carlo stock
Perf.: $1250   **Exc.:** $900   **VGood:** $800
With straight stock
Perf.: $1250   **Exc.:** $900   **VGood:** $800

Winchester Model 101 Trap Single Barrel
Same specs as Model 101 except 12-ga.; 32″, 34″ single over barrel; Full choke; trap-style Monte Carlo stock. Introduced 1967; discontinued 1971.
Perf.: $900   **Exc.:** $600   **VGood:** $500

Winchester Model 101 Waterfowl
Same specs as Model 101 except 12-ga.; 3″ chambers; 32″ barrels; WinChokes. Introduced 1981; discontinued 1982.
Perf.: $1350   **Exc.:** $1000   **VGood:** $900

WINCHESTER MODEL 101 DIAMOND GRADE SKEET
Over/under; boxlock; 12-, 20-, 28-ga., .410; 27 1/2″ barrels; WinChokes; tapered, elevated vent rib; hand-checkered select walnut Skeet-style stock, beavertail forearm; auto ejectors; contoured single selective trigger; combo barrel selector, safety; diamond-pattern engraved satin-finished receiver. Made in Japan. Introduced 1982; discontinued 1987.
Perf.: $1750   **Exc.:** $1300   **VGood:** $1100

Winchester Model 101 Diamond Grade Trap
Same specs as Model 101 Diamond Grade Skeet except 12-ga.; 30″, 32″ barrels; hand-checkered select walnut trap-style stock, beavertail forearm. Introduced 1982; discontinued 1987.
Perf.: $1750   **Exc.:** $1300   **VGood:** $1100

Winchester Model 101 Diamond Grade Trap Combo
Same specs as Model 101 Diamond Grade Skeet except 12-ga.; two-barrel set with 30″ or 32″ over/under barrels and 32″ or 34″ single barrel; hand-checkered select walnut trap-style stock, beavertail forearm. Introduced 1982; discontinued 1987.
Perf.: $2600   **Exc.:** $2100   **VGood:** $1900

Winchester Model 101 Diamond Grade Trap Single
Same specs as Model 101 Diamond Grade Skeet except 12-ga.; 32″, 34″ single barrel; hand-checkered select walnut trap-style stock, beavertail forearm. Introduced 1982; discontinued 1987.
Perf.: $2000   **Exc.:** $1650   **VGood:** $1450

WINCHESTER MODEL 101 XTR
Over/under; boxlock; 12-, 20-ga.; 26″, 28″, 30″ barrels; WinChokes; vent rib; metal bead front sight; cut-checkered American walnut pistol-grip stock, beavertail forearm; chrome-plated chambers, bores; manual safety, barrel selector; single selective trigger; auto ejectors; hand engraving; replaced Model 101. Introduced 1978; discontinued 1987.
Perf.: $1850   **Exc.:** $1400   **VGood:** $1200

Winchester Model 101 XTR American Flyer Live Bird
Same specs as Model 101 XTR except 12-ga.; 2 3/4″ chambers; 28″, 29 1/2″ barrels; under barrel fitted for internal WinChokes, over barrel has Extra-Full choke; competition vent-rib; full fancy American walnut stock; blued receiver, gold wire border inlays; matte finish on top of receiver; marketed in luggage-type case. Made in Japan. Introduced 1987; discontinued 1987.
Perf.: $2450   **Exc.:** $1900   **VGood:** $1700

Winchester Model 101 XTR American Flyer Live Bird Combo Set
Same specs as Model 101 XTR except 2 3/4″

chambers; two-barrel set with 28″, 29 1/2″ barrels; under barrel fitted for internal WinChokes, over barrel has Extra-Full choke; competition vent rib; full fancy American walnut stock; blued receiver, gold wire border inlays; matte finish on top of receiver. Made in Japan. Introduced 1987; discontinued 1987.
Perf.: $3500   **Exc.:** $2900   **VGood:** $2600

Winchester Model 101 XTR Pigeon Grade Featherweight
Same specs as Model 101 XTR except 12-, 20-ga.; 25 1/2″ Improved Cylinder/Improved Modified barrels; checkered walnut straight-grip stock. Introduced 1983; discontinued 1987.
Perf.: $1400   **Exc.:** $950   **VGood:** $850

Winchester Model 101 XTR Pigeon Grade Lightweight
Same specs as Model 101 XTR except 12-, 20-, 28-ga., .410; 26″ to 32″ barrels; hand-checkered French walnut stock; knurled non-slip trigger; alloy construction; hand-engraved satin-finished receiver; marketed in hard case. Made in Japan. Introduced, 1983; discontinued 1987.
Perf.: $1800   **Exc.:** $1350   **VGood:** $1200

Winchester Model 101 XTR Pigeon Grade Skeet
Same specs as Model 101 XTR except 12-, 20-, 28-ga., .410; select checkered Skeet-style stock. Introduced 1980; discontinued 1987.
Perf.: $1200   **Exc.:** $950   **VGood:** $850

Winchester Model 101 XTR Pigeon Grade Trap
Same specs as Model 101 XTR except 12-ga.; select checkered trap-style stock. Introduced 1980; discontinued 1987.
Perf.: $1200   **Exc.:** $950   **VGood:** $850

Winchester Model 101 XTR Waterfowl

Winchester Model 501 Grand European

Winchester Xpert Model 96

Winchester Xpert Model 96 Skeet

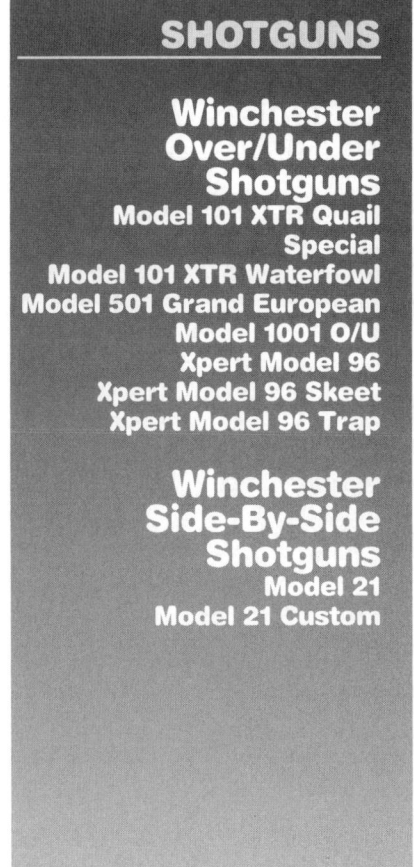

SHOTGUNS

Winchester Over/Under Shotguns
Model 101 XTR Quail Special
Model 101 XTR Waterfowl
Model 501 Grand European
Model 1001 O/U
Xpert Model 96
Xpert Model 96 Skeet
Xpert Model 96 Trap

Winchester Side-By-Side Shotguns
Model 21
Model 21 Custom

Winchester Model 501 Skeet

Winchester Model 501 Trap

Winchester Model 21

Winchester Model 101 XTR Quail Special
Same specs as Model 101 XTR except 12-, 20-, 28-ga., .410; 25 1/2″ vent-rib barrels; Winchokes; checkered American walnut staight-grip stock; silvered engraved receicer. Made in Japan. Introduced 1987; discontinued 1988.

12-ga., .410 models
Perf.: $1850   **Exc.:** $1400   **VGood:** $1200
20-ga. model
Perf.: $2300   **Exc.:** $1700   **VGood:** $1450
28-ga. model
Perf.: $3000   **Exc.:** $2000   **VGood:** $1800

Winchester Model XTR Waterfowl
Same specs as Model 101 XTR except 25 1/2″, 27″, 28″ barrels; weighs 8 1/4 lbs.; vent-rib; bead front and middle sights; hand-checkered French walnut stock, forend. Made in Japan. Introduced 1981; discontinued 1982.
Perf.: $900   **Exc.:** $775   **VGood:** $575

WINCHESTER MODEL 501 GRAND EUROPEAN
Over/under; 12-, 20-ga.; 27″, 30″, 32″ barrels; tapered vent rib; hand-rubbed oil-finish American walnut stock; trap model has Monte Carlo or regular stock; Skeet version has rosewood buttplate; engine-turned breech interior; selective auto ejectors; chromed bores; engraved silvered receiver. Made in Japan. Introduced 1981; discontinued 1986.
Perf.: $1450   **Exc.:** $1150   **VGood:** $950

WINCHESTER MODEL 1001 O/U
Over/under; 12-ga.; 3″ chamber; 28″ barrel; Imp. Cyl., Mod., Imp. Mod., Skeet WinPlus; 45″ overall length; weighs 7 lbs.; select walnut stock; checkered grip, forend; single selective trigger; automatic ejectors; wide vent rib; matte finished receiver top; blued receiver with scroll engraving. Although carried in the 1993 catalog and advertised for a year, the Model 1001 never reached full production status because of technical and quality control difficulties in Italy. Both Field and Sporting Clays versions were planned. Some guns reached consumers, but in late 1994 all were recalled and the project was cancelled.
Note: Since none are in the marketplace, no sales performance can be established and no values are given.

WINCHESTER XPERT MODEL 96
Over/under; boxlock; 12-, 20-ga.; 3″ chambers; 26″, 28″, 30″ barrels; vent rib; depending on gauge, barrel length, weighs 6 1/4 to 8 1/4 lbs.; checkered walnut pistol-grip stock, forend; similar to Model 101. Made in Japan. Introduced 1976; discontinued 1981.
Exc.: $800   **VGood:** $600   **Good:** $500

Winchester Xpert Model 96 Skeet
Same specs as Xpert Model 96 except 2 3/4″ chambers; 27″ barrels; Skeet choke; checkered walnut Skeet-style stock. Made in Japan. Introduced 1976; discontinued 1981.
Exc.: $825   **VGood:** $625   **Good:** $500

Winchester Xpert Model 96 Trap
Same specs as Xpert Model 96 except 12-ga.; 2 3/4″ chambers; 30″ barrels; checkered walnut straight or Monte Carlo trap-style stock; recoil pad. Made in Japan. Introduced 1976; discontinued 1981.
Exc.: $775   **VGood:** $575   **Good:** $495

WINCHESTER SIDE-BY-SIDE SHOTGUNS

WINCHESTER MODEL 21
Side-by-side; boxlock; hammerless; 12-, 16-, 20-, 28-ga., .410; 20″, 28″, 30″, 32″ (12-ga. only) barrels; raised matted or vent rib; Full, Improved Modified, Modified, Improved Cylinder, Skeet chokes; checkered walnut straight or pistol-grip stock, regular or beavertail forend; automatic safety; early models (1931-1944) have double triggers and extractors, single trigger and ejectors optional; later models (1945-1959) have selective single trigger and selective ejectors. Introduced 1931; discontinued 1959.
With double triggers, extractors
Exc.: $1700   **VGood:** $1100   **Good:** $800
With double triggers, selective ejectors
Exc.: $2200   **VGood:** $1700   **Good:** $1450
With single selective trigger, extractors
Exc.: $1900   **VGood:** $1400   **Good:** $1100
With single selective trigger, selective ejectors
Exc.: $2500   **VGood:** $2000   **Good:** $1750
.410 model
Exc.: $30,000   **VGood:** $20,000 **Good:** $14,000
28-ga. model
Exc.: $20,000   **VGood:** $14,000 **Good:** $9000

Winchester Model 21 Custom
Same specs as Model 21 except hand-checkered fancy American walnut stock, forearm; hand-honed internal parts; optional carving, engraving and gold inlays. Introduced 1960; discontinued 1982.
Perf.: $7500   **Exc.:** $5500   **VGood:** $4500

SHOTGUNS

Winchester Side-By-Side Shotguns
Model 21 Duck
Model 21 Grand American
Model 21 Pigeon
Model 21 Skeet
Model 21 Trap
Model 23 Classic
Model 23 Custom
Model 23 XTR
Model 23 XTR Golden Quail
Model 23 XTR Heavy Duck
Model 23 XTR Light Duck
Model 23 XTR Pigeon Grade
Model 23 XTR Pigeon Grade Lightweight
Model 23 XTR Two-Barrel Set Model 24

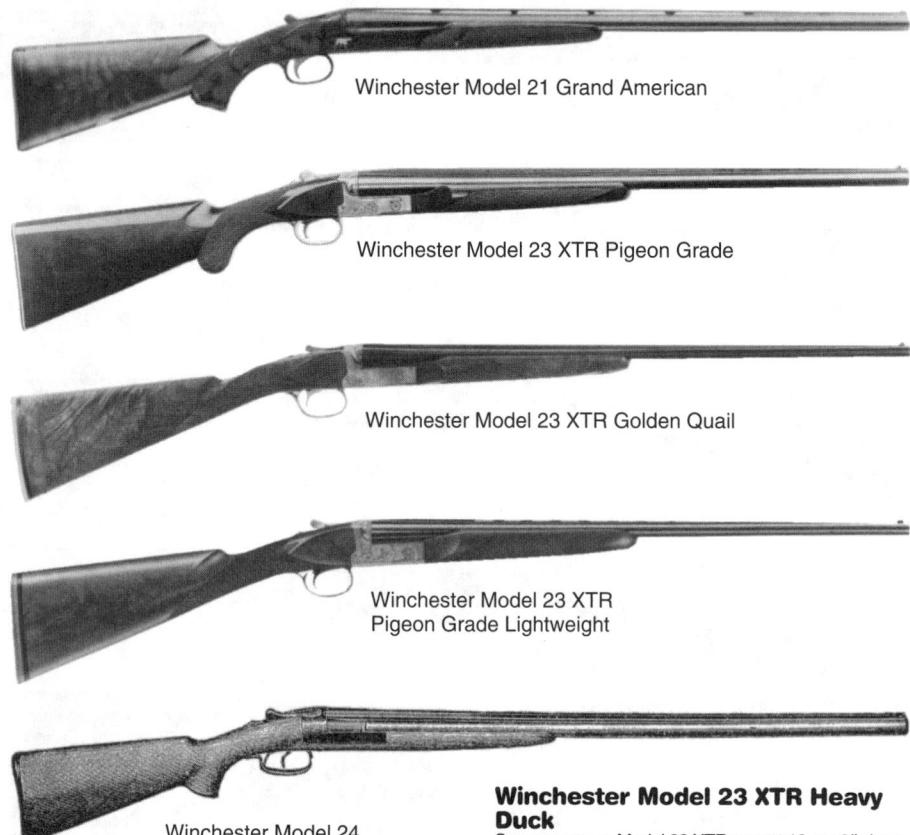

Winchester Model 21 Grand American

Winchester Model 23 XTR Pigeon Grade

Winchester Model 23 XTR Golden Quail

Winchester Model 23 XTR Pigeon Grade Lightweight

Winchester Model 24

Winchester Model 21 Duck
Same specs as Model 21 except 12-ga.; 3″ chamber; 30″, 32″ barrels; Full choke; checkered walnut pistol-grip stock, beavertail forearm; selective ejection; selective single trigger; recoil pad. Introduced 1940; discontinued 1954.
With matted rib
   Exc.: $2500   **VGood:** $2000   **Good:** $1750
With vent rib
   Exc.: $2800   **VGood:** $2300   **Good:** $2000

Winchester Model 21 Grand American
Same specs as Model 21 except two barrel sets; hand-checkered fancy American walnut stock, forearm; hand-honed internal parts; optional carving, engraving and gold inlays. Introduced 1960; discontinued 1982.
   Perf.: $18,750 **Exc.:** $17,000   **VGood:** $14,000

Winchester Model 21 Pigeon
Same specs as Model 21 except hand-checkered fancy American walnut stock, forearm; hand-honed internal parts; optional carving, engraving and gold inlays. Introduced 1960; discontinued 1982.
   Perf.: $10,500 **Exc.:** $8500   **VGood:** $7000

Winchester Model 21 Skeet
Same specs as Model 21 except 26″, 28″ barrels; Skeet chokes; red bead front sight; checkered French walnut stock, beavertail forearm; selective single trigger, selective ejection; non-auto safety. Introduced 1933; discontinued 1958.
With matted rib
   Exc.: $1900   **VGood:** $1600   **Good:** $1300
With vent rib
   Exc.: $2100   **VGood:** $1800   **Good:** $1500

Winchester Model 21 Trap
Same specs as Model 21 except 30″, 32″ barrels; Full choke; checkered walnut pistol-grip or straight stock, beavertail forearm; selective trigger; non-auto safety; selective ejection. Introduced 1932; discontinued 1940.

With matted rib
   Exc.: $1900   **VGood:** $1600   **Good:** $1300
With vent rib
   Exc.: $2100   **VGood:** $1800   **Good:** $1500

WINCHESTER MODEL 23 CLASSIC
Side-by-side; 12-, 20-, 28-ga., .410. 26″ vent-rib barrels; deluxe hand-checkered walnut stock, beavertail forend; auto safety; auto ejectors; single selective trigger; solid recoil pad; gold inlay on bottom of receiver; gold initial plate in stock; ebony inlay in forend; blued, engraved receiver. Made in Japan. Introduced 1986; discontinued 1988.
12-, 20-ga.
   Perf.: $1600   **Exc.:** $1200   **VGood:** $1000
28-ga., .410
   Perf.: $2100   **Exc.:** $1500   **VGood:** $1250

Winchester Model 23 Custom
Same specs as Model 23 Classic except 12-ga.; 27″ WinChoke barrels; chrome-lined bores, chambers for steel shot; high-luster blued receiver, no engraving; marketed in luggage-style case. Made in Japan. Introduced 1986; discontinued 1988.
   Perf.: $2000   **Exc.:** $1200   **VGood:** $1000

WINCHESTER MODEL 23 XTR
Side-by-side; 12-, 20-ga.; 3″ chambers; 25 1/2″, 26″, 28″, 30″ barrels; tapered vent rib; WinChokes; tapered vent rib; cut-checkered high-luster American walnut stock, beavertail forend; mechanical single trigger; selective ejectors; silver-gray satin finish on receiver, top lever, trigger guard; fine-line scroll engraving. Made in Japan. Introduced 1978; discontinued 1986.
   Perf.: $1000   **Exc.:** $700   **VGood:** $600

Winchester Model 23 XTR Golden Quail
Same specs as Model 23 XTR except different chamberings produced each year; 12- (1986), 20- (1984), 28-ga. (1985), .410 (1987); 25 1/2″ solid-rib barrel; fixed chokes; English-style straight-grip stock, beavertail forearm; engraving; gold inlays. Only 500 made each year. Introduced 1984; discontinued 1987.
   Perf.: $1500   **Exc.:** $1000   **VGood:** $900

Winchester Model 23 XTR Heavy Duck
Same specs as Model 23 XTR except 12-ga; 3″ chambers; 30″ Full/Extra-Full barrels; select American walnut stock, beavertail forend; blued receivers, barrels; marketed in hard case. Only 500 made. Made in Japan. Introduced 1983; discontinued 1984.
   Perf.: $1500   **Exc.:** $1000   **VGood:** $900

Winchester Model 23 XTR Light Duck
Same specs as Model 23 XTR except 20-ga.; 3″ chambers; 28″ Full/Full barrels; select American walnut stock, beavertail forend; blued receiver, barrels; marketed in hard case. Only 500 made. Made in Japan. Introduced 1984; discontinued 1985.
   Perf.: $1500   **Exc.:** $1000   **VGood:** $900

Winchester Model 23 XTR Pigeon Grade
Same specs as Model 23 XTR except higher grade of wood, checkering and engraving; marketed in hard case. Made in Japan. Introduced, 1983; discontinued 1986.
   Perf.: $1550   **Exc.:** $1050   **VGood:** $950

Winchester Model 23 XTR Pigeon Grade Lightweight
Same specs as Model 23 XTR except 25 1/2″ barrels; fixed or choke tubes; English-style straight-grip stock, thin semi-beavertail forend; engraved bird scene. Marketed in hard case. Made in Japan. Introduced 1982; discontinued 1988.
   Perf.: $1200   **Exc.:** $900   **VGood:** $800

Winchester Model 23 XTR Two-Barrel Set
Same specs as Model 23 XTR except 20-, 28-ga.; two-barrel set in each gauge of 25 1/2″ barrel, each fitted with its own full fancy American walnut semi-beavertail forend; marketed in handmade leather luggage-style carrying case. Only 500 made. Introduced 1986; discontinued 1986
   Perf.: $3900   **Exc.:** $3100   **VGood:** $2800

WINCHESTER MODEL 24
Side-by-side; hammerless; takedown; 12-, 16-, 20-ga.; 26″ Improved/Modified, 28″ Modified/Full or Improved/Modified, 30″ Modified/Full (12-ga. only) barrels; uncheckered walnut straight or pistol-grip stock, semi-beavertail forearm; composition buttplate; double triggers; automatic ejectors. Introduced 1940; discontinued 1957.
12-, 16-ga. model.
   Exc.: $375   **VGood:** $275   **Good:** $200
20-ga. model
   Exc.: $525   **VGood:** $425   **Good:** $350

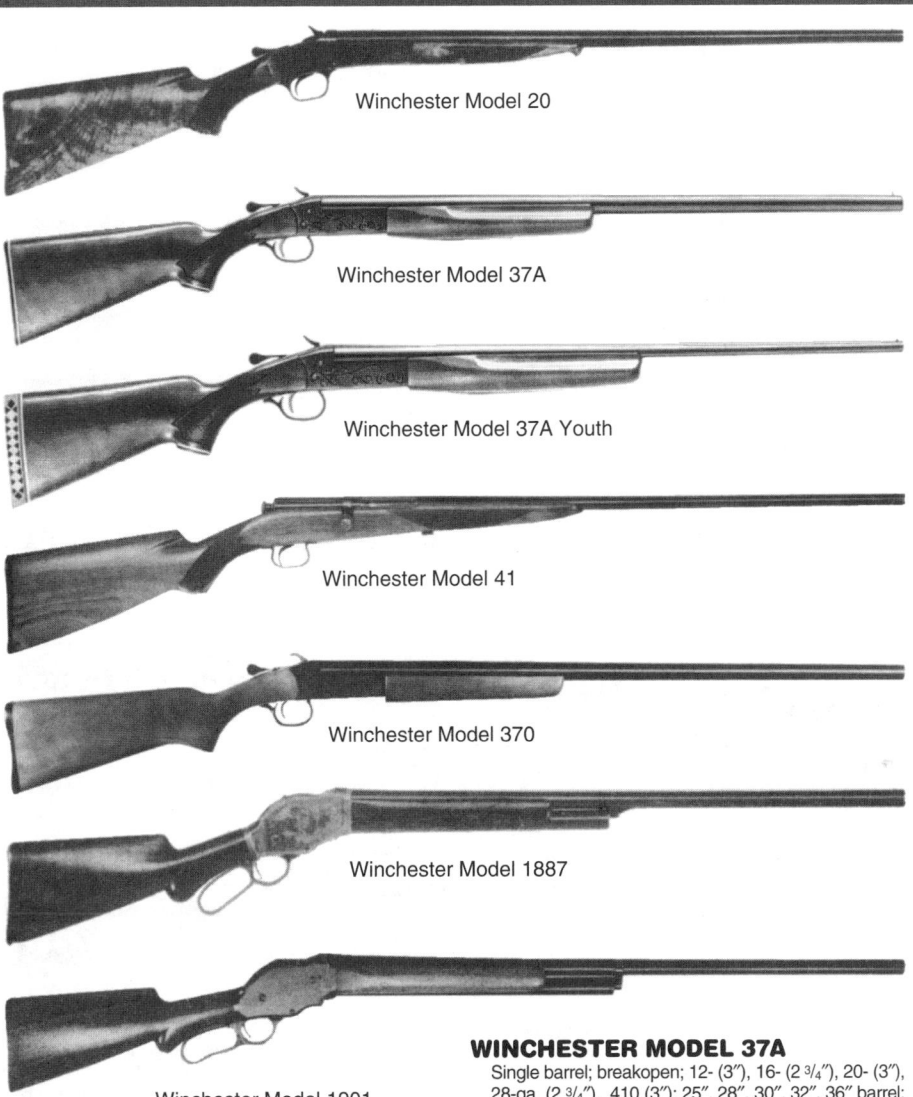

Winchester Model 20

Winchester Model 37A

Winchester Model 37A Youth

Winchester Model 41

Winchester Model 370

Winchester Model 1887

Winchester Model 1901

SHOTGUNS

Winchester Single Shot & Lever-Action Shotguns
Model 20
Model 36
Model 37
Model 37A
Model 37A Youth
Model 41
Model 370
Model 370 Youth
Model 1885 Single Shot
Model 1887
Model 1887 Deluxe
Model 1901

WINCHESTER SINGLE SHOT & LEVER-ACTION SHOTGUNS

WINCHESTER MODEL 20
Single shot; exposed hammer; takedown; .410; 2 1/2" chamber; 26" Full-choke barrel; checkered pistol-grip stock, forearm. Introduced 1919; discontinued 1924.
Exc.: $375 **VGood:** $250 **Good:** $190

WINCHESTER MODEL 36
Single shot; bolt action; takedown; 9mm short shot, 9mm long shot, 9mm ball cartridges interchangeably; 18" round barrel; one-piece plain wood stock, forearm; special trigger guard forms pistol grip; composition buttplate; cocks by pulling rearward on knurled firing pin head, same mechanism used in some Winchester single-shot rifles; guns were not serialized. Introduced 1920; discontinued 1927.
Exc.: $475 **VGood:** $375 **Good:** $300

WINCHESTER MODEL 37
Single shot; breakopen; semi-hammerless; 12-, 16-, 20-, 28-ga.; .410; 26", 28", 30", 32" barrel; Full, Modified, Cylinder choke; metal bead front sight; unchecked American walnut pistol-grip stock, semi-beavertail forarm; automatic ejector; checkered compostion buttplate. Introduced 1936; discontinued 1968.
12-, 16-, 20-ga. model
Exc.: $250 **VGood:** $175 **Good:** $125
28-ga. model
Exc.: $750 **VGood:** $550 **Good:** $400
.410 model
Exc.: $275 **VGood:** $200 **Good:** $100

WINCHESTER MODEL 37A
Single barrel; breakopen; 12- (3"), 16- (2 3/4"), 20- (3"), 28-ga. (2 3/4"), .410 (3"); 25", 28", 30", 32", 36" barrel; Full choke; checkered walnut-stained hardwood pistol-grip stock; concave hammer spur; white spacer between grip cap and buttplate; gold trigger; engraving. Introduced 1973; discontinued 1980.
12-, 16-, 20-ga. model
Perf.: $140 **Exc.:** $100 **VGood:** $80
28-ga., .410 model
Perf.: $190 **Exc.:** $140 **VGood:** $120

Winchester Model 37A Youth
Same specs as Model 37A except 20-ga., .410; 26" barrel; Improved Modified, Full chokes; youth-sized hardwood stock. Introduced 1973; discontinued 1980.
Perf.: $180 **Exc.:** $110 **VGood:** $90

WINCHESTER MODEL 41
Single shot; bolt action; takedown; .410; 2 1/2" (pre-1933), 3" (post-1933) chamber; 24" full-choke barrel; uncheckered one-piece walnut pistol-grip or straight-grip stock; checkering on special order; guns were not numbered serially. Introduced 1920; discontinued 1934.
Exc.: $300 **VGood:** $200 **Good:** $150

WINCHESTER MODEL 370
Single barrel; breakopen; 12- (3"), 16- (2 3/4"), 20- (3"), 28-ga. (2 3/4"), .410 (3"); 26", 28", 30", 32", 36" barrel; Full choke; weight varies with gauges and barrel lengths, from 5 1/2 to 6 1/4 lbs.; bead front sight; plain American hardwood pistol-grip stock, forearm; hard-rubber buttplate; auto ejectors. Introduced 1968; discontinued 1973, replaced by Model 37A.
12-, 16-, 20-ga. model
Perf.: $125 **Exc.:** $100 **VGood:** $80
28-ga., .410 model
Perf.: $175 **Exc.:** $140 **VGood:** $120

Winchester Model 370 Youth
Same specs as Model 370 except 20-ga., .410; 26" Improved Modified barrel; youth-sized hardwood stock; rubber recoil pad. Introduced 1969; discontinued 1973.
Perf.: $160 **Exc.:** $110 **VGood:** $90

WINCHESTER MODEL 1885 SINGLE SHOT
Falling block; takedown or solid frame; 20-ga.; 3" chamber; 26" nickel-steel barrel; extra barrels or matted top barrels optional; Full choke standard; Modified or Cylinder bore optional; uncheckered walnut pistol-grip stock; shotgun version of Model 1885 Hi-Wall rifle. Introduced 1913; discontinued 1918.
Exc.: $3000 **VGood:** $2300 **Good:** $1800

WINCHESTER MODEL 1887
Lever action; solid frame; 10-, 12-ga.; 4-shot tubular magazine; 30", 32" barrel; Full choke; uncheckered American walnut half-pistol-grip stock, forend. Introduced 1887; discontinued 1901.
Exc.: $1650 **VGood:** $1100 **Good:** $800

Winchester Model 1887 Deluxe
Same specs as Model 1887 except Damascus barrel; hand-checkered stock, forend. Introduced 1887; discontinued 1901.
Exc.: $2400 **VGood:** $1850 **Good:** $1400

WINCHESTER MODEL 1901
Lever action; solid frame; 10-ga.; 4-shot tube magazine; 30", 32" barrel; Full choke; uncheckered American walnut half-pistol-grip stock, forend; replaced Model 1887 in line with internal redesign features. Introduced 1901; discontinued 1920.
Exc.: $1100 **VGood:** $800 **Good:** $700

SHOTGUNS

Woodward
Best Quality Double
Best Quality Over/Under
Single Barrel Trap

Zabala
Double

Pietro Zanoletti
Model 2000 Field

Zoli
Delfino
Golden Snipe
Silver Falcon
Silver Fox
Uplander
Woodsman
Z90 Mono-Trap Model
Z90 Skeet Model
Z90 Sporting Clays
Z90 Trap

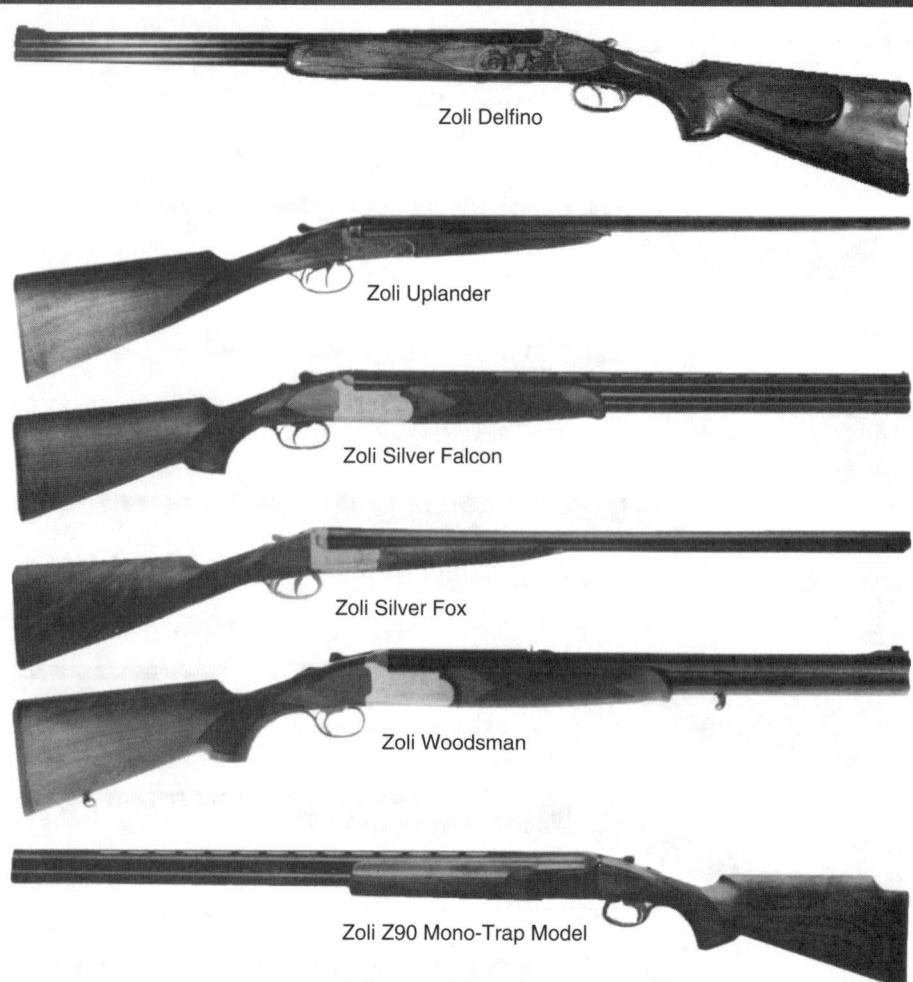

Zoli Delfino

Zoli Uplander

Zoli Silver Falcon

Zoli Silver Fox

Zoli Woodsman

Zoli Z90 Mono-Trap Model

WOODWARD BEST QUALITY DOUBLE
Side-by-side; sidelock; built to customer order in any standard gauge, barrel length, choke; double or single trigger; automatic ejectors; produced in field, wildfowl, Skeet or trap configurations. Made in England. Manufactured prior to WWII.
With single trigger
    Perf.: $28,500 **Exc.:** $21,000   **VGood:** $19,500
With double trigger
    Perf.: $27,500 **Exc.:** $20,000   **VGood:** $18,500

WOODWARD BEST QUALITY OVER/UNDER
Over/under; sidelock; built to customer order in any standard gauge, barrel length, choke; plain barrel or vent rib; double or single trigger; auto ejectors. Made in England. Introduced 1909; discontinued WWII.
With single triggers
    Perf.: $34,000 **Exc.:** $25,000   **VGood:** $20,500
With double triggers
    Perf.: $33,000 **Exc.:** $24,000   **VGood:** $19,000

WOODWARD SINGLE BARREL TRAP
Single shot; sidelock; 12-ga.; built to customer order in any standard barrel length, choke; plain barrel or vent rib; single trigger; auto ejectors. Made in England. Manufactured prior to WWII.
    Perf.: $13,000 **Exc.:** $9000   **VGood:** $8000

ZABALA DOUBLE
Side-by-side; 10-, 12-, 20-ga., .410; 26", 28", 30", 32" barrels; raised matted solid rib; metal bead front sight; hand-checkered French walnut pistol-grip stock, plastic-finished beavertail forend; double triggers; front trigger hinged; hand-engraved action, blued finish. Made in Spain. Introduced 1980; discontinued 1982.
    Perf.: $475   **Exc.:** $395   **VGood:** $325

PIETRO ZANOLETTI MODEL 2000 FIELD
Over/under; boxlock; 12-ga.; 28" barrels; gold bead front sight; checkered European walnut pistol-grip stock, forend; auto ejectors; double triggers; engraved receiver. Made in Italy. Introduced 1984; no longer in production.
    Perf.: $500   **Exc.:** $385   **VGood:** $325

ZOLI DELFINO
Over/under; 12-, 20-ga.; 3" chambers; 28" barrels; vent rib; hand-checkered European walnut pistol-grip stock,

cheekpiece; chromed bores; double triggers; auto sliding safety; color case-hardened receiver; light engraving. Made in Italy. Introduced 1980; discontinued 1989.
    Perf.: $650   **Exc.:** $560   **VGood:** $490

ZOLI GOLDEN SNIPE
Over/under; boxlock; 12-, 20-ga.; 26", 28", 30" vent-rib barrels; standard choke combos; hand-checkered European walnut pistol-grip stock, forend; single trigger; auto ejectors; engraving. No longer in production.
    Perf.: $700   **Exc.:** $560   **VGood:** $490

ZOLI SILVER FALCON
Over/under; boxlock; 12-, 20-ga.; 3" chambers; 26", 28" barrels; standard choke combos or Full choke tubes; Turkish Circassian walnut stock; single selective trigger, auto ejectors; silvered finish, floral engraving on frame. Made in Italy. Introduced 1989; discontinued 1990.
    Perf.: $750   **Exc.:** $600   **VGood:** $525

ZOLI SILVER FOX
Side-by-side; boxlock; 12-, 20-ga.; 3" chambers; 26", 28" barrels; standard choke combos; select Turkish Circassian walnut straight-grip stock, splinter forend; solid recoil pad; single trigger; selective ejectors; engraved, silvered receiver. Made in Italy. Introduced 1989; discontinued 1990.
    Perf.: $1750   **Exc.:** $1425   **VGood:** $1250

ZOLI UPLANDER
Side-by-side; boxlock; 12-, 20-ga.; 3" chambers; 25" Improved Cylinder/Modified barrels; hand-checkered Turkish Circassian walnut straight English-style stock, splinter forend; single trigger; auto ejectors; color case-hardened frame. Introduced 1989; discontinued 1990.
    Perf.: $975   **Exc.:** $875   **VGood:** $775

ZOLI WOODSMAN
Over/under; boxlock; 12-ga.; 3" chambers; 23" vent-rib barrels for rifled slugs; five choke tubes; rifle sights on

raised rib; skip-line-checkered, oil-finished Turkish Circassian walnut stock; rubber buttpad; single selective trigger; auto selective ejectors; with or without sling swivels; blued, engraved frame. Made in Italy. Introduced 1989; discontinued 1990.
    Perf.: $1300   **Exc.:** $1050   **VGood:** $900

ZOLI Z90 MONO-TRAP MODEL
Single barrel; breakopen; 12-ga.; 2 3/4" chamber; 32", 34" single barrel; choke tubes; raised vent rib; two sight beads; checkered Turkish Circassian walnut Monte Carlo pistol-grip stock, forearm; single selective trigger; selective auto ejectors; recoil pad; matte blue finish on receiver. Made in Italy. Introduced 1989; discontinued 1990.
    Perf.: $1600   **Exc.:** $1350   **VGood:** $1100

Zoli Z90 Skeet Model
Same specs as Z90 Mono-Trap except over/under; boxlock; 28" barrels; checkered Turkish Circassian walnut Skeet-style stock, forearm. Made in Italy. Introduced 1989; discontinued 1990.
    Perf.: $1650   **Exc.:** $1350   **VGood:** $1100

Zoli Z90 Sporting Clays
Same specs as Z90 Mono-Trap except over/under; sidelock; 28" barrels; weighs 7 1/4 lbs.; checkered Turkish Circassian walnut pistol-grip stock, forend; Schnabel forend tip; single selective trigger; selective auto ejectors; solid rubber buttpad; silvered, engraved frame. Made in Italy. Introduced 1989; discontinued 1990.
    Perf.: $1350   **Exc.:** $1100   **VGood:** $950

Zoli Z90 Trap
Same specs as Z90 Mono-Trap except over/under; boxlock; 29 1/2", 32" barrels; step-type vent rib; vent center rib; weighs 8 1/2 lbs.; adjustable single selective trigger. Made in Italy. Introduced 1989; discontinued 1990.
    Perf.: $1750   **Exc.:** $1350   **VGood:** $1100

MODERN GUN VALUES

COMMEMORATIVES

Commemoratives & Limited Editions Introduction

EVERY TIME I develop or adopt a theory that might become a full-scale rule that works, something goes astray, and it's back to the old drawing board! Let me state that this has come to be particularly true for me regarding commemorative and limited edition firearms. A few years ago, I was firmly convinced that the values of such arms were tied directly to the price of gold. When gold went up, so did prices on commemoratives, et al. When the value of the precious yellow metal dropped, so did prices on such firearms.

That theory still seems to hold some degree of promise. However, the price of gold fluctuates little these days, and the values of commemoratives and limited editions seem to be more closely bound to supply and demand than to gold, silver or even diamonds. There also are fluctuations and variations in values on a geographic plane. This factor would seem to have a good deal more to do with the state of the economy in a specific area than the model or type of firearm involved—or what event it is meant to commemorate.

There are exceptions to that theory, too. For example, in any instance I have seen, a commemorative Colt Single Action Army invariably will bring a better price in Dallas, Texas, than it will in Portland, Maine. In this situation, nostalgia, local pride and the wish to have something connected with the history of the area all come into play.

If any one firearm has a Texas history, it has to be the SAA. Back in the days when I was in the business of writing low-budget Western movies, the producers would do whatever they could to get the word, Texas, in the title of the film, even if they had to move history a few states geographically. They had found that by so-doing, they could get enough play dates in the Lone Star State alone to cover their production costs! The same thinking tends to apply these days to some facets of the commemorative biz: Where will they sell?

But to get a feel for all this, we have to go back to the 1960s when Colt started what threatened to become the Commemorative of the Month approach, followed closely by Winchester. It has been said that staffs of the two companies sat up nights trying to find worthy bits of history that deserved commemoration...or could be publicized as deserving such commemoration!

It also should be pointed out that, during that era, not much attention was paid to numbers. When Winchester turned out an NRA commemorative, they seemed to assume that every member of the National Rifle Association would willingly pay for one of these rifles. Wrong! Before the supply was absorbed, this particular model was being sold at giveaway prices in discount stores. Original price on the NRA Centennial rifle was listed at $149.95.

Now, though, 25 years later, enough of these rifles have been lost, ruined, broken or simply fired that a virginal, unfired example is worth in the neighborhood of $375 to $400.

In 1981, Winchester issued 50,000 John Wayne commemorative Model 94s. Price was $600 and they were a glut on the market for more than a decade. Suddenly, though, age is giving that particular model stature, and it's bringing in the neighborhood of $900 per copy, fresh and in the original packing box. In recent years, both Colt and Winchester have issued few commemorative arms, unless they are making special runs for someone else. One of the leaders in marketing these has been Cherry's, the gun shop in Greensboro, North Carolina, that boasts of being the "world's leading commemorative gun dealer". Winchester has continued to turn out commemoratives in low numbers, usually fewer than 500 per issue, for this organization. Made in such limited numbers, the law of supply and demand sets in and Cherry's is able to put collector-value prices on these few. Other organizations which have followed this limited-production trend have been the American Historical Foundation and America

Remembers, the latter an organization that markets under several names.

The American Historical Foundation began by turning out replica arms commemorating early history, but in more recent annums has spent a lot of the production and advertising budgets in selling replicas of military armament used in World War II and the various wars and campaigns conducted since. For example, if you want a Thompson submachine gun that fires semi-auto only and commemorates the campaigns in Vietnam, they have it.

At the opposite extreme, they also have a commemorative honoring Teddy Roosevelt, which is alleged to be a gussied-up copy of the old soldier's personal handgun. Of course, Teddy's original was practical. The replica is all loaded up with silver, gold and engraving. Like Cherry's, the American Historical Society works in low production numbers and respectably high prices, even offering time payment plans to customers.

America Remembers has taken a similar approach, but seems to depend more upon personalities than history in selling their commemoratives, special tribute and limited edition guns. They, for example, have done well with versions of the Colt Single Action Army devoted to screen star Tom Mix, deceased now for more than half a century; country singer George Jones; Gene Autry; Roy Rogers and Hopalong Cassidy. Most of these runs are priced at well over $1,500 per gun, but usually are produced in lots of no more than 500 and even as few as 100 units. Some are made by Colt, but these are in the higher priced strata, for the retail price of a new standard Colt SAA today is $1200-plus, without the decorative touches. It should come as no surprise that some of these frontier-oriented commemoratives are manufactured, engraved and plated in Italy by such gunmakers as Uberti. While many commemoratives have not gained in great value over recent years and some have even lost, the short runs just described seem to hold their value and increase steadily. One reason for the continued growing value of the Colt-made Single Action Army revolver commemoratives has been the ever-increasing cost of buying a new standard model of that particular gun. The Colt SAA revolvers now are assembled one at a time in the Colt custom shop, retailing at the aforementioned $1200-plus price.

An example of what is happening is reflected in the values of the Colt Bat Masterson commemorative SAA, of which 500 were released in 1967. At that time, this particular commemorative sold for $180. Today, that same gun is worth a minimum of $1300! It probably would not be worth that much were not the cost of today's standard SAA as high as it is. Reflecting upon all this, it may be that, if one wants to deal in commemoratives or limited edition firearms, the profits are dependent upon numbers, relating to supply and demand—and the economic changes brought about simply by time.

If you have any doubts concerning the supply/demand factor, take note of the fact that the one millionth Sturm, Ruger Mark II was designated as a commemorative, carrying the appropriate 1,000,000 serial number. Totally engraved, gripped in ivory and inlaid with gold, this gun became the ultimate limited edition. It sold in 1980 for $27,200! At this writing, the American Custom Gunmakers Guild has just completed another one-of-a-kind, the John Amber Commemorative Rifle, to be raffled in early 1996. The gun was created by several members of the guild, each lending his particular specialty to completion of this highly decorated single shot. The raffle for which this gun was built is expected to bring in some $60,000!

It should be pointed out that the guns under discussion all are in "as new" condition and unfired. Once a commemorative or limited edition gun has passed a bullet through its barrel, much of the value dies. That's just the way it is!

Jack Lewis

1960 Marlin 90th Anniversary Model

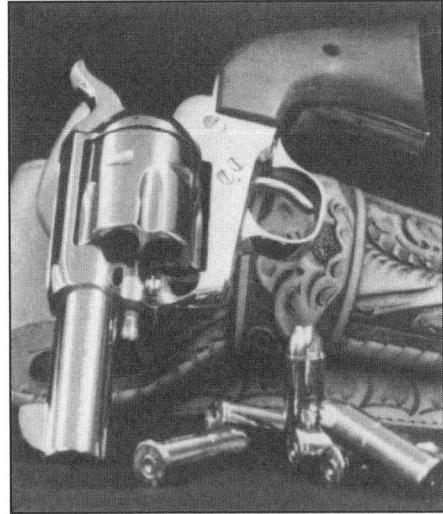
1961 Colt Sheriff's Model

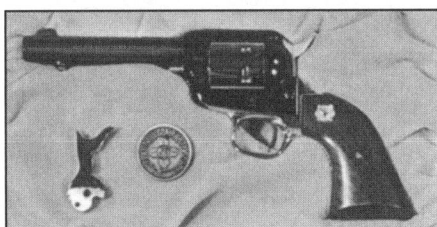
1962 Colt New Mexico Golden Anniversary

COMMEMORATIVES

1960
Marlin 90th Anniversary Model

1961
Colt 125th Anniversary Model
Colt Civil War Centennial
Colt Geneseo Anniversary
Colt Kansas Statehood
Centennial
Colt Pony Express Centennial
Colt Sheriff's Model

1962
Colt Columbus, Ohio,
Sesquicentennial
Colt Fort Findlay, Ohio,
Sesquicentennial
Colt New Mexico Golden
Anniversary
Colt Rock Island Arsenal
Centennial
Colt West Virginia Statehood
Centennial Single-Action
Army

1960

MARLIN 90TH ANNIVERSARY MODEL

Model 39A rifle commemorating firm's ninth decade. Has same general specs as Golden 39A, except for chrome-plated barrel, action; stock forearm of hand-checkered select American walnut; squirrel figure carved on right side of butt stock. Manufactured 1960; only 500 made.
Market Value: $995

1961

COLT 125TH ANNIVERSARY MODEL

Single-Action Army revolver; "125th Anniversary—SAA Model .45 Cal." on left side of 7 1/2" barrel; 45 Colt only; varnished walnut grips; gold-plated Colt medallions; gold-plated hammer, trigger, trigger guard; balance blued. Originally cased in red velvet-lined box; 7390 made in 1961 only. **Issue Price: $150.**
Market Value: $1395

COLT CIVIL WAR CENTENNIAL

Single shot replica of Colt Model 1860 Army revolver; "Civil War Centennial Model—.22 Caliber Short" on left side of 6" barrel; 22 Short only; varnished walnut grips; gold-plated Colt medallions; gold-plated frame, backstrap, trigger guard assembly, balance blued; originally in leatherette case; 24,114 made in 1961 only. **Issue Price: $32.50.**
Market Value: $175

COLT GENESEO ANNIVERSARY

No. 4 Derringer replica; "1836—Geneseo Anniversary Model—1961" on left side of 2 1/2" barrel; 22 Short only; walnut grips; gold plated in entirety; originally cased in velvet/satin-lined box; made especially for Cherry's Sporting Goods, Geneseo, Illinois; 104 made in 1961 only. **Issue Price: $27.50.**
Market Value: $650

COLT KANSAS STATEHOOD CENTENNIAL

Frontier Scout revolver; "1861—Kansas Centennial—1961" on left side of 4 3/4" barrel; 22 LR only; walnut grips; no medallions; gold-plated in entirety; originally cased in velvet-lined box with Kansas state seal inlaid in lid; made in 1961 only. **Issue Price: $75.**
Market Value: $425

COLT PONY EXPRESS CENTENNIAL

Frontier Scout revolver; "1860-61—Russell, Majors and Waddell/Pony Express Centennial Model—1960-61" on left side of 4 3/4" barrel; 22 LR only; varnished walnut grips; gold-plated Colt medallions; gold plated in entirety; originally cased in rosewood box with gold-plated centennial medallion in lid; 1007 made in 1961 only. **Issue Price: $80.**
Market Value: $450

COLT SHERIFF'S MODEL

Single-Action Army revolver; made exclusively for Centennial Arms Corp.; "Colt Sheriff's Model .45" on left side of 3" barrel; 45 Colt only; walnut grips without medallions; 25 made with nickel finish; 478 blued; made only in 1961. **Issue Price: Nickel, $139.50; Blued, $150.**
Nickel
Market Value: $5000
Blued
Market Value: $1995

1962

COLT COLUMBUS, OHIO, SESQUICENTENNIAL

Frontier Scout revolver; "1812—Columbus Sesquicentennial—1962" on left side of 4 3/4" barrel; 22 LR only; varnished walnut grips with gold-plated medallions; gold-plated in entirety; originally cased in velvet/satin-lined walnut case; 200 made in 1962 only. **Issue Price: $100.**
Market Value: $550

COLT FORT FINDLAY, OHIO, SESQUICENTENNIAL

Frontier Scout revolver; "1812—Fort Findlay Sesquicentennial—1962" on left side of 4 3/4", barrel; 22 LR, 22 WMR; varnished walnut grips; gold-plated in entirety; originally cased in red velvet/satin-lined walnut box; 110 made in 1962 only. **Issue Price: $89.50.**
Market Value: $650
Cased Pair, 22 LR, 22 WMR (20 made in 1962)
Market Value: $2500

COLT NEW MEXICO GOLDEN ANNIVERSARY

Frontier Scout revolver; "1912—New Mexico Golden Anniversary—1962" wiped in gold on left side of 4 3/4" barrel; 22 LR only; varnished walnut grips; gold-plated medallions; barrel, frame, base pin screw, ejector rod, rod tube, tube plug and screw, bolt and trigger, hammer screws blued; balance gold-plated; originally cased in redwood box with yellow satin/velvet lining; 1000 made in 1962 only. **Issue Price: $79.95.**
Market Value: $425

COLT ROCK ISLAND ARSENAL CENTENNIAL

Single shot version of Colt Model 1860 Army revolver; "1862—Rock Island Arsenal Centennial Model—1962" on left side of 6" barrel; 22 Short only; varnished walnut grips; blued finish; originally in blue and gray leatherette case; 550 made in 1962 only. **Issue Price: $38.50.**
Market Value: $250

COLT WEST VIRGINIA STATEHOOD CENTENNIAL

Frontier Scout revolver; "1863—West Virginia Centennial—1963" wiped in gold on left side of 4 3/4" barrel; 22 LR only; pearlite grips; gold-plated medallions; blued, with gold-plated backstrap, trigger guard assembly and screws, stock screw; originally cased in blonde wood box with gold velvet/satin lining; 3452 made in 1962 only. **Issue Price: $75.**
Market Value: $425

Colt West Virginia Statehood Centennial Single-Action Army

Same legend on barrel as 22 version; 5 1/2" barrel; 45 Colt only; same blue/gold finish as Scout version; same type of casing; 600 made in 1963 only. **Issue Price: $150.**
Market Value: $1395

1963
Colt Arizona Territorial Centennial
Colt Arizona Territorial Centennial Single-Action Army
Colt Battle of Gettysburg Centennial
Colt Carolina Charter Tercentenary
Colt Carolina Charter Tercentenary 22/45 Combo Set
Colt Fort McPherson, Nebraska, Centennial
Colt Fort Stephenson, Ohio, Sesquicentennial
Colt General John Hunt Morgan Indian Raid
Colt H. Cook 1 of 100
Colt Idaho Territorial Centennial

1964
Colt California Gold Rush Commemorative
Colt California Gold Rush Single-Action Army
Colt Cherry's Sporting Goods 35th Anniversary

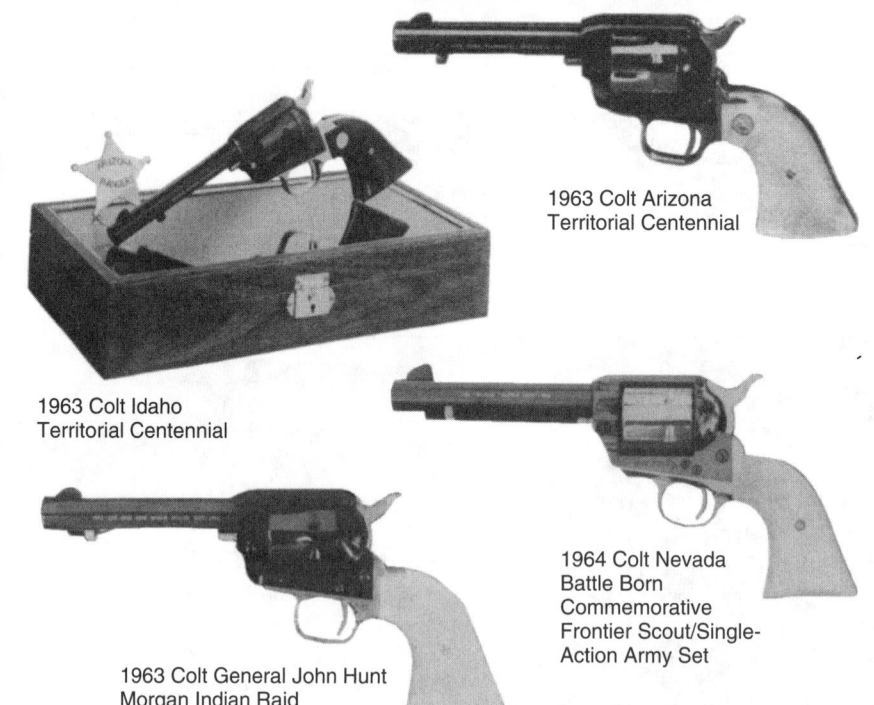

1963 Colt Arizona Territorial Centennial

1963 Colt Idaho Territorial Centennial

1963 Colt General John Hunt Morgan Indian Raid

1964 Colt Nevada Battle Born Commemorative Frontier Scout/Single-Action Army Set

1963

COLT ARIZONA TERRITORIAL CENTENNIAL
Frontier Scout revolver "1863—Arizona Territorial Centennial—1963" wiped in gold on left side of 4 3/4" barrel; 22 LR only; pearlite grips, gold-plated medallions; gold plated, with blue barrel, frame, base pin screw, ejector rod, rod tube, tube plug and screw, bolt and trigger screws, hammer and hammer screw; originally cased in blonde-finished box with yellow velvet/satin lining; 5355 made in 1963 only. **Issue Price: $75.**
   Market Value: $425

Colt Arizona Territorial Centennial Single-Action Army
Same as Colt Arizona Territorial Centennial Frontier Scout with same legend on barrel; 5 1/2" barrel; 45 Colt only; same blue/gold-plated finish; same type case; 1280 made in 1963 only. **Issue Price: $150.**
   Market Value: $1395

COLT BATTLE OF GETTYSBURG CENTENNIAL
Frontier Scout revolver "1863—Battle of Gettysburg Centennial—1963" wiped in gold on left side of 4 3/4" barrel; 22 LR only; walnut grips, gold-plated medallions; gold plated with blued barrel, frame, base pin screw, ejector rod tube, tube plug and screw, and bolt, trigger and hammer screws; originally cased in blonde-finished wood with yellow velvet in bottom, blue satin in lid; 1019 made in 1963 only. **Issue Price: $89.95.**
   Market Value: $425

COLT CAROLINA CHARTER TERCENTENARY
Frontier Scout revolver "1663—Carolina Charter Tercentenary—1963" wiped in gold on left side of 4 3/4" barrel; 22 LR only; walnut grips; gold-plated medallions; gold-plated, with barrel, frame, cylinder, base pin screw, ejector rod, rod tube, tube plug, tube screw, bolt, trigger and hammer screws blued; originally cased in blonde-finished box with yellow velvet/satin lining; 300 made in 1963 only. **Issue Price: $75.**
   Market Value: $425

Colt Carolina Charter Tercentenary 22/45 Combo Set
Includes Frontier Scout described above and Single-Action Army revolver, with same legend on 5 1/2" barrel, 45 Colt only; same finish on grips as Frontier version; larger case to fit both guns; 251 sets made in 1963 only. **Issue Price: $240.**
   Market Value: $1595

COLT FORT MCPHERSON, NEBRASKA, CENTENNIAL
No. 4 Derringer replica; "Fort McPherson/1863—Centennial—1963" wiped in gold on left side of 2 1/2" barrel; 22 Short only; ivorylite grips, no medallions; gold-plated with blued barrel, bolt, trigger screw, hammer and screw, trigger and stock screw; originally cased in walnut-finished box, with gold velvet/satin lining; 300 made in 1963 only. **Issue Price: $28.95.**
   Market Value: $400

COLT FORT STEPHENSON, OHIO, SESQUICENTENNIAL
Frontier Scout revolver "1813—Fort Stephenson Sesquicentennial—1963" wiped in silver on left side of 4 3/4" barrel; 22 LR only; laminated rosewood grips, nickel-plated medallions; nickel-plated finish, with blued barrel, frame, base pin screw, ejector rod, rod tube, tube plug and screw, bolt and trigger and hammer screws; originally cased in blonde-finished wood, with yellow velvet/satin lining; 200 made only in 1963. **Issue Price: $75.**
   Market Value: $550

COLT GENERAL JOHN HUNT MORGAN INDIAN RAID
Frontier Scout revolver; "1863—Gen. John Hunt Morgan Indian Raid—1963" wiped in gold on left side of 4 3/4" barrel; 22 LR; pearlite grips, gold-plated medallions; gold-plated with blued frame, barrel, cylinder, base pin screw, ejector rod, rod tube, tube plug and tube screw and bolt and trigger screw; originally cased

in blonde-finished wood, with gold velvet/satin lining; 100 made in 1963 only. **Issue Price: $74.50.**
   Market Value: $650

COLT H. COOK 1 OF 100
Frontier Scout/Single-Action Army revolvers; sold as set; "H. Cook 1 of 100" on left side of barrels; Scout has 4 3/4" barrel, 22 LR only; SA Army has 7 1/2" barrel, 45 Colt only; pearlite grips; nickel-plated medallions; both nickel-plated with blued frame, base pin, trigger and hammer screws; originally cased in silver-colored box with blue satin/velvet lining; 100 sets made in 1963 only for H. Cook Sporting Goods, Albuquerque, N.M. **Issue Price: $275.**
   Market Value: $1795

COLT IDAHO TERRITORIAL CENTENNIAL
Frontier Scout revolver; "1863—Idaho Territorial Centennial—1963" wiped in silver on left side of 4 3/4" barrel; 22 LR only; pearlite grips; nickel-plated medallions; nickel-plated with blue frame, barrel, base pin screw, ejector rod tube, tube plug and screw, and bolt, trigger and hammer screws; originally cased in blonde-finished wood, with gold velvet/satin lining; 902 made in 1963 only. **Issue Price: $75**
   Market Value: $425

1964

COLT CALIFORNIA GOLD RUSH COMMEMORATIVE
Frontier Scout revolver; "California Gold Rush Model" on left side of 4 3/4" barrel; 22 LR only; ivorylite grips; gold-plated medallions; gold-plated in entirety; originally cased in blonde wood box; blue velvet lining in bottom, gold in lid; 500 made in 1964 only. **Issue Price: $79.50**
   Market Value: $450

Colt California Gold Rush Single-Action Army
Same barrel legend as Frontier Scout version; 5 1/2" barrel; 45 Colt only; same finish, grips, casing; 130 made in 1966 only. **Issue Price: $175.**
   Market Value: $1250 to $1500

COLT CHERRY'S SPORTING GOODS 35TH ANNIVERSARY
Frontier Scout/Single-Action Army revolvers, sold as set; "1929—Cherry's Sporting Goods 1964" on left side of barrel; Scout has 4 3/4" barrel, 22 LR only; SA Army has 4 3/4" barrel, 45 Colt only; both have laminated

rosewood grips; gold-plated medallions; gold-plated in entirety; originally cased in embossed black leatherette, with black velvet/satin lining; 100 sets made in 1964 only. **Issue Price: $275.**
    Market Value: $1695

COLT CHAMIZAL TREATY COMMEMORATIVE
Frontier Scout revolver; "1867 Chamizal Treaty—1964" wiped in gold on left side of 4 ³/₄" barrel; 22 LR only; pearlite grips; gold-plated medallions; gold-plated finish; blued frame, barrel, ejector rod, ejector tube, rod plug and screw, base pin and base pin screw and hammer, trigger and bolt screws; originally cased in blonde-finished wood; yellow velvet/satin lining; 450 made in 1964. **Issue Price: $85.**
    Market Value: $425

Colt Chamizal Treaty Single-Action Army
Same legend on 5 ¹/₂" barrel; 45 Colt only; same grips, finish as Frontier Scout version; same type of case; 50 made in 1964. **Issue Price: $170.**
    Market Value: $1495

Colt Chamizal Treaty Frontier Scout/Single-Action Army Combo
Includes the two revolvers described above in one oversize case; 50 pairs made in 1965. **Issue Price: $280.**
    Market Value: $1995

COLT COL. SAM COLT SESQUICENTENNIAL PRESENTATION
Single-Action Army revolver; "1815—Col. Sam Colt Sesquicentennial Model—1964" on left side of 7 ¹/₂" barrel; 45 Colt only; rosewood grips; roll-engraved scene on cylinder; nickel-plated medallions; silver-plated finish, with blued frame, barrel, ejector rod tube and screw, hammer and trigger; originally cased in varnished walnut box with 12 dummy nickel-plated cartridges in cartridge block; burgundy velvet lining; 4750 made in 1964 only. **Issue Price: $225.**
    Market Value: $1395

Colt Col. Sam Colt Sesquicentennial Deluxe
Same specs as Col. Sam Colt Sesquicentennial Presentation model except hand-fitted rosewood grips with escutcheons rather than medallions; hand-engraved cylinder; case has plate marked "1 of 200"; 200 made in 1964 only. **Issue Price: $500.**
    Market Value: $2500

Colt Col. Sam Colt Sesquicentennial Custom Deluxe
Same specs as Col. Sam Colt Sesquicentennial Deluxe except facsimile of Samuel Colt's signature engraved on backstrap; lid of case engraved with "1 of 50"; name of purchaser engraved when requested; 50 made in 1965. **Issue Price: $1000.**
    Market Value: $4000

COLT GENERAL HOOD CENTENNIAL
Frontier Scout revolver; "1864—General Hood's Tennessee Campaign—1964" on left side of 4 ³/₄" barrel; 22 LR only; laminated rosewood grips, gold-plated medallions; gold-plated finish, except for blued trigger, hammer, base pin, ejector rod, rod head and screw and screws for base pin, hammer, trigger, backstrap and trigger guard; originally cased in blonde-finished wood box with green velvet/satin lining; 1503 made in 1964 only. **Issue Price: $75.**
    Market Value: $425

COLT MONTANA TERRITORY CENTENNIAL
Frontier Scout revolver; "1864—Montana Territory Centennial—1964" on left side of barrel; "1889—Diamond Jubilee Statehood—1964" on right side; both markings wiped in gold; 4 ³/₄" barrel; 22 LR only; rosewood or pearlite grips; gold-plated medallions; gold-plated finish, except for blued barrel, frame, base pin screw, cylinder, ejector rod, rod tube, tube plug and tube screw, bolt, and trigger and hammer screws; originally cased in walnut-finished box with red velvet/satin lining; 2300 made in 1964 only. **Issue Price: $75.**
    Market Value: $425

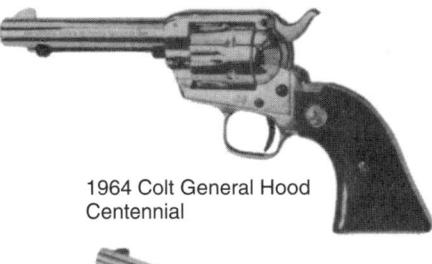
1964 Colt General Hood Centennial

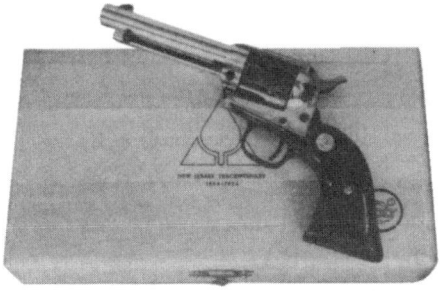
1964 Colt New Jersey Tercentenary

1964 Colt California Gold Rush Commemorative

Colt Montana Territory Centennial Single-Action Army
Same barrel markings as Frontier Scout version; 7 ¹/₂" barrel, 45 Colt only; same grips, finish, except frame is color case-hardened; same casing as Frontier Scout; 851 made in 1964 only. **Issue Price: $150.**
    Market Value: $1395

COLT NEVADA STATEHOOD CENTENNIAL
Frontier Scout revolver "1864—Nevada Centennial—1964" wiped in silver on left side of 4 ³/₄" barrel; 22 LR only; pearlite grips; nickel-plated medallions; nickel-plated finish, with blued barrel, frame, base pin screw, cylinder ejector rod, rod tube, tube plug and tube screw, hammer, bolt, trigger screws; originally cased in gray-finished wood with blue velvet-lined bottom, silver satin-lined lid; 3984 made in 1964 only. **Issue Price: $75.**
    Market Value: $425

Colt Nevada Statehood Centennial Single-Action Army
Same legend on barrel as Frontier Scout; 5 ¹/₂" barrel; 45 Colt only; grips, medallions, finish identical to Scout; same casing motif; 1688 made in 1964 only. **Issue Price: $150.**
    Market Value: $1395

Colt Nevada State Centennial Frontier Scout Single-Action Army Set
Includes the two handguns described above in oversized case; 189 standard sets were made, plus 577 sets featuring extra engraved cylinders; made in 1964 only. **Issue Price: $240 ($350 with extra engraved cylinders).**
Standard Set
    Market Value: $1595
With extra engraved cylinders
    Market Value: $1695

COLT NEVADA BATTLE BORN COMMEMORATIVE
Frontier Scout revolver; "1864—Nevada 'Battle Born'—1964" wiped in silver on left side of 4 ³/₄" barrel; 22 LR only; pearlite grips; nickel-plated medallions; nickel-plated, with blued frame, barrel, base pin screw, ejector rod, tube, tube plug and screw, bolt, trigger and hammer screws; cased in blue-finished wood box, with blue velvet/satin lining; 981 made in 1964 only. **Issue Price: $85.**
    Market Value: $425

Colt Nevada Battle Born Commemorative Single-Action Army
Same legend on barrel as Frontier Scout version; 5 ¹/₂" barrel; 45 Colt only; same grips, finish, casing as Frontier Scout; 80 made in 1964 only. **Issue Price: $175.**
    Market Value: $1495

Colt Nevada Battle Born Commemorative Frontier Scout/Single-Action Army Set
Includes the two handguns previously described in oversize case; 20 sets were made in 1964 only. **Issue Price: $265.**
    Market Value: $2595

COLT NEW JERSEY TERCENTENARY
Frontier Scout revolver; "1664—New Jersey Tercentenary—1964" on left side of barrel; 4 ³/₄" barrel; 22 LR only; laminated rosewood grips; nickel-plated medallions; blued finish, with nickel-plated barrel, frame, ejector rod tube, tube plug and screw; originally cased in blonde-finished box with blue velvet lining in bottom, silver satin in lid; 1001 made in 1964 only. **Issue Price: $75.**
    Market Value: $425

Colt New Jersey Tercentenary Single-Action Army
Same legend on barrel as Frontier Scout; 5¹/₂" barrel; 45 Colt only; grips, medallions, finish the same as on Frontier Scout version; same casing; 250 made in 1964 only. **Issue Price: $150.**
    Market Value: $1395

COMMEMORATIVES

1964
Colt Pony Express Presentation
Colt St. Louis Bicentennial
Colt St. Louis Bicentennial Single-Action Army
Colt St. Louis Bicentennial Frontier Scout/Single-Action Army Set
Colt Wichita Commemorative
Colt Wyatt Earp Buntline
Colt Wyoming Diamond Jubilee
Ithaca Model 48 St. Louis Bicentennial
Remington Montana Centennial
Winchester Wyoming Diamond Jubilee Commemorative

1965
Colt Appomattox Centennial
Colt Appomattox Centennial Frontier Single-Action Army
Colt Appomattox Centennial Frontier Scout/Single-Action Army Combo
Colt Colorado Gold Rush Commemorative
Colt Dodge City Commemorative
Colt Forty-Niner Miner
Colt General Meade Campaign Commemorative
Colt General Meade Campaign Single-Action Army Revolver

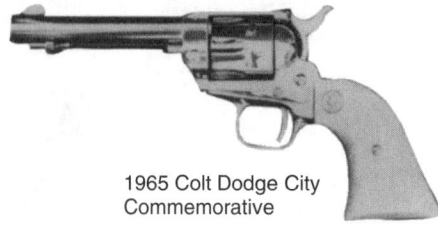

1965 Colt Dodge City Commemorative

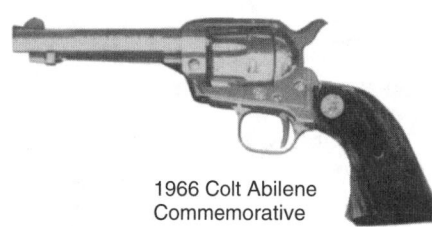

1966 Colt Abilene Commemorative

Colt St. Louis Bicentennial Frontier Scout/Single-Action Army Set
Includes the two handguns described above in oversize case; 200 sets made in 1964 only. **Issue Price: $240.**
  Market Value: $1595

COLT WICHITA COMMEMORATIVE
Frontier Scout revolver; "1864—Kansas Series- Wichita—1964" wiped in silver on left side of 4 3/4" barrel; 22 LR only; gold-plated medallions; gold-plated in entirety; originally cased in blonde-finished wood; lined with red velvet/satin; 500 made in 1964 only. **Issue Price: $85.**
  Market Value: $425

COLT WYATT EARP BUNTLINE
Single-Action Army revolver; "Wyatt Earp Buntline Special" on left side of 12" barrel; 45 Colt only; laminated black rosewood grips; gold-plated medallions; gold-plated in entirety; originally cased in black-finished wood, lined with green velvet/satin; 150 made only in 1964. **Issue Price: $250.**
  Market Value: $2500

COLT WYOMING DIAMOND JUBILEE
Frontier Scout revolver "1890—Wyoming Diamond Jubilee—1965" on left side of barrel; 4 3/4" barrel; 22 LR only; rosewood grips, nickel-plated medallions; nickel-plated finish, except for blued barrel, frame, ejector rod, rod tube, tube plug and plug screw; cased in blond-finished box, with blue velvet bottom lining, silver satin-lined lid; 2357 made in 1964 only. **Issue Price: $75.**
  Market Value: $425

ITHACA MODEL 48 ST. LOUIS BICENTENNIAL
Lever-action; single shot; hand-operated rebounding hammer; 22 LR, 22 Long, 22 Short; 18" barrel; Western carbine-style straight stock; open rear sight, ramp front. Only 200 manufactured in 1964. **Issue Price: $34.95.**
  Market Value: $195

REMINGTON MONTANA CENTENNIAL
Model 600 carbine; bolt action; 6mm Rem. only; deviates from standard Model 600 specs only in better walnut; commemorative medallion inlaid into the stock; barrel inscription reads, "1889-1964/75th Anniversary"; 1000 made in 1964 only. **Issue Price: $124.95.**
  Market Value: $495

WINCHESTER WYOMING DIAMOND JUBILEE COMMEMORATIVE
Model 94 carbine; 30-30 only; 1500 made, distributed exclusively by Billings Hardware Co.; same as standard M94, except for color case-hardened, engraved receiver, commemorative inscription on barrel; brass saddle ring, loading gate; state medallion embedded in stock; made only in 1964. **Issue Price: $100.**
  Market Value: $1295

1964 (cont.)

COLT PONY EXPRESS PRESENTATION
Single-Action Army revolver; "Russell, Majors and Waddell—Pony Express Presentation Model" on left side of barrel; Various Pony Express stop markings on backstrap; 7 1/2" barrel; 45 Colt; walnut grips; nickel-plated medallions; nickel-plated in entirety; originally cased in walnut-finished wood with transparent Lucite lid; lined with burgundy velvet; 1004 made in 1964 only. **Issue Price: $250.**
  Market Value: $1395

COLT ST. LOUIS BICENTENNIAL
Frontier Scout revolver; "1764—St. Louis Bicentennial—1964" wiped in gold on left side of 4 3/4" barrel; 22 LR only; laminated rosewood grips; gold-plated medallions; gold-plated, except for blued frame, barrel, cylinder, ejector rod, rod tube, tube plug and screw; non-fluted cylinder: originally cased in blonde-finished wood box; yellow velvet/satin lining; 802 made in 1964 only. **Issue Price: $75.**
  Market Value: $425

Colt St. Louis Bicentennial Single-Action Army
Same legend on barrel as Frontier Scout version; 5 1/2" barrel; 45 Colt only; same grips, medallions, finish, casing as Scout version; 200 made in 1964 only. **Issue Price: $150.**
  Market Value: $1395

1965

COLT APPOMATTOX CENTENNIAL
Frontier Scout revolver; "1865—Appomattox Commemorative Model—1965" wiped in silver on left side of 4 3/4" barrel; 22 LR only; laminated rosewood grips; nickel-plated medallions; nickel-plated finish, with blued barrel, frame, backstrap and trigger guard screws, ejector rod tube, tube plug and tube screw; originally cased in blonde-finished wood lined with blue velvet on bottom, gray satin in lid; 1001 made in 1965 only. **Issue Price: $75.**
  Market Value: $425

Colt Appomattox Centennial Frontier Single-Action Army
Same legend on 5 1/2" barrel; 45 Colt only; grips, finish, casing the same as for Frontier Scout version; 250 made in 1965. **Issue Price: $150.**
  Market Value: $1395

Colt Appomattox Centennial Frontier Scout/Single-Action Army Combo
Same specs as the two revolvers described above in one oversize case; 250 sets made in 1965 only. **Issue Price: $240.**
  Market Value: $1595

COLT COLORADO GOLD RUSH COMMEMORATIVE
Frontier Scout revolver; "1858—Colorado Gold Rush—1878" wiped in silver on left side of 4 3/4" barrel; 22 LR only; laminated rosewood grips; nickel-plated medallions; gold-plated finish, with nickel-plated hammer, base pin and screw, ejector rod head, hammer and trigger screws, trigger, grip screw; originally cased in blonde-finished wood; black velvet/satin lining; 1350 made in 1965 only. **Issue Price: $85.**
  Market Value: $450

COLT DODGE CITY COMMEMORATIVE
Frontier Scout revolver; "1864—Kansas Series- Dodge City—1964" wiped in silver on left side of 4 3/4" barrel; 22 LR only; ivorylite grips; gold-plated medallions; gold-plated finish, with blued base pin and screw, ejector rod, ejector rod head, bolt and trigger screw, hammer and hammer screw, trigger; originally cased in blonde-finished wood; lined with kelly green velvet/satin; 500 made in 1965 only. **Issue Price: $85.**
  Market Value: $425

COLT FORTY-NINER MINER
Frontier Scout revolver: "The '49er Miner" wiped in gold on left side of 4 3/4" barrel; 22 LR only; laminated rosewood grips; gold-plated medallions; gold-plated finish with blued barrel, frame, backstrap and trigger guard assembly, ejector rod, tube and tube plug, ejector tube screw; originally cased in walnut-finished wood; lined with velvet in bottom, blue satin in lid; 500 made only in 1965. **Issue Price: $85.**
  Market Value: $425

COLT GENERAL MEADE CAMPAIGN COMMEMORATIVE
Frontier Scout revolver; "Gen. Meade Pennsylvania Campaign Model" wiped in gold on left side of 4 3/4" barrel; 22 LR only; ivorylite grips, gold-plated medallions; gold-plated finish; blued frame, barrel, cylinder, ejector rod tube, tube plug and screw, hammer and trigger screws; originally cased in walnut-finished wood; blue velvet lining in bottom, gold satin in lid; 1197 made in 1965 only. **Issue Price: $75.**
  Market Value: $425

Colt General Meade Campaign Single-Action Army Revolver
Same legend on the 5 1/2" barrel; 45 Colt only; same finish, casing as Frontier Scout version; 200 made in 1965 only. **Issue Price: $165.**
  Market Value: $1395

COLT OLD FORT DES MOINES RECONSTRUCTION COMMEMORATIVE

Frontier Scout revolver; "Reconstruction of Old Fort Des Moines" wiped in silver on left side of 4 3/4″ barrel; 22 LR only; pearlite grips; gold-plated medallions; gold-plated in entirety; originally cased in white-finished wood; royal purple velvet lining in bottom; white satin in lid; 700 made in 1965 only. **Issue Price: $89.95.**
    Market Value: $450

Colt Old Fort Des Moines Reconstruction Single-Action Army

Same legend on 5 1/2″ barrel; 45 Colt only; grips, finish the same as on Frontier Scout version; same casing; 100 made in 1965 only. **Issue Price: $169.95.**
    Market Value: $1395

Colt Old Fort Des Moines Frontier Scout/Single-Action Army Combo

Same specs as the two revolvers described above, in one oversize case; 100 sets made in 1965 only. **Issue Price: $289.95.**
    Market Value: $1695

COLT OREGON TRAIL COMMEMORATIVE

Frontier Scout revolver; "Oregon Trail Model," wiped in gold, on left side of 4 3/4″ barrel; 22 LR only; pearlite grips; gold-plated medallions; blued finish with gold-plated backstrap and trigger guard assembly and screws, hammer, trigger and screws, base pin, base pin screw and ejector rod head; originally cased in blonde-finished wood; lined with blue velvet in bottom, gold satin in lid; 1995 made only in 1965. **Issue Price: $75.**
    Market Value: $425

COLT JOAQUIN MURRIETTA 1 OF 100

Frontier Scout/Single-Action Army combo; both have "Joaquin Murrieta 1 of 100" on left side of barrels; Scout has 4 3/4″ barrel, 22 LR only; SAA has 5 1/2″ barrel, 45 Colt only; grips on both are pearlite, with gold-plated medallions; finish for both is gold-plated with blued barrels, frames, ejector rod tubes; originally in one oversize case of walnut-finished wood; blue velvet/satin lining; 100 sets made in 1965 only. **Issue Price: $350.**
    Market Value: $1695

COLT ST. AUGUSTINE QUADRICENTENNIAL

Frontier Scout revolver; "1565—St. Augustine Quadricentennial—1965" wiped in gold on left side of 4 3/4″ barrel; 22 LR only; pearlite grips; gold-plated medallions; gold-plated finish, with blued barrel, base pin, ejector rod, tube, tube plug and screw, frame, hammer and trigger screws, backstrap and trigger guard assembly and screws; cased in blonde-finished wood; gold velvet/satin lining; 500 made in 1965 only. **Issue Price: $85.**
    Market Value: $450

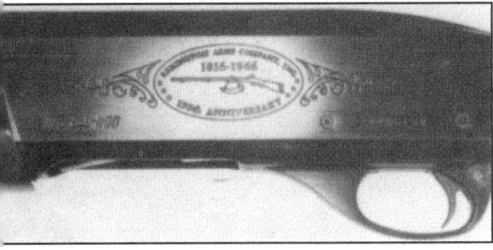

1966 Remington 150th Anniversary Model 1100SA

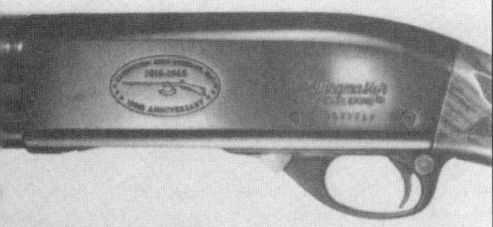

1966 Remington 150th Anniversary Model 870SA

1966

COLT ABERCROMBIE & FITCH TRAILBLAZER

New Frontier single-action Army revolver: "Abercrombie & Fitch Co." wiped in gold on left side of 7 1/2″ barrel; 45 Colt only; rosewood grips; gold-plated medallions; gold-plated finish, blued barrel, cylinder, hammer, sights, ejector rod tube, ejector rod screw, case-hardened frame; roll-engraved, nonfluted cylinder; originally cased in varnished American walnut with brass-framed glass cover; bottom lined with crushed blue velvet; 200 made in 1966 with "New York" marked on butt, 100 with "Chicago" butt marking; 200 with "San Francisco" butt marking. **Issue Price: $275.**
    Market Value: $1295

COLT ABILENE COMMEMORATIVE

Frontier Scout revolver; "1866—Kansas Series Abilene—1966" wiped in silver on left side of 4 3/4″ barrel; 22 LR only; laminated rosewood grips; gold-plated medallions; gold-plated in entirety; originally cased in blonde-finished wood; blue velvet/satin lining; 500 made in 1966 only. **Issue Price: $95.**
    Market Value: $425

COLT DAKOTA TERRITORY COMMEMORATIVE

Frontier Scout revolver; "1861—Dakota Territory—1889" wiped in gold on left side of 4 3/4″ barrel; 22 LR only; laminated rosewood grips; gold-plated medallions; blued finish with gold-plated backstrap and trigger guard assembly and screws, ejector rod and head, base pin, trigger, hammer, stock screw; originally cased in blonde-finished wood; red velvet/satin lining; 1000 made in 1966 only. **Issue Price: $85.**
    Market Value: $425

COLT INDIANA SESQUICENTENNIAL

Frontier Scout revolver; "1816—Indiana Sesquicentennial—1966" wiped in gold on left side of 4 3/4″ barrel; 22 LR only; pearlite grips; gold-plated medallions; blued finish, with gold-plated backstrap and trigger guard assembly, base pin and screw, ejector rod head, cylinder, bolt and trigger screw, hammer and hammer screw, trigger, stock screw; originally cased in blonde-finished wood; bottom lined with gold velvet, lid with blue satin; 1500 made in 1966 only. **Issue Price: $85.**
    Market Value: $425

COLT OKLAHOMA TERRITORY COMMEMORATIVE

Frontier Scout revolver; "1890—Oklahoma Diamond Jubilee—1965" wiped in gold on left side of 4 3/4″ barrel; 22 LR only; laminated rosewood grips; gold-plated medallions; blued finish with gold-plated backstrap and trigger guard assembly and screws, cylinder, ejector rod head, base pin and screw, bolt and trigger; cased in blonde-finished wood; red velvet/satin lining; 1343 made only in 1966. **Issue Price: $85.**
    Market Value: $425

REMINGTON 150TH ANNIVERSARY MODEL 552A

Semi-automatic rifle; 22 LR, 22 Long, 22 Short; same specs as standard Model 552 except stamp-engraved legend on left side of receiver: "Remington Arms Company Inc., 1816-1966, 150th Anniversary," with corporate logo; 1000 made in 1966 only. **Issue Price: $58.**
    Market Value: $275

REMINGTON 150TH ANNIVERSARY MODEL 572A

Slide-action rifle; 22 Short, 22 Long, 22 LR; same specs as standard model 572 except stamp-engraved legend on left side of receiver; "Remington Arms Company, Inc., 1816-1966, 150th Anniversary," with corporate logo; 1000 made in 1966 only. **Issue Price: $60.**
    Market Value: $275

REMINGTON 150TH ANNIVERSARY MODEL 742 ADL

Semi-automatic rifle; 30-06 only; impressed basketweave checkering; has same specs as standard 742 ADL, except for stamp-engraved legend on left side of receiver; "Remington Arms Company Inc., 1816-1966, 150th Anniversary" with corporate logo; 1000 made in 1966 only. **Issue Price: $150.**
    Market Value: $400

REMINGTON 150TH ANNIVERSARY MODEL 760 ADL

Pump-action rifle; 30-06 only; same specs as standard 760 BDL Deluxe model except stamp-engraved legend on left side of receiver: "Remington Arms Company Inc., 1816-1966, 150th Anniversary," with corporate logo; 1000 made in 1966 only. **Issue Price: $135.**
    Market Value: $395

REMINGTON 150TH ANNIVERSARY MODEL 870 SA

Slide-action Skeet gun; 12-ga. only; 26″ barrel; vent rib; specs the same as standard Model 870, except for stamp-engraved legend on left side of receiver: "Remington Arms Company, Inc., 1816-1966, 150th Anniversary" with corporate logo; 1000 made in 1966 only. **Issue Price: $130.**
    Market Value: $450

Remington 150th Anniversary Model 870 TB

Same specs as Remington 150th Anniversary Model 870 SA except recoil pad; 30″ barrel; trap stock; same stamp-engraved legend on receiver; 1000 made in 1966 only. **Issue Price: $165.**
    Market Value: $600

REMINGTON 150TH ANNIVERSARY MODEL 1100 SA

Semi-automatic Skeet shotgun; 12-ga. only; 26″ barrel; vent rib; specs the same as standard Model 1100, except for stamp-engraved legend on left side of receiver: "Remington Arms Company, Inc., 1816-1966, 150th Anniversary" with corporate logo; 1000 made in 1966 only. **Issue Price: $185.**
    Market Value: $600

1966 (cont.)

Remington 150th Anniversary Model 1100 TB
Same specs as Remington 150th Anniversary Model 1100 SA except recoil pad; 30″ barrel; trap stock; same stamp-engraved legend on receiver; 1000 made in 1966 only. **Issue Price: $220.**
Market Value: $650

REMINGTON 150TH ANNIVERSARY NYLON 66
Semi-automatic; 22 LR; same specs as standard Nylon 66 Apache Black model except stamp-engraved legend on left side of receiver "Remington Arms Company Inc., 1816-1966, 150th Anniversary," with corporate logo; 1000 made in 1966 only. **Issue Price: $50.**
Market Value: $175

WINCHESTER NEBRASKA CENTENNIAL COMMEMORATIVE
Model 94 carbine; 30-30 only; same as standard M94 except gold-plated loading gate, buttplate, rear barrel band, hammer; commemorative inscription on barrel, medallion in stock; only 2500 made and distributed only in Nebraska; made only in 1966. **Issue Price: $100.**
Market Value: $1295

WINCHESTER CENTENNIAL '66 COMMEMORATIVE
Model 94; rifle and carbine versions commemorate Winchester's 100th anniversary; produced in 1966 only; 100,478 were made; 30-30 only; rifle version with 26″ half-octagon barrel; full-length 8-shot magazine; gold-plated forearm cap, receiver; post front sight, open rear; walnut stock forearm with epoxy finish; saddle ring; brass buttplate; commemorative inscription on barrel and top tang. Carbine differs only in shorter forearm; 20″ barrel; 6-shot magazine. **Issue Price: $125.**
Rifle, Carbine
Market Value: $450 to $425
Matched set with consecutive serial numbers
Market Value: $925

1967

COLT ALAMO COMMEMORATIVE
Frontier Scout revolver; "Alamo Model," flanked by stars, wiped in gold on left side of 4 ³/₄″ barrel; 22 LR only; ivorylite grips, with inlaid gold-plated Texas star below screw on left grip. Gold-plated finish; blued barrel, frame, ejector rod tube, tube plug and screw; originally cased in blonde-finished wood box; blue velvet/satin lining; 4250 made in 1967 only. **Issue Price: $85.**
Market Value: $425

Colt Alamo Commemorative Single-Action Army
Same legend on barrel; same grips, finish, but with blued barrel, frame and ejector rod tube and tube screw; same casing; 750 made in 1967 only. **Issue Price: $165.**
Market Value: $1395

Colt Alamo Commemorative Frontier Scout/Single-Action Army Combo
Includes two revolvers described above in one oversize case; 250 sets made in 1967 only. **Issue Price: $265.**
Market Value: $1595

COLT BAT MASTERSON
Frontier Scout revolver; "Lawman Series—Bat Masterson" on left side of 4 ³/₄″ barrel; 22 LR only; checkered rubber eagle grips; nickel-plated finish; cased originally in black leatherette; red velvet/satin lining; 3000 made in 1967 only. **Issue Price: $90.**
Market Value: $450

Colt Bat Masterson Single-Action Army
Same legend on 4 ³/₄″ barrel; 45 Colt only; grips, finish, casing are the same as for Frontier Scout version; 500 made in 1967 only. **Issue Price: $180.**
Market Value: $1500

COLT CHATEAU THIERRY COMMEMORATIVE
Semi-automatic; Model 1911 A1; "1917 World War I Commemorative 1967" on right side of slide; roll-engraved scene on left depicting WWI battle; 5″ barrel; 45 auto; checkered walnut grips; inlaid commemorative medallions; left grip inlaid with Chateau Thierry battle bar; blued finish with slide scene, serial number, banner, Colt markings wiped in gold; several features including no trigger finger relief cuts, non-grooved trigger, safety lever, adapted from original M1911 design; Standard model cased in olive drab box; Deluxe and Custom models have oiled, waxed teak cases; Deluxe model case inscribed "One of Seventy-Five/Deluxe Engraved/Chateau Thierry Commemoratives"; Custom model case inscribed "One of Twenty-Five/Custom Engraved/Chateau Thierry Commemoratives"; gun bears gold-filled signature of A.A White, engraver; 7400 Standard versions made in 1967-68, 75 Deluxe, 25 Custom. **Issue Price: Standard, $200; Deluxe, $500; Custom, $1000.**
Standard
Market Value: $795
Deluxe
Market Value: $1350
Custom
Market Value: $2750

COLT CHISHOLM TRAIL COMMEMORATIVE
Frontier Scout revolver; "1867—Kansas Series—Chisholm Trail—1967" wiped with silver on left side of 4 ³/₄″ barrel; 22 LR; pearlite grips; nickel-plated medallions; blued finish, with nickel-plated backstrap and trigger guard assembly and screws, trigger, hammer base pin, ejector rod head, stock screw; originally cased in blonde-finished wood, gold velvet/satin lining; 500 made in 1967 only. **Issue Price: $100.**
Market Value: $425

COLT COFFEYVILLE COMMEMORATIVE
Frontier Scout revolver; "1866—Kansas Series—Coffeyville—1966" wiped in silver on left side of 4 ³/₄″ barrel; 22 LR only; walnut grips; gold-plated medallions; gold-plated finish; blued backstrap and trigger guard assembly, base pin and screw, ejector rod, ejector rod head, hammer and hammer screw, trigger; originally cased in blonde-finished wood, black velvet/satin lining; 500 made in 1967 only. **Issue Price: $95.**
Market Value: $425

REMINGTON CANADIAN CENTENNIAL
Model 742 rifle; semi-automatic; 30-06 only; same as standard model except impressed checkering on pistol grip; left side of receiver is engraved with maple leaves, special insignia, "1867-1967—Canadian Centennial Gun," wiped in white; serial number is preceded by letter C; 1000 made in 1967 only. **Issue Price: $119.95.**
Market Value: $400

RUGER CANADIAN CENTENNIAL 10/22
Standard model of 22 rimfire rifle with silver commemorative medal set in the stock; top of the receiver is engraved with a design composed of the Canadian Exposition symbol, branches of the Canadian maple leaf and the words, "Canadian Centennial Guns." Issued in 1967; 4430 made. **Issue Price: $99.50.**
Market Value: $350

WINCHESTER ALASKAN PURCHASE CENTENNIAL
Model 94 rifle; sold only in Alaska; receiver engraved in 19th Century filigree for "antique" appeal; centered in stock is the official Alaskan Purchase centennial medallion with totem pole symbol of the state; barrel is 26″, with magazine capacity of 8 shots; other facets are standard of Model 94. Introduced, 1967. **Issue Price: $125.**
Market Value: $1495

WINCHESTER CANADIAN CENTENNIAL
Model 64; action obviously is the Model 94; not to be confused with Winchester's Model 64 boys rifle discontinued in 1963. Canadian commemorative is in 30-30 caliber; octagonal 26″ rifle or 20″ carbine barrel; black-chromed receiver engraved with maple leaf motif; forearm tip black-chromed; straight stock finished with "antique gloss." Both versions have a dovetail bead post front sight, buckhorn rear. Carbine comes with saddle ring; 6-shot magazine; the rifle comes with 8-shot magazine. Gold-filled inscription on barrel reads, "Canadian Centennial 1867-1967." Introduced in 1967. **Issue Price: Rifle or Carbine, $125; Matching Set with consecutive serial numbers, $275.**
Rifle
Market Value: $450
Carbine
Market Value: $425
Matched Set
Market Value: $925

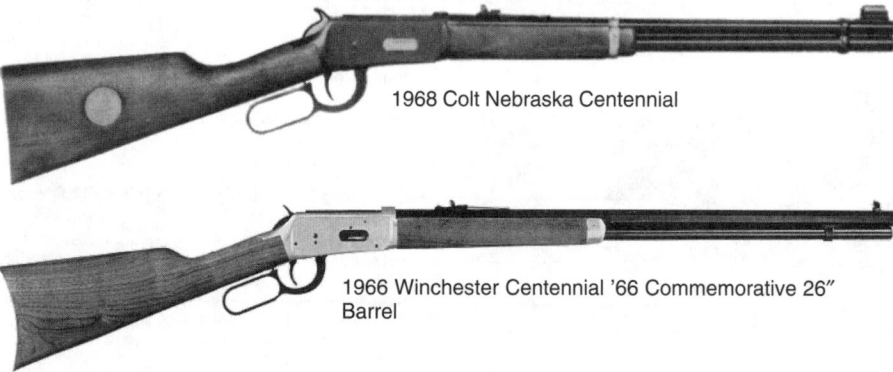

1968 Colt Nebraska Centennial

1966 Winchester Centennial '66 Commemorative 26″ Barrel

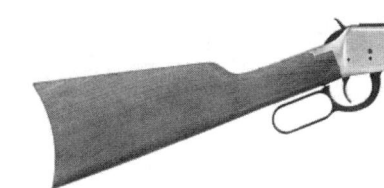
1966 Winchester Centennial '66

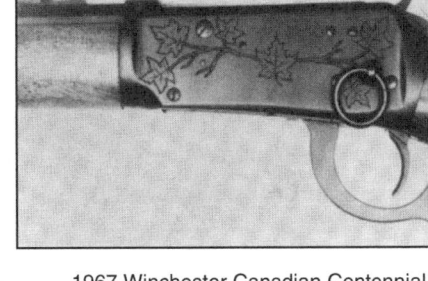
1967 Winchester Canadian Centennial

1967 Colt Bat Masterson

1968 Colt Pat Garrett
Commemorative

COMMEMORATIVES

1968
Colt Belleau Wood
Commemorative
Colt Gen. Nathan Bedford
Forrest
Colt Nebraska Centennial
Colt Pat Garrett
Commemorative
Colt Pat Garrett Single-Action
Army Revolver
Colt Pawnee Trail
Commemorative
Colt Santa Fe Trail
Commemorative
Franchi Centennial
Winchester Buffalo Bill
Commemorative
Winchester Illinois
Sesquicentennial

1968

COLT BELLEAU WOOD COMMEMORATIVE

Semi-automatic; Model 1911 A1; "1917 World War I Commemorative 1967" on right side of slide; roll engraved scene on left side of machine gun battle; 5″ barrel; 45 Auto only; rosewood grips inlaid with commemorative medallions; left grip inlaid with Belleau Wood battle bar; blued finish; slide scene, serial number, banner, Colt markings wiped in gold on Standard model; Deluxe version has slide, frame hand engraved, serial numbers gold-inlaid; Custom has more elaborate engraving; the same features of 1911 model adapted to Chateau Thierry model are incorporated; cases are same as Chateau Thierry model, with brass plate for Deluxe engraved "One of Seventy Five/Deluxe Engraved/Belleau Wood Commemorative"; plate on Custom model reads "One of Twenty-Five/Custom Engraved/Belleau Wood Commemoratives"; production began in 1968, with 7400 Standard types, 75 Deluxe, 25 Custom. **Issue Price: Standard, $200; Deluxe, $500; Custom, $1000.**
Standard
    Market Value: $795
Deluxe
    Market Value: $1350
Custom
    Market Value: $2750

COLT GEN. NATHAN BEDFORD FORREST

Frontier Scout revolver; "General Nathan Bedford Forrest" on left side of 4 3/4″ barrel; 22 LR only; laminated rosewood grips; gold-plated medallions; gold-plated finish; blued cylinder, backstrap and trigger guard assembly; originally cased in dark brown leatherette; red velvet/satin lining; 3000 made in 1968-69. **Issue Price: $110.**
    Market Value: $425

COLT NEBRASKA CENTENNIAL

Frontier Scout revolver; "1867—Nebraska Centennial—1967" on left side of 4 3/4″ barrel; 22 LR; pearlite grips; gold-plated barrel, frame, hammer, trigger, ejector rod head, stock screw; originally cased in blonde-finished wood; lined with blue velvet in bottom, gold satin in lid; 7001 made in 1968 only. **Issue Price: $100.**
    Market Value: $425

COLT PAT GARRETT COMMEMORATIVE

Frontier Scout revolver; "Lawman Series Pat Garrett" on right side of 4 3/4″ barrel; 22 LR only; pearlite grips; gold-plated medallions; gold-plated finish; nickel-plated barrel, frame, backstrap and trigger guard assembly, ejector rod; loading gate is gold-plated; originally cased in black leatherette with gold velvet/satin lining; 3000 made in 1968 only. **Issue Price: $110.**
    Market Value: $450

1968 Colt Belleau Wood Commemorative

Colt Pat Garrett Single-Action Army Revolver

Same barrel legend; 5 1/2″ barrel; 45 Colt only; same grips, finish, casing as Frontier Scout version; 500 made in 1968. **Issue Price: $200.**
    Market Value: $1495

COLT PAWNEE TRAIL COMMEMORATIVE

Frontier Scout revolver; "1868-Kansas Series—Pawnee Trail—1968" wiped in silver on left side of 4 3/4″ barrel; 22 LR; laminated rosewood grips; nickel-plated medallions; blued finish; nickel-plated backstrap and trigger guard assembly and screws, cylinder, base pin, ejector rod head, trigger hammer, stock screw; originally cased in blonde-finished wood; lined with blue velvet in bottom, silver satin in lid; 501 made in 1968. **Issue Price: $110.**
    Market Value: $425

COLT SANTA FE TRAIL COMMEMORATIVE

Frontier Scout revolver; "Kansas Series—Santa Fe Trail—1968" wiped in silver on left side of 4 3/4″ barrel; 22 LR; ivorylite grips; nickel-plated medallions; blued finish with nickel-plated backstrap and trigger guard assembly and screws, hammer, trigger, stock screw, base pin, ejector rod head; originally cased in blonde-finished wood; green velvet/satin lining; 501 made in 1968-69. **Issue Price: $120.**
    Market Value: $425

FRANCHI CENTENNIAL

Semi-automatic take-down rifle; 22 LR only; commemorates 1868-1968 centennial of S.A. Luigi Franchi; centennial seal engraved on receiver; 21″ barrel; 11-shot buttstock magazine; hand-checkered European walnut stock forearm; open rear sight, gold bead front on ramp. Deluxe model has better grade wood, fully engraved receiver. Made only in 1968. **Issue Price: Standard, $86.95; Deluxe, $124.95.**
Standard
    Market Value: $325 to $350
    Deluxe
    Market Value: $450 to $475

WINCHESTER BUFFALO BILL COMMEMORATIVE

Model 94; available with either 20″ or 26″ barrel, both with bead post front sights, semi-buckhorn rear sights. Hammer, trigger, loading gate, forearm tip, saddle ring, crescent buttplate nickel-plated. Barrel, tang inscribed respectively, "Buffalo Bill Commemorative" and "W.F. Cody—Chief of Scouts." Receiver embellished with scrollwork. American walnut stock with embedded Buffalo Bill Memorial Assn. medallion; rifle with 8-shot tubular magazine, carbine with 6-shot magazine. Introduced, 1968. **Issue Price: Rifle, Carbine, $129.95 Rifle marked "1 of 300", $1000.**
Rifle
    Market Value: $450
Rifle marked "1 of 300"
    Market Value: $2650
Carbine
    Market Value: $425

WINCHESTER ILLINOIS SESQUICENTENNIAL

Model 94; "Land of Lincoln," and a profile of Lincoln engraved on the receiver, with gold-filled inscription on barrel, "Illinois Sesquicentennial, 1818-1968"; gold-plated metal buttplate, trigger, loading gate and saddle ring. Official souvenir medallion embedded in walnut stock. First state commemorative to be sold outside the celebrating state by Winchester. Introduced in 1968. **Issue Price: $110**
    Market Value: $395

COMMEMORATIVES

1969

Colt Alabama Sesquicentennial
Colt Arkansas Territorial Sesquicentennial
Colt California Bicentennial
Colt Fort Larned Commemorative
Colt Golden Spike
Colt Wild Bill Hickok Commemorative
Colt Wild Bill Hickok Commemorative Single-Action Army
Colt Meuse Argonne Commemorative
Colt Second Battle of The Marne Commemorative
Colt Shawnee Trail Commemorative
Colt Texas Ranger Commemorative

1969

COLT ALABAMA SESQUICENTENNIAL

Frontier Scout revolver; "1819—Alabama Sesquicentennial—1969" on left side of 4 3/4" barrel; 22 LR only; ivorylite grips; gold-plated medallions; gold-plated finish; nickel-plated loading gate, cylinder, ejector rod, rod head, and tube, base pin and screw, bolt and trigger guard assembly screws, hammer and screw, trigger; originally cased in red leatherette covered wood box; white velvet lining in bottom, red satin in lid; 3001 made in 1969. **Issue Price: $110.**
   Market Value: $425

COLT ARKANSAS TERRITORIAL SESQUICENTENNIAL

Frontier Scout revolver; "1819—Arkansas Territory Sesquicentennial—1969" on left side of 4 3/4" barrel; 22 LR only; laminated rosewood grips; gold-plated medallions; blued frame, backstrap and trigger guard assembly, ejector rod head; gold-plated stock screw nut; originally cased in blonde-finished basswood; red velvet/satin lining; 3500 made; production began in 1969. **Issue Price: $110.**
   Market Value: $425

COLT CALIFORNIA BICENTENNIAL

Frontier Scout revolver; "1769—California Bicentennial—1969" on left side of 6" barrel; 22 LR only; laminated rosewood grips; gold-plated medallions; gold-plated finish; all screws nickel-plated, except base pin, grip screws; hammer, trigger also nickel-plated; originally cased in California redwood; black velvet/satin lining; 5000 made in 1969-70. **Issue Price: $135.**
   Market Value: $425

COLT FORT LARNED COMMEMORATIVE

Frontier Scout revolver; "1869—Kansas Series-Fort Larned—1969" on left side of 4 3/4" barrel; 22 LR; pearlite grips; nickel-plated medallions; nickel-plated finish; blued backstrap and trigger guard assembly, base pin and screw, cylinder, ejector rod head and tube screw, hammer and stock screw, bolt and trigger screw; originally cased in blonde finished wood; blue velvet lining in bottom; silver satin in lid; 500 made in 1969-70. **Issue Price: $120.**
   Market Value: $425

COLT GOLDEN SPIKE

Frontier Scout revolver; "1869—Golden Spike—1969" on right side of 6" barrel; standard barrel markings on left; both wiped in gold; 22 LR only; sand-blasted walnut-stained fir grips; gold-plated medallions; gold-plated finish; blued barrel, frame, backstrap and trigger guard assembly and ejector tube plug and screw; originally cased in hand-stained, embossed simulated mahogany; 11,000 made in 1969. **Issue Price: $135.**
   Market Value: $450

COLT WILD BILL HICKOK COMMEMORATIVE

Frontier Scout revolver; "Lawman Series—Wild Bill Hickok" wiped in silver on right side of 6" barrel; 22 LR only; nonfluted cylinder; pearlite grips; nickel-plated medallions; nickel-plated finish; blued barrel, frame, ejector tube screw; originally cased in black leatherette covered box; bottom lined in blue velvet, lid in silver satin; 3000 made, production began in 1969. **Issue Price: $116.60.**
   Market Value: $450

Colt Wild Bill Hickok Commemorative Single-Action Army

Same legend on 7 1/2" barrel; 45 Colt only; same finish as Frontier Scout version, except for nickel-plated loading gate; same casing; 500 made, production beginning in 1969. **Issue Price: $220.**
   Market Value: $1495

COLT MEUSE ARGONNE COMMEMORATIVE

Semi-automatic; Model 1911A1; "1917 World War I Commemorative 1967" on right side of slide; left has roll-engraved charge on pillbox on Standard; slides, frames on Deluxe, Custom models are hand engraved, serial numbers inlaid in gold; Custom model is more elaborately engraved, inlaid; 5" barrel, 45 Auto only; varnished crotch walnut grips; inlaid commemorative medallions left grip inlaid with Meuse Argonne battle bar; blued finish; engraving, numbers et. al., gold wiped on Standard model; same case as earlier WWI Commemoratives; brass plate for Deluxe reads "One of Seventy-Five/ Deluxe Engraved/Meuse Argonne Commemoratives"; plate on Custom case is inscribed "One of Seventy Five/Custom Engraved/Meuse Argonne Commemoratives"; production began in 1969; 7400 Standard, 75 Deluxe, 25 Custom. **Issue Price: Standard, $220; Deluxe, $500; Custom, $1000.**
Standard
   Market Value: $495
Deluxe
   Market Value: $1350
Custom
   Market Value: $2750

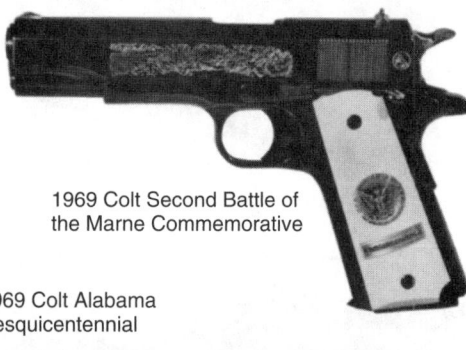

1969 Colt Second Battle of the Marne Commemorative

1969 Colt Alabama Sesquicentennial

1969 Colt Meuse Argonne Commemorative

COLT SECOND BATTLE OF THE MARNE COMMEMORATIVE

Semi-automatic; Model 1911A1; "1917 World War I Commemorative 1967" on right side of slide; roll-engraved combat scene on left side of slide; 5" barrel; 45 Auto; white French holly grips; inlaid commemorative medallions; left grip inlaid with 2nd Battle of the Marne battle bar; blue finish, with slide engraving, serial number on Standard, banner, other markings wiped in gold; Deluxe and Custom models are hand engraved, with serial numbers gold inlaid; work on Custom model is in greater detail; cases are same as others in series, except Deluxe case has brass plate inscribed "One of Seventy-Five/Deluxe Engraved/2nd Battle of the Marne Commemorative"; Custom case has same type of plate inscribed "One of Twenty-Five/Custom Engraved/2nd Battle of the Marne Commemorative"; 7400 Standard guns made in 1969; 75 Deluxe; 25 Custom. **Issue Price: Standard, $220; Deluxe, $500; Custom, $1000.**
Standard
   Market Value: $795
Deluxe
   Market Value: $1350
Custom
   Market Value: $2750

COLT SHAWNEE TRAIL COMMEMORATIVE

Frontier Scout revolver; "1869—Kansas Series—Shawnee Trail 1969" wiped in silver on left side of 4 3/4" barrel; 22 LR only; laminated rosewood grips; nickel-plated medallions; blued finish; nickel-plated backstrap and trigger guard assembly and screws, cylinder, base pin, ejector rod head, hammer, trigger and stock screw; originally cased in blonde-finished wood; red velvet/satin lining; 501 made in 1969 only. **Issue Price: $120.**
   Market Value: $425

COLT TEXAS RANGER COMMEMORATIVE

Single-Action Army revolver; "Texas Ranger Commemoratives/One Riot-One Ranger" wiped in silver on left side of barrel; "Texas Rangers" roll engraved on backstrap; sterling silver star, wreath on top of backstrap behind hammer; YO Ranch brand stamped on bottom of backstrap; 7 1/2" barrel; 45 Colt only; Standard model has rosewood grips, silver miniature Ranger badge inlaid in left grip; blued finish; case-hardened frame; nickel-plated trigger guard, base pin and screw, ejector rod and head, ejector tube screw; gold-plated stock screw, stock escutcheons, medallions. First 200 are custom models, with finish decoration to customer's desires at increasing prices; custom finished guns had deluxe engraved serial numbers, ivory

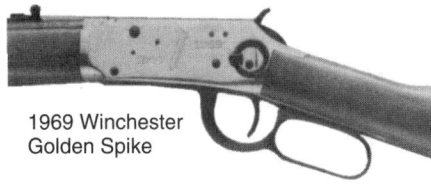

1969 Winchester
Golden Spike

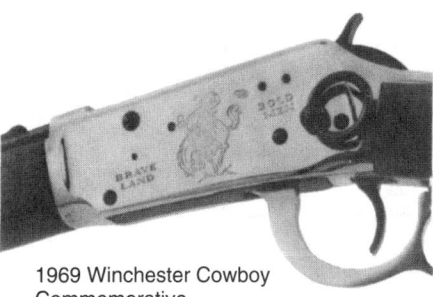

1969 Winchester Cowboy
Commemorative

grips with star inlay; originally cased in special hand-rubbed box with drawers, glass top; red velvet lining; 200 Custom, 800 Standard guns made; production began in 1969. **Issue Price: Custom, varying with customers desires; Standard, $650.**
Standard
  Market Value: $2250
Grade III
  Market Value: $5000
    Grade II
    Market Value: $5500
      Grade I
      Market Value: $6000

WINCHESTER COWBOY COMMEMORATIVE

Model 94; receiver, upper and lower tang, lever, barrel bands are nickel-plated; butt plate is stainless steel, with tang that extends over top of stock for square comb look; stock is straight grip with extended forearm of American walnut; embedded in right side of stock is medallion of cowboy roping a steer; etched on left side of receiver, "Brave Land—Bold Men." Opposite side is engraved with coiled lariat, spurs; barrel is 20″, carrying "Cowboy Commemorative"; upper tang has inscription, "Winchester Model 1894." Has adjustable semi-buckhorn rear sight, blued saddle ring; in 30-30 only. Introduced in 1969. **Issue Price: $125; Marked "1 of 300", $1000.**
  Market Value: $450
Marked "1 of 300"
  Market Value: $2650

WINCHESTER GOLDEN SPIKE

Model 94; features 20″ barrel with twin barrel bands plated in yellow gold; yellow gold receiver, engraved with decorative scrolled border on right side, inscribed on other side with railroad spike flanked by dates, 1869 and 1969. Barrel carries "Golden Spike Commemorative" inscription; upper tang bears words "Oceans United By Rail." Buttstock, forearm are straight-line design of satin-finished American walnut, with fluted comb. Inset in stock is centennial medallion of engines of Central Pacific, Union Pacific meeting on May 10, 1869. It has straight brass buttplate, blued saddle ring; chambered for 30-30; weight is 7 lbs. Introduced in 1969. **Issue Price: $119.95.**
  Market Value: $450

WINCHESTER THEODORE ROOSEVELT COMMEMORATIVE

Model 94 rifle and carbine; made in 1969 only; 49,505 manufactured; 30-30 only; rifle has 26″ octagonal barrel; 6-shot half-magazine; forearm cap, upper tang, receiver plated with white gold; receiver engraved with American eagle, "26th President 1901-1909," Roosevelt's facsimile signature; contoured lever, half pistol grip; medallion in stock. Carbine differs from rifle in shorter forearm, full-length 6-shot tubular magazine; 20″ barrel. **Issue Price: $125.**
Rifle
  Market Value: $450
Carbine
  Market Value: $425
Matched set with consecutive serial numbers
  Market Value: $925

1970

COLT FORT HAYS COMMEMORATIVE

Frontier Scout revolver; "1870—Fort Hays—1970" wiped in silver on left side of 4 3/4″ barrel; 22 LR only; hard rubber grips; nickel-plated finish; blued barrel, backstrap and trigger guard assembly screws, cylinder, base pin screw, ejector tube screw, bolt and trigger screw, hammer screw, trigger; originally cased in blonde finished wood; bottom lined with blue velvet, gold satin in lid; 500 made in 1970. **Issue Price: $130.**
  Market Value: $425

COLT FORT RILEY COMMEMORATIVE

Frontier Scout revolver; "1870—Kansas Series—Fort Riley—1970" wiped in black on left side of 4 3/4″ barrel; 22 LR only; ivorylite grips; nickel-plated medallions; nickel-plated finish; blued backstrap and trigger guard assembly, cylinder, base pin and screw, ejector rod head and tube screw, bolt and trigger screw, hammer and screw, trigger, stock screw; originally cased in blonde-finished wood; black velvet/satin lining; 500 made in 1970. **Issue Price: $130.**
  Market Value: $425

COLT MAINE SESQUICENTENNIAL

Frontier Scout revolver; "1820—Maine Sesquicentennial—1970" on left side of 4 3/4″ barrel; 22 LR only non-fluted cylinder; pearlite grips; gold-plated medallions; gold-plated finish; nickel-plated backstrap and trigger guard assembly, cylinder, base pin screw, hammer and hammer screw, ejector rod, ejector rod head, ejector tube screw, bolt and trigger screw; originally cased in natural knotty white pine; lined with royal blue velvet in bottom; light blue satin in lid; 3000 made in 1970. **Issue Price: $120.**
  Market Value: $425

COLT MISSOURI TERRITORIAL SESQUICENTENNIAL

Frontier Scout revolver; "1820-Missouri Sesquicentennial—1970" wiped in gold on left side of 4 3/4″ barrel; 22 LR only; walnut grips; gold-plated medallions; blued finish; gold-plated cylinder, loading gate, base pin, ejector rod head, ejector tube, tube screw, bolt and trigger screw, hammer, trigger, stock screw, top backstrap screws; originally cased in natural finish willow, lined in red velvet; 3000 made in 1970. **Issue Price: $125.**
  Market Value: $425

Colt Missouri Territorial Sesquicentennial Single-Action Army

Same legend on the 5 1/2″ barrel; 45 Colt only; grips, medallions, finish and plating are same as Frontier Scout version, except for case-hardened frame, loading gate; same casing; 900 made; production started in 1970. **Issue Price: $220.**
  Market Value: $1395

COLT WORLD WAR II/EUROPEAN THEATER

Semi-automatic; Model 1911A1; slide is marked "World War II Commemorative/European Theater of Operations" on left side; right side is roll-engraved with major sites of activity; 5″ barrel; 45 Auto only; bird's-eye maple grips; gold-plated medallions; nickel-plated finish in entirety; originally cased in oak box with oak cartridge block; lid removable; seven dummy cartridges included; infantry blue velvet lining; 11,500 made; production began in 1970. **Issue Price: $250.**
  Market Value: $795

COLT WORLD WAR II/PACIFIC THEATER

Semi-automatic; Model 1911A1; slide is marked "World War II Commemorative/Pacific Theater of Operations" on right side; left side roll-engraved with names of ten major battle areas; both sides of slide bordered in roll-marked palm leaf design; 5″ barrel; 45 Auto only; Brazilian rosewood grips; gold-plated medallions; nickel-plated in entirety; originally cased in Obichee wood; light green velvet lining; seven nickel-plated dummy cartridges in cartridge block; 11,500 made; production began in 1970. **Issue Price: $250.**
  Market Value: $795

1970 Colt Missouri Territorial Sesquicentennial

1970 Colt World War II/
Pacific Theater

1970 Colt World War II/
European Theater

COMMEMORATIVES

1970
Colt Wyatt Earp
Commemorative
Colt Wyatt Earp Single-Action Army
Marlin Model 39 Century Ltd.
Marlin Centennial Matched Pair
Savage Anniversary Model 1895
Winchester Lone Star Commemorative

1971
Colt Fort Scott Commemorative
Colt NRA Centennial Commemorative
Colt NRA Centennial Commemorative

1970 (cont.)

COLT WYATT EARP COMMEMORATIVE
Frontier Scout revolver; "Lawman Series—Wyatt Earp" on right side of barrel; standard model markings on left side; 12" Buntline barrel; 22 LR only; walnut grips; nickel-plated medallions blued finish; nickel-plated barrel, cylinder, ejector tube plug, ejector tube screw, rod head, base pin and base pin screw, hammer, trigger and backstrap and trigger guard assembly; originally cased in black leatherette-covered box; bottom lined with burgundy velvet; lid with red satin; 3000 made; production started in 1970. **Issue Price: $125.**
   Market Value: $495

Colt Wyatt Earp Single-Action Army
Same legend on barrel, but wiped in silver; 16 1/8" barrel; 45 Colt only; same grips, medallions as Frontier Scout version; blued finish; case-hardened frame; nickel-plated hammer, trigger, base pin, base pin cross latch assembly; same casing as Frontier Scout; 500 made; production began in 1970. **Issue Price: $395.**
   Market Value: $2500

MARLIN MODEL 39 CENTURY LTD.
Marking the Marlin Centennial, 1870 to 1970; same specs as standard Model 39A except for square lever; fancy walnut straight-grip uncheckered stock; forearm; 20" octagonal barrel; brass forearm cap; nameplate inset in stock buttplate. Produced only in 1970. **Issue Price: $125.**
   Market Value: $425

MARLIN CENTENNIAL MATCHED PAIR
Combines presentation-grade Model 336 centerfire, rimfire Model 39, in luggage-type case; matching serial numbers, for additional collector value. Both rifles have fancy walnut straight-grip stocks, forearms; brass forearm caps, brass buttplates, engraved receivers with inlaid medallions. Model 336 is chambered for 30-30 only; Model 39 for 22 LR, 22 Long, 22 Short. Only 1000 sets were manufactured in 1970. **Issue Price: $750.**
   Market Value: $1000

SAVAGE ANNIVERSARY MODEL 1895
Replica of Savage Model 1895; hammerless lever-action; marks 75th anniversary of Savage Arms Corp. (1895-1970); 308 Win. only; 24" octagon barrel; 5-shot rotary magazine; engraved receiver, brass-plated lever; brass buttplate; brass medallion inlaid in stock; uncheckered walnut straight grip stock, Schnabel-type forearm. Made only in 1970; 9999 produced. **Issue Price: $195.**
   Market Value: $495

WINCHESTER LONE STAR COMMEMORATIVE
Model 94; produced in rifle version with 26" barrel and carbine with 20" length. Receiver, upper and lower tang, lever, forearm cap, magazine tube cap all are gold-plated; buttplate is crescent shaped, solid brass. Stocks are American walnut with half pistol grip, fluted comb; commemorative medal with faces of Sam Houston, Stephen F. Austin, William Travis, Jim Bowie and Davy Crockett is inset in right side of stock. Left side of receiver is engraved with star and dates, 1845, 1970; both sides are bordered with series of stars; barrel carries inscription, "Lone Star Commemorative." Upper tang has "Under Six Flags," referring to banners of Spain, France, Mexico, Texas Republic, Confederacy and United States, which have flown over territory. It has bead post front sight, semi-buckhorn rear, plus saddle ring. Introduced in 1970. **Issue Price: Rifle, Carbine, $140; Matched Set with consecutive serial numbers $305.**
Carbine
   Market Value: $425
Rifle
   Market Value: $450
Matched set
   Market Value: $925

1971

COLT FORT SCOTT COMMEMORATIVE
Frontier Scout revolver; "1871 Kansas Series—Fort Scott—1971" on left side of 4 3/4" barrel; 22 LR only; checkered rubber, eagle-style grips; nickel-plated finish; blued barrel, cylinder, base pin screw, ejector tube screw, bolt and trigger screw, hammer, hammer screw, trigger; originally cased in blonde-finished wood; gold velvet/satin lining; 500 made in 1971. **Issue Price: $130.**
   Market Value: $425

COLT NRA CENTENNIAL COMMEMORATIVE
Single-Action Army; "1871 NRA Centennial 1971" wiped in gold on left side of 4 3/4", 5 1/2", or 7 1/2" barrels; 357 Magnum, 45 Colt; Goncalo Alves grips; gold-plated NRA medallion inlays; blued finish; case-hardened frame; nickel-silver grip screw escutcheons; originally cased in walnut, with inlaid NRA plate; gold velvet/satin lining; 2412 357 Magnums, 4131 45 Colts made; production began in 1971. **Issue Price: $250.**
357 Magnum
   Market Value: $1295
45 Colt
   Market Value: $1495

COLT NRA CENTENNIAL COMMEMORATIVE
Semi-automatic; Gold Cup National Match model; "1871 NRA Centennial 1971/The First 100 Years of Service/.45 Automatic Caliber" wiped in gold on left side of slide; MK IV barrel; Eliason rear sight; 5" barrel; 45 Auto only; checkered walnut grips; gold-plated NRA medallion inlays; blued; same type of case as NRA commemorative SAA; 2500 made; production began in 1971. **Issue Price: $250.**
   Market Value: $1295

1970 Savage Anniversary Model 1895

1971 Harrington & Richardson Anniversary Model 1873

1971 Colt NRA Centennial Commemorative Single Action

1971 Colt NRA Centennial Commemorative Autoloader

1971 Marlin 39A Article II

1971 Savage Model 71

1971 Winchester National Rifle Association Centennial Model

COMMEMORATIVES

1971

Harrington & Richardson
Anniversary Model 1873
Marlin 39A Article II
Marlin Article II Carbine
Savage Model 71
Winchester National Rifle
Association Centennial Model

1972

High Standard Benner
Commemorative
Colt Florida Territorial
Sesquicentennial
Marlin Model 336 Zane Grey

1973

Churchill One of One Thousand
Colt Arizona Ranger
Commemorative
Colt Peacemaker Centennial

HARRINGTON & RICHARDSON ANNIVERSARY MODEL 1873

Replica of Officer's Model 1873 Trapdoor Springfield commemorating 100th anniversary of H&R (1871-1971); single shot action; 45-70 only; 26″ barrel; engraved receiver, breech block, hammer, lock plate, buttplate; hand-checkered walnut stock with inlaid brass commemorative plate; peep rear sight, blade front; ramrod. Made only in 1971. Production limited to 10,000.
Market Value: $595

MARLIN 39A ARTICLE II

Same general specs as Model 39A; commemorates National Rifle Association Centennial, 1871-1971. Medallion with legend "The Right to Bear Arms" set on blued receiver; 24″ octagonal barrel; tube magazine holds 19 LR, 21 Long, 25 Short; fancy uncheckered walnut pistol-grip stock forearm; brass buttplate, forearm cap. Produced only in 1971. **Issue Price:** $135.
Market Value: $375

Marlin Article II Carbine

Same specs as Marlin 39A Article II rifle except straight grip stock; square lever; shorter magazine; reduced capacity; 20″ octagonal barrel. Produced only in 1971.
Issue Price: $135; Cased Set, $750.
Carbine
Market Value: $375
Cased Set
Market Value: $900

SAVAGE MODEL 71

Single shot lever-action; replica of Stevens Favorite, issued as commemorative to Joshua Stevens, founder of Stevens Arms Co.; 22 LR only; 22″ octagonal barrel; brass-plated hammer, lever, uncheckered straight-grip stock, Schnabel forearm; brass commemorative medallion inlaid in stock; brass butt plate; open rear sight, brass blade front. Made in 1971; only 10,000 produced. **Issue Price:** $75.
Market Value: $295

WINCHESTER NATIONAL RIFLE ASSOCIATION CENTENNIAL MODEL

Introduced in two versions: musket and rifle, both on Model 94 actions; musket resembles Model 1895 NRA musket with military lever to meet requirements for NRA match competition at turn of century; has 26″ tapered round barrel; full-length American walnut forearm; black-chromed steel buttplate; rear sight has calibrated folding rear leaf, blade front sight; 7-shot magazine. Rifle model resembles Model 64, also made on 94 action with 5-shot half-magazine; 24″ tapered, round barrel; hooded ramp and bead post front sight; adjustable semi-buckhorn rear sight; contoured lever, blued steel forearm cap. Both models are 30-30; have quick-detachable sling swivels; receivers are black-chromed steel; NRA seal in silver-colored metal is set in right side of stocks; left side of receivers inscribed appropriately with "NRA Centennial Musket" or "NRA Centennial Rifle." Both were introduced in 1971. **Issue Price: $149.95; Matched set with consecutive serial numbers, $325.**
Musket
Market Value: $425
Rifle
Market Value: $425
Cased set
Market Value: $900

1972

COLT FLORIDA TERRITORIAL SESQUICENTENNIAL

Frontier Scout revolver; "1822—Florida Territory—1972" on left side of 4 3/4″ barrel; 22 LR only; cypress wood grips; gold-plated medallions, blued finish; case-hardened frame, loading gate; gold-plated base pin, base pin screw, ejector rod head and screws, hammer, trigger and trigger screws; originally cased in cypress wood box; gold velvet/satin lining; 2001 made; production began in 1972. **Issue Price:** $125.
Market Value: $425

HIGH STANDARD BENNER COMMEMORATIVE

Super Military Trophy auto; 22 Long only; 5 1/2″ bull barrel; checkered American walnut thumbrest grip; micro rear sight; grip angle is that of the 1911 Government model; engraved with the five-ring Olympic insignia to denote the twentieth anniversary of Huelet O. "Joe" Benner's Olympic Gold Medal win; 1000 guns made; serial numbers, 1 to 1000. Made in 1972. **Issue Price: $550.**
Market Value: $1450

MARLIN MODEL 336 ZANE GREY

Same specs as Model 336A except 30-30 only; 22″ octagonal barrel. Commemorates centennial of Zane Grey's birth, 1872 to 1972; commemorative medallion attached to receiver; selected uncheckered walnut pistol grip stock forearm; brass forearm cap, buttplate; 10,000 produced with special serial numbers, ZG1 through ZG10,000. Produced only in 1972. **Issue Price:** $150.
Market Value: $350

1972 High Standard Benner Commemorative

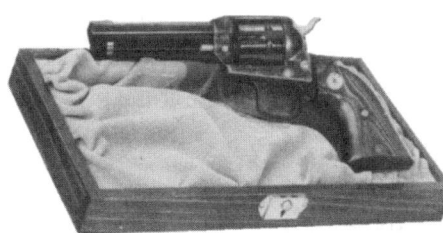

1972 Colt Florida Territorial Sesquicentennial

1973

CHURCHILL ONE OF ONE THOUSAND

Bolt-action; 270 Win., 7mm Rem. Magnum, 308, 30-06, 300 Win. Magnum, 375 H&H Magnum, 458 Win. Magnum; 5-shot magazine in standard calibers, 3-shot in magnum; made on Mauser-type action; classic French walnut stock; hand-checkered pistol-grip, forearm; recoil pad, cartridge trap in butt; sling swivels. Manufactured in England in 1973 to commemorate Interarms' 20th anniversary. Only 1000 made.
Market Value: $2500

COLT ARIZONA RANGER COMMEMORATIVE

Frontier Scout revolver; "Arizona Ranger Commemorative" on left side of 4 3/4″ barrel; 22 LR only; laminated rosewood grips; nickel-plated medallions; blued finish; case-hardened frame; nickel-plated backstrap and trigger guard assembly, hammer trigger, base pin, base pin assembly, screw for backstrap/trigger guard assembly, grips; originally cased in walnut with glass window lid; replica Arizona Ranger badge included in case; lined with maroon velvet; 3001 made; production began in 1973. **Issue Price:** $135.
Market Value: $425

COLT PEACEMAKER CENTENNIAL

Single-Action Army revolver, Frontier Six Shooter configuration; "The Frontier Six-Shooter" etched on left side of barrel, "1873 Peacemaker Centennial 1973" roll-marked on right side; 7 1/2″ barrel; 44-40 only; checkered rubber eagle-style grips; nickel-plated in entirety; originally cased in leather-covered wood box; brown velvet lining; 1500 made; production began in 1973. **Issue Price: $300.**
Market Value: $1495

COMMEMORATIVES

1973
Colt Peacemaker Centennial in 45 Colt Peacemaker
Colt Peacemaker Centennial 45 Colt/44-40 Combo
Harrington & Richardson Custer Memorial Edition
Smith & Wesson Texas Ranger Commemorative
Remington 1973 Ducks Unlimited Commemoratives
Sako Golden Anniversary Model
Stoeger Limited Edition Luger Carbine
Winchester Texas Ranger Commemorative

1974
Charles Daly Wildlife Commemorative
Mossberg Model 472 One in 5000
Remington 1974 Ducks Unlimited Commemorative

1973 Remington 1973 Ducks Unlimited Commemorative

1973 Sako Golden Anniversary Model

1973 Winchester Texas Ranger Commemorative

1974 Charles Daly Wildlife Commemorative

1973 (cont.)

Colt Peacemaker Centennial 45 Colt
Configuration has "1873 Peacemaker Centennial 1973" roll marked on left side of 7 1/2" barrel; 45 Colt only; one-piece varnished walnut grip; blued finish; case-hardened frame, hammer; originally cased in oiled walnut with brass-framed glass cover; maroon velvet lining; 1500 made; production began in 1973. **Issue Price: $300.**
  Market Value: $1495

Colt Peacemaker Centennial 45 Colt/44-40 Combo
Includes the two revolvers described above in oversize case of walnut-stained willow; lined with dark maroon velvet; matching serial numbers on guns; 500 sets made in 1973. **Issue Price: $625.**
  Market Value: $2995

HARRINGTON & RICHARDSON CUSTER MEMORIAL EDITION
Replica of blackpowder 1873 Springfield carbine; 54-caliber; inlaid with gold; heavy engraving; fancy-grade walnut stock; marketed in mahogany case. Each rifle bears the name of a cavalryman who died in the Battle of the Little Big Horn. Manufactured, 1973, in two versions; 25 Officers' Model, 243 Enlisted Men's Model. Officer's Model, **Issue Price:$3000; Enlisted Men's Model, $2000.**
Officer's Model
  Market Value: $3995
Enlisted Men's Model
  Market Value: $1995

REMINGTON 1973 DUCKS UNLIMITED COMMEMORATIVES
Model 1100 autoloading shotgun; 12-ga. only; 30" barrel; Full choke; vent rib. Other specs the same as standard Model 1100, except that serial number is preceded by DU; Ducks Unlimited medallion, surrounded by gilded scrollwork is attached to left side of receiver; 500 made in 1973 only. **Issue Price: $230.**
  Market Value: $495

SAKO GOLDEN ANNIVERSARY MODEL
Same specs as Sako long-action Deluxe sporter; 7mm Rem. Magnum only; floorplate, trigger guard, receiver feature gold oak leaf, acorn decoration; hand-checkered select European walnut stock hand-carved oak leaf pattern. Commemorates firm's 50th anniversary only; 1000 made in 1973.
  Market Value: $995

SMITH & WESSON TEXAS RANGER COMMEMORATIVE
Model 19 357 Combat Magnum; 4" barrel; sideplate stamped with Texas Ranger commemorative seal; uncheckered Goncalo Alves stocks; marketed with specially designed Bowie-type knife in presentation case. Commemorated the 150th anniversary of the Texas Rangers. Reported 8000 sets made in 1973.
  Market Value: $595

STOEGER LIMITED EDITION LUGER CARBINE
Semi-automatic; 22 LR; 11" barrel; forged aluminum frame; walnut pistol grips and forearm. Marketed in a black leatherette, red velvet-lined case. Due to lack of buttstock was classified as a handgun and was legal. Reported 300 made in 1973. Original price unknown.
  Market Value: $450

WINCHESTER TEXAS RANGER COMMEMORATIVE
Model 94; features stock forearm of semi-fancy walnut, with the buttstock having square comb, metal butt plate. Chambered in 30-30; 6-shot tube magazine; a facsimile of Texas Ranger star badge is embedded in the stock; saddle ring is included. Of standard grade, only 4850 were released in April 1973, all of them in the state of Texas. Another 150 Special Edition guns, at $1000 each, were released, in presentation cases, only to the Texas Ranger Association. These were hand-checkered, with full fancy walnut stocks, barrel and receiver highly polished; 4-shot magazine; 16" barrel; weighs 6 lbs.; standard model weighs 7 lbs.; 20" barrel. With Special Edition guns, commemorative star is mounted inside the presentation case instead of in the stock. Also introduced April 1973. Standard model, **Issue Price: $134.95; Special Edition, $1000.**
Standard model
  Market Value: $695
Special Edition
  Market Value: $2650

1974

CHARLES DALY WILDLIFE COMMEMORATIVE
Over/under; 12-ga.; trap and Skeet models only; same general specs as Diamond Grade over/under; fine scrollwork on left side of receiver, duck scene engraved on right side. Manufactured in Japan 1974. Reported 500 guns made.
  Market Value: $795

MOSSBERG MODEL 472 ONE IN 5000
Same specs as Model 472 Brush Gun except brass buttplate; barrel bands and saddle ring; gold-plated trigger; bright blued finish; Indian scenes etched on receiver; select walnut stock and forearm. Limited edition manufactured, 1974 only; numbered 1 through 5,000.
  Market Value: $295

REMINGTON 1974 DUCKS UNLIMITED COMMEMORATIVE
Model 870 pump-action, with gilded scroll-engraved receiver; special serial numbers; DU color medallion set in receiver. Made only in 1974 for auction by DU.
  Market Value: $395

1974 Remington 1974 Ducks Unlimited Commemorative

1973 Colt Peacemaker Centennial in 45 Colt Peacemaker

1975

MOSSBERG DUCK STAMP COMMEMORATIVE

Same specs as the Model 500DSPR Pigeon Grade 12-ga. magnum heavy duck gun; features a heavy 30" vent-rib barrel; Full choke only; receiver carries an etching of a wood duck. Gun was marketed with a special wall plaque commemorating, with the shotgun, the Migratory Bird Hunting Stamp Program. Only 1000 made in 1975.
Market Value: $375

RUGER COLORADO CENTENNIAL

Single-Six revolver; 22 LR; stainless steel grip frame; rosewood grip panels; historic scene roll engraved on cylinder; 6 1/2" barrel; inscribed to signify its purpose as a centennial commemorative; marketed in walnut presentation case; 15,000 made in 1975. **Issue Price: $250.**
Market Value: $350

1976

BROWNING BICENTENNIAL 78

Single shot Model 78; 45-70; same specs as standard model, except for bison and eagle engraved on receiver; scroll engraving on lever, both ends of barrel, buttplate, top of receiver; high-grade walnut stock forearm. Manufactured in Japan. Marketed with engraved hunting knife, commemorative medallion, alder presentation case. Gun and knife serial numbers match, beginning with 1776. Only 1000 sets made. Manufactured only in 1976. **Issue Price: $150.**
Market Value: $1850

BROWNING BICENTENNIAL SUPERPOSED

Over/under; 12-ga.; same basic specs as standard Superposed shotgun, but sideplates engraved, gold-inlaid turkey-hunting scene on right side, U.S. flag, bald eagle on left. State markings are in gold on blue background; hand-checkered American walnut straight-grip stock, Schnabel forearm; marketed in velvet-lined walnut presentation case. Only 51 made; one for each state and District of Columbia. Manufactured in Belgium, 1976.
Market Value: $8995

COLT BICENTENNIAL SET

Includes Colt SAA revolver, Python revolver and 3rd Model Dragoon revolver, with accessories; all have rosewood stocks, matching roll-engraved unfluted cylinders, blued finish, silver medallion bearing the Seal of the United States; Dragoon has silver grip frame; all revolvers in set have matching serial numbers, 0001 through 1776. Marketed with deluxe three-drawer walnut presentation case, reproduction volume of *Armsmear*. Made only in 1976.
Market Value: $2995

ITHACA BICENTENNIAL MODEL 37

Slide-action; 12-ga.; basic specs of Model 37 Supreme, except for Bicentennial design etched into receiver; serialized USA 0001 to USA 1976; full fancy walnut stock slide handle. Only 1976 made in 1976. Marketed in presentation case.
Market Value: $595

MAG-NA-PORT MARK V LIMITED EDITION

Revolver; built on Ruger Super Blackhawk; 44 Magnum; 5" barrel; jeweled, plated hammer and trigger; smoothed trigger pull; front sight altered with red insert; satin nickel backstrap, trigger guard, ejector rod housing, center pin; deluxe blue finish; Mag-na-ported barrel. Marketed in presentation case; made in 1976. **Issue Price: $395.**
Market Value: $3500 to $3700

REMINGTON BICENTENNIAL NYLON 66

Rifle; semi-automatic; 22 LR; same specs as standard Nylon 66 except specially marked with eagle, shield flanked with scrollwork and underlined with "1776-1976"; dates are gilded; Mohawk brown stock only; 12,000 made. **Issue Price: $84.95.**
Market Value: $175

1975 Ruger Colorado Centennial

REMINGTON BICENTENNIAL MODEL 742

Same specs as standard Model 742 Woodsmaster except Bicentennial commemorative inscription etched on receiver; different checkering pattern. Manufactured 1976 only.
Market Value: $395

REMINGTON BICENTENNIAL MODEL 760

Same specs as standard Model 760 Gamemaster except Bicentennial commemorative inscription etched on receiver; different checkering pattern. Manufactured 1976 only.
Market Value: $375

WICKLIFFE '76 COMMEMORATIVE

Single shot; same specs as '76 Deluxe model except filled etching on sidewalls of receiver: 26" barrel only; U.S. silver dollar inlaid in stock; marketed in presentation case. Manufactured 1976 only. Only 100 made.
Market Value: $750

WINCHESTER BICENTENNIAL '76 CARBINE

Model 94; same specs as standard model except 30-30 only; engraved antique silver finish; American eagle on left side of receiver, "76" encircled with thirteen stars on right side; engraved on right side of barrel is legend "Bicentennial 1776-1976." Originally marketed with wooden gun rack with simulated deer antlers, gold colored identification plate. Reported 20,000 made in 1976. **Issue Price: $325.**
Market Value: $595

COMMEMORATIVES

1975
Ruger Colorado Centennial
Mossberg Duck Stamp
Commemorative

1976
Browning Bicentennial 78
Browning Bicentennial
Superposed
Colt Bicentennial Set
Ithaca Bicentennial Model 37
Mag-Na-Port Mark V Limited
Edition
Remington Bicentennial
Nylon 66
Remington Bicentennial
Model 742
Remington Bicentennial
Model 760
Wickliffe Commemorative
Winchester Bicentennial '76
Carbine

1976 Mag-Na-Port Mark V Limited Edition

1976 Remington Bicentennial Nylon 66

1976 Remington Bicentennial Model 742

1976 Remington Bicentennial Model 760

1976 Winchester Bicentennial '76 Carbine

COMMEMORATIVES

1977
Winchester Wells Fargo
Model 94
Winchester Cheyenne
Commemorative
Colt Second Amendment
Commemorative
Smith & Wesson 125th
Anniversary
Smith & Wesson Deluxe
Edition 125th Anniversary
Ithaca Ducks Unlimited
Commemorative
Mag-Na-Port Custom Six
Limited Edition
Mag-Na-Port Tomahawk
Limited Edition

1978
Browning Centennial
Hi-Power
Colt U.S. Cavalry
Commemorative
Winchester Antlered Game
Commemorative
Winchester Legendary Lawman

1976 Browning Bicentennial 78

1977 Ithaca Ducks
Unlimited Commemorative

1977 Smith & Wesson
125th Anniversary

1977 Colt Second
Amendment Commemorative

1978 Browning Centennial Hi-Power

1977

COLT SECOND AMENDMENT COMMEMORATIVE
Peacemaker Buntline revolver; 22 rimfire; 7 1/2″ barrel bears inscription "The Right To Keep And Bear Arms"; polished nickel-plated barrel, frame, ejector rod assembly, hammer, trigger; blued cylinder, backstrap, trigger guard; black pearlite stocks; fluted cylinder; specially serial numbered; marketed in special presentation case, carrying reproduction copy of Second Amendment to the Constitution. Reported 3000 made in 1977. **Issue Price: $194.95.**
   Market Value: $425

ITHACA DUCKS UNLIMITED COMMEMORATIVE
Limited edition of Model 37 Featherlight pump action; commemorates DU 40th anniversary; 12-ga.; 30″ Full choke barrel; vent rib; recoil pad; Raybar front sight; commemorative grip cap; receiver engraved with DU anniversary logo, banner commemorating occasion. Also made in high-grade custom version with more elaborate etching, hand-checkered full fancy American walnut stock, custom-fitted carrying case. Reported 5000 made in 1977. Standard commemorative **Issue Price: $255; Custom model, $600.**
Standard
   Market Value: $400
Custom
   Market Value: $600

MAG-NA-PORT CUSTOM SIX LIMITED EDITION
Revolver; built on Ruger Super Blackhawk; 44 Magnum; 6″ barrel; same specs as Mag-na-port Mark V except yellow front sight insert, gold-outlined rear sight blade; gold inlay. 50 made in 1977. **Issue Price: $350.**
   Market Value: $2250

MAG-NA-PORT TOMAHAWK LIMITED EDITION
Revolver; built on Ruger Blackhawk; 44 Magnum; Mag-na-ported 4 3/4″ barrel; Metalife SS finish; smoothed trigger pull; red front sight insert; white outline rear sight; hammer engraved with Tomahawk logo; topstrap engraved with Mag-na-port and Tomahawk logos; serial numbered 1 to 200; 200 made in 1977. Marketed in carrying case. **Issue Price: $495.**
   Market Value: $2000

SMITH & WESSON 125TH ANNIVERSARY
Model 25 revolver; 45 Colt; 6 1/2″ barrel; blued finish; Goncalo Alves stocks; "Smith & Wesson 125th Anniversary" gold-filled on barrel; sideplate has gold-filled anniversary seal; marketed in case bearing nickel-silver anniversary seal. Included is book, *125 Years With Smith & Wesson.* Reported 10,000 issued in 1977.
   Market Value: $495

Smith & Wesson Deluxe Edition 125th Anniversary
Same specs as Smith & Wesson 125th Anniversary except Class A engraving; ivory stocks; anniversary medallion on box is sterling silver; book leather-bound. Reported 50 issued in 1977.
   Market Value: $1495

WINCHESTER WELLS FARGO MODEL 94
Same specs as standard Model 94 except 30-30 only; antique silver-finish engraved receiver; nickel-silver stagecoach medallion inset in buttstock; checkered fancy American walnut stock forearm; curved buttplate. Reported 20,000 made in 1977.
   Market Value: $495

WINCHESTER CHEYENNE COMMEMORATIVE
Lever-action rifle; same specs as standard Model 94 carbine except 44-40; made specifically for the Canadian market. Reported 11,220 made in 1977. **Issue Price: $375.**
   Market Value: $795

1978

BROWNING CENTENNIAL HI-POWER
Commemorates Browning's centennial anniversary; same specs as standard 9mm Hi-Power except oil-finished hand-checkered walnut stocks; Browning medallion inset on both sides; chrome finish; has centennial inscription with date hand-engraved on side; gold-plated trigger; fixed sights. Issued in fitted walnut case with red velvet lining. Only 3500 produced in 1978, with serial #1878D-0001 through 1878D-3500. **Issue Price: $495.**
   Market Value: $695

COLT U.S. CAVALRY COMMEMORATIVE
Based on 1860 Army design; commemorates 200th anniversary of U.S. Cavalry, 1777 to 1977; blued barrel, hammer assembly, cylinder, backstrap, trigger; frame, hammer color case-hardened; brass trigger guard; one-piece walnut stocks; naval engagement scene roll marked on nonfluted cylinder; marketed with detachable walnut shoulder stock, accessories, in oiled American walnut presentation case. Reported 3000 units manufactured 1978. **Issue Price: $995.**
   Market Value: $1250

WINCHESTER ANTLERED GAME COMMEMORATIVE
Built on Winchester 94 action but with polished 20″ barrel; gold-colored inscription reading "Antlered Game;" gold-plated lever; tang and barrel bands match blue of receiver; 30-30 only. Total of 19,999 made in 1978. **Issue Price: $374.95.**
   Market Value: $495

WINCHESTER LEGENDARY LAWMAN
Model 94 Carbine; 30-30 only; same specs as standard model except 16″ barrel; full-length tube magazine; antique silver-finish barrel bands; right side of barrel bears silver-colored inscription, "Legendary Lawman"; extended forearm, straight-grip stock; nickel-silver medallion set in buttstock features sheriff standing on Western street. Reported 20,000 manufactured in 1978. **Issue Price: $375.**
   Market Value: $495

1979

COLT NED BUNTLINE COMMEMORATIVE

Single-action revolver; 45 Colt; 12" barrel; built on New Frontier SAA frame; adjustable rear sight; nickel-plated; black composite rubber grips; marketed in custom presentation case with six nickel-plated 45 cartridges. Reported 3000 manufactured in 1979. **Issue Price: $900.**

Market Value: $895

MAG-NA-PORT BACKPACKER LIMITED EDITION

Revolver; built on Charter Arms Bulldog; 44 Special; no front sight; Mag-na-ported 1.875" barrel; dehorned, anti-snag hammer; Pachmayr grips; Metalife SS finish; engraved with Mag-na-port logo and Backpacker 1 to 250 serial number; 250 made in 1979. Marketed in carrying case. **Issue Price: $295.**

Market Value: $400

MAG-NA-PORT BULLSEYE LIMITED EDITION

Semi-automatic; built on Ruger MKI Target Model; 22 LR; Mag-na-ported 5 1/2" barrel; gold-outlined rear sight; Clark target trigger; standard black Ruger grips; Metalife SS finish; Mag-na-port logo on end of bolt; lettering, logo gold inlaid; gold bullseye logo on receiver; serial numbered 001-200; 200 made in 1979. Marketed in wild poplar case. **Issue Price: $395.**

Market Value: $700

MAG-NA-PORT CLASSIC LIMITED EDITION

Revolver; built on Ruger Super Blackhawk; 44 Magnum; Mag-na-ported 7 1/2" barrel; yellow insert front sight, gold outline rear; satin nickel finish on backstrap, ejector rod; deluxe blued cylinder, frame, barrel; jeweled, gold-plated hammer, trigger; polished, gold-plated center pin release, ejector rod; "Mag-na-port Classic" engraved, gold inlaid on topstrap; antique grips with gold grip screw; custom tune-up; other lettering, logos gold inlaid; carries Mag-na-port serial numbers 1 through 250; 250 made in 1979. Hammer, trigger were seal-locked to ensure unfired condition. Marketed in case of wild poplar with Mag-na-port logo. **Issue Price: $590.**

Market Value: $2500

MAG-NA-PORT SAFARI #1 CAPE BUFFALO

Revolver; built on Ruger Super Blackhawk; 44 Magnum; 6" barrel; Metalife finish; Mag-na-port vent process; yellow front insert on sight; Omega Maverick rear sight; smoothed action; Mag-na-port logo and serial numbers 001 through 200 engraved; stainless steel ejector rod housing; gold-plated hammer, trigger, center pin, center pin release, ejector rod and grip screws; antiqued factory grips inlaid with 44 Magnum case head; Cape buffalos engraved on cylinder. Only 200 made in 1979. **Issue Price: $995.**

Market Value: $1500

WINCHESTER LEGENDARY FRONTIERSMAN

Model 94 Carbine; receiver decorated with scenes of the old frontier; silver-plated finish on receiver; polished, blued barrel, finger lever, hammer and trigger; forearm and straight-grip stock of semi-fancy American walnut; cut checkering; "Legendary Frontiersman" in silver on right side of barrel. Reported 19,999 made in 1979. **Issue Price: $549.95 each.**

Market Value: $495

WINCHESTER LIMITED EDITION II

Rifle; built on standard Model 94 carbine action but with gold-plated receiver with etched game scenes on each side, gold-plated hammer, lever; rest of metal bright blued; top quality fancy walnut stock forend. Only 1500 made in 1979. **Issue Price: $1750.**

Market Value: $1395

WINCHESTER SET OF 1000 COLLECTOR ISSUE

Combines Model 94 in 30-30 and Model 9422 in 22 WMR; both with game scene engraved on receiver; levers, receivers, barrel bands are gold-plated. Marketed in red velvet-lined wooden case, with brass hardware, lock key. Only 1000 sets made in 1979. **Issue Price: $3000 per set.**

Market Value: $2250

1979 Winchester Legendary Frontiersman

1979 Mag-Na-Port Safari #1 Cape Buffalo

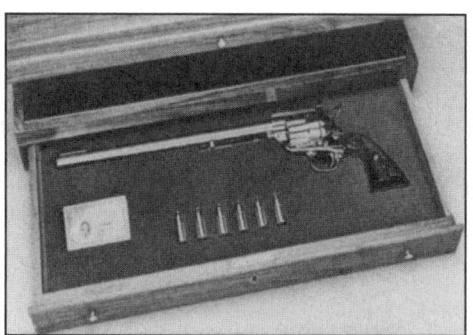

1979 Colt Ned Buntline Commemorative

COMMEMORATIVES

1979
Colt Ned Buntline Commemorative
Mag-Na-Port Backpacker Limited Edition
Mag-Na-Port Bullseye Limited Edition
Mag-Na-Port Classic Limited Edition
Mag-Na-Port Safari #1 Cape Buffalo
Winchester Legendary Frontiersman
Winchester Limited Edition II
Winchester Set of 1000 Collector Issue

1979 Mag-Na-Port Bullseye Limited Edition

1979 Winchester Set of 1000 Collector Issue

1979 Winchester Limited Edition II

1980
Ithaca Model 37 2500 Series Centennial
Mag-Na-Port Safari #2 Elephant
Wickliffe Big Game Commemorative
Winchester Oliver Winchester Commemorative

1981
Browning Waterfowl/Mallard Limited Edition
Colt John M. Browning Commemorative
Mag-Na-Port Safari #3 Lion
Remington 1981 Ducks Unlimited Special
Remington 1981 Classic
Remington Atlantic DU Commemorative
Remington Chesapeake
Remington Model 1100 Collectors Edition
Winchester John Wayne Commemorative

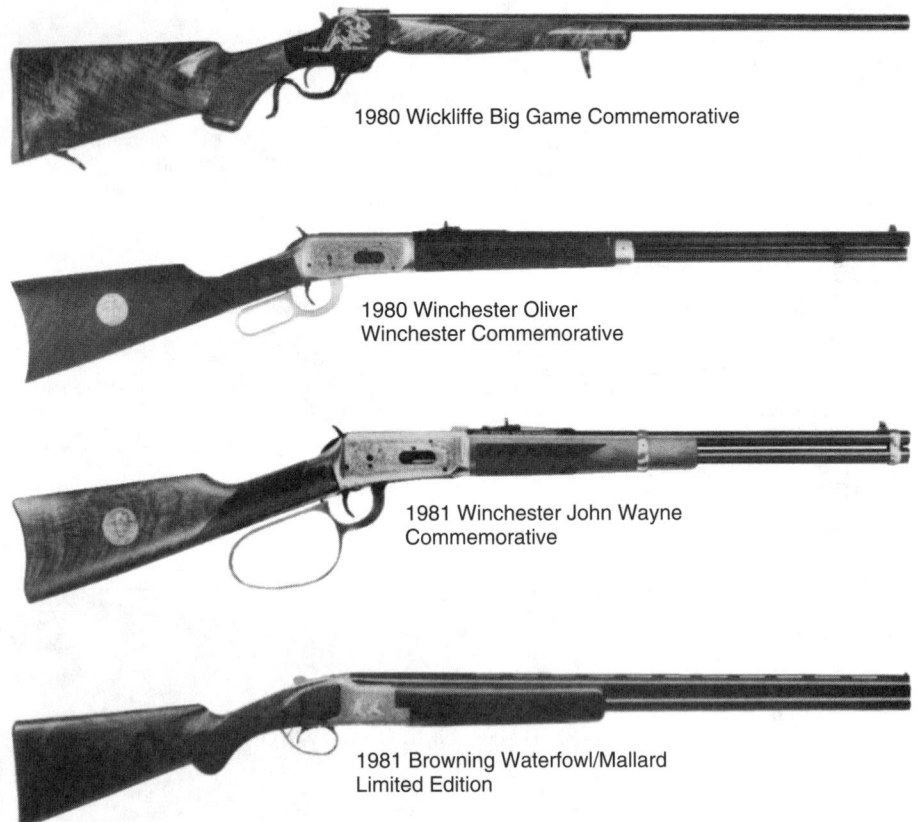

1980 Wickliffe Big Game Commemorative

1980 Winchester Oliver Winchester Commemorative

1981 Winchester John Wayne Commemorative

1981 Browning Waterfowl/Mallard Limited Edition

1980

ITHACA MODEL 37 2500 SERIES CENTENNIAL
Pump-action shotgun; 12-ga. only; etched receiver; antiqued parts; deluxe walnut stock; silver-plated parts. Commemorated Ithaca's 100th anniversary. Unknown number made, 1977.
Market Value: $595

MAG-NA-PORT SAFARI #2 ELEPHANT
Revolver; built on Ruger Super Blackhawk; 44 Magnum; 7 1/2" barrel, with custom inverted crown; includes Mag-na-port venting, satin and bright nickel finish; jeweled and nickeled hammer, trigger; custom rosewood grips with inlaid ivory scrimshaw of elephant; engraved elephant on cylinder; scroll engraving at muzzle; Mag-na-port logo on topstrap; smoothed action; white outline rear sight, red front sight insert. Marketed in walnut presentation case inlaid with ivory elephant scrimshaw. Only 200 made in 1980. **Issue Price: $1395.**
Market Value: $1800

WICKLIFFE BIG GAME COMMEMORATIVE
Single shot; same specs as Wickliffe Model '76 except glossy stock finish; gold-filled receiver etchings, chambered for 338 Win. Magnum only. Only 200 made. Manufactured 1980 only.
Market Value: $550

WINCHESTER OLIVER WINCHESTER COMMEMORATIVE
Model 94; 38-55 Win. only; gold commemorative plaque featuring Oliver Winchester medallion inlaid in American walnut stock; gold-plated, engraved receiver, forend cap. Reported 19,999 made in 1980. **Issue Price: $595**
Market Value: $695

1981

BROWNING WATERFOWL/MALLARD LIMITED EDITION
Over/under; 12-ga.; 28" Modified/Full barrels; other specs same as Lightning Superposed model except each gun is inscribed in gold with the Latin scientific name for the mallard, has gold mallard inlaid in receiver, with two ducks on bottom, one on trigger guard; grayed, engraved receiver; French walnut stock forend; 24 lpi checkering; oil-finished hand-rubbed wood; marketed in velvet-lined black walnut case. Only 500 made in 1981. **Issue Price: $7000.**
Market Value: $4995

COLT JOHN M. BROWNING COMMEMORATIVE
Model 1911; 45 Auto; standard model to commemorate the 70th anniversary of the model's existence. Gold inlay on right side of slide announces the reason for manufacture, with eagle, scrollwork and Colt stallion; right side of slide also features extensive scrollwork; blued hammer and trigger, hand-checkered walnut grips. Reported 3000 made in 1981. **Issue Price: $1100.**
Market Value: $995

MAG-NA-PORT SAFARI #3 LION
Revolver; built on Ruger Blackhawk; 44 Magnum; Mag-na-ported 5 1/2" barrel; highly polished blue-black finish; Omega Maverick rear sight, white front sight insert; silver inlay of Mag-na-port logo; custom grip with ivory scrimshaw of lion; lion engraving on cylinder and "Shumba" engraved on backstrap. Marketed in hand-crafted presentation case. Only 200 made in 1981. **Issue Price: $995.**
Market Value: $1550

REMINGTON 1981 DUCKS UNLIMITED SPECIAL
Model 1100 LT-20; 26" Improved Cylinder, vent-rib barrel; 2 3/4" chamber; left side of receiver panel carries words "Ducks Unlimited Special" in script lettering. Right receiver panel has DU mallard head logo with scrollwork. Right side of buttstock has laser-etched reproduction of DU crest. Only 2400 made in 1981, with serial numbers 0001-DU81 to 2400-DU81, for auction by DU.
Market Value: $495

REMINGTON 1981 CLASSIC
Bolt-action Model 700; 7mm Mauser only; 24" barrel; 44 1/2" overall length; cut-checkered American walnut stock of Classic design; satin finish on wood; no sights; hinged floorplate; rubber recoil pad; sling swivel studs. Limited production in this caliber only in 1981. **Issue Price: $364.95.**
Market Value: $425

REMINGTON ATLANTIC DU COMMEMORATIVE
Model 1100 12-ga. magnum; 32" Full-choke, vent-rib barrel; right side of receiver is embossed with "Ducks Unlimited" and DU mallard head symbol; left side carries the words, "The Atlantic," surrounded by scroll markings. Made only in 1981; carries special DU serial numbers. **Issue Price: $552.**
Market Value: $495

REMINGTON CHESAPEAKE
Ducks Unlimited commemorative for 1981; Model 1100; 12-ga. magnum; 30" barrel; Full choke; select American walnut stock; cut checkering; recoil pad; gold-colored trigger; ivory bead front sight; left side of receiver decorated with plaque of flying duck scrollwork engraved gold-filled legend; furnished with foam-lined hard carrying case. Only 2400 made for DU auctions with opening bid of $950.
Market Value: $495

REMINGTON MODEL 1100 COLLECTORS EDITION
Same specs as standard Model 1100 except deep-relief etching, gold highlights, richly figured walnut stock positive cut checkering. Marketed with certificate showing serial number. Only 3000 made in 1981.
Market Value: $895

WINCHESTER JOHN WAYNE COMMEMORATIVE
Model 94 carbine; 32-40 Win. only; pewter-plated receiver; nickel-silver likeness of Wayne inlaid in American walnut stock; cut checkering; receiver engraved with cattle drive scene on one side, stagecoach under attack by Indians on other; scenes are edged with titles of Wayne's films. Reported 50,000 made in 1981, some marketed by U.S. Repeating Arms. **Issue Price: $600.**
Market Value: $995

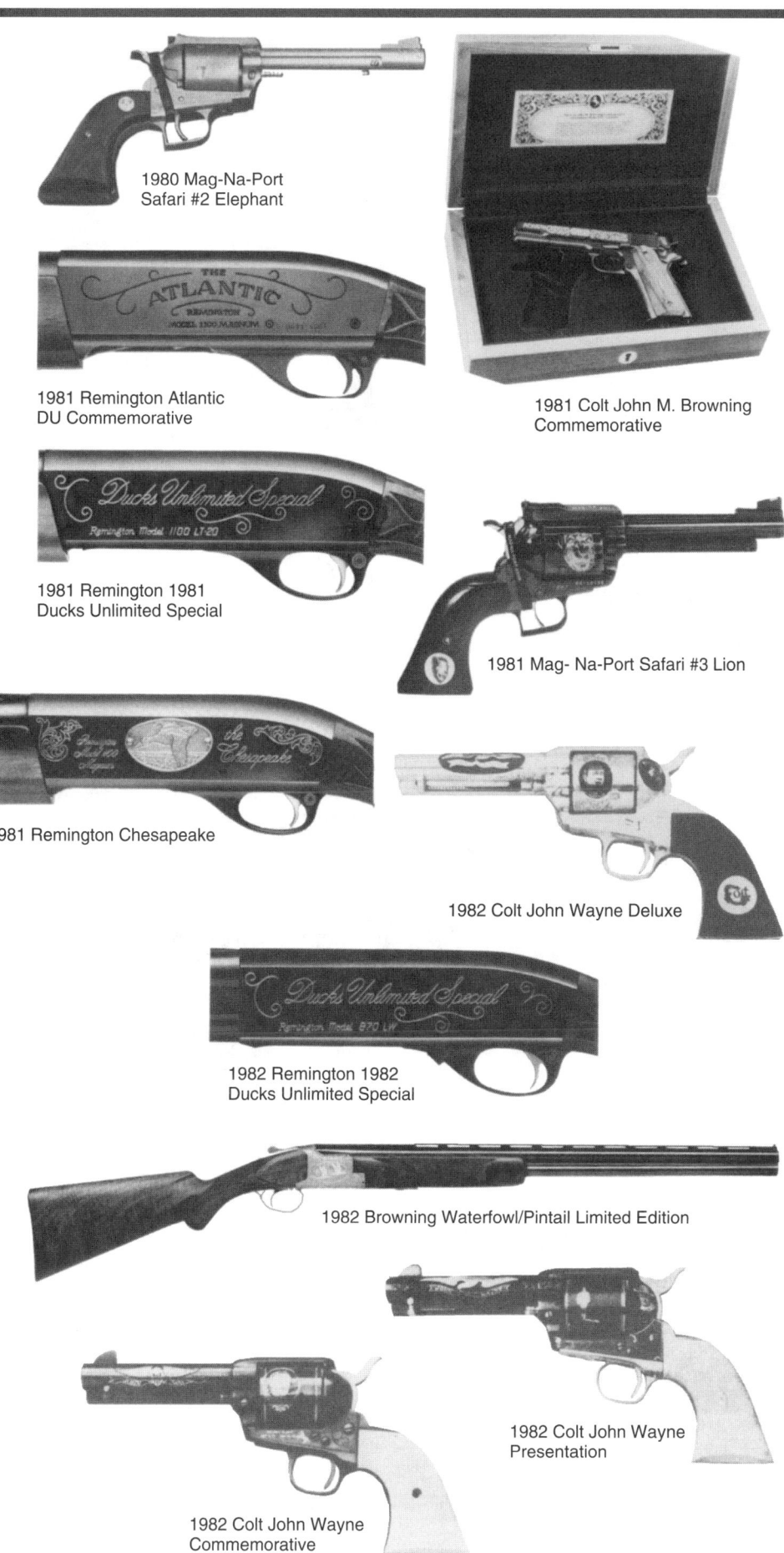

1980 Mag-Na-Port
Safari #2 Elephant

1981 Remington Atlantic
DU Commemorative

1981 Colt John M. Browning
Commemorative

1981 Remington 1981
Ducks Unlimited Special

1981 Mag- Na-Port Safari #3 Lion

981 Remington Chesapeake

1982 Colt John Wayne Deluxe

1982 Remington 1982
Ducks Unlimited Special

1982 Browning Waterfowl/Pintail Limited Edition

1982 Colt John Wayne
Presentation

1982 Colt John Wayne
Commemorative

1982

Browning Waterfowl/Pintail
Limited Edition
Colt John Wayne
Commemorative
Colt John Wayne Deluxe
Colt John Wayne
Presentation
Remington 1982 Classic

1982

BROWNING WATERFOWL/PINTAIL LIMITED EDITION

Over/under; 12-ga.; 28″ Modified/Full barrels; same specs as 1981 Mallard Limited Edition except hand-sculptured inlays of pintails in flight against background of cattails; gold-inlaid head of pintail on trigger; inscribed in gold are "American Pintail" and the scientific name, "Anas Acuta." Stock is of French walnut, with checkered butt. Marketed in velvet-lined black walnut case. Only 500 made in Belgium in 1982. **Issue Price: $8800.**
Market Value: $4995

COLT JOHN WAYNE COMMEMORATIVE

Single-action; 45 Colt; ivory grips; etched, gold-plated portrait of Wayne on cylinder, eagle and shield on barrel with name; gold-filled Wayne signature on backstrap; marketed in presentation case. Reported 3100 made. **Issue Price: $2995.**
Market Value: $1995

Colt John Wayne Deluxe

Same specs as John Wayne Commemorative except silver-plated, hand-engraved finish with 18-karat gold inlaid motif; two-piece ebony grips with ivory insert, gold-plated image of Wayne on horseback etched on right side of cylinder. Only 500 made. **Issue Price: $10,000.**
Market Value: $7500

Colt John Wayne Presentation

Same specs as John Wayne Commemorative except blued finish, gold plating, 24-karat gold inlays, two-piece checkered ivory grips with gold inlay. Only 100 made. **Issue Price: $20,000.**
Market Value: $12,000

REMINGTON 1982 CLASSIC

Same specs as the 1981 Remington Classic except 257 Roberts; limited production of approximately 7000 in this caliber only in 1981. **Issue Price: $381.95.**
Market Value: $400

COMMEMORATIVES

1982
Remington 1982 Ducks Unlimited Special
Remington Model Four Collectors Edition
Remington River DU Commemoratives
Ruger Model I Limited Edition
Winchester Annie Oakley 22

1983
Browning Waterfowl/Black Duck Limited Edition
Colt Buffalo Bill Wild West Show Commemorative
Mag-Na-Port Alaskan Series

1982 Ruger Model 1 Limited Edition

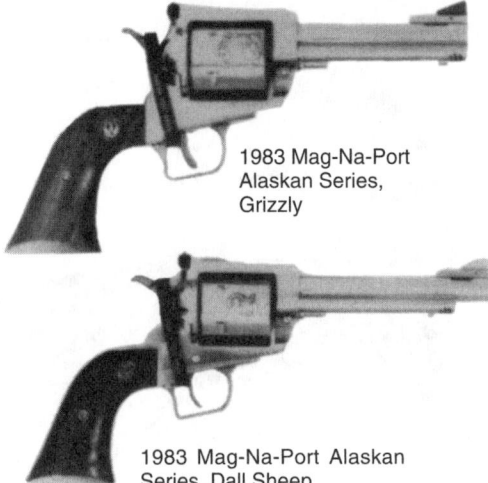

1983 Mag-Na-Port Alaskan Series, Grizzly

1983 Mag-Na-Port Alaskan Series, Dall Sheep

1983 Mag-Na-Port Alaskan Series, Caribou

1983 Mag-Na-Port Alaskan Series, Moose

1983 Colt Buffalo Bill Wild West Show Commemorative

1982 Remington River DU Commemorative

1982 Winchester Annie Oakley 22

1982 (cont.)

REMINGTON 1982 DU SPECIAL
Model 870; 20-ga. lightweight; 2 3/4" chamber, 26" Improved Cylinder barrel; decorated with gold-filled scrollwork inscribed with mallard logo, words, "Ducks Unlimited Special"; has gold-colored trigger, ivory bead front sight. Only 3000 made; serial numbers are 0001-DU82 to 3000-DU82. **Issue Price: $550.**
   Market Value: $475

REMINGTON MODEL FOUR COLLECTORS EDITION
Same specs as standard Model Four semi-automatic except etched receiver; 24K gold inlays; all metal parts with high-luster finish; 30-06 only. Only 1500 made in 1982.
   Market Value: $995

REMINGTON RIVER DU COMMEMORATIVE
Model 670; 12-ga. magnum; 30" Full-choke, vent-rib barrel; dedicated to the Mississippi Flyway; left receiver panel has engraved bronze medallion of mallard duck flanked by script lettering of gun's model and commemorative title, "The River," set off by scroll markings filled in gold color. Special serial numbers are DU82-0001 to DU82-3000. Only 3000 made. **Issue Price: $425.**
   Market Value: $495

RUGER MODEL I LIMITED EDITION
Mark I Target Model semi-automatic pistol of stainless steel introduced January 1, 1982, to mark the end of production of the standard model. Pistol was marketed in the same type of wooden case and printed inner box in which the original Ruger auto was shipped at the beginning of production in 1949. Each pistol was accompanied by an exact reproduction of the original automatic pistol brochure. Barrel of each pistol was marked "1 of 5000" and the receiver roll marked with a replica of the signature of the inventor, William B. Ruger. Only 5000 made in 1982. **Issue Price: $435.**
   Market Value: $495

WINCHESTER ANNIE OAKLEY 22
Rifle; 22 LR, 22 WMR; 20 1/2" barrel; based on Model 9422; barrel inscription gold inlaid; receiver roll engraved on right side with portrait of Annie Oakley, on left side with Oakley standing in saddle, shooting from moving horse; available with sling swivels, saddle ring on special order. Reported 6000 manufactured. **Issue Price: $699.**
   Market Value: $695

1983

BROWNING WATERFOWL/BLACK DUCK LIMITED EDITION
Over/under; 12-ga.; 28" Modified/Full barrels; same specs as 1981 Mallard Limited Edition except 24-karat black ducks inlaid in receiver; words, "Black Duck" and Latin designation "Anas Rubripes," engraved in gold within gold banners. Marketed in velvet-lined black walnut case. Only 500 made in Belgium in 1983. **Issue Price: $8800.**
   Market Value: $4995

COLT BUFFALO BILL WILD WEST SHOW COMMEMORATIVE
Single-Action Army; 45; 4 3/4" barrel etched with gold-plated figure of Colonel Cody on unfluted cylinder, cowboys, Indians and buffalo on barrel with commemorative inscription; two-piece walnut grips; blued finish, color case-hardened frame; marketed in cherry-stained hardwood case. Only 500 made. **Issue Price: $1349.95.**
   Market Value: $1595

MAG-NA-PORT ALASKAN SERIES
Revolver; built on stainless steel Ruger Super Blackhawk; 44 Magnum; four variations in the series; rear sight is original Omega white outline; front sight is C-More by Magnum Sales, with different colored inserts; barrels are Mag-na-ported with Mag-na-port logo etched on each cylinder. "Alaskan Series" is etched into the barrel, "One of 200" into the topstrap, with appropriate animal's profile also etched into the cylinder; each gun was delivered with Mag-na-port seal as guarantee the hammer had never been in cocked position. Each gun marketed in walnut presentation case with appropriate animal's head branded into wood. Introduced, 1983.

1983 Winchester Chief Crazy Horse Commemorative

1984 Winchester 9422 Eagle Scout Commemorative

Mag-Na-Port Alaskan Grizzly

Motif included 50 of the 200 guns in series; barrel had been cut to 4 5/8" and recrowned; C-More front sight carried a green insert. **Issue Price: $695.**
  Market Value: $850

Mag-Na-Port Alaskan Dall Sheep

Same as Grizzly with 50 guns issued except barrel cut to 5 1/2", recrowned; replacement C-More front sight carries pink insert of high-contrast DuPont acetal. **Issue Price: $695.**
  Market Value: $850

Mag-Na-Port Alaskan Caribou

Issue includes 50 guns with 6 1/2" recrowned barrel; insert of C-More sight is blaze orange in color. **Issue Price: $695.**
  Market Value: $850

Mag-Na-Port Alaskan Moose

Comprises 50 guns with standard 7 1/2" Blackhawk barrel; C-More sight insert is bright yellow. **Issue Price: $695.**
  Market Value: $850

MAG-NA-PORT SAFARI #4 RHINO

Revolver; built on stainless steel Ruger Super Blackhawk; 44 Magnum; 4 5/8" Mag-na-ported barrel; velvet hone finish; gold-plated hammer, trigger, ejector rod, center pin, center pin release; Omega white-outline rear sight, red insert front sight; rhino head hand-engraved on cylinder; Mag-na-port serial numbers 001 through 200 hand-engraved on frame. Marketed in walnut presentation case with overlay of African continent and hand-painted head of rhino. Only 200 made in 1983. **Issue Price: $895.**
  Market Value: $1450

REMINGTON 1983 CLASSIC

Same specs as Classic limited issue of previous years except 300 H&H Magnum only. Approximately 3500 made in this caliber only in 1983. **Issue Price: $421.95.**
  Market Value: $475

REMINGTON MISSISSIPPI DU COMMEMORATIVE

Model 870 12-ga. magnum; 32" Full-choke, vent-rib barrel; right side of receiver embossed with words, "Ducks Unlimited," DU mallard head symbol; left side of receiver carries title of gun, "The Mississippi," with scrollwork. **Issue Price: $454.**
  Market Value: $450

WINCHESTER CHIEF CRAZY HORSE COMMEMORATIVE

Rifle; built on Model 94 action; 38-55 Win.; 24" barrel; stock decorated with brass tacks; medallion in stock symbolizes united Sioux tribes; Chief Crazy Horse inscription engraved in barrel; receiver engraved, gold filled on both sides with Indian scenes, names of Sioux tribes in English, Lakota Sioux; saddle ring. Reported 19,999 made. **Issue Price: $600.**
  Market Value: $595

1984

AMERICA REMEMBERS BUFFALO BILL CENTENNIAL

Replica of Colt Model 1860 Army blackpowder 44 revolver; 8" barrel; extensive gold etching of Western scenes; bonded ivory stocks and powder flask; brass accessories; marketed in lined display case. Reported 2500 manufactured in Virginia gun factory, 1984. **Issue Price: $1950.**
  Market Value: $895

AMERICAN HISTORICAL FOUNDATION KOREAN WAR COMMEMORATIVE THOMPSON

Semi-automatic; 45 ACP; same specs as standard Model 1928 Thompson except highly polished walnut buttstock, pistol grip, forend; gold-plated activator knob, Cutts compensator, rear sight base, trigger, sling swivels; commemorative medallions set in buttstock, pistol grip; special serial numbers, KW0001 through KW1500. Reported 1500 made by Auto Ordnance, 1984. **Issue Price: $1,195.**
  Market Value: $1295

AMERICAN HISTORICAL FOUNDATION M1 GARAND WORLD WAR II COMMEMORATIVE

Reproduction of WWII combat rifle; select-grade walnut stock; highly blued finish; gold plating on smaller parts. Reported 2500 made by Springfield Armory in Illinois under special order; carried serial numbers WW0001 through WW2500. Marketed, 1984, with walnut display case. **Issue Price: $1695.**
  Market Value: $1195

COLT KIT CARSON COMMEMORATIVE

Single action; 22 rimfire; based on Colt New Frontier model; 6" barrel; blued, color case-hardened frame; ivory-colored stocks; gold artwork in Western motif on barrel; marketed in custom cherry case with hardbound copy of *Kit Carson's Own Story*; 950 made. **Issue Price: $549**
  Market Value: $450

COLT USA EDITION

Single-action revolver; 44-40; 7 1/2" barrel; fluted cylinder; checkered ivory stocks with fleur-de-lis pattern; gold inlaid frame borders, outline of state or capital city on recoil shield; gold-inlaid rampant colt on loading gate; gold-inlaid "USA" on left side of barrel; two gold-inlaid barrel bands; gold-inlaid ejector tube band; silver- and gold-inlaid stars; only 100 made. **Issue Price: $4995**
  Market Value: $5000

COLT THEODORE ROOSEVELT COMMEMORATIVE

Single action; based on Colt SAA; 7 1/2" barrel; 44-40; hand-fitted ivory stocks; backstrap bears gold TR monogram; barrel, cylinder hand engraved; marketed in oak presentation case; 500 made. **Issue Price: $1695**
  Market Value: $1895

1984 Texas Longhorn South Texas Army Limited Edition

REMINGTON 1984 CLASSIC

Same specs as earlier Classic Limited issues except 250-3000 Savage. Approximately 2500 made in this caliber in 1984. **Issue Price: $420.80.**
  Market Value: $475

TEXAS LONGHORN SOUTH TEXAS ARMY LIMITED EDITION

Revolver; single action; all centerfire pistol calibers; 6-shot cylinder; 4 3/4" barrel; grooved topstrap rear sight, blade front; one-piece fancy walnut stocks; loading gate, ejector housing on left side of the gun; cylinder rotates to left; color case-hardened frame; high polish blue finish; music wire coil springs; hand made; "One of One Thousand" engraved on barrel; marketed in glass-covered display case. Only 1000 made.
  Market Value: $1495

WEATHERBY 1984 MARK V OLYMPIC COMMEMORATIVE

Rifle; bolt-action; 300 Weatherby; same specs as standard Mark V except accents gold-plated; top-grade walnut stock with inlaid star. Reported 1000 made in Japan, 1984. **Issue Price: $2000.**
  Market Value: $1000

COMMEMORATIVES

1984
Winchester 9422 Eagle Scout Commemorative
Winchester 9422 Boy Scout Commemorative
Winchester Grouse Model
Winchester 94 Set

1985
American Historical Foundation WWII Victory Special Edition M1 Carbine
Colt Texas Sesquicentennial
Colt Sesquicentennial Premier
Remington 1985 Classic
Ruger YR-25 Carbine
SIG Model 226 125th Anniversary Commemorative
Winchester ATA Hall of Fame Commemorative

1986
American Historical Foundation Vietnam War Commemorative Thompson
American Historical Foundation M1 Garand Korean War Commemorative
Browning Big Horn Sheep Issue
Browning Model 1886 Montana Centennial

1984 Winchester 9422 Boy Scout Commemorative

1984 Winchester Grouse Model

1985 Colt Texas Sesquicentennial

WINCHESTER/COLT 94 SET
Both guns 44-40 Win.; carbine with 20″ barrel; tube magazine; 38 1/8″ overall length; follows basic Model 94 specs except horse-and-rider trademark, WC monogram etched in gold on left side; gold-etched portrait of Oliver Winchester on right; right side of barrel has rendering of original Winchester factory; additional gold scrollwork; crescent buttplate; deep-cut spade checkering; semi-gloss finish on American walnut woodwork. Colt is Peacemaker SAA model with backstrap bearing Sam Colt signature in gold; WC monogram is gold-etched on left side of cylinder; barrel bears serpentine Colt logo in gold; right side of barrel has gold etching of original Colt factory; gold-plated scrollwork; oil-finished American walnut grips; 7 1/2″ barrel; 4440 sets made. **Issue Price: $3995**
Market Value: $2250

1985

AMERICAN HISTORICAL FOUNDATION WWII VICTORY SPECIAL EDITION M1 CARBINE
Semi-automatic; 30 Carbine; same specs as M1 Carbine standard military version except highly polished; American walnut stock; mirror-polished barrel, receiver; gold-plated rear sight, windage knob, front sight, trigger, magazine release, barrel bands, safety, slide stop; special serial numbers are WW0001 through WW2500. Reported 2,500 made under special contract by Iver Johnson Arms, 1985. **Issue Price: $695.**
Market Value: $595

COLT TEXAS SESQUICENTENNIAL
Single action; 45 Colt; built on SAA Sheriff's Model; 4″ barrel; Texas legend, dates inlaid on barrel in gold; gold star in shield on cylinder; marketed in French-fitted oak presentation case; 1000 made. **Issue Price: $1836**
Market Value: $1495

Colt Sesquicentennial Premier
Same specs as Texas Sesquicentennial; built on SAA design; 45 Colt; 4 3/4″ barrel; 24-karat gold-plated trigger guard, backstrap; engraved gold-inlaid scenes of Texas history, including the Alamo; scrimshawed ivory grips; marketed in four-sided glass presentation case; 75 made. **Issue Price: $7995**
Market Value: $5000

REMINGTON 1985 CLASSIC
Same specs as earlier versions except 350 Rem. Magnum. Approximately 6500 made in this caliber only in 1985. **Issue Price: $474.70.**
Market Value: $475

RUGER YR-25 CARBINE
Semi-automatic; 44 Magnum; same specs as the standard Ruger Model 44 except medallion in stock commemorating the 25th anniversary of the carbine's manufacture; the last year the model was made. Made only in 1985.
Market Value: $450

SIG MODEL 226 15TH ANNIVERSARY COMMEMORATIVE
Same specs as Model 226 except deep black non-reflective finish; ornately carved wooden grips; trigger, hammer, magazine release, decocking lever and slide stop all are gold-plated, all stampings on slide are gold-

filled. Slide marking reads, "125 Jahre SIG Waffen 1860-1985"; marketed in leather presentation case. Only 200 imported into U.S. in 1985. **Issue Price: $2500.**
Market Value: $1195

WINCHESTER ATA HALL OF FAME COMMEMORATIVE
Combo trap set; 12-ga.; 34″ single-barrel; 30″ vent-rib over/under with Winchoke barrels; engraved with Amateur Trapshooting Assn. Hall of Fame logo; serial numbered HF1 through HF250; first 60 were auction items at 1985 Grand American; marketed in fitted luggage style case; only 250 made in 1985 in Japan by Winchester-Olin. **Issue Price: $2795.**
Market Value: $2750

1986

AMERICAN HISTORICAL FOUNDATION VIETNAM WAR COMMEMORATIVE THOMPSON
Semi-automatic; 45 ACP; basic specs of 1928 Thompson submachine gun except highly polished blued finish; gold-plated rear sight, front sight and Cutts compensator, trigger, sling swivels; American walnut stock, forearm, with inset medallions; receiver carries gold-filled commemorative inscriptions, serial number, issuing organization, Thompson patent numbers. Reported 1500 made by Auto Ordnance Corp.,on special order, 1986. **Issue Price: $1,295.**
Market Value: $1295

AMERICAN HISTORICAL FOUNDATION M1 GARAND KOREAN WAR COMMEMORATIVE
Semi-automatic; 30-06; same specs as M1 Garand military version except Fajen select American walnut stock; gold-plated front sight, rear sight base, trigger, windage knob, elevation knob, stacking swivel, sling swivels, safety; barrel, receiver and other metal parts mirror polished; gold-plated medallion inset in stock. Reported 2500 made, 1986, serial numbers KW0001 through KW2500. **Issue Price: $1895.**
Market Value: $1195

BROWNING BIG HORN SHEEP ISSUE
Same specs as standard Browning A-Bolt rifle except 270 Win. only; checkered stock is high-grade walnut with high-gloss finish; has brass spacers under grip cap, recoil pad; deep relief engraving; big horn displayed in 24-karat gold. Only 600 made in Japan. Introduced, 1986.
Market Value: $1250

BROWNING MODEL 1886 MONTANA CENTENNIAL
Same specs as 1886 High Grade lever-action rifle except different stock design; engraving on receiver reads "Montana Centennial". Only 1,000 made in Japan, 1986. **Issue Price: $995.**
Market Value: $1795

1984 (cont.)

WINCHESTER 9422 EAGLE SCOUT COMMEMORATIVE
Same specs as Model 9422 XTR except receiver deep etched, plated in gold; left side has Boy Scout law, right side has Boy Scout oath; frame has "1910-1985" inscription; checkered stock, forend, high-luster finish; Eagle Scout medallion embedded in right side of stock; crescent steel buttplate; gold-plated forend cap; jeweled bolt. Marketed in oak presentation case; 1000 made. **Issue Price: $1710**
Market Value: $3000

Winchester 9422 Boy Scout Commemorative
Same specs as Model 9422 except receiver roll-engraved, plated in antique pewter; frame carries "1910-1985" anniversary inscription, Boy Scout oath, law; lever has engraved frieze of scouting knots; "Boy Scouts of America" inscribed on barrel; medallion embedded in stock; 15,000 made. **Issue Price: $495.**
Market Value: $595

WINCHESTER GROUSE MODEL
Limited edition over/under based on Model 101 configuration; custom built for the Ruffed Grouse Society; 20-ga. only; engraved scene of grouse on action; fancy walnut woodwork. Only 225 made in 1984.
Market Value: $1450

COLT 150TH ANNIVERSARY COMMEMORATIVE

Single Action Army revolver with 10" barrel, royal blue finish, authentic old-style frame with "B" engraving coverage, Goncalo Alves grips; anniversary logo is etched in 24K gold on backstrap, plus gold-etched signature panel. Marketed in an oak case with silver blue velvet lining; top of lid branded "1836-Colt-1986"; inside of the lid has inset 150th anniversary medallion; 1000 made with serial numbers AM-0001 through AM-1000. Made in 1986 only. **Issue Price: $1595.**
Market Value: $1795

COLT 150TH ANNIVERSARY ENGRAVING SAMPLER

Single Action Army; 4 3/4" barrel; blued or nickel finish; carries four styles of hand engraving; Henshaw, Nimschke, Helfricht and Colt Contemporary, with each style contained on one part of the gun; ivory grips are scrimshawed with the names of the four patterns. Made only in 1986. **Blued, Issue Price: $1612.95; Nickel, $1731.50.**
Market Value: $2500 to $3000 Depending on finish, caliber and barrel length.

COLT 150TH ANNIVERSARY DOUBLE DIAMOND

Two-gun set of stainless steel, including the Python Ultimate with 6" barrel and a polished Officers Model 45 ACP automatic. Both guns carry matching serial numbers and rosewood grips as well as the Double Diamond logo. Marketed in a cherrywood presentation case with a framed glass lid; lining is black velvet. Serial numbers extend from 0001-DD through 1000-DD. Only 1000 sets made in 1986. **Issue Price: $1574.95.**
Market Value: $1795

COLT WHITETAILER II

King Cobra AA3080; 357 Magnum; 8" barrel, vented rib; red insert in front sight ramp, fully-adjustable white outline rear; rubber combat-style grips; transfer bar safety; brushed stainless steel finish; barrel marked on left side, "Whitetailer II"; furnished with 1 1/2-4x variable Burris scope, with brushed aluminum finish; Millett satin nickel mounts. Marketed in soft sided custom carrying case. Produced in 1986; 1000 made. **Issue Price: $807.95.**
Market Value: $1095

KIMBER BROWNELL

Rifle; bolt-action; 22 rimfire; honors stockmaker Len Brownell. Has high-grade, full-length, Mannlicher-type stock. Reported 500 made, 1986.
Market Value: $1500

REMINGTON 1986 CLASSIC

No changes in the basic design except 264 Win. Magnum only. **Issue Price: $509.85.**
Market Value: $495

TEXAS LONGHORN SESQUICENTENNIAL MODEL

Revolver; single action; same specs as South Texas Army Limited Edition except with 3/4-coverage engraving, antique gold, nickel finish; one-piece ivory stocks. Marketed in hand-made walnut presentation case. Only 150 made in 1986.
Market Value: $2500

WINCHESTER 120TH ANNIVERSARY LIMITED EDITION

Model 94 carbine; 44-40 only; anniversary medallion in left side of receiver, with horse-and-rider trademark; right side of receiver carries gold-etched portrait of Oliver Winchester; signature is on the tang; right side of barrel has rendering of original Winchester factory; gold-plated scrollwork on barrel and receiver; magazine cap, front sight blade are gold-plated; has hoop-type lever, crescent buttplate. Only 1000 made in 1986. Serial numbers are WRA0001 to WRA1000. **Issue Price: $950.**
Market Value: $895

WINCHESTER SILVER ANNIVERSARY MODEL 101

Special issue in 12-ga. only; 28" barrels, vent rib. Each gun marketed in a custom leather carrying case, complete with snap caps and silver initial plate. Only 101 made in 1986. **Issue Price: $5200.**
Market Value: $2995

1987

AMERICAN DERRINGER TEXAS COMMEMORATIVE

Same specs as Model 1 Derringer except solid brass frame; stainless steel barrel, stag grips. Made in 38 Special, 44-40, 44 American, 45 Colt. Produced in 1987.
Market Value: $295

AMERICAN HISTORICAL FOUNDATION GENERAL PATTON COMMEMORATIVE

Single-action revolver; 45 Colt; 4 3/4" barrel; same general specs as Colt Single Action Army but made in Italy by Uberti; all exposed metal silver-plated; engraved frame, cylinder, barrel; faux ivory grips with Patton initials. Serial numbered P0001 through P2500. Reported 2500 made, 1987. **Issue Price: $1495.**
Market Value: $1495

AMERICAN HISTORICAL FOUNDATION M14 VIETNAM COMMEMORATIVE

Semi-automatic rifle; 7.62mm NATO; version of maker's M1A; 22" barrel; highly polished blued metal; some gold-plated parts; select walnut stock. Reported 1500 in Army motif, 1500 in Marine motif made by Springfield Armoury, 1987. **Issue Price: $1600.**
Market Value: $1195

AMERICAN HISTORICAL FOUNDATION VIETNAM WAR COMMEMORATIVE THOMPSON

Semi-automatic; 45 ACP; polished lacquer American walnut stock; same general specs as standard Model 1928 except gold-plated small parts; roll-engraved; gold-filled legend on receiver. Reported 1500 made by Auto Ordnance with special serial numbers, 1987. **Issue Price: $1295.**
Market Value: $1295

BROWNING PRONGHORN ANTELOPE ISSUE

Same specs as standard Browning A-Bolt except 243 Win. only; extensive detailed engraving; pronghorn in gold on each side of receiver; stock has skip-line checkering, high-gloss finish; brass spacers under pistol-grip cap, recoil pad. Only 500 made in Japan. Introduced, 1987.
Market Value: $1100

F.I.E. YELLOW ROSE LIMITED EDITION

Same specs as the standard model except polymer grips scrimshawed with map of Texas, Texas state flag, yellow rose with green leaves; other scrimshawed depictions available. Marketed in French-fitted American walnut presentation case. Manufactured in 1987.
Market Value: $200

KIMBER CENTENNIAL

Bolt-action rifle commemorates the 100th anniversary of the 22 Long Rifle cartridge; tastefully engraved; match barrel; select walnut stock; skeleton buttplate. Reported 100 manufactured in 1987 with serial numbers C-1 through C-100.
Market Value: $2600

PARKER B GRADE LIMITED EDITION

Replica of Parker Bros. side-by-side shotgun; 12-, 20- and 28-ga.; extensive engraving follows original pattern. Reported 100 made in Japan, 1987, specifically for Parker Reproductions. **Issue Price: $3975.**
Market Value: $4800

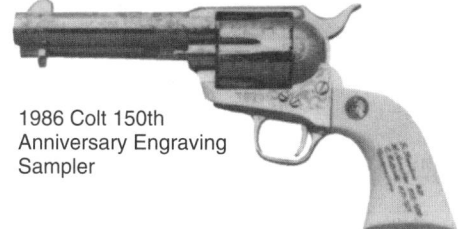

1986 Colt 150th Anniversary Engraving Sampler

REMINGTON 1987 CLASSIC

Same specs as other Model 700s in Classic Series except 338 Win. Magnum only. Approximately 3000 made in this caliber only in 1987. **Issue Price: $421.**
Market Value: $450

WALTHER 100TH YEAR COMMEMORATIVE

Semi-automatic; 9mm Para.; same basic specs as P-5 model except heavy engraving. Marketed in walnut presentation case. Number made, original price unknown. Introduced, 1987.
Market Value: $2350

WINCHESTER MODEL 70 XTR GOLDEN ANNIVERSARY

Similar to sporter magnum except select walnut classic-style stock old-style swivel bases, steel floorplate, grip cap, trigger guard; hand-engraving on barrel, receiver, magazine cover, trigger guard; inscription on barrel, "The Rifleman's Rifle 1937-1987"; hand engraved receiver, floorplate, trigger guard, grip cap, crossbolt; chambered for 300 Win. Magnum only; 500 made. **Issue Price: $939.**
Market Value: $950

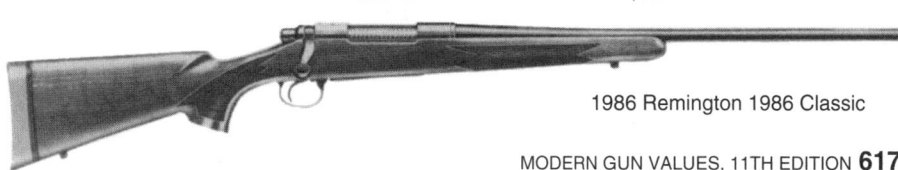

1986 Remington 1986 Classic

COMMEMORATIVES

1988
American Historical Foundation
American Armed Forces Uzi
Commemorative
American Historical Foundation
Commemorative Texas Paterson
American Historical Foundation
M14 Vietnam Commemorative
American Historical Foundation
Law Enforcement
Commemorative Thompson
American Historical Foundation
Teddy Roosevelt
Commemorative
American Historical Foundation
U.S. Army Commemorative
American Historical Foundation
Vietnam War M16 Commemorative
America Remembers U.S.
Cavalry Model
Browning Model 12 Limited
Edition
Colt Heirloom Limited Edition
Colt U.S. Marshals
Bicentennial Commemorative
Krico 720 Limited Edition
Remington 1988 Classic

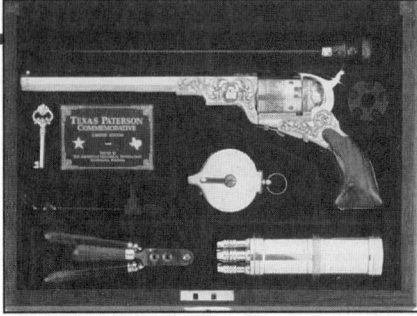

1988 American Historical Foundation
Commemorative Texas Paterson

1988 American Historical Foundation M14
Vietnam Commemorative

1988 American Historical Foundation
Teddy Roosevelt Commemorative

1988 Colt Heirloom
Limited Edition

1988

AMERICA REMEMBERS U.S. CAVALRY MODEL
Limited edition blackpowder Colt Model 1860 replica; 44 blackpowder revolver; 8" barrel; gold-etched scene on cylinder; stag grips. Marketed in presentation case with replica brass cavalry buckle. Reported 2500 made in U.S., 1988. **Issue Price: $1650.**
   Market Value: $895

AMERICAN HISTORICAL FOUNDATION AMERICAN ARMED FORCES UZI COMMEMORATIVE
Semi-automatic; 9mm Para.; basic specs of standard Uzi except highly polished blued finish; gold-plated magazine release, barrel ring nut, pistol-grip bushing, sear and trigger pivot pins, trigger, stock screws and nuts, sling keepers and cocking knob; furniture-finished woodstock carrying medallion with words, "The American Armed Forces Around The World." Marketed with gold-plated sight-adjustment tool. Reported 1500 made under special Israeli Military Industries contract, 1988. **Issue Price: $2195.**
   Market Value: $1500

AMERICAN HISTORICAL FOUNDATION COMMEMORATIVE TEXAS PATERSON
Five-shot revolver; blackpowder; 36-caliber; part of Samuel Colt Golden Tribute Collection, commemorates 150th anniversary of gun's production; gold-plated, hand-engraved in Italy by Pedersoli; hand-fitted burled walnut grips; 9" octagonal barrel. Marketed with combination tool, extra cylinder. Reported 950 made, 1988. **Issue Price: $1495.**
   Market Value: $1495

AMERICAN HISTORICAL FOUNDATION M14 VIETNAM COMMEMORATIVE
Semi-automatic; 7.62mm NATO; same specs as M14 military issue except high-quality Fajen stock; gold-plated components; patriotic inscriptions etched and gold-filled on operating rod and receiver. Made in two configurations,

500 Deluxe Museum Edition and 500 Collector Edition by Federal Arms under special contract, 1988, with special serial numbers. Deluxe Museum Edition, **Issue Price: $1,595; Collector Edition, $1,395.**
Deluxe Museum Edition
   Market Value: $1195
Collector Edition
   Market Value: $995

AMERICAN HISTORICAL FOUNDATION LAW ENFORCEMENT COMMEMORATIVE THOMPSON
Semi-automatic; 45 ACP; same specs as standard 1928 Thompson subgun except highly polished American walnut buttstock, grip and vertical foregrip; gold-plated frontsight, Cutts compensator; sling swivels. Receiver roll-engraved, gold-filled legend; police shield medallion in buttstock, rear grip carries medallion with police motto. Marketed with night stick carrying same serial number as gun. Reported 1500 made on special order by Auto Ordnance, 1988. **Issue Price: $1595.**
   Market Value: $1295

AMERICAN HISTORICAL FOUNDATION TEDDY ROOSEVELT COMMEMORATIVE
Revolver; 45 Colt; same specs as Colt Single Action Army except 7 1/2" barrel; gold-plated ejector rod housing, trigger, hammer and cylinder; all other external metal parts sterling silver-plated; ivory/polymer grips, with buffalo head raised on right panel; TR monogram on left; extensively engraved in Nimaschke pattern of Roosevelt's original SAA. Reported 750 made in Italy by Uberti, 1988. **Issue Price: $1,995**
   Market Value: $1495

AMERICAN HISTORICAL FOUNDATION U.S. ARMY COMMEMORATIVE
Semi-automatic; 45 ACP; same specs as standard Model 1911A1 except ten gold-plated parts; slide etched, gold-filled with U.S. Army seal, legend; Herrett burled walnut grips with Army medallion. Reported 1911 made on special order by Auto Ordnance, 1988. **Issue Price: $995**
   Market Value: $750

AMERICAN HISTORICAL FOUNDATION VIETNAM WAR M16 COMMEMORATIVE
Semi-automatic; 5.56mm; same basic specs as M16 military issue except high-gloss black finish; gold-plated flash suppressor, trigger, selector lever, bolt catch, rear sight windage knob, take-down pins, sling swivels; pistol grip and buttstock carry commemorative medallions. Marketed with adjustable black leather military-type sling. Reported 1,500 made by Colt on special order, 1988. **Issue Price: $1995.**
   Market Value: $1795

BROWNING MODEL 12 LIMITED EDITION
Slide-action shotgun; 20-, 28-ga.; 2 3/4" chamber; 26" barrel; reproduction of the Winchester Model 12; has cut-checkered walnut stock; high-post floating rib, grooved sight plane; crossbolt safety in trigger guard; polished blue finish. Limited edition; 8500 Grade I, 4000 Grade V guns made in Japan in 1988. Imported by Browning. **Issue Price: Grade I, 20-ga., $734.95; Grade I, 28-ga., $771.95; Grade V, 20-ga., $1187; Grade V, 28-ga., $1246.**
Mint Grade I
   Market Value: $650
Mint Grade V
   Market Value: $995

COLT HEIRLOOM LIMITED EDITION
Officers Model 1911A1; 45 ACP; produced in Colt Custom Shop on individual order, with serial numbers to order, combining numbers with family names or initials; full mirror finish; Combat Commander hammer; extended trigger; wide combat grip safety; ambidextrous safety; jeweled barrel, hammer and trigger; ivory grips; special grip medallion. Marketed in mahogany presentation case. Made in 1988-89. **Issue Price: $1500.**
   Market Value: $1395

COLT U.S. MARSHALS BICENTENNIAL COMMEMORATIVE
Single Action Army revolver; 45 Colt; smooth walnut grips; 7 1/2" barrel; blued finish; U.S. Marshal medallion set in left grip panel; on left side of the barrel, marking is "U.S. Marshals Bicentennial 1789-1989"; on right side, legend reads "One of Five Hundred." Marketed in glass-topped case, with French blue velvet lining; U.S. Marshal's badge included. Produced for U.S. Marshals Foundation; 500 made in 1988. **Issue Price: $775.**
   Market Value: $1495

KRICO 720 LIMITED EDITION
Bolt-action rifle; 270; 20 1/2" barrel; full-length Mannlicher-type stock; gold-plated trim and scrollwork on metal parts; serial number inlaid in gold. Made in Germany. Imported by Beeman Precision Arms, 1988. **Issue Price: $2375.**
   Market Value: $2300

REMINGTON 1988 CLASSIC
Same specs as earlier rifles in the series except 35 Whelen. Approximately 7000 made in this caliber in 1988 only. **Issue Price: $440.**
   Market Value: $495

1989 Colt Wyoming Centennial

1989 American Historical Foundation
Browning Model 1885 Deerhunter
Limited Edition

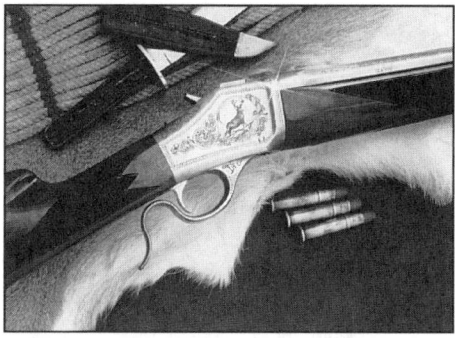

1988 Browning Model 53
Limited Edition

1989 Mossberg National Wild Turkey
Federation Limited Edition

1989

AMERICAN HISTORICAL FOUNDATION BROWNING MODEL 1885 DEERHUNTER LIMITED EDITION

Lever-action; 45-70; same specs as original Winchester 1885 except has complete engraving coverage of receiver featuring a deer in a wreath; hand-finished walnut stock, forearm. Reported 100 made, 1989. **Issue Price: $2775.**
   Market Value: $1750

AMERICAN HISTORICAL FOUNDATION VIETNAM WAR COMMEMORATIVE

Semi-automtatic; 45 ACP; same specs as standard Model 1911A1 except gold-plated hammer, trigger, safety, slide stop, magazine catch, magazine catch lock and grip screws; slide engraved with Oriental dragons, carries bannered legend; American oak grips with medallion. Reported 2500 made by Auto Ordnance on special order, 1989. **Issue Price: $1095.**
   Market Value: $795

BROWNING WYOMING CENTENNIAL

Model 1885 lever-action rifle; 25-06; same basic specs as standard model except tapered octagon barrel, No. 5 walnut stock, Wyoming state seal, centennial logo inlaid in gold. Reported 2000 made in 1989. Marketed by Sports, Inc. **Issue Price: $1200.**
   Market Value: $1200

COLT BEVERLY HILLS POLICE COMMEMORATIVE

Colt Combat Government Model; 45 ACP; commemorates the 75th anniversary of California's Beverly Hills Police Department; gold-plated scene on both sides of slide includes city hall, police car, motorcycle officer foot patrolman; city insignia on rear of slide in gold; city seal inset in grips; gold-bannered commemorative buckle included; special serial numbers, BHPD001 to 150; 150 guns made in 1989. Marketed in red velvet-lined presentation case. **Issue Price: $785.**
   Market Value: $950

COLT JOE FOSS LIMITED EDITION

Government Model auto; 45 ACP; blued finish; fixed sights; smooth walnut grips; left side of the slide covers Foss' life through WWII with gold-plated figures of farmhouse, Marine Corps seal, Marine aviator's wings, cannon and Grumman aircraft; at rear of the slide is bordered rampant colt. On the slide's right side, the gold-plated figures represent Foss' post-WWII career, including South Dakota capitol, a hunter, Mount Rushmore figures, and an F16 fighter aircraft, with words "A Tribute To Joe Foss" Serial numbers are JF0001 through JF2500; 2500 made in 1989. Marketed in walnut presentation case with red velvet lining. Included are brass plate and medallion. **Issue Price: $1375.**
   Market Value: $995

COLT SNAKE EYES SET

Colt Python models; 357 Magnum; two guns in set; 2 1/2" barrels; one gun has blued finish, other is stainless steel; Snake Eyes name appears on left side of barrels; ivory-like grips with scrimshaw snake eyes dice on left side, royal flush poker hand on right; matching serial numbers EYES001 through EYES500 positioned on butt of each gun; glass-front presentation case is etched with gold letters, trim; interior of case is lined with green felt; interior lighting; Colt-marked poker chips, playing cards are French-fitted. 500 sets made in 1989. **Issue Price: $3500 per set.**
   Market Value: $1995

COLT WYOMING CENTENNIAL

Single Action Army revolver; 45 Colt; 4 3/4" barrel; rosewood custom grips; gold-plated rampant colt medallion; left side of barrel carries roll-marked gold-washed logo with "1890 Wyoming Centennial." Unfluted cylinder carries gold-inlaid state seal on one side, centennial logo on the other. Only 200 made in 1989. Marketed in French-fitted walnut case with commemorative plate by Sports, Inc., Lewiston, Montana. **Issue Price: $1250.**
   Market Value: $1395

MOSSBERG NATIONAL WILD TURKEY FEDERATION LIMITED EDITION

Same specs as standard Mossberg Model 835 Ulti-Mag 12-ga. pump-action shotgun, except camo finish; Wild Turkey Federation medallion inlaid in the stock. Marketed with a 10-pack of Federal 3 1/2" turkey-load shotshells. Made in 1989. **Issue Price: $477.**
   Market Value: $295

REMINGTON 1989 CLASSIC

Same specs as earlier caliber chamberings in the series except 300 Weatherby Magnum. Approximately 6000 made in this caliber only. **Issue Price: $485.**
   Market Value: $495

1990

AMERICA REMEMBERS FREDERICK REMINGTON MODEL 1886

Revolver; honors the famed Western artist's 100th anniversary as an associate of the National Academy of Design; 44 blackpowder 1886 Army replica; gold etching on 8" barrel, trigger, cylinder frame and gripstraps. Marketed with cased accessories. Reported 1000 made, 1990. **Issue Price: $1500.**
   Market Value: $750

AMERICA REMEMBERS LIMITED EDITION TEXAS RANGER DRAGOON

Blackpowder 44 replica revolver; 8" barrel; silverplating on gripstraps, trigger guard and cylinder; colorcase-hardened frame, loading lever; gold etched barrel, frame. Marketed in case with accessories. Reported 1000 made in U.S., 1990. **Issue Price: $1585.**
   Market Value: $995

AMERICA REMEMBERS ROY ROGERS COWBOY EDITION

Single-Action Army 45 Colt revolver; 4 3/4" barrel; gold-plated cylinder; gold etchings on barrel, gripstrap, frame; stag grips. Marketed in presentation case. Reported 2500 made, 1990. **Issue Price: $1550.**
   Market Value: $1495

American Remembers Roy Rogers Premier Edition

Same general specs as Roy Rogers Cowboy Edition except extensive inlays and heavy engraving. Reported 250 made, 1990. **Issue Price: $4500.**
   Market Value: $2995

COMMEMORATIVES

1990

America Remembers Roy Rogers Premier Edition
Browning Model 53 Limited Edition
Colt EL Presidente
Kimber 10th Anniversary Issue
Remington 1990 Classic
Smith & Wesson Model 28 Magnaclassic
Smith & Wesson Model 629 Magnaclassic
Winchester Wyoming Centennial Commemorative

1991

America Remembers Interpol Colt SAA
American Historical Foundation WWII Commemorative Hi-Power
Glock Desert Storm Commemorative
Iver Johnson 50th Anniversary M-1 Carbine
Remington Model 11-87 175th Anniversary Commemorative
Remington 1991 Classic
Remington 7400 175th Anniversary Rifle
Savage Model 110WLE Limited Edition
Winchester 125th Anniversary Commemorative Model 84

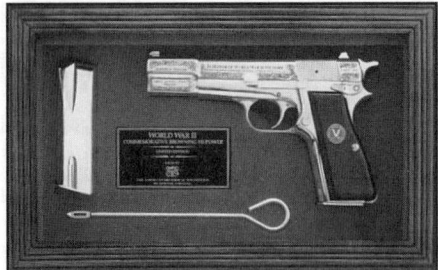

1991 American Historical Foundation WWII Commemorative Hi-Power

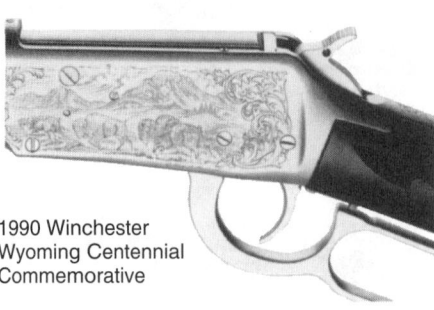

1990 Winchester Wyoming Centennial Commemorative

1991 Winchester 125th Anniversary Commemorative Model 94

1990 (cont.)

AMERICA REMEMBERS ROYAL ARMORIES DRAGOON

Replica of 44 blackpowder 2nd Model Dragoon revolver made in England for Sam Colt; 7 1/2″ barrel; hand-engraved case-hardened action; walnut grips with sterling silver plaque. Reported 1000 made, 1990. **Issue Price: $2450.**
Market Value: $1995

BROWNING MODEL 53 LIMITED EDITION

Rifle; 32-20; 7-shot half-length magazine; 22″ round, tapered barrel; 39.5″ overall length; weighs 6.8 lbs.; post bead front sight, open adjustable rear; cut-checkered select walnut stock, forend; metal grip cap; blued finish. Only 5000 produced in Japan. Based on Winchester Model 92 design. Introduced, 1990; dropped, 1991. **Issue Price: $675.**
Market Value: $595

COLT EL PRESIDENTE

Dressed-up version of the Colt Government Model 38 Super; highly polished Ultimate stainless finish; Pearlex grips; "El Presidente" and rampant colt are engraved on left side of the slide. Only 350 made in 1990 for Lew Horton Distributing. Marketed in special soft case with Colt logo. **Issue Price: $800.**
Market Value: $995

KIMBER 10TH ANNIVERSARY ISSUE

Bolt-action rifle; 223 Remington; French walnut stock with shadow cheekpiece, slim forend; 22″ barrel; round-top receiver; steel buttplate; scope mounts furnished. Introduced, 1990 in extremely limited numbers.
Market Value: $1600

REMINGTON 1990 CLASSIC

Same specs as previous 700 Classic models except 25-06 only; 4-shot magazine; 24″ barrel. Made only in 1990. **Issue Price: $520.**
Market Value: $495

SMITH & WESSON MODEL 28 MAGNACLASSIC

Same specs as standard Model 29 except 7 1/2″ full lug barrel; highly polished finish; target hammer, trigger; ergonomic Goncalo Alves grips with carnuba finish. Marketed in cherrywood case with gold-embossed leather lid, fitted velvet interior, black leather sight box, sight assortment, test target, certificate of authenticity; factory accurized; barrel laser-etched, "1 of 3000." 1550 were made in 1990. **Issue Price: $999.**
Market Value: $995

Smith & Wesson Model 629 MagnaClassic

Same specs as Model 29 MagnaClassic except stainless steel; same markings; marketed with same accessory package. 1450 were made in 1990. **Issue Price: $999.**
Market Value: $995

WINCHESTER WYOMING CENTENNIAL COMMEMORATIVE

Model 94 30-30; features coin-type finish on receiver, barrel bands; receiver engraved both sides with Wyoming scenes; left side carries engraved banners, "1890-1990" and "Wyoming Centennial"; 20″ barrel; full buckhorn Marble sight; special serial numbers; only 999 made. Marketed exclusively by Cherry's. **Issue Price: $895.**
Market Value: $1095

1991

AMERICA REMEMBERS INTERPOL COLT SAA

Single-action revolver; 45 Colt; 4 3/4″ barrel; engraving replica of that on Kornblath Interpol Model; silver-plated, with faux ivory grips; marketed in book-like leather case. Only 154 manufactured by Colt on special order, 1991. **Issue Price: $4500.**
Market Value: $2995

AMERICAN HISTORICAL FOUNDATION WWII COMMEMORATIVE HI-POWER

Semi-automatic; 9mm Para.; same specs as standard Hi-Power model except overall gold plating; carries deep-etched inscriptions, scrollwork; custom-finished walnut grips; inset with victory medallion. Marketed with extra magazine and cleaning rod, both gold-plated. Reported 500 made on special order by Browning 1991. **Issue Price: $1795.**
Market Value: $1295

GLOCK DESERT STORM COMMEMORATIVE

Built on the Glock 17 9mm Parabellum configuration; legend on the right side of the slide reads, "Operation Desert Storm January 16 - February 17, 1991"; on the left side is "New World Order" and the Glock 17 logo. On top of the slide are listed the 30 countries of the United Nations that took part in the campaign. Marketed with a field knife in a glass case bearing the Desert Storm shield. Only 1000 made in 1991. **Issue Price: $900.**
Market Value: $900

IVER JOHNSON 50TH ANNIVERSARY M-1 CARBINE

Semi-automatic; 30 U.S. Carbine; same specs as standard version except deluxe American walnut stock; red, white and blue American enameled flag embedded in stock; gold-filled, roll engraving with words "50th Anniversary 1941-1991" on slide; Parkerized finish. Introduced, 1991; still available. **Issue Price: $384.95.**
Market Value: $375

REMINGTON 1991 CLASSIC

Same Model 700 specs as other rifles in the annual series except 7mm Weatherby Magnum only; 4-shot magazine. Produced only in 1991.
Market Value: $495

REMINGTON 7400 175TH ANNIVERSARY RIFLE

Same specs as the standard Remington Model 7400 except receiver engraved with company's 175th Anniversary scroll design, American eagle; high-gloss American walnut stock; deep-cut checkering; 30-06 only. Introduced, 1991; dropped, 1992.
Market Value: $350

REMINGTON MODEL 11-87 175TH ANNIVERSARY COMMEMORATIVE

Same specs as the Model 11-87 Premier except engraved receiver carries anniversary design with American eagle; walnut stock, forend with high-gloss finish; 28″ vent-rib barrel; 12-ga. only. Marketed with Rem Choke tubes. Made in 1991. **Issue Price: $618.**
Market Value: $450

SAVAGE MODEL 110WLE LIMITED EDITION

Same specs as Model 110G except 250-3000, 300 Savage, 7x57; high-luster fancy-grade American walnut stock, cut checkering, swivel studs, recoil pad, polished barrel; Savage logo is laser-etched on bolt. Marketed with gun lock, shooting glasses, ear plugs, sight-in target. Introduced, 1991; still available.
Market Value: $350

WINCHESTER 125TH ANNIVERSARY COMMEMORATIVE MODEL 84

Rifle; 30-30; features John Ulrich-style, full-coverage engraving; right side has 24-karat gold relief bull moose; left side has gold figures of buck and doe; 20″ octagon barrel, lever are engraved, gold inlaid; full buckhorn rear sight; hand-checkered American walnut stock; marketed in French-fitted walnut presentation case exclusively by Cherry's. Only 125 made in 1991. **Issue Price: $4995.**
Market Value: $5500

1992

AMERICAN HISTORICAL FOUNDATION 40TH ANNIVERSARY RUGER COMMEMORATIVE

Semi-automatic; 22 LR; same specs as Ruger Mark II except faux ivory grips with red Ruger medallion inset in right grip, black medallion in left grip; gold-plated trigger, trigger pivot, hammer pivot, grip-screws, bolt handle, bolt stop, scrollwork; 40th Anniversary inset in barrel in gold. Reported 950 made, 1992. **Issue Price: $995.**
Market Value: $495

AMERICAN HISTORICAL FOUNDATION 50TH ANNIVERSARY WW II COLT

Semi-automatic; 45 ACP; same basic specs as Colt Model 1911A1; blued finish; faux ivory grips; issued in honor of six major battles in Pacific, six in Europe; explanatory legend etched in gold on slide; grips carry historical quotations, details of campaigns, rampant colt insignia. Reported 250 pistols made by Colt, 1992, for each of the twelve campaigns; guns numbered 001 through 250 with prefix appropriate to each campaign. Marketed in OD-lined presentation case with serially numbered Basic Field Manual. **Issue Price: $995.**
Market Value: $995

AMERICAN HISTORICAL SOCIETY SAMUEL COLT GOLDEN TRIBUTE BUNTLINE

Revolver; 45 Colt; same basic specs of Colt Single Action Army model except 12″ barrel; all metal parts gold-plated; heavy engraving in Texas scroll pattern; hand-fitted walnut grips. Marketed in walnut wall display case with glass lid, brass hardware. Reported 950 made in Italy by Uberti, 1992. **Issue Price: $2195.**
Market Value: $1495

AMERICA REMEMBERS COWBOY HALL OF FAME REVOLVER

Single-action revolver; 45 Colt; 4³/₄″ barrel; hand-fitted stag grips; blued steel barrel, trigger guard, gripstraps; gold-plated frame; gold-etched scenes on cylinder. Marketed in velvet-lined display case with glass lid. Marketed in conjunction with the Cowboy Hall of Fame, Oklahoma City. Reported 1000 manufactured in 1992. **Issue Price: $1600.**
Market Value: $1600

AMERICA REMEMBERS EISENHOWER COMMEMORATIVE

Semi-automatic; 45 ACP; Springfield Armory-made Model 19911A1; 5″ barrel; 7-shot magazine; checkered walnut grips with inset bronze medallions; blued slide carries gold decoration. Marketed in velvet-lined display case with Eisenhower's signature etched in walnut lid; framed photo of Ike autographed by his son, John. Reported 2500 made, 1992. **Issue Price: $1675.**
Market Value: $995

AMERICA REMEMBERS MEL TORME COLT SAA

Single-action revolver; 45 Colt; 5 ¹/₂″ barrel; mother of pearl grips; silver-inlaid engraving; gold-inlaid highlights. Marketed in leather-bound display case bearing gold-embossed signature of singer/gun fancier Torme. Reported 100 manufactured by Colt on special contract, 1992. **Issue Price: $4500**
Market Value: $2995

AMERICA REMEMBERS RICHARD PETTY SILVER EDITION

Single-action revolver; 45 Colt; blued steel; sterling silver decoration; faux ivory grips with incised signature. Reported 1000 made by Colt in 1992. **Issue Price: $1495**
Market Value: $1650 to $1700

SPRINGFIELD ARMORY GULF VICTORY SPECIAL EDITION

Semi-automatic; 45 ACP; same specs as Model 1911A1 except decoration; highly blued finish; gold plating on slide, small parts; marketed in padded, embroidered carrying case with jacket patch, window decal and medallion. Number made unknown. **Issue Price: $869.**
Market Value: $775

WINCHESTER ARAPAHO COMMEMORATIVE

Lever-action carbine; 30-30; same specs as standard Model 94 carbine except varnished select walnut stock, decoration; lever, receiver, barrel bands gold-plated, other metal parts blued; right side of the receiver etched with scene of Arapaho life, left side carries scene of buffalo hunt with a banner reading, "The Arapaho Commemorative." Reported 500 made exclusively for Cherry's by USRAC/Winchester, 1992. **Issue Price: $895.**
Market Value: $1095

WINCHESTER KENTUCKY BICENTENNIAL COMMEMORATIVE

Lever-action carbine; 30-30; same specs as standard Model 94 except receiver engraved in Italy, then bone charcoal case-colored; checkered stock, forearm. Banner on left side of receiver reads, "Kentucky Bicentennial 1792-1992"; racehorse etchings also featured. On right side is a Kentucky horse barn. Reported 500 made by USRA/Winchester exclusively for Cherry's, 1992. **Issue Price: $995**
Market Value: $1095

WINCHESTER ONTARIO CONSERVATION COMMEMORATIVE

Lever-action carbine; 30-30; same specs as standard Model 94 except pewter finish of some parts; fancy walnut stock, decoration; right receiver panel carries Ontario game scene and portrait of conservation officer; left panel carries etching of moose, beavers in wilderness setting; replica of conservation officer's badge is set in stock. Reported 500 made by USRA/Winchester exclusively for Cherry's, 1992. **Issue Price: $795.**
Market Value: $1095

1992 Springfield Armory Gulf Victory Special Edition

1992 Winchester Kentucky Bicentennial Commemorative

COMMEMORATIVES

1993

America Remembers
Hpoalong Cassidy Cowboy
Edition
America Remembers
Hopalong Cassidy Premier
Edition
America Remembers George
Jones SAA
America Remembers "Will
Penny" Charlton Heston
Tribute
America Remembers Don't
Give Up The Ship Model
America Remembers
American Eagle Colt
American Historical Foundation
Special Edition M16 Vietnam
Tribute
American Historical Foundation
Showcase Edition Colt M16
American Historical Foundation
General Robert E. Lee Henry
Repeater
American Historical Foundation
Special Edition S&W Model
629 Hunter
Winchester NEZ PERC
Commemorative

1993

AMERICA REMEMBERS AMERICAN EAGLE COLT
Semi-automatic; 45 ACP; Colt-made Model 1911A1; 5″ barrel; blued finish; gold/silver bald eagle etched on each side; gold-plated safety, sights, hammer, grip screws. Marketed in blue velvet-lined walnut-case with American flag and brass identification plate. Reported 2500 made, 1993. **Issue Price: $1950.**
    Market Value: $1950 to $2000

AMERICA REMEMBERS DON'T GIVE UP THE SHIP MODEL
Semi-automatic; 45 ACP; Colt-made Model 1911A1; 4″ barrel; blued finish; gold-etched on the slide are USS Constitution, USS Ticonderoga, plus appropriate legend; rosewood grips with Colt medallion. Reported 1997 made, 1993. **Issue Price: $1570.**
    Market Value: $995

AMERICA REMEMBERS GEORGE JONES SAA
Single-action revolver; 45 Colt; blued steel finish, decorated with gold inlays of music scores; Jones' name on barrel; bonded pearl grips. Reported 950 made, 1993. **Issue Price: $1575.**
    Market Value: $1495

AMERICA REMEMBERS HOPALONG CASSIDY COWBOY EDITION
Single-action revolver; 45 Colt; 5 1/2″ barrel; grips of black water buffalo horn; blued finish; frame, barrel, cylinder decorated with sterling silver. Marketed in velvet-lined cherrywood display case. Reported 950 manufactured, 1993. **Issue Price: $1675.**
    Market Value: $1650

America Remembers Hopalong Premiere Edition
Same specs as Cowboy Edition except in addition to silver decoration it has deep engraving, gold inlays and a leather case embossed with William Boyd's monogram. Only 100 made, 1993. **Issue Price: $4500.**
    Market Value: $2995

AMERICA REMEMBERS "WILL PENNY" CHARLTON HESTON TRIBUTE
Single-action revolver; 45 Colt; 5 1/2″ barrel; blued steel; silver and gold decoration on metal, including cattledrive scene. Minted gold portrait medallion of Heston inset in single-piece walnut grips. Reported 500 manufactured in 1993. **Issue Price: $1850.**
    Market Value: $995

AMERICAN HISTORICAL FOUNDATION GENERAL ROBERT E. LEE HENRY REPEATER
Lever-action rifle; 44-40; 44″ overall length; same specs as original Henry repeating rifle except receiver scrolled with dogwood, cotton flowers, magnolias; high-gloss blued finish; gold-plated trigger, follower, front sight blade and lifter; oil-finished walnut stock; left side of receiver with gold-filled Confederate seal, right side with portrait of Lee and Confederate flag. Reported 250 made in Italy by Uberti, 1993. **Issue Price: $2495.**
    Market Value: $1500

AMERICAN HISTORICAL FOUNDATION SHOWCASE EDITION COLT M16
Semi-auto; 5.56mm; same specs as standard Colt M16 military version except receiver, sights, barrel mirror polished or glass-beaded, then nickel-plated; plastic stock has heavily textured finish; sterling silver plaque inset in stock reads "The Showcase Edition Colt One of One Hundred"; same legend etched on magazine well, surrounding rampant colt. Reported 100 with special serial numbers made by Colt, 1993. **Issue Price: $1995.**
    Market Value: $1795

AMERICAN HISTORICAL FOUNDATION SPECIAL EDITION M16 VIETNAM TRIBUTE
Semi-auto M16; 5.56mm; HBAR match configuration; heavy barrel; custom-textured buttstock, pistol grip and hand guard; gold-plated trigger, magazine release, safety, pivot pins, windage knob, elevation knob, flash suppressor, sling swivels; medallion in pistol grip displays Vietnam ribbon and legend. Reported 1500 made, 1993, on special order by Colt, serial numbers VI0001 through VI1500. **Issue Price: $1995.**
    Market Value: $1795

AMERICAN HISTORICAL FOUNDATION SPECIAL EDITION S&W MODEL 629 HUNTER
Revolver; 44 Magnum; same specs as standard Model 629 except 6″ Mag-Na-Ported barrel with integral scope base, variable underweight; heavily engraved; finger-grooved Goncalo Alves grips. Marketed with Tasco scope mounts, 2x Nikon pistol scope, S&W Performance Center shooting bag, Tasco shooting glasses, Silencio ear plugs. Only 50 made by S&W under special order, 1993. **Issue Price: $3795.**
    Market Value: $1795

WINCHESTER NEZ PERC COMMEMORATIVE
Lever-action carbine; 30-30; same specs as standard Model 94 except for decoration; nickel-finished receiver, lever, barrel bands; other parts blued; left side of the receiver etched with bust of Chief Joseph, map of tribe's travels; right side features Nez Perc encampment against mountain background. Reported 600 made by USRAC/Winchester on special order, 1993. Serial numbered NEZ001 through NEZ600. **Issue Price: $950.**
    Market Value: $1095

1994 American Historical Foundation
Smith & Wesson Bank Note Limited Edition

1994 American Historical Foundation
Colt Heritage Edition SAA

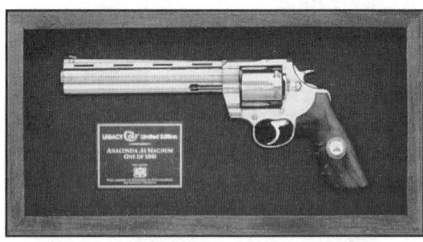

1994 American Historical Foundation
Limited Edition Colt Anaconda

1993 American Historical Foundation
Showcase Edition Colt M16

1994

AMERICA REMEMBERS AMERICAN INDIAN TRIBUTE CARBINE

Rifle; 45 Colt; same specs as Winchester Model 94 Wrangler except gold-plated loop-style lever; gold-plated barrel bands, action; right side of receiver with buffalo highlighted in nickel, Indian-head penny mounted ahead of ejection port; left side features buffalo hunt in nickel, with buffalo-head nickel adjoining. Reported 300 made by USRA, 1994. **Issue Price: $1295.**
Market Value: $1195

AMERICA REMEMBERS ARMY AIR FORCES TRIBUTE

Semi-automatic; 45 ACP; Colt-made Model 1911A1; decoration etched on left side of the slide, including Air Corps wings and Master of the Sky legend, nickel-plated; remainder of slide gold-plated; right side carries insignia of all sixteen Army Air combat units of the World War II era; gold-plated safety, sights and grip screws; rosewood custom grips inset with Colt medallion. Reported 500 made, 1994. **Issue Price: $1500.**
Market Value: $995

AMERICA REMEMBERS GETTYSBURG COMMEMORATIVE

Revolver; 44-caliber replica of Colt 1860 Army Model; 8" barrel; blued finish; select walnut grips; cylinder etched with panorama of Pickett's charge; barrel marked with gold, Gettysburg, Pennsylvania, July 1863 and line from Lincoln's address. Marketed in velvet-lined walnut display case; inset leather lid carries silhouette of a general, with "Gettysburg 1863" stamped in gold. Reported 1863 made, 1994. **Issue Price: $1350.**
Market Value: $995

AMERICA REMEMBERS JOHNNY CASH PATERSON

Revolver; 36-caliber blackpowder Colt Paterson replica; 12 1/2" overall length; bonded pearl grips; blued finish; decorated with 24-karat gold; titles of Cash's hit songs on barrel. Marketed in black-finished hardwood display case bearing Cash's signature. Included was a copy of Cash's biography. Reported 1000 made, 1994. **Issue Price: $1350.**
Market Value: $995

AMERICA REMEMBERS MARINE PACIFIC TRIBUTE

Semi-automatic; 45 ACP; Model 1911A1 design; 5" barrel; gold-plated slide, safety, grip screws, sights; slide etched with WWII battles on right side, appropriate legend and Iwo Jima flag raising on left; custom rosewood grips with gold-plated Colt medallion; 500 made by Colt, 1994. **Issue Price: $1500.**
Market Value: $995

AMERICA REMEMBERS PACIFIC NAVY TRIBUTE

Semi-automatic; 45 ACP; Model 1911A1 design; 5" barrel; gold-plated slide, safety, grip screws; slide etched with appropriate inscriptions, classes of ships that participated in the Pacific in World War II; custom-checkered rosewood grips inset with U.S. Navy seal. Reported 500 made on special order by Auto Ordnance,1944. **Issue Price: $1500.**
Market Value: $995

AMERICA REMEMBERS WWII GOLDEN ANNIVERSARY VICTORY TRIBUTE

Semi-automatic; Model 1911A1 design; 5" barrel, gold-plated slide, safety, sights, grip screws; slide etched with regimental badges of units under Eisenhower's command on D-Day; rosewood grips inset with gold commemorative medallion. Reported 500 made on special order by Colt, 1994. **Issue Price: $1500.**
Market Value: $1995

AMERICAN HISTORICAL FOUNDATION 2ND AMENDMENT BROWNING HI-POWER

Semi-automatic; 40 S&W; dedicated to the right to bear arms; same specs as standard Browning Hi-Power except silver plating overall, with gold-plated hammer, trigger, magazine release, safety, slide release, grip screws; scroll-etched frame, slide; polished rosewood grips, 2nd Amendment medallion inset in left grip. Reported 500 made on special order by Browning, 1994. **Issue Price: $1995.**
Market Value: $1295

AMERICAN HISTORICAL FOUNDATION COLT HERITAGE EDITION SAA

Revolver; 45 Colt; 4 3/4" barrel; honors Sam Colt and company history; same specs as standard Single Action Army model except heavy scroll engraving on frame barrel, cylinder, hammer; gold inlaid decoration, including rampant colt on recoil shield, company insignia on loading gate, bronc rider on frame, Colt 45 Single Action Army bannered in gold inlay on left side of barrel; walnut grips; Colt grip medallions. Marketed in stand-up glass/walnut case. Reported 250 made by Colt, 1995. **Issue Price: $2995.**
Market Value: $2995

AMERICAN HISTORICAL FOUNDATION LIMITED EDITION COLT ANACONDA

Double-action revolver; 44 Magnum; 8" barrel; same specs as standard Anaconda except major parts plated with black and gold titanium; identifying legend, rampant colt trademark etched in left side of barrel, gold-filled; polished rosewood grips inset with special Colt medallion. Reported 1000 made, 1995; serial numbers LE0001 through LE1000. **Issue Price: $1595.**
Market Value: $995

AMERICAN HISTORICAL FOUNDATION SMITH & WESSON BANK NOTE LIMITED EDITION

Revolver; 45 ACP; stainless steel construction; same specs as S&W Model 625; 3" barrel, underlug; left side of frame has "45 hand-cut, surrounded by Bank Note engraving pattern; right side hand-engraved, "Bank Note .45"; etched on right side of barrel ".45 Cal. Model of 1989"; custom Hogue Bill Jordan grips of patented synthetic. Reported 100 produced, 1995. **Issue Price: $2195.**
Market Value: $1295

MARLIN CENTURY LIMITED MODEL 1894

Lever-action rifle; 44-40; 12-shot tube magazine; 24" tapered octagonal barrel; 40 7/8" overall length; weighs 6 1/2 lbs.; checkered semi-fancy American black walnut stock, forearm; brass buttplate; carbine front sight, adjustable semi-buckhorn rear; lever, bolt, lever engraved in Italy. Commemorates 100th anniversary of model. Reported 2500 made, 1994. **Issue Price: $1087.**
Market Value: $950

PEDERSOLI CREEDMORE COMMEMORATIVE

Single shot rifle; 45-70; 34" round/octagon barrel; hand-checkered American walnut stock; hand-fitted German silver forend cap; color case-hardened receiver, buttplate; tasteful engraving; hooded front sight, long Vernier tang rear; weighs 12 lbs., 2oz. Reported 280 made on special order for Cherry's, 1995. **Issue Price: $1495.**
Market Value: $1795

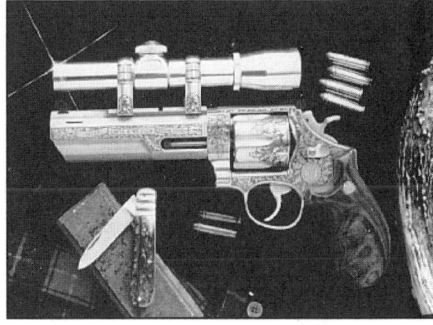

1993 American Historical Foundation Special Edition S&W Model 629 Hunter

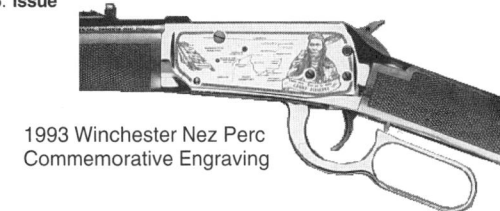

1993 Winchester Nez Perc Commemorative Engraving

1993 Winchester Nez Perc Commemorative

COMMEMORATIVES

1995

America Remembers 7th
Cavalry Tribute
America Remembers Ameri-
can Indian Tribute
America Remembers Tom
Mix Classic Edition
America Remembers Tom
Mix
Premier Edition
America Remembers Buffalo
Bill Sesquicentennial
America Remembers Navajo
Code Talkers Tribute
America Remembers Chuck
Yeager Tribute
America Remembers V-J Day
Tribute
America Remembers The
Confederate Enfield Limited
Edition
America Remembers Roy
Rogers Tribute

1995 America Remembers
Tom Mix Classic Edition

1995 America Remembers
Navajo Code Talkers Tribute

1995 America Remembers
American Indian Tribute

1995

AMERICA REMEMBERS 7TH CAVALRY TRIBUTE

Single-action revolver; 45 Colt; 7 1/2″ barrel; blued finish; 24-karat gold etchings of crossed sabers on each side of barrel, unit designation, mixed cavalry and Indian symbols on cylinder; Custer portrait laser-carved in faux ivory grips. Marketed in walnut presentation case with replica cavalry belt buckle. Reported 500 made, 1995. **Issue Price:** $1500
Market Value: $1495

AMERICA REMEMBERS AMERICAN INDIAN TRIBUTE

Single-action revolver; 45 Colt; 4 3/4″ barrel; genuine hand-fitted stag grips; blued finish; decorated with gold, nickel; left side of frame with Indian battle scene etched in gold; on right side etching of two bison. Individual edition number is stamped on right side of barrel, "The American Indian" on left. Marketed as a companion piece to 1994 issue American Indian Carbine in glass-topped cherrywood case, with brass plaque. Reported 300 made, 1995. **Issue Price:** $1495
Market Value: $1495

AMERICA REMEMBERS BUFFALO BILL SESQUICENTENNIAL

Single-action revolver; 45 Colt; 4 3/4″ barrel; stag grips; blued finish; left side of frame with Buffalo Bill's Wild West etched in gold, right side carries American bison, Chief Scout Cody in Cheyenne territory; cylinder with gold-etched art of Cody's career. Issue commemorates 150th anniversary of his birth. Reported 500 made, 1995. **Issue Price:** $1500.
Market Value: $1495

AMERICAN REMEMBERS CHUCK YEAGER TRIBUTE

Semi-automatic; 45 ACP; 5″ barrel; checkered rosewood grips; silver-plated with gold and blued decorations; left side carries etching of breaking the sound barrier and Yeager's signature; right side portrays subject as WWII fighter pilot with P-51 and Messerschmitt aircraft. Marketed in blue velvet-lined walnut case with signature laser-carved on lid. Reported 500 made on special order by Colt, 1995. **Issue Price:** $1750
Market Value: $995

AMERICA REMEMBERS NAVAJO CODE TALKERS TRIBUTE

Semi-automatic; 45 ACP; Colt-made Model 1911A1; 5″ barrel; blued finish; stippled slide, with engraved Navajo Code Talkers Commemorative in banner; ironwood grips inlaid with silver, red coral and turquoise, silver figures of code talkers and Iwo Jima flag raising. Marketed in ironwood case with David Yellowhorse custom-made knife featuring stippled, gold-plated bolsters, handle of silver, turquoise, red coral and ironwood inlaid with flag-raising. Reported 300 sets made, 1995. **Issue Price:** $3000.
Market Value: $1500

AMERICA REMEMBERS NORTH AMERICAN RODEO TRIBUTE

Lever-action carbine; 30-30; loop-style lever; 16″ barrel; other specs the same as standard Model 94 Winchester except for gold-finished barrel bands, gold etching on trigger, trigger guard, hammer. On left side of receiver, "The Great North American Rodeo" in gold banner, steer wrestling and bronc riding in gold against a nickel background; right side has bull rider and barrel racer featured in gold, nickel. Reported 500 made by USRA/Winchester, 1995. **Issue Price:** $1600.
Market Value: $1195

AMERICA REMEMBERS ROY ROGERS WINCHESTER TRIBUTE

Lever-action rifle; 30-30; same specs as standard Model 94 Winchester except jeweled hammer and breech-bolt, gold-plated lever, barrel bands and front sight hood; left side of receiver features portrait of Rogers in gold and nickel; right side with Rogers on rearing horse, six-guns leveled, also in nickel and gold. Reported 300 made on special order by USRA/Winchester, 1995. **Issue Price:** $2100.
Market Value: $1195

AMERICA REMEMBERS TOM MIX CLASSIC EDITION

Single-action revolver; 45 Colt; blued finish; 5 1/2″ barrel with TM brand in 24-karat gold; Mix brand also on each chamber of cylinder; gold signature on backstrap, portrait of Mix on recoil shield, repro of monument erected at the sight of his death on left; faux ivory grips carry Mix name, Stetson hat and TM brand. Marketed in walnut case with red velvet lining; included is Tom Mix comic book, circa 1948-1953. Reported 200 made by Uberti in Italy, 1995. **Issue Price:** $1395.
Market Value: $995

America Remembers Tom Mix Premier Edition

Same specs as the Classic except barrel carries gold-filled edition numbers 1 through 50. Each gun also furnished with an original-edition Tom Mix Big Little Book of the 1930s. Fifty guns made by Colt, 1995. **Issue Price:** $2595.
Market Value: $1750

AMERICA REMEMBERS THE CONFEDERATE ENFIELD LIMITED EDITION

Single shot muzzle-loading rifle; 58 caliber; overall length, 55″; barrel length, 39″; European walnut stock; blued finish; gold-finished buttplate, hammer and lock, trigger guard, barrel bands, barrel design. Replica of Confederate belt buckle inlaid in right side of stock. Marketed in wall-mount display case with Confederate gray lining and uniform buttons of the eleven Rebel states as decoration. Reported 500 made, 1995. **Issue Price:** $1695.
Market Value: $1250

AMERICA REMEMBERS V-J DAY TRIBUTE

Semi-automatic; 45 ACP; Model 1911A1; 5″ barrel; checkered rosewood grips inset with gold Colt medallion; blued finish, except for safety, hammer

1995 America Remembers
The Confederate Enfield Limited Edition

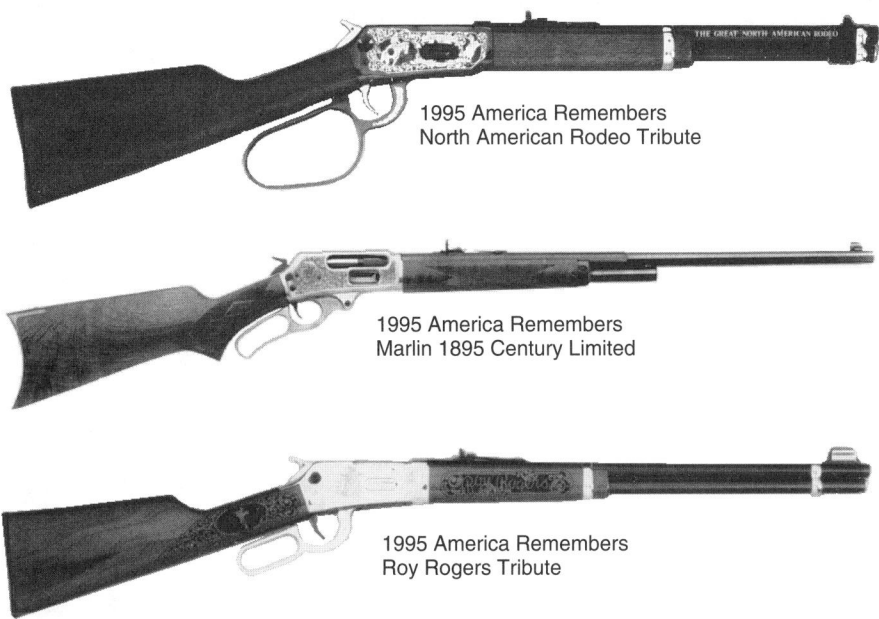

1995 America Remembers
North American Rodeo Tribute

1995 America Remembers
Marlin 1895 Century Limited

1995 America Remembers
Roy Rogers Tribute

COMMEMORATIVES

1995
America Remembers North
American Rodeo Tribute
American Historical
Foundation Second
Amendment S&W Sigma
American Historical Foundation
Limited Edition Colt Gold Cup
American Historical
Foundation World War II 50th
Anniversary Thompson
Marlin Bat Masterson
Commemorative
Marlin 1895 Century Limited
Winchester Florida
Sesquicentennial
Commemorative

and grip screws; decoration on slide; top of slide carries engraved legend, gold seal; left side with gold portrait of Admiral Nimitz, with dive bomber, battleship; right side with Douglas MacArthur in victory wreath with his words at surrender ceremony. Reported 250 made, 1995. **Issue Price: $1495.**
    Market Value: $995

AMERICAN HISTORICAL FOUNDATION LIMITED EDITION COLT GOLD CUP
Semi-automatic; 45 ACP; same specs as standard Colt Gold Cup but all exterior surfaces gold-plated, except backstrap. Banner carries legend, "Gold Cup Edition" and "One of 100". Reported 100 made, 1995. **Issue Price: $2195**
    Market Value: $1495

AMERICAN HISTORICAL FOUNDATION SECOND AMENDMENT S&W SIGMA
Semi-automatic; 40 S&W; 4 1/2″ barrel; same specs as standard Sigma model except gold-filled etched scrollwork on slide, special commemorative medallion inset in grip; special serial numbers. Reported 250 made, 1995. **Issue Price: $1995.**
    Market Value: $750

AMERICAN HISTORICAL FOUNDATION WORLD WAR II 50TH ANNIVERSARY THOMPSON
Semi-automatic; 45 ACP; gold-etched receiver follows military acorn motif; gold-plated trigger, rear sight base, Cutts compensator, actuator knob, sling swivels; American walnut stock inset with three medallions; special serial numbers. Reported 500 made by Auto Ordnance on special order, 1995. **Issue Price: $2495.**
    Market Value: $1295

MARLIN 1895 CENTURY LIMITED
Lever-action; 45-70; commemorates company's 125th anniversary, 100th anniversary of model; same specs as original 1895 model except black walnut stock; French gray-finished receiver, etched with centennial dates; inletted steel buttplate. Unknown number made, 1995. **Issue Price: $1105.**
    Market Value: $995

MARLIN BAT MASTERSON COMMEMORATIVE
Lever-action carbine; 44 Special/44 Magnum; same specs as standard Marlin Model 1894 except decoration; receiver and bolt engraved in Italy; left receiver panel engraved with old Dodge City street scene with portrait of Masterson inset; right panel carries the names of towns where he enforced the law, with his derby, badge, a buffalo skull and other artwork; gold-plated action, other metal parts blued. Reported 500 made exclusively for Cherry's by Marlin, 1995. **Issue Price: $895.**
    Market Value: $895

WINCHESTER FLORIDA SESQUICENTENNIAL COMMEMORATIVE
Lever-action carbine; same specs as standard Model 94 except decoration; lever, receiver, barrel bands gold-plated; barrel hammer, bolt, magazine tube deep-luster blued; left receiver panel engraved with state seal, state bird and Kennedy Space Center launch; right panel features Everglades with alligator, crane and airboat; checkered semi-fancy French walnut stock; special serial numbers FL001 through FL500. Reported 500 made exclusively for Cherry's, 1995. **Issue Price: $1195.**
    Market Value: $1195

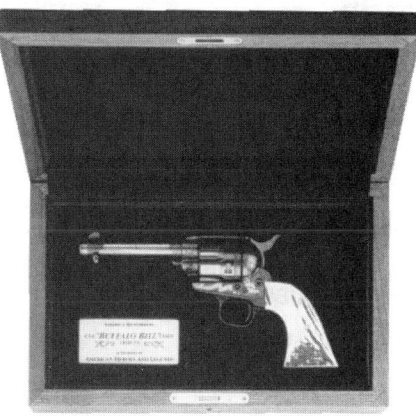

1995 America Remembers
Buffalo Bill Sesquicentennial

UNITED STATES

ALABAMA

Alabama Gun Collectors Assn.
Secretary, P.O. Box 70965, Tuscaloosa, AL 35407

ALASKA

Alaska Gun Collectors Assn., Inc.
C.W. Floyd, Pres., 5240 Little Tree, Anchorage, AK 99507

ARIZONA

Arizona Arms Assn.
Don DeBusk, President, 4837 Bryce Ave., Glendale, AZ 85301

CALIFORNIA

California Cartridge Collectors Assn.
Rick Montgomery, 1729 Christina, Stockton, CA 95204/209-463-7216 evs.

California Waterfowl Assn.
4630 Northgate Blvd., #150, Sacramento, CA 95834

Greater Calif. Arms & Collectors Assn.
Donald L. Bullock, 8291 Carburton St., Long Beach, CA 90808-3302

Los Angeles Gun Ctg. Collectors Assn.
F.H. Ruffra, 20810 Amie Ave., Apt. #9, Torrance, CA 90503

Stock Gun Players Assn.
6038 Appian Way, Long Beach, CA, 90803

COLORADO

Colorado Gun Collectors Assn.
L.E.(Bud) Greenwald, 2553 S. Quitman St., Denver, CO 80219/303-935-3850

Rocky Mountain Cartridge Collectors Assn.
John Roth, P.O. Box 757, Conifer, CO 80433

CONNECTICUT

Ye Connecticut Gun Guild, Inc.
Dick Fraser, P.O. Box 425, Windsor, CT 06095

FLORIDA

Unified Sportsmen of Florida
P.O. Box 6565, Tallahassee, FL 32314

GEORGIA

Georgia Arms Collectors Assn., Inc.
Michael Kindberg, President, P.O. Box 277, Alpharetta, GA 30239-0277

ILLINOIS

Illinois State Rifle Assn.
P.O. Box 637, Chatsworth, IL 60921

Mississippi Valley Gun & Cartridge Coll. Assn.
Bob Filbert, P.O. Box 61, Port Byron, IL 61275/309-523-2593

Sauk Trail Gun Collectors
Gordell M. Matson, P.O. Box 1113, Milan, IL 61264

Wabash Valley Gun Collectors Assn., Inc.
Roger L. Dorsett, 2601 Willow Rd., Urbana, IL 61801/217-384-7302

INDIANA

Indiana State Rifle & Pistol Assn.
Thos. Glancy, P.O. Box 552, Chesterton, IN 46304

Southern Indiana Gun Collectors Assn., Inc.
Sheila McClary, 309 W. Monroe St., Boonville, IN 47601/812-897-3742

IOWA

Beaver Creek Plainsmen Inc.
Steve Murphy, Secy., P.O. Box 298, Bondurant, IA 50035

Central States Gun Collectors Assn.
Dennis Greischar, Box 841, Mason City, IA 50402-0841

KANSAS

Kansas Cartridge Collectors Assn.
Bob Linder, Box 84, Plainville, KS 67663

KENTUCKY

Kentuckiana Arms Collectors Assn.
Charles Billips, President, Box 1776, Louisville, KY 40201

Kentucky Gun Collectors Assn., Inc.
Ruth Johnson, Box 64, Owensboro, KY 42302/502-729-4197

LOUISIANA

Washitaw River Renegades
Sandra Rushing, P.O. Box 256, Main St., Grayson, LA 71435

MARYLAND

Baltimore Antique Arms Assn.
Mr. Cillo, 1034 Main St., Darlington, MD 21304

MASSACHUSETTS

Bay Colony Weapons Collectors, Inc.
John Brandt, Box 111, Hingham, MA 02043

Massachusetts Arms Collectors
Bruce E. Skinner, P.O. Box 31, No. Carver, MA 02355/508-866-5259

MICHIGAN

Association for the Study and Research of .22 Caliber Rimfire Cartridges
George Kass, 4512 Nakoma Dr., Okemos, MI 48864

MINNESOTA

Sioux Empire Cartridge Collectors Assn.
Bob Cameron, 14597 Glendale Ave. SE, Prior Lake, MN 55372

MISSISSIPPI

Mississippi Gun Collectors Assn.
Jack E. Swinney, P.O. Box 16323, Hattiesburg, MS 39402

MISSOURI

Greater St. Louis Cartridge Collectors Assn.
Don MacChesney, 634 Scottsdale Rd., Kirkwood, MO 63122-1109

Mineral Belt Gun Collectors Assn.
D.F. Saunders, 1110 Cleveland Ave., Monett, MO 65708

Missouri Valley Arms Collectors Assn., Inc.
L.P Brammer II, Membership Secy., P.O. Box 33033, Kansas City, MO 64114

MONTANA

Montana Arms Collectors Assn.
Dean E. Yearout, Sr., Exec. Secy., 1516 21st Ave. S., Great Falls, MT 59405

Weapons Collectors Society of Montana
R.G. Schipf, Ex. Secy., 3100 Bancroft St., Missoula, MT 59801/406-728-2995

NEBRASKA

Nebraska Cartridge Collectors Club
Gary Muckel, P.O. Box 84442, Lincoln, NE 68501

NEW HAMPSHIRE

New Hampshire Arms Collectors, Inc.
James Stamatelos, Secy., P.O. Box 5, Cambridge, MA 02139

NEW JERSEY

Englishtown Benchrest Shooters Assn.
Michael Toth, 64 Cooke Ave., Carteret, NJ 07008

Jersey Shore Antique Arms Collectors
Joe Sisia, P.O. Box 100, Bayville, NJ 08721-0100

New Jersey Arms Collectors Club, Inc.
Angus Laidlaw, Vice President, 230 Valley Rd., Montclair, NJ 07042/201-746-0939; e-mail: acclaid-law@juno.com

NEW YORK

Iroquois Arms Collectors Assn.
Bonnie Robinson, Show Secy., P.O. Box 142, Ransomville, NY 14131/716-791-4096

Mid-State Arms Coll. & Shooters Club
Jack Ackerman, 24 S. Mountain Terr., Binghamton, NY 13903

NORTH CAROLINA

North Carolina Gun Collectors Assn.
Jerry Ledford, 3231-7th St. Dr. NE, Hickory, NC 28601

OHIO

Ohio Gun Collectors Assn.
P.O. Box 9007, Maumee, OH 43537-9007/419-897-0861; Fax:419-897-0860

Shotshell Historical and Collectors Society
Madeline Bruemmer, 3886 Dawley Rd., Ravenna, OH 44266

The Stark Gun Collectors, Inc.
William I. Gann, 5666 Waynesburg Dr., Waynesburg, OH 44688

OREGON

Oregon Arms Collectors Assn., Inc.
Phil Bailey, P.O. Box 13000-A, Portland, OR 97213-0017/503-281-6864; off.:503-281-0918

Oregon Cartridge Collectors Assn.
Boyd Northrup, P.O. Box 285, Rhododendron, OR 97049

PENNSYLVANIA

Presque Isle Gun Collectors Assn.
James Welch, 156 E. 37 St., Erie, PA 16504

SOUTH CAROLINA

Belton Gun Club, Inc.
Attn. Secretary
P.O. Box 126, Belton, SC 29627, (864) 369-6767

Gun Owners of South Carolina
Membership Div.: William Strozier, Secretary, P.O. Box 70, Johns Island, SC

29457-0070/803-762-3240; Fax:803-795-0711; e-mail:76053.222@compuserve.com

SOUTH DAKOTA

Dakota Territory Gun Coll. Assn., Inc.
Curt Carter, Castlewood, SD 57223

TENNESSEE

Smoky Mountain Gun Coll. Assn., Inc.
Hugh W. Yabro, President, P.O. Box 23225, Knoxville, TN 37933

Tennessee Gun Collectors Assn., Inc.
M.H. Parks, 3556 Pleasant Valley Rd., Nashville, TN 37204-3419

TEXAS

Houston Gun Collectors Assn., Inc.
P.O. Box 741429, Houston, TX 77274-1429

Texas Cartridge Collectors Assn., Inc.
Robert Mellichamp, Memb. Contact, 907 Shirkmere, Houston, TX 77008/713-869-0558

Texas Gun Collectors Assn.
Bob Eder, Pres., P.O. Box 12067, El Paso, TX 79913/915-584-8183

Texas State Rifle Assn.
1131 Rockingham Dr., Suite 101, Richardson, TX 75080-4326

VIRGINIA

Virginia Gun Collectors Assn., Inc.
Addison Hurst, Secy., 38802 Charlestown Height, Waterford, VA 20197/540-882-3543

WASHINGTON

Association of Cartridge Collectors on the Pacific Northwest
Robert Jardin, 14214 Meadowlark Drive KPN, Gig Harbor, WA 98329

Washington Arms Collectors, Inc.
Joyce Boss, P.O. Box 389, Renton, WA, 98057-0389/206-255-8410

WISCONSIN

Great Lakes Arms Collectors Assn., Inc.
Edward C. Warnke, 2913 Woodridge Lane, Waukesha, WI 53188

Wisconsin Gun Collectors Assn., Inc.
Lulita Zellmer, P.O. Box 181, Sussex, WI 53089

WYOMING

Wyoming Weapons Collectors
P.O. Box 284, Laramie, WY 82073/307-745-4652 or 745-9530

NATIONAL ORGANIZATIONS

Amateur Trapshooting Assn.
David D. Bopp, Exec. Director, 601 W. National Rd., Vandalia, OH 45377/937-898-4638; Fax:937-898-5472

American Airgun Field Target Assn.
5911 Cherokee Ave., Tampa, FL 33604

American Coon Hunters Assn.
Opal Johnston, P.O. Cadet, Route 1, Box 492, Old Mines, MO 63630

American Custom Gunmakers Guild
Jan Billeb, Exec. Director, P.O. Box 812, Burlington, IA 52601-0812/319-752-6114 (Phone or Fax)

American Defense Preparedness Assn.
Two Colonial Place, 2101 Wilson Blvd., Suite 400, Arlington, VA 22201-3061

American Paintball League
P.O. Box 3561, Johnson City, TN 37602/800-541-9169

American Pistolsmiths Guild
Alex B. Hamilton, Pres., 1449 Blue Crest Lane, San Antonio, TX 78232/210-494-3063

American Police Pistol & Rifle Assn.
3801 Biscayne Blvd., Miami, FL 33137

American Single Shot Rifle Assn.
Gary Staup, Secy., 709 Carolyn Dr., Delphos, OH 45833/419-692-3866. Website: www.assra.com

American Society of Arms Collectors
George E. Weatherly, P.O. Box 2567, Waxahachie, TX 75165

American Tactical Shooting Assn.(A.T.S.A.)
c/o Skip Gochenour, 2600 N. Third St., Harrisburg, PA 17110/717-233-0402; Fax:717-233-5340

Association of Firearm and Tool Mark Examiners
Lannie G. Emanuel, Secy., Southwest Institute of Forensic Sciences, P.O. Box 35728, Dallas, TX 75235/214-920-5979; Fax:214-920-5928; Membership Secy., Ann D. Jones, VA Div. of Forensic Science, P.O. Box 999, Richmond, VA 23208/804-786-4706; Fax:804-371-8328

Boone & Crockett Club
250 Station Dr., Missoula, MT 59801-2753

Browning Collectors Assn.
Secretary:Scherrie L. Brennac, 2749 Keith Dr., Villa Ridge, MO 63089/314-742-0571

The Cast Bullet Assn., Inc.
Ralland J. Fortier, Editor, 4103 Foxcraft Dr., Traverse City, MI 49684

Citizens Committee for the Right to Keep and Bear Arms
Natl. Hq., Liberty Park, 12500 NE Tenth Pl., Bellevue, WA 98005

Colt Collectors Assn.
25000 Highland Way, Los Gatos, CA 95030/408-353-2658.

Ducks Unlimited, Inc.
Natl. Headquarters, One Waterfowl Way, Memphis, TN 38120/901-758-3937

Fifty Caliber Shooters Assn.
PO Box 111, Monroe UT 84754-0111

Firearms Coalition/Neal Knox Associates
Box 6537, Silver Spring, MD 20906/301-871-3006

Firearms Engravers Guild of America
Rex C. Pedersen, Secy., 511 N. Rath Ave., Lundington, MI 49431/616-845-7695(Phone and Fax)

Foundation for North American Wild Sheep
720 Allen Ave., Cody, WY 82414-3402/web site: http://iigi.com/os/non/fnaws/fnaw s.htm; e-mail: fnaws@wyoming.com

Freedom Arms Collectors Assn.
P.O. Box 160302, Miami, FL 33116-0302

Garand Collectors Assn.
P.O. Box 181, Richmond, KY 40475

Golden Eagle Collectors Assn. (G.E.C.A.)
Chris Showler, 11144 Slate Creek Rd., Grass Valley, CA 95945

Gun Owners of America
8001 Forbes Place, Suite 102, Springfield, VA 22151/703-321-8585

Handgun Hunters International
J.D. Jones, Director, P.O. Box 357 MAG, Bloomingdale, OH 43910

Harrington & Richardson Gun Coll. Assn.
George L. Cardet, 330 S.W. 27th Ave., Suite 603, Miami, FL 33135

High Standard Collectors' Assn.
John J. Stimson, Jr., Pres., 540 W. 92nd St., Indianapolis, IN 46260

Hopkins & Allen Arms & Memorabilia Society (HAAMS)
P.O. Box 187, 1309 Pamela Circle, Delphos, OH 45833

International Ammunition Association, Inc.
C.R. Punnett, Secy., 8 Hillock Lane, Chadds Ford, PA 19317/610-358-1285; Fax:610-358-1560

International Benchrest Shooters
Joan Borden, RR1, Box 250BB, Springville, PA 18844/717-965-2366

International Blackpowder Hunting Assn.
P.O. Box 1180, Glenrock, WY 82637/307-436-9817

IHMSA (Intl. Handgun Metallic Silhouette Assn.)
PO Box 368, Burlington, IA 52601 Website: www.ihmsa.cor

International Society of Mauser Arms Collectors
Michael Kindberg, Pres., P.O. Box 277, Alpharetta, GA 30239-0277

Jews for the Preservation of Firearms Ownership (JPFO)
2872 S. Wentworth Ave., Milwaukee, WI 53207/414-769-0760; Fax:414-483-8435

The Mannlicher Collectors Assn.
Membership Office: P.O. Box1249, The Dalles, Oregon 97058

Marlin Firearms Collectors Assn., Ltd.
Dick Paterson, Secy., 407 Lincoln Bldg., 44 Main St., Champaign, IL 61820

Merwin Hulbert Association,
2503 Kentwood Ct., High Point, NC 27265

Miniature Arms Collectors/ Makers Society, Ltd.
Ralph Koebbeman, Pres., 4910 Kilburn Ave., Rockford, IL 61101/815-964-2569

M1 Carbine Collectors Assn. (M1-CCA)
623 Apaloosa Ln., Gardner-ville, NV 89410-7840

National Association of Buckskinners (NAB)
Territorial Dispatch—1800s Historical Publication, 4701 Marion St., Suite 324, Live-

stock Exchange Bldg., Denver, CO 80216/303-297-9671

The National Association of Derringer Collectors
P.O. Box 20572, San Jose, CA 95160

National Assn. of Federally Licensed Firearms Dealers
Andrew Molchan, 2455 E. Sunrise, Ft. Lauderdale, FL 33304

National Association to Keep and Bear Arms
P.O. Box 78336, Seattle, WA 98178

National Automatic Pistol Collectors Assn.
Tom Knox, P.O. Box 15738, Tower Grove Station, St. Louis, MO 63163

National Bench Rest Shooters Assn., Inc.
Pat Ferrell, 2835 Guilford Lane, Oklahoma City, OK 73120-4404/405-842-9585; Fax: 405-842-9575

National Muzzle Loading Rifle Assn.
Box 67, Friendship, IN 47021 / 812-667-5131. Website: www.nmlra@nmlra.org

National Professional Paintball League (NPPL)
540 Main St., Mount Kisco, NY 10549/914-241-7400

National Reloading Manufacturers Assn.
One Centerpointe Dr., Suite 300, Lake Oswego, OR 97035

National Rifle Assn. of America
11250 Waples Mill Rd., Fair-fax, VA 22030 / 703-267-1000. Website: www.nra.org

National Shooting Sports Foundation, Inc.
Robert T. Delfay, President, Flintlock Ridge Office Center, 11 Mile Hill Rd., Newtown, CT 06470-2359/203-426-1320; FAX: 203-426-1087

National Skeet Shooting Assn.
Dan Snyuder, Director, 5931 Roft Road, San Antonio, TX 78253-9261/800-877-5338. Website: nssa-nsca.com

National Sporting Clays Association
Ann Myers, Director, 5931 Roft Road, San Antonio, TX 78253-9261/800-877-5338. Website: nssa-nsca.com

National Wild Turkey Federation, Inc.
P.O. Box 530, 770 Augusta Rd., Edgefield, SC 29824

North American Hunting Club
P.O. Box 3401, Minnetonka, MN 55343/612-936-9333; Fax: 612-936-9755

North American Paintball Referees Association (NAPRA)
584 Cestaric Dr., Milpitas, CA 95035

North-South Skirmish Assn., Inc.
Stevan F. Meserve, Exec. Secretary, 507 N. Brighton Court, Sterling, VA 20164-3919

Remington Society of America
Gordon Fosburg, Secretary, 11900 North Brinton Road, Lake, MI 48623

Rocky Mountain Elk Foundation
P.O. Box 8249, Missoula, MT 59807-8249/406-523-4500; Fax: 406-523-4581 Website: www.rmef.org

Ruger Collector's Assn., Inc.
P.O. Box 240, Greens Farms, CT 06436

Safari Club International
4800 W. Gates Pass Rd., Tucson, AZ 85745/520-620-1220

Sako Collectors Assn., Inc.
Jim Lutes, 202 N. Locust, Whitewater, KS 67154

Second Amendment Foundation
James Madison Building, 12500 NE 10th Pl., Bellevue, WA 98005

Single Action Shooting Society (SASS)
23255-A La Palma Avenue, Yorba Linda, CA 92887/714-694-1800; FAX: 714-694-1815/ email: sasseot@aol.com Web-site: www.sassnet.com

Smith & Wesson Collectors Assn.
Cally Pletl, Admin. Asst.,PO Box 444, Afton, NY 13730

The Society of American Bayonet Collectors
P.O. Box 234, East Islip, NY 11730-0234

Southern California Schuetzen Society
Dean Lillard, 34657 Ave. E., Yucaipa, CA 92399

Sporting Arms and Ammunition Manufacturers' Institute (SAAMI)
Flintlock Ridge Office Center, 11 Mile Hill Rd., Newtown, CT 06470-2359/203-426-4358; FAX: 203-426-1087

Sporting Clays of America (SCA)
Ron L. Blosser, Pres., 9257 Buckeye Rd., Sugar Grove, OH 43155-9632/614-746-8334; Fax: 614-746-8605

The Thompson/Center Assn.
Joe Wright, President, Box 792, Northboro, MA 01532/508-845-6960

U.S. Practical Shooting Assn./IPSC
Dave Thomas, P.O. Box 811, Sedro Woolley, WA 98284/360-855-2245

U.S. Revolver Assn.
Brian J. Barer, 40 Larchmont Ave., Taunton, MA 02780/508-824-4836

U.S. Shooting Team
U.S. Olympic Shooting Center, One Olympic Plaza, Colorado Springs, CO 80909/719-578-4670

The Varmint Hunters Assn., Inc.
Box 759, Pierre, SD 57501/ Member Services 800-528-4868

Weatherby Collectors Assn., Inc.
P.O. Box 888, Ozark, MO 65721

The Wildcatters
P.O. Box 170, Greenville, WI 54942

Winchester Arms Collectors Assn.
P.O. Box 230, Brownsboro, TX 75756/903-852-4027

The Women's Shooting Sports Foundation (WSSF)
4620 Edison Avenue, Ste. C, Colorado Springs, CO 80915/719-638-1299; FAX: 719-638-1271/e-mail: wssf@worldnet.att.net

ARGENTINA

Asociacion Argentina de Coleccionistas de Armes y Municiones
Castilla de Correos No. 28,

Succursal I B, 1401 Buenos Aires, Republica Argentina

AUSTRALIA

Antique & Historical Arms Collectors of Australia
P.O. Box 5654, GCMC Queensland 9726, Australia

The Arms Collector's Guild of Queensland Inc.
Ian Skennerton, P.O. Box 433, Ashmore City 4214, Queensland, Australia

Australian Cartridge Collectors Assn., Inc.
Bob Bennett, 126 Landscape Dr., E. Doncaster 3109, Victoria, Ausrtalia

Sporting Shooters Assn. of Australia, Inc.
P.O. Box 2066, Kent Town, SA 5071, Australia

CANADA

ALBERTA

Canadian Historical Arms Society
P.O. Box 901, Edmonton, Alb., Canada T5J 2L8

National Firearms Assn.
Natl. Hq: P.O. Box 1779, Edmonton, Alb., Canada T5J 2P1

BRITISH COLUMBIA

The Historical Arms Collectors of B.C. (Canada)
Harry Moon, Pres., P.O. Box 50117, South Slope RPO, Burnaby, BC V5J 5G3, Can-ada/604-438-0950; Fax:604-277-3646

ONTARIO

Association of Canadian Cartridge Collectors
Monica Wright, RR 1, Mill-grove, ON, LOR IVO, Canada

Tri-County Antique Arms Fair
P.O. Box 122, RR #1, North Lancaster, Ont., Canada K0C 1Z0

EUROPE

BELGIUM

European Cartridge Research Assn.
Graham Irving, 21 Rue Schaltin, 4900 Spa, Bel-gium/32.87.77.43.40; Fax:32.87.77.27.51

CZECHOSLOVAKIA

Spolecnost Pro Studium Naboju (Czech Cartridge Research Assn.)
JUDr. Jaroslav Bubak, Pod Homolko 1439, 26601 Beroun 2, Czech Republic

DENMARK

Aquila Dansk Jagtpatron Historic Forening (Danish Historical Cartridge Collectors Club)
Steen Elgaard Møller, Ulriksdalsvej 7, 4840 Nr. Alslev, Denmark 10045-53846218; Fax:00455 384 6209

ENGLAND

Arms and Armour Society
Hon. Secretary A. Dove, P.O. Box 10232, London, 5W19 2ZD, England

Dutch Paintball Federation
Aceville Publ., Castle House 97 High Street, Colchester, Essex C01 1TH, England/011-44-206-564840

European Paintball Sports Foundation
c/o Aceville Publ., Castle House 97 High St., Colches-ter, Essex, C01 1TH, England

Historical Breechloading Smallarms Assn.
D.J. Penn M.A., Secy., P.O. Box 12778, London SE1 6BX, England. Journal and newslet-ter are $23 a yr., including air-mail.

National Rifle Assn.
(Great Britain) Bisley Camp, Brookwood, Woking Surrey GU24 OPB, England/01483.797777; Fax: 014730686275

United Kingdom Cartridge Club
Ian Southgate, 20 Millfield, Elmley Castle, Nr. Pershore, Worcestershire, WR10 3HR, England

FRANCE

STAC-Western Co.
3 Ave. Paul Doumer (N.311); 78360 Montesson, France/01.30.53-43-65; Fax: 01.30.53.19.10

GERMANY

Bund Deutscher Sports-chützen e.v. (BDS)
Borsigallee 10, 53125 Bonn 1, Germany

Deutscher Schützenbund
Lahnstrasse 120, 65195 Wiesbaden, Germany

SPAIN

Asociacion Espanola de Coleccionistas de Cartu-chos (A.E.C.C.)
Secretary: Apdo. Correos No. 1086, 2880-Alcala de Henares (Madrid), Spain. President: Apdo. Correos No. 682, 50080 Zaragoza, Spain

SWEDEN

Scandinavian Ammunition Research Assn.
Box 107, 77622 Hedemora, Sweden

NEW ZEALAND

New Zealand Cartridge Collectors Club
Terry Castle, 70 Tiraumea Dr., Pakuranga, Auckland, New Zealand

New Zealand Deerstalkers Assn.
P.O. Box 6514 TE ARO, Wellington, New Zealand

SOUTH AFRICA

Historical Firearms Soc. of South Africa
P.O. Box 145, 7725 Newlands, Republic of South Africa

Republic of South Africa Cartridge Collectors Assn.
Arno Klee, 20 Eugene St., Malanshof Randburg, Gauteng 2194, Republic of South Africa

S.A.A.C.A. (Southern Africa Arms and Ammunition Assn.)
Gauteng Office: P.O. Box 7597, Weltevreden Park, 1715, Republic of South Africa/011-679-1151; Fax: 011-679-1131; e-mail: saaaca@iafrica.com. Kwa-Zulu Natal office: P.O. Box 4065, Northway, Kwazulu-Natal 4065, Republic of South Africa

SAGA (S.A. Gunowners' Assn.)
P.O. Box 35203, Northway, Kwazulu-Natal 4065, Republic of South Africa

THE COLLECTOR'S LIBRARY

FOR COLLECTOR ♦ HUNTER ♦ SHOOTER ♦ OUTDOORSMAN

IMPORTANT NOTICE TO BOOK BUYERS

Books listed here may be bought from Ray Riling Arms Books Co., 6844 Gorsten St., P.O. Box 18925, Philadelphia, PA 19119, Phone 215/438-2456; FAX: 215-438-5395. E-Mail: sales@rayrilingarms-books.com. Joe Riling is the researcher and compiler of "The Arms Library" and a seller of gun books for over 32 years. The Riling stock includes books classic and modern, many hard-to-find items, and many not obtainable elsewhere. These pages list a portion of the current stock. They offer prompt, complete service, with delayed shipments occurring only on out-of-print or out-of-stock books.

Visit our web site at **www.rayrilingarmsbooks.com** and order all of your favorite titles on line from our secure site.

NOTICE FOR ALL CUSTOMERS: Remittance in U.S. funds must accompany all orders. For your convience we now accept Visa, Mastercard & American Express. For Shipments in the U.S. add $7.00 for the 1st book and $2.00 for each additional book for postage and insurance. Minimum

order $10.00. International Orders add $13.00 for the 1st book and $5.00 for each additional book. All International orders are shipped at the buyer's risk unless an additional $5 for insurance is included. USPS does not offer insurance to all countries unless shipped Air-Mail please e-mail or call for pricing.

Payments in excess of order or for "Backorders" are credited or fully refunded at request. Books "As-Ordered" are not returnable except by permission and a handling charge on these of 10% or $2.00 per book which ever is greater is deducted from refund or credit. Only Pennsylvania customers must include current sales tax.

A full variety of arms books also available from Rutgers Book Center, 127 Raritan Ave., Highland Park, NJ 08904/908-545-4344; FAX: 908-545-6686 or I.D.S.A. Books, 1324 Stratford Drive, Piqua, OH 45356/937-773-4203; FAX: 937-778-1922.

COLLECTORS

A Glossary of the Construction, Decoration and Use of Arms and Armor in All Countries and in All Times. By George Cameron Stone., Dover Publishing, New York 1999. Softcover. $39.95

An exhaustive study of arms and armor in all countries through recorded history - from the stone age up to the second world war. With over 4500 Black & White Illustrations. This Dover edition is an unabridged republication of the work originally published in 1934 by the Southworth Press, Portland MA. A new Introduction has been specially prepared for this edition.

Accoutrements of the United States Infantry, Riflemen, and Dragoons 1834-1839. by R.T. Huntington, Historical Arms Series No. 20. Canada: Museum Restoration. 58 pp. illus. Softcover. $8.95

Although the 1841 edition of the U.S. Ordnance Manual provides ample information on the equipment that was in use during the 1840s, it is evident that the patterns of equipment that it describes were not introduced until 1838 or 1839. This guide is intended to fill this gap in our knowledge by providing an overview of what we now know about the accoutrements that were issued to the regular infantryman, rifleman, and dragoon, in the 1830's with excursions into earlier and later years.

Age of the Gunfighter; Men and Weapons on the Frontier 1840-1900, by Joseph G. Rosa, University of Oklahoma Press, Norman, OK, 1999. 192 pp., illustrated. Paper covers. $21.95

Stories of gunfighters and their encounters and detailed descriptions of virtually every firearm used in the old West.

Air Guns, by Eldon G. Wolff, Duckett's Publishing Co., Tempe, AZ, 1997. 204 pp., illus Paper covers. $35.00

Historical reference covering many makers, European and American guns, canes and more.

Allied and Enemy Aircraft: May 1918; Not to be Taken from the Front Lines, Historical Arms Series No. 27. Canada: Museum Restoration. Softcover. $8.95

The basis for this title is a very rare identification manual published by the French government in 1918 that illustrated 60 aircraft with three or more views: French, English American, German, Italian, and Belgian, which might have been seen over the trenches of France. Each is describe in a text translated from the original French. This is probably the most complete collection of illustrations of WW1 aircraft which has survived.

American Beauty; The Prewar Colt National Match Government Model Pistol, by Timothy J. Mullin, Collector Grade Publications, Cobourg, Ontario, Canada. 72 pp., illustrated. $34.95

Includes over 150 serial numbers, and 20 spectacular color photos of factory engraved guns and other authenticated upgrades, including rare "double-carved" ivory grips.

The American Military Saddle, 1776-1945, by R. Stephen Dorsey & Kenneth L. McPheeters, Collector's Library, Eugene, OR, 1999. 400 pp., illustrated. $59.95

The most complete coverage of the subject ever written on the American Military Saddle. Nearly 1000 actual photos and official drawings, from the major public and private collections in the U.S. and Great Britain.

American Police Collectibles; Dark Lanterns and Other Curious Devices, by Matthew G. Forte, Turn of the Century Publishers, Upper Montclair, NJ, 1999. 248 pp., illustrated. $24.95

For collectors of police memorabilia (handcuffs, police dark lanterns, mechanical and chain nippers, rattles, billy clubs and nightsticks) and police historians.

Ammunition; Small Arms, Grenades, and Projected Munitions, by Greenhill Publishing. 144 pp., Illustrated. $22.95 The best concise guide to modern ammunition available today. Covers ammo for small arms, grenades, and projected munitions. 144 pp., Illustrated. As NEW – Hardcover.

Antique Guns, the Collector's Guide, 2nd Edition, edited by John Traister, Stoeger Publishing Co., So. Hackensack, NJ, 1994. 320 pp., illus. Paper covers. $19.95

Covers a vast spectrum of pre-1900 firearms: those manufactured by U.S. gunmakers as well as Canadian, French, German, Belgian, Spanish and other foreign firms.

Arming the Glorious Cause; Weapons of the Second War for Independence, by James B. Whisker, Daniel D. Hartzler and Larry W. Tantz, Old Bedford Village Press, Bedford, PA., 1998. 175 pp., illustrated. $45.00

A photographic study of Confederate weapons.

Arms & Accoutrements of the Mounted Police 1873-1973, by Roger F. Phillips and Donald J. Klancher, Museum Restoration Service, Ont., Canada, 1982. 224 pp., illus. $49.95

A definitive history of the revolvers, rifles, machine guns, cannons, ammunition, swords, etc. used by the NWMP, the RNWMP and the RCMP during the first 100 years of the Force.

Arms and Armor In Antiquity and The Middle Ages. By Charles Boutell, Combined Books Inc., PA 1996. 296 pp., w/ b/w illus. Also a descriptive Notice of Modern Weapons. Translated from the French of M.P. Lacombe, and with a Preface, Notes, and One Additional Chapter on Arms and Armour in England. $14.95

Arms and Armor in the Art Institute of Chicago. By Waltler J. Karcheski, Bulfinch, New York 1999. 128 pp., 103 color photos, 12 black & white illustrations. $50.00

The George F. Harding Collection of arms and armor is the most visited installation at the Art Institute of Chicago - a testament to the enduring appeal of swords, muskets and the other paraphernalia of medieval and early modern war. Organized both chronologically and by type of weapon, this book captures the best of this astonishing collection in 115 striking photographs - most in color - accompanied by illuminating text. Here are intricately filigreed breastplates and ivory-handled crossbows, samurai katana and Toledo-steel scimitars, elaborately decorated maces and beautifully carved flintlocks - a treat for anyone who has ever been beguiled by arms, armor and the age of chivalry.

Arms and Armor in Colonial America 1526-1783. by Harold Peterson, Dover Publishing, New York, 2000. 350 pages with over 300 illustrations, index, bibliography & appendix. Softcover. $29.95

Over 200 years of firearms, ammunition, equipment & edged weapons.

Arms and Armor: The Cleveland Museum of Art. By Stephen N. Fliegel, Abrams, New York, 1998. 172 color photos, 17 halftones. 181 pages. $49.50

Intense look at the culture of the warrior and hunter, with an intriguing discussion of the decorative arts found on weapons and armor, set against the background of political and social history. Also provides information on the evolution of armor, together with manufacture and decoration, and weapons as technology and art.

Arms and Equipment of the Civil War, by Jack Coggins, Barnes & Noble, Rockleight, N.J., 1999. 160 pp., illustrated. $12.98

This unique encyclopedia provides a new perspective on the war. It provides lively explanations of how ingenious new weapons spelled victory or defeat for both sides. Aided by more than 500 illustrations and on-the-scene comments by Union and Confederate soldiers.

Arms Makers of Colonial America, by James B. Whisker, Selinsgrove, PA:, 1992: Susquehanna University Press. 1st edition. 217 pages, illustrated. $45.00

A comprehensively documented historical survey of the broad spectrum of arms makers in America who were active before 1783.

Arms Makers of Maryland, by Daniel D. Hartzler, George Shumway, York, PA, 1975. 200 pp., illus. $50.00

A thorough study of the gunsmiths of Maryland who worked during the late 18th and early 19th centuries.

Arms Makers of Pennsylvania, by James B. Whisker, Selinsgrove, PA:: Susquehanna Univ. Press, 1990. 1st edition. 218 pages, illustrated in black and white and color. $45.00

Concentrates primarily on the cottage industry gunsmiths & gun makers who worked in the Keystone State from it's early years through 1900.

Arms Makers of Western Pennsylvania, by James B. Whisker, Old Bedford Village Press. 1st edition. This deluxe hard bound edition has 176 pages, $45.00
Printed on fine coated paper, with many large photographs, and detailed text describing the period, lives, tools, and artistry of the Arms Makers of Western Pennsylvania.

Arsenal Of Freedom: The Springfield Armory 1890-1948, by Lt. Col. William Brophy, Andrew Mowbray, Inc., Lincoln, RI,1997. 20 pgs of photos. 400 pages. As New - Softcover. $29.95
A year by year account drawn from official records. Packed with reports, charts, tables, line drawings, and 20 page photo section.

Artistic Ingredients of the Longrifle, by George Shumway Publisher, 1989 102 pp., with 94 illus. $20.00
After a brief review of Pennsylvania-German folk art and architecture, to establish the artistic environment in which the longrifle was made, the author demonstrates that the sophisticated rococo decoration on the many of the finer longrifles is comparable to the best rococo work of Philadelphia cabinet makers and silversmiths.

The Art of Gun Engraving, by Claude Gaier and Pietro Sabatti, Knickerbocker Press, N.Y., 1999. 160 pp., illustrated. $34.95
The richness and detail lavished on early firearms represents a craftsmanship nearly vanished. Beginning with crossbows in the 100's, hunting scenes, portraits, or mythological themes are intricately depicted within a few square inches of etched metal. The full-color photos contained herein recaptures this lost art with exquisite detail.

Astra Automatic Pistols, by Leonardo M. Antaris, FIRAC Publishing Co., Sterling, CO, 1989. 248 pp., illus. $55.00
Charts, tables, serial ranges, etc. The definitive work on Astra pistols.

Basic Documents on U.S. Martial Arms, commentary by Col. B. R. Lewis, reissue by Ray Riling, Phila., PA, 1956 and 1960. *Rifle Musket Model 1855.*
The first issue rifle of musket caliber, a muzzle loader equipped with the Maynard Primer, 32 pp. *Rifle Musket Model 1863.* The typical Union muzzle-loader of the Civil War, 26 pp. *Breech-Loading Rifle Musket Model 1866.* The first of our 50-caliber breechloading rifles, 12 pp. *Remington Navy Rifle Model 1870.* A commercial type breech-loader made at Springfield, 16 pp. *Lee Straight Pull Navy Rifle Model 1895.* A magazine cartridge arm of 6mm caliber. 23 pp. *Breech-Loading Arms (five models)* 27 pp. *Ward-Burton Rifle Musket 1871-16* pp. Each $10.00.

Battle Weapons of the American Revolution, by George C. Neuman, Scurlock Publishing Co., Texarkana, TX, 2001. 400 pp. Illus. Softcovers. $34.95
The most extensive photographic collection of Revolutionary War weapons ever in one volume. More than 1,600 photos of over 500 muskets, rifles, swords, bayonets, knives and other arms used by both sides in America's War for Independence.

The Bedford County Rifle and Its Makers, by George Shumway. 40pp. illustrated, Softcover. $10.00
The authors study of the graceful and distinctive muzzle-loading rifles made in Bedford County, Pennsylvania. Stands as a milestone on the long path to the understanding of America's longrifles.

Behold the Longrifle Again, by James B. Whisker, Old Bedford Village Press, Bedford, PA, 1997. 176 pp., illus. $45.00
Excellent reference work for the collector profusely illustrated with photographs of some of the finest Kentucky rifles showing front and back profiles and overall view.

The Belgian Rattlesnake; The Lewis Automatic Machine Gun, by William M. Easterly, Collector Grade Publications, Cobourg, Ontario, Canada, 1998. 584 pp., illustrated. $79.95
The most complete account ever published on the life and times of Colonel Isaac Newton Lewis and his crowning invention, the Lewis Automatic machine gun.

Beretta Automatic Pistols, by J.B. Wood, Stackpole Books, Harrisburg, PA, 1985. 192 pp., illus. $24.95
Only English-language book devoted to the Beretta line. Includes all important models.

The Big Guns, Civil War Siege, Seacoast, and Naval Cannon, by Edwin Olmstead, Wayne E. Stark, and Spencer C. Tucker, Museum Restoration Service, Bloomfield, Ontario, Canada, 1997. 360 pp., illustrated. $80.00
This book is designed to identify and record the heavy guns available to both sides by the end of the Civil War.

Birmingham Gunmakers, by Douglas Tate, Safari Press, Inc., Huntington Beach, CA, 1997. 300 pp., illus. $50.00
An invaluable work for anybody interested in the fine sporting arms crafted in this famous British gunmakers' city.

Blue Book of Gun Values, 22nd Edition, edited by S.P. Fjestad, Blue Book Publications, Inc. Minneapolis, MN 2001. $34.95
This new 22nd Edition simply contains more firearms values and information than any other single publication. Expanded to over 1,600 pages featuring over 100,000 firearms prices, the new Blue Book of Gun Values also contains over _ million words of text – no other book is even close! Most of the information contained in this publication is simply not available anywhere else, for any price!

Blue Book of Modern Black Powder Values, by Dennis Adler, Blue Book Publications, Inc. Minneapolis, MN 2000. 200 pp., illustrated. 41 color photos. Softcover. $14.95

This new title contains more up-to-date black powder values and related information than any other single publication. With 120 pages, this new book will keep you up to date on modern black powder models and prices, including most makes & models introduced this year! .

The Blunderbuss 1500-1900, by James D. Forman, Historical Arms Series No. 32. Canada: Museum Restoration, 1994. An excellent and authoritative booklet giving tons of information on the Blunderbuss, a very neglected subject. 40 pages, illustrated. Softcover. $8.95

Boarders Away I: With Steel-Edged Weapons & Polearms, by William Gilkerson, Andrew Mowbray, Inc. Publishers, Lincoln, RI, 1993. 331 pages. $48.00
Contains the essential 24 page chapter 'War at Sea' which sets the historical and practical context for the arms discussed. Includes chapters on, Early Naval Weapons, Boarding Axes, Cutlasses, Officers Fighting Swords and Dirks, and weapons at hand of Random Mayhem.

Boarders Away, Volume II: Firearms of the Age of Fighting Sail, by William Gilkerson, Andrew Mowbray, Inc. Publishers, Lincoln, RI, 1993. 331 pp., illus. $65.00
Covers the pistols, muskets, combustibles and small cannon used aboard American and European fighting ships, 1626-1826.

The Book of Colt Firearms, by R. L. Wilson, Blue Book Publications, Inc, Minneapolis, MN, 1993. 616 pp., illus. $158.00
A complete Colt library in a single volume. In over 1,250.000 words, over 1,250 black and white and 67 color photographs, this mammoth work tells the Colt story from 1832 through the present.

Boothroyd's Revised Directory Of British Gunmakers, by Geoffrey Boothroyd, Long Beach, CA: Safari Press, 2000. Revised edition. 412pp, photos. $39.95
Over a 30 year period Geoffrey Boothroyd has accumulated information on just about every sporting gun maker that ever has existed in the British Isles from 1850 onward. In this magnificent reference work he has placed all the gun makers he has found over the years (over 1000 entries) in an alphabetical listing with as much information as he has been able to unearth. One of the best reference sources on all British makers (including Wales, Scotland and Ireland) in which you can find data on the most obscure as well as the most famous. Contains starting date of the business, addresses, proprietors, what they made and how long they operated with other interesting details for the collector of fine British guns.

Boston's Gun Bible, by Boston T. Party, Ignacio, CO: Javelin Press, August 2000. Expanded Edition.Softcover. $28.00
This mammoth guide for gun owners everywhere is a completely updated and expanded edition (more than 500 new pages!) of Boston T. Party's classic Boston on Guns and Courage. Pulling no punches, Boston gives new advice on which shoulder weapons and handguns to buy and why before exploring such topics as why you should consider not getting a concealed carry permit, what guns and gear will likely be outlawed next, how to spend within your budget, why you should go to a quality defensive shooting academy now, which guns and gadgets are inferior and why, how to stay off illegal government gun registration lists, how to spot an undercover agent trying to entrap law-abiding gun owners and much more.

Breech-Loading Carbines of the United States Civil War Period, by Brig. Gen. John Pitman, Armory Publications, Tacoma, WA, 1987. 94 pp., illus. $29.95
The first in a series of previously unpublished manuscripts originated by the late Brigadier General John Putnam. Exploded drawings showing parts actual size follow each sectioned illustration.

The Breech-Loading Single-Shot Rifle, by Major Ned H. Roberts and Kenneth L. Waters, Wolfe Publishing Co., Prescott, AZ, 1995. 333 pp., illus. $28.50
A comprehensive and complete history of the evolution of the Schutzen and single-shot rifle.

The Bren Gun Saga, by Thomas B. Dugelby, Collector Grade Publications, Cobourg, Ontario, Canada, 1999, revised and expanded edition. 406 pp., illustrated. $65.95
A modern, definitive book on the Bren in this revised expanded edition, which in terms of numbers of pages and illustrations is nearly twice the size of the original.

British Board of Ordnance Small Arms Contractors 1689-1840, by De Witt Bailey, Rhyl, England: W. S. Curtis, 2000. 150 pp. $18.00
Thirty years of research in the Archives of the Ordnance Board in London has identified more than 600 of these suppliers. The names of many can be found marking the regulation firearms of the period. In the study, the contractors are identified both alphabetically and under a combination of their date period together with their specialist trade.

The British Enfield Rifles, Volume 1, The SMLE Mk I and Mk III Rifles, by Charles R. Stratton, North Cape Pub. Tustin, CA, 1997. 150 pp., illus. Paper covers. $16.95
A systematic and thorough examination on a part-by-part basis of the famous British battle rifle that endured for nearly 70 years as the British Army's number one battle rifle.

British Enfield Rifles, Volume 2, No.4 and No.5 Rifles, by Charles R. Stratton, North Cape Publications, Tustin, CA, 1999. 150 pp., illustrated. Paper covers. $16.95
The historical background for the development of both rifles describing each variation and an explanation of all the "marks", "numbers" and codes found on most parts.

British Enfield Rifles, Volume 4, The Pattern 1914 and U. S. Model 1917 Rifles, by Charles R. Stratton, North Cape Publications, Tustin, CA, 2000. Paper covers. $16.95

One of the lease know American and British collectible military rifles is analyzed on a part by part basis. All markings and codes, refurbishment procedures and WW 2 upgrade are included as are the various sniper rifle versions.

The British Falling Block Breechloading Rifle from 1865, by Jonathan Kirton, Tom Rowe Books, Maynardsville, TN, 2nd edition, 1997. 380 pp., illus. $70.00

Expanded 2nd edition of a comprehensive work on the British falling block rifle.

British Gun Engraving, by Douglas Tate, Safari Press, Inc., Huntington Beach, CA, 1999. 240 pp., illustrated. Limited, signed and numbered edition, in a slipcase. $80.00

A historic and photographic record of the last two centuries.

British Service Rifles and Carbines 1888-1900, by Alan M. Petrillo, Excaliber Publications, Latham, NY, 1994. 72 pp., illus, Paper covers. $11.95

A complete review of the Lee-Metford and Lee-Enfield rifles and carbines.

British Single Shot Rifles, Volume 1, Alexander Henry, by Wal Winfer, Tom Rowe, Manardville, TN, 1998, 200 pp., illus. $50.00

Detailed Study of the single shot rifles made by Henry. Illustrated with hundreds of photographs and drawings.

British Single Shot Rifles Volume 2, George Gibbs, by Wal Winfer, Tom Rowe, Maynardville, TN, 1998. 177 pp., illus. $50.00

Detailed study of the Farqurharson as made by Gibbs. Hundreds of photos.

British Single Shot Rifles, Volume 3, Jeffery, by Wal Winfer, Rowe Publications, Rochester, N.Y., 1999. 260 pp., illustrated. $60.00

The Farquharsen as made by Jeffery and his competitors, Holland & Holland, Bland, Westley, Manton, etc. Large section on the development of nitro cartridges including the .600.

British Single Shot Rifles, Vol. 4; Westley Richards, by Wal Winfer, Rowe Publications, Rochester, N.Y., 2000. 265 pages, illustrated, photos. $60.00

In his 4th volume Winfer covers a detailed study of the Westley Richards single shot rifles, including Monkey Tails, Improved Martini, 1872,1873, 1878,1881, 1897 Falling Blocks. He also covers Westley Richards Cartridges, History and Reloading information.

British Small Arms Ammunition, 1864-1938 (Other than .303 inch), by Peter Labbett, Armory Publications, Seattle, WA. 1993, 358 pages, illus. Fourcolor dust jacket. $79.00

A study of British military rifle, handgun, machine gun, and aiming tube ammunition through 1 inch from 1864 to 1938. Photo-illustrated including the firearms that chambered the cartridges.

The British Soldier's Firearms from Smoothbore to Rifled Arms, 1850-1864, by Dr. C.H. Roads, R&R Books, Livonia, NY, 1994. 332 pp., illus. $49.00

A reprint of the classic text covering the development of British military hand and shoulder firearms in the crucial years between 1850 and 1864.

British Sporting Guns & Rifles, compiled by George Hoyem, Armory Publications, Coeur d'Alene, ID, 1997. 1024 pp., illus. In two volumes. $250.00

Eighteen old sporting firearms trade catalogs and a rare book reproduced with their color covers in a limited, signed and numbered edition.

Browning Dates of Manufacture, compiled by George Madis, Art and Reference House, Brownsboro, TX, 1989. 48 pp. $10.00

Gives the date codes and product codes for all models from 1824 to the present.

Browning Sporting Arms of Distinction 1903-1992, by Matt Eastman, Matt Eastman Publications, Fitzgerald, GA, 1995. 450 pp., illus. $49.95

The most recognized publication on Browning sporting arms; covers all models.

Buffalo Bill's Wild West: An American Legend, by R.L. Wilson and Greg Martine, Random House, N.Y., 1999. 3,167 pp., illustrated. $60.00

Over 225 color plates and 160 black-and-white illustrations, with in-depth text and captions, the colorful arms, posters, photos, costumes, saddles, accoutrement are brought to life.

Bullard Arms, by G. Scott Jamieson, The Boston Mills Press, Ontario, Canada, 1989. 244 pp., illus. $35.00

The story of a mechanical genius whose rifles and cartridges were the equal to any made in America in the 1880s.

Burning Powder, compiled by Major D.B. Wesson, Wolfe Publishing Company, Prescott, AZ, 1992. 110 pp. Soft cover. $10.95

A rare booklet from 1932 for Smith & Wesson collectors.

The Burnside Breech Loading Carbines, by Edward A. Hull, Andrew Mowbray, Inc., Lincoln, RI, 1986. 95 pp., illus. $16.00

No. 1 in the "Man at Arms Monograph Series." A model-by-model historical/technical examination of one of the most widely used cavalry weapons of the American Civil War based upon important and previously unpublished research.

Camouflage Uniforms of European and NATO Armies; 1945 to the Present, by J. F. Borsarello, Atglen, PA: Schiffer Publications. Over 290 color and b/w photographs, 120 pages. Softcover. $29.95

This full-color book covers nearly all of the NATO, and other European armies' camouflaged uniforms, and not only shows and explains the many patterns, but also their efficacy of design. Described and illustrated are the variety of materials tested in over forty different armies, and includes the history of obsolete trial tests from 1945 to the present time. More than two hundred patterns have been manufactured since World War II using various landscapes and seasonal colors for their look. The Vietnam and Gulf Wars, African or South American events, as well as recent Yugoslavian independence wars have been used as experimental terrains to test a variety of patterns. This book provides a superb reference for the historian, re-enactor, designer, and modeler.

Camouflage Uniforms of the Waffen-SS A Photographic Reference, by Michael Beaver, Schiffer Publishing, Atglen, PA. Over 1,000 color and b/w photographs and illustrations, 296 pages. $69.95

Finally a book that unveils the shroud of mystery surrounding Waffen-SS camouflage clothing. Illustrated here, both in full color and in contemporary black and white photographs, this unparalleled look at Waffen-SS combat troops and their camouflage clothing will benefit both the historian and collector.

Canadian Gunsmiths from 1608: A Checklist of Tradesmen, by John Belton, Historical Arms Series No. 29. Canada: Museum Restoration, 1992. 40 pp., 17 illustrations. Softcover. $8.95

This Checklist is a greatly expanded version of HAS No. 14, listing the names, occupation, location, and dates of more than 1,500 men and women who worked as gunmakers, gunsmiths, armorers, gun merchants, gun patent holders, and a few other gun related trades. A collection of contemporary gunsmiths' letterhead have been provided to add color and depth to the study.

Cap Guns, by James Dundas, Schiffer Publishing, Atglen, PA, 1996. 160 pp., illus. Paper covers. $29.95

Over 600 full-color photos of cap guns and gun accessories with a current value guide.

Carbines of the Civil War, by John D. McAulay, Pioneer Press, Union City, TN, 1981. 123 pp., illus. Paper covers. $12.95

A guide for the student and collector of the colorful arms used by the Federal cavalry.

Carbines of the U.S. Cavalry 1861-1905, by John D. McAulay, Andrew Mowbray Publishers, Lincoln, RI, 1996. $35.00

Covers the crucial use of carbines from the beginning of the Civil War to the end of the cavalry carbine era in 1905.

Cartridge Carbines of the British Army, by Alan M. Petrillo, Excalibur Publications, Latham, NY, 1998. 72 pp., illustrated. Paper covers. $11.95

Begins with the Snider-Enfield which was the first regulation cartridge carbine introduced in 1866 and ends with the .303 caliber No.5, Mark 1 Enfield.

Cartridge Catalogues, compiled by George Hoyem, Armory Publications, Coeur d'Alene, ID., 1997. 504 pp., illus. $125.00

Fourteen old ammunition makers' and designers' catalogs reproduced with their color covers in a limited, signed and numbered edition. Completely revised edition of the general purpose reference work for which collectors, police, scientists and laymen reach first for answers to cartridge identification questions. Available October, 1996.

Cartridge Reloading Tools of the Past, by R.H. Chamberlain and Tom Quigley, Tom Quigley, Castle Rock, WA, 1998. 167 pp., illustrated. Paper covers. $25.00

A detailed treatment of the extensive Winchester and Ideal lines of handloading tools and bulletmolds plus Remington, Marlin, Ballard, Browning and many others.

Cartridges for Collectors, by Fred Datig, Pioneer Press, Union City, TN, 1999. In three volumes of 176 pp. each. Vol.1 (Centerfire); Vol.2 (Rimfire and Misc.) types; Vol.3 (Additional Rimfire, Centerfire, and Plastic.). All illustrations are shown in full-scale drawings. Volume 1, softcover only, $19.95. Volumes 2 & 3, Hardcover $19.95

Civil War Arms Makers and Their Contracts, edited by Stuart C. Mowbray and Jennifer Heroux, Andrew Mowbray Publishing, Lincoln, RI, 1998. 595 pp. $39.50

A facsimile reprint of the Report by the Commissioner of Ordnance and Ordnance Stores, 1862.

Civil War Arms Purchases and Deliveries, edited by Stuart C. Mowbray, Andrew Mowbray Publishing, Lincoln, RI, 1998. 300pp., illus. $39.50

A facsimile reprint of the master list of Civil War weapons purchases and deliveries including Small Arms, Cannon, Ordnance and Projectiles.

Civil War Breech Loading Rifles, by John D. McAulay, Andrew Mowbray, Inc., Lincoln, RI, 1991. 144 pp., illus. Paper covers. $15.00

All the major breech-loading rifles of the Civil War and most, if not all, of the obscure types are detailed, illustrated and set in their historical context.

Civil War Cartridge Boxes of the Union Infantryman, by Paul Johnson, Andrew Mowbray, Inc., Lincoln, RI, 1998. 352 pp., illustrated. $45.00

There were four patterns of infantry cartridge boxes used by Union forces during the Civil War. The author describes the development and subsequent pattern changes to these cartridge boxes.

Civil War Commanders, by Dean Thomas, Thomas Publications, Gettysburg, PA. 1998. 72 pages, illustrated, photos. Paper Covers. $9.95
    138 photographs and capsule biographies of Union and Confederate officers. A convenient personalities reference guide.

Civil War Firearms, by Joseph G. Bilby, Combined Books, Conshohocken, PA, 1996. 252 pp., illus. $34.95
    A unique work combining background data on each firearm including its battlefield use, and a guide to collecting and firing surviving relics and modern reproductions.

Civil War Guns, by William B. Edwards, Thomas Publications, Gettysburg, PA, 1997. 444 pp., illus. $40.00
    The complete story of Federal and Confederate small arms; design, manufacture, identifications, procurement issue, employment, effectiveness, and postwar disposal by the recognized expert.

Civil War Infantryman: In Camp, On the March, And in Battle, by Dean Thomas, Thomas Publications, Gettysburg, PA. 1998. 72 pages, illustrated, Softcovers. $12.95
    Uses first-hand accounts to shed some light on the "common soldier" of the Civil War from enlistment to muster-out, including camp, marching, rations, equipment, fighting, and more.

Civil War Pistols, by John D. McAulay, Andrew Mowbray Inc., Lincoln, RI, 1992. 166 pp., illus. $38.50
    A survey of the handguns used during the American Civil War.

Civil War Sharps Carbines and Rifles, by Earl J. Coates and John D. McAulay, Thomas Publications, Gettysburg, PA, 1996. 108 pp., illus. Paper covers. $12.95
    Traces the history and development of the firearms including short histories of specific serial numbers and the soldiers who received them.

Civil War Small Arms of the U.S. Navy and Marine Corps, by John D. McAulay, Mowbray Publishing, Lincoln, RI, 1999. 186 pp., illustrated. $39.00
    The first reliable and comprehensive guide to the firearms and edged weapons of the Civil War Navy and Marine Corps.

The W.F. Cody Buffalo Bill Collector's Guide with Values, by James W. Wojtowicz, Collector Books, Paducah, KY, 1998. 271 pp., illustrated. $24.95
    A profusion of colorful collectibles including lithographs, programs, photographs, books, medals, sheet music, guns, etc. and today's values.

Col. Burton's Spiller & Burr Revolver, by Matthew W. Norman, Mercer University Press, Macon, GA, 1997. 152 pp., illus. $22.95
    A remarkable archival research project on the arm together with a comprehensive story of the establishment and running of the factory.

Collector's Guide to Colt .45 Service Pistols Models of 1911 and 1911A1, Enlarged and revised edition. Clawson Publications, Fort Wayne, IN, 1998. 130 pp., illustrated. $45.00
    From 1911 to the end of production in 1945 with complete military identification including all contractors.

A Collector's Guide to United States Combat Shotguns, by Bruce N. Canfield, Andrew Mowbray Inc., Lincoln, RI, 1992. 184 pp., illus. Paper covers. $24.00
    This book provides full coverage of combat shotguns, from the earliest examples right up to the Gulf War and beyond.

A Collector's Guide to Winchester in the Service, by Bruce N. Canfield, Andrew Mowbray, Inc., Lincoln, RI, 1991. 192 pp., illus. Paper covers. $22.00
    The firearms produced by Winchester for the national defense. From Hotchkiss to the M14, each firearm is examined and illustrated.

A Collector's Guide to the '03 Springfield, by Bruce N. Canfield, Andrew Mowbray Inc, Lincoln, RI, 1989. 160 pp., illus. Paper covers. $22.00
    A comprehensive guide follows the '03 through its unparalleled tenure of service. Covers all of the interesting variations, modifications and accessories of this highly collectible military rifle.

Collector's Illustrated Encyclopedia of the American Revolution, by George C. Neumann and Frank J. Kravic, Rebel Publishing Co., Inc., Texarkana, TX, 1989. 286 pp., illus. $36.95
    A showcase of more than 2,300 artifacts made, worn, and used by those who fought in the War for Independence.

Colonial Frontier Guns, by T.M. Hamilton, Pioneer Press, Union City, TN, 1988. 176 pp., illus. Paper covers. $17.50
    A complete study of early flint muskets of this country.

Colt: An American Legend, by R.L. Wilson, Artabras, New York, 1997. 406 pages, fully illustrated, most in color. $60.00
    A reprint of the commemorative album celebrates 150 years of the guns of Samuel Colt and the manufacturing empire he built, with expert discussion of every model ever produced, the innovations of each model and variants, updated model and serial number charts and magnificent photographic showcases of the weapons.

The Colt Armory, by Ellsworth Grant, Manat-Arms Bookshelf, Lincoln, RI, 1996. 232 pp., illus. $35.00
    A history of Colt's Manufacturing Company.

Colt Blackpowder Reproductions & Replica: A Collector's and Shooter's Guide, by Dennis Miller, Blue Book Publications, Minneapolis, MN, 1999. 288 pp., illustrated. Paper covers. $29.95
    The first book on this important subject, and a must for the investor, collector, and shooter.

Colt Heritage, by R.L. Wilson, Simon & Schuster, 1979. 358 pp., illus. $75.00
    The official history of Colt firearms 1836 to the present.

Colt Memorabilia Price Guide, by John Ogle, Krause Publications, Iola, WI, 1998. 256 pp., illus. Paper covers. $29.95
    The first book ever compiled about the vast array of non-gun merchandise produced by Sam Colt's companies, and other companies using the Colt name.

The Colt Model 1905 Automatic Pistol, by John Potocki, Andrew Mowbray Publishing, Lincoln, RI, 1998. 191 pp., illus. $28.00
    Covers all aspects of the Colt Model 1905 Automatic Pistol, from its invention by the legendary John Browning to its numerous production variations.

Colt Peacemaker British Model, by Keith Cochran, Cochran Publishing Co., Rapid City, SD, 1989. 160 pp., illus. $35.00
    Covers those revolvers Colt squeezed in while completing a large order of revolvers for the U.S. Cavalry in early 1874, to those magnificent cased target revolvers used in the pistol competitions at Bisley Commons in the 1890s.

Colt Peacemaker Encyclopedia, by Keith Cochran, Keith Cochran, Rapid City, SD, 1986. 434 pp., illus. $65.00
    A must book for the Peacemaker collector.

Colt Peacemaker Encyclopedia, Volume 2, by Keith Cochran, Cochran Publishing Co., SD, 1992. 416 pp., illus. $60.00
    Included in this volume are extensive notes on engraved, inscribed, historical and noted revolvers, as well as those revolvers used by outlaws, lawmen, movie and television stars.

Colt Percussion Accoutrements 1834-1873, by Robin Rapley, Robin Rapley, Newport Beach, CA, 1994. 432 pp., illus. Paper covers. $39.95
    The complete collector's guide to the identification of Colt percussion accoutrements; including Colt conversions and their values.

Colt Pocket Hammerless Pistols, by Dr. John W. Brunner, Phillips Publications, Williamstown, NJ, 1998. 212 pp., illustrated. $59.95
    You will never again have to question a .25, .32 or .380 with this well illustrated, definitive reference guide at hand.

Colt Revolvers and the Tower of London, by Joseph G. Rosa, Royal Armouries of the Tower of London, London, England, 1988. 72 pp., illus. Soft covers. $15.00
    Details the story of Colt in London through the early cartridge period.

Colt Rifles and Muskets from 1847-1870, by Herbert Houze, Krause Publications, Iola, WI, 1996. 192 pp., illus. $34.95
    Discover previously unknown Colt models along with an extensive list of production figures for all models.

Colt's SAA Post War Models, by George Garton, The Gun Room Press, Highland Park, NJ, 1995. 166 pp., illus. $39.95
    Complete facts on the post-war Single Action Army revolvers. Information on calibers, production numbers and variations taken from factory records.

Colt Single Action Army Revolvers: The Legend, the Romance and the Rivals, by "Doc" O'Meara, Krause Publications, Iola, WI, 2000. 160 pp., illustrated with 250 photos in b&w and a 16 page color section. $34.95
    Production figures, serial numbers by year, and rarities.

Colt Single Action Army Revolvers and Alterations, by C. Kenneth Moore, Mowbray Publishers, Lincoln, RI, 1999. 112 pp., illustrated. $35.00
    A comprehensive history of the revolvers that collectors call "Artillery Models." These are the most historical of all S.A.A. Colts, and this new book covers all the details.

Colt Single Action Army Revolvers and the London Agency, by C. Kenneth Moore, Andrew Mowbray Publishers, Lincoln, RI, 1990. 144 pp., illus. $35.00
    Drawing on vast documentary sources, this work chronicles the relationship between the London Agency and the Hartford home office.

The Colt U.S. General Officers' Pistols, by Horace Greeley IV, Andrew Mowbray Inc., Lincoln, RI, 1990. 199 pp., illus. $38.00
    These unique weapons, issued as a badge of rank to General Officers in the U.S. Army from WWII onward, remain highly personal artifacts of the military leaders who carried them. Includes serial numbers and dates of issue.

Colts from the William M. Locke Collection, by Frank Sellers, Andrew Mowbray Publishers, Lincoln, RI, 1996. 192 pp., illus. $55.00
    This important book illustrates all of the famous Locke Colts, with captions by arms authority Frank Sellers.

Colt's Dates of Manufacture 1837-1978, by R.L. Wilson, published by Maurie Albert, Coburg, Australia; N.A. distributor I.D.S.A. Books, Hamilton, OH, 1983. 61 pp. $6.00
    An invaluable pocket guide to the dates of manufacture of Colt firearms up to 1978.

Colt's 100th Anniversary Firearms Manual 1836-1936: A Century of Achievement, Wolfe Publishing Co., Prescott, AZ, 1992. 100 pp., illus. Paper covers. $12.95
    Originally published by the Colt Patent Firearms Co., this booklet covers the history, manufacturing procedures and the guns of the first 100 years of the genius of Samuel Colt.

Colt's Pocket '49: Its Evolution Including the Baby Dragoon and Wells Fargo, by Robert Jordan and Darrow Watt, Privately Printed, Loma Mar, CA 2000. 304 pages, with 984 color photos, illus. Beautify bound in a deep blue leather like case. $125.00
    Detailed information on all models and covers engraving, cases, accoutrements, holsters, fakes, and much more. Included is a summary booklet containing information such as serial numbers, production ranges & identifying photos. This book is a masterpiece on its subject.

Complete Guide to all United States Military Medals 1939 to Present, by Colonel Frank C. Foster, Medals of America Press, Fountain Inn, SC, 2000. 121 pp.,.illustrated, photos. $29.95
Complete criteria for every Army, Navy, Marines, Air Force, Coast Guard, and Merchant Marine awards since 1939. All decorations, service medals, and ribbons shown in full-color and accompanied by dates and campaigns as well as detailed descriptions on proper wear and display.

Complete Guide to the M1 Garand and the M1 Carbine, by Bruce N. Canfield, 2nd printing, Andrew Mowbray Inc., Lincoln, RI, 1999. 296 pp., illus. $39.50
Expanded and updated coverage of both the M1 Garand and the M1 Carbine, with more than twice as much information as the author's previous book on this topic.

The Complete Guide to U.S. Infantry Weapons of the First War, by Bruce Canfield, Andrew Mowbray, Publisher, Lincoln, RI, 2000. 304 pp., illus. $39.95
The definitive study of the U.S. Infantry weapons used in WW1.

The Complete Guide to U.S. Infantry Weapons of World War Two, by Bruce Canfield, Andrew Mowbray, Publisher, Lincoln, RI, 1995. 303 pp., illus. $39.95
A definitive work on the weapons used by the United States Armed Forces in WWII.

A Concise Guide to the Artillery at Gettysburg, by Gregory Coco, Thomas Publications, Gettysburg, PA, 1998. 96 pp., illus. Paper Covers. $10.00
Coco's tenth book on Gettysburg is a beginner's guide to artillery and its use at the battle. It covers the artillery batteries describing the types of cannons, shells, fuses, etc.using interesting narrative and human interest stories.

Cooey Firearms, Made in Canada 1919-1979, by John A. Belton, Museum Restoration, Canada, 1998. 36pp., with 46 illus. Paper Covers. $8.95
More than 6 million rifles and at least 67 models, were made by this small Canadian riflemaker. They have been identified from the first 'Cooey Canuck' through the last variations made by the 'Winchester-Cooey'. Each is described and most are illustrated in this first book on The Cooey.

Cowboy Collectibles and Western Memorabilia, by Bob Bell and Edward Vebell, Schiffer Publishing, Atglen, PA, 1992. 160 pp., illus. Paper covers. $29.95
The exciting era of the cowboy and the wild west collectibles including rifles, pistols, gun rigs, etc.

Cowboy Culture: The Last Frontier of American Antiques, by Michael Friedman, Schiffer Publishing, Ltd., West Chester, PA, 1992. 300 pp., illustrated.
Covers the artful aspects of the old west, the antiques and collectibles. Illustrated with clear color plates of over 1,000 items such as spurs, boots, guns, saddles etc.

Cowboy and Gunfighter Collectible, by Bill Mackin, Mountain Press Publishing Co., Missoula, MT, 1995. 178 pp., illus. Paper covers. $25.00
A photographic encyclopedia with price guide and makers' index.

Cowboys and the Trappings of the Old West, by William Manns and Elizabeth Clair Flood, Zon International Publishing Co., Santa Fe, NM, 1997, 1st edition. 224 pp., illustrated. $45.00
A pictorial celebration of the cowboys dress and trappings.

Cowboy Hero Cap Pistols, by Rudy D'Angelo, Antique Trader Books, Dubuque, IA, 1998. 196 pp., illus. Paper covers. $34.95
Aimed at collectors of cap pistols created and named for famous film and television cowboy heroes, this in-depth guide hits all the marks. Current values are given.

Custom Firearms Engraving, by Tom Turpin, Krause Publications, Iola, WI, 1999. 208 pp., illustrated. $49.95
Over 200 four-color photos with more than 75 master engravers profiled. Engravers Directory with addresses in the U.S. and abroad.

The Decorations, Medals, Ribbons, Badges and Insignia of the United States Army; World War 2 to Present, by Col. Frank C. Foster, Medals of America Press, Fountain Inn, SC. 2001. 145 pp., illustrated. $29.95
The most complete guide to United States Army medals, ribbons, rank, insignia had patches from WWII to the present day. Each medal and insignia shown in full color. Includes listing of respective criteria and campaigns.

The Decorations, Medals, Ribbons, Badges and Insignia of the United States Navy; World War 2 to Present, by James G. Thompson, Medals of America Press, Fountain Inn, SC. 2000. 123 pages, illustrated. $29.95
The most complete guide to United States Navy medals, ribbons, rank, insignia had patches from WWII to the present day. Each medal and insignia shown in full color. Includes listing of respective criteria and campaigns.

The Derringer in America, Volume 1, The Percussion Period, by R.L. Wilson and L.D. Eberhart, Andrew Mowbray Inc., Lincoln, RI, 1985. 271 pp., illus. $48.00
A long awaited book on the American percussion derringer.

The Derringer in America, Volume 2, The Cartridge Period, by L.D. Eberhart and R.L. Wilson, Andrew Mowbray Inc., Publishers, Lincoln, RI, 1993. 284 pp., illus. $65.00
Comprehensive coverage of cartridge derringers organized alphabetically by maker. Includes all types of derringers known to the authors to have been offered to the American market.

The Devil's Paintbrush: Sir Hiram Maxim's Gun, by Dolf Goldsmith, 3rd Edition, expanded and revised, Collector Grade Publications, Toronto, Canada, 2000. 384 pp., illus. $79.95
The classic work on the world's first true automatic machine gun.

Dr. Josephus Requa Civil War Dentist and the Billinghurst-Requa Volley Gun, by John M. Hyson, Jr., & Margaret Requa DeFrancisco, Museum Restoration Service, Bloomfield, Ont., Canada, 1999. 36 pp., illus. Paper covers. $8.95
The story of the inventor of the first practical rapid-fire gun to be used during the American Civil War.

The Duck Stamp Story, by Eric Jay Dolin and Bob Dumaine, Krause Publications, Iola, WI, 2000. 208 pp., illustrated with color throughout. Paper covers. $29.95; Hardbound. $49.95
Detailed information on the value and rarity of every federal duck stamp. Outstanding art and illustrations.

The Dutch Luger (Parabellum) A Complete History, by Bas J. Martens and Guus de Vries, Ironside International Publishers, Inc., Alexandria, VA, 1995. 268 pp., illus. $49.95
The history of the Luger in the Netherlands. An extensive description of the Dutch pistol and trials and the different models of the Luger in the Dutch service.

The Eagle on U.S. Firearms, by John W. Jordan, Pioneer Press, Union City, TN, 1992. 140 pp., illus. Paper covers. $17.50
Stylized eagles have been stamped on government owned or manufactured firearms in the U.S. since the beginning of our country. This book lists and illustrates these various eagles in an informative and refreshing manner.

Encyclopedia of Rifles & Handguns; A Comprehensive Guide to Firearms, edited by Sean Connolly, Chartwell Books, Inc., Edison, NJ., 1996. 160 pp., illustrated. $26.00
A lavishly illustrated book providing a comprehensive history of military and civilian personal firepower.

Eprouvettes: A Comprehensive Study of Early Devices for the Testing of Gunpowder, by R.T.W. Kempers, Royal Armouries Museum, Leeds, England, 1999. 352 pp., illustrated with 240 black & white and 28 color plates. $125.00
The first comprehensive study of eprouvettes ever attempted in a single volume.

European Firearms in Swedish Castles, by Kaa Wennberg, Bohuslaningens Boktryckeri AB, Uddevalla, Sweden, 1986. 156 pp., illus. $50.00
The famous collection of Count Keller, the Ettersburg Castle collection, and others. English text.

European Sporting Cartridges, Part 1, by W.B. Dixon, Armory Publications, Inc., Coeur d'Alene, ID, 1997. 250 pp., illus. $63.00
Photographs and drawings of over 550 centerfire cartridge case types in 1,300 illustrations produced in German and Austria from 1875 to 1995.

European Sporting Cartridges, Part 2, by W.B. Dixon, Armory Publications, Inc., Coeur d'Alene, ID, 2000. 240 pp., illus. $63.00
An illustrated history of centerfire hunting and target cartridges produced in Czechoslovakia, Switzerland, Norway, Sweden, Finland, Russia, Italy, Denmark, Belguim from 1875 to 1998. Adds 50 specimens to volume 1 (German-Austria). Also, illustrates 40 small arms magazine experiments during the late 19th Century, and includes the English-Language export ammunition catalogue of Kovo (Povaszke Strojarne), Prague, Czeck. from the 1930's.

Fifteen Years in the Hawken Lode, by John D. Baird, The Gun Room Press, Highland Park, NJ, 1976. 120 pp., illus. $24.95.
A collection of thoughts and observations gained from many years of intensive study of the guns from the shop of the Hawken brothers.

'51 Colt Navies, by Nathan L. Swayze, The Gun Room Press, Highland Park, NJ, 1993. 243 pp., illus. $59.95
The Model 1851 Colt Navy, its variations and markings.

Fighting Iron, by Art Gogan, Andrew Mowbray, Inc., Lincoln, R.I., 1999. 176 pp., illustrated. $28.00
It doesn't matter whether you collect guns, swords, bayonets or accoutrement— sooner or later you realize that it all comes down to the metal. If you don't understand the metal you don't understand your collection.

Fine Colts, The Dr. Joseph A. Murphy Collection, by R.L. Wilson, Sheffield Marketing Associates, Inc., Doylestown, PA, 1999. 258 pp., illustrated. Limited edition signed and numbered. $99.00
This lavish new work covers exquisite, deluxe and rare Colt arms from Paterson and other percussion revolvers to the cartridge period and up through modern times.

Firearms, by Derek Avery, Desert Publications, El Dorado, AR, 1999. 95 pp., illustrated. $9.95
The firearms included in this book are by necessity only a selection, but nevertheless one that represents the best and most famous weapons seen since the Second World War.

Firearms and Tackle Memorabilia, by John Delph, Schiffer Publishing, Ltd., West Chester, PA, 1991. 124 pp., illus. $39.95
A collector's guide to signs and posters, calendars, trade cards, boxes, envelopes, and other highly sought after memorabilia. With a value guide.

Firearms of the American West 1803-1865, Volume 1, by Louis A. Garavaglia and Charles Worman, University of Colorado Press, Niwot, CO, 1998. 402 pp., illustrated. $59.95
Traces the development and uses of firearms on the frontier during this period.

Firearms of the American West 1866-1894, by Louis A. Garavaglia and Charles G. Worman, University of Colorado Press, Niwot, CO, 1998. 416 pp., illus. $59.95

A monumental work that offers both technical information on all of the important firearms used in the West during this period and a highly entertaining history of how they were used, who used them, and why.

Firearms from Europe, by David Noe, Larry W. Yantz, Dr. James B. Whisker, Rowe Publications, Rochester, N.Y., 1999. 192 pp., illustrated. $45.00
A history and description of firearms imported during the American Civil War by the United States of America and the Confederate States of America.

Firepower from Abroad, by Wiley Sword, Andrew Mowbray Publishing, Lincoln, R.I., 2000. 120 pp., illustrated. $23.00
The Confederate Enfield and the LeMat revolver and how they reached the Confederate market.

Flayderman's Guide to Antique American Firearms and Their Values, 7th Edition, edited by Norm Flayderman, DBI books, a division of Krause Publications, Iola, WI, 1998. 656 pp., illus. Paper covers. $32.95
A completely updated and new edition with more than 3,600 models and variants extensively described with all marks and specifications necessary for quick identification.

The FN-FAL Rifle, etal, by Duncan Long, Paladin Press, Boulder, CO, 1999. 144 pp., illustrated. Paper covers. $18.95
Detailed descriptions of the basic models produced by Fabrique Nationale and the myriad variants that evolved as a result of the firearms universal acceptance.

The .45-70 Springfield, by Joe Poyer and Craig Riesch, North Cape Publications, Tustin, CA, 1996. 150 pp., illus. Paper covers. $16.95
A revised and expanded second edition of a best-selling reference work organized by serial number and date of production to aid the collector in identifying popular "Trapdoor" rifles and carbines.

The French 1935 Pistols, by Eugene Medlin and Colin Doane, Eugene Medlin, El Paso, TX, 1995. 172 pp., illus. Paper covers. $25.95
The development and identification of successive models, fakes and variants, holsters and accessories, and serial numbers by dates of production.

Freund & Bro. Pioneer Gunmakers to the West, by F.J. Pablo Balentine, Graphic Publishers, Newport Beach, CA, 1997. 380 pp., illustrated $69.95
The story of Frank W. and George Freund, skilled German gunsmiths who plied their trade on the Western American frontier during the final three decades of the nineteenth century.

From the Kingdom of Lilliput: The Miniature Firearms of David Kucer, by K. Corey Keeble and The Making of Miniatures, by David Kucer, Museum Restoration Service, Ontario, Canada, 1994. 51 pp., illus, $25.00
An overview of the subject of miniatures in general combined with an outline by the artist himself on the way he makes a miniature firearm.

Frontier Pistols and Revolvers, by Dominique Venner, Book Sales Inc., Edison, N.J., 1998. 144 pp., illus. $19.95
Colt, Smith & Wesson, Remington and other early-brand revolvers which tamed the American frontier are shown amid vintage photographs, etchings and paintings to evoke the wild West.

The Fusil de Tulole in New France, 1691-1741, by Russel Bouchard, Museum Restorations Service, Bloomfield, Ontario, Canada, 1997. 36 pp., illus. Paper covers. $8.95
The development of the company and the identification of their arms.

Game Guns & Rifles: Percussion to Hammerless Ejector in Britain, by Richard Akehurst, Trafalgar Square, N. Pomfret, VT, 1993. 192 pp., illus. $39.95
Long considered a classic this important reprint covers the period of British gunmaking between 1830-1900.

The Gas Trap Garand, by Billy Pyle, Collector Grade Publications, Cobourg, Ontario, Canada, 1999 316 pp., illustrated. $59.95
The in-depth story of the rarest Garands of them all, the initial 80 Model Shop rifles made under the personal supervision of John Garand himself in 1934 and 1935, and the first 50,000 plus production "gas trap" M1's manufactured at Springfield Armory between August, 1937 and August, 1940.

George Schreyer, Sr. and Jr., Gunmakers of Hanover, Pennsylvania, by George Shumway, George Shumway Publishers, York, PA, 1990. 160pp., illus. $50.00
This monograph is a detailed photographic study of almost all known surviving long rifles and smoothbore guns made by highly regarded gunsmiths George Schreyer, Sr. and Jr.

The German Assault Rifle 1935-1945, by Peter R. Senich, Paladin Press, Boulder, CO, 1987. 328 pp., illus. $60.00
A complete review of machine carbines, machine pistols and assault rifles employed by Hitler's Wehrmacht during WWII.

The German K98k Rifle, 1934-1945: The Backbone of the Wehrmacht, by Richard D. Law, Collector Grade Publications, Toronto, Canada, 1993. 336 pp., illus. $69.95
The most comprehensive study ever published on the 14,000,000 bolt-action K98k rifles produced in Germany between 1934 and 1945.

German Machineguns, by Daniel D. Musgrave, Revised edition, Ironside International Publishers, Inc. Alexandria, VA, 1992. 586 pp., 650 illus. $49.95
The most definitive book ever written on German machineguns. Covers the introduction and development of machineguns in Germany from 1899 to the rearmament period after WWII.

German Military Rifles and Machine Pistols, 1871-1945, by Hans Dieter Gotz, Schiffer Publishing Co., West Chester, PA, 1990. 245 pp., illus. $35.00
This book portrays in words and pictures the development of the modern German weapons and their ammunition including the scarcely known experimental types.

The German MP40 Maschinenpistole, by Frank Iannamico, Moose Lake Publishing, Harmony, ME, 1999. 185 pp., illustrated. Paper covers. $19.95
The history, development and use of this famous gun of World War 2.

German 7.9mm Military Ammunition, by Daniel W. Kent, Daniel W. Kent, Ann Arbor, MI, 1991. 244 pp., illus. $35.00
The long-awaited revised edition of a classic among books devoted to ammunition.

The Golden Age of Remington, by Robert W.D. Ball, Krause publications, Iola, WI, 1995. 194 pp., illus. $29.95
For Remington collectors or firearms historians, this book provides a pictorial history of Remington through World War I. Includes value guide.

The Government Models, by William H.D. Goddard, Andrew Mowbray Publishing, Lincoln, RI, 1998. 296 pp., illustrated. $58.50
The most authoritative source on the development of the Colt model of 1911.

Grasshoppers and Butterflies, by Adrian B. Caruana, Museum Restoration Service, Alexandria, Bay, N.Y. 1999. 32 pp., illustrated. Paper covers. $8.95
No.39 in the Historical Arms Series. The light 3 pounders of Pattison and Townsend.

The Greener Story, by Graham Greener, Quiller Press, London, England, 2000. 256 pp., illustrated with 32 pages of color photos. $64.50
W.W. Greener, his family history, inventions, guns, patents, and more.

A Guide to American Trade Catalogs 1744-1900, by Lawrence B. Romaine, Dover Publications, New York, NY. 422 pp., illus. Paper covers. $12.95

A Guide to Ballard Breechloaders, by George J. Layman, Pioneer Press, Union City, TN, 1997. 261 pp., illus. Paper covers. $19.95
Documents the saga of this fine rifle from the first models made by Ball & Williams of Worchester, to its production by the Marlin Firearms Co, to the cessation of 19th century manufacture in 1891, and finally to the modern reproductions made in the 1990's.

A Guide to the Maynard Breechloader, by George J. Layman, George J. Layman, Ayer, MA, 1993. 125 pp., illus. Paper covers. $11.95
The first book dedicated entirely to the Maynard family of breech-loading firearms. Coverage of the arms is given from the 1850s through the 1880s.

A Guide to U. S. Army Dress Helmets 1872-1904, by Kasal and Moore, North Cape Publications, 2000. 88 pp., illus. Paper covers. $15.95
This thorough study provides a complete description of the Model 1872 & 1881 dress helmets worn by the U.S. Army. Including all components from bodies to plates to plumes & shoulder cords and tells how to differentiate the originals from reproductions. Extensively illustrated with photographs, '8 pages in full color' of complete helmets and their components.

Gun Collecting, by Geoffrey Boothroyd, Sportsman's Press, London, 1989. 208 pp., illus. $29.95
The most comprehensive list of 19th century British gunmakers and gunsmiths ever published.

Gunmakers of London 1350-1850, by Howard L. Blackmore, George Shumway Publisher, York, PA, 1986. 222 pp., illus. $35.00
A listing of all the known workmen of gun making in the first 500 years, plus a history of the guilds, cutlers, armourers, founders, blacksmiths, etc. 260 gunmarks are illustrated.

Gunmakers of London Supplement 1350-1850, by Howard L. Blackmore, Museum Restoration Service, Alexandria Bay, NY, 1999. 156 pp., illustrated. $60.00
Begins with an introductory chapter on "foreign" gunmakers followed by records of all the new information found about previously unidentified armourers, gunmakers and gunsmiths.

The Guns that Won the West: Firearms of the American Frontier, 1865-1898, by John Walter, Stackpole Books, Inc., Mechanicsburg, PA.,1999. 256 pp., illustrated. $34.95
Here is the story of the wide range of firearms from pistols to rifles used by plainsmen and settlers, gamblers, native Americans and the U.S. Army.

Gunsmiths of Illinois, by Curtis L. Johnson, George Shumway Publishers, York, PA, 1995. 160 pp., illus. $50.00
Genealogical information is provided for nearly one thousand gunsmiths. Contains hundreds of illustrations of rifles and other guns, of handmade origin, from Illinois.

The Gunsmiths of Manhattan, 1625-1900: A Checklist of Tradesmen, by Michael H. Lewis, Museum Restoration Service, Bloomfield, Ont., Canada, 1991. 40 pp., illus. Paper covers. $8.95
This listing of more than 700 men in the arms trade in New York City prior to about the end of the 19th century will provide a guide for identification and further research.

The Guns of Dagenham: Lanchester, Patchett, Sterling, by Peter Laidler and David Howroyd, Collector Grade Publications, Inc., Cobourg, Ont., Canada, 1995. 310 pp., illus. $39.95
An in-depth history of the small arms made by the Sterling Company of Dagenham, Essex, England, from 1940 until Sterling was purchased by British Aerospace in 1989 and closed.

Guns of the Western Indian War, by R. Stephen Dorsey, Collector's Library, Eugene, OR, 1997. 220 pp., illus. Paper covers. $30.00

The full story of the guns and ammunition that made western history in the turbulent period of 1865-1890.

Gun Powder Cans & Kegs, by Ted & David Bacyk and Tom Rowe, Rowe Publications, Rochester, NY, 1999. 150 pp., illus. $65.00

The first book devoted to powder tins and kegs. All cans and kegs in full color. With a price guide and rarity scale.

The Guns of Remington: Historic Firearms Spanning Two Centuries, compiled by Howard M. Madaus, Biplane Productions, Publisher, in cooperation with Buffalo Bill Historical Center, Cody, WY, 1998. 352 pp., illustrated with over 800 color photographs. $79.95

A complete catalog of the firearms in the exhibition, "It Never Failed Me: The Arms & Art of Remington Arms Company" at the Buffalo Bill Historical Center, Cody, Wyoming.

Gun Tools, Their History and Identification, by James B. Shaffer, Lee A. Rutledge and R. Stephen Dorsey, Collector's Library, Eugene, OR, 1992. 375 pp., illus. $30.00

Written history of foreign and domestic gun tools from the flintlock period to WWII.

Gun Tools, Their History and Identifications, Volume 2, by Stephen Dorsey and James B. Shaffer, Collectors' Library, Eugene, OR, 1997. 396 pp., illus. Paper covers. $30.00

Gun tools from the Royal Armouries Museum in England, Pattern Room, Royal Ordnance Reference Collection in Nottingham and from major private collections.

Gunsmiths of the Carolinas 1660-1870, by Daniel D. Hartzler and James B. Whisker, Old Bedford Village Press, Bedford, PA, 1998. 176 pp., illustrated. $40.00

This deluxe hard bound edition of 176 pages is printed on fine coated paper, with about 90 pages of large photographs of fine longrifles from the Carolinas, and about 90 pages of detailed research on the gunsmiths who created the highly prized and highly collectable longrifles. Dedicated to serious students of original Kentucky rifles, who may seldom encounter fine longrifles from the Carolinas.

Gunsmiths of Maryland, by Daniel D. Hartzler and James B. Whisker, Old Bedford Village Press, Bedford, PA, 1998. 208 pp., illustrated. $45.00

Covers firelock Colonial period through the breech-loading patent models. Featuring longrifles.

Gunsmiths of Virginia, by Daniel D. Hartzler and James B. Whisker, Old Bedford Village Press, Bedford, PA, 1992. 206 pp., illustrated. $45.00

A photographic study of American longrifles.

Gunsmiths of West Virginia, by Daniel D. Hartzler and James B. Whisker, Old Bedford Village Press, Bedford, PA, 1998. 176 pp., illustrated. $40.00

A photographic study of American longrifles.

Gunsmiths of York County, Pennsylvania, by Daniel D. Hartzler and James B. Whisker, Old Bedford Village Press, Bedford, PA, 1998. 160 pp., illustrated. $40.00

160 pages of photographs and research notes on the longrifles and gunsmiths of York County, Pennsylvania. Many longrifle collectors and gun builders have noticed that York County style rifles tend to be more formal in artistic decoration than some other schools of style. Patriotic themes, and folk art were popular design elements.

Hall's Military Breechloaders, by Peter A. Schmidt, Andrew Mowbray Publishers, Lincoln, RI, 1996. 232 pp., illus. $55.00

The whole story behind these bold and innovative firearms.

The Handgun, by Geoffrey Boothroyd, David and Charles, North Pomfret, VT, 1989. 566 pp., illus. $60.00

Every chapter deals with an important period in handgun history from the 14th century to the present.

Handgun of Military Rifle Marks 1866-1950, by Richard A. Hoffman and Noel P. Schott, Mapleleaf Militaria Publishing, St. Louis, MO, 1999, second edition. 60 pp., illustrated. Paper covers. $20.00

An illustrated guide to identifying military rifle and marks.

Handguns & Rifles: The Finest Weapons from Around the World, by Ian Hogg, Random House Value Publishing, Inc., N.Y., 1999. 128 pp., illustrated. $18.98

The serious gun collector will welcome this fully illustrated examination of international handguns and rifles. Each entry covers the history of the weapon, what purpose it serves, and its advantages and disadvantages.

The Hawken Rifle: Its Place in History, by Charles E. Hanson, Jr., The Fur Press, Chadron, NE, 1979. 104 pp., illus. Paper covers. $15.00

A definitive work on this famous rifle.

Hawken Rifles, The Mountain Man's Choice, by John D. Baird, The Gun Room Press, Highland Park, NJ, 1976. 95 pp., illus. $29.95

Covers the rifles developed for the Western fur trade. Numerous specimens are described and shown in photographs.

High Standard: A Collector's Guide to the Hamden & Hartford Target Pistols, by Tom Dance, Andrew Mowbray, Inc., Lincoln, RI, 1991. 192 pp., illus. Paper covers. $24.00

From Citation to Supermatic, all of the production models and specials made from 1951 to 1984 are covered according to model number or series.

Historic Pistols: The American Martial Flintlock 1760-1845, by Samuel E. Smith & Edwin W. Bitter, The Gun Room Press, Highland Park, NJ, 1986. 353 pp., illus. $45.00

Covers over 70 makers and 163 models of American martial arms.

Historical Hartford Hardware, by William W. Dalrymple, Colt Collector Press, Rapid City, SD, 1976. 42 pp., illus. Paper covers. $10.00

Historically associated Colt revolvers.

The History and Development of Small Arms Ammunition, Volume 2, by George A. Hoyem, Armory Publications, Oceanside, CA, 1991. 303 pp., illus. $65.00

Covers the blackpowder military centerfire rifle, carbine, machine gun and volley gun ammunition used in 28 nations and dominions, together with the firearms that chambered them.

The History and Development of Small Arms Ammunition, Volume 4, by George A. Hoyem, Armory Publications, Seattle, WA, 1998. 200 pp., illustrated $65.00

A comprehensive book on American black powder and early smokeless rifle cartridges.

The History of Colt Firearms, by Dean Boorman, Lyons Press, New York, NY, 2001. 144 pp., illus. $29.95

Discover the fascinating story of the world's most famous revolver, complete with more than 150 stunning full-color photographs.

History of Modern U.S. Military Small Arms Ammunition. Volume 1, 1880-1939, revised by F.W. Hackley, W.H. Woodin and E.L. Scranton, Thomas Publications, Gettysburg, PA, 1998. 328 pp., illus. $49.95

This revised edition incorporates all publicly available information concerning military small arms ammunition for the period 1880 through 1939 in a single volume.

History of Modern U.S. Military Small Arms Ammunition, Volume 2, 1940-1945 by F.W. Hackley, W.H. Woodin and E.L. Scranton. Gun Room Press, Highland Park, NJ. 300 + pages, illustrated. $39.95

Based on decades of original research conducted at the National Archives, numerous military, public and private museums and libraries, as well as individual collections, this edition incorporates all publicly available information concerning military small arms ammunition for the period 1940 through 1945.

The History of Winchester Rifles, by Dean Boorman, Lyons Press, New York, NY, 2001. 144 pp., illus. $29.95

A captivating and wonderfully photographed history of one of the most legendary names in gun lore. 150 full-color photos.

The History of Winchester Firearms 1866-1992, sixth edition, updated, expanded, and revised by Thomas Henshaw, New Win Publishing, Clinton, NJ, 1993. 280 pp., illus. $27.95

This classic is the standard reference for all collectors and others seeking the facts about any Winchester firearm, old or new.

History of Winchester Repeating Arms Company, by Herbert G. Houze, Krause Publications, Iola, WI, 1994. 800 pp., illus. $50.00

The complete Winchester history from 1856-1981.

Honour Bound: The Chauchat Machine Rifle, by Gerard Demaison and Yves Buffetaut, Collector Grade Publications, Inc., Cobourg, Ont., Canada, 1995. $39.95

The story of the CSRG (Chauchat) machine rifle, the most manufactured automatic weapon of World War One.

Hopkins & Allen Revolvers & Pistols, by Charles E. Carder, Avil Onze Publishing, Delphos, OH, 1998, illustrated. Paper covers. $24.95

Covers over 165 photos, graphics and patent drawings.

How to Buy and Sell Used Guns, by John Traister, Stoeger Publishing Co., So. Hackensack, NJ, 1984. 192 pp., illus. Paper covers. $10.95

A new guide to buying and selling guns.

Hunting Weapons From the Middle Ages to the Twentieth Century, by Howard L. Blackmore, Dover Publications, Meneola, NY, 2000. 480 pp., illustrated. Paper covers. $16.95

Dealing mainly with the different classes of weapons used in sport—swords, spears, crossbows, guns, and rifles—from the Middle Ages until the present day.

Identification Manual on the .303 British Service Cartridge, No. 1-Ball Ammunition, by B.A. Temple, I.D.S.A. Books, Piqua, OH, 1986. 84 pp., 57 illus. $12.50

Identification Manual on the .303 British Service Cartridge, No. 2-Blank Ammunition, by B.A. Temple, I.D.S.A. Books, Piqua, OH, 1986. 95 pp., 59 illus. $12.50

Identification Manual on the .303 British Service Cartridge, No. 3-Special Purpose Ammunition, by B.A. Temple, I.D.S.A. Books, Piqua, OH, 1987. 82 pp., 49 illus. $12.50

Identification Manual on the .303 British Service Cartridge, No. 4-Dummy Cartridges Henry 1869-c.1900, by B.A. Temple, I.D.S.A. Books, Piqua, OH, 1988. 84 pp., 70 illus. $12.50

Identification Manual on the .303 British Service Cartridge, No. 5-Dummy Cartridges (2), by B.A. Temple, I.D.S.A. Books, Piqua, OH, 1994. 78 pp. $12.50

The Illustrated Book of Guns, by David Miller, Salamander Books, N.Y., N.Y., 2000. 304 pp., illustrated in color. $34.95

An illustrated directory of over 1,000 military and sporting firearms.

The Illustrated Encyclopedia of Civil War Collectibles, by Chuck Lawliss, Henry Holt and Co., New York, NY, 1997. 316 pp., illus. Paper covers. $22.95
  A comprehensive guide to Union and Confederate arms, equipment, uniforms, and other memorabilia.

Illustrations of United States Military Arms 1776-1903 and Their Inspector's Marks, compiled by Turner Kirkland, Pioneer Press, Union City, TN, 1988. 37 pp., illus. Paper covers. $7.00.
  Reprinted from the 1949 Bannerman catalog. Valuable information for both the advanced and beginning collector.

Indian War Cartridge Pouches, Boxes and Carbine Boots, by R. Stephen Dorsey, Collector's Library, Eugene, OR, 1993. 156 pp., illus. Paper Covers. $20.00
  The key reference work to the cartridge pouches, boxes, carbine sockets and boots of the Indian War period 1865-1890.

An Introduction to the Civil War Small Arms, by Earl J. Coates and Dean S. Thomas, Thomas Publishing Co., Gettysburg, PA, 1990. 96 pp., illus. Paper covers. $10.00
  The small arms carried by the individual soldier during the Civil War.

Japanese Rifles of World War Two, by Duncan O. McCollum, Excalibur Publications, Latham, NY, 1996. 64 pp., illus. Paper covers. $18.95
  A sweeping view of the rifles and carbines that made up Japan's arsenal during the conflict.

Kalashnikov Arms, compiled by Alexei Nedelin, Design Military Parade, Ltd., Moscow, Russia, 1997. 240 pp., illus. $49.95
  Weapons versions stored in the St. Petersburg Military Historical Museum of Artillery, Engineer Troops and Communications and in the Izhmash JSC.

Kalashnikov "Machine Pistols, Assault Rifles, and Machine Guns, 1945 to the Present", by John Walter, Paladin Press, Boulder, CO, 1999, hardcover, photos, illus., 146 pp. $22.95
  This exhaustive work published by Greenhill Military Manuals features a gun-by-gun directory of Kalashnikov variants. Technical specifications and illustrations are provided throughout, along with details of sights, bayonets, markings and ammunition. A must for the serious collector and historian.

The Kentucky Pistol, by Roy Chandler and James Whisker, Old Bedford Village Press, Bedford, PA, 1997. 225 pp., illus. $60.00
  A photographic study of Kentucky pistols from famous collections.

The Kentucky Rifle, by Captain John G.W. Dillin, George Shumway Publisher, York, PA, 1993. 221 pp., illus. $50.00
  This well-known book was the first attempt to tell the story of the American longrifle. This edition retains the original text and illustrations with supplemental footnotes provided by Dr. George Shumway.

Know Your Broomhandle Mausers, by R.J. Berger, Blacksmith Corp., Southport, CT, 1985. 96 pp., illus. Paper covers. $12.95
  An interesting story on the big Mauser pistol and its variations.

Krag Rifles, by William S. Brophy, The Gun Room Press, Highland Park, NJ, 1980. 200 pp., illus. $35.00
  The first comprehensive work detailing the evolution and various models, both military and civilian.

The Krieghoff Parabellum, by Randall Gibson, Midland, TX, 1988. 279 pp., illus. $40.00
  A comprehensive text pertaining to the Lugers manufactured by H. Krieghoff Waffenfabrik.

Las Pistolas Espanolas Tipo "Mauser," by Artemio Mortera Perez, Quiron Ediciones, Valladolid, Spain, 1998. 71 pp., illustrated. Paper covers. $34.95
  This book covers in detail Spanish machine pistols and C96 copies made in Spain. Covers all Astra "Mauser" pistol series and the complete line of Beistegui C96 type pistols. Spanish text.

Law Enforcement Memorabilia Price and Identification Guide, by Monty McCord, DBI Books a division of Krause Publications, Inc. Iola, WI, 1999. 208 pp., illustrated. Paper covers. $19.95
  An invaluable reference to the growing wave of law enforcement collectors. Hundreds of items are covered from miniature vehicles to clothes, patches, and restraints.

Legendary Sporting Guns, by Eric Joly, Abbeville Press, New York, N.Y., 1999. 228 pp., illustrated. $65.00
  A survey of hunting through the ages and relates how many different types of firearms were created and refined for use afield.

Legends and Reality of the AK, by Val Shilin and Charlie Cutshaw, Paladen Press, Boulder, CO, 2000. 192 pp., illustrated. Paper covers. $35.00
  A behind-the-scenes look at history, design and impact of the Kalashnikov family of weapons.

LeMat, the Man, the Gun, by Valmore J. Forgett and Alain F. and Marie-Antoinette Serpette, Navy Arms Co., Ridgefield, NJ, 1996. 218 pp., illus. $49.95
  The first definitive study of the Confederate revolvers invention, development and delivery by Francois Alexandre LeMat.

Les Pistolets Automatiques Francaise 1890-1990, by Jean Huon, Combined Books, Inc., Conshohocken, PA, 1997. 160 pp., illus. French text. $34.95

French automatic pistols from the earliest experiments through the World Wars and Indo-China to modern security forces.

Levine's Guide to Knives And Their Values, 4th Edition, by Bernard Levine, DBI Books, a division of Krause Publications, Iola, WI, 1997. 512 pp., illus. Paper covers. $27.95
  All the basic tools for identifying, valuing and collecting folding and fixed blade knives.

The Light 6-Pounder Battalion Gun of 1776, by Adrian Caruana, Museum Restoration Service, Bloomfield, Ontario, Canada, 2001. 76 pp., illus. Paper covers. $8.95

The London Gun Trade, 1850-1920, by Joyce E. Gooding, Museum Restoration Service, Bloomfield, Ontario, Canada, 2001. 48 pp., illus. Paper covers. $8.95
  Names, dates and locations of London gunmakers working between 1850 and 1920 are listed. Compiled from the original Kelly's Post Office Directories of the City of London.

The London Gunmakers and the English Duelling Pistol, 1770-1830, by Keith R. Dill, Museum Restoration Service, Bloomfield, Ontario, Canada, 1997. 36 pp., illus. Paper covers. $8.95
  Ten gunmakers made London one of the major gunmaking centers of the world. This book examines how the design and construction of their pistols contributed to that reputation and how these characteristics may be used to date flintlock arms.

Longrifles of North Carolina, by John Bivens, George Shumway Publisher, York, PA, 1988. 256 pp., illus. $50.00
  Covers art and evolution of the rifle, immigration and trade movements. Committee of Safety gunsmiths, characteristics of the North Carolina rifle.

Longrifles of Pennsylvania, Volume 1, Jefferson, Clarion & Elk Counties, by Russel H. Harringer, George Shumway Publisher, York, PA, 1984. 200 pp., illus. $50.00
  First in series that will treat in great detail the longrifles and gunsmiths of Pennsylvania.

The Luger Handbook, by Aarron Davis, Krause Publications, Iola, WI, 1997. 112 pp., illus. Paper covers. $9.95
  Quick reference to classify Luger models and variations with complete details including proofmarks.

Lugers at Random, by Charles Kenyon, Jr., Handgun Press, Glenview, IL, 1990. 420 pp., illus. $59.95
  A new printing of this classic, comprehensive reference for all Luger collectors.

The Luger Story, by John Walter, Stackpole Books, Mechanicsburg, PA, 2001. 256 pp., illus. Paper Covers $29.95
  The standard history of the world's most famous handgun.

M1 Carbine, by Larry Ruth, Gun room Press, Highland Park, NJ, 1987. 291 pp., illus. Paper $19.95
  The origin, development, manufacture and use of this famous carbine of World War II.

The M1 Carbine: Owner's Guide, by Scott A. Duff, Scott A. Duff, Export, PA, 1997. 126 pp., illus. Paper covers. $19.95
  This book answers the questions M1 owners most often ask concerning maintenance activities not encounter by military users.

The M1 Garand: Owner's Guide, by Scott A. Duff, Scott A. Duff, Export, PA, 1998. 132 pp., illus. Paper covers. $19.95
  This book answers the questions M1 owners most often ask concerning maintenance activities not encounter by military users.

The M1 Garand Serial Numbers and Data Sheets, by Scott A. Duff, Export, PA, 1995. 101 pp., illus. Paper covers. $11.95
  Provides the reader with serial numbers related to dates of manufacture and a large sampling of data sheets to aid in identification or restoration.

The M1 Garand 1936 to 1957, by Joe Poyer and Craig Riesch, North Cape Publications, Tustin, CA, 1996. 216 pp., illus. Paper covers. $19.95
  Describes the entire range of M1 Garand production in text and quick-scan charts.

The M1 Garand: Post World War, by Scott A. Duff, Scott A. Duff, Export, PA, 1990. 139 pp., illus. Soft covers. $19.95
  A detailed account of the activities at Springfield Armory through this period. International Harvester, H&R, Korean War production and quantities delivered. Serial numbers.

The M1 Garand: World War 2, by Scott A. Duff, Scott A. Duff, Export, PA, 1993. 210 pp., illus. Paper covers. $39.95
  The most comprehensive study available to the collector and historian on the M1 Garand of World War II.

Maine Made Guns and Their Makers, by Dwight B. Demeritt Jr., Maine State Museum, Augusta, ME, 1998. 209 pp., illustrated. $55.00
  An authoritative, biographical study of Maine gunsmiths.

Marlin Firearms: A History of the Guns and the Company That Made Them, by Lt. Col. William S. Brophy, USAR, Ret., Stackpole Books, Harrisburg, PA, 1989. 672 pp., illus. $75.00
  The definitive book on the Marlin Firearms Co. and their products.

Martini-Henry .450 Rifles & Carbines, by Dennis Lewis, Excalibur Publications, Latham, NY, 1996. 72 pp., illus. Paper covers. $11.95
  The stories of the rifles and carbines that were the mainstay of the British soldier through the Victorian wars.

Mauser Bolt Rifles, by Ludwig Olson, F. Brownell & Son, Inc., Montezuma, IA, 1999. 364 pp., illus. $59.95
The most complete, detailed, authoritative and comprehensive work ever done on Mauser bolt rifles. Completely revised deluxe 3rd edition.

Mauser Military Rifles of the World, 2nd Edition, by Robert Ball, Krause Publications, Iola, WI, 2000. 304 pp., illustrated with 1,000 b&w photos and a 48 page color section. $44.95
This 2nd edition brings more than 100 new photos of these historic rifles and the wars in which they were carried.

Mauser Smallbores Sporting, Target and Training Rifles, by Jon Speed, Collector Grade Publications, Cobourg, Ontario, Canada 1998. 349 pp., illustrated. $67.50
A history of all the smallbore sporting, target and training rifles produced by the legendary Mauser-Werke of Obendorf Am Neckar.

Military Holsters of World War 2, by Eugene J. Bender, Rowe Publications, Rochester, NY, 1998. 200 pp., illustrated. $45.00
A revised edition with a new price guide of the most definitive book on this subject.

Military Pistols of Japan, by Fred L. Honeycutt, Jr., Julin Books, Palm Beach Gardens, FL, 1997. 168 pp., illus. $42.00
Covers every aspect of military pistol production in Japan through WWII.

The Military Remington Rolling Block Rifle, by George Layman, Pioneer Press, TN, 1998. 146 pp., illus. Paper covers. $24.95
A standard reference for those with an interest in the Remington rolling block family of firearms.

Military Rifles of Japan, 5th Edition, by F.L. Honeycutt, Julin Books, Lake Park, FL, 1999. 208 pp., illus. $42.00
A new revised and updated edition. Includes the early Murata-period markings, etc.

Military Small Arms Data Book, by Ian V. Hogg, Stackpole Books, Mechanicsburg, PA, 1999. $44.95. 336 pp., illustrated.
Data on more than 1,500 weapons. Covers a vast range of weapons from pistols to anti-tank rifles. Essential data, 1870-2000, in one volume.

Modern Beretta Firearms, by Gene Gangarosa, Jr., Stoeger Publishing Co., So. Hackensack, NJ, 1994. 288 pp., illus. Paper covers. $16.95
Traces all models of modern Beretta pistols, rifles, machine guns and combat shotguns.

Modern Gun Values, The Gun Digest Book of, 10th Edition, by the Editors of Gun Digest, DBI Books, a division of Krause Publications, Iola, WI., 1996. 560 pp. illus. Paper covers. $21.95
Greatly updated and expanded edition describing and valuing over 7,000 firearms manufactured from 1900 to 1996. The standard for valuing modern firearms.

Modern Gun Identification & Value Guide, 13th Edition, by Russell and Steve Quertermous, Collector Books, Paducah, KY, 1998. 504 pp., illus. Paper covers. $14.95
Features current values for over 2,500 models of rifles, shotguns and handguns, with over 1,800 illustrations.

More Single Shot Rifles, by James C. Grant, The Gun Room Press, Highland Park, NJ, 1976. 324 pp., illus. $35.00
Details the guns made by Frank Wesson, Milt Farrow, Holden, Borchardt, Stevens, Remington, Winchester, Ballard and Peabody-Martini.

Mortimer, the Gunmakers, 1753-1923, by H. Lee Munson, Andrew Mowbray Inc., Lincoln, RI, 1992. 320 pp., illus. $65.00
Seen through a single, dominant, English gunmaking dynasty this fascinating study provides a window into the classical era of firearms artistry.

The Mosin-Nagant Rifle, by Terence W. Lapin, North Cape Publications, Tustin, CA, 1998. 30 pp., illustrated. Paper covers. $19.95
The first ever complete book on the Mosin-Nagant rifle written in English. Covers every variation.

The Navy Luger, by Joachim Gortz and John Walter, Handgun Press, Glenview, IL, 1988. 128 pp., illus. $24.95
The 9mm Pistol 1904 and the Imperial German Navy. A concise illustrated history.

The New World of Russian Small Arms and Ammunition, by Charlie Cutshaw, Paladin Press, Boulder, CO, 1998. 160 pp., illustrated. $42.95
Detailed descriptions, specifications and first-class illustrations of the AN-94, PSS silent pistol, Bizon SMG, Saifa-12 tactical shotgun, the GP-25 grenade launcher and more cutting edge Russian weapons.

The Number 5 Jungle Carbine, by Alan M. Petrillo, Excalibur Publications, Latham, NY, 1994. 32 pp., illus. Paper covers. $7.95
A comprehensive treatment of the rifle that collectors have come to call the "Jungle Carbine"—the Lee-Enfield Number 5, Mark 1.

The '03 Era: When Smokeless Revolutionized U.S. Riflery, by Clark S. Campbell, Collector Grade Publications, Inc., Ontario, Canada, 1994. 334 pp., illus. $44.50
A much-expanded version of Campbell's The '03 Springfields, representing forty years of in-depth research into "all things '03."

Observations on Colt's Second Contract, November 2, 1847, by G. Maxwell Longfield and David T. Basnett, Museum Restoration Service, Bloomfield, Ontario, Canada, 1997. 36 pp., illus. Paper covers. $6.95
This study traces the history and the construction of the Second Model Colt Dragoon supplied in 1848 to the U.S. Cavalry.

Official Guide to Gunmarks, 3rd Edition, by Robert H. Balderson, House of Collectibles, New York, NY, 1996. 367 pp., illus. Paper covers. $15.00

Identifies manufacturers' marks that appear on American and foreign pistols, rifles and shotguns.

Official Price Guide to Gun Collecting, by R.L. Wilson, Ballantine/House of Collectibles, New York, NY, 1998. 450 pp., illus. Paper covers. $21.50
Covers more than 30,000 prices from Colt revolvers to Winchester rifles and shotguns to German Lugers and British sporting rifles and game guns.

Official Price Guide to Military Collectibles, 6th Edition, by Richard J. Austin, Random House, Inc., New York, NY, 1998. 200 pp., illus. Paper cover. $20.00
Covers weapons and other collectibles from wars of the distant and recent past. More than 4,000 prices are listed. Illustrated with 400 black & white photos plus a full-color insert.

The Official Soviet SVD Manual, by Major James F. Gebhardt (Ret.) Paladin Press, Boulder, CO, 1999. 112 pp., illustrated. Paper covers. $15.00
Operating instructions for the 7.62mm Dragunov, the first Russian rifle developed from scratch specifically for sniping.

Old Gunsights: A Collector's Guide, 1850 to 2000, by Nicholas Stroebel, Krause Publications, Iola, WI, 1998. 320 pp., illus. Paper covers. $29.95
An in-depth and comprehensive examination of old gunsights and the rifles on which they were used to get accurate feel for prices in this expanding market.

Old Rifle scopes, by Nicholas Stroebel, Krause Publications, Iola, WI, 2000. 400 pp., illustrated. Paper covers. $31.95
This comprehensive collector's guide takes aim at more than 120 scope makers and 60 mount makers and features photos and current market values for 300 scopes and mounts manufactured from 1950-1985.

The P-08 Parabellum Luger Automatic Pistol, edited by J. David McFarland, Desert Publications, Cornville, AZ, 1982. 20 pp., illus. Paper covers. $11.95
Covers every facet of the Luger, plus a listing of all known Luger models.

Packing Iron, by Richard C. Rattenbury, Zon International Publishing, Millwood, NY, 1993. 216 pp., illus. $45.00
The best book yet produced on pistol holsters and rifle scabbards. Over 300 variations of holster and scabbards are illustrated in large, clear plates.

Parabellum: A Technical History of Swiss Lugers, by Vittorio Bobba, Priuli & Verlucca, Editori, Torino, Italy, 1996. Italian and English text. Illustrated. $100.00

Patents for Inventions, Class 119 (Small Arms), 1855-1930. British Patent Office, Armory Publications, Oceanside, CA, 1993. 7 volume set. $250.00
Contains 7980 abridged patent descriptions and their sectioned line drawings, plus a 37-page alphabetical index of the patentees.

Pattern Dates for British Ordnance Small Arms, 1718-1783, by DeWitt Bailey, Thomas Publications, Gettysburg, PA, 1997. 116 pp., illus. Paper covers. $20.00
The weapons discussed in this work are those carried by troops sent to North America between 1737 and 1783, or shipped to them as replacement arms while in America.

The Pitman Notes on U.S. Martial Small Arms and Ammunition, 1776-1933, Volume 2, Revolvers and Automatic Pistols, by Brig. Gen. John Pitman, Thomas Publications, Gettysburg, PA, 1990. 192 pp., illus. $29.95
A most important primary source of information on United States military small arms and ammunition.

The Plains Rifle, by Charles Hanson, Gun Room Press, Highland Park, NJ, 1989. 169 pp., illus. $35.00
All rifles that were made with the plainsman in mind, including pistols.

Powder and Ball Small Arms, by Martin Pegler, Windrow & Green, London, 1998. 128 pp., illus. $39.95
Part of the new "Live Firing Classic Weapons" series featuring full color photos of experienced shooters dressed in authentic costumes handling, loading and firing historic weapons.

The Powder Flask Book, by Ray Riling, R&R Books, Livonia, NY, 1993. 514 pp., illus. $69.95.
The complete book on flasks of the 19th century. Exactly scaled pictures of 1,600 flasks are illustrated.

Proud Promise: French Autoloading Rifles, 1898-1979, by Jean Huon, Collector Grade Publications, Inc., Cobourg, Ont., Canada, 1995. 216 pp., illus. $39.95
The author has finally set the record straight about the importance of French contributions to modern arms design.

E. C. Prudhomme's Gun Engraving Review, by E. C. Prudhomme, R&R Books, Livonia, NY, 1994. 164 pp., illus. $60.00
As a source for engravers and collectors, this book is an indispensable guide to styles and techniques of the world's foremost engravers.

Purdey Gun and Rifle Makers: The Definitive History, by Donald Dallas, Quiller Press, London, 2000. 245 pp., illus. Color throughout. $100.00
A limited edition of 3,000 copies. Signed and Numbered. With a PURDEY book plate.

Reloading Tools, Sights and Telescopes for Single Shot Rifles, by Gerald O. Kelver, Brighton, CO, 1982. 163 pp., illus. Paper covers. $13.95
A listing of most of the famous makers of reloading tools, sights and telescopes with a brief description of the products they manufactured.

The Remington-Lee Rifle, by Eugene F. Myszkowski, Excalibur Publications, Latham, NY, 1995. 100 pp., illus. Paper covers. $22.50
Features detailed descriptions, including serial number ranges, of each model from the first Lee Magazine Rifle produced for the U.S. Navy to the last Remington-Lee Small Bores shipped to the Cuban Rural Guard.

Revolvers of the British Services 1854-1954, by W.H.J. Chamberlain and A.W.F. Taylerson, Museum Restoration Service, Ottawa, Canada, 1989. 80 pp., illus. $27.50
Covers the types issued among many of the United Kingdom's naval, land or air services.

Rhode Island Arms Makers & Gunsmiths, by William O. Archibald, Andrew Mowbray, Inc., Lincoln, RI, 1990. 108 pp., illus. $16.50
A serious and informative study of an important area of American arms making.

Rifles of the World, by Oliver Achard, Chartwell Books, Inc., Edison, NJ, 141 pp., illus. $24.95
A unique insight into the world of long guns, not just rifles, but also shotguns, carbines and all the usual multi-barreled guns that once were so popular with European hunters, especially in Germany and Austria.

The Rock Island '03, by C.S. Ferris, C.S. Ferris, Arvada, CO, 1993. 58 pp., illus. Paper covers. $12.50
A monograph of interest to the collector or historian concentrating on the U.S. M1903 rifle made by the less publicized of our two producing facilities.

Round Ball to Rimfire, Vol. 1, by Dean Thomas, Thomas Publications, Gettysburg, PA, 1997. 144 pp., illus. $40.00
The first of a two-volume set of the most complete history and guide for all small arms ammunition used in the Civil War. The information includes data from research and development to the arsenals that created it.

Ruger and his Guns, by R.L. Wilson, Simon & Schuster, New York, NY, 1996. 358 pp., illus. $65.00
A history of the man, the company and their firearms.

Russell M. Catron and His Pistols, by Warren H. Buxton, Ucross Books, Los Alamos, NM, 1998. 224 pp., illustrated. Paper covers. $49.50
An unknown American firearms inventor and manufacturer of the mid twentieth century. Military, commercial, ammunition.

The SAFN-49 and The FAL, by Joe Poyer and Dr. Richard Feirman, North Cape Publications, Tustin, CA, 1998. 160 pp., illus. Paper covers. $14.95
The first complete overview of the SAFN-49 battle rifle, from its pre-World War 2 beginnings to its military service in countries as diverse as the Belgian Congo and Argentina. The FAL was "light" version of the SAFN-49 and it became the Free World's most adopted battle rifle.

Sam Colt's Own Record 1847, by John Parsons, Wolfe Publishing Co., Prescott, AZ, 1992. 167 pp., illus. $24.50
Chronologically presented, the correspondence published here completes the account of the manufacture, in 1847, of the Walker Model Colt revolver.

J. P. Sauer & Sohn, Sauer "Dein Waffenkamerad" Volume 2, by Cate & Krause, Walsworth Publishing, Chattanooga, TN, 2000. 440 pp., illus. $79.00
A historical study of Sauer automatic pistols. This new volume contains a great deal of new knowledge that has surfaced about the firm J.P. Sauer. You will find new photos, documentation, serial number ranges and historical facts which will expand the knowledge and interest in the oldest and best of the German firearms companies.

Scottish Firearms, by Claude Blair and Robert Woosnam-Savage, Museum Restoration Service, Bloomfield, Ont., Canada, 1995. 52 pp., illus. Paper covers. $8.95
This revision of the first book devoted entirely to Scottish firearms is supplemented by a register of surviving Scottish long guns.

The Scottish Pistol, by Martin Kelvin. Fairleigh Dickinson University Press, Dist. By Associated University Presses, Cranbury, NJ, 1997. 256 pp., illus. $49.50
The Scottish pistol, its history, manufacture and design.

Sharps Firearms, by Frank Seller, Frank M. Seller, Denver, CO, 1998. 358 pp., illus. $55.00
Traces the development of Sharps firearms with full range of guns made including all martial variations.

Simeon North: First Official Pistol Maker of the United States, by S. North and R. North, The Gun Room Press, Highland Park, NJ, 1972. 207 pp., illus. $15.95
Reprint of the rare first edition.

The SKS Carbine, by Steve Kehaya and Joe Poyer, North Cape Publications, Tustin, CA, 1997. 150 pp., illus. Paper covers. $16.95
The first comprehensive examination of a major historical firearm used through the Vietnam conflict to the diamond fields of Angola.

The SKS Type 45 Carbines, by Duncan Long, Desert Publications, El Dorado, AZ, 1992. 110 pp., illus. $19.95
Covers the history and practical aspects of operating, maintaining and modifying this abundantly available rifle.

Smith & Wesson 1857-1945, by Robert J. Neal and Roy G. Jinks, R&R Books, Livonia, NY, 1996. 434 pp., illus. $50.00
The bible for all existing and aspiring Smith & Wesson collectors.

Sniper Variations of the German K98k Rifle, by Richard D. Law, Collector Grade Publications, Ontario, Canada, 1997. 240 pp., illus. $47.50
Volume 2 of "Backbone of the Wehrmacht" the author's in-depth study of the German K98k rifle. This volume concentrates on the telescopic-sighted rifle of choice for most German snipers during World War 2.

Southern Derringers of the Mississippi Valley, by Turner Kirkland, Pioneer Press, Tenn., 1971. 80 pp., illus., paper covers. $4.00
A guide for the collector, and a much-needed study.

Soviet Russian Postwar Military Pistols and Cartridges, by Fred A. Datig, Handgun Press, Glenview, IL, 1988. 152 pp., illus. $29.95
Thoroughly researched, this definitive sourcebook covers the development and adoption of the Makarov, Stechkin and the new PSM pistols. Also included in this source book is coverage on Russian clandestine weapons and pistol cartridges.

Soviet Russian Tokarev "TT" Pistols and Cartridges 1929-1953, by Fred Datig, Graphic Publishers, Santa Ana, CA, 1993. 168 pp., illus. $39.95
Details of rare arms and their accessories are shown in hundreds of photos. It also contains a complete bibliography and index.

Soviet Small-Arms and Ammunition, by David Bolotin, Handgun Press, Glenview, IL, 1996. 264 pp., illus. $49.95
An authoritative and complete book on Soviet small arms.

Sporting Collectibles, by Jim and Vivian Karsnitz, Schiffer Publishing Ltd., West Chester, PA, 1992. 160 pp., illus. Paper covers. $29.95
The fascinating world of hunting related collectibles presented in an informative text.

The Springfield 1903 Rifles, by Lt. Col. William S. Brophy, USAR, Ret., Stackpole Books Inc., Harrisburg, PA, 1985. 608 pp., illus. $75.00
The illustrated, documented story of the design, development, and production of all the models, appendages, and accessories.

Springfield Armory Shoulder Weapons 1795-1968, by Robert W.D. Ball, Antique Trader Books, Dubuque, IA, 1998. 264 pp., illus. $34.95
This book documents the 255 basic models of rifles, including test and trial rifles, produced by the Springfield Armory. It features the entire history of rifles and carbines manufactured at the Armory, the development of each weapon with specific operating characteristics and procedures.

Springfield Model 1903 Service Rifle Production and Alteration, 1905-1910, by C.S. Ferris and John Beard, Arvada, CO, 1995. 66 pp., illus. Paper covers. $12.50
A highly recommended work for any serious student of the Springfield Model 1903 rifle.

Springfield Shoulder Arms 1795-1865, by Claud E. Fuller, S. & S. Firearms, Glendale, NY, 1996. 76 pp., illus. Paper covers. $17.95
Exact reprint of the scarce 1930 edition of one of the most definitive works on Springfield flintlock and percussion muskets ever published.

Standard Catalog of Firearms, 11th Edition, by Ned Schwing, Krause Publications, Iola, WI, 2001.1328 Pages, illustrated. 6,000+ b&w photos plus a 16-page color section. Paper covers. $32.95
This is the largest, most comprehensive and best-selling firearm book of all time! And this year's edition is a blockbuster for both shooters and firearm collectors. More than 12,000 firearms are listed and priced in up to six grades of condition. That's almost 80,000 prices! Gun enthusiasts will love the new full-color section of photos highlighting the finest firearms sold at auction this past year –including the new record for an American historical firearm: $684,000!

Standard Catalog of Winchester, 1st Edition, edited by David D. Kowalski, Krause Publications, Iola, WI, 2000. 704 pp., illustrated with 2,000 B&W photos and 75 color photos. Paper covers. $39.95
This book identifies and values more than 5,000 collectibles, including firearms, cartridges shotshells, fishing tackle, sporting goods and tools manufactured by Winchester Repeating Arms Co.

Steel Canvas: The Art of American Arms, by R.L. Wilson, Random House, NY, 1995, 384 pp., illus. $65.00
Presented here for the first time is the breathtaking panorama of America's extraordinary engravers and embellishers of arms, from the 1700s to modern times.

Stevens Pistols & Pocket Rifles, by K.L. Cope, Museum Restoration Service, Alexandria Bay, NY, 1992. 114 pp., illus. $24.50
This is the story of the guns and the man who designed them and the company which he founded to make them.

A Study of Colt Conversions and Other Percussion Revolvers, by R. Bruce McDowell, Krause Publications, Iola, WI, 1997. 464 pp., illus. $39.95
The ultimate reference detailing Colt revolvers that have been converted from percussion to cartridge.

The Sumptuous Flaske, by Herbert G. Houze, Andrew Mowbray, Inc., Lincoln, RI, 1989. 158 pp., illus. Soft covers. $35.00
Catalog of a recent show at the Buffalo Bill Historical Center bringing together some of the finest European and American powder flasks of the 16th to 19th centuries.

The Swedish Mauser Rifles, by Steve Kehaya and Joe Poyer, North Cape Publications, Tustin, CA, 1999. 267 pp., illustrated. Paper covers. $19.95
Every known variation of the Swedish Mauser carbine and rifle is described including all match and target rifles and all sniper versions. Includes serial number and production data.

Televisions Cowboys, Gunfighters & Cap Pistols, by Rudy A. D'Angelo, Antique Trader Books, Norfolk, VA, 1999. 287 pp., illustrated in color and black and white. Paper covers. $31.95
Over 850 beautifully photographed color and black and white images of cap guns, actors, and the characters they portrayed in the "Golden Age of TV Westerns". With accurate descriptions and current values.

Thompson: The American Legend, by Tracie L. Hill, Collector Grade Publications, Ontario, Canada, 1996. 584 pp., illus. $85.00
The story of the first American submachine gun. All models are featured and discussed.

Toys That Shoot and Other Neat Stuff, by James Dundas, Schiffer Books, Atglen, PA, 1999. 112 pp., illustrated. Paper covers. $24.95

Shooting toys from the twentieth century, especially 1920's to 1960's, in over 420 color photographs of BB guns, cap shooters, marble shooters, squirt guns and more. Complete with a price guide.

The Trapdoor Springfield, by M.D. Waite and B.D. Ernst, The Gun Room Press, Highland Park, NJ, 1983. 250 pp., illus. $39.95
The first comprehensive book on the famous standard military rifle of the 1873-92 period.

Treasures of the Moscow Kremlin: Arsenal of the Russian Tsars, A Royal Armories and the Moscow Kremlin exhibition. HM Tower of London 13, June 1998 to 11 September, 1998. BAS Printers, Over Wallop, Hampshire, England. xxii plus 192 pp. over 180 color illustrations. Text in English and Russian. $65.00
For this exhibition catalog each of the 94 objects on display are photographed and described in detail to provide a most informative record of this important exhibition.

U.S. Breech-Loading Rifles and Carbines, Cal. 45, by Gen. John Pitman, Thomas Publications, Gettysburg, PA, 1992. 192 pp., illus. $29.95
The third volume in the Pitman Notes on U.S. Martial Small Arms and Ammunition, 1776-1933. This book centers on the "Trapdoor Springfield" models.

U.S. Handguns of World War 2: The Secondary Pistols and Revolvers, by Charles W. Pate, Andrew Mowbray, Inc., Lincoln, RI, 1998. 515 pp., illus. $39.00
This indispensable new book covers all of the American military handguns of World War 2 except for the M1911A1 Colt automatic.

United States Martial Flintlocks, by Robert M. Reilly, Mowbray Publishing Co., Lincoln, RI, 1997. 264 pp., illus. $40.00
A comprehensive history of American flintlock longarms and handguns (mostly military) c. 1775 to c. 1840.

U.S. Martial Single Shot Pistols, by Daniel D. Hartzler and James B. Whisker, Old Bedford Village Pess, Bedford, PA, 1998. 128 pp., illus. $45.00
A photographic chronicle of military and semi-martial pistols supplied to the U.S. Government and the several States.

U.S. Military Arms Dates of Manufacture from 1795, by George Madis, David Madis, Dallas, TX, 1989. 64 pp. Soft covers. $6.00
Lists all U.S. military arms of collector interest alphabetically, covering about 250 models.

U.S. Military Small Arms 1816-1865, by Robert M. Reilly, The Gun Room Press, Highland Park, NJ, 1983. 270 pp., illus. $39.95
Covers every known type of primary and secondary martial firearms used by Federal forces.

U.S. M1 Carbines: Wartime Production, by Craig Riesch, North Cape Publications, Tustin, CA, 1994. 72 pp., illus. Paper covers. $16.95
Presents only verifiable and accurate information. Each part of the M1 Carbine is discussed fully in its own section; including markings and finishes.

U.S. Naval Handguns, 1808-1911, by Fredrick R. Winter, Andrew Mowbray Publishers, Lincoln, RI, 1990. 128 pp., illus. $26.00
The story of U.S. Naval Handguns spans an entire century—included are sections on each of the important naval handguns within the period.

Walther: A German Legend, by Manfred Kersten, Safari Press, Inc., Huntington Beach, CA, 2000. 400 pp., illustrated. $85.00
This comprehensive book covers, in rich detail, all aspects of the company and its guns, including an illustrious and rich history, the WW2 years, all the pistols (models 1 through 9), the P-38, P-88, the long guns, .22 rifles, centerfires, Wehrmacht guns, and even a gun that could shoot around a corner.

Walther Pistols: Models 1 Through P99, Factory Variations and Copies, by Dieter H. Marschall, Ucross Books, Los Alamos, NM. 2000. 140 pages, with 140 b & w illustrations, index. Paper Covers. $19.95
This is the English translation, revised and updated, of the highly successful and widely acclaimed German language edition. This book provides the collector with a reference guide and overview of the entire line of the Walther military, police, and self-defense pistols from the very first to the very latest. Models 1-9, PP, PPK, MP, AP, HP, P.38, P1, P4, P38K, P5, P88, P99 and the Manurhin models. Variations, where issued, serial ranges, calibers, marks, proofs, logos, and design aspects in an astonishing quantity and variety are crammed into this very well researched and highly regarded work.

The Walther Handgun Story: A Collector's and Shooter's Guide, by Gene Gangarosa, Steiger Publications, 1999. 300., illustrated. Paper covers. $21.95
Covers the entire history of the Walther empire. Illustrated with over 250 photos.

Walther P-38 Pistol, by Maj. George Nonte, Desert Publications, Cornville, AZ, 1982. 100 pp., illus. Paper covers. $11.95
Complete volume on one of the most famous handguns to come out of WWII. All models covered.

Walther Models PP & PPK, 1929-1945 – Volume 1, by James L. Rankin, Coral Gables, FL, 1974. 142 pp., illus. $40.00
Complete coverage on the subject as to finish, proofmarks and Nazi Party inscriptions.

Walther Volume II, Engraved, Presentation and Standard Models, by James L. Rankin, J.L. Rankin, Coral Gables, FL, 1977. 112 pp., illus. $40.00
The new Walther book on embellished versions and standard models. Has 88 photographs, including many color plates.

Walther, Volume III, 1908-1980, by James L. Rankin, Coral Gables, FL, 1981. 226 pp., illus. $40.00
Covers all models of Walther handguns from 1908 to date, includes holsters, grips and magazines.

Winchester: An American Legend, by R.L. Wilson, Random House, New York, NY, 1991. 403 pp., illus. $65.00
The official history of Winchester firearms from 1849 to the present.

Winchester Bolt Action Military & Sporting Rifles 1877 to 1937, by Herbert G. Houze, Andrew Mowbray Publishing, Lincoln, RI, 1998. 295 pp., illus. $45.00
Winchester was the first American arms maker to commercially manufacture a bolt action repeating rifle, and this book tells the exciting story of these Winchester bolt actions.

The Winchester Book, by George Madis, David Madis Gun Book Distributor, Dallas, TX, 1986. 650 pp., illus. $49.50
A new, revised 25th anniversary edition of this classic book on Winchester firearms. Complete serial ranges have been added.

Winchester Dates of Manufacture 1849-1984, by George Madis, Art & Reference House, Brownsboro, TX, 1984. 59 pp. $9.95
A most useful work, compiled from records of the Winchester factory.

Winchester Engraving, by R.L. Wilson, Beinfeld Books, Springs, CA, 1989. 500 pp., illus. $135.00
A classic reference work of value to all arms collectors.

The Winchester Handbook, by George Madis, Art & Reference House, Lancaster, TX, 1982. 287 pp., illus. $24.95
The complete line of Winchester guns, with dates of manufacture, serial numbers, etc.

The Winchester-Lee Rifle, by Eugene Myszkowski, Excalibur Publications, Tucson, AZ 2000. 96 pp., illustrated. Paper Covers. $22.95
The development of the Lee Straight Pull, the cartridge and the approval for military use. Covers details of the inventor and memorabilia of Winchester-Lee related material.

Winchester Lever Action Repeating Firearms, Vol. 1, The Models of 1866, 1873 and 1876, by Arthur Pirkie, North Cape Publications, Tustin, CA, 1995. 112 pp., illus. Paper covers. $19.95
Complete, part-by-part description, including dimensions, finishes, markings and variations throughout the production run of these fine, collectible guns.

Winchester Lever Action Repeating Rifles, Vol. 2, The Models of 1886 and 1892, by Arthur Pirkie, North Cape Publications, Tustin, CA, 1996. 150 pp., illus. Paper covers. $19.95
Describes each model on a part-by-part basis by serial number range complete with finishes, markings and changes.

Winchester Lever Action Repeating Rifles, Volume 3, The Model of 1894, by Arthur Pirkie, North Cape Publications, Tustin, CA, 1998. 150 pp., illus. Paper covers. $19.95
The first book ever to provide a detailed description of the Model 1894 rifle and carbine.

The Winchester Lever Legacy, by Clyde "Snooky" Williamson, Buffalo Press, Zachary, LA, 1988. 664 pp., illustrated. $75.00
A book on reloading for the different calibers of the Winchester lever action rifle.

The Winchester Model 94: The First 100 Years, by Robert C. Renneberg, Krause Publications, Iola, WI, 1991. 208 pp., illus. $34.95
Covers the design and evolution from the early years up to the many different editions that exist today.

Winchester Rarities, by Webster, Krause Publications, Iola, WI., 2000. 208 pp., with over 800 color photos, illus. $49.95
This book details the rarest of the rare; the one-of-a-kind items and the advertising pieces from years gone by. With nearly 800 full color photos and detailed pricing provided by experts in the field, this book gives collectors and enthusiasts everything they need.

Winchester Shotguns and Shotshells, by Ronald W. Stadt, Krause Publications, Iola, WI, 1995. 256 pp., illus. $34.95
The definitive book on collectible Winchester shotguns and shotshells manufactured through 1961.

The Winchester Single-Shot- Volume 1; A History and Analysis, by John Campbell, Andrew Mowbray, Inc., Lincoln RI, 1995. 272 pp., illus. $55.00
Covers every important aspect of this highly-collectible firearm.

The Winchester Single-Shot- Volume 2; Old Secrets and New Discoveries, by John Campbell, Andrew Mowbray, Inc., Lincoln RI, 2000. 280 pp., illus. $55.00
An exciting follow-up to the classic first volume.

Winchester Slide-Action Rifles, Volume 1: Model 1890 & 1906, by Ned Schwing, Krause Publications, Iola, WI, 1992. 352 pp., illus. $39.95
First book length treatment of models 1890 & 1906 with over 50 charts and tables showing significant new information about caliber style and rarity.

Winchester Slide-Action Rifles, Volume 2: Model 61 & Model 62, by Ned Schwing, Krause Publications, Iola, WI, 1993. 256 pp., illus. $34.95
A complete historic look into the Model 61 and the Model 62. These favorite slide-action guns receive a thorough presentation which takes you to the factory to explore receivers, barrels, markings, stocks, stampings and engraving in complete detail.

Winchester's North West Mounted Police Carbines and other Model 1876 Data, by Lewis E. Yearout, The author, Great Falls, MT, 1999. 224 pp., illustrated. Paper covers. $38.00
An impressive accumulation of the facts on the Model 1876, with particular emphasis on those purchased for the North West Mounted Police.

Worldwide Webley and the Harrington and Richardson Connection, by Stephen Cuthbertson, Ballista Publishing and Distributing Ltd., Gabriola Island, Canada, 1999. 259 pp., illus. $50.00
A masterpiece of scholarship. Over 350 photographs plus 75 original documents, patent drawings, and advertisements accompany the text.

MODERN GUN VALUES READER INTEREST SURVEY

1. **Check all that apply:** Why did you buy this book?
   - ☐ For general information
   - ☐ I'm an active "guntrader" and I need current valuations
   - ☐ I'm an active collector of specific brands/types of firearms

2. **Check all that apply:** In which type of firearm do you have the greatest collecting/ trading interest?

   RIFLES:
   - ☐ Single shot
   - ☐ Bolt action
   - ☐ Autoloaders
   - ☐ Lever action
   - ☐ Slide action
   - ☐ Muzzleloaders

   HANDGUNS:
   - ☐ Single shot
   - ☐ Revolvers
   - ☐ Autoloaders

   SHOTGUNS:
   - ☐ Lever action
   - ☐ Over & Under
   - ☐ Autoloaders
   - ☐ Side-by-Side
   - ☐ Slide action

   OTHER: _____

3. **Check all that apply:** Which of these arms price guides do you regularly buy?
   - ☐ Blue Book of Gun Values
   - ☐ Flayderman's Guide to Antique American Firearms
   - ☐ Gun Trader's Guide
   - ☐ MODERN GUN VALUES (this book)
   - ☐ Standard Catalog of Firearms
   - ☐ Other:_____
   - _____
   - ☐ None of these

4. *Including this one,* how many different price guides do you typically purchase in a year?

   ☐☐ Number of different price guides typically purchased in a year

5. **Check all that apply:** With regard to firearms you acquire, which of the following do you personally perform?
   - ☐ Cleaning
   - ☐ Repairs (of any magnitude)
   - ☐ Assembly and/or disassembly
   - ☐ Test firing

6. **Check all that apply:** Which of the following magazines do you read regularly; i.e., at least three out of every four issues?
   - ☐ Arms Collecting
   - ☐ Guns & Ammo
   - ☐ Double Gun Journal
   - ☐ Guns
   - ☐ Shooting Sportsman
   - ☐ Shooting Times
   - ☐ Sporting Classics
   - ☐ The Gun Report
   - ☐ Man At Arms
   - ☐ Other: _____

7. **Check all that apply:** Which of the following did you read in this 11th Edition of MODERN GUN VALUES?

   ARTICLES:
   - ☐ Inspection Guide to Used Guns, by Patrick Sweeney
   - ☐ Evaluation of Arms Condition (color section), by Tom Turpin
   - ☐ An Illustrated Guide to Condition, by Joseph Schroeder
   - ☐ A Beginner's Guide to Collecting Colt SAAs, by John Luchsinger
   - ☐ Buying Used Custom Rifles, by John Luchsinger

   ASSEMBLY/DISASSEMBLY:
   - ☐ Russian SKS Carbine
   - ☐ Colt Police Positive Revolver

   ARMS PRICING SECTIONS:
   - ☐ GUNDEX
   - ☐ Handgun Models & Pricing
   - ☐ Rifle Models & Pricing
   - ☐ Shotgun Models & Pricing
   - ☐ Commemorative Arms Models & Pricing

   REFERENCE SECTION:
   - ☐ Arms Associations
   - ☐ Arms Collector's Library

8. Any additional comments? _____
   _____

Thank you for filling out the MODERN GUN VALUES Reader Interest Survey. Please return your completed survey no later than JUNE 14, 2002. This offer expires on JUNE 14, 2002.

YOU RECEIVE <u>FOUR</u> <u>FREE</u> <u>ISSUES</u> OF GUN LIST MAGAZINE FOR TAKING THE TIME TO COMPLETE OUR SURVEY!**

☐ Please send me my gift — 4 issues of GUN LIST magazine (a retail value of $22.00). GUN LIST is the leading national newspaper for people who buy and sell quality firearms. *If you are already a subscriber,* your subscription will extended by four additional issues. Thank you for your help.

Name: ☐☐☐☐☐☐☐☐☐☐☐☐☐☐☐☐☐☐☐☐☐☐☐☐☐☐☐☐

Address: ☐☐☐☐☐☐☐☐☐☐☐☐☐☐☐☐☐☐☐☐☐☐☐☐☐☐☐

City: ☐☐☐☐☐☐☐☐☐☐☐☐☐☐☐☐☐☐☐☐☐☐☐☐☐☐☐☐

State: ☐☐   Zip Code: ☐☐☐☐☐   Phone: ☐☐☐–☐☐☐–☐☐☐☐

** PLEASE NOTE: You must complete ALL of the survey questions and return this by June 14, 2002 to qualify to receive your free gift (4 issues of GUN LIST). ABA42D

Tape Here   Tape Here

|·|···|·||·|·||·|·|··||·||·|·||·|··|·|··||·|·|·||·|··|||